Organic Chemistry

A Mechanistic Approach

Organic Chemistry

A Mechanistic Approach

Penny Chaloner

CRC Press
Taylor & Francis Group
Boca Raton London New York

CRC Press is an imprint of the
Taylor & Francis Group, an **informa** business

CRC Press
Taylor & Francis Group
6000 Broken Sound Parkway NW, Suite 300
Boca Raton, FL 33487-2742

Printed and bound in India by Replika Press Pvt. Ltd.

Printed on acid-free paper
Version Date: 20140911

International Standard Book Number-13: 978-1-4822-0690-6 (Hardback)

Library of Congress Cataloging-in-Publication Data

Chaloner, Penny A.
Organic chemistry : a mechanistic approach / Penny Chaloner.
pages cm
"A CRC title."
Includes bibliographical references and index.
ISBN 978-1-4822-0690-6 (hardcover : alk. paper)
1. Chemistry, Organic--Textbooks. 2. Chemical reactions--Textbooks. I. Title.

QD251.3.C43 2014
547--dc23 2013050348

Visit the Taylor & Francis Web site at
http://www.taylorandfrancis.com

and the CRC Press Web site at
http://www.crcpress.com

Contents

Preface

TO THE INSTRUCTOR

Traditional organic chemistry courses receive bad press from students for two reasons. First, they are "sold" as hard courses—I vividly recall being told that in the United States that the point of organic chemists was to keep unsuitable people out of medical schools, because (unlike practitioners and instructors in most other areas) "we still have real standards and we still fail people." Second, organic chemistry courses are perceived by students as requiring a great deal of rote learning—this latter of course being totally untrue. Those new to organic chemistry need to learn the language—the nomenclature, the irregular verbs—but beyond that, a basic mechanistic understanding will allow even beginning students to approach the subject without fear and even with enjoyment!

Organic chemistry teaching is bedeviled by the problem that many students in any sophomore class would really prefer to be elsewhere—they see themselves as biologists, chemical engineers, or premeds, not as aspirant chemists. So as instructors we can either turn somersaults to try to make the students see relevance to their particular discipline—not always easy at an elementary level—or we can try to convert these students into seeing the inherent beauty of the subject and divert them into the infinitely satisfying practice of chemistry.

In this book, I hope to convince students that learning organic chemistry does not need to be a tedious repetition of "preps and props" of the various categories of compounds but can be a deep understanding of underlying principles that can then be applied to a wide range of as yet unknown problems and systems. As they go along, they will also learn a way of thinking and analysis that will serve them well across many academic disciplines.

I have made a conscious effort to include some of the more modern synthetic approaches—particularly those leading to enantioselective syntheses such as Sharpless dihydroxylation and epoxidation. Later chapters deal with the biological, environmental, industrial, and forensic aspects of the subject.

Supplemental materials for instructors are available from the publisher. Please contact orders@taylorandfrancis.com. The PowerPoint slides are designed for a large lecture theater in which all the screen is visible everywhere. There are more slides than are likely to be used—where there are multiple examples, most instructors will want to select. There is also a Word and a PowerPoint version of the solutions to the problems in the text, the latter being designed for use in section teaching. Early chapters have laboriously detailed solutions and explanations, but these reduce as the student feels more comfortable with the basic material.

Each chapter has extensive review problems to solve—any student who works through these thoroughly will become very competent. In many chapters, basic material is reviewed at the start of the problems, particularly nomenclature and spectroscopy. For most chapters, there is also a set of additional challenging problems—mainly for chemistry majors, perhaps.

TO THE STUDENT

Many students approach organic chemistry courses with trepidation. There is a lot of learning to do, isn't there, and the instructors set very difficult problems. No subject can be mastered without effort—this applies in all areas. You would not expect to become competent in speaking a language without learning the irregular verbs, in playing piano without practicing, or running a marathon without training. The most important training you can do in organic chemistry is to practice solving problems—with which each chapter of this book is extensively provided. Generally, the first few are easy and for review—rather like your warm up before running or scales before Chopin.

While some students will go on to further study of chemistry, for others this is a somewhat unloved requirement for their major. I'd like to think that even those will find topics to interest them here and will find that the problem-solving approach to the subject serves them well in disciplines beyond chemistry.

Feel free to share things you have found difficult. Constructive feedback is welcome. If you have found something tough, maybe others have too, and some additional explanation can be given.

Good luck!

Acknowledgments

I am extremely grateful to the many friends and colleagues listed below who have read various parts of this text and suggested several improvements. Without them, the writing would have been much more difficult and the outcome not as good. Particular thanks go to Professor Jim Hanson, who bravely struggled through every chapter. I must also thank many classes of students to whom I have taught this material; their enthusiasm and hard work has been inspiring, and their comments very enlightening. Any remaining errors are, of course, my own.

Individual figures and photographs with which I had help are acknowledged in the text. Special thanks go to Alexandria Rogerson and Iain Swan, who produced all the NMR spectra. Many thanks to

Gwen Baumann
Robin Bedford
Norman Billingham
John Brown
Martin Carter
George Fleet
Jim Hanson
Peter Knowles
Gerry Lawless
Cheryl Marlowe
Tim Metham
Bob Moss
Göran Nillson
Phil Parsons
Ken Seddon
Graham Sewell
Liz Sommerville
John Spencer
John Turner
Eddie Viseaux

Author

Penny Chaloner received her first degree and PhD from the University of Cambridge before moving to the University of Oxford as a research fellow. She then spent a period in the United States teaching at Rutgers University and Harvard Summer School, where she learned that organic chemistry, which had always been the easiest topic for English students, was regarded quite differently elsewhere!

Chaloner returned to a faculty position at the University of Sussex, remaining there for 27 years. During this time, she taught a wide range of courses, but had a long-standing interest in international exchange programs, and taught organic chemistry in the American style to premedical students for many years. Retiring in 2010, she has since had a visiting position at Imperial College, and writes extensively from home.

Review of Atomic and Molecular Structure

1

1.1 INTRODUCTION

Organic chemistry was originally considered to be the chemistry of living things; organic chemicals came from plants and animals, while inorganic ones came from rocks. However, the discovery of the conversion of ammonium isocyanate, on heating, into urea (Figure 1.1) first connected the whole of chemistry together. The starting material is definitely inorganic, but the product urea would be classed as an organic molecule. In the nineteenth century (and indeed today), it was isolated from natural sources.

More organic than inorganic compounds are known, firstly because carbon forms strong bonds to other carbon atoms. Although carbon is not unique in this, it is much the most successful element in the periodic table in this respect. Silicon, below carbon in the periodic table, does not readily form Si–Si bonds (it much prefers to form –Si–O–Si–O– chains) so the silicon life form beloved of science fiction writers stays firmly in the realm of fiction. Carbon also forms multiple (double and triple) bonds to other carbon atoms and forms strong and/or multiple bonds to most nonmetallic elements and to some metals.

The Weeping Angels are amongst Dr Who's scariest adversaries; although they are never specifically defined as a silicon life form, they do spend most of their time as stone

It has been possible to analyze organic molecules since the nineteenth century, and although the technology has been radically altered, the underlying principle of modern microanalysis is the same as was used then. What has changed is the availability of very high accuracy balances that can weigh reliably to ±0.001 mg. Consider the nineteenth-century version of the analysis of ethanol. A 4.6 g sample of ethanol is burned in an excess (XS is a common abbreviation for excess) of

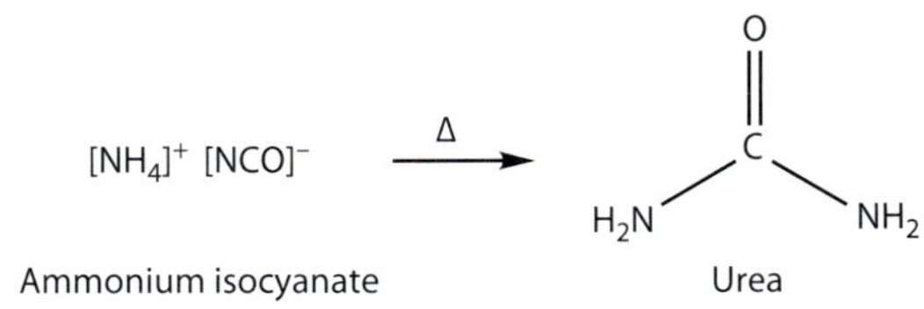

FIGURE 1.1 Conversion of ammonium isocyanate to urea. *Note*: I hope you are familiar with the use of a capital Greek delta, Δ, to indicate heat. New symbols and acronyms will be explained the first time they are used, but they are also collected together in Appendix A.

oxygen to give 8.8 g of carbon dioxide and 5.4 g of water (Equation 1.1). I hope that you can easily convert these figures into percentage composition and formula for ethanol (a quick reminder in the following text, but if any of this seems unfamiliar, it's probably sensible to review the relevant parts of your general chemistry course notes now).

$$\underset{4.6\text{ g}}{CH_3CH_2OH} + \underset{XS}{O_2} \rightarrow \underset{8.8\text{ g}}{2CO_2} + \underset{5.4\text{ g}}{3H_2O} \tag{1.1}$$

Quick Reminder on Analysis

Of the 8.8 g CO_2, 12/(12 + {2 × 16}) = 27.3 % is carbon.
Thus, 4.6 g ethanol yielded 2.4 g C.
Of the 5.4 g H_2O, 2/18 = 11.1 % is hydrogen.
Thus, 4.6 g ethanol yielded 0.6 g H.

We can now calculate the percentages of C, H, and O in ethanol:
C = 2.4/4.6 = 52.2 %
H = 0.6/4.6 = 13.0 %
The remainder is oxygen, 100 − 52.2 − 13 = 34.8 %.

To convert these percentages to an empirical formula, we need to divide by the atomic weights:
Relative number of atoms of carbon = 52.2/12 = 4.35
Relative number of atoms of hydrogen = 13.0/1 = 13
Relative number of atoms of oxygen = 34.8/16 = 2.175

We now divide each of these by the smallest number:
So #C = 4.35/2.175 = 2
#H = 13/2.175 = 5.98
#O = 2.175/2.175 = 1

Thus, our empirical formula for ethanol is C_2H_6O.

Analysis of a compound is normally considered to indicate that it is pure if the analysis figures are within 0.3 % of the correct figure, from the formula. Although you will see microanalyses quoted to two decimal places, this is unrealistic, and good-quality journals don't expect it. The only time you should see a second decimal place is when the number in it is a 5.

The purity of an organic compound is established by a good-quality microanalysis, but there is also a mass spectrometric technique, accurate mass determination, that is widely used, and which we will discuss in Chapter 5. There is also the expectation that the compound, if solid, will have a narrow melting range, and various spectroscopic techniques will not indicate any impurities.

PROBLEM 1.1

Ethylene glycol, used as antifreeze, contains only C, H, and O. Combustion analysis of a 2.346 mg sample yields 2.042 mg H_2O and 3.327 mg CO_2. What is the empirical formula? What is the molecular formula if ethylene glycol has a molecular weight of 62.0 amu*?

Solution

3.327 mg CO_2 contains (12/44) × 3.327 = 0.907 mg C
2.042 mg H_2O contains (2/18) × 2.042 = 0.227 mg H
Thus, % C in ethylene glycol = (0.907/2.346) × 100 = 38.7 %
% H in ethylene glycol = (0.227/2.346) × 100 = 9.7 %
Thus, % oxygen = 100 – 38.7 – 9.7 = 51.6 %
Relative number of atoms C = 38.7/12 = 3.225
Relative number of atoms H = 9.7/1 = 9.7
Relative number atoms O = 51.6/16 = 3.225
Dividing through by 3.225,
#C = 1
#H = 3
#O = 1
Empirical formula = CH_3O
C + 3H + O = 12 + 3 + 16 = 31 mass units
Molecular weight = 62
Molecular formula = $C_2H_6O_2$

1.2 ATOMIC AND MOLECULAR STRUCTURE

1.2.1 ORBITALS OF THE HYDROGEN ATOM

In this section, there will be quite a few statements that I do not intend to justify in detail. You will probably have covered much of it in a general chemistry course (and may later go into much more detail in a physical chemistry course), but as organic chemists, we are interested in the results, rather than the detail of the theory. We know that the hydrogen atom consists of one proton, positively charged, and one electron, negatively charged. Early ideas of the structure of hydrogen involved the electron "orbiting" the proton, rather like a bead on a wire. Niels Bohr (Noble Prize for Physics 1922) restricted the electron to a series of orbits, each associated with a specific energy level, in order to explain the spectra obtained from atomic hydrogen. Electrons could not have just any energy; the energy is quantized. Transitions of the electron between these levels could produce, or absorb, electromagnetic radiation. Because electrons spin on their own axes, either clockwise or counterclockwise, two electrons per orbit were allowed, with opposing spins. (Electrons don't really know about clocks, ☺, but there are still two ways to spin on a single axis).

FIGURE 1.2 1s orbital of the hydrogen atom.

Erwin Schrödinger (Nobel Prize for Physics 1933) took things further, and much of our modern view of atomic structure derives from his ideas. He described electrons in terms of a wave function, ψ; an electron can be considered to have wave as well as particle character, because of its very small size (sometimes called the wave–particle duality). The probability of finding an electron in a particular place is proportional to the value of ψ^2 at that point in space. The solution of Schrödinger's wave equation for the hydrogen atom has the electron in the lowest available energy level, which we call the 1s orbital. This is generally represented as a sphere (Figure 1.2).

FIGURE 1.3 Cutaway diagram of the hydrogen 2s orbital.

The number 1 is the *principal quantum number*, designated as *n*. *n* may have integer values, 1,2,3,4, *etc*., and the energy of the electron increases as *n* increases. The letter s tells us the orbital shape, which comes from the *angular momentum quantum number*, *l*. *l* can have any integral value from 0 to $n - 1$, so if $n = 1$, then $l = 0$, and this describes a spherical 1s orbital. When $n = 2$ and $l = 0$, this describes a larger, spherical 2s orbital (Figure 1.3; the distinct colors represent different signs of the wave function ψ). If $n = 2$ and $l = 1$, this describes a dumbbell-shaped 2p orbital (Figure 1.4).

FIGURE 1.4 2p orbital.

* amu, atomic mass unit.

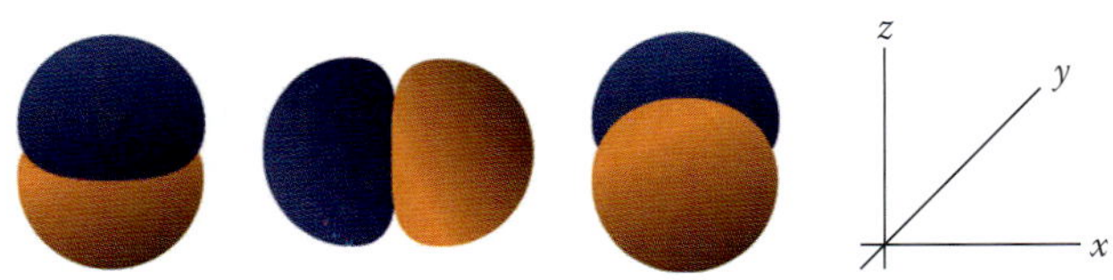

FIGURE 1.5 The three 2p orbitals.

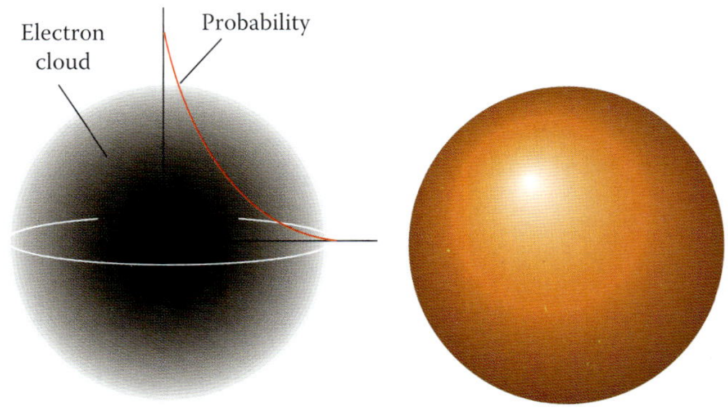

FIGURE 1.6 The 1s orbital.

The magnetic quantum number m_l can have values from $+l$ to $-l$, including zero. Thus, for $n = 2$, $l = 1$, $m_l = -1, 0$, or $+1$. Thus, we have three p orbitals, which are at 90 ° to each other and are conventionally labeled p_z, p_x, and p_y (Figure 1.5). The *spin quantum number s* (sometimes m_s) can have just two values, +1/2 and −1/2, which we can associate with clockwise and counterclockwise spins of the electron.

Conventionally, we draw s orbitals as spheres—but what exactly do we mean by this? In Figure 1.6, there is a diagrammatic representation of the electron cloud and the function ψ^2 plotted against distance from the nucleus. The electron is now shown as a cloud of varying density. One convention is that there is a 95 % probability that the 1s electron is within the sphere that we draw, which would then be a sphere 1.7 Å* from the nucleus. However, another convention is that the sphere represents the most likely distance of the electron from the nucleus, a much smaller number.

We can plot the probability of finding the electron at a particular distance from the nucleus against the distance r. To do this, the function we actually plot is $4\pi r^2\psi^2$. We need to take account of the fact that the larger r, the larger the area of the sphere that has a particular value for ψ.† For the 1s orbital, this is shown in Figure 1.7. Close to the nucleus, ψ and ψ^2 are very large, but r is vanishingly small. At large distances from the nucleus, r is large but ψ and ψ^2 are very small. This gives a maximum at 0.529 Å. Figures 1.8 through 1.10 show ψ and probability for the 2s, 3s, and 4s orbitals. The 2s orbital is larger than the 1s orbital but still spherical. There are two regions of high probability for the electron to be found, with that further from the nucleus being the more important. Figure 1.11 shows the probabilities for the 1s, 2s, 3s, and 4s orbitals on the same scale.

The p orbitals are dumbbell shaped; there are two "lobes" separated by a nodal plane or node. At the node, the sign of the wave function changes (Figure 1.12). ψ and ψ^2 are zero, and there is no probability of finding the electron at the node. In this figure, in most drawing programs, and

* Å Angstrom, 10^{-10} m. Not strictly an SI unit, but because of its similarity in magnitude to atom sizes and bond lengths, it's convenient to use it when describing these.

† Volume of a sphere with radius r is $(4/3)\pi r^3$. Volume of a sphere with radius $r + \delta r$ is $(4/3)\pi(r + \delta r)^3$. The volume of the shell between the two spheres is therefore $4\pi r^2\delta r$ + terms in δr^2.

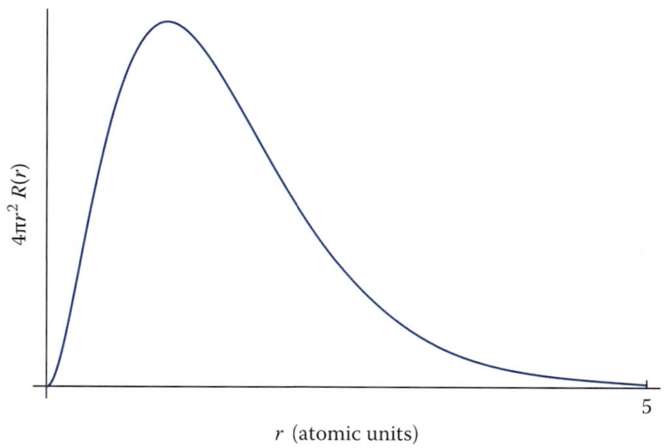

FIGURE 1.7 Probability of electron position for the hydrogen 1s orbital. (Thanks to Prof. Peter Knowles for this figure.)

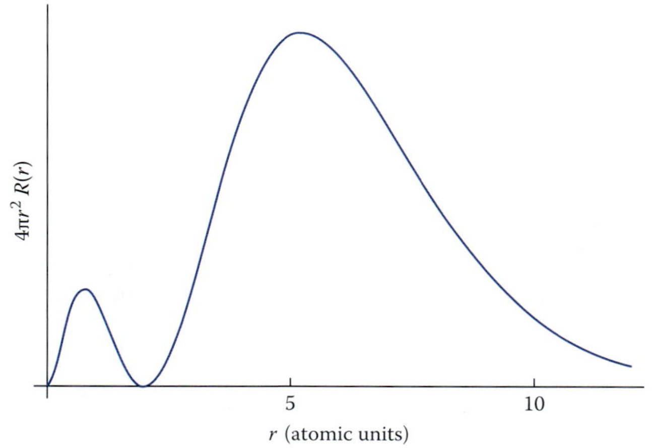

FIGURE 1.8 Probability of the electron position for the hydrogen 2s orbital. (Thanks to Prof. Peter Knowles for this figure.)

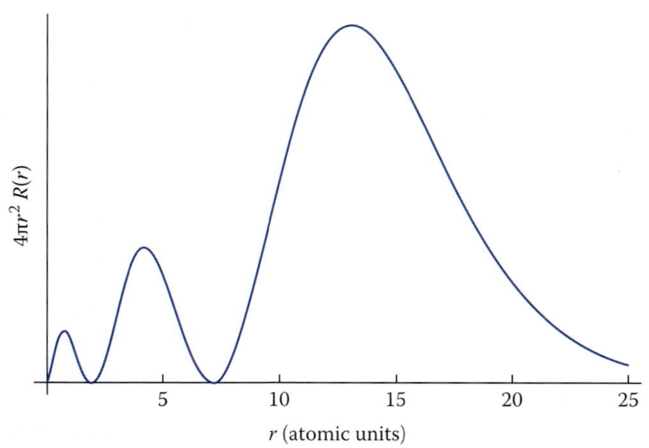

FIGURE 1.9 Probability for the electron position for the hydrogen 3s orbital. (Thanks to Prof. Peter Knowles for this figure.)

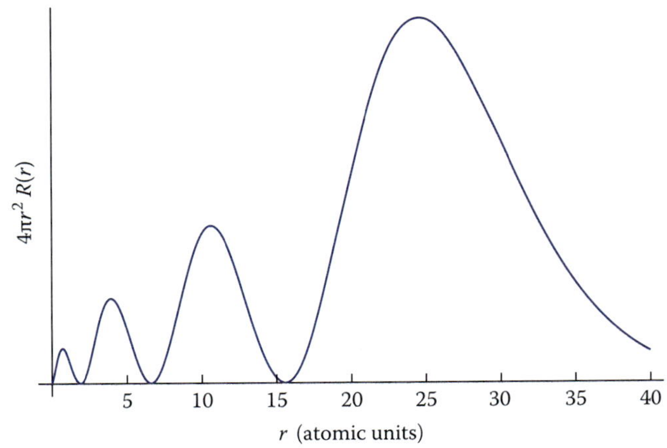

FIGURE 1.10 Probability for the electron position for the hydrogen 4s orbital. (Thanks to Prof. Peter Knowles for this figure.)

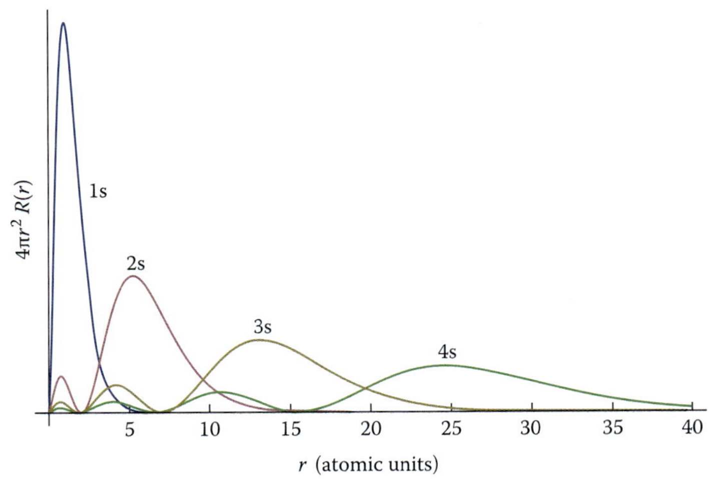

FIGURE 1.11 Probabilities for 1s, 2s, 3s, and 4s orbitals. (Thanks to Prof. Peter Knowles for this figure.)

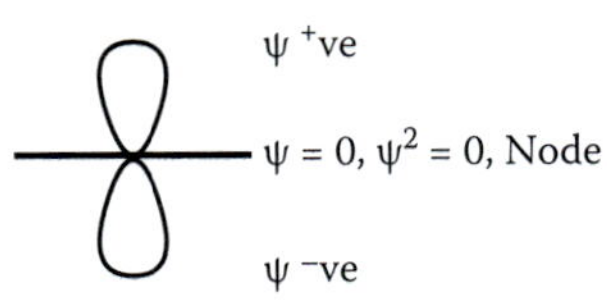

FIGURE 1.12 p Orbital.

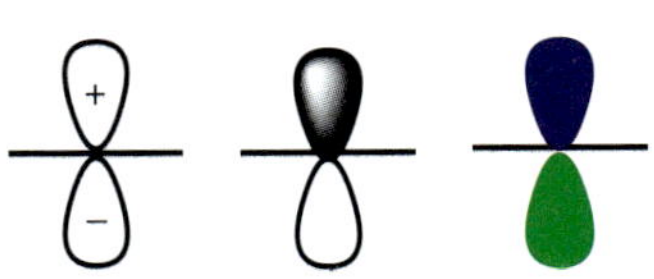

FIGURE 1.13 Common representations of p orbitals.

when we draw this by hand, we tend to draw the lobes of p orbitals as significantly "slimmer" than they actually are (Figures 1.4 and 1.5 show calculated orbitals at "real" sizes), but this is not problematic as we generally only need to see the phase of the orbital. You will see various types of representations of such orbitals in texts (Figure 1.13); I will avoid the one where the phases are marked as + and − on the lobes; it's easy to confuse these with charges, when it actually represents the sign of the wave function.

When the principal quantum number, n, is 3, we have a 3s (l = 0), three 3p (l = 1), and five 3d (l = 2) orbitals. All of the five 3d orbitals have the same energy in the hydrogen atom, despite having rather different shapes (Figure 1.14).

1.2.2 MULTIELECTRON ATOMS

Putting electrons into orbitals in multielectron atoms is governed by three rules, the Aufbau principle, the Pauli exclusion principle, and Hund's rule. The Aufbau, or building-up, principle tells us to put the electrons in the lowest-energy orbital that is available. The Pauli principle restricts the contents of the orbital to two electrons, with spins, s, +1/2 and −1/2. Hund's rule of maximum multiplicity (the law of antisocial electrons...☺) means that where there is more than one orbital of equivalent energy, the electrons distribute between them in order to keep apart.

FIGURE 1.14 Set of five 3d orbitals, from left to right $m = -2, -1, 0, 1, 2$.

TABLE 1.1 Configurations of the Elements of the First Row of the Periodic Table

Element	1s	2s	$2p_x$	$2p_y$	$2p_z$	Configuration	Comments
H	↑					$1s^1$	
He	↑↓					$1s^2$	Filled shell, noble gas
Li	↑↓	↑				$1s^2\ 2s^1$	Outer electron is easily ionized
Be	↑↓	↑↓				$1s^2\ 2s^2$	
B	↑↓	↑↓	↑			$1s^2\ 2s^2\ 2p_x^1$	
C	↑↓	↑↓	↑	↑		$1s^2\ 2s^2\ 2p_x^1\ 2p_y^1$	
N	↑↓	↑↓	↑	↑	↑	$1s^2\ 2s^2\ 2p_x^1\ 2p_y^1\ 2p_z^1$	Hund's rule
O	↑↓	↑↓	↑↓	↑	↑	$1s^2\ 2s^2\ 2p_x^2\ 2p_y^1\ 2p_z^1$	
F	↑↓	↑↓	↑↓	↑↓	↑	$1s^2\ 2s^2\ 2p_x^2\ 2p_y^2\ 2p_z^1$	
Ne	↑↓	↑↓	↑↓	↑↓	↑↓	$1s^2\ 2s^2\ 2p_x^2\ 2p_y^2\ 2p_z^2$	Filled shell, noble gas

Imagine this as seats on a bus—when the English board the bus, each person sits in a new seat, rather than next to someone who is already there. Only when every seat is half full are book bags put on the floor, and the other half of the seat is occupied. Table 1.1 shows the electron configurations of the first row of the periodic table. The electrons are represented by arrows, spin +1/2 up, and spin −1/2 down.

PROBLEM 1.2

Write the electronic configuration for the ground state of each of the following (show the contents of p_x, p_y, and p_z orbitals where appropriate):

(a) C (b) O^- (c) Cl^- (d) Mg^{2+} (e) P

PROBLEM 1.3

Which element corresponds with each of the following electronic configurations?

(a) $1s^22s^22p_x^12p_y^12p_z^1$
(b) $1s^22s^22p_x^22p_y^12p_z^1$
(c) $1s^22s^22p^63s^2$

Solutions

1.2 (a) C $1s^22s^22p_x^12p_y^1$
(b) O^- $1s^22s^22p_x^22p_y^12p_z^1$
(c) Cl^- $1s^22s^22p^63s^23p_x^23p_y^23p_z^2$
(d) Mg^{2+} $1s^22s^22p^6$
(e) P $1s^22s^22p^63s^23p_x^13p_y^13p_z^1$

1.3 (a) $1s^22s^22p_x^12p_y^12p_z^1$ N
(b) $1s^22s^22p_x^12p_y^12p_z^1$ O
(c) $1s^22s^22p^63s^2$ Mg

1.3 BONDING

To a first approximation, we can say that only the valence shell, the outermost shell, electrons are used in the formation of chemical bonds. The simplest bonds to understand are those between ionized species, for example, in lithium fluoride. We know that atoms or ions with filled shells of electrons are very stable; group 18 elements, the noble gases, with such filled shells, were long thought to be completely chemically inert, because they were so reluctant to form bonds. In lithium fluoride, lithium has lost one electron to give Li^+ with the helium configuration, and fluorine has gained one to give F^-, so that it has the neon configuration. Then lithium fluoride is formed by electrostatic attraction between the two ions, and it is a high melting-point solid.

However, many atoms, especially carbon, form bonds with like atoms (*e.g.*, O_2, F_2, N_2, $H_3C–CH_3$, and ethane); clearly, these do not involve ions and electron transfer. In these molecules, the atomic orbitals are combined to give molecular orbitals. This is described as linear combination of atomic orbitals (LCAO). At a crude and simplistic level, we could imagine pushing together two 1s orbitals in hydrogen, in phase, to form what we describe as a σ-bond (Figure 1.15).

However, this is not the whole story. Mathematically, the formation of the molecular orbitals involves adding and subtracting the two atomic orbital wave functions, so that *two* molecular orbitals are produced. The new wave functions, ψ_A and ψ_B, are the molecular orbitals. The number of molecular orbitals produced must equal the number of atomic orbitals from which they were produced. The orbitals (real calculated ones) are shown in Figure 1.16, where energy increases up the page. When we put electrons into molecular orbitals, the same considerations we used with atomic orbitals, Aufbau, Pauli, and Hund, apply.

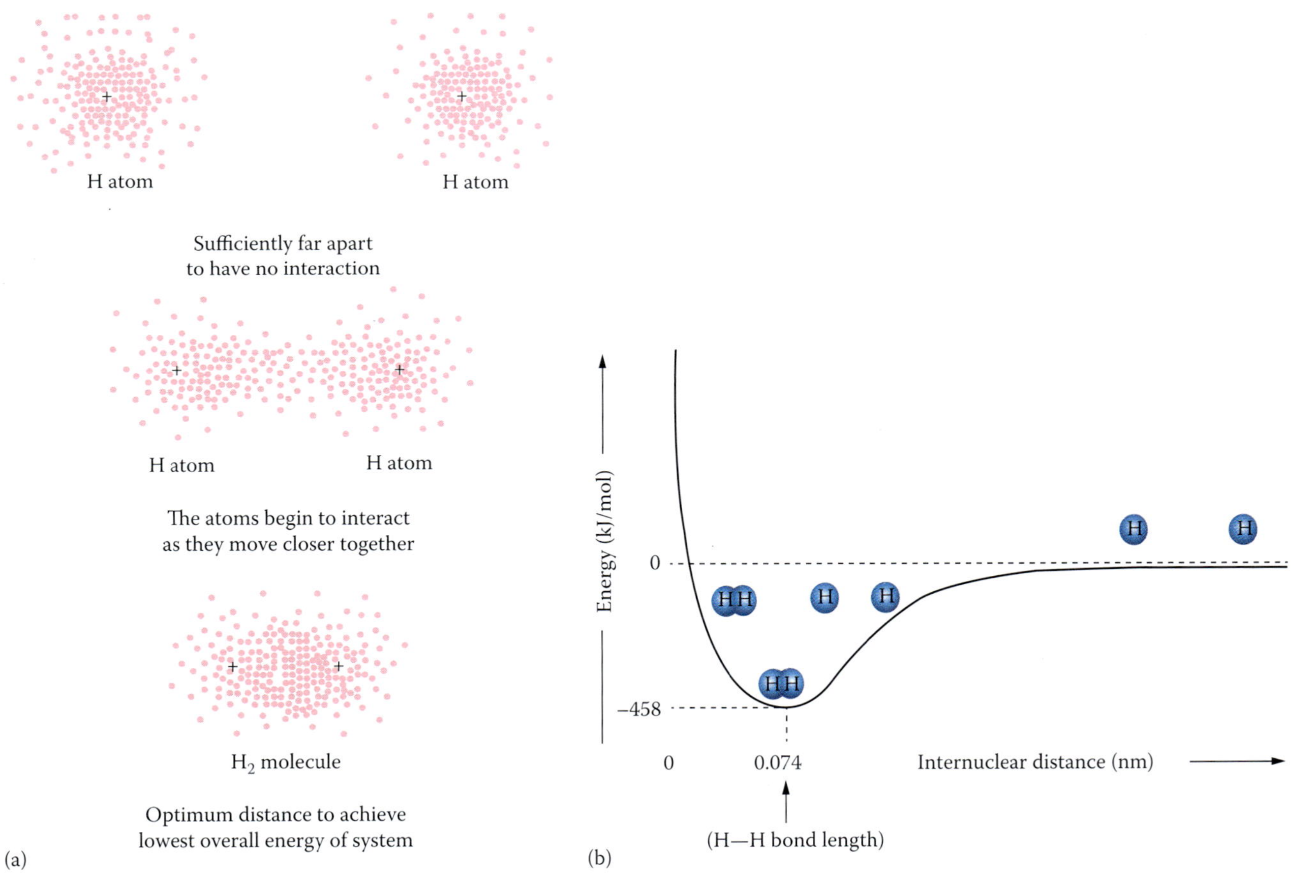

FIGURE 1.15 Formation of a hydrogen molecule. (a) Hydrogen is orbitals combine to form the bond in H_2. (b) Plot of energy against internuclear distance for the formation of the H_2 molecule.

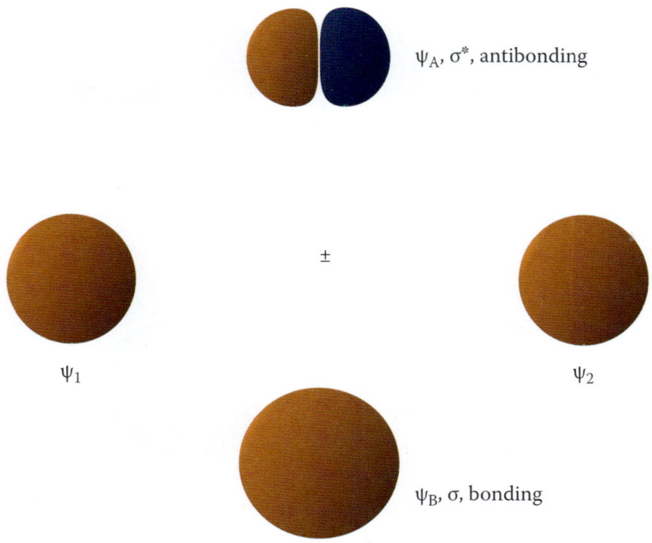

FIGURE 1.16 Bonding and antibonding orbitals of the hydrogen molecule.

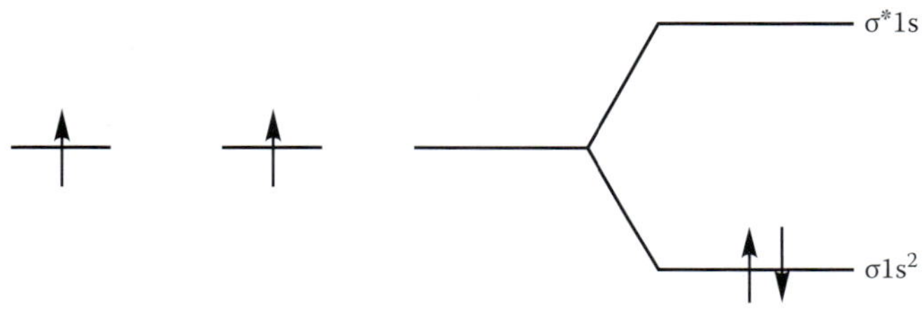

FIGURE 1.17 Molecular orbitals of H_2.

The bond between the hydrogen atoms in H_2 is called a σ-bond; σ refers to its cylindrical symmetry. Both of the electrons, one from each hydrogen atom, go into the lower-energy, bonding, orbital (Figure 1.17). The electronic configuration of the hydrogen molecule is written as $\sigma_{1s}{}^2$; there are two electrons in the bonding orbital, and none in the antibonding orbital, which is designated as σ*. If we were to consider making He_2, a diatomic helium molecule, we would need to accommodate four electrons, and the molecule would be described as $\sigma_{1s}{}^2\,\sigma^{*}_{1s}{}^2$. Thus, we have one bond and one "antibond" that effectively cancel out. There is no energetic advantage for helium to form molecules, and it does not do so; helium is a monatomic gas.

If we want to make a carbon–hydrogen bond, we can consider bringing together a 2p orbital of carbon and the 1s orbital of hydrogen to form a σ-bond (Figure 1.18, which shows only the bonding molecular orbital). We could consider making a carbon–carbon bond by bringing together 2p orbitals nose to nose, as in Figure 1.19 (again, only the bonding orbital is shown, and we have to use the *same* 2p orbital on each carbon atom, conventionally for organic chemists, the p_x). Notice the cylindrical symmetry of both of these σ-bonds.

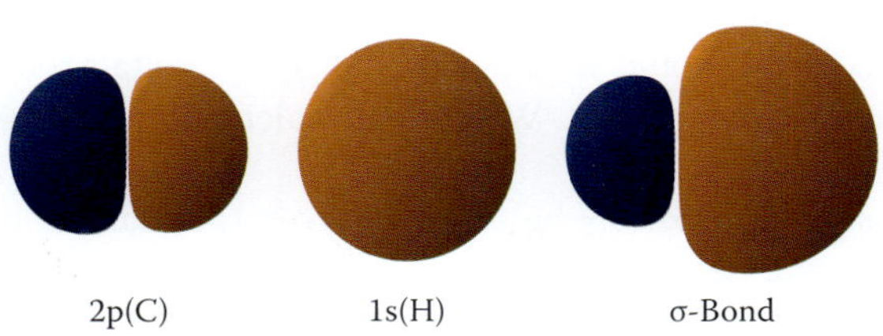

FIGURE 1.18 How can we form a C–H bond?

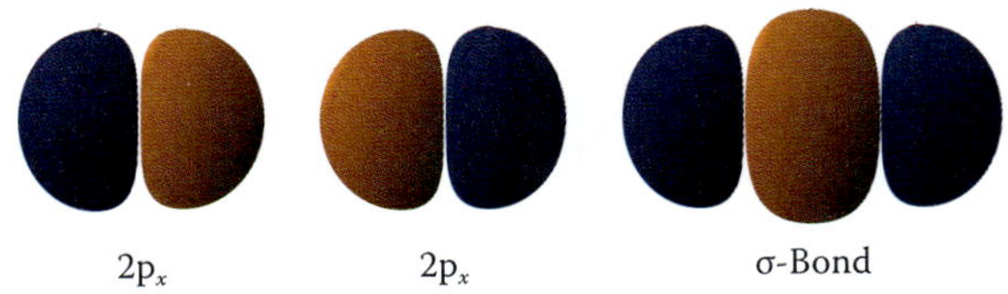

FIGURE 1.19 How can we form a C–C bond?

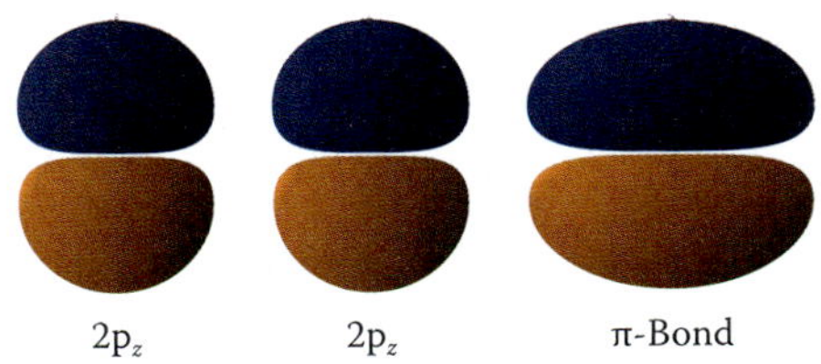

FIGURE 1.20 Formation of a π-bond.

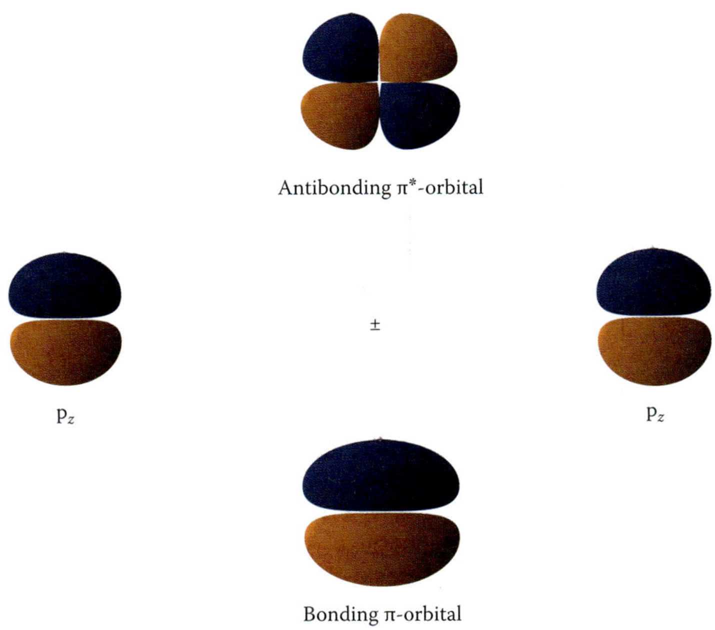

FIGURE 1.21 Formation of π- and π*-orbitals from $2p_z$ orbitals.

If we bring together p orbitals side by side, we get a new type of bond, a π-bond. The orbital overlap is less good. π-Molecular orbitals, π-bonds, do not have cylindrical symmetry (Figure 1.20; conventionally, organic chemists prefer to use p_z to make π-bonds; the plane of a diagram is normally the *xz* plane); if you were to take a stroll around the outside of the bond at any point along the internuclear axis, you would see a phase change. π-Orbitals retain the node that was in the p orbitals from which they were built. We can also calculate the appearance of the π*-orbital (Figure 1.21).

However, this analysis can't possibly be correct. Given that carbon only has two half-filled p orbitals, it could only make two bonds to hydrogen, and they would be at right angles to each other. But we know from the experiment that carbon is 4 valent, and its simplest compound, methane, CH_4, has a perfectly tetrahedral structure. We'll solve this conundrum in Chapter 2.

REVIEW PROBLEMS

1. Caproic acid, responsible for the distinctive smell of old socks, contains C, H, and O. On combustion in an excess of oxygen, 0.45 g caproic acid gives 0.418 g water and 1.023 g carbon dioxide. What is the empirical formula of caproic acid? The relative molecular mass is 116.2; what is the molecular formula?
2. One form of vitamin A, retinol, has the following percentage composition. Derive its empirical formula.
   83.9 % C
   10.5 % H
   5.6 % O
3. The alkaloid morphine has been shown to contain only C, H, N, and O. Morphine contains, according to elemental analysis, 71.6 % C, 6.7 % H, and 4.9 % N by weight. What is the empirical formula of morphine?
4. Write the electronic configuration for the ground state of each of the following species (show the contents of p_x, p_y, and p_z orbitals where appropriate):
   (a) Li^+
   (b) N
   (c) S
   (d) Ga
   (e) F^-
   (f) Cl
   (g) K
   (h) Br
5. Which element corresponds with each of the following electronic configurations?
   (a) $1s^2 2s^2$
   (b) $1s^2 2s^2 2p^6$
6. The density of oak wood is 0.72 g cm^{-3}. Assume oak wood has the formula $C_6H_{12}O_6$ and calculate the mass of water produced when a dry log 12 cm × 14 cm × 25 cm is burned.
7. What are the principal and orbital angular momentum quantum numbers for each of the following orbitals?
   (a) 2s
   (b) 6f
8. Why does N have a less favorable electron affinity than its neighbors on either side in the periodic table, carbon and oxygen?
9. Cytochrome c is an iron-containing enzyme found in the cells of all aerobic organisms. If cytochrome c is 0.43 % Fe by weight, what is the minimum molecular weight of cytochrome c?
10. Calculate the % composition of Tylenol, $C_8H_9NO_2$.
11. Explain clearly, in terms of molecular orbitals, why the molecule He_2 does not exist but $[He_2]^+$ does.

Alkanes and Friends

Structure, Bonding, Properties, and Nomenclature

2

2.1 INTRODUCTION

At the end of the last chapter, we had learned how to combine atomic orbitals into covalent bonds for simple systems—two s orbitals in molecular hydrogen and an s and a p orbital to make a C–H bond, and p orbitals to make either σ- or π-bonds. Unfortunately, these unsophisticated pictures do not allow us to understand even the simplest organic molecule, methane. Carbon is four-valent, and the methane molecule has been long known, and convincingly shown, to be a perfect tetrahedron, but none of the orbitals we have seen so far are directed toward the corners of a tetrahedron. In order to make the right orbitals, we will need to mix, or "hybridize," s and p orbitals. Organic chemists need a model that reliably produces and predicts what we need to understand about organic molecules and their reactions. Physical chemistry involves more serious mathematics and needs other parameters; it may be more accurately descriptive but it is less usable for the complex molecules involved in organic chemistry. It's probably best to think of hybridization not so much as something that *happens* (as though the molecule has a little think, and then decides which orbitals to mix together) ... ☺ but as something that *has happened* in a ground-state molecule.

2.2 HYBRIDIZATION IN METHANE, AMMONIA, AND WATER

What orbitals are available to make the four identical C–H bonds in methane? The four hydrogen atoms provide four 1s orbitals. The configuration of elemental carbon can be written as

$$\mathrm{C} \quad 1s^2\, 2s^2\, 2p_x^{\ 1}\, 2p_y^{\ 1}\, 2p_z^{\ 0}$$

First, we promote one electron from the 2s to the $2p_z$ orbital to give

$$\mathrm{C} \quad 1s^2\, 2s^1\, 2p_x^{\ 1}\, 2p_y^{\ 1}\, 2p_z^{\ 1}$$

This gives us four unpaired electrons that can be paired up with the single electrons in the hydrogen 1s orbitals—but the orbitals are not identical, nor are they pointing in the right directions. So we will mix, or hybridize, the 2s, $2p_x$, $2p_y$, and $2p_z$ orbitals to give four *identical* hybrid orbitals, called sp^3 orbitals. These point toward the corners of a regular tetrahedron—why do we get a tetrahedron? To say that it comes from complex mathematics that we don't need to understand is, of course, a cop-out (though true). This is consistent with VSEPR*—a tetrahedral arrangement of electron pairs minimizes repulsion. Those of you who are mathematically inclined can demonstrate using vector algebra that a tetrahedron is defined by four corners of a cube (the proof is not trivial†) and can thus calculate the tetrahedral angle as 109.5 ° from first principles.

Although we don't always draw sp^3 orbitals in the form that matches physical reality, organic chemistry is very visual and it's useful to be aware of what they actually look like (Figure 2.1). Each of these sp^3 orbitals combines with a 1s orbital to give a σ-bond (Figure 2.2). This is the conventional drawing, but we should not forget that each of the sp^3 orbitals has a "back" lobe that we are not showing and that every time we combine atomic orbitals, we generate not only a bonding but also an antibonding orbital.

* Valence shell electron pair repulsion.

† G.H. Duffey, *J. Chem. Educ.*, **67** (1990) 35.

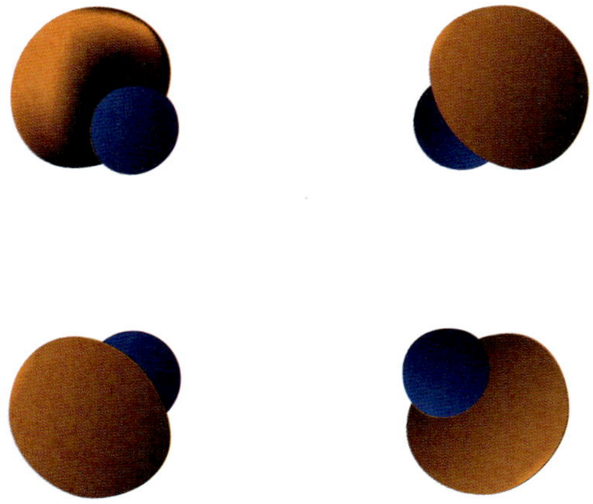

FIGURE 2.1 sp^3 orbitals.

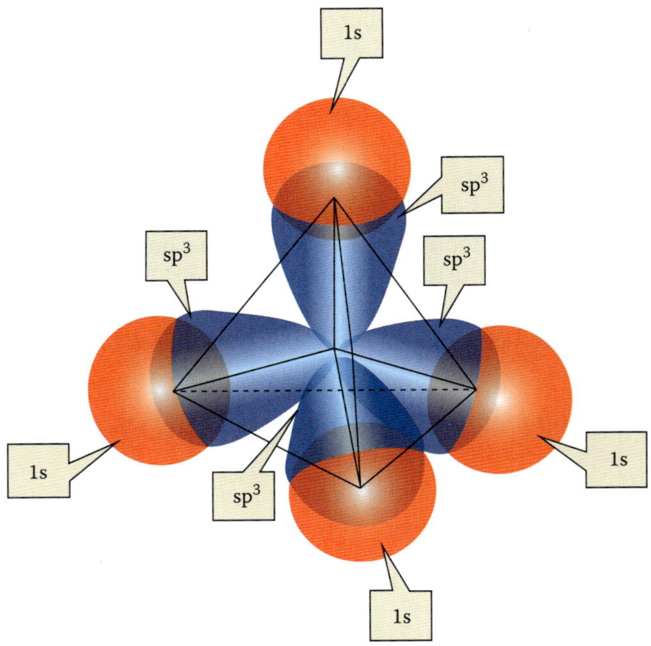

FIGURE 2.2 sp^3 hybridization and bonding in methane.

Ammonia and water also have structures based on sp^3 hybridization at the central atom. We know that ammonia is a base and that it has a nonbonding or lone pair of electrons. The electronic configuration of nitrogen is

$$\text{N} \qquad 1s^2\ 2s^2\ 2p_x^{\ 1}\ 2p_y^{\ 1}\ 2p_z^{\ 1}$$

One might initially think that there are three unpaired electrons, and these could make the three bonds by pairing up with the 1s electrons of hydrogen. But these orbitals are pointing in the wrong directions—electron diffraction measurements in the gas phase tell us that ammonia has a pyramidal shape. Electron counting, and VSEPR, tells us that there are four electron pairs around nitrogen in ammonia, so the shape should be approximately tetrahedral. Thus, we also need sp^3 hybridization for ammonia. If we consider filling the four sp^3 orbitals (Figure 2.3), then there are three unpaired electrons and one lone or nonbonding pair. The unpaired electrons pair up with 1s electrons of hydrogen to make σ-bonds, and the lone pair of electrons is in the sp^3 orbital. VSEPR tells us that lone pairs have a greater repulsive effect than bonding pairs, so the HNH angles are "squeezed down" from 109.5 ° to 107 ° (**2.1**).

Moving on to water, oxygen has the following electronic configuration:

$$\text{O} \qquad 1s^2\ 2s^2\ 2p_x^{\ 2}\ 2p_y^{\ 1}\ 2p_z^{\ 1}$$

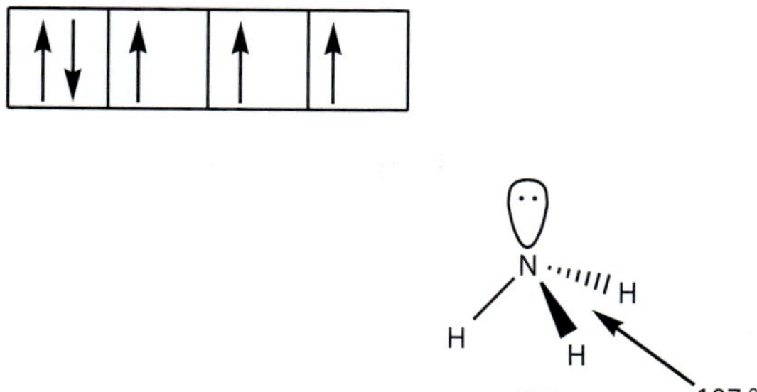

2.1 107 °

FIGURE 2.3 Paired and unpaired electrons in the sp³ orbitals of nitrogen in ammonia.

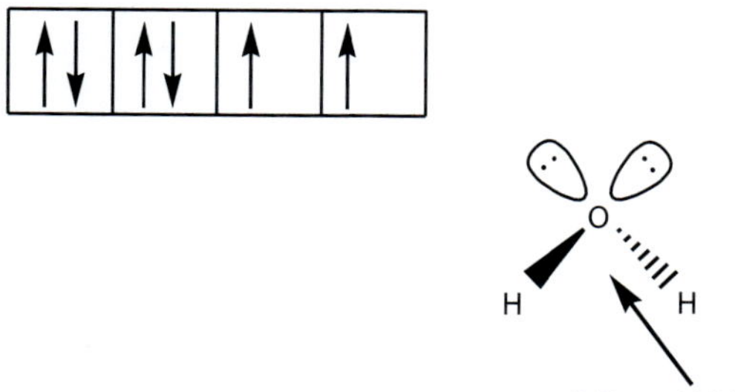

2.2 105 °

FIGURE 2.4 Paired and unpaired electrons in the sp³ orbitals of oxygen in water.

There are two unpaired electrons to make the two bonds to hydrogen, but the orbitals they occupy point in the wrong direction for the known structure of water. Again, we need to make four $2p^3$ hybrid orbitals, of which two will be filled (the lone pairs) and two half filled (to make the bonds with the 1s of hydrogen) (Figure 2.4). VSEPR tells us that the lone pairs repel more than the bonding pairs, so in the structure of water, the HOH angle is 105 ° (**2.2**).

You may be wondering about the way in which the bonds in ammonia and water have been drawn. The use of what are called wedge and hashed bonds indicates stereochemistry, the 3-dimensional shape of the molecule. The solid wedge indicates that the bond is coming out of the plane of the paper, toward the viewer, and the hashed wedge that the bond is behind the plane of the diagram, away from the viewer. We will be using this notation throughout the text—it is a way for us to show the 3-dimensional shape of a molecule projected onto a 2-dimensional page.

The structures of saturated (no multiple bonds) species related to methane, ammonia, and water can usually be deduced from these three basic structures. So CH_3Cl, chloromethane, is like methane and approximately tetrahedral (**2.3**). Note that this will not be a perfect tetrahedron like methane—chlorine is much larger than hydrogen, and much more electronegative, so a C–H bond is not identical to a C–Cl bond. Look carefully at how this is drawn—everyone develops their own style, but for a tetrahedron, two bonds should be in the plane of the page, one in front and one behind. In methanol, CH_3OH, the carbon of the CH_3 is like methane, and the OH is like water. So we can see methanol either as a substituted methane (**2.4a**) or as a substituted water molecule (**2.4b**). Similarly, we can view methylamine, CH_3NH_2, either as a substituted methane (**2.5a**) or as a substituted ammonia (**2.5b**).

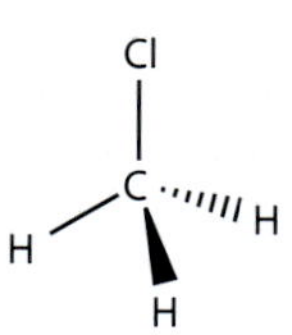

2.3

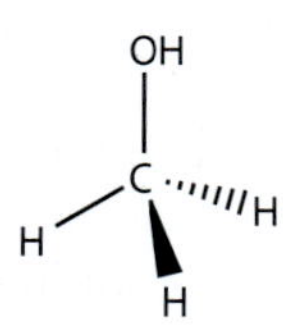

2.4a

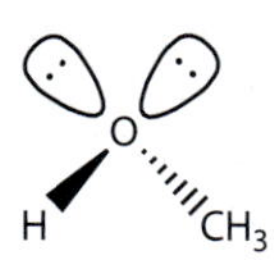

2.4b

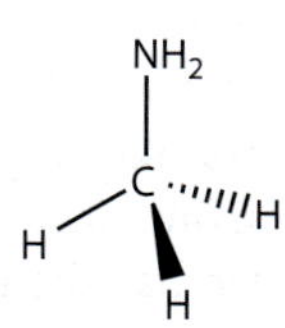

2.5a

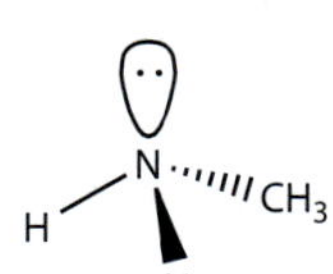

2.5b

PROBLEM 2.1

Each of the following molecules contains one or more sp[3] hybridized atom. Sketch the structures you predict, focusing in turn on each of the C, N, and O atoms:

CF_3Cl CH_3OCH_3 $(CH_3)_2NH$

Solution

CF_3Cl will have threefold symmetry—all the fluorines are identical, but the bond to chlorine will be different—chlorine is larger but less electronegative than fluorine. Structural studies give ClCF = 110 ° and FCF = 108.6 °.

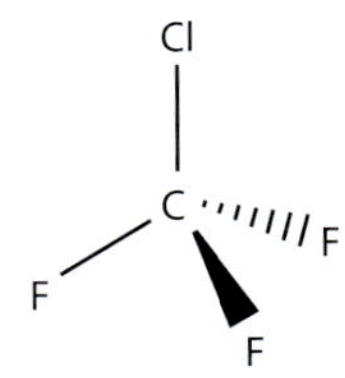

CH_3OCH_3 can be viewed either as a substituted methane or as a disubstituted water. As with water, the COC angle will be about 105 °.

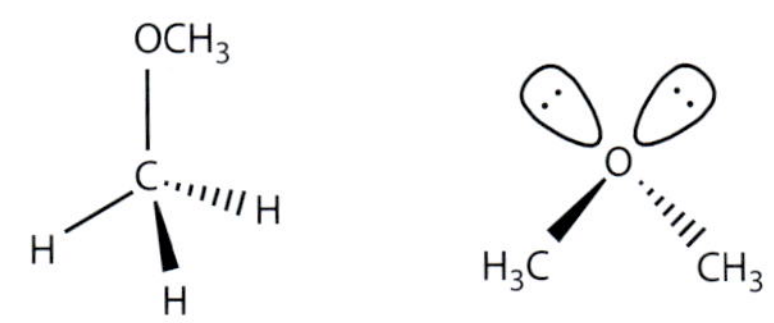

In Me_2NH, two of the hydrogens of ammonia have been replaced by methyl groups—again, we can look at the structure as either a substituted methane or a substituted ammonia. The bond angles at nitrogen, according to VSEPR, will be about 107 °.

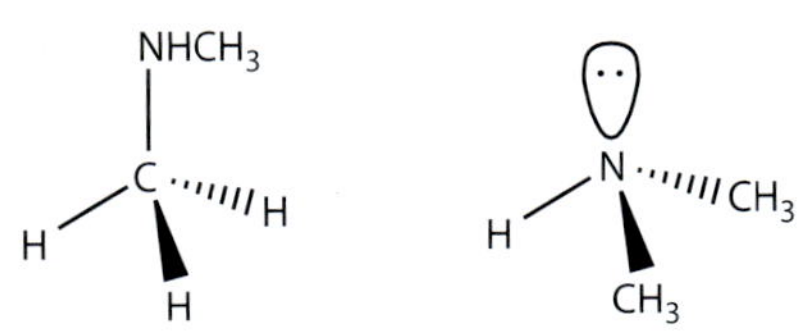

2.3 DRAWING MOLECULES

Organic molecules are three-dimensional, so organic chemists need to have methods of drawing in two dimensions that reflect the original structures. For example, it is possible to focus on one specific bond. If we consider ethane, C_2H_6, and look at it from the side, we see the view shown in **2.6**. The carbon–carbon bond and one carbon–hydrogen bond at each atom are in the plane of the paper, with one hydrogen in front and one behind the plane at each carbon. This is called a wedge form. If we now move around the molecule about 45 °, then an oblique view, called a saw-horse form, is seen. There are two "extreme" forms of this type of diagram, shown as **2.7a** and **2.7b**. In **2.7a**, all the C–H bonds are eclipsed or parallel—hence, we call this the eclipsed form. In **2.7b**, the C–H bonds on the adjacent carbons are as far apart as we can make them—this is called the staggered form. Free rotation is possible about the carbon–carbon single bond—every intermediate form is also accessible. Intuitively, you might expect the staggered form to be the more stable one—and indeed it is, something we will explore in detail in Chapter 7. Notice that in these drawings we show the two carbon atoms simply as vertices. As we progress to larger molecules, it's important to keep diagrams uncluttered, and many of the types of representations we use do not show every atom. Finally, we move further around the molecule so that we are looking along the carbon–carbon bond—**2.8a** and **2.8b** show, respectively, the eclipsed and staggered forms of ethane. Here, again, we don't show the carbon atom specifically—the two aligned carbon atoms are represented by the circle. These are called Newman projections.

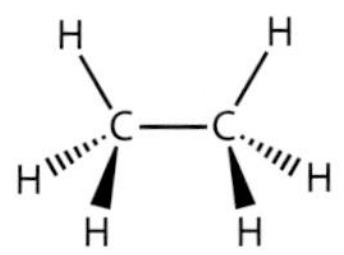

2.6

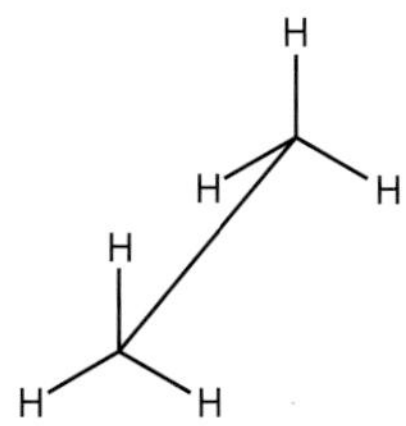

2.7a

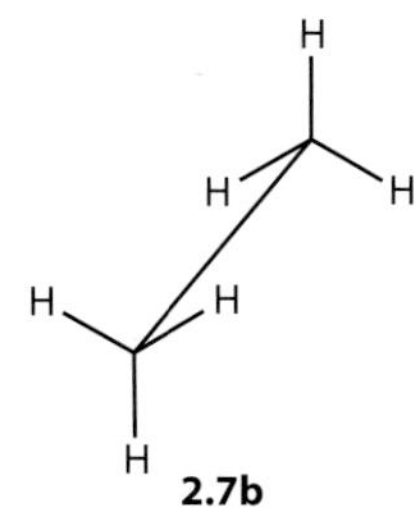

2.7b

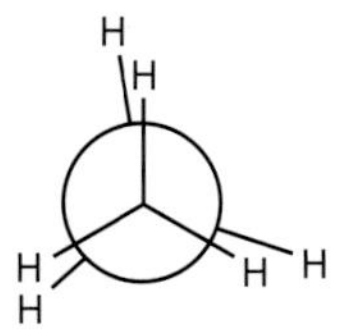

2.8a

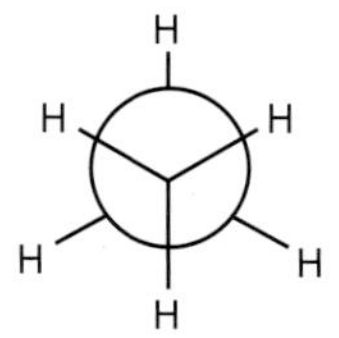

2.8b

Sometimes, a line drawing of a molecule, without showing any three-dimensional geometry, is used, as in **2.9**. It's not ideal as it implies that the bonds around carbon point toward the corner of a square rather than a tetrahedron, but it is a layout you will encounter in your reading. The stereochemistry of the molecule may not be relevant to the reaction being studied, or we simply wish a shorthand representation for a part of the molecule that remains unchanged.

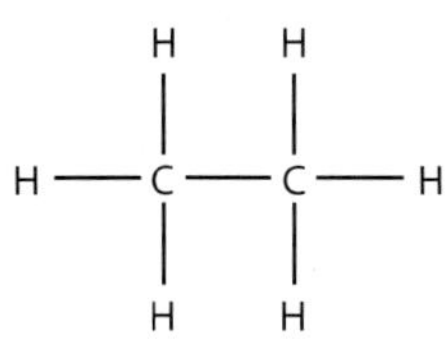

2.9

The representations generally used by organic chemists are even simpler, however—they can be described as stick drawings. These show *only* carbon–carbon bonds, with no hydrogen or carbon atoms (other atoms, oxygen, nitrogen, *etc.*, are shown). Although most students initially find this idea alarming, it is the only practicable way to represent large molecules. So **2.10a** is the shorthand way of writing **2.10b**. Although this is difficult to begin with, a little practice and you will soon find it's both quicker and easier than writing out every atom. Don't forget that carbon will always be four-valent and any bonds not shown at a particular vertex are bonds to hydrogen.

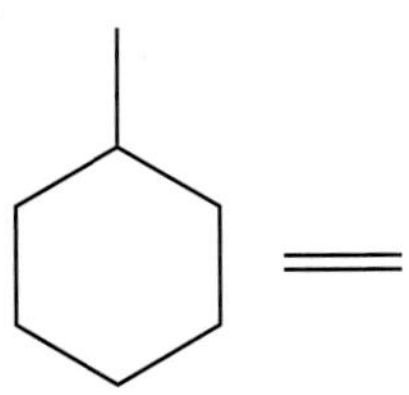

2.10a

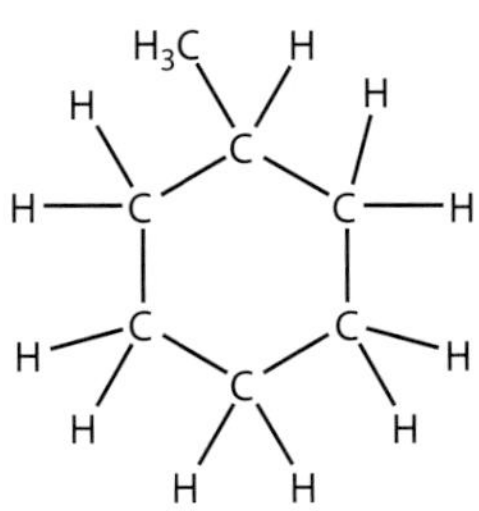

2.10b

Key Points from Sections 2.1 through 2.3

- The structures of methane, ammonia, and water are all based on a tetrahedral arrangement of electron pairs around the central atom, with the central atoms being sp^3 hybridized.
- Methane has four bonding pairs of electrons, ammonia three bonding pairs and one lone pair, and water two bonding pairs and two lone pairs. Exact shapes may be predicted by VSEPR.
- The structures of many molecules can be derived as substituted versions of methane, ammonia, and water.
- There are various two-dimensional representations of three-dimensional molecules—wedge, sawhorse, and Newman representations.
- Complex organic molecules are represented as stick drawings, with hydrogen atoms omitted, and carbon atoms represented by vertices.

2.4 INTRODUCTION TO ALKANES

Organic chemists tend to have a limited interest in alkanes, despite their being some of the most abundant organic molecules on the planet. Abundant yes, reactive no—their most important reaction is combustion. Much of the energy used in the developed world derives from the burning of hydrocarbons, in one form or another. Alkanes are the simplest of hydrocarbons, with a general formula C_nH_{2n+2} and no multiple bonds. They are nonpolar and largely insoluble in water. The lower molecular weight alkanes are gaseous, the next group are nonpolar liquids, and the highest members of the family are oils or waxes. Boiling point and melting point increase with molecular weight. The first members of the family are given in Table 2.1.

How are these and other compounds to be named? With some 60 million compounds recorded into the Chemical Abstracts Services (CAS) database by June 2011, there has to be a system. The CAS registry has been in existence for 40 years, and it took 33 years to register the first 10 million compounds. The last 10 million were registered between September 2009 and June 2011. In the early years of modern organic chemistry, names of new compounds tended to reflect the origins of the materials—for example, lactic acid (**2.11**) is isolated from milk (Latin, *lactis*, of milk). Others were named on the basis of a whim of the discoverer—barbituric acid (the parent compound for many sleeping pills and a class of antiepileptic drugs) (**2.12**) was supposedly discovered on the feast of St. Barbara. Yet other molecules are named for their shape, such as cubane (**2.13**) or buckminsterfullerene* (**2.14**) named for its

TABLE 2.1 Low-Molecular-Weight Alkanes

Name	Formula	BPt (°C)	Comments
Methane	CH_4	−164	Main constituent of natural gas, marsh gas, firedamp. An important greenhouse gas
Ethane	C_2H_6	−89	1 %–6 % of natural gas, from which it is isolated. Thought to exist as lakes and precipitation on Titan (one of the moons of Saturn)
Propane	C_3H_8	−42	Main constituent of LPG, widely used in fuel for heating, portable stoves, barbecues
Butane	C_4H_{10}	−0.5	Fuel for outdoor cooking, cigarette lighters. Propellant in aerosols
Pentane	C_5H_{12}	36	Blowing agent for polystyrene foam

LPG, liquefied petroleum gas.

* Image courtesy of Prof. Ken Seddon, Queen's University, Belfast.

resemblance to the geodesic domes designed by the architect Richard Buckminster Fuller. Blue dyes recently isolated in Verona were named Juliet Blue and Romeo Blue.

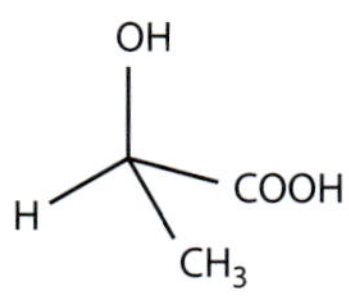

2.11

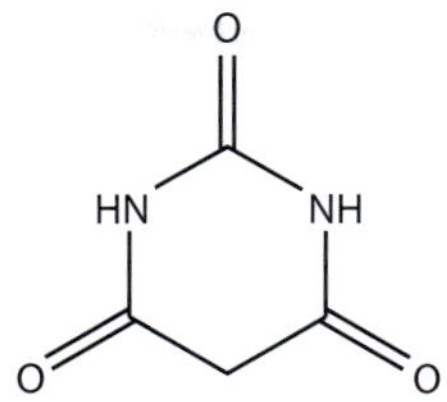

2.12

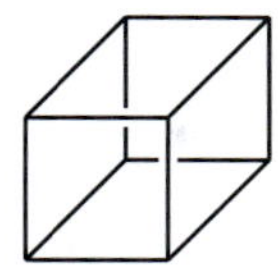

2.13

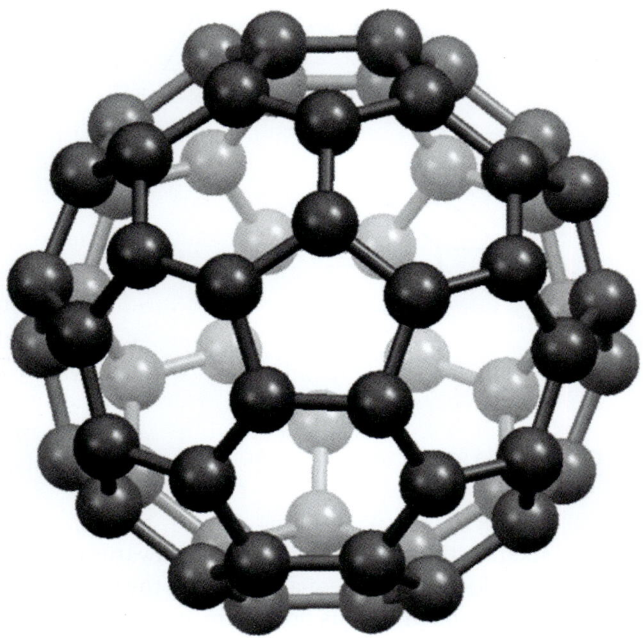

2.14

You might believe that organic chemists don't have a sense of humor, but there are quite a few jokes in nomenclature (Figure 2.5). With the exception of fenestrane (Latin, *fenestra*, window), these molecules or their derivatives have all been synthesized. Fenestrane can be made but only when we have at least one 5-membered ring in the structure. Compound **2.15** is called a ladderane, and surprisingly, it's a natural product, found in anaerobic ammonia-oxidizing bacteria and synthesized by Corey.

2.4.1 FORMAL NOMENCLATURE OF ALKANES

Formal nomenclature is not an exciting subject—no one could pretend otherwise, even when it's enlivened by stories of the origins of the names. However, it is important and something that you will need to keep working on throughout the course. It's analogous to learning the regular and irregular verbs of a language; you may learn much other vocabulary, but if you can't conjugate "to be," "to have," and "to go," which are irregular in most languages, you will always make mistakes. The rules for formal nomenclature are defined (in all languages, not just English) by the International Union of Pure and Applied Chemistry (IUPAC). However, many common names are so well entrenched that they are still widely used, so you will need to recognize these, and they will be discussed as we meet the various organic functional groups. A complete list of common names that you should know (at the very least because this is how the bottles in your lab course will be labeled) is in Appendix B.

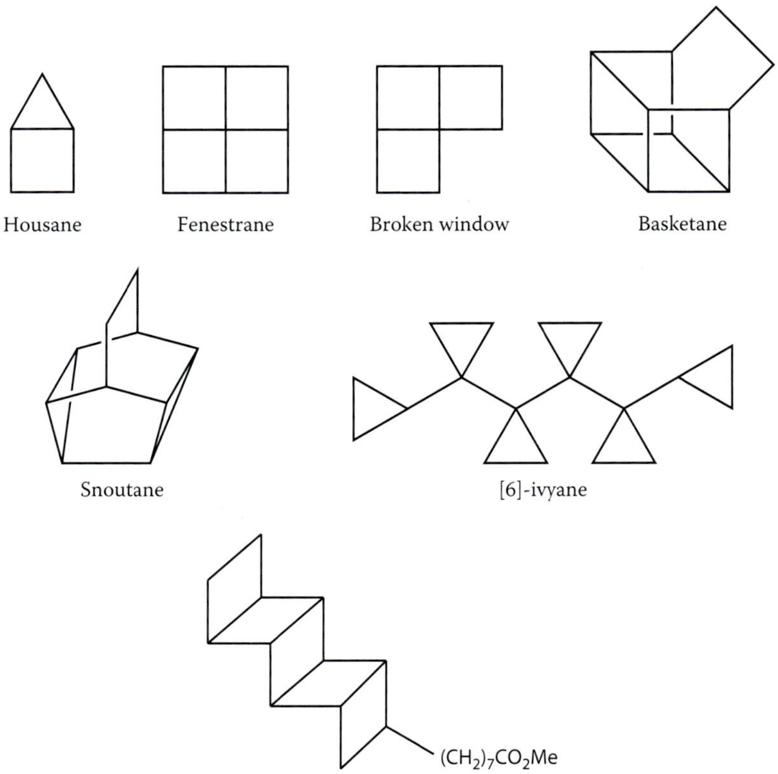

2.15

FIGURE 2.5 Organic chemists *do* have a sense of humor.

TABLE 2.2 Nomenclature of Alkanes

Formula of Linear Alkane	Name
CH_4	Methane
C_2H_6	Ethane
C_3H_8	Propane
C_4H_{10}	Butane
C_5H_{12}	Pentane
C_6H_{14}	Hexane
C_7H_{16}	Heptane
C_8H_{18}	Octane
C_9H_{20}	Nonane
$C_{10}H_{22}$	Decane
$C_{11}H_{24}$	Undecane
$C_{12}H_{26}$	Dodecane

Returning to the simple alkanes, the names from pentane onwards derive from the number of carbon atoms. For now, we'll just look at the linear alkanes, and the names are shown in Table 2.2. Each name is composed of a *root*, which tells us how many carbon atoms we have, plus a *suffix*, which tells us what type of molecule it is. For alkanes, our suffix is "-ane." Most of the prefixes are Greek in origin, and many will be familiar from other words that you know. You would not easily deduce the names of very long alkanes, but most will be understandable, once seen. For example, $C_{74}H_{150}$ is tetraheptacontane; tetra- gives us the 4 and hepta- the 7, and contane is used for the decades between 30 and 100.

2.4.2 STRUCTURAL ISOMERS AND FORMAL NOMENCLATURE

With methane, ethane, and propane, there is only one way the molecule can be put together that conforms to the requirements of valency. However, when we reach butane, there are two possibilities. The carbon atoms can be arranged in a linear four-carbon chain (**2.16**) or a branched chain (**2.17**). These were initially called *n*-butane, normal butane, and iso-butane. We describe these as structural isomers—the carbon atoms are joined together in a different pattern. Notice that I am using the stick drawings; if you draw out the same structures with all the atoms there, you will quickly see why these are better!

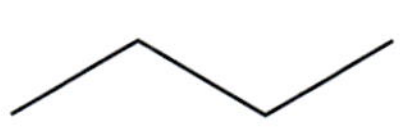

2.16, *n*-butane

2.17, *iso*-butane

With pentane, three structures are possible (**2.18**, **2.19**, **2.20**), *n*-pentane (*normal* pentane), *iso*-pentane, and *neo*-pentane, the new pentane. However, moving on to higher alkanes, clearly, a different system is needed—new prefixes cannot be endlessly invented. If we want to name **2.21**, we need to use the IUPAC system.

2.18

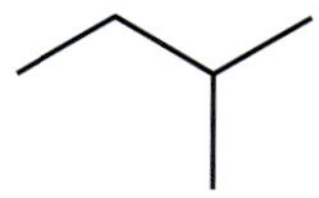

2.19

2.20

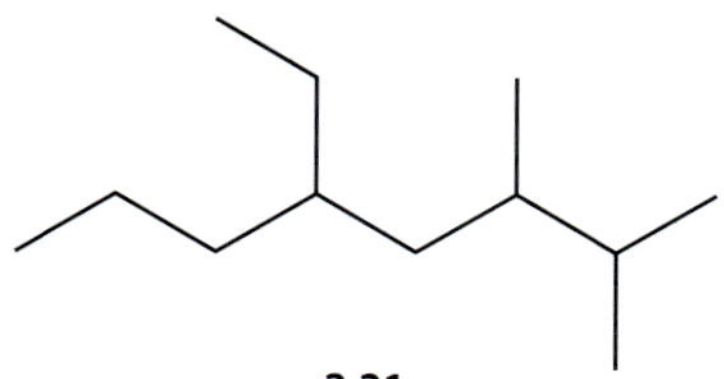

2.21

The first rule is to locate the longest continuous chain of carbon atoms. Don't be fooled by the drawing—although the longest chain will often be near to horizontal across the page, testers do like to check that you are paying attention, by drawing the molecule in an unconventional way. In this case, the longest chain is eight carbons long. So we will name this molecule as an octane. The numbering should start from the end that gives the substituents on the lowest-numbered carbon atom. So **2.21a** is correct, and **2.21b** is wrong. In **2.21a**, the substituents are on carbons 2, 3, and 5, whereas in **2.21b**, they are attached to carbons 4, 6, and 7. Finally, we name the substituents and say where they are. The suffix for an alkyl substituent is "-yl" so $-CH_3$ is methyl, $-CH_2CH_3$ is ethyl, *etc.* (Table 2.3). There are two substituents in the table that don't quite fit with the "purest" version of IUPAC nomenclature, *iso*-propyl and *tert*-butyl. However, these terms are very common and widely used, so it's important to recognize them. Finally, we order the substituents alphabetically and indicate by using di-, tri-, *etc.*, if there is more than one of a particular type. The alphabetic ordering is not the most rigid rule, because it may well come out differently in different languages, and provided that the right number is attached to each substituent, there is no ambiguity, even if they are not correctly alphabetized. So **2.21** is 2,3-dimethyl-5-ethyloctane.

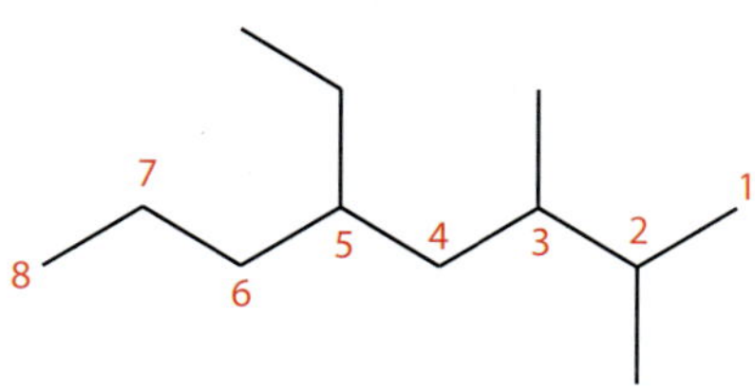

2.21a ✓

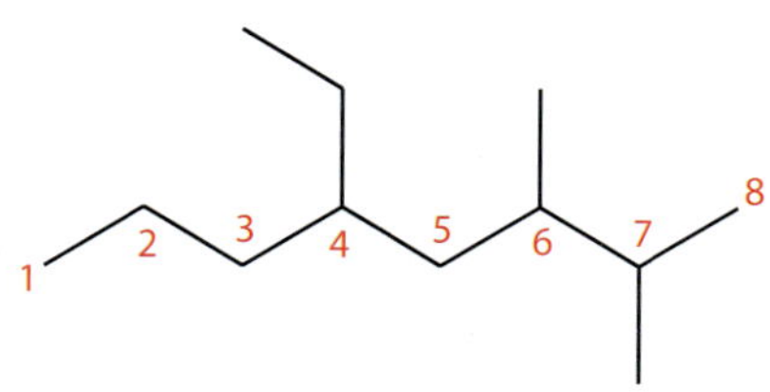

2.21b ×

TABLE 2.3 Nomenclature of Substituents

Formula	Name	Abbreviation
$-CH_3$	Methyl	Me
$-CH_2CH_3$	Ethyl	Et
$-CH_2CH_2CH_3$	*n*-Propyl	*n*-Pr
$-CH(CH_3)_2$	*iso*-Propyl	*i*-Pr
$-CH_2CH_2CH_2CH_3$	*n*-Butyl	*n*-Bu
$-CH_2CH(CH_3)_2$	*iso*-Butyl	*i*-Bu
$-C(CH_3)_3$	*tert*-Butyl	*t*-Bu

How do we deal with a substituent that is itself branched? **2.22** provides an example. We first note that the chain has 12 carbons, so it will be named as a dodecane, and we number from the right, so that the substituent is on C-6 (it would be on C-7 if we numbered from the left). So this will be 6-(???)dodecane. In naming the substituent, we find the longest chain working outwards from the main chain—in this case three carbons long, so this is a propyl group. Finally, we add in the substituents on that propyl chain to get 6-(1,1-dimethylpropyl)dodecane. Students often worry about the location of the parentheses—you will get it right if you bear in mind that their purpose is simply to make things clear.

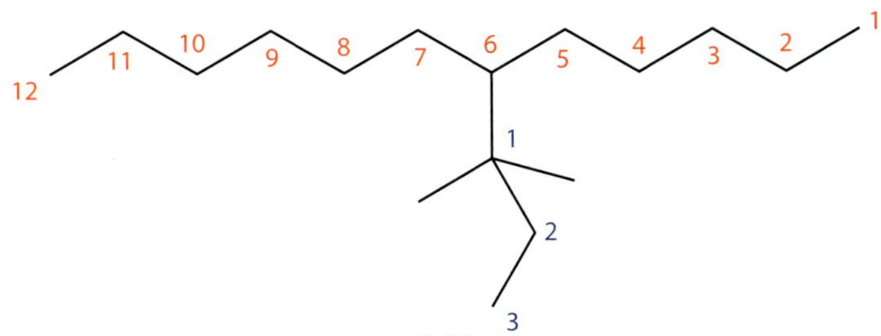

2.22

Digression: Nonpolar (Hydrophobic) Interactions

Molecules (and atoms) are attracted to one another—otherwise solids and liquids would not hold together. All the types of forces between molecules are properly referred to as van der Waals forces, although this term is also sometimes used to refer specifically to dispersion forces. Table 2.4 lists the main forces between molecules (although there is increasing interest in other types of interactions specific to particular types of molecules, especially in the solid state).

Hydrocarbons are clearly not ionic or polar and have no capacity for hydrogen bonding. Their interactions are therefore dominated by London dispersion forces. These can be considered to arise from the fact that electrons in all molecules are in constant motion; thus, at any given instant, they may be distributed such that the molecule has an instantaneous dipole moment. This instantaneous dipole induces a dipole in a nearby molecule and is then attracted to it (Figure 2.6).

Dispersion forces increase with molecular mass; heavier molecules have more electrons and there is more fluctuation in partial charges. Thus, for noble gases, melting point and boiling point both increase with atomic weight. Figure 2.7 shows that although argon is monatomic, the interaction between two argon atoms is attractive until they are less than 3.8 Å apart. Molecular hydrogen has a low boiling point of 20 K—it has only two electrons. Nitrogen and oxygen have more electrons and higher boiling points. In the halogen group, at room temperature, fluorine and chlorine are gases, bromine is a liquid, and iodine is a solid (Figure 2.8).

TABLE 2.4 Interionic and Intermolecular Forces

Type of Interaction	Typical Energy ($kJ\ mol^{-1}$)	Interacting Species
Ion–ion	250	Ions only
Dipole–dipole	0.3–2	Polar molecules
London dispersion forces	2	All molecules
Hydrogen bonding	20	Mainly N, O, F—the link is a hydrogen atom

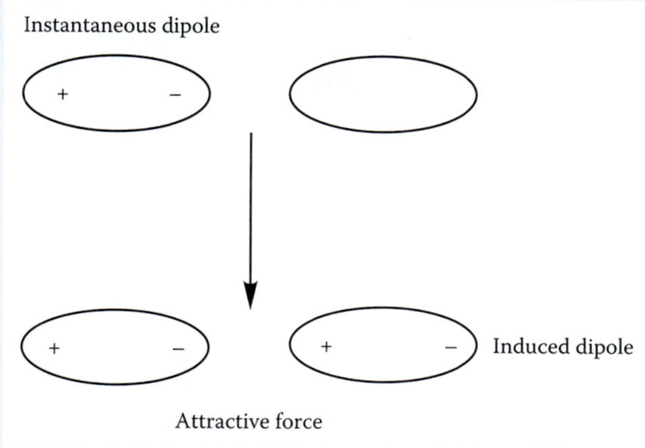

FIGURE 2.6 Instantaneous and induced dipoles attract.

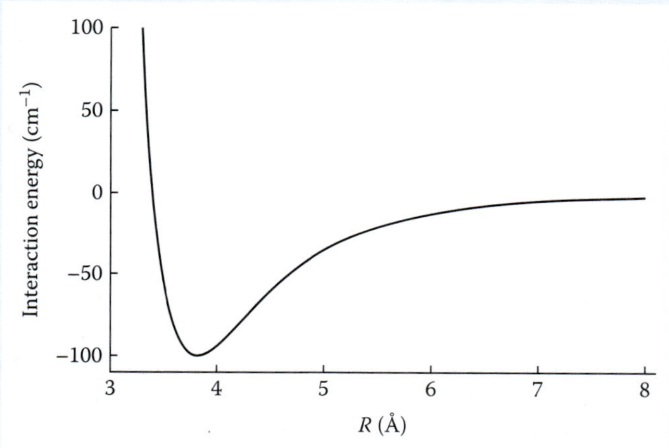

FIGURE 2.7 Energy diagram for the approach of two argon atoms. (Thanks to Prof. Peter Knowles for this figure.)

FIGURE 2.8 Elemental halogens.

TABLE 2.5 Melting and Boiling Points of Nonpolar Compounds

Substance	MPt (°C)	BPt (°C)	Substance	MPt (°C)	BPt (°C)
He	−272[a]	−268.9	F_2	−217	−188
Ne	−248.6	−246.1	Cl_2	−101.5	−34
Ar	−189.4	−187.8	Br_2	−7	59
Kr	−157.35	−153.35	I_2	113.7	183–6[b]
Xe	−111.75	−108.15	At_2	302	337
Rn	−71.15	−61.85	CH_4	−182	−162
H_2	−259.1	−252.9	CF_4	−184	−128
N_2	−209.85	−195.75	CCl_4	−23	77
O_2	−218.35	−182.95	CBr_4	95	190

[a] 2.5 MPa pressure.
[b] Sublimes.

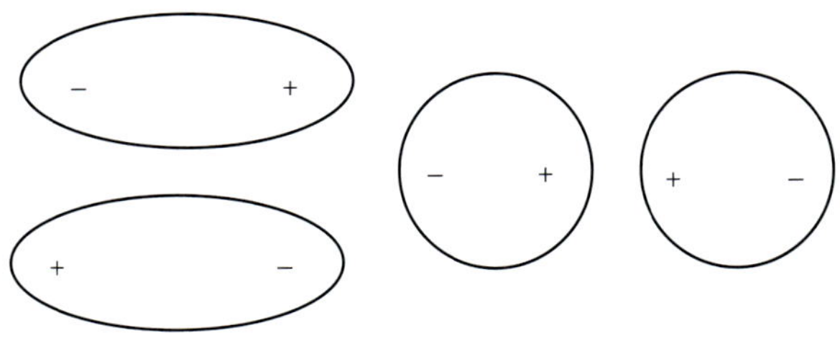

FIGURE 2.9 Dispersion forces depend on molecular shape.

The same arguments can be applied to simple organic molecules, CX_4, that have no dipole (Table 2.5). Note that in each series the boiling points are characteristically monotonically increasing, but there are some oddities in the melting points. This is usually explained by invoking "crystal packing forces"—a convenient catchall phrase that includes crystal morphology and longer-range interactions between molecules.*

Dispersion forces also depend on molecular shape. For example, the boiling points of *n*-pentane and *neo*-pentane are 36 °C and 10 °C, respectively. With rod-shaped molecules, the dipoles are able to get closer to each other and have a stronger interaction than with essentially spherical entities (Figure 2.9).

* In principle, at a given temperature, there should be a single crystalline form for any molecule that has the lowest energy. However, in practice, the kinetics of crystallization may give a different form or polymorph, for a wide range of reasons, often not well understood. Different polymorphs may have different melting points and solubility properties, and using the correct polymorph of medicinally active species is critical in their pharmacodynamics.

2.4.3 HALOALKANES

When an alkane is substituted with one or more halogen atoms, then the halogens are specified as substituents, and the "halo" substituents take precedence in numbering the chain over alkyl substituents. Halogens as substituents are named as chloro-, bromo-, *etc.* Thus, **2.23** is 3-bromoheptane, and **2.24** is 2-chloro-7-methyloctane.

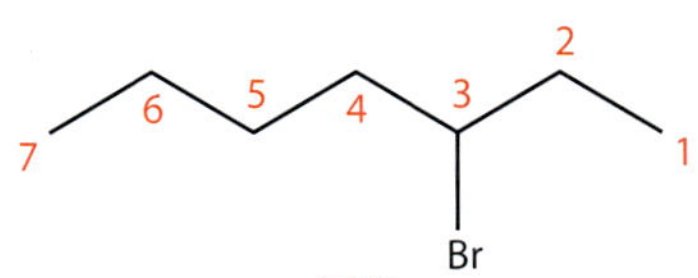

2.23

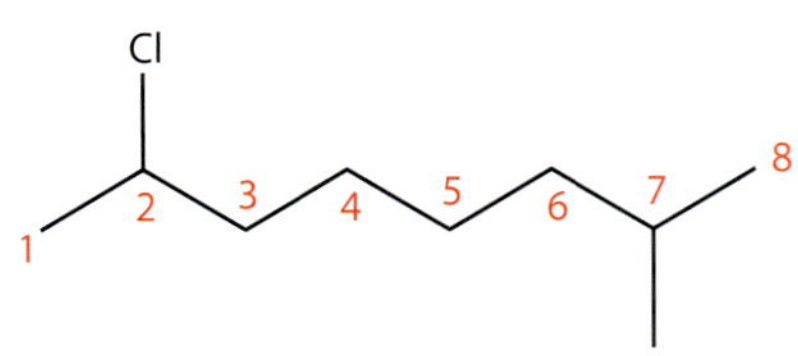

2.24

Digression: Interactions of Polar Molecules

Polar molecules such as haloalkanes have permanent dipoles—because the carbon–halogen bond is polar (halogens are more electronegative than carbon, and hence electron density is attracted to them). Permanent dipoles attract each other (Figure 2.10). The strength of the attractive force depends on the size of the dipole and the shapes of the molecules. Generally, molecules with permanent dipole moments have higher melting and boiling points than those without them. This effect is readily seen with the fluorinated methanes (Table 2.6). The main attractive forces are between the slightly positive hydrogen atoms and the lone pairs on fluorine. The possible interactions are maximized when there are equal numbers of hydrogens and fluorines.

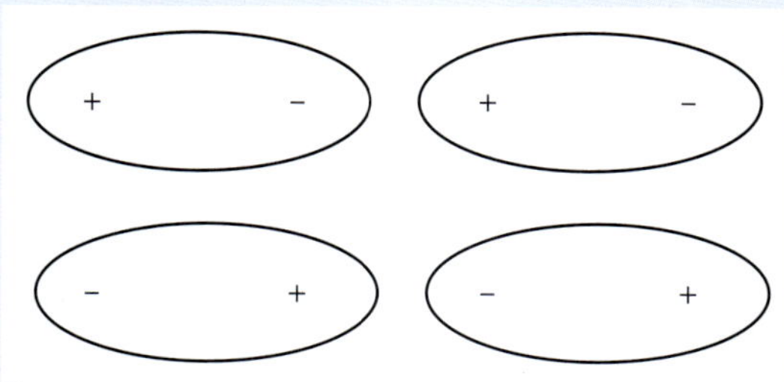

FIGURE 2.10 Attraction of permanent dipoles.

TABLE 2.6 Physical Properties of Fluorinated Methanes

Molecule	CH_4	CH_3F	CH_2F_2	CHF_3	CF_4
Molecular weight	16	34	52	70	88
Melting point (°C)	−182	−115	−136[a]	−163	−184
Boiling point (°C)	−162	−78	−52	−84	−128

[a] Value may be unreliable; all reports say this is a glass rather than a crystalline material.

PROBLEM 2.2

For each of the following compounds, give a name according to the IUPAC nomenclature rules*:

(a)

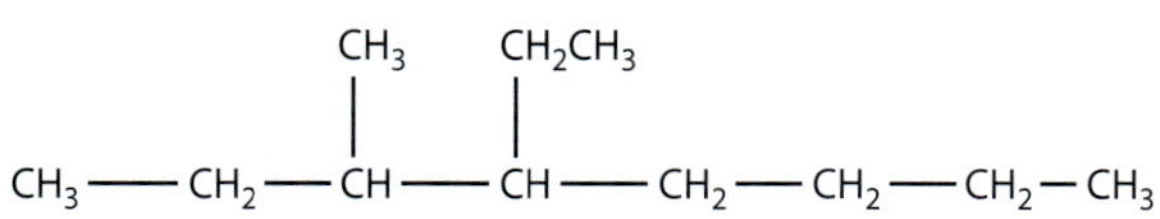

(b)

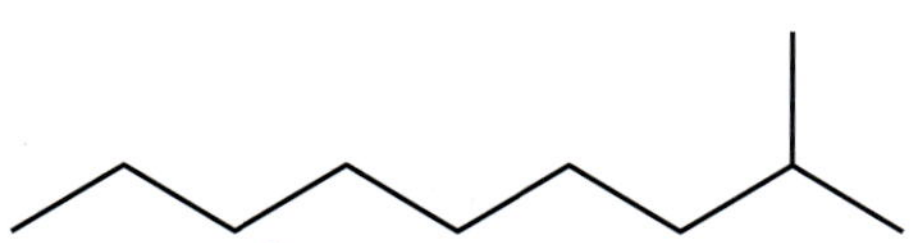

(c)

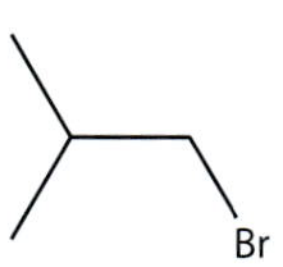

* The difference in layout between the examples is deliberate—you will meet many ways of writing molecules, some good and some less so, and you need to be able to understand all of them.

PROBLEM 2.3

For each of the following compounds, draw a structural formula, using stick-type diagrams:

(a) 2-Bromo-2-methylpropane
(b) 2,2,4-Trimethylpentane
(c) 2,3-Dimethyl-4-ethyloctane

Solutions

2.2(a) We first need to number the chain so that the substituents have the lowest numbers—in this case from the left:

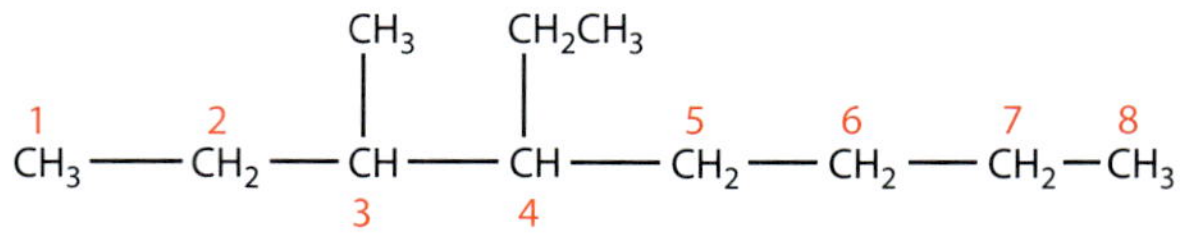

So this is 4-ethyl-3-methyloctane.

2.2(b) This time we will number from the right:

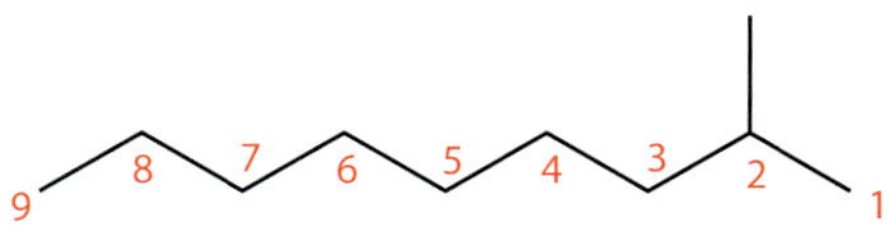

So this is 2-methylnonane.

2.2(c) This problem is trickier. The longest continuous chain is three carbons long, but there are two chains of three carbons that we could select:

Because of the priority of the bromine atom, we select the red chain and number it so that the carbon bearing the priority bromine atom is C-1:

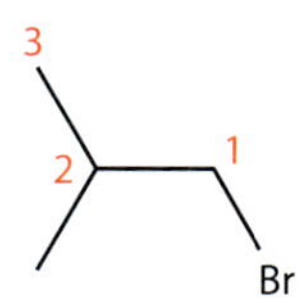

So this is 1-bromo-2-methylpropane.

2.3(a) First, draw and number propane (it's almost always best to start at the end of the name and get the basic skeleton first):

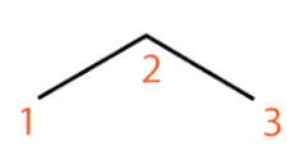

We now add the substituents at the 2 position:

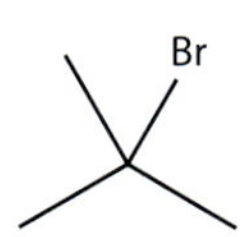

2.3(b) First, draw and number pentane:

Then add the substituents:

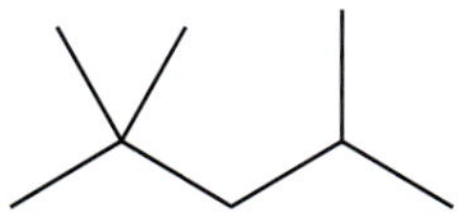

2.3(c) Draw and number octane:

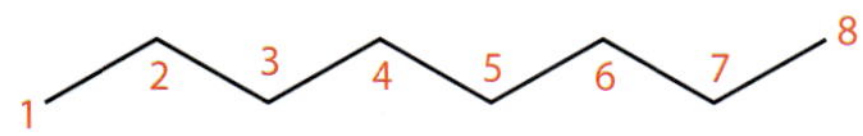

Add the substituents:

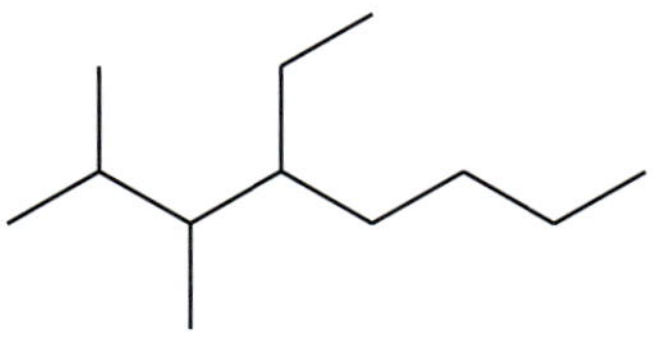

2.4.4 CYCLOALKANES

Cycloalkanes contain a carbocyclic ring, and because there are no methyl group "ends" to the molecule, they have the general formula C_nH_{2n}. The names still derive from the number of carbon atoms, but now the whole name has the prefix "cyclo-" to indicate the ring structure. So **2.25** is cyclopropane, and **2.26** is cyclohexane. Substituted cycloalkanes are named as derivatives of the ring, so **2.27** is methylcyclohexane. If there is more than one substituent on the ring, then we need to state their relative positions. One of the substituents' positions will be designated as carbon 1, and the ring will then be numbered so that the other(s) is on the lowest-numbered carbon atoms. So **2.28** is 1,2-dimethylcyclohexane.

2.25

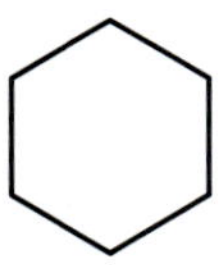

2.26

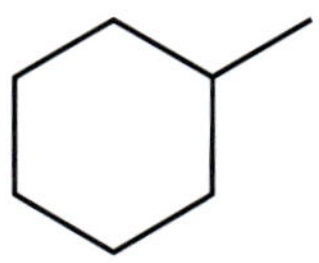

2.27

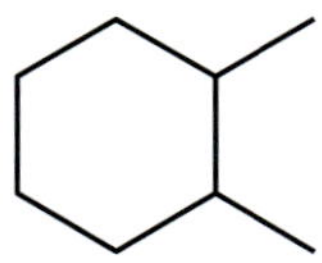

2.28

One more specification needs to be added with cycloalkanes—if there is more than one substituent, are the two on the same or on opposite sides of the ring? We will defer considering what the real three-dimensional shapes of various-sized rings are for the present, returning to this in Chapter 7. We describe two groups on the same "side" of a ring as being *cis* and those on opposite sides as being *trans*. These terms are taken from Latin and are normally italicized. So **2.29** is *cis*-1,2-dimethylcyclohexane, and **2.30** is *trans*-1,3-dimethylcyclopentane.

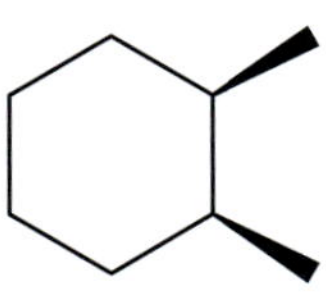

2.29

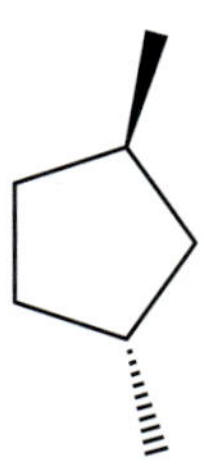

2.30

PROBLEM 2.4

Give a structure for each of the following molecules:

(a) *trans*-1,3-Dimethylcyclobutane
(b) *cis*-1,4-Dichlorocyclohexane

PROBLEM 2.5

Give a name for the following molecules, according to IUPAC rules:

(a)

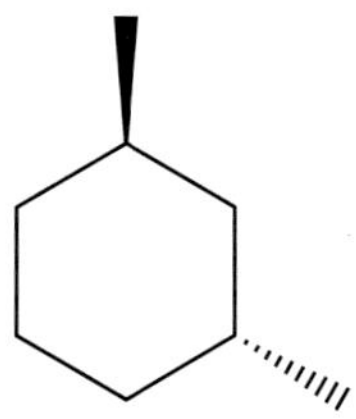

(b)

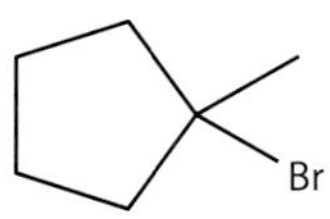

Solutions

2.4(a)

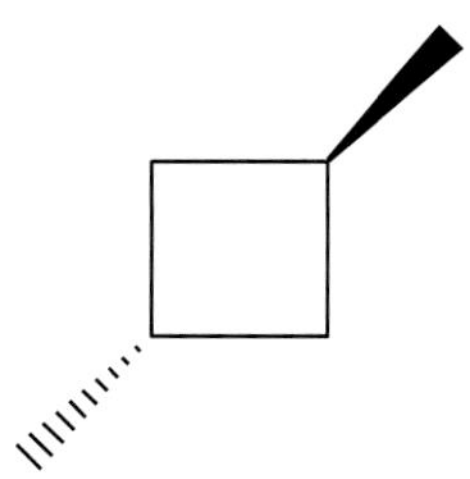

2.4(b)

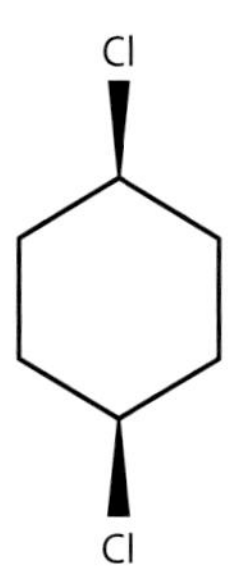

2.5(a) *trans*-1,3-Dimethylcyclohexane
2.5(b) 1-Bromo-1-methylcyclopentane

Focus on Hydrocarbons as Fuels

Throughout the developed world, most of the energy used comes from fossil hydrocarbon fuels, gas, coal, and oil. Coal is largely carbon with small amounts of hydrogen, nitrogen, oxygen, and sulfur. The carbon is mostly in the form of benzene rings, rather than more saturated hydrocarbons. Natural gas, as used in fuel, is mostly methane, with up to 20 % of other hydrocarbons, mostly ethane. Purification of extracted gas removes a range of other impurities. Since methane itself has no odor, small amounts of an unpleasant smelling sulfur compound are added, so that leakage will be readily noticed. As extracted, oil consists of a very complex mixture of hydrocarbons, with a wide range of boiling temperatures. Table 2.7 shows some of the factions, their common names, and primary uses. Many different sets of descriptions and definitions exist now and some have changed over time.

Gasoline is sold with an "octane" rating—for automobiles, generally between 85 and 100. A gasoline with an octane rating of 100 is defined to have the same antiknock properties as 2,2,4-trimethylpentane (one of many isomers for formula C_8H_{18}), and a gasoline with an octane rating of zero has the same antiknock properties as *n*-heptane. Everything in between is defined in terms of the appropriate mixture of the two compounds. The octane rating does not refer to the energy obtained from combustion but to the ability of the fuel to burn in a controlled manner. The higher the octane rating, the higher the activation energy for combustion. Octane numbers greater than 100 are possible, because some additives have higher octane ratings. For example, toluene has an octane rating of 111 and methanol and ethanol about 108. These can be added to lower octane number

TABLE 2.7 Fractions of Saturated Hydrocarbons

Number of Carbon Atoms	Boiling Range (°C)	Common Name	Uses, Comments
1–4	<40	Petroleum gas	LPG for stoves and some cars.
5–9	40–100	Ligroin, naphtha	Ligroin is primarily used as a solvent in chemical manufacture and contains 55 % alkanes and 30 % cycloalkanes. Sometimes referred to as petroleum ether, but it is NOT an ether. Naphtha should not be confused with the solid aromatic compound naphthalene.
5–12	40–205	Gasoline	Automobile fuel. Exact composition varies internationally and with climate.
10–18	150–300	Kerosene, paraffin	Fuel for jet engines and tractors. Some uses as heating oil. Contains significant amounts of aromatic hydrocarbons. Paraffin is the term used in the United Kingdom and Ireland.
>12	250–350	Diesel, gas oil	Fuel for automobiles and heavier vehicles. Ultralow-sulfur diesel fuel is the norm in Europe and North America for auto use, and the higher costs relate to sulfur removal. About 75 % alkanes, 25 % aromatic hydrocarbons.
20–50	300–370	Lubricating oil	Lubrication. Contains significant amounts of aromatic hydrocarbons.
20–70	200–600	Fuel oil	Industrial fuel. Contains significant amounts of aromatic hydrocarbons. Many grades exist, and the lower grades are kerosene and diesel.
>70	>600	Residues	

fuels to improve their antiknock properties—for example, in racing fuel and fuel for piston-engine aircraft (avgas). In the developed world, the classic antiknock additive, tetraethylead, is banned from automobile fuel, because of its environmental toxicity, but it is still used, although at lower levels than in the past, in avgas.

Supermarine Spitfire

My car

Key Points from Section 2.4

- Alkanes are hydrocarbons containing no multiple bonds. They are nonpolar and insoluble in water, and their most important reaction is combustion.
- Nomenclature of organic compounds follows IUPAC rules, and names consist of a prefix indicating the number of carbon atoms, and a suffix to indicate the type of compound.
- In naming alkanes, the prefix derives from the longest chain of carbon atoms; the suffix is "-ane." The chain is numbered so that substituents are on the lowest-numbered carbon atom. Substituents are named with the suffix "-yl" for carbon substituents and as halo (*e.g.*, in chloromethane) for halogens.
- Alkanes with a ring structure take the prefix "cyclo-" followed by the number of the carbon atoms in the ring (*e.g.*, cyclobutane). Rings are numbered so the substituents are on the lowest-numbered carbon and, whether they are on the same or opposite sides of the ring, are indicated by *cis* or *trans*.

2.5 STABILITY, STRAIN, AND PROPERTIES OF ALKANES AND CYCLOALKANES

Although we have not yet studied the shapes of cycloalkanes in detail, there are some simple observations that we can make. Cyclopropane and cyclobutane must be strained molecules—the carbon atoms are sp^3 hybridized but clearly can't attain a CCC angle of 109.5 °. So these cannot be very happy molecules ☹. Table 2.8 shows the strain in the various cycloalkanes. In cyclopropane and cyclobutane, most of the strain is ***angle strain*** resulting from their inability to attain true tetrahedral angles. The other cycloalkanes are not planar (in fact cyclobutane is not quite planar, but the distortion is small), and hence the tetrahedral angles can be accommodated. Notice that the six-membered ring is unstrained, and this gives rise to a theme that you will see repeated throughout your studies. Six-membered rings are the easiest to make, and the most stable of ring structures, in organic chemistry. What may be more surprising is the increase in strain in the 7–12 carbon range, usually referred to as medium rings. With rings with more than 13 carbon atoms, there is no strain.

Where did these figures come from? We get information about the relative stabilities of compounds by looking at their heats of formation or heats of combustion, the latter being simpler to determine experimentally. All alkanes burn in excess oxygen according to Equation 2.1. The greater the number of carbon and hydrogen atoms, the greater the heat of combustion. You might imagine that all isomeric alkanes burn with the same heat of combustion, but this is not so. The strengths of the O=O, C=O, and O–H bonds (*i.e.*, the energy content of the products) are the same in all cases, so these differences must reflect genuine differences in the stability of the original alkane. Table 2.9 shows the heats of combustion of some simple alkanes.

TABLE 2.8 Strain Energy in Cycloalkanes

Cycloalkane	Strain Energy (kJ mol^{-1})
Cyclopropane	115.5
Cyclobutane	110.5
Cyclopentane	27.2
Cyclohexane	0
Cycloheptane	26.4
Cyclooctane	40.2
Cyclononane	52.7
Cyclodecane	50.2
Cycloundecane	46
Cyclododecane	10

(2.1) $$C_nH_{2n+2} + XS\ O_2 \rightarrow nCO_2 + (n+1)\ H_2O + Heat$$

We can see the obvious conclusion that more carbon means more heat, but the differences between the isomeric butanes and the isomeric pentanes are instructive. In each case, the more branched the isomer, the less energy we get on burning it. Since the products are identical each time, this means that the branched isomers are lower in energy than the linear ones. This is quite general for alkanes and indeed quite a number of other compound classes.

If we consider the cycloalkanes, the heat of combustion per carbon atom (they are composed only of identical CH_2 groups) is a good measure of stability (Table 2.10). The more energy that is generated by combustion, the more strained the ring, as the strain energy is completely released on combustion.

TABLE 2.9 Heats of Combustion of Lower Alkanes

Alkane	$\Delta H_{combustion}$ (kJ mol^{-1})
CH_4	891
C_2H_6	1560
C_3H_8	2219
	2877
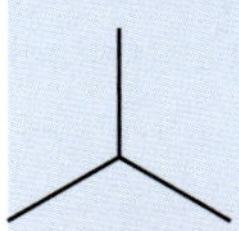	2868
	3535
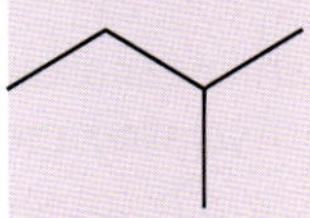	3529
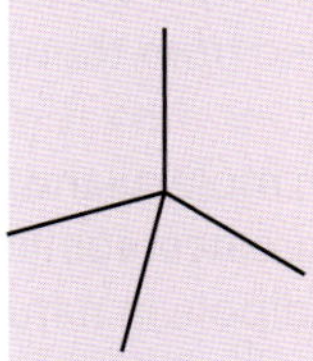	3514

TABLE 2.10 Heats of Combustion of Cycloalkanes per Carbon Atom

Cycloalkane	$\Delta H_{combustion}/n$ (kJ mol^{-1})
Cyclopropane	697
Cyclobutane	680.1
Cyclopentane	658.2
Cyclohexane	655
Cycloheptane	655.2
Cyclooctane	658.2
Cyclodecane	660

n is the number of carbon atoms in the ring.

TABLE 2.11 Physical Properties of Alkanes

Alkane	MPt (°C)	BPt (°C)
Methane	−183	−162
Ethane	−184	−88
Propane	−188	−42
Butane	−137	0
iso-Butane(2-methylpropane)	−133	−11
Pentane	−130	36
iso-Pentane (2-methylbutane)	−60	28
neo-Pentane (2,2-dimethylpropane)	−18	9
Hexane	−95	69
iso-Hexane (2-methylpentane)	−153	61
neo-Hexane (2,2-dimethylbutane)	−100	49
Octane	−53	126
Decane	−30	174
Hexadecane	18	281
Triacontane (C_{30})	66	550

Alkanes, as was indicated earlier, have physical properties that depend on both their molecular weight and three-dimensional shape. Table 2.11 gives the melting and boiling points of various alkanes. Considering first only the linear alkanes, there is a clear monotonic increase in boiling point with molecular weight. Although melting points also generally increase, this is a much less regular progression, because it depends on the form in which the material crystallizes. It is interesting to compare the isomers of C_4, C_5, and C_6 hydrocarbons. In each case, branching lowers the boiling point of the liquid. Boiling points depend on the energy required to overcome the intermolecular forces in the liquid state—for hydrocarbons, these are dispersion forces. These forces depend on the molecules' ability to interact, and these interactions are maximized when the molecules can stack in the liquid phase. Substituents disrupt this stacking, so interactions, and hence boiling points, are lowered.

Key Points from Section 2.5

- Branched alkanes are thermodynamically more stable than linear alkanes.
- Small-ring cycloalkanes are destabilized by angle strain. Cyclohexane is strain free, but medium-ring alkanes are more strained.
- Stabilities of alkanes (and other compounds) can be derived from thermodynamic parameters such as heat of combustion or heat of formation.
- Melting points and boiling points of alkanes increase with molecular weight; the intermolecular interactions are dispersion forces, and these are greatest for linear, unsubstituted species. Intermolecular forces in haloalkanes are dipole–dipole in nature.

REVIEW PROBLEMS

1. 0.543 g of a molecule containing C, H, and O is burned in air to give 1.039 g CO_2 and 0.6369 g H_2O. Calculate its empirical formula.
2. Vitamin C has the following percentage composition. Derive its empirical formula:
   40.9 % C
   4.5 % H
   54.5 % O
3. Give a systematic name for each of the following molecules:
   (a)
   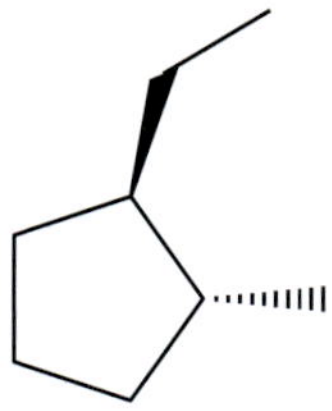
   (b)
   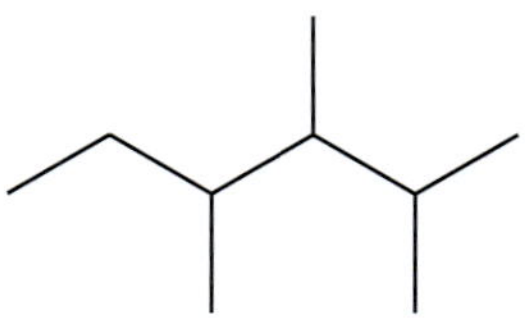
   (c)

   (d)
   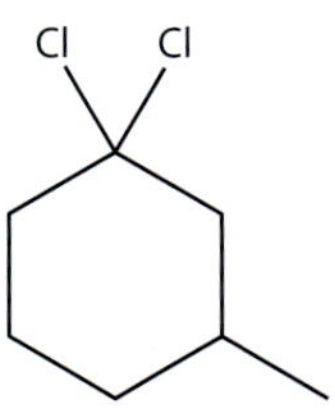

   (e)
   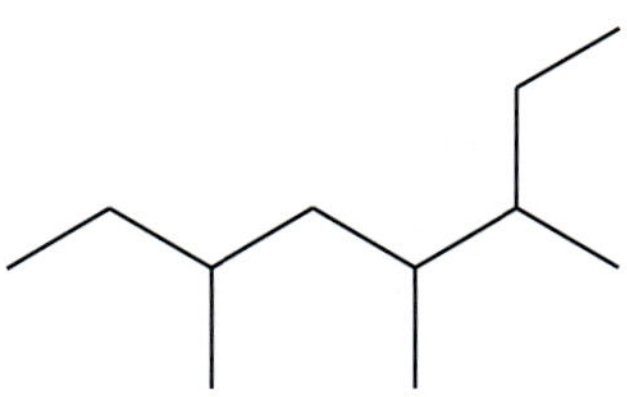
   (f)
   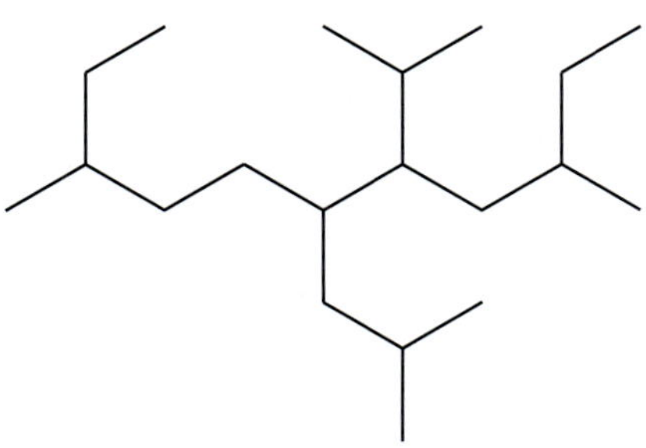
   (g)
   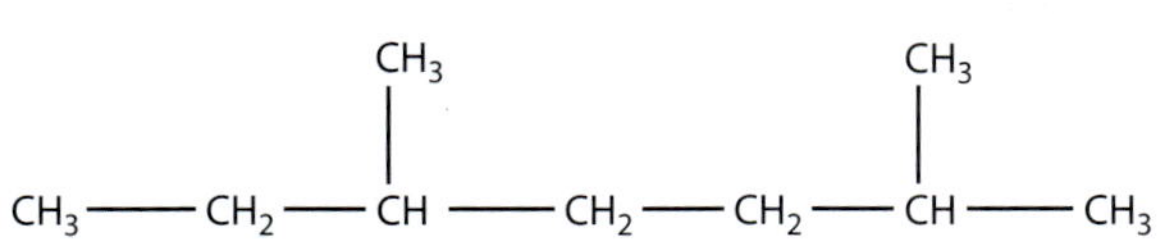

(h)

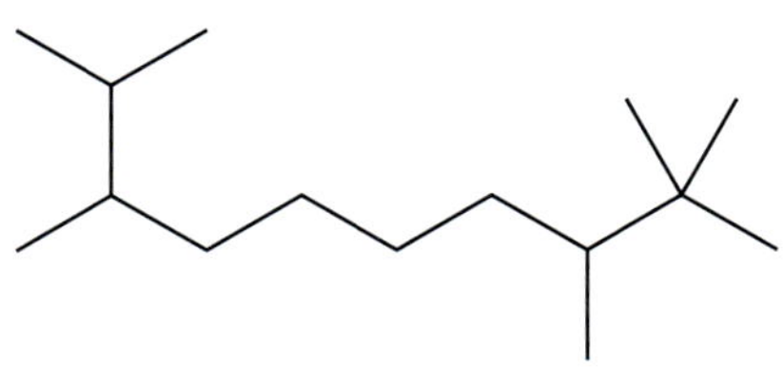

(i)

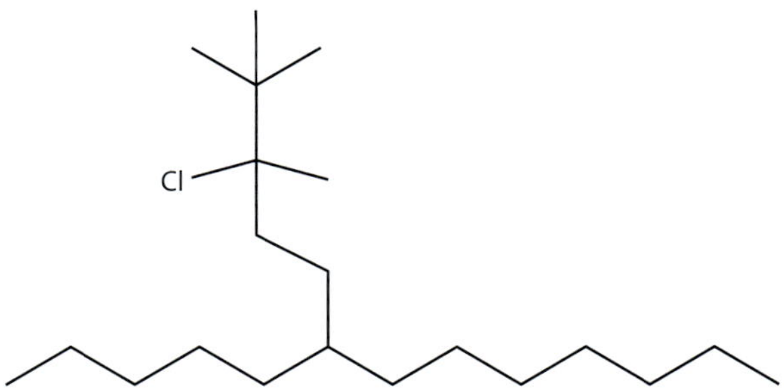

(j)

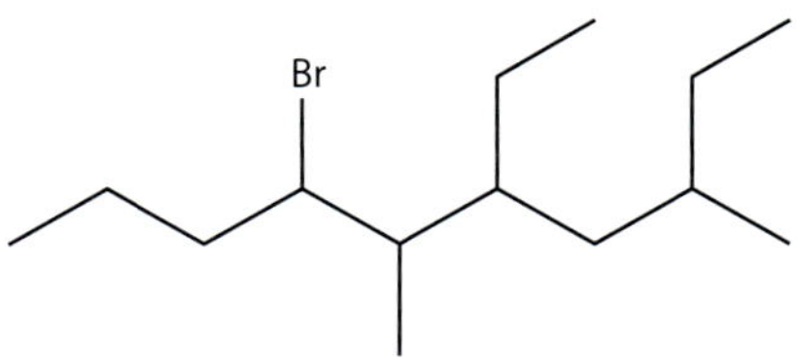

4. Draw a stick structure for each of the following molecules:
   (a) 2-Methylheptane
   (b) Ethylcyclopropane
   (c) 2,3,4-Trimethylheptane
   (d) 2,8-Dimethyl-4-ethyl-3-(1-methylethyl)decane
   (e) 4-Methyl-2,6-diphenylheptane
   (f) *cis*-1-Ethyl-2-propylcyclohexane
   (g) 2,4,6-Trimethylnonane
   (h) *trans*-1,4-Dichlorocyclohexane
   (i) 6-Methyl-4(1-methylpropyl)nonane
   (j) Bromocyclobutane
   (k) 2,3,4-Trimethyloctane
   (l) *cis*-1-Bromo-3-methylcyclopentane
5. What hybridization and what shape would you predict for each of the following molecules (consider nitrogen and oxygen as well as carbon)?

$$CH_2Cl_2 \qquad (CH_3)_3N \qquad CH_3CH_2OCH_2CH_3$$

6. For the following pair of molecules, predict which would have the higher boiling point. Give a reason.

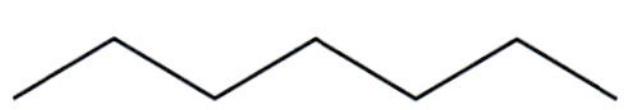

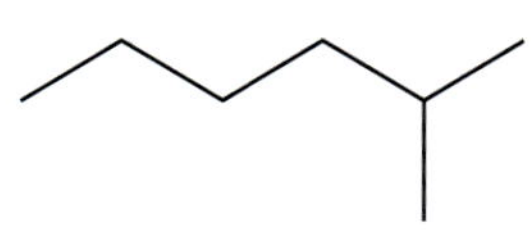

7. What are the most important kinds of intermolecular forces in the following atoms/molecules?
   (a) Krypton
   (b) Tetrafluoromethane
   (c) Hydrogen fluoride
   (d) Butane
8. Give an example of each of the following:
   (a) A diatomic molecule with a permanent dipole moment
   (b) A diatomic molecule without a permanent dipole moment
   (c) An sp^3 hybridized molecule with a permanent dipole moment
   (d) A triatomic molecule with a permanent dipole moment

Alkenes, Alkynes, and Aromatic Compounds

3

3.1 ALKENES

3.1.1 BONDING IN ALKENES

The next group of hydrocarbons that we will study is the alkenes (old name olefins), compounds containing one or more carbon–carbon double bonds. The first member of the class is ethene (**3.1**) in which there is a double bond between the two carbon atoms. The old name for this compound was ethylene, and this is still widely used in the polymer industry. There are two carbon–carbon bonds along the same direction—so clearly, we cannot use sp^3 hybrid orbitals, which we used to make alkanes, for this molecule.

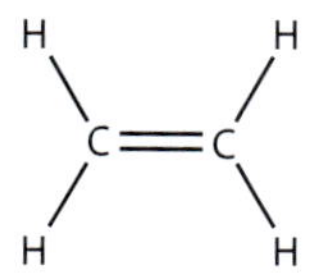

3.1

In order to describe a carbon–carbon double bond, we need to return to the electronic configuration of carbon:

$$\mathrm{C} \quad 1s^2\,2s^2\,2p_x^{\,1}\,2p_y^{\,1}\,2p_z^{\,0}$$

As before, we formally promote one electron to obtain

$$\mathrm{C} \quad 1s^2\,2s^1\,2p_x^{\,1}\,2p_y^{\,1}\,2p_z^{\,1}$$

This time, we set aside the $2p_z$ orbital, which we will use to make a π-bond, and take the 2s, $2p_x$, and $2p_y$ orbitals to make three $2p^2$ hybrid orbitals. These sp^2 hybrids point toward the corners of an equilateral triangle (by VSEPR or some serious mathematics), with angles of 120° between them. The calculated orbitals are shown in Figure 3.1. We can now use these orbitals with the 1s orbitals from hydrogen to make the σ-bonds of ethene (**3.2**). The σ-bonds to hydrogen are each made from a hydrogen 1s and the sp^2 orbital. The carbon–carbon σ-bond is formed from two sp^2 orbitals, coming together effectively nose to nose. Since we used s, p_x, and p_y to make these hybrid orbitals, the σ-framework must be planar in the *xy* plane (Figure 3.2).

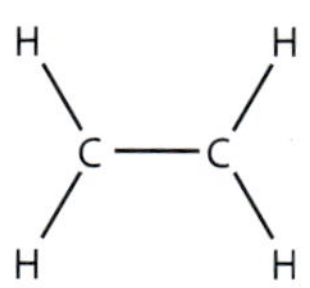

3.2, σ-framework of ethene

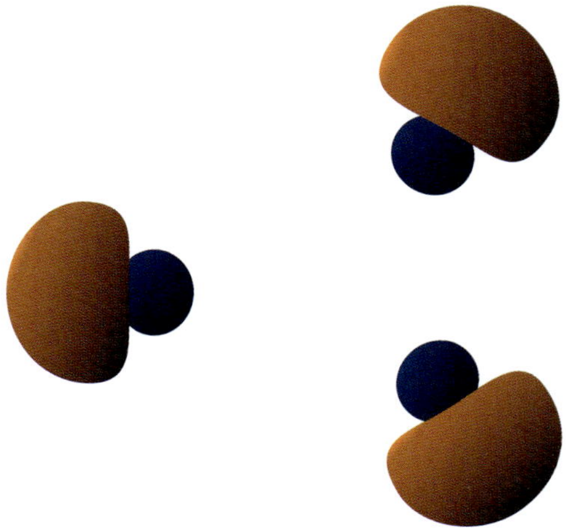

FIGURE 3.1 Calculated sp^2 orbitals.

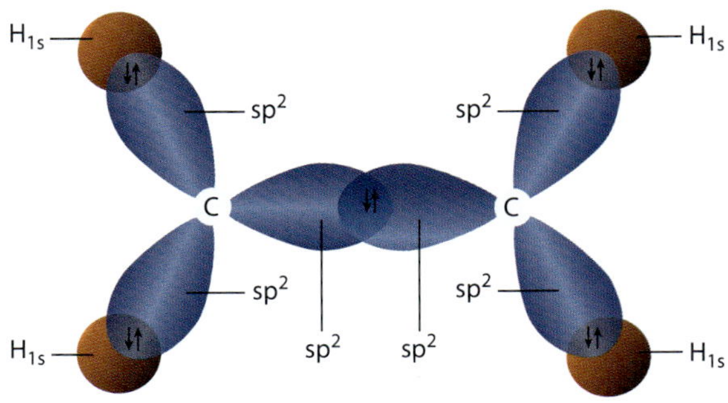

FIGURE 3.2 Sigma bond framework of ethene, showing orbitals.

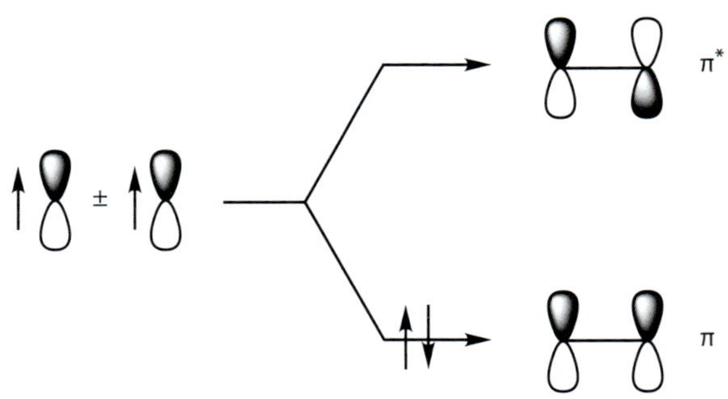

FIGURE 3.3 π- and π^*-orbitals of ethene.

On each carbon atom, we are left with a $2p_z$ orbital containing one electron. The orbitals are brought together side to side (they are perpendicular to the plane of the rest of the molecule) and combined to give a π- and a π^*-orbital (Figure 3.3). We have two electrons to accommodate, and these are both in the π-orbital. Notice that in both **3.2** and **3.3**, we are not drawing the orbitals as they really are; the stylized versions make it easier to see what is happening. Figure 3.4 shows the orbitals as they really are.

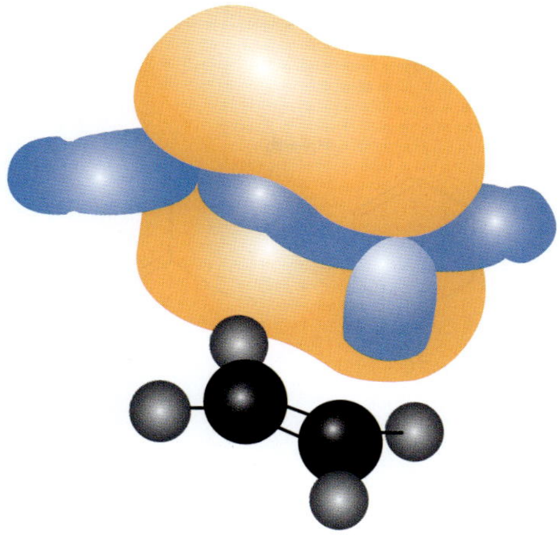

FIGURE 3.4 Bonding orbitals of ethene.

TABLE 3.1 Bond Strengths and Lengths in Hydrocarbons

Bond	Bond Strength (kJ mol^{-1})	C–C Distance (Å)	Bonds
C–C	347	1.54	σ
C=C	610	1.34	$\sigma + \pi$
C=C	836	1.20	$\sigma + \pi + \pi$

The double bond in ethene thus consists of one sigma bond, derived from sp^2 hybrid orbitals, and one π-bond, formed from p_z orbitals. Double bonds are stronger and shorter than single bonds, but they are not twice as strong—the $\sigma + \pi$ double bond is stronger than one σ-bond, but not as strong as two σ-bonds (Table 3.1). The presence of the π-bond has another important consequence for alkene structure—it is not possible to rotate freely about a carbon–carbon double bond, since any rotation would break the π-bond (the p_z orbitals would no longer be aligned). When alkenes react, they do so by using the electrons of the π-bond, which are also more accessible to reagents. A carbon–carbon single bond is always one σ-bond, a double bond one σ-bond + one π-bond, and a triple bond one σ-bond + two π-bonds.

One other term is useful when discussing unsaturated hydrocarbons—and it will be helpful to us when we are trying to determine their structure. If a linear molecule has two double bonds in the chain, then it is said to have two "degrees of unsaturation." Other equivalent terms are double bond equivalents (DBEs) or the index of hydrogen deficiency. Relative to a saturated hydrocarbon, for each double bond or ring, two hydrogens are missing from the formula that an alkane with that number of carbon atoms would possess (Table 3.2). As far as the formula is concerned, the effect of a ring is the same as that of a double bond—but there will usually be other evidence from spectroscopy or reactivity to tell whether the DBE is a multiple bond or a ring structure.

How is this concept useful? Let us say that we have isolated a hydrocarbon of molecular formula C_6H_8. If this had been an acyclic alkane, we would have expected $2n + 2$ hydrogens, in this case 14. So we are six hydrogens "short." So we must have three rings or double bonds, for example, **3.3**, **3.4**, or **3.5**. We can generalize this to say that if we have n carbon atoms and m hydrogen atoms, then (DBE) = $\{(2n + 2) - m\}/2$. Halogens count as if they were hydrogens,

TABLE 3.2 Hydrocarbon Formulae

Class of Compound	Formula
Alkane	C_nH_{2n+2}
Cycloalkane	C_nH_{2n}
Alkene	C_nH_{2n}
Cycloalkene	C_nH_{2n-2}
Alkyne	C_nH_{2n-2}

so for C_6H_5Cl, DBE = {(2 × 6) + 2 − 6}2 = 4. A likely structure would be chlorobenzene (**3.6**) with three double bonds and one ring.

3.3

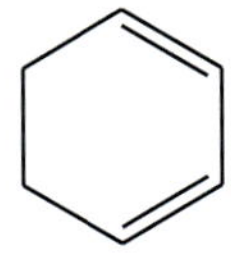

3.4

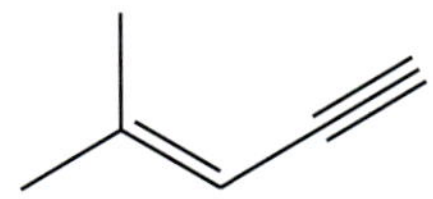

3.5

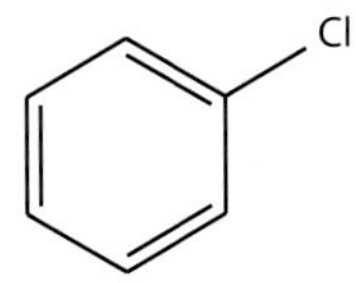

3.6

PROBLEM 3.1

Calculate the DBE for the formula C_8H_{14} and suggest two possible structures for the formula.

Solution

$$\text{DBE} = \frac{\{(2\times 8)+2-14\}}{2} = 2$$

There are many possible structures such as

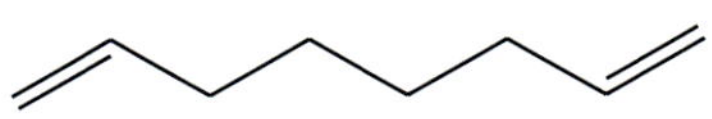

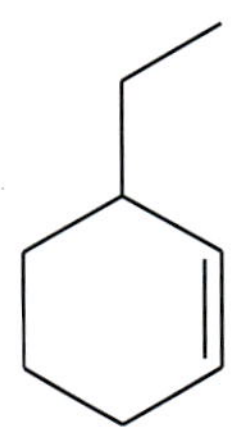

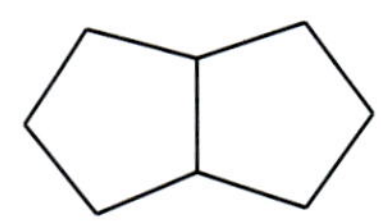

Focus on Ethene in Fruit Ripening

You probably think of ethene as being a major industrial chemical, and indeed it is the most manufactured organic chemical on the planet. But it is also a very important plant hormone—it regulates the ripening of many fruits, bud opening of flowers, and the fall of leaves. If you have ever grown tomatoes, you will know that at the end of the summer, there are always some green ones left on the vine. If you are not a fan of chutney, salsa, or fried green tomatoes, you can ripen these by putting them in a paper bag in a kitchen drawer with a ripe tomato. And you will have heard some version of the saying "one bad apple spoils the whole bushel." Although farmers have known the tricks to ripening fruit at the proper time for millennia (the ancient Egyptians ripened dates and the Chinese pears this way), it was only in the twentieth century that ethene was identified as the regulator of these processes. In the early twentieth century, lemons were picked green and ripened to yellow in kerosene-heated sheds. When more modern heating methods were introduced, the lemons stopped ripening—it was in fact traces of ethene in the vapor from the kerosene, rather than the heat, that had caused the ripening.

Unripe lemon

Ripe lemon

Ripening tomatoes. (Thanks to Paul Deards for this image.)

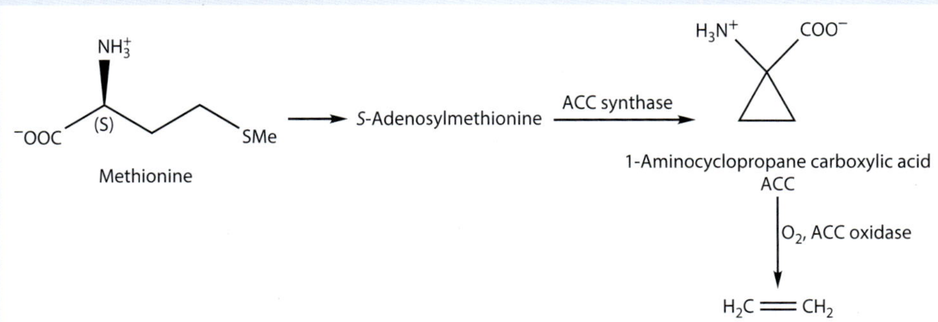

FIGURE 3.5 Biosynthesis of ethene by plants.

Ethene is produced by plants naturally by the sequence shown in Figure 3.5. Normally, it regulates fruit ripening, but ethene is also produced in response to damage to the plant tissue (hence the bad apple spoiling the bushel, because it causes the rest of the fruit to overripen) or environmental stress. Apples, avocados, bananas, melons, peaches, pears, and tomatoes all generate ethene and are ripened by it. Cherries and blueberries neither produce any significant amount nor are they affected by it. Ethene is widely used commercially to ripen tomatoes, pears, and bananas after harvesting. Although the practice is banned in the developed world, some countries do use calcium carbide, CaC_2, in ripening; with moisture, this generates ethyne, which also stimulates ripening. 1-Methylcyclopropene has been shown to block the ethene receptor, and this may be used to slow ripening and hence extend the shelf life of fruit and flowers.

3.1.2 NOMENCLATURE OF ALKENES

In IUPAC nomenclature, the suffix for the class of alkenes is "-ene." Some of the lower members of the class are shown in Figure 3.6. In working out the formal name of the alkene, we now have two new factors to consider—where is the double bond and what is the geometry about the double bond? Remember this geometry is fixed, as there is no rotation about the double bond. While the *cis*- and *trans*-designations will serve for simple alkenes such as 2-butene, we will need a more sophisticated set of rules for larger, more complex molecules. For example, how would compound **3.7** be named?

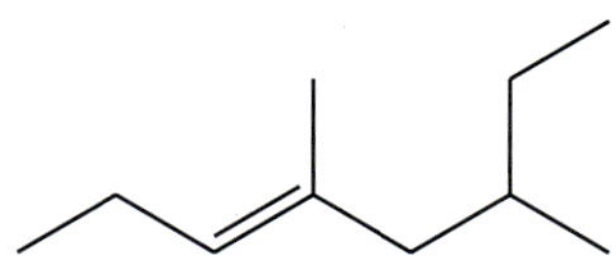

3.7

We still choose the longest chain, but this time it *must* contain the double bond, and the chain is numbered so that the double bond has the lowest indicative number as in **3.8**. So this is a 3-octene (it's only necessary to give the number of one of the carbon atoms of the double bond—the other is then obvious), and when we add in the substituents, it's 4,6-dimethyl-3-octene. Notice that the longest chain here does not run across the page. If we had started at the other end of the molecule, then the double bond would have been between C-5 and C-6, rather than C-3 and C-4. We'll return to the issue of geometry about double bonds shortly.

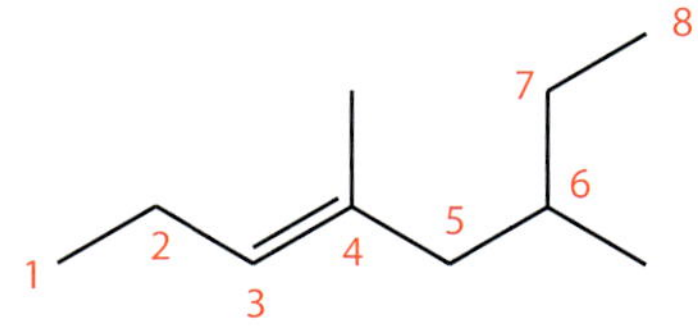

3.8

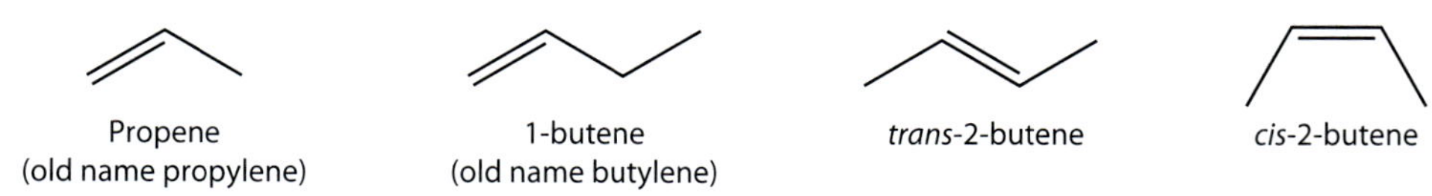

FIGURE 3.6 Simple alkenes.

When naming cycloalkenes, the numbering starts with the double bond as C-1 and C-2 and then goes around the ring to place the substituents on the lowest numbered carbon atoms. Thus, structure **3.9a** is correctly numbered, and structure **3.9b** is not. This molecule would be named as *cis*-3,4-dimethylcyclohexene.

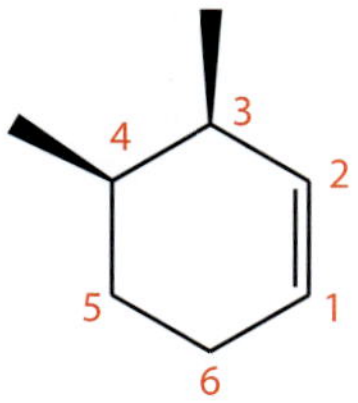

3.9a✓

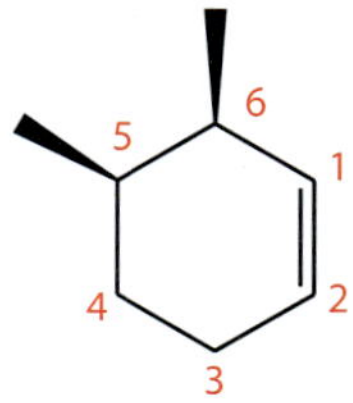

3.9b×

If there is more than one double bond in the molecule, then they are described as diene, triene, *etc*. Thus, **3.10** is 1,4-pentadiene and **3.11** is 1,3,7-octatriene.

3.10

3.11

PROBLEM 3.2

Provide a line drawing for each of the following molecules:

(a) 6-Chloro-2,4-dimethyl-1,4-octadiene
(b) 1,5-Cyclooctadiene

PROBLEM 3.3

Give a systematic name for each of the following compounds:

(a)

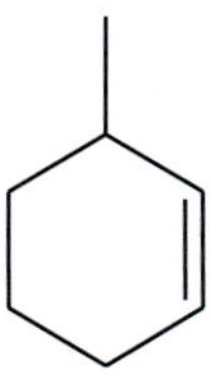

(b)

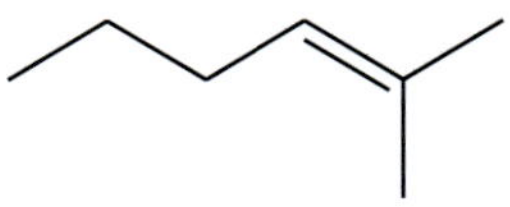

Solutions

3.2(a) First draw and number 1,4-octadiene:

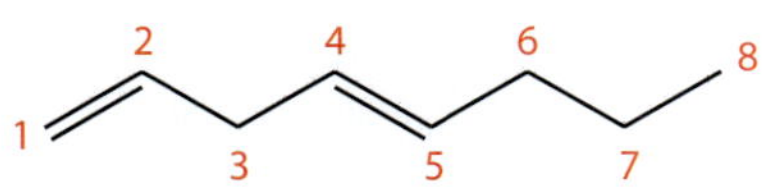

1,4-octadiene

We can then place the substituents at the appropriate positions to give

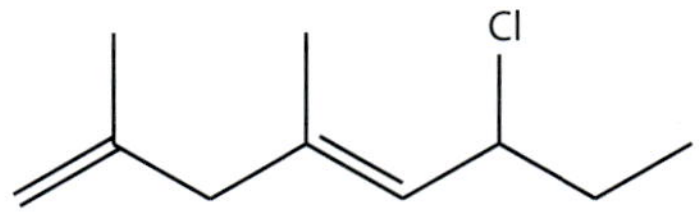

3.2(b)

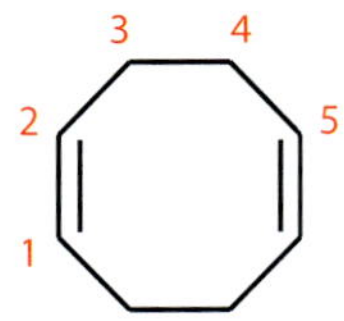

3.3(a) Start by ignoring the substituent, then this will be cyclohexene. In order to get the proper numbering for the molecule, the carbons of the double bond must be C-1 and C-2, and we number the ring counterclockwise so that the methyl substituent is at C-3.

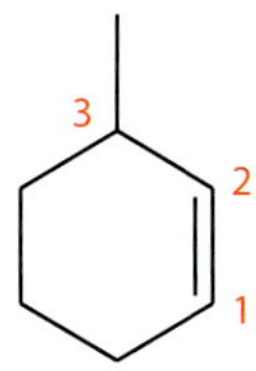

So this is 3-methylcyclohexene. There is no need to specify the location of the double bond—the requirements of the numbering scheme mandate its being between C-1 and C-2.

3.3(b) The longest chain has six atoms, and we number from the right, so that the double bond is between C-2 and C-3.

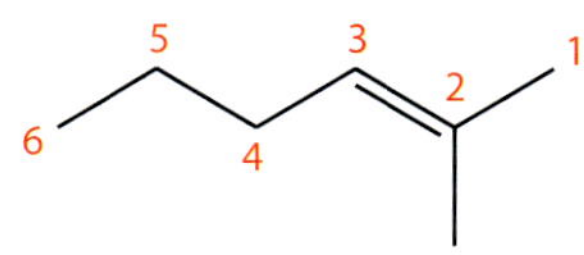

So this is 2-methyl-2-hexene.

3.1.3 GEOMETRIC ISOMERISM IN ALKENES

As we have just seen, 2-butene exists in *cis*- and *trans*-forms—the methyl groups may be on the same side (*cis*) or opposite sides (*trans*) of the double bond. These are usually described as geometric isomers and are distinct from the structural isomers discussed in the last chapter. All the same bonds are present, but the directional property of the bonds means that they are distinct compounds with different chemical and physical properties. This is a subclass of stereo-isomers, which we will discuss in more detail in Chapter 7. The terms *cis* and *trans* are adequate for simple structures with just two substituents on the double bond, but how would we describe the geometry in **3.12**?

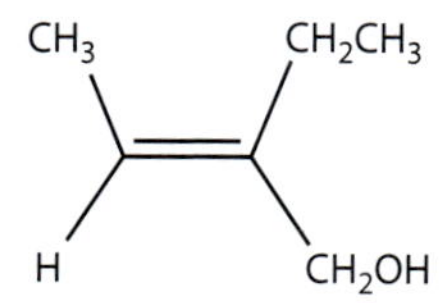

3.12

First, we "rank" the substituents at each end of the double bond, and to do this, we use a method called the Cahn–Ingold–Prelog system. This is a very powerful method, designed to cope with almost any eventuality, and is used in ranking groups to describe stereochemistry in a range of compounds, not only alkenes. *All* the rules are listed here, with examples for the most important; for now, simply skip the last two, which will seem obscure, until we return to them later:

- Atoms are ranked in order of atomic number, highest atomic number first:
  Br > Cl > C > H > lone pair of electrons
- If the first atom is the same, continue down the chain until a difference is found:
  $-CH_2CH_3 > CH_3$ because C > H
  $-CH(CH_3)_2 > -CH_2CH_3$ because two carbons outrank one carbon
- Multiple bonds are treated as two bonds, with a phantom atom (in parentheses below) added in.

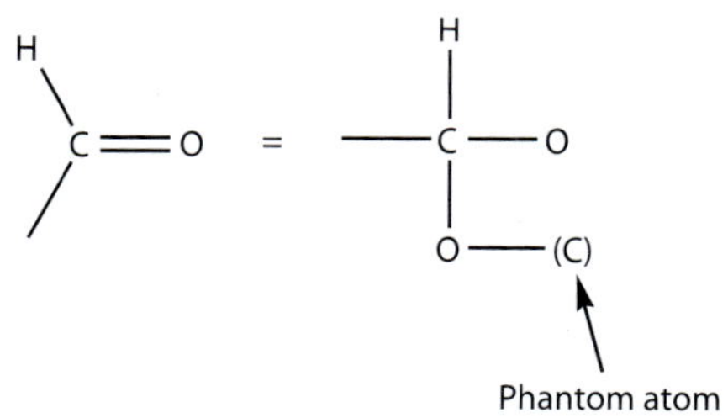

  $-CHO > -CH_2OH$, because two bonds to oxygen outrank one bond to oxygen
- Heavier isotopes outrank lighter isotopes
  T > D > H (tritium, deuterium, protium, or ^3H, ^2H, ^1H)
- *Z* > *E*
- *R* > *S*

So in **3.12**, CH_3 > H and $CH_2OH > CH_2CH_3$. In this molecule, the higher priority groups are on opposite sides of the double bond. The geometry is designated *E*, from the German word *entgegen*, which means apart. Looking at another example (**3.13**), the phenyl ring outranks the methyl group, and the nitrogen outranks the carbon. So the higher priority groups are on the same side of the double bond, and we designate this *Z*, from the German *zusammen*, together.

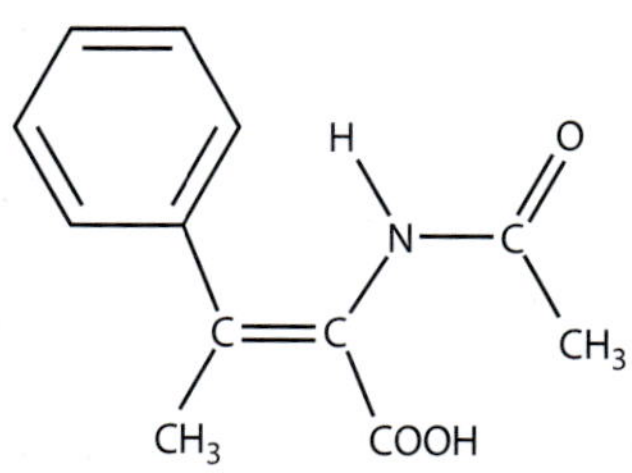

3.13

PROBLEM 3.4

For each of the following sets of substituents, rank them according to the Cahn–Ingold–Prelog system:

(a) –H, –D, $-CH_3$, –OH, $-N(CH_3)_2$
(b) $-CH_3$, $-CH_2CH_3$, $-CH(CH_3)_2$, $-CH_2CH_2CH_3$

PROBLEM 3.5

Designate the double bond geometry as *Z* or *E* for each of the following molecules:

(a)

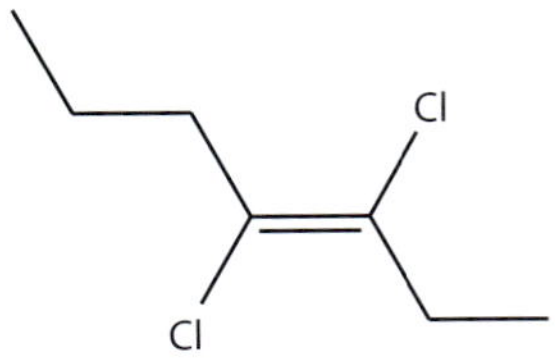

(b)

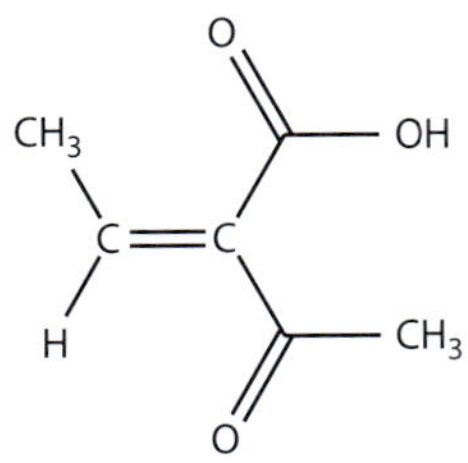

Solutions

3.4(a) –OH > $-N(CH_3)_2$ > $-CH_3$ > –D > –H
3.4(b) $-CH(CH_3)_2$ > $-CH_2CH_2CH_3$ > $-CH_2CH_3$ > $-CH_3$
3.5(a) At each end of the double bond, chlorine takes precedence over carbon so the priority groups are on opposite sides of the double bond, and we would designate this as *E*.
3.5(b) CH_3 outranks hydrogen and –C(=O)OH outranks $-C(=O)CH_3$ because oxygen outranks carbon. So the priority groups are on the same side of the molecule, and this will be designated as *Z*.

3.1.4 STABILITY OF ALKENES AND THE EFFECT OF CONJUGATION

As with alkanes, we can estimate alkene stability from molar heats of combustion, bearing in mind that more carbon always means more heat. The heats of combustion of the molecules with formula C_4H_8 are given in Table 3.3. We remember from the previous chapter that the greater the magnitude of the heat of combustion, the less stable the species. We can draw two conclusions from this table. First, *trans*-alkenes are more stable than *cis*-alkenes. Intuitively, we would have

TABLE 3.3 Molar Heats of Combustion of Alkenes of Formula C_4H_8

Alkene	Structure	ΔH_c (kJ mol^{-1})
1-Butene		−2716.8
cis-2-Butene	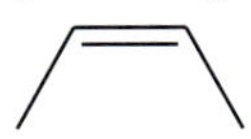	−2709.8
trans-2-Butene	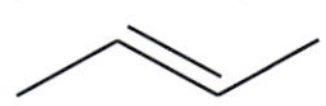	−2706.6
2-Methylpropene		−2699.5

expected this—it is a steric effect, because the energy is lower when the larger substituents are kept apart. The second conclusion is less intuitively obvious—the more substituted an alkene, the more stable it is. This is attributed to a phenomenon called *hyperconjugation*, involving interaction between a C–H σ-bond and the π-bond.*

Heats of hydrogenation of alkenes also reflect their stability. If we consider the linear butenes, all three are hydrogenated in the presence of a platinum catalyst to give butane (Figure 3.7). The reactions are all exothermic, and the heats of hydrogenation are given in Table 3.4.

When double and single bonds alternate in a molecule, then we describe the bonds, and the molecule, as being *conjugated*. Although here we will be concerned with carbon–carbon double bonds, conjugation can involve any sort of multiple bond—examples of conjugated species are given in Figure 3.8 and non-conjugated multiple bonds, which do not interact, in Figure 3.9.

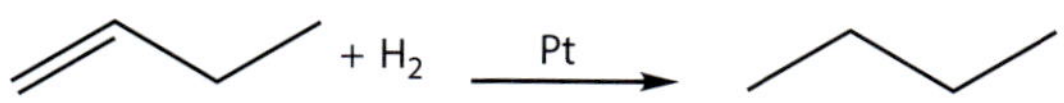

FIGURE 3.7 Hydrogenation of 1-butene.

TABLE 3.4 Heats of Hydrogenation of Linear Butenes

Alkene	Structure	$\Delta H_{hydrogenation}$ (kJ mol^{-1})
1-Butene		−126.5
cis-2-Butene		−119.6
trans-2-Butene		−115.5

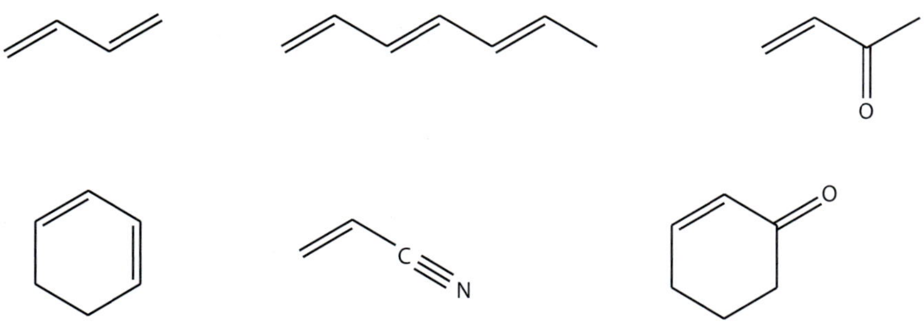

FIGURE 3.8 Molecules containing conjugated multiple bonds.

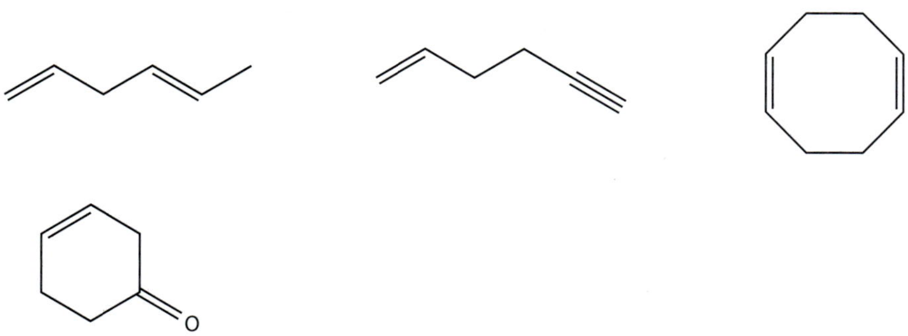

FIGURE 3.9 Molecules containing non-conjugated, non-interacting, multiple bonds.

* Most students don't believe in hyperconjugation until they have the theoretical background to do serious calculations—so for now, just accept that there is a reason for the effect and that the effect is real.

PROBLEM 3.6

In which of the following molecules (a–e) would you describe the double bonds as conjugated?

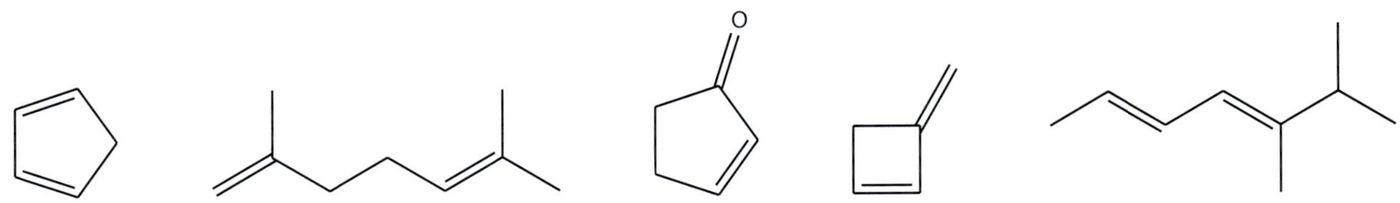

(a) (b) (c) (d) (e)

Solution a, c, d, e

Both 1,3-pentadiene (conjugated) and 1,4-pentadiene (non-conjugated) are hydrogenated in the presence of a platinum catalyst to give pentane (Figure 3.10), but the heats of hydrogenation differ significantly. From these data, we can conclude that conjugation lowers energy and increases the stability of a molecule. The difference in the heats of hydrogenation reflects the difference in stability between the conjugated and the non-conjugated diene—shown schematically in Figure 3.11.

There is other physical evidence that shows that conjugation has a profound effect on the properties of a molecule (see also Section 3.2.1). The lowest-energy state of butadiene is planar, despite the fact that there should be free rotation about the C2–C3 single bond, and steric hindrance is not minimized in the planar form. Also, this C2–C3 single bond is *shorter* than one might expect for a single bond, 1.454 Å, rather than the expected 1.47 Å (a bond between an sp^2 and an sp^3 carbon is always shorter than a single bond between two sp^3 carbon atoms, because the sp^2 orbital is less extended). The double bonds of butadiene (1.338 Å) are slightly longer than those in ethene (1.3305 Å).

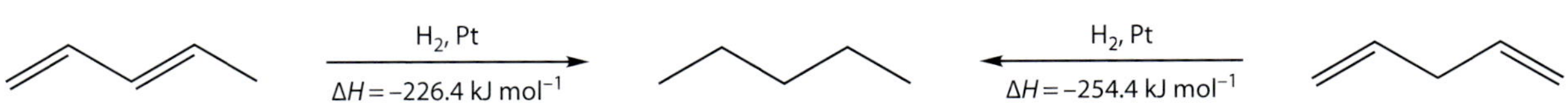

FIGURE 3.10 Hydrogenation of 1,3- and 1,4-pentadiene.

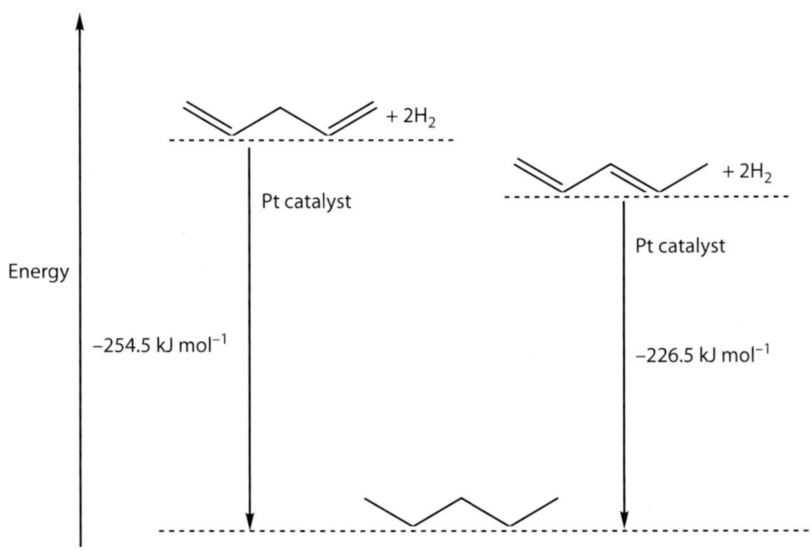

FIGURE 3.11 Energetics of pentadiene hydrogenation.

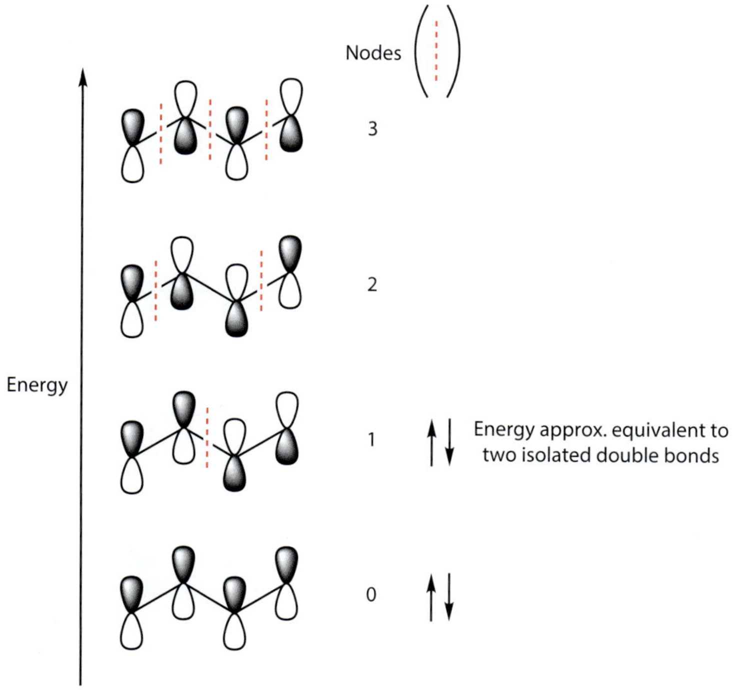

FIGURE 3.12 Simplified picture of the molecular orbitals of butadiene.

Although, as always, there is some complex mathematics lurking in the background, we can explain conjugation in terms of a relatively simple molecular orbital picture. For butadiene, all the carbon atoms are sp^2 hybridized, and we use these sp^2 hybrid orbitals to make the C–H and C–C σ-bonds. There are then four p_z orbitals and four electrons to make the π-orbitals. In ethene, we combined two p_z orbitals to give two molecular orbitals, bonding and antibonding, π and π*. With butadiene, we will combine all four p_z orbitals to form four molecular orbitals, shown in Figure 3.12. When drawing such schemes for yourself, you will find it easiest to start with the best, and worst, arrangements for overlap—these are always the easiest to draw. Count the nodes in the most antibonding orbital and the missing ones in between will have successively fewer until the lowest-energy orbital has no nodes (other than the one in the plane of the molecule derived from the node in all p orbitals). For butadiene, the lowest-energy orbital is strongly bonding; the next one up is bonding, about equivalent to isolated double bonds; and the third and fourth orbitals are successively more antibonding. There are four electrons to be accommodated, and these will be in the lowest two, the bonding, orbitals. By considering the molecular orbitals as belonging to the conjugated system as a whole, rather than to specific double bonds, the overall energy is decreased.

PROBLEM 3.7

By analogy with butadiene, draw a schematic of the six π-molecular orbitals of hexatriene, showing the phases of the p_z orbitals and nodes.

Solution

The following are key points in approaching the problem:

- Don't worry about the shape of the molecule—focus on the orbitals.
- Draw the most bonding and the most antibonding orbitals first—these are always the easiest.
- The most bonding orbital has no nodes, the most antibonding five nodes. The intervening ones will have 1,2,3,4, respectively.
- If there is an odd number of nodes, one will always be at the midpoint of the molecule.
- Nodes must always be distributed symmetrically about the midpoint of the molecule.

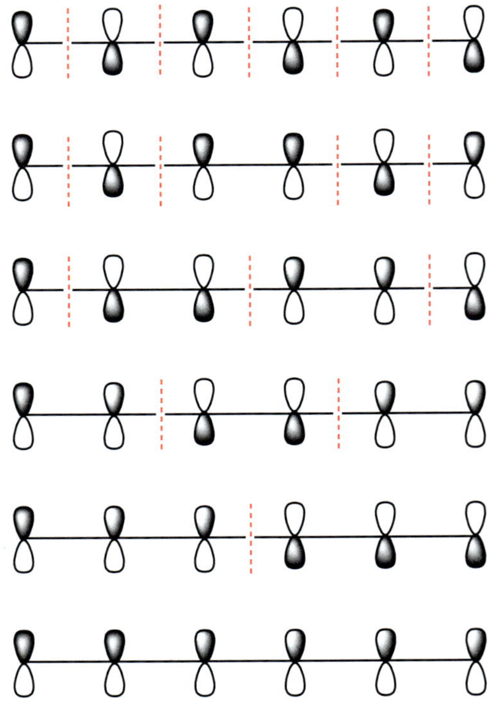

Focus on Naturally Occurring Polyenes

A number of long-chain polyenes give rise to the colors in fruits and vegetables. β-Carotene (**3.14**) is the red-orange pigment in carrots, pumpkin, sweet potato, mango, papaya, cantaloupe, spinach, green beans, peas, and lettuce. Its purpose in plants is to protect plant cells against excess ultraviolet (UV) light. You may be surprised to realize that there are orange pigments in many green leafy vegetables—but in Chapter 5, we will look at a technique that allows us to demonstrate this. The much stronger chlorophyll pigments mask the yellow to orange shades, as they do in the leaves of most trees in summer. When the leaves stop chlorophyll production, in the fall, the yellow and orange pigments may be seen. β-Carotene is needed in the diet to make vitamin A. Retinol and retinal (components of vitamin A, **3.15** and **3.16**) are important light-absorbing molecules in the retina. The yellow color of fat in humans and other omnivores such as chickens arises because omnivores are relatively inefficient in converting carotenes to vitamin A, and thus, some carotene is stored. Herbivores process carotenes better, and obligate carnivores, such as cats and ferrets, not at all.

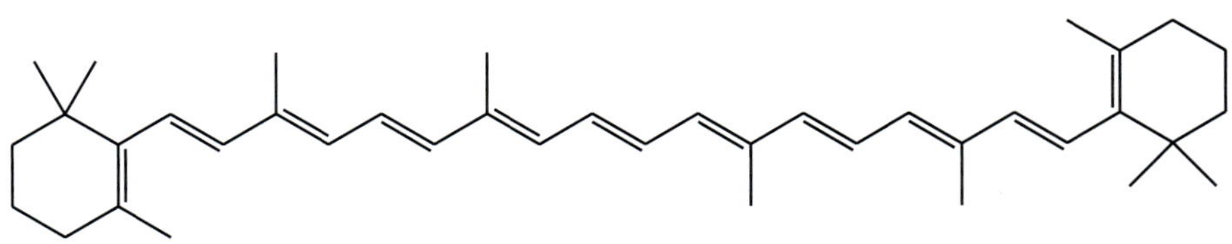

3.14

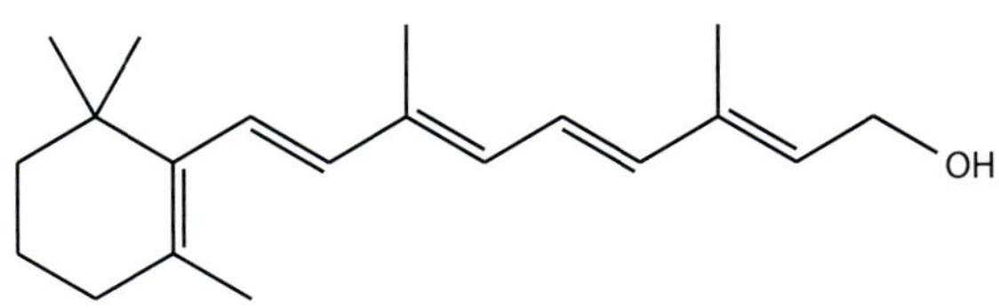

3.15

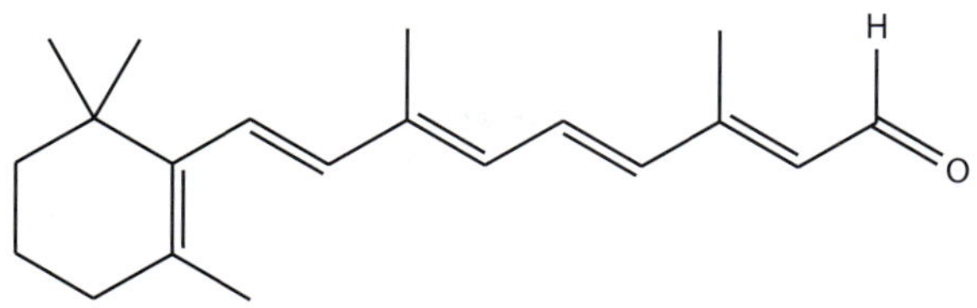

3.16

Carotene-containing fruits and vegetables

The bright red pigment found in tomatoes (also in carrots, watermelons, papaya) is lycopene (**3.17**). It is key intermediate in the biosynthesis of carotenes (using an enzyme called lycopene cyclase), but lycopene from the diet cannot be converted to vitamin A. Both carotene and lycopene are known to be antioxidants and are supposed to retard the aging process.

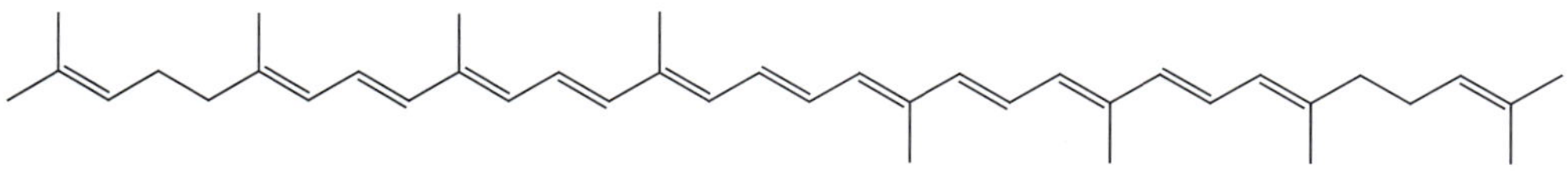

3.17

Lycopene-containing fruits and vegetables

Key Points from Section 3.1

- Alkenes are compounds containing carbon–carbon double bonds. The double bond consists of one σ- and one π-bond. The σ-bond is formed from two sp^2 hybrid orbitals, and the π-bond is formed from two p_z orbitals. Alkenes are planar. The more substituted an alkene, the more stable it is.
- In naming alkenes, we select the longest carbon chain that contains the double bond and number it so that the double bond has the lowest possible number. Substituents are then located and specified.
- The stereochemistry of alkenes is specified by ranking the substituents at each end of the double bond according to the Cahn–Ingold–Prelog system. When the higher-ranking substituents are on the same side of the double bond, the stereochemistry is *Z*. When they are on opposite sides of the double bond, the stereochemistry is *E*.
- Alkenes with alternating double and single bonds are described as conjugated; conjugation significantly lowers energy and increases stability.

3.2 AROMATIC HYDROCARBONS

The formula of benzene as C_6H_6 was established early in the nineteenth century, but its structure was long problematic. Structures such as **3.18** could be written, but the reactions of benzene are quite unlike those of alkenes or alkynes. Benzene is extremely stable, and the characteristic reaction is substitution (**3.1**) rather than the addition typical of alkenes (**3.2**):

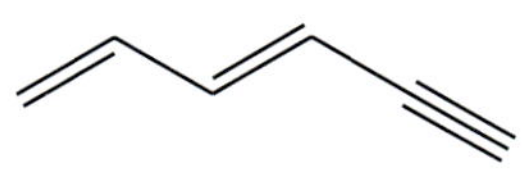

3.18

$$C_6H_6 + Br_2 \rightarrow C_6H_5Br + HBr \qquad \text{Substitution} \tag{3.1}$$

$$C_2H_4 + Br_2 \rightarrow C_2H_4Br_2 \qquad \text{Addition} \tag{3.2}$$

The structure of benzene as a regular hexagon with alternating double and single bonds is attributed to Kekulé, who claims to have fallen asleep on a horse-drawn bus and dreamt of the molecule as snakes of atoms:

> Atoms were gamboling before my eyes, twisting and twining in snakelike motion. But look! One of the snakes has seized its own tail ….
>
> **Kekulé (1865)**

From this, he proposed the cyclic structure with alternating double and single bonds. Although Kekulé is credited with this idea, this may be mistaken; in 1861, Josef Loschmidt had published ring structures for benzene and its derivatives (Figure 3.13), and all the evidence suggests that Kekulé had read this. Both Kekulé and Loschmidt have been commemorated for this work on special issue stamps.

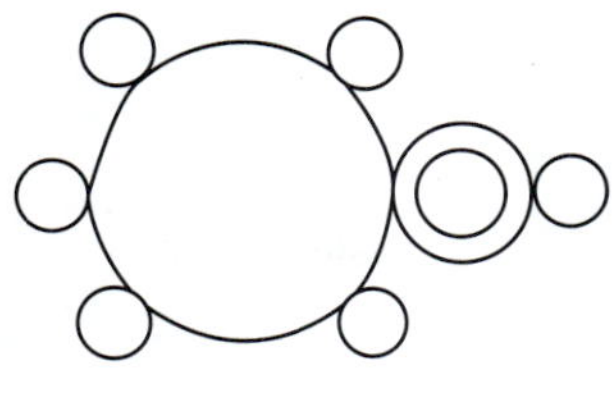

Phenol

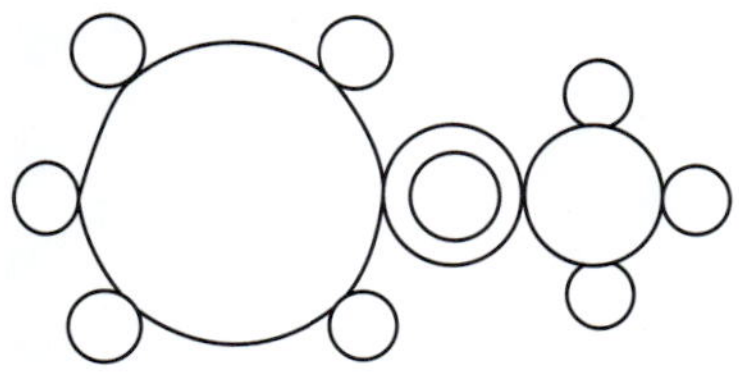

Anisole

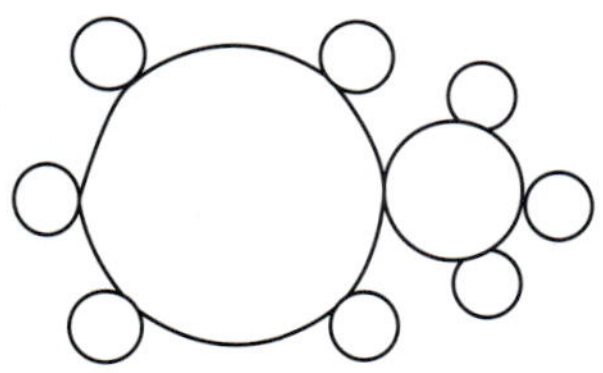

Toluene

FIGURE 3.13 Loschmidt structures of aromatic compounds, 1861.

Commemorative stamps for Kekulé and Loschmidt

3.2.1 BONDING IN AROMATIC COMPOUNDS

Kekulé's original proposal for the structure of benzene involved a ring with alternating double and single bonds, but it was quickly clear that there were some problems with this idea. For example, there was only one 1,2-dibromobenzene—the isomer with the two bromines at either end of a double bond (**3.19**) was exactly the same as the isomer with them at opposite ends of a single bond (**3.20**). Modern structure determination shows that all the carbon–carbon bonds in benzene are identical and 1.4 Å in length (intermediate between carbon–carbon single bonds at 1.54 Å and carbon–carbon double bonds at 1.34 Å). This suggests that the carbon–carbon bonds in benzene correspond to a bond order

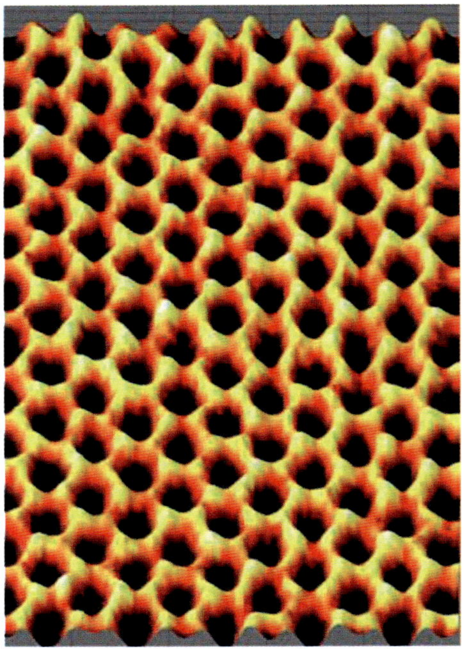

FIGURE 3.14 Image of a single suspended sheet of graphene taken with TEAM 0.5, showing individual carbon atoms (yellow) on the honeycomb lattice. (Courtesy of Steven Pennycook.)

of about 1.5. The problem is resolved by considering two *resonance forms* **3.21a** and **3.21b**. The true structure of benzene lies halfway between these two extremes. Note that this is not an equilibrium, with the two forms interchanging—the structure of benzene is halfway between the two *all the time*. So the two resonance forms are connected not with an equilibrium arrow but with a new symbol, a double-headed arrow. The perfectly hexagonal structure of benzene has been confirmed in recent years by various new techniques in microscopy—a recent image is shown in Figure 3.14.

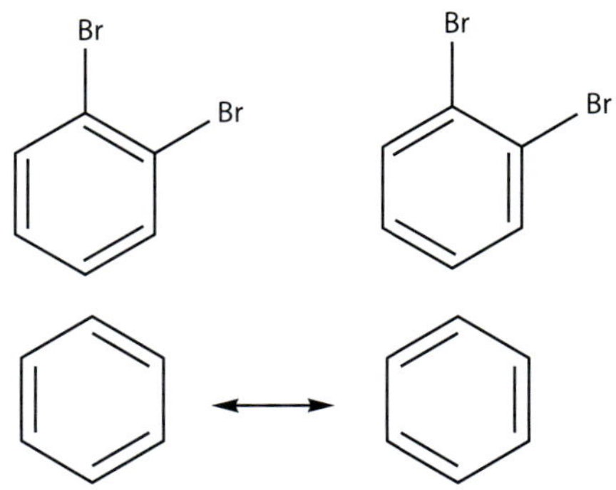

3.21a **3.21b**

The energy the molecule gains by this resonance is described as the resonance energy. This is estimated from heats of hydrogenation for cyclohexene and benzene. When cyclohexene is reduced to cyclohexane (Figure 3.15), 121 kJ mol^{-1} of energy is released. Thus, we might expect three times as much energy to be released from the hydrogenation of benzene. So we expect a hypothetical cyclohexatriene to yield 363 kJ mol^{-1} on complete hydrogenation, but only 206 kJ mol^{-1} is actually obtained. Therefore, benzene is more stable than a hypothetical cyclohexatriene by 157 kJ mol^{-1}, the resonance energy of benzene. This is shown schematically in Figure 3.16.*

We can account for this large difference in energies by looking at the relevant molecular orbitals. The σ-skeleton of benzene is constructed by using sp^2 hybrid orbitals at each carbon

* You may notice in passing that conjugation in 1,3-cyclohexadiene lowers its energy by about 10 kJ mol^{-1} relative to two isolated double bonds.

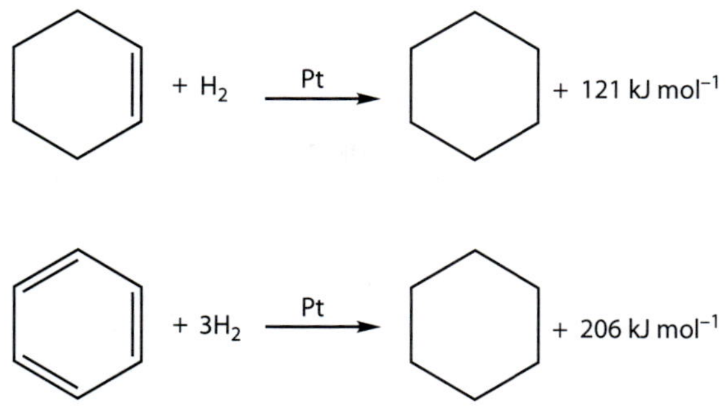

FIGURE 3.15 Hydrogenation of cyclohexene and benzene.

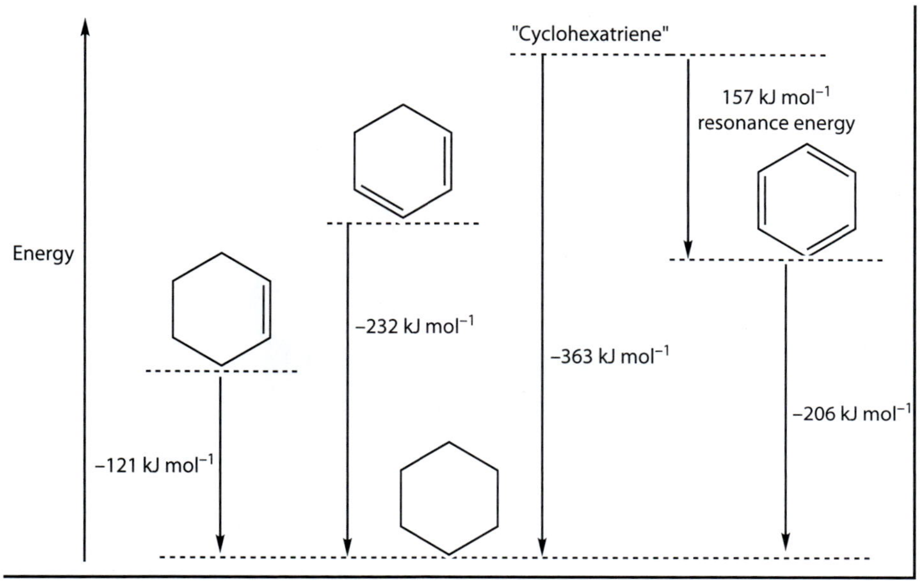

FIGURE 3.16 Energy diagram for benzene and cyclohexene hydrogenation.

atom to form the C–C and C–H σ-bonds. This leaves us with six p_z orbitals and six electrons. One simplistic way of looking at this is to say that these orbitals interact to form an annulus of electron density above and below the plane of the six-membered ring. This leads to the common shorthand drawing of benzene as **3.22**. This is fine as shorthand as long as nothing is happening in the benzene ring but should be avoided otherwise. An energy diagram of the molecular orbitals of benzene is given in Figure 3.17, where six molecular orbitals have been constructed from the six p_z orbitals. All of the electrons can be placed in bonding orbitals, and there is considerable energy gain (the resonance energy) relative to considering isolated double bonds.

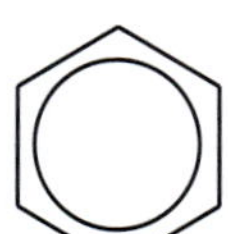

3.22

In the lowest-energy orbital, all the p_z orbitals interact in phase (Figures 3.18 and 3.19). Putting in one node can be done in two ways, either through atoms or through bonds, and these two molecular orbitals (Figure 3.20) are the other two bonding orbitals. Despite their different appearance, they do have the same energy. For the antibonding orbitals of benzene, two (of identical energy) have two nodes (Figure 3.21), and the final one, the most antibonding, has three nodes (Figure 3.22).

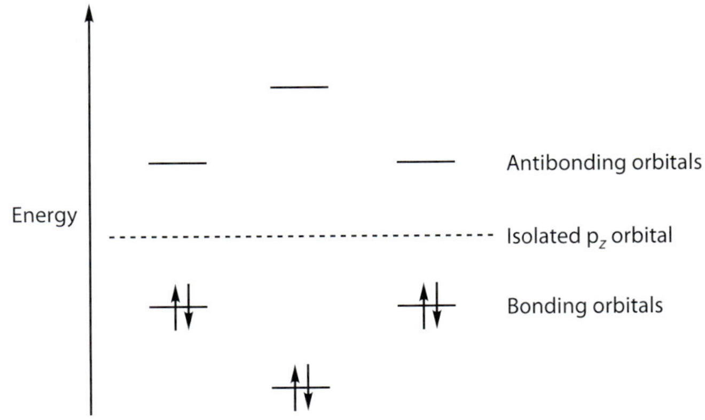

FIGURE 3.17 Energy diagram of the molecular orbitals of benzene.

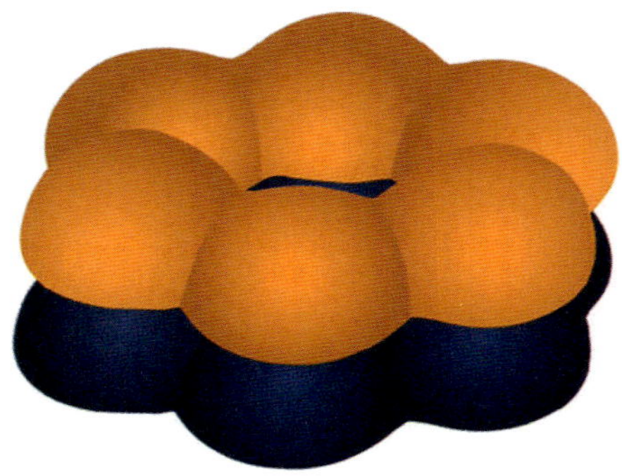

FIGURE 3.18 Lowest-energy π-molecular orbital of benzene (oblique view).

Given what we have learned about benzene, we might expect other cyclic conjugated polyenes to behave in the same way. But this is not so—why is benzene very stable and cyclobutadiene (**3.23**) very unstable? Cyclobutadiene can be isolated only at 4 K, in a solid argon matrix. It is not square but rectangular with real double and single bonds with different lengths. If we examine the energy levels of the molecular orbitals of cyclobutadiene (Figure 3.23), Hund's rule mandates that once the lowest-energy orbital is filled, one electron must go into each of the next two with

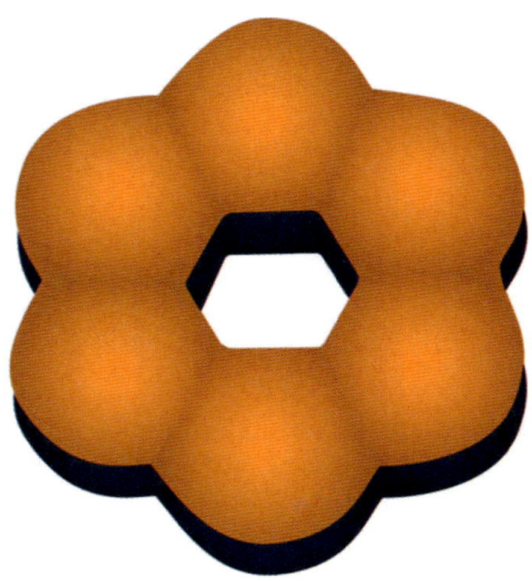

FIGURE 3.19 Lowest-energy π-molecular orbital of benzene (top down view).

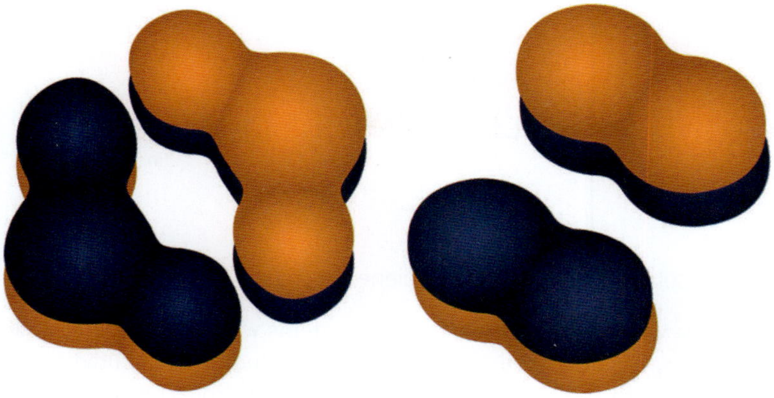

FIGURE 3.20 Bonding molecular orbitals of benzene with one node.

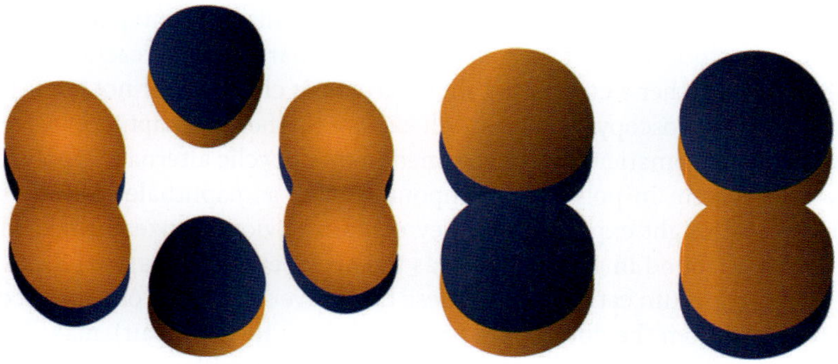

FIGURE 3.21 Molecular orbitals of benzene containing two nodes.

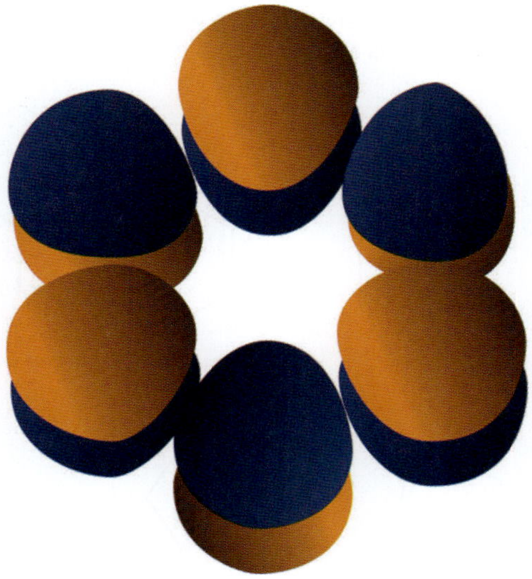

FIGURE 3.22 Molecular orbital of benzene containing three nodes.

the same energy. So not only is little achieved energetically by delocalization, but there would also be unpaired electrons, making the molecule very reactive.

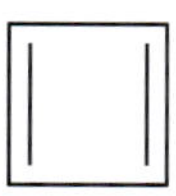

3.23

The stability or otherwise of monocyclic alternating polyenes has been codified as the Hückel rule. If there are $4n + 2$ π-electrons, the system will be stable, and we describe it as aromatic.

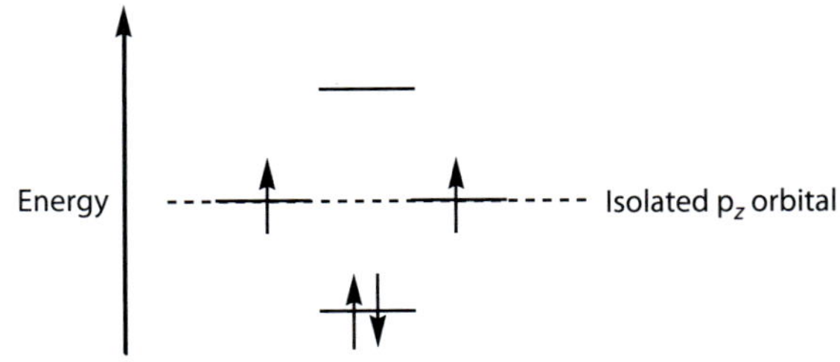

FIGURE 3.23 Molecular orbitals of cyclobutadiene.

This includes benzene with six π-electrons but also systems with 2, 10, 14, *etc.*, π-electrons. If the ring has 4*n* π-electrons, it is described as antiaromatic—including cyclobutadiene with four electrons but also systems with 8, 12, 16, *etc.*, π-electrons. What properties are specific to aromatic compounds? We have already seen that the additional stability is reflected in heats of hydrogenation, and this is also true for other thermodynamic parameters such as heat of combustion or heat of formation. Aromatic molecules are planar, or nearly so—this is essential for proper interaction of the p_z orbitals. Substitution, rather than addition, is the characteristic reaction. The easiest practical measurement of whether a compound shows aromatic character or not comes from nuclear magnetic resonance spectroscopy, which we will explore further in Chapter 6.

The phenomenon of aromaticity is not confined to monocyclic alternating polyenes; aromatic properties are also evident in polycyclic compounds such as naphthalene (**3.24**) and phenanthrene (**3.25**). This we might expect, since they contain obvious benzene rings. However, aromatic character is also found in anionic (such as the cyclopentadienyl anion, **3.26**) and cationic species (such as the tropylium cation, **3.27**). These both have six π-electrons (in the cyclopentadienyl anion, this is four from the double bonds and two from the lone pair), making them Hückel compliant systems, and the electrons are fully delocalized. We will explore the resonance forms of these systems more fully later in this chapter. The energy levels for the molecular orbitals of the cyclopentadienyl system are shown in Figure 3.24. This shows why the cyclopentadienyl cation (**3.28**), a 4 π-system, will be profoundly unstable—little energy gain and unpaired electrons. The tropylium cation is rather stable; **3.29** is a stable isolable salt.

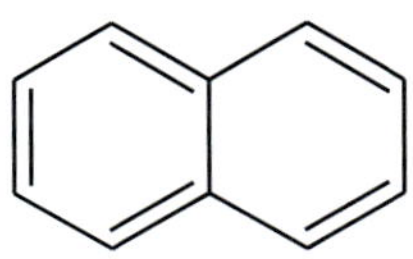

3.24, naphthalene

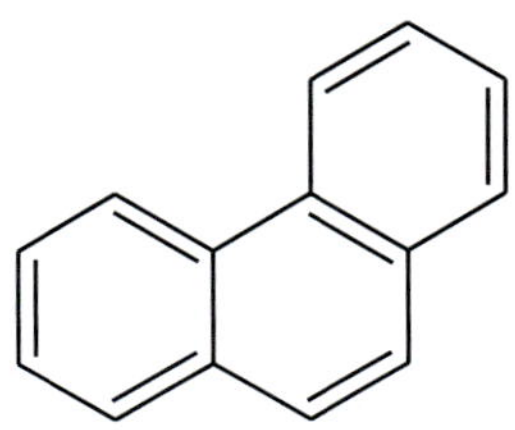

3.25, phenanthrene

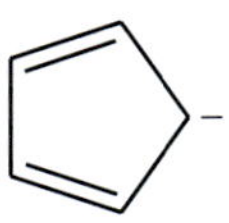

3.26

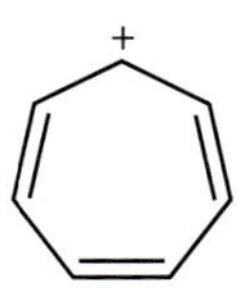

3.27

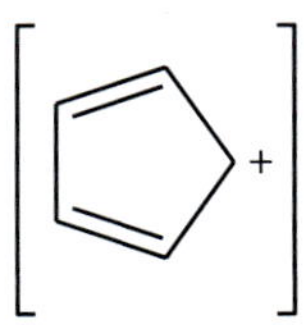

3.28

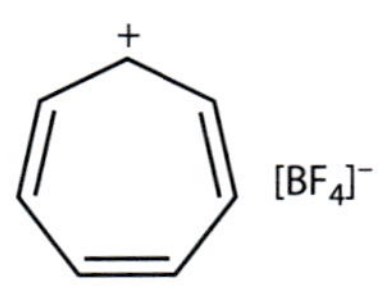

3.29

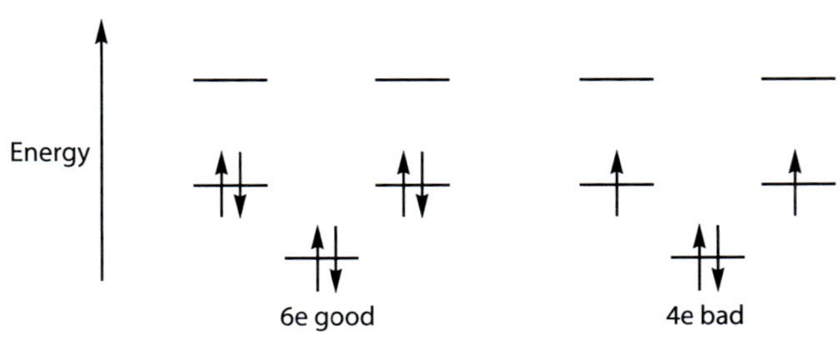

FIGURE 3.24 Energies of the molecular orbitals in the cyclopentadienyl system.

In the cyclopentadienyl anion, the aromatic sextet of electrons is made up of four π-electrons from the two double bonds and a lone pair of electrons on carbon. But the lone pair of electrons need not be associated with a carbon atom. Lone pairs of electrons on oxygen, sulfur and nitrogen can also form a part of an aromatic system. Although the delocalization of the lone pair is less in **3.30**, **3.31**, and **3.32** (furan, pyrrole, and thiophene) than in the cyclopentadienyl anion, there is still some aromatic character and a significant resonance energy. Like benzene, these compounds generally undergo substitution rather than addition, and they are planar. The hybridization of the heteroatom (a term used for any noncarbon atom) is not the same as we would expect in water or ammonia. One lone pair of electrons in furan is in a p_z orbital and interacts with the rest of the π-system, and one is in an sp^2 orbital in the plane of the ring (**3.33** shows the direction and interactions of the lone pairs on oxygen in furan). As we will see later (Chapter 12), this has a major effect on the properties and reactivity of these molecules.

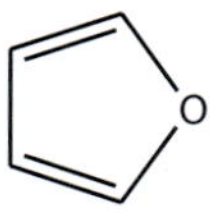

3.30, furan

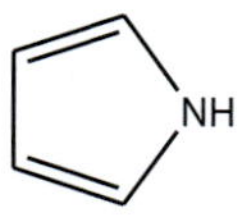

3.31, pyrrole

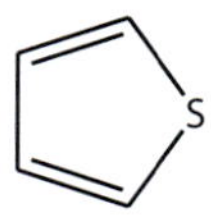

3.32, thiophene

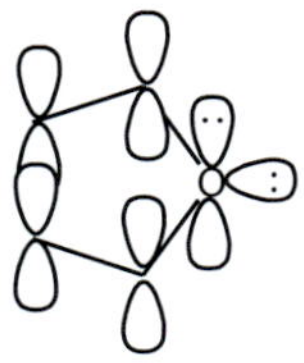

3.33

PROBLEM 3.8

How many π-electrons are involved in resonance in each of the following molecules? Classify each as aromatic, antiaromatic, or nonaromatic.

(a)

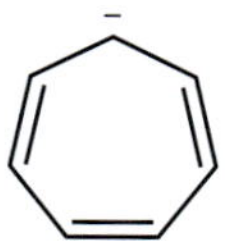

(b)

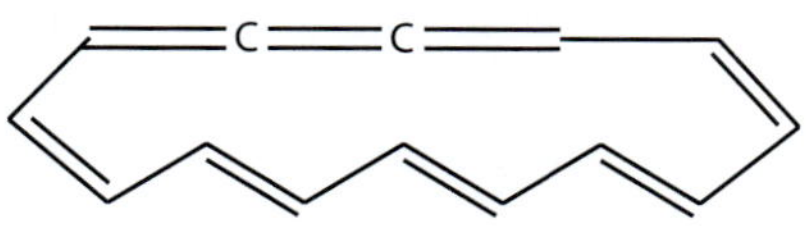

(c)

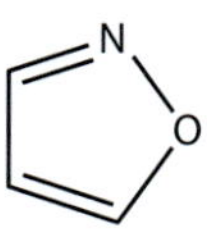

Solutions

(a) Eight π-electrons, antiaromatic.
(b) This is a type of molecule we have not seen before—it has double bonds that are not conjugated but adjacent. The class of compounds is allene, when there are two such bonds, or cumulene, when there are more. The simple alkene conjugated double bonds are easy to count up—5 double bonds, 10 electrons. But we cannot simply count all the electrons in the π-system of the cumulene, because, geometrically, they will not all be able to participate in the conjugated system. When we consider the orbitals of the first double bond of the sequence, we make the π-bond using the p_z orbitals on each carbon atom.

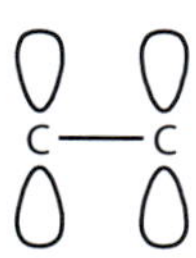

But the p_z orbital on C-2 has now been used and the only option to make the next π-bond is to use p_y, and we must use p_y on both C-2 and C-3 for the overlap we need to make the bond. The p_z orbitals are shown in red and the p_y in blue.

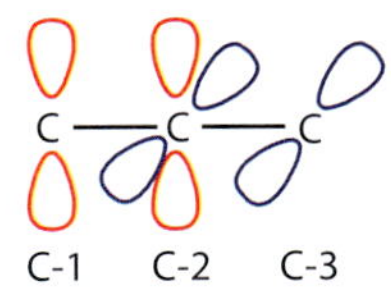

To make the third π-bond, we have to return to p_z orbitals, as the p_y at C-3 has already been used.

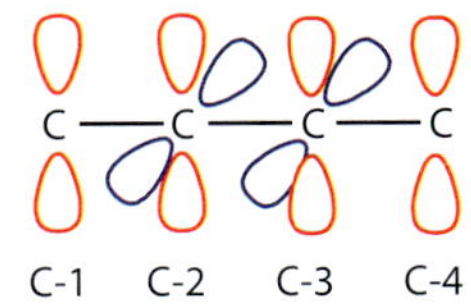

In terms of the conjugation of the entire π-system, we can only count the electrons in bonds made from p_z orbitals. The electrons in the bond made from p_y orbitals cannot participate in the system as it is at 90 ° to the other participating orbitals. Thus, in this molecule, we have a 14 π-system, and this is Hückel rule compliant and this will be an aromatic system.

(c) This molecule is an analog of furan, has six electrons (four from double bonds and two from the oxygen lone pair), and will show aromatic character. Don't be distracted by the second lone pair on oxygen or the lone pair on nitrogen. These atoms both are sp^2 hybridized and have lone pairs in the plane of the ring at 90 ° to the π-system.

3.2.2 NOMENCLATURE OF BENZENE DERIVATIVES

Many benzene derivatives have been known for well over a century and hence have common names that are generally used in preference to the formal systematic names. Some of the common monosubstituted benzenes with "common" names are shown in Figure 3.25. It is difficult to predict whether these names will continue to be used in the long term (phenol used to be called carbolic acid, so names are not fixed forever); toluene, aniline, and phenol show no signs of disappearing, styrene is very important to the polymer industry, but anisole is starting to fade.

When there are two substituents, their relative positions must be specified; thus, **3.34**, **3.35**, and **3.36** are named as 1,2-, 1,3-, and 1,4-dimethylbenzene. The common name for dimethylbenzene is xylene, and given its wide use as an industrial solvent, this name is likely to survive

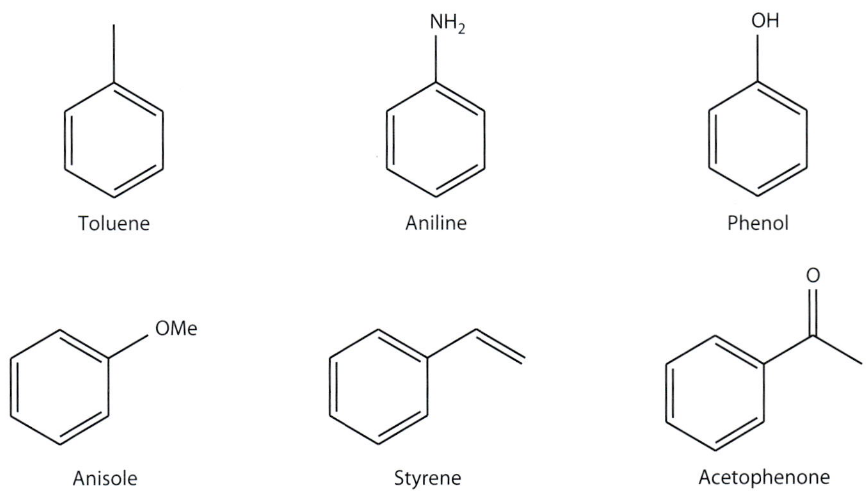

FIGURE 3.25 Some common names of monosubstituted benzene derivatives.

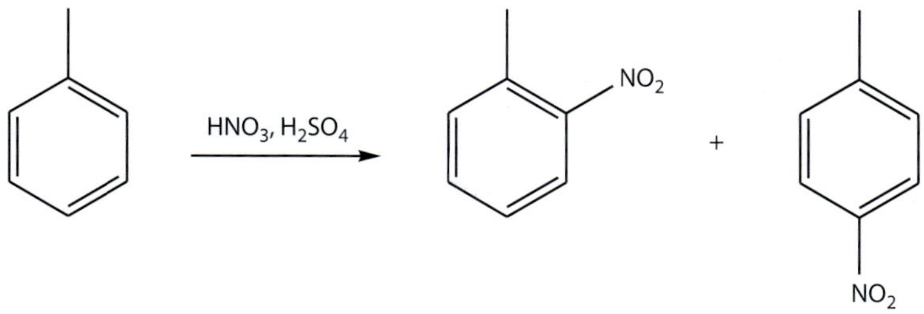

FIGURE 3.26 Nitration of toluene.

for some time. However, there is a parallel system for describing the relationships between substituents on aromatic rings. For a 1,2-relationship, the substituents are *ortho*, for 1,3— *meta*, and for 1,4— *para*. So **3.34**, **3.35**, and **3.36** could also be described as *ortho*-xylene, *meta*-xylene, and *para*-xylene. These terms are always italicized in typescript and are used descriptively as well as in nomenclature. So in **3.37**, we might say that the two nitro groups are *meta* to each other, though this would not be part of the name. In Figure 3.26, we would say that substitution had occurred *ortho* and *para* to the methyl group.

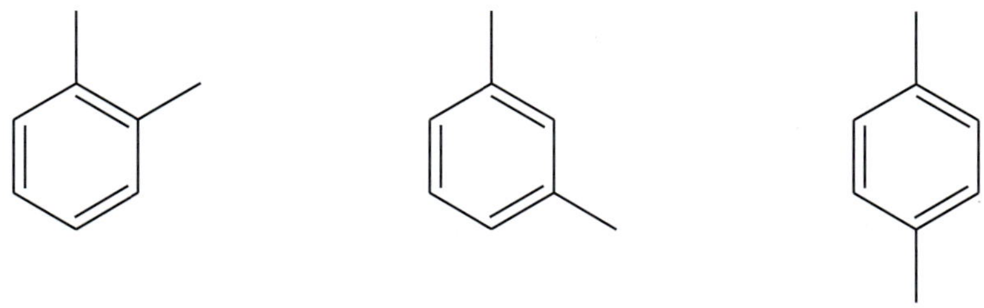

3.34, 1,2-dimethylbenzene *ortho*-xylene

3.35, 1,3-dimethylbenzene *meta*-xylene

3.36, 1,4-dimethylbenzene *para*-xylene

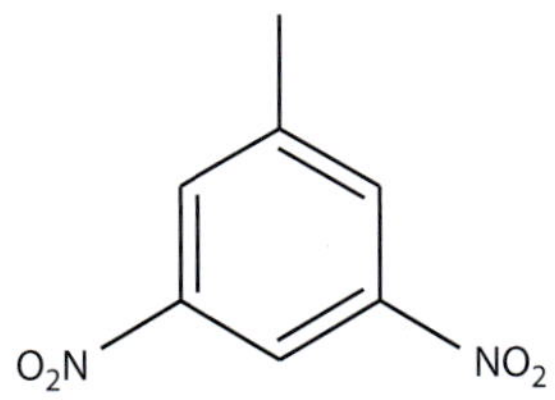

3.37

With more substituents, we assign C-1 as the carbon atom in the ring bearing the priority substituent and then number the ring so that the other substituents are on the lowest numbered carbons. Thus, **3.38** is 3,4-dimethylphenol. If a benzene ring is being described as a substituent of something else, the word phenyl is used. Thus, **3.39** would be named as triphenylmethane. Phenyl is often abbreviated as Ph, or even the Greek letter ϕ. Be careful not to confuse this with benzyl, which means $PhCH_2$– (Bn), and benzoyl, which means PhC(=O)– (Bz).

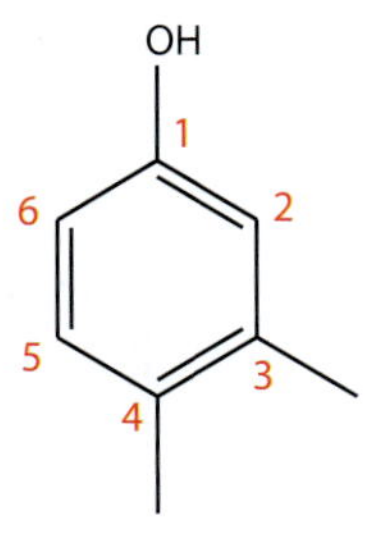

3.38

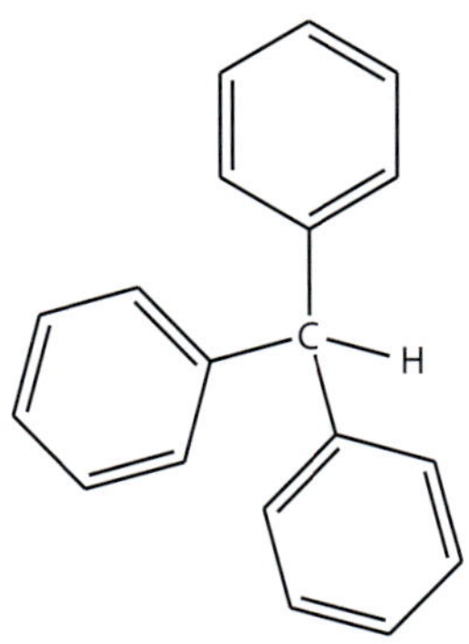

3.39

PROBLEM 3.9

Draw line structures for each of the following molecules:

(a) 3-Amino-4-methylphenol
(b) *m*-Dichlorobenzene
(c) 1,2,3,4,5-Pentamethylbenzene

PROBLEM 3.10

Give systematic names for the following molecules:

(a)

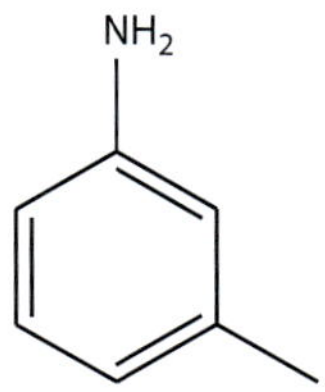

(b)

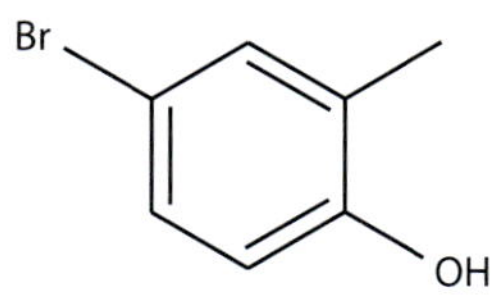

(c)

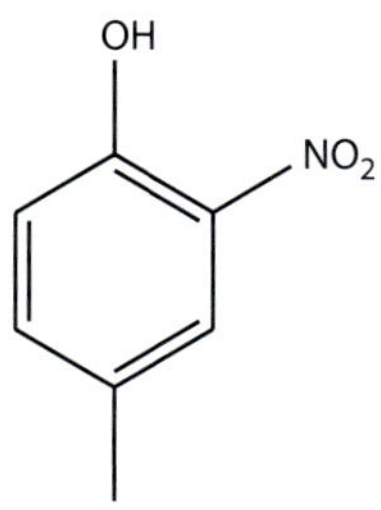

Solutions

3.9(a) First, draw and number phenol:

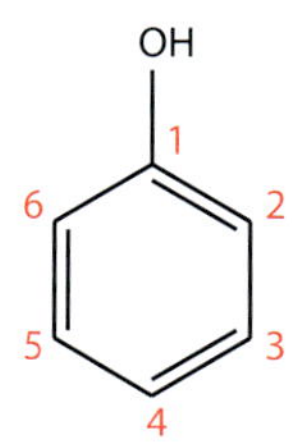

We can now add the substituents at the appropriate positions. The new one here is "amino," which is how we describe the $-NH_2$ group as a substituent:

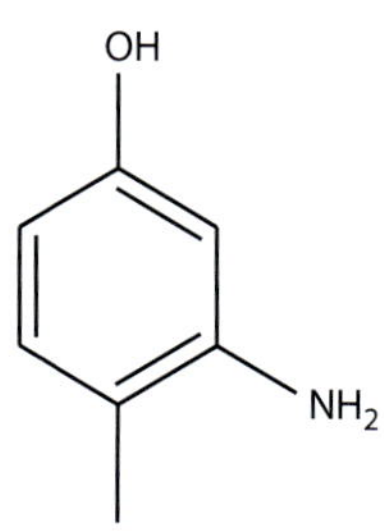

3.9(b) Remember that *meta* means that the two substituents are in a 1,3-relationship:

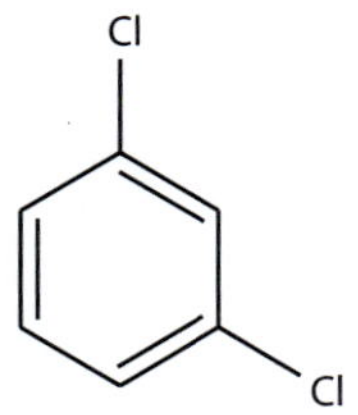

3.9(c) No worries about where to put the substituents this time, as all but one position is substituted:

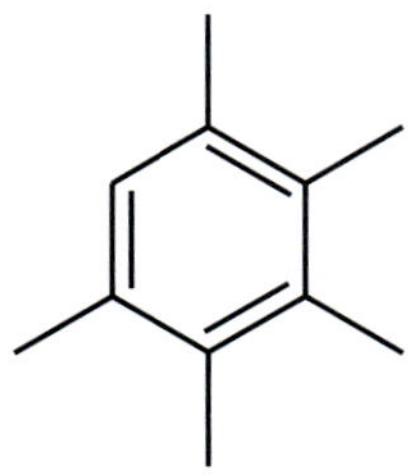

3.10(a) This will be named as an aniline, and numbering from the carbon bearing the amino group, the methyl substituent is on C-3. So this is 3-methylaniline.

3.10(b) Although not in the orientation we are used to (remember different authors will have different styles or wish to emphasize different aspects of a particular molecule), this will be named as a phenol. The carbon bearing the –OH is designated as C-1, and the ring will be numbered counterclockwise so that the substituents are on the lowest numbered carbon atoms. So this is 4-bromo-2-methylphenol.

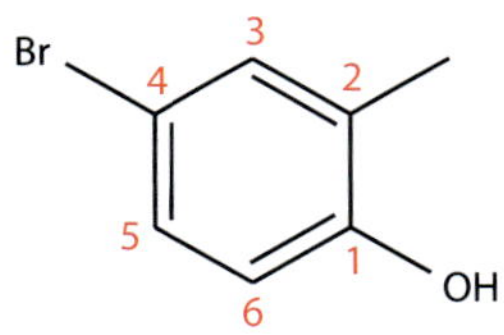

3.10(c) This is another phenol, and this time the numbering is clockwise. The $-NO_2$ substituent is described as a nitro group, so this is 4-methyl-2-nitrophenol.

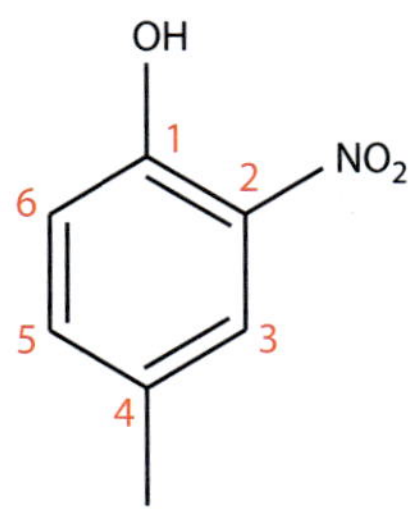

3.3 ALKYNES: HYDROCARBONS CONTAINING TRIPLE BONDS

Hydrocarbons containing a triple bond are described as alkynes (old name acetylenes). The first member of the class, H–C≡C–H, is ethyne, and the suffix for this class is "-yne." The old name for ethyne is acetylene, and this will doubtless continue to be used as long as we use oxyacetylene welding and cutting equipment. As usual, we start with the configuration of carbon

$$\text{C} \quad 1s^2\ 2s^2\ 2p_x^{\ 1}\ 2p_y^{\ 1}\ 2p_z^{\ 0}$$

and promote one electron from the 2s to the $2p_z$ orbital to give

$$\text{C} \quad 1s^2\,2s^1\,2p_x^{\,1}\,2p_y^{\,1}\,2p_z^{\,1}$$

Now we hybridize the 2s and the $2p_x$ orbitals to form two sp hybrid orbitals (Figure 3.27). These are used to form the C–H and C–C σ-bonds. The arrangement is linear, as only the directionality of the p_x orbital is involved. On each carbon atom, we are left with one electron each in the $2p_y$ and $2p_z$ orbitals, and these are used to form two π-bonds. The bonding is shown in Figure 3.28—the π-electrons almost surround the σ-bond in a lumpy annulus of electron density.

The method for naming alkynes, with a general formula of C_nH_{2n-2}, is essentially the same as that for alkenes. Figure 3.29 shows the lower members of the family. Notice that the triple bond is always drawn as linear. Find the longest chain that contains the triple bond, and number the chain so that the triple bond has the lowest number. Then add substituents as before. So **3.40** is 6-methyl-3-octyne. When there is both an alkene double bond and an alkyne triple bond in the molecule, as in **3.41**, the alkene takes precedence in nomenclature. So **3.41** is 1-heptene-3-yne.

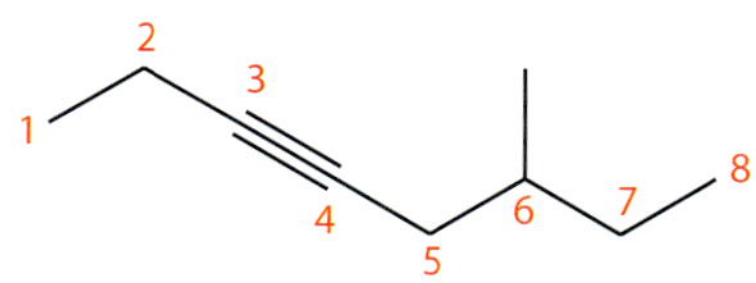

3.40

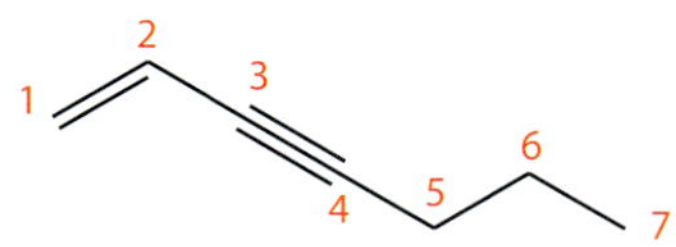

3.41

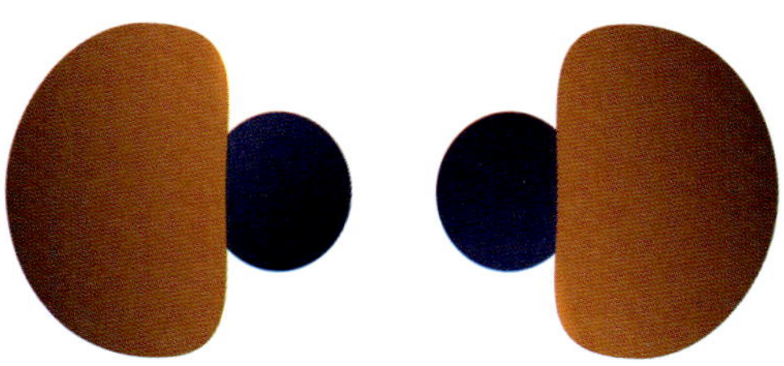

FIGURE 3.27 sp hybrid orbitals.

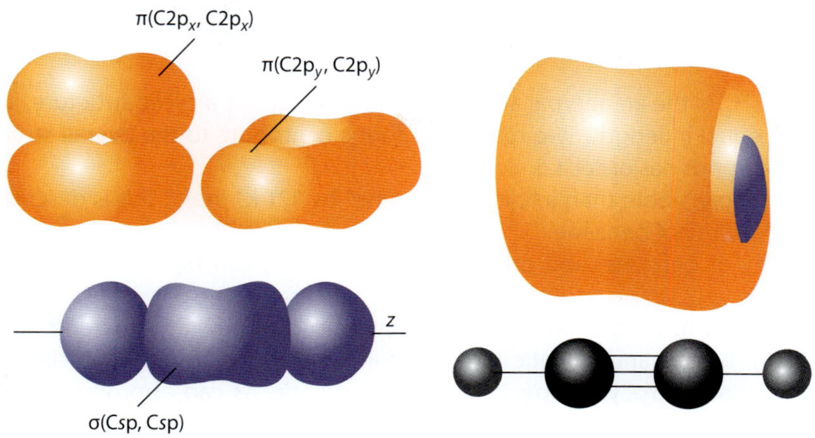

FIGURE 3.28 Bonding in ethyne.

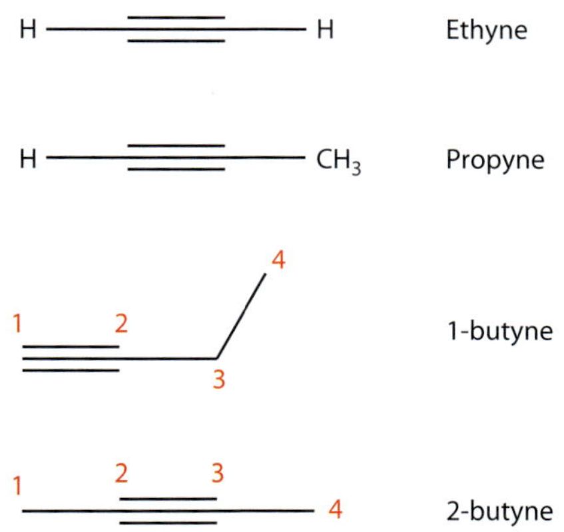

FIGURE 3.29 The lower alkynes.

PROBLEM 3.11

Give a systematic name for $CH_3C{\equiv}CCH(CH_3)CH_2CH_3$.

PROBLEM 3.12

Draw a line structure for 4-methyl-1-heptyne.

Solutions

3.11 The condensed method of writing this formula is not as informative as a stick diagram, so first convert it to that, number the structure so that the triple bond has the lowest number, then add the substituents.

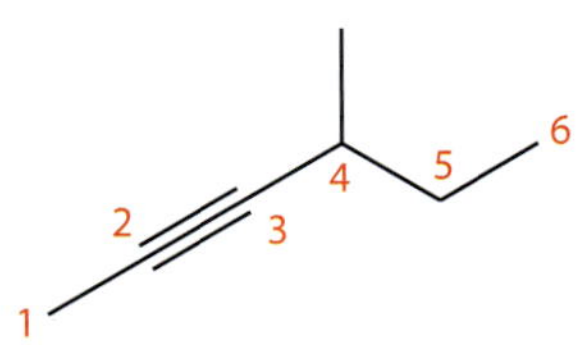

So this is 4-methyl-2-hexyne.

3.12 First, draw and number 1-heptyne, then add the substituent:

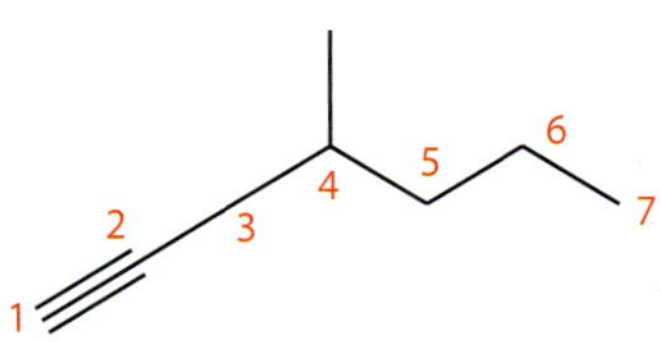

Key Points from Sections 3.2 and 3.3

- Benzene, C_6H_6, is the archetypal aromatic compound. It has a cyclic structure, and all the bonds are equal in length. Two resonance forms contribute to the overall structure and each of the bonds has order 1.5.
- Benzene is more stable than a "cyclohexatriene" by 121 kJ mol^{-1}; this is described as the resonance energy of benzene.
- The stability of benzene, and other aromatic systems, can be explained in molecular orbital terms.
- Species are said to be aromatic if they are cyclic and fully conjugated and have $4n + 2$ π-electrons. They are antiaromatic if they have $4n$ π-electrons. This is the Hückel rule.
- Polycyclic and heterocyclic species may be aromatic, as can appropriate anions and cations.
- Benzene derivatives are named as substituted benzene rings, although many trivial names are also used. When substituents are in 1,2- 1,3-, or 1,4-relationships, they may be described as *ortho*, *meta*, and *para*, respectively.
- Alkynes contain carbon–carbon triple bonds consisting of one σ- and two π-bonds. The two carbon atoms of the triple bond are sp hybridized and the alkyne unit is linear.
- In naming alkynes, choose the longest chain containing the triple bond and number it so that the triple bond has the lowest number. Then locate and name substituents.

3.4 RESONANCE

The structure of benzene is an average or hybrid of two resonance forms, as we saw in Section 3.2.1; there is no equilibrium, the true structure is halfway between the two resonance forms. We now need to have some way of representing the movement of electrons between the two resonance forms, and this is the curly, or curved, arrow (**3.42**). This is a profoundly important symbol that we will use throughout later chapters, and it represents the movement of an electron pair from the origin of the arrow to its tip. It is very important to draw the arrow accurately—something that is easier to achieve by hand (because of the infinite flexibility available) than with a computer program. So we draw the arrows to turn one resonance from of benzene into the other in Figure 3.30—essentially the π-bonds, which represent an electron pair, are being moved.

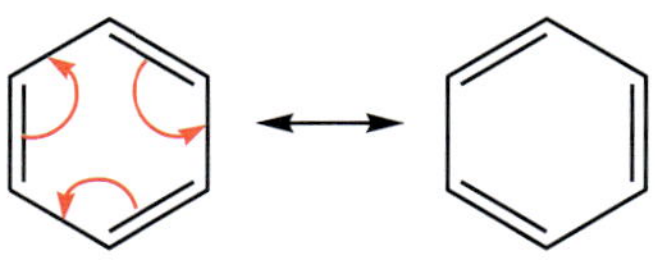

FIGURE 3.30 Resonance forms of benzene.

3.42

Benzene is a straightforward example of resonance, because one set of π-bonds is simply turned into another. We'll now look at a more complicated example, the allyl anion. This has two main resonance forms shown as **3.43a** and **3.43b**; other than the labeling of the atoms, these are identical. A similar pair of resonance forms is available for the anion of a carboxylic acid (**3.44a** and **3.44b**), and you could compare this with some resonance structures such as carbonate (**3.45**) and nitrate (**3.46**) from general chemistry. The key feature of all these structures is that they indicate which atoms are identical. Despite the formalism of where bonds are drawn, the two oxygen atoms of a carboxylate anion are absolutely identical in every way—similarly for the oxygens of nitrate or carbonate.

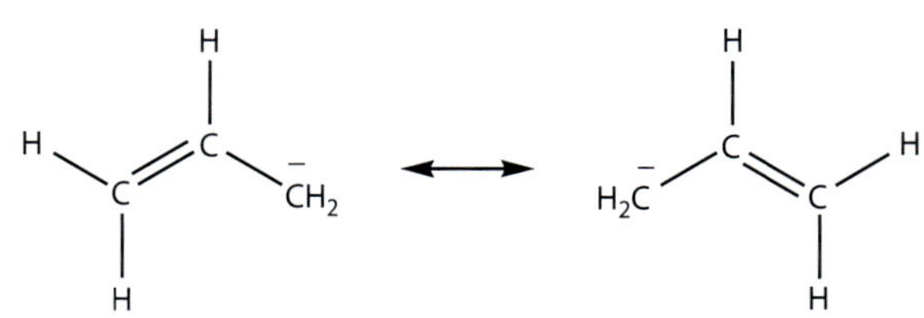

3.43a **3.43b**

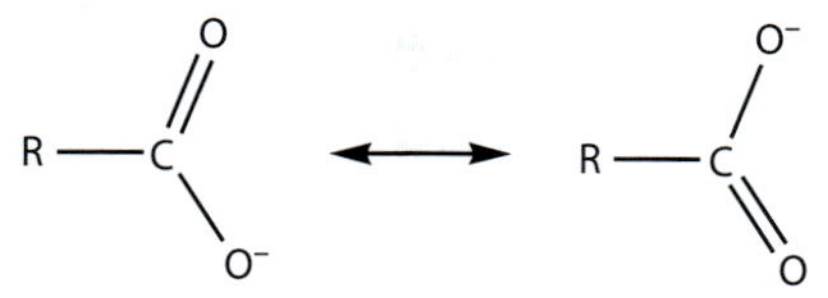

3.44a **3.44b**

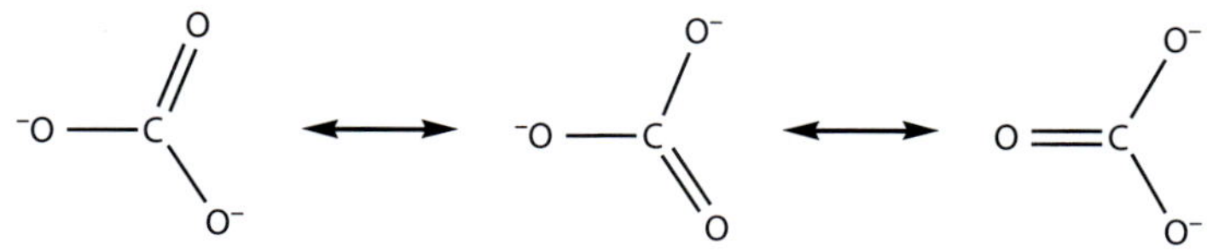

3.45a **3.45b** **3.45c**

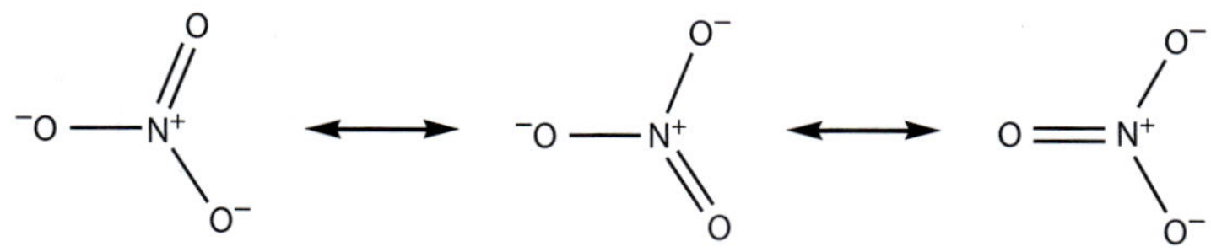

3.46a **3.46b** **3.46c**

In each case, we can show how we obtain one resonance structure from the other, using curly arrows to move the electrons. A lone pair of electrons on carbon or oxygen is moveable in just the same way as an electron pair that forms a π-bond (Figure 3.31). Notice that these two processes are essentially identical—a lone pair of electrons is moved to become a π-bond, and the electrons of a π-bond become a lone pair. The fact that different atoms are involved is incidental to the *process* of generating resonance forms; the curly arrows needed will always look the same.

Key Points: A Series of Simple Rules Will Help You to Write Resonance Forms

- Resonance forms involve only the movement of electrons, not atoms.
- There are the same number of electrons, paired and unpaired, in all valid resonance forms; charge is conserved.
- Equivalent representations contribute equally, and nonequivalent ones unequally.
- Forms with eight electrons at each atom contribute more than those that do not have eight electrons.
- Once the octet rule has been applied, forms with negative charges on electronegative atoms contribute more than forms with positive charges on electronegative atoms.

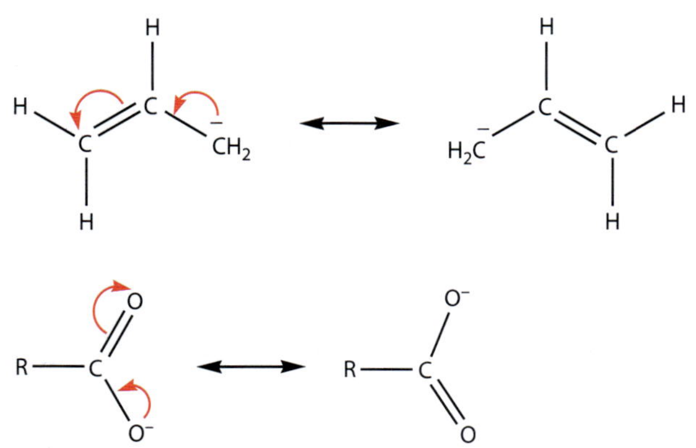

FIGURE 3.31 Connecting resonance forms using curly arrows.

Some examples will illustrate the concept of resonance. The two forms of benzene or the allyl and carboxylate ions are identical, and both contribute equally. However, if we consider the two forms of the enolate anion (**3.47**), although the movement of electrons is just the same as for an allyl system, the two forms are no longer equivalent and will not contribute equally. Intuitively, we would expect that the form with the negative charge on oxygen will contribute more to the overall structure than that with the negative charge on carbon (though when we come to investigate the chemistry of these species, in Chapter 17, there are complications). If we consider ketones (**3.48**), species with a carbon oxygen double bond, then the doubly bonded form, where all the atoms have octets, is the most important one. In the two charge separated forms, the species with a negative charge on oxygen and a positive one on carbon is more important than the one with the charges the other way round—and is an important indicator of the way in which ketones react with positively and negatively charged species. We will explore the concepts and uses of resonance forms in more detail in the next chapter.

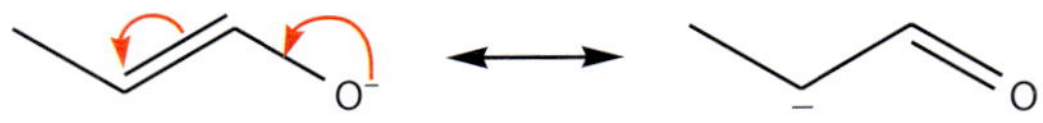

3.47

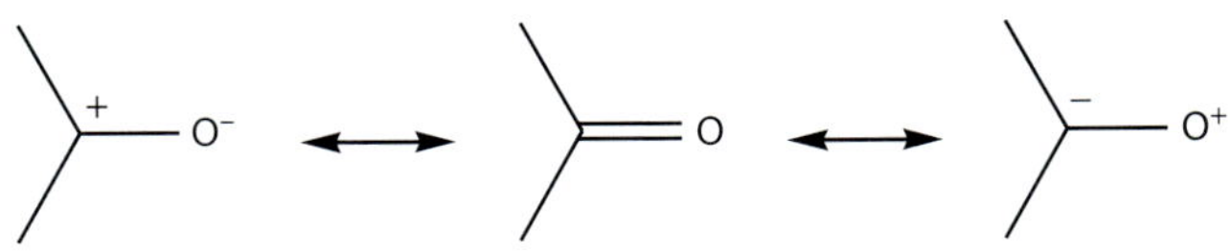

3.48

PROBLEM 3.13

Write down one resonance form for each of the following molecules:

(a)

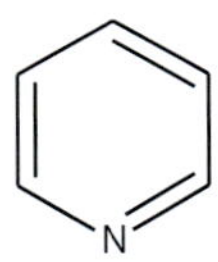

(b)

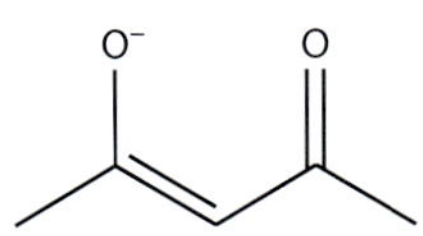

(c)

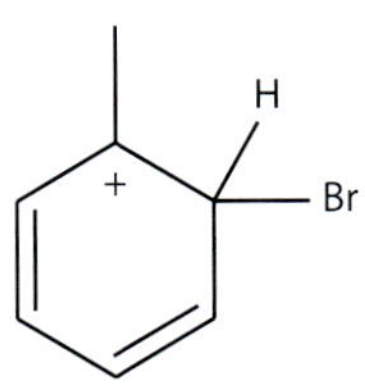

Solutions

(a) The nitrogen makes no difference here, as its lone pair is in an sp^2 orbital in the plane of the ring. Pyridine can be treated in just the same way as benzene:

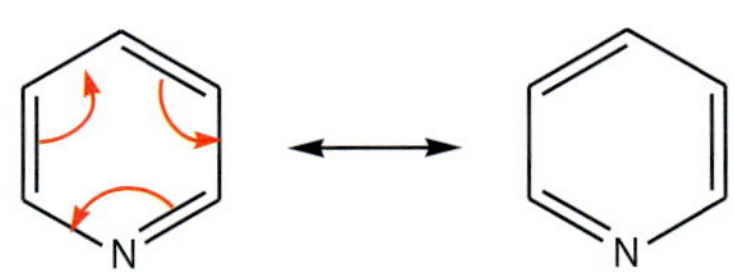

(b) Despite the oxygen atoms, this is essentially the same problem as writing the resonance form of an allyl anion (the "allyl" unit is in blue in the following):

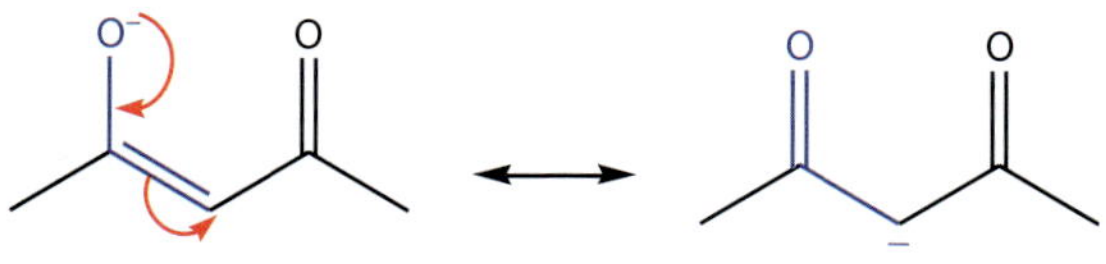

The process can be continued so as to put the negative change on the other oxygen atom:

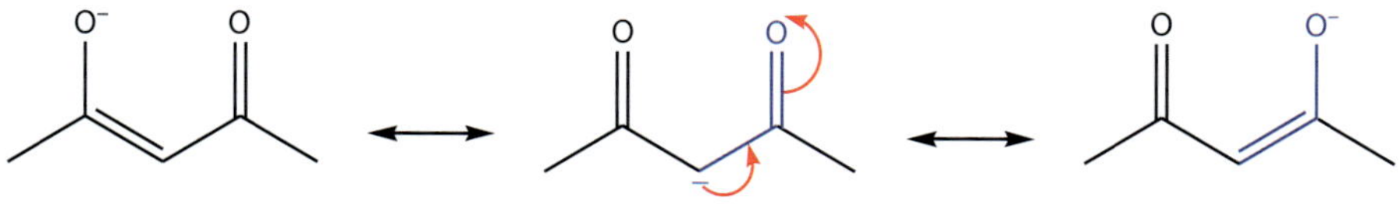

(c) In this problem, we need to consider allylic cations rather than allylic anions, but writing resonance forms can still be seen in terms of allyl units:

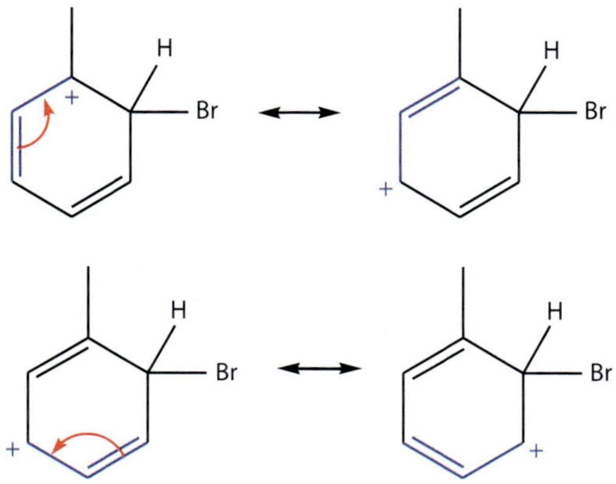

The resonance forms of the allyl anion and cation can also be explained in molecular orbital terms. The σ-framework is formed from sp² orbitals to make the C–C and C–H σ-bonds. What are left are three p_z orbitals, one on each atom, and four electrons, two from the lone pair (the negative charge) and two from the π-bond. The three p_z atomic orbitals must now be combined to produce three molecular orbitals, spanning the three atoms. The lowest-energy molecular orbital, ψ_1, has all the p_z orbitals interacting in phase, with no additional nodes. Inevitably, the highest energy orbital, ψ_3, has them all out of phase, and there are two nodes. So we should conclude that ψ_2 should have one node. Because the anion is symmetric, this must pass through the central atom of the allyl group. It is neither bonding nor antibonding and is generally described as nonbonding (Figure 3.32). The four electrons are in ψ_1 and ψ_2. This corresponds with our picture from the resonance forms—there is more electron density on carbon atoms 1 and 3 than carbon atom 2. Obviously, the real physical chemistry of the system and the relevant mathematics are more complex, but this picture will help us to understand the reactivity of the anion. The allyl cation (Figure 3.33) has two fewer electrons, but the orbitals available are the same. Figure 3.34 shows the calculated orbitals with all the proper coefficients, but it does not add greatly to the understanding of reactivity of these species.

We can now return briefly to the cyclopentadienyl anion and show *how* this system is fully delocalized. The trick of writing all the correct resonance forms is to focus on one isolated allyl system at a time—writing the two resonance forms for this is straightforward and can be continued onward around the ring (Figure 3.35). Remember that the ability to delocalize alone does not make a molecule aromatic—it must also fulfill the Hückel condition of $4n + 2$ π-electrons. We can write delocalized forms of the cyclopentadienyl cation (Figure 3.36), but this is still an antiaromatic system and very unstable.

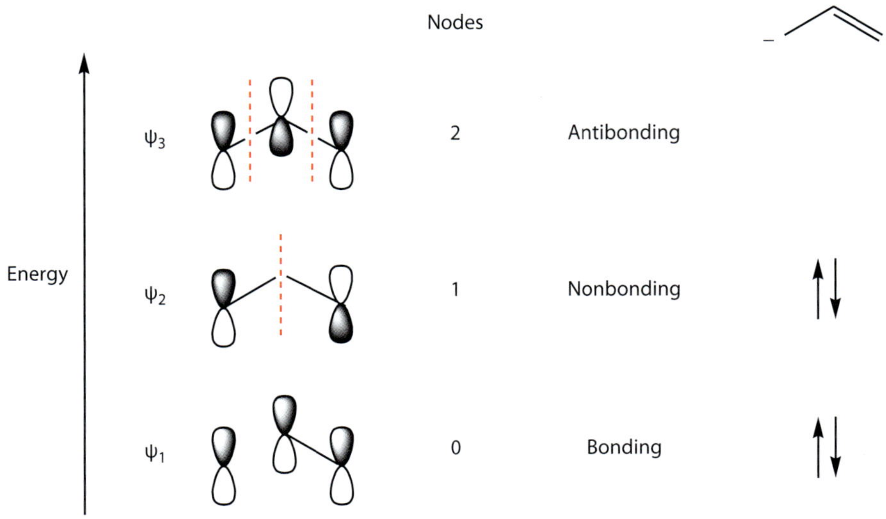

FIGURE 3.32 Molecular orbitals of the allyl anion.

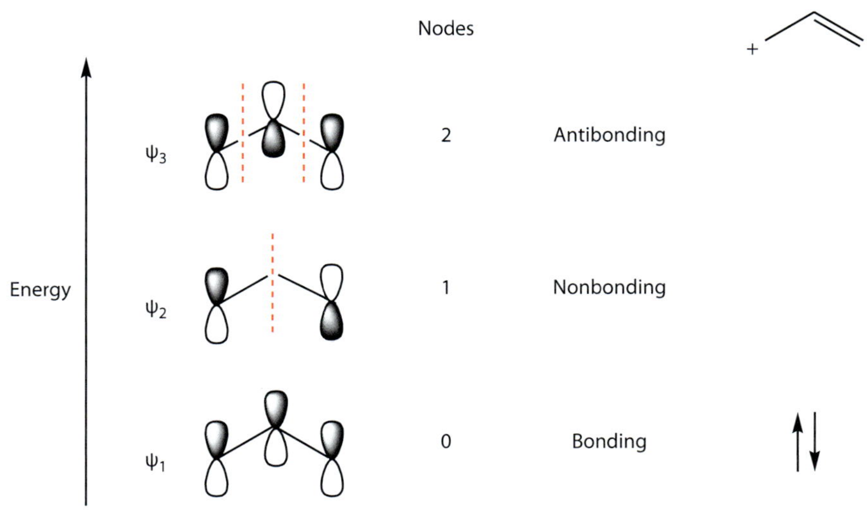

FIGURE 3.33 Molecular orbitals of the allyl cation.

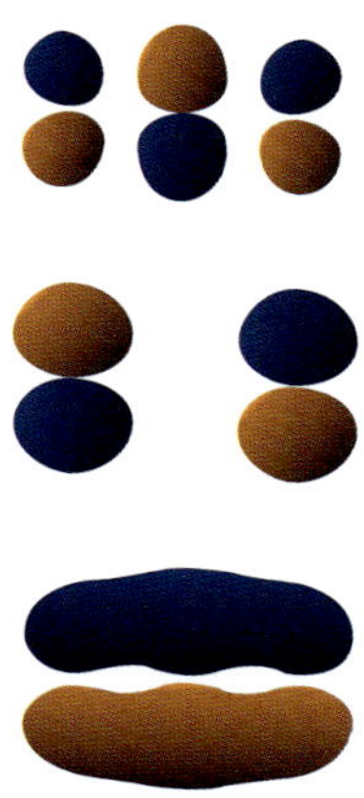

FIGURE 3.34 Calculated molecular orbitals of the allyl system.

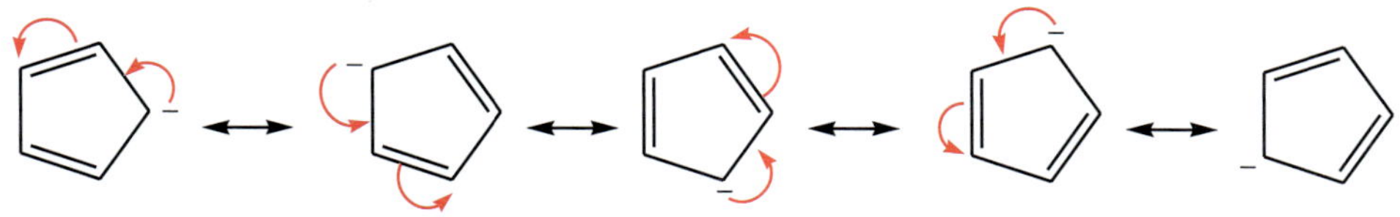

FIGURE 3.35 Resonance forms of the stable cyclopentadienyl anion.

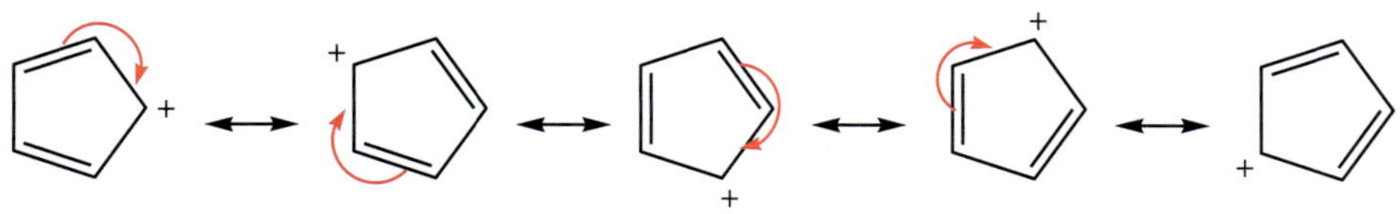

FIGURE 3.36 Resonance forms of the unstable cyclopentadienyl cation.

REVIEW PROBLEMS

1. Draw structures for each of the following molecules:
   (a) 5-Methyl-*cis*-2-hexene
   (b) 6-Chloro-7-methyl-2-octyne
   (c) 4-*tert*-Butyl-2-phenyloctane
   (d) 2-Bromo-4-chloro-6-fluorophenol
   (e) 4-Methyl-*trans*-2-heptene
   (f) 1,4-Cyclohexadiene
   (g) 2-Methyl-1-hexene-4-yne
   (h) 1,3,5-Tribromobenzene
   (i) 2-Methylaniline
   (j) *trans*-2-Hexene-5-yne
   (k) 3-Ethylphenol
   (l) 4-Methyl-*trans*-2-hexene
   (m) 3-Phenyl-1-butyne
   (n) 8-Chloro-3-ethyl-7-methyl-1,5-decadiene
   (o) *p*-Dichlorobenzene
2. Give a systematic name for each of the following compounds; pay attention to stereochemistry where relevant:
   (a)

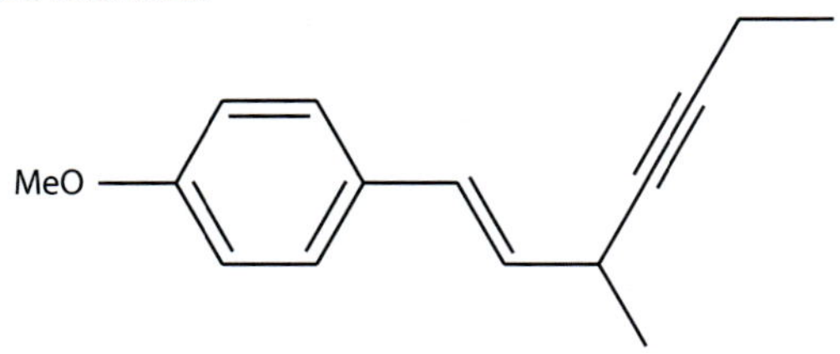

   (b)

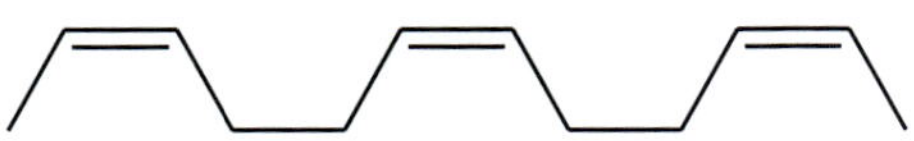

   (c)

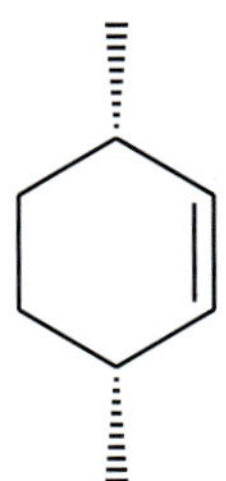

(d)

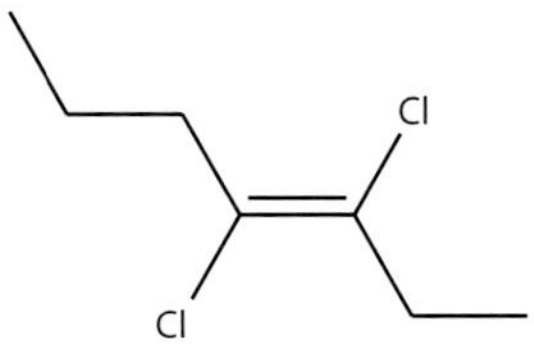

(e)

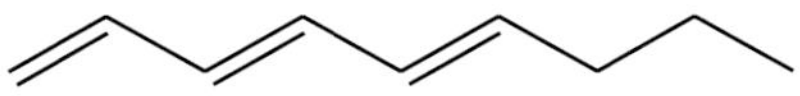

(f)

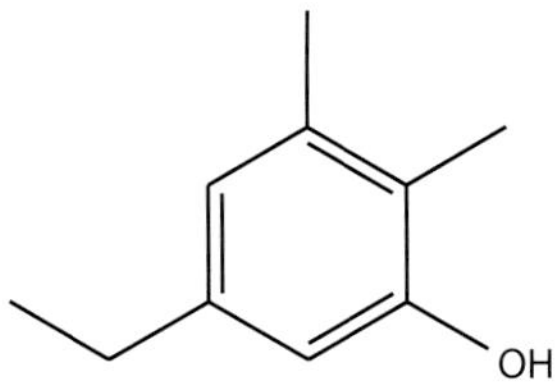

(g)

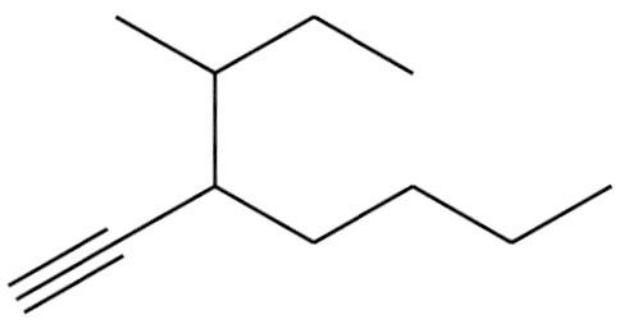

(h)

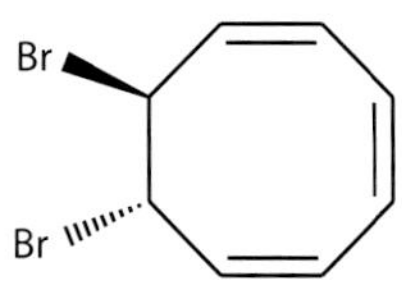

(i)

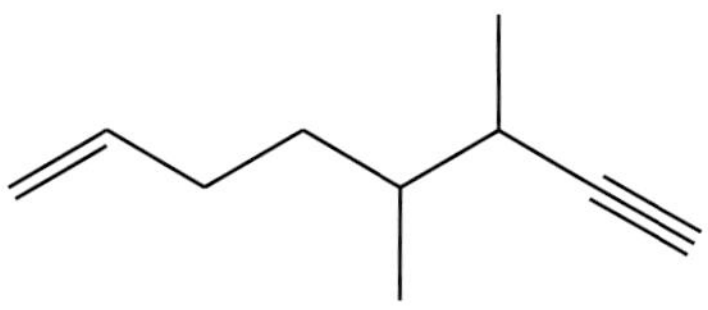

(j)

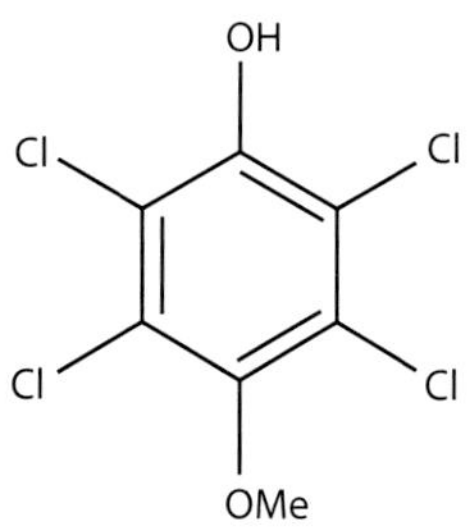

Drosopholin A

Found in wild boar meat, from mushrooms that the boars eat.

3. Write down one resonance form for each of the following molecules:

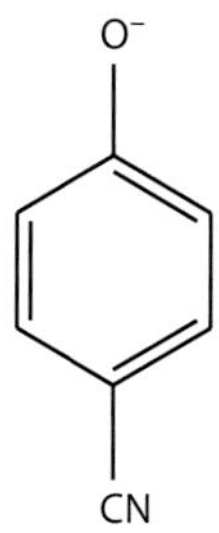

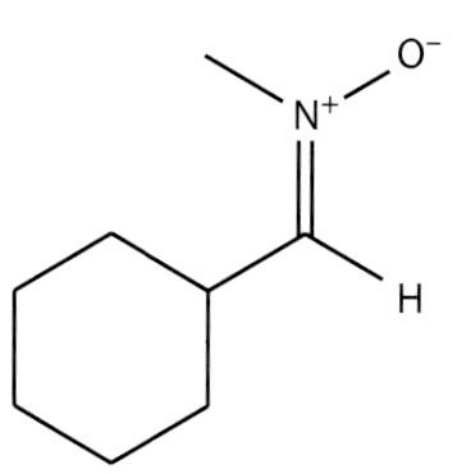

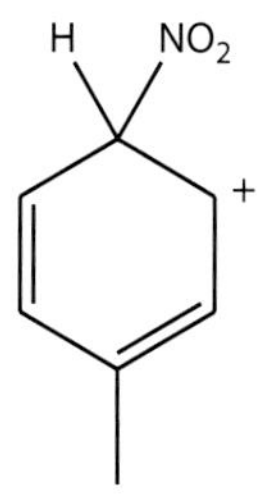

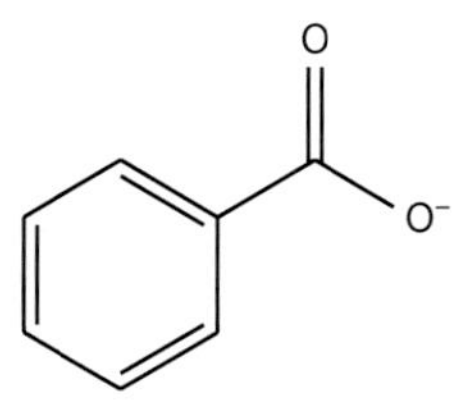

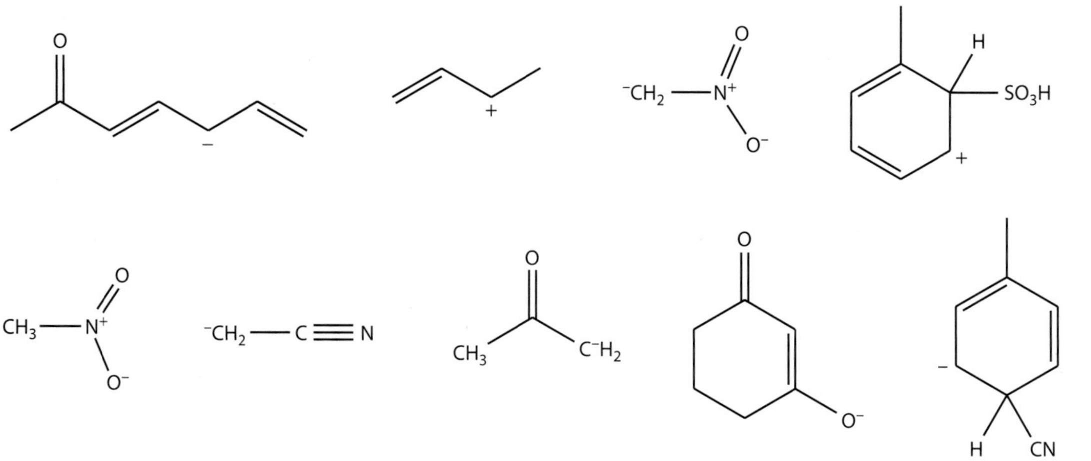

4. How many π-electrons are involved in resonance in each of the following molecules? Which are aromatic or antiaromatic?

(a)

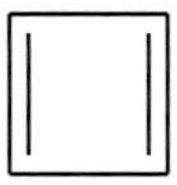

(b)

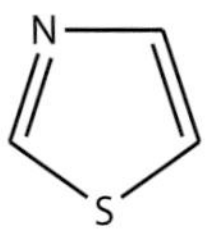

(c)

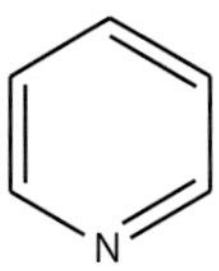

(d)

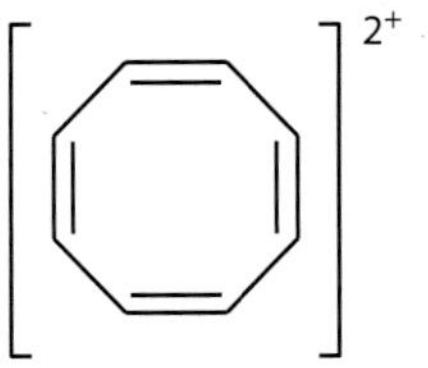

(e)

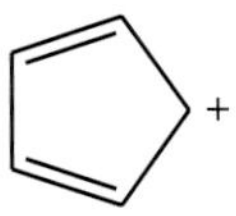

(f)

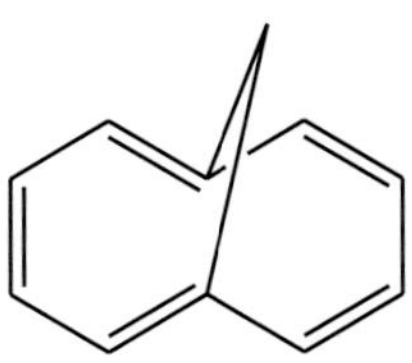

(g)

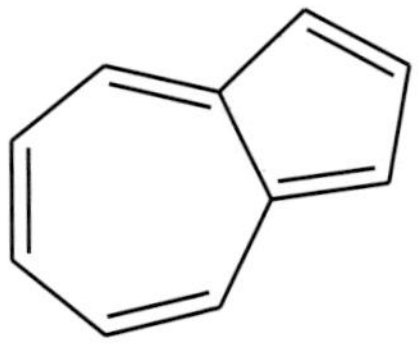

(h)

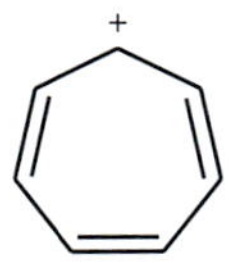

(i)

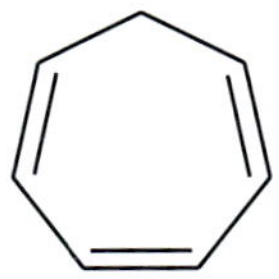

(j)

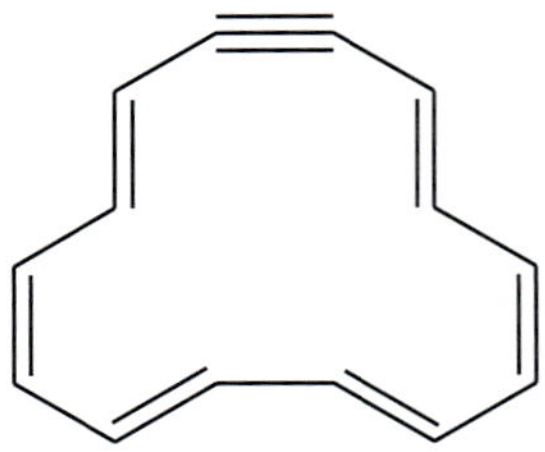

(k)

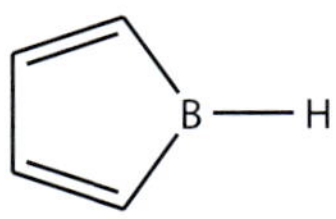

(l)

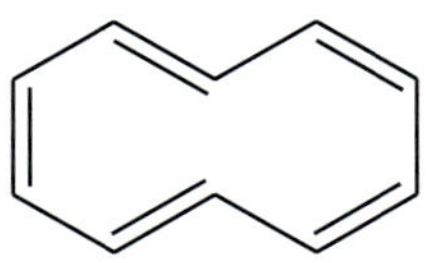

5. What are the most important intermolecular forces in the following:
   (a) Pyridine
   (b) Benzene
   (c) Chloroethane
6. Assign as *E* or *Z* the double bond in each of the following molecules:

(a)

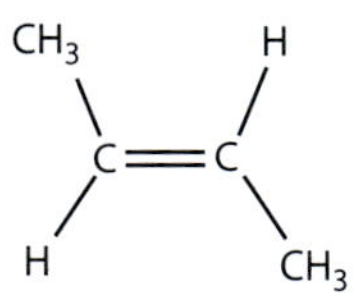

(b)

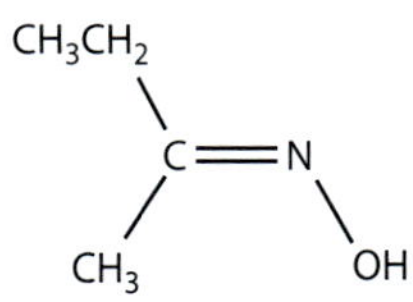

(c)

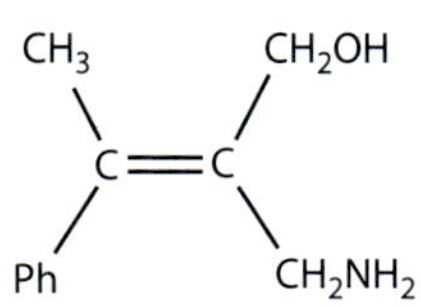

(d)

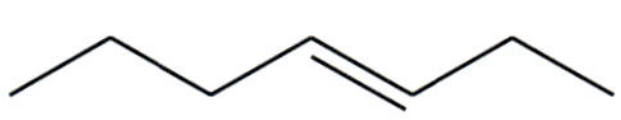

(e) 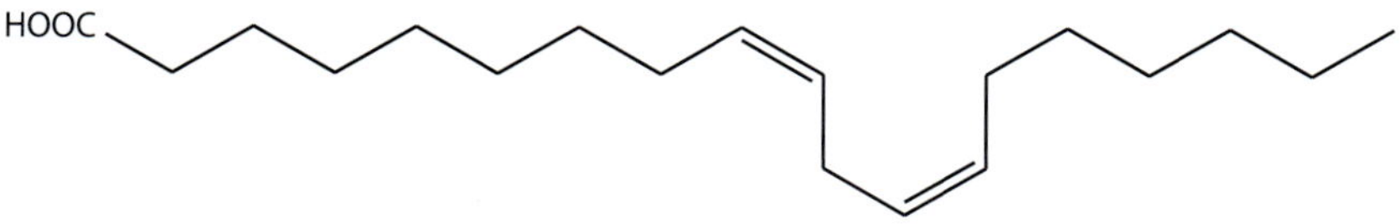

(f) 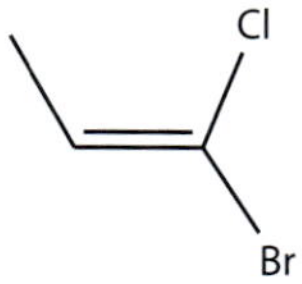

7. Place in order according to the Cahn–Ingold–Prelog system each of the following sets of substituents:
   (a) –H, –D, –CH_3, –F, –NH_2
   (b) –H, –CH_3, –Ph, –CH_2CH_3, –$CH_2CH(CH_3)_2$
   (c) –H, –CH_3, –CH_2CH_3, –CH=CH_2–$C(CH_3)_3$
   (d) –H, –F, –OH, –OCH_3
   (e) –H, –Cl, –OH, –OCH_3
8. From a consideration of molecular orbital theory, deduce the three-dimensional structure of allene (1,2-propadiene), $H_2C{=}C{=}CH_2$.
9. Draw all the acyclic structural isomers of C_5H_{10}. Which of these exhibit *cis–trans*-isomerism?
10. Give an example of each of the following:
    (a) A disubstituted benzene, which has no permanent dipole moment
    (b) An alkene with a permanent dipole moment
11. Which of the following pairs of molecules would you expect to have the higher boiling point?
    (a) Benzene and pyridine
    (b) 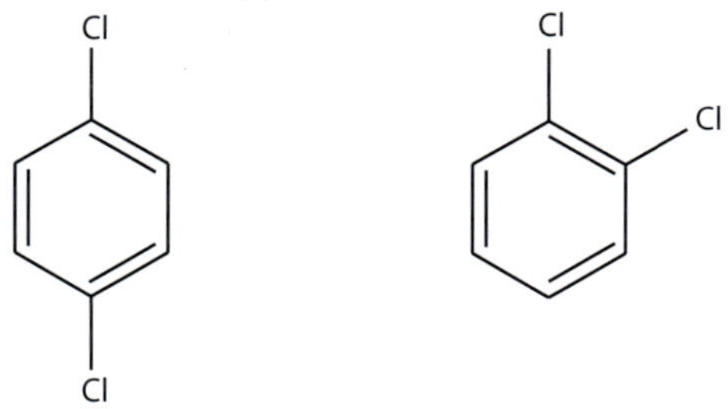

Heteroatom-Containing Functional Groups

4

I promise this is the last chapter on formal nomenclature....

4.1 COMPOUNDS CONTAINING sp^3 NITROGEN ATOMS

Amines are the class of compounds obtained when the hydrogen atoms of ammonia are replaced by carbon functionality. We describe amines as primary, secondary, and tertiary according to the number of the hydrogens that are replaced. Thus, RNH_2 is a primary amine, R_2NH is a secondary amine, and R_3N is a tertiary amine. If all of the hydrogen atoms of the ammonium ion, $[NH_4]^+$, are replaced by alkyl groups, this is described as quaternary ammonium salt, $[R_4N]^+X^-$.

In formal International Union of Pure and Applied Chemistry (IUPAC) nomenclature, CH_3NH_2 would be called methanamine—the suffix for amines is "-amine." However, this is one IUPAC rule that has not "caught on" in the wider Anglophone chemical community, and in English, CH_3NH_2 is almost invariably called methylamine. Me_2NH is dimethylamine and Me_3N is trimethylamine. If the amine is substituted by two different groups, then both substituent names are used, so that MeNHEt is ethylmethylamine. Some common names are also generated by naming amines as amino alkanes. Thus, **4.1** could be called 2-heptylamine or 2-aminoheptane and **4.2** might be named as 2-heptyldimethylamine or 2-dimethylaminoheptane. No one would pretend that this lack of clarity is desirable; unfortunately, for the present, it seems inevitable. The important point to remember is that when you are naming an amine, you need to be as unambiguous as possible.

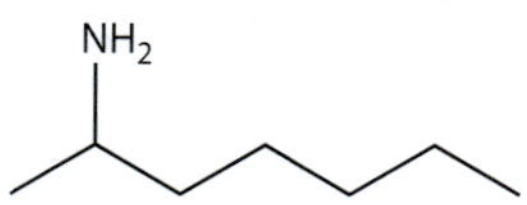

4.1

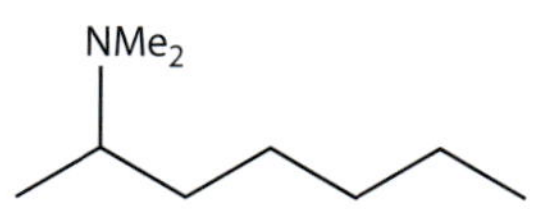

4.2

As we saw in Chapter 2, ammonia has a pyramidal structure with a lone pair of electrons that is associated with its basicity (**4.3**). The arrow next to the molecule shows the direction of the resultant molecular dipole moment. One of the earliest pieces of evidence for the structure of ammonia was that it has a dipole moment, symbol μ. The N–H bond is polarized such that there is a small positive charge on hydrogen and a small negative charge on nitrogen. The resultant molecular dipole when these three bond dipoles are summed lies along the threefold symmetry axis of the molecule. For those of you who have studied vector algebra, this is a vector sum.

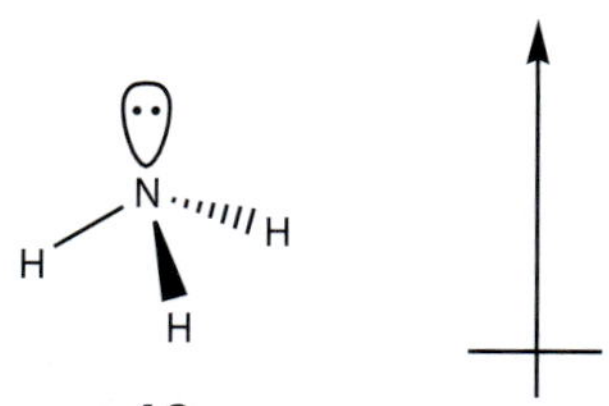

4.3

Quick Digression on Dipole Moments and Hydrogen Bonds

All bonds between different elements have a bond dipole, as the atoms are of differing electronegativity. For example, a carbon–chlorine bond is always polarized toward chlorine, which bears a small negative charge, because the Pauling electronegativity of carbon is 2.55 and that of chlorine 3.16. The greater the difference in electronegativity of the connected atoms, the greater the dipole moment of the bond. Note, however, that dipole moment is a molecular phenomenon, which is measured, whereas the Pauling electronegativity scale is a construct that has been devised to accommodate this and other physical phenomena. Polar bonds do not exist in isolation in organic chemistry; they are part of more complex molecules. Formally, vector algebra on bond dipoles will give us the molecular dipole, but in most simple cases, common sense will suffice. And these molecular dipoles will tell us something about the molecular structure. Thus, two of the dichloroethenes have a dipole moment; the third does not (Figure 4.1). The unit used for molecular dipoles is Debye (in honor of Peter Debye, who first studied this phenomenon) abbreviated D.

Because dipoles are attracted to each other (the positive "end" of one dipole is attracted to the negative "end" of another), compounds with permanent dipole moments tend to have higher melting and boiling points than those that do not. Thus, 1,2-dichlorobenzene, which has a dipole moment, boils at 180 °C, while 1,4-dichlorobenzene, which has no dipole, boils at 174 °C.

Hydrogen bonding can be regarded as a "super" version of a dipolar interaction, but when it is present, it dominates all other intermolecular attractions. We start with a strongly dipolar bond between hydrogen and an electronegative element, usually fluorine, oxygen, or nitrogen. All of these X–H bonds are polarized so that there is a partial negative charge on X and a partial positive charge on H. Then there is a strong interaction between the hydrogen of one molecule and the lone pair on X in another. It is hydrogen bonding that gives water its unusual properties and gives rise to the abnormal boiling points of hydrogen fluoride, water, and ammonia (Table 4.1). In Group 14, where no hydrogen bonding

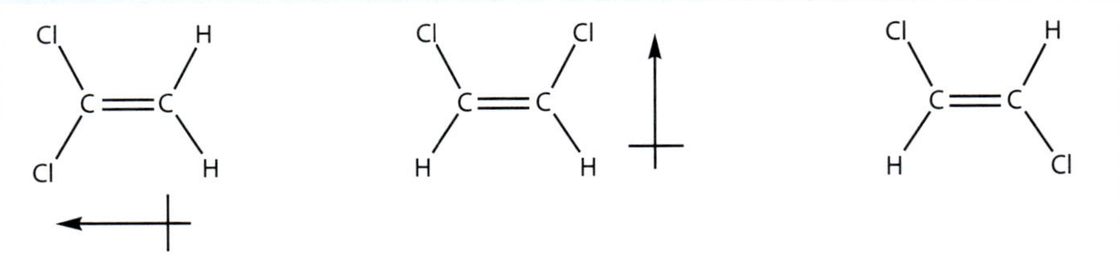

FIGURE 4.1 Dipole moments of dichloroethenes.

TABLE 4.1 Boiling Points of Elemental Hydrides

	BPt (°C)		BPt (°C)		BPt (°C)		BPt (°C)
CH_4	−162	NH_3	−33	H_2O	100	HF	19.5
SiH_4	−112	PH_3	−88	H_2S	−60	HCl	−85
GeH_4	−88	AsH_3	−62.5	H_2Se	−41	HBr	−66
SnH_4	−52	SbH_3	−17	H_2Te	−2	HI	−34

occurs, the boiling points increase down the group as one would expect. In groups 15, 16, and 17, this is also true—**except** for the first member of the group, where hydrogen bonding is so important.

Amines with N–H bonds will hydrogen bond in just the same way as does ammonia (**4.4**). The hydrogen bonds in amines are not as strong as those in water and alcohols but are still significant. The strength of hydrogen bonds reflects the electronegativity of the atom to which the hydrogen is attached and increases in the order N < O < F. Boiling points are thus higher for a non-hydrogen-bonded molecule with a comparable molecular weight. Hydrogen bonding is also important between molecules and solvents; thus, lower amines are water soluble, because they form H-bonds to water (**4.5**).

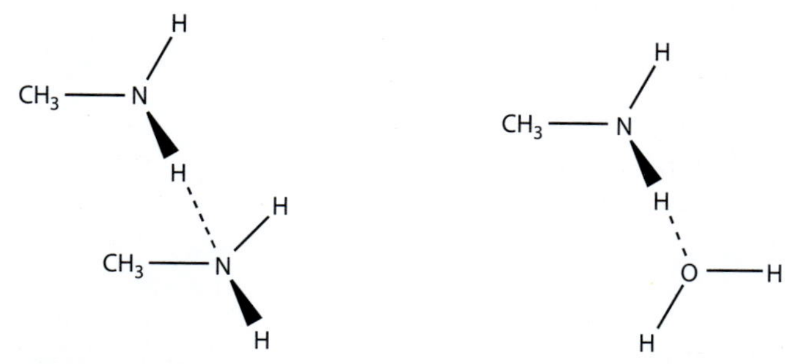

4.4 **4.5**

The lone pair of electrons on ammonia and amines is associated with their basicity; you should be familiar with the reaction of ammonia with a proton (Figure 4.2). We can show what happens schematically by using the curly arrow; notice that the arrow starts from the lone pair of electrons on the amine and finishes in the gap between the nitrogen and the proton—where the new bond is being formed.

Because of the analogy with ammonia, you should feel comfortable with the idea of amines as bases. The pK_a value for a protonated amine, $[R_3NH]^+$, is about 9, so it is a stronger acid than water, but not as strong as H_3O^+. However, amines may also act as acids and can be deprotonated. For a typical secondary amine, the pK_a is 36 (Equation 4.1). We will explore in more detail the acidity and basicity of all classes of compounds in Chapter 8.

$$R_2NH \rightleftharpoons R_2N^- + H^+ \tag{4.1}$$

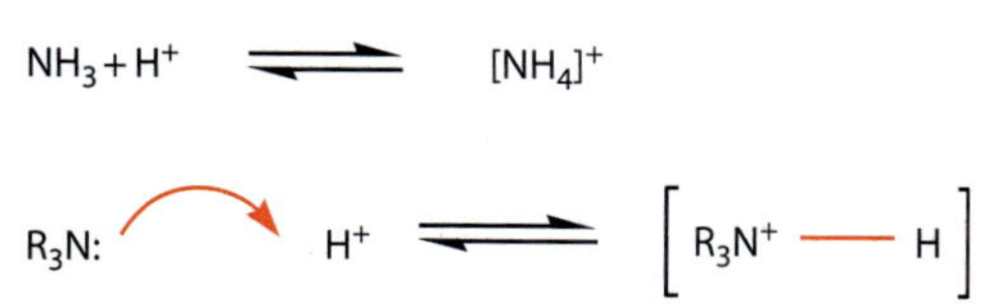

FIGURE 4.2 Protonation of ammonia and amines.

Focus on Amines: From Stinks to Stimulants

The active ingredient of smelling salts, previously called *sal volatile* (because it produced a reaction) or spirits of hartshorn (because ammonium salts were produced by distillation of extracts of deer hooves and antlers), is gaseous ammonia. It was widely used to revive people who had fainted and is still occasionally used on semiconscious boxers. The odor is highly unpleasant and triggers a strong breathing reflex. Most of the volatile lower amines have unpleasant, fishy, odors; individuals with the genetic disorder fish odor syndrome (trimethylaminuria) lack the enzyme to oxidize and hence metabolize trimethylamine:

Revival with smelling salts in Victorian England

The diamines, putrescine and cadaverine, are even more unpleasant smelling (Figure 4.3). They are formed in rotting animal and fish tissue by the breakdown of amino acids and their odor is associated with these, but they are also found in urine and semen, and they are part of major metabolic pathways. Recently, the production of putrescine in

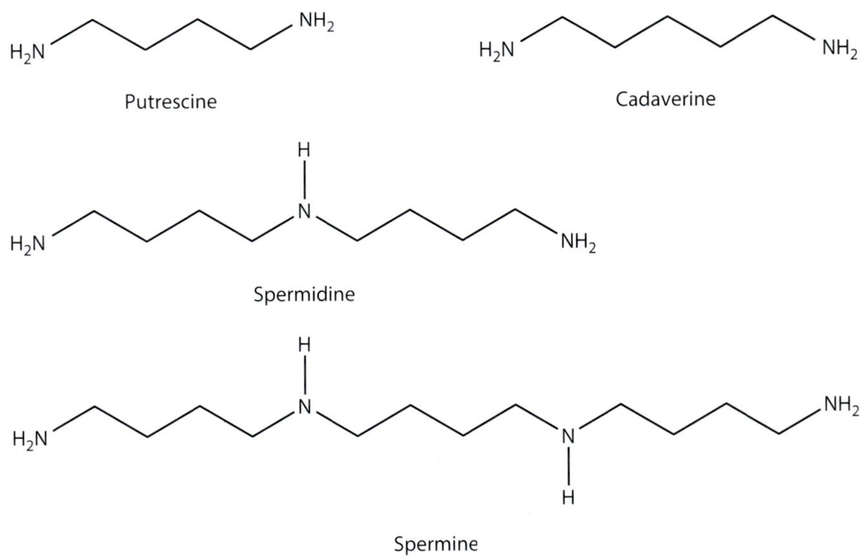

FIGURE 4.3 Some naturally occurring polyamines.

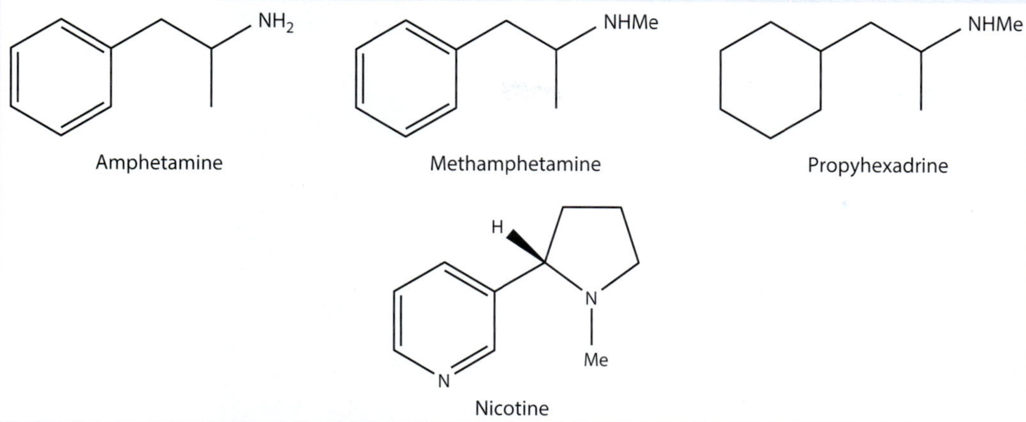

FIGURE 4.4 Medicinal and psychoactive amines.

the brain after an epileptic seizure has been shown to protect the brain against further seizures. Putrescine is also the biological precursor of the polyamines spermine and spermidine, both of which were originally isolated from human sperm (spermine as early as 1678). Spermidine inhibits the enzyme neuronal nitric oxide synthase, and spermine has been shown to induce changes in DNA conformation.

Many amines have biological effects, and they are widely used both medicinally and as recreational drugs. For example, amphetamine (Figure 4.4) is used to treat attention deficit hyperactivity disorder (ADHD) and narcolepsy (trade name Adderall®) and by various countries' military to maintain alertness during long missions. Until 1965, it was used in decongestant inhalers, at this point being replaced by propylhexedrine (benzedrex inhalers). The street drug "speed" (used to describe various amphetamines) was banned in the United States in 1971, because of its addictive potential. The use of methamphetamine, "ice" or "crystal meth," was restricted in the 1940s, but it is still occasionally prescribed. Nicotine contains both a pyridine ring and a tertiary amine and is isolated from a number of plants of the Solanaceae family. Commercial tobacco contains 0.6 %–3 % nicotine. Not only is nicotine as addictive as heroin, but it is also toxic—the lethal dose for an adult is 40–60 mg.

4.2 MOLECULES CONTAINING sp² NITROGEN ATOMS

In the last chapter, we met a few aromatic molecules containing sp² nitrogen atoms, such as pyridine, **4.6**; pyrrole, **4.7**; and imidazole, **4.8**. In pyridine, there is a lone pair of electrons at nitrogen, in an sp² orbital in the plane of the ring, and a singly occupied p_z orbital is part of a benzenoid π-system. In pyrrole, both the electrons of the lone pair at nitrogen are required to make up the aromatic sextet, so they must be in a p_z orbital with the N–C and N–H bonds made with sp² hybrid orbitals. In imidazole, the NH nitrogen is like the NH of pyrrole, with the lone pair a part of the aromatic sextet, and the other nitrogen is sp² hybrid, with an sp² lone pair in the plane of the ring and a p_z orbital with one electron contributing to the π-system.

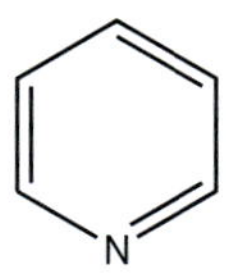

4.6

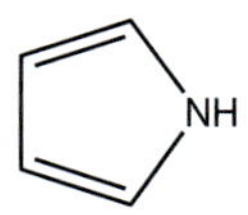

4.7

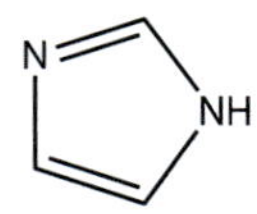

4.8

The other classes of compounds containing sp² nitrogen atoms that we will meet are shown in Figure 4.5, imines, oximes, hydrazones, and semicarbazones. In terms of the types of orbitals used in the bonding, all the classes are the same as far as the carbon–nitrogen double bond is concerned, so we'll look at the oxime as an example. Some parts of this are straightforward and already known to us. The carbon of the double bond is sp² hybridized and forms three σ-bonds to

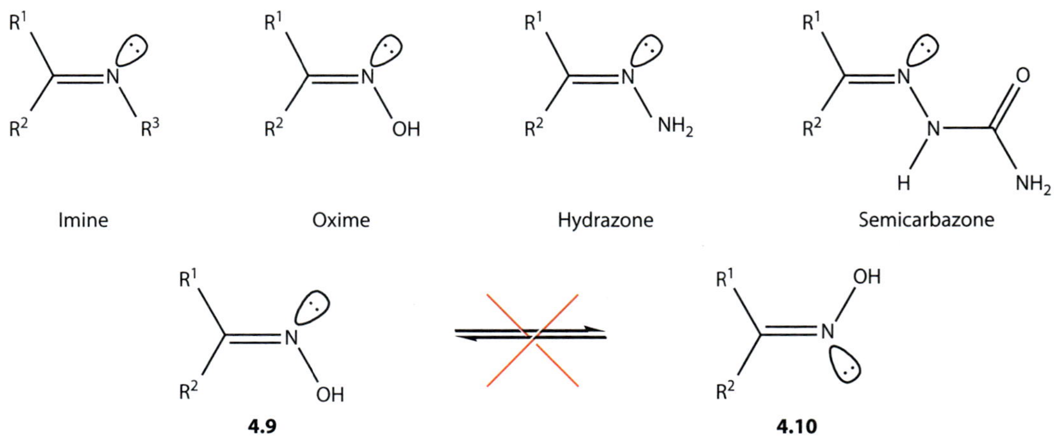

FIGURE 4.5 Classes of compounds containing sp² nitrogen atoms.

the substituents R^1, R^2, and nitrogen. The oxygen atom is sp^3 hybridized, with two sp^3 lone pairs and two σ-bonds to hydrogen and nitrogen. The ground state configuration of nitrogen is

$$\mathrm{N} \quad 1s^2\ 2s^2\ 2p_x^{\ 1}\ 2p_y^{\ 1}\ 2p_z^{\ 1}$$

As we did with carbon when it formed a double bond, we will hybridize the s, p_x, and p_y orbitals to give three sp^2 hybrid orbitals, reserving the p_z containing one electron to form the π-bond to carbon. The sp^2 hybrid orbitals must accommodate four electrons so one is filled and two are half-filled. The half-filled orbitals are used to make σ-bonds to carbon and oxygen. The filled sp^2 orbital is the lone pair at nitrogen, and its presence gives rise to the geometry of the molecule. We know that sp^2 orbitals point toward the corners of an equilateral triangle, so the molecules are all "bent." This means that when there are two different substituents at carbon, oximes may exist as geometric isomers (*e.g.*, **4.9** and **4.10**, where R^1 and R^2 are different) (Figure 4.5). These isomers have different physical and chemical properties and do not readily interconvert. While recognizing that substituents will have an effect, sp^2 lone pairs, in this type of compound, are more "s-like" and less extended than sp^3 lone pairs in ammonia and amines. They are therefore less basic.

4.3 MOLECULES WITH sp-HYBRIDIZED NITROGEN

We shall meet only two classes of compounds in this group, and only one is at all common. This is the class nitrile, or cyanide, R–C≡N. Both the carbon and the nitrogen of the CN group are sp hybridized. At nitrogen, as at carbon, the s and $2p_x$ orbitals are hybridized, and the $2p_y$ and $2p_z$ half-filled orbitals form π-bonds with the corresponding orbitals at the sp-hybridized carbon atom. One of the sp hybrid orbitals is filled, an sp lone pair, and the half-filled one forms the σ-bond to carbon (Figure 4.6).

There are two commonly used methods of naming nitriles. The first involves the entire carbon chain, including the carbon atom of the nitrile group, as the root, with the suffix nitrile. Thus, **4.11** would be named as pentane nitrile. In simple cases, the carbon of the nitrile is labeled as C-1, so **4.12** is 3-methylhexane nitrile. When the CN group is a substituent on a ring, it is named as a substituent with the suffix "-carbonitrile." Thus, **4.13** is cyclohexane carbonitrile. However, the cyanide can also be treated as a substituent—in this system, **4.11** is 1-butylcyanide (note the carbon of the CN is now not included in the longest chain), **4.12** is 2-methyl-1-pentylcyanide, and **4.13** is cyclohexylcyanide. Confusing, certainly, but usually not ambiguous, whichever system is used. Two important common names that you need to know are acetonitrile, CH_3CN, and benzonitrile, PhCN.

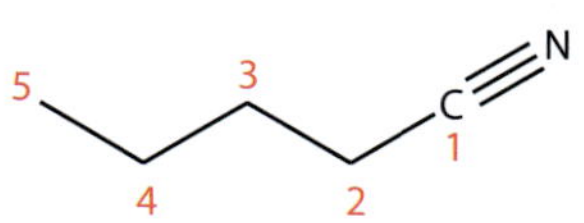

4.11

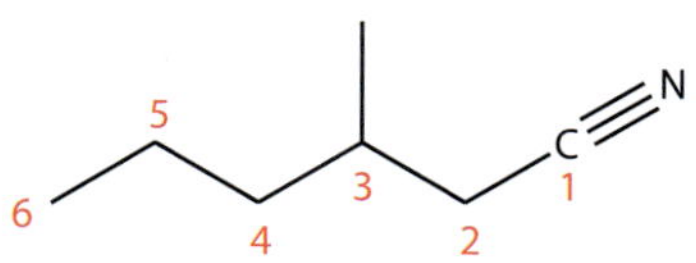

4.12

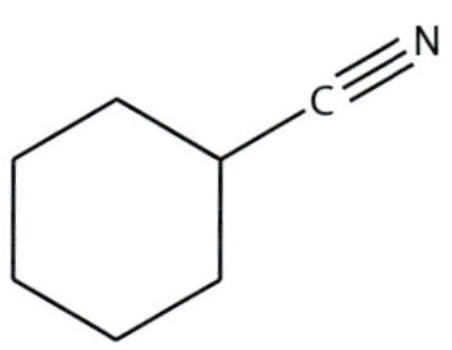

4.13

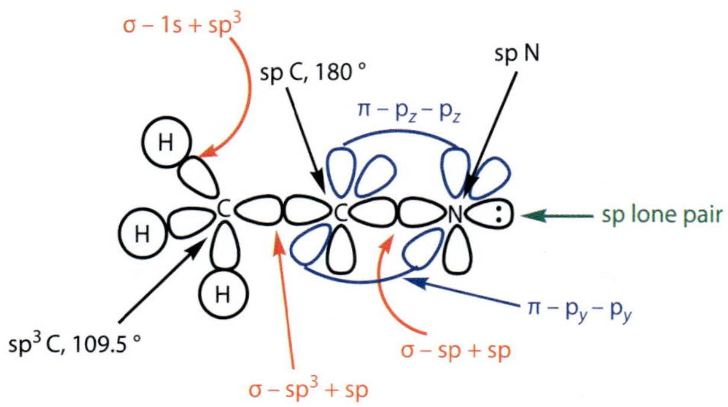

FIGURE 4.6 Bonding in MeCN. While both lobes of p orbitals are shown, only the major lobe of sp^3, sp^2, or sp orbitals is shown.

TABLE 4.2 Melting and Boiling Points of Nitriles and Alkynes

# 1st Row Atoms	Alkyne	MPt (°C)	BPt (°C)	Nitrile	MPt (°C)	BPt (°C)
3	Propyne	−103	−23	Propanenitrile	−93	97
4	1-Butyne	−126	10	Butanenitrile	−112	118
5	1-Pentyne	−106	40	Pentanenitrile	−96	141
6	1-Hexyne	−131	71	Hexanenitrile	−89	162
7	1-Heptyne	−81	99	Heptanenitrile	−64	180
8	1-Octyne	−77	126	Octanenitrile	−53	199
9	1-Nonyne	−50	150	Nonanenitrile	−49	224
10	1-Decyne	−44	174	Decanenitrile	−14	243
11	1-Undecyne	−25	196	Undecanenitrile	−6	253
12	1-Dodecyne	−19	215	Dodecanenitrile	4	277

Another example of intermolecular interactions between polar molecules is provided by a comparison of the melting points and boiling points of 1-alkynes and nitriles (Table 4.2). The boiling points of the nitriles are systematically higher than those of the 1-alkynes of comparable molecular weight—to separate the molecules, energy must energy must be supplied to overcome dipolar interactions. The melting points are more similar. For a dipolar solid, any given molecule is surrounded by a large array of other molecules (not just one or two, as in the liquid phase). There are both attractive and repulsive interactions between the dipoles, which almost cancel out.*

PROBLEM 4.1

Give a line structure for each of the following molecules:

(a) Diethylamine
(b) Benzonitrile

PROBLEM 4.2

Give a systematic name for each of the following molecules:

(a)

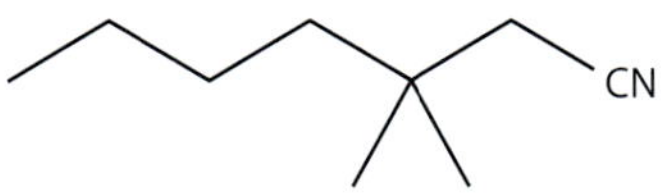

(b)

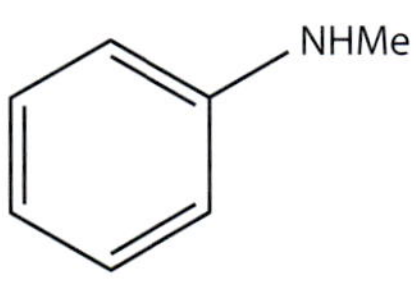

PROBLEM 4.3

For the molecule $MeNH_2$, generate an orbital representation. Label bonds as σ or π, and state which orbitals were used to make them (s, p, sp, sp^2, sp^3). Mark empty orbitals and lone pairs, and state their type.

* You may remember considering something similar if you have studied lattice energies of ionic solids and discussed the Madelung constant.

PROBLEM 4.4

For each of the following pairs of compounds, state which you think will have the higher boiling point, giving a reason for your conclusion:

(a) Benzene or pyrimidine
(b) Methylethylamine and triethylamine

Solutions

4.1(a)

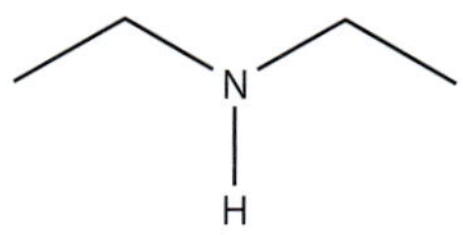

4.1(b)

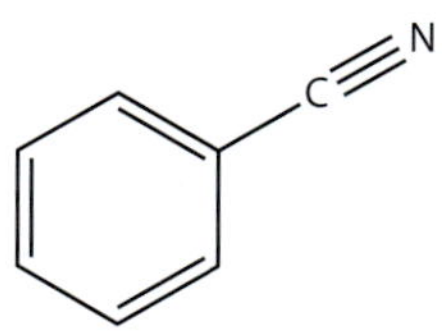

4.2(a) 3,3-Dimethylheptane nitrile
4.2(b) *N*-Methylaniline
4.3

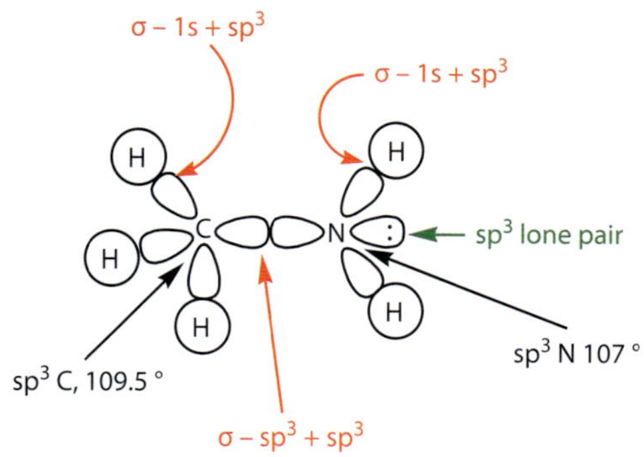

4.4(a) Pyrimidine—it has a dipole moment.
4.4(b) Methylethylamine—this can act as both an H-bond donor and an acceptor. Triethylamine can only act as an H-bond acceptor.

Key Points from Sections 4.1 through 4.3

- Compounds containing sp^3 nitrogen atoms are named as amines. IUPAC mandates methanamine for $MeNH_2$; common usage is methylamine.
- Amines can form hydrogen bonds both to other similar molecules and to appropriate solvents.
- Amines, like ammonia, may be protonated to give ammonium salts. With more difficulty, they may also be deprotonated.
- sp^2 Nitrogen atoms are found in aromatic compounds such as pyridine (where the lone pair is at right angles to the plane of the aromatic system) and pyrrole, where the lone pair, in an sp^2 orbital, is part of the aromatic system.
- sp^2 Nitrogen atoms form double bonds with carbon in imines, oximes, hydrazones, and semicarbazones.
- Compounds with a carbon nitrogen triple bond contain sp hybrid nitrogen atoms and are named as nitriles or cyanides.

4.4 COMPOUNDS CONTAINING sp³ OXYGEN

4.4.1 ALCOHOLS

Alcohols have a general formula ROH, and the bonding is similar to that in water—one of the hydrogens has been replaced by a carbon substituent. The hybridization at oxygen is sp^3, and there are two lone pairs of electrons (**4.14**). As with water, hydrogen bonding is very important. The lower alcohols are all liquids at room temperature and are miscible with water (otherwise, we would have no martinis(!), nor many perfumes or cleaning products). They are often divided into three groups, primary, secondary, and tertiary, with generic formulae RCH_2OH, R_2CHOH, and R_3COH, respectively. Their reactivity varies with structure.

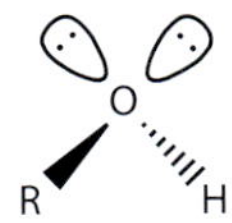

4.14

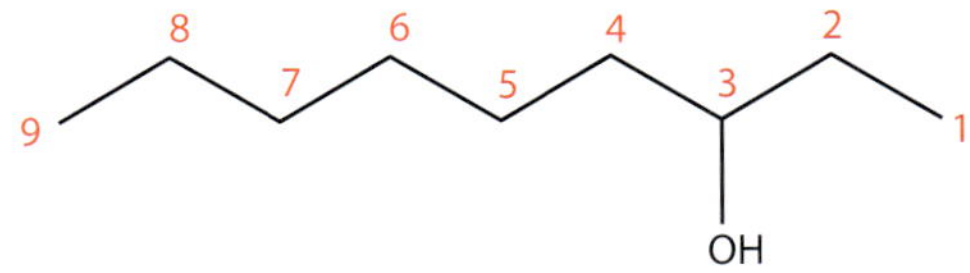

4.15

The suffix for alcohols is "-ol," attached to the name of the alkane (or alkene or alkyne) from which it is derived. Thus, CH_3OH is methanol; CH_3CH_2OH is ethanol. Table 4.3 gives some names, properties, and uses of the lower alcohols. For more complex examples such as **4.15**, the

TABLE 4.3 Properties of Some Common Alcohols

Structure	Systematic Name	Common Name(s)	BPt (°C)	Uses, Comments
CH_3OH	Methanol	Methyl alcohol, wood alcohol	65	Originally from distillation of wood. Toxic, used to denature ethanol for industrial use. Used as a fuel additive. Miscible with water.
CH_3CH_2OH	Ethanol	Ethyl alcohol, grain alcohol	78	Originally produced from sugar fermentation. Used in beverages, paints, perfumes. Antiseptic, used in hand sanitizers. Miscible with water.
$CH_3CH_2CH_2OH$	1-Propanol	Propyl alcohol, *n*-propanol	97–98	Miscible with water.
$(CH_3)_2CHOH$	2-Propanol	*Iso*-Propanol, IPA	82.5	Rubbing alcohol or surgical spirit is a 70 % solution in water, used to cool skin and disinfect. Used in deicing fluids. A slurry with solid CO_2 is used as a cooling bath for a temperature of −78 °C. Miscible with water.
$CH_2{=}CHCH_2OH$	2-Propene-1-ol	Allyl alcohol	97	Miscible with water, moderately toxic.
$CH_3CH_2CH_2CH_2OH$	1-Butanol	Butyl alcohol, *n*-butanol	118	Perfumery, extraction of essential oils. Water soluble.
$(CH_3)_2CHCH_2OH$	2-Methyl-1-propanol	*Iso*-Butanol	108	Additive in ink, paint, polishes. Water soluble.
$CH_3CH_2CH(CH_3)OH$	2-Butanol	*Sec*-Butanol	99	Industrial solvent. Water soluble.
$(CH_3)_3COH$	2-Methyl-2-propanol	*Tert*-Butanol	82	MPt, 25 °C. Additive in fuel and paint stripper. Water soluble.
$HOCH_2CH_2OH$	Ethane-1,2-diol	Ethylene glycol	197	Solution in water used as antifreeze; lowest freezing point, −51 °C, at 70 % ethylene glycol. Polymer precursor. Sweet taste, but rather toxic.
$HOCH_2CH(OH)CH_2OH$	Propane-1,2,3-triol	Glycerol	290	Miscible with water. Used as an additive in pharmaceuticals, food, and personal care products. Key component of fats, triglycerides.

rule is to look for the longest chain that includes the OH-bearing carbon atom and then number so that the hydroxyl group is on the lowest-numbered carbon atom. Thus, **4.15** is 3-nonanol. You should never describe it as 3-hydroxynonane; this is tolerated for amines, but not for alcohols. Taking another example, **4.16**, the answer will be more obvious if we redraw the molecule before numbering it—this is 3-ethyl-3-pentanol. If there is more than one –OH group in the molecule, the terms diol, triol, *etc.*, are used. So **4.17** is 2,3-butanediol.

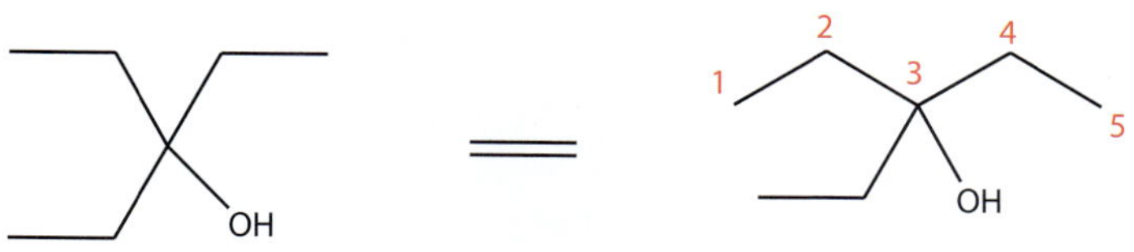

4.16

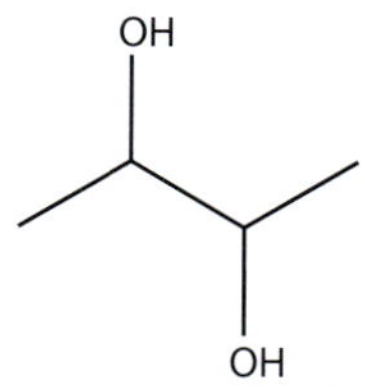

4.17

Where the molecule contains both a double bond and an OH group, the OH group takes precedence in numbering the chain (at the end of this section on nomenclature, there will be a complete list of priorities). So **4.18** is named as 2-propene-1-ol. Its common name is allyl alcohol. We have already met phenol, **4.19**—when there is more than one OH group or where the OH group does not have the highest priority and is named as a substituent, we use the term hydroxy. So, **4.20** is 1,3-dihydroxybenzene.

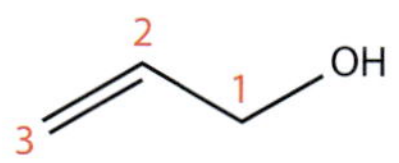

4.18

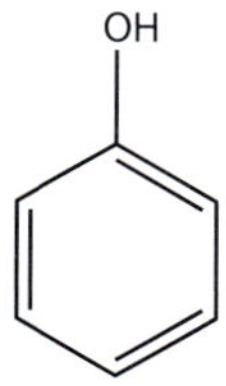

4.19

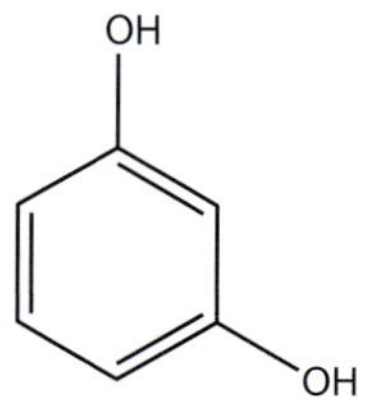

4.20

Focus on Proof Spirit and Moonshine

Ethanol is the oldest recreational drug, and the fermentation of sugar to alcohol was probably one of the first chemistry experiments ever undertaken. The term proof spirit originated in England in the seventeenth century and was defined in terms of the minimum alcohol content that when gunpowder was soaked with it, in a silver spoon, it would still allow the gunpowder to burn. Proof spirit had to have at least 57.15 % alcohol by volume, and this was defined as 100 proof. On this scale, used until 1980 in the United Kingdom (when the European Union [EU] standard was introduced), 100 % alcohol was 175 proof. From 1740 to 1816, a Clarke's hydrometer was used to measure the specific gravity of the liquid, and after 1816, this was replaced by a Sikes hydrometer. In the United States, the proof is twice the percentage alcohol by volume; on this scale, 95 % alcohol, the highest percentage that can be obtained by distillation (alcohol and water co-distill as a 95:5 azeotropic mixture) is 190 proof.

The term moonshine originated in Britain in the early eighteenth century as a term for pursuits undertaken at night—including illegal distillation. It is now most commonly used for illegally distilled corn whiskey in parts of the United States, most notably in the South, the Appalachian region, and parts of Atlantic Canada. The production of moonshine increased dramatically during the prohibition era, 1920–1933, and was in decline by the 1950s. Its production is reported to increase again during every recession, including the most recent, and the largest moonshine bust in the United States was in 2001, involving

1.5 million gallons of alcohol.* Moonshiners discovered that if they powered their vehicles with alcohol, they went faster, starting the sport of stock car racing.

Moonshine still, 1922

Almost every nation has some form of home-distilled spirits available, either legally or, more commonly, illegally. In Ireland, this is *poitín* (Anglicized as poteen or potcheen), the word deriving from the diminutive of the Irish *pota*, pot, as it is made in small domestic pot stills. The materials used are malted barley or potatoes. Since 1989, the manufacture of *poitín* for sale has been licensed, and Irish *poitín* has geographical indicative status from the EU (so no one outside Ireland can manufacture it and call it *poitín*). Traditionally, the first portion or noggin of the distillate is poured on the ground "for the fairy folk," but this has an important practical application as well. This forerun contains toxic methanol (10 cm^3 causes blindness, and 30 cm^3 is lethal) and acetone. Much of Latin America produces *chicha*, while in Eastern Europe, the most common home distillates are fruit brandies such as *slivovice* (Czech, plum brandy) or *rakia* (Serbia). *Witblits*, white lightning, produced in South Africa from grapes, is said to be comparable to Italian grappa.

Poitin

Rakia

Chicha

* If you want to know more, try to find a copy of the out of print title *Me and My Likker*, Marvin (popcorn) Sutton (1999).

4.4.2 ETHERS

Ethers are the class of compounds in which both of the hydrogen atoms of water have been replaced by carbon substituents. The bonding is the same as in water or in alcohols; the oxygen atom is sp^3 hybridized, with two lone pairs of electrons. However, these molecules differ from water and alcohols in that they are not hydrogen-bond donors; they thus have lower boiling points and are at best sparingly soluble in water. The nomenclature is not complex, but there is no single system that is used universally. Where the substituents are simple alkyl groups, then these are usually used in the nomenclature. Thus, **4.21** is diethyl ether (commonly referred to just as ether, as it is a very widely used solvent), and **4.22** is ethyl methyl ether. However, a more systematic approach can also be useful. In this, the entire chain is named as an alkane, with a CH_2 substituted for the oxygen. Then the oxygen is "acknowledged" with the term "oxa" and its position. Thus, **4.21** is then 3-oxapentane and **4.22** 2-oxabutane.

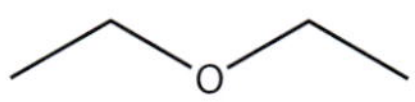

4.21

Diethyl ether
3-Oxapentane

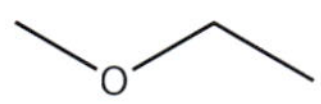

4.22

Methyl ethyl ether
2-Oxabutane

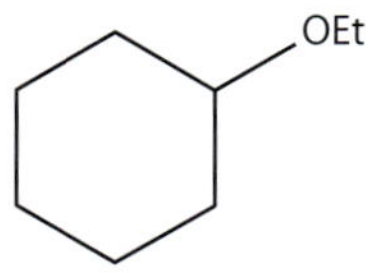

4.23

Cyclohexyl ethyl ether
Ethoxycyclohexane

When one of the substituents is a ring, as in **4.23**, another method is common. This could be named as cyclohexyl ethyl ether, but can also be described as ethoxycyclohexane. One rather special class of ethers is the epoxides, or oxiranes, such as **4.24** and **4.25**. There are two methods by which they can be named—the first refers to the alkene from which they are generally synthesized plus the word oxide. Thus, **4.24** is ethene oxide and **4.25** cyclohexene oxide. Alternatively, the molecule may be named as an alkane, and then the word epoxy, and the location of the three-membered ring added. In this case, the names would be ethane epoxide and cyclohexane-1,2-epoxide.

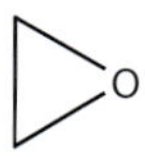

4.24

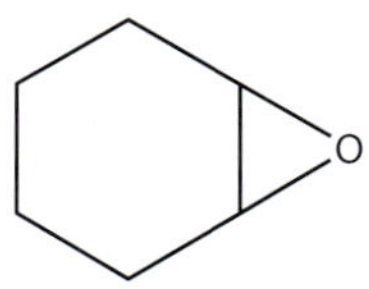

4.25

Digression: More about Hydrogen Bonds

Where is the proton in a hydrogen bond? One might think that it would be halfway between the two atoms connected by the hydrogen bond. With one exception, the anion $[HF_2]^-$, this is not the case. The state when the hydrogen is exactly halfway between the two atoms is an energy maximum and represents the transition state for the transfer of a proton from one atom to the other (Figure 4.7). The best arrangement for hydrogen bonding, and the best

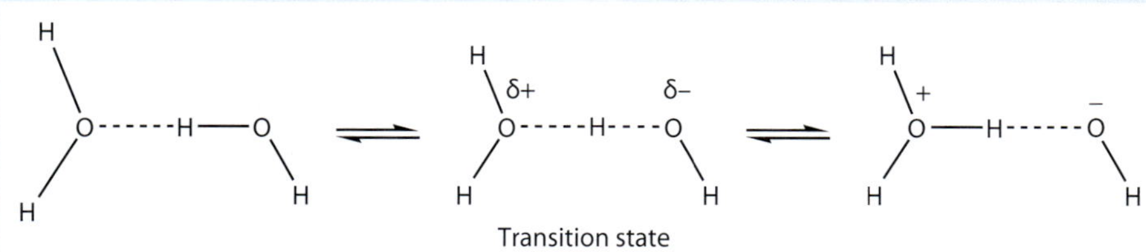

FIGURE 4.7 Transfer of a proton between water molecules.

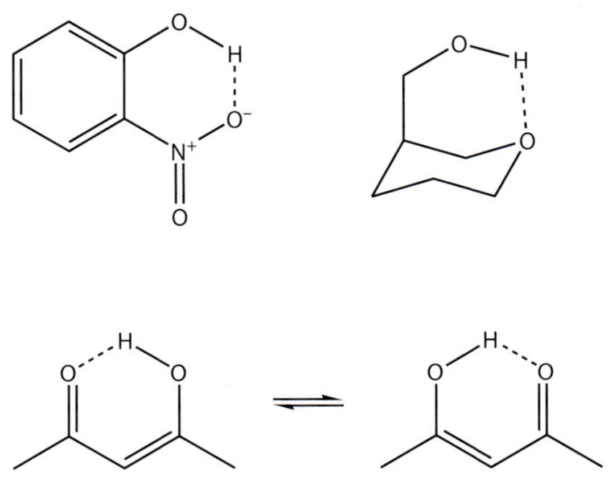

FIGURE 4.8 Molecules with intramolecular hydrogen bonds.

transition state for proton transfer, is linear, but this is not always possible. A number of molecules, such as *ortho*-nitrophenol (Figure 4.8), are stabilized by intramolecular hydrogen bonding, and linearity here would be impossible.

The solubility of a material in water is often related to its ability to hydrogen bond to it. Water can act either as an H-bond donor or an H-bond acceptor. With acetone, **4.26**, and tetrahydrofuran (THF), **4.27**, water acts only as a donor, and both are miscible with water. For comparison, the solubility of cyclopentane in water is 0.01 %.

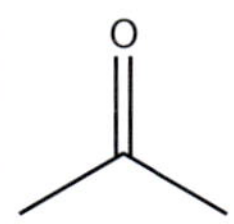

4.26

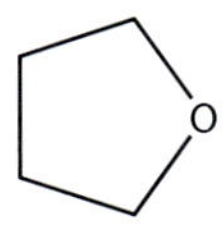

4.27

PROBLEM 4.5

Give a systematic name for each of the following molecules:

(a) $CH_3CH_2CH(CH_3)CH_2CH_2OH$

(b)

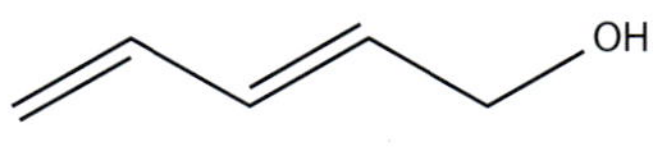

(c)

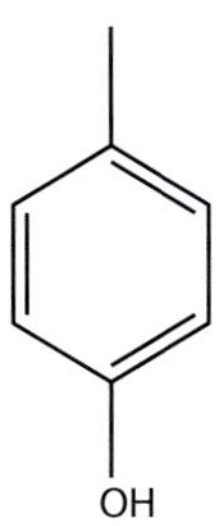

PROBLEM 4.6

For methanol, CH_3OH, generate an orbital representation. Label bonds as σ or π, and state which orbitals were used to make them (s, p, sp, sp^2, sp^3). Mark empty orbitals and lone pairs, and state their type.

PROBLEM 4.7

For each of the following pairs of molecules, predict which would have the higher boiling point, giving a reason:

(a) Methoxybenzene and benzyl alcohol
(b) Ethylamine or ethanol
(c) Hexanol or ethane-1,2-diol

Solutions

4.5(a) The molecule is numbered from the carbon bearing the –OH group, and the longest chain has five carbons. So this will be named as a 1-pentanol. There is a methyl substituent at C-3, hence 3-methyl-1-pentanol.

4.5(b) Again, the molecule is numbered from the carbon bearing the –OH group, and there are five carbons in the chain. The two double bonds start at C-2 and C-4, hence 2,4-pentadiene-1-ol.

4.5(c) This will be named as a phenol, hence 4-methylphenol.

4.6

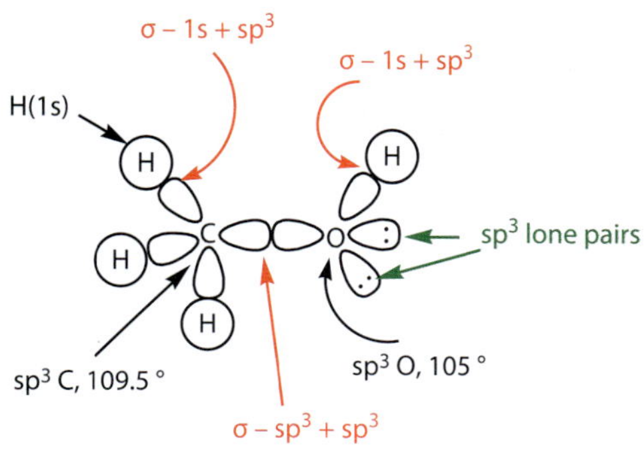

4.7(a) Benzyl alcohol—can form hydrogen bonds.
4.7(b) Ethanol—oxygen forms stronger hydrogen bonds than nitrogen.
4.7(c) Ethane-1,2-diol—two OH groups mean more potential for H-bonding.

4.4.3 CROWN ETHERS, IONOPHORES, AND THE SOLVATION OF METAL IONS

Many small cations, especially those of the first two groups of the periodic table, are attracted to lone pair electron density on a heteroatom, usually nitrogen or oxygen, in much the same way that protons are. This is the reason that a salt such as sodium chloride dissolves in water; the ions are solvated (Figure 4.9).

Such interactions are even better when one metal cation can interact with several lone pairs on heteroatoms in the same molecule. Species that can do this are known as bidentate, tridentate,

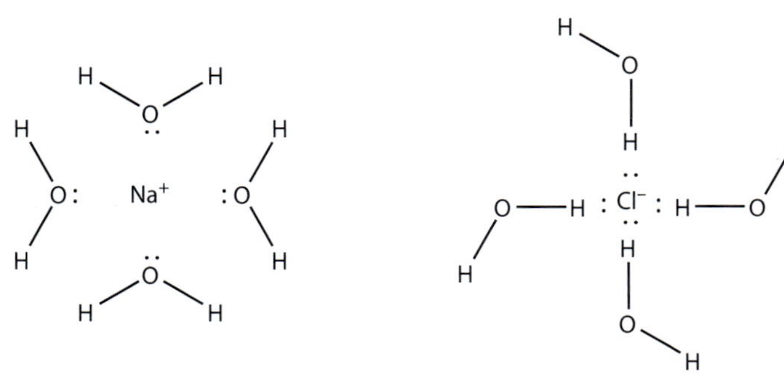

FIGURE 4.9 Solvation of sodium chloride in water.

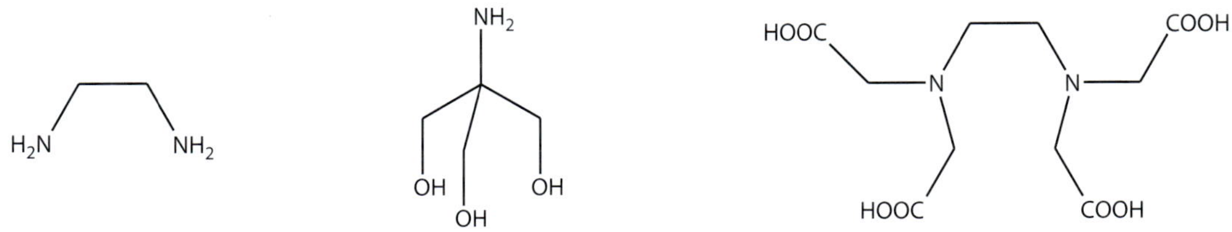

Ethylene diamine

tris (Hydroxymethyl)-aminomethane (TRIS)

Ethylene diamine tetraacetic acid (EDTA)

FIGURE 4.10 Molecules that can act as multidentate ligands.

or polydentate ligands (Greek πολυς, *polus*, many; Latin *dentatus*, toothed), and the process of multiple interactions of this type is called chelation (Greek χηλή, *chelē*, claw). Such materials (examples in Figure 4.10) are often used to solubilize metal ions in media in which they would not ordinarily dissolve.

Naturally occurring lipid-soluble molecules that are able to transport ions across cell membranes are called ionophores (ion carriers, Greek φέρω, *phero*, to carry), and they play numerous critical roles in biology. They encase metal ions in a mainly hydrocarbon "shell," and this allows them to pass through lipid membranes. Sometimes, as with the puffer fish toxin, tetrodotoxin, **4.28**, this means that they are very toxic, as they upset the normal ion balance within cells. In other cases, as with amphotericin B, **4.29**, there is a medicinal use; although it is quite toxic, it can be used in low doses to treat the protozoan infection leishmaniasis.

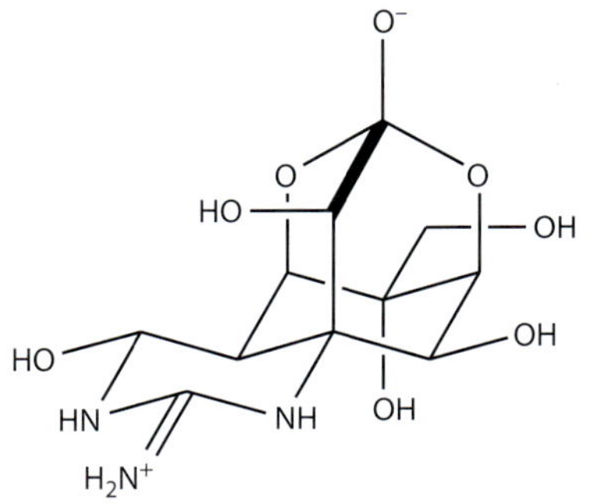

4.28, tetrodotoxin

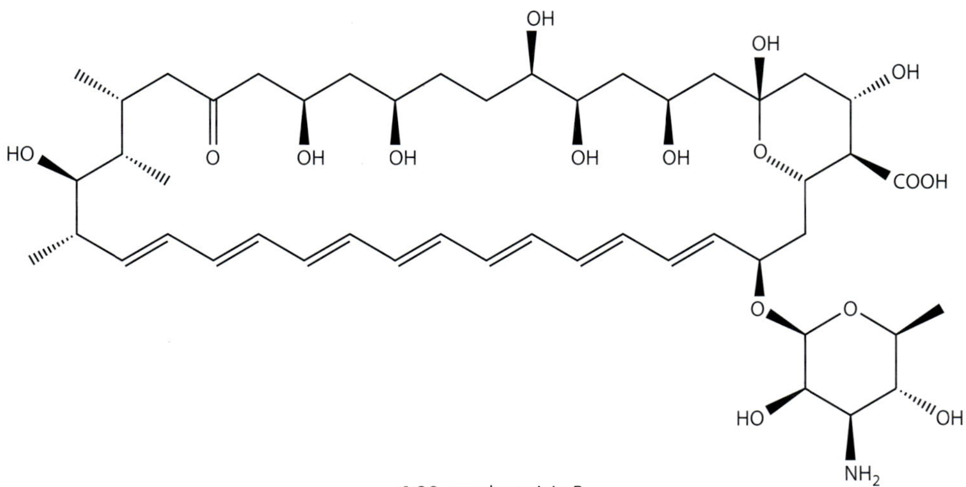

4.29, amphotericin B

Focus on Tetrodotoxin

Tetrodotoxin is found in puffer fish, porcupine fish, ocean sunfish, and some species of toads and newts; its purpose is to deter predators. It is a neurotoxin, and there is no antidote. Less than 1 mg will kill an adult. Symptoms start with numbness of the lips and tongue, and death occurs by paralysis of the diaphragm. There are some 30–40 poisonings per annum reported in Japan, mostly from fishermen eating their own catch. "Fugu" restaurants, where the fish is served as a highly prized delicacy, are strictly regulated so that the most toxic parts of the fish are discarded. Recent research has shown that puffer fish bred in captivity do not produce tetrodotoxin (and in some parts of the world, farmed fish is now the main source for restaurants, and hence safe, whatever they may advertise); the material is thought to originate in the wild from symbiotic bacteria.

Puffer fish

There are also numerous synthetic cyclic ionophores, which can transport metals ions into nonpolar solvents. Their effectiveness depends on the "fit" of the metal ion into the cavity at the center of the ligand. The best known examples are the crown ethers, for example, 18-crown-6 (18-C-6), **4.30**. This is an ionophore with an 18-membered ring containing 6 ether-type oxygen atoms. The cavity is the right size to encapsulate K^+. In the presence of 18-C-6, the powerful oxidant potassium permanganate can be dissolved in a nonpolar solvent such as benzene (so-called purple benzene), thus enabling it to be a very useful reagent for the oxidation of hydrophobic substrates. Li^+ is best complexed with 12-crown-4, **4.31** and Na^+ with 15-crown-5, **4.32**. The name crown ether derives from the fact that in the crown ether–metal complex, the ring adopts a conformation reminiscent of the sort of crowns found in Christmas crackers.* These were originally discovered by Charles Pedersen, and many

* Christmas crackers are popular in the United Kingdom and Ireland and some Commonwealth countries; they consist of a cardboard tube wrapped in paper, which is pulled with a small bang like a cap gun, at the start of a holiday meal. Inside is a gift and a rather foolish looking paper crown, which is supposed to be worn for the rest of the meal.

variants were developed by Jean-Marie Lehn and Donald Cram—these three scientists shared the Nobel Prize for chemistry for this work in 1987. We will return to the uses of the compounds in Chapter 23, on catalysis.

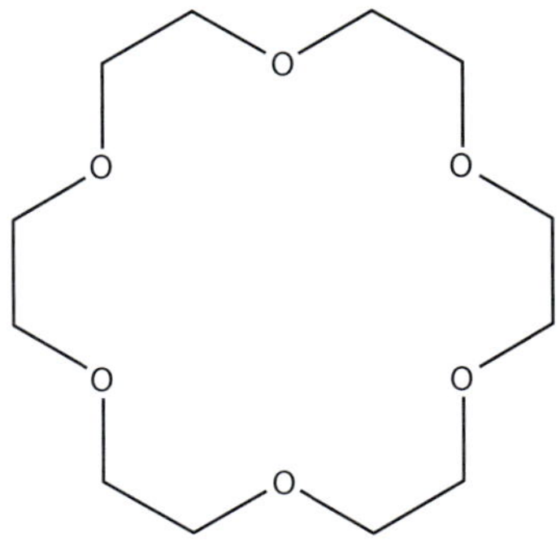

4.30, 18-crown-6

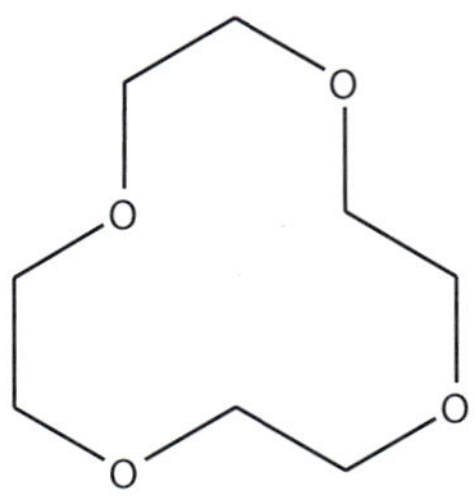

4.31, 12-crown-4

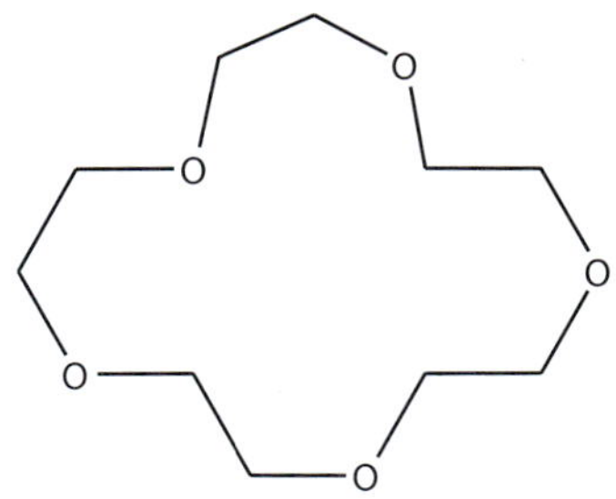

4.32, 15-crown-5

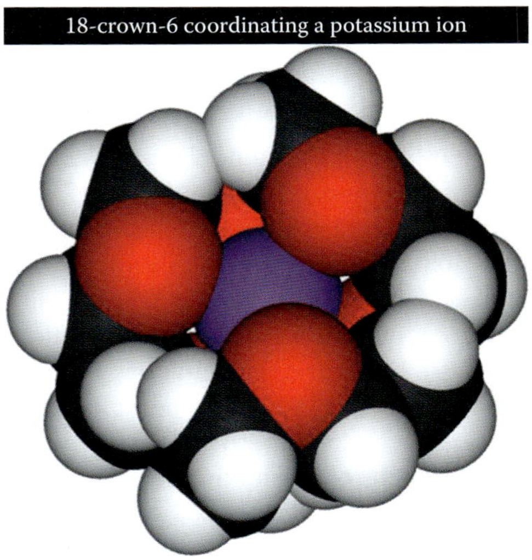

PROBLEM 4.8

Give an example of each of the following:

(a) An ionophore used as an antibiotic (structure essential)
(b) A crown ether
(c) A tridentate ligand

Solution

Many examples of each are possible:

(a) Nonactin has a strong preference for the complexation of potassium over sodium, thus making bacterial cells leak K^+.

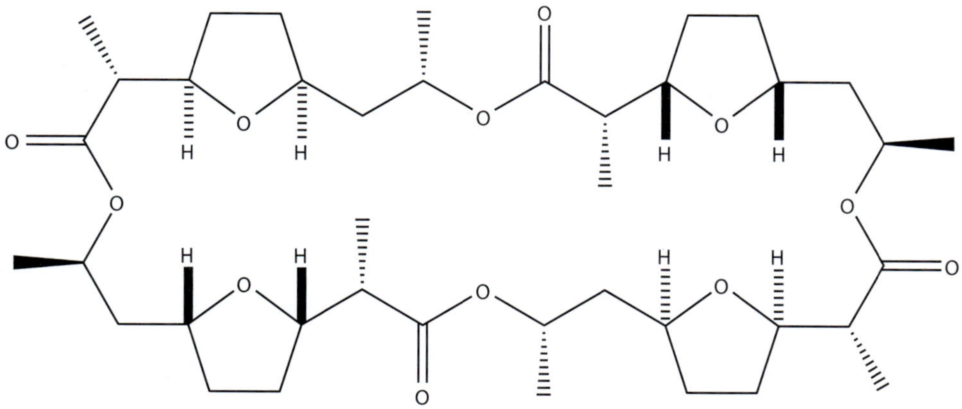

Nonactin

(b) 18-C-6

(c)

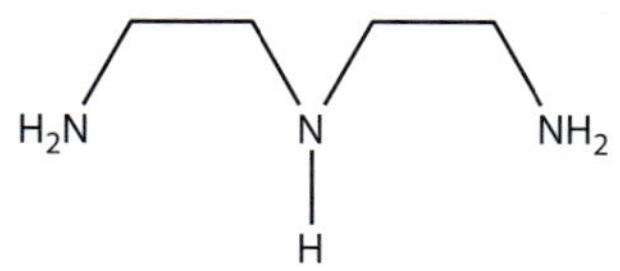

Key Points from Section 4.4

- Alcohols are named from the longest chain bearing the OH group. The suffix is -ol. Several schemes are available for naming ethers.
- Intermolecular interactions in alcohols are dominated by hydrogen bonding, so that they have higher boiling points than hydrocarbons or amines of comparable molecular weight. Like water, they can act as H-bond donors or H-bond acceptors.
- Lower alcohols are miscible with water, because of their hydrogen-bonding ability.
- Metal ions can be solvated by interaction with lone pairs on oxygen (or nitrogen); the solubilization is enhanced if there are multiple sites in a molecule for binding—the chelate effect.
- Naturally occurring and synthetic ionophores have many sites for metal coordination and are able to transport metal ions across lipid membranes.
- Crown ethers are an important class of synthetic ionophore; they consist of cyclic polymers of the (CH_2CH_2O) unit, and their size can be designed to encapsulate ions of a specific radius.

4.5 COMPOUNDS CONTAINING sp² OXYGEN

All the classes of compounds that will be discussed here contain a carbon–oxygen double bond, called a carbonyl group (**4.33**). They are classified and named according to the substituents on the carbon atom of the carbonyl, but the carbon–oxygen bonding is the same in each case. Carbon is sp^2 hybrid and uses its p_z orbital to make a π-bond to oxygen. Thus the p_z orbital at oxygen must be used to complete this π-bond, and the oxygen is also sp^2 hybridized. One of the sp^2 hybrid orbitals is half full and makes the σ-bond to carbon, and the other two are filled—thus, there are two lone pairs at oxygen. When we consider the reactivity of the carbonyl group, these lone pairs will be crucial in the ready protonation of most types of carbonyl compounds.

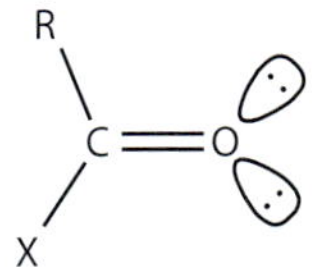

4.33

4.5.1 ALDEHYDES

Aldehydes are the class of carbonyl compounds where one substituent is an alkyl, alkenyl, or aryl group, and the other is a hydrogen atom, **4.34**. The suffix for aldehydes is "-al" added to the hydrocarbon skeleton name, so if R = CH_3, **4.34** would be ethanal (the old name, which is still commonly used, was acetaldehyde). In formal nomenclature, the aldehyde carbon is designated as C-1. When R = H, the name is methanal, but this compound is almost invariably known by its common name, formaldehyde. With more complex structures, the longest chain starting at C-1 is discerned, numbered, and the substituents put in as usual. So **4.35** is 2,3-dimethylpentanal.

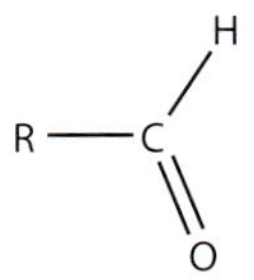

4.34

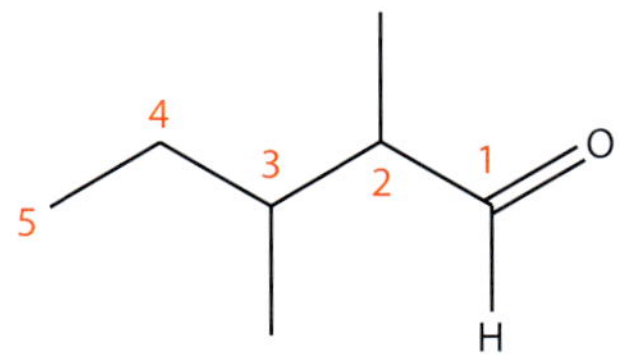

4.35

PROBLEM 4.9

Give a line structure for 3-methyl-5-bromooctanal.

Solution

First, draw and number octanal (it is not always necessary to draw out the whole carbonyl structure; –CHO is usually sufficient), remembering that the carbon of the carbonyl is C-1:

Then add the substituents at the appropriate positions:

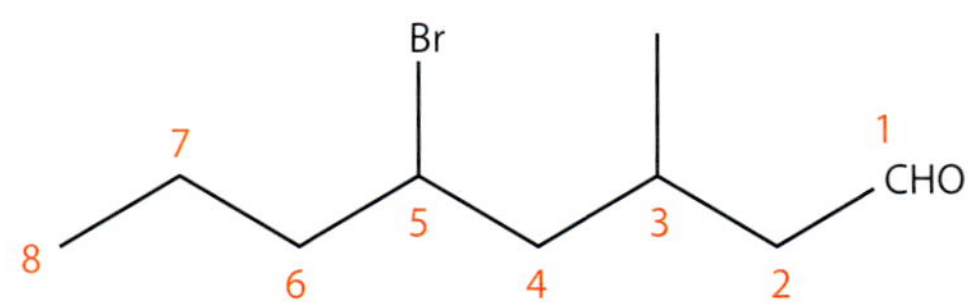

When the –CHO group is a substituent on a ring, we use the term carbaldehyde. Thus, **4.36** is cyclohexane carbaldehyde and **4.37** is benzene carbaldehyde, usually contracted to benzaldehyde:

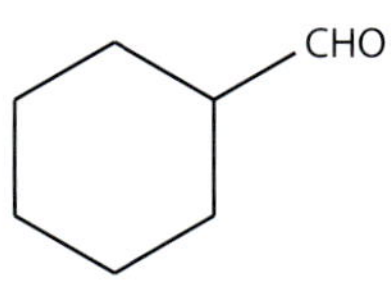

4.36

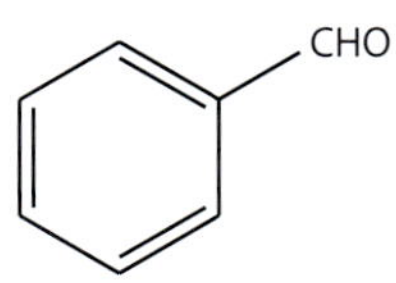

4.37

Lower aldehydes are generally water soluble; although they cannot act as hydrogen-bond donors, they are excellent hydrogen-bond acceptors. The solution of formaldehyde in water is referred to as formalin and is widely used for the preservation of biological specimens.

4.5.2 KETONES

In this class of compounds, both the substituents on the carbonyl group are carbon—alkyl, aryl, alkenyl, *etc.* The suffix is "-one." So **4.38** is propanone (common name, invariably used, is acetone), and **4.39** is 2-pentanone. The longest chain used in the nomenclature must contain the carbon atom of the carbonyl group and is numbered so that the carbonyl has the lowest number it can. Where there are two carbonyl groups, the molecule is named as a dione; thus, **4.40** is butane-2,3-dione.

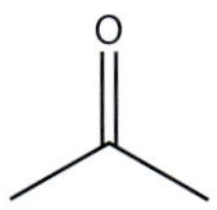

4.38

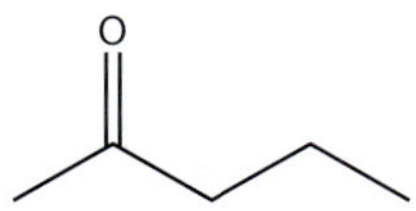

4.39

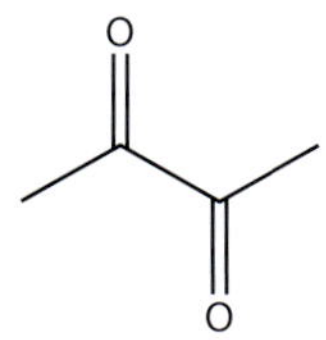

4.40

If we consider **4.41**, where there is both a ketone and an aldehyde functional group, the aldehyde takes precedence, and the compound is named as an aldehyde. The keto group is now named as an "oxo" substituent—thus, 4-oxopentanal.

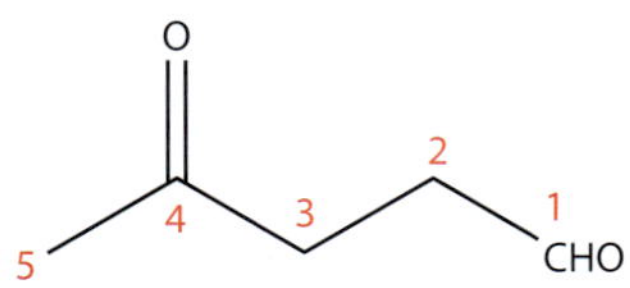

4.41

As with aldehydes, the lower ketones are soluble in water because they can act as hydrogen-bond acceptors. Acetone is very widely used as a water-miscible solvent.

PROBLEM 4.10

Give a line structure for each of the following molecules:

(a) 2,3,4-Trimethyl-2-cyclohexenone
(b) 2-Chloro-5,5-dimethyl-4-nonanone

Solutions

(a) If this molecule was simply 2-cyclohexenone, we would draw a six-membered ring with the carbonyl carbon as C-1 and the double bond between C-2 and C-3:

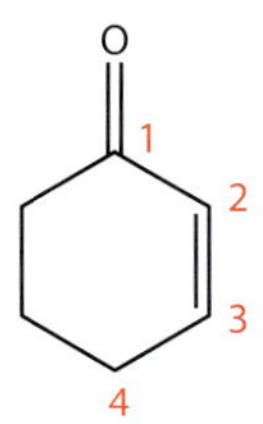

Then we add the methyl substituents at the appropriate positions:

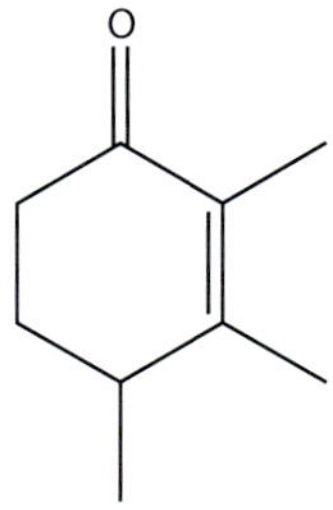

(b) Draw a nine-carbon chain, number it, and add a carbonyl group at C-4:

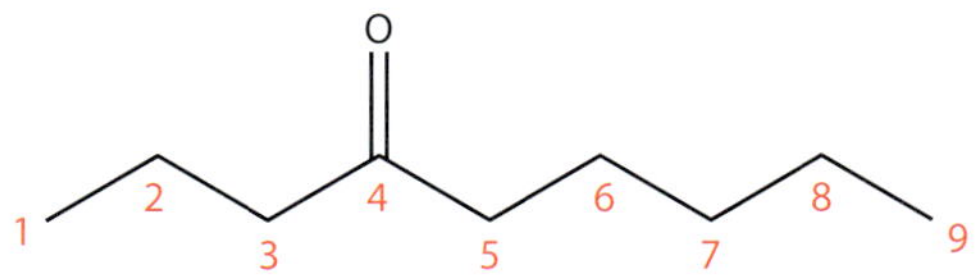

Then add the substituents at the appropriate positions:

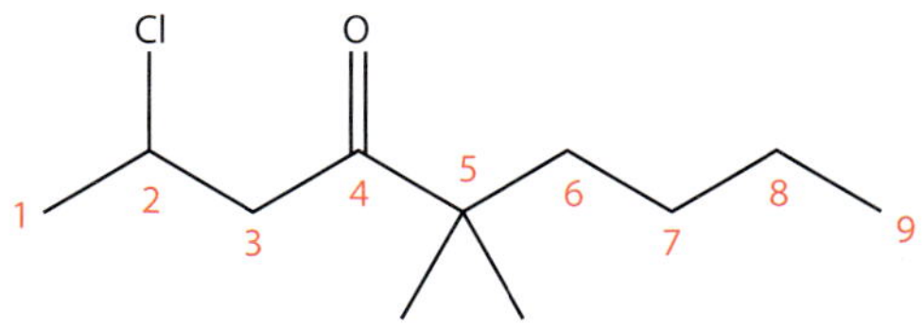

4.5.3 CARBOXYLIC ACIDS

In carboxylic acids, one of the substituents at the carbonyl group is carbon (or H in HCOOH), and the other is an OH group. The suffix is "-oic acid," and the carboxyl carbon atom is always C-1. Thus, **4.42** is 3-ethyl-2-methyl hexanoic acid. Methanoic (**4.43**), ethanoic (**4.44**), and benzene carboxylic acid (**4.45**) acids generally go by their common names, formic (from Latin, *formica*, ant—it was originally obtained from red ants) acid, acetic (Latin, *acetum*, vinegar) acid, and benzoic acid. When the formula of the acid is written in text, it can appear as CH_3COOH, CH_3CO_2H, or $CH_3C(=O)OH$. All of these mean the same molecule and simply reflect a stylistic difference on the part of the author.

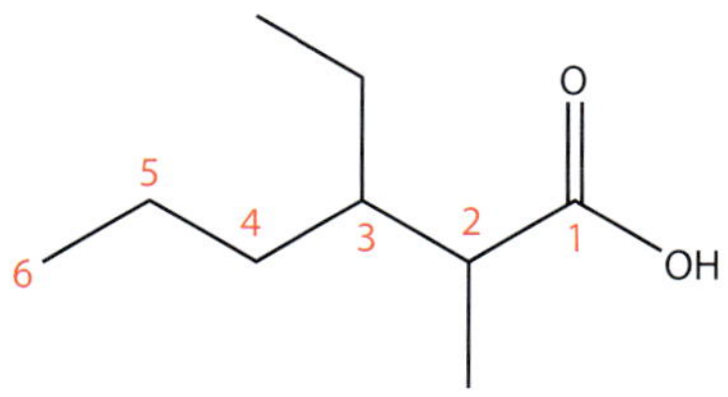

4.42

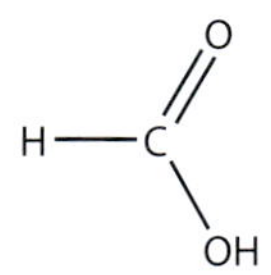

4.43

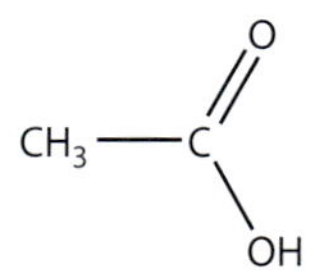

4.44

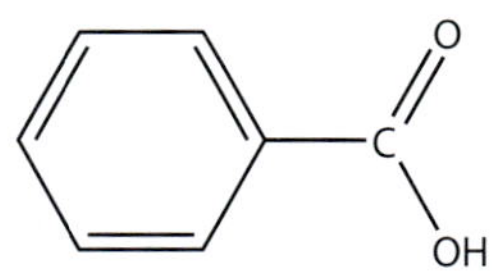

4.45

Carboxylic acids hydrogen bond strongly to water—they can act as both H-bond donors and acceptors (**4.46**), and the lower acids are miscible with water. In nonpolar solvents, such as benzene, they self-associate through hydrogen bonding to give dimers (**4.47**).

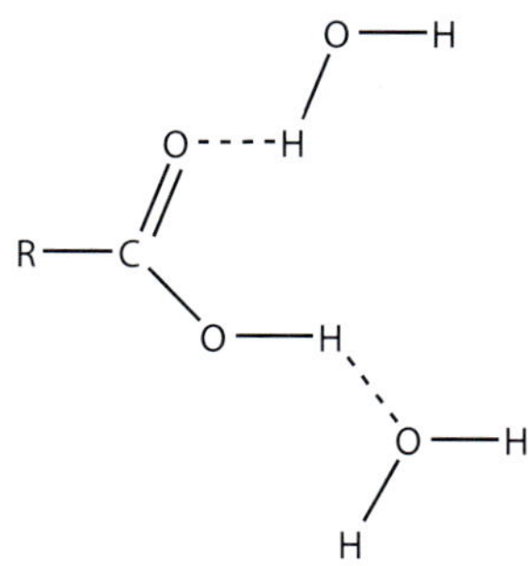

4.46

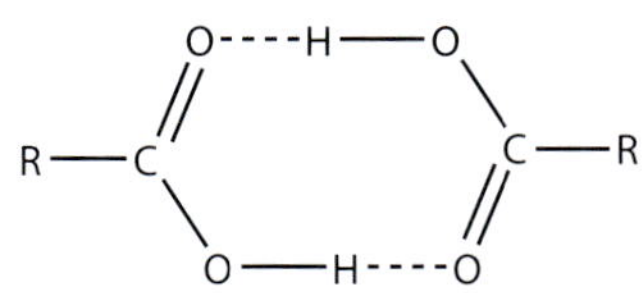

4.47

PROBLEM 4.11

Give systematic names for each of the following molecules:

(a)

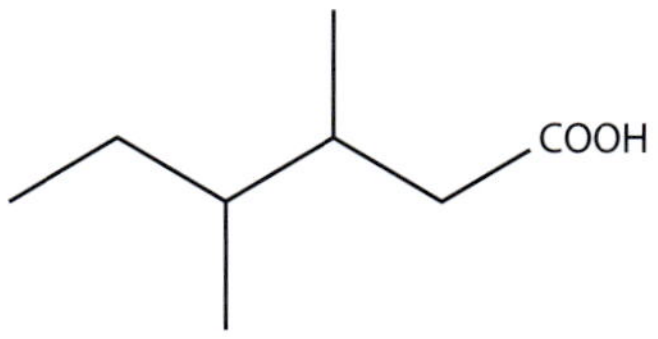

(b)

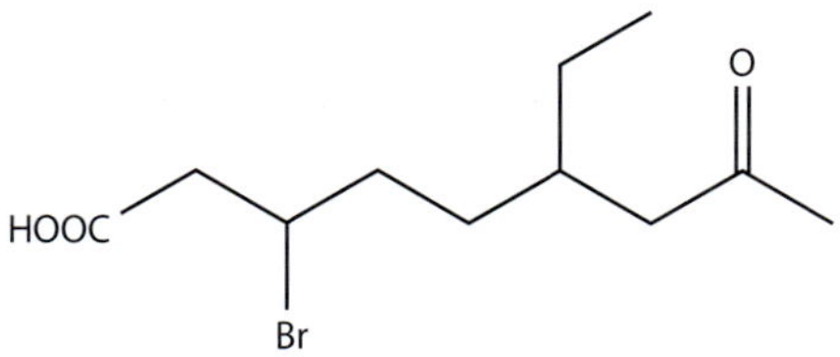

(c)

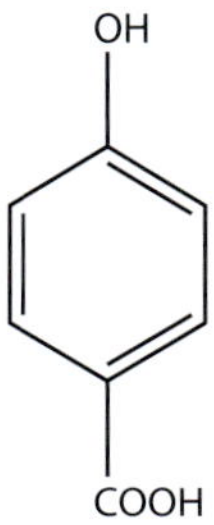

Solutions

(a) The longest chain here is fairly obvious and has six carbon atoms, so this is a hexanoic acid. Add the substituents, thus 3,4-dimethylhexanoic acid. The commonest error in problems of this type is not to start the numbering scheme on the carboxylic acid carbon atom.

(b) Despite the presence of a ketone, this will be named as a carboxylic acid, so the chain is numbered from the –COOH group.

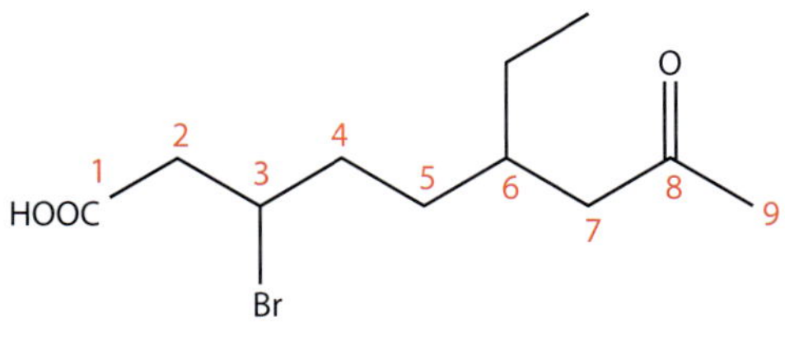

So this is 3-bromo-6-ethyl-8-oxononanoic acid.

(c) This will be named as a benzoic acid, 4-hydroxybenzoic acid, as the carboxyl group takes precedence over the –OH.

4.5.4 CARBOXYLATE ESTERS

Esters (**4.48**) have one carbon substituent at the carbonyl and one –OR group. They are formed from acids and alcohols (Figure 4.11), and the name of the ester is a combination of the acid and alcohol names. The name of the alcohol used the form of a substituent, methyl, ethyl, phenyl, *etc.*, and the acid part now takes the suffix "-oate," to indicate the ester functionality. Thus **4.49** is methyl propanoate. The most common error that students make in naming esters is that they don't recognize the functional group and try to name them as substituted ketones. If you are finding the nomenclature of these difficult, it's often helpful to separate the ester into the acid and alcohol from which it was made. **4.50** was made from ethanoic (acetic) acid and 1-pentanol; hence, the name is 1-pentyl ethanoate or 1-pentyl acetate. In text, esters may be written as

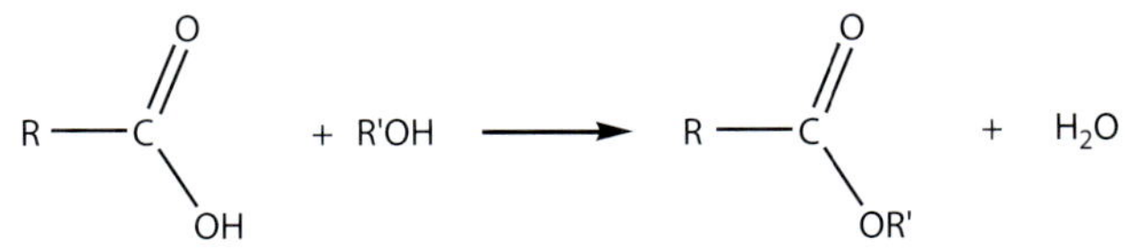

4.48

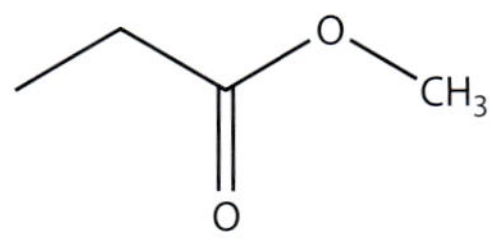

4.49

FIGURE 4.11 Formation of an ester.

RCOOEt, RCO_2Et, or RC(=O)OEt—these all mean exactly the same thing. Lower esters have some solubility in water, because of their ability to act as hydrogen-bond acceptors, but higher esters are largely insoluble. Many esters have sweet and pleasant odors.

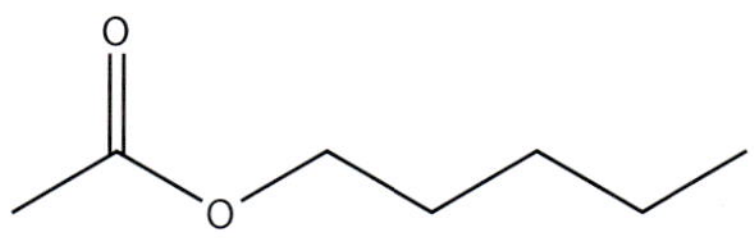

4.50

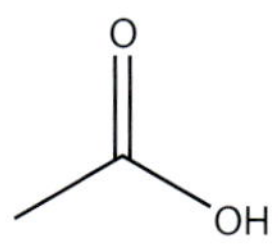

Ethanoic acid

1-Pentanol

PROBLEM 4.12

Give a line drawing of each of the following molecules:

(a) 1-Propyl-3-oxobutanoate
(b) Methyl 2-aminobenzoate, a component of grape flavor

Solutions

(a)

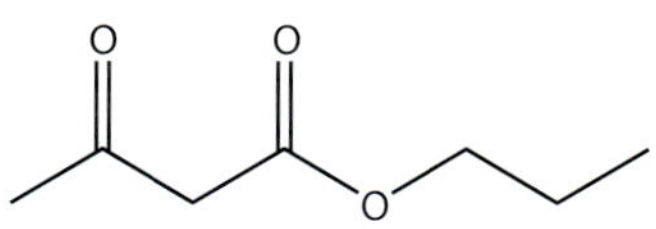

(b)

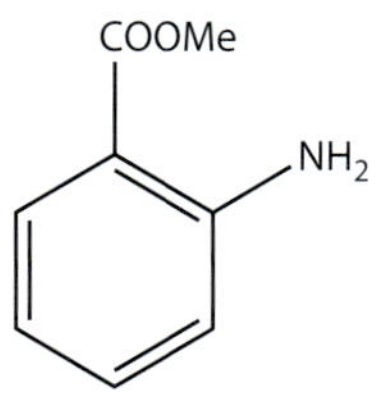

4.5.5 AMIDES

We describe a compound as an amide when the OH of the carboxylic acid is replaced by an $-NH_2$ or $-NR_2$ group. Thus, **4.51** is ethanamide and **4.52** *N*-methyl ethanamide. In the latter, the capitalized, italicized *N* indicates that the substituent is at nitrogen. Other substituents are named normally, from their position on the longest chain starting with the carbonyl group as C-1. Thus, **4.53** is 4,*N*-dimethylhexanamide.

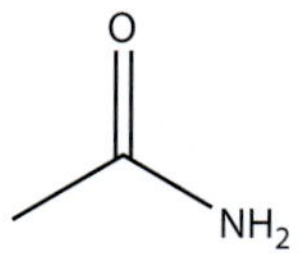

4.51

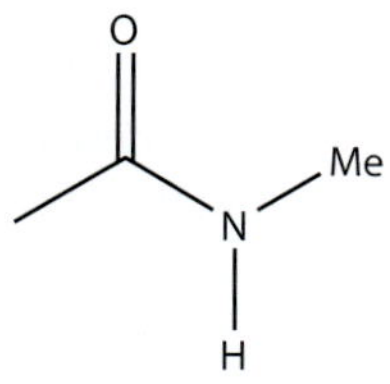

4.52

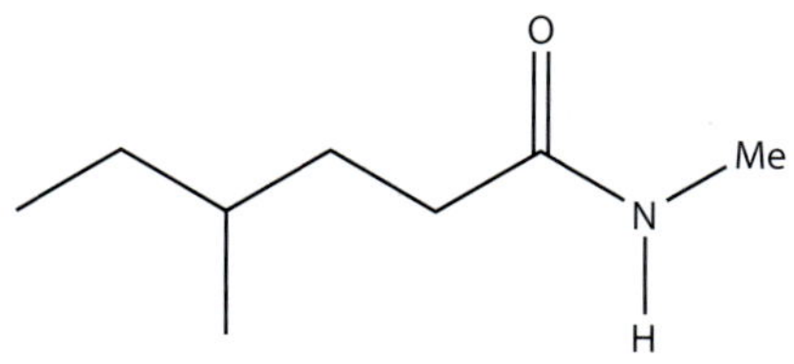

4.53

Amides are both good H-bond donors and acceptors (**4.54**) and are extensively associated in both the solid state and solution. Many have high melting and boiling points because of the large number of hydrogen bonds that have to be broken in these phase changes.

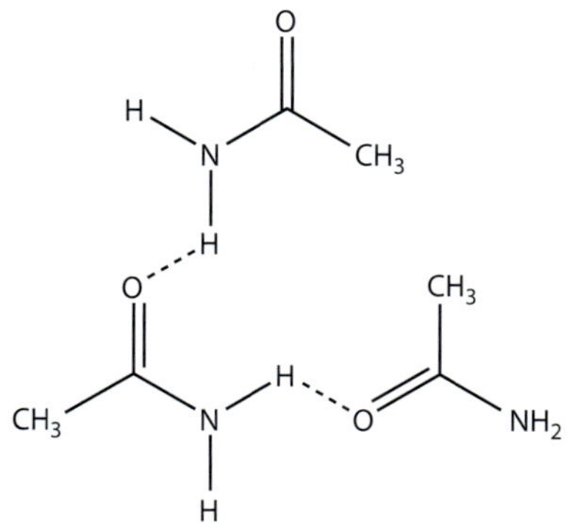

4.54

Additional resonance forms are also very important in the structure of amides—and ultimately affect the structure of many biomolecules, especially proteins, which contain amide functionalities. As we saw in the last chapter, with any carbonyl group, there is a contribution from a resonance form that has a positive charge at carbon and a negative charge at oxygen (**4.55a**). We can now take this one step further. In **4.55b**, the charge-separated resonance form, not all the atoms have octets—hence it contributes less to the overall structure. However, it is now next to a nitrogen atom with a lone pair of electrons that it can share with the positive center, giving **4.55c**. **4.55c** is a *very* important contributor to the overall structure of amides—all the atoms now have octets of electrons. The key feature of the contribution of **4.55c** is that amides are *planar*—all four atoms attached to the carbon of the carbonyl and the nitrogen atoms are in the same plane. While this double bond does not have the stereochemical integrity of an alkene—*cis*- and *trans*-isomers cannot be separated—there is a significant, and measurable, barrier to rotation about the carbon–nitrogen double bond. In **4.56**, the rotation about the C=N is slow enough that the two methyl groups are distinguishable by NMR spectroscopy (see Chapter 6).

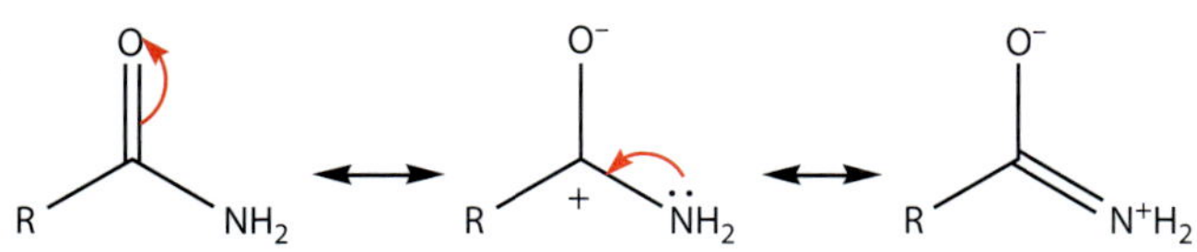

4.55a **4.55b** **4.55c**

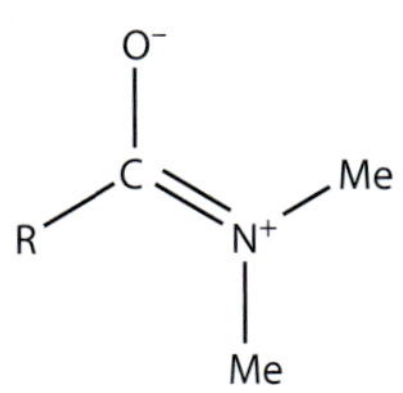

4.56

4.5.6 ACYL HALIDES AND ANHYDRIDES

In acyl halides, the –OH of the carboxylic acid is replaced by a halogen atom. While derivatives of all the halogens are known, chlorides are much the most commonly used. The suffix here is "-oyl chloride." Thus, **4.57** is ethanoyl chloride, **4.58** is 2-methylbutanoyl chloride (the carbonyl is C-1, as always), and **4.59** is benzoyl chloride. Although acyl halides could act as H-bond acceptors, they react with water (and other protic solvents) rather than dissolving and are among the most reactive of all carbonyl derivatives (see Chapter 15). Anhydrides, as their name might suggest, may be considered formally as being formed from two moles of acid, with the loss of a molecule of water. The two carbonyl groups of what were formally the acids are linked by a single oxygen atom (**4.60**). The names derive from the name of the parent acid; thus **4.60** is ethanoic (acetic) anhydride and **4.61** is benzoic ethanoic anhydride. Like acyl chlorides, anhydrides react with water and protic solvents rather than dissolve in them.

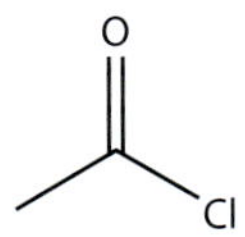

4.57

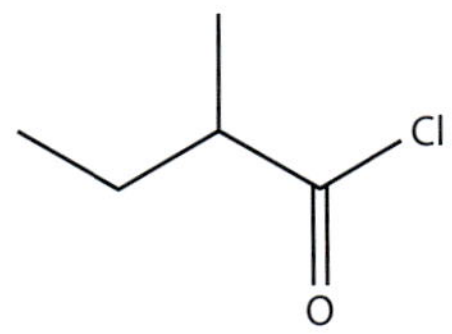

4.58

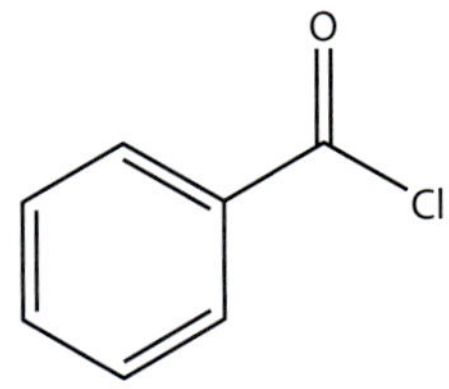

4.59

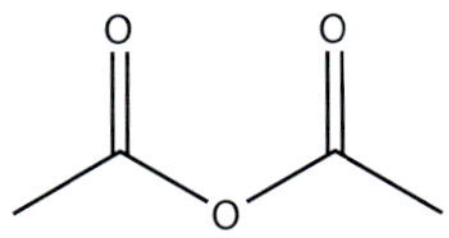

4.60

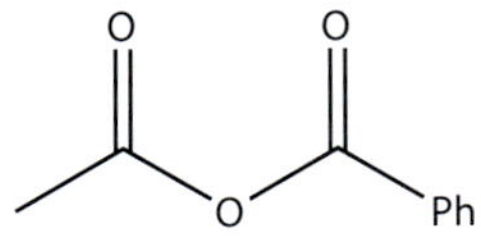

4.61

PROBLEM 4.13

Give a systematic name for each of the following compounds:

(a)

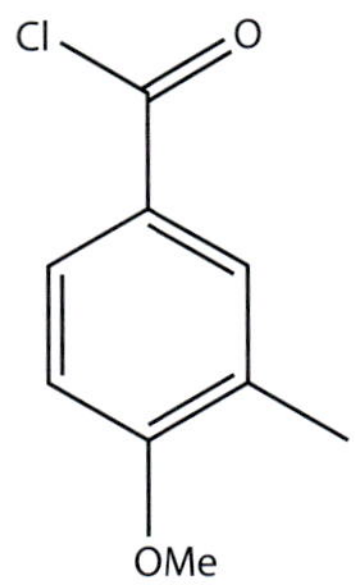

(b)

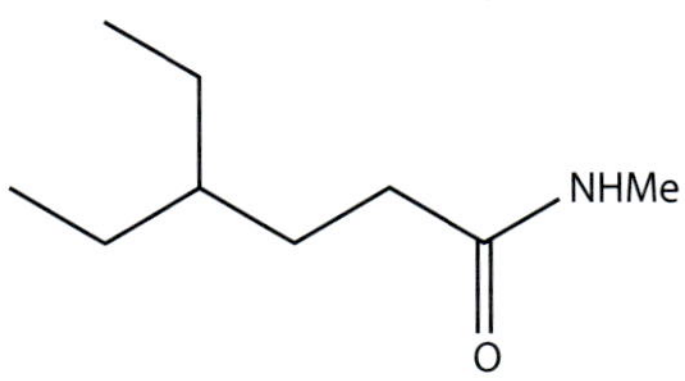

PROBLEM 4.14

For acetyl chloride, $CH_3C(=O)Cl$, generate an orbital representation. Label bonds as σ or π, and state which orbitals were used to make them (s, p, sp, sp^2, sp^3). Mark empty orbitals and lone pairs, and state their type.

PROBLEM 4.15

For each of the following pairs of molecules, predict which would have the higher boiling point, giving your reason:

(a) Butanoic acid and ethyl acetate
(b) Cyclohexane and cyclohexanone
(c) Acetone and acetic acid

Solutions

4.13(a) The numbering of the benzene ring starts at the carbon bearing the acid chloride group and goes counterclockwise so that the substituents are on the lowest-numbered atoms, hence 4-methoxy-3-methylbenzoyl chloride.

4.13(b) The key point here is to recognize the amide functional group. The amide carbonyl is designated as C-1, and the longest chain has six carbon atoms, so this is a substituted hexanamide. The substituents are an ethyl group at C-4 and a methyl group on nitrogen. Hence, this is *N*-methyl-4-ethylhexanamide.

4.14

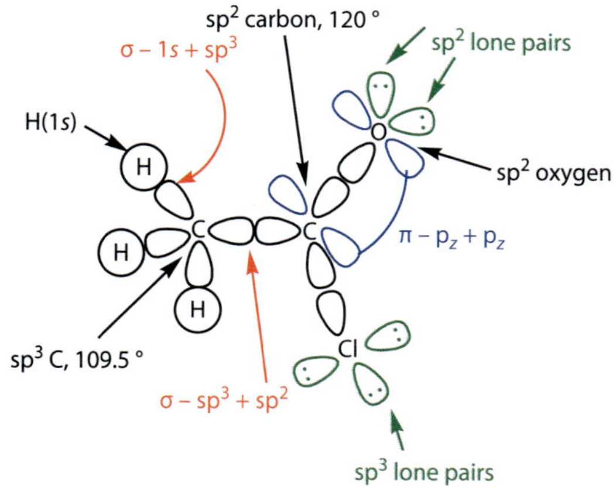

4.15(a) Butanoic acid—molecules are held together by strong hydrogen bonds.
4.15(b) Cyclohexanone—polar molecules have higher boiling points than nonpolar ones.
4.15(c) Acetic acid—molecules are held together by hydrogen bonds.

Key Points from Section 4.5

- All carbonyl compounds involve a polar carbon–oxygen double bond, and both atoms are sp^2 hybridized. The oxygen atom bears two lone pairs of electrons.
- Aldehydes have a formula RC(=O)H (also written RCHO) and a nomenclature suffix -al. The carbonyl carbon atom is C-1.
- Ketones have a formula $R_2C{=}O$ and a nomenclature suffix -one. The chain is numbered to give the carbonyl carbon atom the lowest number.
- Acid derivatives include carboxylic acids, RC(=O)OH, suffix -oic acid, esters RC(=O)OR′, suffix -oate; amides, $R(C{=}O)NH_2$ or $RC({=}O)NR'_2$, suffix, -amide, acid chlorides, RC(=O)Cl, suffix -oyl chloride; anhydrides RC(=O)OC(=O)R′, suffix -oic anhydride.
- All carbonyl compounds can act as hydrogen-bond acceptors. Carboxylic acids and amides are also excellent hydrogen-bond donors.

TABLE 4.4 Summary and Priorities in Nomenclature

Class of Compound	Suffix	Name When a Substituent
Carboxylic acid	-oic acid	n/a
Anhydrides	-oic anhydride	n/a
Esters	-oate	Monoesters of diacids are usually named as such—for example, butane dioic acid monomethyl ester
Acyl halides	-oyl halide	n/a
Amides	-amide	-carboxamide
Nitriles	-nitrile	-cyano
Aldehydes	-al	-oxo, -carbaldehyde on ring systems
Ketones	-one	-oxo
Alcohols	-ol	-hydroxy
Thiols	-thiol (but note that PhSH and analogs are named as the phenol, but thio added, thus thiophenol.)	-mercapto
Amines	-amine (but rarely used)	-amino
Ethers	Ether	-oxa (for location of oxygen)
Alkenes	-ene	-ene
Alkynes	-yne	-yne
Halides	n/a	Chloro, bromo, *etc.*
Nitro compounds	n/a	Nitro
Alkanes	-ane	n/a

4.6 PRIORITIES IN NOMENCLATURE

We have mostly so far looked at molecules containing just one functional group—but of course, many really interesting compounds contain more than one. We need to have a system that allows us to prioritize. If a molecule contains both a carboxylic acid and a ketone, do we name it as an acid or as a ketone? The IUPAC rules tell us to name it as an acid—so **4.62** is 4-oxoheptanoic acid. Table 4.4 summarizes the ordering of groups and how they are named when they are a substituent on a molecule containing a higher priority group. In general, you will see that the more highly oxidized the functional group, the higher up the table it appears. There is neither any prospect, nor any need, to learn all the IUPAC rules in one go (they run to several volumes, and there are various classes we have not dealt with, because we will encounter them infrequently), and later chapters will give review problems to remind you of the way we name the molecules that chapter deals with.

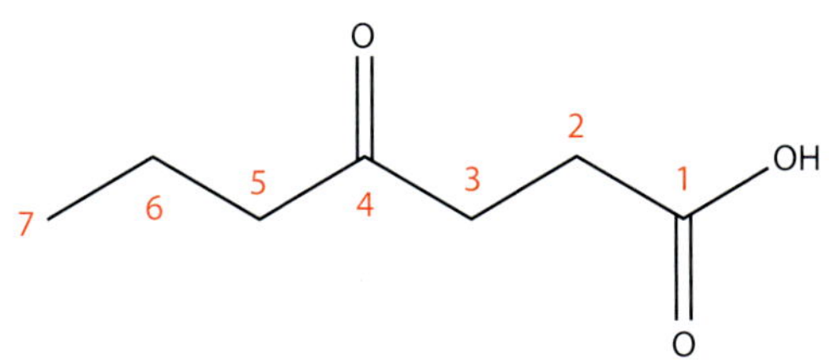

4.62

4.7 STRUCTURE AND BONDING IN SOME HIGHLY REACTIVE SPECIES

Although these species do not necessarily contain heteroatoms, the structures and bonding in some highly reactive species are most conveniently considered here.

When carbon–carbon bonds are broken, there are two ways in which this can occur. The first is described as homolytic cleavage, where one of the electrons of the carbon–carbon bond

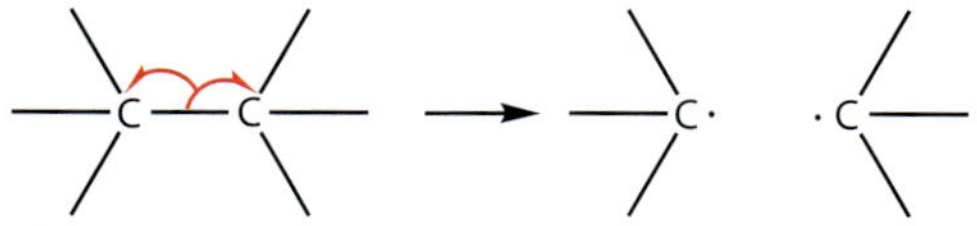

FIGURE 4.12 Homolytic cleavage of a carbon–carbon bond.

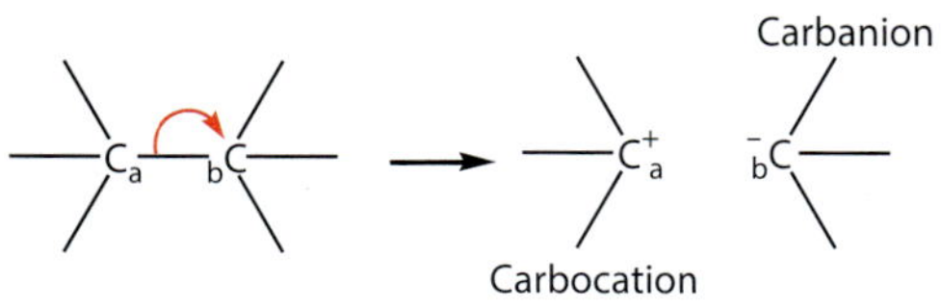

FIGURE 4.13 Heterolytic cleavage of a carbon–carbon bond.

is given to each of the carbon atoms (Figure 4.12). This forms species with a single electron at carbon, described as a radical or free radical. Notice the new symbol here—sometimes called a fishhook arrow. This single-headed arrow denotes the movement of a single electron. The radicals produced generally have very high reactivity.

Alternatively, we can break the carbon–carbon bond by heterolytic cleavage, to give one negatively charged and one positively charged carbon atom (Figure 4.13). Negatively charged carbon species are called **carbanions** and positively charged carbon species are called **carbocations**. These are two more examples of reactive intermediates.

4.7.1 CARBOCATIONS

In a carbocation, there are only six electrons around the carbon atom, so the structure will be trigonal planar (VSEPR), with the three pairs of electrons in sp^2 hybrid orbitals. The p_z orbital is empty. Thus, the structure of $[CH_3]^+$ is **4.63**. Carbocations are stabilized by electron donors and destabilized by electron-withdrawing groups. Groups able to donate electron density through the attached σ-bonds are primarily alkyl groups, the so-called σ-donors. This means that the more alkyl groups there are attached to a carbocation, the more stable it will be (Figure 4.14). Conjugation can also stabilize a carbocation—essentially, the more sites over which the positive charge can be delocalized, the more stable the species will be. We have already met the allyl cation, which is stabilized by resonance (**4.64**). The same can be said of the benzyl cation, **4.65**. If we look at the benzyl cation side-on (**4.66**), it is easy to see how the empty p_z orbital of the cation can interact with the π-system of the benzene ring. The effects are additive—the more benzene rings, the greater the stability, and $[Ph_3C][BF_4]$ is a stable, isolable salt.

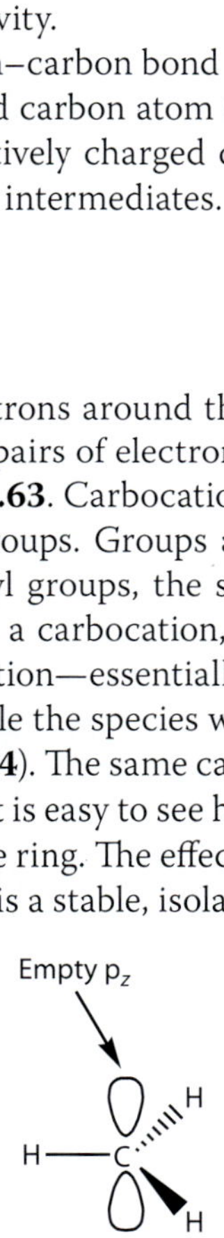

4.63

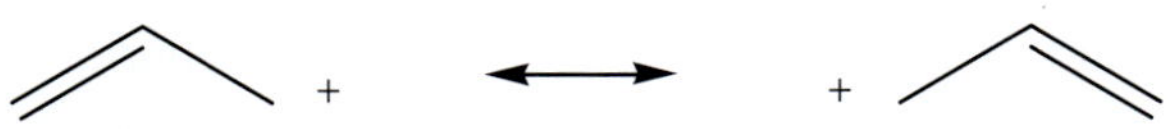

4.64

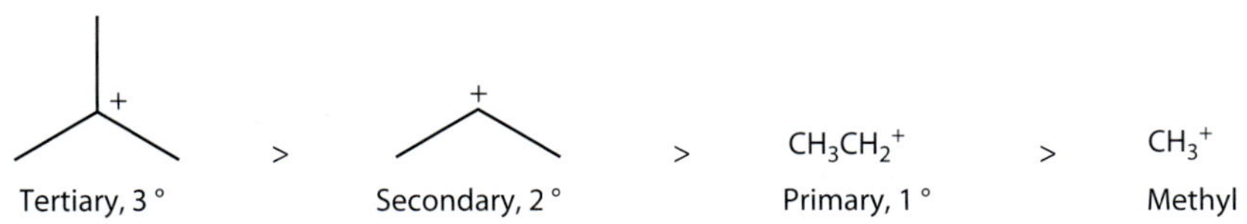

FIGURE 4.14 Relative stability of simple carbocations.

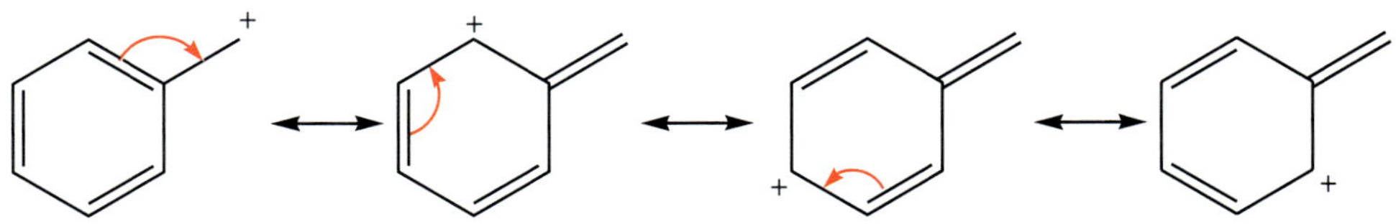

4.65

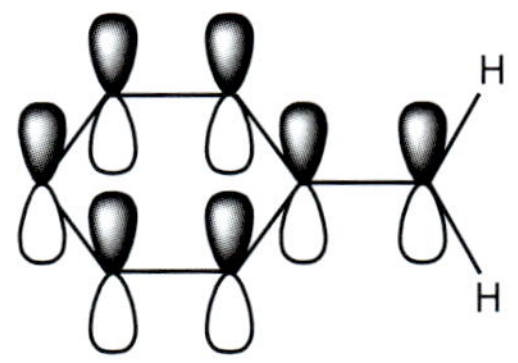

4.66

The type of resonance that we see in the allyl and benzyl cations is called π-delocalization. Cations can also be stabilized by sharing a lone pair on an adjacent atom—called *n*-delocalization (*n* is for nonbonding electrons). Examples are provided by **4.67** and **4.68**, and the resonance structures with the charge on the heteroatom are very significant, since in these, all the atoms have octets of electrons.

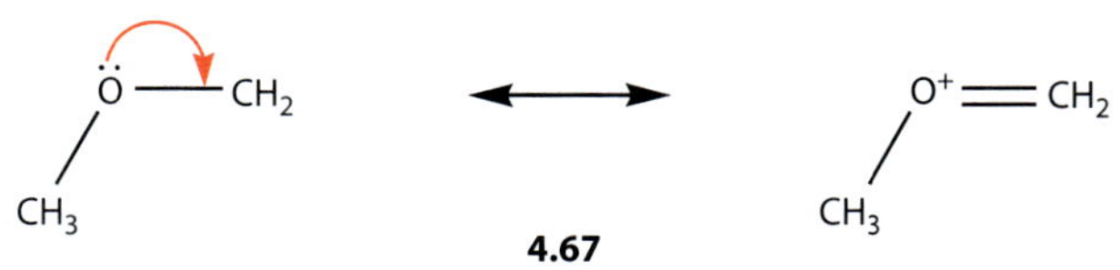

4.67

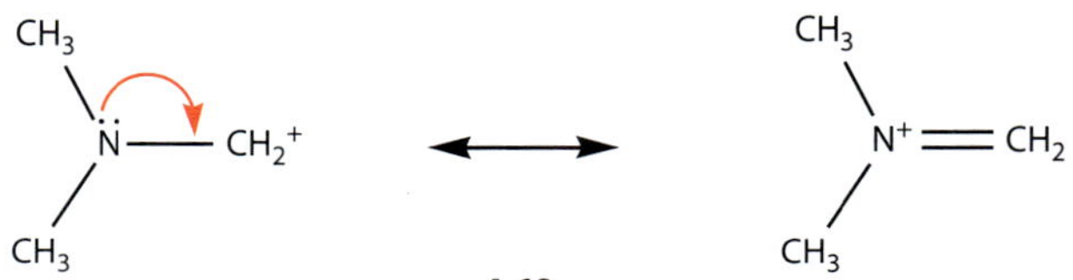

4.68

Cations that are constrained not to be planar are not generally stable. The energy required to deform a carbocation toward a pyramidal structure with an empty sp^3 orbital is some 100 kJ mol^{-1}. So in **4.69**, the cation is constrained not to be planar, and this cation cannot be prepared. And as we have seen in the previous chapter, cations with aromatic character, such as the tropylium ion, **4.70**, are especially stable, while those with antiaromatic character, such as the cyclopentadienyl cation, **4.71**, are especially unstable.

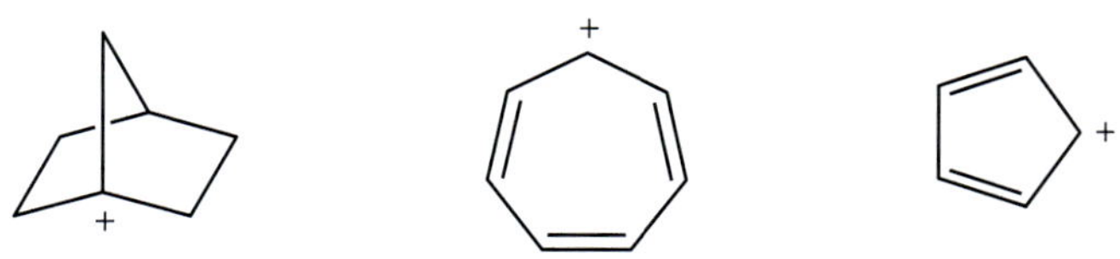

4.69 **4.70** **4.71**

We can see the stability of carbocations manifested in an important reaction that leads to their formation. Alcohols are a useful source of carbocations, by heterolytic cleavage of the carbon–oxygen bond. But before the carbon–oxygen bond cleavage is possible, the oxygen must be protonated (Figure 4.15); the lone pair of electrons is used to form a new oxygen–hydrogen bond. This is analogous to the protonation of water to give the hydronium ion, H_3O^+. The ease of the heterolytic reaction depends on the stability of the carbocation produced (water is the other product in every case). So the reaction to give RCH_2^+ is difficult, because the primary carbocation is rather unstable, but that to give R_3C^+ is easy, because the tertiary carbocation is much more stable. Other reactions that are relatively easy because the cation produced is stable are shown in Figure 4.16.

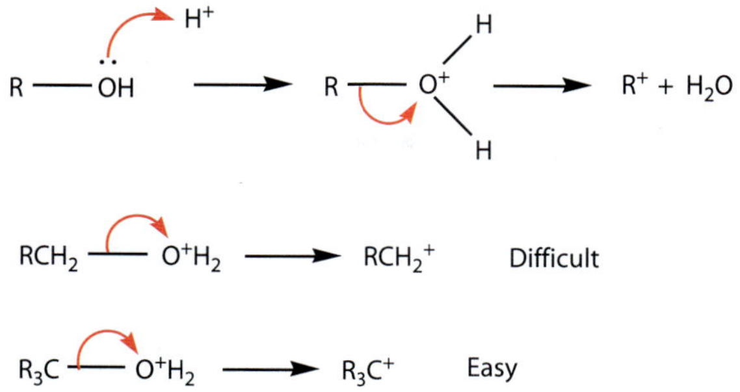

FIGURE 4.15 Formation of carbocations from alcohols.

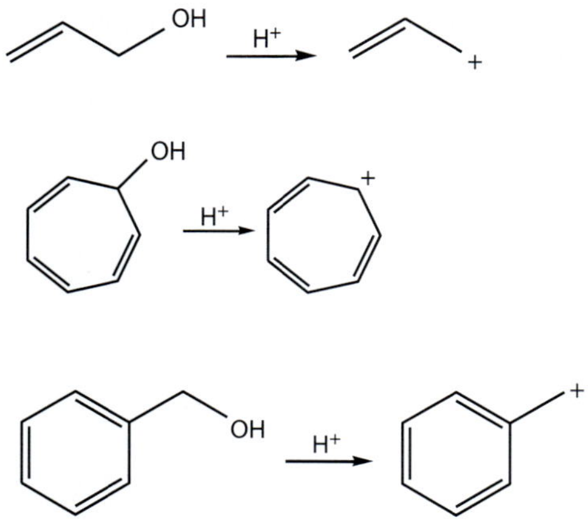

FIGURE 4.16 Production of relatively stable carbocations from alcohols.

PROBLEM 4.16

Place the cations shown in the following in order of their relative stability:

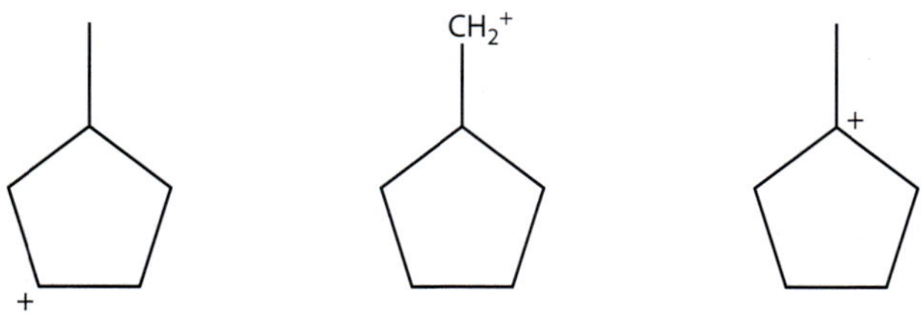

Solution

The tertiary cation is the most stable, followed by the secondary, and finally the primary:

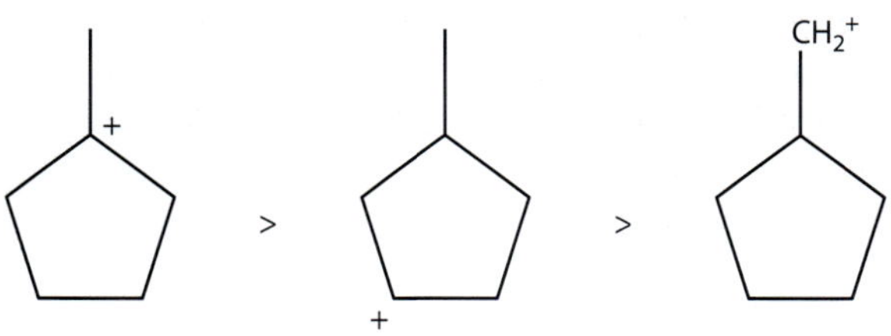

Focus on Some Very Stable Cations

One quantitative approach to determining the stability of a carbocation is to establish the equilibrium constant for reaction 4.2; the lower the equilibrium constant, the more stable the carbocation. This is more usually expressed logarithmically, as the function $pK_{R}+$—the more positive this is, the more stable the carbocation. Some values are given in Table 4.5:

$$R^{+} + H_2O \rightleftharpoons ROH + H^{+} \tag{4.2}$$

TABLE 4.5 $pK_{R}+$ Values for Carbocations

Cation	$pK_{R}+$	Comments
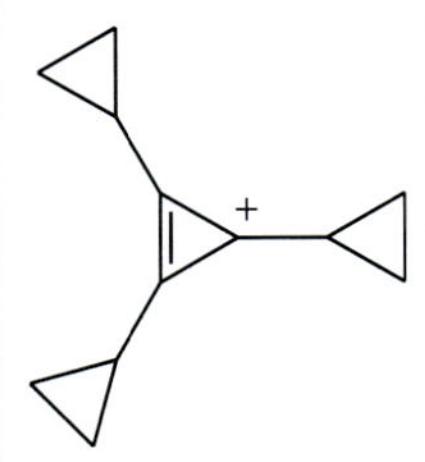	+9.7	Aromatic 2π-electron system
$(4\text{-}Me_2NC_6H_4)_3C^{+}$	+9.36	Gentian violet or crystal violet
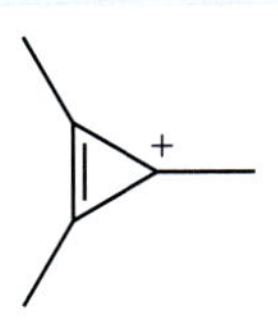	+7.8	Aromatic 2π-electron system
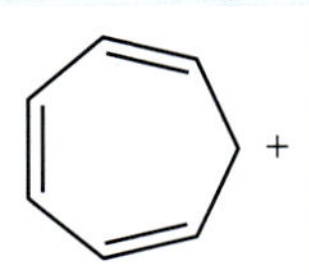	+4.7	Aromatic 6π-electron system
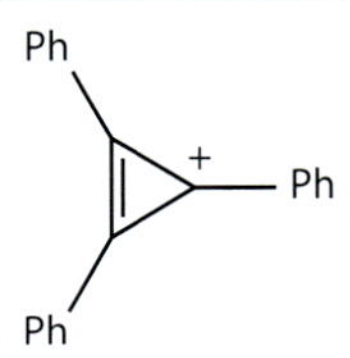	+3.1	Aromatic 2π-electron system
$(4\text{-}MeOC_6H_4)_3C^{+}$	0.82	
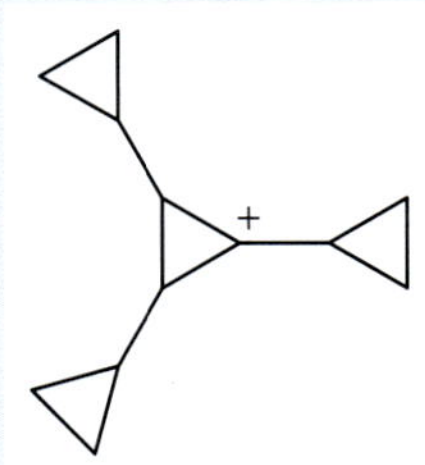	−2.3	
$(4\text{-}MeC_6H_4)_3C^{+}$	−3.56	
Ph_3C^{+}	−6.63	
$(4\text{-}ClC_6H_4)_3C^{+}$	−7.74	
$(3\text{-}ClC_6H_4)_3C^{+}$	−11	
Ph_2CH^{+}	−11.3	Secondary carbocation
Me_3C^{+}	−15.5	
$(4\text{-}O_2NC_6H_4)_3C^{+}$	−16.27	

Source: Data taken from Carey and Sundberg, *Advanced Organic Chemistry Part A*, 5th edn., Springer, 2007, p. 427.

A number of observations here will strike you as unusual. Cyclopropane rings stabilize carbocations very effectively, better than methyl groups and better than phenyl rings. The 2π aromatic cations seem to be more stable than the 6π systems. That electron donors such as MeO, NMe_2, and methyl stabilize cations and electron-withdrawing nitro groups destabilize them will not be unexpected, but why does the position of a chlorine atom on the ring matter so much? (Remember, this is a log scale so the difference in equilibrium constant is >1000.) Why does the –NMe_2 group have a much greater effect than the –OMe group in stabilizing the carbocation? We can't yet answer all of these questions, but many of the reasons will become clear as we learn more about the availability of electrons from various functional groups (see Chapter 8).

Triarylmethyl cations have proved to be important dyes. Gentian, or crystal, violet, **4.72**, is particularly stable because all of the –NMe_2 groups are involved in the stabilization of the cationic center. It is used as a dye for paper and in inks, as a histological stain, an antiseptic, and to develop latent fingerprints and shoeprints. A related compound is leucomalachite green. **4.73** itself is colorless, but when it reacts with hydrogen peroxide, in the presence of mammalian blood, it is converted first to the alcohol, **4.74**, and then to the green cation, **4.75**. This dye is used to identify blood spatter in crime scene investigation. Malachite green has also been used as a biological stain and as a dye for silk, leather, and paper.

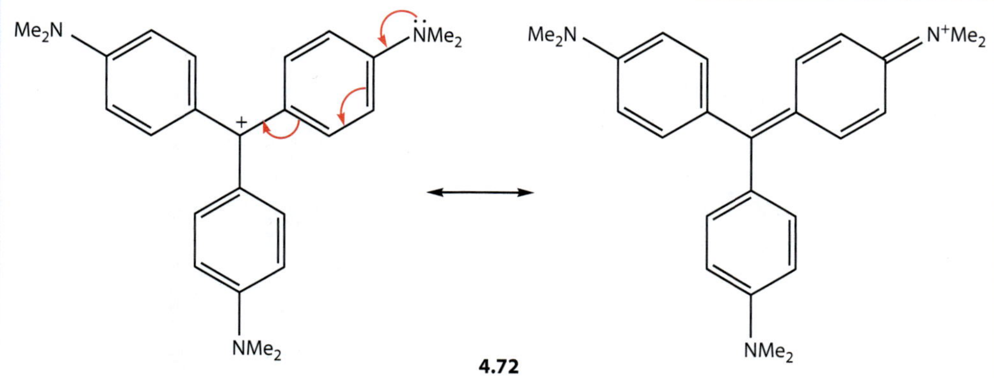

4.72

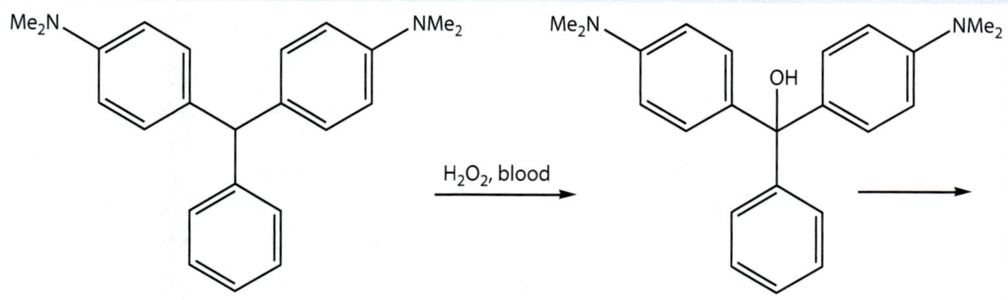

4.73 **4.74**

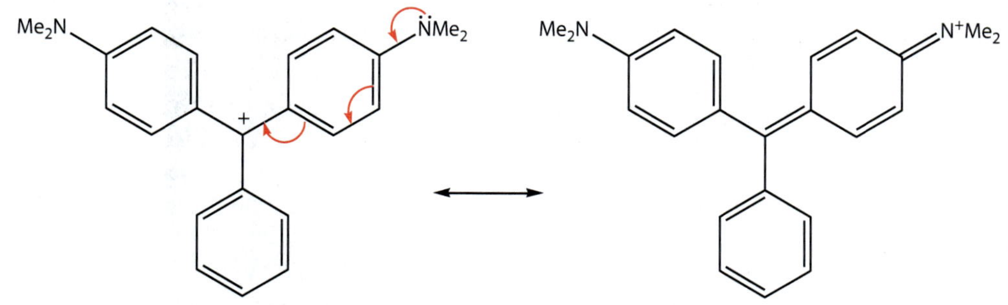

4.75

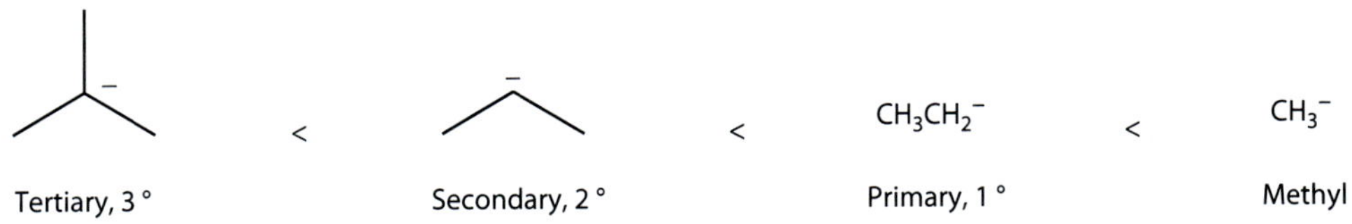

FIGURE 4.17 Stabilities of simple carbanions.

4.7.2 CARBANIONS

Carbanions are isoelectronic with ammonia and have the same structure—sp^3 hybridization, pyramidal geometry, and sp^3 lone pair, which we will see is associated with much of the reactivity of carbanions. The structure of $[CH_3]^-$ is **4.76**, and as with the carbocations, there is significant resistance to any distortion from this geometry. Carbanions are stabilized by inductive withdrawers of electrons—so $[CCl_3]^-$ is relatively stable. The effect of inductive donors such as alkyl groups is, predictably, exactly the opposite of their effect on carbocations—so, tertiary carbanions are the least stable, and primary carbanions are the most stable (Figure 4.17).

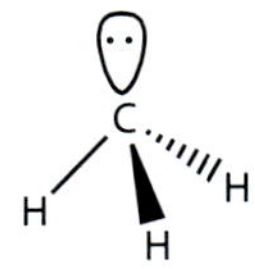

4.76

Carbanions, like carbocations, may be stabilized by resonance delocalization; the allyl and benzyl anions are relatively stable. The enolate anion, **4.77**, is even more so, since in one of the resonance forms, the negative charge is located on the electronegative oxygen atom. The stability of carbanions will be very important to our discussions of the pK_a values for organic compounds in Chapter 8.

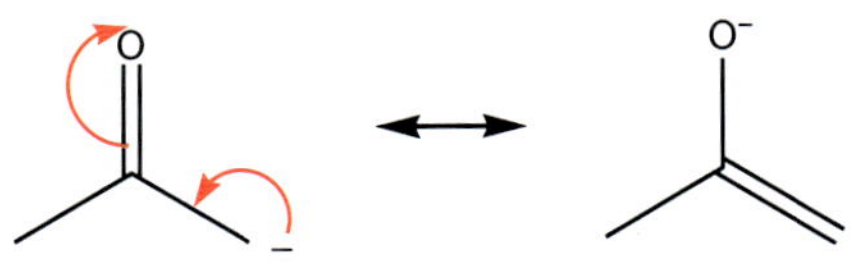

4.77

PROBLEM 4.17

For the carbanion $[CCl_3]^-$, generate an orbital representation. Label bonds as σ or π, and state which orbitals were used to make them (s, p, sp, sp^2, sp^3). Mark empty orbitals and lone pairs, and state their type.

Solution

The structure is based on that of CH_3^-, though this is considerably more stable. Only one C–Cl bond is shown in full. Hybridization at C is sp^3, with angles approx. 107°. Whether the orbitals of chlorine are hybridized or not is an interesting but complex question, and not one needs to trouble us in organic chemistry.

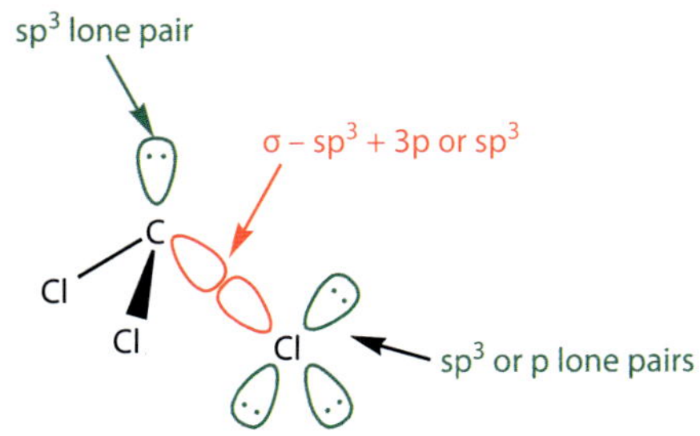

4.7.3 CARBON-CENTERED RADICALS

Carbon radicals have seven electrons around the central carbon atom—thus three electron pairs and one half-filled orbital. This is a situation where VSEPR does not give a straightforward result, and you might expect that the structure of radicals would be somewhere in between that for the carbocation and the carbanion. In fact, the structure is much more like the carbocation, and radicals are generally formally written as sp^2 hybridized and planar with the odd electron in the p_z orbital (**4.78**). However, radicals are much more tolerant than either carbanions or carbocations of changes in molecular geometry. Pyramidalized radicals such as **4.79** can be generated, although as we have already noted, the related carbocation cannot be made.

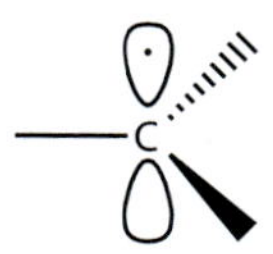

4.78

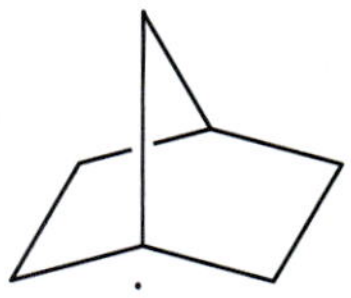

4.79

Radical stability is, generally, related to carbocation stability. Thus, tertiary radicals are the most stable, and primary ones the least so. As with carbanions and carbocations, resonance delocalization increases stability; the allylic and benzylic radicals are more stable than alkyl radicals. Notice that in **4.80** and **4.81**, we use single-headed arrows to move one electron at a time in generating the resonance forms.

4.80

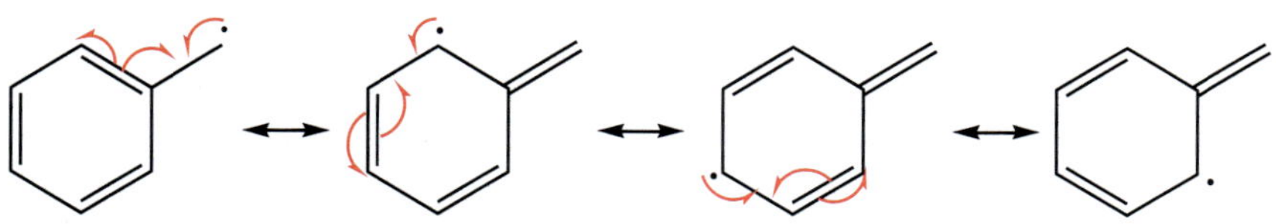

4.81

Key Points from Section 4.7

- Carbocations are three-coordinate sp^2 carbon species. They are planar and resistant to deformation.
- Carbocations are stabilized by inductive donors of electron density such as alkyl groups, by resonance involving π-electrons or nonbonding electrons (lone pairs) or by aromatic character.
- Carbanions are pyramidal sp^3 species, isoelectronic with ammonia. They have a lone pair of electrons and, like ammonia, invert rapidly.
- Carbanions are stabilized by inductive withdrawers of electron density such as halogens, but are destabilized by inductive donors such as alkyl groups. They are stabilized by resonance with π-electrons and aromatic character.
- Carbon-centered radicals are 7-electron species, formed by bond homolysis. They are planar and sp^2 hybridized, with the odd electron in the p_z orbital, but the barrier to deformation is relatively low. Their stability broadly follows that of carbocations.

4.8 MORE RESONANCE

It's useful at this point to review how our ideas about resonance have been extended, particularly with atoms other than carbon. For example, with ketones, the charge-separated resonance form with the negative charge on oxygen is very important (**4.82**) and will lead us to an understanding of the chemistry of this group of compounds. If we now consider **4.83a**, we would first write a resonance form with the positive charge on carbon and the negative charge on oxygen, **4.83b**. But if we look at this form, we can see an allyl cation unit, and we can write an additional form **4.83c**. This suggests that there are two possible sites for this molecule to react with a negatively charged species. This is indeed the case as Figure 4.18 demonstrates. The negatively charged cyanide ion may attack at the carbonyl carbon atom or at the more remote site where positive charge may also be localized in the resonance form.

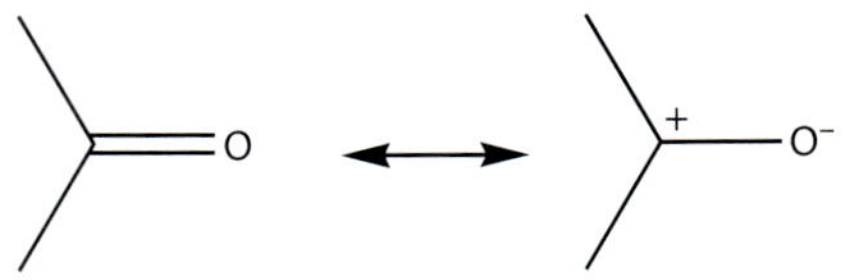

4.82

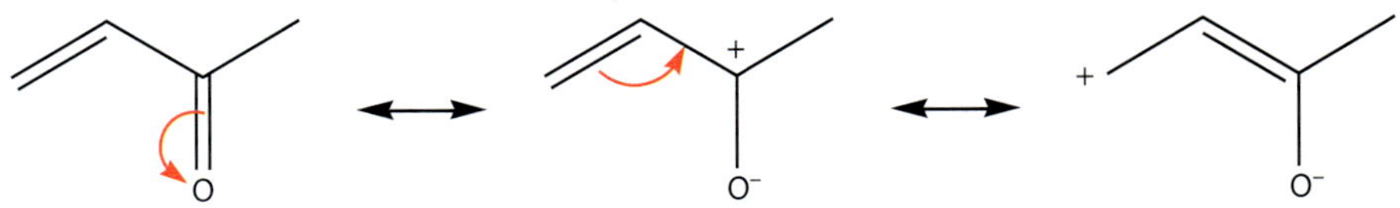

4.83a **4.83b** **4.83c**

Almost all problems involving the writing of resonance forms can be reduced to five motifs—aromatic resonance, the allyl anion, the allyl cation, charge separation in double bonds, and sharing lone pairs with a positive center, shown in Figure 4.19. Generally, it does not matter what the atoms are; the resonance forms will be similar. But knowing what the atoms are will allow you to attach greater or less importance to a particular form.

Problems that appear to be more complex can usually be broken down into these five motifs, provided that you isolate and focus on one bit of the molecule at a time. Until you have had more practice, looking at a small part of the molecule at a time is a good strategy.

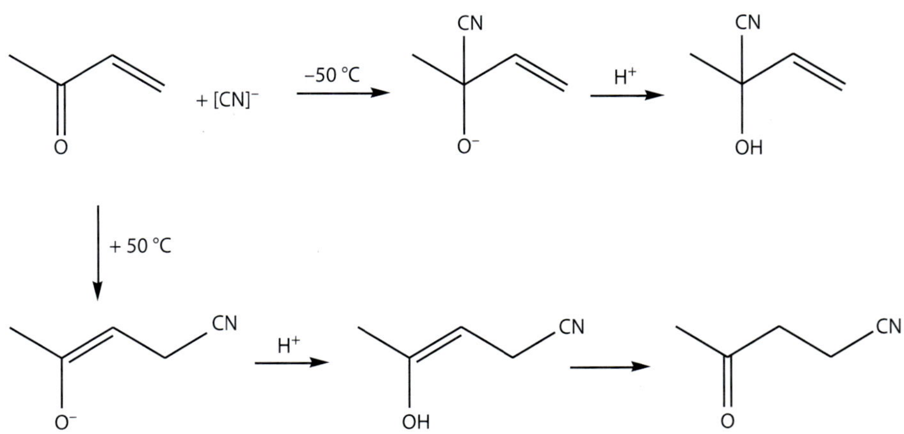

FIGURE 4.18 Addition of cyanide to **4.83**.

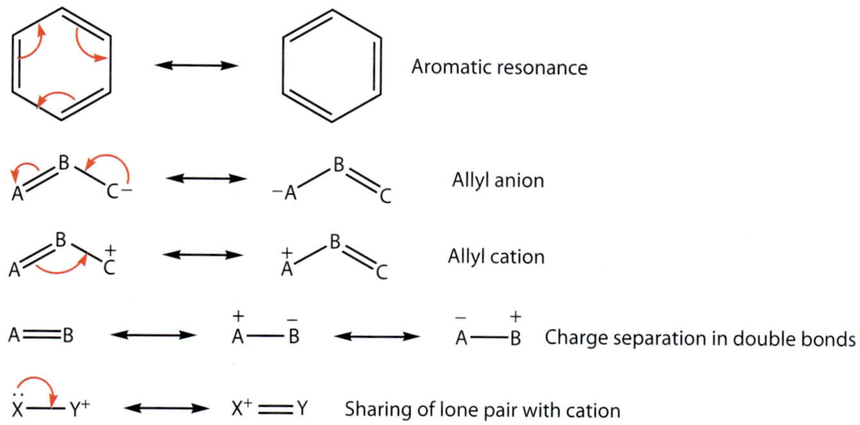

FIGURE 4.19 Key points from Section 4.8—motifs for resonance.

REVIEW PROBLEMS

1. For each of the following molecules, draw a line structure:
   (a) Ethyl-2-amino-5-methylhexanoate
   (b) 2-Ethyl-5,8-dimethylnonanoyl chloride
   (c) 2-Oxohexanoic acid
   (d) 4-(2-Bromopropyl)-3-oxooctanal
   (e) Hexylamine
   (f) 4,5-Dimethyl-3-oxohexanoic acid
   (g) 2-Bromo-4-chloro-6-fluorophenol
   (h) *N*,3-Dimethylhexanamide
   (i) 3,5-Dimethoxybenzoyl chloride
   (j) 3-Methylbenzoic acid
   (k) 2-Methyl-3-nonanone
   (l) 3,7-Dimethyl-5-oxooctanal
   (m) 3-Methylhexanal
   (n) Ethyl 3-oxobutanoate
   (o) 2,2-Dibromohexane dioic acid
   (p) 2-Methyl-2-butanol
   (q) 4-Methylhept-3-ene-2-one
   (r) Methyl pentanoate (a component of apple flavor)
   (s) 4,4-Dimethylpentanal
2. Give a systematic name for each of the following compounds:
   (a)

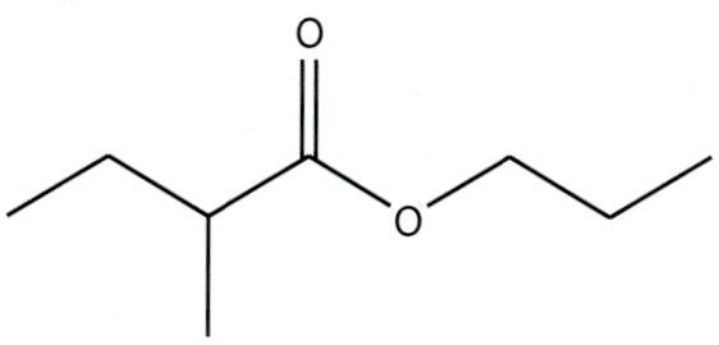

   (b)

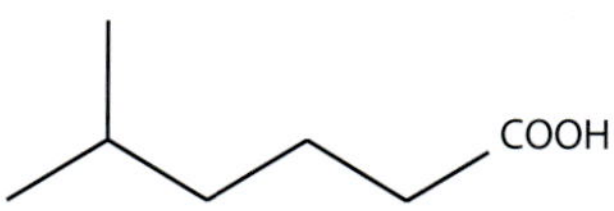

   (c)

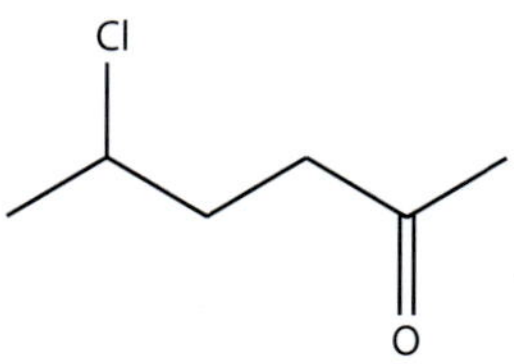

(d)

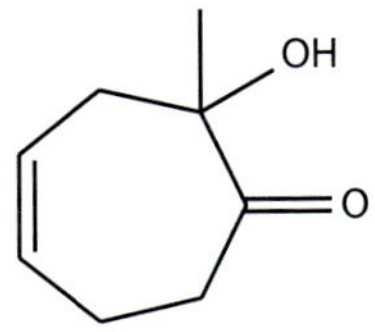

(e)

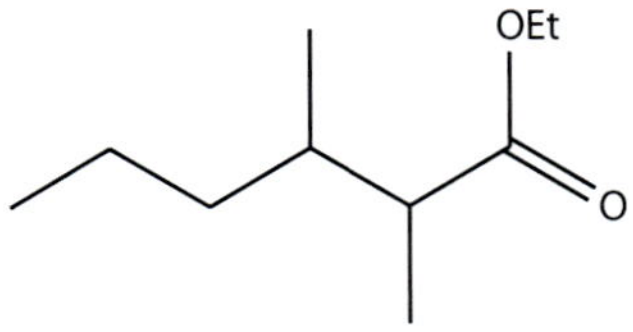

(f) $CH_3CH_2CH_2COOCH_2CH_3$

(g)

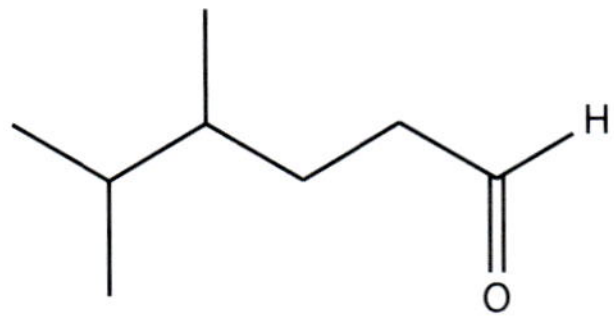

(h)

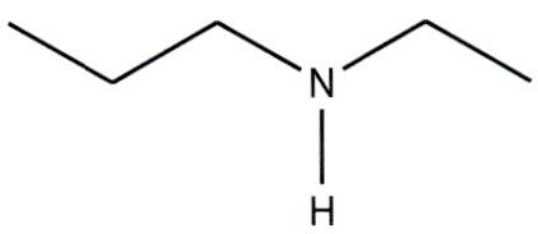

(i)

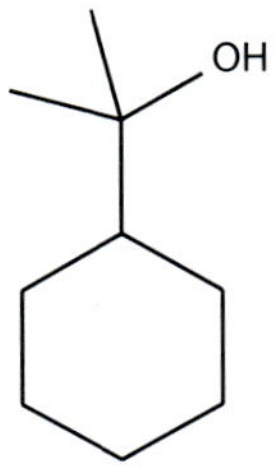

(j)

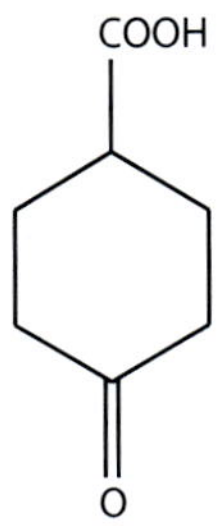

(k)

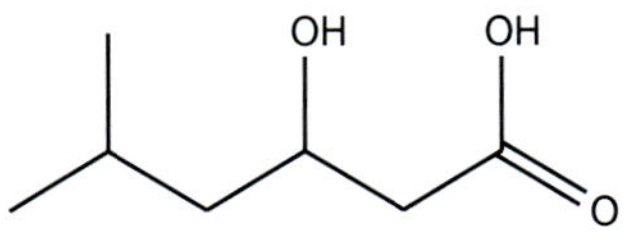

(l)

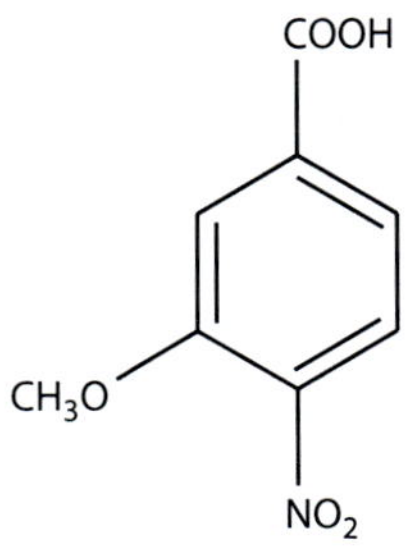

(m)

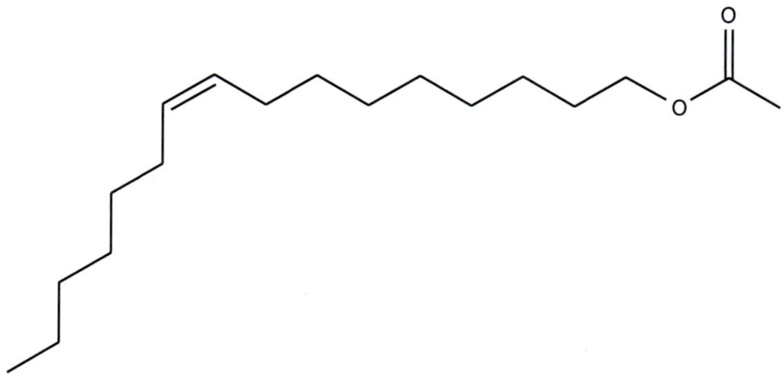

(n)

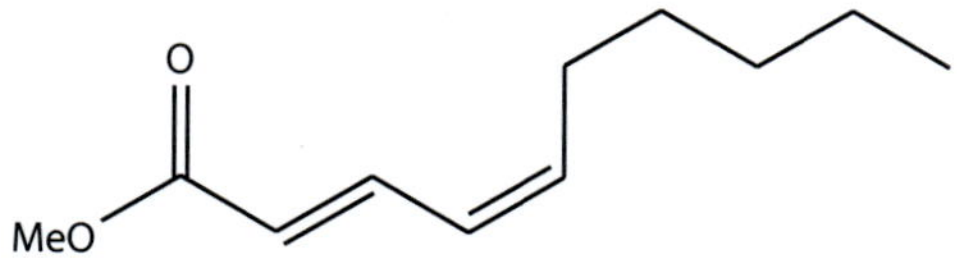

(o)

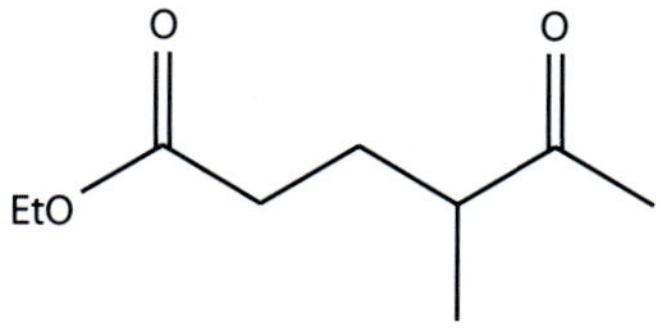

(p)

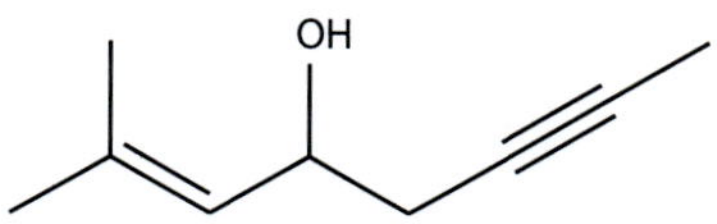

(q)

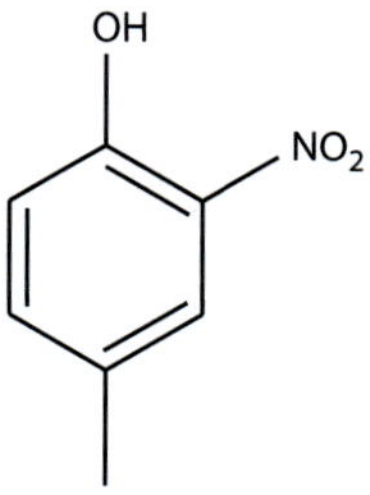

(r)

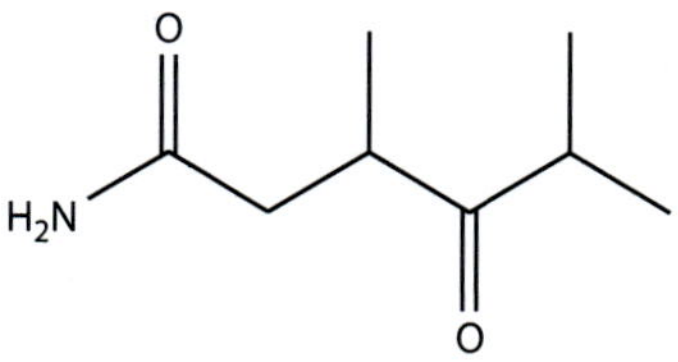

(s)

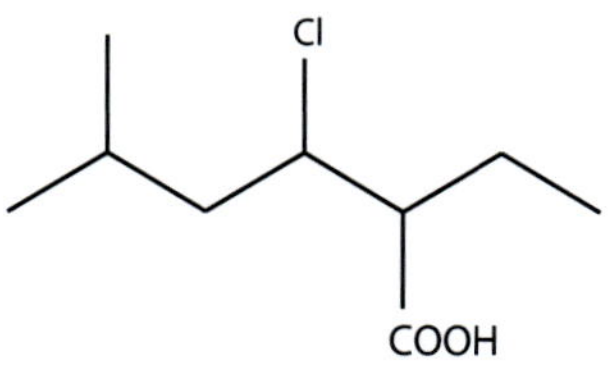

3. For the following species, indicate the hybridization of first row elements, and state what shape you expect that each molecule will have:

H_3O^+, O_3, CO, CH_3CN, CO_2, CH_3NC, $[BH_4]^-$, H_2O

4. For the following molecules, generate orbital representations. Label bonds as σ or π, and state which orbitals were used to make them (s, p, sp, sp^2, sp^3). Mark empty orbitals and lone pairs, and state their type:
   (a) $H_2C{=}O$
   (b) $CH_3CH_2^+$
   (c) $CH_2{=}NOH$
   (d) $[NH_2]^-$
   (e) HCOOMe
   (f) NH_2OH
   (g) $(CH_3)_2C{=}O$
5. Indicate the direction of the molecular dipole for

CH_3Cl CO_2 CCl_4 CO $Br_3CCH_2CBr_3$

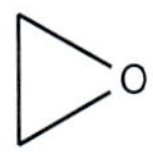

You will need to start by giving a three-dimensional structure for each molecule, showing clearly any empty orbitals, formal charges, or lone pairs.
6. Write down one reasonable resonance form for each of the following structures:
   (a)

(b)

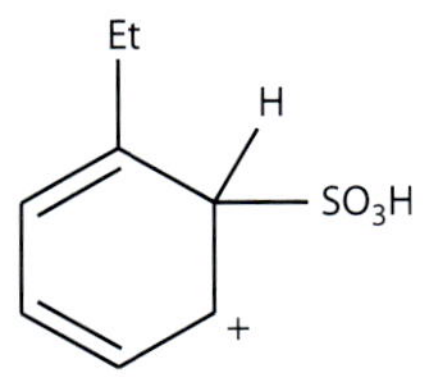

(c)

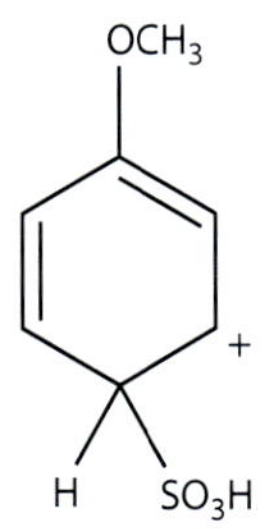

(d)

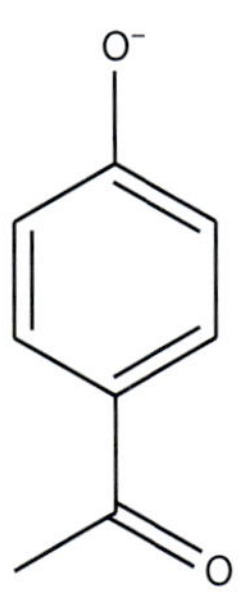

(e)

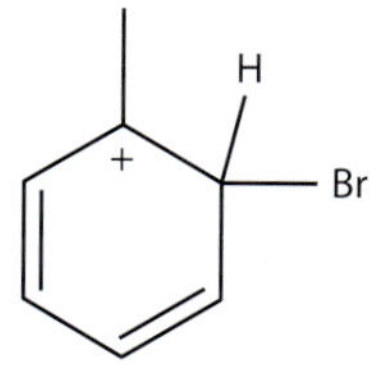

(f)

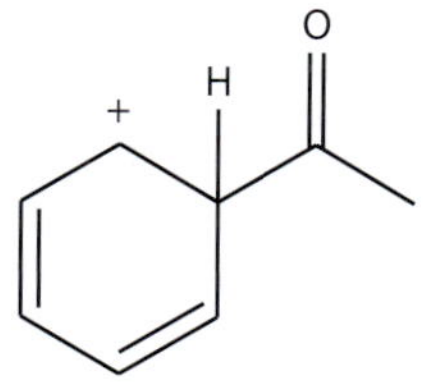

(g)

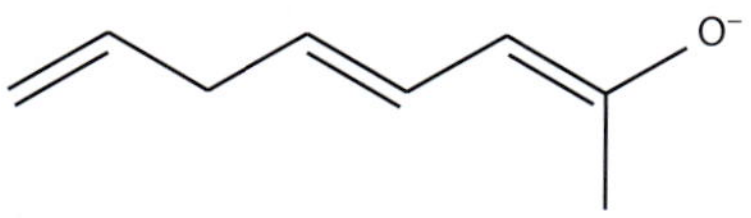

(h)

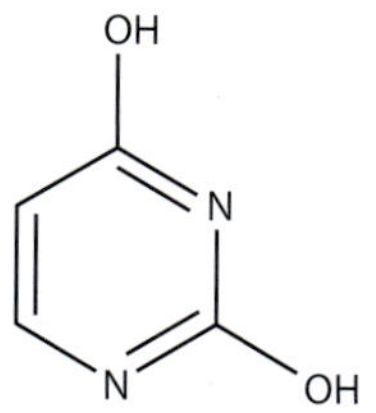

7. For each of the following pairs of compounds, state which you would expect to have the higher boiling point. Give a reason:
   (a) 1-Hexylamine and 1-hexanol
   (b) $CH_3CH_2CH_2OH$ and $CH_3C(=O)OH$
   (c) $CH_3CH_2CH_2CH_3$ and $CH_3CH_2OCH_3$
   (d)

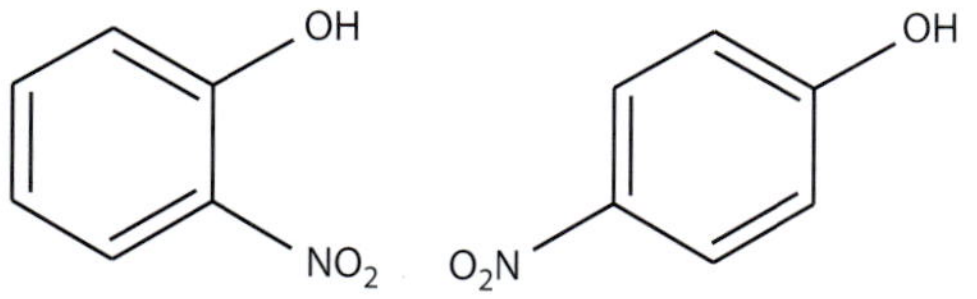

   (e)

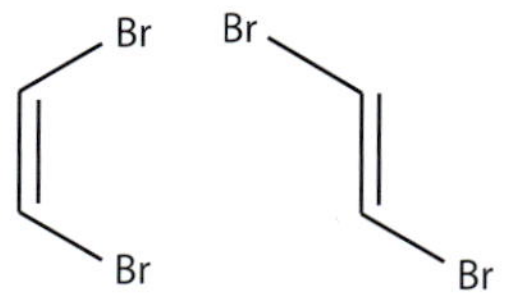

   (f)

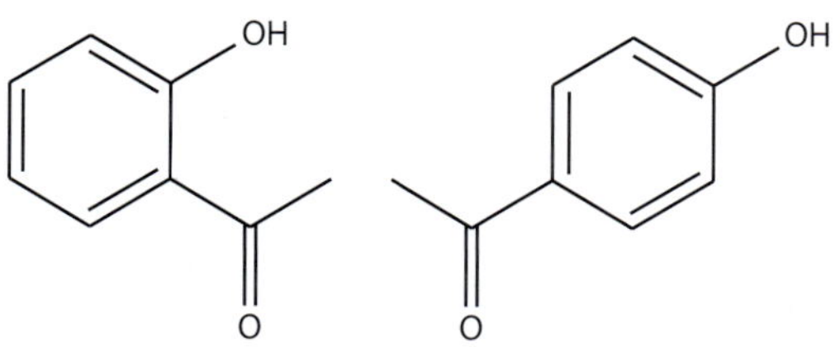

   (g) 2-Butanol and 2-butanone
8. Give an example of each of the following:
   (a) A lactone
   (b) A cyclic amide (lactam)
   (c) A molecule with an sp^2 lone pair at nitrogen

(d) An alkene with *Z*-geometry
(e) A molecule with an sp lone pair of electrons
(f) An *E*-alkene with a dipole moment
(g) A bidentate ligand
(h) A linear triatomic molecule with no permanent dipole moment

9. The common name for the compound shown in the following is oleamide. Give a systematic name according to IUPAC rules:

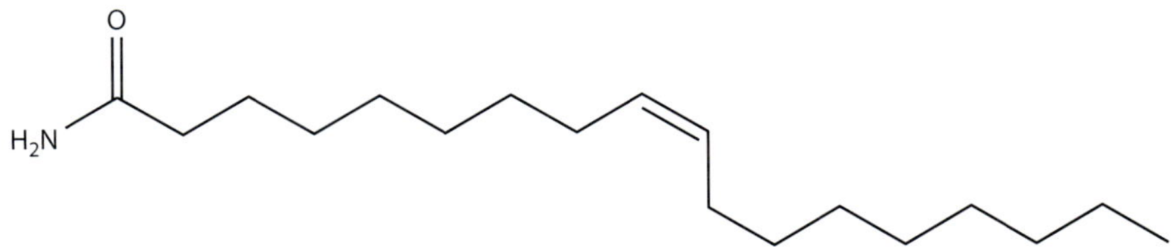

10. Give a systematic name for **4.84**, which is the sex pheromone used by both female elephants and some female moths:

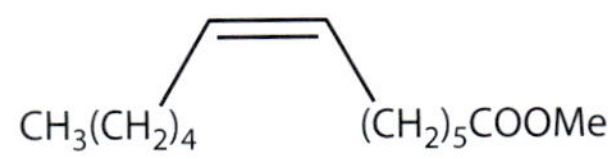

4.84

11. Give a systematic name for **4.85**, which is the attractant of the silk moth, *Bombyx mori*:

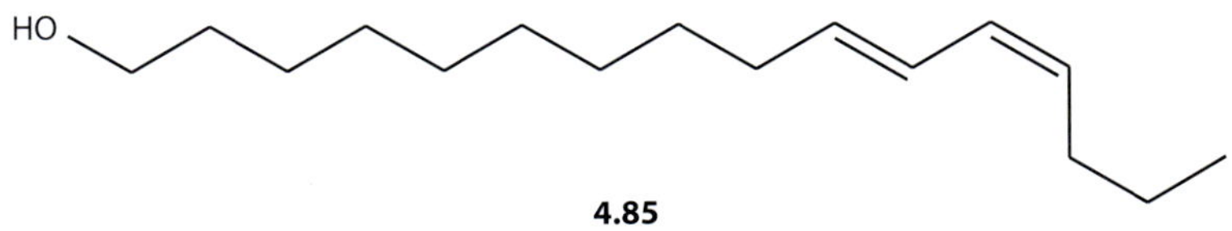

4.85

12. What are the most important kinds of intramolecular forces in
    (a) $PhCH_2OH$
    (b) Aniline
    (c) 2-Chloroethanol
    (d) Xenon
    (e) Trichloromethane

13. Which of the following pairs of compounds would you expect to be more soluble in water? Why?
    (a) $[Bu_4N]Br$ $[(C_{16}H_{33})_4N]Br$
    (b) $CH_3(CH_2)_3NH_2$ $CH_3(CH_2)_3NMe_2$
    (c) Diethyl ether Butane
    (d) 3-Pentanol 3-Pentanone
    (e) Butanoic acid Ethyl acetate

14. Give a specific example of each of the following:
    (a) A molecule that forms intramolecular hydrogen bonds
    (b) A tetradentate ligand
    (c) A synthetic ionophore

MORE CHALLENGING PROBLEMS

1. For each of the following molecules, draw a line structure:
    (a) 3-Bromo-1-ethyl-4-*iso*-propylcyclooctanol
    (b) *N*,4-Dimethyl-3-oxohexanamide
    (c) 2-Dimethylamino-3-pentanol

2. Give a systematic name for each of the following molecules:

(a)

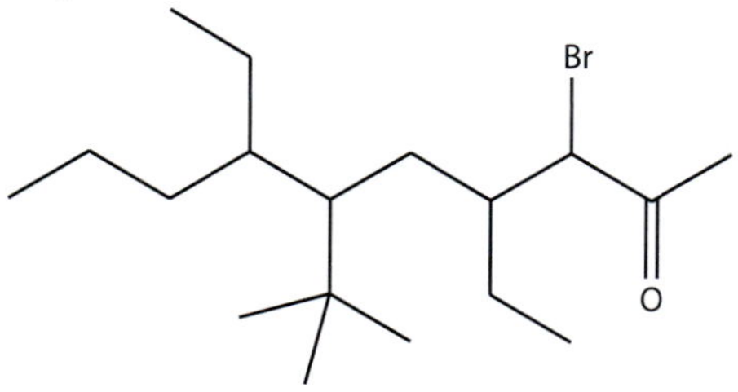

(b)

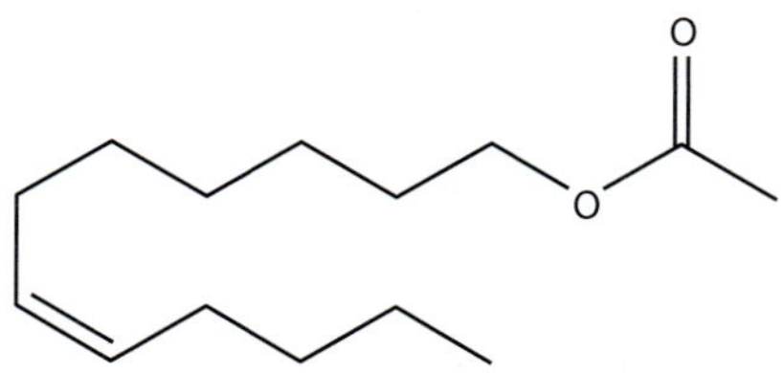

(c)

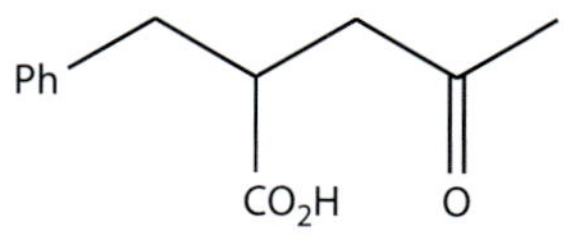

3. For the following molecules, generate orbital representations. Label bonds as σ or π, and state which orbitals were used to make them (s, p, sp, sp^2, sp^3). Mark empty orbitals and lone pairs, and state their type:

(a) $CH_3C(=O)OCH_3$

(b) CH_3NHOH

(c) $CH_2=NCH_3$

(d)

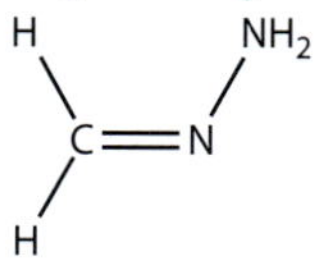

4. Draw the molecular orbitals of ketene, **4.86**. Indicate the hybridization of each atom. What is the relationship between the planes that contain the C=O and C=C π-bonds?

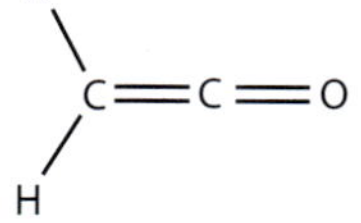

4.86

5. The structure of one of the degradation products of the antibiotic vermiculene is shown in the following. Give a systematic name:

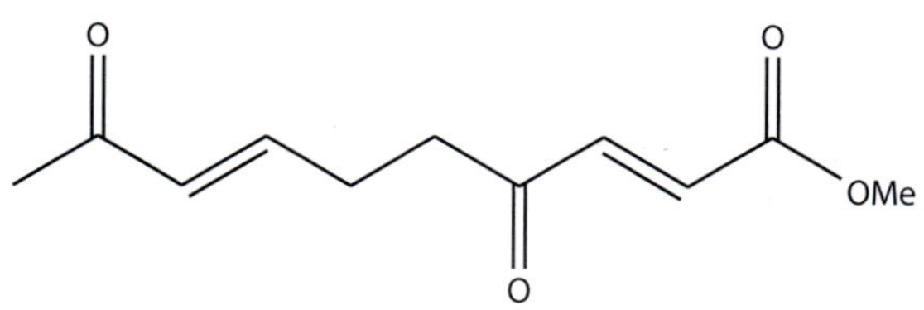

6. Consider the molecules **4.87** and **4.88** shown in the following. **4.87** is rather unstable, but **4.88** is stable. Explain.

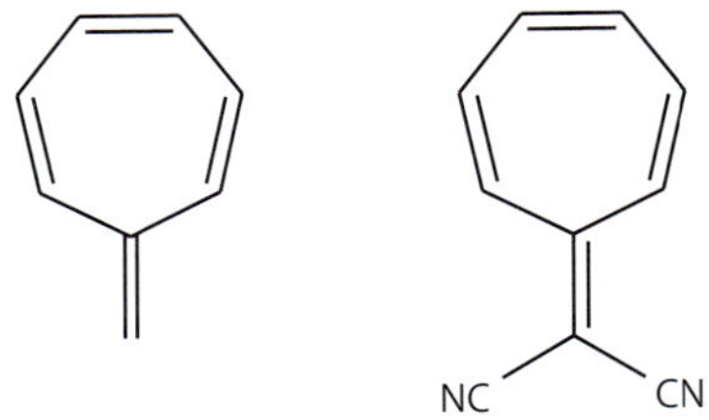

4.87 **4.88**

7. Explain why tetrahydrofuran, **4.89**, is water miscible, while diethyl ether is only sparingly soluble in water.

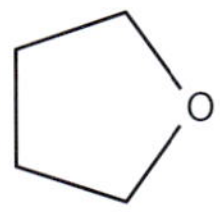

4.89

8. Draw a resonance form of **4.90** in which there is no charge separation:

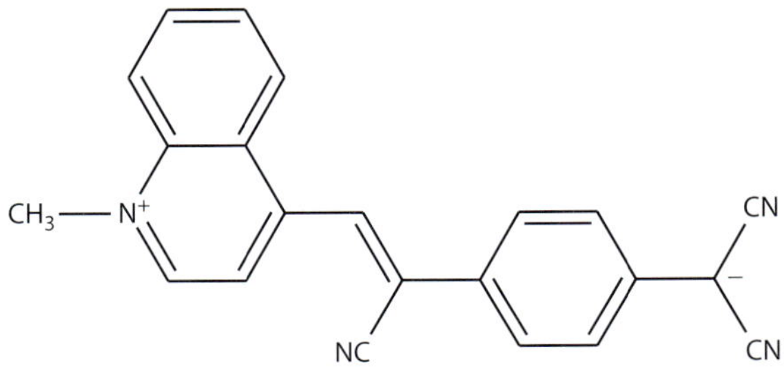

4.90

9. Which would you expect to be more basic, RNH_2 or RC≡N? Give a reason.
10. One of the three isomers (*ortho*, *meta*, and *para*) of hydroxybenzaldehyde is much less soluble in water and has a lower boiling point than the other two. Identify this isomer and explain why its properties are different from those of the other two isomers.

Purification and Identification of Organic Compounds 1

5

5.1 INTRODUCTION

This chapter will begin with a quick review of the techniques that can be used to purify organic compounds, which will be covered in more detail in your lab classes. Then it will investigate how the materials that have been purified can be identified, using various spectroscopic techniques. Chapter 5 will deal with mass spectrometry, infrared (IR), and UV/visible spectroscopy, with Chapter 6 devoted to NMR spectroscopy. Not only gross structure but also details such as stereochemistry are available from such techniques.

5.2 PURIFICATION OF ORGANIC COMPOUNDS

We need to have good techniques for purification of organic compounds because when they are synthesized, or extracted from natural materials, they are not always in a pure form. For example, when a chemical reaction is designed to turn A into B, it rarely does so with 100 % efficiency. The material we obtain may contain by-products, solvents and reagents, and unreacted starting material. Extracting materials from natural sources can give mixtures of bewildering complexity, with large numbers of components. How can we separate and characterize the material that we want?

5.2.1 CRYSTALLIZATION

Probably all of you are familiar with the beauty and fun of crystal growing. Although crystals are formed by both organic and inorganic materials (and by pure elements), those you may be most familiar with are inorganic materials such as copper sulfate and alum and those that adorn jewelry. Sadly, I cannot give you a simple recipe to grow diamonds. However, organic materials can also form fine crystals. Crystals are regular in appearance reflecting the internal regularity of their structure, and the way in which they grow generally precludes the incorporation of other molecules (except occasionally solvent molecules). Thus, if you can grow good crystals, then your sample is very likely to be pure. Many chemists use a final crystallization for purification, even when a sample has been separated from impurities by other means.

Inorganic crystals: Copper sulfate crystal

Alum crystals

Crystals have long been valued:
The Steinmetz Pink diamond

Emerald studded dagger from the Topkapi Museum, Istanbul

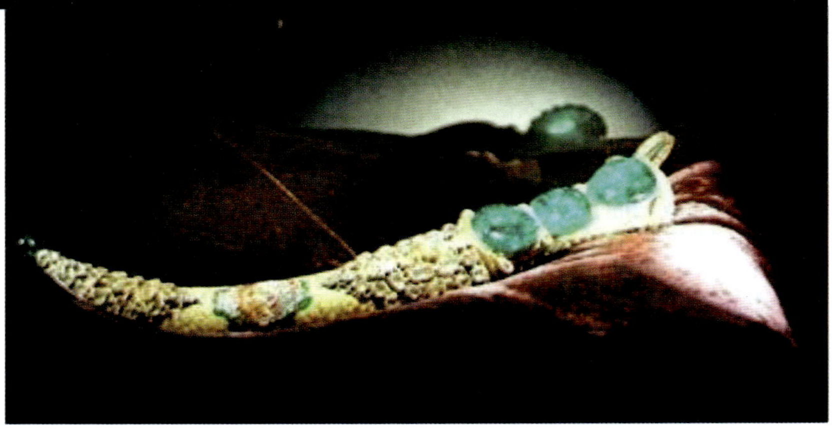

Crystals of organic compounds: Sucrose

Protein crystal: Bovine insulin

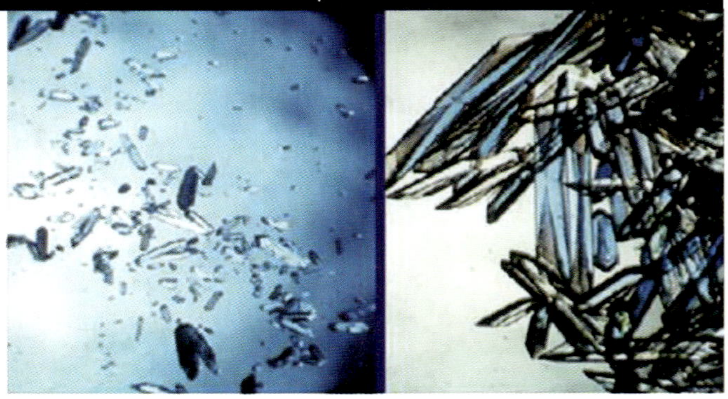

One simple method of crystallizing organic compounds is to find a solvent where the material is soluble when hot but insoluble when cold. A classic example of this method is benzoic acid, which is soluble in hot, but not in cold, water. Insoluble impurities are removed by hot gravity filtration; the hot solution is filtered through a fluted filter paper into a heated flask (Figure 5.1). The flask into which the material is filtered needs to contain some hot solvent, so that the funnel and the flask are kept warm, and crystals do not form prematurely. The filter paper is fluted to maximize its surface area and to minimize the contact with the cooler funnel. This technique works well when any impurities are insoluble in the hot solution or soluble in the solvent at any temperature.

Finding a single solvent for the hot gravity filtration technique to work well is not always possible. Sometimes better results are obtained using a mixture of solvents—the solute is very much more soluble in one of the solvents than in the other. In this case, the solute is dissolved in a small amount of the solvent in which it is soluble, and then the other solvent is added until faint turbidity (cloudiness) is detected and the material is then set aside (often in the fridge or an ice

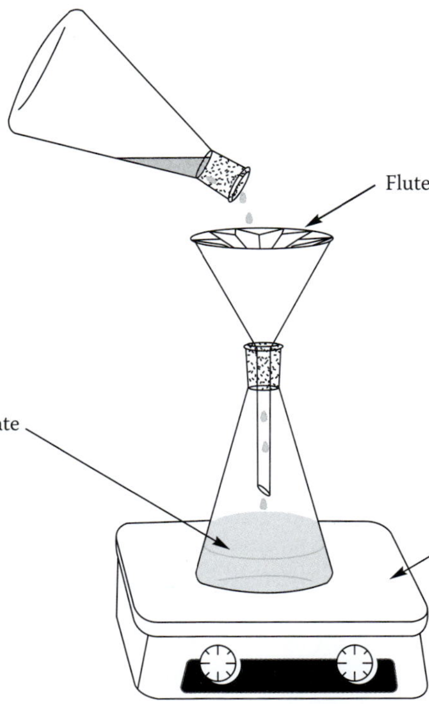

FIGURE 5.1 Hot gravity filtration.

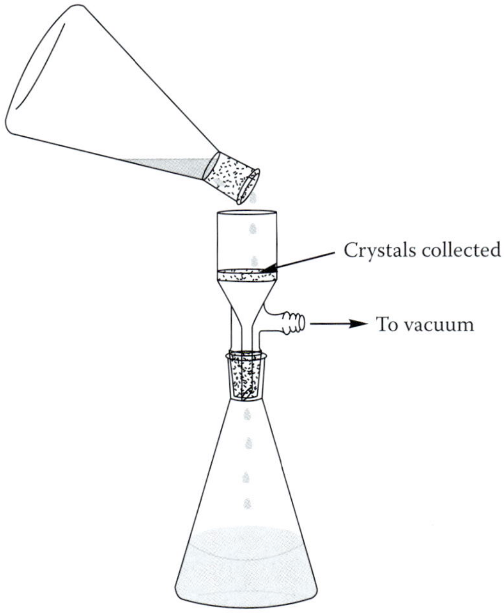

FIGURE 5.2 Collection of crystals on a glass frit.

bath) for crystallization to be completed. Where the solvents have significantly different densities (dichloromethane, density 1.3 g cm^{-3}, and hexane, density 0.655 g cm^{-3} are a good pairing), a good technique is to layer the hexane on top of the dichloromethane by cautiously adding it down the side of the vessel and allowing the crystals to grow at the interface as the two solvents slowly diffuse into one another. It's generally the case that if you want large and beautiful crystals (*e.g.*, if you want to determine a structure by x-ray crystallography), you need to grow them slowly.

Once crystals are obtained, by any technique, they are collected by filtration, using either filter paper on a Buchner funnel or a glass frit (Figure 5.2). A vacuum is used to suck the mother liquor away from the crystals, and they may be washed with a small amount of cooled solvent (so that they do not redissolve). In general, it is wiser to dry crystals as much as possible in this way and then place them in a desiccator to remove the last traces of the solvent. Unless you have an oven with a very reliable thermostat, and you already know the melting point of the crystals, drying them hurriedly in an oven has its risks!

5.2.2 DISTILLATION AND SUBLIMATION

Many organic compounds are not solid at room temperature, and hence cannot easily be crystallized. Although it is possible to grow, and collect, crystals at very low temperatures, this is not technically simple and not commonly accomplished. If your compound is a liquid, then it is usually purified by distillation. Simple distillation (Figure 5.3) is successful if the desired compound is the only one in the mixture that is volatile. However, this is not invariably the case, and components of a mixture need to be separated by a process called fractionation, effectively by boiling point. This is usually done on a glass surface, the length of the column being determined by the degree to which the boiling points of the materials are different (Figure 5.4). Fractions are collected according to their boiling points, so that the desired material can be isolated. For many organic liquids, it is not possible to distill them at atmospheric pressure; they would decompose before they boiled. In these cases, distillation must be carried out at reduced pressure, so that the boiling point is reduced. The vacuums used vary from the 15 mmHg generally provided by a simple water pump to the 0.01 mmHg available from oil diffusion pumps.

The process of sublimation is the conversion from the solid to the gas phase, without the substance melting. This is possible only for a relatively small proportion of all organic compounds

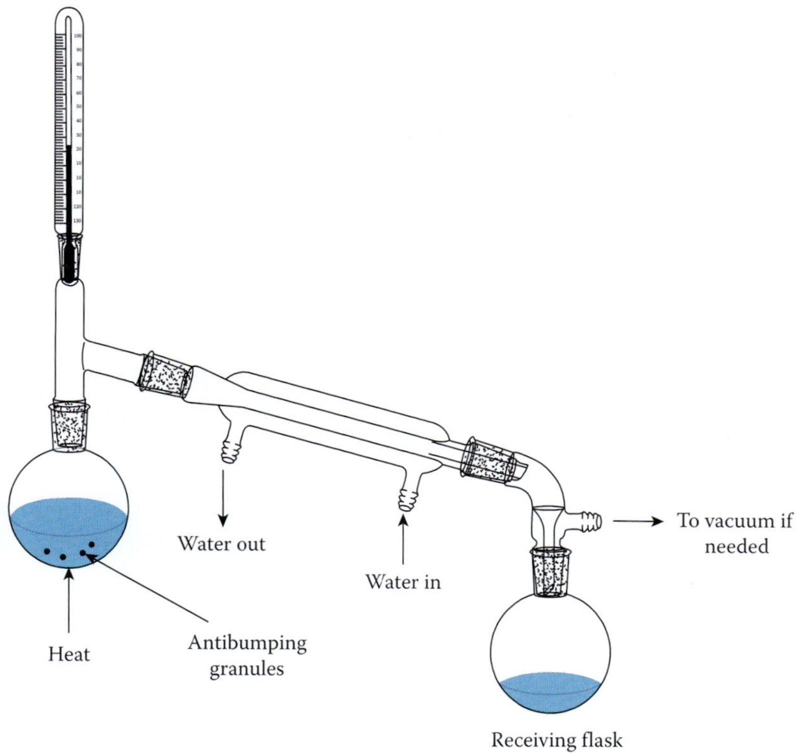

FIGURE 5.3 Simple distillation.

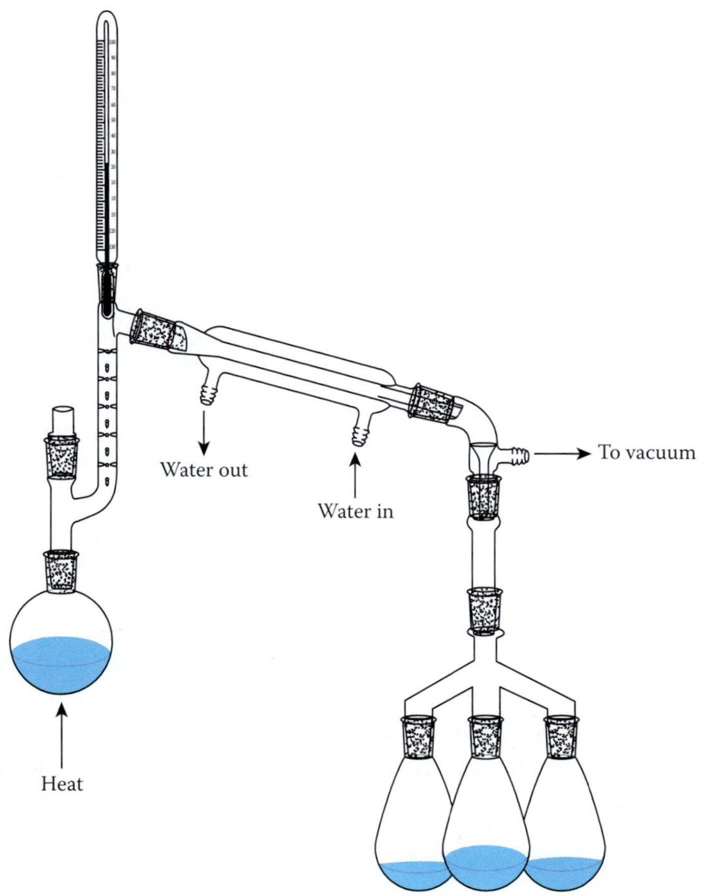

FIGURE 5.4 Fractional distillation of liquids.

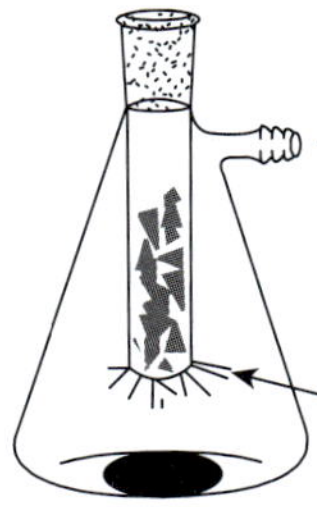

FIGURE 5.5 Simple sublimation apparatus.

(the reasons for this are complex, involving a study of the phase diagrams of the materials), so that when it works, it is an excellent method for separating a material that will sublime from those that will not. Few materials will sublime at atmospheric pressure; water and iodine are important exceptions. Generally, the material to be purified is heated under vacuum, and the sublimate is collected on a "cold finger," cooled with water, ice, or solid CO_2 (Figure 5.5). Examples of organic compounds that can be purified by sublimation include naphthalene, camphor, benzoic acid, and caffeine.

PROBLEM 5.1

List five criteria that should be used in selecting a solvent for recrystallization.

Solution

- Material should be soluble in hot solvent.
- Material should not be soluble in cold solvent.
- Material should not react with the solvent.
- Impurities should be either insoluble in hot or cold solvent or fully soluble in hot or cold solvent.
- Solvent should have a boiling point lower than the melting point of the solid (it is otherwise difficult to tell if it has dissolved or melted).
- Volatile solvents are generally preferred as they are easier to remove and hence dry the crystals.

Key Points from Sections 5.1, 5.2.1, and 5.2.2

- Crystallization is an excellent method of purification of organic compounds. Insoluble impurities may be removed by hot gravity filtration. Mixed solvents may be used when a single solvent is not suitable.
- Crystals are regular in shape because of the internal regularity of their structure.
- Crystal size depends on the rate of growth.
- Some solids may be purified by sublimation.
- Liquids are often purified by distillation; fractionation of the distillate may be needed, and the process is often carried out under reduced pressure.

5.2.3 CHROMATOGRAPHY

The techniques discussed previously work best with compounds that are already reasonably pure; they would not generally succeed in separating and purifying a natural product extract, containing many compounds in comparable amounts. The word chromatography comes from the Greek for colored writing (χρῶμα, color, and γράφειν, to write), and it refers to a number of related techniques for the separation of mixtures into pure compounds.

TABLE 5.1 Common Types of Chromatography

Type of Chromatography	Stationary Phase	Mobile Phase	Scale	Uses, Comments
Gas chromatography	SiO_2 usually coated with an involatile liquid	Inert gas	µg–mg	Almost always analytical; fractions may be analyzed by techniques such as mass spectrometry in single instruments. Old name was gas liquid chromatography, and older texts may refer to GLC.
Liquid chromatography	SiO_2, sometimes with a range of modifications	Solvent	mg–kg	May be analytical or preparative. Fractions of the analyte may be passed directly into other spectroscopic techniques for structure determination. Can be used, with an appropriate chiral stationary phase, to separate enantiomers (see Chapter 7). Usually takes place under high pressures, hence HPLC, which is used to indicate both high-pressure and higher-performance liquid chromatography.
Column chromatography	SiO_2 or Al_2O_3	Solvent	mg–kg	Usually preparative. Passage of the mobile phase may be speeded up by use of pressure/vacuum, in "flash" chromatography.
Thin layer chromatography (TLC)	SiO_2 or Al_2O_3	Solvent	µg–mg	Almost always analytical, although thicker layers of stationary phase have been used preparatively. Used to monitor the progress of reactions and to determine appropriate solvents for preparative chromatography.

All of these involve putting a mixture onto a stationary phase, such as paper, silica, or alumina, and then allowing a mobile phase to flow over it. Different materials "stick" to the stationary phase to different extents and hence move more or less rapidly with the mobile phase. The process of "sticking" is not a simple one if thoroughly analyzed (but we don't need to understand the theory completely to use the technique); on silica, part of the process involves interaction with surface –OH groups, and part a partition between the solvent and water adsorbed onto the silica. With alumina, the adsorption component is more important. Table 5.1 details the main sorts of chromatography used by organic chemists for separation of mixtures. Other types of chromatography exist; for example, polymer chemists determine molecular weights of materials by their rate of passage through a stationary phase with particular sized "holes," ion exchange chromatography is used to analyze a range of charged species, and amino acids are sometimes separated either on paper or on Sephadex®, a cross-linked gel.

Thin-layer chromatography (TLC) is used to monitor the progress of an organic reaction—has the starting material all disappeared, and is there a single product or more than one? It is quick, easy, and inexpensive, so repeating it at regular intervals through a reaction is no problem. It is also used to monitor the fractions obtained from column chromatography to determine which of the many fractions collected contain the desired material. Silica or alumina is bonded to a plastic or aluminum foil sheet (glass slides can also be made but are now less common as they are more expensive and more fragile). TLC can also be used to check that a particular material is identical with a known compound of which you have a sample or that a particular material is completely pure. The plate, with a spot of a mixture of compounds, is placed in a vessel containing a small amount of an appropriate solvent (Figure 5.6), so that the solvent does not cover the spot of material to be analyzed. The solvent then moves up the plate by capillary action, and the various components of the mixture move at varying rates, so that they are separated (Figure 5.7). We define the term R_f, retardation factor, as the distance moved by a particular material, divided by the distance moved by the solvent front. Although different materials generally appear in the same order on the plate, the precise R_f depends on the reaction conditions to some extent, so if one is using TLC to check that two compounds are identical, then it is wise to run them side by side on the same plate, so that all the conditions of the process are identical. In this example, colored materials are separated, but most organic compounds are colorless or close to it. In these cases, we need another method of visualization. In most cases, this is done with UV light (Figure 5.8). The silica is impregnated with a fluorescent compound, so under UV, the

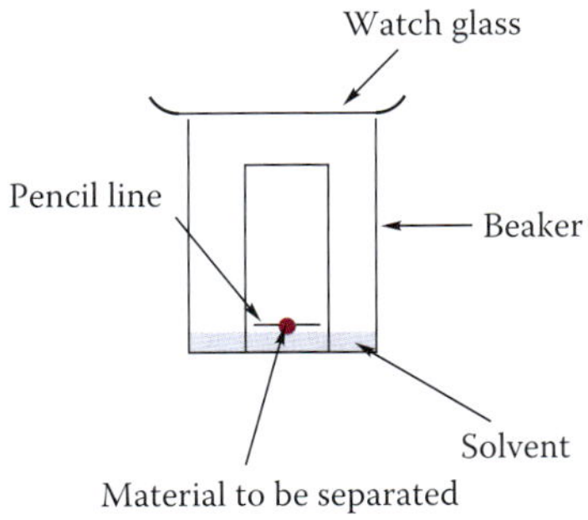

FIGURE 5.6 TLC—at the start of the experiment.

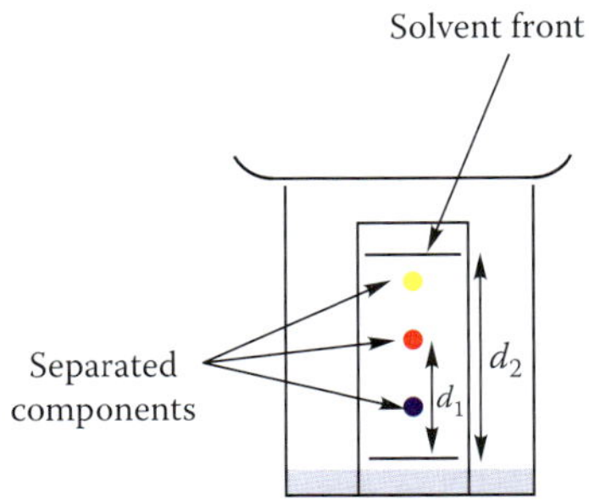

FIGURE 5.7 TLC—after 5–10 min.

FIGURE 5.8 Visualization of spots on a TLC plate using UV light.

plate glows green. However, where there is a spot of material, the fluorescence is quenched, and the spot appears dark. Iodine has also been used to visualize spots; it is absorbed, reversibly, by most organic compounds (Figure 5.9).

Column chromatography relies on the same principles as TLC. Material is put on the top of a column of silica or alumina, and the solvent is allowed to percolate through the stationary phase by gravity, taking with it the components of the mixture. However, the scale of this process is quite different, and it is normally used preparatively. Depending on the ease with which the materials are separated, up to 1 g of a mixture can be separated on 100 g of the stationary phase. Figures 5.10 and 5.11 show the setup and the experiment in progress. The sand is added to the top and bottom of the column to protect the surface of the stationary phase. The bottom of the

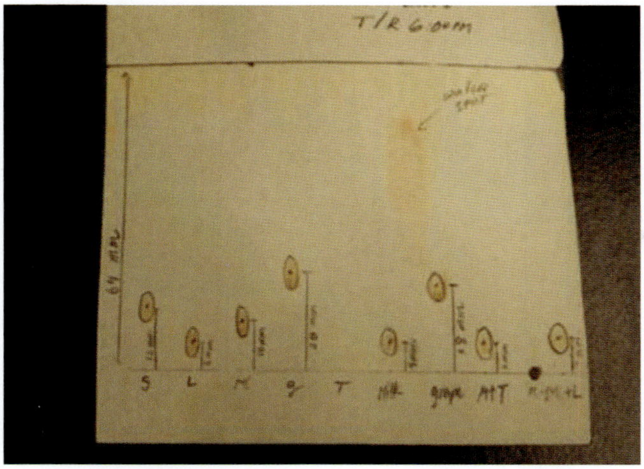

FIGURE 5.9 Visualization of TLC plate using iodine.

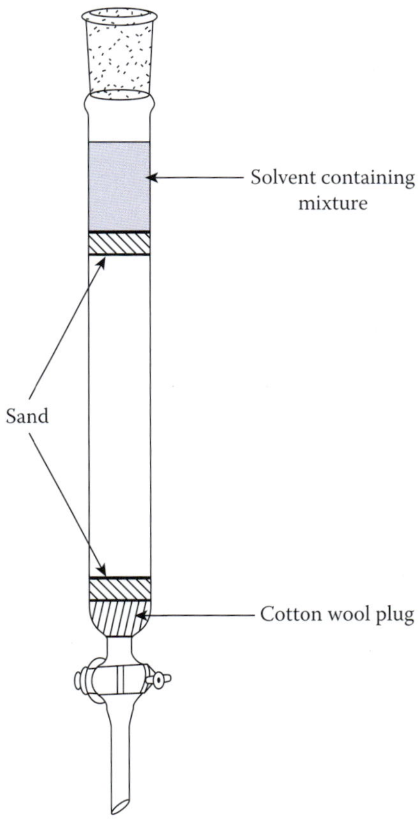

FIGURE 5.10 Column chromatography setup.

column is sealed with a glass frit or, failing that, a plug of cotton wool. The stationary phase is normally packed as a slurry with the solvent that will initially be used to elute the column. This requires some care and patience; the column must be exactly vertical, and the material should be allowed to settle under gravity for the best packing. The better the column is initially packed, the better the separation. As the solvent is eluted from the column, "fractions" are collected and labeled, to be tested by TLC to see if they contain the required materials (as before, most organic compounds are colorless, so fractions of fixed volume are collected for later testing). Automated versions with a continuous solvent supply, and automatic fraction collection, are available. Sometimes it is possible to elute all the required material using a single solvent, but

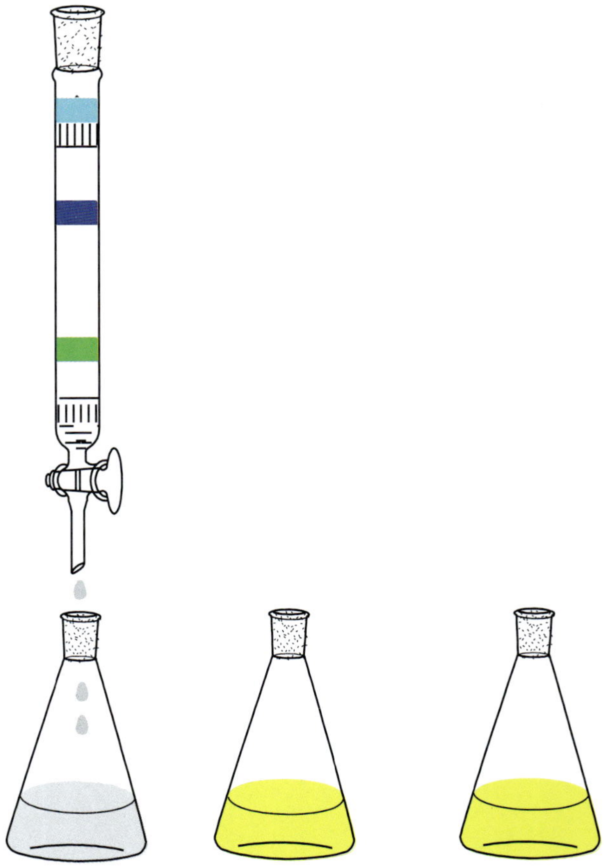

FIGURE 5.11 Column chromatography in progress.

on other occasions, it will be necessary to increase the solvent polarity during the experiment to ensure that all the material is recovered. This should be done gradually, and with care—the heats of mixing of the solvents can make the interface quite warm, and formation of vapor bubbles makes packing uneven and separation poorer. Because silica and alumina have surface hydroxyl groups, they are *polar*. Therefore, the least polar molecules bind least well and are eluted first. More polar molecules are bound better, and a more polar solvent is needed to elute them.

The resolution of a column (or TLC plate) depends on the surface area of the solid phase and hence on the size of the silica or alumina particles. Small particles give a larger surface area, and hence better resolution, but the flow of liquid through the column is slower. Size is described in terms of "mesh size"—higher mesh size means better resolution but slower elution. Not only were graduate students never the most patient of folk, but leaving materials on a chromatography column for long periods of time can mean that some compounds decompose (hydrolysis of esters and oxidations are common); neither silica nor alumina is strictly chemically inert. Initial efforts at forcing solvent through faster using gas pressure developed into a technique called flash chromatography, where the liquid is forced through the column by pressure and/or suction. For gravity columns, a 70–230 mesh material is used, while flash columns are packed dry and use 230–400 mesh material. A schematic of a pressure-driven system is shown in Figure 5.12.

There are limits, however, to the pressures that a glass column can tolerate. This leads us to high-pressure (or high-performance) liquid chromatography (HPLC) in which liquid is forced through a steel-encased column by a pump. This is a powerful, but relatively expensive, technique and is highly automated, both for analytical and preparative use. Taking TLCs of the many fractions that HPLC produces would be prohibitively slow, so the system needs its own techniques to detect that material other than solvent being eluted. One way to do this is by refractive index measurements; this depends on the fact that the refractive index of a solution is different from that of pure solvent. Alternatively, the UV absorption of the material is monitored. HPLC is used for a wide range of analytical purposes (*e.g.*, Figure 5.13) and also preparatively for fine chemicals:

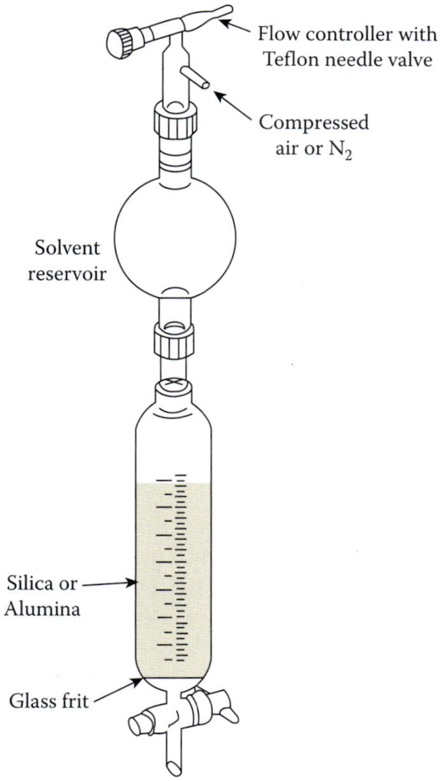

FIGURE 5.12 Pressure driven flash chromatography.

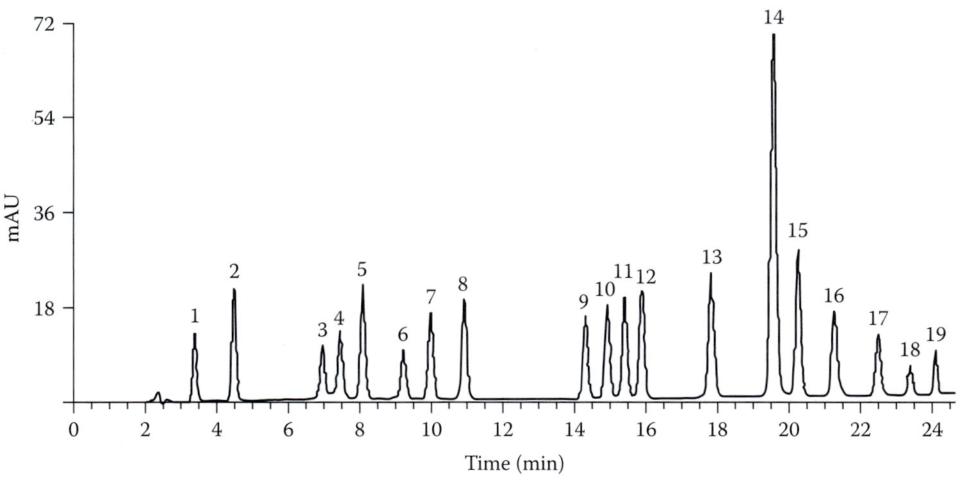

FIGURE 5.13 HPLC separation of opiates. (1) Normorphine, (2) morphine, (3) norcodeine, (4) naloxone, (5) codeine, (6) hydrocodeine, (7) 6-monoacetylmorphine, (8) ethylmorphine, (9) acetyldihydrocodeine, (10) thebacone, (11) acetylcodeine, (12) heroin, (13) internal standard, (14) papaverine, (15) dextromethorphan, (16) noscapine, (17) buprenorphine, (18) normethadone, and (19) methadone. (From Dams, R. et al., *J. Chromatogr. A* 896(1–2), 311, 2000, Figure 2a.)

In gas chromatography (schematic in Figure 5.14), the solid phase is usually silica coated with an involatile silicone polymer. Columns may be 10–100 m in length, so separation is usually excellent. Materials are eluted in approximate order of their boiling points (so this only works if they are reasonably volatile and do not decompose readily on heating). Detection of material being eluted relies on a thermal conductivity detector or flame ionization detector. The thermal conductivity detector depends on the fact that the conductivity of gas plus eluent is different from that of gas

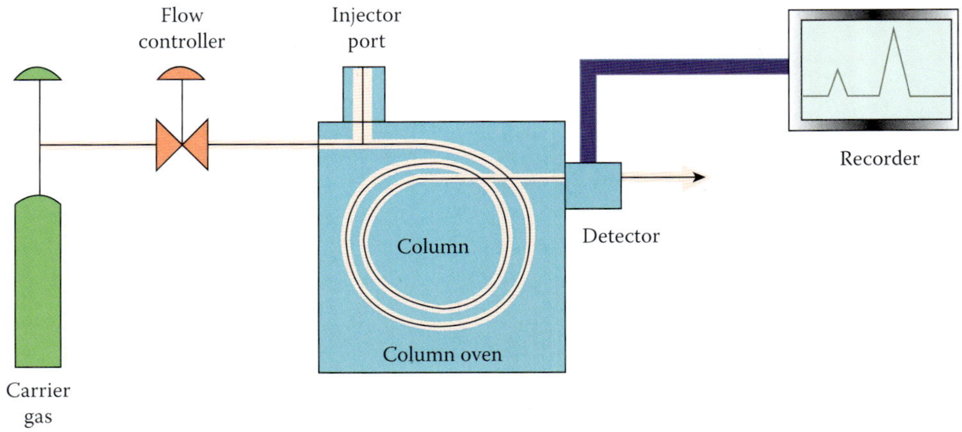

FIGURE 5.14 Schematic of a gas chromatograph.

alone. In flame ionization detectors, the eluent gas is burned and the ions from the eluent detected. Increasingly, both HPLC and gas chromatography (GC) systems are connected to mass spectrometers (and other types of spectrometer) so that the spectrum of each fraction can be determined automatically.

PROBLEM 5.2

What will be the result of the following errors in TLC technique?

(a) Too much sample applied
(b) Solvent of too high polarity
(c) Solvent pool in developing jar too deep
(d) Forgetting to remove the plate when the solvent has reached the top of the plate

Solution

(a) The plate will be overloaded, and the spots may merge.
(b) All the material will move to the top of the plate, and separation will be poor.
(c) The material will wash off into the solvent and not be carried up the plate.
(d) Materials will continue to move up the plate, and separation will deteriorate with time. It will not be possible to calculate R_f values.

Key Points from Section 5.2.3

- The types of chromatography used to analyze and purify organic compounds are shown in Table 5.1.
- TLC is used to analyze mixtures, check purity, and monitor the progress of reactions.
- Column chromatography is mostly a preparative technique that is able to handle larger amounts of material. Efficiency of separation depends on the surface area of the stationary phase. The rate of elution can be increased by pressure/vacuum techniques, as in flash chromatography.
- HPLC columns are long and narrow and made of steel, so that much higher pressures can be employed. Fraction collection and analysis is automated, and the technique can be used both preparatively and analytically.
- In GC, the mobile phase is gaseous, with a carrier gas and the test material. Materials are eluted in a general order of their boiling points.
- Both GC and HPLC can be coupled to spectroscopic techniques for analysis of the materials eluting. Mass spectrometry is most commonly coupled.

5.3 MASS SPECTROMETRY

Mass spectrometry differs from the other spectroscopic techniques we will discuss in that it does not involve electromagnetic radiation—essentially it weighs molecules. Many organic compounds, even those we might not ordinarily think of as volatile, can be made to pass into the vapor phase if the pressure is low enough. Once a molecule is in the gas phase, it can be ionized, using a variety of techniques, the simplest (and oldest) of which is to bombard it with a stream of high-energy electrons. This expels a single electron from the test molecule to give a species we describe as a radical cation. It is a cation, because it has a single positive charge, and it is a radical because it has an unpaired electron (Equation 5.1). In this equation, the arrow under the electron indicates that it is high in energy and fast moving. $M^{+\bullet}$ is called the molecular ion,* and we often put it in parentheses, as $[C_2H_6]^{+\bullet}$, to indicate that we are not quite sure where either the charge or the unpaired electron will be located. A mass spectrometer is designed only to detect charged particles, and the assumption is made that singly charged particles form the majority of the material. Figure 5.15 shows a schematic of a simple mass spectrometer. The particles are accelerated through a voltage and then enter a magnetic field, strength B. They are deflected in the magnetic field describing an arc of radius r, where $r^2 = 2(m/z)V/B^2$. m/z is the ratio of the mass to charge of the particle, and we normally only worry about species for which $z = 1$. Light particles are deflected more than heavy ones. In order for species of different mass to strike the collector, the collector is not moved, but the voltage is varied. Modern spectrometers use high-powered superconducting magnets, and there are many new ionization techniques that allow ions to be produced by gentler means (they are often called soft ionization techniques), and directly from solids such as peptides and proteins, which would not withstand the older methods.

(5.1) $$\underset{\longrightarrow}{e^-} + M \longrightarrow M^{+\bullet} + 2e^-$$

Mass spectrometry can do more than simply determine the molecular weight of a compound. A high-resolution instrument can determine it to three or more decimal places, and this allows the formula of the molecule to be determined (extensive tables exist). Although we often think

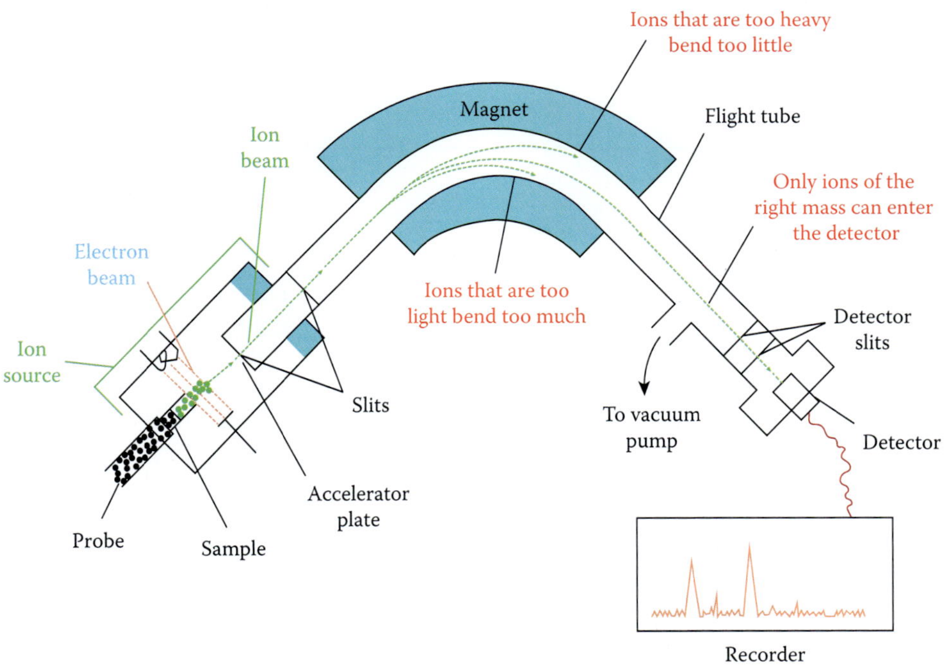

FIGURE 5.15 Schematic of a mass spectrometer.

* Generally pronounced as "M plus dot."

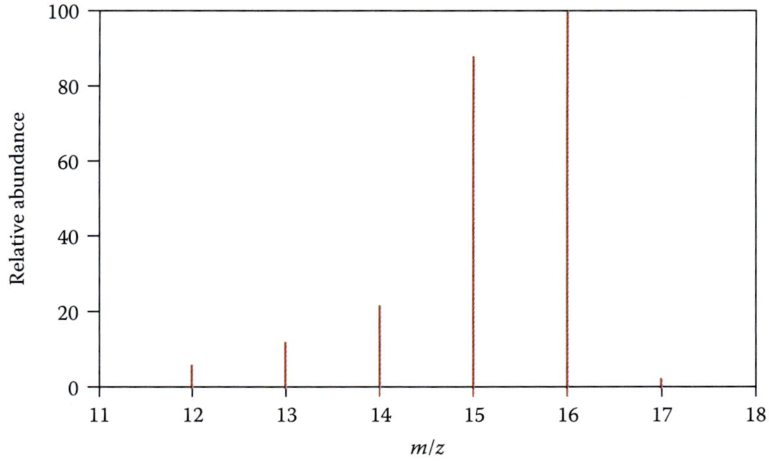

FIGURE 5.16 Mass spectrum of methane.

of atoms as having integral masses, this is not exactly true. For example, C_2H_4, CO, and N_2 all have nominal molecular weights of 28, but more accurately, these are 28.0312, 28.006, and 27.994, and these can be readily resolved. Isotope peaks can also help quickly to identify particular elemental components of a compound. Both carbon and hydrogen have "heavy" isotopes, ^{13}C (1.1 %) and deuterium (^2H, 0.015 %). So when we inspect the mass spectrum of methane (Figure 5.16), we see a small peak at 17 mass units due to $^{13}CH_4$. In this case, the peak is hard to see, but with larger molecules, the chance of finding at least one ^{13}C in the molecule increases, so for a C_{30} molecule, the peak at M + 1 mass units would be about one-third the size that of the molecular ion.

It is particularly easy to recognize molecules containing chlorine or bromine. Chlorine has two isotopes, ^{35}Cl and ^{37}Cl, in the ratio of about 3:1, and bromine has two isotopes, ^{79}Br and ^{81}Br, in approximately 1:1 ratio. The appearance of mass spectra of molecules containing either chlorine or bromine is very characteristic (Figures 5.17 through 5.19). In Figure 5.17, there are two molecular ions, mass 108 and 110. Notice that there is also a cluster of peaks at 29 mass units and below, which clearly do not contain bromine; these arise from fragments of the original molecule, with the peak at 29 deriving from the ethyl cation, $[C_2H_5]^+$. In Figure 5.18, the chlorine-containing molecular ions are seen at *m/z* 80 and 78, but there

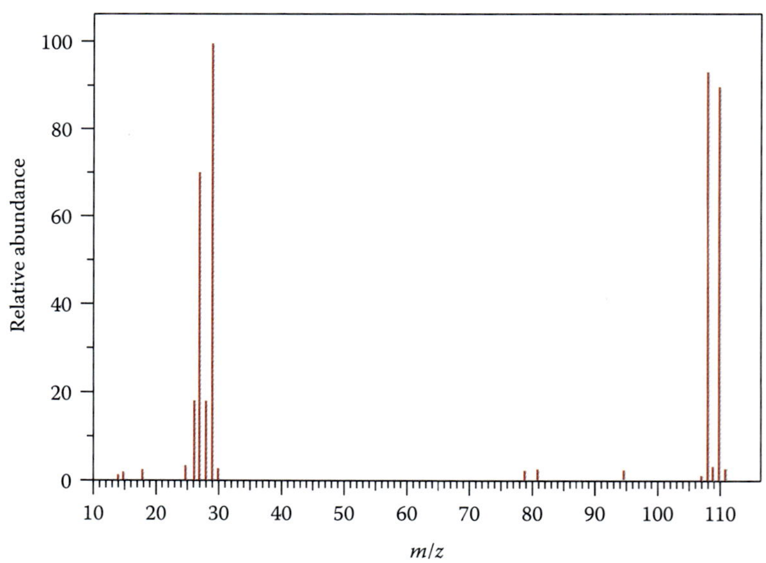

FIGURE 5.17 Mass spectrum of bromoethane.

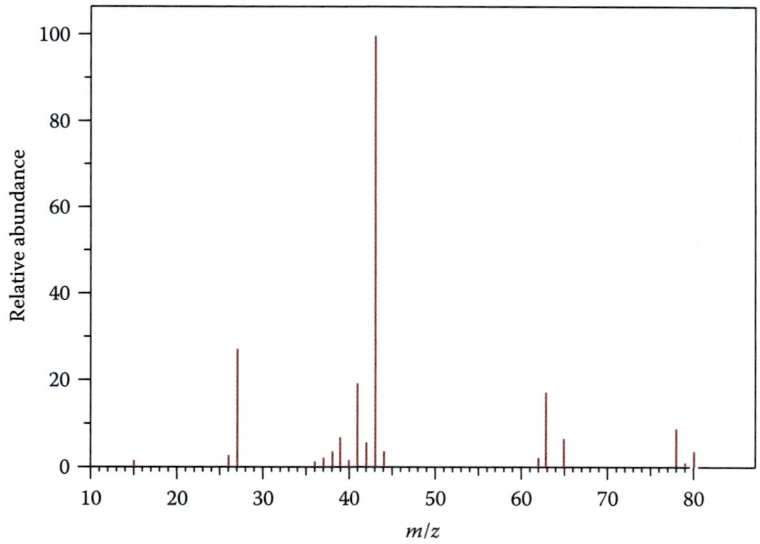

FIGURE 5.18 Mass spectrum of 2-chloropropane.

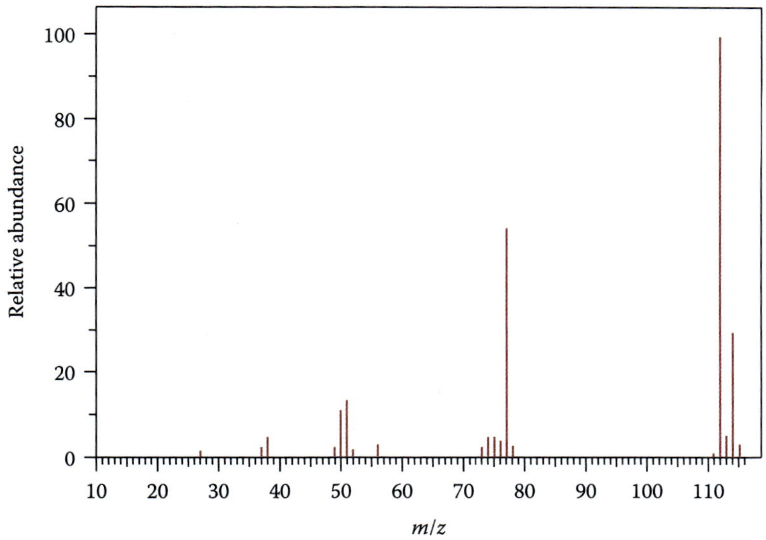

FIGURE 5.19 Mass spectrum of chlorobenzene.

is also a chlorine-containing fragment at 65/63 mass units; this is attributed to a fragment that has lost a methyl group, formula C_2H_4Cl. In the mass spectrum of chlorobenzene (Figure 5.19), the molecular ions are seen at 112 and 114 mass units; notice that the ^{13}C peaks are now much more visible. There is a strong peak at *m/z* 77, for $[C_6H_5]^+$—this will become quite familiar, as it is present in the spectrum of most compounds containing a monosubstituted benzene ring.

The "nitrogen rule" states that if a molecule has an odd number of nitrogen atoms, it will have an odd molecular weight. The rule derives from the fact that nitrogen is unique among common elements/isotopes in having an odd valency but an even atomic number. Figure 5.20 shows the mass spectrum of aniline, with M^+ at *m/z* 93.

In all of these spectra, we have seen peaks for parts or fragments of the molecules. This is because the molecular ion is generated with a lot of excess energy—enough to break bonds within the molecule. However, the fragmentation is not random but generally takes place to give

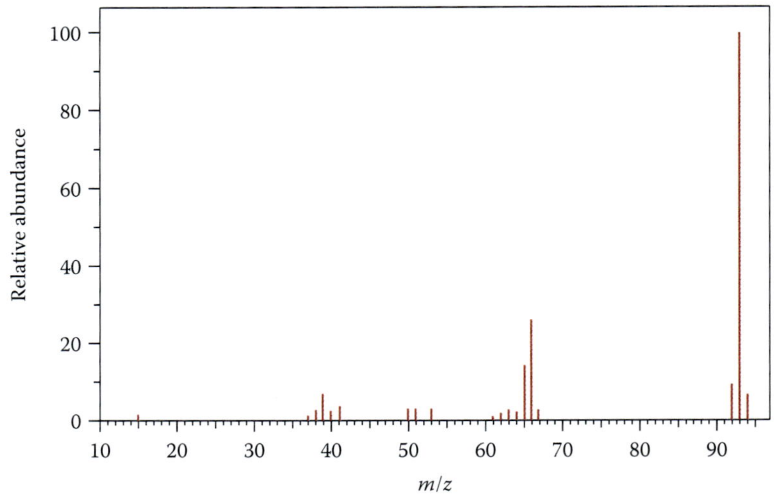

FIGURE 5.20 Mass spectrum of aniline.

TABLE 5.2 Mass Spectrometric Fragmentations

Mass Lost	Group Lost	Significance
15	CH_3	
16	NH_2	$RCONH_2$
17	OH	ROH
18	H_2O	Alcohol
28	C_2H_4	ArC(=O)Et, RCOOEt, $RC(=O)CH_2CH_2CH_3$
29	CHO	Aldehyde
29	C_2H_5	RC(=O)Et
31	OCH_3	RCOOMe, ArOMe
44	CO_2	RCOOH

stable cations. Some losses can be easily spotted—Table 5.2 gives some of these and their significance in terms of the type of molecule involved.

Considering the spectra of the isomeric compounds 3,3-dimethylhexane and *n*-octane (Figures 5.21 and 5.22) will illustrate some of the general principles. The molecular ion for each should be at *m/z* 114, but the molecular ion is not seen for the branched compound.

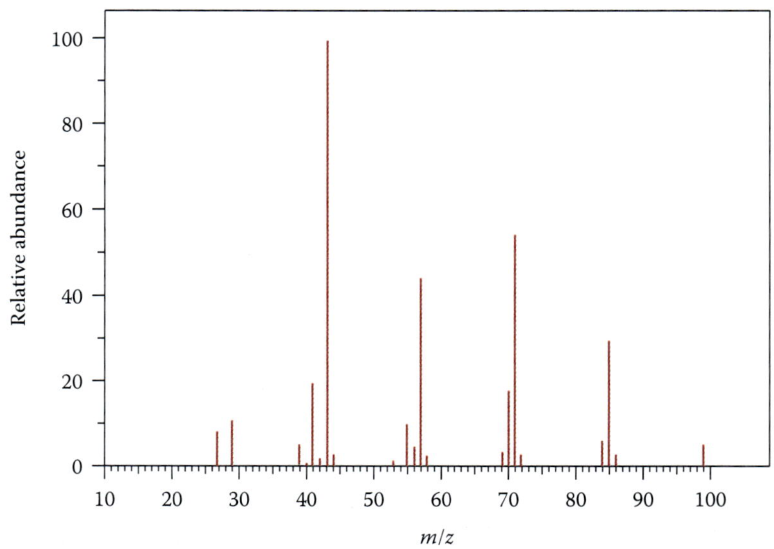

FIGURE 5.21 Mass spectrum of 3,3-dimethylhexane.

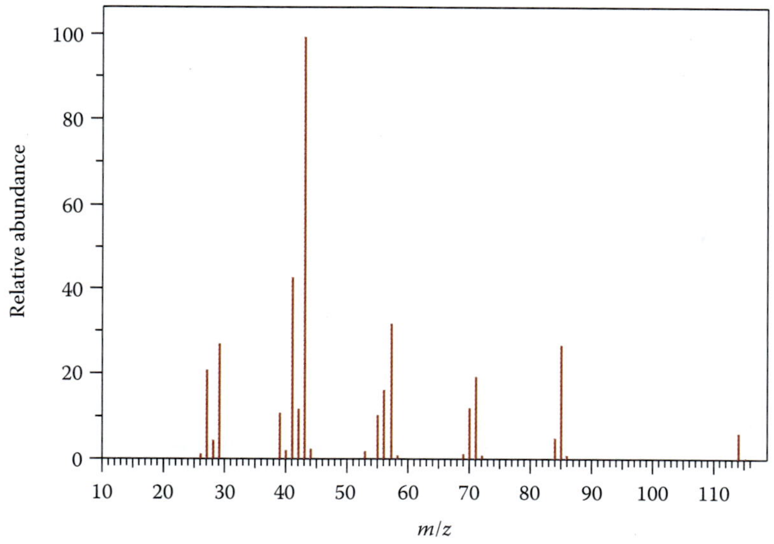

FIGURE 5.22 Mass spectrum of *n*-octane.

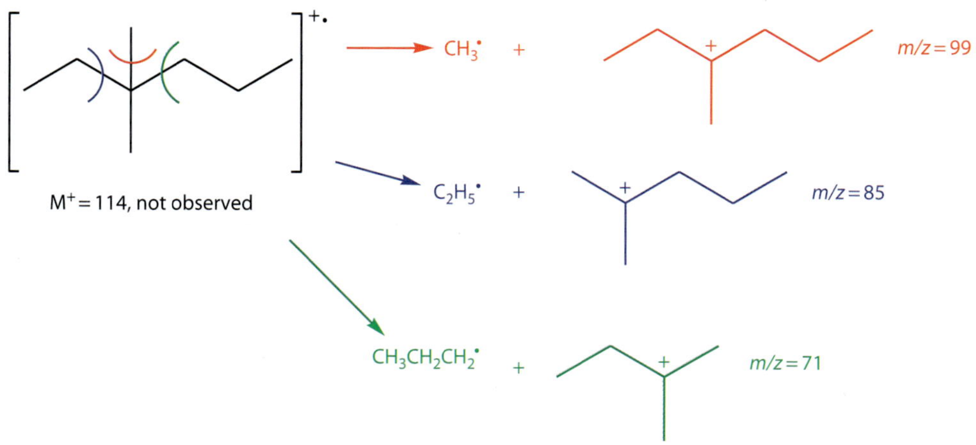

FIGURE 5.23 Fragmentation of 3,3-dimethylhexane.

Instead, the first peak is at *m/z* 99—it has lost a methyl group. There are other strong peaks at *m/z* 85, 71, 57, and 43. We can account for these as in Figure 5.23—in each of the primary fragmentations, a stable tertiary carbocation is formed (remember for carbocations, tertiary is more stable than secondary, which is more stable than primary), and this process is so favorable that the molecular ion is not seen at all. The peaks at 57 and 43 are $[Me_3C]^+$ and $[Me_2CH]^+$, which are readily produced by further fragmentation of the first produced cations. For *n*-octane, there is no readily available path for fragmentation to a tertiary cation, so the molecular ion persists; the fragments derive from breaks at random points of the chain. We can conclude that fragmentation will usually occur at branch points of the molecule. Note that these various fragmentations can occur concurrently—one sequence of fragmentation may not account for all the peaks observed.

Carbocations can also be stabilized by a heteroatom that bears a lone pair of electrons able to donate into the empty p_z orbital (Figure 5.24). Thus, in mass spectrometry, the fragmentation generally occurs such that the cation formed can be stabilized by the heteroatom. The mass spectrum of 2-methyl-2-butanol (Figure 5.25) illustrates. The molecular ion (M^+ would be 88) is not seen as there are good routes to stable cations. The peaks at *m/z* 73 and 59 represent the loss of methyl and ethyl radicals to give cations stabilized by lone pair

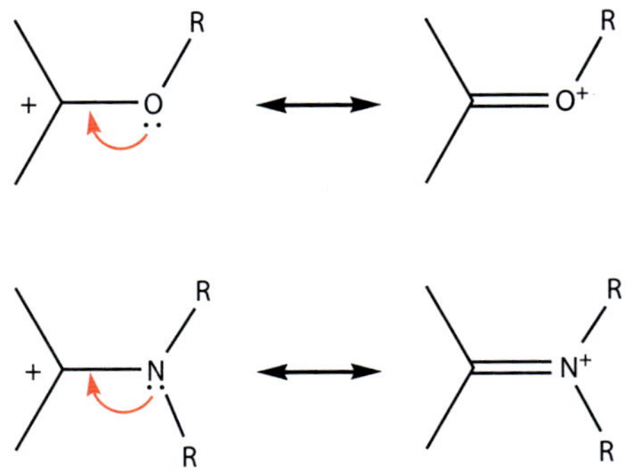

FIGURE 5.24 Stabilization of carbocations by lone pair donation.

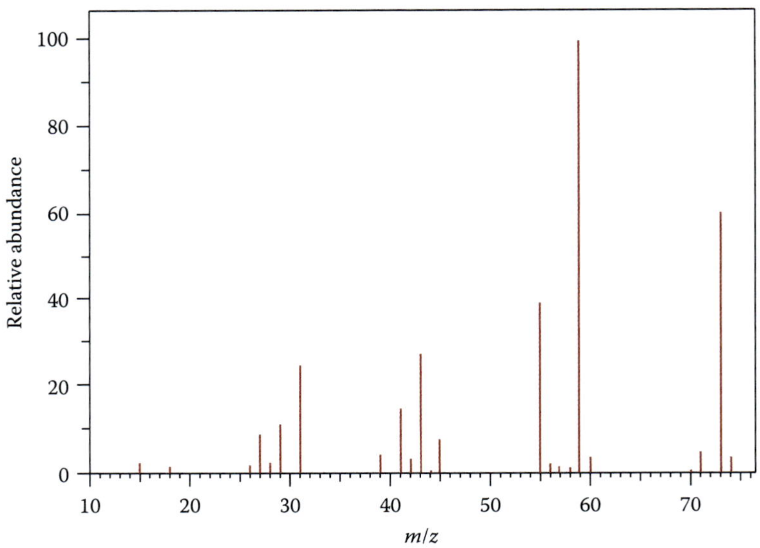

FIGURE 5.25 Mass spectrum of 2-methyl-2-butanol.

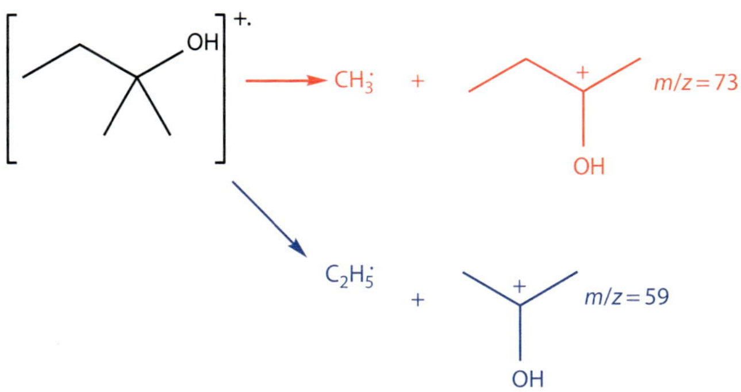

FIGURE 5.26 Fragmentation of 2-methyl-2-butanol.

donation from oxygen (Figure 5.26). Notice that there is only a very small peak at *m/z* 71 for $[CH_3CH_2C(CH_3)_2]^+$; a tertiary cation with stabilization from a lone pair is more stable than a simple tertiary cation.

Lone pair stabilization is also responsible for the pattern of fragmentation of ketones, which involves cleavage on either side of the carbonyl group. Thus, the two main fragmentations of 3-heptanone give peaks at *m/z* 85 and 57 (Figure 5.27). The species thus formed are called acylium ions, and the resonance forms are shown in Figure 5.28. Acylium ions readily lose 28 mass units, carbon monoxide, to give simple carbocations.

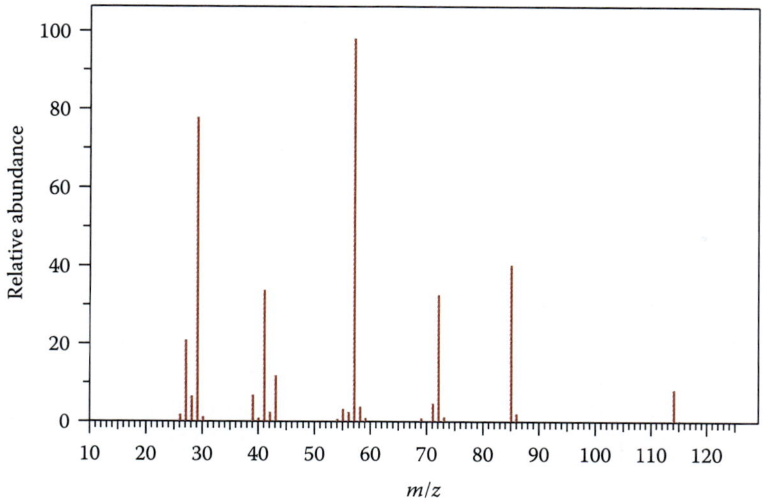

FIGURE 5.27 Mass spectrum of 3-heptanone.

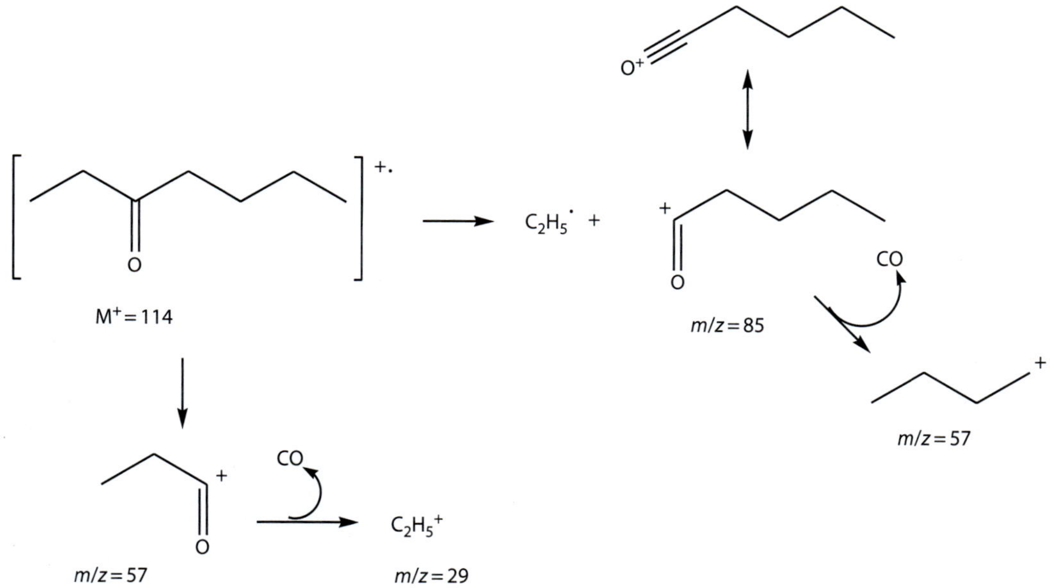

FIGURE 5.28 Fragmentation of 3-heptanone.

PROBLEM 5.3

The mass spectra of the compounds **5.1** and **5.2** both showed molecular ions at *m/z* = 134. One showed peaks at *m/z* 119, 91, 77, and 43, and the other at 105, 77, 57, and 29. Which is which?

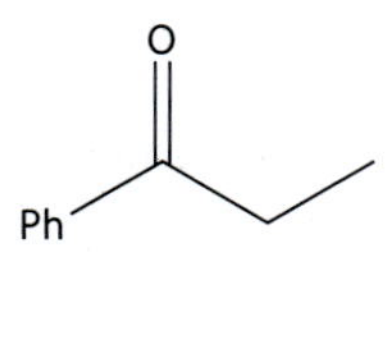

5.1

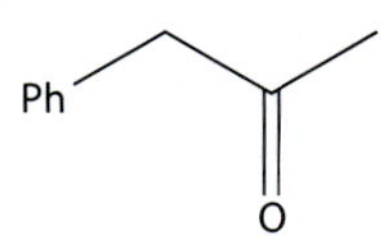

5.2

PROBLEM 5.4

The molecular ions for polyhalogenated compounds are complicated. Taking the natural abundance of ^{81}Br and ^{79}Br as 50 % each and the abundance of ^{35}Cl and ^{37}Cl as 75 % and 25 %, respectively, calculate the masses of the molecular ions of compounds of the following formulae. What are the relative amounts of each ion?

(a) CH_2Cl_2
(b) CH_2Br_2

Solutions

5.3 We expect ketones to fragment by breaking the bond next to the carbonyl group. This is followed by loss of CO (28 mass units). So for **5.1**, we would expect fragments $[PhCO]^+$, Ph^+, $[CH_3CH_2CO]^+$, and $[CH_3CH_2]^+$ with masses 105, 77, 57, and 29. For **5.2**, we would expect $[PhCH_2CO]^+$, $[PhCH_2]^+$, and $[CH_3CO]^+$ with masses 119, 91, and 43. Although not formed by the initial fragmentation, the peak at 77 mass units is clearly a phenyl ring, presumably deriving from a fragmentation of $[PhCH_2]^+$.

5.4 This is really a simple probability question.

(a) Probability of $CH_2{}^{35}Cl_2 = 3/4 \times 3/4$ Mass = 84
Probability of $CH_2{}^{35}Cl^{37}Cl = 3/4 \times 1/4$ Mass = 86
Probability of $CH_2{}^{37}Cl^{35}Cl = 1/4 \times 3/4$ Mass = 86
Probability of $CH_2{}^{37}Cl_2 = 1/4 \times 1/4$ Mass = 88
So the ratio of the peaks at 84, 86, and 88 is 9:6:1 (or 56.25 %, 37.5 %, and 6.25 %).

(b) Probability of $CH_2{}^{79}Br_2 = 1/2 \times 1/2$ Mass = 172
Probability of $CH_2{}^{79}Br^{81}Br = 1/2 \times 1/2$ Mass = 174
Probability of $CH_2{}^{81}Br^{79}Br = 1/2 \times 1/2$ Mass = 174
Probability of $CH_2{}^{81}Br_2 = 1/2 \times 1/2$ Mass = 176
So the ratio of the peaks at 172, 174, and 176 mass units is 1:2:1 (or 25 %, 50 %, and 25 %).

Focus on Stable Isotope Mass Spectrometry

In literature tables of isotopic abundances, the values given are averages; however, this does not mean that every sample containing carbon contains exactly 1.1 % of ^{13}C or that all water contains exactly 0.015 % of heavy water, D_2O. For example, there is a higher abundance of ^{13}C in marine sediments and a lower abundance in crude oil. Although the differences in isotopic abundances are small, usually less than 0.1 atom %, they are measurable using specially designed mass spectrometers. The δ-notation (Equation 5.2) expresses the differences in units per 1000, relative to a standard. R_s is the measured ratio of the heavier isotope over the lighter one in the sample, and R_{std} is the measured isotope ratio in the standard, δ being defined as zero for the standard. The amount of ^{13}C is defined in terms of Pee Dee Belemnite (PDB), which is derived from a marine fossil from the Cretaceous period, *Belemnitella Americana*, from the Pee Dee formation in South Carolina (natural sources are now exhausted, so Vienna PDB, a virtual material, is now used as standard). This has a higher $^{13}C/^{12}C$ ratio (0.0112372) than most natural substances, and thus most δ ^{13}C values are negative. Oxygen and hydrogen isotopes are defined relative to either standard mean ocean water (SMOW) or standard light Antarctic precipitation (SLAP). Nitrogen is reported relative to atmospheric air ($^{15}N/^{14}N$ = 0.003676) and sulfur relative to Cañon Diablo triolite ($^{34}S/^{32}S$ = 0.045005). Stable isotope profiling has been used for ecological and environmental research, archaeological research, detection of food fraud and the profiling of murder victims, and the victims of mass disasters.

$$\delta_S = \left(\frac{R_s - R_{std}}{R_{std}}\right) \times 1000\,[‰] \qquad (5.2)$$

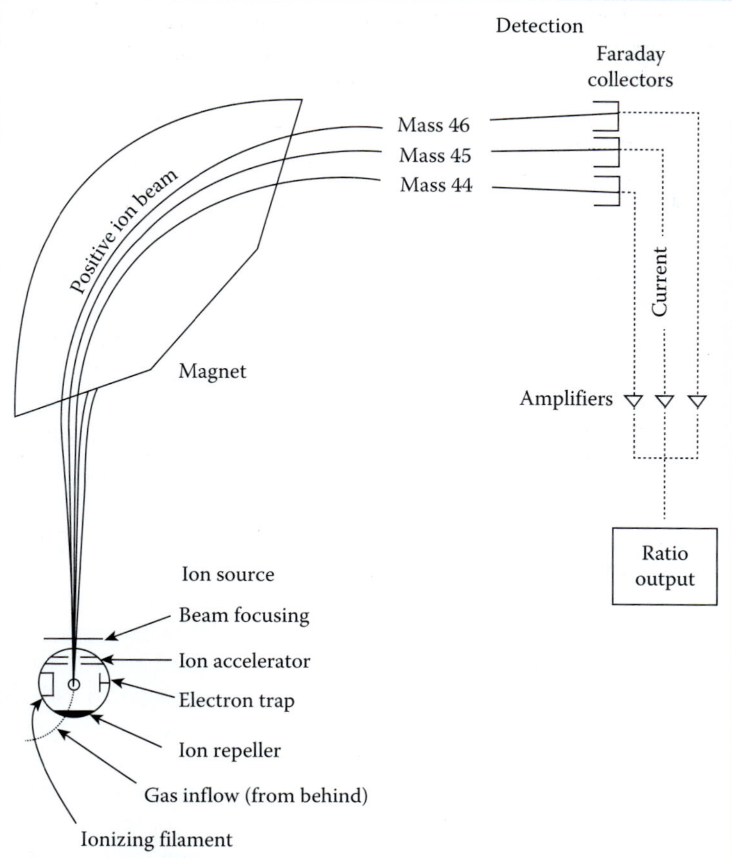

FIGURE 5.29 Schematic of an isotope ratio mass spectrometer.

Special mass spectrometers are needed to measure the small variations encountered in this field. An isotope ratio mass spectrometer gives accurate and precise measurement of isotopic abundance at the expense of the flexibility of scanning mass spectrometry. The material for analysis is converted to a gas (CO_2, N_2, SO_2) in such a way that the isotope ratio of the gas is representative of the ratio in the parent material. Once in the instrument, the molecular ions are separated by a magnetic field, filtered, and focused onto a dedicated detector (Figure 5.29). For CO_2, only masses of 44 ($^{12}C^{16}O_2$), 45 ($^{13}C^{16}O_2$), and 46 ($^{12}C^{18}O^{16}O$) need to be considered; the chances of having more than one less abundant isotope in the molecule are very small.

Why might stable isotope ratios differ in different materials? Some of the effects derive from thermodynamics and are associated with differences in IR absorption, molar volume, vapor pressure, and boiling and melting points. Others derive from primary kinetic isotope effects (KIEs). We will discuss these further in Chapter 8, but bonds to heavier and lighter isotopes may be broken at different rates in chemical and biological processes, because the bonds have different strengths.

^{13}C is fractionated in nature because different plants fix CO_2 by two distinct pathways. The Calvin–Benson or C3 pathway produces 3-phosphoglycerate (see **5.3**), and the Hatch–Slack or C4 pathway produces the C_4 diacid oxaloacetate (see **5.4**). Carbon isotopes are fractionated strongly and differently by the two processes. Glucose from C3 plants has $\delta\,^{13}C = -25‰$, and for C4 plants, this is −11‰. Most plants in Europe and North America are C3 photosynthesizers, with C4 plants found in hot dry climates, as the Hatch–Slack process uses water more efficiently. However, there are exceptions, including some major crops such as maize (corn), sorghum, millet, and sugarcane. In North America, these form a major part of animal feed, but this is not the case in Europe. Most sweetened items in North America use either sugarcane or corn syrup, but beet sugar is more important in Europe. So the ^{13}C isotope profile would be expected to be different in North Americans and Europeans.

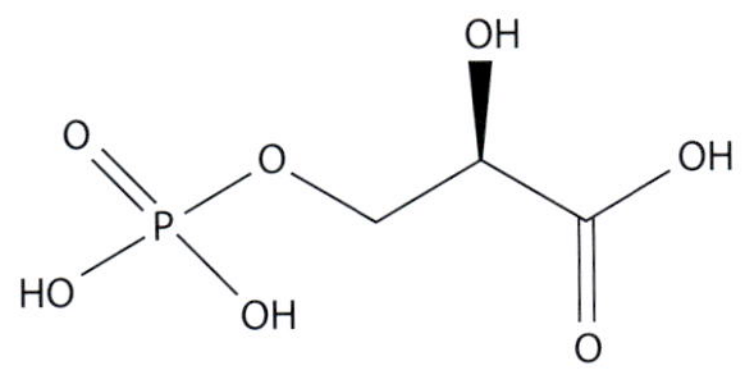

5.3, 3-phosphoglyceric acid

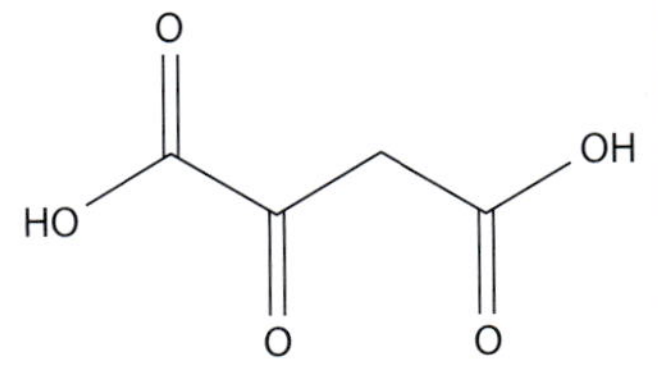

5.4, oxaloacetic acid

^{2}H and ^{18}O concentrations in hair reflect drinking water. Drinking water in early childhood is also the main source of ^{18}O in the dental enamel, bioapatite (teeth, unlike bones, are not remodeled over time). The concentrations of the heavier isotopes vary with latitude, distance from the sea, and altitude. This is a physical mass discrimination because of differences in rates of evaporation, condensation, and precipitation. Meteoric water (the technical name for water from the sky) has $\delta\ ^{2}H$ varying from +20‰ to −230‰. Less negative values are typical of coastal/equatorial regions, and more negative values of high latitudes, altitudes, and inland areas. Good data are available* (*e.g.*, Figure 5.30).

These techniques have been used in forensic science to identify murder victims. For example, in Ireland, the headless body of a male 35–40 years old was discovered. Skin color and physical characteristics suggested that he was of African/Afro-Caribbean or African-American ancestry. Stable isotope ratios in his hair, bones, and teeth were compared with those of people whose geographic origin and life history were known. From the hair, it was established that he had lived in Ireland for at least 9–12 months before his death. The femur was then tested; while ribs remodel in about 10 years, long bones such as the femur take about 20 years. The inner part of the bone reflects the most recent history, and the $^{18}O/^{16}O$ profile corresponded to local drinking water. However, the outer, older part of the femur suggested water with a δ value of

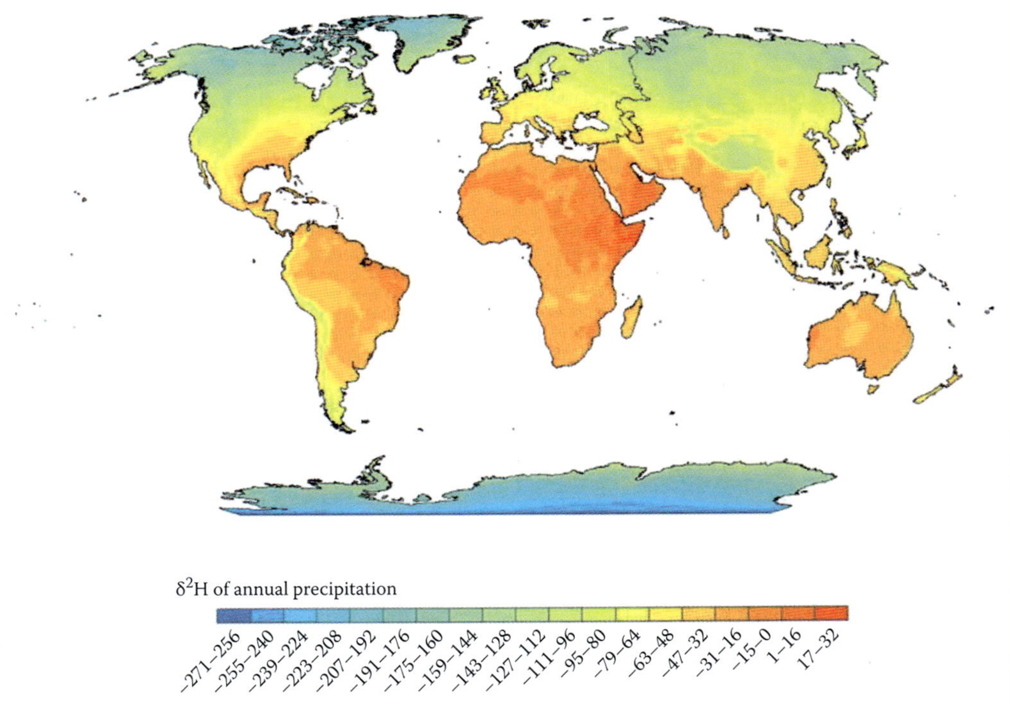

FIGURE 5.30 Deuterium isotope distribution in annual precipitation.

* http://wateriso.eas.purdue.edu/waterisotopes/; http://isohis.iaea.org.

δ −1.73‰. This signature is unusual and distinctive to coastal regions near the equator. Taken together, the results indicated that the victim came to Ireland some 6–8 years before his death, and the best fit for the old bone was the Horn of Africa. It was eventually established that the date was correct and that he originated from an area near the Kenya/Somalia border.

Other uses for stable isotope profiling have included establishing the origin of lamb meat and whether it was in fact grass fed as claimed, establishing the authenticity or adulteration of wine, tequila, honey, vanilla extracts, and royal jelly, finding the origins of drugs and explosives and climate change research. The techniques are also widely used in archaeological research to establish where individuals originated, where they lived and died, and patterns of migration. Information concerning their diet and agricultural practice can be inferred as well as the origins of the metal ores used in early toolmaking.

Key Points from Section 5.3

- Mass spectrometry is a technique for determination of the molecular weight of compounds. High-resolution instruments can measure this to three or four decimal places, and such results can be correlated with known data to give a molecular formula.
- Molecules in the gas phase are ionized and then passed through a magnetic field to separate species with different masses.
- Molecules containing chlorine or bromine can readily be identified as these halogens have two common isotopes in a known ratio.
- The nitrogen rule indicates that molecules with an odd molecular weight contain an odd number of nitrogen atoms.
- Fragmentation of the initially formed ions occurs in the mass spectrometer; the fragmentations give, where available, stable cations, either tertiary or stabilized by donation from a lone pair of electrons on a heteroatom.
- Fragmentation of carbonyl compounds generally occurs by breaking one of the bonds to the carbonyl group to give an acylium ion. Acylium ions readily lose CO.

5.4 UV/VISIBLE SPECTROSCOPY

All the other techniques that will be discussed here for the identification of organic compounds involve spectroscopy in various parts of the electromagnetic spectrum (Figure 5.31). We will specifically be looking at UV/visible spectroscopy, IR spectroscopy, and NMR spectroscopy, which uses frequencies in the radio region. From the basic relationships (5.3) and (5.4), we know that frequency and wavelength are inversely related, and the energy of the radiation increases with frequency. Absorptions in UV spectroscopy are described in terms of their wavelength in nanometers (abbreviated nm, 10^{-9} m, from Greek νᾶνος, dwarf). The range 100–200 nm is called the vacuum UV and is not accessible with normal laboratory instruments. The UV region is 200–400 nm, and (at least for humans) the visible region is 400–800 nm. In the UV/visible region, we are observing electronic transitions from bonding to antibonding orbitals. In energy terms, the visible region corresponds to 150–300 kJ mol^{-1} and the accessible UV region 300–600 kJ mol^{-1}. Note that these energy values are similar to those that bring about reactions:

$$c = \lambda \nu \tag{5.3}$$

$$E = h\nu \tag{5.4}$$

where

λ is the wavelength
ν is the frequency
c is the velocity of light
E is the energy
h is Planck's constant

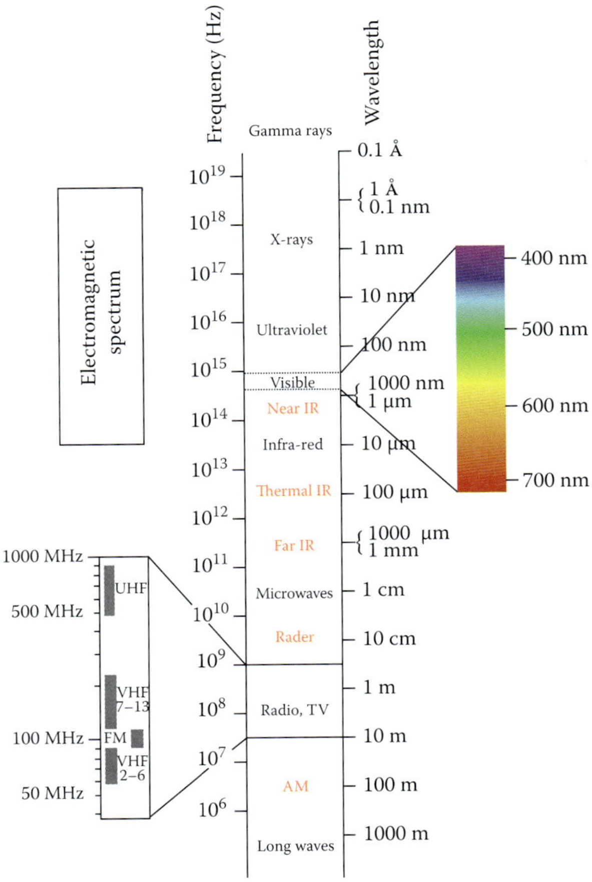

FIGURE 5.31 The electromagnetic spectrum.

Students sometimes ask "why do we still have to learn this?" No one really assigns structures from these spectra any more, surely? That's certainly true, though the availability of UV spectroscopy preceded IR or NMR spectroscopy by many years. However, UV spectroscopy is still important in other ways. For example, UV absorption is about 1000 times easier to detect per mole than NMR absorption. It is used in the following reactions where the absorbance of the product is different from that of the starting materials, and the timescale is fast, because it is so sensitive. It is widely used to study kinetics in biochemistry. Much quantitative analysis in organic chemistry uses HPLC with a UV detector, and it is the sensitivity of UV spectroscopy that makes HPLC such a sensitive technique. Uses have been as varied as checking the cleaning process for a production vessel to urinalysis for anabolic steroids in drug testing in athletics.

The basic equation of UV/visible spectroscopy is the Beer–Lambert law (5.5). A simpler way of expressing this is (5.6), where A is absorbance and c is concentration in moles per liter—what this says is that the amount of light absorbed depends on the path length, and the amount of the absorbing material—hardly surprising!

(5.5) $$\varepsilon = \frac{M}{cl}\log_{10}\left(\frac{I_o}{I_t}\right)$$

where

- ε is the molar extinction coefficient, a property of the material
- M is the molecular weight
- l is the path in cm
- c is the concentration in g L^{-1}
- I_o is the incident light
- I_t is the transmitted light

(5.6) $$A = \varepsilon cl$$

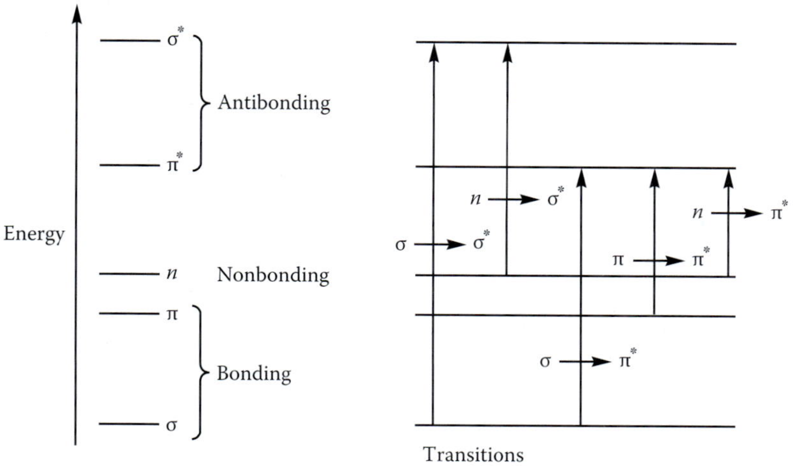

FIGURE 5.32 Transitions in UV spectroscopy.

The transitions that we can observe in the UV spectrum are shown in Figure 5.32. The only ones that will concern us in practice are the π to π^* and n to π^* transitions. Simple σ to σ^* transitions are invariably too high in energy and fall beyond 200 nm. Even the π to π^* transition of ethene is just out of range at 190 nm. However, for conjugated systems, the absorption is both more intense and at longer wavelength (Table 5.3). The appearance of the spectra is more complex than Table 5.3 would indicate; Figure 5.33 shows some real spectra of polyenes. The reasons for the "fine structure" observed relate

TABLE 5.3 π **to** π^* **Transitions in Conjugated Polyenes**

Polyene	λ_{max} (nm)	ε
$H_2C{=}CH{-}CH{=}CH_2$	217	18,000
$H_2C{=}CH{-}CH{=}CH{-}CH{=}CH_2$	258	35,000
$H_2C{=}CH{-}CH{=}CH{-}CH{=}CH{-}CH{=}CH_2$	296	77,000
$H_2C{=}CH{-}CH{=}CH{-}CH{=}CH{-}CH{=}CH{-}CH{=}CH_2$	334	118,000
$CH_3{-}(CH{=}CH)_6CH_3$	380	146,500

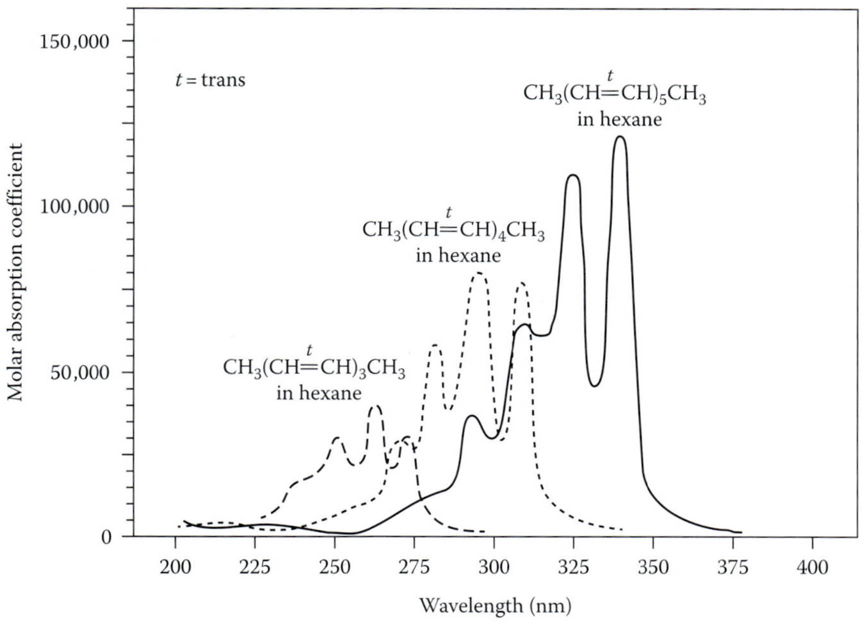

FIGURE 5.33 UV spectra of the polyenes $CH_3(CH{=}CH)_nCH_3$ for n = 3, 4, and 5. (Spectra taken from Naylor, T. and Whiting, M.C., *J. Chem. Soc.*, 3042, 1955.)

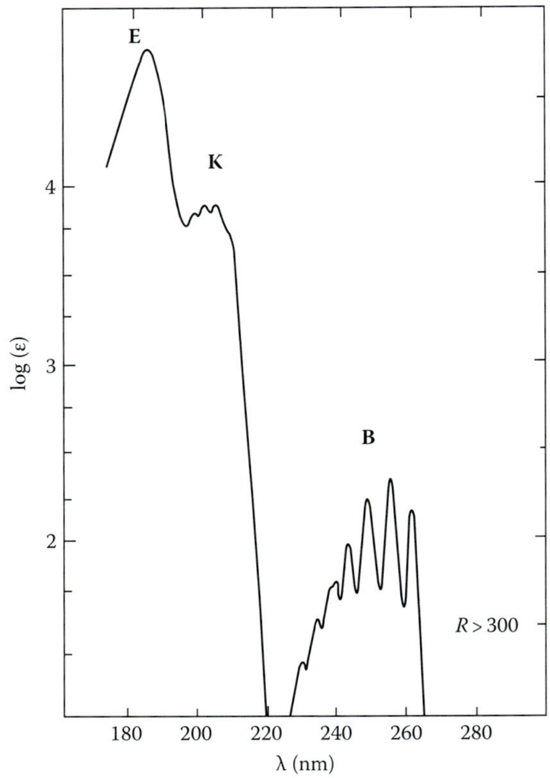

FIGURE 5.34 UV spectrum of benzene in heptane solvent.

to the fact that the molecules are also vibrating and rotating; when an electronic transition occurs, this happens too rapidly for the nuclei to move, but after the transition, they do adjust to minimize the total energy in the new electronic configuration. The theory is well understood, and you will study it if you take a more advanced physical chemistry course. It's also well understood why the gap between the highest occupied and lowest unoccupied molecular orbital (HOMO and LUMO) diminishes as the length of the conjugated system increases (again a topic for a more advanced physical or theoretical course), but the simple take-home message is that both λ_{max} and ε increase with conjugation.

Since benzene is highly conjugated, it would be expected to have an interesting UV spectrum—and indeed it does (Figure 5.34). There are three main bands, marked as E, K, and B (don't worry about the labels), with ε values of approximately 40,000, 2,000, and 250, respectively. Further conjugation, with additional aromatic rings, double bonds, or atoms bearing lone pairs, increases the values of both λ_{max} and ε. The strong UV spectrum of benzene has an important implication for parties given by chemistry graduate students. The so-called "absolute" ethanol contains benzene and it is very toxic. And 95 % ethanol is also toxic, but only in rather larger quantities. If a chemist offers you homemade punch, ask to see the UV spectrum of the ethanol before you take it.

UV spectra are also important in other compounds containing double bonds, particularly ketones. For a simple ketone, the π to π* transition is at 187 nm; although the ε value is high, typically 10,000, it is out of range of normal observation. The *n* to π* transition (from the lone pair of electrons on oxygen to the π* orbital of the double bond) lies between 270 and 300 nm, but the ε value is much lower, typically 20. More useful spectra are obtained from enones such as **5.5**. The π to π* transition is now in range at 219 nm and is a strong band (ε 7100). The *n* to π* transition is 324 nm but is still weak (ε 27).

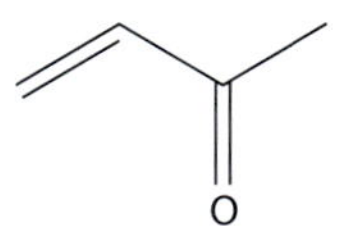

5.5

For an organic compound to absorb in the visible region (λ_{max} > 400 nm), and hence to be colored, significant conjugation is necessary. The color observed is what is not absorbed. So if blue is absorbed,

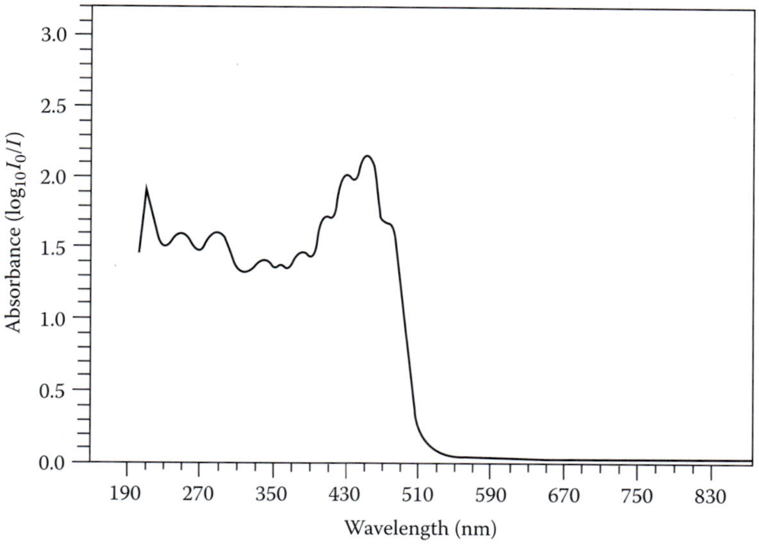

FIGURE 5.35 UV/Visible spectrum of β-carotene.

the compound will appear red or yellow, and if red/yellow is absorbed, the compound appears blue/green. Very-long-chain polyenes, such as carotene, **5.6**, have absorption that has been shifted out of the UV into the visible (blue) region of the spectrum (Figure 5.35). Carotene appears bright orange. Acid/base indicators also show color changes in the visible part of the spectrum, where changes in pH cause substantial alterations in chemical structure and hence in UV/visible spectra.

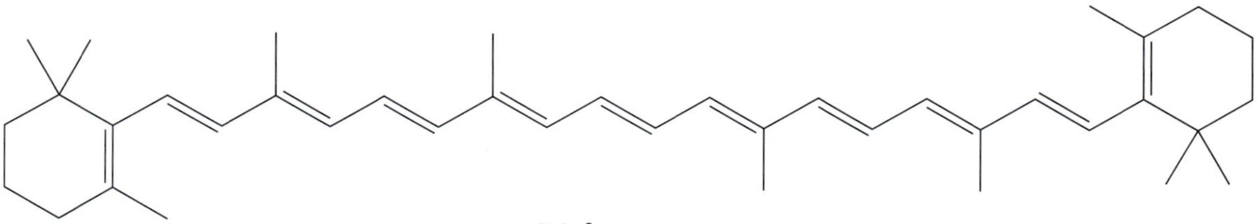

5.6, β-carotene

PROBLEM 5.5

The molar absorptivity for cytosine is 6100 M^{-1} cm^{-1} at 220 nm. If the absorbance of a sample is measured to be 0.2, for a 1 cm path length, what is the concentration?

PROBLEM 5.6

Rank the following compounds in order of decreasing λ_{max}.

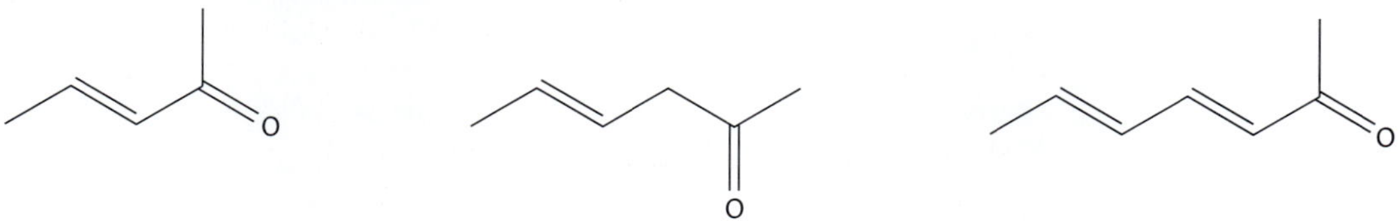

Solutions

5.5 This is a Beer's law problem

$$A = \varepsilon c l$$

where
- A is the absorbance
- c is the concentration in moles per liter
- l is the path in cm

Putting in the numerical values

$$0.2 = (6100\ \mathrm{M}^{-1}\ \mathrm{cm}^{-1}) \times c \times 1\ \mathrm{cm}$$

$$c = \left(\frac{0.2}{6100}\right)\mathrm{M} = 3.3 \times 10^{-5}\ \mathrm{M}$$

5.6 The more conjugated the molecule, the longer the wavelength for absorption. So

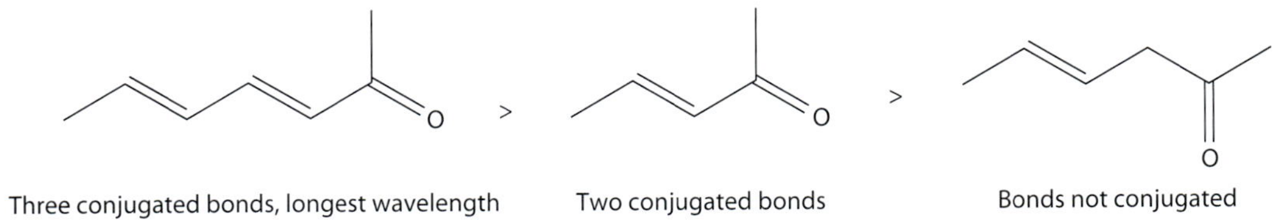

Focus: What Does Visible Really Mean?

Vision depends on light-sensitive cells in the retina, called rods and cones. The rods are sensitive to low light levels, but not to color, and there are various types of cone cell that are sensitive to specific light wavelengths. We class as visible light wavelengths from 400 to 700 nm, but this is simply the range of light to which humans (and old world primates) are sensitive. Other nonprimate species see the world in rather different colors. In the human eye, there are three types of cones (hence, we are classed as trichromats, Figure 5.36), responding to frequencies centered at 430, 545, and 570 nm. The cones for the blue part of the spectrum are less numerous than the other two types, and there is some evidence that blue color perception diminishes with age. In humans, and most terrestrial animals, the rods, used for night vision, are most sensitive at about 500 nm; slightly surprisingly, this is not optimized for moonlight.

The evolution of color vision suggests that multicellular organisms evolved monochromatic photoreceptors some 800 million years ago. The frequency to which these were sensitive is not clear; arguments have been made for values of 360 nm (in the modern UV) and 625 nm. By 450 million years, BP, evolution led to a four-cone system (tetrachromats) with sensitivity maxima at 360, 430, 560, and 625 nm. So the ancestors of all vertebrate species at one time had UV vision. Mammals evolved about 208 million years BP, but by 125 million years BP, they had lost both the UV and red sensors (maybe they were largely nocturnal during the age

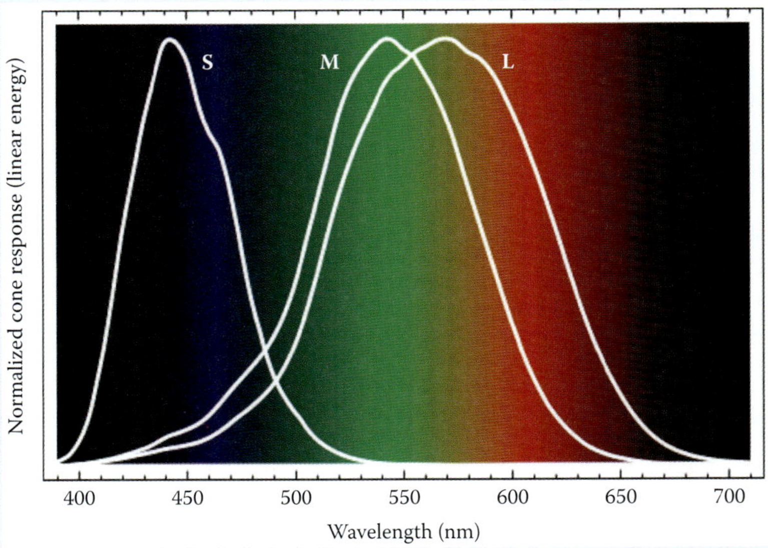

FIGURE 5.36 Frequency response of the three types of cones in the human eye.

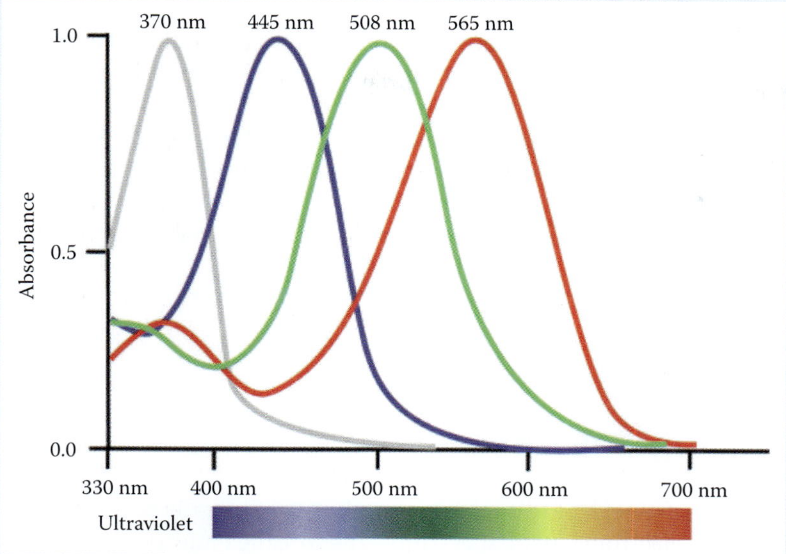

FIGURE 5.37 Absorbances of visual pigments of birds.

of the dinosaurs?); the common ancestor of all placental mammals was a dichromat. This is still true in most modern mammals—dogs have two types of cones receptive at about 420 and 570 nm, and this type of vision is also the norm for felines and ungulates. Some 40–50 million years ago, some primates split the yellow cone of their ancestors into two new types, blue/green and yellow/green. Humans and most old world primates have this adaptation, but it is rarer in new world primates, suggesting that the two adaptations occurred independently.

However, a wide range of species retained their four-color vision during evolution, and the UV vision is important to them in mate selection, foraging, and communication. This includes most birds (Figure 5.37), fish, amphibians, and insects. Most vertebrate UV sensitivity peaks around 360 nm, as in ancestral species, although some bird species have a receptor that peaks about 390 nm. Birds and some lizards use their UV vision to distinguish gender in species where male and female appear identical to the human eye. Many berries show high UV reflection, in contrast to leaves and bark that reflect little UV, allowing birds to forage more effectively. Many flowers have very distinctive patterns in the UV to assist in attracting pollinating species.* Interestingly, honeybees have only three types of cone, receptive at 544, 436, and 344 nm—they see UV light, but not the red end of the spectrum. Many fish have UV receptors and an array of skin pigments that reflect UV—hunters can connect with prey species.

Only a few mammals have any level of perception in the UV region. Some marsupials are known to be trichromats, their receptors being in the red, blue, and UV regions. Some rodents also have UV vision, which seems to be a specific adaptation to detect urine trails; rats have two types of cones with maximum efficacy at 359 and 510 nm. Recently, it has been established that reindeer can also use UV light; UV light is strongly reflected by snow but largely absorbed by the reindeer's main food source, lichen. Unlike most mammalian corneas (UV light damages both cornea and retina, so corneal pigments exclude it), up to 50 % of UV light is admitted, and the reindeer retina is sensitive to it. However, although it is not yet fully understood, the physiology of the process is different from that of other vertebrates; reindeer do not have a UV-specific receptor.†

Extension of vision into the IR region is rare. Rattlesnakes have IR detectors to allow them to locate prey, and some types of beetles and butterflies use perception of the near IR to detect forest fires. Some IR sensitivity has been observed in goldfish and mantis shrimp, but the reason for this adaptation is less obvious.

* Look at http://www.naturfotograf.com/UV_flowers_list.html, for some wonderful UV photography of flowers.

† C. Hogg, M. Neveu, K. Stokkan, L. Folkow, P. Cottril, R. Douglas, D.M. Hunt, and G. Jeffery, *J. Exp. Biol.*, **214**(12) (2011) 2014–2019.

Key Points from Section 5.4

- UV/visible spectroscopy involves absorption of electromagnetic radiation exciting electronic transitions between bonding (or nonbonding) and antibonding molecular orbitals. The most important transitions are *n* to π^* and π to π^*.
- Absorption is governed by the Beer–Lambert law, which relates absorption to concentration, path length, and the extinction coefficient ε, which is a property of the molecule.
- λ_{max}, the wavelength for the maximum absorption is measured in nanometers, and both it and ε increase with conjugation in molecules.

5.5 INFRARED SPECTROSCOPY

5.5.1 INTRODUCTION AND THEORY

When organic chemists speak of IR spectroscopy, they are generally referring only to a portion of the IR range of the electromagnetic spectrum—the so-called mid-IR with frequencies 4,000–400 cm^{-1} (the full range is 14,000–10 cm^{-1}). This is a longer wavelength and a lower frequency than the UV/visible region. The usual unit for IR spectroscopy is wavenumbers, cm^{-1}. You may wonder why UV spectroscopy uses a wavelength unit and IR spectroscopy uses a frequency unit, but there is no simple answer to this question—it's just how the techniques developed. In the early stages of the development of IR spectroscopy, wavelengths, in μm, were sometimes used (2.5–25 μm for the usual region of interest), but this is now rarely seen. The energy of the IR region of the electromagnetic spectrum corresponds to molecular vibrations, the bending and stretching of bonds. Although you will learn more about IR spectroscopy by running your own spectra in a lab course, this section aims to have you recognize single, double, and triple bonds, bonds to hydrogen, and some important functional groups. You should also know in a general way which bands will be intense.

Samples for IR spectroscopy may be solids, liquids, or gases, though organic chemists rarely encounter gaseous samples. If the compound you are analyzing is a liquid, a drop is placed between two sodium chloride plates (Figure 5.38). Glass is not transparent in the IR region of the spectrum, so all plates and cells for IR spectroscopy are made from highly polished sodium chloride. Therefore, your samples must be dry—otherwise the moisture dissolves the surface of the plate and they then have to be repolished. If the sample is a solid, it may be made into a paste by grinding it (in a mortar of nonabsorbent material such as agate) with a liquid hydrocarbon mixture called Nujol, which is then applied to the sodium chloride plates. Alternatively, it may be ground with potassium bromide and pressed into a thin translucent

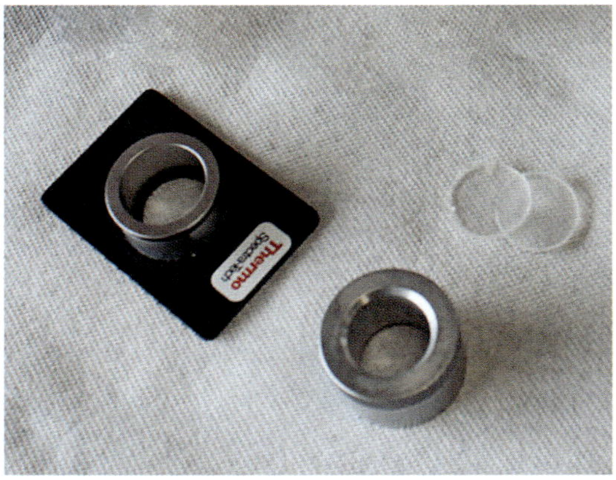

FIGURE 5.38 Sodium chloride plates and holder. (From Derry, P. and Barron, A., IR sample preparation: A practical guide, http://cnx.org/content/m43564/1.1/.)

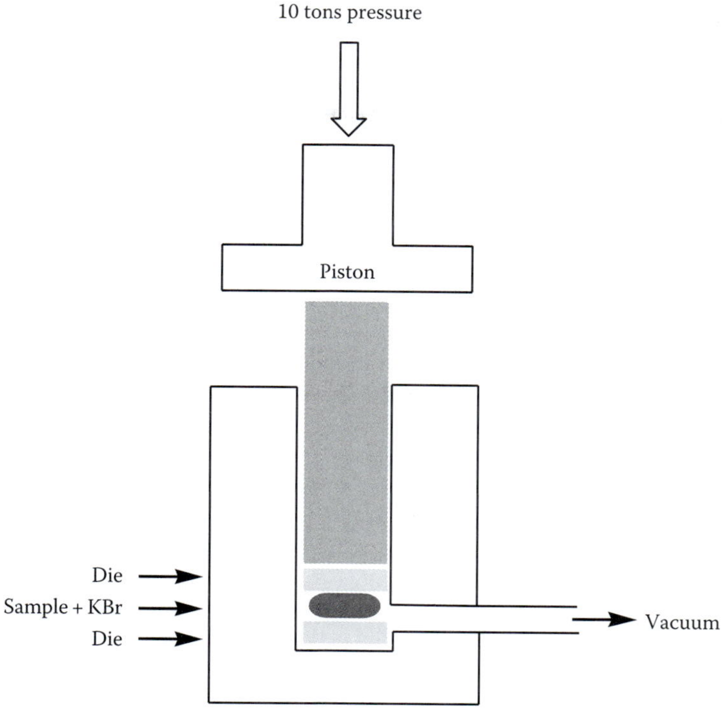

FIGURE 5.39 Hydraulic press to make KBr discs.

disc with a hydraulic press (Figure 5.39). The most recent technique uses attenuated total reflectance, where a solid material is pressed on the surface of a crystal for analysis. The full range of IR radiation is then passed through the sample, and the detector measures which frequencies have been absorbed. The system requires that a background scan is run, so that the frequencies from air may be subtracted (although neither oxygen nor nitrogen absorbs in the IR, both water and CO_2 do). Almost all modern spectrometers use a Fourier transform technique for data processing; while the electronics of the system are not immediately relevant here, the technique allows multiple scans of the sample to be added, allowing the identification of weak samples.

Agate pestle and mortar

We can exemplify the type of molecular vibrations we are studying in the IR spectrum by looking at simple systems. For example, in CO_2 (Figure 5.40), there are two ways in which the molecule can stretch, either symmetrically or antisymmetrically, and two bending modes at right

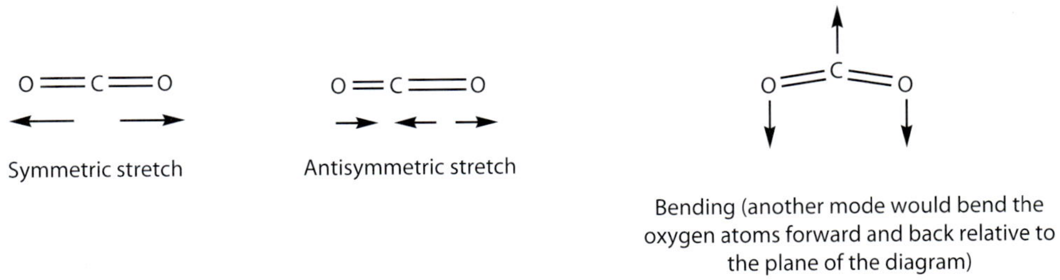

FIGURE 5.40 Vibrational modes of CO_2.

angles to each other. Figure 5.41 shows the vibrational modes for a CH_2 group, which at first sight seem alarmingly complex, especially when you think about how many CH_2 groups a molecule might have.

Stretching frequencies are generally at higher energy (higher frequency) than bending modes and are more characteristic of the functional groups of the molecule. All but the simplest molecules have spectra that are too complex for us to analyze all the bands in the IR region, but fortunately, we can extract useful information without doing that. Functional groups have *characteristic frequencies* that do not depend very much on the nature of the rest of the molecule. Figure 5.42 shows some typical frequencies of particular functional groups—looking at the IR spectra is an exercise in identifying the functional groups present in the molecule.

In considering stretching vibrations, we can view the bond as a spring; physics tells us that the frequency is given by Equation 5.7, where ν is frequency, k is the force constant or stiffness of the spring, and $\bar{m}$ (pronounced m-bar), the reduced mass, is $m_1m_2/(m_1 + m_2)$.* Some simple illustrations

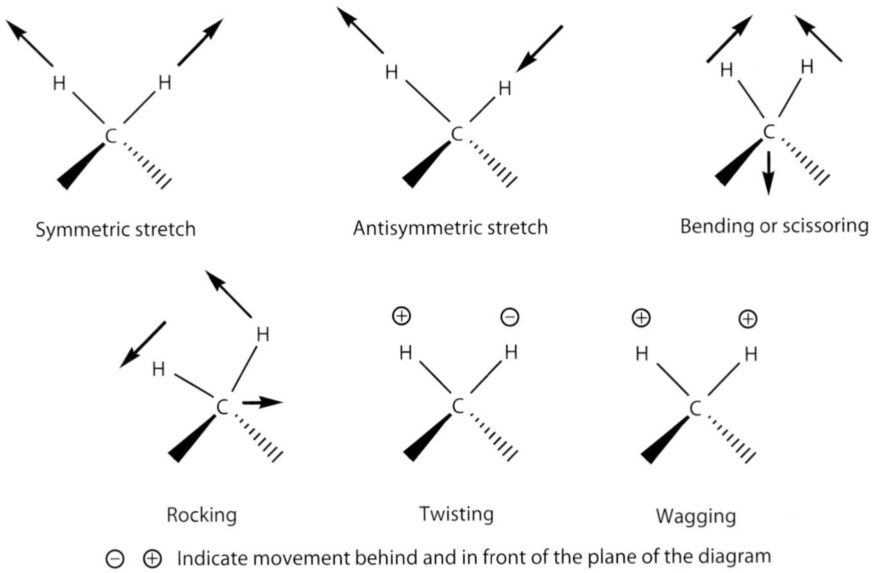

FIGURE 5.41 Vibrational modes of a CH_2 group.

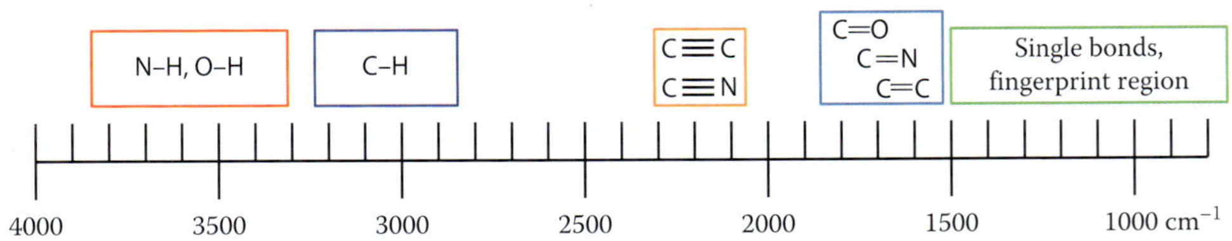

FIGURE 5.42 Characteristic frequencies in the IR spectrum.

* m_1 and m_2 are the masses of the atoms joined by the bond.

TABLE 5.4 Some IR Stretching Frequencies

Bond	ν (cm^{-1})	Bond	ν (cm^{-1})
C–I	500	C–O	1100
C–Br	500–700	C=O	1600–1800
C–Cl	600–800	C≡O	2200
C–F	1000–1400		

show that the heavier the atoms, the lower the frequency, but the stronger the bond, the higher the frequency (Table 5.4). In comparing the halogens, the main effect is the change in the reduced mass, although the force constant of the bond also changes. In comparing carbon–oxygen single, double, and triple bonds, the reduced mass is identical, but the force constant varies significantly.

(5.7) $$\nu = \frac{1}{2\pi}\sqrt{\frac{k}{\bar{m}}}$$

A good illustration is provided by comparing the spectra of $CHCl_3$ and $CDCl_3$. The C–H stretching frequency in $CHCl_3$ is 3020 cm^{-1} (Figure 5.43). If we assume that the C–H and C–D bond strengths are the same (in fact not quite true—the C–D bond is stronger, as we will see in Chapter 8), we can then use the reduced mass to predict where the C–D stretch should be seen (5.8). ν_{CD} = 3020/1.36 = 2220 cm^{-1}. The observed frequency (Figure 5.44) is 2260 cm^{-1}, which is an excellent agreement for a fairly crude model, with one known approximation ignored.

(5.8) $$\frac{\nu_{CH}}{\nu_{CD}} = \frac{\sqrt{(k/\bar{m})(CH)}}{\sqrt{(k/\bar{m})(CD)}} = \sqrt{\frac{\bar{m}(CD)}{\bar{m}(CH)}} = \sqrt{\frac{2\times12\times13}{1\times12\times14}} \sim 1.36$$

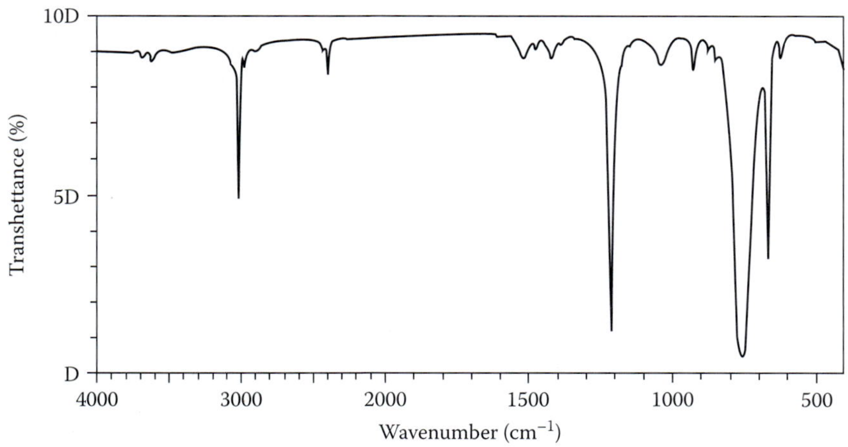

3683	84	1216	11
3618	84	1047	84
3019	47	1041	84
2976	81	929	81
2400	79	760	4
1520	84	871	31
1423	84	627	84

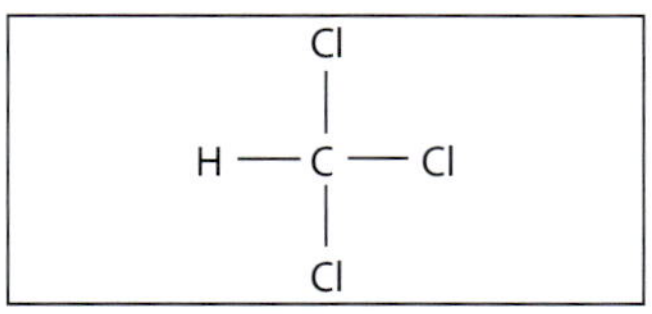

FIGURE 5.43 IR spectrum of chloroform, $CHCl_3$.

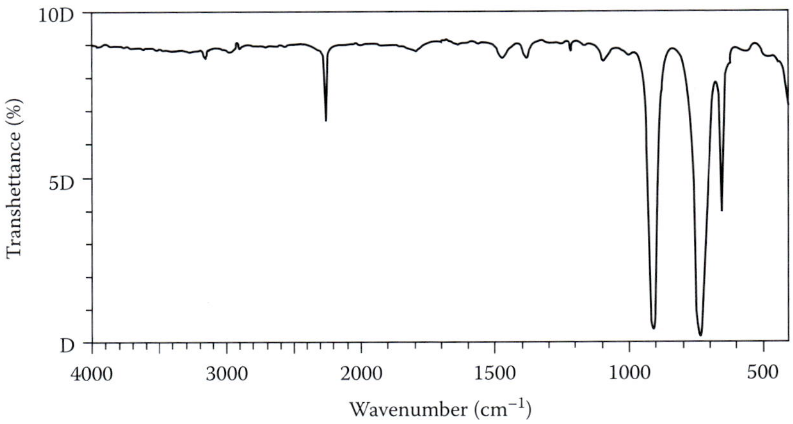

3164	86	662	39
2254	68		
1473	86		
1381	86		
1096	84		
907	7		
736	4		

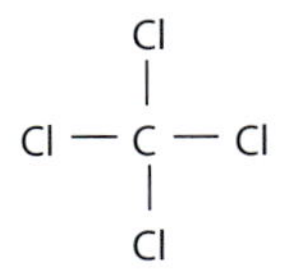

FIGURE 5.44 IR spectrum of deuterochloroform $CDCl_3$.

5.5.2 REGIONS OF THE INFRARED SPECTRUM

Although relatively few molecules contain triple bonds, the stretching frequencies are always between 2100 and 2300 cm^{-1}, a region in which there is little else, so they are easy to assign. Carbon–carbon triple bonds in alkynes give rise to bands around 2200 cm^{-1}, usually weak in intensity. Carbon–nitrogen triple bonds in nitriles give rise to bands around 2230 cm^{-1}, and these are usually intense (*e.g.*, Figure 5.45). Strong absorption in the IR is associated with a large change in the molecular dipole over the course of the vibration. This *selection rule* explains why neither molecular oxygen nor molecular nitrogen absorbs in the IR; the molecules have no dipoles, and no amount of stretching (which is all a diatomic molecule can do) will generate one. However, for most complex organic molecules, there is at least a small change in the dipole over a vibration, so there is almost always a band, even if it is a weak one. Dipole moments depend on both charge separation and the distance between the charges. The charge separation does not change over the course of a stretching vibration, but the distance between the atoms bearing the small charges does. Since nitrogen is more electronegative than carbon, in the carbon–nitrogen triple bond, there is a small positive charge on carbon and a small negative charge on nitrogen—so the dipole will change as the bond is stretched. However, for a carbon–carbon triple bond, this will only be true if one of the atoms bears an electron-donating or electron-withdrawing group, and even then the dipole may be quite small. Very weak bands are the norm, except for 1-alkynes.

The absorption for carbon–carbon double bonds is in the region 1620–1680 cm^{-1}, and this is also weak unless the alkene is substituted with an electron-withdrawing or electron-donating group. Conjugated double bonds absorb at the lower end of the range, and benzene derivatives show one, two, or three bands at about 1500, 1580, and 1600 cm^{-1} (Figure 5.46). You may be surprised to see the two bands between 1800 and 2000 cm^{-1}; these are overtone/combination bands from the strong C–H bending vibrations. The patterns of such absorptions are characteristic of the substitution pattern of the benzene ring and were at one time used to elucidate this, but this is now invariably done more easily and reliably by NMR spectroscopy, as we will see in Chapter 6.

Stretching frequencies of carbonyl groups are, when they are present, usually the dominant feature of an IR spectrum. The bond has a substantial dipole moment; hence, the absorption is intense.

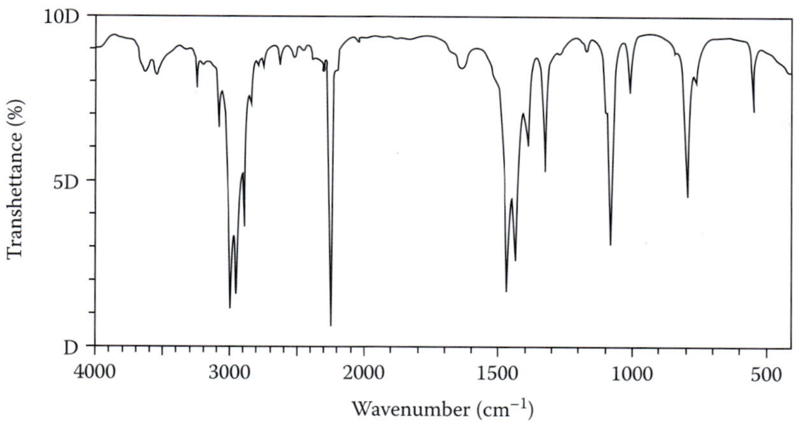

3632	79	2844	70	2299	81	1320	60	646	68
3548	79	2787	81	2248	4	1170	86		
3249	74	2749	81	2198	81	1075	29		
3082	64	2624	81	1639	81	1006	74		
2997	10	2521	84	1464	14	836	84		
2952	13	2461	86	1432	23	786	49		
2891	34	2460	86	1387	68	767	77		

FIGURE 5.45 IR spectrum of propanenitrile.

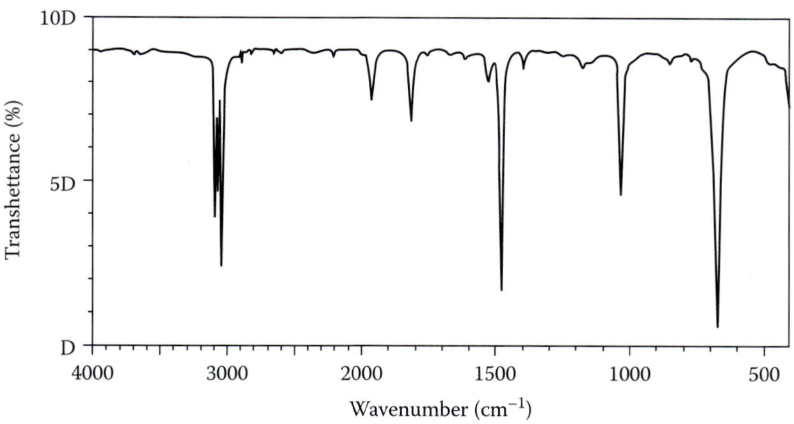

3091	42	1393	84
3072	49	1176	86
3036	27	1038	49
1961	77	674	4
1815	70		
1528	81		
1479	20		

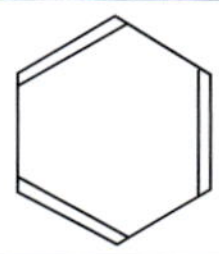

FIGURE 5.46 IR spectrum of benzene.

It's fortunate that even small variations in the position of the band can be interpreted in terms of the type of carbonyl group (Table 5.5). Anhydrides show two bands because the carbonyl vibrations are coupled, and they can vibrate in phase or out of phase (**5.7**, **5.8**); these are called the symmetric and antisymmetric stretches. The carbonyl stretching frequency can also be used to establish ring size if the carbonyl carbon atom is part of the ring. Cyclohexanone shows an entirely normal stretching frequency of 1716 cm^{-1}; the values for cyclopentanone, cyclobutanone, and cyclopropanone are, respectively, 1748, 1783, and 1850 cm^{-1}. Similar differences are noted for lactones (cyclic esters) and lactams (cyclic amides).

TABLE 5.5 Carbonyl Stretching Frequencies

Compound Class	Formula	ν (cm^{-1})
Acid chlorides	RC(=O)Cl	1790–1815
Acid anhydrides	RC(=O)OC(=O)R	~1820 and 1760
Esters	RC(=O)OR′	1750–1730
Aldehydes	RC(=O)H	1740–1720
Carboxylic acids	RC(=O)OH	1725–1700
Ketone	RC(=O)R′	1725–1700
Amide	RC(=O)NH$_2$	1680–1640

5.7, symmetric stretching mode

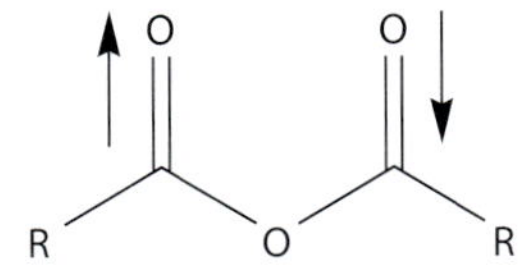

5.8, antisymmetric stretching mode

At first sight, the low value for the amide carbonyl might appear puzzling; the reason is that the carbon–oxygen double bond is weakened by the contribution of the resonance form in which that bond is single (**5.9**). Remember that the contribution of that resonance form also means that amides are planar, with important consequences for the structures of peptides and proteins. Because of the electronegativity of oxygen, this type of resonance form is less important for esters.

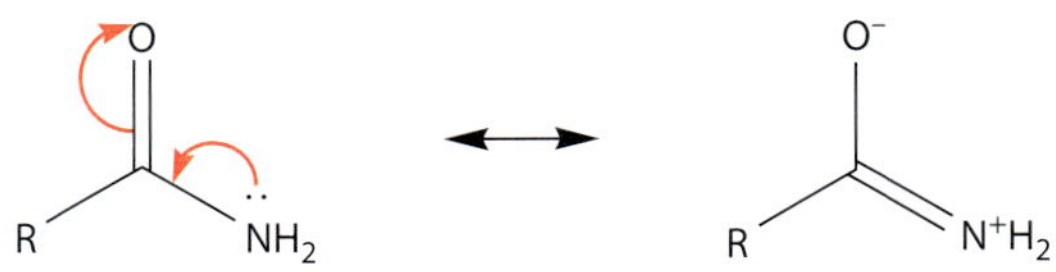

5.9

Conjugation of the carbonyl group, either with double bonds or an aromatic ring, reduces the stretching frequency of the carbonyl group, usually by 25–45 cm^{-1}. This is related to the contributions of the additional resonance forms available. All carbonyl groups have an important resonance form in which there is a negative charge on oxygen and a positive charge on carbon; as we will see in later chapters, this is a good predictor of their reactivity (Figure 5.47). However, with a conjugated carbonyl group, we must also consider an additional resonance form, **5.10**. This additional resonance form also has a carbon–oxygen single bond, and its contribution leads to weakening of the carbon–oxygen bond and hence a lowering of the stretching frequency.

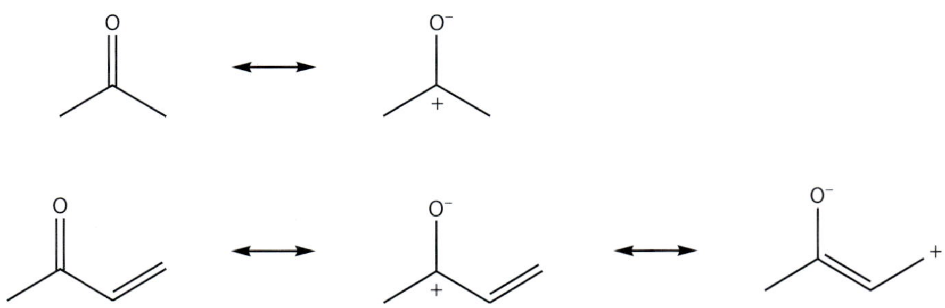

5.10

FIGURE 5.47 Resonance forms of carbonyl compounds and α,β-unsaturated carbonyl compounds.

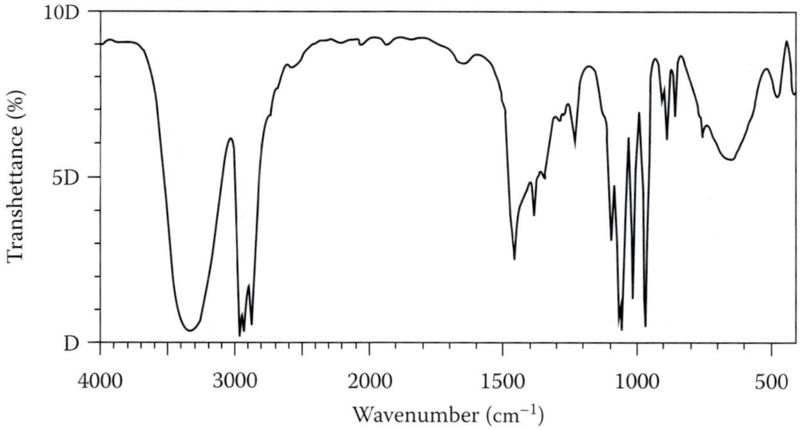

3333	6	1383	39	1017	16
2963	4	1346	50	969	8
2938	6	1293	68	905	72
2878	8	1236	60	888	62
1656	84	1100	32	858	68
1851	84	1069	9	755	82
1466	26	1066	6	479	74

CH_3—CH_2—CH_2—OH

FIGURE 5.48 IR spectrum of neat propanol.

Single bonds, other than those to hydrogen, are generally difficult to assign, falling in the fingerprint region of the spectrum, 1400–800 cm^{-1}. The fingerprint region is so called because two samples of the same material will give the same spectrum, but it will not usually be possible to assign every band. For example, the C–N stretch in simple alkylamines, R-NH_2, is in the range 1020–1150 cm^{-1}, with the value for an aromatic amine, Ar-NH_2, being 1250–1360 cm^{-1}. So these should be readily distinguishable? True, but unfortunately, the C–O single bonds of esters absorb between 1180 and 1300 cm^{-1} and those of ethers in the range 1050–1270 cm^{-1}. While we may be able to assign these bands, once the structure of the compound is known, they are unreliable diagnostically. This region is useful for the identification of a molecule, where a comparison sample is available, but usually not to establish the presence of particular functional groups.

Hydrogen is a very light atom, so we would expect the vibrations of bonds to hydrogen to have a high frequency—and X–H stretches are well away from the rest of the spectrum. Free, nonhydrogen-bonded O–H groups show a stretching frequency around 3600 cm^{-1}, with the band being sharp. Hydrogen-bonded O–H groups show a much broader band in the range 3600–3200 cm^{-1}. These absorptions are invariably intense, because of the large dipole moment of the O–H bond. How do we know which is which? Figures 5.48 and 5.49 show the IR spectrum of propanol, respectively, as a dilute solution in tetrachloromethane and as a neat liquid and as a dilute solution in tetrachloromethane. In the neat liquid, we would expect very extensive hydrogen bonding, and only the band due to the H-bonded OH is seen. However, diluted in CCl_4 (which is inert for this purpose), two bands are seen. The sharp band at higher frequency is assigned to non-H-bonded OH; in a dilute solution, not every propanol will be able to hook up.

Carboxylic acids always show strongly hydrogen-bonded O–H groups; this is a very broad and characteristic band. Even in dilute solution in a nonpolar solvent, the sociable carboxylic acids go around in pairs (**5.11**).

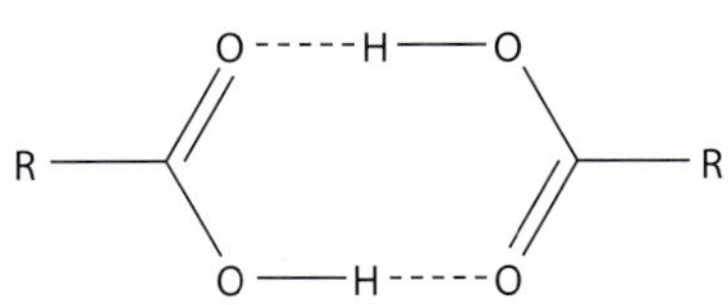

5.11

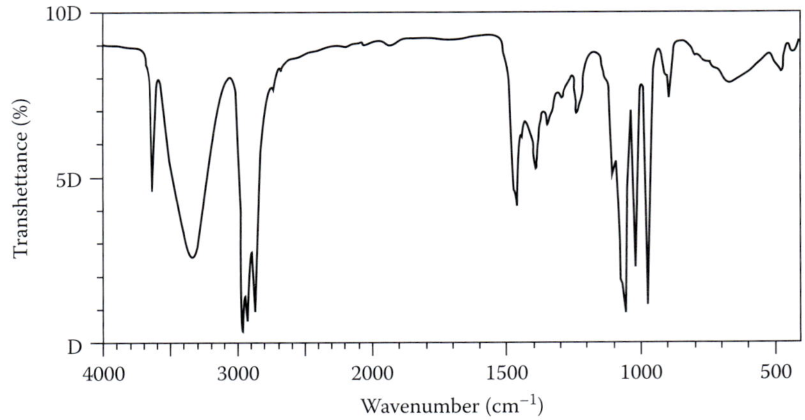

3637	46	1387	62	967	12
3340	26	1345	66	899	79
2964	4	1291	72	887	72
2937	8	1231	68	661	77
2878	10	1097	50	655	77
1465	44	1054	10	473	81
1466	41	1016	23	426	86

CH_3—CH_2—CH_2—OH

FIGURE 5.49 IR spectrum of solution of 1-propanol in tetrachloromethane.

The stretching frequency of N–H bonds is usually in the range 3300–3500 cm^{-1}, and the absorption is less intense than for OH (the N–H dipole is smaller). Primary amines, RNH_2, show two bands for symmetric and antisymmetric stretches (**5.12**) and R_2NH shows a single N–H absorption. The N–H absorptions in amides, which are strongly hydrogen bonded (**5.13**), are in the range 3200–3050 cm^{-1}.

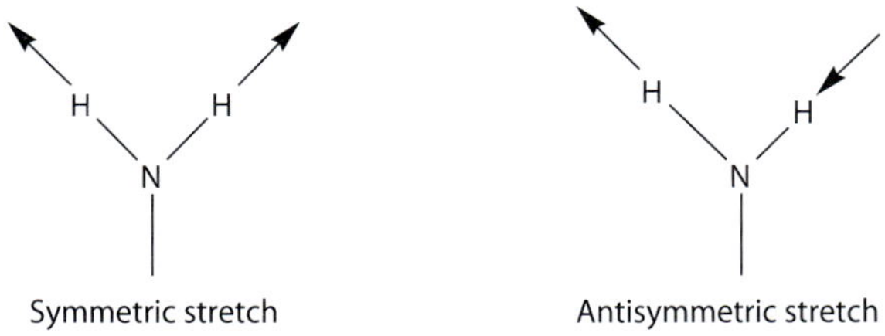

5.12

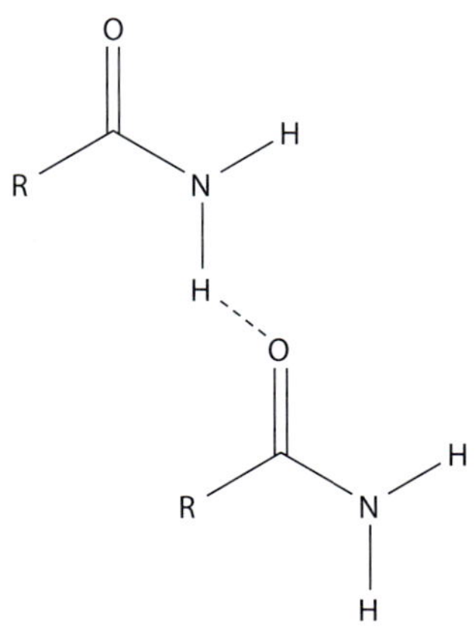

5.13

TABLE 5.6 C–H Absorption in the IR Spectrum

Bond Type	ν (cm^{-1})	Bond Strength (kJ mol^{-1})
sp^3 C–H	2850–2960	411.6
sp^2 C–H (alkenes)	3010–3090	432.6
sp C–H	3300	525
sp^2 C–H in RC(=O)H	2850, 2750	368–377

It might not be immediately obvious that C–H stretching vibrations have much to tell us in terms of organic structures—most organic compounds have C–H bonds! However, the bond strengths of sp^3, sp^2, and sp C–H are sufficiently different to be reflected in different IR absorptions (Table 5.6). You should be intrigued that there are usually two bands for the stretching of the C–H bond in aldehydes, when there is obviously no possibility for symmetric and antisymmetric vibrations. The explanation is a complex one, involving Fermi resonance between the C–H stretch and an overtone (double the frequency of the main band) of the C–H bending vibration in aldehydes at 1390 cm^{-1}, and is beyond the scope of this course. Many of the bands arising from CH_2 and CH_3 groups have structure, in that there are symmetric and antisymmetric vibrations, but it is rarely possible to interpret these in detail. The most important benchmark is at 3000 cm^{-1}; almost no sp^3 C–H stretches are above this value and almost no sp^2 or sp C–H stretches below it.

Examination of typical spectra of a number of functional group containing molecules will allow us to see what we should be looking for in these spectra. Remember when running your own spectra that what you will see is the desired compound plus any residual solvent or impurities; "real" spectra are often imperfect. Figures 5.50 and 5.51 show the spectra of the isomeric compounds, *n*-octane and 2-methyl heptane.

Since these molecules are both simple hydrocarbons with mostly the same bonds, the two spectra are very similar. There are various C–H stretching vibrations just below 3000 cm^{-1} and C–H bending vibrations between 1300 and 1500 cm^{-1}. We should note, however, that the fingerprint regions of the spectra, though similar, are not identical. Next, consider Figures 5.52 and 5.53 with the spectra of 1-octene and 1-octyne. In Figure 5.52, notice the band for an sp^2 C–H stretch above 3000 cm^{-1} and the C=C stretch at 1640 cm^{-1}. In the spectrum of 1-octyne, we note a very prominent and sharp sp C–H stretch at 3315 cm^{-1}, the C≡C stretch

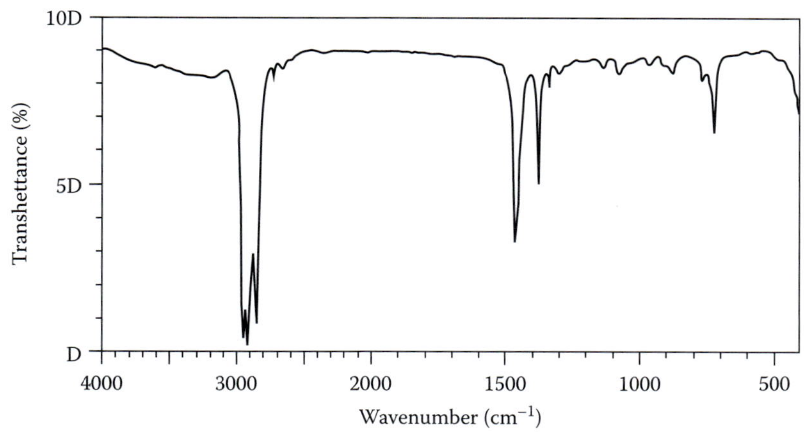

2968	6	1461	38	879	84
2925	4	1379	52	766	81
2874	16	1340	81	722	86
2867	11	1300	84		
2733	84	1135	86		
2870	86	1080	84		
1468	34	887	84		

FIGURE 5.50 IR spectrum of octane.

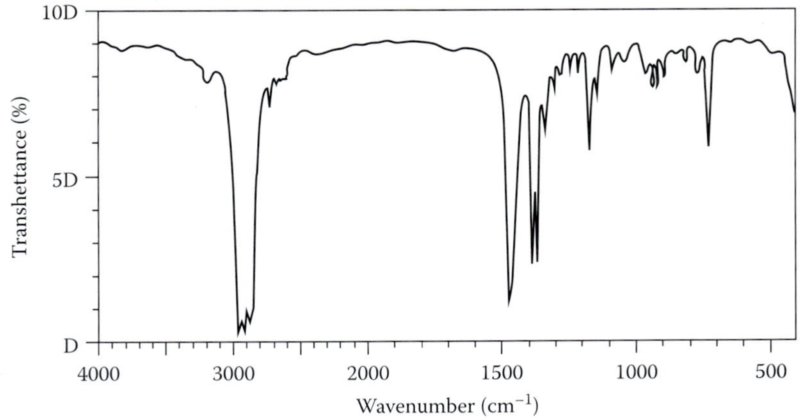

3182	79	2674	79	1282	81	1044	84	814	84
2958	4	1469	14	1276	81	1039	84	767	81
2926	5	1384	26	1238	84	961	81	758	84
2873	8	1379	33	1209	81	966	81	726	60
2857	10	1367	26	1171	58	933	77		
2730	72	1339	66	1143	74	918	77		
2720	72	1302	78	1084	84	893	79		

CH_3—CH—$(CH_2)_4$—CH_3
|
CH_3

FIGURE 5.51 IR spectrum of 2-methylheptane.

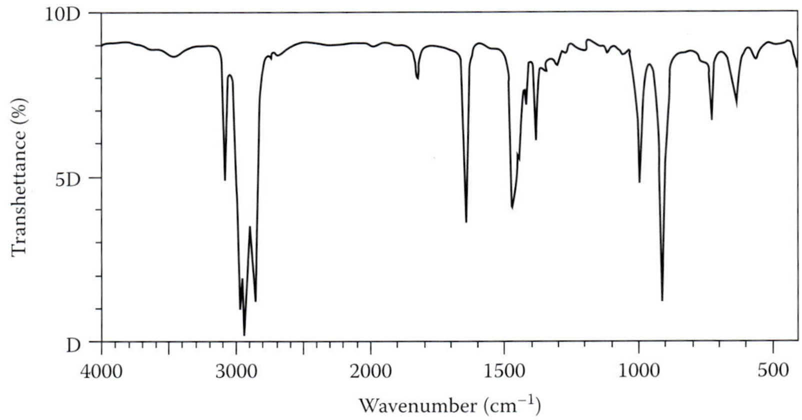

3079	60	1642	37	994	49
2998	57	1468	42	910	14
2959	12	1460	44	725	88
2928	4	1416	72	633	72
2874	23	1379	62	558	86
2857	13	1344	81		
1823	79	1299	84		

CH_2═CH—$(CH_2)_5$—CH_3

FIGURE 5.52 IR spectrum of 1-octene.

at 2120 cm^{-1} and the sp C–H bending vibration at 620 cm^{-1}. This demonstrates that the C≡C stretch is stronger in 1-alkynes than in internal alkynes—the two "ends" of the triple bond are more different. Figure 5.54 shows the IR spectrum of propylbenzene—note that there are both sp^2 and sp^3 C–H stretches, above and below 3000 cm^{-1}. A typical aromatic pattern of C=C stretches at 1500, 1580, and 1600 cm^{-1} is also notable as well as sp^2 and sp^3 C–H bending vibrations.

The spectra of hydrocarbons can be informative, but they are more notable for what they lack (absorptions attributable to functional groups) than what they demonstrate. Figure 5.55

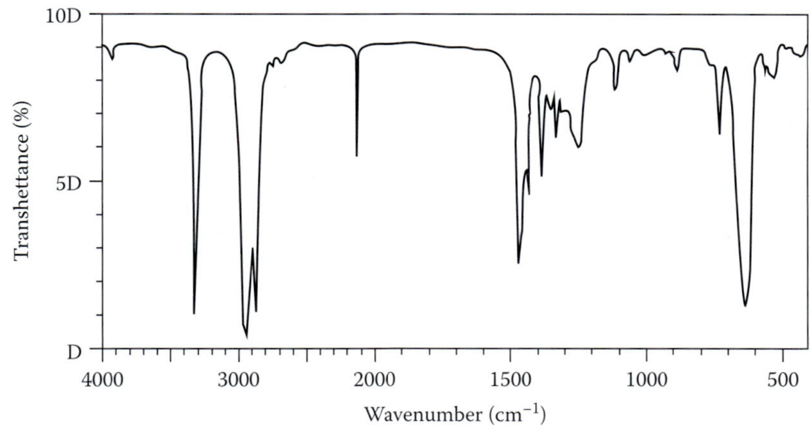

3926	86	2673	86	1327	64
3315	10	2120	58	1242	60
2959	7	1468	26	1113	77
2933	4	1461	30	886	81
2874	15	1432	46	725	64
2861	10	1380	52	830	13
2734	84	1346	72	628	81

$CH_3—(CH_2)_5—C{\equiv}CH$

FIGURE 5.53 IR spectrum of 1-octyne.

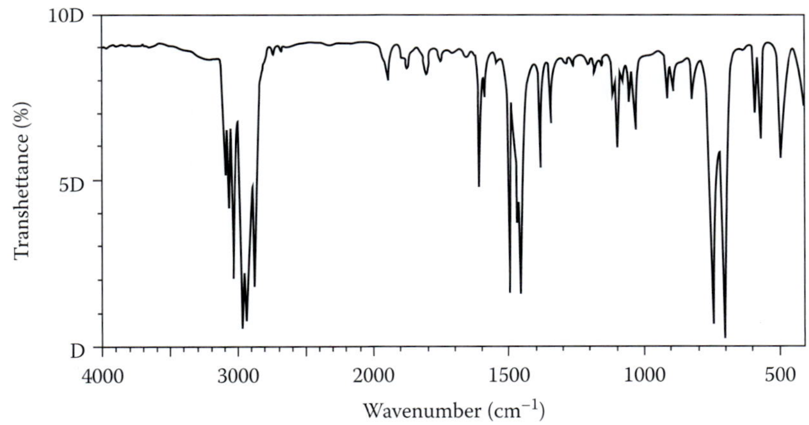

3108	77	2873	20	1466	41	1111	77	819	74
3085	52	1941	79	1453	18	1095	62	743	9
3064	43	1801	81	1379	55	1075	81	698	4
3028	22	1604	49	1341	70	1060	74	687	70
3002	62	1584	74	1258	84	1031	66	565	62
2960	8	1538	84	1203	84	909	74	490	57
2931	10	1496	18	1178	81	888	77	481	66

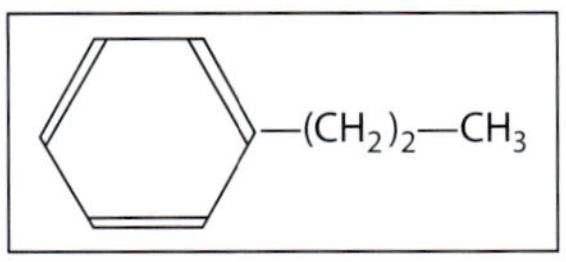

FIGURE 5.54 IR spectrum of propylbenzene.

shows the spectrum of cyclopentanol, run on the neat liquid. There is no mistaking that this is an alcohol, characterized by a very strong, broad O–H stretch; as one would expect in a pure liquid, the O–H groups are strongly hydrogen bonded. In Figure 5.56, the spectrum of propylamine shows the expected symmetric and antisymmetric stretches of the NH_2 group. It is worth noting the difference in intensity of these bands compared with that of the OH—although they occur in broadly the same region of the spectrum, they are generally easy to distinguish for this reason. Figure 5.57 shows dipropylamine, with a single N–H stretch. Even

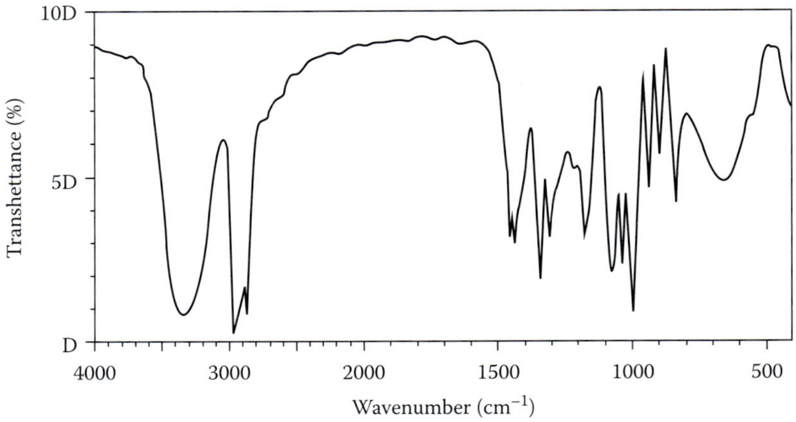

3346	10	1308	34	837	47
3335	10	1174	34	664	49
2961	4	1073	29	858	49
2873	12	1034	26		
1451	34	994	12		
1438	32	934	50		
1341	23	894	68		

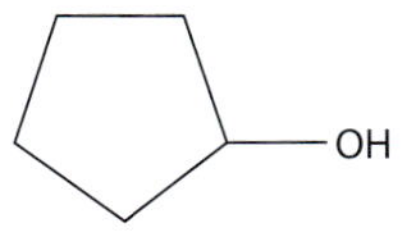

FIGURE 5.55 IR spectrum of cyclopentanol.

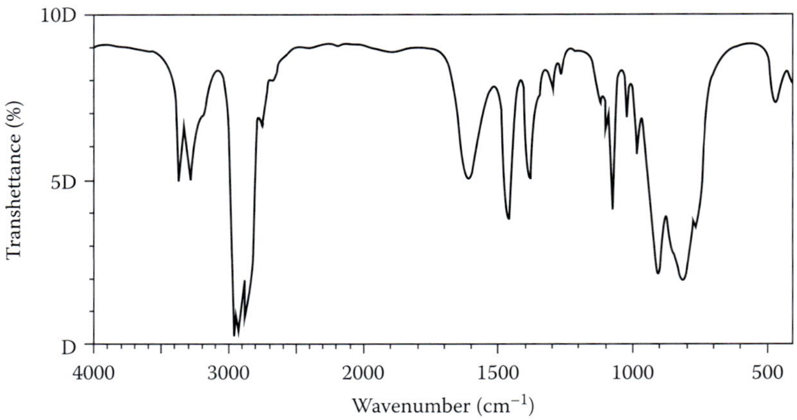

3369	62	1607	60	1119	74	763	36
3291	52	1465	35	1095	66	466	74
2959	4	1460	38	1072	42		
2930	6	1388	62	1019	70		
2876	10	1381	52	980	58		
2861	13	1298	79	904	22		
2766	66	1268	81	812	20		

FIGURE 5.56 IR spectrum of propylamine, $CH_3CH_2CH_2NH_2$.

allowing for hydrogen bonding, this absorption seems unusually broad, and the sample on which it was run was probably wet, so this is partly an O–H peak. This is an important point when you look at spectra you run on your own samples. In undergraduate teaching labs, there is not always enough time for compounds to be thoroughly dried, and some samples will still probably be slightly wet. You should not interpret a weak O–H peak as necessarily implying that you have an alcohol; if the molecule is an alcohol, the O–H peak will be one of the strongest in the spectrum.

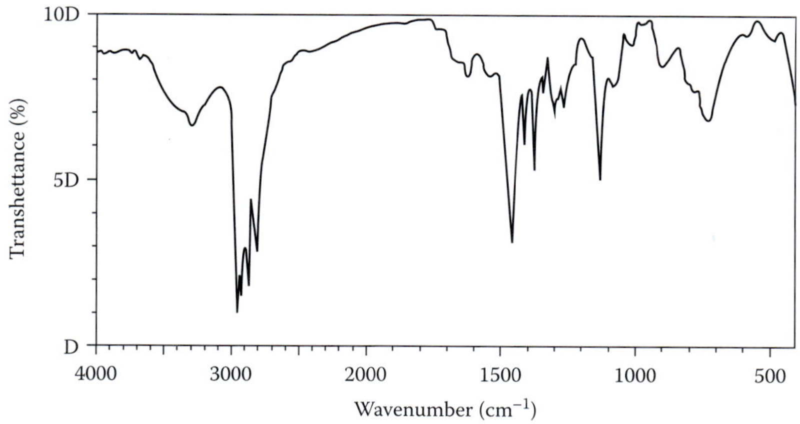

3682	79	1646	74	1266	64	899	77
3292	58	1540	74	1130	43	730	60
2959	4	1460	25	1086	70	485	84
2934	7	1411	63	1081	72		
2876	10	1379	47	1074	72		
2811	21	1341	70	1019	84		
1624	74	1301	64	1014	84		

CH_3—$(CH_2)_2$—NH—$(CH_2)_2$—CH_3

FIGURE 5.57 IR spectrum of dipropylamine.

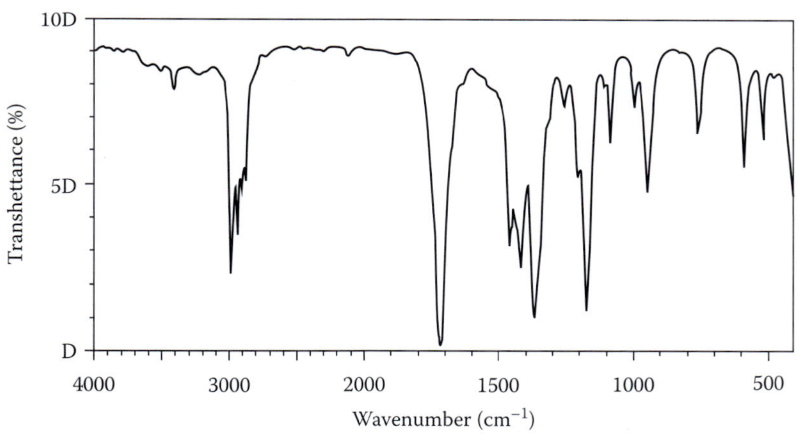

3611	84	1718	4	1173	14
3416	79	1461	34	1087	64
2981	26	1454	38	996	74
2964	44	1417	27	946	49
2940	37	1366	13	761	66
2909	50	1257	74	590	57
2883	62	1206	63	617	66

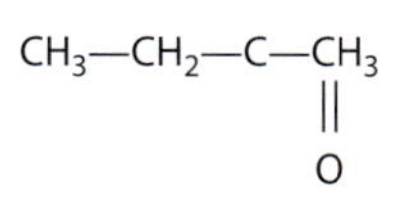

FIGURE 5.58 IR spectrum of 2-butanone.

Figure 5.58 shows the spectrum of 2-butanone, a typical ketone. The most prominent feature of the spectrum is the carbonyl stretching peak at 1718 cm^{-1}. The small peak in the O–H region of the spectrum should be disregarded; the sample is slightly wet, and it is often particularly difficult to dry lower aldehydes and ketones. Figure 5.59 shows the IR spectrum of butanal, again slightly wet. The carbonyl is, as always, the most prominent feature of the

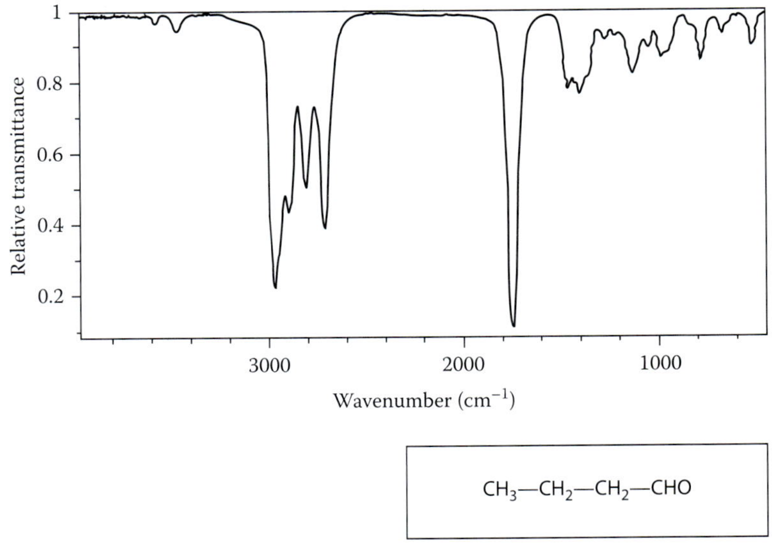

FIGURE 5.59 IR spectrum of butanal.

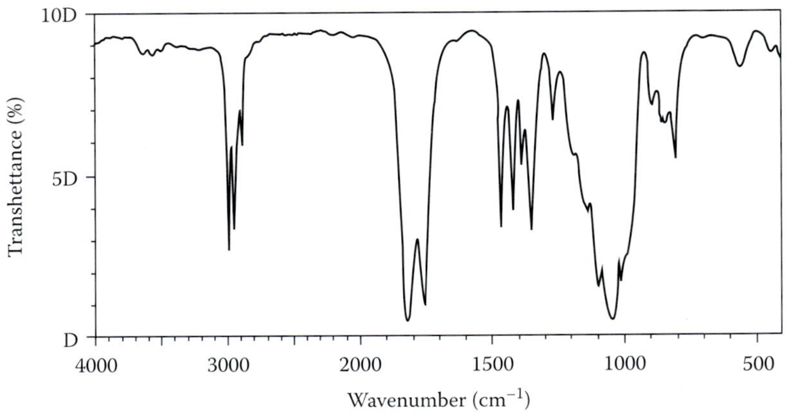

3635	84	1464	34	1042	6	806	63
3559	84	1418	39	1009	17	559	79
2989	26	1385	52	890	88	442	84
2948	33	1349	34	867	64		
2889	58	1267	66	851	64		
1820	4	1138	38	845	84		
1762	10	1096	16	840	64		

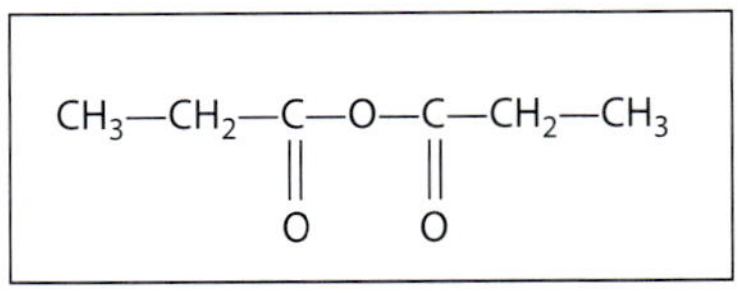

FIGURE 5.60 IR spectrum of propanoic anhydride.

spectrum, but you should also notice the two characteristic C–H stretching bands of the aldehyde C–H. Generally, the presence or absence of these absorptions will allow you to distinguish between aldehydes and ketones.

Figure 5.60 shows the IR spectrum of propanoic anhydride; the most notable features are the symmetric and antisymmetric carbonyl stretches. In Figure 5.61, the spectrum of propanoic acid, there are two features that identify this as a carboxylic acid. The strong carbonyl absorption is obvious, but the shape and size of the O–H stretch is very characteristic of acids. It is exceptionally strong and broad and effectively swamps and obscures the C–H stretches beneath it. Both this and the broad carbonyl absorption may be attributed to hydrogen bonding.

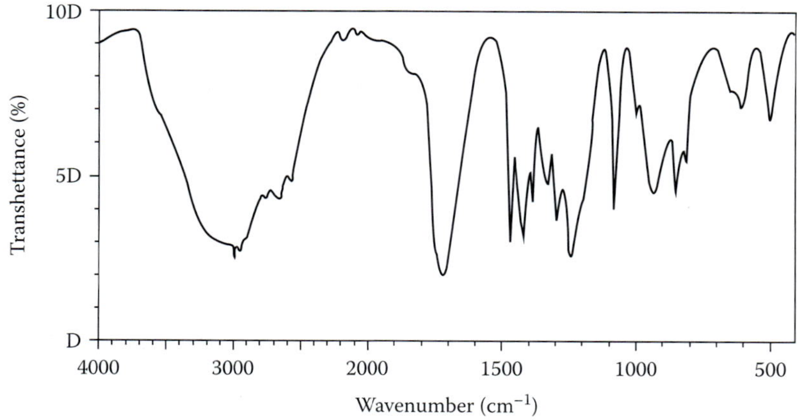

2986	10	1384	32	848	36
2948	12	1327	37	811	44
2860	31	1291	25	814	86
2666	38	1240	12	605	64
1716	4	1080	29	501	60
1467	18	995	82		
1416	19	933	34		

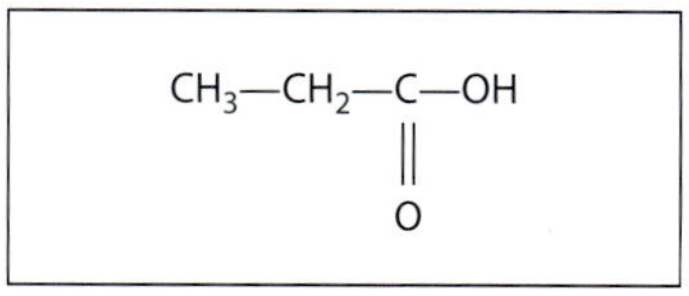

FIGURE 5.61 IR spectrum of propanoic acid.

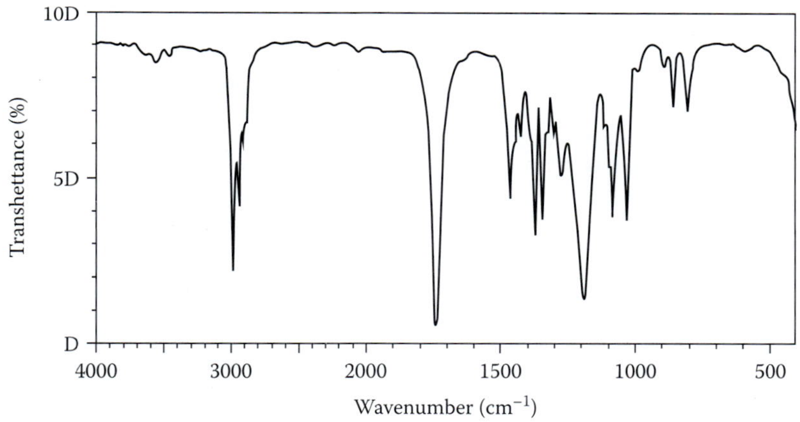

3664	84	1427	64	1035	37
2985	23	1372	34	991	81
2945	44	1347	38	895	84
2910	60	1301	64	860	72
1740	4	1275	52	807	72
1465	47	1192	12		
1461	60	1084	38		

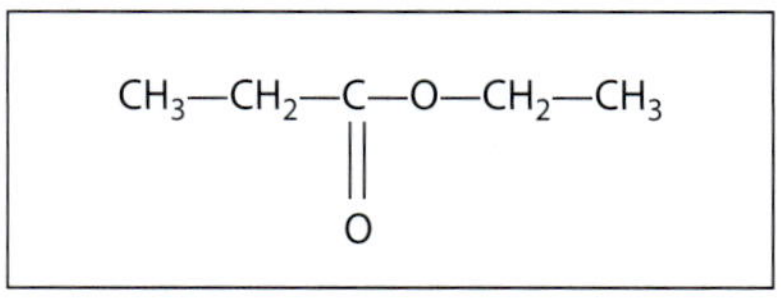

FIGURE 5.62 IR spectrum of ethyl propanoate.

Figure 5.62 shows the spectrum of ethyl propanoate; the carbonyl stretch at 1740 cm^{-1} is the most prominent feature. The spectrum of Figure 5.63, of octanamide, shows the expected symmetric and antisymmetric N–H stretches; the intensity of the bands is increased by the hydrogen bonding characteristic of the molecule. The carbonyl band might appear to be split, but this is not the case; the other absorption is an N–H bending vibration.

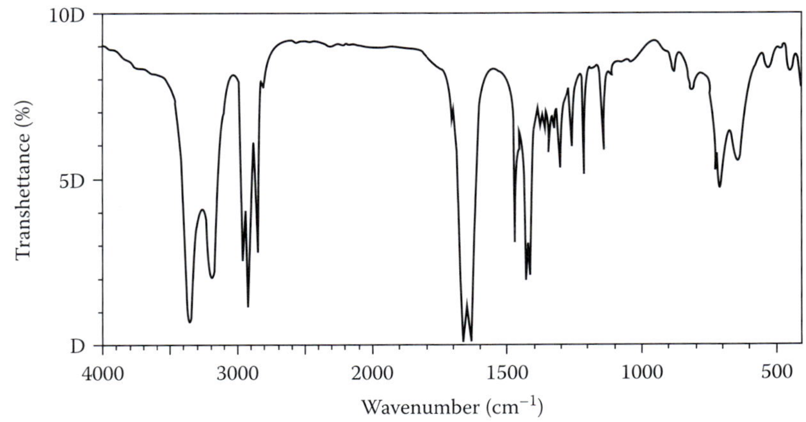

3362	9	2811	77	1376	68	1140	60	706	49
3190	22	1703	68	1363	66	1112	81	641	57
2956	26	1661	5	1341	58	881	81	520	84
2936	18	1633	4	1326	66	872	84	442	81
2923	14	1469	35	1304	55	810	77		
2871	44	1428	22	1259	82	798	79		
2866	30	1413	29	1214	63	724	69		

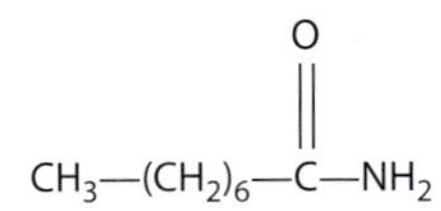

FIGURE 5.63 IR spectrum of octanamide.

Focus on Molecules in the Cosmos

The IR portion of the electromagnetic spectrum is able to penetrate heavy dust around newly formed stars, so that the IR spectroscopy allows observations that would not be possible optically. IR spectroscopy can provide fingerprints of molecules in galaxies far, far away. Interstellar material accounts for some 20 %–30 % of the mass of the universe, but only 0.1 % of this is elements such as carbon, nitrogen, and oxygen (most is hydrogen, with some helium). Much of these elements are ejected from old or dying stars, but dense interstellar clouds are also the birthplaces of stars and planetary systems. Both airborne and space observatories have been used to collect IR spectroscopic data. All types of spectra may be Doppler shifted on observation because distant galaxies are receding from us.

More than 100 molecules (including radicals and cations) have been identified from the IR spectra of materials in interstellar media. Some of these are produced in space even at the low temperatures that prevail there, but others are thought to be produced on the surfaces of dust or ice particles. Poly-ynes such as **5.14–5.18** are detected in the region of red giant stars and are thought to be formed in regions where hydrogen is scarcer than usual. These poly-ynes are probably the precursors of the many polyaromatic hydrocarbons (PAHs) common in cosmic dust (Figure 5.64). Conspicuous IR emission bands are seen at 3030, 1610, 1280, 1150, and 890 cm^{-1}; the molecules are excited in regions of high UV light intensity and emit in the IR.

5.14

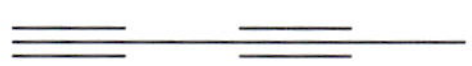

5.15

5.16

5.17

5.18

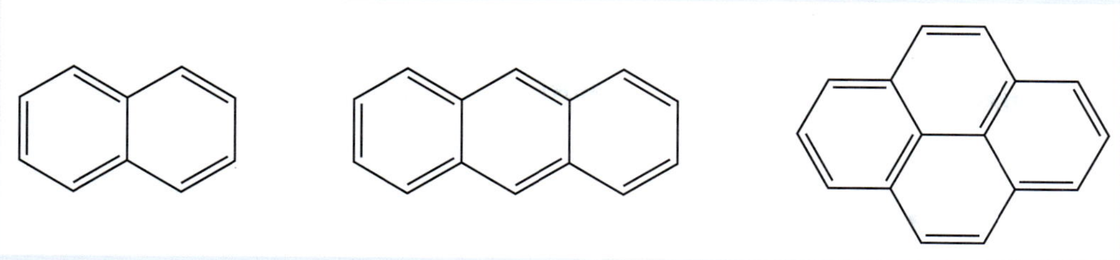

FIGURE 5.64 PAHs detected in cosmic dust.

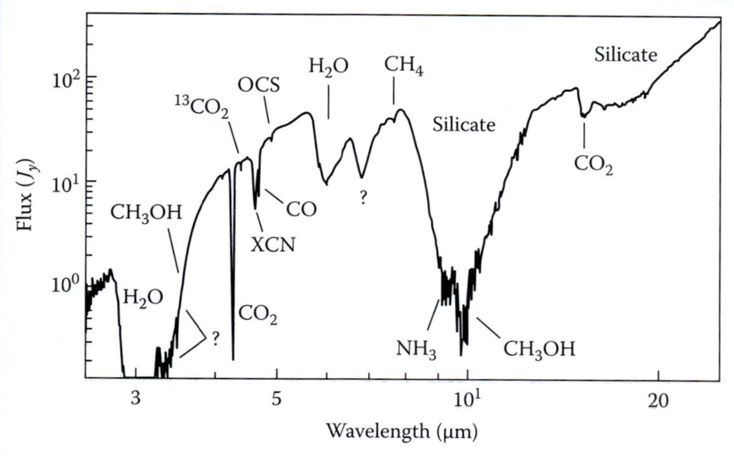

FIGURE 5.65 IR spectrum of light toward protostar W33A.

A wide variety of molecules containing oxygen and nitrogen have also been detected including CO, CH_4, $H_2C{=}O$, MeOH, $Me_2C{=}O$, HCOOH, C_2H_4, C_2H_6, HCOOMe, EtOH, HCOOEt, $H_2C{=}C{=}O$, CH_3CHO, CH_3COOH, $HOCH_2CH_2OH$, CH_3NH_2, CH_3CN, CH_3CH_2CN, H_2NCHO, HC(=O)CN, $CH_3C({=}O)NH_2$, and various species where a terminal alkyne is replaced by a cyanide group. Figure 5.65 shows the IR spectrum looking toward the protostar W33A*; numerous small molecules can be identified.

* E.L Gibb, D.C.B. Whittet, W.A. Schutte, A.C.A. Boogert, J.E. Chiar, P. Ehrenfreund, P.A. Gerakines, A.G.G.M. Tielens, E.F. van Dishoeck, and O. Kerkhof, *Astrophys. J.*, **536** (2000) 347–356.

PROBLEM 5.7

(a) Give an example of a compound which would show a strong IR absorption at 1715 cm^{-1} and a peak in its mass spectrum at M-15.
(b) How would you distinguish the compounds PhC≡N and PhC≡CH by IR spectroscopy? What other spectroscopic method would be useful?
(c) Explain why the C=C stretch in the IR spectrum of allyl alcohol is stronger than that in propene.

Solutions

(a) Any methyl ketone, for example, 2-pentanone.
(b) The easiest way to do this would be to see which has the sp C–H stretch at approximately 3300 cm^{-1}; this will be present in the alkyne but absent in the nitrile. The triple-bond stretching band for the more polar nitrile group should be more intense than that for the alkyne triple bond, but a qualitative distinction is less easy to assess than a presence/absence dichotomy. Mass spectrometry would also distinguish the compounds.
(c) In allyl alcohol, there is a polar group attached to the double bond, and this increases the bond dipole, relative to a hydrocarbon. Since the selection rule for IR spectroscopy requires a change in dipole over the molecular vibration under consideration, the C=C stretch will be a more intense absorption if the bond bears polar substituents.

Key Points from Section 5.5

- In IR spectroscopy, electromagnetic radiation is absorbed at energies that correspond to molecular stretching and bending vibrations. Absorptions are described in terms of a frequency unit, the wavenumber (cm^{-1}), and the range of interest is 400–4000 cm^{-1}.
- Samples may be solids, liquids, or gases. Glass is not transparent to IR radiation, so sample cells are made of polished NaCl or KBr.
- The frequency of bond stretching may be modeled approximately using Hooke's law for mechanical springs. Higher frequencies are associated with strong bonds and/or light atoms. Stretching frequencies are generally higher than bending frequencies.
- Stretching and bending frequencies for particular functional groups are characteristic of the group and depend only to a limited extent on the rest of the molecule. Many tables of data exist for correlation.
- The selection rule for IR spectroscopy is that for absorption to occur, the molecular dipole must change during the course of the vibration. Thus, simple diatomic molecules such as O_2 do not absorb. The most intense absorptions involve polar functional groups—carbonyl groups absorb more strongly than alkenes, and nitriles more strongly than alkynes.
- The absorptions of carbonyl groups are all very strong and may be analyzed in detail as to the type of carbonyl group present.
- Hydrogen bonding, for O–H and N–H groups, results in broadened signals and slightly lowered frequencies.
- Other than bonds to hydrogen, the stretches of single bonds are difficult to assign specifically. They mainly occur in the fingerprint region and are useful in establishing that two materials are identical.

REVIEW PROBLEMS

Note that this is a shorter than usual selection of problems, but these topics will be reviewed again in the next chapter (and indeed in many subsequent ones), when they are combined with NMR spectroscopy in the elucidation of organic structures:

1. The molecular ions for polyhalogenated compounds are complicated. Taking the natural abundance of ^{81}Br and ^{79}Br as 50 % each and the abundance of ^{35}Cl and ^{37}Cl as 75 % and 25 %, respectively, calculate the masses of the molecular ions of compounds of the following formulae. What are the relative amounts of each ion?
   (a) $CHCl_3$
   (b) CH_2BrCl
2. Give an example of each of the following:
   (a) A molecule that would show an IR absorption at 1715 cm^{-1} and a loss of 43 mass units in the mass spectrum
   (b) A molecule that would show an IR absorption at 1715 cm^{-1} and a peak in the mass spectrum at M-29
   (c) A compound that would show strong IR absorptions at 1670 and 1630 cm^{-1}
   (d) A compound that shows two medium intensity IR absorptions in the region of 3250 cm^{-1}
   (e) A compound that shows no IR bands between 2500 and 3000 cm^{-1}, several closely spaced bands between 3000 and 3100 cm^{-1}, and a strong broad band at 3400 cm^{-1}
3. An 0.018 μM solution of 1,3,5-hexatriene has an absorbance, $A = 0.63$, with a path length of 1 cm. What is the molar absorptivity of 1,3,5-hexatriene?
4. Define or describe each of the following terms as applied to recrystallization (not all discussed in this chapter but will be useful to you in the lab course!):
   Solution recrystallization
   Mixed solvents
   Solvent selection
   Decolorization
   Hot gravity filtration
   Vacuum filtration
   Fluted filter paper
   Mother liquor
   Filtrate
   Solute
   Seeding
   Percent recovery
5. In hot gravity filtration, what might happen if the filter funnel is not preheated before the solution is poured through it?
6. When you have collected one batch of crystals, a second crop of crystals may sometimes be obtained by concentrating the filtrate under vacuum and cooling. Why is this crop of crystals probably less pure than the first crop?
7. The solubility of benzoic acid in water is 2.65 g in 100 cm^3 at 79 °C and 0.29 g in 100 cm^3 at 20 °C. What is the minimum volume of water in which 1 g of benzoic acid can be dissolved at 79 °C?
8. When you are collecting your crystals by vacuum filtration, the solvent used to wash them is generally specified to be *cold*. Why?
9. How would you use IR spectroscopy to distinguish between the following pairs of compounds?

(a) $(CH_3)_3N$ and $CH_3CH_2NHCH_3$

(b)

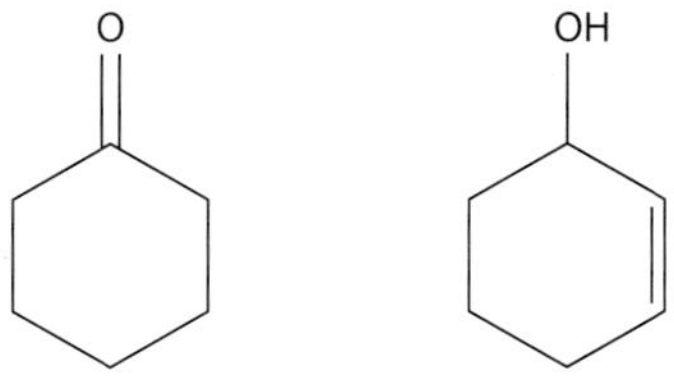

(c) $CH_3CH_2CH_2COOH$ and $CH_3CH_2COOCH_3$

(d)

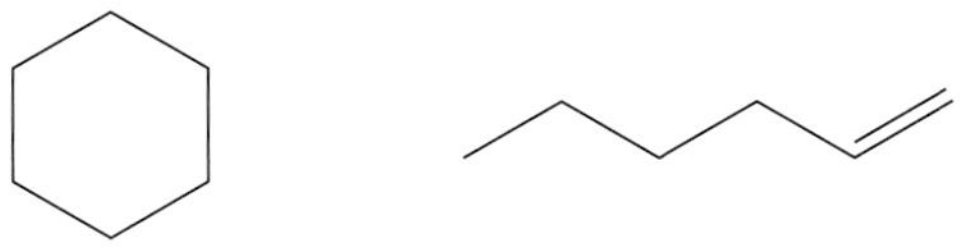

(e) CH_3CH_2COOH and $HOCH_2CH_2C(=O)H$

(f) $CH_2=CHOCH_3$ and CH_3CH_2CHO

(g)

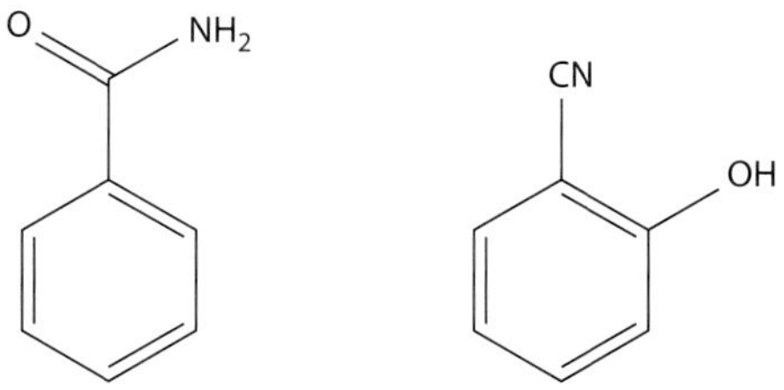

10. Write down as many isomeric structures as you can with the formula C_3H_6O. If the compound you have shows a strong IR absorption at 1715 cm^{-1}, what can you conclude concerning your sample?
11. Oxidation of 1-butanol to give butanal produces a sample that is mostly butanal but that contains small amounts of an impurity with peaks in its mass spectrum at *m/z* 88 and 72. What is the likely impurity?
12. How would you follow the reduction of cyclohexanone to cyclohexanol, using IR spectroscopy?
13. Arrange the following compounds in order of their λ_{max} in their UV spectra:

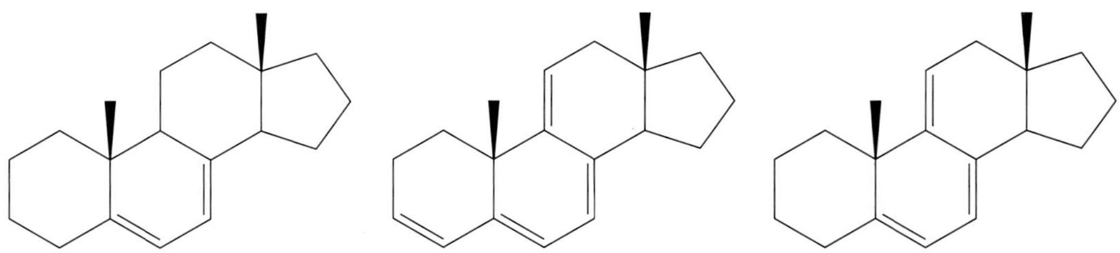

14. A compound containing only C, H, and O gave 80 % C and 6.7 % H on elemental analysis. The mass spectrum showed m/z = 120, 105, 77, 51, and 43. The IR spectrum showed a peak at ν_{max} = 1690 cm^{-1}. Deduce a structure and explain the data given.
15. How would you follow the reduction of phenylethanone to 1-phenylethanol by $Na[BH_4]$ using IR spectroscopy?

16. The following UV spectroscopic maxima have been measured. What conclusion can you draw about the effect of adding methyl groups to the diene? Approximately what effect does each added alkyl group have?

Compound	λ_{max} (nm)
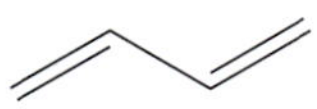	217
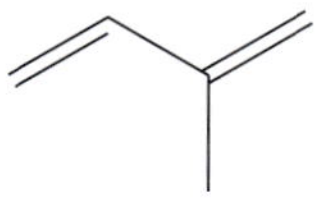	220
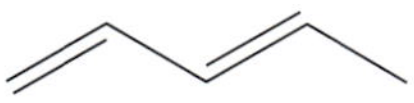	223
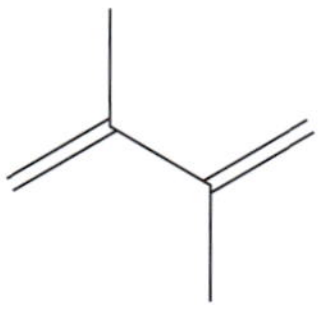	226
	227
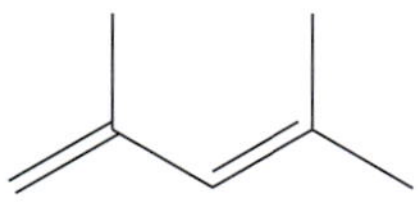	232
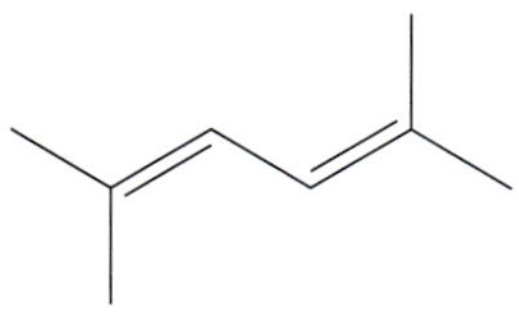	240

17. The IR spectrum of a linear hydrocarbon with formula C_6H_{10} shows strong bands at 3350, 2120, and 615 cm^{-1}. Deduce its structure and assign the IR stretching frequencies.
18. Explain clearly why 2-hexanol gives rise to a mass spectrum with key peaks at m/z = 102, 84, and 45.
19. Anhydrides, RCOOCOR′, show two carbonyl bands between 1600 and 1850 cm^{-1} in their IR spectra, while ketones, RCOR′, show only one. Explain.
20. Rank the following compounds in order of decreasing λ_{max}:

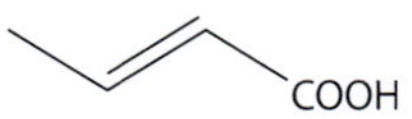

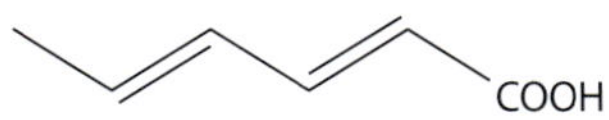

21. The base (most intense) peak in the mass spectrum of pentanal has m/z = 44. Propose a structure and account for its formation.

MORE CHALLENGING PROBLEMS

1. Describe two methods for isolating a desired organic compound from plant material.
2. A compound, $C_5H_{10}O$, has an IR maximum at 1715 cm^{-1} and shows peaks in its mass spectrum at 86, 71, 58, and 43 mass units. Draw its structure and explain the mass spectrometric data.

3. The compound trimyristin, **5.19**, can, unusually, be isolated from ground nutmeg by ether extraction (its concentration and solubility are high). It is purified by recrystallization from acetone:

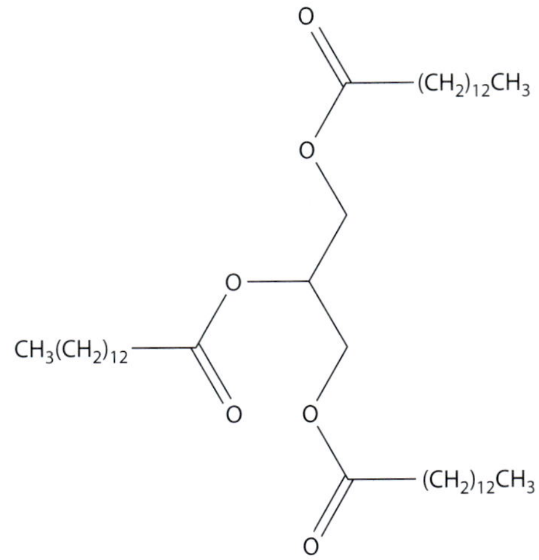

5.19

(a) Why is the residual nutmeg to be removed from the ether extract by gravity filtration rather than vacuum filtration?
(b) Why is a fluted filter paper rather than a plain filter paper used?
(c) A certain plant material known to contain mainly trimyristin and tripalmitin (look up the structure of this compound), in approximately equal amounts, was extracted with diethyl ether. After removal of the diethyl ether from the extract, an oil remained, which was difficult to crystallize. Explain.

4. The mosquito attractant oct-1-ene-3-ol is prepared by reduction of oct-1-ene-3-one. Suggest THREE ways of monitoring the progress of the reaction and describe what you would observe in each case.
5. Metabolism of drugs by the body often depends on the species. The molecule coumarin, an anticoagulant, is metabolized to a toxic compound in rats but is safe in humans. How would the UV and IR spectra differentiate between these liver metabolites, shown in the following?

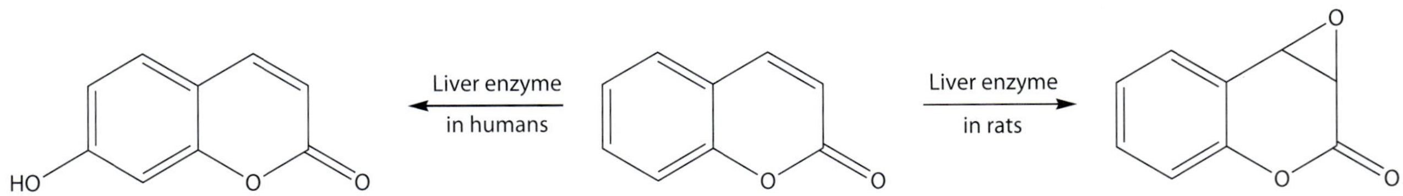

6. The stretching frequencies for some carbonyl groups are given in the following; explain why the stretching frequency for a particular class of compound increases as the ring size decreases.

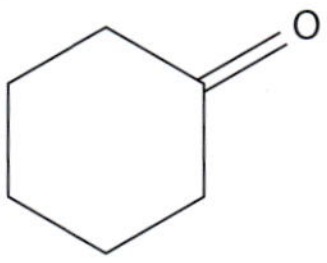

1716 cm^{-1}

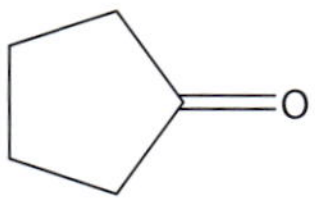

1748 cm^{-1}

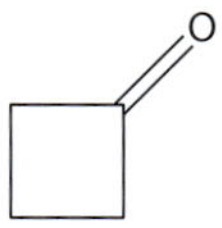

1783 cm^{-1}

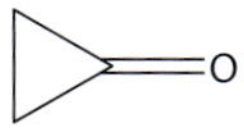

1850 cm^{-1}

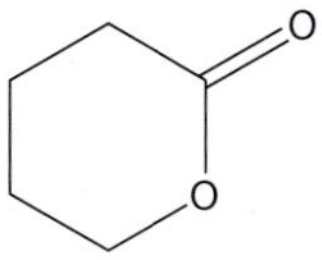

1740 cm^{-1}

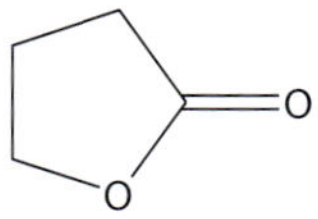

1765 cm^{-1}

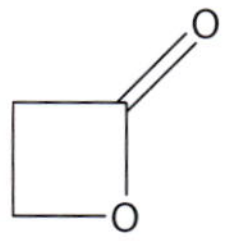

1840 cm^{-1}

Identification of Organic Compounds 2

Nuclear Magnetic Resonance Spectroscopy

6

6.1 INTRODUCTION

The nuclei of some atoms have a magnetic moment, designated by the symbol I. I is quantized and can have values 0, 1/2, 1, 3/2, *etc.* Many elements important to organic chemists have isotopes with spin 1/2, which is associated with odd atomic mass—and this provides us with the simplest and most common NMR spectroscopy experiment (Table 6.1). The NMR experiment is not possible if a nucleus does not have a magnetic moment, so there are no NMR spectra of the common isotopes of carbon or oxygen. Although the most common isotope of nitrogen has a magnetic moment, the signals are very broad, and this method is rarely used. Unless a material is enriched with ^{17}O, its low isotopic abundance makes the experiment challenging. We will mainly discuss ^{1}H and ^{13}C NMR spectra, as these are of the most use in determining the structures of organic molecules. Nuclei of isotopes of many other elements have appropriate spins and are extensively used in inorganic chemistry. The gyromagnetic ratio is a fundamental characteristic of the particular isotope and provides a measure of the sensitivity of the nucleus in the NMR experiment; the higher the value, the more sensitive the nucleus, and the more speedily the experiment can be performed.

A nucleus with a nuclear spin can be visualized as a small bar magnet. This is placed in a powerful magnetic field, H_0 (some texts use B_0). The nuclear magnetic moment can be aligned with or against the external field (Figure 6.1); these are described as spin states of +1/2 and −1/2, respectively, and the lower energy state is that in which the nuclear spin is aligned parallel to the external applied field. The energy difference between the two spin states corresponds to absorption in the radio-frequency (RF) region (Figure 6.2). This is a small energy difference, and at room temperature, both spin states are roughly equally populated.*

The fundamental relationship between magnetic field and transition frequency is given by Equation 6.1, where ν is the frequency in Hz, γ is the gyromagnetic ratio, and H is the magnetic field precisely at the nucleus under observation. This is not the same as H_0; H is given by the sum of H_0 and all the local magnetic fields generated by the movement of electrons (6.2). For ν = 300 MHz, H is approx. 7 T for protons. Early NMR spectrometers used a magnetic field generated by a large iron electromagnetic and scanned through RFs to establish which were absorbed. However, all modern spectrometers use a Fourier transform process.

(6.1) $$\nu = \frac{\gamma H}{2\pi}$$

(6.2) $$H = H_0 + \Sigma H_{local}$$

Samples for NMR spectroscopy are dissolved in a suitable solvent, usually $CDCl_3$ or other deuterated solvent, and placed in a sample tube, generally 5 mm in diameter for proton spectra, but occasionally 10 mm in diameter for ^{13}C and other less abundant nuclei. The sample tube is then placed in a spinner and spun in the magnetic field. This ensures that the molecules are tumbling

* You may have previously encountered the Boltzmann distribution in general chemistry; this allows the calculation of the relative populations of the two states.
Antiparallel/parallel = $e^{(-2\mu H/kbT)}$ where $\mu = \gamma I h/2\pi$, kb is the Boltzmann constant, h is Planck's constant, and I is the nuclear spin. For protons, the calculation yields a ratio of 0.999999999999525 for a 7 T instrument.

TABLE 6.1 Nuclear Spins of Some Common Nuclei

Isotope	Natural Abundance (%)	Spin (*I*)	Gyromagnetic Ratio[a] (γ)
^{1}H	99.98	1/2	26.753
^{2}H	0.015	1	4.107
^{3}H	~0	1/2	28.545
^{10}B	19.9	3	2.88
^{11}B	80.1	3/2	8.56
^{12}C	98.9	0	—
^{13}C	1.1	1/2	6.728
^{14}N	99.64	1	1.933
^{15}N	0.36	1/2	−27.116
^{16}O	99.76	0	—
^{17}O	0.038	5/2	−36.264
^{18}O	0.2	0	—
^{19}F	100	1/2	25.179
^{29}Si	4.7	1/2	−5.319
^{31}P	100	1/2	10.840

[a] γ in units of 10^7 rad T^{-1} s^{-1} (radian per tesla per second).

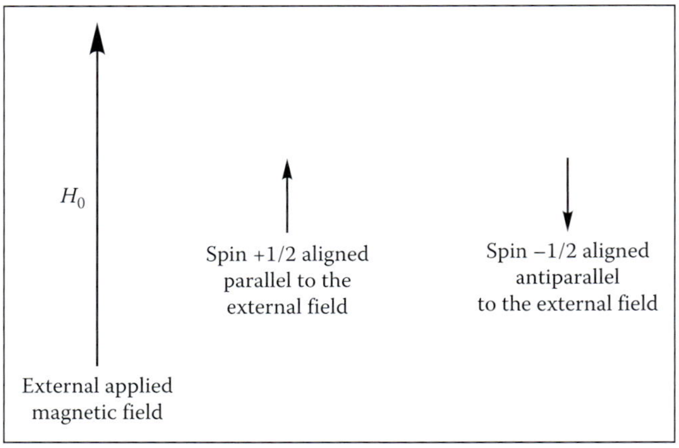

FIGURE 6.1 Spin states in NMR spectroscopy.

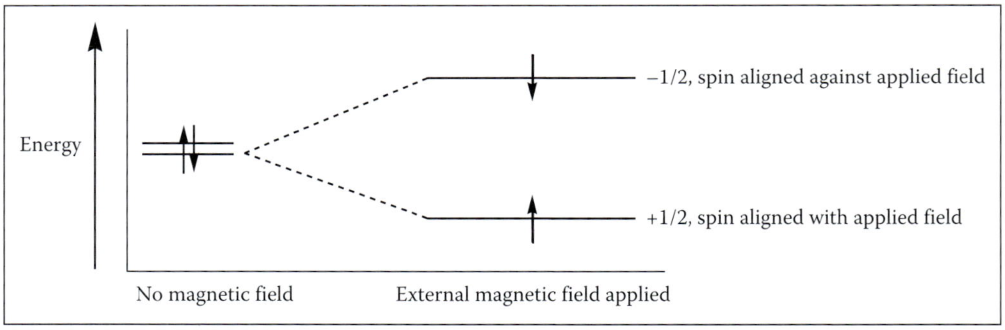

FIGURE 6.2 Splitting of nuclear spin energy levels in a magnetic field.

rapidly, and randomly, relative to the external magnetic field. While spectra can be obtained on solid samples, the technique required is different, and the signals are usually significantly broader. The use of a deuterated solvent is essential; the deuterium frequency (which is quite different from that of protons because of its lower gyromagnetic ratio) is used to calibrate the spectrum. Since solvent is much more abundant than solute, the signals from any protons in the solvent would swamp the signals from the solute that we need to observe. In practice, the production of solvents with total deuteration would be prohibitively expensive; standard deuterochloroform is

99.8 atom% pure. This means that in dilute solutions, a small peak due to $CHCl_3$ will be observed, and you need to be aware of this in interpreting spectra.

Sample tube for ^{1}H NMR spectroscopy

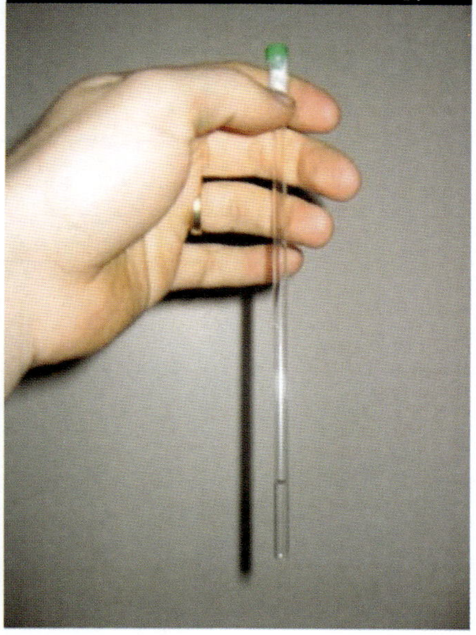

6.2 CHEMICAL SHIFTS

Parameters in NMR spectra are not given in frequency or wavelength terms because these would depend on the magnetic field at which the spectrometer operates—and the experiment is only useful if the signals for a molecule using one spectrometer are directly comparable to those obtained on any other. In order to establish a useful scale, a reference compound was chosen (a reference has been established for each NMR active nucleus). For proton spectra, this is tetramethylsilane, TMS, **6.1**. Why was this particular compound chosen? It is inexpensive, readily available, and chemically largely inert—hence not likely to react with samples. It's also volatile, so it can be readily evaporated from samples and material recovered after the experiment. Although TMS is nowadays rarely added directly to samples, with the scale being established from the deuterium resonance of the deuterated solvent, in the early days of the technique this was routine, so these features were important. Most importantly, the signal for TMS is at one end of the range of values shown by protons in "ordinary" organic compounds. The parameter δ, the chemical shift, is defined in Equation 6.3. Δν is the difference in frequency between the signal of interest and the signal for TMS, and ν is the operating frequency of the spectrometer in Hz. Thus, the value of δ, the chemical shift, of TMS is defined as zero.

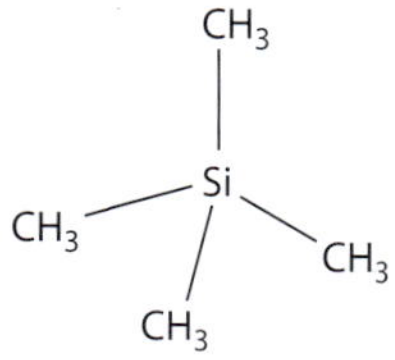

6.1

$$\delta = \frac{\Delta\nu}{\nu} \times 10^6 \quad (6.3)$$

By custom, "TMS," or rather δ = 0, is positioned at the right of any spectrum, and δ increases to the left (Figure 6.3). If Figure 6.3 is a spectrum from a 300 MHz spectrometer, then

$$\delta \text{ for the signal of interest} = (900/300 \times 10^6) \times 10^6 = 3 \text{ ppm}.$$

δ is dimensionless, and its value is independent of the spectrometer frequency. For protons, most signals are in the range δ 0–10. The ^{13}C resonance of TMS is used as the reference compound for ^{13}C NMR spectra, but the range of values of δ for ^{13}C is much greater—between 0 and 220 ppm (Figure 6.4).

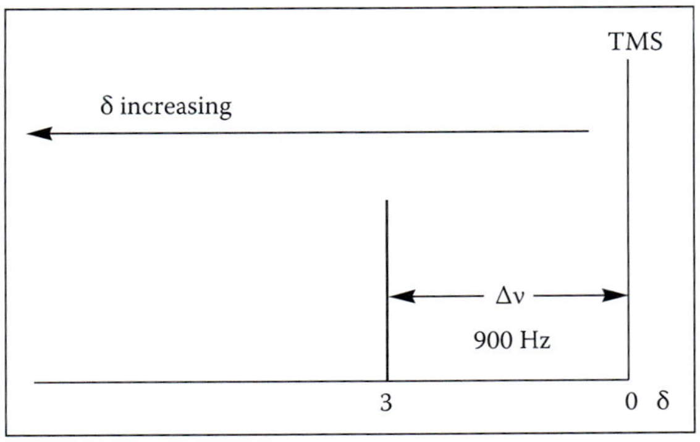

FIGURE 6.3 The δ-scale for proton NMR spectra.

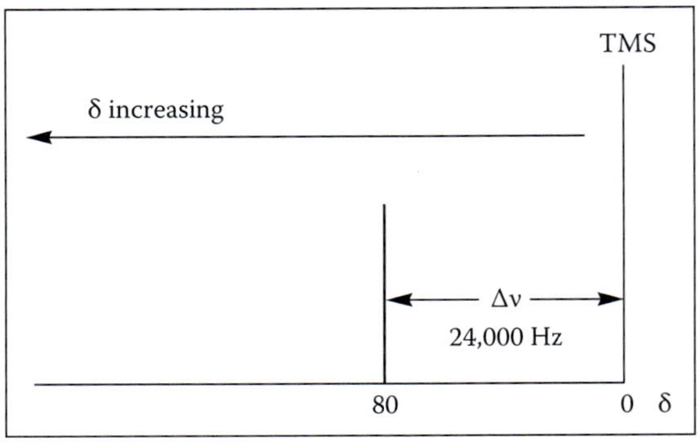

FIGURE 6.4 The δ-scale for ^{13}C NMR spectra.

PROBLEM 6.1

Using a 500 MHz NMR spectrometer, how many Hz downfield from TMS is a chemical shift of 6.5 ppm?

Solution

We need simply to plug the numbers into Equation 6.3; Thus $\Delta\nu = 6.5 \times 500 = 3250$ Hz.

As we noted earlier, the RF of the NMR transition depends on the magnetic field at the nucleus of interest. This has two components, the external applied field and the local fields generated by the movement of electrons. So H, and hence ν, are specific to nuclei and characteristic of their environment. In accord with Le Chatelier's principle, the local fields *oppose* the applied field, H_0. We say that the nucleus is *shielded* from the applied field. This is shown schematically for a single proton in Figure 6.5. While the physics of the process is elegant and fascinating, we don't need to understand it in detail to use the technique. We need to remember only that shielding generally decreases the value of δ, and we will have tables of well-established values to guide us in assigning particular peaks to protons of specific types.

All modern NMR spectrometers operate using the Fourier transform principle. A short (~0.1 s), but intense, RF pulse is applied to the sample, covering all the frequencies that correspond to the energy required to flip protons from the lower to higher energy state (Figure 6.2). After this, all the nuclei are in the excited state. Then we watch the RFs emitted as the nuclei return to the equilibrium state. A "free induction decay" (Figure 6.6) is obtained, and a black box of mathematics and electronics converts this to the spectrum. The data are repeatedly collected and added together to obtain a spectrum with a good signal to noise ratio. Spectrometers currently in use range from operating frequencies of 250 MHz up to 1 GHz, though the highest field instruments are exclusively used for investigation of complex biological molecules. All of the magnets involved are superconducting, operating at 4 K, and all that can be seen from the outside is the Dewar jacket containing liquid nitrogen, used to slow the evaporation of the liquid helium required to maintain the temperature required for the superconducting magnet to operate. While there have

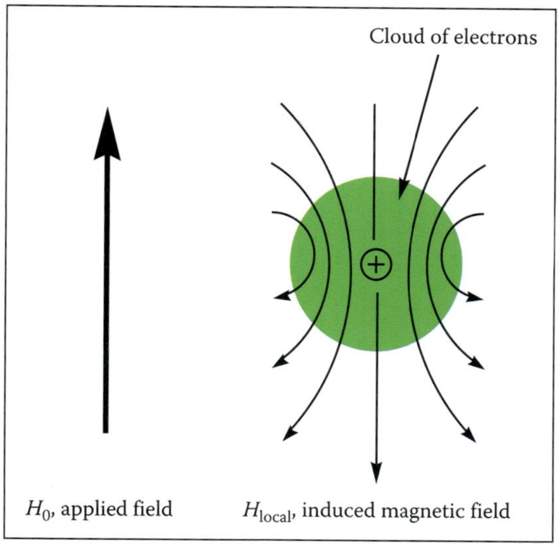

FIGURE 6.5 Shielding of an isolated nucleus by the movement of the surrounding electron cloud.

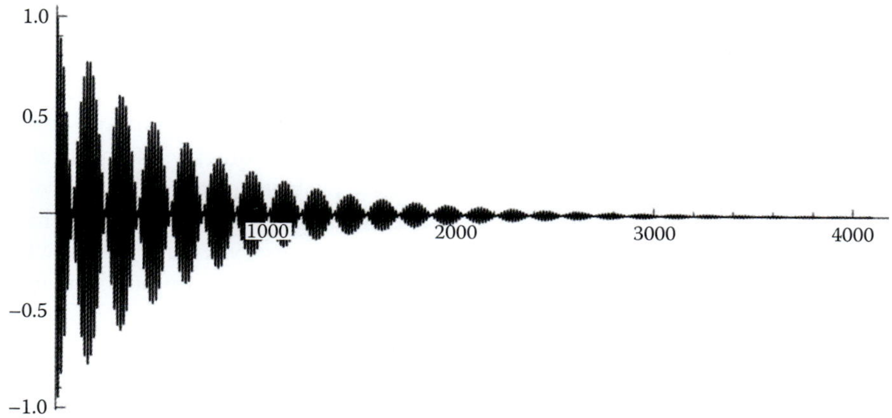

FIGURE 6.6 Typical free induction decay.

been improvements in recent years, there is often a lot of stray magnetic field in the vicinity of the instruments; metal objects, watches, credit cards, magnetic storage media, and anyone with a heart pacemaker should give them a wide berth.

950 MHz NMR spectrometer. (Photo courtesy of Jason Schnell, Department of Biochemistry, University of Oxford, Oxford, U.K., www.bioch.ox.ac.uk/aspsite/index.asp?pageid=651.)

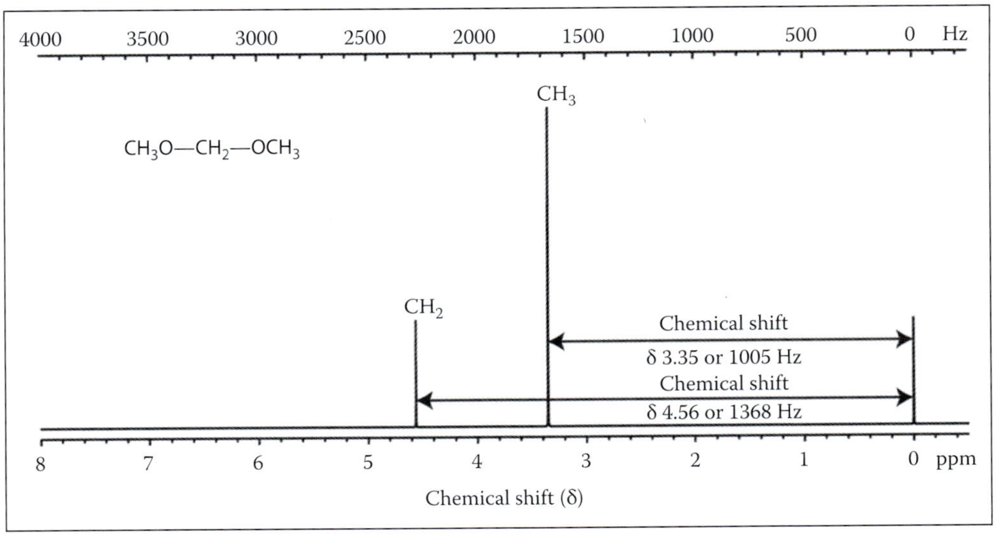

FIGURE 6.7 ^{1}H NMR spectrum of dimethoxymethane. The lower axis is the chemical shift scale in terms of δ (ppm) and the upper scale gives it in frequency units. TMS has been added to the sample and the signal is seen at δ 0.

Although the theory of NMR spectroscopy is complex, the applications in organic chemistry are both powerful and relatively straightforward and require little understanding of the theory. They do, however, require experience in order to assign spectra quickly and accurately. Many later chapters have some NMR problems in the review problems, and you will probably also be expected to assign spectra as a part of your practical course; this is not a topic that you can become expert in all at once, and long-term practice is necessary. Figure 6.7 shows a simple spectrum for dimethoxymethane. There are two different proton environments, hence two signals. The size of the signals reflects the relative abundance of the two types of proton; there are six identical protons in the methyl groups that are distinct from the CH_2 group. Electronic integration of the area under the peaks (Figure 6.8) gives us the correct ratio of protons of each type.

We can make some generalizations about the chemical shifts of protons in particular environments, and many tables of data (such as Table 6.2) exist. Table 6.2 demonstrates that, all else being equal, the more hydrogen atoms per carbon, the greater the shielding and the smaller the value of δ. δ-values increase in a series methyl < methylene < methine. Attaching

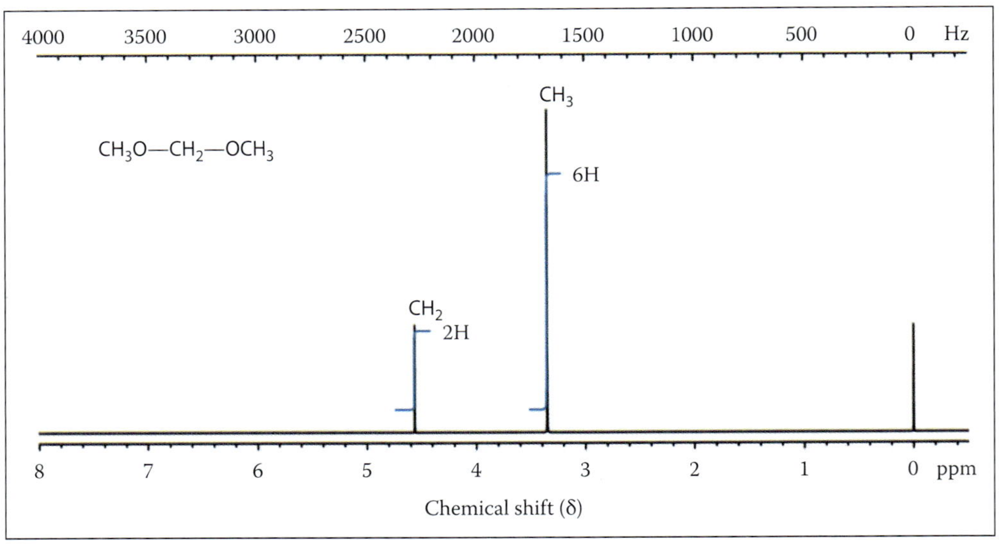

FIGURE 6.8 ^{1}H NMR spectrum of dimethoxymethane with integration. The integration is the blue superimposed curve for each signal. The vertical height of the integral tells us how many protons of each type there are. Only a ratio can be obtained this way.

TABLE 6.2 Typical Chemical Shifts of Protons Attached to sp^3 Carbon Atoms

Methyl Protons	δ	Methylene Protons	δ	Methine Protons	δ
CH_3–alkyl	0.7–1.6	–CH_2–alkyl	1.4	–CH–alkyl	1.5
CH_3–aryl	2.3	–CH_2–aryl	2.3–2.7	–CH–aryl	3.0
CH_3–C(=O)alkyl	2.1	–CH_2–C(=O)alkyl	2.4	–CH–C(=O)alkyl	2.7
CH_3–C(=O)aryl	2.6	–CH_2–C(=O)aryl	2.9	–CH–C(=O)aryl	3.3
CH_3–C(=O)Oalkyl	2.0	–CH_2–C(=O)Oalkyl	2.3	–CH–C(=O)Oalkyl	2.6
CH_3–C(=O)Oaryl	2.4	–CH_2–C(=O)Oaryl	2.6	–CH–C(=O)Oaryl	2.8
CH_3–C(=O)Cl	2.7	–CH_2–C(=O)Cl	2.9	–CH–C(=O)Cl	2.9
CH_3–C(=O)NH_2	1.8	–CH_2–C(=O)NH_2	2.3	–CH–(C=O)NH_2	2.4
CH_3–CN	2.0	–CH_2CN	2.3	–CH–CN	2.7
CH_3–F	4.1	–CH_2–F	4.3		
CH_3–Cl	3.0	–CH_2–Cl	3.6	–CH–Cl	4.2
CH_3–Br	2.7	–CH_2–Br	3.3–3.5	–CH–Br	4.3
CH_3–I	2.1	–CH_2–I	3.2	–CH–I	4.3
CH_3–OH	3.4	–CH_2–OH	3.6	–CH–OH	3.9
CH_3–Oalkyl	3.3	–CH_2–Oalkyl	3.4	–CH–Oalkyl	3.7
CH_3–Oaryl	3.8	–CH_2–OAryl	4.3	–CH–Oaryl	4.5
CH_3–OC(=O)alkyl	3.7	–CH_2–OC(=O)aryl	4.1	–CH–OC(=O)alkyl	4.8
CH_3–OC(=O)aryl	4.1	–CH_2–OC(=O)aryl	4.4	–CH–OC(=O)aryl	5.0
		–CH_2–NH_2	2.6	–CH–NH_2	3.1
CH_3–NHalkyl	2.3	–CH_2–NHalkyl	2.6	–CH–NHalkyl	2.7
CH_3–NHaryl	3.0	–CH_2–NHaryl	3.3		
CH_3–NHC(=O)alkyl	2.9	–CH_2–NHC(=O)alkyl	3.3	–CH–NHC(=O)alkyl	3.9

electronegative atoms to a carbon pulls electron density toward the electronegative atom, reduces shielding and increases the value of δ. The effect is roughly proportional to the electronegativity of the attached atom and is additive. Thus, the chemical shifts for the protons in $CHCl_3$, CH_2Cl_2, and CH_3Cl are 7.3, 6.5, and 3.0, respectively, compared with 0.9 for the nonpolar ethane. This also accounts for the difference in the chemical shift of the two types of proton in dimethoxymethane—the CH_2 is attached to two electronegative oxygen atoms, but the CH_3 is attached to only one. Table 6.3 gives the chemical shifts of the residual protons in common deuterated solvents. The choice of solvent will depend on solubility and stability of the compound in various solvents, the temperature at which the experiment is to be run, the predicted shifts of the compound (which should be different from that of the solvent, especially if solubility is low or limited material is available), and, of course, cost of the deuterated solvent. Although very large variances are rare, there are differences in chemicals shifts for a compound in different solvents, especially where there is the potential for some interaction between the solvent and the compound you are observing. Aromatic rings can form stacks with benzene or toluene as solvent, acids and alcohols may act as H-bond donors, and pyridine, acetone, and DMSO are H-bond acceptors.

For hydrogens attached to sp^2 and sp carbon atoms, there are some other considerations. With alkenes, hydrogen atoms in the plane of the carbon–carbon double bond are deshielded, and δ-values are typically 4–6 ppm. For aldehyde hydrogens, directly attached to the carbon of a carbonyl group, the deshielding is very strong, and δ-values of 8–10 are typical. We have already noted in Table 6.2 the electron-withdrawing effect of carbonyl groups, which increases the chemical shift of a methyl group from about 1 to about 2 ppm. As we mentioned in Section 3.2.1, there are some special features of the NMR spectra of aromatic compounds such as benzene. In benzene, there is an annulus of electron density above and below the plane of the six-membered ring and the electrons in this circulate (it's usually called a ring current) so that the induced magnetic field inside the ring opposes the applied field (Figure 6.9). However, since the lines of magnetic

TABLE 6.3 Chemical Shifts of Residual Protons in Common Deuterated Solvents

Solvent	Formula	MPt (°C)	BPt (°C)	δ (ppm)	Comments
Chloroform	$CHCl_3$	−63	61	7.26	Inexpensive, most commonly used.
Acetic acid	CH_3COOH	17	118	11.65 2.04	
Acetone	$(CH_3)_2C{=}O$	−94	56	2.06	Signal often seen in spectra in other solvents because acetone is commonly used to wash NMR tubes.
Acetonitrile	$CH_3C{\equiv}N$	−45	82	1.94	
Benzene	C_6H_6	6	80	7.16	Not suitable for low-temperature work.
Cyclohexane	C_6H_{12}	7	81	1.38	Not suitable for low-temperature work.
Dichloromethane	CH_2Cl_2	−97	40	5.32	
Dimethyl sulfoxide	$(CH_3)_2S{=}O$	19	191	2.50	Difficult to obtain completely dry.
1,4-Dioxane	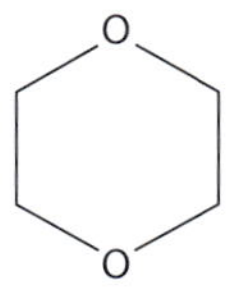	12	101	3.53	Difficult to obtain completely dry.
Methanol	CH_3OH	−97	65	3.3 4.8	Position of the signal for the OH peak is strongly temperature dependent; the distance between the two peaks has been used to calibrate temperature in variable temperature experiments.
Nitromethane	CH_3NO_2	−29	101	4.33	
Pyridine	C_5H_5N	−41	115	7.19 7.55 8.71	
Tetrahydrofuran	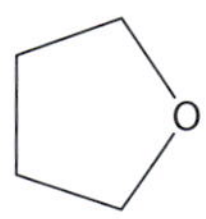	−108	66	1.73 3.58	
Toluene	$C_6H_5CH_3$	−95	111	2.3 7.04 7.06 7.14	Excellent for variable temperature experiments; long liquid range.
Trifluoroacetic acid	CF_3COOH	−15	74	11.5	
Water	H_2O	0	100	4.70	The δ-value is strongly temperature and pH dependent.

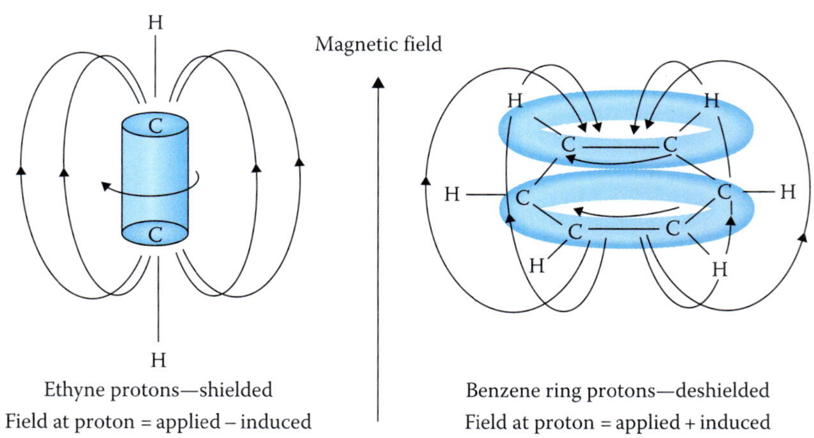

Shielding and deshielding of protons in ethyne and benzene

FIGURE 6.9 Ring currents in alkynes and aromatic compounds.

force must be continuous (don't worry if the physics of this is not clear to you—just use the result) outside the ring, around the protons attached to it, the induced magnetic field reinforces the applied field. Thus, protons attached to the outside of an aromatic ring are deshielded relative to alkene protons, and most resonances are in the range δ 6–8. The value of δ for benzene is 7.16 ppm. If one considers a ring current in the π-electrons of an alkyne (remember that the two π-bonds form an almost continuous annulus of electron density around the carbon–carbon bond), this induces a field opposing the applied field inside the electron ring and one that reinforces the external field outside the ring. However, this time, the protons that we want to observe are *inside* the ring current, so they are shielded from the applied field and δ is reduced, typically to 2.2–2.8 ppm.

PROBLEM 6.2

Assign each of the signals in the following NMR spectra:

(a)

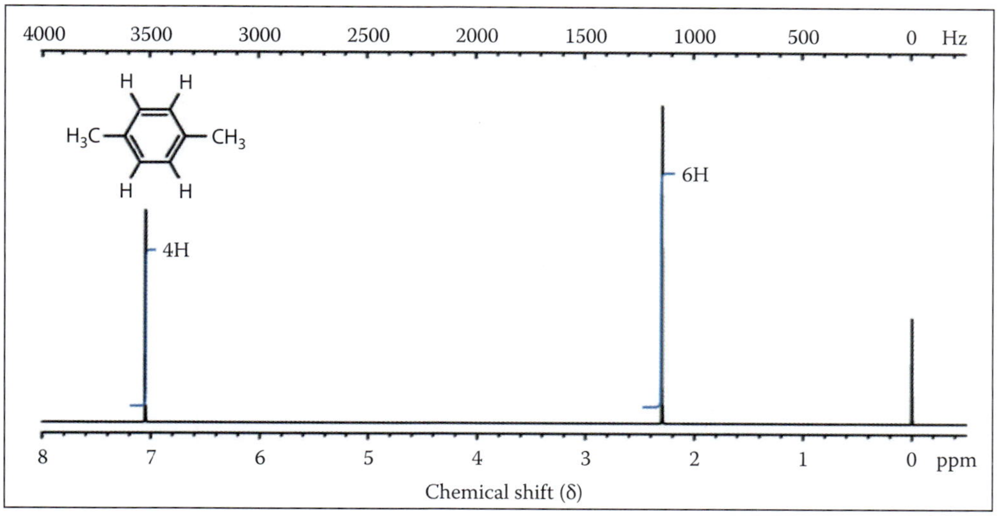

(b)

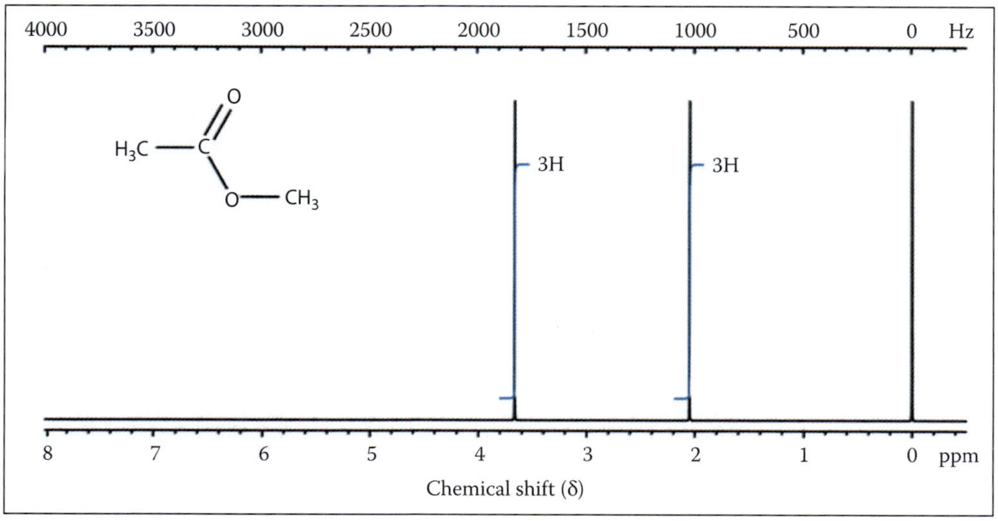

(c)

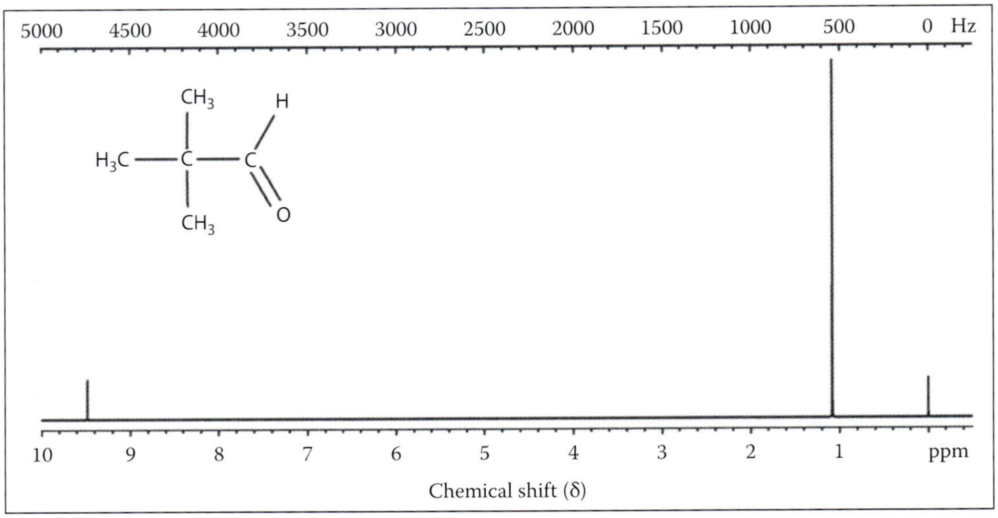

(d)

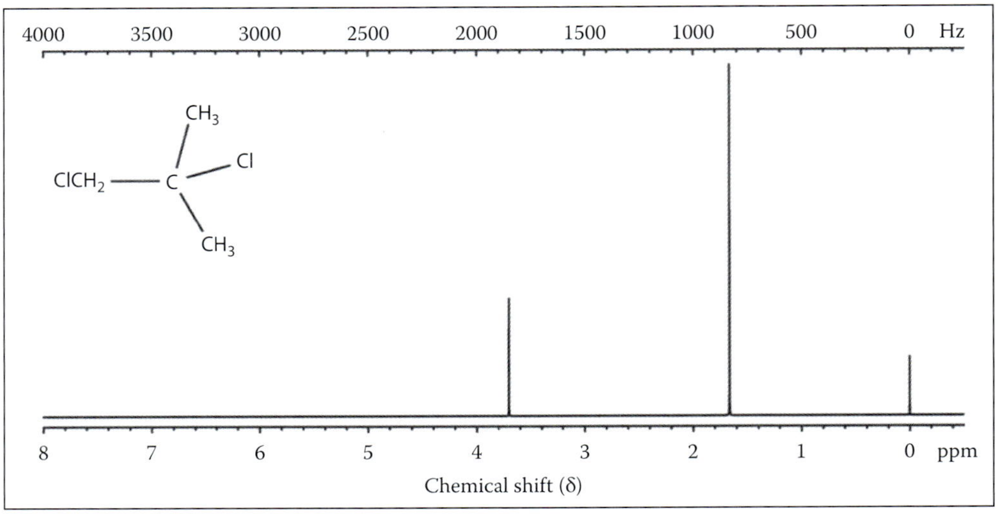

Solutions

(a) There are two types of protons in this molecule, six in the two methyl groups and four on the aromatic ring. The values of 2.3 and 7.1 for their chemical shifts are consistent with this. δ for the methyl group is 2.3 ppm and δ for the aromatic hydrogen atoms is 7.1 ppm. The integration is correctly 3:2 (6:4).

(b) There are two distinct methyl groups. The one with δ 3.8 is the *O*-methyl group and that with δ 2.1 is the methyl group attached to the carbonyl.

(c) Again there are two types of hydrogen atom. The three methyl groups are all identical, integrating to nine hydrogens at δ 1.1. The single aldehyde hydrogen is strongly deshielded by the electron-withdrawing carbonyl group at δ 8.4.

(d) Again there are two types of hydrogen atom, two identical methyl groups and a CH_2. The CH_2 is directly attached to a chlorine atom and the chemical shift is δ 3.8 ppm. The two methyl groups show a chemical shift of δ 1.6, which is somewhat higher than expected for a simple methyl group attached to another sp^3 carbon atom. This increase in δ arises from the electron-withdrawing effect of the chlorine—although it is not directly attached to the methyl groups, its inductive effect is transmitted, albeit less strongly, through more than one single bond.

Key Points from Sections 6.1 and 6.2

- Certain nuclei, including ^1H and ^{13}C, have magnetic moments and can be aligned parallel or antiparallel to an external magnetic field. The energy for the transition between the two states corresponds to EM radiation in the RF region.
- The energy for the transition depends on the magnetic field at a particular nucleus, and this in turn depends on the chemical environment of the nucleus.
- Chemical shift is a characteristic for a particular proton and is defined relative to the signal from TMS.
- Most proton chemical shifts lie between δ 0 and 10, and most ^{13}C chemical shifts between δ 0 and 200.
- Extensive tables exist correlating chemical shift with proton environment. δ increases with the proximity of electronegative atoms and is higher for protons attached to sp^2 carbon atoms.

6.3 SPIN–SPIN COUPLING

6.3.1 SIMPLE ALIPHATIC SYSTEMS

So far, all the spectra we have examined have been relatively simple ones; they have shown a single line for each type of proton in the molecule. However, this is relatively unusual. Let's look at the proton NMR spectrum of 1,1,2-tribromo-3,3-dimethylbutane, **6.2** (Figure 6.10). The three identical methyl groups give rise to a single signal at δ 1.2. However, each of the other two non-identical protons gives rise to a signal that has been split into two—this is described as a doublet. Why are these signals split? It is because they are adjacent to each other and hence "coupled."

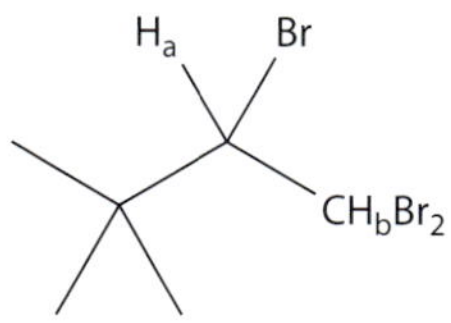

6.2

Suppose that H_a is emitting RF as it falls to the lower energy state. How can H_b affect it? H_b can be aligned either parallel or antiparallel to the external field. If it is aligned with H_0, then H_a "feels" a little more external field. If H_b is aligned against H_0, then H_a "feels" less field. This means that we

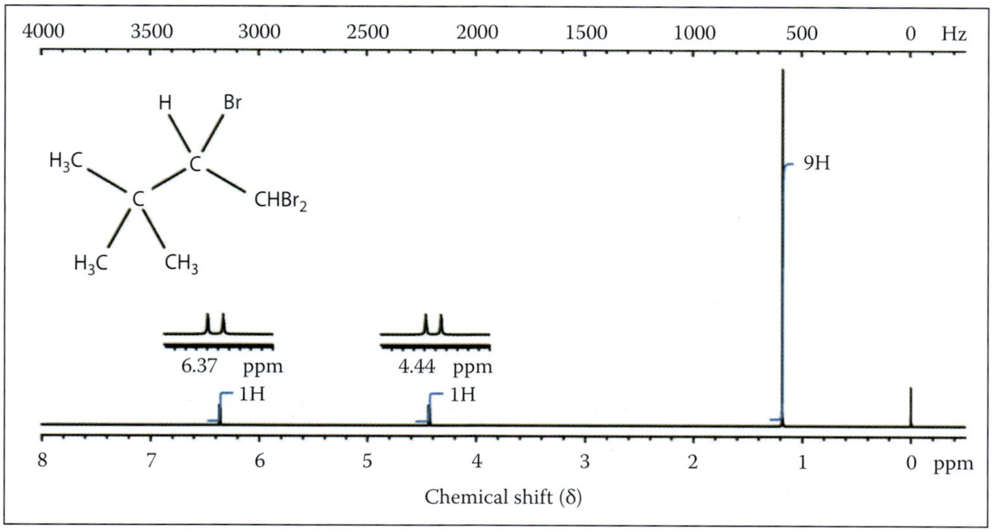

FIGURE 6.10 Proton NMR spectrum of 1,1,2-tribromo-3,3-dimethylbutane.

see a doublet for H_a. The lines are equal in intensity because the upper and lower energy states are approximately equally populated. The center of the doublet is δ, the chemical shift of the unperturbed signal. If we consider H_b emitting, then it will be affected in the same way by the two possible states of H_a—so we would expect another doublet. The energy of the interaction is the same in either direction, so the splitting of the two signals is identical. This is called the coupling constant, symbol J, and is measured in Hz. For proton–proton coupling in organic molecules, the value of J is in the range 0–20 Hz, and it is independent of the spectrometer operating frequency. For a system such as **6.3**, we can write in full $^3J_{ab}$. The superscript "3" indicates the number of bonds between the nuclei that are coupled, and the subscript "ab" indicates which hydrogens are involved in the coupling. Three-bond coupling is also called vicinal coupling (Latin *vicinus*, neighboring, near).

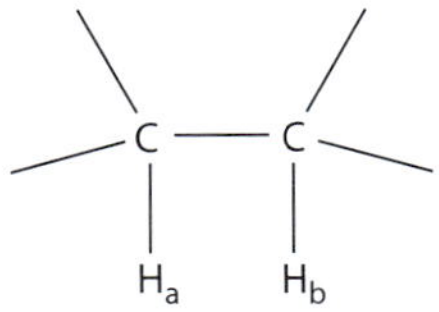

6.3

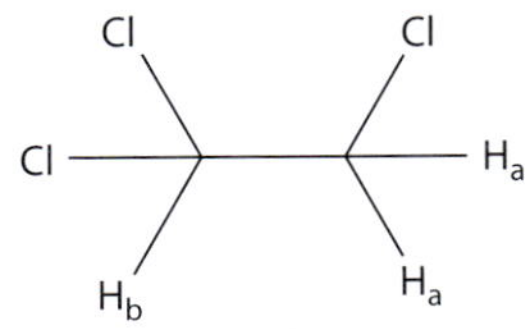

6.4

If we now consider the proton NMR spectrum of 1,1,2-trichloroethane, **6.4** (Figure 6.11), we can identify and assign the signals from H_a and H_b from the δ-values and the integration. H_b appears as a 1:2:1 triplet and H_a as a 1:1 doublet (Figure 6.12 shows an expansion). When H_a is

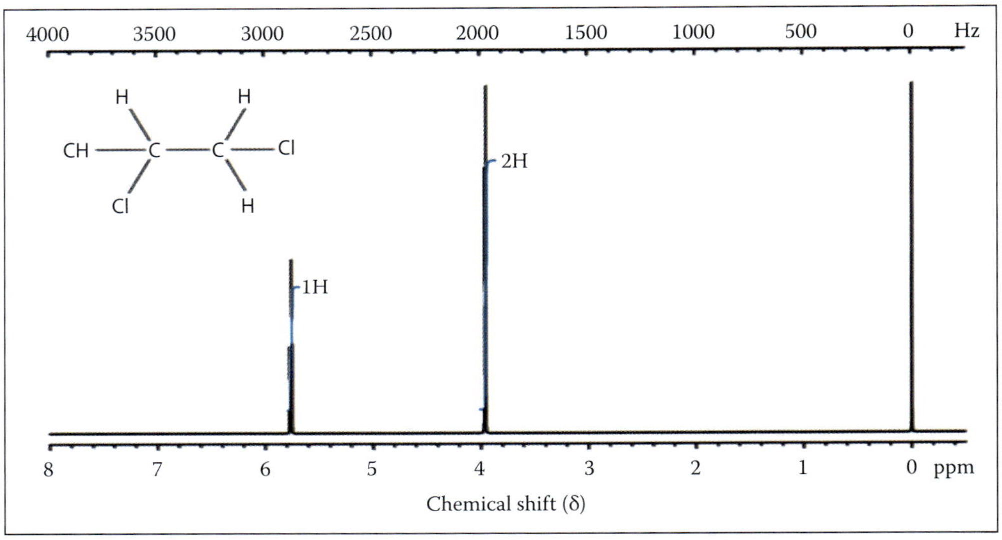

FIGURE 6.11 Proton NMR spectrum of 1,1,2-trichloroethane.

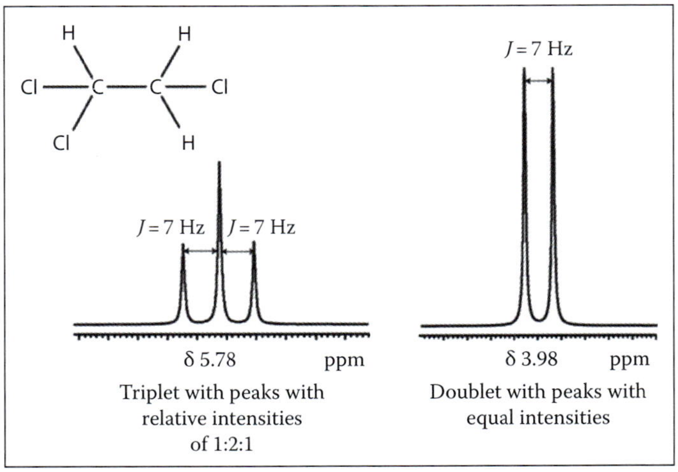

FIGURE 6.12 Expansion of the proton NMR spectrum of 1,1,2-trichloroethane.

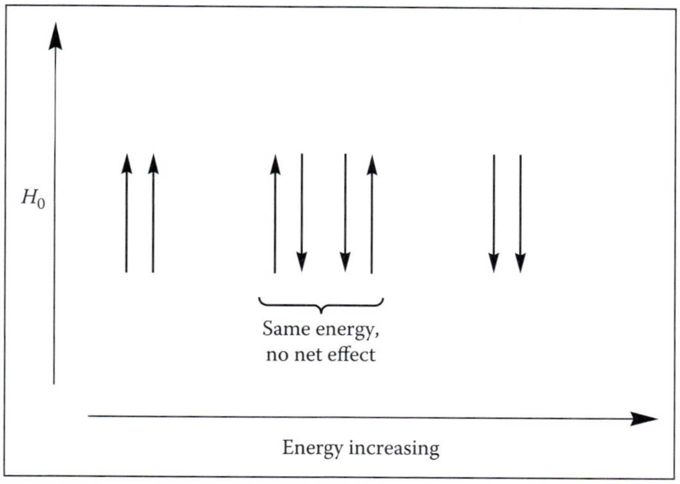

FIGURE 6.13 Schematic of spin states in an A_2B system.

emitting, H_b has two possible orientations—hence, we see a doublet. What is the effect of the two H_a nuclei on H_b? We must consider all the possible spin states (Figure 6.13). There are three possible energy states, hence three lines, which is referred to as a triplet.

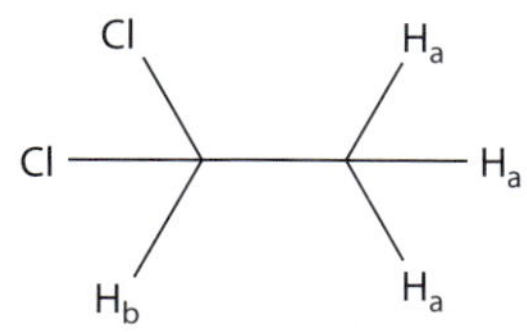

6.5

Next, we will consider 1,1-dichloroethane, **6.5** (Figure 6.14). As before, H_a is split into a doublet by H_b. The effect of the three H_a nuclei on H_b is shown in Figure 6.15, considering all the possibilities as having equal probabilities. This means that we should (and do) see four lines in the ratio 1:3:3:1, which is described as a quartet (Figure 6.16). By now, you might be beginning to suspect that there is a "rule" lurking here. You would be correct, and it's generally referred to as the $n + 1$ rule. This states that *n* ***equivalent*** protons on adjacent atoms split a signal into $n + 1$ lines. Note that equivalent protons do not split each other. The relative heights of the peaks may be obtained from Pascal's triangle (Figure 6.17). In simple cases, *J* is the line separation and can be measured from either of a pair of coupled signals. The rule only works when all the couplings involved are essentially equal; when they are very different, we see a doublet of doublets rather than a triplet (*vide infra*).

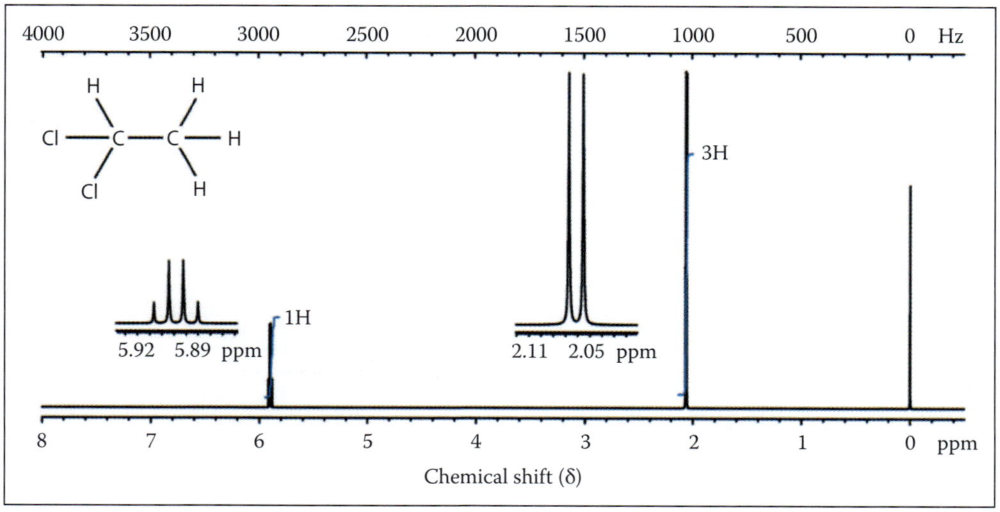

FIGURE 6.14 Proton NMR spectrum of 1,1-dichloroethane.

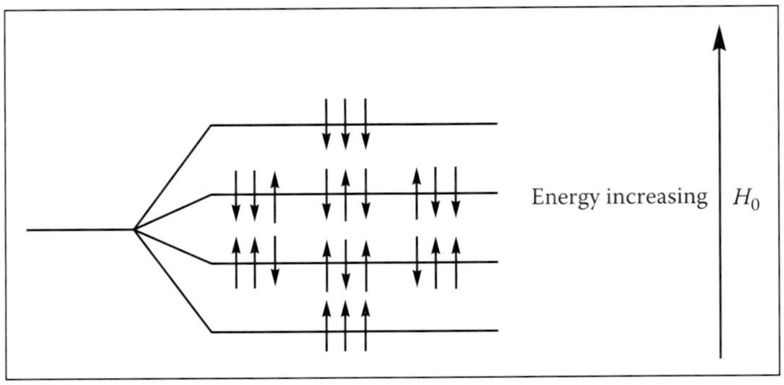

FIGURE 6.15 Schematic of spin states in an A_3B system.

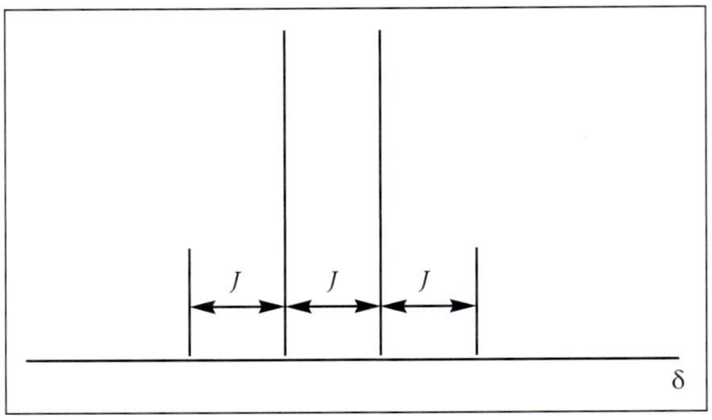

FIGURE 6.16 Quartet in an NMR spectrum from coupling to three equivalent protons.

1

1 1

1 2 1

1 3 3 1

1 4 6 4 1

1 5 10 10 5 1

1 6 15 20 15 6 1

FIGURE 6.17 Pascal's triangle.

Three more examples will illustrate the principle. Figure 6.18 shows the spectrum of bromoethane, with expansions of each of the peaks. The methyl group at δ 1.67 is a triplet because it is next to a carbon bearing two protons, and the methylene signal at δ 3.43 is a quartet because it is next to a carbon bearing three protons. The spectrum of 1,3-dichloropropane in Figure 6.19 shows that it is the total number of protons on adjacent carbon atoms that is important—they do not have to be all attached to the same carbon atom. The two CH_2 groups bearing chlorine atoms may be distinguished from the central CH_2 by both their chemical shift and relative integration. They are next to a CH_2 group, so they appear as a four-proton triplet. The CH_2 group in the middle of the chain has four identical protons on adjacent carbon atoms, two on each side. Thus, the signal is split into five lines in the ratio 1:4:6:4:1. In 2-bromopropane (Figure 6.20), the signal for the six methyl protons appears at δ 1.75 and is a doublet coupled to the unique proton at C-2. The signal for the lone proton is at δ 4.3 (because of the inductive effect of the bromine) and is split into a septet by the six adjacent protons. The smallest outermost peaks can be difficult to see, as the ratio of the peak heights is 1:6:15:20:15:6:1.

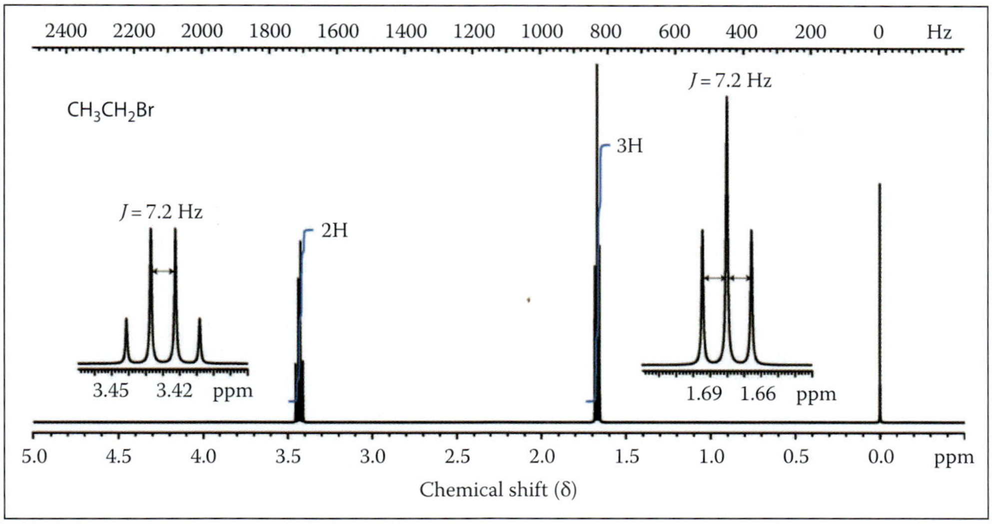

FIGURE 6.18 Proton NMR spectrum of bromoethane.

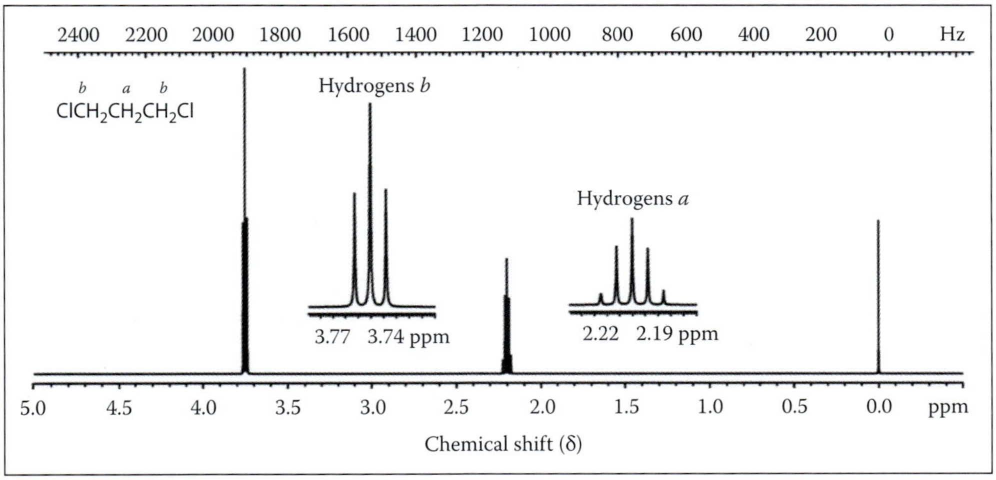

FIGURE 6.19 Proton NMR spectrum of 1,3-dichloropropane.

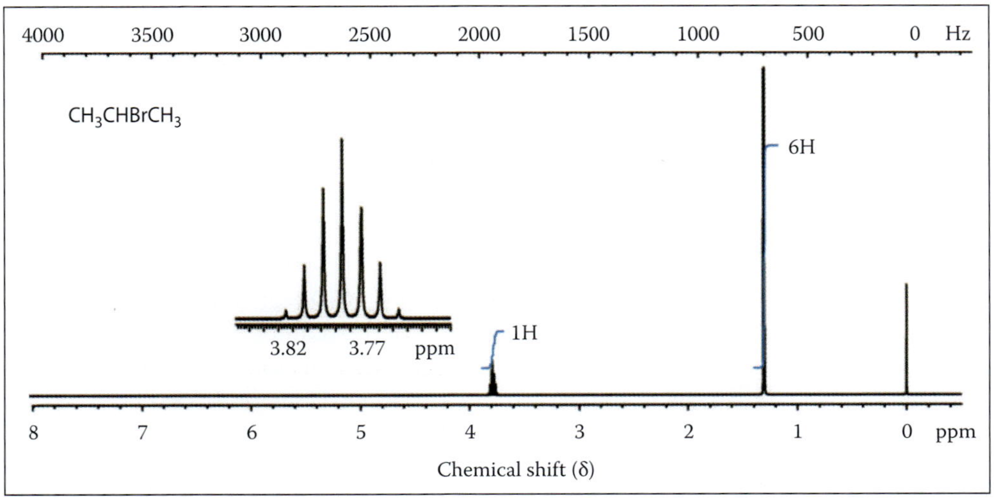

FIGURE 6.20 Proton NMR spectrum of 2-bromopropane.

Key Points from Section 6.3.1

- Equivalent protons do not couple to one another. Also, nuclei with the same chemical shift do not exhibit spin–spin coupling. They may be coupled, but the splitting cannot be directly observed.
- Nuclei separated by three or fewer bonds will usually be spin coupled and will show mutual splitting of their signals (*i.e.*, the same value of *J*), provided that their chemical shifts are different.
- The magnitude of the spin–spin coupling depends on a range of factors and is given by the coupling constant, *J* expressed in Hz. For hydrogens attached to sp^3 carbon atoms, vicinal coupling is usually 5–8 Hz.
- *J* is the same for both coupling partners and is independent of the operating frequency of the spectrometer.
- The splitting pattern for a given nucleus can be predicted using the $n + 1$ rule, where n is the number of neighboring spin coupled nuclei with similar or the same *J* values.

PROBLEM 6.3

How would you distinguish between compounds (a)–(d) on the basis of NMR spectroscopy?

(a) 1,4-Dichlorobutane
(b) 1,3-Dichlorobutane
(c) 2,3-Dichlorobutane
(d) 1,2-Dichloro-2-methylpropane

Solution

It is important to note in this problem that you are NOT being asked to predict the whole spectra of these compounds or indeed numerical values of chemical shifts or coupling constants. The two key features to focus on are symmetry and the multiplicity of the signal for the protons attached to the same carbon as the chlorine—these will be well separated from the rest of the spectrum and hence easily distinguished. However, the question is also designed to make sure that you can draw each of the molecules from its name:

(a) 1,4-Dichlorobutane

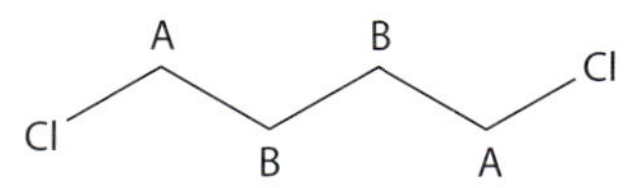

There are only two types of protons here, labeled A and B on the structure. There are four A-type protons and four B-type protons. The A-type protons will give rise to a triplet (δ 3.58) and the B-type also to a triplet (δ 1.94; remember the B protons cannot split each other).

(b) 1,3-Dichlorobutane

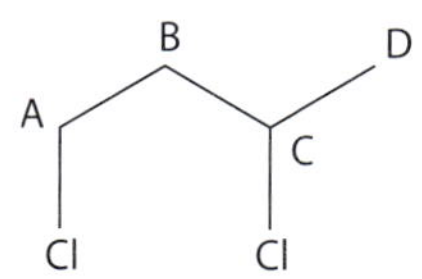

There are four distinct types of protons here, with A:B:C:D = 2:2:1:3. The signals due to A, C, and D should be simple. The chemical shift of the A and C protons will be in the range 3.5–4.5 ppm (actually δ 3.96 for A and δ 4.25 for C). The signal for A will be a triplet; that for C will be more complex. The methyl group D should be easy to assign and a doublet (δ 1.56, $J = 6.5$ Hz). The signal for the two B protons will be a complex multiplet. For reasons that will become clear in the next chapter, the two B protons are not strictly equivalent (though they are very similar) and will be coupled both to each other and to the protons at A and C.

(c) 2,3-Dichlorobutane

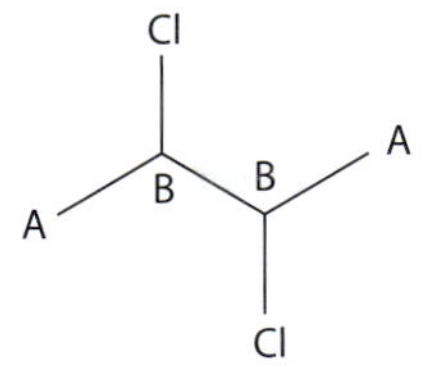

This molecule also has some symmetry, with two types of proton, and an A:B ratio of 3:1. We expect a quartet for B with δ in the range 3.5–4.5 (actually δ 4.1) and a doublet for A (δ 1.6).

(d) 1,2-Dichloro-2-methylpropane

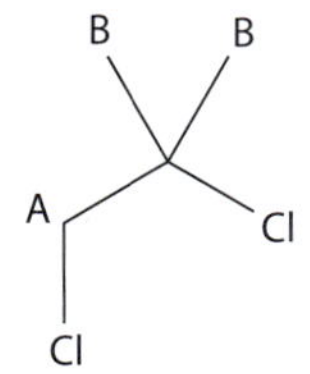

There are two types of proton here with A:B = 2:6. They are not coupled as they are not attached to adjacent carbon atoms, so will appear as two singlets (δ 3.7, 1.7 ppm).

Looking back at the analysis, there is clearly overspecification in our arguments—we could have distinguished (a) and (b), simply by looking at integrations:

(a) 1:1
(b) 2:2:1:3
(c) 3:1
(d) 3:1

(c) and (d) are easily distinguished by the splitting patterns.

6.3.2 SPIN–SPIN COUPLING INVOLVING PROTONS ATTACHED TO sp^2 CARBON ATOMS

The magnitude of spin–spin coupling between nonequivalent protons on a carbon–carbon double bond is one of the most definitive ways in which we can establish the geometry about the double bond. If the two protons are *cis*, the coupling is in the range 7–10 Hz, and if they are *trans* the coupling is 12–18 Hz. The two ranges do not overlap, and the distinction is regarded as diagnostic of double-bond geometry. This is exemplified by the spectra of *E*- and *Z*-3-chloropropenoic acid (Figure 6.21); notice the larger couplings for the *trans*-protons. Since the bonds in a benzene ring resemble alkenes, we might expect that the vicinal couplings in aromatic systems would be similar to those for *cis*-alkenes, and this is indeed the case, with couplings of *ortho*-protons in the range 6–10 Hz. This is exemplified by the proton NMR spectrum of 4-bromonitrobenzene (Figure 6.22).

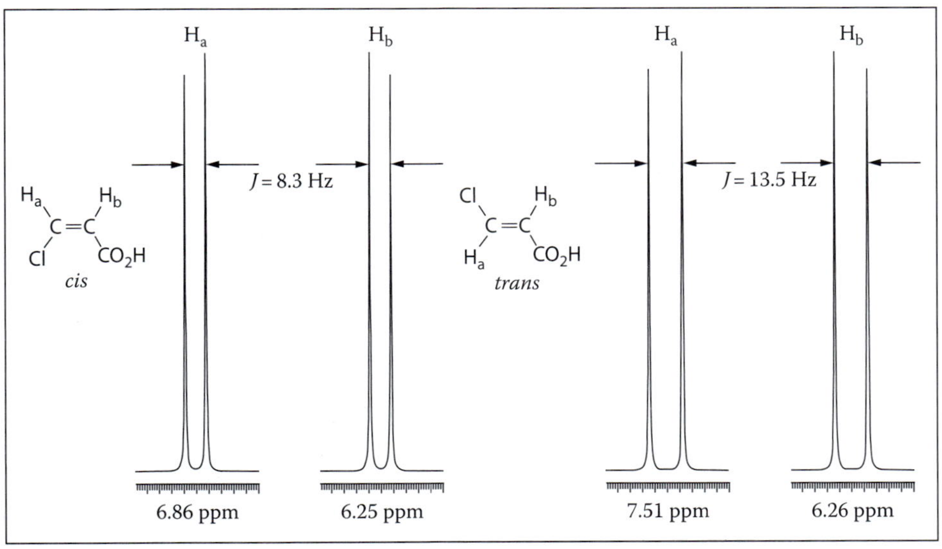

FIGURE 6.21 Sections of the proton NMR spectrum of *E* and *Z*-3-chloropropenoic acid.

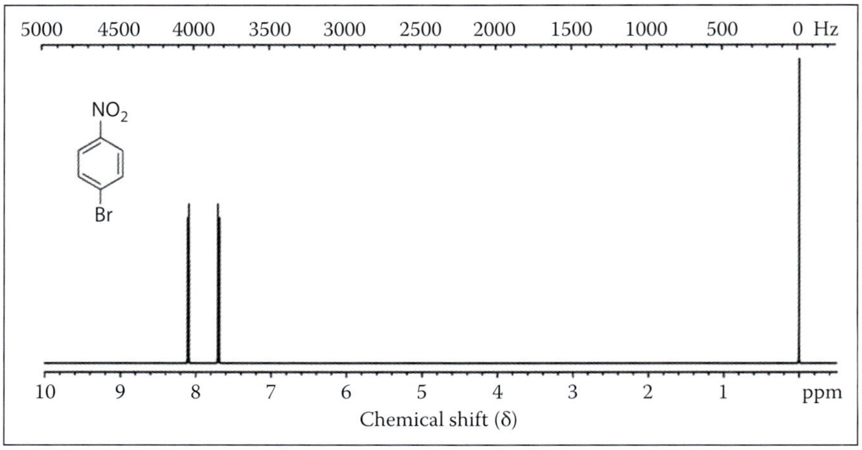

FIGURE 6.22 Proton NMR spectrum of 4-bromonitrobenzene.

The four aromatic protons are equivalent in pairs and each is a doublet. In both allylic and aromatic systems, it is sometimes possible to observe longer-range couplings. For example, Figure 6.23 shows the proton NMR spectrum of *E*-2-butenoic acid, and the expansions show that the proton at C-2, as well as that at C-3, is coupled to the protons of the methyl group. Similarly *meta*-couplings (0–3 Hz) can be clearly seen in the proton NMR spectrum of salicylic acid (Figure 6.24). Ranges of couplings are summarized in Table 6.4, including some that we will meet later.

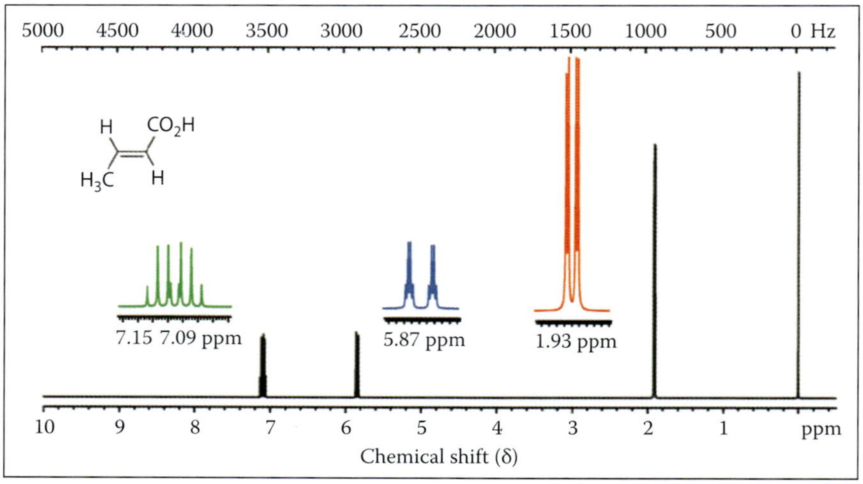

FIGURE 6.23 Proton NMR spectrum of *E*-2-butenoic acid with expansions. Note that the OH peak is at approx. δ 12 and not shown here.

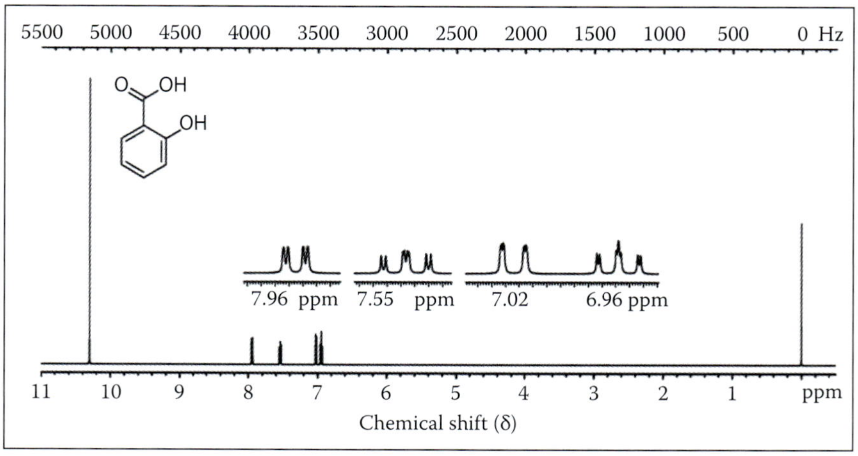

FIGURE 6.24 Proton NMR spectrum of salicylic acid.

TABLE 6.4 Proton–Proton Spin–Spin Coupling Constants

Coupling Type	Typical Structure	Range (Hz)	Comments
^3J	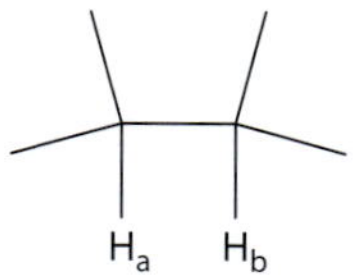	5–8	Magnitude depends on dihedral angle[a] and can occasionally be very small.
^3J	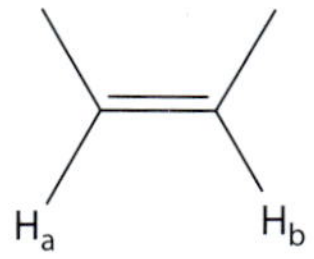	6–10	Diagnostic of double-bond geometry.
^3J	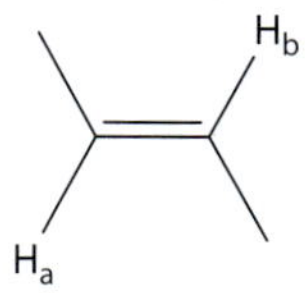	12–18	Diagnostic of double-bond geometry.
^3J	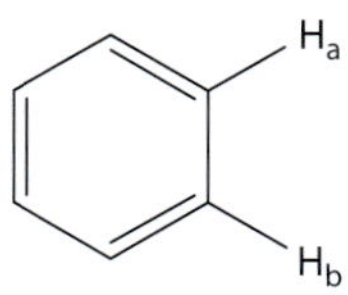	6–10	
^4J	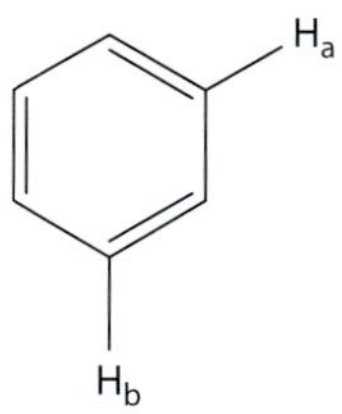	0–3	Long-range coupling, not always observed.
^4J	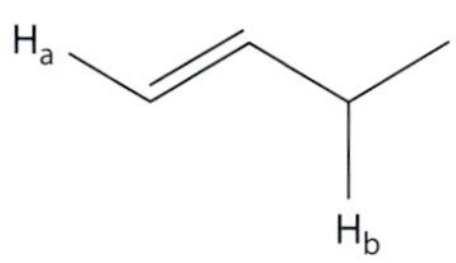	0–3	Long-range coupling, not always observed.
^2J	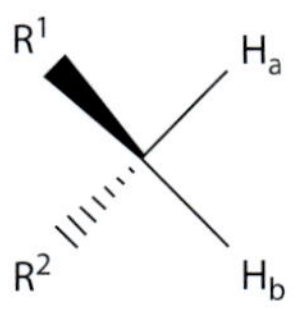	10–15	Geminal coupling; for H_a and H_b to be distinguishable, R^1 and R^2 must be different, and one must contain a chiral center or a rigid group.
^2J	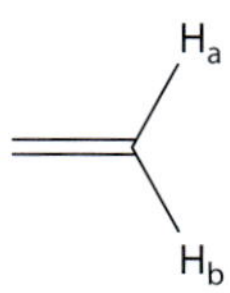	0–3	Geminal coupling.

[a] The dihedral angle is defined as the angle between the bonds to hydrogen, when viewed along the carbon–carbon bond. This is most easily seen in a Newman projection.

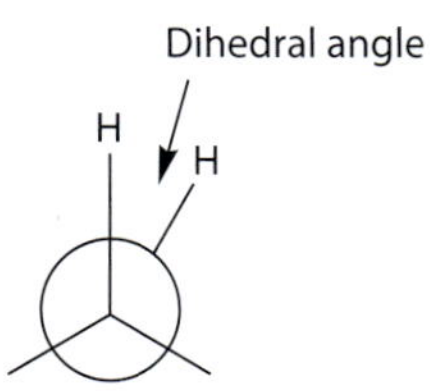

Focus: Tips on Analysis of the Splitting Patterns for Benzene Rings

When you are asked to identify a compound containing a benzene ring, with various substituents, you first need to check how many aromatic protons there are—and hence how many positions are substituted. In most cases, this should be easy from integration (though in a few compounds, there may be some overlap with alkene protons). You then need to decide the relative positions of the various substituents, and this is most easily done by looking at the splitting patterns of the protons for the various possibilities. In a monosubstituted benzene ring, there are three types of protons in the ratio 2:2:1 (**6.6**). Assuming we can neglect any long-range couplings (and we can certainly distinguish them), H_a is a doublet (coupled to H_b). H_b is a triplet (coupled to H_a and H_c) and H_c is a triplet (coupled to both H_bs).

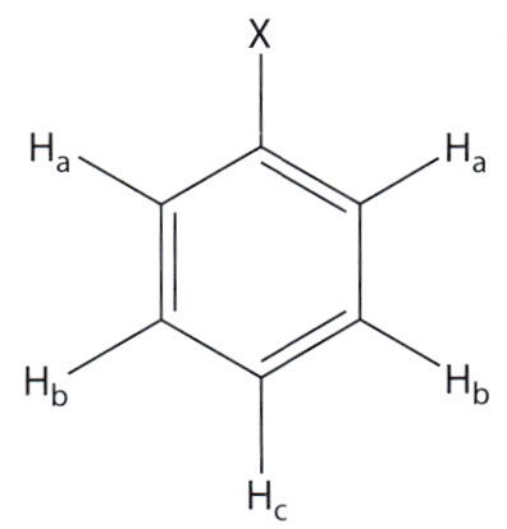

6.6

You can conduct the same analysis for more substituted compounds. This is shown in Figure 6.25 for benzene rings substituted with two identical or nonidentical substituents. Remember that no coupling can be observed between identical protons, so although the H_b protons in 1,2-$C_6H_4X_2$ have two neighbors, one of these is an identical H_b proton and has to be ignored. Similar analyses can be undertaken for tri- and tetrasubstituted rings.

The spectra of *ortho-*, *meta-*, and *para-*bromonitrobenzene will be illustrated (Figures 6.26 through 6.28). The *para-*isomer is most easily distinguished by its symmetry—two doublets.

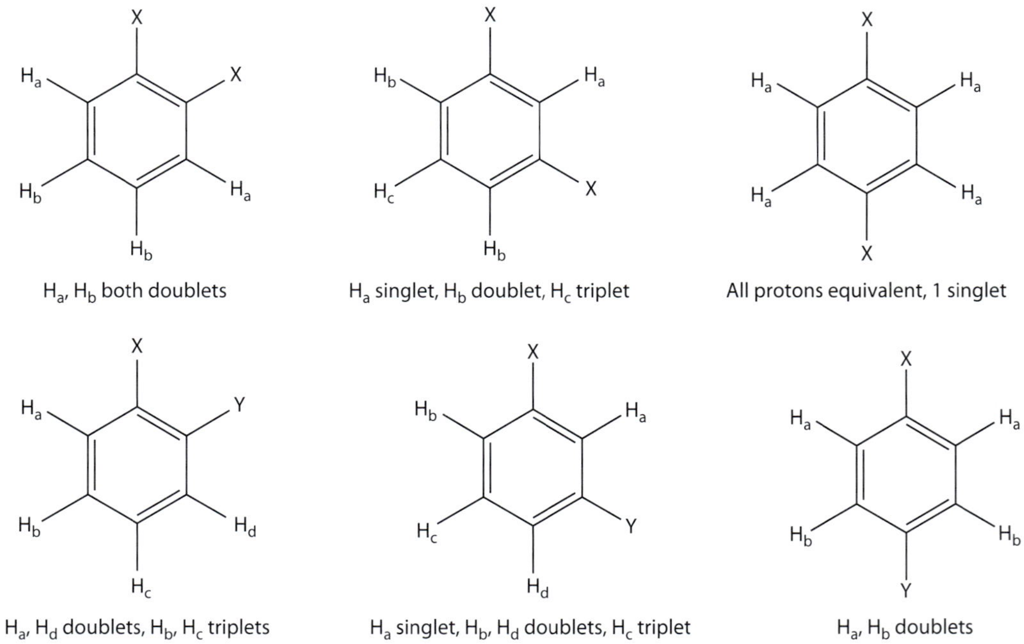

FIGURE 6.25 Splitting patterns in the proton NMR spectra of polysubstituted aromatic compounds.

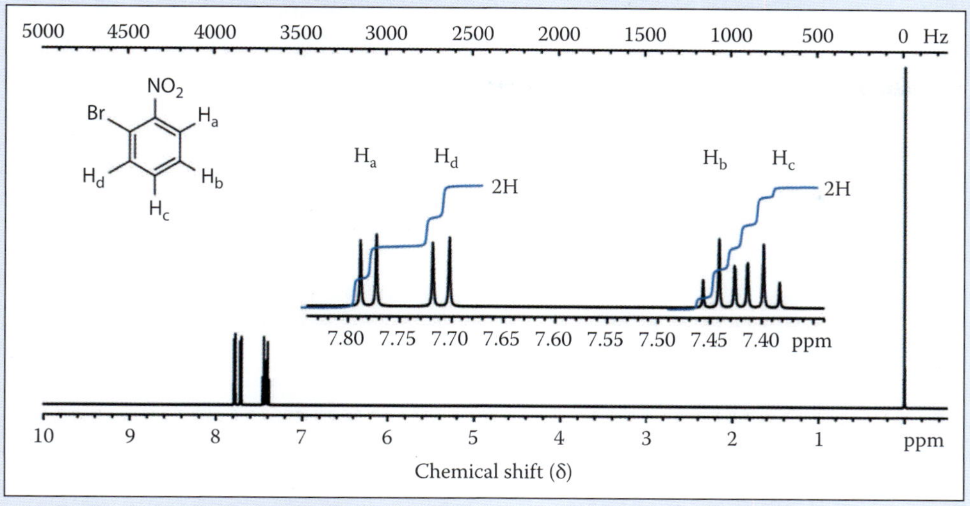

FIGURE 6.26 Proton NMR spectrum of *ortho*-bromonitrobenzene.

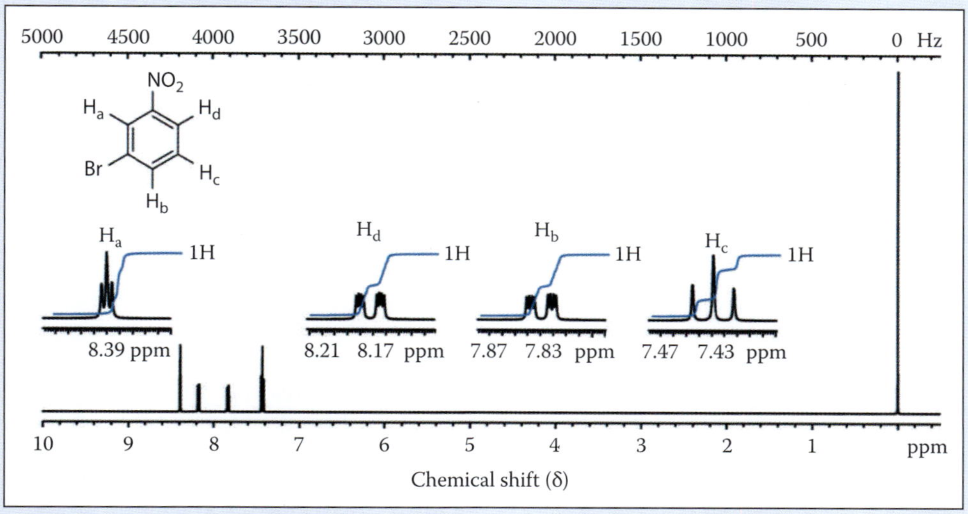

FIGURE 6.27 Proton NMR spectrum of *meta*-bromonitrobenzene. Note that H_b and H_d show *meta*-couplings both to each other and to H_a.

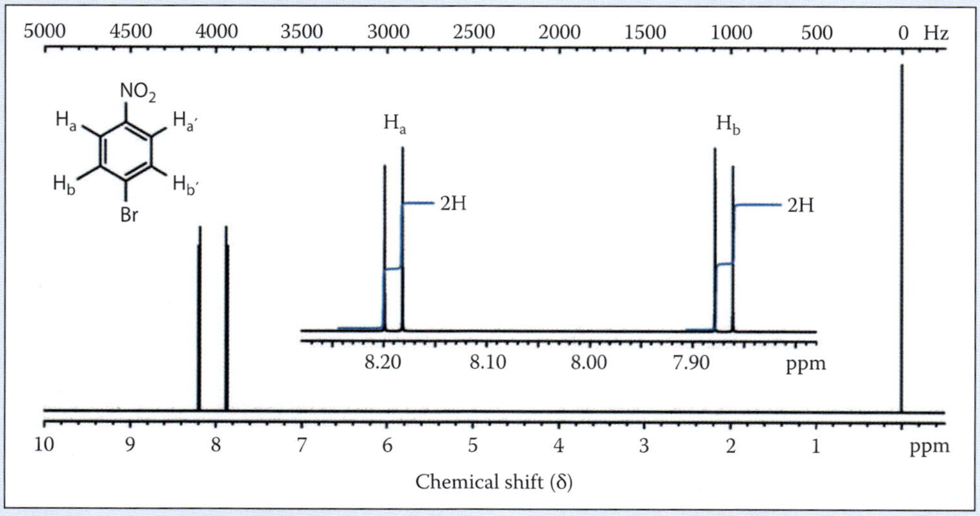

FIGURE 6.28 Proton NMR spectrum of *para*-bromonitrobenzene.

For the *ortho*-isomer, we expect two doublets and two triplets; while the two doublets are easily distinguished, the triplets require expansion to clarify. This points up a common issue with the proton spectra of aromatic compounds; the fact that two signals are inherently different does not invariably mean that we will be able to easily observe the difference. The range available for these signals is relatively small, and overlaps are common. The spectrum of the *meta*-isomer looks complex, but this is mainly due to the fact that the long-range couplings can be observed. If these are ignored, there is one singlet, two doublets, and one triplet, just as predicted.

* H_a is a triplet because of a long-range *meta*-coupling to H_b and H_d. H_b and H_d each show a *meta*-coupling to H_a and also a *meta*-coupling to each other. The values of J are different so the signals appear as doublets of doublets of doublets rather than doublets of triplets.

PROBLEM 6.4

Three isomeric compounds, C_8H_8O, **6.7**, **6.8**, and **6.9**, give rise to the following NMR spectra. Identify the compounds and explain how they were distinguished:

	δ		δ		δ
6.7	9.95 (s, 1H)	**6.8**	7.95 (d, 2H)	**6.9**	9.95 (s, 1H)
	7.8 (d, 2H)		7.5 (m, 3H)		7.65 (d, 1H)
	7.3 (d, 2H)		2.6 (s, 3H)		7.6 (s, 1H)
	2.4 (s, 3H)				7.4 (d, 1H)
					7.3 (t, 1H)
					2.4 (s, 3H)

Solution

Both the formulae and the chemical shifts of some of the protons indicate that these are all aromatic compounds. The DBE is 5 for all three, which indicates that there is one degree of unsaturation that is not accounted for by the benzene ring (three double bonds plus ring). Given the presence of an oxygen atom, this is likely to be a carbonyl group. The singlet peaks at δ 9.95 in **6.7** and **6.9** can readily be assigned as an aldehyde proton. **6.8** lacks such a signal and must therefore be a ketone; the only possible structure is PhC(=O)CH_3. The five aromatic protons are also indicative of a monosubstituted benzene ring; the signals from the protons at the *meta*- and *para*-positions overlap, which is not uncommon. **6.7** and **6.9** are therefore isomers of methylbenzaldehyde; the signals at δ 2.4 are assigned to a methyl group attached to an aromatic ring. **6.7** is easily identified as the *para*-isomer by its symmetry; there are only two types of aromatic protons. All four aromatic protons would be distinct in either the *ortho*- or *meta*-isomers. With the *ortho*-isomer, we expect two doublets and two triplets, whereas the *meta*-isomer would give one singlet, two doublets, and a triplet. Hence, **6.9** is the *meta*-isomer.

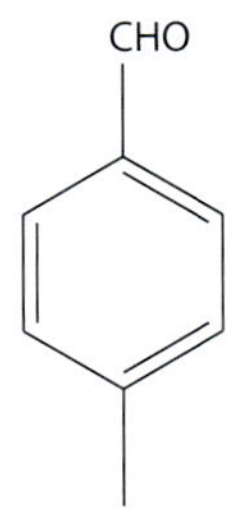

6.7

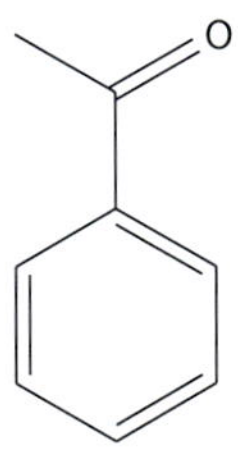

6.8

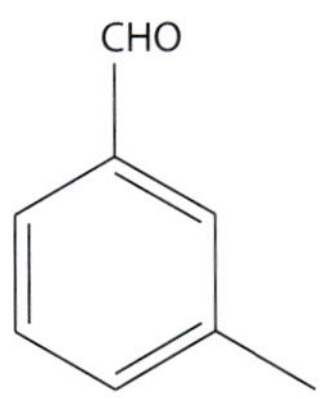

6.9

The observation that couplings across *cis*- and *trans*-alkenes are not identical means that the spectra of monosubstituted alkenes will not be simple—although all the protons that are coupling are on adjacent carbon atoms, they are no longer equivalent. If we consider a generic

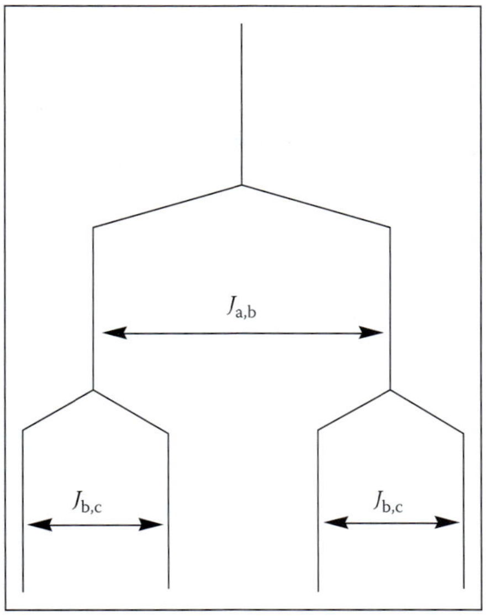

FIGURE 6.29 Splitting of H_b in **6.10**.

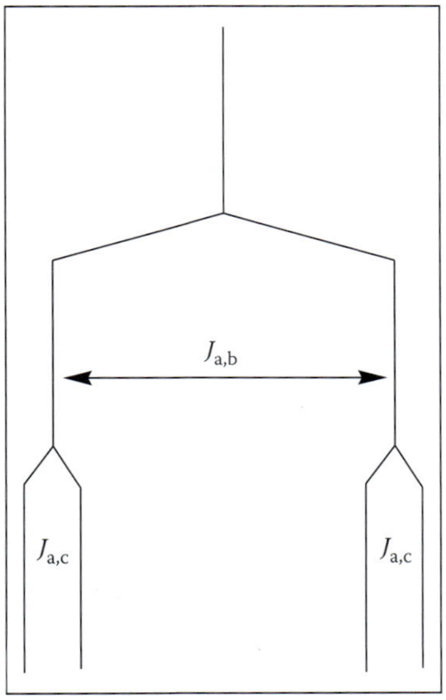

FIGURE 6.30 Splitting of H_a in **6.10**.

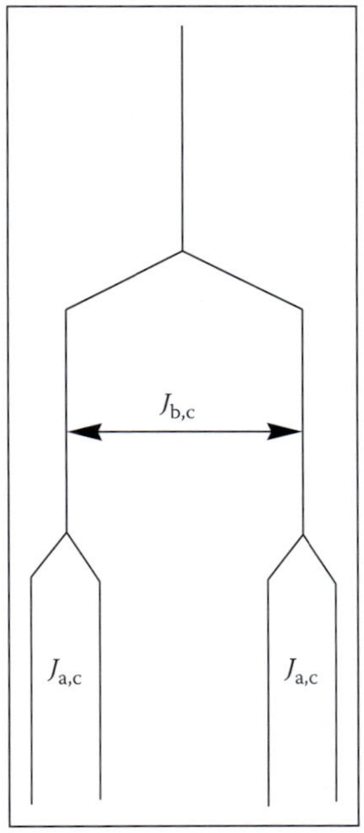

FIGURE 6.31 Splitting of H_c in **6.10**.

monosubstituted alkene, **6.10**, the atoms H_a, H_b, and H_c will all be quite distinct and J_{ab} = 12–18 Hz, J_{bc} = 7–10 Hz, and J_{ac} = 0–2 Hz (J_{ab} and J_{bc} are vicinal couplings and J_{ac} is a geminal coupling). So what would we expect to see for the signal assigned to H_b? There is a large coupling to H_a and a smaller one to H_c. In these systems, we consider the largest coupling first. H_b will be split into two peaks by H_a. Next, each of these two peaks will be split into two by H_c. Thus, we have four lines, and we describe this as a doublet of doublets (Figure 6.29). We can apply similar considerations to H_a and H_c (Figures 6.30 and 6.31).

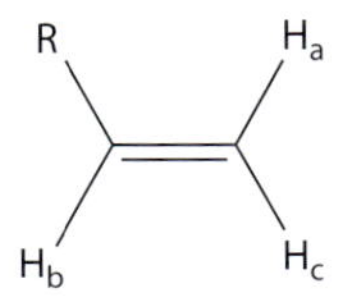

6.10

Now we can examine the spectrum of a real example, the proton NMR spectrum of **6.11** (Figure 6.32). For this particular example, J_{cd} = 14 Hz, J_{bd} = 6 Hz, and J_{bc} = 2 Hz. The nine-proton

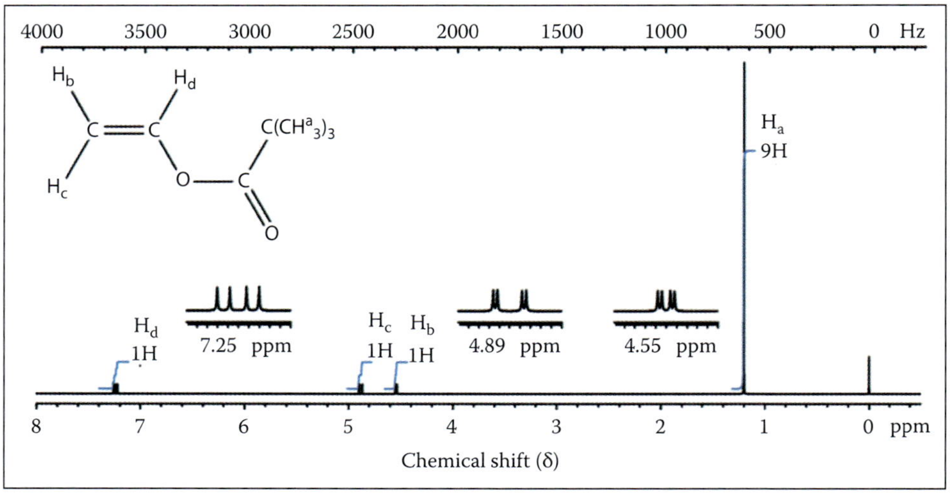

FIGURE 6.32 Proton NMR spectrum of vinyl 2,2-dimethylpropanoate.

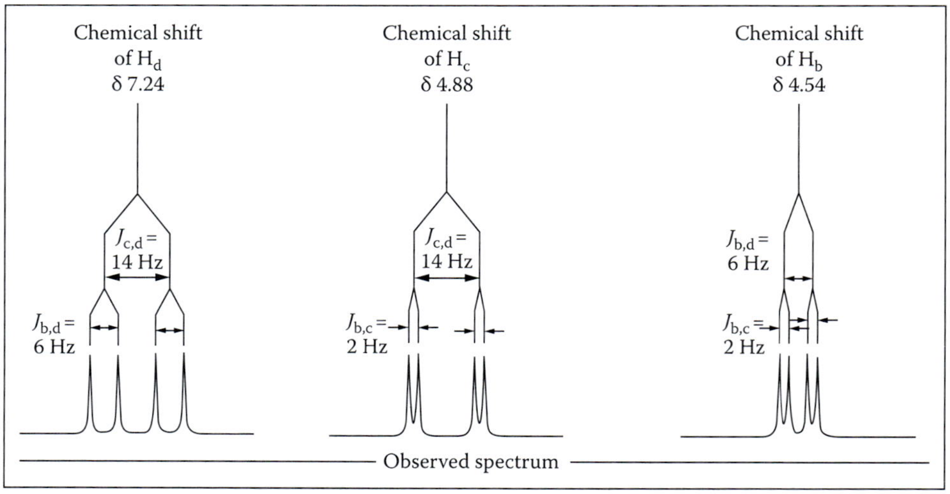

FIGURE 6.33 Schematic splitting diagram for the vinyl protons of vinyl 2,2-dimethylpropanoate.

singlet for H_a is at δ 1.2, and the patterns for the three vinyl protons are essentially as described previously (Figure 6.33).

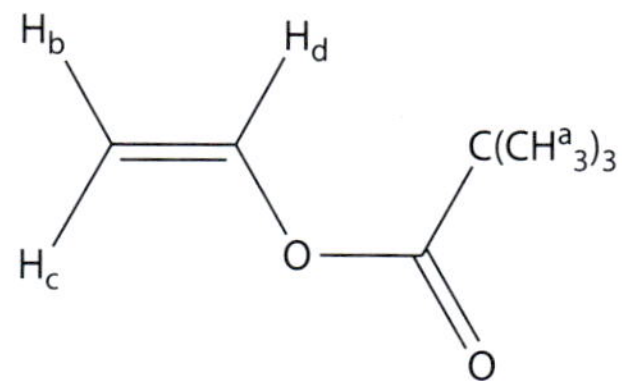

6.11

PROBLEM 6.5

The compound $C_6H_{10}O$ is insoluble in aqueous sodium carbonate solution and shows strong bands in its IR spectrum at 1680 and 1630 cm^{-1}. Its ^{1}H NMR spectrum is detailed in the following. Identify it and explain the spectrum:

δ 9.5 (d, J = 7 Hz, 1H)
6.9 (d of t, J = 15, 6 Hz, 1H)
6.1 (d of d, J = 15, 7 Hz, 1H)
2.3 (q, J = 6 Hz, 2H)
1.5 (m, 2H)
1.0 (t, J = 6.5 Hz, 3H)

Solution

With this type of problem, there is no simple answer to the question of "where do I start?" You should always start with the information that *you* can readily interpret, then work through the rest, but this will not give the same order of events for everyone. When you have a structure, it is good practice to go back over all the data, to make sure that everything fits and that you have not become enamored of one answer early on and bent other data to fit.

DBE = 2. Insolubility in sodium bicarbonate solution indicates that this is not a carboxylic acid (there is no OH peak in the IR spectrum either). The IR peaks at 1680 and 1630 cm^{-1} suggest a carbonyl conjugated with a double bond (sneaking ahead to the NMR spectrum, there's an aldehyde proton and various alkene-type protons).

The 3H singlet at δ 1 is a methyl group, and it is next to a CH_2 (split into a triplet). The coupling constant we are given doesn't match anything else, so the best option for the next group along is the CH_2 multiplet. This CH_2 must have a proton bearing carbon on its other side, as it is not simply a quartet from coupling to the methyl group.

The ^1H signal at δ 9.5 must be an aldehyde, and it is coupled to one proton—that at δ 6.1. The two protons at δ 6.1 and 6.9 must be attached to an alkene (from their chemical shifts), and they are coupled to each other by 15 Hz, indicating that they are *trans* to each other. We now have a clear view of two parts of the molecule, **6.12** and **6.13** (chemical shifts in red, coupling constants in blue):

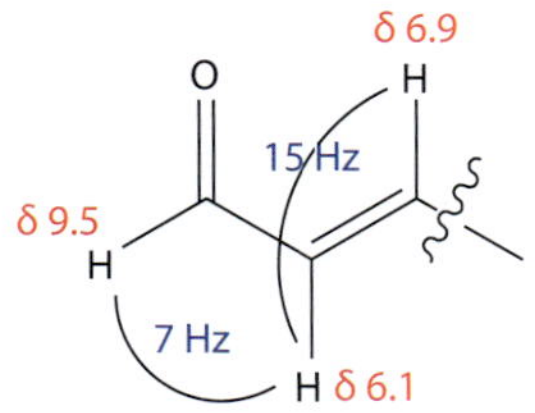

6.12

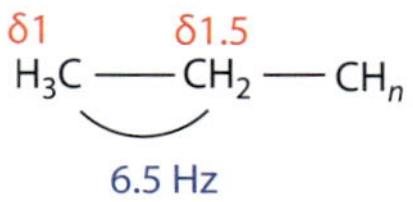

6.13

We have now solved the problem, since we have "found" all of the molecule except one CH_2 group—and there is only one place this can go, joining up the two segments (**6.14**):

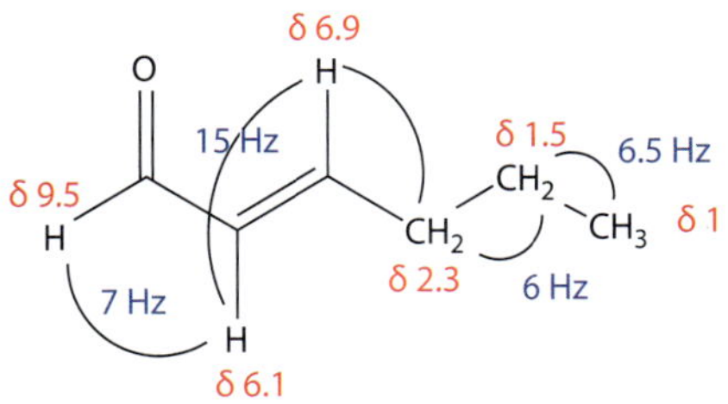

6.14

The CH_2 group at δ 2.3 is a quartet because it is coupled to both the adjacent CH and the adjacent CH_2 groups; fortuitously, the coupling constants are roughly the same. The signal at δ 6.1 is a doublet of doublets, with a large coupling to the proton *trans* to it across the double bond and a smaller one to the aldehyde proton. The signal at δ 6.9 is a doublet of triplets—a large coupling to the proton *trans* to it across the double bond and a smaller one to the adjacent CH_2 group.

When solving this type of problem in a test, it's important that you read the question carefully and answer the question actually asked. If all you need is the structure, get there as fast as is consistent with getting it right, but don't spend time on explaining your assignment of every parameter. On the other hand, if you are asked to assign or explain the spectrum, make sure that you do this clearly—probably half or more of the marks for the question will be allocated for that. Conversely, with that type of question, always assign as much as you can even if you can't get to the solution—you should get some partial credit.

6.3.3 SPIN–SPIN DECOUPLING

Because the range of values of *J* is small, in a complex molecule it may not always be easy to assign the pairings. Additional information can be obtained using ***spin decoupling***. The technique involves irradiating one resonance specifically; this signal is then not observed and all the couplings to other atoms disappear as well (Figure 6.34). A practical example is provided by the spectrum of 1-nitropropane (Figure 6.35). A triplet at δ 4.35 is assigned to the CH_2 next to the nitro group; it is a triplet because it is next to a CH_2. The signal at δ 2.04 is assigned to the protons at C-2; this is split into six lines because it is coupled both to the CH_3 and the other CH_2 groups. Finally, the signal at δ 1.02, the methyl group, is coupled to the adjacent methylene and is hence a triplet. The coupling constants here are very similar, 7.2 and 7.5 Hz. Figures 6.36 through 6.38 show the effects of irradiating in turn each signal. When the methyl group is irradiated the adjacent methylene group appears as a simple triplet—it is now coupled only to C-1H_2. Irradiating C-2H_2 collapses both the other methylene group and the methyl group to singlets—they are both coupled to it. Irradiation of C-1H_2 collapses the other methylene to a quartet, as it is now only coupled to the methyl group.

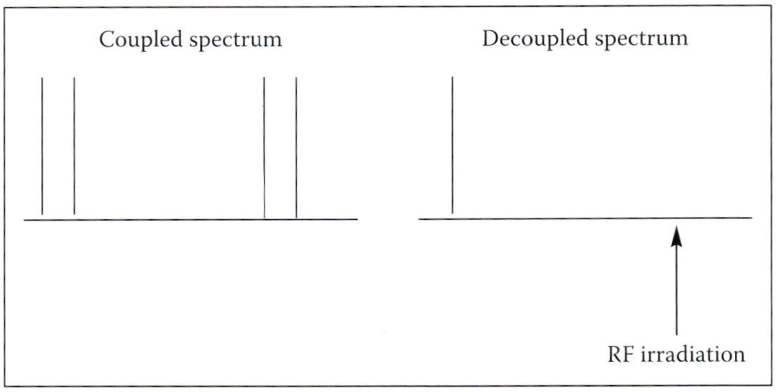

FIGURE 6.34 Decoupling of a simple two-proton system.

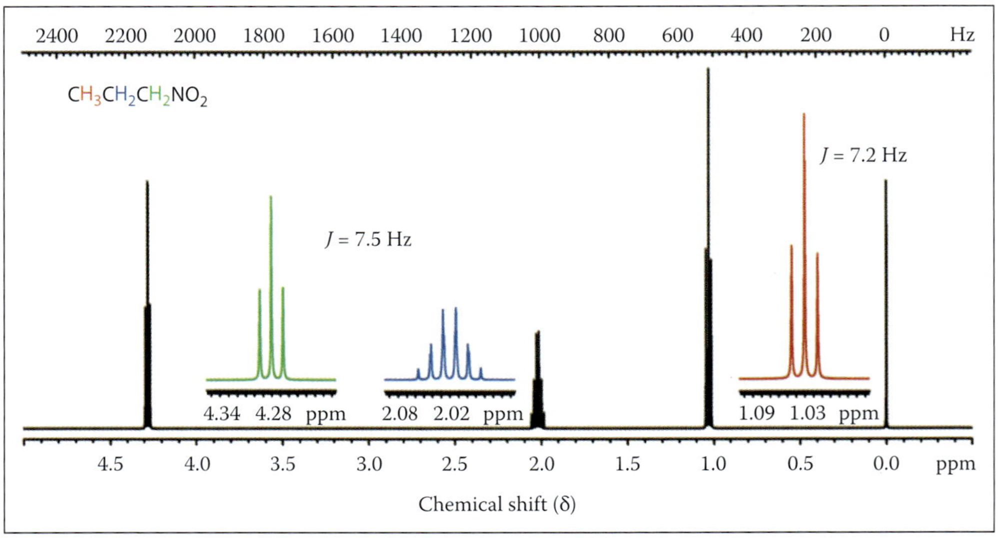

FIGURE 6.35 Proton NMR spectrum of 1-nitropropane.

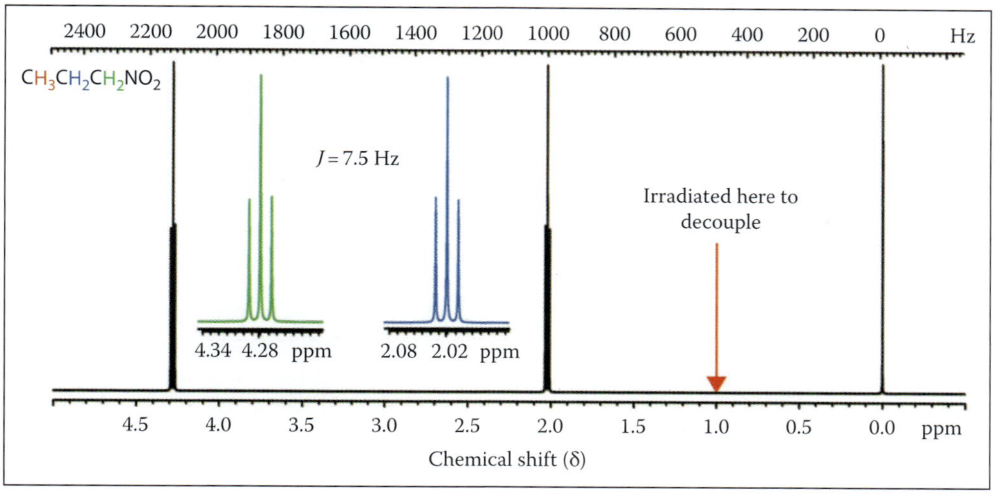

FIGURE 6.36 Proton NMR spectrum of 1-nitropropane with the methyl group irradiated.

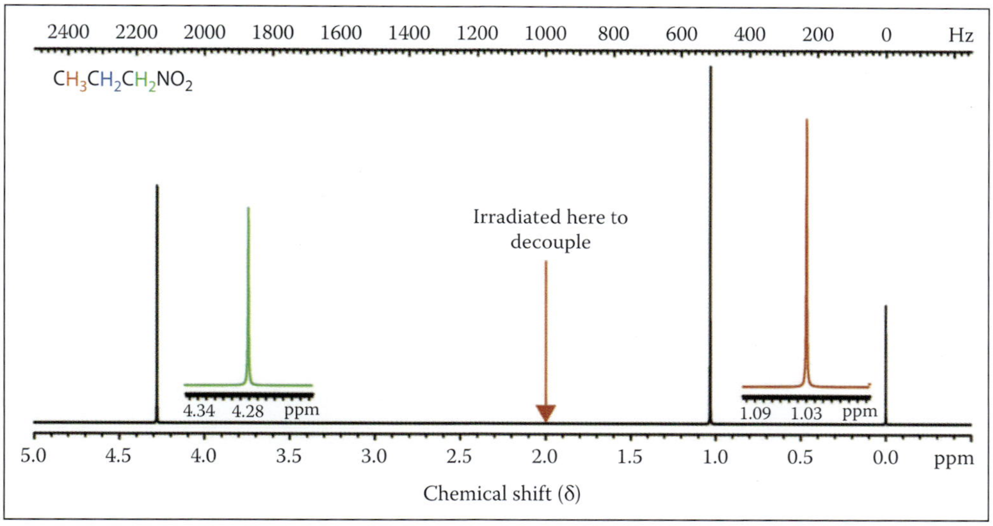

FIGURE 6.37 Proton NMR spectrum of 1-nitropropane with the C-2H_2 group irradiated.

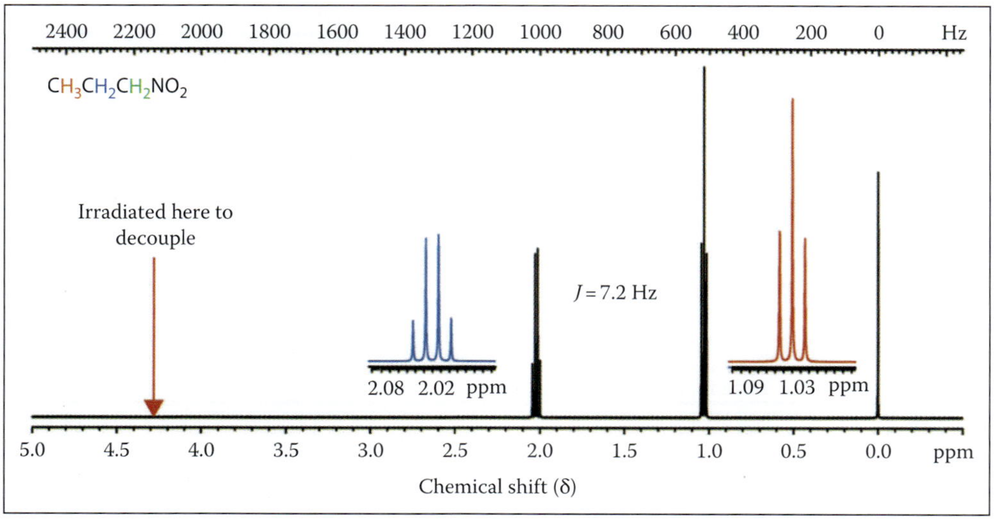

FIGURE 6.38 Proton NMR spectrum of 1-nitropropane with the C-1H_2 group irradiated.

6.4 PROTONS ATTACHED TO OXYGEN AND NITROGEN

The signals assigned to OH and NH groups are almost always rather broad, and their chemical shifts are not particularly easy to predict, but they are fortunately rather easy to identify. The protons attached to carboxylic acids invariably appear as very broad signals, in the region δ 10–13 (*e.g.*, Figure 6.39). Since there are rarely any other signals in this region, they are easy to identify; the only problems arise when the signal is so broad that it is missed completely. The breadth of the signal is attributed to the wide range of hydrogen-bonded forms existing in dynamic equilibrium in solution. However, the chemical shifts of the OH groups of simple alcohols are much more variable, ranging from δ 2 to 7. All types of –OH protons can be identified by shaking the sample with D_2O. The equilibrium of reaction (6.4) is established and is far to the right because D_2O is added in excess. The signal due to the original OH group disappears, and a signal for HOD, at a new position, appears. While deuterium NMR spectra can be acquired, the frequencies involved are quite different. Amines, RNH_2 and R_2NH, also exchange deuterium with D_2O, but the reaction is generally a little slower. The signals of almost all hydrogens attached to nitrogen atoms, whether they exchange with anything external or not, are somewhat broad. This is because ^{14}N has a nuclear quadrupole; the theory of this is beyond the scope of this course, but the broadening effect may be noticeable to you in looking at the spectra of nitrogen-containing compounds. The only exception is when the nitrogen atom is in a symmetric environment as in $[R_4N]^+$.

(6.4)

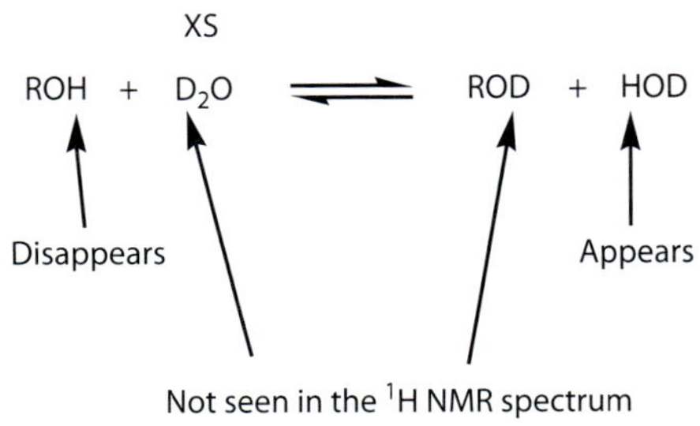

Under normal operating conditions, it is unusual to see any coupling between the protons of a hydroxyl group and protons on the carbon atom to which it is attached. This is exemplified by the proton NMR spectrum of ethanol under "ordinary" conditions compared

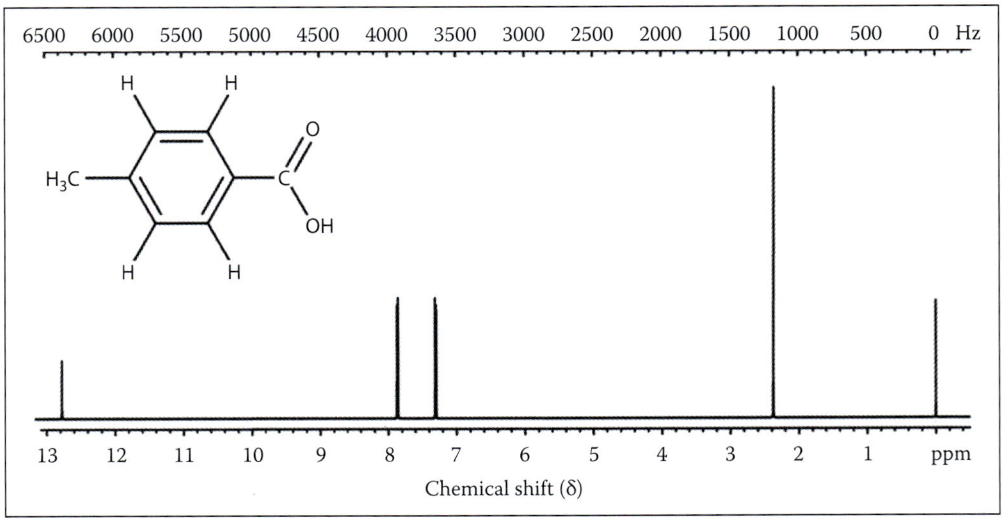

FIGURE 6.39 Proton NMR spectrum of 4-methylbenzoic acid.

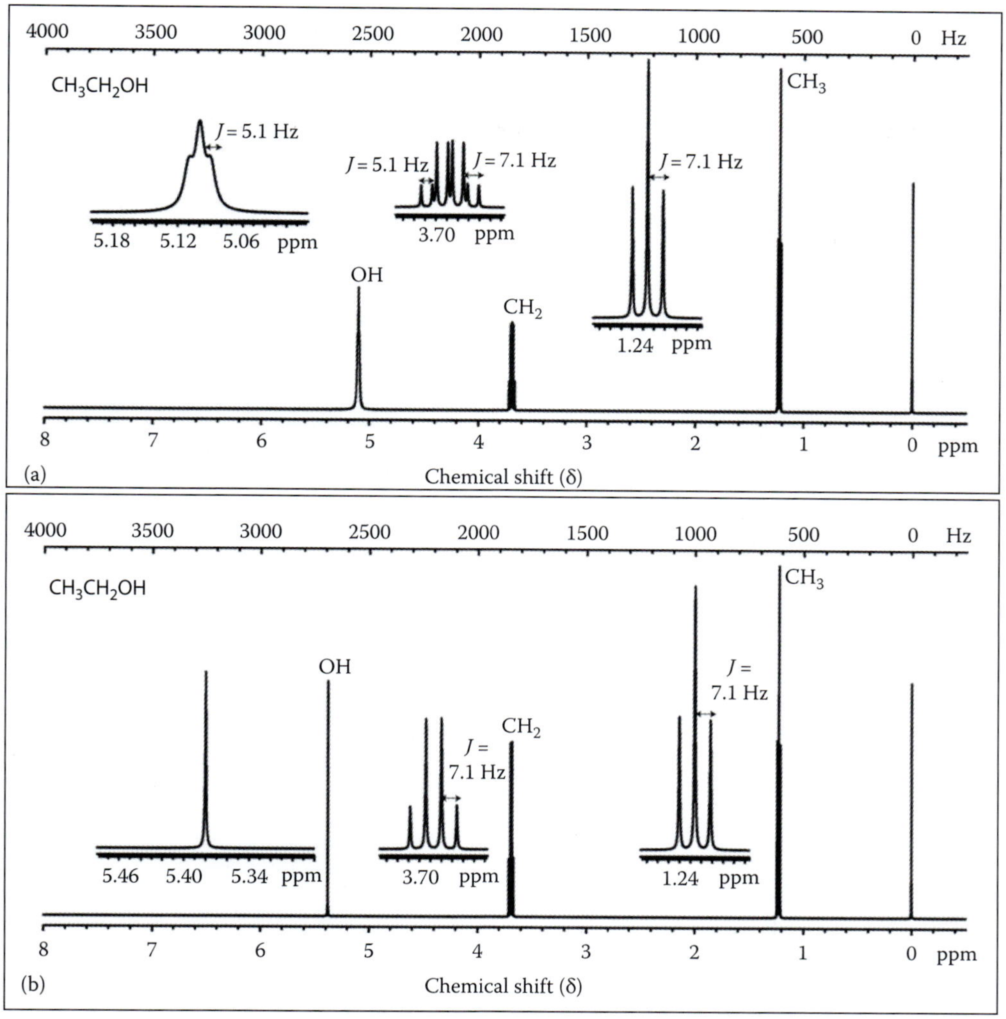

FIGURE 6.40 Proton NMR spectra of (a) dry and (b) "wet" ethanol.

with a spectrum of superdry and acid-free ethanol (Figure 6.40). In very dry ethanol, all the couplings are seen. The OH group is a triplet (albeit a broad one), and the CH_2 group is a complex multiplet because the coupling between the CH_2 and the CH_3 is rather different from that between CH_2 and OH. The methyl group is a simple triplet. However, with "wet" ethanol, the OH signal is a singlet and the CH_2 and CH_3 signals are similar to those of any other ethyl group, a quartet and a triplet. In wet ethanol, the OH protons exchange very rapidly between ethanol molecules, and the incoming proton will have a 50 % probability of having the opposite spin state to that of the departing one. Thus, each CH_2 effectively "sees" its OH protons constantly changing between spin states with every exchange; this is sometimes described as chemical decoupling. Notice that the position of the OH signal has also changed; quite generally, the position of OH signals depends on pH, concentration, and the extent of hydrogen bonding.

Two examples will illustrate. The proton spectrum of 2-methyl-3-butyne-2-ol is shown in Figure 6.41, and the spectrum after the sample has been shaken with D_2O in Figure 6.42. In the original spectrum, it is not easy to tell which of the two one-proton singlets is which, but they are unequivocally assigned after the exchange with deuterium. The same is true for 4-hydroxy-4-methyl-2-pentanone (Figures 6.43 and 6.44). It is of interest that the chemical shifts for the OH peak in these two compounds are quite different. In the hydroxy ketone, there is the opportunity for the formation of a hydrogen bond between the alcohol and the ketone; such hydrogen bonding has a significant effect on the chemical shift.

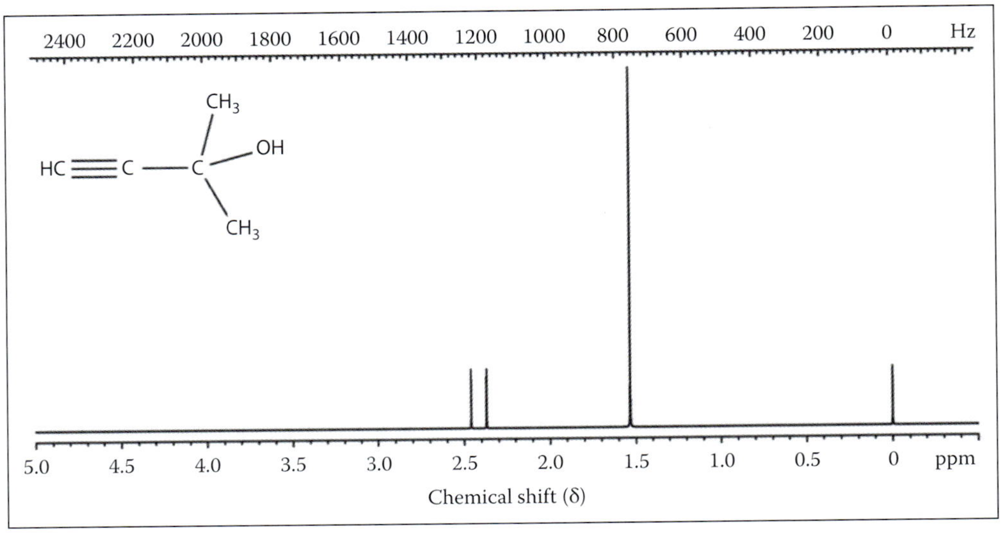

FIGURE 6.41 Proton NMR spectrum of 2-methyl-2-butyn-2-ol.

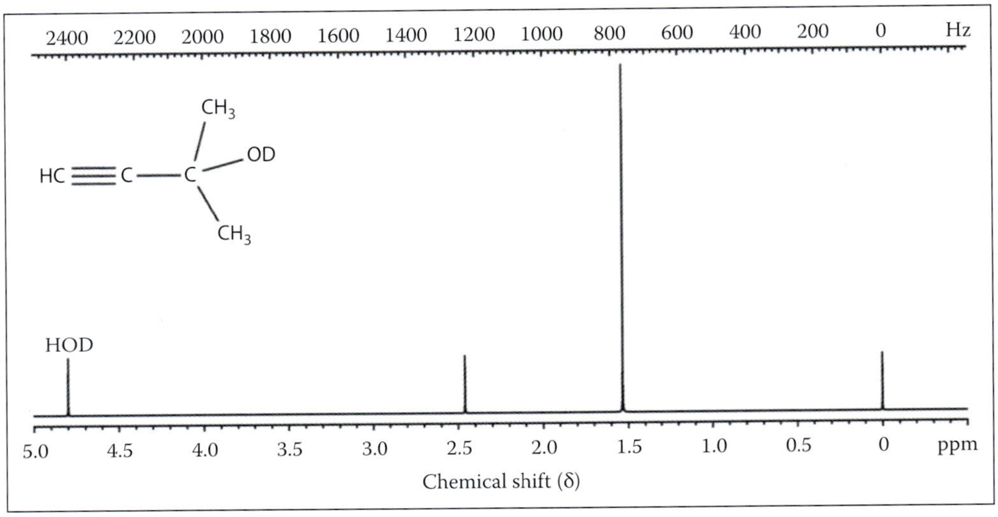

FIGURE 6.42 Proton NMR spectrum of 2-methyl-2-butyn-2-ol after D_2O shake.

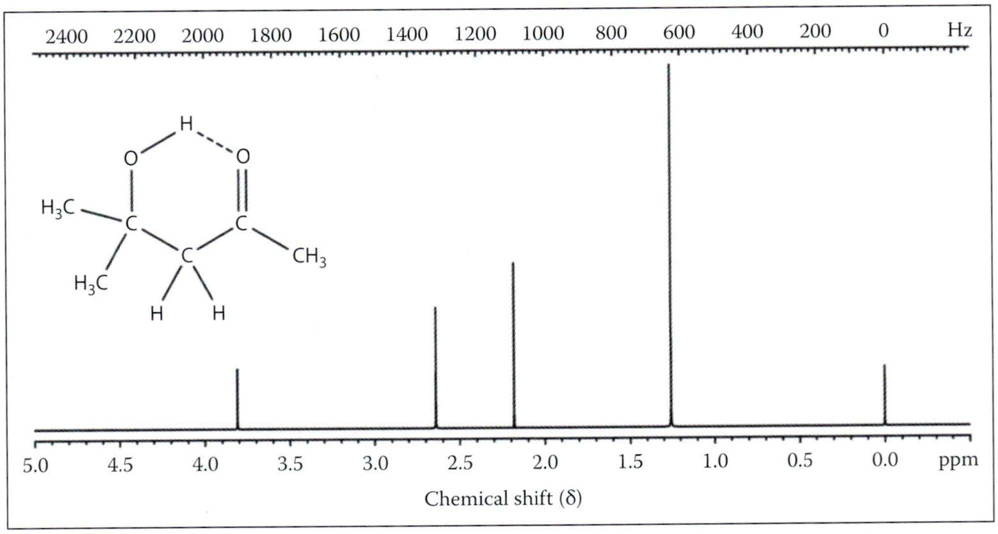

FIGURE 6.43 Proton NMR spectrum of 4-hydroxy-4-methyl-2-pentanone.

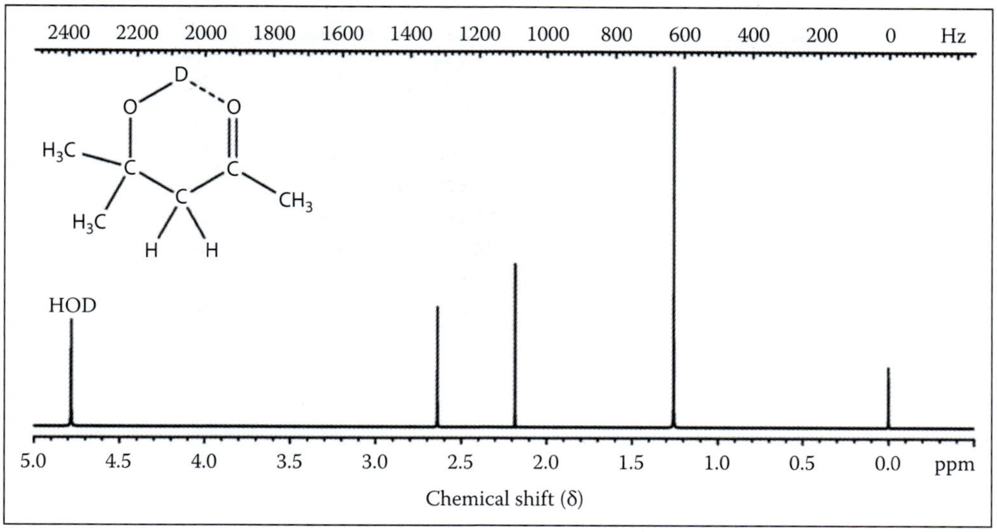

FIGURE 6.44 Proton NMR spectrum of 4-hydroxy-4-methyl-2-pentanone after a D_2O shake.

Focus on OH Chemical Shifts: pK_a, H-Bonding, and Temperature

Thus far, we have considered the chemical shifts of OH protons in carboxylic acids and alcohols, acids δ 10–13 and alcohols δ 2–7. Phenols are intermediate between these two (δ 4–8), as indeed they are in terms of acidity. This will be discussed more fully in Chapter 8, but for now the approx. pK_as of acids, phenols, and alcohols are 5, 10, and 15, respectively. Hence, it would seem that the more acidic the proton, the higher the chemical shift. However, this is not the whole story—the extent of hydrogen bonding is critical in determining the observed chemical shift for alcohols and phenols (carboxylic acids are hydrogen bonded under all conditions of temperature and solvation).

The proton NMR spectrum of methanol at 30 °C shows two signals, both slightly broadened singlets at δ 4.8 and 3.35 ppm. As the temperature of the sample is decreased, the rate of exchange of the OH protons is slowed until at −45 °C, the signal at δ 3.35 is a doublet, and the OH proton appears as a quartet at δ 5.5. The difference between the two chemical shifts shows a linear dependence on temperature and was used to establish probe temperature in older NMR spectrometers where accurate external calibration was unavailable. At lower temperature, there is more hydrogen bonding, as well as a slower rate of exchange. The factors affecting the rate of exchange are multiple and include the sample temperature, the chosen solvent, the concentration, and pH of the solution.

The importance of hydrogen bonding is also demonstrated by studying the proton NMR spectra of dilute methanol solutions (the solution is dilute so that the methanol cannot hydrogen bond with itself) in solvents that have more or less hydrogen-bonding ability (Table 6.5). The chemical shift of the methyl group varies little; small changes between solvents are common, and there is no need to interpret them. However, the chemical shift of the OH group changes rather dramatically between $CDCl_3$, which has no capacity to act as a hydrogen bond acceptor, and DMSO, which is an excellent hydrogen bond acceptor. In some cases, the use of DMSO as a solvent can also "restore" the coupling between the OH proton and protons on the adjacent carbon atom.

The effect of hydrogen bonding on chemical shift is also highlighted by the variation in the chemical shift of the OH proton in phenol with concentration—the higher the concentration, the greater the hydrogen bonding (Table 6.6). Where intramolecular hydrogen bonding is inevitable, as in **6.15**, high values of δ (12.25 ppm) are noted. The *meta*- and *para*-isomers have more normal chemical shifts (δ 7.44 and 8.69 ppm, respectively) for the phenolic proton.

TABLE 6.5 Proton NMR Spectrum of Methanol in Other Solvents

Solvent	$CDCl_3$	CD_3CN	$(CD_3)_2C{=}O$	$(CD_3)_2S{=}O$
δ (CH_3)	3.4	3.28	3.31	3.16
δ (OH)	1.10	2.16	3.12	4.01

TABLE 6.6 Proton Chemical Shift of the OH Proton in Phenol

Concentration (% in $CDCl_3$)	100	20	10	5	2	1
δ (OH)	7.45	6.8	6.45	5.95	4.90	4.35

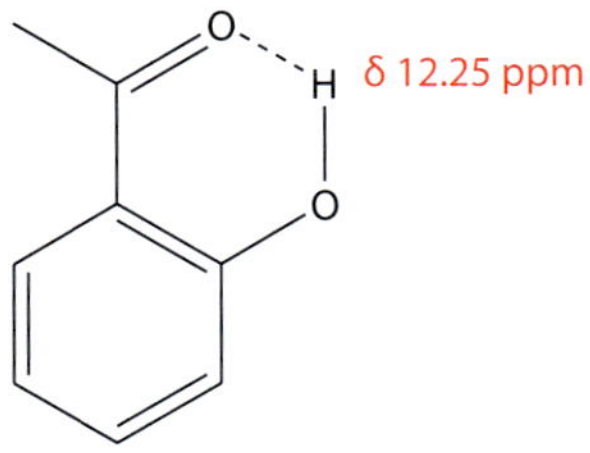

6.15

PROBLEM 6.6

Two isomeric compounds, **6.16** and **6.17**, have the formula $C_8H_{10}O_2$ and show a molecular ion M^+ at ***m/z*** 138. The main fragment ions are as follows:

6.16 121, 76
6.17 121, 107

Their IR spectra show key bands as follows:

6.16 3352 cm^{-1} (br)
6.17 3326 cm^{-1} (br)

Their ^1H NMR spectra are as follows:

6.16 δ 7.3 (d, 2H, J = 7 Hz)
6.8 (d, 2H, J = 7 Hz)
4.44 (s, 2H)
3.7 (s, 3H)
3.55 (s, 1H, exchanges with D_2O)

6.17 δ 7.67 (s, 1H, exchanges with D_2O)
7.2 (d, 2H, J = 7 Hz)
6.8 (d, 2H, J = 7 Hz)
4.53 (s, 2H)
3.5 (s, 3H)

Deduce the structures for **6.16** and **6.17**, explaining clearly how you differentiate between them.

Solution

Both the IR spectra and the proton NMR spectra clearly show that both of these compounds contain a single exchangeable OH group. However, these OH groups have very distinct chemical shifts—perhaps a hint that one is a phenol and one an aliphatic alcohol? Looking at the pattern of the aromatic protons, both **6.16** and **6.17** contain *para*-disubstituted benzene rings. With the OH groups, this accounts for C_6H_5O of the molecular formula, leaving "C_2H_5O" to find. Both the proton NMR spectra show signals for one CH_3 and one CH_2 group—and these are not adjacent, as they are not coupled. In both compounds, the chemical shift tells us that the CH_2 and CH_3 groups are bonded to oxygen (there is no sign of a carbonyl absorption in the IR spectrum, so the other oxygen must be part of an ether). There are now only two ways that we can fit the pieces of the molecule together, 6.x and 6.y:

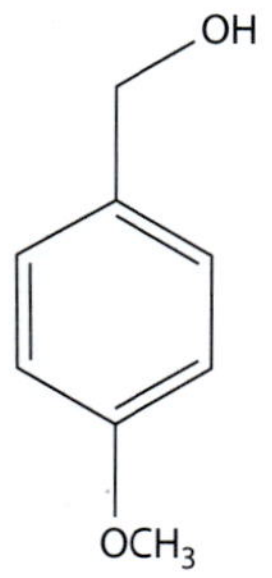

6.x

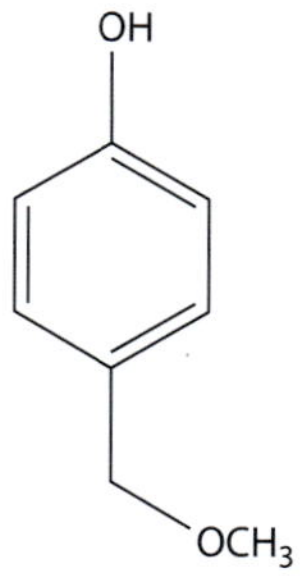

6.y

Distinguishing between these two structures spectroscopically is not trivial. The chemical shifts of the CH_2 and CH_3 groups are similar, as would be expected, given the similarity of their environments. The values of the chemical shifts of the aromatic protons, equally, are so similar as to be uninformative. The biggest difference is in the chemical shifts of the OH peaks. Since shift is roughly correlated with pK_a, we would expect a higher shift for the phenol than for the alcohol and would assign **6.16** as 6.x and **6.17** as 6.y. Chemically, it would be easy to distinguish the two—the phenol would be expected to have a pK_a of about 10 and would be soluble in sodium hydroxide solution. The benzyl alcohol would have a pK_a of about 15 and would not dissolve in sodium hydroxide solution.

Focus on Magnetic Resonance Imaging

Magnetic resonance imaging ([MRI] sometimes magnetic resonance tomography) has been developed over the last 20 years as a method of studying the human body that involves no ionizing radiation and is hence thought to be extremely safe. Paul Lauterbur and Peter Mansfield received the Nobel Prize for this work in 2003. The human body contains a very large fraction of water, and it is the protons in the water molecules that are studied. The main, high-field, magnet is used, with an RF pulse, to align some of the proton spins in the same direction. The gradient magnets (much lower fields) are effectively used to produce an additional magnetic field superimposed on the first, which mimics the effect of the chemical shift such that H_2O molecules in different parts of the body have different chemical shifts. The gradients can be manipulated around the axis of the patient's body, and the resulting images produce virtual slices through the area of the scanned body. Where there appears to be a lot of water, the image will be bright, whereas where there is little water, as in bone, it will be dark. While water may be in different environments in different tissues, you might be tempted to ask how different can these be—how is any distinction made? The differences arise from the water density in different tissues, different chemical shifts in different environments, and from the rates at which the proton spins return to equilibrium

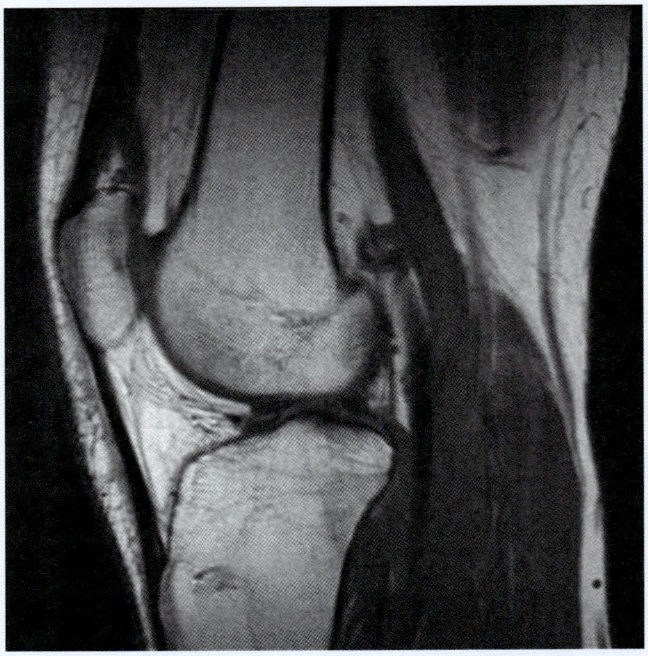

FIGURE 6.45 MRI scan of a human knee.

(these are called relaxation rates; often the patient may be given a solution of a proton relaxation reagent to improve the image). The technique is particularly valuable in detecting tumor masses, and the technology is constantly improving. Early scanners were narrow, slow, and claustrophobic, but the time for image acquisition has now been reduced and so-called open scanners have been developed. All remain, however, extremely noisy and cannot be used for patients with any type of ferromagnetic implant (such as a heart pacemaker or cochlear implant). Figure 6.45 shows an MRI image of a knee.

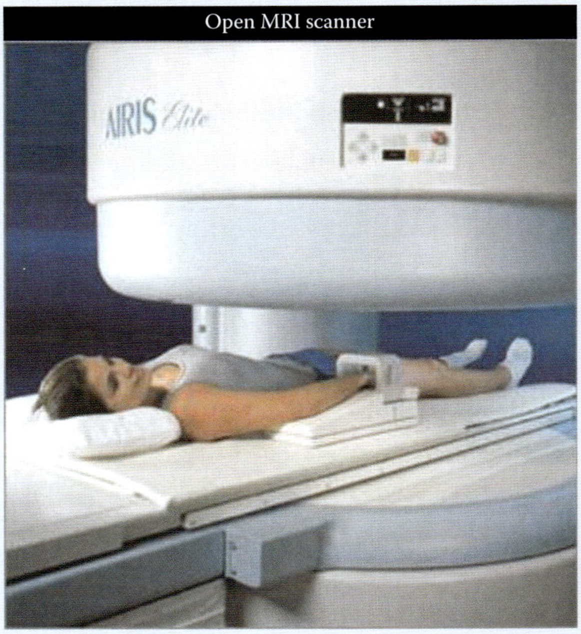

Key Points from Sections 6.3.2, 6.3.3, and 6.4

- *cis*- and *trans*-relationships between protons attached to alkenes can be distinguished by the magnitude of the *J* values, 6–10 Hz for *cis*-protons and 12–18 Hz for *trans*-protons.
- Couplings between vicinal protons on benzene rings are similar to those for *cis*-alkenes, in the range 7–10 Hz. Longer-range couplings are sometimes observed.
- Coupling patterns in alkene systems may be quite complex; they can be predicted by starting with the largest coupling constant for a particular proton and proceeding successively to the smaller ones.
- In spin–spin decoupling, *all* the couplings to a particular proton may be removed/suppressed by irradiating with RF corresponding to the resonance frequency for that proton.
- Chemical shifts, and the observation or otherwise of couplings, for protons attached to oxygen (and nitrogen) depend on the extent of proton exchange and hydrogen bonding, which in turn depend on temperature, solvent, pH, and pK_a of the proton in question.
- OH and NH peaks can be identified by shaking the sample with D_2O—they then disappear.

6.5 ^{13}C NMR SPECTROSCOPY

Ordinary carbon contains only 1.1 % of ^{13}C, so the acquisition of ^{13}C NMR spectra is clearly more difficult than for protons, which are almost 100 % abundant. The gyromagnetic ratio of ^{13}C is also one quarter that for a proton. These two factors combine to produce a nucleus that is over 5000 times more difficult to observe. Most ^{13}C atoms in organic molecules have hydrogens bonded to them—and hence coupled to them, thus further reducing the intensity of individual signals. One-bond C–H coupling is in the range 130–270 Hz, and two- and three-bond couplings are also significant in magnitude, so a spectrum could be very complex indeed. Thus, to obtain ^{13}C NMR spectra, even with modern instrumentation, it is often necessary to obtain more scans to get acceptable signal to noise in comparison with a proton spectrum.

Despite the difficulties, ^{13}C NMR spectra represent a powerful addition to our armory of techniques for structure determination, and there are some advantages. $^{13}C^{13}C$ coupling can be neglected, as only one molecule in 10^4 has two adjacent ^{13}C nuclei. Also, the chemical shift range is greater (Figure 6.46), so overlap is rare. Even for quite complex molecules, we can make a first-order assumption that the number of signals is the same as the number of distinct carbon atoms.

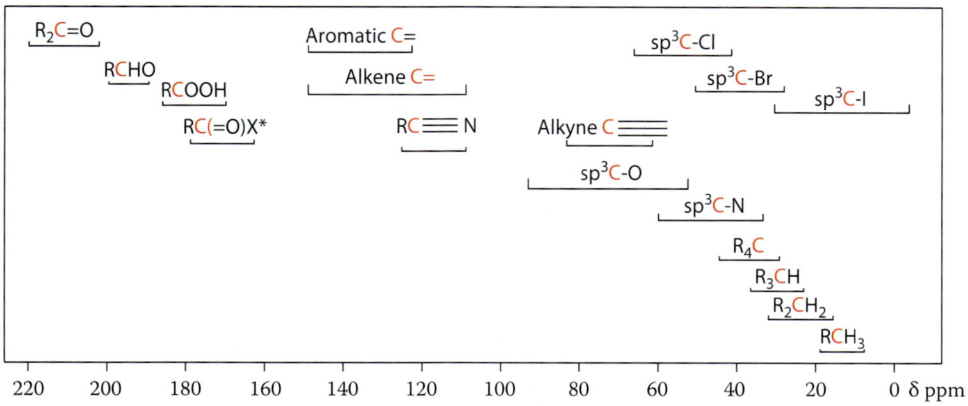

FIGURE 6.46 ^{13}C chemical shift ranges. *X = Cl, OR, NH_2, NHR.

Because ^{13}C is coupled both to directly attached, and more distant protons, it is normal to decouple all the protons as the spectrum is acquired. Technically, this involves irradiation of the proton spectrum with a band of appropriate RFs covering the entire proton chemical shift range; the practical result is that a single line is seen for each ^{13}C resonance. There are many data for ^{13}C chemical shifts, and the factors determining them are reasonably well understood. There exist computer programs that allow prediction of the chemical shifts of many carbon atoms in simple (and some more complex) molecules. Although the theory of the subject is reasonably well developed, most programs rely on empirical processes—what is the effect of adding a methyl group or a halogen at a particular position relative to the carbon atom under study?

One important difference from proton NMR spectroscopy is that the size of the signal does not give information as to how many of a particular type of carbon atom it represents. Many factors affect signal intensity, including how the spectrum was run. While the theory is beyond the scope of this course, we can make an important generalization; the decoupling process results in the observation that carbon atoms with directly attached protons produce more intense signals than those that have none. This is exemplified by the ^{13}C NMR spectrum of 4-amino-3-methylbenzoic acid, **6.18** (Figure 6.47). Although each of the signals can be assigned to a particular carbon atom, they are not all of the same intensity, and the quaternary carbon atoms are lower in intensity than those with attached protons.

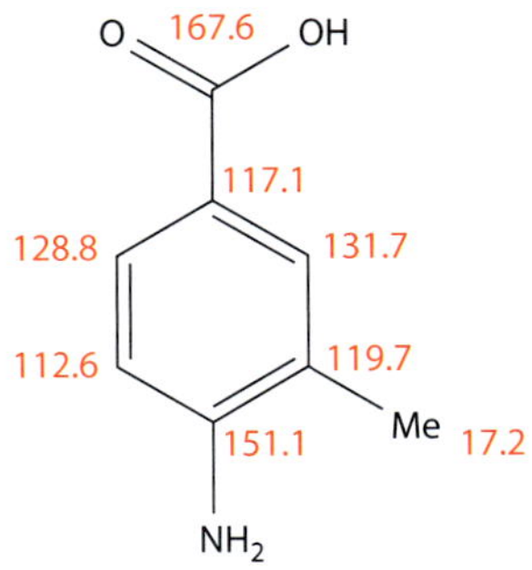

6.17, with assigned chemical shifts (δ values)

In some spectra, especially of dilute solutions in $CDCl_3$, you will be able to see the signal for the carbon atom of the solvent. This appears at δ 77 and is a 1:1:1 triplet, J = 32 Hz. This triplet reflects the coupling to deuterium, which has a spin of 1. This means that there are three possible spin states for deuterium, +1, 0, and −1—hence three lines. More complex patterns are obtained from solvents containing multiple deuterium atoms, but fortunately the signals are generally weak.*

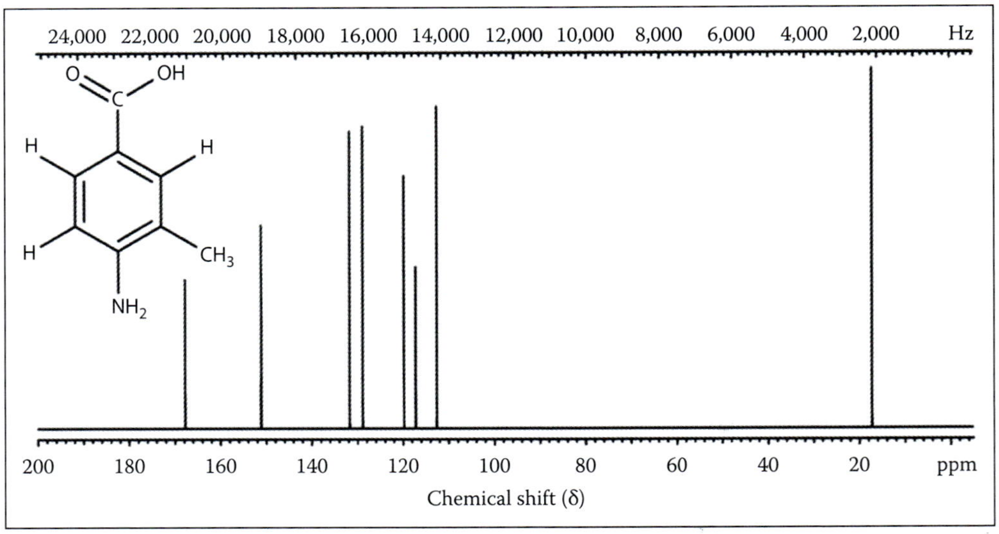

FIGURE 6.47 ^{13}C NMR spectrum of 4-amino-3-methylbenzoic acid, **6.17.**

* Deuterium does not relax carbon as well as do protons; relaxation effects are very important in NMR spectroscopy, but you will have to wait for a more advanced course for the full details.

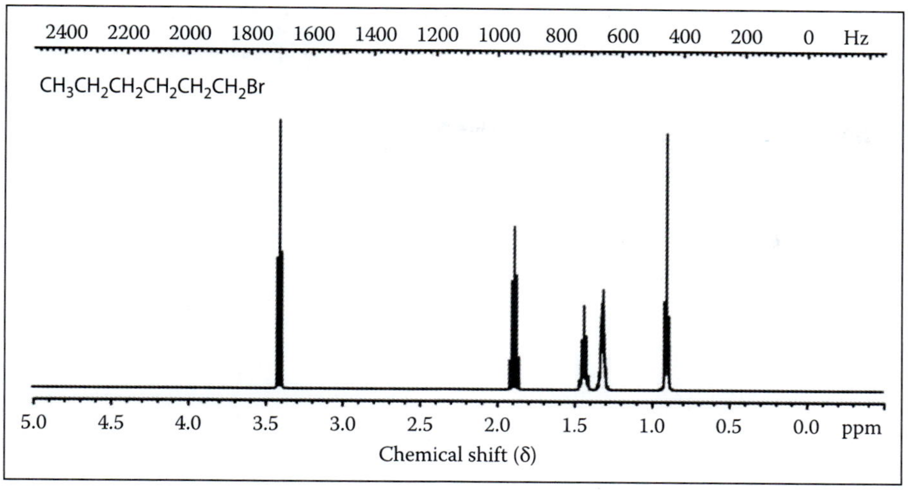

FIGURE 6.48 Proton NMR spectrum of 1-bromohexane (500 MHz).

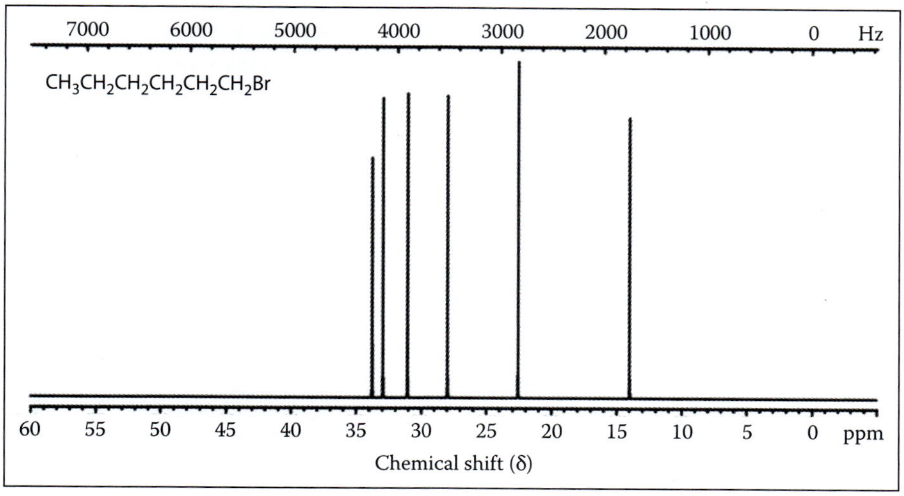

FIGURE 6.49 Carbon NMR spectrum of 1-bromohexane (125 MHz).

The advantages of ^{13}C NMR spectroscopy are readily exemplified by comparing the proton and carbon NMR spectra of 1-bromohexane (Figures 6.48 and 6.49). The proton spectrum can be analyzed, and the signals assigned, but this is not a trivial process; the spectrum is complex, and there are significant overlaps. By contrast, the carbon spectrum is clear; all six carbons are readily distinguished and assigned.

PROBLEM 6.7

How would you distinguish 2,3-dimethylphenol and 2,6 dimethylphenol using ^{13}C NMR spectroscopy?

Solution

This problem is about the symmetry of the molecules. In 2,3-dimethylphenol, all the carbon atoms are distinct. In 2,6-dimethylphenol, the molecule is symmetric, and there are only four types of aromatic carbon atoms and a single type of methyl group (δ ^{13}C values are given on the structures):

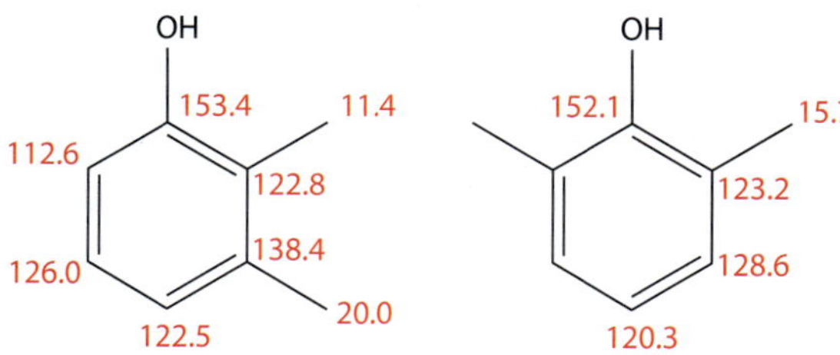

Focus on Phosphorus and Fluorine NMR Spectroscopy

Both ^{19}F and ^{31}P nuclei are 100 % abundant in natural material and are hence good candidates for the NMR spectroscopic experiment. Fluorine has a high gyromagnetic ratio and 83 % of the sensitivity of proton NMR spectroscopy. Phosphorus NMR spectroscopy is 6.6×10^{-3} as sensitive as the ^{1}H experiment, but this is still significantly more sensitive (by a factor of 38) than ^{13}C NMR spectroscopy. The other advantage for these two nuclei is that generally any impurities in the samples (and whatever solvent we choose) will not contain the relevant nucleus, so they are particularly suitable for observing biological systems, or reaction mixtures. Factors beyond the scope of this course means that integration is reliable for ^{19}F, but not for ^{31}P NMR spectra.

Because there are nine electrons around fluorine (rather than the single electron associated with protons), the ^{19}F chemical shift is much more sensitive to its environment than is that for ^{1}H. The reference compound is $CFCl_3$ and chemical shifts range from −540 (UF_6) to +415 (HF). Fluorine has been particularly widely used to study biological systems; since there is essentially no fluorine in biological systems, the only signals are from the molecules added to probe the system. Fluorinated analogues of many biological molecules are now commercially available, and some, such as 5-fluorouracil (**6.19**), have found medicinal applications. The dispersion of fluorine chemical shifts means that when a fluorinated amino acid such as a derivative of phenylalanine or tyrosine (**6.20, 6.21**) is incorporated in a protein, the fluorine signals for the amino acids at different sites can usually all be distinguished, even though there is substantial overlap between the proton signals from the same residues. ^{19}F NMR spectroscopy is often used to study the interaction of small fluorinated molecules with receptors; the large chemical shift changes from relatively small changes in environment are again useful here.

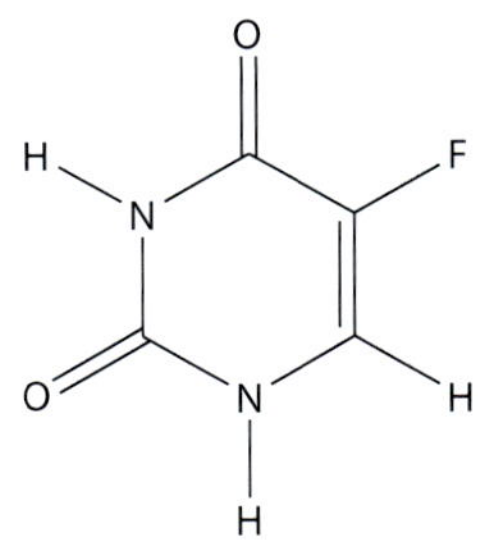

6.19

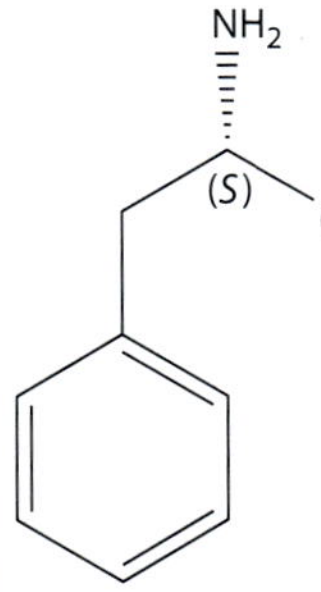

6.20

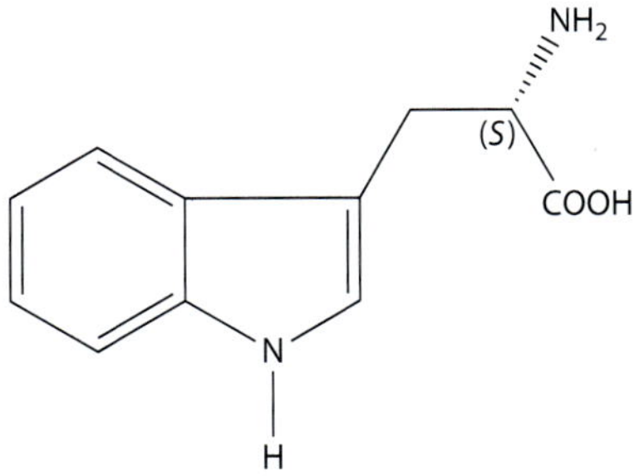

6.21

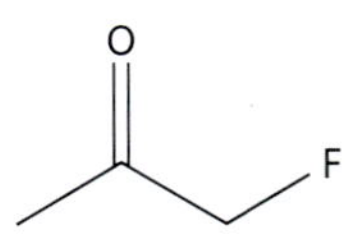

6.22

^{19}F NMR spectra can be run with or without proton decoupling; fluorine–fluorine coupling constants are generally significantly larger than proton–proton couplings, and long-range couplings are common, but many general trends are similar (Table 6.7). Fluorine

TABLE 6.7 Fluorine–Fluorine Coupling Constants in NMR Spectroscopy

Type of Coupling	Typical Structure	J (Hz)
^2J (geminal, sp^3 C)	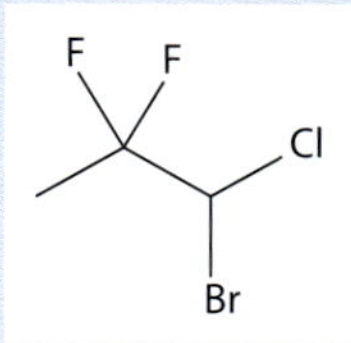	40–370
^2J (geminal, sp^2 C)	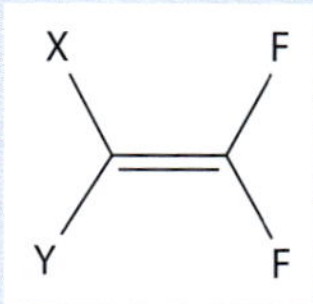	0–110
^3J (vicinal, sp^3 C)	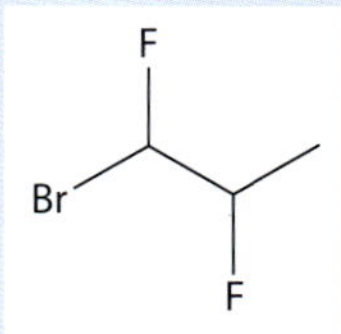	0–45
^3J (vicinal, sp^2 C, *cis*)	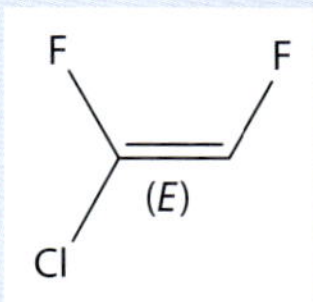	0–60
^3J (vicinal, sp^2 C, *trans*)	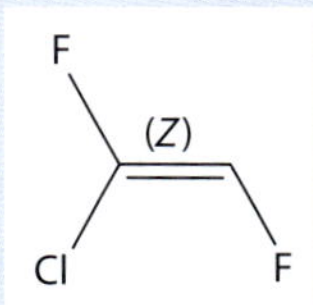	100–150
^3J (vicinal, aromatic)	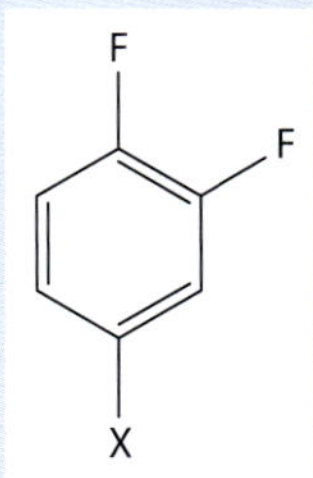	18–35
^4J (long range, aromatic)	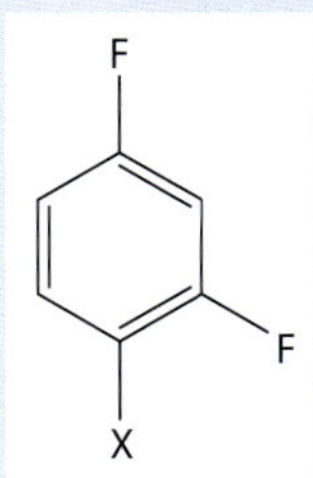	0–15
^5J (long range, aromatic)	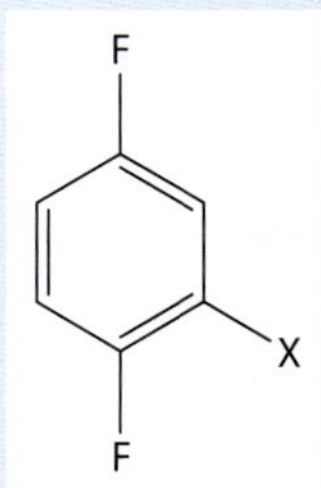	4–16

couplings are also seen in proton and carbon spectra. In fluoroacetone, **6.22**, $^{2}J_{HF}$ = 48 Hz and $^{4}J_{HF}$ = 4.3 Hz. In the ^{13}C spectrum of the same molecule, $^{1}J_{CF}$ = 185 Hz and $^{2}J_{CF}$ = 20 Hz. All these couplings are strongly dependent on molecular geometry. Modern techniques allow both fluorine and protons to be decoupled simultaneously from carbon, but it is rarely necessary.

In ^{31}P NMR spectroscopy, the reference compound is 85 % H_3PO_4, phosphoric acid, and the range of chemical shifts runs from +250 to −180 ppm. ^{31}P NMR spectra are usually run with proton decoupling; as $^{1}J_{PH}$ is very large (691 Hz in $(CH_3CH_2O)_2P\text{-}H$), this requires significantly more power than for ^{13}C. Longer-range couplings are also significant ($^{3}J_{PH}$ in $(CH_3CH_2O)_2P\text{-}H$ is 8.9 Hz) and can be readily seen in the proton spectra.

Two areas of research have particularly benefited from the use of ^{31}P NMR spectroscopy. The first is the *in situ* study of biological systems. High-energy phosphates such as ATP can be studied in living tissue using a coil on the surface of the skin. The other area of interest is a study of the structures of metal phosphine complexes such as Wilkinson's catalyst (**6.23**, we will meet this again in Chapter 23). The complex is square planar, and there are two types of phosphorus ligand—two are *trans* to one another, and the other is *trans* to chlorine. These are coupled to each other, so one triplet and one doublet, but also to rhodium, which is a 100 % spin half nucleus (Figure 6.50).

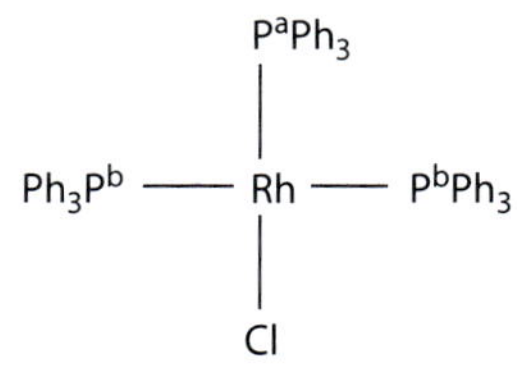

6.23

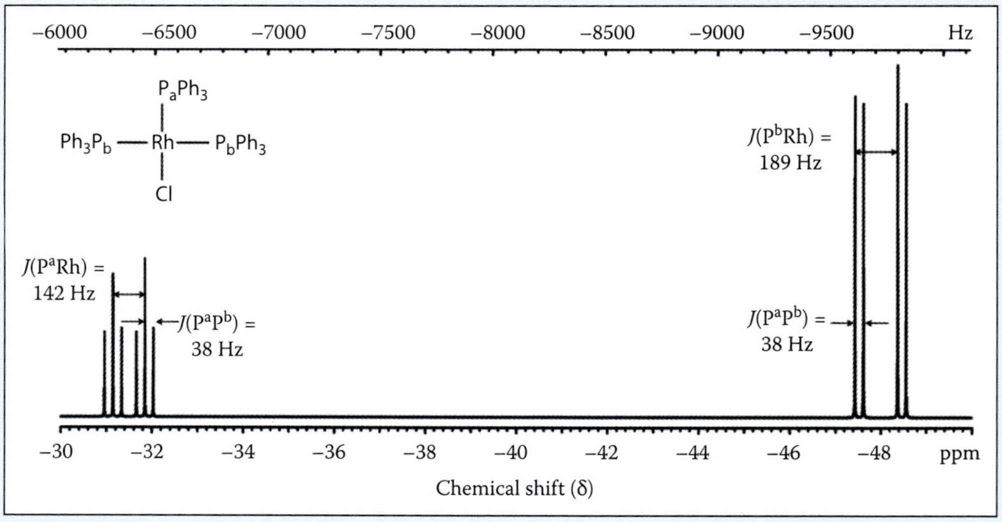

FIGURE 6.50 ^{31}P NMR spectrum of Wilkinson's complex.

Key Points from Section 6.5

- Because of the inherent insensitivity of the ^{13}C nucleus and its low natural abundance, many scans are often needed to obtain a ^{13}C NMR spectrum.
- ^{13}C NMR spectra are generally run with complete decoupling from both attached and more remote protons.
- Overlap in ^{13}C NMR spectra is rare, and the number of signals generally corresponds to the number of distinct carbon atoms in the molecule.
- Integration of standard ^{13}C NMR spectra is not meaningful, but in general signals from carbon atoms with directly attached protons are more intense than those from quaternary carbon atoms.

REVIEW PROBLEMS

1. Using a 600 MHz NMR spectrometer, how many Hz downfield from TMS is a chemical shift of 5.5 ppm?
2. Suggest the structure of a compound with each of the following spectroscopic characteristics. Predict the approximate chemical shifts for the protons in the compound you suggest:
   (a) C_2H_6O with one ^1H and one ^{13}C resonance
   (b) C_3H_7Cl, with two ^1H and two ^{13}C resonances
   (c) C_3H_9N with one ^1H and one ^{13}C resonance
   (d) An alcohol, $C_4H_{10}O$, with two ^1H and two ^{13}C resonances
3. Would tetraethylsilane, $(CH_3CH_2)_4Si$, be a good standard for ^1H NMR spectroscopy? If not, why?
4. Predict the ^1H NMR spectrum, including multiplet patterns, for each of the following molecules. Ignore possible long-range couplings:
   (a) 3-Pentanone
   (b) *E*-Dimethyl butenedioate
   (c) Propanal
   (d) 4-Propylnitrobenzene
   (e) 1-Butyne
   (f) 3-Methylphenol
   (g) Pyridine
   (h) 2-Pentanone
   (i) Butanoic acid
   (j)

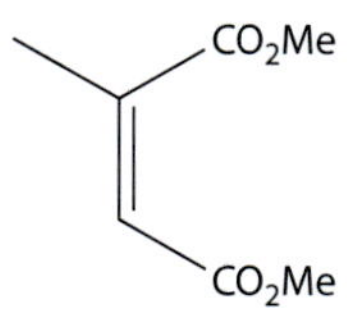

   (k)

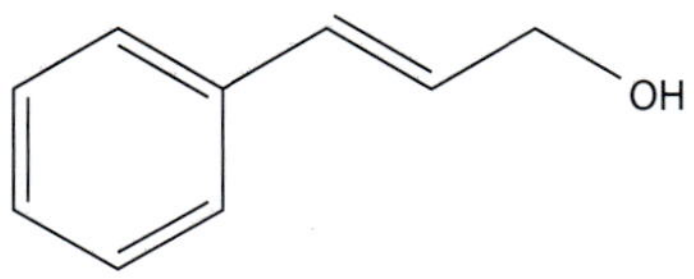

5. How would you distinguish spectroscopically between each of the following pairs of compounds? Give as much detail as you can.

(a)

(b)

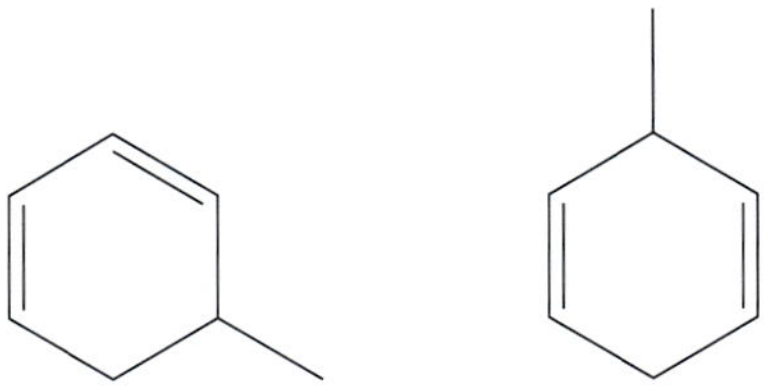

(c)

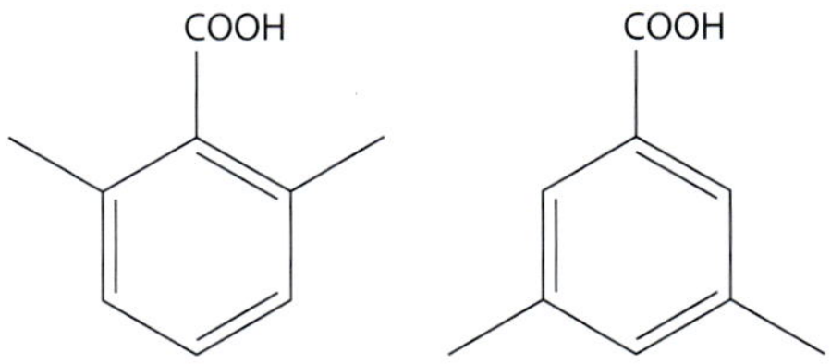

(d)

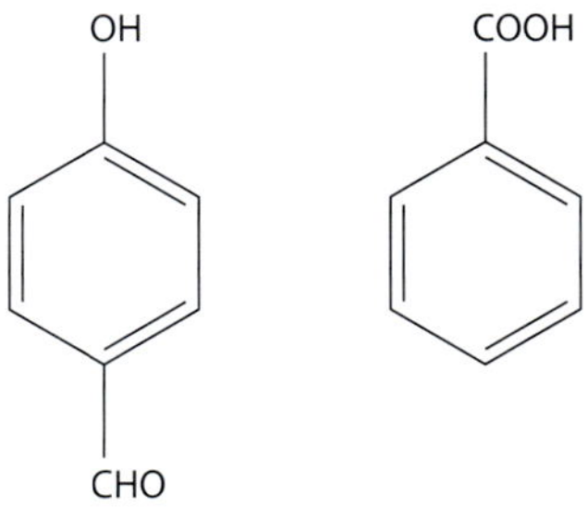

(e)

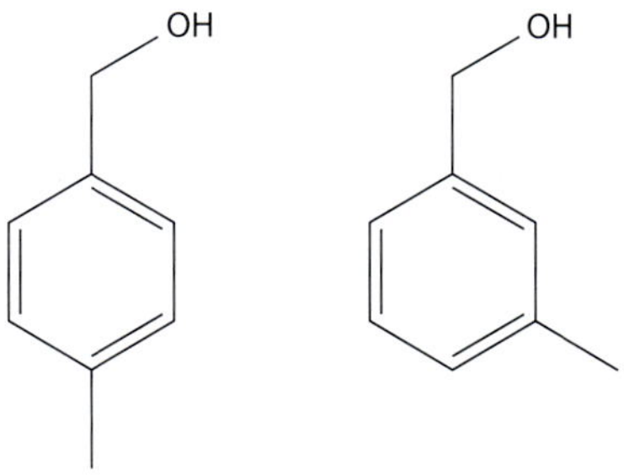

(f)

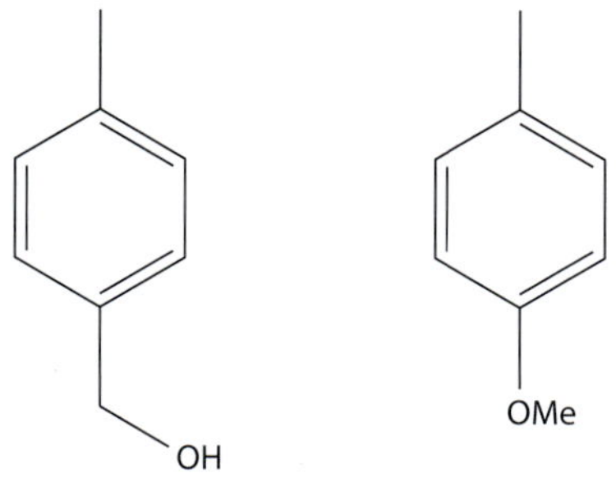

(g)

6. Identify each of the following compounds using the spectroscopic data given, and assign the spectra:
   (a) A compound C_8H_8O shows a strong absorption in its IR spectrum at 1703 cm^{-1} and strong peaks in the mass spectrum at 119 and 91 mass units. Its ^1H NMR spectrum shows the following signals:

   δ 2.4, s, 3H
   7.2, d, 2H
   7.9, d, 2H
   10.0, s, 1H

   (b) A compound $C_7H_{12}O_3$ shows strong absorptions in its IR spectrum at 1735 and 1715 cm^{-1}. Its ^1H NMR spectrum shows the following signals:

   δ 1.3, t, 3H
   2.1, s, 3H
   2.6, t, 2H
   2.8, t, 2H
   4.3, q, 2H

   (c) A compound $C_{10}H_{12}O$ shows peaks in its mass spectrum at m/z = 148, 105, and 77 and shows a strong peak in its IR spectrum at 1685 cm^{-1}. Its ^1H NMR spectrum shows peaks at

   δ 1.05 (t, 3H)
   1.80 (sextet, 2H)
   2.95 (t, 2H)
   7.50 (m, 3H)
   7.98 (d, 2H)

(d) A compound $C_9H_{10}O_2$ shows peaks in its mass spectrum at *m/z* = 150, 135, 107, and 77 and shows a strong peak in its IR spectrum at 1670 cm^{-1}. Its ^1H NMR spectrum shows peaks at

δ 2.6 (s, 3H)
3.9 (s, 3H)
7.0 (d, *J* = 7 Hz, 2H)
7.9 (d, *J* = 7 Hz, 2H)

(e) A compound $C_7H_6BrNO_2$ shows the following ^1H NMR spectrum:

δ 2.5, s, 3H
7.4, d, *J* = 7 Hz, 1H
8.4, d, *J* = 2 Hz, 1H

(f) A compound $C_7H_{12}O$ shows peaks in its IR spectrum at 1670, 1630 cm^{-1}, and no peaks above 3000 cm^{-1}. Its mass spectrum shows peak at *m/z* = 112, 97, 55, and 43. Its ^1H NMR spectrum has peaks at

δ 0.95, t, 3H
1.51, m, 2H
2.22, m, 2H
2.25, s, 3H
6.08, d, *J* = 16 Hz, 1H
6.80, d of t, *J* = 16, 6.8 Hz, 1H

7. Four isomeric compounds of formula $C_8H_{10}O$, **6.24**, **6.25**, **6.26**, and **6.27** showed the following spectroscopic data. Identify them, assign the spectra, explain your reasoning clearly, and comment on any features of interest:

6.24	IR (CCl_4 solution)	3616 cm^{-1} (sharp, s)
	IR (Nujol)	3262 cm^{-1} (br, s)
	Mass spectrum	*m/z* 122, 107
	^1H NMR spectrum	δ 2.15 (s, 3H), 2.27 (s, 3H) 4.76 (s, 1H, exchanges with D_2O) 6.6 (d, 1H) 6.8 (d, 1H) 7.0 (t, 1H)
	^{13}C NMR spectrum	δ 11.4, 20.0, 112.9, 122.5, 122.8, 126.0, 138.4, 153.0
6.25	IR (CCl_4 solution)	No peaks above 3100 cm^{-1}
	Mass spectrum	*m/z* 122, 107, 91, 76
	^1H NMR spectrum	δ 2.21 (s, 3H), 3.76 (s, 3H) 6.8 (d, 1H) 6.9 (t, 1H) 7.0 (t, 1H) 7.1 (t, 1H)
	^{13}C NMR spectrum	δ 16.2, 55.12, 110.0, 120.4, 126.7, 126.9, 130.7, 157.9
6.26	IR (CCl_4 solution)	3613 cm^{-1} (sharp, m) 3481 cm^{-1} (br, m)
	IR (liq. film)	3420 cm^{-1} (br, s)
	Mass spectrum	*m/z* 122, 107
	^1H NMR spectrum	δ 1.2 (t, 3H), 2.6 (q, 2H) 4.8 (s, 1H, exchanges with D_2O) 6.8 (d, 1H) 6.9 (t, 1H) 7.1 (t, 1H) 7.2 (d, 1H)
	^{13}C NMR spectrum	δ 14.0, 22.9, 115.0, 121.0, 127.0, 129.4, 130.3, 153.2
6.27	IR (CCl_4 solution)	3620 cm^{-1} (sharp, s) 3489 cm^{-1} (br, m)
	IR (liq. film)	3339 cm^{-1} (br, s)
	Mass spectrum	*m/z* 122, 91
	^1H NMR spectrum	δ 2.01 (s, 1H, exchanges with D_2O) 2.8 (t, 2H *J* = 7 Hz), 3.8 (t, 2H *J* = 7 Hz) 7.1–7.4 (m, 5H)
	^{13}C NMR spectrum	δ 39.1, 63.5, 126.0, 128.5, 129, 138.6

MORE CHALLENGING PROBLEMS

1. The ^{1}H NMR spectrum of hydrocarbon **6.28** shows three multiplets centered at 9.55, 8.43, and −5.43 in the ratio 2:2:1. Explain (think carefully about the aromaticity or otherwise of this compound and whether various π-bonds are formed from p_z or p_y orbitals).

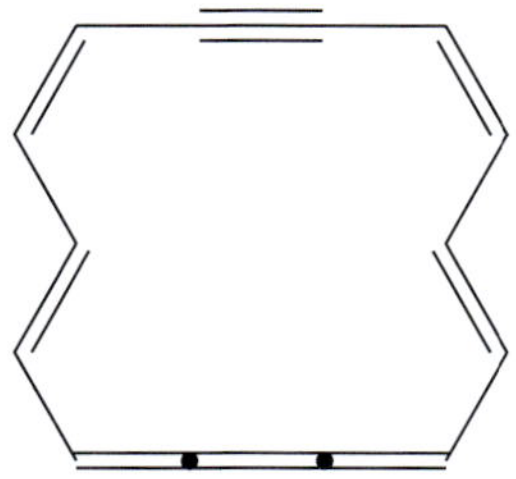

6.28

2. In the ^{1}H NMR spectrum of **6.29**, 10 protons give signals in the region δ 8.2–8.7, while the signals due to the CH_3 groups are at δ −4.25. Explain.

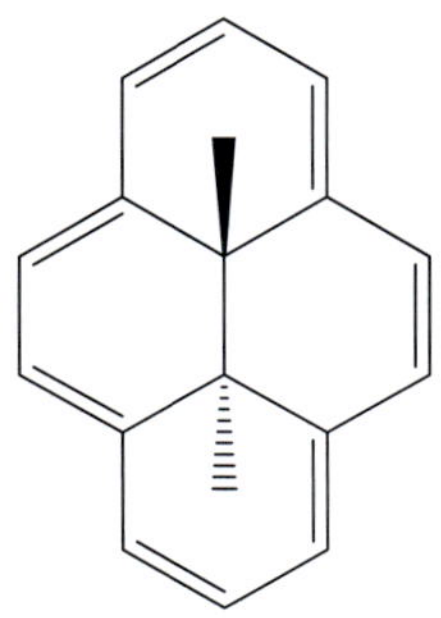

6.29

3. Identify each of the following compounds using the spectroscopic data given and assign the spectra.

(a) $C_{11}H_{20}O$

Mass spectrum *m/z*	168 (M$^+$), 126, 111, 83, 57
IR spectrum	3020, 2990, 1680, 1620 cm^{-1}
^{1}H NMR spectrum	δ 0.9 (d, *J* = 8 Hz, 6H)
	1.1 (s, 9H)
	1.5 (nonet, *J* = 8 Hz, 1H)
	2.3 (d, *J* = 8 Hz, 2H)
	6.1 (d, *J* = 18 Hz, 1H)
	6.9 (d, *J* = 18 Hz, 1H)

(b) $C_7H_{10}O_4$

IR spectrum	1740, 1720, 1670 cm^{-1}
Mass spectrum	*m/z* = 158 (M$^+$), 143, 127, 99
^{1}H NMR spectrum	δ 3.4 s 2H
	3.8 s 3H
	3.9 s 3H
	5.7 d *J* = 1 Hz 1H
	6.3 d *J* = 1 Hz 1H
^{13}C NMR spectrum	δ 174.2, 168.1, 129.0 128.1 (CH_2), 53.2 (CH_3), 51.7 (CH_3) 34.6 (CH_2)

The indicators in parentheses for the ^{13}C spectrum have been obtained using a technique that allows the determination of the number of hydrogens attached to a particular carbon atom; those not designated have no attached hydrogens.

(c) $C_7H_{12}O$

IR spectrum	δ 3350 (br, s), 3310 (sharp, s), 2096 (m) cm^{-1}
^1H NMR spectrum	δ 0.92 (t, 3H) 1.64–1.11 (m, 4H), 1.69 (m, 2H) 2.38 (s, 1H, exchanges with D_2O) 2.46 (s, 1H) 4.36 (t, 1H)
^{13}C NMR spectrum	δ 14.0, 22.4, 27.3, 37.4, 62.2, 72.8, 85.3

4. How would you distinguish spectroscopically between the following isomeric compounds? Be as economical as you can—enough detail to get it right, but no more.

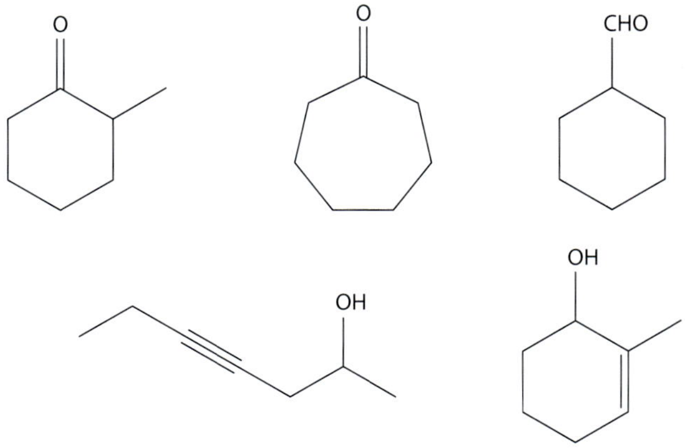

5. How would you use NMR spectroscopy to distinguish between the following isomeric alcohols?

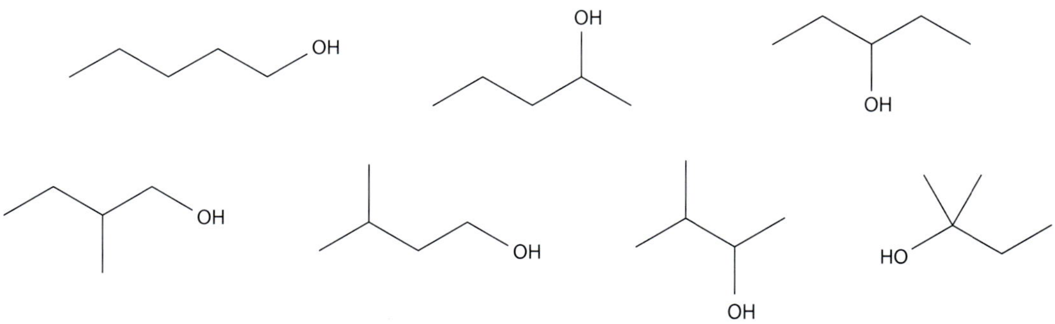

Stereochemistry

7

7.1 INTRODUCTION

We have previously discussed the existence of isomers of organic compounds—compounds that have the same molecular formula but a different construction. Isomers can be divided into two types—structural isomers and stereoisomers. For example, *n*-butane (**7.1**) and 2-methyl propane (**7.2**) are *structural* isomers—the atoms are joined together in a different way. Structural isomers are described as having a different *constitution*.

7.1

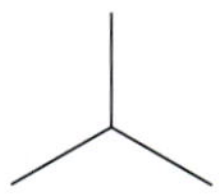

7.2

Stereoisomers have the same constitution, with all the same atoms joined together by the same type of bonds but a different *shape*. The existence of stereoisomers depends on the theory of organic structure and the fact that chemical bonds have directional properties. Stereoisomers can be further divided into two mutually exclusive groups: *enantiomers* and *diastereoisomers* (sometimes elided as diastereomers). Enantiomers are distinguishable molecules related as object and mirror image. Diastereoisomers are all other stereoisomers.

Although there are many computer programs allowing you to generate molecules on screen and move them around in space, physical models are also useful in understanding stereochemistry. "Ball and stick" models are useful to get an overall picture of a molecule, including the relative sizes of the atoms, but they tend to be quite expensive (Figure 7.1). Skeletal models allow you to see just the bonds in a molecule but give no idea of steric effects. The advantage is that they are generally quite inexpensive (Figure 7.2). "Space-filling" models focus attention on the atoms, rather than the bonds; they are excellent for studying steric effects, giving a clearer view of atom size (Figure 7.3). These are sometimes known as Corey, Pauling, Koltun (CPK) models where atoms are represented as overlapping spheres, with radii genuinely representative of atomic radii. They are frequently used to study biological interactions, such as the docking of drugs at receptors. They are less good for studying stereochemical relationships and are very expensive. It's rare nowadays to build these, but they are often computer generated.

As we saw in previous chapters, we need to be able to draw a two-dimensional representation of a three-dimensional structure. It is usual to focus on a particular carbon–carbon bond of interest. A view from a direction perpendicular to the bond of interest is called a wedge representation (**7.3a** and **7.3b**). At each carbon atom, two bonds are in the plane of the drawing: one, represented by a wedge, is in front of the plane, and another, described as a hashed wedge, (or colloquially as "a hash" or "hashed") is behind the plane. **7.3a** and **7.3b** represent the eclipsed and staggered forms of ethane, which we discussed earlier in Chapter 2. You should recognize that whether computer or hand drawn, the angles and appearances of such diagrams can vary—without necessarily being incorrect. For your own understanding, you should develop a good style and stick with it.

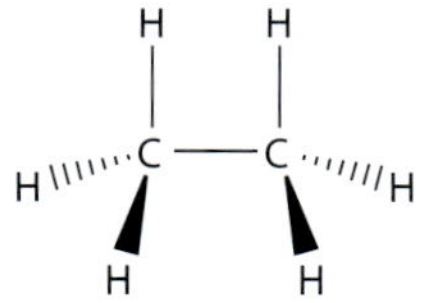

7.3a

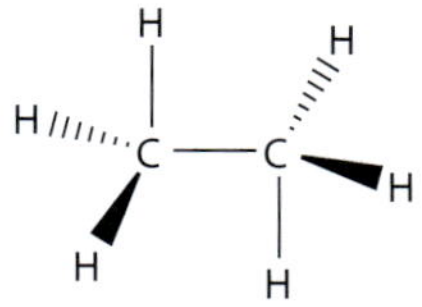

7.3b

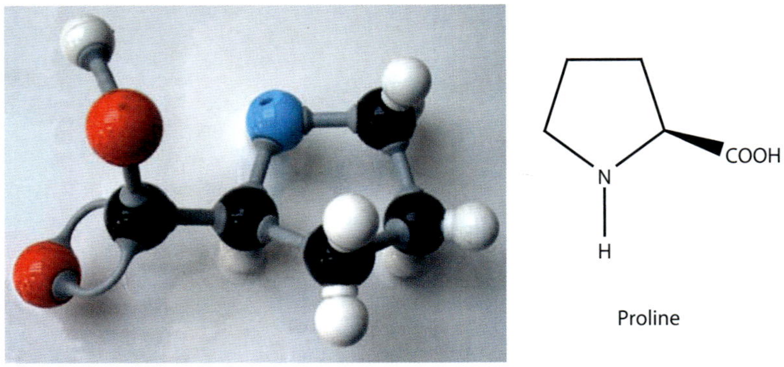

FIGURE 7.1 Ball and stick model of proline.

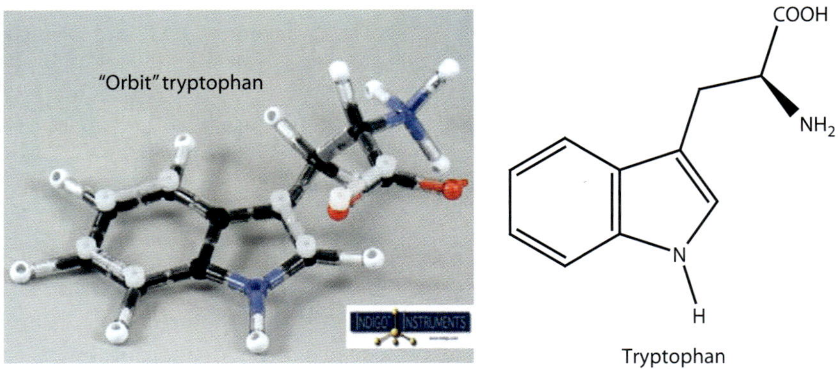

FIGURE 7.2 Skeletal model of tryptophan.

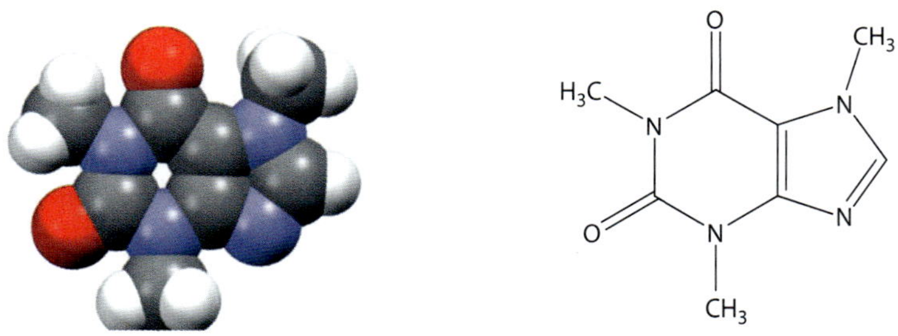

Caffeine

FIGURE 7.3 Computer-generated space-filling model of caffeine. (Thanks to Prof. Ken Seddon, Queens University, Belfast, U.K.)

In larger and more complex structures, the stereochemical arrangements of substituents are often shown by wedges and hashes relative to a "linear" carbon chain (really a horizontal orientated zigzag). **7.4** shows the usual representation, which you will meet in many later reactions. In this type of diagram, the hydrogen atoms are not shown, but their position is implied. Generally, the substituents are the key to the reactivity of the compound, so the diagram is kept simple by focusing on them.

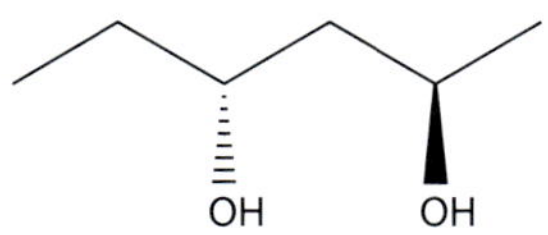

7.4

An oblique or sawhorse view of the molecule is at 45 ° to the bond of interest—again we can have either eclipsed or staggered conformations of the molecule (**7.5a** and **7.5b**), as well as any conformation in between these. In Newman projections, we look down the bond of interest, **7.6a** and **7.6b**:

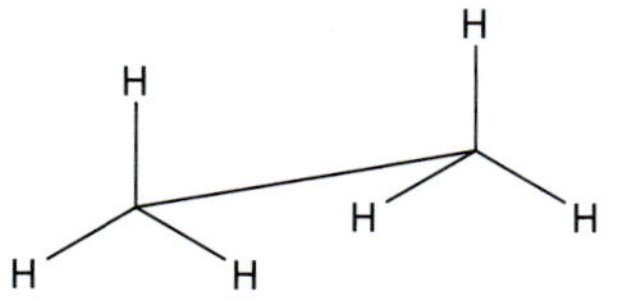

7.5a

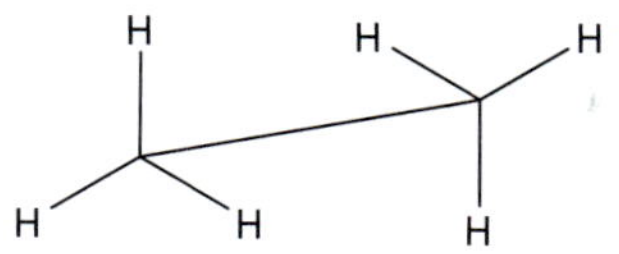

7.5b

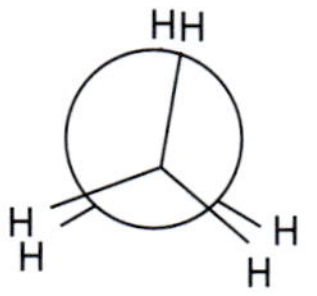

7.6a

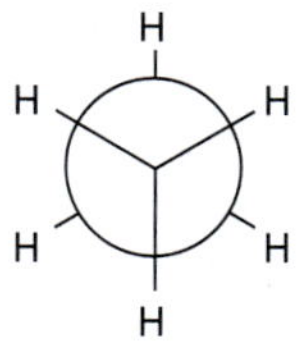

7.6b

PROBLEM 7.1

(a) Draw a sawhorse representation of ethanol in the staggered conformation.
(b) Draw a sawhorse diagram of the eclipsed conformation of $(CH_3)_2CHCH_2CH_3$, focusing on the C2 to C3 bond.

PROBLEM 7.2

(a) Draw a Newman projection of the staggered form of propane.
(b) Draw a Newman projection of the eclipsed conformation of ethane-1,2-diol.

Solutions

7.1 For (a), there is only one possible solution, but for (b), there are two options, depending on which groups are eclipsed; the second conformer, where the methyl groups are as far apart as possible, is lower in energy:

(a)

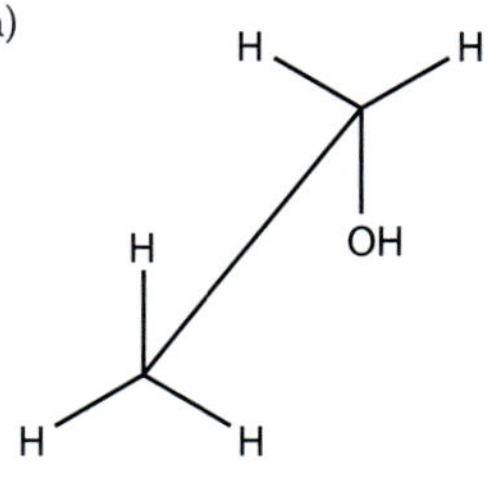

(b)

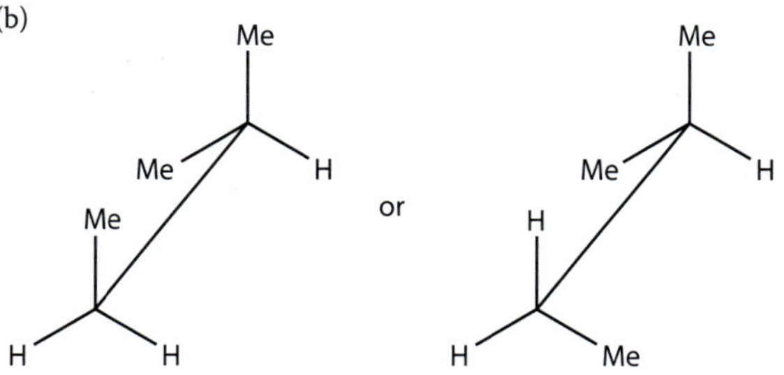

7.2 Again for (a), there is a single solution, but for (b), there are two options, depending on whether OH eclipses hydrogen or the other OH. Although both are technically correct, the less hindered version is the more likely:

(a)

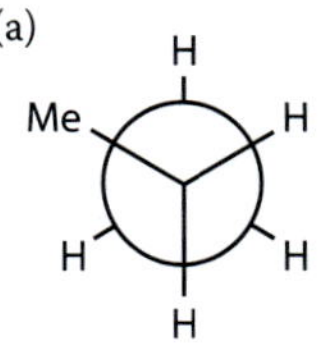

(b)

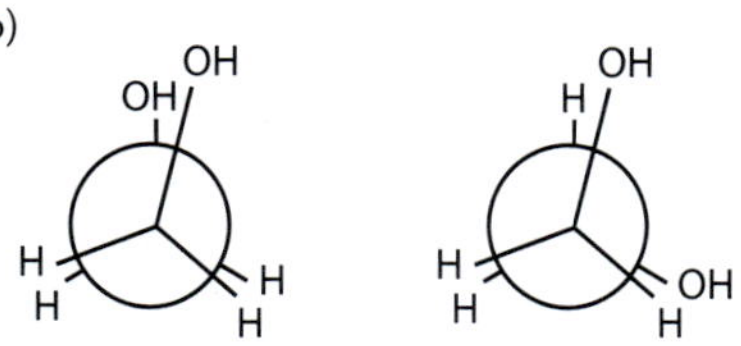

7.2 CONFORMATIONS OF ALKANES

The various versions of ethane that we can generate by rotation around the carbon–carbon bond are *conformations*, momentary arrangements of atoms in space interconverted by rotation about single bonds. Let's consider the energy changes when we rotate about the carbon–carbon bond of ethane (Figure 7.4). The highest energy conformation is the eclipsed form, and the lowest energy conformation is the staggered form. The difference in energy between the two is relatively small, 12 kJ mol^{-1}. There are conflicting theories about the true origin of the energy difference, but steric factors are certainly important.

If we think about the C2–C3 bond of butane, we now have different eclipsed and staggered conformations, depending on whether the methyl groups eclipse each other or hydrogens (Figure 7.5). In one of the eclipsed forms, hydrogens eclipse methyl groups, but in the other, the two methyl groups are eclipsed. Clearly the second is higher in energy. Similarly, the anti-staggered conformation with the methyl groups as far apart as possible will be lower in energy than the one where they are adjacent. The one where they are adjacent is described as a gauche conformation. Figure 7.6 shows the energy changes associated with this rotation.

PROBLEM 7.3

(a) Draw a Newman projection of the anti-staggered conformation of 1,2-dibromoethane.
(b) Draw a sawhorse representation of the gauche staggered form of 1,2-dibromoethane.

Solutions

(a)

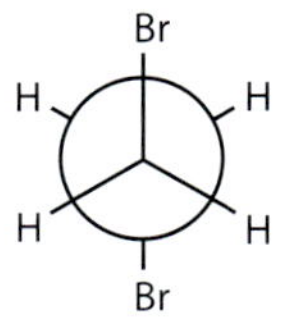

(b)

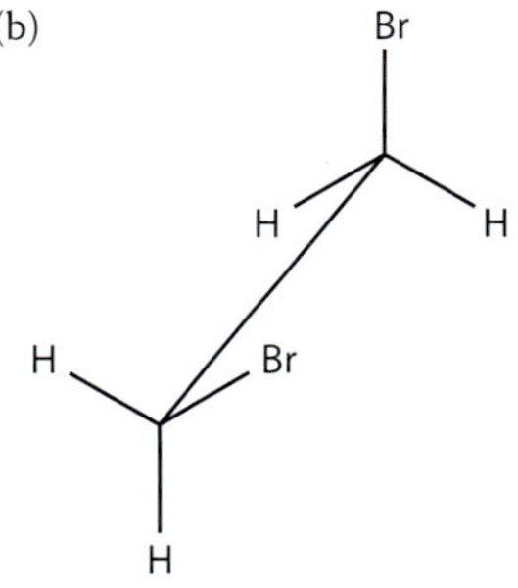

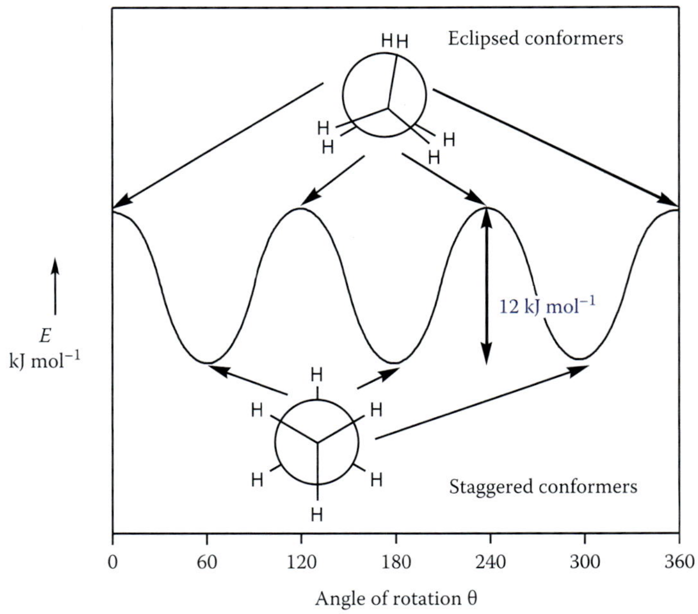

FIGURE 7.4 Changes in energy on rotation about the single bond of ethane.

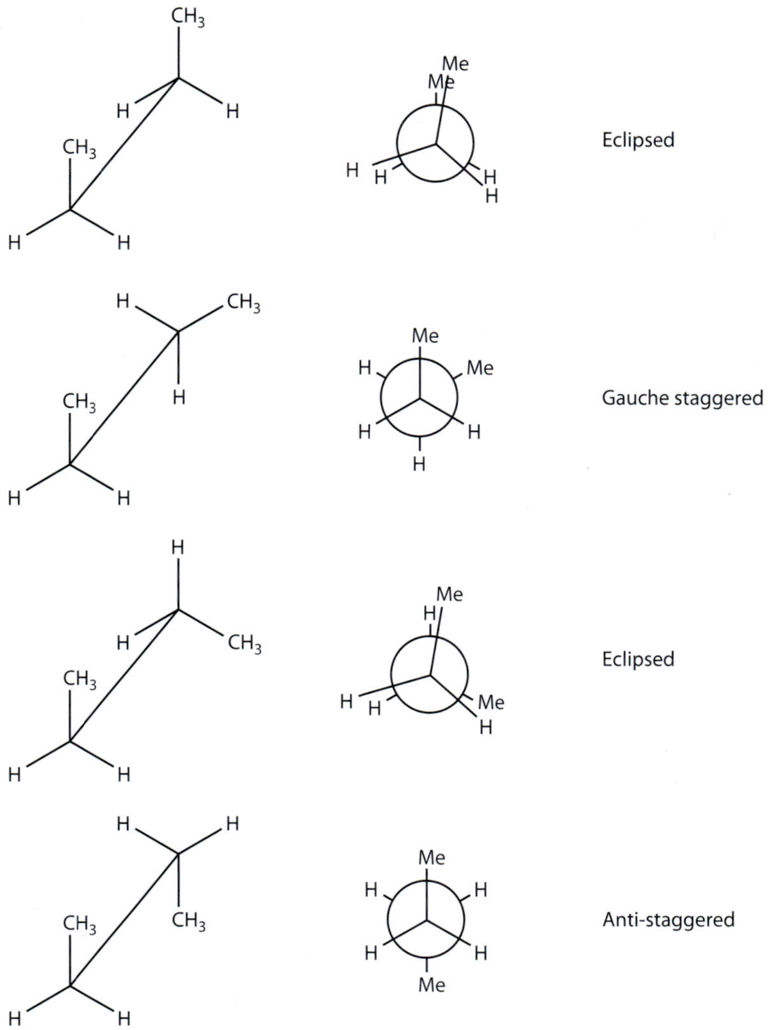

FIGURE 7.5 Conformations of butane.

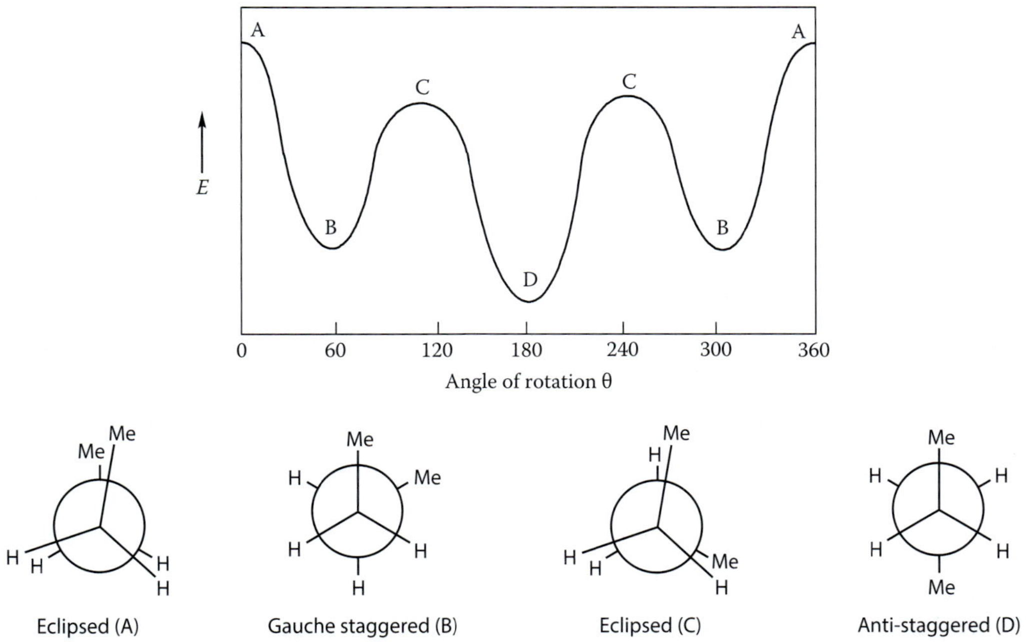

FIGURE 7.6 Energy changes induced by the rotation about the C2–C3 bond of butane.

Key Points from Sections 7.1 and 7.2

- Enantiomers are molecules that are not superimposable on their mirror images. Diastereoisomers are all other stereoisomers.
- The two-dimensional representations of three-dimensional molecules include wedge, sawhorse, and Newman projections.
- Conformations are instantaneous arrangements of bonds in a molecule, which can generally be interconverted by rotation about single bonds.
- For alkanes, conformations where bonds are eclipsed are higher in energy than those where the bonds are staggered, though differences in energy are generally less than 20 kJ mol^{-1}.
- The origins of such energy differences are primarily steric, and the differences increase with the sizes of the substituents involved.

7.3 CYCLOALKANES

7.3.1 CYCLOPROPANE

As we noted in Chapter 2, there is considerable angle strain in cyclopropane and cyclobutane, because the angles in the ring are, by necessity, rather different from the 109.5 ° optimum for sp^3 carbon atoms. In cyclopropane, all the angles between the carbon–carbon internuclear directions *must* be 60 °. In reality, the bonds/orbitals of the carbon–carbon bonds lie "outside" the line joining the centers of the atoms (Figure 7.7). They are sometimes referred to as "banana bonds" and are relatively weak, despite the fact that the internuclear distance is rather short (1.51 Å *vs.* 1.54 Å for a regular single bond between two sp^3 carbon atoms). The angles between the bonds, as opposed to the internuclear vectors, are about 104 °. An additional type of strain suffered by cyclopropane is eclipsing strain. We noted that in ethane, the energy was higher when the C–H bonds were eclipsed than when they were staggered. The same is true in cycloalkanes, and in cyclopropane, all the C–H bonds must be eclipsed (**7.7**).

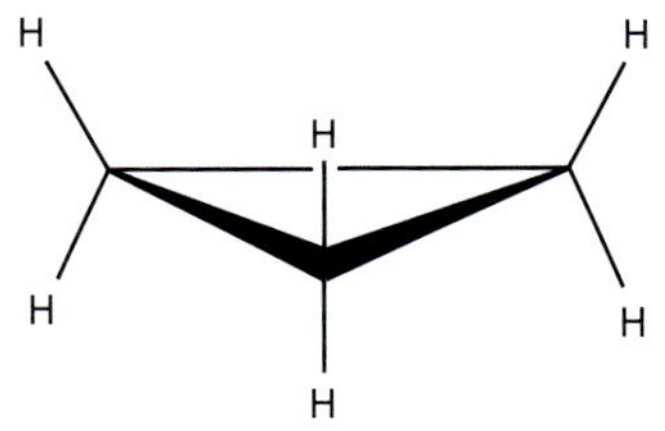

7.7

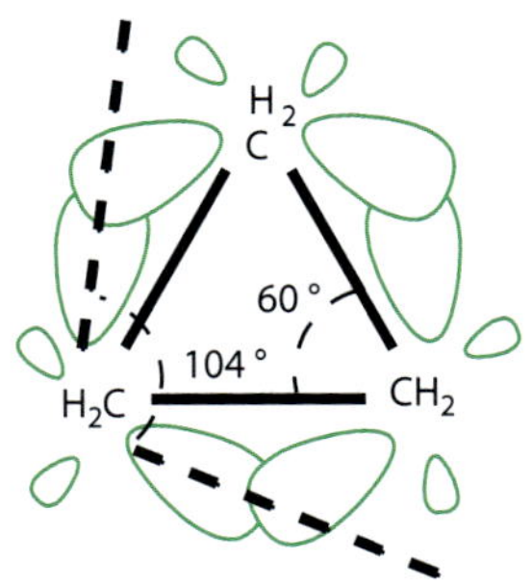

FIGURE 7.7 Bonding in cyclopropane.

7.3.2 CYCLOBUTANE

If we assumed cyclobutane to be planar, then the angles between the bonds would be 90 ° rather than the optimal 109.5 °, and all the C–H bonds would be eclipsed (**7.8**). In practice, it is not quite as bad as this—the ring flexes slightly (the interplanar angle in **7.9** is about 25 °) to relieve some of both angle and eclipsing strain (**7.9**).

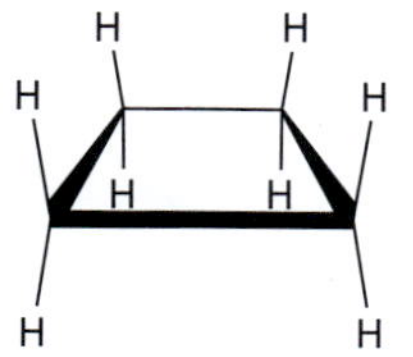

7.8, disfavored form

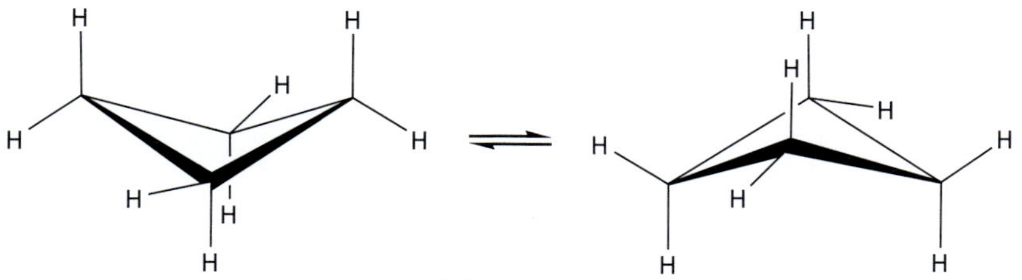

7.9, favored form

7.3.3 CYCLOPENTANE

In cyclopentane, there is little angle strain—the internal angle of a pentagon is 108 ° and the ring is very flexible. However, if the molecule were planar, there would be considerable eclipsing strain. The lowest energy form is described as an envelope conformation (**7.10**), but the ring is both dynamic and floppy, with each atom taking a turn at being the "flap" of the envelope.

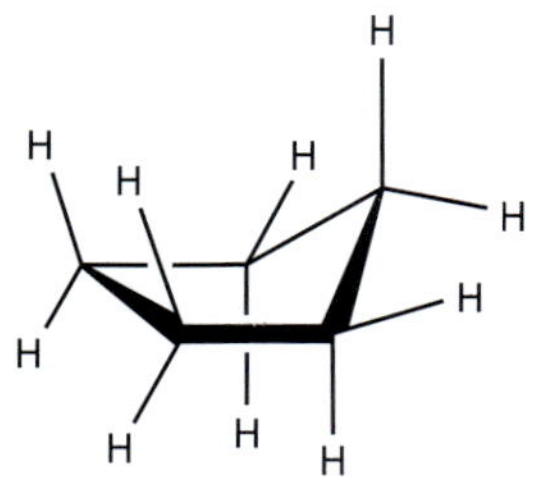

7.10

7.3.4 SIX-MEMBERED RINGS

Cyclohexane is unstrained. In its lowest energy chair conformation (**7.11**, Figure 7.8), all the angles for the sp^3 carbon atoms are 109.5 °, and all of the bonds are perfectly staggered. Another possible conformation of cyclohexane is a boat, **7.12**, which is some 27 kJ mol^{-1} higher in energy than the chair form, partly at least due to what is described as the bowsprit interaction. In the twist-boat conformation, **7.13**, the bowsprit interaction is partially relieved. Figure 7.9 shows an energy diagram for the interconversion of the various conformations of cyclohexane. Note that the boat conformation is an energy maximum and hence can only be a transition state between the two half boats.

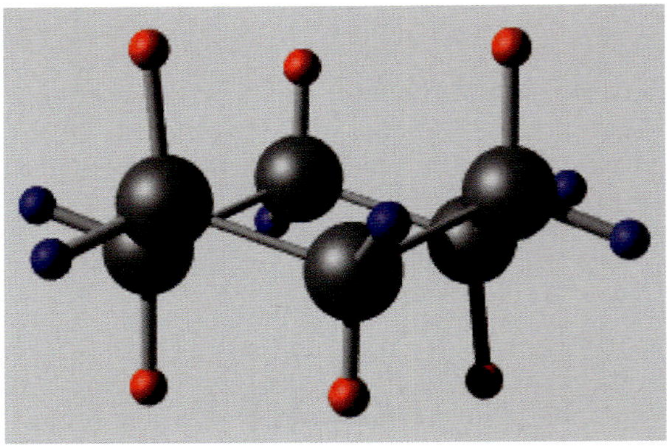

FIGURE 7.8 Ball and stick model of the chair form of cyclohexane. (Thanks to Dr. John Turner, University of Sussex, East Sussex, U.K.)

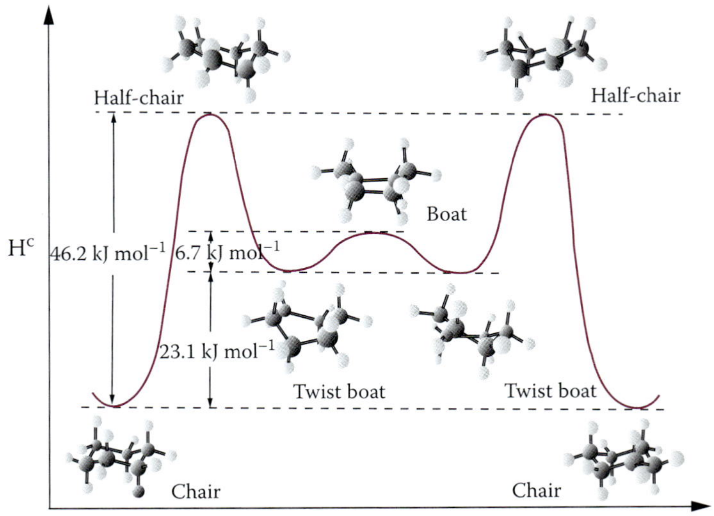

FIGURE 7.9 Energy profile for the interconversion of various conformations of cyclohexane.

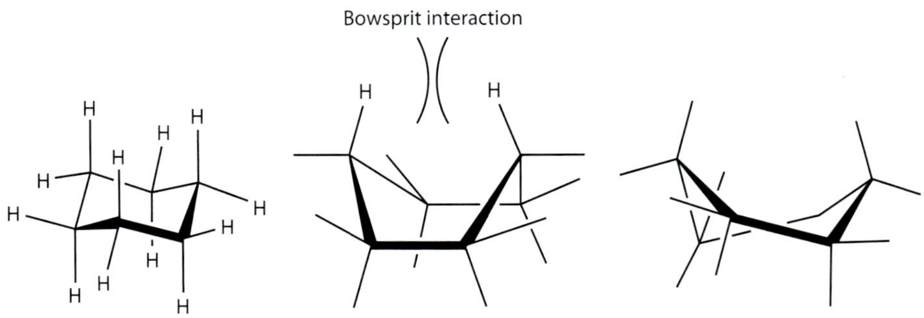

7.11 **7.12** **7.13**

When the chair form of a cyclohexane is "inverted" or "flipped," to get the alternative chair form, *via* the boat form transition state, all the axial substituents become equatorial, and all the equatorial substituents become axial (Figure 7.10). The terms axial and equatorial refer to the orientation of bonds relative to the mean plane of the ring. Axial bonds are above or below the plane of the ring, and when you are drawing these, make sure that they are always strictly vertical. Equatorial substituents are closer to the mean plane and should be drawn approximately parallel to the appropriate ring bond. Imagine rotating the left-hand carbon of **7.14** upward, to give the boat, **7.15**. The flipped chair, **7.16**, is obtained by rotating the right-hand carbon of the boat downward. Thus, Ha and He have changed places in **7.16**. Although this is easiest to see for

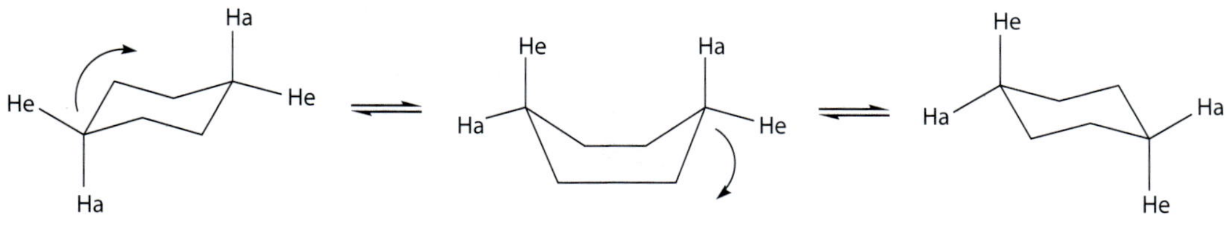

7.14 7.15 7.16

FIGURE 7.10 Inversion of a cyclohexane ring.

the outer edges of the molecule as drawn, this is true for all the hydrogens—it's simply a matter of the direction from which we observe.

If the cyclohexane ring bears a nonhydrogen substituent, it may occupy an axial or an equatorial position, and these two forms, **7.17a** and **7.17b**, will be in equilibrium. The equatorial site is preferred for steric reasons, deriving from 1,3-diaxial interactions (**7.17c**) and from gauche interactions (Figure 7.11). If we look along the red-colored bond in **7.17a**, we see the Newman projection of **7.18**—with a gauche interaction between the X and the CH_2 of the ring. The energy of this interaction is likely to be comparable with the X/Me gauche interaction in the 1-substituted propane, **7.19**. By contrast in the equatorial series, **7.17b**, the Newman projection, **7.20**, shows that the substituent and the ring carbon are anti to one another—the lowest energy conformation, comparable to the *anti*-arrangement of substituents in **7.21**.

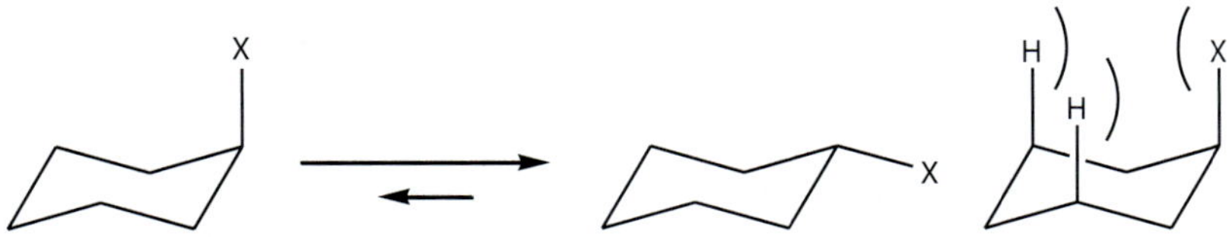

7.17a 7.17b 7.17c

The position of the equilibrium depends mainly on the size of the substituent; for example, the *tert*-butyl group, $-CMe_3$, and the trimethylammonium group, $-N^+Me_3$, never occupy axial positions. The *tert*-butyl group is generally said to "lock" the ring in a particular conformation, but with most other substituents, there is some axial conformer at equilibrium. Table 7.1 shows some of the thermodynamic data for a range of substituents for the equilibrium of Figure 7.12.

What happens when there is more than one substituent on a cyclohexane ring—for example, in the *cis*- and *trans*-isomers **7.22** and **7.23**? How are these flat drawings, which are very clear for

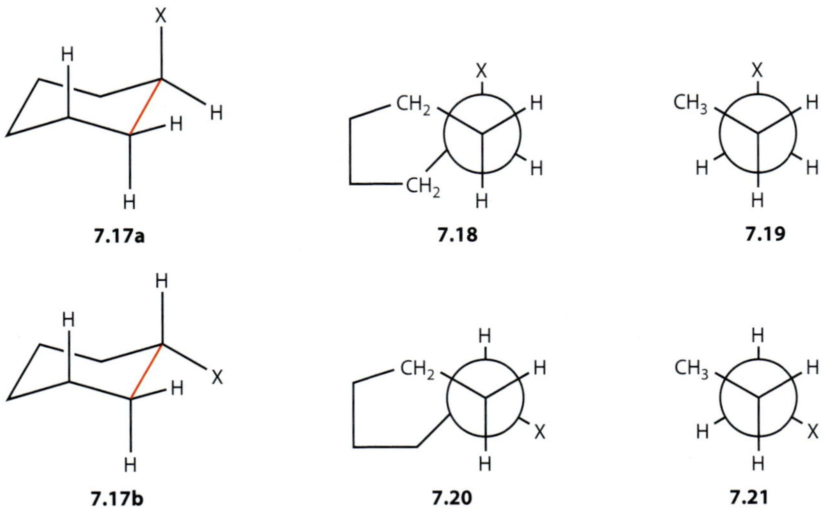

FIGURE 7.11 Gauche interactions in a substituted cyclohexane.

TABLE 7.1 Proportions of Axial Isomers in Monosubstituted Cyclohexanes

X	ΔG (kJ mol⁻¹)[a]	K_{eq}	% Axial
H	0	1	50
Me	7.5	20	5
Et	7.5	20	5
i-Pr	8.8	33	3
t-Bu	>19	>76	<0.01
F	1	1.5	40
Cl, Br, I	2	2.2	31
OH	3.8	4.5	18

[a] The ΔG value is the difference in energy between the molecule with the substituent axial and that with it equatorial.

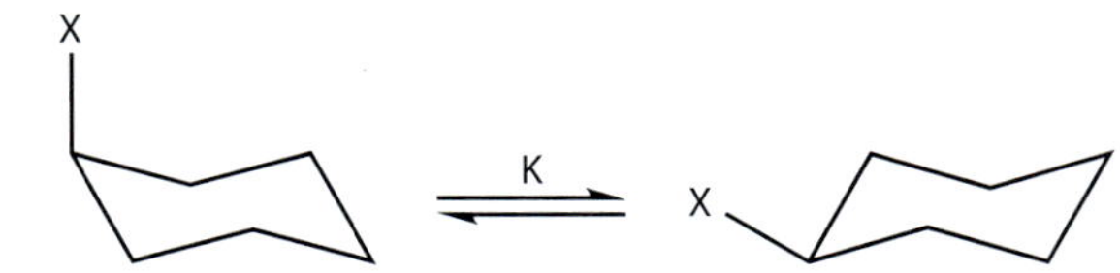

FIGURE 7.12 Equilibrium between conformers of monosubstituted cyclohexanes.

a study of symmetry, but far from the real three-dimensional structures, turned into chair forms? Relying on intuition, or the "obvious," will not invariably give the right answer. For both isomers, two chair forms are possible, and it is only after we have drawn these correctly that we will be in a position to evaluate which is likely to be the more stable. A good tip in these drawings is to regard the wedge and hash symbols as "up" or "down" relative to the mean plane of the ring. Thus, there are two conformers of the *cis*-isomer, both with one substituent equatorial and one axial. Plainly, these are identical energetically (Figure 7.13). Two chair forms can also be drawn for the *trans*-isomer, but these are very different. It's very obvious that the diaxial form is indeed *trans*, but it's less intuitively obvious that the diequatorial form is also *trans*. This is where the tip of looking at "up" and "down" is helpful—although the two equatorial substituents don't look to be on opposite sides of the bond, one

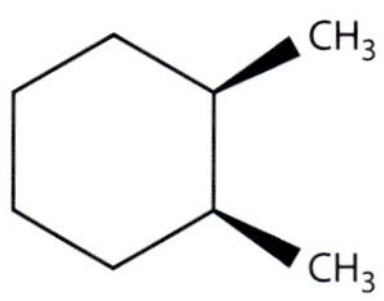

7.22, *cis*

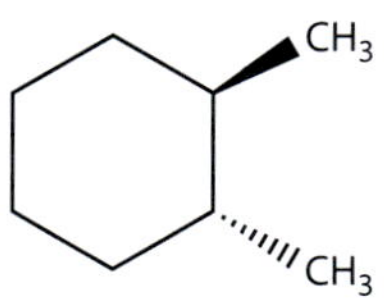

7.23, *trans*

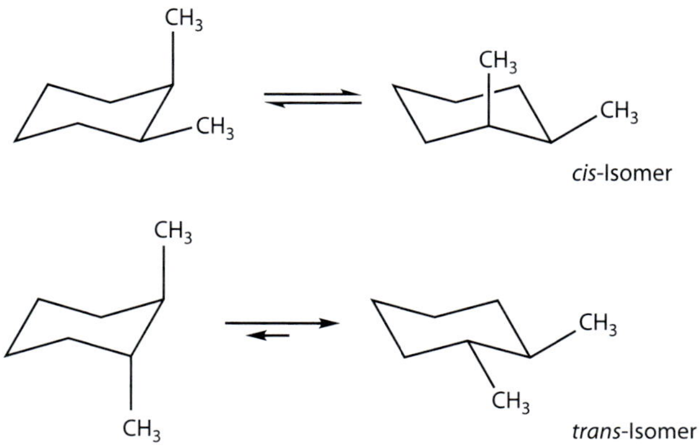

FIGURE 7.13 Conformational equilibria in 1,2-disubstituted cyclohexanes.

is definitely "up" and the other "down" relative to the mean plane of the ring. However, when you are working out how to draw more complex examples, it is often easier to focus on the axial substituents (even if these are hydrogens that you have to draw in), as it is much easier to see whether these are *cis* or *trans*.

PROBLEM 7.4

Draw both chair forms of **7.24**, and predict which will be the more stable:

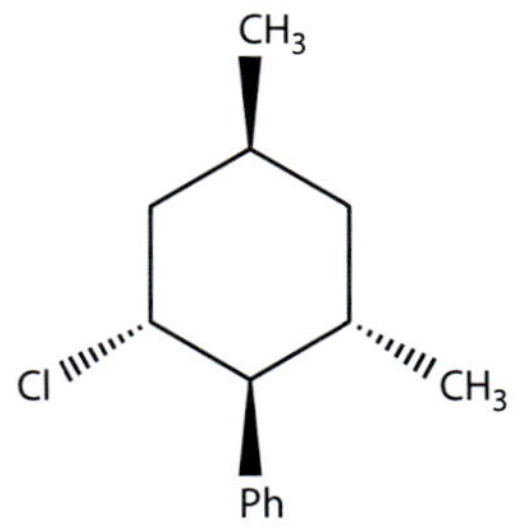

7.24

7.25 **7.26**

Solution

Many students find it helpful to develop their own convention as to where to start in drawing these rings; some orientations are easier to draw, and the solutions easier to follow, than others. For right-handers, a good option is to start with the right-handers' ring (**7.25**) and start at the top right and then go round the ring counterclockwise. Left-handers will probably find it easier to start with **7.26**. Thus, we get

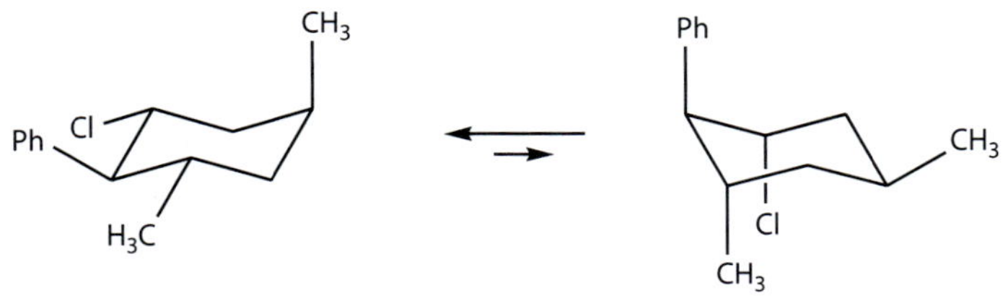

The equilibrium lies far to the left, toward the conformation where three of the four substituents are equatorial.

When two cyclohexane rings are fused as in **7.27** or **7.28**, they go by the common name of decalin, a contraction of the old name of the compounds, decahydronaphthalene. The structure of the *trans*-decalin is both easier to describe and to draw; two chair cyclohexanes are fused so that the bonds of the second ring are equatorial with respect to the first and *vice versa* (**7.29**). Neither of these cyclohexane rings can invert, as this would mean putting the bonds of the other ring into two axial positions—and the four-carbon chain is not long enough to allow this. This structure is one we will encounter later in many natural products, particularly steroids. By contrast, the *cis*-decalin has rings fused with one axial and equatorial bond, and the rings can invert. Although more flexible than *trans*-decalin, it is less thermodynamically stable (some 11.3 kJ mol^{-1}), this being mainly attributed to greater steric interactions between the hydrogens on the "underside" (as drawn, **7.30**) of the molecule.

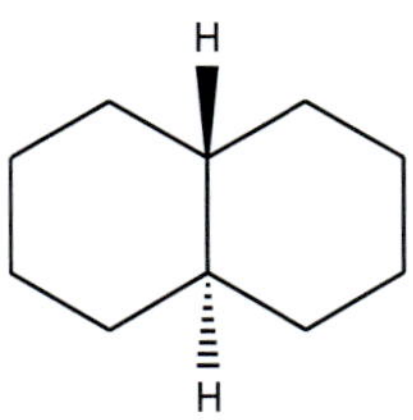

7.27

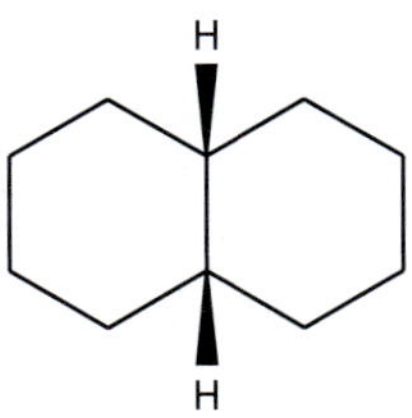

7.28

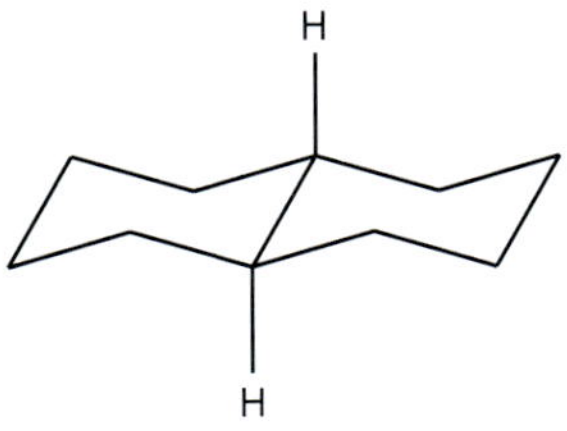

7.29

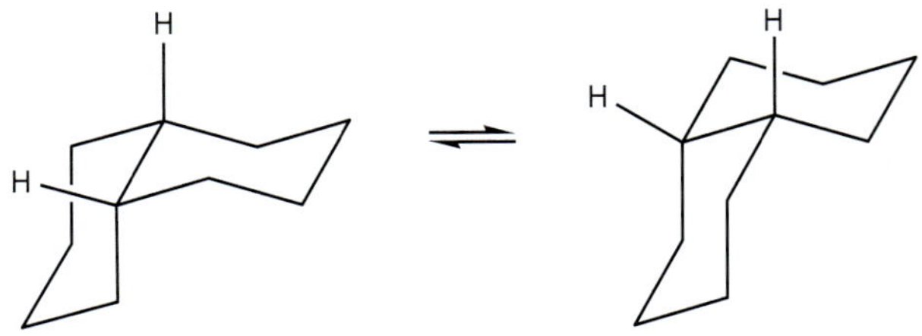

7.30

Introducing one or more sp² carbon atoms into the six-membered ring introduces more strain, since these atoms require 120 ° angles. Any addition reaction that converts the system to a saturated cyclohexane will tend to be more favorable than for a comparable acyclic system. Thus, cyclohexanone is a little more susceptible to addition reactions than acetone. However, in cyclohexanone, two 1,3-diaxial interactions are removed (**7.31**, **7.32**). This means that for substituted cyclohexanones, the axial conformation is less unfavorable than for a related cyclohexane. As noted earlier, the difference in energy between axial and equatorial conformations for methyl cyclohexane is 7.5 kJ mol^{-1}, and there is only about 5 % of the axial conformer at equilibrium. For 3-methylcyclohexanone, the energy difference is only 2.9 kJ mol^{-1}, and at equilibrium, there is 25 % of the axial isomer (**7.33**).

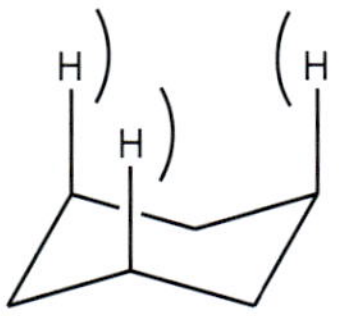

7.31

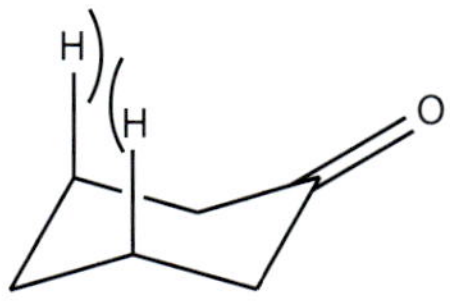

7.32

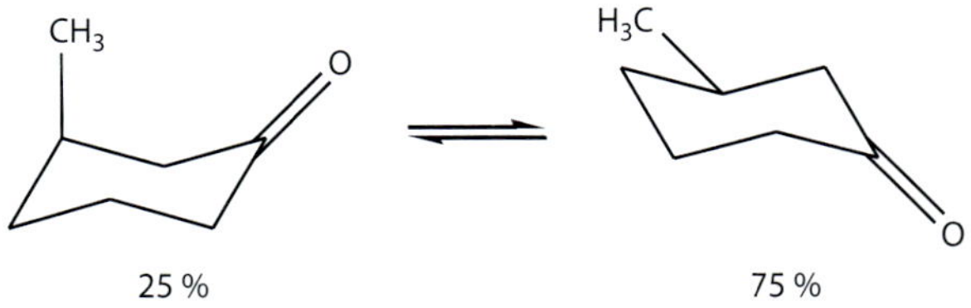

7.33

In cyclohexenes, two sp² carbon atoms must be accommodated in the ring, and the conformation is thus rather different. Two half-chair conformations, **7.34** and **7.35**, are in equilibrium, with an energy barrier of 22.2 kJ mol^{-1} to ring inversion.

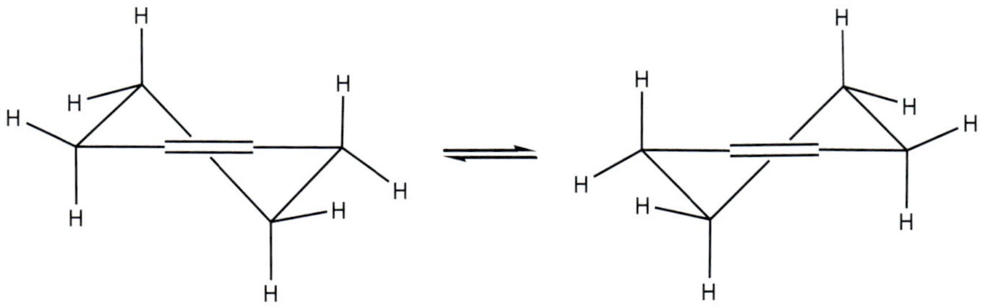

7.34

7.35

7.3.5 MEDIUM AND LARGE RINGS

When we consider the geometries and strain energies of medium rings, it may at first seem surprising that strain increases in 7–12-membered rings. Although these are much more flexible structures, it is difficult completely to avoid eclipsing strain. And we must now consider another form of strain—transannular strain. This is most easily illustrated in cyclodecane, **7.36**. Some of the hydrogens must now be *inside* the ring, and clearly these will interact across the ring. Very large rings, >13 atoms, are regarded as being strain free. Many conformations are energetically accessible; there is no angle strain, no transannular interactions, and no more gauche interactions than any open-chain compound would experience.

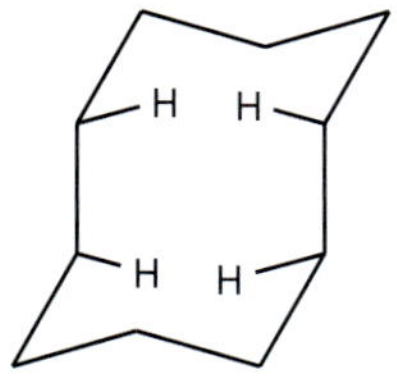

7.36

PROBLEM 7.5

Draw each of the chair conformations of *myo*-inositol. Predict which form will be more stable, giving your reasons:

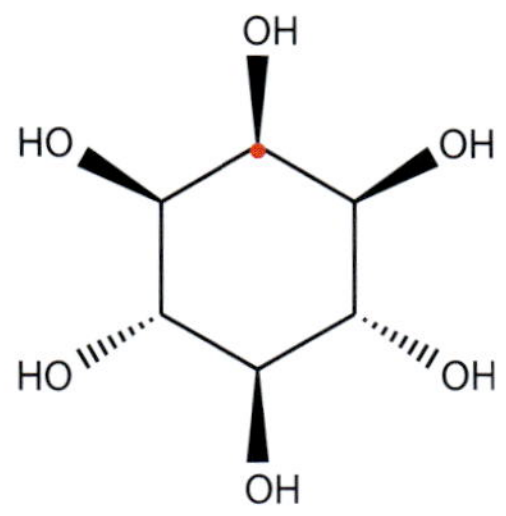

myo-Inositol

Solution

In drawing the representations, the "top" carbon of the flat structure (identified by a red dot) is placed at the top right of the left-hand structure and is hence at the bottom right of the second one. The first structure has five of its six substituents equatorial and will clearly be the more stable (though perhaps less than would be expected given the opportunities for hydrogen bonding afforded by the axial orientation).

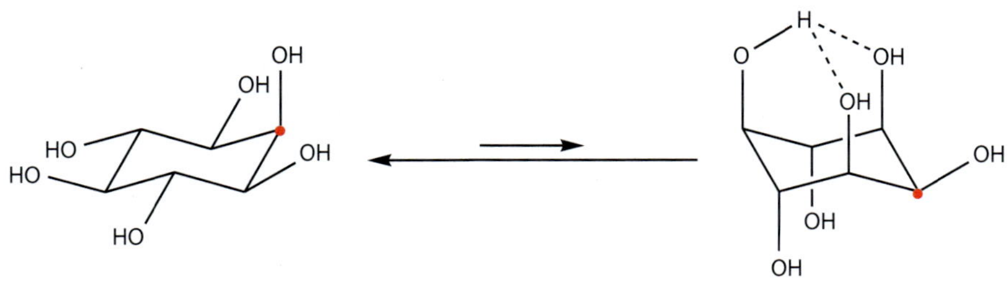

7.3.6 RING SYNTHESIS

How do the considerations of ring strain we have studied affect ring formation from acyclic precursors? Ring strain is reflected in thermodynamic parameters for ring closure—how stable is the new ring? We would predict that six-membered rings would be best, followed by five-membered rings. Three-membered and four-membered rings, and medium rings, would be predicted to be difficult to make. There should be no thermodynamic barriers to making unstrained large rings.

However, kinetic factors are also important. How likely is it that the two ends of a chain that we need to bring together will actually meet one another? This is an entropic argument—the shorter the chain, the more likely it is that the two "ends" will be in appropriate proximity. From this argument, the smaller the ring, the easier it should be to make. Putting the two factors together, it is found that it is relatively easy to make three-, five-, and six-membered rings, but more difficult to make rings of other sizes. Three-membered rings are formed rapidly, but they are not invariably stable and may undergo other reactions. Five-membered rings tend to be formed more rapidly than six-membered rings, where the two processes are in competition, but six-membered rings are decidedly more stable.

Key Points from Section 7.3

- Cyclopropane and cyclobutane exhibit considerable angle and eclipsing strain. Cyclopropane relieves this by forming banana bonds and cyclobutane by flexing from the planar.
- Cyclopentane adopts a flexible envelope conformation.
- Cyclohexane adopts a chair conformation. C–H bonds are described as being axial, perpendicular to the mean plane of the ring, or equatorial, close to the mean plane of the ring. The chair can invert, rendering all axial substituents equatorial and all equatorial substituents axial.
- In substituted cyclohexanes, the chair conformation with the most and the largest substituents equatorial is the most stable; this is attributed to gauche and 1,3-diaxial interactions. The difference in energy between axial and equatorially substituted forms is dependent mainly on the size of the substituent.
- *tert*-Butyl and trialkylammonium substituents always occupy equatorial sites on cyclohexanes.
- Medium rings are less thermodynamically stable than five- or six-membered rings because of eclipsing and transannular strain. Rings with >13 atoms are generally unstrained.
- A combination of thermodynamic and kinetic factors means that three-, five- and six-membered rings are easy to prepare from acyclic precursors.

7.4 CHIRALITY

The word chirality comes from the Greek χείρ, kheir, hand—and most of nature, both macroscopic and microscopic, is chiral or handed. Almost all humans are either right-handed or left-handed (a very small number are fully ambidextrous), and there is evidence, albeit some of it controversial, that other members of the animal kingdom also have a preference for one "hand" over the other. Helical structures are formed in nature from the microscopic in DNA to the macroscopic in gastropod shells. Over 90 % of modern gastropod shells are "right-handed," a few species are exclusively "left-handed," and even fewer have equal numbers of righties and lefties. Even climbing plants coil in a particular way (the majority are right-handed).

> The fragrant honeysuckle spirals clockwise to the sun,
> And many other creepers do the same.
> But some climb anti-clockwise, the bindweed does, for one,
> Or Convolvulus, to give her proper name.*

* http://www.members.optusnet.com.au/pennywyatt/Interests/FlandersSwann/DropOfaHat/At the Drop of a Hat10.html.

Modern gastropod shells.
Left, sinistral shell of *Neptunea angulata*;
right, dextral shell of *Neptunea despecta*

Turret or screw shells

Necklace made from Cuban mountain snail shells

A molecule is said to be chiral if it is not superimposable on its mirror image, like your right and left hands. Molecules that will exhibit chirality are required to have a "stereogenic unit," a grouping within the molecule that is a focus of stereoisomerism; at least one of these must be present in any enantiomer.* For a molecule to be chiral, three conditions must be fulfilled. There must be no plane of symmetry in the molecule, no center of symmetry, and no low-energy path for the interconversion of the mirror images. At this point, we need a...

Short Digression on Symmetry

We need to review briefly what is meant by the terms "plane of symmetry" and "center of symmetry." A plane of symmetry is defined mathematically as a plane that divides a three-dimensional object into two halves that are exact mirror images of each other. For example, a cube can be divided into mirror image halves in many different ways (it has nine planes of symmetry, such as **7.37**). If we consider a molecule such as benzene, which has very high symmetry, it has seven planes of symmetry. Three pass through the centers of bonds (**7.38**), three pass through atoms (**7.39**), and the final one is the plane of the molecule itself (not shown). Although it is often easier to identify planes of symmetry that bisect bonds, those bisecting atoms are equally common. Most molecules, of course, have much lower symmetry. For example, 2-propanol, **7.40**, has just one plane of symmetry. The molecule is drawn in a slightly unconventional orientation to show this more clearly; the plane of symmetry is the plane of the paper, with one methyl group pointing toward us, one pointing away, and the other two groups in the plane of the paper.

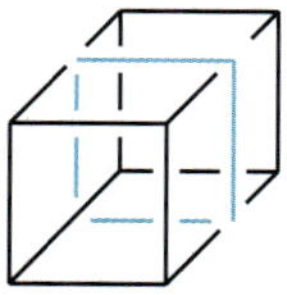

7.37

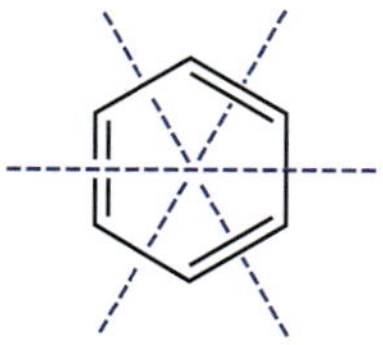

7.38

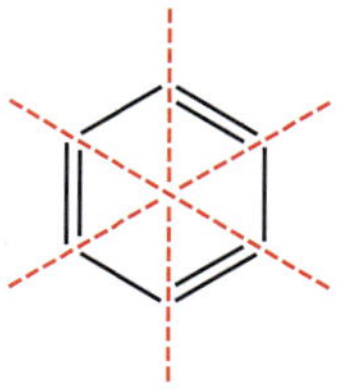

7.39

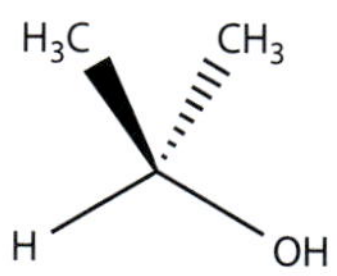

7.40

In terms of identifying symmetry elements that mean that a molecule is achiral (*i.e.*, not chiral), planes are much the most common element. However, a few achiral molecules have no plane of symmetry but do have a center of symmetry. Many have both symmetry elements—for example, benzene. A center of symmetry is a point in the molecule such that for any atom in the molecule, there is another identical atom, at the opposite end

* However, the presence of a stereogenic unit does not require the corresponding chemical species to be chiral, as we will see later when we discuss *meso*-compounds.

of the "diameter" through the center of symmetry. It's often useful to imagine the molecule inside a sphere. Consider *cis*- and *trans*-1,3-dimethylcyclobutane, **7.41** and **7.42**. In the *cis*-isomer, there are two planes of symmetry, through each pair of vertices. In the *trans*-isomer, there is one plane of symmetry, through the methyl-bearing vertices and a center of symmetry (the red dot). **7.43** is an example of the small group of molecules that have a center of symmetry, but no plane of symmetry. Notice that we are not worrying about the conformation of this molecule in determining its symmetry—many molecules have temporary *conformations* in which there is no plane or center of symmetry, but the definition of a conformation tells us that these are only instantaneous arrangements of atoms in space, constantly changing by low-energy rotations about single bonds.

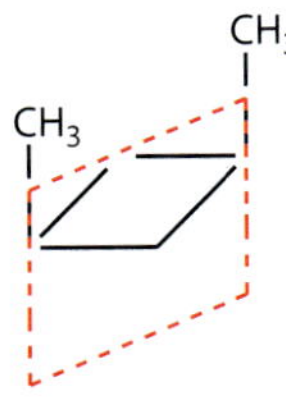

7.41a

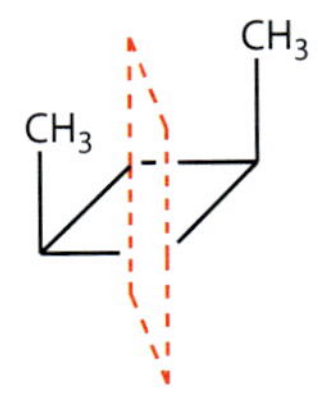

7.41b

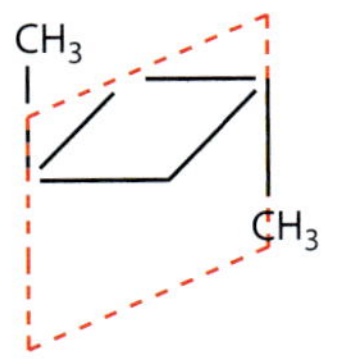

7.42a

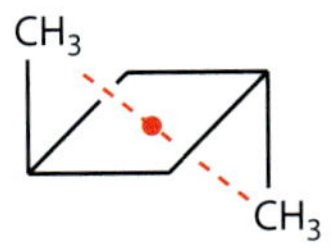

7.42b

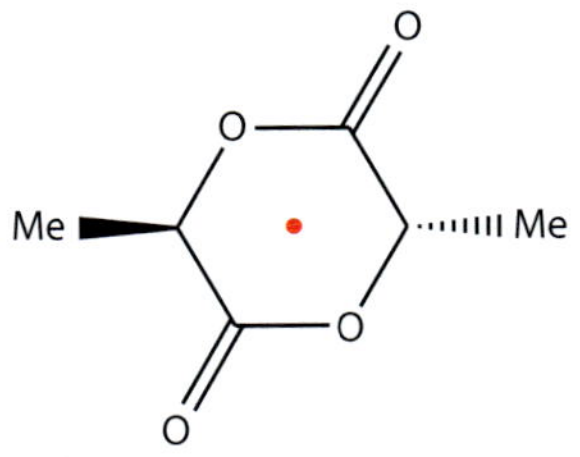

7.43

How do we detect in practice whether a molecule is chiral? It is not a seashell where we can simply look at it. We detect chirality of molecules by the ability of their solutions to rotate the plane of polarization of plane-polarized light. Plane-polarized light will be a familiar topic to those of you who have previously studied physics, but fortunately, it's not necessary for us to understand the details to use the phenomenon. The solution rotates the angle of polarization through an angle α; conventionally clockwise rotations are regarded as positive and counterclockwise ones as negative. A schematic diagram of a polarimeter is shown in Figure 7.14. Although manual polarimeters are still made, most modern instruments are automatic and simply give printouts of the results. A specific rotation, characteristic of the material being tested, is defined by Equation 7.1. The specific rotation may vary slightly with wavelength and temperature; hence, these are part of the specification for $[\alpha]$. Polarimeters are usually designed to operate using the bright orange light of the sodium D-line, and we then refer to $[\alpha]_D$. Specific rotations up to several thousand degrees are known; where this is suspected, successive dilutions of the sample will allow the correct value to be obtained. It might seem a little odd to measure the path length in decimeters, but this is not problematic, as all polarimeters are built containing a cell exactly 1 dm in length. The solvent, and the concentration, is

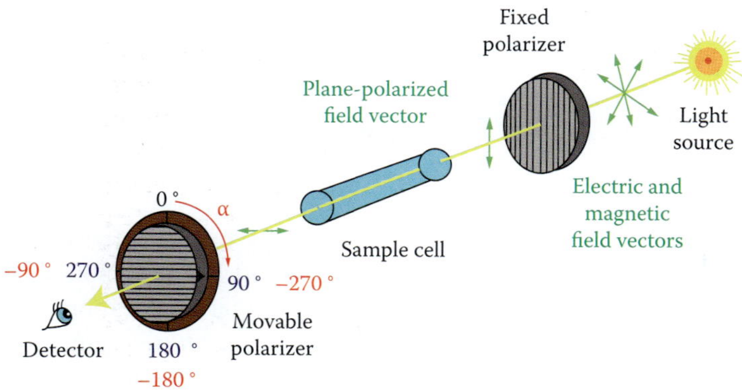

FIGURE 7.14 Schematic of a polarimeter.

always specified for polarimetric measurements; variation is not large but may be measurable, especially when the molecules under test are capable of forming strong hydrogen bonds or aggregating in some way:

(7.1)
$$[\alpha]_\lambda^T = \frac{\alpha}{ld}$$

where
- [α] is the specific rotation, characteristic of the material
- α is the measured angle of rotation
- *T* is the temperature
- λ is the wavelength of light used
- *l* is the length of path in decimeters
- *d* is the concentration in g cm^{-3}

In simple cases, mirror-image molecules, enantiomers, have equal and opposite rotations. For example, lactic acid, $CH_3CH(OH)COOH$, has rotations at 20 °C of +3.82 ° and −3.82 ° for its two enantiomers, in water as solvent (optical rotation in this case does vary with both solvent, concentration, and pH). Lactic acid contains an asymmetric carbon atom (sometimes referred to as a chiral center) as its stereogenic unit, where four different groups are attached to a single carbon atom. Asymmetric carbon atoms (and atoms of other elements) are the commonest form of stereogenic unit. The enantiomers of lactic acid (**7.44**) are related as object and nonsuperimposable mirror image. Although in most cases you will find it easiest to write a mirror-image molecule by reflecting it across a mirror that is formally vertical, there is nothing absolute about this—horizontal or plane of the paper mirrors also works fine, and another way of drawing the enantiomers of lactic acid is shown in **7.45**.

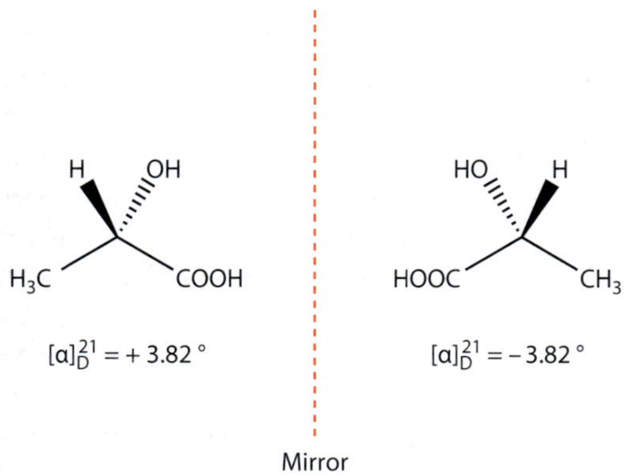

7.44

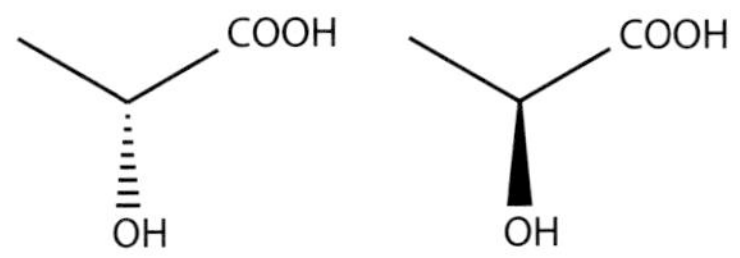

7.45

Chirality is a necessary, but not sufficient, condition for optical activity to be measured. If a solution contains equal numbers of "right-handed" and "left-handed" molecules, they rotate the plane-polarized light to an equal and opposite extent, so that no net rotation is observed. This mixture is called a racemate, or racemic mixture (Latin, *racemus*, a bunch of grapes, because Pasteur first isolated racemic tartaric acid from wine), and is sometimes designated (±) to indicate that there are equal amounts of (+) and (−) material. Between an equal mixture and a pure enantiomer, every other mixture is also possible. We therefore define the term "optical purity" or enantiomer excess (ee, **7.2**). If we have a mixture of 75 % of the (+)-enantiomer and 25 % of the (−)-enantiomer, we say that the optical purity is 50 %. Although the physical properties, other than optical rotation, of the two enantiomers are identical, it is not invariably the case that a racemate has exactly the same physical properties as a pure enantiomer. For example, either pure enantiomer of lactic acid melts at 53 °C but a racemic mixture melts at 16.4 °C.

$$ee = \frac{\text{Rotation of the sample}}{\text{Rotation of pure material}} \times 100 \qquad (7.2)$$

PROBLEM 7.6

For each of the following molecules, state whether or not they are chiral. If they are, draw the two enantiomers as mirror images. If they are not chiral, indicate the plane or center of symmetry:

(a) 2-Butanol
(b) *cis*-1,2-Dibromocyclobutane
(c) *trans*-1,4-Dibromocyclohexane

Solutions

(a) Chiral:

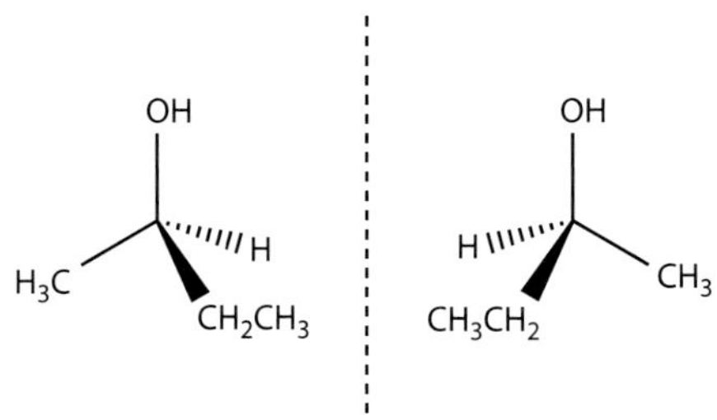

(b) Plane of symmetry, so despite the fact that there are two asymmetric carbon atoms, this is achiral:

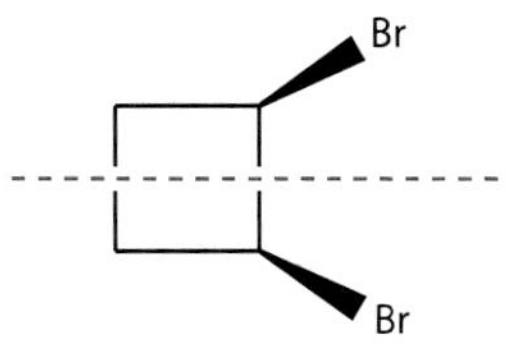

(c) Plane of symmetry and center of symmetry, achiral:

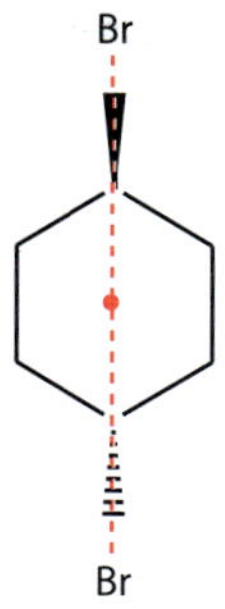

Focus on the Origins of Natural Chirality and Asymmetry*

From molecules to mollusks, nature tends toward chirality. Why? Studies in this field encompass physics, chemistry, and biology, and the chemistry is probably the best understood area. In physics, it has been known since the 1950s that for one of the four fundamental forces of the universe, the weak nuclear interaction, parity, is not conserved. Essentially this means that the universe is inherently slightly "left-handed" and that the energy of the two enantiomers of chiral molecules is very slightly different. But this energy difference is so tiny that it's hard to understand how it could translate into a preponderance of one chiral form of, for example, an amino acid, over the other. Although there are many theories as to how the weak nuclear force could imprint chirality on the universe, two have the most currency. If amino acids are subjected to circularly polarized light (about 17 % of stellar radiation emitted in some areas of the universe where stars form), crystal growth of one enantiomer of certain amino acids is favored. Grinding of these enriched crystals (a proxy for evolution) under conditions in which they can racemize gives chiral material.† This theory has the disadvantage that although it is clear that prebiotic molecules can be formed in interstellar space, they are certainly not crystalline in that medium. Another theory relates to the denaturation (unwinding) of left- or right-handed polymers of amino acids in water. The pH at which this occurs is reported to be higher for the nonnatural material by 0.2–0.3 units. This was attributed to the interaction of the magnetic fields of the enantiomers with the nuclear spin states of the hydrogen atoms of water (the so-called *ortho*-isomer, with identical spins, predominates by a ratio of 3:1 at room temperature). The effect disappeared in D_2O, since deuterium has a nuclear spin of 1 rather than 1/2.‡

The connection between the asymmetry of the physics of the universe and the production of chiral molecules may well exist, but the mechanisms for it remain tenuous. Another theory suggests that initial symmetry breaking in molecular terms may have been random, rather than deterministic. By chance, an aggregate of molecules had a small preponderance of one enantiomer over the other, and this was then amplified over time. Up to 60 amino acids have been isolated from carbonaceous meteorites, and in all of these, the *S*-amino acids are in slight preponderance (begging another question as to how this comes about). Were these amino acids, of extraterrestrial origin, the start of chirality

* *Origin of Chirality in the Molecules of Life*, Albert Guijaro and M. Yus, 2008, RSC.

† C. Viedma, J.E. Ortiz, T. de Torres, T. Izumi, and D.G. Blackmond, *J. Am. Chem. Soc.*, **130** (2008) 15274–15275; W.L. Noorduin, A.A.C. Bode, M. van der Meijden, H. Meekes, A.F. van Etteger, W.J.P. van Enckevort, P.C.M. Christanen, B. Kaptein, R.M. Kellog, T. Rasing, and E. Vleig, *Nat. Chem.*, **1** (2009) 729–732; S.P. Fletcher, *Nature: Chem.*, **1** (2009) 692–693.

‡ Y. Scolnik, I. Portnaya, U. Cogan, S. Tal, R. Haimovitz, M. Fridkin, A.C. Elitzur, D.W. Deamer, and M. Shinitzky, *Phys. Chem. Chem. Phys.*, **8** (2006) 333–339.

on earth? However it began, once any preponderance of one enantiomer over the other is established, it tends to increase—and there are well established and understood chemical and physical processes to do this. Although it is not known what processes were involved on the prebiotic earth, there are sufficient examples established to be confident that such processes could, and did, exist. They do not need to be efficient; a time scale of billions of years is available. For example, crystallization of the amino acid proline (**7.46**) can raise its enantiomeric purity from a few percent to 100 %. If serine (**7.47**) crystals with an ee of 3 % are sublimed, the vapor has an ee of 95 % and can be condensed into an almost enantiopure solid (in the vapor phase, serine forms octamers, which only fit together if they all have the same chirality).

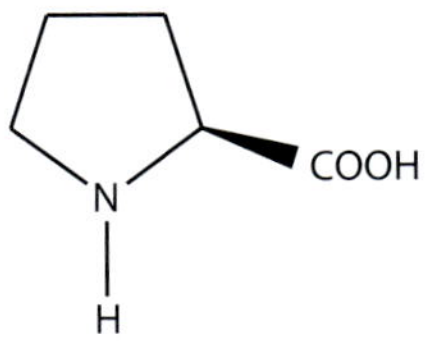

7.46, *S*-Proline

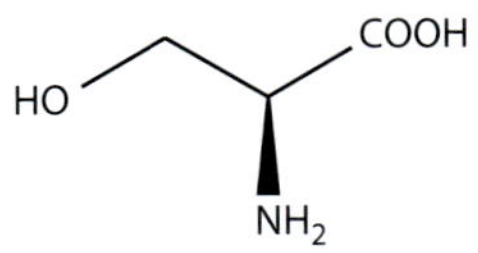

7.47, *S*-Serine

There are also chemical processes that allow a small imbalance in enantiomers to be amplified. The best known is the Soai asymmetric amplification, where the chiral product of a process catalyzes the formation of material with higher enantiomer excess in the next cycle, and this has achieved 85 % enantiomer excesses from material initially enriched by only 0.1 %.* In another process, racemic glyceraldehyde (2,3-dihydroxypropanal, *vide infra*) and 2-aminooxazole (**7.48**) react in the presence of 1 % enriched proline to give almost enantiopure RNA precursors such as **7.49**.†

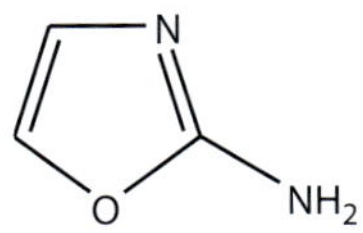

7.48

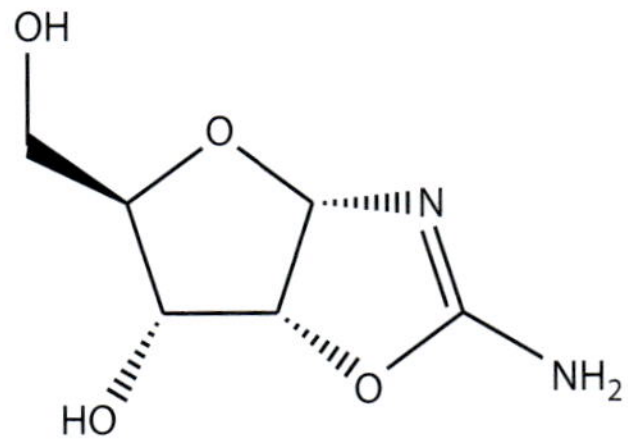

7.49

How, if at all, does the chirality of molecules determine the morphological symmetry or asymmetry of organisms? This dates almost to the first appearance of life on earth, established from both body fossils and trace fossils.‡ Chirality is heritable, and genes have been identified that are related to lack of symmetry, but there is as yet very limited evidence for any strong connection with molecular chirality.

* T. Kawasaki, Y. Matsumura, T. Tsutsumi, K. Suzuki, M. Ito, and K. Soai, *Science*, **324** (2009) 492–495.
† J. Hein, E. Tse, and D. Blackmond, *Nat. Chem.*, **3** (2011) 704–706.
‡ L.A. Babcock, *Eur. Rev.*, **13**(suppl. 2) (2005) 135–143.

7.4.1 ABSOLUTE CONFIGURATION

The absolute configuration of a molecule is a formal method of describing it as "right-handed" or "left-handed." The simplest question to ask involves a compound containing one asymmetric carbon atom, with four different groups around it, as the stereogenic unit. The first step is to order the groups according to the Cahn–Ingold–Prelog system that we met in Section 3.1.3. We number them 1–4. Then view the molecule along the carbon-4 bond from the side remote from group 4. If when this is done the sequence 1 → 2 → 3 is clockwise, we designate the molecule *R* (Latin, *rectus*, right; chirality designations are normally italicized). If the 1 → 2 → 3 sequence is counterclockwise, then the molecule is designated as *S* (Latin, *sinister*, left). This can be illustrated using 2-chlorobutane, **7.50**. The Cahn–Ingold–Prelog rules give us the sequence Cl > CH_2CH_3 > CH_3 > H. So we need to look down the C–H bond along the direction of the arrow. Think of the bond from carbon to group 4 as a steering column, and the other groups as the spokes of a classic design steering wheel. For many of you, it will be immediately obvious that the 1 → 2 → 3 sequence is clockwise and the molecule is *R*. Others will find it helpful to build a model and turn it around physically, to do this sort of problem. However, good spatial perception is not a gift enjoyed by everyone nor is it one that can easily be acquired, simply by working at it. We will explore some simple strategies that will help with problem solving.

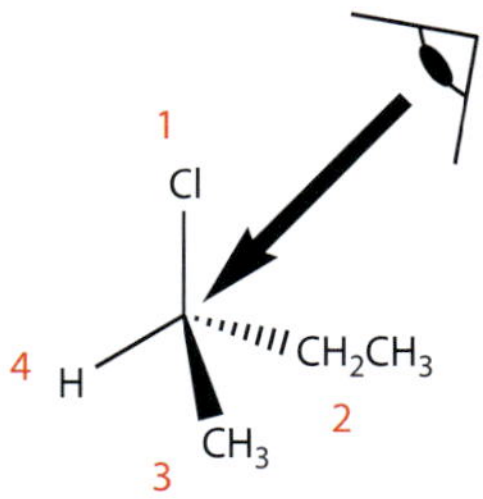

7.50

Taking another example, **7.51**, we order the groups OH > CHO > CH_2OH > H, and look along the C–H bond. Again the 1 → 2 → 3 sequence is clockwise, and the molecule is *R*. For those of you to whom this is obvious, move on, but for those of you finding this difficult, two strategies may be helpful. The first is to imagine this as a physical model and "hold it" by the CHO group. Then rotate the other three groups around the C–CHO axis, counterclockwise, until the hydrogen actually *is* pointing away from you, behind the plane of the paper as in **7.52**. Note that **7.52** is *exactly* the same molecule as **7.51**—but now the 1 → 2 → 3 clockwise sequence is now much more apparent.

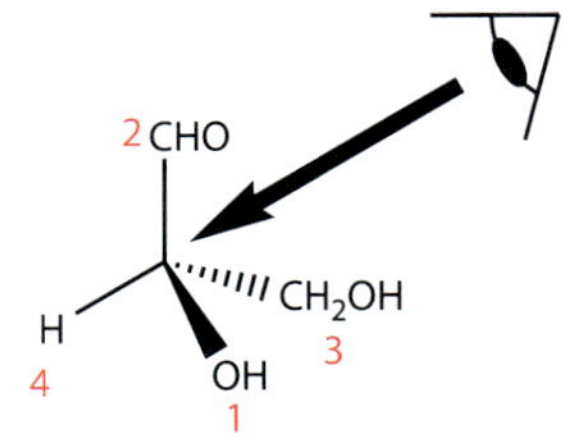

7.51

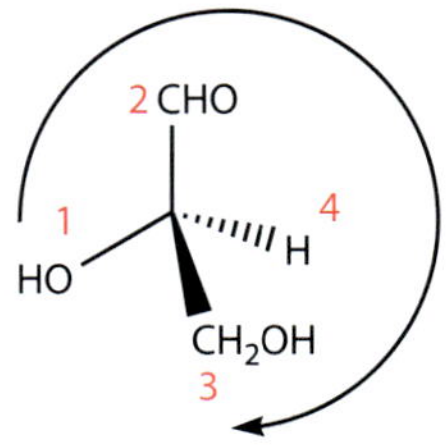

7.52

An alternative strategy is one that involves swapping over two groups in the diagram so that group number 4 is actually pointing away from the observer. *For a single asymmetric carbon atom, every time a pairwise swap is made, whatever type of representation is being used, the product is the enantiomer of the starting molecule.* This is a slow, clumsy, and mechanical process, but it requires no facility with depth or spatial perception, and it is reliable. If we apply this to **7.51**, we would swap the hydrogen and the CH_2OH groups to give the enantiomer of the starting material, ent-**7.51**. The 1 → 2 → 3 track is now clearly counterclockwise, so ent-**7.51** is *S* and **7.51** is therefore *R*. There is no magical one-size-fits-all strategy that will work for every student; try out problems, and stick to what works for you.*

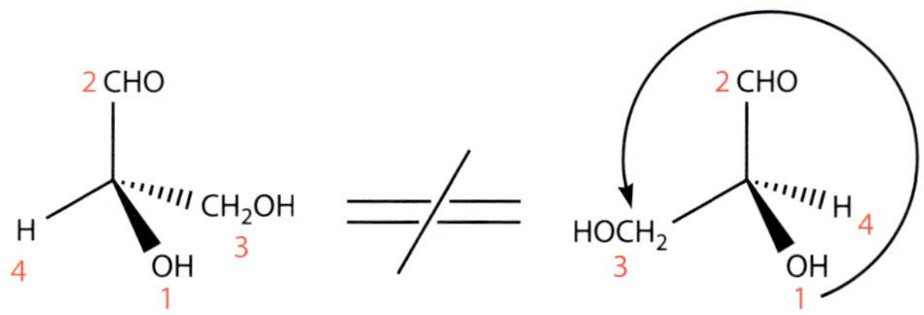

7.51 ent-**7.51**

The existence of stereochemical problems was recognized by the late nineteenth century, particularly by Fischer, studying sugars. Compound **7.51**, 2,3,-dihydroxypropanal, which has the common name glyceraldehyde, is a key molecule in the development of stereochemical theory. At that time (and indeed for over 50 years afterward), there was no *practical* way to determine the absolute configuration of a molecule. The sign of the optical rotation is *not* correlated with absolute configuration. Fischer guessed that the form of glyceraldehyde with a positive rotation was what we now call the *R*-enantiomer, and all subsequent designations involved chemical correlation, ultimately with glyceraldehyde. Reactions that are used for chemical correlation must be those that either do not affect the chiral center at all or affect it in a completely predictable way. Fortunately, when an x-ray crystallographic technique that allowed the determination of absolute configuration was developed in the 1950s, Fischer was proved to have guessed correctly.

* This is an area where students with even mild dyslexia tend to have difficulty, especially if they have more general difficulties with right/left and clockwise/counterclockwise. Wearing an old-fashioned dial-based watch in a test is generally permitted and is an excellent reminder of the directions of clockwise and counterclockwise. ☺

PROBLEM 7.7

Study molecules shown in the following. Then for each of them,

(i) Assign priorities to the four groups around the asymmetric carbon atom according to the Cahn–Ingold–Prelog system.
(ii) Assign an absolute configuration.
(iii) Give a systematic name for the molecule.

(a)

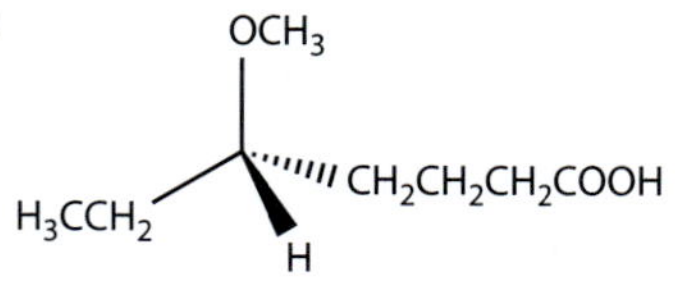

(b)

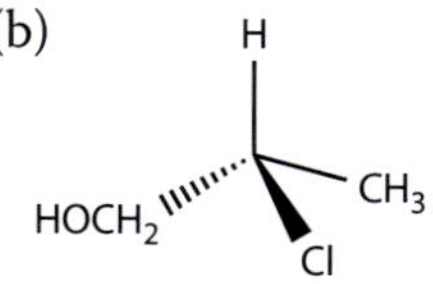

PROBLEM 7.8

Assign an absolute configuration to each of the chiral centers in the anti-AIDS drug, 3′-azido-2′-deoxythymidine, AZT:

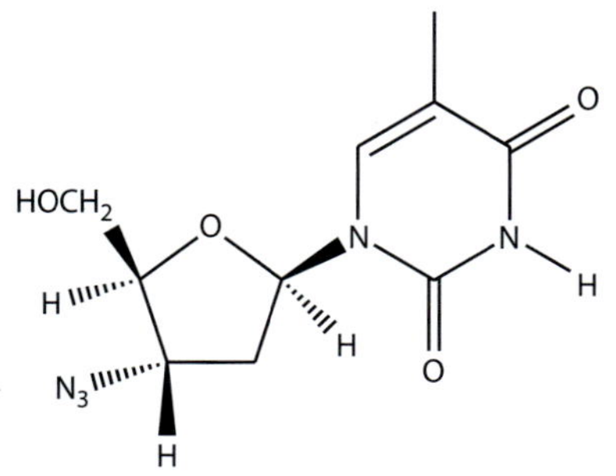

Solutions

7.7(a)
(i) $OCH_3 > CH_2CH_2CH_2COOH > CH_2CH_3 > H$
(ii) *S*
(iii) *S*-5-Methoxyheptanoic acid

7.7(b)
(i) $Cl > CH_2OH > CH_3 > H$
(ii) *R*
(iii) *R*-2-Chloro-1-propanol

7.8 There are three chiral centers to be considered, numbered on the figure as follows (the proper numbering and nomenclature need not detain us here):
At position 1, O > N > C > H, thus *R*.
At position 2, $O > CH_2OH > CHN_3 > H$, thus *S*.
At position 3, $N_3 > CO > CH_2 > H$, thus *S*.

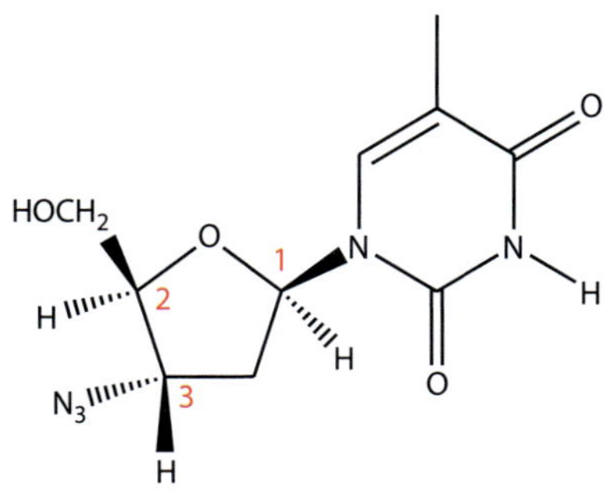

7.4.2 MOLECULES WITH MORE THAN ONE CHIRAL CENTER

Consider **7.53** and its enantiomer ent-**7.53**. These are clearly related as object and mirror image and, in the enantiomer, chirality, is reversed at each of the two asymmetric carbon atoms. What is the relationship between these two molecules and the enantiomeric pair **7.54** and ent-**7.54**? Remembering what we said about stereoisomers at the start of the chapter, the relationship between **7.53** and **7.54** is that of diastereoisomers—all relationships that are not enantiomeric are diastereoisomeric. Diastereoisomers are different compounds, with different chemical and physical properties. Enantiomers have the same physical properties other than their ability to rotate the plane of plane-polarized light and the same chemical reactivity unless they are reacting with a chiral reagent.

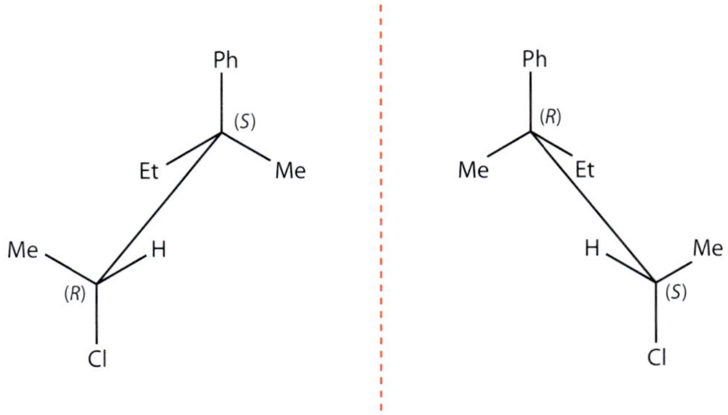

7.53 ent-**7.53**

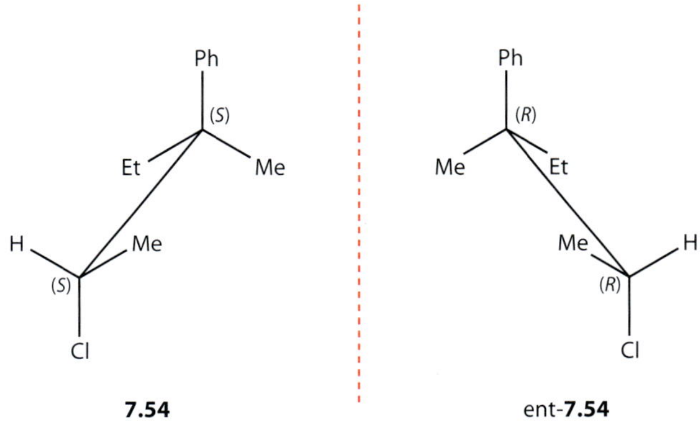

7.54 ent-**7.54**

The difference in physical properties of diastereoisomers suggests a route that we might use to separate enantiomers (a process described as *resolution*)—we will convert them, by some reversible process, into separable diastereoisomers. If we consider the reduction of **7.55** using sodium borohydride (we'll meet this reaction in Chapter 14), the ketone is planar and can be approached from either face by Na[BH_4]. These approaches are equally likely, so the product obtained will be racemic. If this alcohol is then reacted with a chiral carboxylic acid (Equation 7.3), then two distinct esters will be formed, with *R*,*R*- and *R*,*S*-stereochemistries. The two esters are diastereoisomers, which may be separated by physical techniques. Once the esters have been separated, then they can be hydrolyzed to obtain the separated alcohols and recover the chiral acid:

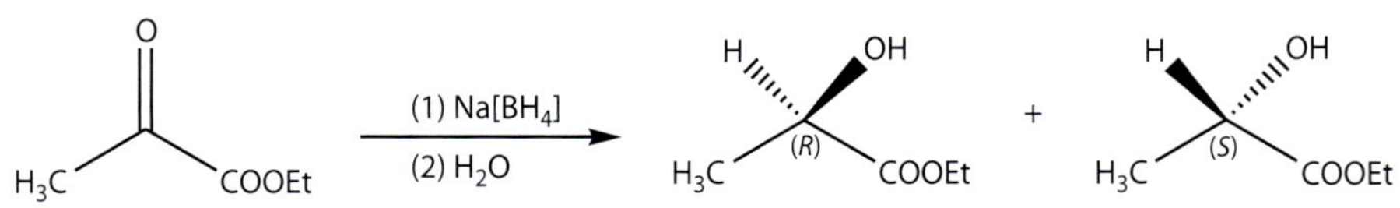

7.55 50 % 50 %

Racemic product

(7.3) $$R,S\text{-alcohol} + R\text{-acid} \rightarrow R,S\text{-ester} + R,R\text{-ester}$$

Separations may also be accomplished using chiral biological reagents, such as enzymes. Most enzymes, since they are composed of chiral amino acids, will accept only one enantiomer of a molecule as their substrate. Thus, the enzyme hog kidney acylase hydrolyzes only the natural enantiomer of amides. We will develop this idea further in Chapter 15, when we consider the reactions of carboxylic acid derivatives in more detail.

In Problem 7.6(b), we noted that *cis*-1,2-dibromocyclobutane was not a chiral compound, because it has a plane of symmetry. However, it does have two asymmetric carbon atoms, each substituted by four different groups, Br, H, CH_2, and CHBr. There are many such compounds, called *meso*-compounds. If we consider 2,3-dibromobutane, **7.56**, there are two chiral centers, and thus, there should, in principle, be 2^2, that is, 4, stereoisomers. However, the "mirror images," **7.57**, are actually identical, and the plane of symmetry should be obvious. Note that for this type of molecule, it is easiest to see if a compound is *meso* when it is drawn in an eclipsed sawhorse form. **7.58** and ent-**7.58** are true enantiomers and not superimposable. Thus, for this compound, there are three, not four, distinguishable stereoisomers.

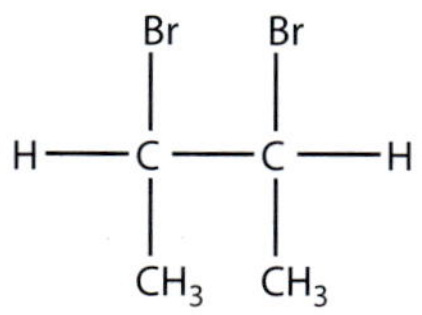

7.56

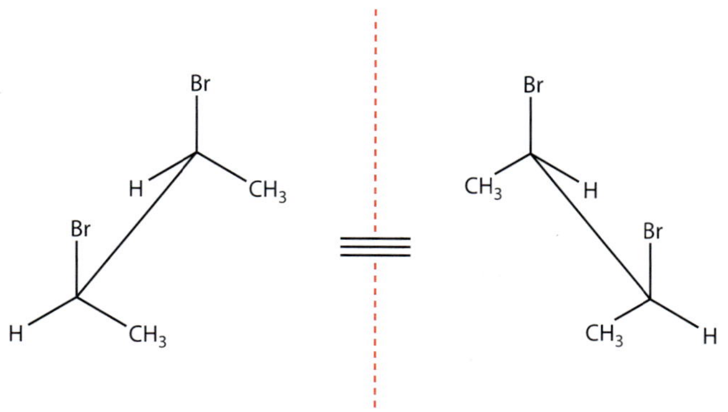

7.57

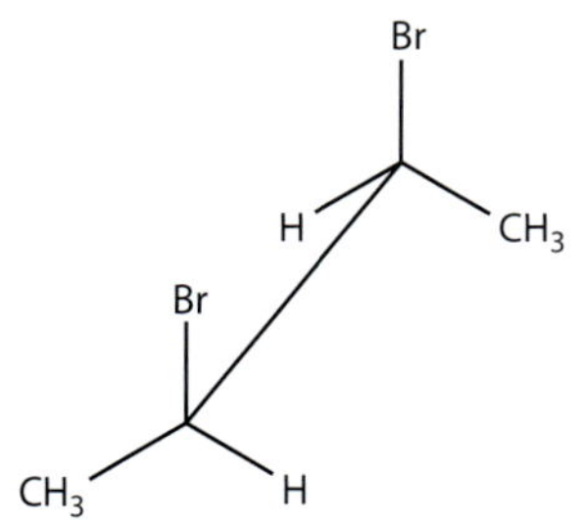

7.58

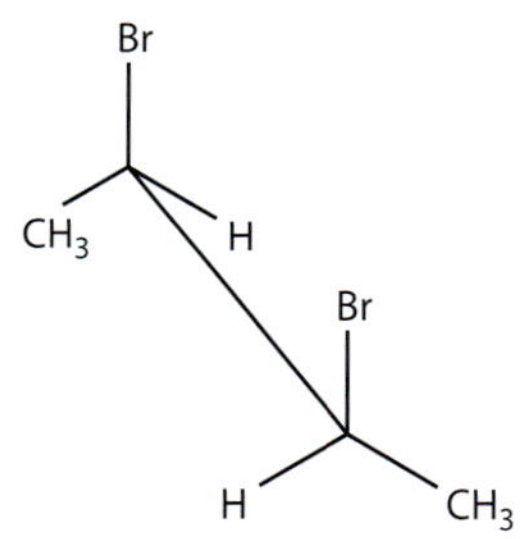

ent-**7.58**

Focus on Enantiomers with Different Biological Roles and Effects

Natural molecules are almost invariably chiral, so all types of biological receptors—taste, odor, and drug receptors—are also chiral. This means that enantiomers interact differently with these receptors. For example, *R*-limonene, **7.59**, is derived from citrus peel and has an odor of oranges. By contrast, *S*-limonene is obtained from Douglas fir needle oil and smells of turpentine/pine. *R*-Carvone, **7.60**, has the odor of spearmint and is extracted from spearmint oil. *S*-Carvone is extracted from dill seeds and smells of caraway. Menthol, **7.61**, is a major component of peppermint oil, but only the natural material, the 1*R*, 2*S*, 5*R*-isomer, has the familiar taste and odor. The scent of sandalwood is typified by the odor of *Z*-β-santalol, **7.62**. The (–)-isomer, as shown, has the odor of sandalwood, but the (+)-isomer is almost odorless.

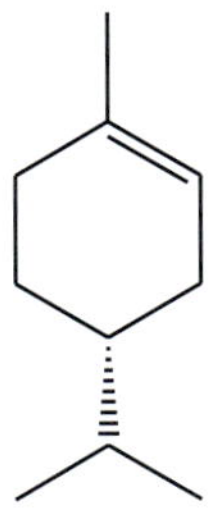

7.59, *R*-limonene

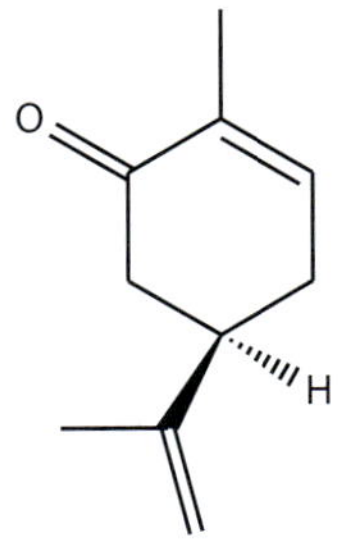

7.60, *R*-carvone

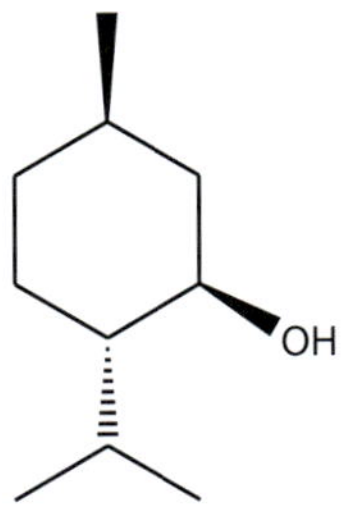

7.61, natural menthol

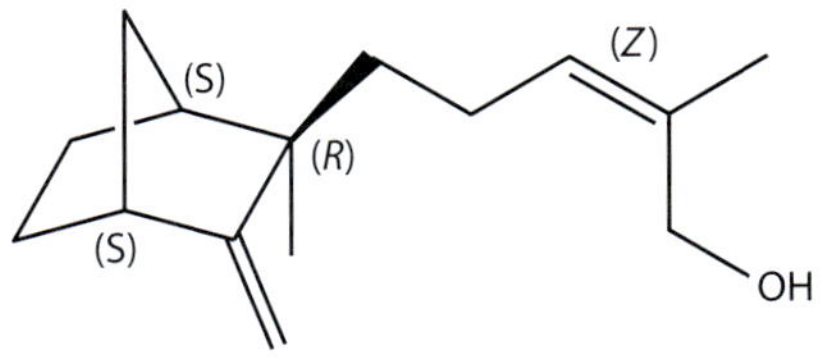

7.62, (–)-*Z*-β-santalol

Citrus peel—a source of *R*-limonene

Douglas Fir needles—a source of *S*-limonene

Sandalwood—source of (–)-*Z*-β-santalol

Recent work on chiral odorants shows that 52 % of all odorants that show chiral discrimination contain six-membered rings, and flexibility in the six-membered ring is broadly associated with a difference in enantiomer scent. Thus, (+)- and (–)-geosmins, **7.63** and **7.64**, differ in the intensity of their odor, but not its nature—they are stereochemically rigid *trans*-decalins. Similarly, (+)-tetrahydronootkatone, **7.65** (which is inflexible), has the same odor as its enantiomer, but 4*R*,4*S*,6*R*-nootkatone, **7.66** (which is flexible), has a grapefruit odor, though its enantiomer has none.*

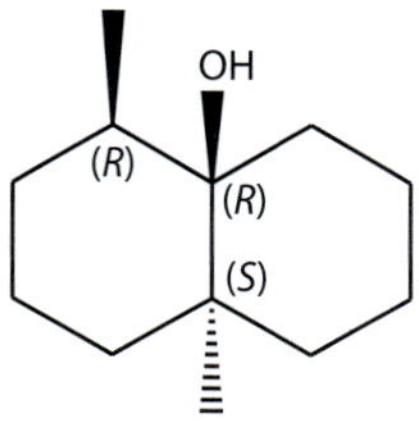

7.63, (+)-geosmin

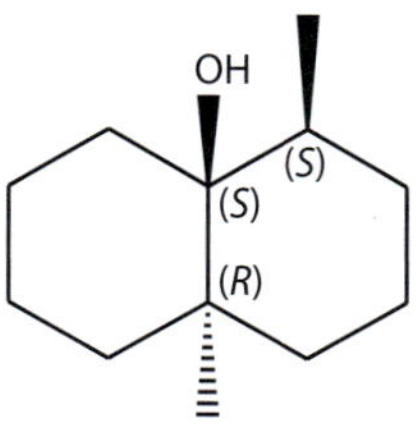

7.64, (–)-geosmin

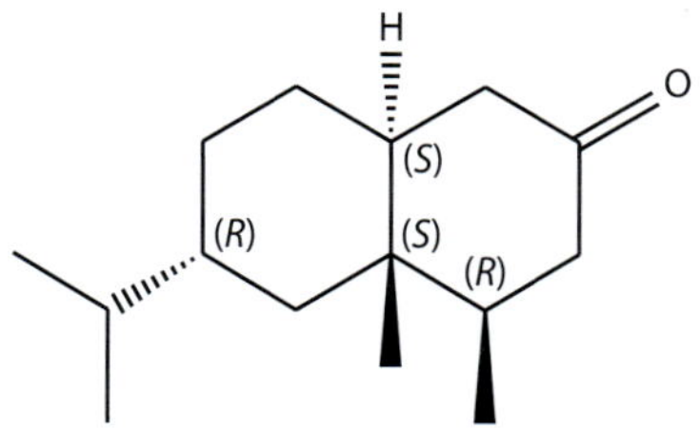

7.65, tetrahydronootkatone

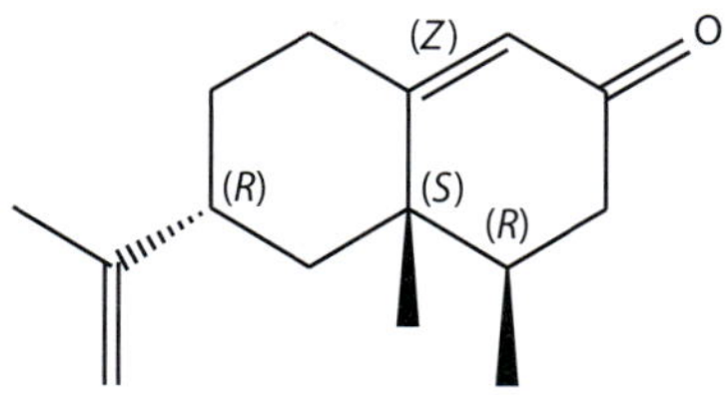

7.66, nootkatone

Taste and smell are strongly correlated; if molecules smell different, then, to most individuals, they will taste different. A simple example is provided by the enantiomers of asparagine, **7.67**, and ent-**7.67**. One enantiomer tastes sweet and the other bitter.

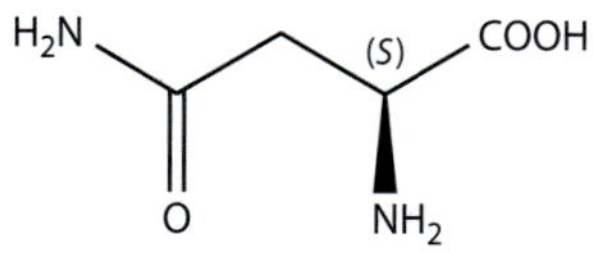

7.67, sweet

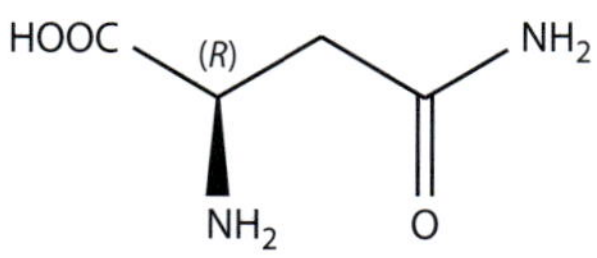

ent-**7.67**, bitter

* J.C. Brookes, A.P. Horsefield, and A.M. Stoneham, *J. R. Soc. Interface*, **6** (2009) 75–86.

Some 50 % of all generally used drugs are chiral compounds, and of these, about half are administered as single enantiomers. Probably the most infamous example of a drug where the enantiomers have different activities is thalidomide, **7.68**. The *R*-enantiomer, as shown, is an effective antinausea agent, but the *S*-enantiomer is a powerful teratogen. It might be thought that administration of the pure enantiomer would have avoided the tragedy that occurred, but this is not so—the two enantiomers can be interconverted in the liver. In some cases, the "wrong" enantiomer is actively toxic; examples include the enantiomers of penicillamine, **7.69** (antiarthritic); naproxen, **7.70** (anti-inflammatory); and L-DOPA, **7.71** (anti-Parkinson's disease). In other cases, the unwanted enantiomer is inactive, but nontoxic; it may still be clinically desirable to administer the chiral compound to reduce the incidence of side effects and the burden on the liver. For ibuprofen, **7.72**, only the *S*-isomer is active, but the *R*-isomer can be slowly converted to *S in vivo*. Only the *S*-isomer of methylDOPA, **7.73**, is an active hypertensive. In some cases, the racemic material may be advantageous. For example, the enantiomers of cicletanine, **7.74**, a diuretic used in treating hypertension, have a synergistic effect. The *S*-isomer decreases the peripheral vascular resistance and hence enhances the antihypertensive effect of the *R*-enantiomer.

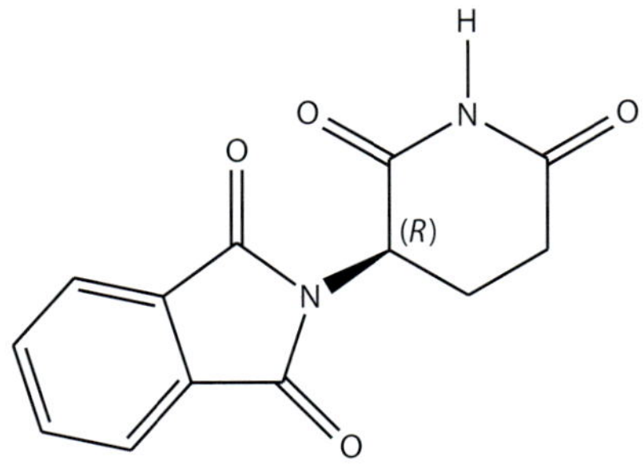

7.68, *R*-thalidomide

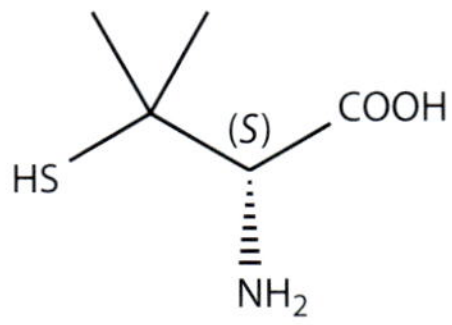

7.69, penicillamine

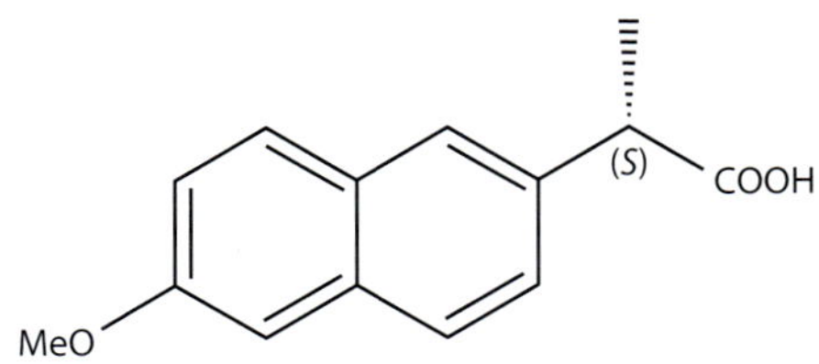

7.70, naproxen

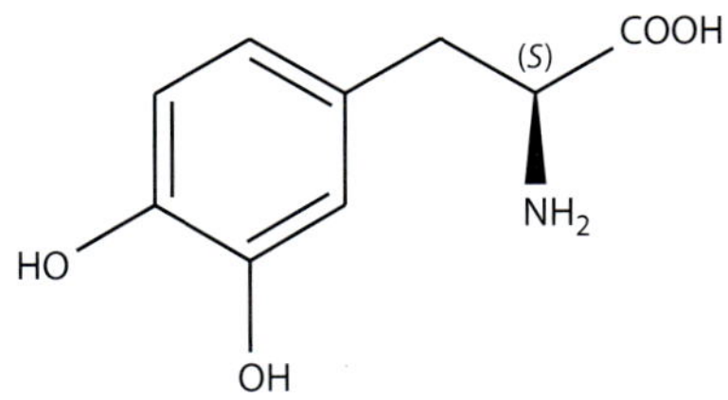

7.71, L-DOPA

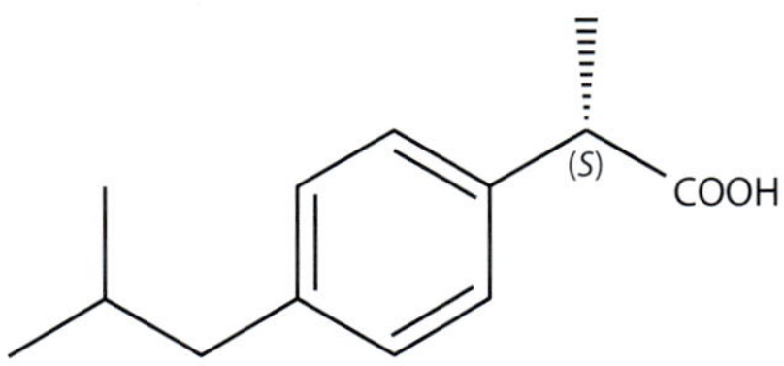

7.72, ibuprofen

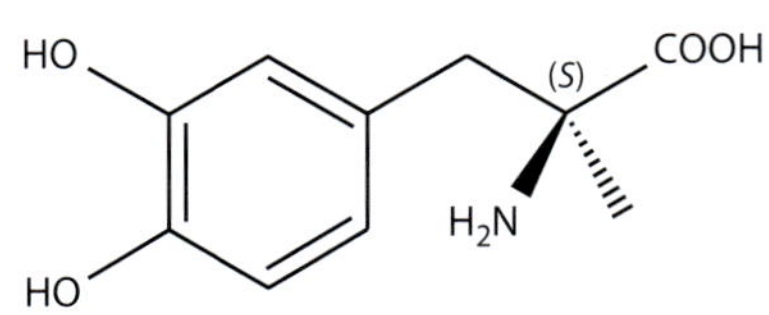

7.73, methylDOPA

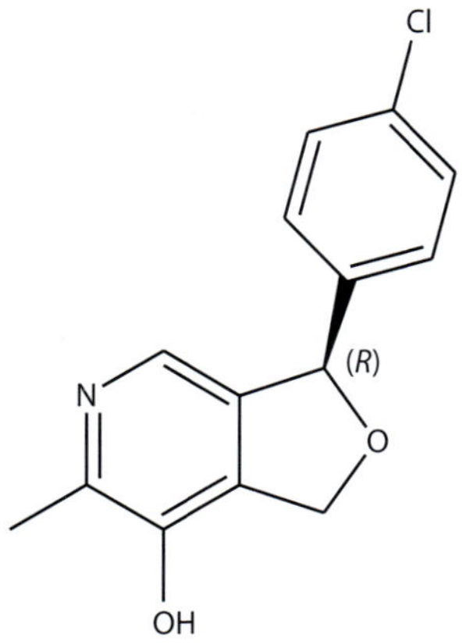

7.74, cicletanine

PROBLEM 7.9

What stereoisomers would you expect for the following compounds? Label double bonds *E* or *Z* and chiral centers *R* or *S*. Distinguish enantiomers and diastereoisomers. Which isomers would you expect to be optically active?

(a) $CF_3CHBrCl$
(b) $CH_3CH{=}CHCH{=}NOH$
(c) $CH_3CHOHCHOHCH_3$

(d)

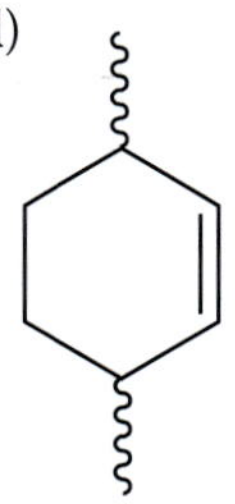

(e)

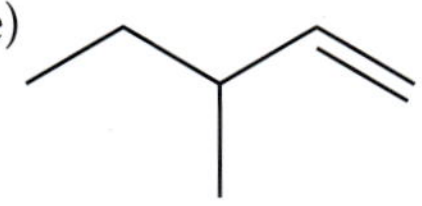

(f)

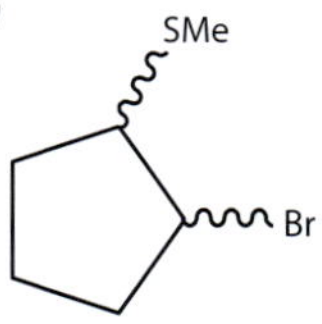

Solutions

(a) There is a single asymmetric carbon atom, so we expect two enantiomers; to make the assignment of absolute stereochemistry as easy as possible, your first drawn structure should have the lowest ranked group, that is, hydrogen, pointing away from you. In terms of ranking, Br > Cl > CF_3 > H:

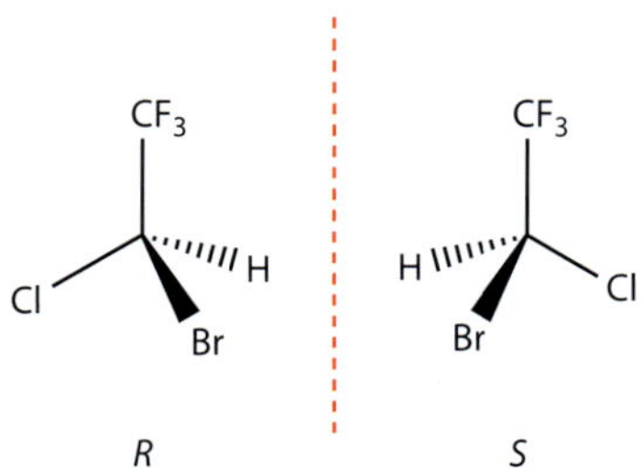

(b) There are four possible stereoisomers, depending on configuration about the double bonds. CH_3 outranks H, C=N outranks H, C=C outranks H, and OH outranks a lone pair. They are all diastereoisomers but could also be described as geometric isomers:

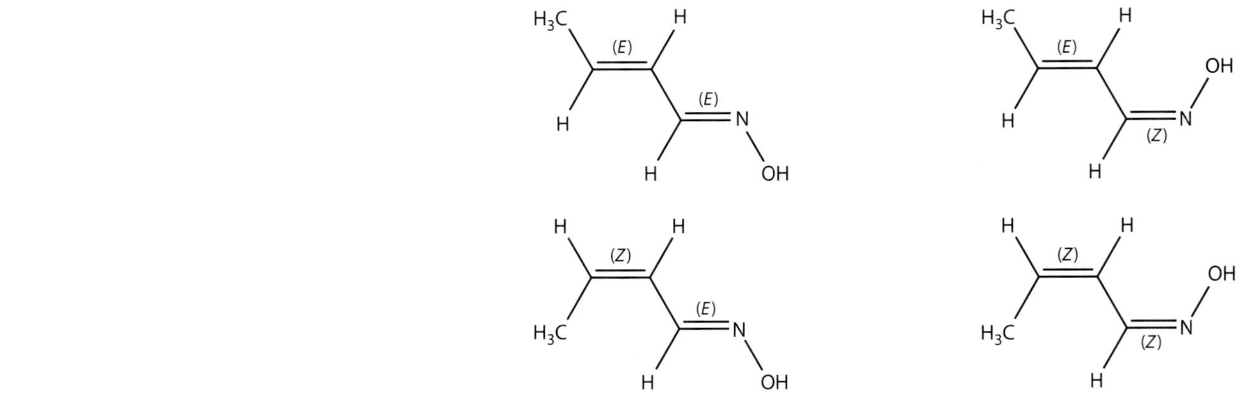

(c) In terms of ranking, OH > CHOH > CH_3 > H. However, as with $CH_3CHBrCHBrCH_3$, there are only three stereoisomers, a pair of enantiomers and a *meso*-diastereoisomer.

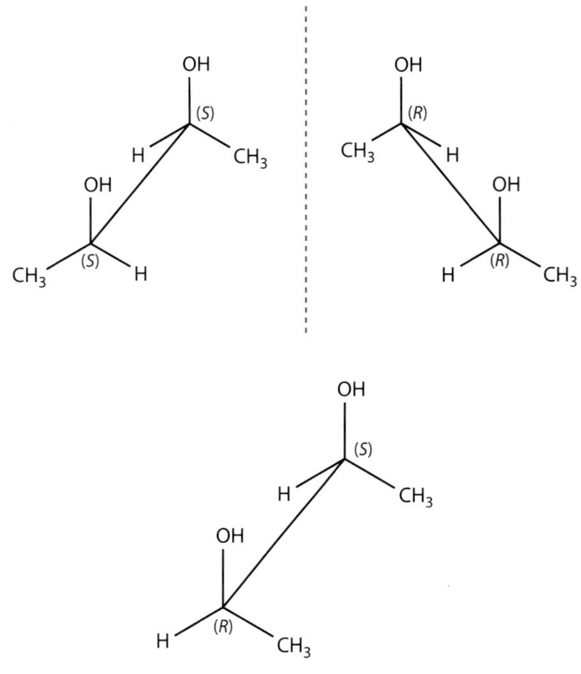

meso-Compound

(d) The methyl groups may be *cis* or *trans* on the ring. The *cis*-isomer is *meso*. The *trans*-isomer is a pair of enantiomers.

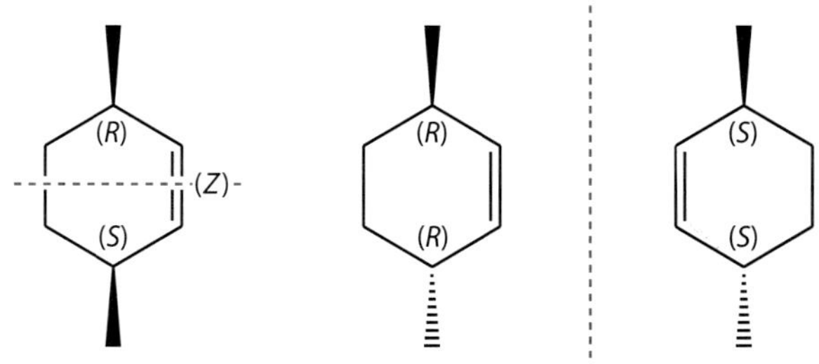

meso Enantiomers

(e) This molecule has a single chiral center, hence two enantiomers:

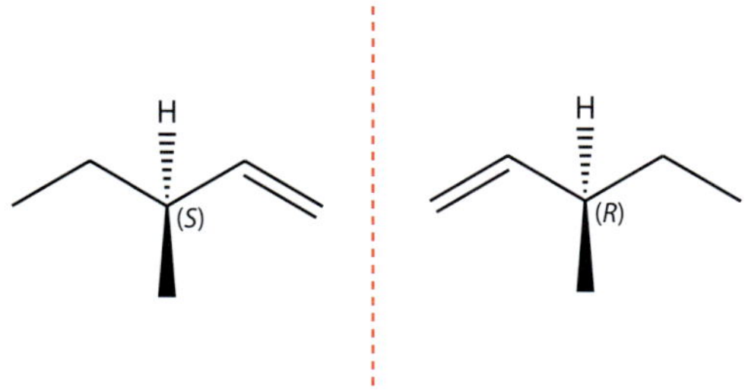

(f) There are two chiral centers, no symmetry (hence no *meso*-compounds). We expect two pairs of enantiomers (which are diastereoisomers of each other):

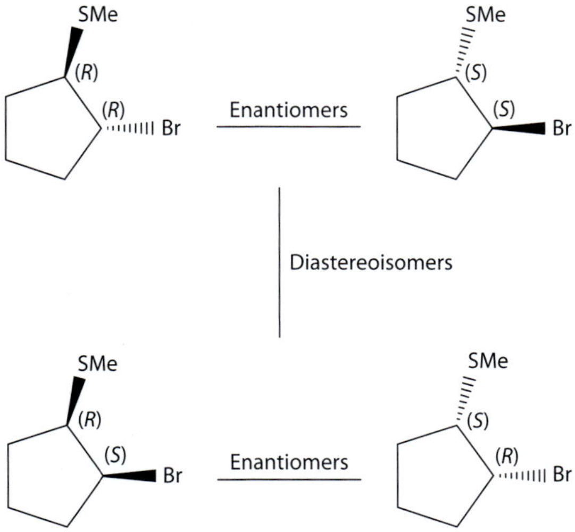

7.4.3 CHIRAL MOLECULES WITH NONCARBON CHIRAL CENTERS

Chirality is a symmetry property, and there is no reason that the chiral center type of stereogenic unit should be restricted to carbon atoms. For example, both phosphines and phosphine oxides can have chiral centers at tetrahedral phosphorus atoms; for the phosphines, the fourth position is occupied by a lone pair of electrons (which takes the lowest priority in the Cahn–Ingold–Prelog rules). We will meet chiral phosphines again in Chapter 23, as their metal complexes can be used to catalyze enantioselective reactions. By VSEPR, sulfoxides also adopt approximately tetrahedral geometry, and can be chiral, as can sulfonium salts, $[R_3S]^+$. In general, the heavier elements of each periodic table group behave similarly to the lighter ones; thus, silanes with four different groups at silicon are also chiral. Some examples are shown in Figure 7.15.

You may be wondering why nitrogen was not mentioned previously. Surely an amine with three different attached groups is chiral, in just the same way as a phosphine would be. The structure of **7.75** is indeed chiral, but at room temperature, inversion at nitrogen is very rapid, the activation energy for inversion being between 20 and 30 kJ mol^{-1}. So even though the individual structures are chiral, it is not possible to separate them. There is a much higher barrier to inversion in chiral phosphines, where typical activation energies would be 130–150 kJ mol^{-1}. However, ammonium salts where all the groups are different are chiral, and many have been resolved (*e.g.*, **7.76**).

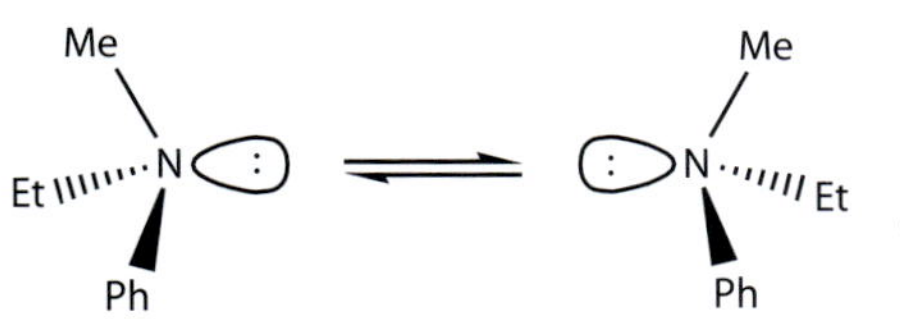

S-**7.75** *R*-**7.75**

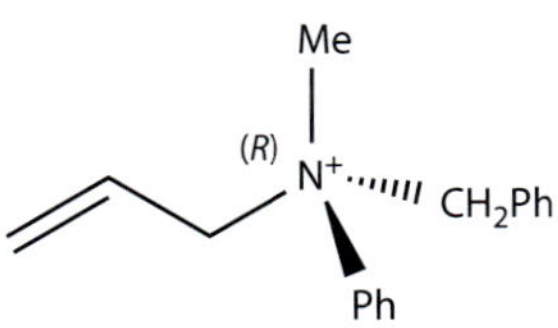

7.76

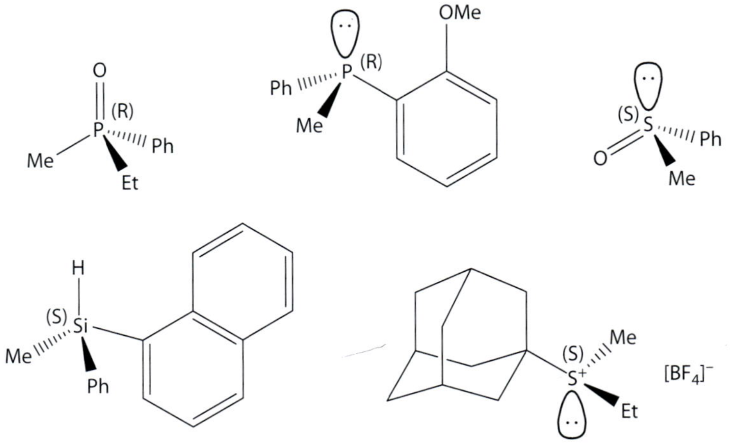

FIGURE 7.15 Chiral centers at atoms other than carbon.

Focus: More on Chirality at Nitrogen

Although the most common "potentially chiral" nitrogen compounds, the amines, are not chiral in practice because they invert so rapidly, recent years have seen an increasing number of chiral nitrogen compounds studied and resolved. The very first of these, Troger's base, **7.77**, was first synthesized in 1887, but the enantiomers were not separated until 1944. Here, the chirality at nitrogen is stable, because the cage structure means that inversion is impossible. Other trivalent nitrogen atoms that do not invert are those in three-membered rings: the aziridines, **7.78**; oxaziridines, such as **7.79**; and the diaziridines, **7.80**. The barrier to inversion at nitrogen in **7.79** is 133.5 kJ mol^{-1} and that for **7.80** 115 kJ mol^{-1}. Why are these nitrogen atoms so difficult to invert? At the transition state for inversion, the hybridization at nitrogen is essentially sp^2—this would like to have angles of 120 °, and in a three-membered ring with 60 ° angles, this imposes too much strain, so that the energy of the transition state is correspondingly raised.

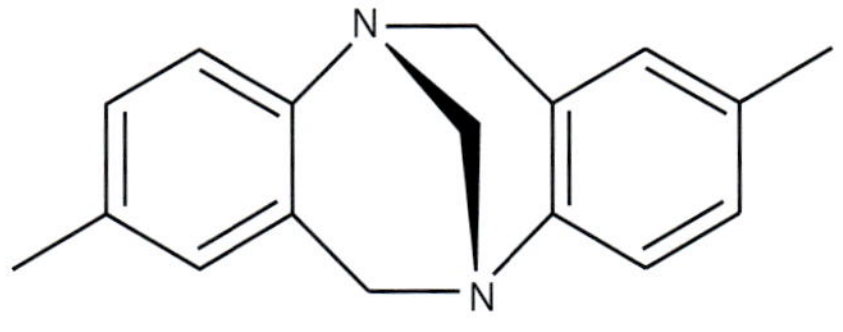

7.77, Troger's base

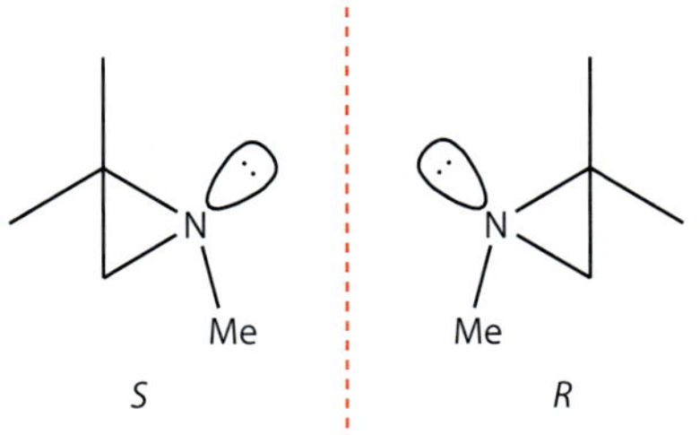

7.78

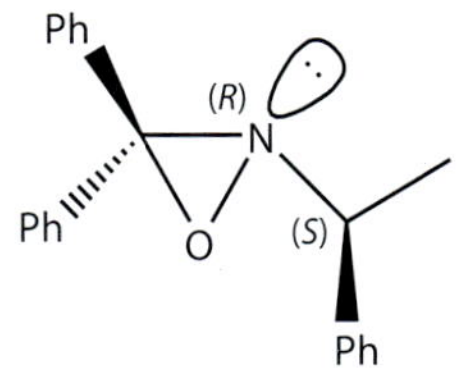

7.79

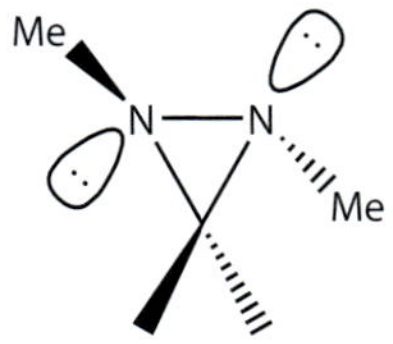

7.80

A few cyclic hydrazines,* such as **7.81** and **7.82**, also have high barriers to inversion at nitrogen, have been resolved, and are stereochemically stable. In these examples, inversion at nitrogen is inhibited by large steric constraints. Finally, amine oxides, obtained from amines using hydrogen peroxide as an oxidant, have four groups around nitrogen and no opportunity to invert. Thus, **7.83** has been resolved.

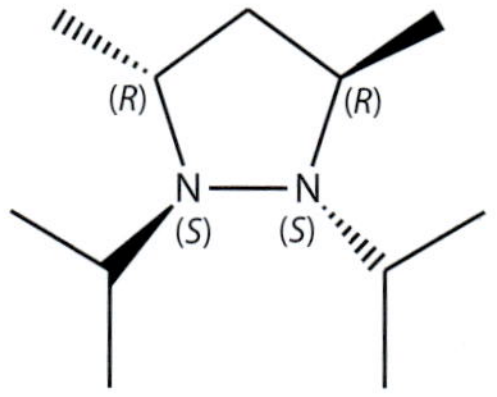

7.81

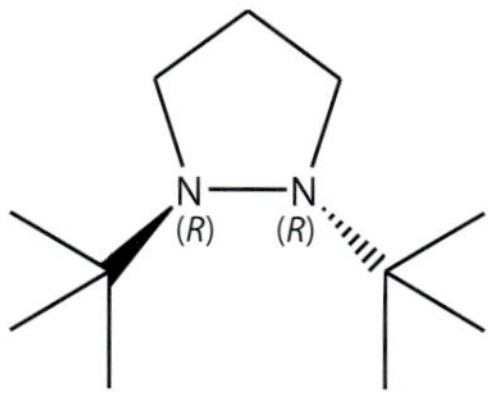

7.82

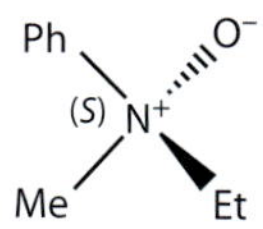

7.83

*Hydrazine is $H_2N–NH_2$.

7.4.4 OTHER CHIRAL COMPOUNDS

There are also classes of organic compounds that are chiral and cannot be superimposed on their mirror images, but which do not contain any classical asymmetric atoms. The stereogenic unit in these molecules is essentially a twisted structure, with either a right- or left-handed twist. The first category comprises the allenes. If you look back to Problem 3.8(b), you should remember that allenes are not planar. If one double bond is formed from p_z atomic orbitals, then the other must be formed from p_y orbitals (**7.84**). So the shape of 2,3-pentadiene, **7.85**, is twisted, and the two mirror images cannot be superimposed.

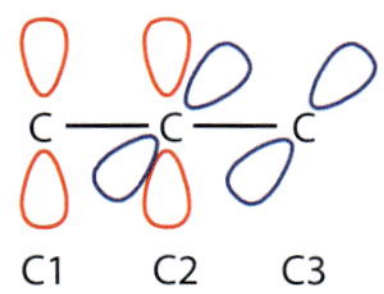

7.84

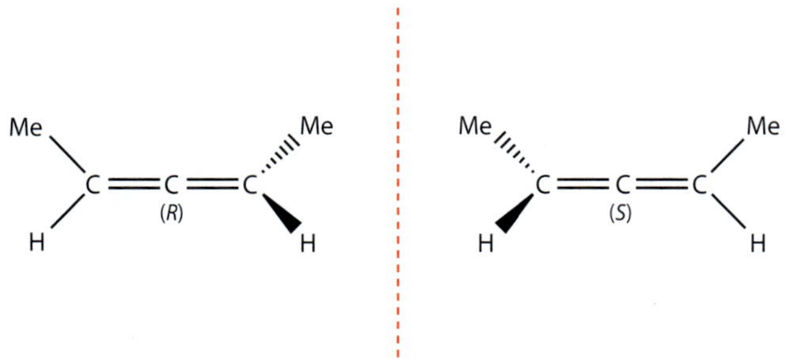

7.85

Biphenyl, Ph–Ph, might be thought to be planar to maximize the conjugation between the aromatic rings. However, this would also maximize the steric interactions between protons at the *ortho*-positions. For the unsubstituted molecule, the angle between the rings in the solid-state crystal structure is 0 °, but this is due to crystal packing forces, and the angle between the rings in the ground state conformation in solution, gas or melt phases, is 44 °. The barrier to rotation for biphenyl is only 6–8 kJ mol^{-1}, but this rises rapidly for biphenyls substituted at the *ortho*-position. To obtain a chiral structure, each ring must be nonsymmetrically substituted, and the substituents must be large enough to restrict rotation about the single bond between the rings. An example is provided by **7.86**. This type of optical isomerism is called atropisomerism and is sometimes defined in terms of the half-life for racemization being greater than 1000 s.

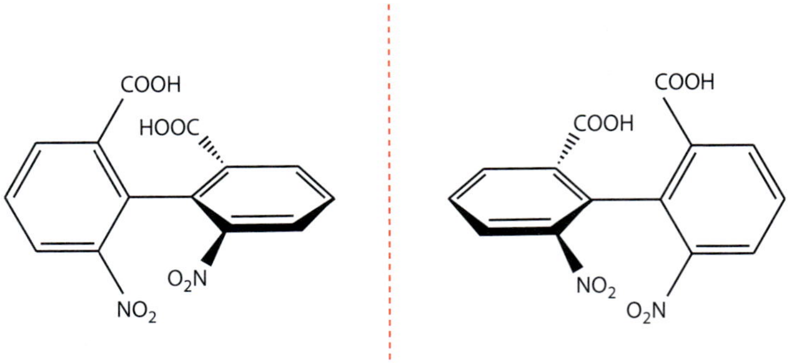

7.86

trans-Cycloalkenes are, in principle, chiral, but in the unsubstituted case, this is important only for *trans*-cyclooctene, **7.87**, which is the smallest ring *trans*-cycloalkene stable at room temperature (it is still 38 kJ mol^{-1} less stable than the *cis*-isomer). Racemization of *trans*-cyclooctene occurs on heating to about 180 °C, with an activation energy of 149 kJ mol^{-1}. Although chiral *trans*-cyclononene has been isolated, racemization is much faster, with an activation energy of 84 kJ mol^{-1}; the half-life for racemization is about 4 min at 0 °C. The comparable activation energy for *trans*-cyclodecene is about 40 kJ mol^{-1}—when the chain linking the ends of the *trans*-double bond is longer, the transition state required for racemization is more easily accessed. One final group of chiral molecules is the helicenes, which, as their name implies, have helical structure. Hexahelicene, **7.88**,* has a specific rotation of several thousand degrees, which is surprisingly sensitive to the wavelength of the light used:

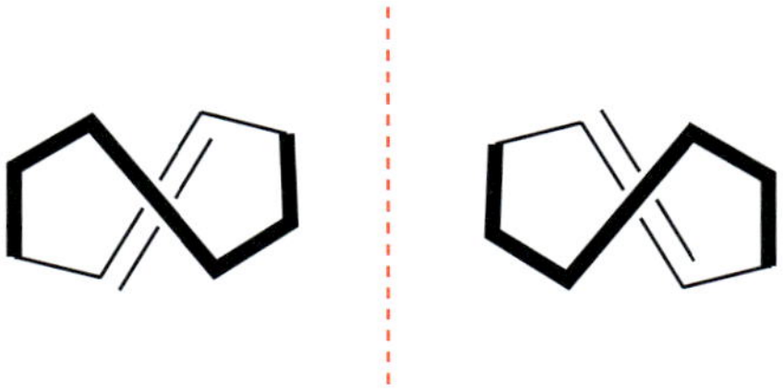

7.87

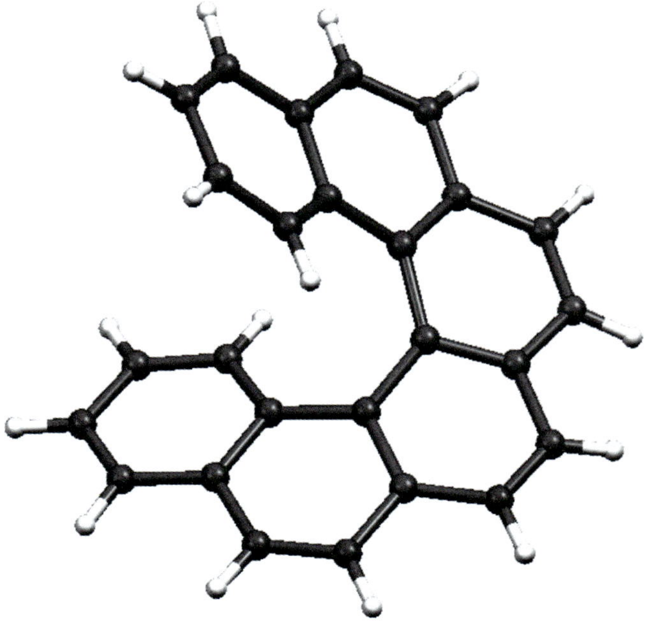

7.88

* Thanks to Prof. Ken Seddon, Queens University Belfast for this image.

PROBLEM 7.10

Assign an absolute configuration to each of the following molecules.

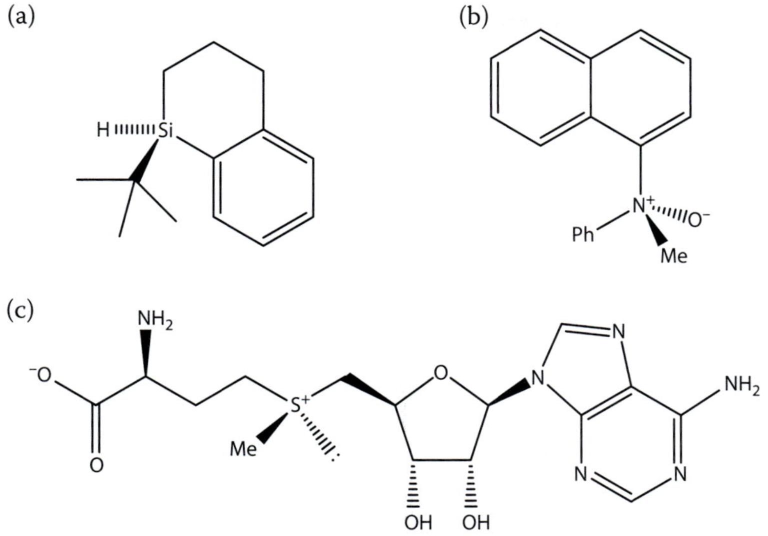

Solutions

(a) C=C > CMe_3 > CH_2 >H; *R*.
(b) O > naphthyl > Ph >Me; *R*.
(c) The molecule is *S*-adenosylmethionine, important in many biological methylations:

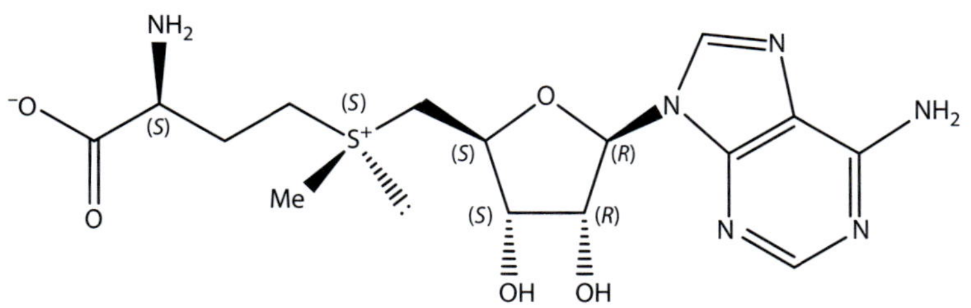

7.5 FISCHER PROJECTIONS

Either by hand or by computer, it's reasonably easy to draw and manipulate "real" stereochemical representations of molecules with two chiral centers, sometimes even with three. However, many biological molecules have many more chiral centers than this, and Fischer devised a way of representing these in a flat form, in which it is easy to see stereochemical relationships and, particularly, to identify *meso*-compounds. Fischer projections consist of a cross motif, **7.89**, and in this, groups b and d are pointing toward us, out of the paper, while groups a and c point away from us, behind the paper.

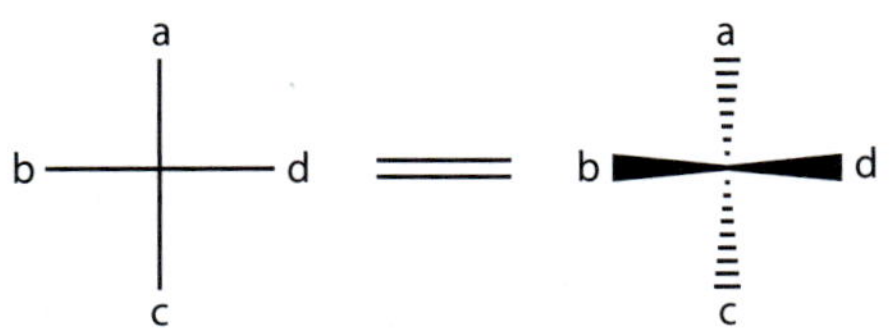

7.89

This type of projection is particularly useful in looking at the stereochemistry of sugars, such as glucose, **7.90**; although it is simply a formalism, and not a likely conformation for the molecule, it is still a useful shorthand. If ribose, **7.91**, is oxidized, **7.92** is obtained. The fact that this is a

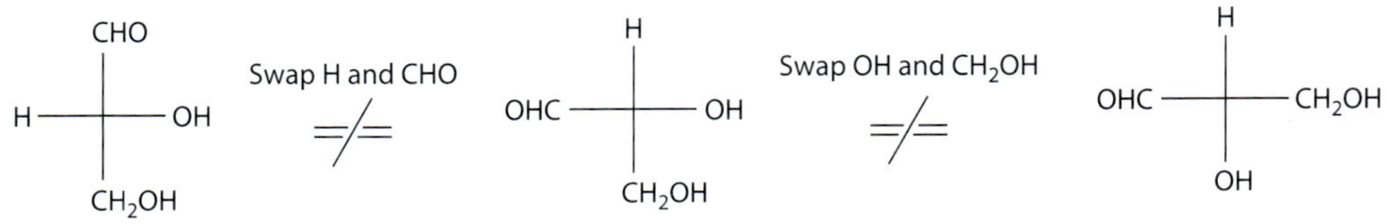

FIGURE 7.16 Manipulation of Fischer projections.

meso-compound, and hence optically inactive, tells us something important about the stereochemistry of the original ribose molecule—there must have been a pseudoplane of symmetry across the central carbon of the molecule.

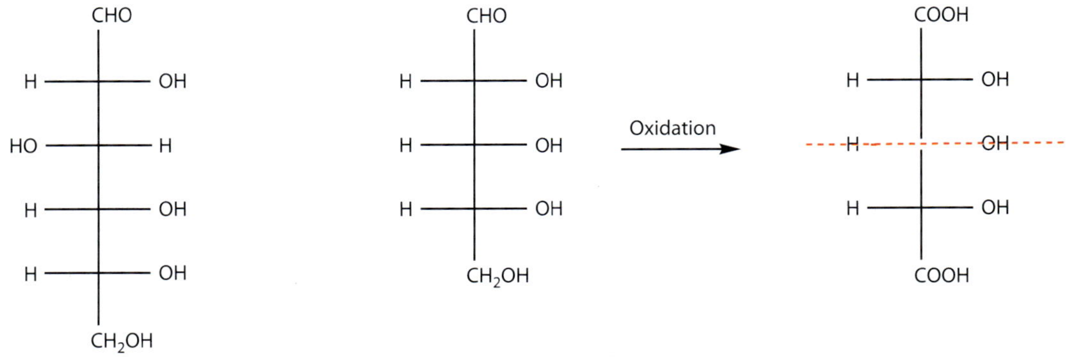

7.90, (+)-glucose **7.91**, ribose **7.92**

Textbooks offer many rules for the manipulation of Fischer projections, but we already know the only one we need—*any pairwise swap, at a single asymmetric carbon atom, reverses the chirality.* Thus, two swaps will get us back to the same configuration we started with, and Figure 7.16 shows how we can prove that **7.93** and **7.95** are identical. While a few of you will be able to look at **7.93** and **7.95** and say it's obvious that they are the same, most find the mechanical process both more comfortable and more reliable.

Assignment of the absolute configuration at a chiral center in a Fischer projection requires that the lowest priority group, by the Cahn–Ingold–Prelog system, be at the top of the diagram. This is essentially the manipulation in Figure 7.16. When H is at the top of the diagram and $1 \rightarrow 2 \rightarrow 3$ is clockwise, then the molecule is *R*. Similarly if $1 \rightarrow 2 \rightarrow 3$ is counterclockwise, the molecule is *S*. For **7.93**, the ordering of the groups is OH > CHO > CH_2OH > H. Thus, **7.94** has *S*-stereochemistry and **7.95** (and hence **7.93**) has *R*-stereochemistry. **7.93** is *R*–*D* (+)-glyceraldehyde (non-IUPAC). We will learn the meaning of the other designation, *D*, when we look formally at sugars in Chapter 17.

PROBLEM 7.11

Assign an absolute configuration for each of the chiral centers of arabinose:

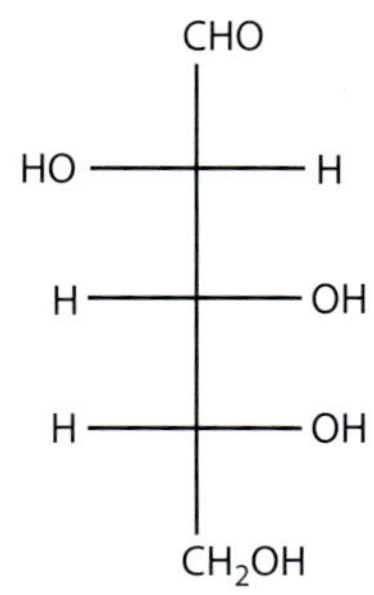

(−)-Arabinose

Solution

This is not quite as easy as for a single asymmetric carbon atom, and we will need to work on each of the centers separately. They are labeled 2, 3, and 4 as follows (in numbering the chain, the aldehyde takes precedence over the alcohols):

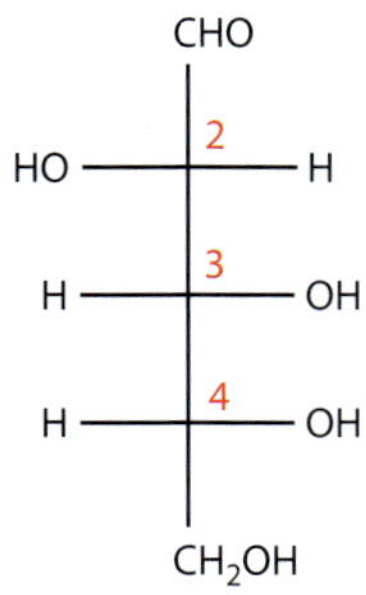

At position 2, OH > CHO > CH(OH) > H, so we have

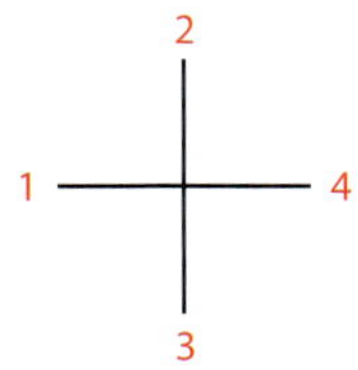

We need to place group 4 at the top of the diagram, by swapping 2 and 4 to give

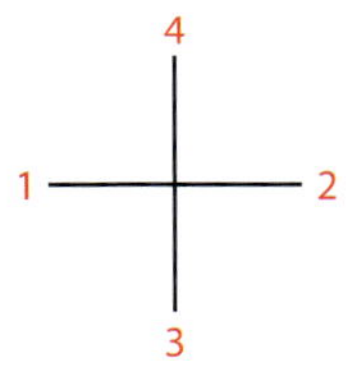

This clearly has *R*-stereochemistry, so the original was *S*.

For position 3, OH > CH(OH)CHO > CH(OH)CH_2OH > H; thus, we have

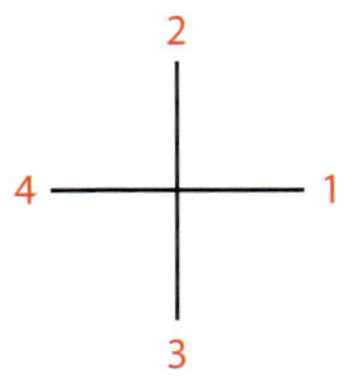

To put group 4 in the correct position, we swap 2 and 4, giving

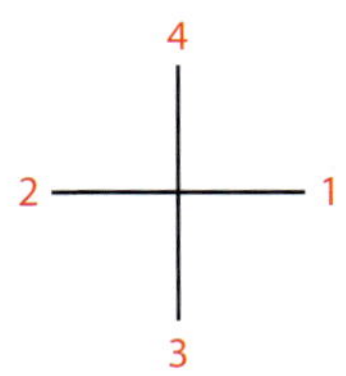

This is clearly *S*, so the original was *R*.

At position 4, HO > CH(OH)C > CH_2OH > H. A similar process to those previously presented gives us *R*-stereochemistry. So overall, we have

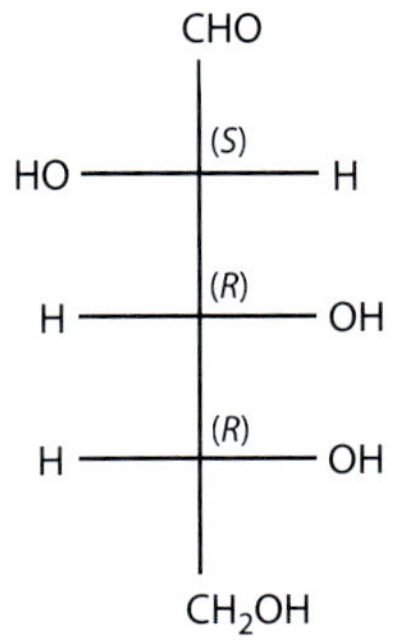

Key Points from Sections 7.4 and 7.5

- Chiral molecules are not superimposable on their mirror images and must contain a stereogenic unit. The simplest chiral organic molecules contain a carbon atom (a chiral center or asymmetric carbon atom) bonded to four distinct groups.
- For a molecule to be chiral, it must lack a plane of symmetry and a center of symmetry, and it must not be possible to interconvert enantiomers by a low-energy process (simple amines and carbanions with three different substituents and a lone pair fail the last test as there is a low-energy process for inversion).
- Chiral molecules are detected by their ability to rotate the plane of plane-polarized light. The specific rotation $[\alpha]_D$ is measured using a polarimeter and is a property specific to the molecule. For molecules with a single chiral carbon atom, the rotations of the enantiomers are equal and opposite.
- The absolute configuration of a chiral center in a molecule is designated *R* or *S*. Substituents are ordered according to the Cahn–Ingold–Prelog scheme, the molecule is viewed along the C-(group 4) bond, and if the 1,2,3 substituents appear in clockwise order, the molecule is *R*.
- Whatever the representation of the molecule, switching any pair of groups at a chiral center gives the enantiomer with the opposite absolute configuration.
- Absolute configurations were originally related by chemical reactions to *R*-glyceraldehyde, the configuration of which was guessed around 1900. In the 1950s, x-ray crystallography showed that the original guess was correct.
- Absolute configuration is not correlated with the sign of the optical rotation.
- When a molecule has *n* chiral centers, there may be up to 2^n stereoisomers. However, some of these may be *meso*-compounds, where the molecule has asymmetric carbon centers but also possesses a plane or center of symmetry and hence is not chiral.
- Racemic mixtures (equal amounts of the two enantiomers) may be resolved by conversion to separable diastereoisomers or by enzymatic methods.
- Chiral centers may be located at noncarbon atoms, including P, S, and Si, as well as many metals. Molecules that lack a specific asymmetric carbon atom may be chiral by virtue of a lack of symmetry—examples are found among allenes, biphenyls, and helicenes.
- Fischer projections are useful to demonstrate stereochemical relationships in molecules with many chiral centers.
- To determine the absolute configuration of a molecule from its Fischer projection, the projection must be manipulated (switches giving opposite enantiomers) so that the lowest priority group is at the top of the diagram.

REVIEW PROBLEMS

1. Explain and give a specific example of the following terms used in stereochemistry:
   (a) Resolution
   (b) *meso*-Compound
   (c) Center of symmetry
   (d) Gauche conformation
   (e) Sawhorse representation
   (f) Eclipsing strain
   (g) Transannular strain
   (h) Plane of symmetry
   (i) Specific rotation
2. Draw a sawhorse representation of
   (a) The eclipsed form of *meso*-2,3-butane diol
   (b) The anti-staggered conformation of *meso*-2,3-dibromobutane, focusing on the C2 to C3 bond
3. Draw a Newman projection of
   (a) Chloroethane in the staggered conformation
   (b) The staggered conformation of 2,3-dimethylbutane, focusing on the C2 to C3 bond
4. Give a structure for each of the following molecules:
   (a) *S*-1-Chloro-2-propanol
   (b) *S*-3-Methyl-1-pentyne-3-ol
   (c) *S*-1-Bromo-3-chloro-2-methylpropane
5. Study the molecules shown in the following. Then for each of them
   (i) Assign priorities to the four groups around the asymmetric carbon atom according to the Cahn–Ingold–Prelog system.
   (ii) Assign an absolute configuration.
   (iii) Give a systematic name for the molecule:

(a)
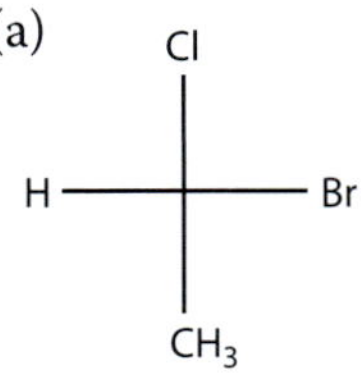

(b)
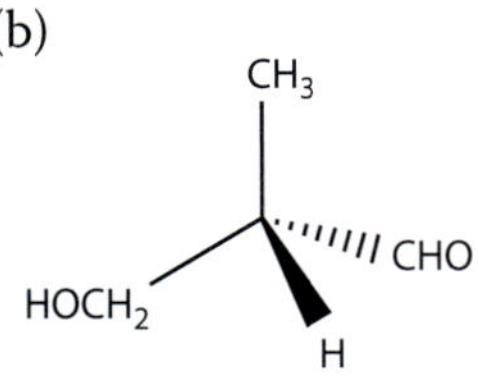

(c)
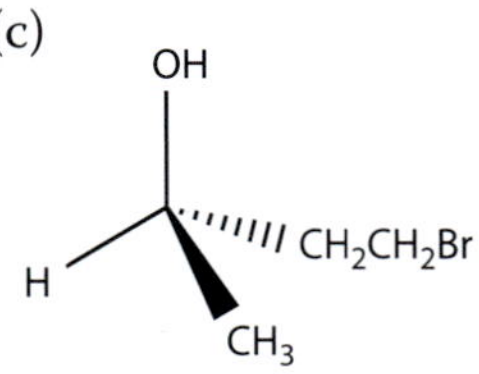

(d)
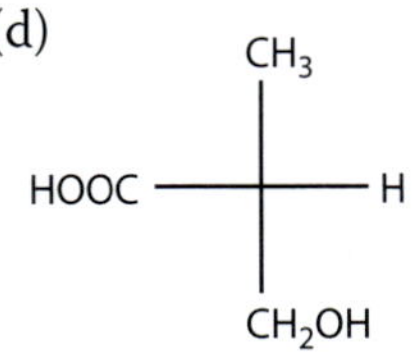

(e)
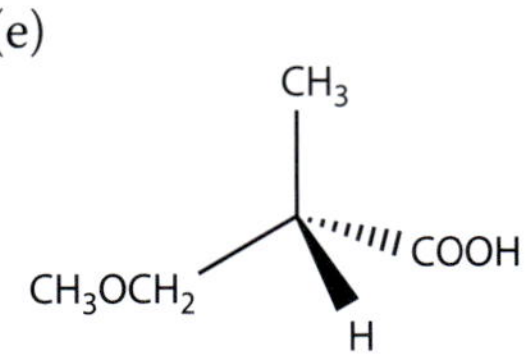

(f)
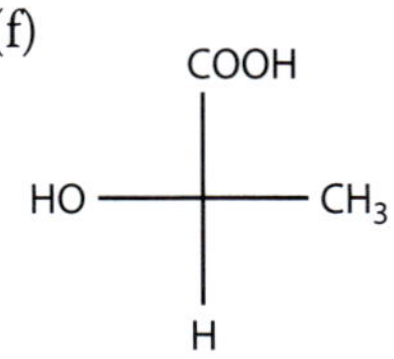

6. Draw each of the chair forms of the following molecules. Predict which will be more stable in each case, and give your reasons:

(a)
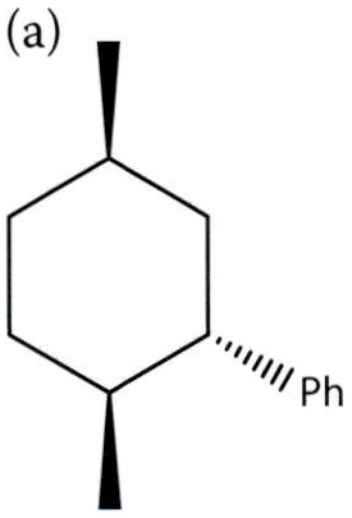

(b)
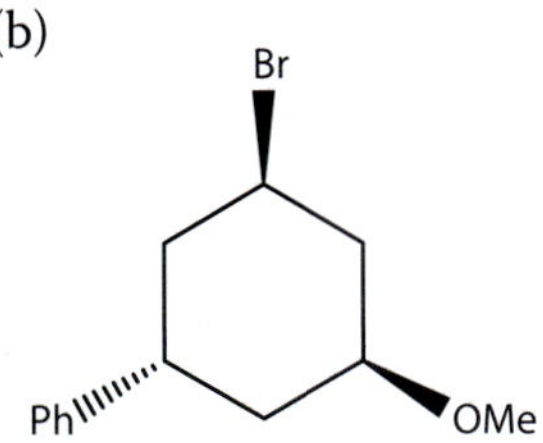

(c)
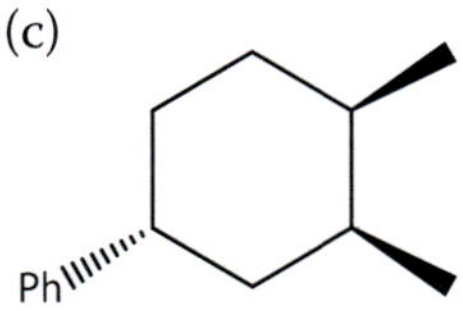

(d)

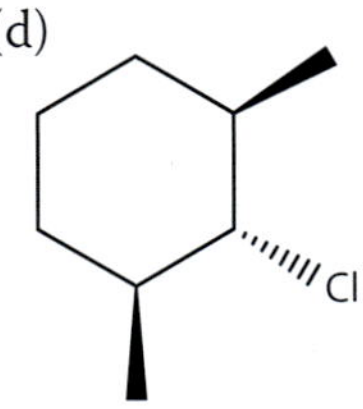

(e)

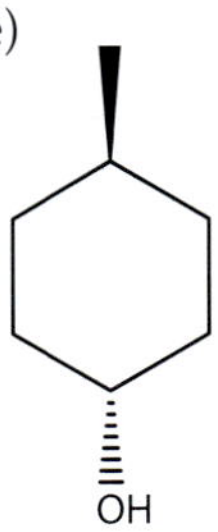

(f)

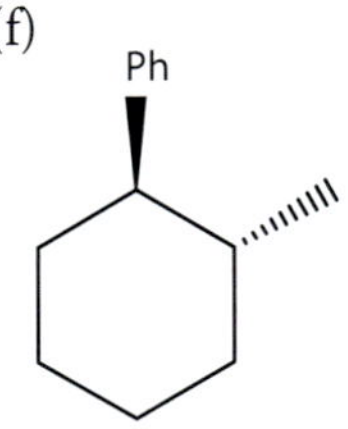

7. Draw tetrahedral representations of the two enantiomers of cysteine, $HSCH_2CH(NH_2)COOH$, and identify them as *R* and *S*.

8. What is the stereochemical relationship between the two molecules shown in the following text? Is either chiral?

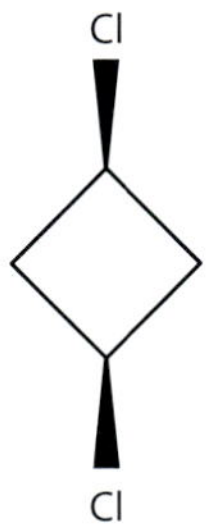

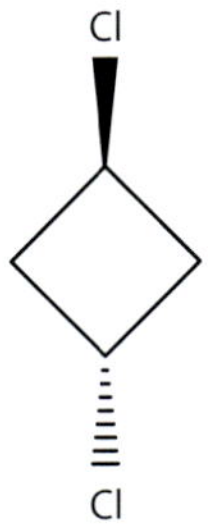

9. If a stereocenter is present in the following molecules, mark it with an asterisk (*).

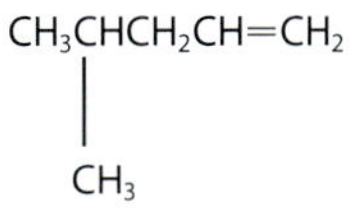

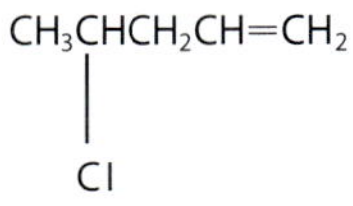

$ClCH_2CH_2CH_2($

10. Which of the following molecules contains a center of symmetry?

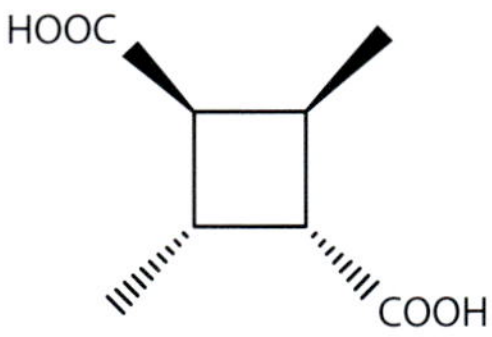

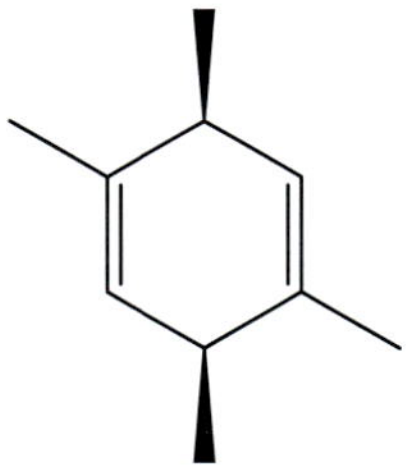

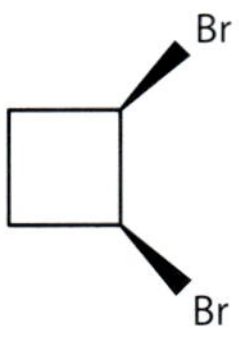

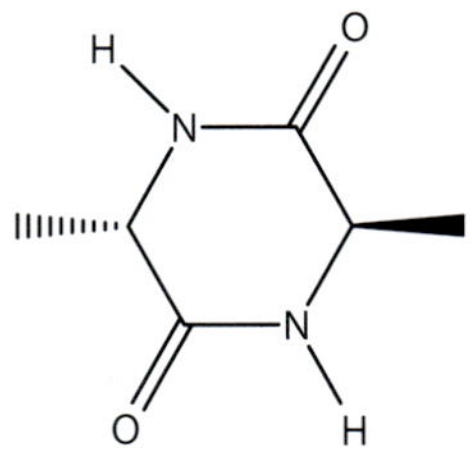

11. Assign an absolute configuration to (−)-amphetamine and (+)-citronellal.

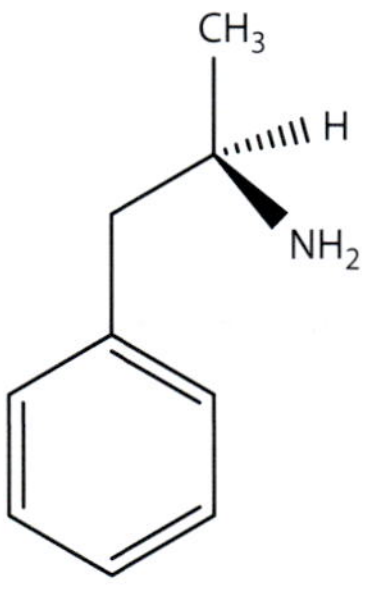

(−)-Amphetamine

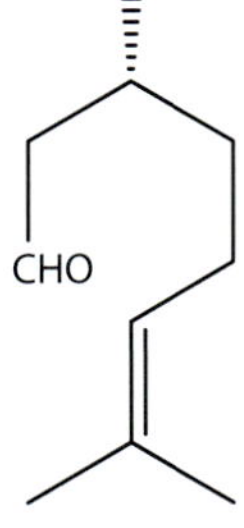

(+)-Citronellal

12. Assign an absolute configuration to each of the chiral centers tartaric acid and the antimigraine drug fluoxetine.

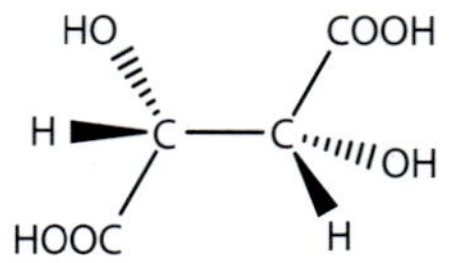

Tartaric acid

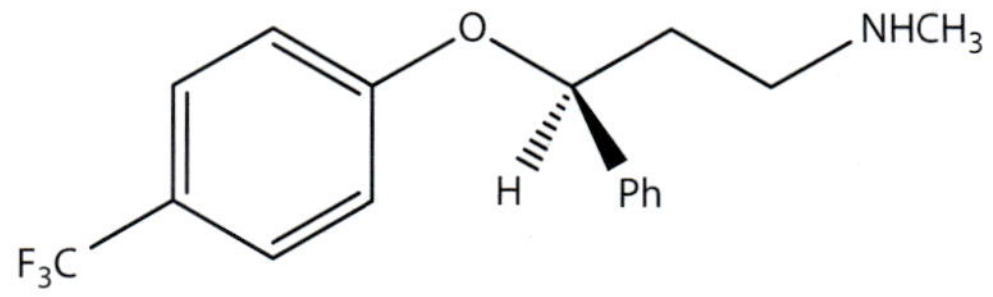

Fluoxetine

13. Assign absolute configurations to nicotine and (–)-limonene.

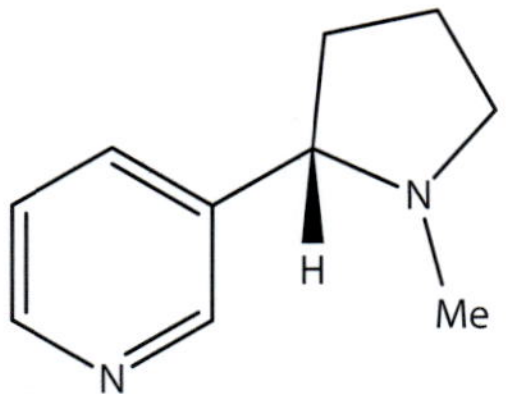

Nicotine

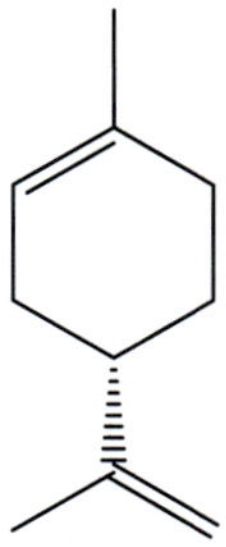

(–)-Limonene

14. The molecules cathinone and cathine are derived from *Catha edulis*, or khat, and have stimulant properties. Assign the absolute configurations of the natural materials.

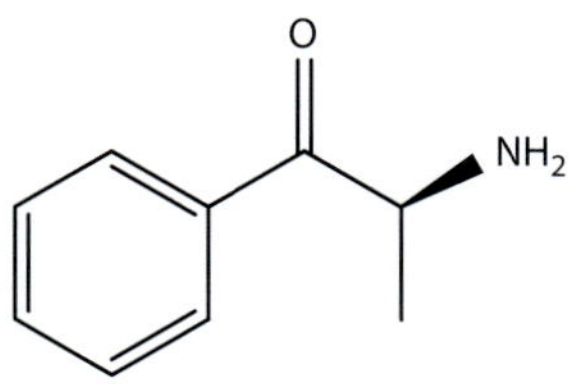

Cathinone

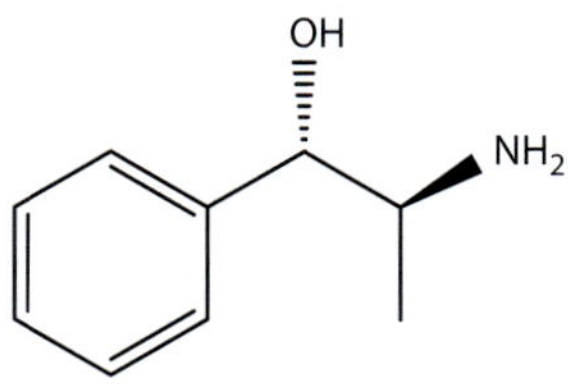

Cathine

15. Draw each of the chair conformations of menthol, a naturally occurring terpene. Predict which form will be more stable, giving your reasons.

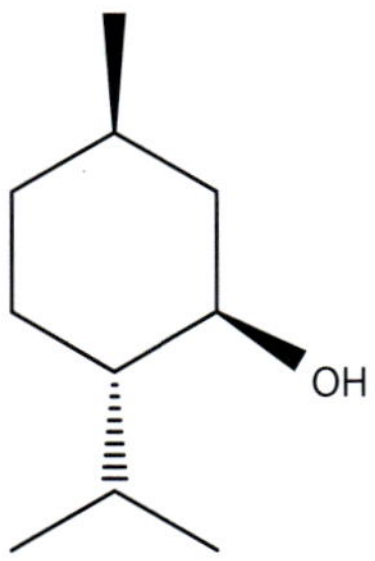

Menthol

16. Draw each of the chair conformations of the sugar (+)-glucopyranose. State which will be more stable, giving your reasons.

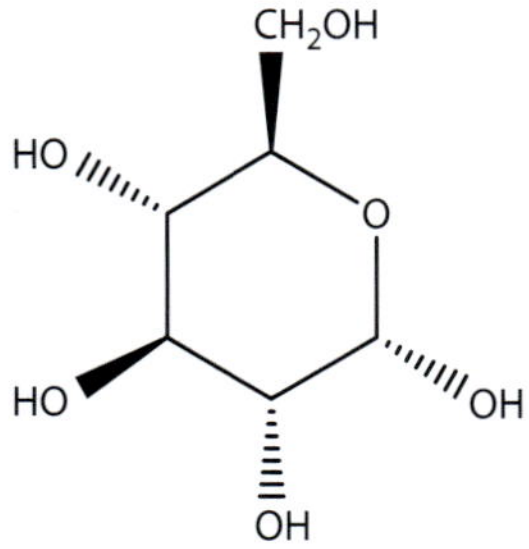

(+)-Glucopyranose

17. The energy cost of putting a *tert*-butyl group in an axial position on a six-membered ring is more than 19 kJ mol^{-1}. However, the comparable value for the $SiMe_3$ group is 10.5 kJ mol^{-1}. Explain this observation.
18. In piperidine, **7.96**, would you expect the N–H bond to be axial or equatorial? Draw both forms.

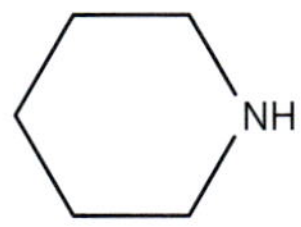

7.96

MORE CHALLENGING PROBLEMS

1. In CCl_4, the OH stretching frequency of **7.97** is independent of concentration, but in **7.98**, it varies considerably, going from 3650 cm^{-1} at 0.03 M to 3300 cm^{-1} at 1.0 M. Explain.

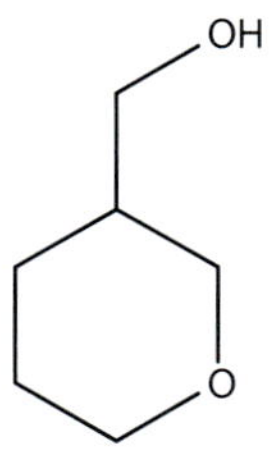

7.97

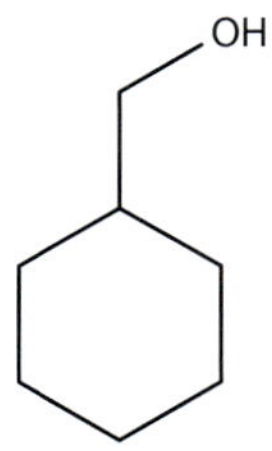

7.98

2. The biphenyls **7.99** and **7.100** were synthesized in optically active form, and the kinetics of their racemization was studied. At –19.8 °C, the rate of racemization, k_H for **7.99**, was 6.48 × 10^{-5} s^{-1}. The rate for **7.100** k_D = 7.7 × 10^{-5} s^{-1}. So k_H/k_D = 0.84. What conclusion would you draw about the relative sizes of H and D? Is this what you would have expected?

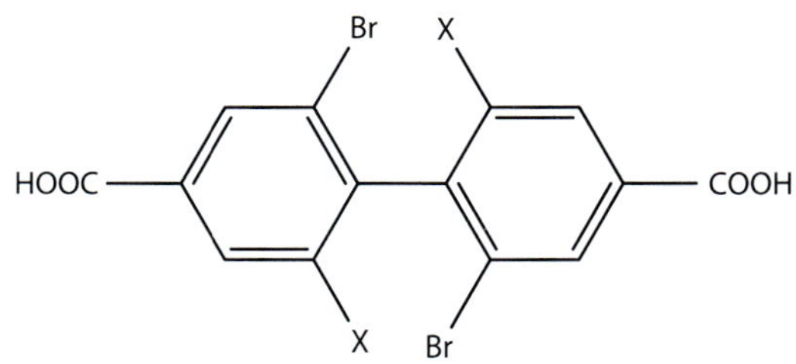

7.99, X = H
7.100, X = D

3. Draw each of the chair conformations of the amino sugar mycosamine, which is a component of many antifungal compounds. Predict which form will be more stable, giving your reasons.

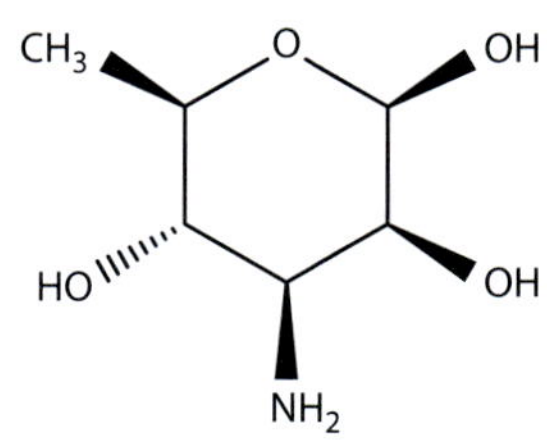

β-Mycosamine

4. Draw both chair conformations of *cis*-1,3-dimethoxycyclohexane. Which is the more stable? Draw analogous structures for *cis*-1,3-dihydroxycyclohexane. Would you expect the relative stability of the conformers to be the same? Give your reasoning.

5. Assign an absolute configuration to (+)-carvone, oil of caraway, and an absolute configuration at each of the chiral centers in ascorbic acid, vitamin C.

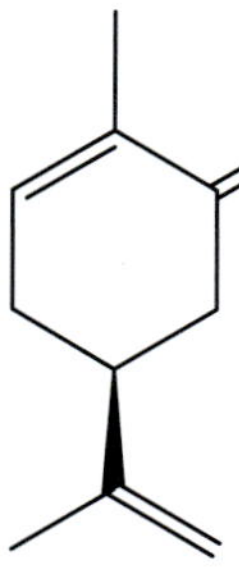

(+)-Carvone
oil of carraway

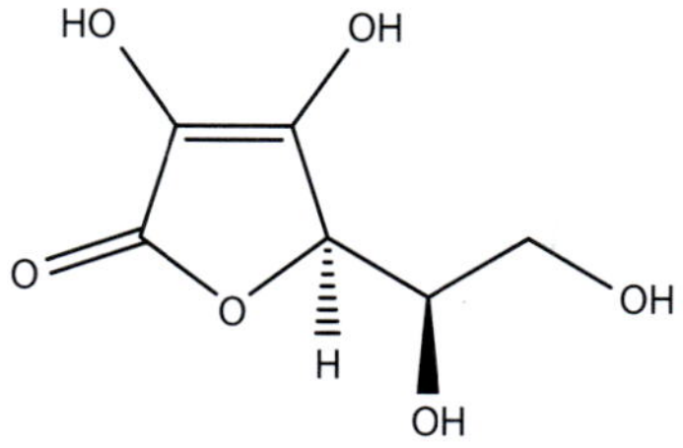

Ascorbic acid
Vitamin C

6. Assign an absolute configuration to each of the chiral centers in the antibiotic chloramphenicol.

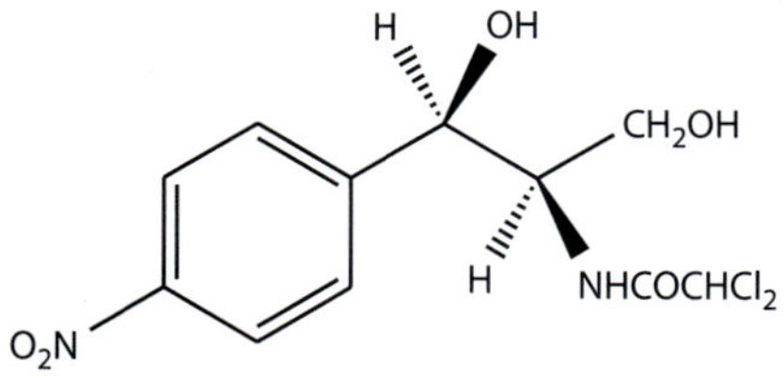

Chloramphenicol

7. Natural products of the general formula $PhCH(OH)CH(NHMe)CH_3$ are isolated from *Ephedra* species and are called ephedrine and pseudoephedrine. They have various physiological effects, including uses as decongestants and appetite suppressants. Draw all four of the possible stereoisomers as Fischer projections, and assign the stereochemistry at each chiral center. Why do ephedrine and pseudoephedrine have different physiological effects? Do the natural and unnatural isomers have different effects? What is the relationship of these molecules to methamphetamine?
8. The energy differences between axial and equatorial conformers for **7.101** are given in the below table in comparison with those for the related cyclohexanes. Explain this observation.

R	Δ*G* (kJ mol⁻¹), 7.101	Δ*G* (kJ mol⁻¹) Cyclohexyl-*R*
H	0	0
Me	3.4	7.5
Et	3.4	7.5
i-Pr	4.2	8.8
t-Bu	5.9	>19

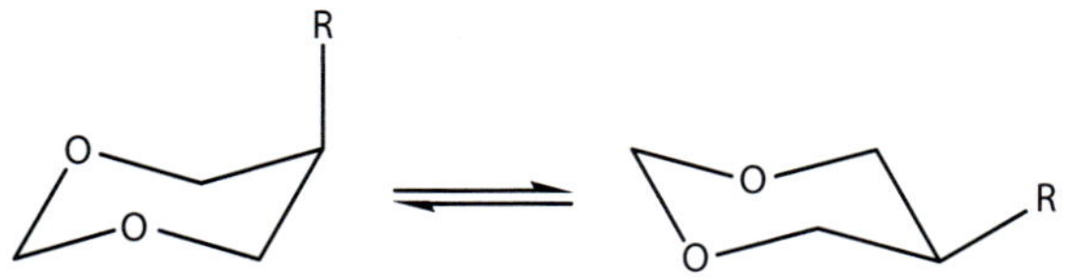

7.101

9. The CNC angle in Me_3N is 110.9 °, whereas the CPC angle in Me_3P is 98.6 °. In $(Me_3C)_3P$, the CPC angle is 109.7 °. Explain.

Introduction to Mechanism

8

8.1 INTRODUCTION

This chapter is a crucial one for developing your understanding of organic chemistry. If you do not understand *how* reactions work, you won't be able to either predict what will happen, or understand what has happened, in any transformation. This would take you back to the dread that many students have of organic chemistry—a large number of reactions that follow no rational pattern but simply have to be memorized. It does not have to be like that, and it shouldn't be. Here, we will study the methods available to us for establishing and understanding organic reaction mechanisms. Electron availability and deficiency, which often determine the directions of reactions, will also be investigated. This will be exemplified by studying the pK_a values of acids and bases—what stabilizes positive and negatively charged species? In Chapter 4, you met the idea of curly arrows to show the movement of electrons in writing resonance forms. Now, and most crucially, we will develop an understanding of how to represent *mechanisms* by the use of curly arrows. If you understand and internalize this material, you will be able to predict what will happen in completely unfamiliar reactions as well as those you have seen before—organic chemistry will become *easy and fun* ☺.

8.2 STUDYING ORGANIC REACTION MECHANISMS

An organic reaction mechanism is a step-by-step description of the movement of electrons, and bond-making and bond-breaking processes, as well as an understanding of the nature of any intermediates that may be a part of the process.

8.2.1 THERMODYNAMICS

We should first think of the energetics of the process; if this is not favorable, then the reaction will not occur naturally. This does not mean that no individual step in the reaction can be endothermic or require the input of energy, but this will need to be compensated by an exothermic step elsewhere. Let's first consider an exothermic one-step process (Figure 8.1); this reaction coordinate is typical for a simple substitution reaction such as the reaction of bromoethane with PhS^-. You should remember from thermodynamics that ΔG tells us if the reaction is possible, and $\Delta G^{\ddagger}$, the free energy of activation energy, tells us how easy or difficult it is to persuade the reaction to proceed. The value of ΔG, or more specifically ΔG^o, tells us where the equilibrium between the starting material and product lies, but not how fast the reaction will get there (Equation 8.1). We would regard a single-step reaction as nonreversible if K, the equilibrium constant, is greater than 10^7; this corresponds to a ΔG^o of -42 kJ mol^{-1}.

(8.1) $$\Delta G^o = -RT \ln K$$

The free energy of the reaction has both enthalpy and entropy components (Equation 8.2); ΔH is related to the strengths of the bonds made and broken and is frequently the major factor involved in determining whether an organic reaction will work. In Chapter 4, we looked at measuring bond strength and how bonds may be broken—homolytically or heterolytically (Figure 8.2). In homolytic cleavage, each of the atoms of the bond being broken acquires one of the two electrons

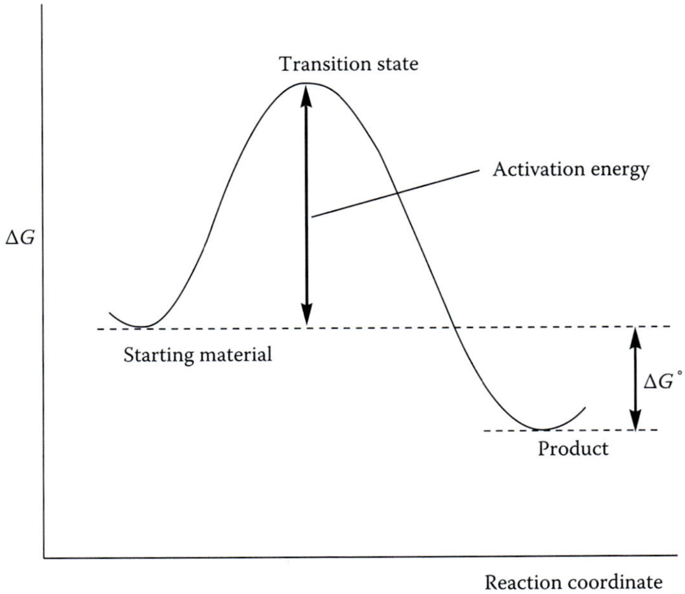

FIGURE 8.1 Energetics of an exothermic one-step process.

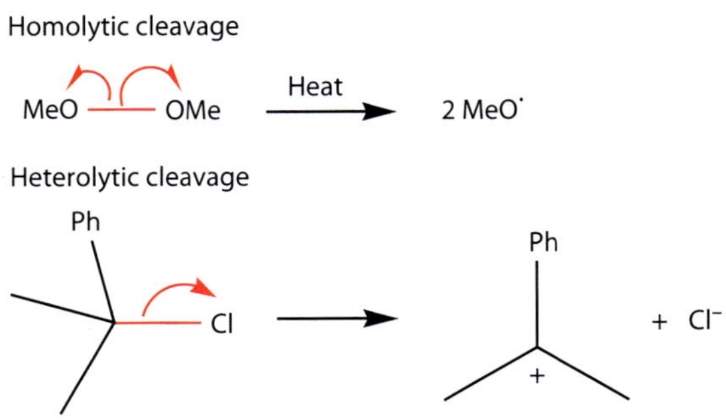

FIGURE 8.2 Homolytic and heterolytic cleavage of bonds.

of the bond, giving radicals. In heterolytic cleavage, both electrons go to one of the two atoms, so that an anion and a cation are formed. We also first met the idea that we use a curly (or curved) arrow to represent the movement of a pair of electrons in bond-making and bond-breaking processes. A single-headed arrow is used to denote the movement of a single electron.

(8.2) $$\Delta G^{\circ} = \Delta H^{\circ} - T\Delta S^{\circ}$$

Quick Review of Arrow Types Used in Organic Chemistry

Arrow / **Meaning**

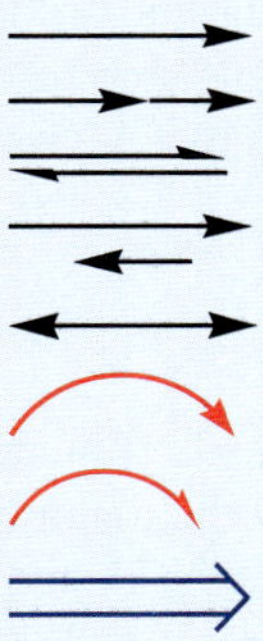

- Reaction goes in the forward direction.
- Indicates more than one reaction step—often similar processes.
- Approximately balanced equilibrium.
- Reaction is an equilibrium, biased to the right.
- Connects resonance forms of a molecule.
- Movement of an electron pair.
- Movement of a single electron.
- Retrosynthetic arrow; will be discussed further in Chapter 19.

If we consider a very straightforward reaction, the burning of methane to give CO_2 and water (8.3), then we can readily prove to ourselves that this is an exothermic reaction, and we can use natural gas to heat our homes and keep ourselves warm in winter.

(8.3) $$CH_4 + 2O_2 \rightarrow CO_2 + 2H_2O$$

We perform a simple bookkeeping operation, looking at the energies of the bonds broken and the bonds made.

Bonds Broken		Bonds Made	
4 × C–H	1660 kJ mol^{-1}	2 × C=O	1500 kJ mol^{-1}
2 × O=O	1000 kJ mol^{-1}	4 × O–H	1756 kJ mol^{-1}
Total	2660 kJ mol^{-1}	*Total*	3256 kJ mol^{-1}

Thus, we are "ahead" by 596 kJ mol^{-1}, as we would have expected. If we look at another example, however, the reaction of chloromethane with water to give methanol and HCl (8.4), things turn out a little differently.

(8.4) $$CH_3Cl + H_2O \rightarrow CH_3OH + HCl$$

Bonds Broken		Bonds Made	
C–Cl	217 kJ mol^{-1}	H–Cl	431 kJ mol^{-1}
O–H	464 kJ mol^{-1}	C–O	359 kJ mol^{-1}
Total	681 kJ mol^{-1}	*Total*	790 kJ mol^{-1}

Thus, the reaction is exothermic by 109 kJ mol^{-1}, but it does not proceed easily at room temperature. Although the thermodynamics are OK, the activation energy is high enough that the reaction is very slow. The rate of the reaction is related to the activation energy by the Arrhenius equation (8.5), where *Ea* is related to the activation energy, $\Delta G^{\ddagger}$.

(8.5) $$\text{Rate} = Ae^{(-Ea/RT)}$$

ΔG^o has two components: enthalpy (ΔH^o) and entropy (ΔS^o). The enthalpy of reaction is related to the energy required to break chemical bonds. ΔS^o relates to the order, or degrees of freedom, in the starting material and the product. For a process where one molecule is split into two, the production of more molecules means that there are more degrees of freedom, and more disorder, so ΔS would be positive. Conversely, if two molecules combine into one, ΔS would be negative (Figure 8.3). While the thermodynamics of many organic reactions is dominated by enthalpy factors, Diels–Alder reactions provide an example of an exceptional type; for reactions such as that in Figure 8.3, the reaction entropy, ΔS^o, is typically −146 JK^{-1} mol^{-1}. This means that the entropy term makes a significant contribution to the free energy change, and many Diels–Alder reactions can be reversed on heating to >400 K.

In another specific category of reactions, entropy is crucially important—the formation of compounds containing rings (these may be all carbon, or contain other atoms, which we refer to as heteroatoms). Intramolecular processes leading to formation of rings are favored by entropy

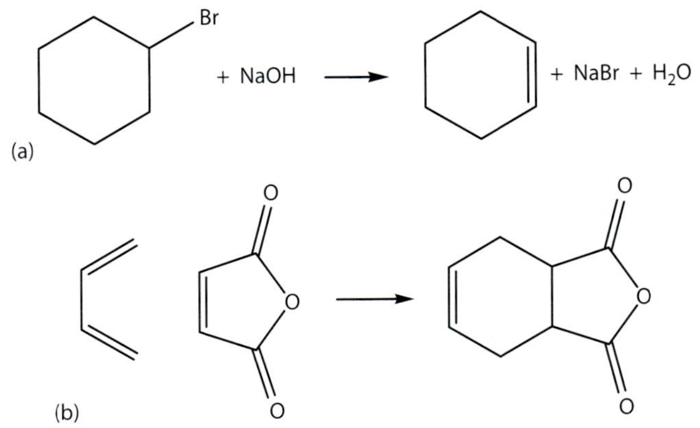

FIGURE 8.3 Processes with positive or negative entropic terms. (a) Two molecules react to give three molecules—entropy increases. (b) Two molecules combine to give one—entropy decreases in the Diels–Alder reaction.

over intermolecular processes, because there is no need for a molecular collision to take place (Figure 8.4). You may be worried by the fact that I have not balanced the equation in Figure 8.4; this is quite common in organic chemistry. The other products are water and chloride ion, which are not of great interest from an organic chemical point of view and, in practical terms, would be discarded in the work-up of the reaction mixture. As we will see in a range of reactions, enthalpic and entropic considerations combine so that the formation of three-, five-, and six-membered rings is easy; rings of other sizes present more challenges.

Many reactions involve more than one step; Figures 8.5 and 8.6 show examples of the reaction coordinate for two-step reactions. The "highest hill" that the reaction has to climb is the rate-determining step (RDS), sometimes called the rate-controlling or the rate-limiting step. It is the slowest step of the reaction process, and its rate determines the overall rate—effectively a kinetic bottleneck. At each maximum on the energy curve, there is a transition state—something all elementary processes have, where the bonds are partially made or partially broken. Because these are energy maxima, they cannot be isolated or observed under any normal conditions,* but a full understanding of the reaction sometimes allows us to say something useful about their structure. These are drawn as in **8.1**, with bonds half made or half broken represented as dotted lines and charges in parentheses to indicate that they are not "full" charges. **8.1** represents a possible transition state for the displacement of one halide ion by another at a methyl group—part of the negative charge is on the chlorine and part on the iodine. In this particular reaction, the transition state is not symmetric, and there is no requirement that it should be—we will meet some transition states that are symmetric, and some that are not, as we study a range of reactions.

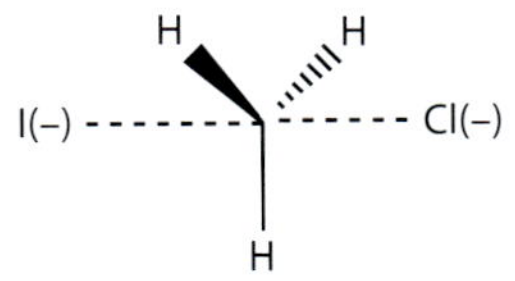

The dashed bonds indicate bond partially made or partially broken
The charges in parentheses indicate partial charges

8.1

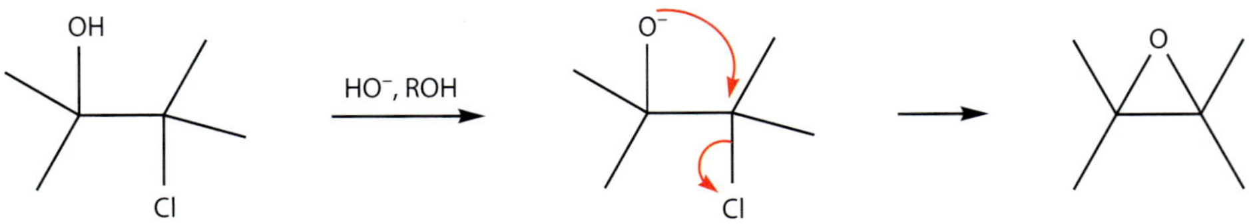

FIGURE 8.4 Intramolecular processes may be favored because they have low entropy of activation.

* A few activated complexes have been observed on the femtosecond scale.

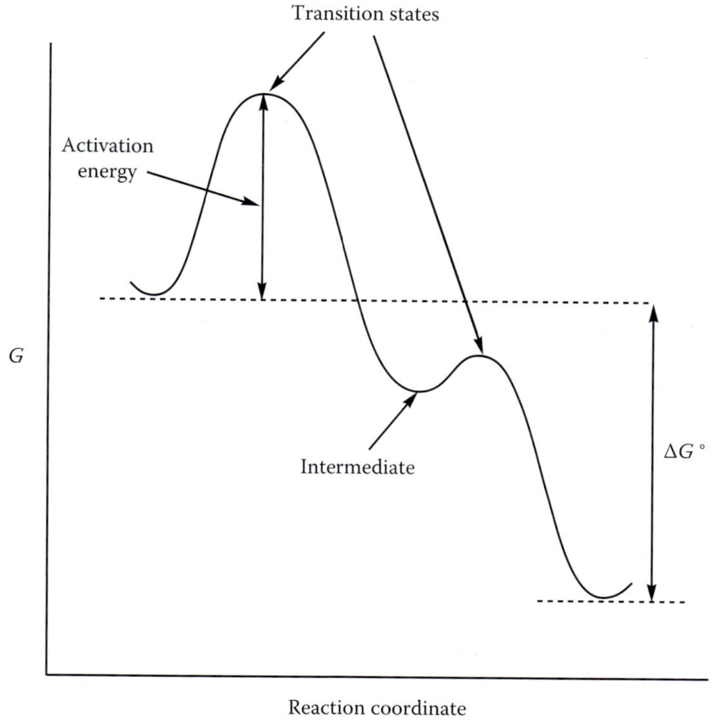

FIGURE 8.5 Energetics of a two-step process where the first step is rate controlling.

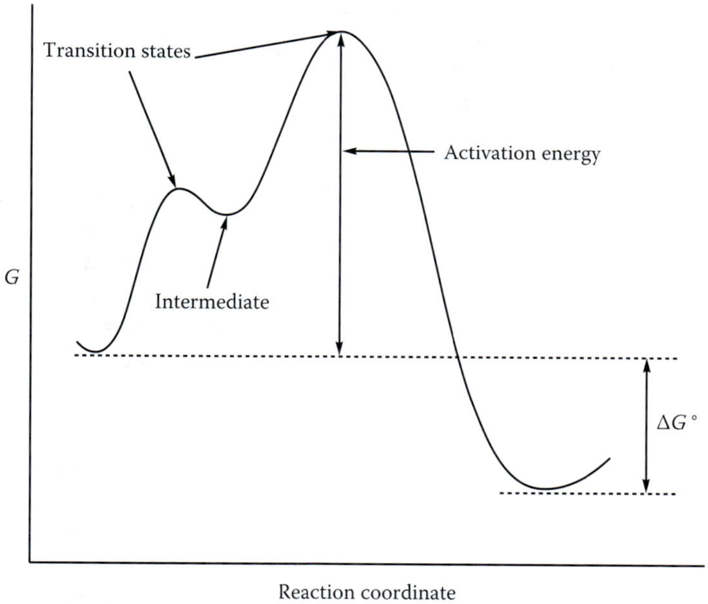

FIGURE 8.6 Energetics of a two-step process where the second step is rate controlling.

In principle, although not always in practice, reaction intermediates may be isolated. They exist at minima on the energy curve and have a definite lifetime, which may be long or short. For example, in the reaction of an excess of bromoethane with methylamine to give the ammonium salt $[Et_3NMe]^+Br^-$ (8.6), the intermediates ethylmethylamine and diethylmethylamine are stable compounds, and if the reaction was stopped at an appropriate time, they could be isolated. Other intermediates, such as the carbocation, **8.2**, are short lived and unstable and cannot be isolated, although they can *sometimes* be observed spectroscopically (Figure 8.7).

$$\underset{XS}{EtBr} + MeNH_2 \rightarrow [Et_3NMe]^+Br^- \qquad (8.6)$$

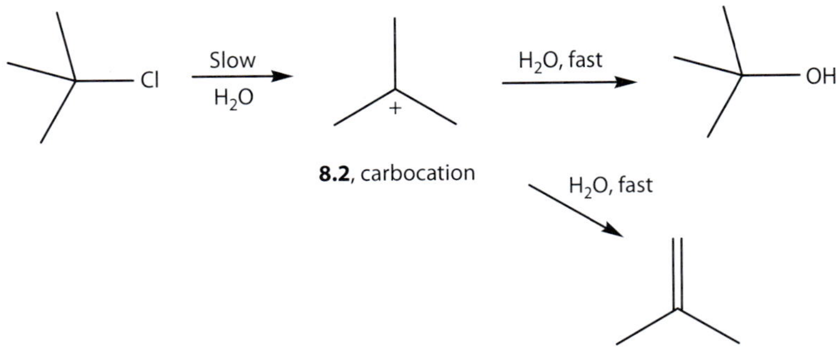

FIGURE 8.7 Some reactions of 2-methyl-2-chloropropane (*tert*-BuCl).

PROBLEM 8.1

In the reactions shown in Figure 8.7, would you expect to be able to observe the carbocation spectroscopically? Explain your conclusion.

Solution

As we noted in Chapter 4, the *tert*-butyl cation would be expected to be relatively stable because of the electron-donating inductive effect of the three methyl groups. The NMR spectrum of this cation has indeed been recorded and salts stable enough for x-ray structure determination have been prepared. However, we would never be able to observe it in either of these processes—it is formed slowly, but consumed rapidly, so its concentration will always be too low to detect.

8.2.2 KINETICS

ΔG^o tells us where a reaction equilibrium lies, in which direction it will proceed, and $\Delta G^\ddagger$ tells us how fast we will get there. The free energy of activation also has enthalpic and entropic components (8.7). We can describe the kinetics of organic reactions, just like any other, in terms of a rate equation. For example, the hydrolysis of iodomethane (8.8) shows the rate (Equation 8.9). Measured kinetics of this type tell us about the whole reaction process, whether one or many steps are involved. Another useful classification is the molecularity of processes, since this is applicable to individual steps in reaction mechanisms. Figure 8.8 shows typical unimolecular and bimolecular reaction steps. In the unimolecular process, a single molecule is involved in the RDS, and in many (but not all) cases, this will lead to the rate depending only on the concentration of substrate. In a bimolecular reaction, two molecules are involved in the RDS, and the rate depends on the concentration of each of them.

$$\Delta G^\ddagger = \Delta H^\ddagger - T\Delta S^\ddagger \quad (8.7)$$

$$CH_3I + HO^- \rightarrow CH_3OH + I^- \quad (8.8)$$

$$\frac{d[CH_3OH]}{dt} = k[CH_3I][HO^-] \quad (8.9)$$

Kinetic studies of reactions will not, generally, tell us absolutely what the reaction mechanism is, but they may well tell us what it is not. Compare the two reactions of Figure 8.9. The kinetics of the substitution of bromomethane are second order, while those for *tert*-butyl bromide are first order. This tells us immediately that the two reactions do not proceed by the same mechanism. In the reaction of *tert*-butyl bromide, the hydroxyl ion must be involved somewhere, because it is there in the product, but it is not involved in the RDS. Thus, we must conclude that there are at least two steps in this process; we are measuring only the rate of the RDS, the kinetic bottleneck. The mechanism proposed is a two-step process, where the RDS is the heterolytic cleavage of the C–Br bond to give a carbocation, which then reacts rapidly with the hydroxyl ion (Figure 8.10). Note that mechanisms may be consistent with experimental data, but such data do not "prove"

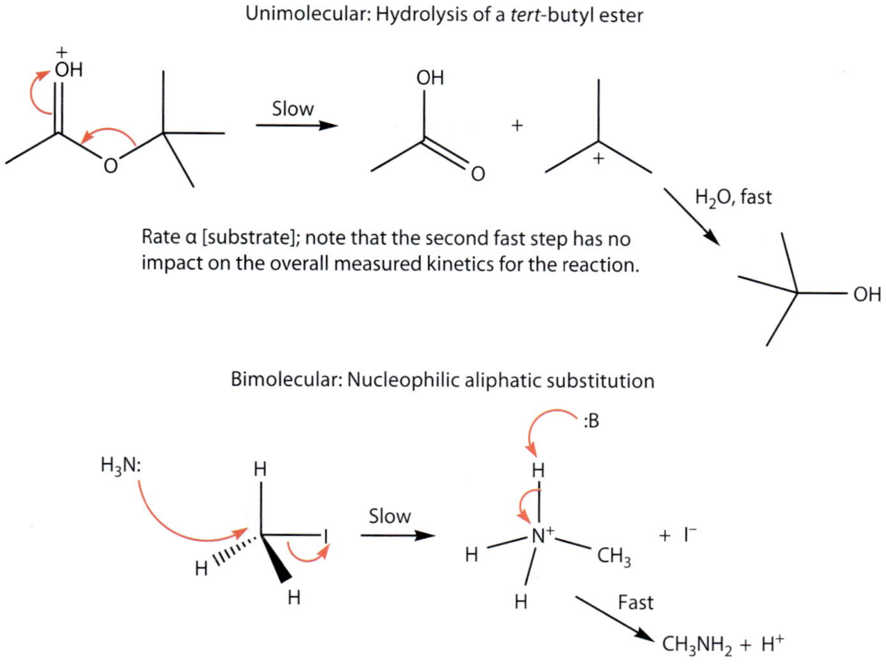

FIGURE 8.8 Unimolecular and bimolecular RDSs.

$CH_3Br + HO^- \rightarrow CH_3OH + Br^-$

$d[CH_3OH]/dt = k[CH_3Br][HO^-]$

$Me_3CBr + HO^- \rightarrow Me_3COH + Br^-$

$d[Me_3COH]/dt = k[Me_3CBr]$

FIGURE 8.9 Kinetics of substitution of alkyl halides.

mechanisms. The rate equation tells us what is involved in the reaction up to and including the RDS, but it is not definitive in revealing the mechanism of the RDS.

8.2.3 STUDY OF PRODUCTS

This may sound a little obvious—we need to look at what the reaction yields in order to understand what has happened. However, there is more information to be gleaned. Is there a single product, or are there some by-products? If there is more than one product, are the proportions of the products the same throughout the course of the reaction? If we consider the reaction of Figure 8.11, two products, an alcohol and an alkene, are formed, and the ratio between the two is constant throughout the reaction. This means that the two are not interconverted during the reaction (*i.e.*, one might be formed; then a subsequent reaction converts it to the other)—they are formed by different, divergent processes.

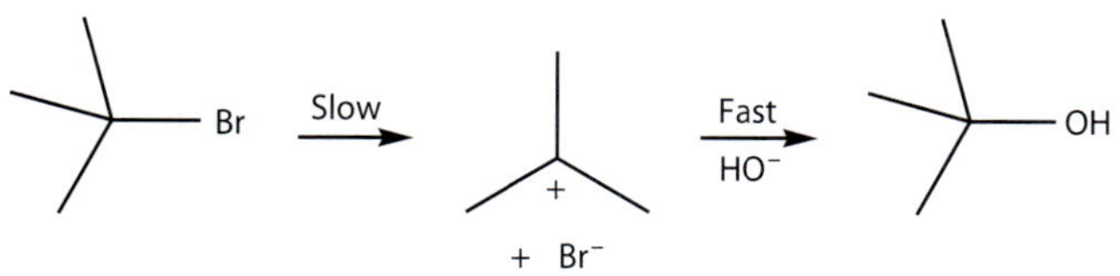

FIGURE 8.10 Proposed mechanism for hydrolysis of *tert*-butyl bromide.

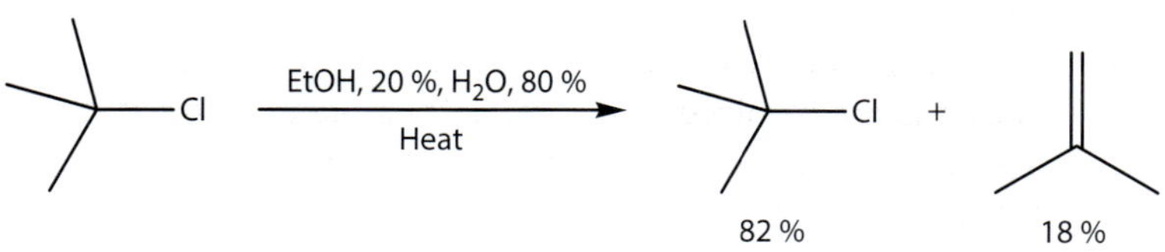

FIGURE 8.11 Hydrolysis and elimination reactions of *tert*-butyl chloride.

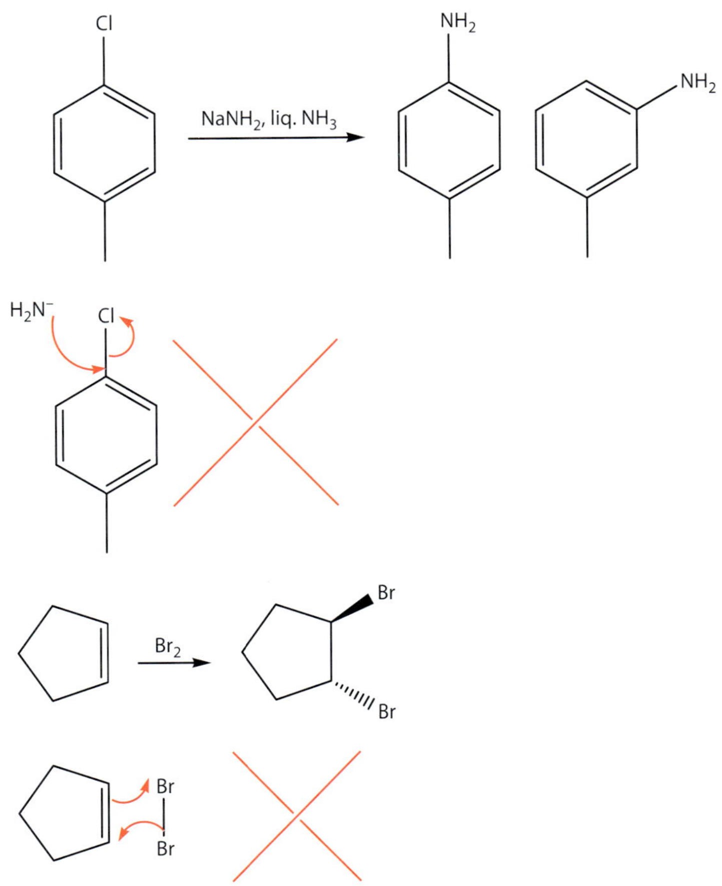

FIGURE 8.12 Product analysis shows what a mechanism can't be.

Figure 8.12 illustrates how a product study may not show what a mechanism is, but it can show what it is not. Reaction of the aryl halide with amide anion in liquid ammonia gives two products, in almost exactly equal amounts. This means that the replacement of chloride by an amino group cannot be a direct displacement, as this could give rise to only one of the products observed. In the second example, the stereochemistry of the product shows that it cannot have been produced in a simple one-step process—as this would give rise to a *cis*-product.

In Figure 8.13, a ketone is brominated at the α-position; the starting material is chiral, but the product is racemic. This tells us that there must have been a planar intermediate in the process, which can be attacked equally easily from either face, so this is not a single-step reaction.

We will encounter examples of reactions where the nature of the product depends on the temperature at which the reaction is conducted. Examples are shown in Figure 8.14. Where this is observed, it means that there are two reaction pathways available. One has a low activation energy and is easily accessed at low temperature, but the reaction product is less

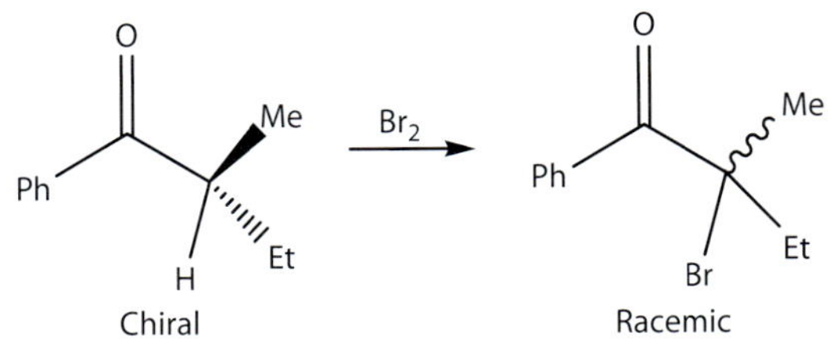

FIGURE 8.13 Bromination of a chiral ketone.

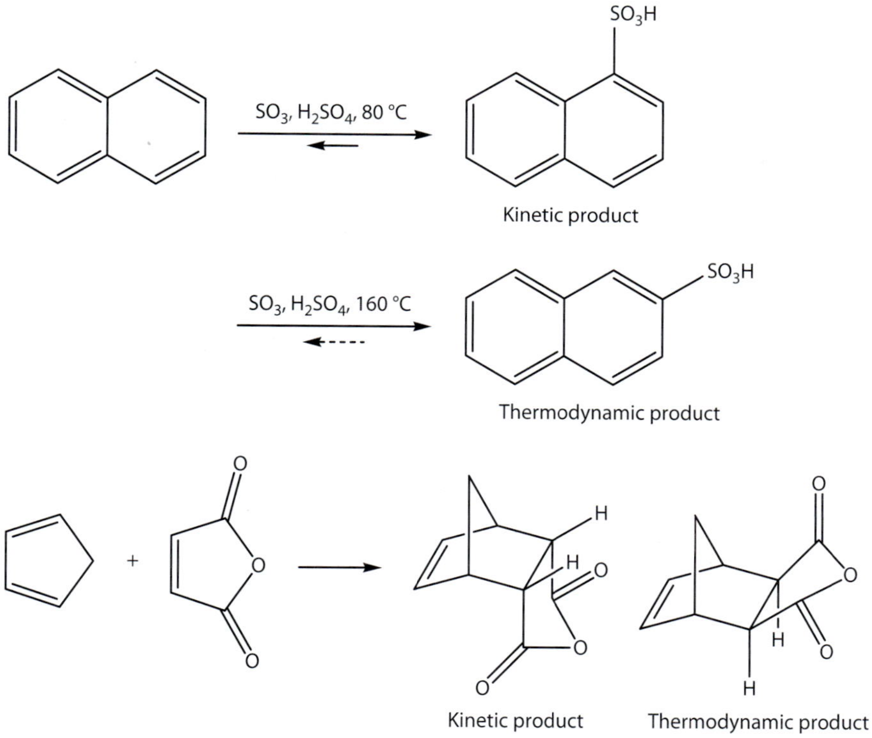

FIGURE 8.14 Reactions capable of giving kinetic or thermodynamic products. In practice, both isomers may be isolated from this reaction, depending on the temperature and duration of the reaction.

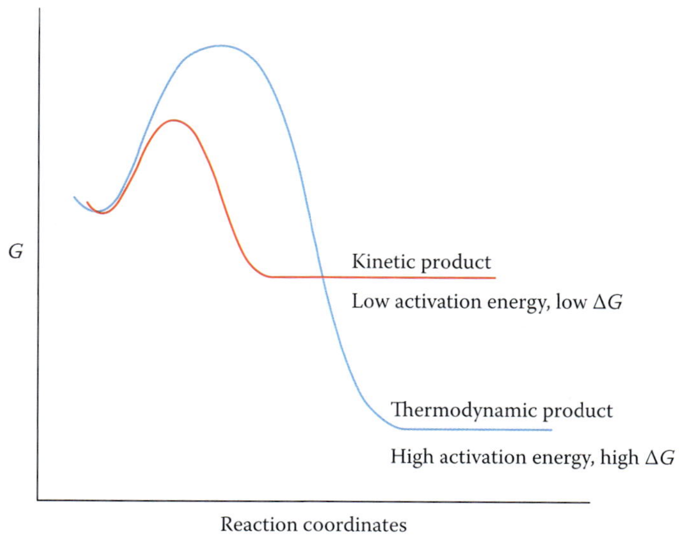

FIGURE 8.15 Energetics of the formation of kinetic and thermodynamic products.

thermodynamically stable. The other has a high activation energy (so it needs more energy to get started), but the product is thermodynamically more stable (Figure 8.15). For each example, in the first case, kinetics controls the outcome, while thermodynamics controls the outcome of the second case. For the turnover from one pathway to the other to be successful, the kinetic pathway must be reversible, which is not always the case; in some reactions, the kinetic product is invariably isolated.

8.2.4 ISOTOPIC LABELING AND KINETIC ISOTOPE EFFECTS

Isotopes are forms of an element in which the number of protons and the atomic number are the same, but the number of neutrons differs. Hydrogen has three well-known isotopes, protium (^1H or more usually just H), deuterium (D or ^2H), and tritium (T or ^3H), with, respectively, no, one, and two neutrons in the nucleus. Tritium is radioactive. Similarly, we have ^{12}C, the normal form, ^{13}C, which is about 1.1 % of natural carbon, and ^{14}C, which is radioactive. Most elements have either stable or radioactive isotopes available (or both), and the chemistry that compounds containing the various isotopes undergo is almost identical. To help with the elucidation of mechanisms, a specific atom in a reactant is labeled with one of the nonstandard isotopes, and its position/distribution is tracked through the reaction sequence. This tracking may involve spectroscopic techniques (mainly mass spectrometry and NMR spectroscopy) or radiological techniques to detect radioactive isotopes. Some simple examples are shown in Figure 8.16. In the first example (the * represents a ^{13}C label), the labeling pattern in the products indicates that there has been some symmetric intermediate, in which the labeled carbon is identical to its neighbor. In the second example, a single product is obtained; this rules out any process in which the allyl group becomes free.

Although carbon and hydrogen isotopes are the most commonly used in mechanistic studies, many other isotopes have also been useful. For example, the reaction of radiolabeled iodide ion with 2-iodooctane established the stereochemistry of some substitution reactions (Section 9.3), and ^{15}N, ^{17}O, and ^{18}O have been used along with carbon and hydrogen isotopes to establish pathways in biosynthesis. The use of labeled compounds is one of the most important tools in establishing biosynthetic pathways, and labeled compounds were extensively used in the elucidation of the conversion of aminolevulinic acid, **8.3**, through porphobilinogen, **8.4**, to protoporphyrin IX, **8.5**, the precursor of heme. The NIH shift* process in the biological hydroxylation of aromatic rings was established initially using tritium-labeled material (Figure 8.17); intermediates are thought to

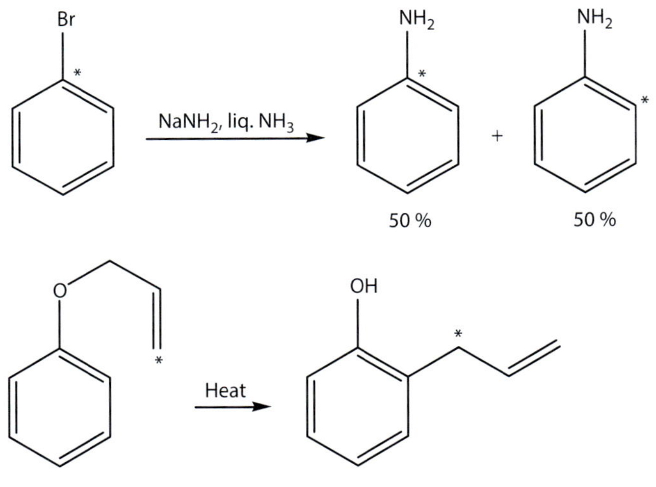

FIGURE 8.16 Labeling techniques in mechanistic studies.

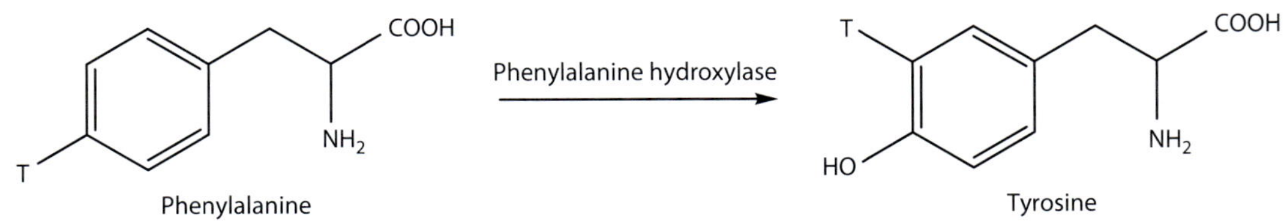

FIGURE 8.17 The NIH shift in the hydroxylation of phenylalanine.

* The name NIH shift derives from the fact that the process was first observed at the National Institute for Health.

include aromatic epoxides. Enzyme mechanisms have been investigated using chiral methyl groups, labeled stereospecifically as –CHDT, and chiral phosphate groups containing ^{16}O, ^{17}O, and ^{18}O.

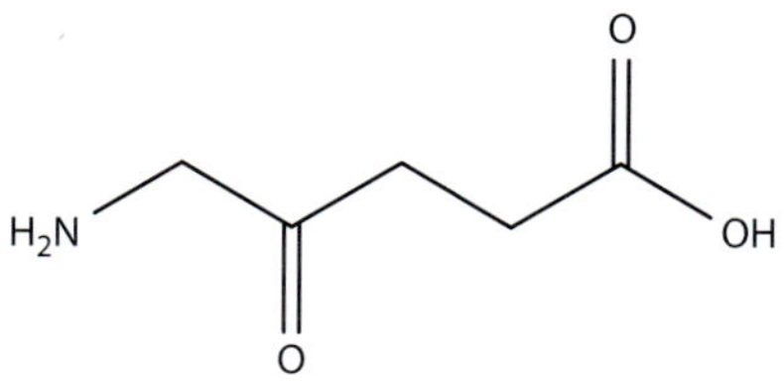

8.3, δ-aminolevulinic acid

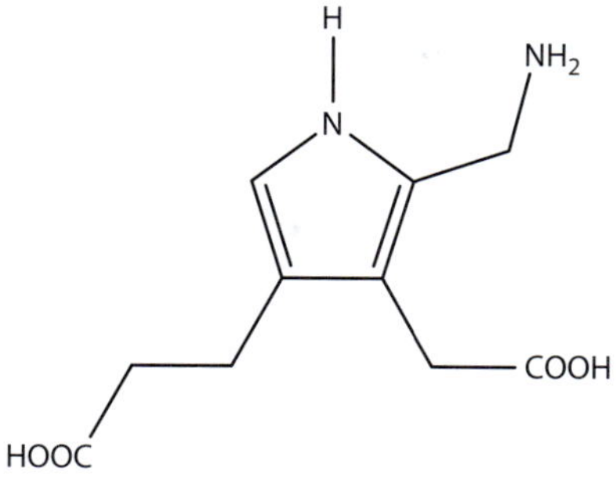

8.4, porphobilinogen

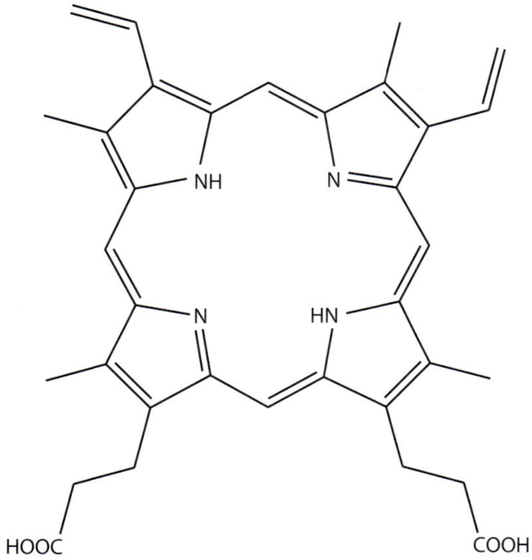

8.5, protoporphyrin IX

Kinetic measurements can tell us which species are involved in a reaction up to and including the RDS—but such measurements tell us little about the detailed structure of the transition state for the RDS. It was stated earlier that the chemistry of materials containing heavy isotopes was essentially the same as that for the standard light isotopes. This is almost true, but the differences will be useful to us. The physical chemistry underlying these observations is beyond our scope here, but the result is that carbon–deuterium bonds are more difficult to break than carbon–hydrogen bonds. The effect exists for all pairs of heavy/light atom bonds, but because for hydrogen, the mass ratio is much greater, the difference is more marked (look back at Section 5.5.1 where the vibration frequencies of C–H and C–D are discussed). The activation energy for the breaking of a carbon–deuterium bond is higher than that for a carbon–protium bond, and this can translate to a kinetic isotope effect (KIE), which is the ratio of the rate for protium (or another light isotope) to that for deuterium (or another heavy isotope), that is, k_H/k_D. To observe a primary KIE of this type, it is necessary that a C–H/C–D bond is broken in the RDS. Some examples will illustrate (Figure 8.18; H* represents either H or D). In the first two examples, a C–H/C–D bond is broken in the RDS, but in the third example, it clearly is not. The magnitude of the KIE for these first two examples is typical of a linear, and roughly symmetric, transition state in which the KIE is maximized. Transition states for proton transfer are not invariably linear; KIEs for bent transition states are usually lower, in the region of 2.5–3.5. KIEs are also lower for nonsymmetric transition states, which resemble either reactants or products. This is a manifestation of the Hammond postulate, which states that transition states proceeding to unstable intermediates can be approximated by the structure of the intermediates. Essentially, if two species, a transition state and an intermediate, are close in energy, then their structures will be similar.

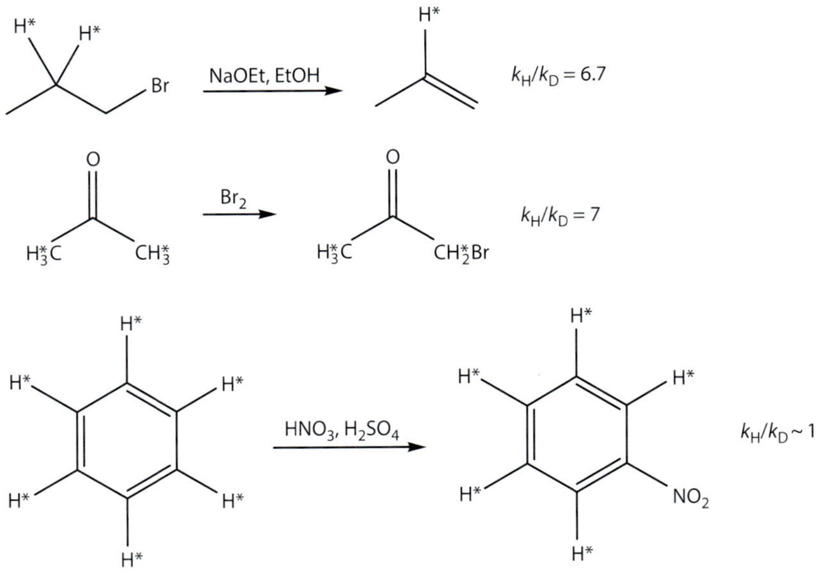

FIGURE 8.18 KIEs indicate whether a C–H bond is broken in the RDS.

8.2.5 STUDY OF INTERMEDIATES

This might be thought to be another rather obvious way to study organic reactions, but it's not quite as easy as one might suppose. Look back at Problem 8.1—there was indeed an intermediate, but this does not mean that it can be seen or studied. This would be true of any intermediate that either is so unstable that it has a very short lifetime or is one that is formed slowly but consumed quickly. However, if an intermediate is formed rapidly but consumed slowly, then it can sometimes be observed spectroscopically, even if it is not stable enough to be isolated. An example is shown in Figure 8.19. The intermediate in this case can be observed by IR spectroscopy.

In an ideal case, we would hope to isolate and characterize intermediates in a reaction, and this was achieved successfully in the reaction of Figure 8.20. The two intermediates shown are reasonably stable and can be isolated. However, this type of study comes with a *caveat*; the fact that a species that *could* be an intermediate is isolated from a reaction mixture does not necessarily mean that it *is* on the main reaction pathway. It is important that you show that the isolated species can be put back under the reaction conditions and proceeds to the final product. If this is not done, then what you isolated may be related to a minor pathway, or be in equilibrium with something on the main pathway, but have been precipitated because it is less soluble than the true intermediate.

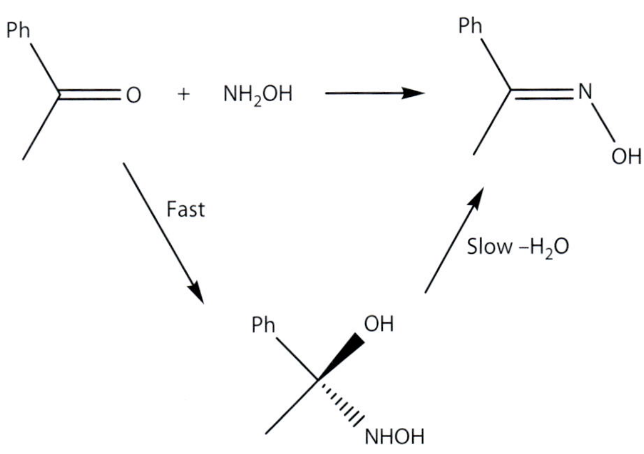

FIGURE 8.19 Observation of an unstable intermediate.

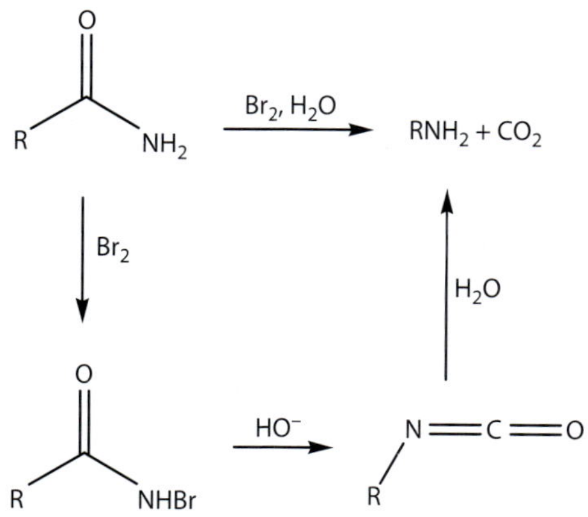

FIGURE 8.20 Sometimes intermediates can be isolated.

Review of Reactive Intermediates and a Little More

In Section 4.7, we encountered the main types of reactive intermediate and the factors that made them more or less stable. We can summarize:

Carbocations

- Planar, sp^2 hybrid, empty p_z orbital.
- High barrier to deformation, so nonplanar cations are difficult to make.
- Stabilized by inductive electron donors, so 3 ° > 2 ° > 1 °.
- Stabilized by sharing electrons with atoms bearing lone pairs (O, N, S).
- Stabilized by resonance; allyl and benzyl cations are stabilized.
- Stable when aromatic (*e.g.*, tropylium ion).
- Destabilized by electron-withdrawing atoms or groups.

Carbanions

- Pyramidal, sp^3 hybridization, lone pair of electrons.
- Stereochemically unstable, invert (like ammonia).
- Destabilized by inductive electron donors, so 1 ° > 2 ° > 3 °.
- Stabilized by electron-withdrawing atoms or groups.
- Stabilized by resonance; allyl, benzyl, and particularly enolate ions are stabilized.
- Stable when aromatic (*e.g.*, cyclopentadienyl anion).

Radicals

- Generally planar with sp^2 hybridization and a single electron in a p_z orbital.
- Barrier to deformation is relatively small; nonplanar radicals can be made, unlike nonplanar cations.
- Stability mostly parallels that of cations (3 ° > 2 ° > 1 °), but overall stability usually lower.
- Stabilized by resonance (*e.g.*, allyl and benzyl radicals).

We will also encounter a few other reactive intermediates. We previously met radical cations formed in the mass spectrometer. They are encountered elsewhere, but rarely. Radical anions are formed by the addition of electrons to double bonds (Figure 8.21), the source of the electron generally being the ionization of an alkali metal ($Na \rightarrow Na^+ + e^-$).

Carbenes

Carbenes are disubstituted carbon atoms bearing two electrons (which may be paired or unpaired). The parent species is methylene, $:CH_2$. The structures of carbenes are based on sp^2 hybridization, and there are two possible arrangements of the electrons within the

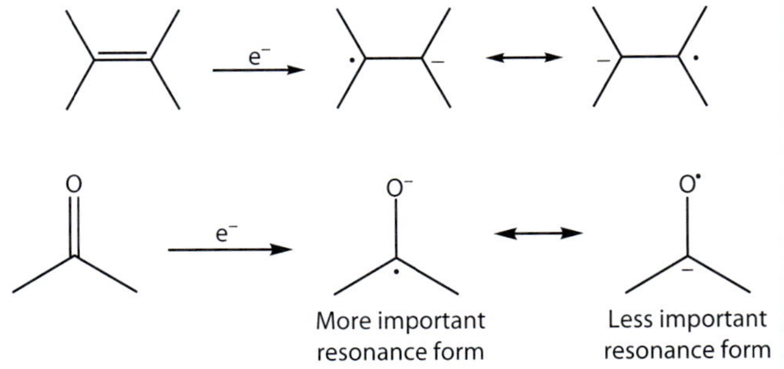

FIGURE 8.21 Formation of radical anions. In this example, the two resonance forms contribute equally.

FIGURE 8.22 Singlet and triplet methylene.

available orbitals (Figure 8.22). If both electrons are in the sp^2 orbital, in the plane of the molecule (as one might expect from VSEPR), this is described as singlet methylene and the HCH angle is closed down to 102 °. This form of methylene undergoes concerted reactions. Alternatively, we can put one electron in each of the p_z and sp^2 orbitals, with their spins parallel. This is described as the triplet form, and the HCH angle is opened up to 140 °. Triplet methylene behaves as a diradical and is lower in energy than the singlet form by about 37.8 kJ mol^{-1}. You will meet both methylene and more stable carbenes such as CCl_2 in Section 11.4.2 on their addition to carbon–carbon multiple bonds.

Key Points from Section 8.2

- Thermodynamic considerations determine whether or not an organic reaction is possible. Enthalpy terms are generally dominant and depend on the strengths of the bonds made and broken.
- Entropy terms are important where there is a choice between inter- and intramolecular processes.
- Bonds may be broken homolytically or, more commonly, heterolytically.
- Rates of reaction depend on the activation energy of the highest energy step, the RDS. Observed kinetics depend on the process before and includes the RDS, but not later steps.
- Molecularity refers to the number of molecules involved in a specific step of the reaction.
- Reactions may be single step or multistep.
- Studying reaction products and by-products gives useful information. Kinetic products are those resulting from the pathway with the lowest activation energy, while thermodynamic products are those with the lowest energy.
- Stable isotopes can be used to track the fate of particular atoms in both chemical and biological mechanisms. KIEs can be used to establish whether a particular C–H bond is broken in the RDS of a reaction.
- Intermediates may be long or short lived; in some cases, they may be observed or even isolated.
- Processes where a racemic product is formed from a chiral starting material are multistep, involving a planar intermediate.
- The rates of formation and consumption of reactive intermediates depend on their stability.

8.3 CLASSIFICATION OF REACTIONS

Organic chemists like to describe all reactions as belonging to six main types—addition, elimination, substitution, condensation, rearrangement, and isomerization. These categories are mainly useful to describe and label processes rather than telling us anything fundamental about the reaction mechanisms. Addition may be exemplified as in Figure 8.23; X=Y may be C=C, C=O, C≡C, C≡N, C=N, *etc.*, and either A–B may be a small molecule such as H_2 or HCN or the process may involve something significantly more complex.

Elimination reactions involve the removal of two atoms or groups from a molecule to give a multiple bond (Figure 8.24). Thus, it is the reverse of the addition reaction.

Substitution reactions involve the replacement of one atom or group by another. The examples of Figure 8.25 display a considerable diversity of mechanisms. Condensation reactions

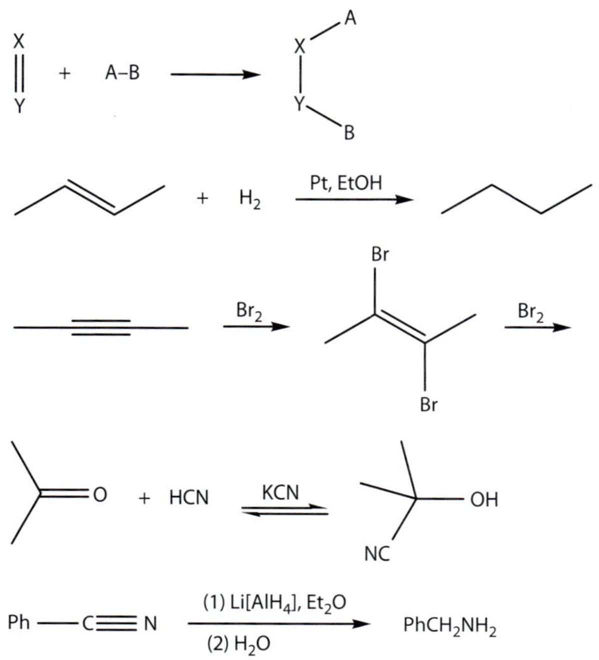

FIGURE 8.23 Addition reactions.

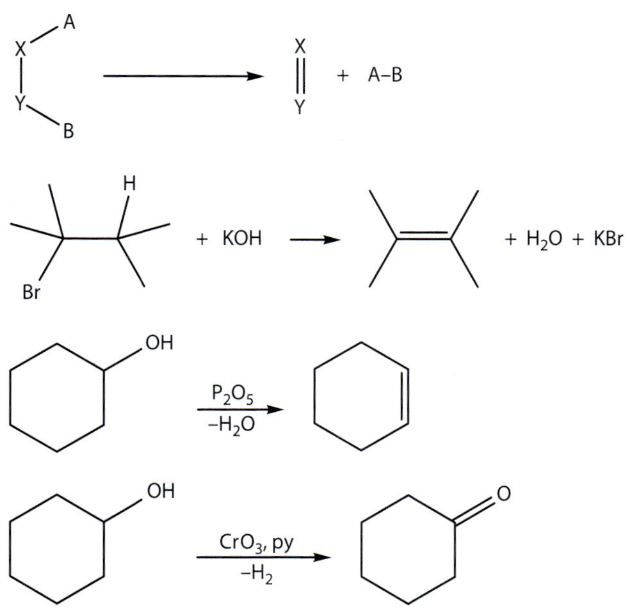

FIGURE 8.24 Elimination reactions.

Nucleophilic aliphatic substitution

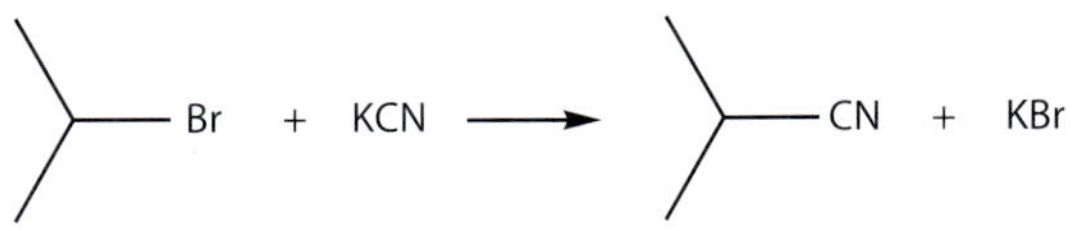

Electrophilic aromatic substitution (this is not a single-step process but involves an addition followed by a proton loss)

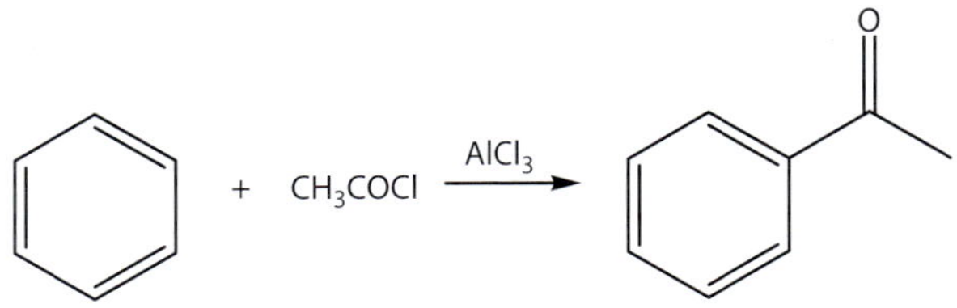

Radical chain substitution

$$CH_4 + Cl_2 \xrightarrow{h\nu} CH_3Cl + HCl$$

FIGURE 8.25 Substitution reaction.

(Figure 8.26) involve the joining together of two molecules with the loss of a small molecule (often water or an alcohol). We will meet many of these reactions when we discuss polymers, as this is the main route to step-growth polymers such as nylon. Rearrangements (Figure 8.27) are a small, but highly diverse, group of reactions in which the carbon skeleton of the molecule is altered. Isomerizations (Figure 8.28) involve moving bonds in a molecule, without rearrangement of the carbon skeleton. The additional description of redox reactions potentially overlaps with any of these main categories; this will generally be obvious when we meet reactions of this type, because the reagent is either an oxidant or a reductant.

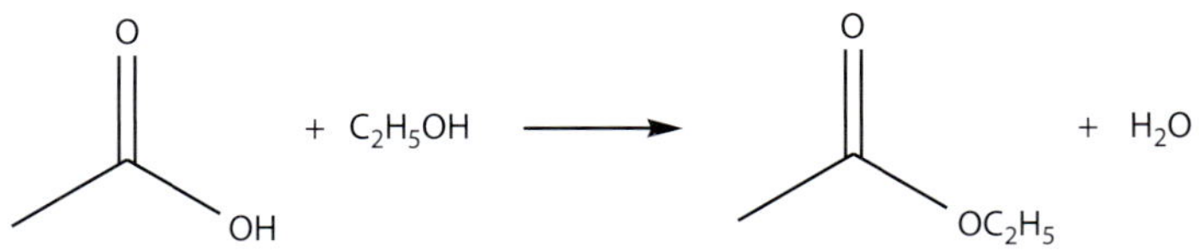

FIGURE 8.26 Condensation reaction.

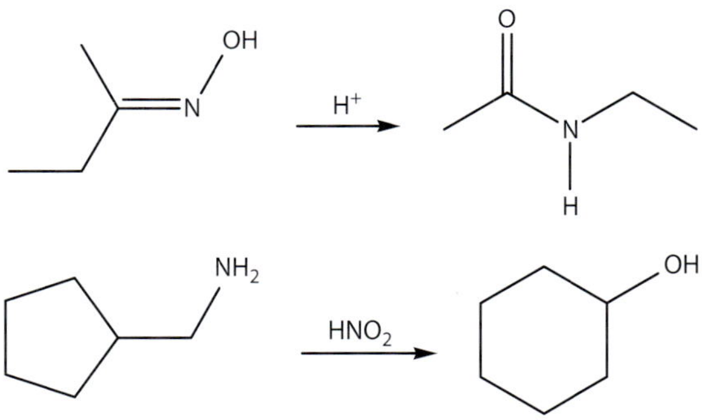

FIGURE 8.27 Rearrangement reactions.

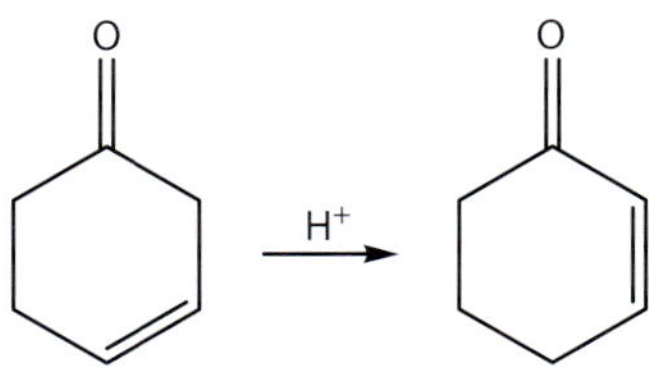

FIGURE 8.28 Isomerization reaction.

PROBLEM 8.2

Classify each of the following reactions as one of substitution, addition, elimination, condensation, rearrangement, or isomerization:

(a)

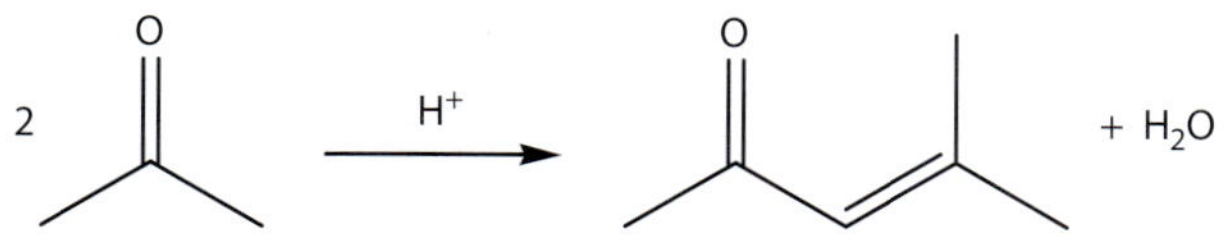

(b) $CH_3C(=O)Cl + H_2O \rightarrow CH_3C(=O)OH + HCl$

(c)

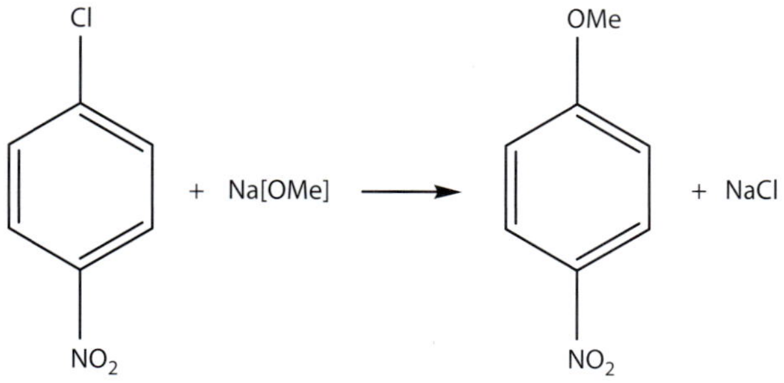

(d)

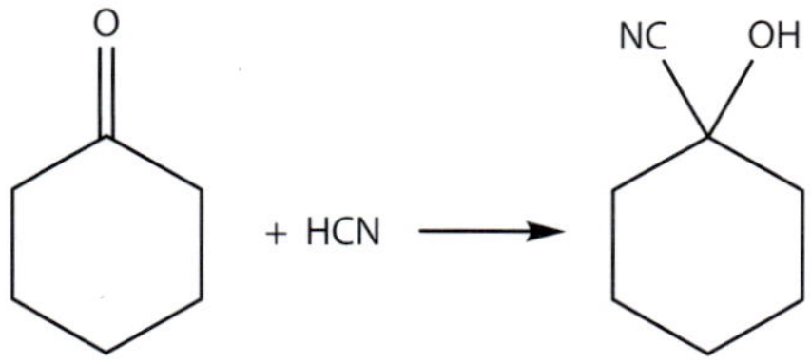

(e)

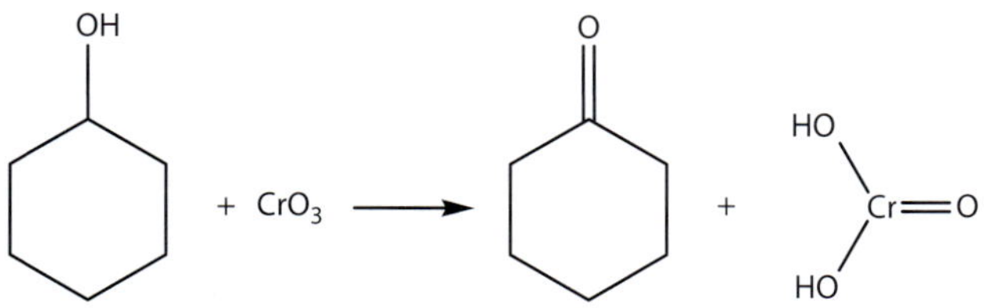

(f)

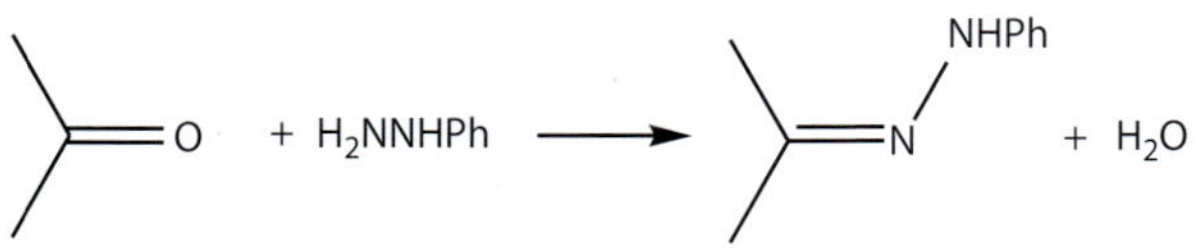

(g)

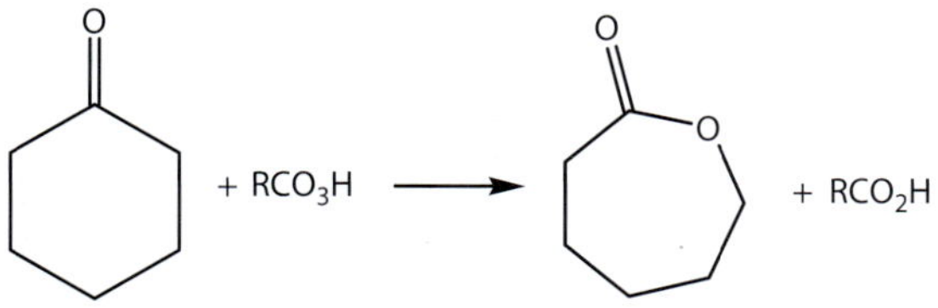

Solutions

(a) Condensation
(b) Substitution
(c) Substitution
(d) Addition
(e) Elimination
(f) Condensation
(g) Rearrangement

8.4 ACIDS AND BASES

Acidity and basicity in organic compounds are related to electron availability and will teach us a lot about resonance forms and the stability of anions and cations. First, it is useful to review pK_a, which is defined as $-\log_{10}K_a$ where K_a is the equilibrium constant for (8.10), $K_a = [H^+][A^-]/[HA]$. The pK_a is the pH at which the acid is exactly half ionized (8.11 through 8.14) and tells us the strength of the acid HA.

(8.10) $$HA \rightleftharpoons H^+ + A$$

(8.11) $$pK_a = -\log K_a$$

(8.12) $$\log(K_a) = \log[H^+] + \log[A^-] - \log[HA]$$

(8.13) $$\text{When } [A^-] = [HA], \quad \log[A^-] - \log[HA] = 0$$

(8.14) $$\log K_a = \log[H^+] = -pH$$

At a pH greater than its pK_a, the acid is ionized as A^-. At a pH less than its pK_a, the acid will exist as HA. The higher the pK_a, the weaker is the acid; the lower the pK_a, the stronger the acid. The pK_a also tells you about the strength of the conjugate base, A^-. The higher the pK_a of the acid HA, the stronger the conjugate base, A^-. The lower the pK_a of HA, the weaker the base. It is important that we remember that for acids and base, we are always looking at equilibria; any comments that we make about the strength of the acid have to be related to thermodynamics, not kinetics. Thus, we should not say "it is easier for HA to lose a proton," but "the anion A^- is stabilized relative to HA."

Many factors affect the acidity and basicity of organic compounds, and we will consider each in turn. The pK_a values are more widely variable than you will have encountered for inorganic acids and bases; it's important to remember that with organic compounds, generally we are not dealing with aqueous solutions. The values for pK_a given in the text are taken from a range of sources; although every effort has been made to give the best available value, you will find that values vary slightly from text to text. Some of the values, especially at the extremes, are difficult to measure, and some quoted data refer to gas phase processes and others to solution data. Tables 8.1 and 8.2 give a fairly comprehensive listing, and you will want to refer back to these in many subsequent chapters. But it is not here to be memorized....

8.4.1 ORBITAL EFFECTS

If we compare the pK_a values of alkanes (typically 50), alkenes (typically 40–45), and alkynes (typically 25), we can attribute the differences to the greater stability of the lone pair in an

TABLE 8.1 pK_a Values (in Water) for Compounds Important in Organic Chemistry Where the Proton Is Attached to a Heteroatom

Acid	Conjugate Base	pK_a	Acid	Conjugate Base	pK_a
$HClO_4$	$[ClO_4]^-$	−10	CH_3OH	$[CH_3O]^-$	15.5
HI	I^-	−10	CH_3CH_2OH	$[CH_3CH_2O]^-$	16
HBr	Br^-	−9	$(CH_3)_2CHOH$	$[(CH_3)_2CHO]^-$	17
HCl	Cl^-	−7	$(CH_3)_3COH$	$[(CH_3)_3CO]^-$	18
HF	F^-	3.2	CF_3CH_2OH	$[CF_3CH_2O]^-$	12.5
HOCl	$[ClO]^-$	7.5	CH_3OOH	$[CH_3OO]^-$	11.5
H_2SO_4	$[HSO_4]^-$	−10	PhSH	$[PhS]^-$	7.8
$[HSO_4]^-$	$[SO_4]^{2-}$	2.0	CH_3CH_2SH	$[CH_3CH_2S]^-$	10.5
HNO_3	$[NO_3]^-$	−1.3	PhOH	$[PhO]^-$	9.99
HNO_2	$[NO_2]^-$	3.29	$3\text{-}O_2NC_6H_4OH$	$[3\text{-}O_2NC_6H_4O]^-$	8.35
H_3PO_4	$[H_2PO_4]^-$	2.1	$4\text{-}O_2NC_6H_4OH$	$[4\text{-}O_2NC_6H_4O]^-$	7.14
$[H_2PO_4]^-$	$[HPO_4]^{2-}$	7.2	$2,4\text{-}(O_2N)_2C_6H_3OH$	$[2,4\text{-}(O_2N)_2C_6H_3O]^-$	3.96
$[HPO_4]^{2-}$	$[PO_4]^{3-}$	12.3	$2,4,6\text{-}(O_2N)_3C_6H_2OH$	$[2,4,6\text{-}(O_2N)_3C_6H_2O]^-$	0.38
$[HN_3]$	$[N_3]^-$	4.72	$[(CH_3)_2OH]^+$	CH_3OCH_3	−3.8
HCN	$[CN]^-$	9.3	$[CH_3OH_2]^+$	CH_3OH	−2.5
H_2CO_3	$[HCO_3]^-$	6.35	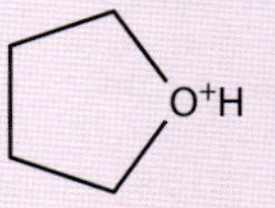	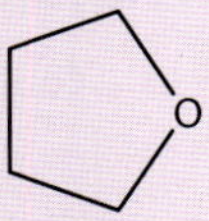	−2.05
$[HCO_3]^-$	$[CO_3]^{2-}$	10.3			
CH_3SO_3H	$[CH_3SO_3]^-$	−2	$[(CH_3)_2C{=}OH]^+$	$(CH_3)_2C{=}O$	−7.3
CF_3SO_3H	$[CF_3SO_3]^-$	−14	$[CH_3C({=}OH)OCH_3]^+$	$CH_3C({=}O)OCH_3$	−6.5
$PhSO_3H$	$[PhSO_3]^-$	−6.5	$[CH_3C({=}OH)OH]^+$	$CH_3C({=}O)OH$	−6.1
HCOOH	$[HCOO]^-$	3.75	$[NH_4]^+$	NH_3	9.21
CH_3COOH	$[CH_3COO]^-$	4.76	NH_3	$[NH_2]^-$	38.0
FCH_2COOH	$[FCH_2COO]^-$	2.66	$[CH_3NH_3]^+$	CH_3NH_2	10.63
$ClCH_2COOH$	$[ClCH_2COO]^-$	2.85	CH_3NH_2	$[CH_3NH]^-$	40
$BrCH_2COOH$	$[BrCH_2COO]^-$	2.69	$[(CH_3)_2NH_2]^+$	$(CH_3)_2NH$	10.66
ICH_2COOH	$[ICH_2COO]^-$	3.12	$[(CH_3)_3NH]^+$	$(CH_3)_3N$	9.76
$Cl_2HCCOOH$	$[Cl_2HCCOO]^-$	1.3	$[PhNH_3]^+$	$PhNH_2$	4.62
Cl_3CCOOH	$[Cl_3CCOO]^-$	0.65	$[PhCH_2NH_3]^+$	$PhCH_2NH_2$	9.34
F_3CCOOH	$[F_3CCOO]^-$	−0.25	$[H_3N^+CH_2COOH]$	H_2NCH_2COOH	9.6
O_2NCH_2COOH	$[O_2NCH_2COO]^-$	1.68	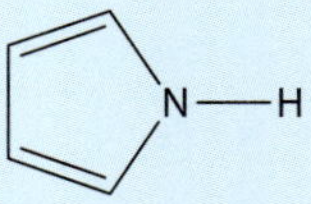	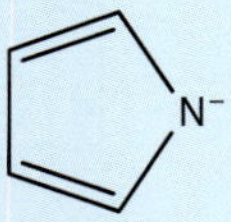	17.0
$NCCH_2COOH$	$[NCCH_2COO]^-$	2.43	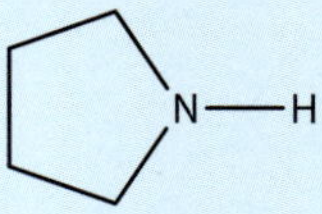	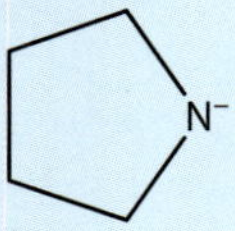	36
PhCOOH	$[PhCOO]^-$	4.2			
$PhCH_2COOH$	$[PhCH_2COO]^-$	4.28			
CyclohexylCOOH	[CyclohexylCOO]$^-$	4.9	$[pyH]^+$	Pyridine	5.2
trans-$CH_3CH{=}CHCOOH$	[*trans*-$CH_3CH{=}CHCOO]^-$	4.69	$[NH_3OH]^+$	NH_2OH	6.0
CH_3CO_3H	$[CH_3CO_3]^-$	8.2	$[H_2NNH_3]^+$	H_2NNH_2	8.2
$[H_2NCH_2COOH]$	$H_2NCH_2COO^-$	2.31	$[PhC{\equiv}NH]^+$	$PhC{\equiv}N$	−10

(*continued*)

TABLE 8.1 (continued) pK_a Values (in Water) for Compounds Important in Organic Chemistry Where the Proton Is Attached to a Heteroatom

Acid	Conjugate Base	pK_a	Acid	Conjugate Base	pK_a
$[H_3O]^+$	H_2O	−1.7	$CH_3C(=O)NH_2$	$[CH_3C(=O)NH]^-$	15.1
H_2O	$[HO]^-$	15.7	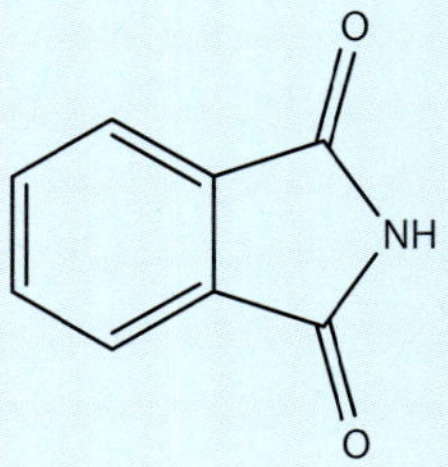	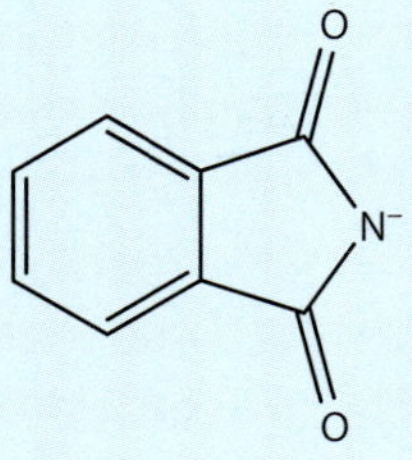	8.3
HOOH	$[HOO]^-$	11.6			
H_2NOH	$[H_2NO]^-$	13.7			
H_2S	$[HS]^-$	7.0			

Note: It will be obvious to you that these values are not measured in water, as many of these anions could not exist in water; we will return to solvent effects shortly.

TABLE 8.2 pK_a Values for Carbon Acids

Acid	Conjugate Base	pK_a	Acid	Conjugate Base	pK_a
$(CH_3)_3CH$	$(CH_3)_3C^-$	53	CH_3NO_2	$[CH_2NO_2]^-$	10.29
$(CH_3)_2CH_2$	$[(CH_3)_2CH]^-$	51	$CH_3C{\equiv}N$	$[CH_2C{\equiv}N]^-$	25
CH_4	$[CH_3]^-$	48–50	$CHCl_3$	$[CCl_3]^-$	25
C_2H_6	$[CH_3CH_2]^-$	48	$CH_3C(=O)CH_3$	$[CH_2C(=O)CH_3]^-$	20
Cyclopropane	$[C_3H_5]^-$	43–46	CH_3CHO	$[CH_2CHO]^-$	17
C_2H_4	$[H_2C=CH]^-$	44	$CH_3C(=O)CHCl_2$	$[CH_3C(=O)CCl_2]^-$	15
C_6H_6	Ph^-	43	$PhC(=O)CH_3$	$[PhC(=O)CH_2]^-$	15.9
$H{-}C{\equiv}C{-}H$	$H{-}C{\equiv}C^-$	25	$CH_3C(=O)OCH_2CH_3$	$[CH_2C(=O)OCH_2CH_3]^-$	24.5
$CH_3CH=CH_2$	$[CH_2=CHCH_2]^-$	43	$CH_3C(=O)N(CH_3)_2$	$[CH_2C(=O)N(CH_3)_2]^-$	30
$PhCH_3$	$[PhCH_2]^-$	41	$CH_3C(=O)CH_2C(=O)CH_3$	$[CH_3C(=O)CHC(=O)CH_3]^-$	9.0
Ph_2CH_2	$[Ph_2CH]^-$	33.5	$CH_3C(=O)CH_2COOEt$	$[CH_3C(=O)CHCOOEt]^-$	11.0
Ph_3CH	Ph_3C^-	31.5	$CH_3C(=O)CH_2CHO$	$[CH_3C(=O)CHCHO]^-$	5.9
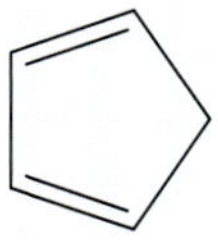	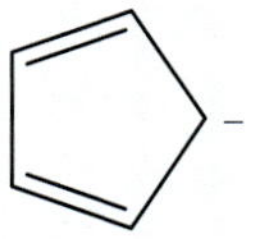	15	$CH_2(COOMe)_2$	$[CH(COOMe)_2]^-$	13.0
			$CH_2(NO_2)_2$	$[CH(NO_2)_2]^-$	4.0
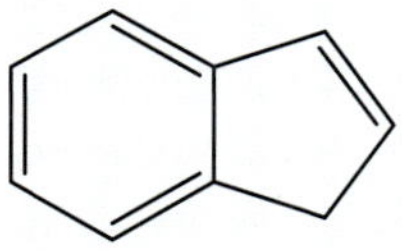	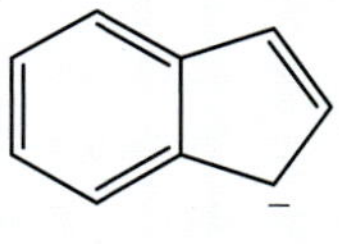	20	$CH_2(CHO)_2$	$[CH(CHO)_2]^-$	5.0
			$CH_2(CN)_2$	$[CH(CN)_2]^-$	11.8

sp orbital relative to an sp^2 or sp^3 orbital. The sp orbital has more s-character and is closer to the nucleus, so the lone pair is "happier" there ☺. Hence, alkynes are stronger acids and much easier to deprotonate than either alkenes or alkanes. The same effect can be seen with nitrogen. $CH_3C{\equiv}N^+H$ is a very strong acid (pK_a −10) compared with either $[pyH]^+$ (pK_a 5.2) or $[CH_3NH_3]^+$ (pK_a 10.63). The lone pairs of electrons in the conjugate bases are in sp, sp^2, and sp^3 orbitals, respectively—the nitrile, with the lone pair close to the nucleus in the sp orbital, is the most stable of the conjugate bases. This group is more often described in terms of the basicity of the uncharged species; so CH_3NH_2 is more basic than pyridine, which is much more basic than acetonitrile.

PROBLEM 8.3

Predict the position of the following equilibrium:

$$CH_3CH_2CH_2CH_2^- + HC{\equiv}CH \rightleftharpoons CH_3CH_2CH_2CH_3 + HC{\equiv}C^-$$

Solution

The equilibrium will lie far to the right (equilibrium constant will exceed 10^{20}). The anion of ethyne is very much more stable than that of butane, because the lone pair is close to the nucleus in an sp orbital. Butyl lithium is often used to deprotonate 1-alkynes to their anions.

8.4.2 ELECTRONEGATIVITY

For similar types of molecules, the more electronegative the atom on which the charge is placed in a conjugate base, the more stable the conjugate base will be. Thus, CH_3O^- is favored over $[CH_3NH]^-$, which is favored over CH_3^-, because oxygen is the most electronegative of the atoms on which the charge is located and carbon the least so. This gives pK_a values of 50, 38, and 15.5 for the pK_a values of methane, ammonia, and methanol, respectively.

PROBLEM 8.4

Place the following compounds in order of increasing acidity:

$$PhCH_3;\; PhOH;\; PhNH_2;\; PhSH$$

Solution

You can always look up the pK_a values; the only one not in the table is aniline, and this can be modeled or checked on line:

$$PhCH_3 - 41;\; PhOH - 9.99;\; PhNH_2 - 30.6;\; PhSH - 7.8$$

The results for aniline, phenol, and toluene are clearly related to the electronegativity of the atom on which the charge resides in the conjugate base; oxygen is more electronegative than nitrogen, which is more electronegative than carbon. The one we need to think about a little more is thiophenol—why is this more acidic than phenol, when sulfur is less electronegative than oxygen? If we check back to Table 8.1, this is not an isolated result; ethanethiol is more acidic than ethanol. So it seems as though our comments on electronegativity need a qualification—it only works when the elements being compared are in the same row of the periodic table. So in the halogen acids, HI is the strongest acid and HF the weakest—again in the opposite direction to electronegativity. The difference between sulfur and oxygen can be explained by two factors. The H–S bond is significantly weaker than the H–O bond (385 and 485 kJ mol^{-1} in H_2S and H_2O, respectively), so its loss is less energetically undesirable. Also, when the negative charge resides on sulfur, this is a larger atom, so the charge is dispersed over a larger surface.

8.4.3 INDUCTIVE EFFECTS

Any atom or group that withdraws electron density through bonds will have the effect of stabilizing an anion. This is easily exemplified in Table 8.1, by the pK_a values of substituted acetic acids. The more halogens are substituted into the methyl group and the more electronegative the halogen substituted, the stronger the acid. Notice that nitro and cyano groups are also clearly strong withdrawers of electron density through their inductive effect. Similar effects are noted with alcohols. 2,2,2-Trifluoroethanol has a pK_a value of 12.5 by comparison with 16 for ethanol. The converse is also true—alkyl groups are inductive donors of electron density, and secondary and tertiary alcohols are consequently weaker acids than primary alcohols.

PROBLEM 8.5

Explain the pK_a values given for the substituted butanoic acids in the following:

Butanoic acid – 4.83
2-Chlorobutanoic acid – 2.86
3-Chlorobutanoic acid – 4.05
4-Chlorobutanoic acid – 4.52
2-Oxobutanoic acid – 2.5

Solution

The pK_a value for butanoic acid is entirely typical of a carboxylic acid. The effect of the chloro substituent is to withdraw electron density inductively, and its influence in stabilizing the carboxylate anion is greatest when it is closest to the site of the anion. The effect of the 2-oxo substituent indicates that this is a stronger electron withdrawer than –CHClR.

8.4.4 DELOCALIZATION

The more the delocalization of the charge in the conjugate base, A^-, the more stable it will be, and hence the more acidic the acid HA. Thus, the pK_a of methanol, where the change is localized on a single oxygen atom, is 15.5, but CH_3COOH, where the charge can be delocalized over two oxygen atoms (Figure 8.29), has a pK_a of 4.76. In CH_3SO_3H, the negative charge in the conjugate base is delocalized over three oxygen atoms, and the pK_a is –2. A similar effect is seen comparing propane and propene. In propane, we have a localized carbanion; in propene, there are two possible resonance forms. The pK_a values are 50 and 35, respectively. Toluene may be compared with methane; the pK_a value of 41 for toluene reflects the possibility of charge delocalization around the aromatic ring in the benzyl anion, $PhCH_2^-$.

In the examples of Figure 8.29, the charges were delocalized onto identical and often equivalent atoms. However, any delocalization is helpful, even where the atoms are not identical. So phenol (acidic enough to cause serious skin burns) has a pK_a value of 10, because the negative charge on the oxygen can be delocalized around the aromatic ring (Figure 8.30). It is considerably more acidic than a regular alcohol, but because no other oxygen atoms are involved, and the aromatic sextet has to be breached to write the resonance forms, it is much less acidic than a carboxylic acid. In general, the anions of phenols are stabilized by inductive withdrawers of electron density attached to the benzene ring and destabilized by inductive donors. Thus, the pK_a of *meta*-nitrophenol is 8.35, reflecting the inductive effect of the nitro group. However, the fact that the nitro group can act as a resonance acceptor of electron density is even more significant. The pK_a of 4-nitrophenol is 7.14—but the nitro group is further from the anion, and hence, any inductive effect will be less. The resonance effect is dominant here (Figure 8.30)—and this is generally the case with the strongly accepting nitro group.

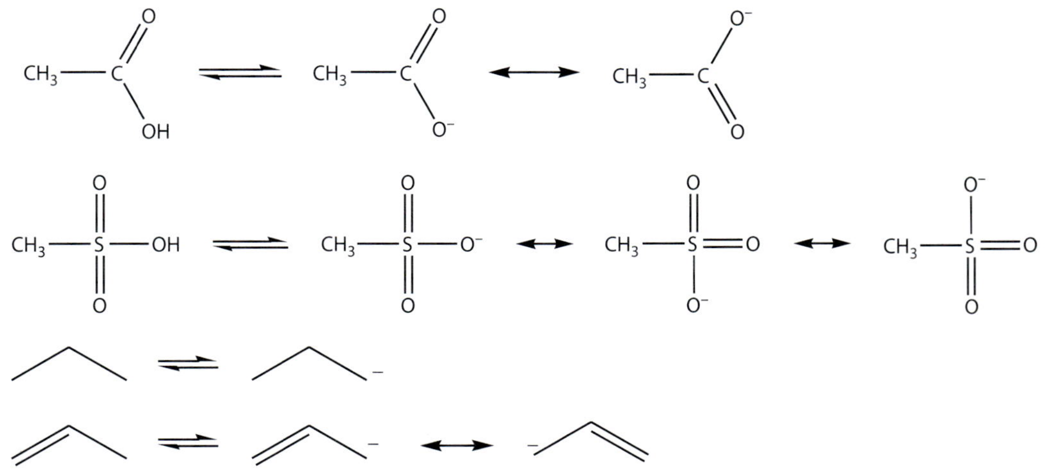

FIGURE 8.29 Delocalization of charge stabilizes conjugate bases.

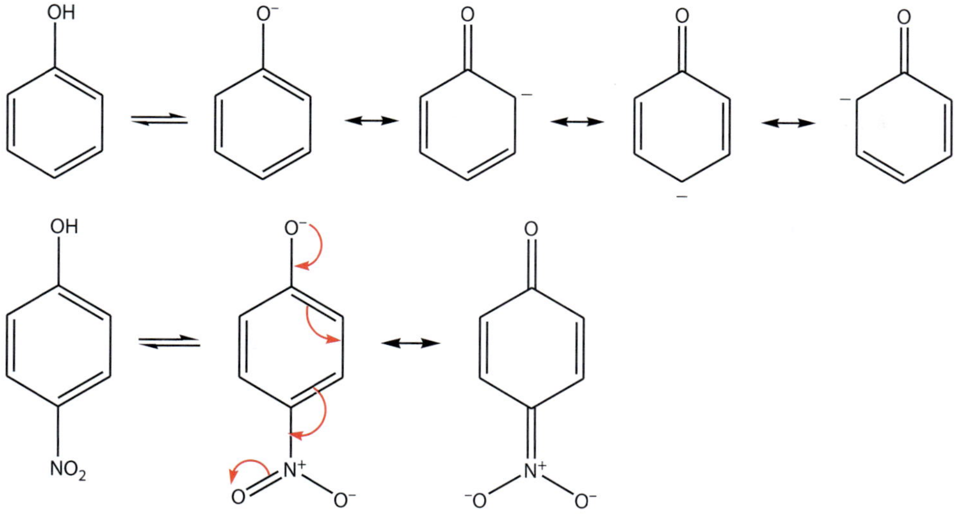

FIGURE 8.30 More delocalization of charge to stabilize conjugate bases.

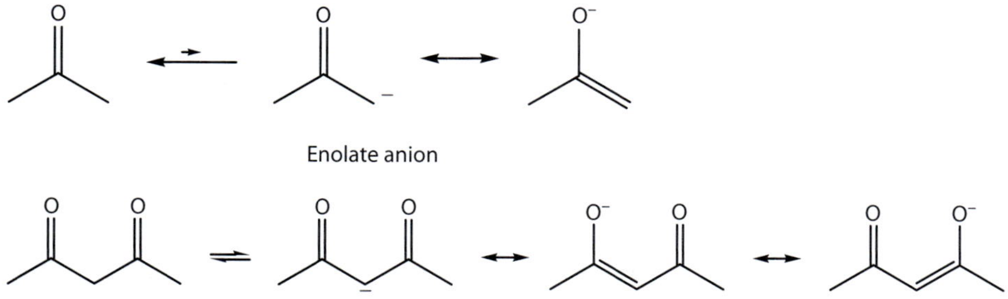

FIGURE 8.31 Anions of carbonyl compounds.

Other important anions stabilized by resonance delocalization are those on carbon atoms attached to carbonyl groups, as in acetone (Figure 8.31). Some of the stabilization of the anion comes from the inductive effect (see Problem 8.5), but much more from delocalization, as the charge may be delocalized onto the electronegative oxygen atom. The acidity of simple carbonyl compounds depends on the other substituent—typically ketones have a pK_a of 20, esters 25, and *N,N*-dialkylamides 30. As you can see in pentane 2,4-dione (pK_a 9), the introduction of a second carbonyl group renders the CH_2 group between the two carbonyls significantly more acidic.

PROBLEM 8.6

Circle the most acidic hydrogen in each of the following molecules:

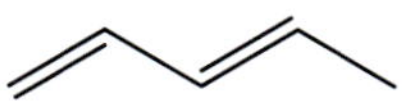

(a)

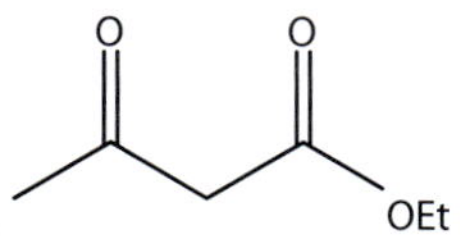

(b)

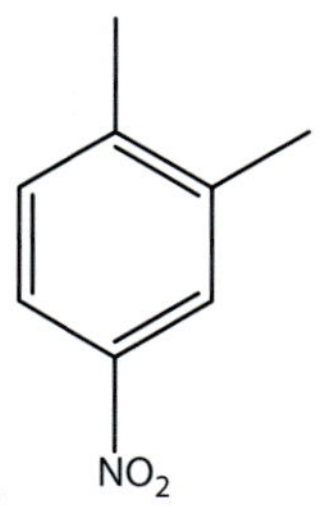

(c)

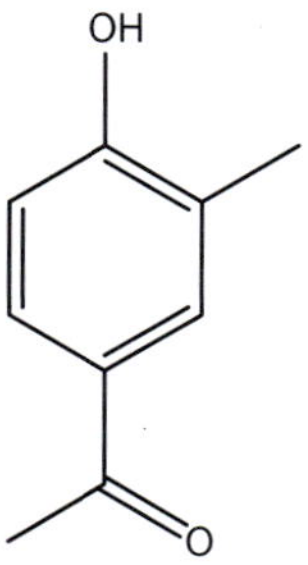

(d)

(a) The protons on the methyl group are the most acidic, as the anion can be delocalized:

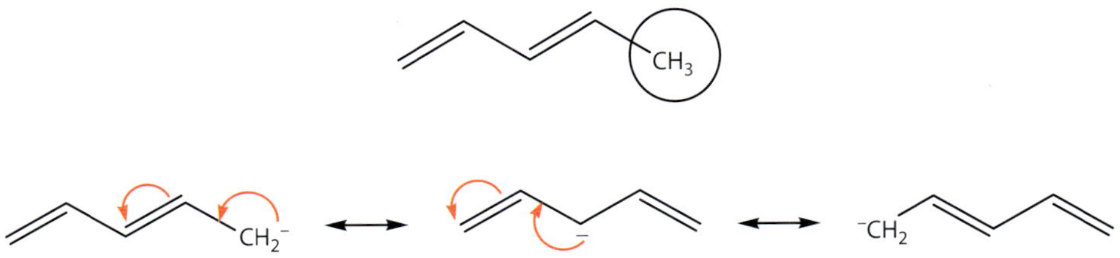

(b) The protons on the CH_2 group between the two carbonyl groups are the most acidic, since the anion can be stabilized by resonance with both carbonyl groups (and benefits from the inductive effect of both carbonyl groups).

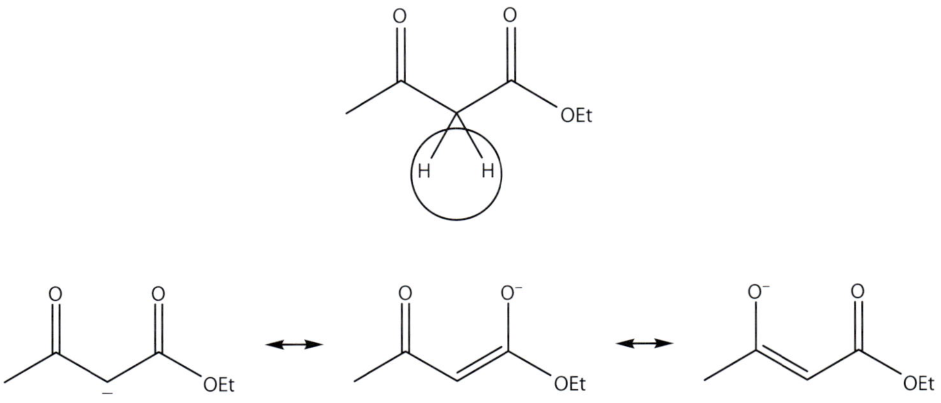

(c) An anion on either of the methyl groups would have resonance overlap with the π-system of the benzene ring, but only the anion *para* to the nitro group would benefit from resonance delocalization into the nitro group.

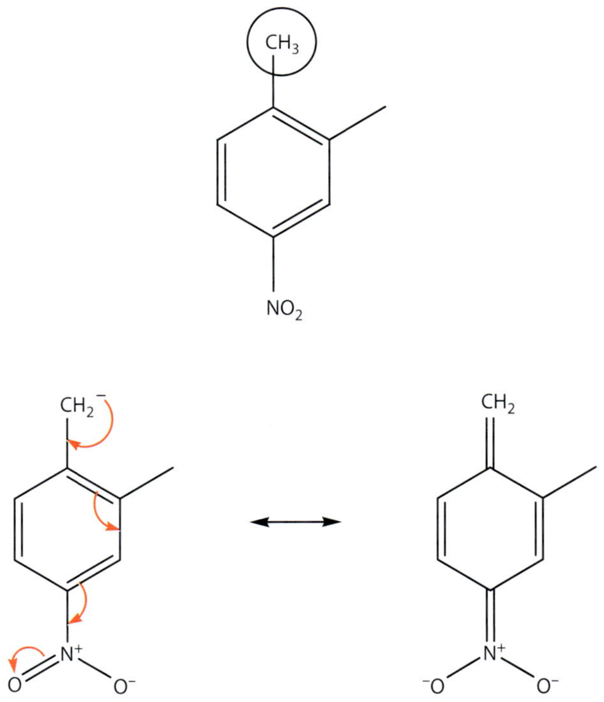

(d) Sorry if this one seems like a trick—but it is important not to miss the obvious!

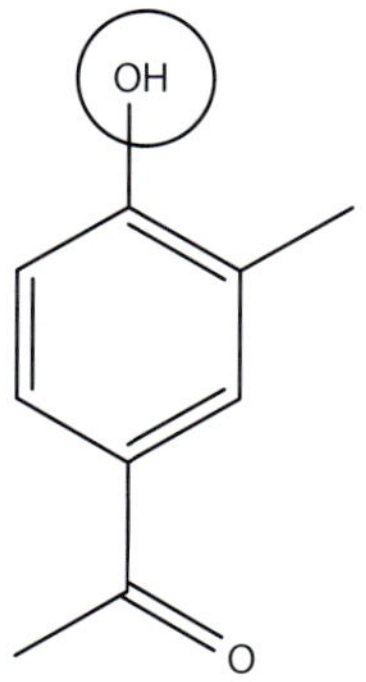

8.4.5 SOLVATION

Although we first think of pK_a, acidity, and basicity in terms of aqueous solvents, it must be clear that many of the anions and cations we have discussed could not even exist in an aqueous medium. Solvents do have a significant effect on pK_a.* In solution, *t*-BuO⁻/*t*-BuOH is a stronger base than MeO⁻/MeOH. However, in the gas phase, methoxide is the stronger base. Methoxide is small and highly solvated, especially in protic solvents, whereas *t*-BuO⁻ is large and poorly solvated (naked, ☺ and hence more keen to reacquire a proton).

8.4.6 HYDROGEN BONDING

Solvents capable of acting as hydrogen bond donors tend to stabilize anions where the charge resides on oxygen or nitrogen. However, intramolecular hydrogen bonding, where it is possible, is even more effective. If we consider the pK_a values for 2- and 4-hydroxybenzoic acid (Figure 8.32), we can see this effect. The values for the 4-hydroxy compound are essentially normal for a phenol and a benzoic acid. However, those for the 2-hydroxy compound are quite unusual. The formation of the monoanion, by loss of the proton from the carboxylic acid, is favorable, because it is strongly intramolecularly hydrogen bonded. However, the stability of this anion means that a second deprotonation is much less favorable, giving two adjacent negatively charged groups attached to the ring.

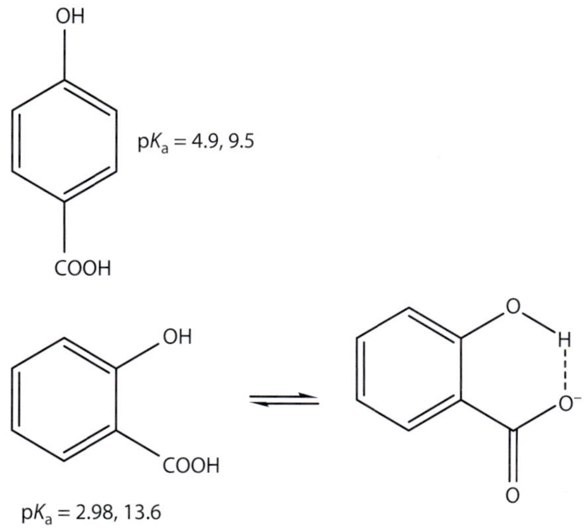

FIGURE 8.32 pK_a values of 2- and 4-hydroxybenzoic acid.

---

* http://www.chem.wisc.edu//areas//reich//pkatable// has a series of values of pK_a values in DMSO.

PROBLEM 8.7

The monoanion of fumaric acid is both a stronger acid and a stronger base than the monoanion of maleic acid. Explain.

Fumaric acid Maleic acid

Solution

The pK_a values for the first and second ionization of fumaric acid are 3.03 and 4.44; the corresponding values for maleic acid are 1.83 and 6.07. Thus, the statement is clearly true. The monoanion to dianion conversion for fumaric acid is more favorable than that for maleic acid, because the second pK_a is lower. The monoanion to diacid conversion is also more favorable for the monoanion of fumaric acid. The monoanion of maleic acid is favorable to form, because there is an intramolecular hydrogen bond, but it is difficult to deprotonate, for exactly the same reason. The hydrogen bonding stabilizes the maleate monoanion against both protonation and deprotonation.

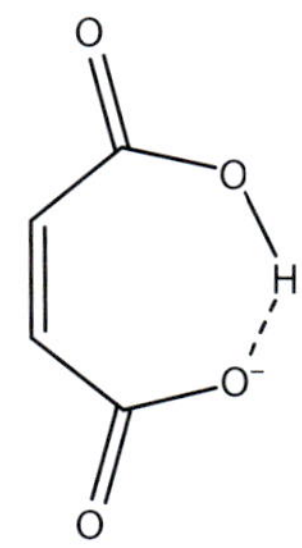

8.6, monoanion of maleic acid

8.4.7 STERIC EFFECTS

If we consider the ionization of cyclopentanone, then the pK_a is approximately 20, similar to the value for acetone. However, it is very difficult to ionize **8.7**; if we draw the enolate anion, the reason becomes clear—the formation of a conventional enolate anion is impossible because the carbon–carbon double bond would not be planar.

The relatively low basicity of aniline relative to aliphatic amines results from the interaction of the lone pair of electrons at nitrogen with the aromatic ring; this can only occur when the system is completely planar as in **8.8**, because the lone pair of electrons at nitrogen needs to be in a p_z orbital that is parallel to the p_z orbitals of the π-system. However, in **8.9**, the NMe_2 group is forced to twist out of the plane of the ring to avoid interaction with the *ortho*-methyl groups; this amine is more basic.

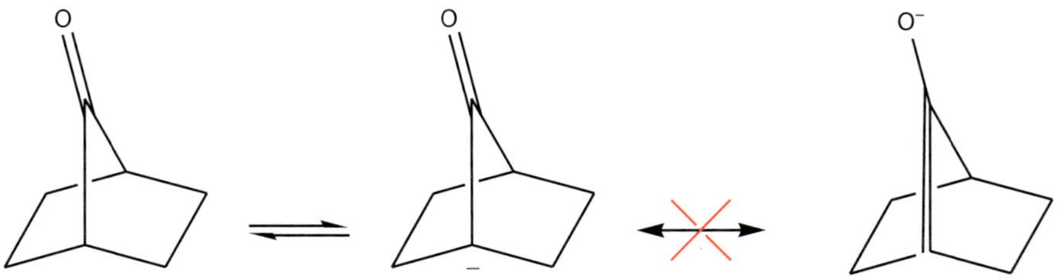

8.7 Geometrically impossible

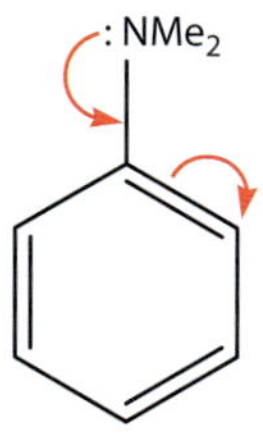

pK$_a$ [PhNHMe$_2$]$^+$ = 5.1

8.8

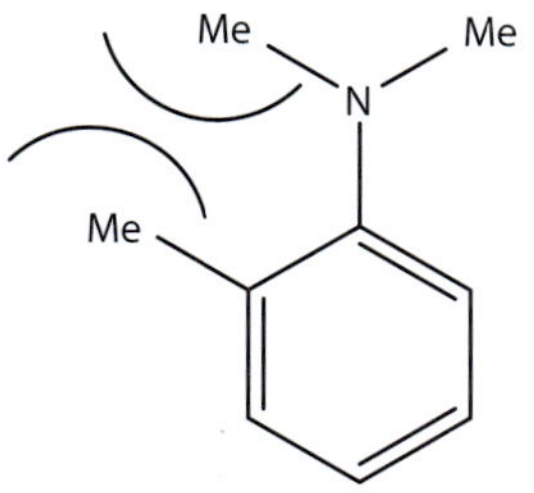

pK$_a$ [ArNHMe$_2$]$^+$ = 5.86

8.9

8.4.8 AROMATICITY

When an anion is aromatic, its formation is particularly favored. Thus, the pK_a of cyclopentadiene is 15, in comparison with a typical value of 35–40 for a simple allylic system. Conversely, when the anion being formed is antiaromatic, as in the ionization of **8.10**, the process is very unfavorable.*

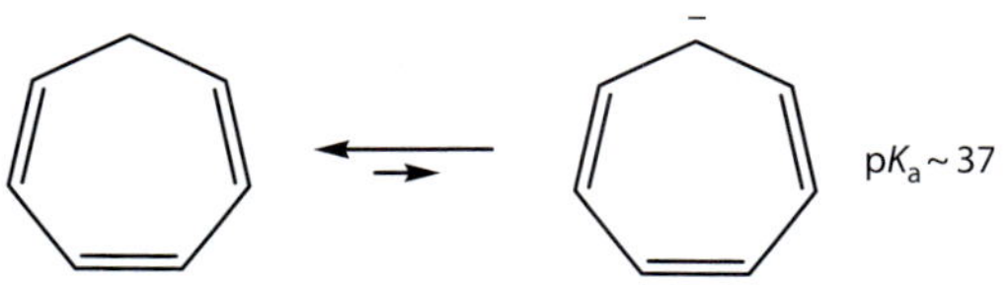

8.10 8-π antiaromatic system

PROBLEM 8.8

For each of the following pairs of compounds, state which will be more acidic, giving a brief reason:

(a)

(b)

* There are two low-energy states of this anion: one planar and antiaromatic and one nonplanar, nonaromatic and less conjugated.

Solutions

(a) Cyclopentadiene is more acidic because the anion formed is a 6π aromatic system.

(b) Protonated pyrrole (pK_a 0.4) is a very strong acid, because the neutral molecule is aromatic. Protonated imidazole is a moderate acid (pK_a 6.95)—both the neutral molecule and the protonated forms have aromatic character, so this molecule gains less by releasing the proton.

Key Points from Sections 8.3 and 8.4

- Organic reactions can be classified as addition, elimination, substitution, condensation, rearrangement, or isomerization. Redox reactions may belong to a number of these categories.
- Acidity and basicity depend on the various factors that influence the position of the equilibrium between an acid HA and its conjugate base A^-. These include orbital effects, electronegativity, inductive effects, resonance effects, solvation hydrogen bonding, steric effects, and aromaticity.

8.5 TAUTOMERS: A SLIGHT DIGRESSION

Tautomerism differs from resonance in that both electrons and hydrogen atoms are moved in order to make the new tautomer, but only electrons may be moved to create a new resonance form. The best way to generate a tautomer of a particular structure is using a thought experiment. First, remove a proton that gives a plausible anion. Write an alternative resonance form of the anion that you produced. Finally, replace the proton on the new resonance form. This is shown in Figure 8.33 for acetone. We make the enolate anion by removing a proton from one of the methyl

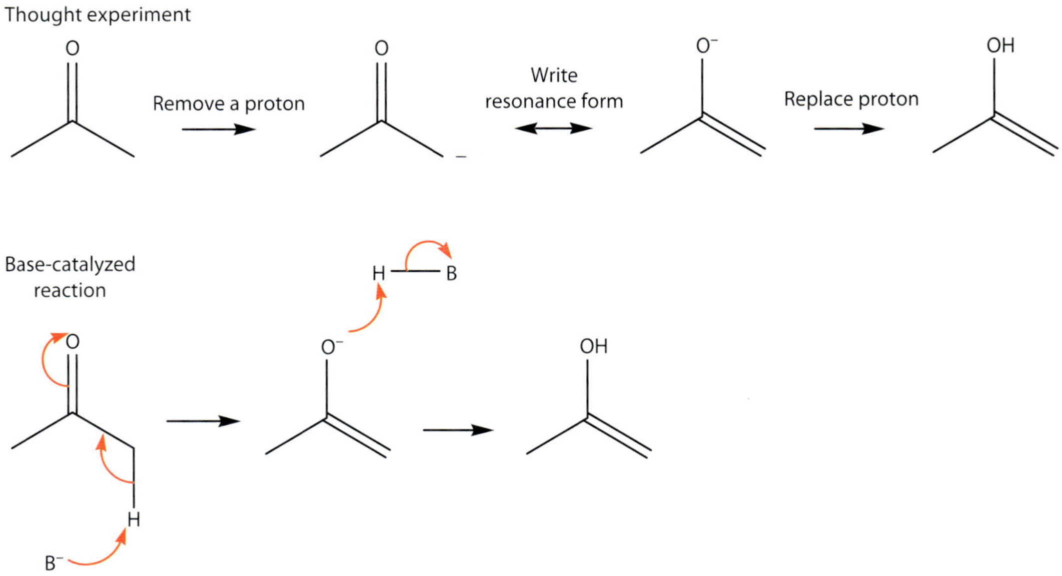

FIGURE 8.33 Tautomerization of acetone.

groups and then write the resonance form with the charge on oxygen. Replacing the proton on oxygen gives the enol tautomer. In practice, ketone/enol tautomerization may be catalyzed by either acid or base, a process that we will study in more detail in Chapter 17. However, the base-catalyzed reaction strongly resembles the "thought experiment" (Figure 8.33). At equilibrium, for a simple ketone such as acetone, there is only 6×10^{-7} % enol at equilibrium, but as we will see later, it is still very important in the reactivity of acetone.

Other carbonyl compounds have more of the enol form present at equilibrium. For example, pentane-2,4-dione, **8.11a**, contains up to 90 % enol, **8.11b**, at equilibrium, depending on the solvent. The enol form is stabilized both by conjugation of the double bonds and hydrogen bonding. An example where one tautomer is completely dominant is phenol, **8.12a**. Although two distinct and sensible keto tautomers, **8.12b** and **8.12c**, can be written, neither of these are present at equilibrium. However, as we will see later, they give a useful clue to the reactivity of phenol with electrophiles. In principle, though not always in practice, tautomers can be separated. Under strictly neutral conditions, and at low temperature, **8.11b** is isolable. Where the carbonyl compound is conjugated with double bonds as in **8.13a**, a more remote proton can be removed, to **8.13b**. Moving the electrons along the chain gives a stable resonance form of the anion, **8.13c**, and replacing the proton gives the enol **8.13d**.

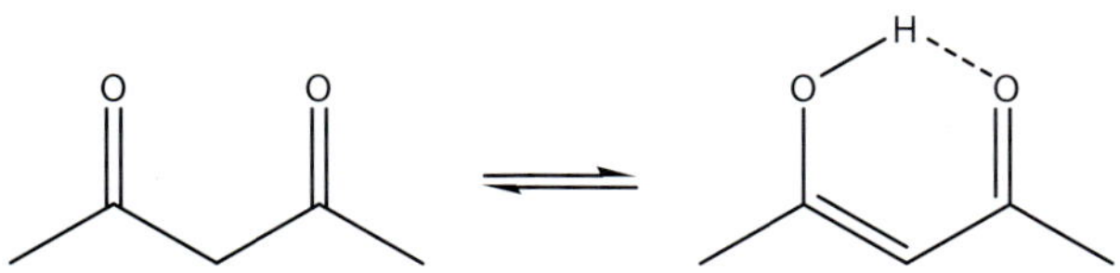

8.11a **8.11b**

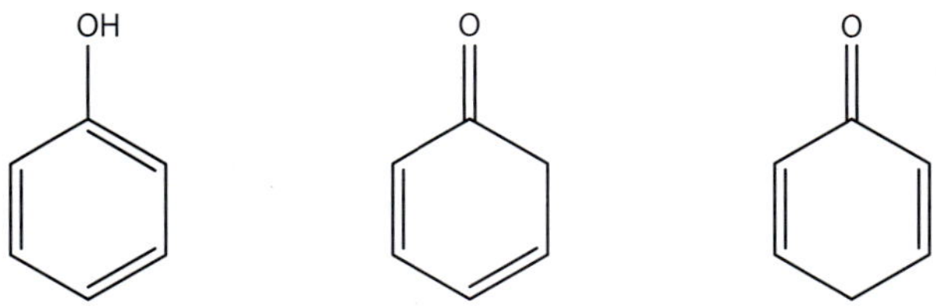

8.12a **8.12b** **8.12c**

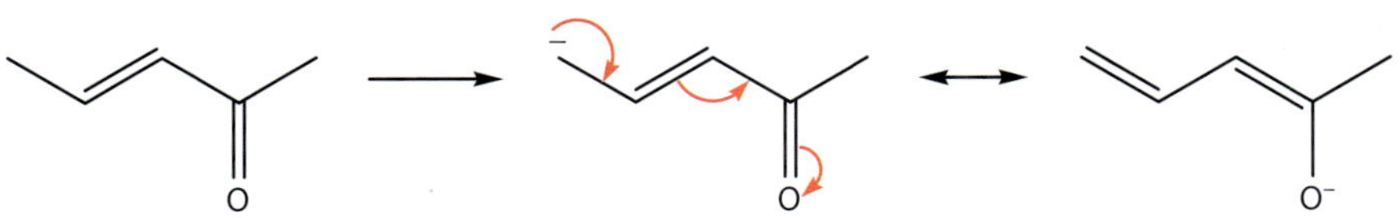

8.13a **8.13b** **8.13c**

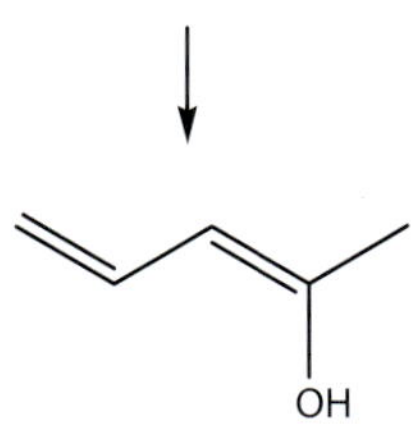

8.13d

There are many other molecules where it is possible to write tautomers. The tautomerism between 2-hydroxypyridine, **8.14a**, and 2-pyridone, **8.14b**, is a model for our understanding of tautomerism in the bases of DNA and RNA. In writing tautomers, if there is an NH or OH attached to a double bond, it is often good to start by removing that proton in your thought experiment. However, removing a proton from carbon can also be useful as in the tautomerization of the nitroso compound, **8.15a**. The tautomer here is described as an oxime, **8.15b**—and if you had started with the oxime, the proton removed would have been that attached to oxygen.

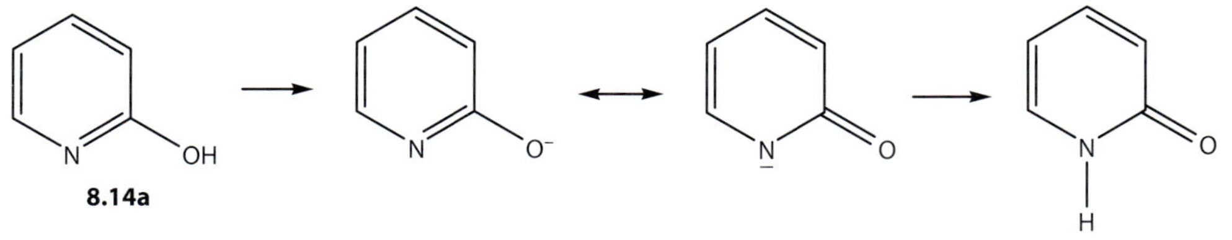

8.14a **8.14b**

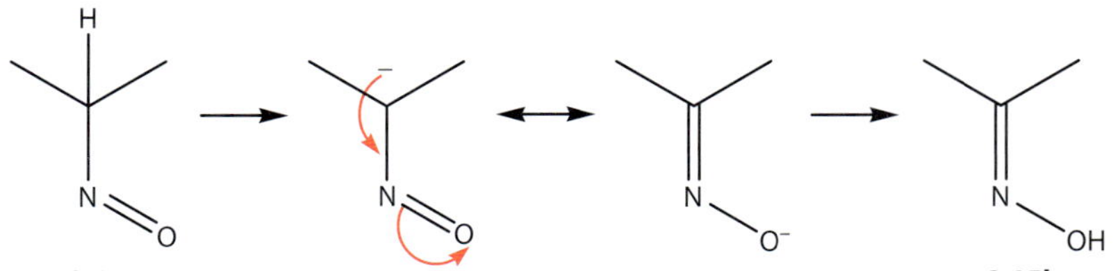

8.15a **8.15b**

PROBLEM 8.9

Give a tautomer for each of the following species and comment on the likely relative stability of the species you produce.

(a)

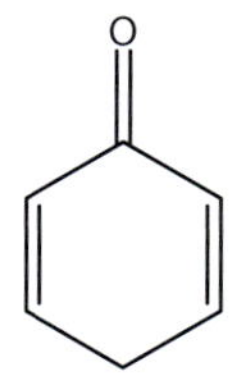

(b)

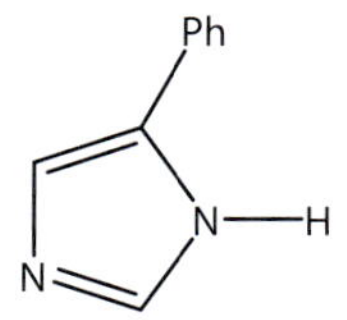

(c)

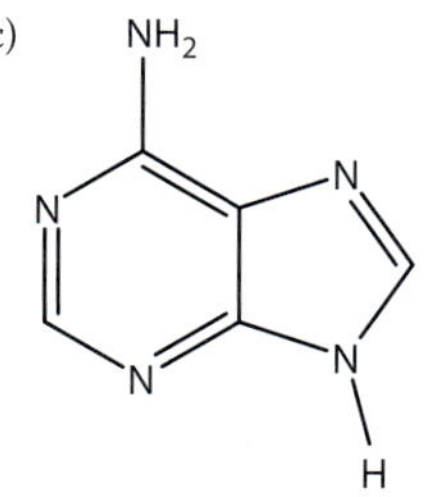

Adenine

(d)

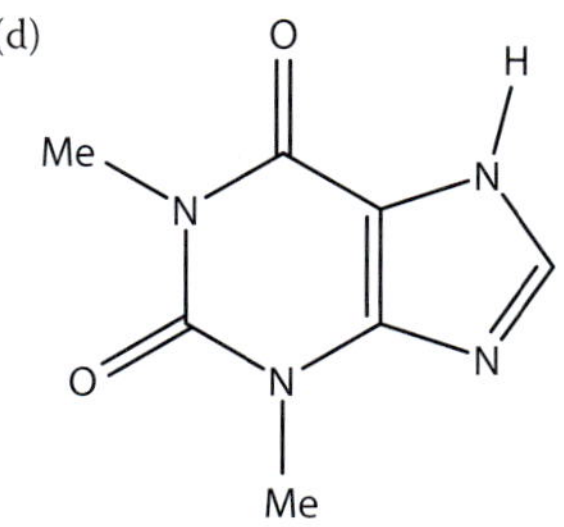

Theophylline

Solutions

(a) This one should be easy; the tautomer is phenol, and the equilibrium is entirely in favor of the aromatic phenol.

(b) Remove the proton on nitrogen, write the resonance form with the charge on the other nitrogen, and replace the proton.

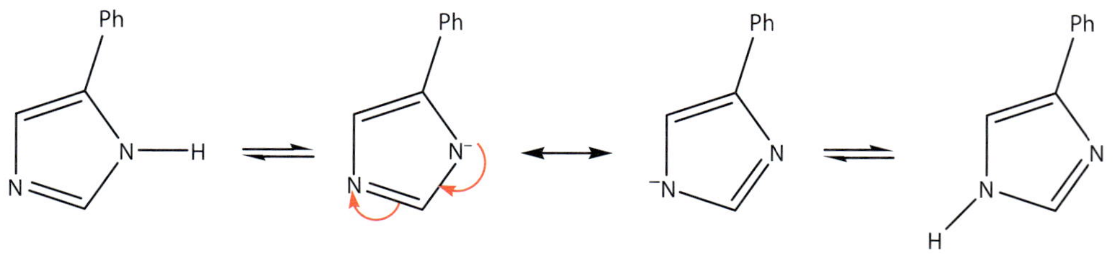

Both tautomers have reasonable structures, and the equilibrium between them is probably fairly equal.

(c) This question will teach some strategy as well as a solution. When the question simply asks for the structure of *a* tautomer, then *always* give the simplest answer, which involves moving the proton between the two nitrogen atoms in the five-membered ring (as in (b)).

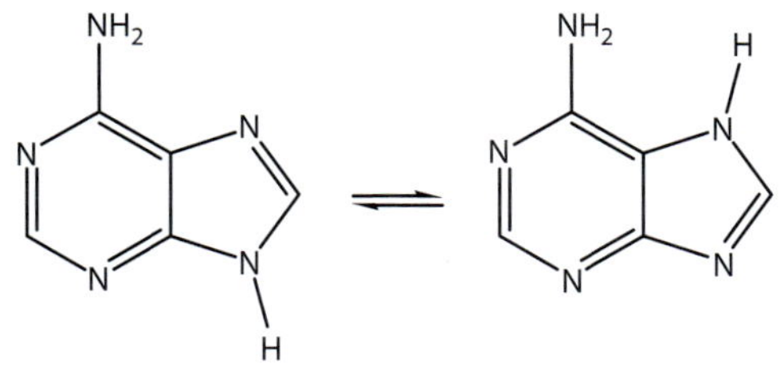

However, if the question also asks for a reason for the relative stability of the two tautomers, taking this solution gives a problem; the reason for the difference between these two tautomers is quite subtle. You might then be better advised to give the imino tautomer of the aromatic amine.

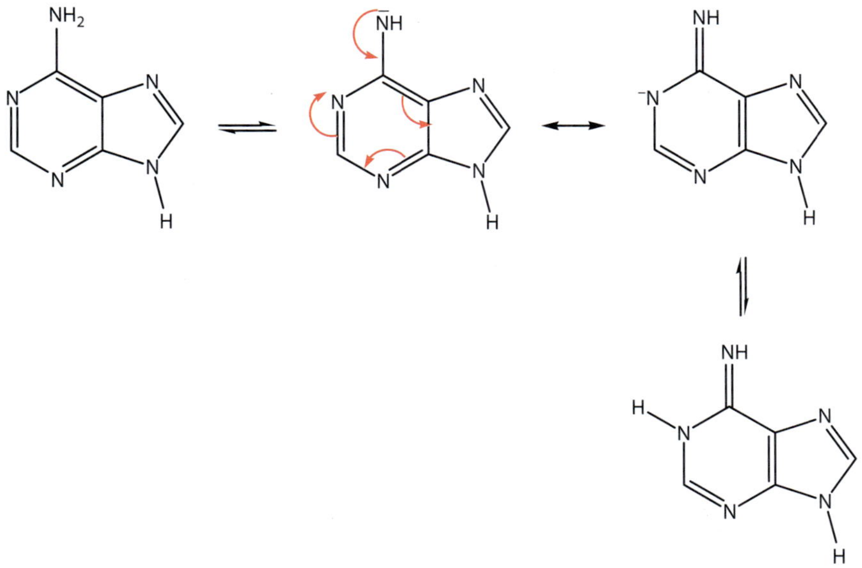

The argument for the amino tautomer is now relatively easy; the strength of the carbon–nitrogen double bond is insufficient to compensate for the loss of aromaticity in the six-membered ring.

(d) For theophylline, there is only one choice—there are no protons on nitrogen in the six-membered ring, so we must move a proton from one nitrogen atom to the other across the five-membered ring.

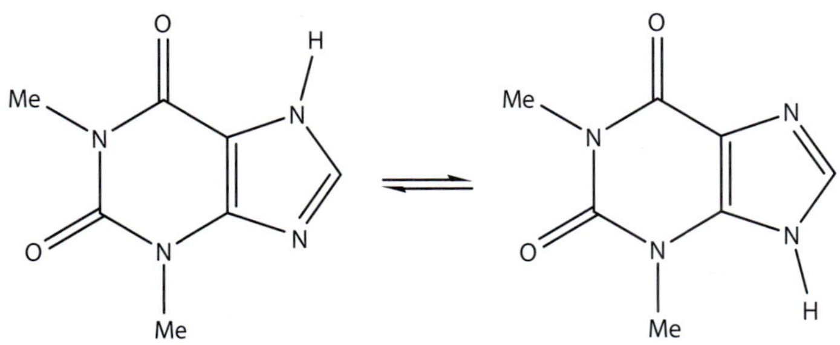

The most plausible explanation for the greater stability of the first tautomer, at this level, is that there is a (long) intramolecular hydrogen bond—but it's not entirely convincing.

Key Points from Section 8.5

- Tautomers are formed by the movement of protons and electrons.
- To generate a tautomer, remove a proton to give an anion, write a resonance form of the anion, and then replace the proton.
- The proportions of particular tautomers at equilibrium depend on the relative stability of the various structures, but a low proportion of a particular tautomer does not rule out its being important in the reactivity of the molecule.

8.6 WRITING MECHANISMS USING CURLY ARROWS

We first encountered the curly (or curved) arrow when we looked at resonance forms, as a method to show the movement of electrons around the molecule to give new resonance forms. I have sneaked a few reaction arrows in since, but we now need to look at these more formally. The curly arrow, **8.16**, means the movement of two electrons from one end of the arrow to the other, and it is very important to draw your arrows carefully and accurately. This is one of the times that the human hand is better than any computer drawing program—computers have a clear view of what curves are allowed and what is not—but the hand has infinite flexibility. Your arrows will start where there is a pair of electrons, on a species that we describe as a nucleophile—a nucleus seeker. Nucleophiles are species that seek either a positive charge or a region of electron deficiency. In principle, any pair of electrons can be used as a nucleophile, but some work much better than others. The arrow should end where a new bond is made or on an atom to which you are giving a pair of electrons.

8.16

One type of nucleophile is a molecule such as water, or ammonia, that has a lone pair of electrons. The simplest of all reactions is protonation (Figure 8.34). In each case, the arrow starts on the lone pair and ends in the gap between the heteroatom and the hydrogen—these electrons are being used to make the new bond, which is shown in red. These should be very familiar reactions—the protonation of ammonia to give the ammonium ion and the protonation of water to give the hydronium ion, H_3O^+. The charge in these species, as you know, is on the central atom—two electrons have been taken from the original neutral species. Note also that these reactions, and all other curly arrow mechanisms, are charge balanced—we started with one positive charge in the reactants, and there is one charge on the product. This may sound obvious, and in these examples, it is, but in more complex mechanisms, it's a useful check that you have done it correctly.

Students sometimes ask whether it is important exactly *where* the lone pair of electrons should be positioned on the heteroatom. Ideally, you might draw out ammonia or water in full and put the lone pair exactly where the structure demands. However, we rarely do this, and given that, the answer is that, within reason, you should put the lone pair where it is clear and convenient. Thus, the reaction of ammonia could have been reasonably written as in Figure 8.35; this means exactly

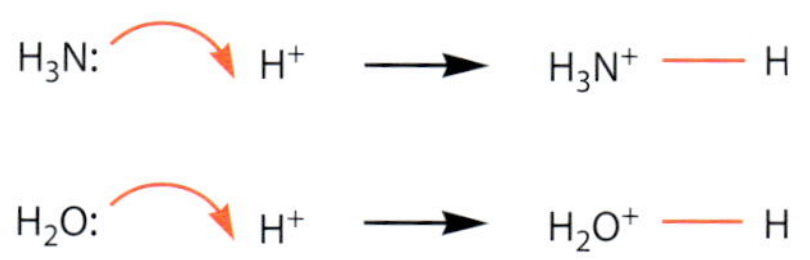

FIGURE 8.34 Protonation of nucleophiles.

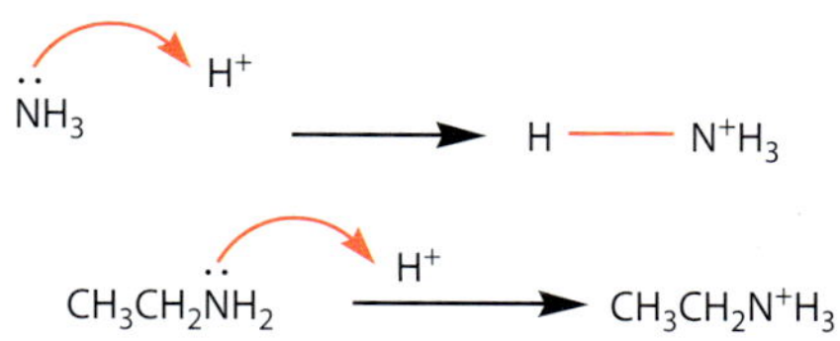

FIGURE 8.35 Drawing the protonation of amines.

the same thing as the version in Figure 8.34. However, if we want to protonate ethylamine, it is easier to put the lone pair above or below the nitrogen; the diagram would come out rather oddly if you tried to put it vertically next to the nitrogen. It's important to do enough practice with simple examples (and have someone check over your answers) to feel confident of the difference between a stylistic variation and a mistake.

PROBLEM 8.10

Draw mechanisms for the protonation of H_2S, CH_3OH, Ph_3P, and diethyl ether.

Solution

All of these molecules have a lone pair of electrons on the heteroatom, and this is what is used to make the new bond to hydrogen (shown in red).

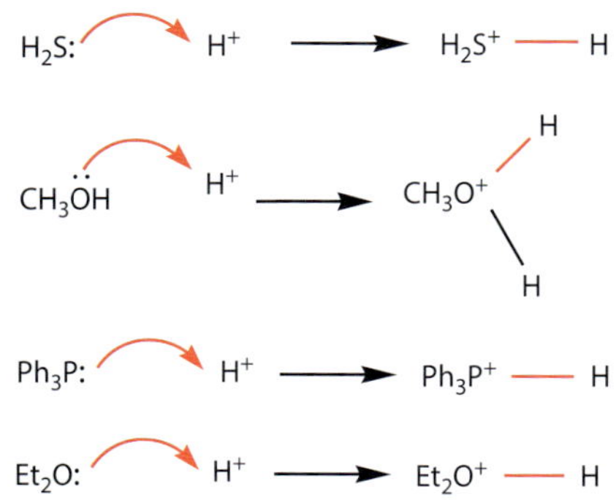

Some nucleophiles are negatively charged, and in these, our arrow will start on the negative charge—this is also a lone pair of electrons not involved in bonding. Many examples are possible; this is how many of the conjugate bases of Tables 8.1 and 8.2 are protonated to give the acid forms (Figure 8.36). Note that in each case, we start with an anion and a cation and end with a neutral molecule; thus, charge balance is preserved.

Bonding as well as nonbonding electrons can act as nucleophiles and can be used to make new bonds. One of the best examples of this involves the π-electrons of alkenes (Figure 8.37). We already know (Chapter 3) that the π-bond in an alkene is weaker than the σ-bond, so that a

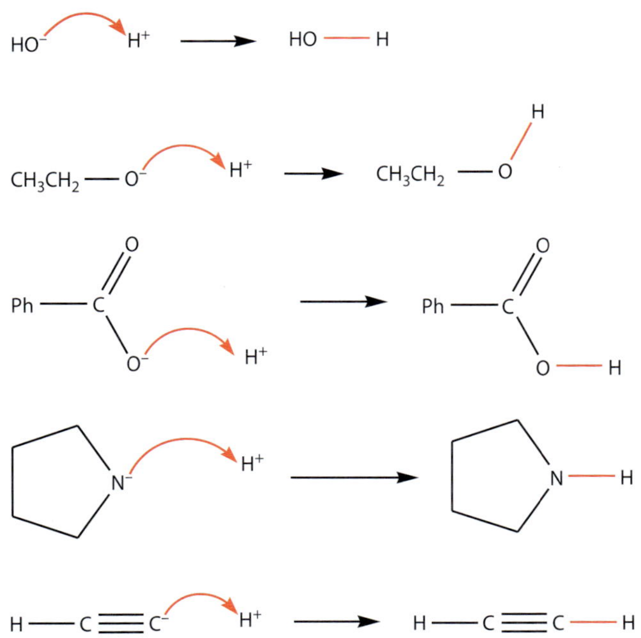

FIGURE 8.36 Protonation of anions.

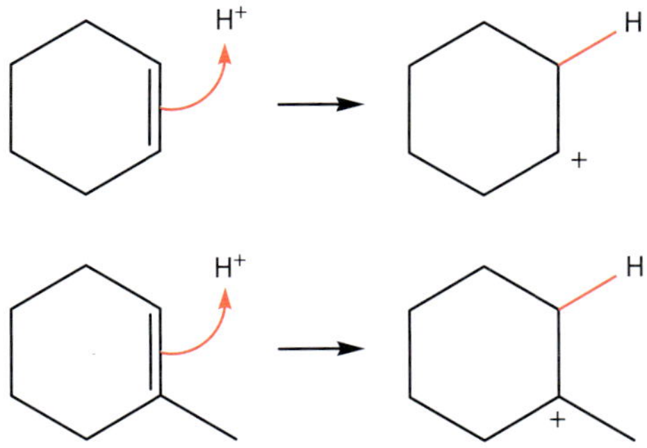

FIGURE 8.37 Using the π-electrons of an alkene as a nucleophile.

process where the π-bond is sacrificed to make a new σ-bond is likely to be reasonably favorable. Thus, the protonation of cyclohexene to give a carbocation is possible, even though the cation is a relatively unstable, high-energy, species. A new bond is made at one of the carbon atoms; the other is left electron deficient, and the positive charge is located there. When the alkene is not symmetric as in 1-methylcyclohexene, we have to decide which "end" of the double bond to protonate. We always do this so that the more stable of the two possible carbocations is formed, in this case the tertiary species.

PROBLEM 8.11

Protonate styrene (1-phenyl-1-propene) to give the more stable carbocation.

Solution

Ideally, when we draw arrows of this type, they should "emerge" from the double bond through the atom at which the new bond is made; this is much easier to achieve with hand drawing than with any computer package. We must consider all features that stabilize carbocations when judging which will be more stable. Both of the cations that can be produced are secondary, but the benzylic cation is much more stable than the alternative, because of the possibilities of resonance forms involving the π-electrons of the aromatic ring.

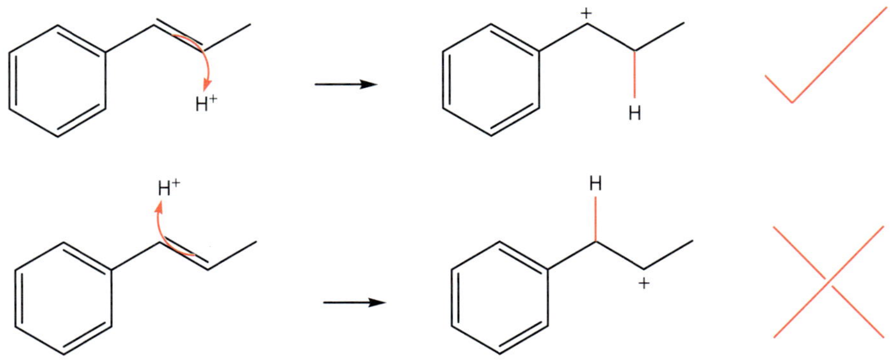

So we have used lone pairs of electrons, negative charges, and double bonds as sources of electrons, or nucleophiles. Each time, we have chosen to react these nucleophiles with a specific cation, the proton, at the other end of our arrow. While this is a common process, and we will meet many examples of it, the reaction also works well with other cations (Figure 8.38).

You should also remember from general chemistry that some molecules are electron deficient, though not charged—these are Lewis acids such as BF_3 or $AlCl_3$. These can also react with nucleophiles (Figure 8.39). The first two examples show Lewis acids reacting with halides to give anions at boron or aluminum. Aluminum hydride is also Lewis acidic (though mainly dimeric at room temperature), and its reaction with hydride ion is the key step of the synthesis of lithium aluminum hydride, a reagent we shall use often in later chapters. Note that in these first three examples, we begin with one anion and one neutral species, and the product is an anion—charge

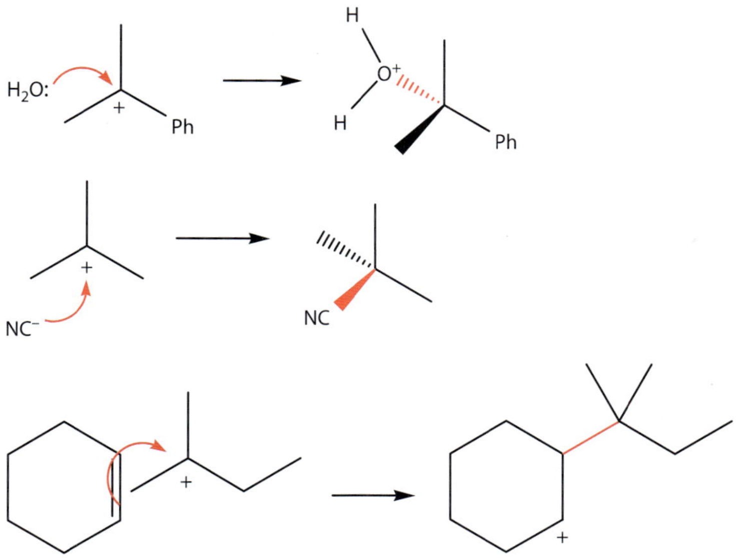

FIGURE 8.38 Reactions of nucleophiles with various cations.

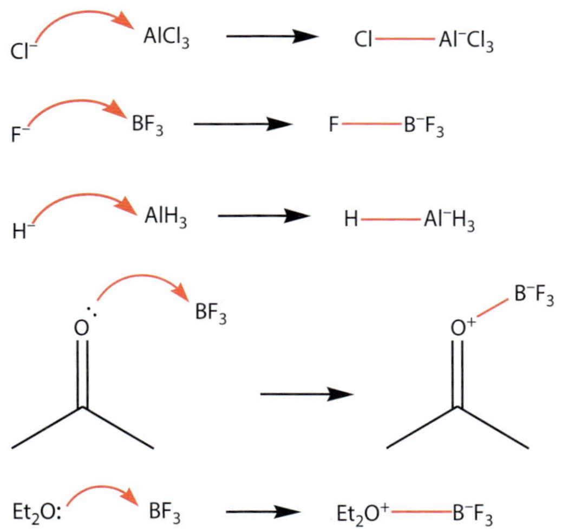

FIGURE 8.39 Reactions of Lewis acids with nucleophiles.

balance is maintained. In the last two examples, we use a species with a lone pair as the nucleophile (both species are neutral)—in each case, our product has one positive and one negative center, so again charge balance is maintained. We describe these cations and Lewis acids as *electrophiles*—electron seekers, and all the reactions we have described here, have involved bond formation from a nucleophile and an electrophile.

So far, all the mechanisms we have looked at have involved the making of new bonds. However, we already know that curly arrows can be used to show the breaking of bonds, most commonly by heterolytic fission (Figure 8.40); the bond being broken is shown in blue. In each of the first three examples, we start with a neutral species and end up with a positive and a balancing negative charge. However, other charge types are possible; the next two examples are the reverse reactions of the protonation of ketones and alkenes, respectively. Notice that in the final example, a new carbon–carbon π-bond is being made as well as the carbon–hydrogen σ-bond being broken.

The ability of some bonds to undergo easy heterolytic fission gives us another source of electrophiles—molecules can be electrophiles by virtue of their ability to break a σ- or π-bond.

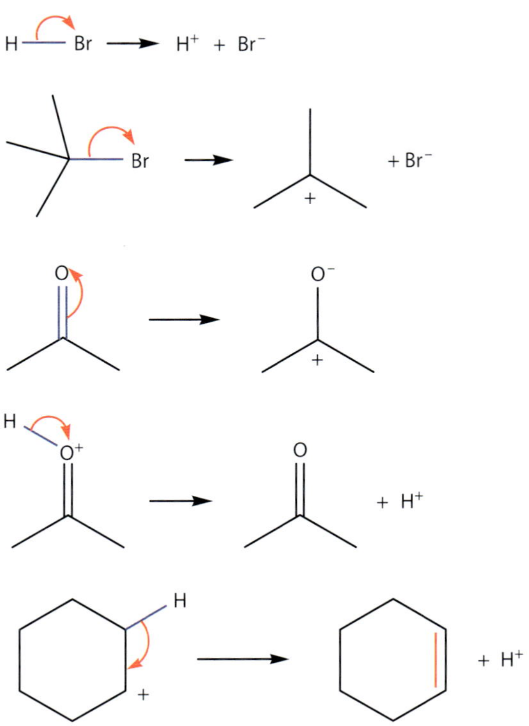

FIGURE 8.40 Heterolytic fission.

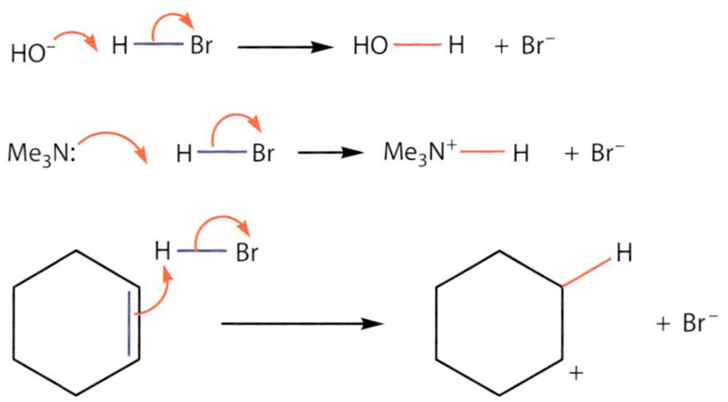

FIGURE 8.41 H–Br as an electrophile.

Thus, HBr is an excellent electrophile (Figure 8.41). This gives us an additional category of electrophile—a molecule with a polar σ-bond. Notice that we were not tempted to break the HBr bond in the opposite direction to give a hydride ion—we know that this is considerably less stable than bromide ion.

Bonds do not need to be as polar as HBr to be effective electrophiles—all that is required is some polarity and that the group being expelled (the leaving group) is stable. This is exemplified by the examples in Figure 8.42. Hydroxyl ion can displace bromide from bromomethane, because the C–Br bond is polar and bromide is a stable anion. The nucleophile attacks the positive end of the polar bond, and an anion or neutral species is expelled. We can assess the ability of a group to leave by looking at the pK_a of the conjugate acid—good leaving groups are the conjugate bases of strong acids. Thus, bromide ion (pK_a HBr = −9) and water (pK_a $[H_3O]^+$ = −1.7) are good leaving groups; hydroxyl ion (pK_a H_2O = 15.7) and cyanide ion (pK_a HCN = 9.3) are poor leaving groups. Occasionally, as in the final example, the group "leaves" but remains attached to the molecule. RO^- is not a good leaving group, but this molecule has an advantage—opening the three-membered ring relieves strain.

A smaller group of electrophiles are molecules that have relatively weak σ-bonds, such as molecular bromine (bond enthalpy is 193 kJ mol^{-1}). Chlorine, bromine, and iodine are the

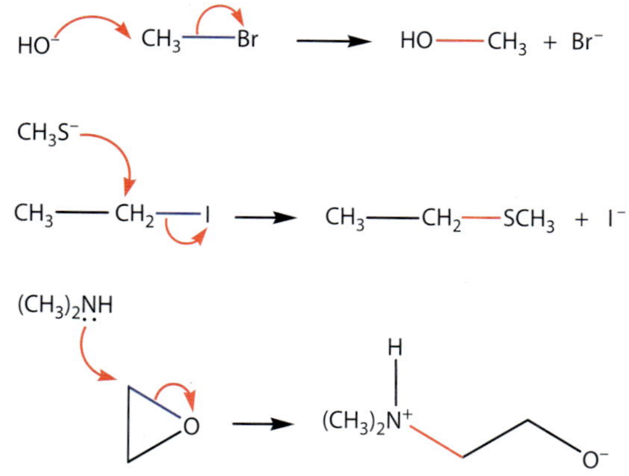

FIGURE 8.42 Molecules with polar σ-bonds as electrophiles.

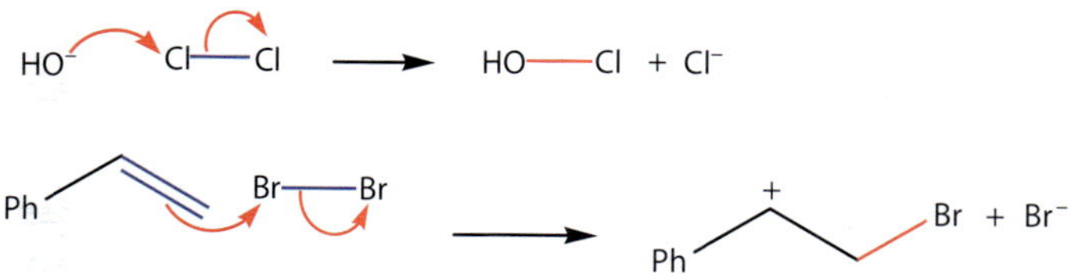

FIGURE 8.43 Electrophiles with weak σ-bonds.

common examples of this type that you will encounter (Figure 8.43). The second example introduces a new question—when styrene is reacted with bromine, which of two possible carbocations will be formed? In this case, fortunately, the answer is easy—the secondary benzylic action is much more stable than the alternative primary cation.

Polarized π-bonds are also good electrophiles—the carbon–oxygen double bond in carbonyl compounds is the most commonly seen, but carbon–nitrogen double and triple bonds are also electrophilic. When a carbonyl group reacts with a nucleophile, the oxygen stays attached, because only the weaker π-bond is broken; the stronger σ-bond is unaffected (Figure 8.44). Where there is a choice between a polarized σ-bond and a polarized π-bond, it is almost always the π-bond that is initially broken (third example of Figure 8.44), as it is weaker and more polarizable.

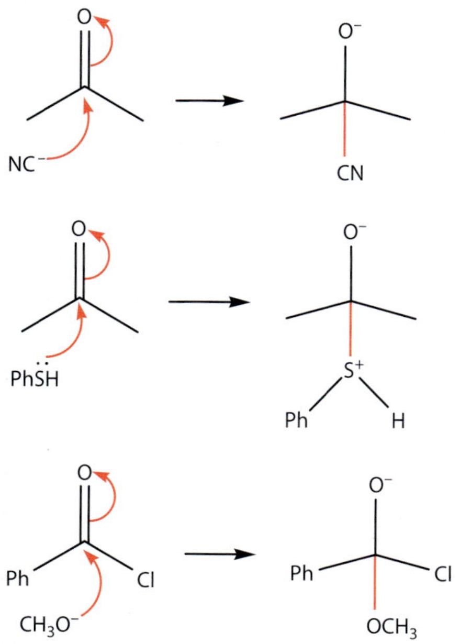

FIGURE 8.44 Reaction of nucleophiles with carbonyl compounds.

Key Points from Section 8.6

- Curly arrows denote the movement of electrons from one end of the arrow to the other. They always start where there is a pair of electrons and end where a bond is formed or on a species being expelled. Arrows may demonstrate the formation or the breaking of a chemical bond.
- Nucleophiles are nucleus (or positive charge)-seeking species and may be anions, lone pairs of electrons, or electrons in π-bonds.
- Electrophiles are electron-seeking species; they may be electron-deficient species such as cations or Lewis acids or may contain a polarized σ-bond, a polarized π-bond, or a weak, polarizable σ-bond.
- In writing mechanisms, it is important to check that electrical charge has been balanced.

PROBLEM 8.12

Provide curly arrows to show the mechanisms of the following reactions:

(a)

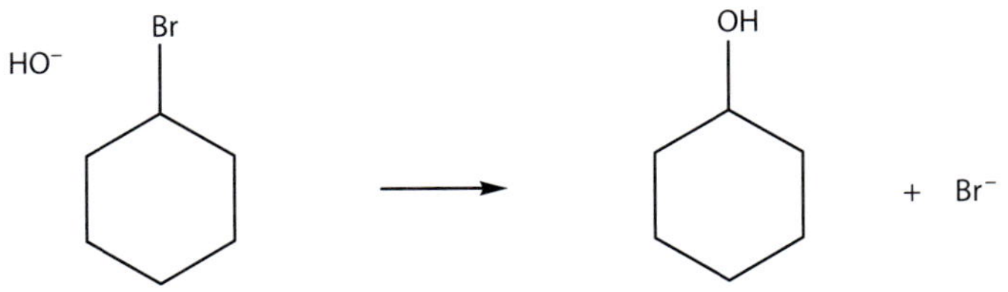

(b)

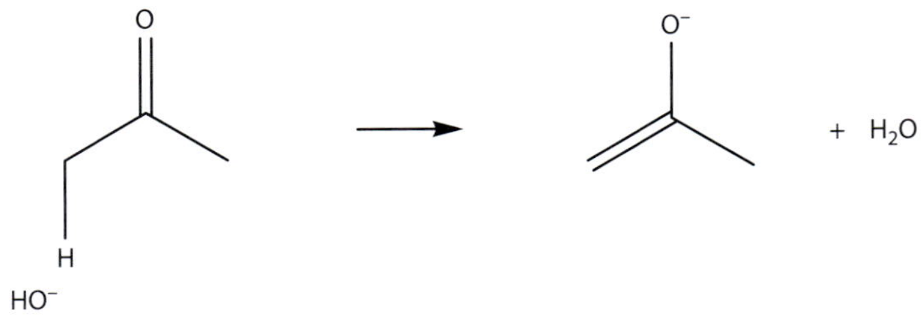

(c)

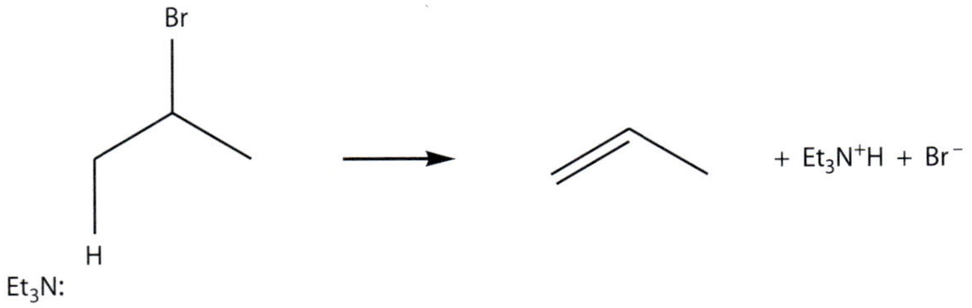

(d)

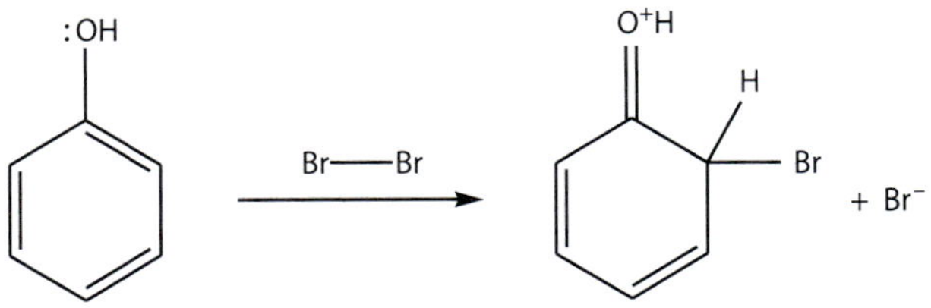

(e)

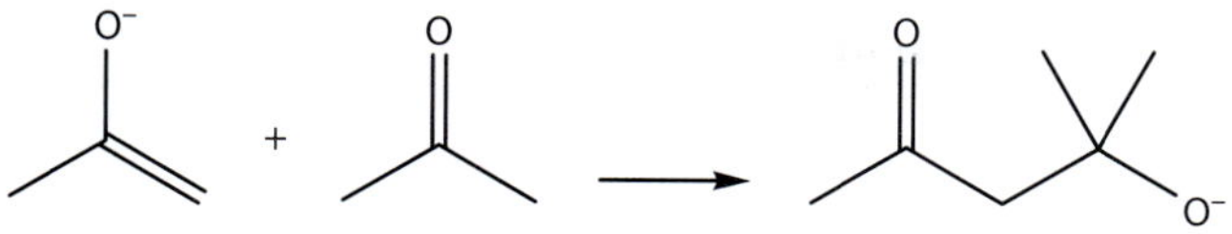

(f)

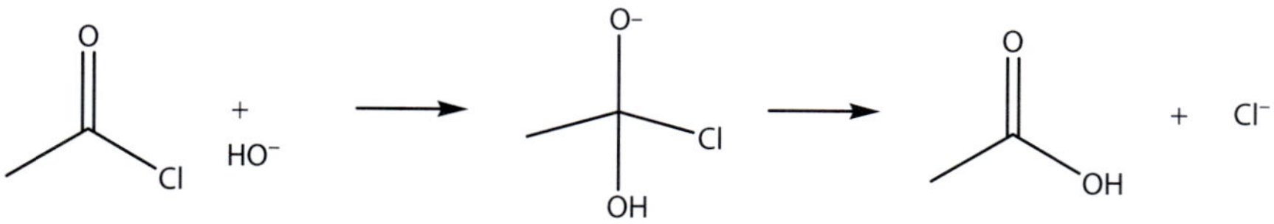

(g)

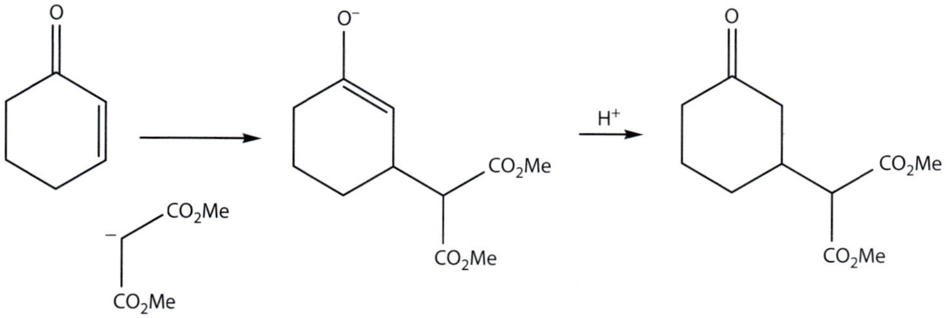

Solutions

Note first that these problems are simple exercises in arrow pushing—the reactions will all be unfamiliar to you at this point. But this is the point of the question—when you can see the products and intermediates, you can usually write a sensible mechanism, even if you are completely unfamiliar with the process. These problems are graded in that the earlier examples have been drawn in a manner that is helpful to you in figuring out the mechanisms, the later ones less so. You won't always get lone pairs drawn in helpfully for you, but with practice, you should know where to put them. The last two examples are multistep processes:

(a)

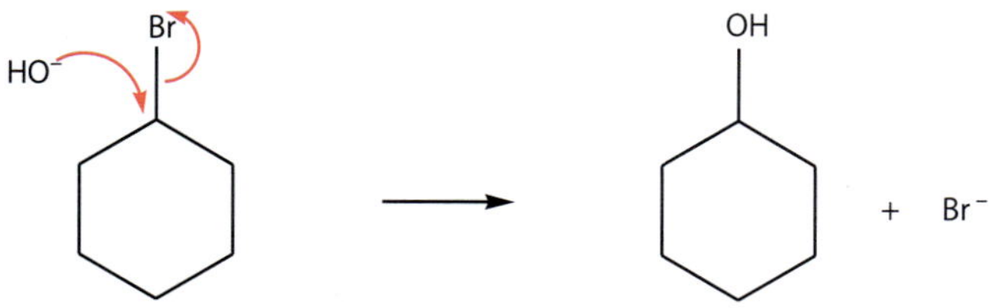

(b)

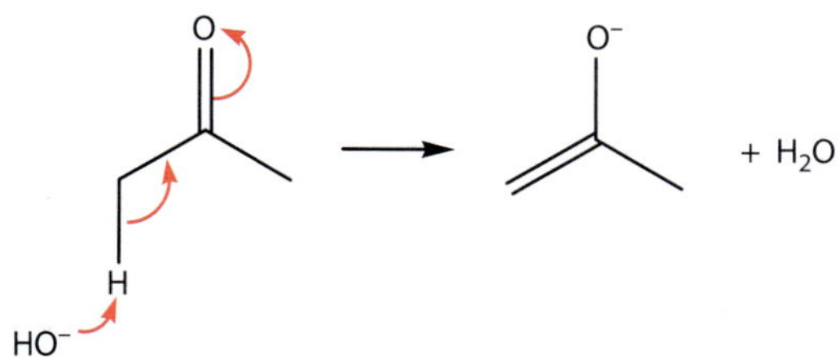

(c)

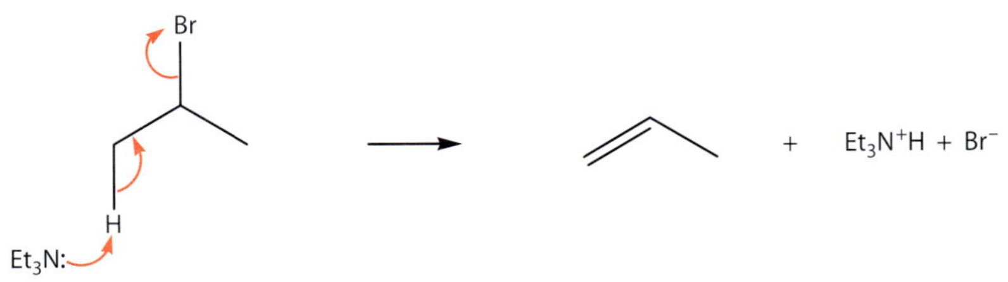

(d)

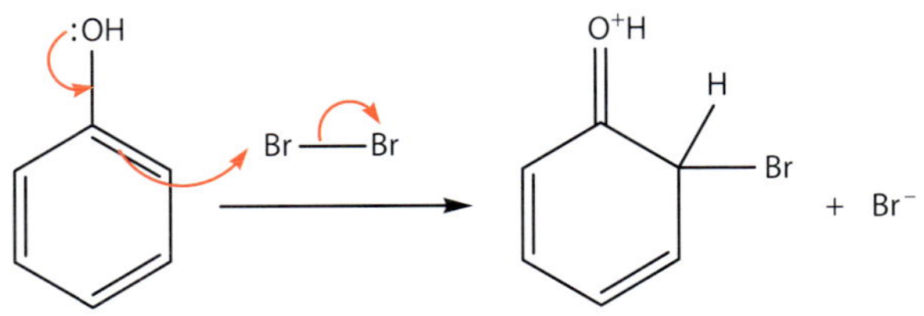

(e) Getting trickier here, as this is no longer written in the most helpful orientation. Look carefully at which bonds have been made. And while it may sound obvious, count the atoms at the start and the end—many errors result from an atom having been gained or lost in the drawing.

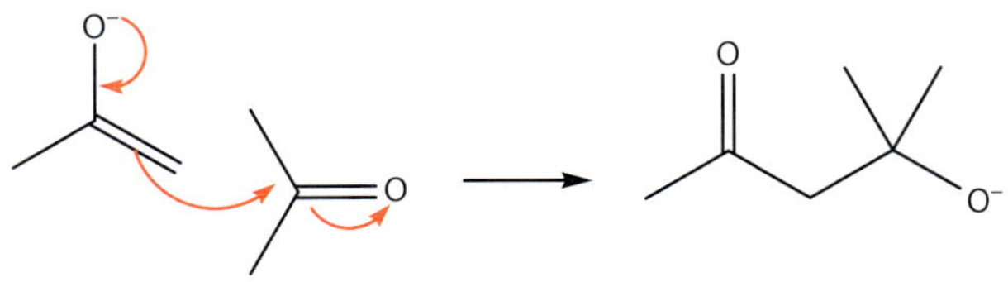

(f) The first step here is one we have seen before—the addition of hydroxyl to a carbon–oxygen double bond to give the sp^3 intermediate. The second step essentially reverses this but with a different (and better) leaving group.

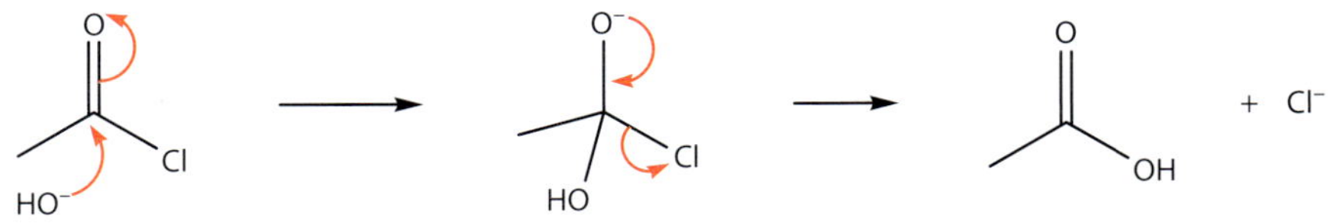

(g)

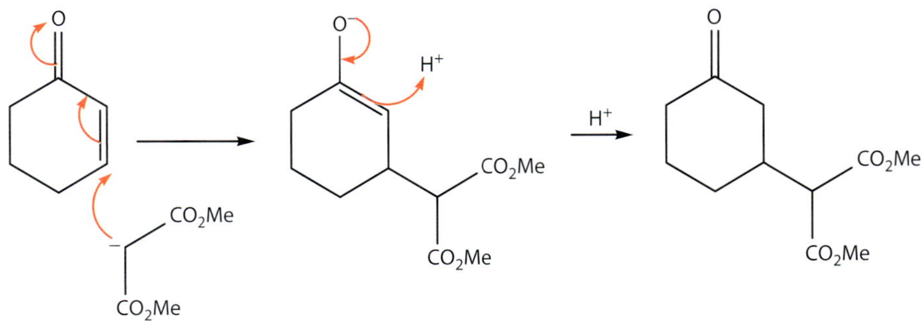

REVIEW PROBLEMS

1. Give the structure of the conjugate base of each of the following species (if there is more than one possible site of ionization, consider the easiest ionization):
   (a) $PhCH_2C(=O)Ph$
   (b) Et_2NH

(c) $H{-}C{\equiv}C{-}CH_2O^-$
(d) $CH_3CH_2CH_2COOMe$

(e)
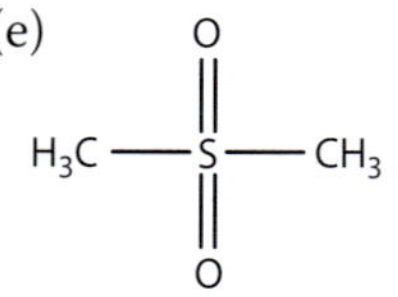

(f)
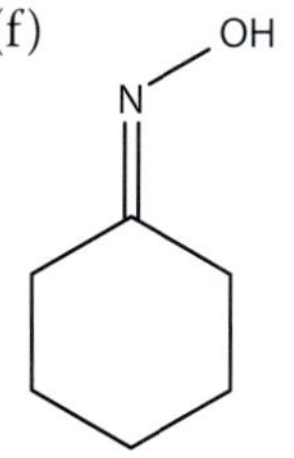

(g)
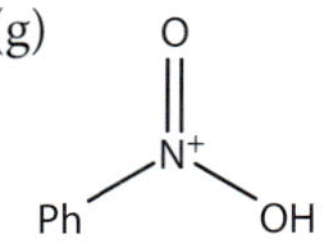

2. Give the structures of the conjugate acid formed by protonation of each of the following species:
(a) Et_3N
(b) pyridine
(c) $CH_3C(=O)NH_2$
(d) H_2NCH_2COOH
(e) CF_3CH_2OH

(f)
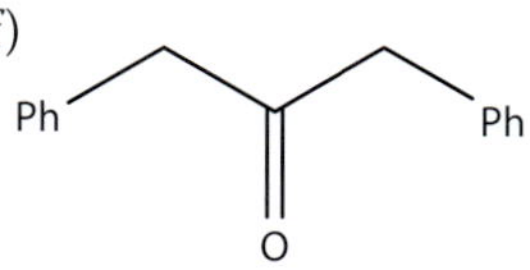

(g)
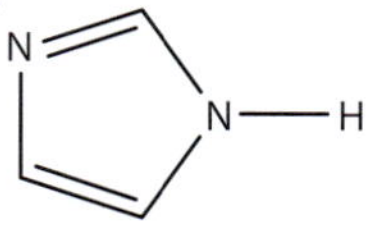

3. Explain the following sets of pK_a values:

(a)
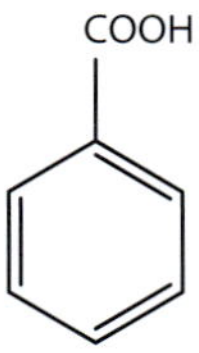
4.2

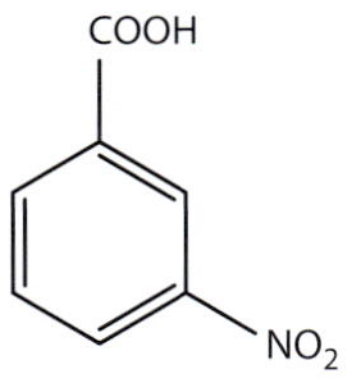
3.5

3.4

(b)
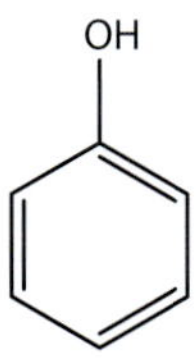
10

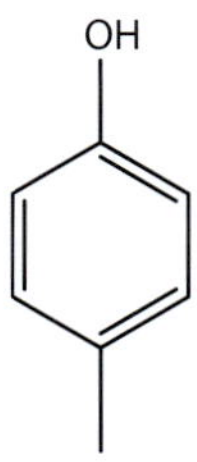
10.2

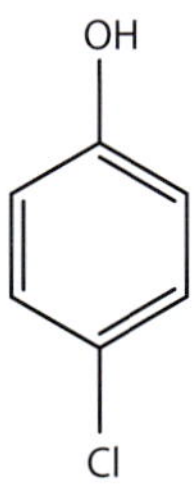
9.38

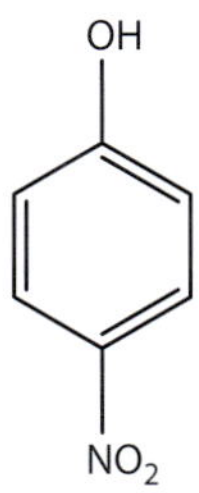
7.17

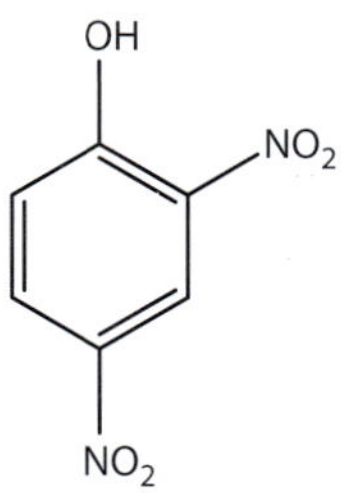
4.02

(c)
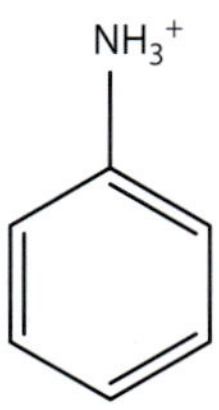
4.62

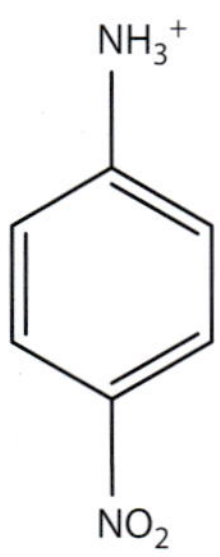
0.92

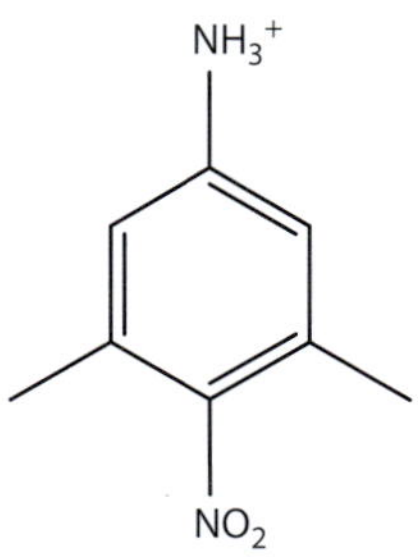
3.0

4. Explain the pK_a values for the conjugate acids of the following amines:

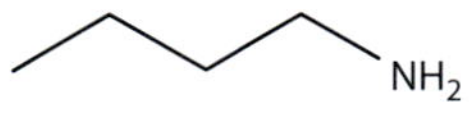

10.6

9.92

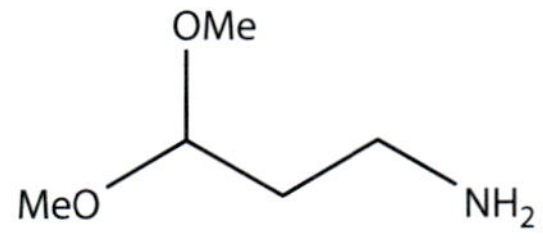

8.54

5. For each of the following pairs of compounds, state which you would expect to be the stronger acid, giving a reason:

(a)

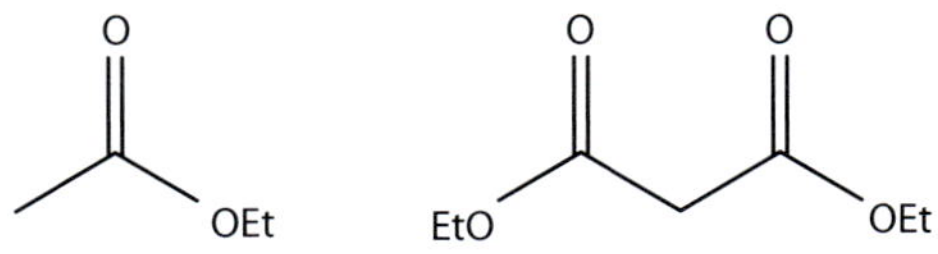

(b)

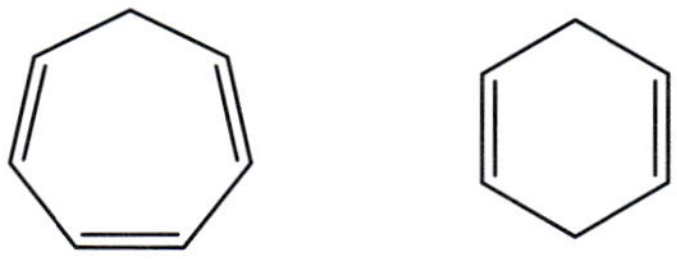

(c)

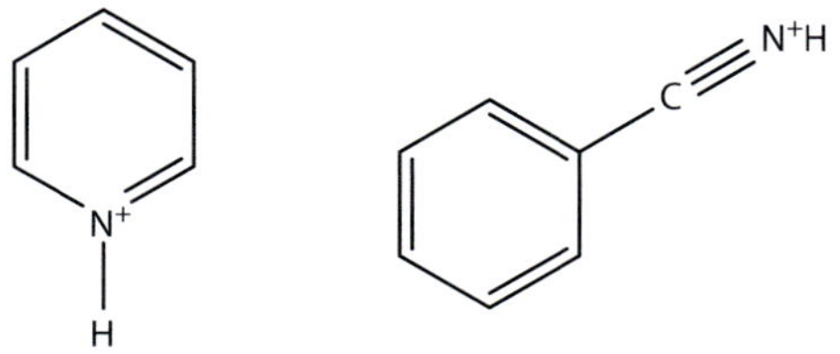

(d)

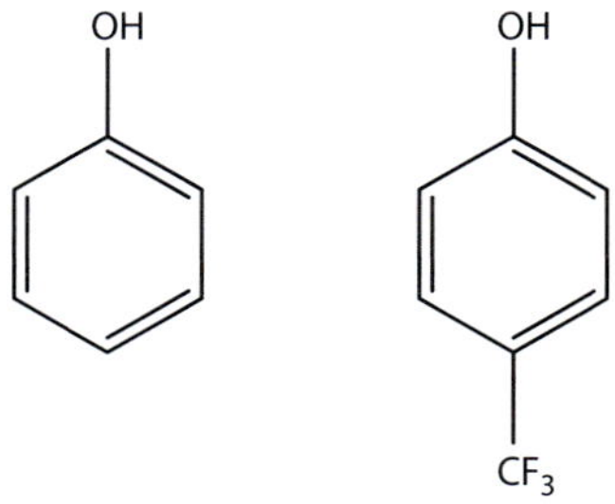

(e)

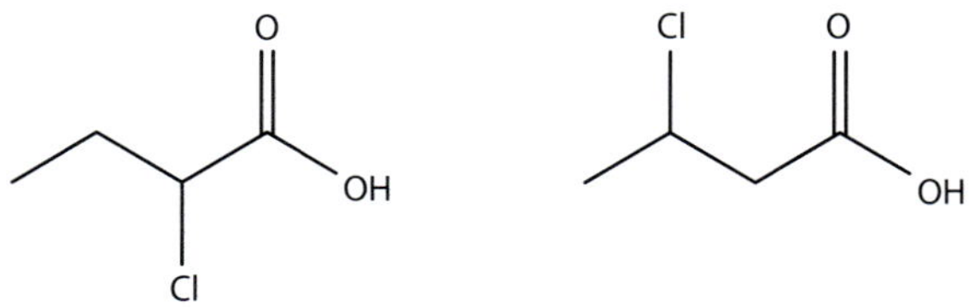

(f) CF_3COOH CFH_2COOH

(g)

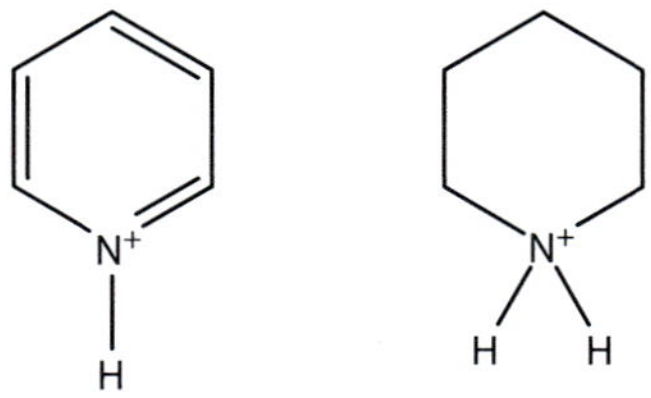

(h)

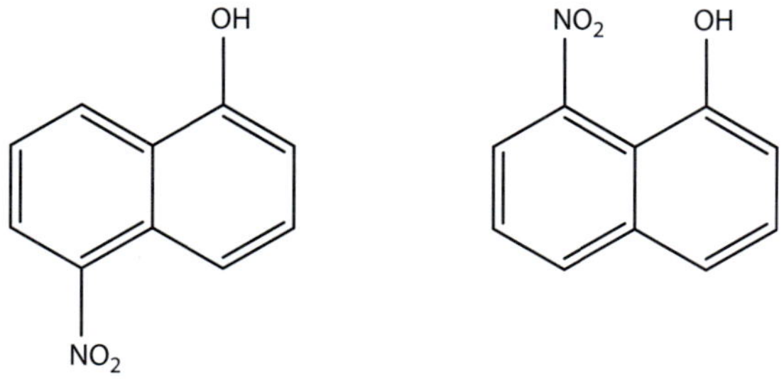

(i)

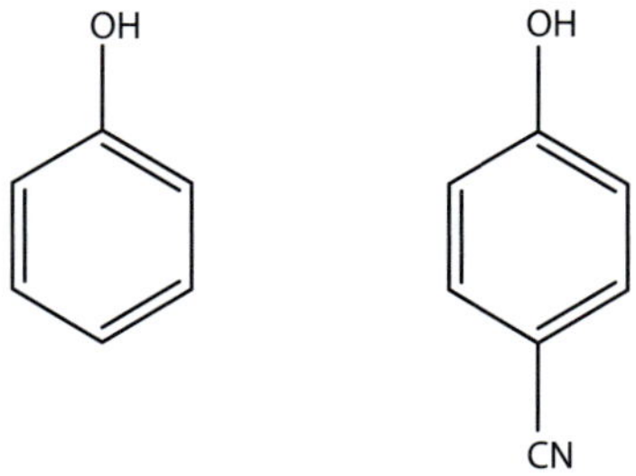

(j)

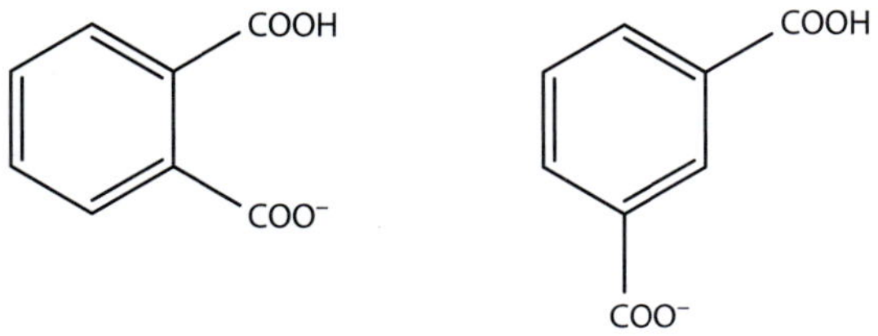

(k)

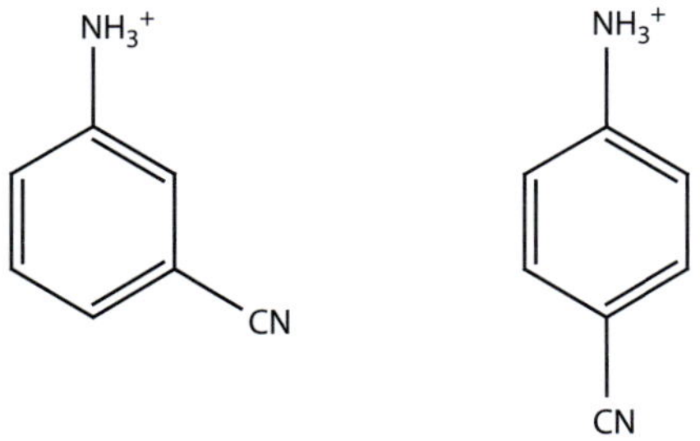

6. Give a tautomer for each of the following species and comment on the likely relative stability of the species you produce:

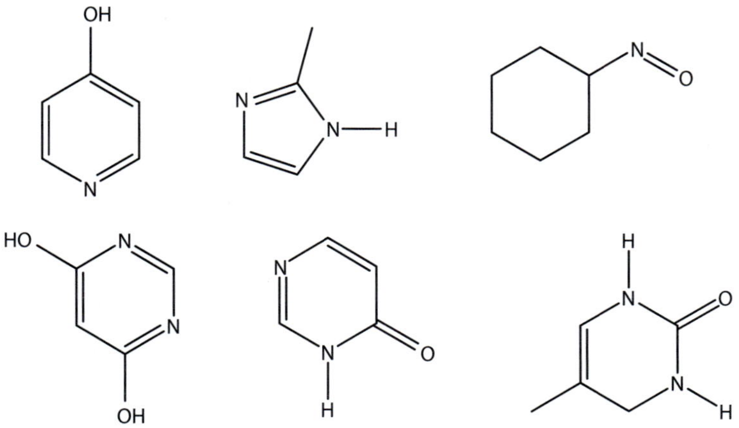

7. In the reaction

$$ROH + H^+ \rightleftharpoons R^+ + H_2O$$

at 25 °C in 0.1 M H_2SO_4, the compounds **8.17**, **8.18**, and **8.19** gave, respectively, 10^{-5} %, 10^{-15} %, and 50 % of $[Ar_3C]^+$ at equilibrium:

Ar_3COH

8.17, Ar = Ph

8.18, Ar =

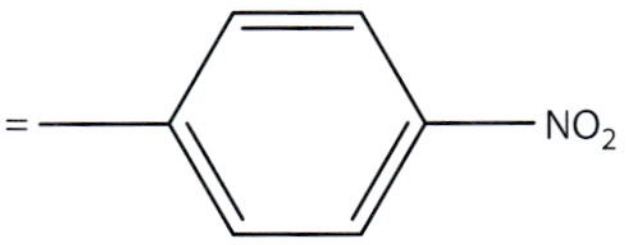

8.19, Ar =

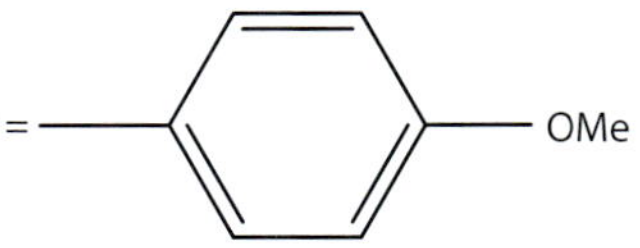

Explain.

8. Predict the position of the following equilibria and explain your reasoning:
(a)

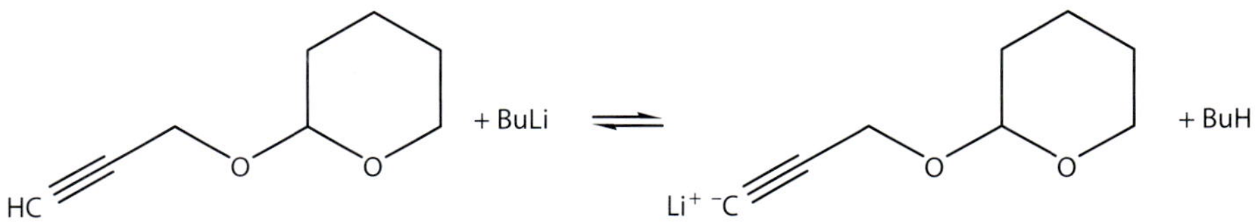

(b)

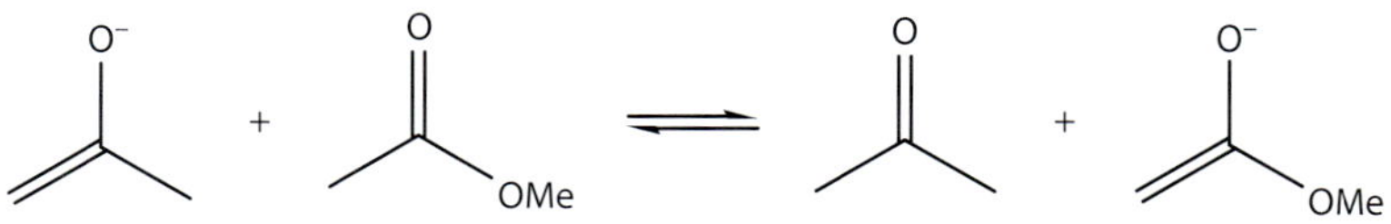

(c)

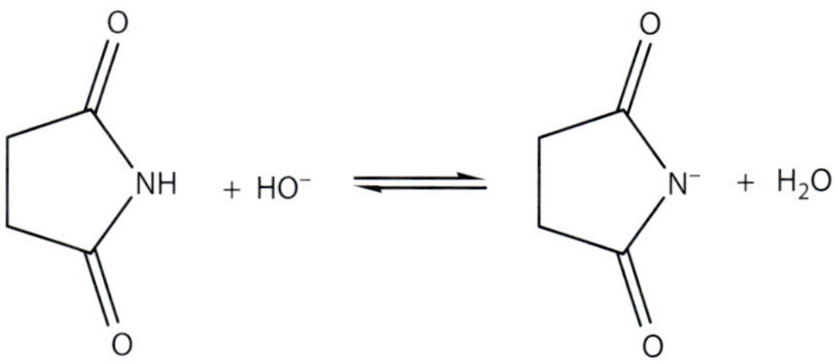

(d)

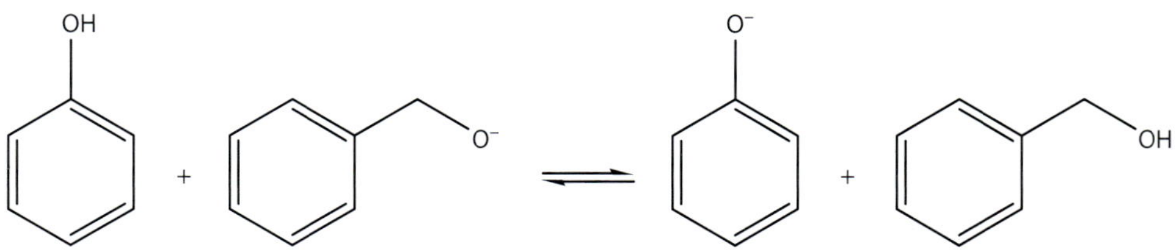

9. Circle the most acidic hydrogen in each of the following molecules:

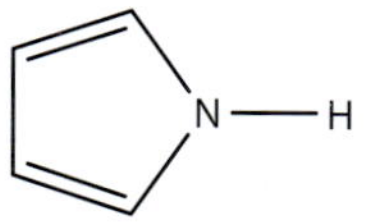

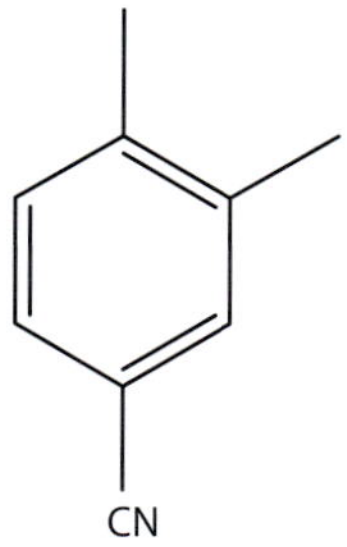

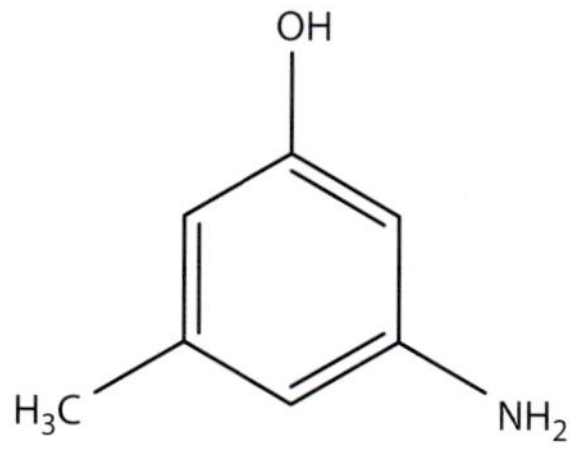

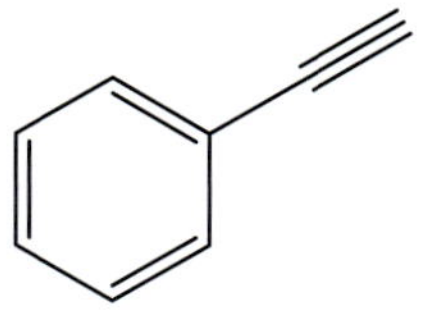

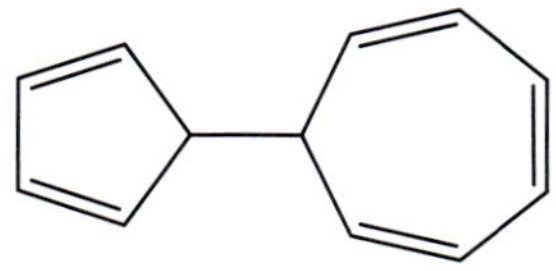

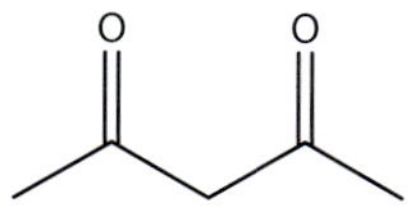

10. Explain the following observations:
   (a) Amines are basic, but amides are essentially neutral.
   (b) 7-Bromo-1,3,5-cycloheptatriene is water soluble.
   (c) Hexanoic acid is soluble in an aqueous solution of $Na[HCO_3]$ but hexanol is not.

11. For each of the following pairs of compounds, state which would be the more basic, and explain your reasons:
   (a) Me_3N $(CF_3)_3N$

   (b)

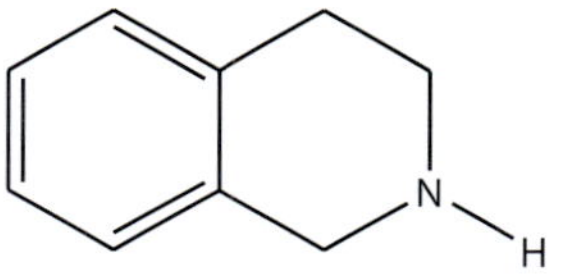

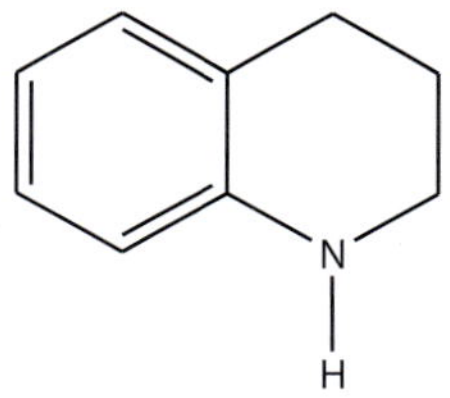

   (c)

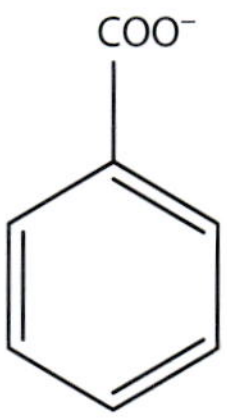

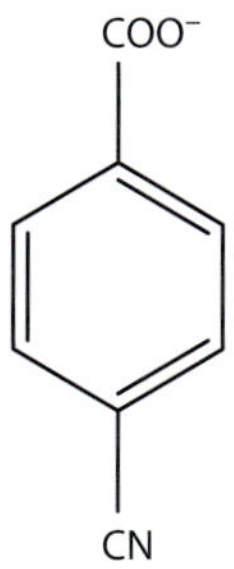

12. Provide curly arrows to show the mechanisms of each of the following processes:

   (a)

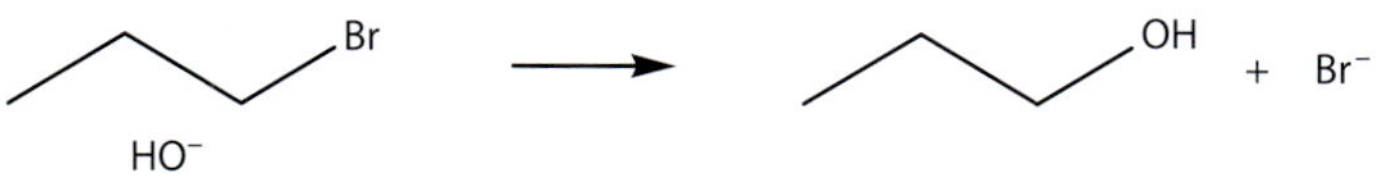

   (b)

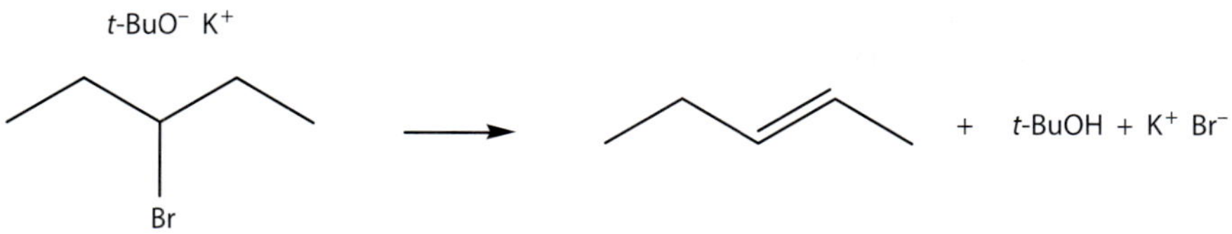

(c)

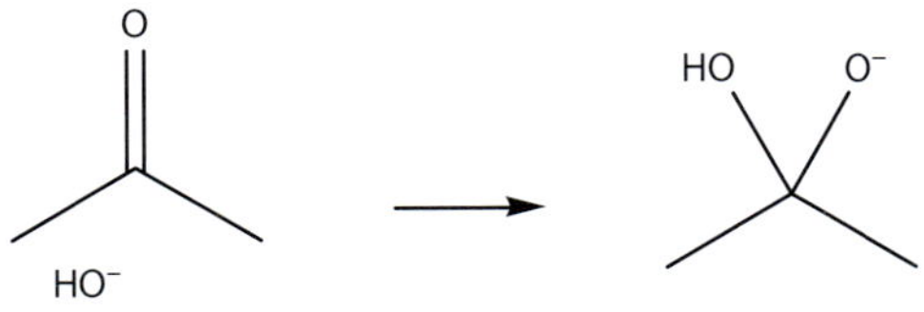

(d)

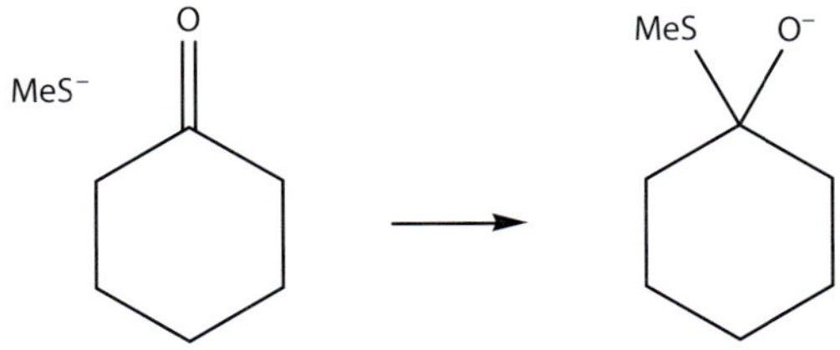

(e) $H_3C{-}Cl + AlCl_3 \rightarrow H_3C^{+}\,{}^{-}AlCl_4$

(f)

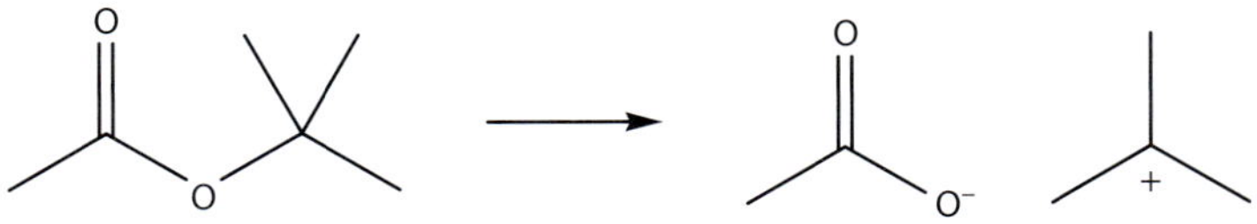

(g)

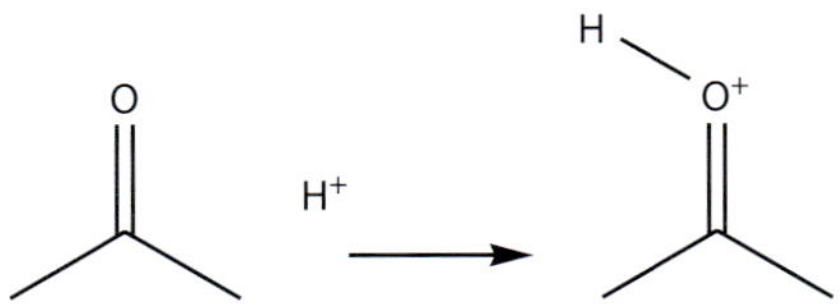

(h)

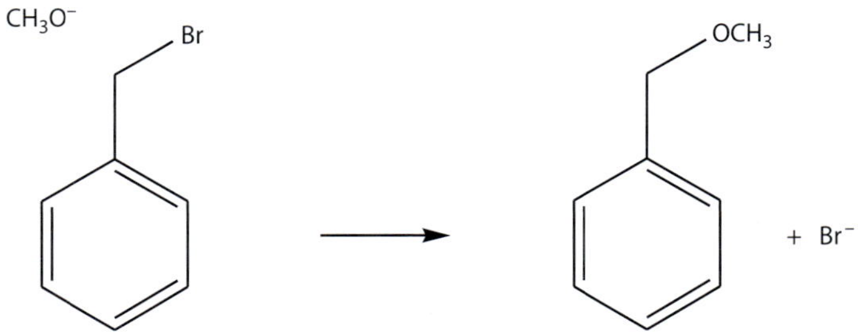

(i)

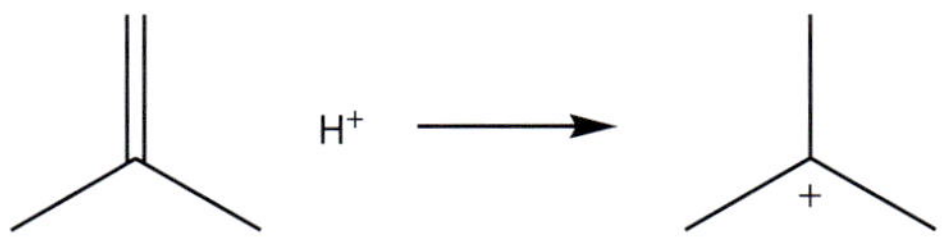

(j)

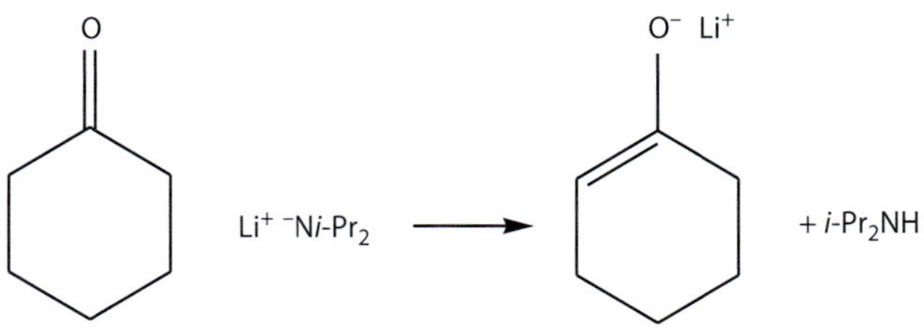

MORE CHALLENGING PROBLEMS

1. Place in relative order of acidity the following groups of compounds:

   (a)

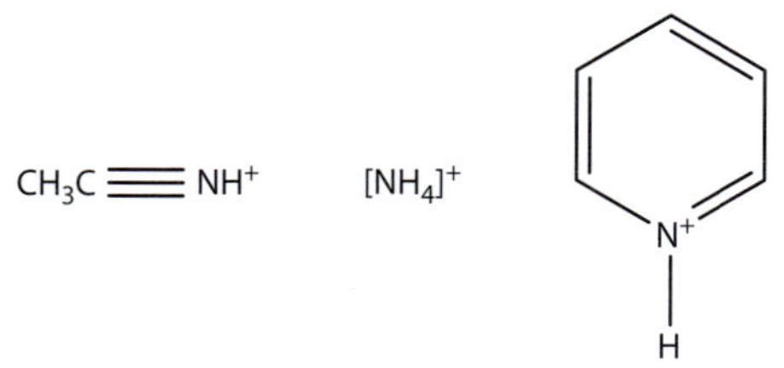

   (b) Propanone ethanoic acid ethanamide

   (c) PhOH, CH_3COOH, EtOH, Me_3CCOOH, Ph_3CH

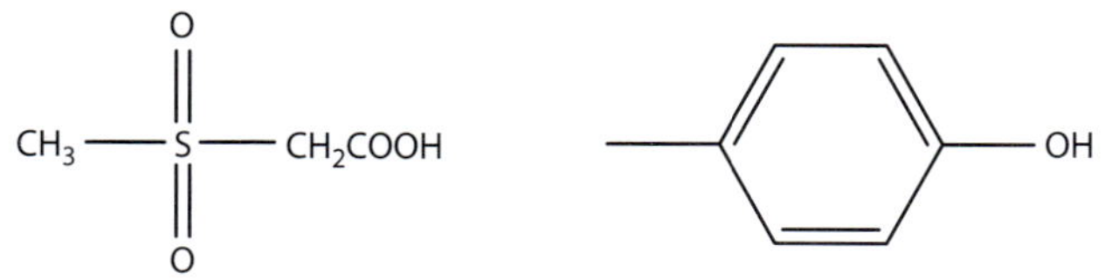

2. The molecule shown in the following is a strong acid with pK_a = 1.3. Bearing in mind that most enols have a pK_a of about 10, why is this so acidic?

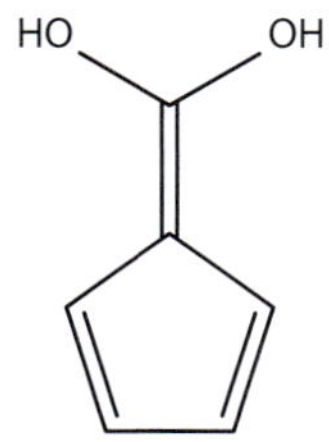

3. Give a tautomer for each of the following species and comment on the likely relative stability of the species you produce:

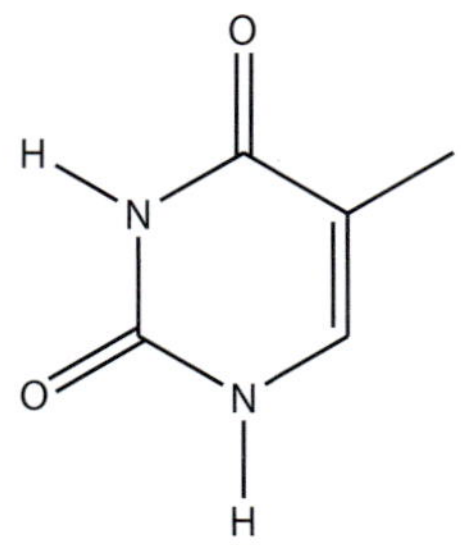
Thymine

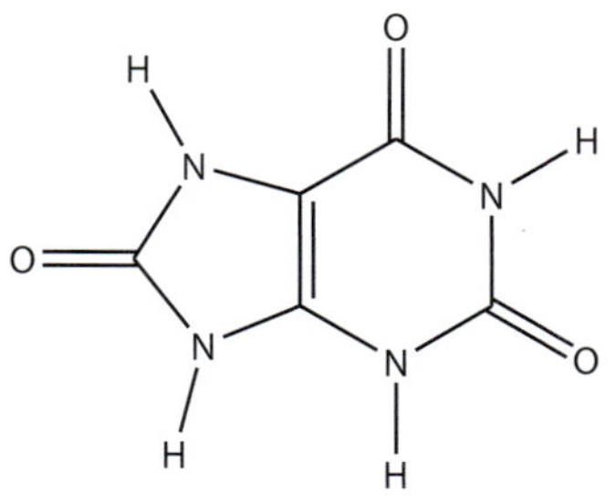
Uric acid

4. The herbicide sulcotrione has a pK_a of 3.13. Which is the most acidic proton, and why is the molecule so acidic?

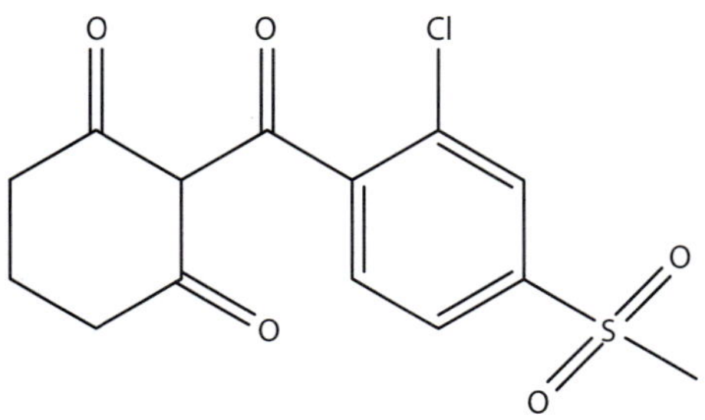

Sulcotrione

5. The amino acid alanine has pK_a values of 2.35 (the COOH group) and 9.69 (the NH_3^+ group). Write down all the possible protonated/deprotonated forms of alanine, from strongly acid to strongly basic solution. In what form would alanine exist at physiological pH (7.2) and at the pH of the human stomach (3.0)? You should use the Henderson–Hasselbalch equation to estimate the proportion of ionization, as appropriate.
6. Provide curly arrows for each of the following reaction sequences:

(a)

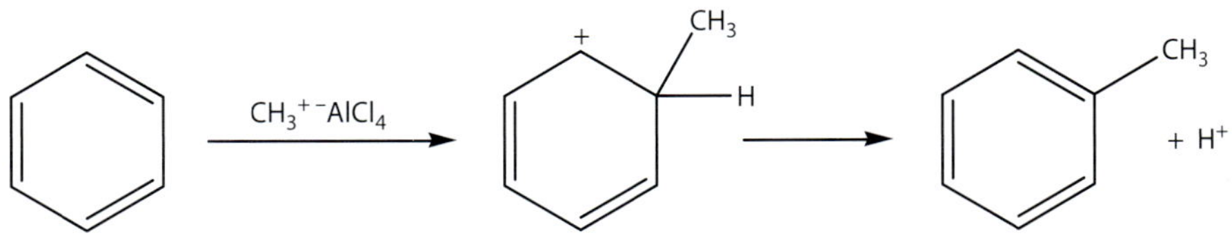

(b)

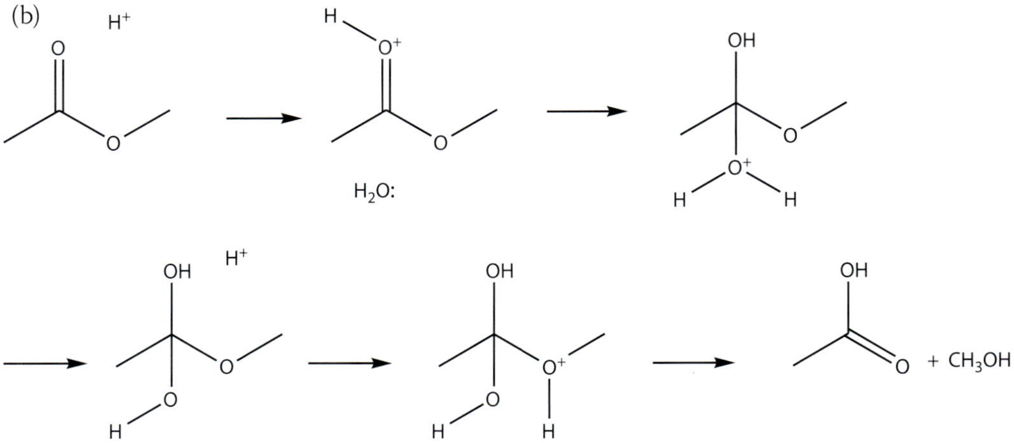

(c)

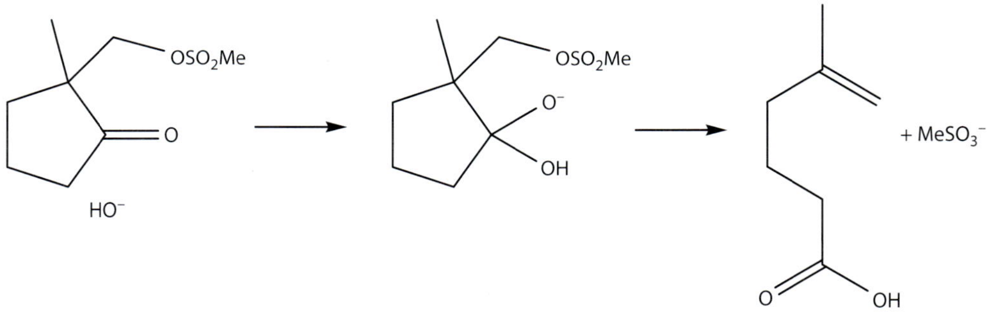

Nucleophilic Substitution at Saturated Carbon

9

9.1 INTRODUCTION

In earlier chapters, we have thought about what molecules are and how we can study their structures and spectra. Now, as we move on to think about what they do, nucleophilic substitution is the first class of reactions that we will consider in detail. The prototype reaction is shown in Figure 9.1. A nucleophile (frequently, but not always, negatively charged) attacks a substrate at a carbon atom that has on it a leaving group. The word substrate is one that we will come across often—it just means the molecule that is involved in a reaction. The leaving group, X, is replaced by the nucleophile. The simplest version of this is that both the nucleophile and the leaving group are negatively charged, but either can be neutral, and we will see examples of all the possible combinations.

9.2 MECHANISTIC TYPES

There are two mechanisms for nucleophilic substitutions, plus one that is rather special and less common. These are designated, and divided, by their kinetics. The first of these is designated as S_N1; this stands for substitution, nucleophilic, unimolecular. The rate of the reaction is proportional only to the concentration of the substrate; only the substrate is involved in the RDS. The reaction involves two steps, the first of which is rate determining. The example shown in Figure 9.2 will illustrate. In the first, rate-determining, step of the reaction, the carbon–bromine bond is broken heterolytically, to give a bromide ion and a carbocation. The carbocation is then attacked by the lone pair of electrons on a water molecule in a fast step, and finally a proton is shed from the oxygen to yield an alcohol.

Note in passing that almost all transfers of protons to and from oxygen and nitrogen are fast, and they are almost never rate determining. The stereochemistry at the start, which has disappeared in the products (the wavy bond indicated unspecified stereochemistry), is deliberate—we'll look at the reasons for this in the next section, but for now, think about the shape of the carbocation intermediate and where it will be attacked.

The energy profile for a typical S_N1 reaction is shown in Figure 9.3. The activation energy for the first step, forming the carbocation, is higher than for the second step, the attack of the nucleophile on the carbocation, so it is this step that is rate determining. Equation 9.1 describes the reaction kinetics. The carbocation is a high-energy intermediate; it is formed slowly but consumed rapidly. Although in principle it may be possible to observe intermediates in multistep reactions, that is not invariably so in practice, and observing most types of carbocation requires rather special conditions.

$$\text{Rate } (S_N1 \text{ reaction}) = k[\text{substrate}] \quad (9.1)$$

The other mechanism for nucleophilic substitution is the S_N2 reaction, substitution, nucleophilic, bimolecular. The rate of the reaction is proportional to both the concentrations of the nucleophile and the substrate, and this is a single-step process. The kinetics are given by

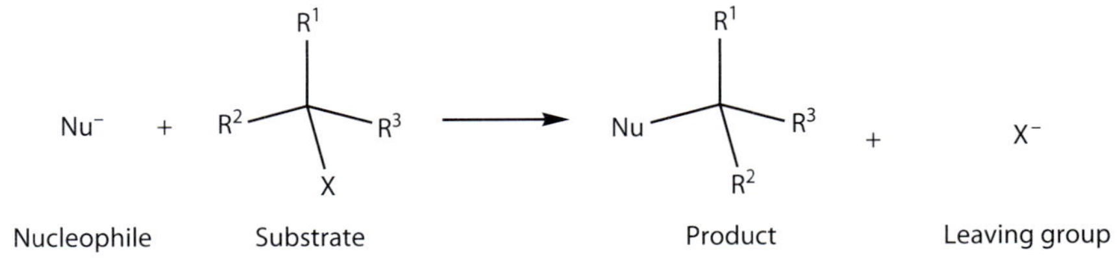

FIGURE 9.1 Prototype substitution reaction at sp^3 carbon atom. R^1, R^2, R^3 may be hydrogen, alkyl or aryl; no particular stereochemical outcome should be inferred at this point.

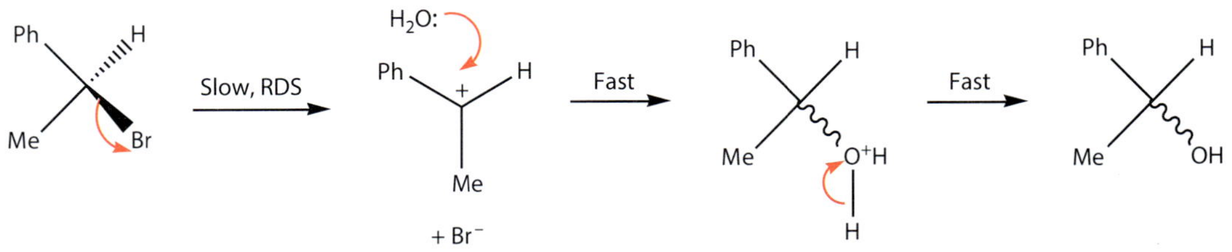

FIGURE 9.2 Mechanism of S_N1 substitution.

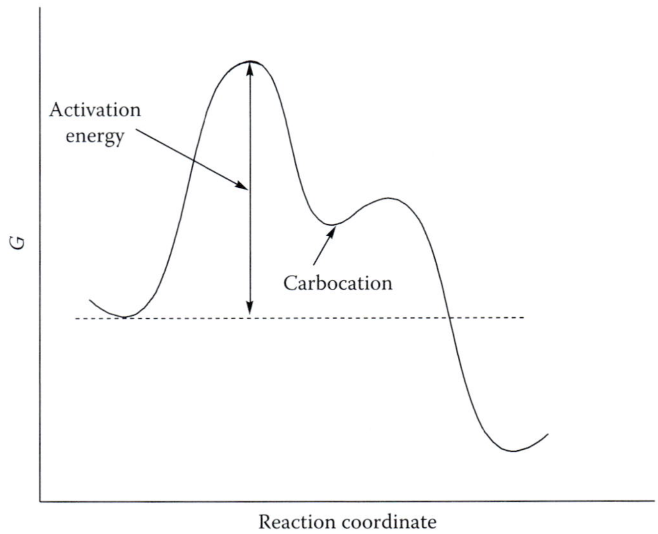

FIGURE 9.3 Energy profile of a typical S_N1 reaction.

Equation 9.2. Figure 9.4 shows a typical process—the new bond to the nucleophile is formed as the bond between the carbon atom and the leaving group is broken. There is no intermediate. We can draw a transition state (**9.1**) for the reaction, where in this case the carbon–oxygen bond is partly made and the carbon–bromine bond partly broken. Remember, however, that we can't really observe transition states, though we can infer something about their structure from various reaction parameters. Although **9.1** is drawn as a fairly symmetric structure, this is conjecture; some transition states are symmetric, while others are not. That the negative charges are in parentheses indicates that they are partial charges—the hydroxyl and bromine have one negative charge between them, but at the moment of the transition state, each atom bears some of it.

$$\text{Rate}\ (S_N2\ \text{reaction}) = k[\text{substrate}][\text{nucleophile}] \qquad (9.2)$$

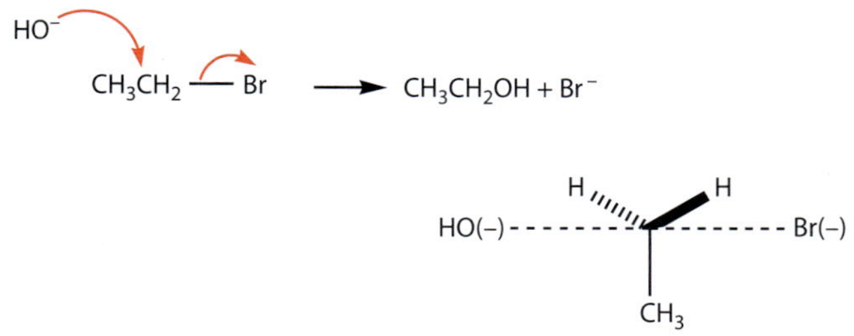

9.1, transition state for an S_N2 reaction

FIGURE 9.4 Mechanism of the S_N2 reaction.

PROBLEM 9.1

Explain why the *rate* of disappearance of Me_3CCl in aqueous solution is unaffected by the presence of $Na[N_3]$, sodium azide, in the solution, but the products include both the azide and the alcohol.

Solution

If the rate of the reaction is unaffected by the concentration of the nucleophile, then the reaction must go by the S_N1 mechanism. So in the RDS of the reaction, the carbon–chlorine bond is broken to give the carbocation. This is not surprising as this is a stable tertiary carbocation. The carbocation is very reactive, so in the second, fast, step, it will be attacked by both azide ion and water:

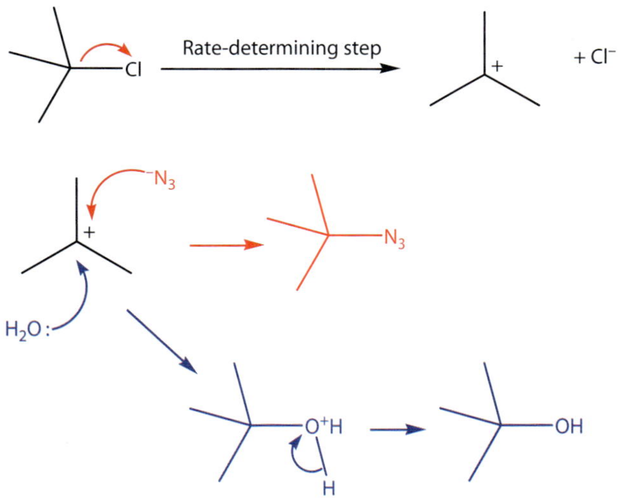

9.3 STEREOCHEMICAL IMPLICATIONS

The intermediate in the S_N1 reaction is a carbocation. As you will remember from Sections 4.7.1 and 8.2.5, carbocations are planar and trigonal, sp^2 hybridized, with an empty p_z orbital. Because they are planar, they can be attacked by nucleophiles from either face, so that the product should be racemic. If we return to the reaction of Figure 9.2, the starting material is chiral, and in Figure 9.5,

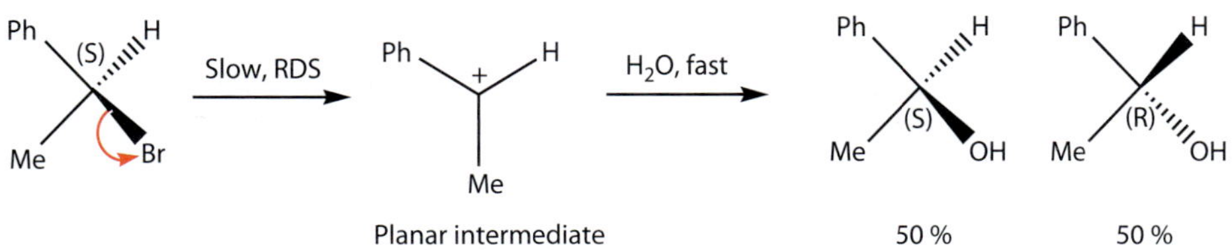

FIGURE 9.5 Stereochemical outcome of an S_N1 reaction.

we look at the reaction of the enantiomer with *S*-stereochemistry. The intermediate carbocation is planar and hence may be attacked with equal likelihood from above or below, leading to equal amounts of *R* and *S* products.

Many S_N1 reactions happen in this way, leading to a racemic product. However, the exact outcome can depend a little on conditions. When the leaving group departs, electrostatic interaction may mean that it does not go very far, so that one face of the planar carbocation is slightly "protected" from the attacking nucleophile. This is called an ion pair effect. When the reaction of Figure 9.5 was repeated with the chloride rather than the bromide as the leaving group, the ratio of enantiomers was 41:59, rather than strictly 50:50. This does not mean that some other mechanism is operating, however, as the reaction kinetics are strictly first order.

If attack from the two faces of our carbocation produces diastereoisomers, rather than enantiomers, then it is likely that we will get both of them, but there is no necessary expectation that they will be formed in equal amounts. Diastereoisomers are different compounds, so the transition states leading to them will have different energies, and hence the processes will occur to different extents. We could also explain this as one "face" of the carbocation being more hindered to the approach of a nucleophile than the other. An example is provided in Figure 9.6; in this S_N1 reaction, two diastereoisomers are formed (notice there is no change in the stereochemistry of the methyl group at the other chiral center—no reaction has happened here to change anything). One diastereoisomer will predominate over the other, and the difference may be small or large.

Other examples of S_N1 reactions are shown in Figure 9.7. In the second example, it's worth noting that it is shown by this experiment that sulfur is a much better nucleophile than oxygen (so $[HS]^-$ captures the cation much more effectively than does methanol or methoxide), something that we will return to shortly.

There is no intermediate in the S_N2 reaction, and the stereochemistry involves inversion of configuration, rather like an umbrella blowing inside out. The nucleophile approaches the substrate along a direction opposite to the carbon-to-leaving group bond that we intend to break, and a symmetric transition state for this reaction would be a planar arrangement of the three groups that are not reacting, with the arriving and departing groups on either side of this plane, as in **9.1**.

What is our evidence for this inversion? There are many S_N2 reactions where inversion occurs (though note that this does not invariably change an *S* molecule for an *R* one or change the sign of the optical rotation; that will depend on the substituents), but the most rigorous evidence comes from a classic experiment in which chiral 2-iodooctane (containing regular,

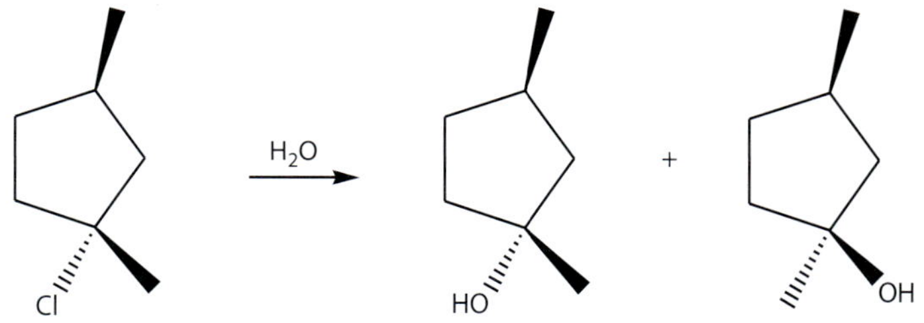

FIGURE 9.6 S_N1 reactions may produce diastereoisomers in unequal amounts.

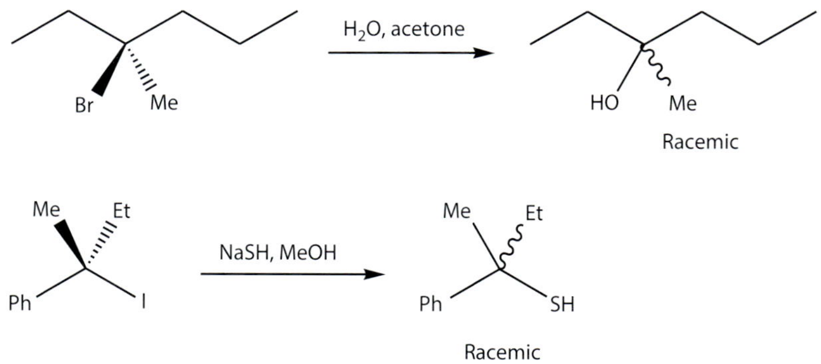

FIGURE 9.7 Typical S_N1 reactions.

nonradioactive [127]iodine) was reacted with radioactive iodide ion, $^{131}I^-$. It was first demonstrated that the reaction was second order—the rate was proportional to both substrate and added iodide ion. It was further observed that the rate of incorporation of radioactive iodine was exactly one half of the rate of racemization. Let's imagine that we start with 10 molecules of *S*-2-iodooctane, and we do just one replacement of iodine by radiolabeled iodine. In principle, we would have three choices, stereochemically, as to the outcome. If the outcome of each individual process was retention, we would still have 10 molecules with *S*-stereochemistry. So there is no racemization and so this is NOT consistent with the experimental observation. If the outcome of each molecular event was racemization, we would have a 50:50 chance of getting an *S*- or an *R*-product. So we would say that we had 9 *S* molecules (the ones that have not reacted) plus 0.5 *S* molecules plus 0.5 *R* molecules. Obviously, we can't really have half molecules, but we don't really have 10 molecules, we have trillions! So the rotation we would measure would be that for 9 *S* molecules—the 0.5 *S* and 0.5 *R* rotations would cancel out. So we have 10 % incorporation and 10 % racemization—again not consistent with the observed experimental result. Finally, if each molecular event results in inversion, after we have done one incorporation, we will have nine *S* molecules and one *R* molecule. The rotations of one *S* molecule and one *R* molecule will cancel, and what will be measured will be the rotation of eight *S* molecules. So now we have 10 % incorporation of radiolabel and 20 % racemization—which is exactly what is observed. This is set out in Table 9.1 and shown diagrammatically in Figure 9.8—the two iodine isotopes are colored differently.

Typical S_N2 reactions are shown in Figure 9.9; both show clean inversion of configuration. Again in the second example, we see that sulfur is a better nucleophile than oxygen.

TABLE 9.1 Outcomes of Various Possible Stereochemistries for S_N2 Reactions

Stereochemistry of the Reaction	10 *S* Molecules Give	% Incorporation Radioactive Iodine	% Racemization	Consistent with Experiment?
Retention	10 *S* molecules	10	0	No
Racemization	9 *S* + 0.5 *S* + 0.5 *R* = 9 *S*	10	10	No
Inversion	9 *S* + 1 *R* = 8 *S*	10	20	Yes

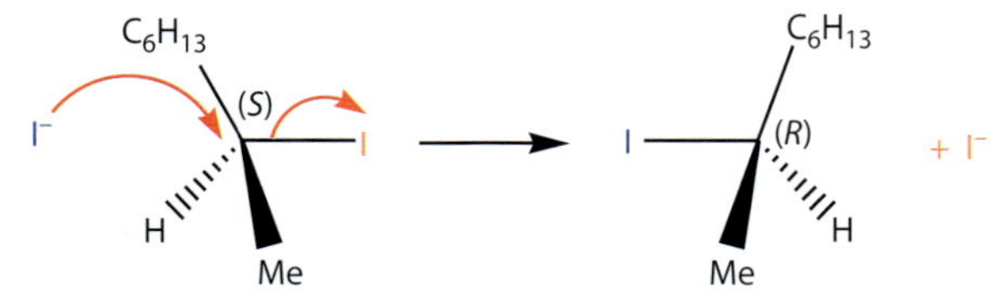

FIGURE 9.8 S_N2 substitution goes with inversion of stereochemistry.

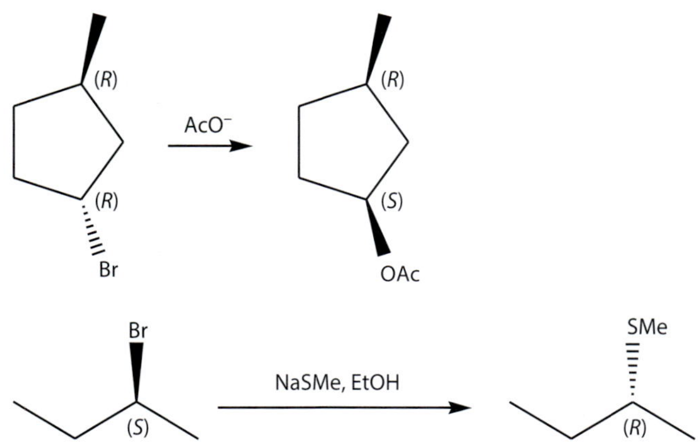

FIGURE 9.9 Typical S_N2 reactions.

PROBLEM 9.2

The optically pure tosylate of 2,2-dimethyl-1-phenylpropan-1-ol ($[\alpha]_D$ = −30.3 °) was heated in acetic acid to give as the major product an acetate with $[\alpha]_D$ = +5.3 °. If the optically pure acetate has $[\alpha]_D$ = +53.6 °, what percentage racemization and what percentage inversion have occurred? Explain this in terms of the mechanism(s) involved:

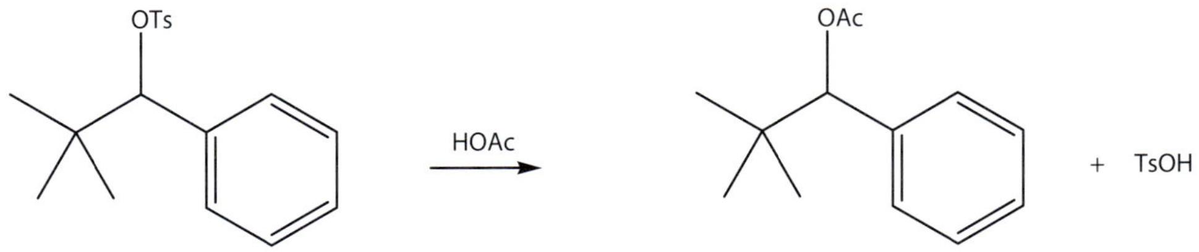

$[\alpha]_D$ = −30.3 °

Observed $[\alpha]_D$ = +5.3 °
$[\alpha]_D$ for optically pure material = +53.6 °

Solution

We should consider both S_N1 and S_N2 mechanisms for the substitution. We might have some preference for S_N1 because of the steric hindrance in the molecule (S_N2 reactions are more sterically demanding because we need to crowd more groups around a single carbon atom) and the fact that the cation is benzylic and hence stabilized by resonance. The product is approximately 10 % optically pure. This means we have 55 % of one enantiomer and 45 % of the other. If we assume that the inverted product comes from an S_N2 reaction, and the racemized product from an S_N1 reaction, we would conclude that our reaction is 90 % S_N1 and 10 % S_N2. However, we should also consider the possibility that the reaction is entirely S_N1, but that there is a small ion pair effect.

PROBLEM 9.3

Assuming that all the following substitutions go by an S_N2 mechanism, predict the products, including stereochemistry, where appropriate:

(a)

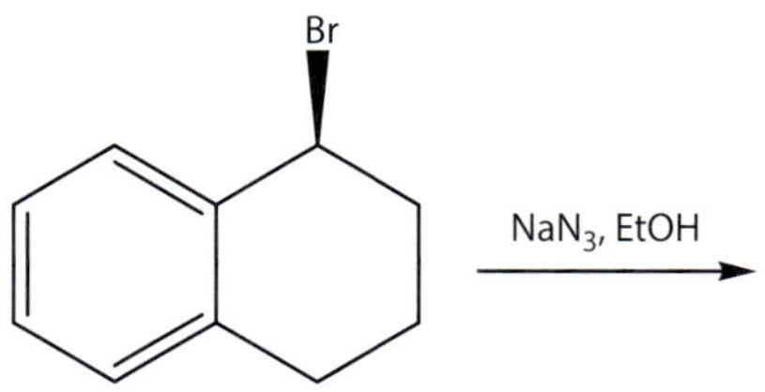

(b)

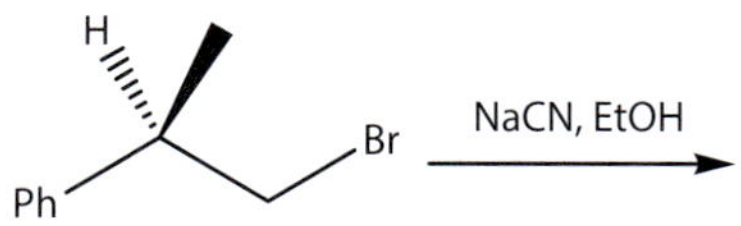

(c)

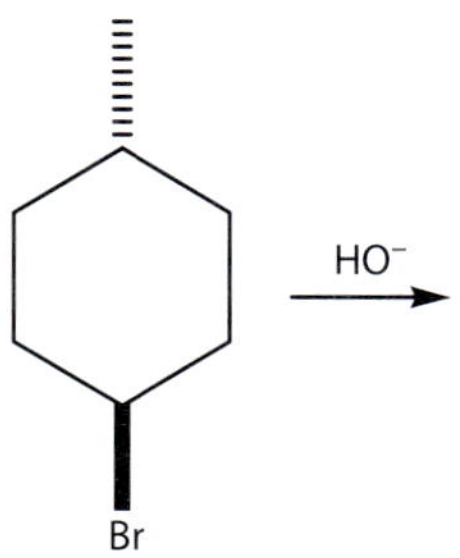

Solutions

(a) S_N2 with inversion gives

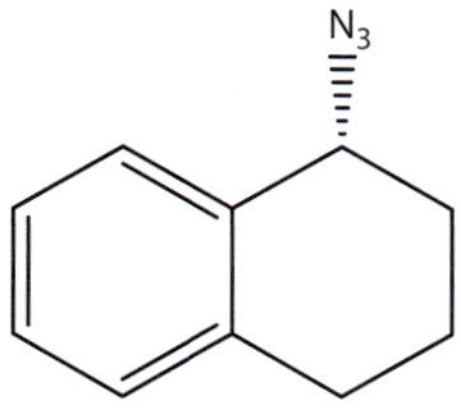

(b) In this case, the stereocenter is remote from the site of the reaction, and nothing is happening there. So the halide is substituted, but the chiral center is unaffected:

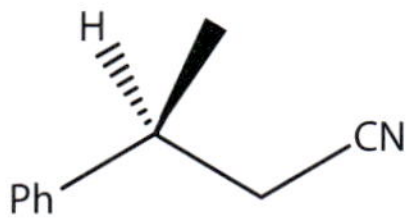

(c) S_N2 with inversion gives

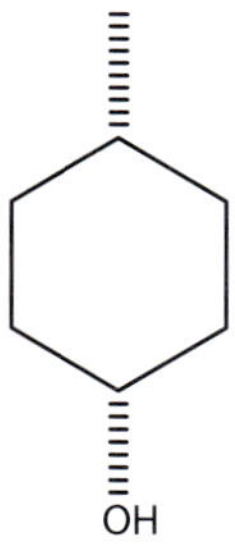

Key Points from Sections 9.1 through 9.3

- There are two common mechanisms for nucleophilic substitution at sp^3 carbon atoms, designated as S_N1 and S_N2. The S_N1 mechanism is a first-order, unimolecular process, the rate of which is dependent only on substrate concentration. The S_N2 process is bimolecular, second order, with rate depending on the concentration of both substrate and nucleophile.
- The S_N1 mechanism is a two-step process with a planar carbocation as the intermediate. The RDS is the formation of the carbocation, which then reacts rapidly with nucleophiles. The planar carbocation intermediate may be attacked from either face resulting in racemization of an initially chiral substrate.
- The S_N2 reaction is a single-step process, which proceeds with inversion of configuration at the reacting center.

9.4 EFFECT OF NUCLEOPHILE

In Figures 9.7 and 9.9, we saw examples where we had either oxygen or sulfur nucleophiles available, with the sulfur nucleophile proving to be more reactive. This brings us to a very important concept—nucleophilicity and basicity are NOT equivalent (alkoxides, RO^-, are significantly more basic than thiolates, RS^-). Basicity derives from thermodynamics; pK_a comes from the equilibrium constant, K_a, and the way in which the proton is added or removed is irrelevant. We are looking only at the relative energy levels of the two sides of the equilibrium. However, nucleophilicity is essentially related to kinetics; how well does a nucleophile react with a range of substrates? Size, solvation, and the nature of the substrate will also be important. As a reference reaction, we usually consider how well a nucleophile will react with a simple, sterically uncongested substrate, methyl iodide, remembering that even this, the simplest possible substrate, is much larger than a proton.

Some trends are common to both nucleophilicity and basicity. For example, the two run in parallel when the attacking atom is the same. Thus, for nucleophilicity,

$$EtO^- > HO^- > PhO^- > [CO_3]^{2-} > AcO^- > H_2O$$

with the respective pK_a values of the protonated species being 16.0, 15.7, 9.95, 10.3, 4.76, and −1.7. Similarly, if the attacking atom is in the same period of the periodic table,

$$H_3C^- > H_2N^- > HO^- > F^-$$

with the pK_a values for methane, ammonia, water, and hydrogen fluoride being, respectively, 50, 36, 15.7, and 3.7. All of the values given are related to aqueous solutions, and the relative nucleophilicities are for substrates in which steric hindrance is minimal.

However, other trends do not parallel basicity. If we consider the halide ions, fluoride is the most basic, and iodide the least so. But in terms of nucleophilicity the reverse is true; iodide is the best nucleophile, and fluoride the least good, Similarly, RO^- is more basic, but much less nucleophilic, than RS^-. Table 9.2 shows some typical relative reactivities in nucleophilic reactions. We can observe some generalizations. The simplest (compare MeOH and MeO^-) is that anions are more nucleophilic than their conjugate acids—not a great surprise. We can also see (compare the halide anions, $PhO^-/PhS^-/PhSe^-$, and Et_3N/Et_3P) that nucleophilicity increases as we go down the periodic table. Finally, nucleophiles that have two bonded heteroatoms seem to have much higher reactivity than their simple relatives. Thus, hydroxylamine is more nucleophilic than ammonia, and the anion of hydrogen peroxide is more nucleophilic than hydroxyl ion. Although the theoretical justification for this is beyond the scope of this course, the observation is important, not confined to their reactions with simple substrates, and something that we will make use of later.

Why are species in which the nucleophilic atom is in the second or third row of the periodic table more nucleophilic than those containing, oxygen, nitrogen, or fluorine? This is related to the polarizability of the atom. With second and third row elements, the outer electrons are further away from the nucleus. Thus, when they approach a charge or dipole, the electron cloud is more easily squashed out of shape, in preparation for the formation of the new bond. We describe these nucleophiles as "soft." Polarizability means lower basicity and higher nucleophilicity. RS^- is more polarizable than RO^-,

TABLE 9.2 Relative Reactivity of Nucleophiles with Iodomethane

Nucleophile	Relative Rate of Reaction
MeOH	1
F^-	500
AcO^-	2×10^4
Cl^-	2.3×10^4
Et_2S	2.2×10^5
NH_3	3.2×10^5
PhO^-	5.6×10^5
Br^-	6.2×10^5
N_3^-	6.3×10^5
MeO^-	1.9×10^6
HO^-	3.2×10^6
NH_2OH	4×10^6
Et_3N	4.6×10^6
CN^-	5×10^6
I^-	2.6×10^7
Et_3P	5.2×10^8
HOO^-	6.3×10^7
PhS^-	8.3×10^9
$PhSe^-$	5×10^{10}

so RS^- is a better nucleophile but a poorer base. Polarizability increases with atom size, decreases with charge. Think of these as Nerf balls or beach balls that are easily deformed. "Hard" nucleophiles, such as RO^-, are more like baseballs or table tennis balls. Iodomethane and molecular bromine are "soft" electrophiles and respond to nucleophilicity. H^+, HX, $R_2S{=}O$ are hard electrophiles and respond to basicity. This is described as HSAB theory, standing for hard and soft acids and bases.

Hard and soft sports balls

Another way to look at this is in molecular orbital terms. Hard nucleophiles have a low-energy highest occupied molecular orbital (HOMO) and usually have a negative charge. Soft nucleophiles have a high-energy HOMO and do not necessarily have a negative charge. Hard electrophiles have a high-energy LUMO and usually have a positive charge. Soft electrophiles have a low-energy LUMO and don't necessarily have a positive charge. Hard–hard reactions are fast because of the Coulombic attraction between the species. Soft–soft reactions are fast because the HOMO of the nucleophile and the LUMO of the electrophile are relatively close in energy and thus interact well. If you want to pursue this elegant theory further, read Fleming, *Molecular Orbitals and Organic Chemistry Reactions*. Table 9.3 classifies both nucleophile/bases and electrophiles/acids as soft, intermediate, and hard.

Steric effects are also important in determining nucleophilicity. The reaction between a base and a proton is sterically completely undemanding. However, an S_N2 reaction is sterically much more challenging—at the transition state (**9.1**), five groups must be accommodated around the central carbon atom. Thus, as well as softness and hardness, we need to consider size. Bulky species, such as Ph_3C^- and *tert*-BuO^-, are good bases but very poor nucleophiles. Triethylamine and quinuclidine, **9.2**, have similar basicity, but quinuclidine is the better nucleophile, because the alkyl groups are "tied back" and out of the way.

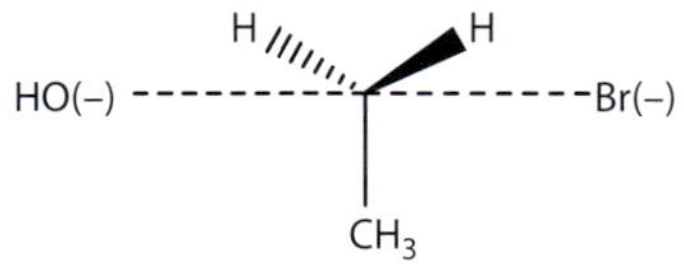

9.1, transition state for an S_N2 reaction

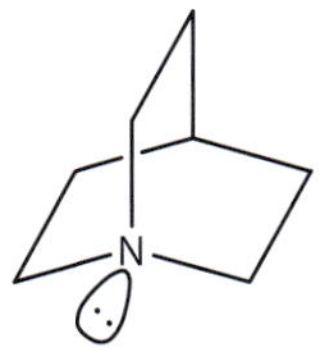

9.2, quinuclidine

TABLE 9.3 Hard and Soft Acids and Bases

	Bases	Acids
Soft	R_2S, RSH, RS^-, I^-, CN^-, arenes, alkenes, R_3P	I_2, Br_2, RSX, RCH_2X (X = halogen, softest with I), BH_3, Cu^+, Ag^+, zero-valent metals
Intermediate	Br^-, N_3^-, $ArNH_2$, pyridine, $[NO_2]^-$	Cu^{2+}, Zn^{2+}, R_3C^+, R_3B, Fe^{2+}
Hard	NH_3, RNH_2, H_2O, HO^-, HOO^-, ROH, RO^-, Cl^-, F^- NO_3^-	HX, Li^+, Na^+, K^+, Mg^{2+}, Ca^{2+}, $AlCl_3$, SO_3, NO_2^+, CO_2, various carbonyl compounds

PROBLEM 9.4

Which of each of the following pairs of compounds is the more nucleophilic?

(a) NH_3 or $[NH_2]^-$
(b) CH_3CH_2OH or PhOH
(c) $(CH_3)_3B$ or $(CH_3)_3N$
(d) $[HO]^-$ or $[HS]^-$
(e)

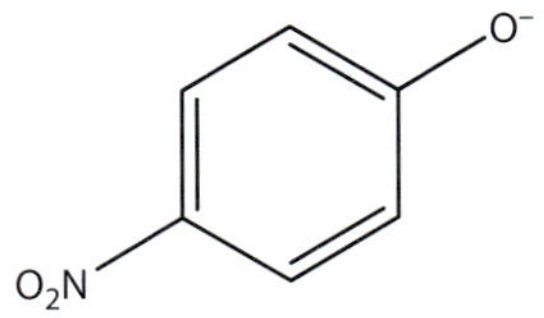

or

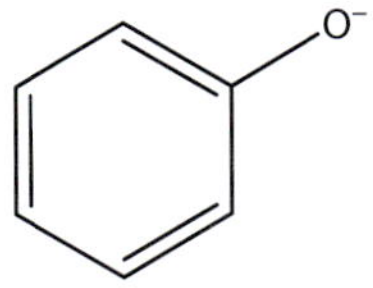

(f)

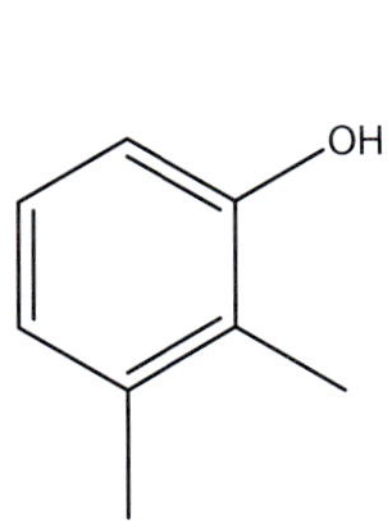

or

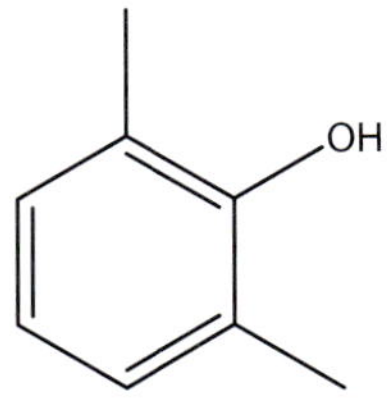

(g) $[HO]^-$ or $[HOO]^-$

Solutions

(a) $[NH_2]^-$; anions are almost always stronger nucleophiles than related neutral species.
(b) CH_3CH_2OH; lone pair is more available in ethanol than in phenol, because in phenol, it interacts with the π-electrons of the benzene ring.
(c) $(CH_3)_3N$; trimethylborane is sp^2 hybridized with an empty p_z orbital and is an electrophile. The lone pair on trimethylamine is a reasonable nucleophile.
(d) $[HS]^-$; sulfur is more polarizable.
(e) PhO^-; the nitro group is strongly electron withdrawing and can delocalize the negative charge away from the oxygen of the phenolate in 4-nitrophenolate.
(f)

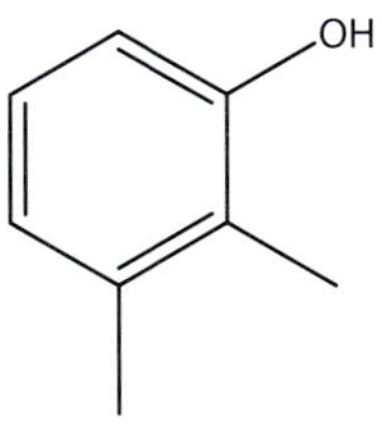

The OH group is less sterically hindered.
(g) $[HOO]^-$; species containing two bound heteroatoms are particularly nucleophilic.

9.5 EFFECT OF THE LEAVING GROUP

There is one key message here—*good leaving groups are the anions/conjugate bases ("happy" anions ☺) of strong acids.* We will return again and again to this concept, in looking at many other reactions. So iodide (pK_a HI = −9), bromide (pK_a HBr = −7), and H_2O (pK_a H_3O^+ = −1.7) are good leaving groups, but cyanide (pK_a HCN = 9.1) and hydroxyl (pK_a H_2O = 15.7) are poor. If you are in doubt as to whether something unfamiliar will be a good leaving group, go back to Chapter 8 and look up the pK_a. In Figure 9.10, you see that iodide is displaced in preference to chloride.

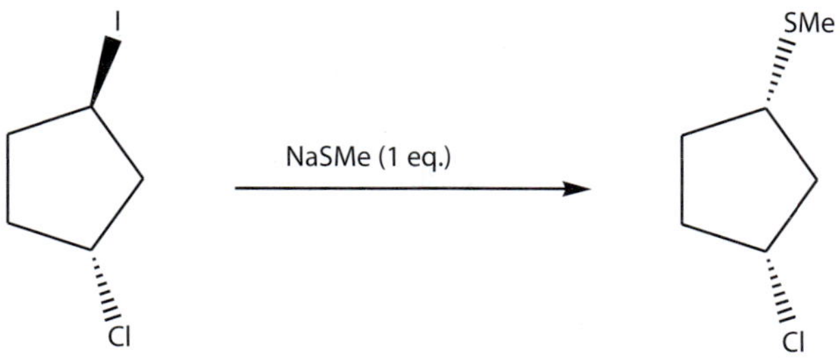

FIGURE 9.10 Selectivity in an S_N2 reaction.

PROBLEM 9.5

Predict the product(s) of each of the following reactions:

(a)

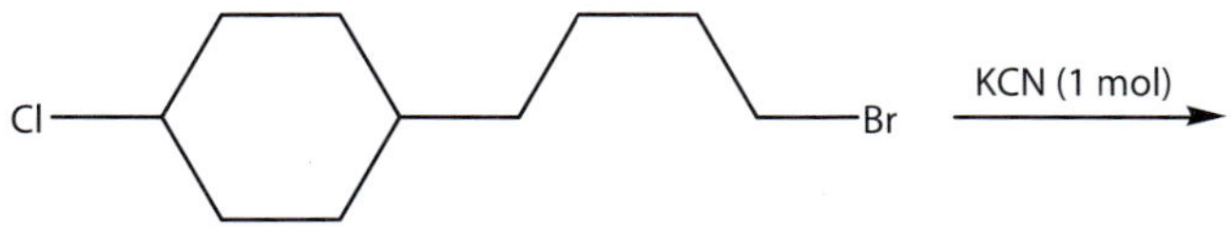

(b)

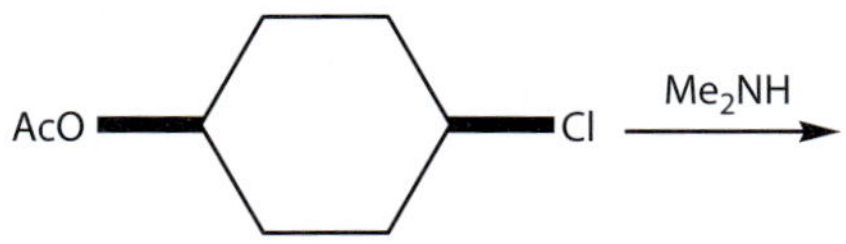

(c)

(d)

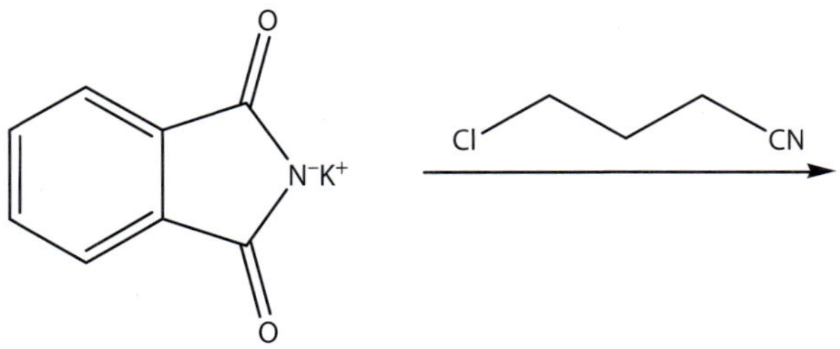

Solutions

(a) Bromide is a better leaving group than chloride (HBr is a stronger acid than HCl). Moreover, the bromide is primary and the chloride secondary; hence, only the bromide is substituted.

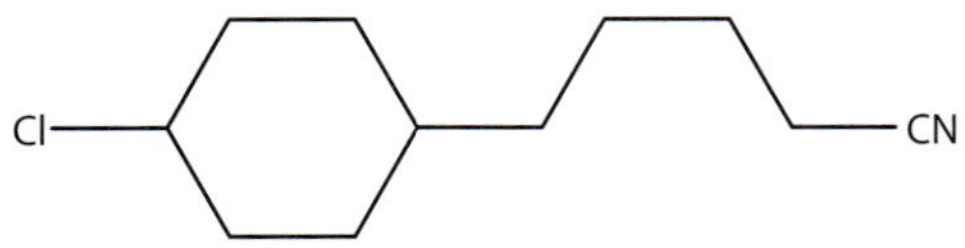

(b) Chloride is a better leaving group than acetate (HCl is a stronger acid than HOAc). It is displaced in the S_N2 process, with inversion of stereochemistry.

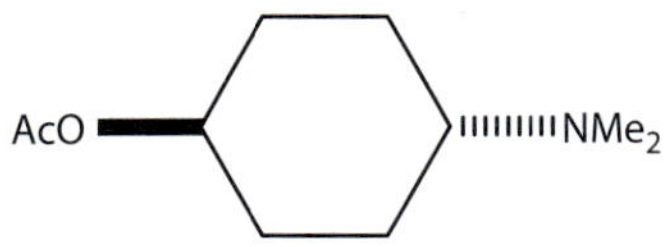

(c) Bromide is a better leaving group than ethoxide (HBr is a stronger acid than ethanol).

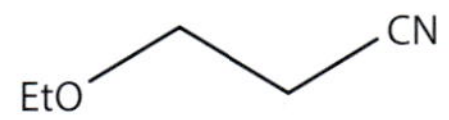

(d) Chloride is a better leaving group than cyanide (HCl is a stronger acid than HCN).

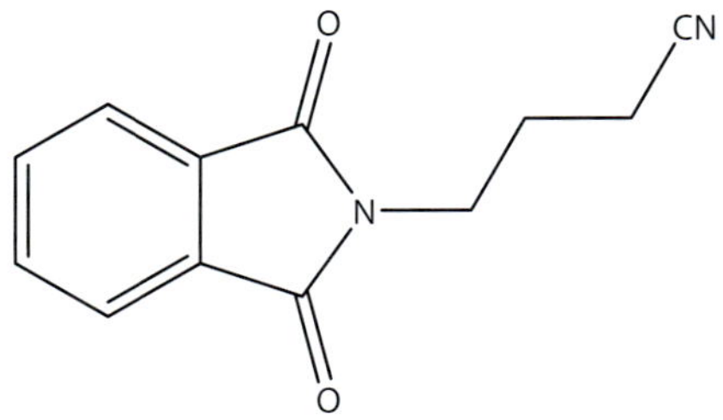

Key Points from Sections 9.4 and 9.5

- Nucleophilicity and basicity parallel each other only when there are no steric effects, and the species compared are within the same row of the periodic table.
- Heavier atoms tend to be more nucleophilic because they are more polarizable; thus, RS^- is more nucleophilic than RO^-.
- In HSAB theory, hard electrophilic centers such as protons and carbonyl groups respond best to hard bases such as hydroxyl ion. Soft, polarizable centers such as MeI or Br_2 respond best to soft nucleophiles such as RS^- and alkenes.
- Good leaving groups are the anions of strong acids.

9.6 EFFECT OF SUBSTRATE STRUCTURE

9.6.1 S_N1 REACTIONS

Ionization of the leaving group to give a carbocation is the RDS in the S_N1 process, so this is the only one we need to consider. A good leaving group leads to easy ionization. The formation of a stable carbocation leads to easy ionization. The Hammond postulate operates; if the carbocation is low in energy, then the transition state leading to it will also be of low energy. This is a good point to go back and review the factors that make carbocations stable (Section 8.2.5). But we can usefully make some generalizations. Tertiary centers, with or without conjugating substituents, will almost always give rise to S_N1 processes. Primary centers, unless conjugated (and not always even then), almost never give S_N1 reactions. Secondary centers may react by either S_N1 or S_N2 mechanisms; it will depend on the exact structure, the nucleophile, and the solvent. So the reaction in Figure 9.11 gives a racemic product; the nucleophile is poor and the cation is secondary, but it is also stabilized by conjugation with the aromatic rings.

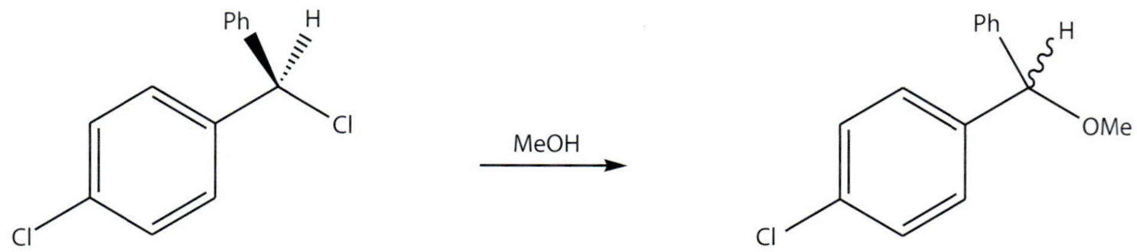

FIGURE 9.11 S_N1 substitution at a secondary benzylic center.

PROBLEM 9.6

Rank the following three reactions in order of the rate at which you expect they will proceed:

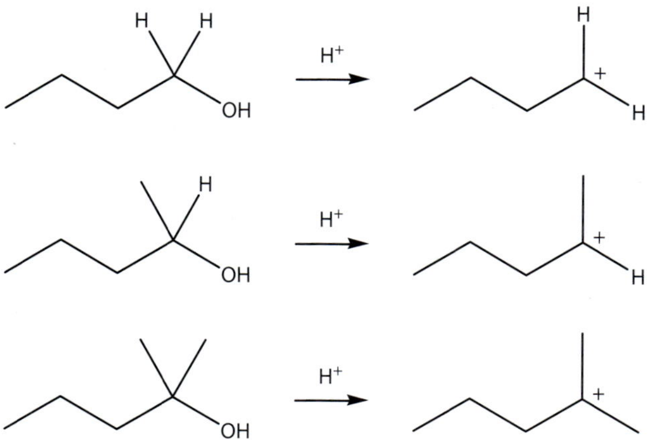

Solution

In each case, the acid protonates the hydroxyl in order to make it into a good leaving group, that is, water. The order is 3 ° > 2 ° > 1 °. Tertiary carbocations are more stable than secondary, in turn more stable than primary.

Focus on Carbocations

We've previously discussed how carbocations may be stabilized—electron-donating alkyl groups, lone pairs of electrons and π-conjugation. Also how they may be destabilized—electron-withdrawing groups, nonplanarity. We can illustrate some of these effects by looking at the relative rates of a reaction such as that in Figure 9.12. The truly dramatic acceleration shown when the carbocation produced is tertiary is obvious. In fact the accelerations for secondary and tertiary cation formation are even greater than is apparent from these rates; the reactions of the methyl and ethyl derivatives are not true S_N1 reactions, and most of the rate measured is that of the S_N2 process that prevails for these substrates. The effect is even more dramatic when the substituents are benzene rings (Table 9.4). The trityl cation, Ph_3C^+, was first prepared in 1901, and its structure was determined crystallographically in the 1960s (Figure 9.13). It can be prepared either by treating Ph_3CCl with $Ag[PF_6]$ (the related fluoroborate and perchlorate salts are also known, and the original structure was done on the perchlorate; although perchlorates frequently give excellent crystals for structure determination, they are not generally to be recommended, because of their propensity to explode!) to give insoluble silver chloride and trityl hexafluorophosphate or from triphenyl methanol and sulfuric acid. The cation is rigorously planar about the central sp^2 carbon atom, but the benzene rings are tilted out of this plane in a propeller shape, to avoid steric interactions between the *ortho*-hydrogen atoms. The shortened C–C bond lengths around (~1.45 Å) the central carbon indicate that there is still resonance stabilization in this structure.

R	**Me**	**Et**	***i*-Pr**	***tert*-Butyl**
Relative rate	1.05	1.0	11.6	1.2×10^6

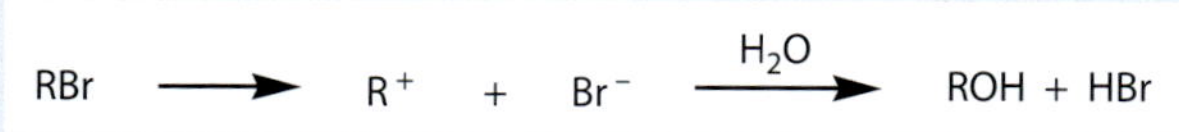

FIGURE 9.12 Relative rates of cation formation.

TABLE 9.4 Relative Rates of Solvolysis of Benzylic Halides

Substrate	MeCl	$PhCH_2Cl$	Ph_2CHCl	Ph_3CCl
Relative rate	⋘1	1	2000	3×10^7

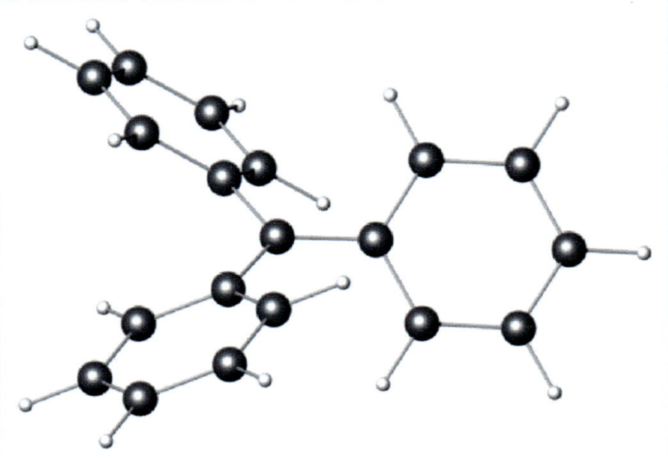

FIGURE 9.13 X-ray crystal structure of the trityl cation. (Thanks to Dr. John Turner for this image.)

There are now many carbocations that have been fully structurally characterized. For example, $[PhCMe_2][SbF_6]$ has been characterized at −124 °C. The cation is close to planar, and the shortening of the bond between the benzene ring and the cationic center (1.41 Å *versus* 1.54 Å for an ordinary carbon–carbon single bond) as well as the bond lengths in the benzene ring indicates that benzylic resonance is important in stabilizing it. The bonds between the cation and the methyl groups are also slightly shortened, indicating some stabilization from hyperconjugation (*vide infra).*

The *tert*-butyl cation may be regarded as the archetype for unconjugated tertiary cations, and it has been structurally characterized as its $[Sb_2F_{11}]$ salt. It is prepared by the reaction of Me_3CF with SbF_5; interestingly, it is also obtained from 1-fluorobutane, 2-fluorobutane, and 2-methyl-1-fluoropropane, indicating that this cation is indeed the lowest energy species available to a four-carbon cation. The carbon–carbon bond lengths range from 1.439 to 1.446 Å, against a normal $sp^3(C)–sp^2(C)$ of 1.54 Å (Figure 9.14). This shortening of the bonds may be attributed to the hyperconjugative stabilization provided by the methyl groups. Estimates of energy required to form various cations by heterolytic cleavage of the related

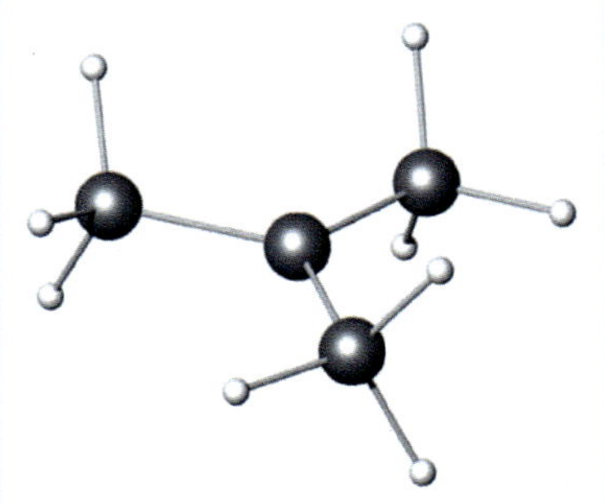

FIGURE 9.14 X-ray structure of the cationic portion of $[Me_3C^+][Sb_2F_{11}]$. (Thanks to Dr. John Turner for this image.)

TABLE 9.5 Energy Required for Heterolytic Dissociation of Gas Phase Hydrocarbons

Ion, R^+	CH_3^+	$CH_3CH_2^+$	$(CH_3)_2CH^+$	$(CH_3)_3C^+$	Ph^+	$H_2C{=}CH^+$	$H_2C{=}CHCH_2^+$	$PhCH_2^+$	$MeC^+({=}O)$
$RH \rightarrow R^+ + H^-$ (kJmol^{-1})	1316	1158	1043	970	1230	1200	1070	996	962

hydrocarbon (from gas phase measurements) are given in Table 9.5; notably this also shows why S_N1 reactions do not occur at sp^2 carbon atoms.

We also noted that carbocations that cannot be planar are generally very difficult to form. This is readily seen in the relative rates of solvolysis of the bromides shown in Figure 9.15. The more nonplanar the cation would be forced to be, the less likely it is to be produced.

The 3,5,7-trimethyladamantyl cation provides an interesting example. Although it appears in its conventional drawing to be rather pyramidal around the positive center, the crystal structure shows distinct flattening at C-1, toward planarity (Figure 9.16; the counterion is $[Sb_2F_{11}]^-$). This is allowed by the hyperconjugative stabilization from the adjacent C—C bonds, effectively a contribution from a no-bond resonance structure (**9.3**). The C$^+$—Cα bonds are shortened to 1.44 Å and the Cα—Cβ bonds elongated to 1.61 Å. In the parent hydrocarbon, these bonds are quite conventional in length, 1.528 and 1.530 Å, respectively.

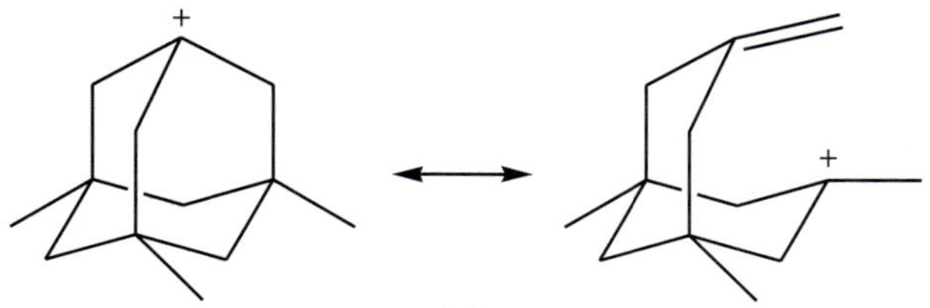

9.3

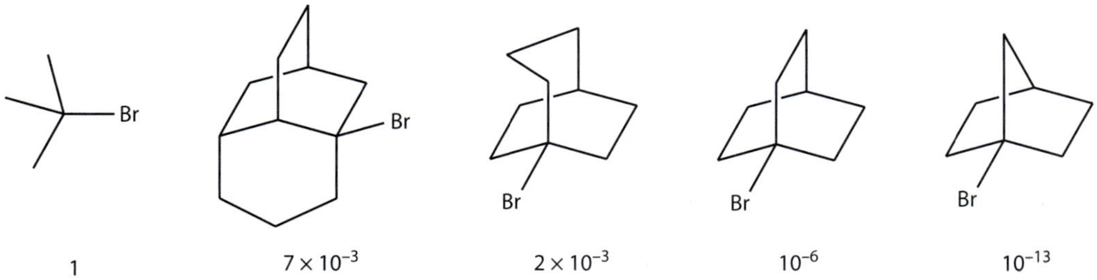

FIGURE 9.15 Relative rates of solvolysis of bicyclic halides.

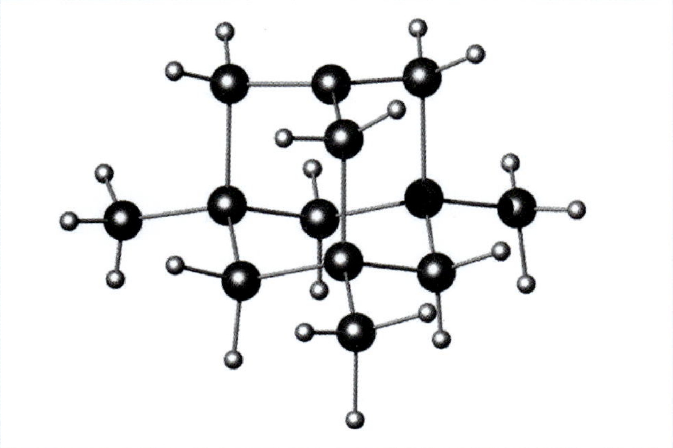

FIGURE 9.16 X-ray crystal structure of 3,5,7-trimethyl adamantyl cation, showing flattening at C-1. (Thanks to Dr. John Turner for this image.)

9.6.2 S_N2 REACTIONS

The main consideration in S_N2 processes is steric, because five atoms/groups must be accommodated around the reacting center at the transition state. Thus, in general, primary centers react fastest and tertiary ones not at all. However, other steric hindrance is also important; 1-chloro-2,2-dimethylpropane, **9.4**, reacts slowly, even though it is primary. In Figure 9.17, the unhindered chloride is selectively substituted, although both are secondary.

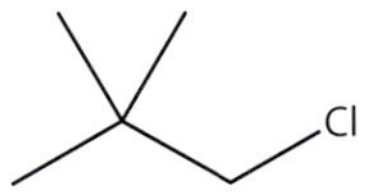

9.4

A Quick Aside on Relative and Absolute Stereochemistry

You should notice that in the structures in Figure 9.17 (and some previous worked examples), the wedge and hash bonds look just slightly different from those we have used previously, in that they do not taper toward one end. This is not accidental. Usually we need to show real absolute stereochemistry, and for this, we use the conventional tapered hash and wedge bonds. However, in these two molecules, this is inappropriate—both are *meso*-compounds, as the molecules have planes of symmetry in the vertical (as drawn) passing through the two heteroatoms, and the "mirror images" would be the same as we would obtain by flipping the molecule over. Where we are showing only *relative* stereochemistry, we will use these linear wedge and hash bonds.

Epoxides may be readily opened in S_N2 processes to give, after work-up, alcohols. The reaction is favorable because of the release of angle strain in the epoxide. In a simple example (Figure 9.18), a *trans* but racemic product is obtained (the starting material is not chiral and

FIGURE 9.17 Steric hindrance inhibits S_N2 reactions.

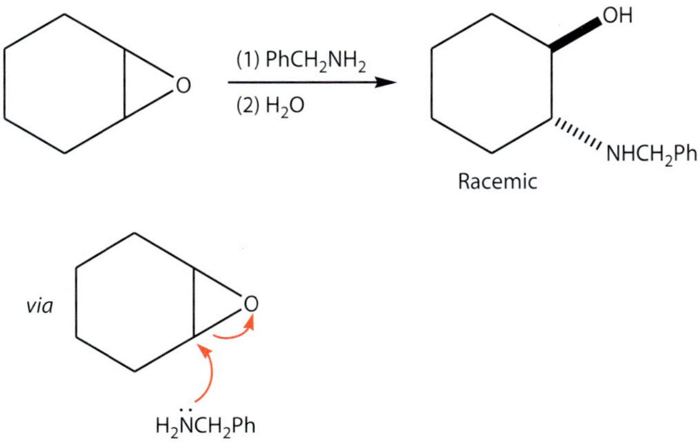

FIGURE 9.18 S_N2 opening of epoxides.

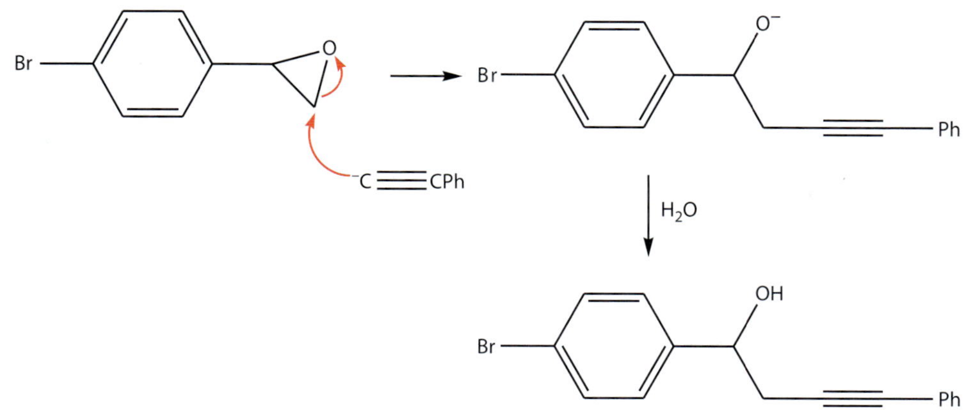

FIGURE 9.19 S_N2 opening of epoxides occurs selectively at the less hindered center.

can be attacked from either face; hence, we use the linear hash and wedge notation). The two carbons of the epoxide are identical, so neither can be specifically favored. However, when the epoxide is nonsymmetric (Figure 9.19), attack occurs specifically at the less substituted, the less hindered, carbon atom. Notice a new nucleophile here, the anion of an alkyne, which can be prepared *in situ* from the alkyne and a strong base such as lithium diisopropylamide (LDA); in some examples, the metal ion may assist the reaction by coordination to the oxygen atom of the epoxide.

There are also electronic effects in systems containing double bonds; substrates **9.5**, **9.6**, and **9.7** all react faster than the saturated compound **9.8**. This is because the transition state in each of these is stabilized by interaction with the p_z orbitals of the π-bond (**9.9**). Allylic systems react about 40 times faster than saturated ones, benzylic ones 120 times faster. Ketones substituted with a leaving group at the α-position react some 50,000 times faster; as well as the orbital overlap effect, the electron-withdrawing effect of the carbonyl groups stabilizes a transition state that is essentially negatively charged.

9.5

9.6

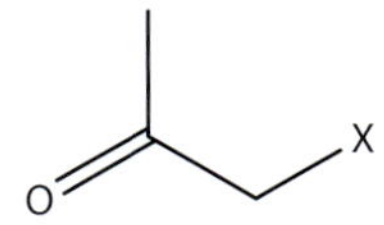

9.7

9.8

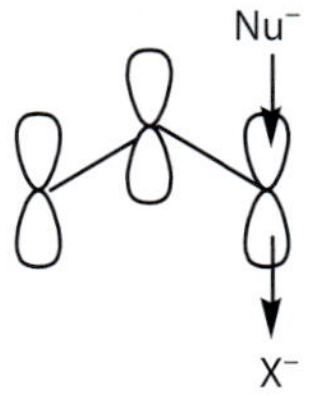

9.9

PROBLEM 9.7

For each of the following sets of molecules, predict the order of reactivity in an S_N2 reaction:

(a)

CH_3I

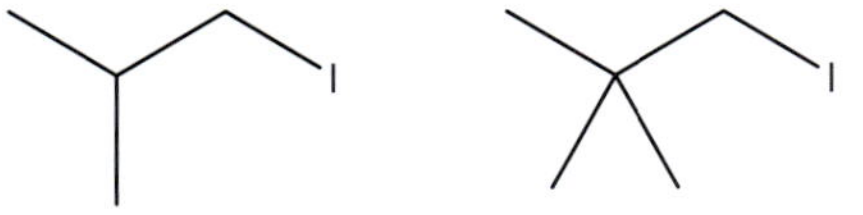

(b)

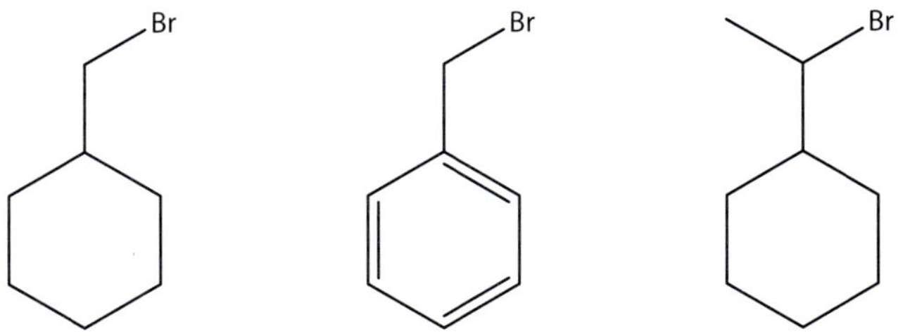

Solution

(a) $CH_3I > (CH_3)_2CH_2I > (CH_3)_3CH_2I$—steric hindrance.
(b) $PhCH_2Br > CyCH_2Br > CyCH(CH_3)Br$, Cy = cyclohexyl; benzylic halides react faster than saturated systems, primary halides react faster than secondary halides.

9.7 EFFECT OF SOLVENTS

Polar solvents solvate ionic species and thus favor ion formation and separation in S_N1 reactions. Part of the effect is due to dielectric constant, which facilitates charge separation (Table 9.6), and part to the ability of the solvent to solvate both anion and cation. Dipolar protic solvents are chosen since they are able to solvate both anions and cations (Figure 9.20). Solvation of cations involves interaction between lone pairs and the positive center and solvation of anions the formation of hydrogen bonds. The solvation of any type of ion increases with its charge and decreases with its size. Typically, S_N1 reactions are carried out in polar solvents such as water and alcohols.

Polar solvents solvate departing leaving groups in S_N2 reactions, but this is not the main effect we need to consider. Polar *protic* solvents also solvate the nucleophile. This lowers the energy of an anionic nucleophile relative to the reaction transition state and hence increases the activation energy and slows the reaction. Think of this as a nucleophile in a thick down winter coat—the smart outfit underneath will not get much attention while the coat is there. The ideal solvent for an S_N2 reaction is polar but aprotic, and S_N2 reactions are typically carried out in solvents such as acetone, acetonitrile, DMSO, or DMF. All of these solvents solvate cations well, so the counterion of our nucleophile is separated from any ion pair. Very reactive nucleophile/electrophile pairs will still react by an S_N2 mechanism in protic solvents, but in more marginal cases, changing the solvent may lead to a change in mechanism.

TABLE 9.6 Relative Rate of Reaction of *tert*-Butyl Chloride in Various Solvents

Solvent	Dielectric Constant	Relative Rate
AcOH	6	1
MeOH	33	3
H_2O	78	150,000

FIGURE 9.20 Solvation of cations and anions by water.

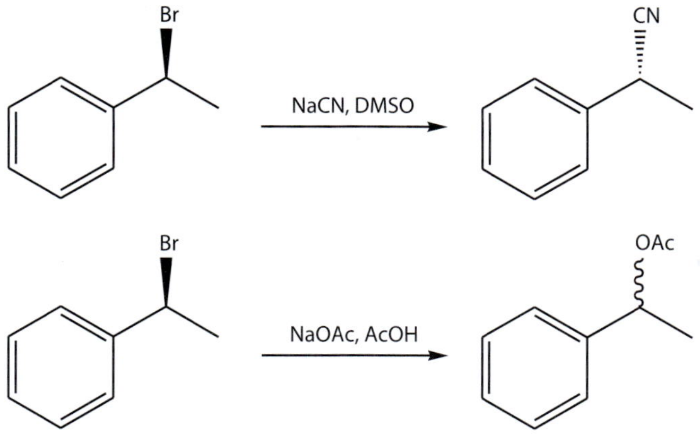

FIGURE 9.21 Mechanism of substitution may depend on the solvent used.

In the first example of Figure 9.21, we have a secondary substrate, with a good nucleophile and an S_N2 friendly solvent; clean inversion of configuration is observed. In the second reaction, we have the same substrate, but a somewhat poorer nucleophile, and a polar, protic solvent. The product acetate is racemic.

9.8 COMPETING REACTIONS

Elimination, the loss of two atoms or groups to form a double bond, competes with both S_N1 and S_N2 reactions (Figure 9.22). This is a sizable topic and gets the next chapter to itself.

We already know that some types of carbocations are more stable than others; tertiary cations are more stable than secondary and primary, cations can be stabilized by resonance (π-conjugation) or by the presence of atoms that can share a lone pair of electrons (*n*-conjugation). Thus, when we form a carbocation in an S_N1 reaction, there exists the possibility that it will rearrange to give a more stable cation. Cations can rearrange by migrating an alkyl group or a hydrogen atom from one carbon to an adjacent positive center. The migrating group takes its electrons with it, so that the atom from which it migrates will then bear the positive charge. Figure 9.23 illustrates. Ionization of the bromide gives a secondary carbocation. This is a fairly stable carbocation, but if a methyl group is migrated along the carbon chain, a more stable, tertiary, cation is obtained. This then reacts with water to give the tertiary alcohol product. At this point, you may be wondering why we don't get the alcohol from attack of the water on the initially formed carbocation. The answer is that you can get either or both alcohols; the relative amounts depend on the relative rates of rearrangement and capture by water. There is no quick mantra that will tell you which reaction is faster. While it's possible to work from analogy with a similar system where the outcome is known, mostly you will only be able to rationalize this rather than predict it.

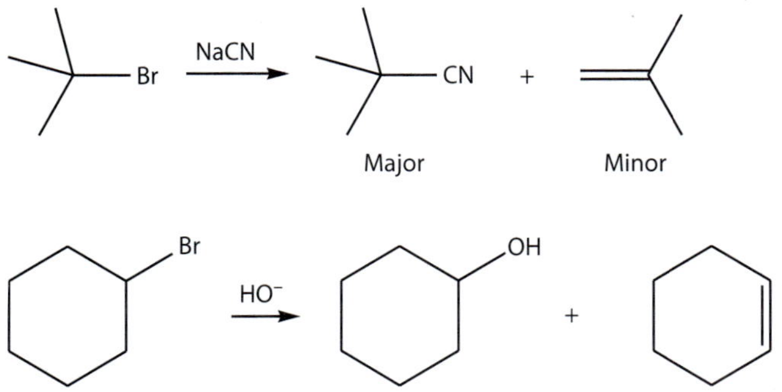

FIGURE 9.22 Elimination may compete with both S_N1 and S_N2 reactions.

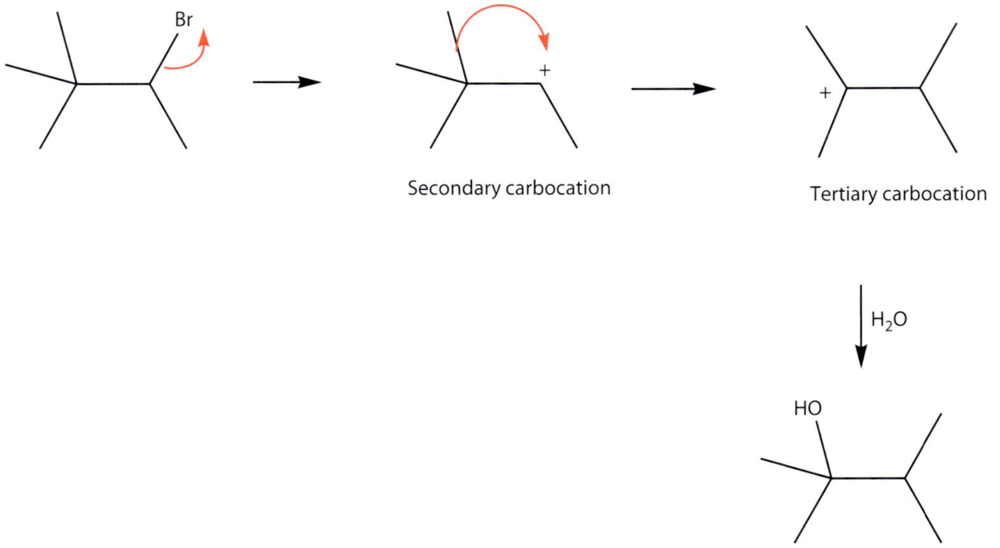

FIGURE 9.23 Rearrangement during S_N1 solvolysis of a secondary halide.

Rearrangement, or lack of it, sometimes tells us what mechanism is operating. Consider compound **9.10**. With hydroxyl ion, it undergoes a pure S_N2 reaction to give **9.11**. No rearranged products are observed. However, under slightly different conditions, with water as the nucleophile, both **9.11** and **9.12** (which predominates) are formed, indicating an S_N1 process. Notice that in this case, the rearrangement that occurs involves migration of a hydride ion (we describe it that way because the hydrogen takes both electrons of the C–H bond along with it). Migrating a methyl group will not help here—we would simply exchange one secondary carbocation for another.

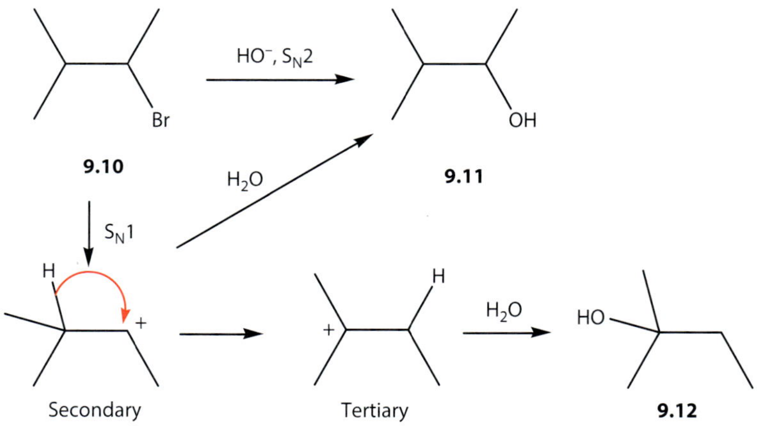

PROBLEM 9.8

Propose a mechanism for this reaction:

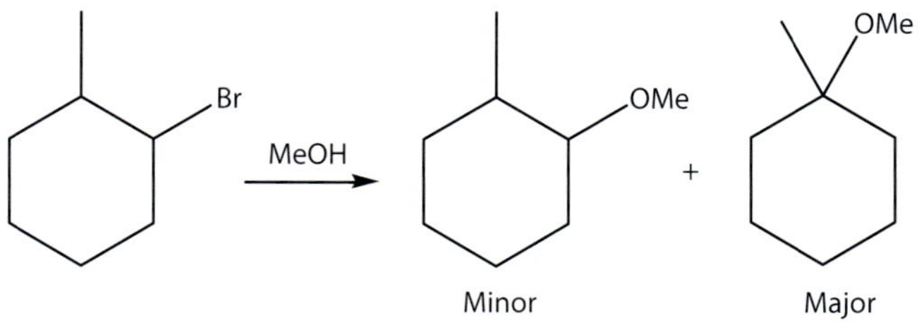

Solution

Since methanol is a poor nucleophile, and the halide is secondary, this is likely to be an S_N1 reaction. The initially formed secondary carbocation can react directly with methanol but also rearranges easily by migration of a hydride to give a tertiary cation. The tertiary cation reacts with methanol to give the major product:

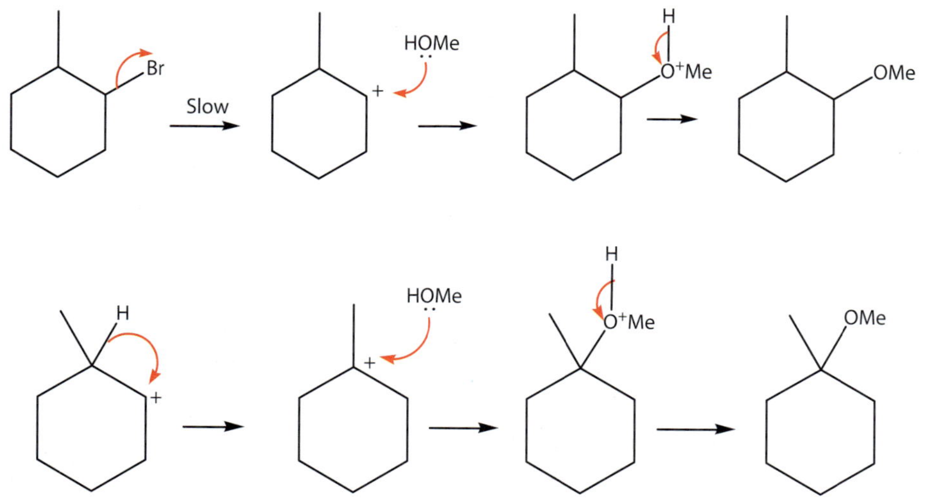

Another class of substrates in which rearrangements commonly occur is allylic derivatives, **9.13** (Figure 9.24). If we do a simple S_N2 displacement on the allyl bromide, the product is obvious. However, another process also involves sensible curly arrows and gives exactly the same product. We describe this as an S_N2' process. In the absence of any substituents, how can we tell which one is operating?

The solution is to label either the hydrogens with deuterium or one of the carbon atoms with ^{13}C (Figure 9.25).

If an allyl halide reacts by an S_N1 mechanism, then resonance will equivalence the two termini, in the absence of substituents (Figure 9.26). The two products differ only in the location of the ^{13}C label and will be formed in equal amounts.

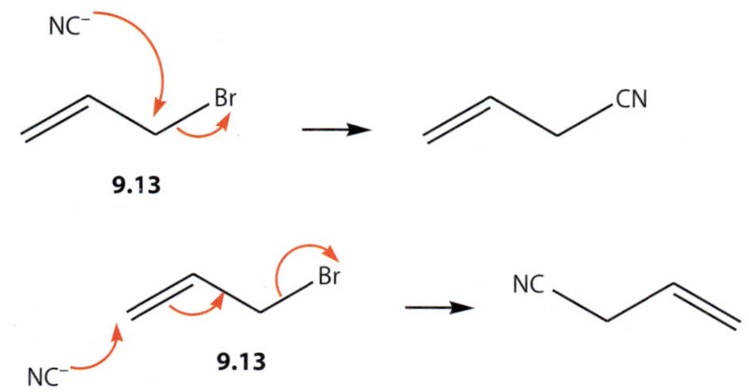

FIGURE 9.24 S_N2 and S_N2' processes in allylic substitution.

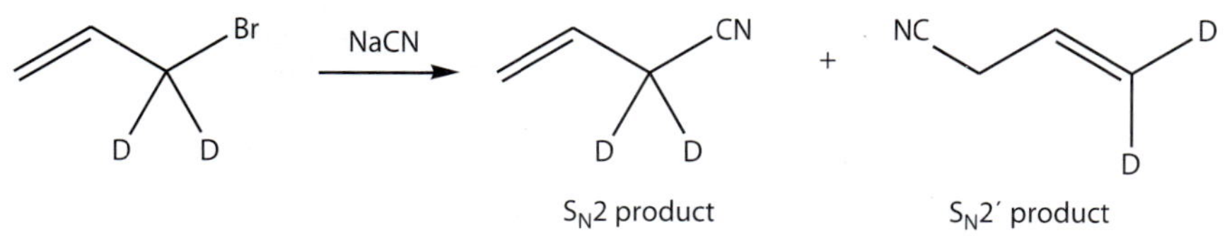

FIGURE 9.25 Deuterium labeling is used to study allylic substitution.

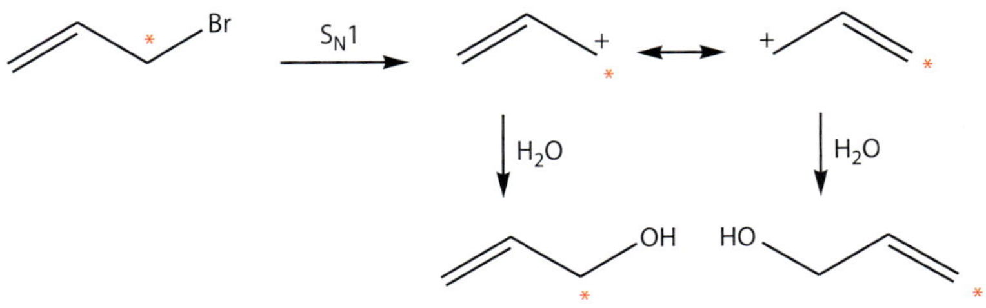

FIGURE 9.26 S_N1 substitution of an allylic system.

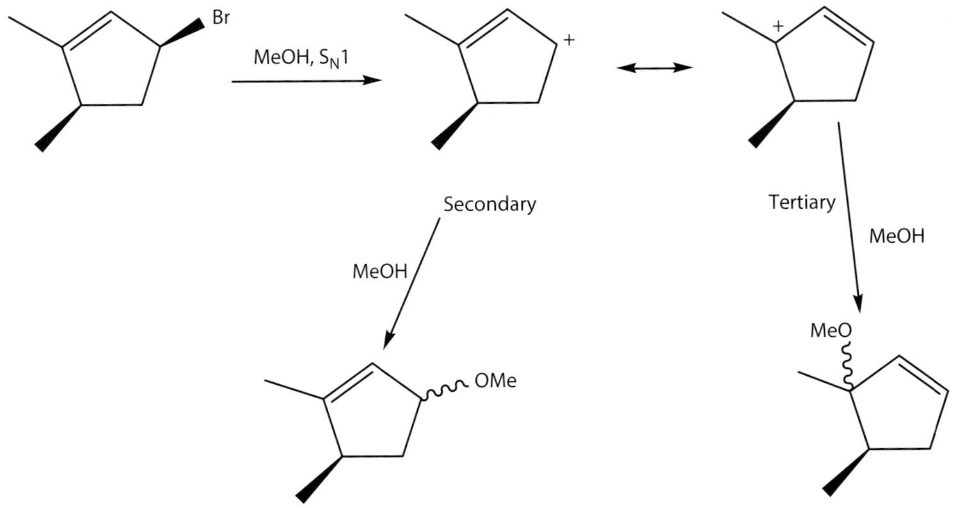

FIGURE 9.27 S_N1 substitution of a nonsymmetrical allylic system.

However, if the two resonance forms of the allylic cation are not equivalent, then the major product will be formed from the one that contributes most to the structure. An example is shown in Figure 9.27. The allylic cation produced in an S_N1 process has two forms, one secondary and one tertiary. The tertiary form is the more important, and the products are largely derived from that. Note that we get two diastereoisomers; the cation is planar and may be attacked from either face.

PROBLEM 9.9

If the substrate in the previous reaction is treated with methoxide ion rather than methanol, then a bimolecular process occurs. What is the likely product?

Solution

Although both S_N2 and S_N2' processes are possible in principle, steric hindrance makes the direct S_N2 reaction much the more likely. It will go with inversion of configuration to give a single product, **9.14**.

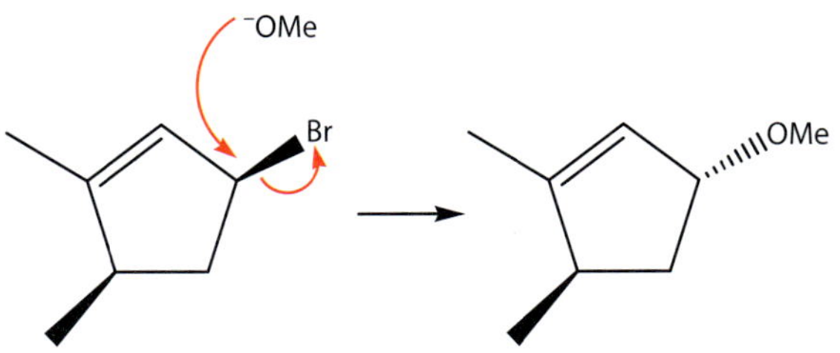

9.14

9.9 COMPETITION BETWEEN S_N1 AND S_N2 PROCESSES

This is a good point to pause and summarize (Table 9.7) what we know about the two processes we have discussed—what factors favor S_N1 and what favors S_N2? Figure 9.28 shows how a change in nucleophile can result in a change of mechanism. This is a reaction that could be either S_N1 or S_N2. The cation formed would be primary, but benzylic, so good, but not great. The leaving group is good, which would help either mechanism. With the good nucleophiles, PhS^- and HO^-, the rate of the reaction is clearly related to the nature of the nucleophile. So the nucleophile is involved in the RDS and these are S_N2 processes. With the weaker nucleophiles, Cl^-, PhO^-, and AcO^-, the reaction rates are much the same. This suggests that here we are seeing the rate of a single process, cation formation, without the involvement of the nucleophile, that is, an S_N1 process.

PROBLEM 9.10

How would we order compounds A–E in terms of their reactivity in

(a) A typical S_N1 hydrolysis in water
(b) A typical S_N2 reaction with iodide ion in acetone:

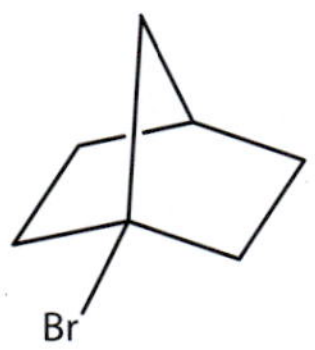

(a)

$PhCH_2Br$
(b)

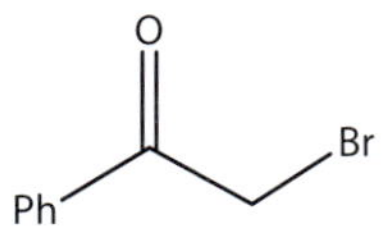

(c)

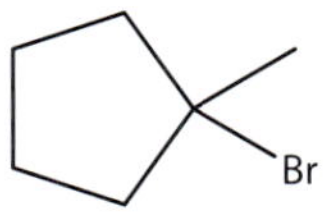

(d)

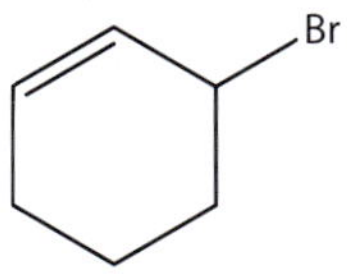

(e)

Solution

(a) D > E > B ≫ A, C (don't react)
Although A is tertiary, the carbocation that would be formed on breaking the carbon–bromine bond could not be planar, and this system never undergoes S_N1 reactions.
(b) C > B > E > D,A (don't react)
Look back at Section 9.6, if you did not get the first three in the right order.

TABLE 9.7 Factors Favoring S_N1 and S_N2 Substitution

Favors S_N1	Favors S_N2
Good, stable carbocation	Low steric hindrance
Good leaving group	Good leaving group
Weak nucleophile (to avoid S_N2)	Good nucleophile
Medium that solvates both anions and cations (typically water, alcohols)	Medium that does not heavily solvate the nucleophile (typically acetone, DMSO)

$$X^- + Ph_2S^+CH_2Ph \rightarrow Ph_2S + PhCH_2X$$

X^-	PhS^-	HO^-	Cl^-	PhO^-	AcO^-
Rate × 10^5	110	74	3.8	4.0	3.9

FIGURE 9.28 Competition between S_N1 and S_N2 processes.

Key Points from Sections 9.6 through 9.9

- Tertiary substrates react by the S_N1 mechanism, primary (unconjugated) substrates by S_N2 processes. Secondary substrates may follow either mechanism, depending on the substrate and the nucleophile. Any type of steric hindrance slows S_N2 reactions.
- Substrates with an adjacent double bond, especially α-haloketones, react more rapidly in S_N2 processes.
- Polar protic solvents favor ionization and hence S_N1 reactions. Polar aprotic solvents, unable to solvate nucleophiles, favor S_N2 processes.
- Elimination processes compete with both types of substitution.
- Carbocation rearrangements compete with S_N1 processes. In allylic systems, allylic transposition may occur in either S_N1 or S_N2 reactions.
- S_N1 reactions are favored by good carbocations, good leaving groups, polar protic solvents, weak nucleophiles. S_N2 reactions are favored by low steric hindrance, good nucleophiles and leaving groups, and polar aprotic solvents.

9.10 APPLICATIONS AND SCOPE OF THE REACTION

Before we discuss specific nucleophiles, leaving groups, and transformations, we should be clear about what this type of reaction will not do. Substituents attached to sp or sp^2 carbon atoms cannot be substituted in this way. S_N1 is precluded by the instability of sp^2 or sp cations. S_N2 is impossible because backside attack is inhibited by the π-system. Thus, in Figure 9.29, only the allylic or benzylic halides are substituted. This is not to say that replacement of vinyl or aryl halides is impossible—simply that it does not happen in the same way, or under the same conditions (see Sections 13.4 and 23.6.3), as substitution at sp^3 carbon atoms.

9.10.1 HALIDE IONS AS NUCLEOPHILES

Halide ions (other than fluoride) are both good nucleophiles and good leaving groups, so we expect that displacement of a leaving group by a halide ion would be good way to prepare haloalkanes. The main substitution reaction used to make haloalkanes involves the loss of water from an alcohol. If we were to consider the reaction of ROH with Br^- as a route to an alkyl bromide, this looks unpromising—hydroxyl ion is a poor leaving group, and bromide ion will not displace it. So we need to make the OH into a good leaving group. There are several ways to do this—protonating it so that the leaving group is water is the simplest. So to convert ROH to RBr, we can use HBr (Figure 9.30). Water is the conjugate base of H_3O^+, which is a strong acid. Examples are shown in Figure 9.31; the first two are S_N2 processes at primary centers, the third involves an S_N1 mechanism.

The same substitution reaction can be accomplished with HCl to give a chloride, but this is more difficult, as chloride is a poorer nucleophile than bromide. Zinc chloride, a Lewis acid, is sometimes added to facilitate the reaction (Figure 9.32).

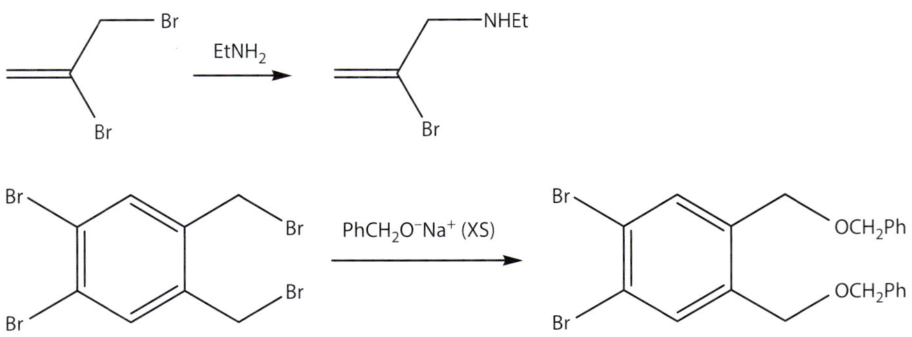

FIGURE 9.29 Neither S_N1 nor S_N2 substitution occurs at sp^2 or sp centers.

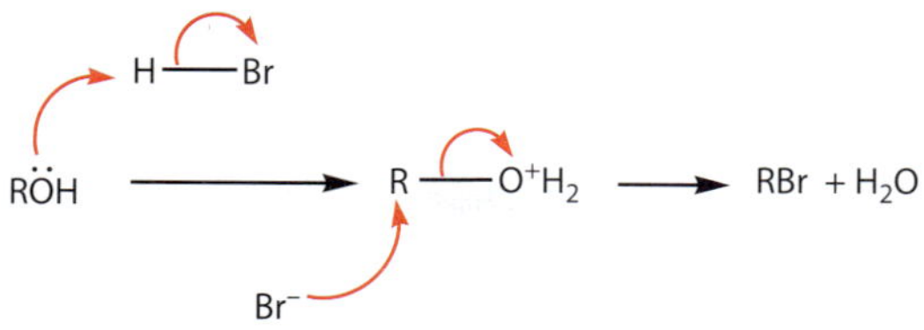

FIGURE 9.30 OH can be converted into a good leaving group by protonation.

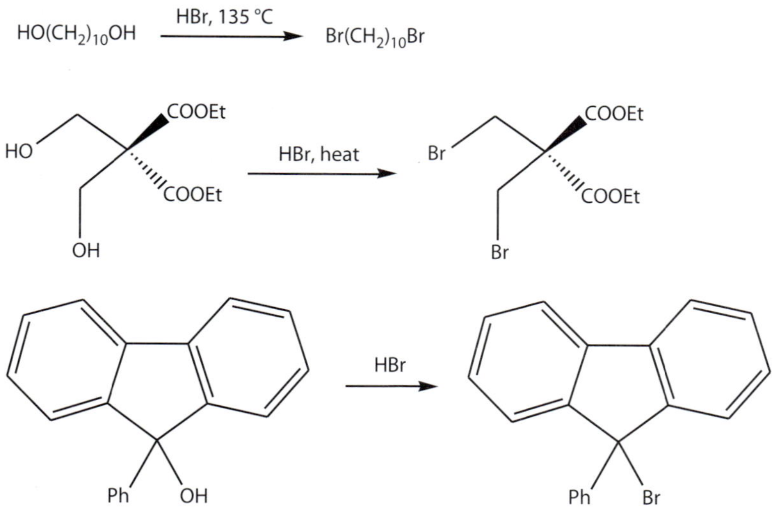

FIGURE 9.31 Conversion of alcohols to bromides with HBr.

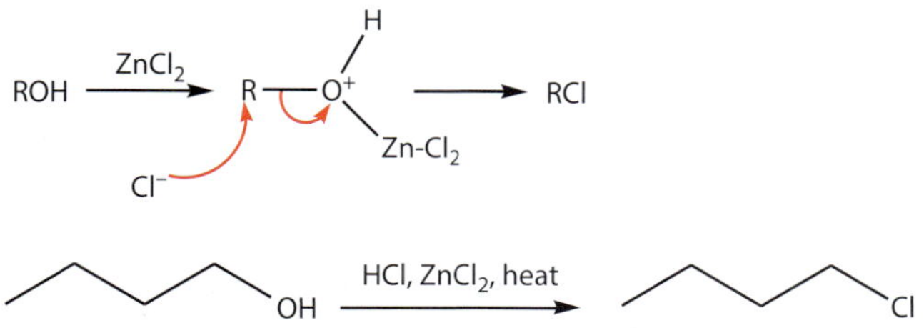

FIGURE 9.32 Conversion of alcohols to chlorides.

Although the reactions with acids are easy, and inexpensive, not all molecules can tolerate the harsh conditions involved. Another method for turning OH into a leaving group, and then displacing it with halide, involves the reaction of the alcohol with PX_3 (X = Cl or Br; PI_3 can be formed and used *in situ*). The lone pair of electrons on oxygen attacks phosphorus, displacing bromide. All three bromides can be displaced. The bromide ions then attack at carbon to give the alkyl bromide and eventually $P(OH)_3$, which rearranges to $HP(=O)(OH)_2$ (Figure 9.33). Because of the S_N2 nature of the final displacement step, this reaction is only useful for primary and secondary alcohols. Some examples are given in Figure 9.34. Of the three halogens, much the most popular to be introduced in this way is bromine; other methodologies are more common for

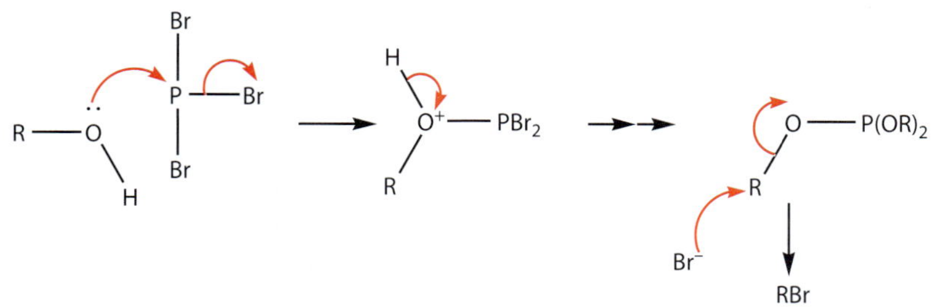

FIGURE 9.33 Mechanism of halogenation of alcohols with PX_3.

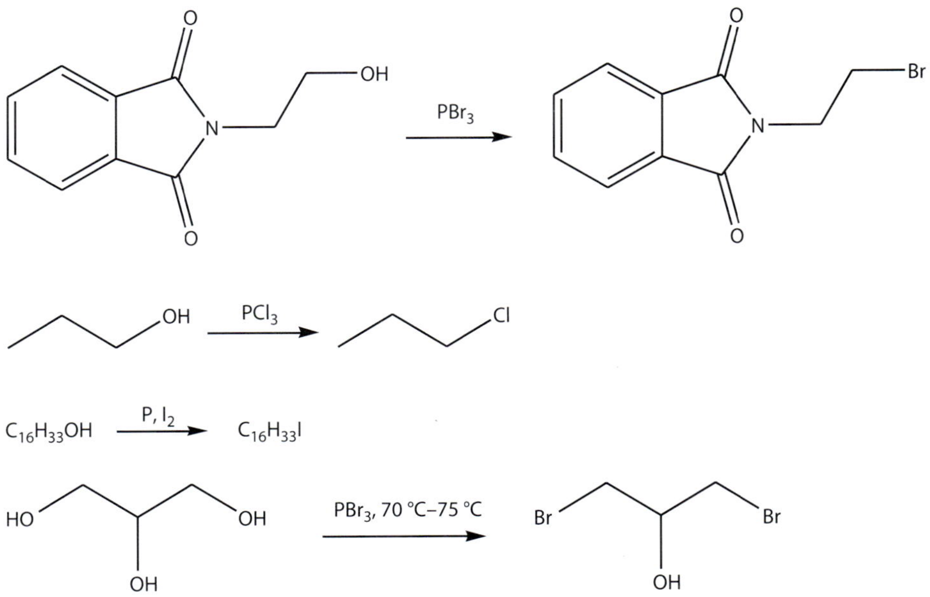

FIGURE 9.34 Halogenation of alcohols using PX_3.

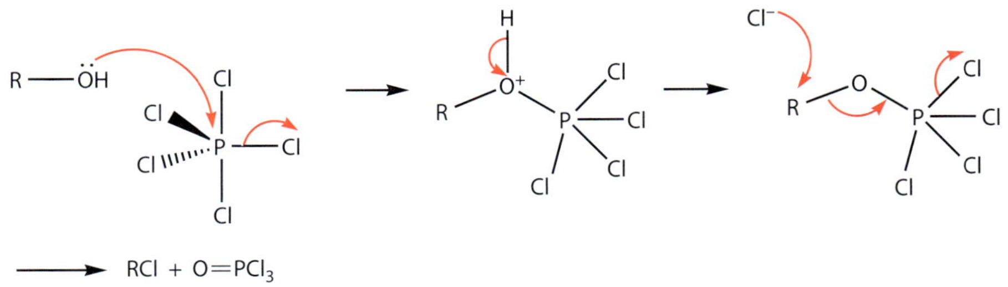

FIGURE 9.35 Mechanism of chlorination of alcohols by PCl_5.

chloride and iodide. The last example is interesting; the primary alcohols can be reacted selectively in the presence of the more hindered, secondary alcohol.

Bromine and chlorine can also be introduced by reaction of alcohols with PX_5; the reaction mechanism is shown in Figure 9.35. The species expelled is $POCl_3$, and the driving force for the process is the strength of the phosphorus–oxygen double bond (460 kJ mol^{-1}).

A method specifically for conversion of alcohols to chlorides involves the use of thionyl chloride ($Cl_2S{=}O$), and this brings us to our third and final mechanism for substitution, the S_Ni process. This stands for substitution, nucleophilic, intramolecular. The method has the advantage of interesting and predictable stereochemistry, and by-products that are all gaseous, and so it is generally preferred over the phosphorus halide–based methods to introduce chloride. The mechanism of the first step of the reaction is shown in Figure 9.36. The alcohol lone pair attacks the sulfur atom of the thionyl chloride, and the π-electrons of the double bond are displaced to oxygen. A proton is lost from the oxygen atom, and the sulfur oxygen double bond is reformed, with loss of chloride ion. The two products of the process are the chlorosulfite ester, in which the OH has been converted into a better leaving group, and hydrogen chloride, formed from the lost proton, and the displaced chloride.

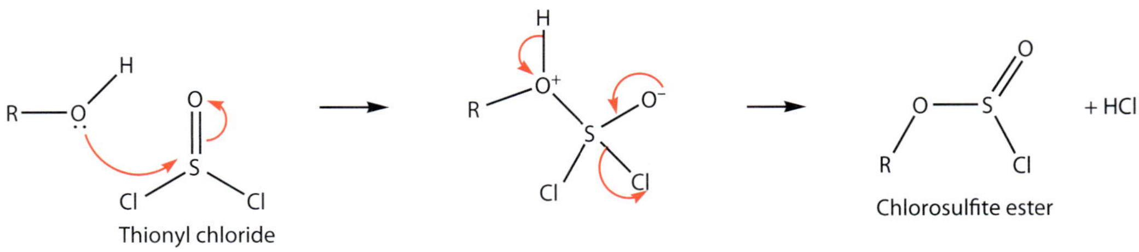

FIGURE 9.36 Conversion of an alcohol to a chlorosulfite ester.

What happens next depends on the solvent in which the reaction is performed. If the reaction takes place in ether, HCl is not dissociated, so there is no free chloride ion to act as a nucleophile. We are dependent on a much slower process, the collapse of the chlorosulfite ester to give a carbocation, and the chlorosulfite anion, which rapidly collapses to SO_2 transferring chloride to the nearest face of the carbocation (the formation of this molecule of gas, with its associated entropy, is the driving force for this whole process). Overall, the process gives an unusual retention of stereochemistry—the cation has no opportunity to rotate before it is attacked (Figure 9.37). This is the SN_i mechanism.

By contrast, if pyridine is present in the solution (other bases can also be used), this reacts with HCl to give the salt pyridinium chloride (Figure 9.38).

Now free chloride ion is available, and the reaction goes by a regular S_N2 process, with inversion of stereochemistry (Figure 9.39). This is the more common version of the reaction, and there are many examples involving primary and secondary alcohols.

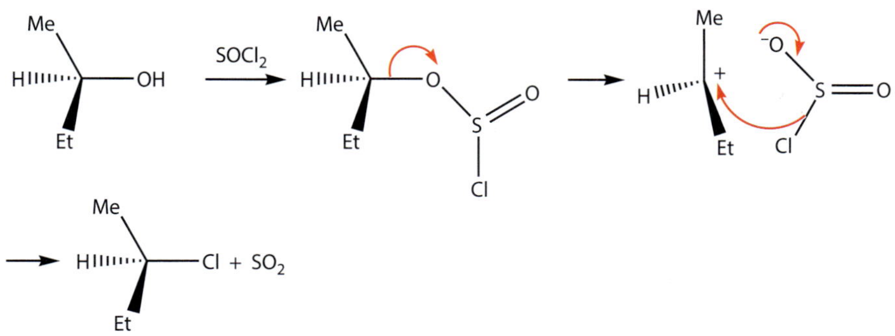

FIGURE 9.37 The S_Ni mechanism for chlorination of alcohols.

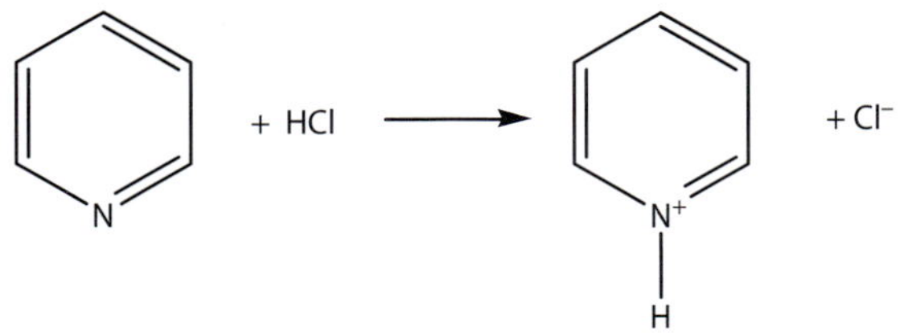

FIGURE 9.38 Formation of pyridine hydrochloride.

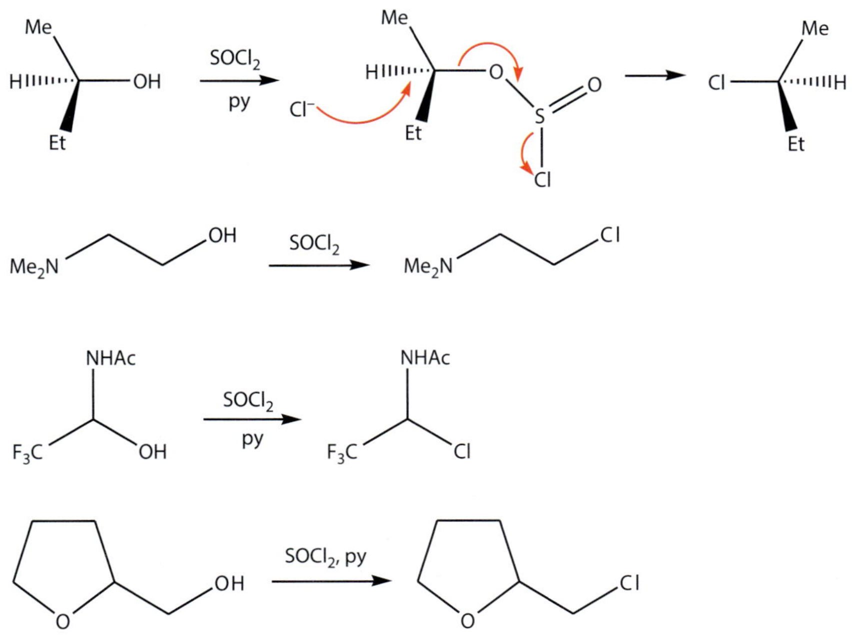

FIGURE 9.39 S_N2 chlorination of alcohols using thionyl chloride in pyridine.

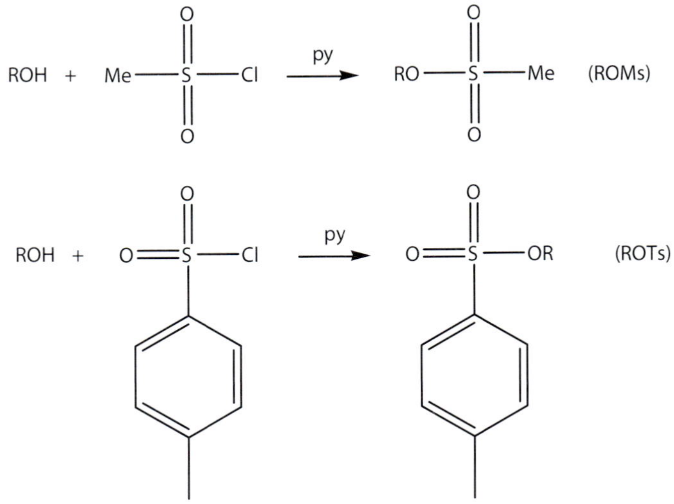

FIGURE 9.40 Formation of mesylates and tosylates.

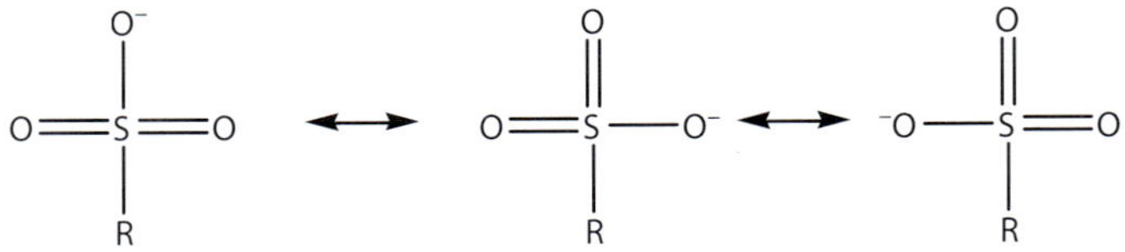

9.15

Our final method of converting OH to a good leaving group is shown in Figure 9.40 with a sulfonyl chloride such as $MeSO_2Cl$ (methane sulfonyl chloride or mesyl chloride) or 4-$MeC_6H_4SO_2Cl$ (toluene sulfonyl chloride or tosyl chloride) to give sulfonate esters that are described, respectively, as mesylates and tosylates. Tosylate and mesylate anions are excellent leaving groups as they are the anions of the strong acids methane sulfonic acid ($pK_a = -1.9$) and toluene sulfonic acid ($pK_a = -2.8$). Why are these sulfonic acids so strong? The anions have several resonance forms where the negative charge is on oxygen (**9.15**).

Tosylates and mesylates are readily displaced by halides; examples are given in Figure 9.41. At this point, you may feel we are rather laboring a point, about conversion of alcohols to halides; however, this lays the groundwork for conversions of alcohols to a range of other functional groups—we will always have the issue that hydroxyl ion is not a good leaving group, and we need to make it into one. All of the conversions we have done here will become familiar themes as we look at other types of displacements.

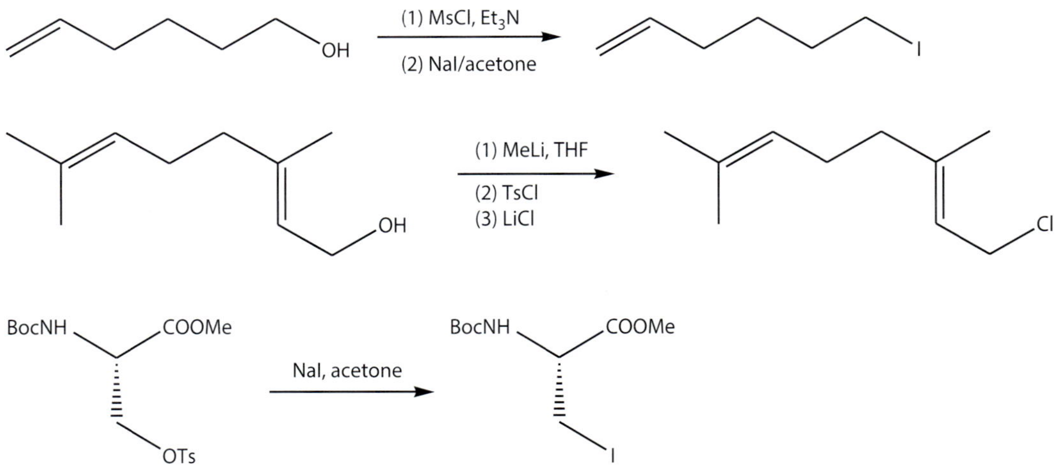

FIGURE 9.41 Displacement of tosylates and mesylates by halide anions. Boc is a protecting group for amino groups—we'll learn about this in Section 22.4.2.

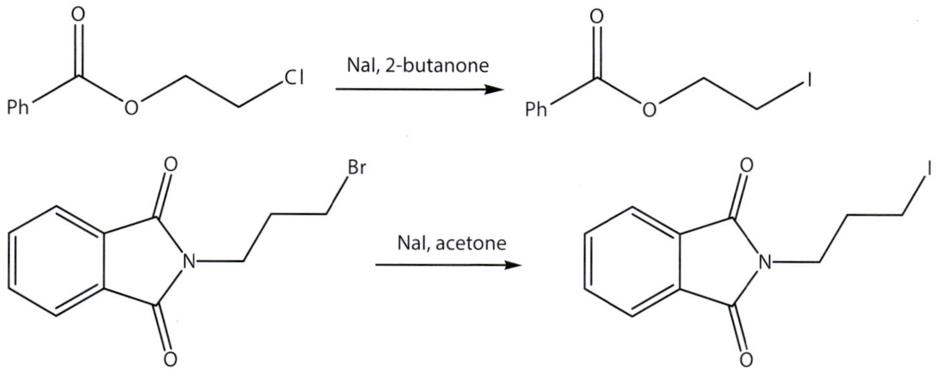

FIGURE 9.42 Formation of iodides by displacement of other halides.

PROBLEM 9.11

Explain the previous reaction, drawing mechanisms, with particular regard to the stereochemical outcome:

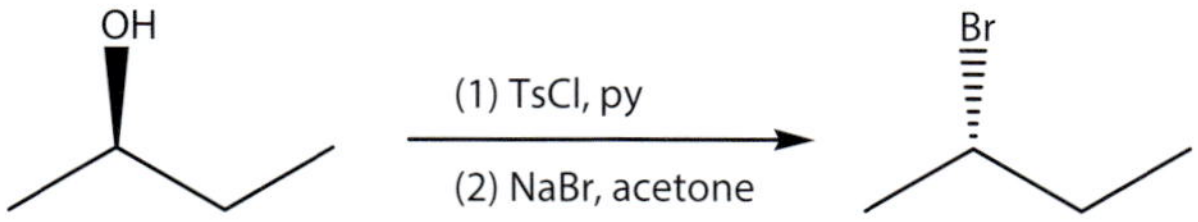

R-2-Butanol   *S*-2-Bromobutane

Solution

The first step of the reaction involves the formation of the tosylate, **9.16**. At this point, no reaction has occurred at the chiral center, so the stereochemistry is unchanged. The second step is an S_N2 reaction, which goes with inversion of configuration.

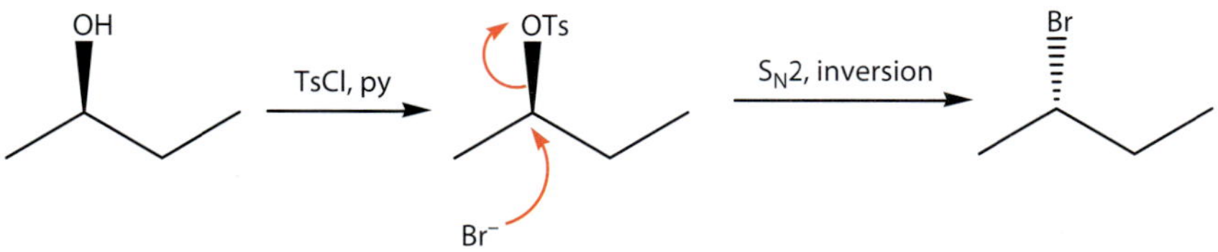

9.16

You may be wondering if halides displace anything other than OH groups that have been modified in some way. These hydroxyl to halide transformations form the bulk of those in practical use with halide nucleophiles. However, there is a small group of reactions where iodide displaces another halide, where the starting material is either readily available or easier to synthesize than the iodide (there is no iodide or bromide equivalent of $SOCl_2$, and red phosphorus/iodine mixtures are difficult to handle, with by-products requiring careful disposal) (Figure 9.42).

PROBLEM 9.12

Predict the product(s) of each of the following reactions:

(a)

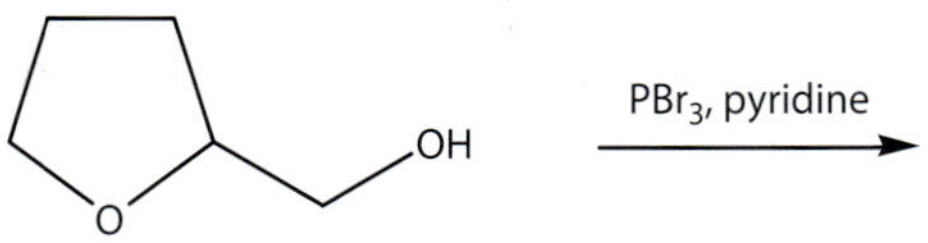

(b) $n\text{-}C_{12}H_{25}OH$ $\xrightarrow{\text{HBr, 100 °C–120 °C}}$

(c)

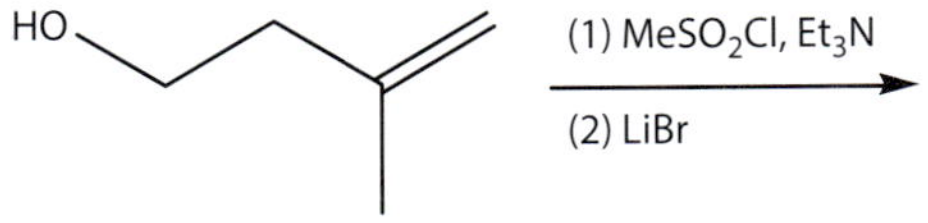

(d)

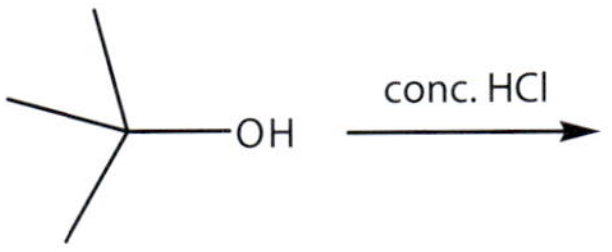

(e)

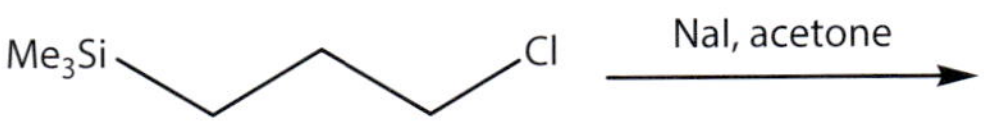

Solution

All of these are straightforward examples, showing the range of reagents that may be used. (a), (b), and (c) are all S_N2 processes; in (c), the mesylate intermediate may be isolated, if necessary. (d) is an S_N1 reaction. In (e), the reaction is S_N2 with the excellent nucleophile, iodide, displacing chloride:

(a)

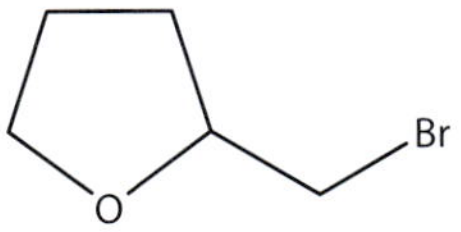

(b) n-$C_{12}H_{25}Br$ (c)

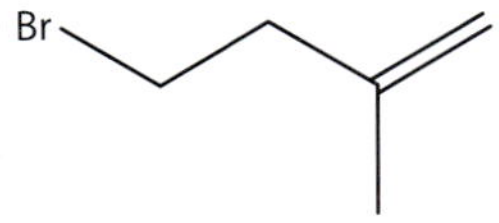

(d)

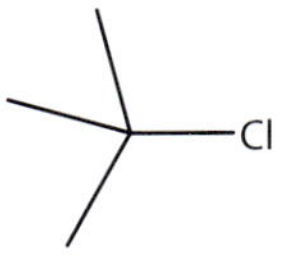

(e)

9.10.2 OXYGEN AND SULFUR AS NUCLEOPHILES

Hydroxyl ion and alkoxyl anions are good nucleophiles, alcohols and water rather less so. The sulfur analogues are much better nucleophiles. Generally speaking, we would choose anions, where we expect an S_N2 reaction, and the neutral molecules for S_N1 reactions at tertiary centers. Thus, primary and secondary halides can be converted into alcohols/ethers, with hydroxyl or alkoxy ions (the first two reactions in Figure 9.43), and tertiary halides with water/alcohols (the third example). In practice, alcohols are rarely prepared this way, since the alkyl halide was often prepared from the corresponding alcohol. The product of the second reaction is 18-crown-6 (18-membered ring with 6 oxygen atoms; see Sections 4.4.3 and 23.4); the base converts the alcohol groups into more nucleophilic alkoxides. All of these processes result in the formation of ethers, and the reaction is called the Williamson synthesis. Elimination reactions compete with the S_N2 substitutions; both elimination and rearrangement compete with S_N1 processes. Sulfur analogues are also known (final example in Figure 9.43); because sulfur is a better nucleophile than oxygen, elimination is less of a problem.

Intramolecular versions of these reactions are common, particularly when three-, five-, or six-membered rings are being formed (Figure 9.44; the bases deprotonate the OH groups to make them into better nucleophiles). Remember that when we consider both enthalpy and entropy, it is easy to make three-, five-, and six-membered rings but more difficult to make others.

Allylic halides can react either by S_N2 or S_N2' processes. The first reaction of Figure 9.45 goes entirely by an S_N2 process, for steric reasons. Notice that the alkenyl halide is unreactive. By contrast, in the second example, the main reaction is S_N2', again for steric reasons.

Carboxylate anions are less good nucleophiles than alkoxides (the charge is spread between the two oxygen atoms by resonance), but they can still displace halides (Figure 9.46). Sometimes, however, the reaction needs a little help, often by using a cesium salt of the carboxylate anion.

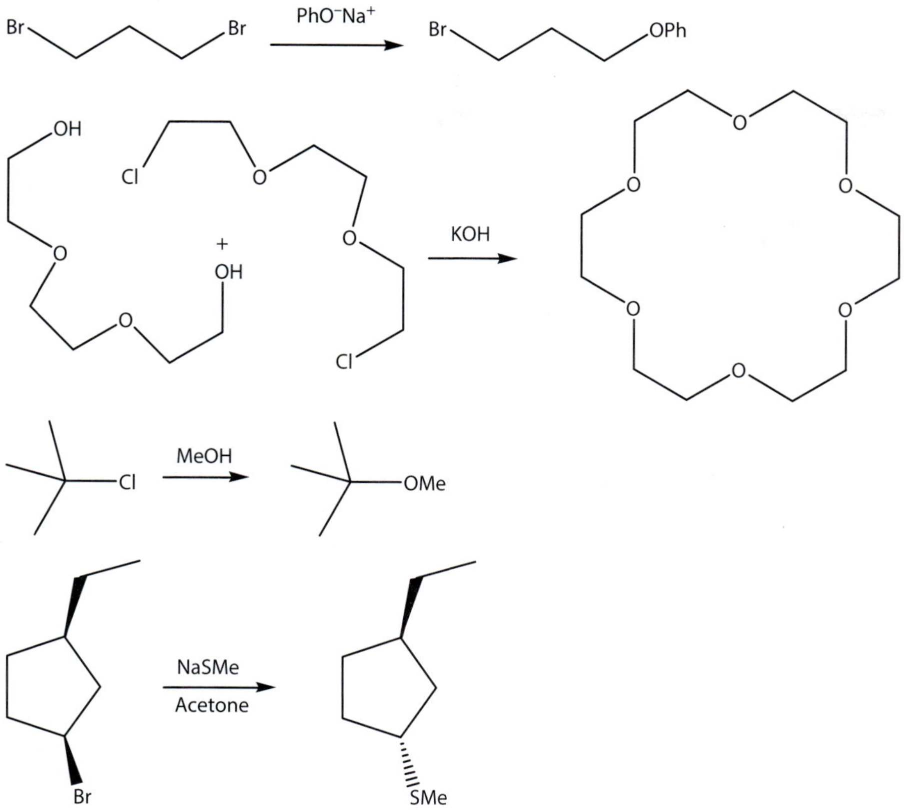

FIGURE 9.43 Williamson syntheses of ethers and thioethers.

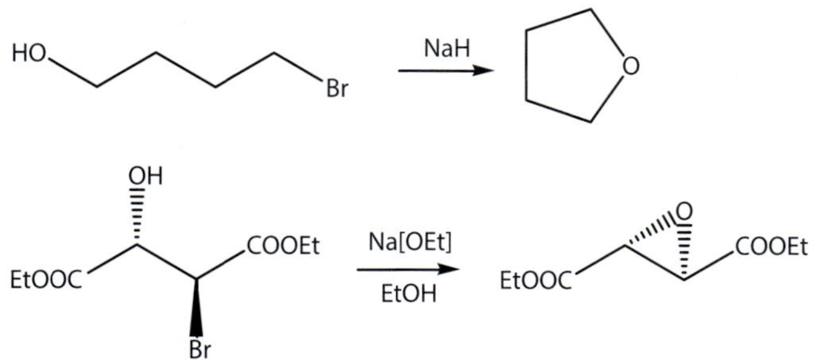

FIGURE 9.44 Intramolecular reactions for ether synthesis.

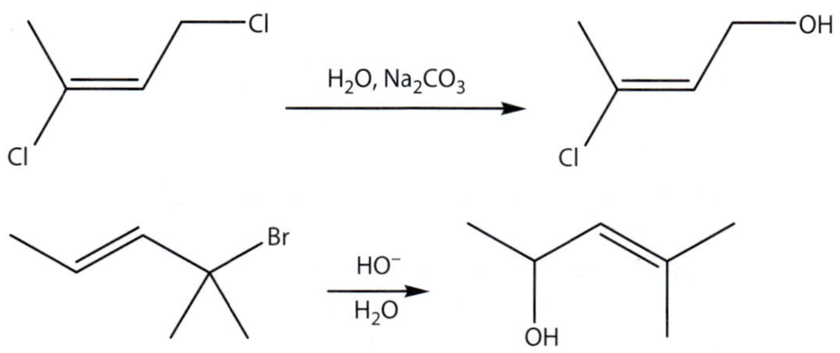

FIGURE 9.45 Substitution of allylic halides by oxygen nucleophiles.

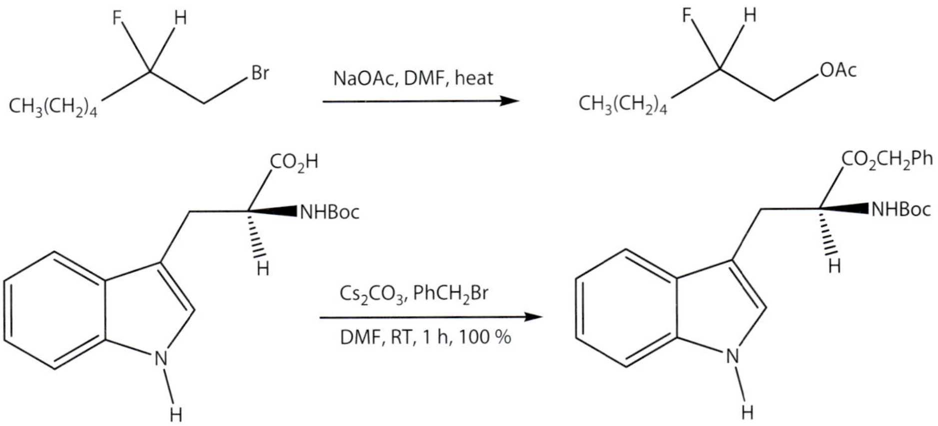

FIGURE 9.46 Displacement of halides by carboxylate anions.

PROBLEM 9.13

Predict the product(s) of each of the following reactions:

(a)

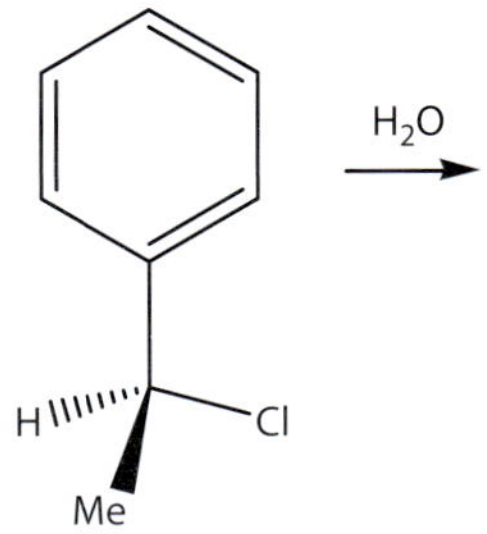

(b)

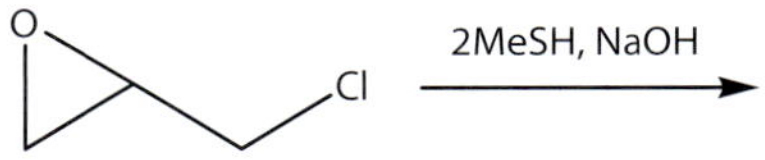

(c)

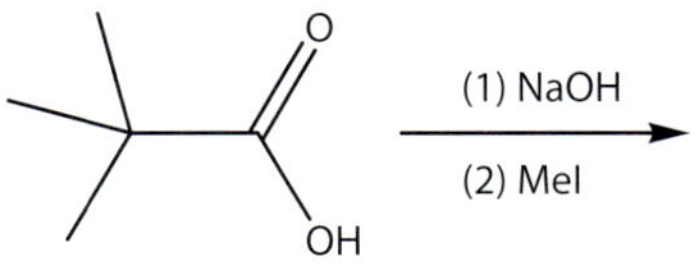

Solutions

(a) Water as nucleophile, and a secondary benzylic center, suggests that this will be an S_N1 reaction. Since the carbocation is planar and may be attacked from either face, a racemic product will be obtained (remember that the wavy bonds indicate random stereochemistry):

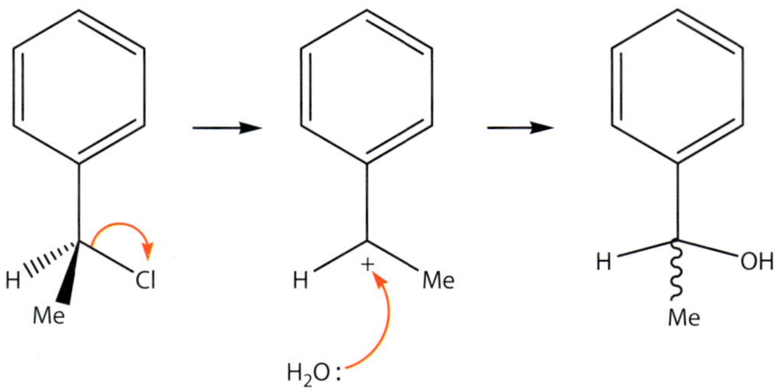

(b) The question states that we have 2 mol of the good nucleophile MeS^- (MeSH and HO^- are in equilibrium, but the thiolate is much the better nucleophile), and we have two electrophilic centers, so we expect both to be substituted. The halide is substituted in an S_N2 process, and the epoxide is attacked, also in an S_N2 reaction, at the less hindered center:

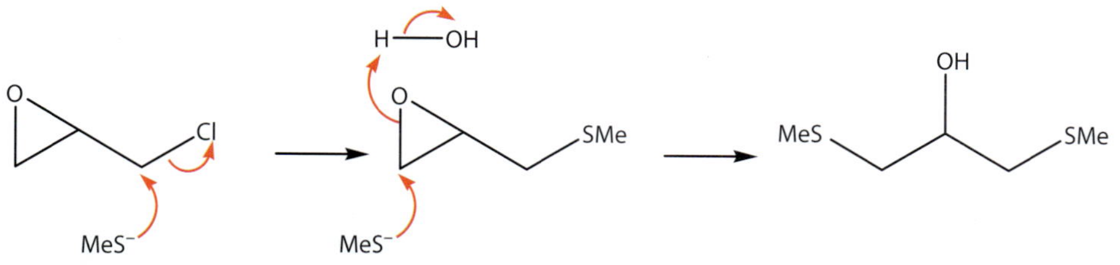

(c) Reaction of the acid with sodium hydroxide produces the anion, and this then substitutes iodomethane to give the ester. This route was chosen because the usual ones (which, as we will see in Chapter 15, involve nucleophilic addition to the carbonyl group) are inhibited by steric hindrance:

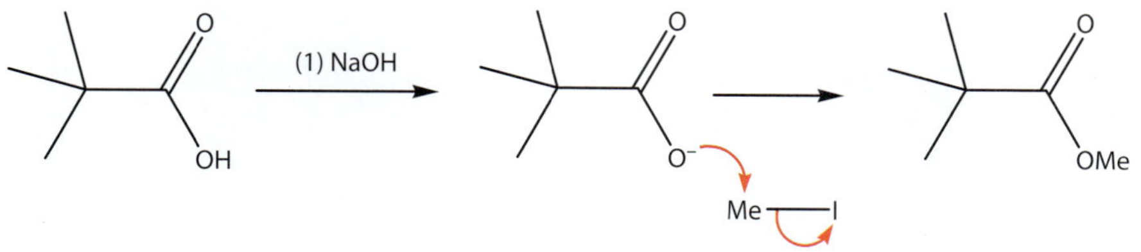

Key Points from Sections 9.10.1 and 9.10.2

- All the halide anions can act as nucleophiles; iodide is the best, fluoride the least good.
- In order for alcohols to be converted to halides, the OH group has to be converted into a good leaving group. This may involve protonation, reaction with PX_3, PX_5, or $SOCl_2$, or conversion to a sulfonate.
- Iodide can displace chloride or bromide readily.
- Alkoxide anions are used in S_N2 reactions with halides to prepare ethers in the Williamson synthesis. Neutral alcohols are preferred in S_N1 processes.
- Thioethers are prepared similarly, but the reaction is easier, and there is less competing elimination.
- Carboxylate anions can displace halides to give esters; the process may be facilitated by the preformation of various metal carboxylate salts.

9.10.3 NITROGEN NUCLEOPHILES

Although ammonia and amines react readily with alkyl halides to give alkylation, the reaction is not always useful synthetically. Consider the reaction of methylamine with iodomethane (Figure 9.47). The product is dimethylamine. Because alkyl groups donate electron density, dimethylamine is *more* nucleophilic than methylamine, so that, as soon as it is formed, it will compete with methylamine for the available iodomethane. So if we restrict the amount of iodomethane, we will get a mixture of products, and if we add an excess of iodomethane, we will have a runaway reaction, to give the quaternary salt tetramethylammonium iodide.

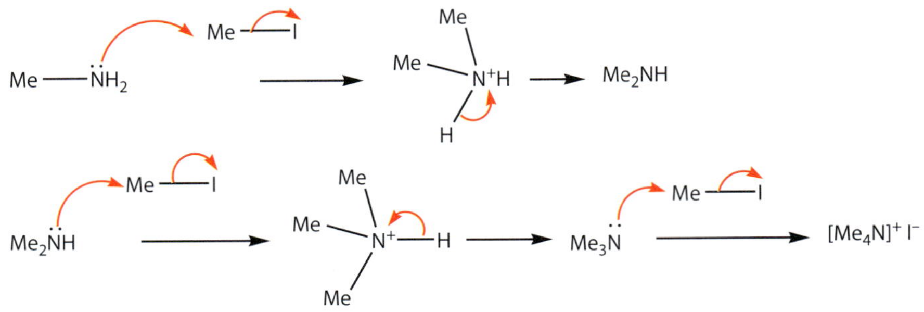

FIGURE 9.47 Reaction of methylamine with iodomethane.

Satisfactory syntheses are achieved, especially for tertiary centers, when one component is rather sterically hindered (Figure 9.48). In the first example, the reaction is exclusively S_N2 rather than S_N2', for steric reasons, and goes with clean inversion of stereochemistry. In the second reaction, the nitrogen is not a particularly good nucleophile, because of the interaction of the lone pair with the π-electrons of the aromatic rings; we need to make it more nucleophilic by deprotonation with sodium hydride. A second substitution would also be sterically disfavored. Alternatively, the synthesis works well when the electrophile bears another electron-withdrawing group such as carboxylate, which renders the first-formed product less, rather than more, electrophilic than the

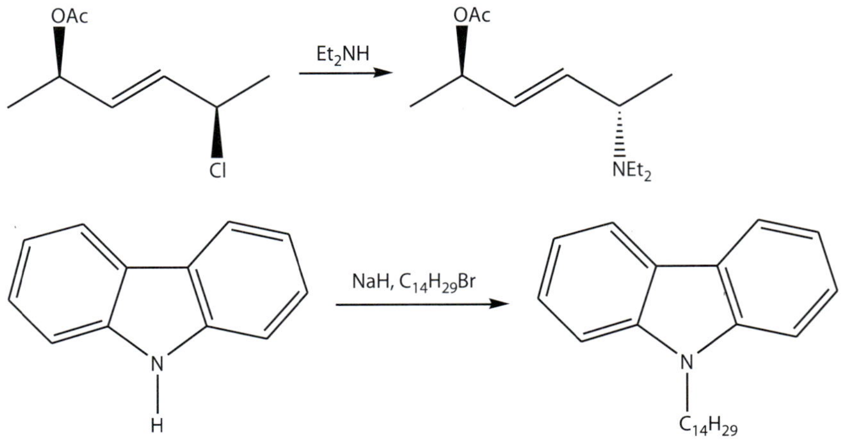

FIGURE 9.48 Overreaction can be prevented by steric hindrance.

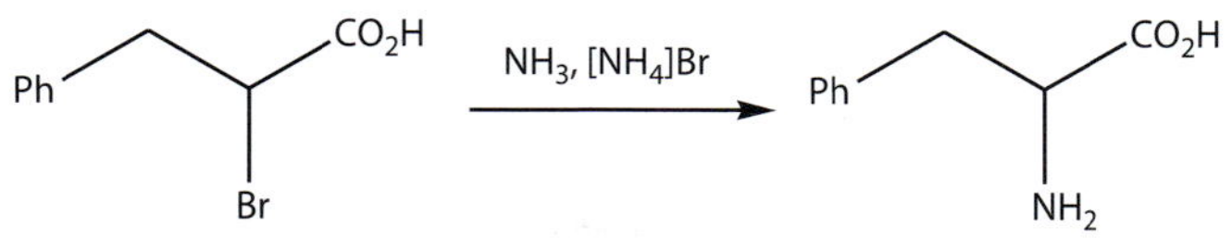

FIGURE 9.49 Clean monosubstitution of α-halocarboxylic acids by ammonia.

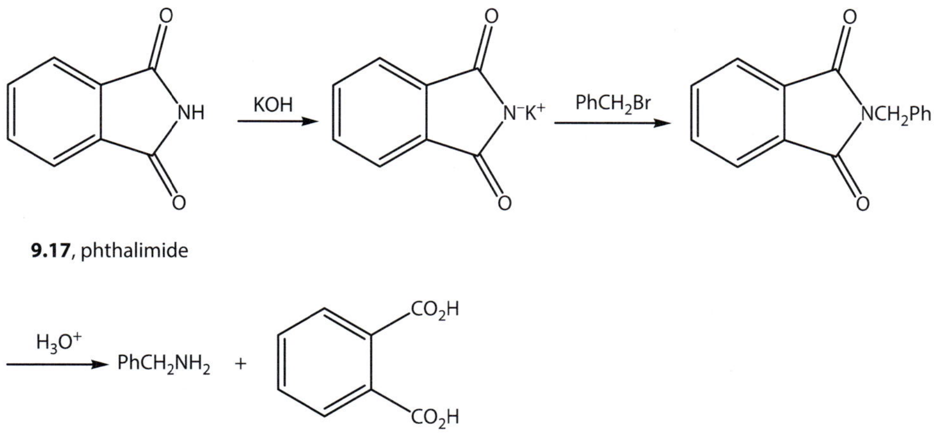

FIGURE 9.50 Gabriel phthalimide synthesis.

starting material (Figure 9.49, the synthesis of (±)-phenylalanine). This latter process has been used for the synthesis of a wide range of amino acids.

Other amine syntheses tend to involve other nitrogen nucleophiles than can later be readily transformed to amines. The Gabriel synthesis uses the anion of phthalimide, **9.17**, as the nucleophile. The pK_a of phthalimide is 8.3, so it is readily deprotonated by KOH; the negative charge is stabilized by delocalization involving the carbonyl groups. This can react only when deprotonated; the imide nitrogen is not nucleophilic. When the nitrogen has been alkylated, hydrolysis releases the amine and phthalic acid (Figure 9.50). Although the phthalic acid could be recycled to phthalimide to give a more "atom efficient" process, this is not invariably economic. The value of this approach is that overalkylation is impossible. The process works well with primary and secondary alkyl halides and sulfonates; because the displacement is an S_N2 process, elimination is the main reaction with tertiary substrates.

Another useful reagent in this context is hexamethylenetetramine, **9.18**. Think of this as a very hindered version of ammonia that will only do a single substitution. In Figure 9.51, it substitutes the benzylic halide (notice that the aryl bromide is unreactive) to give an intermediate that can be hydrolyzed to a primary amine (the other products are formaldehyde and ammonia). We will meet the mechanism for this hydrolysis in Chapter 14.

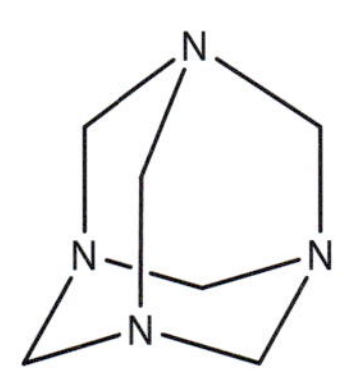

9.18, hexamethylene tetramine

The final nitrogen nucleophile that we will consider is the azide ion, $[N_3]^-$. The most important resonance form is $^-N{=}N^+{=}N^-$, and it is isoelectronic with CO_2. All azides, either organic or inorganic, should be regarded as potentially explosive; decomposing sodium azide provides the explosive release of nitrogen in car air bags, and lead azide is used as a detonator to ignite secondary explosives. Azide is the conjugate base of hydrazoic acid, HN_3, which has

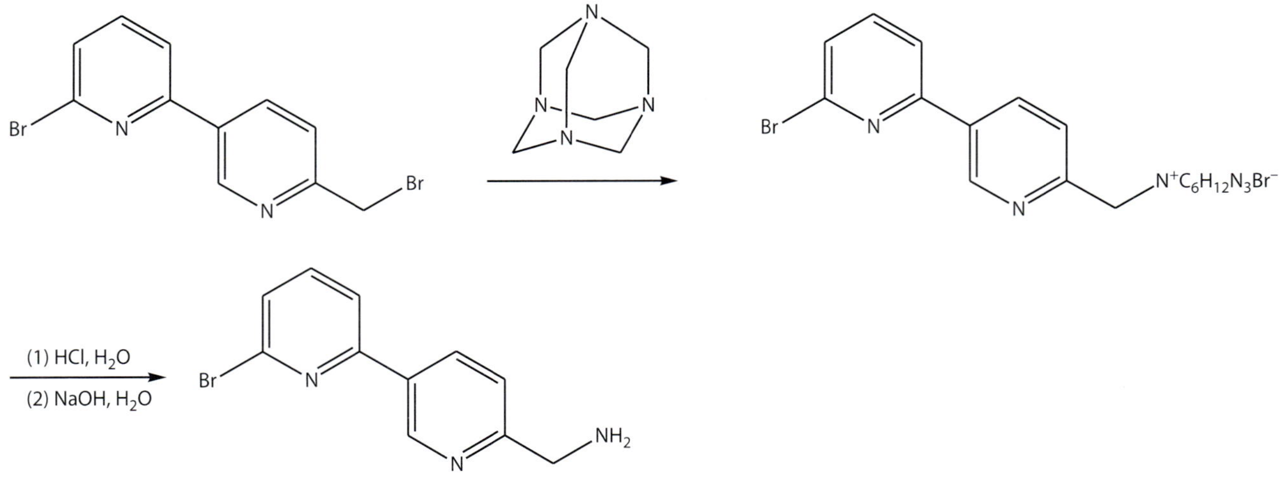

FIGURE 9.51 Substitution using hexamethylene tetramine as a proxy for ammonia—the Délepine reaction.

a pK_a of 4.6, comparable to acetic acid. So this will not be a particularly strong base, or good leaving group, but its nucleophilicity is comparable to that of bromide. Its value in synthesis is as a surrogate, more properly a synthon, or synthetic equivalent of $[NH_2]^-$. It undergoes S_N2 reactions readily at either primary or secondary centers (Figure 9.52), with reliable inversion of stereochemistry. A few S_N1 reactions at tertiary centers are known, as in Figure 9.53.

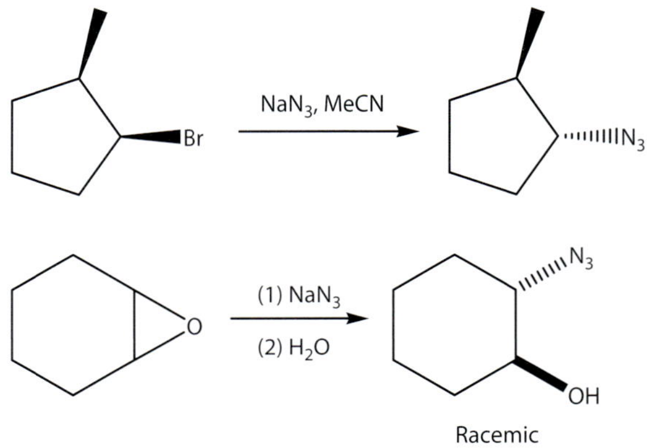

FIGURE 9.52 Azide is an excellent nucleophile.

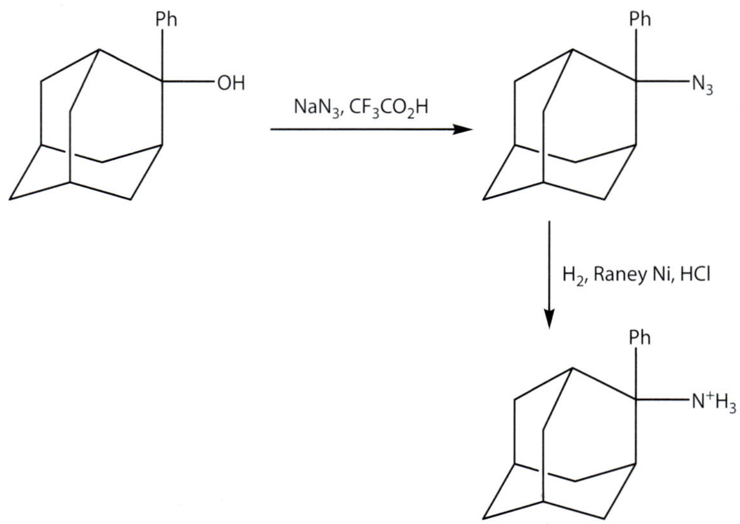

FIGURE 9.53 Azide may also be used as a nucleophile in S_N1 reactions.

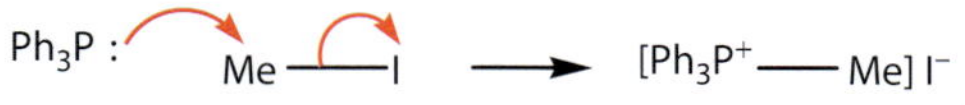

FIGURE 9.54 Phosphines are excellent nucleophiles.

Many methods of reduction from azide to amine are available: hydrogen with various metal catalysts, lithium aluminum hydride, or cobalt(II) chloride/sodium borohydride.

We noted earlier that phosphorus is a better nucleophile than nitrogen, and one particular reaction (Figure 9.54) will be very important to us when we come to study carbonyl chemistry. The phosphonium salt produced is a precursor of a Wittig reagent that is used to convert carbonyl compounds to alkenes.

PROBLEM 9.14

Predict the product(s) of each of the following reactions:

(a)

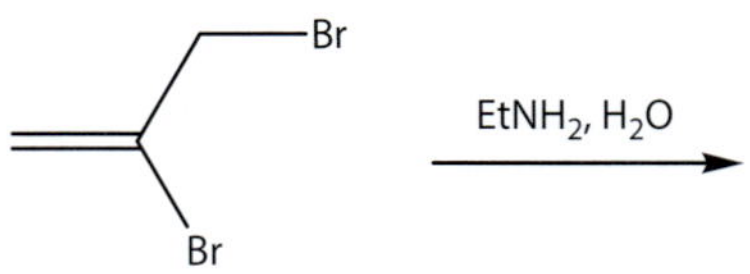

(b)

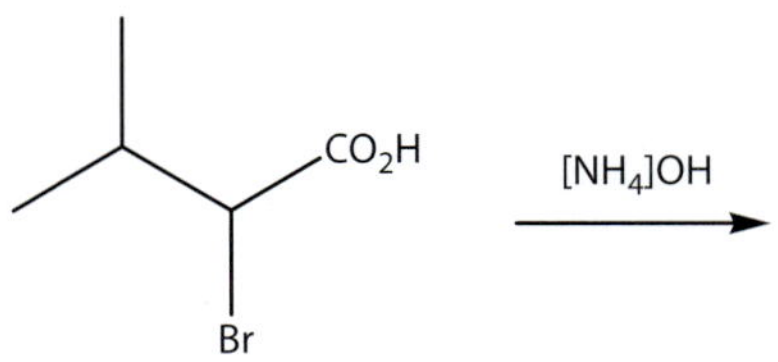

(c)

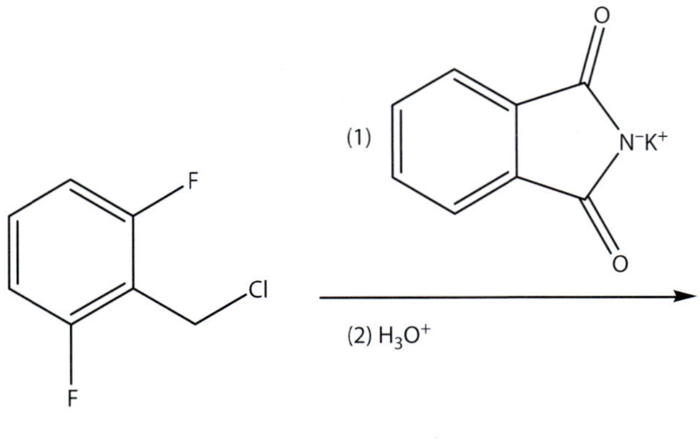

(d)

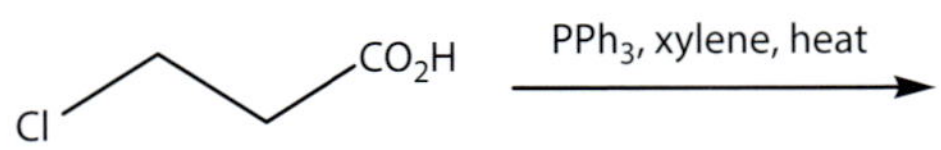

Solutions

(a) Only the allylic halide is substituted; the high reactivity of the substrate is important in the success of the reaction. Either an S_N2 or an S_N2'mechanism might be operating; the structure of the product would be the same irrespective of which predominated:

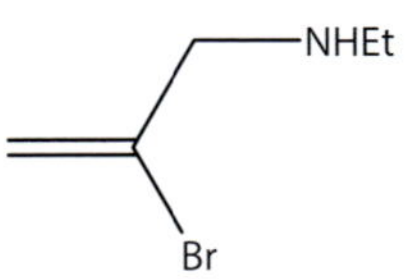

(b) This amino acid synthesis (of (±)-valine) is one of those where after substitution, the new amine is less nucleophilic than the starting material, so overalkylation is not a problem:

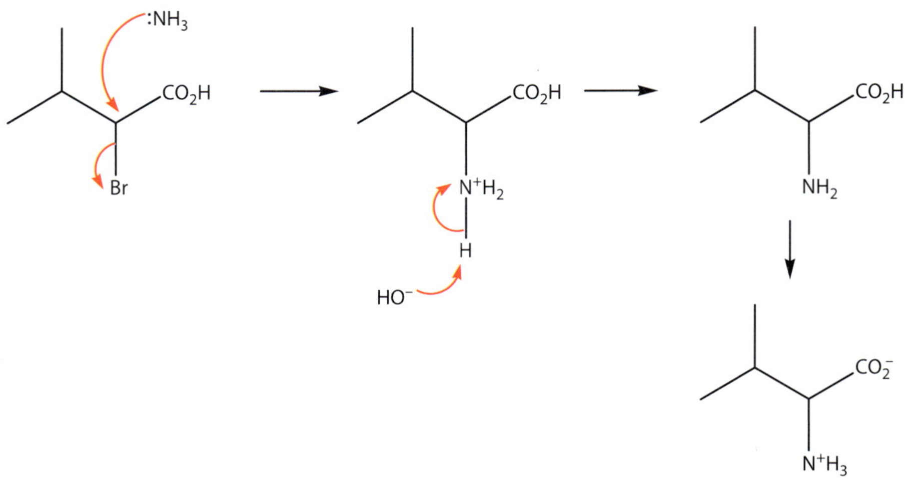

(c) This is a standard phthalimide synthesis. Only the benzylic chloride is reactive, giving

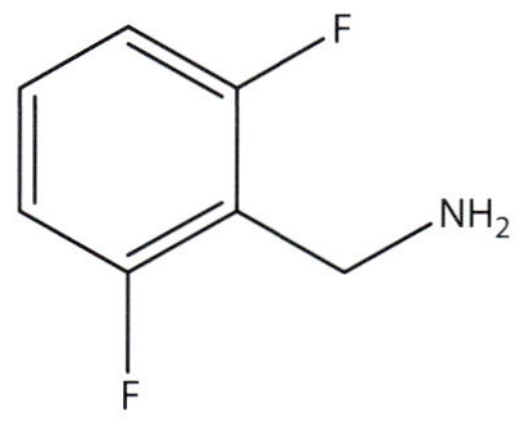

(d) Phosphorus displaces chloride to give the phosphonium salt:

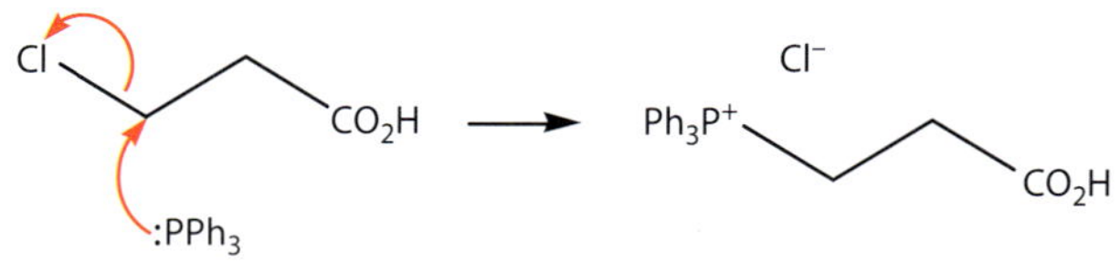

Focus on Biological Substitution Reactions

S-Adenosylmethionine, SAM, **9.19**, is the co-substrate involved in a wide range of biological methyl group transfers. Over 40 reactions are known, including those of proteins, nucleic acids, and lipids. Nucleophiles attack at the methyl group bonded to sulfur, so that the leaving group is a neutral thioether, *S*-adenosylhomocysteine. The cycle is completed by hydrolysis of *S*-adenosylhomocysteine to homocysteine and adenosine, followed by remethylation of the amino acid by cobalamin (vitamin B_{12}) using one of the

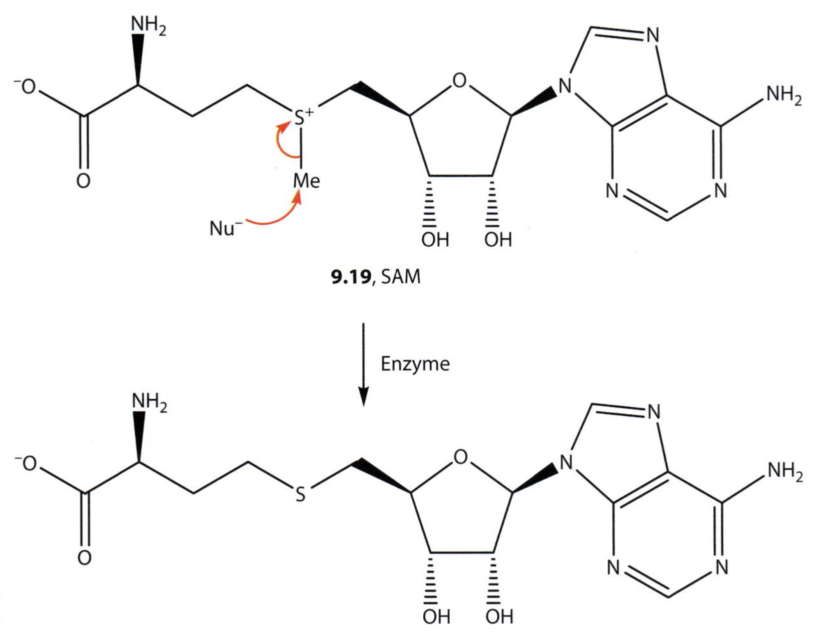

FIGURE 9.55 Methylation using SAM.

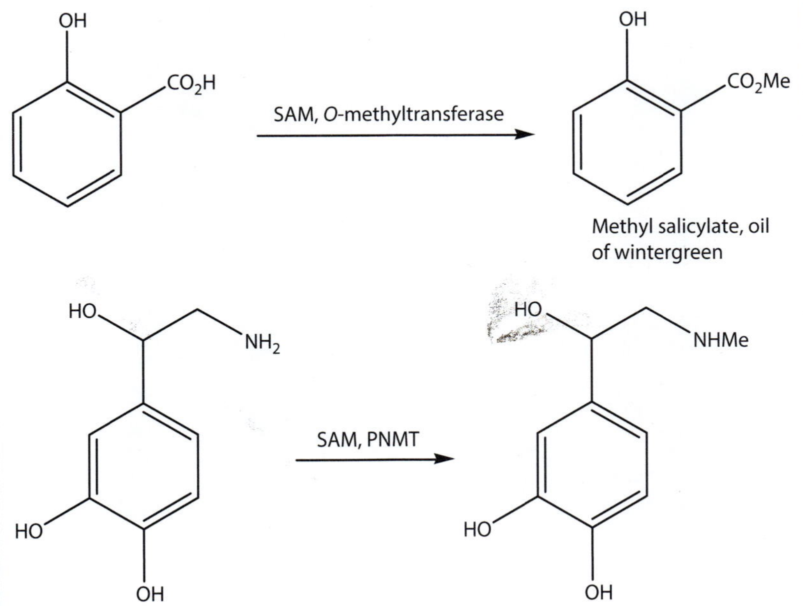

FIGURE 9.56 Biological methylations using SAM.

methionine synthase enzymes and 5-methyltetrahydrofolate as the ultimate source of the methyl group (Figure 9.55).

A wide range of alcohols and amines can be methylated using SAM in S_N2 processes. For example, salicylic acid is converted to methyl salicylate (oil of wintergreen) and norepinephrine to epinephrine (adrenaline) (Figure 9.56, PNMT is the enzyme

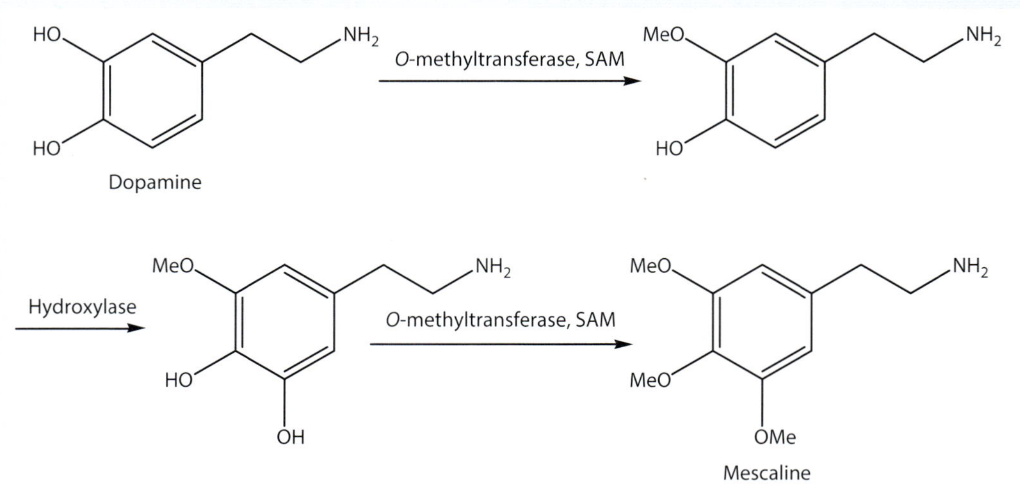

FIGURE 9.57 Biosynthesis of mescaline from dopamine.

phenylethanolamine *N*-methyltransferase). SAM also provides the methyl groups for the biosynthesis of the psychoactive principle of the peyote cactus, mescaline, from the neurotransmitter dopamine (Figure 9.57).

Another important biological molecule that makes use of the high nucleophilicity of thiols is glutathione, **9.20**. A series of enzymes known as glutathione *S*-transferases bind toxic foreign material to glutathione, so that they become more water soluble and can be excreted. One of these is the toxic metabolite of aflatoxin B1, **9.21**—found in moldy nuts and one of the most potent carcinogens and mutagens known. Aflatoxin undergoes oxidation mediated by cytochrome P450, to the epoxide, which exerts its mutagenic effect as it is opened by guanine. The sulfur of glutathione successfully competes with guanine for the epoxide, transforming it into a glutathione "conjugate," which is relatively water soluble and can be excreted.

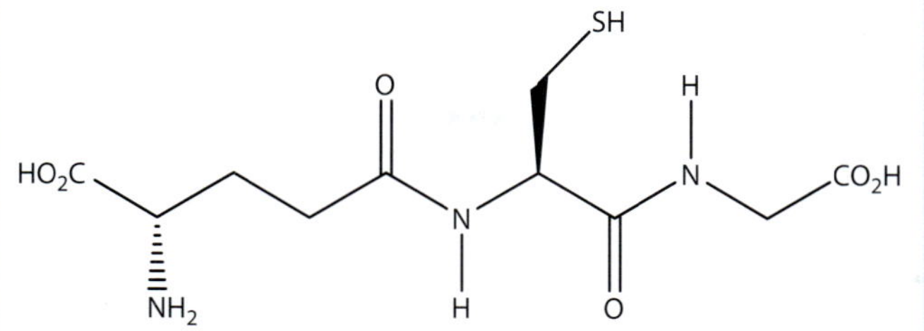

9.20, glutathione

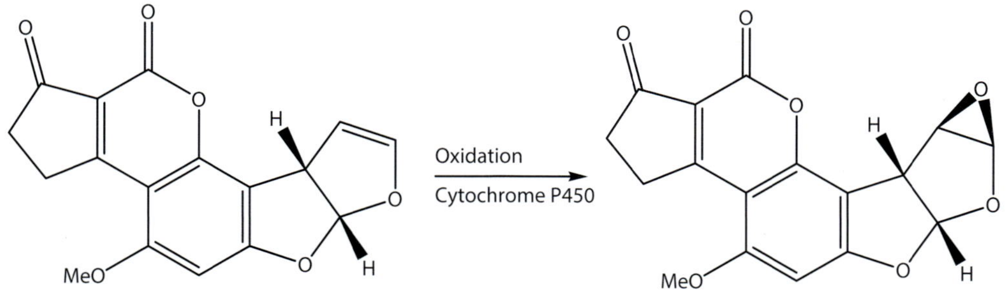

9.21, aflatoxin B1

9.10.4 CARBON NUCLEOPHILES

The cyanide ion is a relatively poor leaving group (the pK_a of HCN is 9.21, so not a strong acid), but it is a good nucleophile, between bromide and iodide in reactivity. Thus, it undergoes good S_N2 reactions at primary and secondary centers (Figure 9.58). The first example shows that bromide is more readily displaced than chloride. In the second reaction, cyanide both opens the three-membered ring, attacking at the less hindered carbon of the epoxide, and displaces chloride. It is rarely used under the more acidic conditions that would be the norm for S_N1 reactions, because of the risk of generating highly toxic hydrogen cyanide gas.

Although there are a number of interesting organic nitriles, the main value of this process lies in the other functionalities into which we can convert the CN group. Reduction with lithium

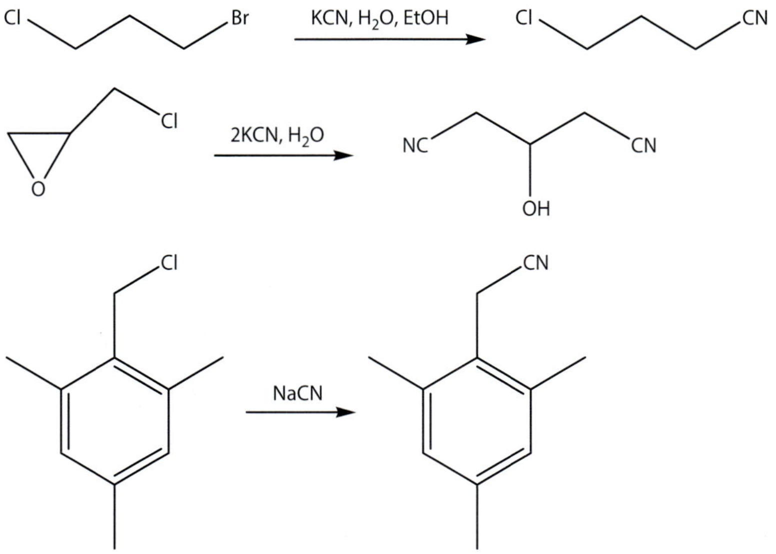

FIGURE 9.58 S_N2 reactions of cyanide ion.

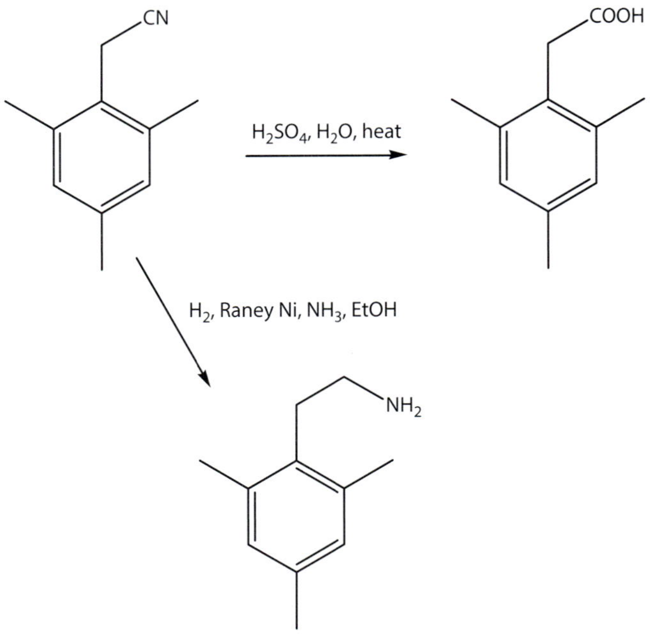

FIGURE 9.59 Transformations of cyanide groups.

aluminum hydride or hydrogenation in the presence of Raney nickel gives a primary amine (see Chapter 11), while hydrolysis gives a carboxylic acid (we will look more closely at the mechanism for this process in Chapter 14) (Figure 9.59).

PROBLEM 9.15

Predict the product of the following reactions. Show important intermediates, but mechanisms are not required:

(a)

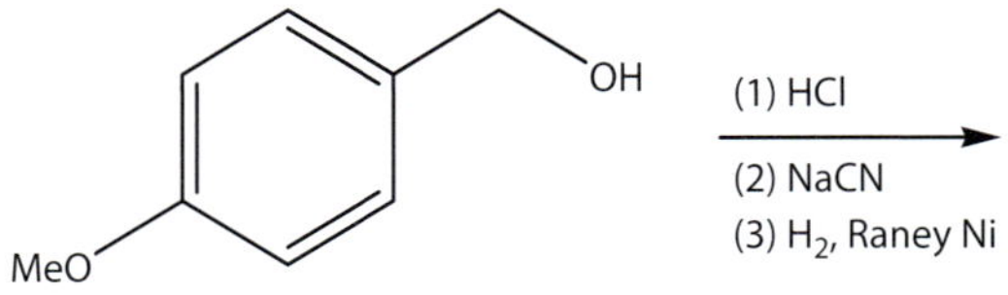

(b) n-$C_{12}H_{25}Br$ $\xrightarrow[\text{(2) KOH, } H_2O]{\text{(1) KCN}}$

Solutions

(a) The first step converts the hydroxyl into a better leaving group, halide; then the halide is replaced by cyanide, which is reduced to an amine in the final step:

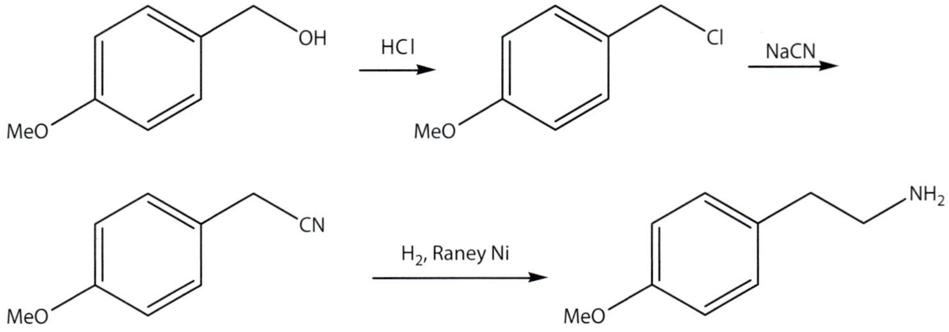

(b) The halide is replaced by cyanide, which is then hydrolyzed to give a carboxylic acid:

$$n\text{-}C_{12}H_{25}Br \xrightarrow{KCN} n\text{-}C_{12}H_{25}CN \xrightarrow{KOH,\ H_2O} n\text{-}C_{12}H_{25}CO_2H$$

As we discovered in Chapter 8, most organic compounds can be deprotonated, if you treat them with a strong enough base. We also learned what kind of effects we needed to consider in assessing the relative stability of carbanions. For hydrocarbons, we looked at orbital effects (so that alkynes are more acidic than alkenes, than alkanes), electron-donating (alkyl groups destabilize carbanions, so that primary carbanions are more stable than secondary and are more stable than tertiary) and electron-withdrawing (such as halogens) groups, and resonance effects. So we know that we can convert ethyne into its anion, using the strong bases sodamide or butyl lithium (Figure 9.60). This anion then reacts well with alkyl halides or sulfonates to give a 1-alkyne. The bromide or iodide (iodides are more reactive, and iodide ion is sometimes added to bromides as a catalyst) must be primary; elimination is the main reaction with secondary and tertiary halides. We can deprotonate this again, with another mole of base, and introduce another substituent. This process is useful in synthesis because it also allows us to prepare the corresponding alkenes, alkanes and methyl ketones—Figure 9.61 gives us a look ahead to Chapter 11.

Most organic compounds can be deprotonated, using an appropriate base, and we will meet examples of this later in the course. However, there is a relatively simple way to make the equivalent of a carbanion, an organometallic compound. Many metals can be persuaded to form σ-bonds to carbon; those we shall meet most often are derivatives of magnesium, Grignard reagents, and organolithium compounds. You may worry about whether the carbon–metal bond is truly covalent or has ionic character (metal$^+$carbon$^-$). Although we can make some generalizations, this isn't a trivial question—it depends on the nature of the metal, the stability or otherwise of the carbanion, and the reaction solvent. However, for many synthetic purposes the nuances will not matter—we can view these organometallic species simply as more or less feisty carbanions. Broadly, the more electropositive the metal, the more ionic the bond, and the more reactive the species. Solvation may assist ionization, but it decreases nucleophilicity.

Grignard reagents, organomagnesium halides, are prepared by the reaction of metallic magnesium with a wide range of organic halides (reaction 9.3). Dry ether type solvents are essential.

FIGURE 9.60 Deprotonation of ethyne.

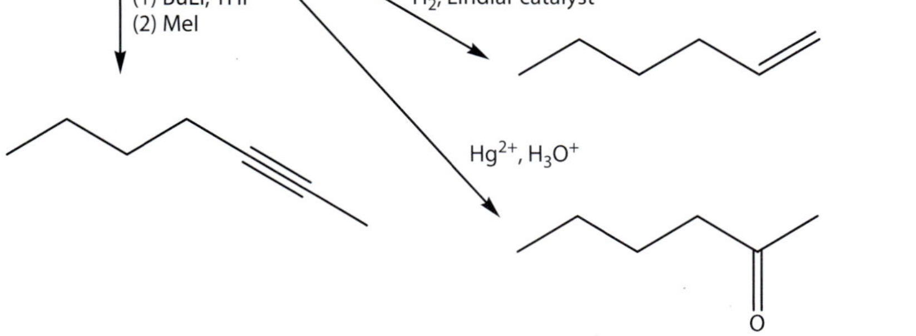

FIGURE 9.61 Alkylation of alkynes and subsequent transformations.

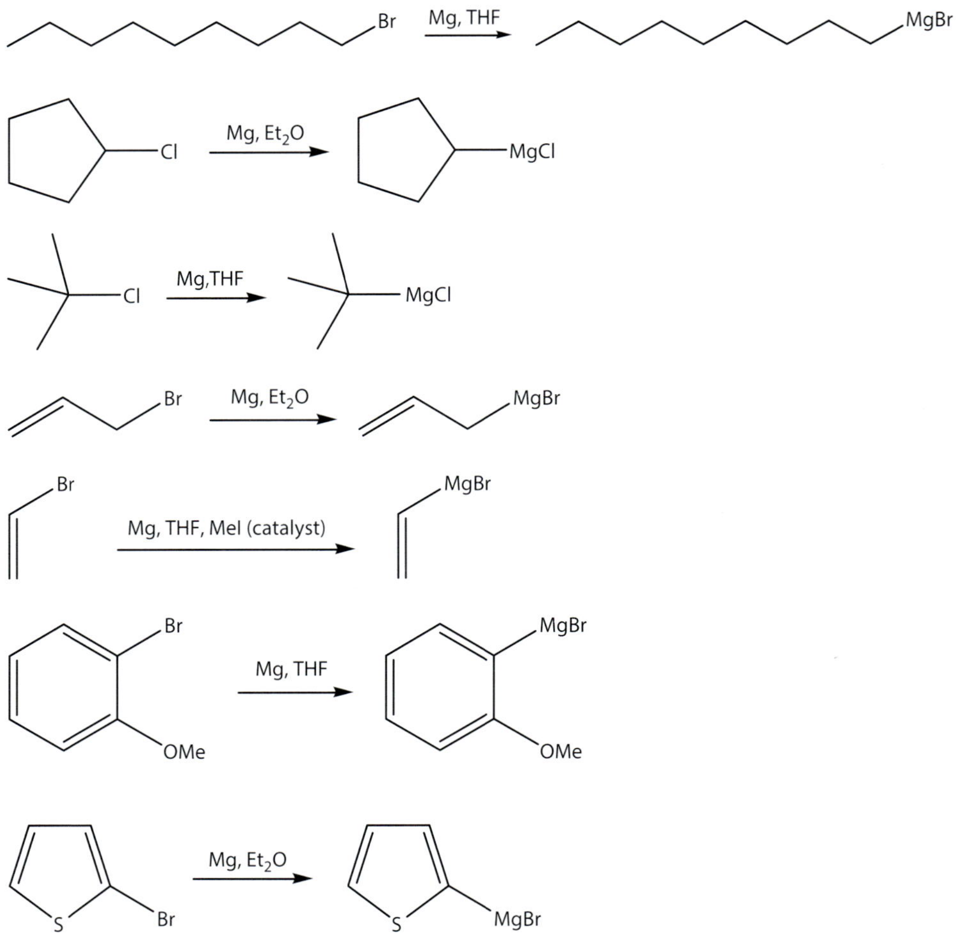

FIGURE 9.62 Preparation of Grignard reagents.

Grignard reagents are strong bases and react with water to give hydrocarbons (reaction 9.4 shows how we can use this to introduce deuterium into molecules), so they must be both prepared and used in anhydrous conditions. Ethereal solvents are essential to solvate the reagent. There are few restrictions on the nature of the group R—primary, secondary, and tertiary alkyl, aryl, alkenyl, allyl, and benzyl halides are all reactive, although rates vary. Bromides and iodides are more reactive than chlorides (although chlorides are generally less expensive), and fluorides are rarely useful. Some examples are given in Figure 9.62.

$$RBr + Mg \xrightarrow{\text{Dry ether or THF}} RMgBr \quad (9.3)$$

$$RMgBr + D_2O \longrightarrow RD + MgBr(OD) \quad (9.4)$$

It would seem an attractive idea that the synthesis of a wide range of hydrocarbons could be accomplished by reacting Grignard reagents with suitable alkyl halides. In an uncatalyzed mode (see also Section 23.6.3), this is a rare reaction and successful only with very reactive halides, generally allylic or benzylic derivatives (Figure 9.63). Notice that in the first example, the vinyl bromide is unreactive. The homocoupling in the second example is a problem in preparing benzylmagnesiumbromide. Generally, the less reactive chloride is preferred, if we want to prepare the Grignard reagent. Dialkyl sulfates, with the excellent sulfate leaving group, also give successful substitutions. However, Grignard reagents are strong bases, and competing elimination of HX is both common, and the main reaction in many cases.

However, Grignard reagents do react well with epoxides, opening them at the less hindered site, in an S_N2 process with clean inversion of configuration. This reaction can be used for

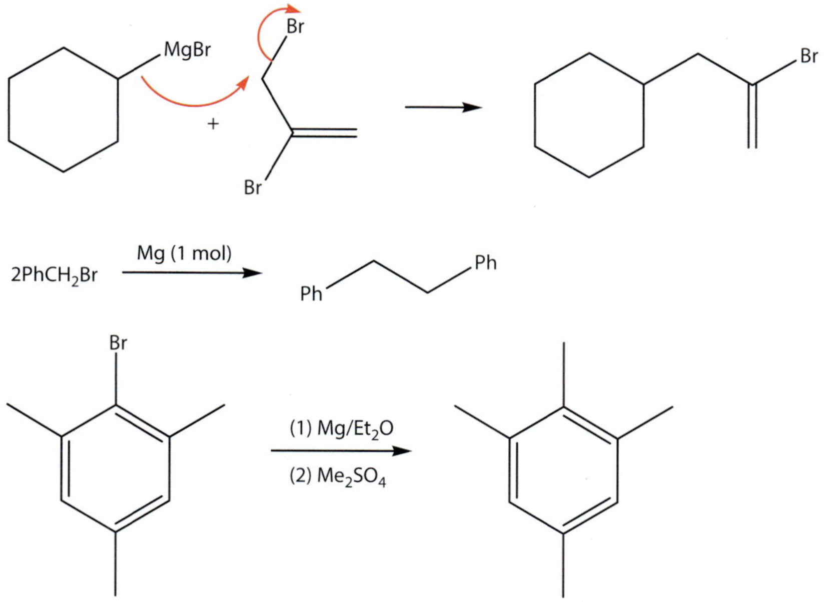

FIGURE 9.63 Synthesis of hydrocarbons by coupling of Grignard reagents with very reactive alkyl halides.

extending a hydrocarbon chain by two carbon atoms (Figure 9.64). The stereoselectivity of the process is clear in the second example.

There are many methods for generating organolithium reagents, but the simplest involves lithium metal and an organic halide (reaction 9.5). Some examples are given in Figure 9.65, and many of the simpler reagents are commercially available.

(9.5) $$RX + 2Li \longrightarrow RLi + RX$$

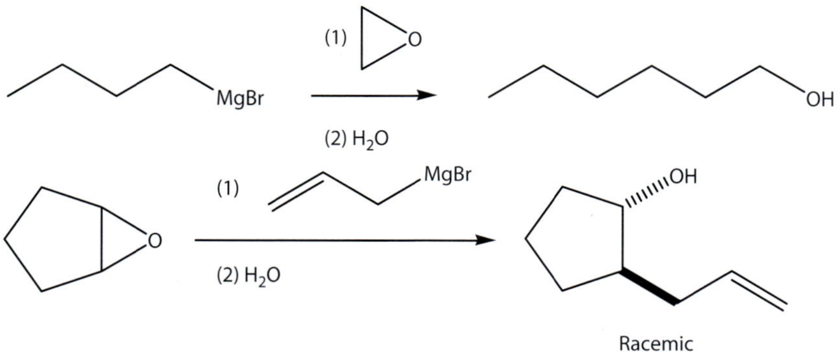

FIGURE 9.64 Reaction of Grignard reagents with epoxides.

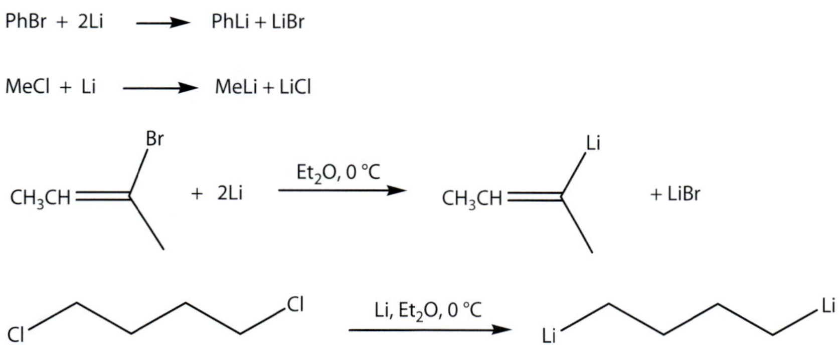

FIGURE 9.65 Preparation of organolithium reagents.

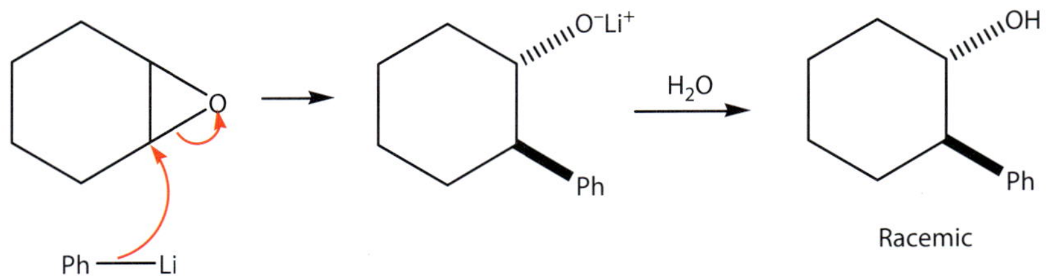

FIGURE 9.66 Reaction of organolithium compounds with epoxides.

Like Grignard reagents, these are strong bases and hence must be used under anhydrous conditions. They generally react well and regio- and stereospecifically with epoxides (Figure 9.66) but poorly with simple halides.

Focus on the Structures of Grignard Reagents and Organolithium Compounds

Preparation of Grignard reagents requires clean magnesium surfaces, dry solvents, and, sometimes, an activator such as molecular iodine or 1,2-dibromoethane. Iodides react most readily, and fluorides scarcely at all, unless special Rieke magnesium is used. This is very finely divided metal, with a high surface area. The mechanism of the reaction involves a single electron transfer at the magnesium surface to give a radical anion $[RX]^{-\bullet}$, which collapses to give the halide anion X^- and the organic radical $R^\bullet$. Mg^+ combines with X^- to give $MgX^\bullet$, and this recombines with the radical to give RMgX (Figure 9.67). Both the mechanism for formation and the fact that Grignard reagents have sufficient carbanionic character to invert rapidly mean that even when a starting halide is chiral, all stereochemical information is lost in the Grignard reagent.

A wide range of Grignard reagents have now been characterized crystallographically. Both monomeric (such as $[Ph_3CMgBr(Et_2O)_2]$, Figure 9.68 and $[MeMgBr(THF)_3]$ Figure 9.69) and dimeric ($[Et_2Mg_2Br_2(i\text{-}Pr_2O)_2]$, Figure 9.70) species are known, but such solid-state structures should be treated with caution in terms of what is present in solution. It can happen that what crystallizes is the least soluble species present in solution, rather than the most abundant. In tetrahydrofuran (THF) solution, RMgX is monomeric for X = Cl, Br, I over a wide

$$RX + Mg \longrightarrow [RX]^{-\bullet} + Mg^+$$

$$[RX]^{-\bullet} \longrightarrow R^\bullet + X^-$$

$$X^- + Mg^+ \longrightarrow [MgX]^\bullet$$

$$[MgX]^\bullet + R^\bullet \longrightarrow RMgX$$

FIGURE 9.67 Mechanism of formation of Grignard reagents.

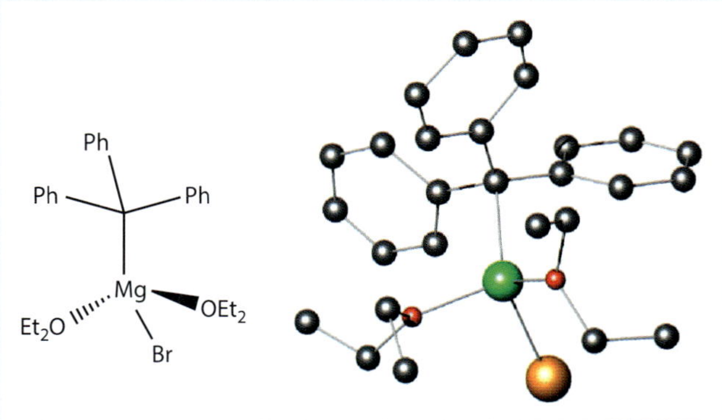

FIGURE 9.68 Structure of $Ph_3CMgBr(Et_2O)_2$. (Thanks to Dr. John Turner for this image.)

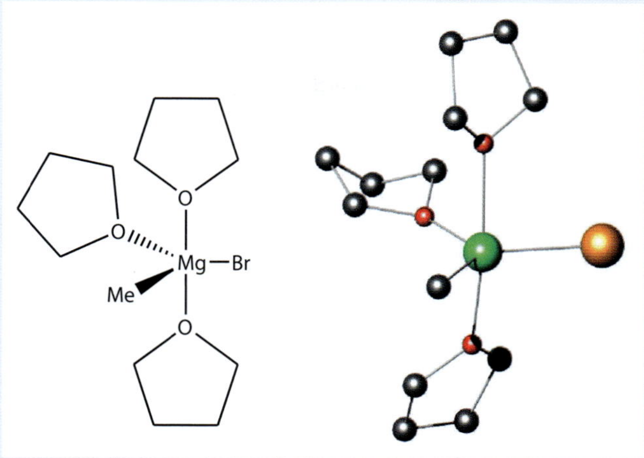

FIGURE 9.69 Structure of the THF solvate of methylmagnesium bromide. (Thanks to Dr. John Turner for this image.)

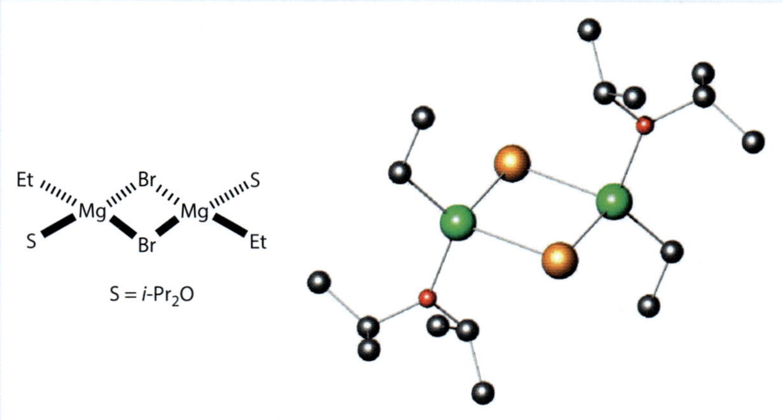

FIGURE 9.70 Structure of di*iso*propyl ether solvate of ethylmagnesium bromide. (Thanks to Dr. John Turner for this image.)

concentration range, but fluorides are dimeric. In diethyl ether solution, chlorides and fluorides are dimers. The situation is more complex with bromides and iodides; at low concentrations, they are monomeric, but at higher concentrations, dimers and higher oligomers are formed:

An additional complication is introduced because Grignard reagents are in equilibrium in solution with dialkylmagnesium compounds, the so-called Schlenk equilibrium (Equation 9.6). The position of the equilibrium depends on the nature of the R group and the solvent. In general, K is very small in hydrocarbons and very large in the presence of amines and generally higher in ether than in THF. The equilibrium can be shifted entirely to the left, to give a pure dialkylmagnesium compound, by addition of dioxane, in which magnesium halides are insoluble. The structure of $(Me_3C)_2Mg$ has been determined. In the solid state, it is dimeric, with bridging and terminal *tert*-butyl groups (Figure 9.71). In solution, the dimeric structure is retained, but there is some ring opening, and the bridging and terminal alkyl groups can be exchanged. Many other dialkylmagnesium compounds, such as Ph_2Mg, have polymeric structures in the solid state.

(9.6) $$R_2Mg + MgX_2 \overset{K}{\rightleftharpoons} 2RMgX$$

Organolithium structures, both in the solid state and in solution, are dominated by oligomeric species. As with organomagnesium compounds, the degree of aggregation

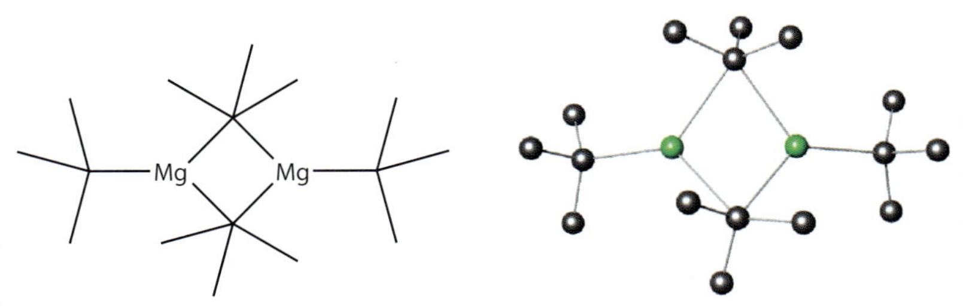

FIGURE 9.71 Structure of di-*tert*-butylmagnesium. (Thanks to Dr. John Turner for this image.)

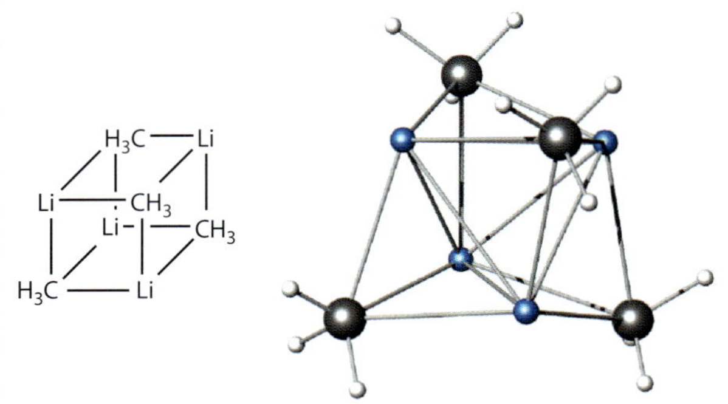

FIGURE 9.72 Solid-state structure of methyllithium. (Thanks to Dr. John Turner for this image.)

depends on steric hindrance in the alkyl groups, concentration, solvent, and presence of base. In the solid state, methyllithium is isolated as a tetramer based on a cubic structure (Figure 9.72). Many reactions of organolithium compounds require the addition of a base such as 1,2-diaminoethane (old name ethylenediamine, abbreviated en) or better *N,N,N,N*-tetramethylethane-1,2-diamine (old name tetramethylethylene diamine, abbreviated tmeda). These disaggregate the oligomers to give the more reactive monomer.

Many of the oligomeric structures that have been previously drawn should be giving you pause for thought—they don't seem to correspond to what you have learned in the past about the valency of various elements. And why don't we get simple anions and cations, when the difference in electronegativity of carbon (2.5) and lithium (0.98) or magnesium (1.31) suggests significant ionic character? The answers to these questions are beyond the scope of this course, but a simple idea that will be helpful is that electron counting still matters. So in many cases, we have only two electrons available to bond—we say that these bridging structures have three-center two-electron bonds.

The final group of carbon nucleophiles, which we will introduce briefly here (we will look at these is much more detail in Chapters 17 and 20), are the anions of dicarbonyl compounds such as pentane 2,4-dione (**9.22**) and diethyl malonate (**9.23**). These have pK_a values, respectively, of 7.0 and 12.9, so they may be readily deprotonated by a simple base such as hydroxyl or ethoxide ion. The anions thus produced react with a range of primary and secondary halides and sulfonates, as well as epoxides (Figure 9.73).

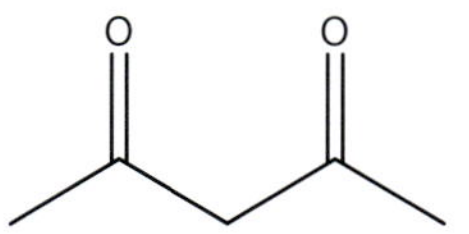

9.22, pentane-2,4-dione

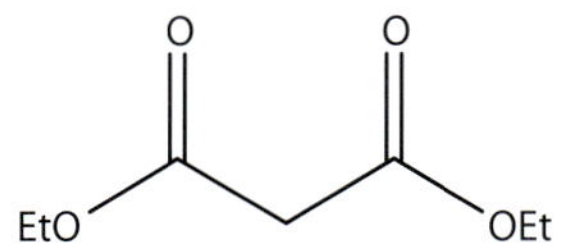

9.23, diethyl malonate

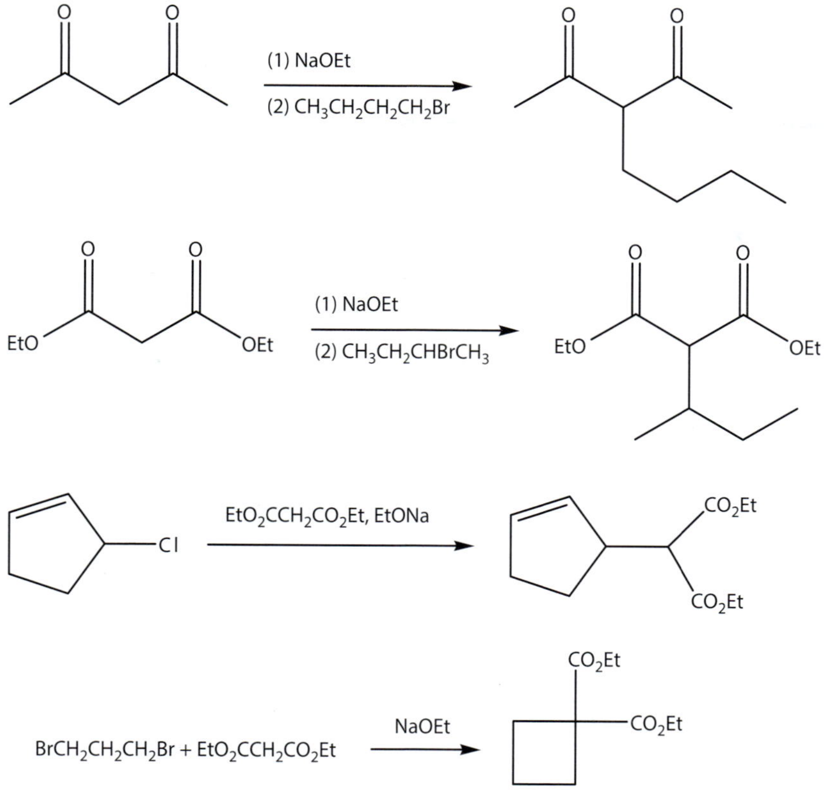

FIGURE 9.73 Anions of β-dicarbonyl compounds as nucleophiles.

Focus on Epoxide Opening Reactions

Epoxides are excellent electrophiles and substrates for S_N2 reactions; the release of strain in the three-membered ring makes them particularly reactive. They react with all the nucleophiles we have discussed, with reliable inversion of stereochemistry (Figure 9.74). Note that the stereochemistry in the products is relative, not absolute—since the starting material is not chiral, racemic mixtures would be obtained (hence, the plain rather than the tapered wedge and hash bonds). The one unfamiliar example here is the reaction with lithium aluminum deuteride, Li[AlD_4]. For the present, think of this as essentially a deuteride anion (or hydride from lithium aluminum hydride), but this reactive molecule will quickly become a good friend to you in reductions and devising syntheses.

When the epoxide substrate is nonsymmetrical, then, in principle, two regioisomers can be formed (Figure 9.75). In both regioisomers, the stereochemistry of the incoming nucleophile and the OH is *trans*; clearly both processes are fundamentally S_N2 in character and go with inversion of stereochemistry. With good nucleophiles, under basic conditions, the reaction is straightforward; as in most S_N2 reactions, attack takes place predominantly or exclusively at the less hindered of the two carbon atoms of the epoxide. Examples are shown in Figure 9.76.

However, in acidic solution, the epoxide may be protonated, and this leads to reactions with more S_N1 character. The protonated epoxide is considered to be partially opened, with one bond elongated, and some cationic character at the more substituted carbon (**9.24**). Under these circumstances, the nucleophile may have a greater attraction to the more positive carbon atom, even though there is more steric hindrance. This is particularly true when the substituent is a benzene ring, which can stabilize the incipient cation by π-resonance.

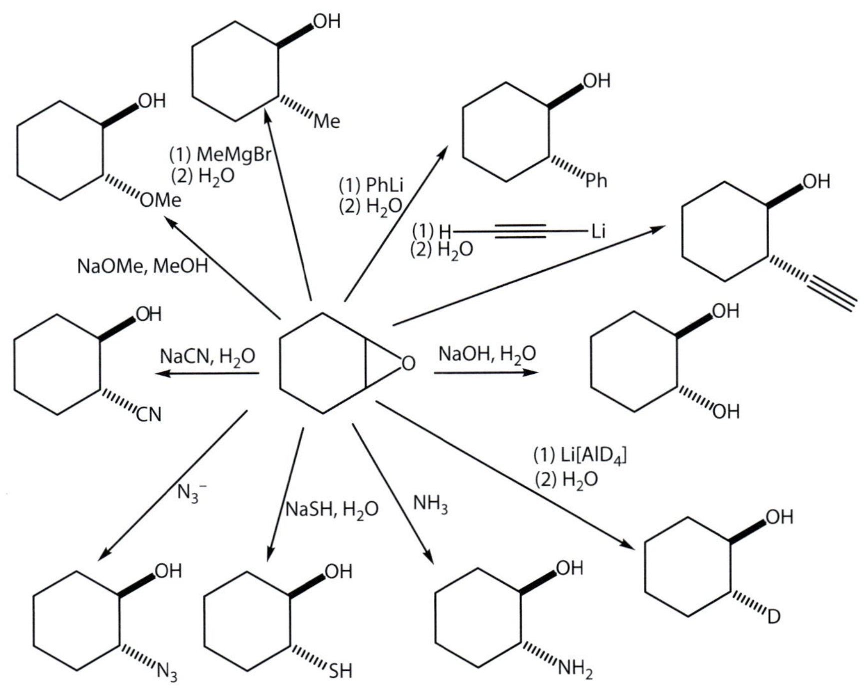

FIGURE 9.74 Reactions of cyclohexene oxide with nucleophiles.

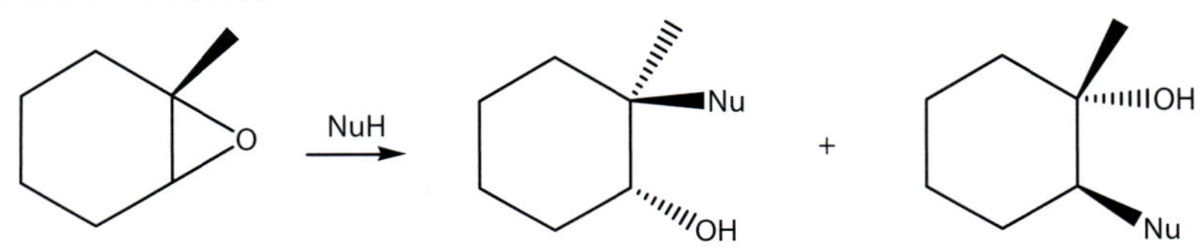

FIGURE 9.75 Ring opening of a nonsymmetrical epoxide in an S_N2 process.

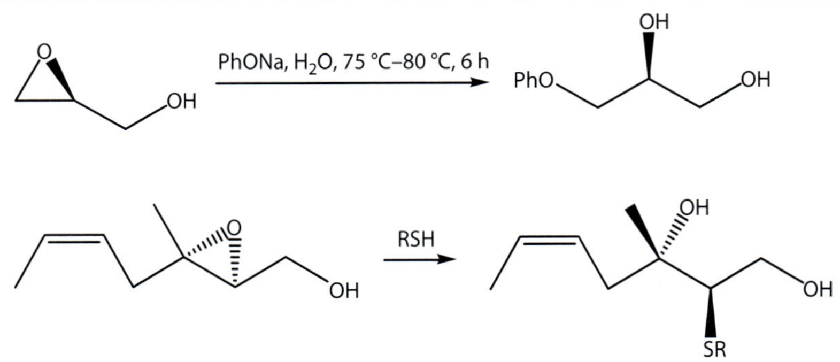

FIGURE 9.76 Ring opening of epoxides in synthesis.

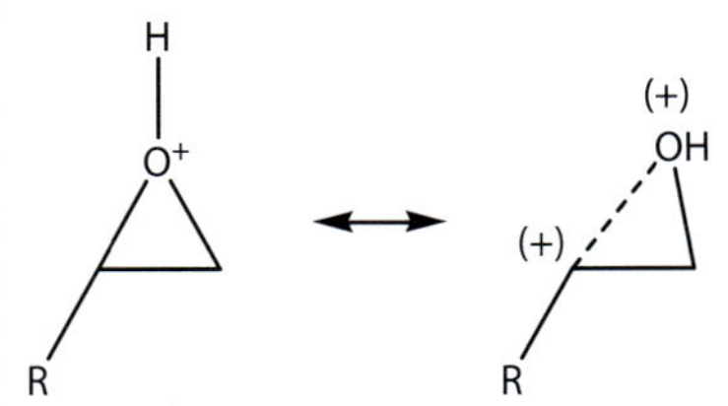

9.24

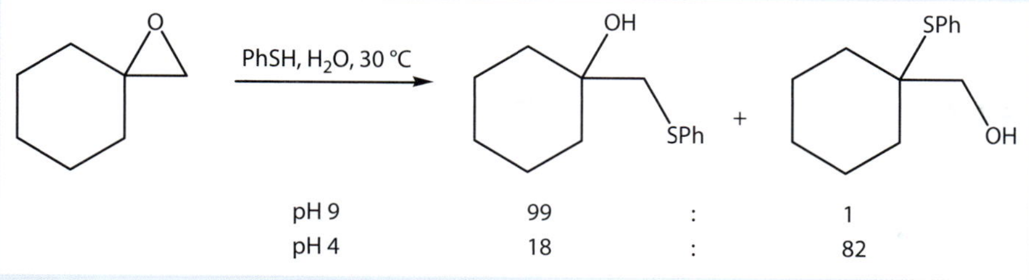

FIGURE 9.77 Effect of pH on regiochemistry of epoxide ring opening.

This is demonstrated by the effect of pH on the epoxide ring opening in Figure 9.77. At pH 9, the nucleophile is PhS^-, and attack is essentially exclusively at the less hindered site. At pH 4, the epoxide is at least partially protonated, and the nucleophile is PhSH. The major product is of attack at the more substituted carbon of the epoxide.

This is the first of a number of processes that I will describe as Alice (in Wonderland) reactions.

> Alice laughed. "There's no use trying," she said, "one can't believe impossible things."
>
> "I dare say you haven't had much practice," said the Queen. "When I was your age, I always did it for half an hour a day. Why, sometimes I've believed as many as six impossible things before breakfast."
>
> **Lewis Carroll**
> *Through the Looking Glass*

We have to believe two mutually contradictory things about our substrate simultaneously. If we think of it as a closed, three-membered ring, we get the stereoselectivity correct. If we think of it as an open carbocation, we get the regiochemistry correct. The truth is somewhere in between, and in most cases, we get predictable stereochemistry *and* regiochemistry.

PROBLEM 9.16

Predict the product(s) of each of the following reactions:

(a)

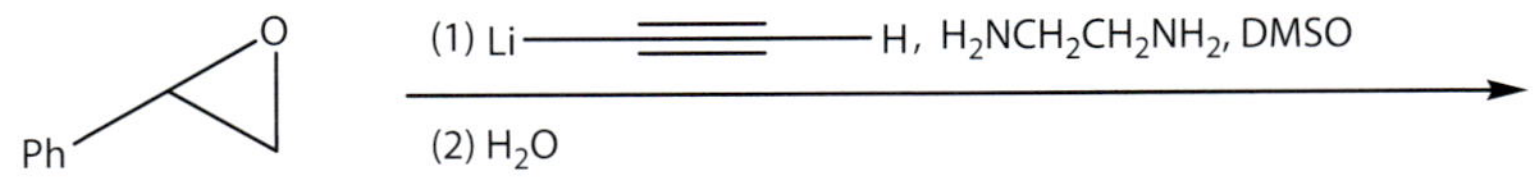

(b)

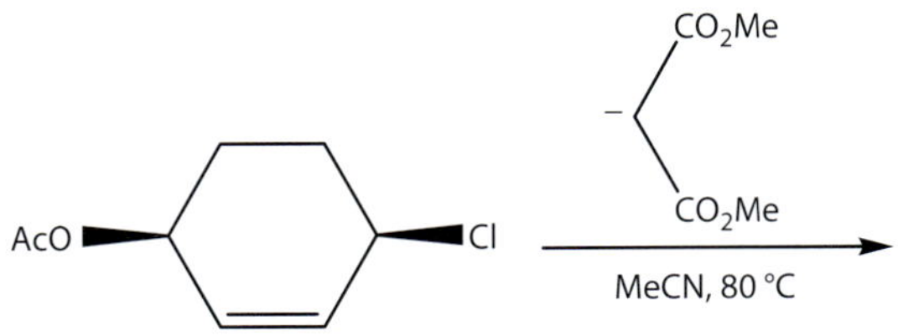

(c)

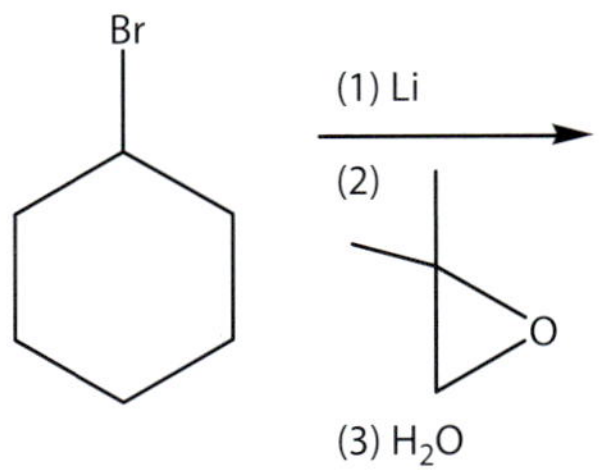

Solutions

(a) This reaction involves S_N2 nucleophilic opening of the epoxide, at the less hindered site. The alkoxide ion formed is protonated in the work-up step.

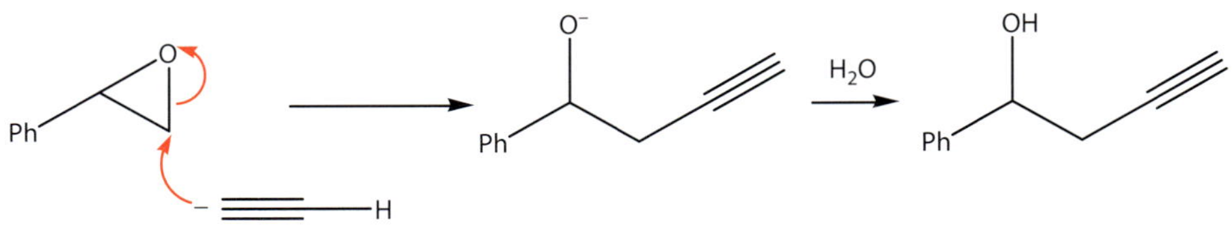

(b) This is an S_N2 substitution, with inversion. S_N2' would be disfavored for steric reasons.

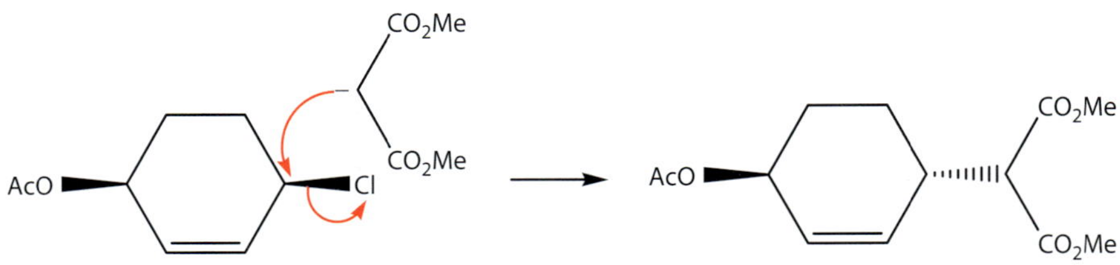

(c) In the first step, the halide is converted to the organolithium compound. This then attacks the epoxide at the less hindered site. In the work-up step, the alkoxide produced is reprotonated.

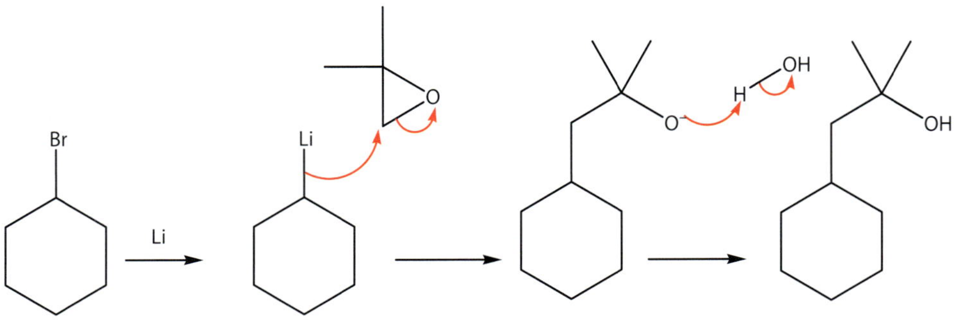

Key Points from Sections 9.10.3 and 9.10.4

- Amines readily displace halides from haloalkanes, but in many cases, this is not synthetically useful because of overalkylation. Useful examples are found where the halide is sterically hindered or bears an electron-withdrawing group.
- Monosubstitution of halides, without overalkylation, may be achieved using phthalimide (Gabriel synthesis) or hexamethylene tetramine (Délepine synthesis).
- The azide anion is an excellent nucleophile and readily displaces a wide range of halides. Azides may be reduced to amines using lithium aluminum hydride or H_2/Raney nickel.
- Cyanide is an excellent nucleophile under S_N2 conditions; the resultant organonitriles may be reduced to amines (H_2/catalyst) or hydrolyzed to carboxylic acids.
- The anions of alkynes are readily generated using strong bases and substitute primary haloalkanes to give more complex alkynes that can be converted to alkenes, alkanes, and ketones.
- A wide range of organic carbanions may be generated using appropriate substrates and bases; many react well with primary and secondary halides and sulfonates.
- Epoxides may be opened by a wide range of nucleophiles in S_N2 processes with reliable inversion of configuration and predictable regiochemistry.

9.11 NEIGHBORING GROUP PARTICIPATION AND INTRAMOLECULAR REACTIONS

We have already seen a range of examples where intramolecular substitution reactions led to ring formation. Although in principle rings of all sizes can be formed by a process such as that of Figure 9.78, the process is much more favorable when the ring is three-, five-, or six-membered. Typically five- and six-membered rings are formed 10^5–10^6 times faster than a related intermolecular reaction, and three-membered rings 100 times faster. Three- and five-membered rings tend to be formed under conditions of kinetic control, and six-membered rings under conditions of thermodynamic control. Examples are shown in Figure 9.79.

All of the rings made previously were built to last—but this is not always the case. In some reactions, a ring is formed temporarily and then reopened. An example is shown in Figure 9.80. When α-chloroacid is hydrolyzed in base, unexpectedly, the reaction goes with retention of configuration. There is no obvious mechanism that we have seen so far to account for this. This cannot be a classic S_Ni process, as it does not involve thionyl chloride. When the acid is treated with base, the first thing that happens is deprotonation—remember that transfer of protons to and from oxygen and nitrogen is very rapid compared with almost any other processes. The carboxylate anion then attacks the back side of the carbon–chlorine bond, in an S_N2 reaction, with inversion of configuration, forming a very unstable strained three-membered ring. This intermediate is called an α-lactone, particularly unstable because of the sp^2 carbon, which wants an angle of 120 ° but gets 60 °. Thus, it is easy for the ring to be reopened by hydroxyl ion, in another S_N2 process, with inversion. Two inversions take us back to the original configuration. In general, retention of configuration in a substitution is a clue that this sort of process, involving two inversions, has occurred. We describe this as neighboring group participation, or anchimeric assistance.

Lone pairs of electrons, on appropriately positioned atoms, can also provide a "neighboring" group to assist a substitution. If we consider the conversion of the alcohol, **9.25**, to a chloride (Figure 9.81), two products are obtained—clearly this is not a simple S_N2 reaction.

In order to make the OH into a leaving group, we must protonate it. Although at this point it could be displaced by chloride, displacement by the lone pair on sulfur to make a three-membered

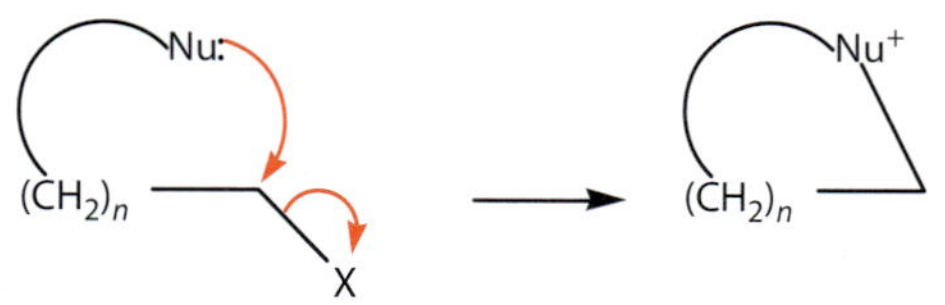

FIGURE 9.78 Intramolecular S_N2 reaction.

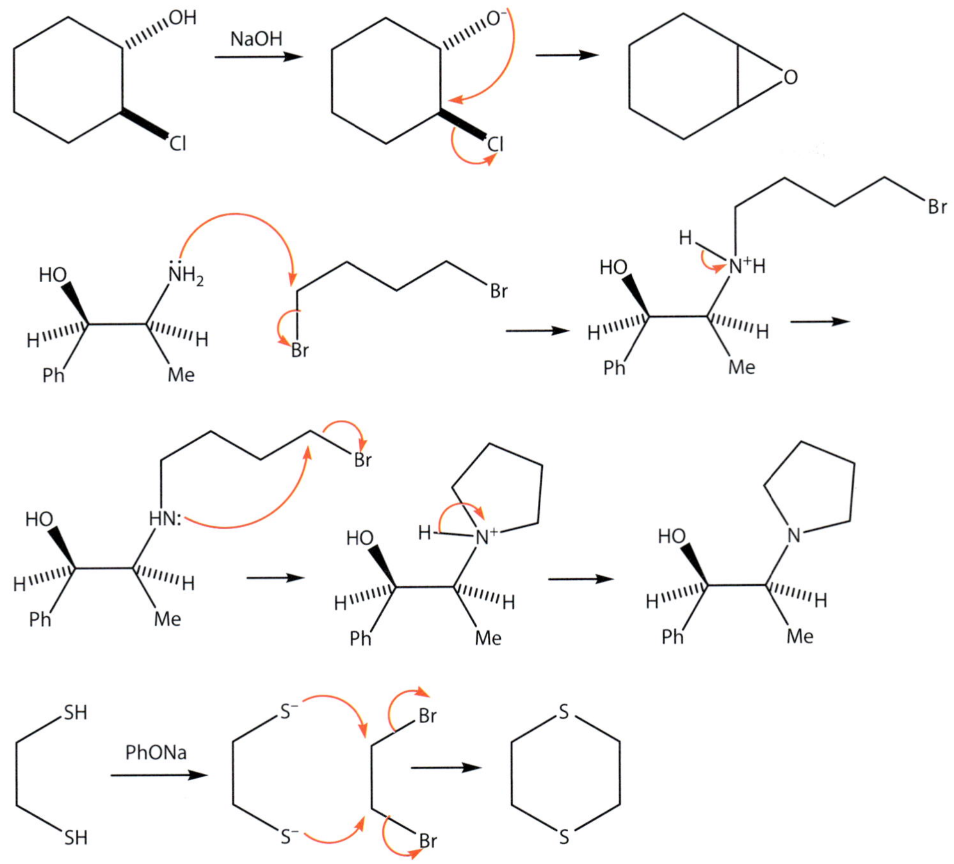

FIGURE 9.79 Intramolecular substitution reactions.

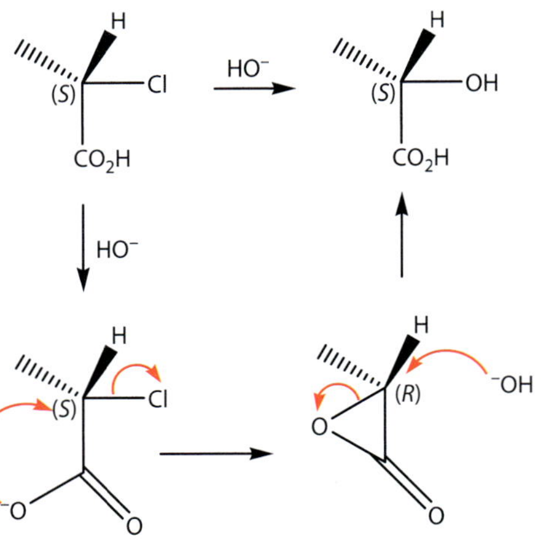

FIGURE 9.80 Retention of configuration involving neighboring group participation.

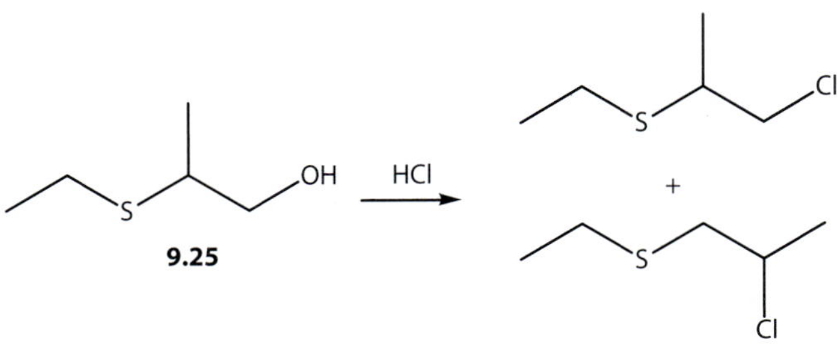

FIGURE 9.81 Neighboring group participation involving sulfur.

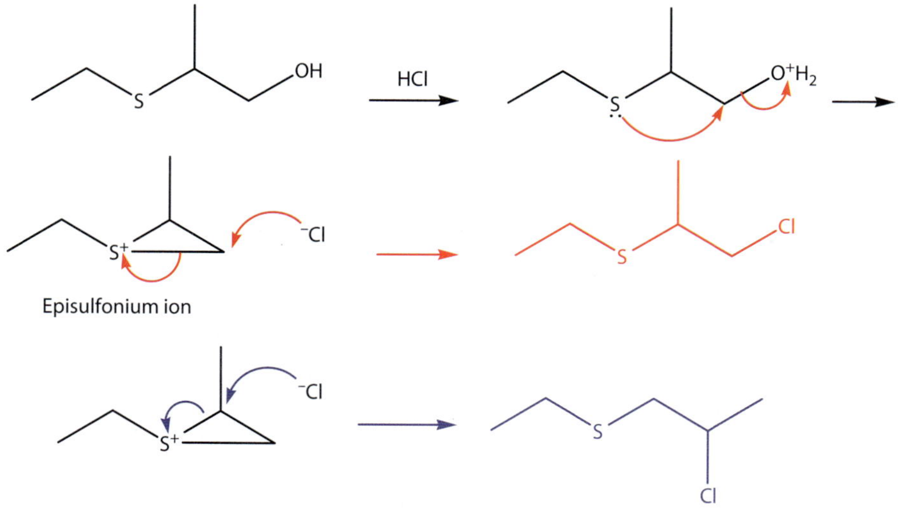

FIGURE 9.82 Mechanism of neighboring group participation by sulfur.

ring, an episulfonium ion, is much faster. This strained three-membered ring can then be opened by chloride, to give the two products (Figure 9.82).

The electrons of a π-bond can also act as a neighboring group. The reaction in Figure 9.83 goes with retention of configuration, and the reaction proceeds 10^{11} faster than for compound **9.26**, without the double bond. The tosylate is protonated and leaves with the assistance of the double bond. Acetic acid then approaches from the side remote from the original double bond, and retention of configuration is observed in the product—again the result of two inversions. You may wonder why we have drawn the intermediate as a delocalized system, rather than as a simple carbocation, such as **9.27**. The answer lies in a study of the effect of methylation of the double bond on the reaction. We would expect that any alkyl substitution of the alkene would make the double bond more electron rich and thus better able to participate in the reaction. One methyl group increases the rate by a factor of 13.3, two methyl groups by a factor of 148. If the cation were localized rather than delocalized, then a second methyl group should have a very minor effect as the cation would be located at a single site. Thus, we can conclude that the cation must be delocalized.

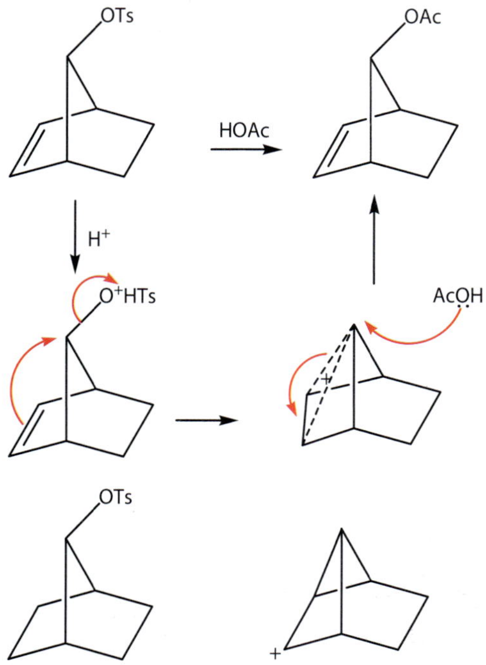

9.26 **9.27**

FIGURE 9.83 Neighboring group participation of a π-bond.

PROBLEM 9.17

Explain the following observations:

(a) On reaction with acetic acid, both the *cis*- and *trans*-isomers **9.28a** and **9.28b** give the same diacetate product. The *trans*-isomer reacts 670 times faster than the *cis*-isomer:

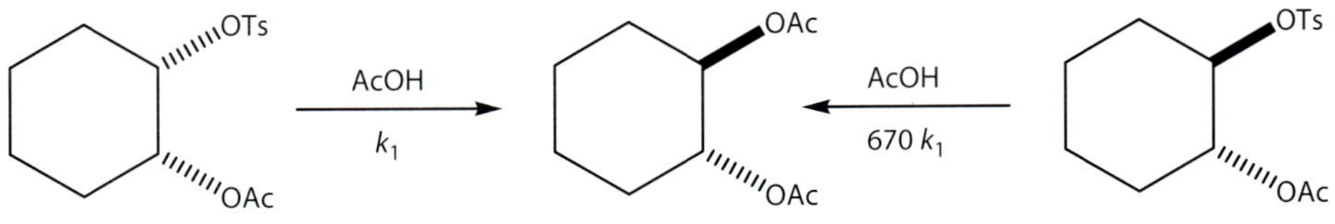

9.28a **9.28b**

(b) The rate of hydrolysis of mustard gas, **9.29**, is first order in mustard gas and independent of [HO⁻]. Suggest a mechanism to account for these kinetic data (remember that primary carbocations do not usually form, even in very ionizing solvents, and that S has lone pairs of electrons):

$$ClCH_2CH_2SCH_2CH_2Cl \xrightarrow{HO^-,\ H_2O} HOCH_2CH_2SCH_2CH_2OH$$

9.29

Solutions

(a) The reaction of the *cis*-isomer appears to be a straightforward S_N2 substitution of the protonated tosylate. With the *trans*-isomer, however, the reaction goes with retention of stereochemistry, which suggests a double inversion. As always with this type of problem, it's crucial to draw out the six-membered rings in their proper chair form and consider both possible chair forms. For the *cis*-isomer, in neither form is the acetate group "behind" the departing tosylate, and hence it cannot assist with its departure. An S_N2 reaction, with inversion of stereochemistry, is the only available option. However, in the *trans*-isomer, the acetate is appropriately positioned to assist the departure of the tosylate, to give the intermediate **9.30**. This reaction forms a five-membered ring, so we would expect it to be very much faster than an intermolecular process. The five-membered ring is then reopened by acetic acid, to give the *trans*-diacetate. We shall meet a number of examples of this type, where two fast reactions outpace a single slow step:

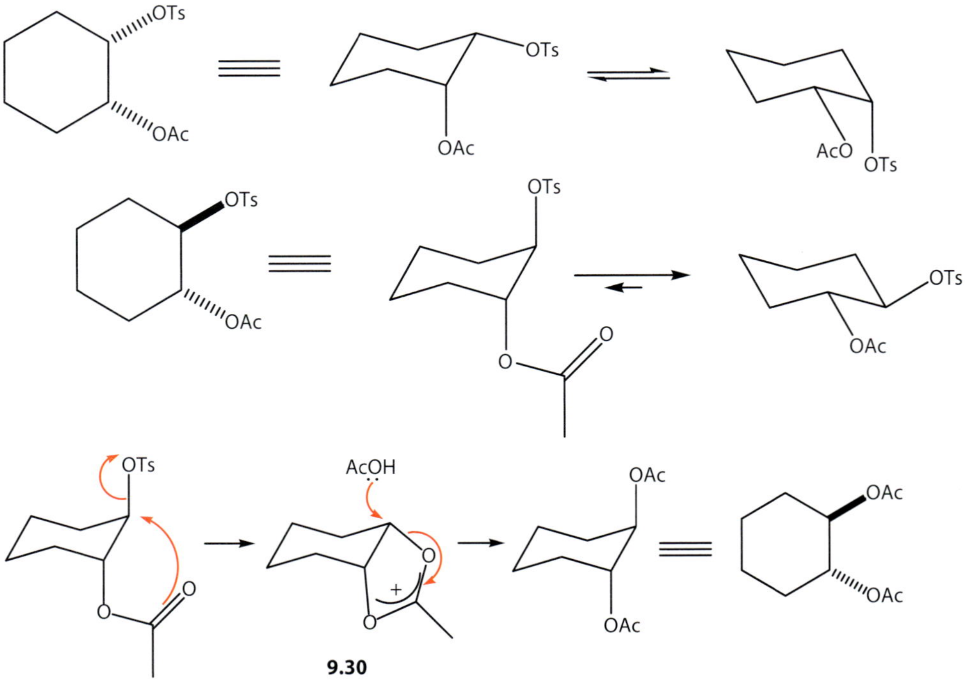

9.30

(b) The reaction cannot be a simple S_N1 process, despite the kinetics; if we are writing down something involving an unconjugated primary carbocation, we can be fairly sure we've got it wrong. The chloride is expelled using the lone pair of electrons on sulfur in an RDS. Only the substrate is involved in this step, so the kinetics are first order. The strained episulfonium ion is then opened by hydroxyl ion to give the hydrolyzed product:

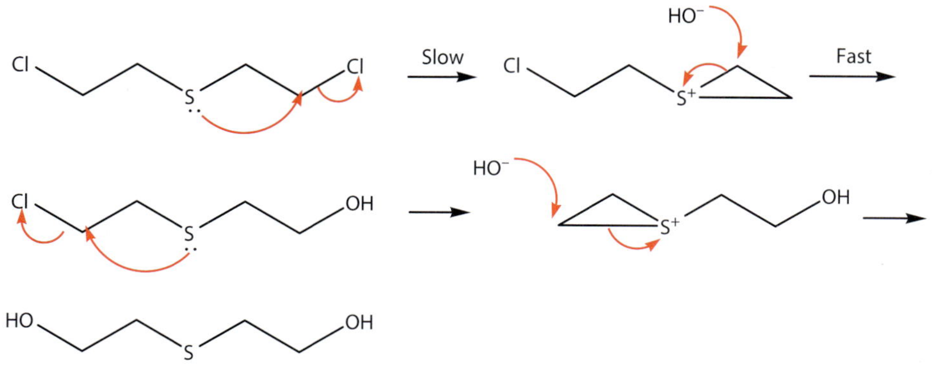

Key Points from Section 9.11

- Intramolecular reactions for three-, five-, or six-membered rings are generally faster than intermolecular reactions.
- Appropriately positioned anions, atoms with lone pairs of electrons to share, and π-bonds can act to assist the departure of a leaving group, forming a temporary ring structure, which is then reopened by an external nucleophile. Such processes can result in retention of stereochemistry *via* two inversions or structural rearrangements.

REVIEW PROBLEMS

1. Give systematic names for each of the following molecules:

(a) 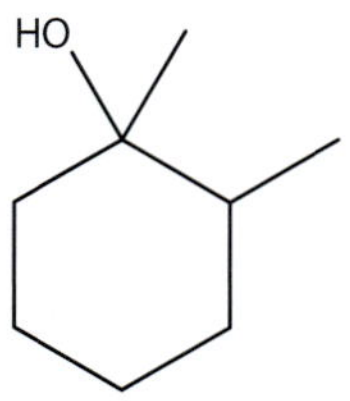

(b) 

(c) 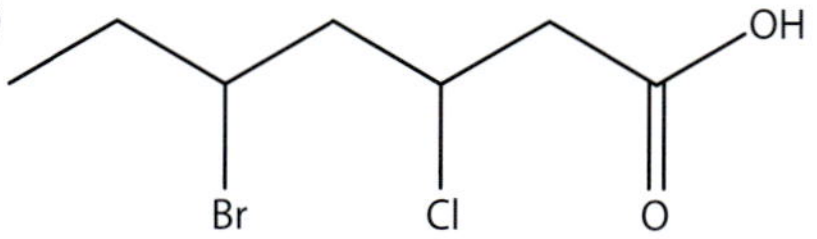

(d) $CH_3CH_2CH(F)COOH$

2. Give structures for each of the following molecules:
(a) 2,2-Dibromohexane dioic acid
(b) Pentane-2,3-diol
(c) 4-Chlorohexanamide
(d) Cyclohexylamine

3. (a) Predict the ^1H NMR spectrum of (i) benzyl alcohol and (ii) $CH_3CH_2CH_2Br$, giving both approximate chemical shifts and relative peak intensities.
   (b) How would you distinguish the following isomers, using only ^1H NMR splitting patterns to guide you?

$$CH_3CH_2CH_2Cl \quad CH_3CH_2CH(Cl)CH_3 \quad (CH_3)_2CHCH_2Cl \quad (CH_3)_3CCl$$

4. Predict the product(s) of each of the following reactions, paying attention to the stereochemistry where appropriate:

(a)

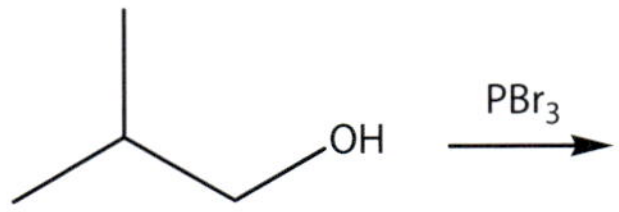

(b)

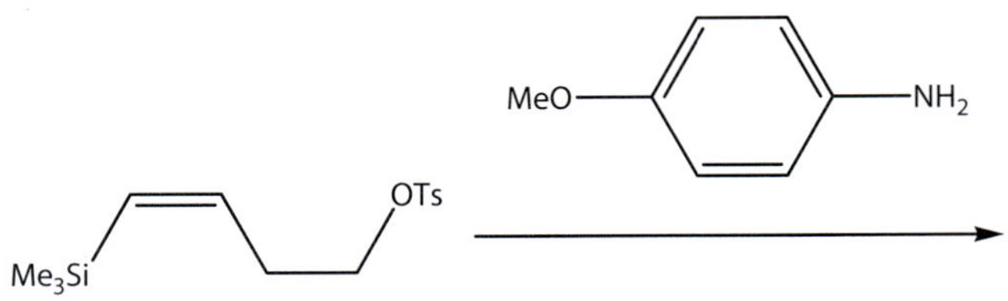

(c)

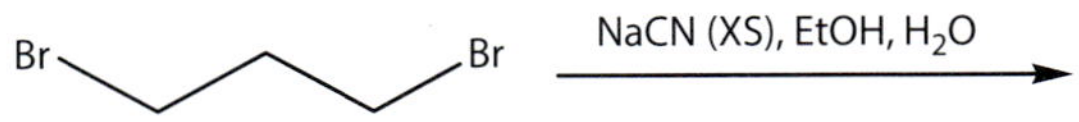

(d)

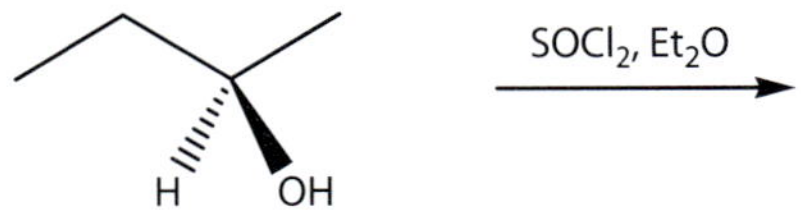

(e)

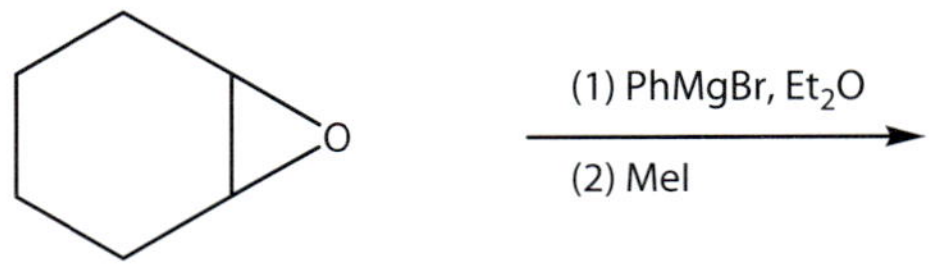

(f)

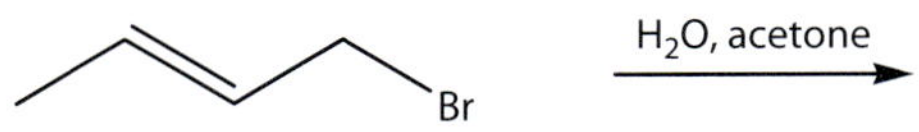

(g)

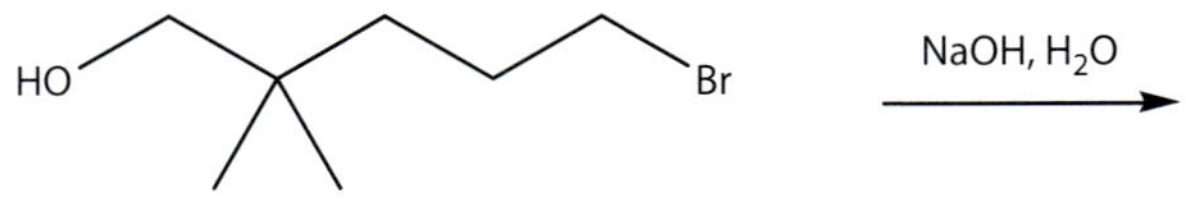

(h)

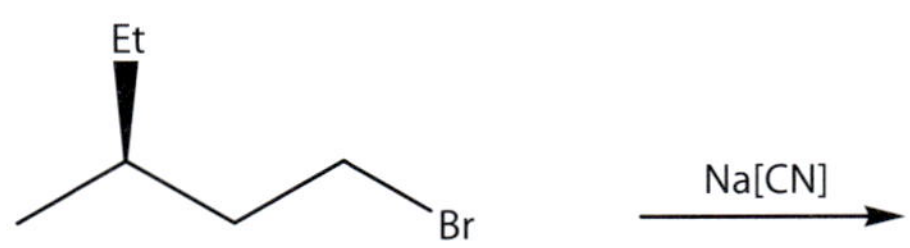

(i)

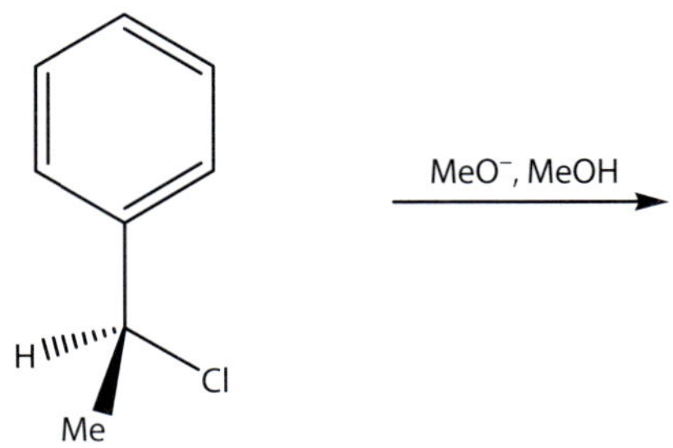

(j)

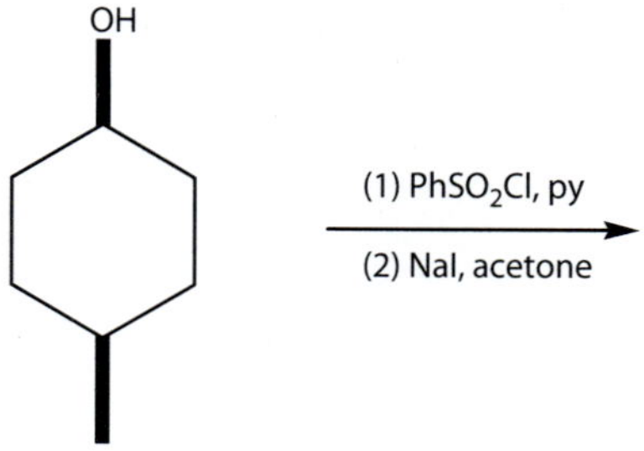
OH
(1) PhSO₂Cl, py
(2) NaI, acetone

(k)

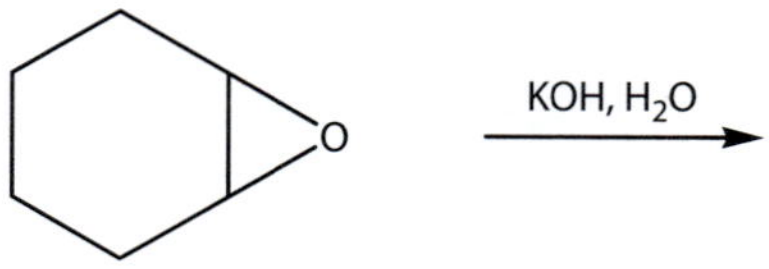
O
KOH, H₂O

(l)

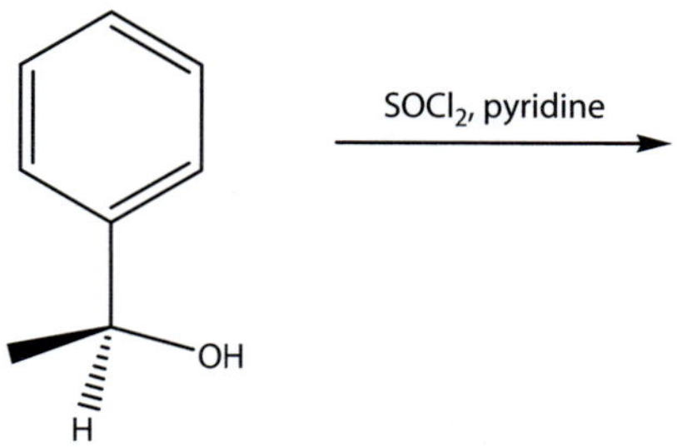
OH
H
SOCl₂, pyridine

(m)

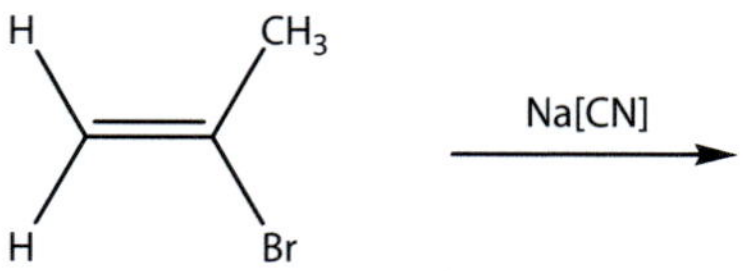
H
CH₃
H
Br
Na[CN]

(n)

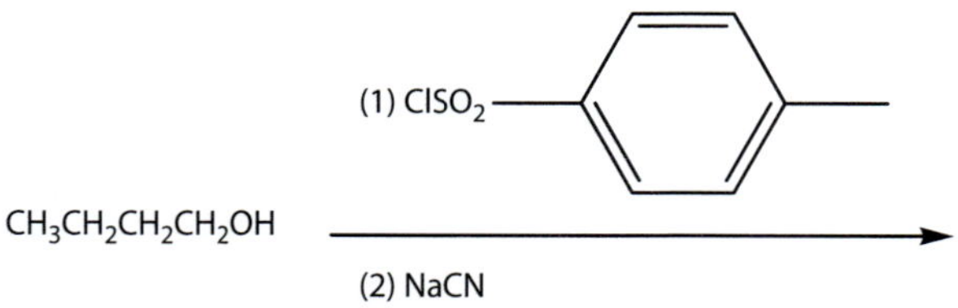
CH₃CH₂CH₂CH₂OH
(1) ClSO₂
(2) NaCN

(o)

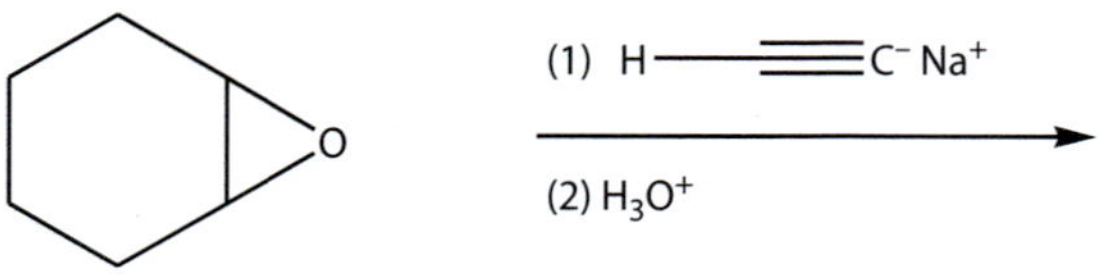
O
(1) H—≡C⁻ Na⁺
(2) H₃O⁺

(p)

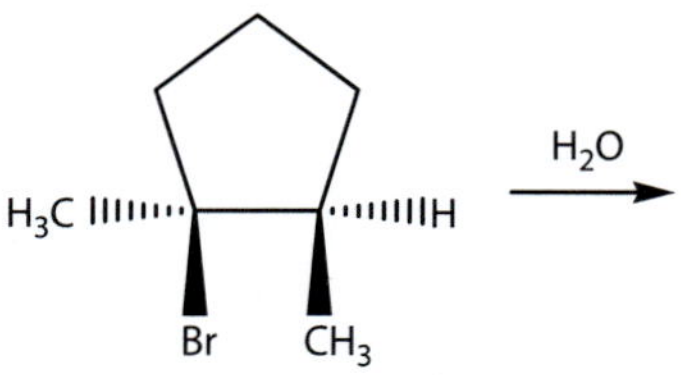
H₃C
H
Br
CH₃
H₂O

5. Rank the following three reactions in order of the rate at which you expect they will proceed, giving a brief explanation.

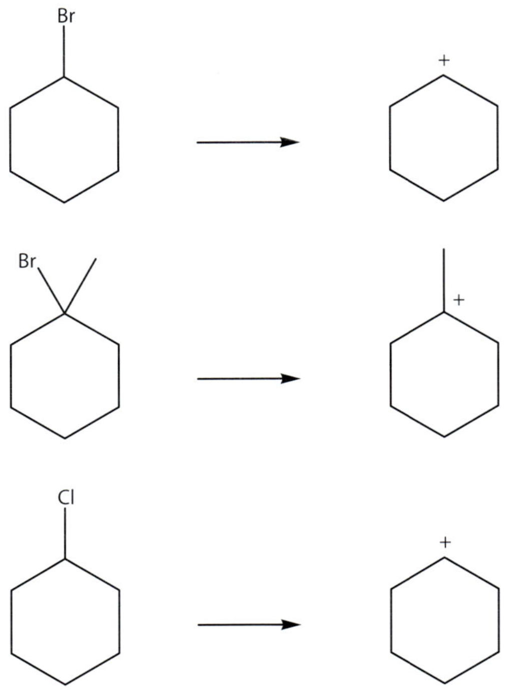

6. For each of the following sets of molecules, predict the order of reactivity in an S_N1 reaction.

(a)

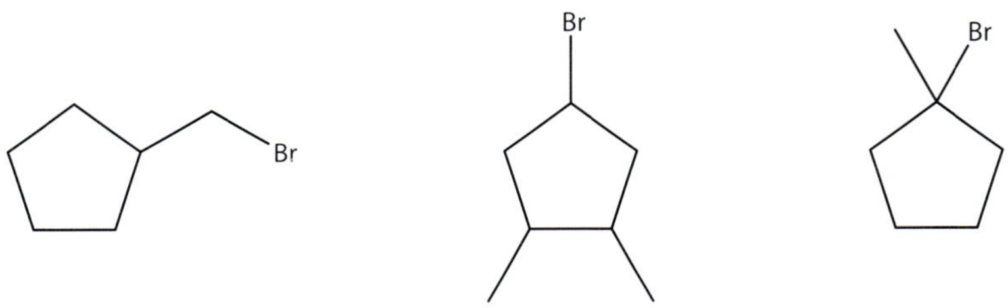

(b)

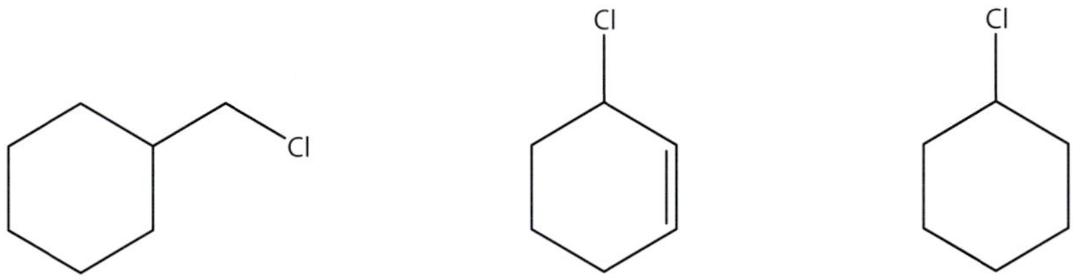

7. Place the following compounds in order with respect to their reactivity in a typical S_N2 reaction, giving brief reasons.

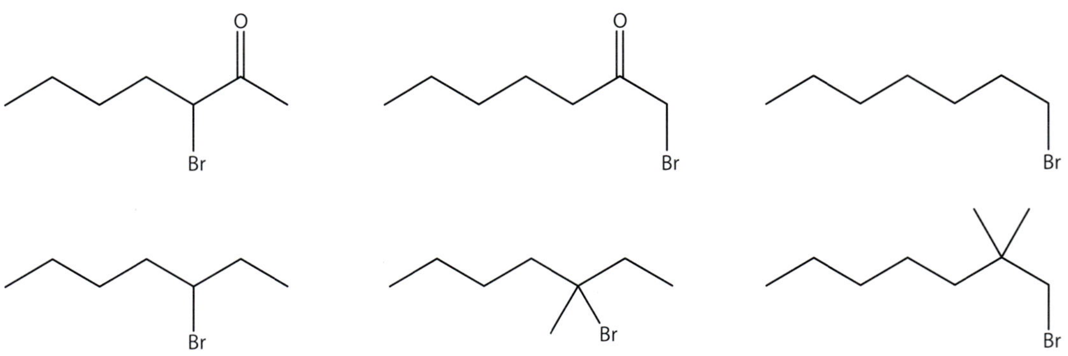

8. Place the following compounds in order with respect to their reactivity in a typical S_N1 reaction, giving brief reasons.

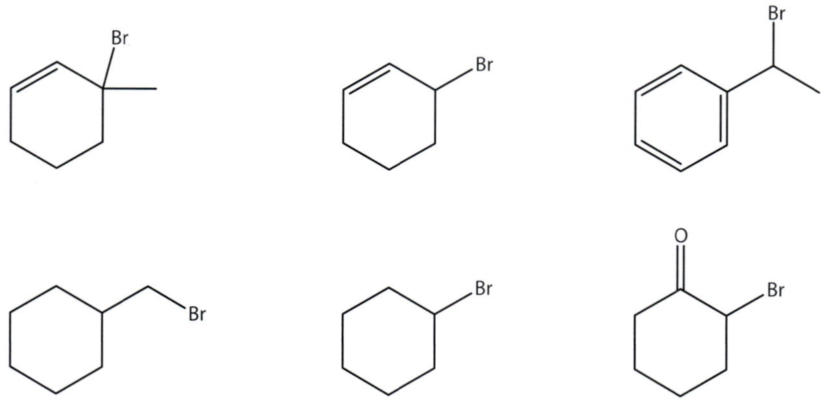

9. Propose a mechanism for each of the following reactions:

(a)

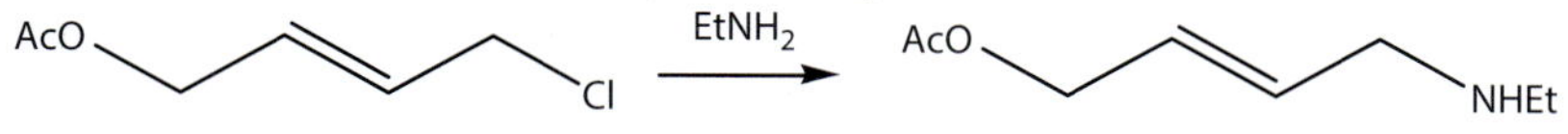

(b)

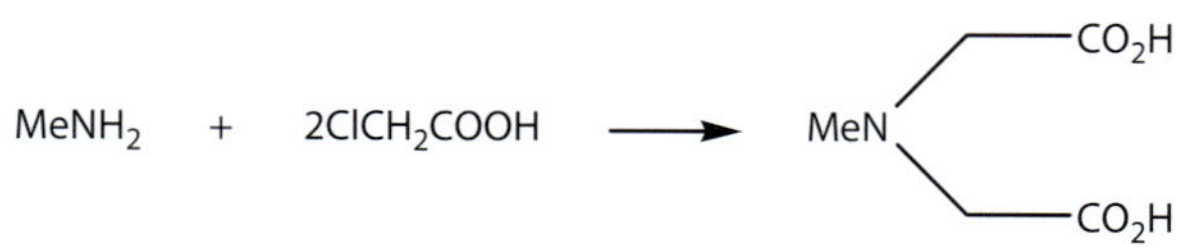

(c)

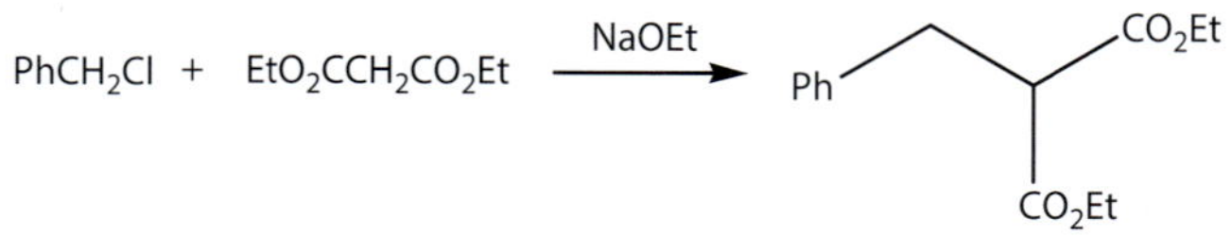

(d) $Et_2O + 2HI \rightarrow 2EtI + H_2O$

(e)

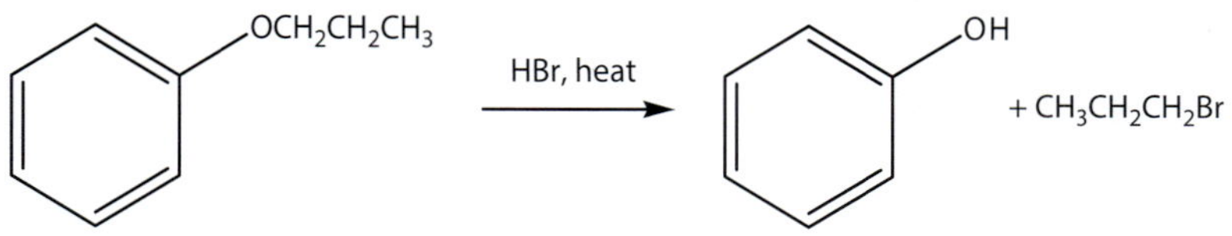

(f)

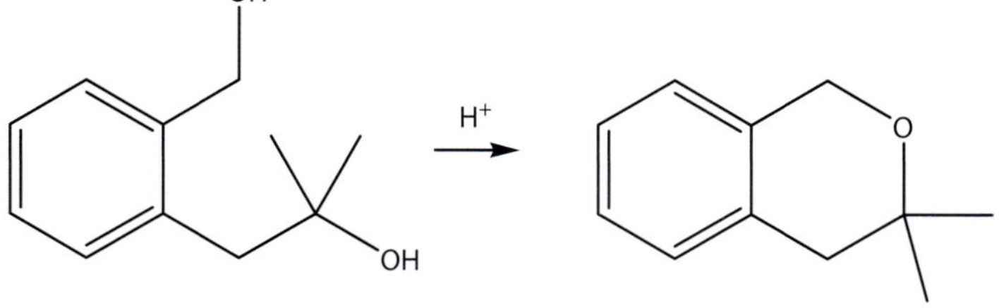

(g)

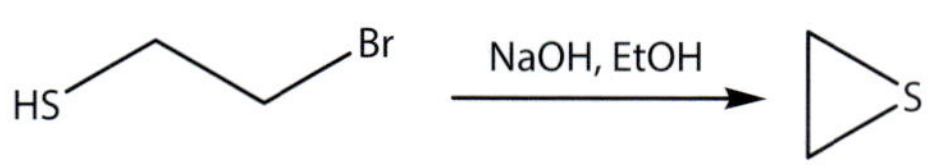

(h)

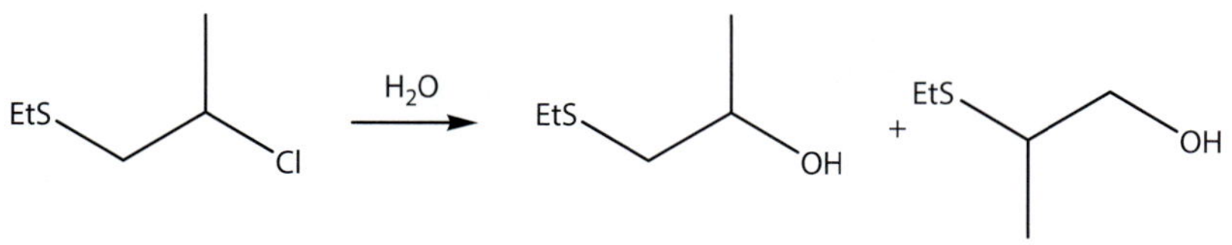

(i)
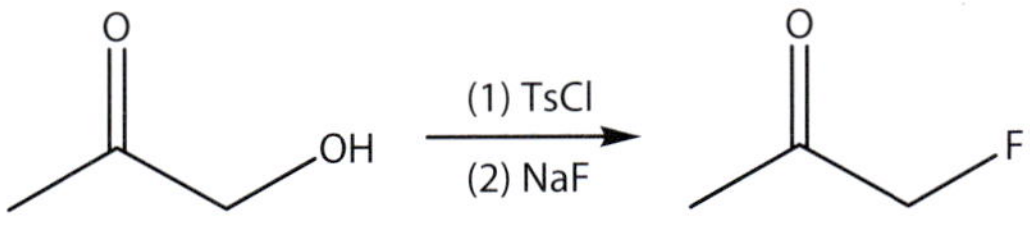

(j)
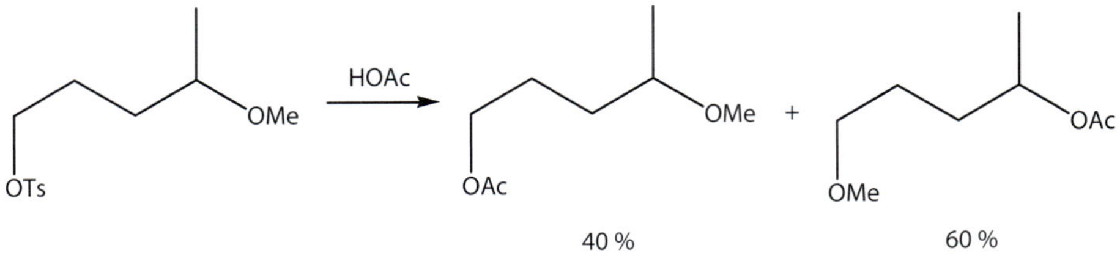

(k)
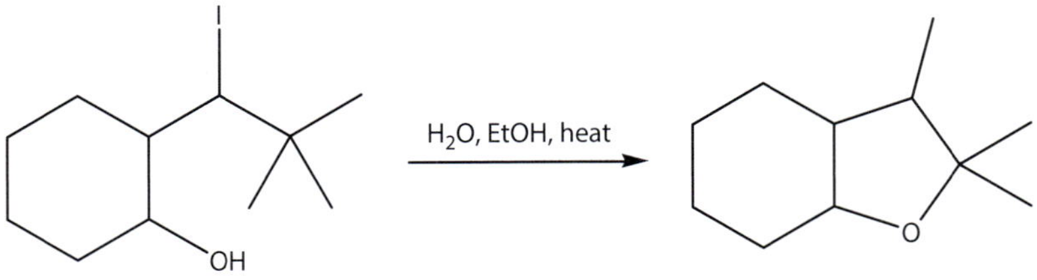

(l)
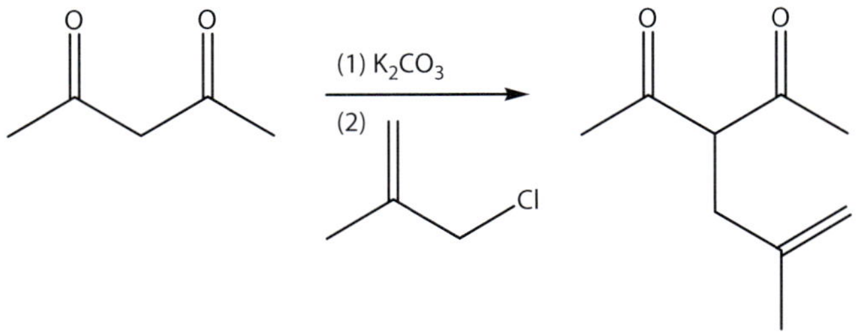

(m)
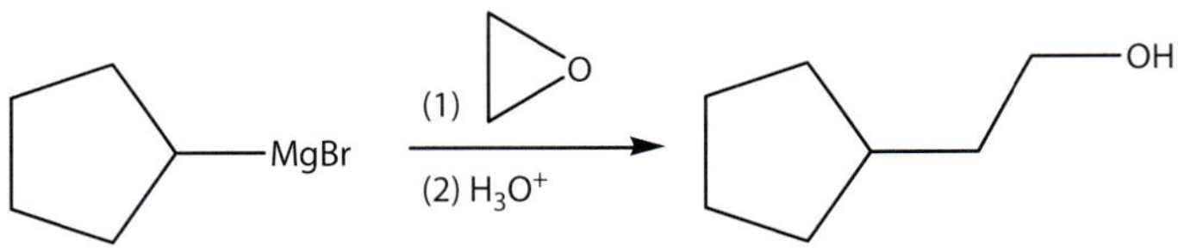

10. Explain the following observations:

(a)
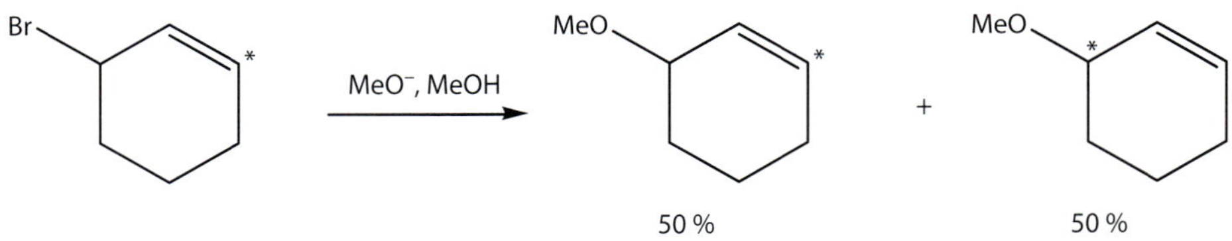

* indicates a ^{14}C label.

(b) Attack of [PhS]⁻ on **9.31** is 58 times faster than the reaction with **9.32**:

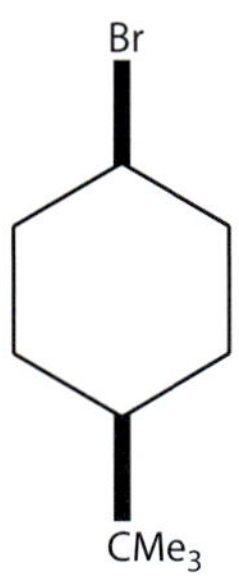

9.31

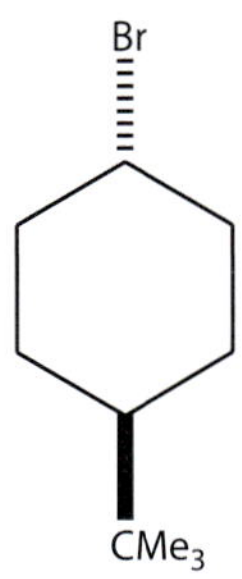

9.32

(c) When Me_3CBr reacts with water at pH 8 to give Me_3COH, the rate is proportional to $[Me_3CBr]$, but for the reaction of Ph_3CBr under the same conditions, the rate is proportional to $[Ph_3CBr][HO^-]$.

(d) The rate of solvolysis of **9.33** in CF_3CH_2OH at 25 °C exceeds that of **9.34** by a factor of 10^{14}:

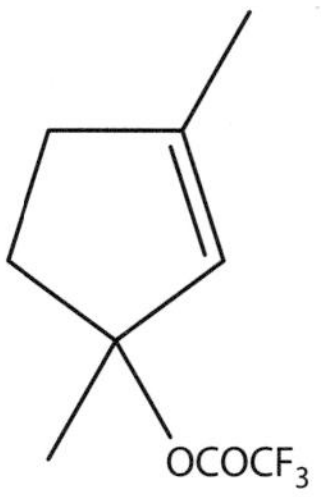

9.33

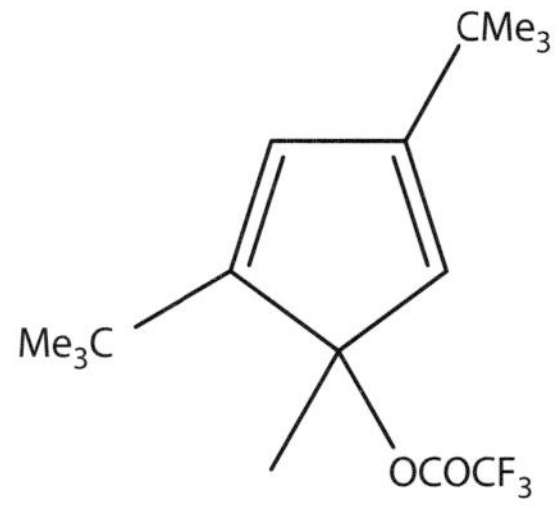

9.34

11. Study the following reactions shown:

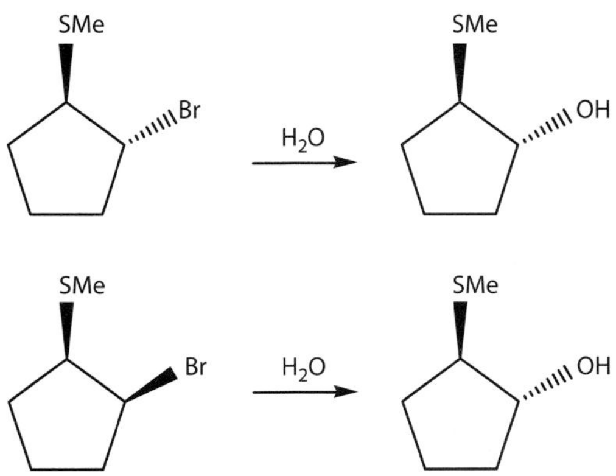

Which would you expect to be the faster? Give your reasons and explain the observed stereochemistry.

12. Identify the compounds **9.35** and **9.36**, and explain their formation with a full mechanism ($MsCl = CH_3SO_2Cl$):

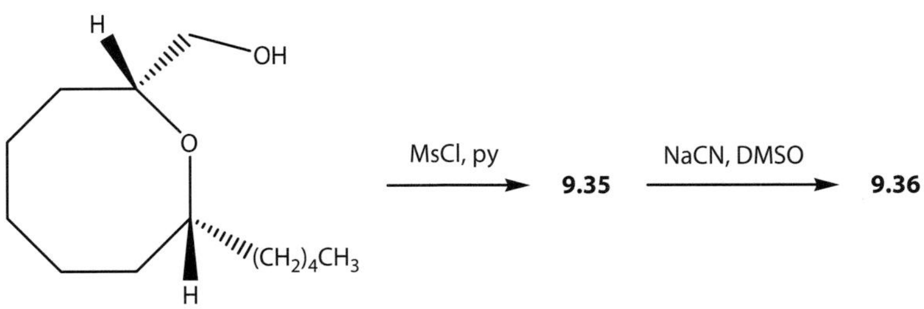

13. Identify the dianion **9.37** in the following sequence shown, and explain why its reaction with bromononane is so selective:

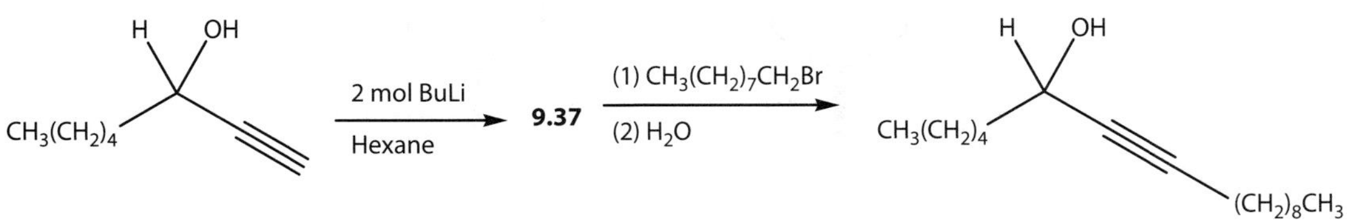

14. When a primary alcohol is treated with TsCl at room temperature in the presence of a base such as pyridine, a tosylate is formed. When the reaction is carried out at higher temperature, an alkyl chloride is often obtained. Why?

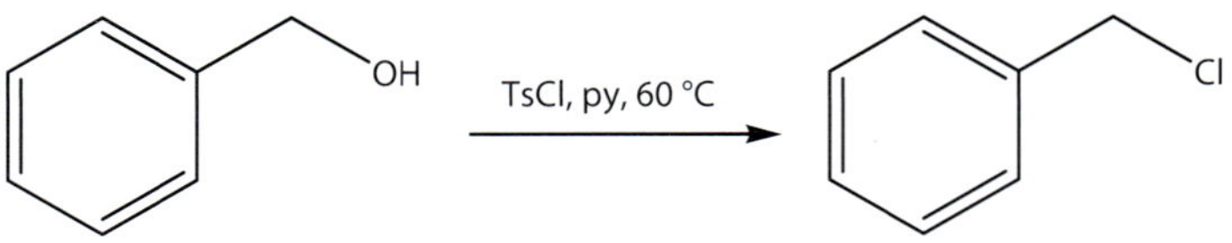

MORE CHALLENGING PROBLEMS

1. Predict the order of reactivity of the following compounds:
   (a) In an S_N2 reaction with $[EtO]^-$/EtOH
   (b) In an S_N1 solvolysis reaction with HCOOH as the solvent

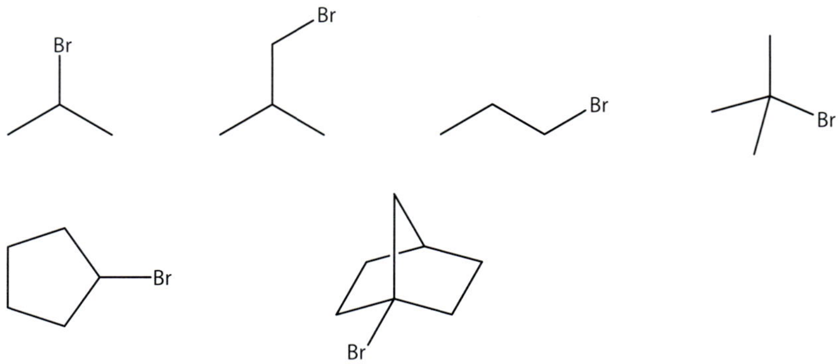

2. Predict the product of each of the following processes:

(a)

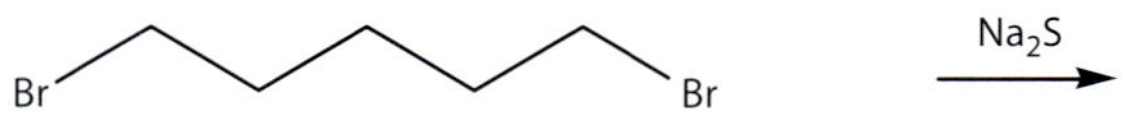

(b)

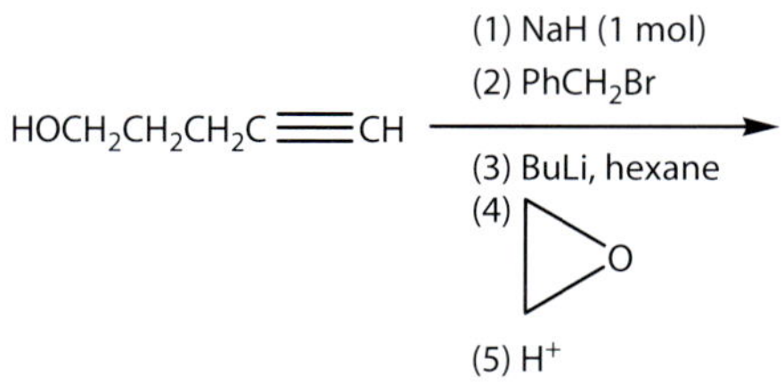

(c)

(d)

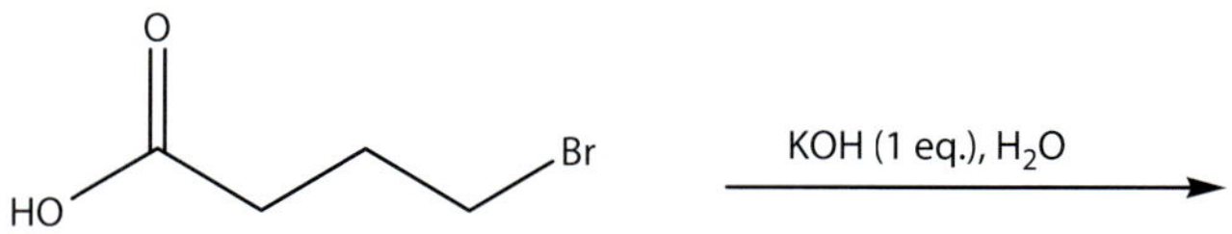

(e)

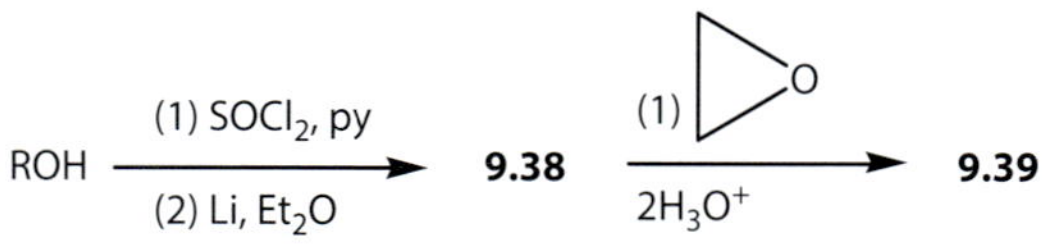

(f)

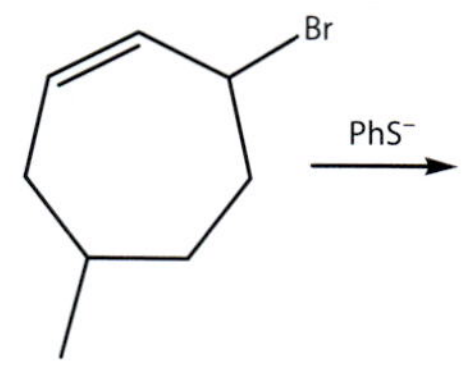

(g)

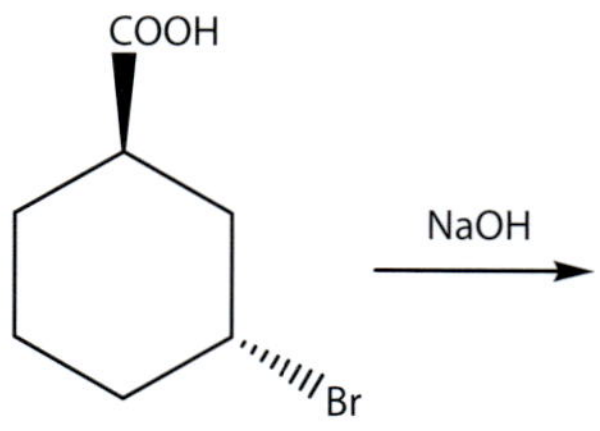

(h)

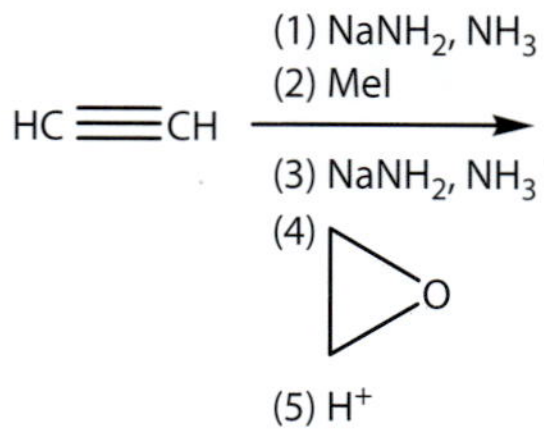

3. Write a mechanism for each of the following processes:

(a)

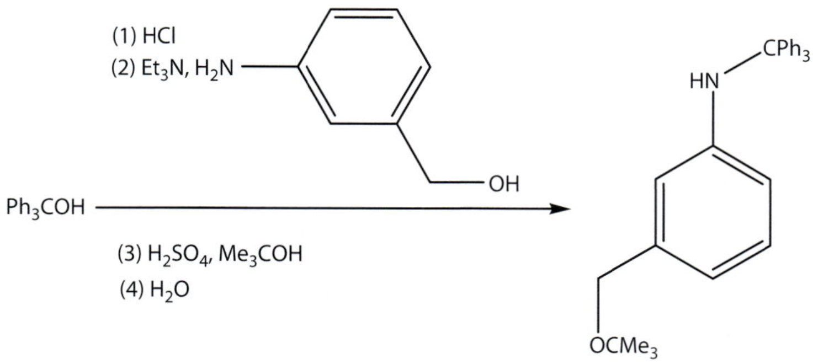

(b)

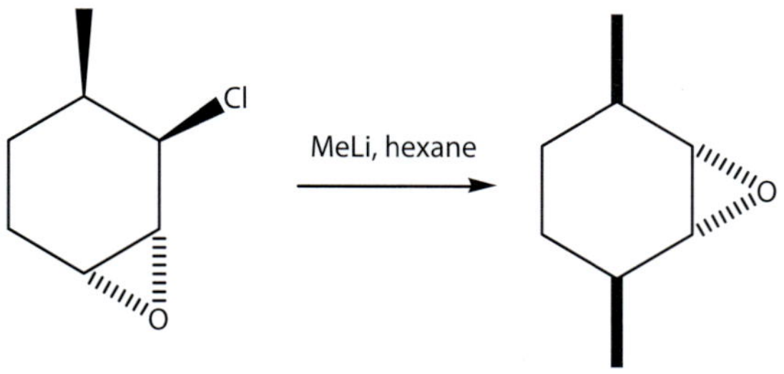

Optically active

meso

(c)

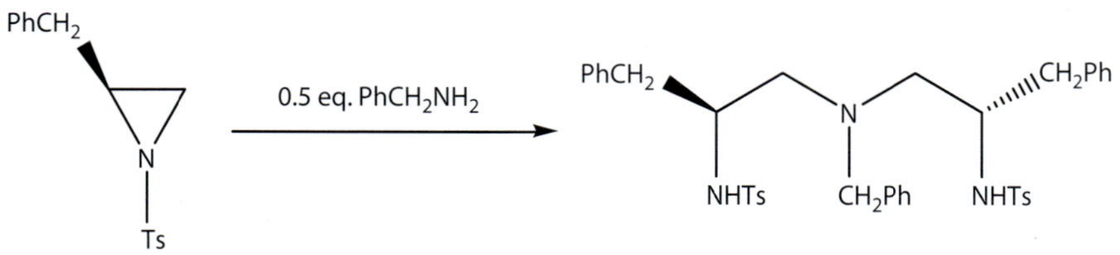

(d)

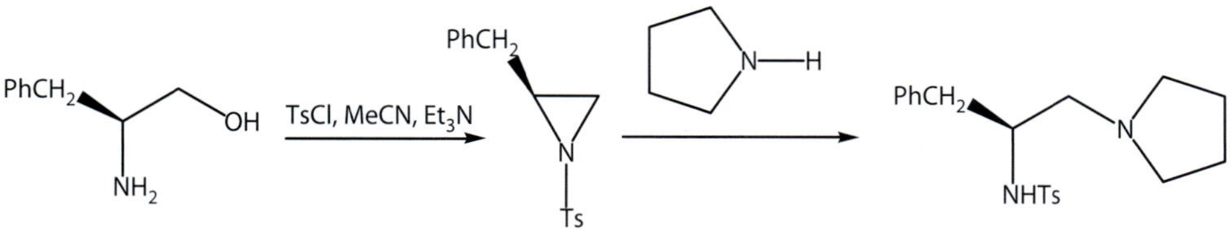

(e)

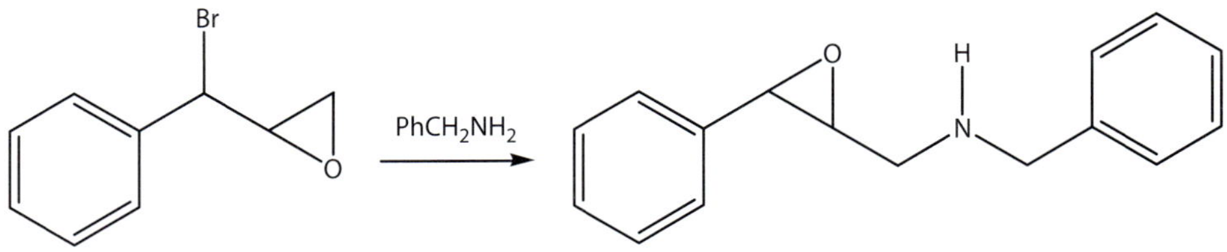

(f)

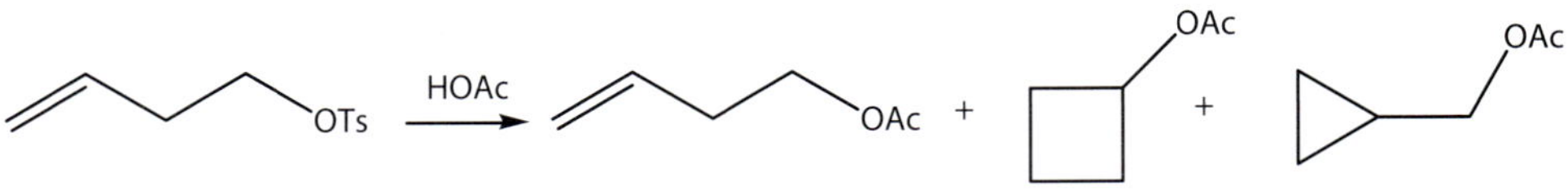

(g)

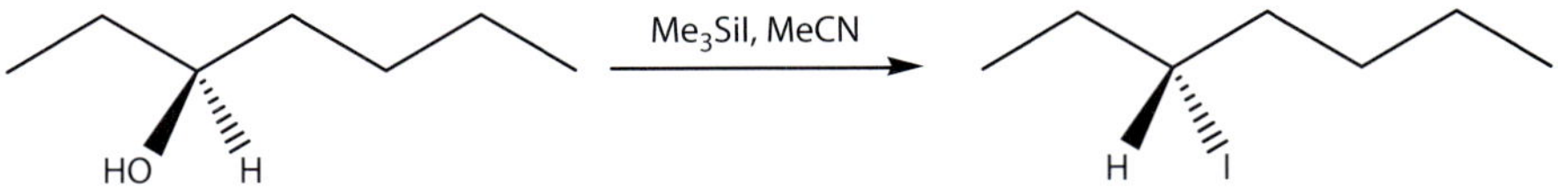

(h)

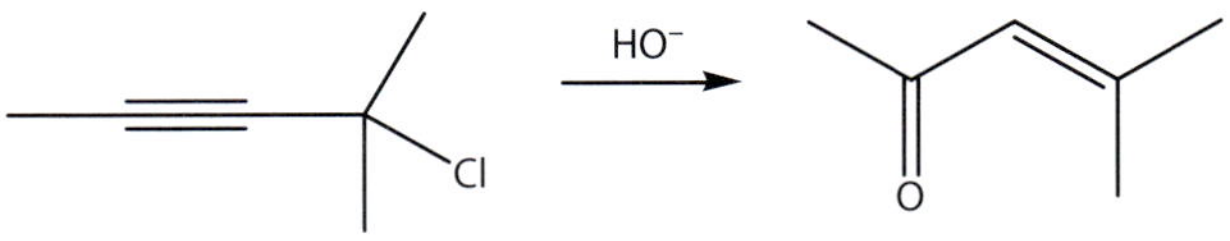

(i)

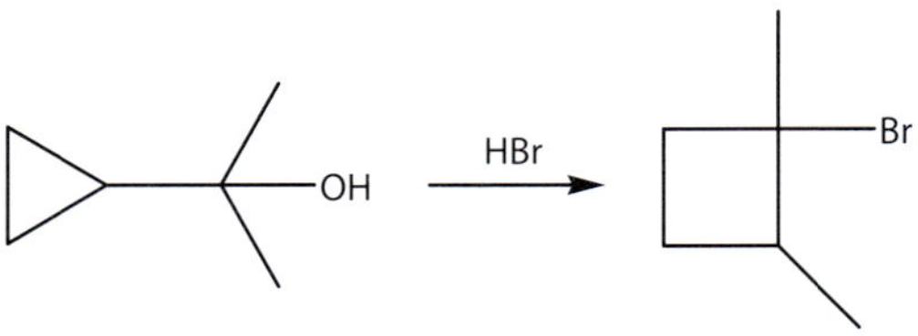

4. Explain the following observations:
   (a) In the S_N2 reaction

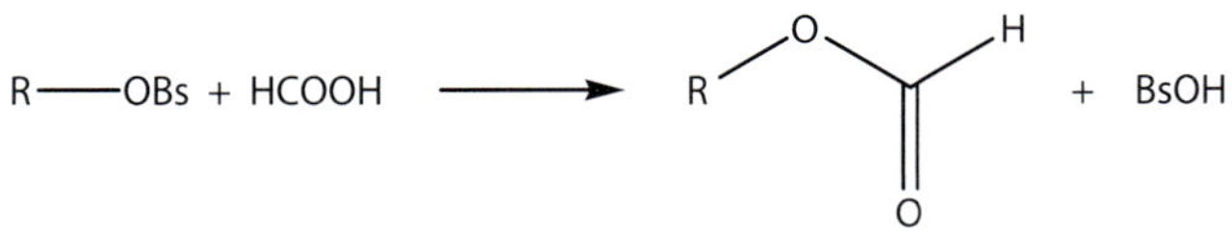

(OBs = $OSO_2C_6H_4$–4-Br—consider it as similar to tosylate)
The relative rates of reaction are given in the following:

	k_{rel} (75 °C)
R = Bu	1
R = $MeOCH_2CH_2$	0.1
R = $MeO(CH_2)_3$	0.33
R = $MeO(CH_2)_4$	46.1
R = $MeO(CH_2)_5$	32.6
R = $MeO(CH_2)_6$	1.13

(b) The rate of substitution of chloride by water as shown in the following depends on the nature of the group X. The rates decrease in the order X = NMe_2 > OMe > Ph > Cl.

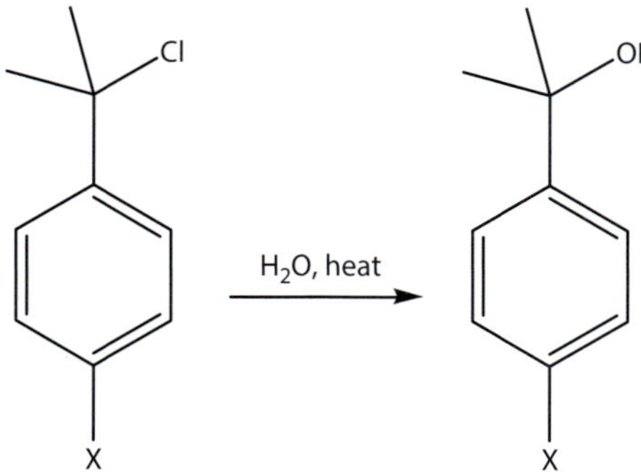

5. Can the molecule **9.40** undergo S_N2 substitution? Can it undergo S_N1 substitution? What would be the product of the solvolysis of **9.40** in methanol? Would **9.40** react faster or more slowly than **9.41** under these conditions? Explain your conclusion.

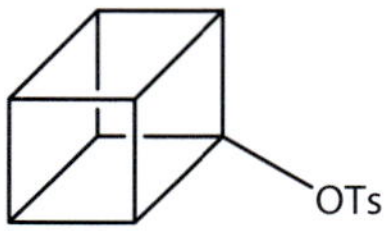

9.40

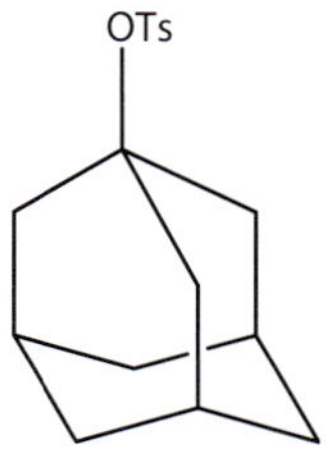

9.41

6. Explain the kinetic data given in the following:

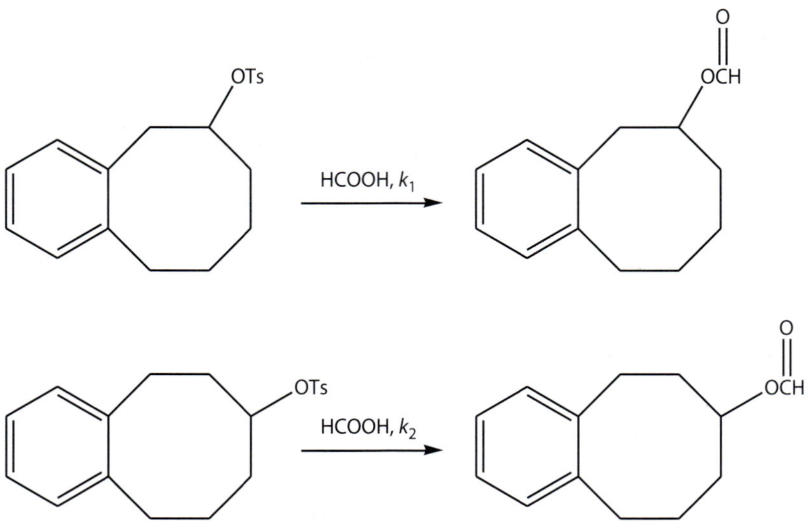

$k_1 = 1000k_2$

7. Propose mechanisms for the following reactions, explaining clearly why the outcomes are different.

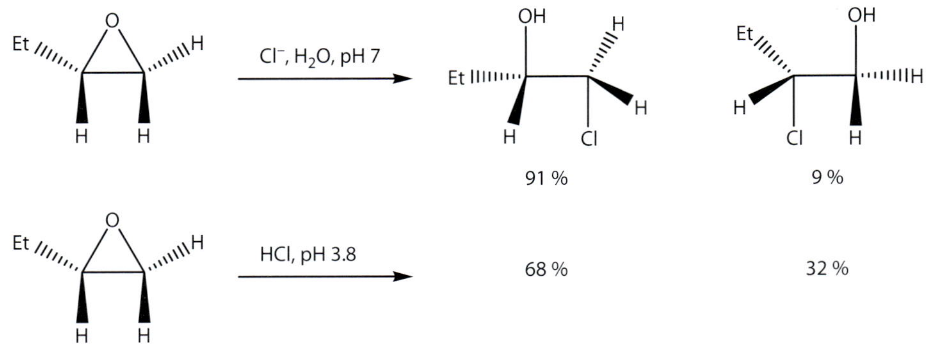

Elimination Reactions

10

10.1 INTRODUCTION

Elimination reactions are those in which we remove two atoms or groups from a molecule to generate a multiple bond. Most of the examples we will discuss involve making a carbon–carbon double bond, but we will also make some triple bonds and a few carbon–heteroatom bonds. We previously saw elimination processes as a "side reaction" of substitution—now, we turn the tables and see substitution as a side reaction of a desired elimination. Although we have good methodologies for inducing reactions to go in one direction or the other, we should recognize that there may always be some competition. In this chapter, we will concentrate initially on eliminations to give alkenes, turning later to alkynes, and multiple bonds involving heteroatoms.

10.2 MECHANISMS

As with substitution, we have two mechanisms that are relatively common and one that is much rarer. The two common mechanisms, E1 and E2, have strong similarities to the S_N1 and S_N2 processes described in the previous chapter and are generally favored by similar conditions to these analogues. The third mechanism, E1cB, is rather different, with no direct analogy in substitution chemistry. We will initially exemplify all the reactions by removal of hydrogen halides, HX, in the presence of base and then explore the scope of the reactions.

10.2.1 E1 ELIMINATION, UNIMOLECULAR

This mechanism (Figure 10.1) has strong similarities to the S_N1 process—indeed, the first RDS is identical. A leaving group is lost, taking the electrons from the bond being broken with it, in a slow step, to give a carbocation. However, the next step is not capture of the carbocation by a nucleophile but a rapid loss of a proton to give an alkene.

The observed kinetics of the reaction are first order, with the rate proportional only to the concentration of the alkyl halide, not to any added base. The RDS is unimolecular, hence the name E1. The analogy to S_N1 is clear. This type of reaction is favored if a good carbocation, usually tertiary or conjugated, is formed, and by a good leaving group. As with S_N1, polar solvents are generally advantageous because they solvate the ions formed in the initial step. Although less crucial, the absence of a strong base is usually desirable—strong bases, as we will see shortly, promote E2 reactions.

10.2.2 E2 ELIMINATION, BIMOLECULAR

This is analogous to the S_N2 reaction; it is a single-step, bimolecular process, involving both the base, B^-, and the substrate (Figure 10.2).

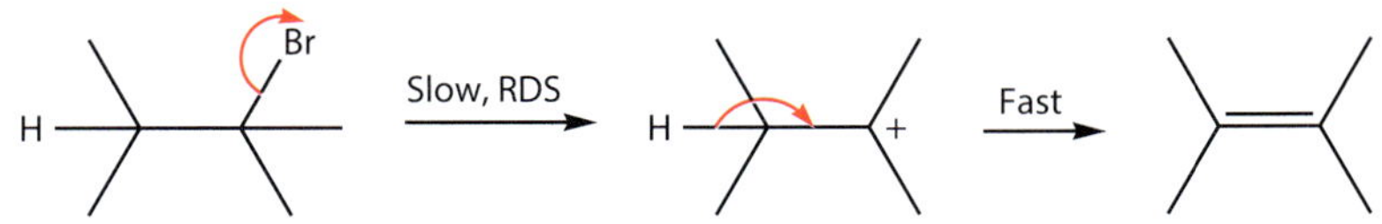

FIGURE 10.1 Mechanism of the E1 elimination process.

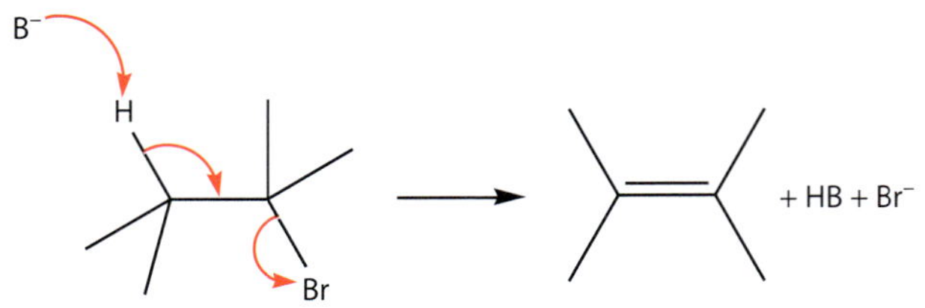

FIGURE 10.2 Mechanism of the E2 elimination process.

The observed kinetics are second order, with the rate proportional to both base and substrate (Equation 10.1). It is generally favored by strong, non-nucleophilic bases such as *tert*-butoxide, low-polarity solvents (among other factors so that the base is not heavily solvated), and appropriate stereochemistry, a point that we will return to in a later section.

$$\text{Rate } \alpha [\text{RX}][\text{Base}] \qquad (10.1)$$

10.2.3 E1cB ELIMINATION, UNIMOLECULAR, CONJUGATE BASE

This relatively rare mechanism is a two-step process, but we still describe it as unimolecular and give it a "1" tag because the RDS is unimolecular. A single species, the carbanion, or conjugate base (hence cB), from the starting material, is involved. The first step is fast and sometimes, but not invariably, reversible. The process involves deprotonation of the substrate to give a carbanion, which must be a relatively stable one, if this mechanism is to operate. This carbanion then slowly loses the leaving group to give the alkene (Figure 10.3).

The rate of the reaction is proportional only to the concentration of the anion. However, the anion concentration in turn depends on the concentration of the substrate and the base since the first fast step is generally an equilibrium, though it may be a highly biased one. So we usually see second-order kinetics as in (**10.1**). This type of reaction is favored by the formation of a stabilized anion, the presence of a strong base, and a poor leaving group.

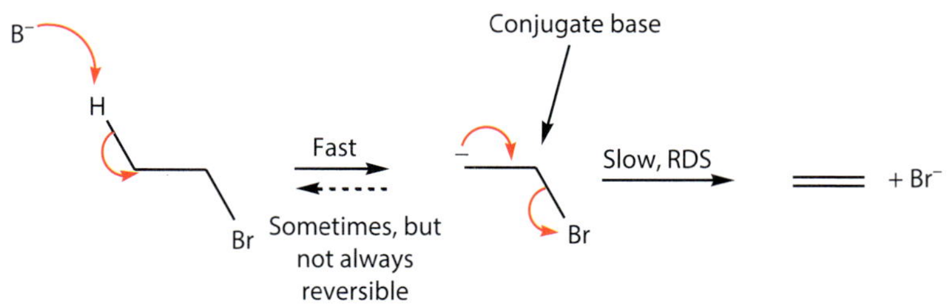

FIGURE 10.3 Mechanism of the E1cB elimination process.

Focus on Kinetics

The first step of an E1cB reaction may be a rapidly established equilibrium, or it may be effectively irreversible. We can consider two apparently similar eliminations of water, under basic conditions (where the OH group is a very poor leaving group), which show rather different kinetics. In the reaction of Figure 10.4, the first step is very rapid, any equilibrium far to the right, and the second step is much slower.

$k_1 \gg k_2$, so the kinetics appear to be first order. By contrast in the reaction of Figure 10.5, the first step is a true equilibrium, so second-order kinetics are observed, with rate dependent both on the concentration of the substrate and the base.

We may note in passing that this tells us that a nitro group is better at stabilizing a carbanion than an adjacent carbonyl group; it is more electron withdrawing.

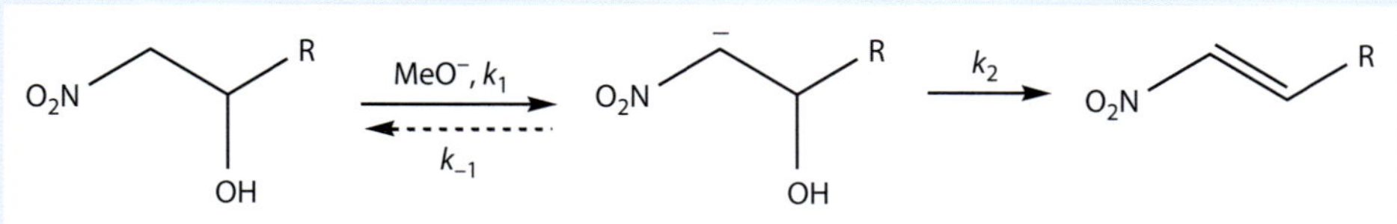

FIGURE 10.4 E1cB reaction showing first-order kinetics.

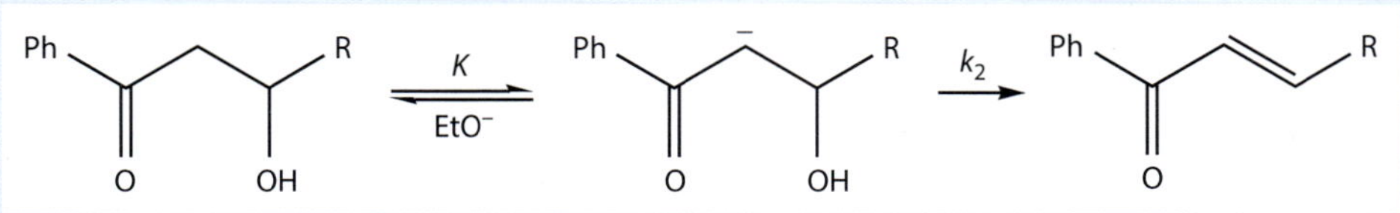

FIGURE 10.5 E1cB reaction showing second-order kinetics.

10.3 STEREOCHEMISTRY

Although stereochemistry for elimination is generally a less critical issue than for substitution, each of the three mechanisms has "requirements" and "preferences" (for E2, a strong preference). Both requirements and preferences will have an influence on which mechanism is dominant and sometimes on whether substitution or elimination predominates.

10.3.1 E1 ELIMINATIONS

The requirement for E1 reactions is that the C—H (or C—X) bond that is broken in the second fast step must be parallel to the empty p_z orbital of the carbocation. These are the two orbitals that are used to make the developing π bond. However, in general, this requirement is easily achieved by rotation around a C—C single bond and moreover can be achieved in more than one orientation, as shown in Figure 10.6. The relative stability of the two conformations of the carbocation will depend on the relative sizes of the groups R^1 to R^4, and the process will generally give predominantly the less hindered, more thermodynamically stable product.

10.3.2 E2 ELIMINATIONS

The E2 mechanism is the "fussiest" of the three we are considering. It is an absolute requirement that the two bonds that are being broken are coplanar, and there is a preference for the two bonds

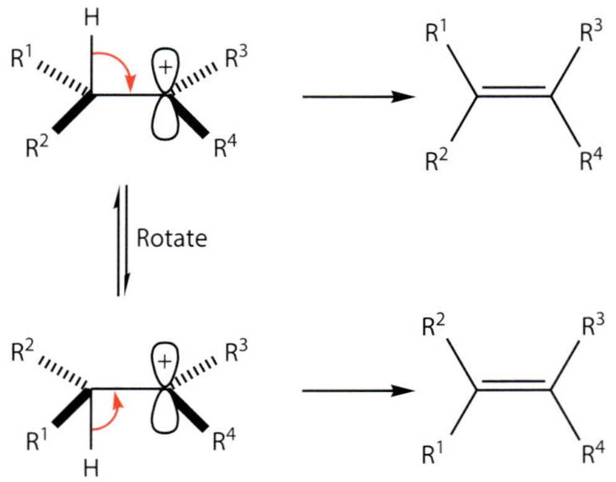

FIGURE 10.6 Stereochemical requirements of the E1 elimination process.

being *trans-antiperiplanar* as in **10.1**. This is just a fancy way of saying that the reaction works best when the two bonds are on opposite sides of the developing double bond.

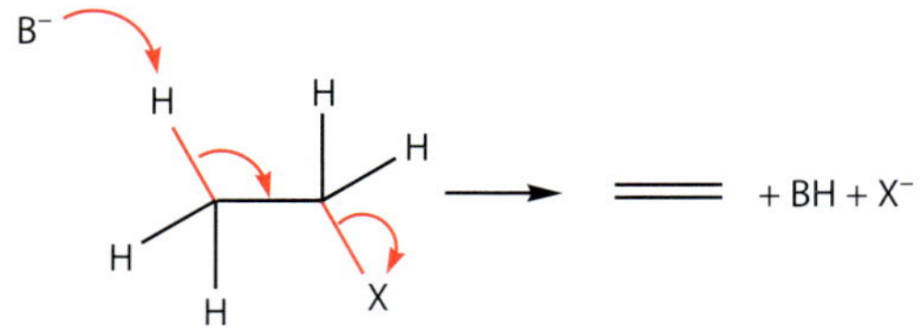

10.1

You will find that the easiest way to determine outcomes is using a good sawhorse diagram. Groups can only be *trans-antiperiplanar* when you have a staggered conformation, and this means that you will often need to rotate about the single bond in your original structure to get the groups in the right positions. Most people find it is easiest to work out what is happening when they draw the conformation where the two groups that are going to be eliminated are in vertical positions.

An example will illustrate. Let us say that we are asked to determine the stereochemistry of the product of elimination of HBr from meso-1,2-dibromo-1,2-diphenylethane. The first step is to draw the *meso*-compound, and as we will remember from Chapter 7, this is most easily done in the eclipsed conformation **10.2**. But this conformation is no good for our elimination reaction,

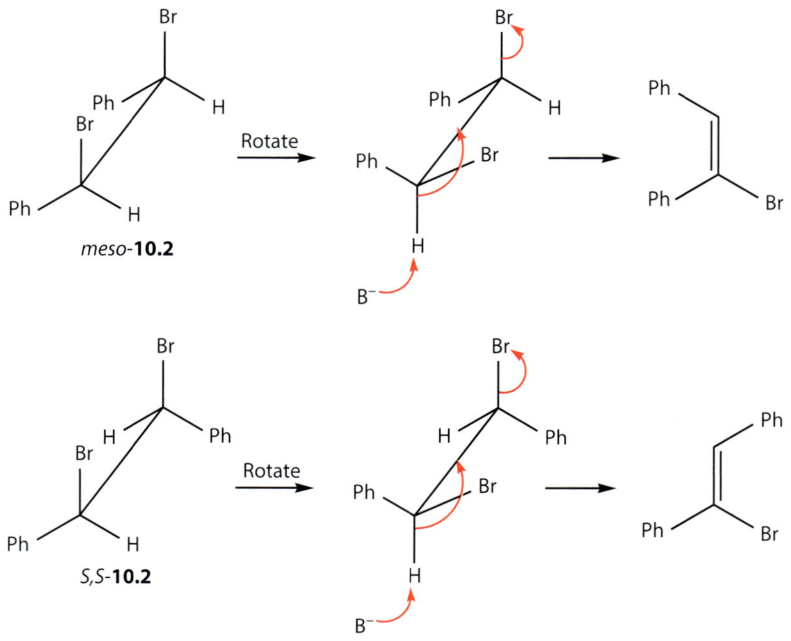

FIGURE 10.7 *Trans*-antiperiplanar elimination of HBr.

so we must rotate about the single bond until a hydrogen and a bromine are clearly opposite to one another. It's usually best to think about this as holding one carbon atom still and rotating the bond to the other. The stereochemistry of the product is then obvious—we get the *E*-isomer (Figure 10.7). If we now look at the same process with the *S,S*-isomer, we also get a completely stereospecific reaction, this time to give the *Z*-isomer. So the outcome of the elimination reaction is completely determined by the stereochemistry of the starting material.

PROBLEM 10.1

Zinc metal gives rise to the elimination of vicinal dibromides, so 1,2-dibromoethane would yield ethene. Predict the product, including the stereochemistry, of the zinc-promoted elimination of bromine from **10.3**.

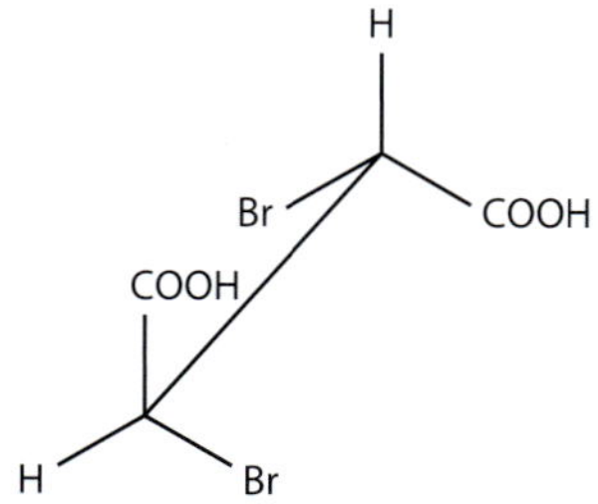

10.3

Solution

The conformation with which we presented in this problem is not appropriate for elimination—we need to have the two bromine atoms *trans-anti* to one another, and what we are given is an eclipsed conformation. We will see most easily and clearly how the stereochemistry will play out if we have the bonds to the two bromines in vertical and *anti* to one another. This involves rotating both carbon atoms, and you might want to do that in two steps. But once done, the stereochemistry of the product becomes obvious:

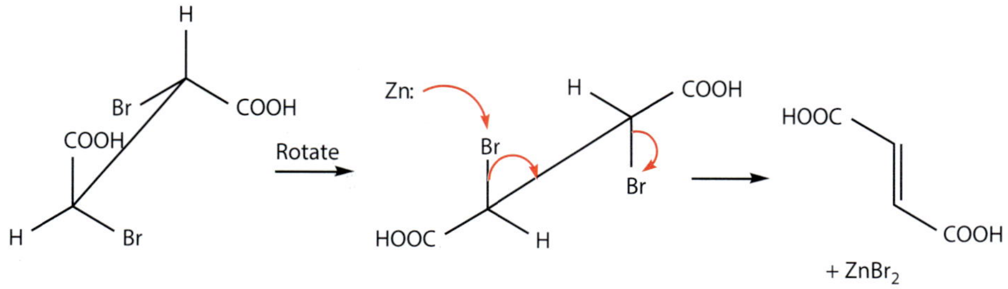

Stereochemical requirements can influence the rate of E2 reactions as well as their outcome. For example, if we consider the elimination of HCl from **10.4**, we will find that it is very slow. If we draw out the molecule in its proper chair conformation, the best conformation has all the substituents (but most especially the very bulky *tert*-butyl group) in the favored equatorial positions. However, no E2 elimination is possible from this conformation—the group *trans-anti* to the chlorine that we want to eliminate is a carbon–carbon bond, not a carbon–hydrogen bond. To undergo elimination, the molecule must "flip" to the highly unfavorable conformation with all the substituents in axial. Then only one elimination is possible, because only one *trans-anti* hydrogen is available—and this will be very slow (Figure 10.8). We would describe the reaction as regiospecific—only one alkene is formed.

By contrast, if we investigate the reaction of **10.5**, the reaction is much faster, and two products are formed. The most stable chair conformation of **10.5** has the chlorine axial (the large *tert*-butyl group, as always, dominates the equilibrium stereochemistry), and two hydrogen atoms are available in *trans-anti* positions for elimination. Thus, two alkenes can be formed; the more stable, the more substituted alkene predominates by a factor of 3:1 (Figure 10.9).

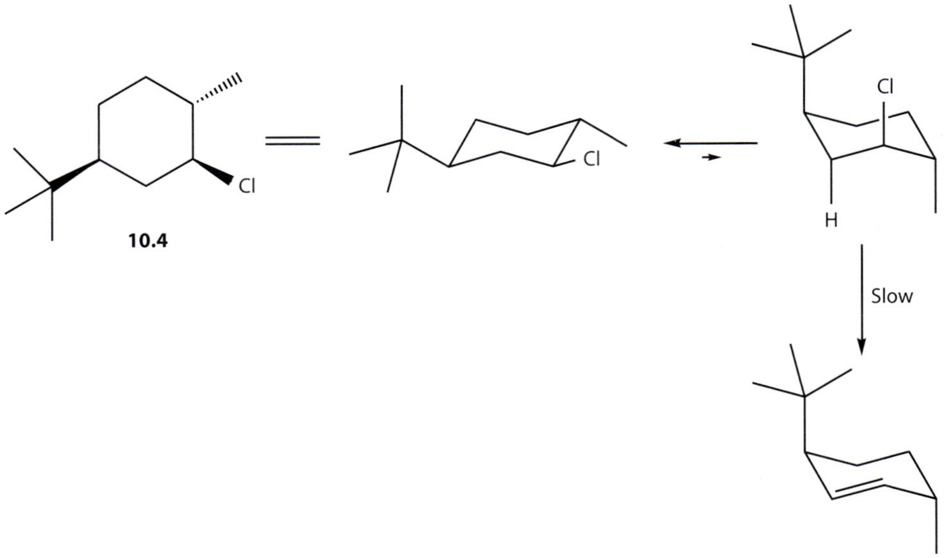

FIGURE 10.8 E2 elimination is slowed by inaccessibility of the required conformation.

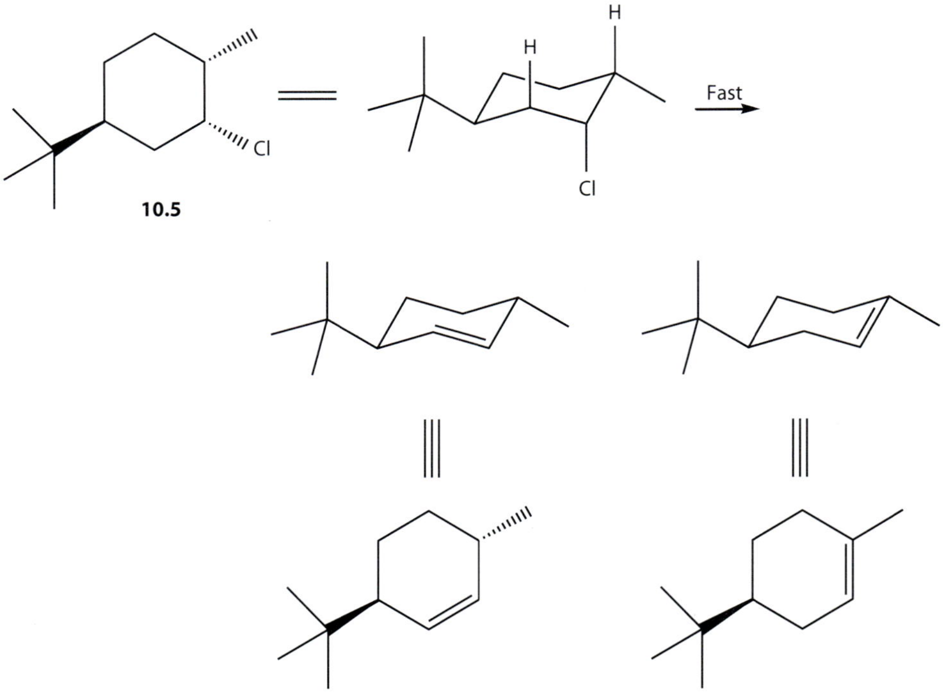

FIGURE 10.9 E2 elimination from a favorable conformation.

If the geometry of the molecule is such that no antiperiplanar arrangement is available, then a *syn*-coplanar arrangement of the departing groups is acceptable. Electronically, *syn*-coplanar eliminations are always possible, but in acyclic systems, they are very unlikely, because they require that the molecule is in an eclipsed conformation, which is disfavored relative to the staggered conformation needed for *trans-anti* elimination. We generally see this type of elimination in rigid polycyclic systems such as **10.6**. No *trans-anti* arrangement of any hydrogen/deuterium with the bromine is possible, and the molecule is inflexible, so

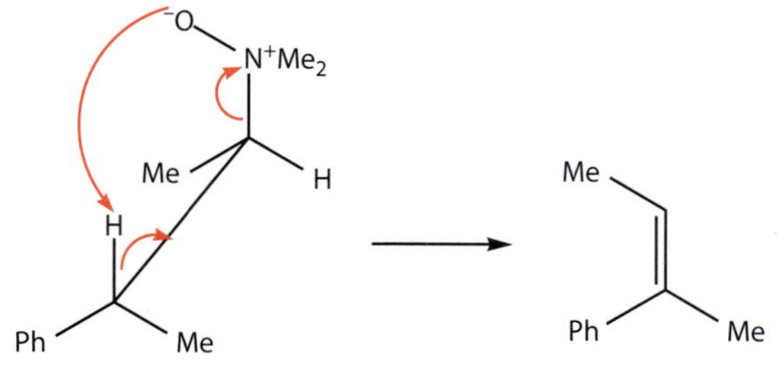

FIGURE 10.10 Thermal elimination of an amine oxide.

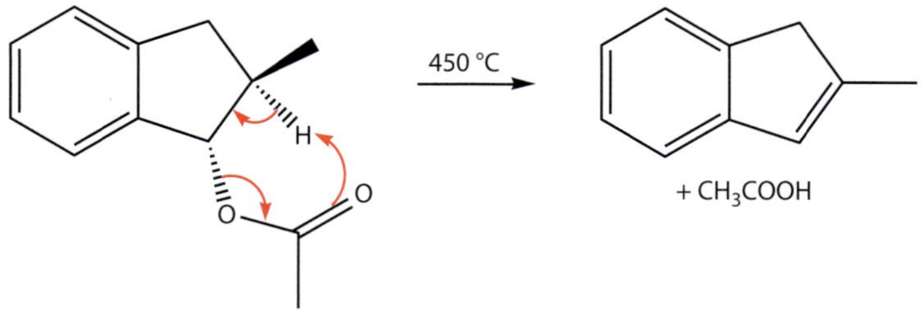

FIGURE 10.11 Pyrolytic elimination of a carboxylic acid from an ester.

there can be no conformational change. Elimination leads to a product called norbornene that is 94 % *de*deuterated (**10.7**).

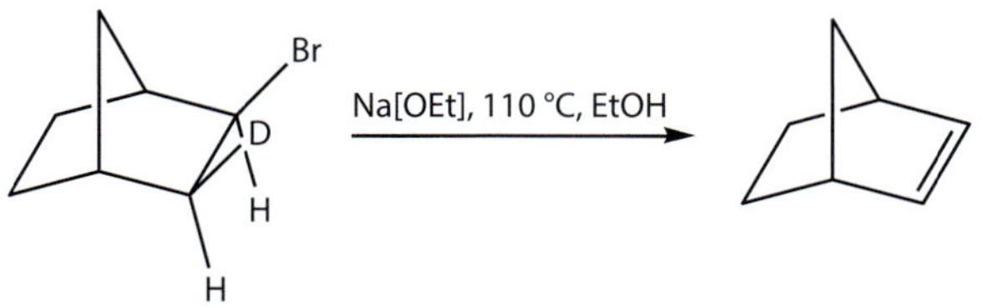

10.6 **10.7**

A few special examples always have *syn*-coplanar geometry because the reactions are intramolecular. Although we will discuss them here because of their mechanistic similarity to E2 processes, they are technically unimolecular and may show first- or second-order kinetics, depending on the reaction type. One example (Figure 10.10) is provided by the thermal elimination of amine oxides (prepared by the reaction of hydrogen peroxide or peracids with amines). In this example, 90 % of the desired compound is formed. The by-product is not the stereoisomeric alkene but the 1-alkene, CH_2=CHCH(Me)Ph, formed by abstraction of a proton from the methyl group. Esters are also pyrolyzed at high temperatures in a similar reaction (Figure 10.11). For both of these processes, it is necessary that a *syn*-coplanar hydrogen is available.

10.3.3 E1cB ELIMINATIONS

In this process, the lone pair of the carbanion must be in the same plane as the C–X bond to be broken in the second RDS (Figure 10.12). The carbanion is pyramidal (review the structure of the methyl anion, CH_3^-, if you have forgotten this), but it inverts very rapidly, so all the original stereochemistry is lost.

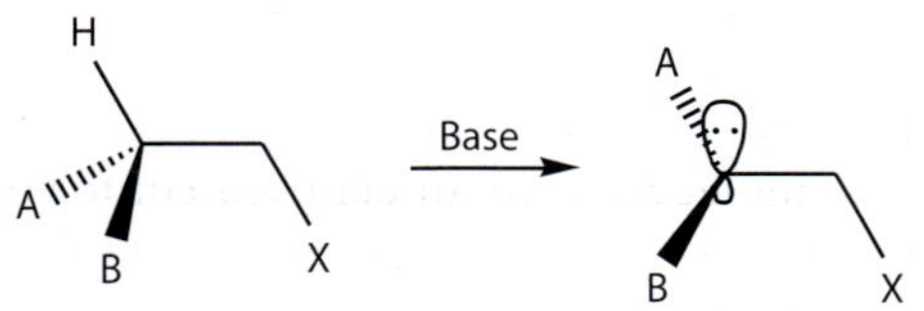

FIGURE 10.12 Stereochemistry of E1cB elimination.

PROBLEM 10.2

Study the reaction shown in the following. Don't worry for the present about the protecting groups on the oxygen atoms; we will meet these in detail later (Chapter 20), and they don't participate in this reaction. The peracid oxidizes the amine to an amine oxide. Write a mechanism for the elimination process and explain why it shows the regiochemistry observed:

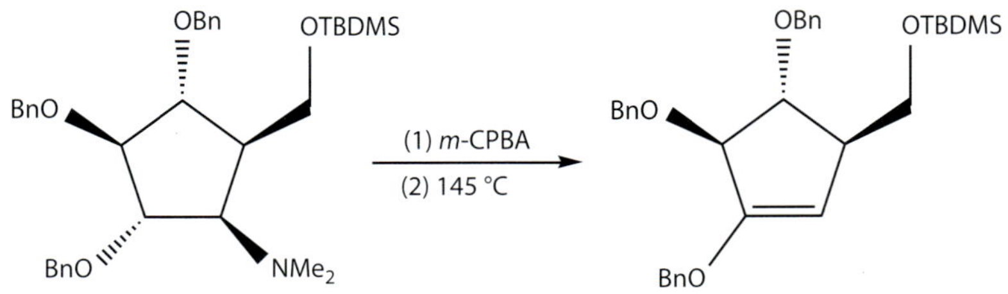

Solution

The first step of the process is oxidation of the amine to the amine oxide. At this stage, nothing is changed about the molecular stereochemistry, because no reaction has occurred at any stereocenter.

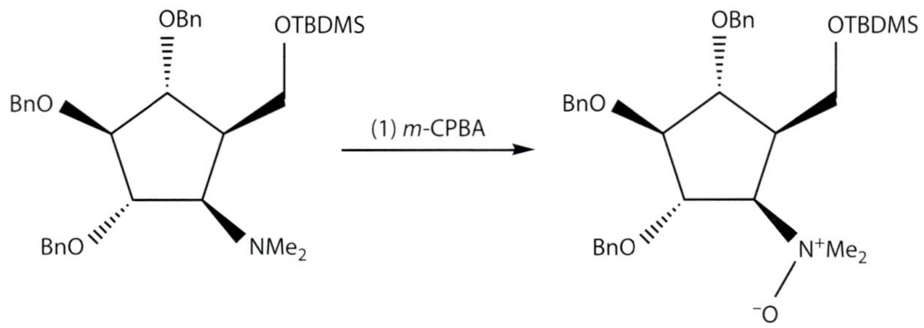

The elimination step is an intramolecular *syn*-elimination. Only one proton is available that is *syn* to the amine oxide; hence, only one regioisomer is formed.

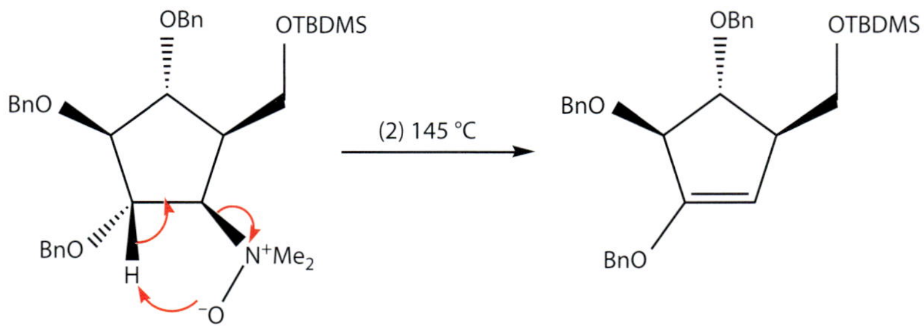

Key Points from Sections 10.1 through 10.3

- E1 elimination, related to S_N1 substitution, is a two-step process involving carbocation formation, followed by loss of (usually) a proton. This is favored by the formation of a stable carbocation.
- E2 elimination is a single-step process, involving simultaneous loss of two atoms or groups. It is required that the bonds to these two moieties are coplanar, and there is a strong preference for a *trans*-antiperiplanar arrangement of the departing groups.
- E1cB elimination is a two-step process; in the first step, a base removes a proton to give a reasonably stable carbanion, and in the second RDS, the leaving group is lost. This mechanism is favored by stable carbanions and poor leaving groups.

10.4 REGIOCHEMISTRY

The term regiochemistry describes which alkene is formed—which "region" of the molecule reacts. In a few cases, there are no options—dehalogenation of 1,2-dibromides using zinc metal or iodide ion, for example (Figure 10.13).

However, in many examples, two regioisomeric products could in principle be produced by a dehydrohalogenation. From 2-bromobutane, we could write sensible curly arrows to obtain *E*- or *Z*-2-butene or 1-butene (Figure 10.14). Which of these products are obtained, under what conditions, and why?

We need to consider the three mechanisms in turn. In the E1 process, the formation of the product from the intermediate carbocation is fast. This means that the stability of the developing alkene is the only factor that we need to consider. The most stable alkene, generally the most substituted one, will predominate. This is sometimes referred to as Zaitsev elimination and is exemplified in Figure 10.15. Either proton could be lost from the intermediate carbocation, and the major product is the trisubstituted rather than the disubstituted alkene.

Where all else is equal, as in the formation of a 2-alkene (the 1-alkene will be a very minor by-product) from the carbocation in Figure 10.16, steric effects are important, and the lowest energy

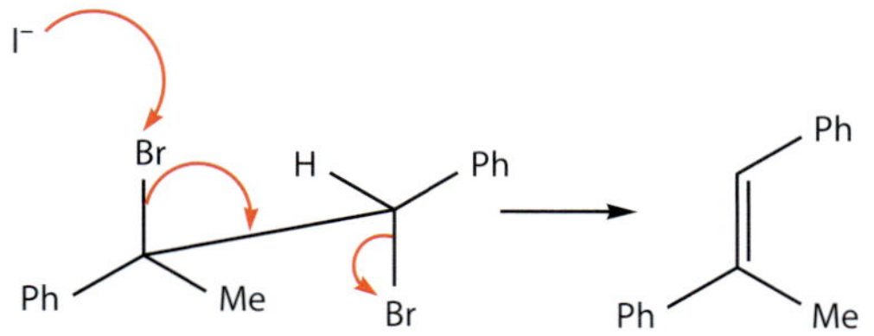

FIGURE 10.13 Iodide ion dehalogenation of a vicinal dibromide.

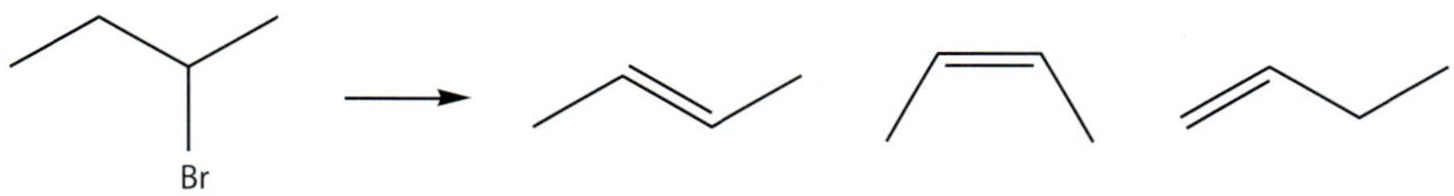

FIGURE 10.14 Dehydrobromination of 2-bromobutane.

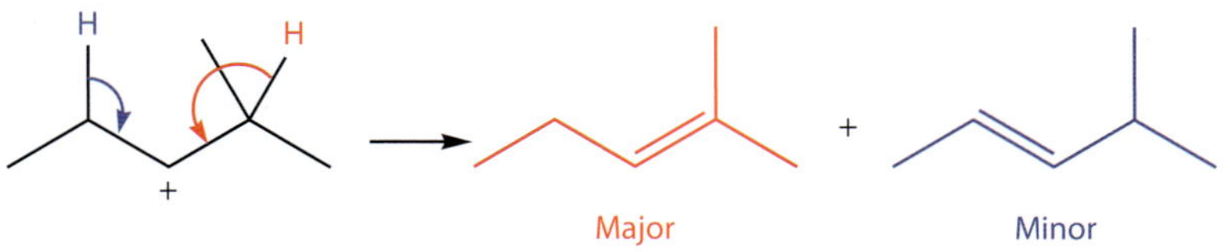

FIGURE 10.15 Regiochemistry of E1 elimination.

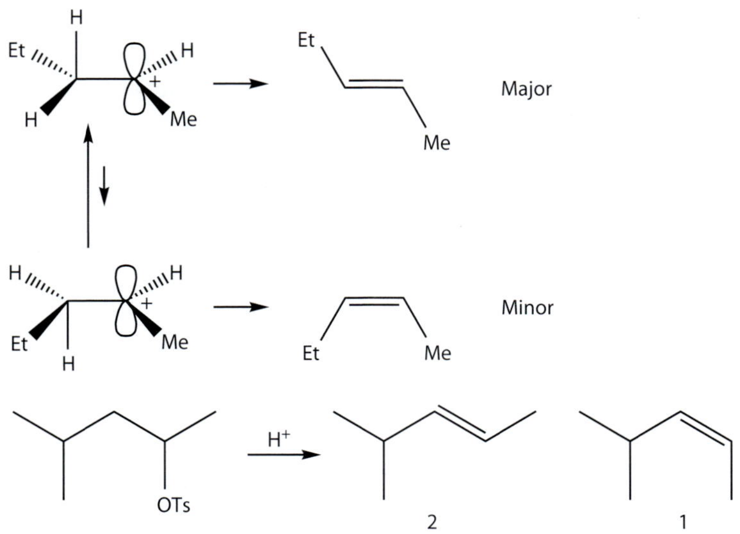

FIGURE 10.16 Steric effects on the regiochemistry of E1 eliminations.

process will be the one from the lowest energy conformation of the carbocation intermediate. *trans*-Products predominate in such reactions. However, the effect is not a large one unless the substituents are very bulky, as the second example demonstrates.

PROBLEM 10.3

Explain the following observation:

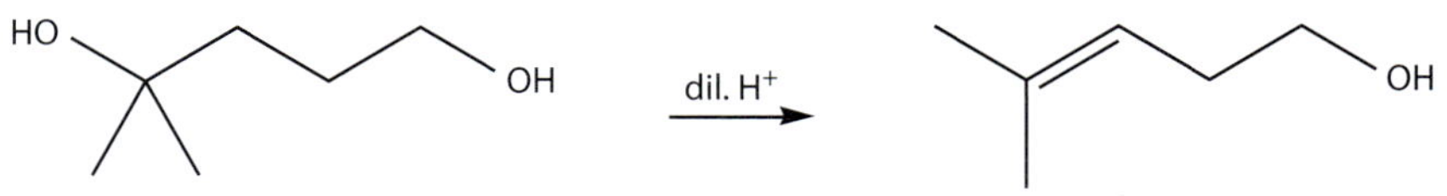

Solution

There are two hydroxyl groups in the molecule, and both of these could be protonated. However, only one molecule of water is eliminated. This reflects the E1 mechanism of the reaction. Only the tertiary alcohol can readily lose water to give a stable tertiary carbocation, and hence, only this undergoes elimination. Although an alternative elimination from this carbocation to give the 1-alkene is theoretically possible, the trisubstituted 2-alkene is more stable and hence is the predominant product:

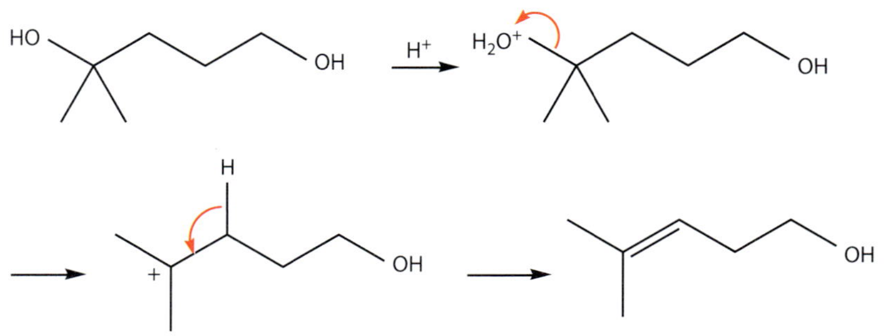

In the E1cB reaction, the regiochemistry is entirely dependent on the ease of formation of the carbanion. Which hydrogen is the most acidic, and which one, when removed, gives the most stable carbanion? We should briefly review carbanion stability. Because alkyl groups are electron donors, primary carbanions are the most stable, and tertiary ones the least:

$$CH_3^- > CH_3CH_2^- > (CH_3)_2CH^- > (CH_3)_3C^-$$

We also need to consider all the other features that stabilize carbanions. Electron-withdrawing atoms such as chlorine will stabilize a carbanion, so $[CCl_3]^-$ is more stable than $[CHCl_2]^-$. More importantly in this context, carbanions can be stabilized by resonance; thus, benzylic, allylic, and enolate anions are significantly more stable than their unconjugated counterparts. The greatest advantage is gained with the enolate anion, as, in one of the resonance forms, the negative charge is located on the electronegative oxygen atom (Figure 10.17).

We can now readily deduce the regiochemistry of elimination for **10.8**; the reaction goes through an E1cB mechanism because the intermediate enolate anion is easily formed.

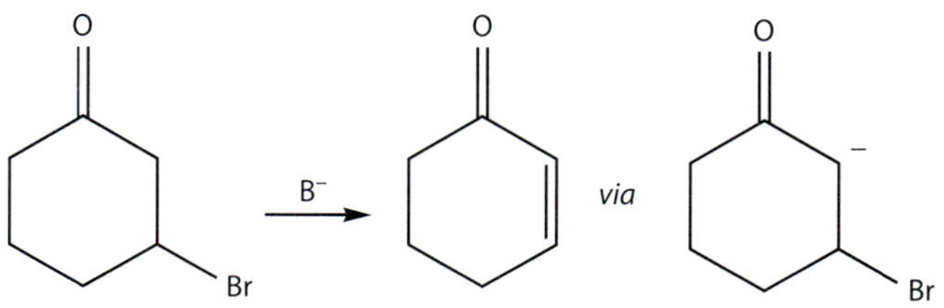

10.8

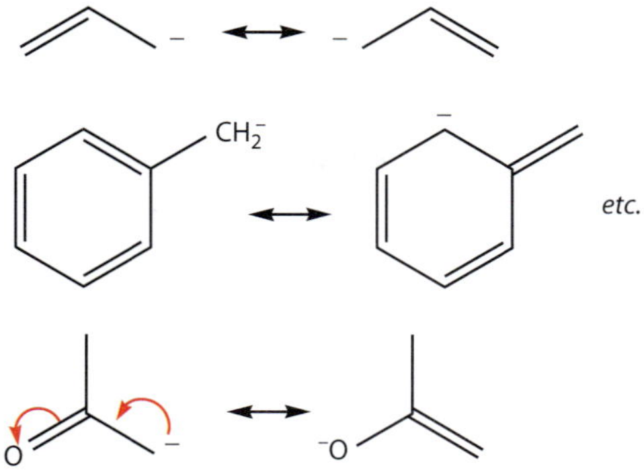

FIGURE 10.17 Stabilization of anions by resonance.

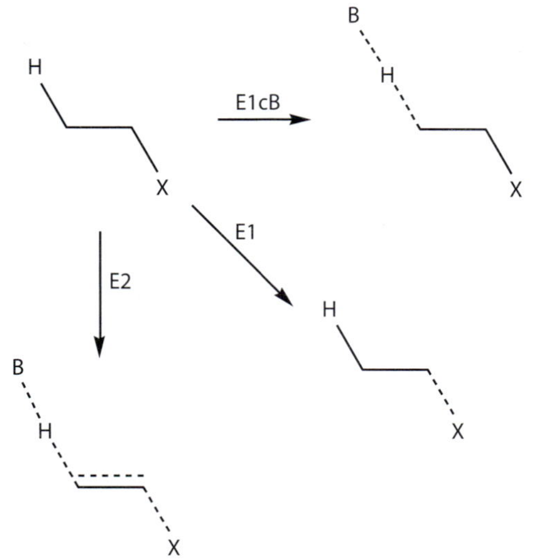

FIGURE 10.18 Transition states for elimination reactions.

You may be wondering why we have looked at the mechanisms in this order. First, the E1 and E1cB processes have a clear and well-defined answer on the subject of regiochemistry, whereas that for E2 requires rather more subtlety. Also, in thinking about what happens in the E2 process, these well-defined answers from the other mechanisms will help our understanding. The transition state for an E2 reaction may be, but is not necessarily, symmetric. If we draw the transition states for each process (Figure 10.18), we see that E2, in bond breaking at least, has some features of both of the other mechanisms.

Broadly speaking, we characterize E2 transition states as being "alkene-like"—approximately symmetric with the development of the new double bond quite advanced at the transition state—or "carbanion-like," which is closer in character to the E1cB process. Reactions with alkene-like transition states are favored when we have a good leaving group, such as bromide or tosylate, and a good base such as ethoxide. Because the transition state is alkene-like, the formation of the most stable alkene is favored, Zaitsev elimination. In passing, we can note that an E2 process that was E1-like would be indistinguishable from this in its outcome; E1 reactions also give predominantly Zaitsev elimination. "Carbanion-like" transition states are favored when the leaving group is poor, and the base strong. In this case, the removal of the proton from the least substituted carbon is favored, as this position would give the best carbanion. Thus, the less substituted alkene would predominate, and this is described as Hofmann elimination.

Some examples will illustrate (Figure 10.19). In the reaction of 2-bromobutane with hydroxyl ion, we have a good leaving group and a good base. The *trans*-2-alkene predominates in the product. However, with 2-fluorohexane, we use a stronger base, methoxide, and have fluoride as the leaving group. As we saw when we considered substitution, fluoride is a poor leaving group. So, in this case, the transition state is carbanion-like, and the major product is the 1-alkene.

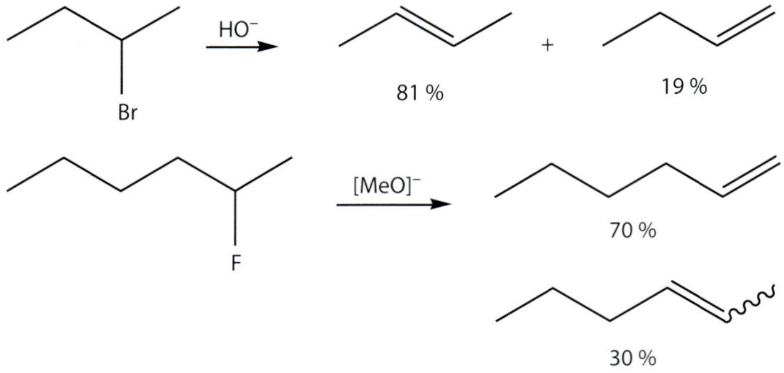

FIGURE 10.19 Regiochemistry of E2 elimination reactions.

So far we have looked mainly at single-atom leaving groups, where steric constraints are limited. However, if either the base or the leaving group is very large, steric hindrance may be an issue. Where this is the case, the base will generally abstract the most accessible proton, usually one that is either terminal or otherwise unhindered. For example, for the reaction of 2-bromo-2-methylbutane (Figure 10.20), more Hofmann product is produced with *t*-BuO⁻ as base, mainly for steric reasons (although it is also a stronger base and might be expected to lead to a more carbanion-like transition state, this is probably a less important effect).

Similarly, in the reaction of 2-substituted pentanes (Figure 10.21), as the size of the leaving group is increased, the proportion of 1-alkene produced also increases; where steric hindrance is a problem, the base abstracts the more accessible, terminal hydrogen atom. The reaction features that favor Zaitsev or Hofmann elimination in E2 processes are summarized in Table 10.1.

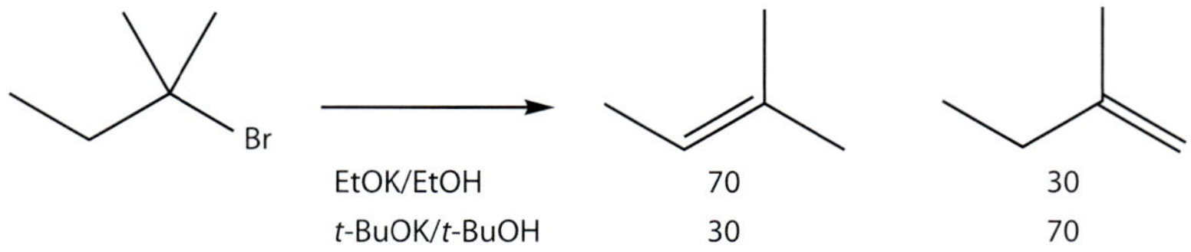

FIGURE 10.20 Steric effects on the regiochemistry of E2 elimination.

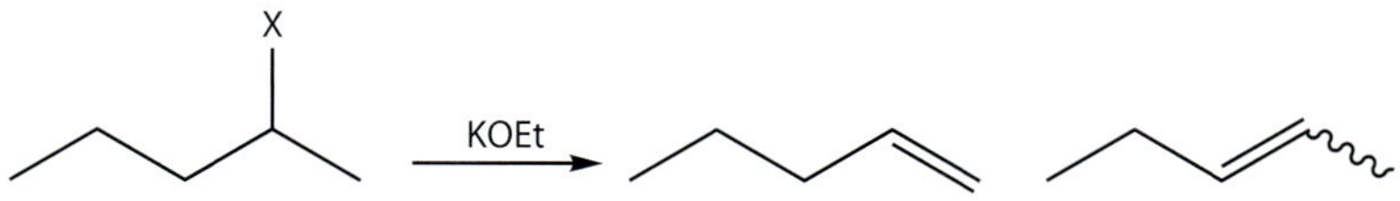

X	Br	OTs	S^+Me_2	SO_2Me	N^+Me_3
% 1-pentene	31	48	87	89	98

FIGURE 10.21 Regiochemistry of E2 elimination from 2-substituted pentanes.

TABLE 10.1 Features of Zaitsev and Hofmann Elimination in E2 Reactions

Hofmann Elimination	Zaitsev Elimination
Strong base	Moderate base
Poor leaving group	Good leaving group
High steric demand from either leaving group or base	Low steric demand
Typical leaving groups F, Me_3N	Typical leaving group Br, I

PROBLEM 10.4

Explain the following observations:

(a)

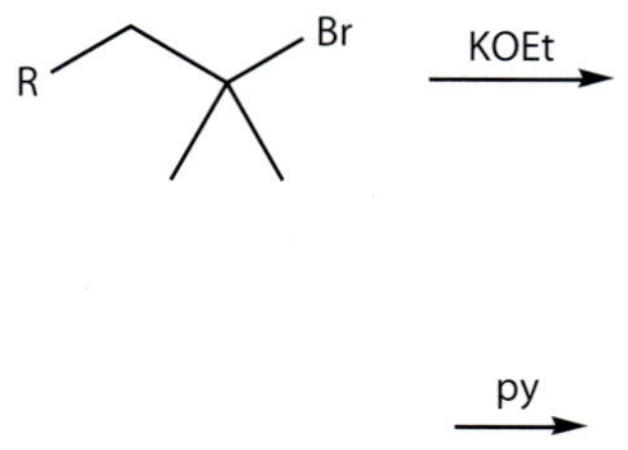

	R	1-alkene	2-alkene
KOEt	Me	30	70
	Et	50	50
	i-Pr	54	46
	t-Bu	84	14
py	Me	25	75
	Et	32	68
	i-Pr	44	56
	t-Bu	70	30

(b)

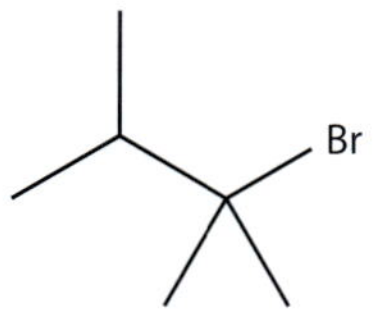

Base	% 1-alkene
EtO^-	21
t-BuO^-	73
$EtMe_2CO^-$	81
Et_3CO^-	92

Solutions

(a) These examples clearly demonstrate the effect of bulk in the substrate on the regioselectivity of the elimination reaction. The larger the group R, the more inclined the base will be to deprotonate at the 1-position, giving the 1-alkene.

(b) This demonstrates the effect of base size on regioselectivity. The more bulky the base, the less likely it is to deprotonate at the more hindered site, so that again, the proportion of the 1-alkene increases.

Focus on Kinetic Isotope Effects (See Section 8.2.4)

The full theory of kinetic isotope effects, and what affects their magnitude, is beyond the scope of this course. At the simplest level, the isotope effect is largest (generally around 7) when the transition state for the reaction is approximately symmetrical with respect to removal of the proton. Thus, E2 reactions should show large isotope effects, but E1 and E1cB reactions should show very small ones. This is slightly complicated by the fact that some E2 reactions have some E1cB character, and the isotope effects here, while not insignificant, tend to be smaller. We can illustrate the continuum of E2 to E1cB character by examining the isotope effects in elimination from **10.9**. We know that poor, bulky, leaving groups give more carbanion character to the elimination reaction, and thus the isotope effect decreases.

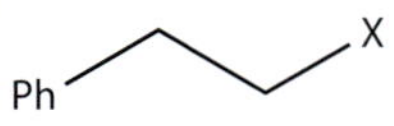

10.9

X	Br	OTs	S^+Me_2	N^+Me_3
k_H/k_D	7.11	5.66	5.07	2.98

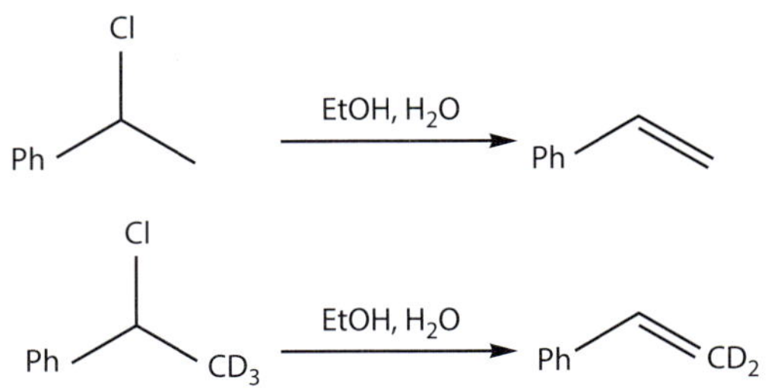

FIGURE 10.22 Minimal kinetic isotope effects in E1 reactions.

If we set up conditions where the E2 reaction is almost completely carbanion-like, as for the elimination of HF from **10.10** using *t*-butoxide as base (poor leaving group, high steric demand), then the isotope effect is reduced still further to 1.2.

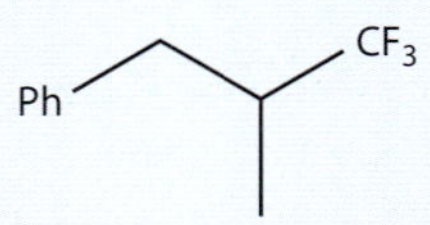

10.10

Moving to reactions that are essentially E1 in character, the study of the pair of reactions in Figure 10.22 gave k_H/k_D = 1.2 (*i.e.*, close to unity). The benzylic cation is reasonably stable, no strong base is present, and the polar solvent is likely to stabilize the ions formed in an E1 process.

10.5 EFFECT OF SUBSTRATE

In E1 reactions, the rate of the reaction depends mainly on carbocation stability, just as we saw for S_N1 reactions. (Remember primary > secondary > tertiary for carbocations, conjugation either with multiple bonds or lone pairs of electrons stabilizes carbocations, and nonplanar carbocations cannot readily be formed). Most of the quantitative data (Table 10.2) for carbocation stability, however, derive from substitution rather than elimination reactions, and even some of these are a little misleading. Rates of substitution of bromomethane and bromoethane tell us little about carbocations; these are purely S_N2 reactions in which no carbocation participates. While good gas phase and superacid data do exist, what they can tell us about reactions in solution is limited. That said, the data do tell us clearly that tertiary cations are the most stable and that conjugation stabilizes carbocations.

We discussed the data from bicyclic systems that tell us about cation stability derived from substitution processes in Chapter 9 (Figure 9.16). Eliminations are implausible in most of these systems because the resulting alkene would be too strained. The instability, and hence non-formation, of alkenes at the bridgeheads of bicyclic systems is described as Bredt's rule. The difficulty of solvolysis reflects the relative strain, and non-planarity, of the carbocation to be formed.

In E2 reactions, the accessibility of ***trans-anti*** or *syn*-coplanar conformations is crucial. If the molecule can't manage this, then another reaction, either an E1 elimination or a substitution, will probably predominate.

TABLE 10.2 Relative Rates for Halide Hydrolysis in Water

Alkyl Bromide	Relative Rate of Hydrolysis in Water	Benzyl Halide	Relative Rate for Hydrolysis in Water
MeBr	1.05	$PhCH_2Cl$	1.0
EtBr	1.0	Ph_2CHCl	2,000
Me_2CHBr	11.6	Ph_3CCl	30,000,000
Me_3CBr	12,000,000		

PROBLEM 10.5

Predict the product(s) of the following reaction:

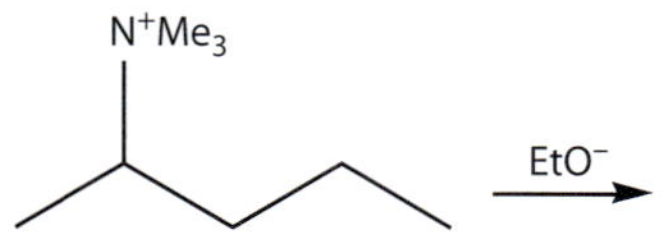

Solution

In this case, we have a bulky, poor leaving group and a reasonably strong base. This indicates that the Hofmann product, the 1-alkene, is likely to predominate in the E2 elimination reaction, although some 2-alkene may be produced as a by-product.

10.6 SUBSTITUTION *VERSUS* ELIMINATION

Both S_N1 and E1 reactions are favored by polar solvents, good leaving groups, the formation of a stable carbocation, and relief of steric strain. Both S_N2 and E2 reactions are favored by nonpolar solvents, good leaving groups, and a relatively uncrowded molecule. So how can we predict what will happen?

The choice between S_N1 and E1 reactions is the more difficult to control, because the rate-limiting step is the same for both of them, and the product is formed in a fast step, with relatively low-energy requirements. In general, crowding favors elimination, because large substituents are further apart in a product with sp^2 hybridization, with 120 ° angles, than in a saturated sp^3 system, with 109.5 ° separations. The presence of a good base favors E1, that of a good nucleophile S_N1. Elimination is favored by higher temperatures. The first example of Figure 10.23 shows a process where E1 and S_N1 are in competition, with substitution being the main outcome. In the second example, elimination predominates; the molecule is more crowded; and by using sulfuric or phosphoric acid as the source of protons, we eliminate any possibility of substitution, as the anions of these acids are very feeble nucleophiles. Note that the more stable alkene is produced, essentially exclusively.

Because the E2 reaction is a single-step process, with precise stereochemical requirements, we can do more to control the outcome. As far as the substrate is concerned, branching at α- and/or β-positions slows substitution more than it does elimination. Because in substitution, there is a requirement for a "five-coordinate" carbon at the transition state, the steric demand of this process is much higher than for elimination. Also, branching at the α- or β-positions stabilizes an alkene-like transition state. As we have previously noted, stereochemistry is critically important, especially to E2 reactions. Elimination will be inhibited if the *trans-anti* arrangement of the atoms or groups to be eliminated is not available or is disfavored. Substitution will be unaffected. If we compare the reactions of **10.11a** and **10.11b** using ethoxide (which can be either a

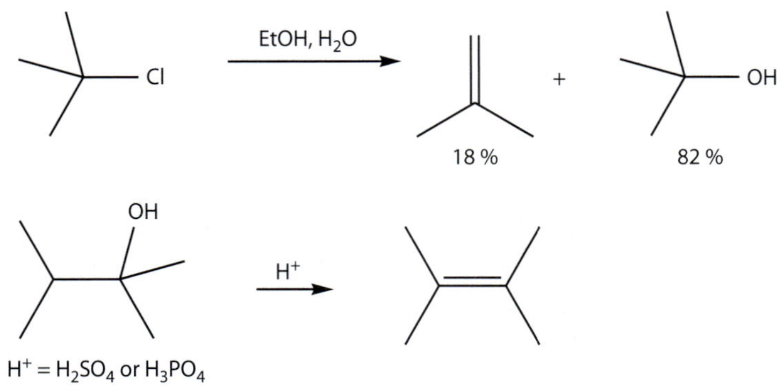

FIGURE 10.23 Competition between S_N1 and E1 reactions.

base or a nucleophile), we find that **10.11a** undergoes *only* substitution by an S_N2 process. The conformation that would be required for elimination would place both of the very bulky groups in axial positions on the six-membered ring, which would be very disfavored. By contrast, **10.11b** undergoes mainly elimination (92 %) under the same conditions, with only 8 % of the S_N2 process occurring. In the six-membered ring, one of the two bulky substituents is forced to occupy an axial position; they are of comparable size, so there is no strong preference. Having the leaving group axial is ideal for an E2 elimination, and removal of the very bulky group reduces steric congestion in the product.

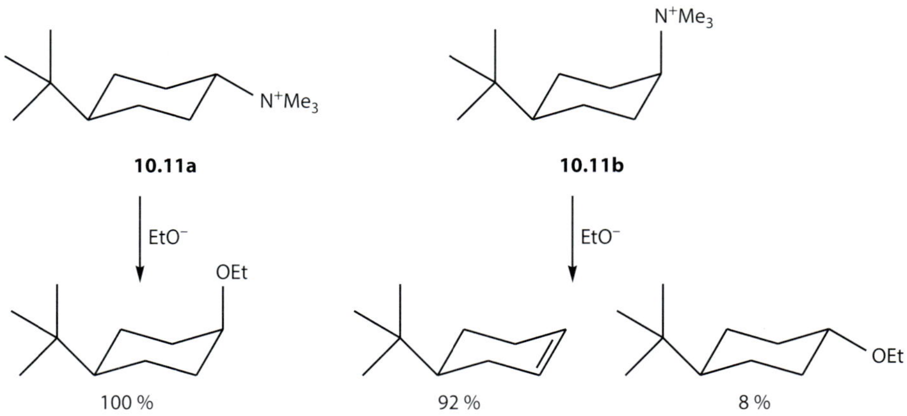

The trickiest issue in deciding whether elimination or substitution will predominate takes us back to the discussion in Chapter 9 (Section 9.4) on whether a particular species is more likely to behave as a nucleophile or as a base. Before going further, review the concept of HSAB. In the contest between elimination and substitution, a good base will favor elimination, and a good nucleophile will favor substitution. And, for a given basicity, the larger the attacker, the more elimination will occur. So in Figure 10.24, we see that PhS^-, an excellent nucleophile (soft and polarizable), reacts with **10.12** by substitution, whereas potassium *tert*-butoxide (hard, bulky, and basic) gives an elimination reaction.

Solvent polarity may also influence the balance between S_N2 and E2 reactions. Consider the two transition states **10.13a**, for elimination, and **10.13b**, for substitution. (This discussion is specific to eliminations/substitutions with negatively charged attacking and leaving groups, as this is the most common type, and there are some differences with other charge types.) Charge dispersion at the transition state is favored by low-polarity solvents. The charge dispersal in the transition state for the E2 process is greater than in the S_N2 process, because the charge is

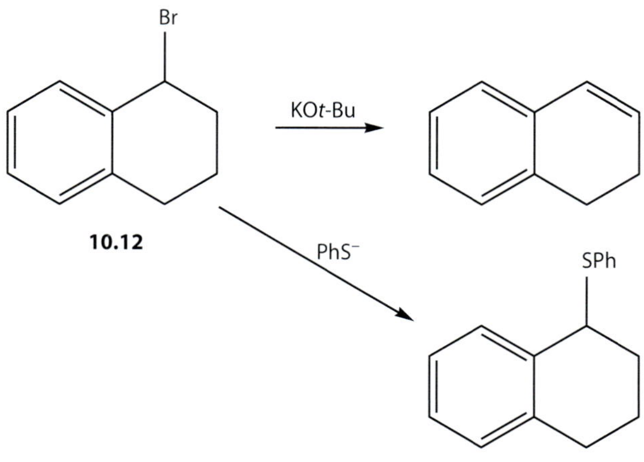

FIGURE 10.24 Competition between substitution and elimination.

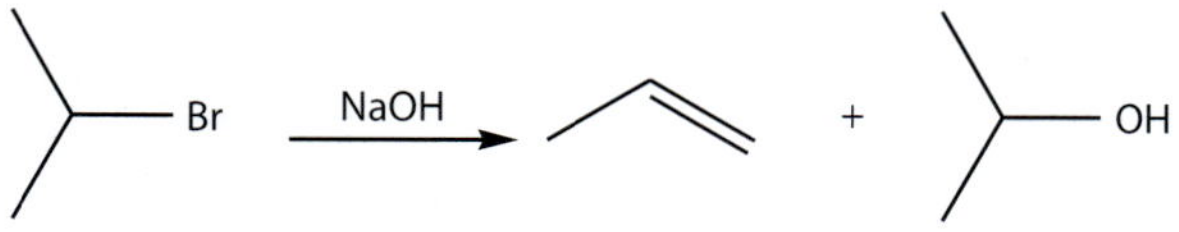

H_2O:EtOH	40:60	20:80	0:100
E_2:SN_2	1.17	1.44	2.45

FIGURE 10.25 Solvent effects on selectivity between S_N2 and E2 reactions.

dispersed over five atoms rather than three. Hence, elimination will be favored over substitution by a low-polarity solvent. We should also note that in a low-polarity solvent, the anion will be less encumbered by solvation.

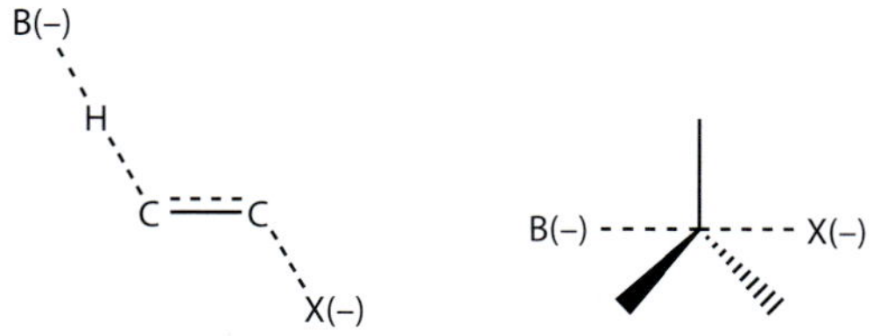

10.13a **10.13b**

An example is provided by the reaction of 2-bromopropane with sodium hydroxide in mixtures of water and ethanol (Figure 10.25). Water is more polar than ethanol, and the ratio of elimination to substitution increases as the water content of the mixture decreases.

PROBLEM 10.6

Explain the following observation:

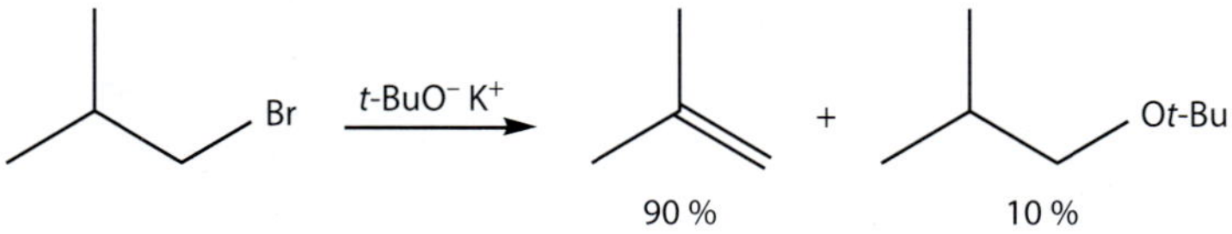

Solution

tert-Butoxide is a strong base but, because of its bulk, a relatively poor nucleophile. So elimination (there is only one possible product) predominates, especially as the molecule is quite congested.

10.7 CARBOCATION REARRANGEMENTS

Since we are producing the same type of carbocations that we did when considering substitution reactions, we can reasonably expect to see the same sorts of rearrangements of those carbocations. Figure 10.26 shows how, in the reaction of a carbocation, the secondary carbocation can rearrange to a tertiary carbocation, a process that is comparable in rate to proton loss.

A more complex example is shown in Figure 10.27. Although a tertiary carbocation is initially formed by water loss, rearrangement to an alternative tertiary carbocation gives a six-membered ring, which is of lower energy than its seven-membered counterpart. Loss of a proton gives the alkene in the final step.

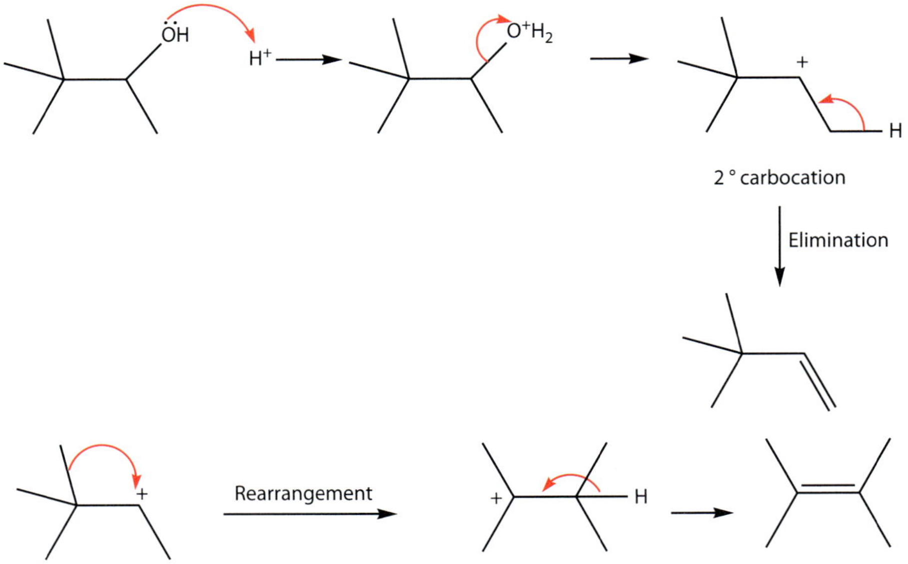

FIGURE 10.26 Carbocation rearrangement in an E1 elimination.

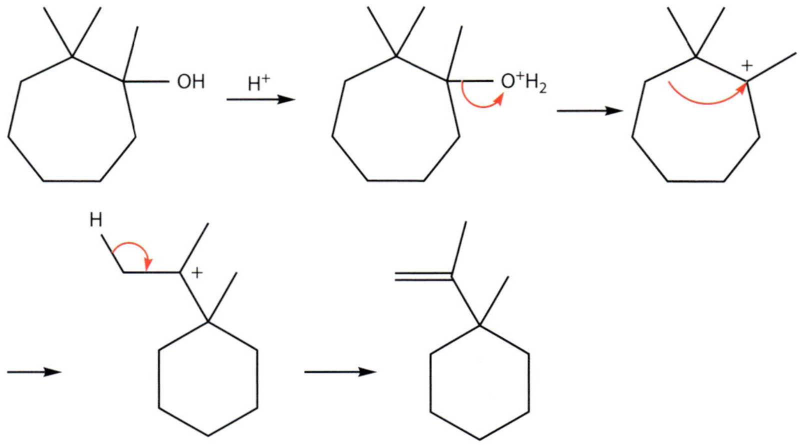

FIGURE 10.27 Carbocation rearrangement in E1 elimination.

PROBLEM 10.7

Why does the carbocation initially formed in the final reaction of Figure 10.28 not rearrange to give a tertiary cation?

Solution

We first need to draw the cation that will be produced after protonation by trifluoroacetic acid (a strong acid) and loss of water.

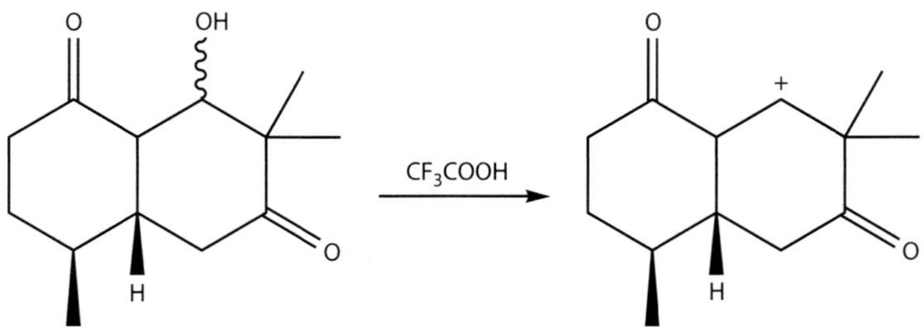

We can then consider migrating a methyl group from the carbon adjacent to the positive charge, which would give a tertiary carbocation.

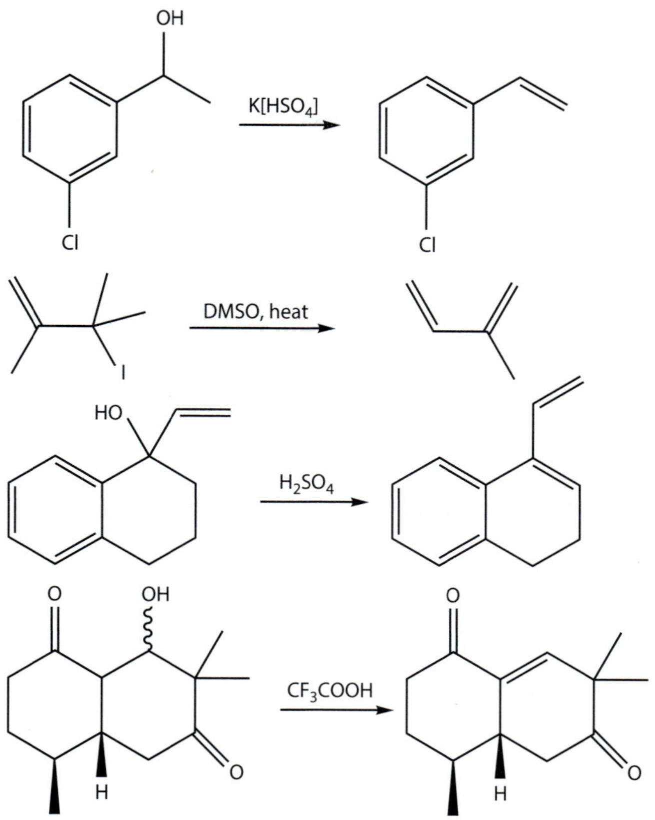

FIGURE 10.28 Applications of E1 elimination in synthesis.

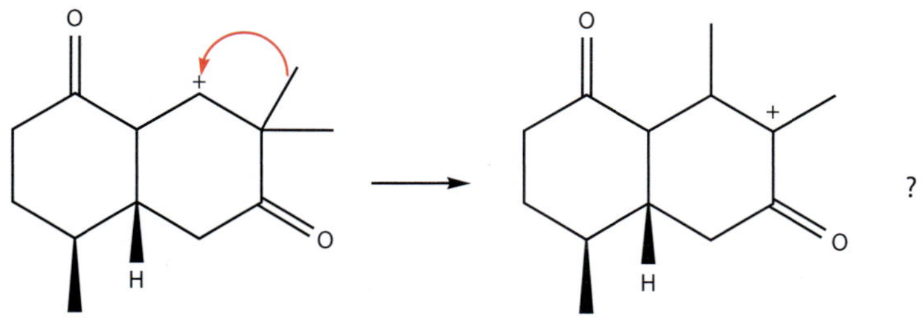

What is wrong with this idea—since it clearly does not happen? Although the new carbocation is tertiary, this particular tertiary carbocation is *not* stable, because the positive charge is adjacent to the strong electron-withdrawing carbonyl group. Moreover, proton loss from the initially formed carbocation would be very rapid as the alkene produced is very stable, being both trisubstituted and conjugated with the carbonyl group.

Key Points from Sections 10.4 through 10.7

- The regiochemistry of E1 reactions is difficult to control; the most stable alkene is generally formed. However, initially formed carbocations may rearrange under the reaction conditions.
- The regiochemistry of E1cB reactions is determined by the location of the most acidic proton.
- The regiochemistry of E2 reactions depends on the nature of the transition state. If this is alkene-like (good leaving group, low steric demand, good base), the most substituted alkene predominates. If it is carbanion-like (very strong base, bulky base and/or leaving group, poor leaving group), the less substituted alkene predominates.
- In competition between substitution and elimination, substitution is favored by good nucleophiles and modest steric demands. Elimination is favored by strong, nonnucleophilic, or bulky bases, low-polarity solvents, and high steric demand.
- Carbocations formed in E1 reactions may rearrange to give more stable species.

10.8 EXAMPLES AND APPLICATIONS

10.8 1 E1 REACTIONS

In synthesis, the most common use of this process is the acid-catalyzed elimination of water from an alcohol to give an alkene. We have already seen a number of syntheses of alcohols; we will meet many more in later chapters. Some examples are shown in Figure 10.28; notice that in most cases, there was only one plausible elimination, and an acid with a nonnucleophilic counterion was used. Try writing the mechanisms for each of these reactions—this is always a good practice.

10.8.2 E2 REACTIONS

Although the process is a common one, many reactions in practice give a mixture of products. While we may influence the outcome by using a more or less strong/bulky base, sometimes, the selectivity may still be too poor for practical application. In Figure 10.29, there is only one possible elimination process.

An illustration of the effect of group size is provided by the elimination of HBr from the esters **10.14** and **10.15** (both are made similarly by a reaction that we will meet in a later chapter). Elimination from the methyl ester, which has low steric demand, gives mainly the 2-alkene, with 10 %–15 % of the 1-alkene. However, using a bulkier ester and a bulkier base, the 1-alkene is produced with >95 % selectivity. In the reaction of **10.16**, there is only one possible elimination product, but S_N2 substitution predominates in this process.

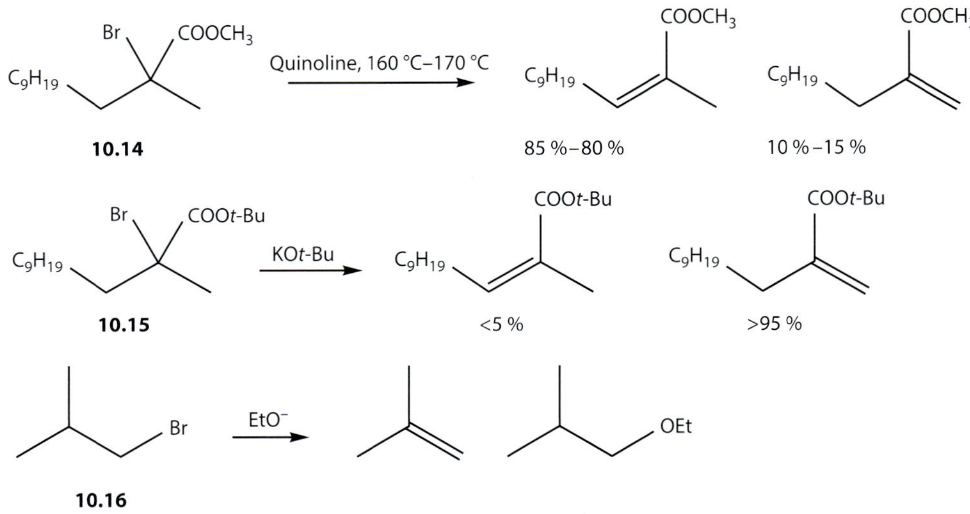

Elimination of an amino group, NR_2, is a difficult problem, whatever mechanism is involved. The typical pK_a of an amine, R_2NH, is 35, so expulsion of R_2N^- is unrealistic. If we need to do this, we must first make the amine into a quaternary ammonium salt, by reaction with methyl iodide.

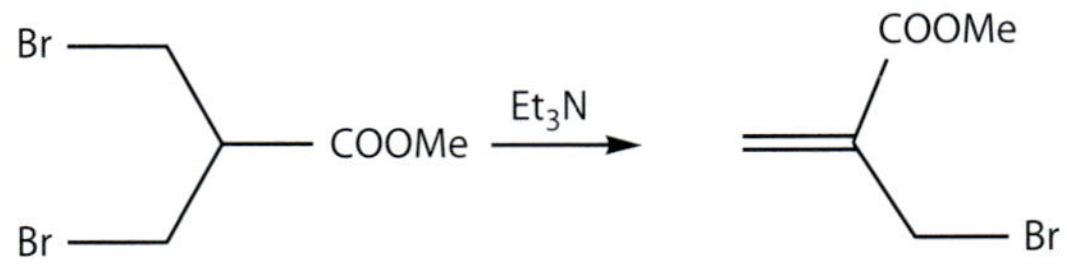

FIGURE 10.29 E2 elimination of HBr.

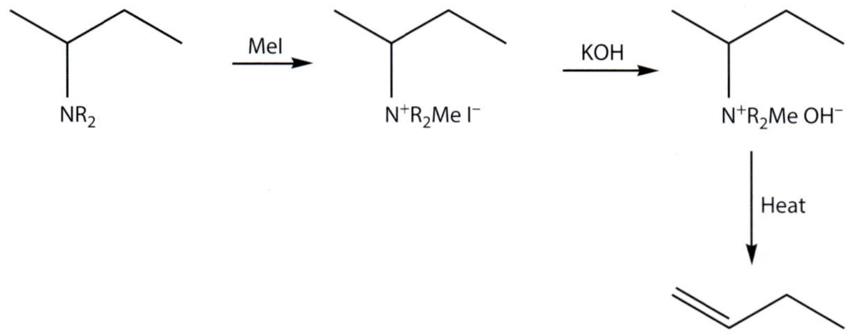

FIGURE 10.30 Elimination of a tertiary amine.

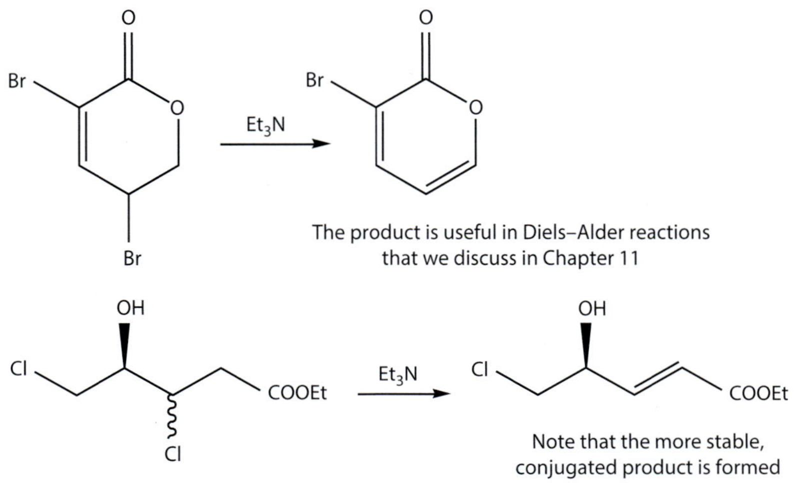

FIGURE 10.31 Examples of E2 elimination reactions in synthesis.

R_3N is not a very good leaving group, but it is possible to expel it, especially when we incorporate the base we intend to use into an intimate ion pair with the ammonium salt. The leaving group is both large and poor, and Hoffmann elimination predominates (Figure 10.30). Some other examples of E2 eliminations that have found uses in synthesis are shown in Figure 10.31.

10.8.3 E1cB REACTIONS

Recall that this mechanism operates when there is a good base but a poor leaving group and when the carbanion produced is stabilized. The first reaction in Figure 10.30 provides an example: triethylamine is a strong base, cyanide is a poor leaving group, and the intermediate carbanion is stabilized by the electron-withdrawing effect of the cyanide substituents. Loss of OH or OR groups in elimination is also often indicative of an E1cB process, as these are poor leaving groups, especially in basic solution. In the second example of Figure 10.32, notice that a very strong base* is needed to carry out the reaction.

Because of the potential for carbanion inversion in the intermediate in E1cB reactions, a mixture of stereoisomers is usually obtained. If the reaction of **10.17** had been strictly an E2 process, only the *Z*-isomer would have been obtained.

* LDA is lithium diisopropylamide, Li^+ $[N\textit{i}\text{-}Pr_2]^-$, a very strong, but non-nucleophilic, base.

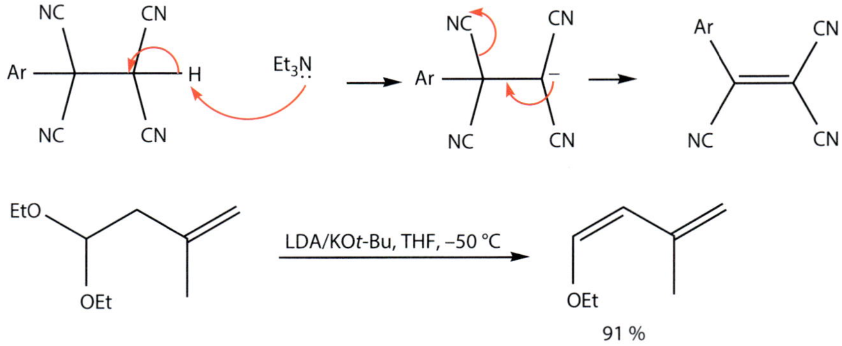

FIGURE 10.32 E1cB elimination reactions.

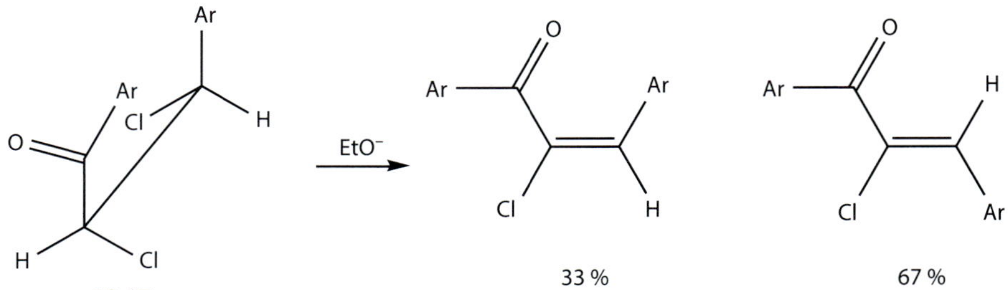

10.17

10.8.4 MIXED OR VARIABLE MECHANISMS

Dehydration of β-hydroxyketones is both a common reaction (we will see in Chapters 17 and 20 why these molecules are easy to prepare) and a very straightforward one. The product obtained is invariably the conjugated one, whatever the mechanism. Acid-catalyzed reactions are common; they are usually E1 with tertiary alcohols and E2 with primary alcohols. Secondary alcohols such as **10.18** may react by either or both mechanisms.

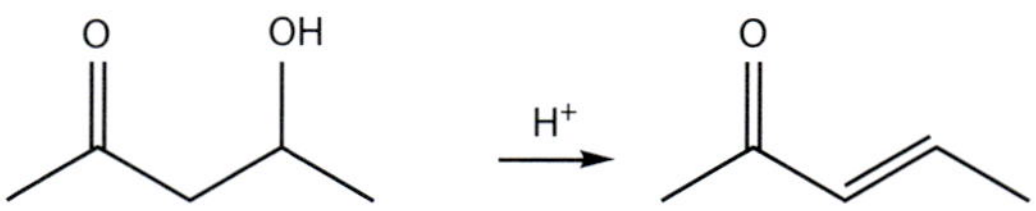

10.18

Dehydration of β-hydroxyketones can also occur under basic conditions (Figure 10.33), the molecule forms a good carbanion (because of the inductive and resonance effect of the carbonyl group), and the leaving group, hydroxyl ion, is poor.

Although our discussion has been dominated by reactions in which HX, where X is a halogen or OH, is eliminated, other examples are possible. In elimination from **10.19**, we might have wanted to eliminate HBr to give the unsaturated acid, but something quite different happens. The proton that is most easily and rapidly removed, by many orders of magnitude, is that on the carboxyl group. So what is lost is carbon dioxide and bromide ion.

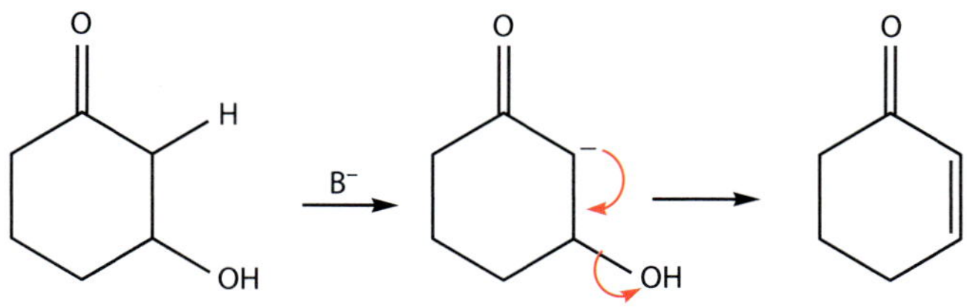

FIGURE 10.33 Dehydration of a β-hydroxyketone by the E1cB mechanism.

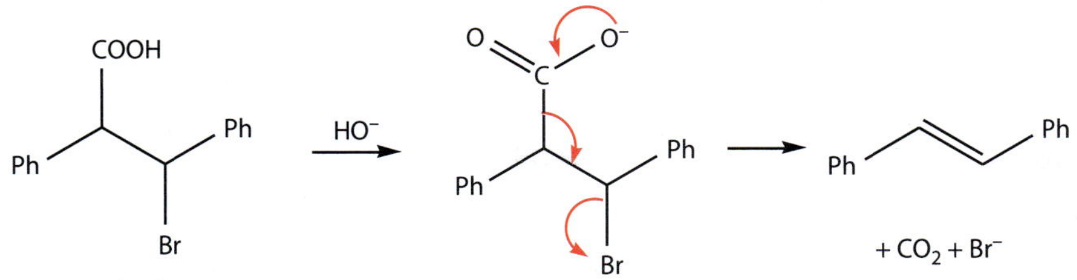

10.19

PROBLEM 10.8

Iodide ion may be used to dehalogenate vicinal dibromides as shown in the following:

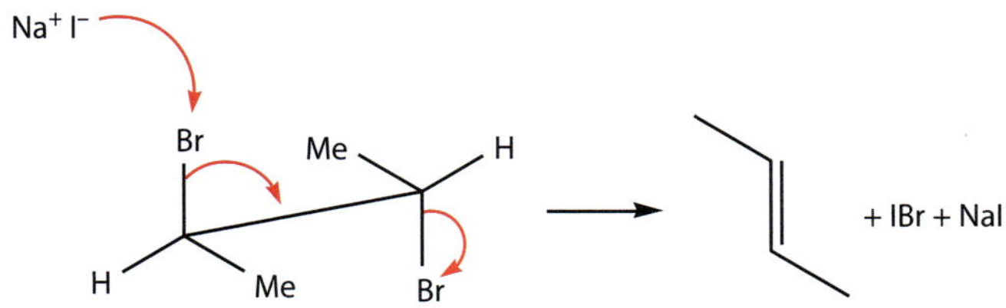

Given this observation, predict the product of the reaction shown as follows:

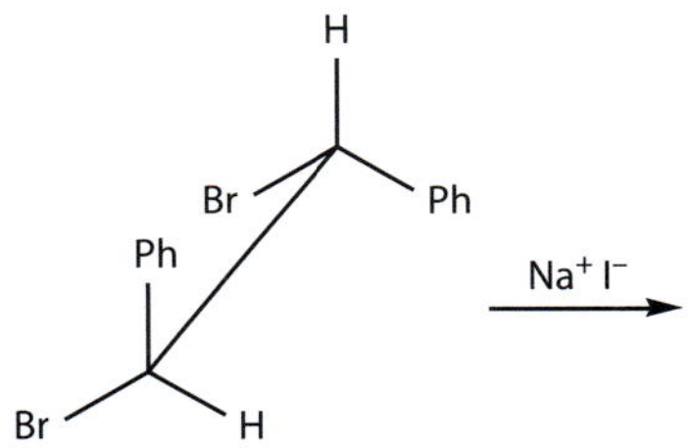

Solution

We will first need to rotate about the single bond in this molecule to put it into the right conformation to do the elimination. The elimination process is then identical to that shown earlier.

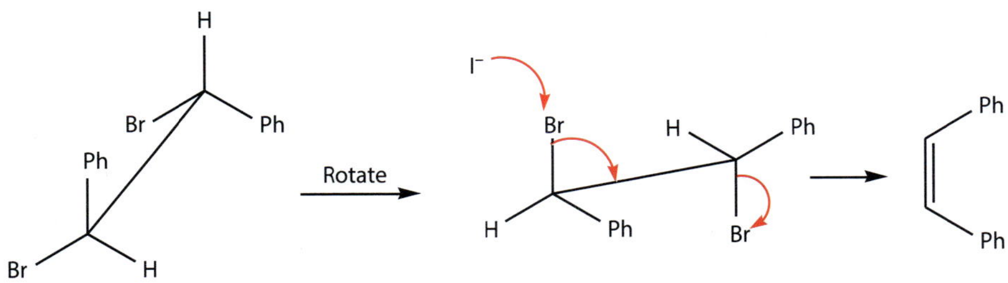

10.9 ELIMINATION TO GIVE ALKYNES

We can begin by thinking about possible elimination mechanisms on the basis of those we know that lead to alkene formation. If we start by losing a leaving group, in an analogy to E1, we produce a vinyl cation, which would then lose a proton to give the alkyne (Figure 10.34). If we consider an E2-type mechanism, then either *syn*-coplanar or *trans*-antiperiplanar processes are possible, depending on the geometry of the starting material. We can also write an E1cB version of the process, involving a vinyl anion.

The E1 process is very unlikely; vinyl cations are quite unstable. Both E2 and E1cB are viable options; vinyl anions are reasonably stable (remember the orbital effect in considering anion stability).

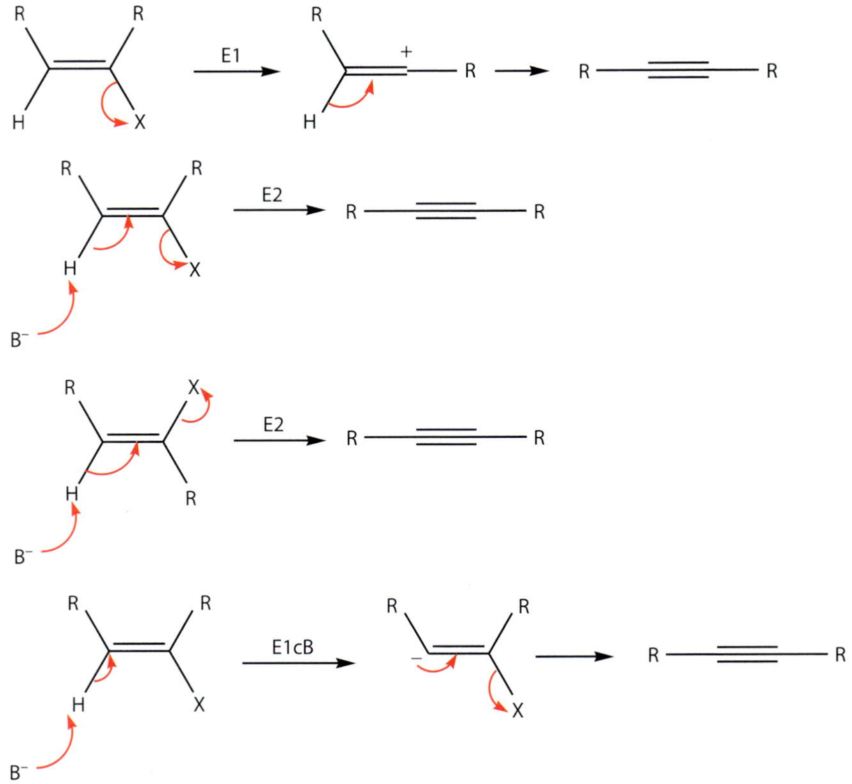

FIGURE 10.34 Possible mechanisms of elimination to give alkynes.

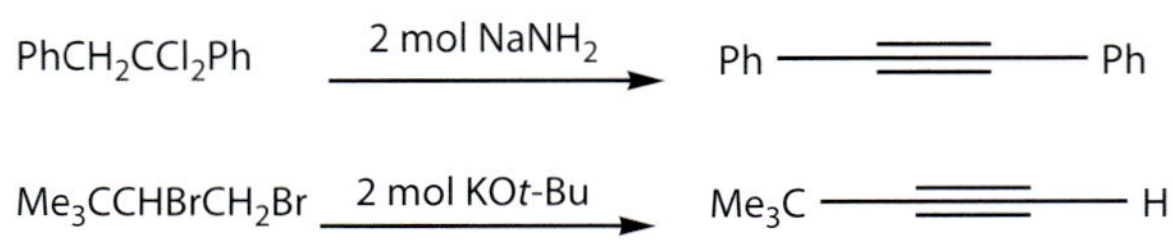

FIGURE 10.35 Dehydrohalogenation reactions to produce alkynes.

However, for any elimination to give an alkyne, the presence of a strong base will be essential, typically sodamide or BuLi. Why should this be? In this process, we have to break an sp^2 C–H (or C–X) bond, and these are significantly stronger than sp^3 C–H bonds. Simple examples are shown in Figure 10.35; in each case, two dehydrohalogenation reactions are occurring. Other examples that have been used to produce intermediates of synthetic value are shown in Figure 10.36.

Sodamide also reacts with bromoarenes by elimination of HBr (Figure 10.37) to give benzyne, a very reactive and unstable intermediate. Mechanistically, the elimination may be E2 or E1cB or somewhere in between. Although it is common, and quite conventional, to write the structure of benzyne with a triple bond, this is not a triple bond of the same type as we find in alkynes. The extra "bond" is formed by weak overlap of two sp^2 orbitals, each with one electron in it (**10.20**). It is probably more correct to think of it as a resonance hybrid of the triply bonded form, diradicals, and charge-separated species. The IR spectrum of isolated benzyne in an inert gas matrix at 8 K has been studied; although assignments have been a matter of discussion, the value of 1846 cm^{-1} for the C–C triple bond stretch is at least approximately correct. This suggests that the "triple bond" is considerably weaker than that in a conventional alkyne where a frequency of 2200 cm^{-1} would be typical.

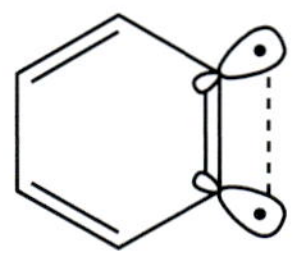

10.20

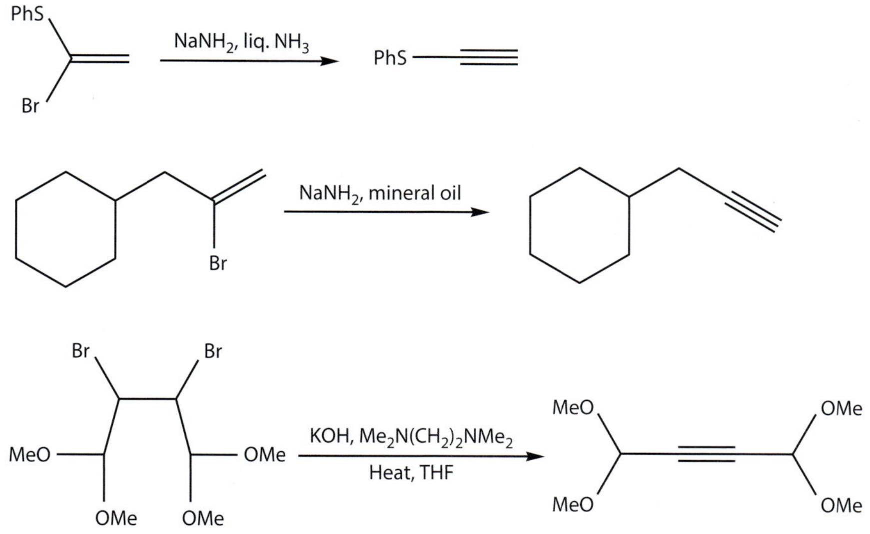

FIGURE 10.36 Synthesis of alkynes as intermediates in total synthesis.

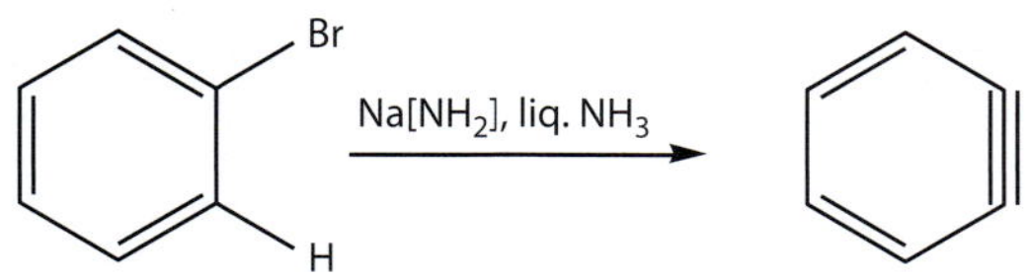

Benzyne

FIGURE 10.37 Elimination of HBr from bromobenzene to give benzyne.

We will meet the reactions of this species in some later chapters—left to itself, it dimerizes and trimerizes; it can undergo cycloaddition with both alkenes and dienes (Section 11.4.6) and add species such as liquid ammonia (Section 13.6) (Figure 10.38).

PROBLEM 10.9

Write a mechanism for the reaction shown:

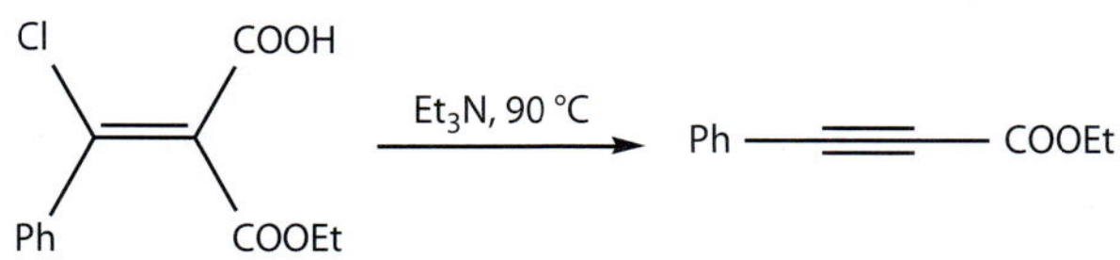

Solution

The two groups that have been eliminated here are carboxylate and chloride ion. The first thing that happens when we add triethylamine to the starting material is the deprotonation of the carboxylic acid to give a carboxylate anion. Loss of CO_2 is always a good reaction, because of the entropy gain from a mole of escaping gas. The elimination has no choice but to be *syn*-coplanar.

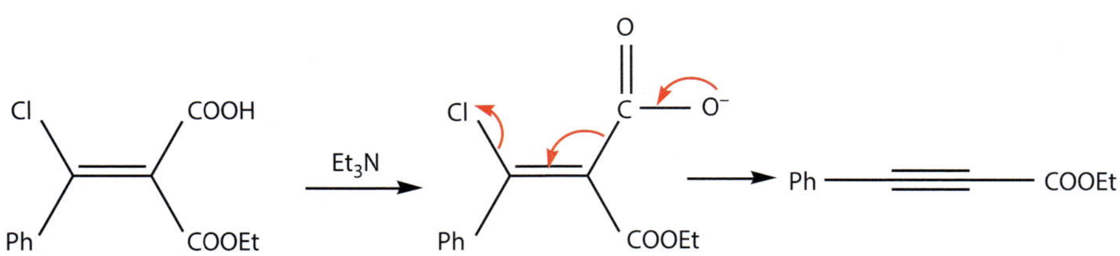

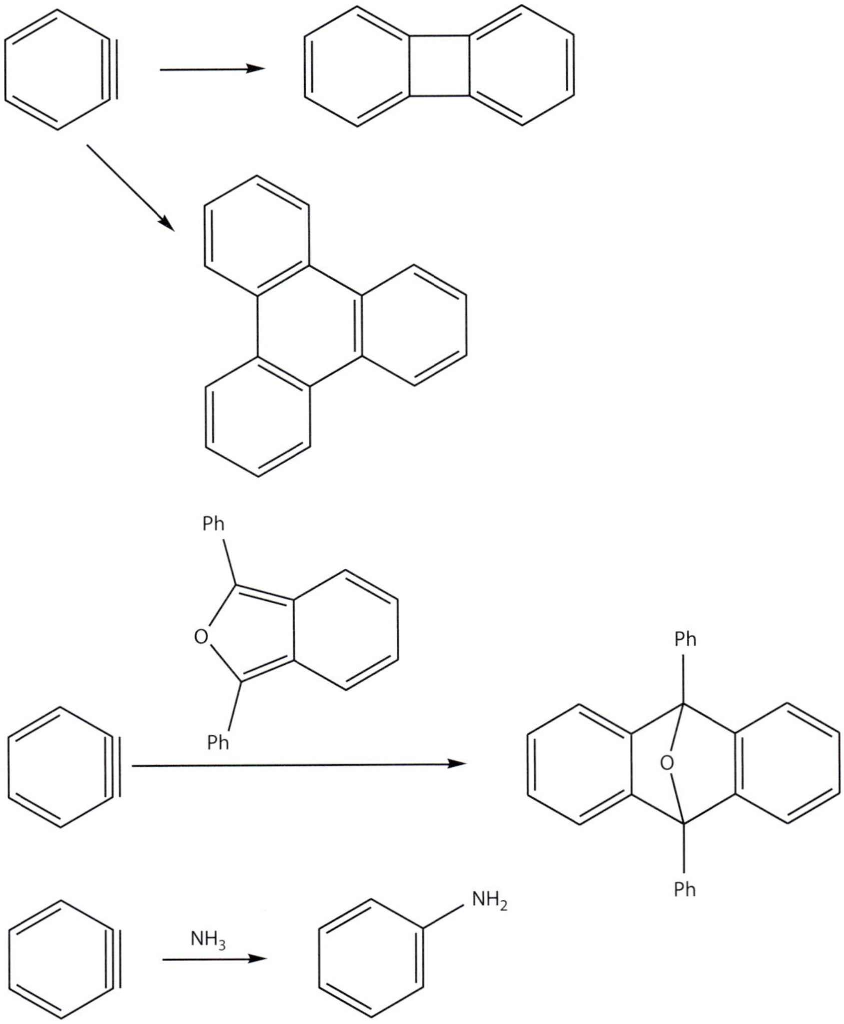

FIGURE 10.38 Some reactions of benzyne.

10.10 OTHER ELIMINATION REACTIONS

While there are many other reactions that are mechanistically eliminations, they are generally discussed more often in their synthetic contexts, so we shall meet them again later on. The most important is elimination to give carbonyl compounds—much more commonly considered as an oxidation of alcohols by chromium(VI). When CrO_3 is dissolved in water, the resulting solution contains various equilibrating species (Figure 10.39). However, for our purposes, we can simply consider the reaction of an alcohol with CrO_3 (Figure 10.40). The lone pair on the hydroxyl attacks at chromium, and a pair of electrons from the Cr=O double bond are displaced to oxygen; this is analogous to what happens when CrO_3 is dissolved in water, and the lone pair of water attacks at chromium. The intermediate is described as a chromate ester. In the elimination step, a proton and chromium are lost, to give a carbon–oxygen double bond.

When we compare the rates of reaction of axial and equatorial hydroxyl groups in six-membered rings, we find that the reactions of the axial alcohols are faster than those of equatorial alcohols. This tells us that it is the second step, the elimination process, that is rate determining; the hydrogen to be removed from the axial alcohol is in the equatorial position and hence more accessible (Figure 10.41).

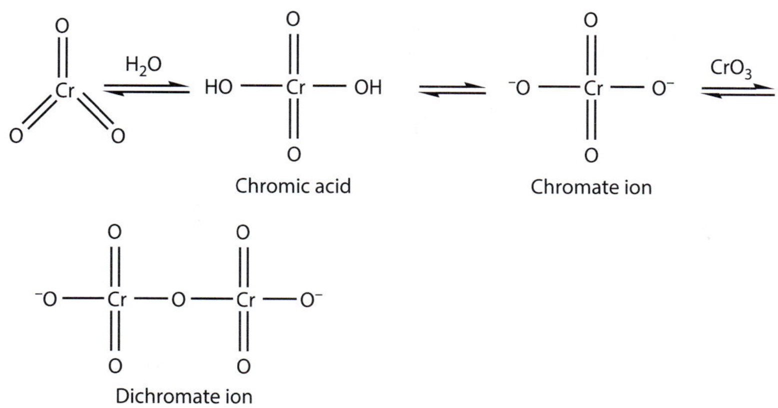

FIGURE 10.39 Aqueous equilibria involving Cr(VI).

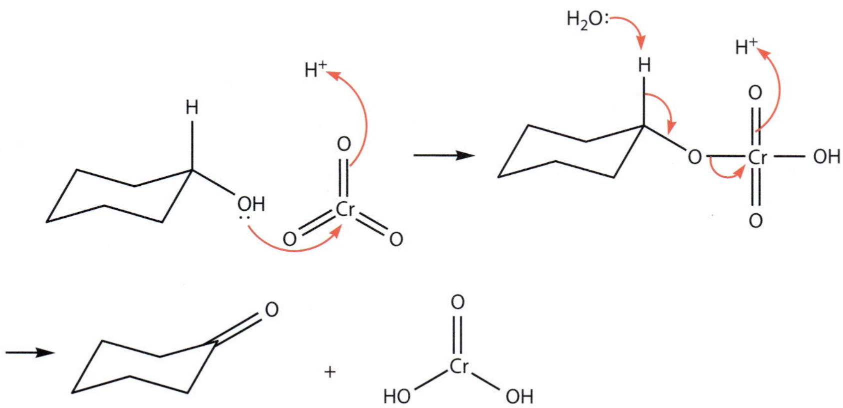

FIGURE 10.40 Cr(VI) oxidation of alcohols.

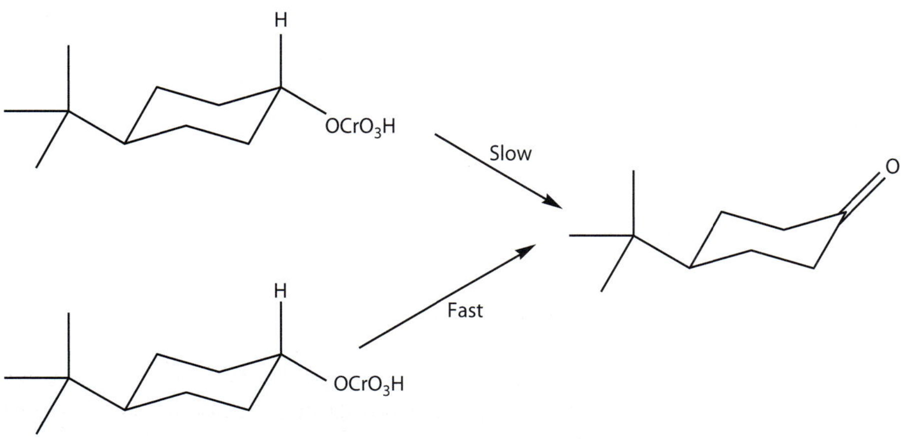

FIGURE 10.41 Chromium(VI) oxidation of axial and equatorial alcohols.

PROBLEM 10.10

Propose mechanisms for the following reactions:

(a)

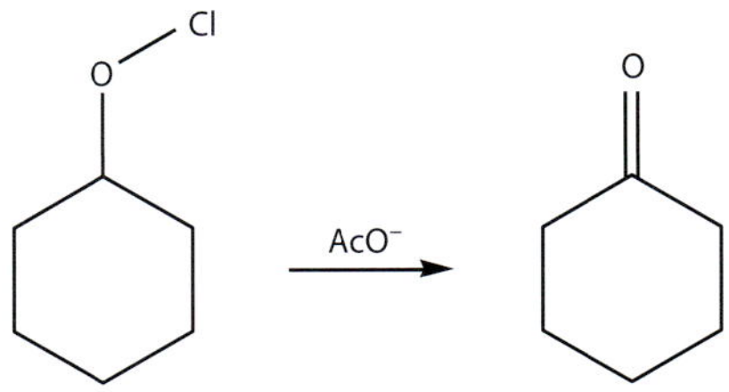

(b)

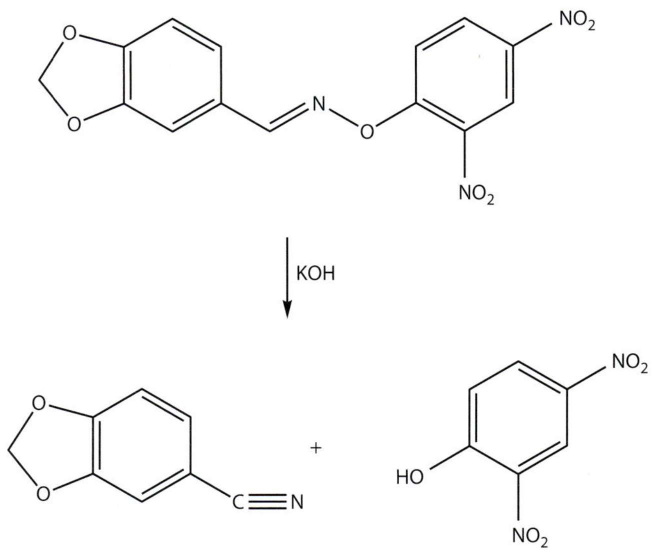

Solution

Your first thought on looking at these problems might be EEK! I have not seen anything like this before. However, as we have said earlier, once you understand the mechanistic principles, you CAN tackle and understand new types of problem. In both cases, once you work out which atoms/groups have been eliminated, the arrow pushing is relatively easy.

(a) In this example, you have removed a hydrogen from carbon and a chlorine from oxygen to make a carbon–oxygen double bond. H^- cannot be a leaving group, so it must be removed as H^+ by the acetate. Cl^- is a reasonable leaving group.

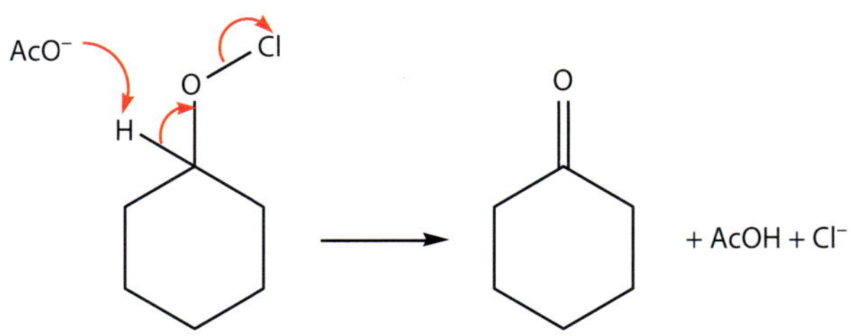

(b) Here, we are making a nitrile, containing a carbon–nitrogen triple bond, by removal of a hydrogen atom, and the anion of 2,4-dinitrophenol. The pK_a of 2,4-dinitrophenol is about 7, so although it's not the greatest leaving group, it is viable. Because of the geometry of the original double bond, this must be a *syn*-coplanar elimination.

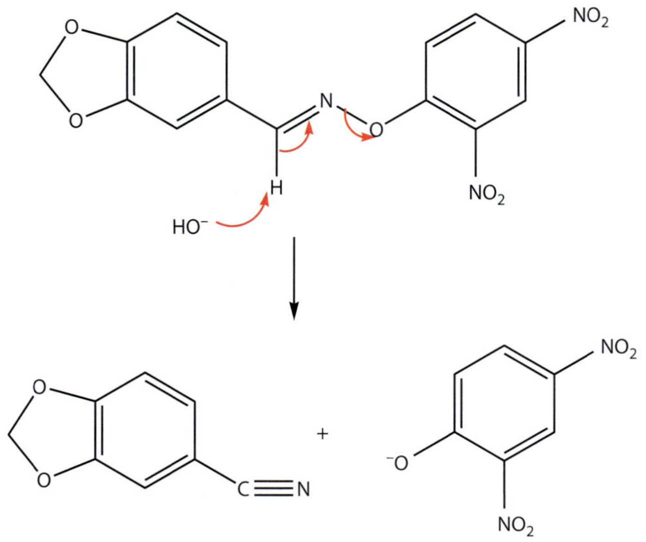

Not so difficult, after all!

Focus on Alcohol Oxidation: A Look Ahead

When we come to study synthesis, we will find that oxidation of alcohols by chromium(VI) compounds is an important reaction. Primary and secondary alcohols can both be oxidized under appropriate conditions, and by controlling the conditions, we can oxidize primary alcohols either to aldehydes or carboxylic acids (Figure 10.42). Pyridinium chlorochromate (PCC) $[pyH]^+[CrO_3Cl]^-$ can be, and often is, used instead of the CrO_3/pyridine combination. Because this reaction is an elimination, it is necessary that there is a hydrogen atom attached to the carbon bearing the OH group to be oxidized. Therefore, tertiary alcohols are not oxidized under these conditions.

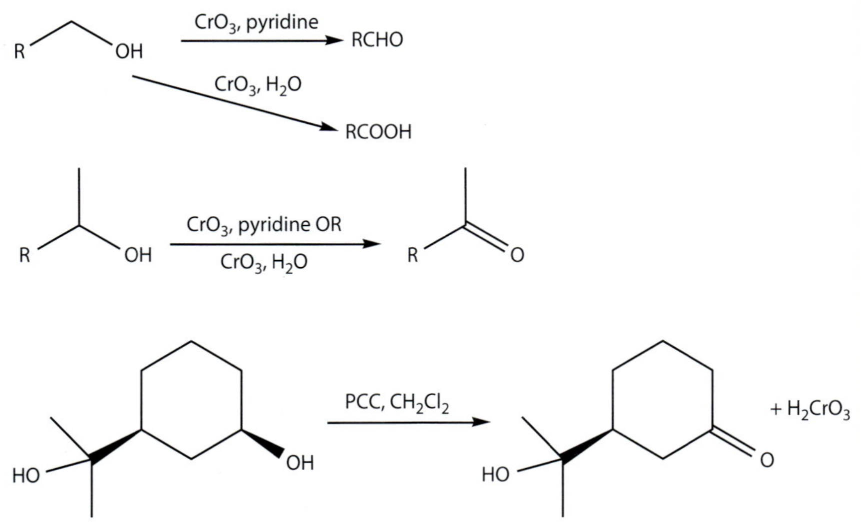

FIGURE 10.42 Oxidation of primary and secondary alcohols using chromium(VI).

Key Points from Sections 10.9 and 10.10

- Alkynes can be formed by elimination reactions, which may be E2 or E1cB in character. Very strong bases are generally required.
- Benzyne may be formed by elimination of HBr from bromobenzene using sodamide as a base. This contains a very weak triple bond, and the molecule is highly reactive, especially toward addition processes.
- Formal elimination of molecular hydrogen from primary and secondary alcohols may be achieved using chromium(VI) reagents.

REVIEW PROBLEMS

1. Give systematic names for the following compounds:

(a)

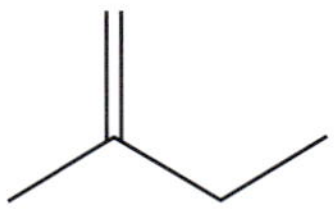

(b)

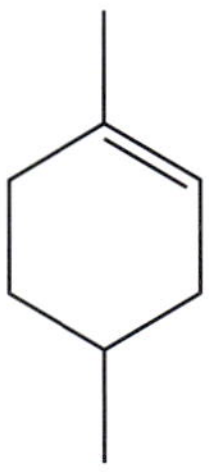

(c)

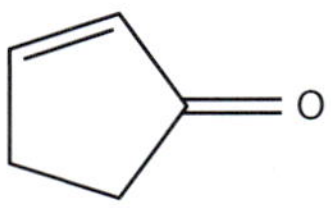

(d)

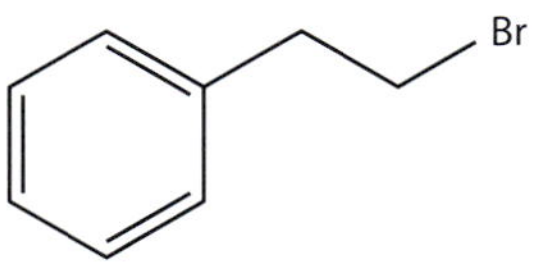

(e)

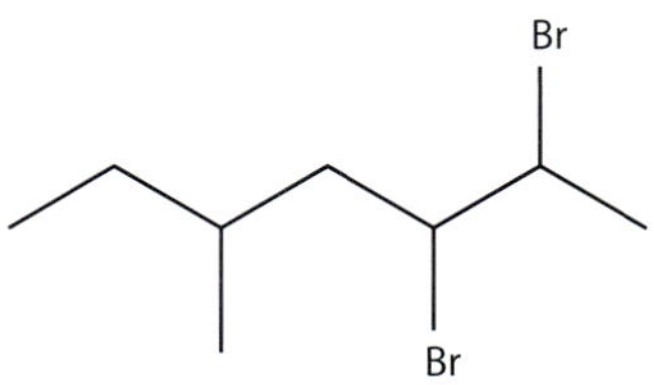

2. Draw structures for each of the following compounds:
   (a) *E*-2-hexene
   (b) 3-Bromocyclohexene
   (c) 2-Methyl-2-pentanol
   (d) 1-Phenylethanol
   (e) 3,3,-Dimethyl-1,4-cycloheptadiene

3. (a) Deduce a structure from the following set of spectroscopic data:

$C_5H_{10}O$

Mass spectrum		*m*/*z* 86 (M^+), 71, 43
IR spectrum		3400, 1610 cm^{-1}
^1H NMR spectrum	δ	1.3 (s, 6H)
		2.2 (s, 1H, exchanges with D_2O)
		4.9 (dd, *J* = 9.5, 1.5 Hz, 1H)
		5.2 (dd, *J* = 17, 1.5 Hz, 1H)
		5.9 (dd, *J* = 17, 9.5 Hz, 1H)

(b) ^1H NMR spectroscopic data are given in the following for various isomers of $C_5H_{10}Br_2$. Deduce their structures:

(i) δ 1.0 (s, 6H), 3.4 (s, 4H)
(ii) δ 1.0 (t, 6H), 2.4 (q, 4H)
(iii) δ 0.9 (d, 6H), 1.5 (m, 1H), 1.85 (t, 2H), 5.3 (t, 1H)
(iv) δ 1.0 (s, 9H), 5.3 (s, 1H)
(v) δ 1.0 (d, 6H), 1.75 (m, 1H), 3.95 (d, 2H), 4.7 (q, 1H)
(vi) δ 1.3 (m, 2H), 1.85 (m, 4H), 3.35 (t, 4H)

4. Predict the product(s) of the following reactions:

(a)

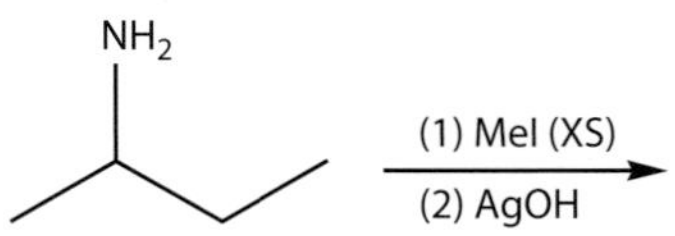

(b)

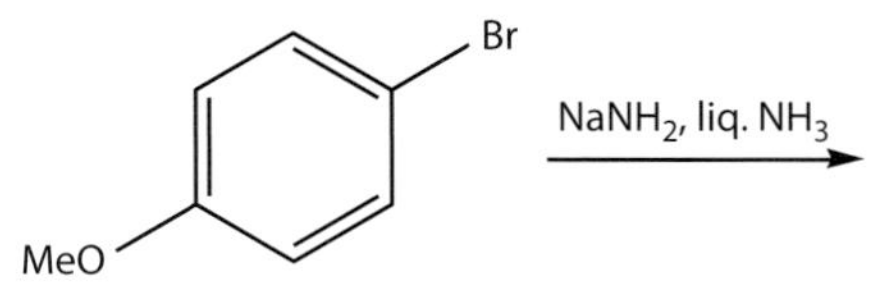

(c)

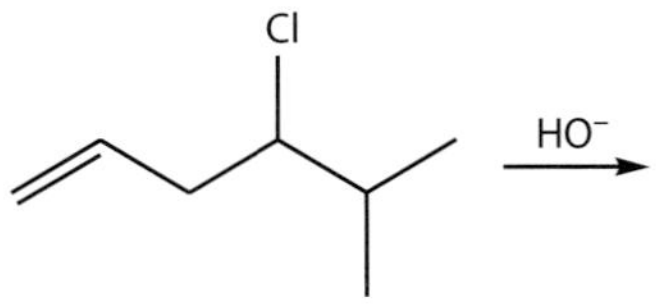

(d)

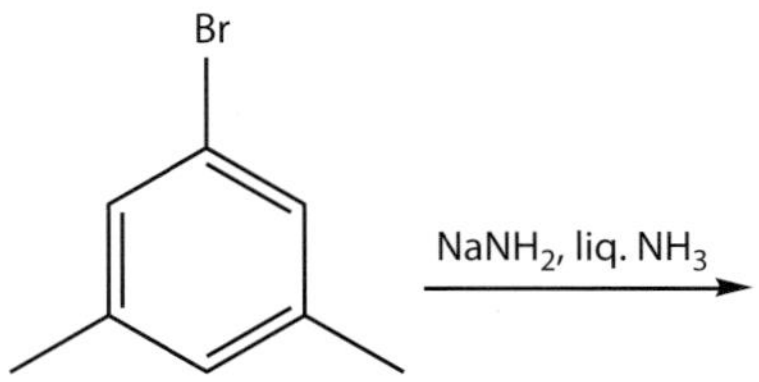

(e)

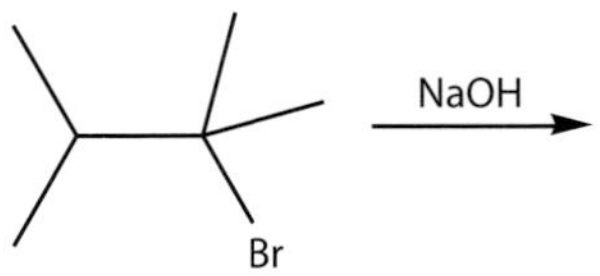

(f)

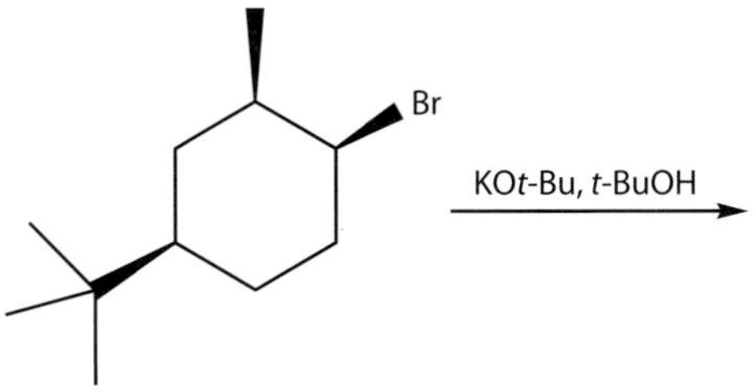

(g)

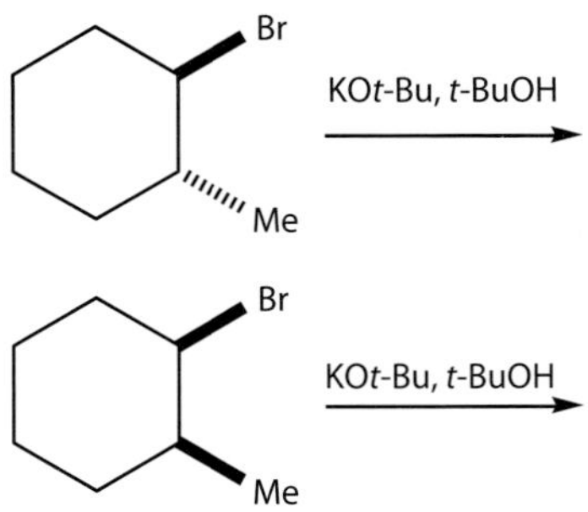

(h)

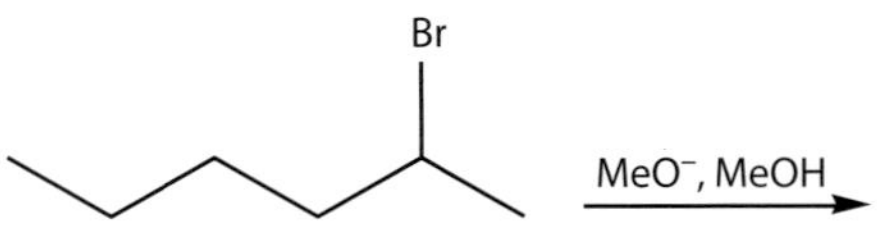

(i)

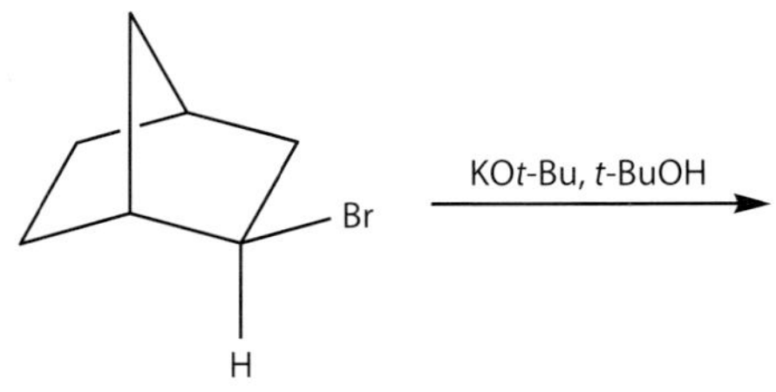

(j)

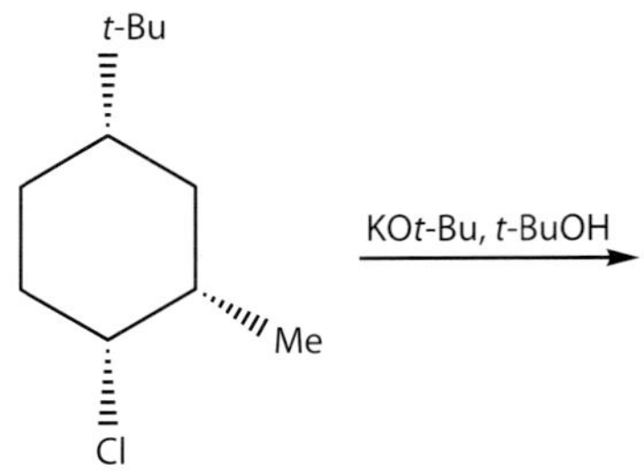

(k)

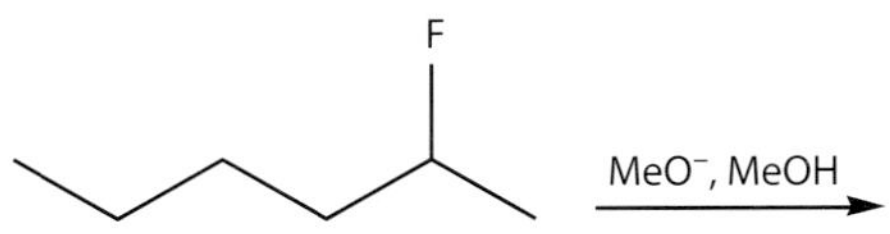

(l)

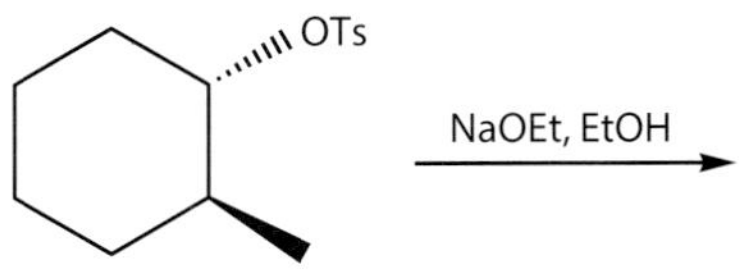

(m)

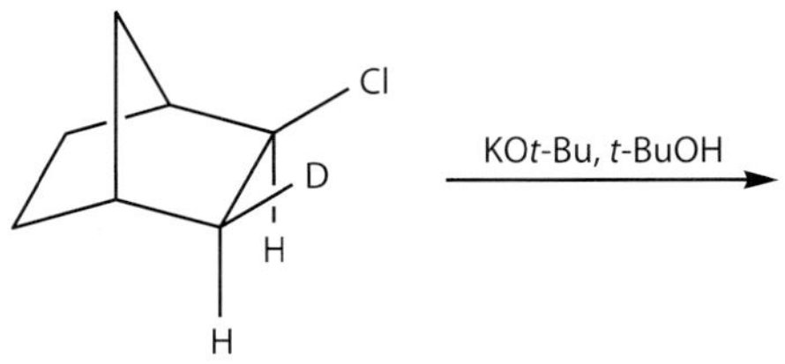

(n)

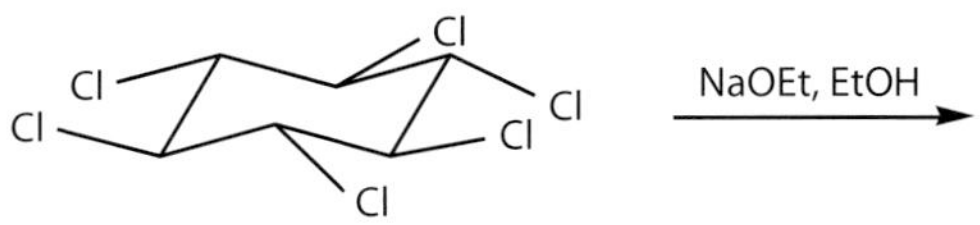

5. Propose a mechanism for each of the following reactions:

(a) *meso*-PhCHBrCHBrPh $\xrightarrow{\text{KO}t\text{–Bu, }t\text{–BuOH}}$ PhCH=CBrPh

Which stereoisomer do you get?

(b)

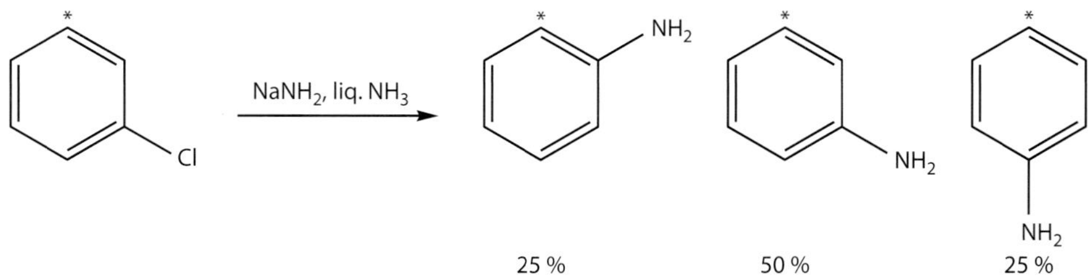

* indicates a ^{13}C label.

(c)

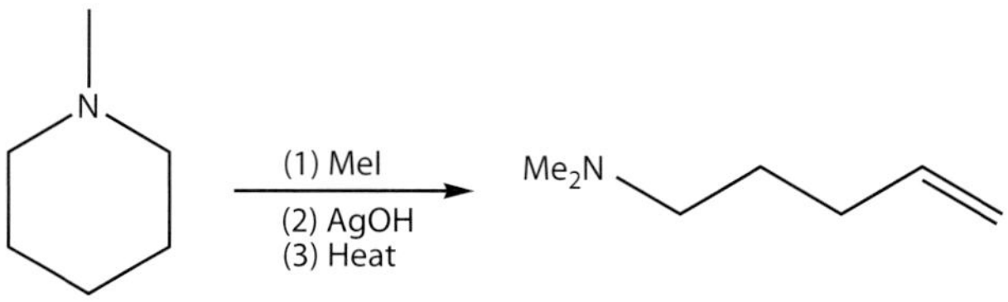

(d)

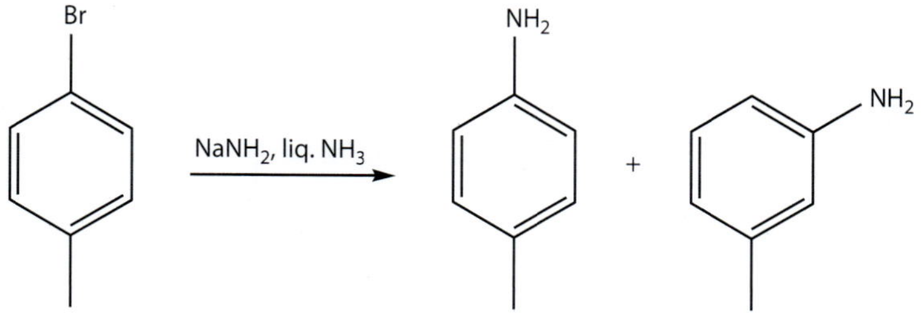
Br
NaNH₂, liq. NH₃
NH₂
+
NH₂

(e)

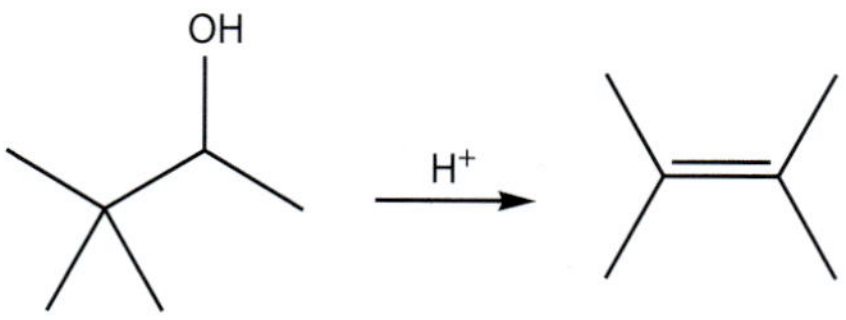
OH
H⁺

(f)

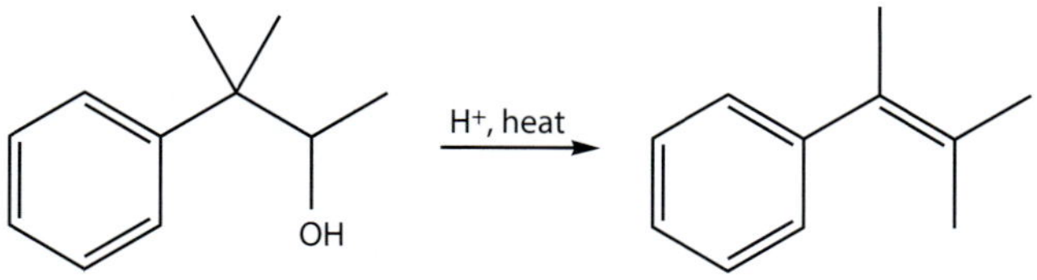
OH
H⁺, heat

(g)

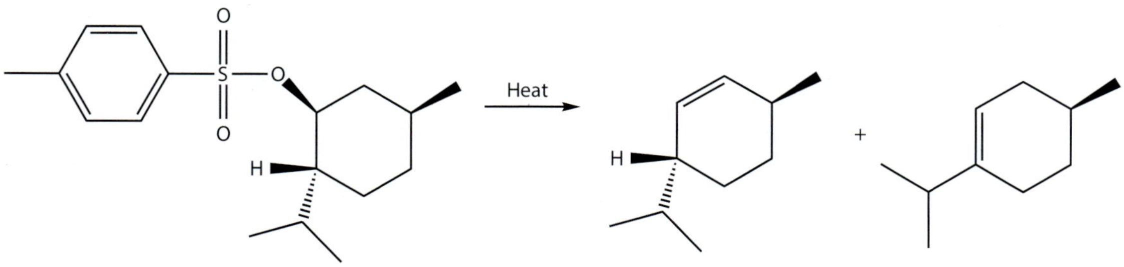
O
S
O
O
H
Heat
H
+

(h)

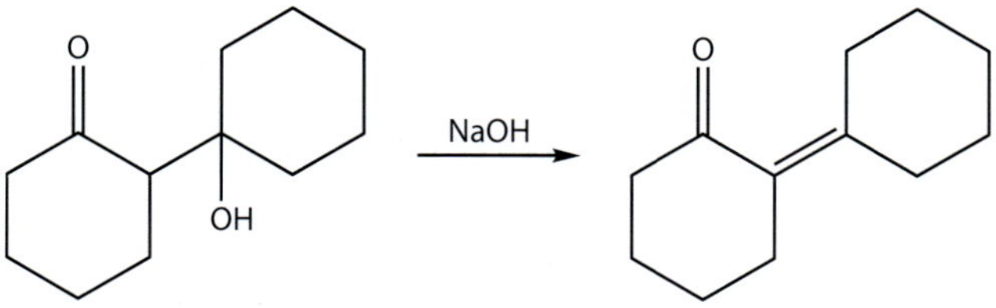
O
OH
NaOH
O

(i)

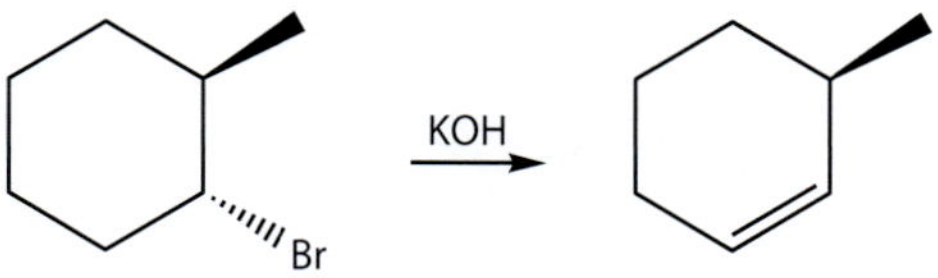
Br
KOH

(j)

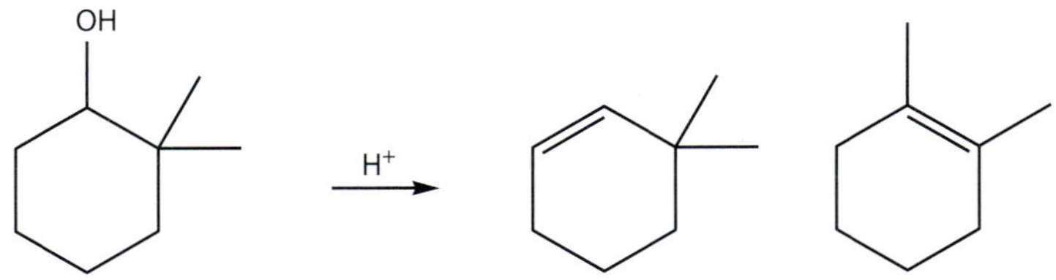

6. One of the molecules in the following undergoes elimination using NaOEt/EtOH much faster than the other. Which is which?

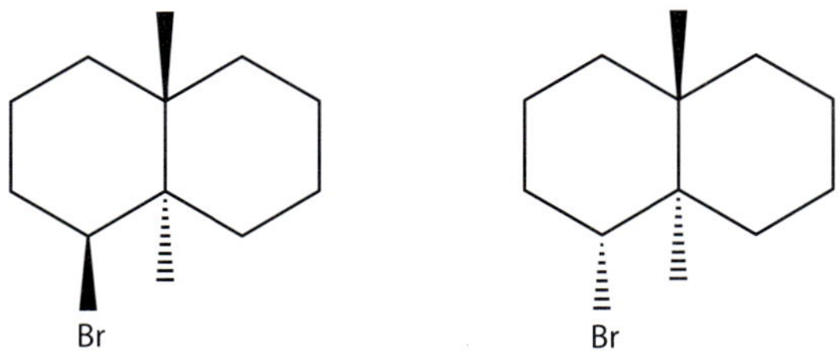

Explain the observations on the deuterated compounds shown as follows:

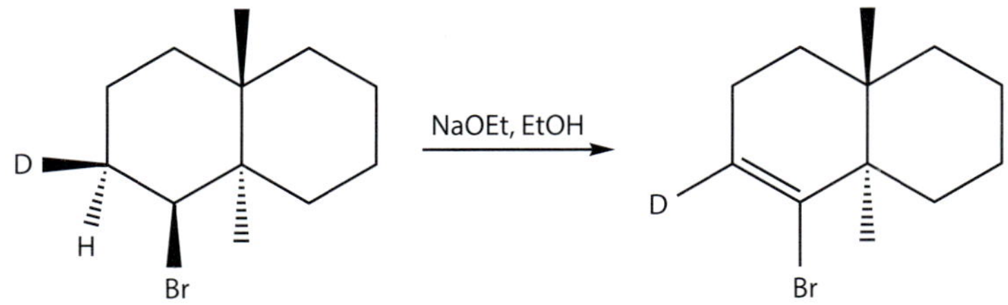

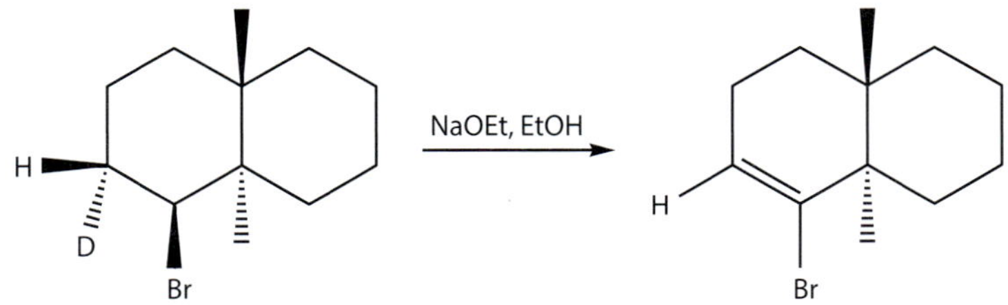

7. Explain the following observations:

(a)

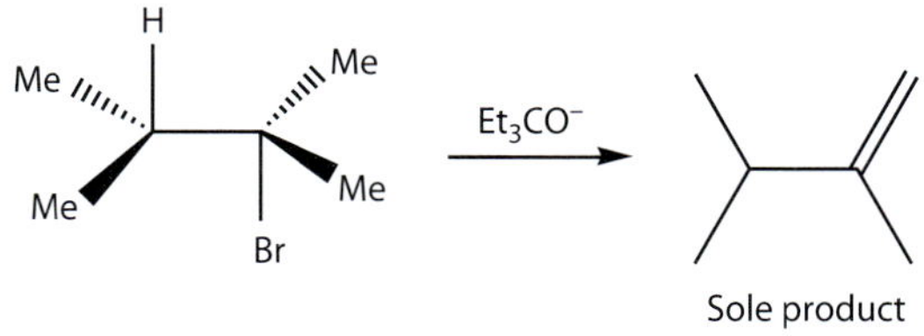

(b) One of these two isomers undergoes HCl elimination about 100 times as fast as the other one. Which isomer is more reactive and why?

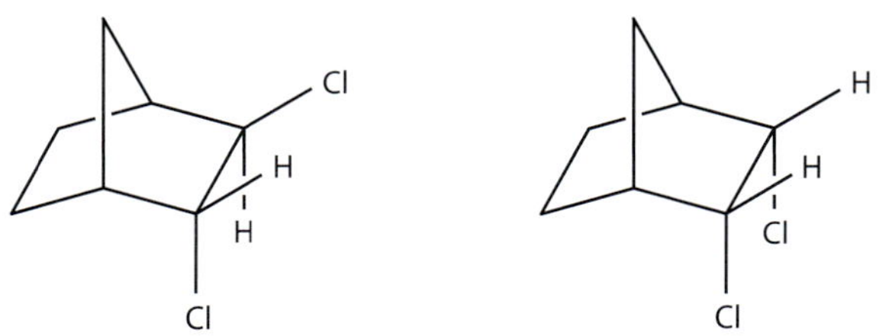

(c) When 2-chloropropane is reacted with ethoxide, the major product is that of elimination, propene, along with a small amount of the substitution product, ethyl *iso*-propyl ether. By contrast, when 2-chloropropane is reacted with acetate ion, only *iso*-propyl acetate is obtained.

(d) Compound **10.21** undergoes base-assisted elimination more slowly than **10.22**.

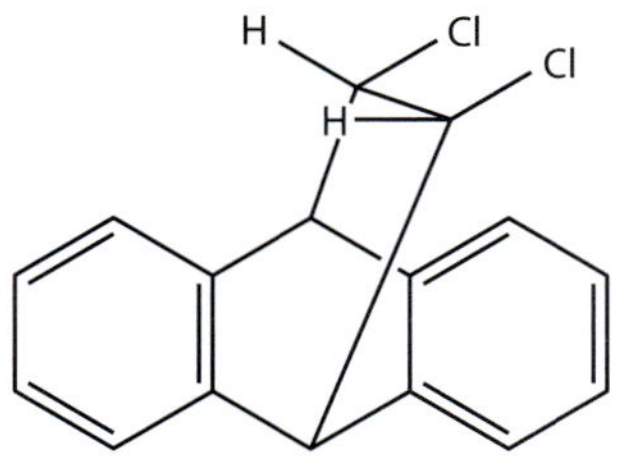

10.21

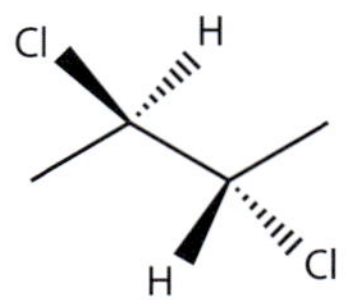

10.22

(e) Cyclohexanol, when warmed with HBr, gives bromocyclohexane. When warmed with sulfuric acid, the product is cyclohexene.

(f) Compound **10.23**, DDT, loses HCl when treated with $[EtO]^-/EtOH$. Substitution of H* by D reduces the reaction rate by a factor of 3.8. The use of EtOT does not introduce T into the unreacted DDT. Remember that C–D bonds are broken more slowly than C–H bonds—this is the kinetic isotope effect. (Hint: Consider each possible mechanism in turn and check whether the experimental data are consistent with it.)

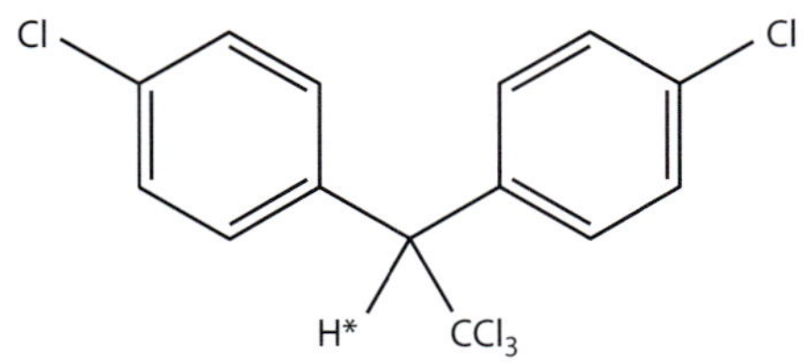

10.23

(g) Explain why elimination of HBr from one of the two isomers shown in the following is rapid, using KOH. The other reacts very slowly. Show a full mechanism.

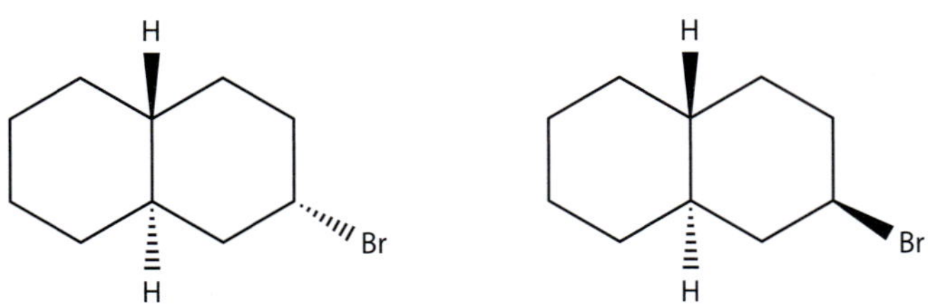

(h)

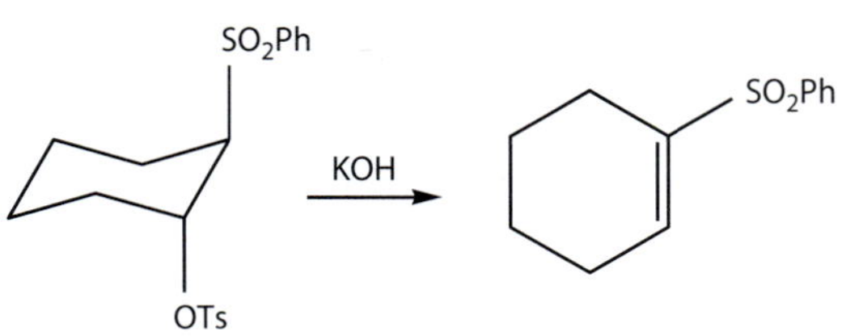

(i) Bromobenzene reacts 20 times faster than chlorobenzene with KNH_2/liq. NH_3.

(j) Compound **10.24** undergoes chromic acid oxidation 10 times faster than compound **10.25**.

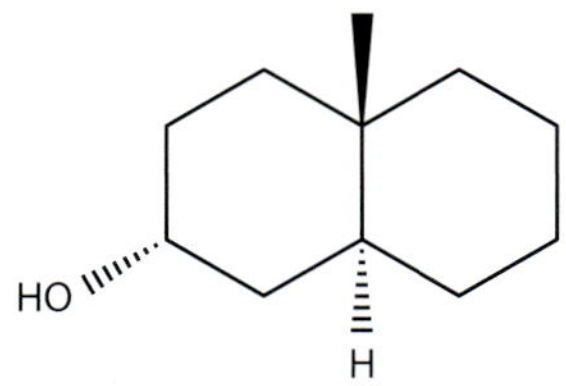

10.24

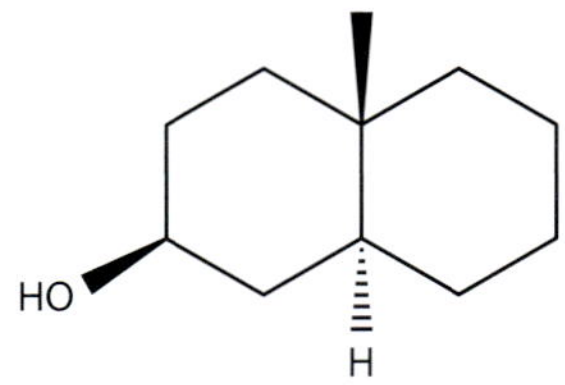

10.25

8. Study the following reaction:

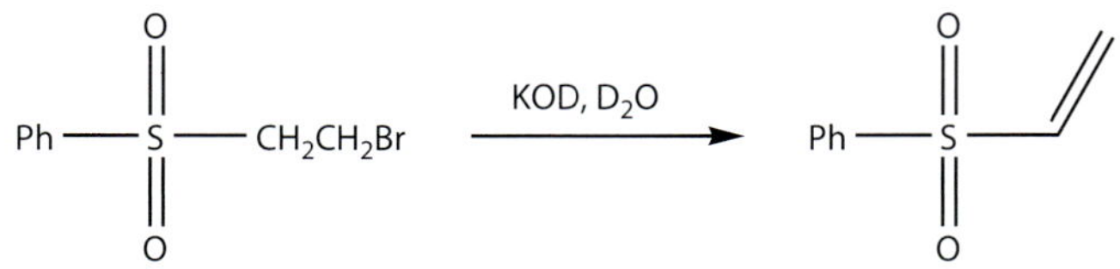

If the starting material is recovered before the reaction is complete, it is found to be deuterium labeled at the CH_2 bonded to the SO_2 group. Account for this with a full reaction mechanism.

MORE CHALLENGING PROBLEMS

1. Predict the product(s) of each of the following reactions:

(a)

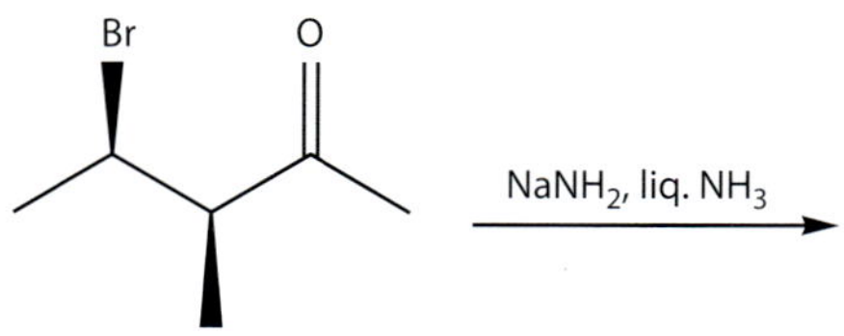

(b)

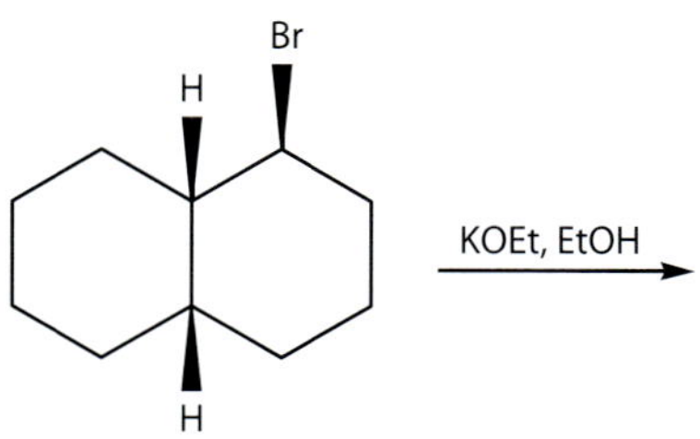

(c)

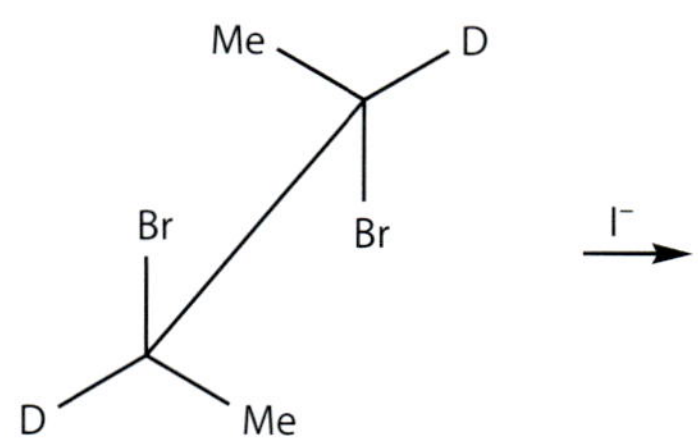

(d)

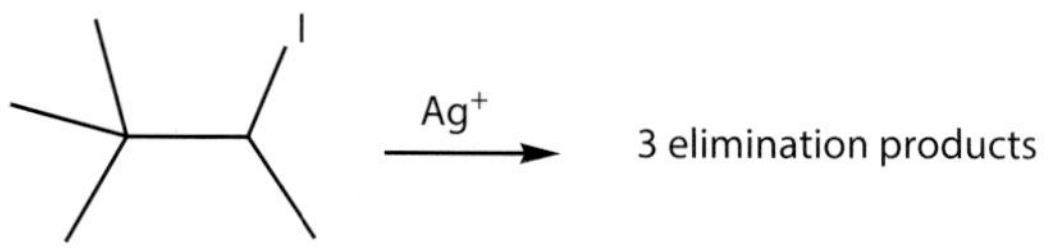

(e)

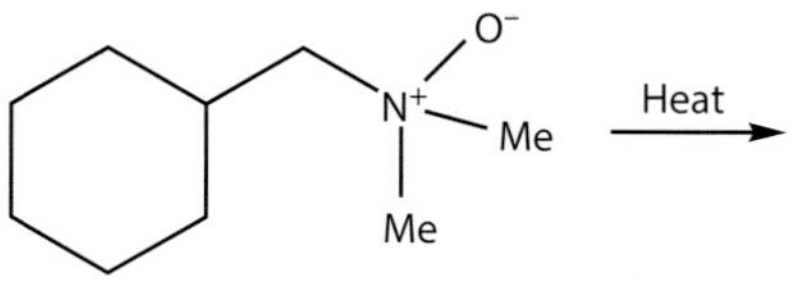

2. Propose a mechanism for each of the following reactions, paying attention to stereochemical considerations as appropriate:

(a)

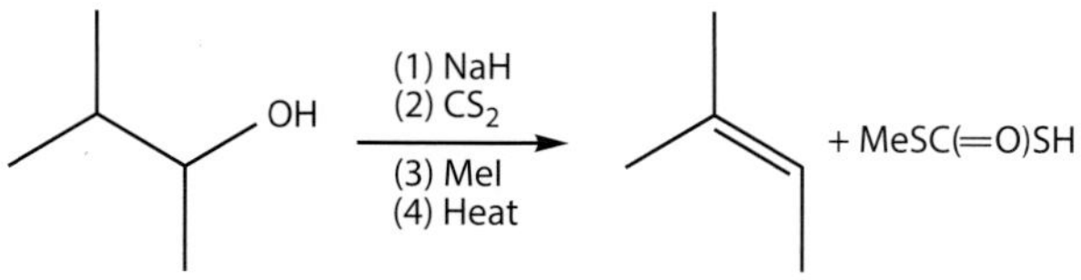

(b)

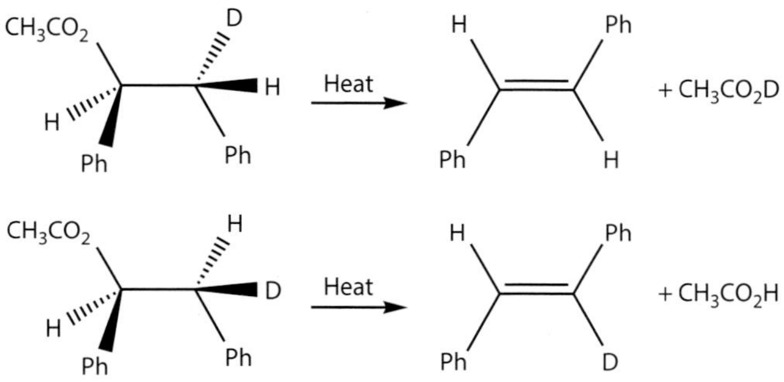

(c)

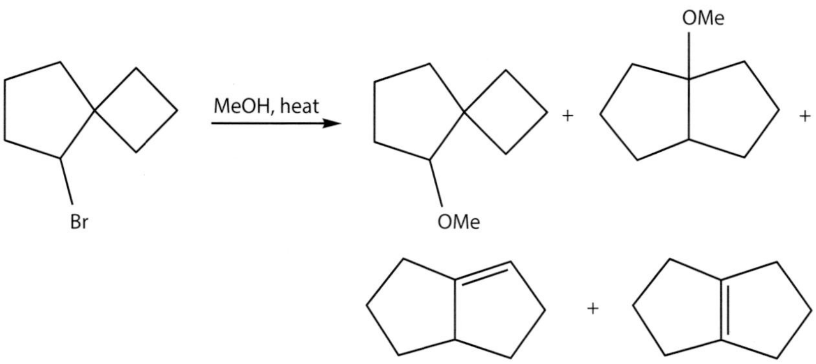

(d)

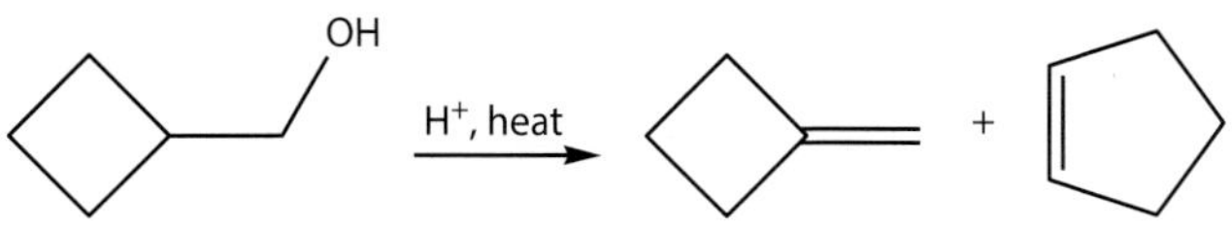

3. Explain each of the following observations:

(a) There are eight diastereoisomers of 1,2,3,4,5,6-hexachlorocyclohexane. One isomer loses HCl in an E2 reaction nearly 1000 times more slowly than any one of the others. Which isomer reacts so slowly and why?

(b) Reaction of *S*-2-bromo-2-fluorobutane with sodium methoxide gave three fluoroalkenes and *S*-2-fluoro-2-methoxybutane.

(c) Treatment of 1-bromo-2-deutero-2-phenylethane with strong base leads to a mixture of deuterated and nondeuterated phenylethenes in which the deuterated product predominates by approximately 7:1.

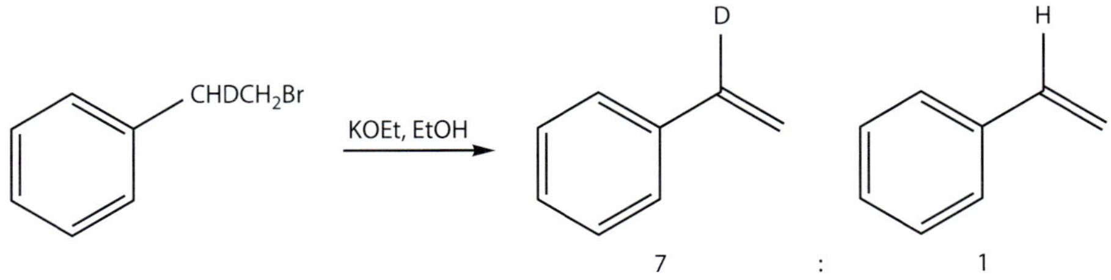

4. Identify the isomeric compounds **10.26** and **10.27** in the following sequence, justifying your assignment by writing a complete mechanism for all the processes involved (you will find this easiest using sawhorse representations).

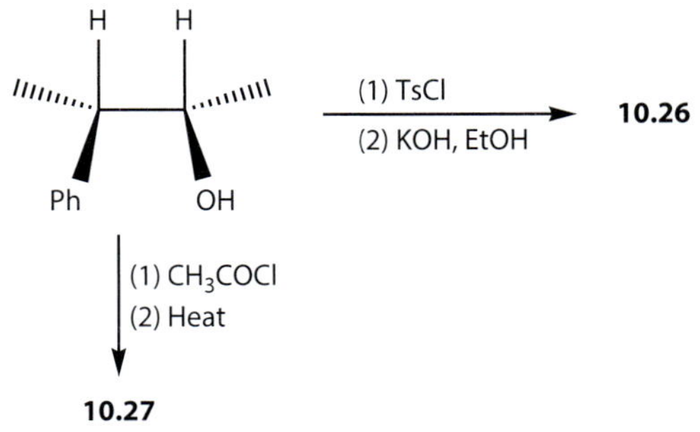

5. The elimination of HF from **10.28** is thought to proceed by a conventional E2 mechanism. **10.28** may be reacted with methyl iodide to give **10.29**. Elimination of HF from **10.29** is thought to proceed by an E1cB reaction, in which anion formation is irreversible. Draw mechanisms for the processes and explain why the reactions proceed by different mechanisms.

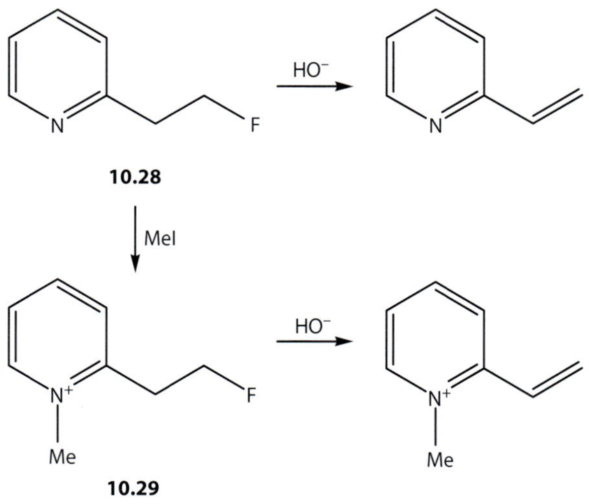

6. The following reaction has been studied by several groups:

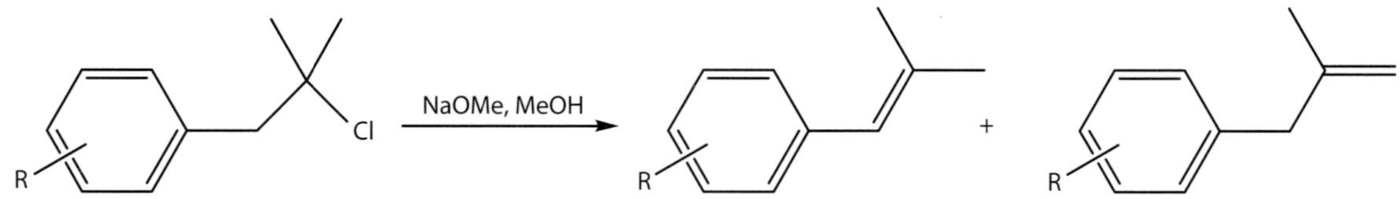

10.30 **10.31** **10.32**

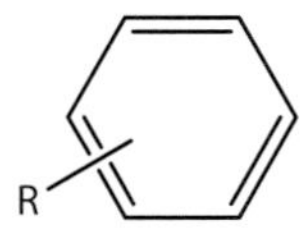

means a substituent on the aromatic ring

e.g., R = 4-Me means

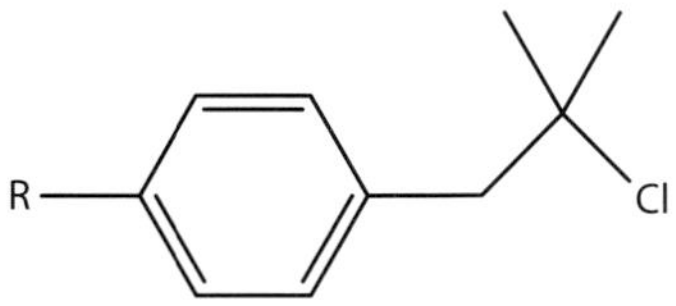

The following observations have been made:

(a) The rate of reaction is given by the expression

$$-\mathrm{d}[X]/\mathrm{d}t = k_1[X] + k_2[X][\mathrm{MeO}^-]$$

(b) When $ArCD_2CMe_2Cl$ is compared with $ArCH_2CMe_2Cl$, the value of k_1 is affected only to a very small degree, but the value of k_2 for the CH_2 compound is 2.6 times as great as for the CD_2 compound.

(c) The nature of R has a small effect on k_1, with electron donors accelerating this part of the reaction.

(d) When R is strongly electron withdrawing (*e.g.*, R = NO_2), k_2 is increased, and the proportion of **10.31** in the products is markedly increased.

Take each of these observations and show how it is consistent with the accepted view of elimination mechanisms.

Addition Reactions

11

11.1 INTRODUCTION

In this chapter, we will be studying addition reactions to carbon–carbon multiple bonds; this is the converse process of the eliminations that we studied in the previous chapter. Addition to carbon–heteroatom multiple bonds is coming up in Chapter 14. Nucleophiles, electrophiles, and radicals can all add across double bonds; first, we will concentrate on electrophiles and radicals, as nucleophiles only add readily when the double bond bears a group (such as a carbonyl, nitro, or nitrile; Chapter 17) capable of accepting electron density. Reactions with electrophiles or radicals add two moieties, atoms or groups, by a stepwise process; the two atoms or groups are not added simultaneously. However, there is another class of reactions where the two new bonds are made simultaneously—these are called concerted reactions.

We should first recall the electronic structure of alkenes, with carbon atoms sp^2 hybridized and a sigma framework of bonds at approximately 120 ° to each other. Above and below the plane of the molecule is a π-orbital, derived from the two remaining p_z orbitals (**11.1**).

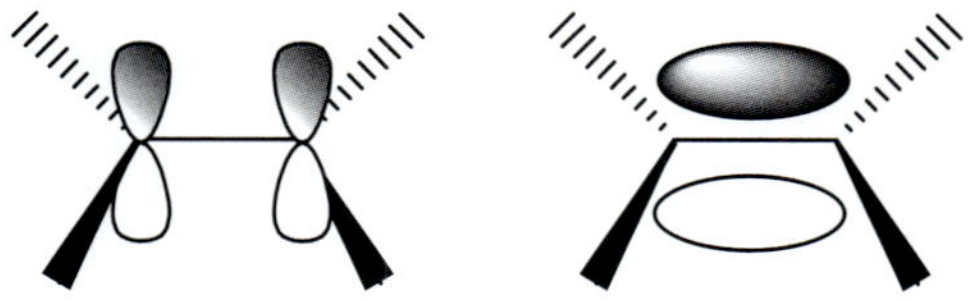

11.1

The addition of two atoms or groups to an alkene is the most important reaction for this type of compound. The two electrons of the π-bond provide two of those needed to make the two new σ-bonds.

11.2 ELECTROPHILIC REACTIONS

11.2.1 REACTION MECHANISM

This reaction is typified by the addition of a hydrogen halide, HX, or water, to an alkene. The process involves two steps, with a carbocation as the intermediate. In the first step (Figure 11.1), an electrophile, for example, a proton, is added to the carbon–carbon bond to form a carbocation. Notice how the curly arrow is drawn—the proton is being added to the "upper" carbon of the double bond and the electrons are "taken away" from the "lower" carbon, leaving it positively charged. In the second step, the counterion, for example, bromide, attacks the carbocation to give a saturated product. Note that we only show one of the new bonds we have made in this example (in red); a carbon–hydrogen bond has also been made.

In this example, it did not matter which carbon initially became bonded to the proton; the same carbocation is obtained whichever one is chosen. However, in many examples, there is a choice—and we always choose to make the more stable of the two possible carbocations. Figure 11.2 illustrates this. Depending on which of the two carbons the proton adds to, we obtain either a tertiary or a secondary carbocation. The reaction that actually occurs is invariably that which forms the more stable, in this case the tertiary, carbocation.

This is often described as Markovnikov addition, and this name is used to define any electrophilic addition that proceeds *via* the most stable carbocation. Many common addition reactions

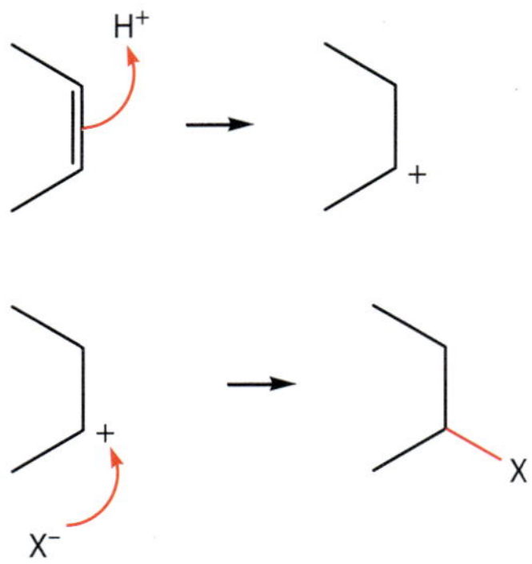

FIGURE 11.1 Mechanism of electrophilic addition of HX to an alkene.

(examples in Figure 11.3) of HBr, HCl, HOCl, and H^+/H_2O proceed in this way. The first example is essentially the one we just looked at, but this time adding HBr. In the next one, we add water, again *via* a tertiary carbocation.

What is our evidence for this two-step process, with the first step being rate determining? The Markovnikov regiochemistry in these addition reactions is completely consistent with carbocation formation. We also observe that alkenes substituted by electron-donating groups react faster with electrophiles; this implies that the first step, the addition of the electrophile, is the RDS. Some relevant examples are shown in Table 11.1.

Few electrophilic addition reactions of HX are stereospecific, or even stereoselective; this is consistent with the intermediacy of a planar carbocation intermediate that can be attacked from either face (Figure 11.4). Here we make the assumption that the proton initially approaches the

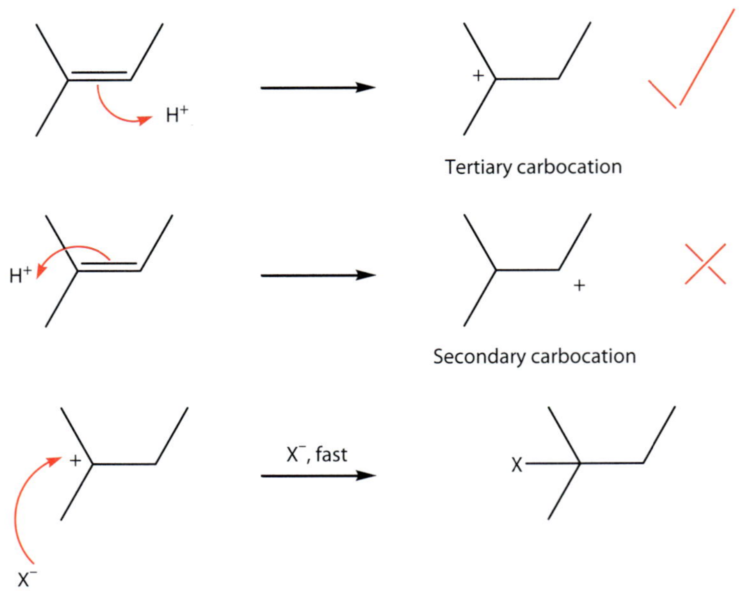

FIGURE 11.2 Addition of HX to a nonsymmetrical alkene.

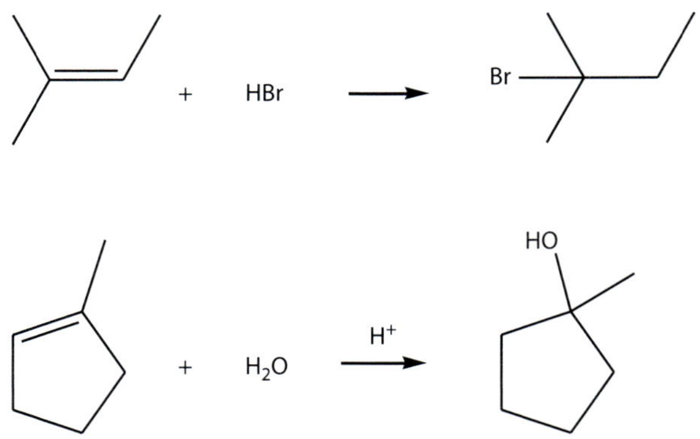

FIGURE 11.3 Examples of Markovnikov addition.

TABLE 11.1 Relative Rate of Acid-Catalyzed Addition of Water to Alkenes

Alkene	Type of Carbocation Formed on Protonation	Relative Rate
$CH_2{=}CH_2$	Primary	1
$CH_2{=}CHCH_3$	Secondary	2×10^6
$(CH_3)_2C{=}CHCH_3$	Tertiary	8.7×10^9
$CH_3OCH{=}CH_2$	Stabilized by resonance donation of oxygen lone pair	5×10^{14}

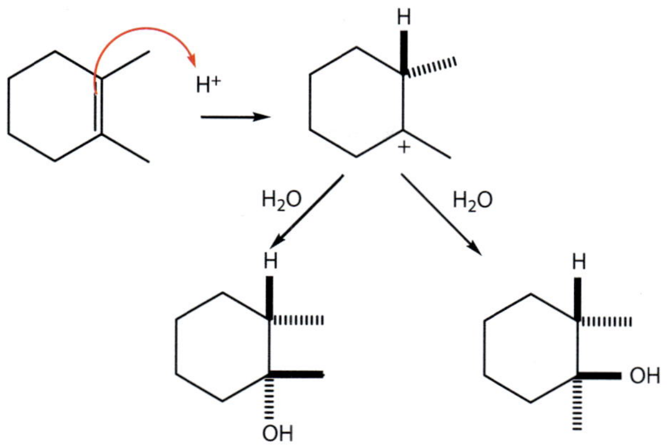

FIGURE 11.4 Electrophilic addition reactions are rarely stereospecific.

top face of the molecule (though both faces are equally likely). When the intermediate carbocation is attacked by water, in a fast step, either face can be attacked, so we get two diastereoisomeric products (and both enantiomers of each will be obtained). Also, the carbocation can be captured by other suitable nucleophiles, as, for example, in Figure 11.5, where the acetic acid solvent competes for the carbocation.

Finally, in electrophilic addition processes, many of the carbocation rearrangements that we previously met in E1 and S_N1 processes are observed (Figure 11.6).

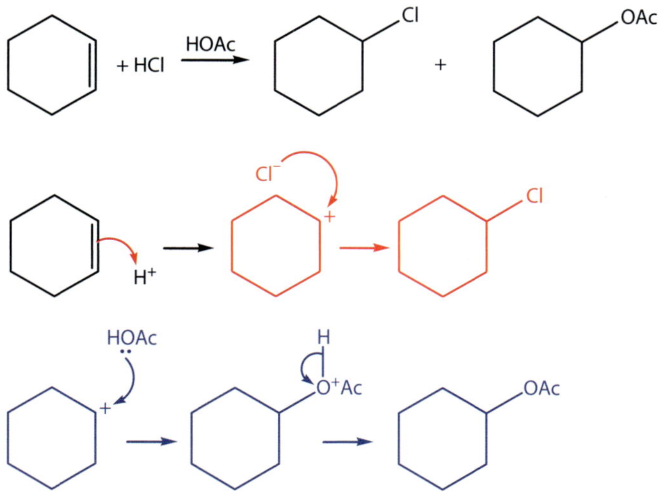

FIGURE 11.5 Intermediate carbocations can be captured by solvent.

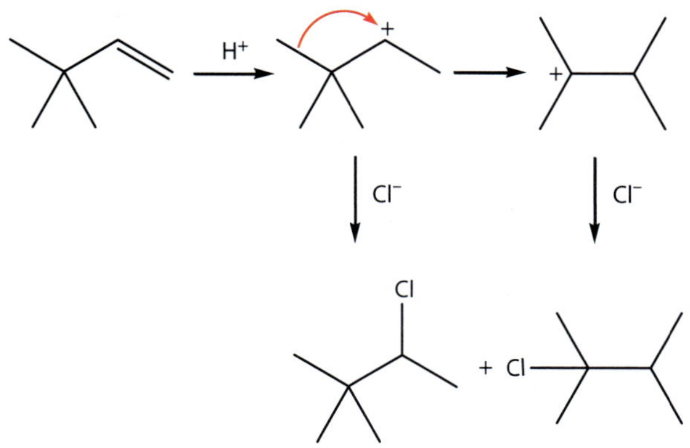

FIGURE 11.6 Carbocation rearrangements in addition reactions.

PROBLEM 11.1

Protonate each of the following alkenes to give the most stable carbocation.

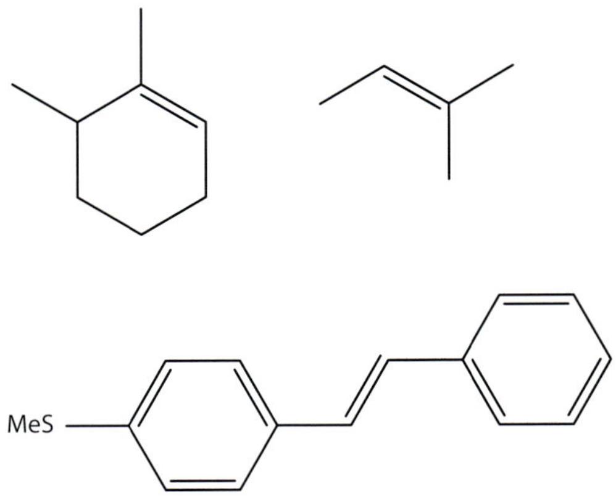

Solution

Protonating the first two substrates is relatively straightforward; in each case, we have the choice between producing a secondary and a tertiary carbocation, so we chose to make the tertiary species:

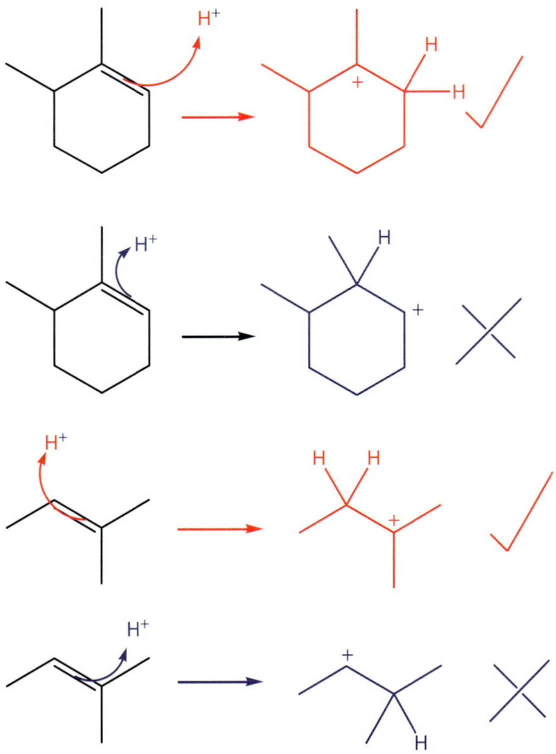

The third example is more challenging, and we will need to remember what we learned about resonance forms. Both of the two possible carbocations are benzylic. The difference between them must be related to the presence of the –SMe group on one of the two rings. We have not met many sulfur compounds, but we do know that sulfur is below oxygen in the periodic table and that it has a lone pair of electrons that it can share with a positive center. In this case, the positive center is some distance away, but we can still write a good resonance form involving the lone pair of electrons at sulfur. Hence, the cation is formed next to the ring bearing the sulfur, because of its ability to share its lone pair through resonance.

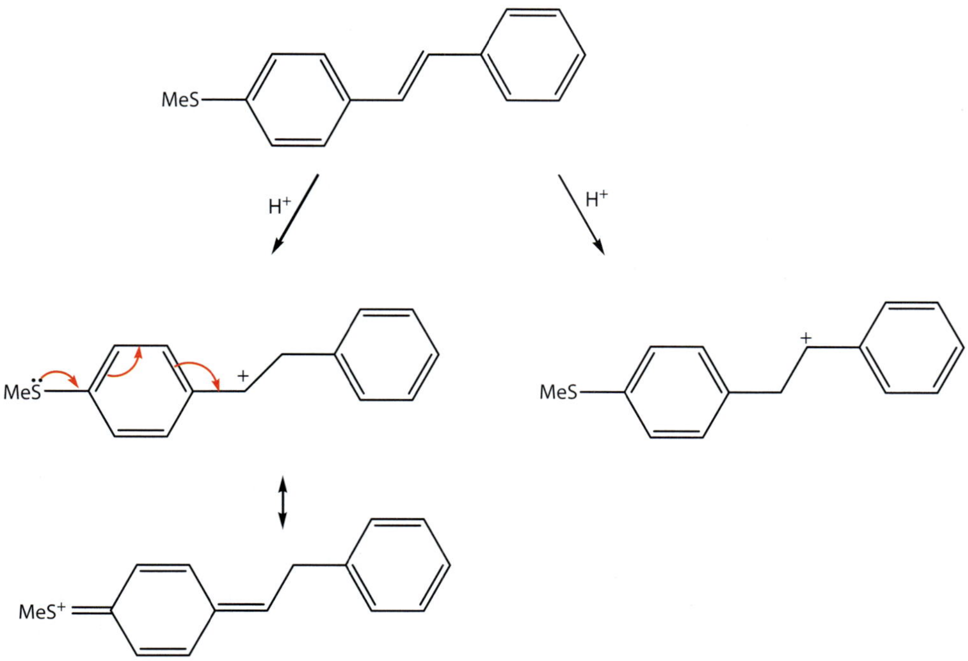

PROBLEM 11.2

Predict the outcome of each of the following reactions.

(a)

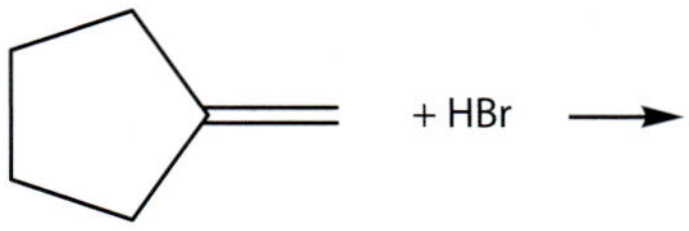

(b)

Solutions

(a) The first step is to protonate the carbon–carbon double bond to give the more stable, tertiary carbocation:

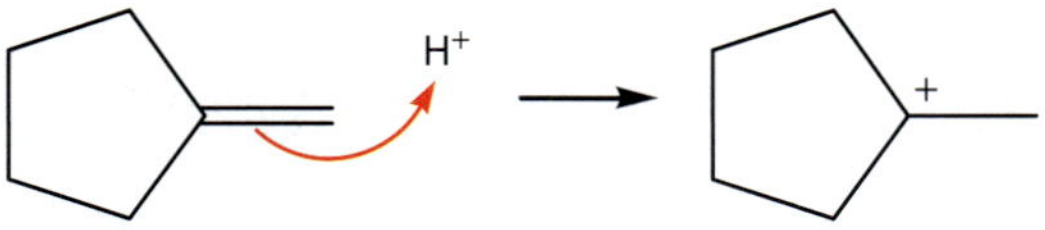

Then the bromide ion attacks the positive center to give the tertiary bromide as the product:

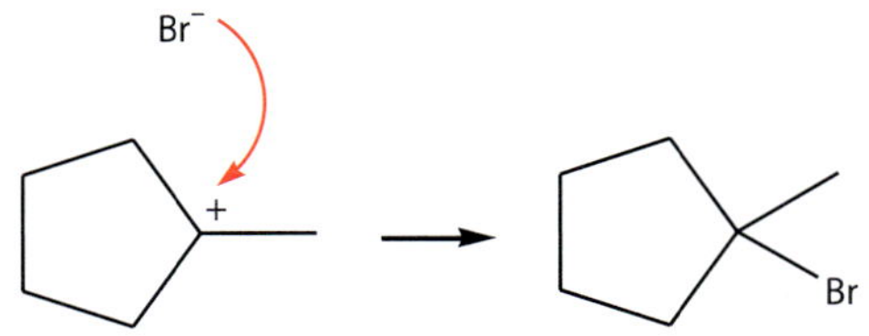

(b) Whichever way we protonate this, we will get a secondary cation, but in one case, the cation will be stabilized by resonance with the benzene ring—so this one is favored. In the second step, the chloride captures the cation to give the final product.

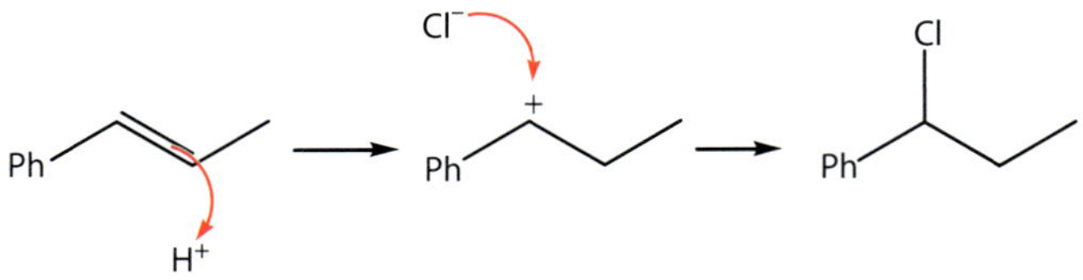

11.2.2 SCOPE OF THE REACTION

The simplest reactions to understand are those adding hydrogen halides; examples are shown in Figure 11.7. All show Markovnikov selectivity. Note that in the third example, a secondary benzylic cation is preferred over a simple secondary cation, as in the problem earlier. The rate of reaction of the hydrogen halides is in the order HI ≫ HBr > HCl > HF. In practice, however, additions of HBr or HCl are the most common. Although the reaction with HI is faster, purification of HI is more difficult, and it is easily oxidized to iodine, leading to a range of side reactions. Notice in the last example how unfavorable it is to place a cationic center next to an electron-withdrawing CF_3 group.

The addition of HF to alkenes is practicable, but the addition reaction is slow and the handling of anhydrous HF is challenging. With anhydrous HF, alkene polymerization is a major

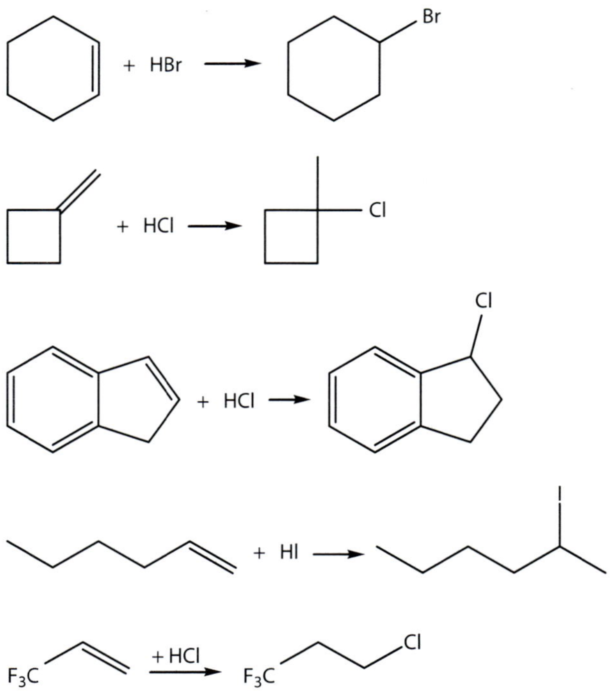

FIGURE 11.7 Addition of HCl or HBr to alkenes.

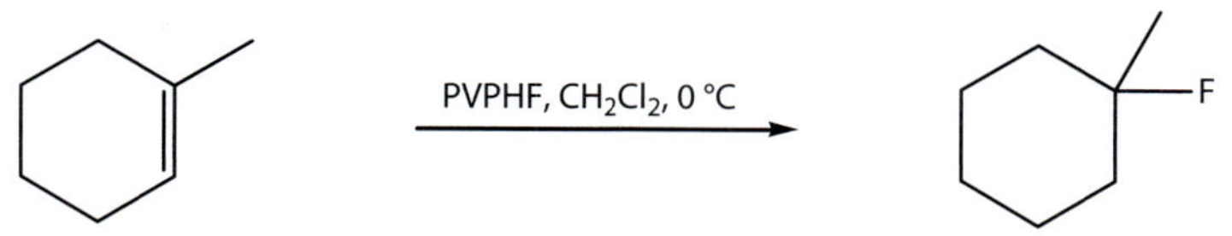

FIGURE 11.8 Addition of "HF" to an alkene.

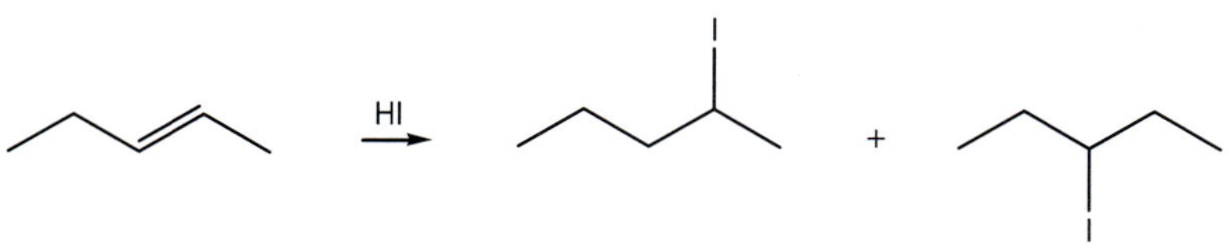

FIGURE 11.9 Addition of HX to alkenes is not regioselective when the intermediate cations are of comparable stability.

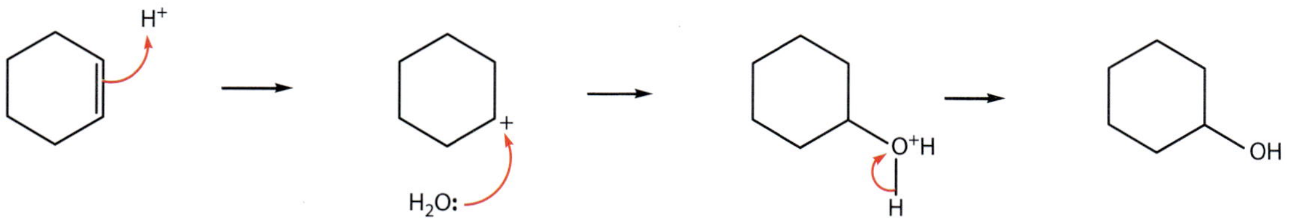

FIGURE 11.10 Acid-catalyzed addition of water to an alkene.

side reaction. A polymeric "version" of HF, polyvinylpyridiniumpolyHF (PVPHF),* has been used successfully (Figure 11.8).

When there is little to choose between the available cation intermediates, mixtures of products are generally obtained, as, for example, in Figure 11.9; both products are formed from secondary carbocation intermediates.

Water alone is not a sufficiently strong acid to protonate an alkene, so all additions of water have to be carried out in the presence of an acid catalyst. Usually a strong acid, with a non-nucleophilic counterion such as sulfate, is used to reduce the risk of by-products from the capture of the carbocation intermediate by the counterion. In very strong sulfuric acid, intermediate sulfates, $ROSO_3H$, can sometimes be isolated, but these are readily hydrolyzed to alcohols. Figure 11.10 shows a simple example. As we noted in Table 11.1, the rate of reaction depends on the stability of the carbocation intermediate. Although the study of this reaction was important in understanding the mechanism of addition reactions, its current use in the synthesis of alcohols is quite limited; there are more selective alternatives, using less harsh conditions.

When the alkene is nonsymmetric, the addition reaction is exclusively Markovnikov and generally nonstereospecific. Some examples are shown in Figure 11.11. The first two examples proceed through a tertiary carbocation. In the third example, we see that alcohols can be added in the same way as water. The last reaction of this group is widely used as the first step of a protection/deprotection sequence for alcohols; we will meet this again later, when we look more closely at synthesis. Notice the role of the lone pair on oxygen stabilizing the adjacent carbocation.

---

*
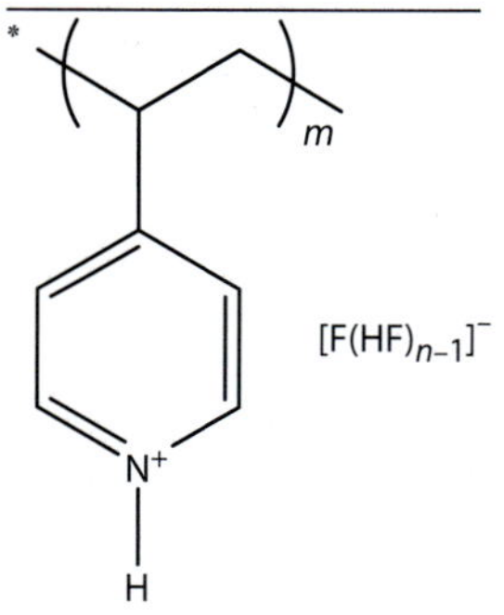

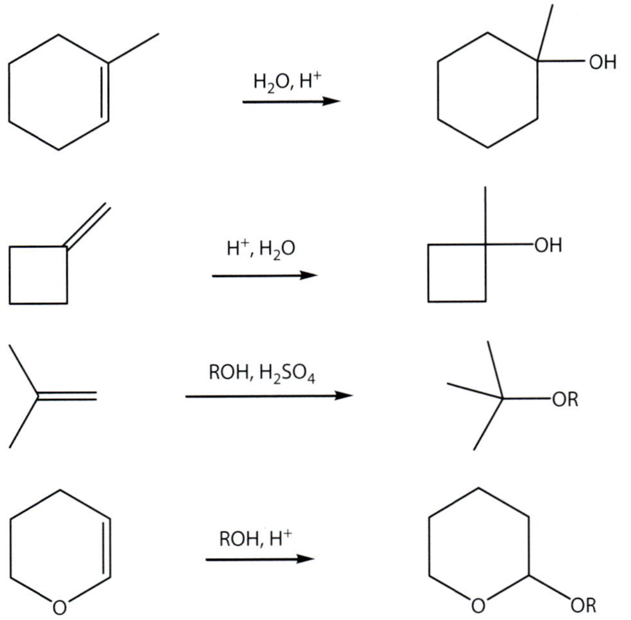

FIGURE 11.11 Addition of water or alcohols to alkenes.

PROBLEM 11.3

Predict the product(s) of the following reaction.

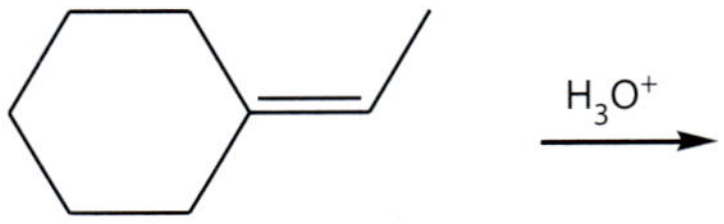

Solution

First, we protonate the double bond to give the more stable tertiary carbocation; then this is attacked by water. The final step is loss of a proton to give the alcohol.

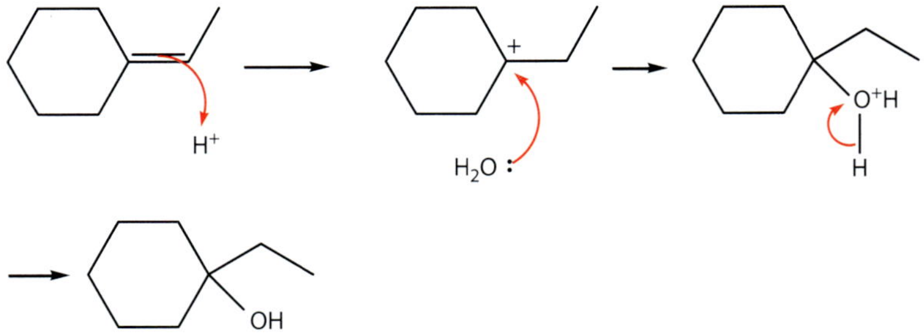

PROBLEM 11.4

The addition of HCl at 0 °C to 3-methyl-1-butene gave a 40:60 mixture of 2-chloro-3-methylbutane and 2-chloro-2-methylbutane. Explain, writing a complete mechanism.

Solution

This is quite a complex problem, as it requires you to translate the names of compounds back into structures, showing that you remember your basic nomenclature (remember, however, that there will probably be some points for this in any grading scheme, so even if you can't complete this sort of problem, you should not score a zero on it). So first, we write out what happens in schematic form:

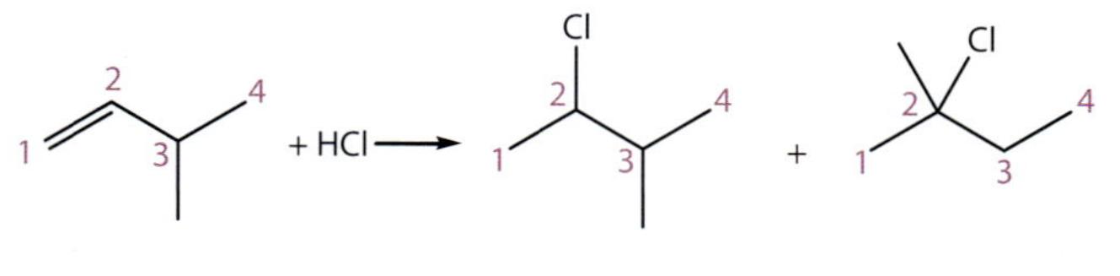

The formation of the first product should give us no problem—it is a simple addition to the original alkene, *via* the more favorable secondary carbocation:

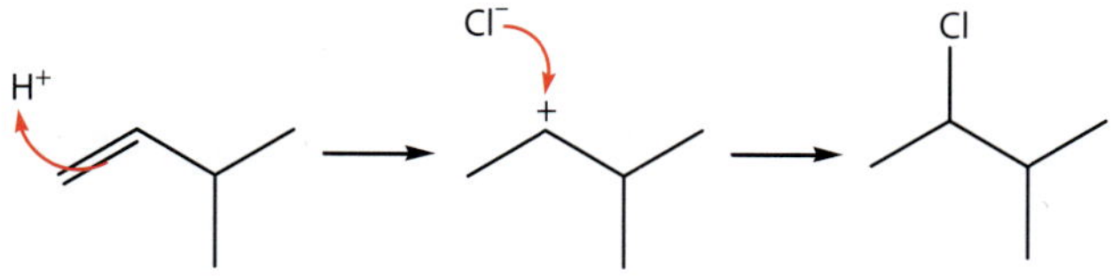

However, in forming the second product, something more fundamental has clearly occurred—a hydrogen atom has migrated (notice that this is not a methyl migration, which would only lead to exactly the same secondary carbocation). Specifically, this has happened to the initially produced carbocation, so that the secondary carbocation is converted into the tertiary one, prior to capture by the chloride ion. This is an example when it's helpful to think about the reaction backward from the end—what carbocation was captured by the chloride to give the second product? That's the one we need to get to, so we do the migration that is required to achieve it.

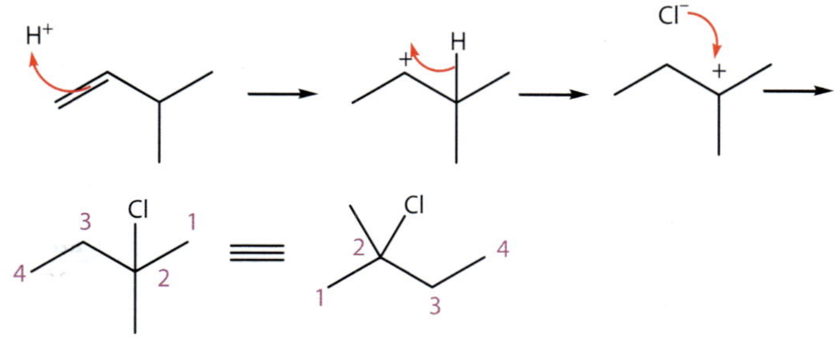

I hope you were not concerned by the fact that the molecule came out "the other way around" from the way we first wrote it. This will happen often, and it's nothing to worry about. In fact, if you are puzzling over a reaction mechanism, it's often a good idea to write the molecule in a different orientation, as sometimes this will clue you into the relationship between the atoms/groups in the starting material and the product.

The addition of halogens to alkenes presents some unusual observations. The addition of bromine to most alkenes is *trans* and stereospecific (Figure 11.12; the product is, however, racemic, hence the straight wedge and hash bonds), but additions of the other halogens give rise to mixtures of products from *syn* and *anti*-addition. How and why does this *trans*-addition occur? It must certainly be a stepwise reaction, as any synchronous process would give rise to *cis*-addition.

This can be explained by the formation of a bromonium ion, **11.2**, in which the bromine atom bridges the two carbons of the alkene. The opening of the cyclic species by the bromide

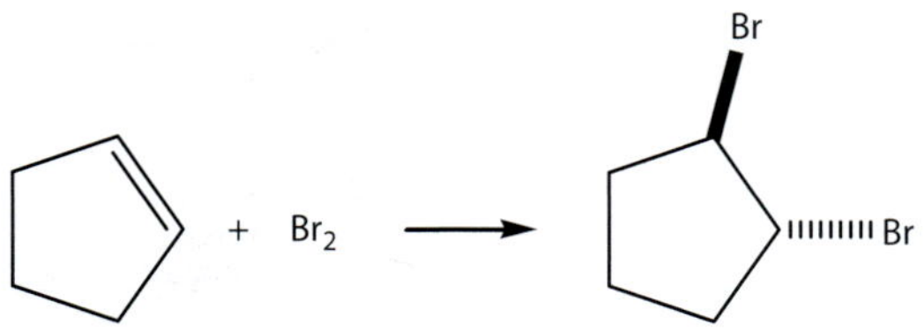

FIGURE 11.12 Addition of molecular bromine to cyclopentene.

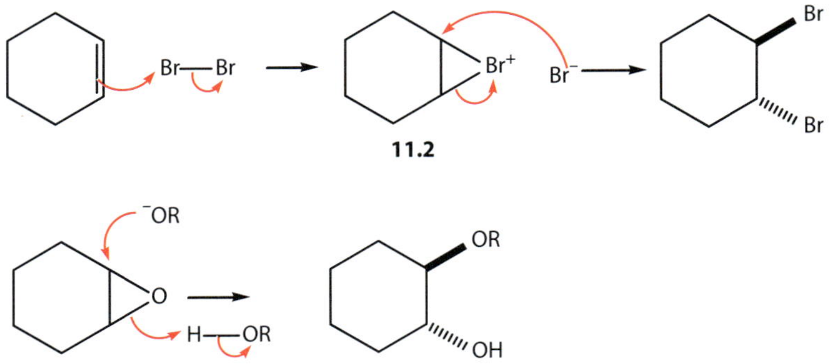

11.2

FIGURE 11.13 S_N2 ring opening of epoxides and bromonium ions.

counterion requires S_N2 backside attack, so that the two bromines end up on opposite sides of the original double bond. This is analogous to a reaction we met earlier when we opened epoxides with a range of nucleophiles (Figure 11.13). In each of these sequences, a racemic mixture would be obtained. In the absence of any substituents, there is no reason that the initial bromonium ion would be formed on one face of the double bond in preference to the other.

Other than this observed stereochemistry, what evidence do we have for the existence of the bromonium ion? In very rare cases, a stable bromonium ion can be isolated. For example, when adamantylideneadamantane reacts with bromine, the stable bromonium ion salt, **11.3**, can be isolated, and its crystal structure has been determined. Any backside approach to this cation by a nucleophile would require skills exceeding those of Luke Skywalker attacking the Death Star—there simply is no space for the attack. The iodine and chlorine analogues of **11.3** have also been prepared and characterized crystallographically. Generally, the addition reactions of chlorine are less stereoselective than for bromine—the atom appears to be not quite large enough to bridge the double bond symmetrically. The addition of fluorine is rarely used because of its high reactivity and difficulty in handling the gas. The addition of iodine gives predominantly *trans*-products, but the reaction is readily reversible.

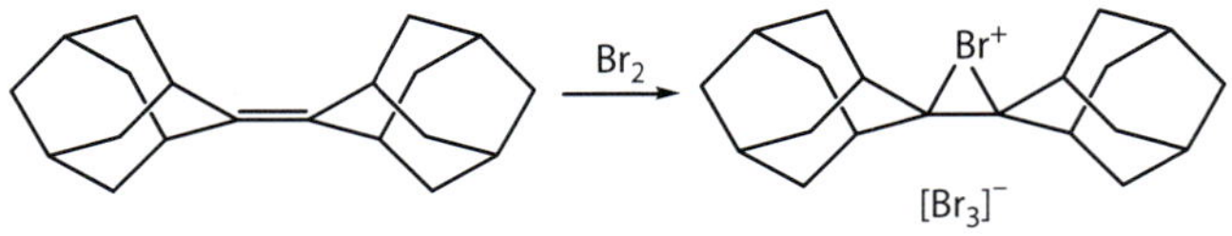

11.3

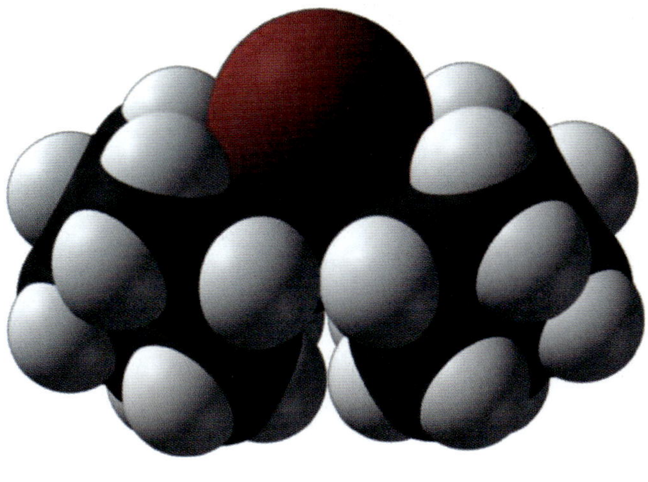

Space-filling model of **11.3**

Synthetically, most of the reactions we shall meet involve the addition of bromine or chlorine; some examples are shown in Figure 11.14. Because the two atoms we are adding are the same, we have no need to worry about regioselectivity here.

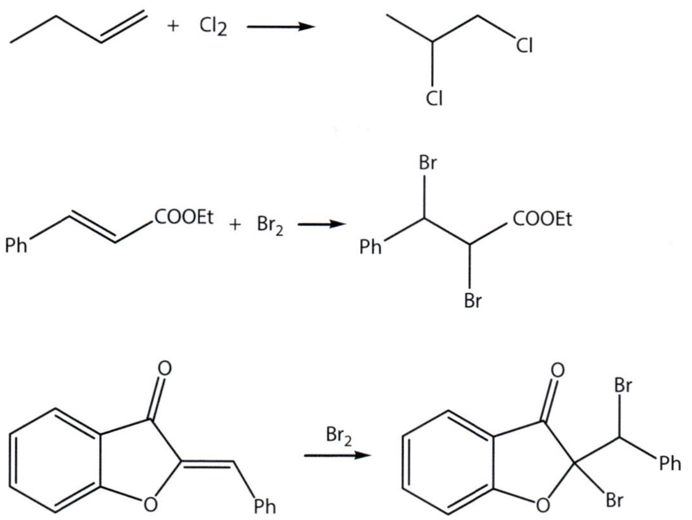

FIGURE 11.14 Addition of halogens to alkenes.

This is a reaction that I'll describe as a "three bears" type of process. Bromine is just the right size, chlorine is too small, and iodine a little too large. We'll meet a number of similar reactions of halogens in other chapters.

We remember that when we studied the addition of hydrogen halides to alkenes, the intermediate carbocation could be captured by "another" nucleophile, such as a nucleophilic solvent. Can the same thing happen with a halonium ion? The answer is yes, it can, and this turns out to be a useful synthetic process. If chlorine or bromine is added in the presence of water (dissolving the halogen in water gives what is sometimes described as chlorine or bromine water), then the major product obtained is the halohydrin (Figure 11.15). Because the reaction involves the opening of a cyclic halonium ion, the stereochemistry of the product is *trans* from *anti*-addition. These "halohydrins" are deprotonated by base and undergo a ring closure reaction to give epoxides.

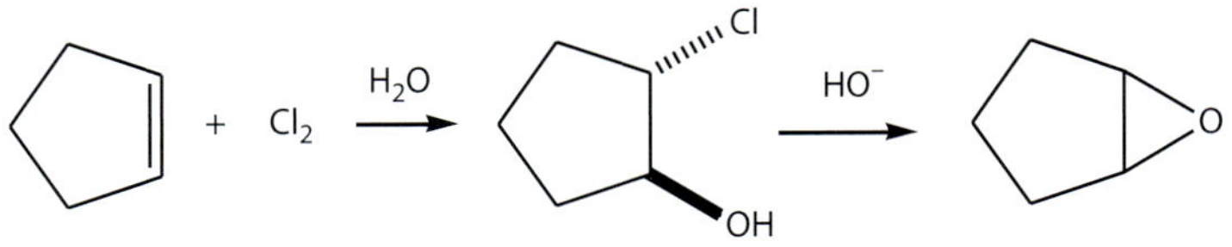

FIGURE 11.15 Addition of chlorine water to an alkene.

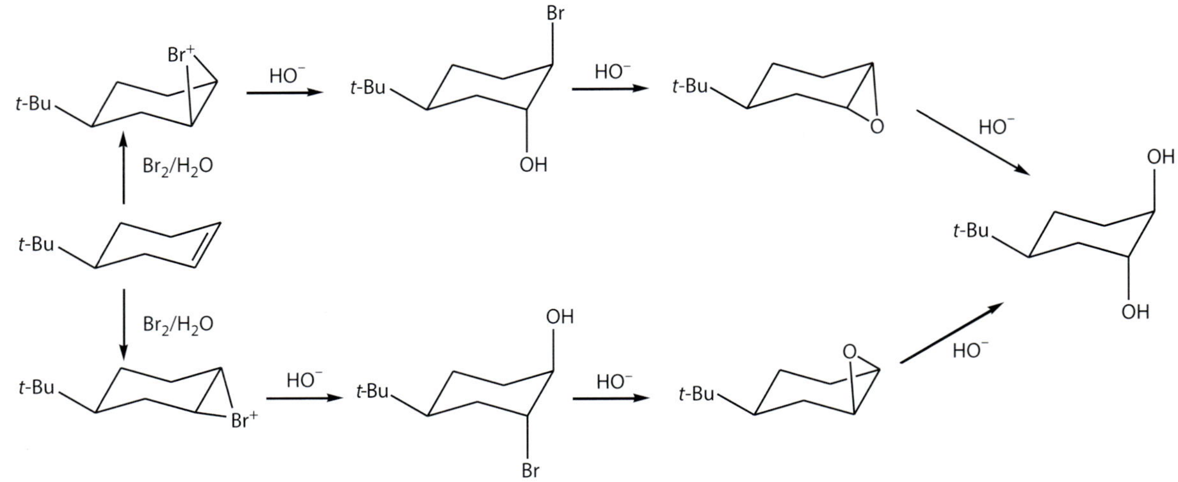

FIGURE 11.16 Formation and opening of halonium ions and epoxides fused to cyclohexane rings.

Examples in six-membered rings show the stereochemical implications of the process clearly (Figure 11.16). By putting a *tert*-butyl substituent on the ring, we lock it against ring inversion. The bromonium ion can be formed on either the top or the bottom face of the double bond. The presence of the *tert*-butyl group may make one face slightly more attractive than the other, but not to the extent of exclusivity. When the bromonium ion is opened by hydroxyl ion in an S_N2 process, the substituents are *trans* and diaxial, in fact ideally set up for the reclosure to the epoxide. The epoxide in turn can be opened to give the *trans*-diol. The key point to take away from this is that S_N2 ring opening of these three-membered rings fused to cyclohexanes gives us *trans*-diaxial geometry and *trans*-diaxial geometry is required to close up to a fused three-membered ring.

So far we have carefully chosen examples where both ends of the double bond are the same—we did not have to worry about regioselectivity. With unsymmetrical alkenes, the attack of water (or an alcohol, these work just as well) is always at the more positive, the more substituted carbon atom (Figure 11.17). But it is still stereospecific.

This is another of my *Alice in Wonderland* reactions—in order to understand one aspect of the reaction, we need to think of it in one way, but in order to understand another aspect, we need to think of it in a different, and contradictory, way. In this process, the regiochemical outcome is determined by thinking of the intermediate as a carbocation and the stereochemical outcome by thinking of it as a closed halonium ion. The truth lies in between—where alkenes are markedly nonsymmetric, the halonium ion has significant carbocation character, as shown in the

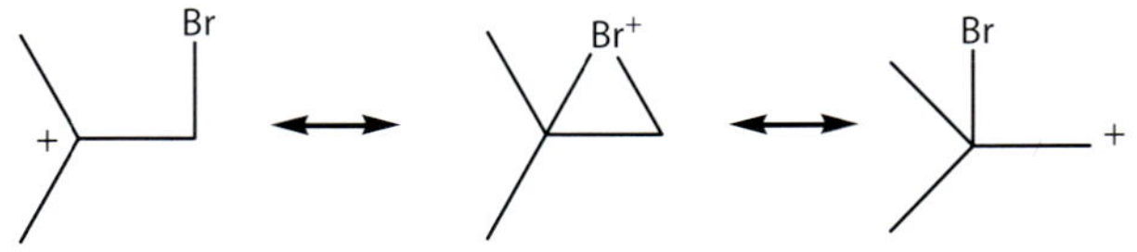

11.4

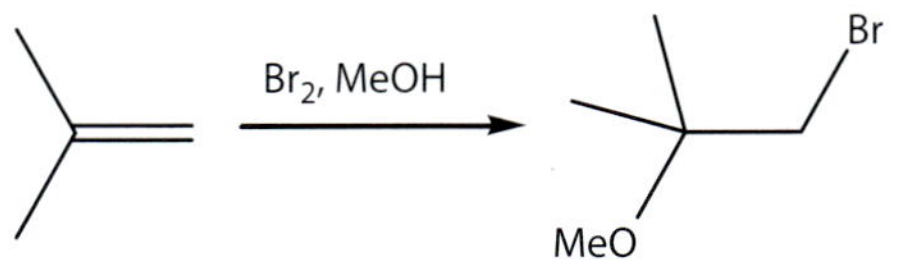

FIGURE 11.17 Regiospecific addition of Br_2/MeOH to an alkene.

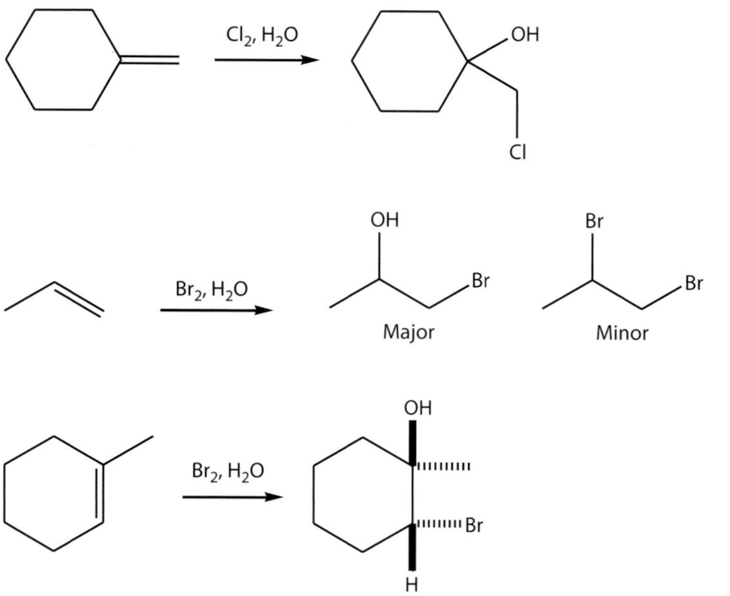

FIGURE 11.18 Addition of halogens to alkenes in the presence of water or alcohols.

resonance forms (**11.4**); the first two forms are important contributors to the overall structure; the third, a primary carbocation, is not. Examples of this type of reaction are shown in Figure 11.18. More recent syntheses often use *N*-bromosuccinimide as a source of electrophilic bromine, to avoid the necessity of handling elemental bromine, but the mechanism is essentially the same.

An alternative method of Markovnikov addition of water to alkenes involves an intermediate rather similar to the bromonium ion. This is a mercurinium ion, formed by the reaction of an alkene with mercury (II) acetate (check out the sizes of mercury and bromine, and you will see why these are so similar). This is often preferred to acid-catalyzed hydration, as conditions

Focus on Halolactonization

So far, we have opened halonium ions with halide ion, water, and alcohols. Let's now consider what would happen if we used a carboxylic acid as our nucleophile. These are generally less nucleophilic than the other species, so to give this the best shot of working, let's put the carboxylic acid in the same molecule as the halonium ion, so that our reaction will be intramolecular (remember that intramolecular processes, especially those leading to five- or six-membered rings, are very much faster than intermolecular processes). Let's consider how **11.5** would react with iodine (Figure 11.19). First, we make an iodonium ion and this is then attacked by the carboxylate group to give a five-ring lactone.

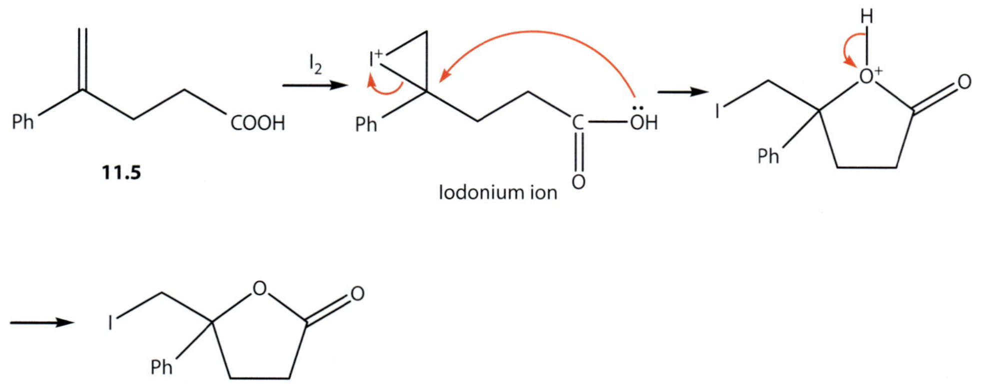

FIGURE 11.19 The mechanism of halolactonization.

You might reasonably wonder why the alternative six-ring lactone, **11.6**, was not formed by the attack of the carboxylic acid at the other end of the iodonium ion. I thought we knew that six-membered rings were more stable than five-membered ones? This is quite true—thermodynamically six-membered rings are more stable, and under thermodynamic conditions, they would predominate. But most halolactonizations proceed under conditions of kinetic control, where five-membered rings are formed more rapidly because the two reacting atoms are closer together.

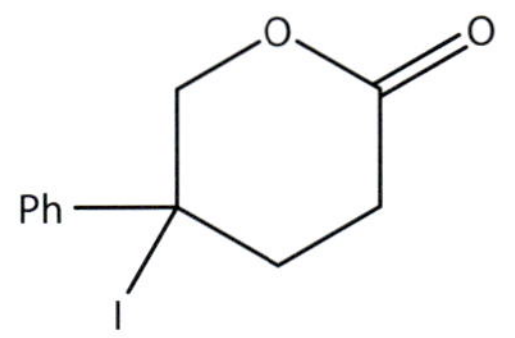

11.6

This powerful reaction has found extensive applications in synthesis. Diastereoselectivity is often very good (the oxygen of the carboxylate attacks the halonium ion from the back side, and the halogen, usually bromine, approaches the less hindered face of the molecule), and complex structures can be assembled relatively easily (Figure 11.20).

Figure 11.21 shows the use of halolactonization in the total synthesis of the antibiotic erythronolide B. In this case, only the formation of the six-ring lactone is possible—if the carboxylate attacked the other end of the bromonium ion, an unfavorable seven-membered ring would have been generated. Notice that for both the key reactions, the bromine and the oxygen of the lactone are on opposite sides of the original double bond.

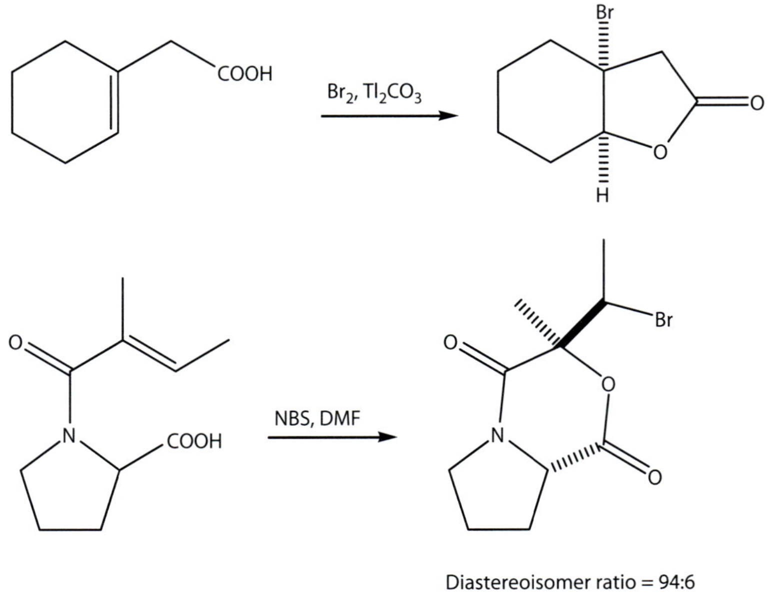

FIGURE 11.20 Halolactonization in synthesis. *Note:* Thallium carbonate is acting as a base to deprotonate the acid.

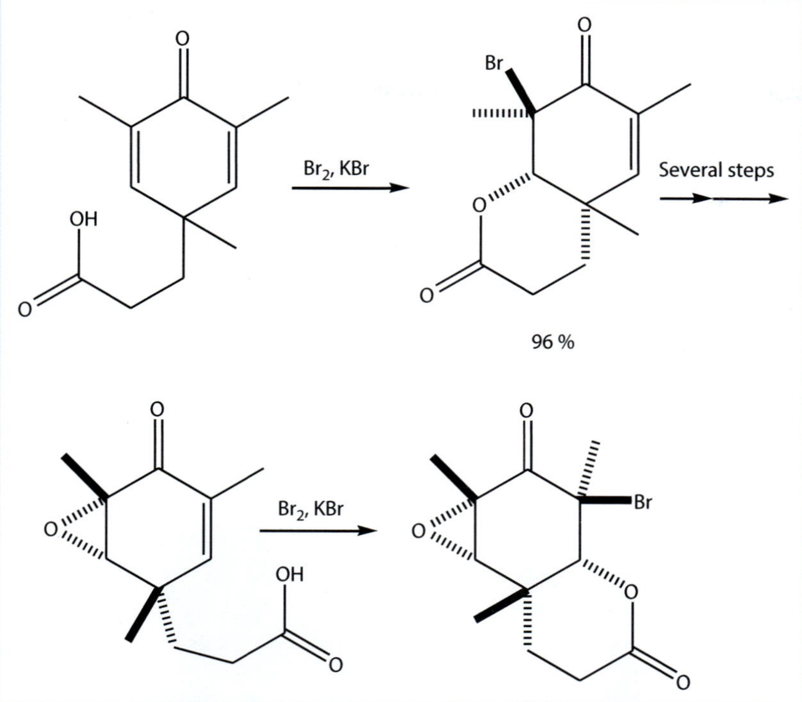

FIGURE 11.21 Halolactonization in the synthesis of erythronolide B.

are milder. The alkene is a soft nucleophile, and mercury is a soft electrophile, whereas the proton is a very hard electrophile. Figure 11.22 shows the mechanism of the reaction—again the intermediate has characteristics of both open and closed resonance forms. The cyclic form of the mercurinium ion tells us about the stereochemistry of ring opening, while the open form tells us what to expect in terms of regiochemistry. Another example is shown in Figure 11.23.

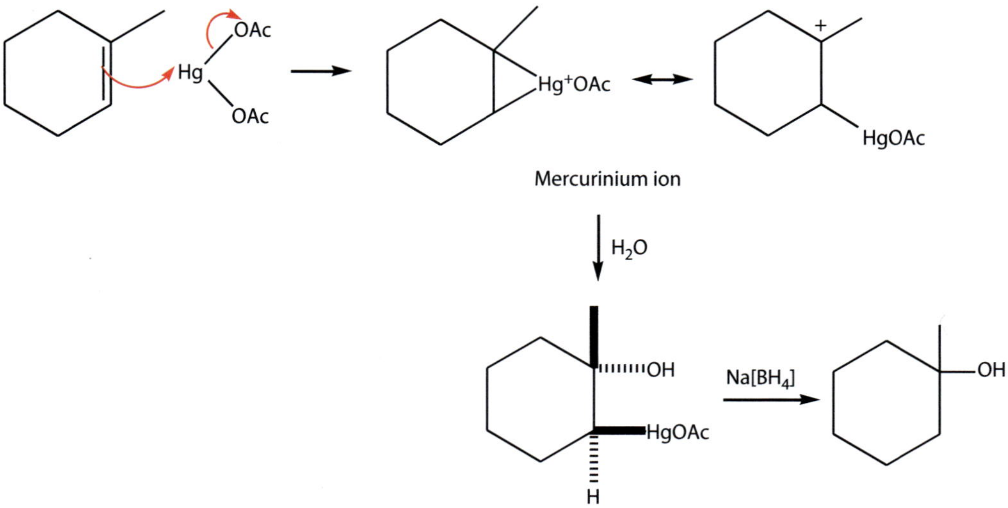

FIGURE 11.22 Mechanism of the oxymercuration reaction.

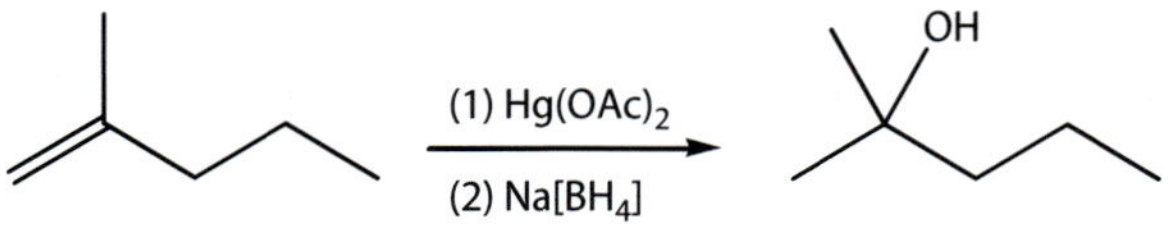

FIGURE 11.23 Oxymercuration in synthesis.

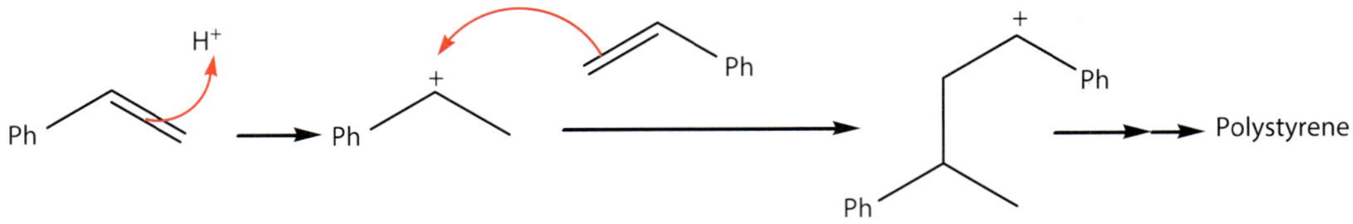

FIGURE 11.24 Cationic polymerization of styrene.

The final example of an electrophile that we will consider here is the carbocation, derived from the alkene itself. For example (Figure 11.24), styrene is readily protonated to give the secondary benzylic carbocation. If no other nucleophile is available, this may be attacked by the π-bond of another molecule of alkene. Continuing with this process ultimately gives a polyalkene, in this case, the familiar polystyrene. We will discuss this type of reaction in more detail in Chapter 21.

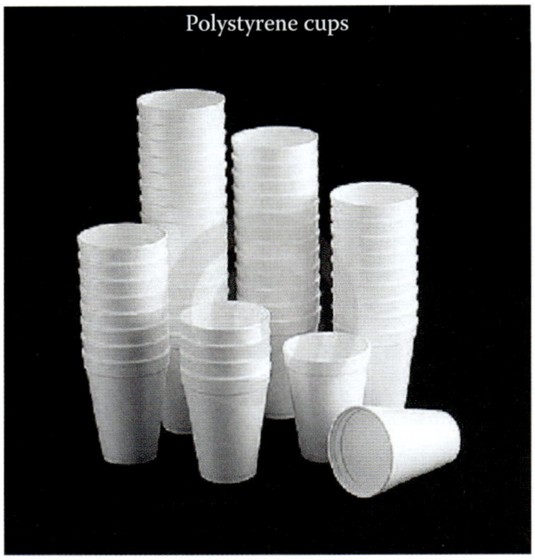

PROBLEM 11.5

Predict the product(s) of the following reactions:

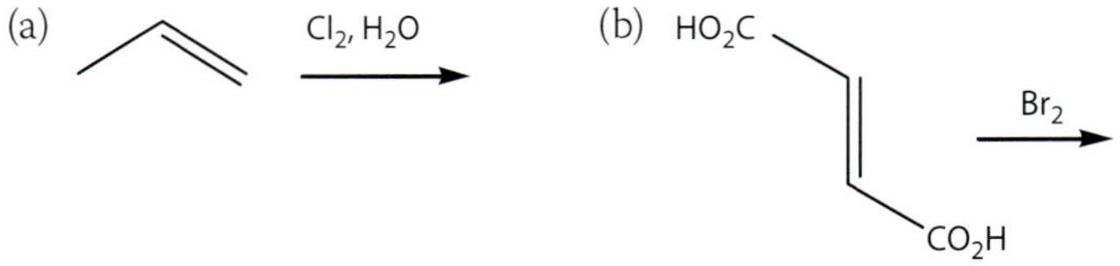

Solutions

(a) This is a relatively straightforward problem. We recognize that we will add one chlorine and one hydroxyl to the double bond, the only question being the regiochemistry (no stereochemistry to worry about). The first step will be to form a chloronium ion, where we must consider both open and closed forms:

Notice that I have only drawn the open form where the cation is secondary—the open primary cation will make no significant contribution to the structure. Now, it is obvious where the water should be added:

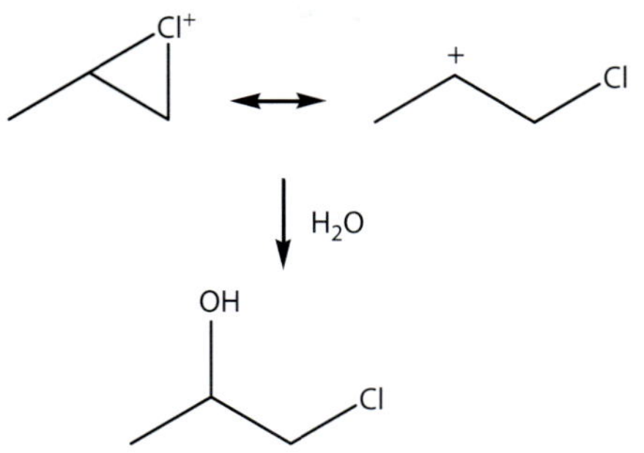

(b) This is a problem where stereochemistry will be an issue, and we will be able to see this most easily if we draw the molecule in a sawhorse form. The first step is the formation of the bromonium ion—we will first assume that bromine attacks the top face, though both are equally likely:

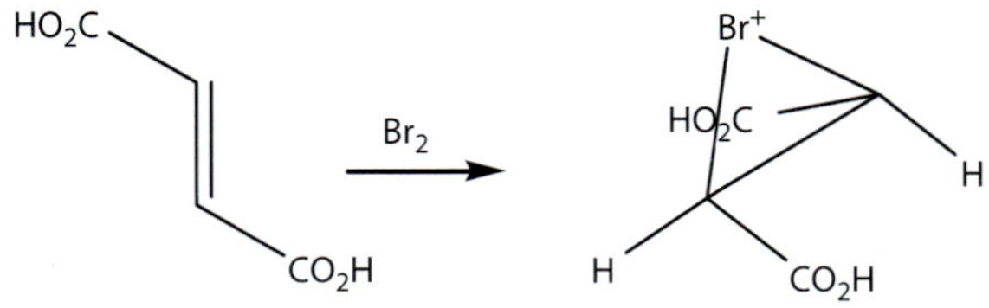

We could then attack the bromonium ion with bromide ion in an S_N2 process at either end:

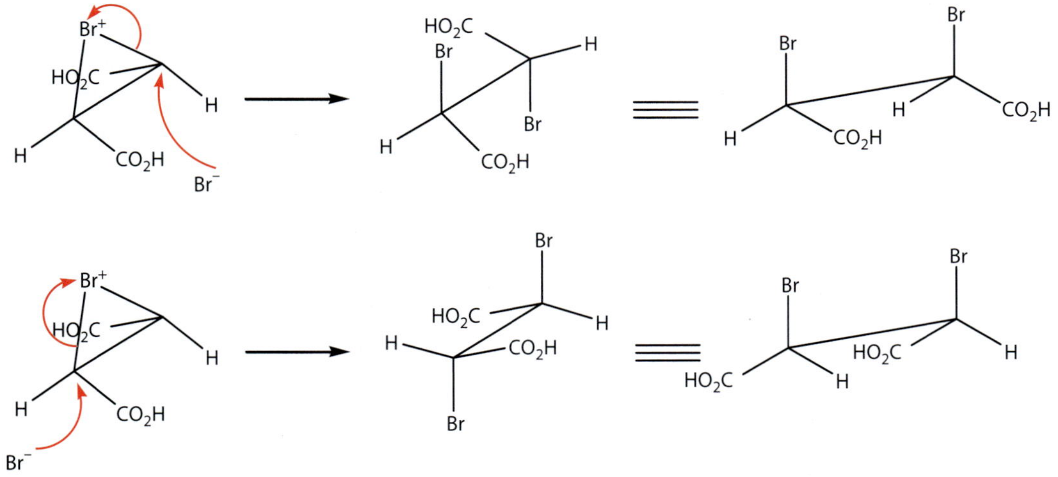

Although the two products come out drawn as mirror images, of course, they are not—they are identical. There is a mirror plane through the molecule, and this is a *meso*-compound. You should prove to yourself that exactly the same result is obtained if we initially attack the underneath face of the double bond with the bromine. Since the carbon–carbon double bond of fumaric acid is rather nonnucleophilic, there may be a radical component to this reaction, but as we shall see in a later section, the reaction will probably still be fairly stereoselective.

11.2.3 ELECTROPHILIC ADDITION TO CONJUGATED DIENES

This section only refers to molecules in which the double bonds alternate and hence interact through the extended molecular orbitals that we saw in Chapter 3. When there are two double

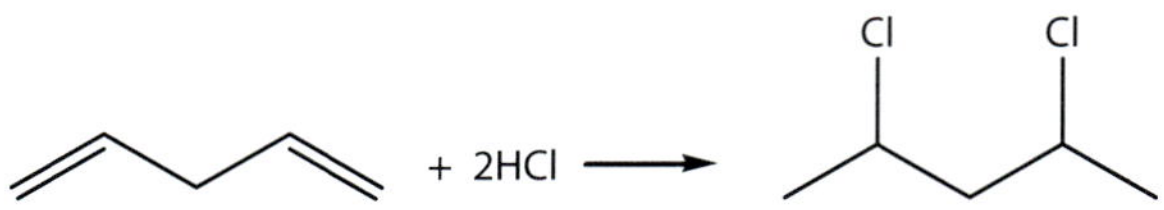

FIGURE 11.25 Addition of chlorine to nonconjugated double bonds.

bonds in a molecule that are not conjugated, they undergo electrophilic addition just as any isolated double bond would—effectively they ignore one another (Figure 11.25).

Let's think about what will happen if we protonate butadiene. We will prefer to make the allylic cation, **11.7**, rather than the isolated primary cation **11.8**. **11.7** has two resonance forms; the secondary cation is the more important contributor to the overall structure. So in adding electrophiles to dienes, we might expect that most of the product would come from the secondary form of the allylic cation.

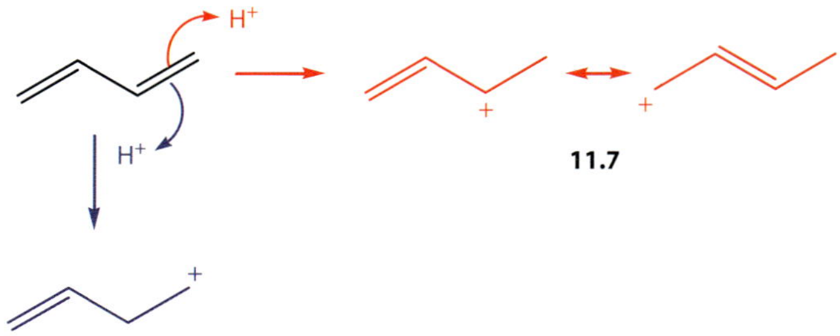

11.7

11.8

If we study the reaction of HBr with butadiene at low temperature (Figure 11.26), this is indeed what happens. 1,2-Addition predominates over 1,4-addition to the extent of 80:20. What we are seeing here reflects the relative importance of the two resonance forms; the temperature is low, and these are the products of kinetic control. At higher temperatures, thermodynamic control may predominate leading to the formation of more 1,4-addition product, as the more substituted alkene formed is more stable. So at +40 °C, 1,4-addition predominates to the extent of 85:15. Also, if we take either of the purified products at +40 °C, we get the same 85:15 mixture of isomers; clearly, this is the position of the thermodynamic equilibrium. Irrespective of temperature, the 1,2-addition product is formed more rapidly, but at low temperature, this is effectively irreversible—insufficient energy is available for the back reaction needed to reform the allylic cation.

The terms 1,2- and 1,4-addition can sometimes seem confusing. They refer to the positions on the original diene in which the new atoms are added. They do not refer to the numbering of the whole molecule, only to the diene fragment that we are reacting. So if we add HBr to 2,4-hexadiene, 1,2- and 1,4-addition products are obtained (Figure 11.27), but this nomenclature does not refer to the final positions on the carbon chain of the added atoms in conventional IUPAC nomenclature.

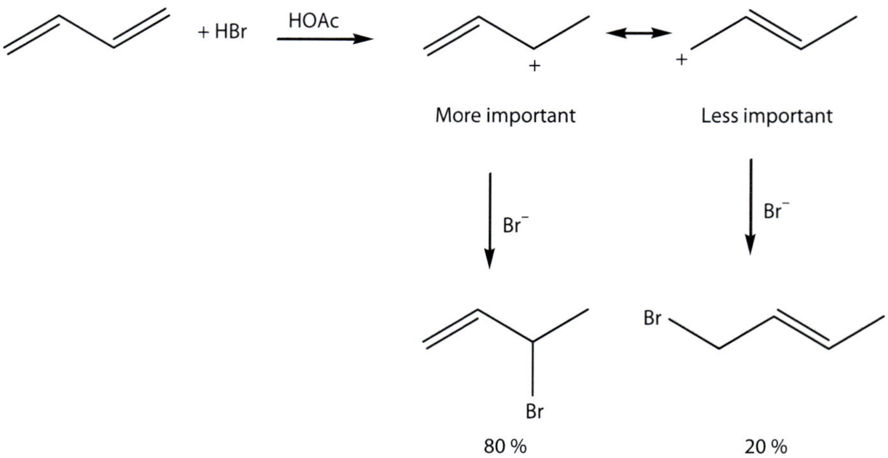

FIGURE 11.26 Addition of HBr to butadiene at low temperature.

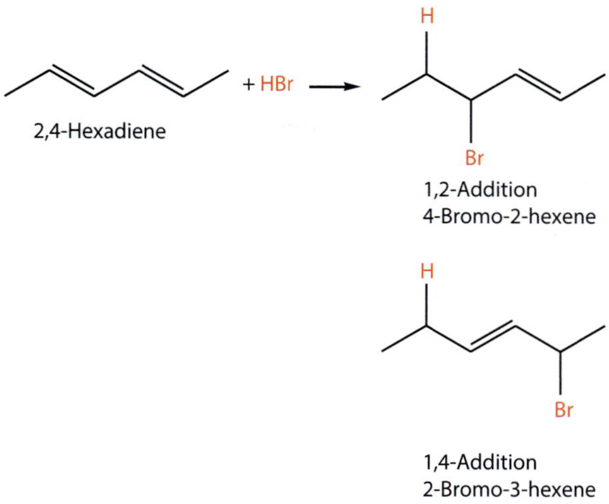

FIGURE 11.27 Addition of HBr to 2,4-hexadiene.

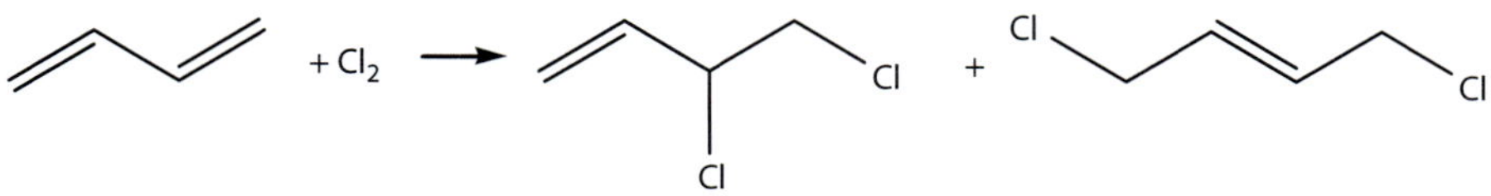

FIGURE 11.28 Addition of chlorine to butadiene.

Halogens also add readily to dienes to give a mixture of 1,2- and 1,4-addition products (Figure 11.28). With butadiene under conditions of kinetic control, the 1,2-addition product predominates, with more of the 1,4-product being formed at higher temperature.

PROBLEM 11.6

Under conditions of thermodynamic control, 1,4-dimethylcyclohexadiene adds bromine to give a mixture of the products shown, with the 1,2-addition product predominating. Explain this observation.

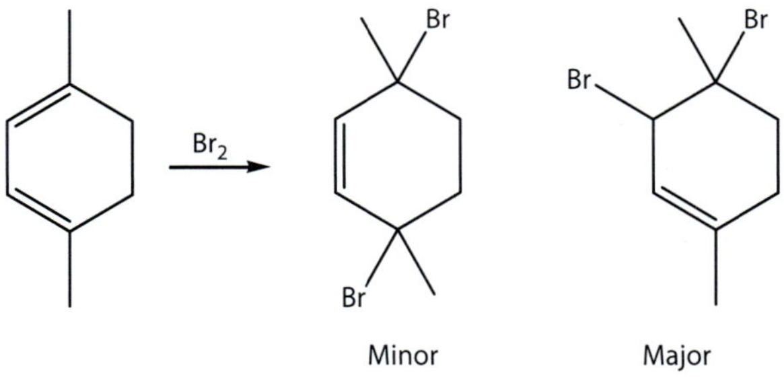

Solution

Our first step must be to react bromine with the diene to form the allyl cation (because the allyl cation is very well stabilized by resonance, the formation of bromonium ions is not favored). As usual, we must consider both resonance forms, and the tertiary allyl cation will contribute more to the overall structure.

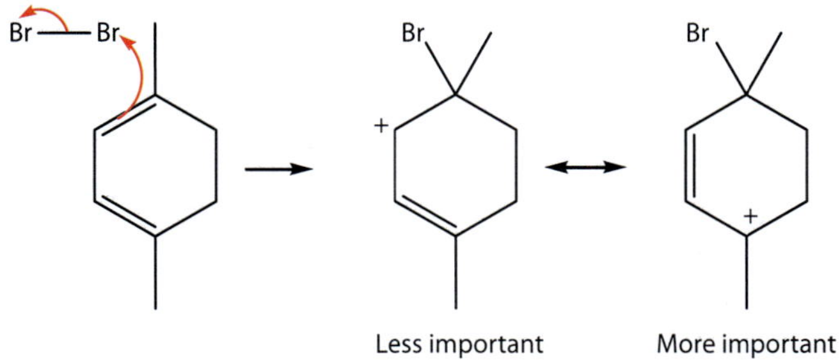

The products of reactions under kinetic control derive from the more important resonance form, so at low temperature, we would expect more of the 1,4-addition product. However, trisubstituted alkenes are more stable than disubstituted ones, so at higher temperature, the 1,2-addition product predominates. This problem points up the pitfalls of simply memorizing that butadiene gives mainly 1,2-addition at low temperature and 1,4-addition at high temperature. This will indeed work for butadiene, but with more complex structures, we may get quite a different (though still entirely predictable) result.

11.2.4 TERPENES: BIOLOGY OLIGOMERIZES DIENES

The class of biomolecules called terpenes comprises a vast range of different types of molecules ranging from mammalian sex hormones to plant essential oils. All of them are formed biosynthetically from one basic building block, isopentenyl diphosphate,* **11.9**, and they are oligomers of 2-methylbutadiene, commonly known as isoprene (**11.10**). An early name for them was isoprenoids.

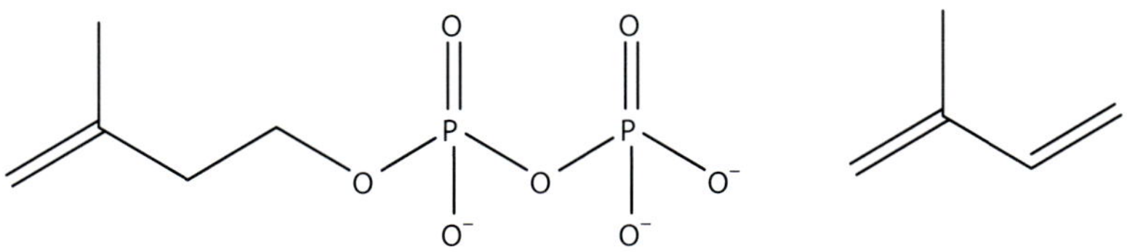

11.9, isopentenyl diphosphate **11.10**, isoprene

All terpenes are composed of C_5 units, but these can be joined together, and functionalized, in many different ways. Head to tail coupling, as in **11.11**, is the commonest motif, but all variants are known.

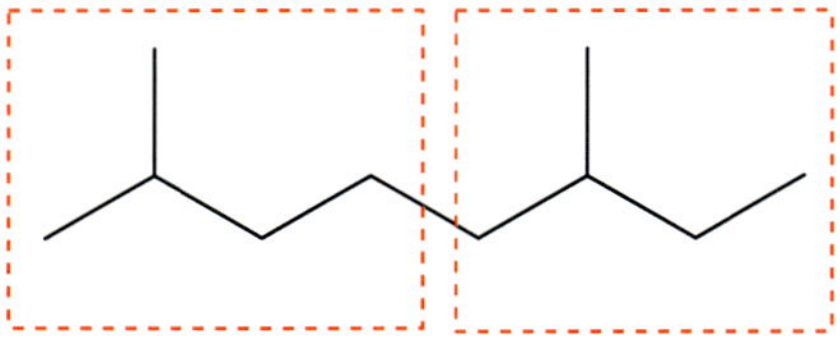

11.11

Monoterpenes contain 2 isoprene units, 10 carbon atoms. Examples include the essential oils citral (lemon grass oil) and menthol (peppermint oil) and grandisol, the sex attractant of the cotton boll weevil (Figure 11.29).

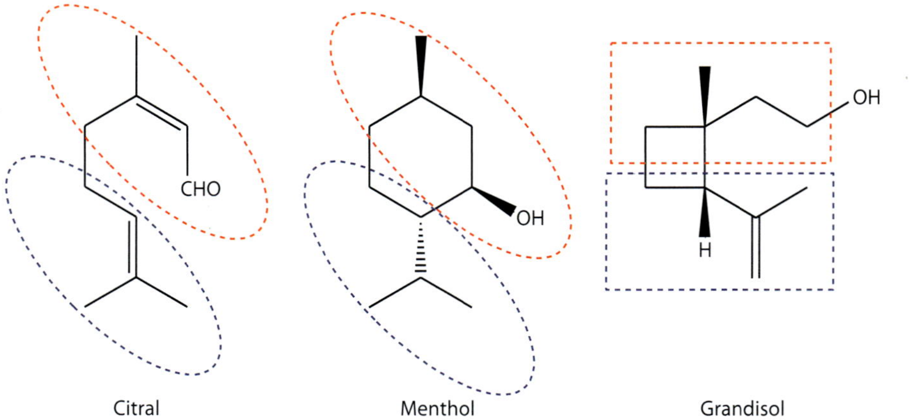

FIGURE 11.29 Structures of some common monoterpenes.

* Older texts will describe this as a pyrophosphate.

Peppermint leaves—The natural source of menthol

Sesquiterpenes contain 15 carbon atoms, 3 isoprene units, diterpenes 20 carbons, and triterpenes 30 carbons. Some examples are shown in Figure 11.30. Humulene, a sesquiterpene, is isolated from hops and gives flavor to beer. Retinol, vitamin A, is a diterpene, important in vision and bone growth. In lanosterol, a triterpene, some of the isoprene units can be easily recognized, but not all of them can be identified; during the biosynthesis, some migrations of methyl groups have occurred. Lanosterol is the biological precursor of all the steroids (Section 16.2.5). All steroids have this four-ring structure (**11.12**), and they include examples of sex hormones and contraceptives.

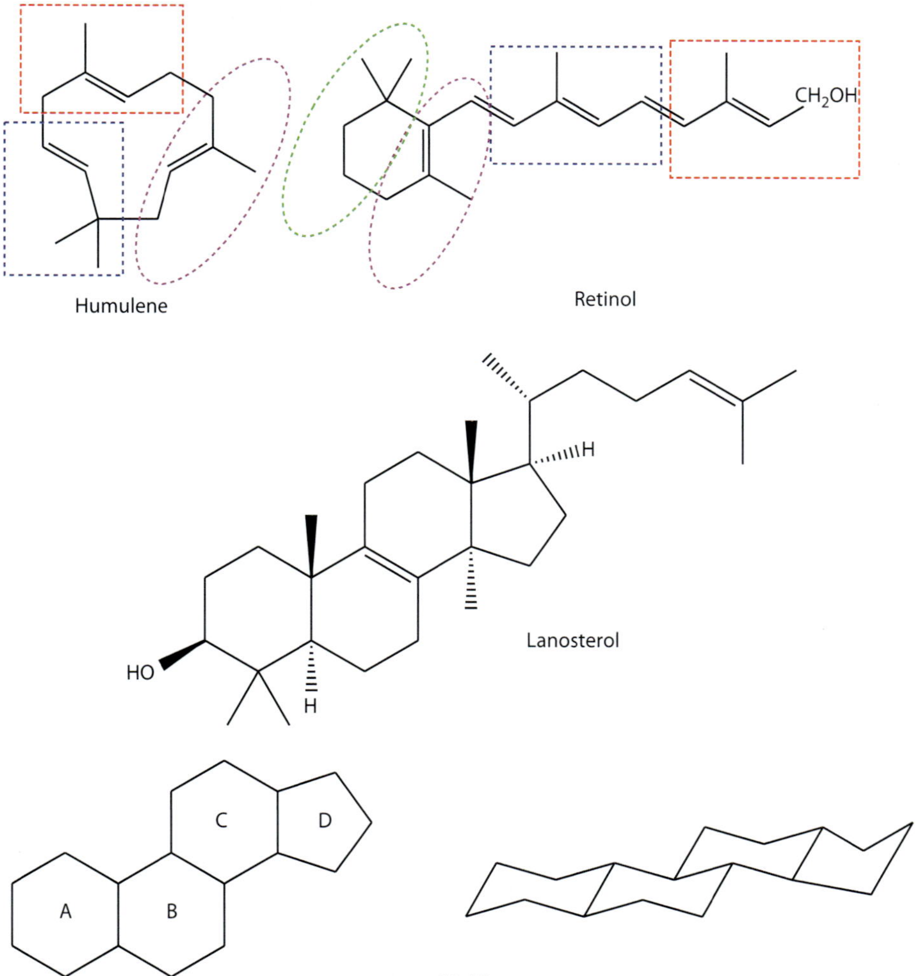

11.12

FIGURE 11.30 Sesquiterpenes, diterpenes, and triterpenes.

As we noted, the key building block for terpenes is isopentenyl diphosphate. The phosphate is an excellent leaving group, as it is the anion of a strong acid (pK_a of phosphoric acid is approx. 2), comparable to a sulfate or a sulfonate. Isopentenyl diphosphate is in equilibrium with dimethylallyl diphosphate. This is an acid-catalyzed reaction resulting from protonation of the double bond, followed by proton loss (Figure 11.31). These two species can couple together in various ways; the formation of geranyl diphosphate is one of the simplest. The alcohol corresponding to geranyl diphosphate is called geraniol; it smells of roses and is important in the perfume industry.

Geraniol may be isolated from palmarosa oil

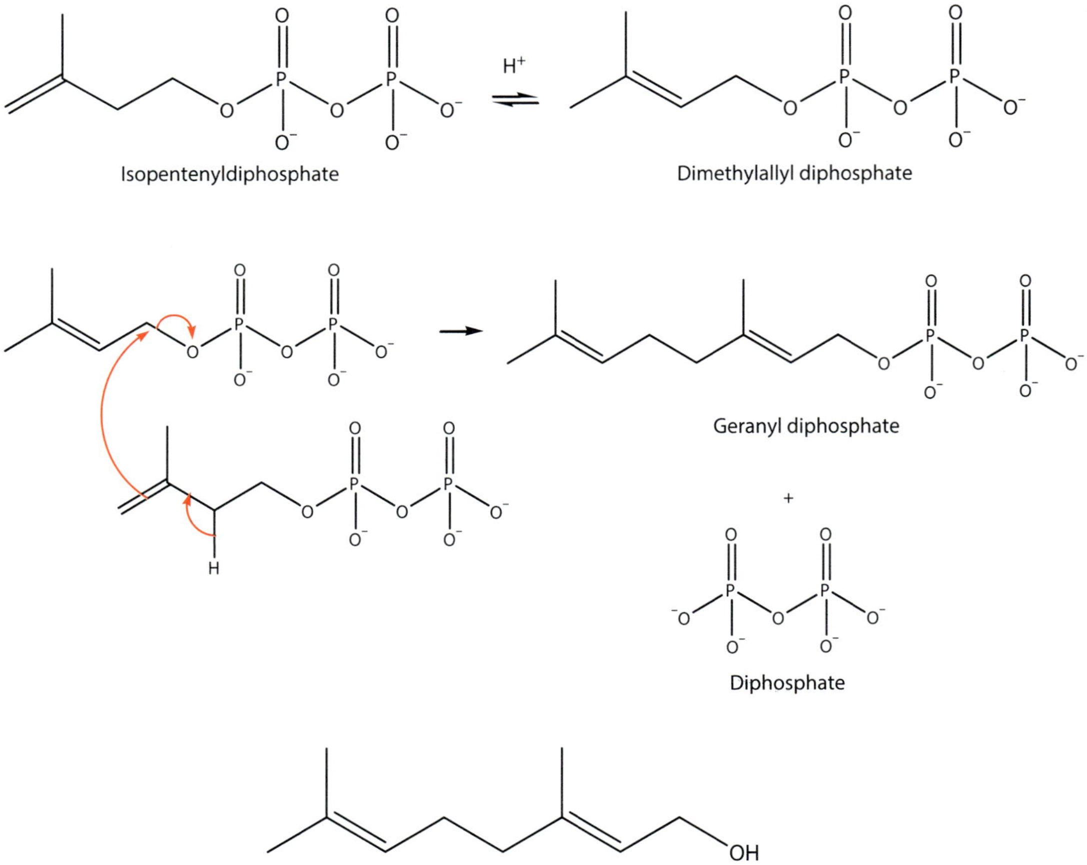

Geraniol, isolated from rose oil, palmarosa oil, and geraniums

FIGURE 11.31 Formation of geranyl diphosphate.

PROBLEM 11.7

Identify the isoprene units in each of the following molecules and classify them as monoterpenes, diterpenes, *etc.*

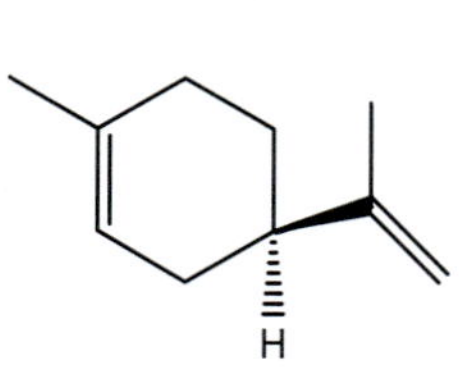

Limonene, from lemons

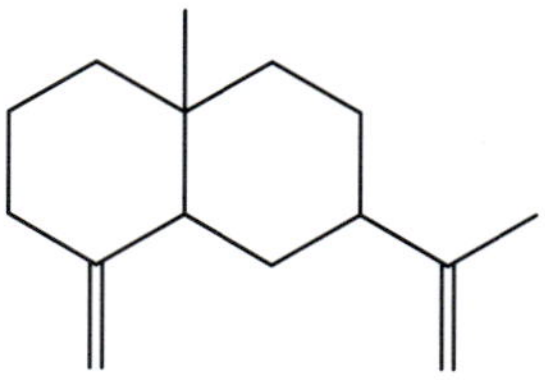

β-Selinene, oil of celery

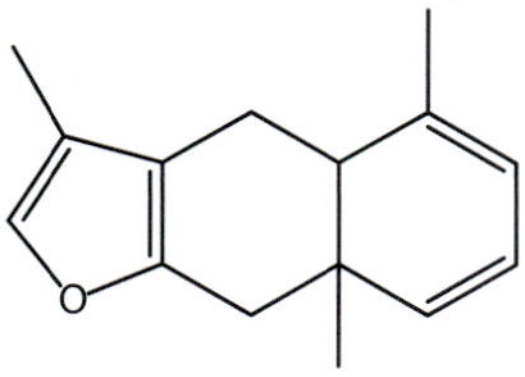

Furanoeudesma-1, 3-diene, isolated from myrrh

Myrrh tree

Solution

With this type of problem, it's best to start at one "end" of the molecule, where you can clearly identify a terminal branch point. Thus, for limonene, the first four atoms of the isoprene unit are obvious, marked in the following with a red x.

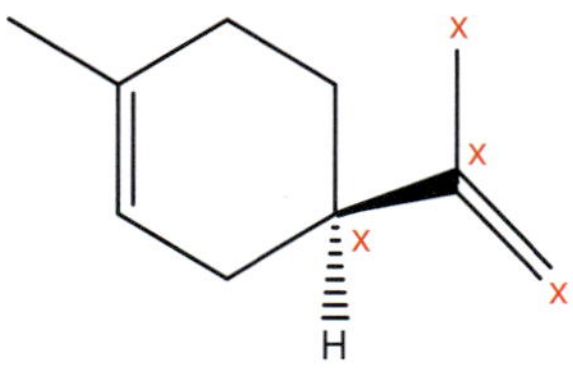

But which way around the ring do we then go? This example is fortunately straightforward—whichever way we go, we will still have a proper isoprene unit remaining. In terms of the biosynthesis, there is a "correct" solution, but we can't discern that without a lot more detail. So we have two possible solutions, and the molecule would be classed as a monoterpene.

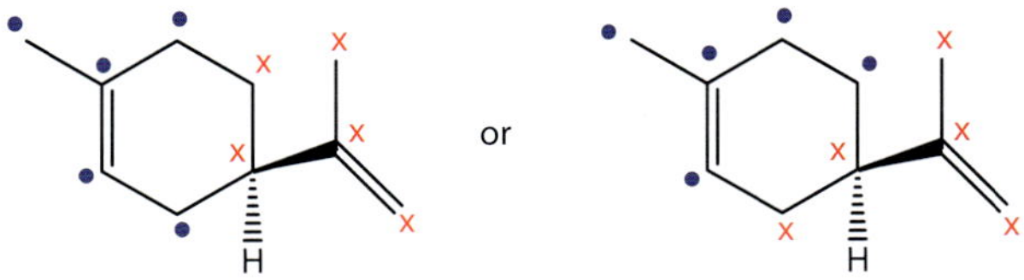

If we adopt the same strategy with β-selinene, we get the same dilemma—which way do we go around the ring?

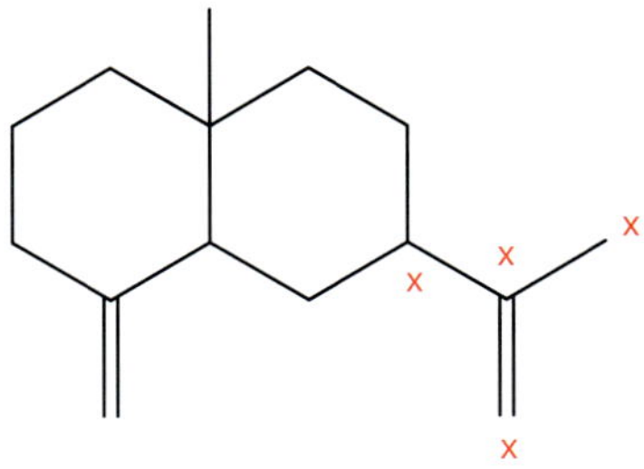

Let's try going "upward," then try to identify the next isoprene unit:

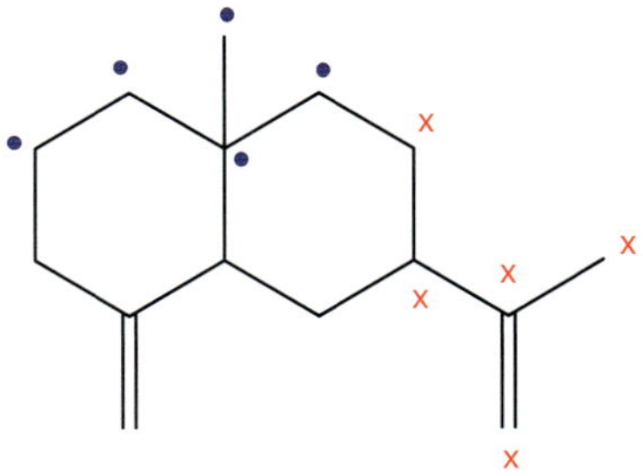

This looks fine, and what we have left is also a proper isoprene unit:

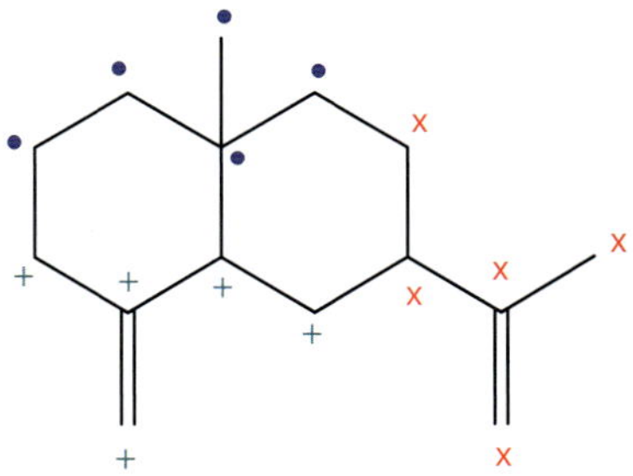

If we had gone the other way, in this case, it would also have worked out OK, but that's not always so. Do be wary of this, and check carefully that you have no five-carbon unit that is linear rather than branched. This has 15 carbons and is a sesquiterpene.

With the myrrh oil, the "end" is a little less easy to see, but the oxygen provides a clue. Again, identifying the first four atoms is easy, but which way should we go next?

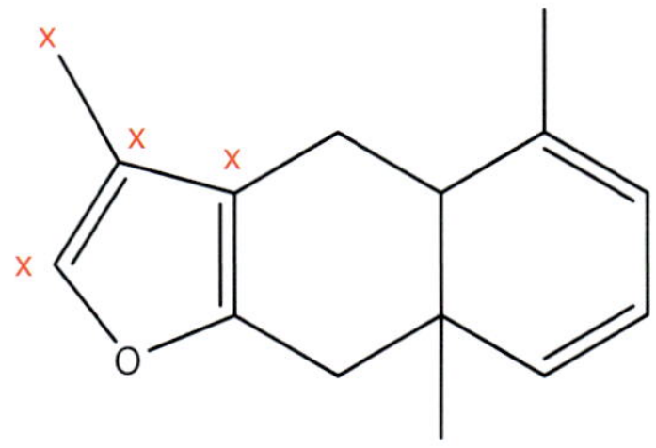

If we go further round the furan ring, we have two options to put in the next isoprene unit, **11.13a** and **11.13b**. Both have three five-carbon units, but **11.13a** is NOT a valid answer. One of the five-carbon units is linear, and so could not have come from isoprene. **11.13b** is fine, and this is a sesquiterpene.

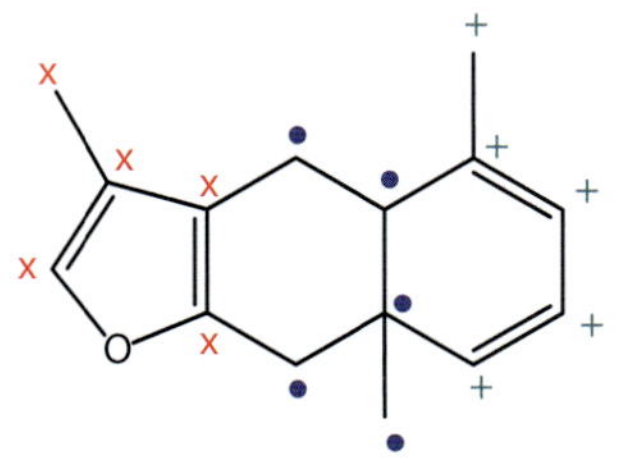

11.13a

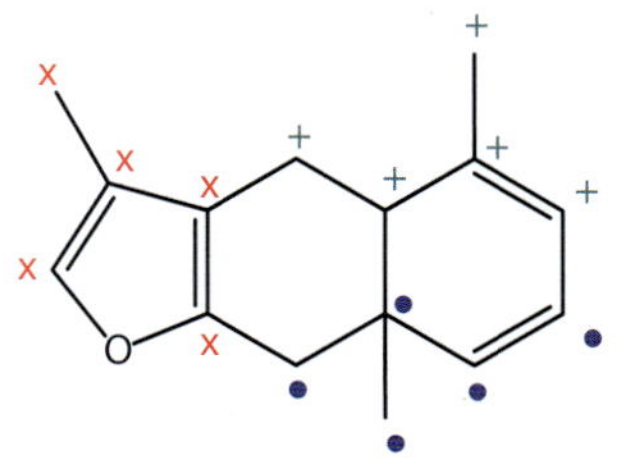

11.13b

If we had headed in the other direction for the fifth carbon of the first unit, we also get two possible options, **11.14a** and **11.14b**. This time, only **11.14a** is the only valid solution; **11.14b** has a linear five-carbon unit. You should note that the "real" answer to these problems with two apparent possibilities has generally been established by biosynthetic studies—but the structure itself does not allow you to do more than guess.

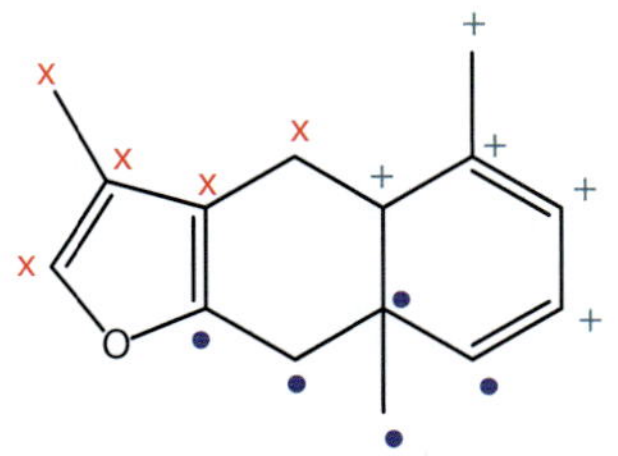

11.14a

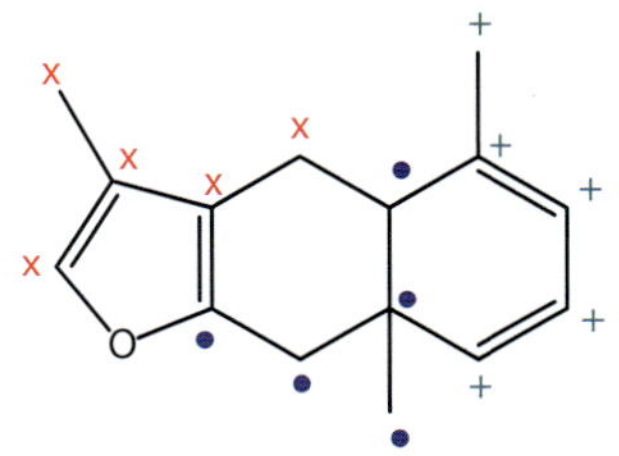

11.14b

11.2.5 ELECTROPHILIC ADDITION TO ALKYNES

If we consider the protonation of 2-butyne by HBr, then we obtain an unstable vinyl cation, **11.15**. This is linear because the carbon atom is sp-hybridized, with an empty p-orbital. Thus, it can be attacked by bromide ion from either side, resulting in the formation of both *E*- and *Z*-alkene products (Figure 11.32). The instability of the vinyl cation intermediate means that alkynes react more slowly with electrophiles than do alkenes. The addition of a second mole of HBr to the alkene generally gives the geminal dibromide; the intermediate carbocation is stabilized by interaction with the lone pair of electrons from bromine (Figure 11.33).

The addition of hydrogen halides to nonsymmetric alkynes follows the Markovnikov rule; vinyl cations are also stabilized by electron-donating alkyl groups (Figure 11.34). The addition of

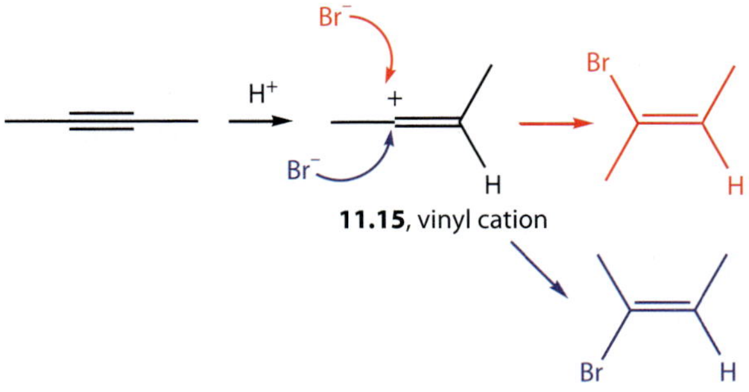

FIGURE 11.32 Mechanism of HBr addition to an alkyne.

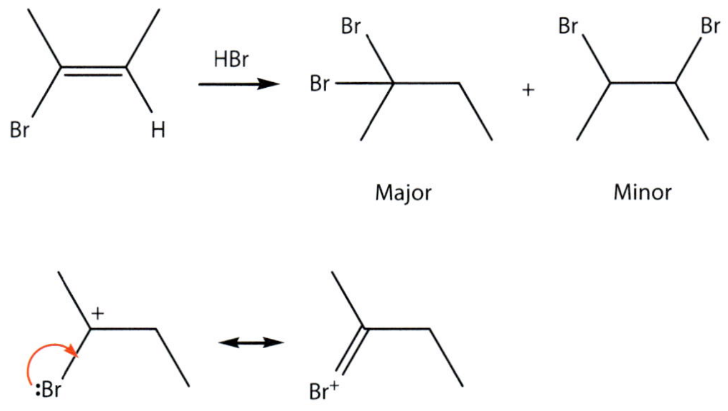

FIGURE 11.33 Addition of HBr to a bromoalkene.

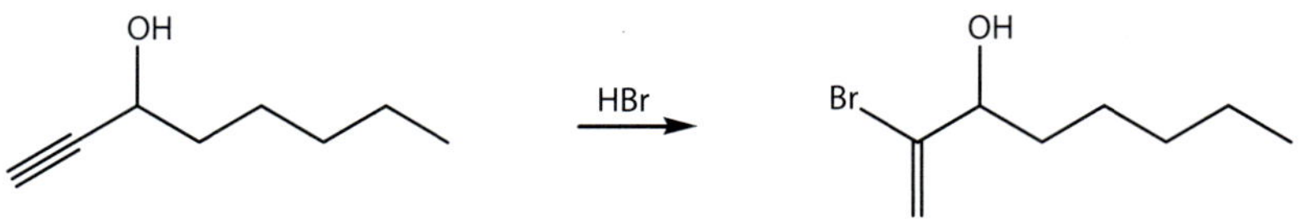

FIGURE 11.34 Markovnikov rule in addition of HBr to an alkyne.

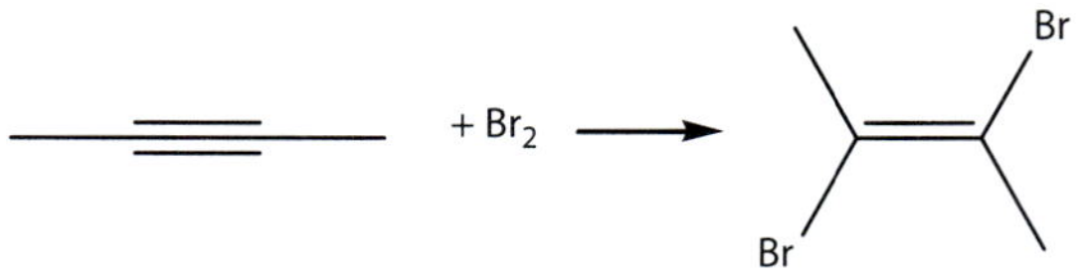

FIGURE 11.35 Addition of bromine to 2-butyne.

FIGURE 11.36 Alkenes are more reactive than alkynes in addition reactions.

bromine, chlorine, or iodine to alkynes gives mainly anti-addition *via* a halonium ion; bromine and iodine reactions are more stereospecific than that with chlorine (Figure 11.35).

Selectivity between alkenes and alkynes is exemplified by the reaction of 1-pentene-4-yne to give mainly 4,5-dibromo-1-pentyne (Figure 11.36). This selectivity derives from the poor stability of vinyl cations.

One of the most useful addition reactions of alkynes is the addition of water. The product is an enol, but, as we already know, in most cases, the equilibrium between ketones and enols is toward the ketone. Thus, we see reactions such as that in Figure 11.37. The symmetry of the first example means that a single product is obtained—it does not matter which way round the water adds, only one product is possible. With less symmetric alkynes, two products are obtained in approximately equal amounts.

The reaction mechanism (Figure 11.38) involves the formation of a mercurinium ion, which is opened by water at the more hindered carbon (the carbon that is most "comfortable" bearing positive charge) to give an enol. Tautomerization and mercury loss give the final product. The reaction is particularly useful for terminal alkynes, which are converted to methyl ketones.

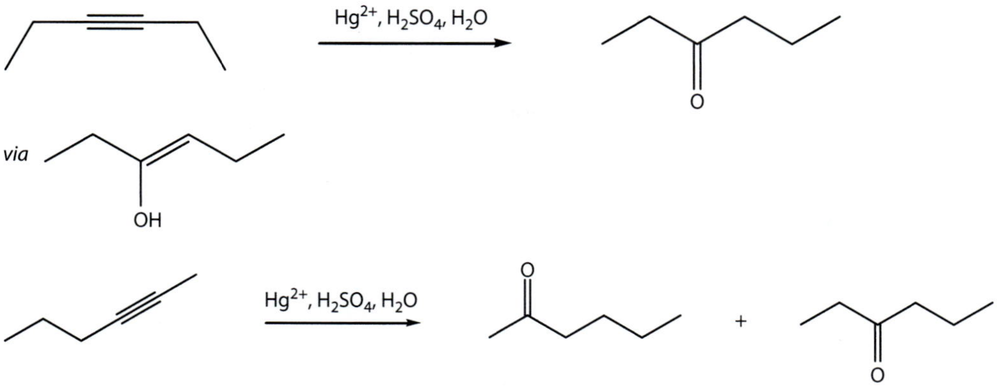

FIGURE 11.37 Addition of water to alkynes.

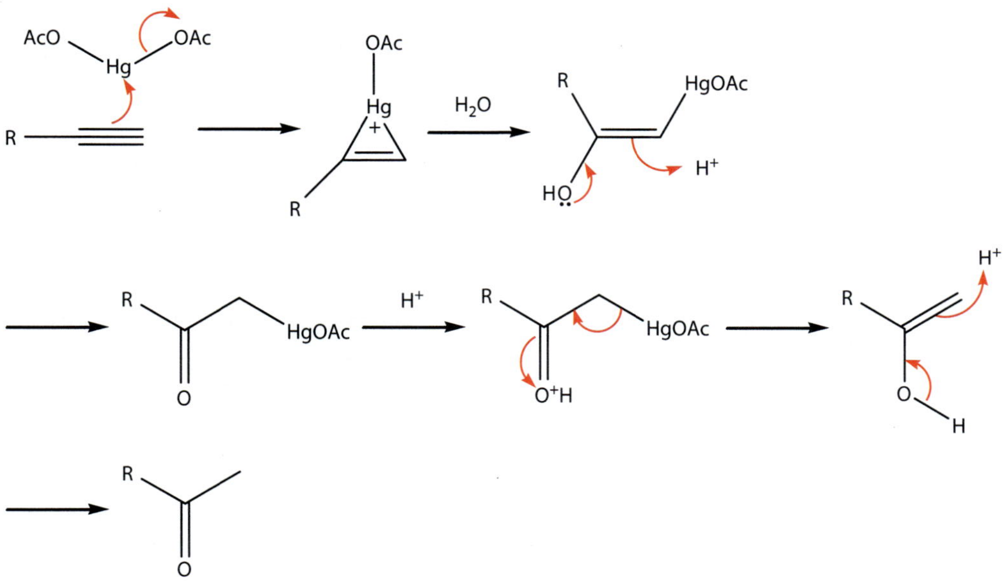

FIGURE 11.38 Mechanism of mercury-catalyzed addition of water to an alkyne.

PROBLEM 11.8

Predict the product(s) of the following reactions.

(a)

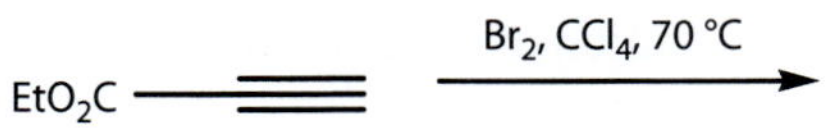

(b)

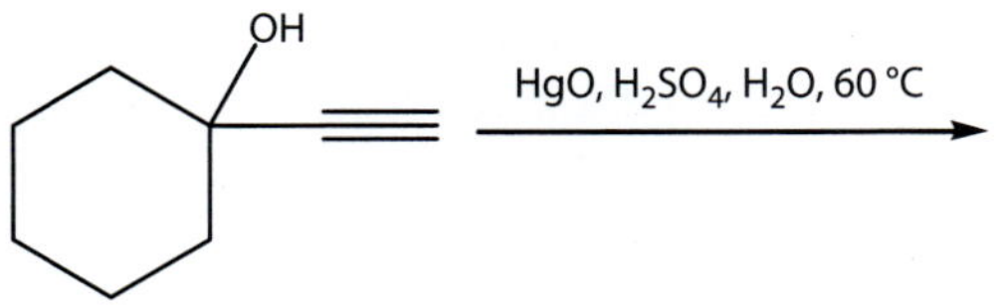

Solutions

(a) This is an addition reaction, and our product will be the dibromoalkene. Although conventionally the *trans*-bromide should predominate, in this particular case, steric considerations led to the formation of a predominant *cis*-dibromide.

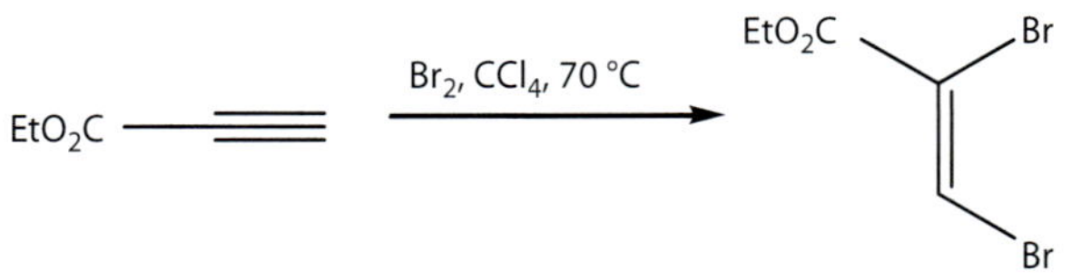

(b) This is an addition of water to the triple bond. It goes *via* the more substituted vinyl cation to give the enol, which tautomerizes to the ketone.

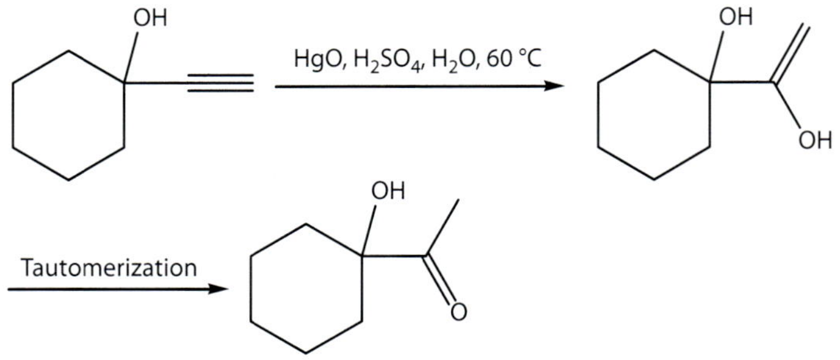

Key Points from Sections 11.1 and 11.2

- Alkenes add electrophiles readily. Most reactions proceed *via* carbocations, and initial addition of the electrophile gives rise to the most stable carbocation (Markovnikov rule).
- The addition of hydrogen halides, and water, proceeds by this mechanism. These reactions are almost invariably regiospecific, but not stereospecific.
- The intermediate carbocation may be captured by other available nucleophiles and may rearrange to give a more stable carbocation, prior to capture by a nucleophile.
- The addition of halogens, most particularly bromine, is stereospecifically *trans*. The reaction of the alkene with the halogen gives a halonium ion, which is opened by the halide ion in an S_N2 process. The halonium ion can also be opened by other competing nucleophiles.
- Electrophilic addition to conjugated dienes proceeds *via* allyl cations. Under conditions of kinetic control, the outcome is determined by the relative stabilities of the resonance forms of the allyl cation. At high temperature, under conditions of thermodynamic control, the most stable alkene is favored.
- The natural products, the terpenes, are formed by the coupling of the five-carbon-unit, isoprene.
- Electrophilic addition to alkynes is slower than to alkenes but follows the same regiochemistry according to the Markovnikov rule. The addition of water is particularly useful, as the enol product formed rapidly tautomerizes to a ketone.

11.3 RADICAL ADDITION REACTIONS

Early investigations of the addition of HBr to alkenes tended to give some of the anti-Markovnikov as well as the expected Markovnikov products. Amounts were variable and depended on the reaction conditions and, in some cases, the care of the experimenter. The cause of this was that atmospheric oxygen was reacting with the HBr to give bromine radicals—and these reacted with alkenes by a different mechanism, and with a different regiochemistry from that for HBr in polar

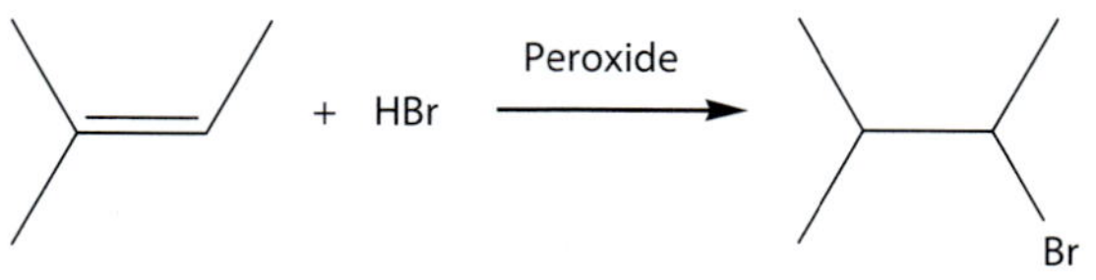

FIGURE 11.39 Radical addition of HBr to 2-methyl-2-butene.

solvents, where the reaction is initiated by H^+. If we deliberately encourage a radical reaction, by adding peroxide, the anti-Markovnikov product is obtained regiospecifically (Figure 11.39).

Although molecular oxygen can initiate radical reactions, more reliable results are obtained with dibenzoyl peroxide, **11.16** (yes, this has a use other than the treatment of acne). On heating to 90 °C–100 °C, this decomposes homolytically to give radicals. This oxygen–oxygen single bond, like all similar ones, is relatively weak (approx. 200 kJ mol^{-1}). The benzoate radical formed is unstable with respect to loss of carbon dioxide, and thus, the active initiating radical is often the phenyl radical, Ph$^•$. This reacts with HBr to abstract hydrogen and give a bromine radical. You might ask why it does not abstract bromine and give a hydrogen radical—the answer lies in the thermodynamics. The hydrogen radical is much less stable than the bromine radical, and the carbon–hydrogen bond (460 kJ mol^{-1}) is much stronger than the carbon–bromine bond (330 kJ mol^{-1}). Notice that in Figure 11.40, all of our arrows are single headed, as we are moving single electrons rather than electron pairs.

The bromine radical then adds to the double bond to give the more stable radical. Radical stability broadly follows carbocation stability, so the stable species is the tertiary rather than the secondary radical (Figure 11.41). This tertiary radical then abstracts a hydrogen from HBr to regenerate the bromine radical. We describe this as a radical chain reaction—once started, the reaction should continue without further initiator. The steps of Figure 11.40 are described

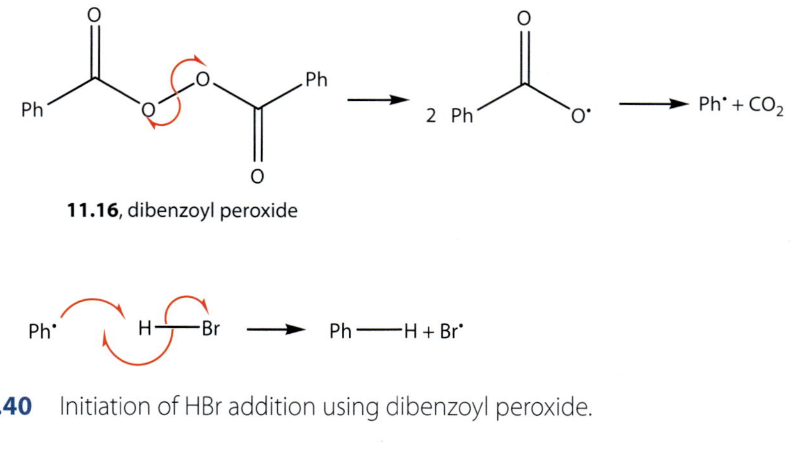

FIGURE 11.40 Initiation of HBr addition using dibenzoyl peroxide.

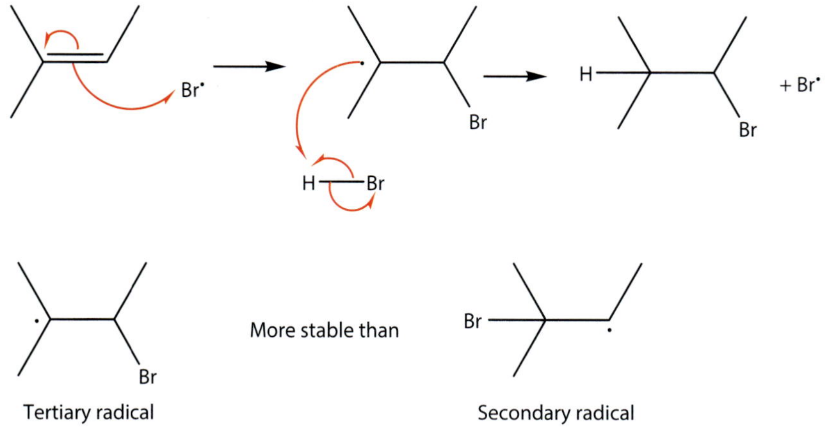

FIGURE 11.41 Radical addition of HBr to an alkene.

as initiation; in principle, we need to do them only once, to set the chain in motion. The steps of Figure 11.41 are described as propagation—a bromine radical goes in, and another is left at the end to repeat the cycle. In practice, radical chain reactions are not "immortal." A number of steps may lead to "termination" of the chain. Two radicals may encounter one another and recombine—for example, two bromine radicals can recombine to give a bromine molecule. Some radicals are lost by collision with the walls of the reaction vessel.

Of all the hydrogen halides, only HBr reacts efficiently in this way. In HF, the hydrogen–fluorine bond is too strong for the initial radical to be formed. With HCl, the second half of the propagation process is very slow; the reaction $CH_3CH^{\bullet}CH_2Cl$ + HCl to $CH_3CH_2CH_2Cl$ and chlorine radical is endothermic. With HI, the first propagation step is very slow; the addition of an iodine radical to an alkene is endothermic.

In radical additions of HBr, the reaction regiochemistry depends on the relative stability of two possible radicals. We remember that radical stability is modeled on carbocation stability, so 3 ° > 2 ° > 1 °. Benzylic and allylic radicals are stabilized by resonance. For example, radical addition of HBr to 1-phenylpropene proceeds *via* the stabilized benzylic radical (Figure 11.42). The reactions are generally stereoselective for *anti*-addition, but not stereospecific. Although the existence of bromonium radicals has been postulated, these are not comparable in stability to bromonium cations.

Unusual, but occasionally useful, reactions involve the addition of thiols or CX_4 (X = Cl, Br) under radical conditions. Examples are shown in Figure 11.43—try writing a mechanism for each of them.

Forming a radical from an alkene can also lead to polymerization (Figure 11.44); we will discuss this process in more detail in Chapter 21.

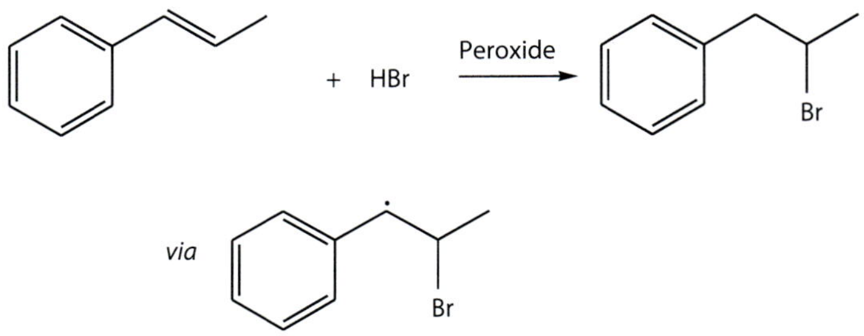

FIGURE 11.42 Addition of HBr to *Z*-1-phenyl-1-propene under radical conditions.

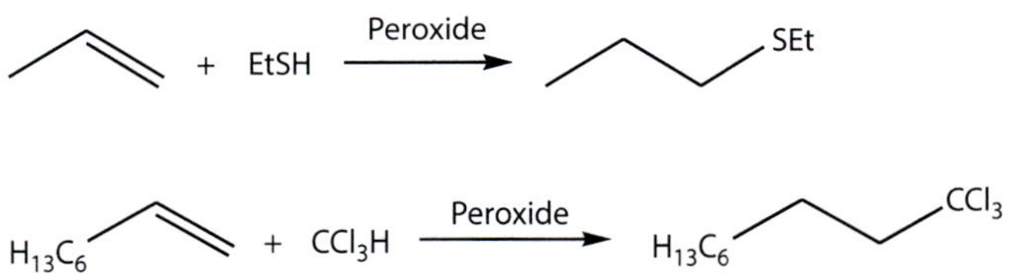

FIGURE 11.43 Other radical additions to alkenes.

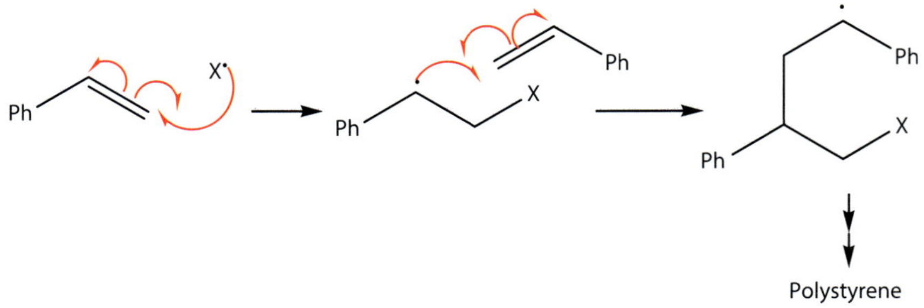

FIGURE 11.44 Radical-initiated polymerization of styrene.

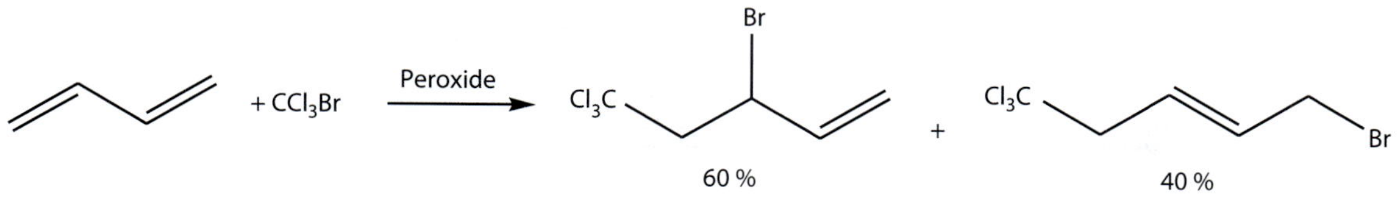

FIGURE 11.45 Radical addition to butadiene.

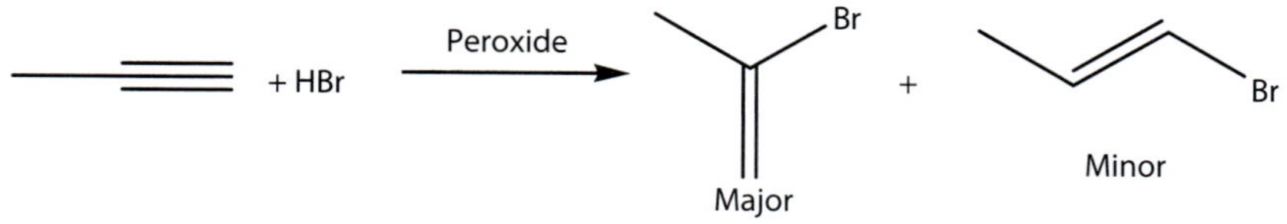

FIGURE 11.46 Radical addition to an alkyne.

Radicals can also be added to dienes; products are generally formed under conditions of kinetic control and reflect the relative stabilities of the resonance forms of the allylic radical. However, radicals are very reactive, and selectivity is often low (Figure 11.45). Radical addition of HBr to alkynes is known, but not widely used (Figure 11.46).

PROBLEM 11.9

Each of the following alkenes can add the thiophenyl radical, PhS$^{\bullet}$, to give a carbon radical product, for example:

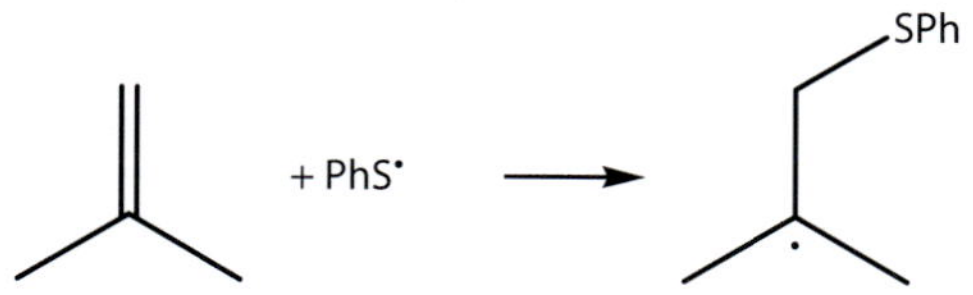

Which radical would be formed by adding PhS$^{\bullet}$ to each of the following?

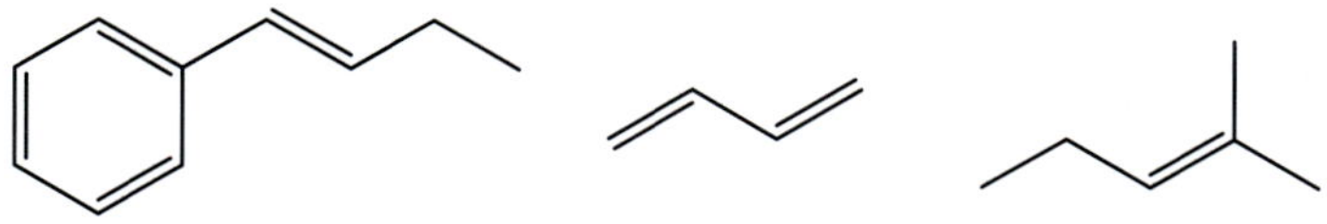

Solution

In the first example, both of the radicals we would obtain are secondary, but the one conjugated with the benzene ring is favored:

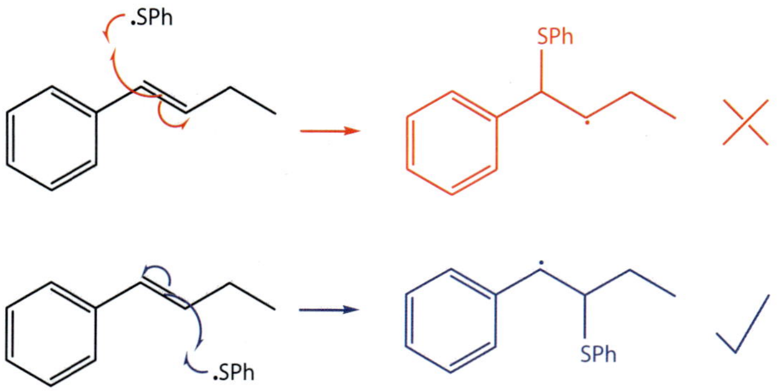

In the second example, we make the stabilized allylic radical, and in the third one, the favored radical is at the tertiary center.

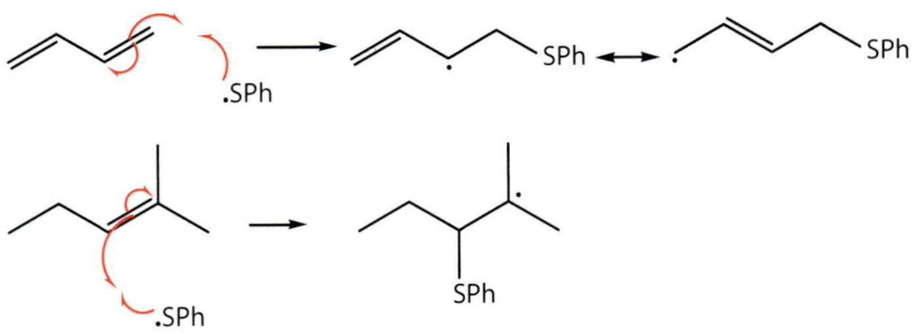

PROBLEM 11.10

Predict the product of the following reaction:

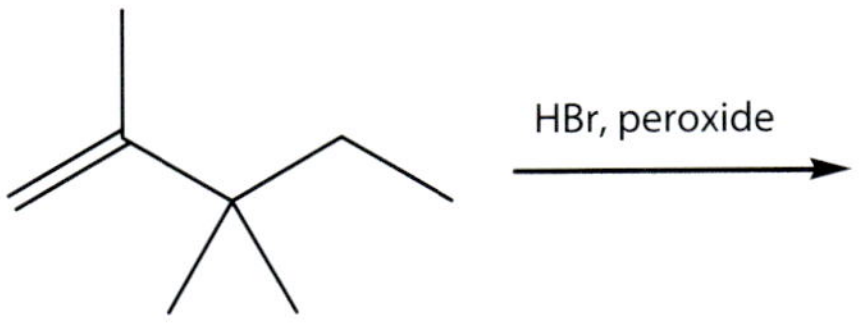

Solution

In the presence of peroxide, we expect anti-Markovnikov addition *via* the more stable tertiary radical, **11.17**.

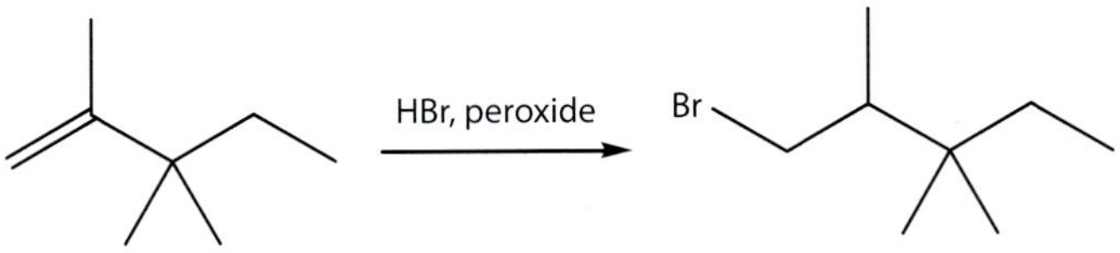

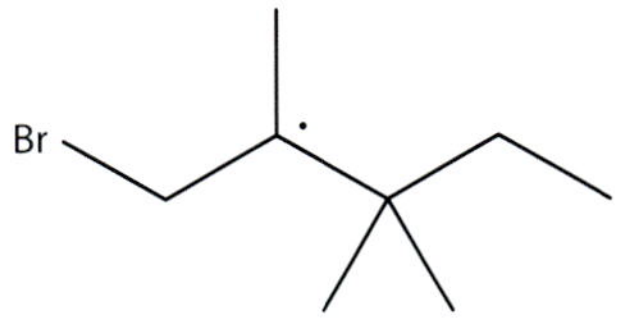

11.17

> **Key Points from Section 11.3**
>
> - In the presence of peroxides, HBr may be added to alkenes in an anti-Markovnikov fashion. The reaction involves a radical chain mechanism and proceeds *via* the most stabilized radical.
> - Comparable reactions of other hydrogen halides are rarely successful.

11.4 MULTICENTER PROCESSES

Although the reactions we shall study here are a diverse group, they have one common feature—two or more bonds are formed simultaneously (Figure 11.47). Overall, both single-step concerted and multistep processes may be involved.

FIGURE 11.47 Synchronous addition of A–B to an alkene.

11.4.1 EPOXIDATION USING PERACIDS

In this process, a single oxygen atom is added to a carbon–carbon double bond to make an epoxide. The oxygen comes from a type of compound we have not met before, a peracid. Percarboxylic acids can, in principle, be made from any carboxylic acid by replacing the –OH with –O–O–H. The driving force for the epoxidation reaction is the weakness of the oxygen–oxygen single bond of the peracid. The mechanism is shown in Figure 11.48. Although the new bonds are made synchronously, the easiest way to think about this reaction is to start with the curly arrow that indicates the use of the electrons of the alkene π-bond to make a new bond between one of the carbon atoms and the oxygen; this emphasizes that the alkene is the nucleophile and the peracid the electrophile. This has implications for the selectivity of the reaction, to which we will return in a moment. Our "next" curly arrow breaks the weak oxygen–oxygen bond, moving the electrons into the carbon–oxygen bond. The π-electrons of the carbonyl are then used to make a new O–H bond, with the electrons of the original O–H closing up the three-membered ring to complete the formation of the epoxide.

The most commonly used peracids are *meta*-chloroperbenzoic acid, MCPBA, **11.18**, and peracetic acid, $MeCO_3H$ (used as a 40 % solution in acetic acid). MCPBA was long considered to be the most convenient peracid, as it is a solid and reasonably stable. Recently, magnesium monoperoxyphthalate (MMPP), **11.19**, has come to be preferred over MCPBA; reactivity in epoxidation is comparable, MMPP is a crystalline solid, and it is more stable than MCPBA. Peracetic acid has to be stored at low temperature, and the peracid content is usually titrated before use, as this decays over time. It is, however, a more powerful oxidant than the aromatic peracids. Pertrifluoroacetic acid, CF_3CO_3H, is a still more powerful oxidant.

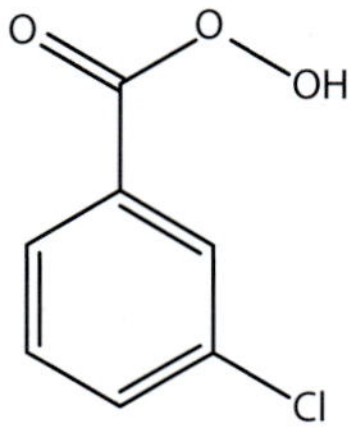

11.18, MCPBA

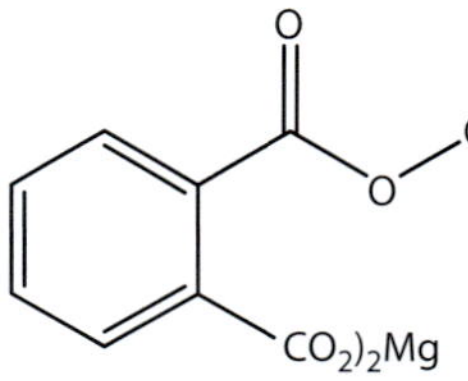

11.19, MMPP

As we noted in the mechanism, peracid epoxidation is an electrophilic process—the π-electrons of the double bond attack the oxygen, breaking the O–O bond. This means that more electron-rich double bonds react more rapidly. In the reaction of **11.20**, the tetrasubstituted double bond reacts faster than the disubstituted ones, giving useful selectivity.

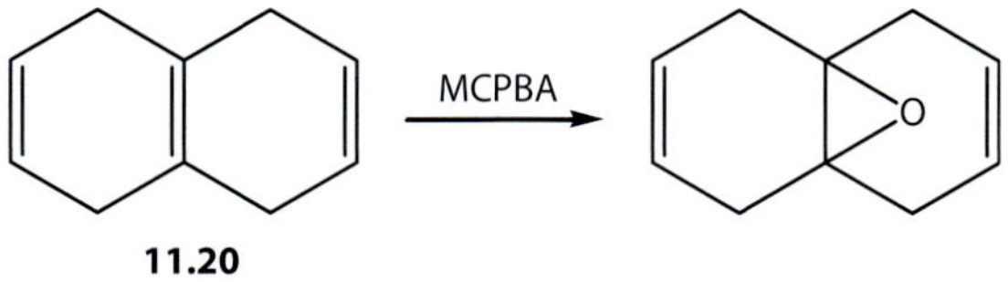

11.20

Most reactions are stereospecific, with the geometry of the original double bond being retained; some examples are shown in Figure 11.49.

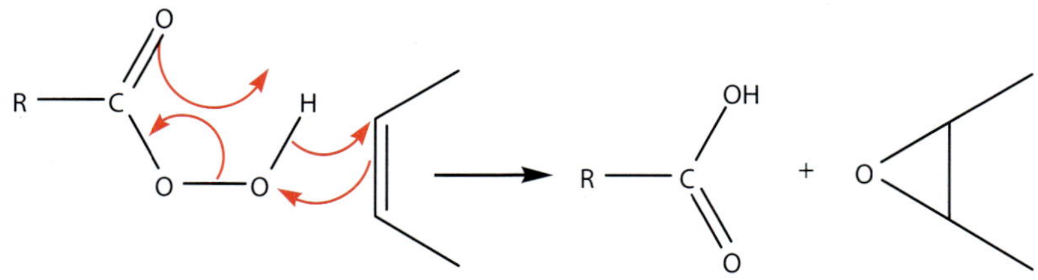

FIGURE 11.48 Mechanism of alkene epoxidation using peracid.

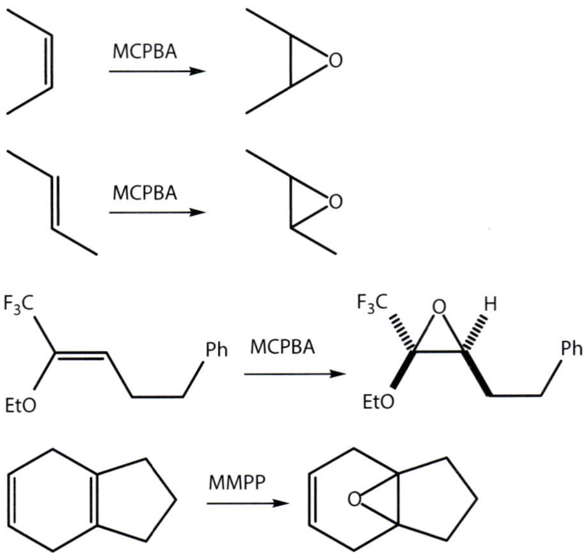

FIGURE 11.49 Peracid-mediated epoxidation of alkenes.

Focus on Sharpless Asymmetric Epoxidation of Allylic Alcohols

Ring opening of epoxides is, generally, stereochemically "reliable"; opening with nucleophiles takes place by an S_N2 mechanism, with clean inversion of configuration. Thus, chiral epoxides are very valuable in synthesis, as they can be opened to give two new stereocenters. Epoxidation with simple peracids may be diastereoselective, but it cannot be enantioselective, as the reagent is achiral. During the 1980s, a new method was developed by Sharpless (his work on enantioselective oxidations was rewarded by the Nobel Prize in 2001) that converts allylic alcohols to chiral epoxides in good chemical yield and usually more than 90 % enantiomer excess. The reagent used is *tert*-BuOOH, with a dialkyl tartrate (**11.21**; review Fischer projections as necessary) as the chiral catalyst. Dimethyl tartrate (DMT), diethyl tartrate (DET), and di-*iso*-propyl tartrate (DIPT) are all available inexpensively in both enantiomeric forms, so both enantiomers of the epoxide can be produced by choosing the correct catalyst. The *R*,*R*-form is the naturally occurring form of tartaric acid, being present in a range of fruits. You may have seen small crystals of potassium bitartrate on wine corks (it's quite harmless).

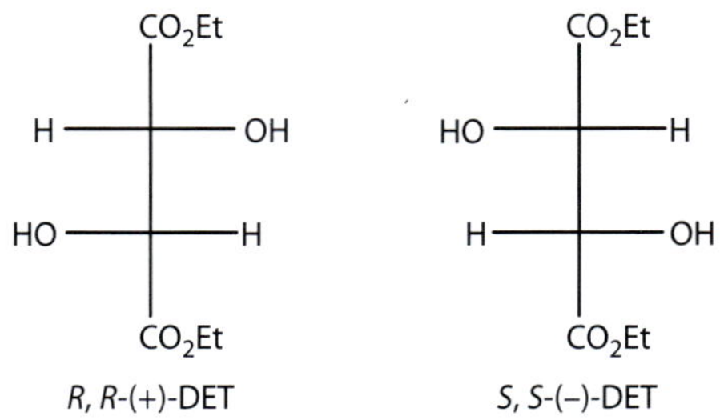

11.20

The process is applicable to a very wide range of allylic alcohol substrates; some examples are shown in Figure 11.50 (remember ee means enantiomer excess, and an ee of 95 % means 97.5 % of one enantiomer and 2.5 % of the other). Notice in the first example that the reaction is not only very enantioselective but also completely selective for the double bond of the allylic alcohol over the other trisubstituted alkene. In the second example, the product was used in a synthesis of disparlure, the sex attractant of the gypsy moth. In the final example, it is rigorously demonstrated that opposite antipodes of catalyst lead to the formation of enantiomeric products.

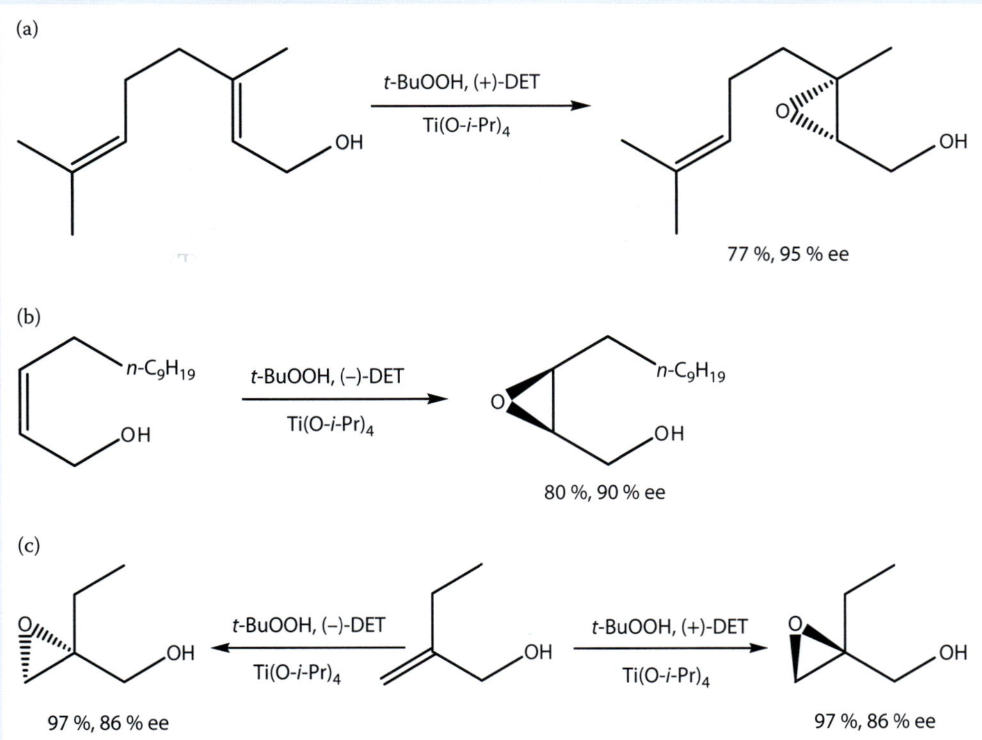

FIGURE 11.50 Sharpless asymmetric epoxidation. (a) 2S, 3S (b) 2S, 3R (c) left-hand molecule R, right-hand molecule, S; CH_2OC outranks CH_2OH.

Gypsy moth trap contains the enantiomerically pure sex attractant, disparlure, formed by Sharpless asymmetric epoxidation

A useful review exercise for you is to assign the absolute configurations of the epoxides produced in Figure 11.50. The solutions are (a) 2*S*, 3*S* (b) 2*S*, 3*R* (c) left-hand molecule *R*, right-hand molecule, *S*; CH_2OC outranks CH_2OH.

The outcome of the reaction is readily predicted using a quadrant method and is largely independent of the nature of the substituents on the double bond. We take a generic allylic alcohol, **11.22**, and position it so that the OH group is in the lower right quadrant, **11.23**. In this orientation (–), *S*,*S*-DET will deliver oxygen from the top (front) face and (+) *R*,*R*-DET will deliver oxygen from the bottom (back) face of the molecule.

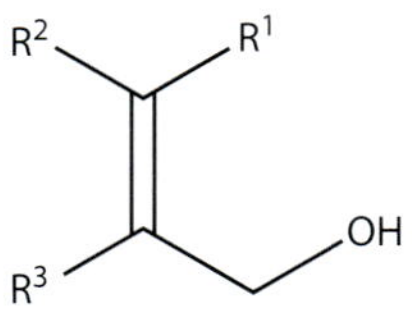

11.22

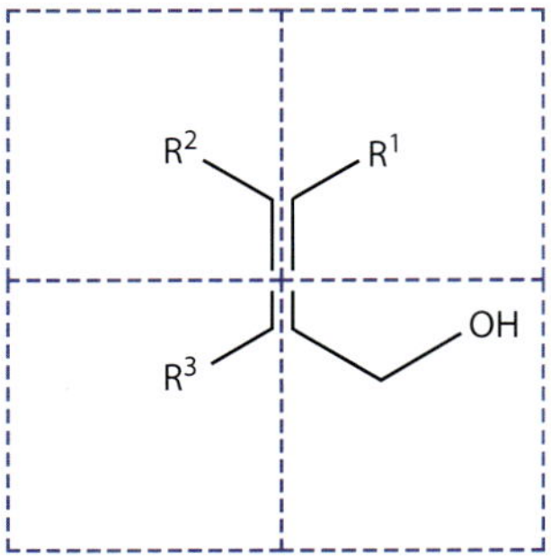

11.23

11.4.2 CARBENE ADDITION

We met the reactive species, methylene, $:CH_2$, in Section 8.2, as one of a number of reactive intermediates. In terms of its electronic configuration, methylene can exist in either a singlet or triplet state. Both have sp^2 hybridization at carbon; they differ in where we put the other two electrons. In singlet methylene, both are placed in the sp^2 hybrid orbital. Because under VSEPR theory lone pairs repel more than bonding pairs, this closes up the HCH angle to 102 °. In the triplet form, one electron is placed in each of the p_z and sp^2 orbitals. Because the sp^2 orbital is now only half full, the

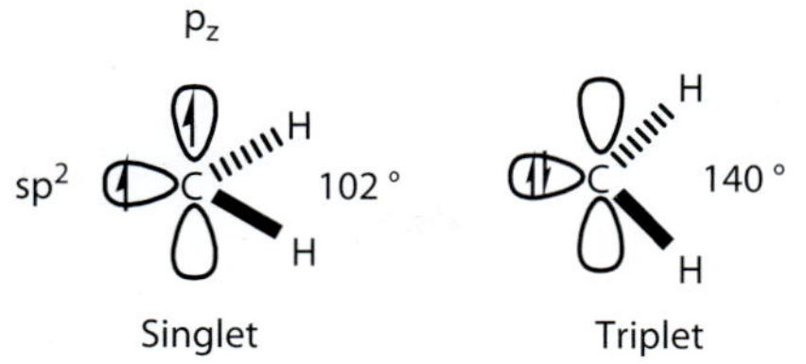

FIGURE 11.51 Singlet and triplet methylene.

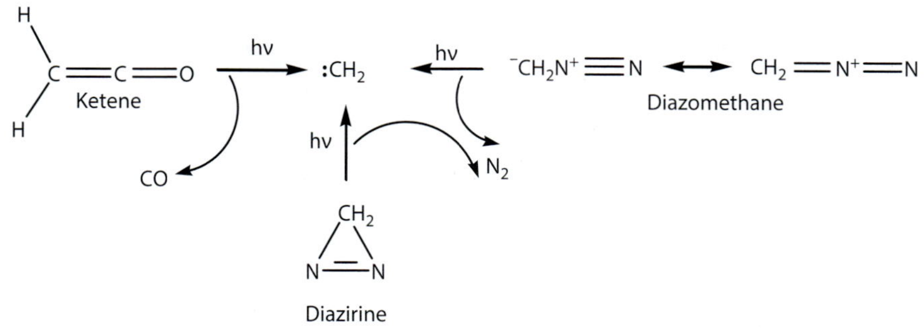

FIGURE 11.52 Generation of methylene.

repulsion is reduced, and the HCH angle is opened up to 140 ° (Figure 11.51). If these singlet and triplet labels are not familiar to you, don't worry; just consider them as labels. If you go on to do PChem next year, you will hear a lot more about them. Both theory and experiment have shown that the singlet state is higher in energy than the triplet state by about 37.8 kJ mol^{-1}.

The generation of methylene (Figure 11.52) almost invariably means using a high energy intermediate and losing a mole of gas, with the concomitant entropy gains. Starting materials include ketene, which loses carbon monoxide, and diazomethane or diazirine, which lose molecular nitrogen. Notice that in each case, we are losing a gas molecule with a *very* strong triple bond.

Free methylene is too energetic (whether singlet or triplet) to be synthetically useful, and it is normally used as a metal complex. Reacting diiodomethane with a zinc/copper couple in ether gives **11.24**, which is stable at room temperature, but behaves as a "toned down" version of methylene. We can write the mechanism as if it were methylene reacting (Figure 11.53). This is an example of a cycloaddition reaction, and we shall meet several more of these later in this chapter. These Simmons–Smith* reactions with double bonds give cyclopropanes, with the stereochemistry of the starting double bond normally being maintained. If there is a polar functional group in the molecule, the carbene is generally added from that side of the molecule (Figure 11.54).

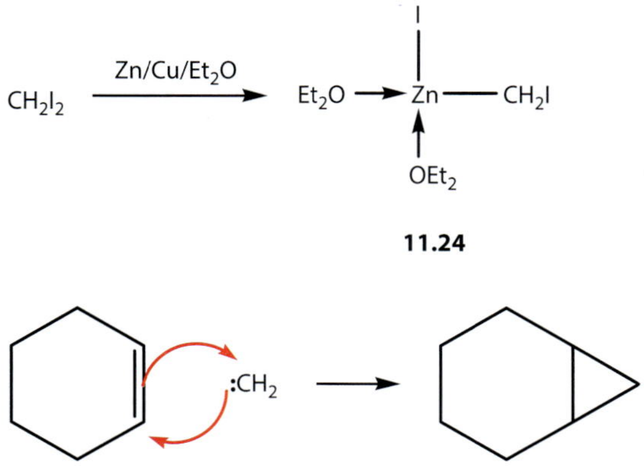

FIGURE 11.53 Generation and use of a methylene equivalent in the Simmons–Smith reaction.

* The name is specific to the reaction with Zn/Cu and CH_2I_2.

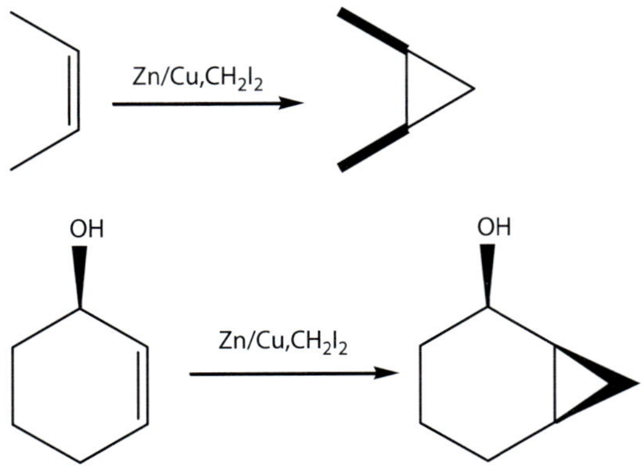

FIGURE 11.54 Simmons–Smith reactions for cyclopropanation of alkenes.

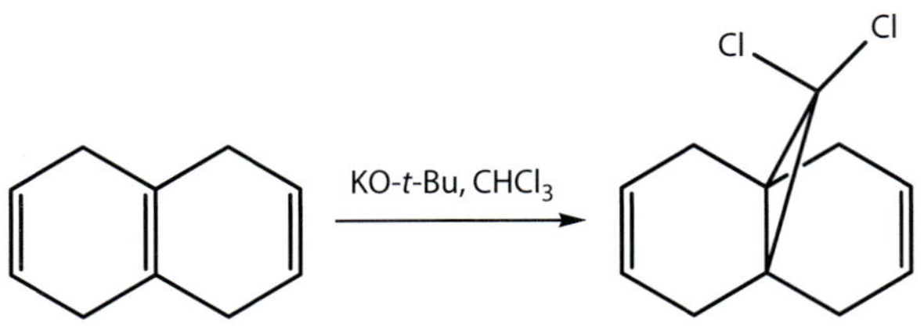

FIGURE 11.55 Dichlorocarbene is electrophilic and reacts selectively with more electron-rich double bonds.

The generation of some less energetic carbenes is easier. For example, when chloroform is treated with the strong base potassium *tert*-butoxide, the $[CCl_3]^-$ anion is formed. This collapses to chloride ion and dichlorocarbene. Like most common carbenes, this is an electrophile, and the most substituted double bond reacts preferentially (Figure 11.55).

PROBLEM 11.11

Predict the product(s) of the following reactions:

(a)

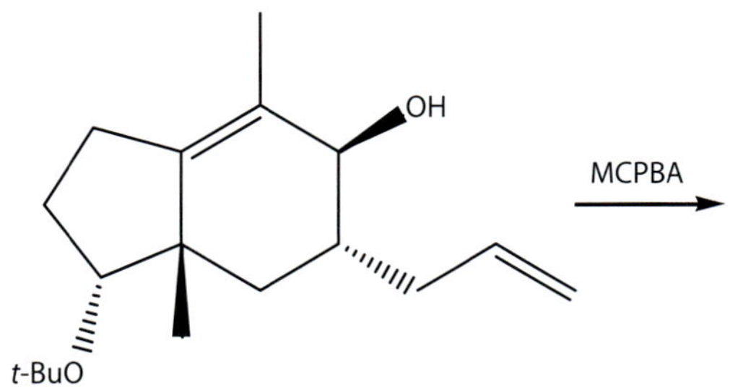

(b)

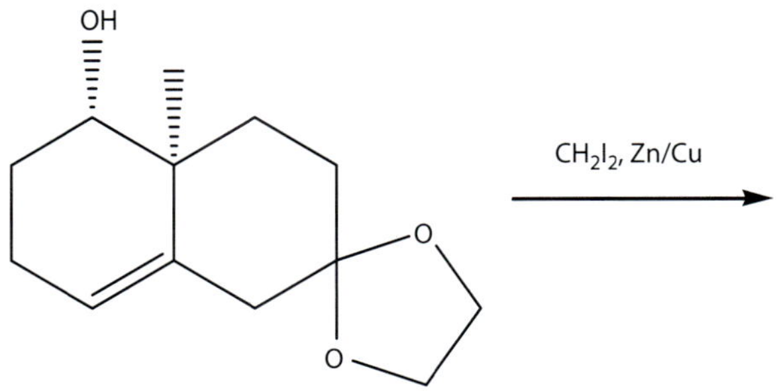

Solution

(a) In principle, we could obtain two diastereomeric products from this process, depending on whether the epoxidation occurs on the top or the bottom face of the trisubstituted double bond. The trisubstituted double bond is *much* more electron rich than the monosubstituted one and reacts much more rapidly. In fact, we are also in luck with the diastereoselectivity—epoxidation occurs from the same face as the hydroxyl group, with excellent selectivity.

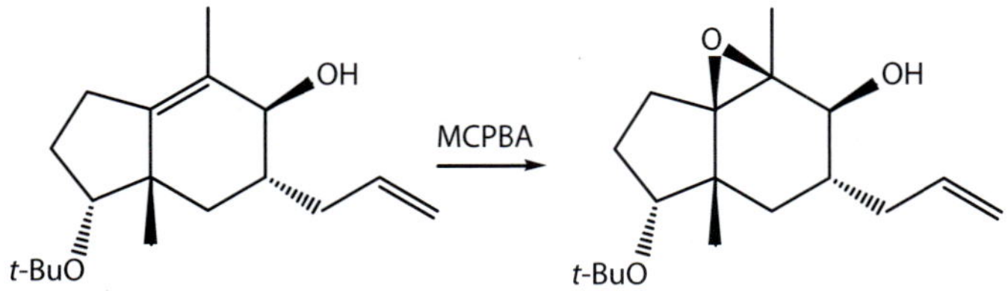

(b) This is another example where our reagent approaches from the face bearing the polar functional group. Cyclopropanation takes place *cis* to the hydroxyl substituent in 76 % yield.

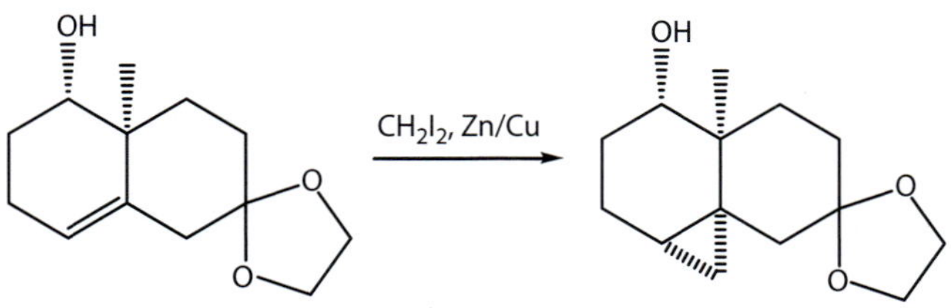

11.4.3 HYDROBORATION

Hydroboration is defined as the addition of borane or one of its derivatives to a multiple bond. It is an enormously versatile reaction synthetically, developed by H.C. Brown in the 1950s and recognized by his Nobel Prize in 1979. We will be looking specifically at the use of a hydroboration–oxidation sequence to accomplish the stereo- and regiospecific anti-Markovnikov addition of water to alkenes.

At room temperature, borane, BH_3, (**11.26**) is in equilibrium with diborane, B_2H_6 (**11.25**), with the equilibrium being strongly in favor of the dimer. Diborane has a very interesting three-center two-electron bond that you will learn more about if you take more inorganic chemistry. In THF solution, there is sufficient of the monomer (THF complexed) that we can sensibly consider our addition reaction being that of monomeric borane. A quick electron count at boron tells us that it has only six electrons, so it must be sp^2 hybridized, trigonal planar in shape, and having an empty p_z orbital. We can thus safely characterize it as an electrophile.

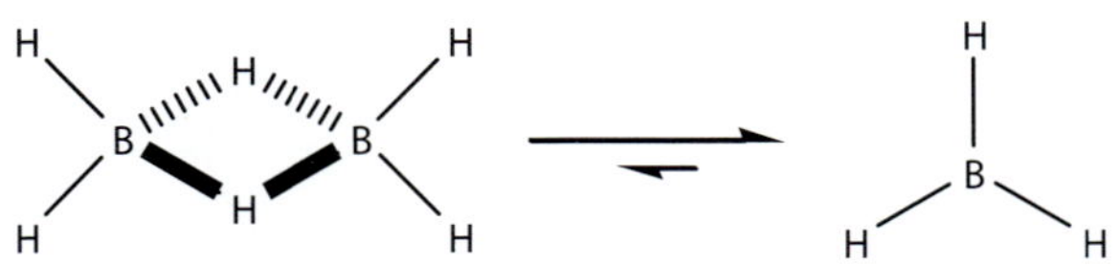

Diborane

11.25

Borane

11.26

If we consider the addition of borane to cyclohexene, then the reaction is concerted and shows *cis*-stereochemistry. Once this has happened, the other B–H bonds can react similarly with further cyclohexene, to give, eventually, a trialkylborane (Figure 11.56).

The electronic structure at boron for the trialkylborane is much the same as that for borane itself—sp^2 hybridization, empty p_z orbital, and electrophilic. Thus, we would expect that it would

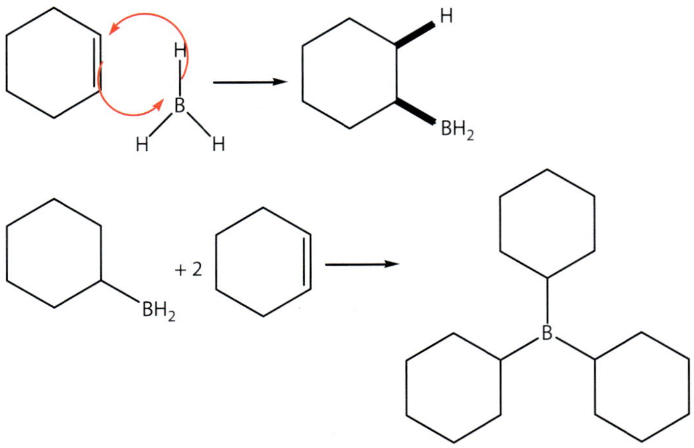

FIGURE 11.56 Addition of borane to cyclohexene.

react readily with nucleophiles, specifically in this case, the anion of hydrogen peroxide. If we treat hydrogen peroxide with base, an equilibrium is established between hydroxyl and hydroperoxyl anions. Fortunately, the hydroperoxyl anion is significantly more nucleophilic than is hydroxyl ion (look back at Table 9.2), so we only need to consider its reaction with the trialkylborane. Thus, HOO^- attacks at boron; boron now has an octet of electrons but also a negative charge. Moreover, we now have a weak oxygen–oxygen bond in our molecule, which will be easy to break. An alkyl group migrates from boron to oxygen, the O–O single bond breaks, and hydroxyl ion is expelled, to give R_2BOR. We can repeat this process twice more to obtain $(RO)_3B$, which is then hydrolyzed to alcohol (Figure 11.57). So, overall, we have added water to the double bond.

Why would we bother with this complex process, when we can accomplish the same result using much less expensive H^+/H_2O? Our first reason is that this reaction, unlike the acid-catalyzed addition of water, is stereospecific. The migration of the alkyl group from boron to oxygen is stereospecific with retention of configuration at the migrating carbon atom. Remember that because the acid-catalyzed reaction involves a carbocation, a mixture of stereoisomers is usually obtained. In the example we used earlier, you can't tell what the stereochemistry was, but if we instead use 1,2-dimethylcyclohexene as substrate, the difference is obvious (Figure 11.58). Hydroboration–oxidation gives stereospecifically *cis*-addition of water.

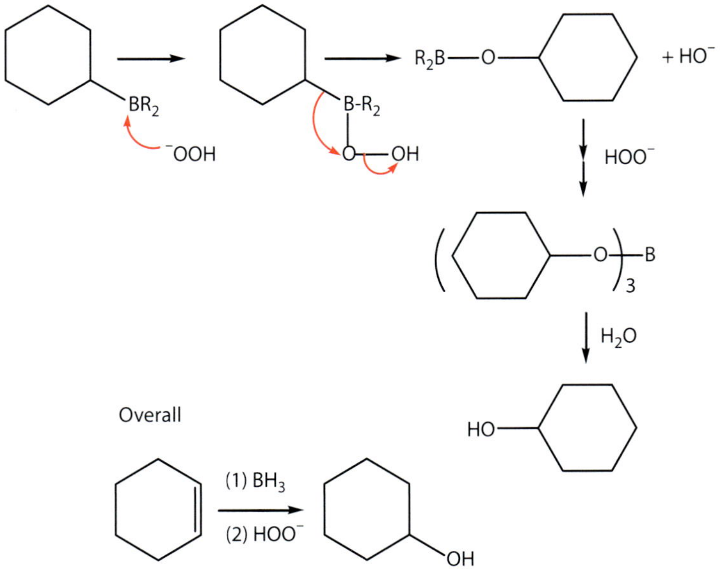

FIGURE 11.57 Reaction of trialkylborane with hydroperoxide anion.

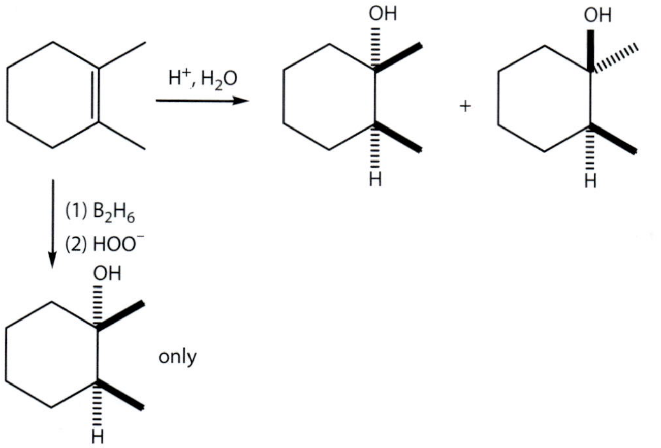

FIGURE 11.58 Addition of water to 1,2-dimethylcyclohexene.

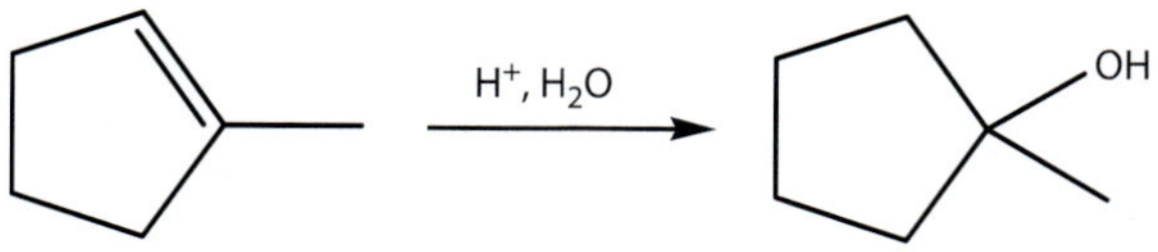

FIGURE 11.59 Regiospecific addition of water to an alkene under acid catalysis.

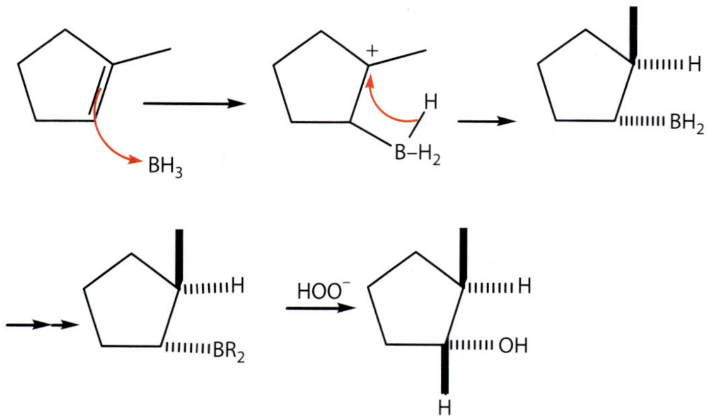

FIGURE 11.60 Anti-Markovnikov addition of water *via* hydroboration–oxidation.

Second, the reaction is regiospecific, and the regiospecificity is opposite to that obtained by acid-catalyzed hydration (or oxymercuration). If we add water to 1-methylcyclopentene in the presence of acid, we obtain 1-methylcyclopentanol, *via* the tertiary carbocation (Figure 11.59).

By contrast, hydroboration gives anti-Markovnikov addition of water (Figure 11.60). If we view the initial attack on borane as an essentially electrophilic process, this would occur at the 2-carbon to give a tertiary carbocation. Then we could envisage transfer of hydride from boron to the tertiary center. We already know that the migration is stereospecific, so we can complete the sequence.

You may already have realized that we are back *Through the Looking Glass*, with Alice. The regiochemistry of the reaction derives from looking at it as an electrophilic process, while the stereochemistry comes from considering it as concerted. The real reaction is both regio- and stereospecific.

Some examples of the use of this process in synthesis are shown in Figure 11.61. Notice in the last example that alkynes react, but with borane itself, regioselectivity is not good. The enols produced tautomerize readily to the relevant ketone.

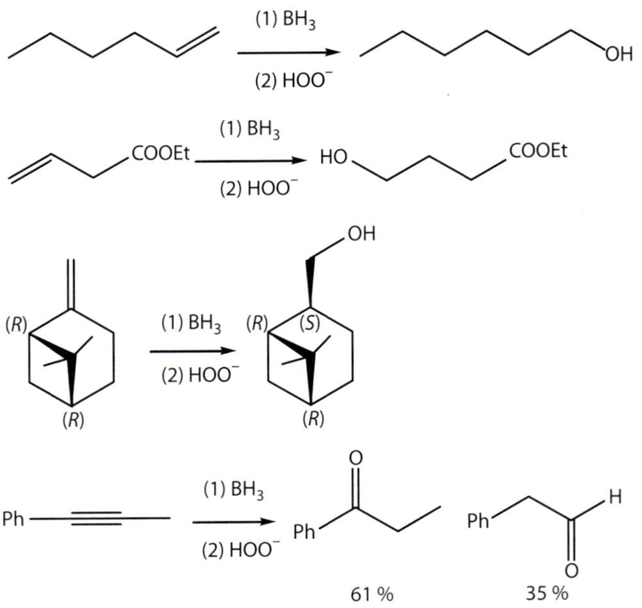

FIGURE 11.61 Hydroboration–oxidation in synthesis.

PROBLEM 11.12

Predict the product(s) of the following reaction, writing a mechanism to justify your answer.

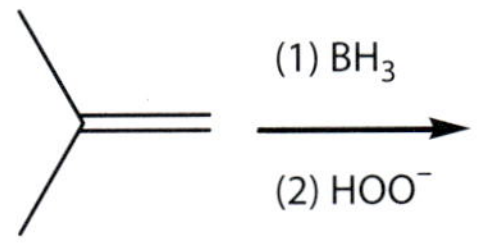

Solution

The addition of borane to the double bond proceeds so that the electrophilic boron is attached to the less substituted carbon atom:

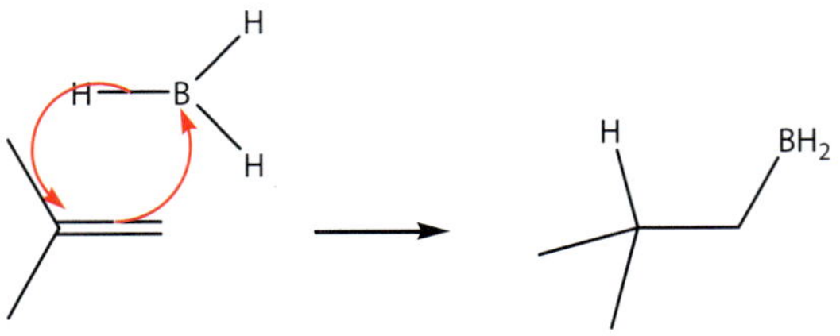

The same process happens with the other two B—H bonds, so that we get a trialkylborane:

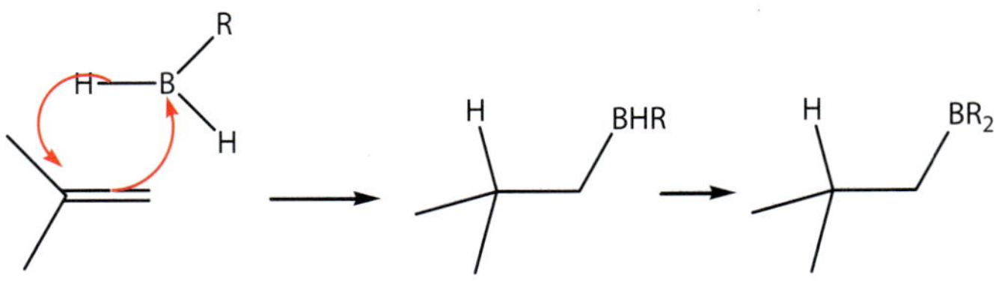

The trialkylborane is next attacked by the hydroperoxide anion (it is still an electrophile with an empty p_z orbital), and the alkyl group migrates from boron to oxygen. Finally, the trialkoxyborane is hydrolyzed by water.

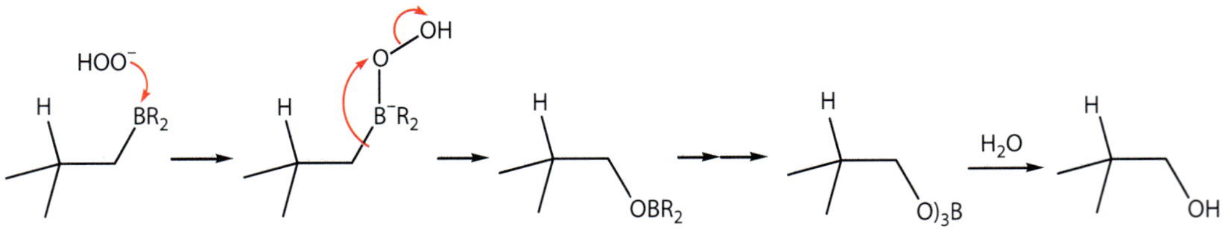

PROBLEM 11.13

How would you accomplish each of the following conversions?

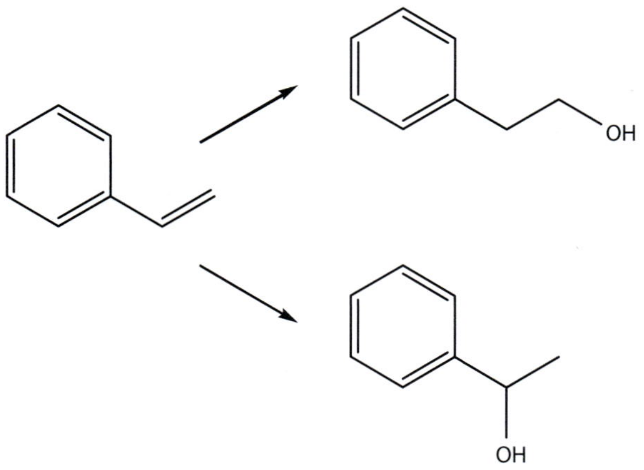

Solution

The conversion to linear alcohol is an anti-Markovnikov addition of water, which is most easily accomplished by hydroboration/oxidation—so our reagents would be (1) BH_3 (2) HOO^-. The branched alcohol is the product of Markovnikov addition of water *via* the stabilized secondary benzylic carbocation, so our reagent is simply aqueous acid.

Focus on Boranes in Organic Synthesis

For many purposes, borane does a great job in terms of both reactivity (high) and selectivity (not always so high). However, there are processes where a less reactive and more selective borane will be preferred, and numerous compounds are available. If borane is reacted with 2,3-dimethyl-2-butene, only a single addition takes place, because of the steric hindrance that would result from a further addition. The product is termed thexylborane (**11.27**). Similarly, disiamylborane (Sia_2BH, **11.28**) is prepared from 2-methyl-2-butene. The number of times the borane reacts is clearly a function of steric hindrance. 9-Borabicyclo[3.3.1]nonane (9-BBNH, **11.29**) is prepared from 1,5-cyclooctadiene, and catecholborane (HBcat, **11.30**) from catechol (1,2-benzenediol) and borane. All of these have been used to improve selectivity for specific reactions—in general, the more hindered the borane, the more selective the reaction.

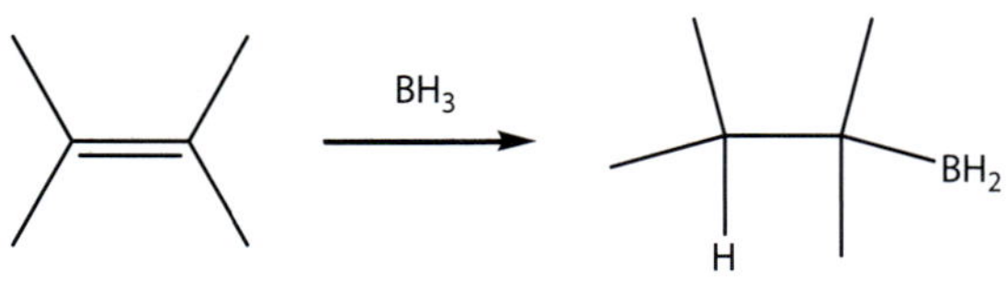

Thexylborane

11.27

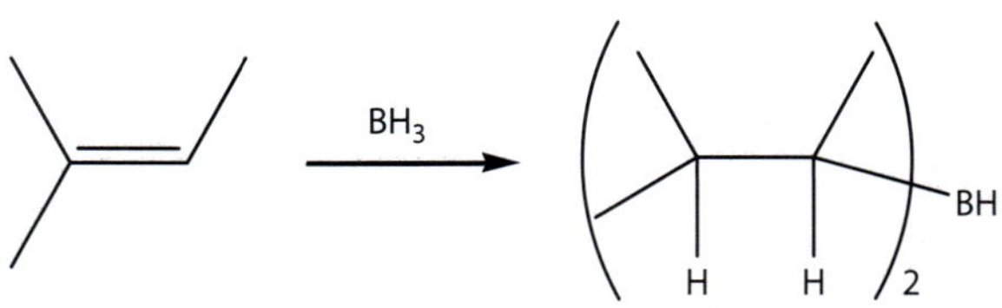

Disamylborane

11.28

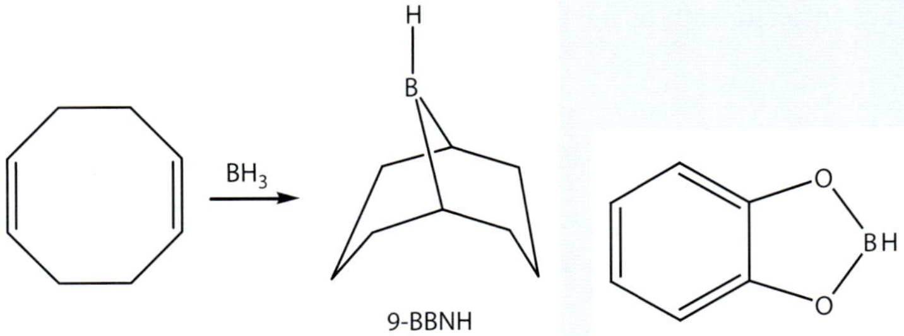

9-BBNH **11.29** **11.30**

Let's consider the hydroboration/oxidation of *Z*-4-methyl-2-pentene, shown in Figure 11.62. Here, the Markovnikov rule, or considerations of the best place for positive charge, will not help us, as both potential positive centers are secondary and similar. To achieve selectivity, we are completely dependent on steric effects—and the more sterically hindered the borane we use, the more selective our reaction is.

We can also use these reagents to react selectively with the least hindered double bond in a molecule; in the reaction of Figure 11.63, only the monosubstituted alkene reacts.

The addition of borane itself to alkynes can be a useful reaction; the enol produced tautomerizes to an aldehyde (Figure 11.64). However, many substrates give significant side products, usually because the alkenylborane intermediate can react with further borane. The use of one of the hindered boranes is usually more successful, and either a carbonyl compound or a *cis*-alkene can be obtained as shown in Figure 11.65. If the borane is protonated rather than oxidized, as in the second example, a hydrocarbon is obtained.

If we start with a chiral alkene such as α-pinene, **11.31**, then we can prepare chiral alcohols. The attack of borane occurs on the less hindered lower face to give diisopinocampheylborane

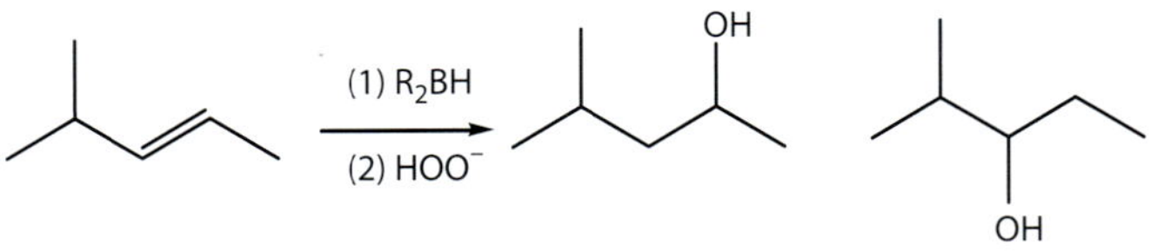

BH_3/THF	57	43
Thexylborane	66	34
Sia$_2$BH	97	3
9-BBNH	99.8	0.2

FIGURE 11.62 Hydroboration–oxidation of *Z*-4-methyl-2-pentene.

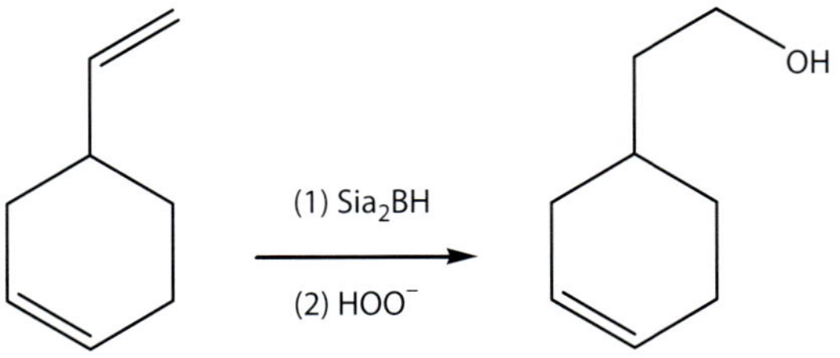

FIGURE 11.63 Selectivity between double bonds in reaction with a hindered borane.

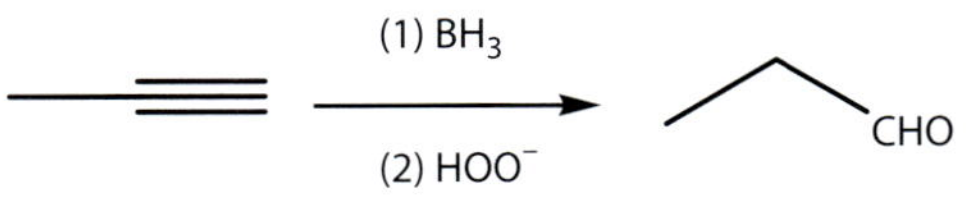

FIGURE 11.64 Hydroboration of an alkyne.

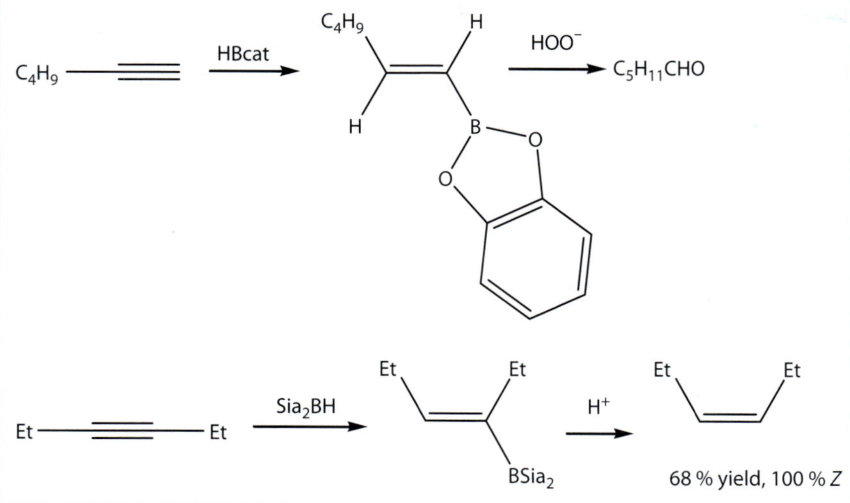

FIGURE 11.65 Reactions of alkynes with hindered boranes.

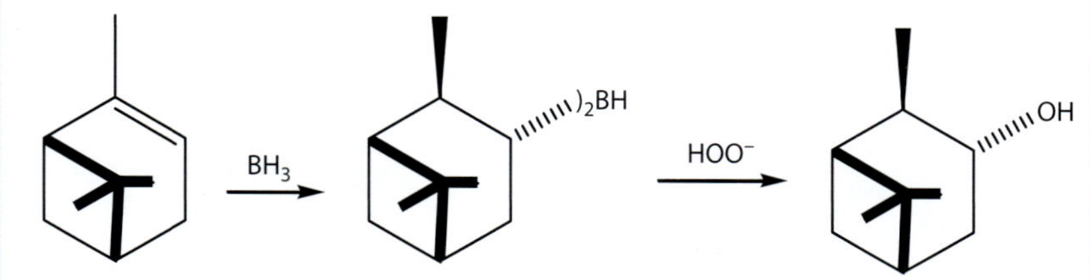

11.31, α-pinene **11.32**, Ipc_2BH **11.33**

FIGURE 11.66 Enantioselective hydroboration using a chiral borane.

(Ipc_2BH, **11.32**). This can itself be converted to the corresponding chiral alcohol, **11.33**. More excitingly, however, Ipc_2BH reacts with other, achiral, alkenes, to give, after oxidative work-up, chiral alcohols with good enantiomeric excess (Figure 11.66).

11.4.4 DIHYDROXYLATION OF ALKENES

What if we wanted to add two OH groups to an alkene to give a 1,2-diol? We have already seen a method for the synthesis of *trans*-diols. First, we convert the alkene to an epoxide using peracid; then we do an S_N2 opening of the epoxide with hydroxyl ion (Figure 11.67).

If we want a *cis*-diol, then both of the OH groups must be added from the same face of the molecule, and this is ideally achieved by a synchronous process. When osmium tetroxide

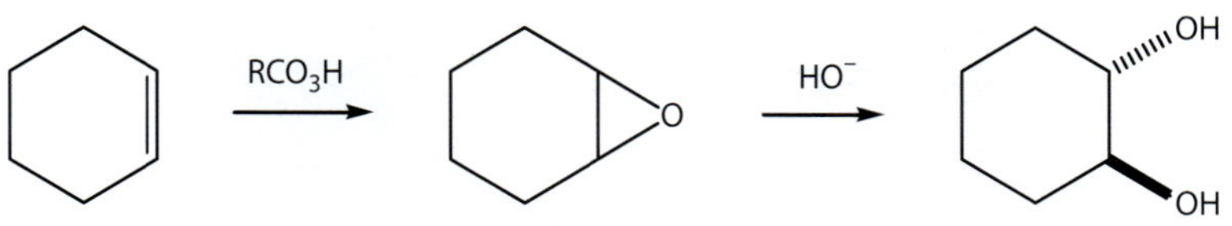

FIGURE 11.67 Synthesis of *trans*-diols.

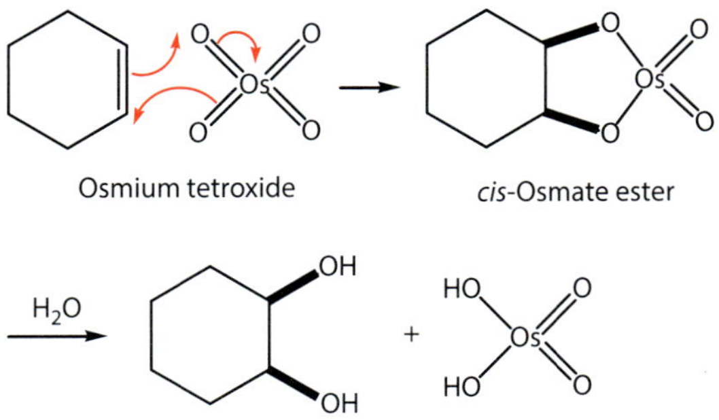

FIGURE 11.68 *cis*-Hydroxylation by osmium tetroxide.

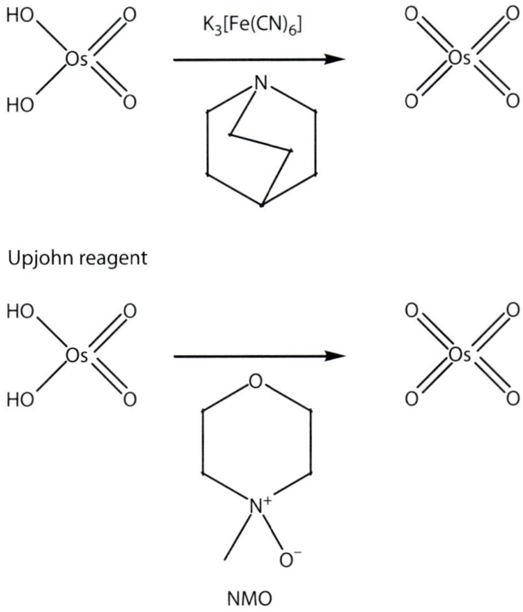

FIGURE 11.69 Catalytic system for the *cis*-dihydroxylation of alkenes.

(osmium(VIII)) is reacted with an alkene, two new C–O bonds are formed, to give a *cis*-osmate ester. This is hydrolyzed by water to give the *cis*-diol and osmium (VI) (Figure 11.68).

Since osmium is both expensive and very toxic, it is now the norm to use a catalytic amount of osmium tetroxide, together with a reoxidizing system (Figure 11.69). An inexpensive reagent, either iron(III), as potassium ferricyanide, or *N*-methylmorpholine *N*-oxide (NMO) is used to reoxidize osmium (VI) to osmium(VIII). Pure osmium tetroxide is colorless, but most samples are described as pale yellow, probably due to OsO_2 (which is brown) as an impurity. It smells terrible—indeed, the name osmium comes from the Greek, osme, όσμή, stench.

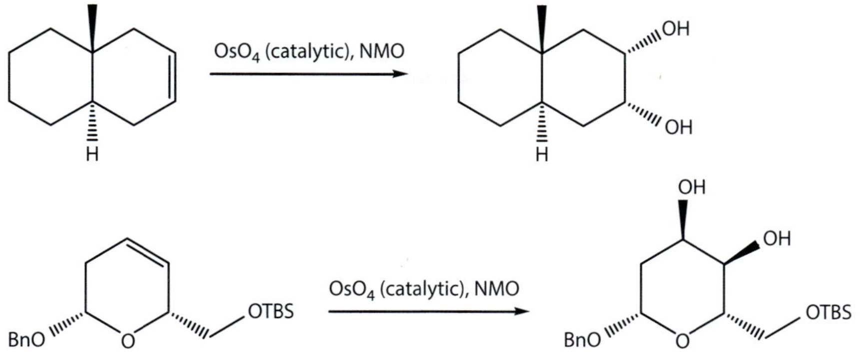

FIGURE 11.70 Applications of dihydroxylation reactions in synthesis.

Examples of uses of the Upjohn system in synthesis are shown in Figure 11.70. Notice that in each case, the reagent approaches from the less hindered face of the molecule.

PROBLEM 11.14

Predict the product(s) of the following reaction.

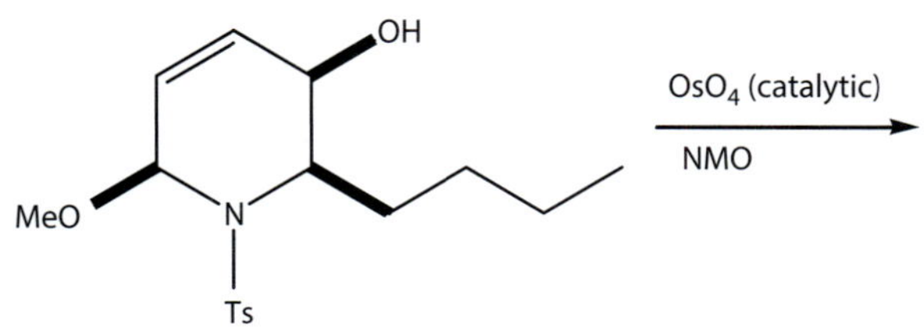

Solution

This is the Upjohn reagent for *cis*-dihydroxylation of alkenes. We need to consider which face of the molecule will react; one is significantly less hindered than the other, and this is the face that reacts, so the product is **11.33**.

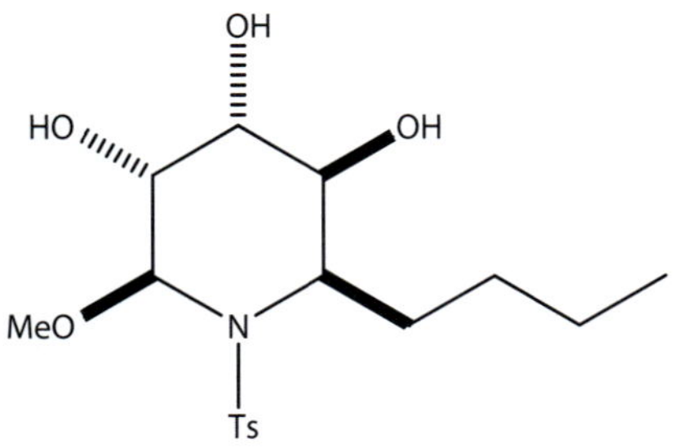

11.33

Focus on Sharpless Asymmetric Dihydroxylation

The work of Sharpless on asymmetric processes led to the development of simple, commercially available reagents to catalyze enantioselective dihydroxylation of a wide range of alkenes. These are marketed as ADmix-α and ADmix-β and are composed of $K_2OsO_4 \cdot 2H_2O$ (osmium (VI), which is oxidized to Os(VIII) *in situ*), ligand, $K_3[Fe(CN)_6]$, potassium carbonate, and $MeSO_2NH_2$ (this accelerates the hydrolysis of the osmate ester, allowing faster reaction at a lower temperature). The two ligands used are derivatives of dihydroquinine and dihydroquinidine, $(DHQ)_2PHAL$ (α, 1,4-*bis*(9-*O*-dihydroquinyl)phthalazine, **11.34**) and $(DHQD)_2PHAL$ (β, 1,4-*bis*(9-*O*-dihydroquinidinyl)phthalazine, **11.35**). The outcome of the reaction is predictable using a quadrant method similar to that we saw for Sharpless epoxidation; if we place the largest substituent in the lower left corner, ADmix-β hydroxylates the top face and ADmix-α the bottom face (**11.36**; the substituents are labeled as L, M, and S, large, medium, and small). Examples of the uses of these systems are shown in Figure 11.71.

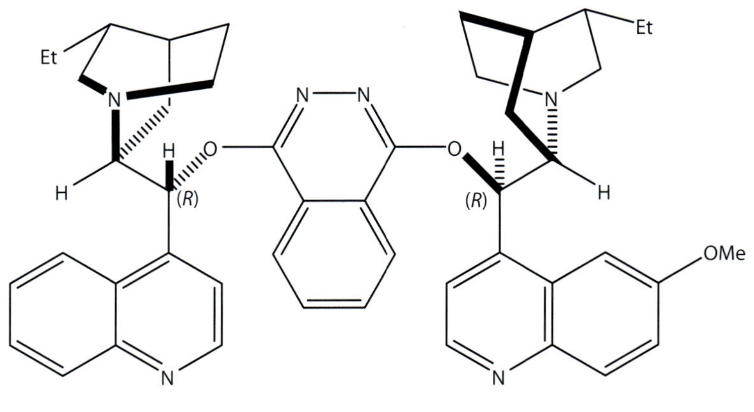

11.34, $(DHQ)_2PHAL$

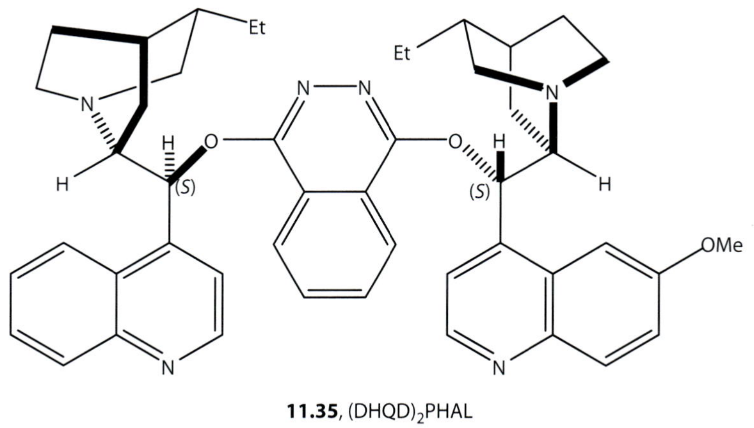

11.35, $(DHQD)_2PHAL$

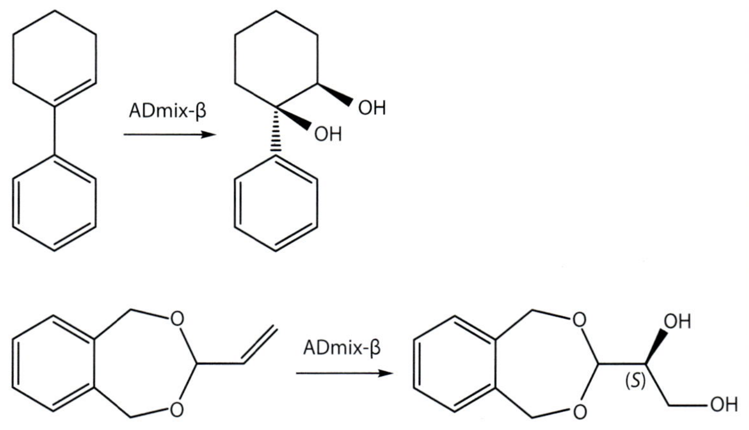

FIGURE 11.71 Uses of asymmetric dihydroxylation in synthesis.

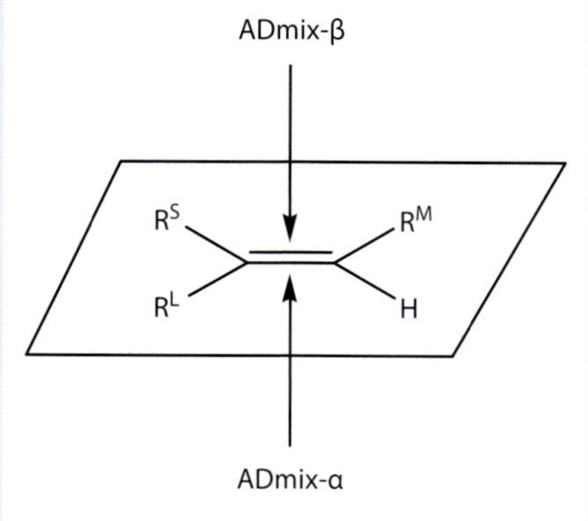

11.36

11.4.5 OZONOLYSIS

The reaction of alkenes with ozone is used to cleave the double bond to two carbonyl compounds. This combined with spectroscopic analysis of the products (mainly mass spectrometry) allows a double bond to be located in a long chain of carbon atoms. As you will remember from Chapter 4, we write ozone in a dipolar form, with a positive charge on the central carbon atom, one oxygen–oxygen double bond, and one negatively charged oxygen. There are two identical resonance forms (**11.37**).

11.37

Ozone reacts with double bonds in a concerted manner to give, initially, a highly explosive molozonide (Figure 11.72). This is a new type of cycloaddition, more specifically cycloaddition of a 1,3-dipole (the resonance forms allow the positioning of the charge at either atom-1 or atom-3). We will meet a number of related examples when we look in more detail at synthesis. The molozonide (sometimes called a primary ozonide) reopens to give a ketone and a hydroperoxide-type anion. This reaction is essentially the reverse of the cycloaddition, but giving a different 1,3-dipole as the product. These reunite to give an ozonide (sometimes secondary ozonide). Although ozonides are also explosive, they are less reactive than the molozonides, and some have been isolated in pure form at low temperature.

The ozonide can be worked up under either oxidative (H_2O_2) or reductive (Zn, $Na[BH_4]$ or Me_2S) conditions. Under oxidative conditions, a diacid is obtained; under reductive conditions, the dialdehyde is isolated (Figure 11.73). Some examples are shown in Figure 11.74.

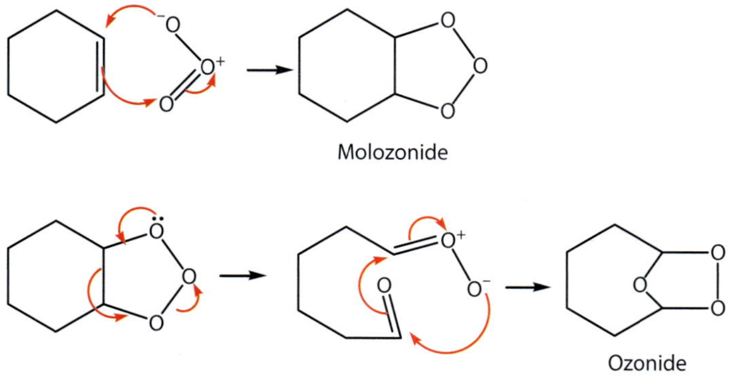

FIGURE 11.72 Reaction of ozone with alkenes to give an ozonide.

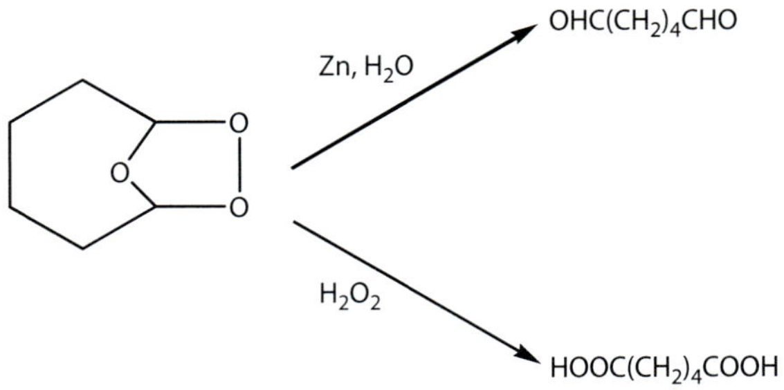

FIGURE 11.73 Oxidative and reductive work-ups for ozonides.

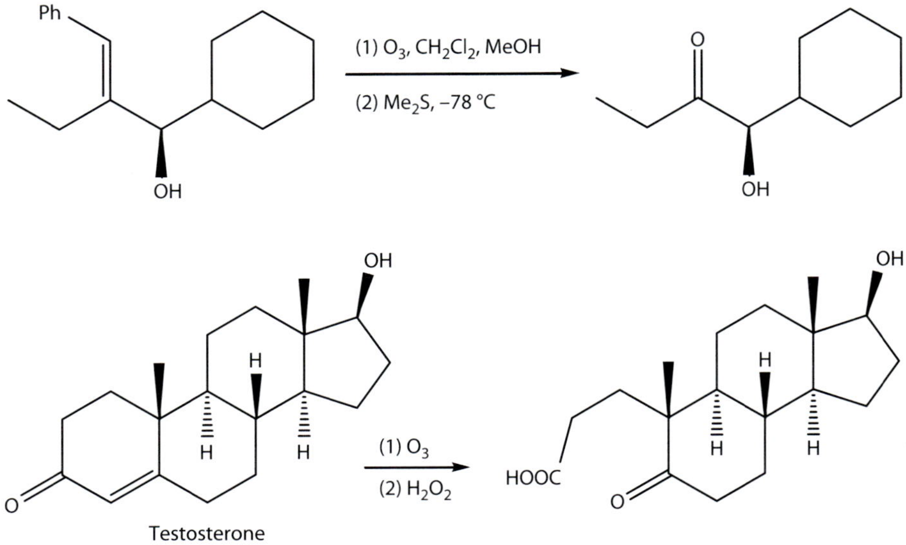

FIGURE 11.74 Ozonolysis in synthesis.

PROBLEM 11.15

A hydrocarbon, **11.38**, formula C_7H_{12}, reacts with ozone followed by dimethylsulfide to give a single ketoaldehyde, **11.39**. **11.38** may be hydrogenated to methylcyclohexane. A reaction of **11.38** with HBr (unfortunately the experimenter let air into the reaction) gave two monobromides, **11.40** and **11.41**. When treated with [EtO]Na/EtOH, **11.40** gave a hydrocarbon **11.42**, which reacted with ozone followed by dimethylsulfide to give a dialdehyde, **11.43**. **11.41** reacted with [EtO]Na/EtOH to give mainly **11.38** and some **11.42**. Identify the species **11.38** to **11.42**, explaining your reasoning.

Solution

The formula, C_7H_{12}, indicates two double bond equivalents. Given that **11.38** can be hydrogenated to methylcyclohexane, one of these is a ring and one a double bond. We know what carbon skeleton we have; the only issue is where is the double bond. The answer to this comes from consideration of the ozonolysis reaction, which gives a ketoaldehyde. First, this means that the double bond must be *in* the ring. If we had methylenecyclohexane, this would ozonize to formaldehyde and cyclohexanone. The fact that we get a ketone means that one end of the original double bond must have been substituted with an alkyl group. This is only true if the double bond is adjacent to the methyl group. Thus we have

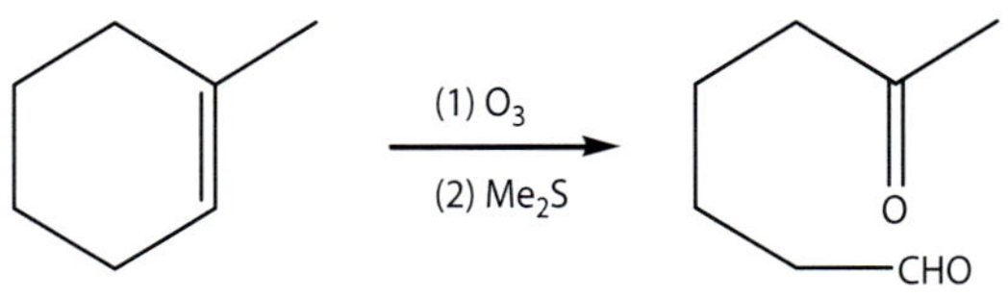

11.38 **11.39**

The key feature of the "careless" reaction with HBr is that if air (oxygen is a diradical and can initiate radical reactions) is present, a radical pathway, leading to anti-Markovnikov products, will predominate. However, this will not be stereospecific, so both *cis*- and *trans*-isomers will be formed:

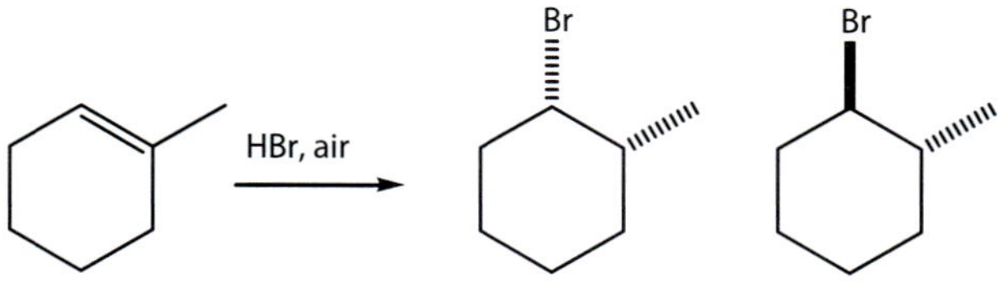

In order to understand what is happening in the elimination reaction, we need to draw proper chair forms for each of the bromides. Taking the *cis*-isomer first, elimination can only occur when the bromine is axial, and two products can be obtained, depending on which of the two available axial protons is removed. As we saw in the previous chapter, the more stable, more substituted alkene usually predominates, in this case 1-methylcyclohexene, **11.38**. In the *trans*-isomer, only one elimination reaction is possible, as there is only one available hydrogen that is *trans anti* to the departing bromine.

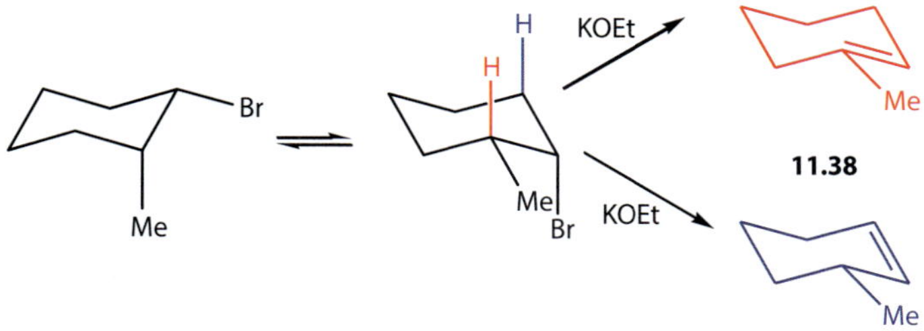

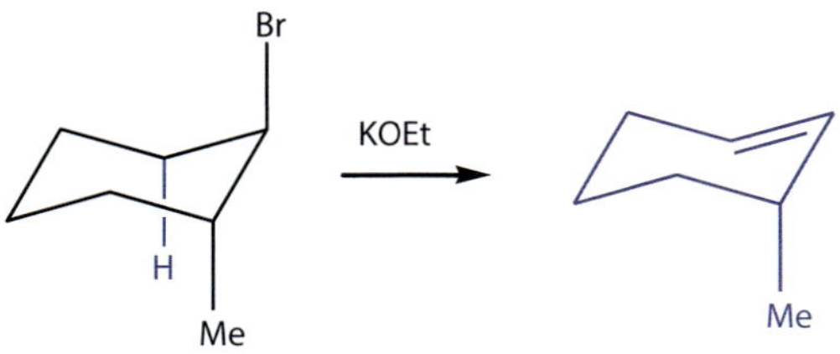

What can we conclude here? **11.40** gave the single product; **11.41** gave two products, mainly **11.38**. Thus, **11.40** is the *trans*-isomer, and **11.41** is the *cis*-isomer. Finally, **11.42**, 3-methylcyclohexene, is ozonized to the dialdehyde **11.43**:

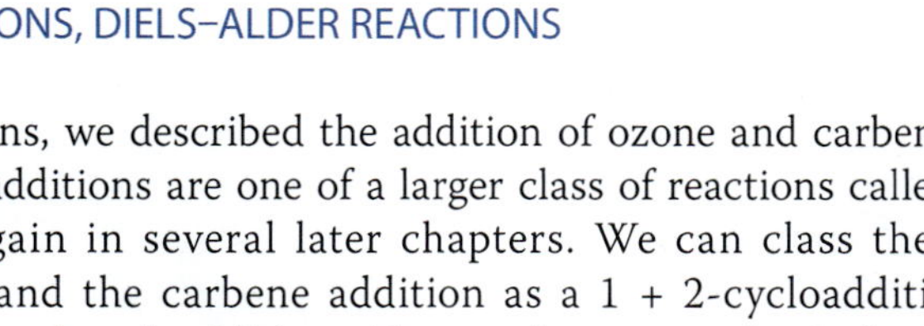

11.42 **11.43**

11.4.6 CYCLOADDITIONS, DIELS–ALDER REACTIONS

In the previous sections, we described the addition of ozone and carbenes to double bonds as cycloadditions. Cycloadditions are one of a larger class of reactions called pericyclic processes that we will meet again in several later chapters. We can class the ozone reaction as a 3 + 2-cycloaddition and the carbene addition as a 1 + 2-cycloaddition. This section will deal with 2 + 2- and 4 + 2-cycloadditions. The numbers come simply from the number of atoms in each component that we are adding together to make a new cyclic compound. The Diels–Alder reaction was first documented in 1928, and Otto Diels and Kurt Alder received the Nobel Prize for their work in 1950. In its simplest form, it involves the addition of an alkene, usually

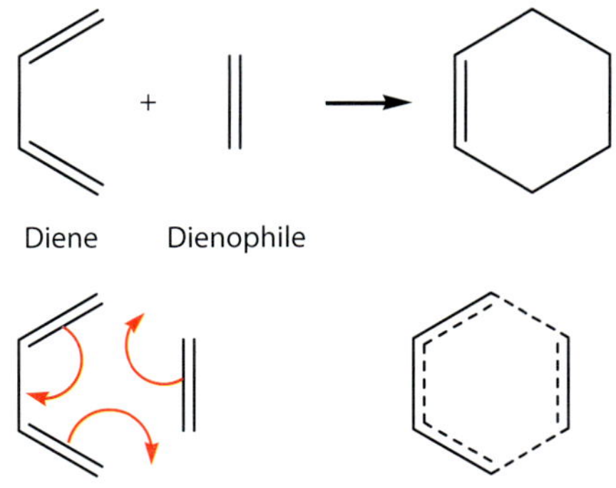

FIGURE 11.75 Diels–Alder reaction.

called the dienophile, and a conjugated diene (Figure 11.75). The new bonds are formed simultaneously, and the transition state is cyclic. In practice, the reaction of ethene and 1,3-butadiene is slow (although the reaction is exothermic; ΔH^o = –167 kJ mol^{-1}, ΔS^o = –182 J °K^{-1} mol^{-1} ΔG^o = –113 kJ mol^{-1}), and harsh conditions of temperature and pressure are required. The favorable enthalpy change arises because we are losing two π-bonds and gaining two stronger σ-bonds. However, because we are going from two molecules to one, degrees of freedom are reduced, and the entropy change is unfavorable. The entropy of activation is also a challenge for the reaction; in order to make the new bonds simultaneously, the molecules must come together in the correct orientation. Moreover, the butadiene must be in the *s-cis* conformation, **11.44**, which is less favorable than the *s-trans*, **11.45**. At room temperature, 96 % of butadiene is in the *s-trans* form, the energy difference between the *s-cis* and *s-trans* forms being 9.6 kJ mol^{-1}.

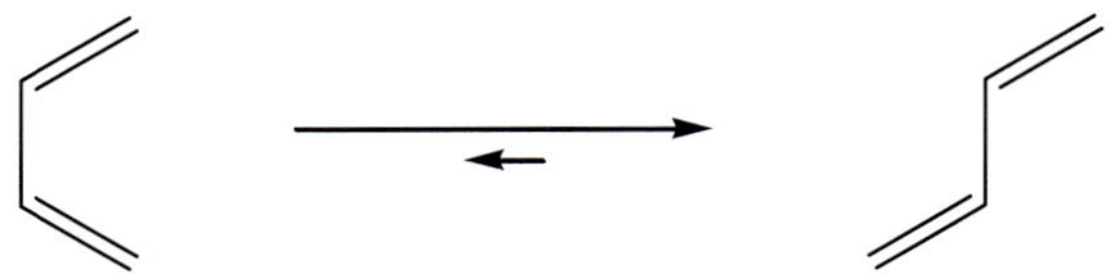

s-cis
11.44

s-trans
11.45

Synthetically, more practical processes are achieved when the dienophile (the alkene) bears at least one electron-withdrawing group such as carbonyl, nitro, or cyanide. For example, butadiene reacts readily with maleic anhydride. Alkynes with appropriate electron-withdrawing groups also react well (Figure 11.76).

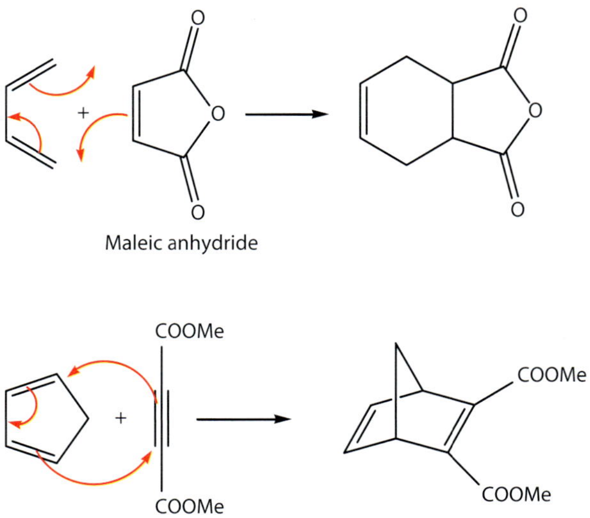

FIGURE 11.76 Practical Diels–Alder reactions.

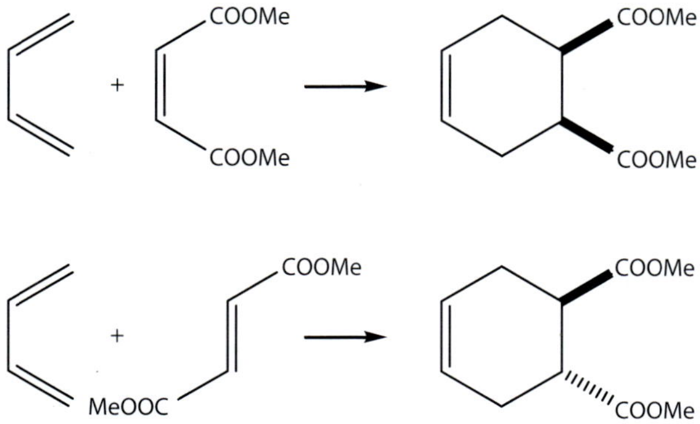

FIGURE 11.77 Diels–Alder reactions are stereospecific.

The reaction is generally stereospecific—as you see in Figure 11.77, the reactions of butadiene with dimethyl fumarate and dimethyl maleate preserve the original geometry of the alkenes in the product.

The reactions of cyclopentadiene are generally faster than those of open-chain dienes, because the diene is held in the *cis*-geometry that is required for the reaction to take place. Relative reactivities of maleic anhydride with various dienes are shown in Figure 11.78. Putting substituents on butadiene increases the rate because of two separate effects. First, the presence of substituents tends to favor the *s-cis* conformation to a greater extent, especially with the bulky *tert*-butyl group. However, the presence of electron-donating substituents on the diene more generally accelerates the reaction. Notice that the reaction with cyclopentadiene is particularly rapid, as it is locked into the favorable *s-cis* conformation.

We might reasonably ask why these 4 + 2-cycloadditions are favorable, when 2 + 2-cycloadditions do not proceed under thermal conditions. We could write apparently sensible curly arrows for 2 + 2-cycloadditions, but they don't readily occur on heating (Figure 11.79).

The answer lies in considering the molecular orbitals that are involved in making the new bonds. The outcome of all pericyclic reactions depends on the symmetry of the orbitals involved. The theory was codified by Woodward and Hoffmann in the late 1960s. Hoffmann

Diene	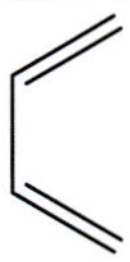	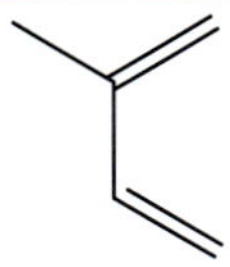	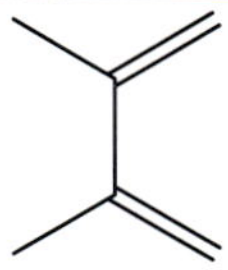	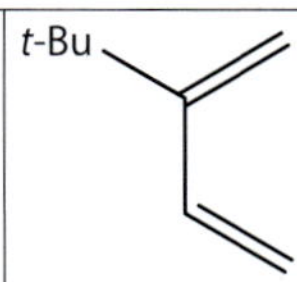	
Relative rate of reaction with maleic anhydride	1	2.3	4.9	27	1350

FIGURE 11.78 Relative reactivity of dienes with maleic anhydride.

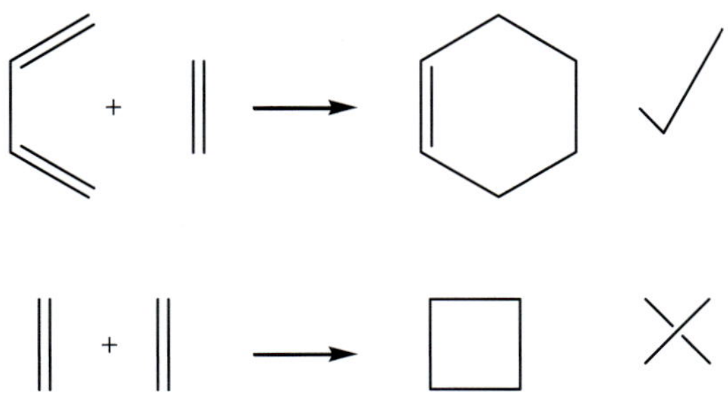

FIGURE 11.79 4 + 2-Cycloadditions occur on heating, but 2 + 2-cycloadditions do not.

received the Nobel Prize for this achievement in 1981; Woodward, who had earlier received the Nobel Prize for his work in synthesis, died in 1979. The theory is elegant and all encompassing, and what we will use here is a very much simplified picture. The HOMO–LUMO approach considers these orbitals as the ones we will use to make the new bonds in a thermal reaction. For a simple alkene, the HOMO is the bonding, π-orbital, and the LUMO the π*, antibonding orbital (Figure 11.80).

If we now consider bringing together one HOMO and one LUMO from two alkene molecules (Figure 11.81), we see immediately that we cannot use these orbitals to make the two new σ-bonds that we require. To get in phase overlap for the orbitals to make two new bonds, we would have a suprafacial process on one alkene (suprafacial simply means that we make our two new bonds on the same face of the molecule) and an antarafacial one on the other (antarafacial means that our new bonds are made on opposite faces of the original molecule). So the only process that is allowed from an orbital point of view is geometrically impossible. We describe the process as disallowed thermally.

The only way we can persuade two alkenes to form a cyclobutane is in a photochemical reaction. When we shine light on the molecule, electrons are promoted from the HOMO to the LUMO, and in photochemical processes we consider LUMO + LUMO (Figure 11.82). All the orbitals we need are now in phase; the reaction is suprafacial on both components and is therefore allowed photochemically.

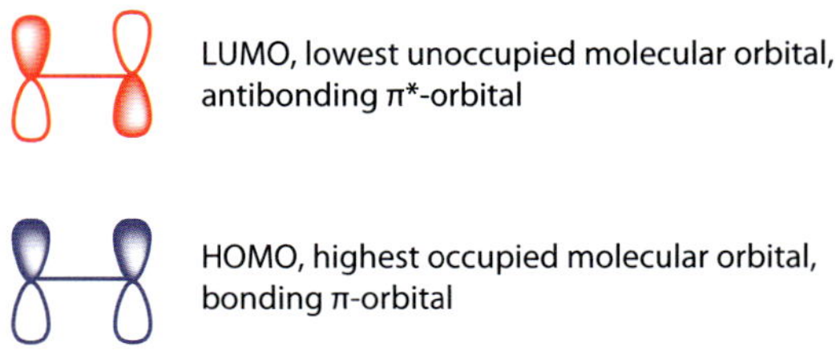

FIGURE 11.80 HOMO and LUMO for a simple alkene.

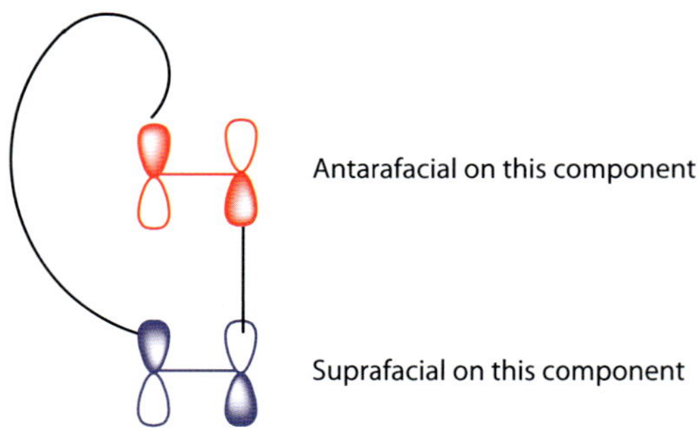

FIGURE 11.81 HOMO + LUMO for a 2 + 2-cycloaddition.

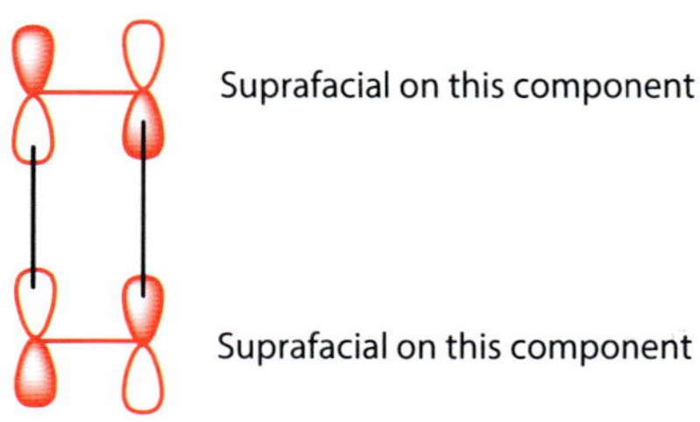

FIGURE 11.82 LUMO + LUMO for a 2 + 2-cycloaddition.

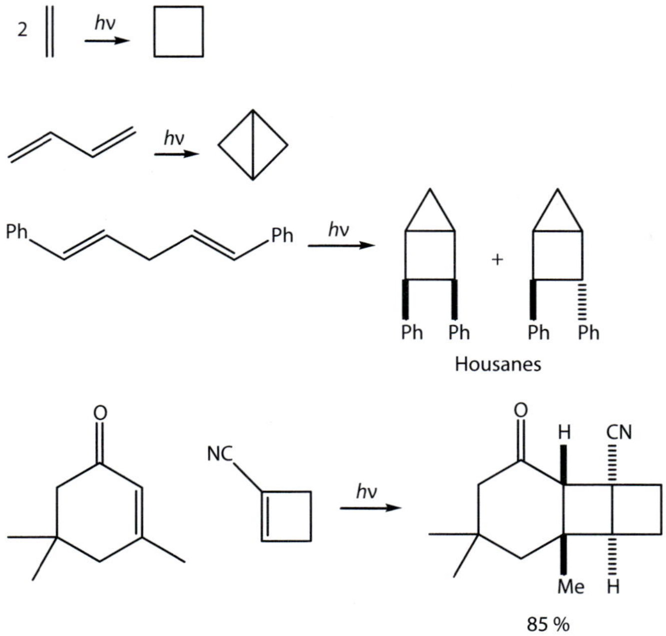

FIGURE 11.83 2 + 2-Cycloadditions.

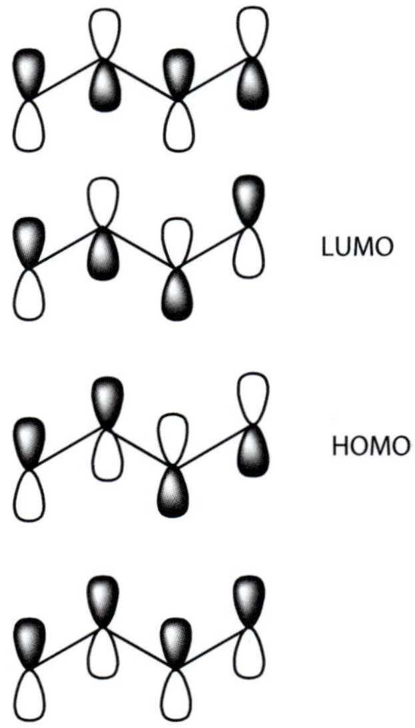

FIGURE 11.84 Molecular orbitals of butadiene.

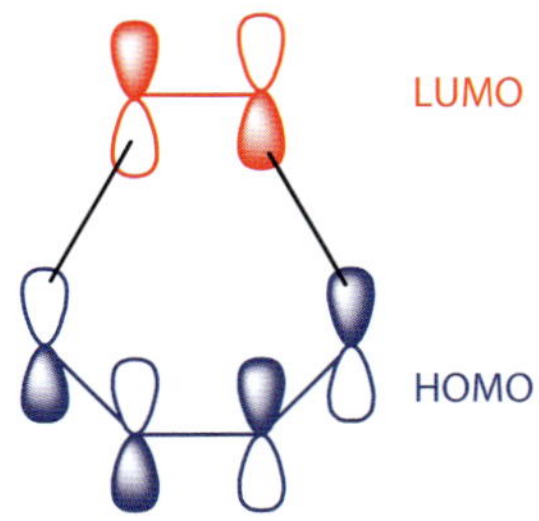

FIGURE 11.85 HOMO + LUMO for 4 + 2-cycloaddition.

Some examples of practical photochemical 2 + 2-cycloadditions are shown in Figure 11.83. The final example should intrigue you—the regioselectivity of the addition is high, if initially surprising. This and other regioselectivities in cycloaddition reactions can be predicted from an understanding of how substituents perturb the energies (though not the symmetry) of the molecular orbitals of the diene and dienophile; there is a good account of this in Fleming, *Molecular Orbitals and Organic Chemical Reactions.*

Turning now to consider the orbital interactions for 4 + 2-cycloadditions, we need to look at the LUMO of the alkene or dienophile and the HOMO of butadiene. So we should review the molecular orbitals of butadiene, which we met first in Chapter 3 (Figure 11.84).

When we now bring together the LUMO of ethene and the HOMO of butadiene, the symmetry of the orbitals is correct for two bonds to be made simultaneously. So the 4 + 2-cycloaddition is suprafacial on both components and is allowed thermally (Figure 11.85).

Both inter- and intramolecular Diels–Alder reactions are widely used in synthesis; some examples are given in Figure 11.86. The first reaction is rapid because the dienophile bears two strong electron-withdrawing groups—you might want to think about what will happen to the product. In the second example, the reaction is intramolecular, *and* the alkene bears an electron-withdrawing group. The third example uses a carbon–oxygen bond as dienophile. Many examples are known where one or more of the carbon atoms of the diene or dienophile are replaced by a heteroatom. That the reaction still works reminds us that the symmetry of molecular orbitals generally does not depend on the atoms involved, only on the p_z orbitals we are using (though relative energy levels will change). The regioselectivity displayed by both this and the final example is reliable, but the reason for it is beyond the scope of this course.

A particular stereochemical issue arises when the diene we use is cyclic. If we consider the reaction of cyclopentadiene with acrolein, $CH_2{=}CHCHO$, then we can envisage the production of two possible stereoisomeric products. In one, the *endo*-product, the substituent from the alkene is on the opposite side of the molecule from the bridge. In the *exo*-product, the substituent is on the same face as the bridge (Figure 11.87).

Under conditions of kinetic control, the *endo*-isomer is generally formed; this is sometimes called the *endo*-rule. The *exo*-product is formed under conditions of thermodynamic control. The *endo*-rule arises from secondary orbital interactions (shown in Figure 11.88), which are beyond the scope of this course.

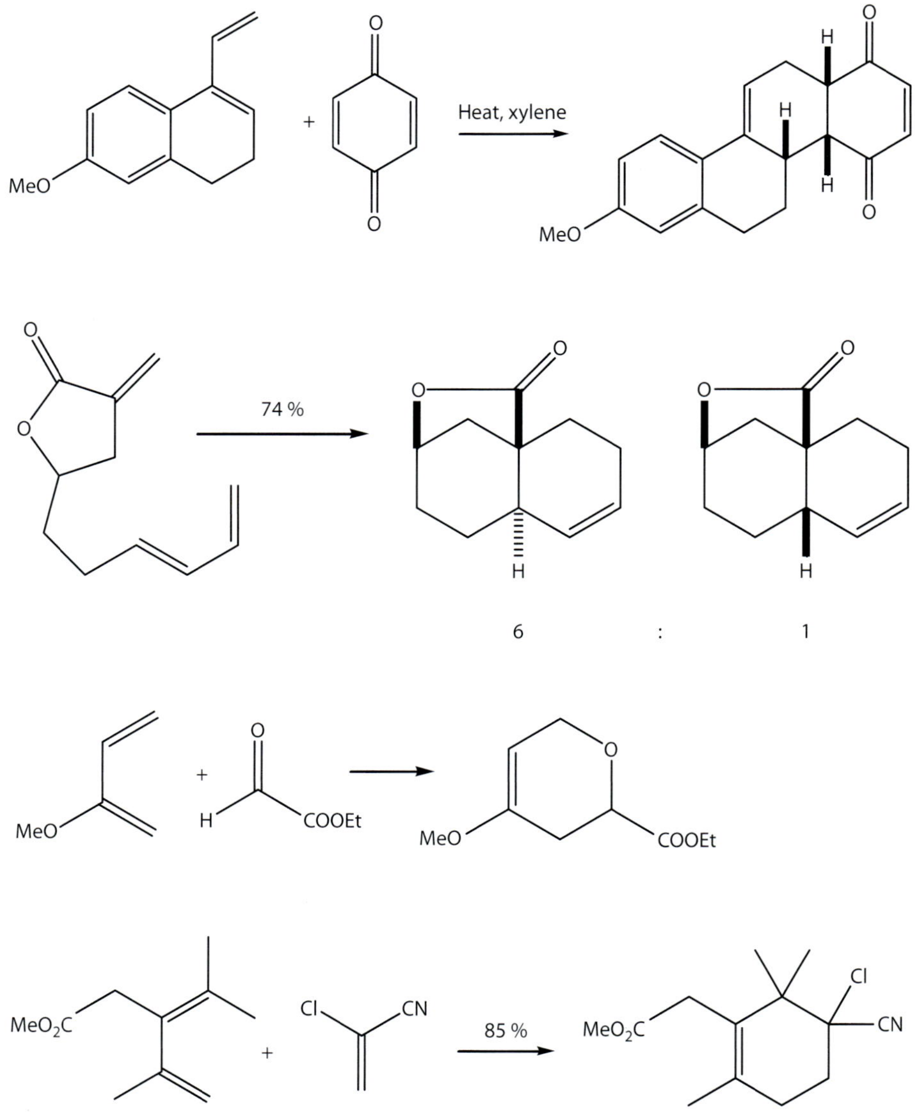

FIGURE 11.86 Diels–Alder reactions in synthesis.

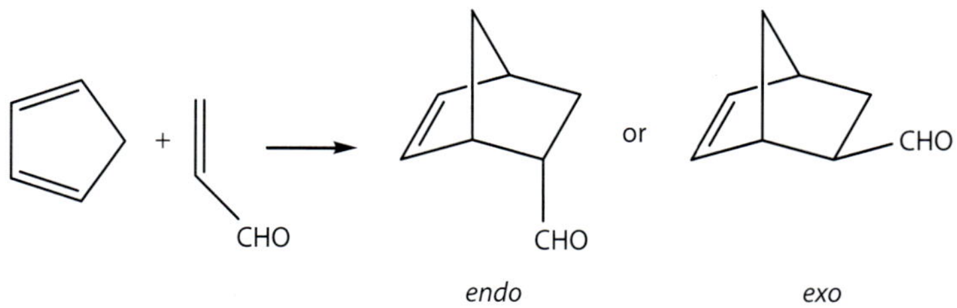

FIGURE 11.87 The *endo*-rule in Diels–Alder reactions of cyclic dienes.

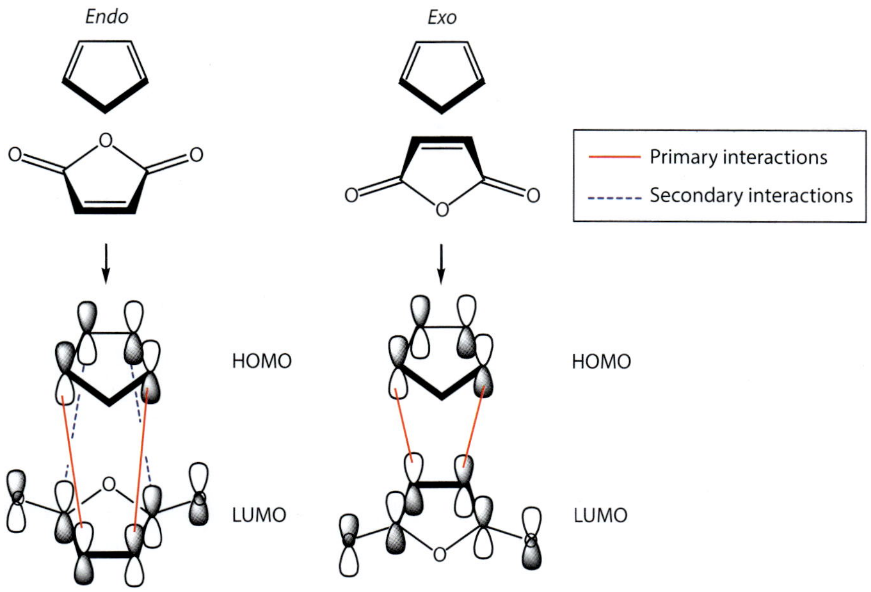

FIGURE 11.88 Secondary orbital interactions lead to the *endo*-rule.

PROBLEM 11.16

Predict the product(s) of the following reactions:

(a)

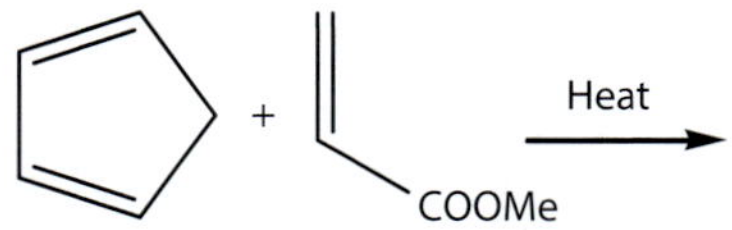

(b)

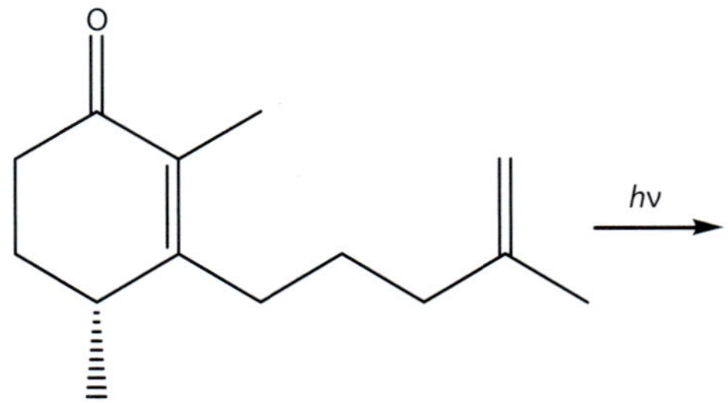

(c)

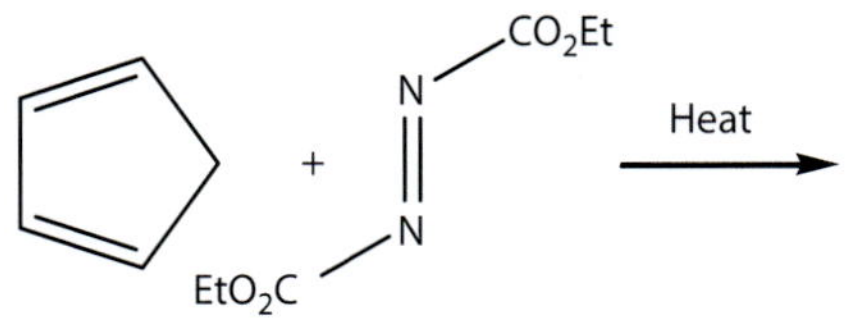

Solutions

(a) This is a Diels–Alder that will follow the *endo*-rule. Because the diene is cyclopentadiene, the reaction goes under mild conditions, giving a 92 % yield over 68 h at room temperature.

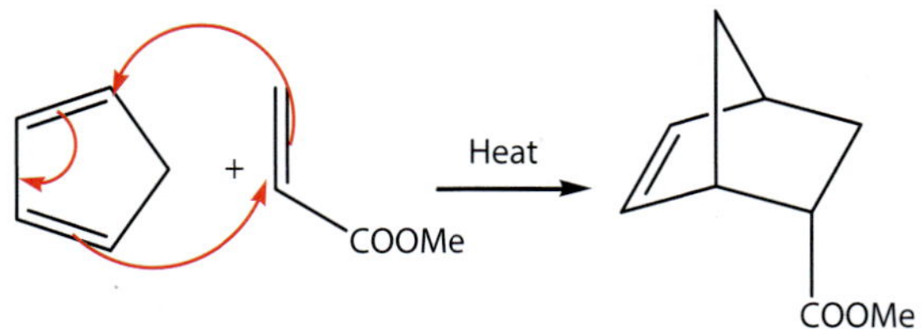

(b) The fact that this is a photochemical reaction suggests that we should be looking for a 2 + 2-cycloaddition to give a cyclobutane. The length of the chain joining the two double bonds means that only one regiochemistry is possible. *cis*-Stereochemistry is required at junctions of four- and six-membered rings, but the other relative stereochemistry obtained is not trivial to predict.

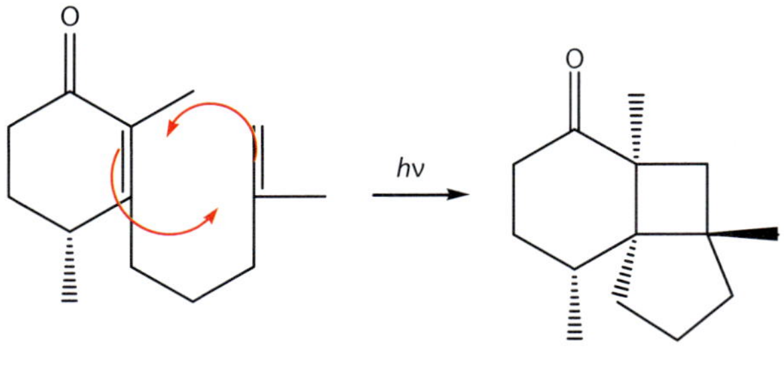

(c) Don't be put off by the nitrogen atoms here—we still have a double bond bearing electron-withdrawing groups, and the 4 + 2-cycloaddition proceeds just the same way as it would for a carbon–carbon double bond bearing the same substituents. The only difference in the product is that the geometry at sp[3] nitrogen atoms is not fixed, as they invert rapidly.

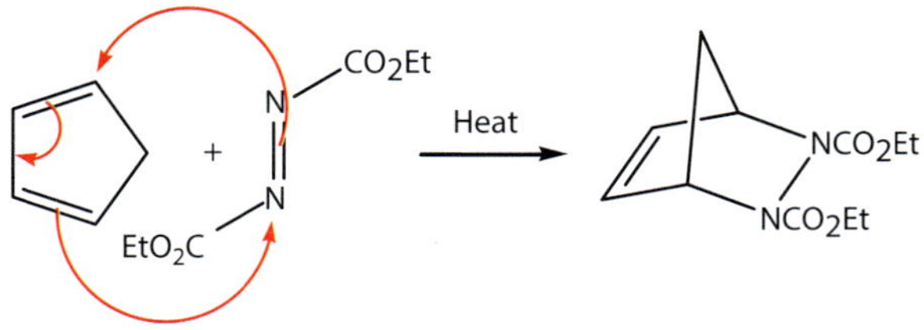

Key Points from Section 11.4

- Alkenes are converted to epoxides using a range of peracids, RCO_3H. The reaction is essentially electrophilic in character, and electron-rich double bonds react most rapidly.
- The addition of carbenes to alkenes, a cycloaddition reaction, affords cyclopropanes. The most common reagent is the Simmons–Smith, CH_2I_2 with a zinc/copper couple.
- The addition of borane to alkenes is stereospecifically *cis* and leads to the formation of trialkylboranes. These may be oxidized to alcohols using the anion of hydrogen peroxide. Overall addition of water is achieved, in a *cis*-stereospecific, anti-Markovnikov manner.
- Hydroboration/oxidation of alkynes gives ketones, after tautomerization of the enol formed.
- *cis*-Dihydroxylation of alkenes is accomplished with catalytic OsO_4 plus an oxidant such as NMO or $K_3[Fe(CN)_6]$. This contrasts with the formation of *trans*-diols by epoxidation of alkenes followed by the opening of the epoxide with hydroxyl ion.
- Ozonolysis of alkenes, involving an initial 3 + 2-cycloaddition of ozone, cleaves the double bond to give carbonyl compounds. Aldehydes are obtained if the work-up is reductive (Zn/H_2O or Me_2S) and acids if it is oxidative (H_2O_2).
- The 4 + 2-cycloaddition of alkenes and dienes occurs on heating to give cyclohexenes. The reaction is most favorable when the alkene bears electron-withdrawing groups.
- The 2 + 2-cycloaddition of alkenes occurs under photochemical conditions to give cyclobutanes.
- The possibility, or otherwise, of cycloaddition reactions is determined by the symmetry of the orbitals involved. For thermal reactions, HOMO + LUMO are considered, with LUMO + LUMO for photochemical reactions.

11.5 HYDROGENATION

We cannot leave the topic of addition to carbon–carbon multiple bonds without considering hydrogenation; it is the reaction we might first think of when we consider the chemistry of alkenes and alkynes. Mechanistically, however, it has little in common with the reactions we have studied to date. Although hydrogenation is thermodynamically favorable (ΔH° for ethene hydrogenation is –136 kJ mol^{-1}), the activation barrier is substantial, and all reactions are carried out in the

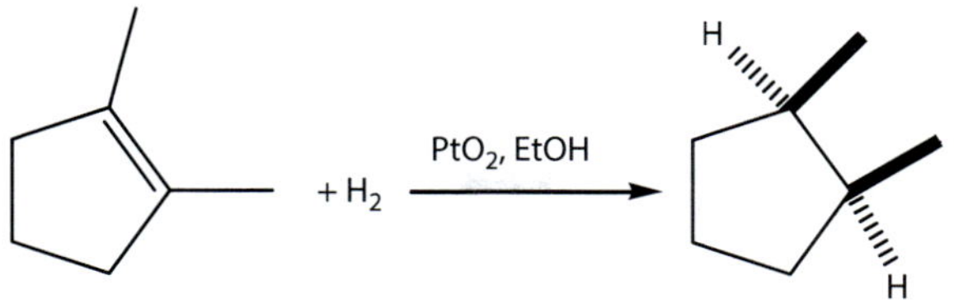

FIGURE 11.89 Hydrogenation of an alkene using a heterogeneous catalyst.

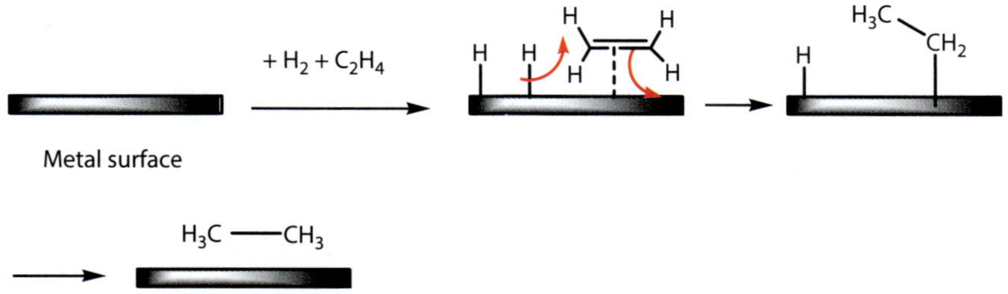

FIGURE 11.90 Mechanism of heterogeneous hydrogenation of alkenes.

presence of heterogeneous or homogeneous catalysts. *cis*-Addition of hydrogen to alkenes (Figure 11.89; note that the platinum is added as its oxide, and this is readily reduced to finely divided metal by the hydrogen) occurs in the presence of a wide range of finely divided metals as catalyst—platinum, palladium, and nickel are the most commonly used. Heterogeneous catalysis has the advantage that the catalyst can be removed and recovered by filtration at the end of the process.

The accepted mechanism for the reaction is that both the hydrogen and the alkene are adsorbed onto the metal surface, and the hydrogen is dissociated to hydrogen atoms, surface bound. The two hydrogen atoms are transferred one at a time to the alkene, with a metal-bound alkyl group as the intermediate. When the second hydrogen has been transferred, the alkane is desorbed (Figure 11.90). Stereoselectivity is generally excellent, because the metal surface can be only on one side of the double bond. Imagine putting a white plate down on a surface covered with ketchup—only one side of it will be covered in ketchup.

If a molecule contains bulky groups, the addition of hydrogen will be mainly or exclusively from the less hindered face, as it is easier for this to be adsorbed on the surface. We can see this in the reactions of Figure 11.91, where the hydrogens that have been added are shown in red.

In the last 30 years, the catalysis of hydrogenation by metal complexes in solution (homogeneous catalysis) has become more popular, because of the specific types of selectivity (and in some cases very high activity) that can be obtained. A preview of some reactions of this type is shown in Figure 11.92; we will return to this topic in more detail in Chapter 23. In the first

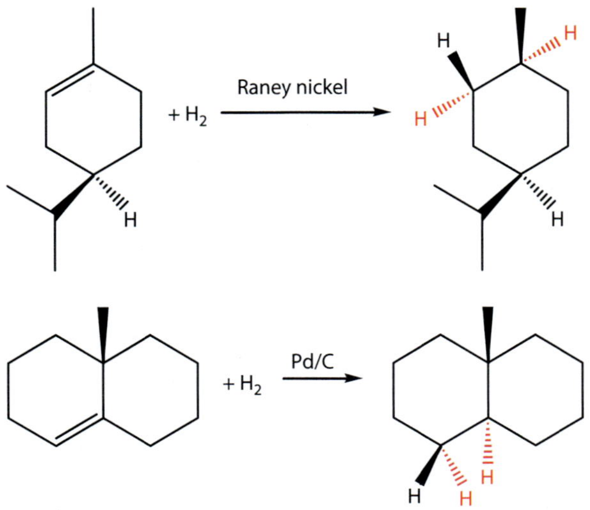

FIGURE 11.91 Addition of hydrogen to alkenes in the presence of a heterogeneous catalyst takes place from the less hindered face of the double bond.

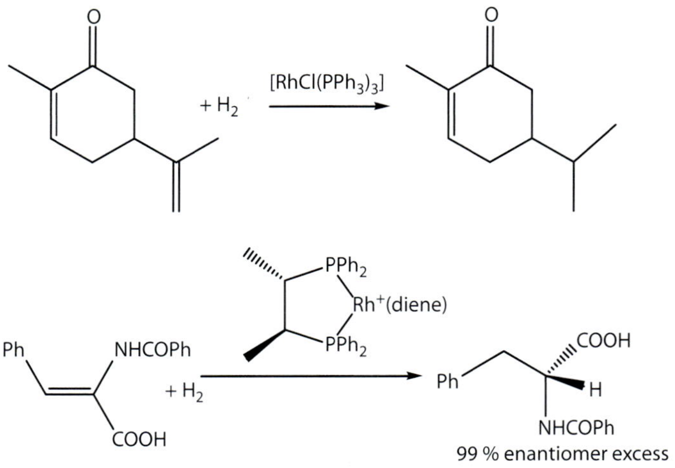

FIGURE 11.92 Homogeneous catalysis of hydrogenation by metal complexes.

example, we are able to reduce a disubstituted double bond in the presence of a trisubstituted one (and a ketone, though selecting alkenes over ketones is not difficult), using tris(triphenylphosphine) rhodium(I) chloride, [RhCl(PPh_3)$_3$], Wilkinson's catalyst. In the second example, a chiral rhodium complex catalyzes an enantioselective reaction. One of the most reactive catalysts ever reported was the lutetium complex **11.46**; this complex turns over 120,000 molecules per molecule of catalyst per hour, at room temperature and pressure, for the reduction of 1-hexene.

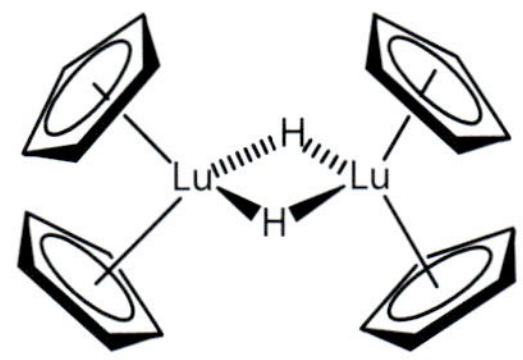

11.46

Most of the catalysts that will reduce alkenes to alkanes will also reduce alkynes to alkanes. However, since alkynes are more easily reduced than alkenes, it is, both in principle and practice, possible to reduce alkynes to *cis*-alkenes, if an appropriate catalyst is used. The most popular catalyst is a deactivated palladium species, called the Lindlar catalyst. This is prepared by precipitating palladium onto calcium carbonate, then deactivating the catalyst with lead acetate and quinoline, **11.47**. However, a range of palladium catalysts with various deactivating additives have also been employed. Examples are given in Figure 11.93. The first example was used in the synthesis of the pigment β-carotene and the second in the preparation of a prostaglandin.

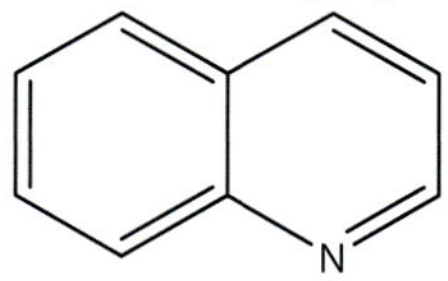

11.47

We can summarize the relative reactivities of various multiple bonds toward standard conditions of hydrogenation as alkyne > alkene > C=N > C=O > arene. This tells us that we can reduce alkynes in the presence of alkenes, and we can hydrogenate alkenes in the presence of ketones and aromatic rings.

What can we do if we want the *trans*-alkene rather than the *cis*-isomer from alkyne reduction? This can be accomplished using a "dissolving metal" reduction. When an alkali metal such as sodium is added to liquid ammonia, it is ionized to give solvated electrons (these are blue, but that's a story for the physical chemists…). One electron is added to the alkyne to give a radical anion (Figure 11.94). Because electrons repel each other, the orbitals containing the lone pair and the odd electron are on opposite sides of the triple bond. The lone pair is protonated by the solvent; then a further electron and proton are added to complete the process. Thus, 4-octyne is cleanly reduced to *trans*-4-octene.

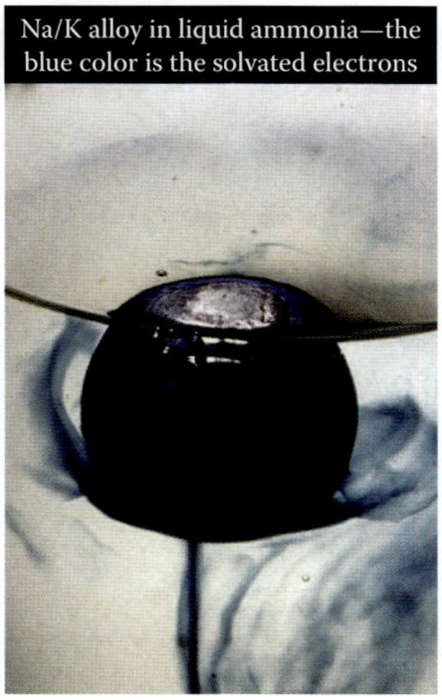

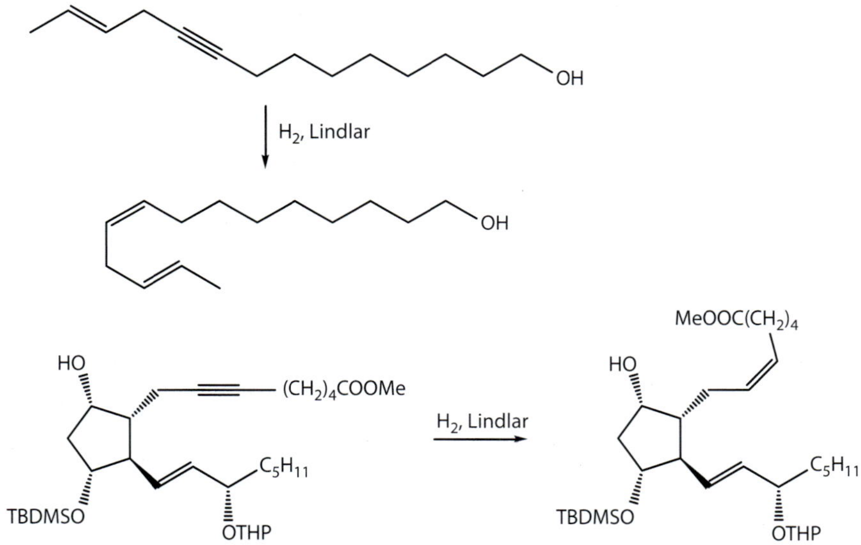

FIGURE 11.93 Partial reduction of alkynes to *cis*-alkenes.

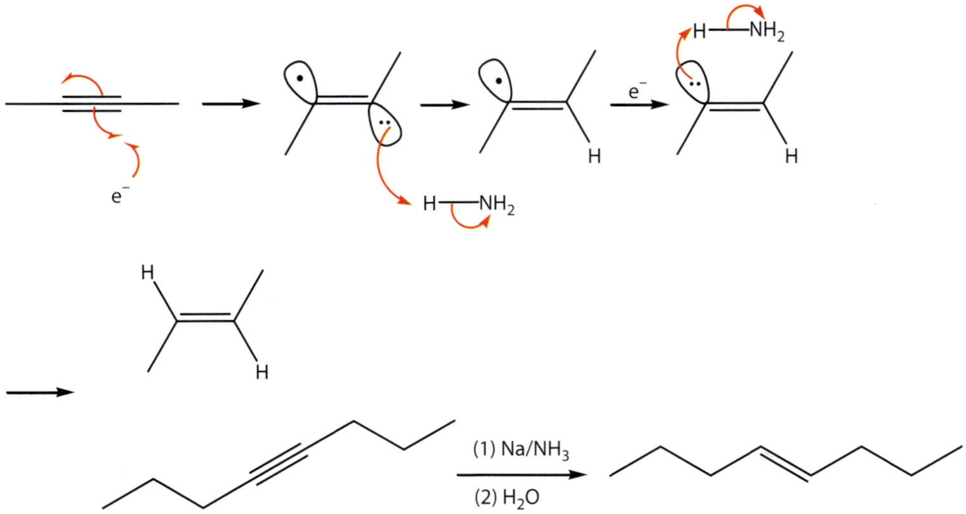

FIGURE 11.94 Dissolving metal reduction of alkynes.

PROBLEM 11.17

Propose reagents/catalysts/conditions for each of the following processes:

(a)

(b)

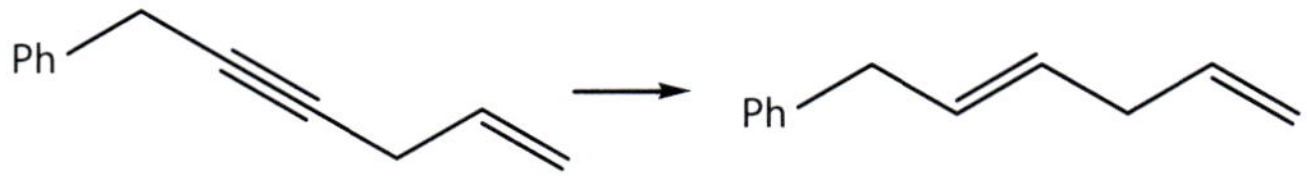

(c)

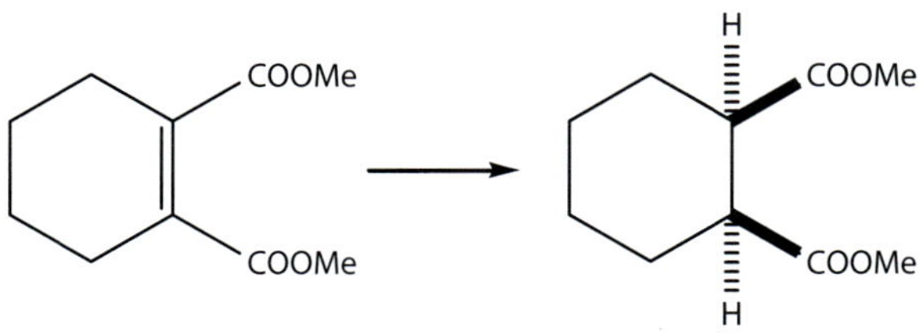

Solutions

(a) Hydrogen with most deactivated palladium catalysts would work here; in fact, the Lindlar catalyst was used.
(b) This is also a fairly standard reaction—an alkali metal, in this case sodium, in liquid ammonia. These reactions are generally worked up with water.
(c) Hydrogen with any of a wide variety of catalysts would be appropriate here; there is little likelihood of reduction of the ester carbonyl groups. In practice, platinum was used as the catalyst.

Focus on Fats: Saturated, Unsaturated, Polyunsaturated, and *trans*

Fats are one of a group of compounds described as lipids. The compounds discussed here are triglycerides, the esters of propane-1,2,3-triol, known as glycerol, with long-chain fatty acids (Figure 11.95). All naturally occurring fatty acids have even numbers of carbon atoms, because they are built up biosynthetically from two-carbon units.

The fatty acids that concern us from a dietary point of view range in length from C_{12} to C_{18}. The acids may contain one or more double bonds; the presence of the double bonds lowers the melting point of the fat, because the chains pack less well. In the C_{18} series (Figure 11.96), we can identify stearic acid (saturated, mainly found in animal fats but also in cocoa butter), oleic acid (one double bond, the triglyceride is the main constituent of olive oil) linoleic acid (two double bonds, one of the "essential" fatty acids that we must consume for good health, found in safflower, sunflower, and corn oil), and linolenic acid (three double bonds, another essential fatty acid, found in canola, soya, and walnuts). In this series (and related ones), the saturated fats are solids, whereas the unsaturated fats are liquids at room temperature. So in order to

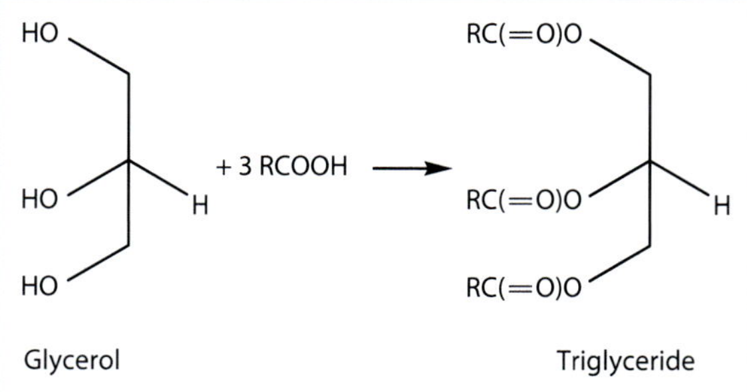

FIGURE 11.95 Formation of triglycerides.

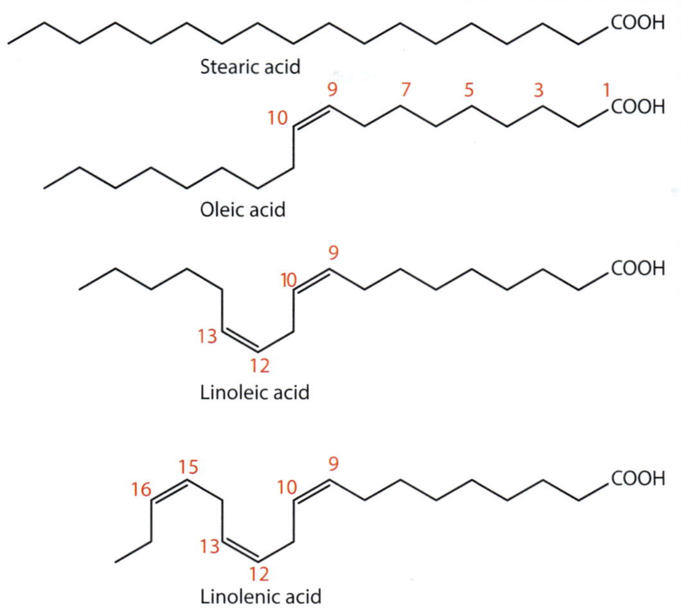

FIGURE 11.96 C_{18} fatty acids.

convert a liquid fat into a solid one—for example, to give a spread or for cake baking—some of the double bonds in a fat must be hydrogenated, essentially to give margarine (Did you know that the yellow color of margarine is usually added artificially, something that was illegal in Ontario, Canada, until 1995?). This is done by controlled hydrogenation over a metal catalyst—because saturated fats are considered to be less healthy than unsaturated ones, it's better for you and better for the manufacturers' marketing to advertise their spread as "rich in polyunsaturates," so hydrogenation was only pursued far enough to get a solid.

The triglyceride of oleic acid is the main constituent of olive oil (Thanks to Camilla Tornoe for this image.)

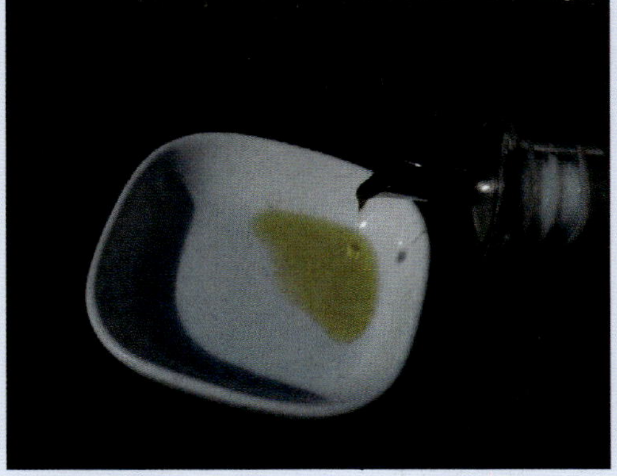

Looking at how the double bonds are arranged in the natural materials, you will see that nature is very good at chemistry. These double bonds are what we call "skipped" double bonds; there is a saturated carbon between them in the chain. They also all have *cis*-stereochemistry. Neither of these situations is thermodynamically that of the lowest energy—double bonds "prefer" to be conjugated and *trans*. So there is a problem, if we wish to "harden" fats by partially hydrogenating them. Unfortunately, when fats are partially hydrogenated chemically (usually a nickel catalyst), thermodynamics tends to take over, and some *trans*-fats are produced. There is significant evidence that consumption of *trans*-fats is unhealthy, with correlations with heart disease, diabetes, and Alzheimer's disease. The response of the food industry, both to these findings and subsequent regulation, has varied widely around the world. In 2007, Crisco, in the United States, reformulated its product to avoid the presence of *trans*-fats. California has banned the use of any *trans*-fats in restaurants. In the United Kingdom, partially hydrogenated fats have long been banned in margarine (though not in other processed foods), and margarine must be fortified with vitamins A and D. Denmark and Switzerland have banned the use of partially hydrogenated fats, containing *trans*-fats, in any food products. Solid margarines are now mostly produced by using a different mixture of vegetable fats, with higher proportions of the more saturated (and semisolid) palm oil and palm kernel oils. The moral of the story is too look VERY carefully at ingredient lists when buying fats; it may look good and taste good, but it does not necessarily do you good.

Key Points from Section 11.5

- Alkenes may be readily hydrogenated over a wide range of metal catalysts, with *cis*-addition of the hydrogen.
- Regio- and stereoselectivity are usually determined by steric hindrance with heterogeneous catalysts, but there is also a wide range of homogeneous catalysts able to produce other selectivity patterns.
- Alkynes may be partially reduced to *cis*-alkenes using hydrogen and a deactivated palladium catalyst and to *trans*-alkenes using Na/liquid ammonia.
- Aromatic rings, carbonyl groups, and nitrile groups are much more resistant to hydrogenation than alkenes and alkynes and are generally unaffected by standard reduction conditions.

REVIEW PROBLEMS

1. Give a structure for each of the following compounds:
   (a) 4-Methyl-*trans*-2-hexene
   (b) 3-Phenyl-1-butyne
   (c) *E*-8-Chloro-3-ethyl-7-methyl-1,5-decadiene
   (d) 1,5-Cyclooctadiene
   (e) *trans*-2-Hexene-5-yne

2. Give a systematic name for each of the following molecules:
   (a)

   (b)

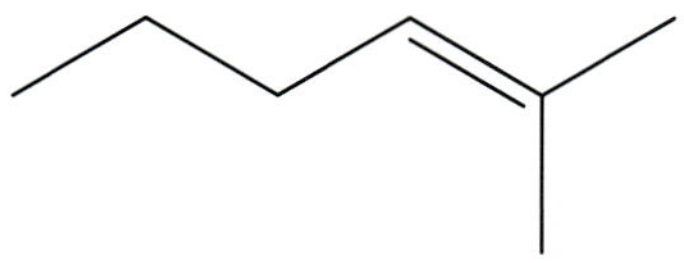

   (c)

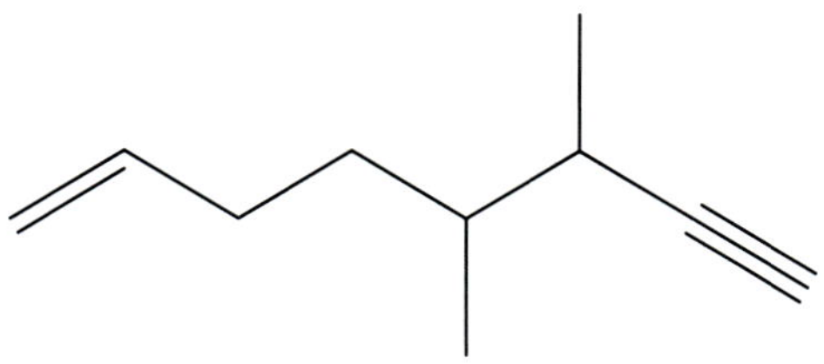

(d)

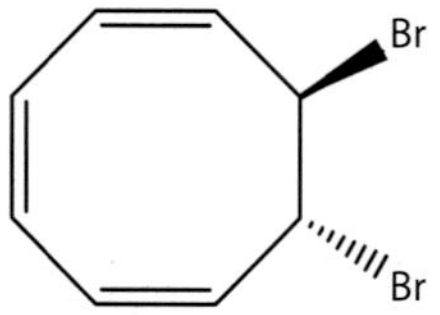

(e)

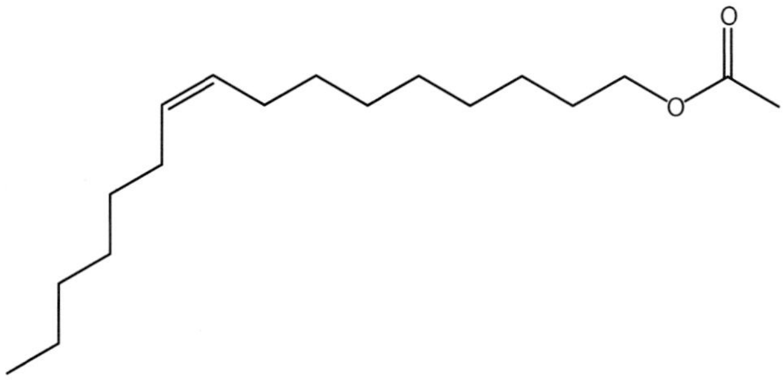

3. Protonate each of the following alkenes to give the most stable carbocation:

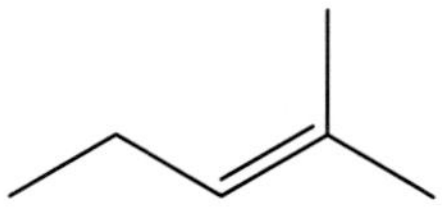
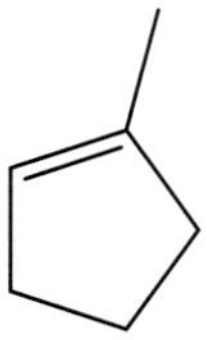
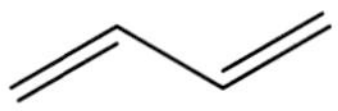
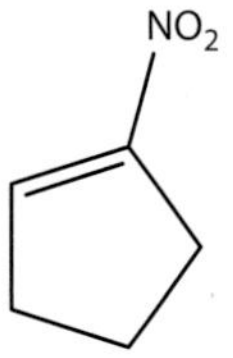
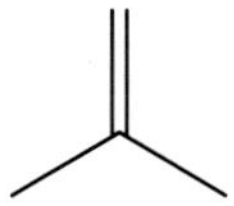
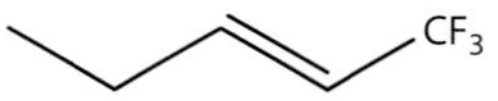

4. Each of the following alkenes can add the thiophenyl radical, PhS, to give a carbon radical product, for example:

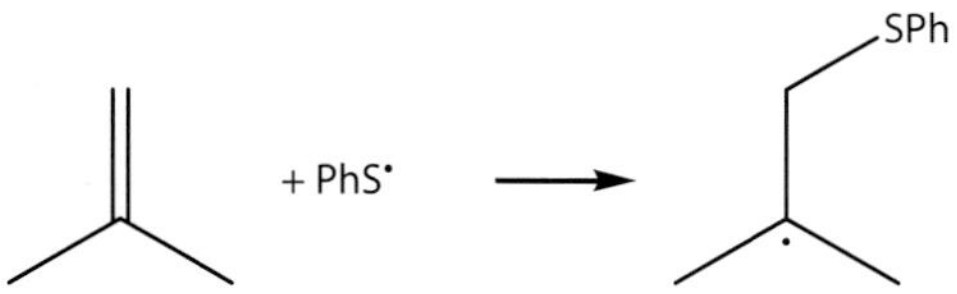

Which radical would be formed by adding PhS· to each of the following?

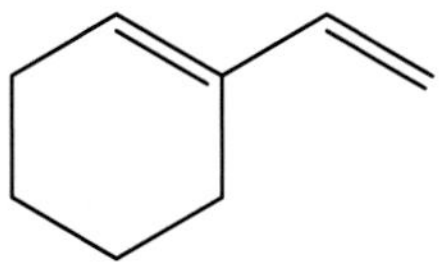
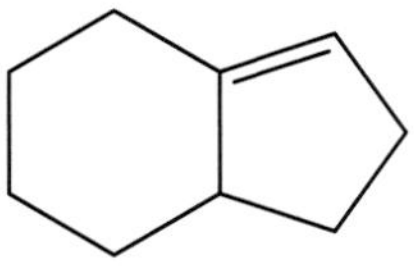
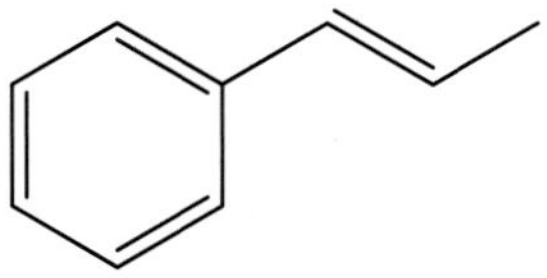

5. Write a mechanism for each of the following reactions:

(a)

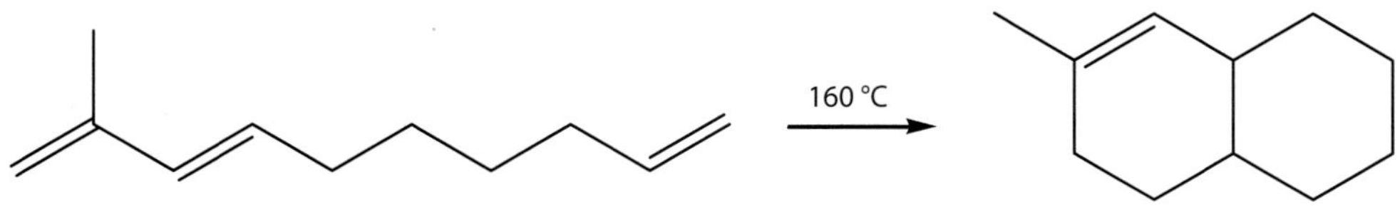

(b)

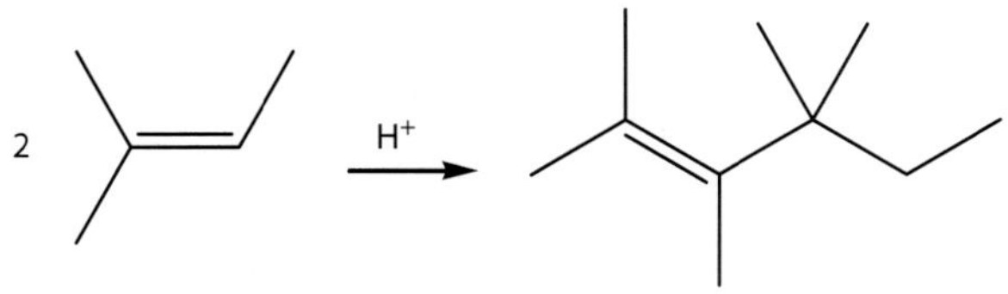

(c)

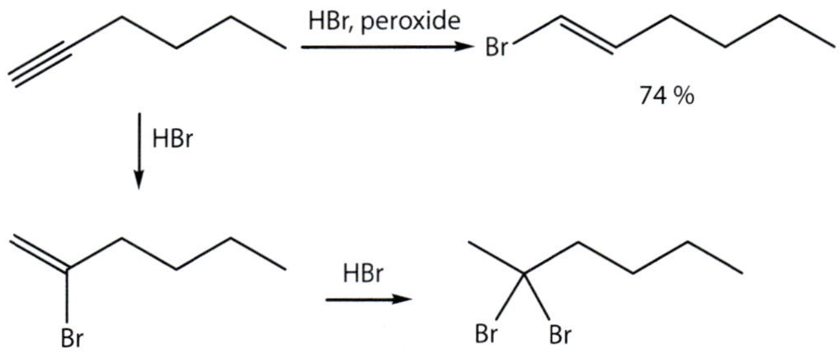
HBr, peroxide
Br
74 %
HBr
Br
HBr
Br
Br

(d)

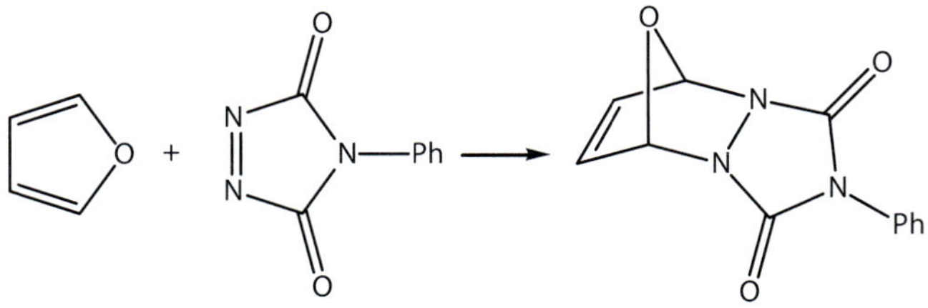
O
N
N
N
Ph
O
+
O
N
N
N
O
O
Ph

(e)

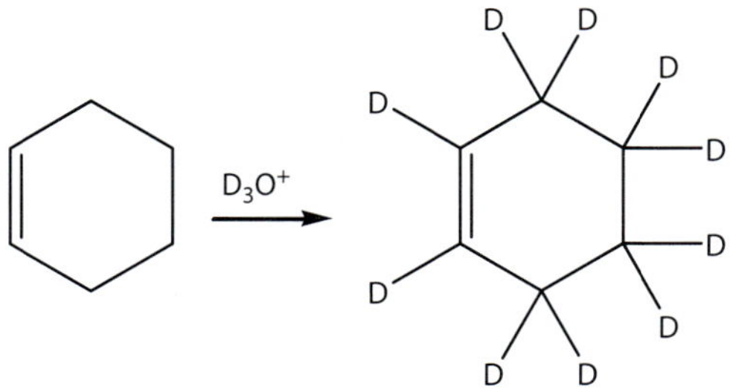
D₃O⁺
D
D
D
D
D
D
D
D
D
D

(f)

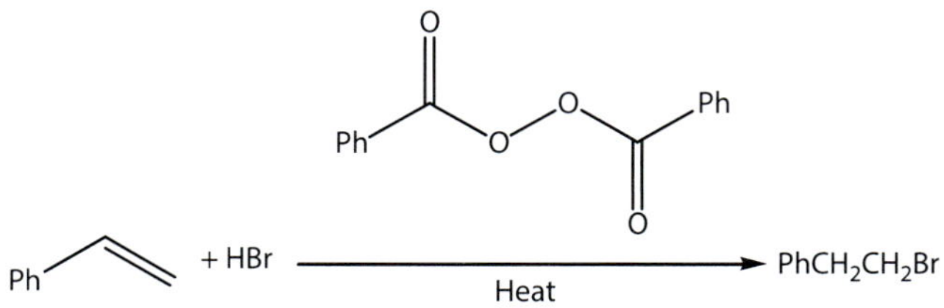
O
Ph
O
O
Ph
O
Ph
+ HBr
Heat
PhCH₂CH₂Br

(g)

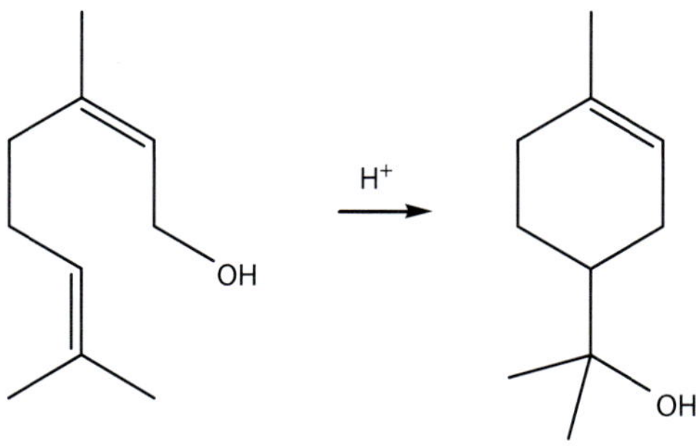
OH
H⁺
OH

(h)

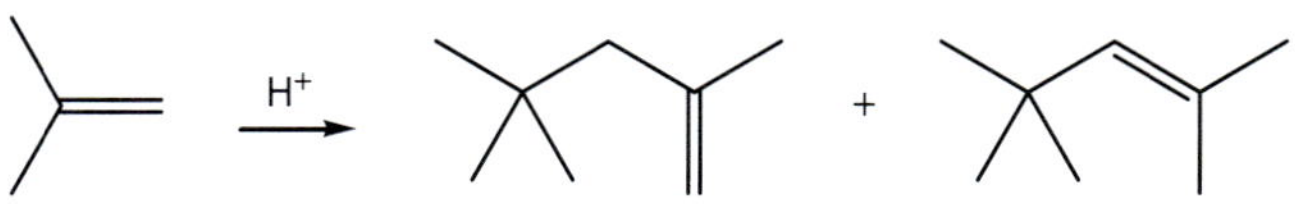
H⁺
+

(i)

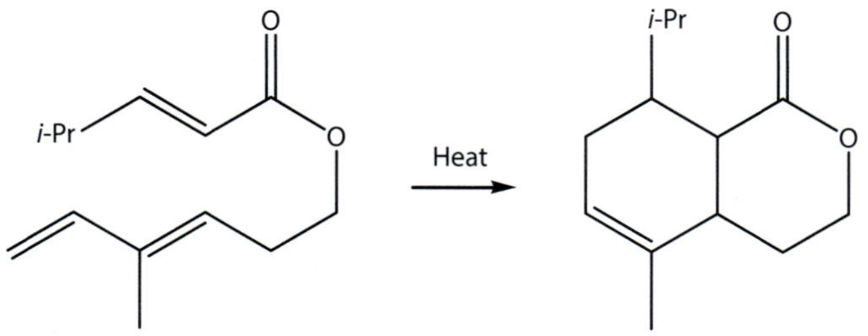

6. Predict the product(s) of each of the following reactions, paying attention to stereochemistry where appropriate:

(a)

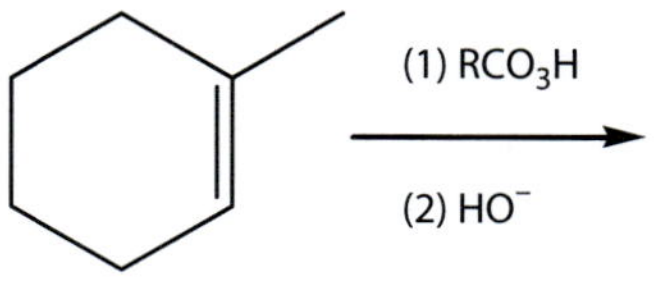

(b)

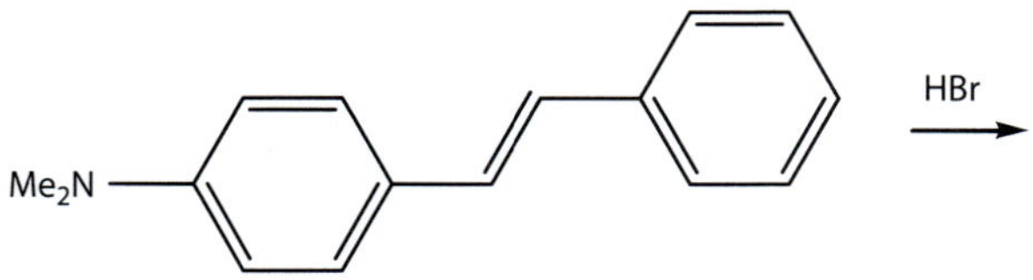

(c)

(d)

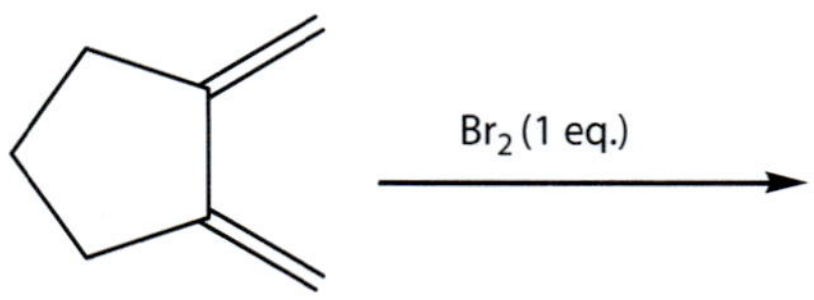

(e)

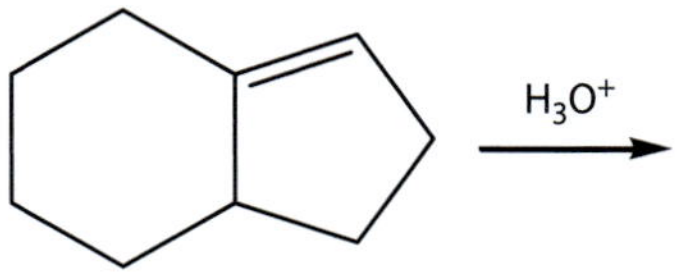

(f)

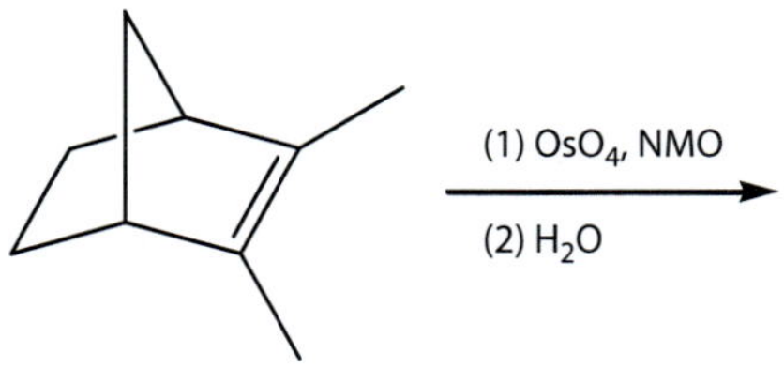

(g)

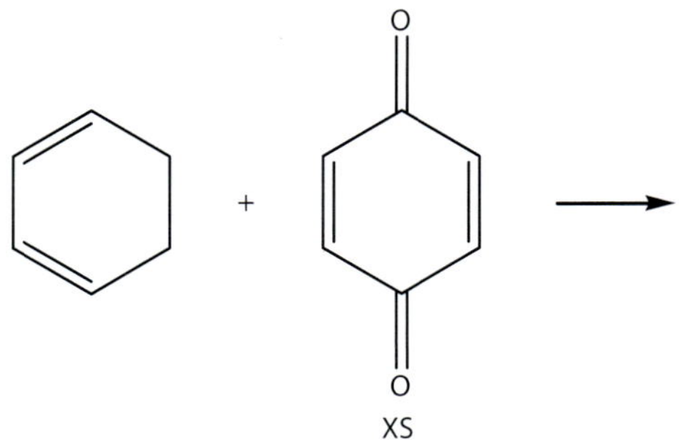
O
+
O
XS

(h)

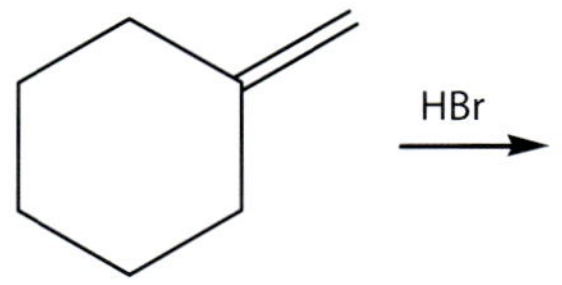
HBr

(i)

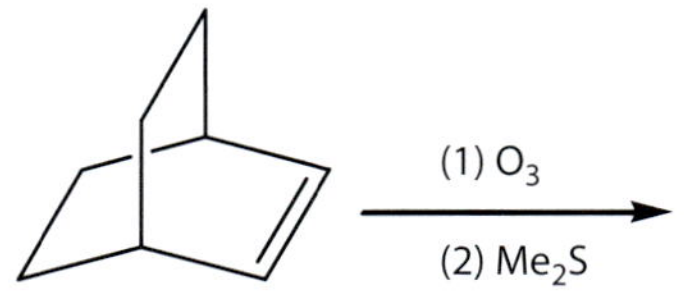
(1) O_3
(2) Me_2S

(j)

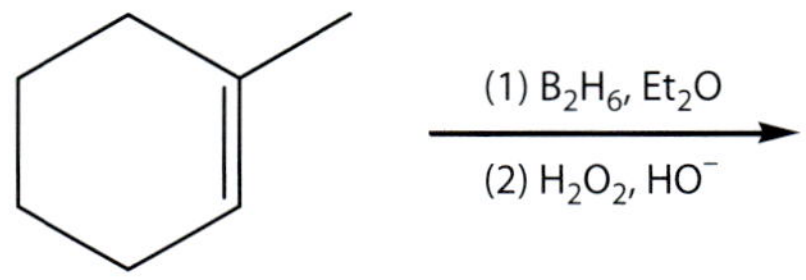
(1) B_2H_6, Et_2O
(2) H_2O_2, HO^-

(k)

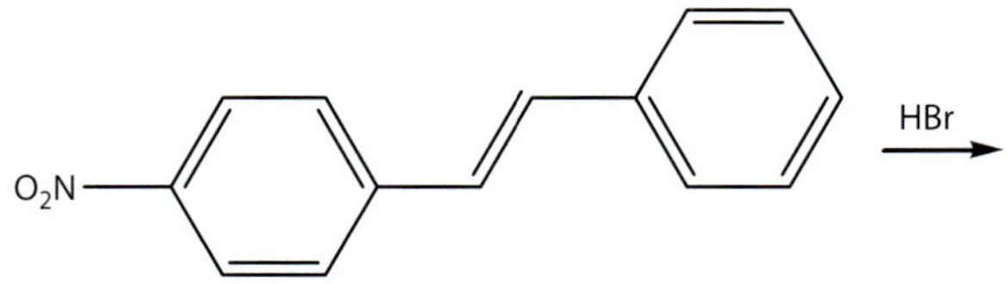
O_2N
HBr

(l)

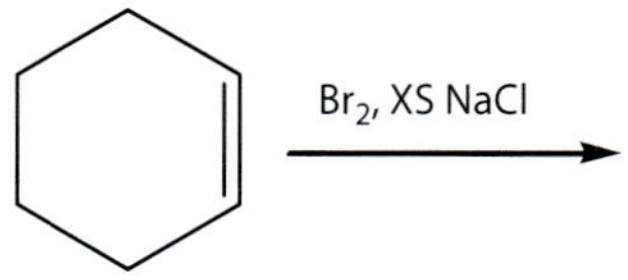
Br_2, XS NaCl

(m)

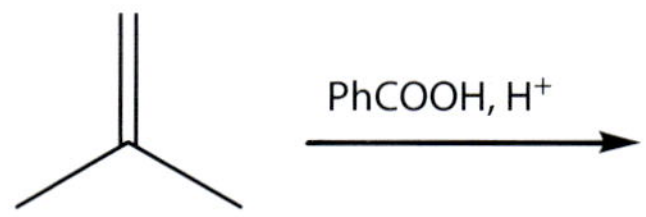
PhCOOH, H^+

(n)

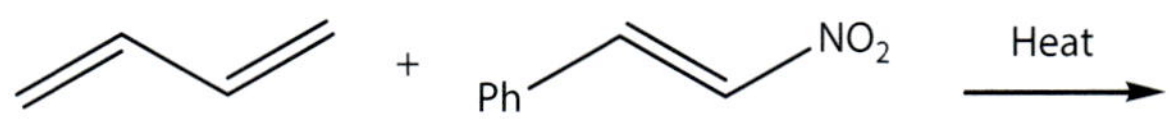
+
Ph
NO_2
Heat

7. How would you carry out each of the following transformations, which may require more than one step?

(a)

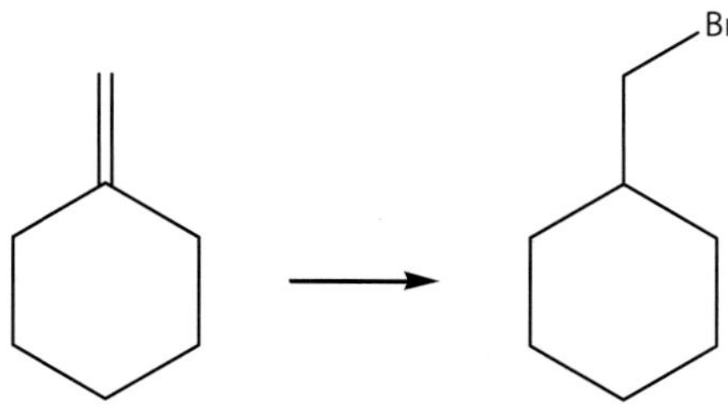

(b)

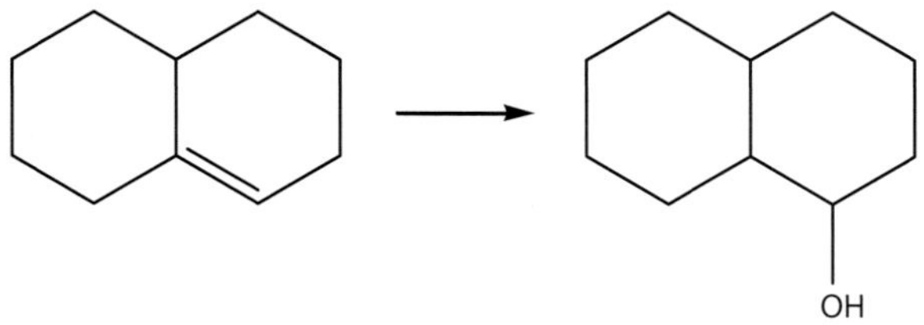

(c)

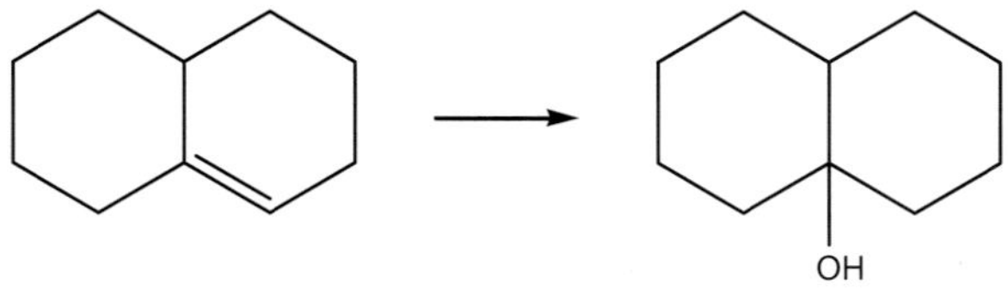

(d)

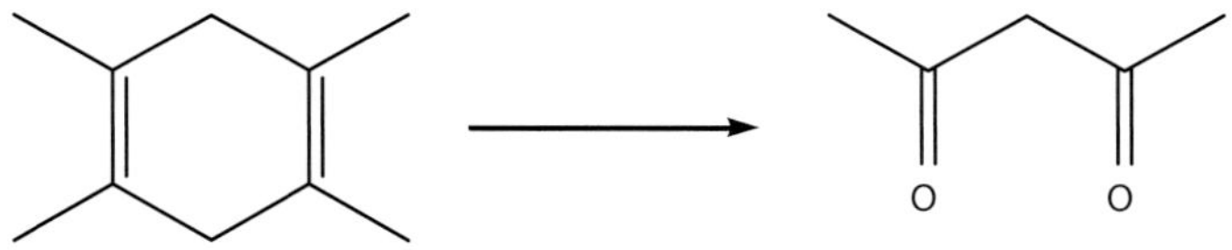

(e)

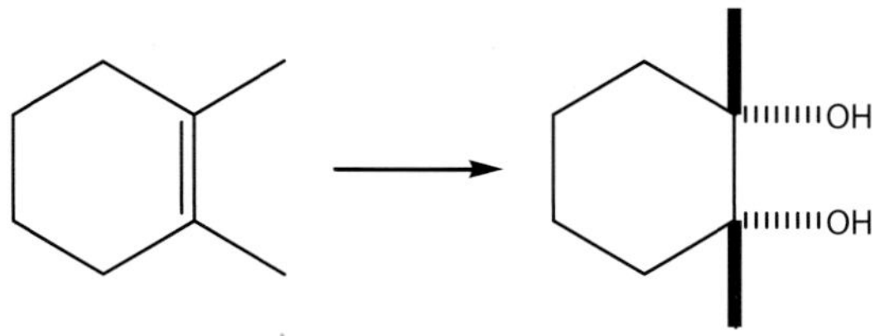

(f)

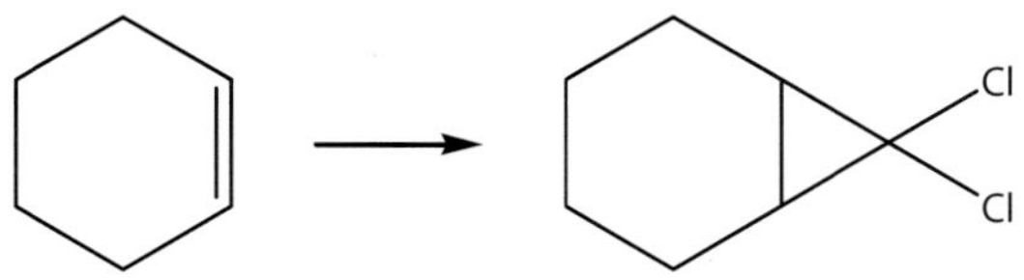

(g)

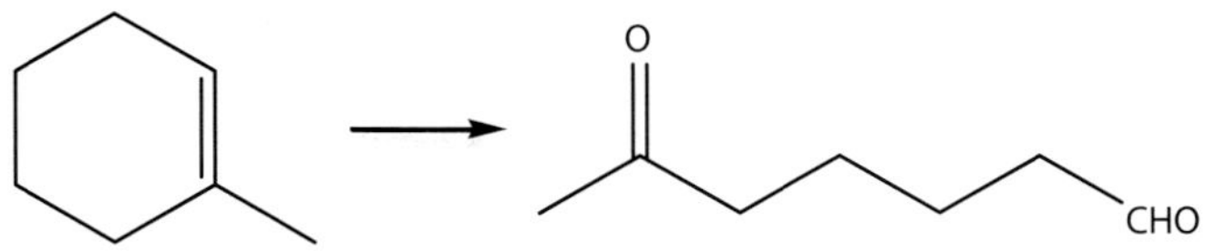

(h)

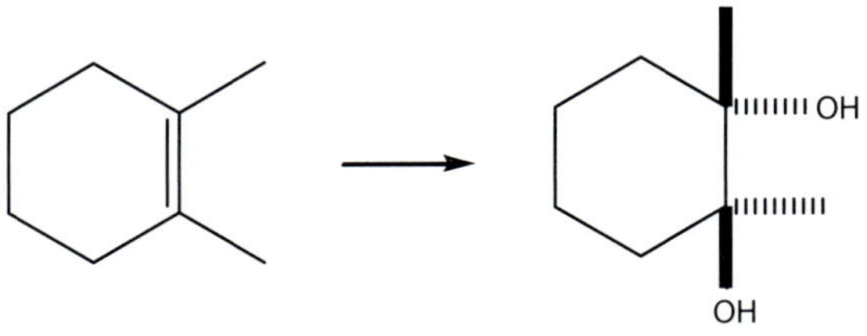

(i)

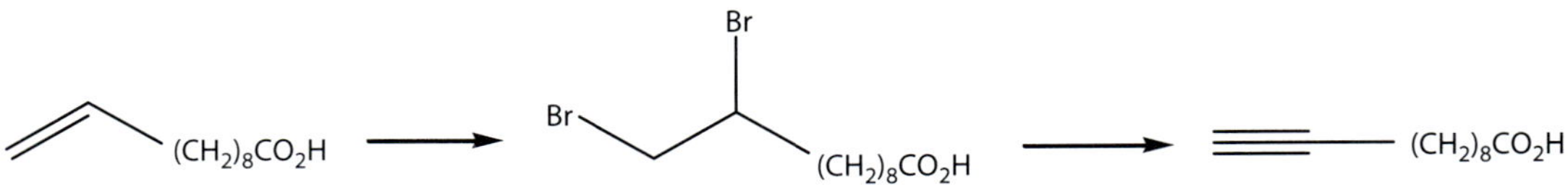

(j)

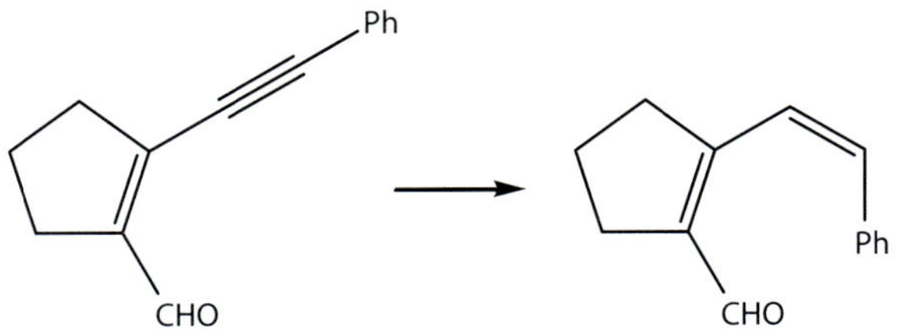

(k)

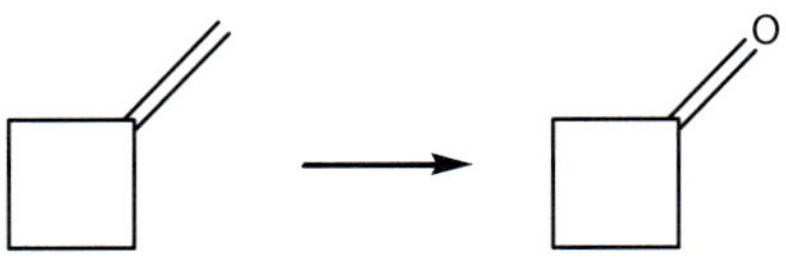

8. What starting materials would you need to prepare each of the following compounds by Diels–Alder reactions?

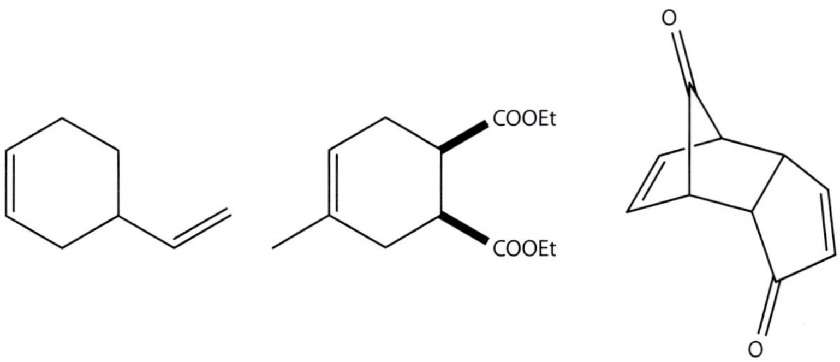

9. Identify the isoprene units in each of the following terpenes and characterize them as monoterpenes, diterpenes, *etc.*

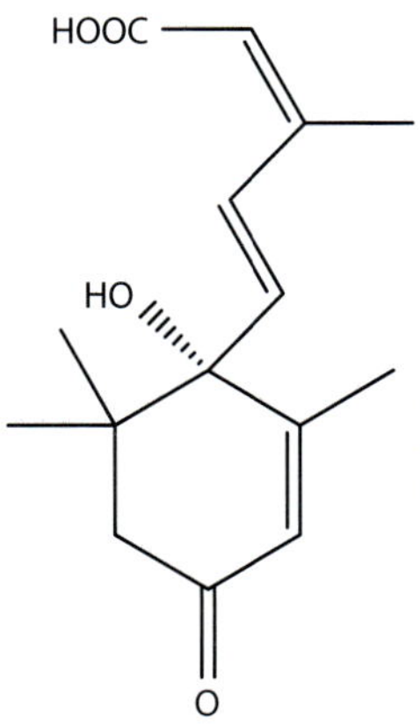

(+) Abscisic acid, a plant hormone controlling fruit ripening and stress response

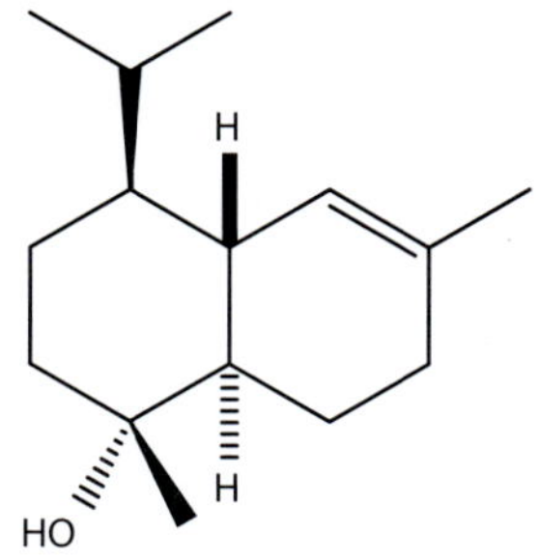

α-Cadinol

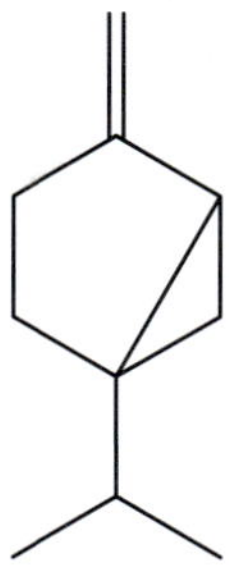

Sabinene

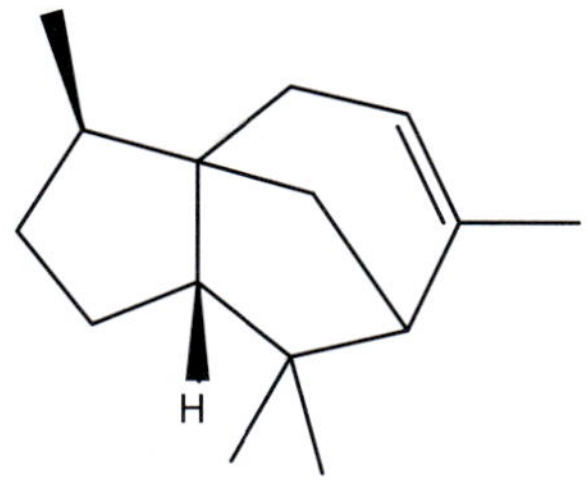

Cedrene

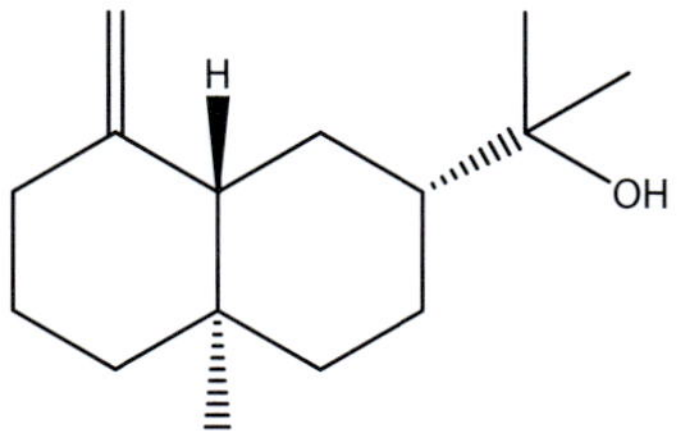

β-Eudesmol (from eucalyptus)

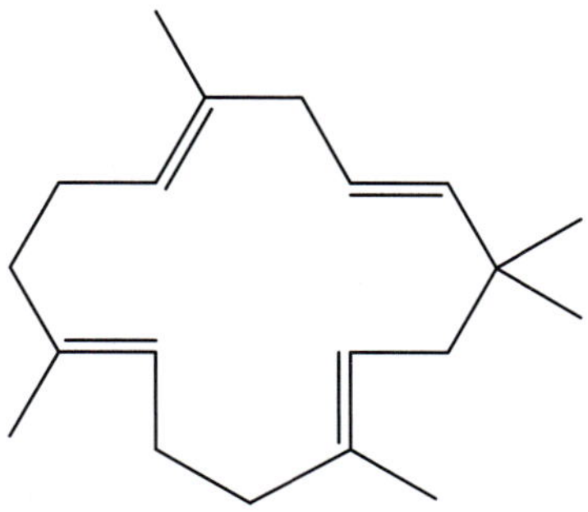

Flexibilene (isolated from a marine coral)

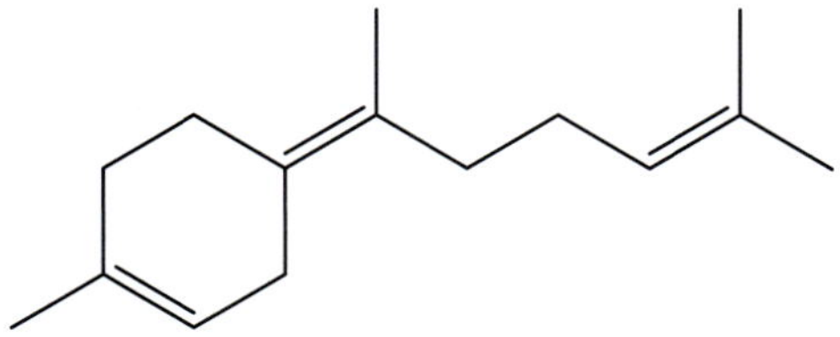

γ-Bisabolene

10. Explain the following observations, providing a full mechanism justifying your conclusion.

   (a) Dihydroxylation of *cis*-2-butene using OsO_4/NMO gives a *meso*-compound, but the reaction of the *trans*-isomer gives a racemate.

   (b)

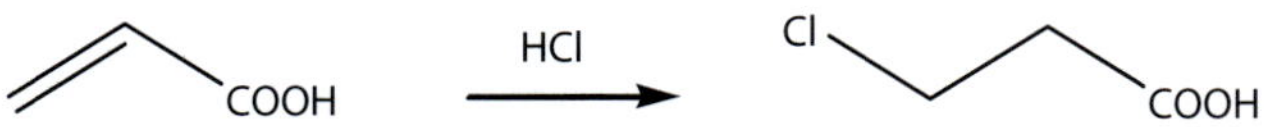

but

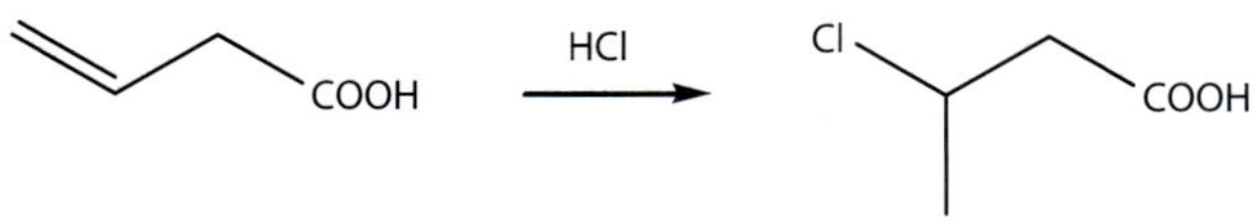

   (c) 1-Alkynes react with bromine and water to give bromoketones, for example:

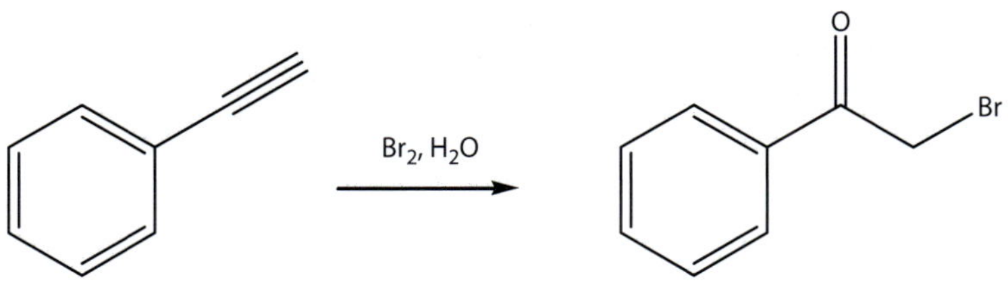

11. The bicyclic alkene, car-3-ene, a constituent of turpentine, undergoes catalytic hydrogenation to give only one of the two possible stereoisomeric products. Explain.

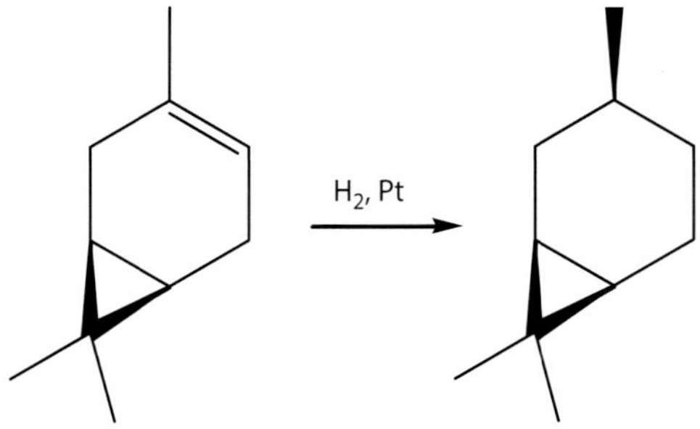

12. The relative rates of reaction of various alkenes with bromine, in methanol as solvent, are given in the following. Comment on these, explaining what it tells us about the nature of the reaction.

Alkene	**Relative Rate of Reaction**
$CH_2{=}CH_2$	3.0
$CH_3CH_2CH_2CH{=}CH_2$	2.9×10^3
Z-$CH_3CH_2CH_2CH{=}CHCH_3$	1.3×10^5
$(CH_3)_2C{=}C(CH_3)_2$	2.8×10^7

13. Both geraniol and its stereoisomer nerol are converted to α-terpineol and terpin in aqueous acid. Nerol reacts faster.

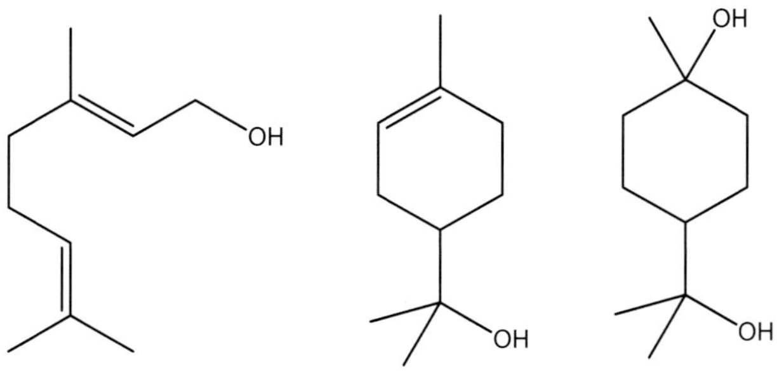

Geraniol α-Terpineol Terpin

(a) Propose a structure for nerol, and explain how nerol and geraniol can interconvert under acidic conditions.

(b) Propose a mechanism for the conversion of geraniol and nerol to terpin and α-terpineol.

14. The reaction of 3,3-dimethyl-1-butene with aqueous acid gives a mixture of two alcohols. Give their structures and account for their formation.

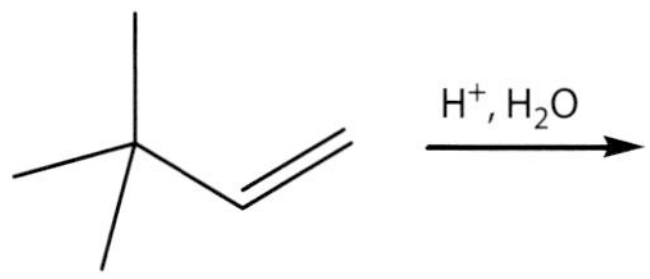

15. Cyclopentadiene, on storage, gives a dimer, which on heating regenerates the monomer. Suggest a structure, and write a mechanism.

16. Explain both the regiochemical and stereochemical outcomes in the following:

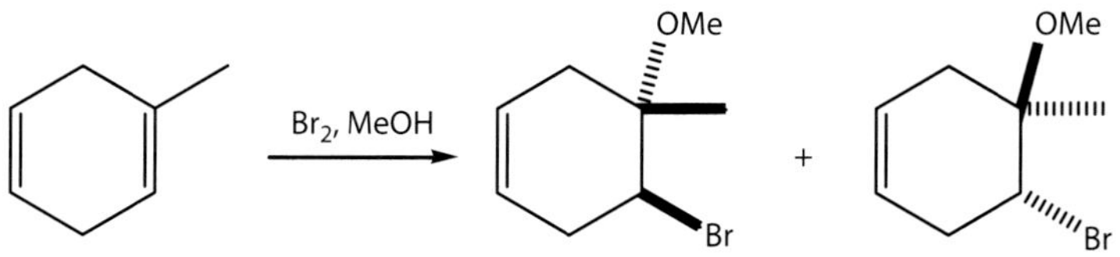

17. When compound **11.48** is reacted with 1 mol of peracetic acid, a single monoepoxide is cleanly formed in 86 % yield. Give a structure for the product and explain the selectivity of the reaction:

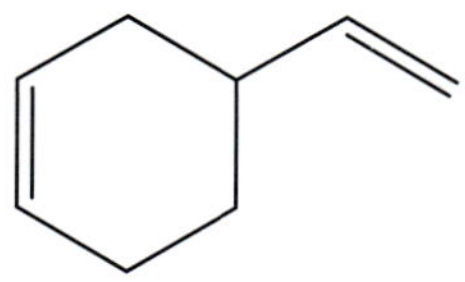

11.48

MORE CHALLENGING PROBLEMS

1. Propose a mechanism for each of the following transformations:

(a)

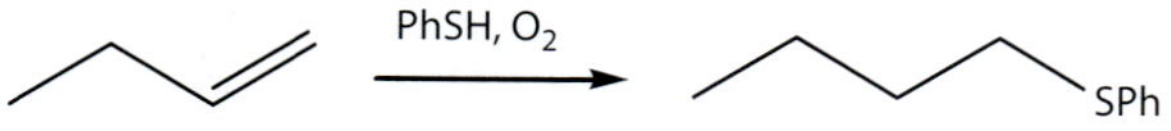

Remember O_2 is a radical initiator.

(b)

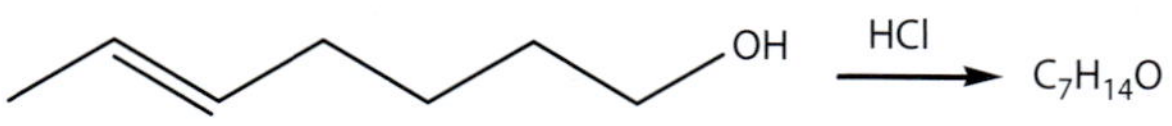

(c)

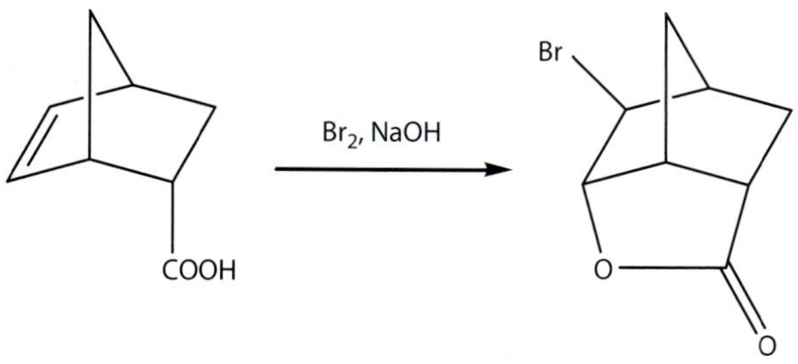

(d)

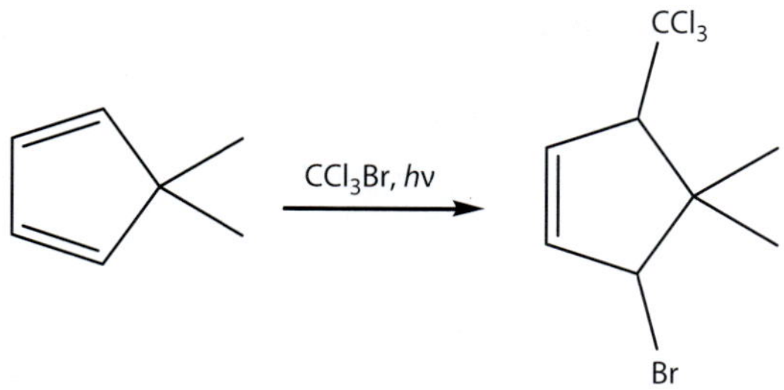

(e)

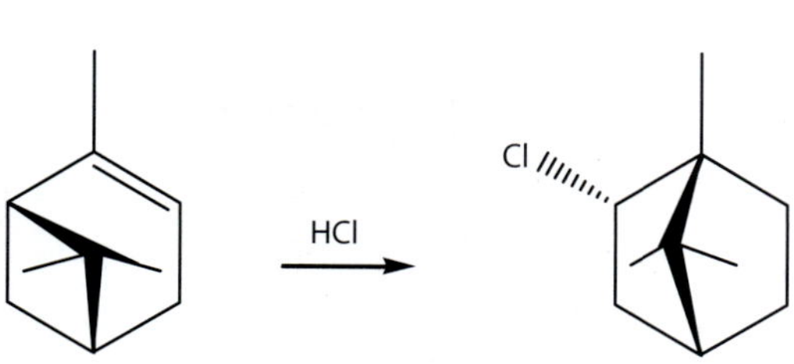

α-Pinene Bornyl chloride

(f)

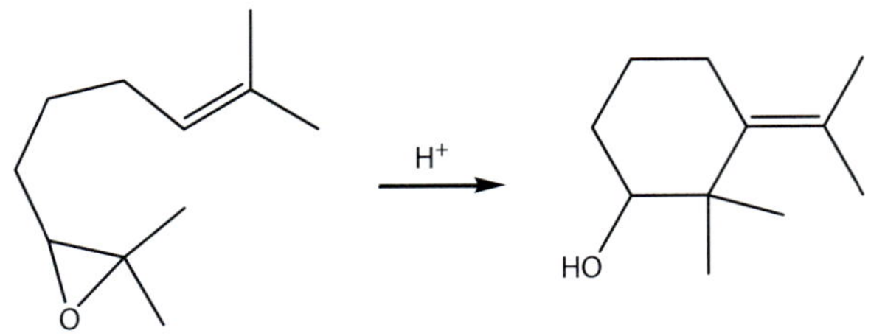

(g)

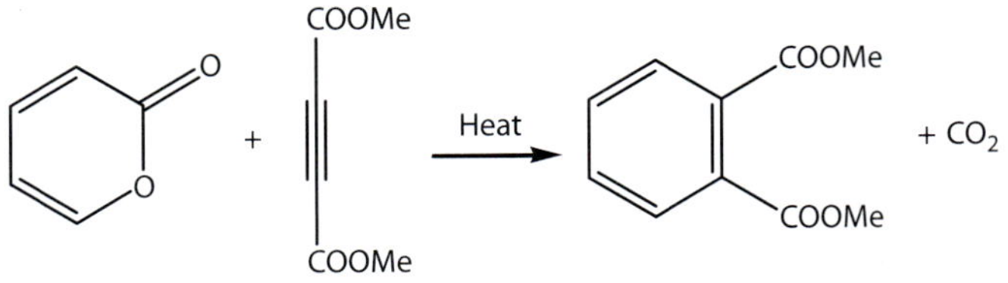

(h)

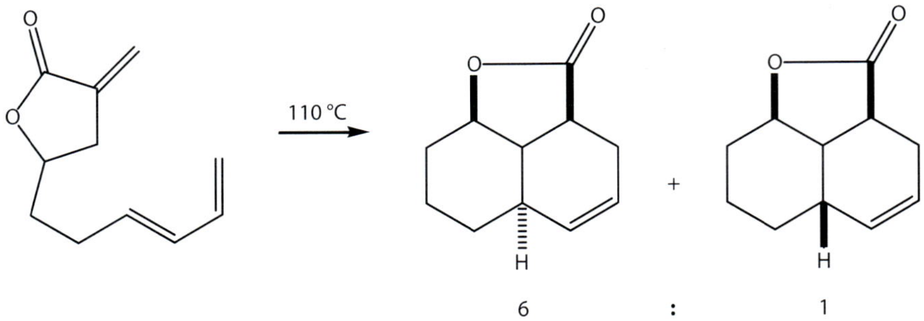

(i)

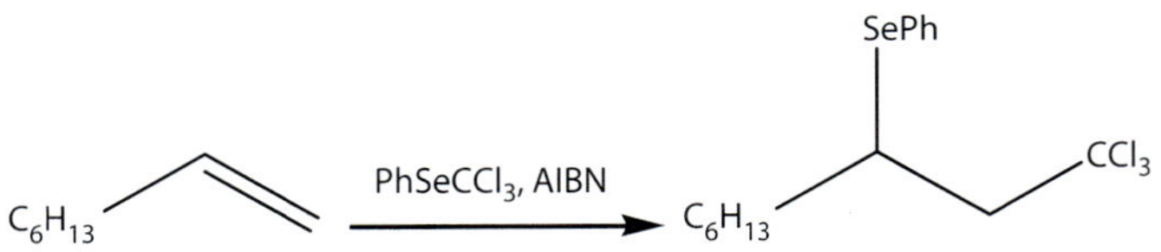

AIBN is a radical initiator.

(j)

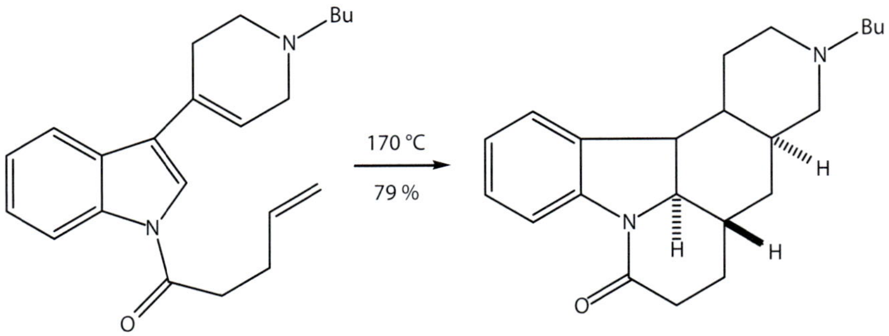

(k)

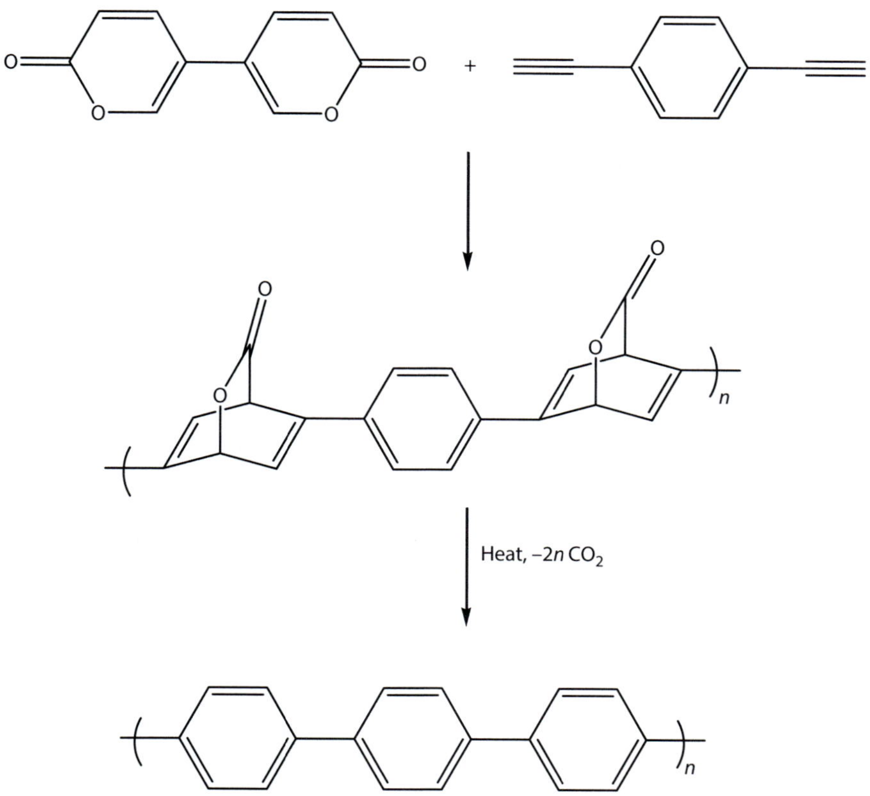

(l)

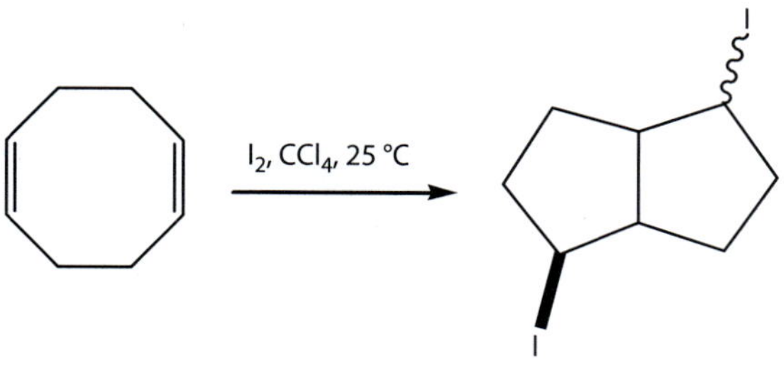

trans:cis = 49:51

2. Predict the products of each of the following reactions:

(a)

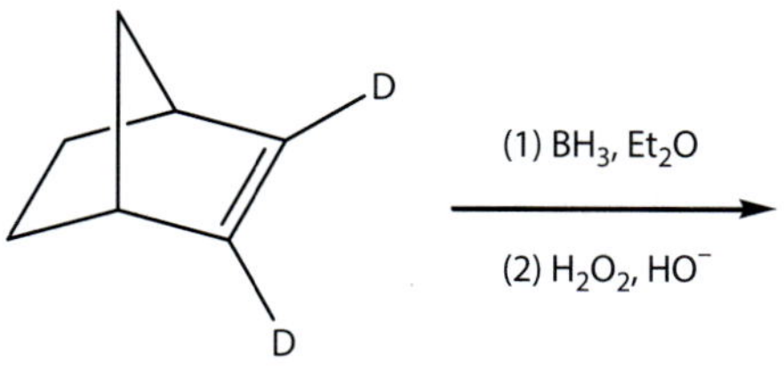

(b)

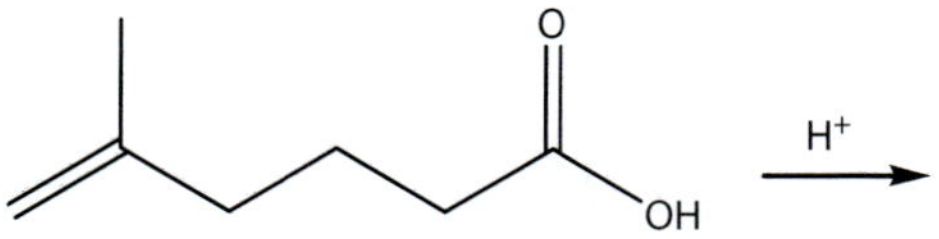

(c)

HCl

(d)

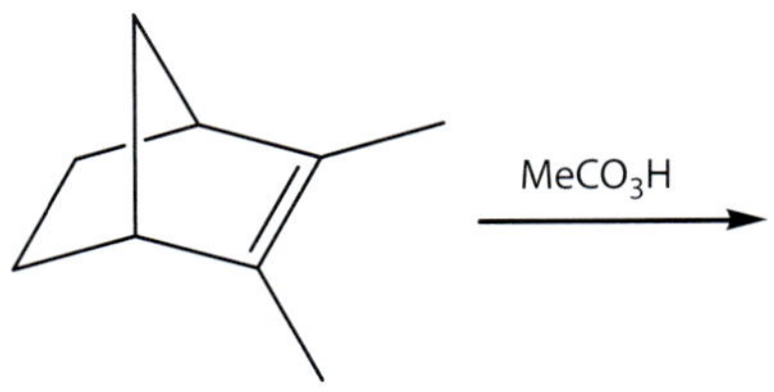
MeCO$_3$H

(e)

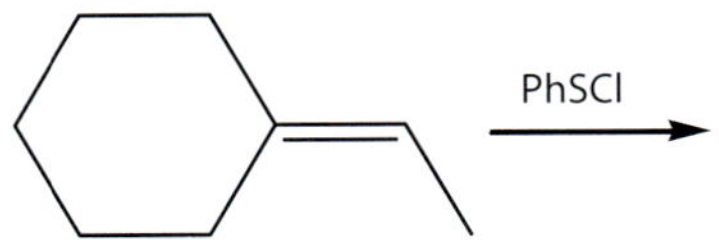
PhSCl

(f)

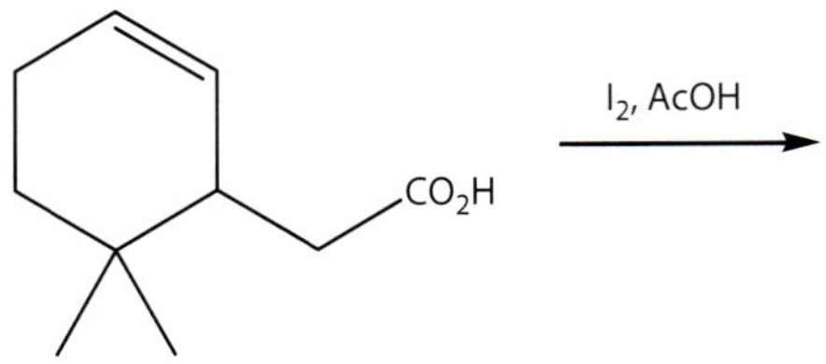
CO$_2$H
I$_2$, AcOH

(g)

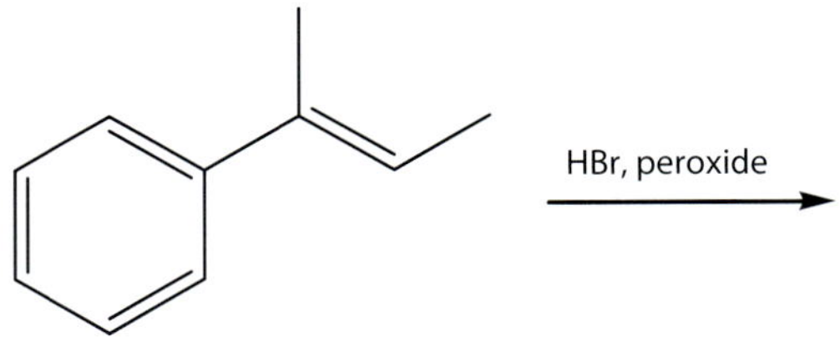
HBr, peroxide

(h)

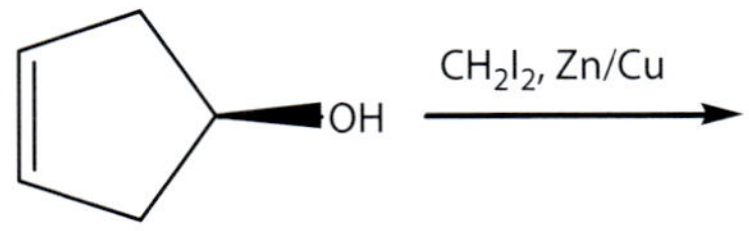
OH
CH$_2$I$_2$, Zn/Cu

(i)

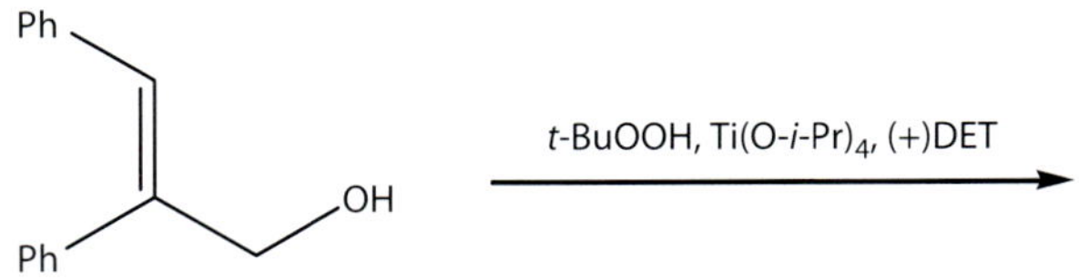
Ph
Ph
OH
t-BuOOH, Ti(O-*i*-Pr)$_4$, (+)DET

(j)

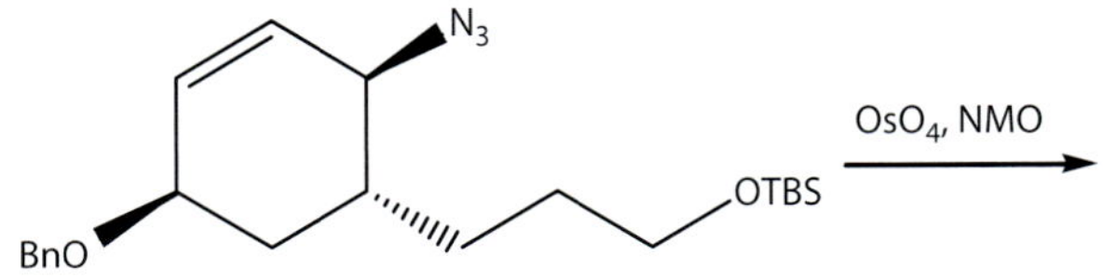
N$_3$
BnO
OTBS
OsO$_4$, NMO

3. How would you accomplish the following conversions, which may require more than one step?

(a)

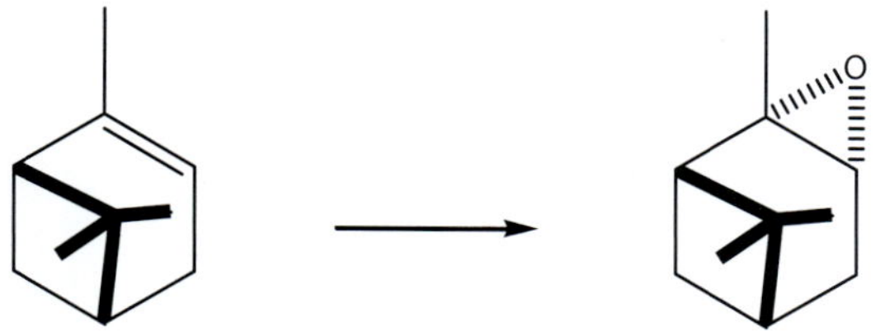

(b)

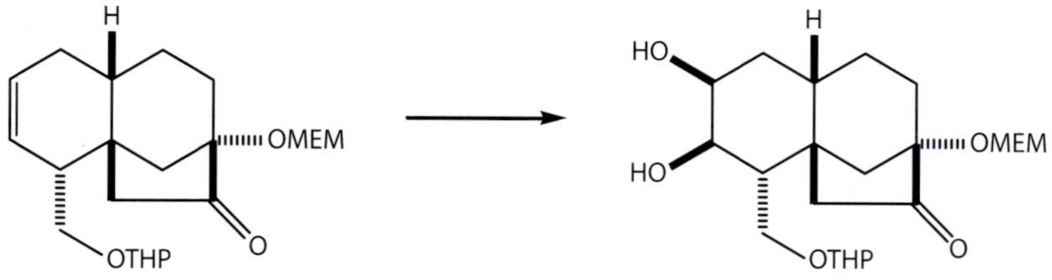

Used in synthesis of Gibberellic acid GA3

(c)

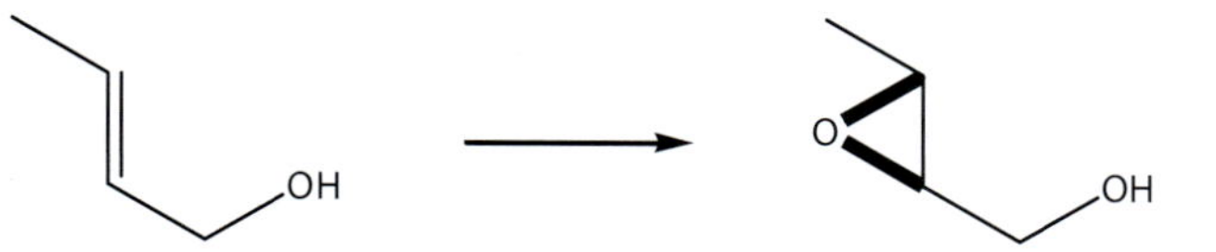

58 %, 95 % ee, used in erythromycin synthesis

(d)

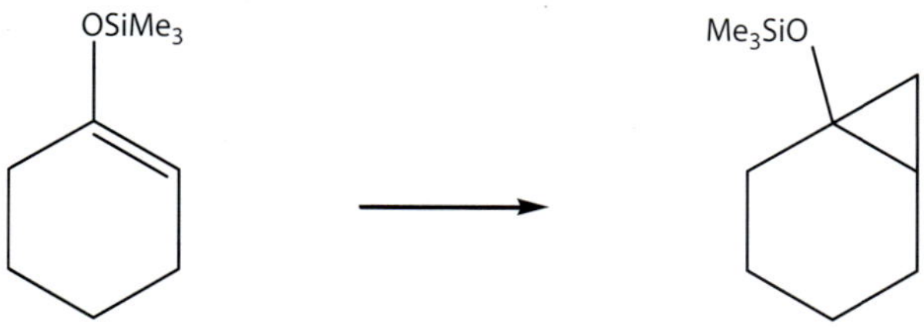

(e)

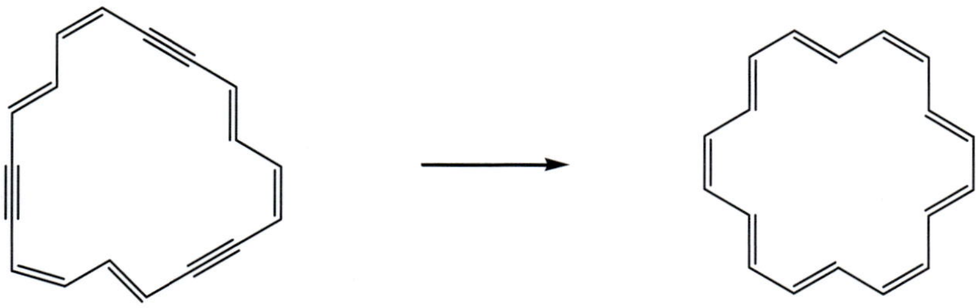

[18]-annulene

(f)

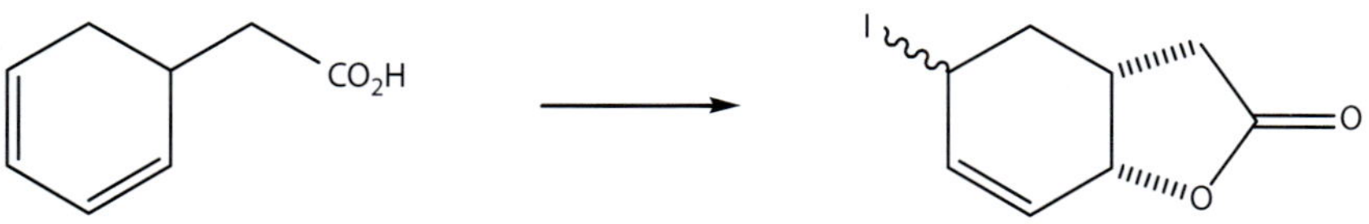

60 %

(g)

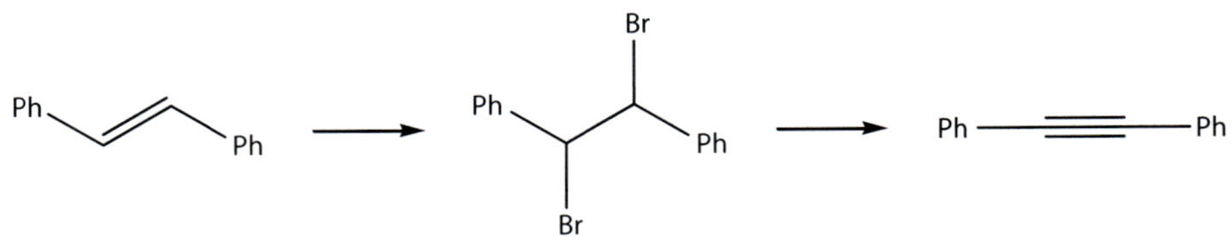

4. When *trans*-2-butene is treated with Br_2, 2,3-dibromobutane is formed. Is this the *meso*- or the racemic compound? Give a detailed mechanism to justify your conclusion.
5. What is the structure of the product of the reaction of cyclopentadiene with CH_2=CHCHO?
   (i) After a short time
   (ii) After a longer time and more intense heat
6. When HBr reacts with 3-methyl-1-butene, the products 2-bromo-3-methylbutane and 2-bromo-2-methylbutane are obtained in approximately equal amounts. Explain, writing a complete mechanism.
7. Under conditions of thermodynamic control, the addition of bromine to 1,3-cyclohexadiene gives 3,6-dibromocyclohexene and 3,4-dibromocyclohexene in comparable amounts. Explain.
8. Chloroethene, the monomer for the production of PVC, was originally prepared by the addition of HCl to ethyne. However, ethyne is very much more expensive than ethene. How could you prepare chloroethene from ethene in an economical process?
9. Study the following sequence, which is the standard preparation of 1,5-oxido-10-annulene.

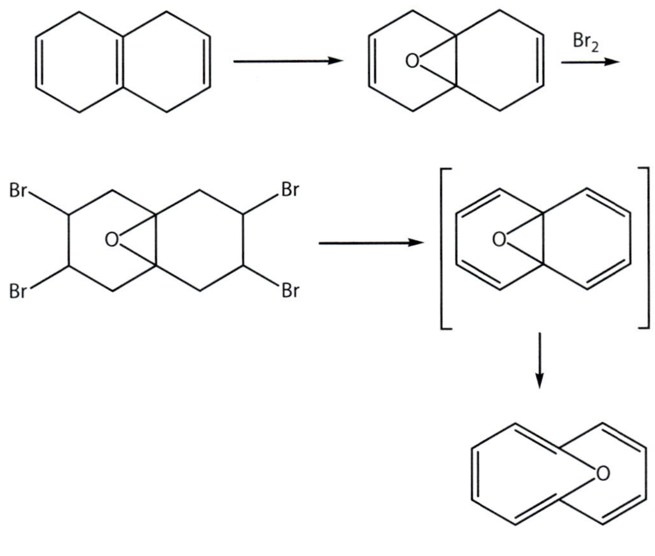

   (a) Which peracid would you choose to accomplish the epoxidation, and why is only one double bond oxidized under these conditions?
   (b) The stereochemistry of the bromine addition is not shown. What is it likely to be?
   (c) Suggest a reagent for dehydrobromination.
   (d) The last step of the reaction is a process we will not meet until Chapter 18. However, you should be able to write plausible arrows for this step.

Electrophilic Aromatic Substitution

12

12.1 INTRODUCTION

In Chapter 11, we saw how we could add atoms or groups across carbon–carbon multiple bonds (Figure 12.1). The double bond was "lost" and two new σ-bonds were formed, at the expense of one σ-bond (the A—B bond) and one π-bond (in the alkene). With few exceptions, this is thermodynamically favorable; σ-bonds are generally stronger than π-bonds.

The first group of reactions of alkenes that we considered was that of electrophilic reagents such as HBr or H_2O, and the first step of all these processes was the addition of an electrophile, usually a proton, to the double bond, to generate a carbocation. This carbocation was then captured by the nucleophilic part of the reagent to give a saturated product (Figure 12.2).

When we consider the reaction of electrophiles with the double bonds of aromatic compounds, the process starts in the same way; the electrophile adds to a double bond in the benzene ring, to give a positively charged species. This is much more stable than the carbocations that we made from simple alkenes, as it has three resonance forms (Figure 12.3).

However, the special stability conferred by aromaticity (review Section 3.2 if you need to) changes what happens next—it is now thermodynamically more favorable for the molecule to lose a proton to return to aromatic character than to add a nucleophile to the intermediate cation. So the typical reaction of benzene with electrophiles may be represented as in Figure 12.4.

The mechanism of the reaction is a two-step process—addition of an electrophile, followed by loss of a proton (Figure 12.5). Almost all the reactions we will meet in this section go by identical mechanisms, so that once we have mastered one, we know them all. The intermediate cation produced is called a Wheland intermediate or a σ-complex.

We need to establish which of the two steps is rate determining—and there is ample evidence to point us in the right direction. The nitrations of benzene, C_6H_6, and hexadeuterobenzene, C_6D_6, go at the same rate. This implies that C—H or C—D bond breaking is not the RDS. (If you need to review kinetic isotope effects, look back at Sections 8.2.4 and 10.4.) Thus, it follows that the addition of the electrophile must be rate determining.

Reactions of benzene with different electrophiles go at very different rates—we would only expect this if the first addition step was rate determining. In almost all the reactions we shall meet, the first step, the addition of the electrophile, is rate determining, and subsequent proton loss is fast. However, in a few very special cases, the intermediate cation has been observed or isolated (Figure 12.6).

In this case, the intermediate is stabilized because all the relevant resonance forms are tertiary cations (Figure 12.7).

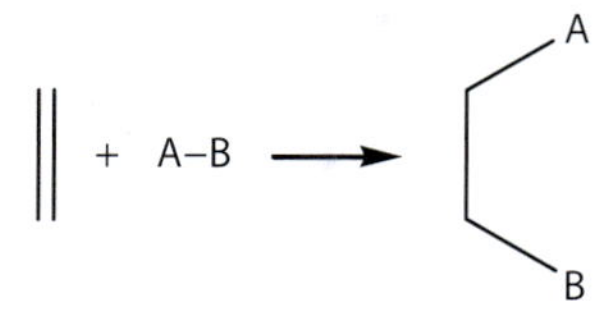

FIGURE 12.1 Addition to carbon–carbon double bond.

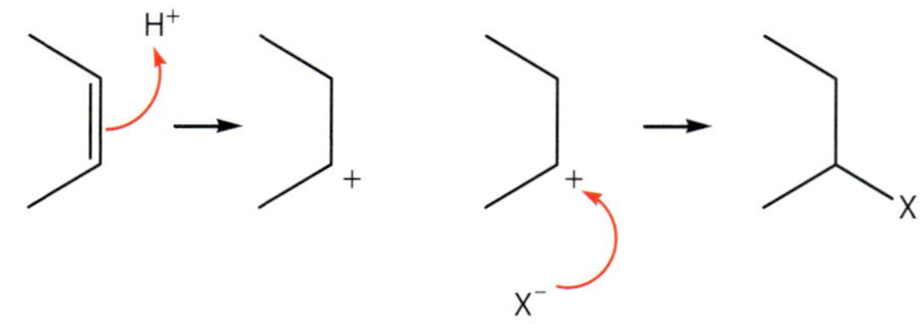

FIGURE 12.2 Addition of HX to alkenes.

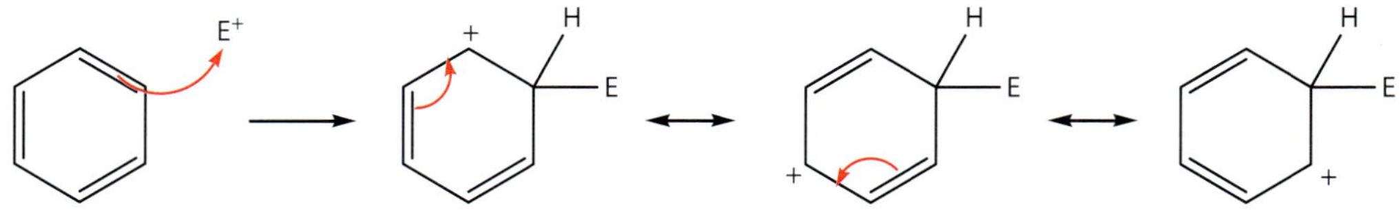

FIGURE 12.3 Reaction of benzene with an electrophile.

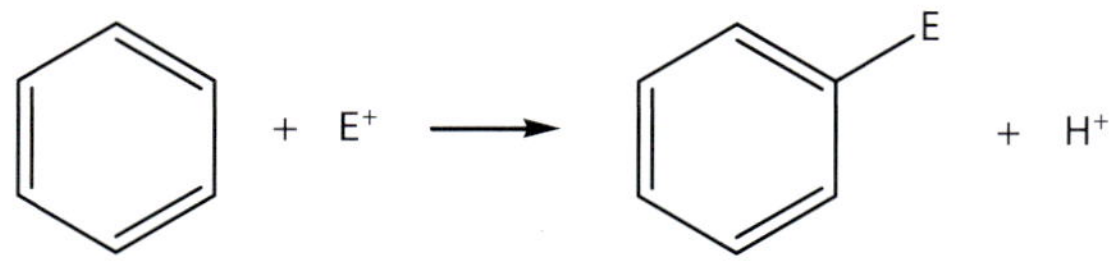

FIGURE 12.4 Electrophilic aromatic substitution. *Note:* E^+ is an electrophile such as H^+, Br^+ (Br_2), $[NO_2]^+$, $[Me_3C]^+$, $[CH_3C(=O)]^+$, *etc.*

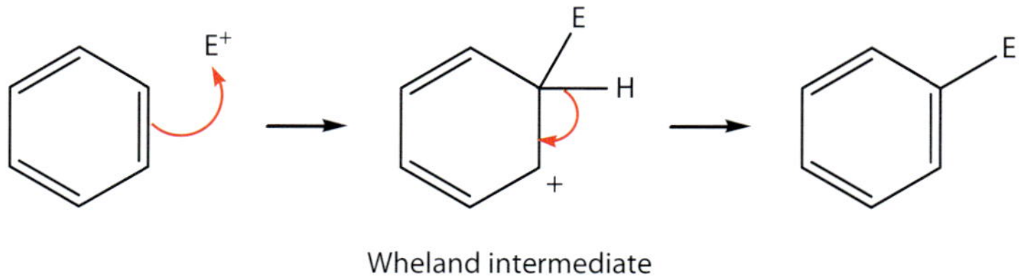

FIGURE 12.5 Mechanism of electrophilic aromatic substitution.

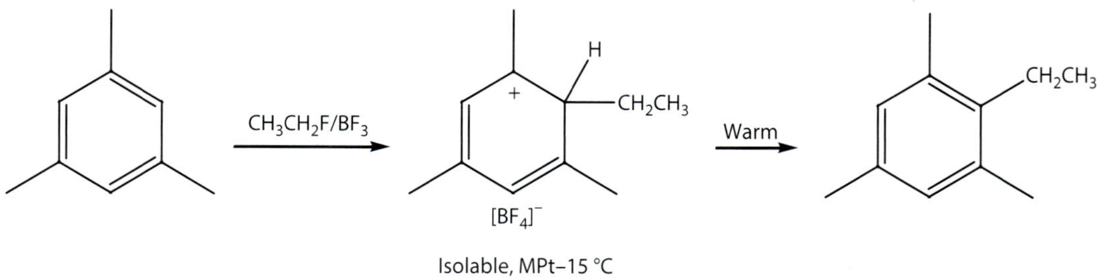

FIGURE 12.6 Alkylation of 1,3,5-trimethylbenzene by fluoroethane in the presence of BF_3.

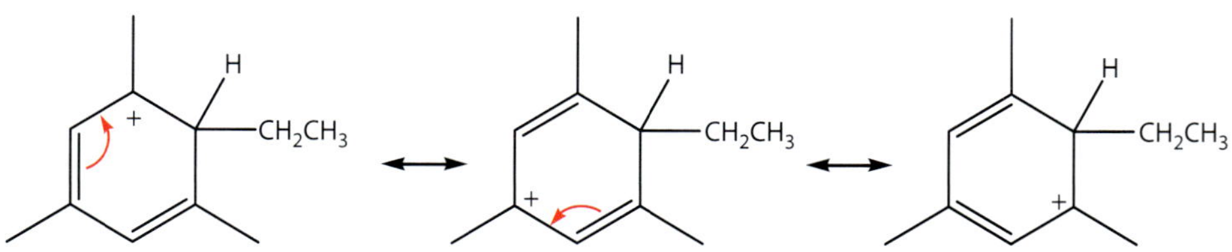

FIGURE 12.7 Resonance forms of a Wheland intermediate.

12.2 ELECTROPHILES

12.2.1 PROTONS

The simplest of all electrophiles is H^+, but exchanging one proton for another is not very interesting! So we use D^+ so that we can see what has happened. In concentrated D_2SO_4, we can readily exchange H and D, and this is the method used to prepare C_6D_6 to be used in mechanistic studies such as the one mentioned earlier, or as a solvent in NMR spectroscopy (Figure 12.8). The mechanism is just as before (Figure 12.8), and the reaction is driven to completion by a mass action effect in the presence of an excess of D_2SO_4.

Notice that this reaction, unusually for electrophilic substitutions, must be reversible; the difference in energy between loss of a proton and a deuterium atom from the Wheland intermediate in Figure 12.8 is small. The reaction can go either way, though we might expect to see some small difference because the C–D bond is stronger than the C–H bond. This is why we drive the reaction to completion using a mass action effect from an excess of D_2SO_4.

12.2.2 CARBOCATIONS

We might next think about using carbocations as electrophiles—but we know that these are not very stable and have to be generated *in situ*. One version of this reaction involves generating a carbocation from a haloalkane using a strong Lewis acid such as aluminum chloride, $AlCl_3$. This is known as the Friedel–Crafts reaction; the yields of many of the simple versions of this process are, however, low (Figure 12.9).

The mechanism of the reaction involves the formation of an ion pair, by abstraction of a halide from a haloalkane by the powerfully Lewis acidic aluminum chloride (Equation 12.1):

$$MeCl + AlCl_3 \rightarrow Me^+[AlCl_4]^- \qquad (12.1)$$

The cation is then attacked by the π-system, and the substitution is completed as usual by proton loss (Figure 12.10).

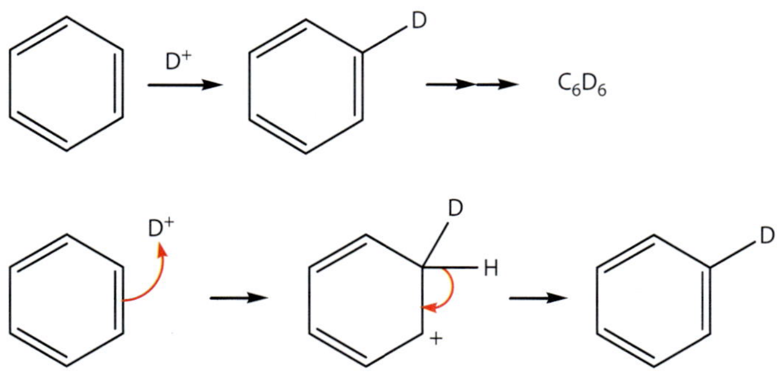

FIGURE 12.8 Deuteration of benzene.

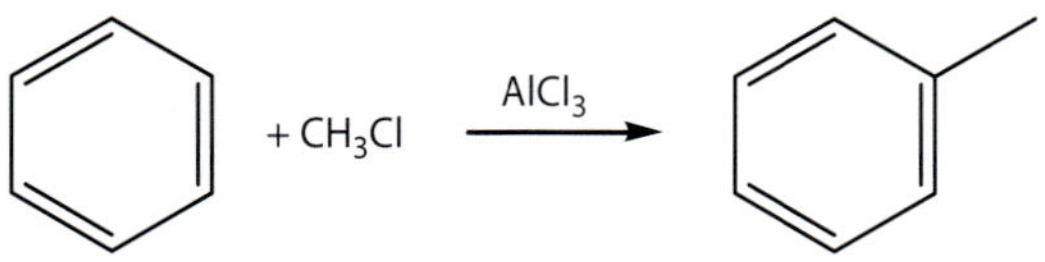

FIGURE 12.9 Friedel–Crafts alkylation of benzene.

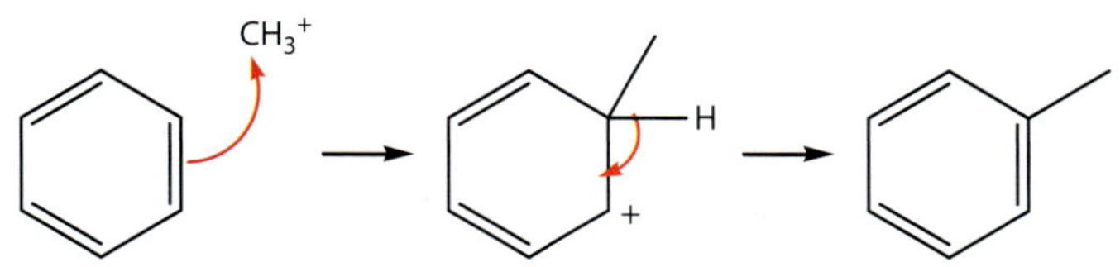

FIGURE 12.10 Mechanism of Friedel–Crafts methylation of benzene.

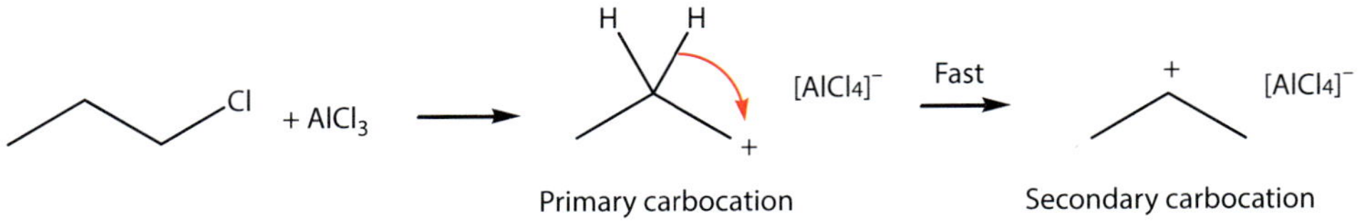

FIGURE 12.11 Rearrangement of a propyl cation.

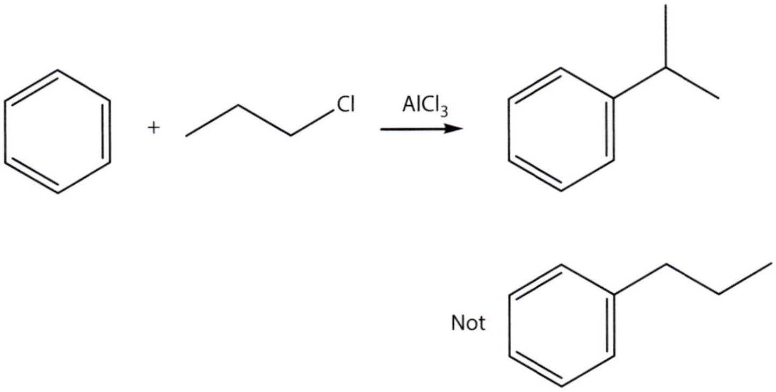

FIGURE 12.12 Friedel–Crafts reaction of benzene with 1-chloropropane.

Why does this reaction not work well? First, as we have seen in substitution, elimination, and addition reactions, carbocations tend to rearrange. This could not happen with a methyl cation, but it happens readily with other primary carbocations. For example, if we try to prepare propylbenzene by the reaction of benzene with 1-chloropropane and aluminum chloride, the main product is *iso*-propylbenzene. The primary cation is formed initially, but it rapidly rearranges by hydride migration to give the more stable secondary cation (Figure 12.11).

This cation then reacts with the benzene ring in the usual way to give *iso*-propylbenzene (Figure 12.12).

The carbocation rearrangement is faster than the reaction of the primary carbocation with benzene. Another example is provided in Figure 12.13. The mechanism of this reaction also involves a rearrangement.

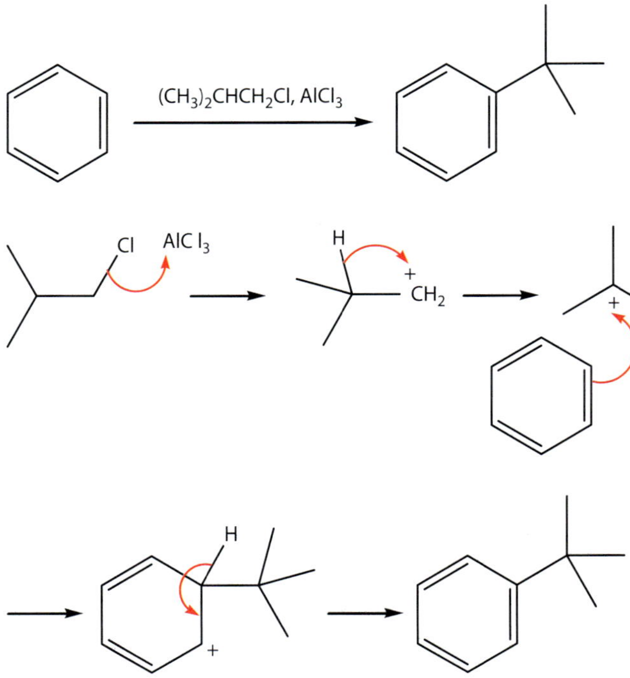

FIGURE 12.13 Friedel–Crafts reaction of 1-chloro-2-methylpropane with benzene.

Another problem with many Friedel–Crafts reactions is that multiple substitutions are both easy and common. Consider the reaction of benzene with chloromethane in the presence of aluminum chloride. Because alkyl groups are electron donors, the aromatic ring of toluene is more electron rich than is benzene—the methyl group donates electron density. Since the ability of an aromatic molecule to react with electrophiles depends on its being electron rich, toluene reacts faster with electrophiles than does benzene. So as soon as any toluene is formed, it will react with $[CH_3]^+[AlCl_4]^-$ faster than the starting material, benzene. So, multiple methylated benzene rings will be obtained (Figure 12.14; we will study the likely relative positions of the groups in a later section).

There are some other limitations. The reaction will not work where the aromatic ring we want to react bears a substituent more electron withdrawing than a halogen. Alkenyl and aryl halides cannot be used as the source of the substituent as the cations produced would be too unstable (remember that these substrates could not form cations for S_N1 reactions either, Chapter 9). Aniline does not react successfully as it forms a strong Lewis acid/Lewis base donor acceptor complex with aluminum chloride and other strong Lewis acids.

At this point, you may be thinking—what is the point of this reaction, which does not seem to work very well? It's often useful for putting tertiary groups onto aromatic rings (Figure 12.15). Here, there is no possibility of any rearrangement, as the cation formed is already tertiary, and multiple reactions are usually inhibited by the size of the *tert*-butyl group.

The reaction also works reasonably well with cations generated from other substrates, as long as these are relatively stable and not prone to rearrangement. For example, the cation in the first example of Figure 12.16 is generated by protonation of a carbon–carbon double bond and the

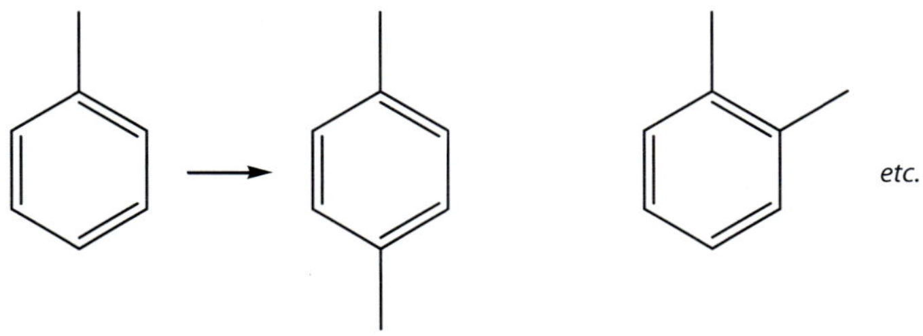

FIGURE 12.14 Multiple alkylations in the Friedel–Crafts reaction.

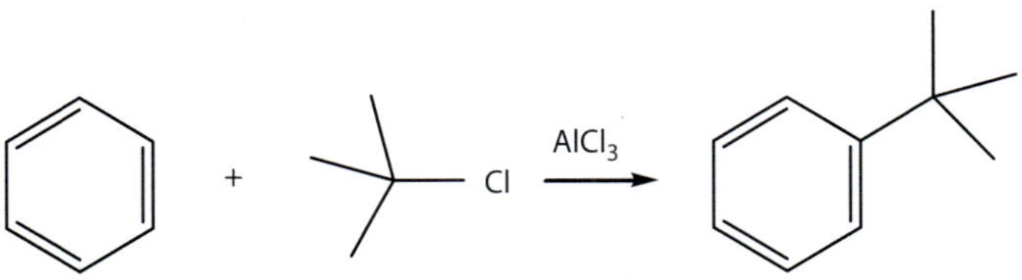

FIGURE 12.15 Friedel–Crafts alkylation with a tertiary halide.

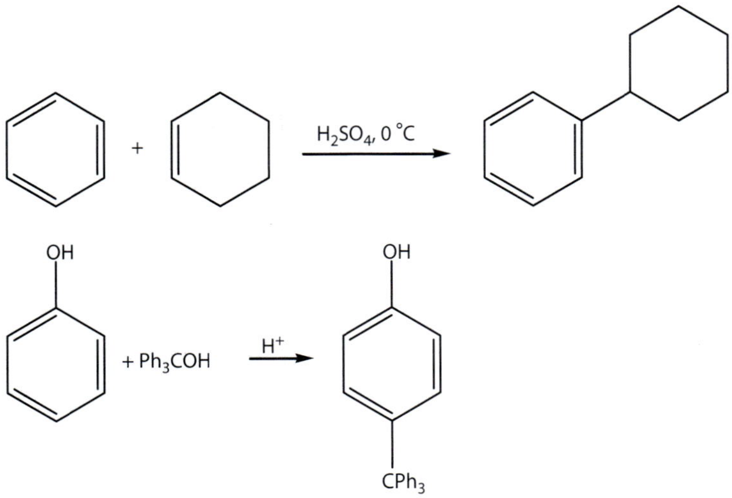

FIGURE 12.16 Other methods of generating cations for the Friedel–Crafts reaction.

second example by protonation of the OH group followed by loss of water. In the second example, the reaction occurs preferentially at the *para*-position because of the size of the substituent (as well as the directing effect of the OH group, to which we shall return shortly).

PROBLEM 12.1

Write down the intermediate cations that react with the benzene ring in the reactions of Figure 12.16.

Solutions

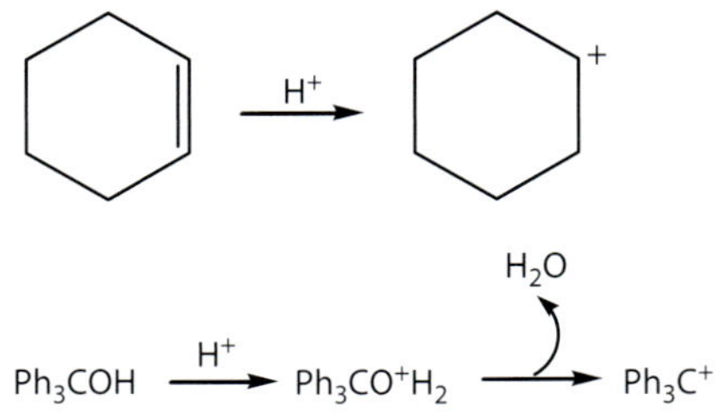

Another version of the Friedel–Crafts reaction that, by contrast, works very well is acylation (Figure 12.17). The reaction mechanism involves the formation of the acylium ion (we met this species briefly earlier when discussing the mass spectrometry of carbonyl compounds, Section 5.3).

PROBLEM 12.2

Write out in full the mechanism for the reaction of CH_3CH_2COCl with benzene in the presence of $AlCl_3$.

Solution

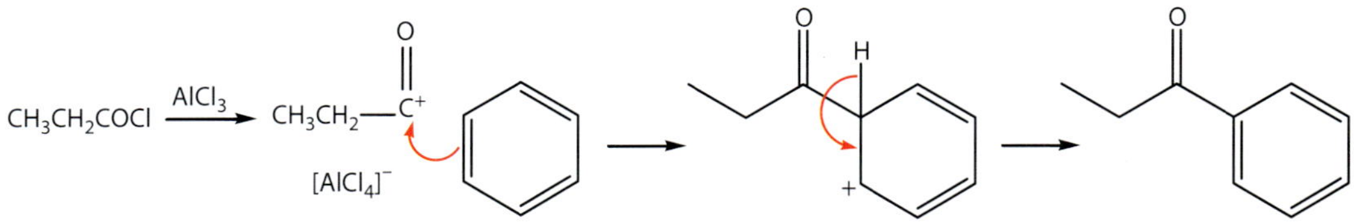

This process has two major advantages over the aliphatic version of the reaction. The acylium ion has no available pathway for rearrangement. Also, since the acyl group is electron withdrawing, the product is **less** reactive than the starting material, and therefore, there is no likelihood of multiple reactions. The success of this reaction also provides us with a pathway to benzene rings substituted with linear alkyl groups—which we could not make by the aliphatic version of the Friedel–Crafts, because the carbocations we needed rearranged. There are a number of ways in which a carbonyl group can be reduced to a CH_2, and we will discuss these in greater detail when we learn about carbonyl chemistry (Section 14.4.1), but for now, we can use the Clemmensen reduction using zinc amalgam and conc. HCl (Figure 12.18).

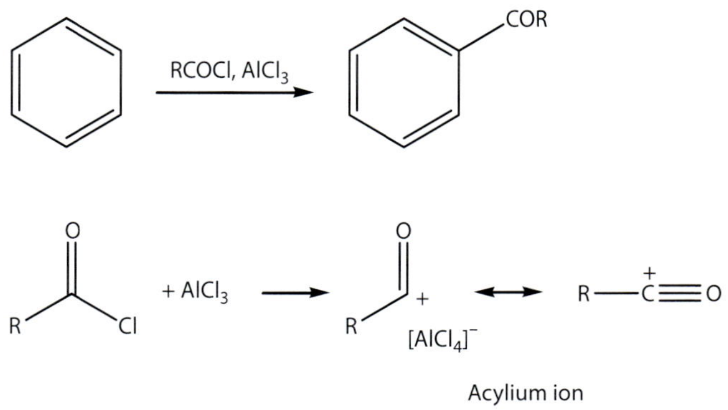

FIGURE 12.17 Friedel–Crafts acylation.

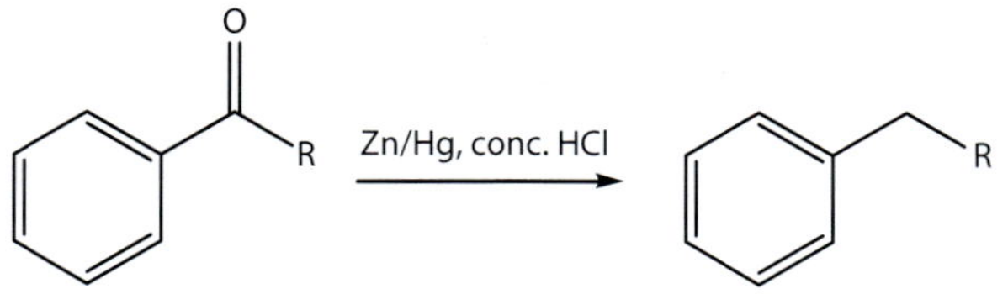

FIGURE 12.18 Clemmensen reduction of aryl ketones.

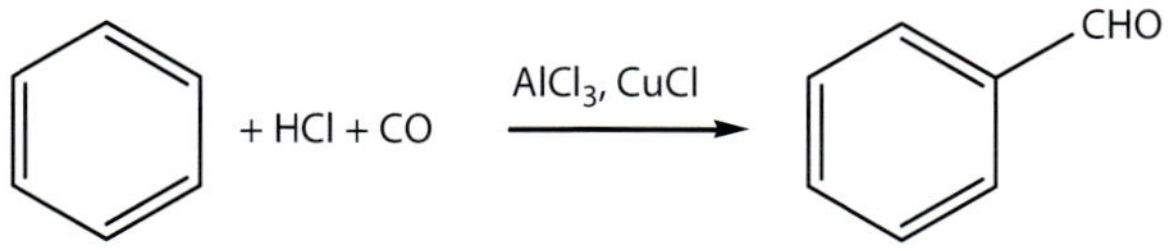

FIGURE 12.19 Gattermann–Koch formylation of benzene.

The Friedel–Crafts acylation reaction works for R = alkyl, alkenyl, or aryl, but it is not a useful route to ArCHO because the required acyl halide, HC(=O)Cl, is not stable. However, we can use a mixture of HCl and CO, in the presence of $AlCl_3$/CuCl—the Gattermann–Koch formylation (Figure 12.19).

12.2.3 HALOGENS

The reaction of benzene with pure bromine is quite slow; most reactions use a halogen carrier such as $FeBr_3$ (Figure 12.20). Unlike the reaction with alkenes, no bromonium ion is involved. The purpose of the halogen carrier is to increase the electrophilicity of molecular bromine. We can represent this as a Lewis acid/Lewis base process.

$$Br_2 + FeBr_3 \rightarrow Br^{(+)} \text{-----} Br^{(-)}FeBr_3$$

Chlorine will react similarly, but molecular fluorine is too reactive—explosions usually result! Iodine can be induced to react with aromatic rings in the presence of $CuCl_2$, but it's not a reaction that is commonly used; we will meet better ways to introduce iodine in the next chapter. This is another of the "three bears" type of reaction; bromine and chlorine are "just right," but fluorine is too reactive and iodine too unreactive. The origin of the effect is thermodynamic rather than steric.

Focus on Thermodynamics

A simple Hess cycle will give us approximate heats of reaction for substitution of benzene by the halogens. The "cost" is the sum of the C–H bond in benzene and the X–X bond in the halogen molecule. The "gain" is the benzene to halogen bond and the H–X bond (Table 12.1).

The reason for our experimental observation is now very obvious. The reaction with fluorine gives rise to a large exotherm, and it is too violent for practical use. The reaction with iodine is thermodynamically unfavorable.

TABLE 12.1 Thermodynamic Parameters for Halogen Addition to Benzene

	Bond Energies in kJ mol^{-1}				
Halogen	**Ph–H**	**X–X**	**Ph–X**	**H–X**	**Energy Gain in Reaction (kJ mol^{-1})**
Fluorine	472	156.9	532.2	570	473.3
Chlorine	472	243	406.3	431.6	122.8
Bromine	472	193.7	351.5	366.3	52.1
Iodine	472	148.4	280.3	298.4	−87

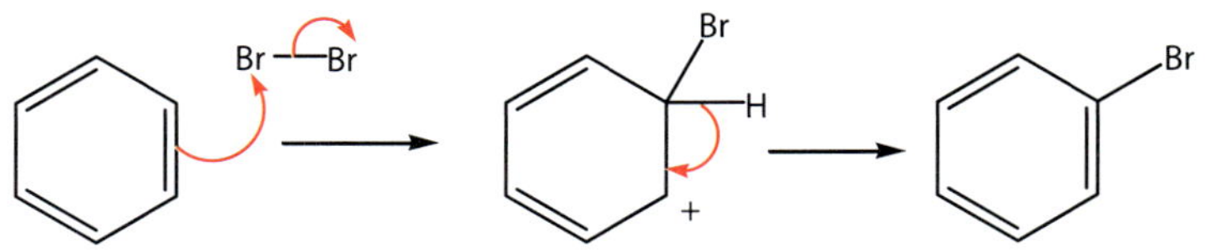

FIGURE 12.20 Bromination of benzene.

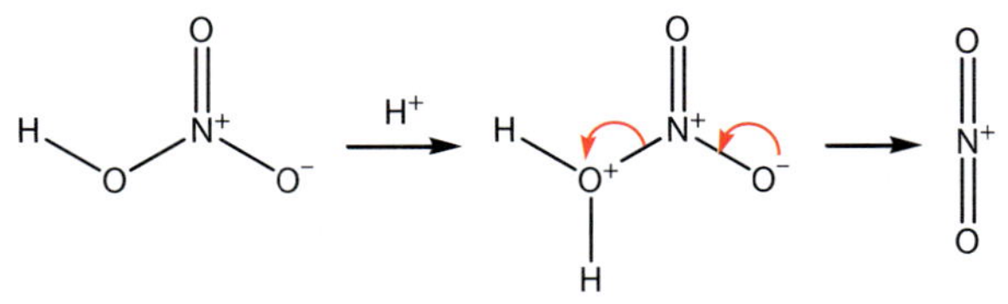

FIGURE 12.21 Reaction of nitric acid with sulfuric acid.

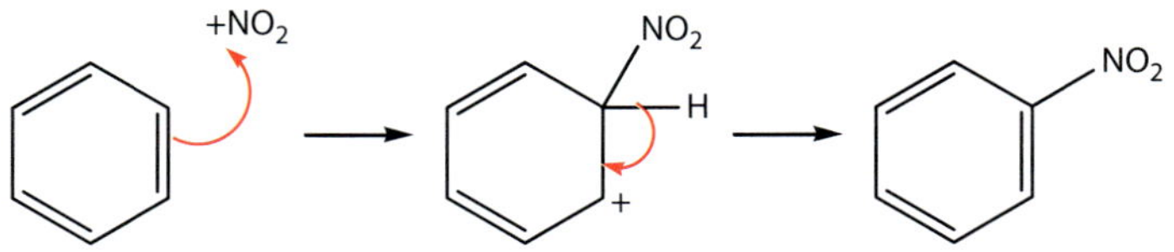

FIGURE 12.22 Mechanism of nitration of benzene.

12.2.4 NITRATION

This reaction was one of the earliest electrophilic substitutions to be developed in the history of aromatic chemistry and dates back to the mid-nineteenth century. The reagent used then (and now) was "nitrating mixture," formed by mixing 1 mol of conc. nitric acid with 2 mol of conc. sulfuric acid. Cryoscopy (depression of freezing point, check back to your general chemistry text on colligative properties) shows that this results in the formation of four ions (Equation 12.2):

$$\text{conc. } HNO_3 + \text{conc. } 2H_2SO_4 \rightarrow [NO_2]^+ + H_3O^+ + 2[HSO_4]^- \quad (12.2)$$

The reaction mechanism involves the protonation of nitric acid by sulfuric acid, followed by loss of water (Figure 12.21).

The product, $[NO_2]^+$, is called the nitronium ion and its presence in nitrating mixture has been demonstrated by Raman spectroscopy. The isolable (but expensive) salt $[NO_2]^+[BF_4]^-$ gives the same nitration reactions as the nitrating mixture. It might initially seem more modern and "greener" to use this salt rather than the strong acid mixture for nitrations, but cost prohibits this on any serious scale. The reaction mechanism for nitration follows exactly all those we have seen so far (Figure 12.22).

12.2.5 SULFONATION

Sulfonation is an oddity among the simple electrophilic substitutions—unlike the other reactions, the first step, addition of SO_3, is reversible, and the rate of the reverse reaction is comparable with proton loss. This means that the reaction overall is also reversible (Figure 12.23).

This can be very useful—because the SO_3H group can be removed, we will be able to use it to block a site on the aromatic ring, while we work on another position, and then remove it when we have finished. If we react the acid with NaOH to give the sodium salt, and heat this in base, phenol is obtained (Figure 12.24).

Sulfonic acids are strong acids, but are generally quite soluble in organic solvents, and are hence often used as acid catalysts for organic reactions. In the substituted example of Figure 12.25,

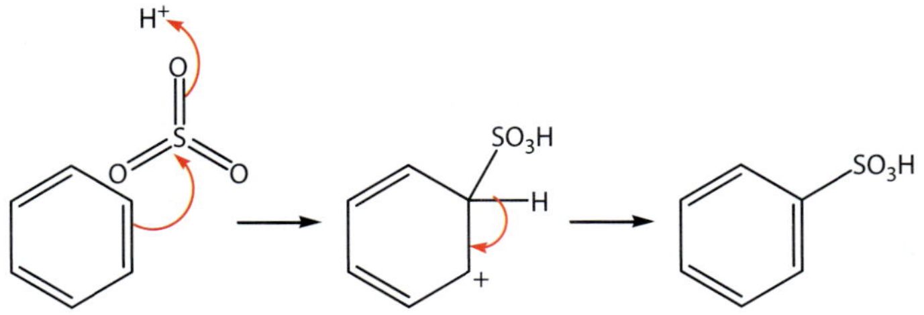

FIGURE 12.23 Mechanism of sulfonation of benzene.

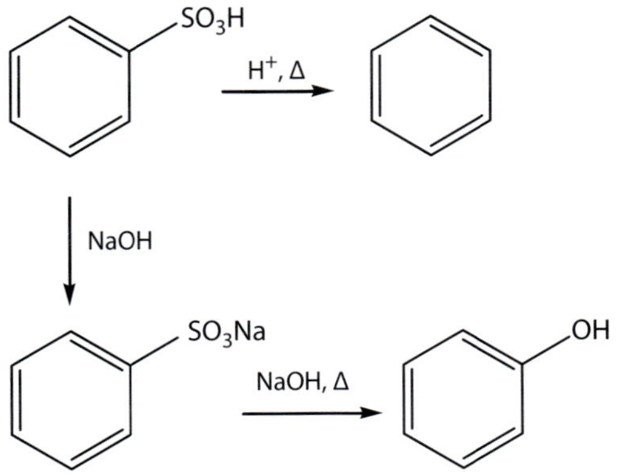

FIGURE 12.24 Reactions of benzene sulfonic acid.

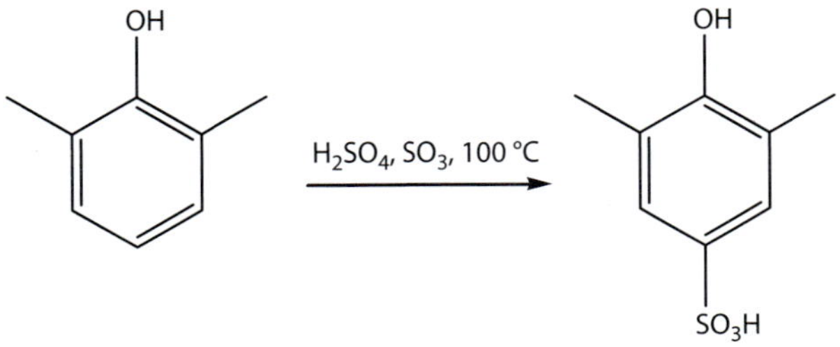

FIGURE 12.25 Sulfonation of a substituted phenol.

reaction occurs exclusively at the *para*-position; we will discuss in detail why this should be in the next sections. In the next chapter, we will explore how this reversibility can be useful to us in the synthesis of aromatic compounds.

The reversibility of sulfonation leads to the possibility that we can obtain different products under different reaction conditions. Under conditions of kinetic control, naphthalene is sulfonated to give naphthalene-1-sulfonic acid, whereas under conditions of thermodynamic control, the more stable naphthalene-2-sulfonic acid is obtained (Figure 12.26).

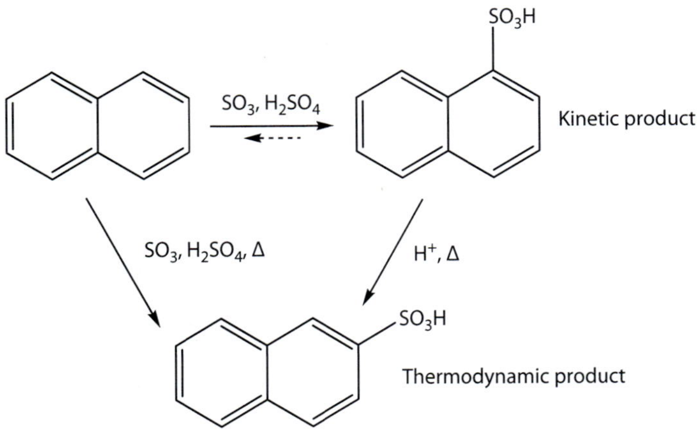

FIGURE 12.26 Sulfonation of naphthalene under thermodynamic and kinetic conditions.

Focus on Biologically Active Phenols

The medicinally active phenol longest known to mankind is salicylic acid, **12.1**. Its analgesic and antipyretic properties were known in the ancient world, and willow bark (*salix* is the Latin word for a willow tree) infusions were documented in Greece in the fifth century BCE. Acetyl salicylic acid, aspirin (**12.2**), was first prepared in the mid-nineteenth century, and, by the turn of the twentieth century, was beginning to replace salicylic acid. The acetyl group acts as a labile protecting group, and the molecule may be termed a "prodrug," which becomes an active species in the acid environment of the stomach, to some extent masking the gastric bleeding caused by salicylic acid. In the twentieth century, aspirin has been one of the most widely used over the counter remedies. Latterly, it has been partially superseded for analgesia by acetaminophen (paracetamol) and ibuprofen, but low-dose aspirin has found new uses as a protectant against heart attacks, strokes, and, most recently, bowel cancer.

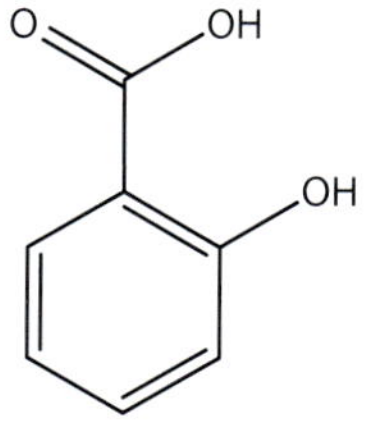

12.1, Salicyclic acid

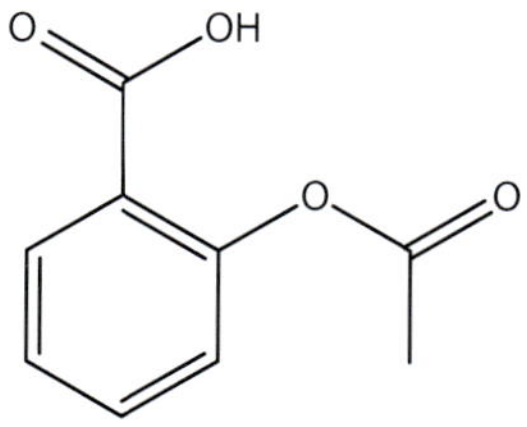

12.2, Acetyl salicyclic acid–aspirin

Aspirin

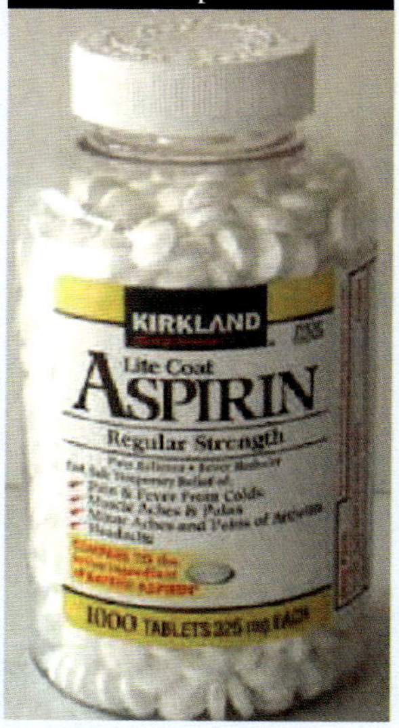

Capsaicin (**12.3**) is the active, "hot," ingredient in chili peppers. It has found uses medicinally in ointments to relieve neuropathic pain such as that in shingles and is widely employed as *oleoresin capsicum* (OC) or pepper spray for personal defense and as a non-lethal agent in riot control. It is a potent lachrymator and causes pain to the eyes, skin, and mucous membranes and temporary breathing difficulties. Its use in warfare is prohibited by various chemical weapons conventions, but it is widely used in law enforcement and, in some jurisdictions, is permitted for self-defense against an individual attacker and/or an animal attack.

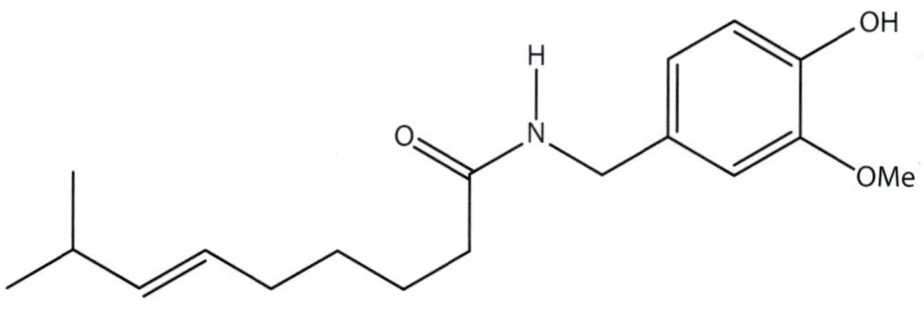

12.3, Capsaicin

Pepper spray

A compound that has attracted wide interest recently is resveratrol, **12.4**. It is found in red grape skins, grape juice, and red wine, and increases longevity in simple organisms such as fruit flies and worms, and also obese mice. Past studies have focused on its effect on enzymes called sirtuins that remove acetyl groups from proteins, but recent work suggests that resveratrol blocks phosphodiesterase enzymes. Like many natural products, it has a number of different effects, and it remains unclear as to which are crucial in its anti-aging effects.

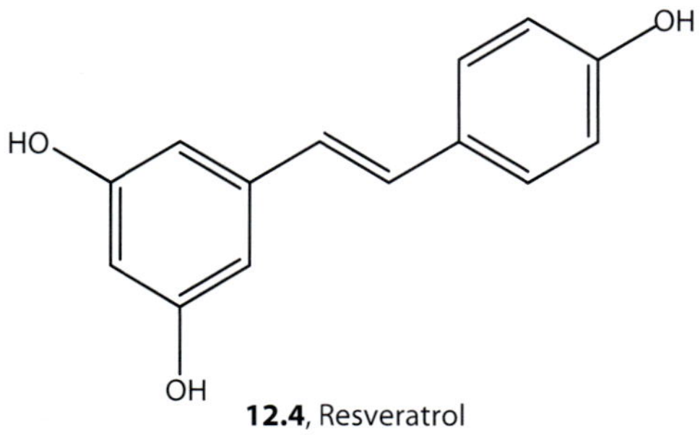

12.4, Resveratrol

Red wine contains resveratrol

Key Points from Sections 12.1 and 12.2

- Reaction of electrophiles with aromatic compounds results in substitution.
- The substitution reaction occurs in two steps, addition of the electrophile to the ring followed by loss of a proton to restore the aromatic sextet. The first step is rate determining, and product determining, in most cases.
- A range of electrophiles may be used (Table 12.2).

TABLE 12.2 Reaction of Electrophiles with Benzene

Reagent	Product	Comments
D^+	$C_6H_{6-n}D_n$	Ratio of hydrogen to deuterium is controlled by mass action; this is the usual synthesis of C_6D_6.
$RCl/AlCl_3$	PhR	Reaction works well only with tertiary alkyl halides (alkenyl and aryl halides are completely unreactive); rearrangement and multiple reactions are otherwise common. Friedel–Crafts reaction.
		Other methods for carbocation generation may be used.
		May not be used on aromatic rings with strongly electron-withdrawing substituents.
$RCOCl/AlCl_3$	PhCOR	Friedel–Crafts acylation is generally more useful than alkylation, as the multiple substitutions do not occur. PhCOR may be reduced to $PhCH_2R$ by the Clemmensen reduction with zinc amalgam and conc. HCl.
		May not be used on aromatic rings with strongly electron-withdrawing substituents.
Cl_2 (/$FeCl_3$)	PhCl	Halogen carrier, $FeCl_3$, is needed, except for very reactive aromatic rings.
Br_2 (/$FeBr_3$)	PhBr	Halogen carrier, $FeBr_3$, is needed, except for very reactive aromatic rings.
		Iodobenzene and fluorobenzene cannot be prepared by this method.
Conc. HNO_3/conc. H_2SO_4 (nitrating mixture)	$PhNO_2$	$[NO_2]^+$ is the active electrophile in the reaction, and $[NO_2][BF_4]$ can also be used as nitrating agent.
SO_3/H_2SO_4	$PhSO_3H$	This reaction is unusual in that it is reversible. Heating $PhSO_3H$ in acid gives benzene and heating its sodium salt with NaOH gives phenol.

PROBLEM 12.3

Predict the product(s) of each of the following processes:

(a)

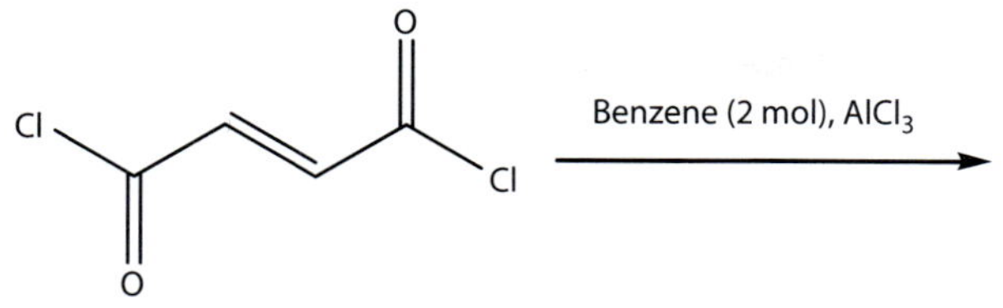

(b)

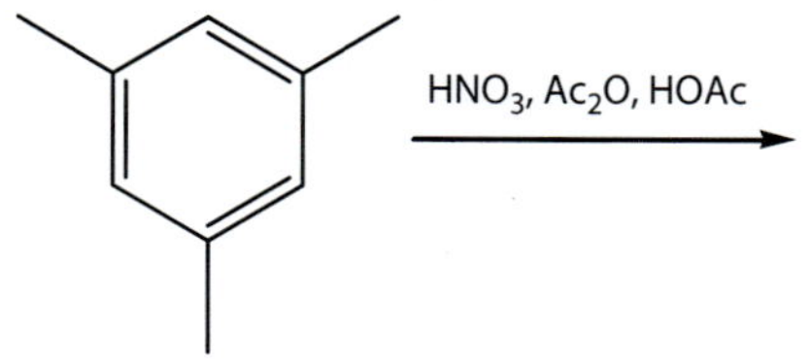

Solutions

(a) This is a Friedel–Crafts acylation, seen from the less usual perspective of the acid chloride. Since 2 mol of benzene are provided, both halides will be substituted.

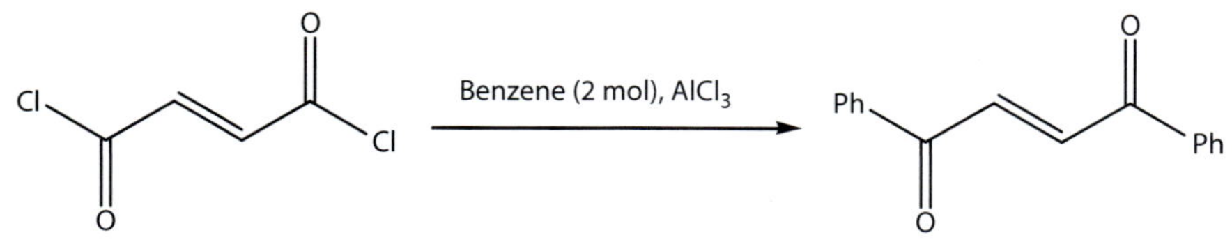

(b) We don't yet know anything about directing effects in aromatic substitutions, but it does not matter in this symmetric molecule. There is only one type of site that can be substituted. As we will see later, nitration deactivates the aromatic ring toward further substitution, so only monosubstitution occurs.

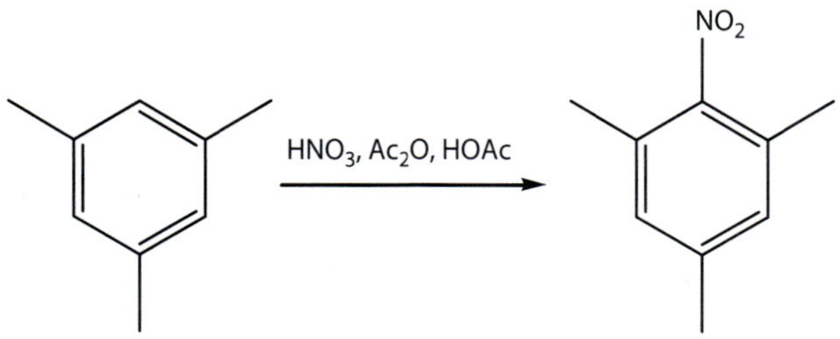

PROBLEM 12.4

Suggest a reagent/conditions for each of the following transformations:

(a)

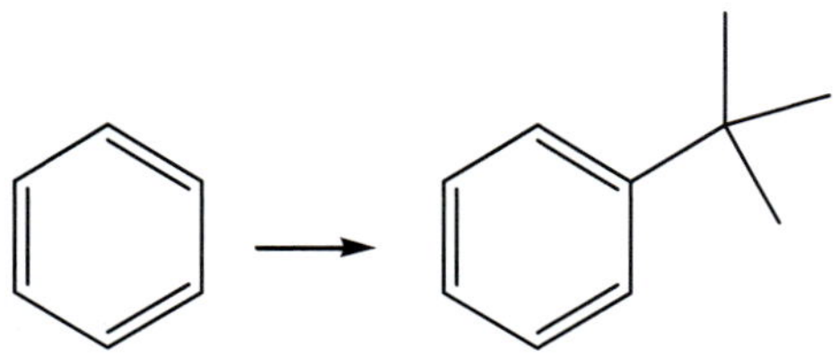

(b)

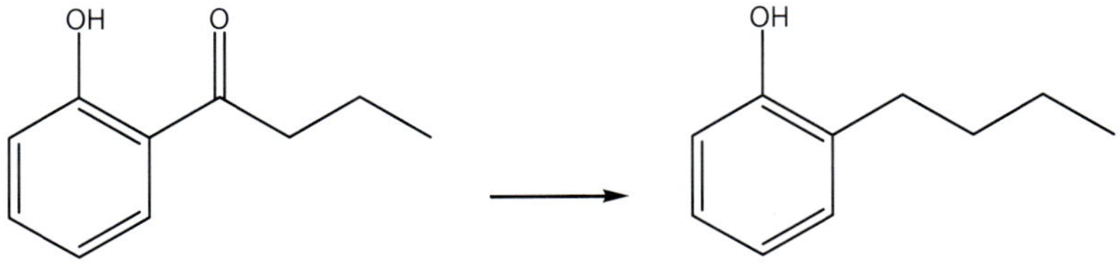

Solutions

(a) A Friedel–Crafts alkylation, $Me_3CCl/AlCl_3$
(b) Clemmensen reduction, Zn/Hg, conc. HCl

12.3 ORIENTATION EFFECTS

12.3.1 INDUCTIVE DONORS

If we consider the mononitration of toluene, we could, in principle, get three products, with the nitro group *ortho*, *meta*, or *para* to the methyl group (Figure 12.27).

Which product or products is/are produced, and can we account for this mechanistically? The first RDS is generally irreversible and hence will also determine which product is formed, so this is the key process. The questions we must ask are whether a particular substituent donates or withdraws electron density and whether this is an inductive or a resonance process. Let's look at what happens when toluene reacts with an electrophile—remembering that the methyl group is an inductive donor of electrons, so we will expect the reaction to be more favorable than for benzene (the aromatic ring is more electron rich). If we consider the three possible resonance forms for the Wheland intermediate formed in the reaction at the *ortho*-position, shown in Figure 12.28, one of these is a tertiary carbocation, and we may expect that this will be the most important resonance form and will contribute most to the stability of the structure.

If we consider the three structures for the intermediate formed in *meta*-substitution (Figure 12.29), we see that none of them is particularly stable—however, we move around charges and bonds; we get no tertiary carbocation. So we would not expect this intermediate to be favorable, relative to that leading to the *ortho*-isomer.

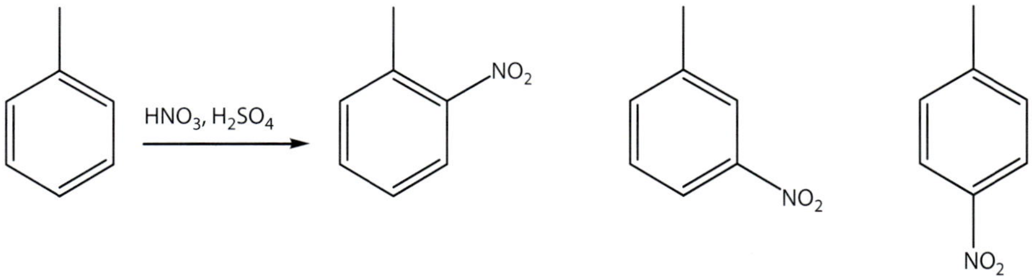

FIGURE 12.27 Regiochemistry of nitration of toluene.

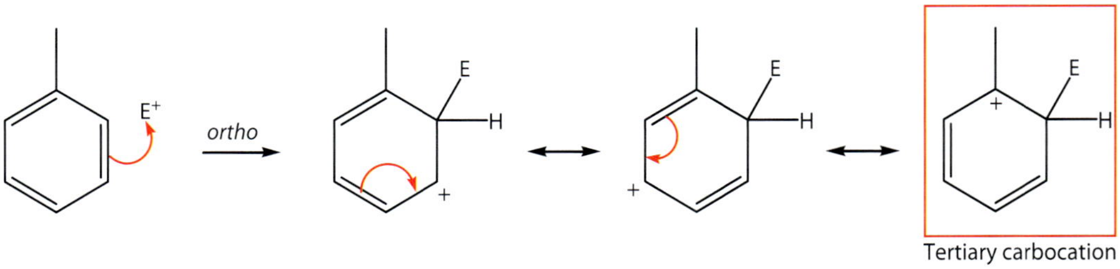

FIGURE 12.28 Resonance forms of the Wheland intermediate in electrophilic substitution of toluene at the *ortho*-position.

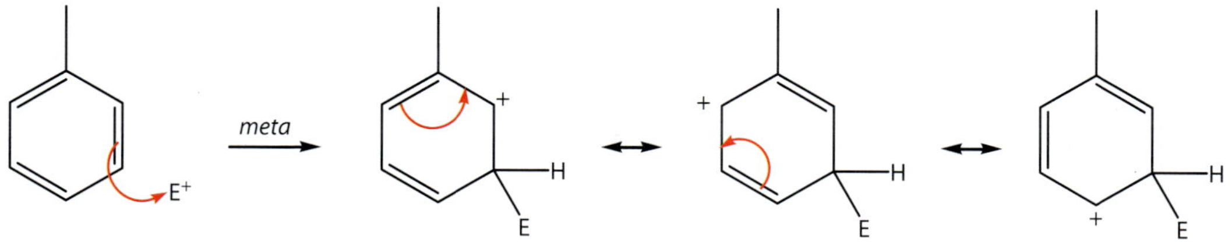

FIGURE 12.29 Resonance forms of the Wheland intermediate in electrophilic substitution of toluene at the *meta*-position.

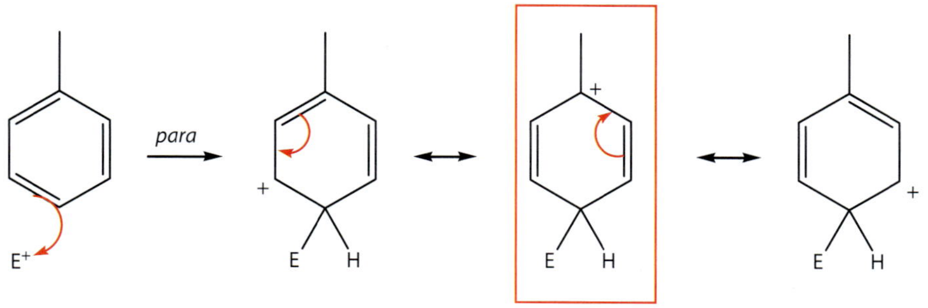

FIGURE 12.30 Resonance forms of the Wheland intermediate in electrophilic substitution of toluene at the *para*-position.

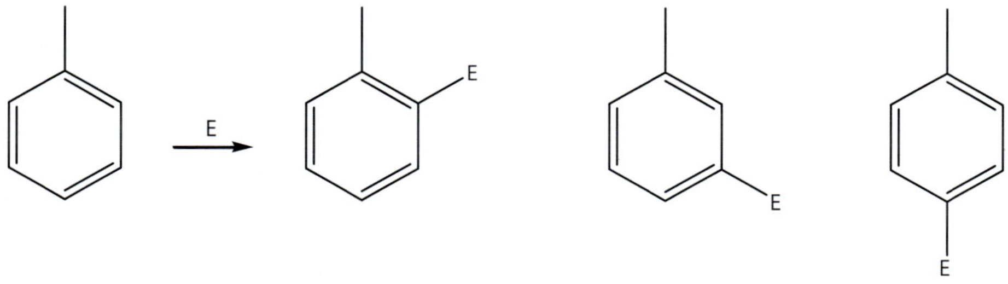

HNO_3, H_2SO_4	59	4	37
Cl_2	60	0	40

FIGURE 12.31 Regiochemistry of toluene substitution.

Moving on to the *para*-isomer, there are again three possible resonance forms, one of which is a favored tertiary carbocation (Figure 12.30).

The outcome is that we would expect to see substitution *ortho* and *para* to the methyl group, and we describe the methyl group as *ortho, para*-directing and activating. This is true of all simple alkyl groups. Alkyl substituents form a group that we describe as *inductive donors.* Because the alkyl groups donate electron density to the ring, alkylbenzenes react about 20 times faster with electrophiles than benzene itself (Figure 12.31).

PROBLEM 12.5

Predict the main product(s) of the following reactions:

(a)

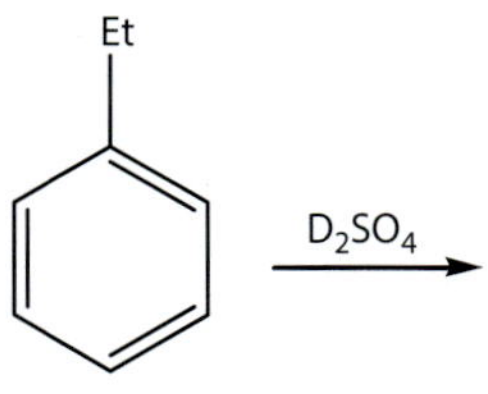

(b)

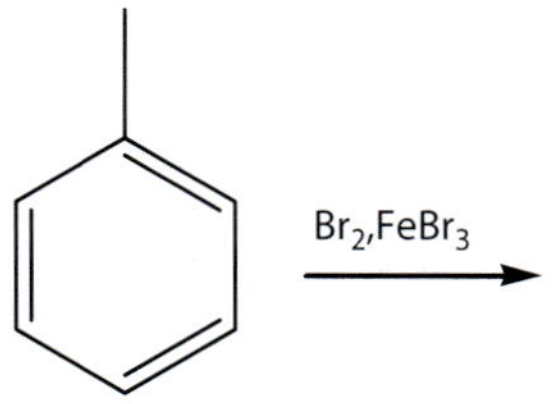

Solutions

(a) Et is *o,p*-directing, so deuteration occurs at those positions.

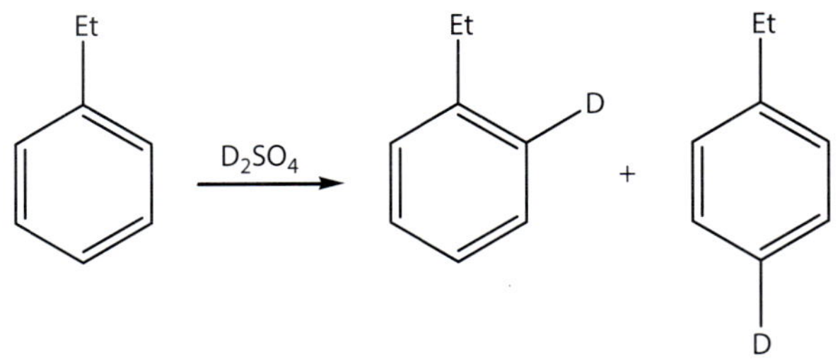

(b) Methyl is also *o,p*-directing. Note that the methyl group is not sufficiently activating that we can dispense with the halogen carrier.

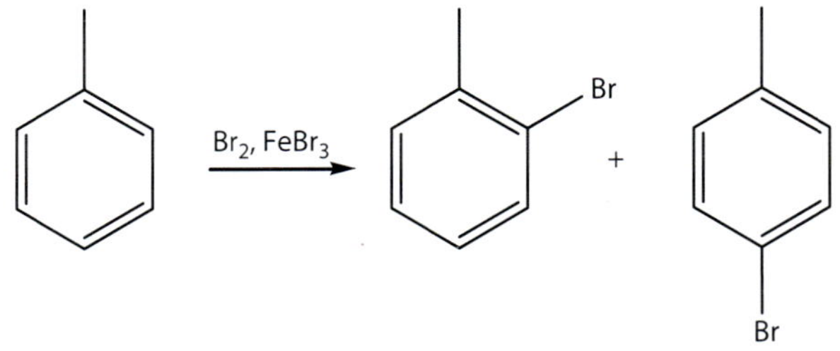

Aromatic substituents, as in biphenyl, **12.5**, are slightly problematic. Experimentally, a phenyl group is weakly activating, and *ortho, para*-directing. We might reasonably have expected that the presence of an additional phenyl ring would confer significant resonance stabilization on any cationic intermediate. However, the lowest energy conformation of biphenyl is far from planar, and the two aromatic rings are not fully conjugated, so this effect is quite small. Conformational analysis of biphenyl suggests that there is an energy minimum when the rings are angled at about 44 ° to each other—a balance between maximizing conjugation and minimizing steric hindrance between the *ortho*-hydrogen atoms.

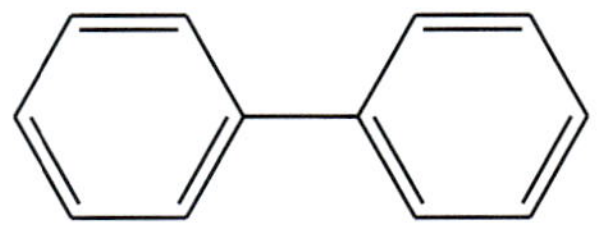

12.5, biphenyl

12.3.2 RESONANCE DONORS

Resonance donors such as OMe and NR_2 are also able to stabilize an adjacent positive charge as in Figure 12.32. We often show this by using the lone pair on the heteroatom as we write the mechanism for the substitution (Figure 12.33). We can do the same for substitution at the *para*-position.

However, if we were to substitute at the *meta*-position, the lone pair on the oxygen atom cannot stabilize the positive charge in the intermediate. As we will see later when we quantify these effects, the *meta*-position is a little less reactive than benzene—while the lone pair of electrons on oxygen makes the OMe group a resonance donor, the electronegativity of oxygen means that it is

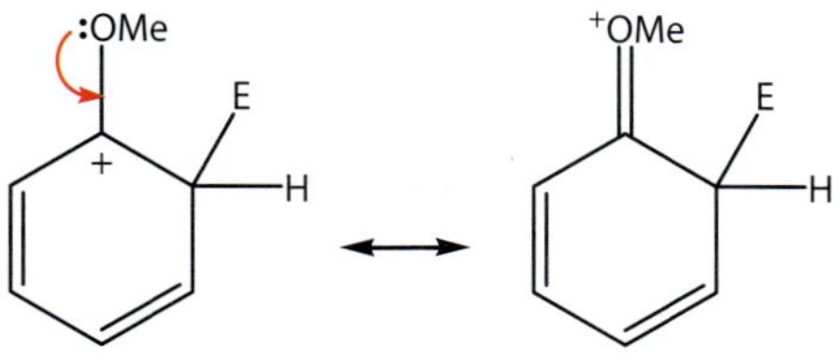

FIGURE 12.32 Lone pairs stabilize a positive center.

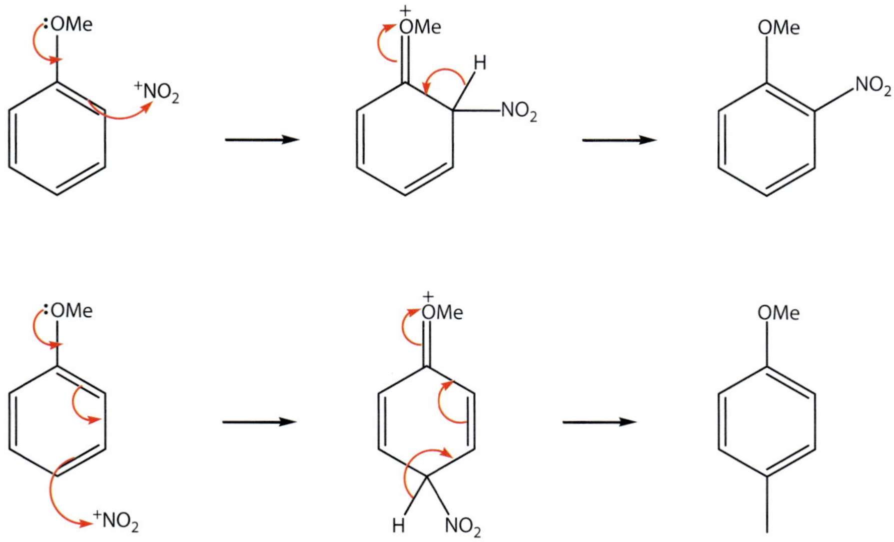

FIGURE 12.33 Mechanism of electrophilic substitution of an aromatic ring bearing a resonance donor.

also an inductive withdrawer of electrons. At the *ortho-* and *para*-positions, the resonance effect substantially outweighs the inductive one; this is quite general. Where there are both inductive and resonance effects, the resonance effects are usually more important in determining the outcome of the reaction.

Resonance donor groups include −OMe, −OH, $-O^-$, −OC(=O)Me, −SH, −SR, $-NH_2$, $-NR_2$, and −NHC(=O)Me. All of these are described as *ortho, para*-directing and are activating. Many of the groups are very strongly activating. For example, if phenol is reacted with molecular bromine, no halogen carrier is needed, and trisubstitution to give 2,4,6-tribromophenol is very rapid (Figure 12.34).

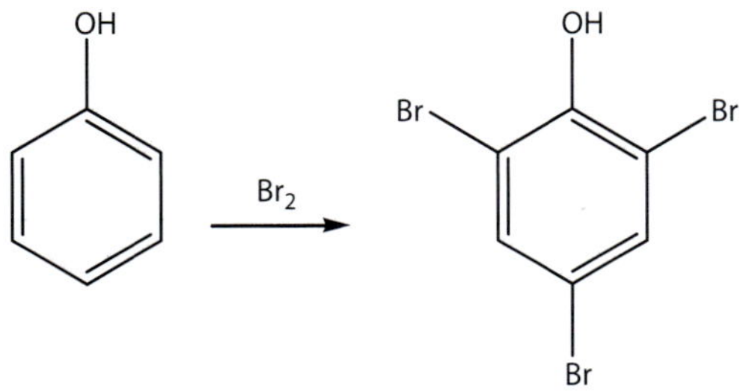

FIGURE 12.34 Trisubstitution of phenol by bromine.

PROBLEM 12.6

Draw a complete mechanism for the tribromination of aniline by molecular bromine.

Solution

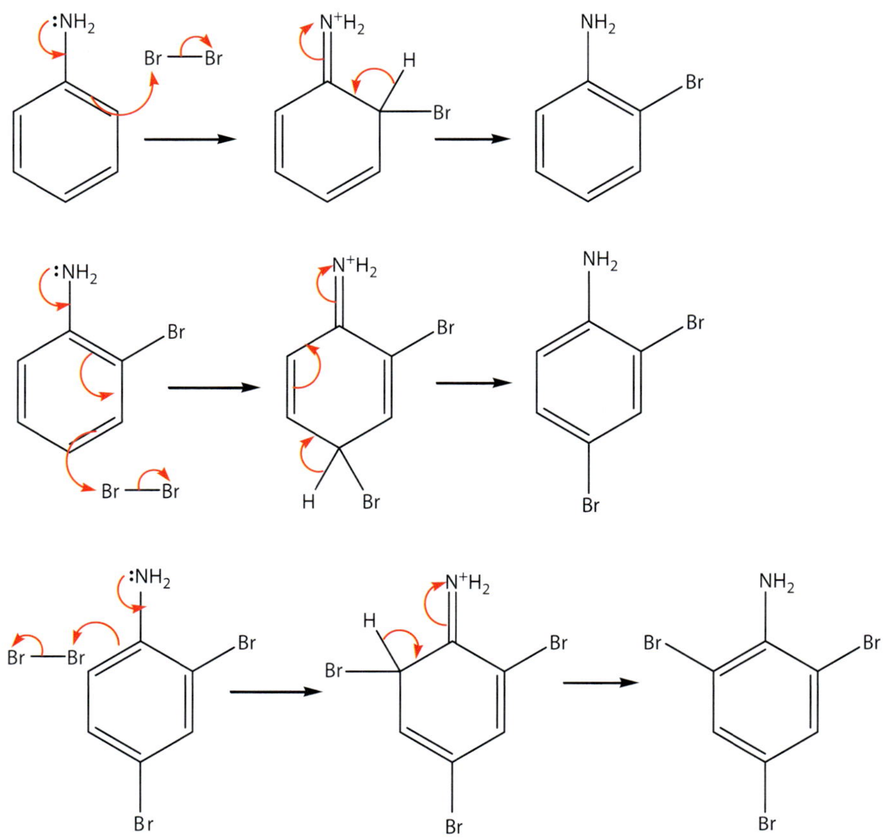

Notice that we do not need a halogen carrier for this reaction to work; the NH_2 group is strongly activating, and it is difficult to stop this reaction at the mono- or disubstitution stage.

12.3.3 RESONANCE ACCEPTORS

Where a benzene ring bears a resonance acceptor group, substitution occurs at the *meta*-position and is slow compared with the reaction of benzene. We describe these substituents as *meta*-directing and deactivating. We can exemplify this by thinking about the reaction of nitrobenzene with electrophiles. The nitro group is a powerful electron withdrawer, both by resonance and induction (Figure 12.35).

In this case, we have two structures (shown enclosed in blue boxes) of the nine possible resonance forms for the intermediates that have the positive charge located adjacent to the electron-withdrawing nitro group. When the substituent was an electron donor, having the positive charge next to the substituent was favorable, but now that it is an electron withdrawer, it is very unfavorable. So the fact that we see *meta*-substitution is not that this is particularly favored but that it is simply *less unfavorable* than the *ortho*- and *para*-alternatives. *Meta*-directing groups include $-CHO$, $-C(=O)R$, $-COOH$, $-COOR$, $-CONH_2$, $-NO_2$, $-SO_3H$, and $-CN$. There is one additional substituent that we may encounter that we should mention here, $-CF_3$. This is strongly electron withdrawing inductively and has no resonance contribution. Thus, it also destabilizes an adjacent positive charge and hence is *meta*-directing and deactivating (Figure 12.36).

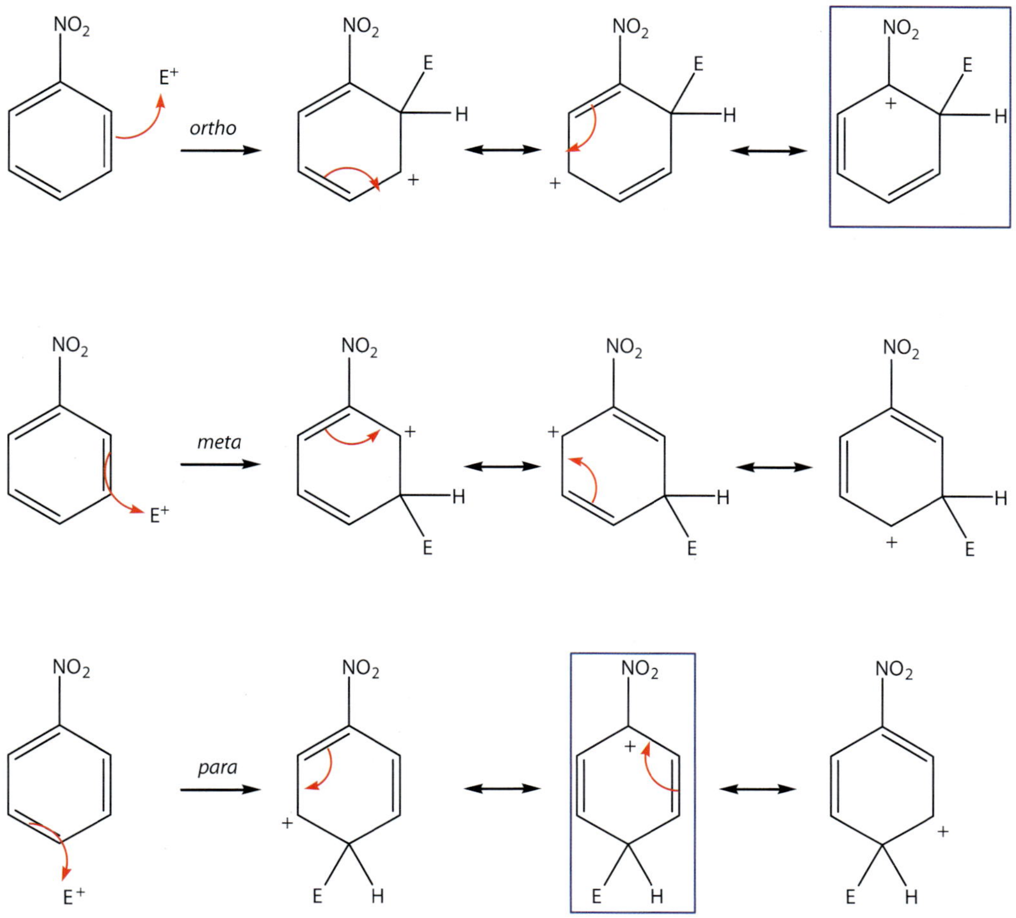

FIGURE 12.35 Resonance structures of the possible intermediates for the electrophilic substitution of nitrobenzene.

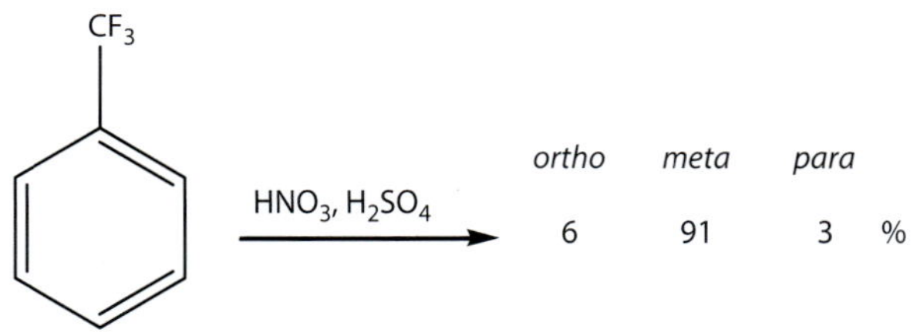

FIGURE 12.36 $-CF_3$ is a *meta*-directing, inductive withdrawer of electron density.

PROBLEM 12.7

Predict the product(s) of the following reactions:

(a)

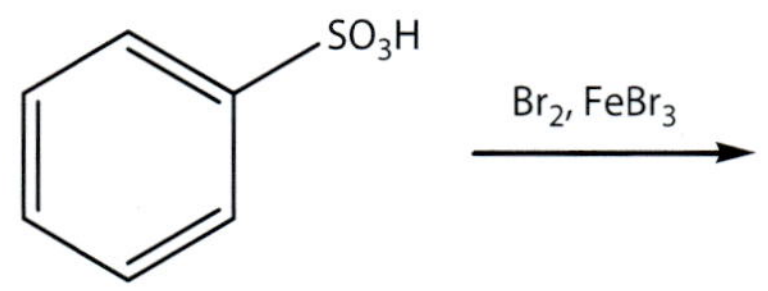

(b)

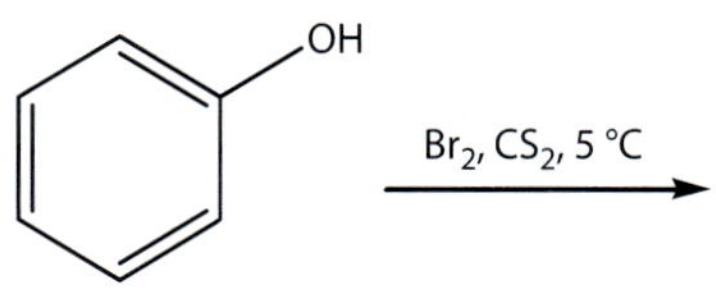

(c)

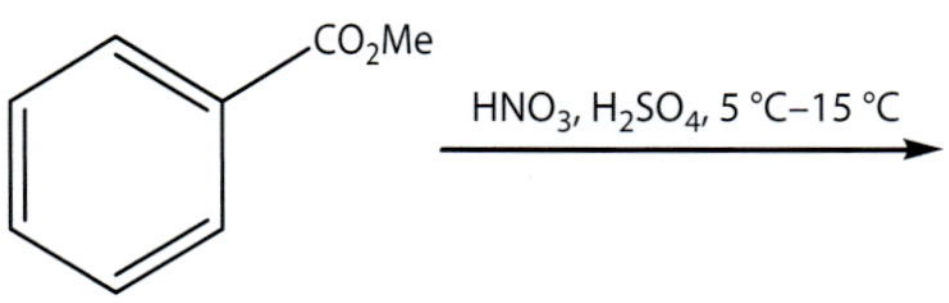

Solutions

(a) The $-SO_3H$ group is *meta*-directing and deactivating.

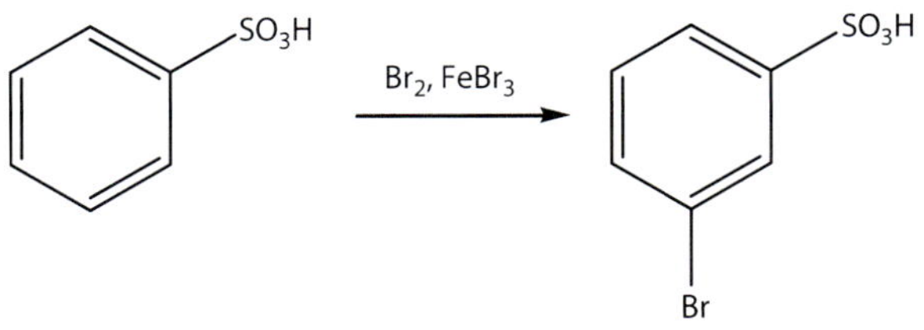

(b) $-OH$ is *o,p*-directing and activating. Under these particular conditions, the main product was the *para*-isomer, obtained in 80 %–84 % yield.

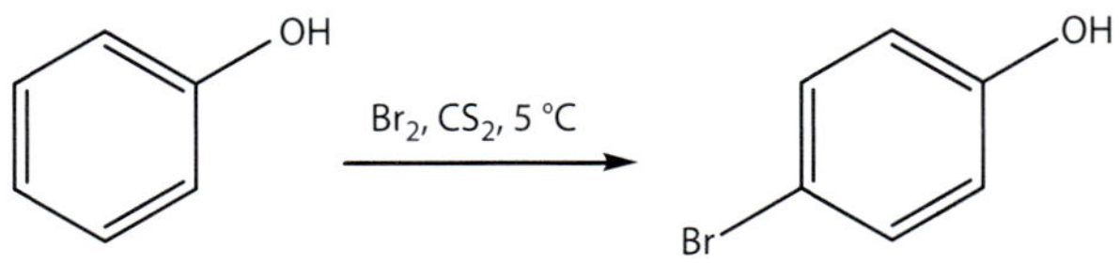

(c) The $-CO_2Me$ group is *meta*-directing and deactivating.

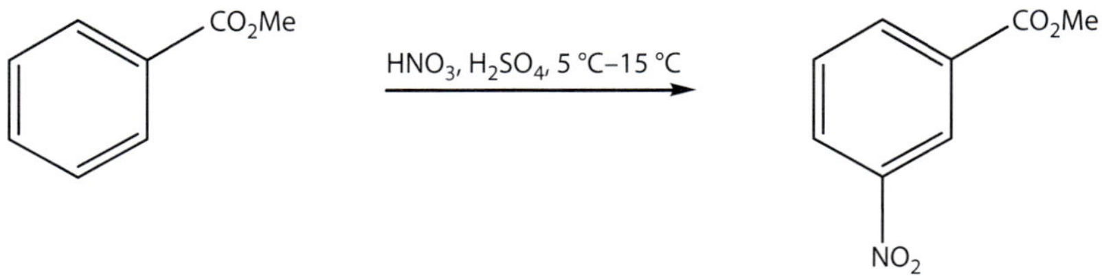

12.3.4 HALOGENS

Halogen substituents do not fit neatly into our previous categories. The experimental observation is that they are *ortho, para*-directing but that the reactions of halobenzenes are slower than for benzene itself. For example, nitration of chlorobenzene (Figure 12.37) gives almost exclusively *ortho*- and *para*-substitution.

We have two effects to consider—inductive and resonance. Halogens are inductive withdrawers of electron density, but they also have lone pairs that can assist in stabilizing a carbocation. Initially, you may feel uncomfortable with the idea that halogens can be resonance donors, but it should not be so surprising, as we have seen it before when we considered the addition of bromine

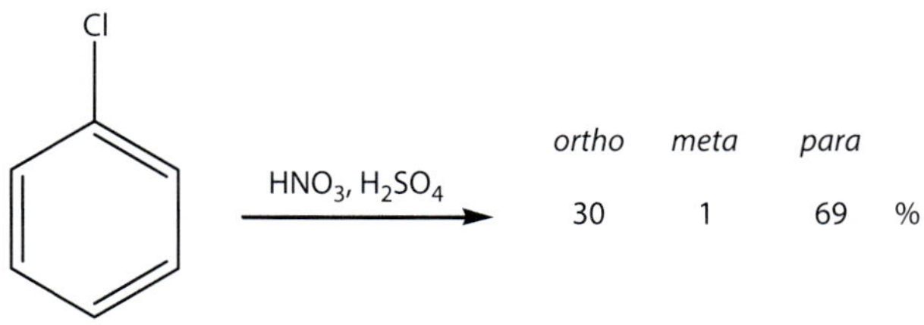

FIGURE 12.37 Nitration of chlorobenzene.

to alkynes (Section 11.2.5). Also, remember that you are very familiar with the sharing of a lone pair on oxygen with an adjacent carbocation, and oxygen is more electronegative than any element other than fluorine.

There is a complex balance here. In terms of electronegativity, fluorine is the most electronegative element and, hence, the strongest inductive withdrawer; the electronegativity of the other halogens decreases down the group. However, the best overlap in terms of resonance donation from the lone pairs on the halogen into the empty p_z orbital of the carbocation occurs when the orbitals are of comparable size, so this is best for fluorine and worst for iodine (you may have discussed this in general chemistry in terms of the structures of the boron trihalides).

Although it may seem counterintuitive, there are ground-state effects that give evidence for this phenomenon. In chlorobenzene, the contribution of resonance forms such as **12.8**, in which there is a positive charge on chlorine, results in a shortened carbon–chlorine bond (1.69 Å *vs.* 1.77 Å in chloroethene and 1.72–1.75 Å in haloalkanes) and a lowered dipole moment.

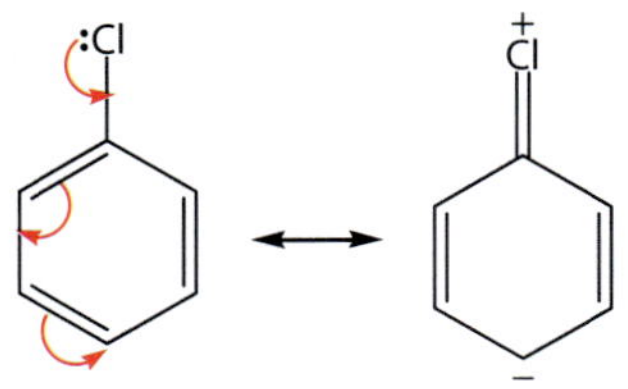

12.8

PROBLEM 12.8

Predict the product(s) of the following reaction:

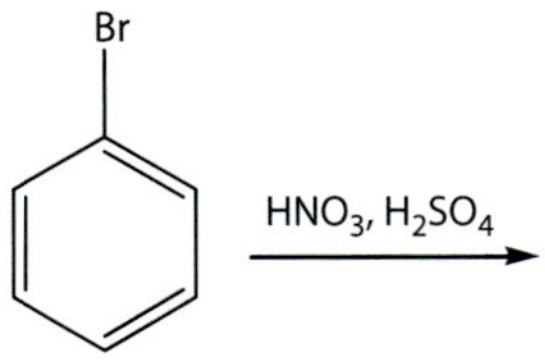

Solution

Br is *o,p*-directing:

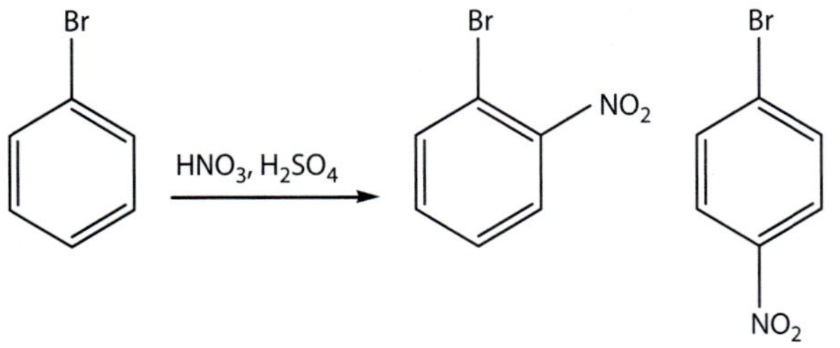

12.3.5 TWO OR MORE GROUPS IN CONFLICT

If we first consider electronic effects, then where groups are in conflict, the more "activating" group "wins." So bromination of 4-methylphenol (Figure 12.38) occurs *ortho* to the OH group. Both the OH and the methyl groups are *ortho, para*-directing and activating, but resonance donation is usually "stronger" than inductive donation.

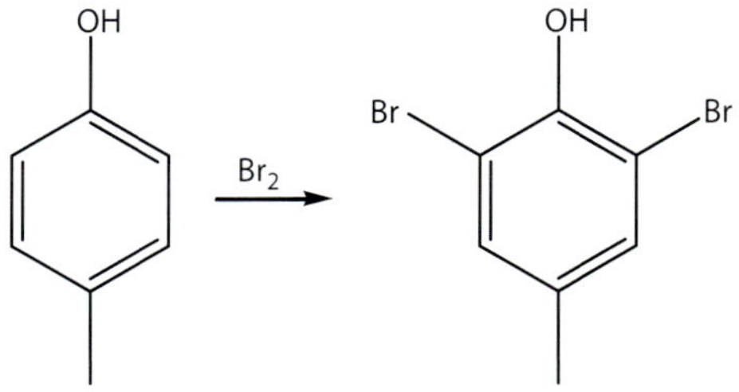

FIGURE 12.38 Electrophilic substitution of 4-methylphenol.

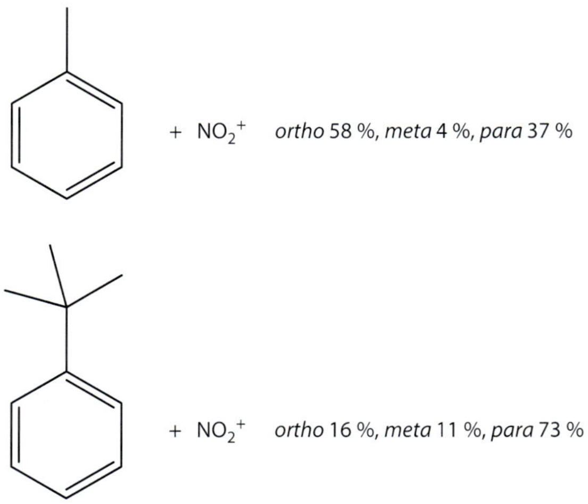

FIGURE 12.39 Size may influence *ortho/para* ratios.

Steric effects can also be important. If we compare the nitration of toluene with that of *tert*-butylbenzene, we get very little substitution *ortho* to the *tert*-butyl group, because of its size (Figure 12.39).

We can look at a more complete table (Table 12.3) of these data; it is clear that as the size of the substituent increases, the more likely substitution is to take place at the *para*-position.

TABLE 12.3 Nitration of Alkyl Benzenes

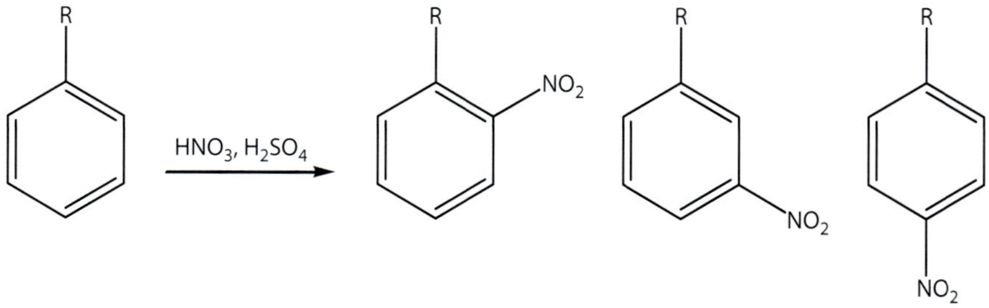

R = Me	58	4	37
Et	45	6	49
i-Pr	30	8	62
t-Bu	16	11	73

PROBLEM 12.9

Predict the major site(s) of electrophilic nitration of each of the following molecules:

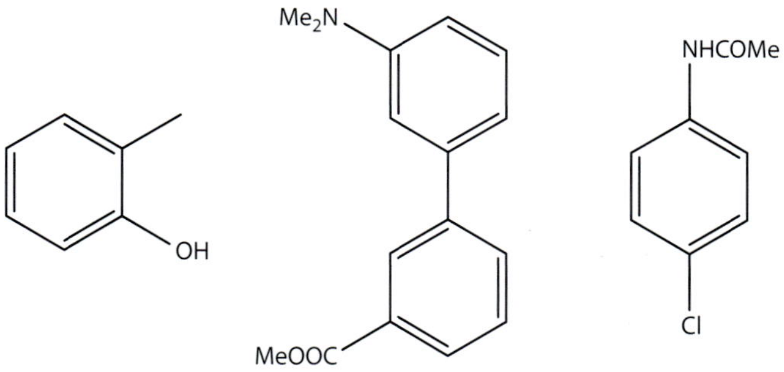

Solutions

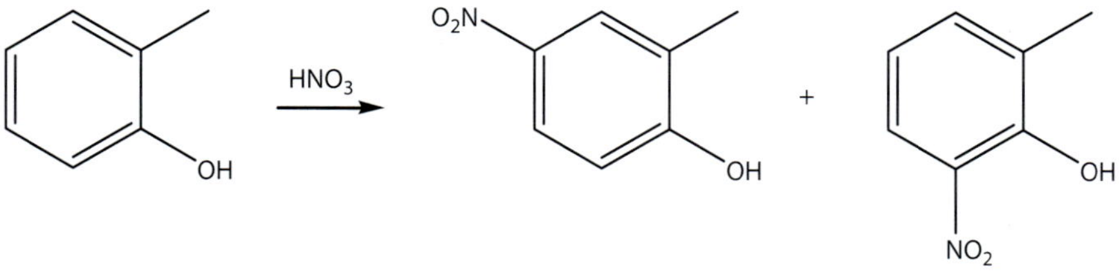

Both groups are *ortho, para*-directing and activating, but the resonance donation of the OH group is stronger than the inductive donation of the methyl group, so its directing effect dominates.

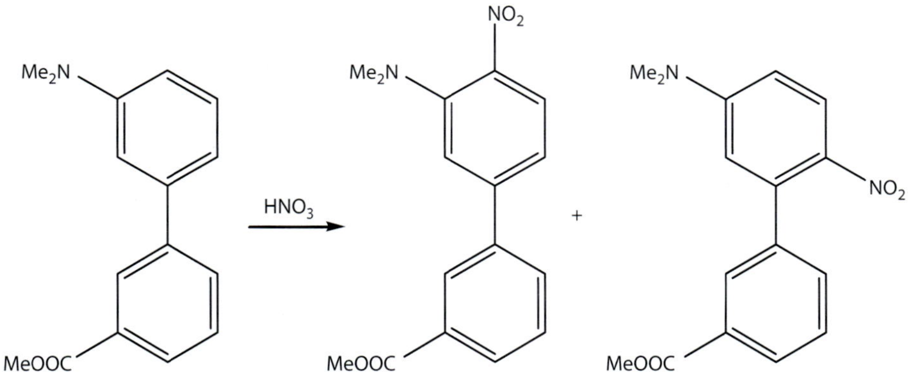

With a biphenyl substrate, we first need to decide which ring is likely to react. Since NMe_2 is a donor and COOMe withdraws electrons, it is the nitrogen-bearing ring that is more electron rich and hence more reactive. NMe_2 is *ortho, para*-directing and activating. The other product of *ortho*-substitution is possible but less likely to be formed, as that position would be relatively sterically hindered.

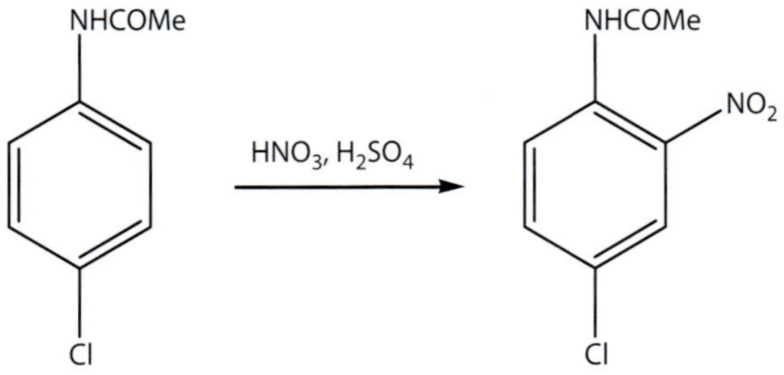

The NHCOMe group is *ortho*, *para*-directing and mildly activating; the chlorine is *ortho*, *para*-directing and deactivating. The activating group dominates the outcome. Notice that because we have one mildly activating and one deactivating group on the ring that we are reacting, we need more vigorous reaction conditions than for the other two examples. Those could be nitrated with dilute nitric acid in acetic acid as solvent, because of the presence of powerful activating groups.

12.3.6 PARTIAL RATE FACTORS

Partial rate factors (PRFs) are a quantitative way of describing directing effects in electrophilic aromatic substitution. If we take the data from Table 12.3, together with the fact that nitration of toluene proceeds 24 times faster than nitration of benzene, we can obtain an estimate of the relative reactivity of a specific site on toluene relative to a single site on benzene. Benzene has six identical protons that can be replaced, whereas toluene has two *ortho*-, two *meta*-, and one *para*-position.

So we calculate the PRF for the *ortho*-position (Equation 12.3):

$$\text{PRF}(ortho) = 0.58 \times 24 \times \frac{6}{2} = 41.8 \tag{12.3}$$

The value 0.58 comes from the fraction of the product that is substituted at the *ortho*-position, 24 from the relative rates of reaction of toluene and benzene, and 6/2 from the number of sites on benzene for substitution relative to the number of *ortho* positions on toluene. Similarly,

$$\text{PRF}(meta) = 0.04 \times 24 \times \frac{6}{2} = 2.88 \tag{12.4}$$

$$\text{PRF}(para) = 0.37 \times 24 \times \frac{6}{1} = 53.3 \tag{12.5}$$

The fact that the value for the *meta*-position is >1 reflects the overall increase in electron density in the aromatic ring; although the intermediate for substitution at that position is not specifically stabilized, the more electron-rich ring in toluene is more nucleophilic than benzene. The greater PRF at the *para*-position is probably steric in origin.

PRFs vary over a very large range—substituted benzenes may be very much more, or less, reactive than benzene. The nitro group is one of the most deactivating substituents, and the PRFs are correspondingly tiny (Figure 12.40).

Notice that these values bear out our conclusions about the directing effect of the nitro group from studying the electronic effects in the resonance forms of the Wheland intermediates for the substitution. The *meta*-position is not "favored"; it is simply the least disfavored option. This means that polynitration of aromatics is likely to be difficult; your lab assignment

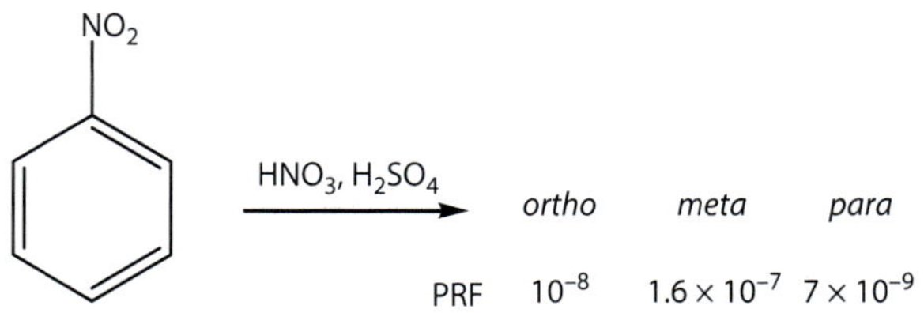

FIGURE 12.40 Partial rate factors for nitration of nitrotoluene.

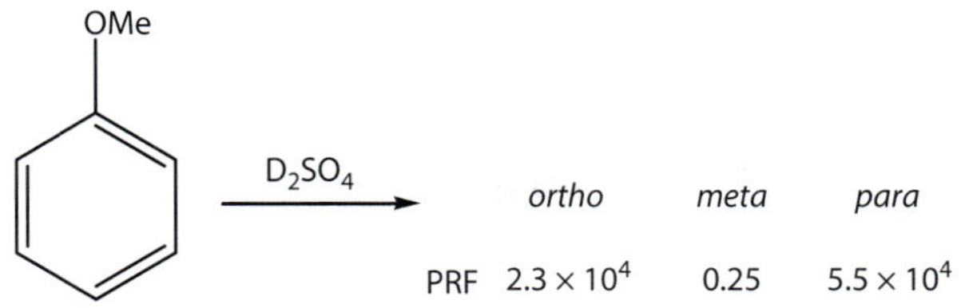

FIGURE 12.41 Partial rate factors for deuteration of anisole.

for this topic will not be the synthesis of trinitrotoluene (TNT) ☺. Although the first nitration of toluene is relatively favorable, the subsequent ones are much less so. Industrially, TNT is prepared in three steps involving isolation of the mononitro and dinitro compounds, with the final nitration step involving an anhydrous mixture of nitric acid and oleum (a mixture of sulfur trioxide and sulfuric acid). By contrast, if we consider the deuteration of anisole, which bears a strongly activating group, we get high PRFs (Figure 12.41).

TNT used in a car bombing in India

The low PRF at the *meta*-position of anisole is noteworthy. At the *meta*-position, there is no resonance effect, so we see the inductive withdrawal effect of the OMe group. At the *ortho*- and *para*-positions, this is masked by the much greater resonance donor effect.

Focus on Partial Rate Factors

Extensive tables are available for a range of substituents—but be wary of using these uncritically. PRFs are frequently very dependent on the exact conditions used for the reaction (Table 12.4). Do not assume that because you have seen the PRFs for nitration of toluene with HNO_3/H_2SO_4, you can apply these to the likely outcome when acetyl nitrate, $CH_3C(=)NO_2$, is used as the nitrating agent. Temperature and solvent are also important and can lead to substantial changes in the outcome. And a close perusal of the chemical literature suggests that different research groups have observed subtly different outcomes even when the reaction parameters are apparently identical.

Although these numbers sometimes appear confusing, they clearly do support the directing effects that we have discussed, and interestingly, many of the numerical differences arise from differences in measured reaction rates, rather than differences in the proportions of products obtained.

TABLE 12.4 Partial Rate Factors in Electrophilic Substitution

Substrate	Reagents/Conditions	PRF (*ortho*)	PRF (*meta*)	PRF (*para*)
Toluene	HNO_3/AcOH/25 °C	49	2.4	70
	HNO_3/$MeNO_2$/25 °C	49	2.5	56
	HNO_3/Ac_2O/0 °C	39	3	51
	$AcONO_2$/AcOH/25 °C	47	3	62
Toluene	Cl_2/AcOH/H_2O	534	—	552
	Cl_2/AcOH	617	5	820
	Cl_2/MeCN	1830	91.1	6250
	HOCl/H_2O/H^+	134	4	82
Toluene	D^+	330	7.2	313
PhF	D^+	0.136	—	1.70
PhCl	D^+	0.035	—	0.161
PhF	HNO_3/H_2SO_4	0.059	—	0.783
PhCl	HNO_3/H_2SO_4	0.03	9.9×10^{-4}	0.14
PhBr	HNO_3/H_2SO_4	0.033	9×10^{-4}	0.11
PhI	HNO_3/H_2SO_4	0.21	0.011	0.648

PROBLEM 12.10

Explain the PRFs for nitration of toluene and *tert*-butyl benzene, given below:

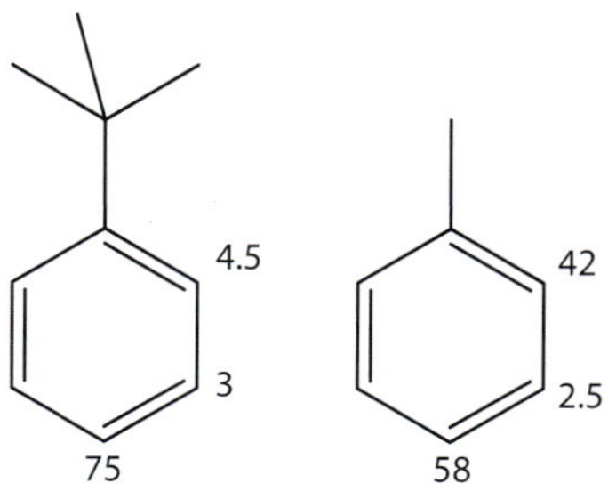

Solution

All alkyl substituents are *ortho, para*-directing and activating, so the high reactivity at the *o*- and *p*-positions of toluene is not surprising. The fact that the reactivity at the *meta*-position is higher than for benzene reflects the electron-rich nature of the ring, because of the inductive donating effect of the alkyl group. The same electronic considerations apply for *tert*-butylbenzene, but in this molecule, the reactivity at the *ortho*-position is markedly reduced by the steric effect of the bulky substituent.

Key Points from Section 12.3

- Existing substituents on an aromatic ring control where any further electrophilic substitution will occur.
- Inductive donors are mildly activating and direct *ortho, para* (alkyl groups).
- Resonance donors are strongly activating and direct *ortho, para* (–OH, –OR, $-O^-$, –OC(=O)R, $-NH_2$, $-NR_2$, NHC(=O)R, –SH, –SR).
- Resonance acceptors are deactivating and direct *meta* (–COOH, –COOR, –C(=O)R, –CHO, $-NO_2$, $-SO_3H$, –CN).
- Inductive withdrawers are deactivating and direct *meta* ($-CF_3$, $-N^+R_3$).
- Halogens are deactivating but direct *ortho, para*.
- If the directing effects of two substituents are in conflict, the more activating substituent has the dominant effect.
- Steric effects may be important in determining *ortho:para* ratios.
- PRFs can be used to quantify directing effects.

12.4 REACTIONS OF POLYCYCLIC AROMATIC COMPOUNDS

Many molecules are known where benzene rings are fused together along a bond. The simplest examples are naphthalene (the main ingredient in mothballs), anthracene, and phenanthrene (**12.10–12.12**).

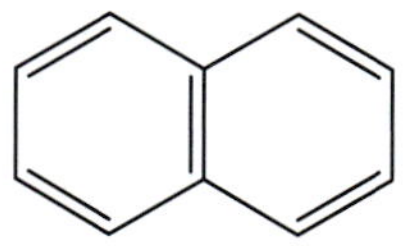

12.10, naphthalene

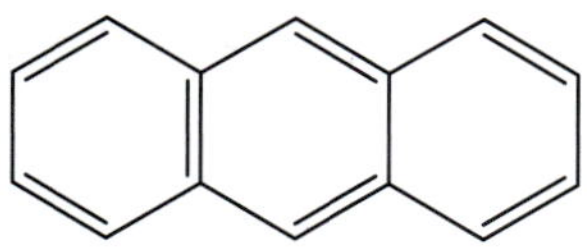

12.11, anthracene

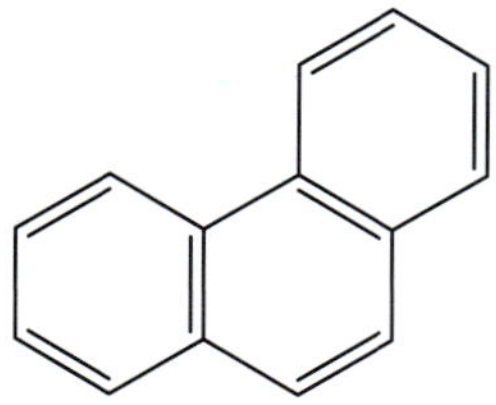

12.12, phenanthrene

One of the most harmful of all the polycyclic aromatic compounds is benzo[*a*]pyrene, **12.13**, which is a constituent of tobacco (and in greater quantities in marijuana) smoke, and was established to be the precarcinogenic compound in the cancers of the scrotum suffered by chimney sweeps in nineteenth-century Britain. *In vivo*, it is catalytically oxidized in the presence of various cytochromes and a hydrolase enzyme to **12.14**, which is both carcinogenic and mutagenic (the epoxide group is susceptible to ring-opening reactions by nucleophiles such as the bases of DNA and proteins).

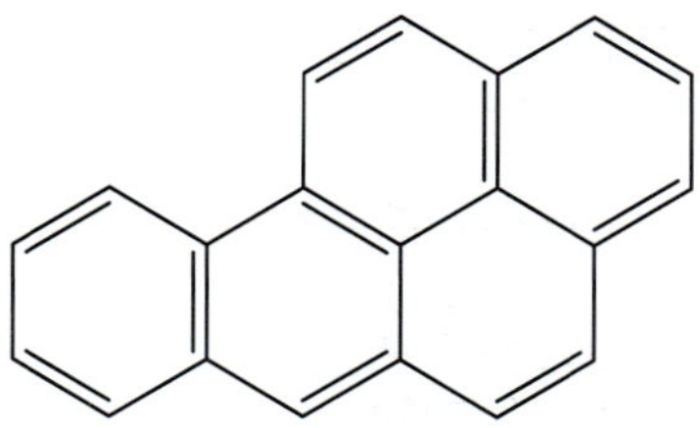

12.13, benzo[*a*]pyrene

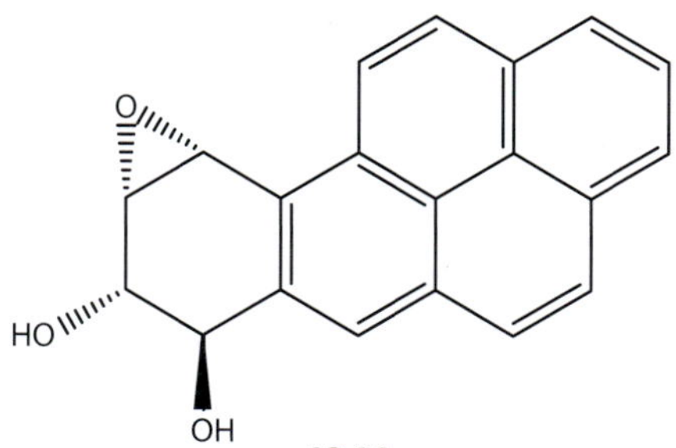

12.14

Polycyclic aromatic hydrocarbons (PAHs) may be petrogenic or pyrogenic. Petrogenic means that they are derived from petroleum products as they are released into the atmosphere. Pyrogenic means that they are formed from combustion, including wood burning, cigarette smoke, burning household waste, and charring steak on a barbecue.

Although naphthalene and indeed most polycyclic aromatic hydrocarbons are usually obtained from coal tar, naphthalene can be synthesized by a rather unusual Friedel–Crafts reaction

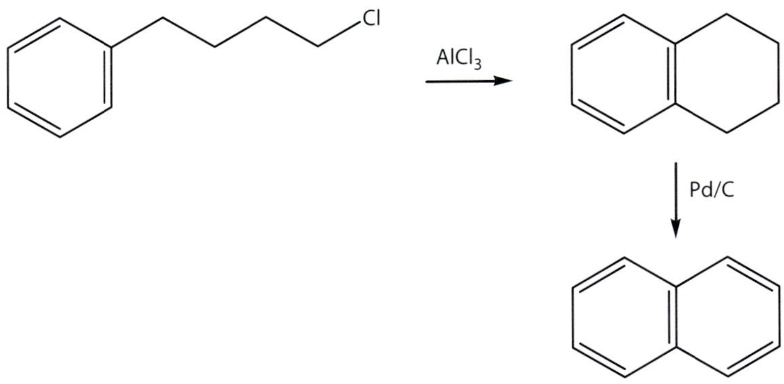

FIGURE 12.42 Synthesis of naphthalene.

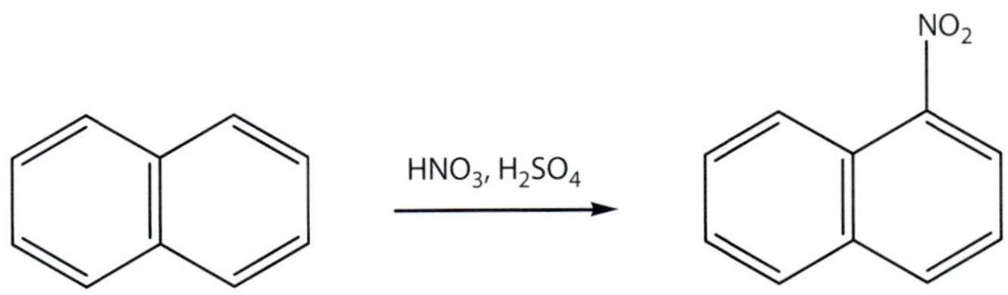

FIGURE 12.43 Nitration of naphthalene.

(Figure 12.42). Although the cation formed in this case is primary and could easily rearrange, the capture by the aromatic ring is greatly accelerated because the reaction is intramolecular, and the ring being formed is six membered. Most derivatives of naphthalene are prepared from naphthalene itself, rather than from substituted versions of this precursor.

The resonance energy of naphthalene is 255 kJ mol^{-1}. We can compare this with benzene, which has a resonance energy of 157 kJ mol^{-1}. Effectively, naphthalene is "less aromatic" than benzene—although the MO theory is beyond the scope of this course, we can see that only one ring at a time can be fully Hückel compliant with six π-electrons. The consequence for the chemistry of naphthalene is that electrophilic substitution is easier than for benzene—less resonance energy is lost in the initial step to form the Wheland intermediate, as one benzene ring will effectively remain intact and fully aromatic. Electrophilic substitution of naphthalene could, in principle, occur at either the 1- or 2-position. In most cases, we see only the kinetic product, which derives from substitution at the 1-position (Figure 12.43).

This is easily explained if we consider the resonance forms of the intermediates (Figure 12.44). When substitution occurs at the 1-position, there are two resonance forms where an intact benzene ring is retained, but for 2-substitution, there is only one of these favorable resonance forms. Hence, electrophilic substitution takes place at the 1-position under conditions of kinetic control. However, we remember that sulfonation is an exception in that it is reversible, and under appropriate conditions, we can obtain the thermodynamic product, naphthalene-2-sulfonic acid (Figure 12.26).

Why is the 2-substitution preferred under thermodynamic conditions? It is less sterically hindered because of the unfavorable *peri*-interaction as in **12.15**.

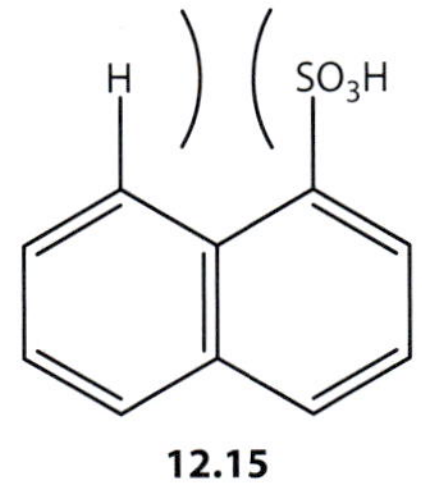

12.15

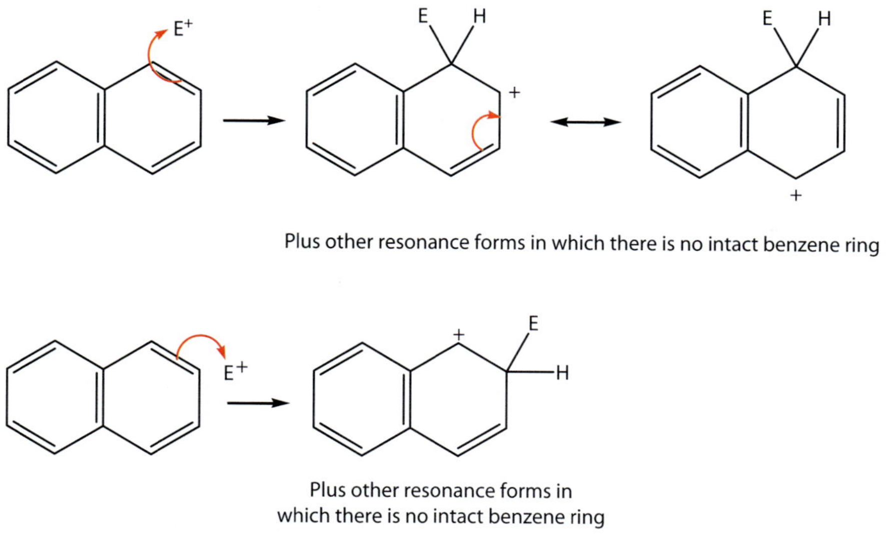

FIGURE 12.44 Mechanism and regiochemistry of electrophilic substitution of naphthalene.

The resonance energy of anthracene, **12.11**, is 351 kJ mol^{-1}; this is only just over twice that for benzene. So we would expect anthracene to be more reactive again than naphthalene. This is born out experimentally; in Figure 12.45, showing the bromination of anthracene, initial reaction is in the center ring, so that in the most important resonance form of the intermediate there are two intact benzene rings. The cost in terms of resonance energy to form this intermediate is low, and indeed, at low temperature, the intermediate reacts with bromide ion to give the addition product **12.16**. On heating, **12.16** loses HBr to give a more conventional substitution product.

Another reaction that demonstrates the similarity in energy between one anthracene and two benzene rings is the Diels–Alder reaction of anthracene with maleic anhydride (Figure 12.46).

Substituent effects in electrophilic substitution of polycyclic compounds are similar to those for benzene, but all the reactions go faster, because there is less loss of resonance energy in the RDS.

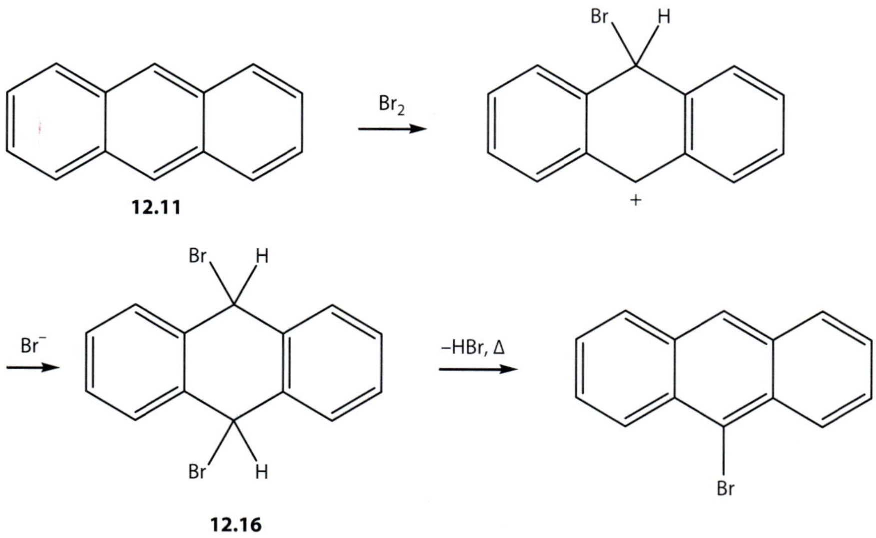

FIGURE 12.45 Bromination of anthracene.

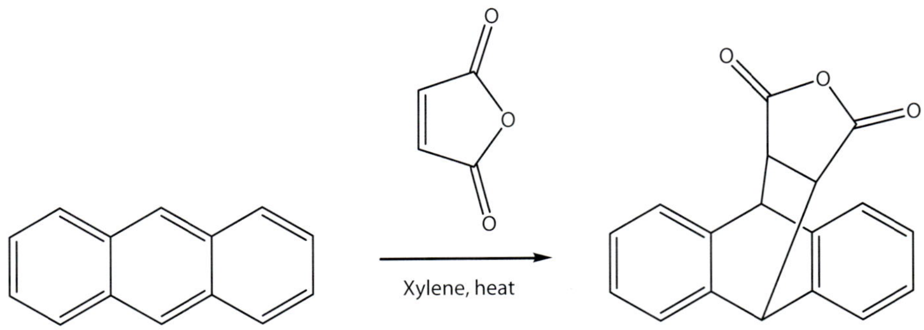

FIGURE 12.46 Diels–Alder reaction of anthracene and maleic anhydride.

PROBLEM 12.11

Predict the product(s) of the electrophilic substitution of **12.17** and **12.18**:

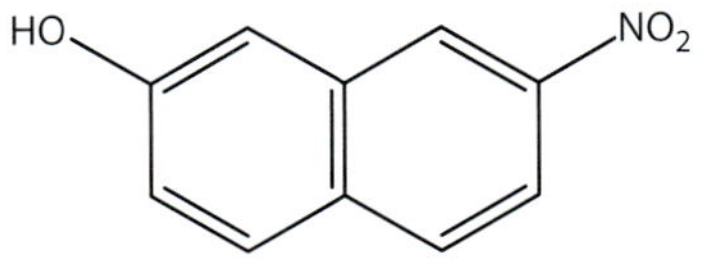

12.17

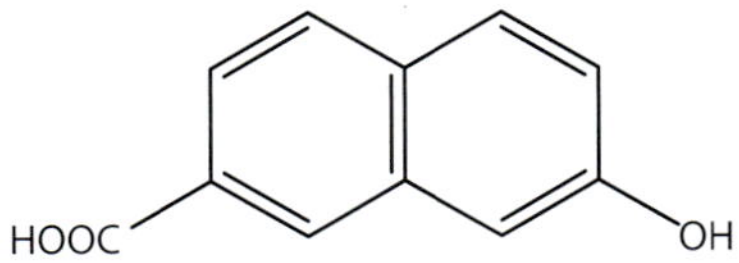

12.18

Solutions

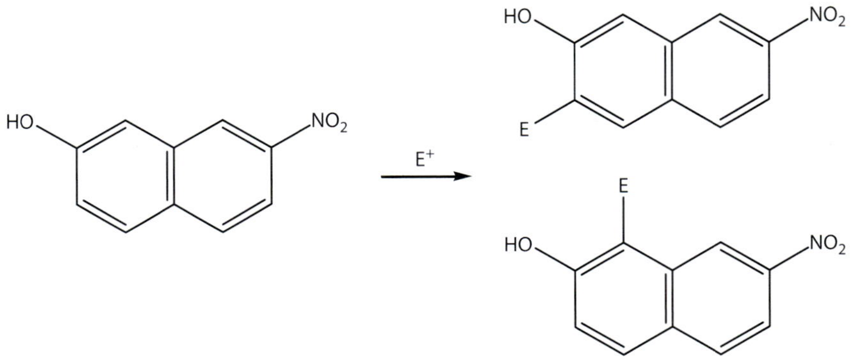

The ring bearing the −OH group is more electron rich and hence more reactive with electrophiles. OH is *ortho, para*-directing and activating, but the *para*-position is blocked by the ring. Given that the 1-position of naphthalene is favored over the 2-position, the second product will probably predominate under conditions of kinetic control.

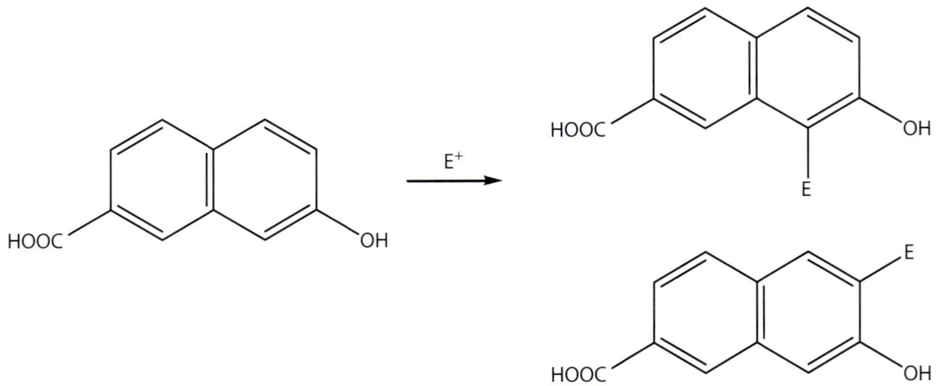

This problem is similar to the previous one. OH is electron donating and activating and COOH is electron withdrawing and deactivating, so the ring bearing the OH group is more electron rich, and hence electrophilic reaction occurs in this ring. OH is *ortho*, *para*-directing, and the naphthalene substitutes preferentially at the 1-position, so the first product will predominate under conditions of kinetic control. In both problems, substitution at the less hindered position would predominate under conditions of thermodynamic control, because of the unfavorable *peri*-interaction in the kinetic products.

Key Points from Section 12.4

- Electrophilic substitution of polycyclic aromatic compounds is generally faster than that of benzene, because they have lower aromatic stabilization.
- The location of the substitution is determined by the loss of aromatic character in the σ-complex; structures where there are "intact" benzene rings are favored.
- Under conditions of kinetic control, naphthalene undergoes electrophilic substitution mainly at the 1-position.
- The directing effects of existing substituents can be predicted from their effects on substitution of benzene.

12.5 HETEROCYCLIC AROMATIC COMPOUNDS

We refer to compounds as heterocycles when there is an atom other than carbon as part of the ring.

12.5.1 PYRIDINE AND RELATED COMPOUNDS

When we replace one carbon atom of benzene or naphthalene with an sp^2 nitrogen atom, we get pyridine, quinoline, and isoquinoline (Figure 12.47).

Acridine is a nitrogen analogue of anthracene, isolated from coal tar. It is known to be a potent carcinogen and mutagen. Pyrimidine and purine are the parent compounds of the bases in nucleic acids (Figure 12.48).

Many other nitrogen-containing heterocycles have biological activity; examples are shown in Figure 12.49.

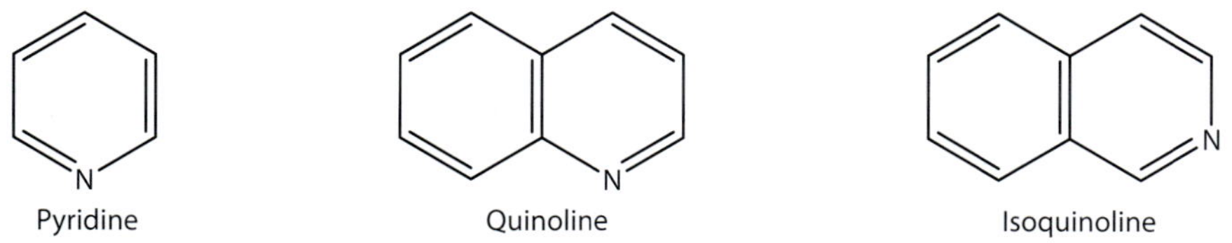

FIGURE 12.47 Pyridine and related heterocycles.

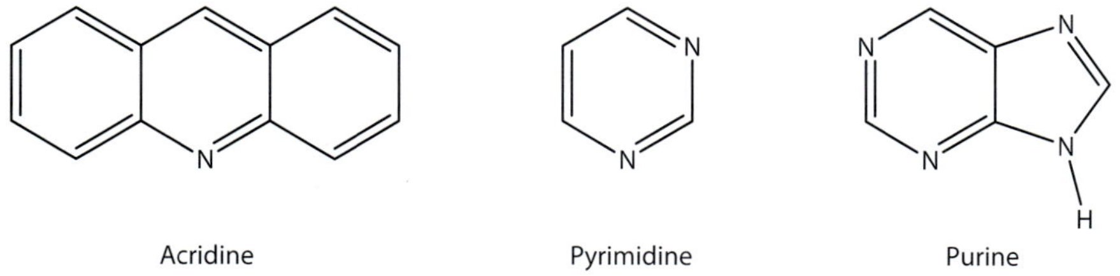

FIGURE 12.48 More pyridine-related heterocycles.

Paraquat is an effective herbicide but is now banned in many countries because of its toxicity

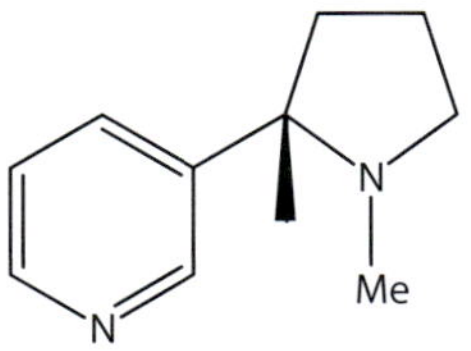

Nicotine

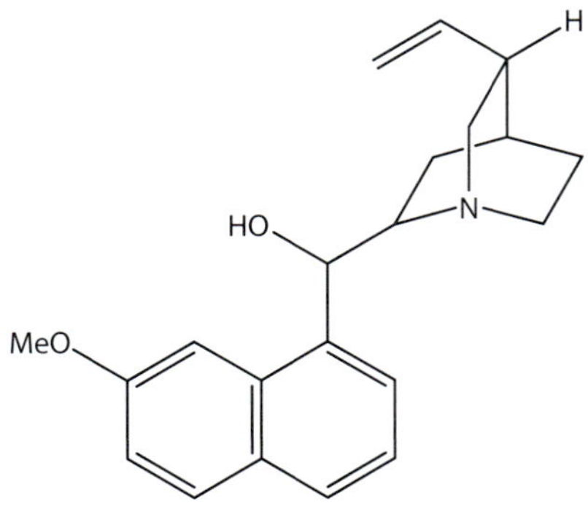

Quinine–anti–malarial

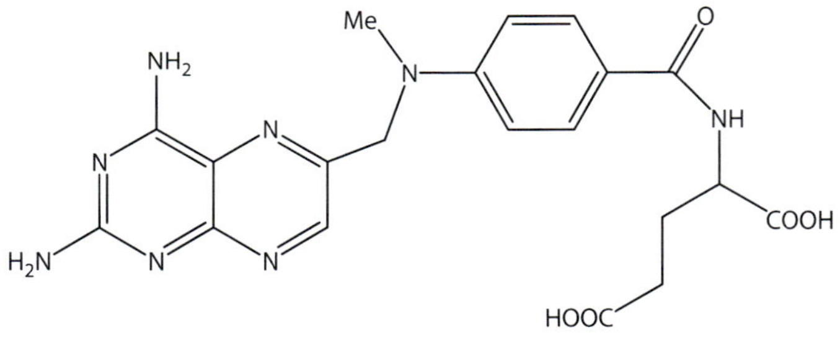

Methotrexate, used in cancer treatment

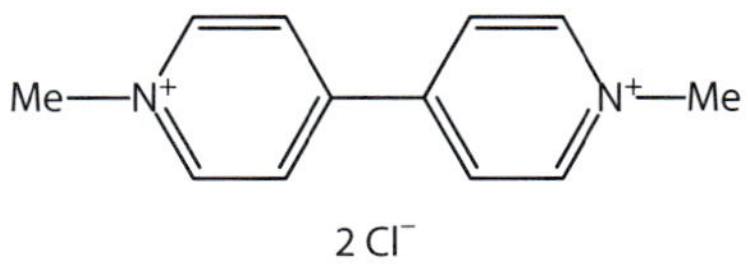

Paraquat–weed killer

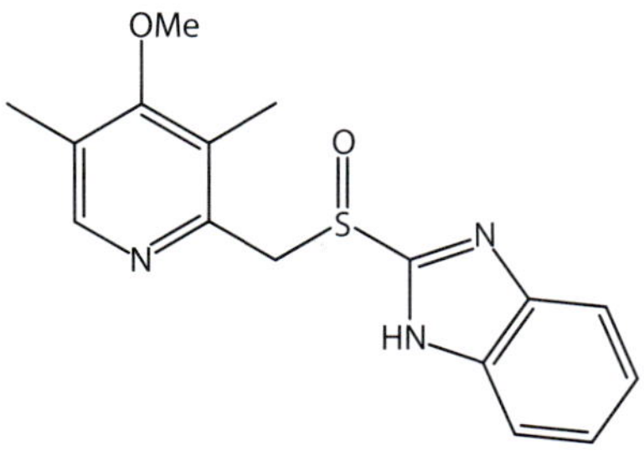

Omeprazole (Losec, prilosec)–antacid, proton pump inhibitor used for ulcer treatment

FIGURE 12.49 Biologically important nitrogen-containing heterocyclic compounds.

How is pyridine different from benzene? First, the electronegativity of the nitrogen atom means that it has a dipole moment, 2.21 D. The bonds are no longer strictly equal in length ($N{-}C_2$ = 1.34 Å, $C_2{-}C_3$ = 1.39 Å, $C_3{-}C_4$ = 1.40 Å; compare C–C in benzene = 1.40 Å). The presence of the nitrogen atom removes the degeneracy from the molecular orbitals of pyridine.* However, the resonance stabilization energy of pyridine is 134 kJ mol^{-1}, quite similar to that of benzene, so any major differences in reactivity will probably not derive from the differences in resonance energy.

The nitrogen atom bears a lone pair, in the plane of the aromatic ring, and it is able to act as a base. Because this is an sp^2 lone pair, pyridine is a less strong base than the sp^3 nitrogen-containing compound, piperidine (Figure 12.50). Pyridinium hydrochloride, formed by the reaction of HCl with pyridine, is relatively insoluble in most organic solvents at room temperature, and pyridine is often used to remove HCl produced as a by-product of organic reactions.

The lone pair on the nitrogen of pyridine also means that it is nucleophilic and can react with a range of electrophiles. Reaction with iodomethane gives *N*-methylpyridinium hydrochloride, and other electrophilic haloalkanes react similarly. Reaction with hydrogen peroxide gives initially the *N*-hydroxy derivative, which is in equilibrium with pyridine *N*-oxide (Figure 12.51).

Pyridine does not undergo electrophilic substitution easily; its reactivity is comparable with that of nitrobenzene. For example (Figure 12.52), sulfonation goes at the 3-position, under very forcing conditions (note that when we describe the position of substituents on a pyridine ring, we number it with the nitrogen atom as 1—this is true for all simple heterocycles).

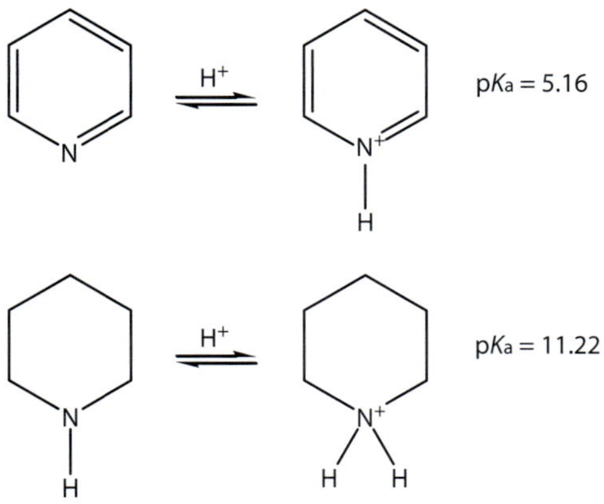

FIGURE 12.50 Acid/base properties of pyridine and piperidine.

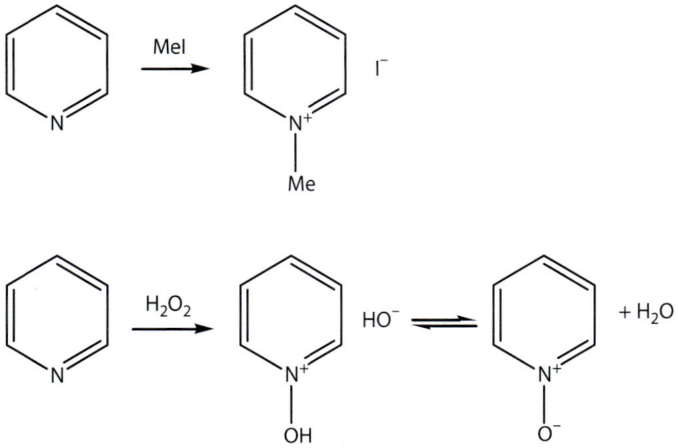

FIGURE 12.51 Reaction of pyridine with methyl iodide or hydrogen peroxide.

* You can find out more about MO theory for pyridine in *The Chemistry of Heterocycles*, Eicher and Hauptmann, 2nd edn., Wiley, 2003.

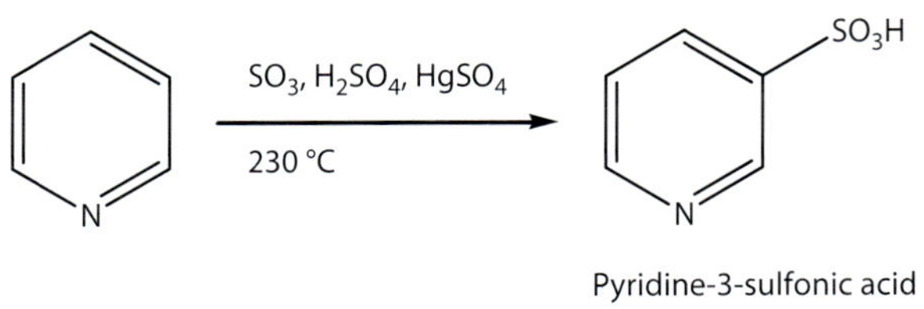

FIGURE 12.52 Sulfonation of pyridine.

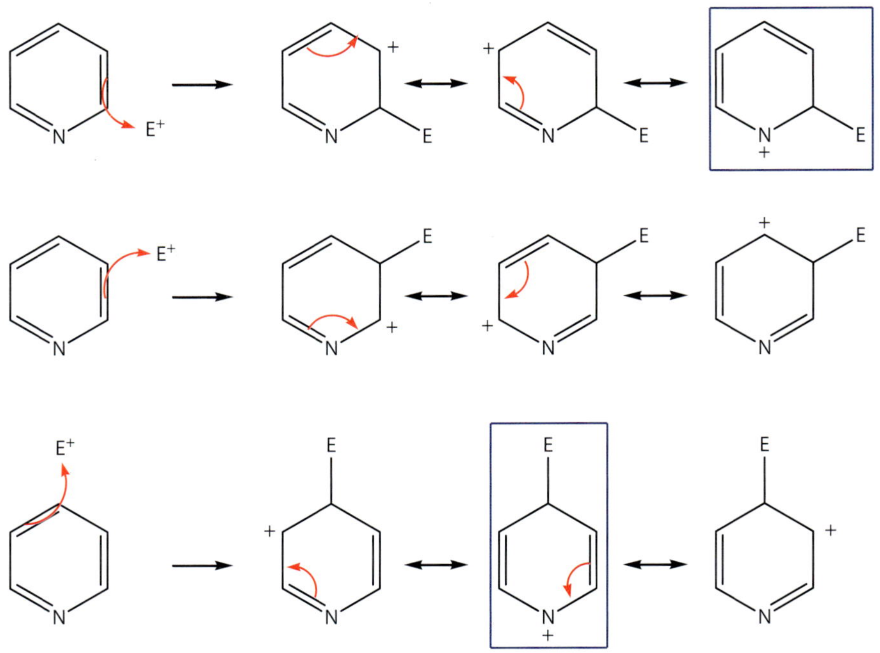

FIGURE 12.53 Resonance forms of possible intermediates produced in electrophilic substitution of pyridine.

This can be explained by the same type of analysis that we used for benzene derivatives; in Figure 12.53, two structures, enclosed in blue boxes, are particularly disfavored, as the nitrogen is positively charged and does not have an octet of electrons. So, as we saw for *meta*-directing, deactivating, substituents on benzene rings, substitution at the 3-position is not favored, but it is the least disfavored of the options available.

12.5.2 ANALOGUES OF THE CYCLOPENTADIENYL ANION

When we studied the idea of aromaticity in Chapter 3, we encountered the cyclopentadienyl anion, **12.19**, which is a charged Hückel system with six π-electrons delocalized over five atoms. The heterocyclic analogies of the cyclopentadienyl anion are furan, thiophene, and pyrrole (Figure 12.54). In each of these, there are four electrons from two π-bonds, and the other two are provided by a lone pair of electrons on the heteroatom. Thus, the lone pair is an integral part of the aromatic sextet of electrons and must occupy an sp^2 orbital parallel to those that form the π-bonds in the ring. For furan and thiophene (though not pyrrole), there is an additional lone pair of electrons in an sp^2 orbital in the plane of the ring; O, N, and S are sp^2 hybridized in all these molecules.

12.19

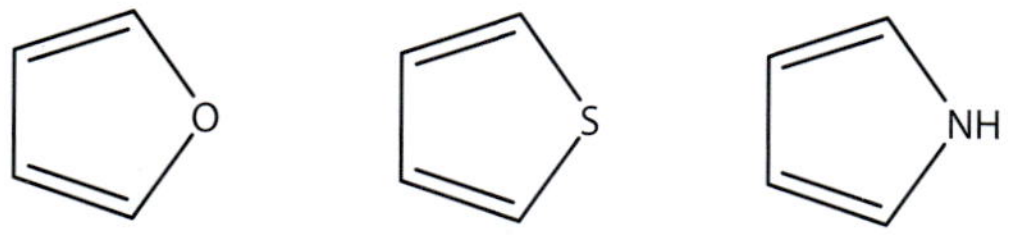

FIGURE 12.54 Heterocyclic analogues of the cyclopentadienyl anion.

Despite the lone pair on nitrogen, pyrrole is a very weak base, because the lone pair is a part of the aromatic system. The pK_a of the cation formed by protonation of pyrrole, **12.20**, is –3.8; the cation is strongly disfavored, as the aromaticity of pyrrole has been lost. However, pyrrole is relatively easy to deprotonate; its pK_a is 17, comparable to that of cyclopentadiene at 16. The anion still has a full octet of electrons at nitrogen, and the aromatic system remains intact.

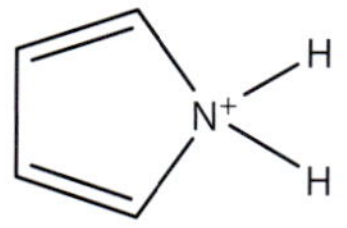

12.20

Pyrrole and its analogues are found widely in biologically important molecules; note that the purine nucleotides that we mentioned earlier contain a pyrrole-type as well as a pyridine-type ring. Pyrrole is a key component of the porphyrins; the 18-π aromatic system protoporphyrin IX, **12.21**, is the precursor of mammalian porphyrins, and there are related structures in heme, chlorophyll, and vitamin B12.

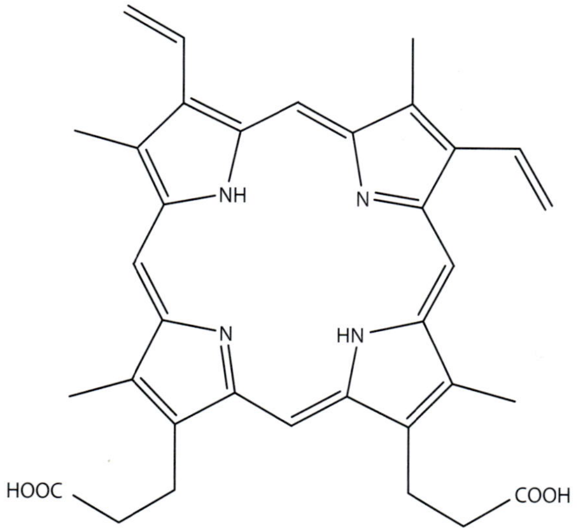

12.21, protoporphyrin IX

Imidazole, **12.22**, is a pyrrole with an additional nitrogen atom at the 3-position (again, we number with the nitrogen atom as 1); we have already met it as a molecule that tautomerizes particularly easily. This ability of picking up a proton at one side and dropping one off at the other is critical to its biological function where the amino acid histidine, **12.23**, is at the active site of protease (and other) enzymes (see Section 23.8).

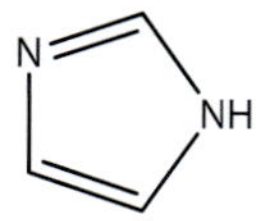

12.22, imidazole

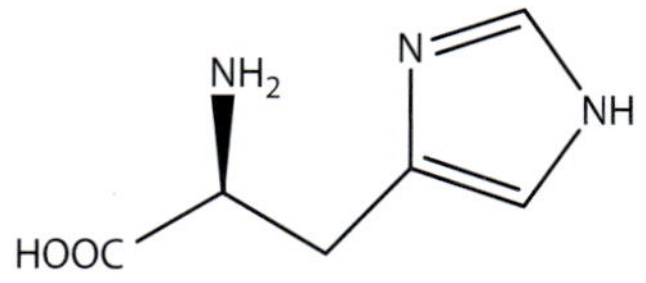

12.23, histidine

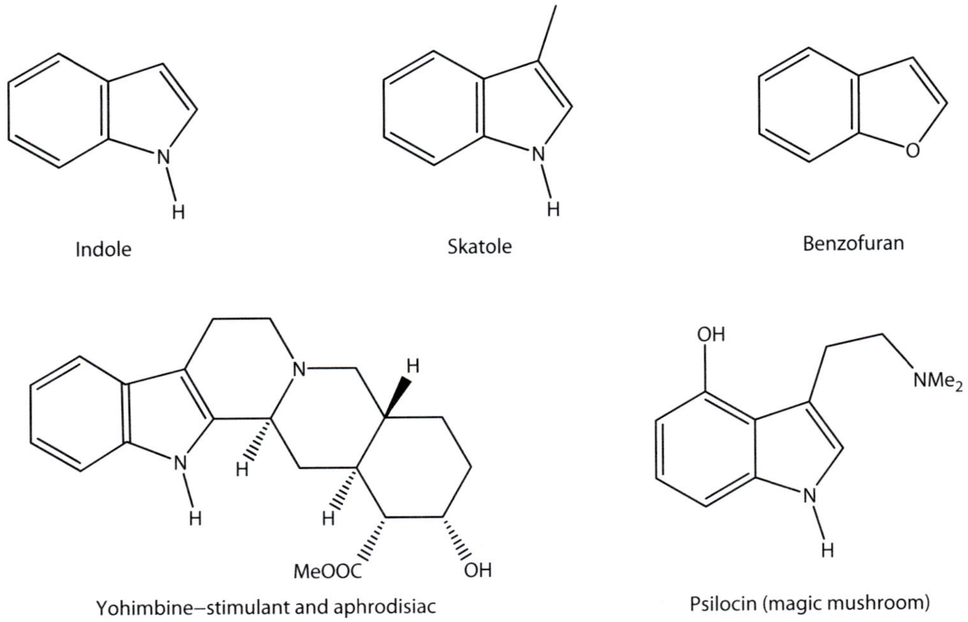

FIGURE 12.55 Indole derivatives and related compounds.

The molecule with pyrrole fused to a benzene ring, indole, is a key component of many alkaloids (Figure 12.55); skatole has a particularly unpleasant smell ☹. Benzofuran is less commonly found biologically and is normally obtained from coal tar.

Furan, thiophene, and pyrrole all have lower resonance energies than benzene (Table 12.5), and they are more reactive toward electrophiles. Even the simplest analysis should convince us that they are more electron rich than benzene—the six π-electrons are distributed over only five atoms.

The relative reactivity with electrophiles of the three 5-ring heterocycles is pyrrole > furan > thiophene > benzene. Although the resonance energy for furan is the lowest, the electronegativity of oxygen reduces its willingness to share electrons or to bear a positive charge in the intermediate. Furan and thiophene undergo electrophilic substitution most readily at the 2-position; the intermediate formed by electrophilic attack here has three viable resonance forms, whereas that from attack at the 3-position has only two (Figure 12.56).

Both pyrrole and furan are very sensitive to the acid conditions usually employed in electrophilic substitution, so control of the reaction conditions is very important. Examples of the mild conditions normally used are shown in Figure 12.57.

Furan, the least aromatic of this group of compounds, undergoes the Diels–Alder reaction readily (Figure 12.58). Although the kinetic product, the *endo*-isomer, can be detected at the early stages of the reaction, the more stable *exo*-isomer is the one isolated. Thiophene is "more" aromatic than furan, so it has more to lose in a Diels–Alder reaction. It undergoes cycloadditions only with great reluctance, and high pressures and temperatures are usually needed. Its reaction with maleic anhydride yields 47 % of the *exo*-adduct in 3 h at 100 °C, under 20 kbar pressure. Pyrrole tends to undergo nucleophile addition to maleic anhydride and other αβ-unsaturated carbonyl compounds (see Chapter 17).

TABLE 12.5 Resonance Energy of Various Heterocyclic Compounds

Molecule	Resonance Energy (kJ mol^{-1})
Benzene	150
Furan	67
Thiophene	121
Pyrrole	88

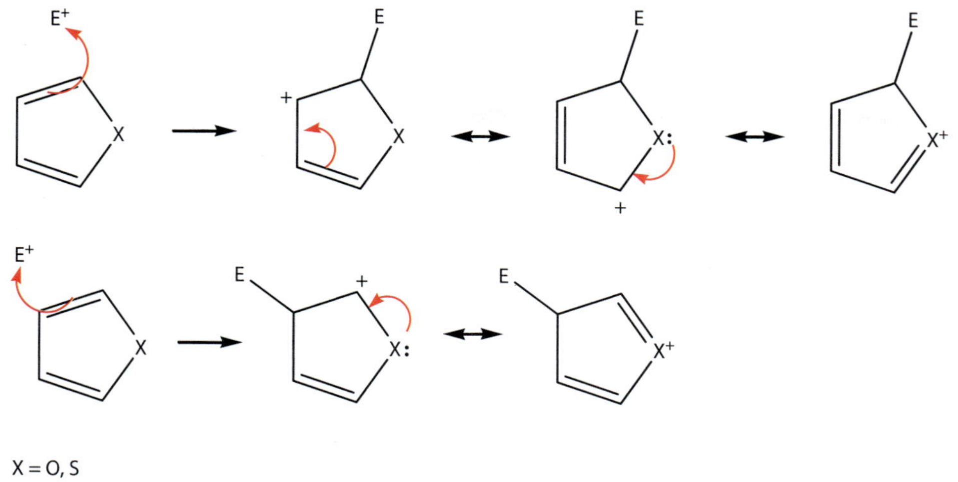

FIGURE 12.56 Electrophilic substitution of five-ring heterocycles.

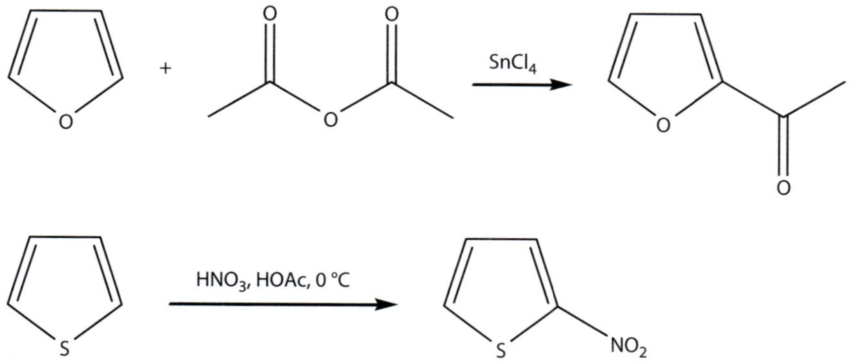

FIGURE 12.57 Typical electrophilic substitutions of furan and thiophene.

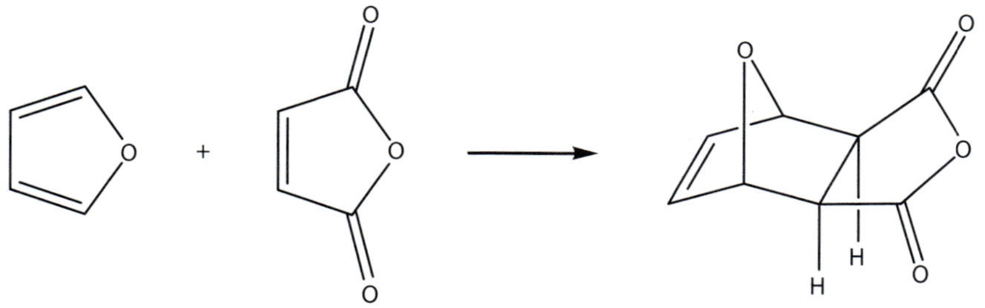

FIGURE 12.58 Diels–Alder reaction of furan.

Key Points from Section 12.5

- The sp^2 lone pair in the plane of the pyridine ring means that it can act both as a base and as a nucleophile.
- Pyridine is relatively unreactive toward electrophilic substitution (comparable with nitrobenzene). It is substituted at the 3-position under forcing conditions.
- The lone pair of electrons on nitrogen in pyrrole is a part of the aromatic sextet, and pyrrole is a very poor base. It is, however, a moderately good acid with a pK_a of 17.
- Pyrrole, furan, and thiophene are electron-rich heterocycles and react readily with electrophiles at the 2-position. Mild conditions are normally used as pyrrole and furan are sensitive to acid.
- Furan is a good substrate for the Diels–Alder reaction, but thiophene reacts only under very forcing conditions. This reflects their relative aromatic stabilization energy.

REVIEW PROBLEMS

1. Give a systematic name for each of the following compounds:

(a)

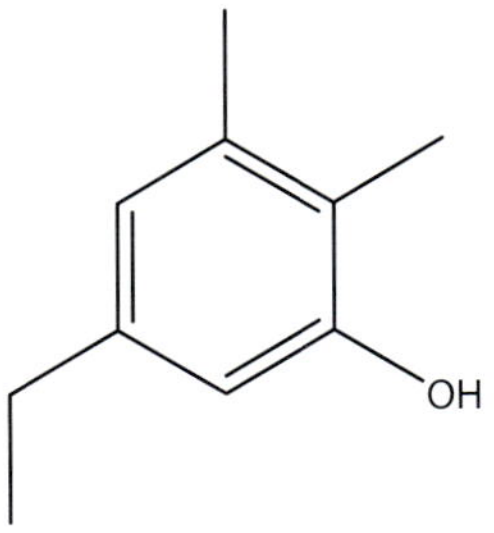

(b)

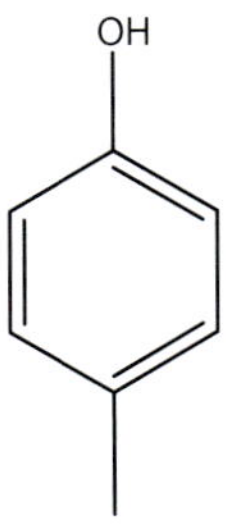

(c)

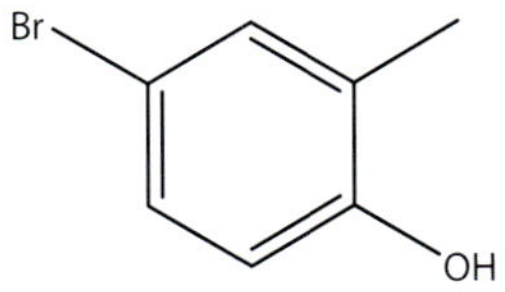

(d)

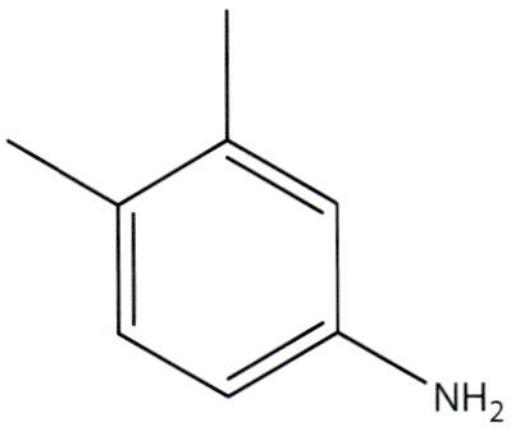

(e)

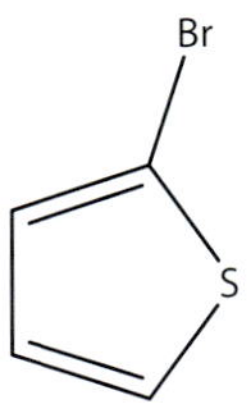

(f)

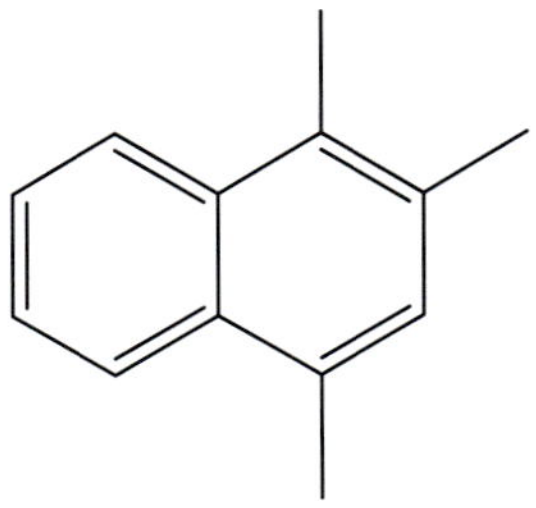

2. Draw a structure for each of the following compounds:
   (a) 3-Ethylphenol
   (b) 1,2,3,4,5-Pentamethylbenzene
   (c) 3-Amino-4-methylphenol
   (d) 1,3,5-Tribromobenzene
   (e) *m*-Dichlorobenzene
   (f) 4-Nitropyridine
   (g) Furan-3-carboxylic acid
   (h) Chrysene
3. How many π-electrons can potentially interact in each of the following? State whether each compound is aromatic, antiaromatic, or nonaromatic:

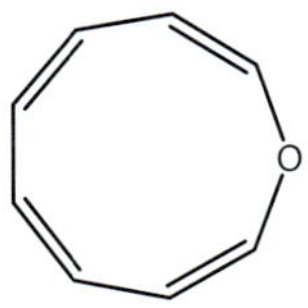

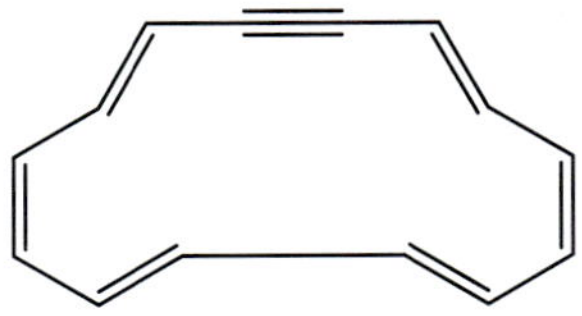

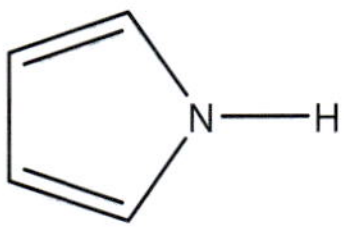

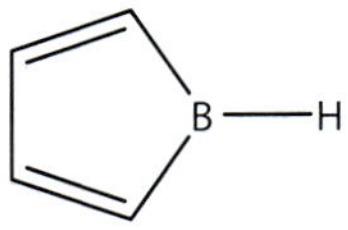

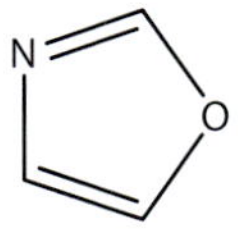

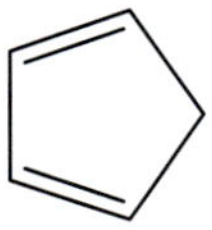

4. The following sets of spectroscopic data are all taken from substituted benzenes. Identify the molecules and explain as many of the data as you can:
   (a) $C_{10}H_{12}O$ shows a strong band in its IR spectrum at 1685 cm^{-1}. Its ^1H NMR spectrum shows the following resonances:

δ		
1.2	d	6H
3.5	Septet	1H
7.9	d	2H
7.5	m	3H

   (b) $C_9H_{10}O_2$ shows peaks in its mass spectrum at m/z = 150, 135, 107, and 77 and shows a strong peak in its IR spectrum at 1670 cm^{-1}. Its ^1H NMR spectrum shows peaks at

δ
2.6 (s, 3H)
3.9 (s, 3H)
7.0 (d, J = 7 Hz, 2H)
7.9 (d, J = 7 Hz, 2H)

   (c) $C_{11}H_{14}O_3$

Mass spectrum	m/z	194 (M^+), 134, 121, 76
IR spectrum		1730 cm^{-1}
^1H NMR spectrum	δ	2.5 (t, J = 8 Hz, 2H)
		2.9 (t, J = 8 Hz, 2H)
		3.7 (s, 3H)
		3.8 (s, 3H)
		6.8 (d, J = 7 Hz, 2H)
		7.2 (d, J = 7 Hz, 2H)

5. Predict the site of electrophilic deuteration of each of the following:

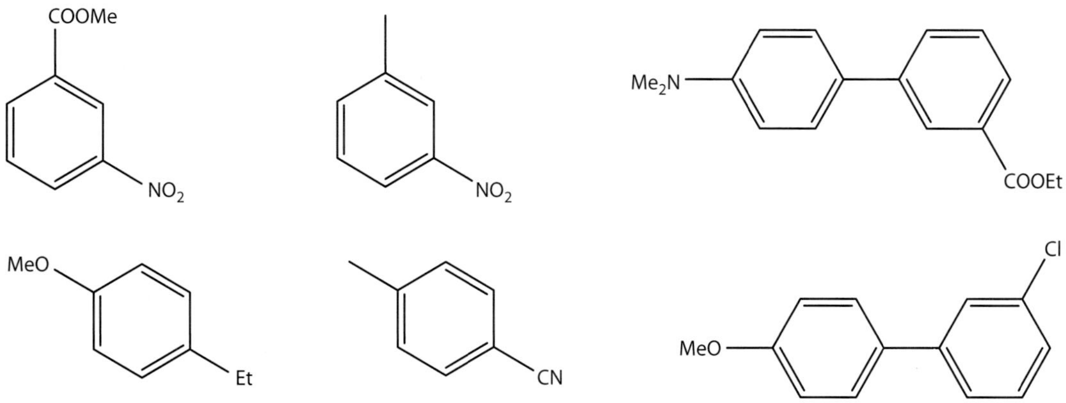

6. For each of the following pairs of compounds shown in the following, state which would react more rapidly with electrophiles. For each compound of the pairs, indicate the principle site(s) of substitution:
   (a)

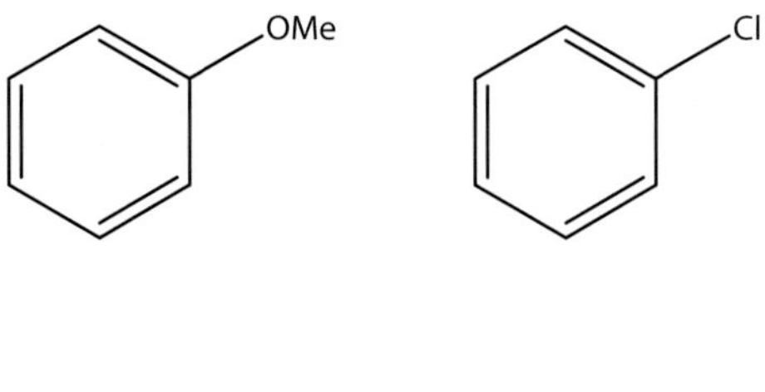

(b)

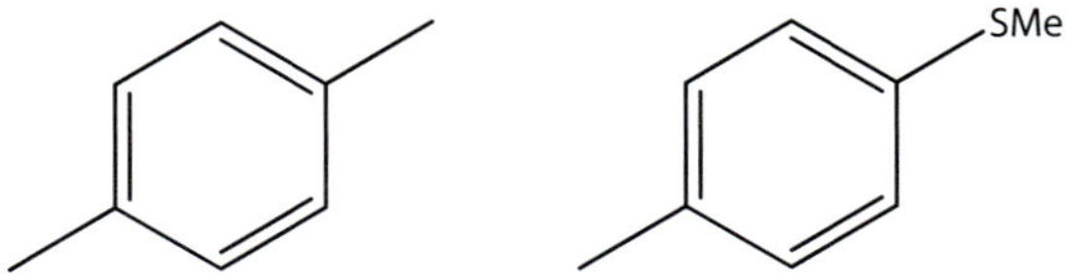

(c)

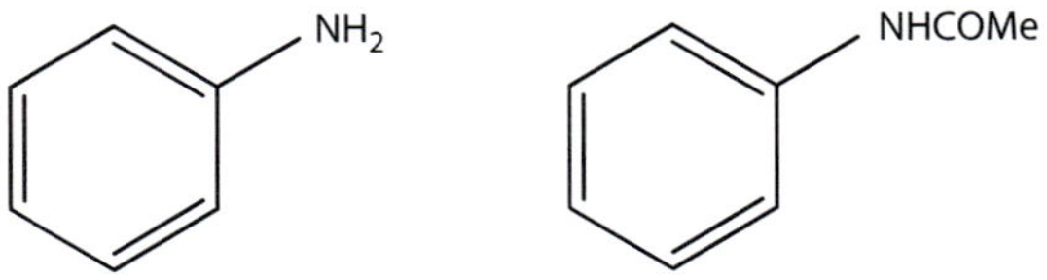

(d)

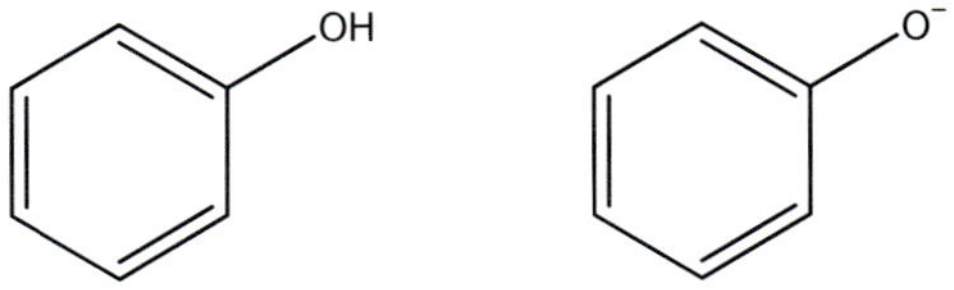

(e)

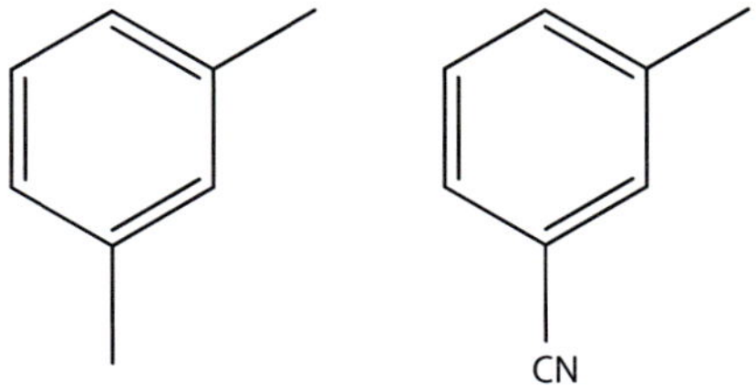

(f)

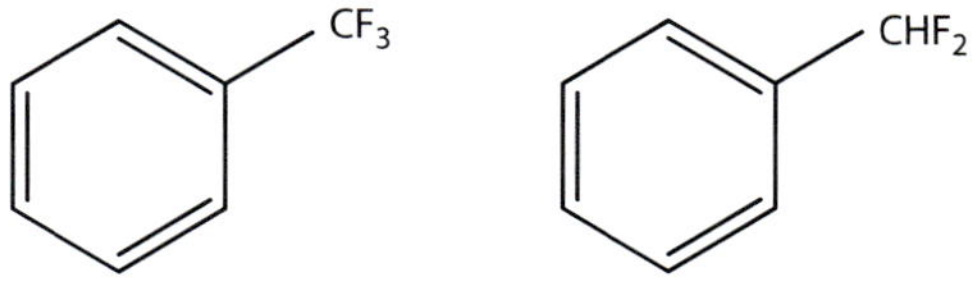

7. Arrange in order of reactivity with MeCOCl/$AlCl_3$:
   PhCOOMe, PhOMe, PhMe
8. Predict the predominant site of electrophilic substitution for each of the following molecules. State whether each is activated or deactivated:

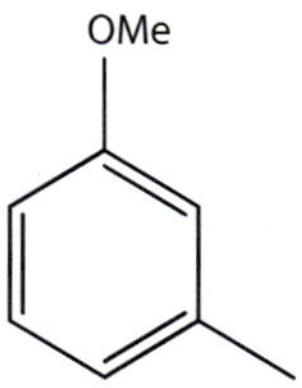

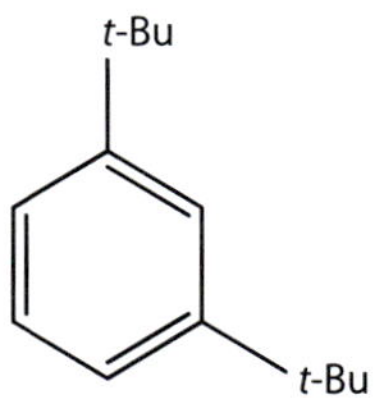

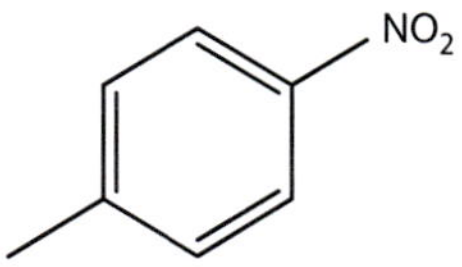

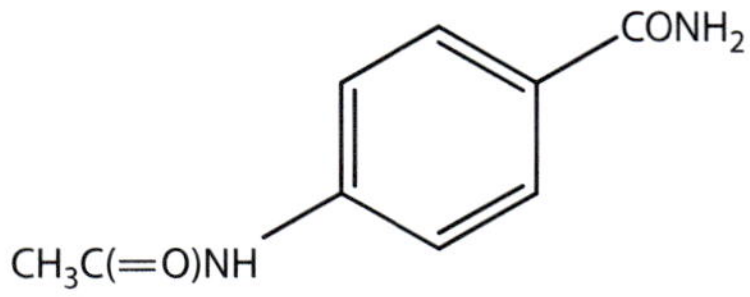

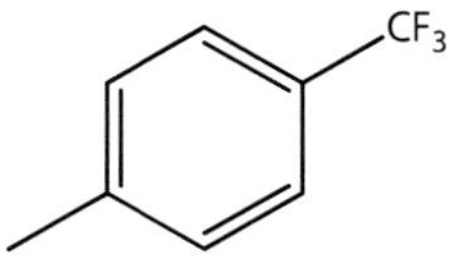

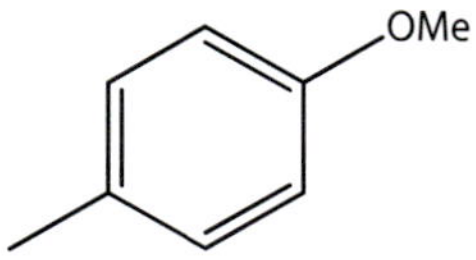

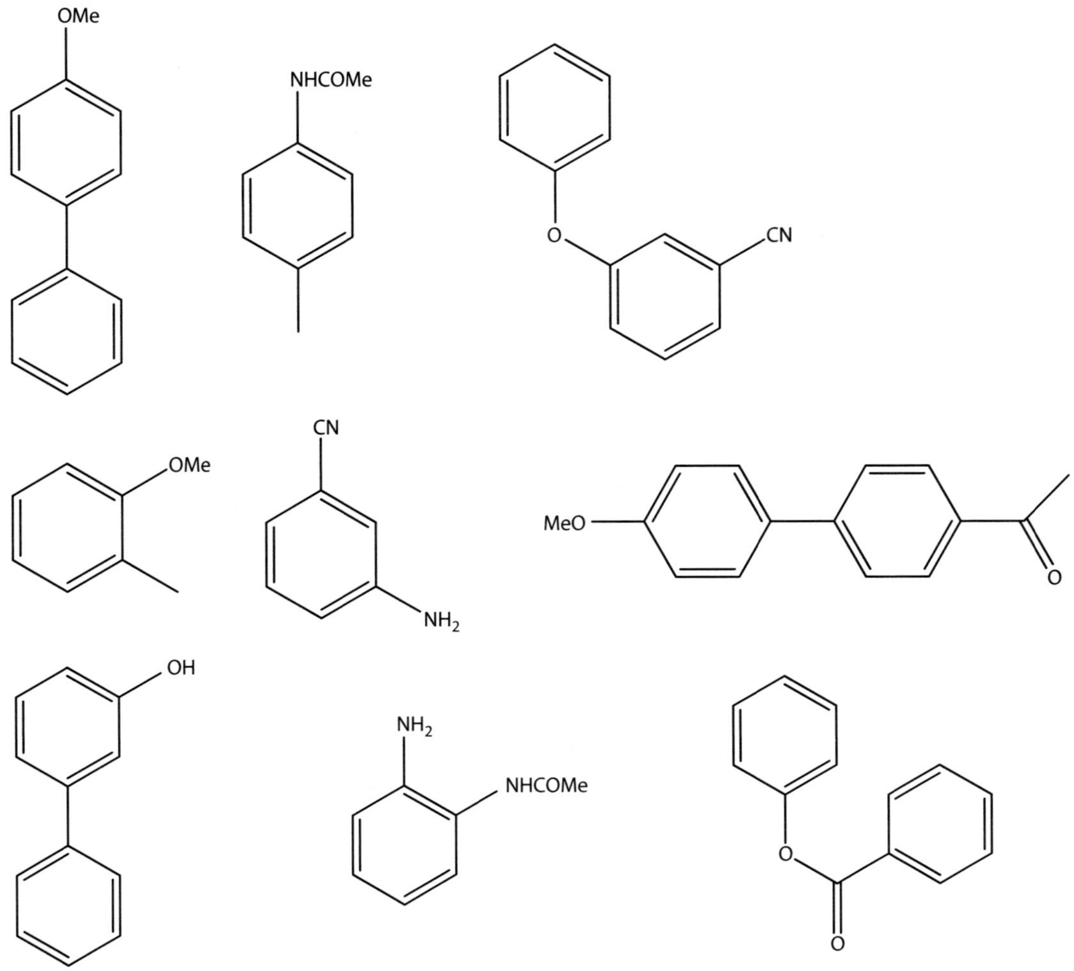

9. Propose a mechanism for each of the following transformations:

(a)

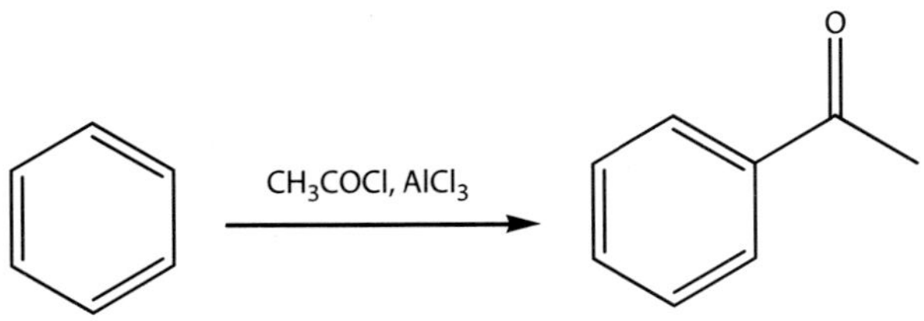

(b)

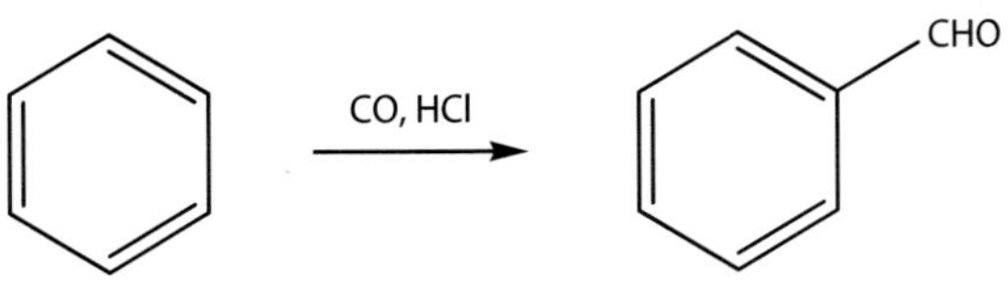

(c)

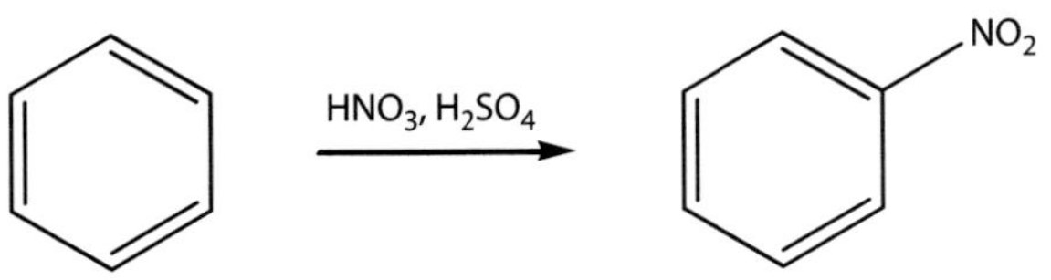

(d)

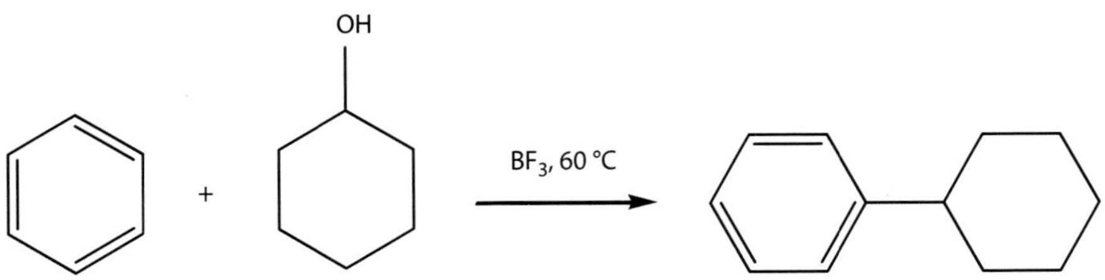

(e)

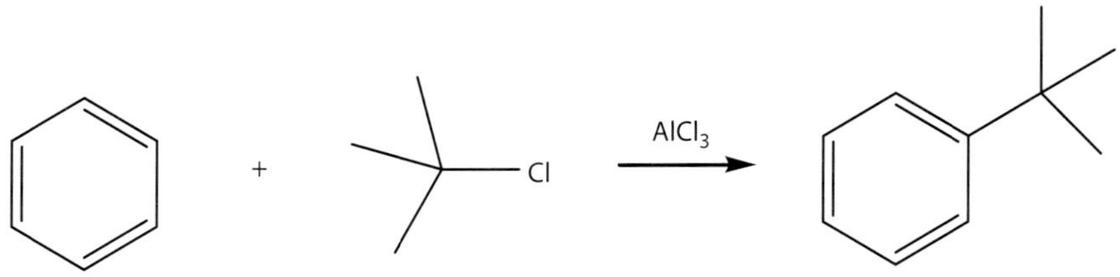

(f)

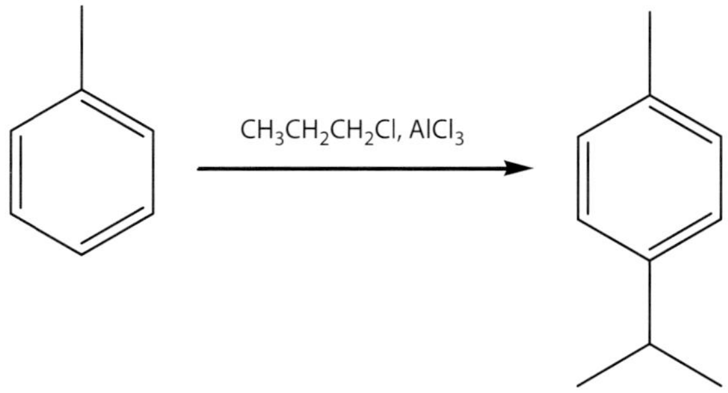

(g)

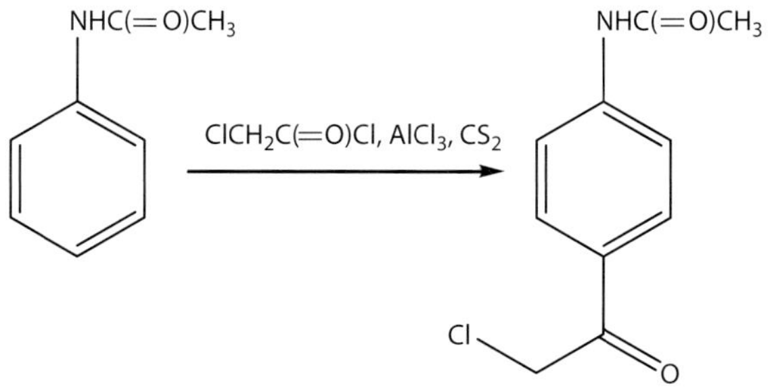

(h)

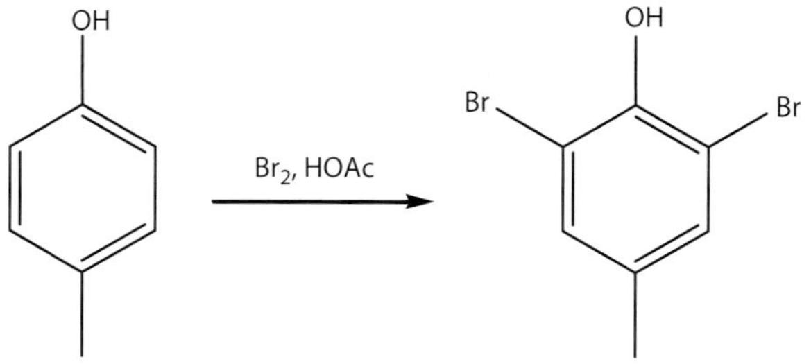

(i)

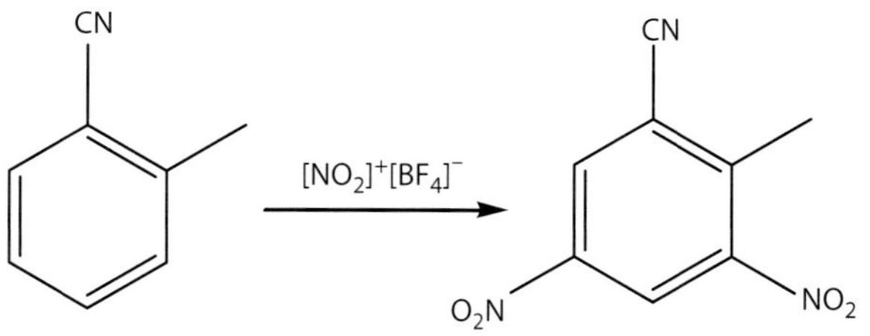

10. Predict the product(s) of each of the following reactions:

(a)

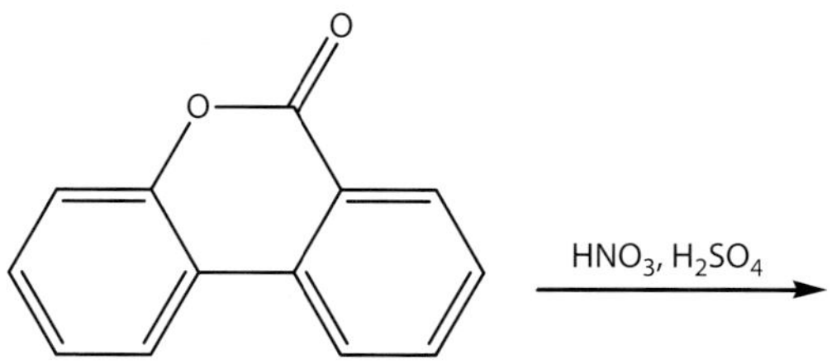

(b)

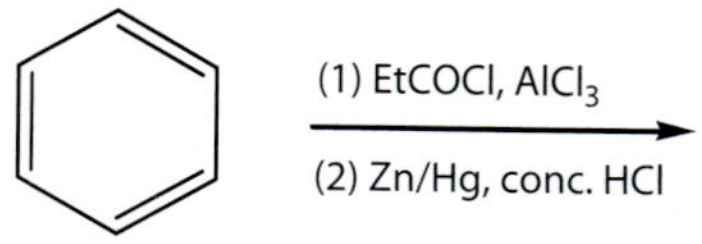

(c)

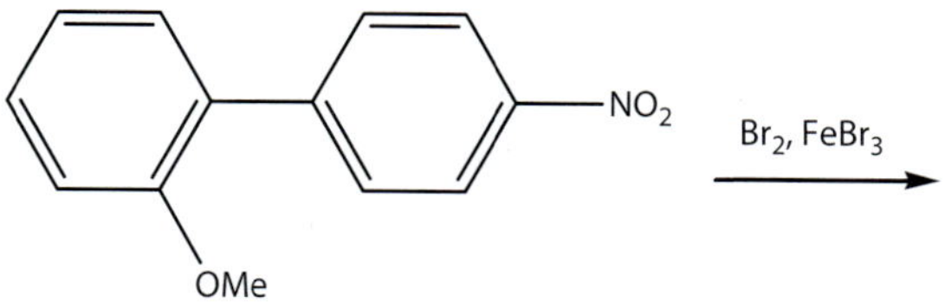

(d)

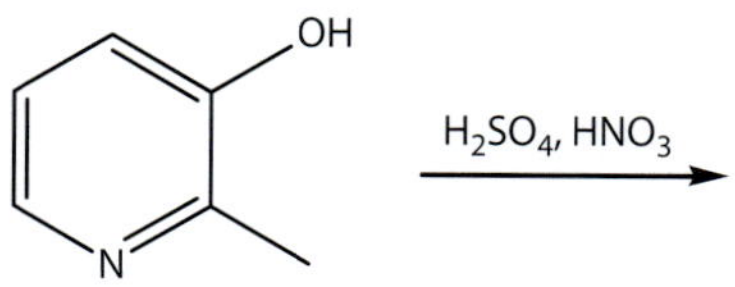

(e)

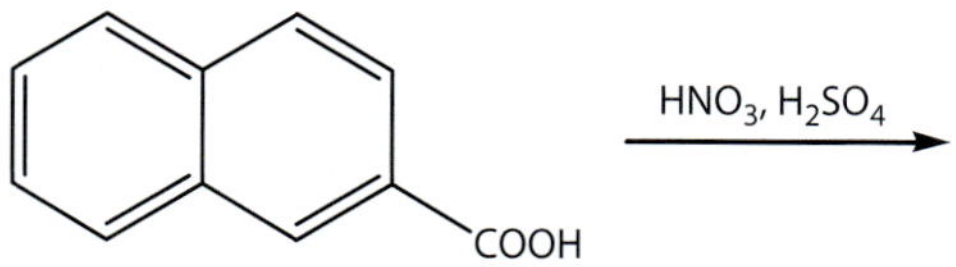

(f)

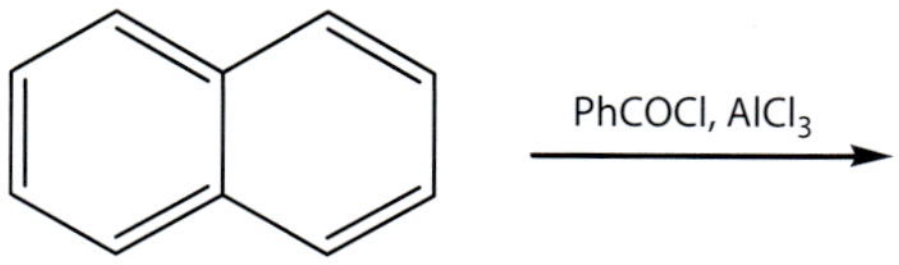

(g)

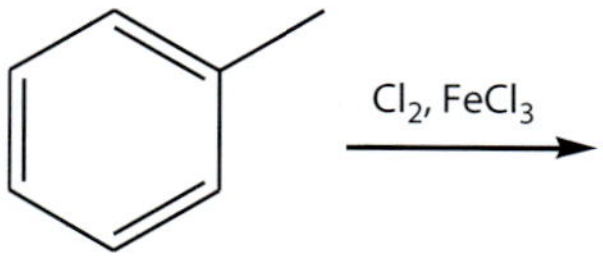

(h)

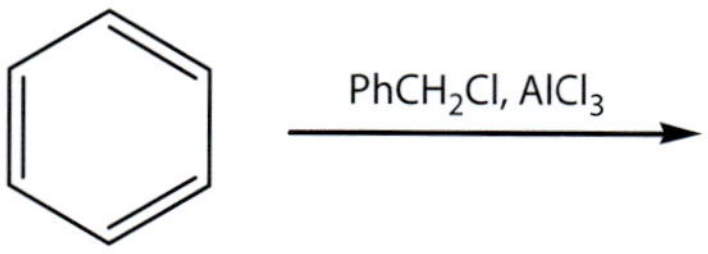

(i)

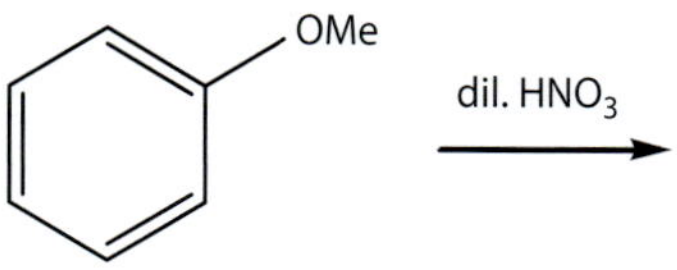

(j)

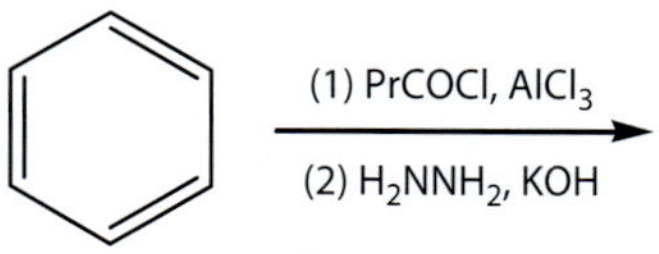

(k)

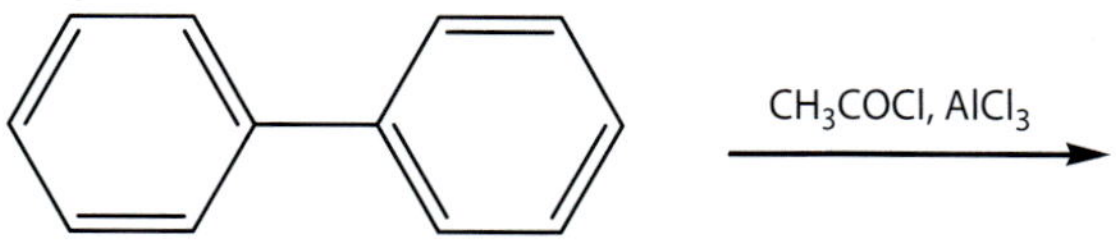

(l)

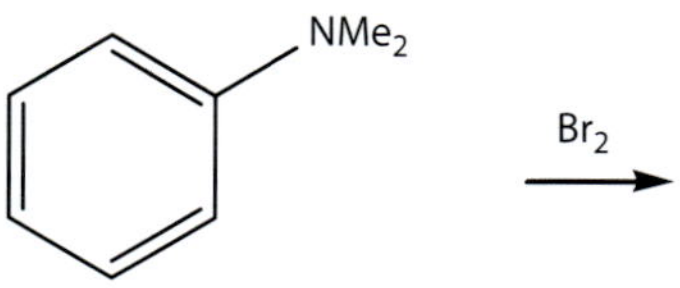

(m)

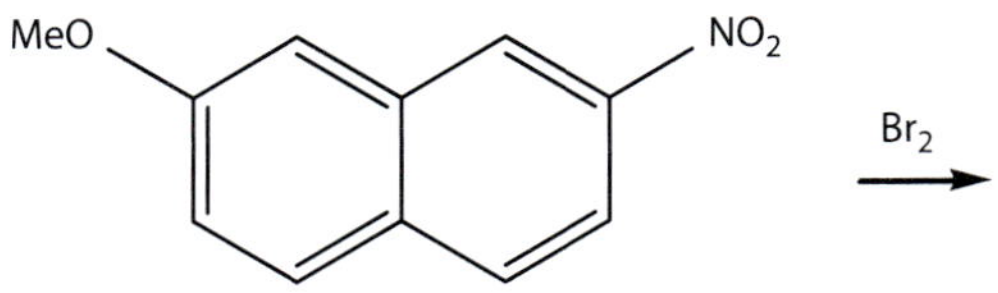

(n)

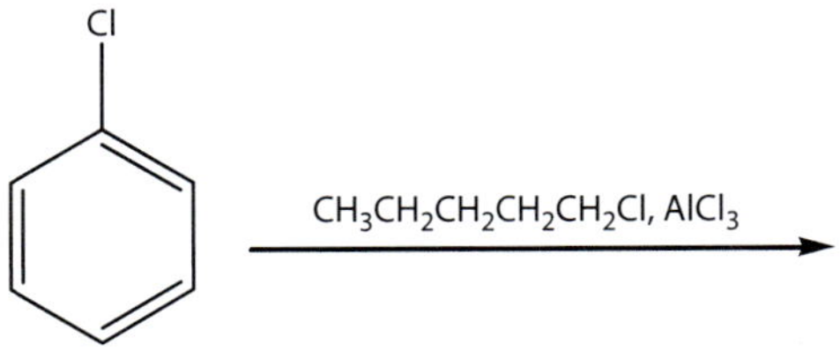

(o)

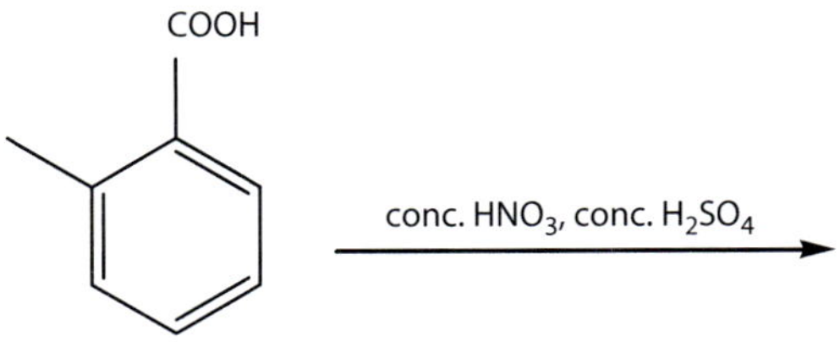

(p)

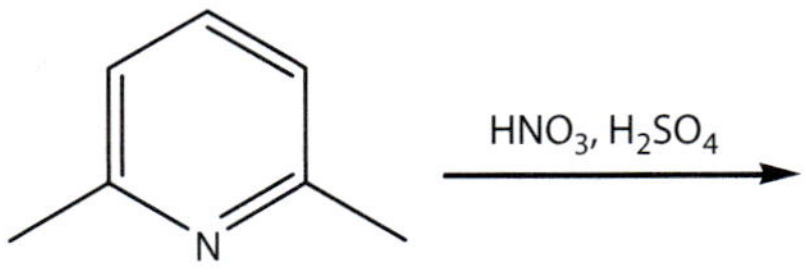

(q)

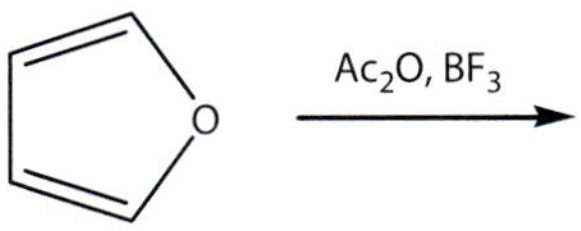

11. How would you carry out each of the following conversions?

(a)

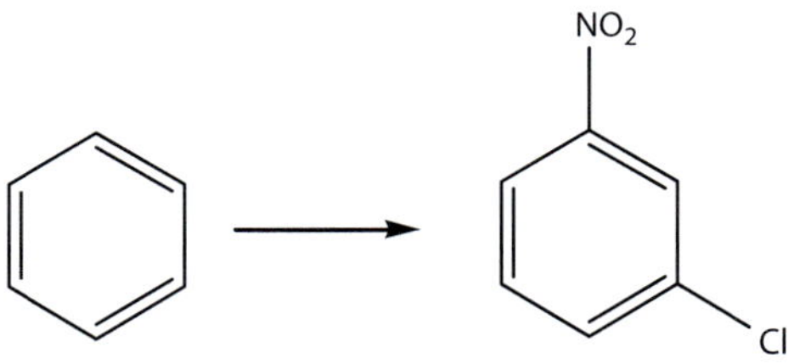

(b)

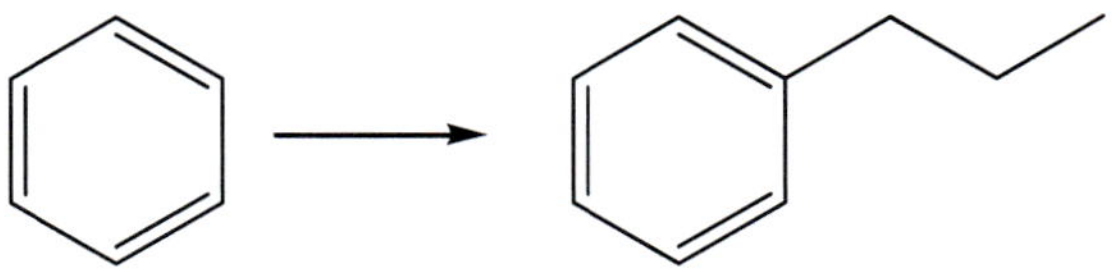

(c)

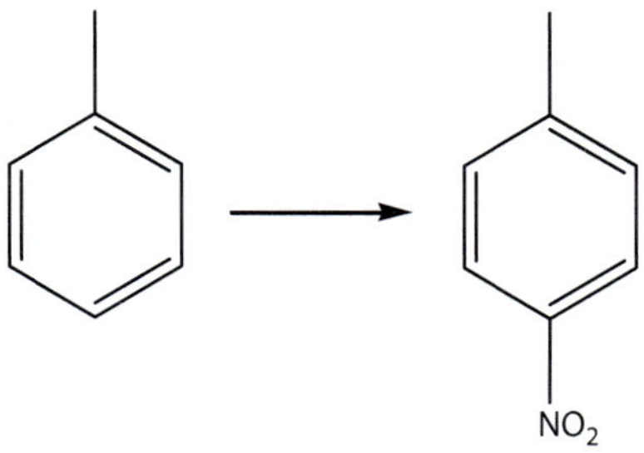

(d)

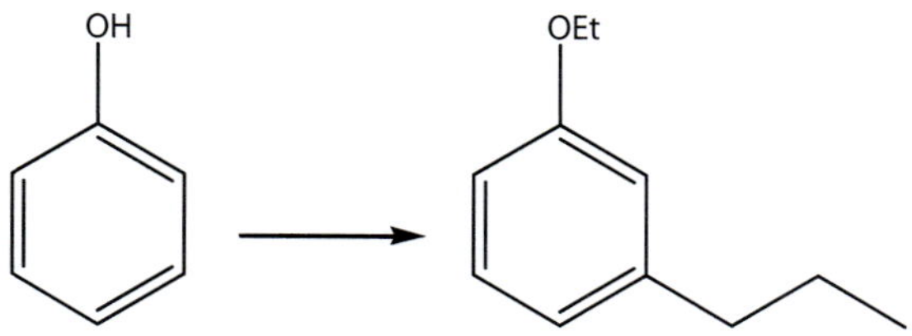

(e)

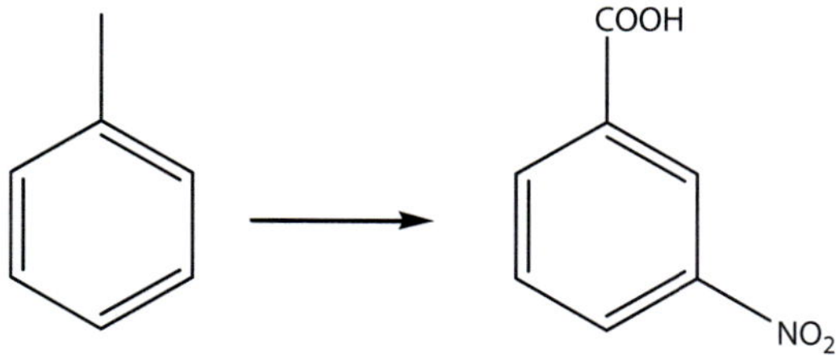

(f)

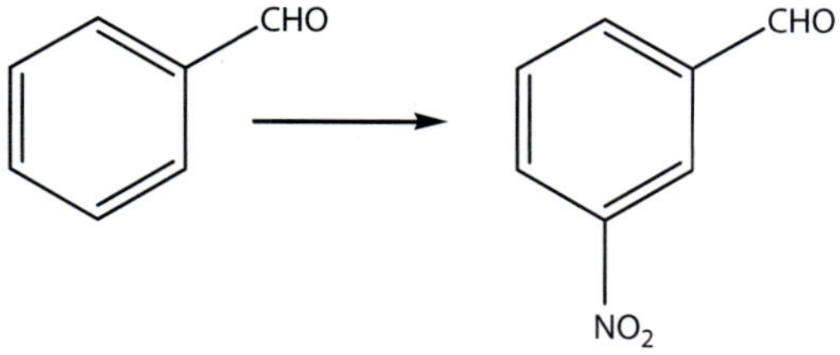

(g)

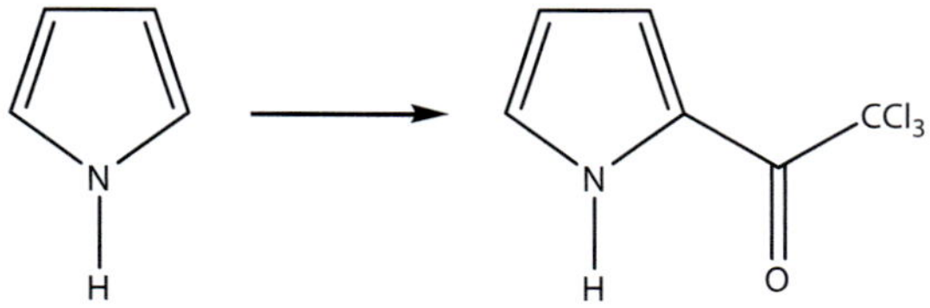

12. In the reaction shown in the following, why do you get the observed product, and why do you not need a catalyst?

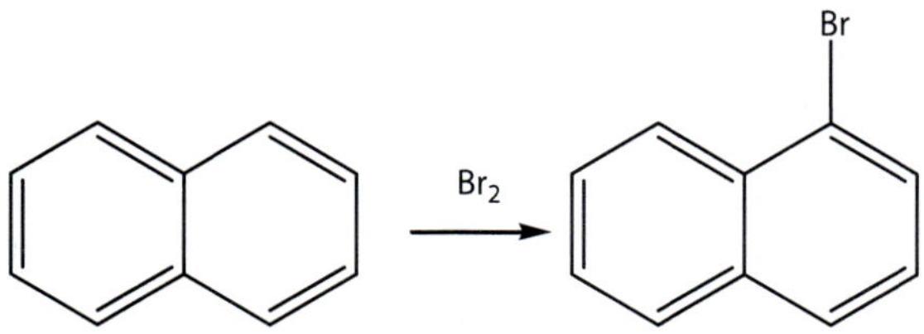

13. Explain the following observations:
    (a) Esters and amides are less activating in electrophilic aromatic substitution than are ethers or amines. Amines are better activators than alcohols.
    (b) Nitrobenzene is often used as a solvent for Friedel–Crafts reactions of other aromatic compounds.
    (c)

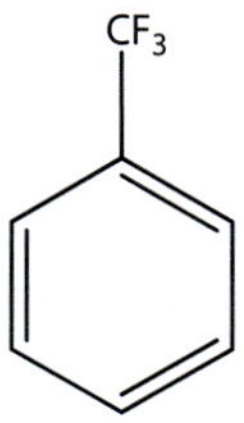

PRFs for nitration
ortho 4.5×10^{-6}
meta 67×10^{-6}
para 4.5×10^{-6}

14. Compare and contrast the reaction with conc. HNO_3/conc. H_2SO_4 of phenyl ethanoate and ethyl benzoate.

MORE CHALLENGING PROBLEMS

1. The ^{1}H NMR spectrum of hydrocarbon **12.24** shows three multiplets centered at δ 9.55, 8.43, and −5.43 in the ratio 2:2:1. Explain. (Think carefully about the aromaticity or otherwise of this compound and whether various π-bonds are formed from p_z or p_y orbitals.)

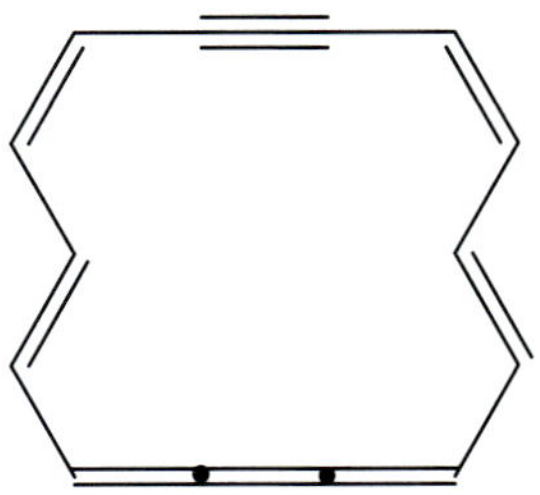

12.24

2. Propose a mechanism, and, where appropriate, explanation, for the following transformations:

(a)

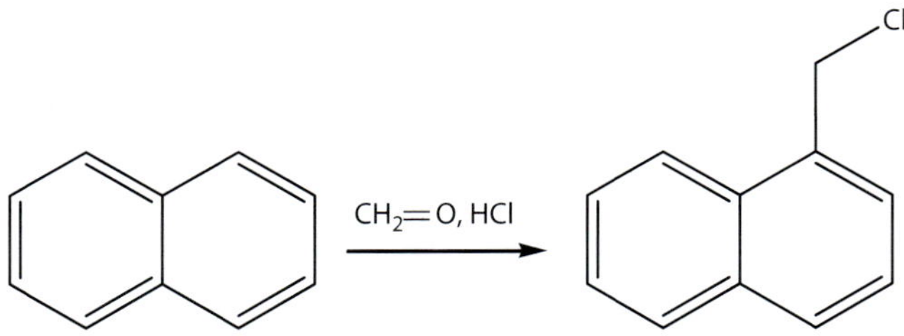

(b)

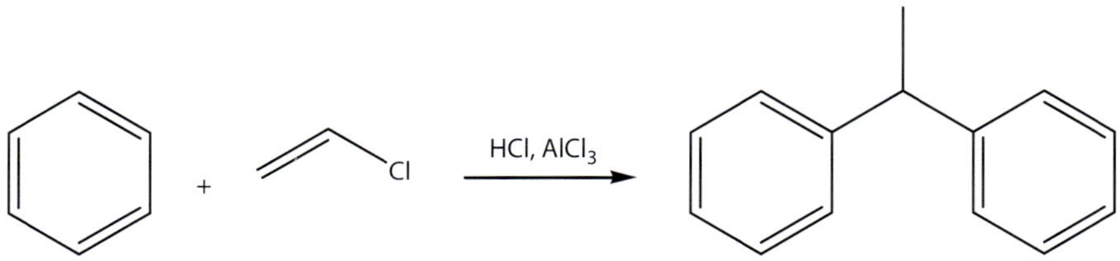

(Hint: this does not involve a vinyl cation, $CH_2{=}CH^+$, and you may find it easier to work backward from the end of the problem.)

(c)

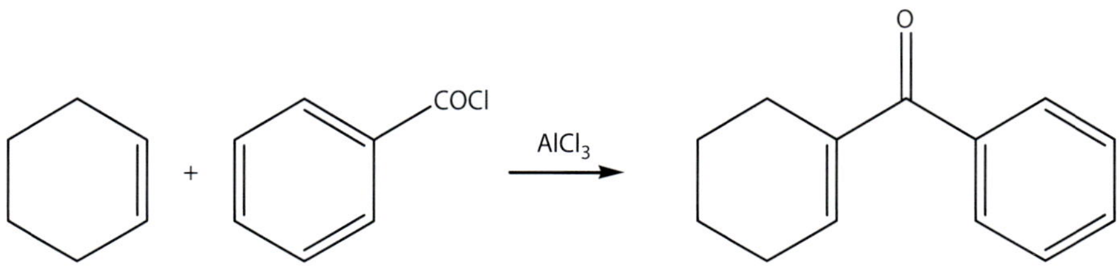

(d)

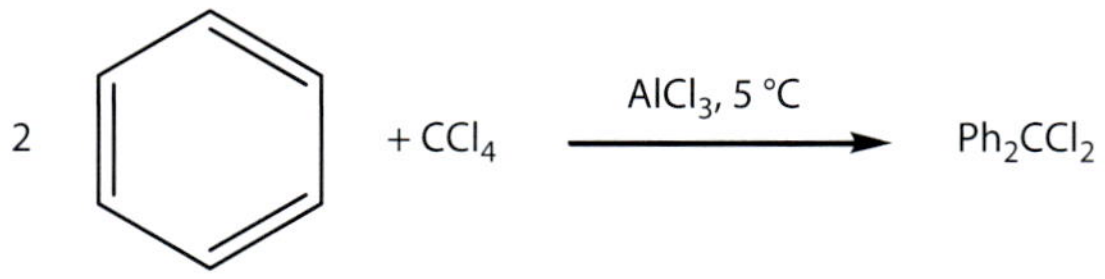

(e)

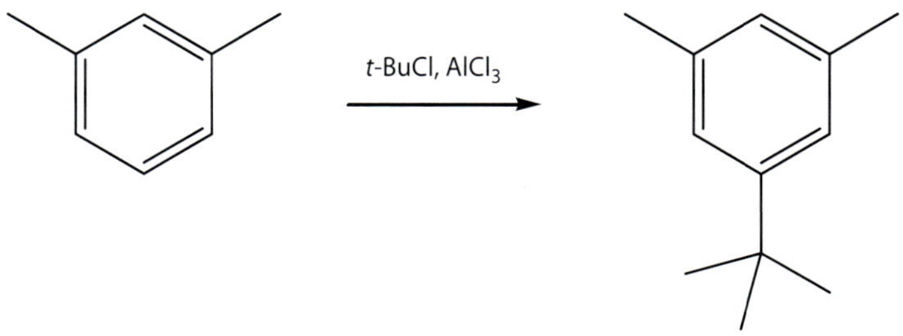

(f)

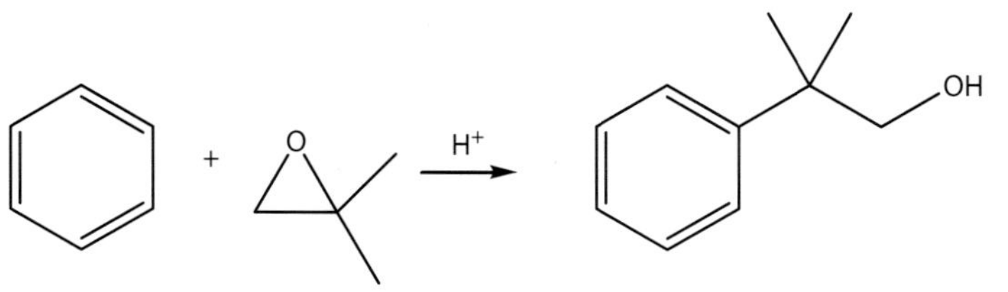

(g)

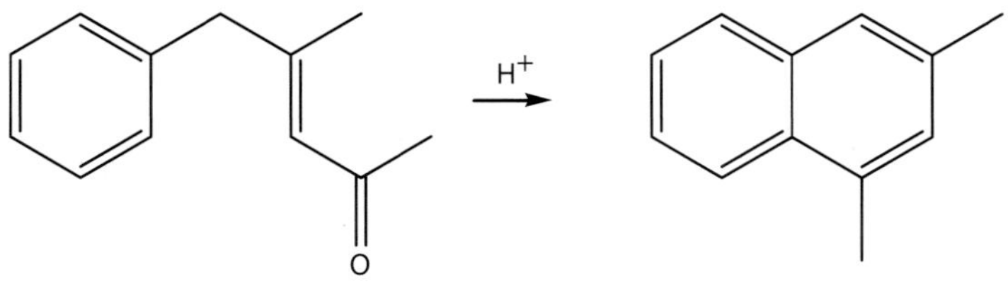

(h)

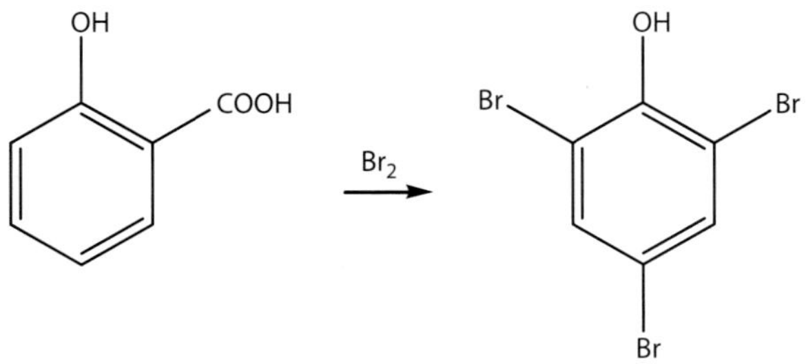

(i)

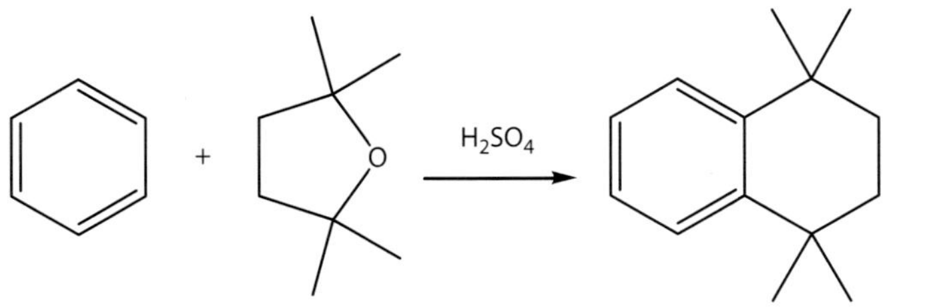

(j)

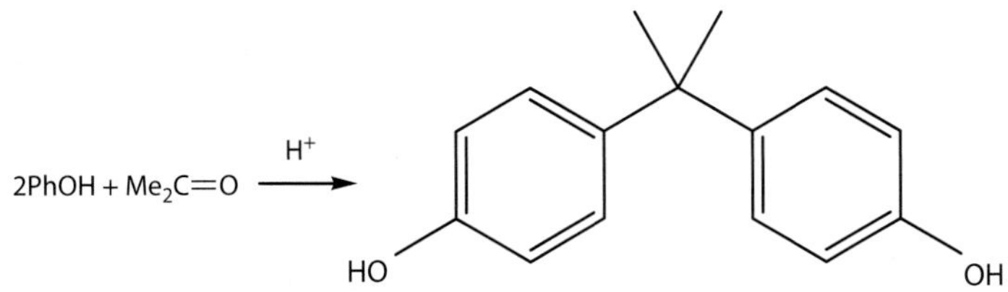

(k)

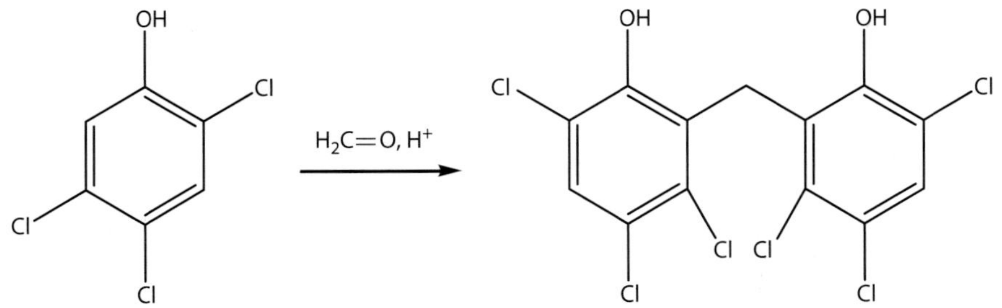

The product is hexachlorophene, used in germicidal soaps.

(l)

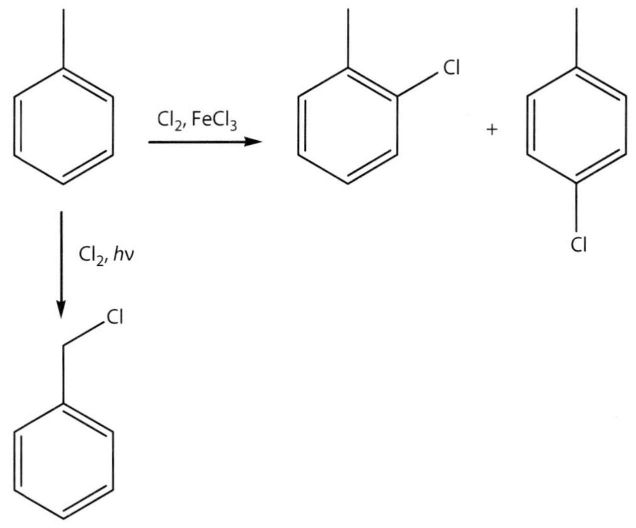

3. Predict the product(s) of the following reactions:

(a)

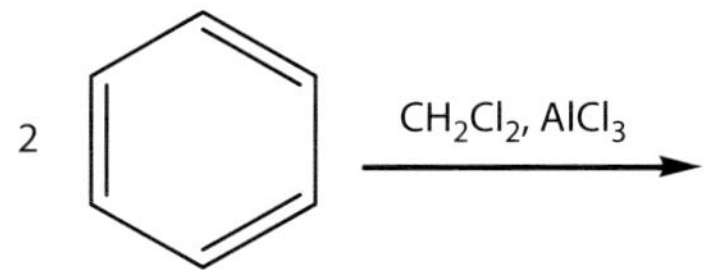

(b)

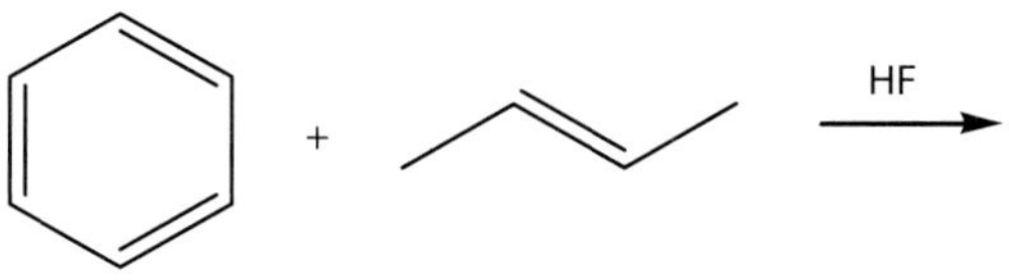

(c)

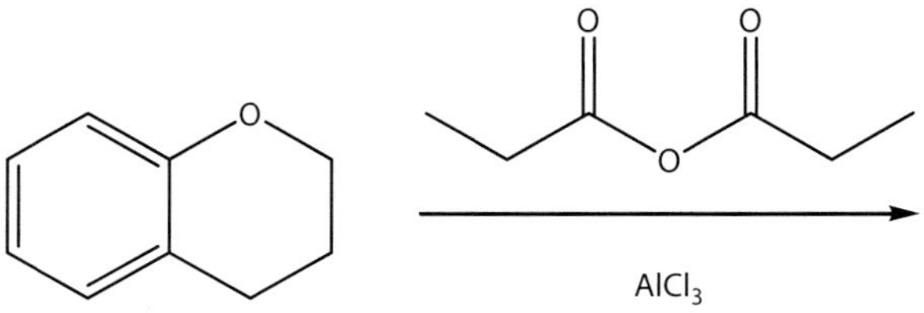

(d)

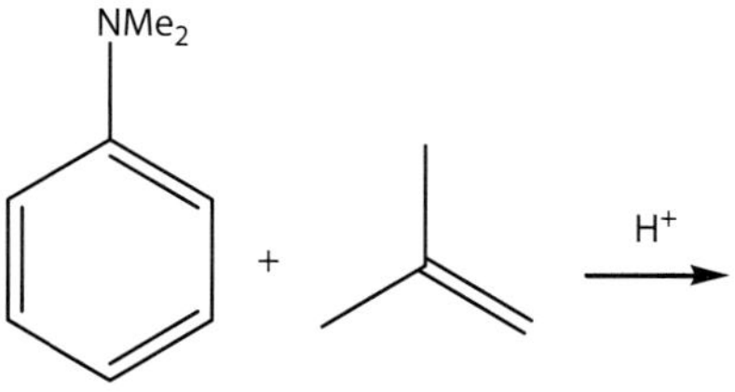

(e)

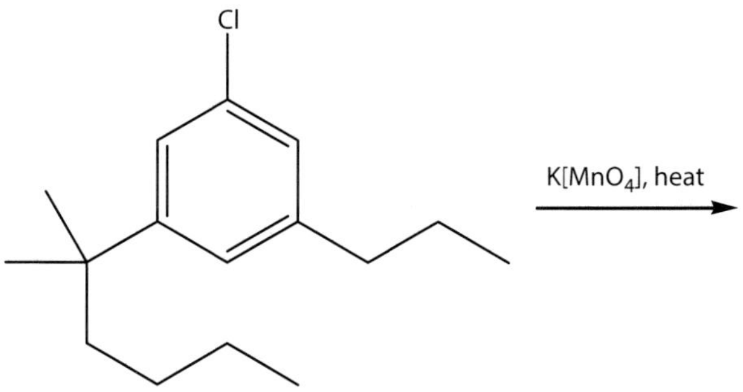

(f)

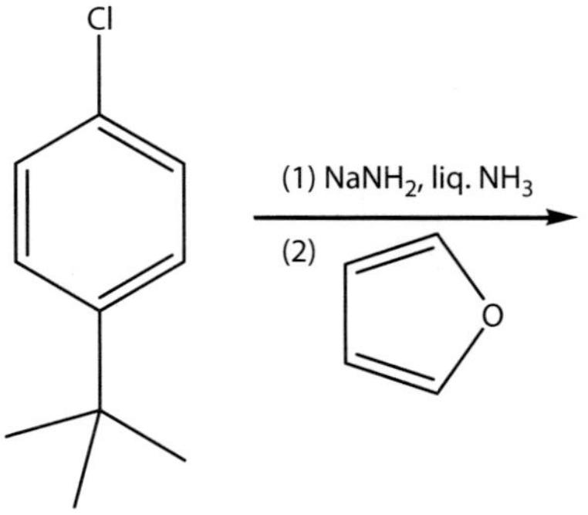

(g)

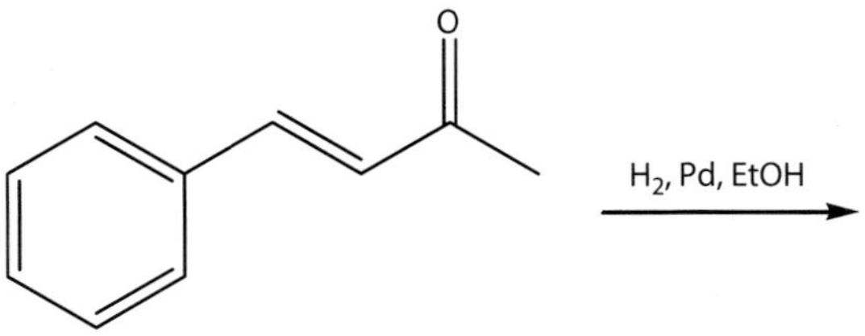

(h)

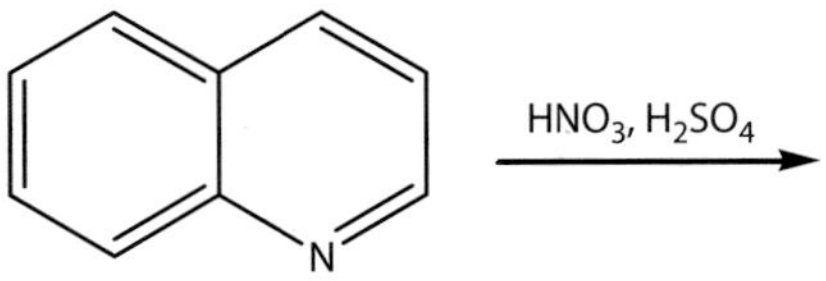

4. Comment on the stereochemical outcome of the Friedel–Crafts reaction of the chiral halide shown in the following.

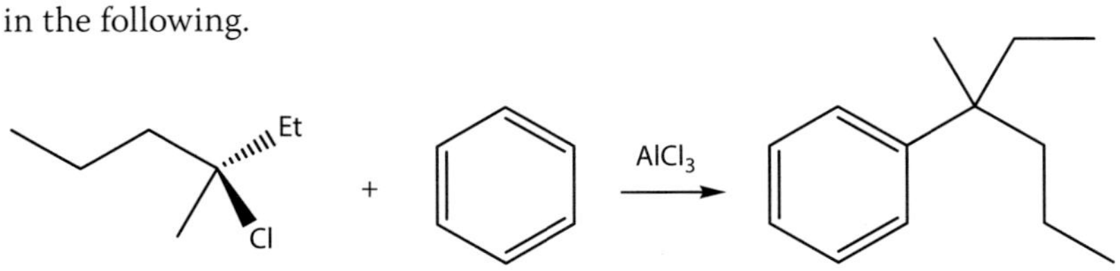

5. Explain the following observations:
   (a) Toluene reacts with 1-chloropropane in the presence of $AlCl_3$ to give **12.25**. 1,3,5-Trimethylbenzene reacts under the same conditions to give **12.26**.

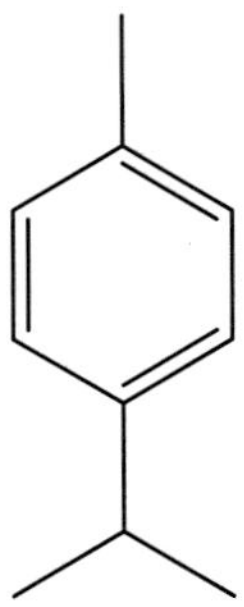

12.25

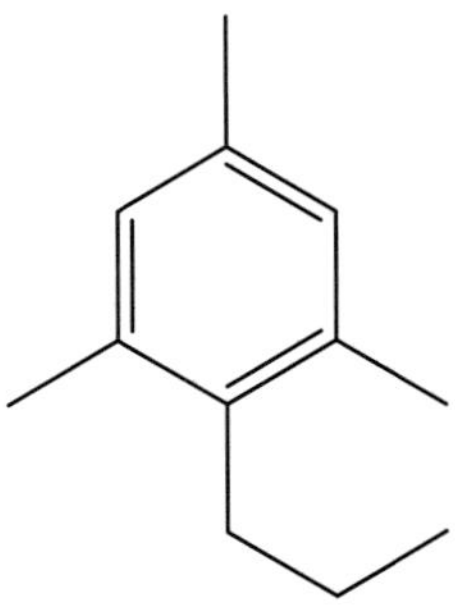

12.26

   (b)

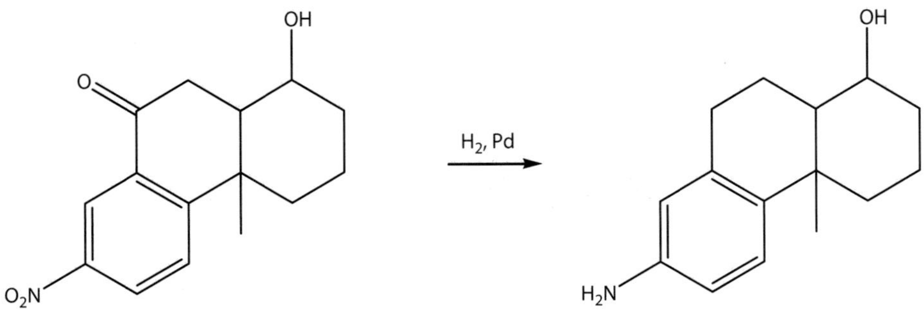

(c) The rates of reaction relative to benzene and the partial rate factors for the nitration (using acetyl nitrate) of various nitrated biphenyls are shown in the following:

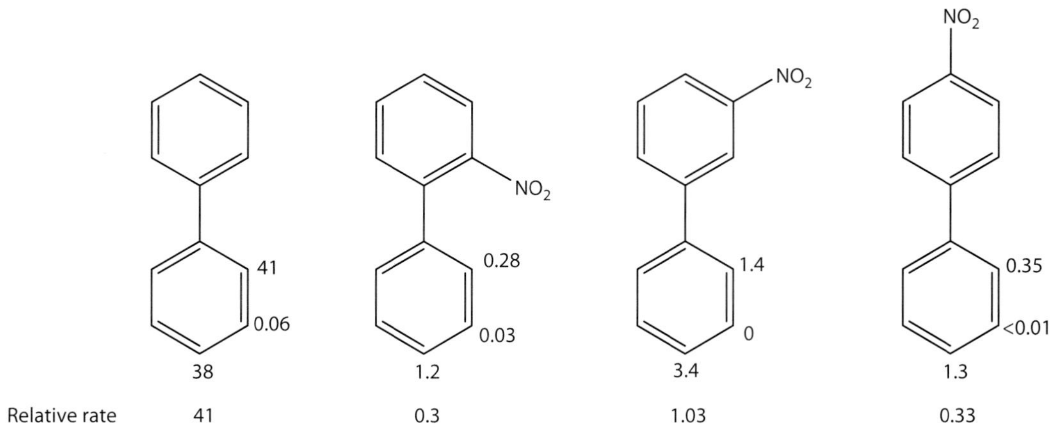

(d) The reaction of fluorene with electrophiles is significantly faster than comparable reactions of biphenyl.

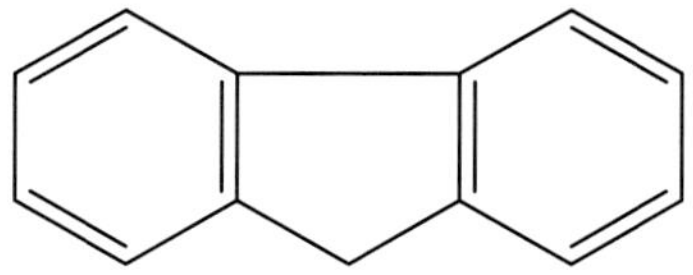

Fluorene

(e) Both 2-chlorothiophene and thiophene-2-carboxylic acid are nitrated predominantly at the 5-position.

(f) Indole reacts with electrophiles predominantly at the 3-position, as shown in the following:

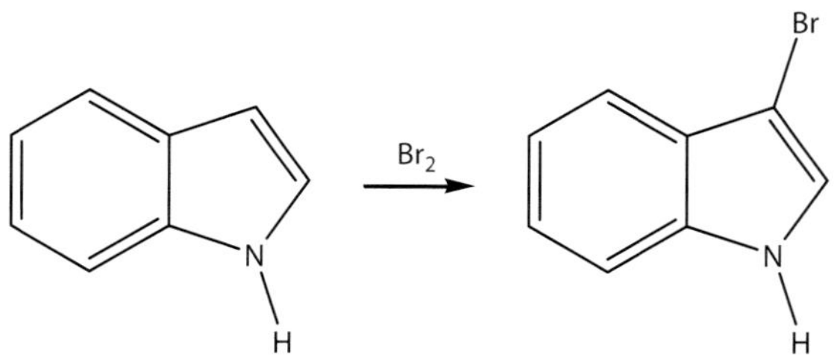

6. Fill in the blanks in the following sequences:

(a)

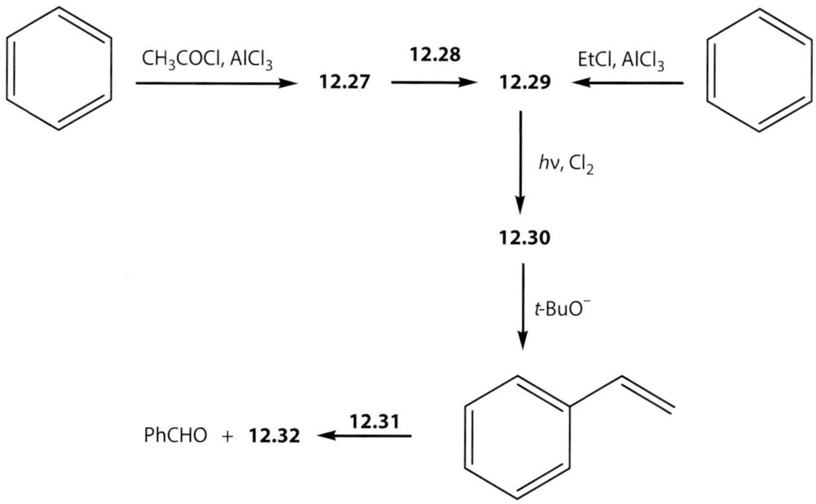

(b)

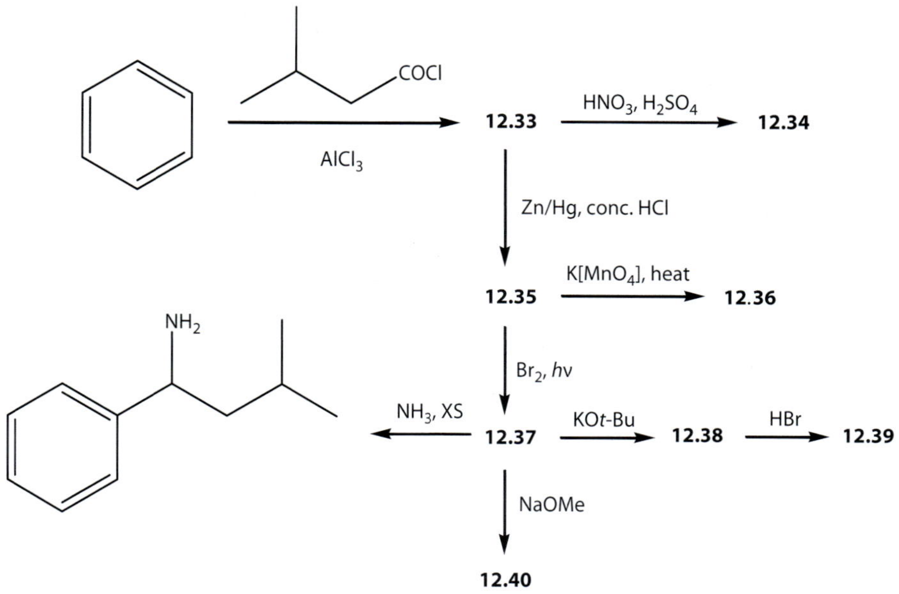

7. How would you carry out the following transformations?

(a)

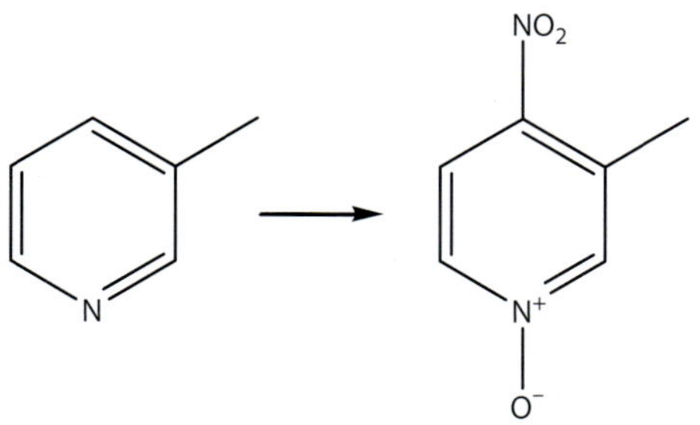

(b)

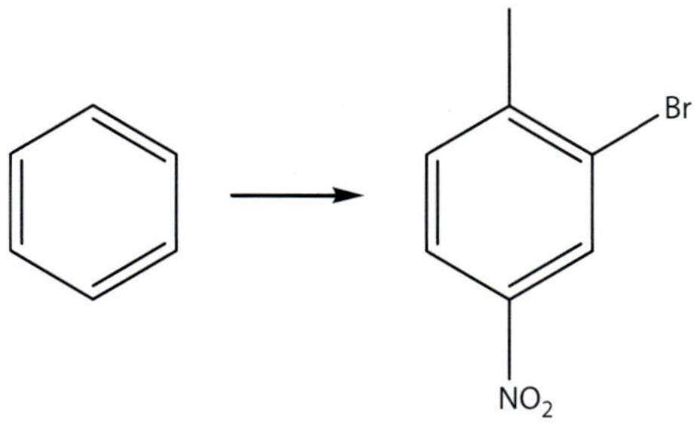

8. Calculate the partial rate factors for nitration of the halobenzenes, given the data in the following:

Halogen	Rate Relative to Benzene	% *ortho*	% *meta*	% *para*
F	0.15	13	0	87
Cl	0.033	30	1	69
Br	0.03	37	1	62
I	0.18	38	2	60

Comment on the outcome.

Nucleophilic Aromatic Substitution and Synthesis of Aromatic Compounds

13

13.1 INTRODUCTION

In the previous chapter, we learned how to substitute aromatic rings with a range of electrophiles. The first part of this chapter will show how one substituent may be exchanged for another, by various mechanisms. This will be followed by a brief section on reactions of side chains/substituents on aromatic rings, and then we will be ready to tackle problems in the synthesis of aromatic compounds.

13.2 UNIMOLECULAR NUCLEOPHILIC SUBSTITUTION OF AROMATIC COMPOUNDS: THE REACTIONS OF DIAZONIUM SALTS WITH SIMPLE NUCLEOPHILES

Nitrous acid, HNO_2, is an unstable compound, which must be prepared and used *in situ*. Cold solutions decompose to nitrogen dioxide, nitric oxide, and water (Equation 13.1), but at higher temperatures, the products are nitric acid, nitric oxide, and water (Equation 13.2):

(13.1) $$2HNO_2 \rightarrow NO_2 + NO + H_2O$$

(13.2) $$3HNO_2 \rightarrow 2NO + HNO_3 + H_2O$$

For our purpose, nitrous acid is usually prepared at 0 °C from sodium nitrite and HCl (Equation 13.3). Its reaction with an aromatic amine (Equation 13.4) at 0 °C gives a diazonium salt, **13.1** (X may be nitrite or chloride), which reacts with a wide range of nucleophiles to give aromatic rings with nucleophilic substituents.

(13.3) $$Na[NO_2] + HCl \rightarrow HNO_2 + H_2O$$

(13.4) $$Ph - NH_2 \xrightarrow{HNO_2,\ 0\ °C} Ph - \overset{+}{N}{\equiv}N\ X^- \xrightarrow{Nu^-} Ph - Nu$$

13.1

Before we look at the mechanism of the reaction, we need to review the structure of nitrous acid. We sometimes write it as HONO, to remind us how the atoms are joined together (**13.2**); the *transoid*-form has been shown to be more stable in the gas phase by infrared spectroscopy.

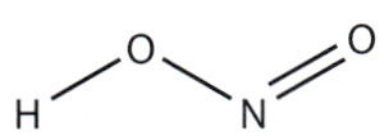

13.2

The first step of the mechanism is attack of the lone pair of electrons of the amine on the N=O double bond. There are then various proton losses and gains, and the loss of water (remember our solution is acidic, so this should be favorable), to give an *N*-nitroso compound, **13.3**. This tautomerizes to **13.4**. Protonation facilitates water loss to give the diazonium salt, which is reasonably stable at 0 °C (Figure 13.1).

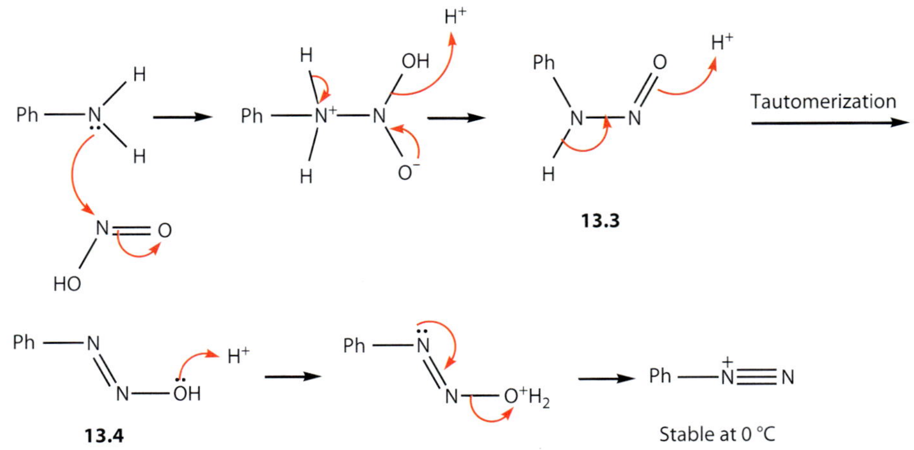

FIGURE 13.1 Mechanism of formation of a diazo compound.

$PhN_2^+ \rightarrow Ph^+ + N_2$

$Ph^+ + Nu^- \rightarrow Ph\text{–}Nu$

FIGURE 13.2 Reaction of an aromatic diazonium salt with nucleophiles.

If we had done this reaction with an aliphatic compound, for example, cyclohexylamine, we would still make the diazonium salt, but aliphatic diazonium salts are much less stable than aromatic ones, and this would decompose with loss of molecular nitrogen at −50 °C. Aromatic diazonium compounds are more stable, because the aryl cation that would be produced by nitrogen loss is at an sp^2 carbon atom, which, as we noted in Chapter 11, is very unfavorable. However, molecular nitrogen is a fabulous leaving group; the N≡N triple bond has lots of enthalpy, and the process also benefits from the entropy of the loss of a mole of gas. So the diazonium salt decomposes slowly to give an aromatic cation, which is then captured by any available nucleophile (Figure 13.2).

This is a unimolecular process, with the RDS being loss of nitrogen (hence, the nomenclature S_N1Ar); the cation is very reactive and is rapidly captured by nucleophiles. This formalism will serve us adequately to use the process for synthesis and to make sensible predictions about what will happen, but we should note in passing that it is probably not true for all nucleophiles; the Sandmeyer reactions of copper halides, and probably some other processes, are actually radical reactions.

What nucleophiles give good results? Water gives phenol; alcohols give phenyl ethers. Potassium iodide gives iodobenzene (remember that we can't easily put iodine onto an aromatic ring by electrophilic substitution; see Section 12.2.3). Copper cyanide gives benzonitrile, PhCN. In all of these cases, there was no simple electrophilic process available to give us these results. This is summarized in Figure 13.3, and some synthetic examples are given in Figure 13.4.

The Sandmeyer reaction involves the replacement of the diazo group by a halogen, usually chlorine or bromine, using copper(I) halides (Figure 13.5). The reaction with CuCN, noted previously, may have the same mechanism (*vide infra*).

We now have two ways to substitute an aromatic ring with chlorine and bromine. The diazonium route also allows us to produce fluoroaromatics, although the process is slightly different. If we convert the initially formed diazonium salt to a tetrafluoroborate salt, by adding HBF_4, we obtain $[PhN_2^+][BF_4]$. Heating this salt gives fluorobenzene (Equation 13.5). The same reaction occurs with the $[PF_6]$ salt. This is among the relatively few simple ways to fluorinate an aromatic ring, and the reaction is called the Schiemann or the Balz–Schiemann reaction. A practical example is given in Figure 13.6.

$$\left[PhN_2^+\right]\left[BF_4^-\right] \xrightarrow{\text{heat}} PhF + N_2 + BF_3 \qquad (13.5)$$

$$PhN_2^+ \xrightarrow{H_2O} PhOH \qquad PhN_2^+ \xrightarrow{KI} PhI$$

$$PhN_2^+ \xrightarrow{ROH} PhOR \qquad PhN_2^+ \xrightarrow{CuCN} PhCN$$

FIGURE 13.3 Nucleophiles that react readily with diazonium salts.

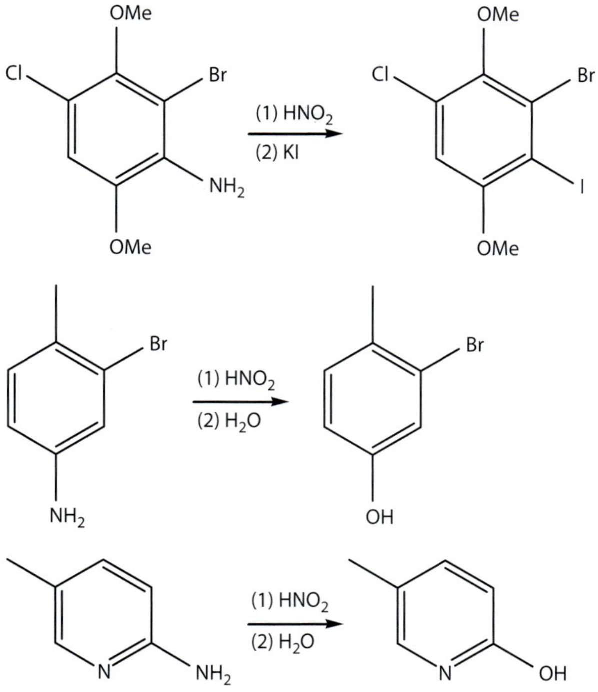

FIGURE 13.4 Synthetic examples of substitution of diazonium salts.

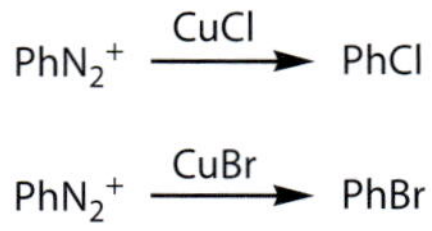

FIGURE 13.5 Sandmeyer reaction.

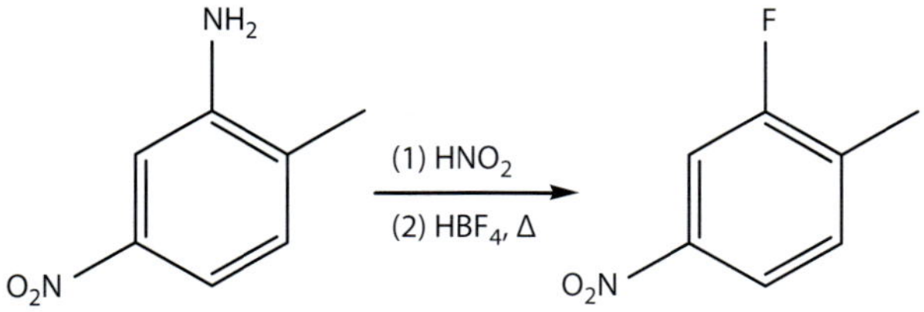

FIGURE 13.6 Fluorination of an aromatic ring.

Focus on the Sandmeyer Reaction

Although it is convenient for us to think about the reactions of diazonium salts as involving nitrogen loss to give a phenyl cation, then attack of a nucleophile on the cation, this is almost certainly not true for the Sandmeyer reaction where the nitrogen is replaced by a halogen from a copper(I) halide. There is a special reason that we use copper salts—the copper(I) and copper(II) oxidation states may be relatively easily interconverted (we will meet this again when we look at copper-catalyzed conjugate addition to enones in Chapter 17). The diazonium salt reacts with Cu(I)X by a one-electron transfer to give a radical and copper(II) (Equation 13.6). This loses molecular nitrogen to give an aryl radical, which is slightly less unstable than an aryl cation (13.7). Finally, the aryl radical reacts with the copper(II) cation to give the aryl halide and regenerate copper(I) (13.8). Some examples of the process in synthesis are shown in Figure 13.7.

(13.6) $$ArN_2^+ + Cu(I)X \rightarrow \left[Cu(II)X\right]^+ + ArN_2^{\bullet}$$

(13.7) $$ArN_2^{\bullet} \rightarrow Ar^{\bullet} + N_2$$

(13.8) $$Ar^{\bullet} + \left[Cu(II)X\right]^+ \rightarrow ArX + Cu(I)$$

$$X = Cl, Br$$

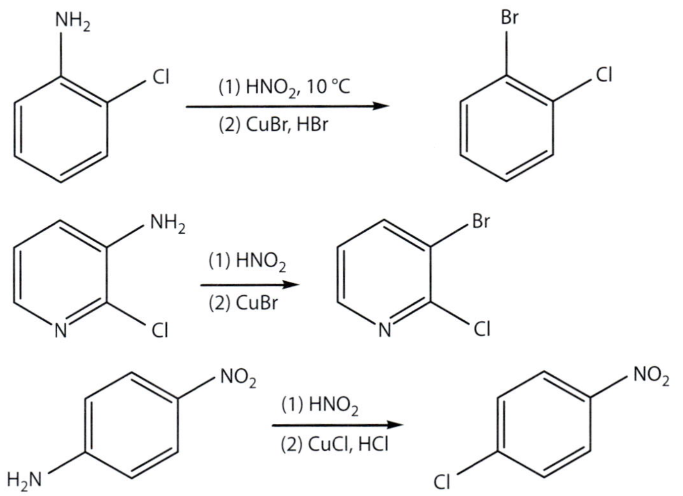

FIGURE 13.7 Sandmeyer reactions in synthesis.

Finally, we can reduce the diazonium salt to the aromatic hydrocarbon using H_3PO_2, hypophosphorous acid. The formula of hypophosphorous acid is more properly written as $H_2P(=O)OH$; it is a monobasic acid and a powerful reducing agent. This reaction (Equation 13.9) almost certainly also goes through a radical mechanism. An example is given in Figure 13.8. Notice our synthetic strategy here—we have used the activating and directing effect of the amino group to introduce three bromine atoms onto the benzene ring and then removed the amino group to obtain our final product. We could not have obtained 2,4,6-tribromobenzoic acid by direct bromination of benzoic acid because the COOH group is *meta*-directing and deactivating. The reaction would have required more forcing conditions, and the bromines would be in the wrong place.

$$PhN_2^+X^- + H_3PO_2 + H_2O \rightarrow PhH + N_2 + H_3PO_3 \tag{13.9}$$

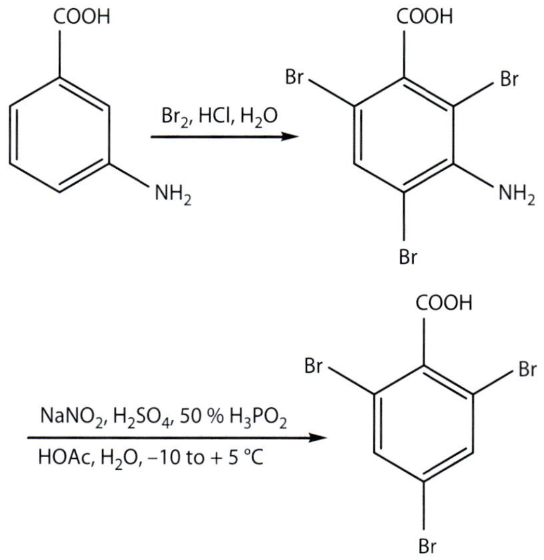

FIGURE 13.8 Removal of an amino group by diazotization followed by reduction.

PROBLEM 13.1

Predict the products of each of the following reaction sequences:

(a)

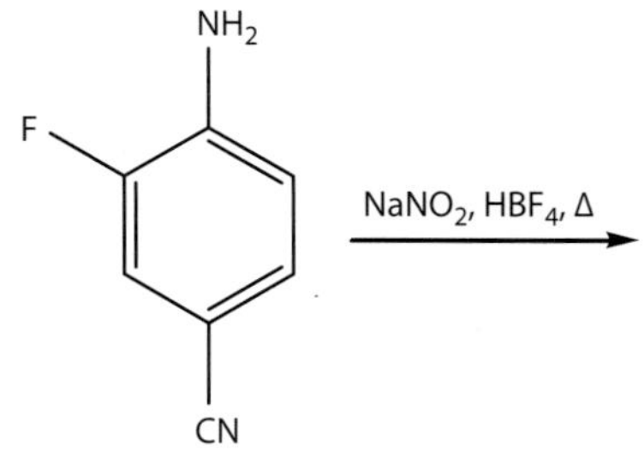

(b)

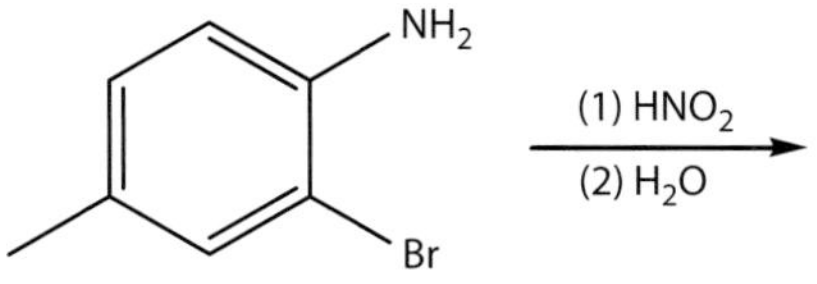

(c)

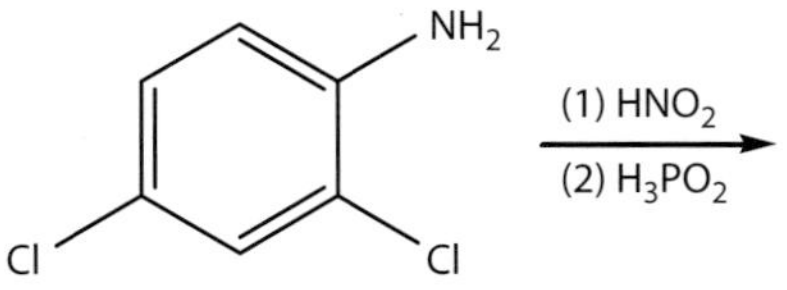

(d)

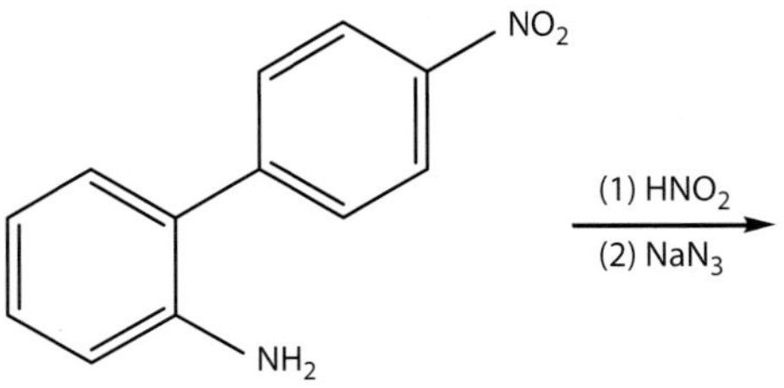

Solutions

(a) This is a standard Balz–Schiemann reaction; the amino group is replaced by a fluorine atom to give

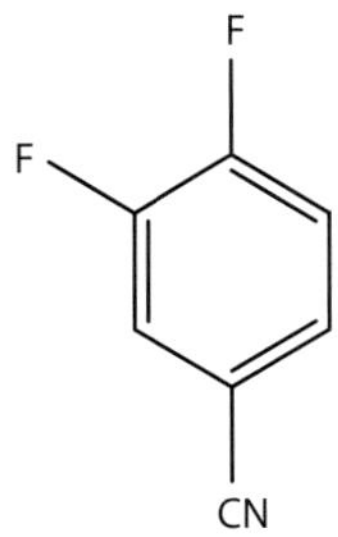

(b) The amino group is replaced by OH to give a phenol:

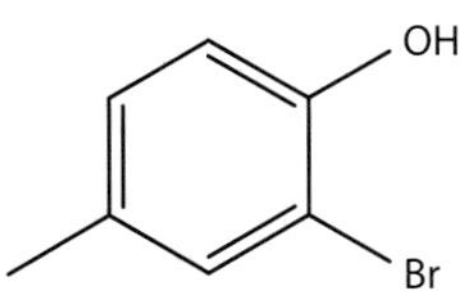

(c) In this process, the amino group is removed completely and replaced by hydrogen to give

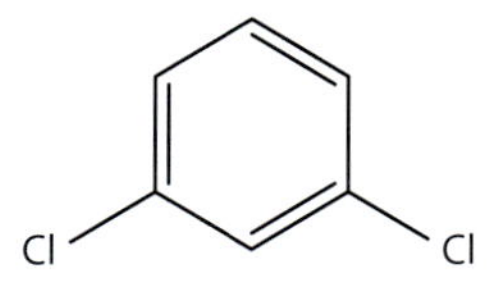

(d) Although we have not seen this nucleophile before, we should be confident that it will react in the same way as the others to give

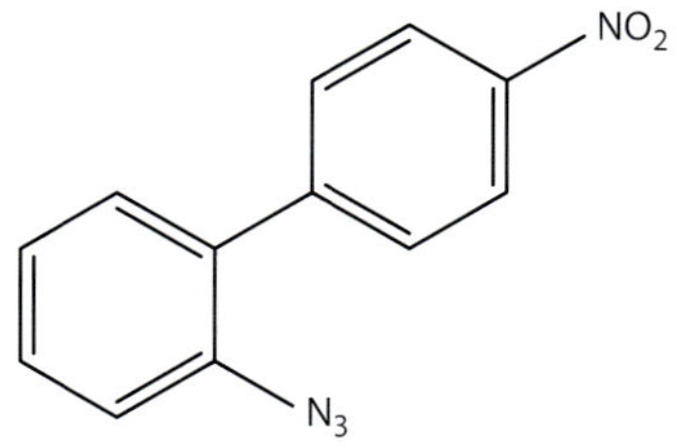

13.3 AZO COUPLING REACTIONS

Diazonium salts also participate in another type of aromatic substitution, azo coupling, where the two nitrogen atoms are retained in the product. They can act as electrophiles toward activated aromatic compounds, which are generally substituted at the *para*-position (Figure 13.9). The arene acting as the nucleophile must be electron rich—bearing OR, NR_2, or a comparable group.

Remember that the geometry of the azo compound must be bent because of the lone pairs on the sp^2 nitrogen atoms. In principle, both *E*- and *Z*-isomers could be formed; the *E*-isomer (**13.5**, shown) is usually the more stable.

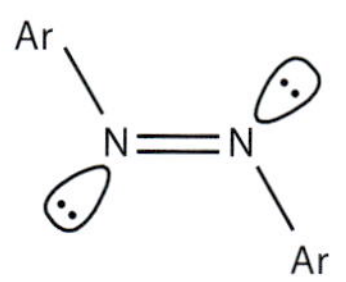

13.5

Azo compounds are important in the dyestuff industry (see Chapter 24) and have found wide applications as acid–base indicators. The colors arise from the extended conjugated system. The synthesis of the indicator methyl red is shown in Figure 13.10.

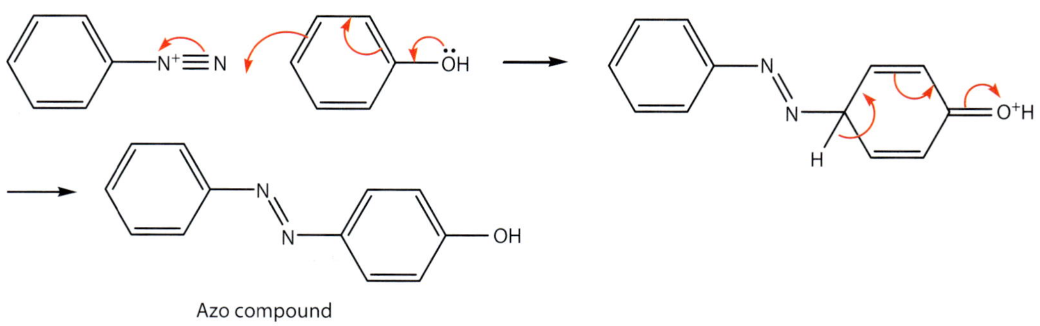

FIGURE 13.9 Azo coupling of phenol.

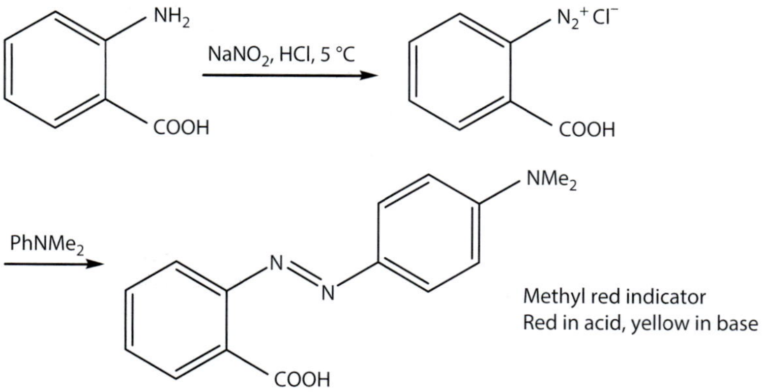

FIGURE 13.10 Synthesis of methyl red by azo coupling.

Focus on Azo Compounds

Azo dyes (as well as the aniline-based dyes alizarin and mauveine) were first prepared in the nineteenth century and today represent at least half of all dyes in use. Their uses range from dyeing of fabrics, inks, leather, and manufactured goods to acid/base indicators, stains used in preparing biological samples, and materials to help with the visualization of fingermarks and biological fluids in forensic science. The most common ones are yellow, red or orange in color, but examples are also known in blue and black.

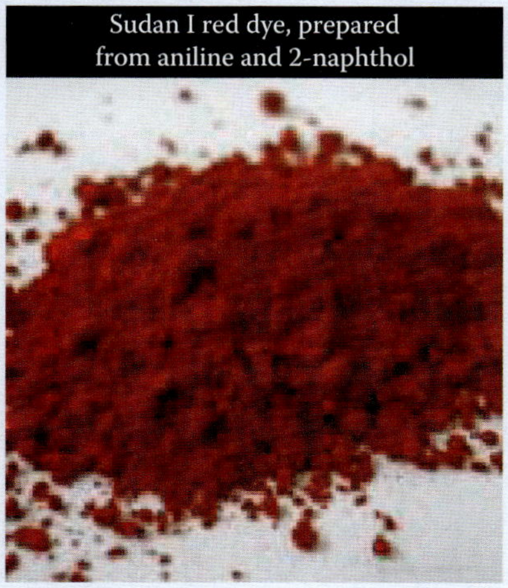

Orange II (also known as acid orange 7), prepared as in Figure 13.11, is used as a fabric dye. Notice that the reaction takes place at the 1-position of the naphthalene that is also *ortho* to the *o,p*-directing hydroxyl substituent.

Amido black 10B and Orange G are used as stains to prepare biological samples, and Direct Blue 12 (Chicago sky blue 6B) has been used to help in the location of the sentinel lymph node in malignant melanoma and early breast cancer. Oil Red O is used to visualize fingerprints, particularly on paper and similar substrates (Figure 13.12).

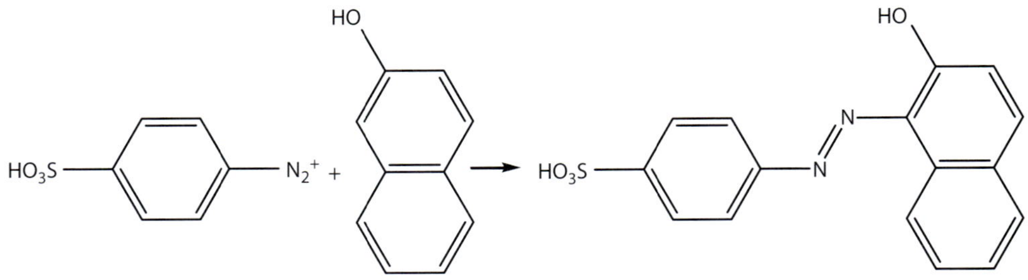

FIGURE 13.11 Preparation of Orange II.

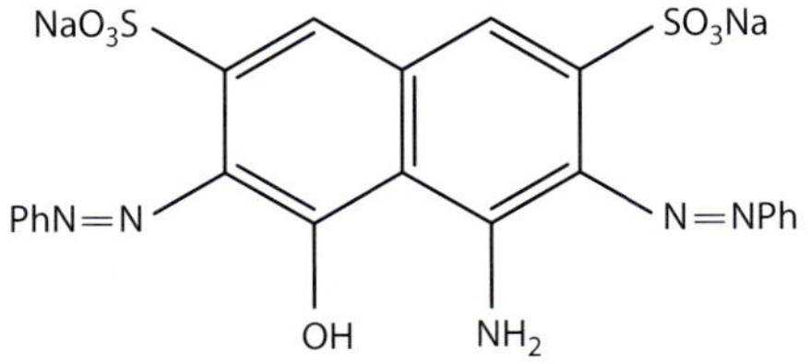

Amido black 10B—used to stain the proteins collagen and reticulin

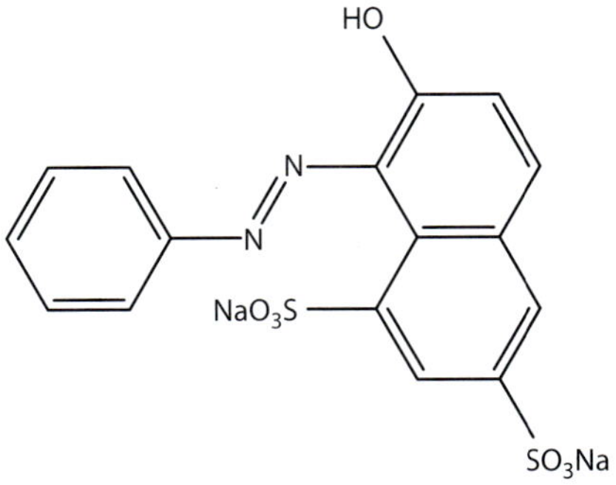

Orange G. Used to stain cells in the pancreas and pituitary gland

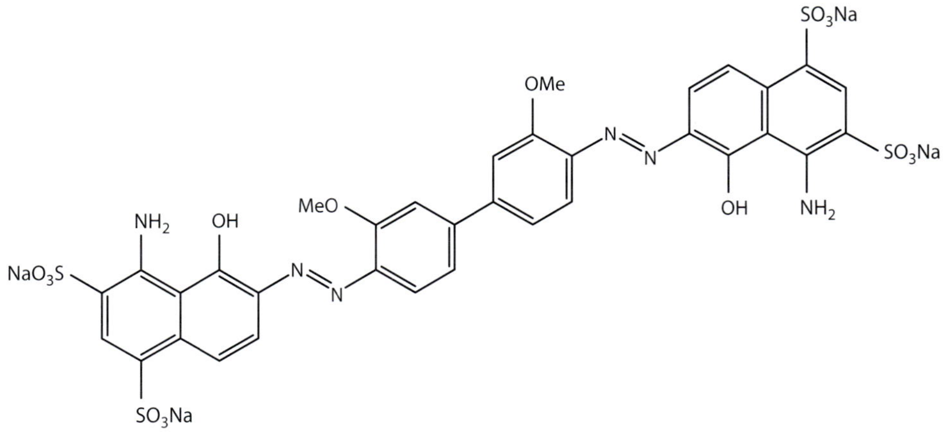

Direct blue 1

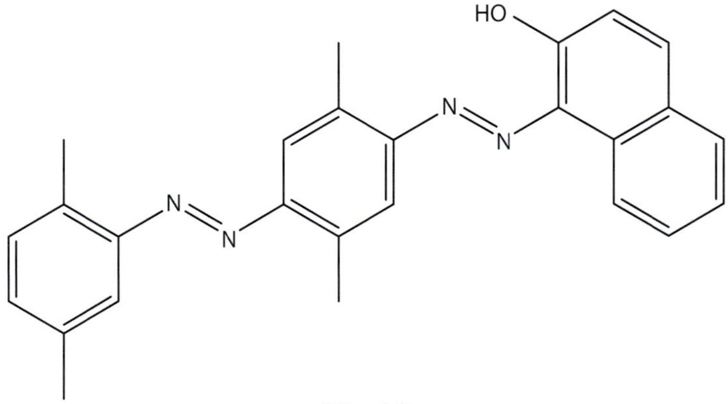

Oil red O

FIGURE 13.12 Azo compounds used as biological stains.

A food and cosmetic colorant that has attracted a good deal of unfavorable attention is tartrazine, **13.6**. It has been established that it causes some allergies and is implicated in worsening asthma. It is also widely believed to be implicated in the development of ADHD in children.

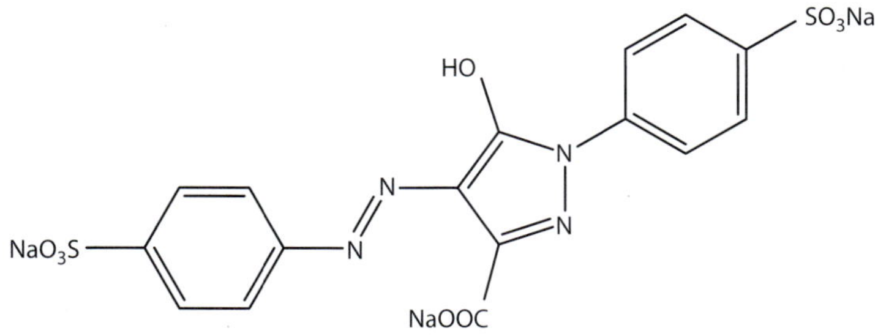

13.6, tartrazine

Tartrazine (Kandlekaz, Perth, Western Australia.)

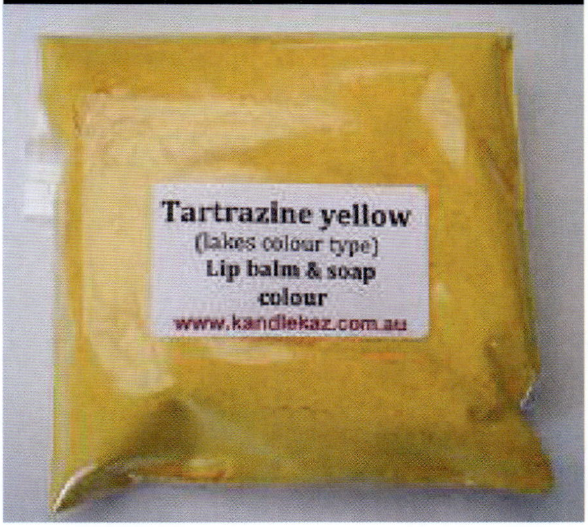

PROBLEM 13.2

Write a full mechanism for the following reaction:

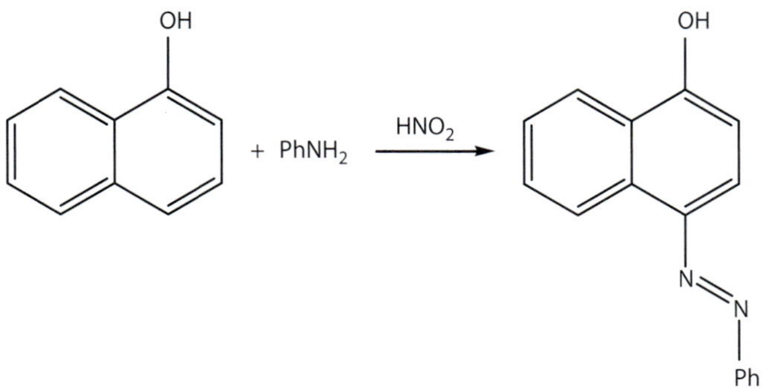

Solution

In the first reaction, we need to convert aniline into a diazonium salt, as shown in Figure 13.1. Then, we react with the 1-napthol at the position *para* to the hydroxyl group. We know that −OH is *o,p*-directing. We might infer that the selectivity for the *para*-position is electronic in origin—naphthalenes prefer to substitute at the position next to the ring junction.

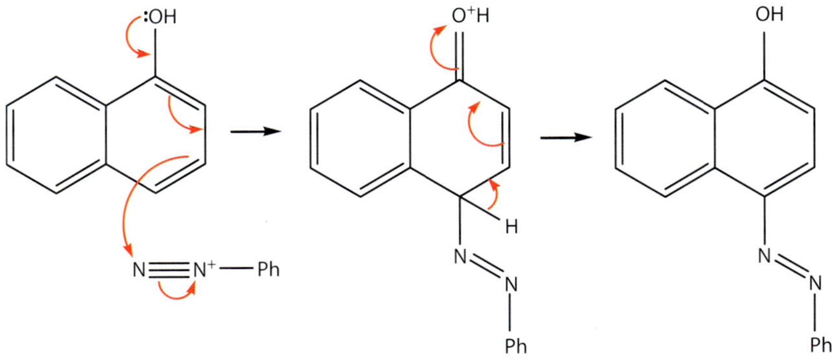

Key Points from Sections 13.1 through 13.3

- Aromatic amines, $ArNH_2$, are converted by nitrous acid to diazonium salts $[ArN_2^+]X^-$. Unlike aliphatic diazonium salts, these are fairly stable at 0 °C.
- Aryl diazonium salts react with a wide range of nucleophiles, Nu, to give substitution yielding Ar–Nu. Although the reaction is usually written as involving a phenyl cation, in a number of cases, a phenyl radical is the more likely intermediate.
- Diazonium salts react with arenes bearing strongly electron-donating groups to give azo compounds (ArN=NAr′), which have found wide applications as dyes.

13.4 NUCLEOPHILIC SUBSTITUTION OF ARYL HALIDES, S_N2Ar

As noted in Chapter 9, conventional nucleophilic substitution reactions of the type described for aliphatic compounds don't occur at sp^2 hybrid carbon atoms. The formation of aryl cations is disfavored because they are unstable, and attack from the backside of the C–halogen bond is rendered impossible by the ring structure. So the substitution of aryl halides is much more difficult, and the mechanism of the S_N2Ar reaction is quite different, involving addition followed by elimination—a two-step process. For example, although substitution of chlorobenzene by hydroxyl ion is possible, conditions are harsh (350 °C, high pressure), and even then, yields are low. The reaction (Figure 13.13) probably involves a benzyne intermediate (see Sections 10.9 and 13.6).

However, haloarenes that also bear electron-withdrawing groups substitute much more readily. The process involves two steps, addition followed by elimination (Figure 13.14). The intermediate, negatively charged, species is called a Meisenheimer complex.

Which of the two steps, addition or elimination, is rate determining? All the evidence suggests that it is the first, the addition, step. When the rates of reaction of arenes bearing different halogens are determined, fluorobenzene reacts the most rapidly and iodobenzene the most slowly. Fluorine is the most electronegative of the halogens, so that a carbon atom bonded to fluorine will have a greater partial positive charge than would be the case with the other halogens. The more positive that carbon atom is, the faster it may be expected to react with nucleophiles. The electron-withdrawing effect of fluorine may also be expected to stabilize the intermediate anion,

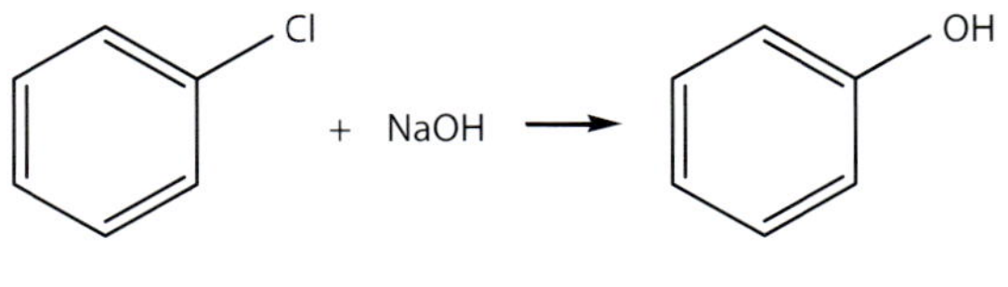

FIGURE 13.13 Substitution of chlorobenzene.

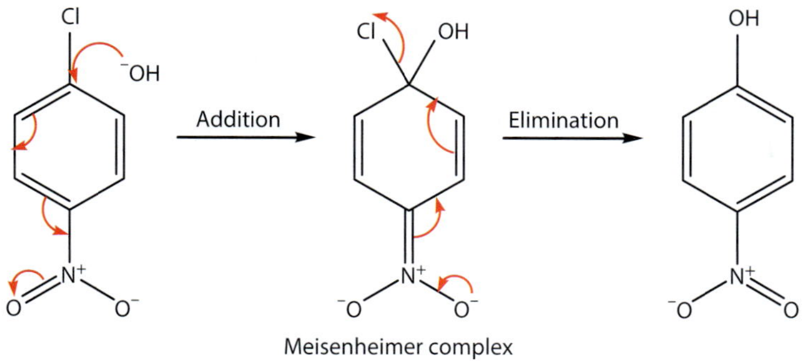

FIGURE 13.14 Substitution of a nitrohaloarene.

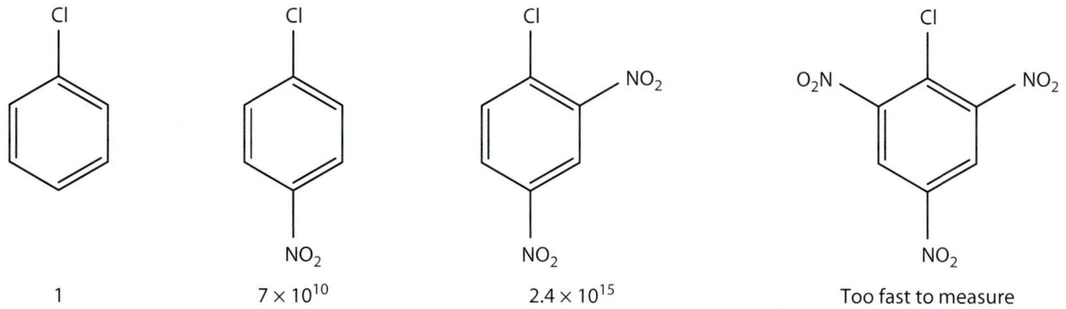

FIGURE 13.15 Relative rates of reaction of nitrated halobenzenes with methoxide ion.

and it is the smallest of the halogens, so steric effects are minimized. Also, fluorine is the poorest leaving group of the halogens, and iodine the best, so the elimination step would be expected to be slow for fluorine and fast with iodine.

The effects of electron-withdrawing groups at the *ortho*- and *para*-positions are additive—more nitro groups means faster substitution (Figure 13.15).

PROBLEM 13.3

Explain the relative reactivity of the 2,4-dinitrohalobenzenes with piperidine:

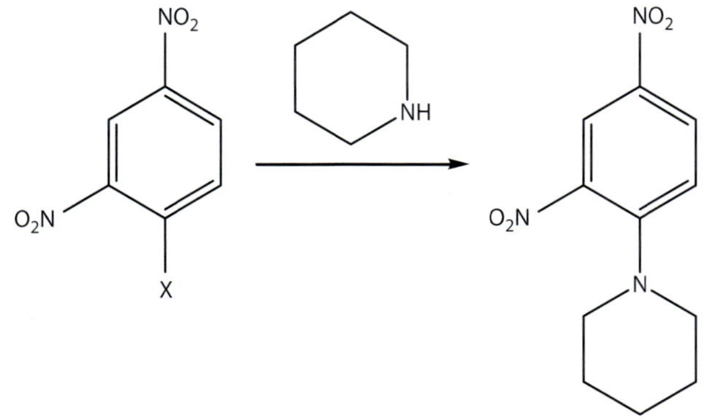

X	F	Cl	Br	I
Relative rate of reaction	3300	4.3	4.3	1

Solution

Fluorine is the most electronegative halogen, and hence, the carbon to which it is attached will be the most positively charged. Its electronegativity also helps to stabilize a negative charge in the intermediate Meisenheimer complex. This means that the rate-controlling step of the reaction is accelerated. The addition of the nucleophile is rate determining, and subsequent loss of halogen to restore aromaticity is rapid, so the relative ability of the halide anions to act as leaving groups is irrelevant.

We should examine why it is that electron-withdrawing groups only accelerate the reaction when they are positioned *ortho* or *para* to the halogen to be substituted. If we first consider the reaction of an unsubstituted haloarene with a nucleophile, then the charge in the Meisenheimer complex produced can only be located on the positions *ortho* and *para* to the position of the attack (Figure 13.16).

We can now see why it is that only when we have groups that act as resonance acceptors at the *ortho*- and *para*-positions can the electron density be delocalized into groups such as nitro, carbonyl, and nitrile. Conversely, electron-donor substituents at these positions will destabilize the Meisenheimer complex. In Figure 13.17, we see how three nitro groups and a relatively poor leaving group result in the formation of a stable Meisenheimer complex. The elimination process is unfavorable because alkoxides are poor leaving groups, especially in basic solution.

This can give rise to some interestingly selective reactions—in Figure 13.18, only the chlorine *para* to the nitro group is substituted by pyrrolidine. Only by attack at the *para*-position can the intermediate Meisenheimer complex be stabilized by delocalization of the charge into the nitro group.

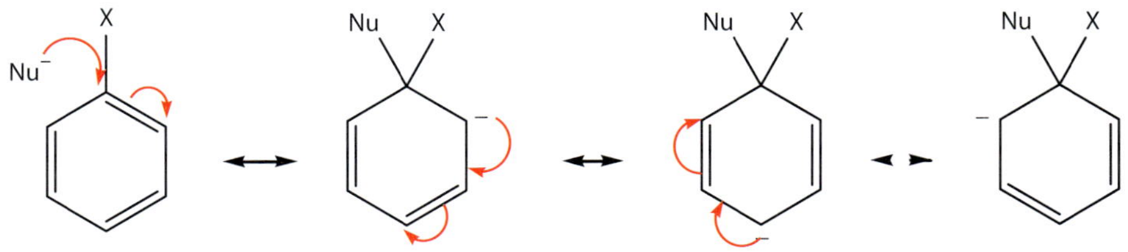

FIGURE 13.16 Resonance forms of a Meisenheimer complex.

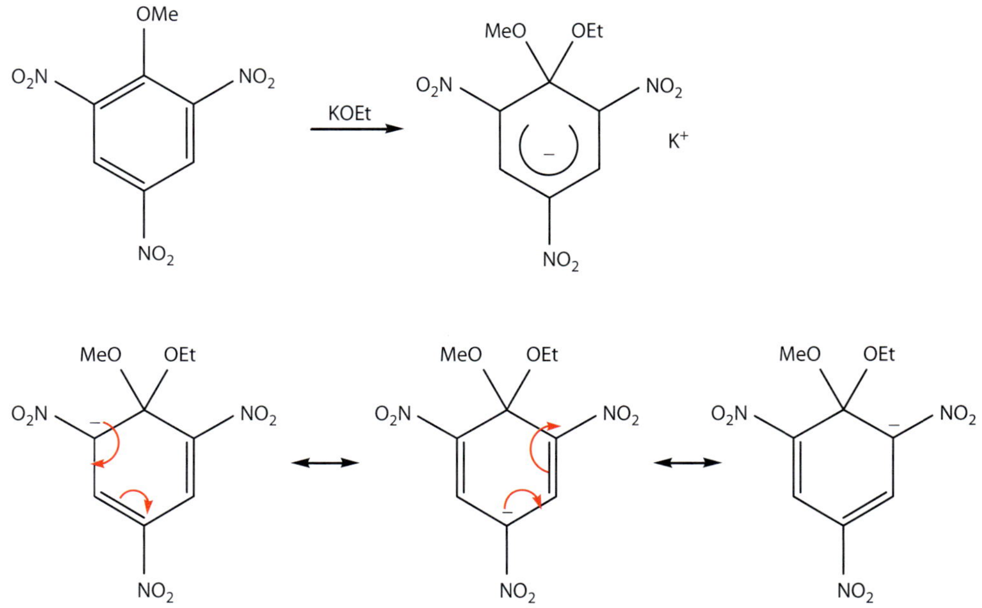

FIGURE 13.17 Formation of a stable Meisenheimer complex.

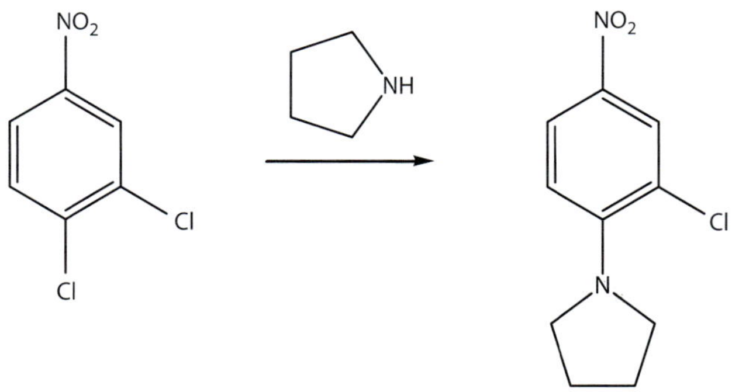

FIGURE 13.18 Selective nucleophilic aromatic substitution.

PROBLEM 13.4

Propose a mechanism for the following reaction:

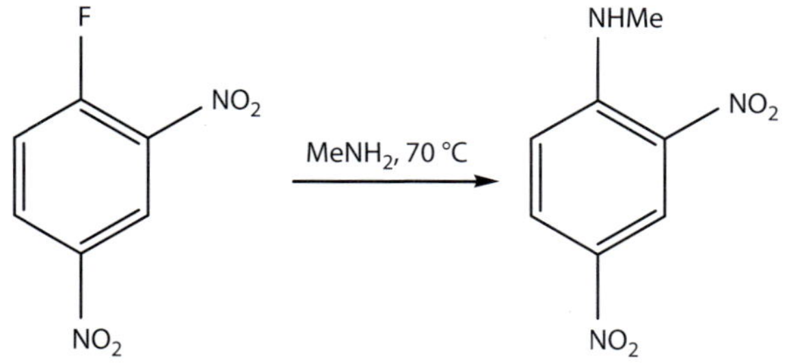

Solution

Attack occurs at the carbon atom bearing the fluorine, and the negative charge is delocalized into both nitro groups (only the resonance form involving the *para*-nitro group is shown). Bringing back the charge from the nitro group allows the expulsion of the fluorine:

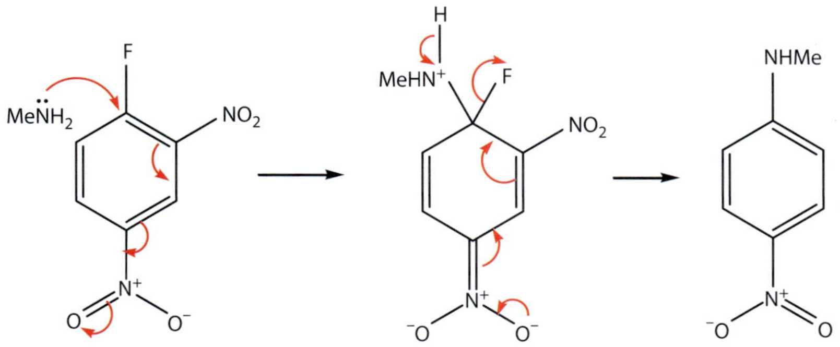

13.5 SUBSTITUTION OF HALOHETEROCYCLES

You will remember that in the last chapter, we said that the reactivity of pyridine was comparable to that of nitrobenzene. So it is not unexpected that 2- and 4-halopyridines can be substituted by nucleophiles (Figure 13.19). In this reaction, we have a resonance form that is particularly favored

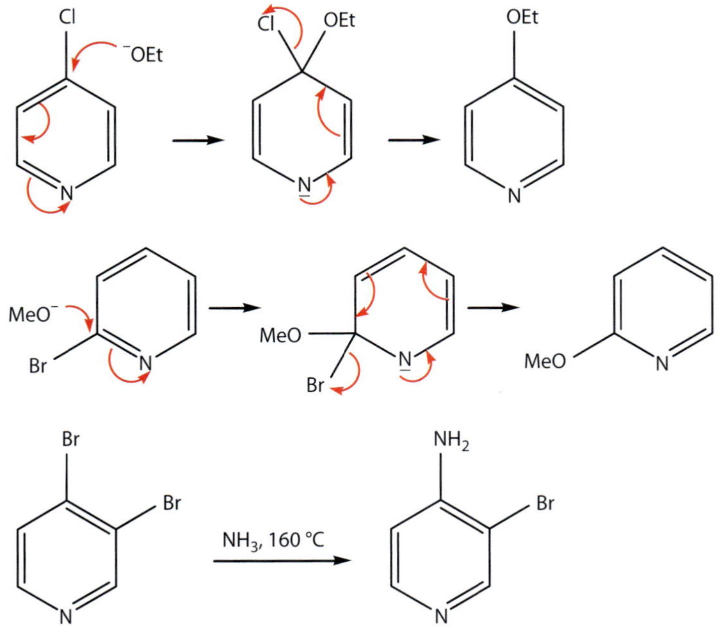

FIGURE 13.19 Mechanism and selectivity of substitution of halopyridines.

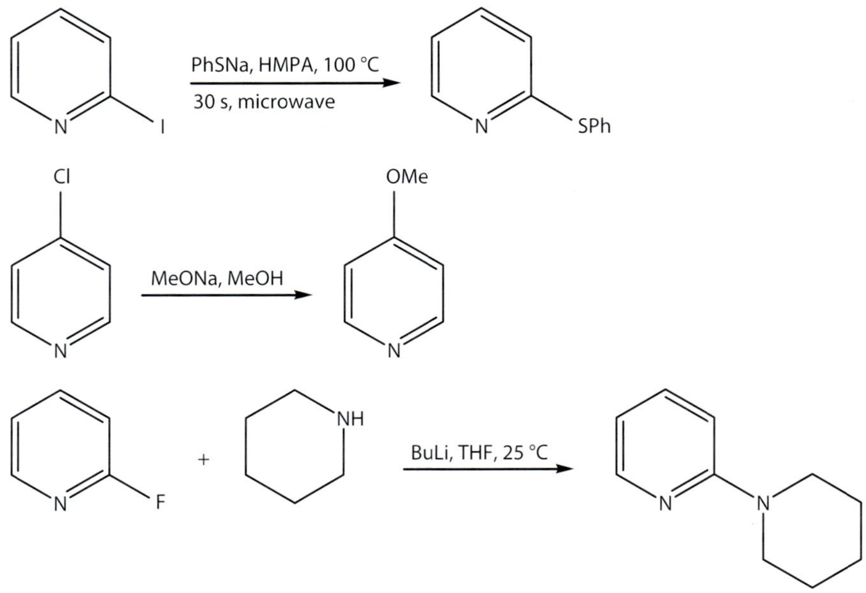

FIGURE 13.20 Substitutions of halopyridines.

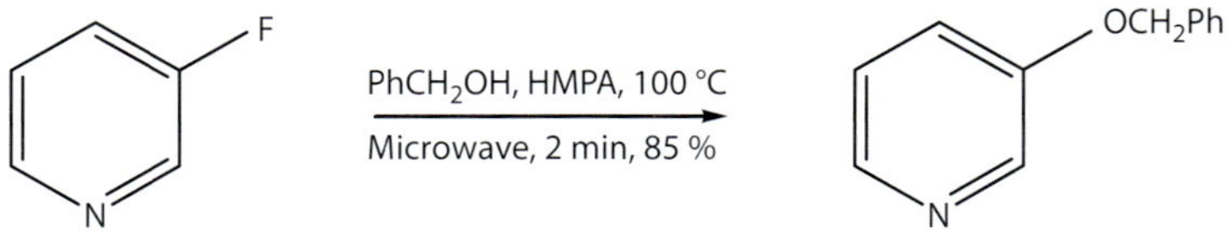

FIGURE 13.21 Substitution of 3-halopyridines using microwave heating.

because the negative charge in the intermediate is located on the electronegative nitrogen atom. The selectivity of the reaction is similar to that noted for halonitrobenzenes; a halogen at the 2- or 4-position is substituted much more rapidly than one at the 3-position. Examples are shown in Figure 13.20.

3-Halopyridines are more reactive than chlorobenzene; if the reaction is microwaved, substitution can occur, although the reaction is slower than for the 2- and 4-halopyridines under comparable conditions (Figure 13.21).

PROBLEM 13.5

Predict the product(s) of the following reaction:

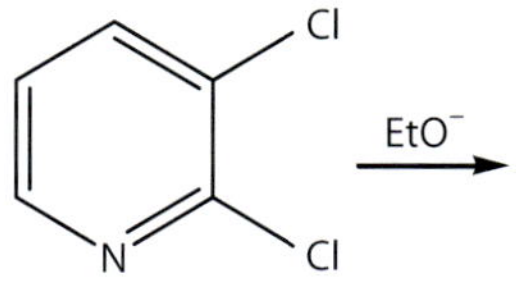

Solution

The chlorine at the 2-position is much more easily substituted, because the charge in the intermediate can be located on the nitrogen atom.

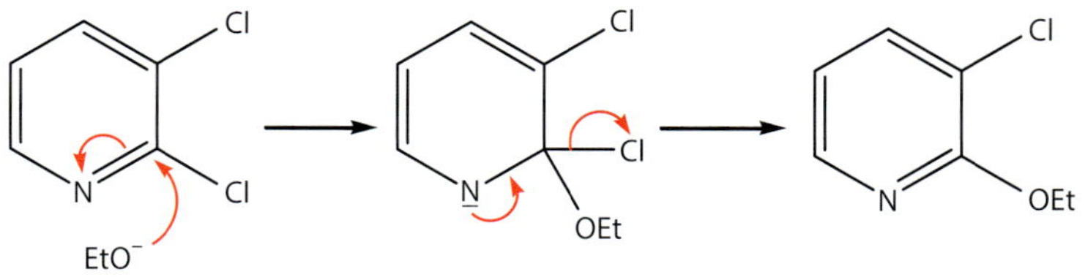

13.6 SUBSTITUTION *VIA* BENZYNE AND RELATED INTERMEDIATES

We have already briefly considered substitution reactions involving benzyne intermediates in Chapter 10, when we focused on the elimination process. To have a substitution reaction, overall we need to have elimination, followed by addition (Figure 13.22). We should note that because the rate-limiting step of the process is elimination, it's crucial that we use a very strong base, such as sodamide, $Na^+[NH_2]^-$.

The reaction mechanism involves an elimination to give the aryne (Figure 13.23). As we noted previously, this can be an E2- or an E1cB-type process, though it is shown as E2 for simplicity. Although it is common in textbooks, and quite conventional, to write benzyne as having a triple bond, this is not a triple bond of the same type that we have in alkynes. The extra "bond" is formed by weak overlap of two sp^2 orbitals, each contributing one electron. It is probably more correct to think of it as a resonance hybrid of the triply bonded form, singlet and triplet diradicals (although all tests for radical character in reactions of benzyne fail), and charge-separated species.

The second step of the process, the addition of the nucleophile, is very rapid and frequently quite unselective as in Figure 13.22—the methyl group has only a slightly electron-releasing effect, so the difference in contribution between the various resonance forms is slight. The mechanism can also be demonstrated, perhaps slightly more rigorously, by the use of a ^{13}C-labeled substrate (Figure 13.24). In this example, there can be no possible effect, either steric or electronic, from a substituent.

Where there are strongly donating or withdrawing substituents on the aromatic ring, it is sometimes possible to predict the outcome by considering the reaction as having some E1cB character (so that the most acidic proton is removed) and then considering the relative

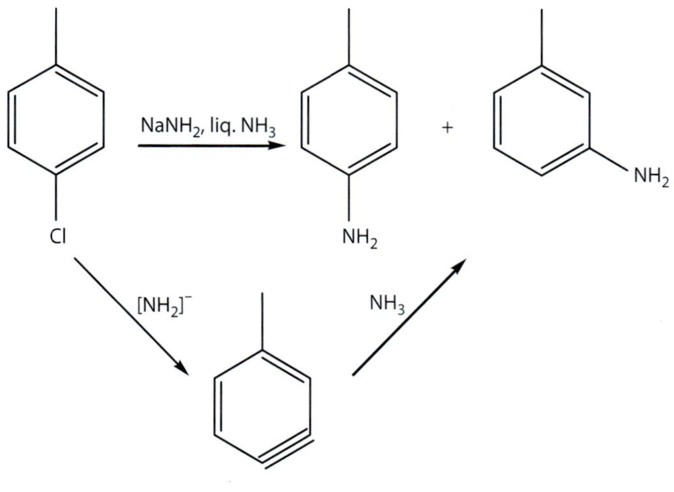

FIGURE 13.22 Substitution of 4-chlorotoluene *via* a benzyne intermediate.

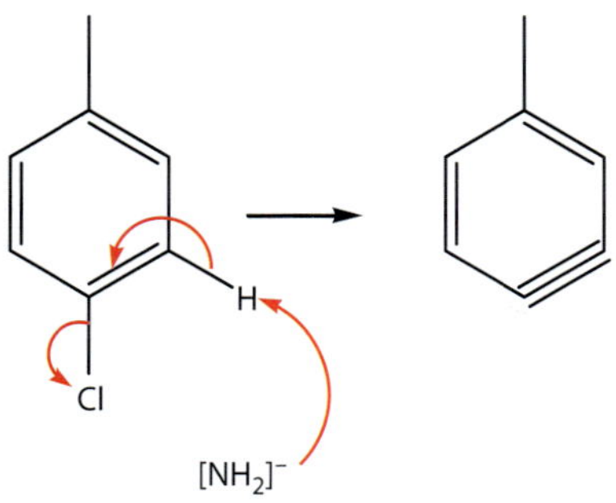

FIGURE 13.23 Mechanism of benzyne formation.

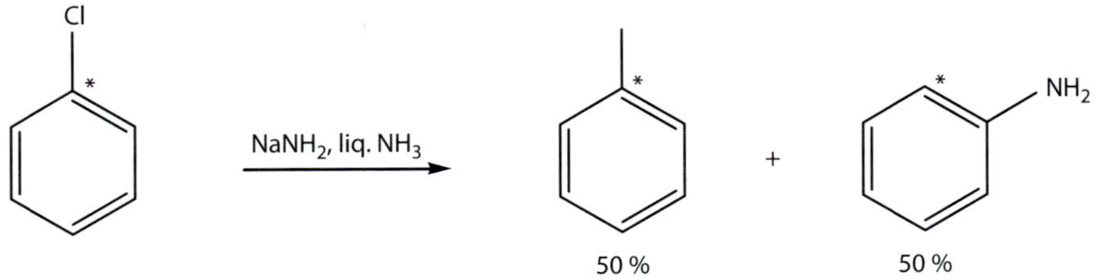

FIGURE 13.24 Substitution of ^{13}C-labeled chlorobenzene.

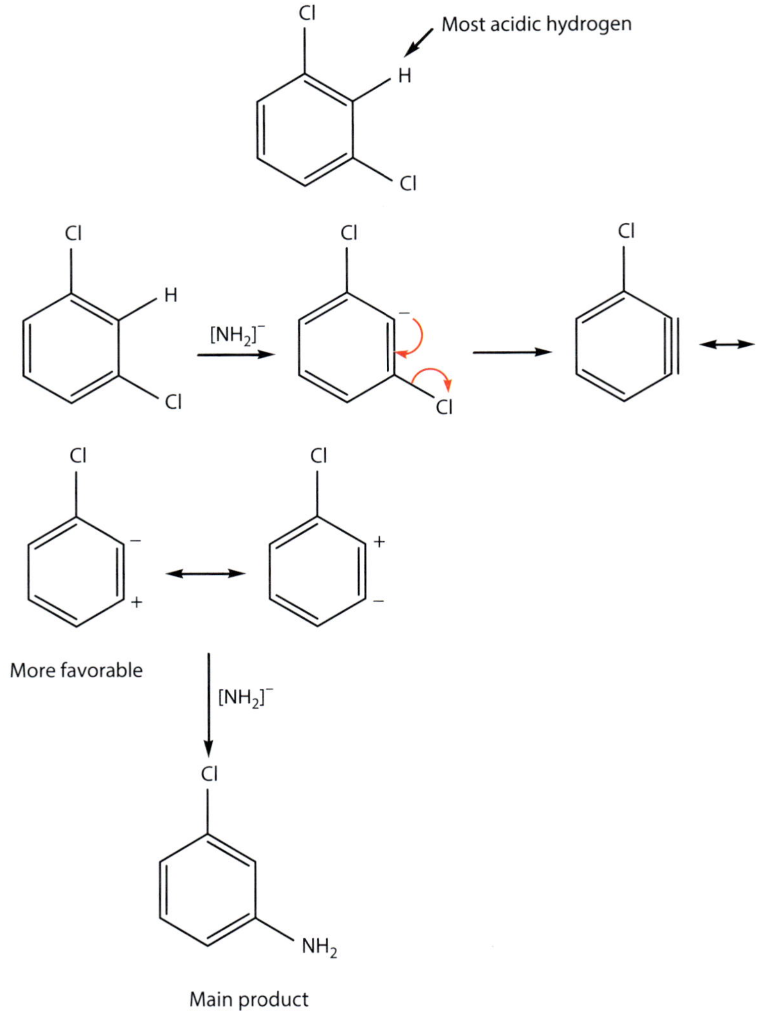

FIGURE 13.25 Regioselectivity in substitution *via* aryne intermediates.

stabilities of the charge-separated resonance forms of the aryne produced (Figure 13.25). The most acidic hydrogen is the one flanked by the two electron-withdrawing chlorine atoms. In the aryne intermediate, the most significant charged resonance form is the one with the negative charge adjacent to chlorine, and it is from this form that most of the product is derived.

Pyridynes can also be produced from halopyridines. In 3-chloropyridine, the hydrogen at the 4-position is significantly more acidic than that at the 2-position (H-4 exchanges with D_2O in base 53 times more rapidly than H-2), so that the main pyridyne produced is that with the

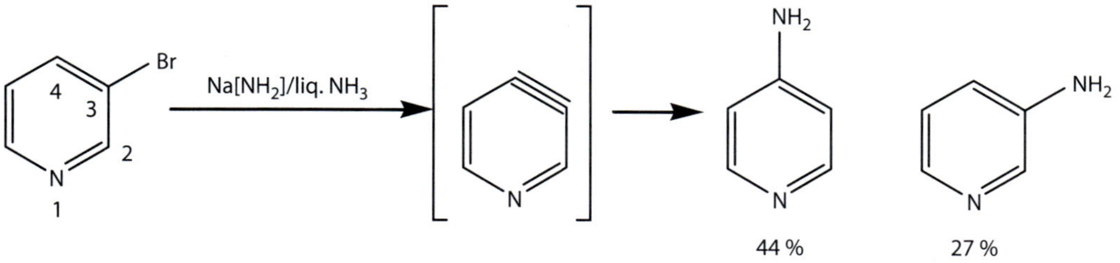

FIGURE 13.26 Substitution *via* pyridyne.

"triple bond" between C-3 and C-4. Thus, the reaction of 3-bromopyridine with $Na[NH_2]$ in liquid ammonia gives a mixture of 3-aminopyridine and 4-aminopyridine (Figure 13.26). The products are obtained in comparable ratio to those from 4-chloropyridine, where 3,4-pyridyne is the only one that can be produced.

PROBLEM 13.6

Explain the following observation, drawing a complete mechanism:

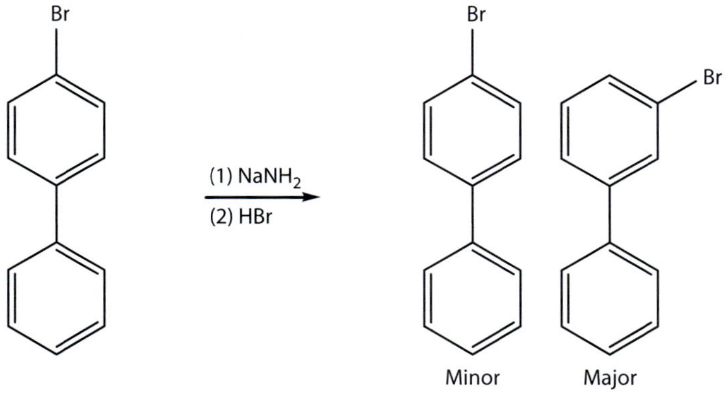

Solution

The two products from this process should tell us that the mechanism of this process involves a benzyne intermediate. Only one benzyne can possibly be produced in the first elimination step:

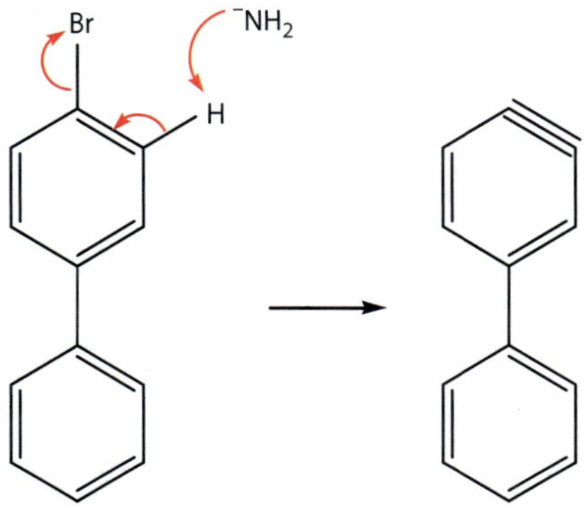

Two charged resonance forms can be written for the benzyne intermediate. Given the outcome, it is clear which one is more important. We can rationalize this by remembering that a phenyl group is a donor of electron density and so is better able to stabilize the positive charge, when it is nearer to it. The process is completed by the reaction of HBr with each of the formal resonance forms.

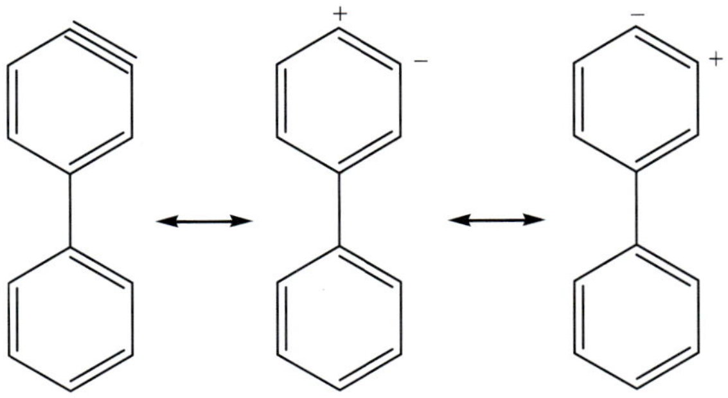

Key Points from Sections 13.4 through 13.6

- Nucleophilic substitution of haloarenes is difficult, and poor reactions are obtained unless the arene bears a group able to accept electrons by resonance.
- Substitution of haloarenes bearing strongly electron-accepting groups proceeds by an addition–elimination mechanism. The addition step is rate determining, and the effect of the electron-withdrawing groups is additive—more electron acceptors means a faster reaction.
- To be effective in delocalizing charge in the intermediate, the electron-withdrawing groups must by *ortho* or *para* to the group we wish to substitute. Conversely, electron donors at the *ortho*- or *para*-positions retard reaction. Groups at the *meta*-position have little effect.
- 2-Halopyridines and 4-halopyridines may also be readily substituted by nucleophiles by an addition–elimination mechanism.
- Halobenzenes and pyridines that are not specifically activated toward substitution by appropriately placed electron-accepting substituents may be substituted by an elimination–addition mechanism involving benzyne and related aryne intermediates; a very strong base is required.

13.7 REACTIONS IN SIDE CHAINS OF AROMATIC COMPOUNDS

13.7.1 REACTIONS OF ALKYLBENZENES

Toluene is readily oxidized to benzoic acid by powerful oxidants such as potassium permanganate or potassium dichromate (Figure 13.27). The reaction also works with benzenes substituted with primary and secondary alkyl groups, but tertiary alkyl groups are not oxidized. The mechanism almost certainly involves radical intermediates. It is compatible with halogen and nitro substituents. In synthetic terms, we have accomplished something that will be useful—for electrophilic substitutions, alkyl groups are *ortho,para*-directing and activating, but carboxyl groups are *meta*-directing and deactivating. This will allow us to use the same initial substituent to accomplish substitution in different places on our aromatic ring.

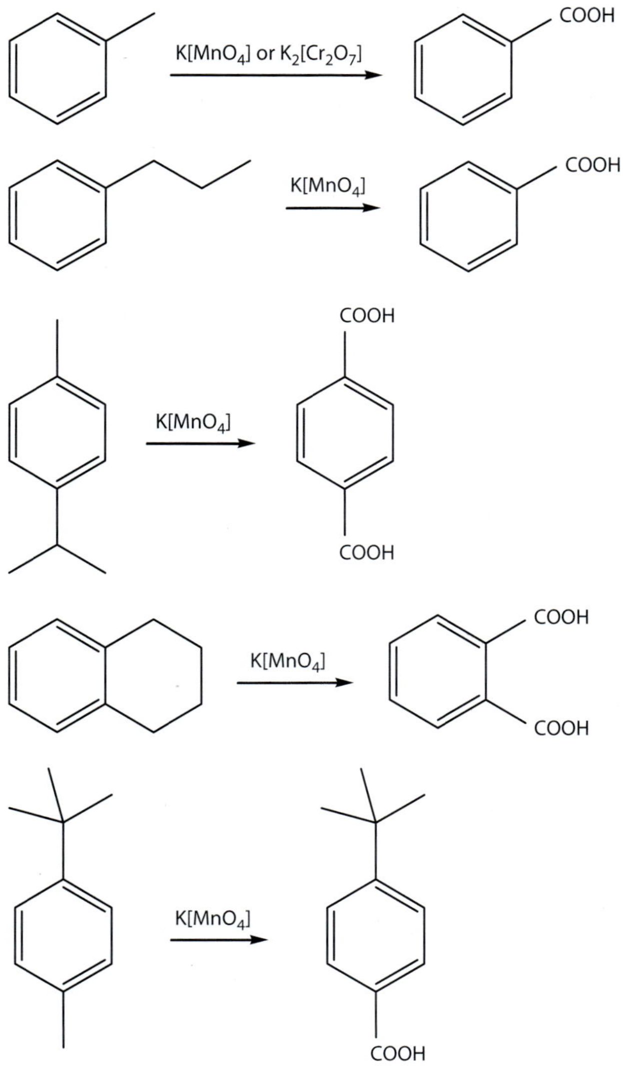

FIGURE 13.27 Oxidation of alkylbenzenes.

Focus on the Industrial Synthesis of Phenol

While the diazonium salt route is probably the most commonly used laboratory synthesis of phenols, phenol itself is manufactured from *iso*-propylbenzene, which has the common name cumene. Cumene is prepared by a Friedel–Crafts reaction between propene and benzene, in the presence of a strong acid such as H_3PO_4 (Figure 13.28). The key intermediate is the 2-propyl cation, formed by protonation of propene.

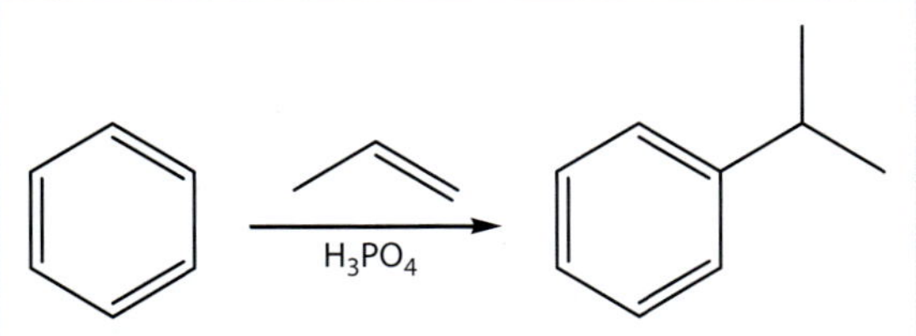

FIGURE 13.28 Preparation of cumene.

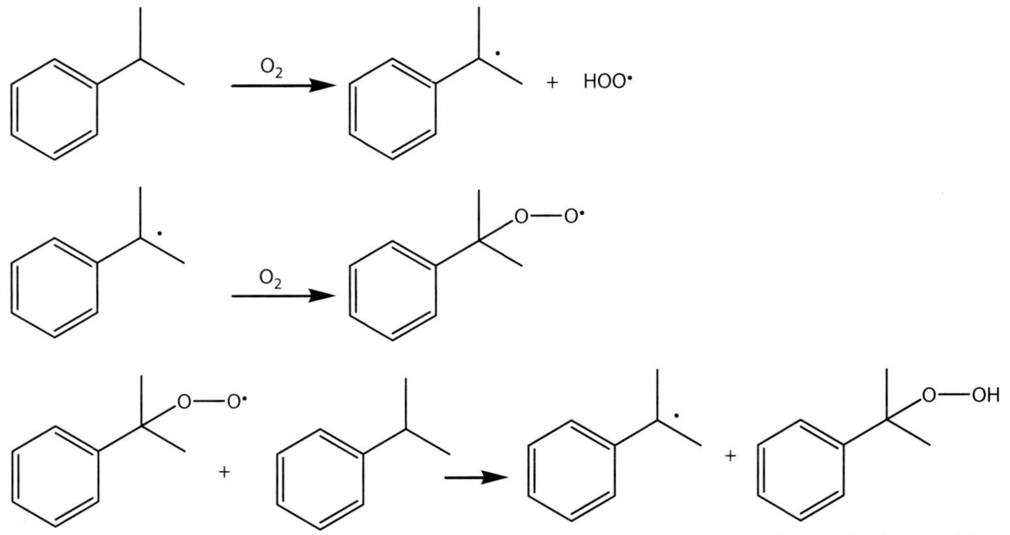

FIGURE 13.29 Atmospheric oxidation of cumene to cumene hydroperoxide.

Cumene is then allowed to react with atmospheric oxygen (which has unpaired electrons and may be considered to be a diradical), which abstracts a hydrogen atom from the benzylic position to give the relatively stable benzylic radical. The radical reacts with molecular oxygen to give the hydroperoxide radical, which in turn abstracts hydrogen from another molecule of cumene (Figure 13.29). This type of atmospheric oxidation of organic compounds is common and is generally called autoxidation. It is one of the processes responsible for the spoilage of foods.

In acid solution, cumene hydroperoxide undergoes a cationic rearrangement. The OH group is protonated, thus made into a leaving group. As water departs, the phenyl ring migrates to oxygen to give a stable tertiary carbocation. The process is driven both by the stability of the carbocation formed and the fact that oxygen–oxygen single bonds are weak and easily broken. We will look in more detail at reactions of this type, involving migration of groups to electron-deficient atoms, in Chapter 18. The cation is then attacked by water and collapses to acetone and phenol (Figure 13.30). The advantage of this process is that the by-product, acetone, also has a commercial value—so we have good atom economy, with all our carbon atoms being usefully employed. Environmentally, however, the use of strong acids is not so desirable, and new catalysts that are more environmentally friendly have recently been developed.

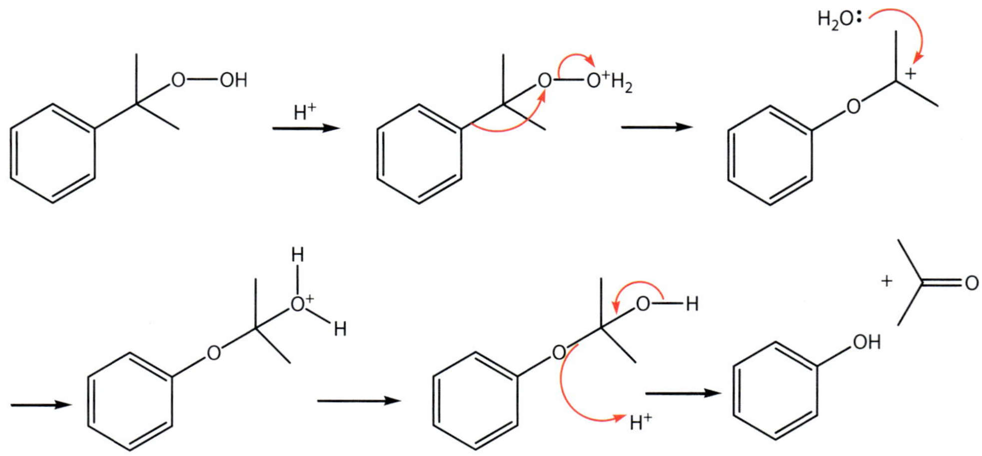

FIGURE 13.30 Acid-catalyzed rearrangement of cumene hydroperoxide.

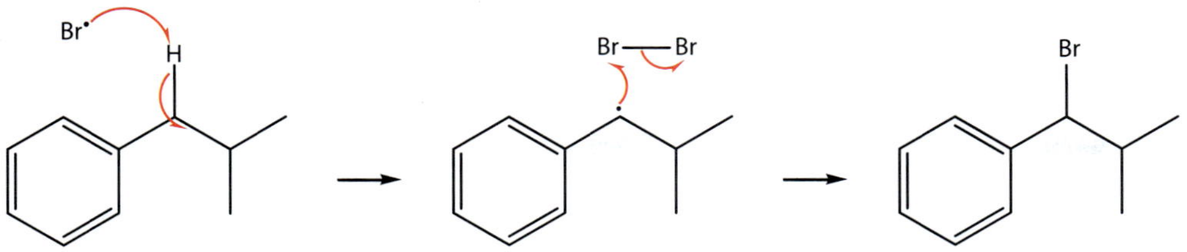

FIGURE 13.31 Radical chain bromination of a benzylic center.

Alkylbenzenes can be halogenated using bromine and light—the light generates bromine radicals from bromine molecules by homolytic fission (Equation 13.10):

$$Br_2 \xrightarrow{hv} 2Br^{\bullet} \tag{13.10}$$

These radicals can then abstract a hydrogen atom from a carbon at the benzylic position to give a favorable benzyl radical. The radical then reacts with a further molecule of bromine to give a brominated product and regenerate a bromine radical (Figure 13.31). This is a radical chain reaction (see **11.3** for radical chain reactions in addition to alkenes); the light is only required to initiate reaction, and the reaction continues for some time, even when the light is switched off. The termination of the chain reaction occurs either when two radicals recombine or when a radical impacts the wall of the reaction vessel.

Under these conditions, the ring is not brominated, and the reaction is quite selective. This is made possible by the fact that the benzylic C–H bond is relatively weak—368 *versus* 436 kJ mol^{-1} for methane. The same process is possible for chlorine but is a little less selective—the chlorine radical is more reactive than the bromine radical, and polychlorination occurs to some extent. The fluorine radical is too reactive for practical use—this is related to the strength of the H–F bond that is formed on hydrogen abstraction (see Section 12.2). The HI bond, by contrast, is weak, and the process of radical formation is endothermic. Many sets of reaction conditions for this process are known—*N*-chlorosuccinimide or *N*-bromosuccinimide in the presence of a radical initiator is also widely used (Figure 13.32). We should recall that the S_N1 or S_N2 substitution of benzyl bromide is very facile, so we can use this process to put a range of substituents into the benzylic position (*e.g.*, as in Figure 13.33).

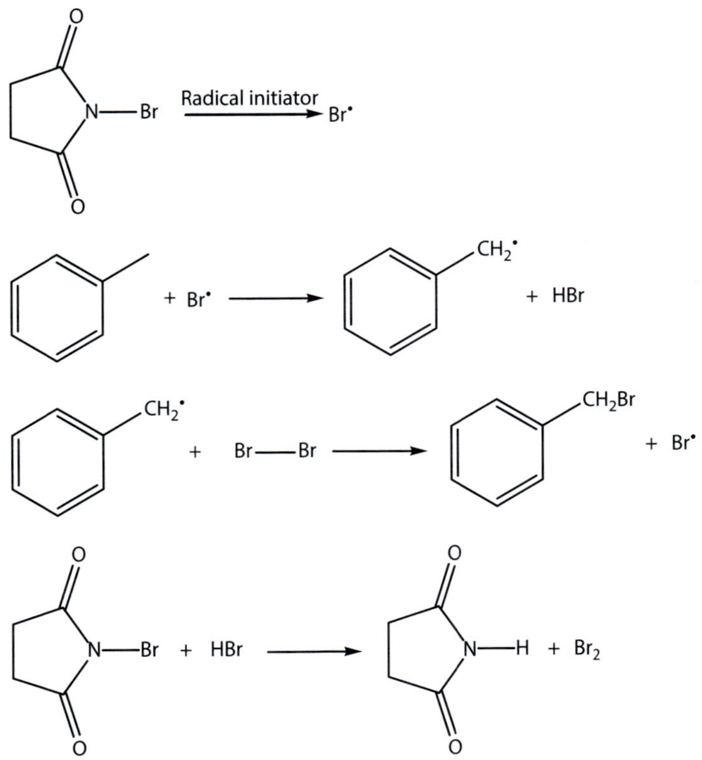

FIGURE 13.32 Benzylic bromination using *N*-bromosuccinimide.

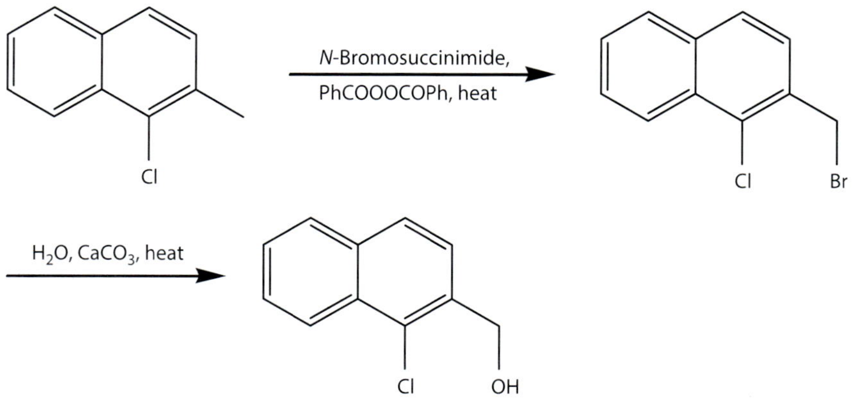

FIGURE 13.33 Radical chain bromination in synthesis.

PROBLEM 13.7

How would you carry out the following conversion, which may require more than one step?

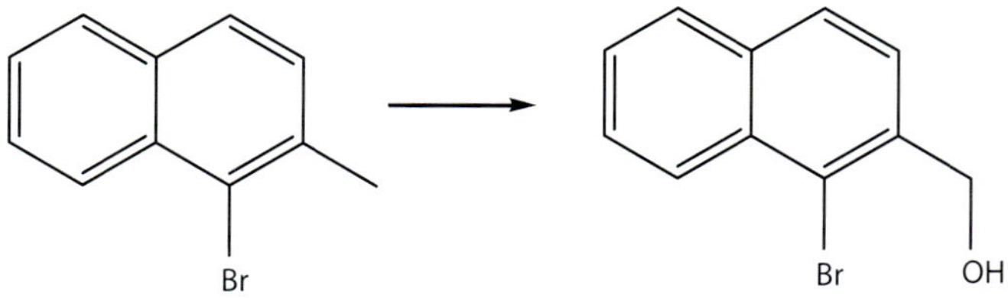

Solution

The first step is the radical bromination of the methyl group. Bromine and a radical initiator would also be a satisfactory reagent. We then remember that S_N2 reactions are easy at benzylic centers, and the second step is accomplished with hydroxyl ion:

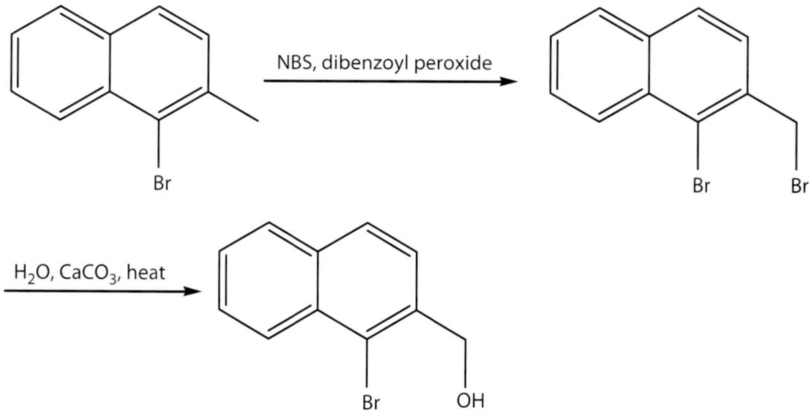

13.7.2 REDUCTION OF NITRO GROUPS

When we discussed the formation of diazonium salts from anilines, noting how useful they are, and the additional scope they give us in synthesis, you might have had a small worry—how do we make the relevant aniline in the first place? So far, we only know how to substitute halides or other leaving groups with amines—there is no simple equivalent of an electrophilic "$[NH_2]^+$." And if we have to start off with a halide, then we are beginning to go round in circles. Fortunately, there is a very simple solution. We can reduce nitro groups to amino groups with tin or iron and HCl (Figure 13.34). There are many experimental protocols available, and precise conditions can be fine-tuned for specific substrates.

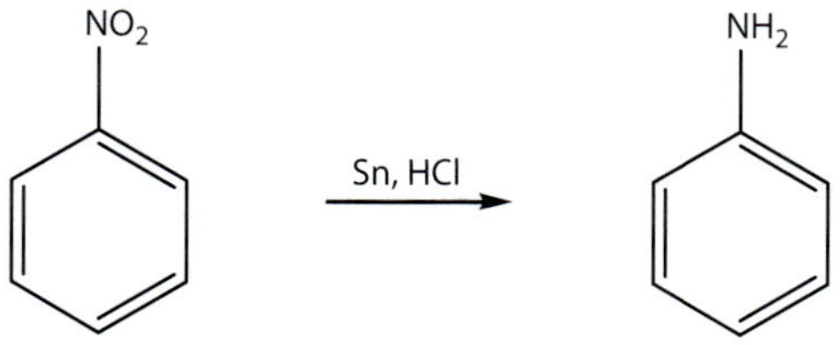

FIGURE 13.34 Reduction of nitroaromatic compounds with tin and HCl.

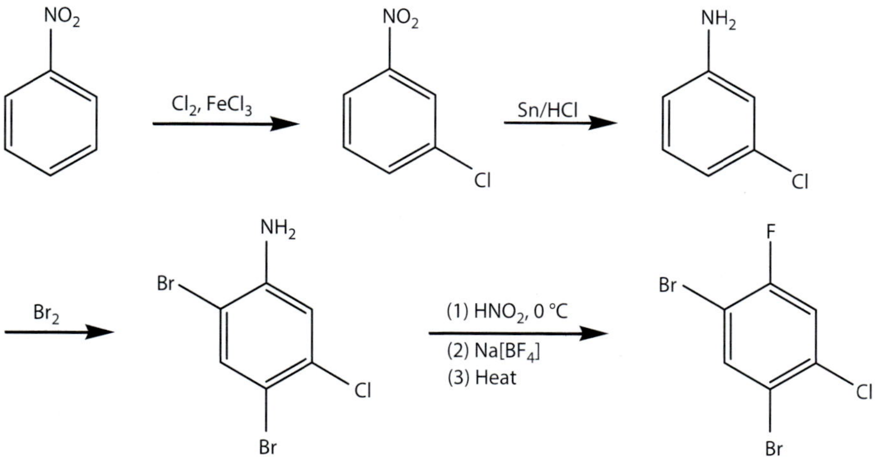

FIGURE 13.35 Use of the different directing effects of the nitro and amino groups in synthesis.

This simple reaction will be very valuable to us in synthesis—the nitro group is readily introduced. In electrophilic substitution, it is deactivating and *meta*-directing, and it assists nucleophilic substitution of good leaving groups at *ortho*- and *para*-positions. The amino group is *ortho, para*-directing and very activating in electrophilic substitution—and *via* reduction of a diazonium salt, it can be removed after we have used it for this purpose. So by reducing nitro groups at an appropriate point in our synthesis, we can put in additional substituents almost anywhere we need to on our aromatic ring (Figure 13.35).

13.7.3 REDUCTION OF ARYL KETONES

We have briefly met one method of reducing a ketone to a hydrocarbon, the Clemmensen reduction, using zinc amalgam and concentrated HCl (Figure 13.36). We note that the reaction works for both aldehydes and ketones. However, the reaction conditions are strongly acidic, so it is not compatible with acid-sensitive functional groups. The reaction is incompatible with nitro groups, which are often also reduced under these conditions.

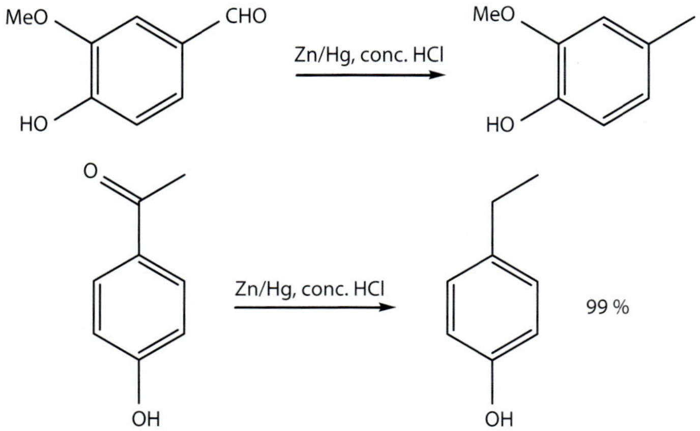

FIGURE 13.36 Clemmensen reduction of aryl aldehydes and ketones.

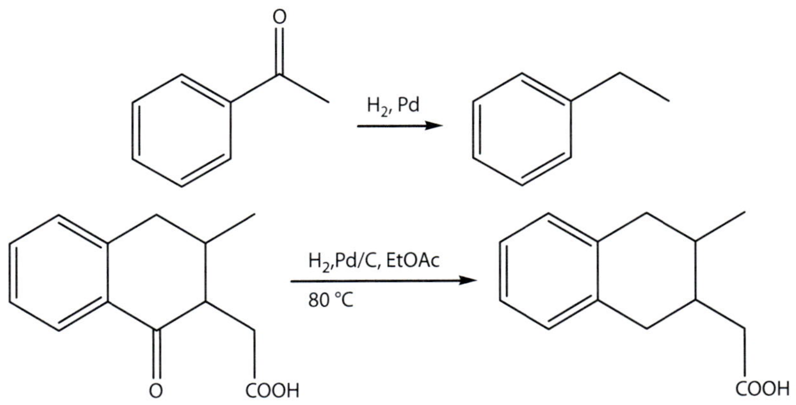

FIGURE 13.37 Hydrogenolysis of aryl ketones.

There are a number of other methods available for the reduction of carbonyl groups to hydrocarbons, specifically the Wolff–Kishner (which operates under strongly basic conditions) and Raney nickel desulfurization of thioacetals. We will return to these and their reaction mechanisms later, when we study carbonyl chemistry in more detail (Chapter 14). Aryl ketones can be reduced to hydrocarbons using molecular hydrogen and a palladium catalyst (Figure 13.37; the process is usually described as hydrogenolysis). This reaction is fairly specific to aryl ketones; aliphatic ketones are not normally reduced. Notice that here also (as with the nitro and amino groups), we have converted a *meta*-directing substituent into one that is *ortho,para*-directing.

13.8 REDUCTION OF BENZENE RINGS

Because benzene is aromatic, it is highly resistant to the reduction of the six-membered ring. Indeed, many reductions of unsaturated side chains may be accomplished, with no risk to the aromatic core (Figure 13.38). Catalytic systems are known that do allow reduction of the benzene ring, but generally, conditions are forcing (Figure 13.39).

Of more interest synthetically is the Birch reduction, which uses an alkali metal in liquid ammonia to provide electrons for reduction to cyclohexadienes. The prototype reaction is shown in Figure 13.40. The reaction mechanism involves the addition of electrons to the aromatic system to give first a radical anion. Unsurprisingly, the negative charge is located at one end of the six-membered ring, and the radical center at the other—we know that electrons repel each other. This is then protonated and a second electron added to give a carbanion, which is again protonated by the added alcohol.

In the reduction of substituted benzene rings, the anionic and radical centers are formed where one might expect—anions are stabilized by electron-withdrawing groups and destabilized by electron-donating groups. Examples of reduction of substituted benzenes are shown in Figure 13.41.

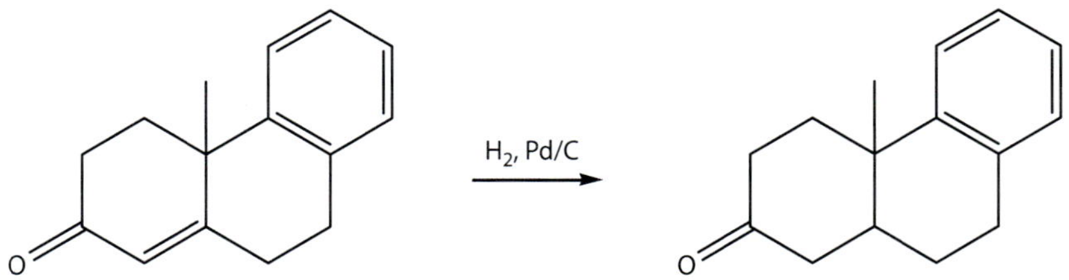

FIGURE 13.38 Selective hydrogenation of an alkene in the presence of an aromatic ring.

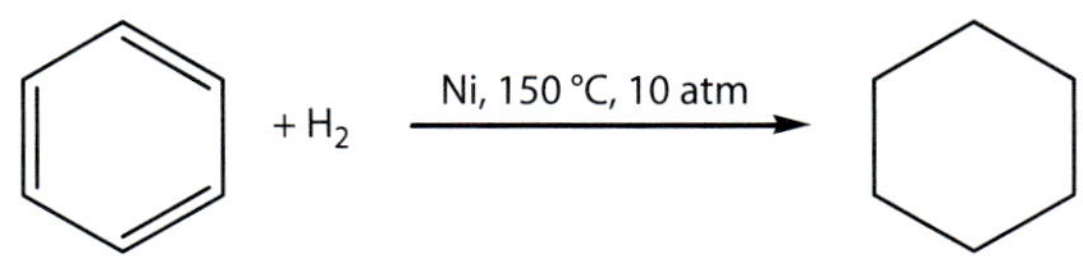

FIGURE 13.39 Hydrogenation of benzene.

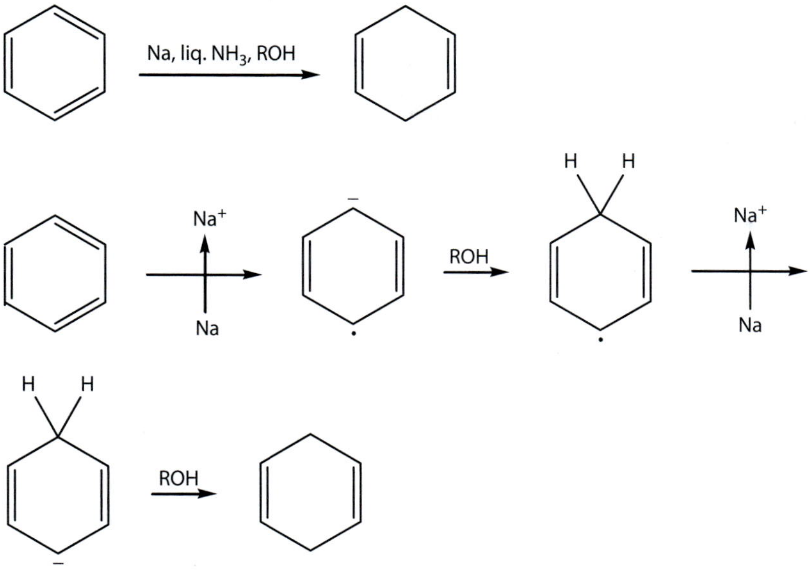

FIGURE 13.40 Mechanism of the Birch reduction.

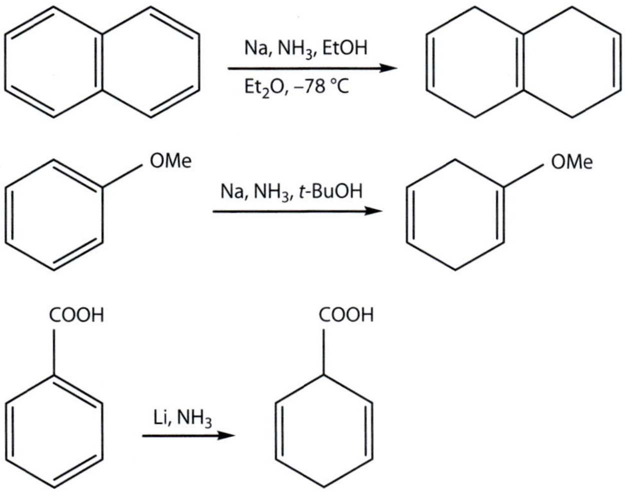

FIGURE 13.41 Examples of the Birch reduction.

Key Points from Sections 13.7 through 13.8

- Primary and secondary alkyl groups attached to benzene rings are oxidized to benzoic acids using potassium permanganate or potassium dichromate. Tertiary alkyl groups are unreactive under these conditions.
- Bromination or chlorination at the benzylic position may be accomplished using the molecular halogen under photolytic conditions or using the *N*-alkylsuccinimide and a radical initiator.
- Nitro groups may be reduced to amino groups using tin and HCl.
- Aryl ketones may be reduced to hydrocarbons by the Clemmensen reduction using zinc amalgam and HCl or by catalytic hydrogenolysis in the presence of palladium metal.
- Dissolving metal reduction of aromatic rings using an alkali metal in liquid ammonia gives 1,4-hexadienes.

PROBLEM 13.8

Suggest reagents for each of the following transformations:

(a)

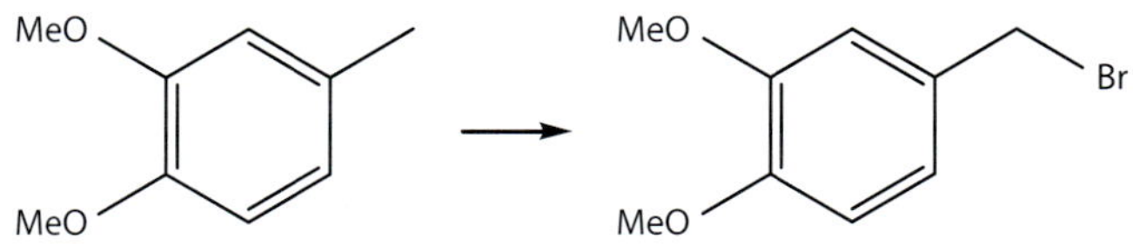

(b)

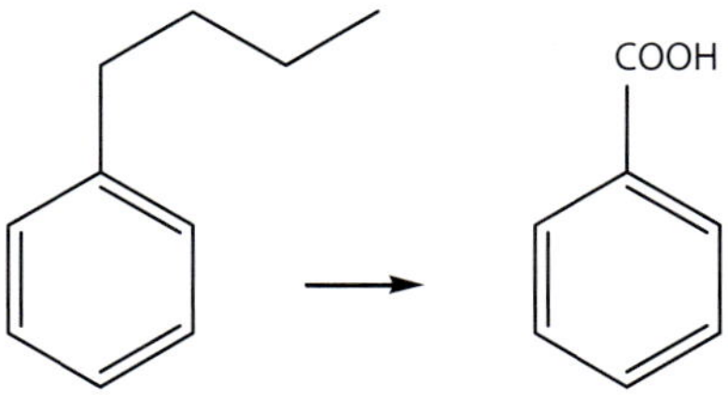

(c)

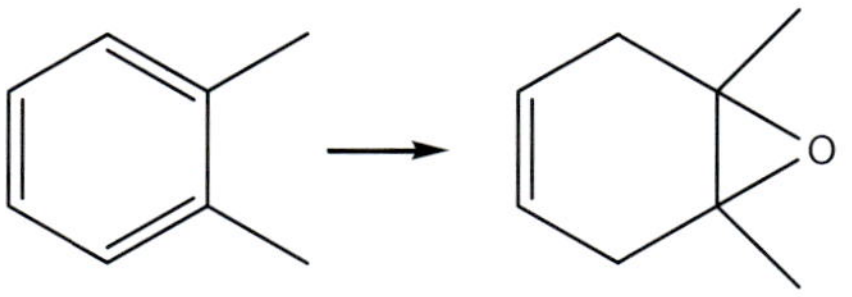

Solutions

(a) A number of reaction conditions involving bromine radicals would work here, but the conditions actually used were *N*-bromosuccinimide, CCl_4, heat, 12 h.
(b) Powerfully oxidizing conditions are required; the usual reagent is $K[MnO_4]$.
(c) In the first step of the reaction, the arene is reduced under Birch conditions to give **13.7**. This can be selectively epoxidized to give the required product. In both steps, the selectivity is as required. The anionic centers in the Birch reduction will be at sites away from the electron-donating methyl groups. In the second step, the more electron-rich double bond reacts selectively with the electrophilic peracid.

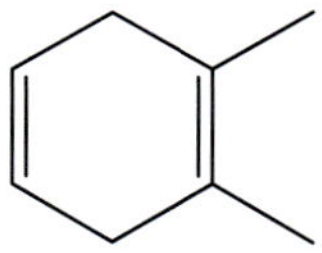

13.7

13.9 CARBON–ARYL BOND FORMATION

Before we start to work on the synthesis of aromatic compounds, it will be useful to summarize and recap all the reactions we have learned that will allow us to attach a carbon side chain to an aromatic ring. Some of these appear earlier in this chapter, or in Chapter 12, some in earlier chapters, and we will also meet a few new ones that we will not study fully until we look at the chemistry of carbonyl groups.

13.9.1 FRIEDEL–CRAFTS ALKYLATION AND ACYLATION (SEE ALSO SECTION 12.2.2)

These reactions involve the electrophilic substitution of an arene by a carbocation or an acylium ion, derived by the reaction of the relevant halide with aluminum chloride or related Lewis acid. The alkylation reactions only give reliable yields when the halide is tertiary (Figure 13.42);

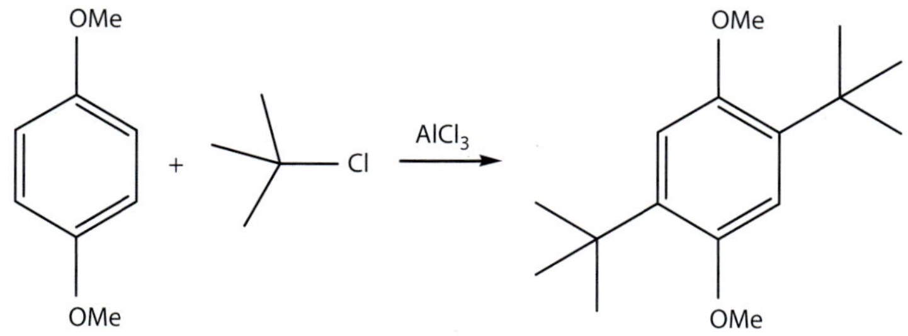

FIGURE 13.42 Friedel–Crafts reaction with *tert*-butyl chloride.

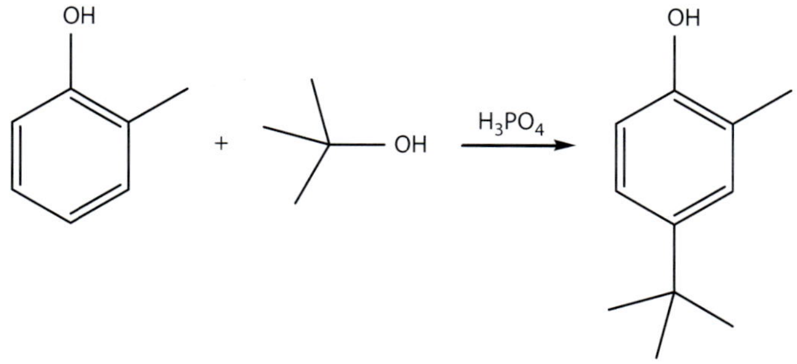

FIGURE 13.43 Alkylation of benzene using *tert*-butanol.

problems with primary and secondary halides include unintended polyalkylation and rearrangement of intermediate carbocations.

Cations may also be generated by other means (*e.g.*, Figure 13.43); any reaction that gives rise to a stable cation will usually give a good result. Notice that the hydroxyl group, which is a strong resonance donor of electrons, dominates the directing effects. For steric reasons, the substitution occurs at the *para*-position.

Friedel–Crafts acylation reactions are generally more reliable than alkylations, and the acyl group can be reduced to a hydrocarbon by the Clemmensen reduction, if necessary. An interesting double substitution is shown in Figure 13.44. Note that once one ring has acylated, it is deactivated to further acylation, and although there is an excess of acylating agent, subsequent reaction occurs only in the other ring.

We might hope to put in an aldehyde group in this way, but that is not as easy as we might imagine. HC(=O)Cl is too unstable to use as a substrate for the Friedel–Crafts reaction. We can use HC(=O)F, but this is also reactive and difficult to handle (Figure 13.45).

A simpler way to accomplish this transformation involves the Gattermann–Koch reaction (Figure 13.46); we may consider this as a way to produce $[HC(=O)]^+$ *in situ*.

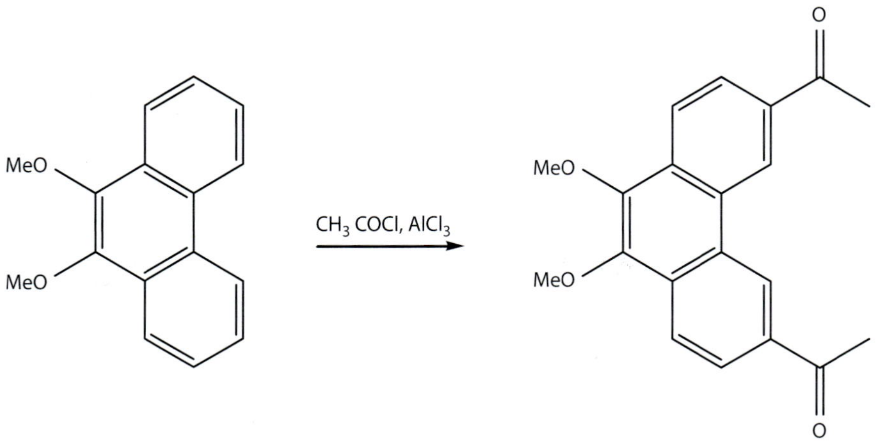

FIGURE 13.44 Friedel–Crafts acylation of a polycyclic compound.

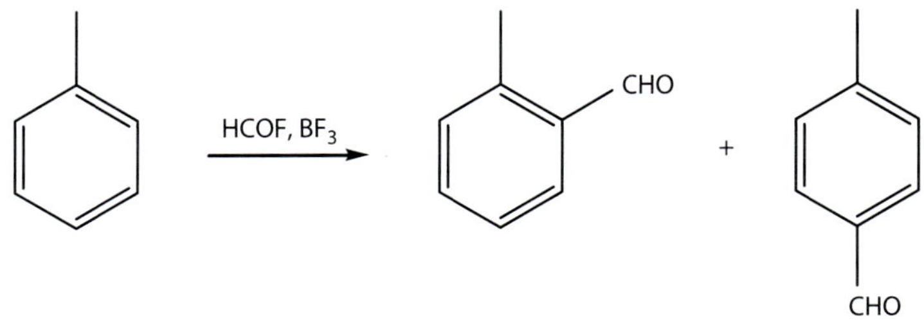

FIGURE 13.45 Friedel–Crafts formylation.

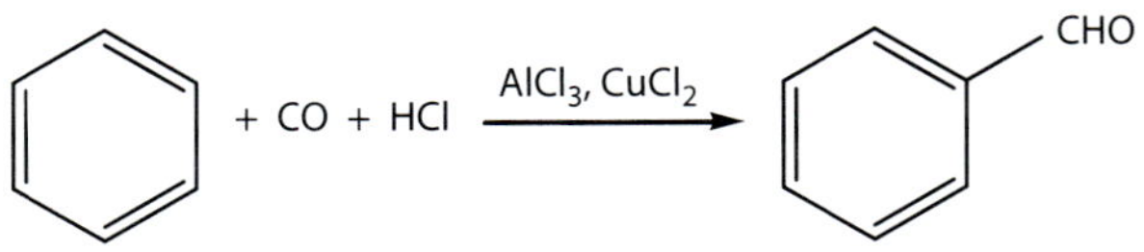

FIGURE 13.46 Gattermann-Koch formylation of benzene.

PROBLEM 13.9

Predict the product(s) of each of the following reactions:

(a)

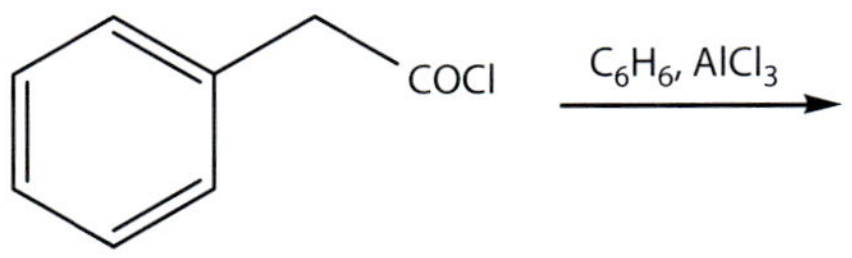

(b)

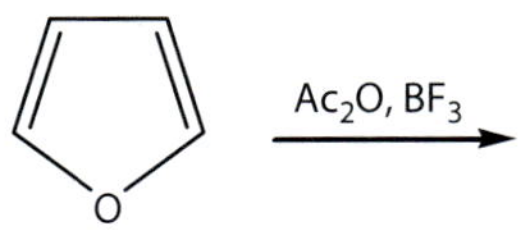

Solutions

(a) This is a Friedel–Crafts acylation.

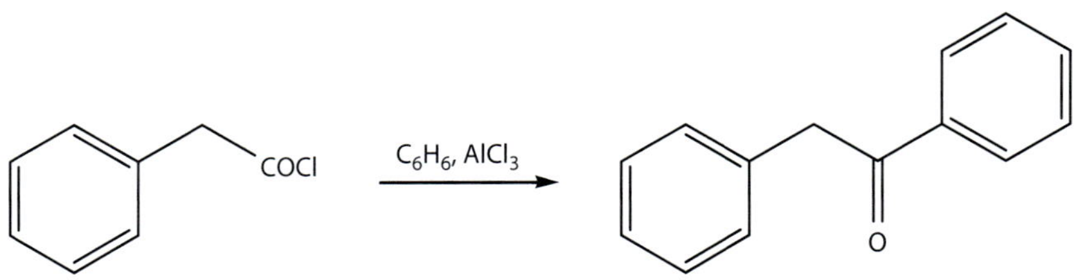

(b) Another Friedel–Crafts, this time on a heterocyclic ring; notice the mild conditions and the selectivity for the 2-position:

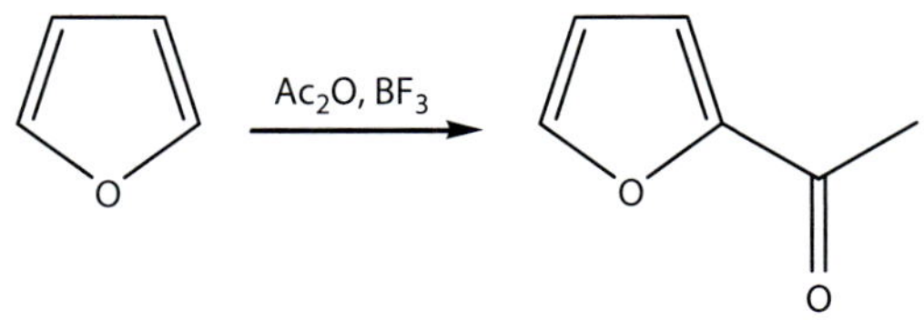

13.9.2 HALOALKYLATION REACTIONS

This is an electrophilic substitution we have not seen before—but we do know the mechanisms for the various components of it (Figure 13.47). The first step is protonation of the formaldehyde to make it into a better electrophile (when we look at carbonyl chemistry in the next chapter, we will find this is a very common process). This is then attacked by the aromatic ring in a conventional electrophilic substitution to give benzyl alcohol. Under the reaction conditions, the benzyl alcohol undergoes an S_N2 substitution to give chloromethylbenzene (Figure 13.48). This process can be

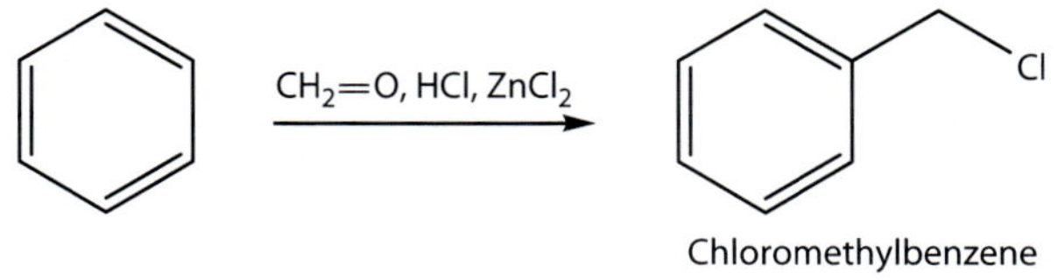

FIGURE 13.47 Haloalkylation of benzene.

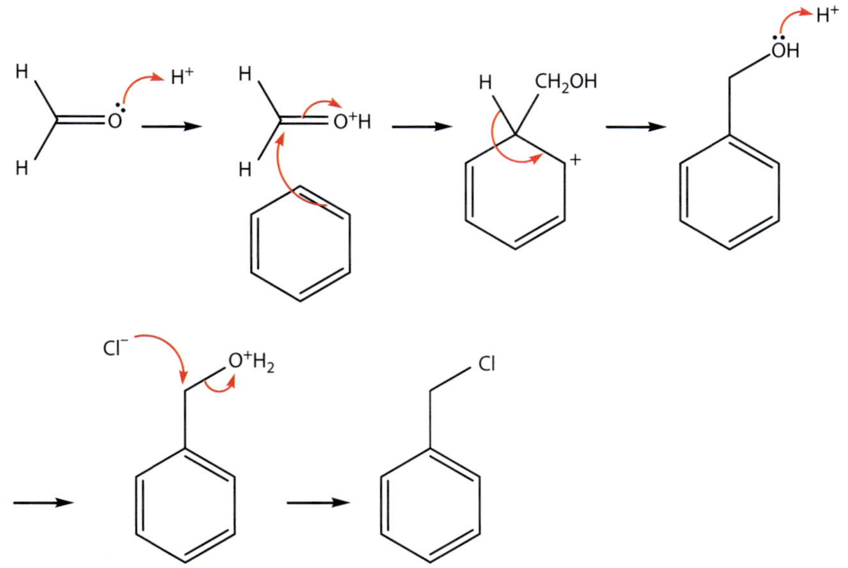

FIGURE 13.48 Mechanism of chloromethylation of benzene.

viewed as complimentary to the radical chain chlorination of toluene (Section 13.7.1) but can be usefully more selective, as there is no mechanism for polyhalogenation. However, if the arene is highly activated, the substitution of the chloromethylarene to give $ArCH_2Ar$ can occur, so this is not a good reaction for phenol or aniline. However, it does work well for deactivated arenes such as nitrobenzene, in contrast to the Friedel–Crafts reaction. In a look ahead, we will meet this reaction later in a very important process in polymer chemistry—it is used to chloromethylate polystyrene as a precursor to a wide range of functionalized polymers (Figure 13.49). The reagent is slightly different to accommodate the solubility properties of the materials, but the mechanism is similar.

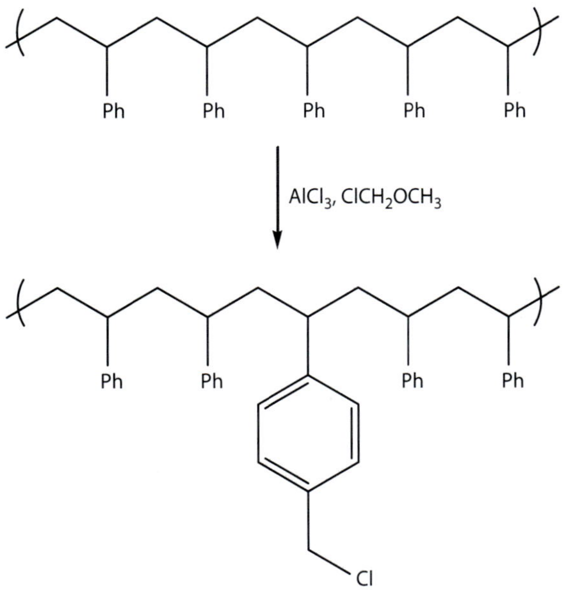

FIGURE 13.49 Chloromethylation of polystyrene.

PROBLEM 13.10

Write a mechanism for the chloromethylation of benzene using the reagents from Figure 13.49 for the chloromethylation of polystyrene.

Solution

Although this is not the most familiar reagent for this process, we do know how organic halides react with $AlCl_3$. The chloride will be abstracted by the strong Lewis acid, to give, in this case, $[CH_3OCH_2]^+$ $[AlCl_4]^-$. This cation then reacts with the aromatic ring.

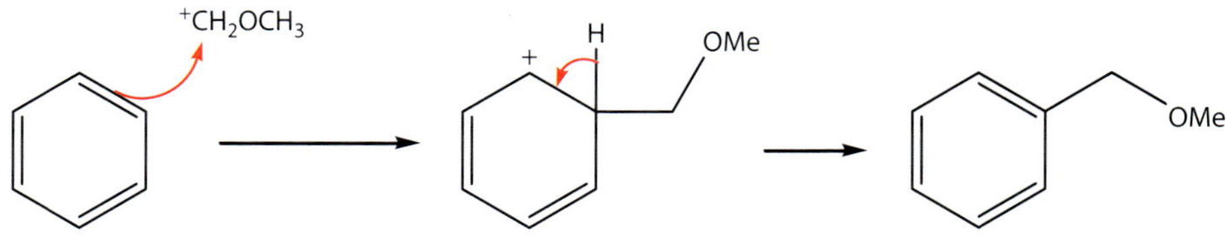

The reaction is completed by the coordination of $AlCl_3$ to the oxygen atom, making it easier to displace in an S_N2 substitution by chloride. The reason for the selectivity for the *para*-position over the *ortho*-site in the reaction of the polymer is clearly steric.

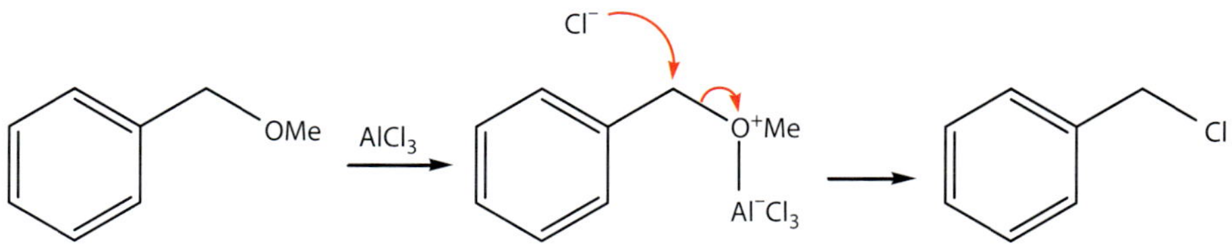

13.9.3 REACTIONS OF GRIGNARD REAGENTS

A proper consideration of these reactions belongs in the next chapter, when we will study carbonyl chemistry, but we have already met the reaction of Grignard reagents with epoxides, and a quick look ahead at this point will give us some useful additional reactions for our syntheses. The main reactions are shown in Figure 13.50; although the mechanisms may be obvious to you, we will defer looking at them in detail until the next chapter. Note that all the reactions require a work-up step with water.

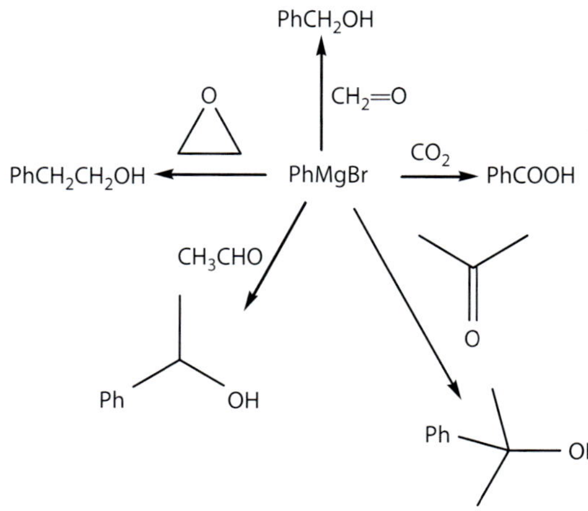

FIGURE 13.50 Key reactions of Grignard reagents.

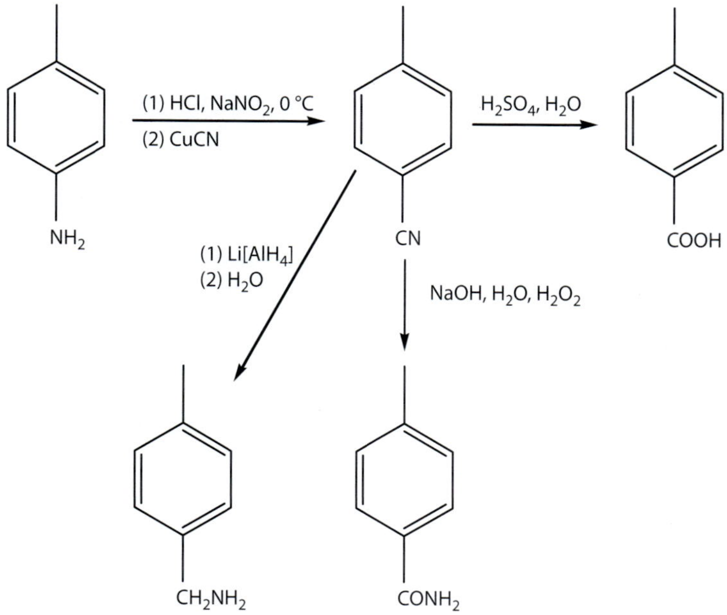

FIGURE 13.51 Useful reactions of aromatic nitriles.

13.9.4 NUCLEOPHILIC SUBSTITUTION BY CYANIDE

A haloarene bearing suitable electron-withdrawing groups may be substituted by cyanide ion. We can also convert diazonium salts to nitriles using CuCN (Figure 13.51). We will return to the mechanistic detail in later chapters, but cyanide groups can be converted to amides, carboxylic acids, or amines (Figure 9.59).

PROBLEM 13.11

Suggest reagents for each of the following transformations, which may require more than one step:

(a)

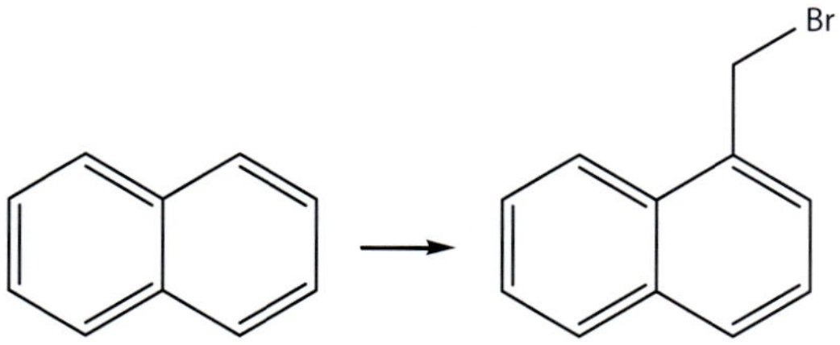

(b)

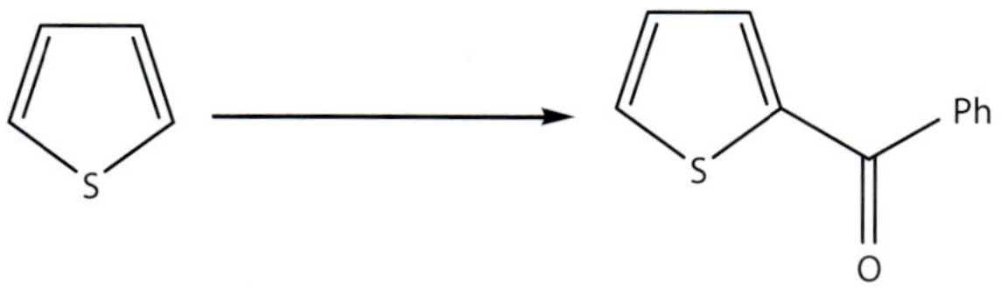

(c)

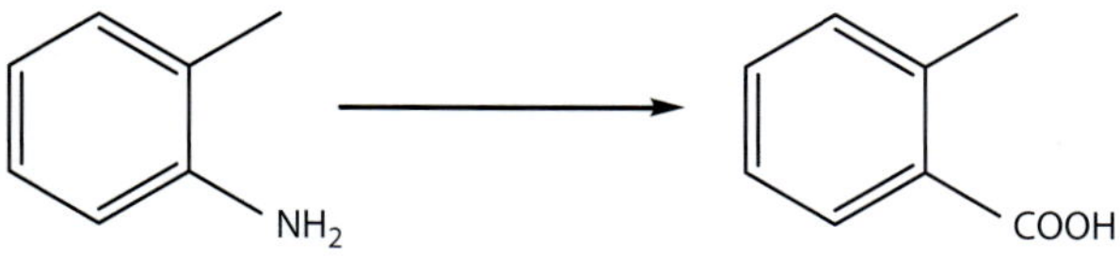

(d)

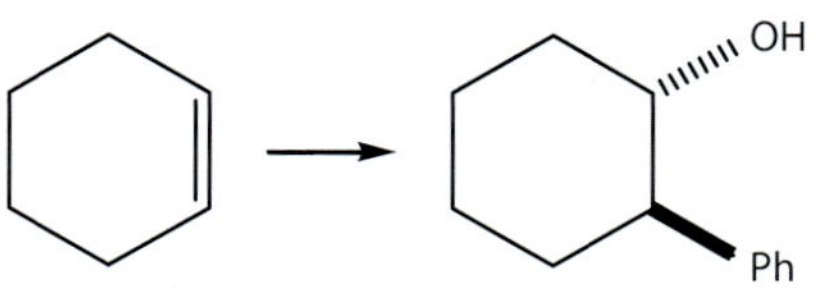

Solutions

(a) Although bromomethylation is a less common reaction than chloromethylation, the methodology is the same, so the reagents will be $CH_2{=}O$ and HBr, with methanol as solvent.
(b) This is a Friedel–Crafts reaction. A range of conditions could be employed; the most successful was PhCOCl, $AlCl_3$, CS_2 solvent.
(c) The key feature of this problem is to recognize that the carboxylic acid was previously a cyanide. Then, the problem is straightforward. In the first step, the amine is converted to the diazonium salt by nitrous acid, and the cyanide is obtained by substitution of the diazonium salt with CuCN. The final step is hydrolysis of the cyanide to the carboxylic acid with H_3O^+.
(d) This problem is a somewhat longer one, but the stereochemistry of the final product gives us a clue as to where to start. *Trans*-stereochemistry of this type is generally achieved when a Grignard reagent (or an organolithium compound) is reacted with an epoxide. So our first step must be to prepare the epoxide, using a peracid.

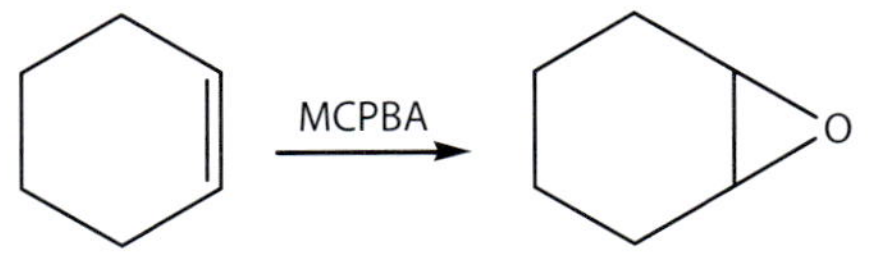

I have used MCPBA, but most peracids would be fine. Bromobenzene is then converted into phenylmagnesium bromide using magnesium in dry ether and opens the epoxide to give the required product:

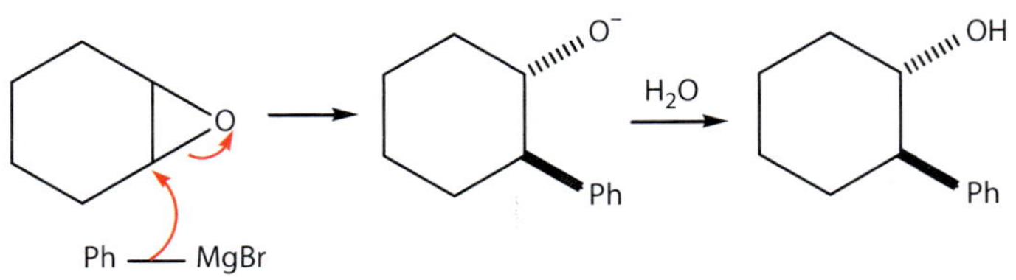

Key Points from Section 13.9

- Friedel–Crafts alkylation and acylation provide useful methods for the formation of carbon to arene bonds. Alkylation reactions work best at tertiary centers. Acyl benzenes may be readily reduced to hydrocarbons (Clemmensen or catalytic hydrogenolysis).
- Haloalkylation, using formaldehyde/HCl, converts arenes to chloromethylarenes.
- Haloarenes may be converted to Grignard reagents with magnesium in dry ether and provide a source of "Ph^-."
- Aryl nitriles can be prepared by the Sandmeyer reaction and can be converted to various carbonyl compounds and amines.

13.10 AROMATIC SYNTHESIS

There are a number of types of problems that you may be asked to tackle in the synthesis of aromatic compounds. The most straightforward of these is of the type "how would you convert A to B?" where the conversion requires only one or two steps. A variant on this is of the type "here is a sequence of reactions, you add the reagents." More challenging is a problem of the type "how would you make A starting from a monosubstituted aromatic compound?" (sometimes specifying a range that may be employed). Finally, you might be asked "how would you make X from readily available starting materials?" This is the most challenging type of problem, not least because the question "what is readily available?" can be puzzling. In a test, you will usually be told what is available to you (this type of question is rarely given in a very open-ended form on a test for this reason—inevitably, it reverts to the slightly less complex type), but it is a useful challenge in practice and homework problems. Often your instructor will give guidance, but a general precept would be that if Aldrich* sell it, it's readily available.

* www.sigmaaldrich.com.

Problem 13.11(d) raised another issue. Often there is more than one correct answer to a synthesis problem. Almost any of the commonly used peracids will epoxidize cyclohexene, and there is no *a priori* way that you can tell which of them will give the best yield. Similarly, the ring opening could have used phenyllithium rather than phenylmagnesium bromide; both would work. A practical synthesis will be designed for high yield, low cost, and as little waste product as possible, but this is frequently a matter of trial and error. I will usually give a solution from the chemical literature, but this does not necessarily mean that your "alternative" is incorrect. Obviously, that does not mean that there are no wrong answers. Sometimes, I will draw attention to straightforward alternatives, but if you feel you have a good but different solution, consult your TA or instructor, who will be able to tell you if it's a reasonable solution, or if it isn't, why it would not work.

Let's try some examples of the simplest type first.

PROBLEM 13.12

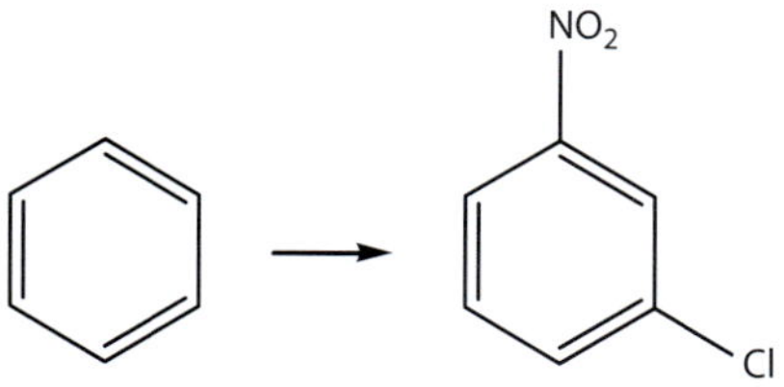

Solution

Chlorine is *ortho,para*-directing; nitro is *meta*-directing. So we must nitrate first. Our reagents are conc. HNO_3/conc. H_2SO_4 for the first step, followed by $Cl_2/FeCl_3$ for the second step. The halogen carrier is essential because the nitro group is very deactivating.

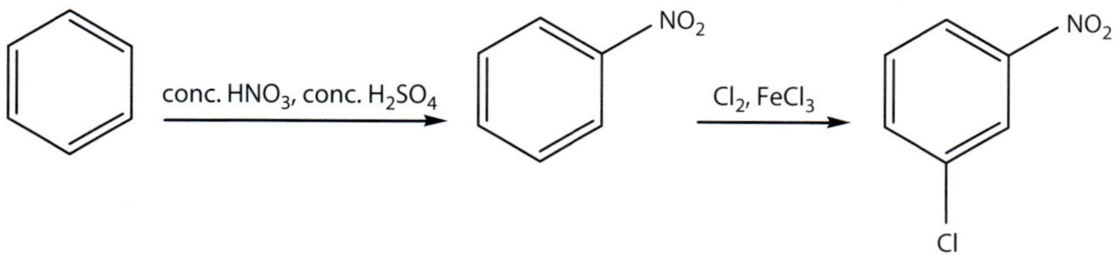

PROBLEM 13.13

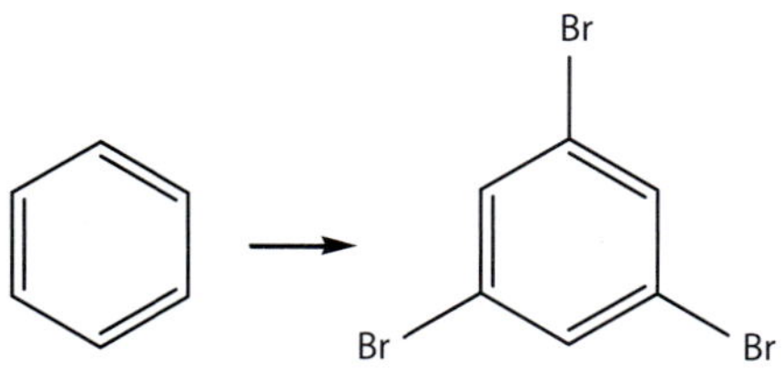

Solution

We note that the bromine atoms are *meta* to one another, but bromine is *ortho,para*-directing. So we will use the directing effect of an amino group (which we will be able to remove at the end of the process) to put the bromines in the right places. So we first nitrate benzene and reduce the nitrobenzene to aniline. Next, we brominate aniline to give tribromoaniline—as long as the bromine is present in excess, this is easy because the amino group is highly activating. Finally, now that we have finished with the amino group, we can remove it by the usual diazotization/reduction sequence.

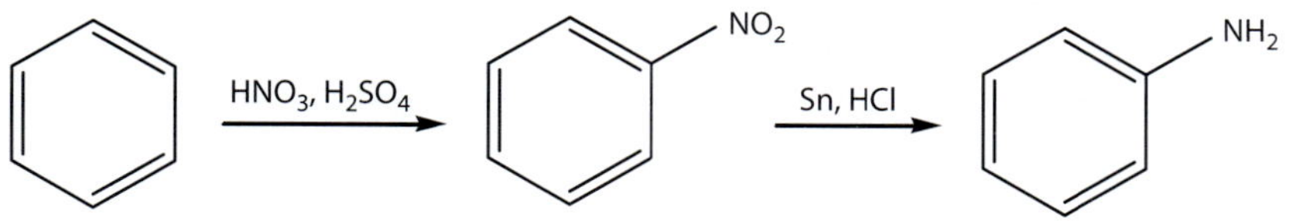

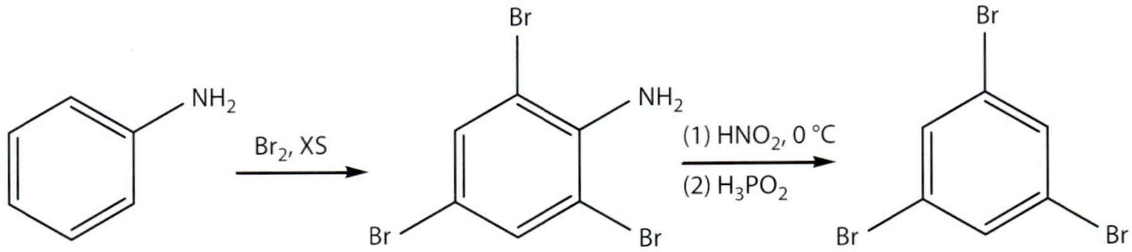

PROBLEM 13.14

How would you accomplish the following conversion?

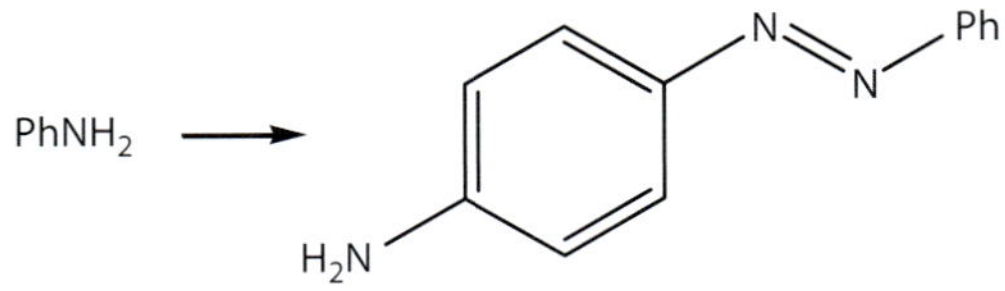

Aniline yellow dye

Solution

We will need first to convert the aniline into the diazonium salt. This then undergoes an azo coupling with unreacted aniline.

$$PhNH_2 \xrightarrow{HNO_2,\ 0\ °C} PhN_2^+$$

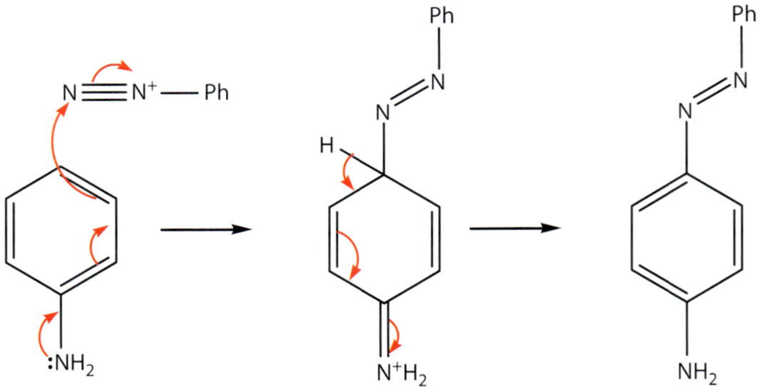

Now for a couple of examples that are more challenging...

PROBLEM 13.15

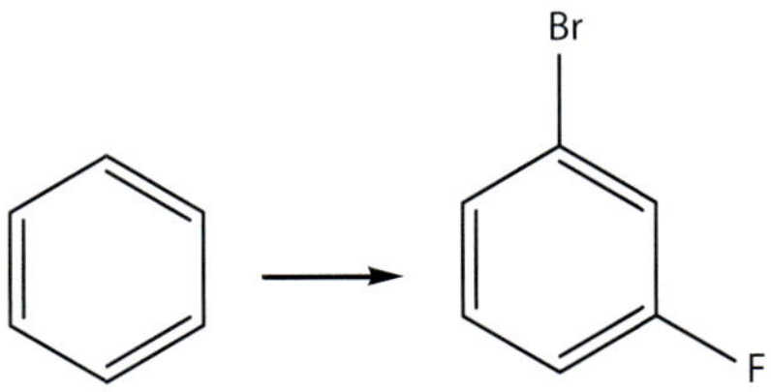

Solution

Both F and Br are *ortho,para*-directing, but in this molecule, they are *meta* to each other. We also know only one way to substitute fluorine onto an aromatic ring, and this is by a diazotization sequence. We remember that the nitro group (which we will convert to a diazo group) is *meta*-directing. So we first nitrate the aromatic ring. Next, we brominate the ring; the bromine is located, correctly, *meta* to the nitro group. The synthesis is completed by the usual sequence to convert the nitro group into the fluoride.

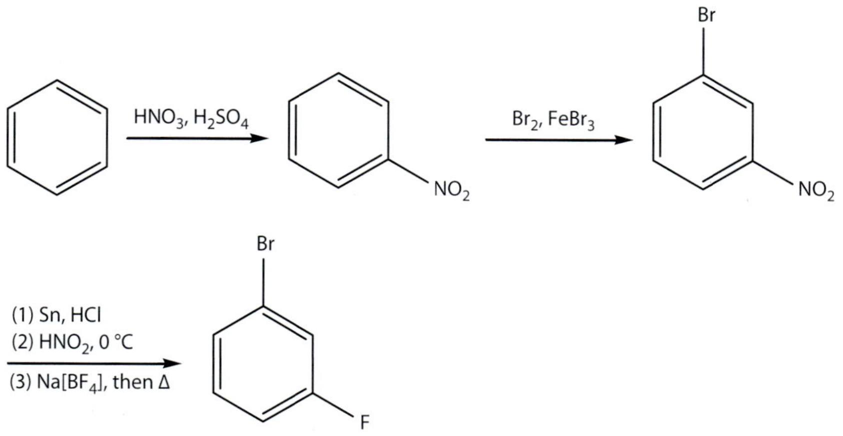

PROBLEM 13.16

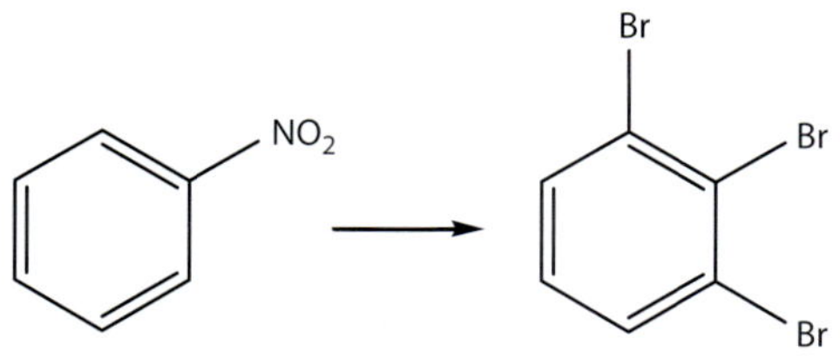

Solution

Putting three substituents next to one another will always be a challenge—in any process where we are relying on a substituent to direct *ortho,para*-substitution, the *para*-position will often be favored, for steric reasons. So we will need to block this position, reversibly, if we are to achieve our goal. We usually achieve this by sulfonation; aniline reacts with sulfuric acid at 180 °C to give sulfanilic acid. So our first step is to reduce nitrobenzene to aniline. Bromination of sulfanilic acid can now only occur at the *ortho*-positions. We can remove the sulfonic acid group in acid and finally convert the amino group to a bromine by diazotization, followed by the Sandmeyer reaction with copper bromide (Figure 13.52).

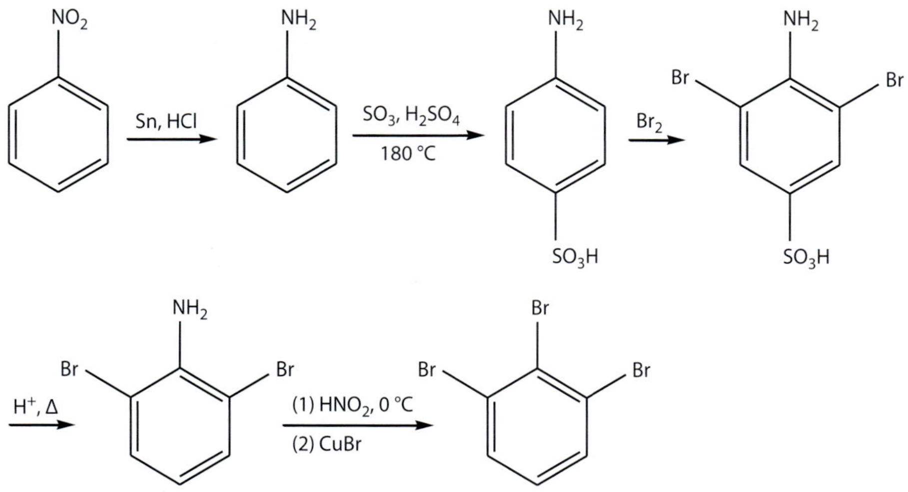

FIGURE 13.52 Solution to Problem 13.16.

Moving on to the second type of problem, let's try this one.

PROBLEM 13.17

Starting from benzene or phenol, devise a synthesis of

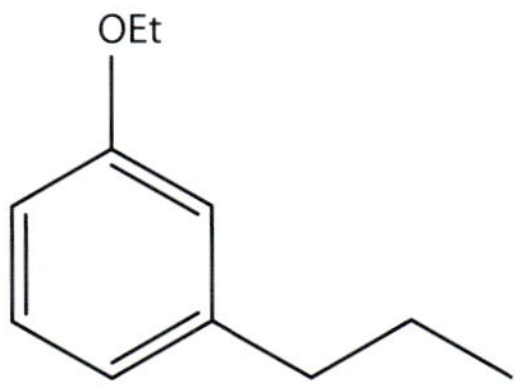

Solution

I have chosen to start from benzene, but there are probably other routes. Remember when doing syntheses, that except for the simplest of processes, it's likely that more than one strategy is viable. In the first steps of the synthesis, we focus on adding the alkyl side chain. We know that we cannot do this by a Friedel–Crafts alkylation reaction, because this would result in a rearrangement, so we use an acylation, as these generally go in good yield. Forming the acyl benzene first gives us another advantage—the directing effect of this group is to the ***meta***-position, which is where we want our other substituent. So our next step is to nitrate. Then, we successively reduce the carbonyl to a hydrocarbon, using the Clemmensen reduction (this might also reduce the nitro group, depending on conditions), and the nitro group to an amine using tin and hydrochloric acid. Finally, the amino group is converted to an ethoxy by diazotization and substitution (Figure 13.53).

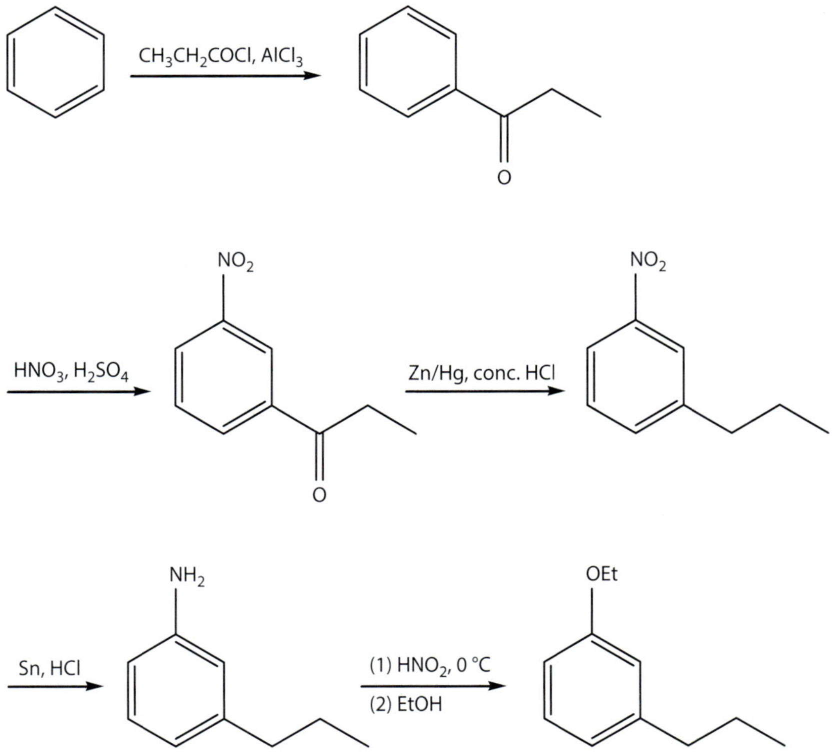

FIGURE 13.53 Solution to Problem 13.17.

PROBLEM 13.18

How would you synthesize the following molecule from a monosubstituted benzene?

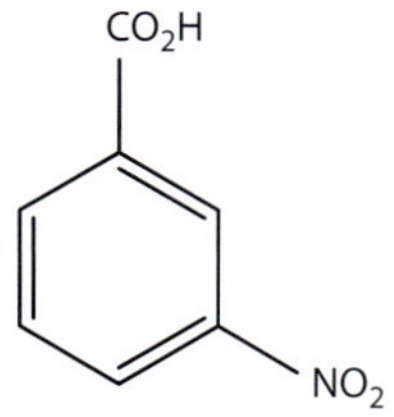

Solution

We note that both COOH and NO_2 are *meta*-directing and deactivating. They are also *meta*- to each other, so at first sight, this is not a challenging problem. In fact, we might think that we have two options—to nitrate benzoic acid or to do a Friedel–Crafts of some sort on nitro benzene. Both benzoic acid and nitrobenzene are readily available. Nitration of benzoic acid will work well.

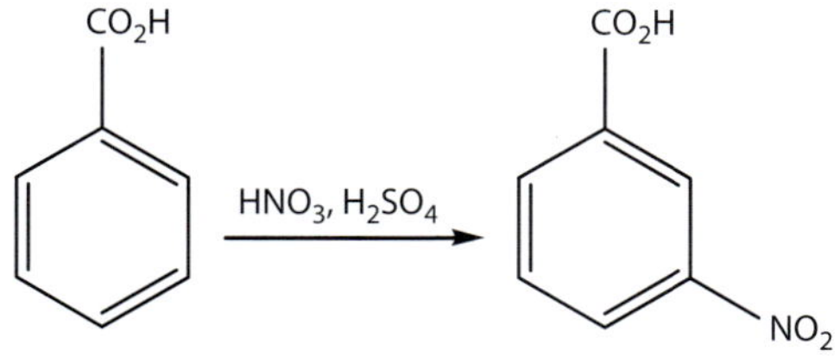

However, our alternate strategy will not be successful—it may look OK on paper, but nitrobenzene is far too deactivated to undergo any type of Friedel–Crafts reaction—indeed, it's often used as a solvent for the reaction of other, more activated, arenes.

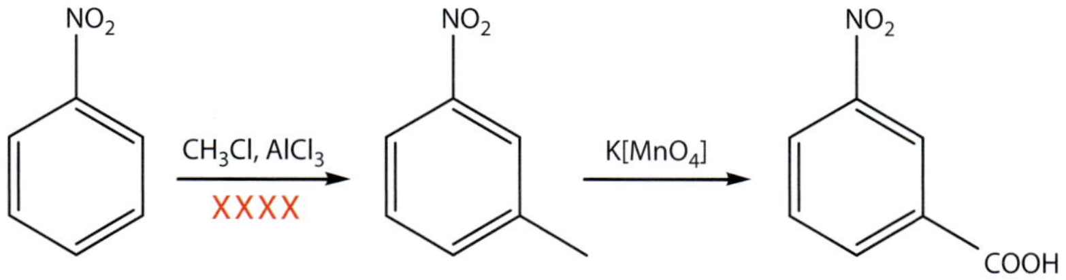

PROBLEM 13.19

How would you synthesize *ortho*-nitrophenol uncontaminated by the *para*-isomer and with as little waste as possible?

Solution

We might first consider simply nitrating phenol to give a mixture of *ortho*- and *para*-isomers and separating them. The reaction is relatively easy, and only dilute acid is needed because of the strong activating effect of the −OH group. These two isomers can be separated by steam distillation (you might like to think about which one would be more volatile and why), but this is not a general phenomenon; for example, it is quite difficult to separate *ortho*- and *para*-nitrotoluene. But it's quite wasteful, as a significant amount of the material will end up as the unwanted *para*-isomer.

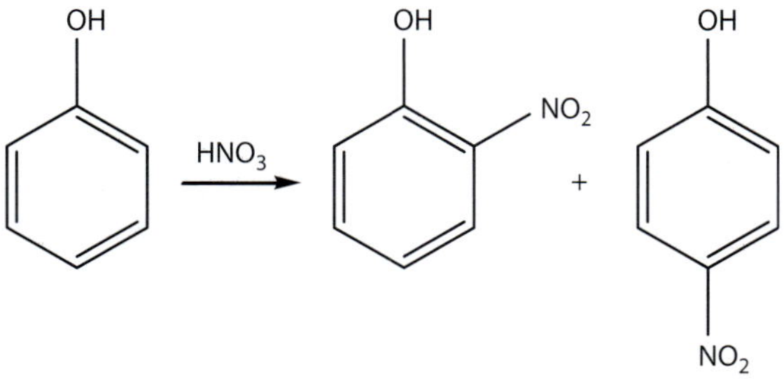

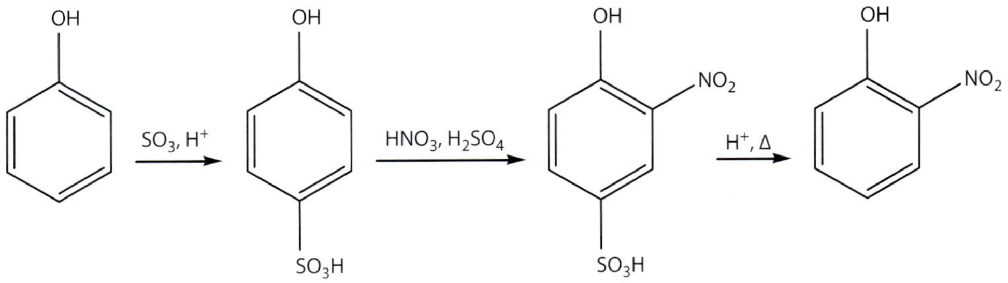

FIGURE 13.54 Synthesis of *ortho*-nitrophenol.

As an alternative, we can "protect" the *para*-position by sulfonating it. We are using the reversibility of the sulfonation reaction here—we will be able to remove it at the end of the reaction. Because of the size of the sulfonate group, it reacts exclusively at the *para*-position. We can then nitrate exclusively at the *ortho*-site and finally remove the sulfonic acid protecting group. Although there are more steps in this process, we do not end up throwing away 40 % of our material (Figure 13.54).

PROBLEM 13.20

How would you synthesis the following molecule, starting from a monosubstituted benzene?

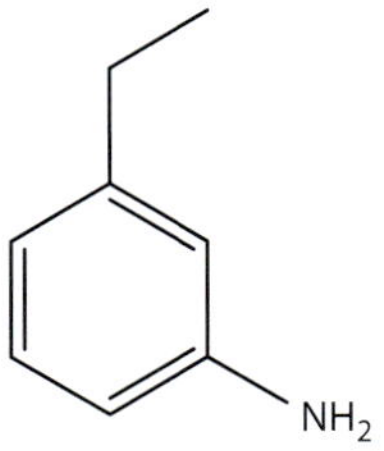

Solution

At first sight, this looks demanding; both of the substituents are *ortho,para*-directing—but they are *meta* to each other. But both of the substituents are usually prepared by reductions—the amino group is usually prepared by the reduction of a nitro group, and the ethyl substituent could be prepared by the reduction of an acyl benzene. So in a new form, our problem would be to make **13.8**:

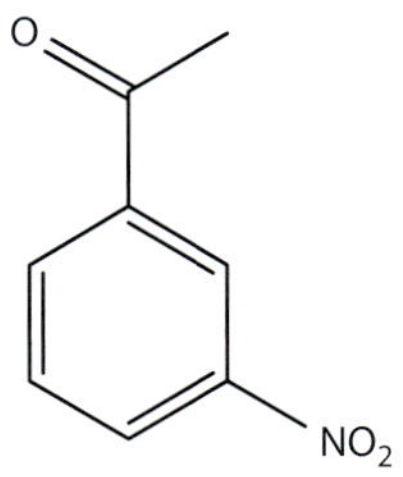

13.8

Acetophenone is readily available (and if it were not, it's simple to make by a Friedel–Crafts acylation reaction), and it is readily nitrated to *meta*-nitroacetophenone. For the reduction of the ketone to a hydrocarbon, I've chosen a new reaction, the Wolff–Kishner reduction, which we

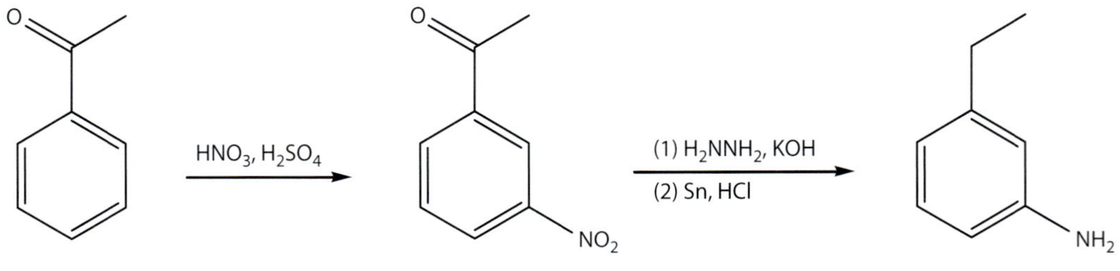

FIGURE 13.55 Synthesis of 3-ethylaniline.

will study in more detail in Chapter 14. The others that we already know, the Clemmensen and palladium-catalyzed hydrogenolysis, would also be effective (Figure 13.55).

We might think about starting from nitrobenzene—but as we noted before, the deactivating effect of the nitro group means that Friedel–Crafts reactions will not work.

PROBLEM 13.21

How would you prepare 2,4-dinitrophenylhydrazine, **13.9**, from benzene?

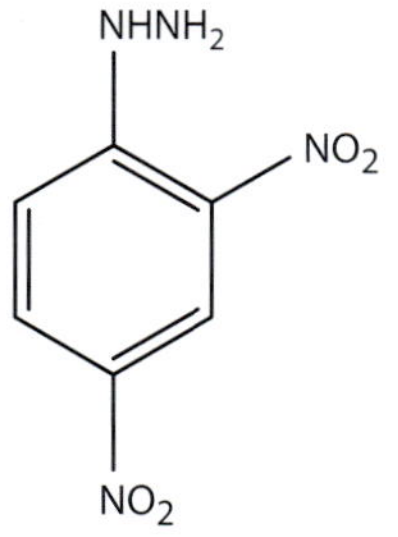

13.9

Solution

Although we have not seen a substituted hydrazine often before, we can make a reasonable inference that hydrazine, H_2NNH_2, is a good nucleophile. So if we approach this problem from the product, we could prepare it by reacting hydrazine with 2,4-dinitrochlorobenzene; we know that the electron-accepting effect of the two nitro groups makes this chloride relatively easy to displace:

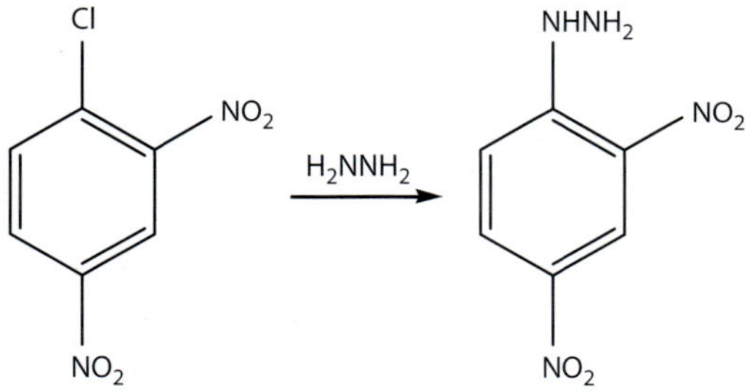

How to prepare 2,4-dinitrochlorobenzene? Chlorine is *o,p*-directing, and deactivating, so we should nitrate chlorobenzene. Chlorobenzene is readily prepared from benzene itself, so the synthesis is now complete (Figure 13.56).

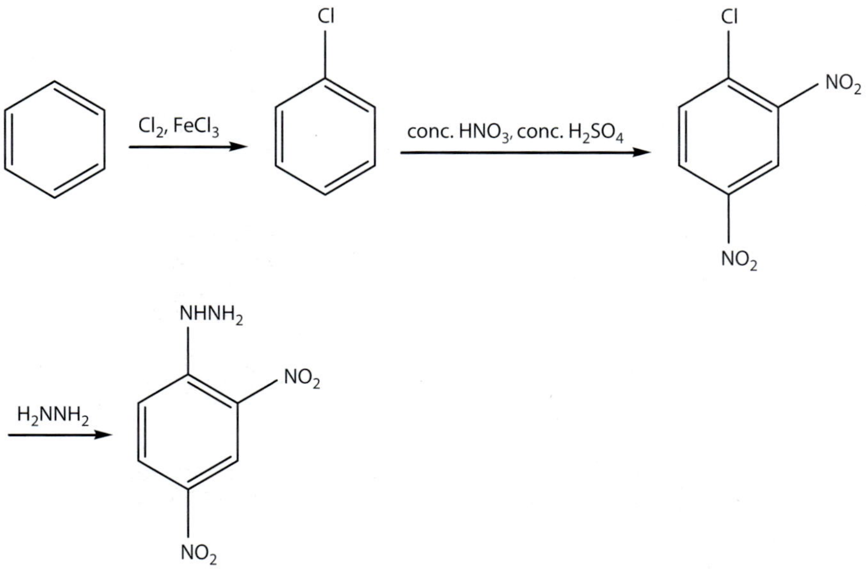

FIGURE 13.56 Synthesis of 2,4-dinitrophenylhydrazine.

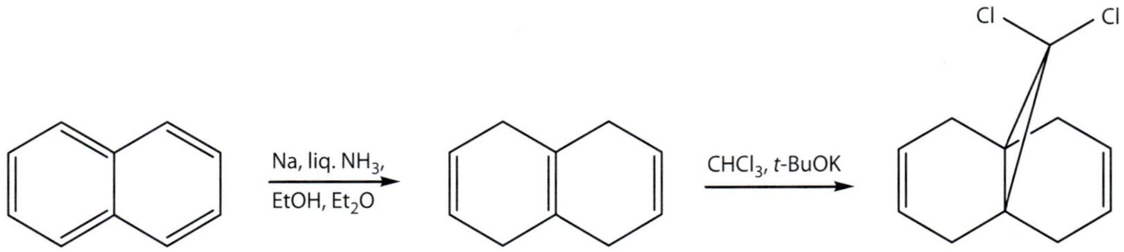

FIGURE 13.57 Solution to Problem 13.22.

PROBLEM 13.22

How would you synthesize **13.10** from naphthalene?

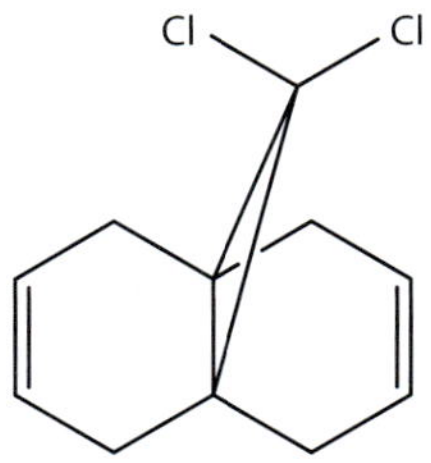

13.10

Solution

Looking at this problem, two things should occur to us. The first is that we can reduce naphthalene in a Birch reduction to something that has fewer double bonds, situated much as these are (look back at Figure 13.41). The second is that we only know one way to make a cyclopropane, and that involves a carbene. The carbene we need here is CCl_2, which is generated by the action of strong base on chloroform. The carbene will react with the appropriate double bond, because the carbene is electrophilic and the most substituted double bond is the most electron rich (Figure 13.57).

REVIEW PROBLEMS

1. Give a systematic name for each of the following compounds:

(a)

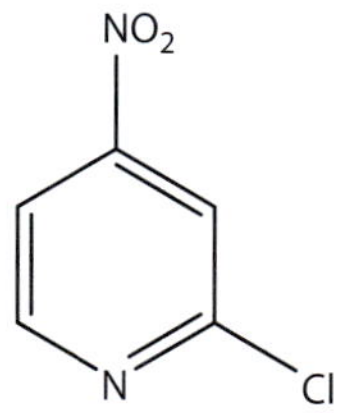

(b)

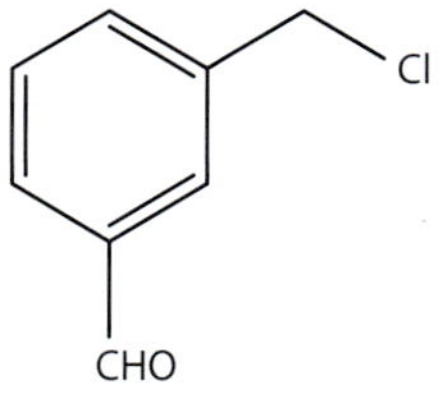

(c)

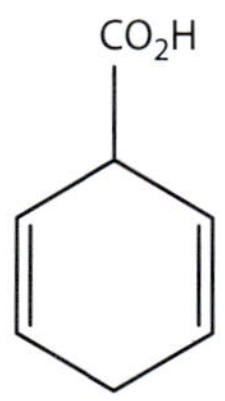

(d)

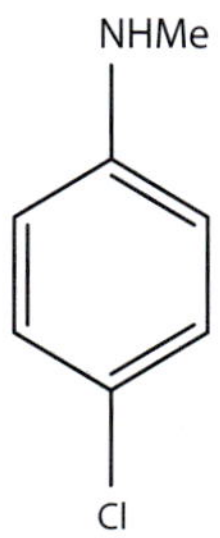

2. Give a structure for each of the following compounds:
   (a) 2-Bromomethylnaphthalene
   (b) 3,4,5-Triiodonitrobenzene
   (c) 1,4-Cyclohexadiene
   (d) 3-Methylthiophene
   (e) 1,2-Dimethoxy-4-chloromethylbenzene
3. Identify the compound for which spectroscopic data are given in the following:
   $C_{10}H_{12}O_3$
   Mass spectrum 180(M^+), 133, 122
   IR spectrum ν_{max} = 1738 cm^{-1}
   ^1H NMR spectrum
   δ 2.06 (3H, s), 3.79 (3H, s), 5.03 (2H, s), 6.88 (d, 2H, J = 8.5 Hz), 7.28 (d, 2H, J = 8.5 Hz)
4. Propose a mechanism for each of the following processes:

(a)

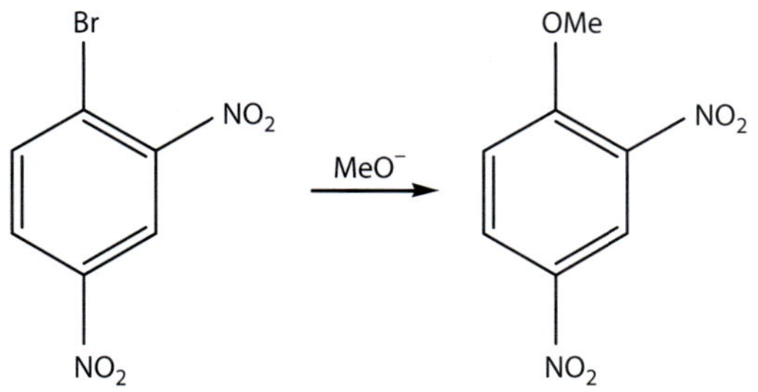

(b)

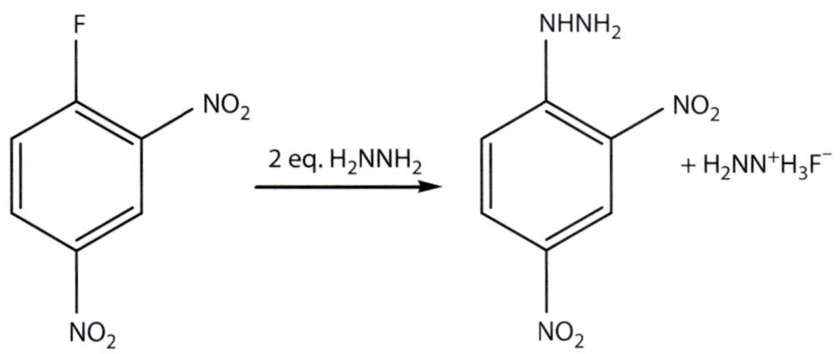

(c)

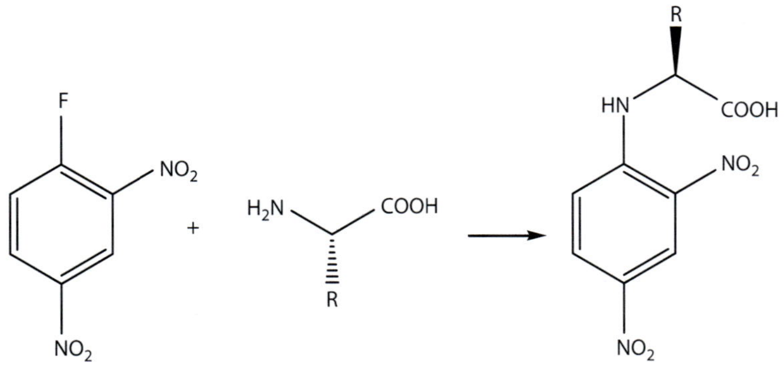

(d)

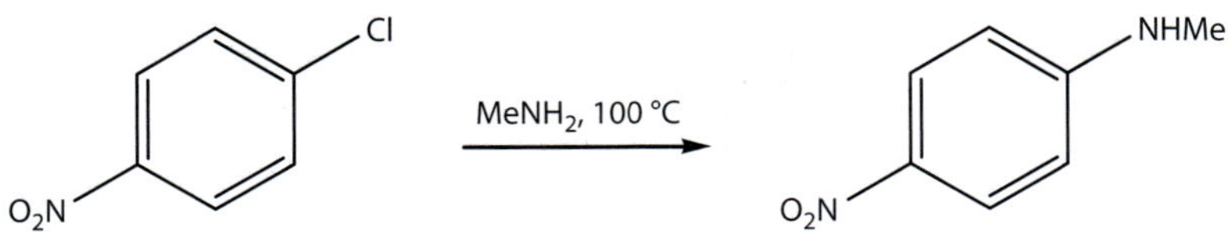

(e)

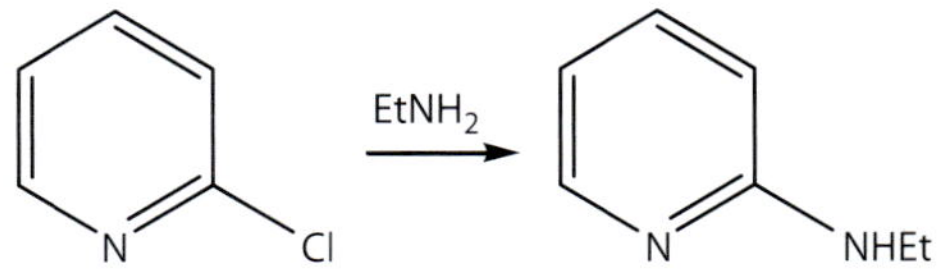
N
Cl
EtNH2
N
NHEt

(f)

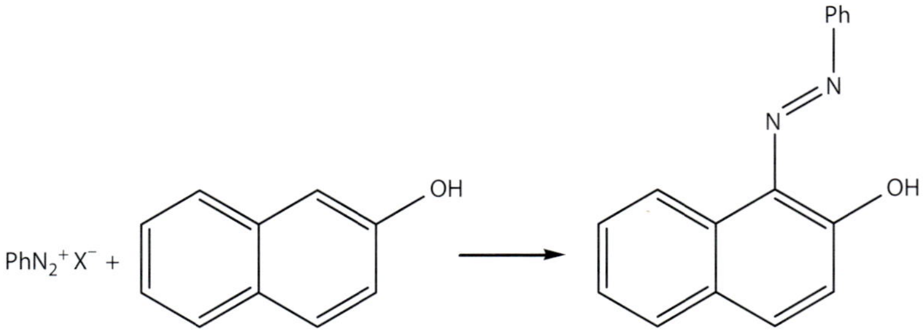
PhN2+ X− +
OH
Ph
N
N
OH

(g)

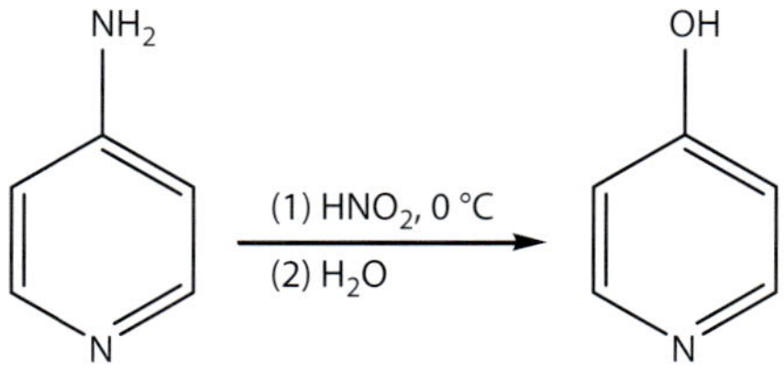
NH2
N
(1) HNO2, 0 °C
(2) H2O
OH
N

(h)

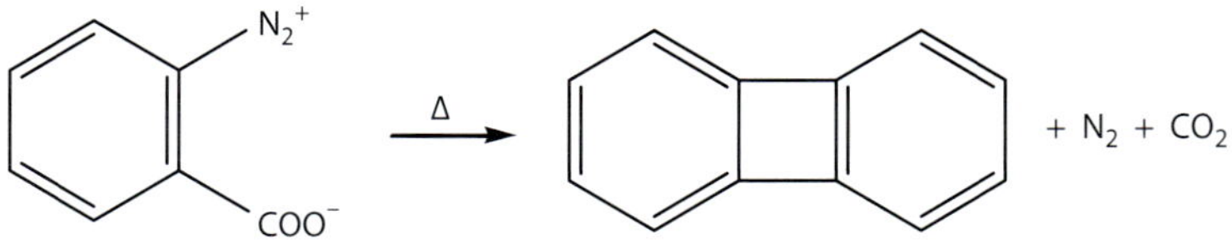
N2+
COO−
Δ
+ N2 + CO2

(i)

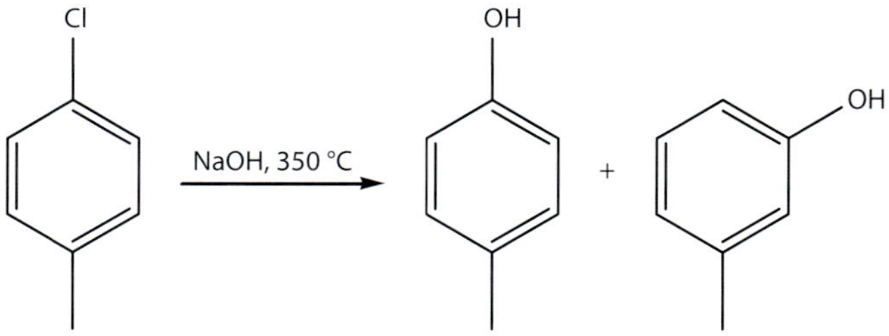
Cl
NaOH, 350 °C
OH
+
OH

(j)

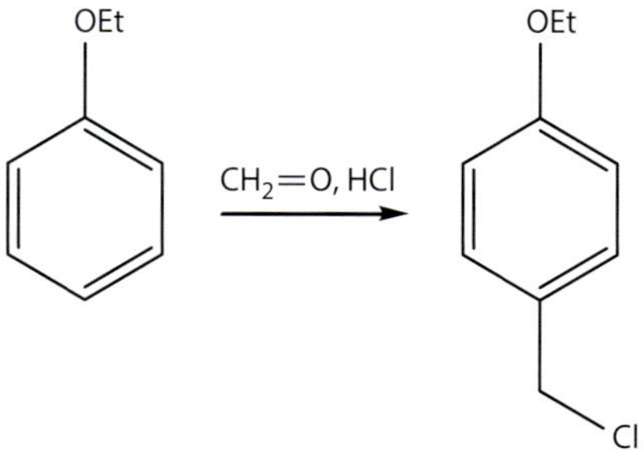
OEt
CH2=O, HCl
OEt
Cl

(k)

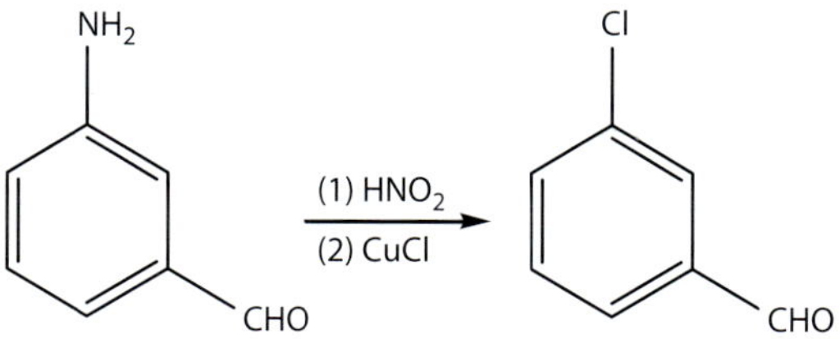
NH2
CHO
(1) HNO2
(2) CuCl
Cl
CHO

(l)

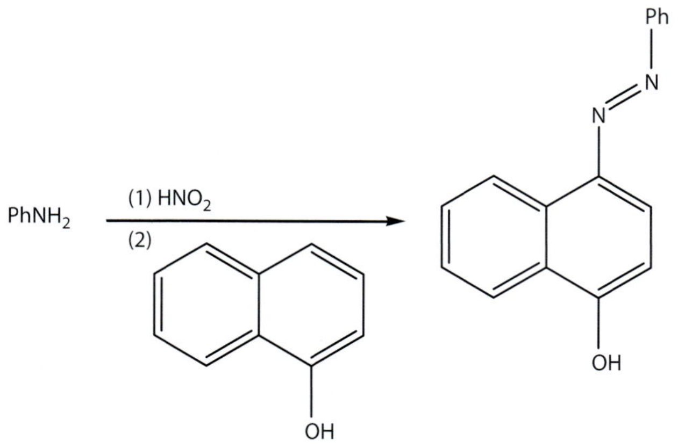

(m)

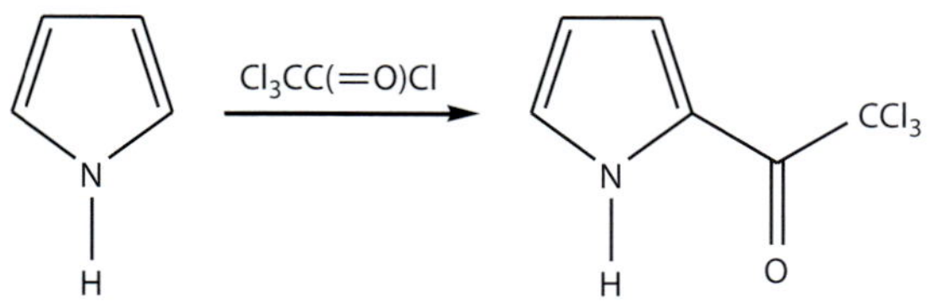

(n)

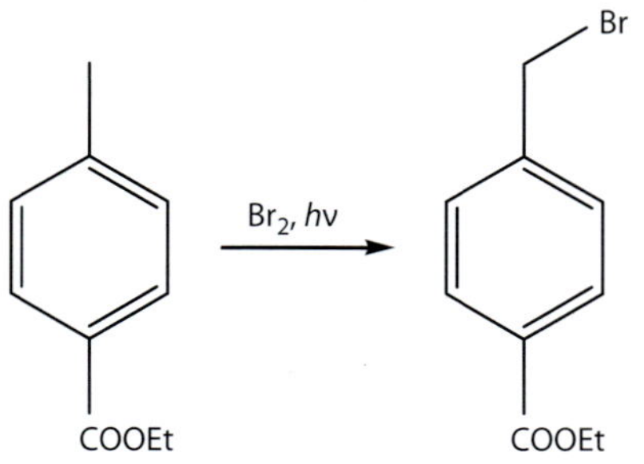

(o)

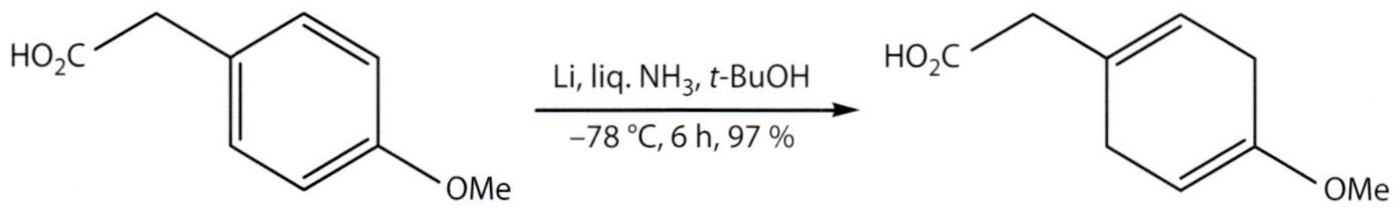

(p)

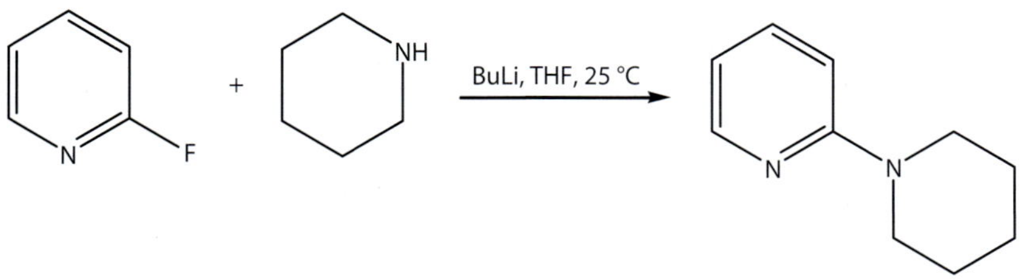

5. Predict the product(s) of each of the following reactions:

(a)

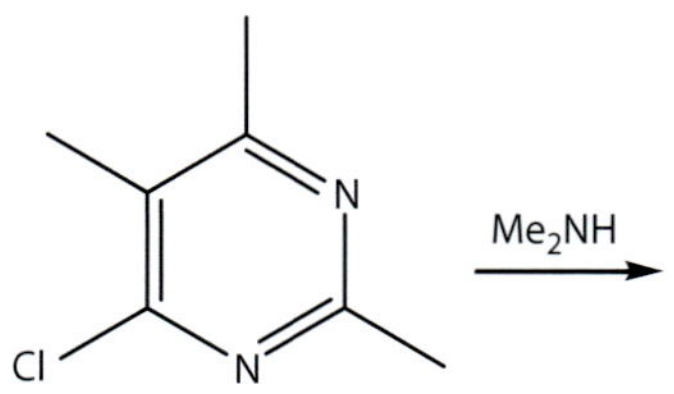

(b)

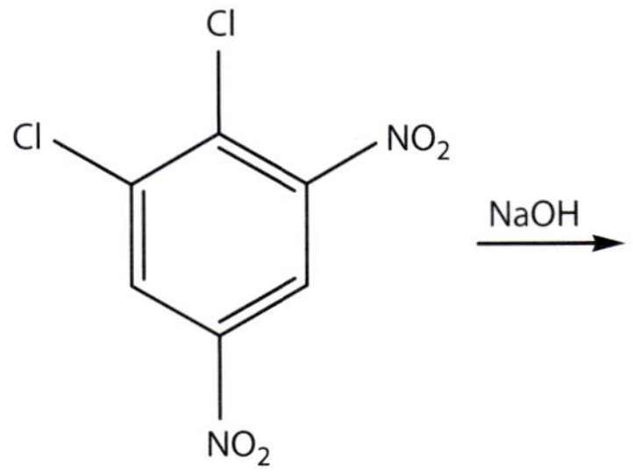
Cl
Cl
NO2
NaOH
NO2

(c)

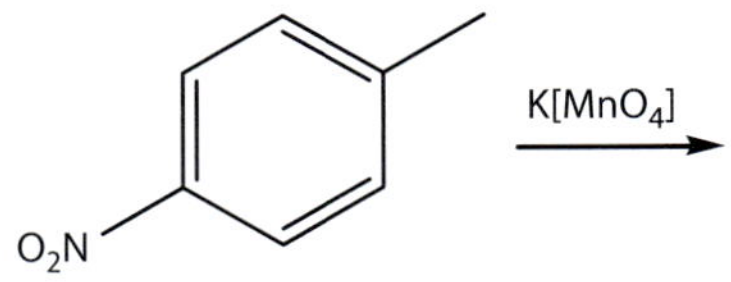
K[MnO4]
O2N

(d)

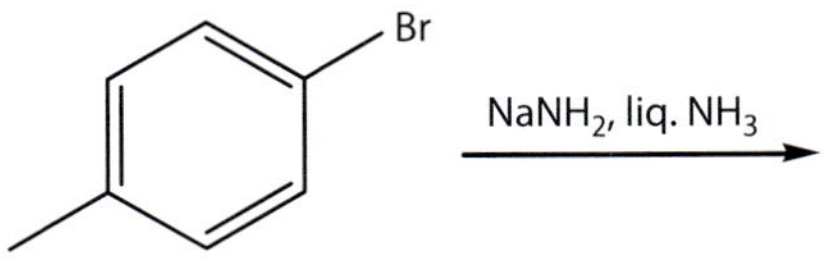
Br
NaNH2, liq. NH3

(e)

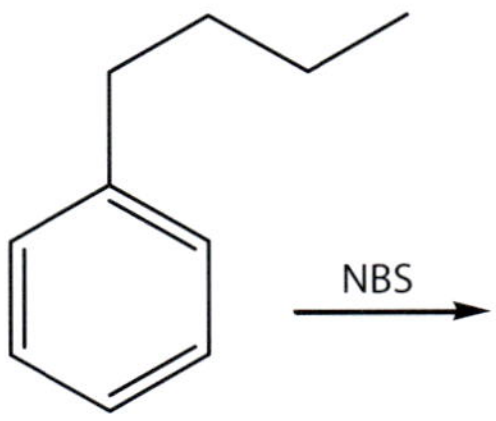
NBS

(f)

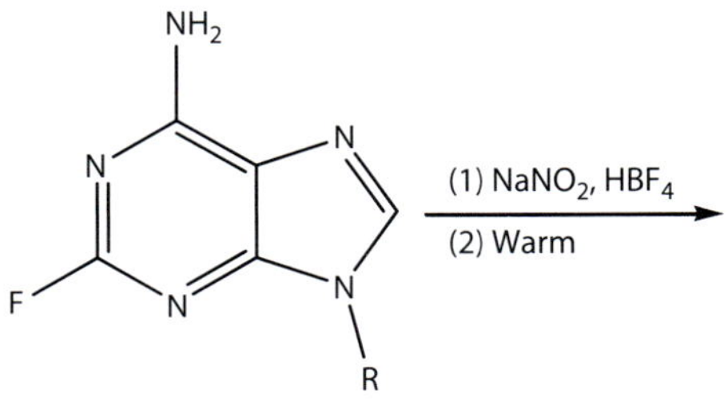
NH2
N
N
F
N
N
R
(1) NaNO2, HBF4
(2) Warm

(g)

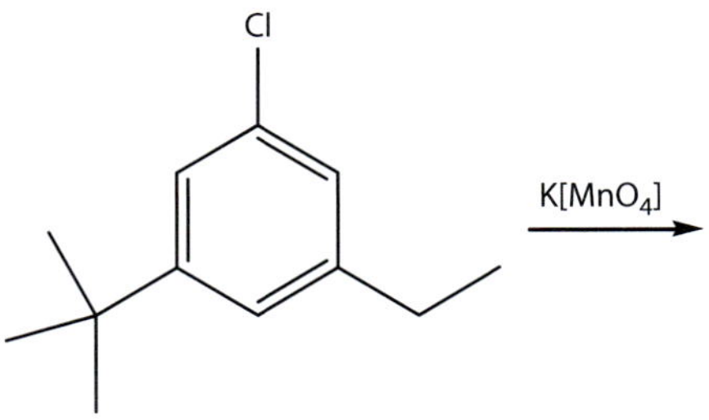
Cl
K[MnO4]

(h)

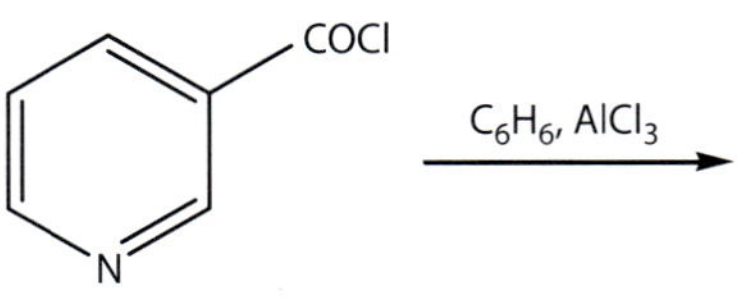
COCl
C6H6, AlCl3
N

(i)

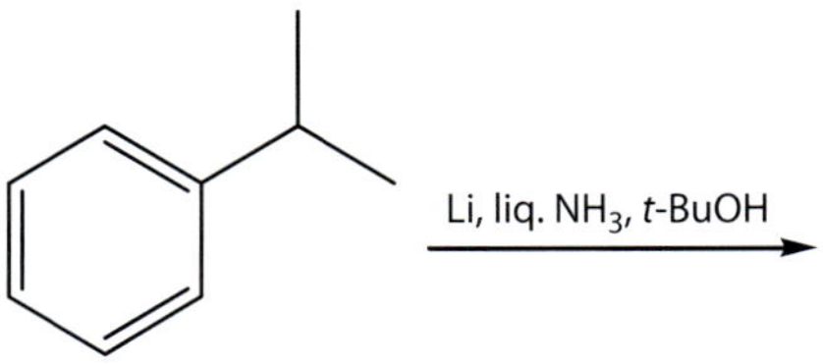
Li, liq. NH3, t-BuOH

(j)

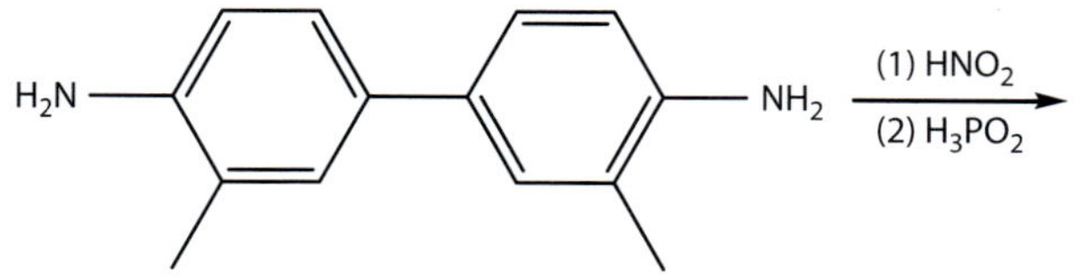

(k)

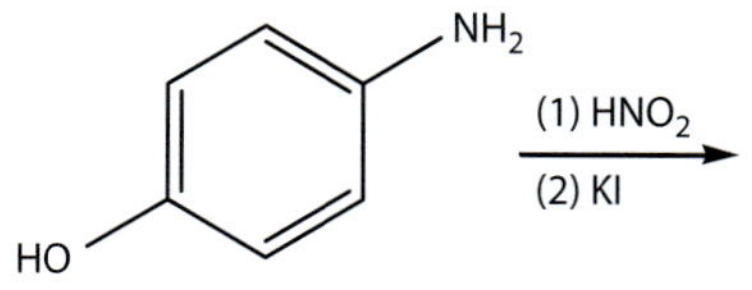

(l)

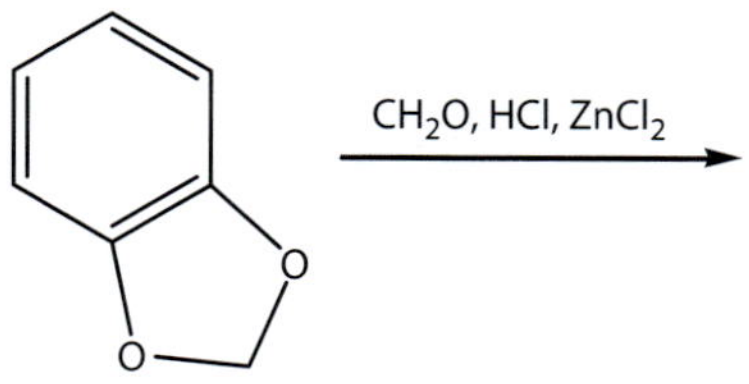

6. How would you carry out each of the following transformations, most of which require more than one step?

(a)

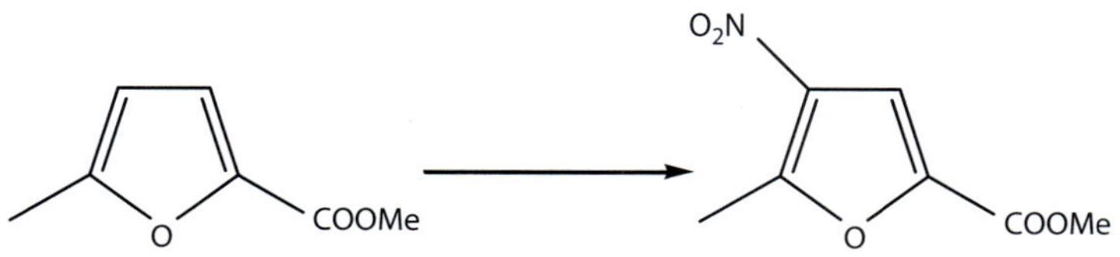

(b)

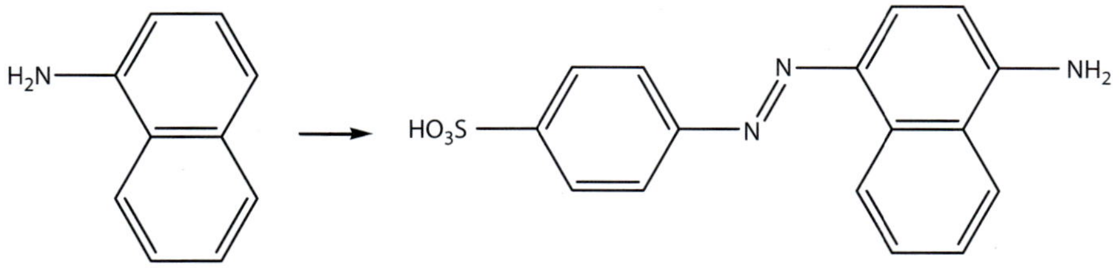

1-Naphthylamine red

(c)

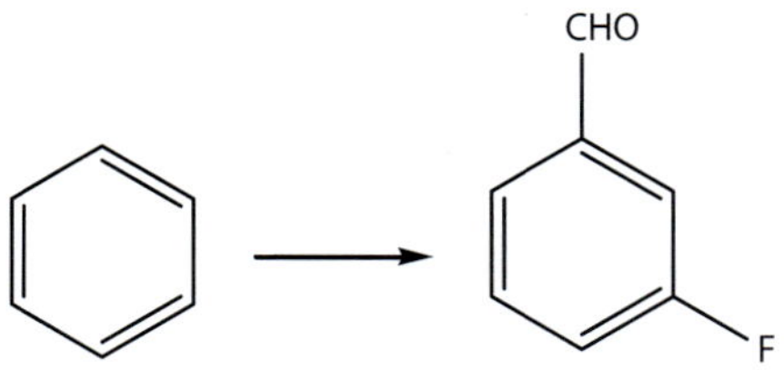

(d)

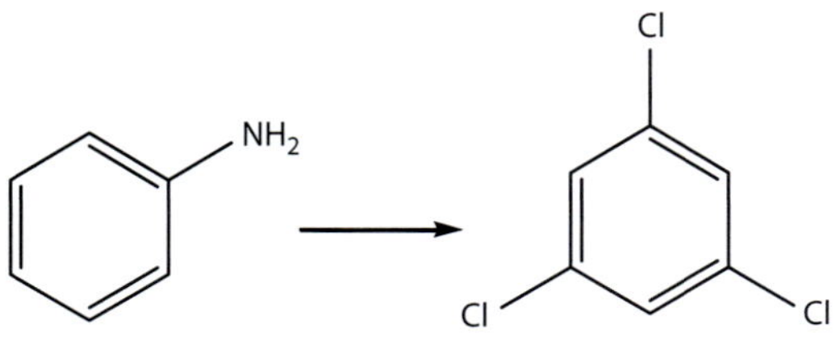

(e)

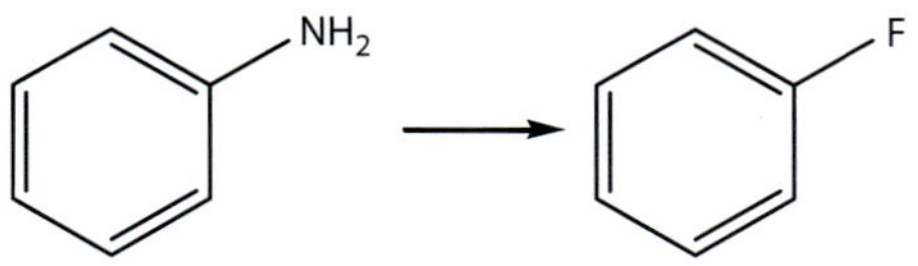

(f)

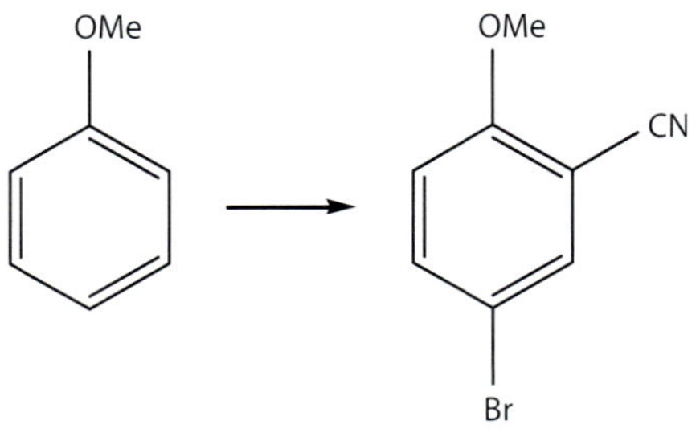

(g)

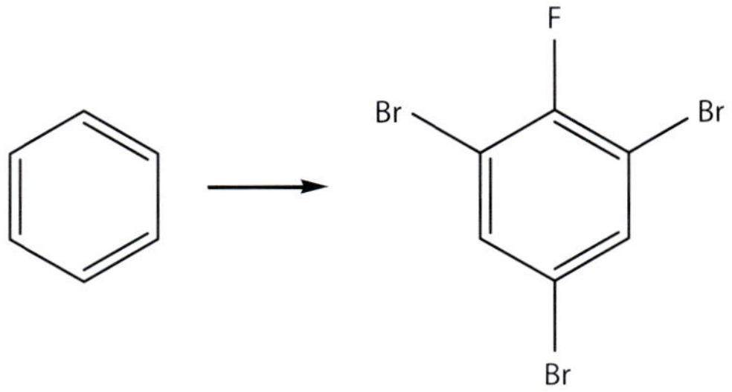

(h)

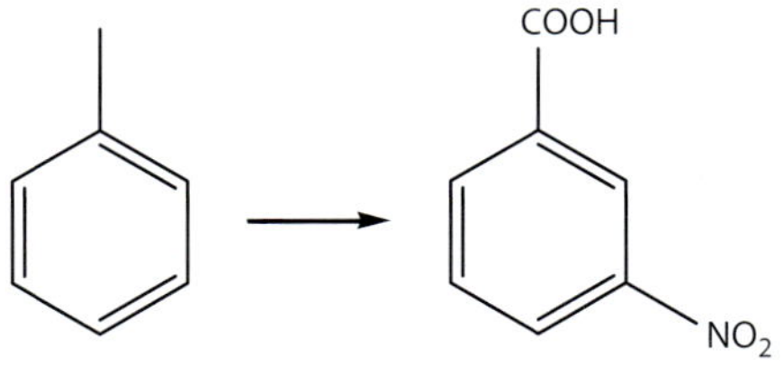

(i)

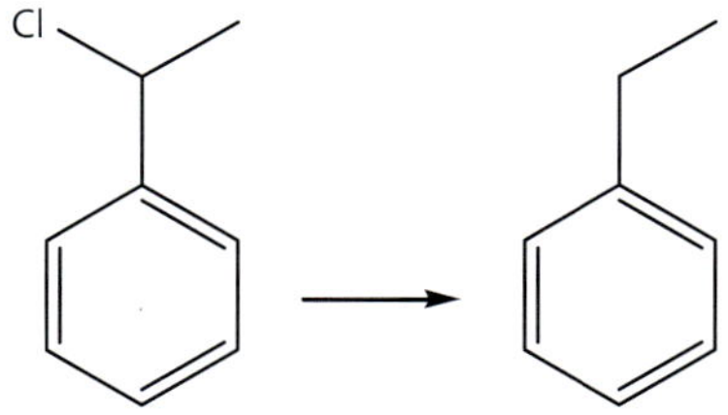

(j)

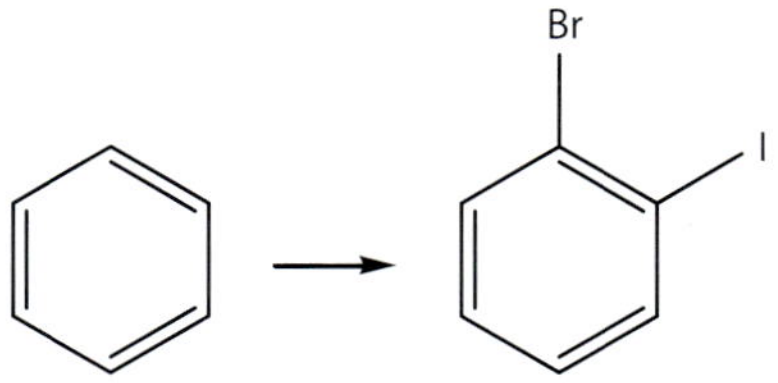

(k)

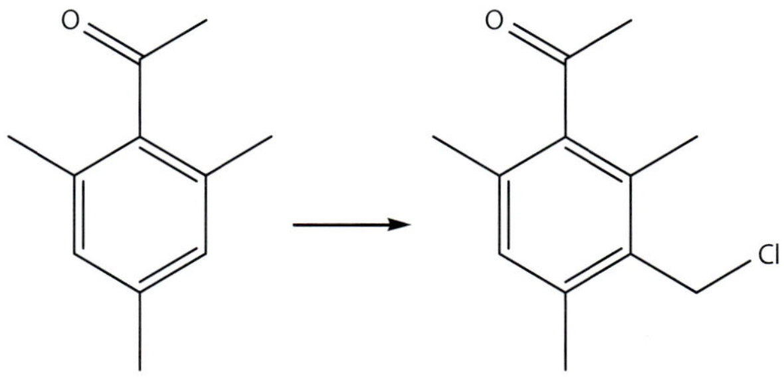

(l)

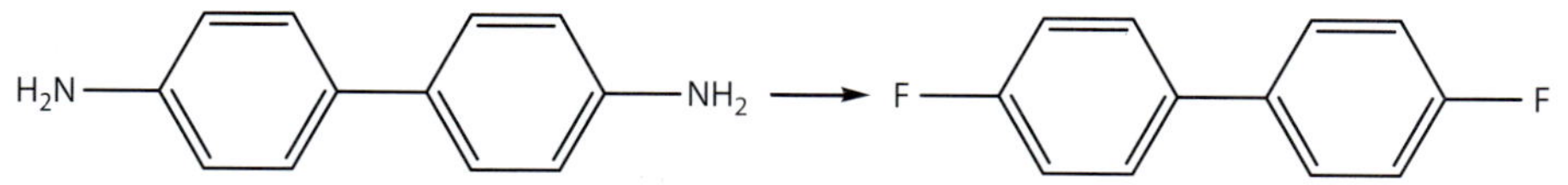

(m)

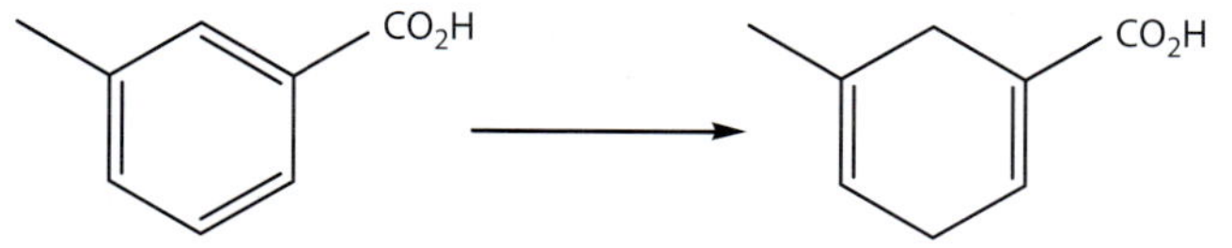

7. Explain the relative rates of substitution of chlorine by methoxide in the following compounds given:

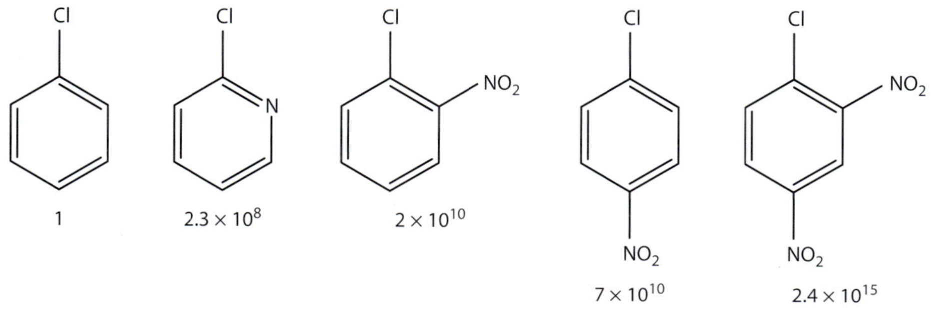

8. Explain the following observations, providing a full mechanism where appropriate:

(a)

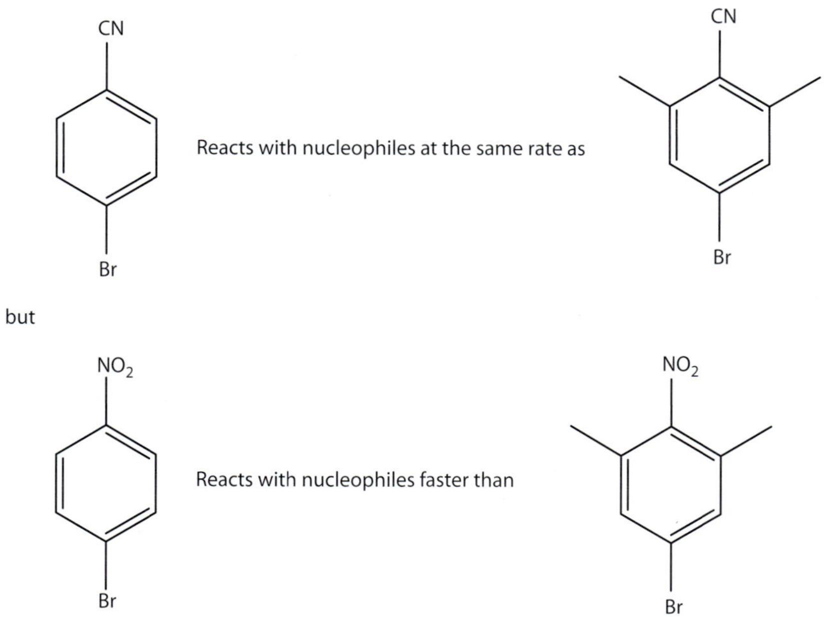

(b) Bromobenzene reacts 20 times faster than chlorobenzene with KNH_2/liq. NH_3.

(c)

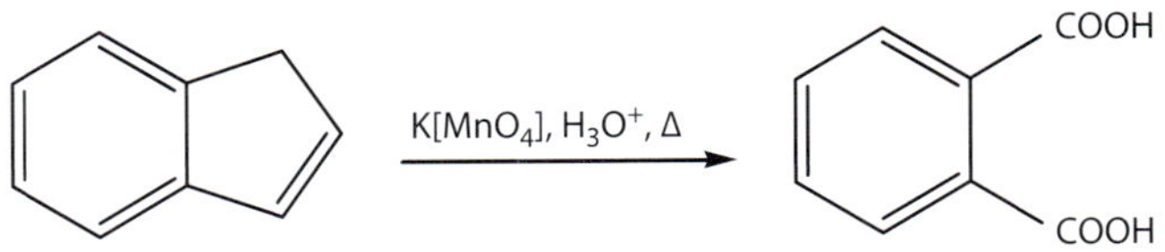

(d) The red form of methyl orange absorbs at 520 nm, whereas the yellow form absorbs at 440 nm.

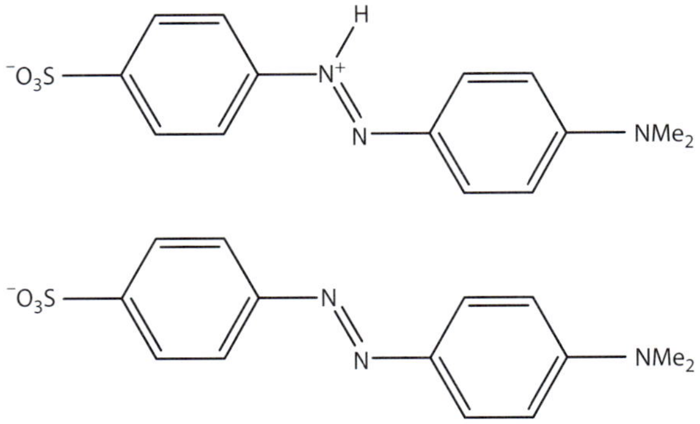

Red form of methyl orange
λ_{max} 520 nm

Yellow form of methyl orange
λ_{max} 440 nm

9. Identify the reagents needed for the synthetic sequences shown in the following:

(a)

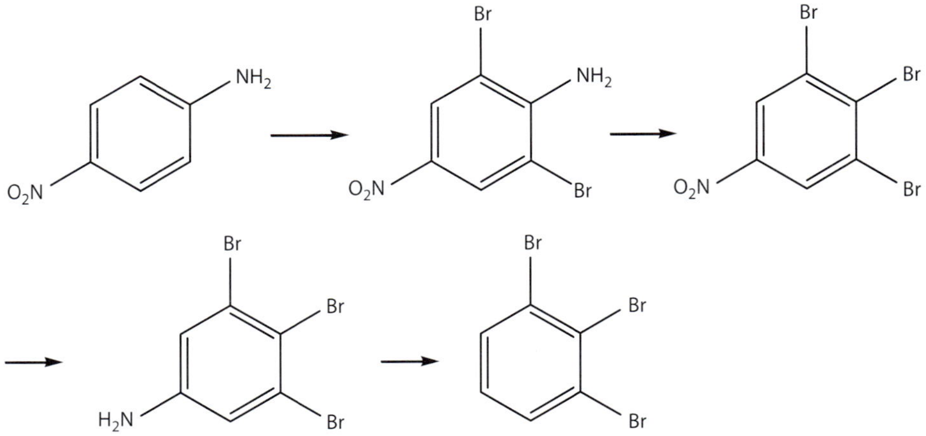

(b)

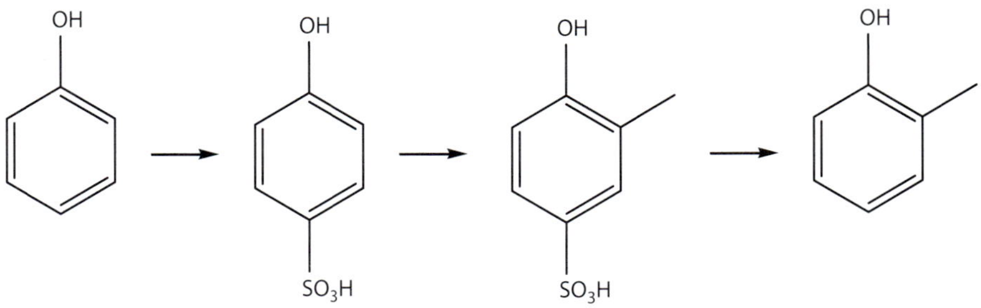

MORE CHALLENGING PROBLEMS

1. Propose a mechanism for each of the following processes:

(a)

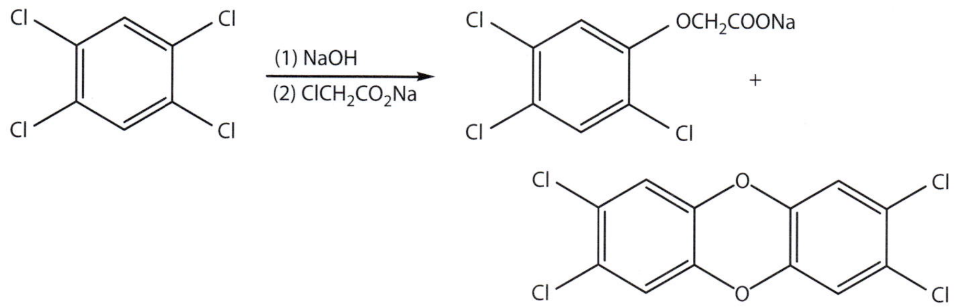

(b)

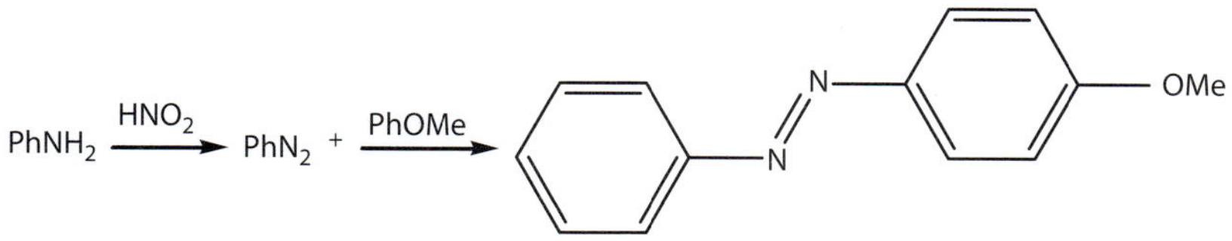

(c)

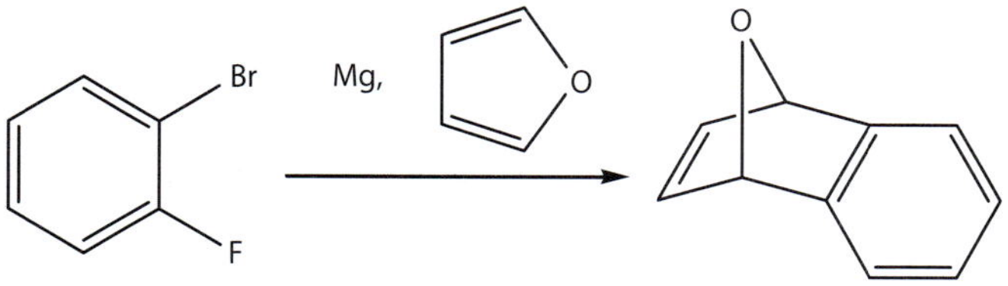

(d)

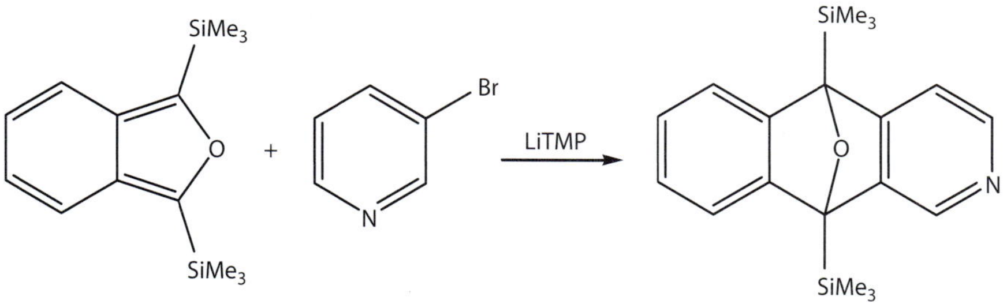

Lithium tetramethylpiperidide, LiTMP, is a strong, nonnucleophilic base.

(e)

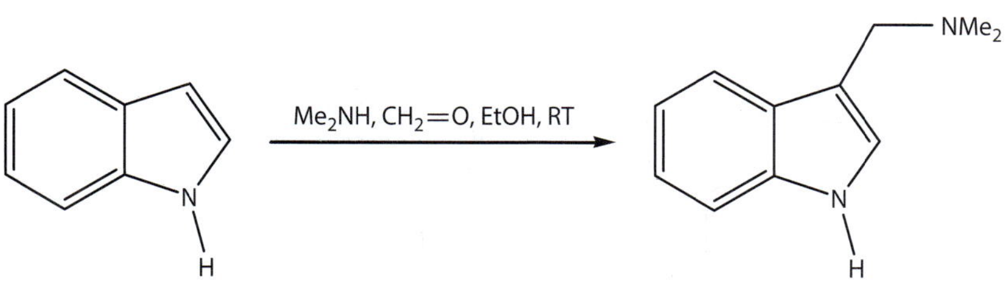

(f)

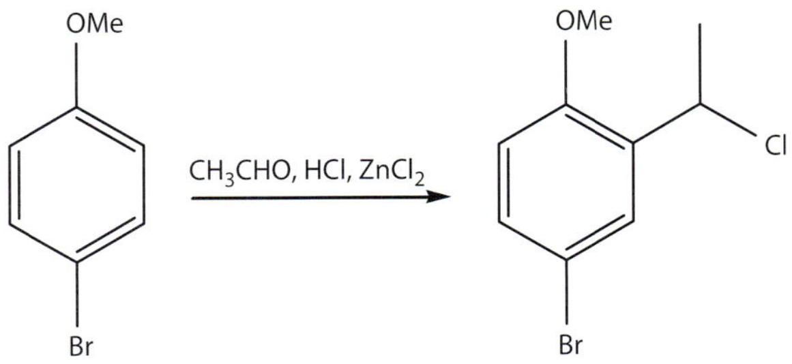

(g)

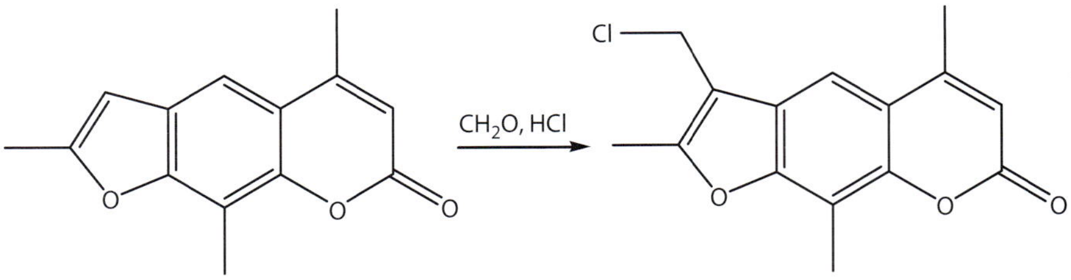

2. Predict the product(s) of each of the following reactions:

(a)

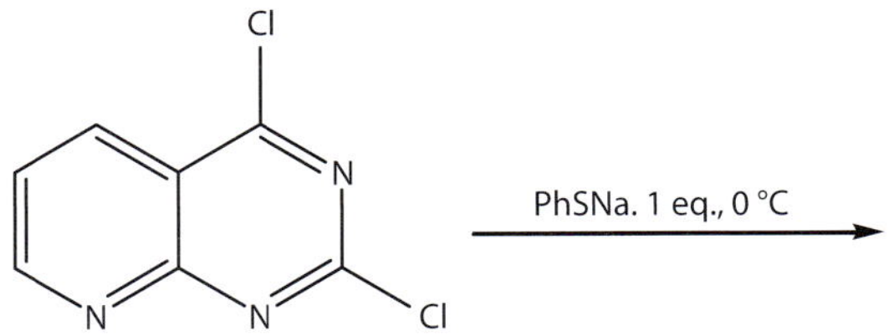

(b)

(c)

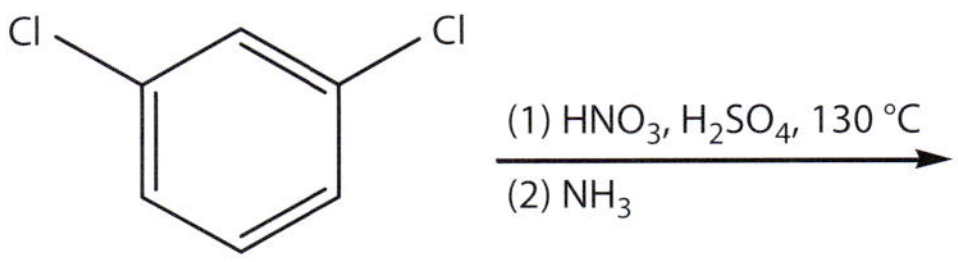

3. How would you carry out each of the following transformations, which will require more than one step?

(a)

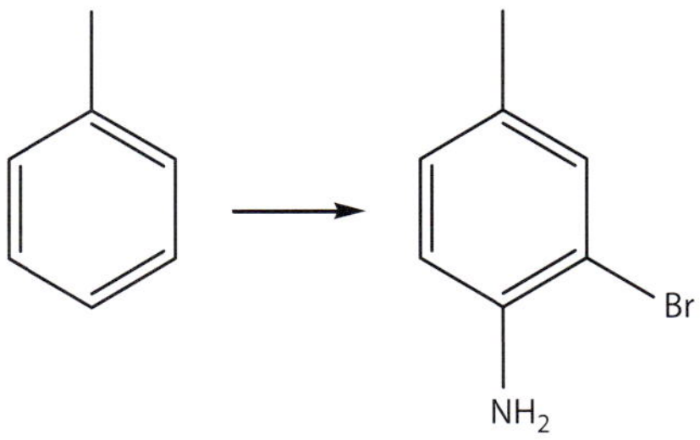

(b)

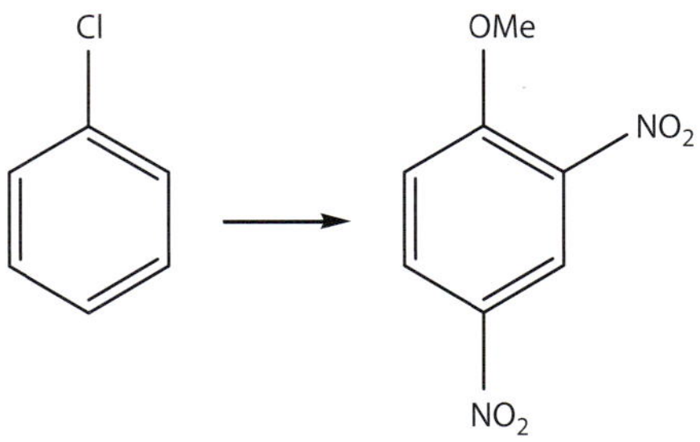

(c)

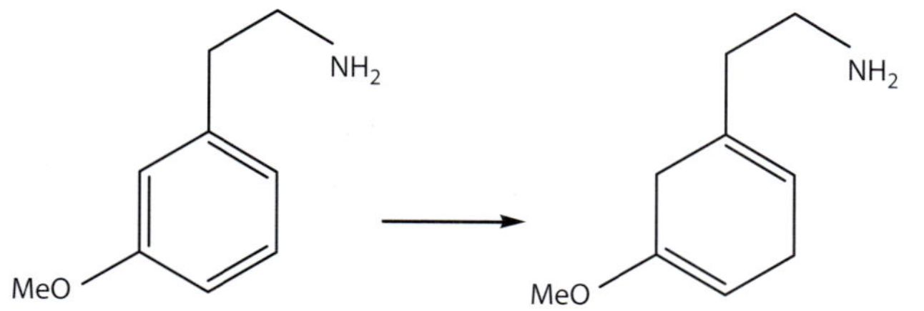

(d)

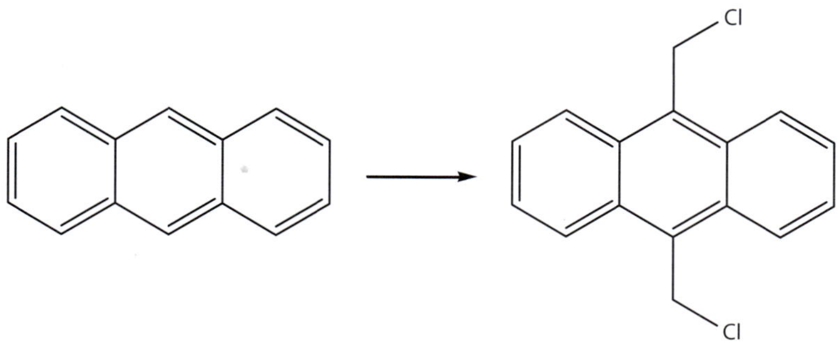

4. Provide reagents for each step in the following reaction scheme, explaining why the transformation shown needs to be done in this way:

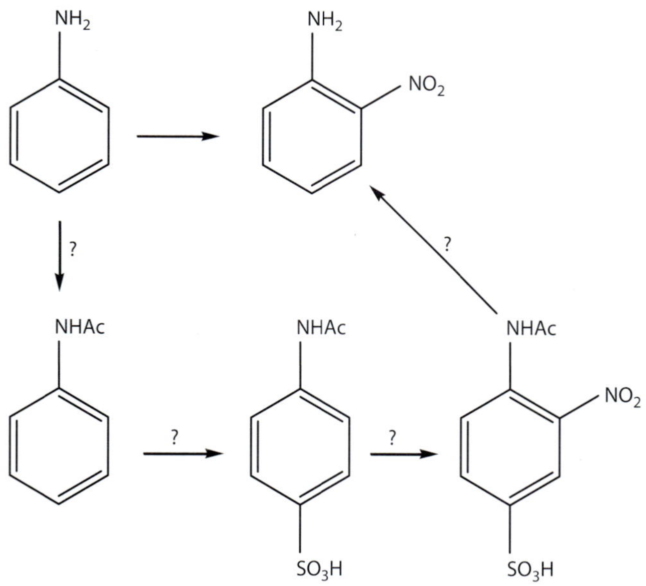

5. Explain the following observations, providing a full mechanism as appropriate:
   (a) If a Birch reduction is worked up with an appropriate electrophile, instead of water, an alkylated product is obtained, as shown in the following reaction:

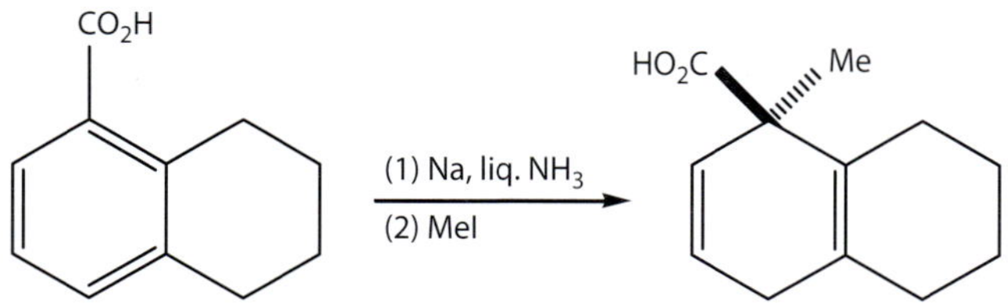

(b) The relative rates of chloromethylation of some aromatic compounds are given:

Compound	Relative Rate
Benzene	1
Toluene	3
1,3-Xylene	24
1,3,5-Trimethylbenzene	600
PhOMe	1,300
3,5-Dimethylanisole	100,000
2,4,6-Trimethylchlorobenzene	2

(c) For the nucleophilic substitution shown in the following, the relative rates for various leaving groups are F:Cl:Br:I = 3300:4.3:4.3:1:

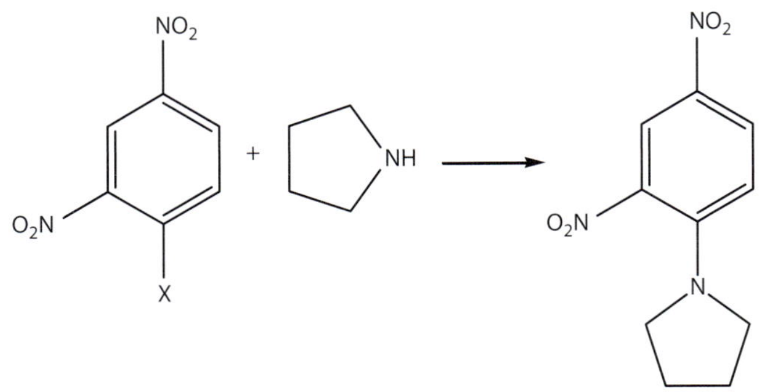

Addition to Carbon–Heteroatom Double Bonds

14

14.1 INTRODUCTION AND REVIEW

Let's take a moment to review the nomenclature of aldehydes and ketones, before we look at their reactions. The suffix for the aldehyde group, CHO, is -al, and the carbonyl carbon atom is always numbered as C-1. The suffix for ketones is -one, and the chain is numbered so that the carbonyl group has the lowest number. So **14.1** is 5-chloro-4-ethyl-2-methyheptanal, and **14.2** is 4-chloro-6,8-dimethyl-3-nonanone. Remember that aldehydes and ketones are described as oxo substituents when there is an acid derivative in the molecule and that the acid derivative will take precedence in the numbering scheme.

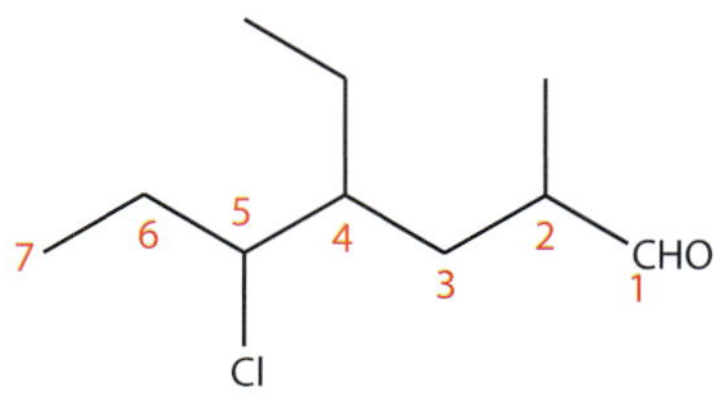

14.1

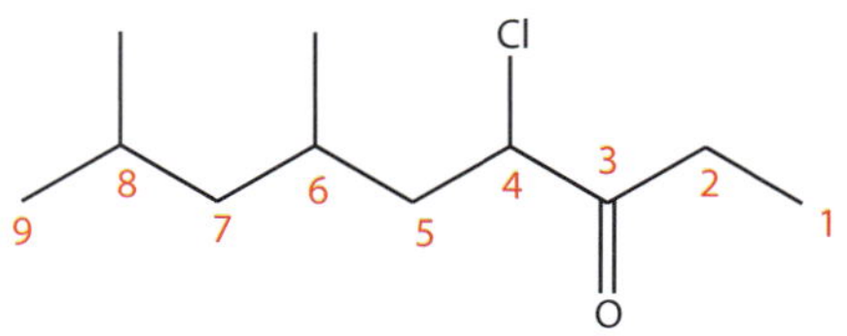

14.2

We have already met a number of reactions that produce aldehydes and ketones. Oxidation of primary or secondary alcohols using chromium(VI) reagents (Figure 14.1, Section 10.10) gives, respectively, aldehydes and ketones. Aldehydes and ketones could be obtained either by hydroboration/oxidation of alkynes and ketones by acid-catalyzed addition of water to alkynes (Figure 14.2, Sections 11.2.2 and 11.4.3). Ozonolysis of double bonds gives aldehydes/ketones as in Figure 14.3 (Section 11.4.5), and Friedel–Crafts acylations give aryl ketones (Figure 14.4, Section 12.2.2).

PROBLEM 14.1

(Review) Propose a mechanism for each of the following reactions:

(a)

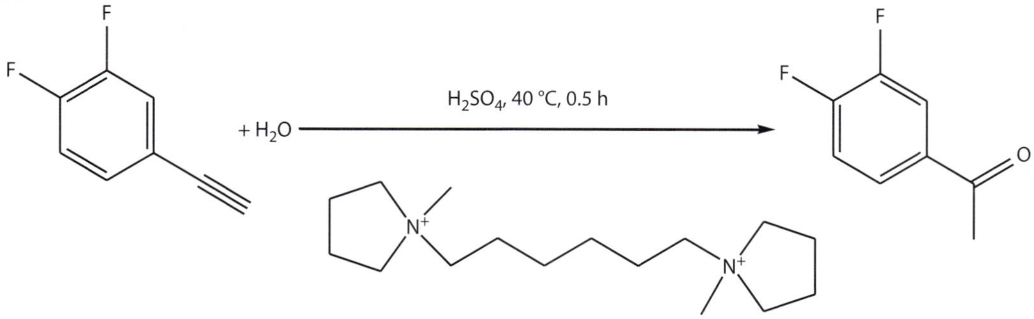

(b)

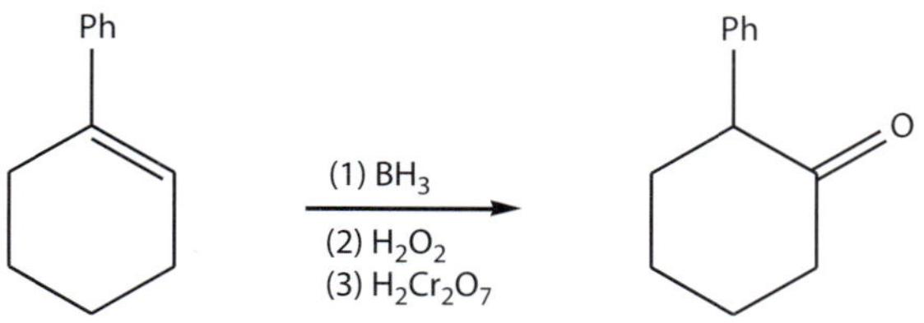

(c)

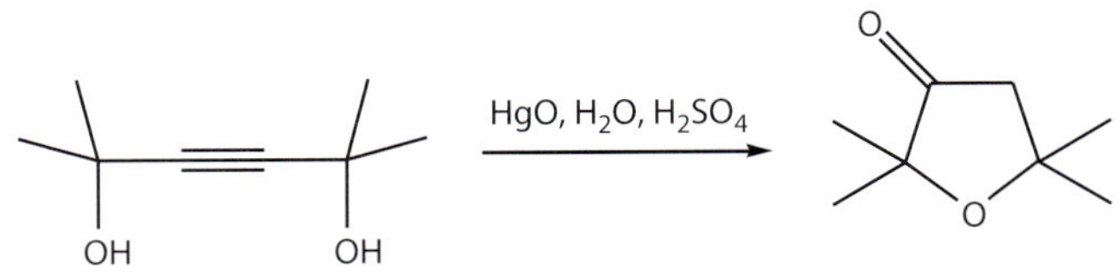

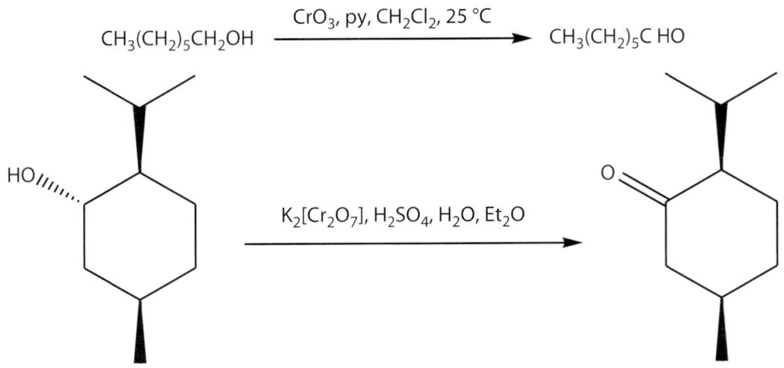

FIGURE 14.1 Chromium(VI) oxidation of primary and secondary alcohols.

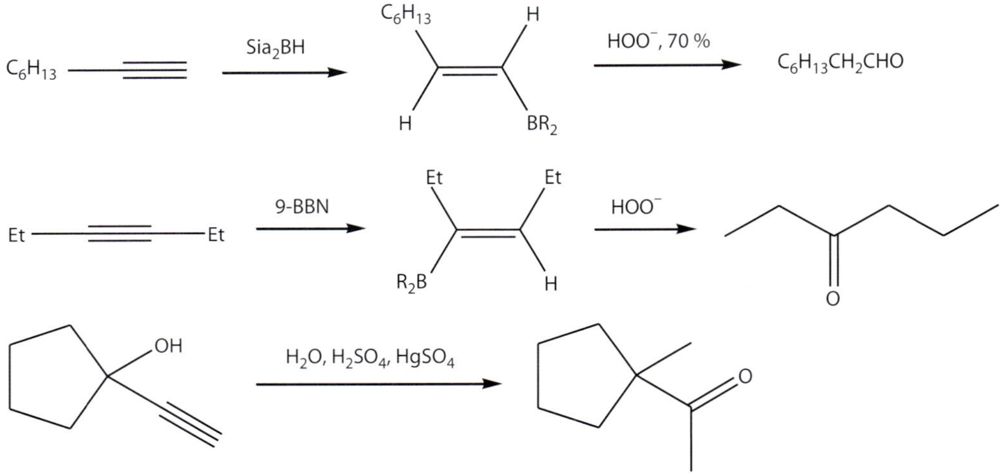

FIGURE 14.2 Carbonyl compounds synthesized by addition of water to alkynes (Sia_2BH is disiamylborane, IUPAC name bis (1,2-dimethylpropyl)borane. 9-BBN is 9-borabicyclo[3.3.1]nonane).

FIGURE 14.3 Carbonyl compounds synthesized by ozonolysis.

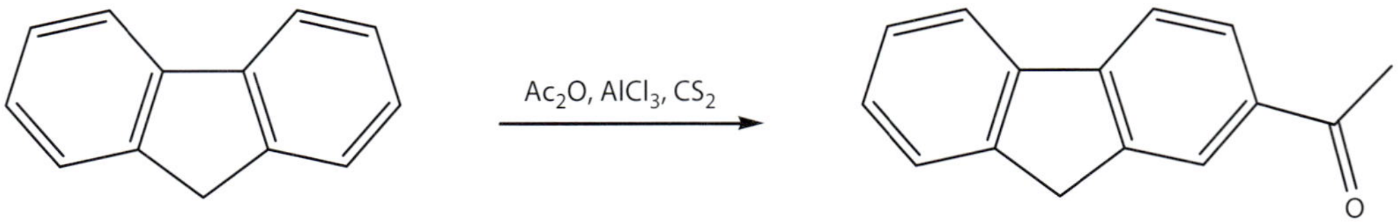

FIGURE 14.4 Friedel–Crafts acylation of arenes to give aryl ketones.

Solutions to Problem 14.1

(a) Although there is none of the usual mercury catalyst here, and we have no real idea (yet) what the diammonium salt is doing, the basic idea of this reaction is straightforward. The alkyne has been hydrated in a Markovnikov addition and then tautomerizes to the ketone:

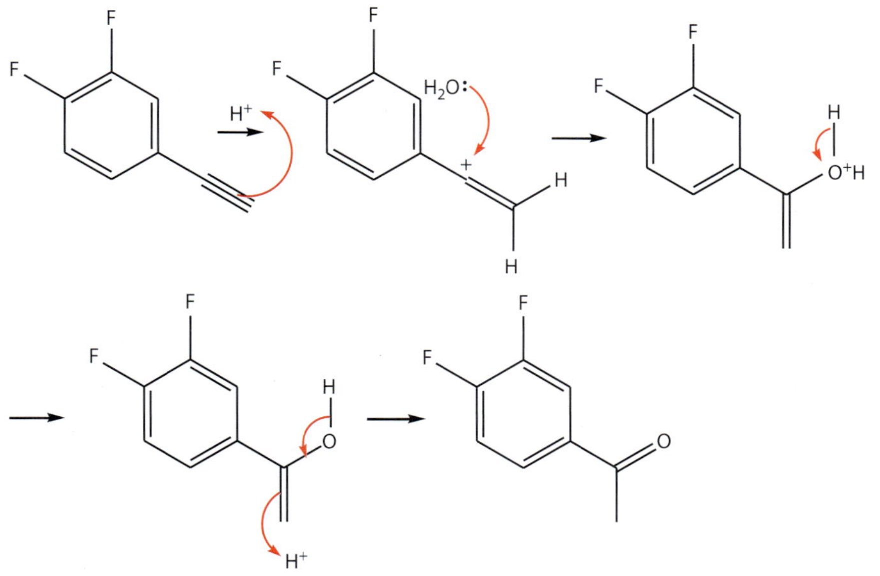

(b) This reaction has two steps—the first is anti-Markovnikov hydration of the double bond, and then the alcohol is oxidized using chromium(VI):

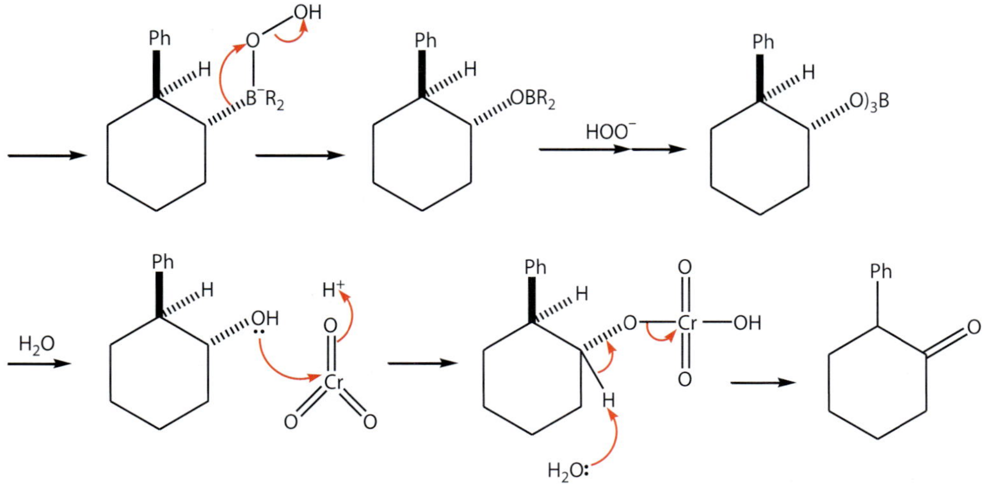

(c) It's reasonably easy here to follow which atom has arrived where, and there is a reaction that we should readily recognize the hydration of an alkyne to give a ketone:

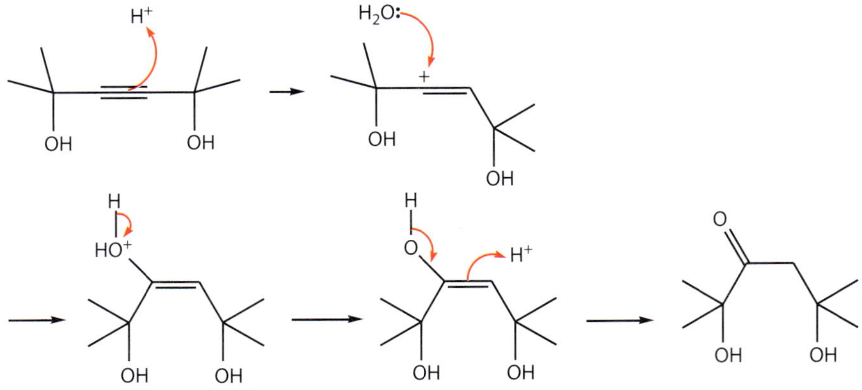

The ring closure step clearly takes place in the acid conditions. We know that protonation of tertiary alcohols followed by water loss is relatively easy, and this gives us a carbocation, ideal for cyclization. Note that we have chosen to protonate the OH group that, when it is lost, will *not* give rise to a carbocation next to the electron-withdrawing carbonyl group. The cyclization is easy because the ring being formed is five-membered:

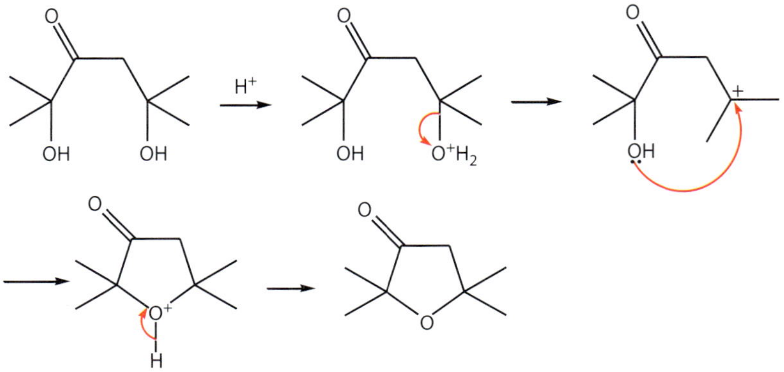

Nucleophilic addition reactions dominate the chemistry of aldehydes and ketones and are largely specific to these types of carbonyl compound and their close relations. The addition of nucleophiles reflects the resonance structures that are important in carbonyl compounds (Figure 14.5). Looking at the charge-separated resonance forms (and this is a useful tip, if you are at any point trying to work out which part of a reagent attacks where—draw out the charge separated form, and it should become clear), we conclude that nucleophiles react at carbon and electrophiles at oxygen. Despite the importance of this charge-separated resonance form, the carbon–oxygen bond is shorter, stronger, and more polar than the carbon–carbon double bond (Figures 14.6 and 14.7). While aldehydes and ketones cannot act as hydrogen-bond donors, and hence do not hydrogen bond among themselves, the carbonyl oxygen is a good hydrogen-bond acceptor, and aldehydes and ketones do hydrogen bond to water, alcohols, amines, amides, and carboxylic acids. The lone pairs on oxygen are attracted to partially positive hydrogen atoms.

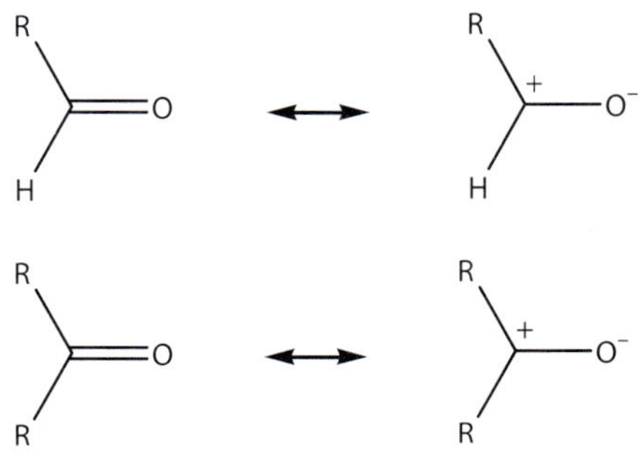

FIGURE 14.5 Resonance forms of aldehydes and ketones.

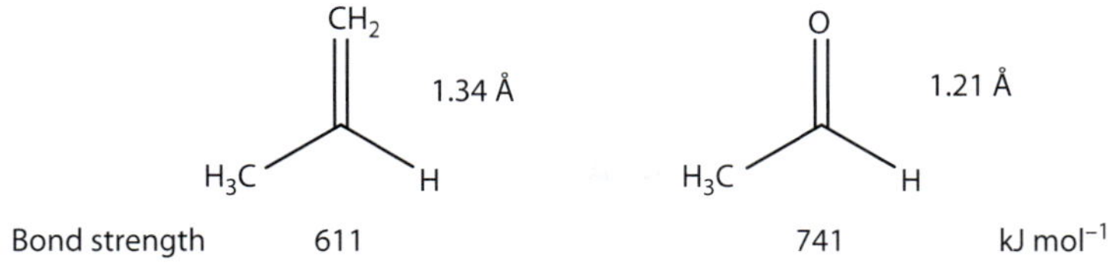

FIGURE 14.6 Comparison of double bonds in alkenes and aldehydes.

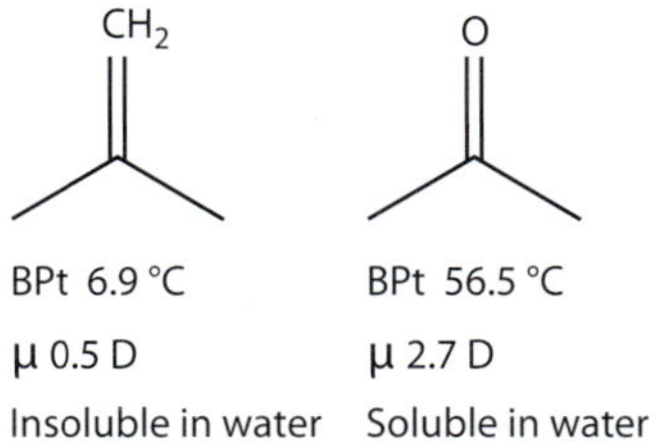

FIGURE 14.7 Comparison of polarity of alkenes and ketones.

Focus on Scent and Flavor

Although not strictly true, it's a commonly told story that Chanel No. 5, launched in the 1920s, was the first perfume based on a synthetic aldehyde fragrance. The perfume contains C_{10}, C_{11}, and C_{12} aldehydes, with the most important component being 2-methylundecanal. Lower aldehydes and ketones generally smell unpleasant. The old name for butanal was butyraldehyde, which comes from the Greek word βούτυρον meaning butter, and the aldehyde smells of, and is an important constituent of, rancid butter. Higher aldehydes are generally pleasantly scented. Hexanal smells of freshly cut grass, and both this and various hexenals are widely used in the perfume industry (they are actually produced by mown grass). *E*-2-Hexenal and *Z*-3-hexenal are referred to as leaf aldehydes, the former apparently being the source of the grassy taste of good Sauvignon blanc wine. Heptanal, octanal, nonanal, decanal, and undecanal have all found use in perfumes having various fruit/floral odors (Figure 14.8). Nonanal, together with exhaled CO_2, has been implicated in attracting mosquitoes. By contrast, citronellal, from the lemon-scented gum tree, is used as an insect repellant.

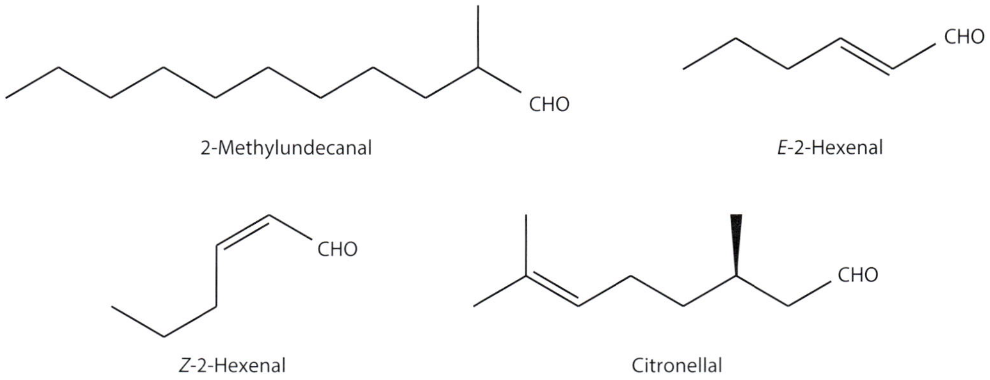

FIGURE 14.8 Some aldehydes used in perfumes.

Lemon-scented gum tree is a source of citronellal

Aldehydes are also important in the flavor and aroma of foods. Benzaldehyde has an almond scent, and many aromatic aldehydes (Figure 14.9) have strong scents and tastes. Cinnamaldehyde is isolated from cinnamon bark and is an important constituent of the taste of coriander, while citral is isolated from lemon grass. All of these materials were first isolated as natural products, but demand now outstrips supply and much vanilla extract is now synthetic or bioengineered.

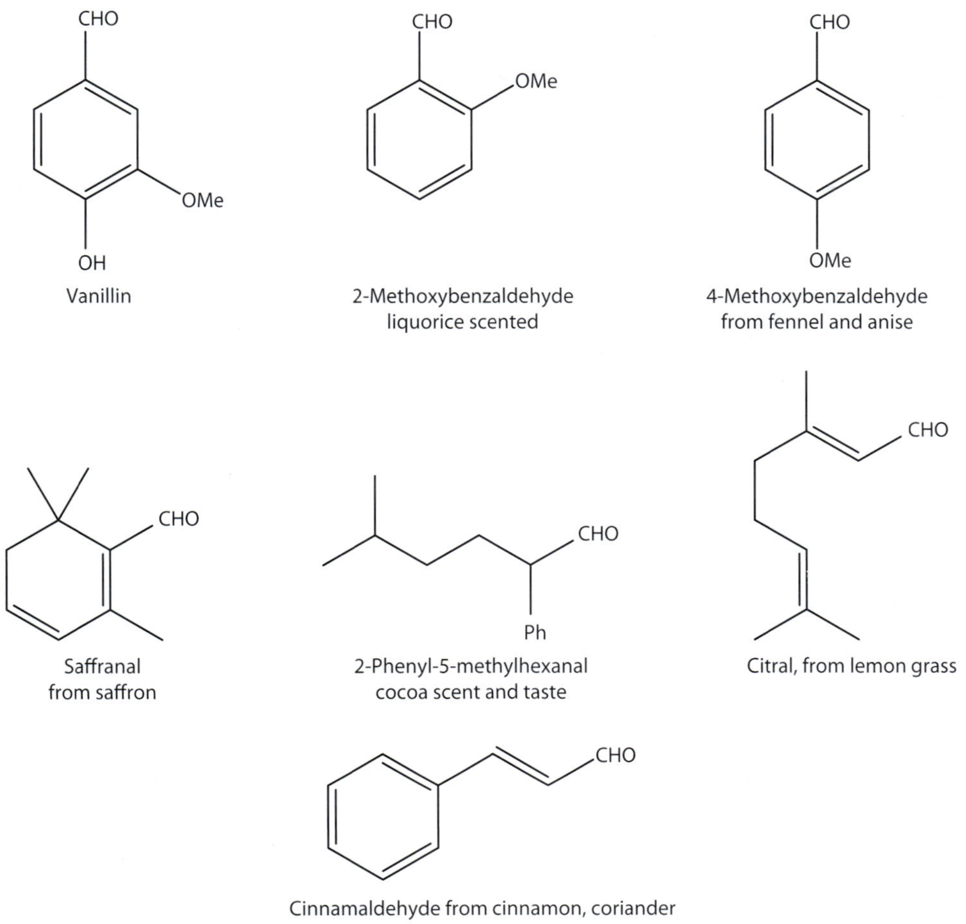

FIGURE 14.9 Aldehydes important in flavor and aroma of food.

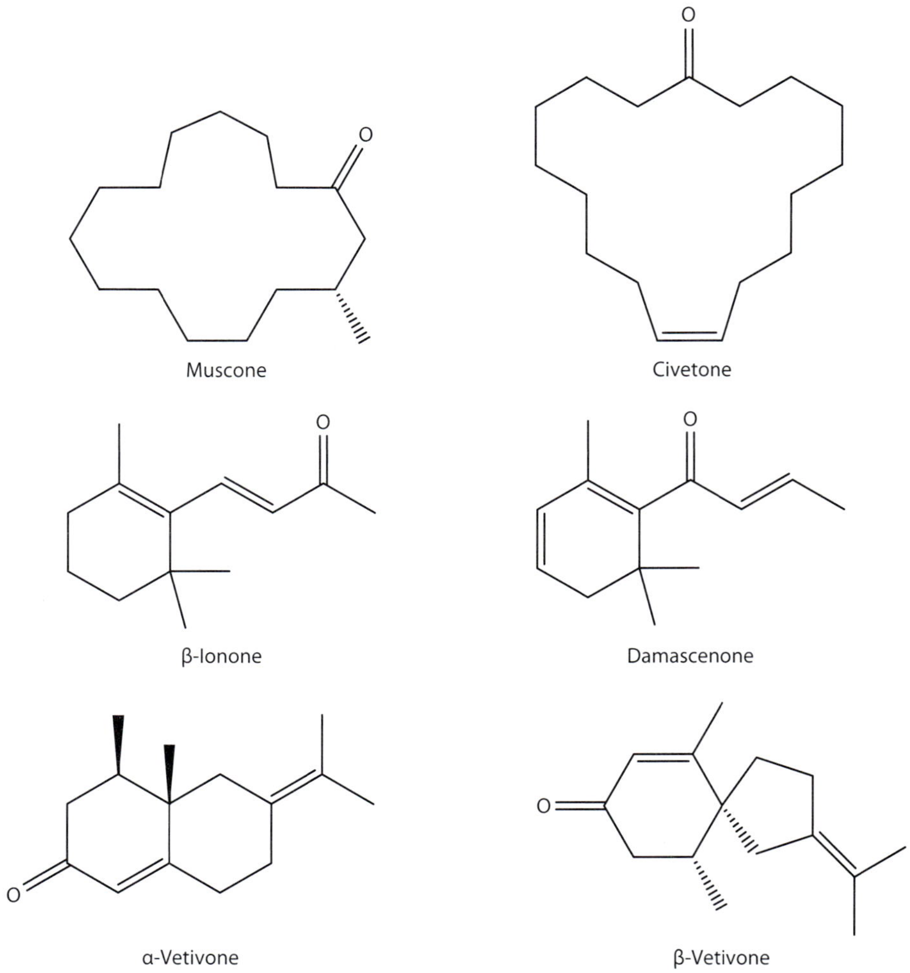

FIGURE 14.10 Ketones important in the fragrance industry.

Many ketones are also important in the perfume industry (Figure 14.10). Natural muscone and civetone, extracted originally from animal scent glands, have been used since ancient times, though most modern material is synthetic. β-Ionone and damascenone have violet/rose scents. Vetiver oil, extracted from a type of grass (the name comes from the Tamil word for grass), is used as a fixative in many perfumes. It is a very complex mixture of materials, including the ketones, α- and β-vetivone.

The molecule carvone (**14.3**) is interesting because the two enantiomers have different plant origins and very different tastes. *R*-Carvone is the main taste constituent of spearmint oil, but *S*-carvone is obtained from caraway seed oil. Camphor (**14.4**) comes from the wood of the camphor laurel and is widely used in Asia in cooking and religious ceremonies.

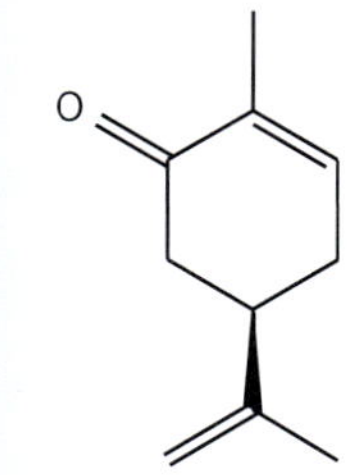

14.3a, *R*-Carvone

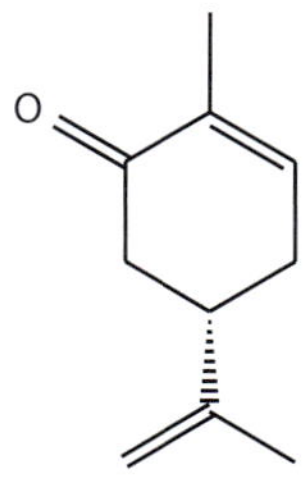

14.3b, *S*-Carvone

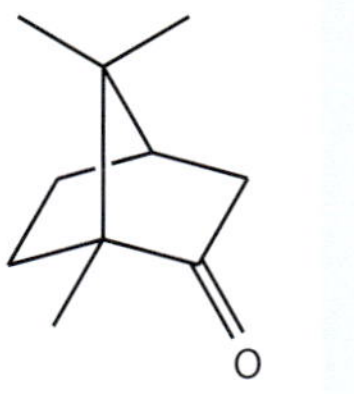

14.4, Camphor

Key Points from Section 14.1

- The suffix for aldehyde nomenclature is -al; the aldehyde carbonyl is always C-1.
- The suffix for ketone nomenclature is -one; the numbering scheme gives the ketone carbonyl the lowest possible number.
- When the aldehyde or ketone is being named as a substituent, the term "oxo" is used.
- Aldehydes can be prepared by oxidation of primary alcohols, hydroboration of 1-alkynes, or ozonolysis of alkenes. Aryl aldehydes can be prepared from the arene with CO/HCl in the presence of $AlCl_3$/CuCl.
- Ketones can be prepared by oxidation of secondary alcohols, hydroboration of internal alkynes, addition of water to alkynes, and ozonolysis of alkenes; aromatic ketones can be prepared by the Friedel–Crafts reaction.

14.2 MECHANISM OF ADDITION REACTIONS

Reactions of nucleophiles with aldehydes and ketones can occur under either acidic* or basic conditions. The prototype for reactions in base generally involves a negatively charged nucleophile, which attacks at the carbon of the carbonyl, with the electrons of the π-bond being displaced to oxygen. The process is completed when the negatively charged oxygen picks up a proton, either from protonated nucleophile or solvent (Figure 14.11). In the acid-catalyzed process (Figure 14.12), the reaction is initiated by protonation of the oxygen of the carbonyl group. The nucleophile (usually neutral) then attacks the carbon atom, and the π-electrons are displaced to oxygen neutralizing the positive charge. The final step is loss of a proton from the nucleophile. You should have noticed that in both processes, base or acid is just a catalyst—the proton or hydroxyl group is regenerated at the end of the reaction. As we will see, this means that the rate of this type of reaction is strongly dependent on solution pH. Under acidic conditions, the carbonyl is rendered more electrophilic by protonation (or coordination to a Lewis acid). Under basic conditions, an attacking species is more likely to be deprotonated and hence more nucleophilic.

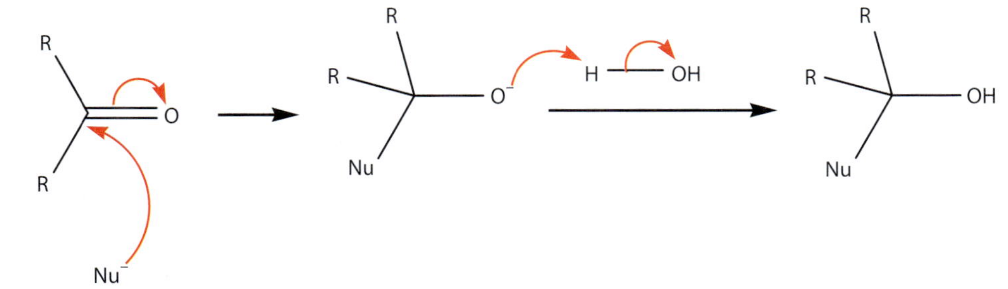

FIGURE 14.11 Mechanism of addition to a carbonyl group under basic conditions.

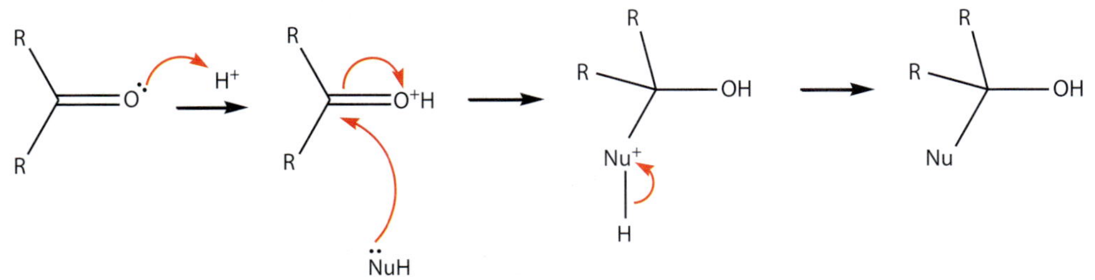

FIGURE 14.12 Mechanism of addition to a carbonyl group under acidic conditions.

* Here, we will consider mainly reactions in the presence of Brønsted acids, but in Chapter 23, we will look in more detail at the reactions that can be accomplished in the presence of Lewis acids.

14.3 REVERSIBLE ADDITION REACTIONS

14.3.1 CYANOHYDRIN FORMATION

Addition of HCN to aldehydes or ketones gives a cyanohydrin (Figure 14.13). However, HCN is not a sufficiently strong acid (pK_a = 9.1) to protonate the carbonyl group to make it into a better electrophile. The dissociation of HCN gives an insufficient concentration of cyanide effectively to attack the carbonyl group. So it is necessary to catalyze the process with potassium cyanide (KCN), which is fully dissociated. The process is reversible, and the position of the equilibrium is dependent on both steric and electronic factors.

If the carbonyl compound is substituted with very bulky groups, as, for example, in **14.5**, less cyanohydrin formation will be observed at equilibrium. At the sp^2 carbon of the ketone (**14.6**), the angle between the groups is 120 °. When the cyanohydrin (**14.7**) is formed, this is contracted to 109.5 °, which is clearly disfavored. In certain other molecules, angle strain is high in the ketone and is partially relieved in the cyanohydrin. For example, cyclopentanone, which has flexibility and internal angles compatible with both the 120 ° requirement of an sp^2 carbon and the 109.5 ° requirement of an sp^3 carbon, gives a proportion of cyanohydrin at equilibrium comparable with unhindered acyclic ketones. If we consider the reaction of cyclopropanone, **14.8**, however, the outcome is quite different. In cyclopropanone, the angle at the carbonyl carbon wants to be 120 °, but it is forced to be 60 °, so the molecule is very unhappy ☹. In the cyanohydrin, the angle at the cyanohydrin carbon wants to be 109.5 °, but it is still forced to be 60 °. The molecule is a little less unhappy 😐, and cyclopropanone is completely converted to its cyanohydrin at equilibrium:

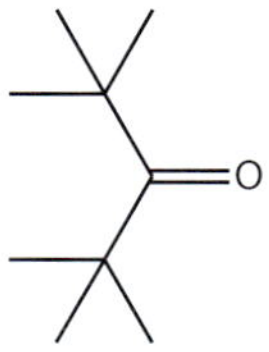

14.5

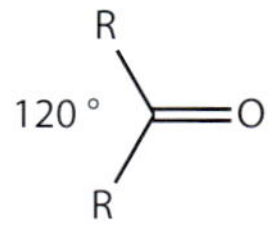

14.6

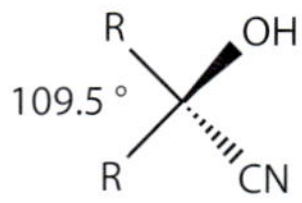

14.7

14.8

Electronic factors are also important. If the starting carbonyl compound is conjugated, the conjugation lowers the overall energy ☺. Conjugation is lost on cyanohydrin formation ☹, so this is disfavored, and **14.9** and **14.10** would give less cyanohydrin at equilibrium than their non-conjugated analogues. Strongly electron-withdrawing groups, such as CF_3, destabilize carbonyl compounds (two or three strong inductive withdrawers of electron density, all pulling on a single

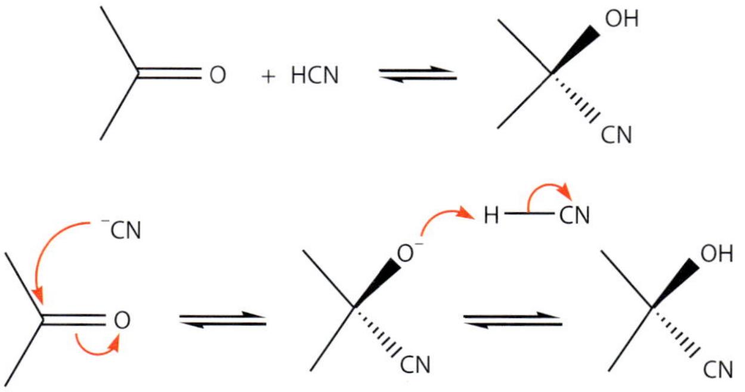

FIGURE 14.13 Cyanohydrin formation.

TABLE 14.1 Equilibrium Constants for Cyanohydrin Formation in Aqueous Solution

Carbonyl Compound	CH_2O	CH_3CHO	Me_2CO	*i*-PrC(=O)*t*-Bu	PhCHO	PhCOMe	$PhCOCF_3$
K (M^{-1})	1.6×10^7	7100	28	0.12	200	0.8	3.4×10^5

carbon atom) more than they do to cyanohydrins. Hexafluoroacetone, **14.11**, is completely converted to its cyanohydrin at equilibrium. Some equilibrium constants for cyanohydrin formation are given in Table 14.1.

14.9

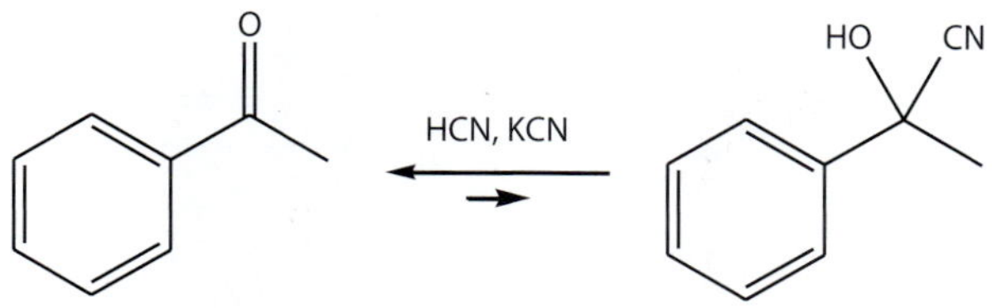

14.10

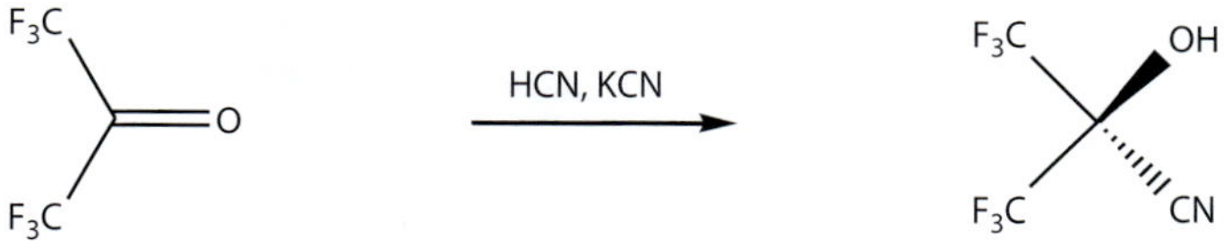

14.11

In practical synthetic terms, formation of cyanohydrins using HCN/KCN works well for aldehydes and for simple aliphatic ketones. Aryl alkyl ketones can be "persuaded" to react, but benzophenone derivatives make poor substrates. The substitution of trimethylsilylcyanide for HCN is useful, as the irreversible formation of the strong silicon–oxygen bond (~460 kJ mol^{-1}) blocks the reverse reaction (Figure 14.14). As we saw with cyanide substitution reactions in Chapter 9, the great value of attaching a cyanide functionality is that it can be transformed into either a carboxylic acid or an amine (Figure 14.15—the silicon–oxygen bond is cleaved in the work-up step of the second example).

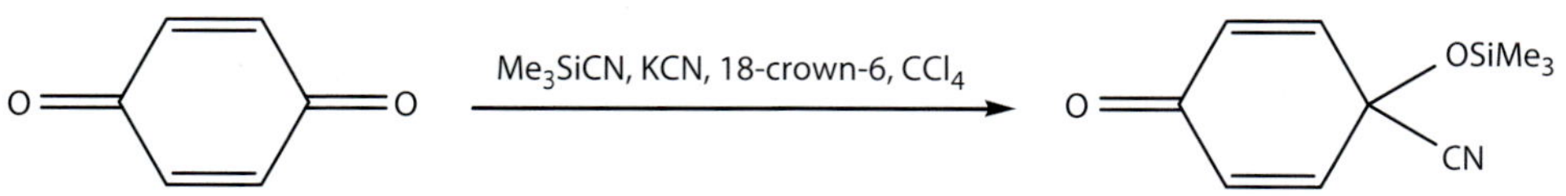

FIGURE 14.14 Reaction of a ketone with trimethylsilyl cyanide (The 18-crown-6 sequesters potassium, breaking up KCN ion pairs, so that the cyanide acts as a better nucleophile).

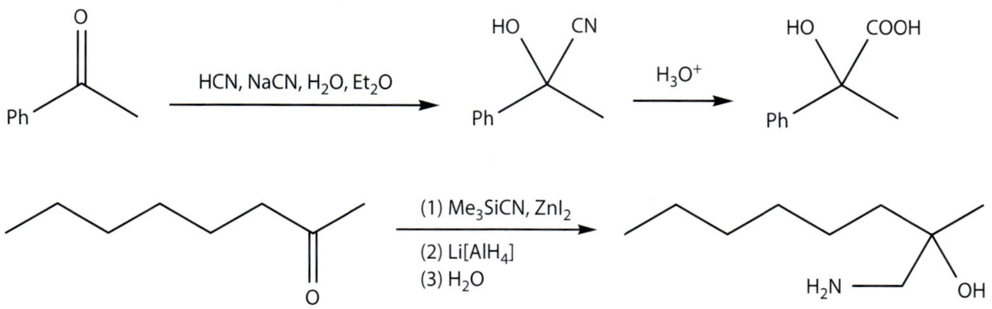

FIGURE 14.15 Synthetic applications of cyanohydrin formation.

Focus on Naturally Occurring Cyanohydrins

Despite, or perhaps because of, the high toxicity of cyanide, cyanohydrins are widely distributed in nature. The cyanohydrin of benzaldehyde, PhCH(OH)CN, is used as a defense mechanism by the millipede *Apheloria corrugata*. When the millipede is attacked, it excretes the cyanohydrin, which is then enzymatically converted to benzaldehyde and HCN, the latter being toxic to the attacker.

Amygdalin, **14.12** (from Greek, ἀμυγδάλη, amygdálē, almond), is a sugar derivative of benzaldehyde cyanohydrin, which can be isolated from the kernels of apricots, peaches, plums, cherries, and almonds. Laetrile, **14.13**, may be derived from this. Both amygdalin and laetrile are degraded, releasing HCN, by the enzyme β-glucosidase, which is naturally present in the human small intestine. It has been claimed that the enzyme is more abundant in cancerous cells and that laetrile is an antitumor compound. There is no evidence that this is so, and while laetrile is still obtainable in some parts of the world, its use is discredited by reputable physicians. It is not only ineffective but also highly toxic.

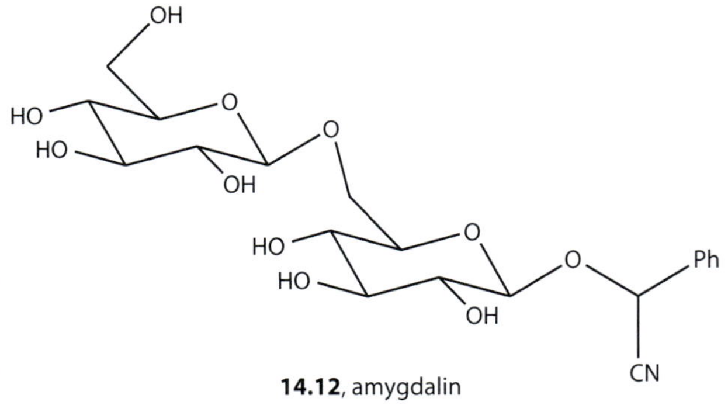

14.12, amygdalin

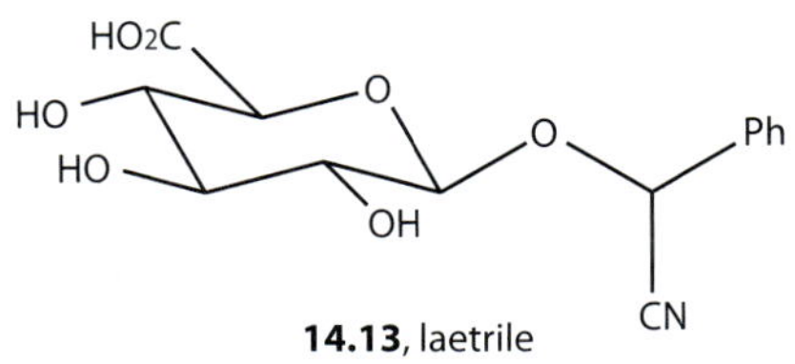

14.13, laetrile

Amygdalin is isolated from apricot kernels

The staple crop cassava, also called manioc, is the third largest carbohydrate source in the world. It originated in South America, but is now grown throughout tropical and subtropical regions. It grows successfully in poor soils and is pest and drought resistant. It contains both the sugar cyanohydrin linamarin (**14.14**) and the enzyme linamarinase that releases HCN from linamarin. There is enough cyanide in cassava root to kill those who consume it, so why does it not poison most of the people who eat it? The cyanide concentration in the food consumed depends on both the variety of the cassava (both sweet, low cyanide and bitter, higher cyanide varieties are known) and how it is prepared. The disease *konzo* was first documented in the 1930s, and later in the 1980s, as a neurotoxic motor disease caused by chronic cyanide poisoning. *Konzo* outbreaks occur when food is scarce, and there is not enough time for the shredded cassava to be sufficiently soaked or fermented (which removes the cyanide), and cooked, or more bitter varieties are consumed.

Cassava in cultivation in the Democratic Republic of Congo

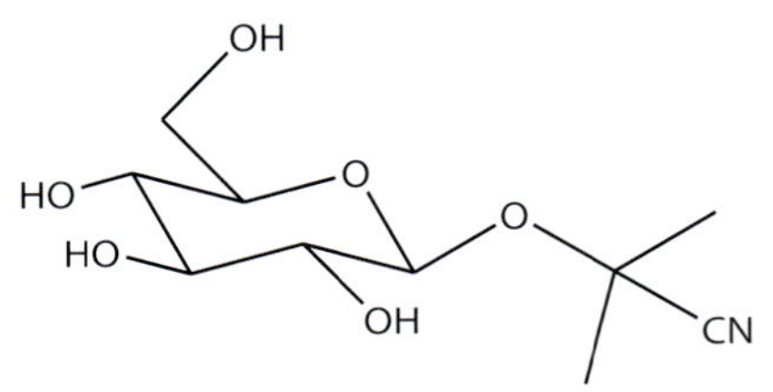

14.14, linamarin

PROBLEM 14.2

Study the reaction sequence in the following and answer the questions given:

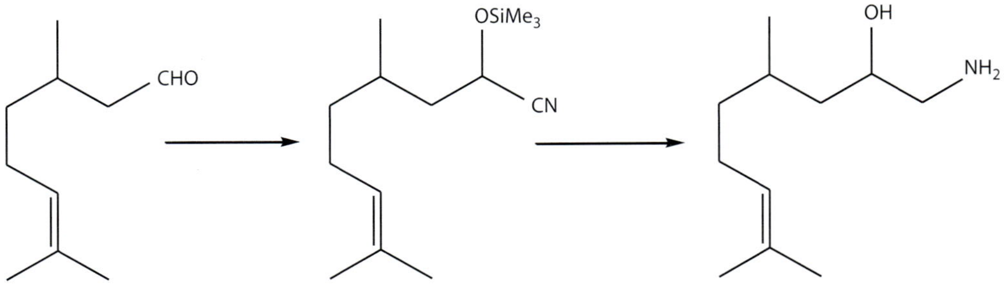

14.15 **14.16** **14.17**

(a) Identify the isoprene units in **14.15**, and characterize it as a mono-, di-, *etc.*, terpene.
(b) What reagent would you use to convert **14.15** to **14.16**?
(c) We have met two reagents for the reduction of nitriles to amines. Name them.
(d) Which would you choose for the conversion of **14.16** to **14.17**?

Solutions

(a) **14.15** is a monoterpene:

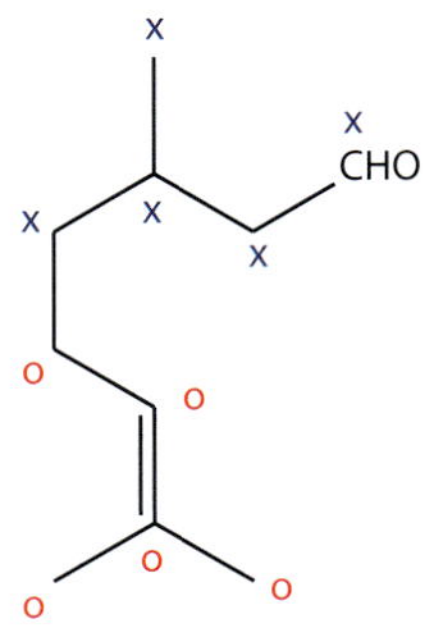

(b) Me_3SiCN.
(c) Cyanides can be reduced using either H_2/metal catalyst or $Li[AlH_4]$ followed by aqueous work-up.
(d) Reduction of **14.16** to **14.17** should be effected by $Li[AlH_4]$. Any metal-catalyzed hydrogenation would also reduce the carbon–carbon double bond. The aqueous work-up of the reduction will also hydrolyze the silyl ether.

PROBLEM 14.3

Explain the equilibrium constants shown in the following for cyanohydrin formation:

$$R_2C{=}O + HCN \xrightleftharpoons{\text{KCN (catalyst)}} R_2C(OH)CN + H_2O$$

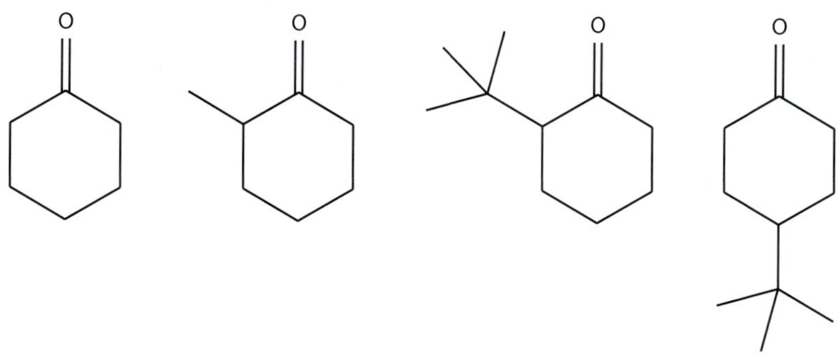

K 1700 930 7.8 1700

Solution

Electronically, these molecules differ little, so this is clearly about steric hindrance. A group at the 2-position of cyclohexanone will interfere sterically with the formation of the cyanohydrin. A methyl group is relatively small, but a *tert*-butyl group has a much larger effect. At the 4-position, the *tert*-butyl group is too far away to exert any steric influence on the equilibrium, so the result is the same as for cyclohexanone itself.

14.3.2 HYDRATES, HEMIACETALS, AND ACETALS

Water adds reversibly to aldehydes and ketones under acid- or base-catalyzed conditions to give hydrates (Figure 14.16). Hydrate formation is generally less favorable than cyanohydrin formation, but the equilibrium is influenced by the same considerations of steric and electronic effects.

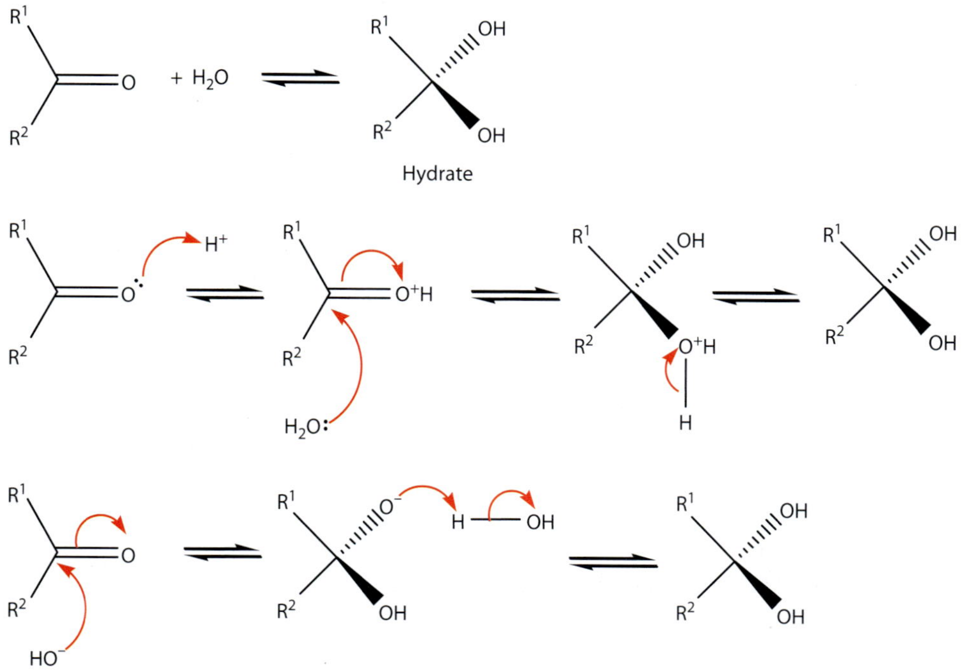

FIGURE 14.16 Addition of water to a ketone occurs under acidic or basic conditions.

Acid and base catalysis of the process are shown. The acid catalysis arises from the fact that protonated ketone is a better electrophile than the ketone and the base catalysis because hydroxyl ion is a better nucleophile than water.

Alcohols may be considered to be substituted versions of water, and alcohols will undergo this type of reaction just as well as water itself—again, the reaction is acid or base catalyzed, and the product is described as a hemiacetal (Figure 14.17; sometimes, only aldehyde derivatives are called hemiacetals, and ketone derivatives hemiketals, but other authors use the term hemiacetals for derivatives of aldehydes or ketones). Again, in acid, the carbonyl group is rendered more electrophilic, while in base, the RO^- anion is more nucleophilic than the alcohol. And whatever alcohols can do as nucleophiles, thiols do at least as well or better (Figure 14.18).*

In considering the equilibrium between carbonyl compounds and their hydrates, we find that much higher proportions of hydrate are obtained with aldehydes (for both steric and electronic reasons), and for most simple ketones, the hydrate cannot be detected (Table 14.2). The ability of most aldehydes to hydrate, at least to some extent, is critical to their mechanism of oxidation with chromium reagents (Section 10.10); it is the hydrate that is the true intermediate in the oxidation. Although the hydrate of acetone is barely detectable, acetone containing ^{16}O in the carbonyl group exchanges with ^{18}O water, *via* low concentrations of the hydrate (Figure 14.19).

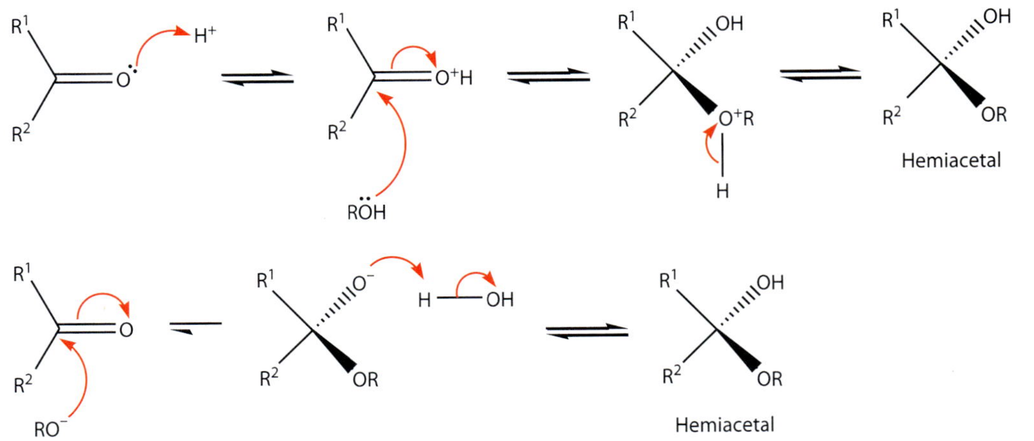

FIGURE 14.17 Acid- and base-catalyzed formation of hemiacetals.

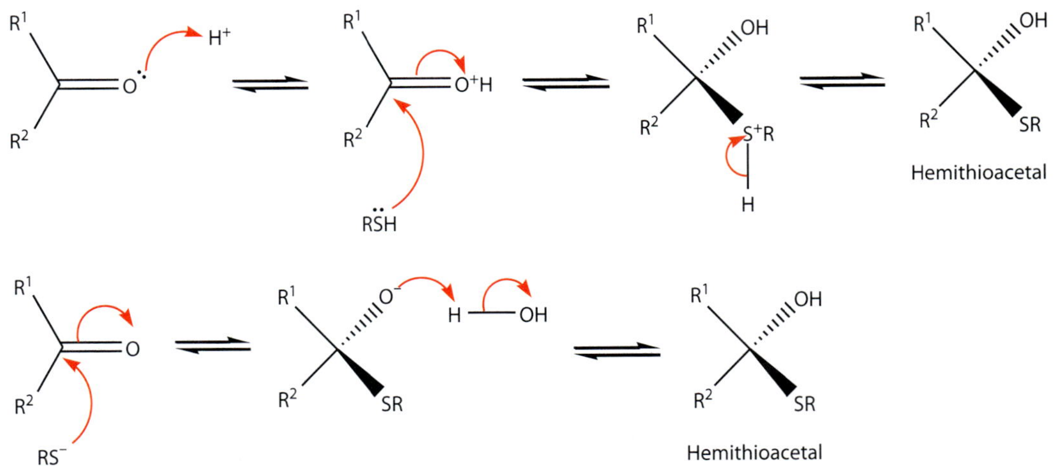

FIGURE 14.18 Acid- and base-catalyzed formation of hemithioacetals.

* We will see in a later chapter that there are some risks of competing reactions when enolizable aldehydes or ketones are treated with either acid or base.

TABLE 14.2 Equilibrium Constants and Proportions of Hydrate at Equilibrium for Aldehydes and Ketones

Carbonyl Compound	*K*	% as Hydrate at Equilibrium
$H_2C{=}O$	~10^4	>99.9
CH_3CHO	1.5	60
Me_3CCHO	0.2	17
$Me_2C{=}O$	0.0014	0.14
CCl_3CHO	500	99.8
$(CF_3)_2C{=}O$	22,000	>99.9
PhCHO	0.008	0.8

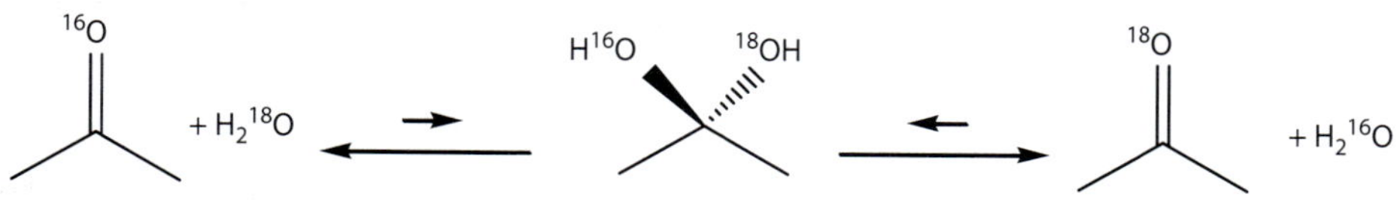

FIGURE 14.19 Exchange of the oxygen atom of acetone with ^{18}O water.

A solution of chloral, trichloroethanal, in aqueous alcohol is known as knockout drops or a "Mickey Finn." The original Mickey Finn was a bartender in Chicago around 1900, who drugged and then robbed his customers. Chloral hydrate has had legitimate uses as an aid to sleep, and in dental anesthesia, and has also been used recreationally. Therapeutic doses of chloral were implicated in the drug overdose death of Anna Nicole Smith in 2007, and Marilyn Monroe was reported to have chloral hydrate in her system when she died.

Rather few hemiacetals are stable. $CF_3CH(OH)OMe$ is a stable crystalline solid once prepared, and **14.18** is stable in the absence of nucleophiles. The important exception to this, however, is when the hemiacetal is cyclic, in a five- or six-membered ring. For **14.19** and **14.20**, there are variable amounts of the cyclic acetal at equilibrium. Taking a look ahead, this phenomenon is crucially important to the structure of sugars such as glucose, **14.21**. As we will see in Chapter 16, glucose exists almost exclusively in the form of two equilibrating six-membered rings (Figure 14.20). When a hemiacetal is formed, a new chiral center is generated, and both diastereoisomers are present in aqueous solution.

14.18

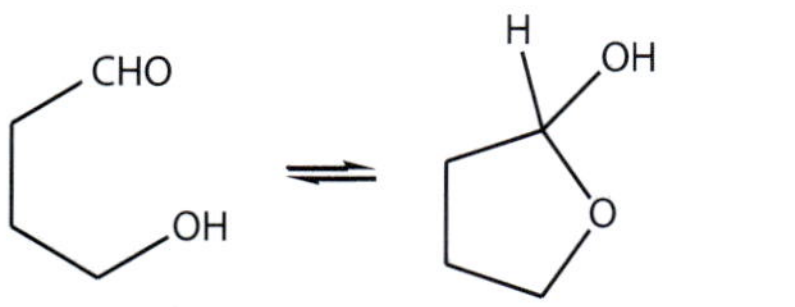

14.19 89 % at equilibrium

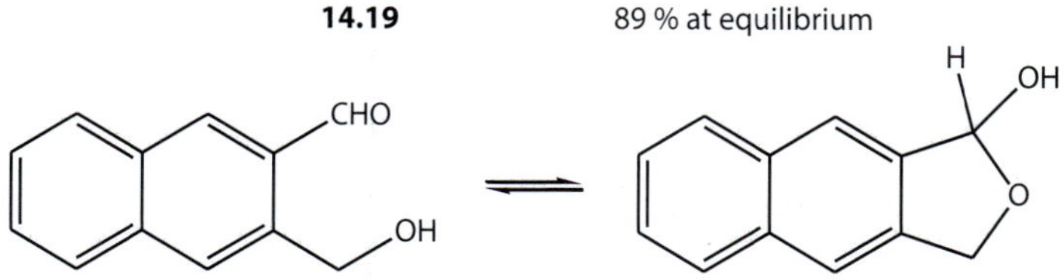

14.20 20 % at equilibrium

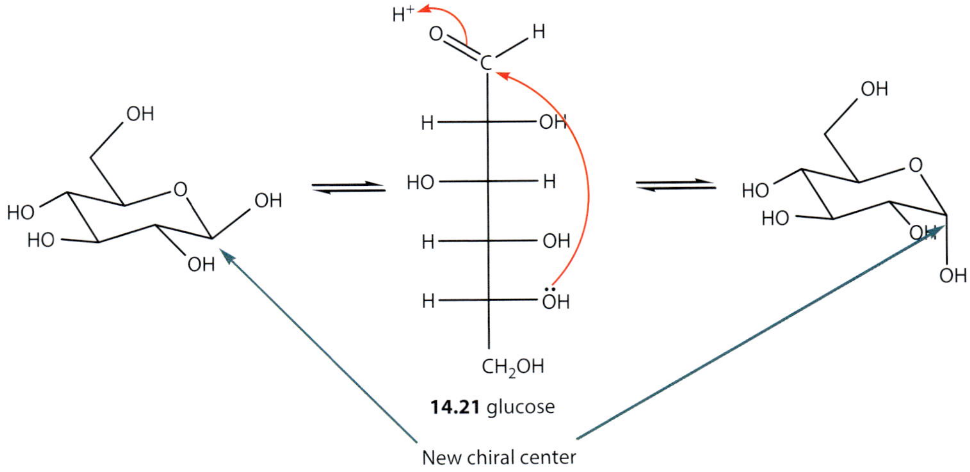

FIGURE 14.20 Cyclization of glucose.

PROBLEM 14.4

Explain the equilibrium data shown in the following table for the hydration of aromatic aldehydes:

$$ArCHO + H_2O \overset{K}{\rightleftharpoons} ArCH(OH)_2$$

Ar	% Hydrate at Equilibrium	*K*
Ph	0.8	0.008
4-ClC_6H_4	1.6	0.016
4-$F_3CC_6H_4$	5.2	0.055
4-$O_2NC_6H_4$	15	0.17
3,5-$(O_2N)_2C_6H_3$	78	2.1

Solution

None of the substituents on the aromatic rings are at the *ortho* positions, so we can safely discount any steric effects. The more electron-withdrawing the substituents, the higher the proportion of hydrate at equilibrium. It is tempting, and true, to say that the electron-withdrawing substituents make the carbon of the carbonyl group more positive and hence more likely to be attacked by a nucleophile. However, the speed of the attack is not what is being measured here; the position of the equilibrium reflects thermodynamics, not kinetics. The correct thermodynamic explanation is that the electron-withdrawing groups destabilize the carbonyl more than they do the hydrate.

14.3.3 ACETALS AND THIOACETALS

The next group of reactions that we will study are reversible, but involve the loss of the original carbonyl oxygen atom. Consider the ethanol hemiacetal of acetone and how it can revert to SM under acid-catalyzed conditions (Figure 14.21). The mechanism as drawn is correct, but should leave us with a question. Why is the OEt rather than the OH protonated in the first step? The answer must be that this is what we needed to do to make the ethanol into a good leaving group (remember, EtO^- is a poor leaving group, because ethanol is not a strong acid). But –OH and –OEt groups differ little in the likelihood that they will be protonated. Let's explore what would happen if we were to protonate the OH instead and have this as our leaving group.

By losing water rather than ethanol from the protonated hemiacetal, we seem to have reached an *impasse*—we can't just drop off an ethyl cation to go back to the ketone. However, if we compare **14.22** with the protonated carbonyl compound, **14.23**, electronically, they are quite similar. So we would expect that **14.20** could be attacked by a nucleophile in the same

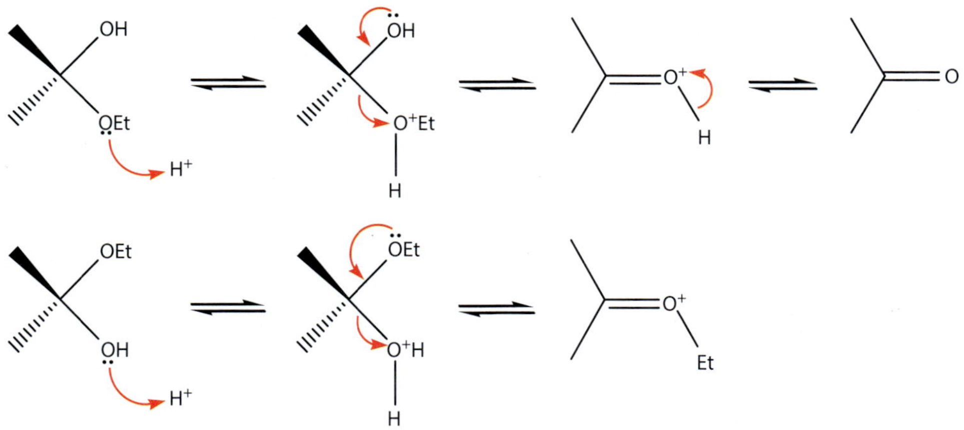

FIGURE 14.21 Reactions of a hemiacetal with acid.

14.22 **14.23**

way as was **14.23**. **14.22** can be attacked by another molecule of ethanol, to give an acetal. The original oxygen of the ketone is gone, replaced by two OEt groups (Figure 14.22). Notice that this reaction <u>requires</u> acid catalysis to turn the OH into a leaving group. While hemiacetals can be produced in either acid or base, formation of acetals can only be accomplished under acid conditions. Hydrolysis of acetals to regenerate the ketone also occurs only in acid; acetals are generally stable to bases and nucleophiles. Figure 14.23 shows the mechanism of formation of a cyclic acetal. Cyclic acetals with five- or six-membered rings are among the easiest to prepare, because of both the product stability and the fact that the rate of a key step is enhanced because it is intramolecular. Two points to remember when drawing mechanisms of this type (there are many similar ones, and once you have mastered the general principle, you should be able to tackle unfamiliar ones). First, whatever group you are expelling should be protonated. HO^-, RO^-, RS^-, R_2N^- are all poor leaving groups here, just as they were back in Chapter 9, when they could not be expelled in nucleophilic substitutions. Second, even though we are making a five-membered ring, any S_N2-type substitution at a tertiary center does <u>not</u> occur. The lone pair of electrons on the other oxygen atom (this also works with other atoms with lone pairs) pushes out the leaving group, then the sp^2 intermediate is attacked by an internal or external nucleophile.

We have said that acetal formation is reversible, so the process shown in Figure 14.24 is an equilibrium. Can we persuade it to go specifically in one direction or the other? It turns out that this is relatively easy, using mass action effects. If we want the reaction to go from right to left, we use aqueous acid as the solvent. If we want to go from left to right, to form the acetal, we use an excess of ethanol, toluene sulfonic acid as a catalyst (this works well because it is

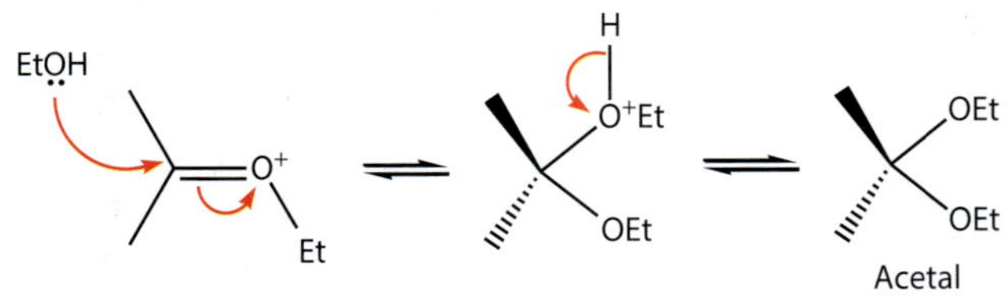

FIGURE 14.22 Acetal formation.

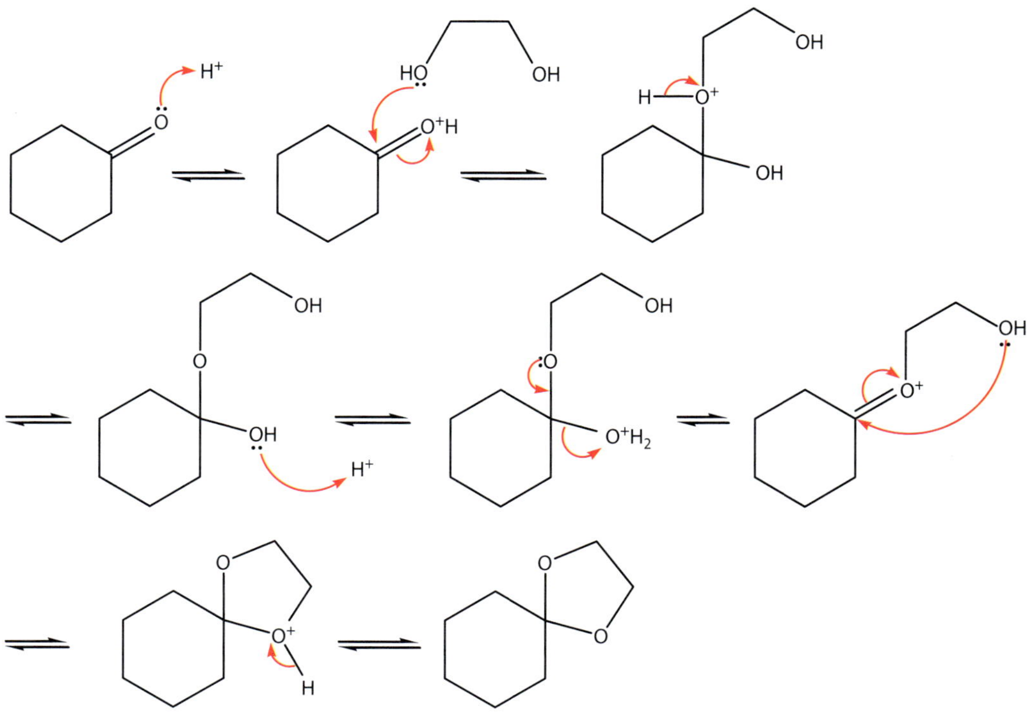

FIGURE 14.23 Mechanism of formation of a cyclic acetal.

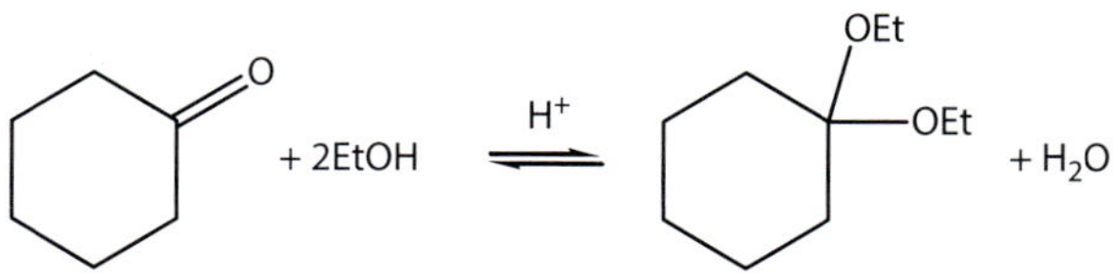

FIGURE 14.24 Reversible acetal formation.

soluble in most organic solvents), and we remove the water as it is formed. This involves the use of a Dean–Stark apparatus (Figure 14.25). The reaction takes place in a round-bottomed flask, in refluxing benzene. As you will know from general chemistry, benzene and water form an azeotropic mixture as they boil. The mixed solvents are condensed into what is essentially a burette, and there, on cooling, the water and benzene separate, with the water being the lower layer. The top layer, the benzene, runs back into the reaction flask. Because the amount of water can be measured (and some of it run off during the process as necessary), then it is clear when the reaction is complete, and it can be stopped when the stoichiometric amount of water has been collected.

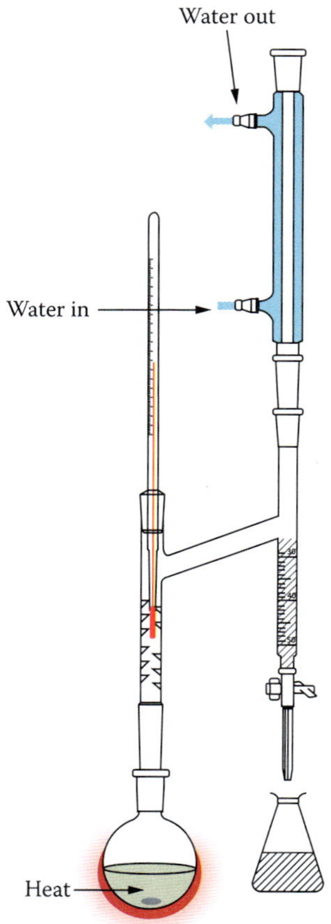

FIGURE 14.25 Dean–Stark apparatus.

PROBLEM 14.5

Thiols react with carbonyls in the same way as do alcohols. Using the reaction of alcohols as your guide, write a mechanism for the reaction in the following, of 2-hexanone:

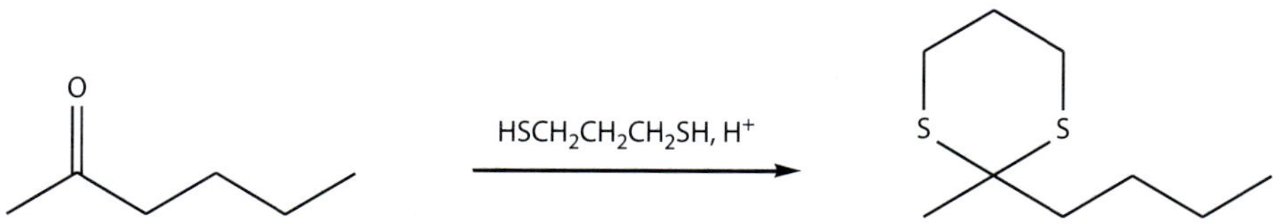

Solution

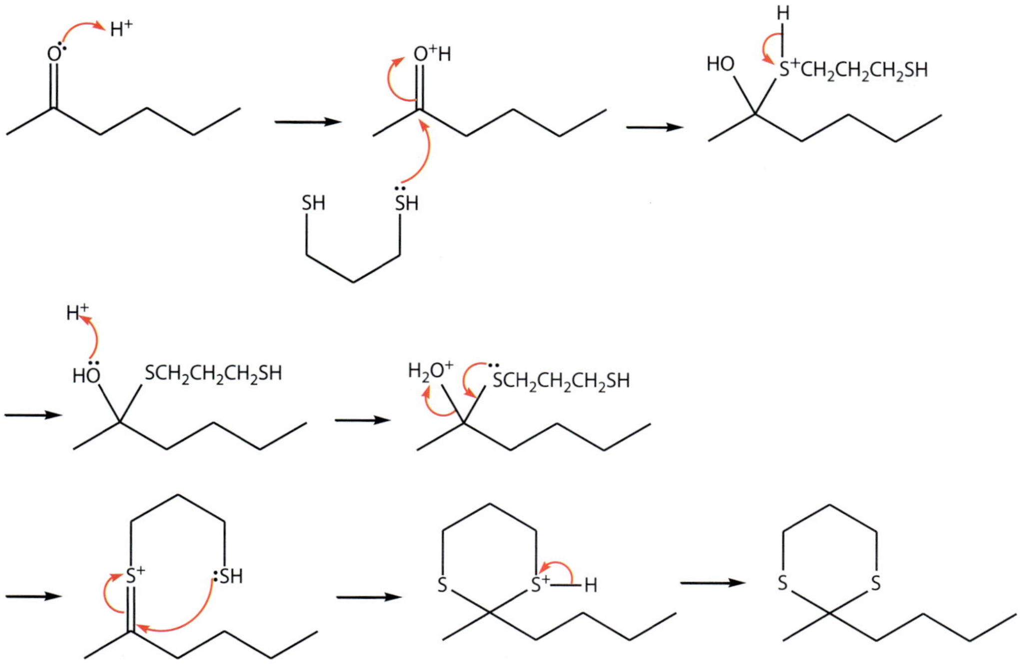

Note that thioacetals are not easily deprotected using aqueous Brønsted acids; sulfur is much less easily protonated than oxygen. The normal conditions for thioacetal hydrolysis employ Hg^{2+} catalysis.

The main use for acetals in organic synthesis is to protect carbonyl compounds, in order to perform a transformation on some other parts of the molecule—with a reagent that would normally react with a carbonyl compound. For example, if our substrate is a ketoester, sodium borohydride, $Na[BH_4]$ reduces the ketone, but not the ester. Lithium aluminum hydride, $Li[AlH_4]$, reduces both the ketone and the ester (Figure 14.26). How could we reduce the ester and not the ketone? There is no simple reagent that will do this, because ketones are more reactive than esters. So we temporarily "turn off" the ketone functionality by conversion into the acetal, so that we can reduce the ester. At the end of that reaction, the acetal is hydrolyzed back to the ketone.

In another synthetic sequence (Figure 14.27), it was necessary to alkylate a 1-alkyne. The strong base required to do this would deprotonate at the carbon atom next to the ketone ($pK_a \sim 20$) rather than at the terminal alkyne ($pK_a \sim 25$). So the ketone is temporarily converted to the acetal and then released once the alkyne has been alkylated.

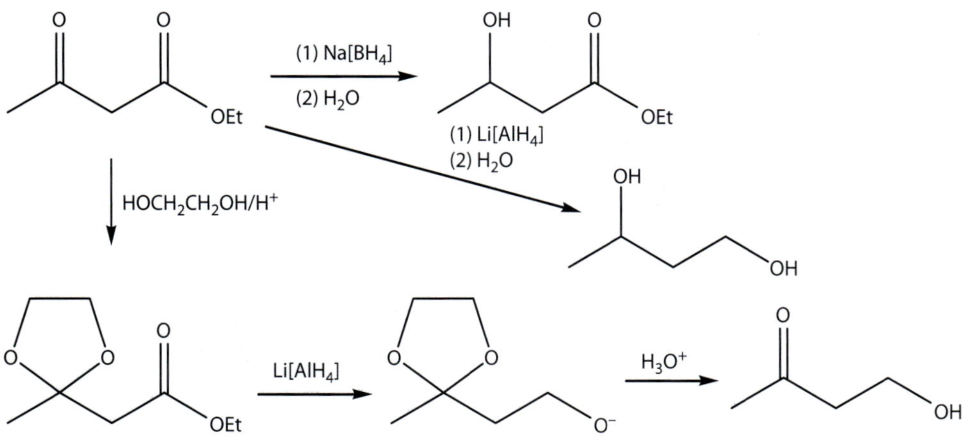

FIGURE 14.26 Reduction of ketoesters.

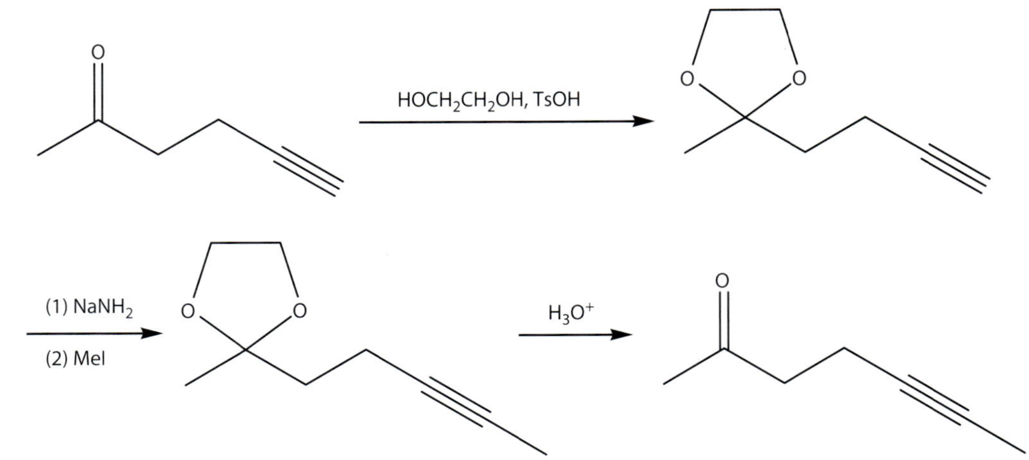

FIGURE 14.27 Use of a cyclic acetal as a protecting group.

PROBLEM 14.6

Write a complete mechanism for the reactions in Figure 14.27.

Solution

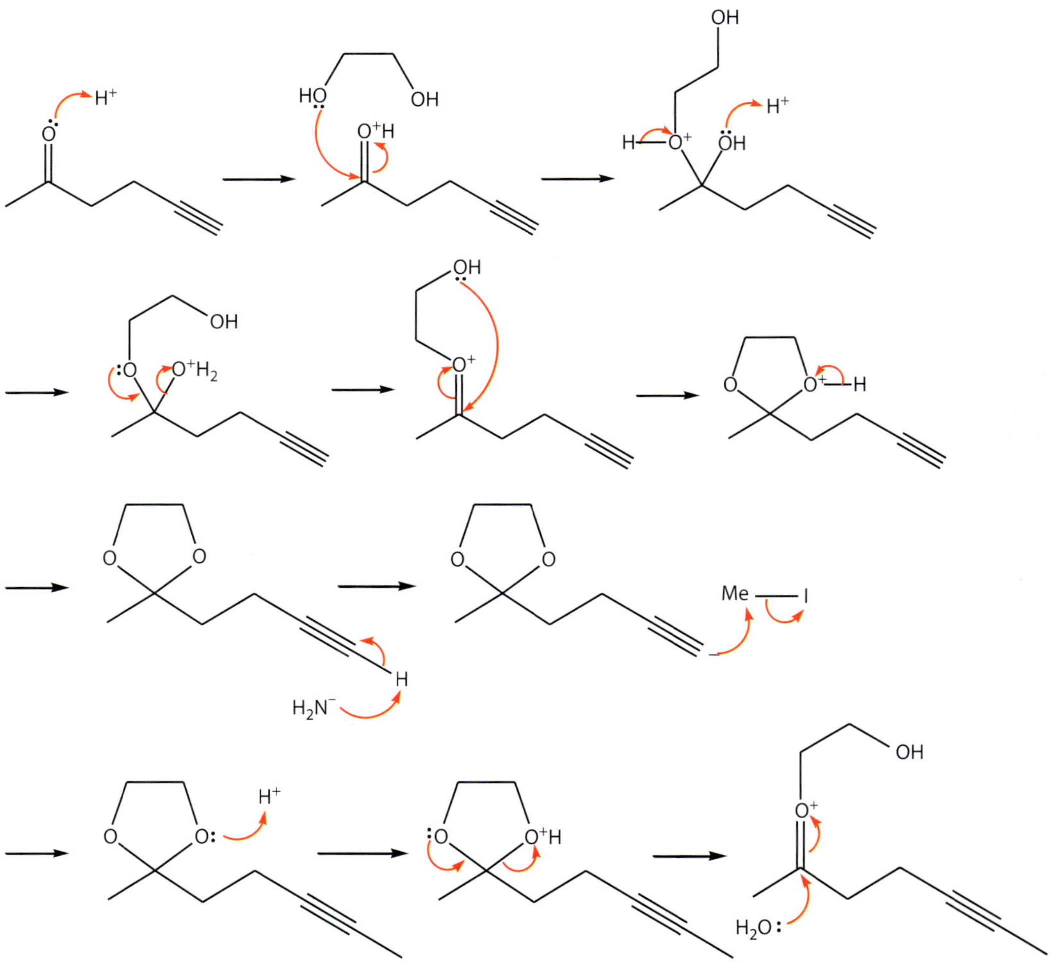

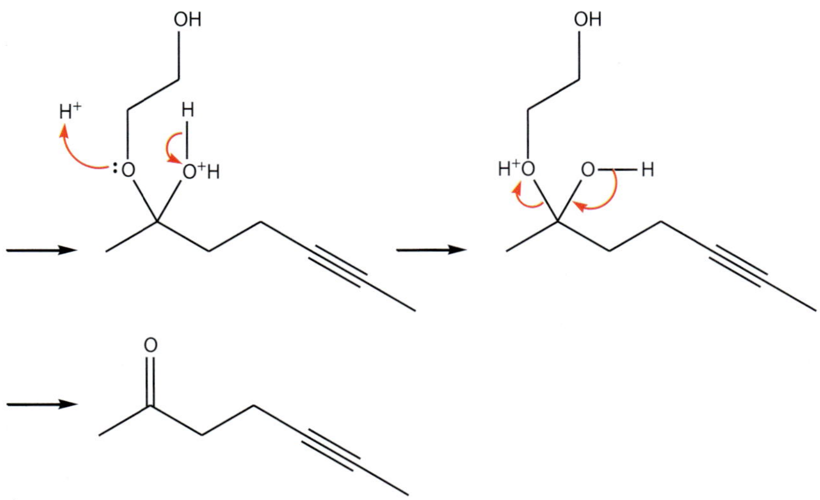

You might have thought that the alkyne could also be hydrated under conditions of aqueous acid, but it was possible to adjust conditions to select for the sequence shown.

14.3.4 IMINES, ENAMINES, OXIMES, AND HYDRAZONES

We know that ammonia is a good nucleophile, so we would expect that it would attack carbonyl compounds. This does happen, but the product, a hemiaminal, is generally unstable (an exception is provided by $(CF_3)_2C(OH)NHR$) and loses water (Figure 14.28) to give an imine. Unsubstituted imines are also quite unstable (again, the derivative of hexafluoroacetone is an exception), but imines substituted at nitrogen are more stable (Figure 14.29). Notice that the structure of all imines and related compounds is bent at nitrogen—although we do not usually show it, there is a lone pair of electrons on the sp^2 nitrogen atom. All of the steps are reversible, and imines are readily hydrolyzed. This is both biologically important, in transaminase reactions, and synthetically useful, as we will see later in this chapter.

Some imine-forming reactions are shown in Figure 14.30—all of these were part of longer synthetic sequences. Imines undergo many of the same types of addition reactions as aldehydes and ketones. A particularly useful example of this is their reaction with cyanide ion, the Strecker reaction (Figure 14.31). When a carbonyl compound is treated with ammonia and sodium or KCN, the ammonia adds to the carbonyl to give an unstable imine. This is attacked by cyanide to give the α-aminocyanide. Since the cyanide can be hydrolyzed to a carboxylic acid, this constitutes a simple amino acid synthesis.

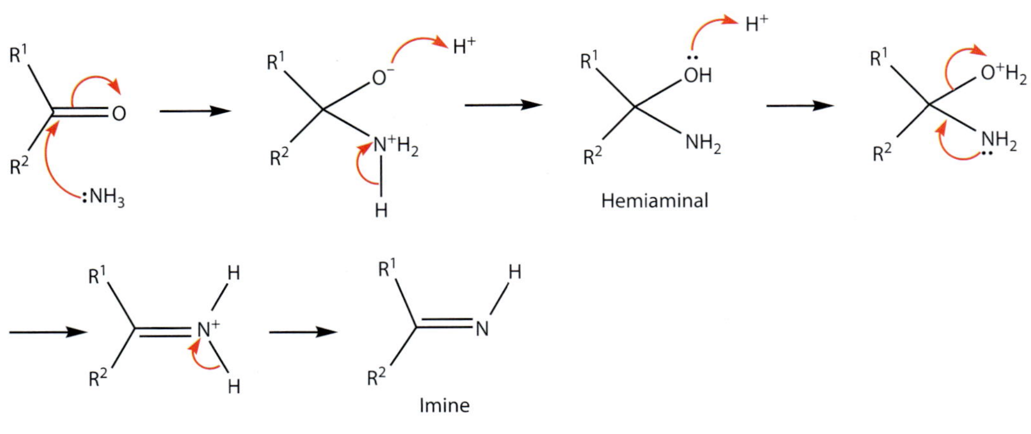

FIGURE 14.28 Mechanism of formation of an unsubstituted imine.

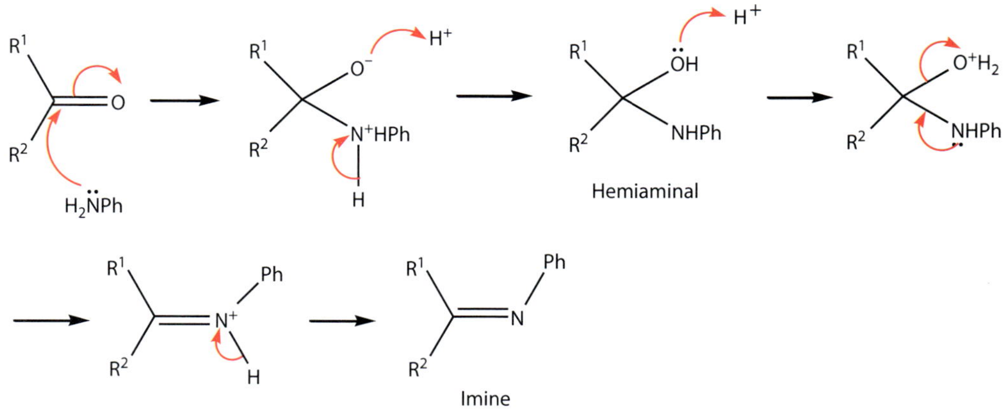

FIGURE 14.29 Mechanism of formation of a substituted imine.

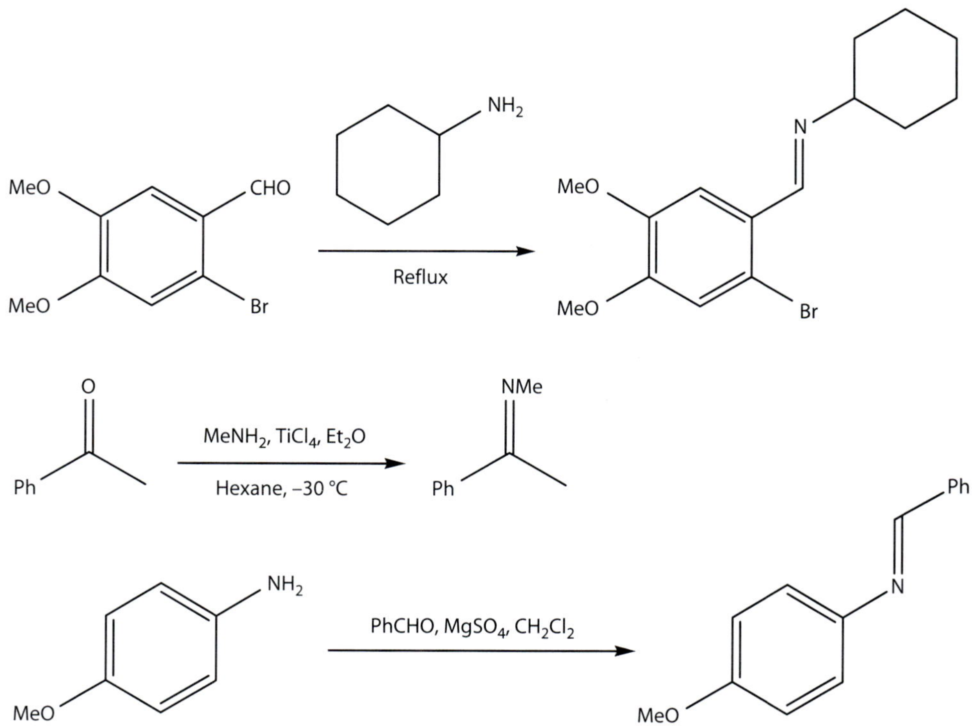

FIGURE 14.30 Imine formation reactions.

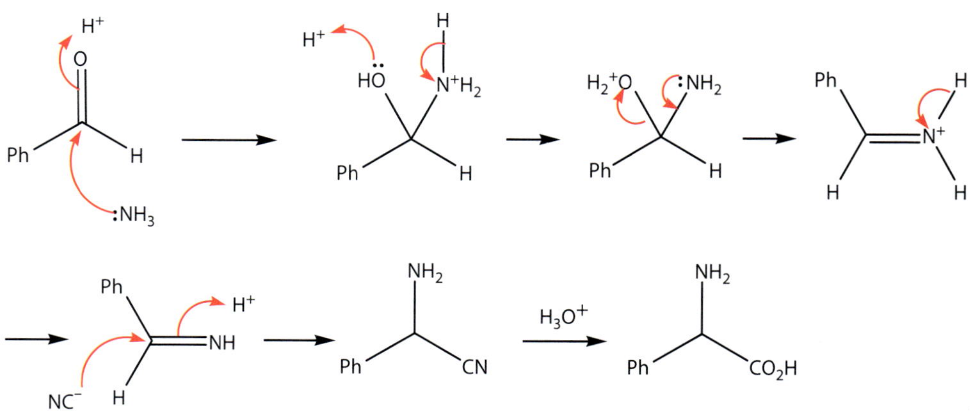

FIGURE 14.31 Strecker synthesis of amino acids.

PROBLEM 14.7

Write down the structures of the imines formed in the following reactions, and assign *E*/*Z* stereochemistry to the carbon–nitrogen double bonds. It is reasonable to assume that the less sterically hindered isomer will be formed.

(a) $PhCHO + MeNH_2 \rightarrow$

(b)

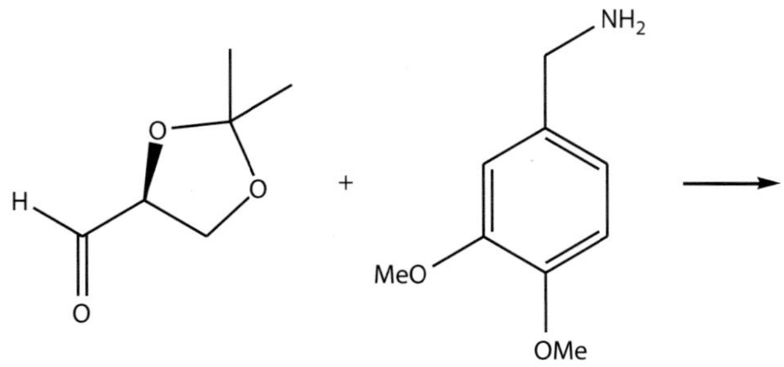

Solutions

Notice that in (b), the acetal is unreactive and stable to the reaction conditions:

(a)

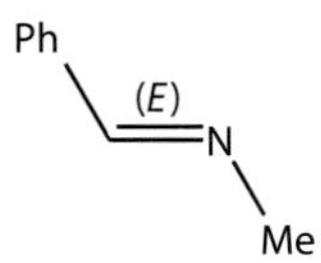

(b)

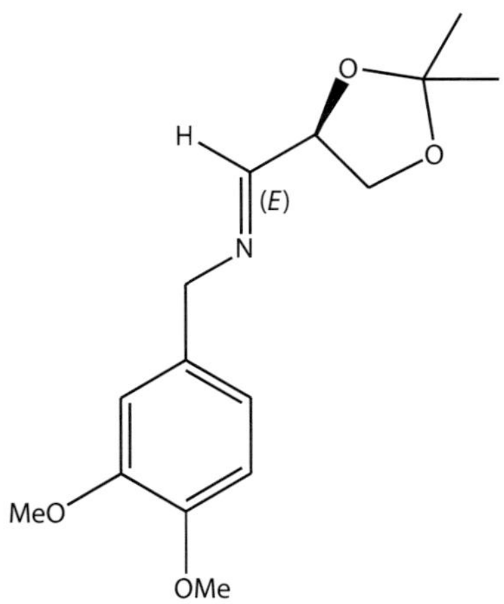

PROBLEM 14.8

In Chapter 9, we used the reagent hexamethylenetetramine, **14.24**, as an alternative for ammonia in nucleophilic substitutions to prepare amines. This reagent is prepared by condensation of ammonia and formaldehyde. Draw out the fundamental reactions that are necessary to build this caged structure. When the nitrogen has substituted an alkyl halide, the product is converted to the amine by acid hydrolysis. Draw out the key steps in this process.

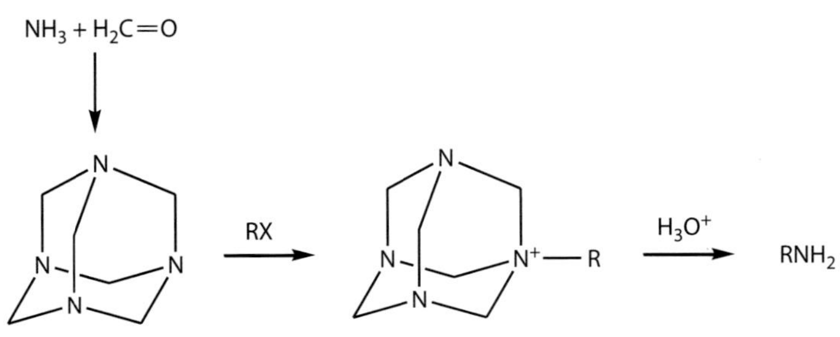

14.24

Solution

In forming the reagent, the key reactions are nucleophilic addition of the amine to formaldehyde, formation of an imine or iminium salt, followed by attack of the ammonia on the carbon nitrogen double bond. Many cyclic and caged structures could be made this way, and there is no inherent way to determine the order of assembly of the cage. The particular structure isolated has the same structure as the very stable adamantane and is highly crystalline.

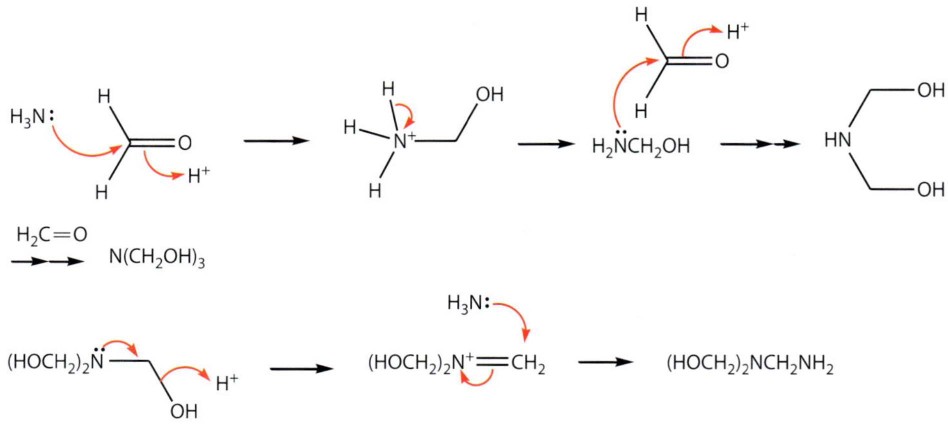

When the nitrogen has been alkylated, and the intermediate is subjected to acid-catalyzed hydrolysis, then the process goes into reverse:

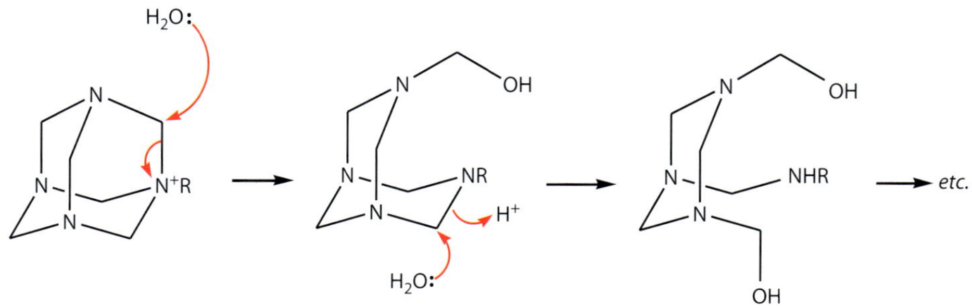

Focus on Transamination

The transamination reaction is important biologically in amino acid metabolism. Simple aldehydes are rare in biological systems and are mostly masked as imines. Biochemists often refer to them as Schiff bases, which are a special class of aldehyde imine where the nitrogen atom is substituted by an alkyl or aryl group. The transamination reaction interconverts amino and carbonyl functionalities (Figure 14.32). The enzymes involved in the process are called transaminases, and they require pyridoxal phosphate as a cofactor.

Although it is superficially attractive to write a mechanism that consists of formation of an imine, tautomerization, and hydrolysis, this is not what happens nor does it explain the need for pyridoxal phosphate in the reaction. The pyridoxal phosphate is bonded to a lysine of the enzyme as an imine (**14.25**). In the transamination reaction, another free amino acid displaces the lysine that is part of the enzyme from pyridoxal, in an imine exchange reaction (Figure 14.33).

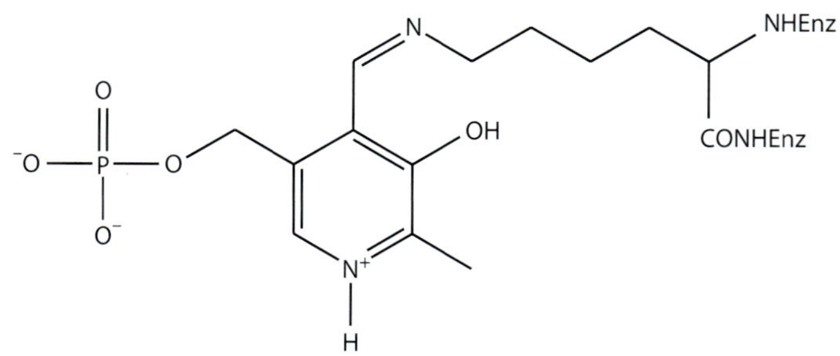

14.25

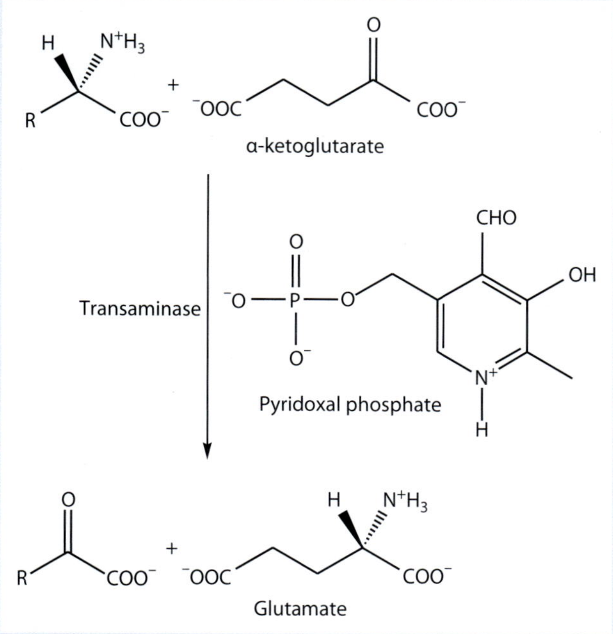

FIGURE 14.32 Transamination reaction.

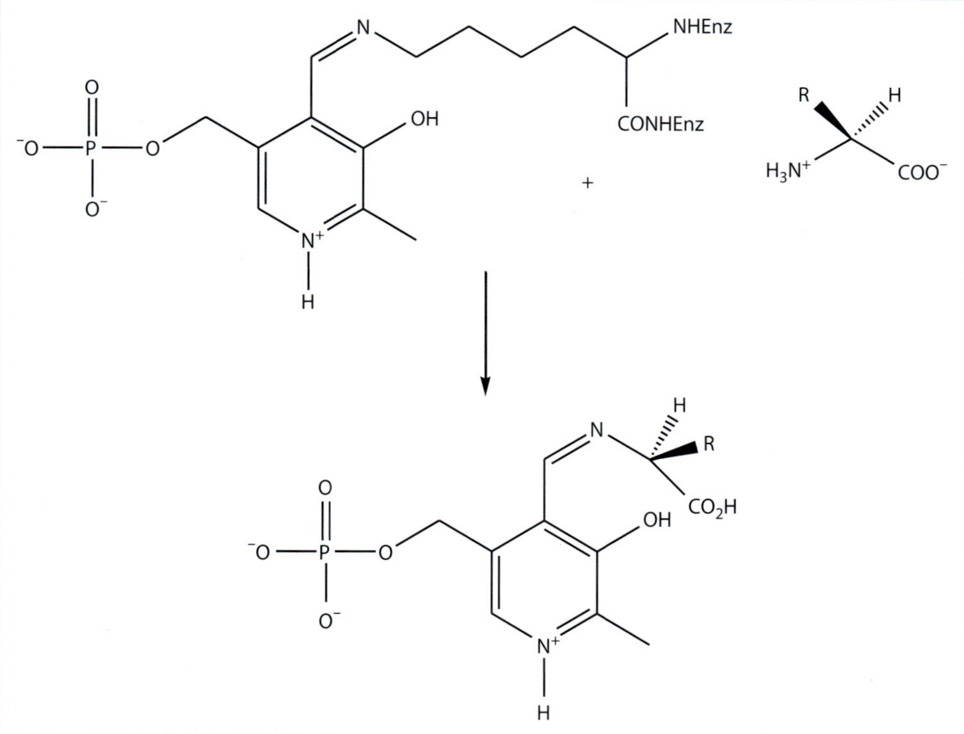

FIGURE 14.33 Imine exchange in the transamination process.

Next, the transaminase enzyme catalyzes the tautomerization of the imine, which is then hydrolyzed to the α-ketoacid and pyridoxamine (Figure 14.34). To reconvert the pyridoxamine to pyridoxal, the reaction runs in the reverse direction, this time converting α-ketoglutarate to glutamate.

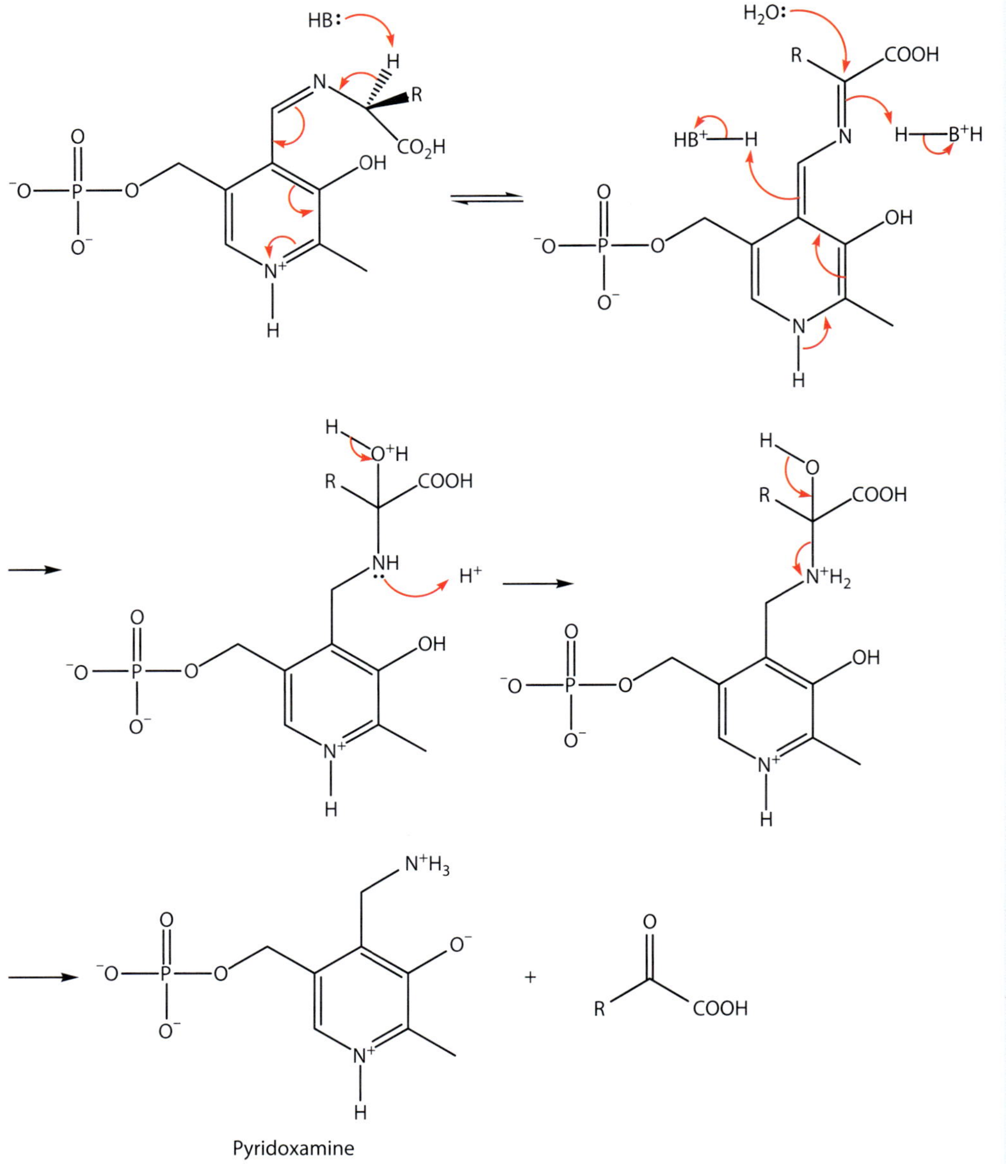

FIGURE 14.34 Mechanism of transamination.

Imines can only be formed from carbonyl compounds and primary amines, but secondary amines are also good nucleophiles and also attack aldehydes and ketones. The initial product (**14.26**) is a hemiaminal, but again this is not stable. The lone pair of electrons on nitrogen is used to expel water, to give an iminium salt. Since this iminium salt does not have a proton on nitrogen to lose, it must lose one from carbon, to form an enamine (Figure 14.35). Enamines are stable and may be distilled (most are liquid) without decomposition in the absence of water. But in aqueous acid, they are hydrolyzed, exactly reversing the steps by which they were formed. This is another equilibrium reaction that we can drive in the required direction by mass action effects. Enamines are prepared using a Dean–Stark apparatus so that water is removed as it is formed, and are hydrolyzed in

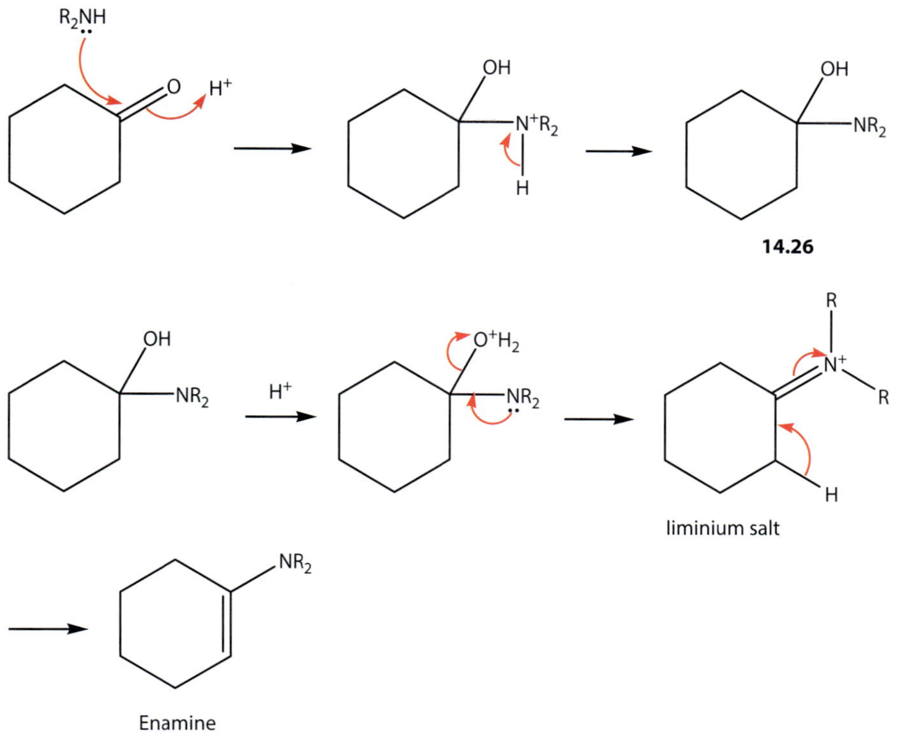

FIGURE 14.35 Mechanism of enamine formation.

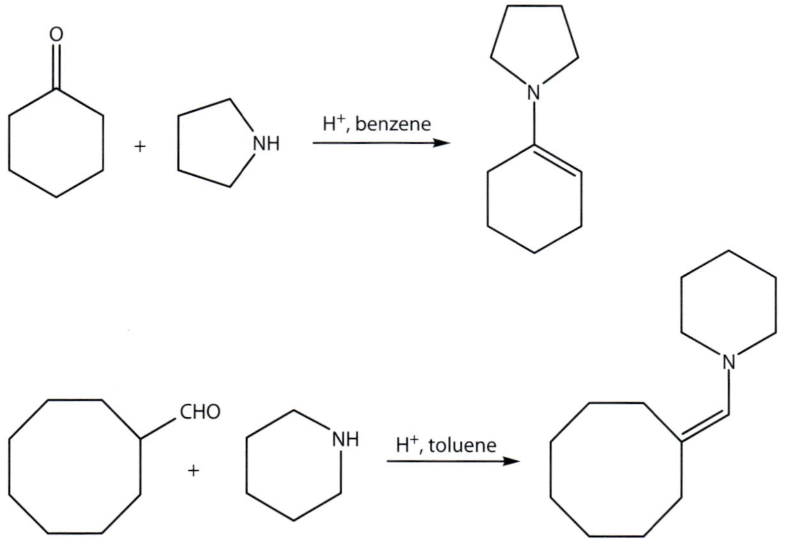

FIGURE 14.36 Preparation of enamines for use in synthesis.

aqueous acid. Figure 14.36 shows some enamine-forming reactions; as with the imines, these reactions are all part of longer synthetic sequences, and most practical examples employ cyclic secondary amines. Note that while imines are electrophilic (like carbonyls), enamines are generally nucleophilic, like enols or enolate anions. We will explore these ideas more fully in Chapters 17 and 20.

Adducts of other amino-containing compounds tend to be more stable and are less easily hydrolyzed. For example, hydroxylamine reacts with ketones to give oximes (Figure 14.37). The first thing to notice about the process is that, although in principle the hydroxylamine has two nucleophilic sites, at oxygen and nitrogen, it is the nitrogen that attacks the carbonyl. Nitrogen is generally more nucleophilic than oxygen, because it is less electronegative, and thus more willing to share its lone pair of electrons. The rate of oxime formation is strongly dependent on solution pH, and this dependence is similar for most other nitrogen nucleophiles we will meet. A plot of rate against pH is a bell-shaped curve with a maximum at approx. pH 4.0. In mildly

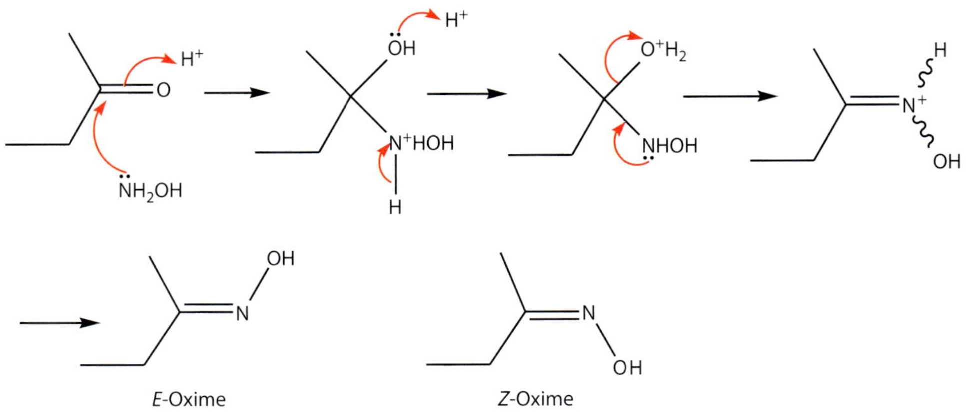

FIGURE 14.37 Mechanism of oxime formation.

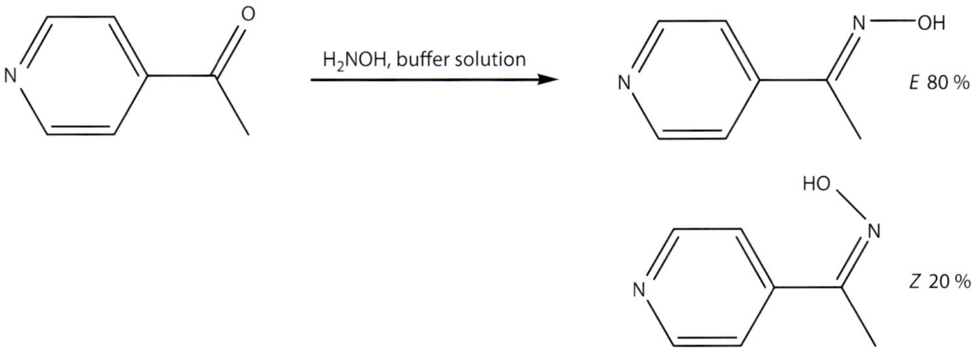

FIGURE 14.38 Oxime formation may be stereoselective.

acidic, neutral, or basic solutions, the loss of water is the rate-limiting step. This depends on protonation of the OH group, to make it into a good leaving group, so we might expect that the reaction would become more rapid as the solution became more acidic. However, in very strong acid, the hydroxylamine itself becomes protonated to give H_3N^+OH, which is not a nucleophile. So in strongly acid solution, the first step, the attack of hydroxylamine on the carbonyl, becomes rate limiting. We'll return to this type of rate/pH dependence when considering catalysis in Chapter 23.

Because the lone pair of electrons on the nitrogen atom is stereochemically active, two oxime isomers can be obtained from an aldehyde or a nonsymmetric ketone. Generally, the less sterically hindered isomer predominates. For example, in the reaction of Figure 14.38, 80 % of the *E*-isomer is obtained, and careful recrystallization isolates a stereochemically pure sample.

The synthetic usefulness of oximes lies in their subsequent transformations. A quick look ahead (Figure 14.39) shows us that they can be reduced to amines, and can rearrange, or undergo elimination to nitriles.

Hydrazine, H_2NNH_2, reacts with aldehydes and ketones in the same way as hydroxylamine, to give compounds called hydrazones. The reaction shows a similar pH dependence. However, depending on the stoichiometry, hydrazine can react twice, using both of the NH_2 groups (Figure 14.40).

A very common derivative of hydrazine, 2,4-dinitrophenylhydrazine, reacts with aldehydes and ketones to give yellow, orange, or red crystalline derivatives (Figure 14.41). Using the substituted hydrazine, only the terminal nitrogen can now react; the other is rendered much less nucleophilic by the electron-withdrawing 2,4-dinitrophenyl group. The reagent was long used as a test for aldehydes and ketones, and these crystalline derivatives were used to characterize liquid ketones. A typical test is shown in Figure 14.42 and the dinitrophenylhydrazone of benzaldehyde in Figure 14.43.

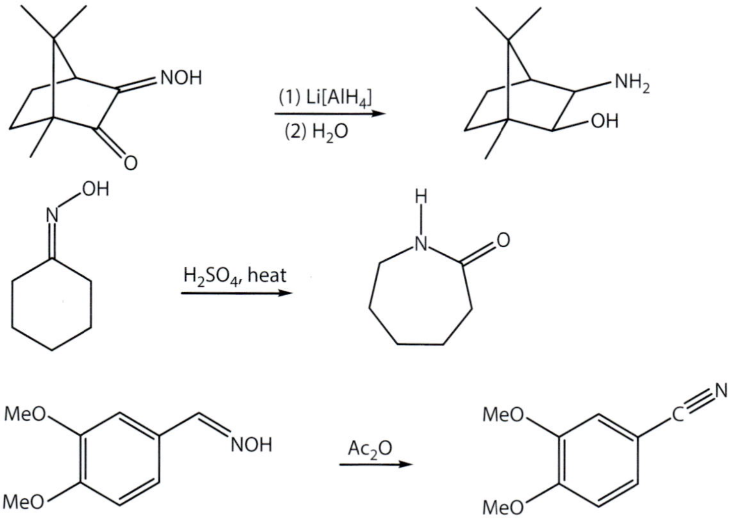

FIGURE 14.39 Uses of oximes in synthesis.

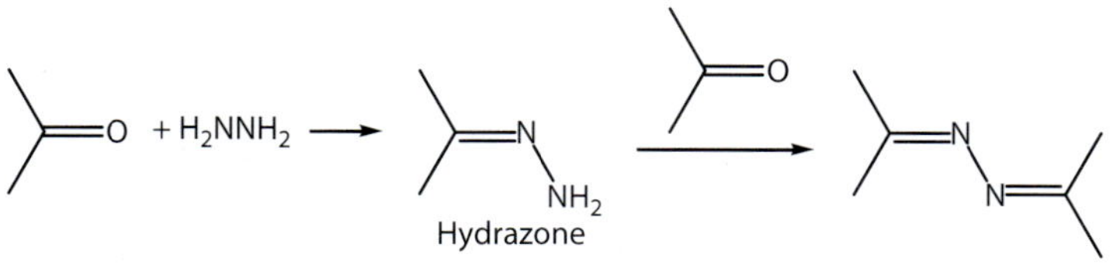

FIGURE 14.40 Reaction of a ketone with hydrazine.

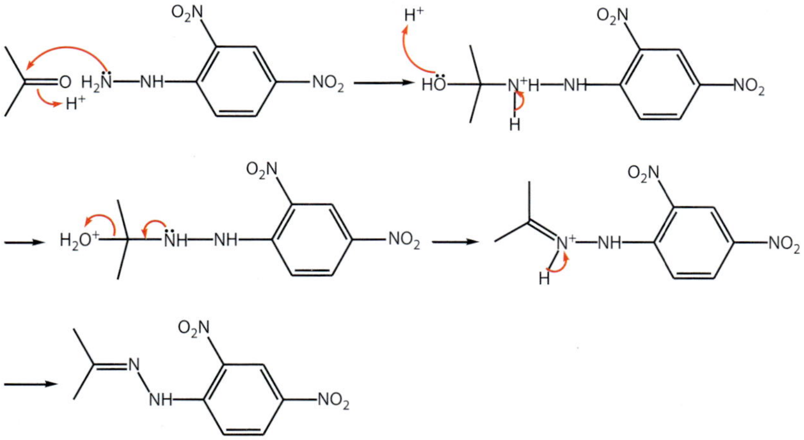

Acetone 2,4-dinitrophenylhydrazone

FIGURE 14.41 Mechanism of reaction of 2,4-dinitrophenylhydrazone with ketones.

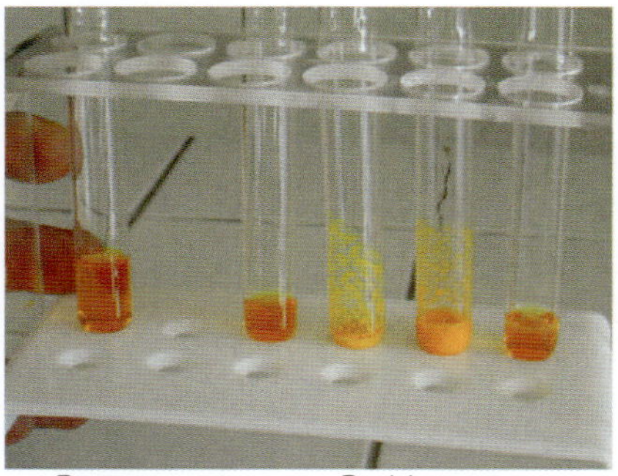

Reagent Positive test

FIGURE 14.42 Testing for ketones using dinitrophenylhydrazine.

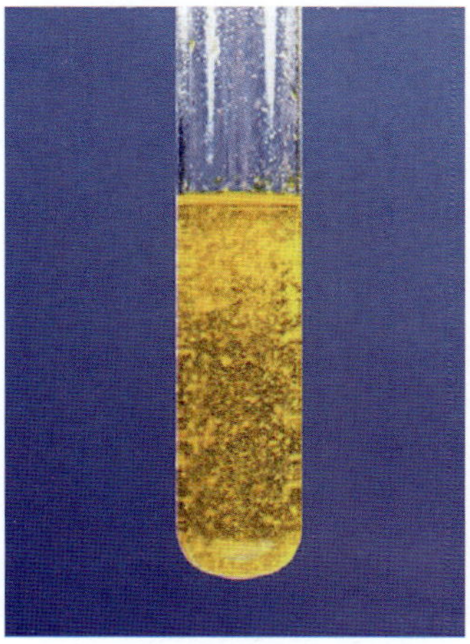

FIGURE 14.43 Ethanal dinitrophenylhydrazone. (From Andrew Lambert/Science.)

PROBLEM 14.9

Predict the product(s) of each of the following reaction sequences:

(a)

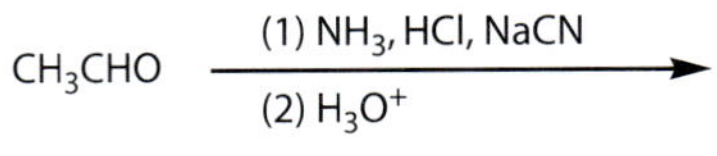

(b)

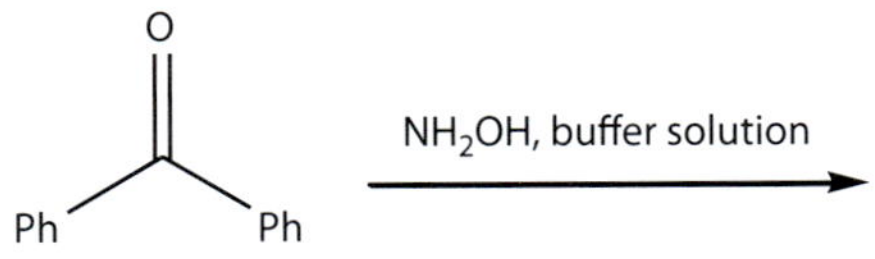

(c)

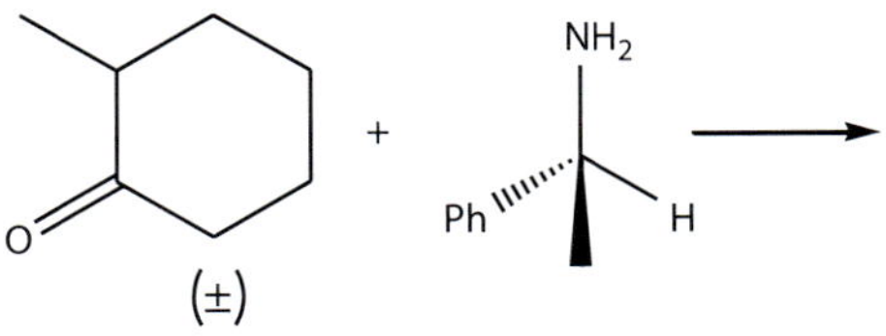

(d)

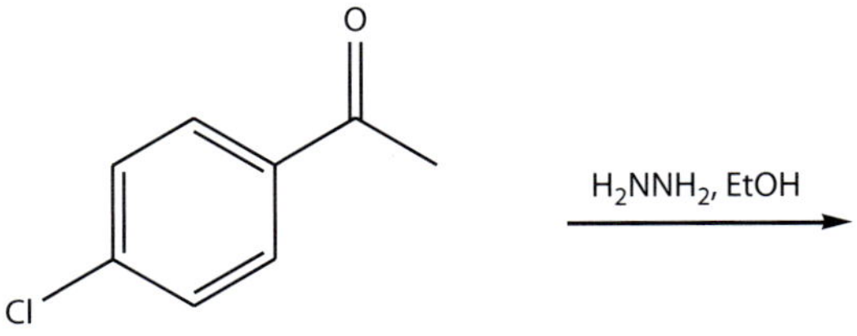

Solutions

All of the reactions shown are fairly standard in type. The first is a Strecker synthesis; the first formed product is the α-aminonitrile, which is then hydrolyzed to an amino acid. The second reaction is a standard oxime synthesis; as the ketone is symmetric, there is no stereochemistry to worry about. In (c), an imine is formed; the stereochemistry about the carbon–nitrogen double bond is determined by steric hindrance. Since the SM is racemic, two diastereoisomers will be formed, hence the wavy bond to the methyl group. In the final example, a hydrazine is prepared; while it's not possible to be certain of the stereochemistry about the C=N double bond, it's logical to assume that the less sterically hindered isomer will predominate:

(a)

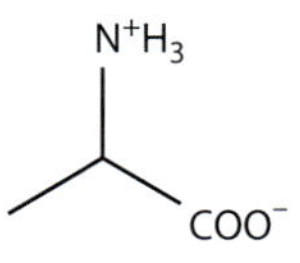

(b)

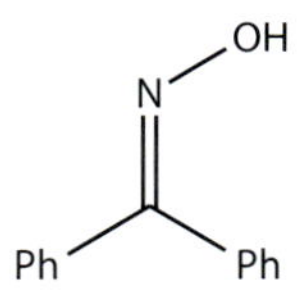

(c)

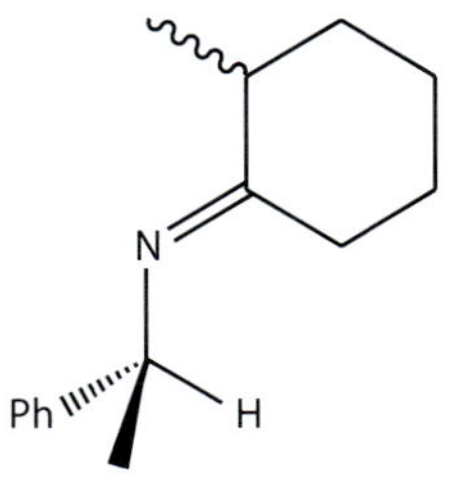

(d)

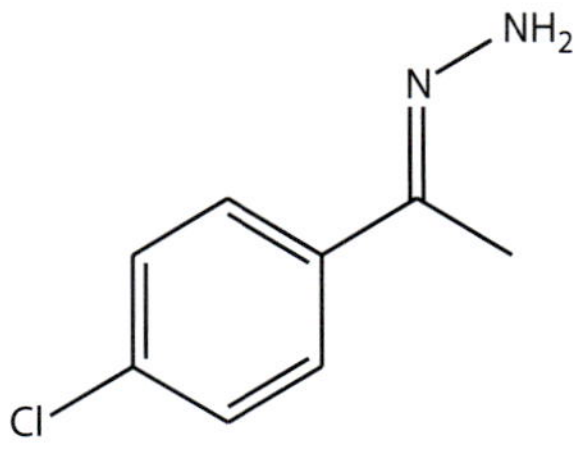

Key Points from Sections 14.2 and 14.3

- Addition to aldehydes and ketones can be catalyzed by either acid or base. In acid solution, the first step is protonation of the carbonyl at oxygen. In base, the first step is attack of the nucleophile at carbon.
- Water and HCN add reversibly to carbonyl compounds to give, respectively, hydrates and cyanohydrins. The position of the equilibrium depends on both electronic and steric factors.
- Alcohols add reversibly to aldehydes and ketones in acid or basic solution to give hemiacetals. The most stable hemiacetals are those where a five- or six-membered ring is formed, by an intramolecular process.
- Alcohols add reversibly to aldehydes and ketones under acid-catalyzed conditions to give acetals (or ketals); these are relatively stable in the absence of water and may be used to protect carbonyl compounds.
- Thiols react in the same way as alcohols, to give hemithioacetals and thioacetals.
- A wide range of compounds containing $-NH_2$ groups add to carbonyl compounds, with loss of water, to give compounds with carbon–nitrogen double bonds. Primary amines are converted to imines, hydroxylamine gives oximes, and hydrazine gives hydrazones. Stability is variable, but all these compounds can be hydrolyzed to regenerate the carbonyl group.
- The formation of dinitrophenylhydrazones has been used to test for, and characterize, aldehydes and ketones.
- Secondary amines react with carbonyl compounds to give enamines, which are stable in the absence of water.

14.4 IRREVERSIBLE ADDITION REACTIONS

Thus far, we have looked mainly at the reactions of oxygen, sulfur, and nitrogen nucleophiles with carbonyl compounds. All of these (and cyanide) have at least the potential to be leaving groups, once they are protonated. However, if we add carbanionic or hydridic nucleophiles to the carbonyl group, there are no circumstances in which these can behave as leaving groups. So additions of this type of nucleophile are nonreversible.

14.4.1 REPLACEMENT OF CARBONYL OXYGEN BY HYDROGEN

In this reaction, we want to remove the carbonyl oxygen and replace it with two hydrogen atoms to give a CH_2 moiety. Remember that when we looked at substituting a benzene ring with a linear alkyl group, this was something that we needed to be able to do (Section 12.2.2). Friedel–Crafts alkylations frequently lead to multiple reactions and rearrangements, but acylations are much smoother. The acyl benzene can then be reduced to the hydrocarbon. The reactions that we met to do this were the Clemmensen reduction, using zinc amalgam and conc. HCl (the mechanism of this reaction is complex, probably involving a carbene bound to the zinc surface), and hydrogenolysis using a palladium catalyst (Figure 14.44). However, the Clemmensen reduction is not suitable for substrates containing acid-sensitive functional groups, and hydrogenolysis works only for ketones next to an aromatic ring. Hydrogenolysis is also unsuitable for substrates with carbon–carbon double bonds elsewhere in the molecule, as these would also be reduced.

The use of a hydrazine intermediate leads to the Wolff–Kishner reduction, which is carried out under basic, rather than acidic, conditions (Figure 14.45). The hydrazone is formed as previously and is then deprotonated at nitrogen by the strongly basic conditions. We can view this step as a type of tautomerism. The terminal NH is again deprotonated, and this time, the molecule fragments to give a carbanion and molecular nitrogen. The carbanion is protonated by the solvent, regenerating the base. The process is driven by the release of molecular nitrogen.

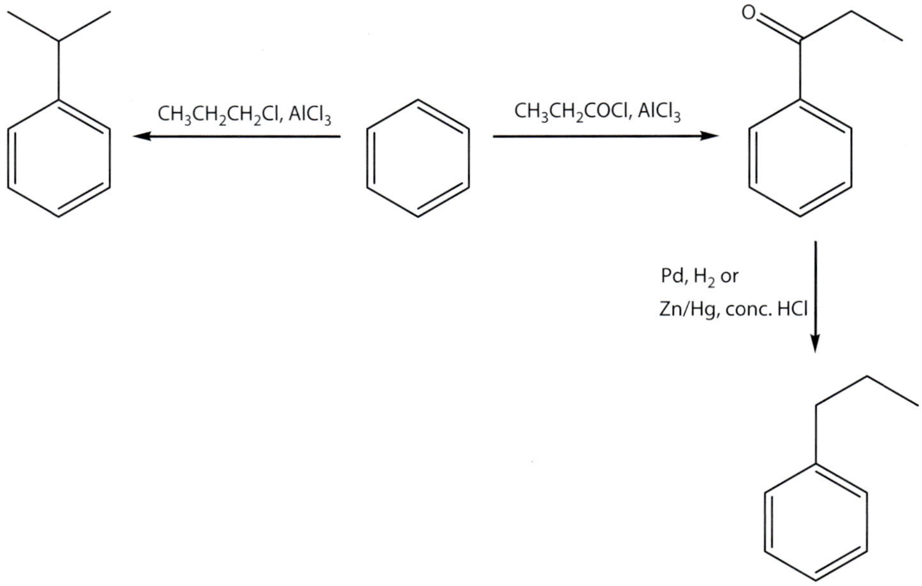

FIGURE 14.44 Friedel–Crafts alkylation and acylation.

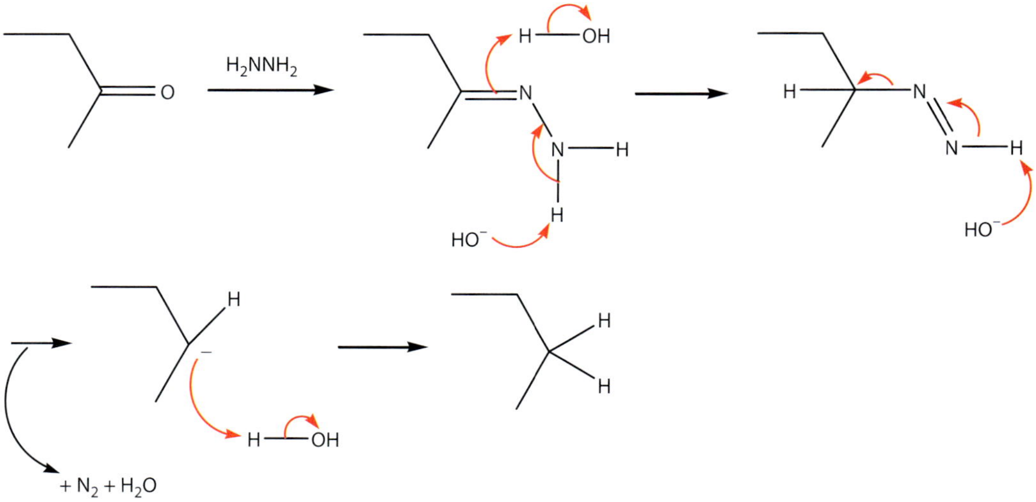

FIGURE 14.45 Mechanism of the Wolff–Kishner reduction.

Examples of the Wolff–Kishner reduction are given in Figure 14.46. In the first example, notice that the ester group is not reduced—esters don't form hydrazones. In the second example, the acetal group survives the conditions of the Wolff Kishner; a Clemmensen reduction would not have succeeded here because of the sensitivity of the acetal to acid conditions. Neither the Clemmensen nor the Wolff–Kishner reductions are generally successful for enones. One final method that may be employed involves the formation of a dithioacetal, followed by treatment with hydrogen/Raney nickel (Figure 14.47). We will explore the affinity for this specially prepared type of nickel for sulfur in Chapter 23.

14.4.2 ADDITION OF HYDRIDE: REDUCTION TO ALCOHOLS OR AMINES

Simple hydrides such as sodium or potassium hydride are too insoluble in organic solvents to be useful in reductions. The compounds sodium borohydride, $Na[BH_4]$, and lithium aluminum hydride, $Li[AlH_4]$, are the usual sources of hydride for carbonyl reduction reactions.

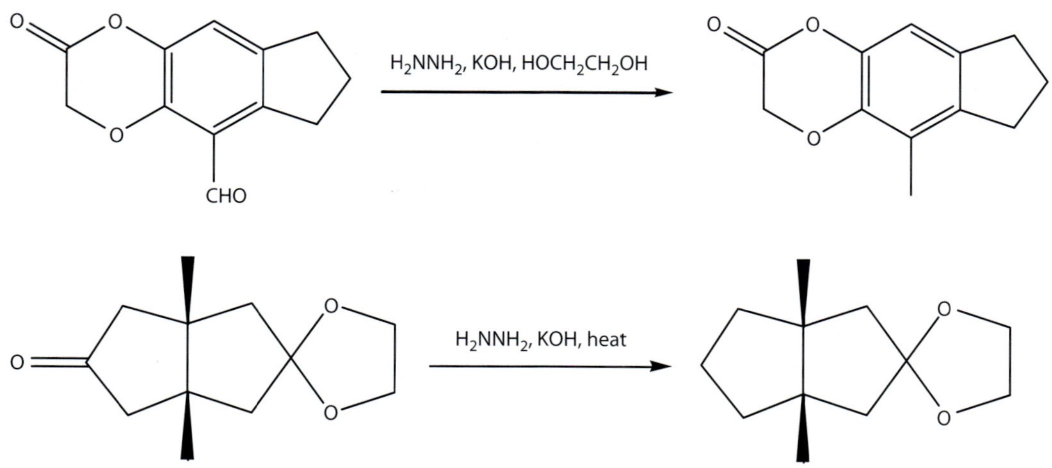

FIGURE 14.46 Uses of the Wolff–Kishner reduction.

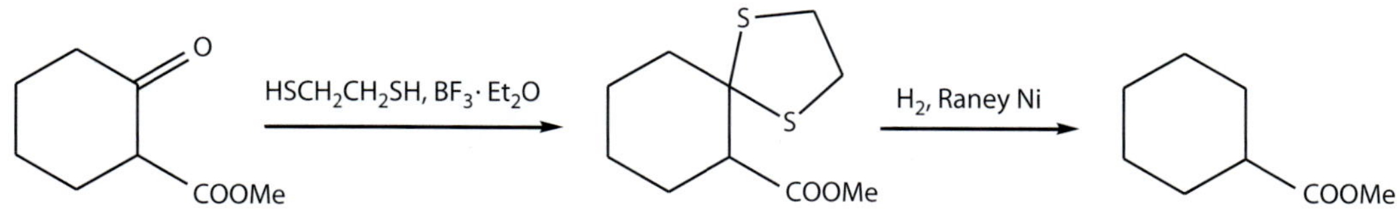

FIGURE 14.47 Desulfurization of a thioacetal using Raney nickel.

Lithium aluminum hydride is prepared from LiH and AlH_3; as a student, it will quickly become one of your best friends, because it will reduce almost everything! Many analogues and substituted versions of these reagents exist, and vary widely in reactivity, both for steric and electronic reasons. Hydride is transferred from $[AlH_4]^-$ to the carbon of the carbonyl, which is initially rendered more electrophilic by interaction with the lithium cation. At the end of the reaction, the alkoxide is protonated to give the alcohol (Figure 14.48). Lithium aluminum hydride is VERY reactive and reacts violently with water or any protic solvent, to give hydrogen—and this frequently ignites in such incidents. So it is important that the solvents used for these reactions are rigorously dry. Sodium borohydride reacts similarly with aldehydes and ketones, but it is much less reactive. Although it does react slowly with alcohols and alkaline water, it can be used in these solvents.

Because the carbon of the carbonyl group is sp^2 hybridized, it is planar and may be attacked from either face with equal likelihood. So, in the reduction of acetophenone, equal amounts of the two enantiomers are produced (Figure 14.49). However, if the two products produced are

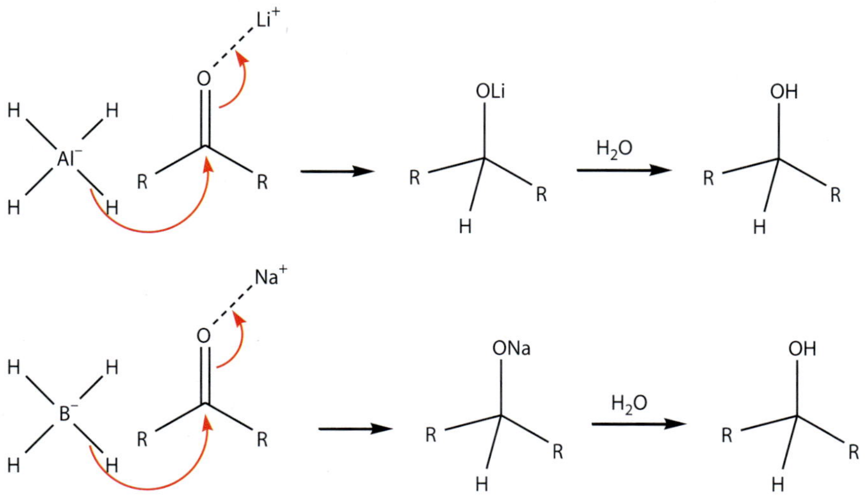

FIGURE 14.48 Addition of hydride to carbonyl compounds.

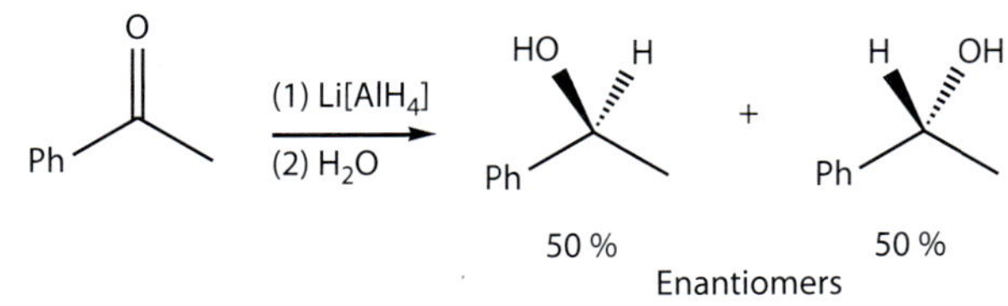

FIGURE 14.49 Reduction of acetophenone with lithium aluminum hydride gives a racemic product.

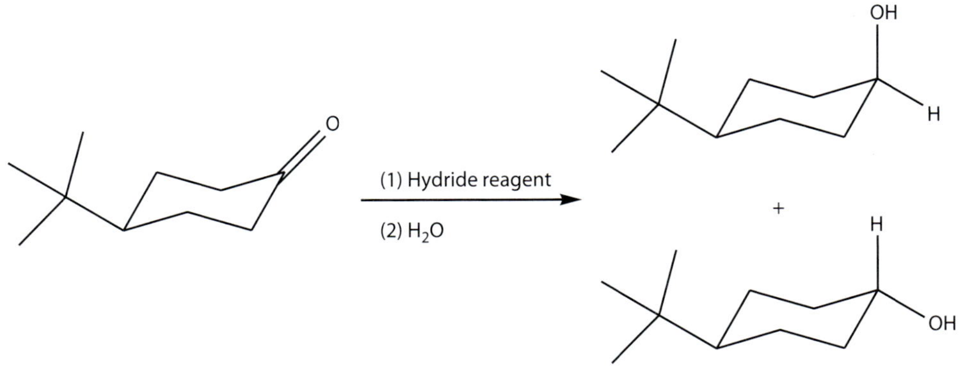

FIGURE 14.50 Diastereomeric reduction products are produced in unequal amounts.

diastereoisomers rather than enantiomers, then they will be produced in unequal amounts. They are different compounds, and we can reasonably expect that the transition states leading to each will be of different energy. If we consider the reaction in Figure 14.50, we can approach a prediction of the reaction stereochemistry in two ways. If we consider the stability of the product, the formation of the equatorial alcohol will be favored; we would describe this as product development control. However, if we consider the approach of the reagent to the carbonyl group, then the approach from the equatorial direction is less hindered; this would give the axial alcohol as the hydride would be equatorial. This is described as steric approach control. The actual outcome for a particular reaction will depend on a number of factors including solvent, temperature, and the nature of the hydride reagent. A substituted hydride-reducing agent such as Li[AlH(O-*t*-Bu)$_3$] is very bulky and would tend to approach from the least hindered direction.

Examples of hydride reduction of aldehydes and ketones are shown in Figure 14.51. Notice in the first reaction that sodium borohydride selectively reduces the aldehyde in the presence of an ester; nitro and nitrile groups are also tolerated. In the second reaction, the borohydride reduction of a ketone, the diastereoselectivity, reflects the steric hindrance to the approach of $[BH_4]^-$; the methyl groups on the bridging carbon atom block the approach from the top face of the molecule. By contrast, little diastereoselectivity is observed in the reduction of 2-hydroxyclobutanone. The final example illustrates the reduction of an α,β-unsaturated ketone.

Can carbon–nitrogen double bonds be reduced by the same reagents as carbon–oxygen double bonds? Yes, but because the carbon–nitrogen bond is less polar than the carbon–oxygen bond, imines and related compounds are somewhat less reactive. Imines can be reduced either with sodium borohydride (Figure 14.52) or lithium aluminum hydride and its derivatives. However, the most commonly used reagent is sodium cyanoborohydride, Na[BH_3CN]. The electron-withdrawing cyanide group makes this a little less reactive than sodium borohydride, and it can be used in slightly acidic solutions, where it is possible for the imine to be protonated to enhance its electrophilicity (Figure 14.53). Notice here that the reagent is selective for the imine; the cyanide is not reduced. Overall, preparation of an imine followed by hydride reduction is a useful synthesis of amines.

Synthesis of amines by this route is also possible using secondary amines, in a one-pot process (Figure 14.54). This reductive amination involves reduction of the electrophilic iminium salt intermediate.

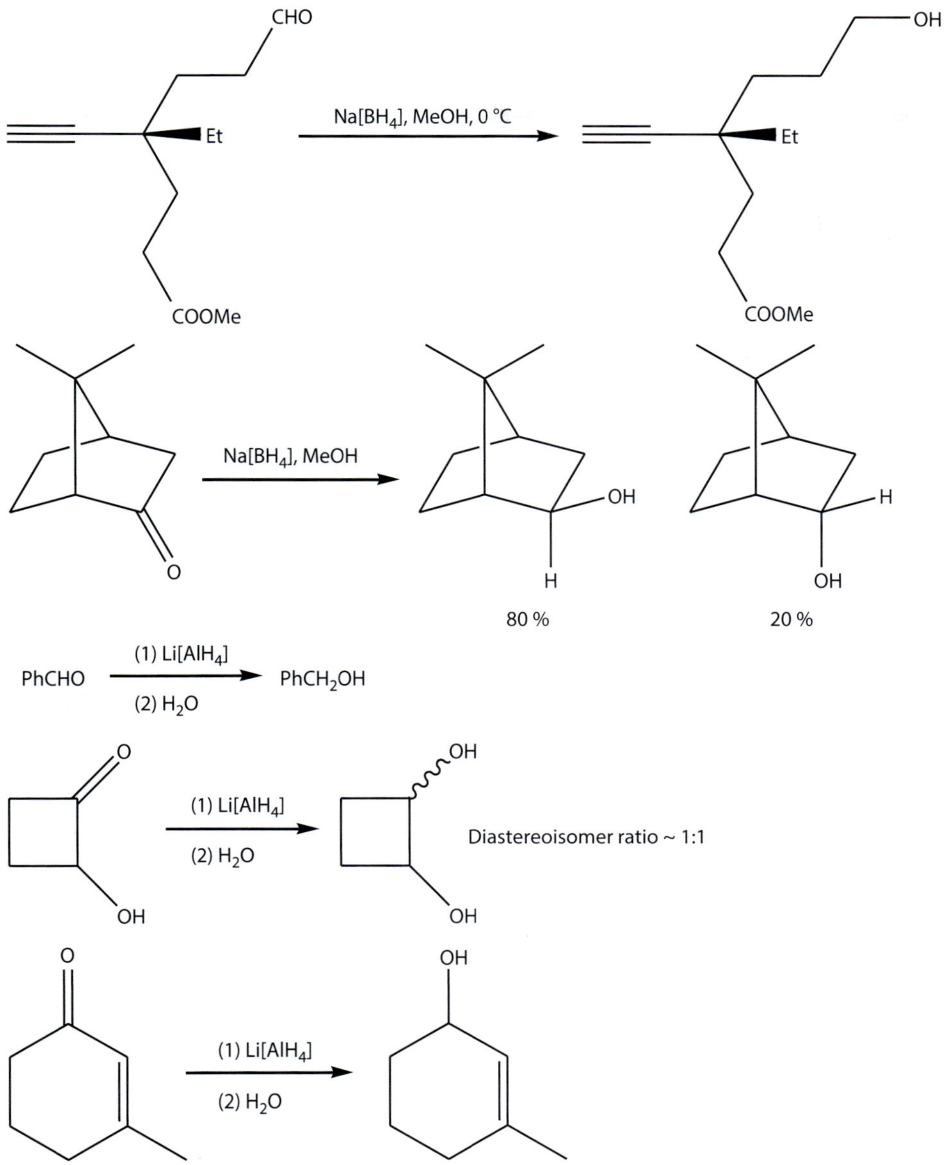

FIGURE 14.51 Hydride reductions of aldehydes and ketones.

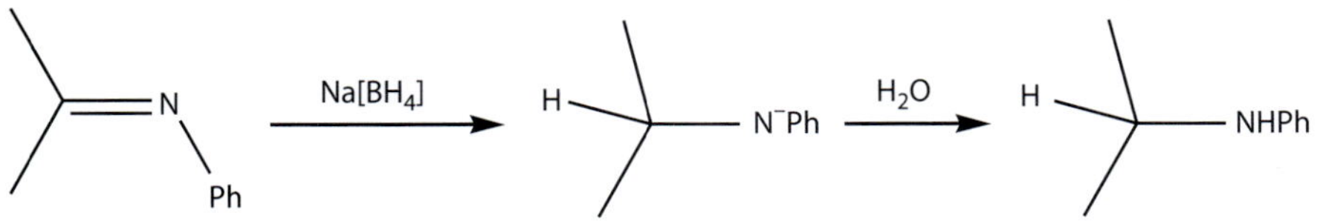

FIGURE 14.52 Imine reduction using sodium borohydride.

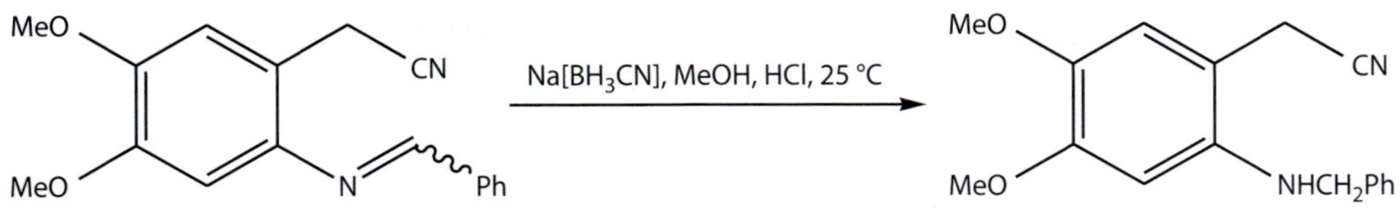

FIGURE 14.53 Imine reduction using sodium cyanoborohydride.

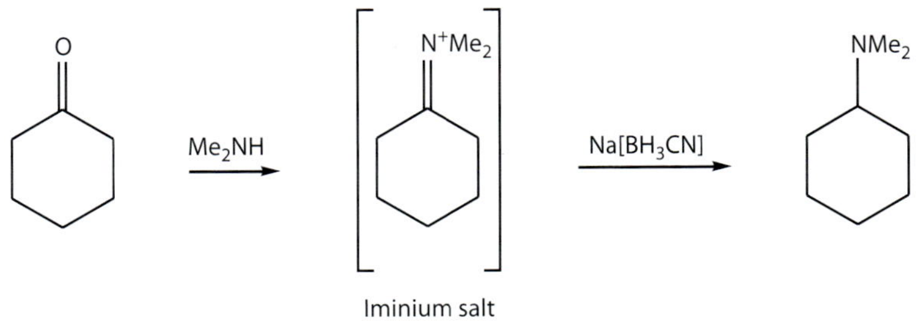

FIGURE 14.54 Reductive amination using a secondary amine.

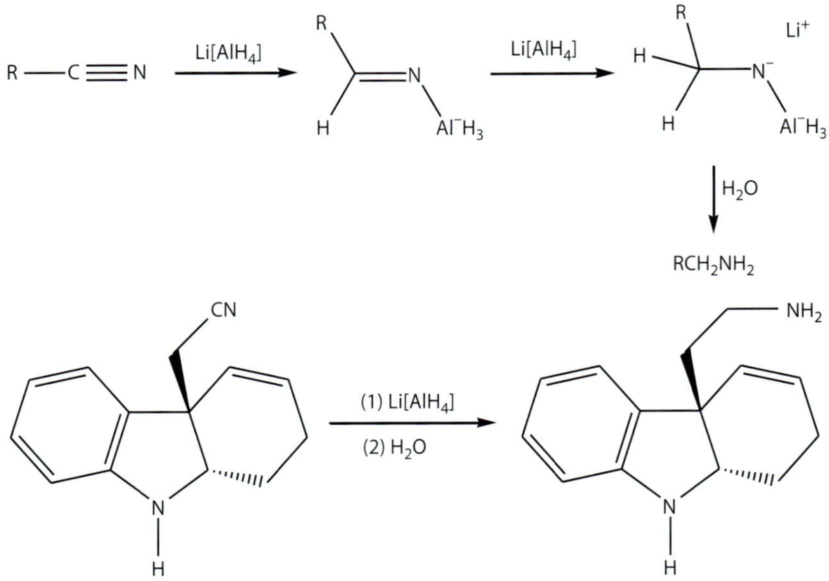

FIGURE 14.55 Reduction of cyanides using lithium aluminum hydride.

Nitriles can also be reduced by hydrides, a useful alternative to the hydrogenations that we saw in Chapter 9, if the molecule contains other double bonds (Figure 14.55). The reaction proceeds *via* an iminium salt.

PROBLEM 14.10

Predict the product(s) of each of the following reactions:

(a)

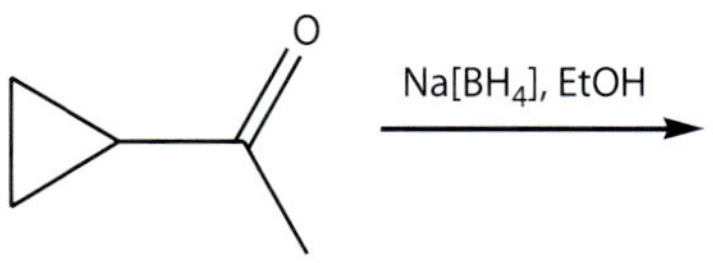

(b)

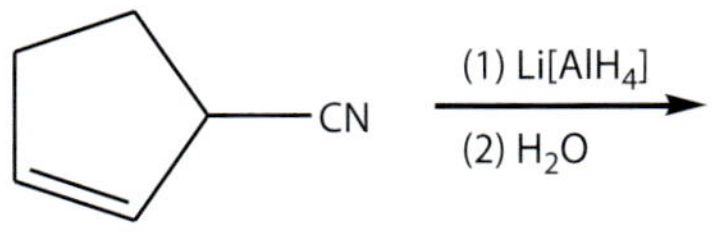

(c)

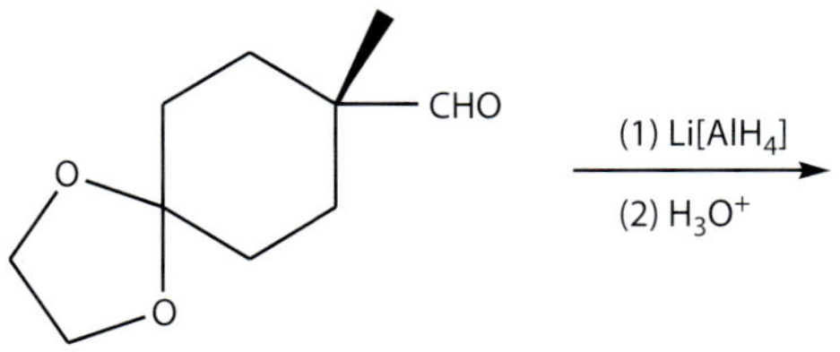

(d)

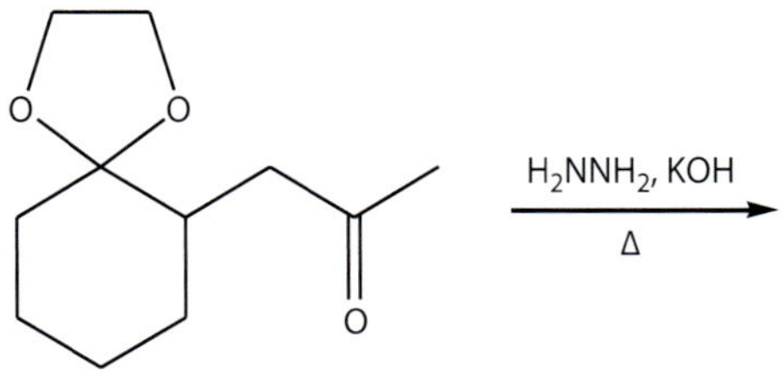

Solutions

Reaction (a) is simply a hydride addition to the carbonyl group, followed by protonation of the alkoxides produced. Notice that in (b), reduction of the nitrile with hydrogen and a metal catalyst would not have been successful as the carbon–carbon double bond would also have been hydrogenated. In (c), the first step is reduction of the aldehyde, but in the work-up with aqueous acid, the acetal is hydrolyzed. In (d), where the reaction remains basic throughout, the acetal is not hydrolyzed:

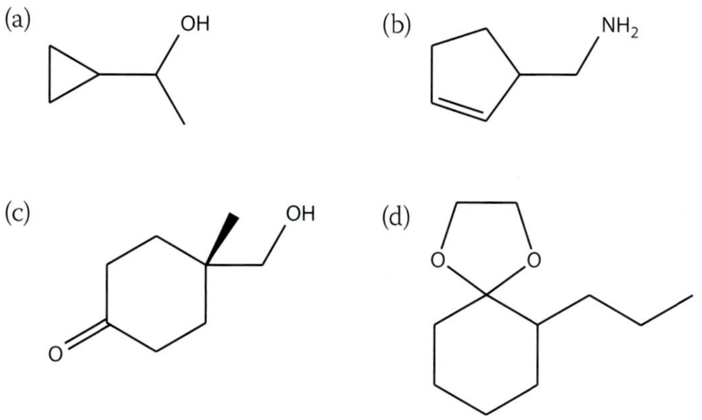

14.4.3 ADDITION OF CARBON NUCLEOPHILES

In Chapter 9, we learned how to make Grignard reagents, and organolithium compounds, and studied their reactions with epoxides (Figure 14.56). In the second reaction, phenylmagnesium bromide is readily prepared from PhBr and magnesium in dry THF.* Grignard reagents behave as

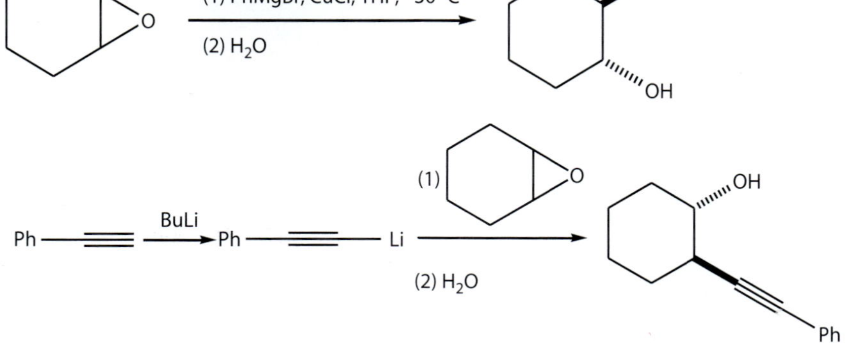

FIGURE 14.56 Preparation of Grignard and organolithium reagents and their reactions with epoxides.

* Grignard reagents can be prepared in either diethyl ether or THF.

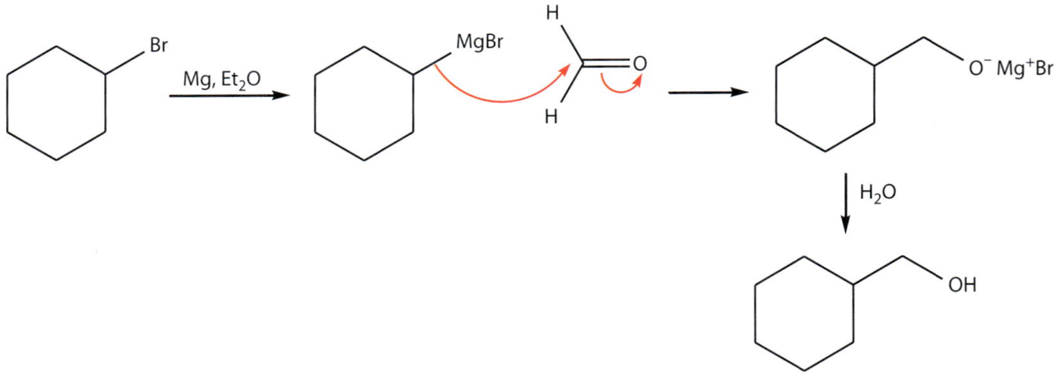

FIGURE 14.57 Reaction of a Grignard reagent with formaldehyde.

though they were carbanions, even though there is no actual dissociation, and they react readily with most types of carbonyl compounds. Figure 14.57 shows the process with formaldehyde; after work-up of the magnesium alcoholate with water, a primary alcohol is obtained, and the original carbon chain has been extended by one carbon atom.

The mechanism of this, and all reactions of Grignard reagents with carbonyl compounds, involves a cyclic transition state, either **14.27** or **14.28**, in which the magnesium atom acts as a Lewis acid toward the oxygen. Reactions with other aldehydes give secondary alcohols and with ketones, tertiary alcohols (Figure 14.58). The dehydration of the product alcohol in the second example to give the anticancer drug tamoxifen is easy and can be achieved under very mild conditions, because the carbocation formed in the E1 process is tertiary and benzylic. When we start to look more formally at synthesis, these reactions will be crucial. One final useful process that we should consider is the reaction of Grignards with carbon dioxide (Figure 14.59), to give, after

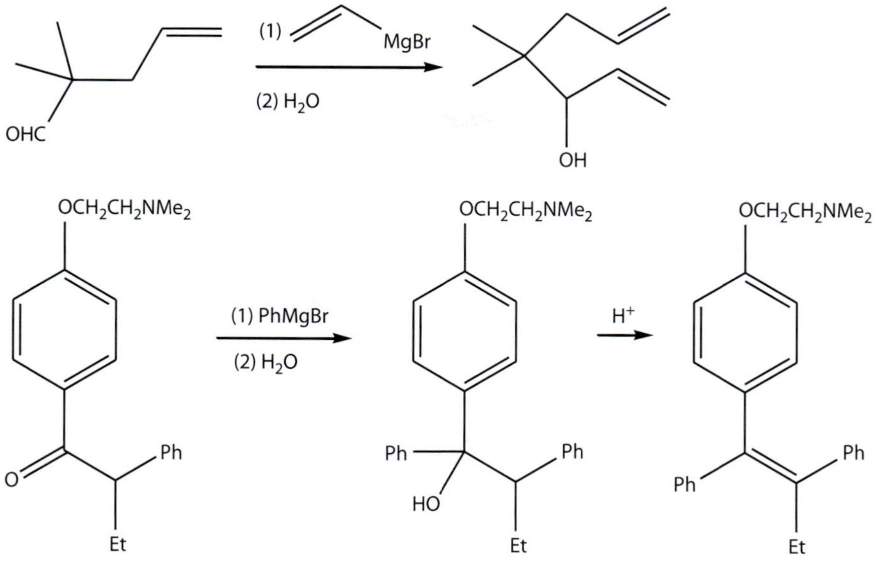

Tamoxifen

FIGURE 14.58 Reaction of Grignard reagents with aldehydes and ketones.

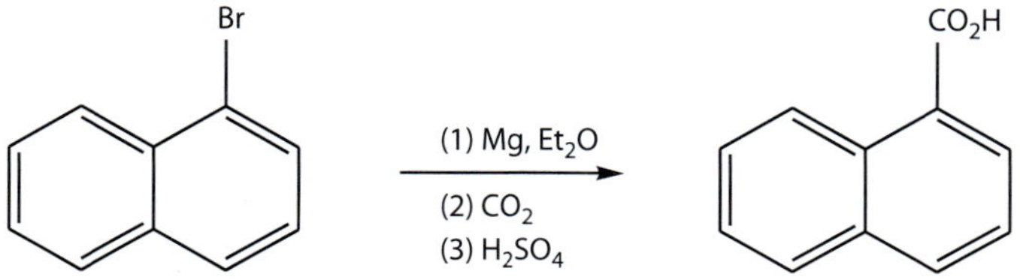

FIGURE 14.59 Reaction of a Grignard reagent with CO_2.

work-up, carboxylic acids; this is another method of extending a molecule by one carbon atom and is also useful for introducing a ^{13}C label into the molecule.

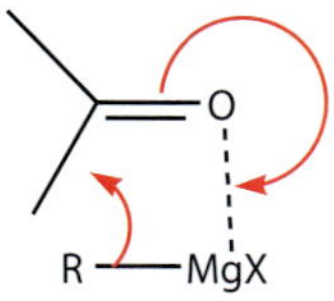

14.27

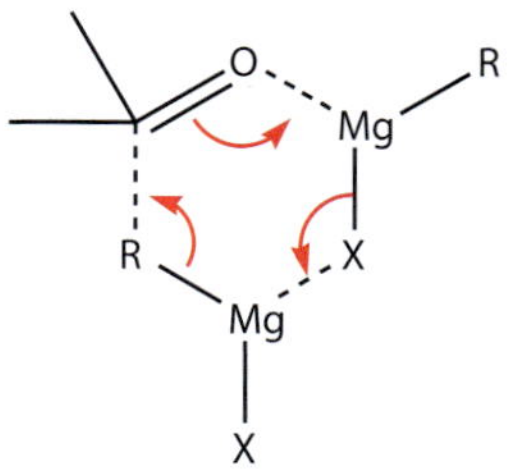

14.28

Grignard reagents also add to carbon–nitrogen double and triple bonds. Reactions with imines (Figure 14.60) give amines with branched substituents. The reaction with nitriles (Figure 14.61) has a different outcome. The first product is a magnesiated imine (Figure 14.62), which does not react with further Grignard reagent. Hydrolysis in the aqueous work-up both protonates this species and converts the product imine to a ketone.

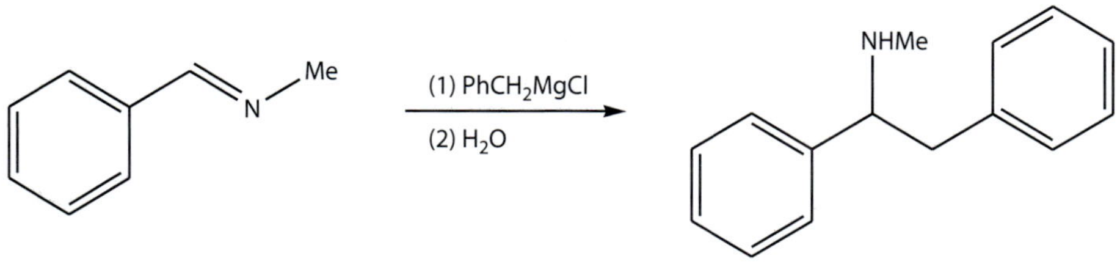

FIGURE 14.60 Reaction of a Grignard reagent with an imine.

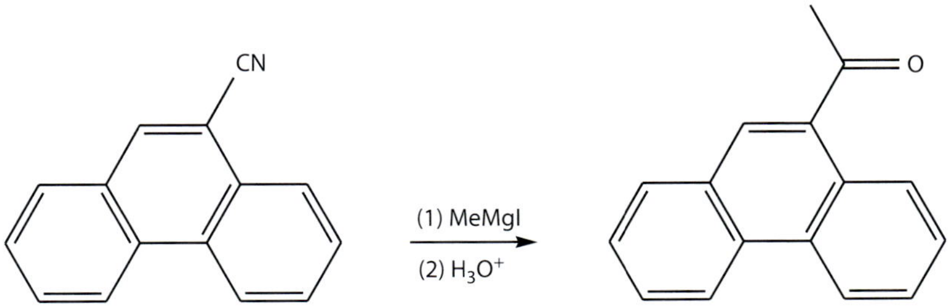

FIGURE 14.61 Reaction of a Grignard reagent with a nitrile.

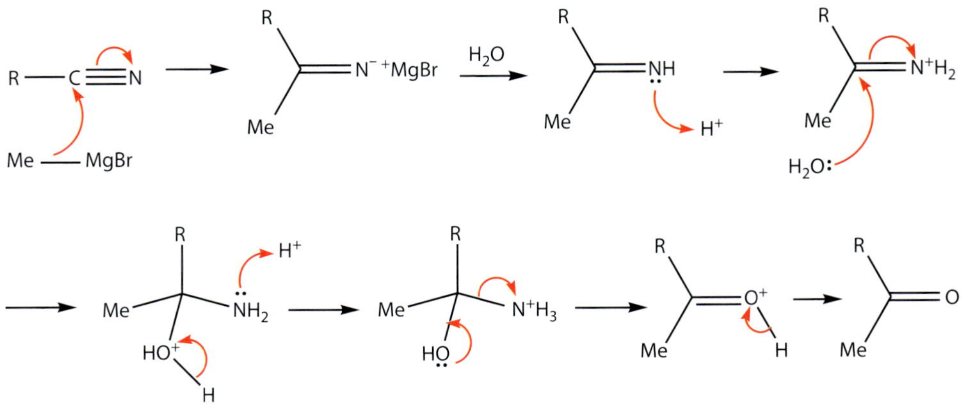

FIGURE 14.62 Mechanism of reaction of Grignard reagents with nitriles.

The reactions of Grignard reagents with a range of substrates are summarized in Table 14.3; the reactions with acid chlorides and esters are included for completeness, and we will look at them in detail in the next chapter.

Other organometallics also react with ketones and imines. For example, the alkyne anions that we prepared in Chapter 8 (remember that the pK_a of an alkyne is about 25, so anions can easily be generated using either $NaNH_2$ or BuLi) add readily to aldehydes and ketones (Figure 14.63). In the second sequence, the alkyne is hydrated to the enol, under acid- and mercury-catalyzed conditions. The enol is then tautomerized to a methyl ketone.

TABLE 14.3 Reactions of Grignard Reagents, RMgX

Substrate	Product	Comments
H_2O	RH	Grignard reagents must be prepared and used in dry solvents to avoid this process.
D_2O	RD	Used to label substrates for mechanistic studies.
CO_2	RCO_2H	Useful for ^{13}C labeling and to prepare very hindered acids.
$H_2C{=}O$	RCH_2OH	Primary alcohol synthesis.
R′CHO	RR′CHOH	Secondary alcohol synthesis.
R′R″C=O	RR′R″COH	Tertiary alcohol synthesis.
R′COCl	$R_2R'COH$	Reaction can be halted at the ketone stage only if either Grignard or substrate is very hindered.
R′COOMe	$R_2R'COH$	Difficult to halt at the ketone stage.
$R^1R^2C{=}NR^3$	$RR^1R^2CNHR^3$	Most useful reactions involve imines of aldehydes.
R′CN	RR′C=O	A few examples are known where the intermediate imine may be isolated.
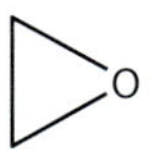	RCH_2CH_2OH	Reaction goes by an S_N2 process, at the less hindered carbon, with inversion of stereochemistry.

Aqueous work-up is assumed for all these reactions.

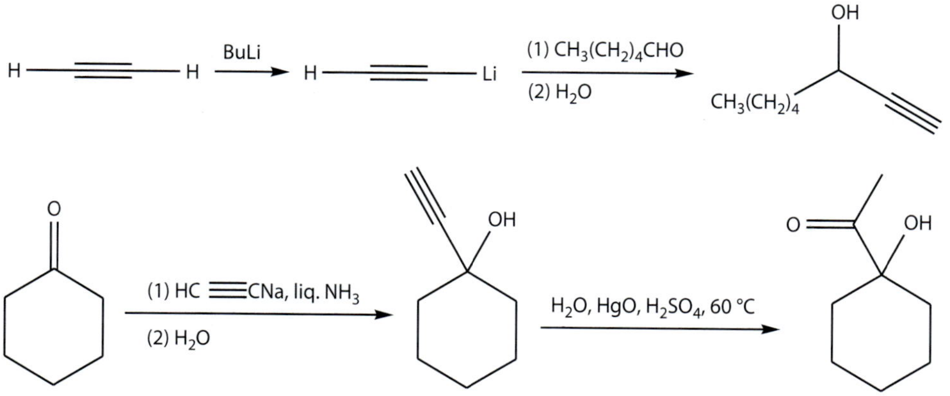

FIGURE 14.63 Addition of alkyne anions to aldehydes and ketones.

Although sometimes more challenging to prepare, organolithium compounds are generally more reactive than Grignard reagents and less prone to eliminations when the organometallic or the substrate is sterically hindered. For example, adamantanone, **14.29**, undergoes mainly reduction to the alcohol with EtMgBr, but ethyl lithium adds cleanly to give **14.30**.

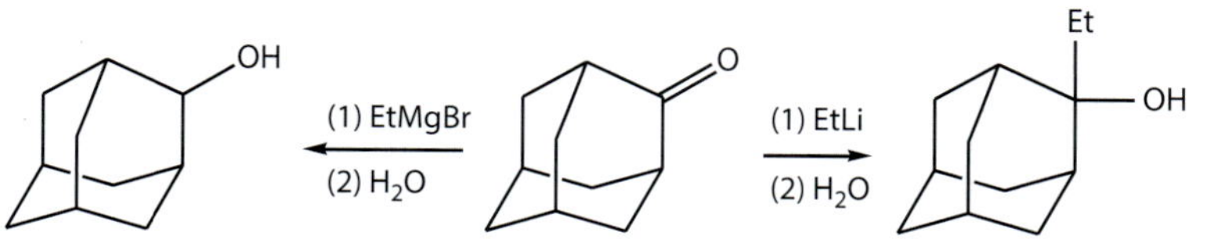

14.29 **14.30**

PROBLEM 14.11

Predict the product(s) of the following reaction sequences:

(a) 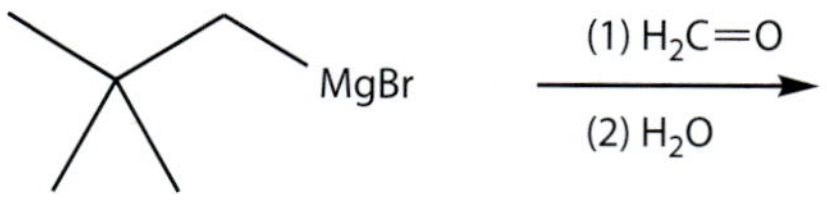

(b) 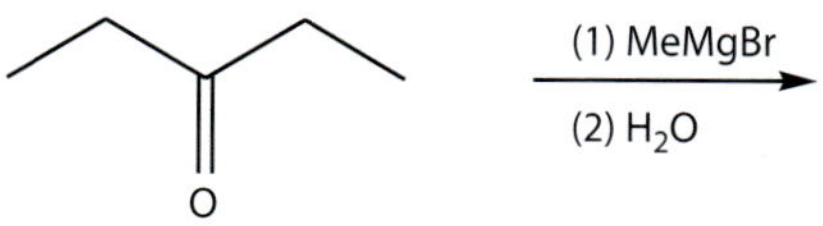

(c) 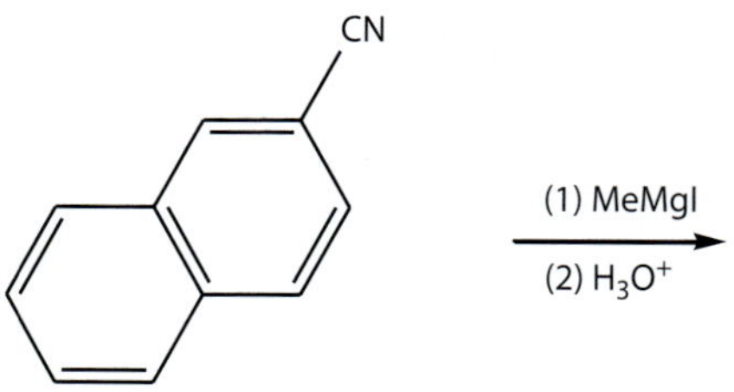

(d) 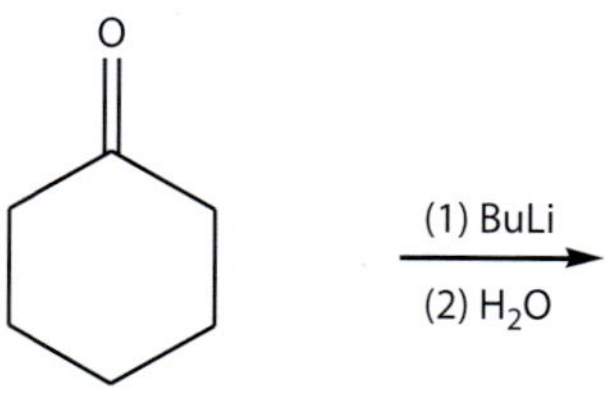

(e) 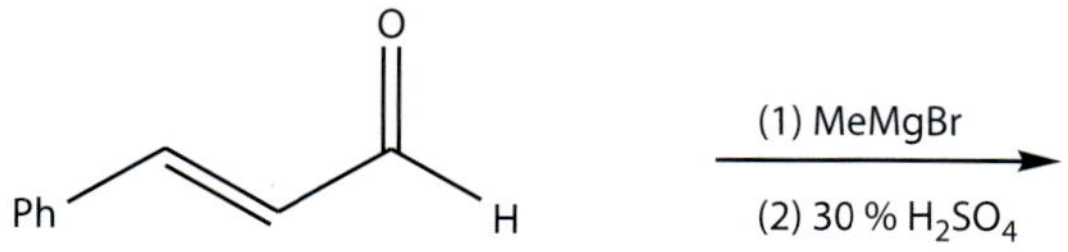

Solutions

The first four reactions are essentially textbook. The organometallic adds to the carbon–oxygen or carbon–nitrogen multiple bond, and then the products are protonated. The work-up steps with the nitrile in (c) also involve hydrolysis of the intermediate imine. The final reaction is a little different. The addition to the carbonyl is exactly as would be predicted, but the work-up here is clearly different. Under such acid conditions, the product alcohol was dehydrated to give a diene.

(a)

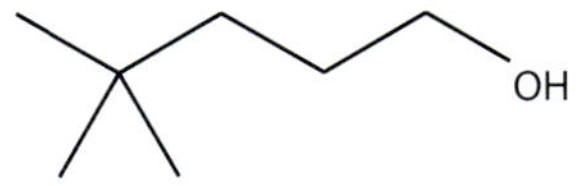

(b)

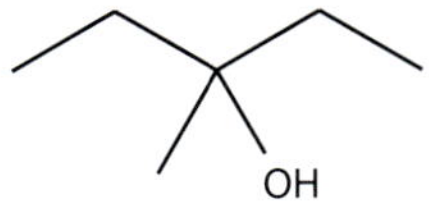

(c)

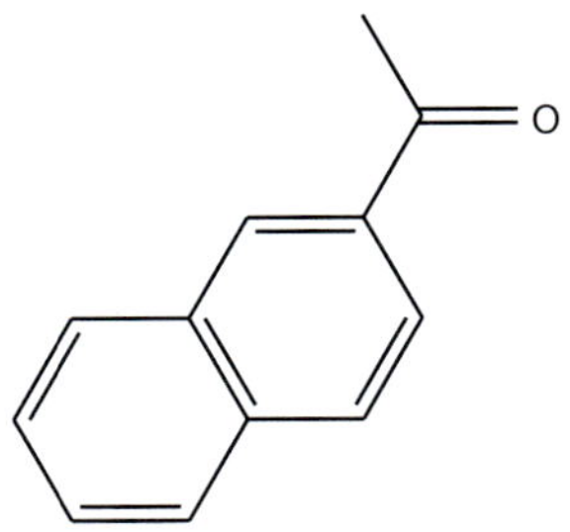

(d)

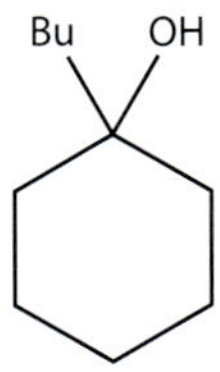

(e)

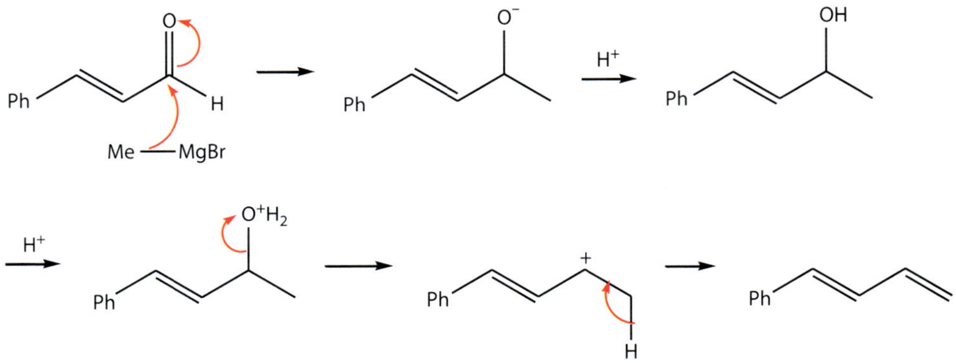

Addition of carbon nucleophiles to carbonyl groups is a key bond-forming reaction in organic synthesis. The alcohols produced are very versatile and can be transformed into a range of other functionalities (halides, ethers, other carbonyl compounds, alkenes, *etc.*). Looking ahead a little, let's now try to approach these reactions from the point of view of the molecule that we want to synthesize and consider which bonds we might want to make and how we would do that. This technique is called retrosynthetic analysis or the disconnection approach. If we consider **14.31** as our target, we will disconnect one of the bonds next to the hydroxyl group to give a Grignard reagent and a ketone. The fat arrow shown is called a retrosynthetic arrow. There are two options for this as we have two distinct groups that are attached to the carbon bearing the OH group. How should we choose between them? In a general case, we might try both, look at costs and see where we get the best practical yield. In this case, however, we get the greatest simplification by reacting benzyl Grignard with acetone. Although this would not be immediately obvious, phenylacetone is rarely used as a substrate; in most jurisdictions, it is a legally controlled substance as it may be used in the manufacture of amphetamines.

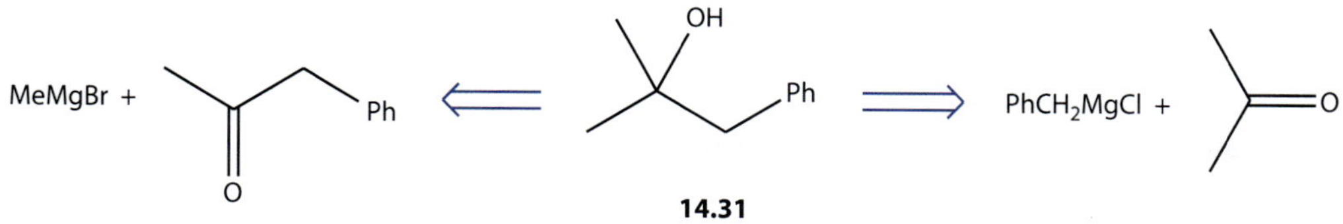

14.31

The object of simplification is a general one—if we consider how we would make **14.32**, disconnection of the methyl group gives us sensible reagents, but disconnection of one of the bonds of the six-membered ring makes things more complicated rather than simpler. So our first general rule is to simplify. We should also use symmetry where it exists—generally, symmetric ketones are more readily available, and less expensive, than non-symmetric ones. Students often worry about the question of what SM is readily available. In many tests, you will be told what you may start with, but if you are doing a homework problem, it's often a good idea to look at the Sigma-Aldrich website* and see if the SM you want is in their catalog.

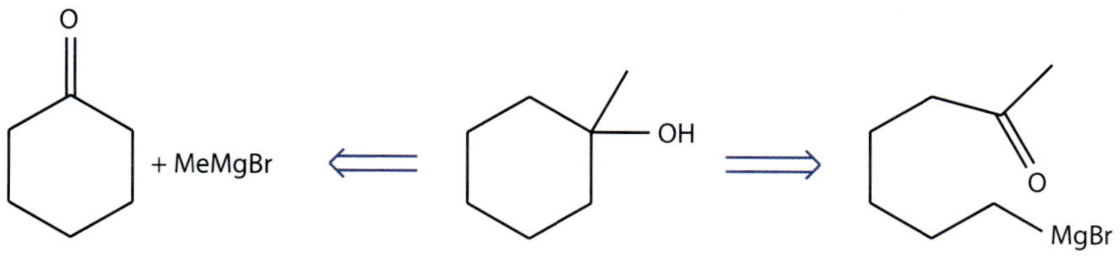

14.32

In Chapter 8, we met the chromium(VI) oxidation of alcohols to aldehydes and ketones (Figure 14.64). We can use the oxidation of secondary alcohols to ketones to construct tertiary alcohols by a two-stage process shown in Figure 14.65.

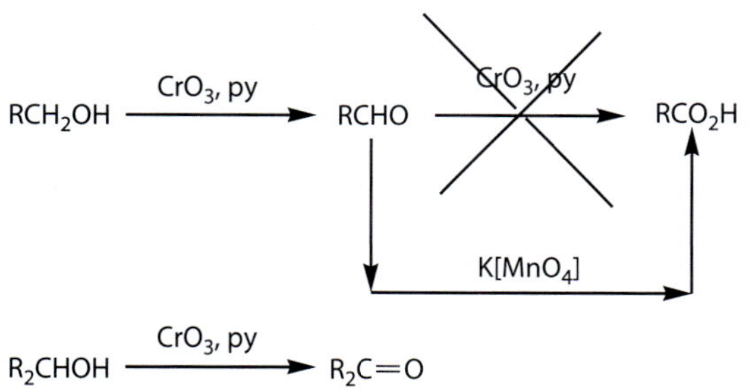

FIGURE 14.64 Oxidation of alcohols.

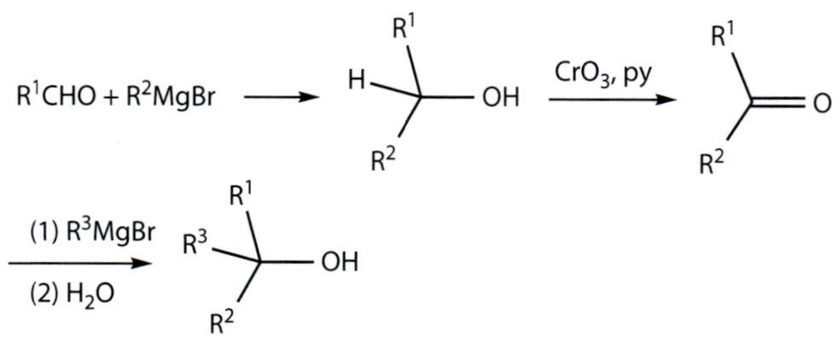

FIGURE 14.65 Grignard reagent-based syntheses.

---

* www.sigmaaldrich.com.

PROBLEM 14.12

Show how you would use Grignard reagents in the synthesis of each of the following alcohols:

(a) 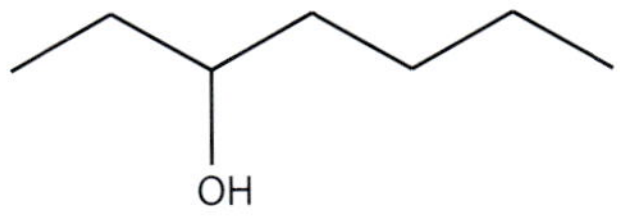

(b) 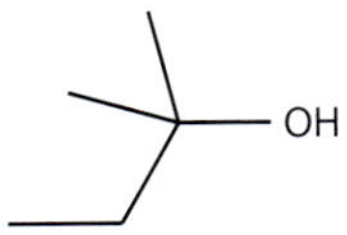

Solution

(a) We must consider which of the two bonds to the carbon bearing the OH group we wish to break, and then write down the reagents that would be needed for the forward Grignard reaction:

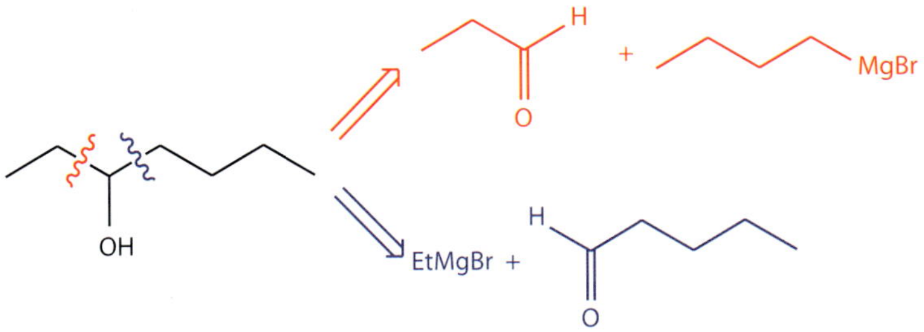

In this example, there is no obvious reason to prefer one route over the other. Propanal, pentanal, bromoethane, and 1-bromobutane are all commercially available. In practice, a decision would be made on the basis of the yields in the two reactions, the costs of the raw materials, and the costs of waste disposal.

(b) Again, there are two possible disconnections, and both forward Grignard reactions would work. In this case, however, there is a clear preference. Acetone is much less expensive than butanone, and this disconnection gives the greater simplification:

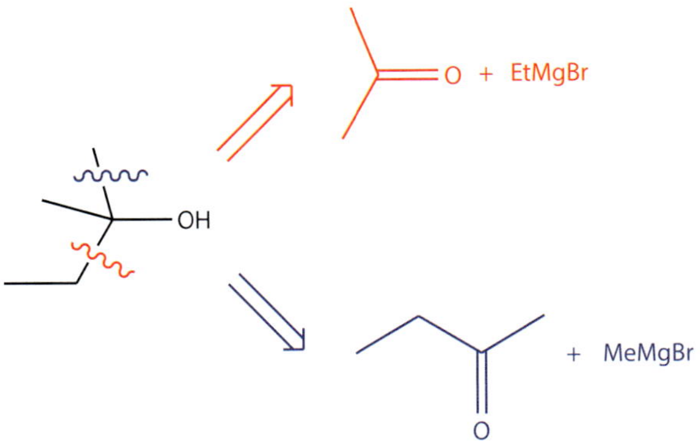

14.4.4 WITTIG REACTIONS

Although the Wittig reaction technically involves a carbon nucleophile, it is sufficiently different from the reactions of simple carbanions to consider it separately. The transformation (Figure 14.66) converts an aldehyde or ketone into an alkene, with complete replacement of the carbonyl oxygen atom. George Wittig received the Nobel Prize in 1979 for the development of this reaction.

The first step of the reaction involves the preparation of a phosphonium salt from triphenylphosphine and an appropriate alkyl halide. Phosphines are excellent nucleophiles, and the salts are stable. The salt is then deprotonated at carbon using a very strong base such as sodium hydride. The species produced is called an ylid—although we will write it in the charged form to draw the reaction mechanism, its stability derives from the doubly bonded form, **14.33**. Because phosphorus has available d orbitals, it can expand its octet to make the double bond. The ylid then attacks the carbonyl compound to give an intermediate betaine. Ring closure gives a four-membered ring, but this is not stable and breaks open to extrude triphenylphosphine oxide, leaving the alkene. The reaction is driven by the formation of the very strong phosphorus–oxygen double bond (~545 kJ mol^{-1}) (Figure 14.67).

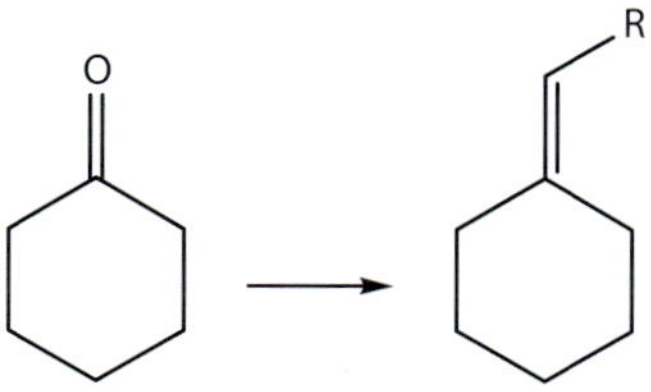

FIGURE 14.66 Wittig reaction.

The reaction works for aldehydes and ketones, aromatic or aliphatic or conjugated, and tolerates functional groups such as alkene, alkyne, hydroxyl, ester, ether, acetal, $ArNO_2$, halogen, and

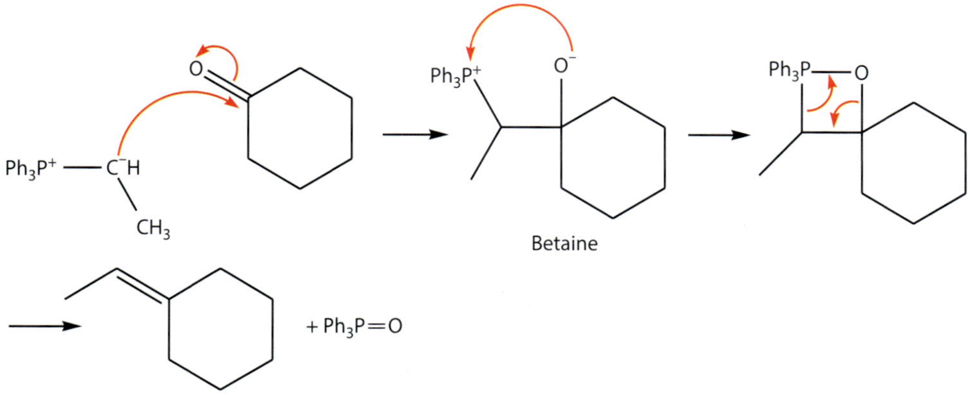

FIGURE 14.67 Mechanism of the Wittig reaction.

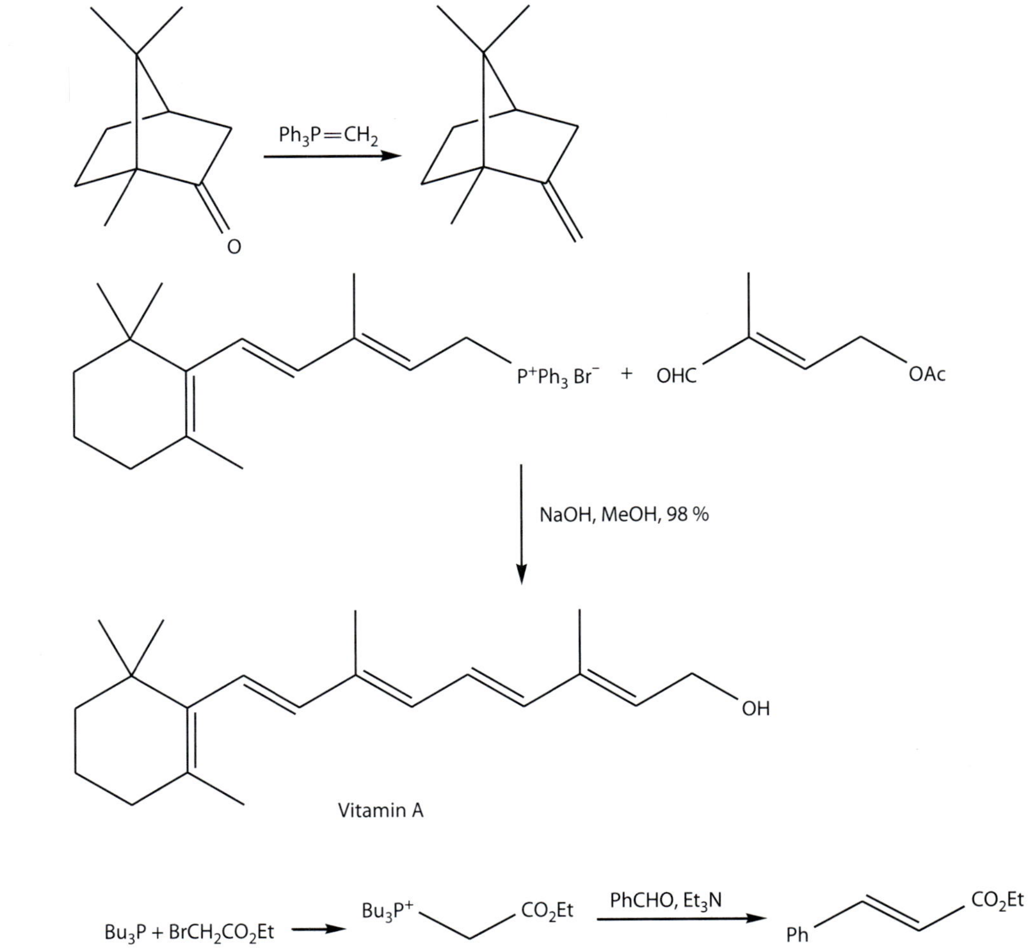

FIGURE 14.68 Synthetic uses of the Wittig reaction.

tertiary amine. Some examples of the use of the reaction in synthesis are shown in Figure 14.68. The second reaction is the industrial synthesis of vitamin A; the acetate ester is hydrolyzed in the work-up. In the final reaction, we see that phosphorus nucleophiles other than triphenylphosphine can be used; the reasons for choices and changes of conditions will be explored when we study synthesis in more depth (Chapter 19).

PROBLEM 14.13

Give the structures of **14.34** and **14.35** in the following sequence. In the second step, why is it possible to use a moderate base like triethylamine, rather than a very strong one such as NaH?

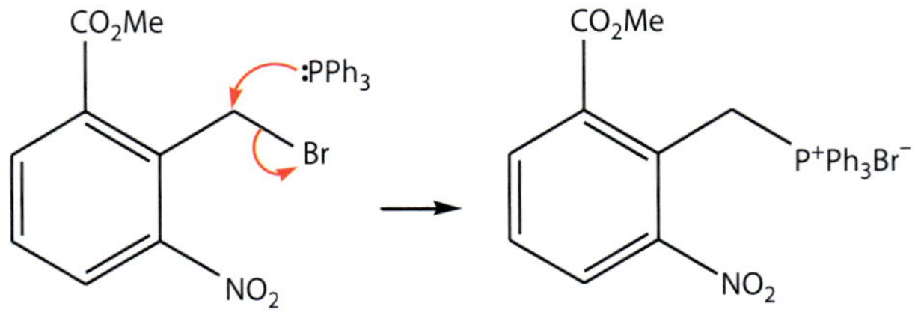

14.34

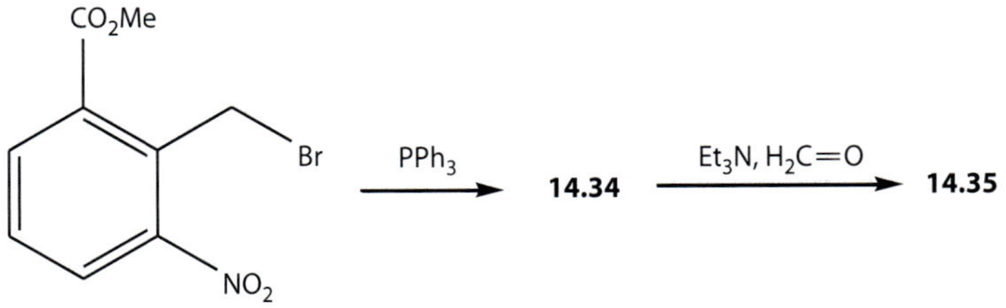

Solution

The first step of the reaction is a nucleophilic displacement of bromide by triphenylphosphine to give a phosphonium salt.

The phosphonium salt is then deprotonated by triethylamine and undergoes the Wittig reaction with formaldehyde. Deprotonation with the modestly basic triethylamine is possible because the anion is stabilized by resonance forms involving both the electron-withdrawing groups on the benzene ring:

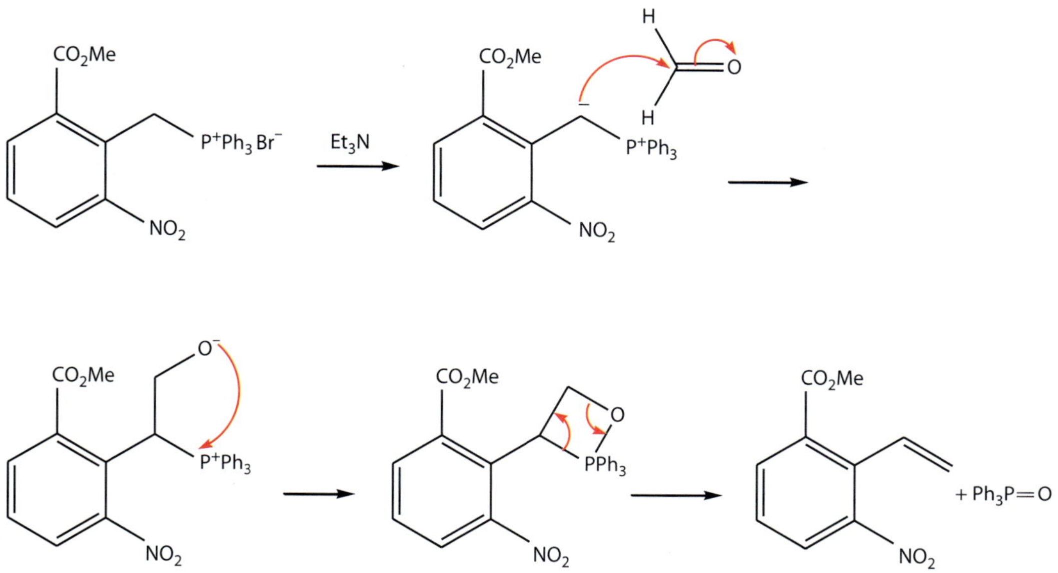

14.35

Key Points from Section 14.4

- Reduction of aldehydes or ketones to hydrocarbons can be accomplished using the Clemmensen reaction (Zn/Hg/HCl), the Wolff–Kishner reaction (hydrazine, KOH), Raney nickel/H_2 reduction of a dithioacetal, or palladium-catalyzed hydrogenolysis (aryl ketones only).
- Hydridic reduction of aldehydes or ketones gives, after work-up, alcohols. Li[AlH_4] is more reactive than Na[BH_4] for this purpose.
- Imines may be reduced to amines using Li[AlH_4] or, more commonly, Na[BH_3(CN)]. Reductive amination of aldehydes and ketones may be accomplished with primary or secondary amines, using either H_2/catalyst or Na[BH_3(CN)] as reductant in a one-pot process.
- Nitriles may be reduced to primary amines using Li[AlH_4] followed by aqueous work-up. This is an alternative to catalytic hydrogenation of the triple bond.
- The reactions of Grignard reagent with various substrates are summarized in Table 14.3. Grignard reagents also add to imines to give amines, after work-up.
- Addition of Grignard reagents to nitriles gives ketones, after work-up.
- Organolithium reagents add to aldehydes and ketones; they are more reactive than Grignard reagents.
- The Wittig reaction, using deprotonated alkyltriphenyl phosphonium salts, converted aldehydes and ketones to alkenes.

REVIEW PROBLEMS

1. Give a structure for
   (a) 3-Chlorobenzaldehyde
   (b) 3,6,6-Trimethyl-2-heptanone
   (c) 2-Methyl-7-bromo-3-nonanone
   (d) *E*-2-Ethyl hex-2-enal
   (e) Cyclodeca-1,6-dione
2. Give a systematic name for

(a)

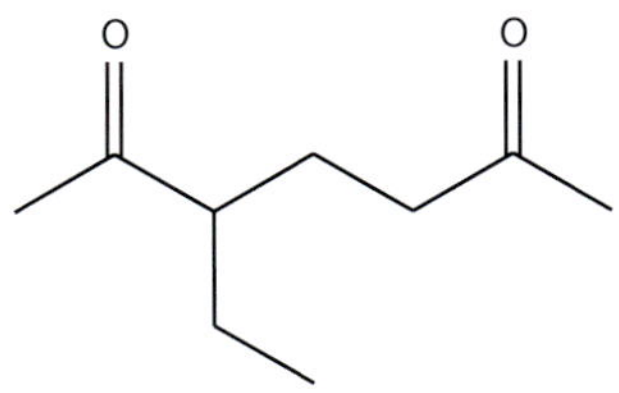

(b)

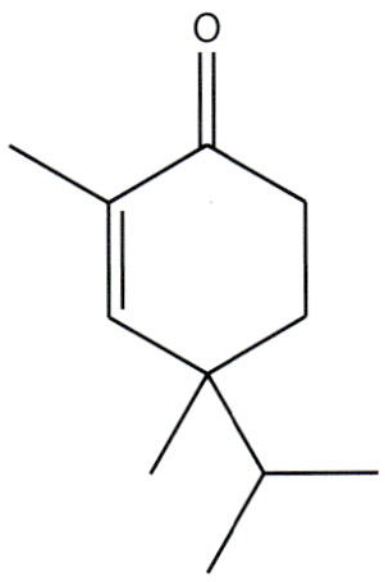

(c)

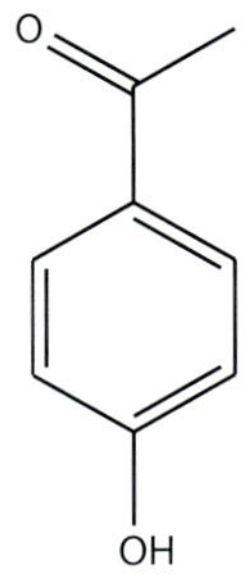

(d)

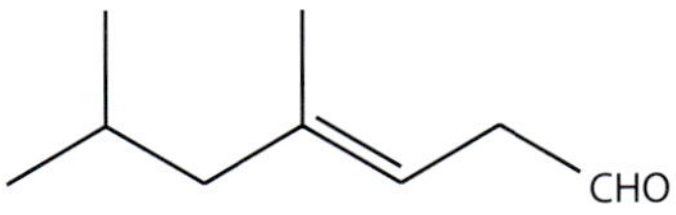

(e)

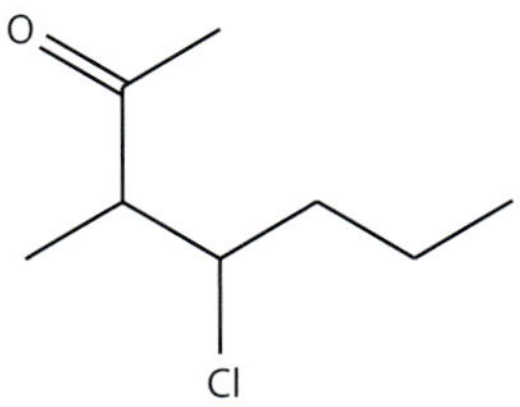

3. Compound **14.36**, $C_6H_{12}O_2$, reacts with dilute H_2SO_4 to give **14.37**, C_4H_8O, which is readily distilled, and **14.38**, $C_2H_6O_2$, which has a high boiling point. **14.37** reacts with MeMgI, followed by aqueous work-up to give **14.39**, $C_5H_{12}O$. Identify compounds **14.36–14.39**, explaining your reasoning.

14.36	IR spectrum 2985, 1070 cm^{-1}	
	Mass spectrum *m/z* 116, 101, 87	
	^1H NMR spectrum δ	0.88 t 3H
		1.22 s 3H
		1.61 q 2H
		3.88 s 4H
14.37	IR spectrum 2985, 1720 cm^{-1}	
	^1H NMR spectrum δ	0.99 t 3H
		2.07 s 3H
		2.42 q 2H
14.38	IR spectrum 3360, 2945, 1085 cm^{-1}	
	^1H NMR spectrum δ	3.63 s 4H
		4.36 s 2H, exchanges with D_2O

4. Propose a mechanism for each of the following processes:

(a)

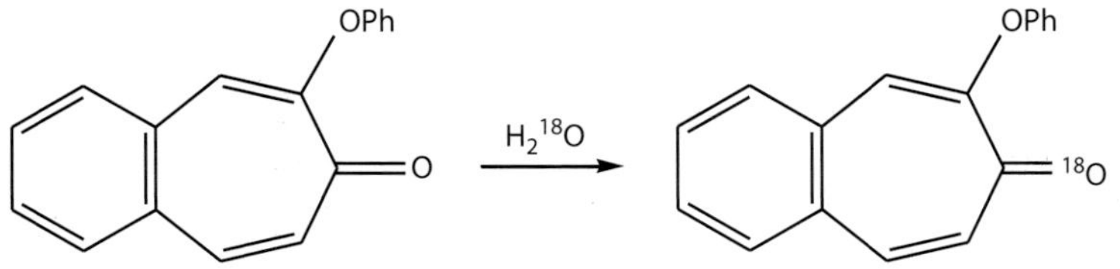

(b)

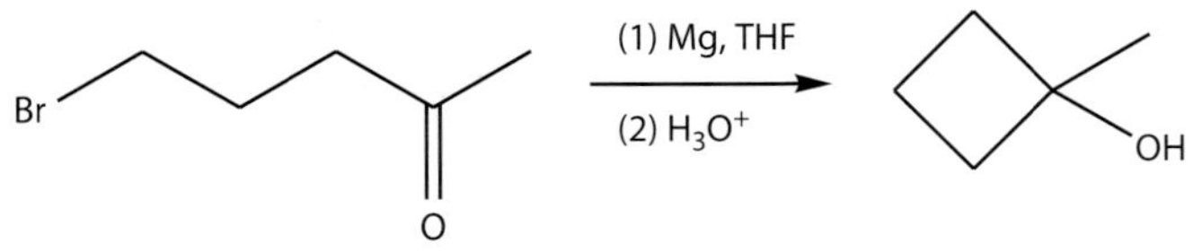

(c)

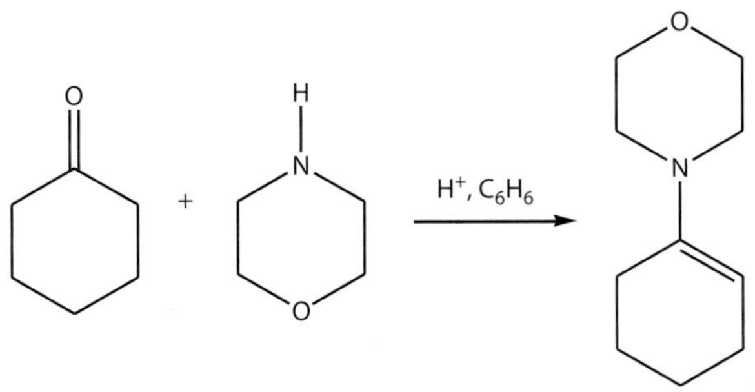

(d)

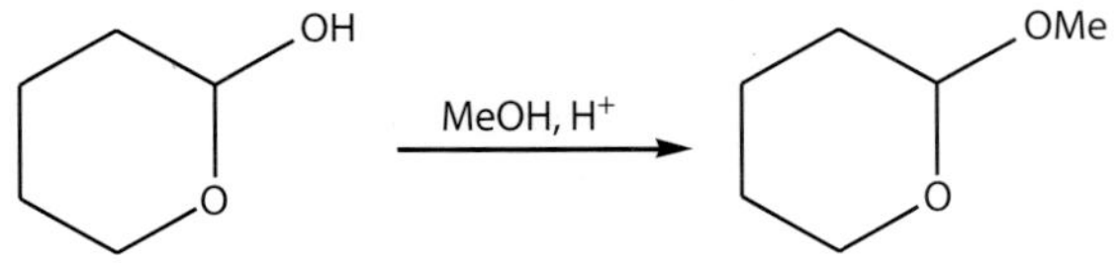

(e)

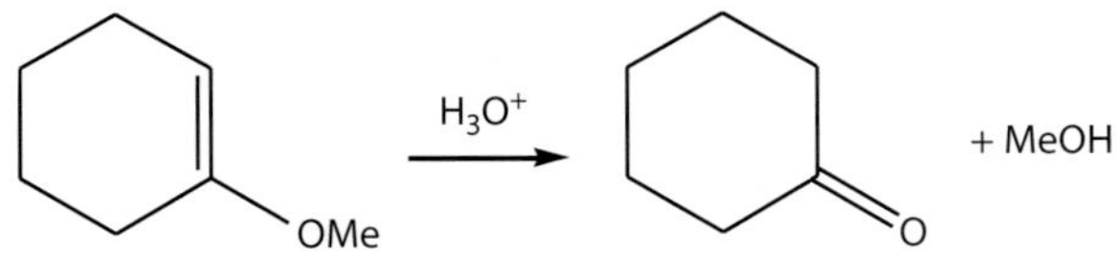

(f)

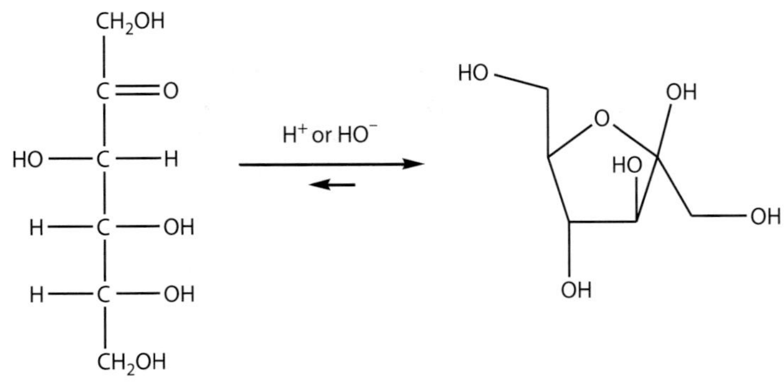

(g)

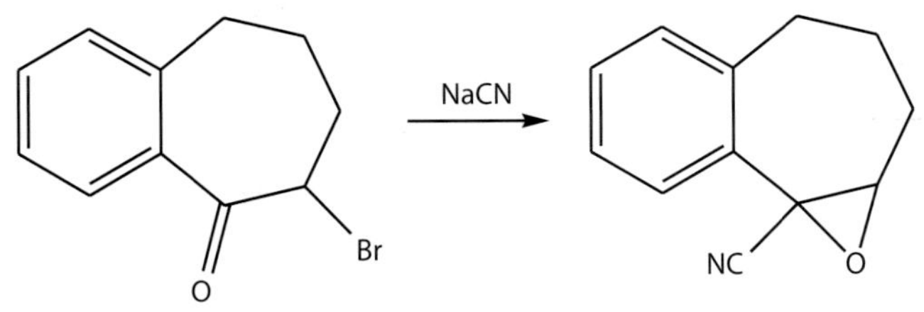

(h)

O
C
R
H
Me3SiCN, catalytic KCN
OSiMe3
C
R
H
NC

(i)

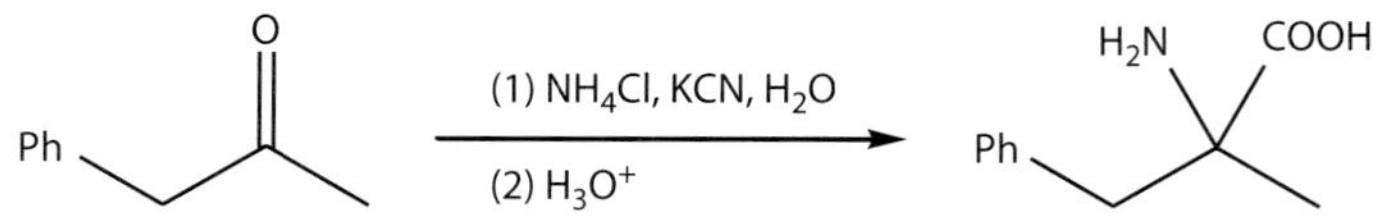
O
Ph
(1) NH4Cl, KCN, H2O
(2) H3O+
H2N
COOH
Ph

(j)

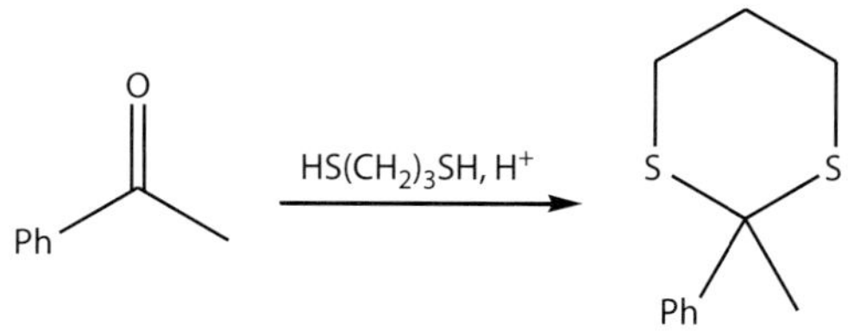
O
Ph
HS(CH2)3SH, H+
S
S
Ph

(k)

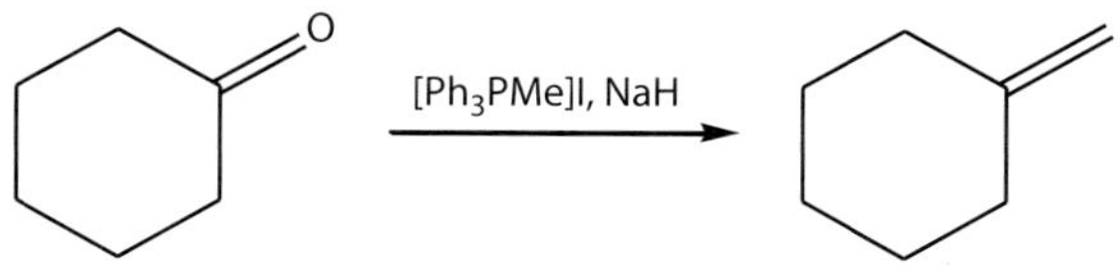
O
[Ph3PMe]I, NaH

(l)

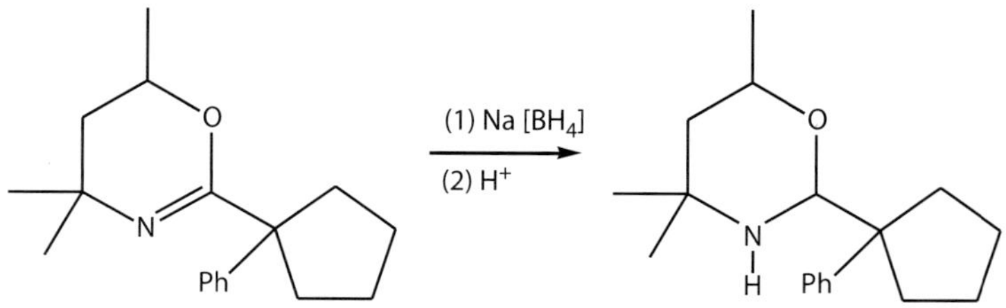
O
N
Ph
(1) Na [BH4]
(2) H+
O
N
H
Ph

(m)

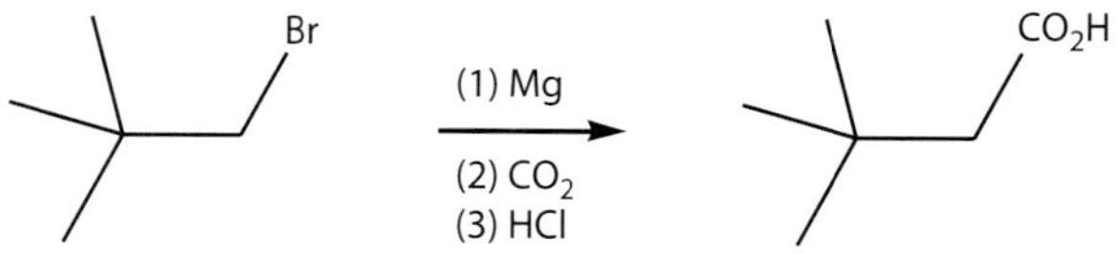
Br
(1) Mg
(2) CO2
(3) HCl
CO2H

(n)

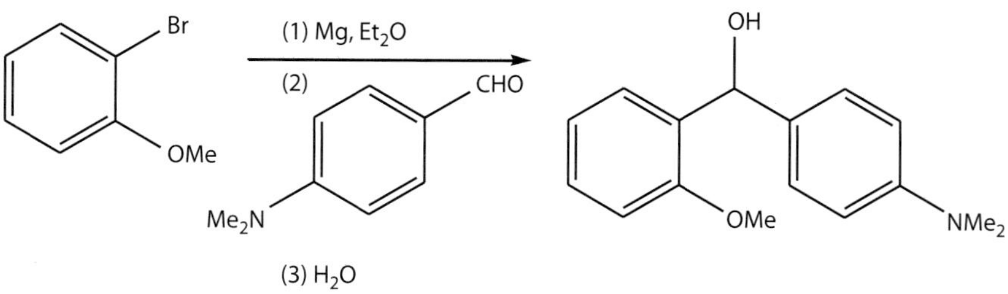
Br
OMe
(1) Mg, Et2O
(2)
CHO
Me2N
(3) H2O
OH
OMe
NMe2

(o)

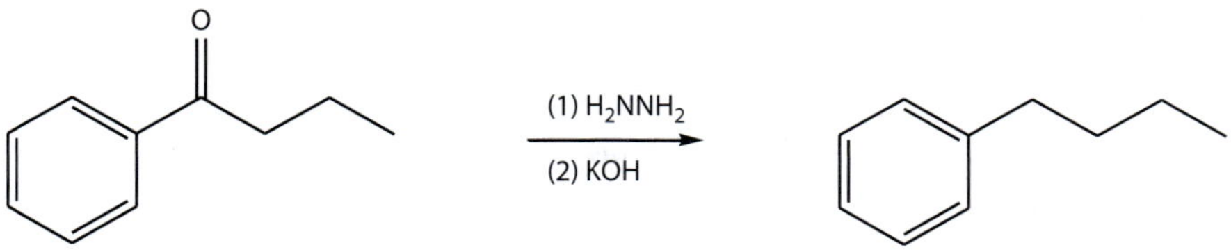

(p)

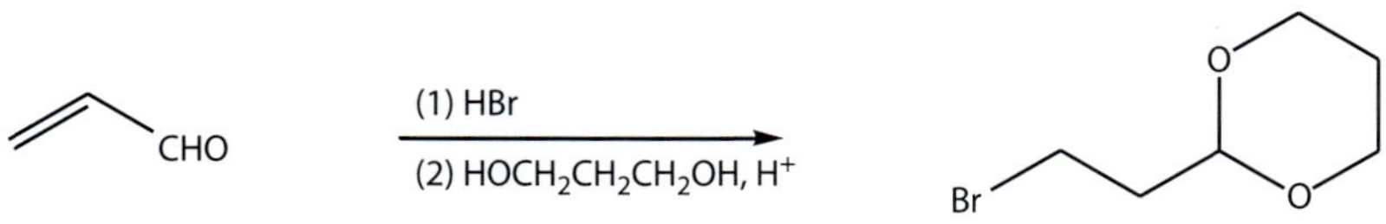

5. How would you accomplish each of the following transformations?

(a)

(b)

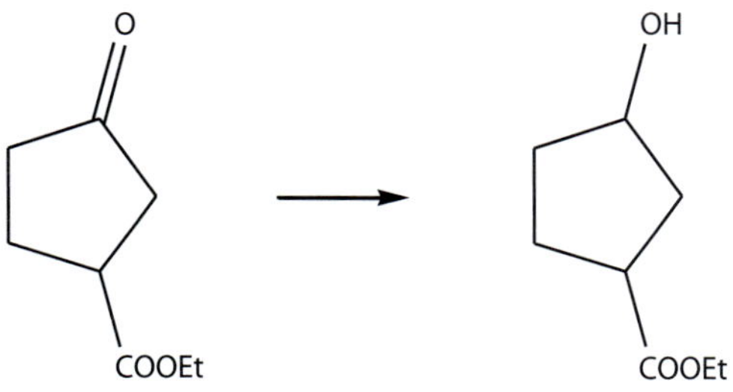

(c) RCl to RD

(d)

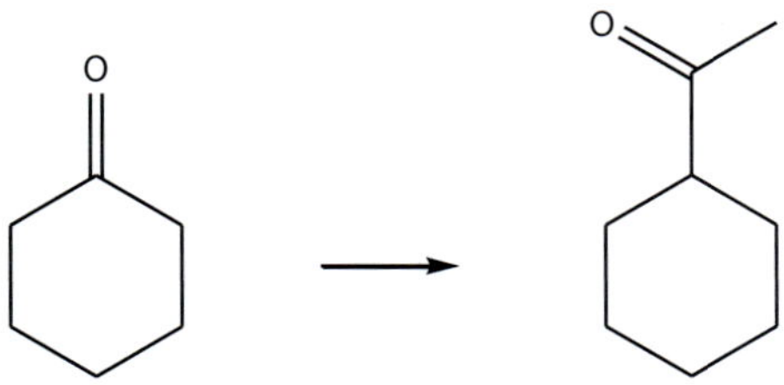

(e)

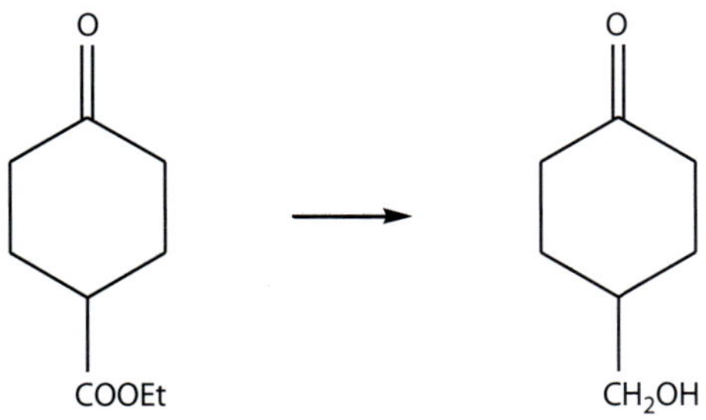

(f)

(g)

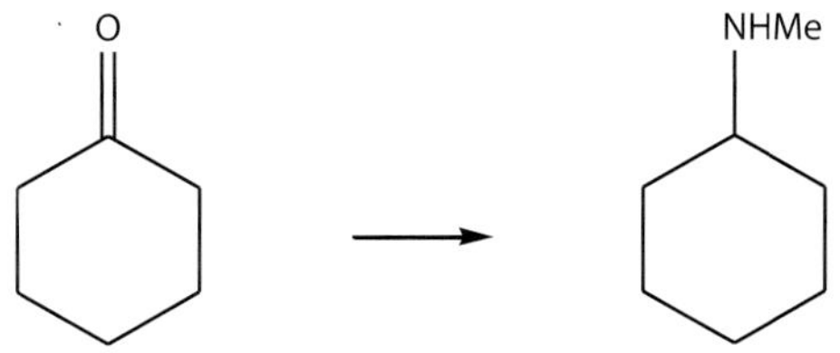
O
NHMe

(h)

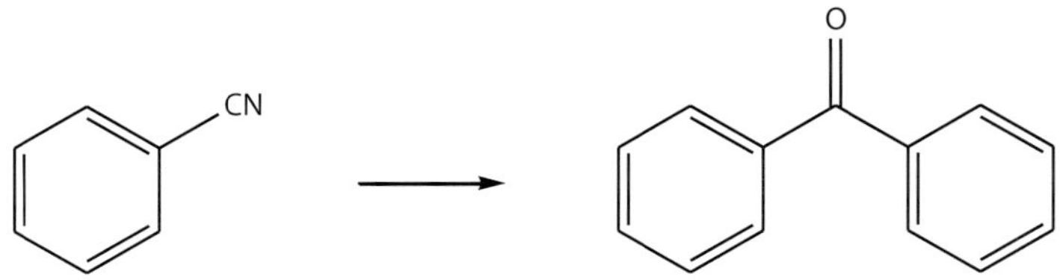
CN
O

(i)

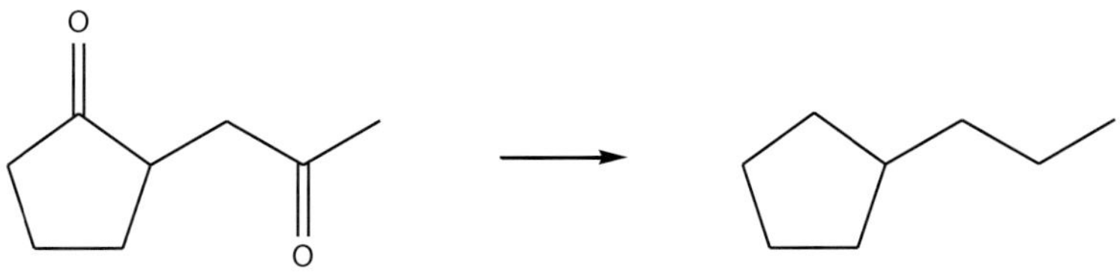
O
O

(j)

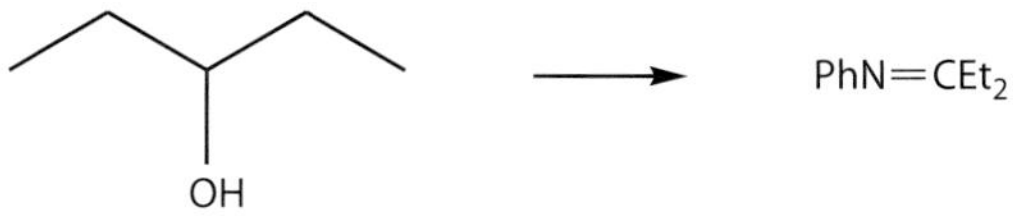
OH
PhN=CEt2

(k)

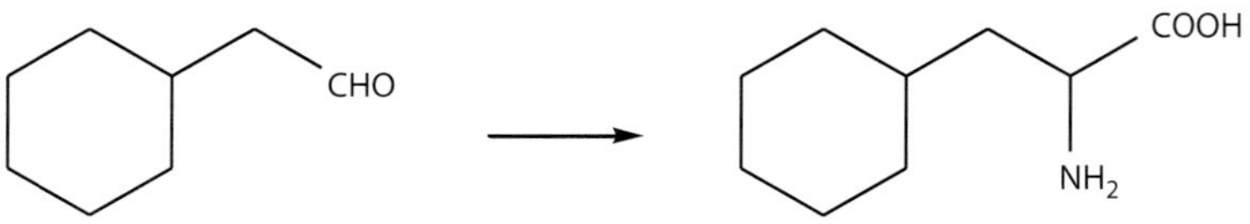
CHO
COOH
NH2

(l)

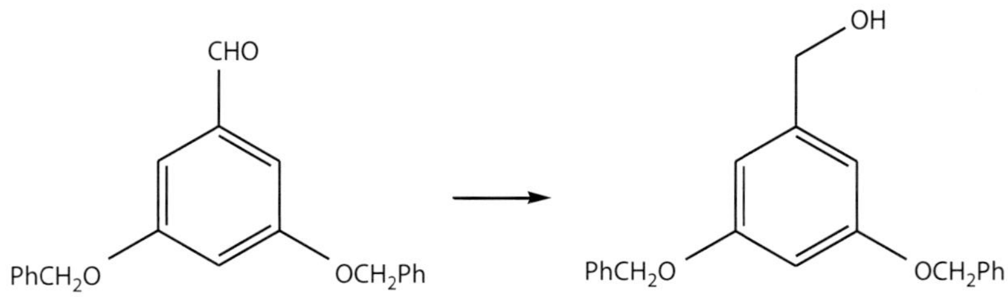
CHO
PhCH2O
OCH2Ph
OH
PhCH2O
OCH2Ph

(m)

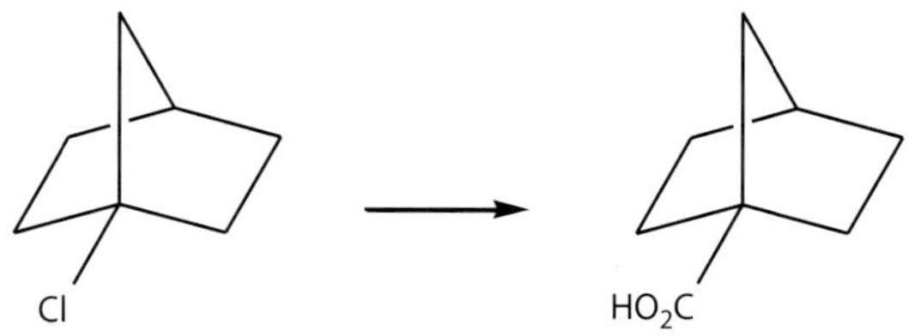
Cl
HO2C

(n)

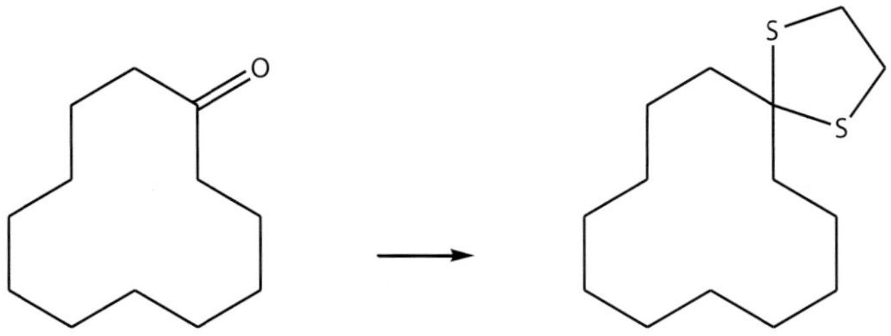

6. Predict the product(s) of each of the following reaction sequences:

(a)

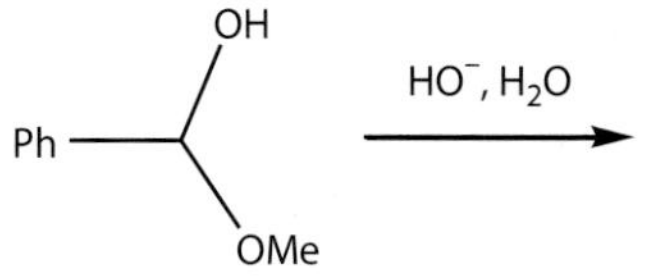

(b)

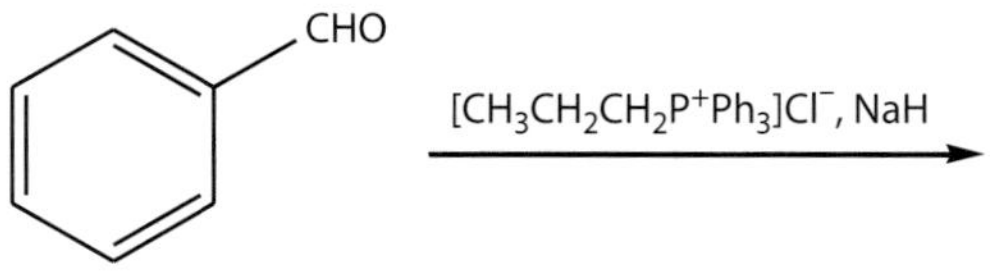

(c)

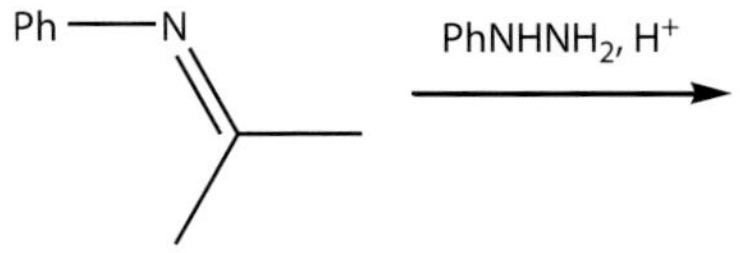

(d)

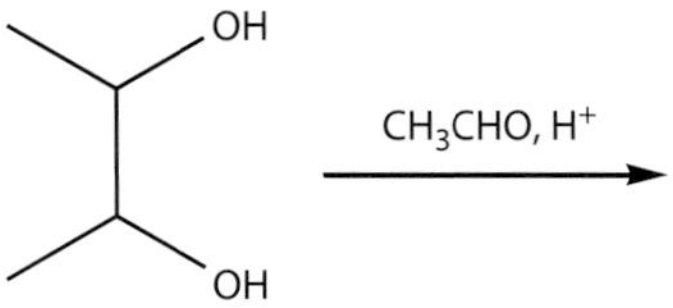

(e)

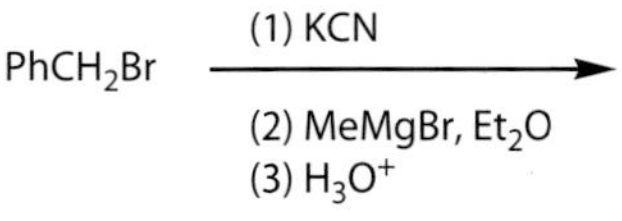

(f)

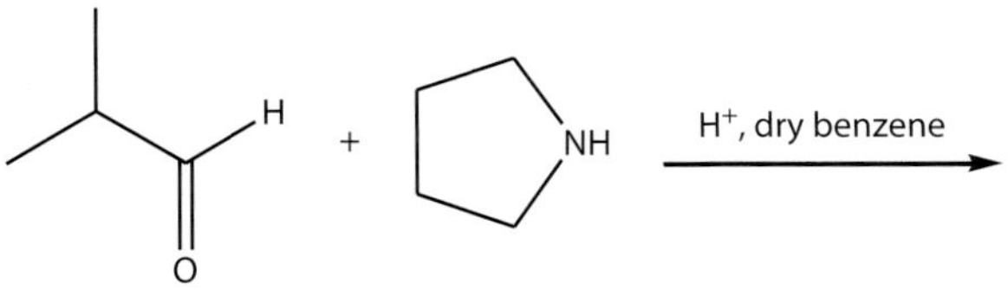

(g)

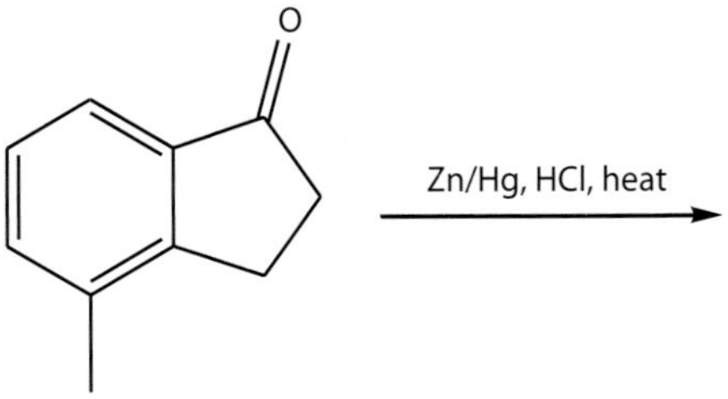

(h)

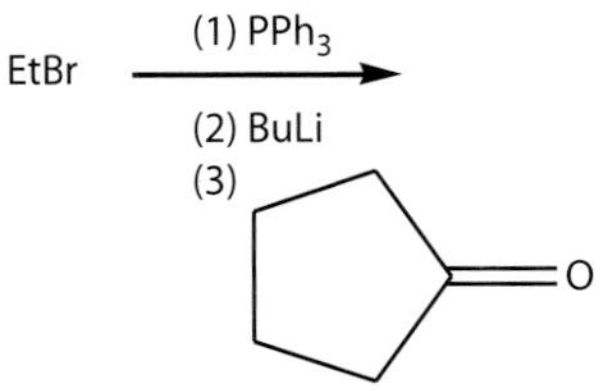

(i)

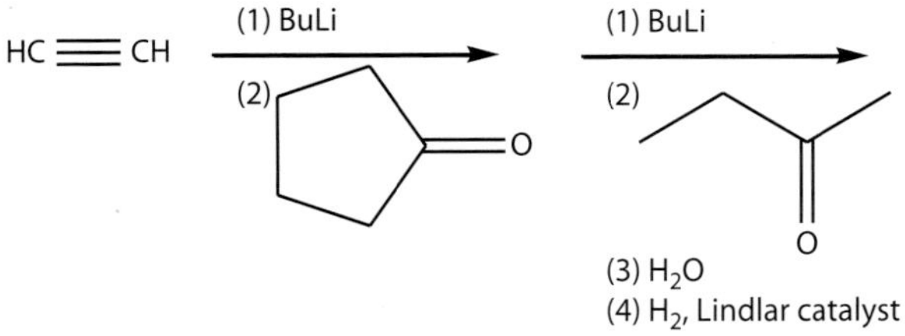

(j)

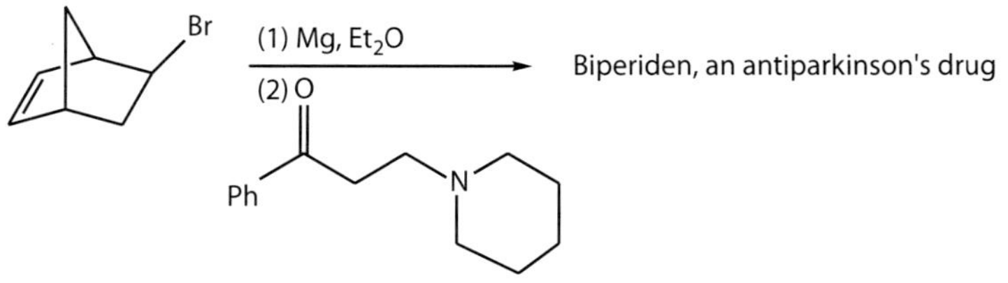

(k)

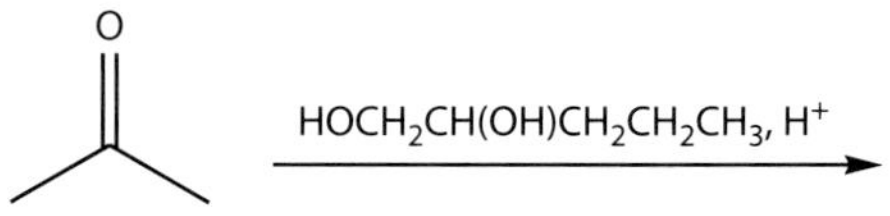

(l)

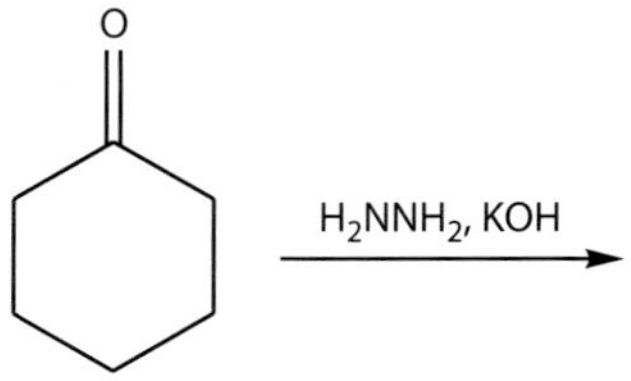

(m)

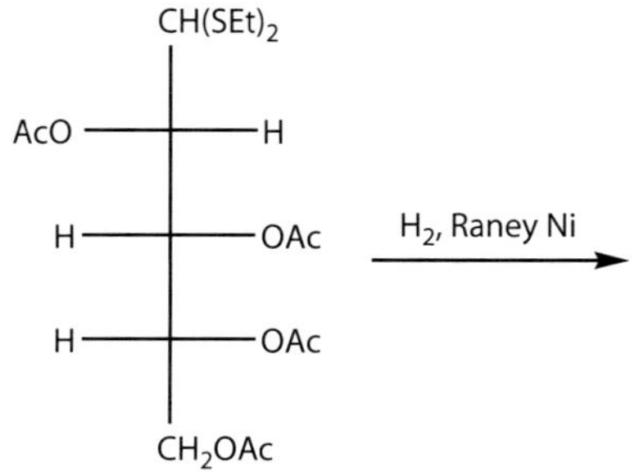

(n)

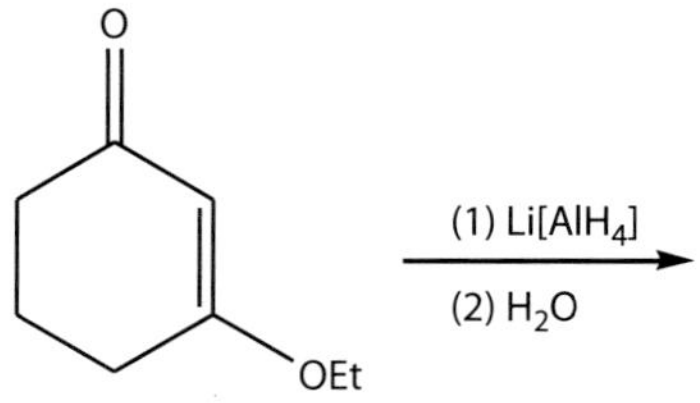

(o)

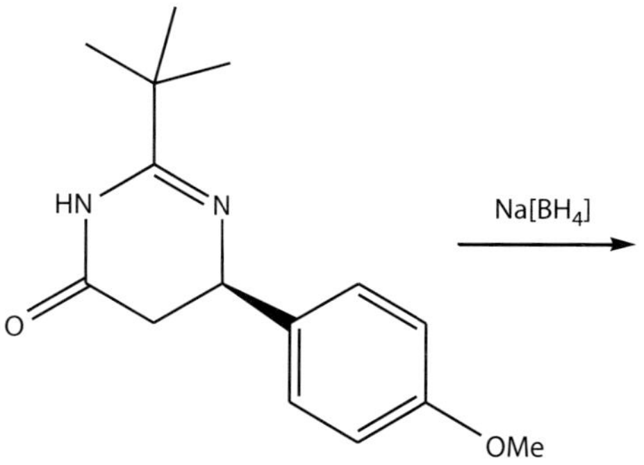

7. Predict the equilibrium constant for each of the following processes, giving a brief reason:

(a) $t\text{-Bu}_2C{=}O + H_2O \rightleftharpoons t\text{-Bu}_2C(OH)_2$

(b)

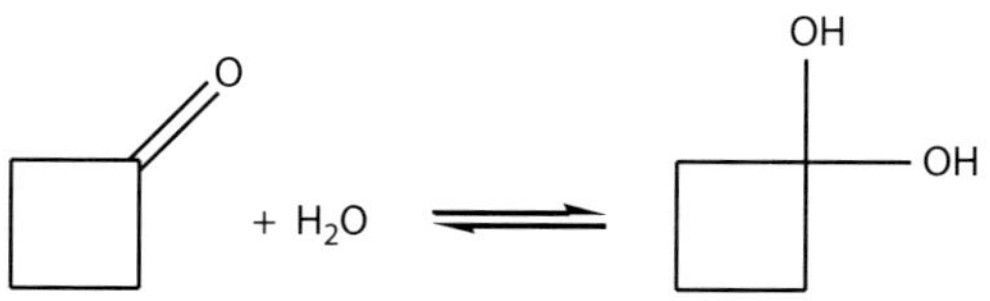

(c) $Ph_2C{=}O + H_2O \rightleftharpoons Ph_2C(OH)_2$

8. Identify the unknown compounds in the following sequence, justifying your conclusions by mechanisms:

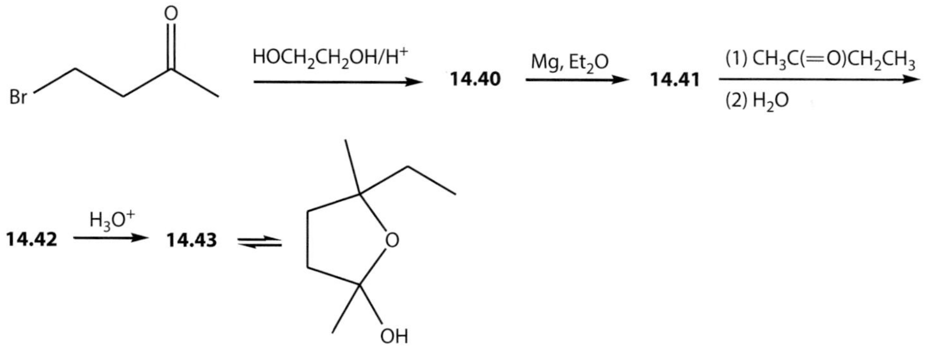

9. Draw the most stable conformation of the molecule **14.44**. Reaction of **14.44** with $Li[AlH_4]$ gave a mixture of two alcohols **14.45** and **14.46**. Draw the chair conformations of **14.45** and **14.46**, and predict which one would predominate, giving your reasons. The compound $Li[AlH(OCMe_3)_3]$ also reduces ketones to alcohols. If this was reacted with **14.44**, what proportions of **14.45** and **14.46** would you expect?

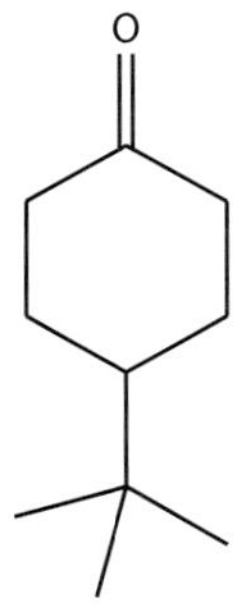

14.44

10. Explain the following observations:
    (a) The rate of formation of acetone semicarbazone from acetone and $H_2NCONHNH_2$ is pH dependent with a maximum at pH 6. You will need to draw a detailed mechanism for the reaction.
    (b) The equilibrium constants for the hydration of the three aldehydes shown in the following:

$H_2C{=}O$

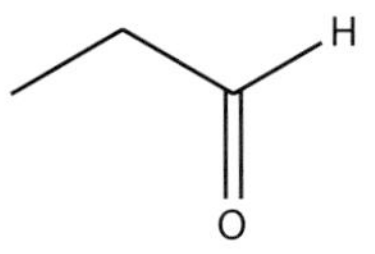

Cl_3CCHO

$H_2C{=}O$	propanal (structure above)	Cl_3CCHO
40	0.7	500

    (c) When *R*-$PhCH_2CH(OH)CHO$ is reacted with KCN/HCN and the products hydrolyzed, unequal amounts of two acids are obtained.
    (d) Hemiacetal formation may be catalyzed by acid or base, but acetal formation can only be catalyzed by acid and not by base.
    (e) The presence of water is essential in the oxidation of aldehydes to carboxylic acids by Cr(VI).

11. Compare and contrast the following:

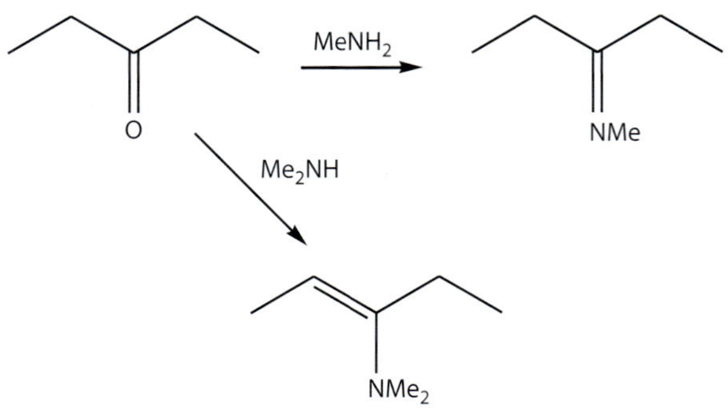

MORE CHALLENGING PROBLEMS

1. Propose a mechanism for each of the following reactions:

(a)

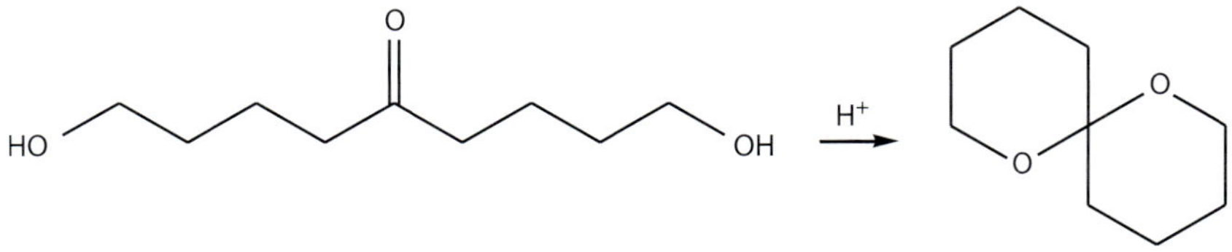

(b)

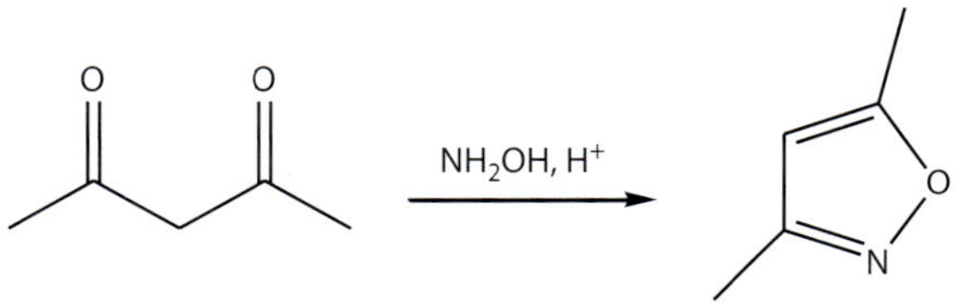

(c)

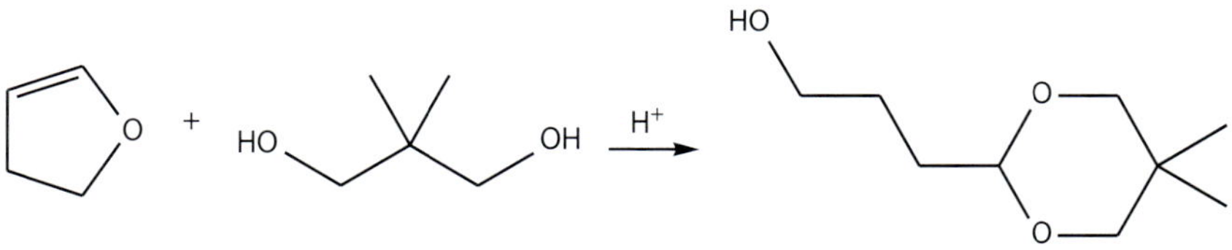

(d)

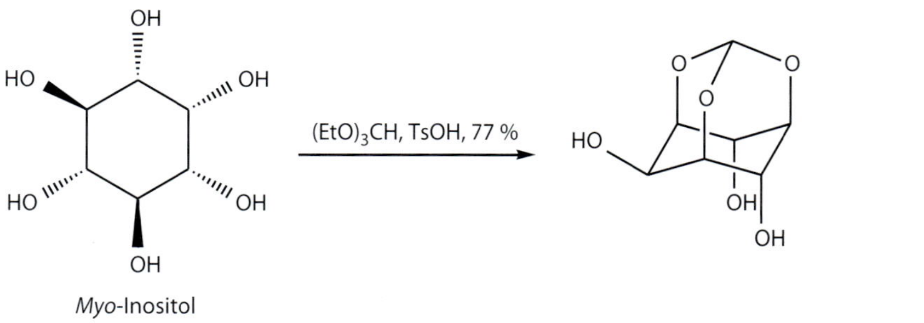

(e)

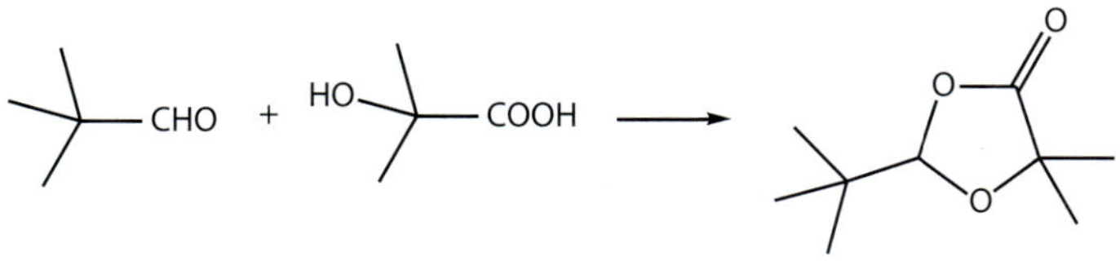

(f)

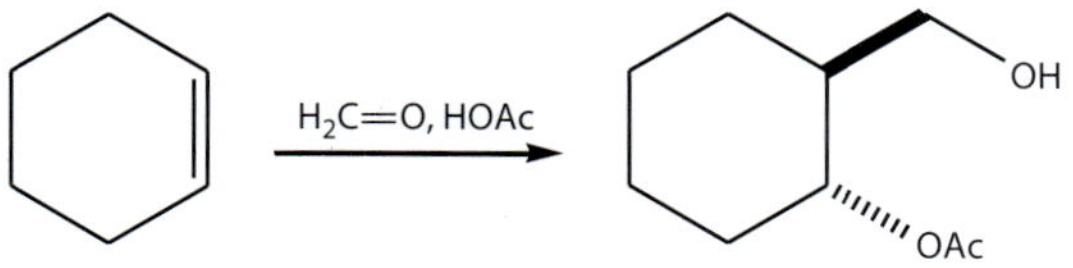

(g)

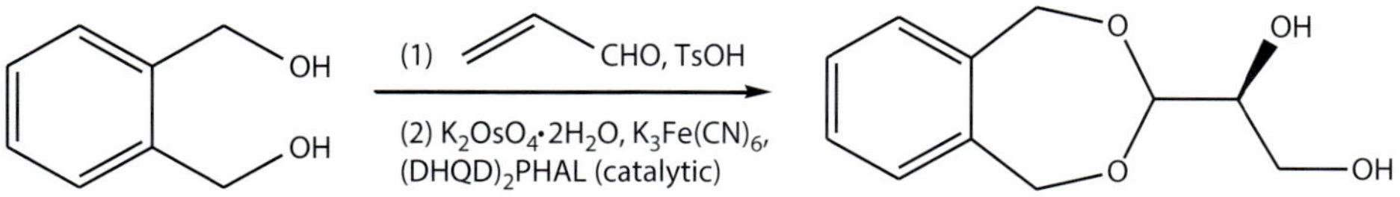

2. How would you accomplish each of the following transformations, which may require more than one step?

(a)

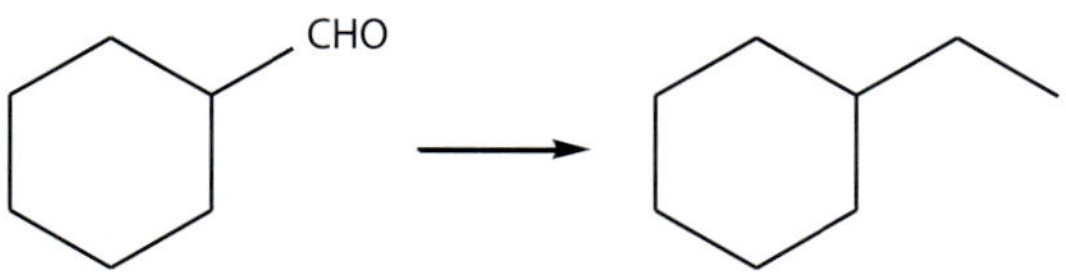

(b)

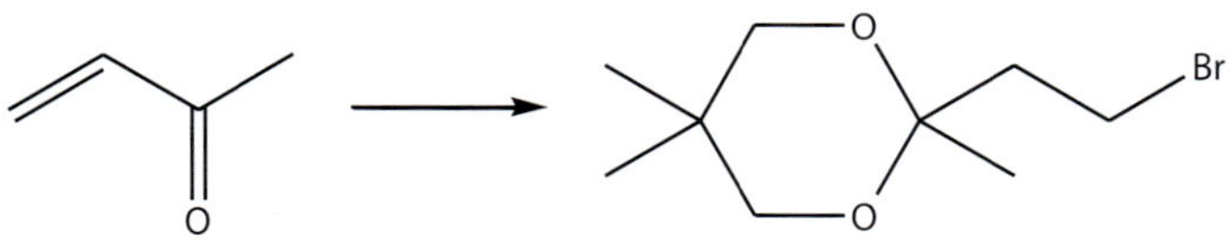

(c)

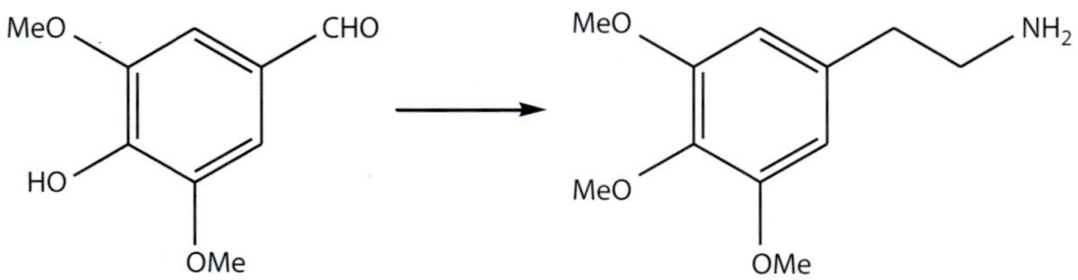

Syringaldehyde

Mescaline

3. Predict the product(s) of the following reactions:

(a)

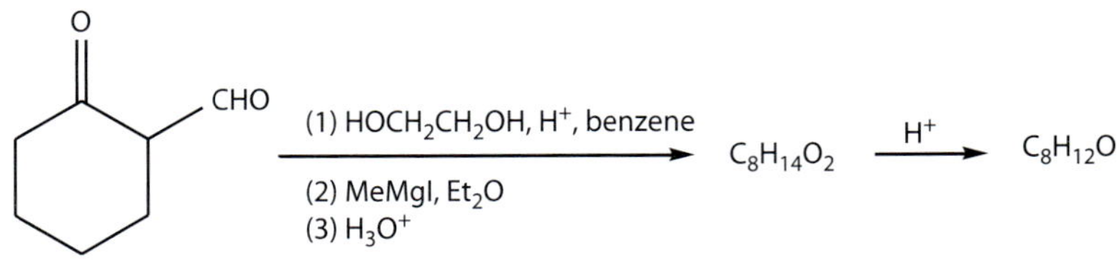

(b) $PhCHO + Ph_3P^+CH_2COOEt \xrightarrow{EtO^-}$

(c)

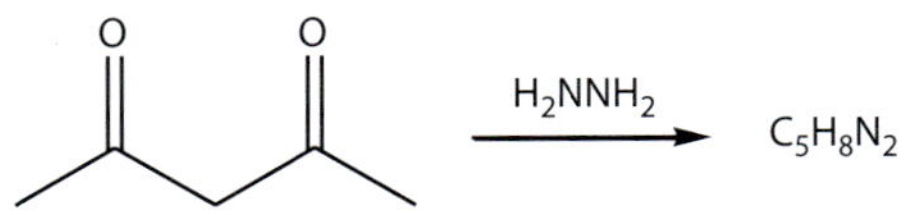

(d)

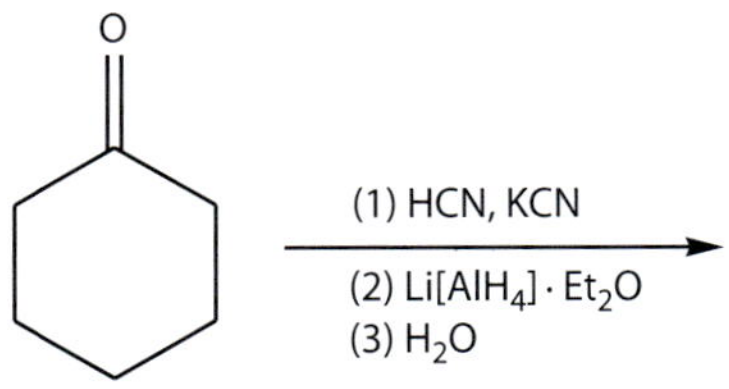

(e)

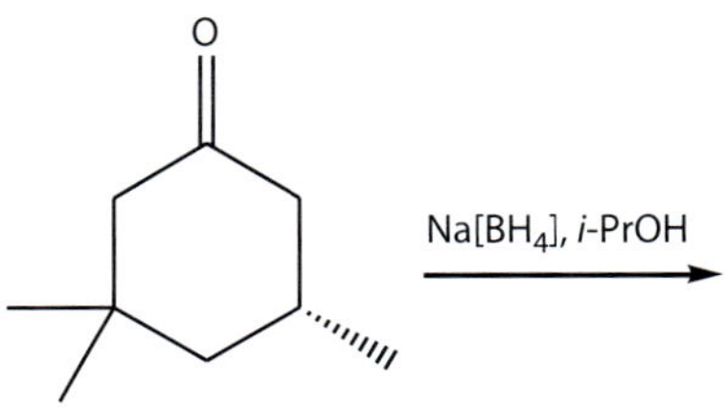

(f)

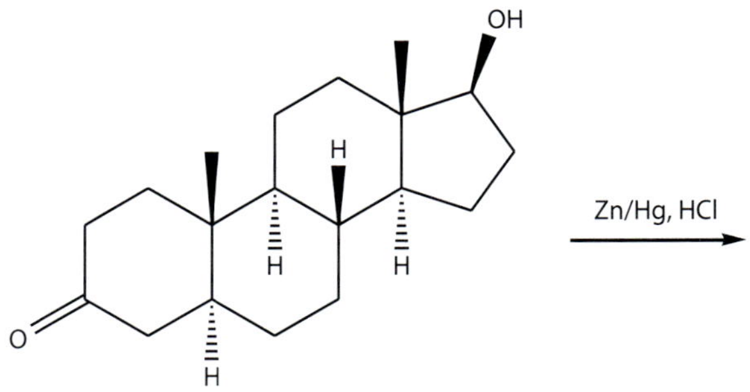

(g)

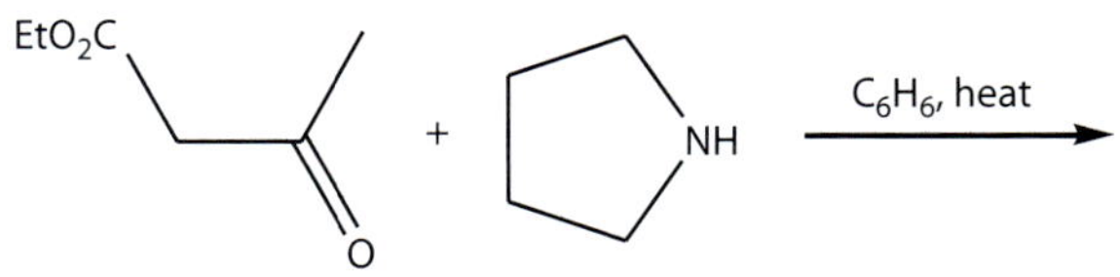

4. Estragole, **14.47**, is a major constituent of tarragon oil. Treatment with HBr in the presence of dibenzoyl peroxide, PhC(=O)–O–O–C(=O)Ph, gives **14.48**. **14.48** reacts with magnesium in ether followed by CO_2 and acidification to give **14.49**. Spectroscopic information on **14.49** is given in the following. Deduce the structure of **14.49** and hence of **14.48** and estragole.

14.49 Mass spectrum $m/z = 194$ (M^+)
IR spectrum 3350 (br, s), 1696 (s) cm^{-1}
^1H NMR spectrum
δ 1.95 (quintet, J = 7 Hz, 2H)
2.37 (t, J = 7 Hz, 2H)
2.59 (t, J = 7 Hz, 2H)
3.74 (s, 3H)
6.75 (d, J = 8 Hz, 2H)
7.05 (d, J = 8 Hz, 2H)
11.6 (s, br, 1H)

5. Explain the following observations:
(a) Warming formaldehyde (methanal) with glycerol (propane-1,2,3-triol) for a few hours gave **14.50** and **14.51** in a 40:60 ratio. If the mixture is left at room temperature for several weeks, the **14.50**/**14.51** ratio is 80:20:

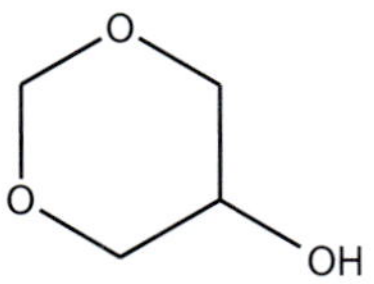

14.50

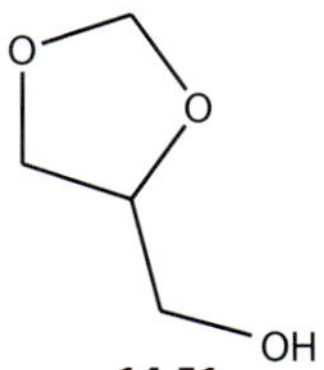

14.51

(b) The hydroxyketone, **14.52**, shows no peaks in its IR spectrum between 1600 and 1800 cm^{-1}, but it does show a broad absorption between 3000 and 3400 cm^{-1}:

14.52

(c) The equilibrium constants, K, for cyanohydrin formation from aromatic carbonyl compounds are given in the following:

Compound	PhCHO	4-ClC_6H_4CHO	4-$O_2NC_6H_4CHO$	4-$MeOC_6H_4CHO$	PhC(=O)Me
K	200	204	1420	32	0.8

(d)

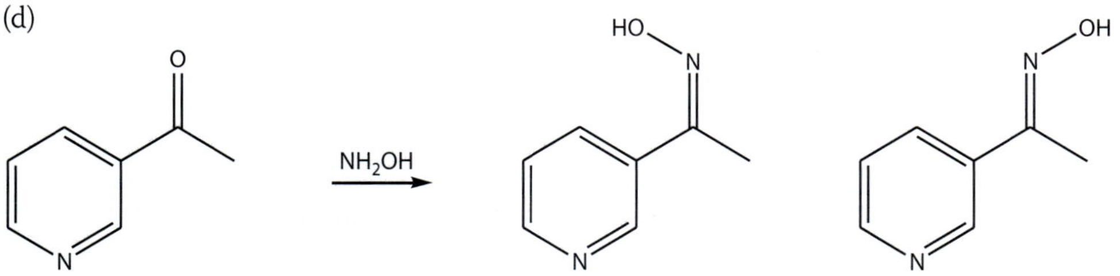

60 °C	97 %	3 %
25 °C	88 %	12 %

6. Suggest reagents for each of the steps in the synthetic sequences shown in the following:

(a)

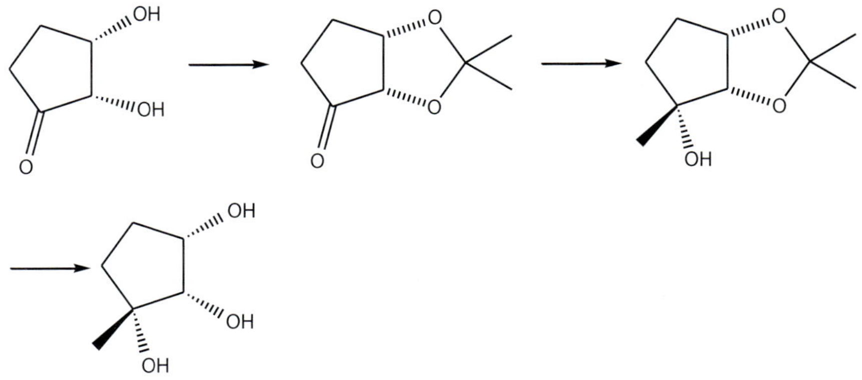

(b)

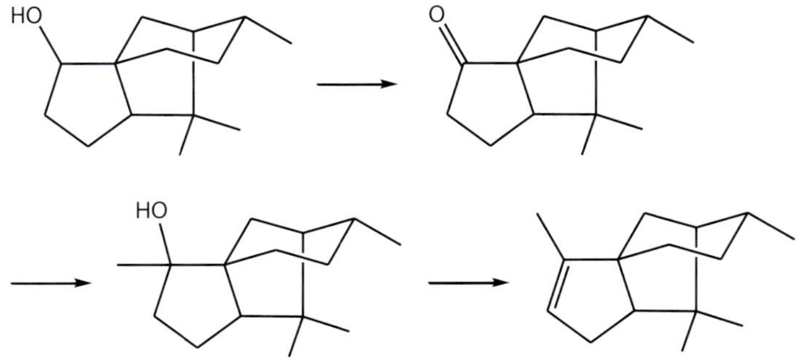

7. When the terpene citral is allowed to react, in the presence of dilute acid, with olivetol, there is an obtained mixture of products including **14.53**, one of the psychoactive components of marijuana. Starting by protonating the aldehyde, and doing an electrophilic substitution of the aromatic ring, shows the steps in a likely mechanism.

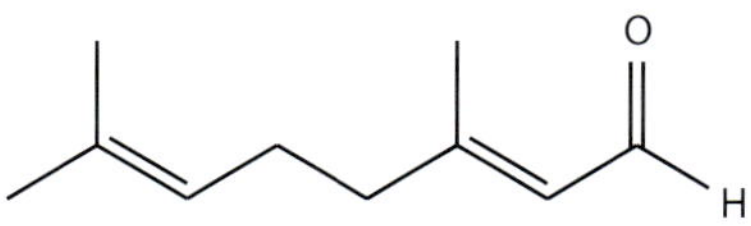

Citral

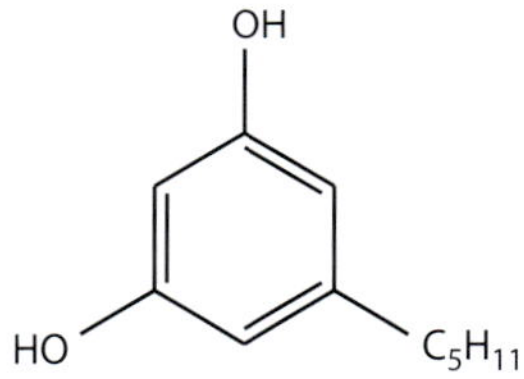

Olivetol

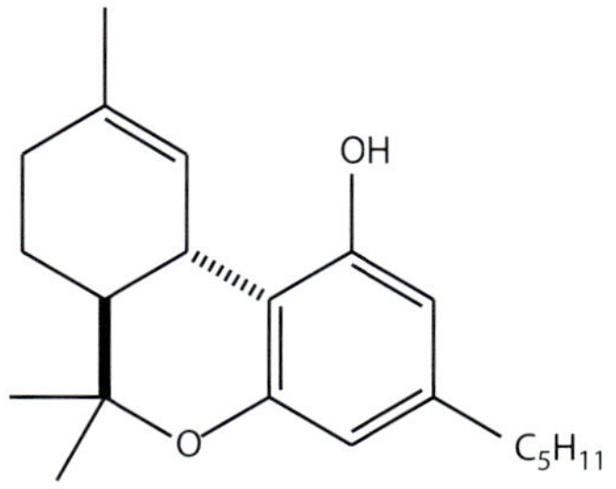

14.53, Δ-1, 3, 4-*trans*-Tetrahydrocannabinol

8. When nitriles, RCN, are hydrogenated in the presence of Raney nickel, the main product is RCH_2NH_2. However, small amounts of RCH_2NHCH_2R are also formed. Write a mechanism to account for the formation of this by-product, and suggest a method for the reduction that would avoid its formation.

9. Study the reaction sequence shown in the following, and answer the questions posed:

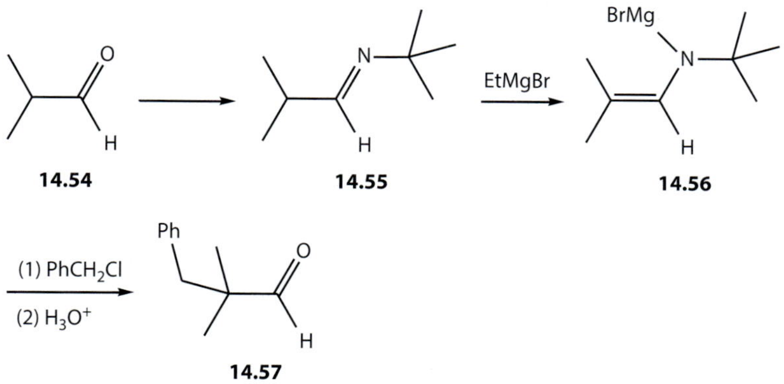

(a) Suggest a reagent/conditions for the conversion of **14.54** to **14.55**.
(b) Why does the Grignard reagent not add to the carbon–nitrogen double bond of **14.55**?
(c) Write a mechanism for the conversion of **14.55** to **14.56**.
(d) Write a mechanism for the conversion of **14.56** to **14.57**.

Substitution of Carbonyl Compounds

Carboxylic Acids and Their Derivatives

15

15.1 INTRODUCTION: REVIEW OF NOMENCLATURE

In this chapter, we will be discussing the chemistry of carboxylic acids, esters, acyl halides, anhydrides, and amides. This is dominated by substitution, where one group is exchanged with another. Substitution is NOT possible for aldehydes and ketones, as you can't displace H^- or R^-—they are hopeless leaving groups. First, let's review some nomenclature. The suffix for carboxylic acids is -oic acid and the carbonyl of the acid is always numbered as C-1. The acid takes precedence over most other functional groups. Some examples are shown in Figure 15.1. Notice that when we have both a ketone and an acid in the molecule, it is named as a carboxylic acid, and the ketone is described as "oxo."

The names of esters are derived from the names of the alcohol and carboxylic acid from which they were made. The name of the alcohol comes first, so esters of methanol begin with methyl, those of ethanol with ethyl, *etc.* The suffix for the acid part of the name is -oate. So we might have ethyl ethanoate. For **15.1**, the alcohol is 3-methyl-1-butanol, and the acid is ethanoic acid. So the ester is named 3-methyl-1-butyl ethanoate. It is banana scented. Similarly, **15.2** is 1-propyl ethanoate and **15.3** ethyl butanoate. If you find this challenging, it's a good strategy to write down the acid and alcohol separately before coming to a conclusion.

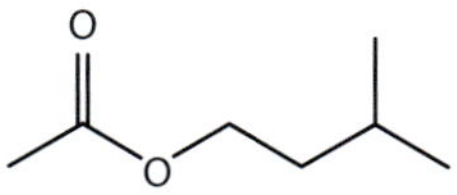

15.1

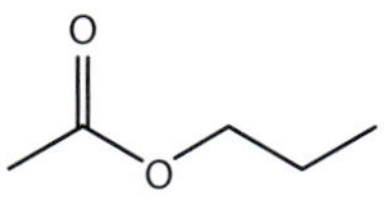

15.2

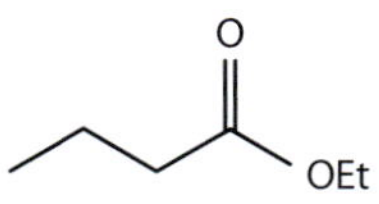

15.3

To name acid chlorides (the same idea applies to other halides, but we will encounter these less frequently), we take the name of the acid and replace the suffix with -oyl chloride. So **15.4** is ethanol chloride (but it mostly goes by its common name of acetyl chloride), **15.5** is 3-butenoyl chloride, and **15.6** is benzoyl chloride. Anhydrides are also named for the acids from which they are derived; thus, **15.7** is ethanoic anhydride (but commonly goes by acetic anhydride) and **15.8** is benzoic pentanoic anhydride.

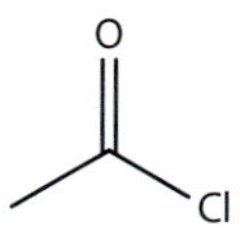

15.4

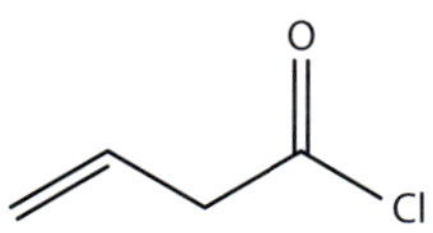

15.5

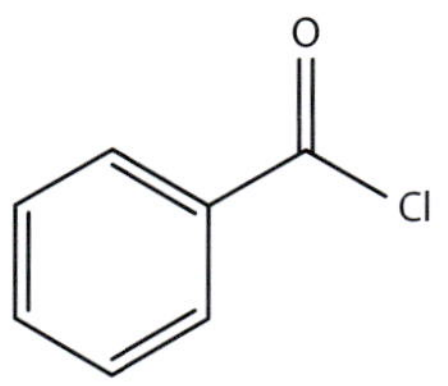

15.6

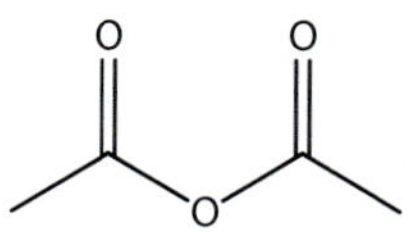

15.7

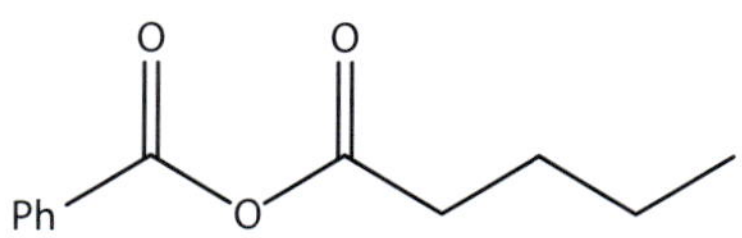

15.8

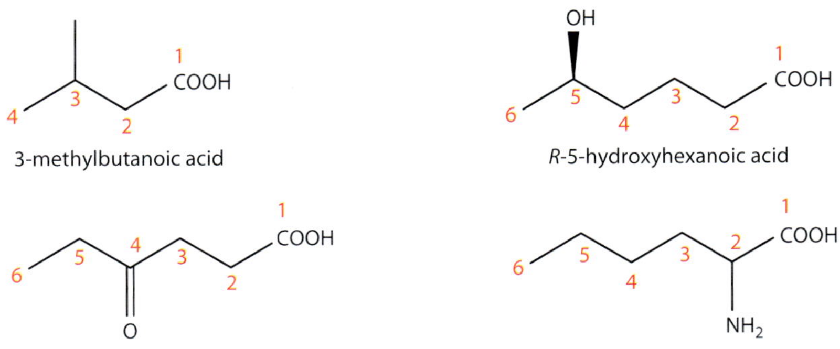

FIGURE 15.1 Nomenclature of carboxylic acids.

The suffix for amides is -amide. If there are substituents on the nitrogen atom (as opposed to those on the carbon chain where they are numbered in the usual way with the carbonyl carbon as C-1), then they are written as *N*-ethyl or *N,N*-dimethyl. So **15.9** is butanamide and **15.10** is 3-ethyl-*N*-methylhexanamide.

15.9 **15.10**

PROBLEM 15.1

(Review) Give a systematic name for each of the following compounds:

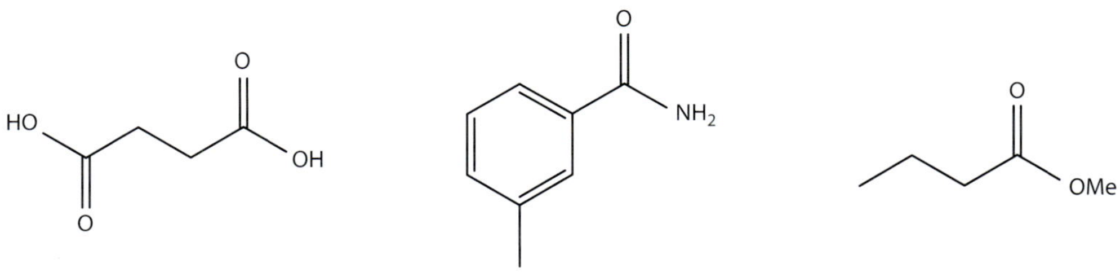

Solution

The first compound is a diacid with four carbon atoms and no substituents—thus butane dioic acid. The common name is succinic acid.

The second compound would be named benzamide if we ignored the methyl group on the benzene ring. To put the substituent in, we need to number the benzene ring from the point that the carbonyl group is attached as 1. Thus, the full name is 3-methylbenzamide. The final compound is an ester formed between methanol and butanoic acid, thus methyl butanoate.

PROBLEM 15.2

Give a structure for each of the following compounds:

(a) Benzoyl chloride
(b) 2-Methyl-*N,N*,-diethylbutanamide
(c) 3-Chloro-5-bromoheptanoic acid

Solution

(a)

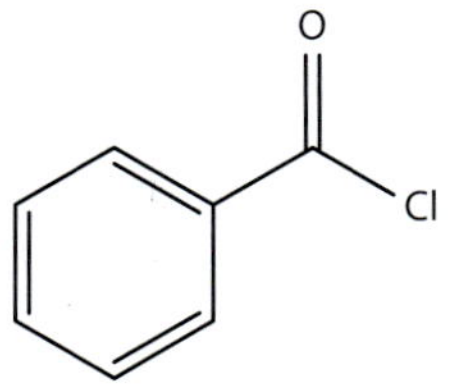

(b) First, draw and number butanamide:

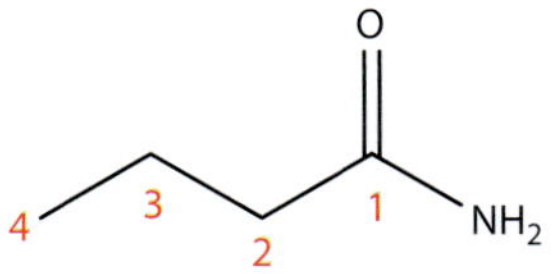

We can now put the substituents in at the appropriate positions:

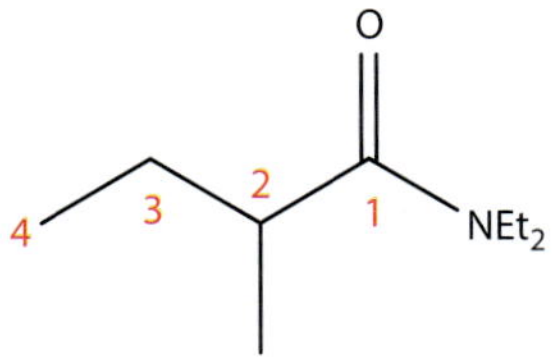

(c) First, draw and number heptanoic acid:

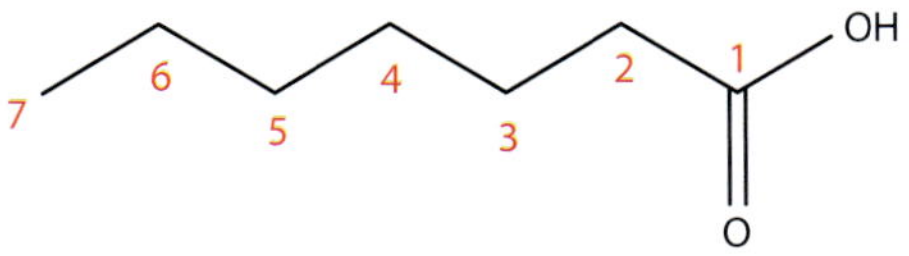

We can now put in the substituents at the appropriate positions:

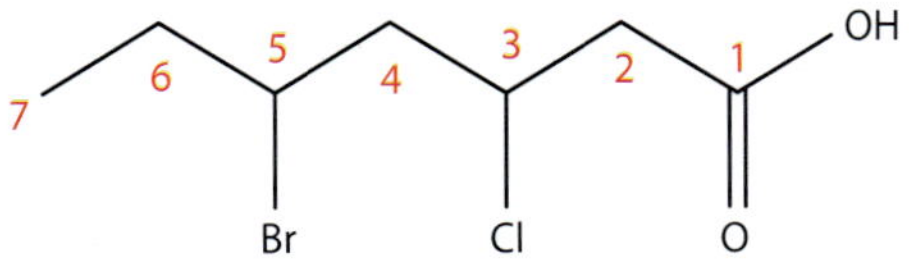

15.2 SUBSTITUTION REACTION

15.2.1 EFFECT OF SUBSTRATE

The reaction that we will be studying in this chapter is shown in Figure 15.2. We know that it cannot be a simple or direct substitution—there are no S_N2 reactions at sp^2 carbon atoms. The mechanism of the reaction is analogous to some we saw in aromatic chemistry—an addition followed by an elimination. The first step is rate determining, and the relative reactivity of various carbonyl compounds depends on the electrophilicity of the carbonyl carbon atom. We saw in the last chapter the effect of inductive donation. Aldehydes are more electrophilic than ketones because alkyl groups donate electron density more than hydrogen does. Thus, the carbon in a ketone was less positive than in an aldehyde. From an inductive point of view, chlorine, oxygen, and nitrogen all withdraw electron density from the carbon of the carbonyl group. We will also meet examples where the initial step is protonation of one of the lone pairs on the oxygen of the carbonyl group, followed by attack by an uncharged nucleophile.

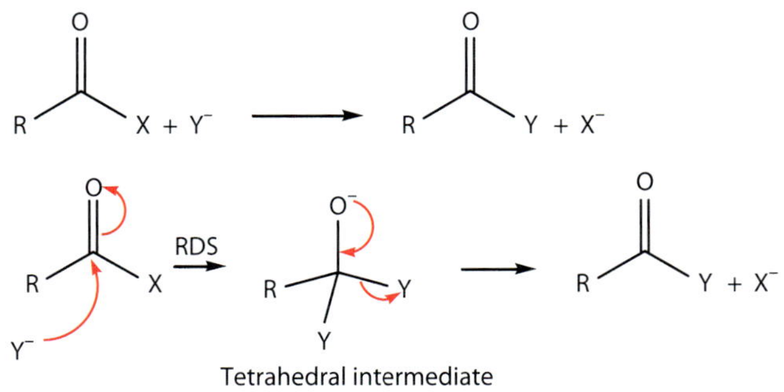

FIGURE 15.2 Substitution at a carbonyl group.

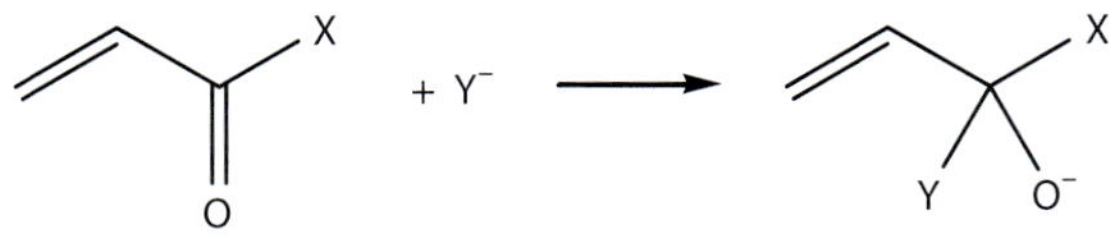

FIGURE 15.3 Formation of a tetrahedral intermediate is disfavored when conjugation is lost.

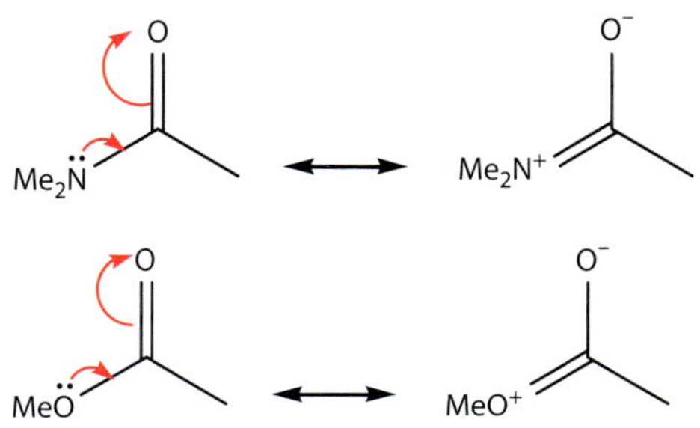

FIGURE 15.4 Resonance forms of esters and amides.

In the last chapter, we noted that when an aryl ketone or an α,β-unsaturated ketone underwent an addition reaction, conjugation was lost, and hence, the process was less favorable than for nonconjugated species. We will see the same outcome with other carbonyl compounds—the RDS is the formation of the tetrahedral intermediate. This also causes loss of the conjugation and so is unfavorable (Figure 15.3).

Another significant effect on reactivity is the ability of the group X to act as a resonance donor. The lone pairs of electrons on both oxygen and nitrogen can be shared with the positive carbon of the carbonyl, resulting in contributions from resonance forms where the carbonyl carbon is less positive and hence less reactive (Figure 15.4). We have already discussed the contribution that the amide resonance form makes to its structure; not only does this mean that amides are planar, but the typical IR stretching frequency of the carbon–oxygen double bond of an amide is one of the lowest of all the carbonyls. The effect is greater for amides than for esters because the relative electronegativities of oxygen and nitrogen mean that nitrogen is more willing to share its lone pair of electrons than is oxygen.

We can summarize all these effects to come up with a sequence for relative reactivity of carbonyl groups (15.1). The carboxylic acid is deliberately omitted, because in general, nucleophiles react with acids to deprotonate them to give the carboxylate anion, $RCOO^-$ and NuH. The anion, being negatively charged, is reluctant to react with nucleophiles. In any given subset, conjugated species will react more slowly than nonconjugated ones. Lists of this sort always need to be qualified by saying that substituents on the R group are not unimportant; compounds containing fluorinated alkyl groups would be expected to react more rapidly than those of unsubstituted alkyls:

$$RC(=O)Cl > RC(=O)OC(=O)R > RC(=O)H > R_2C=O > RC(=O)OR' > RC(=O)NH_2 \quad (15.1)$$

Focus on Structures of Carbonyl Compounds

When discussing carbonyl reactivity, we frequently make arguments that depend on inductive and resonance effects to explain our observations. But, other than conclusions justified by outcome, do we have any physical evidence to support these assertions? Fortunately, there is such evidence, from two sources: structural studies and IR spectroscopy. If we first examine the length of the carbon–oxygen double bond, the variation is quite small, but it is in the right direction to explain the variation in carbon–oxygen stretching frequency (Table 15.1). The shorter the bond, the stronger the bond and the higher the stretching frequency. Carboxylic acids and unsubstituted amides have been deliberately excluded because hydrogen bonding has a significant effect for these compounds. The substituted amide is an outlier—because the resonance form with a single bond between carbon and oxygen is so important, the carbon–oxygen double bond is long and weaker than for other types of carbonyl compound.

The resonance effects that we adduced in amides and esters should also be visible in the lengths of the bond between the heteroatom and the carbon of the carbonyl, and this is indeed the case; they are significantly shorter than normal C–X single bonds (Table 15.2). It is clear that there is no significant resonance donation from chlorine and that donation from the lone pair on nitrogen is greater than that from oxygen.

The best comparative model for an amide would be a species in which all the same bonds are present, but the lone pair of electrons on nitrogen is not able to overlap with the π-bond of the carbonyl. Models are provided by the so-called twisted amides, of which the simplest unsubstituted example, quinuclidone, **15.11**, was synthesized only in 2006. The amide itself is too unstable for it to be possible to determine its structure directly, but its HBF_4 salt has been characterized. The carbon–oxygen bond is short (1.192 Å) and strong (ν_{max} = 1822 cm^{-1}), and the carbon (of the carbonyl)–nitrogen single bond is long, 1.50 Å. Protonation occurs at nitrogen in contrast to other amides, where protonation is invariably at oxygen. Calculations on the unprotonated species predict a C=O bond length of 1.183 Å and an N–(C=O) bond length of 1.433 Å. These are among the longest carbon–nitrogen single bonds ever characterized. Quinuclidone, unlike other amines, is comparable in basicity to a secondary or tertiary amine; since the lone pair is not involved with the carbonyl group, it is available to capture a proton. The compound is extremely unstable, being hydrolyzed in D_2O with a half-life less than 15 s at room temperature. The structure of the β-lactam ring in penicillin (general structure **15.12**) allows its relatively easy hydrolysis and is probably implicated in its mode of action. It has been postulated

TABLE 15.1 Correlation between Carbon–Oxygen Bond Lengths and IR Stretching Frequencies for $CH_3C(=O)X$

X	Cl	OMe	H	Me	NMe_2
r(C=O), Å	1.181	1.207	1.204	1.212	1.214
ν(C=O), cm^{-1}	1800	1735	1730	1715	1662

TABLE 15.2 Comparison of C–X Distances in Carbonyl Derivatives

X	Cl	OMe	NR_2
r(C–X) in RC(=O)X, Å	1.78	1.33	1.35
r(C–X) in alkyl-X, Å	1.77	1.40	1.46

that protease enzymes, which catalyze the hydrolysis of peptide bonds *in vivo*, bind the amide in such a way that it becomes twisted, with reduced conjugation, and is hence more susceptible to nucleophilic attack.

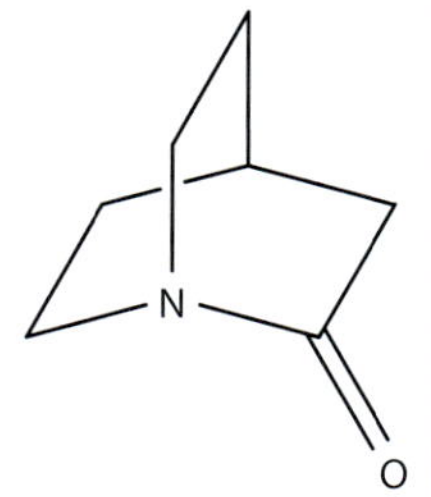

15.11

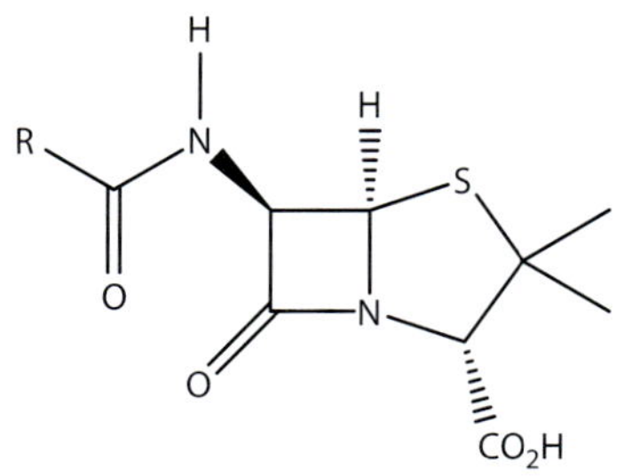

15.12

15.2.2 EFFECT OF LEAVING GROUP

How does the ability of the group X in our substrate to be a leaving group affect the outcome of these reactions? Consider the reaction in Figure 15.5. Because the carbon of the carbonyl group is rather positive, attack of the hydroxyl on it will be fast. Looking at the tetrahedral intermediate, we would instinctively expel chloride to give the product, rather than hydroxyl to return to starting material. We know that chloride is a good leaving group because it is the anion of HCl, a strong acid, but hydroxyl is a poor leaving group because water is a weak acid. Good leaving groups are the anions of strong acids and we can rank the usual substituents at the carbonyl in order of their leaving groups ability as $Cl^- > RCOO^- > HO^- > RO^- > R_2N^-$. This limits which substitution reactions are possible; a good leaving group cannot displace a poor one. Thus, the first two examples in Figure 15.6 are possible, while the final one is not.

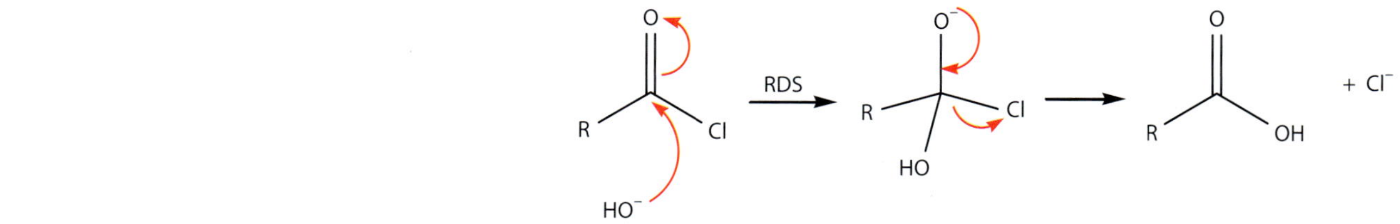

FIGURE 15.5 Reaction of an acid chloride with base.

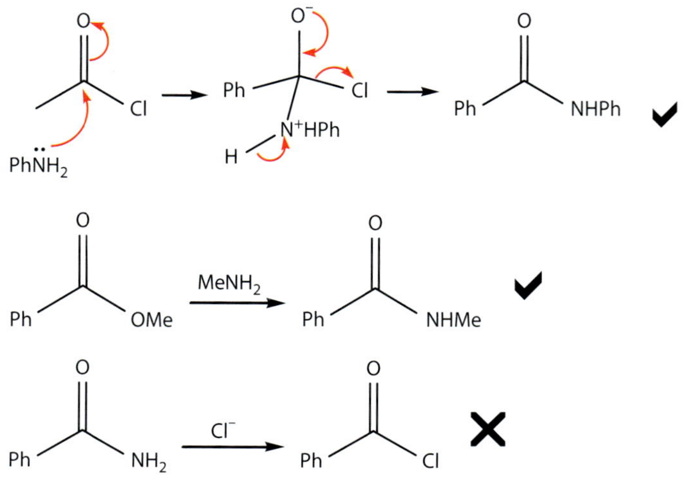

FIGURE 15.6 Availability of reactions depends on the leaving group.

PROBLEM 15.3

Explain the following reaction scheme:

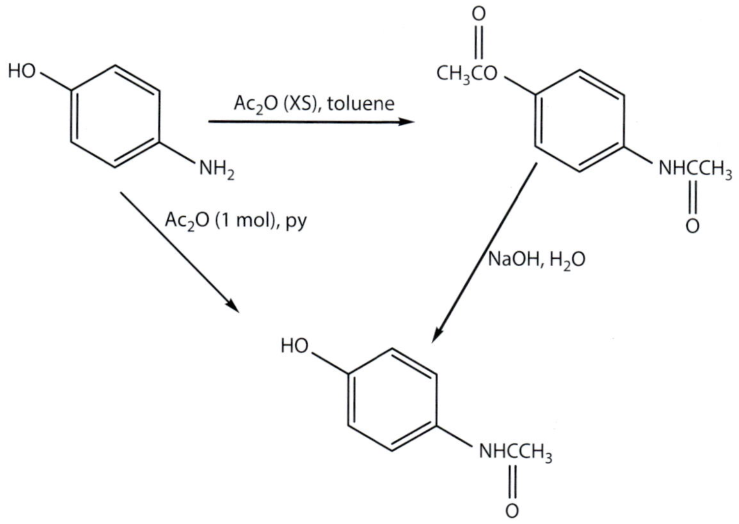

Acetaminophen (paracetamol)

Solution

In the diacetylation, both hydroxyl and amino are able to displace an acetate leaving group from acetic anhydride, because acetic acid is a stronger acid than either aniline or phenol:

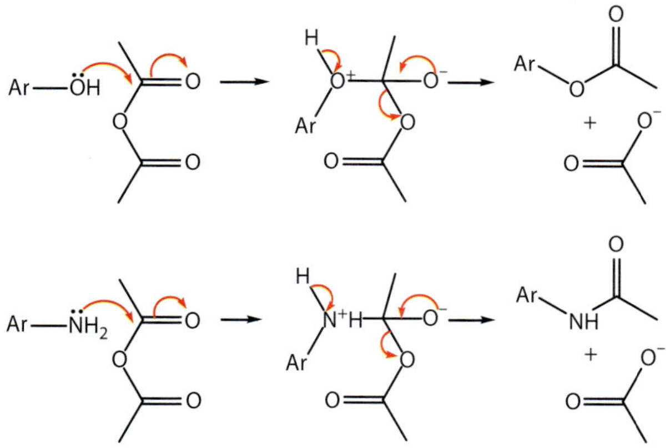

If only one mole of acetic anhydride is available, we need to consider which is the best available nucleophile, and an amino group beats hydroxy, because nitrogen is less electronegative and more willing to share its electrons. In considering the hydrolysis, the ester is more electrophilic than the amide, so more easily attacked. Also, particularly in basic solution, the $[NH_2]^-$ is a very poor leaving group.

Key Points from Sections 15.1 and 15.2

- The suffixes for carboxylic acids, amides, acid anhydrides, and acyl halides are, respectively, -oic acid, -amide, -oic anhydride, and -oyl chloride.
- The names of esters involve the alcohol and the acid from which they are prepared, thus ethyl propanoate from ethanol and propanoic acid.
- Substitution of acid derivatives occurs through an addition–elimination mechanism, *via* a tetrahedral intermediate.
- The leaving group ability of the substituent determines which reactions are possible and which are not.
- Relative reactivities are in the order

$$RCOCl > (RCO)_2O > RCHO > R_2CO > RCOSR > RCOOR \gg RCONH_2$$

15.3 OXYGEN NUCLEOPHILES

We should consider three types of oxygen nucleophile—water, alcohols, and carboxylic acids. Starting with the simplest, water, we can consider its reaction with acid chlorides, anhydrides, esters, and amides. Remember that water will not easily displace ammonia or an amine, so we will need to make some special effort to make amides hydrolyze. In each case, the product will be the carboxylic acid. The reactions of Figure 15.7 with a chloride and anhydride, respectively, are relatively straightforward mechanistically. On a practical note, notice that the reaction with acid chlorides generates choking and toxic fumes of HCl. Pyridine is generally added to such reactions to mop up the HCl as pyridinium hydrochloride (**15.10**), which precipitates and can be removed by filtration. Because acid chlorides are so reactive, they react with the moisture in the air to give HCl, so they should always not only be used in a fume hood but stored there as well.

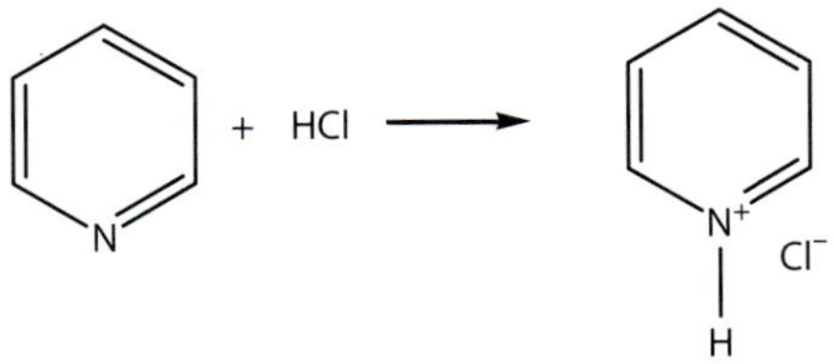

15.10, Pyridinium hydrochloride

Notice that there is a new symbol here—a double-headed arrow. This is a shorthand way of saying that the electrons move out to oxygen to give a tetrahedral intermediate, then return to expel the leaving group (Figure 15.8). Use this symbol if you are comfortable with it, as it does save time, but it will never be wrong to write out the two stages in full.

The other oxygen nucleophiles, alcohols and carboxylic acids, also react with acid chlorides (Figure 15.9) and anhydrides (Figure 15.10). In the second reaction of Figure 15.10, we see that we can prepare

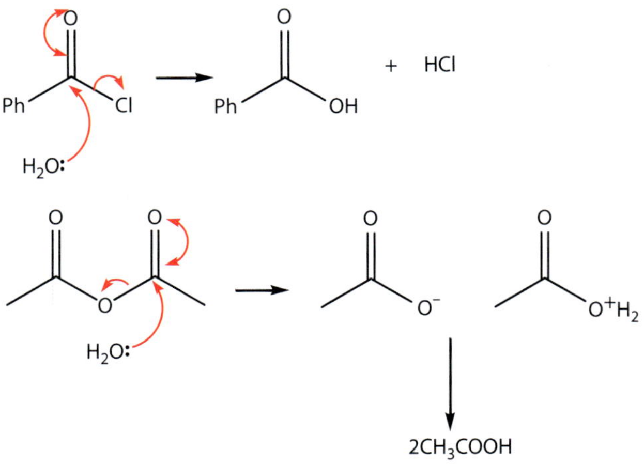

FIGURE 15.7 Hydrolysis of acid chlorides and anhydrides.

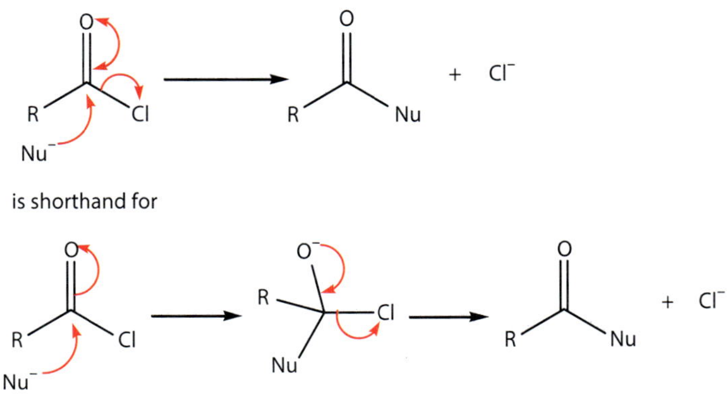

FIGURE 15.8 The double-headed arrow.

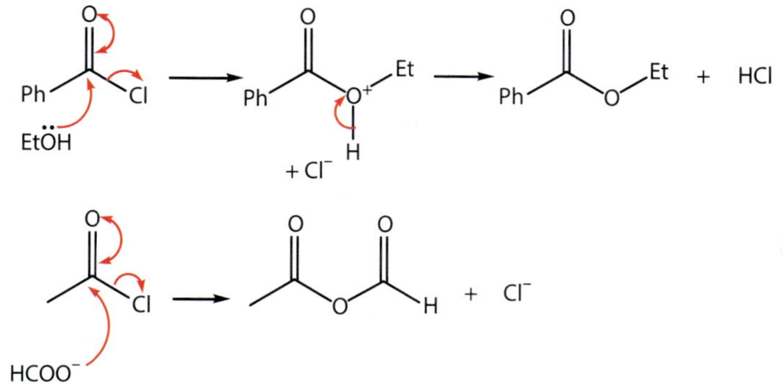

FIGURE 15.9 Reactions of acid chlorides with alcohols or carboxylic acids.

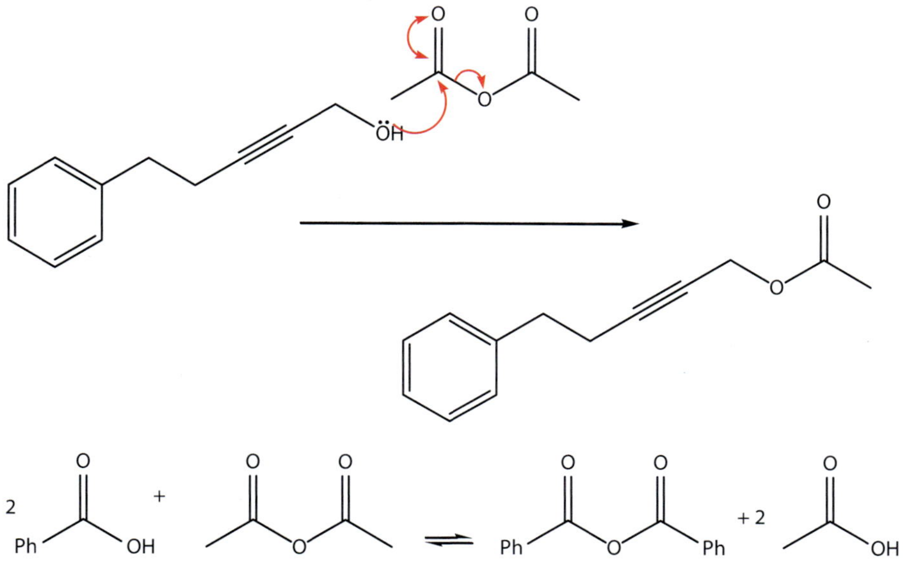

FIGURE 15.10 Reactions of anhydrides with alcohols or carboxylic acids.

other anhydrides from acetic anhydride and the relevant carboxylic acid. The equilibrium is driven to the right by removing acetic acid (the most volatile component of the mixture) by distillation, as it is formed. Although esters can be prepared from the relevant acid and alcohol, methods using anhydrides or acid chlorides are often preferred because of the higher reactivity of these substrates. Some examples that were used in longer synthetic sequences are shown in Figure 15.11.

Reactions with either acid chlorides or anhydrides show some sensitivity to steric hindrance. Although primary, secondary, and tertiary alcohols do all react, the rates are sufficiently different that selectivity between them can be achieved, as shown in Figure 15.12.

The reaction of water with esters, hydrolysis, is the reverse of the reaction by which they were made, the condensation reaction, as shown in Equation 15.2. Hydrolysis of esters can be carried out under either acid- or base-catalyzed conditions. Under basic conditions (Figure 15.13), the RDS is the attack of hydroxide ion on the ester carbonyl to give a tetrahedral intermediate. This collapses rapidly, to lose ethoxide. This would appear to break our rules for what reactions are, or are not, possible, because hydroxide is a better leaving group than ethoxide. But this does work, for two reasons. The mass action effect is on our side, as water is in large excess. Also, the final step of the reaction is an irreversible one; ethoxide (pK_a of ethanol is 16) deprotonates the carboxylic acid (pK_a typically 5). This final step means that the process in base is not fully reversible; esters may be hydrolyzed in base but cannot be synthesized in base. The alcohol is not a sufficiently good nucleophile to attack the carboxylate anion, which is the major species present in base.

$$\text{RCOOH} + \text{R'OH} \rightleftharpoons \text{RCOOR'} + \text{H}_2\text{O} \tag{15.2}$$

Under conditions of acid catalysis, the esterification/hydrolysis reaction is fully reversible (Figure 15.14). Mass action is still important, but we now have much better leaving groups

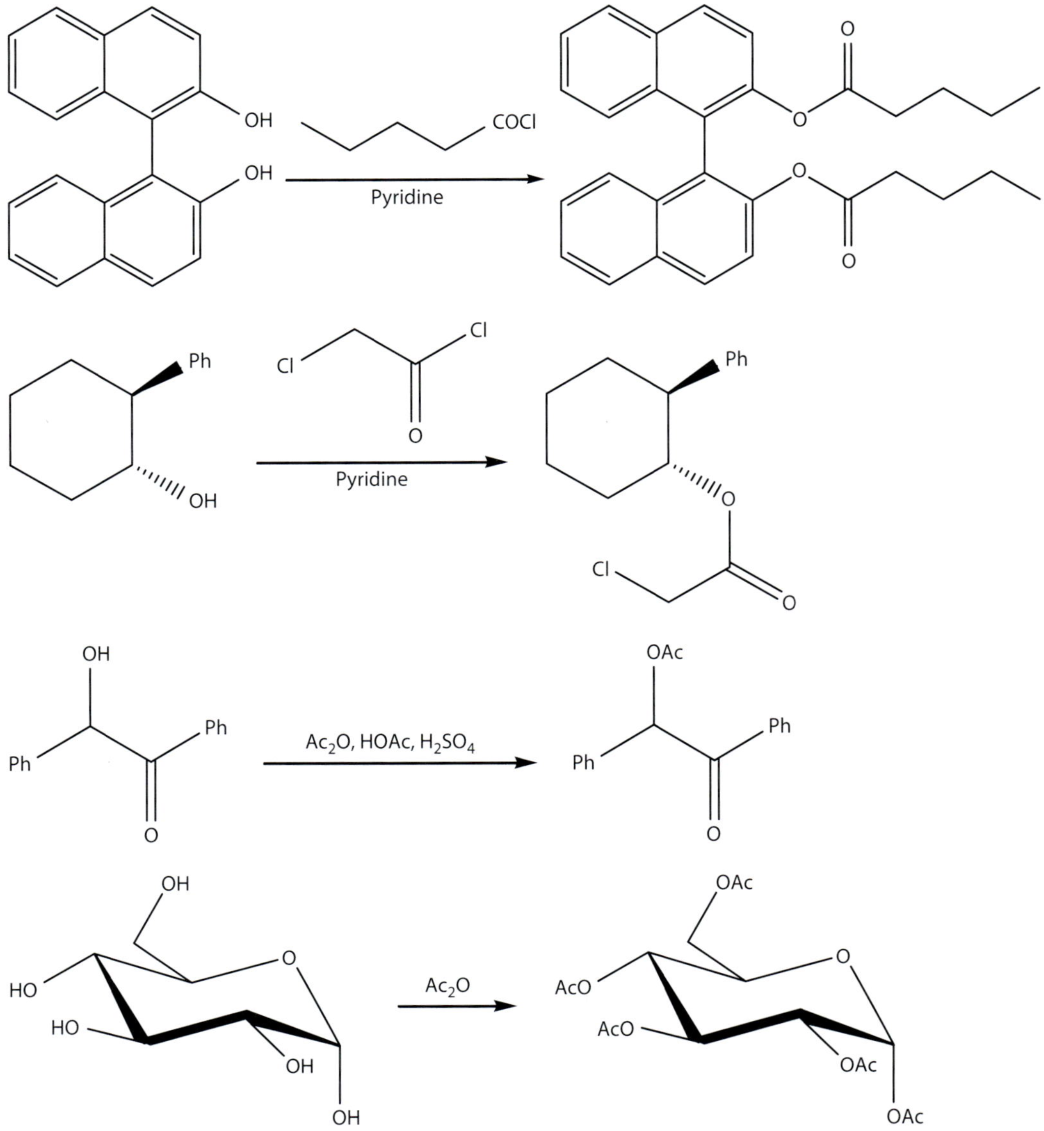

FIGURE 15.11 Ester forming reactions in synthesis.

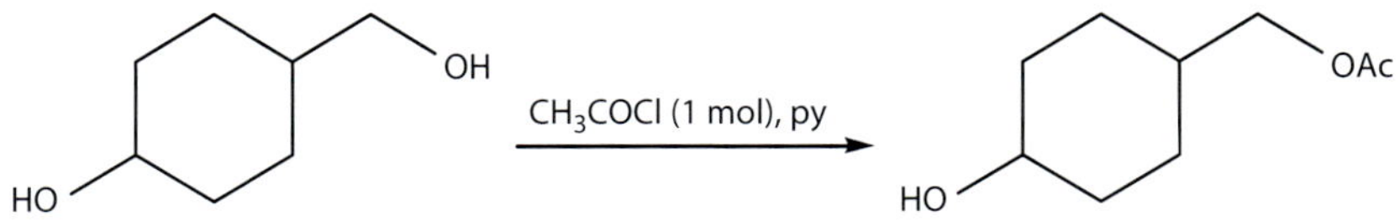

FIGURE 15.12 Reactions with acid chlorides are sensitive to steric hindrance.

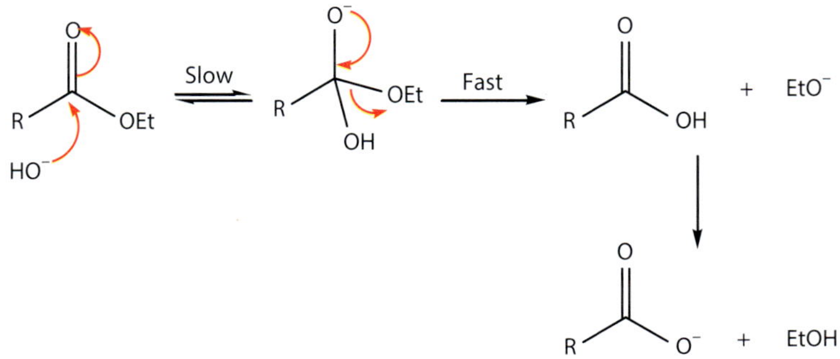

FIGURE 15.13 Base-catalyzed hydrolysis of esters.

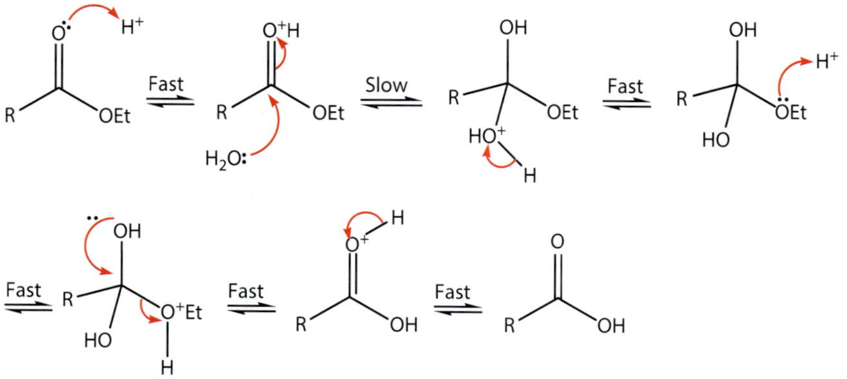

FIGURE 15.14 Acid-catalyzed hydrolysis of esters.

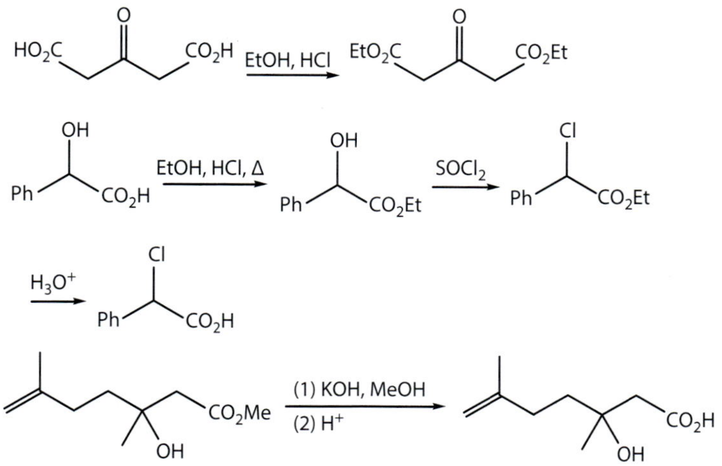

FIGURE 15.15 Esterifications and hydrolysis of esters in synthesis.

(alcohols rather than alkoxide anions). Esters are usually prepared using the alcohols as solvent and are hydrolyzed in aqueous solution. In terms of writing the mechanism, you might wonder why the initial protonation is on the carbonyl oxygen atom; the evidence for this involved ^{1}H and ^{13}C NMR spectroscopy. Some esterifications from synthetic schemes using this methodology are shown in Figure 15.15. Notice in the second reaction that we must convert the acid to an ester before we can do a substitution on the aliphatic OH group. Conditions for ester hydrolysis are normally mild enough that there will be no S_N2 substitution at the chloride. In the third example, the hydrolysis is carried out under basic conditions and acidification at the end of the reaction is gentle; any strongly acidic conditions would be likely to result in elimination of water from the tertiary alcohol.

Focus on Esters: Fruit, Fragrance, and Flavor

Most esters have a pleasant scent, and many are constituents of the scent and flavor of fruits. True natural flavors are very complex, with many components; artificial fruit flavorings may have several components but are much less complex. Our perception of taste depends not only on the taste receptors on the tongue (which are limited to salt, sweet, bitter, sour, and umami) but to a great extent on the aroma of the food. This is why food does not taste good, or indeed much at all, when you have a bad cold. Some examples are given in Table 15.3—you did not think that the rum flavor in Rum 'n' Raisin ice cream actually involved the use of the spirit, did you?

Although esters are quite commonly used in scratch and sniff adverts (they are microencapsulated, and your fingernail breaks open the capsules) and air fresheners, they are not widely used in perfumery. The moisture in human skin results in hydrolysis of the esters; the alcohol portion may largely evaporate, but the acids remain, and many have a most unpleasant scent; butanoic acid is the main scent from rancid butter.

TABLE 15.3 Esters Important in Taste and Fragrance

Ester	Aroma/Taste	Comment/Structure
Ethyl formate	Rum, raspberry, lemon	
1-Propyl acetate	Gummy worms, pears	
1-Butyl acetate	Apple	
1-Pentyl acetate	Pear, banana	
1-Hexyl acetate	Apple, cherry, pear, pineapple, floral	
1-Octyl acetate	Oranges	
2-Methylbutyl acetate	Candy, citrus, floral	$CH_3C(=O)OCH_2CHMeCH_2CH_3$
3-Methylbutyl acetate	Banana	$CH_3C(=O)OCH_2CH_2CH(CH_3)_2$; this is also the alarm pheromone of the honey bee—don't eat bananas around beehives.
Benzyl acetate	Jasmine, peach, banana, strawberry, pear, apple	
Linalyl acetate	Bergamot, lavender	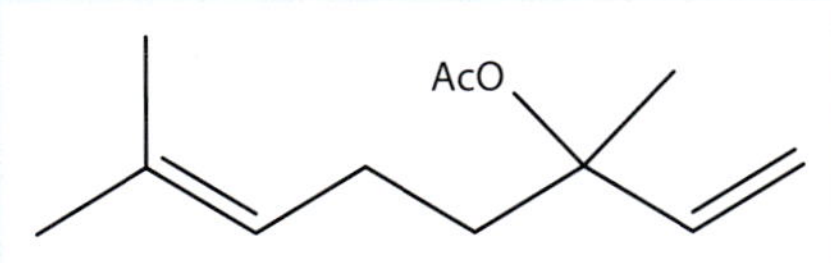
2-Methylpropyl propanoate	Rum	$CH_3CH_2C(=O)OCH_2CH(CH_3)_2$
3-Methylbutyl propanoate	Juicy fruit flavor	$CH_3CH_2C(=O)OCH_2CH_2CH(CH_3)_2$
Methyl butanoate	Apple, pineapple, strawberry	
Ethyl butanoate	Pineapple, cherry, rum, butterscotch, apple, banana, pineapple	
1-Butyl butanoate	Pineapple	
1-Pentyl butanoate	Apricot	
Benzyl butanoate	Floral	
3-Methylbutyl butanoate	Apple	$CH_3CH_2CH_2C(=O)OCH_2CH_2CH(CH_3)_2$
S-Methyl 2-methylbutanoate	Apple, strawberry	$CH_3CH_2CH(CH_3)COOCH_3$
Ethyl 2-methylbutanoate	Cantaloupe	$CH_3CH_2CH(CH_3)COOCH_2CH_3$
Ethyl 3-hydroxybutanoate	Blueberry	$CH_3CH(OH)CH_2COOCH_2CH_2$
1-Butyl 2-butenoate	Blueberry	$CH_3CH=CHC(=O)OCH_2CH_2CH_2CH_3$
Ethyl pentanoate	Green apple	
1-Pentyl pentanoate	Apple	
3-Methylbutyl pentanoate	Apple	$CH_3CH_2CH_2CH_2C(=O)OCH_2CH_2CH(CH_3)_2$
Ethyl hexanoate	Apple, banana	
Allyl hexanoate	Pineapple	
Ethyl heptanoate	Apricot, cherry, grape, raspberry	
Ethyl *E*,*Z*-2,4-decadieneoate	Pear	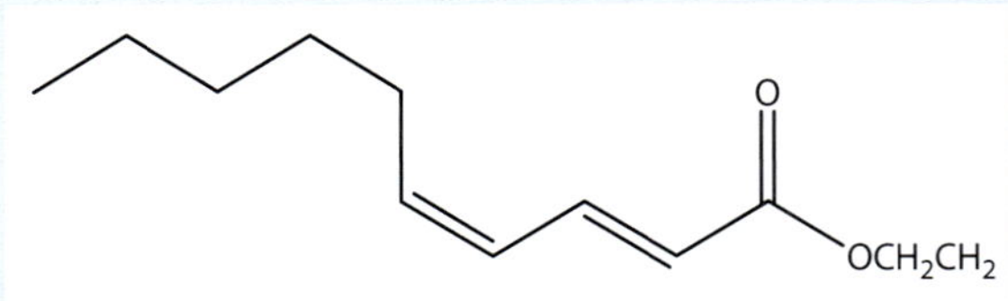

TABLE 15.3 (continued) Esters Important in Taste and Fragrance

Ester	Aroma/Taste	Comment/Structure
Ethyl phenyl acetate	Apple, strawberry, honey	$PhCH_2C(=O)OCH_2CH_3$
E-Methyl cinnamate	Strawberry	$PhCH=CHCOOCH_3$
E-Ethyl cinnamate	Cinnamon	$PhCH=CHCOOCH_2CH_3$
Ethyl safronate	Blueberry, apple, rose	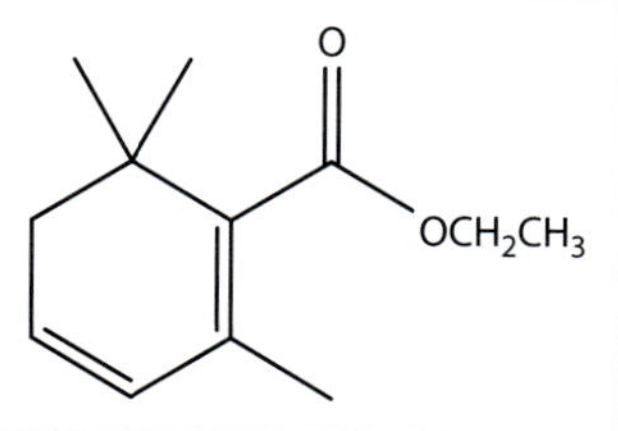
Methyl anthranilate	Grape, apple	Flavor of purple Kool-Aid, bird repellant. 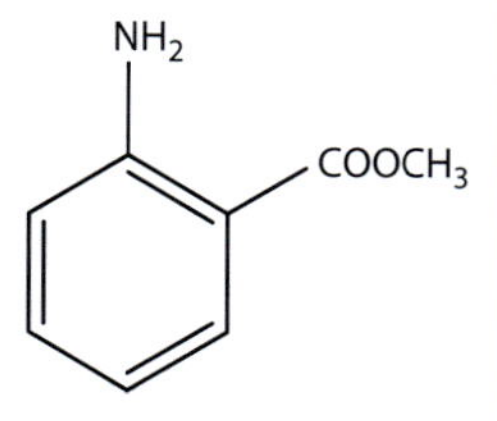

Hydrolysis of amides is more problematic than that of esters, because of the difficulty of expelling amines or ammonia, both of which are poor leaving groups, much less good than water. Initial attack of nucleophiles on the carbonyl group of an amide is also slow, because the resonance stabilization of the amide is lost in forming the tetrahedral intermediate. However, it is an important process, as it is a model for biological peptide hydrolysis. Because amines are poor leaving groups, we would favor acid-catalyzed conditions, and the reaction mechanism is shown in Figure 15.16. The mechanism is similar to that for esters, and the process is only possible because of the irreversible proton transfer between the products. Base-catalyzed reactions are much less favorable and work best when the amine is gaseous, and the reaction can be driven by a significant entropy gain and irreversible gas loss. In Figure 15.17, the loss of $[NH_2]^-$ is very unfavorable, but the reaction is driven by the escape of gaseous ammonia. Some practical applications in synthesis are shown in Figure 15.18. In the second example, the aromatic amine is temporarily converted to the amide, to allow the nitration to proceed without risk of protonating the amine to give $-N^+H_3$, which is a ***meta***-directing, deactivating group (remember that the amino group is ***ortho, para***-directing and strongly activating).

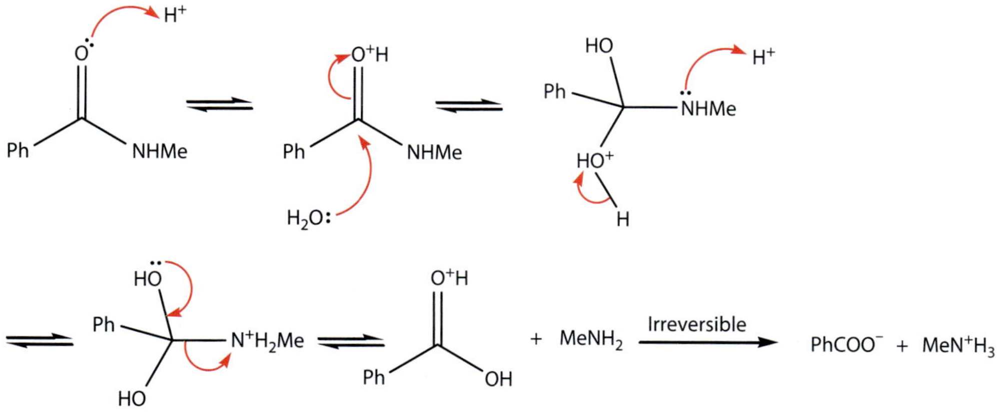

FIGURE 15.16 Mechanism of acid-catalyzed hydrolysis of amides.

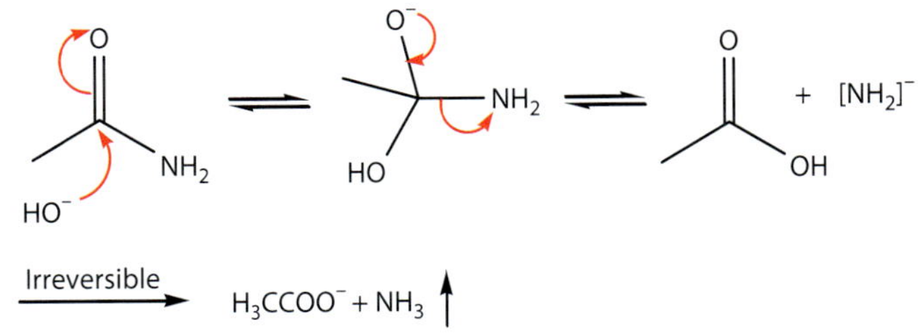

FIGURE 15.17 Mechanism of base-catalyzed hydrolysis of unsubstituted amides.

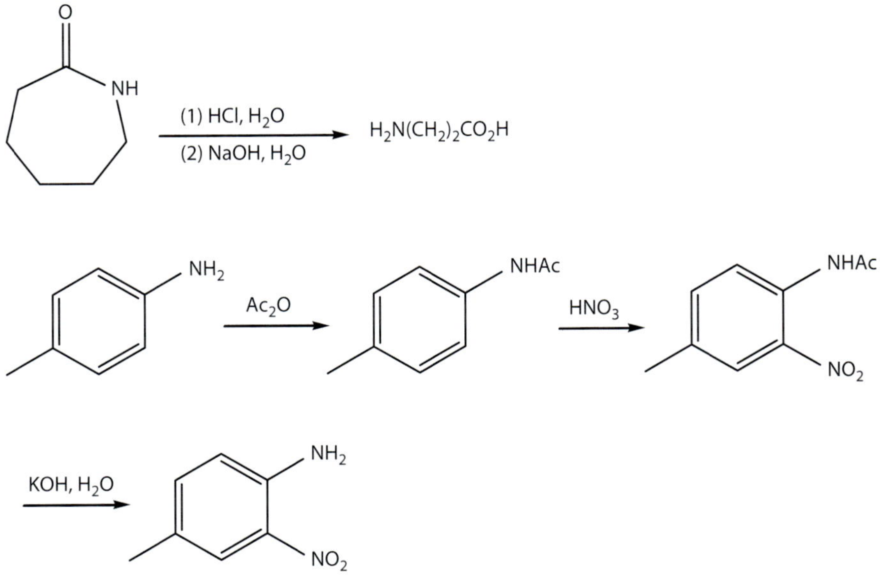

FIGURE 15.18 Practical applications of amide hydrolysis in synthesis.

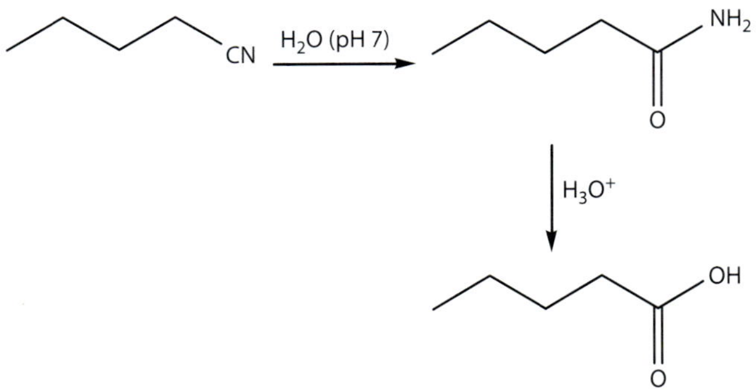

FIGURE 15.19 Selectivity in nitrile hydrolysis.

The difficulty of hydrolyzing amides in basic solution allows for some selectivity in the hydrolysis of nitriles (Figure 15.19). In neutral or weakly basic solutions, the amide is isolated, but under more forcing, and acidic, conditions, the carboxylic acid is obtained (Figure 15.20). Some examples of nitrile hydrolysis in synthesis are shown in Figure 15.21. In the first example, both the ester and the nitrile are hydrolyzed. In the second example, the reaction can be stopped at the amide stage.

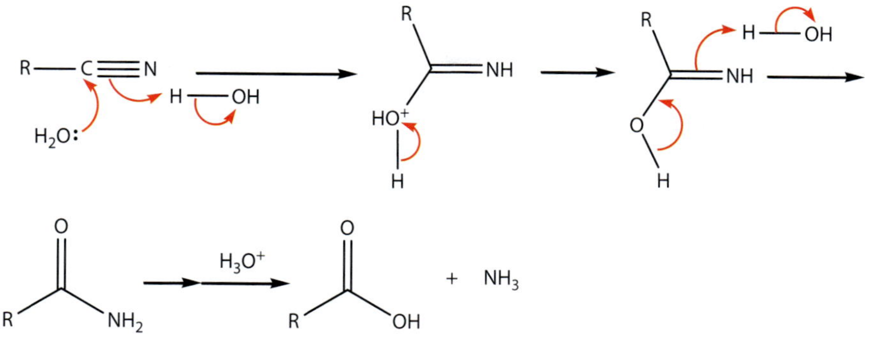

FIGURE 15.20 Mechanism of hydrolysis of nitriles.

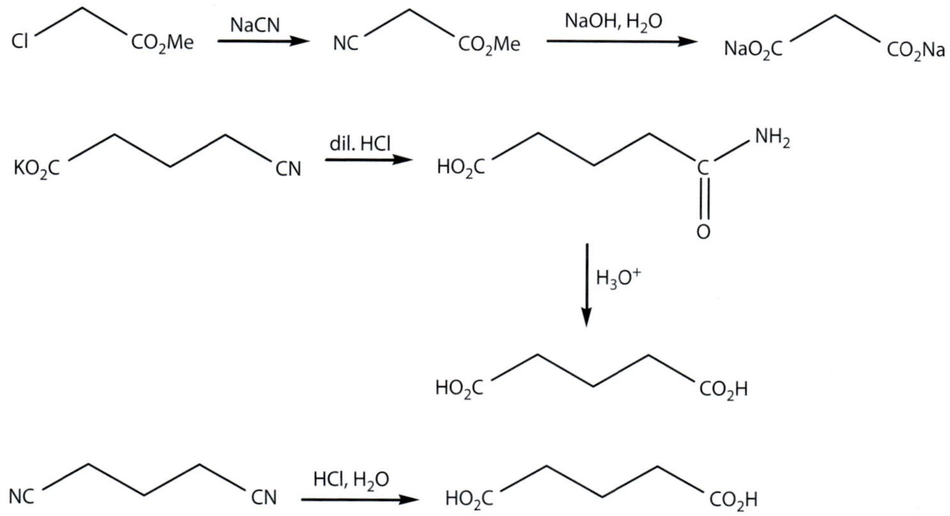

FIGURE 15.21 Nitrile hydrolysis in synthesis.

Focus on Enantioselective Enzymatic Hydrolysis of Esters and Amides

In Chapter 7, we briefly mentioned that there were a number of biological methods for resolution of racemic mixtures of chiral compounds. Because enzymes, biological catalysts, are composed of chiral amino acids, they will generally accept only the natural enantiomer of a compound as their substrate. Some enzymes are very specific as to their substrate, but others can be used with whole classes of substrate, even where these are not very close in overall structure to their natural substrates. The simplest form of resolution involves a racemate, where one enantiomer is transformed by the enzyme and the other is not. This is called kinetic resolution, since it depends on the relative rates of enzymatic catalysis of the reactions of one enantiomer over the other. Our first example is shown in Figure 15.22. When the epoxide is opened by the Grignard reagent, only the *trans*-product is formed, but this is a racemic mixture. The alcohol is converted to an ester, which is then hydrolyzed enzymatically using a lipase enzyme from *Pseudomonas fluorescens*. Notice the very mild conditions for

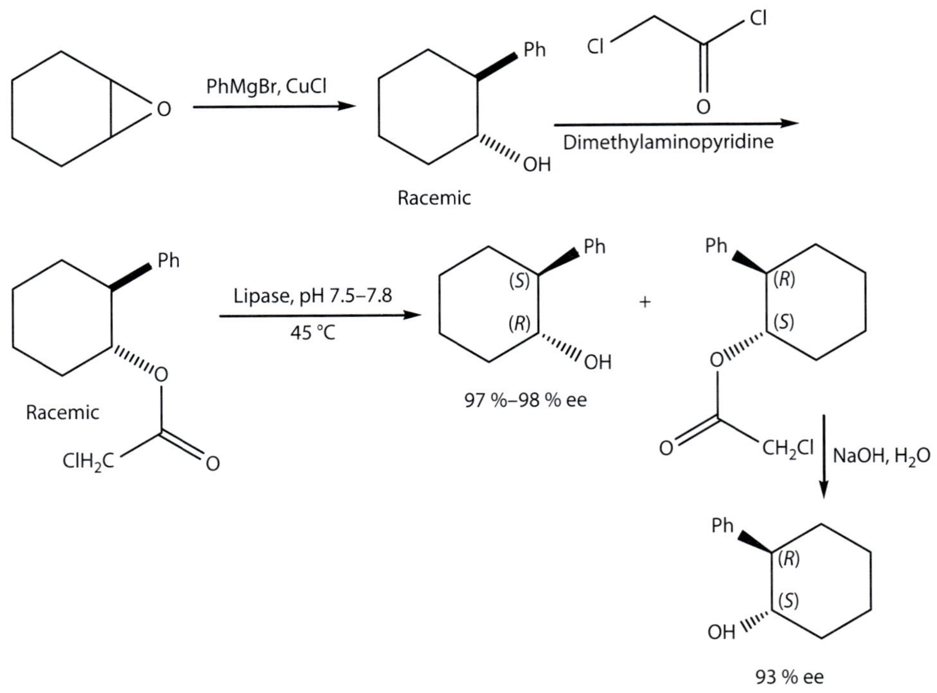

FIGURE 15.22 Kinetic resolution by ester hydrolysis.

the reaction. One enantiomer of the ester is readily hydrolyzed, and the alcohol and the ester can be separated chromatographically. Chemical hydrolysis of the ester yields the other enantiomer of the alcohol. Both chiral alcohols are obtained in good optical yield.

Another example, with an even more obviously non-natural substrate, is shown in Figure 15.23. Chiral binaphthol has been widely used as an addend to promote enantioselective reactions, to prepare other chiral catalysts, and to design chiral ion receptors. In this example, a simple recrystallization of the ester, obtained in 92 % enantiomer excess, gave material with 99 % enantiomer excess. The enzyme is cholesterol esterase, but a crude and inexpensive form was used, bovine pancreas acetone powder.

One of the disadvantages of this process is that, generally, only one of the enantiomers is really useful, so processes where the unwanted isomer can be recycled by racemization are useful. An example is shown in Figure 15.24, which can be used for a range of amino acid amides. The unnatural, unhydrolyzed, amide is converted to an imine, and under strongly basic conditions, this is racemized, and then the imine is hydrolyzed.

A final example shows a process called desymmetrization (Figure 15.25). We start with a *meso*-compound (remember this is a molecule that contains asymmetric carbon atoms but is not chiral, because it has a plane of symmetry) with two identical functional groups. Under enzymatic hydrolysis, only one of these reacts, so a chiral compound is produced, and all the material is used, without any need to recycle. The enzyme used in this reaction is electric eel acetylcholinesterase (EEAc). Pig liver esterase is also commonly used, as a relatively crude extract is inexpensive and gives good results. An example is shown in Figure 15.26, in the synthesis of *R*-mevalonolactone, important in the biosynthesis of terpenes and steroids.

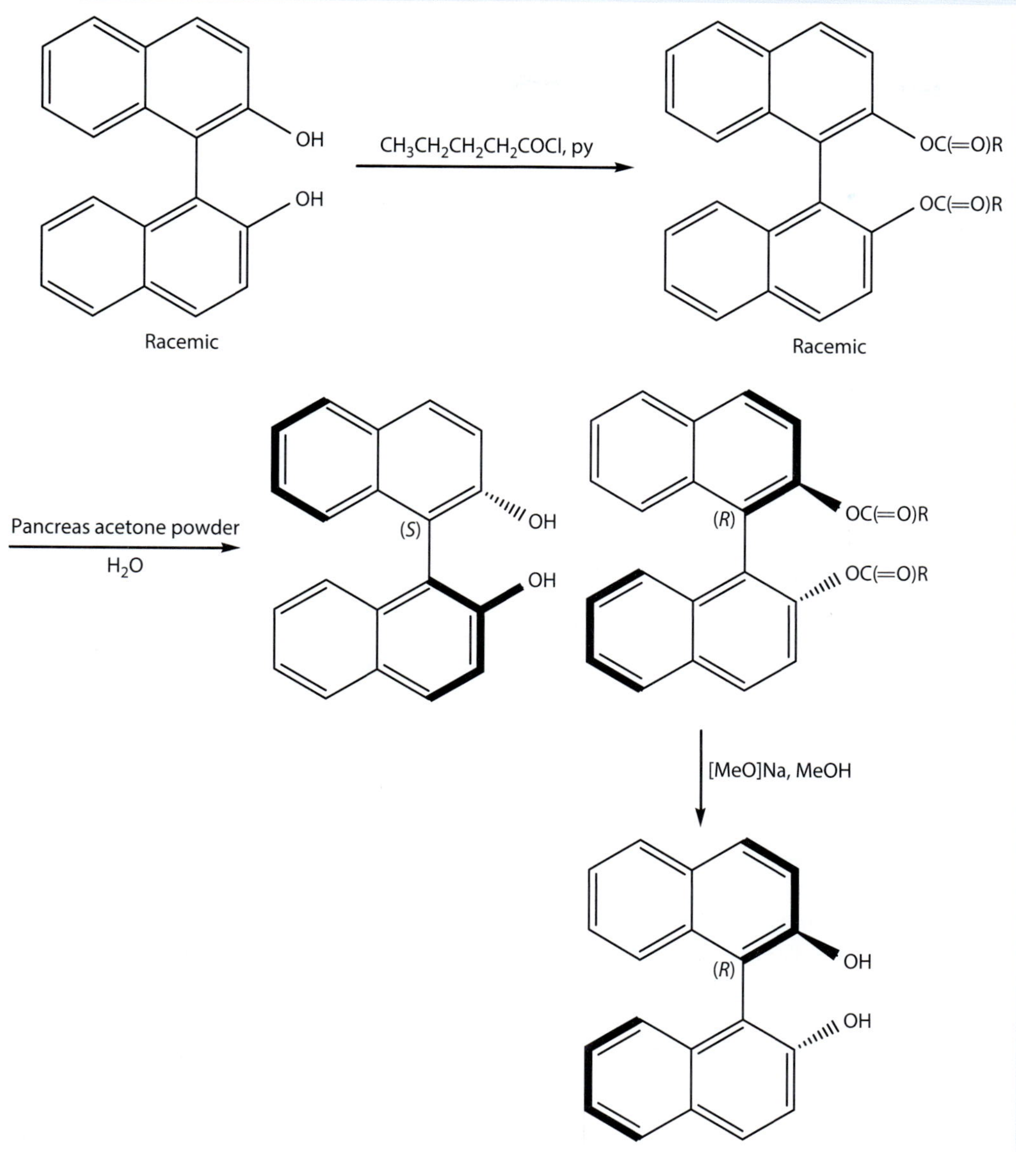

FIGURE 15.23 Kinetic resolution of a binaphthol ester.

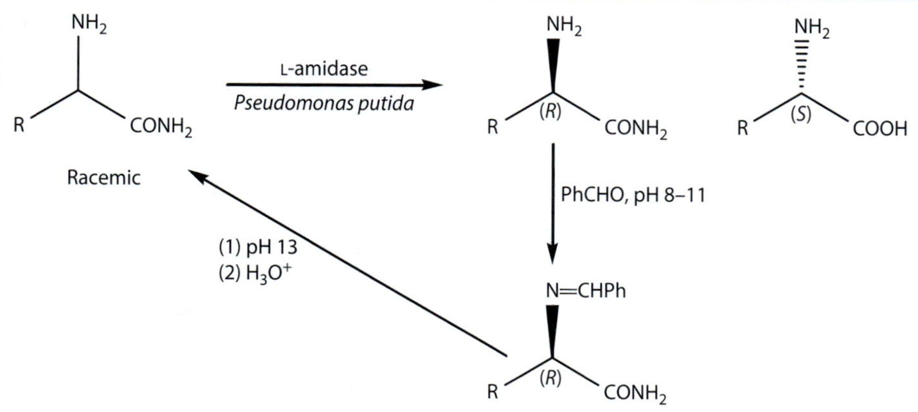

FIGURE 15.24 Kinetic resolution of an amide, with recycling.

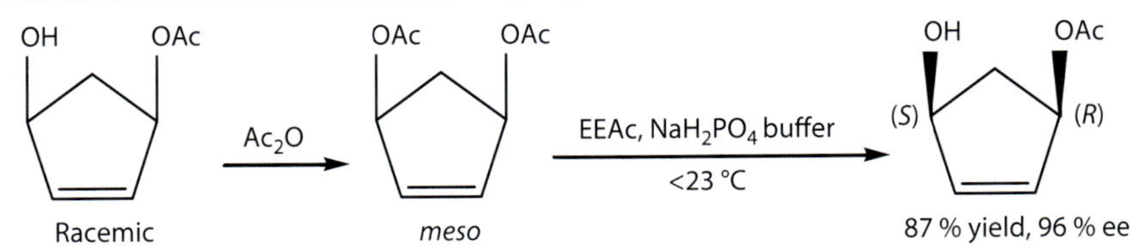

FIGURE 15.25 Enzymatic desymmetrization.

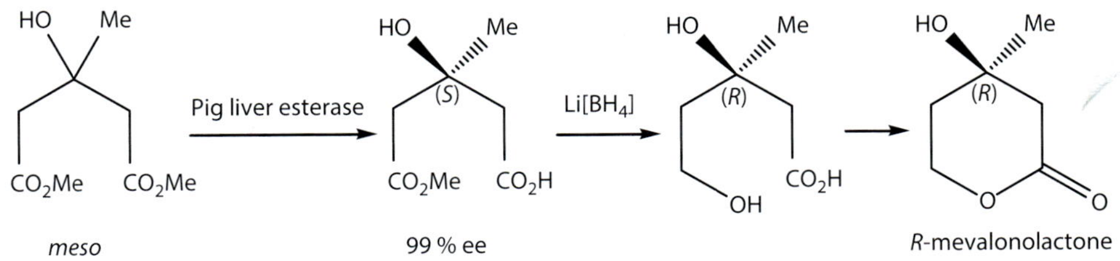

FIGURE 15.26 Asymmetric synthesis of *R*-mevalonolactone.

Key Points from Section 15.3

- Esters, amides, acid chlorides, acid anhydrides, and nitriles can all be hydrolyzed to carboxylic acids, but rates vary widely.
- Esters can be prepared by reaction of alcohols with carboxylic acids under acidic catalysis but can be hydrolyzed in either acid or basic solution. They can also be prepared by reaction of alcohols with acid chlorides or anhydrides.
- Hydrolysis of amides is usually carried out in acid solution, but unsubstituted amides can be hydrolyzed in base under forcing conditions.
- Hydrolysis of nitriles involves an amide intermediate, and under appropriate mild conditions, the amide can be selectively isolated.
- Reaction of carboxylic acids with acid chlorides yields anhydrides. These can also be prepared by anhydride exchange.

PROBLEM 15.4

Predict the product(s) of each of the following reactions:

(a)

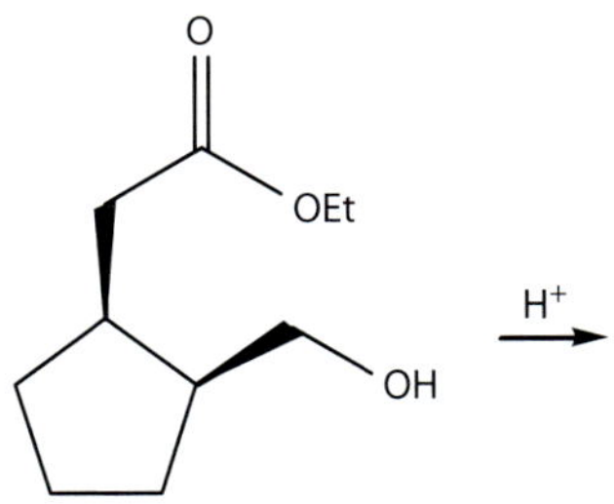

(b)

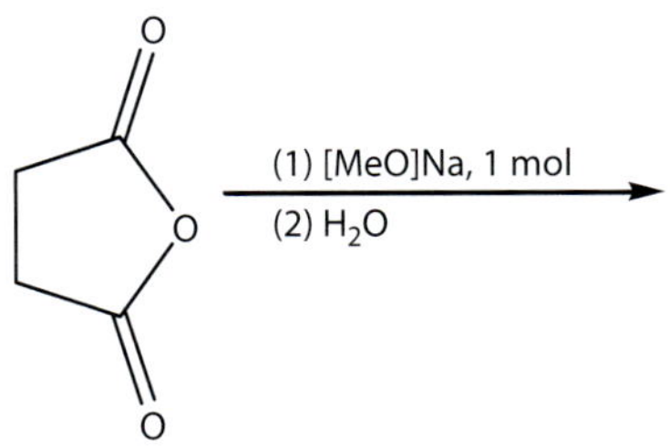

(c)

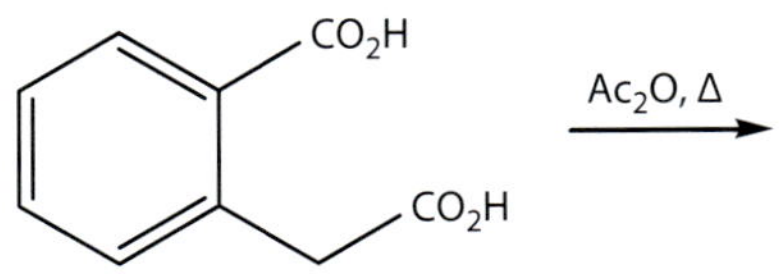

Solutions

(a) This type of reaction is called a transesterification with the intramolecular hydroxyl group displacing ethoxyl. The reaction is successful because a favorable six-membered ring is being formed:

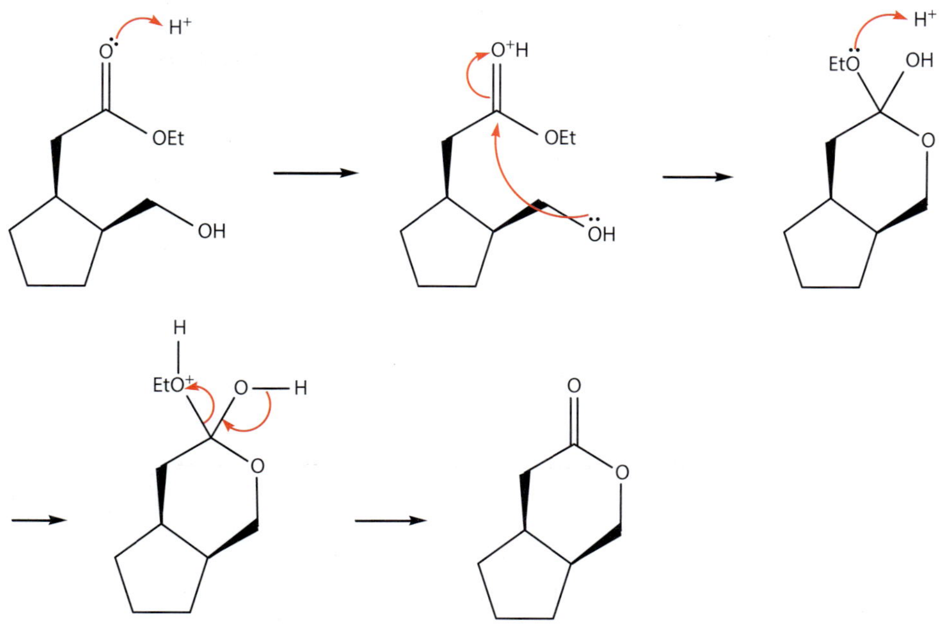

(b) In this reaction, the methoxyl ion attacks the carbonyl of the anhydride, displacing carboxylate. Because at this point the solution is basic, no further reaction can occur. Not only is only 1 mol of the methoxyl available, but it is unlikely to attack a negatively charged carboxylate anion. Aqueous work-up gives the molecule with one acid and one ester group:

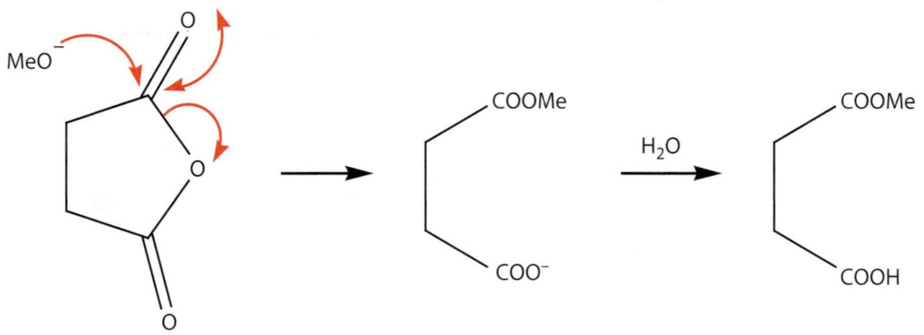

(c) This is a fairly standard reaction, where acetic anhydride is used to prepare another anhydride, with acetic acid being distilled out of the mixture as it is formed. In this case, the reaction is particularly easy as it is intramolecular, and a six-membered ring is formed:

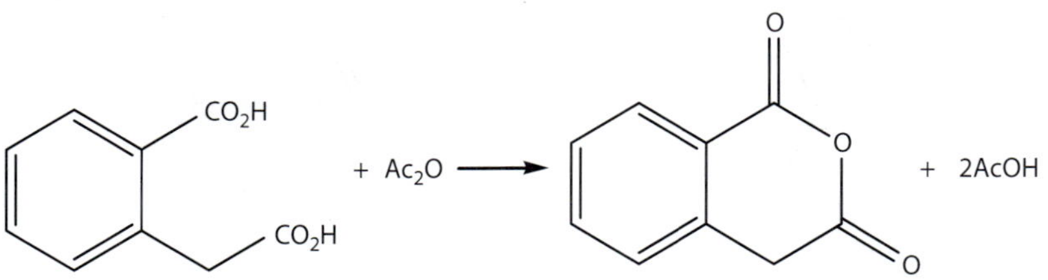

PROBLEM 15.5

Write a mechanism for each of the following reactions:

(a) $PhCO_2Et + H_2O \xrightarrow{HO^-} PhCOO^- + EtOH$

(b)

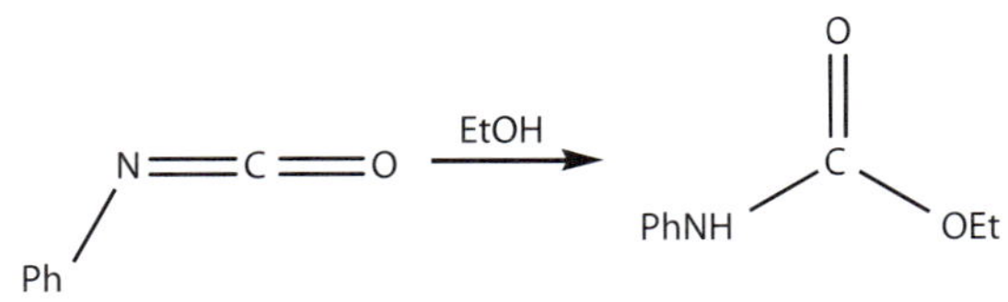

(c)

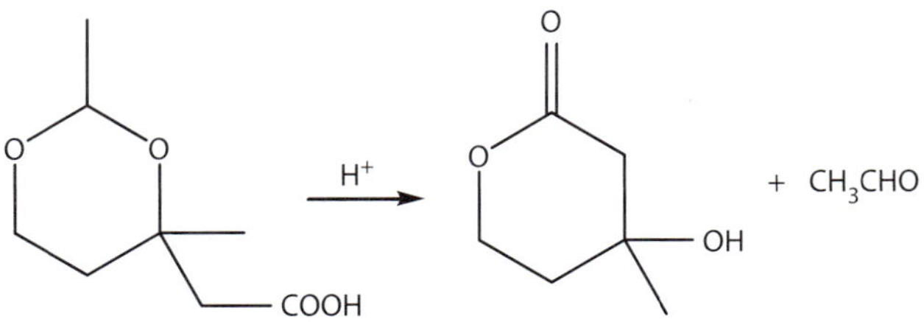

Solutions

(a) This is a standard bookwork problem; if you are having difficulty, look back at Figure 15.13:

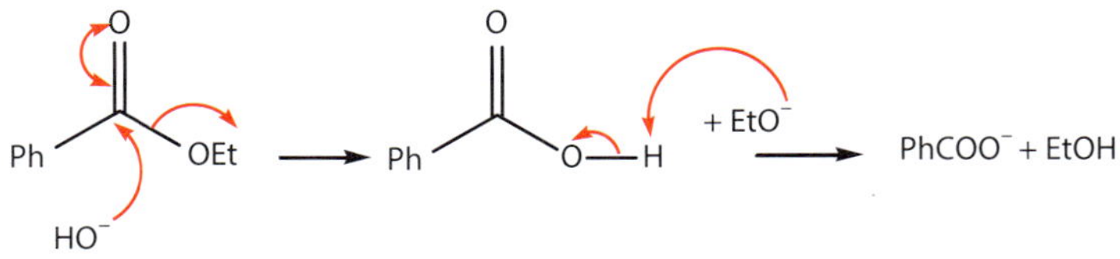

(b) This is a less familiar example, but once we attack the carbon of the isocyanate (simply think of it as a carbonyl and look where the OEt has become attached), the solution becomes clear:

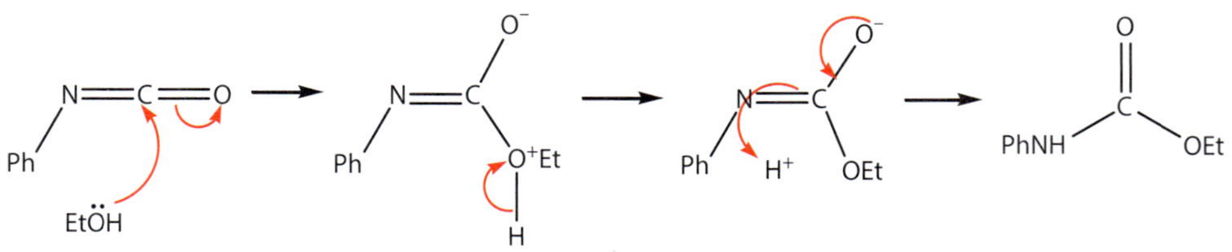

(c) This problem is more challenging, because there is no obvious starting point; the only reagent is acid, and there are a number of oxygen atoms that we could think about protonating. The clue lies in the second product, acetaldehyde. This formed part of the initial acetal, so we should start by hydrolyzing the acetal. Except, oops, we are not given any water! Nonetheless, we start as if we were going to hydrolyze the acetal and see where this leads us:

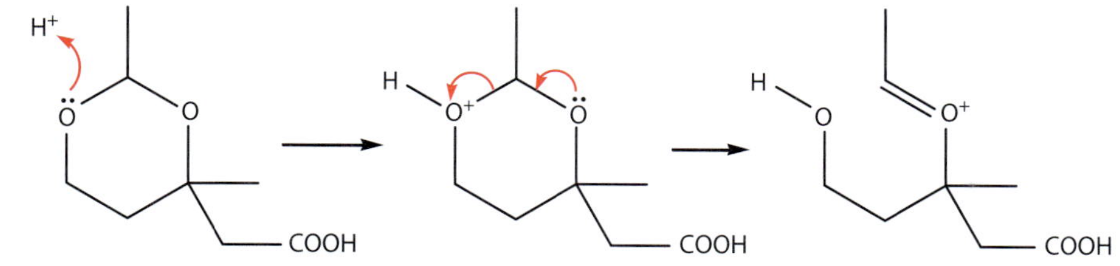

At this point, we can see how the ester can be formed, from the alcohol and the carboxylic acid—and this generates a mole of water that we can use to complete the acetal hydrolysis:

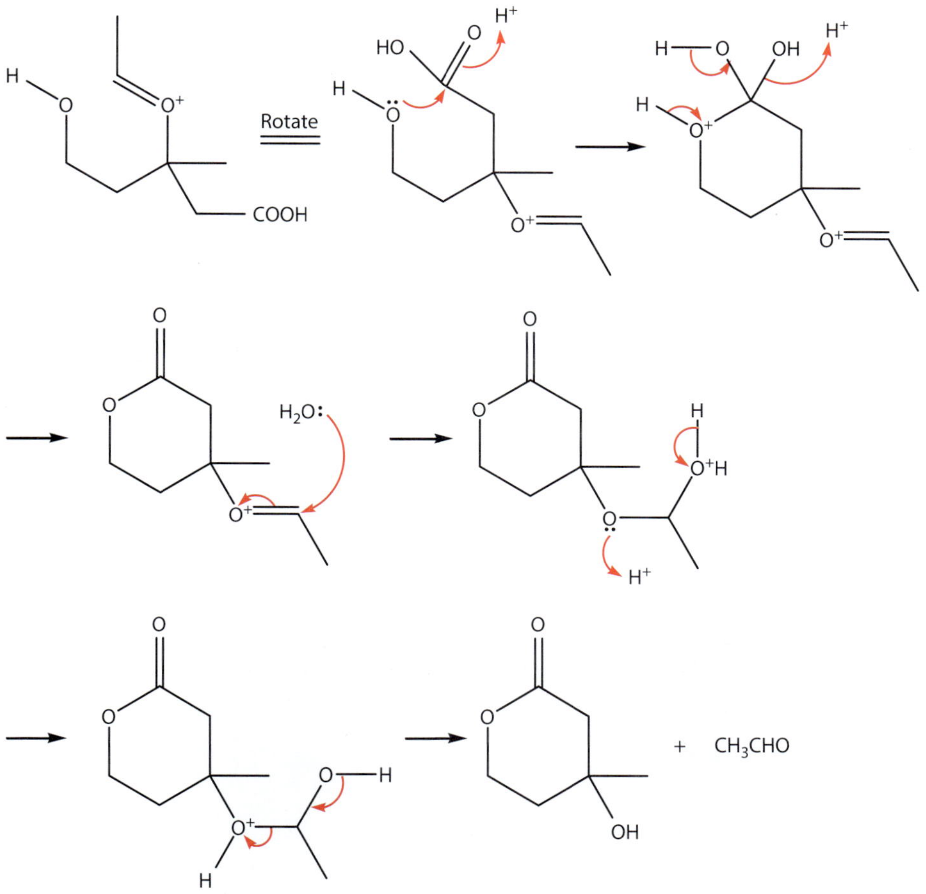

15.4 NITROGEN NUCLEOPHILES

Amides can, in principle, be prepared by thermal dehydration of ammonium salts of carboxylic acids, such as $[NH_4]^+[RCOO]^-$ or $[RNH_3]^+[RCOO]^-$, but this route is rarely used in practice, because of the reversibility of the process. Most strategies depend on the reaction of amines with carbonyl derivatives where there is a better leaving group than OH attached to the carbonyl.

Amines and ammonia are good nucleophiles and are rather hard to displace, once a tetrahedral intermediate has been formed. The synthesis of amides from acid chlorides and amines has been widely applied (Figure 15.27). Notice in the second example that it is the nitrogen rather than the oxygen of the amino alcohol that acts as the nucleophile, and in the third example that the acid chloride is more reactive toward substitution than the aliphatic bromide.

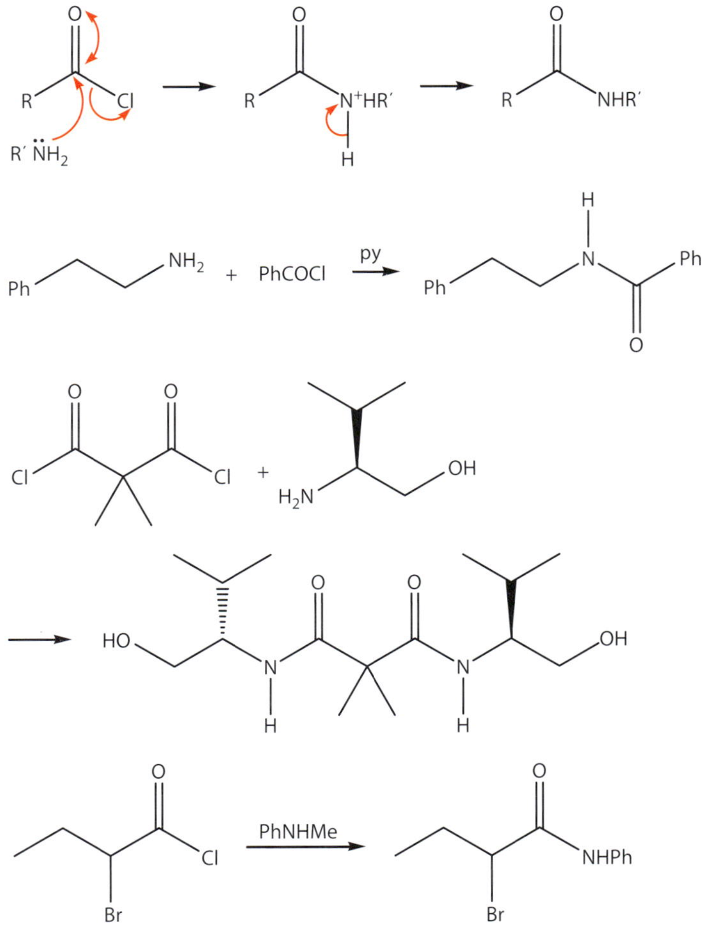

FIGURE 15.27 Synthesis of amides by reaction of ammonia and amines with acid chlorides.

The facility and generality of the substitution of acid chlorides by amines should worry us a little—up to now, we have not prepared any acid chlorides. And given that chloride is the best leaving group we have seen, how will we do so, as chloride will not displace any of the groups we have so far seen attached to the carbonyl? We will come back to this problem in the next section, but there are alternatives for amide synthesis. Amines will displace carboxylate groups from anhydrides and alcohols from esters. Although these reactions are not as rapid as with acid chlorides, fewer toxic materials are used or produced, so they can be a desirable alterative. Examples using anhydrides are shown in Figure 15.28 and using esters in Figure 15.29. In Figure 15.28, the second example shows again that nitrogen is a better nucleophile than oxygen. In the first example of Figure 15.29, we see the same phenomenon; the hydroxyester does not polymerize, but the OPh of the ester can be displaced by the amino group.

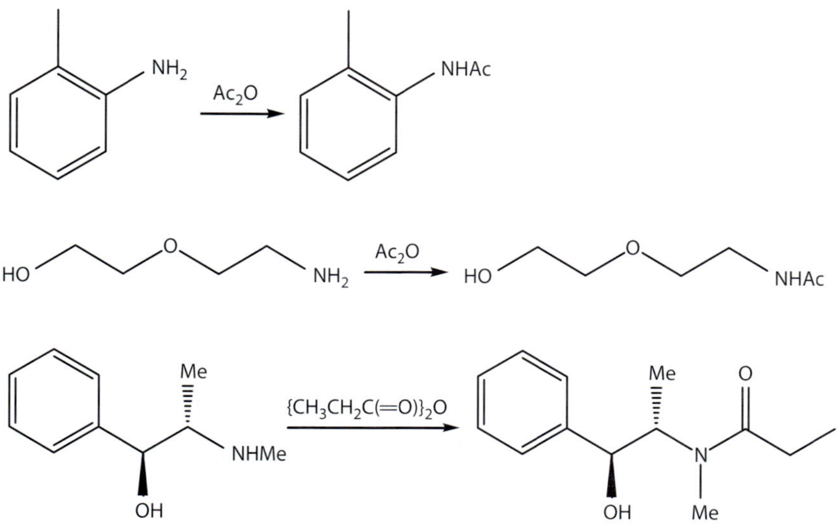

FIGURE 15.28 Synthesis of amides using anhydrides.

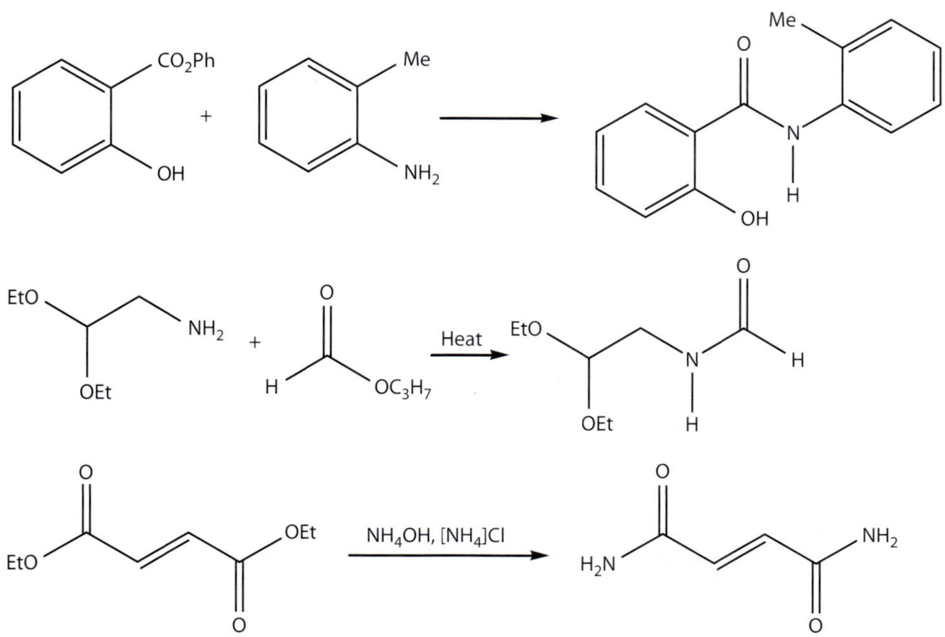

FIGURE 15.29 Synthesis of amides using esters.

PROBLEM 15.6

Write mechanisms for the final example in each of Figures 15.27 through 15.29.

Solution

Figure 15.27

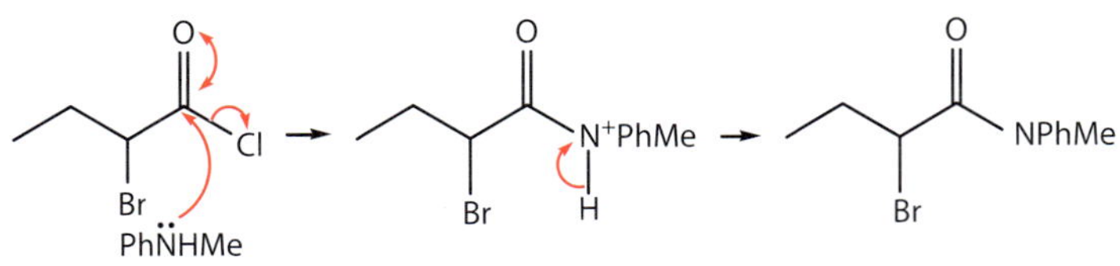

Figure 15.28

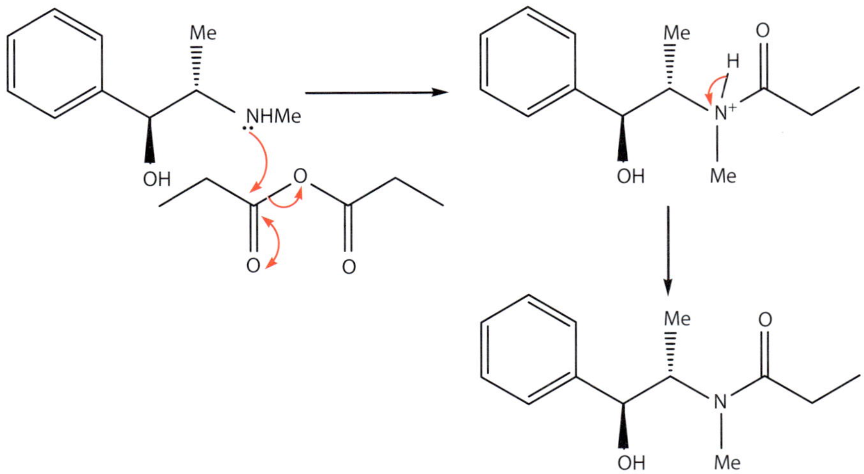

Figure 15.29

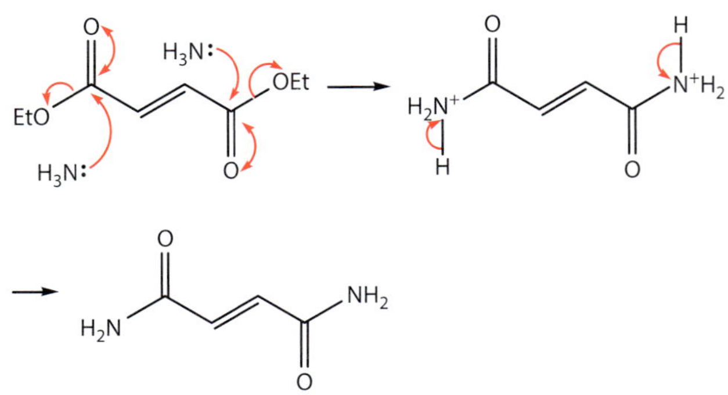

I hope that these problems convince you that these displacements are essentially all the same mechanistically and give you the confidence to tackle less familiar examples.

15.5 SYNTHESIS OF ACID CHLORIDES

In order to prepare acid chlorides, we will need to have a group attached to the carbonyl group that is a better leaving group than chloride, so that it, X, rather than chloride is lost from a tetrahedral intermediate such as **15.13**. We have already met some likely candidates when we converted alcohols to chlorides. Reaction of an aliphatic OH group with thionyl chloride, $SOCl_2$, gave a chlorosulfite ester, which was an excellent leaving group, because of the entropy gain in losing a mole of gaseous SO_2. The mechanism of the reaction with carboxylic acids is shown in

Figure 15.30. The advantage of this reaction, which is used very widely, is that the by-products are gaseous, so isolation of the product is usually straightforward. The disadvantage is that both starting reagent and gaseous products are toxic and unpleasant to use. Some uses in synthesis are shown in Figure 15.31. In the first example, the indirect method of ester preparation is preferred to direct reaction with ethanol, because the substrate is so hindered. In the second, the acid chloride is used for a Friedel–Crafts acylation, and in the final example, the differential reactivity of an acid chloride and an ester is used to prepare a species with one amide and one ester group:

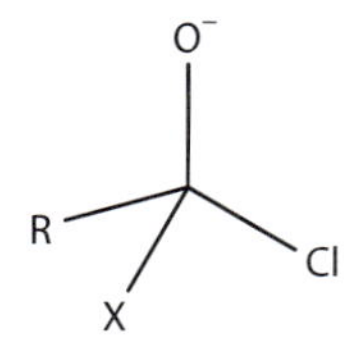

15.13

A slightly milder reagent that has been used to prepare acid chlorides is oxalyl chloride, **15.14**. This also has the advantage that the by-products are gaseous. The reaction can be catalyzed by a trace of dimethylformamide. The reaction and some examples of use in synthesis are shown in Figure 15.32. In the second example, an unusual acyl azide was prepared; the ability of azide to displace chloride should not be surprising, as it was also an excellent nucleophile

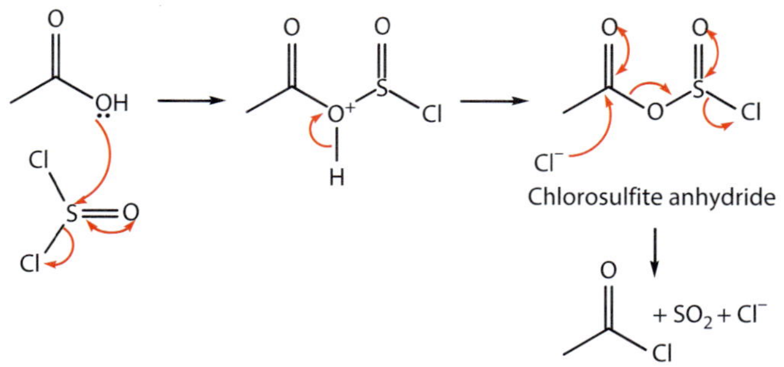

FIGURE 15.30 Mechanism of acid chloride synthesis using thionyl chloride.

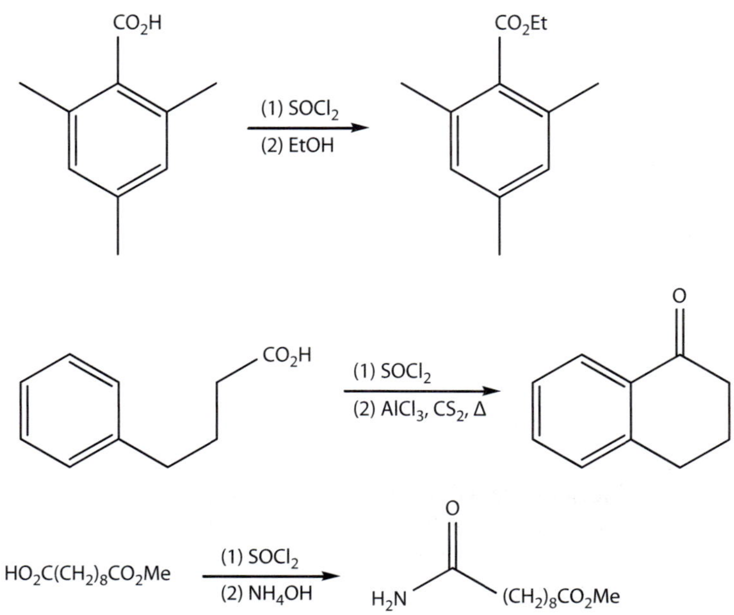

FIGURE 15.31 Preparation and use of acid chlorides.

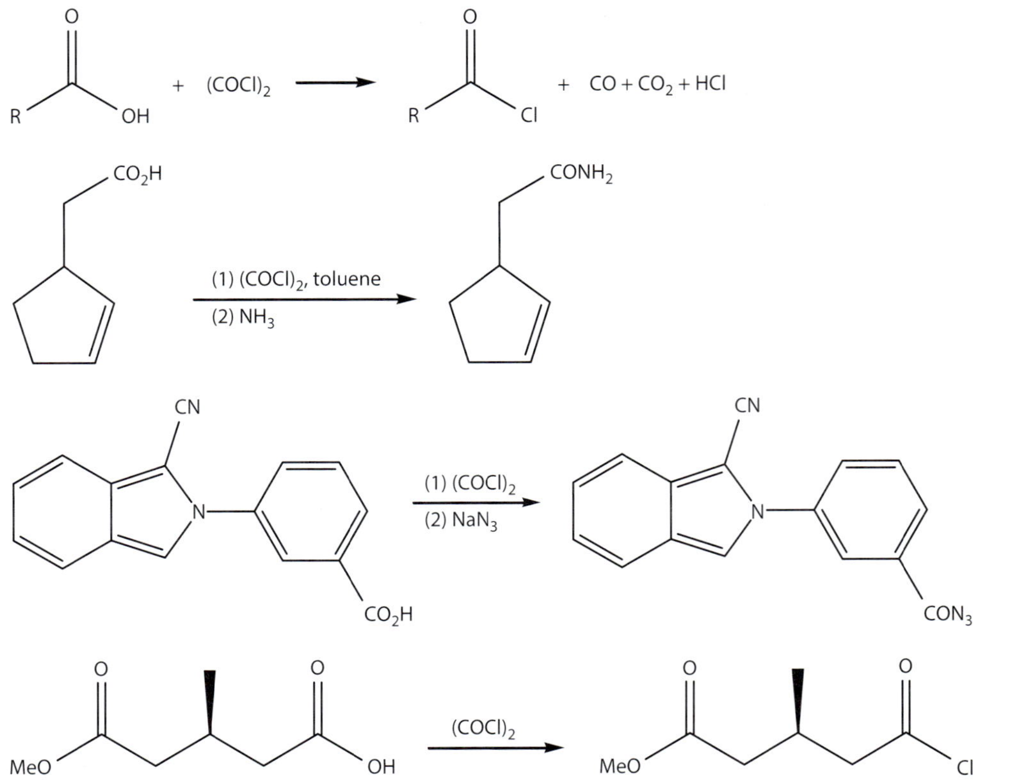

FIGURE 15.32 Preparation and use in synthesis of acid chlorides using oxalyl chloride.

and poor leaving group in aliphatic substitution reactions. In the final example, the conversion was achieved with good retention of stereochemistry in the resolved compound; using $SOCl_2$ resulted in some racemization.

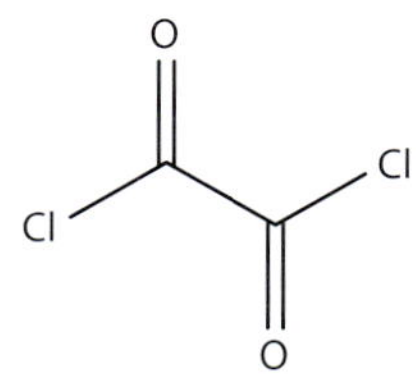

15.14

Phosphorus halides, such as PCl_5 (Figure 15.33), also effect this transformation but tend to be less popular, because phosphorus-containing by-products must be separated and disposed of. In this case, the reaction is driven by the favorable formation of a strong phosphorus–oxygen double bond (529 kJ mol^{-1}).

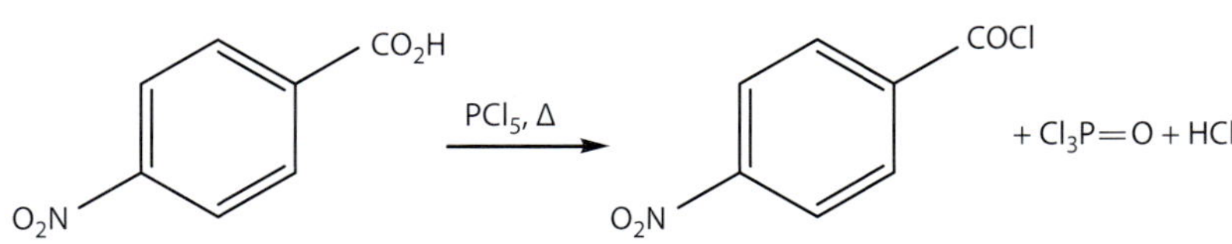

FIGURE 15.33 Formation of acid chlorides using PCl_5.

PROBLEM 15.7

Based on your understanding of the chlorination of simple alcohols with PCl_5, propose a mechanism for the conversion in Figure 15.33.

Solution

The first step of the reaction is the displacement of chloride from PCl_5 by the acid. The displaced chloride will then attack the carbonyl, expelling $Cl_3P{=}O$ and chloride:

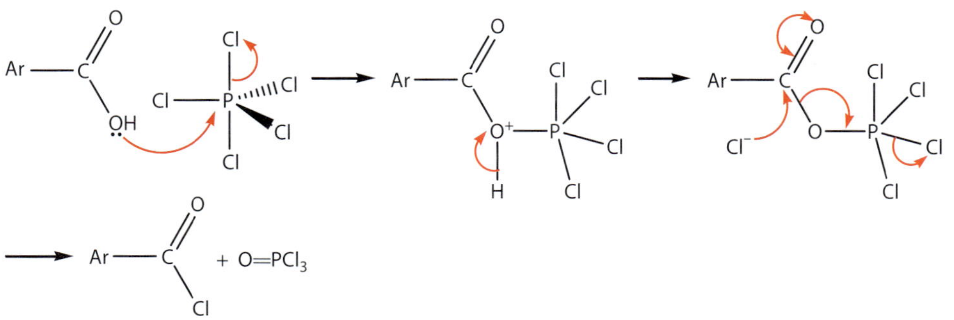

> **Key Points from Sections 15.4 and 15.5**
>
> - Amides can be synthesized by reaction of amines with acid chlorides, anhydrides, or esters.
> - Acid chlorides can be prepared from carboxylic acids using thionyl chloride, $SOCl_2$, oxalyl chloride $(COCl)_2$, or PCl_5.

15.6 HYDRIDE AS NUCLEOPHILE

Consider the reaction of lithium aluminum hydride with an ester (Figure 15.34). The hydride will attack the carbonyl to give a tetrahedral intermediate, just as with aldehydes and ketones. However, this tetrahedral intermediate does not wait around to be protonated but loses ethoxide to give an aldehyde. We already know how lithium aluminum hydride reacts with aldehydes, and our final product, after work-up, is a primary alcohol.

It is not possible to stop the reaction at the stage of the aldehyde, because the aldehyde is rapidly reduced by lithium aluminum hydride. The same is true for acid chlorides, anhydrides, and even carboxylic acids. In this last case, the acid is first converted to its lithium salt, but lithium aluminum hydride is a strong enough reducing agent to attack even this negatively charged species (Figure 15.35).

Reduction of esters to aldehydes can be achieved using diisobutylaluminum hydride (DIBAH), $(Me_2CHCH_2)_2AlH$. The reaction is sensitive to conditions, and to achieve a good yield, a single equivalent of the reducing agent and low temperatures are required (Figure 15.36). The selectivity is thought to derive from the relative stability of the hemiacetal anion at low temperature. Selective reduction of acid chlorides to aldehydes can be accomplished using the very hindered reducing agent (size matters ☺), Li[Al(O-*t*-Bu)$_3$H], which can be prepared *in situ* from lithium aluminum hydride and *t*-butanol (Figure 15.37).

Sodium borohydride, which was useful for aldehyde and ketone reduction, is less commonly used for other acid derivatives. While it will reduce acid chlorides or anhydrides to primary

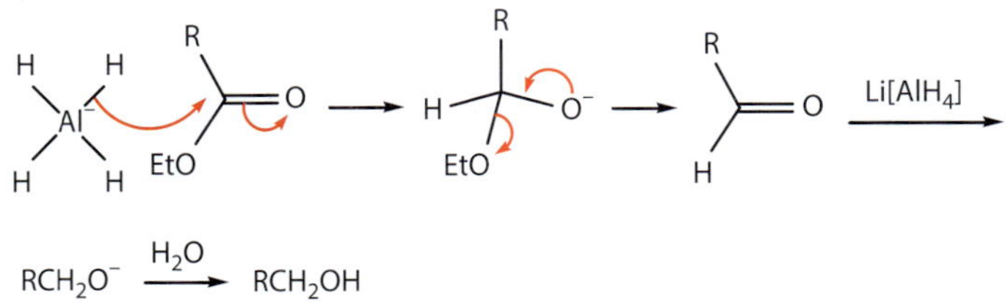

FIGURE 15.34 Reaction of lithium aluminum hydride with an ester.

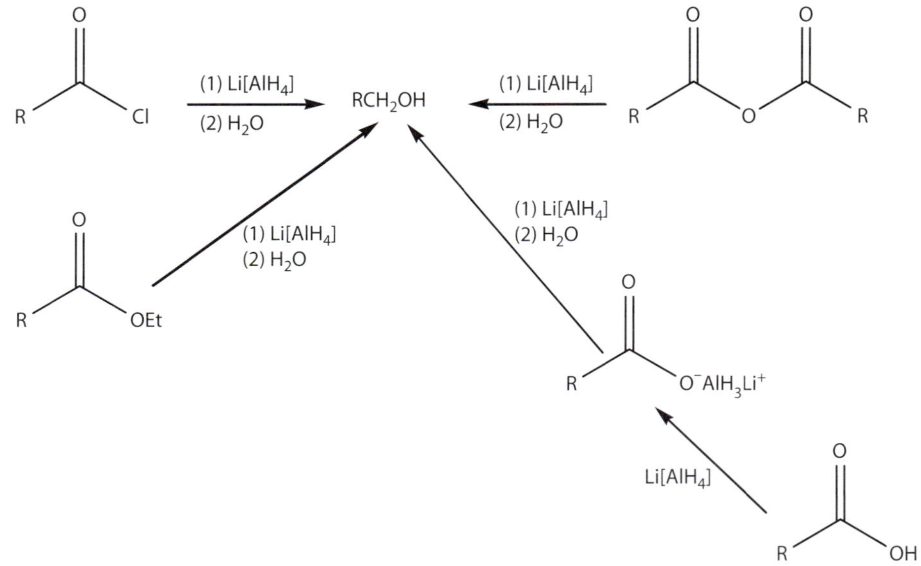

FIGURE 15.35 Reduction of some acid derivatives by lithium aluminum hydride.

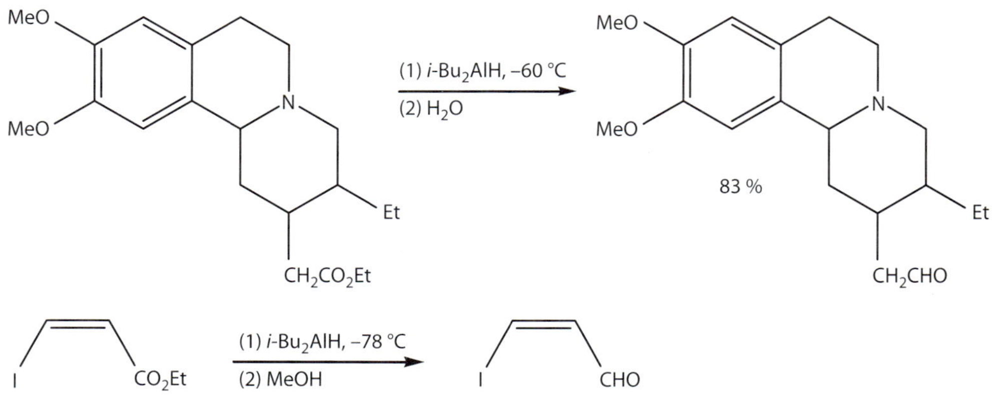

FIGURE 15.36 Reduction of esters to aldehydes using DIBAH.

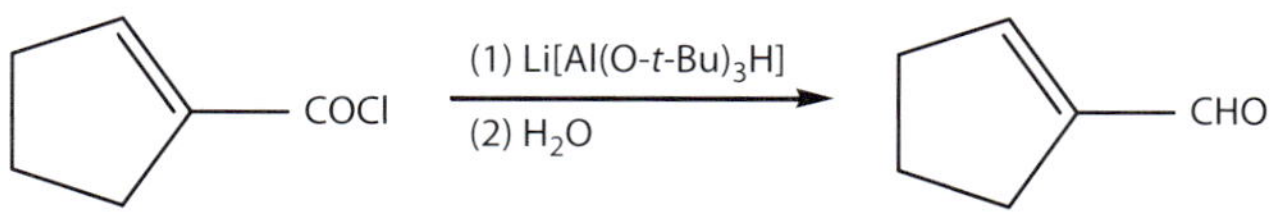

FIGURE 15.37 Reduction of acid chlorides to aldehydes using Li[AlH(O-*t*-Bu)$_3$].

alcohols, it does not react with esters, amides, or carboxylic acids. Notice that in the first example of Figure 15.38, the acetate is not reduced. Lithium borohydride reduces esters but not amides or carboxylic acids. The greater reactivity of the lithium salt is attributed to the greater Lewis acidity of the lithium ion relative to sodium. In the second example of Figure 15.38, we see that, usefully, it does not normally reduce amides or nitriles.

You may have noticed that in all the discussion so far, one class of compounds has been largely omitted, amides. While most reducing agents don't react with amides, our good friend lithium aluminum hydride does react—but the outcome is different from the reactions we have seen so far (Figure 15.39). When the first tetrahedral intermediate is produced, the expulsion of $[NH_2]^-$ is strongly disfavored; instead, the negatively charged oxygen hooks up with the Lewis acidic AlH_3. This intermediate reacts with further $[AlH_4]^-$ to give a species with two negative charges, which expels $Li_2[OAlH_3]$ to give an imine. The imine is then reduced in the conventional way to give, after work-up, an amine. A modified route may operate for substituted amides and must do so for those with two alkyl substituents at nitrogen (Figure 15.40). Since amides are readily prepared from acid chlorides or esters and amines, this sequence represents an excellent general synthesis of substituted amines. Some examples in synthesis are shown in Figure 15.41.

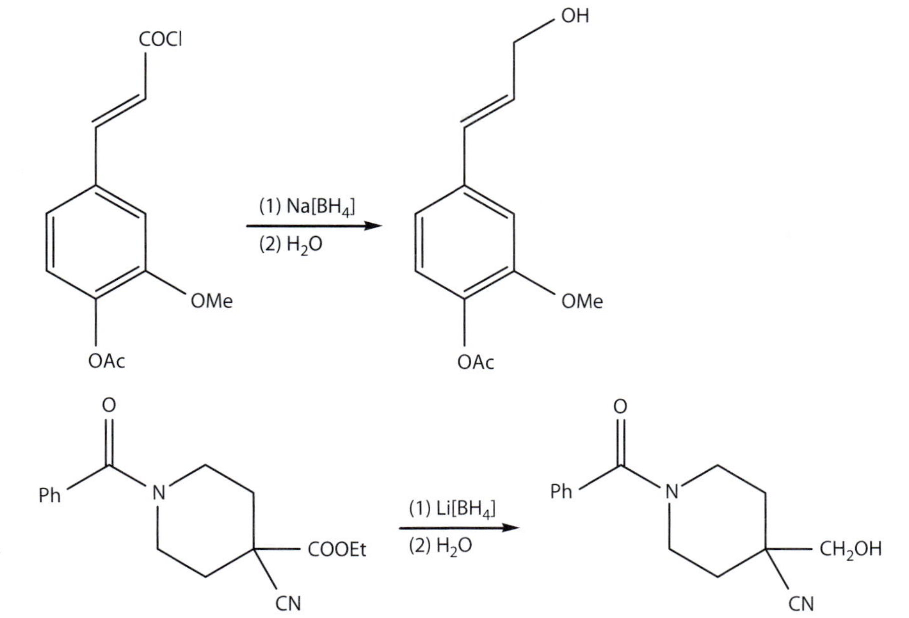

FIGURE 15.38 Reduction of acid derivatives by borohydride reagents.

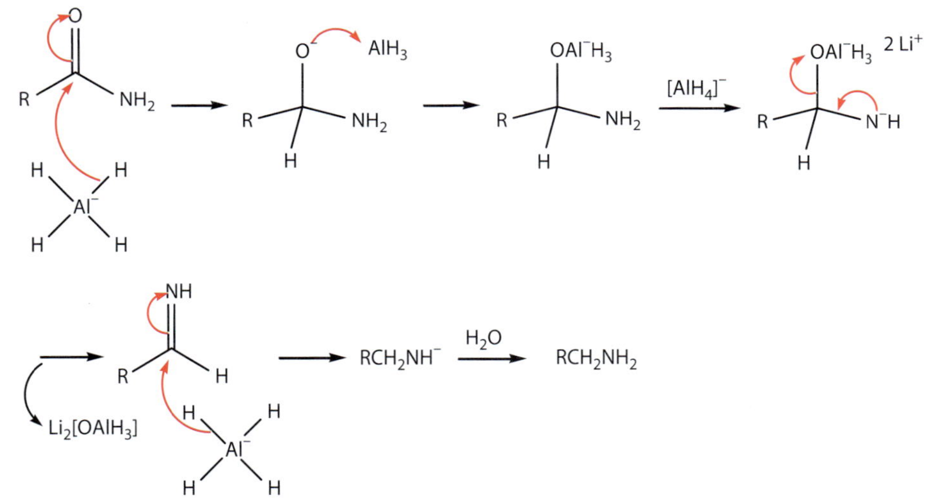

FIGURE 15.39 Mechanism of reduction of unsubstituted amides by lithium aluminum hydride.

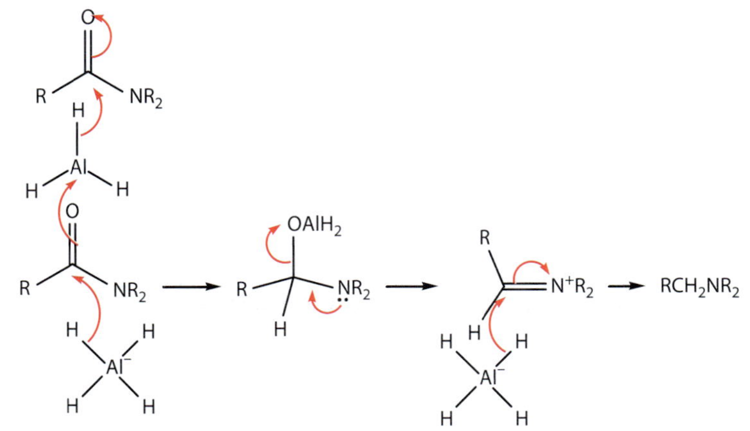

FIGURE 15.40 Lithium aluminum hydride reduction of substituted amides.

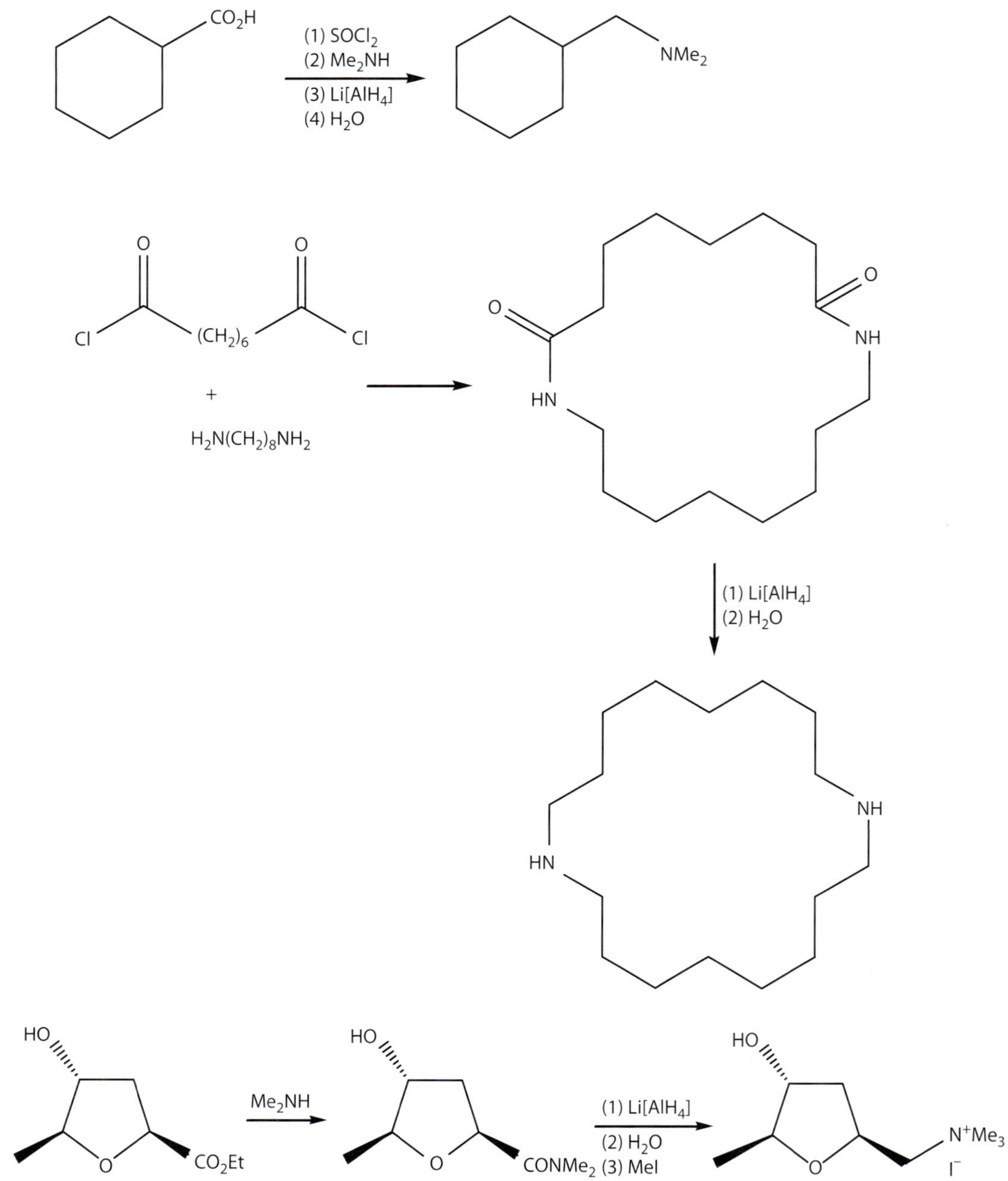

FIGURE 15.41 Amine synthesis from amides.

PROBLEM 15.8

Suggest an appropriate reducing agent for each of the following transformations:

(a)

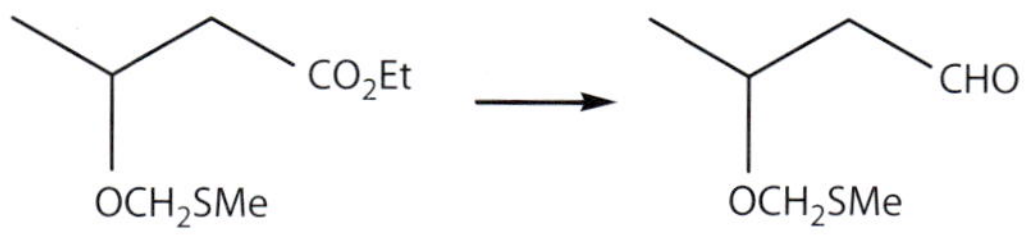

(b)

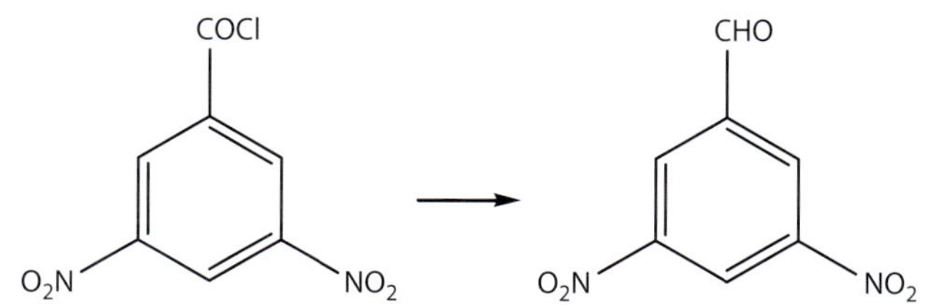

(c)

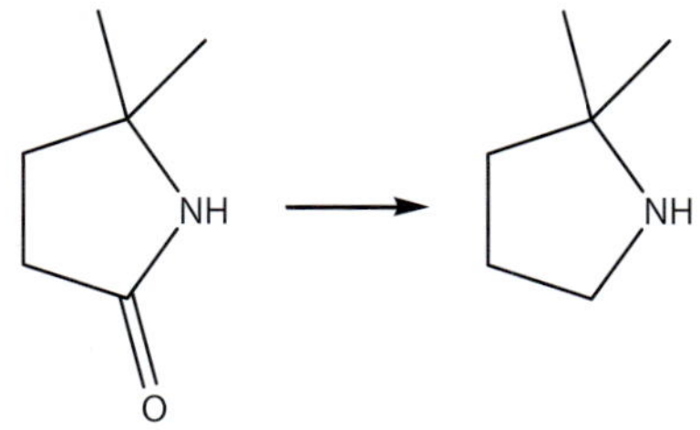

(d)

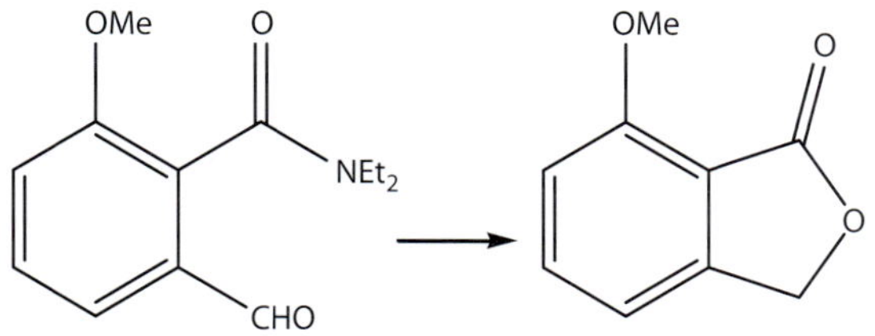

Solutions

(a) *i*-Bu_2AlH, −78 °C; DIBAH at low temperature is specific for reduction of esters to aldehydes.

(b) Li[AlH(O-*t*-Bu)$_3$]; the bulk of the reagent means that it is possible to stop the reaction at the aldehyde stage. Nitro groups are not reduced, and esters, acetals, nitriles, and lactones are also tolerated.

(c) Li[AlH_4] is the only reducing agent that we know that will reduce amides to amines.

(d) This one is a little more challenging, as clearly something happened after the reduction took place. The only way that the lactone can have been formed is if the aldehyde was reduced to an alcohol, and then cyclization displaced the NEt_2 moiety. Ordinarily, this would not be an easy reaction, but in this case, it is possible because a favorable five-membered ring is formed. So we need a reagent that will reduce the aldehyde but not the amide or lactone; Na[BH_4] is suitable. In order to facilitate the cyclization, the work-up should be acidic:

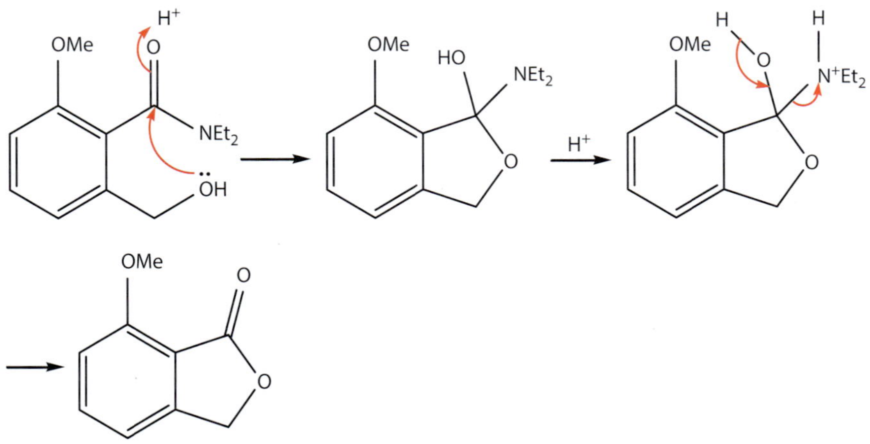

15.7 CARBON NUCLEOPHILES

Most organometallic compounds will react in some way with carbonyl groups, but in this section, we will confine the discussion to magnesium (Grignard reagents are very important), lithium, cadmium, zinc, and copper.

15.7.1 GRIGNARD REAGENTS

Looking back at Figure 15.34 gives us a clue as to what we should expect in adding Grignard reagents to esters (Figure 15.42). The Grignard attacks the carbonyl group, to give a tetrahedral

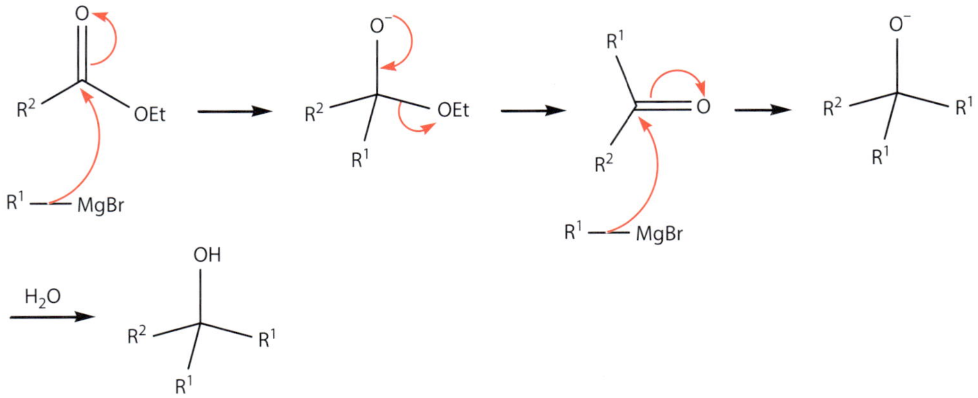

FIGURE 15.42 Reaction of a Grignard reagent with an ester.

intermediate, which expels ethoxide to give a ketone. The reaction is difficult, if not impossible, to stop at this stage, as the ketone is more electrophilic and reactive than the starting ester. The final product is a tertiary alcohol, with two of the substituents derived from the Grignard reagent. Although less commonly used (more expensive and harder to prepare), acid chlorides and anhydrides react in the same way. Amides are unreactive, and carboxylic acids simply protonate the Grignard reagent to give a hydrocarbon and a carboxylate anion.

The reaction is widely used in synthesis, and it is common to find that the product tertiary alcohol is readily dehydrated to give an alkene. Some examples are shown in Figure 15.43. In the first example, the Grignard is prepared from 2-bromonapthalene (Np = 2-naphthyl). The hydroxyl groups of dimethyl tartrate must be protected as an acetal (with which Grignard reagents do not react), before the Grignard addition, as they would otherwise protonate the organomagnesium compound to give naphthalene. In the second example, the product is

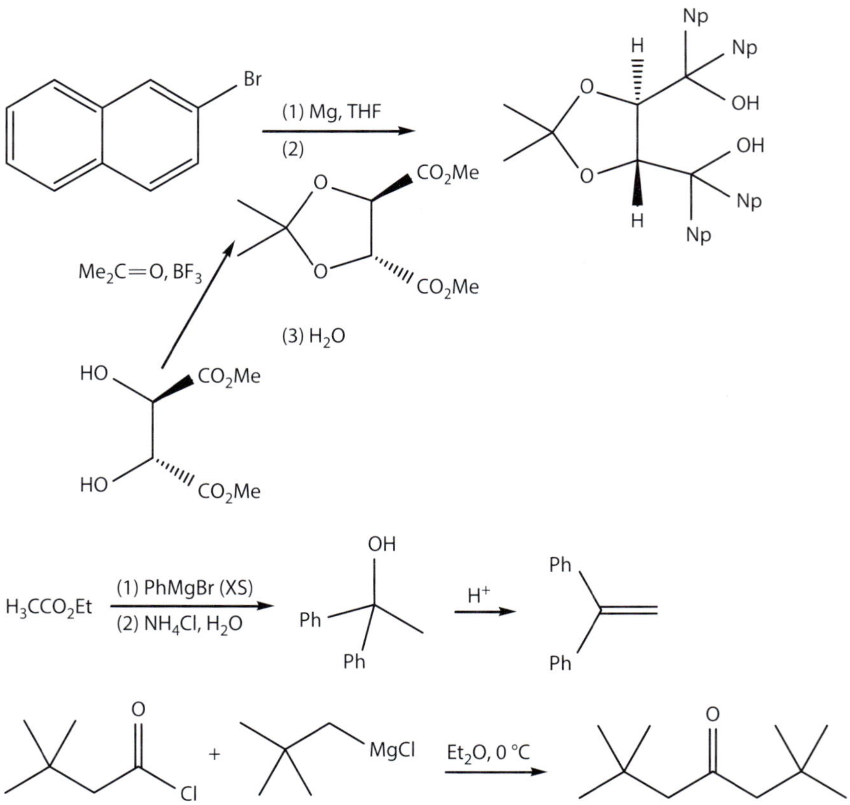

FIGURE 15.43 Grignard reactions with acid derivatives.

readily dehydrated to give an alkene. The final reaction is a rare example of a process in which the reaction can be stopped at the ketone stage. The conditions are carefully controlled, the chloride is more reactive than the ketone, and the steric hindrance is such that addition to the ketone is disfavored.

15.7.2 ORGANOLITHIUM COMPOUNDS

Alkyl and aryl lithium reagents react with esters, acyl halides, anhydrides, and even carboxylic acids (Figure 15.44). However, except for the reaction with acids, Grignard reagents are generally preferred as they are less expensive and easier to handle. In the reaction with a carboxylic acid, 1 mol of the organolithium is "wasted" in deprotonating the acid. But the organolithium is sufficiently nucleophilic to add to the carboxylate anion, to give a dilithium salt. On work-up of this, a ketone hydrate is obtained, and in most cases, this loses water to give the ketone. A practical application is given in the final example. If organolithium reagents are reacted with amines not substituted at nitrogen, then the main reaction is deprotonation of the $-NH_2$ or $-NHR$ group. With $RC(=O)NR'_2$, addition can occur, and ketones or alcohols may be obtained, depending on substituents and reaction conditions.

15.7.3 ORGANOCADMIUM COMPOUNDS

The use of organocadmium compounds, R_2Cd, lies in the fact that they react only with acid chlorides and no other carbonyl compounds. Thus, acid chlorides can be converted to ketones with no risk of further reaction. Organocadmium compounds are prepared (Equation 15.3) by reaction of cadmium chloride with a Grignard reagent, and the reactions only work well when R is aryl or primary alkyl. An example is provided in Figure 15.45.

(15.3) $$2RMgBr + CdCl_2 \rightarrow R_2Cd + MgCl_2 + MgBr_2$$

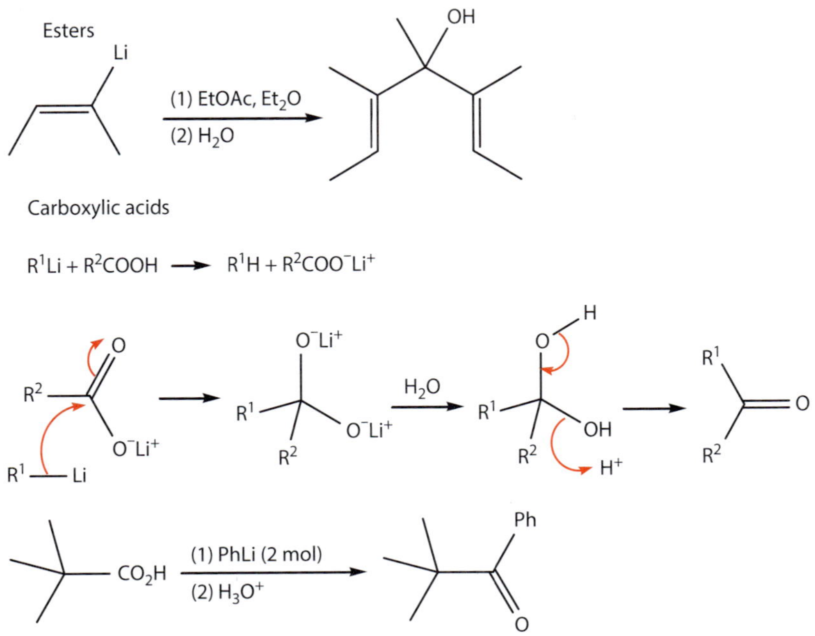

FIGURE 15.44 Reactions of organolithium compounds with acid derivatives.

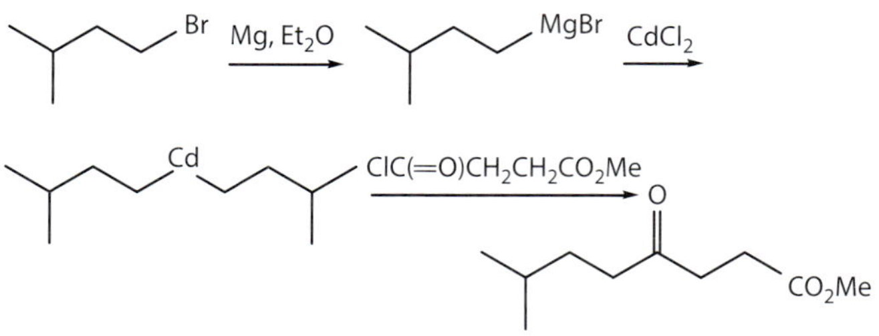

FIGURE 15.45 Preparation and use of an organocadmium compound.

15.7.4 ORGANOZINC COMPOUNDS: THE REFORMATSKII REACTION

Organozinc compounds react with acid chlorides, aldehydes, and ketones but not esters or amides (carboxylic acids are deprotonated to give a hydrocarbon and $RCOO^-$). The classic Reformatskii reaction involves the formation of an organozinc compound by reaction of an α-haloester with zinc powder. The crystal structures of the derivatives of the ethyl and *t*-butyl esters of bromoacetic acid have been determined. The reagent exists in the solid state as a stabilized dimer, **15.15**. Although we would get the correct answer by considering the reagent in solution as a simple organometallic analogous to a Grignard reagent, it can also be considered to behave as a zinc enolate, **15.16**, with the zinc associated with the oxygen rather than carbon (more on this in Chapter 17). The enolate adds to aldehydes and ketones, to give, after work-up, β-hydroxyesters. These are dehydrated under mild conditions to yield the α,β-unsaturated ester (Figure 15.46). Notice that this process would not be possible using magnesium; Grignard reagents react with esters, and a polymer would be obtained. Examples of the reaction are given in Figure 15.47. In the second, more substituted, example, a good ratio of diastereoisomers is obtained.

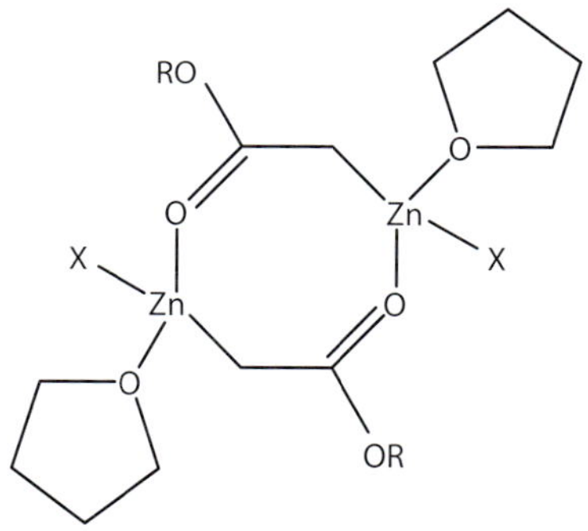

15.15

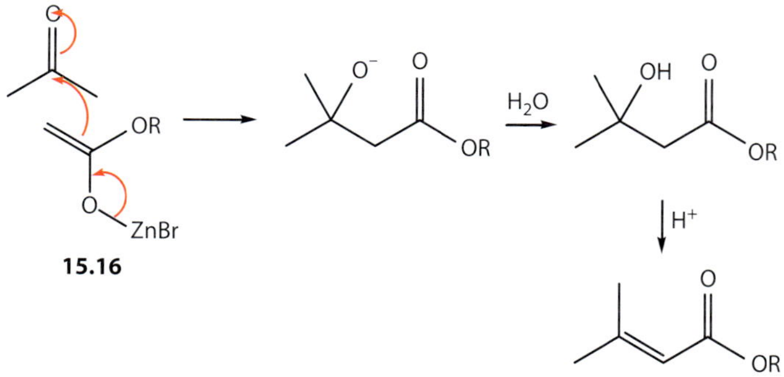

15.16

FIGURE 15.46 Mechanism of the Reformatskii reaction.

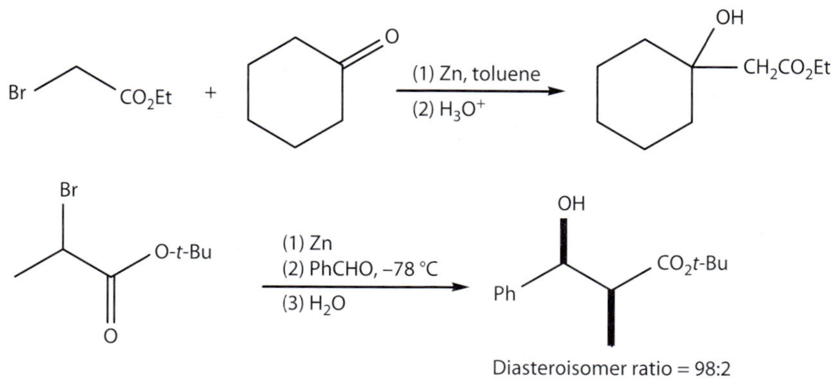

FIGURE 15.47 Examples of Reformatskii reactions in synthesis.

15.7.5 ORGANOCOPPER REAGENTS

While we shall meet organocopper reagents again in another context, here we will simply discuss their reactions with acyl halides. Dialkylcopperlithium reagents, sometimes called Gilman reagents, are prepared by the reaction of alkyl lithiums with copper(I) iodide. The structures of these reagents are complex, and the formulation that we give them reflects only their stoichiometry. Aryl and primary alkyl reagents are stable for several hours at 0 °C, but secondary and tertiary alkyl derivatives decompose above −20 °C. These reagents are less reactive, and more selective than Grignard reagents, and react with acid chlorides to give ketones (Figure 15.48). While we can write a simple mechanism for the process (Figure 15.49), it is probably not strictly accurate.

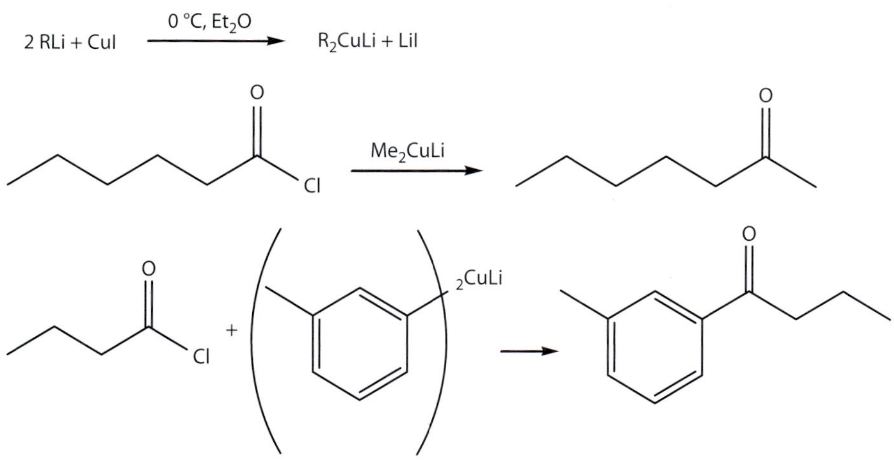

FIGURE 15.48 Preparation of organocopper compounds and their reaction with acyl halides.

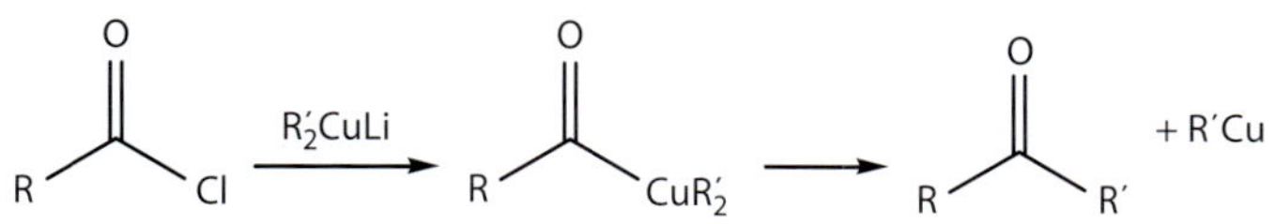

FIGURE 15.49 Mechanism of reaction of organocopper compounds with acyl halides.

PROBLEM 15.9

Suggest reagents for each of the following transformations; some may require more than one step:

(a)

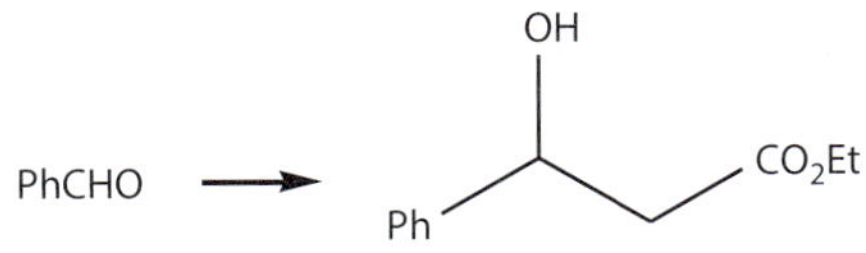

(b)

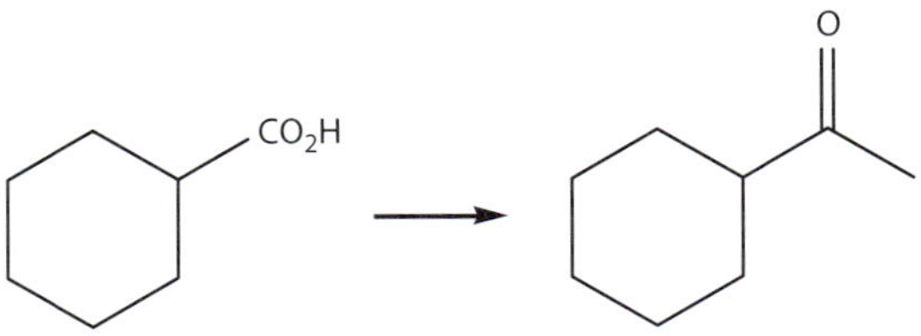

(c)

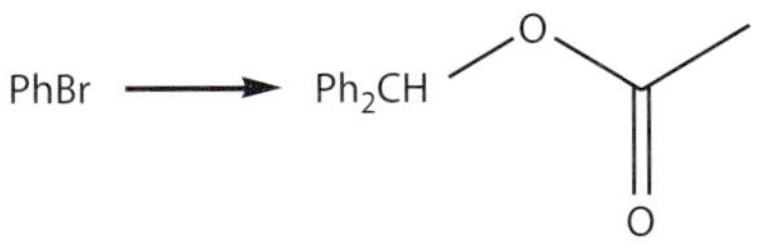

(d)

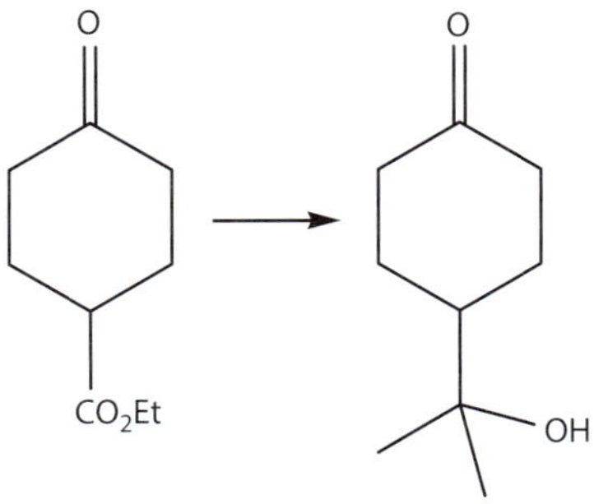

Solutions

(a) This is a classic Reformatskii reaction, using zinc metal and $BrCH_2CO_2Et$.

(b) The quickest route here involves an organolithium reagent, the only one of the common organometallics to react with carboxylic acids. Two moles of methyl lithium followed by an acidic work-up could be used. In the practical reaction, the carboxylate was first deprotonated by the less expensive base, lithium hydride. An acceptable alternative solution would be to convert the acid to the acid chloride with $SOCl_2$ then react the acid chloride with either Me_2Cd or Me_2CuLi.

(c) In this reaction, we should first strip off the acetate group; it is essentially decorative and can be added at the end. We need to make the alcohol, Ph_2CHOH. The first step is to convert the bromobenzene to the Grignard reagent with magnesium in dry ether, and then we react this with any formate ester. It's probably unnecessary to isolate the alcohol; work-up with acetic anhydride or acetyl chloride should give the required product:

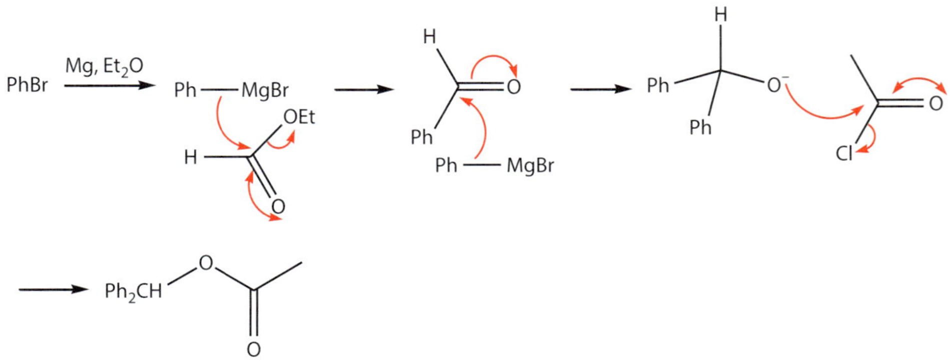

(d) This reaction presents a problem. We need a Grignard reagent to react with the ester but not the ketone. But the ketone is more reactive than the ester. So we must first protect the ketone as an acetal, complete the Grignard reaction, and then hydrolyze the acetal:

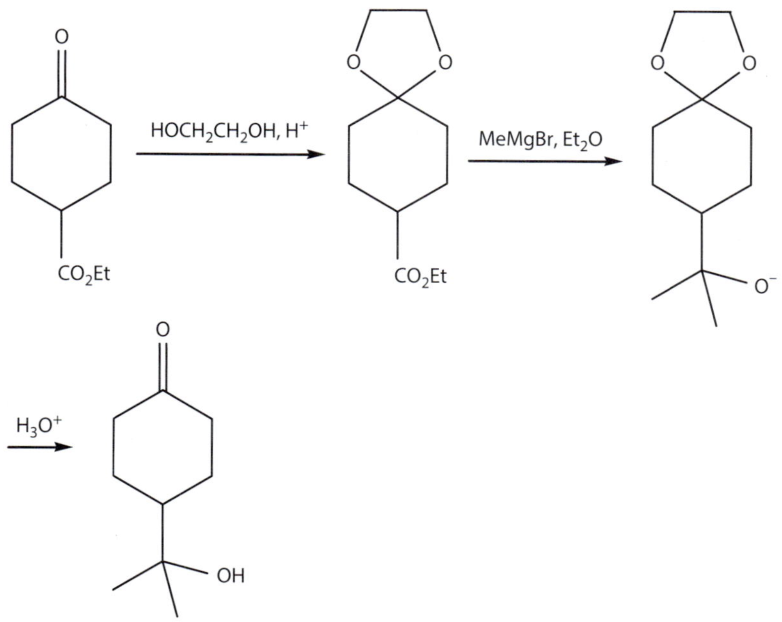

Key Points from Sections 15.6 and 15.7

- Carboxylic acids, esters, acid chlorides, and acid anhydrides can all be reduced to primary alcohols by lithium aluminum hydride.
- Esters are reduced to aldehydes by DIBAH at low temperature.
- Acid chlorides are reduced to aldehydes by Li[AlH(O-*t*-Bu)$_3$].
- Amides and nitriles are reduced to amines by lithium aluminum hydride.
- Sodium borohydride reduces acid chlorides and anhydrides to primary alcohols but does not react with esters or amides.
- Grignard reagents react with esters, acid chlorides, and anhydrides to give tertiary alcohols with two of the attached groups identical. They are protonated by carboxylic acids.
- Organolithium compounds react with esters, acid chlorides, and anhydrides in the same way as Grignard reagents. Reaction with carboxylic acids gives ketones, but 2 mol of organolithium are required.
- Organozinc compounds react with acid chlorides, aldehydes, and ketones but not with esters, and useful reactions can be accomplished with the zinc derivatives of α-haloesters.
- Acid chlorides can be converted to ketones with organocadmium (R_2Cd) or organocopper reagents (R_2CuLi).

REVIEW PROBLEMS

1. Give systematic names for the following compounds:

(a)

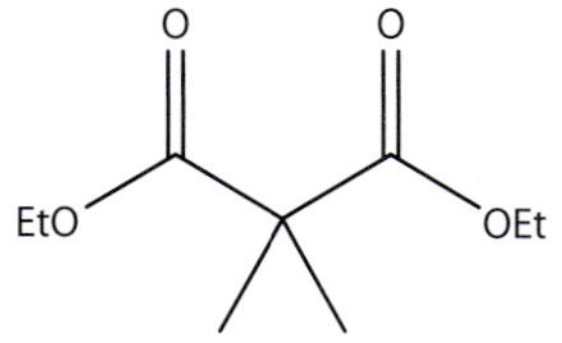

(b)

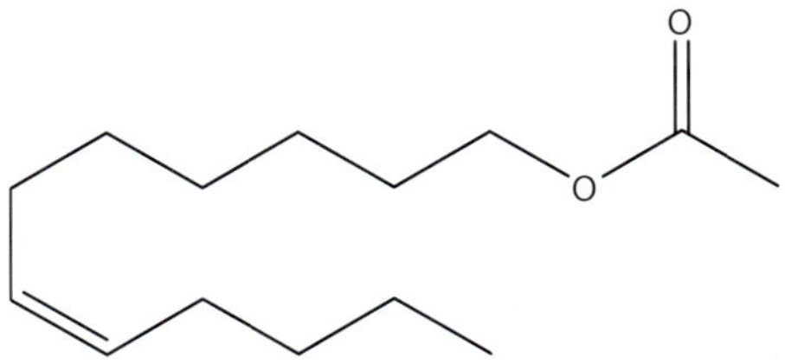

(c)

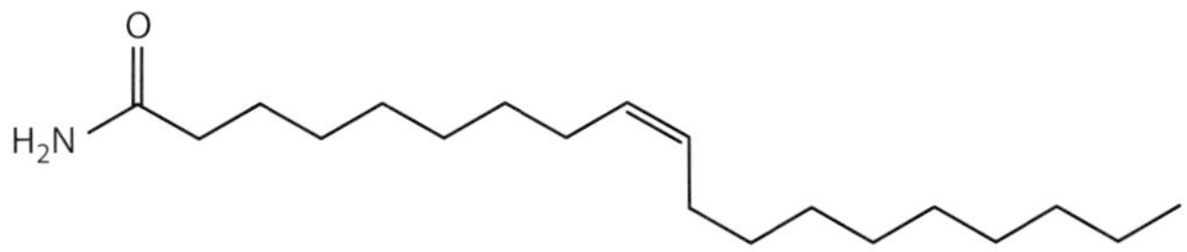

Common name is oleamide.

2. Give structures for
   (a) 1-Hexyl ethanoate
   (b) *trans*-3-Phenylpropenoic acid
   (c) 2-Oxohexanoic acid
3. Three isomeric compounds, **15.17**, **15.18**, and **15.19**, have the formula $C_9H_{10}O_2$, and all show a molecular ion, M^+, at *m*/*z* 150. The main fragment ions are

15.17	135, 107
15.18	108, 107
15.19	119, 91

Their IR spectra show key bands as follows:

15.17	1670 cm^{-1}
15.18	1760 cm^{-1}
15.19	1724 cm^{-1}

Their ^1H NMR spectra are as follows:

15.17	δ	8.0 (d, 2H, *J* = 7 Hz)
		6.95 (d, 2H, *J* = 7 Hz)
		3.90 (s, 3H)
		2.55 (s, 3H)
15.18	δ	7.2 (d, 2H, *J* = 7 Hz)
		7.0 (d, 2H, *J* = 7 Hz)
		2.46 (s, 3H)
		2.35 (s, 3H)
15.19	δ	7.95 (d, 2H, *J* = 7 Hz)
		7.26 (d, 2H, *J* = 7 Hz)
		3.92 (s, 3H)
		2.47 (s, 3H)

Deduce the structures for **15.17**, **15.18**, and **15.19**, explaining clearly how you differentiate between them.

4. Explain the following pK_a values for substituted butanoic acids:

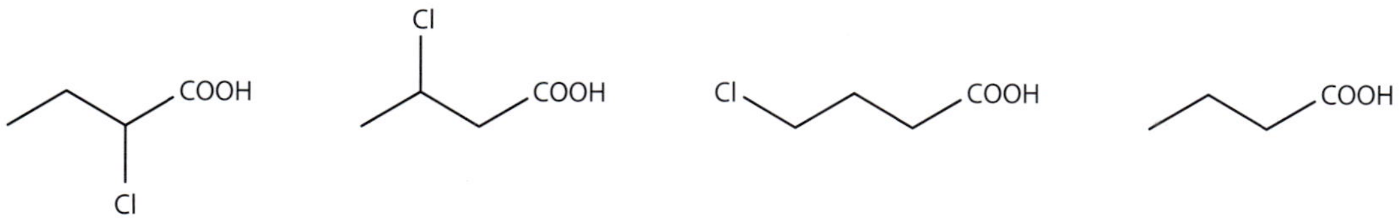

pK_a 2.83 3.98 4.52 4.82

5. Propose a mechanism for each of the following processes:

(a)

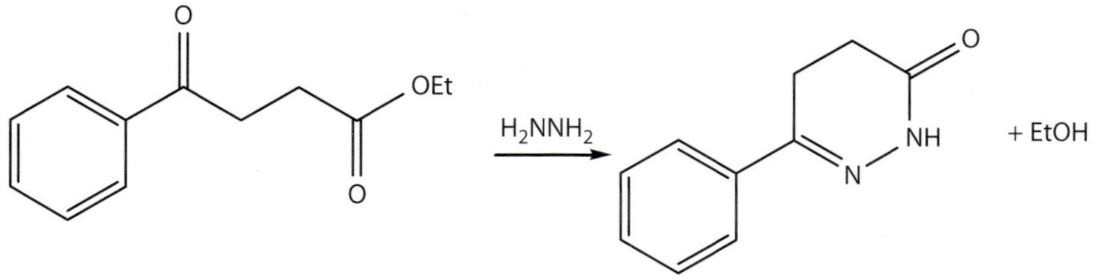

(b) $PhC(OMe)_3 \xrightarrow{H_3O^+} PhCOOH + 3MeOH$

(c)

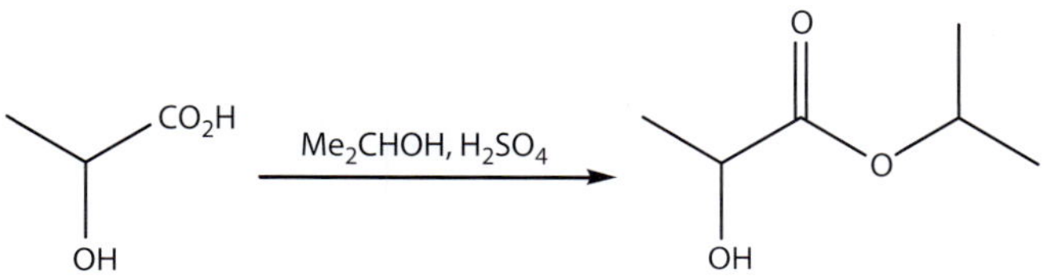

(d)

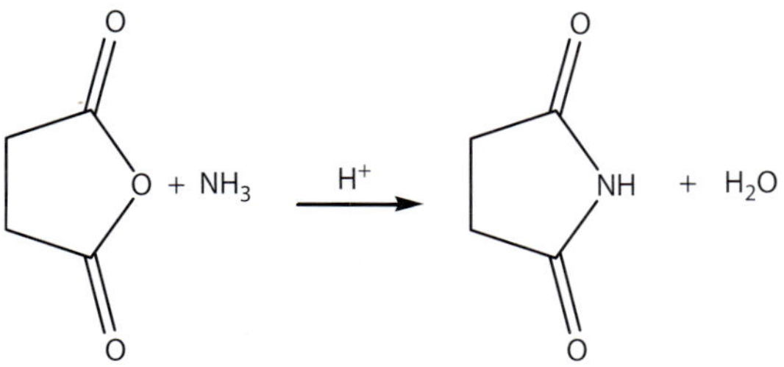

(e)

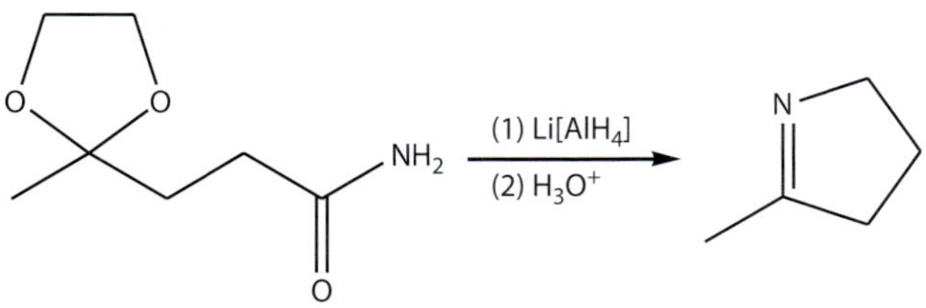

(f)

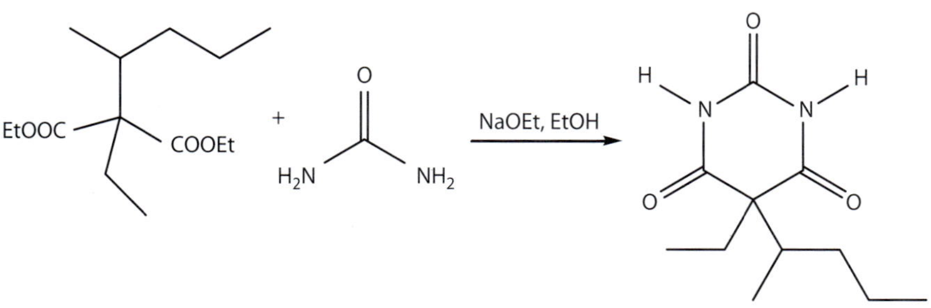

Nembutal

(g)

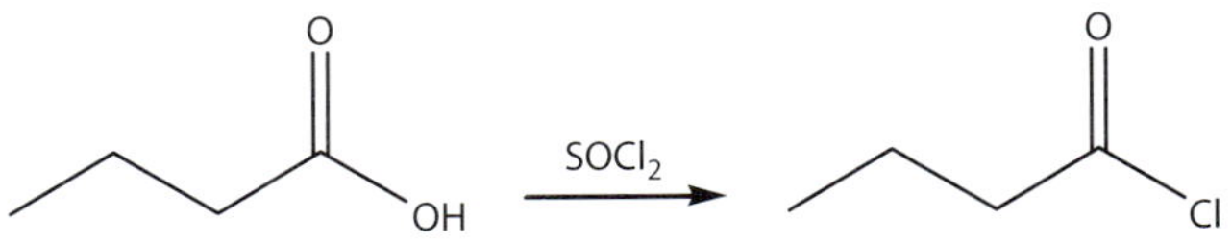
O
OH
$SOCl_2$
O
Cl

(h)

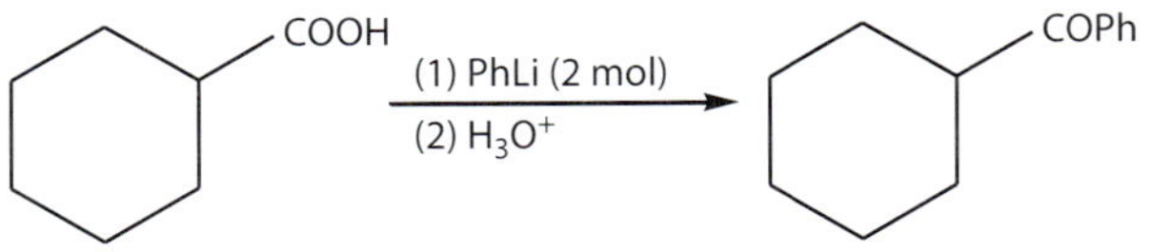
COOH
(1) PhLi (2 mol)
(2) H_3O^+
COPh

(i)

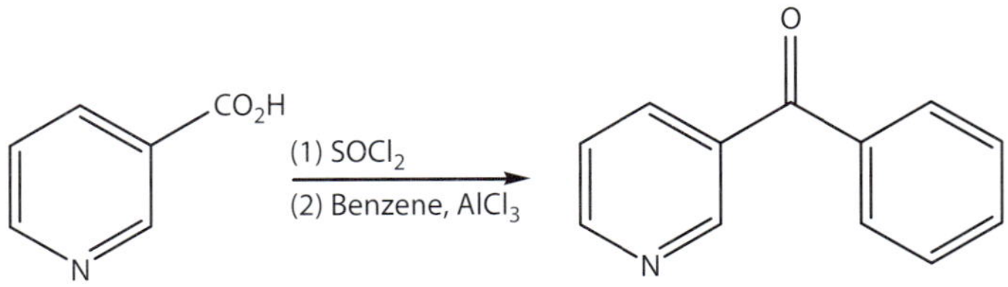
CO_2H
N
(1) $SOCl_2$
(2) Benzene, $AlCl_3$
O
N

(j)

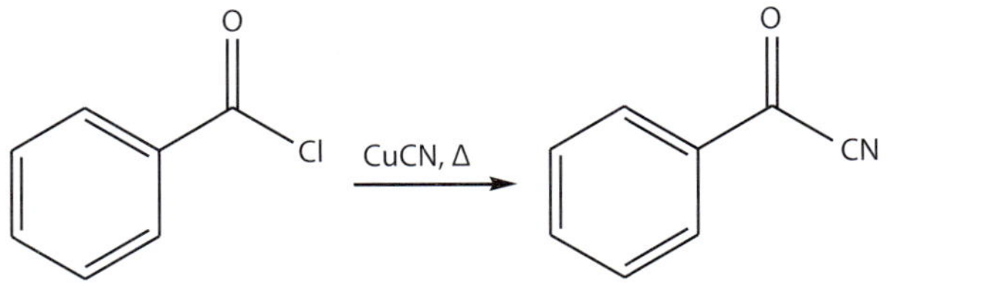
O
Cl
CuCN, Δ
O
CN

(k)

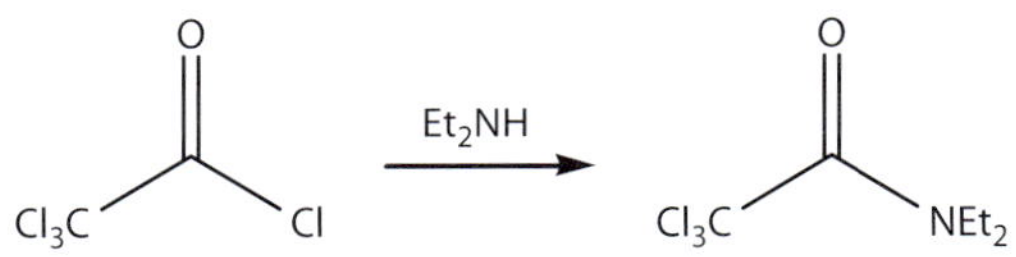
O
Cl_3C
Cl
Et_2NH
O
Cl_3C
NEt_2

(l)

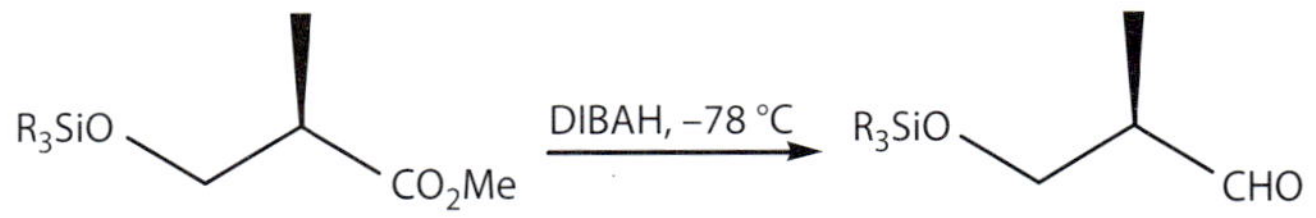
R_3SiO
CO_2Me
DIBAH, –78 °C
R_3SiO
CHO

(m)

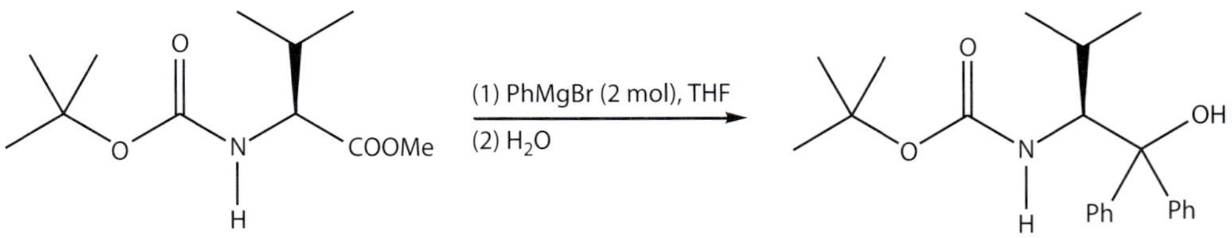
O
O
N
H
COOMe
(1) PhMgBr (2 mol), THF
(2) H_2O
O
O
N
H
OH
Ph
Ph

(n)

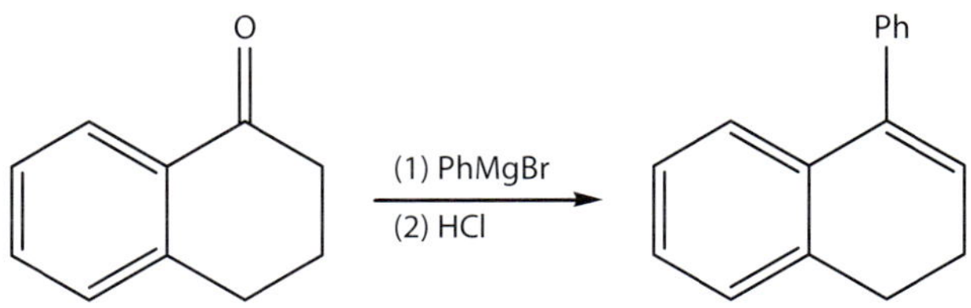
O
(1) PhMgBr
(2) HCl
Ph

(o)

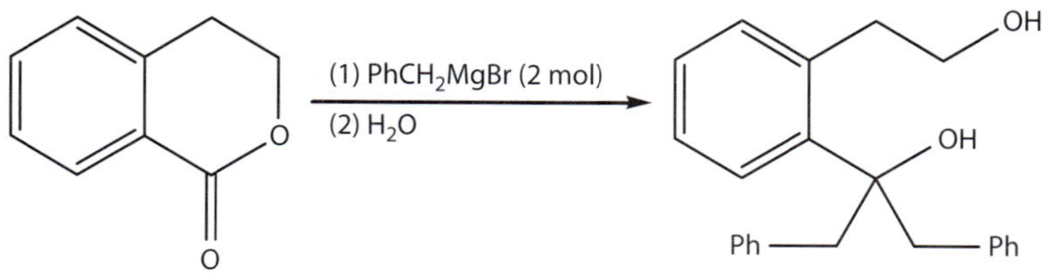

(p)

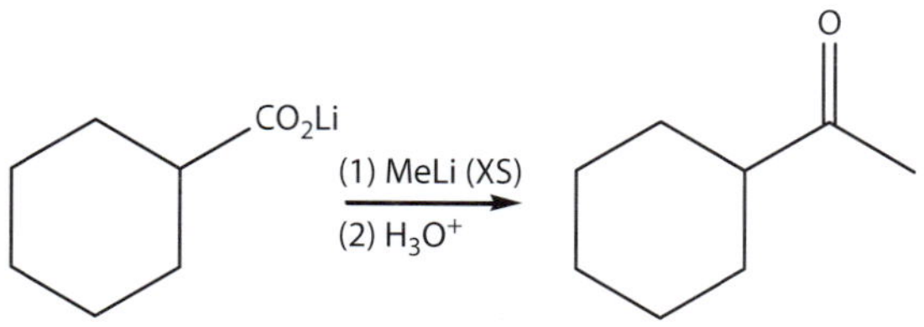

(q) $PhCN + H_2O \xrightarrow{H^+} PhCOOH + NH_3$

6. Predict the outcome of each of the following processes:

(a)

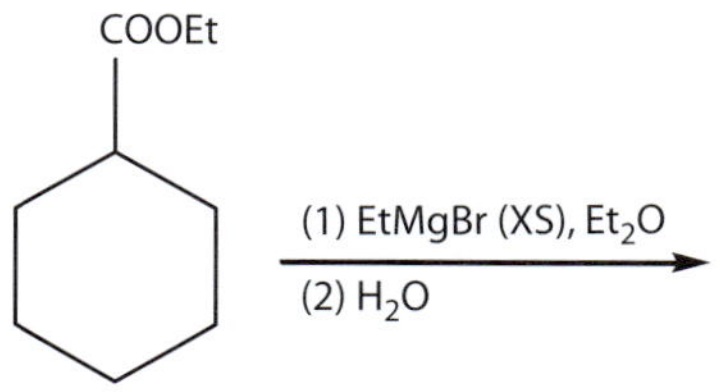

(b)

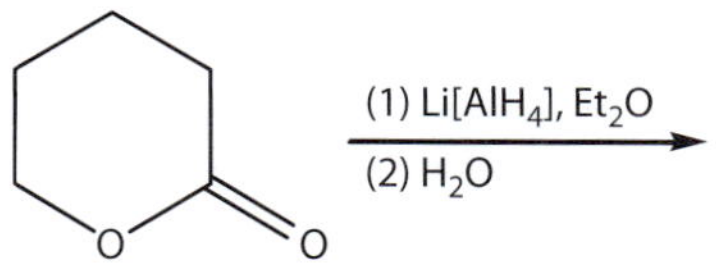

(c)

(d)

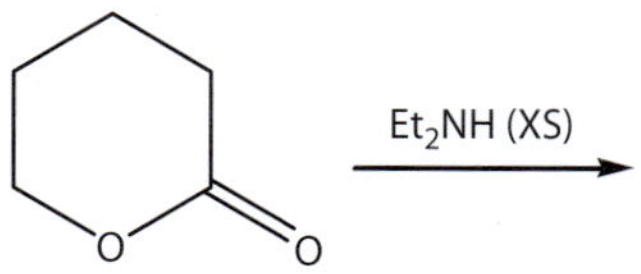

(e) $CH_3CN \xrightarrow{MeLi}$ **15.20** $\xrightarrow{H_3O^+}$ **15.21**

(f)

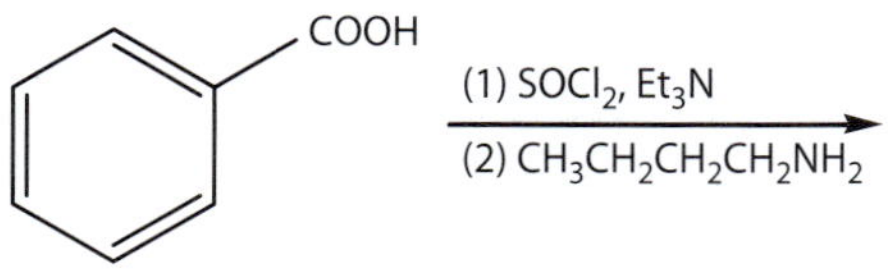

(g)

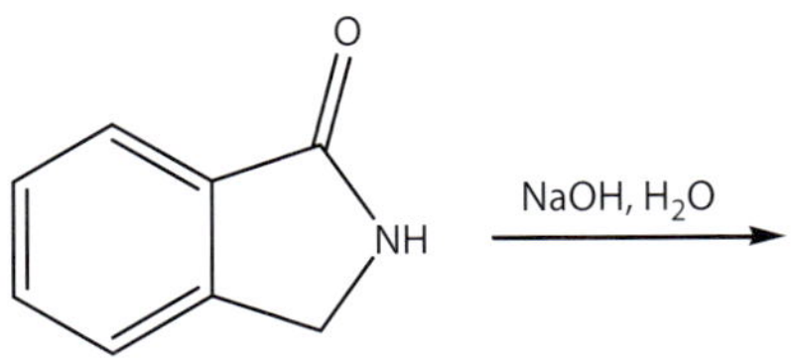

(h)

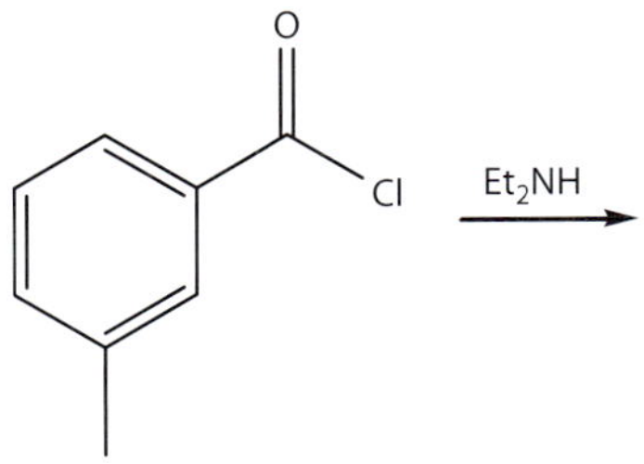

(i)

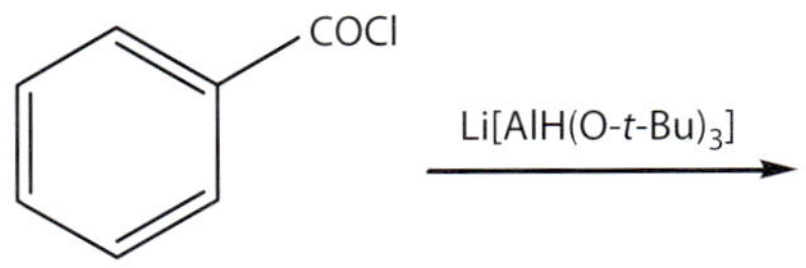

(j) $CH_3(CH_2)_{10}COOMe \xrightarrow[\text{(2) } H_3O^+]{\text{(1) DIBAH, toluene, –78 °C}}$

(k)

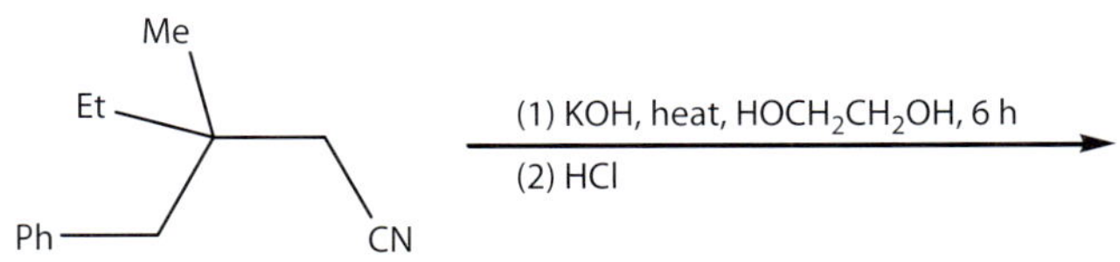

(l)

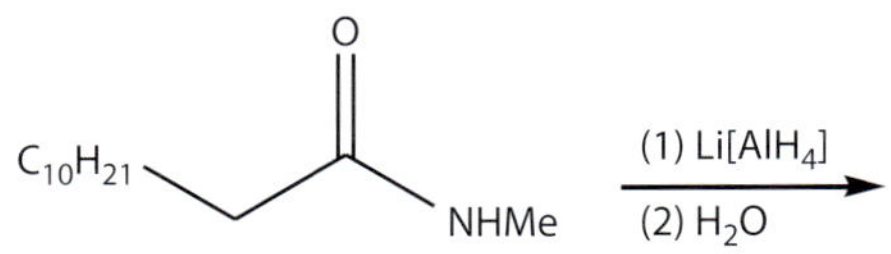

(m)

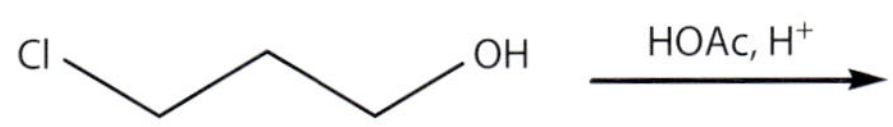

(n)

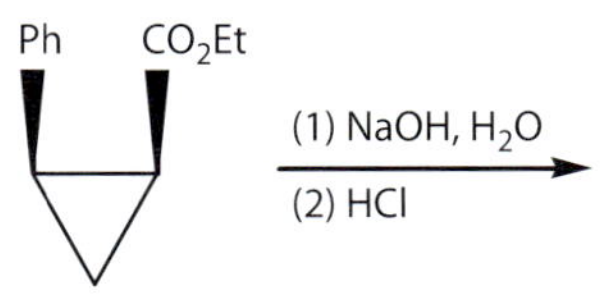

(o)

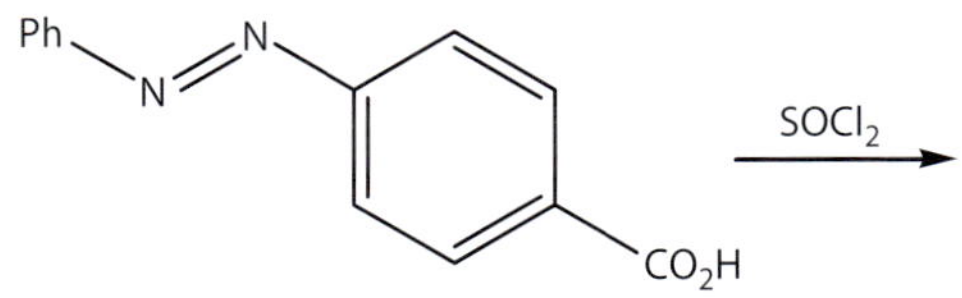

(p) 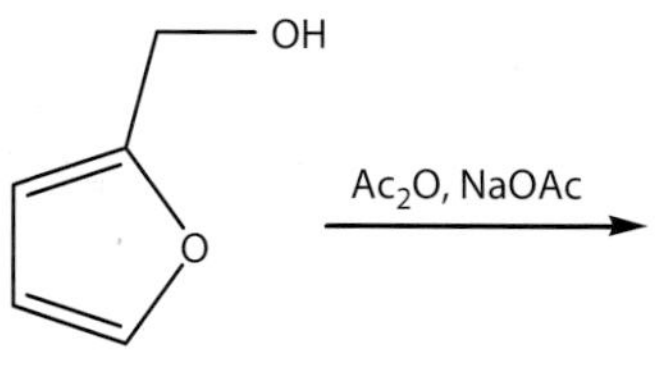

(q) 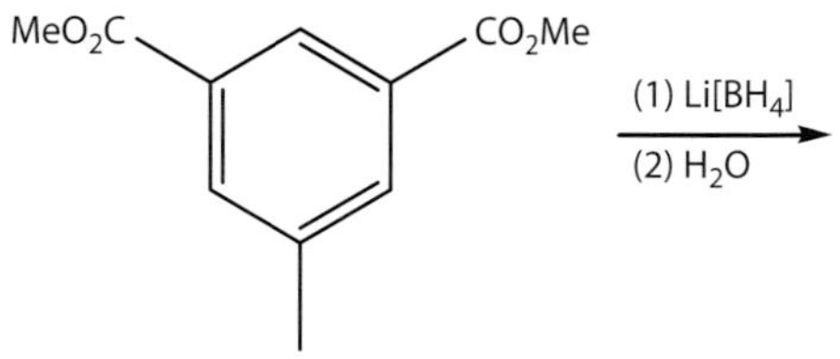

7. How would you accomplish each of the following transformation that may require more than one step?
   (a) PhCOCl to PhCOMe
   (b) 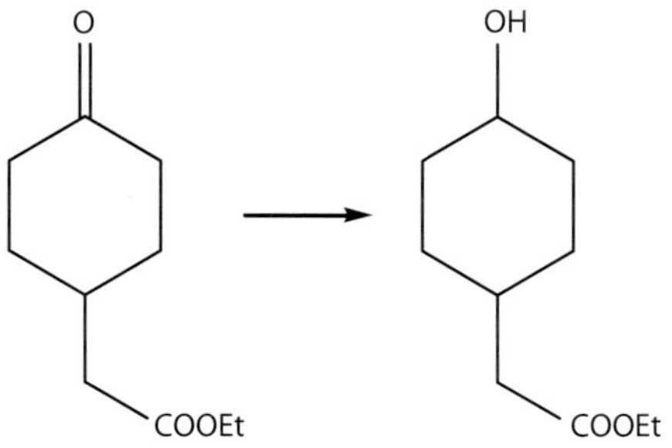

   (c) 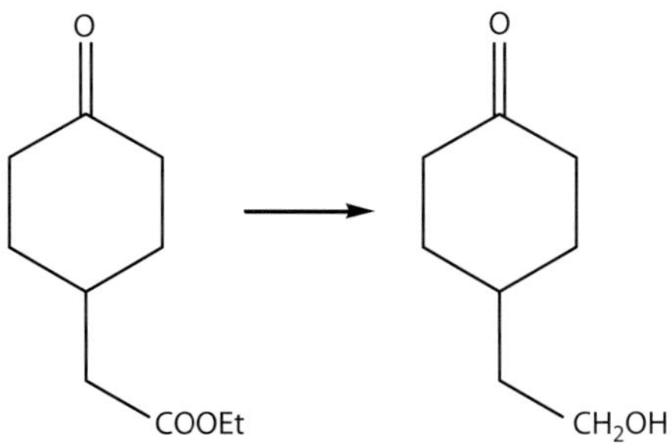

   (d) 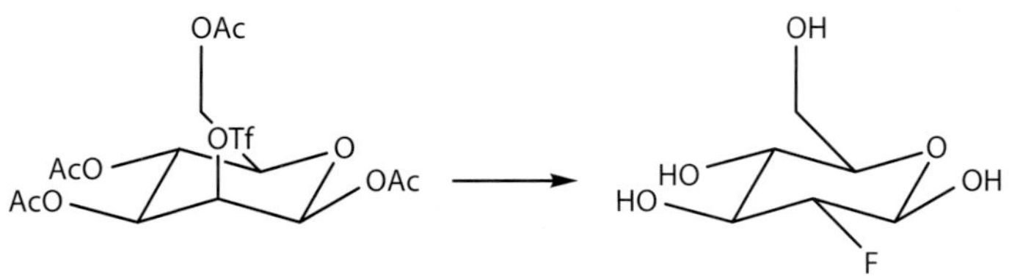

   Remember that triflate is a good leaving group.

   (e) 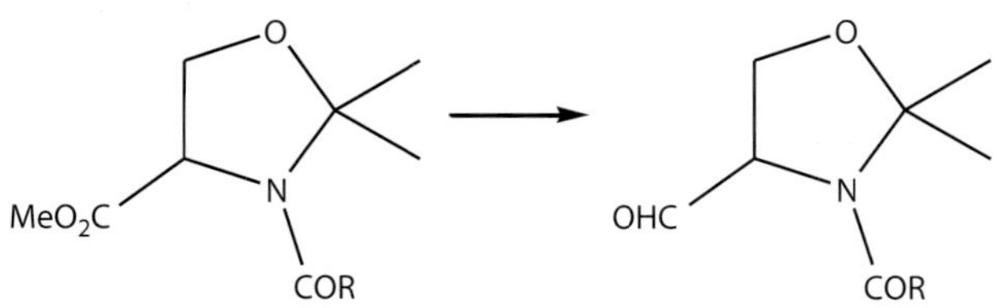

(f)

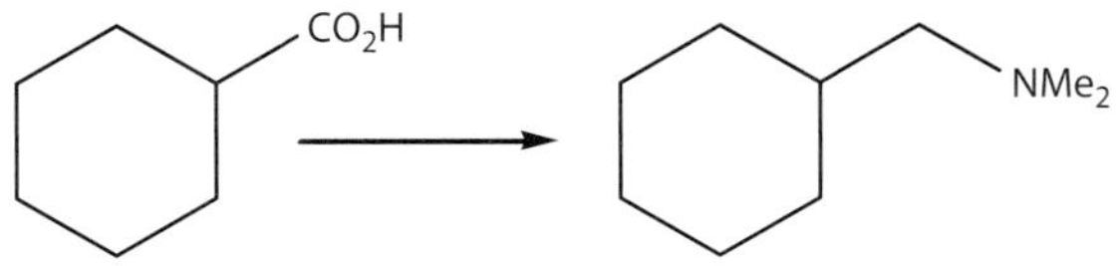

(g)

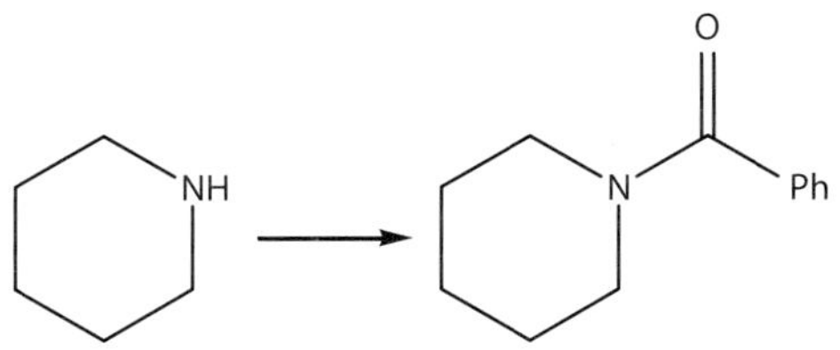

(h)

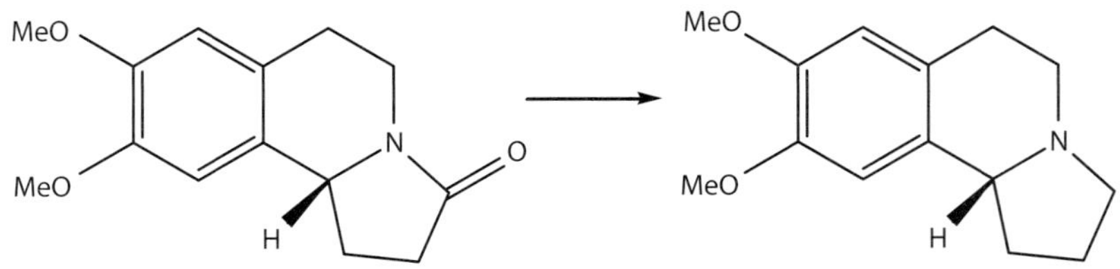

(i)

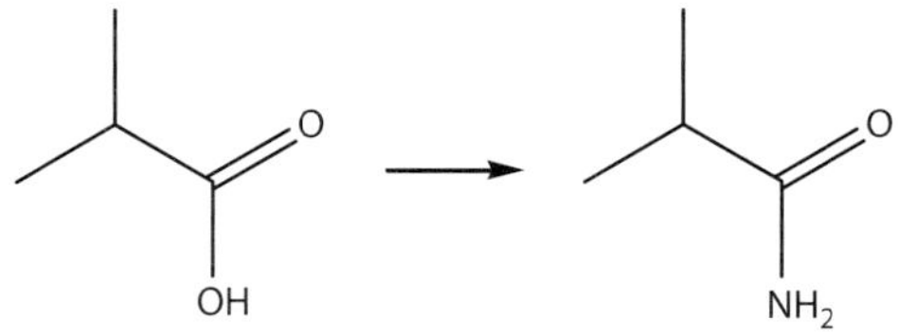

(j) $HO_2C(CH_2)_9CO_2H \rightarrow MeO_2C(CH_2)_9CO_2Me$

(k)

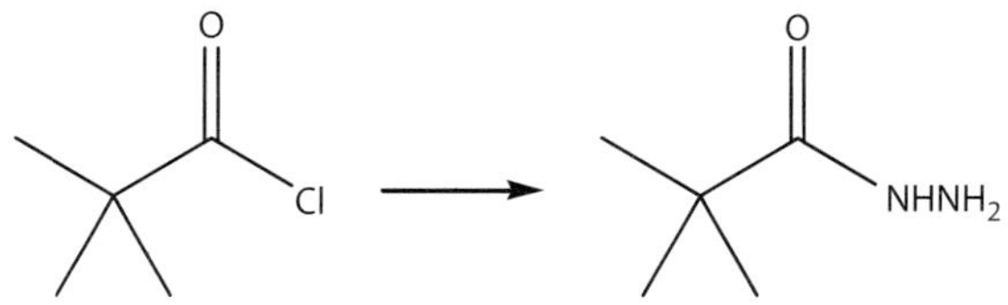

(l)

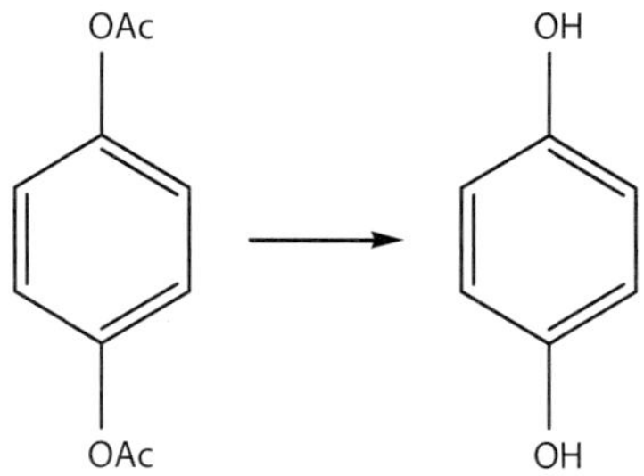

(m)

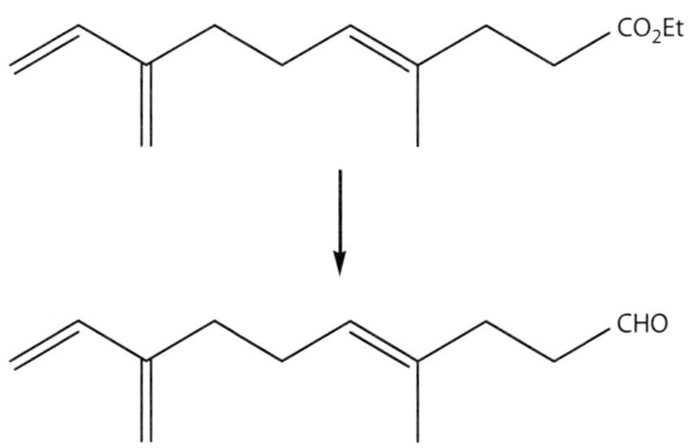

8. Hydroxy acids may be cyclized under acidic conditions to give lactones. The position of the equilibrium varies with the ring size, as shown in the following. Draw a mechanism for the reaction and comment on the proportions of acid and lactone:

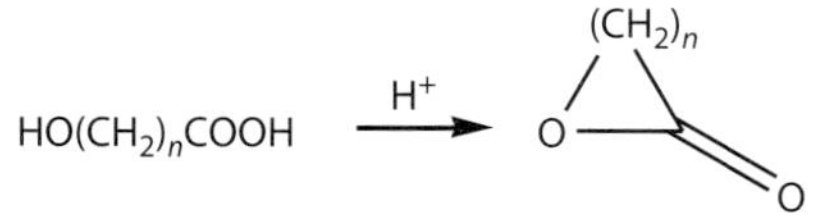

Ring Size	% Acid	% Lactone
4	100	0
5	27	73
6	91	9
7	100	0

9. Explain the following observations:
   (a) Methyl 3-nitrobenzoate is hydrolyzed in aqueous basic solution more rapidly than is methyl benzoate.
   (b) Attempted esterification of 2,4,6-trimethylbenzoic acid with MeOH/HCl is unsuccessful. The ester can be obtained using $SOCl_2$ followed by methanol or with conc. H_2SO_4 followed by methanol. The methyl ester is resistant to hydrolysis in either dilute aqueous acid or base but can be hydrolyzed by using conc. H_2SO_4 followed by water.
   (c) Methyl ethanoate is hydrolyzed in $H_2{}^{18}O$ under basic catalysis. The rate is proportional to $[HO^-]$ and [MeCOOMe]. If MeCOOMe is isolated before the reaction is complete, it is found to contain MeC(=^{18}O)OMe.
   (d) The compound γ-hydroxybutyrate (GHB) is widely used as a club drug and has been implicated in date rape incidents. Its possession is illegal in many jurisdictions. However, there is a substance, γ-butyrolactone, which is subject to fewer restrictions and is now being widely used. How are these substances connected?

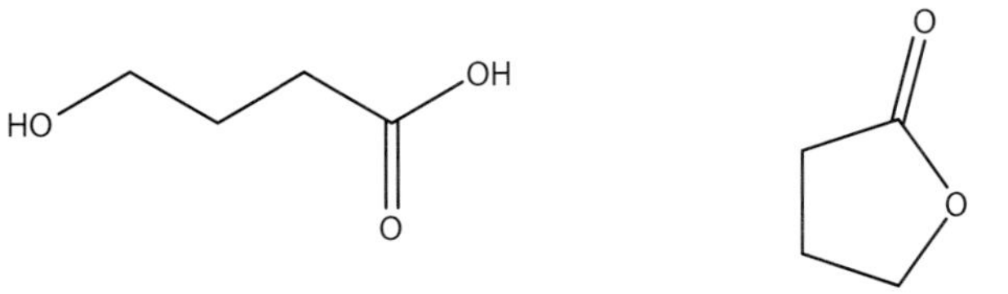

GHB γ-butyrolactone

   (e) The reaction shown in the following is successful, but with EtMgBr, only reduction is observed:

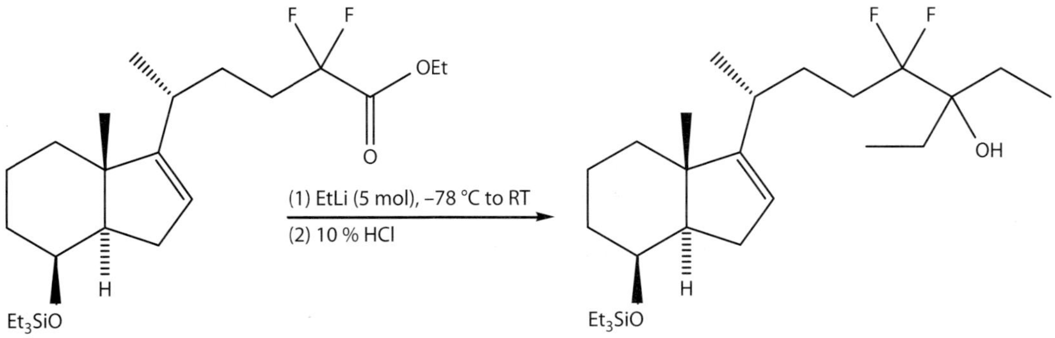

10. Study the reaction sequence shown in the following and suggest reagents for each step:

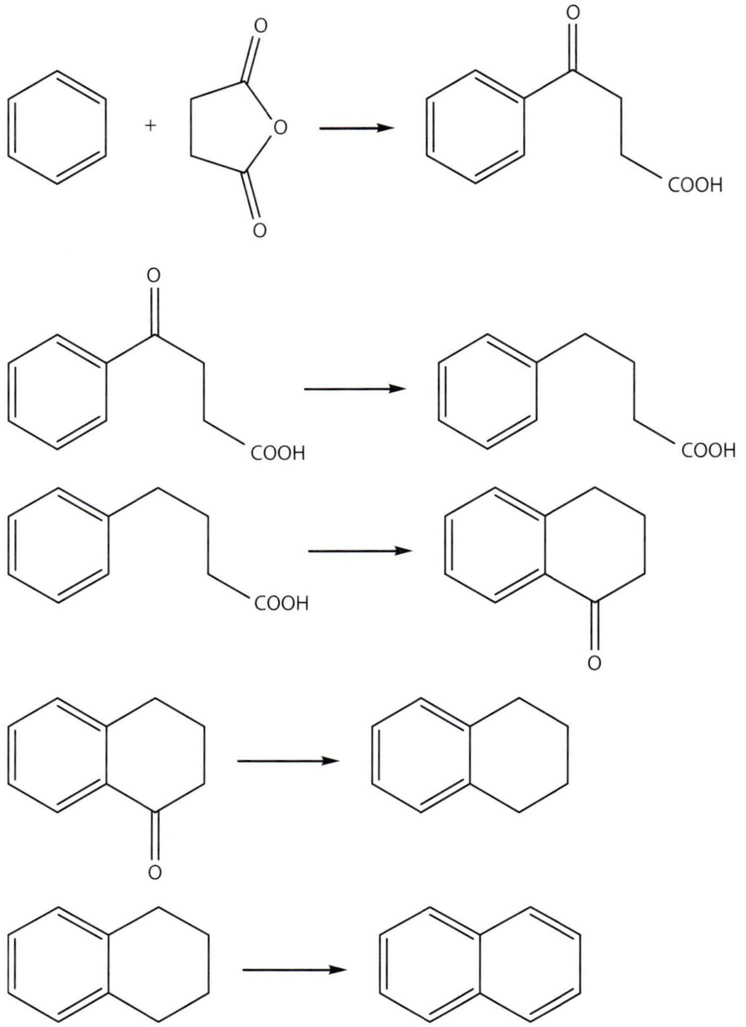

MORE CHALLENGING PROBLEMS

1. Propose a mechanism for each of the following processes:

(a)

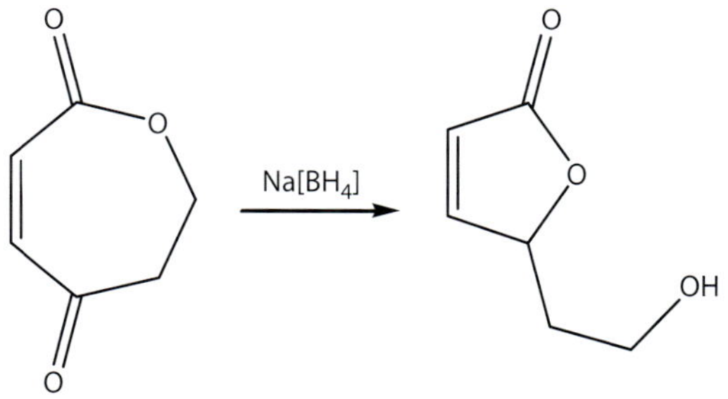

(b)

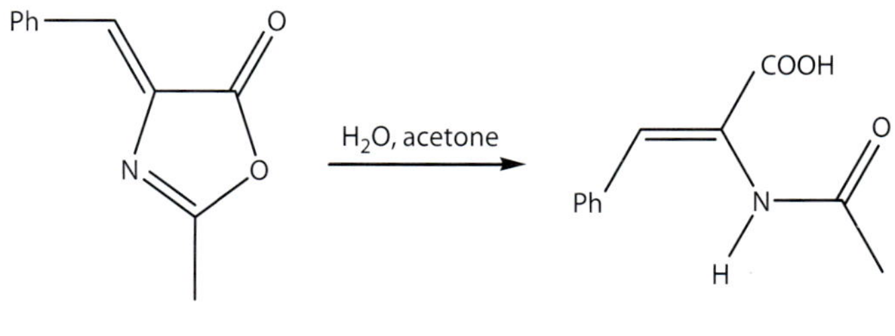

(c)

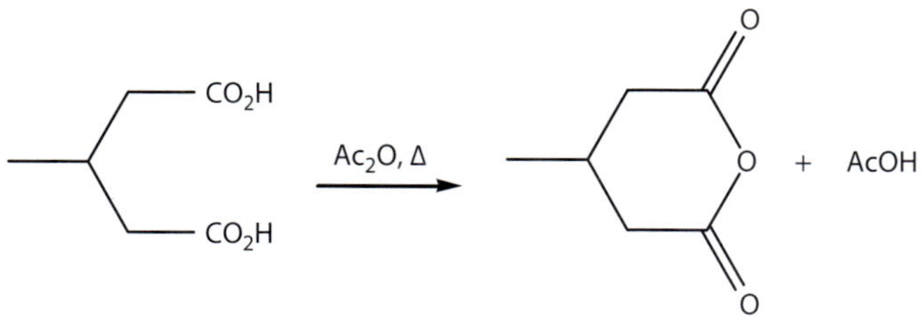
CO2H
CO2H
Ac2O, Δ
O
O
O
+ AcOH

(d)

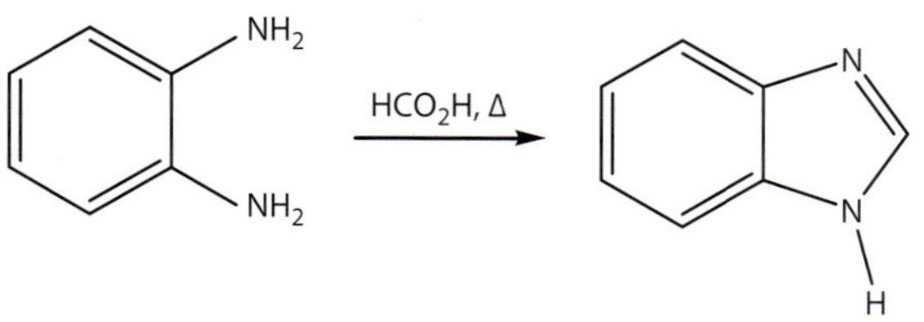
NH2
NH2
HCO2H, Δ
N
N
H

(e)

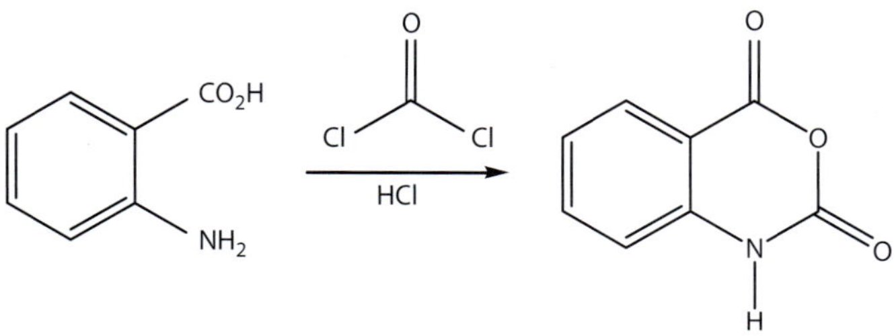
CO2H
NH2
O
Cl
Cl
HCl
O
O
N
O
H

(f)

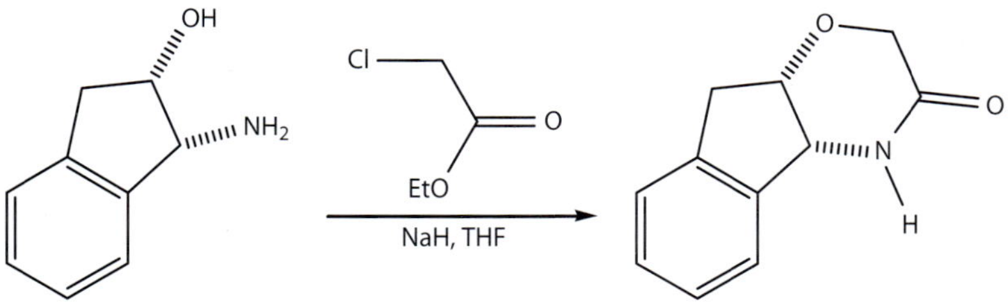
OH
NH2
Cl
O
EtO
NaH, THF
O
O
N
H

(g)

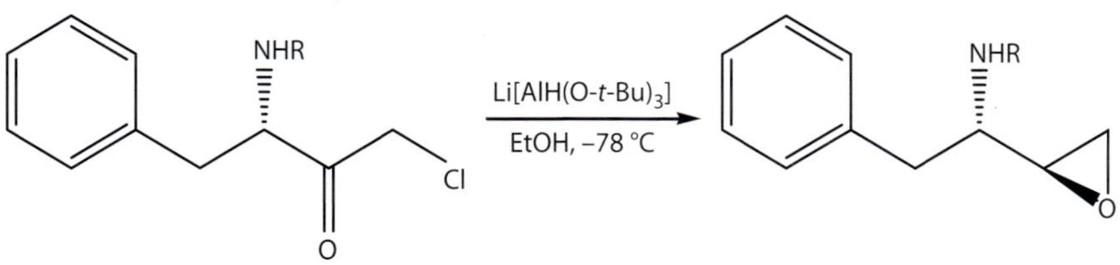
NHR
Cl
O
Li[AlH(O-t-Bu)3]
EtOH, –78 °C
NHR
O

(h)

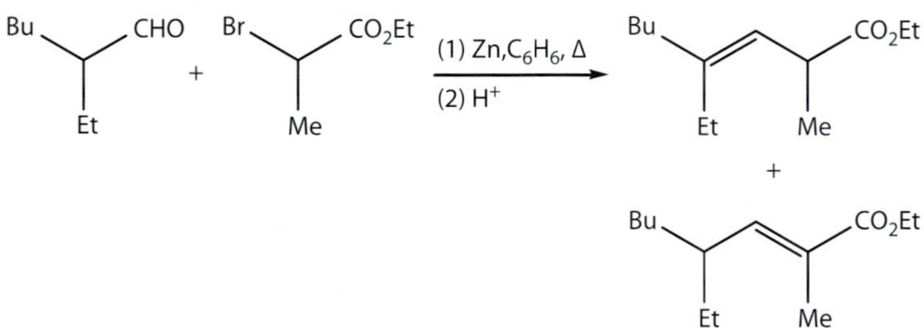
Bu
CHO
Et
+
Br
CO2Et
Me
(1) Zn,C6H6, Δ
(2) H+
Bu
CO2Et
Et
Me
+
Bu
CO2Et
Et
Me

(i)

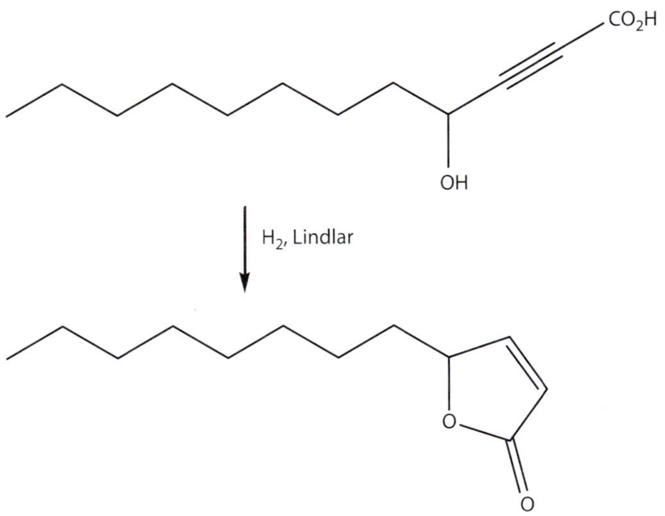

2. Predict the product(s) of the following reaction sequences:

(a)

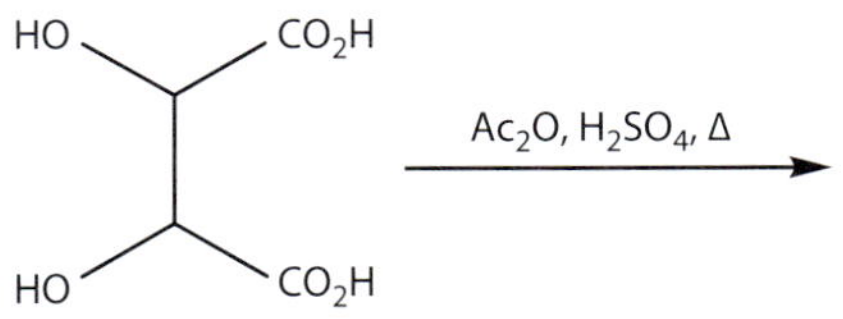

(b)

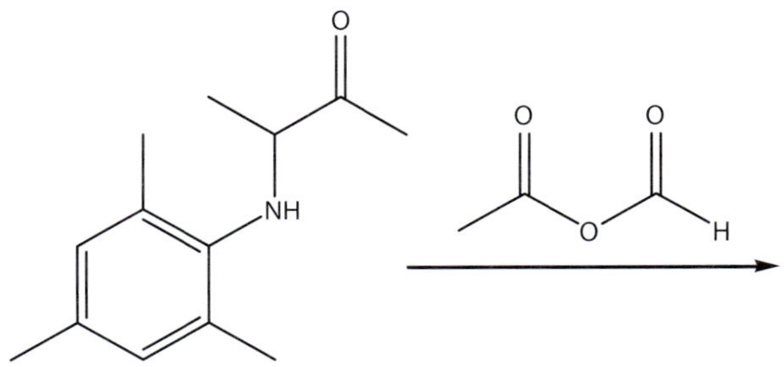

(c)

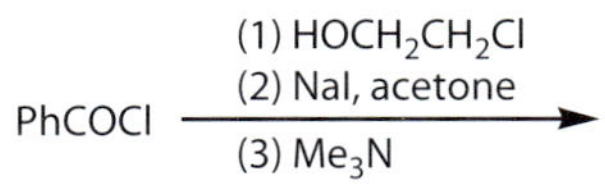

(d)

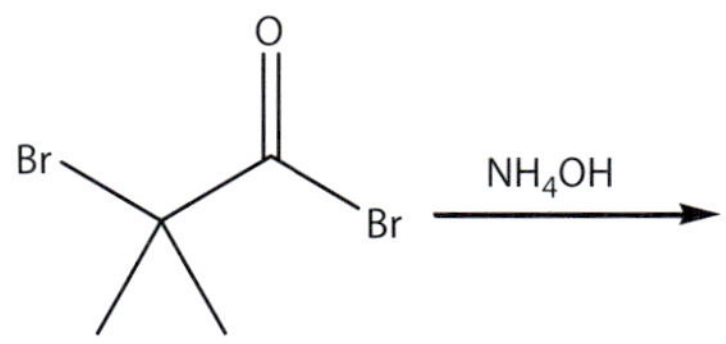

(e)

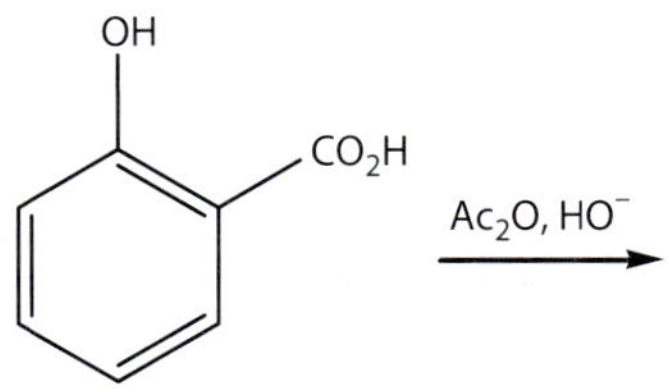

This is a key step in aspirin synthesis.

3. How would you accomplish each of the following transformations, which may require more than one step?

(a)

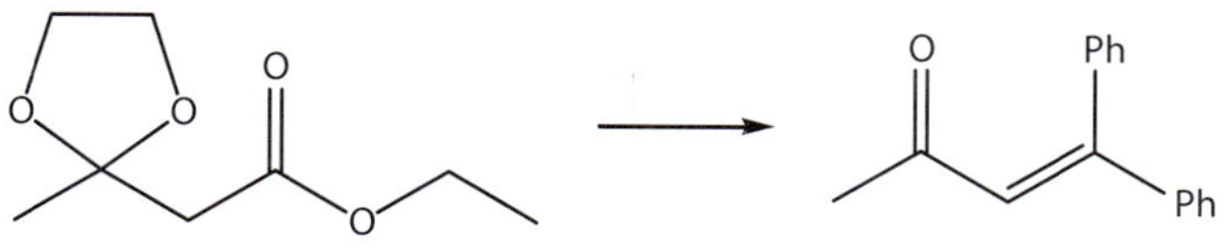

(b)

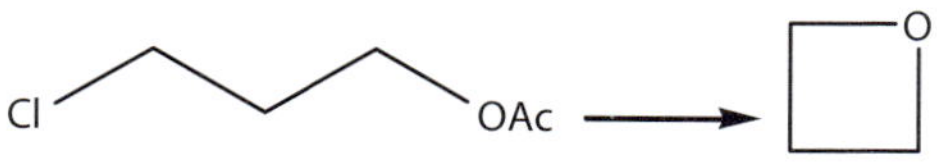

(c)

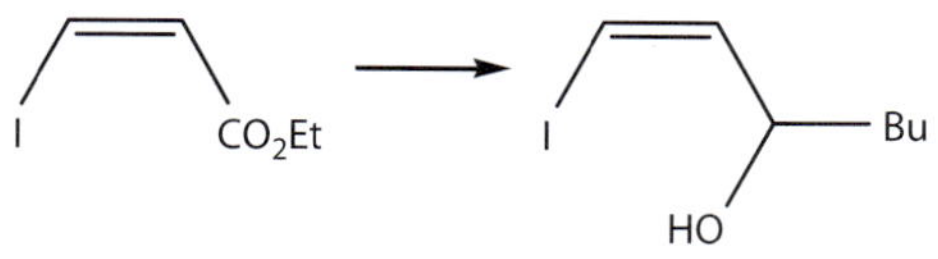

(d)

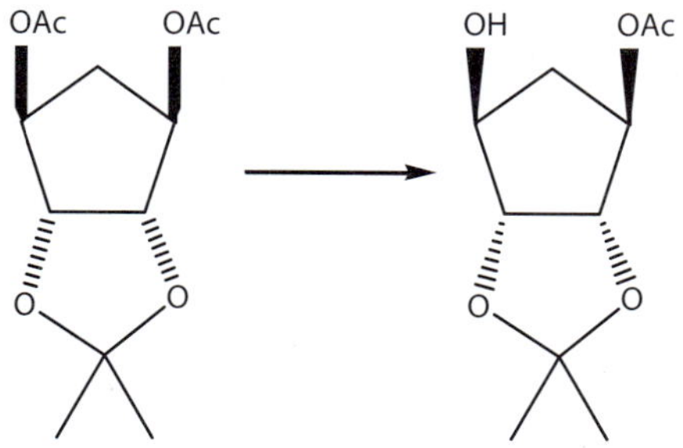

(e)

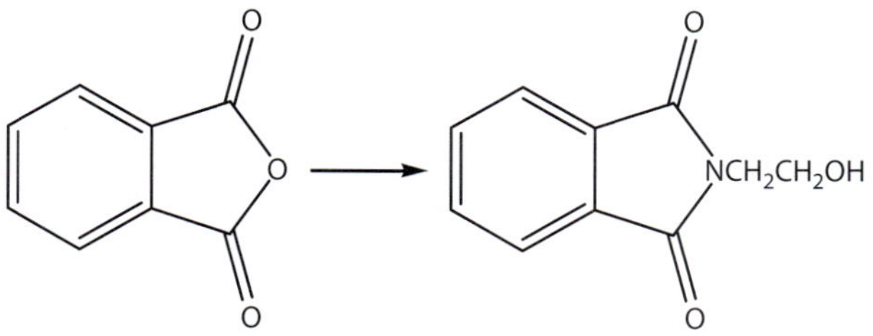

4. Explain the following observations:
   (a) Methyl esters can be prepared cleanly (the only other product is molecular nitrogen) from carboxylic acids and the reactive species diazomethane, the structure of which is given below:

$$^{-}CH_2 - N^{+} \equiv N$$

   (b) Compound **15.22** is hydrolyzed 10^8 times faster than normal amides:

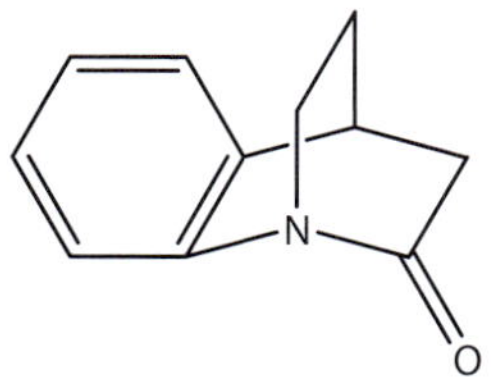

15.22

   (c) The acid-catalyzed hydrolysis of esters of *tert*-butanol follows a different mechanism from that of esters of less hindered alcohols. If the hydrolysis is carried out in O^{18}-labeled water, then O^{18} is found in both the carboxylic acid and the *tert*-butanol produced. *tert*-Butyl esters can be synthesized from either *tert*-butanol or 2-methylpropene under acid-catalyzed conditions.

5. Fill in the blanks in the following sequence:

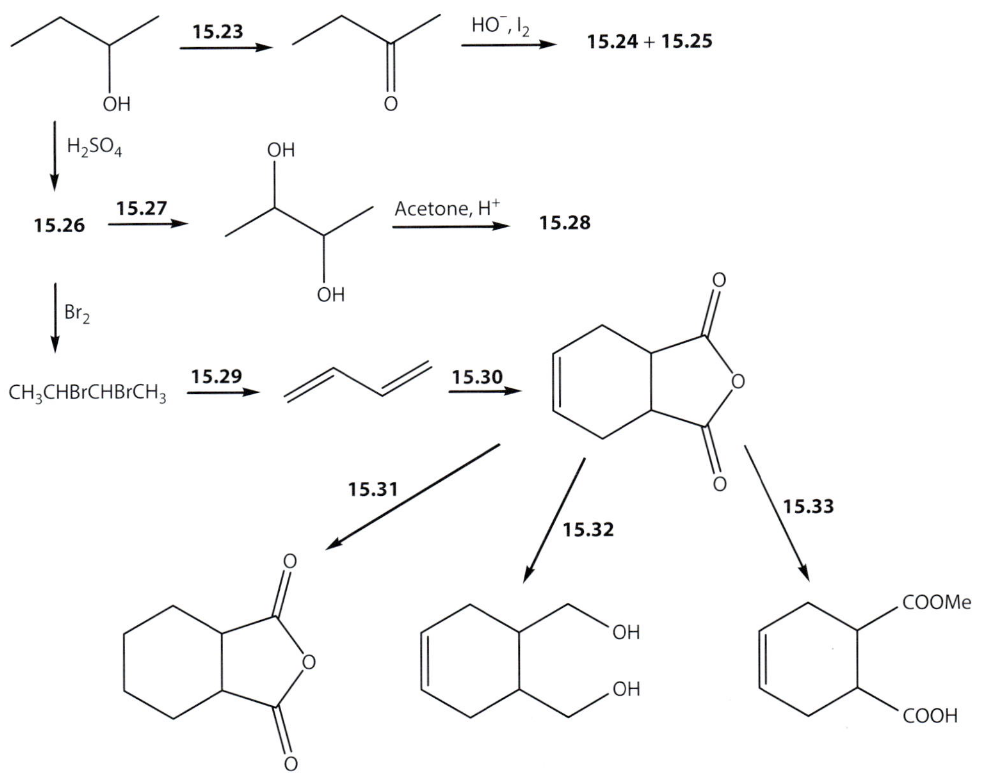

6. The synthesis of the barbiturate veronal is shown in the following:

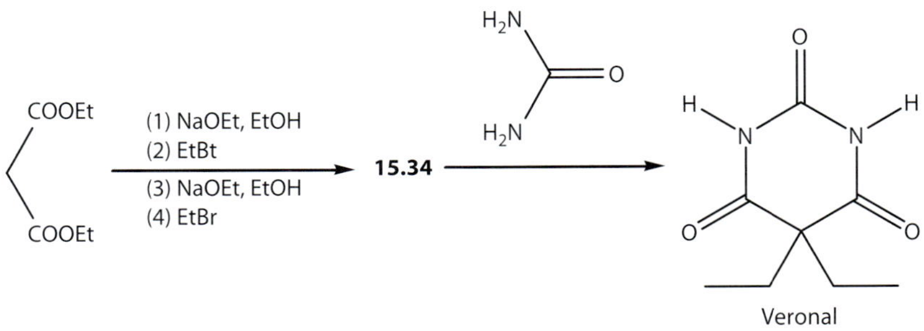

Write a mechanism for the process, identify **15.34**, and show how you could similarly synthesize Nembutal and Seconal:

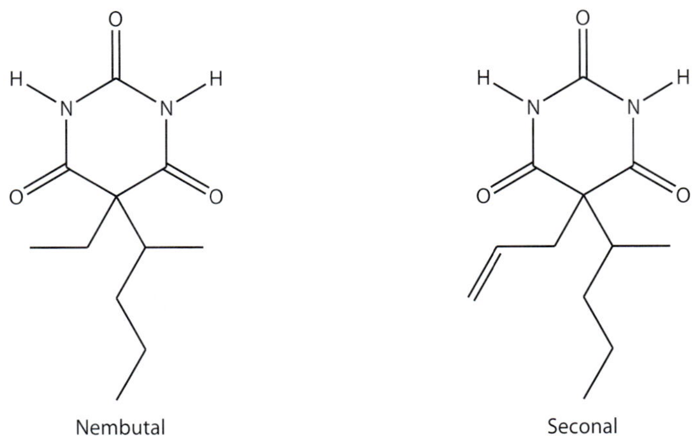

7. Give structures for each of the compounds in this sequence. Only the structures are required, but you may obtain partial credit for mechanisms heading in the right direction:

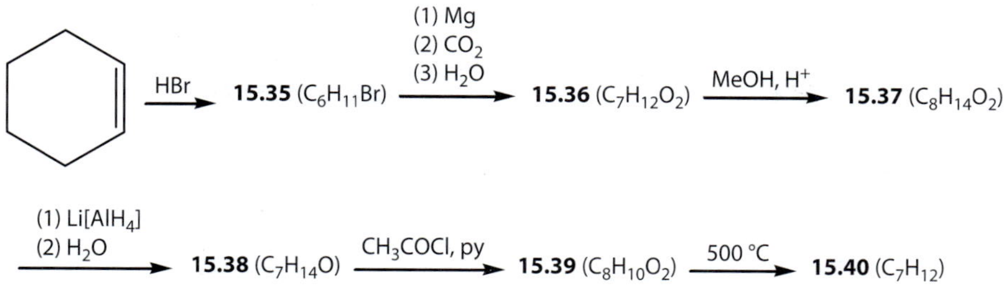

8. Another reagent that can be used to promote the direct coupling of carboxylic acids and amines to give amides is $B(OH)_3$. The first step in the process is assumed to be the formation of an intermediate such as **15.41**, a mixed anhydride between the carboxylic acid and the boric acid:

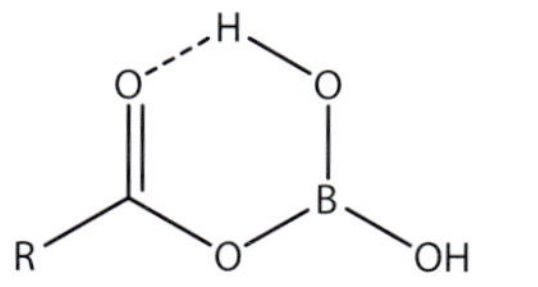

15.41

Explain why formation of this intermediate facilitates the reaction and write a complete mechanism for the amide formation shown:

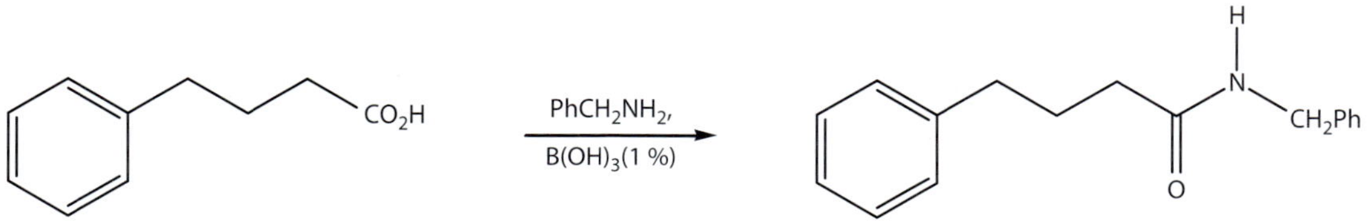

9. Explain the terms kinetic resolution, and desymmetrization, in the context of ester hydrolysis. Classify each of the following reactions, as to which type they are. In the second example, explain why the enantiomer excesses of the two products vary with the extent of conversion:

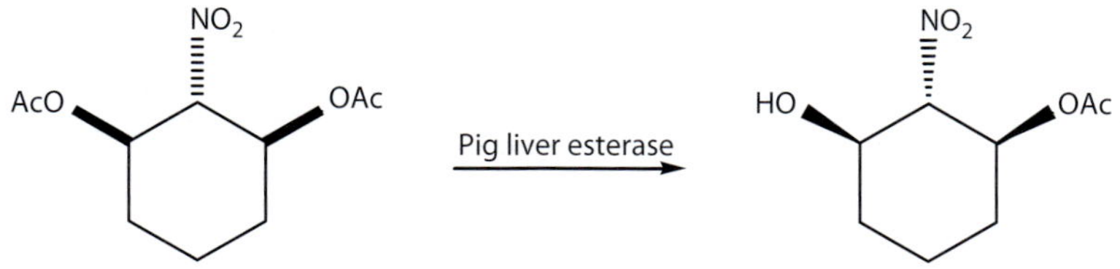

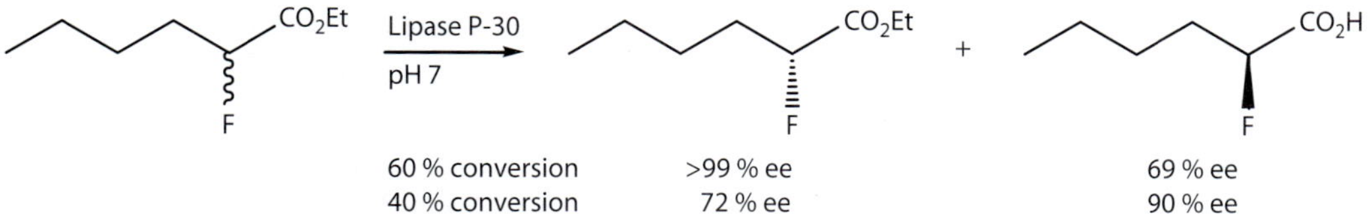

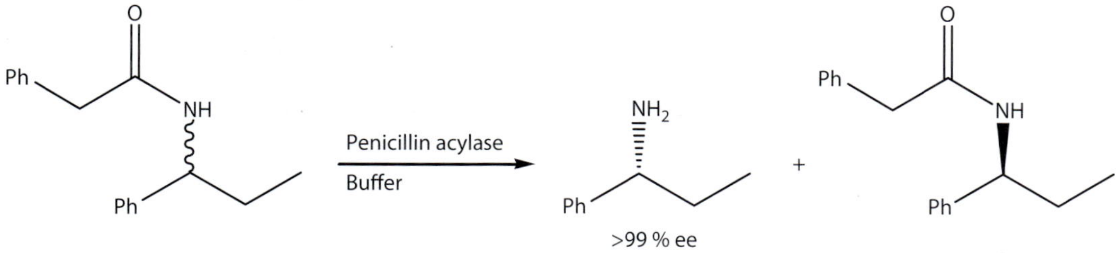

Lipids and Carbohydrates

16

16.1 INTRODUCTION

This chapter deals with two classes of natural products, lipids and carbohydrates; the chemistry involved is applied carbonyl and alcohol chemistry. Although the molecules we will meet are often larger and appear more complex than those we discussed in previous chapters, what happens to them is essentially the same as with the simpler molecules. We will meet some new reagents, especially in carbohydrate chemistry, but the need for these mostly reflects a difference in the solubility of the substrates. Most organic molecules are soluble in organic solvents, but carbohydrates have limited solubility in organic solvents and are very soluble in water. However, the *processes* they undergo have not changed.

16.2 LIPIDS

The term lipid technically means fat, but it is used to describe quite a wide range of related compounds, including fatty acids, steroids, prostaglandins, lipoproteins, sphingolipids, and phospholipids. Fatty acids and triglycerides are energy sources and used for energy storage as well as cell membrane construction. Steroids and prostaglandins fulfill many biological functions as chemical messengers. Some of these molecules, particularly the fatty acids, have a single polar "head group" and a long hydrocarbon "tail." Despite the head group, they are generally more soluble in nonpolar than in polar media and have surfactant properties, explored in more detail in Section 16.2.1. They are sometimes called amphiphiles, as there is one part of the molecule, the head group, that is best solubilized in water and one part, the hydrocarbon tail, that is more compatible with nonpolar solvents (Figure 16.1). This allows the formation of a range of structures including monolayers, bilayers, micelles, and vesicles, depending on the molecule, the solvent, and the concentration.

16.2.1 SURFACTANTS

Figure 16.1 shows a typical surfactant molecule, with a polar head group and a nonpolar tail. How these molecules arrange themselves depends on the medium to which they are added and their concentration. For example, if you put a drop of a surfactant onto the surface of water, it may form a monolayer where the head groups are in the aqueous phase, and the hydrocarbon chains are side by side in the air (Figure 16.2).

When the concentration of surfactant is higher, a micelle is formed; this has a hydrocarbon-like interior and a polar surface. It is a fluid structure, with molecules arriving and leaving all the time. Figure 16.3 shows the cross section of a typical micelle.

As the concentration of a surfactant in water increases, there is a critical point, called the critical micelle concentration (CMC), at which micelles form, and there is, at this point, a sharp change in physical properties such as viscosity and refractive index (Figure 16.4). As the solution becomes more concentrated, other phases are formed including cylindrical micelles and bilayers. In a neat surfactant phase, the main structure is a bilayer; bilayers are commonly found when

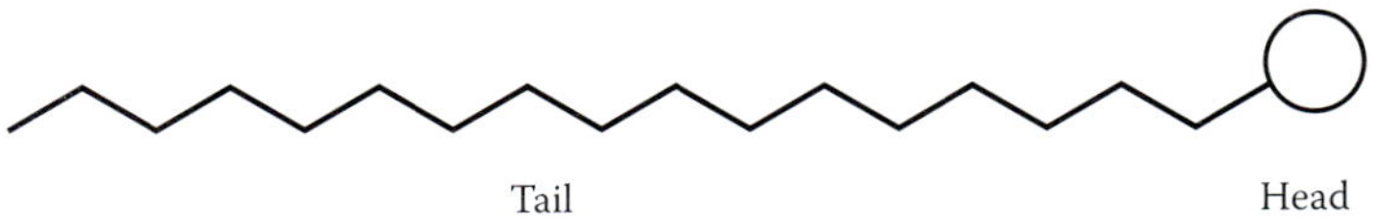

FIGURE 16.1 Surfactant molecule.

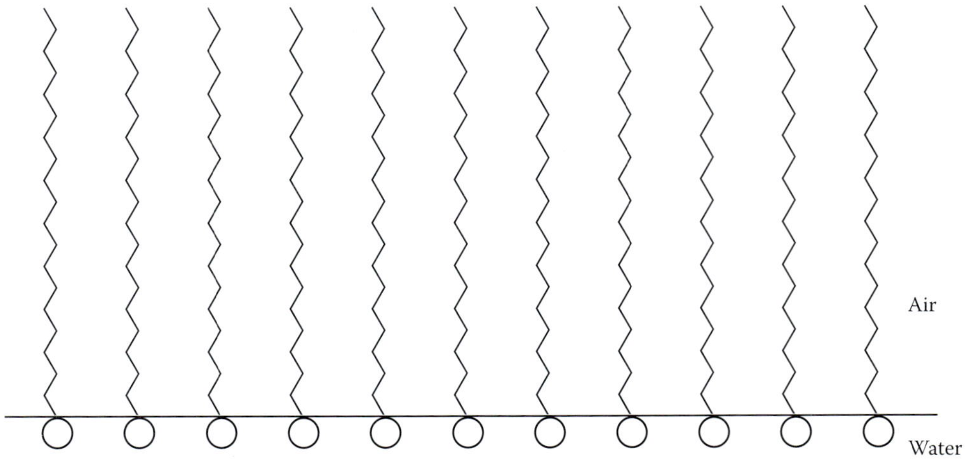

FIGURE 16.2 Monolayer of a surfactant on water.

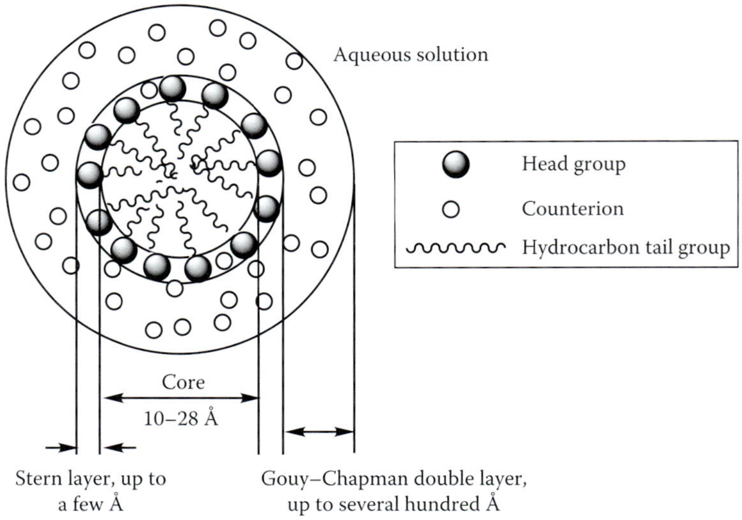

FIGURE 16.3 Schematic of a spherical ionic micelle.

the surfactant head group bears two hydrocarbon tails as in a phospholipid such as **16.1**. Such bilayers are involved in the basic structure of all cell membranes (Figure 16.5).

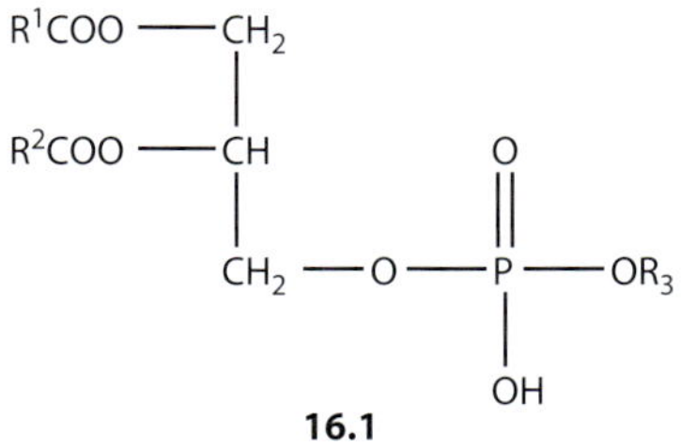

16.1

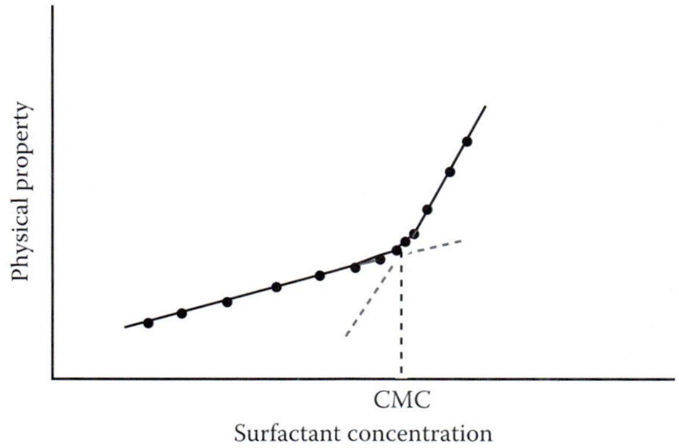

FIGURE 16.4 Formation of micelles gives rise to a discontinuity in a plot of physical properties against concentration.

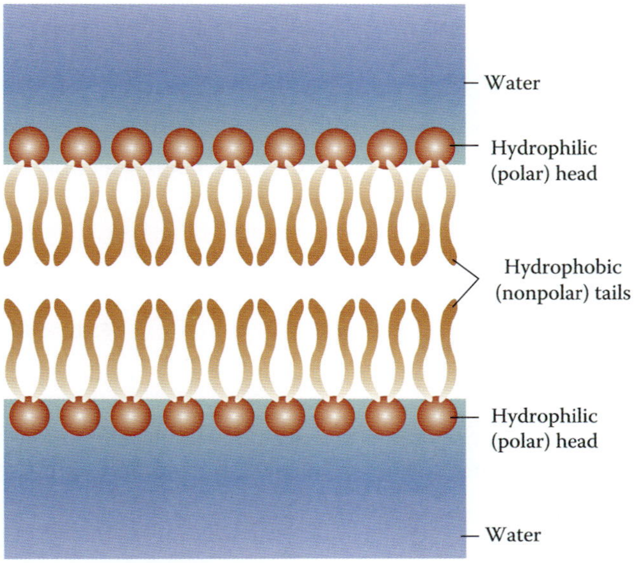

FIGURE 16.5 Bilayer structure of cell membranes.

Bilayers can fold in on themselves to form vesicles (Figure 16.6), where there is an aqueous phase both inside and outside the vesicle. Reactions inside the vesicle may be different from those in the bulk solvent, and vesicles are much longer lived than micelles. Certain types of vesicle have found uses in drug delivery.

16.2.2 FATTY ACIDS AND TRIGLYCERIDES

Triglycerides are the glycerol esters of long-chain fatty acids (Figure 16.7). In some triglycerides, all the fatty acids are the same—trimyristin, **16.2**, which is readily isolated from nutmeg in good quantity, is one such. However, more commonly triglycerides contain a variety of fatty acids. Triglycerides are hydrolyzed, either partially or completely, by enzymes called lipases. The metabolism of triglycerides provides twice as much energy as the same amount of carbohydrate or protein. We must consider the issue of nomenclature. Myristic acid is formally called tetradecanoic acid in the IUPAC system, but this is an area where common usage uses common names (not always even consistently), so it is important to recognize these. The acids commonly found in fats range from C_{12} to C_{18}. Most fatty acids produced by animals and plants are linear and have even numbers of carbon atoms because of their method of biosynthesis (see Chapter 17 for the Claisen condensation of acetyl coenzyme A). Bacteria can synthesize branched acids and those with odd

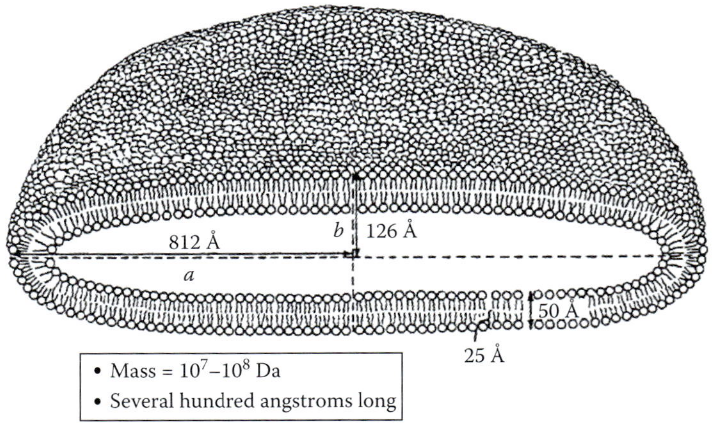

FIGURE 16.6 Schematic representation of a unilamellar vesicle. *Note*: Unilamellar means single walled; vesicles can form more complex structures with one inside another. (From Fendler, J.H., *Acc. Chem. Res.*, 13, 8, 1980, Fig. 2.)

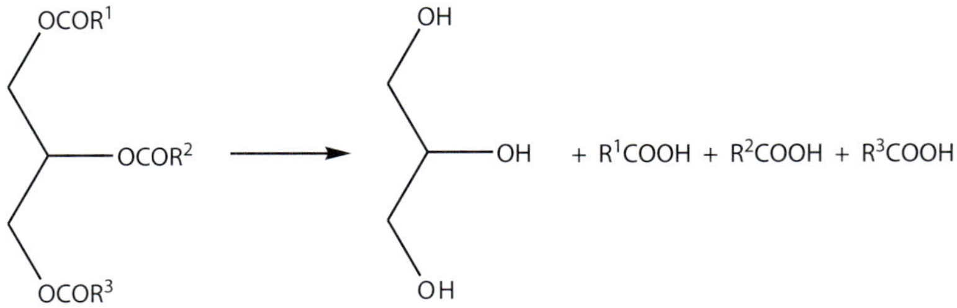

FIGURE 16.7 Hydrolysis of triglycerides.

numbers of carbon atoms, so these are sometimes found in fats from ruminants, having been synthesized by the bacteria in the rumen. Table 16.1 shows the common saturated fatty acids.

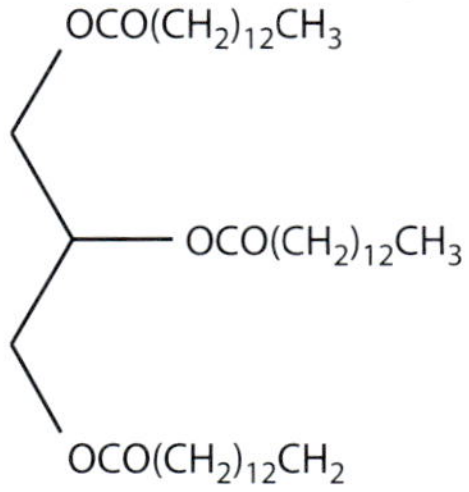

16.2, trimyristin

As we discussed briefly in Section 11.5, ratios of saturated and unsaturated fatty acids are of great concern from a dietary point of view. Because the introduction of *cis*-double bonds into the long chain causes a "kink," meaning that the long hydrocarbon chains pack together less well, they have lower melting points than saturated fats and are generally oils at room temperature. Most naturally occurring unsaturated acids have *cis*-double bonds, and where there is more than one, the "skipped" arrangement (as in **16.3**, linolenic acid) is the norm. Another term you may have heard in describing fatty acids is ω-3 (or some other number). This term describes where you would find the first double bond if you were going down the chain from the hydrocarbon end—in other words, in the opposite direction from IUPAC numbering. ω is the last letter of the Greek alphabet and hence a designation for the "end" of the molecule. So linolenic acid is an ω-3 fatty acid. Mammals require, but cannot synthesize, ω-3 or ω-6 fatty acids (they can put double bonds into the chain up to C-9 but not beyond, so ω-9 acids are not considered to be essential), so linolenic and linoleic acids are described as "essential fatty acids"; they must be obtained from the diet. Strictly, these are the only "essential" acids as they can be converted *in vivo* to arachidonic acid,

TABLE 16.1 Common Saturated Fatty Acids

Common Name	Formula	IUPAC Name	Commonly Found in	Uses
Capric acid	$CH_3(CH_2)_8CO_2H$	Decanoic acid	Coconut oil, palm kernel oil, milk	Food additive, cosmetics
Lauric acid	$CH_3(CH_2)_{10}CO_2H$	Dodecanoic acid	Coconut oil, palm kernel oil, milk	Soap, detergents, cosmetics
Myristic acid	$CH_3(CH_2)_{12}CO_2H$	Tetradecanoic acid	Nutmeg, palm kernel oil, coconut oil, butterfat, spermaceti	Isopropyl myristate is used in medical and cosmetic preparations for good skin penetration. Perfume and food additive
Palmitic acid	$CH_3(CH_2)_{14}CO_2H$	Hexadecanoic acid	Palm oil, coconut oil, butter, cheese, milk, meat	Soap (sodium palmitate), cosmetics, food additives, napalm
Stearic acid	$CH_3(CH_2)_{16}CO_2H$	Octadecanoic acid	Many animal and vegetable fats, especially cocoa butter and shea butter	Plastics, cosmetics, soap (sodium stearate), candles, lubricants
Arachidic acid	$CH_3(CH_2)_{18}CO_2H$	Eicosanidic acid	Peanut oil	Soaps, plastics, waxes, lubricants
Behenic acid	$CH_3(CH_2)_{20}CO_2H$	Docosanoic acid	Ben tree oil, rapeseed, peanut oils	Cosmetics, plasticizers, waxes
Lignoceric acid	$CH_3(CH_2)_{22}CO_2H$	Tetracosanoic acid	Wood tar, rice bran wax	Constituent of cerebrosides and sphingomyelin
Cerotic acid	$CH_3(CH_2)_{24}CO_2H$	Hexacosanoic acid	Beeswax, carnauba wax	Cosmetics, wax polishes

eicosapentaenoic acid (EPA), and DHA. This is not a particularly efficient process, however, and diets rich in these higher fatty acids are recommended for good health. Cats lack the enzymes to convert linolenic acid to these longer-chain acids and must obtain these from the diet; hence, they are described as obligate carnivores, as these acids cannot be obtained from plants. Some common mono- and polyunsaturated fatty acids are detailed in Table 16.2.

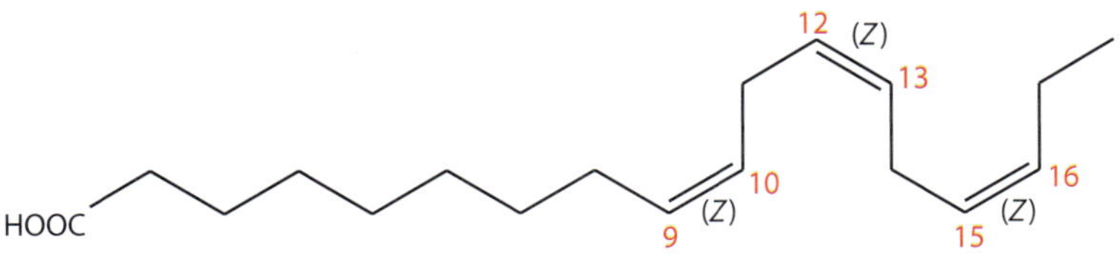

16.3, linolenic acid

The main fatty acid in olive oil is oleic acid. (Thanks to Camilla Tornoe for this photo.)

TABLE 16.2 Unsaturated Fatty Acids

Common Name	Formula	IUPAC Name	ω	Found in	Used for
Myristoleic acid	*cis*-$CH_3(CH_2)_3CH{=}CH(CH_2)_7CO_2H$	*Z*-Tetradec-9-enoic acid	5	Seed oil from ***Myristaceae*** species	
Palmitoleic acid	*cis*-$CH_3(CH_2)_5CH{=}CH(CH_2)_7CO_2H$	*Z*-Hexadec-9-enoic acid	7	Macadamia oil, sea buckthorn oil, human adipose tissue	
Oleic acid	*cis*-$CH_3(CH_2)_7CH{=}CH(CH_2)_7CO_2H$	*Z*-Octadec-9-enoic acid	9	Olive oil, pecan, peanut, grape seed, soya, sesame oil, animal fats, human adipose tissue	Pharmaceuticals, food additives, cosmetics
Linoleic acid	*cis,cis*-$CH_3(CH_2)_4CH{=}CHCH_2CH{=}CH(CH_2)_7CO_2H$	*Z,Z*-9,12-Octadecadienoic acid	6	Poppy seed, safflower, sunflower, corn oils	Soaps, emulsifiers, cosmetics
Linolenic acid	*cis,cis,cis*-$CH_3CH_2CH{=}CHCH_2CH{=}CHCH_2CH{=}CH(CH_2)_7CO_2H$	*Z,Z,Z*-9,12,15-Octadecatrienoic acid	3	Rapeseed, soya, flax, walnuts, hemp, chia sage	Food additive, oil paint, varnishes
Arachidonic acid	$CH_3(CH_2)_4(CH{=}CHCH_2)_4CH_2CH_2CO_2H$	5*Z*,8*Z*,11*Z*,14*Z*-5,8,11,14-Eicosatetraenoic acid	6	Brain, muscle, liver	Biological precursor of prostaglandins
all-cis-Eicosapentaenoic acid (EPA), timnodonic acid	$CH_3CH_2(CH{=}CHCH_2)_5CH_2CH_2CO_2H$	5*Z*,8*Z*,11*Z*,14*Z*,17*Z*-Eicosa-5,8,11,14,17-pentaenoic acid	3	Oily fish—cod, herring, mackerel, salmon, sardine	Anti-inflammatory, benefits to mental health, and coronary artery disease
all-cis-Docosahexaenoic acid (DHA), cervonic acid	$CH_3CH_2(CH{=}CHCH_2)_6CH_2CH_2CO_2H$	4*Z*,7*Z*,10*Z*,13*Z*,16*Z*,19*Z*-Docosa-4,7,10,13,16,19-hexaenoic acid	3	Cold water fatty ocean fish, brain phospholipids, retina	Food additive

Salmon are an excellent source of omega-3 fatty acids

PROBLEM 16.1

An optically active triglyceride, when completely hydrolyzed, yields twice as much stearic as palmitic acid. What is its structure?

Solution

Two of the ester groups are formed with stearic and one with palmitic acid. In order to have an asymmetric carbon atom, the structure must be the one with lower symmetry:

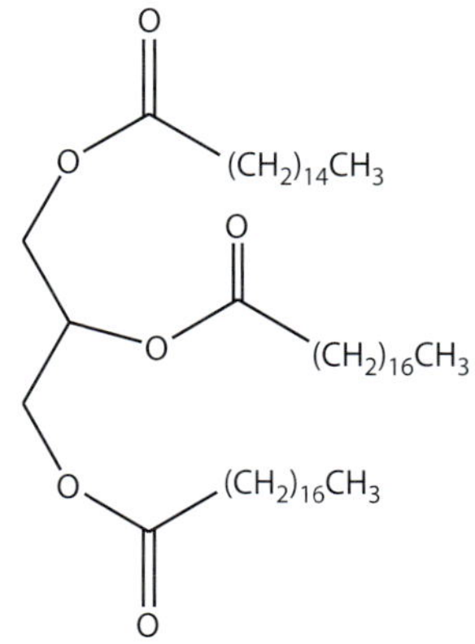

16.2.3 PROSTAGLANDINS

Prostaglandins, together with prostacyclins and thromboxanes, make up the class of eicosanoids, biochemically active molecules all derived biosynthetically from arachidonic acid. All are functionalized derivatives of prostanoic acid, **16.4**. Various prostaglandins regulate inflammation, platelet aggregation and disaggregation, vasodilation, calcium transport, cell growth, and smooth muscle contraction. An outline of the biosynthesis is shown in Figure 16.8. The first chemical syntheses of $PGF_{2\alpha}$ and PGE_2 were reported by Corey in 1969, and in 1971, it was established that aspirin (and other nonsteroidal anti-inflammatory compounds such as ibuprofen and naproxen) inhibits prostaglandin production by blocking the cyclooxygenase enzymes.

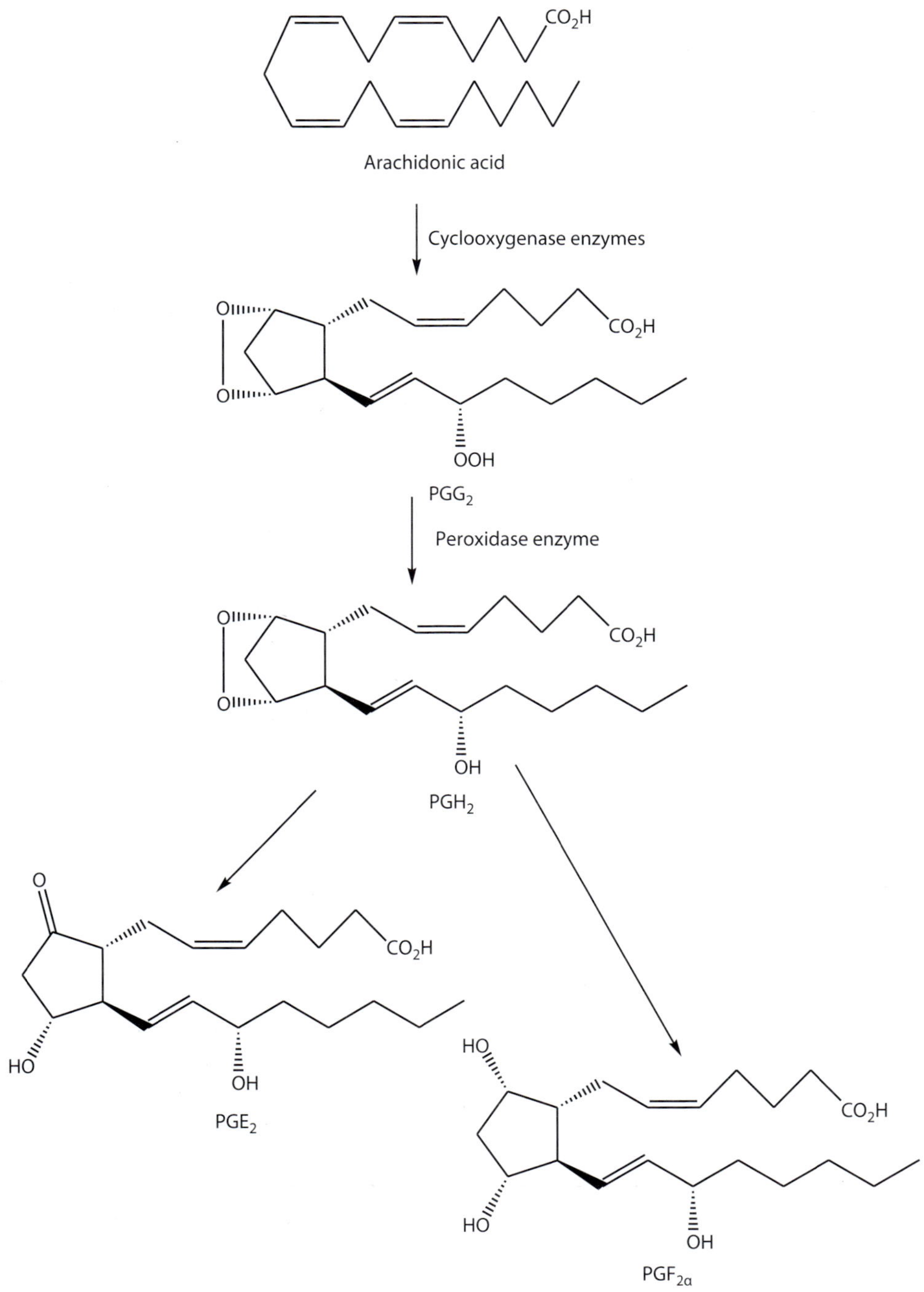

FIGURE 16.8 Prostaglandin biosynthesis.

The synthesis was a challenging one; the stereochemistry of the double bonds and the four or five chiral centers is critical to the activity.

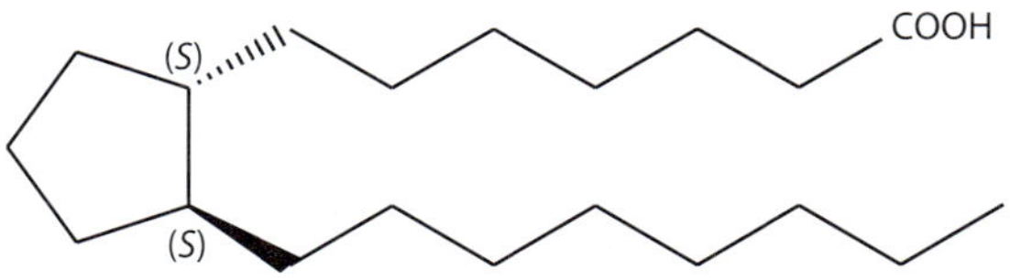

16.4

Prostaglandins have found widespread uses in medicine because of their control of physiological and biochemical functions. A number of these are detailed in Table 16.3.

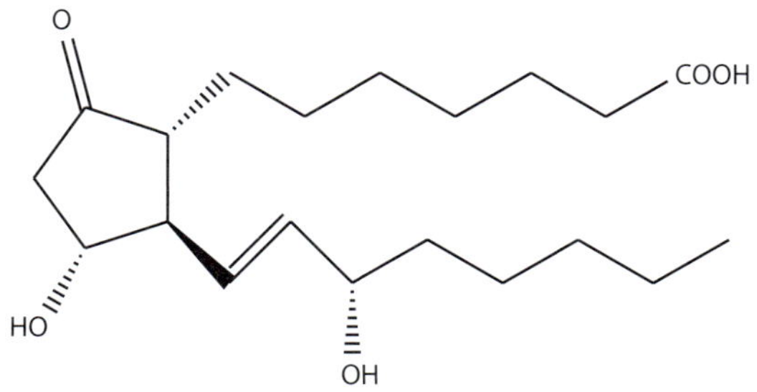

16.5, PGE_1

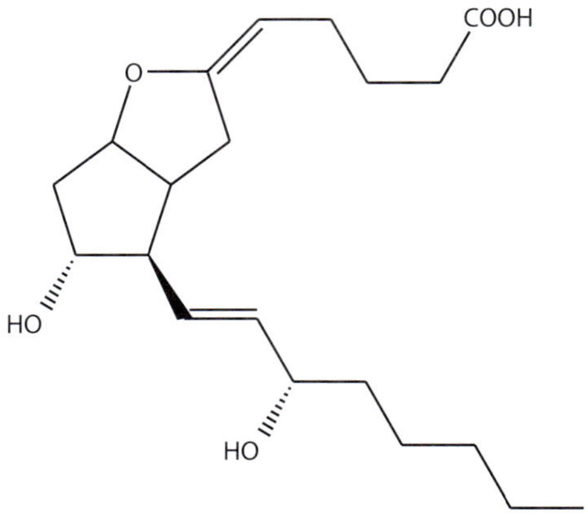

16.6, prostacyclin I_2

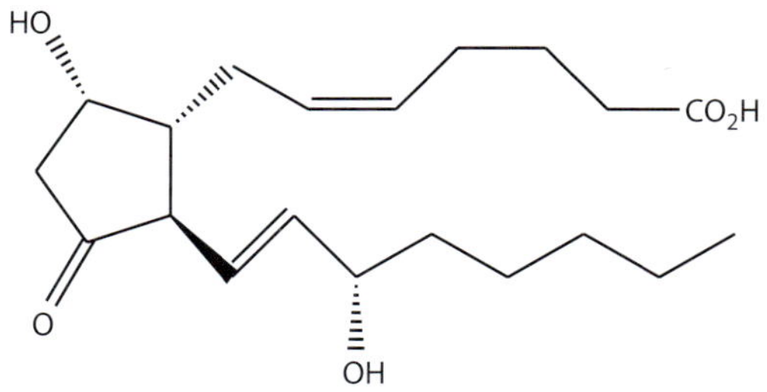

16.7, PGD_2

TABLE 16.3 Uses of Prostaglandins

Prostaglandin	Physiological Function	Medicinal Use
PGE_2	Bronchoconstriction, gastric mucus secretion Uterus contraction, GI tract smooth muscle contraction and relaxation	Initiation and acceleration of labor
PGE_1, **16.5**	Vasodilation	Alprostadil, treatment of erectile dysfunction
$PGF_{2\alpha}$	Bronchoconstriction, uterus contraction	Induction of labor, abortifacient
Prostacyclin I_2, **16.6**	Vasodilation, inhibition of clotting, bronchodilation	Epoprostenol sodium for primary pulmonary hypotension
PGD_2, **16.7**	Found in mast cells, critical in development of allergic asthma	

PROBLEM 16.2

Study the synthetic sequence in the following, which is part of a longer prostaglandin synthesis, and answer the questions posed:

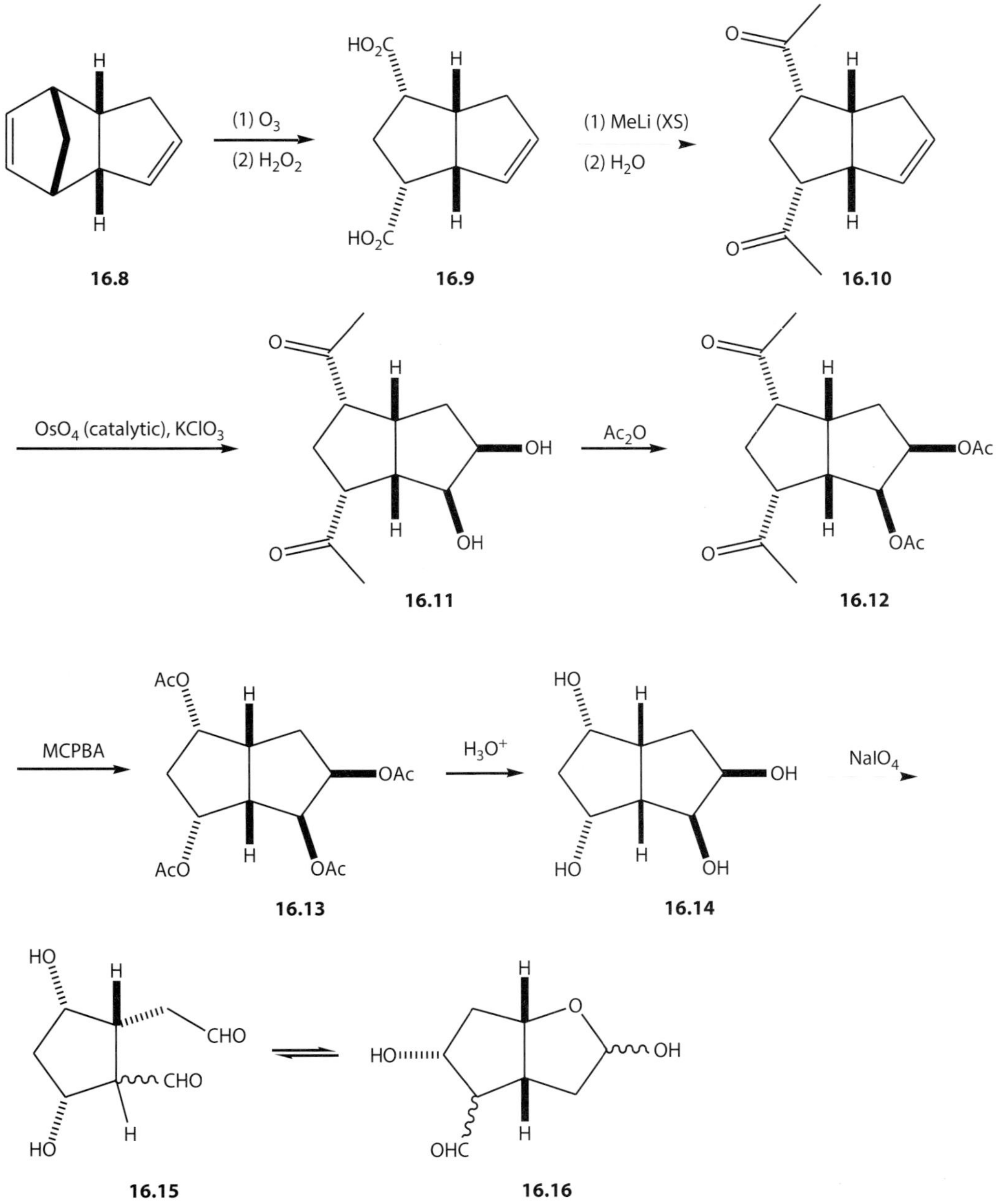

(a) **16.8** is the product of a Diels–Alder reaction. What species act as the diene and the dienophile?

(b) In the conversion of **16.8** to **16.9**, can you foresee a problem with the selectivity of the reaction?

(c) In the conversion of **16.8** to **16.9**, what would have the product been if the intermediate ozonide has been reacted with Me_2S rather than H_2O_2?

(d) Write a mechanism for the reaction of a carboxylic acid with methyl lithium as in the transformation of **16.9** to **16.10**.

(e) What selectivity problem might you anticipate in the conversion of **16.10** to **16.11**?
(f) Write a mechanism representative of the transformation of **16.13** to **16.14**.
(g) Write a mechanism for the interconversion of **16.15** and **16.16**.

Solution

(a) This is the product of the reaction of two molecules of cyclopentadiene; one acts as the diene and the other as the dienophile. The stereochemistry of the product derives from the *endo*-rule:

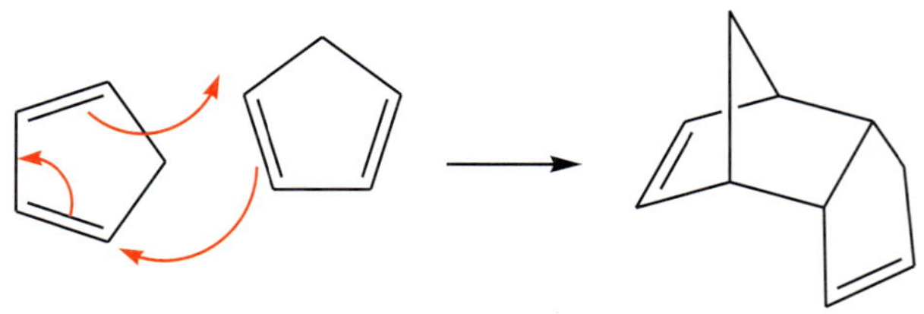

(b) Only one of the double bonds is reacting with ozone, probably because it is the more strained.
(c) With a reductive rather than an oxidative work-up, a dialdehyde would be obtained.
(d)

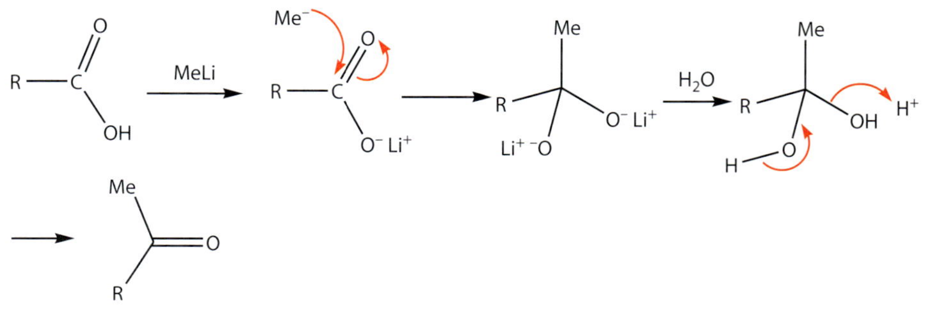

(e) This is an issue of which face will be attacked by the OsO_4—we know that the two OH groups will be *cis*- to each other, but will they be up or down?
(f)

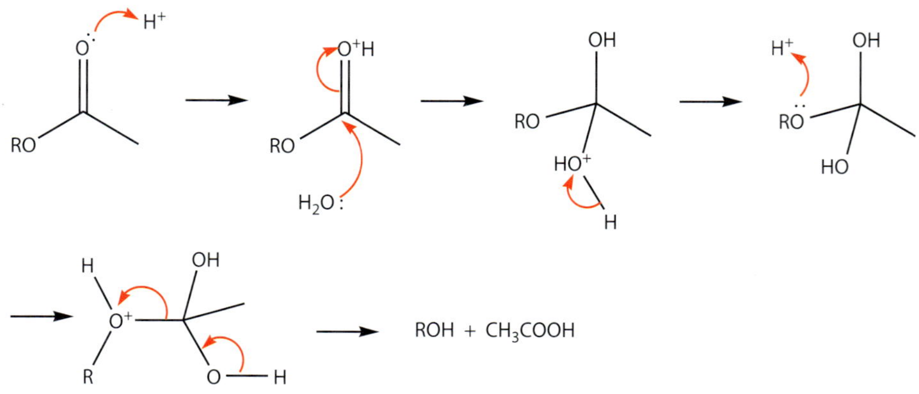

(g) This is quite tricky; an alcohol has clearly attacked an aldehyde, but which? The reaction proceeds as shown because this is the only way to make a five-membered ring, which is not part of a bicyclic system. Other reactions may happen, but all of the processes forming a hemiacetal are reversible, so we will end up with the most stable one:

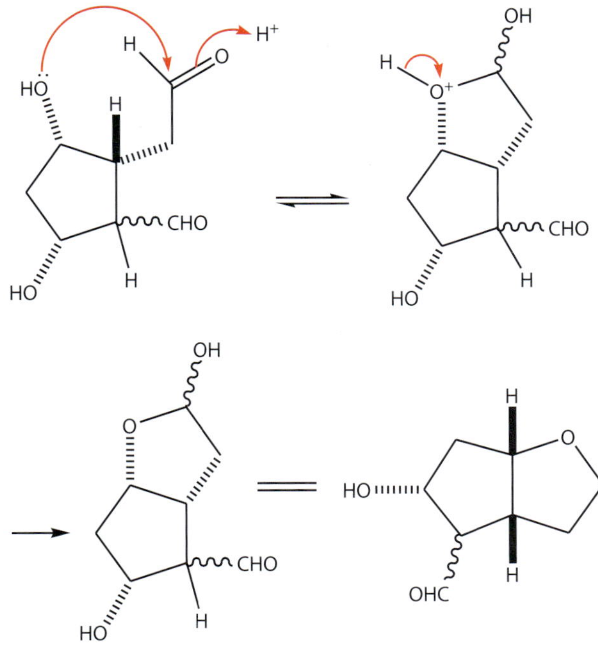

16.2.4 WAXES

The word wax is derived from the Anglo-Saxon *weax*, meaning beeswax. But there is no simple definition of a wax. In a material science sense, it is defined by its properties as a material that softens at or near room temperature, melting above 45 °C to give a low-viscosity liquid. These criteria include many synthetic waxes or waxes derived from petrochemicals. Organic chemists are generally more concerned with naturally occurring waxes, where the main constituents are esters formed between long-chain fatty acids (C_{16}–C_{36}) and long-chain alcohols (C_{16}–C_{36}). These materials are insoluble in water, and most are quite resistant to hydrolysis. All natural waxes are mixtures of compounds, mostly esters. Some are very complex—in lanolin, derived from sheep wool, esters from 200 acids and 100 alcohols have been identified. One of the most common alcohols in lanolin is lanosterol, a steroidal alcohol. In some other examples, one or two components are predominant. Some examples, together with their uses, are shown in Table 16.4.

Beeswax candles

TABLE 16.4 Naturally Occurring Waxes

Common Name	Origin	Main Components	Uses
Beeswax	Bees	Melissyl palmitate, $CH_3(CH_2)_{14}COO(CH_2)_{29}CH_3$. Cerotic acid.	Candles, furniture polish, surf wax, cosmetics
Chinese wax	*Ceroplastes ceriferus*	Ceryl lignocerate, montanyl lignocerate, melissyl lignocerate, ceryl cerotate, ceryl montanate.	Candles, polishes, lubricant, Chinese traditional medicine
Spermaceti	Sperm whale	Cetyl palmitate, $CH_3(CH_2)_{14}COO(CH_2)_{15}CH_3$.	Cosmetics, candles, pharmaceuticals
Lanolin	Sheep wool	Very complex mixture. Forty percent of the fatty acids are α-hydroxy acids. Some sterol esters.	Lubricants, cosmetics, water repellants
Bayberry wax	*Myrica faya*	Esters of lauric, myristic, palmitic acid.	Candles, distinctive fragrance
Carnauba wax	Carnauba palm, *Copernicia prunifera confera*	Melissyl cerotate, $CH_3(CH_2)_{24}COO(CH_2)_{29}CH_3$. ω-Hydroxy acids, C_{26}–C_{30}.	Coatings, car polish, boat wax, shoe polish, surf wax, cosmetics, paper coatings
Rice bran wax	Rice bran	Esters of palmitic, behenic, lignoceric, and higher acids with cetyl alcohol, melissyl alcohol.	Cosmetics, paper and textile coating, waterproofing
Jojoba oil	Jojoba, *Simmondsia chinensis*	C_{18}, C_{20} (70 %), C_{22} acids (mostly containing double bonds, meaning an MPt of 7 °C) esters with C_{20}, C_{22} alcohols.	Cosmetics, fungicide, oxidation resistant
Castor wax	Castor oil tree	Hydrogenation product of the triglyceride of ricinoleic acid, *cis*-$CH_3(CH_2)_5CH(OH)CH_2CH{=}CH(CH_2)_7CO_2H$.	Cosmetics, polishes, lubrication

Cetyl alcohol, $CH_3(CH_2)_{15}OH$; melissyl alcohol, $CH_3(CH_2)_{29}OH$ (also myricyl alcohol); ceryl alcohol, $CH_3(CH_2)_{25}OH$; montanyl alcohol, $CH_3(CH_2)_{27}OH$; montanic acid, $CH_3(CH_2)_{26}COOH$.

Lipsticks contain pigments, oils, and waxes

PROBLEM 16.3

Compound **16.17** is one of a class of compounds called an estolide and can be isolated from the waxy coating on some conifer species. Which compounds would be obtained by complete hydrolysis of this molecule? Give systematic names for each.

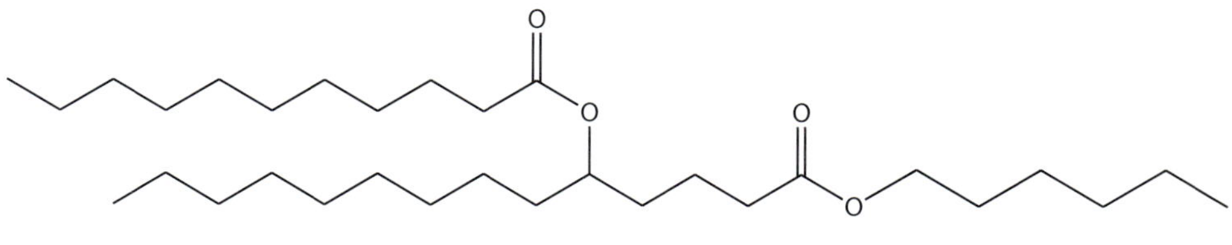

16.17

Solution

We would obtain one linear carboxylic acid, one hydroxy acid, and one alcohol:

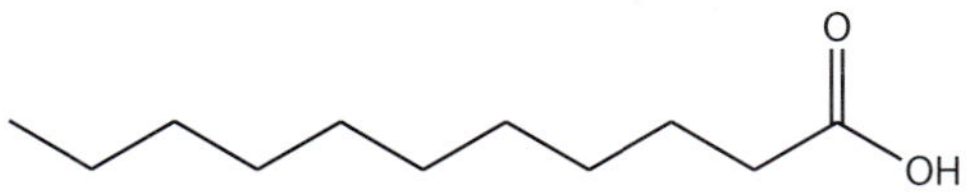

Undecanoic acid

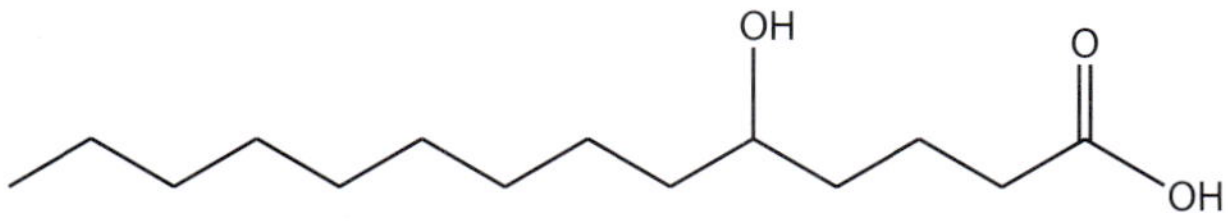

5-hydroxytetradecanoic acid

1-hexanol

16.2.5 STEROIDS

Steroids are lipids that (according to IUPAC) possess the skeleton of cyclopenta[a]phenanthrene or a skeleton derived from this by one or more bond scissions, ring expansions, or contractions. The four rings are referred to as A, B, C, and D as shown. The basic structure is that of cholestane, **16.18**. Sterols have a hydroxyl group at C-3. The stereochemistry at C-5 is variable—steroids with the hydrogen below the plane of the paper are described as α, those with it above β. Steroids are synthesized by animals, plants, and fungi and fulfill a wide range of roles, from sex hormones, through controllers of inflammation, to plant defense compounds. Numerous synthetic steroids have been prepared, and many have useful medicinal properties.

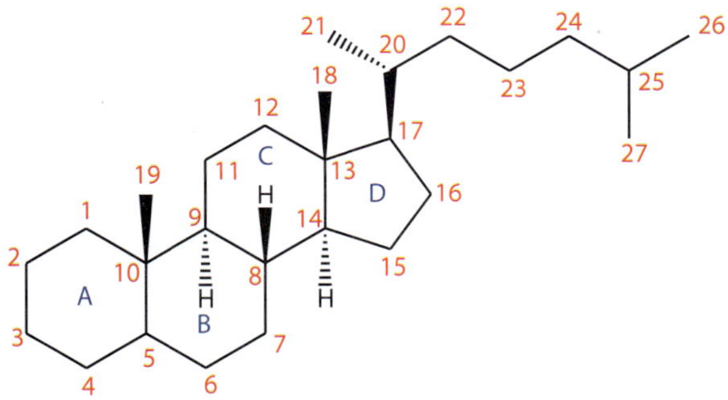

16.18

All steroids are biosynthesized from the triterpene squalene, which is converted to an epoxide by the enzyme squalene 2,3-epoxidase. In mammals, this is converted to lanosterol by 2,3-oxidosqualene–lanosterol cyclase. You should find the first step of the process fairly easy to follow, as it is simply a cascade of electrophilic additions to double bonds. The next step is a cascade of rearrangements—come back and look at that again, once you have studied rearrangements in Chapter 18, as at this point, it may seem a bit mysterious (Figure 16.9). There are 7 stereocenters in lanosterol, which gives a potential for 128 stereoisomers, but a single stereoisomer is produced by the enzymatic reaction.

Lanosterol is converted into cholesterol, **16.19**, in 19 (!) steps catalyzed by enzymes in the endoplasmic reticulum. In the liver, cholesterol is converted into cholic acid, **16.20**, and other bile acids. In various other tissues, it is converted *via* pregnenolone, **16.21**, to all the steroid hormones. Formally, there are five classes of these:

- Progestins—precursors of the others
- Mineral corticoids—includes aldosterone, and regulate ion balance in the kidneys
- Glucocorticoids—promote glucose synthesis (gluconeogenesis) and suppress inflammation
- Androgens—male sex hormones
- Estrogens—female sex hormones, generally formed from one of the androgens by an aromatase enzyme (Figure 16.10)

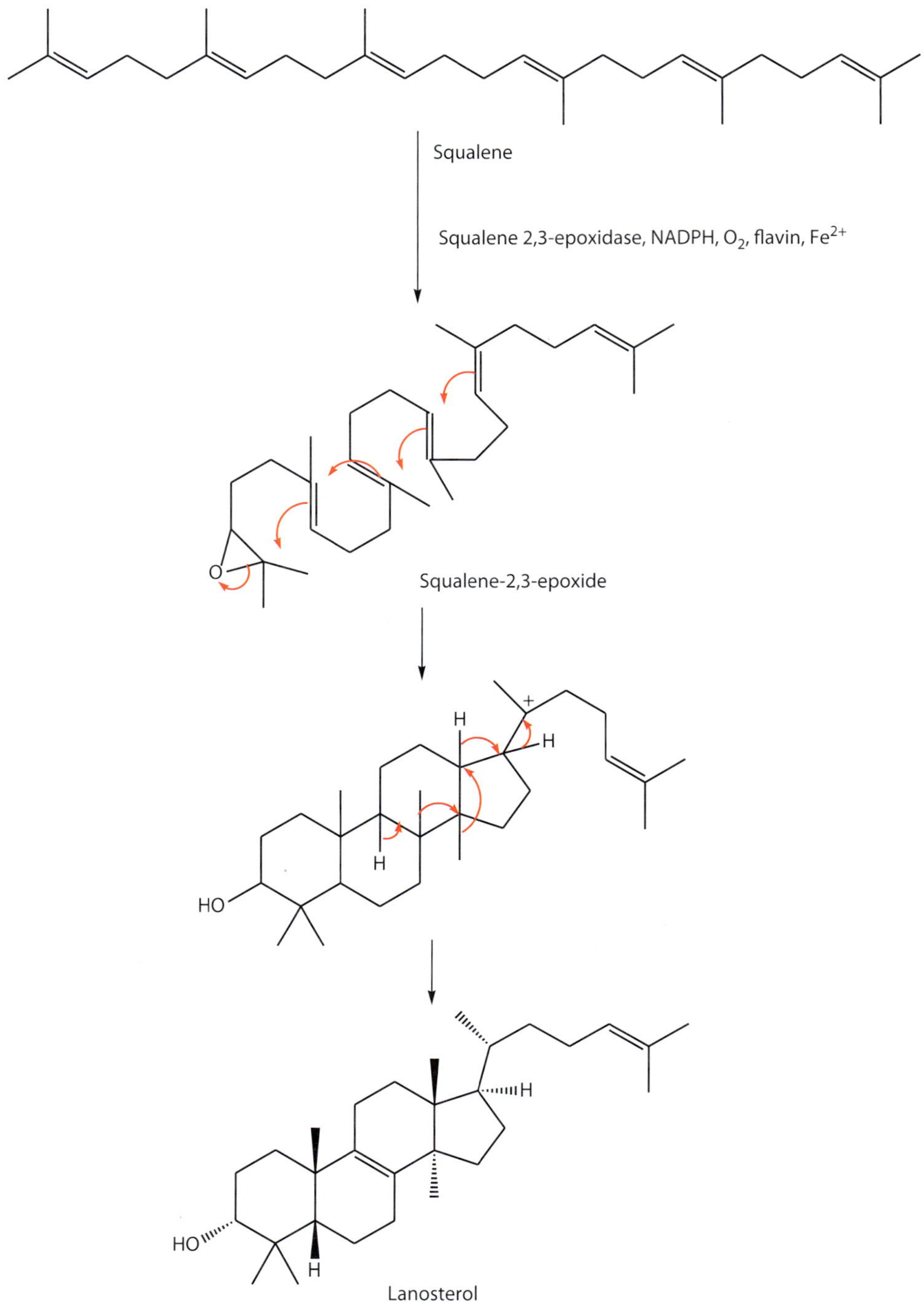

FIGURE 16.9 Biosynthesis of lanosterol in mammals and fungi.

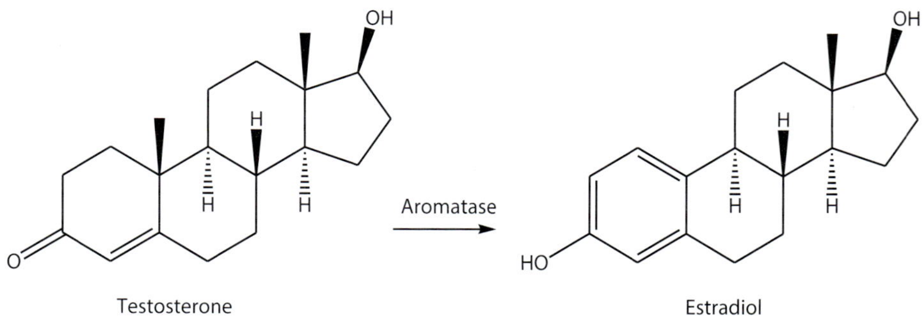

FIGURE 16.10 Conversion of testosterone to estradiol.

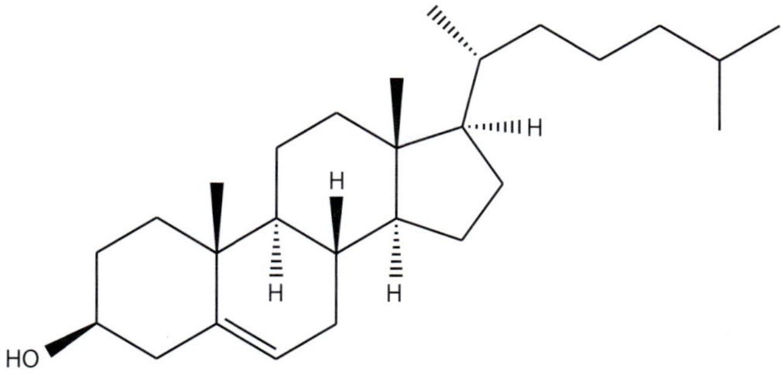

16.19, cholesterol

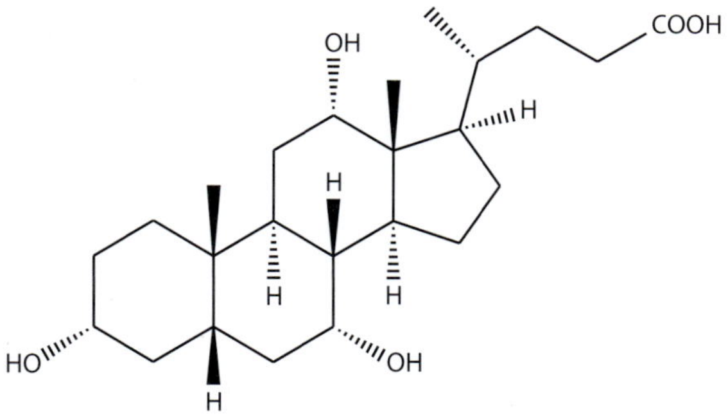

16.20, cholic acid

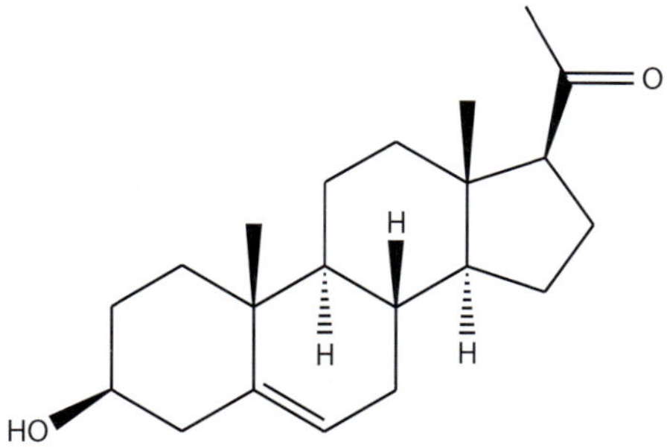

16.21, pregnenolone

Table 16.5 details some of the uses and functions of the steroid hormones, but this list is far from comprehensive—steroids turn up everywhere in control of physiology and in medicinal chemistry.

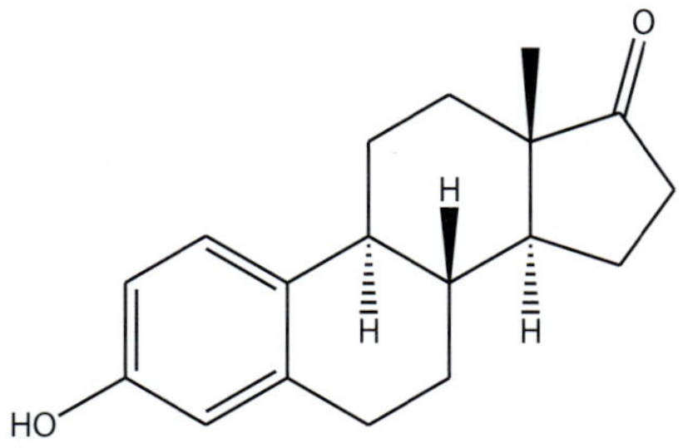

16.22, estrone

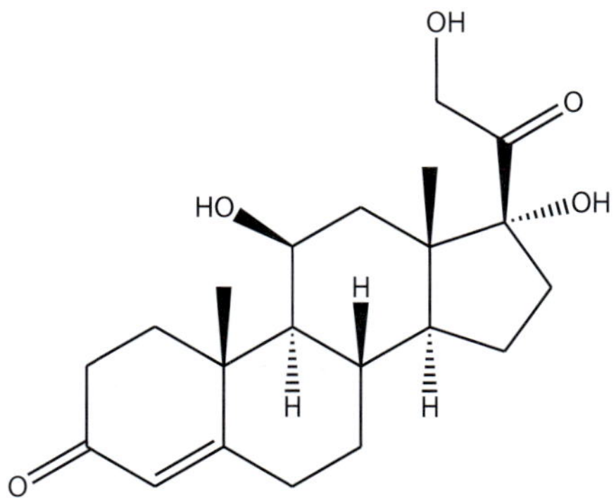

16.23, cortisol

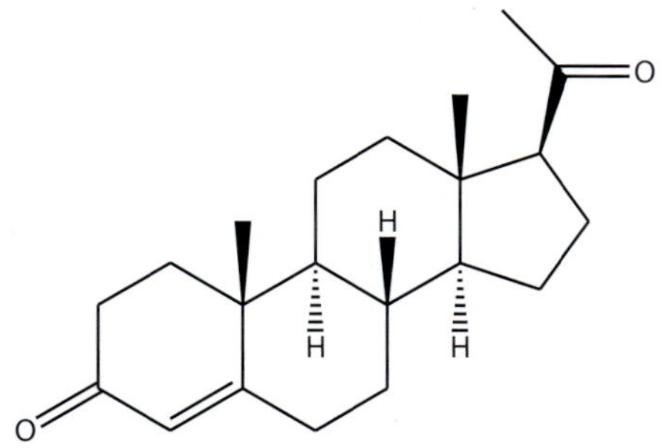

16.24, progesterone

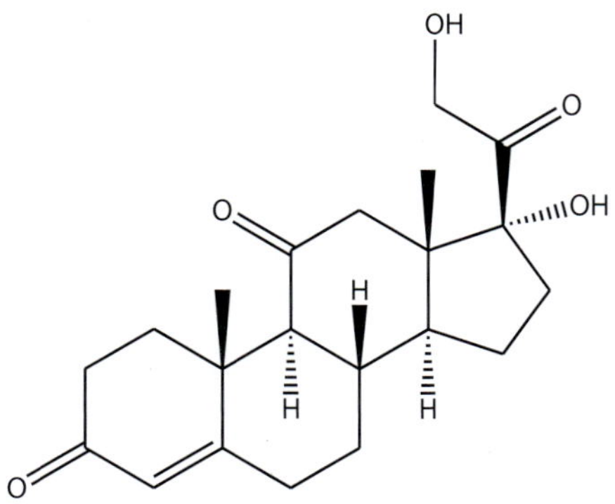

16.25, cortisone

TABLE 16.5 Uses and Functions of Steroid Hormones

Name	Natural or Synthetic?	Function	Use
Estradiol	Natural	Female sex hormone	HRT[a]
Estrone, **16.22**	Natural	Female sex hormone	
Testosterone	Natural	Male sex hormone Anabolic steroid	TRT,[b] hypogonadism treatment
Cortisol, **16.23**	Natural	Increases glucose and glycogen production	Immunosuppressive, used in treating anaphylaxis
Cortisone, **16.25**	Natural		Anti-inflammatory, treatments for rheumatoid arthritis, asthma, skin allergies
Progesterone, **16.24**	Natural	Supports pregnancy	Support of pregnancy, treatment of male to female transsexuals
Prednisolone	Synthetic		Anti-inflammatory, treatment of asthma, allergic reactions, immunosuppressant
Dexamethasone	Synthetic		Anti-inflammatory, immunosuppressant, 20–30 times as potent as cortisol, 4–5 times more effective than prednisolone. Treatment of rheumatoid arthritis, multiple myeloma
17-α-Ethynylestradiol, **16.26**	Synthetic		Oral contraceptive
Depo-Provera, **16.27**	Synthetic		Long-term injectable contraceptive, management of endometriosis

[a] HRT, Hormone replacement therapy.
[b] TRT, Testosterone replacement therapy.

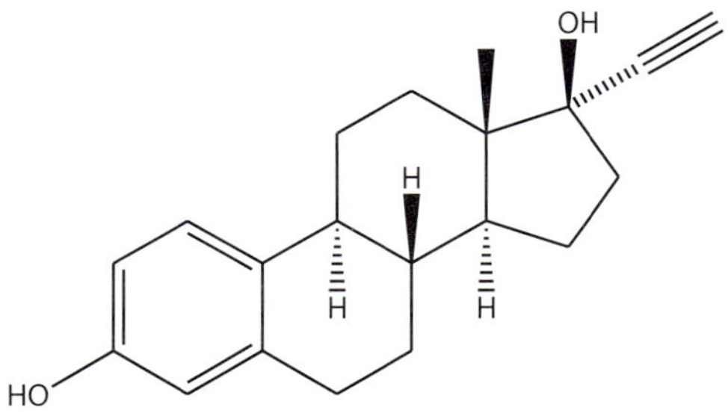

16.26, 17-ethynyl estradiol

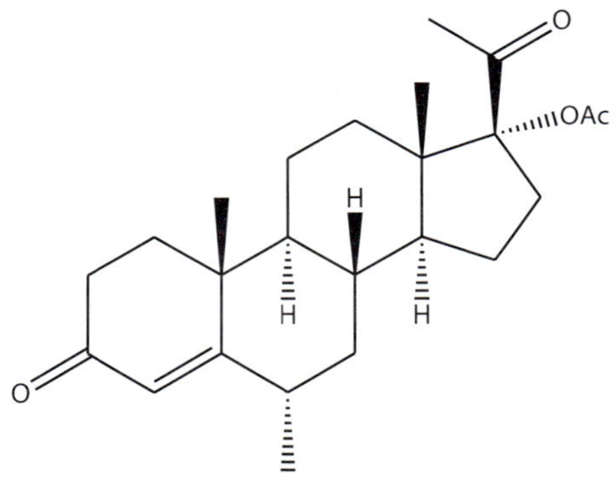

16.27, Depo-Provera

In plants, the squalene initially cyclizes in the same way, but the fate of the intermediate cation is different, as it is converted to cycloartenol, **16.28**, rather than lanosterol. Cycloartenol is converted on to a range of phytosterols, phytostanols, and related compounds, depending on the plant species involved. Some examples are given in Figure 16.11.

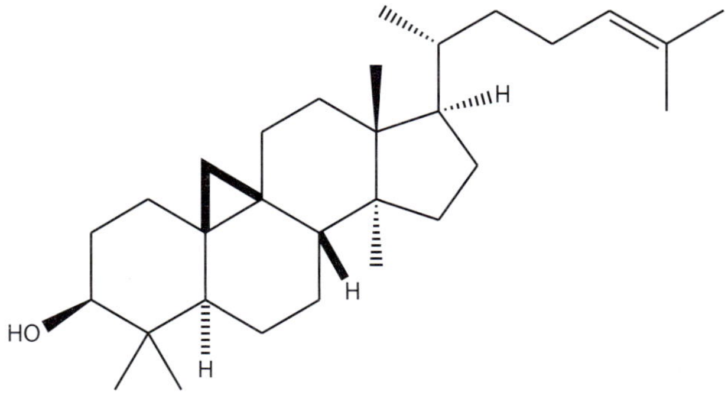

16.28, cycloartenol

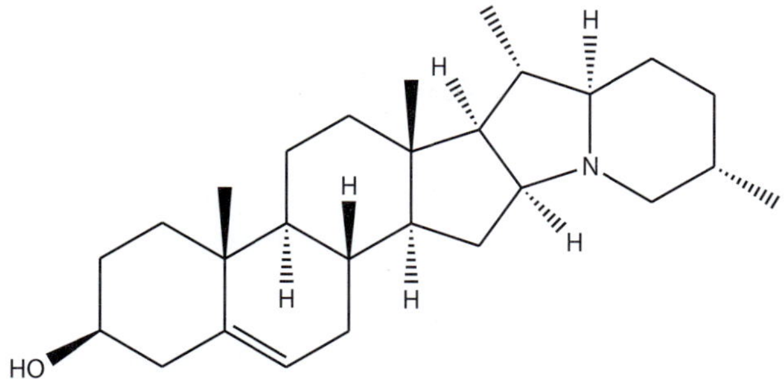

Solanidine, toxin from green potato tubers

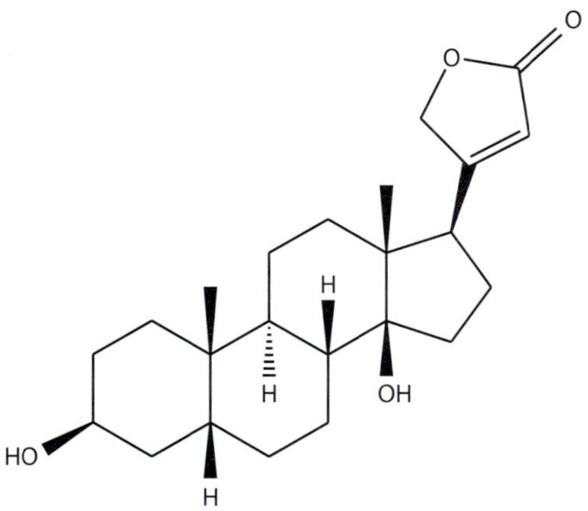

Digitoxigenin, cardenolide from the hydrolysis of digitoxin

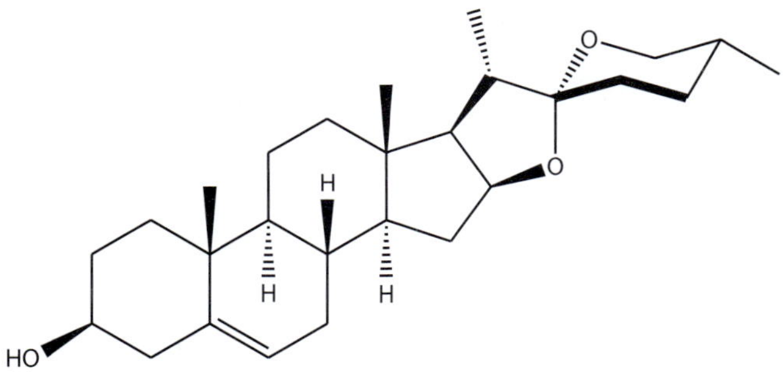

Diosgenin, isolated from Mexican yams and used as precursor of progesterone in its semisynthesis as an early oral contraceptive

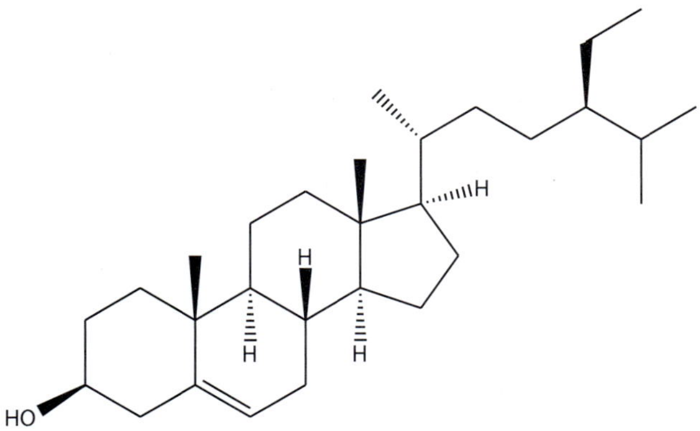

β-sitosterol, isolated from pumkin seeds, pecans, corn and soya oils, reduces blood cholesterol, used in treatment of benign prostate enlargement

FIGURE 16.11 Plant-derived steroids.

Focus on Anabolic Steroids in Sport

Anabolic steroids include the naturally occurring testosterone (Figure 16.10), dihydrotestosterone, **16.29**, and various mimics of these, both natural and synthetic. Testosterone is marketed for oral use in the form of esters with brand names such as *Andriol* or *Sustanon*. Anabolic steroids have the effect of increasing protein synthesis and hence muscle mass, increasing strength, faster recovery from injury, and decreased body fat. Harmful effects can include heart disease, liver damage, high blood pressure, gynecomastia, excessive aggression, enlarged prostate and testicular atrophy in males, and menstrual disorders and masculinized features in females. Most of these steroids also have androgenizing, or masculinizing, effects, resulting in increased body hair and deepening of the voice. The use of these steroids is banned by almost all athletic bodies, but they remain the drugs most commonly detected in athletes. Many anabolic steroids have legitimate medical use in human and veterinary medicine, and different countries have adopted a wide range of legal attitudes to their possession, use, and distribution. If you search the web for the term anabolic steroid, some 90 % of the first 100 entries will be sites offering to sell you the drugs, and most of the remainder are sites urging you not to use them. Studies have indicated that the largest number of users are noncompetitive athletes (who will never be tested), using them primarily for cosmetic purposes.

Steroids undergo many biotransformations, mostly in the liver, and urinalysis generally tests for metabolites of the drugs, which can often be detected for a month after use. Individuals vary widely, and naturally, in their levels of testosterone, so the primary test is for the ratio of testosterone to *epi*-testosterone, **16.30**. **16.30** differs from testosterone in the stereochemistry of the OH group on the five-membered ring. For most people, the ratio is 1:1; in testing, a ratio >6 is taken to be evidence of steroid abuse. However, this has been circumvented by athletes also administering *epi*-testosterone to reduce the ratio to an acceptable level. The case of the British middle-distance runner Diane Modahl is interesting. In 1994, she tested positive for testosterone use with an extraordinary testosterone: *epi*-testosterone ratio of 42, and she was convicted and banned. The ban was lifted in 1995 when it was established that the urine sample had been improperly stored, at 35 °C. Many bacteria are capable of interconverting testosterone and *epi*-testosterone, and at 35 °C (instead of the appropriate 0 °C), these would have been likely to multiply rapidly.

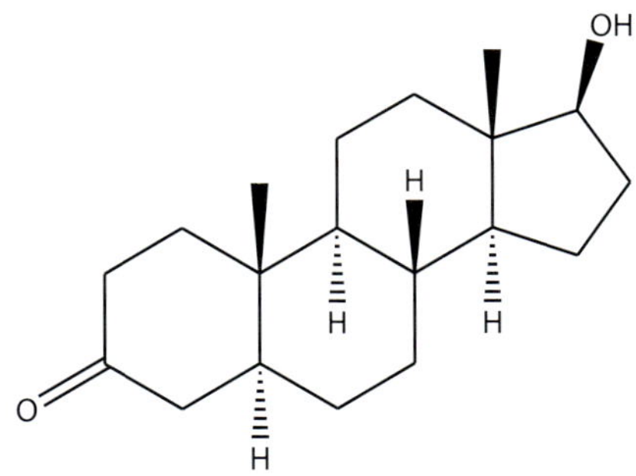

16.29, dihydrotestosterone

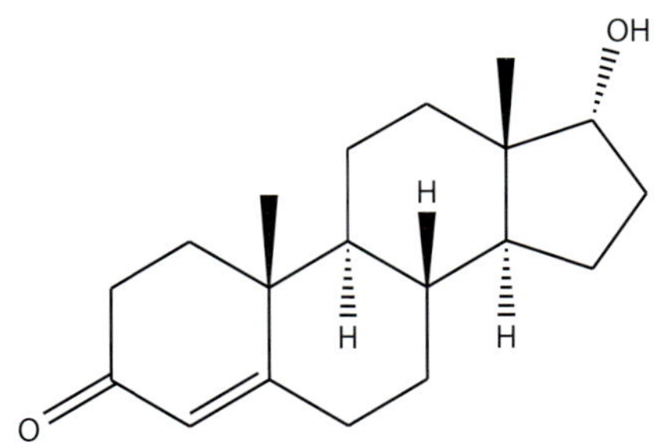

16.30, *epi*-testosterone

A new method for establishing whether testosterone is endogenous (produced by the athlete's own body) or exogenous (synthetic, administered) involves the proportions of ^{13}C and ^{12}C isotopes in the testosterone. Synthetic testosterone is really "semisynthetic"; the precursor phytosterols are isolated from yams or soya. Terrestrial plants fix CO_2 by two different photosynthetic pathways. The Calvin–Benson or C3 pathway produces 3-phosphoglycerate (3PGA), while the Hatch–Slack or C4 process produces the C4 diacid oxaloacetate. Carbon isotopes are fractionated strongly, and differently, by the two processes. Most plants, including soya, use the C3 mechanism, but a number of important food crops including maize, sugar cane, sorghum, and millet are C4 photosynthesizers.

C3 photosynthesizers produce metabolites with less ^{13}C than do C4 plants. Thus, a low level of ^{13}C in testosterone probably means that it is synthetic.

Other anabolic steroids that have been widely used in athletics include stanozolol, **16.31**, trenbolone, **16.32**, and nandrolone, **16.33**. Stanozolol is used in veterinary medicine and in treatments for anemia and is marketed as *Winstrol*. The Canadian sprinter, Ben Johnson, was banned for its use after the Seoul Olympics in 1988. Trenbolone is a veterinary drug used to increase muscle growth and maintain weight en route to slaughter in cattle. Nandrolone testing has been particularly controversial, with a number of high-profile athletes being convicted of its use and later being cleared. The International Olympic Committee (IOC) limit on the nandrolone metabolite 19-norandrosterone in urine is 2 ng cm^{-3}, and there is speculation that levels just a little above this can arise innocently through natural processes.

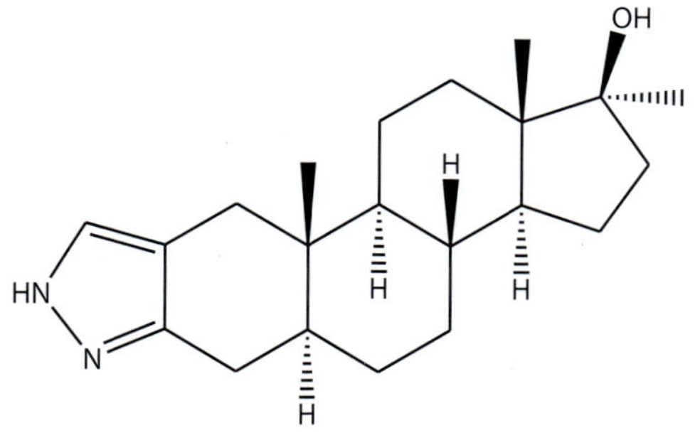

16.31, stanozolol

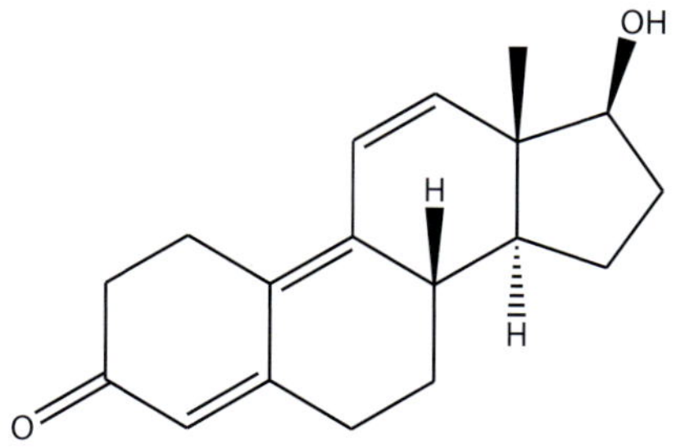

16.32, trenbolone

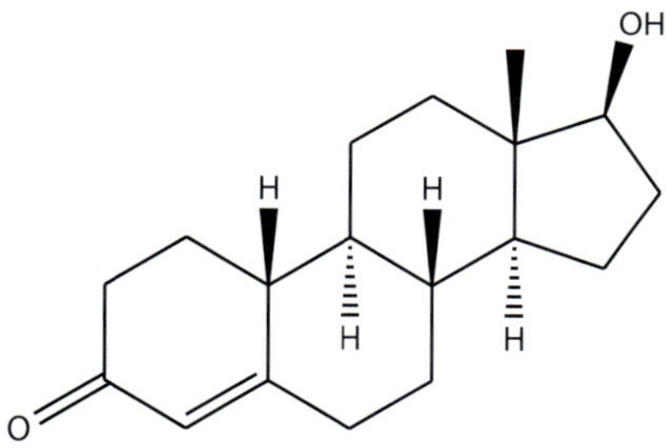

16.33, nandrolone

Each time a new technique for concealing doping, or a new drug, appears, those testing for it have to catch up. The design drug tetrahydrogestrinone (THG), **16.34**, known as "the Clear," provides a useful example. It is prepared by the reduction of the alkyne in gestrinone (**16.35**), which is a synthetic steroid used in the treatment of endometriosis, but THG itself is neither produced naturally nor used medicinally. In 2003, reporters obtained a sample of THG, and with this in hand, it was quickly possible to ban its use and devise an appropriate test. However, the material had been in use at least since 2000, as it was used by the sprinter Marion Jones prior to the 2000 Sydney Olympics. But it's only possible to test for something if you know what it is!

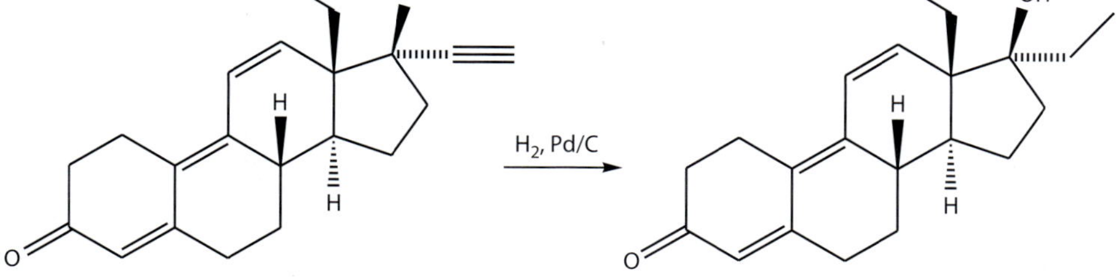

16.35, gestrinone

16.34, tetrahydrogestrinone

Key Points from Section 16.2

- Triglycerides, important compounds for the storage of fats, are the esters of glycerol (propane-1,2,3-triol) and long-chain fatty acids.
- The fatty acids involved in triglycerides are all linear and have even numbers of carbon atoms. The most common acids are C_{12}–C_{20}.
- The number and location of the double bonds in unsaturated fatty acids are important to their role in nutrition and biological function.
- Prostaglandins are biosynthesized from the polyunsaturated fatty acid, arachidonic acid; they have a wide range of physiological functions and medicinal uses.
- Waxes are esters of long-chain fatty acids (C_6–C_{36}) with long-chain alcohols (C_{16}–C_{36}).
- Steroids are polycyclic compounds containing three six-membered and one five-membered ring in a specific, fused arrangement.
- In animals, steroids are biosynthesized from the triterpene squalene, *via* lanosterol. Plant steroids are also biosynthesized from squalene but *via* cycloartenol.
- Naturally occurring steroids include male and female sex hormones. Synthetic steroids have been used as anti-inflammatory agents, immunosuppressives, and contraceptives.

PROBLEM 16.4

A section of the Johnson–Christiansen synthesis of estrone is shown in the following. Study it and answer the questions posed:

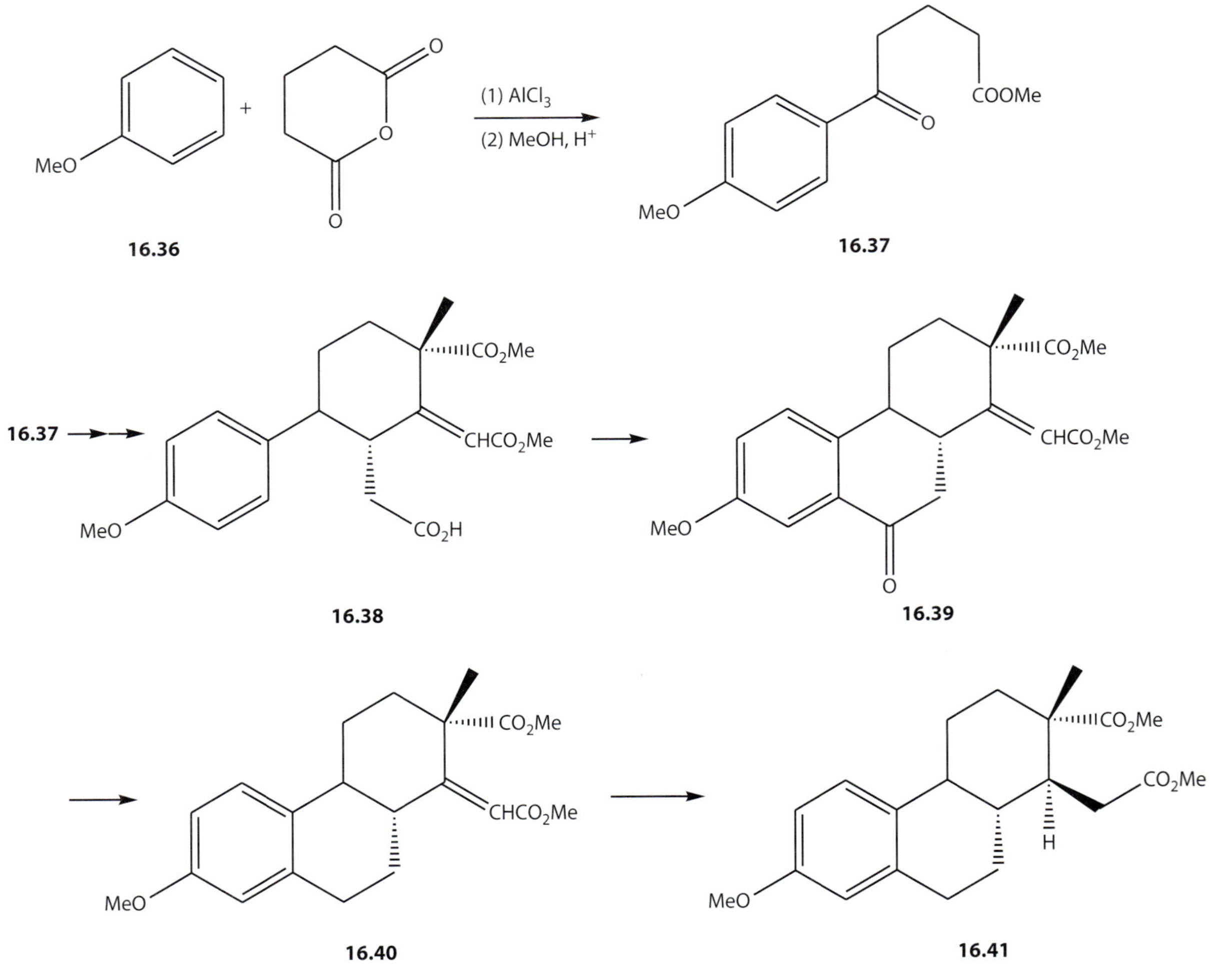

(a) Write a mechanism for each step of the conversion of **16.36** to **16.37**.
(b) What reagents would you use for the conversion of **16.38** to **16.39** (two steps needed)?
(c) What reagents would you use for the conversion of **16.39** to **16.40**?
(d) What reagents would you use for the conversion of **16.40** to **16.41**?

Solutions

(a) This is Friedel–Crafts acylation, followed by esterification. MeO is *ortho, para*-directing and activating; presumably, the selectivity for the *para*-position is steric in origin:

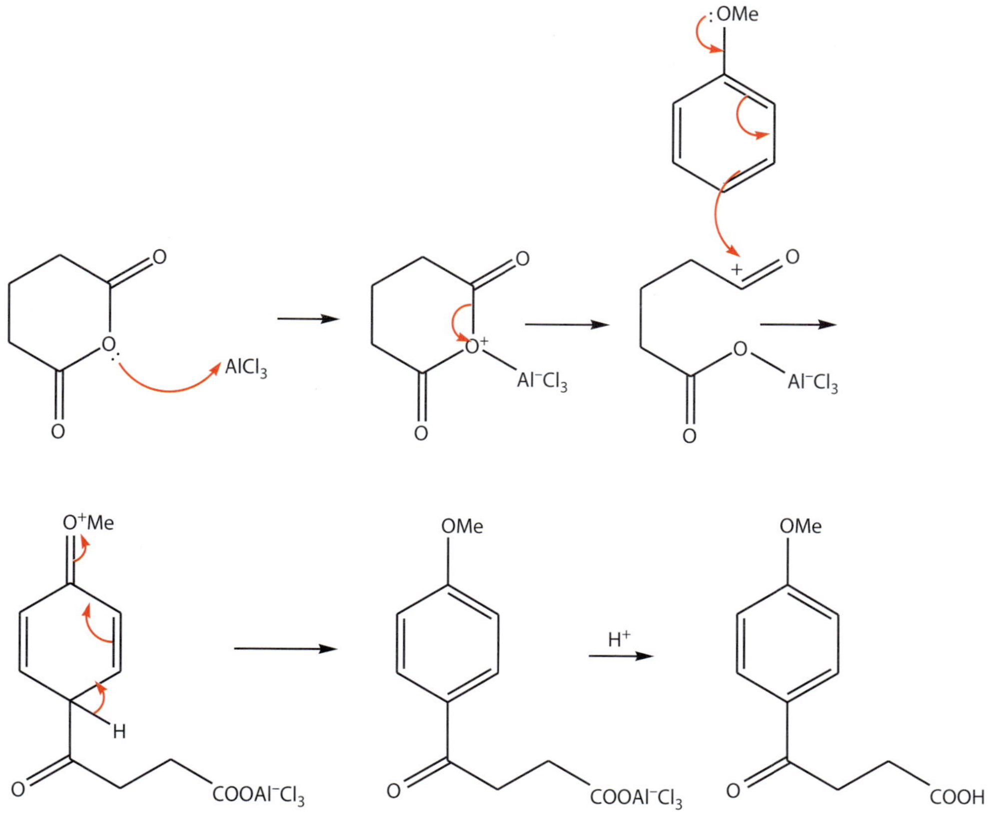

The ester formation is quite standard:

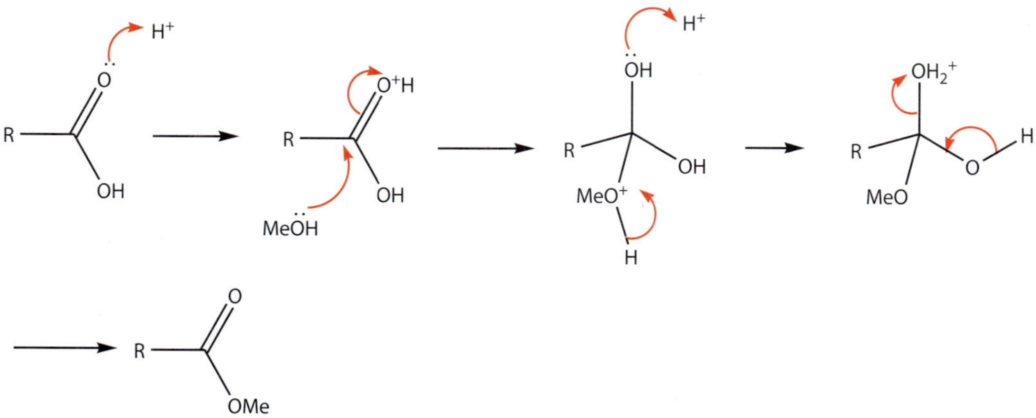

(b) This is another Friedel–Crafts acylation. The acid was converted to the acid chloride with $SOCl_2$, and then the acylation completed using $AlCl_3$.
(c) Hydrogenation using a Pd/C catalyst is specific for the complete reduction of carbonyl groups next to aromatic rings.
(d) Hydrogen with a Pd/$SrCO_3$ catalyst was used, but there are other catalysts that would also have worked.

16.3 CARBOHYDRATES

16.3.1 INTRODUCTION

Carbohydrates are biosynthesized in plants from CO_2 and water, using sunlight and chlorophyll, in the process described as photosynthesis. Oxygen is evolved. Many carbohydrates are involved in providing energy through nutrition, but they are also found in nucleic acids, cellulose, and sugar antibiotics. Formation of water-soluble sugar "conjugates" is implicated in converting various toxins to soluble derivatives for excretion and in administration of drug molecules in a biocompatible form. Carbohydrates were initially classed as compounds with molecular formula $C_n(H_2O)_m$ such as glucose, $C_6H_{12}O_6$, **16.42**. There are many carbohydrates with such formula, but we also now class as carbohydrates many related compounds with more variable compositions. There is a unique system of nomenclature and various drawing conventions specific to carbohydrates. The suffix for all types of sugars is -ose. Species with aldehyde functional groups are called aldoses (like glucose) and with ketone functional groups as ketoses (like D-fructose, **16.43**). Further information about the compound type comes from a number, the number of carbon atoms, *viz.*, tri-, tetra-, penta-, and hexa-. Thus, glucose would be described as an aldohexose.

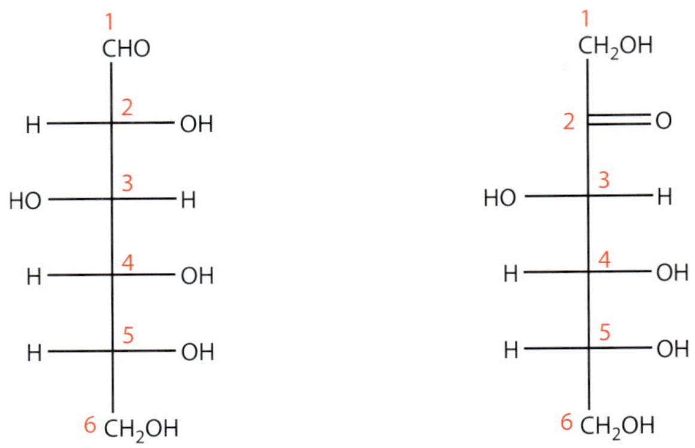

16.42, D-glucose   **16.43**, D-fructose

We will also find that some of the reagents we need to perform familiar transformations on carbohydrates are different from those to which we are accustomed. Most of the compounds we have considered to date are soluble in organic solvents rather than aqueous ones, but carbohydrates, with their many –OH groups, are very soluble in water, but not in nonpolar organic solvents.

16.3.2 TRIOSE SUGARS

The simplest of all sugars are the trioses, which have three carbon atoms. When we studied stereochemistry in Chapter 7, we met D-glyceraldehyde, **16.44**, as the reference compound for stereochemical designations. D-Glyceraldehyde, shown here as its Fischer projection, is the *R*-enantiomer. At the time when Fischer was working, there was no way to establish this, so Fischer guessed the absolute configuration, and fortunately, he was correct. This is a good point to review some stereochemistry, especially the manipulation of Fischer projections, as these are ubiquitous in sugar chemistry. Structure **16.44** also introduces important conventions about drawing sugar structures. The Fischer projection is oriented so that the most oxidized carbon is at the top of the diagram. Thus, the aldehyde is at the top, and the primary alcohol at the bottom. If, after obeying this convention in the Fischer projection, the OH on the physically lowest, but highest numbered, carbon atom is to the right, then this is described as a D-sugar. Almost all naturally occurring sugars are D-sugars. Glyceraldehyde is described as an aldose sugar because it contains an aldehyde group. Ketose sugars contain a ketone group at C2. The three-carbon ketose sugar is 1,3-dihydroxyacetone, **16.45**—almost unique among sugars in being achiral. Both of these trioses are involved in carbohydrate metabolism.

Emil Fischer, 1852–1919; Nobel Prize 1902

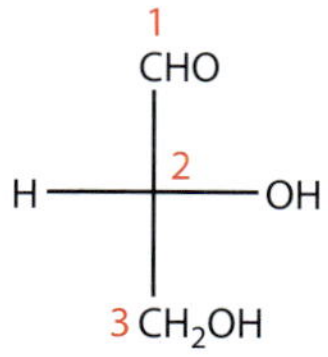

16.44, *R*, D-glyceraldehyde

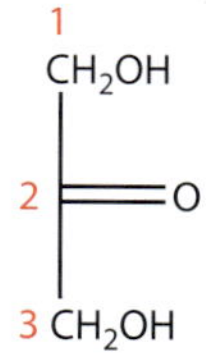

16.45, 1,3-dihydroxy acetone

Quick Review of the Fischer Projections

Fischer projections are a method of showing stereochemistry and particularly the relationship between chiral centers, with uncomplicated drawings. Thus, **16.46a** is represented as **16.46b**. For students with good spatial perception, some of the relationships described will be obvious, but this will not be true for everyone. If you can simply "see" what is happening, that's great—but for many people, a slow, mechanical, pedestrian approach, which will always work, is helpful. How can these projections be manipulated? There are many rules in other textbooks, but we can make do with just one. *Any pairwise swap reverses the absolute configuration at an asymmetric carbon atom.* For those of you able to "translate" immediately between projection and three-dimensional structure, it may be obvious that **16.47a** and **16.47c** are identical, but the swapping system will let the rest of us catch up.

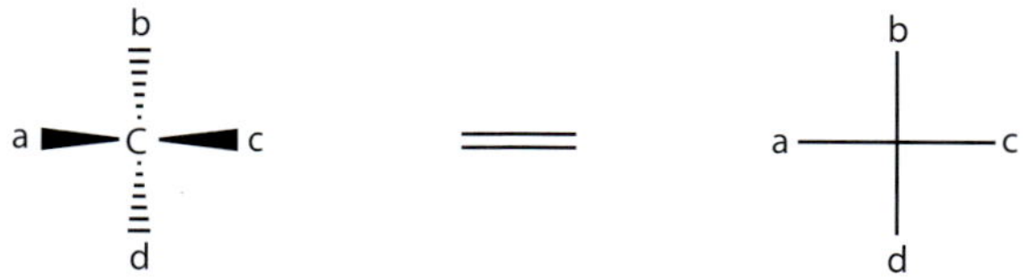

16.46a **16.46b**

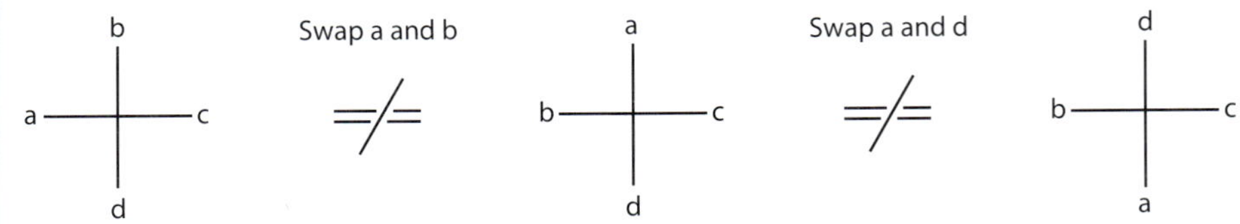

16.47a **16.47b** **16.47c**

If we need to assign an absolute configuration from a Fischer projection, we need to put the lowest priority (under the Cahn–Ingold–Prelog system) group at the "top" of the diagram. Then we track around the other three groups. If the track is clockwise, the molecule is *R*; if counterclockwise, it is *S*. We can apply this to our prototype molecule, glyceraldehyde. We order the substituents by the Cahn–Ingold–Prelog system, as OH > CHO > CH_2OH > H, as in **16.48a**. But the lowest ranked group is not at the top of the diagram, and we need to put it there, which we do by a pairwise swap to **16.48b**. In **16.48b**, the track from OH to CHO to CH_2OH is counterclockwise, so **16.48b** has *S* absolute configuration. Since our pairwise swap reversed the absolute configuration, it follows that the original molecule is *R*-glyceraldehyde.

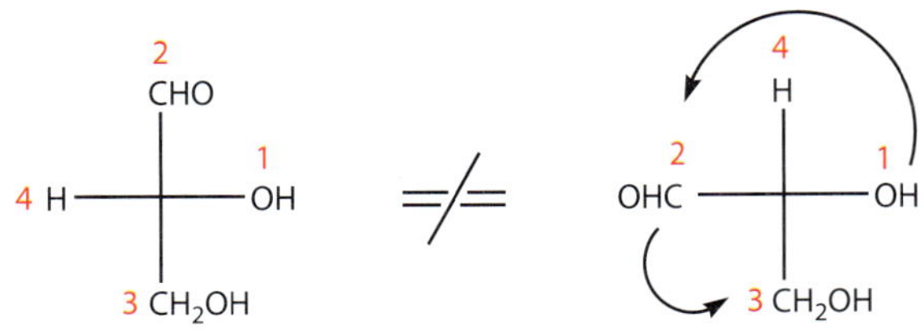

16.48a **16.48b**

PROBLEM 16.5

What is the stereochemical relationship between each of the following pairs of compounds (enantiomers, diastereoisomers or identical)?

(a)

(b)

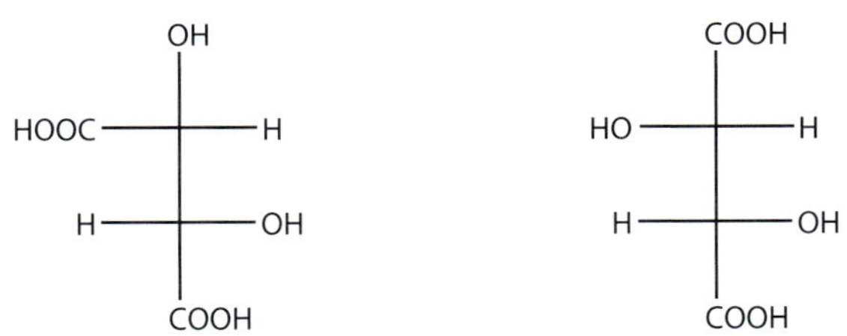

Solutions

(a) If you can simply "see" the answer to this, that's great. Making a model and looking at it from different directions is also a good approach. The third, and more pedestrian, approach is to assign the absolute configuration at each center and see if they are the same.

Under the Cahn–Ingold–Prelog system, OH > COOH > CH_2OH > H. In the three-dimensional drawing, H is pointing toward us, so we need to swap it with the COOH to get the opposite enantiomer. Our new structure is clearly *S*, so the starting material must have been *R*.

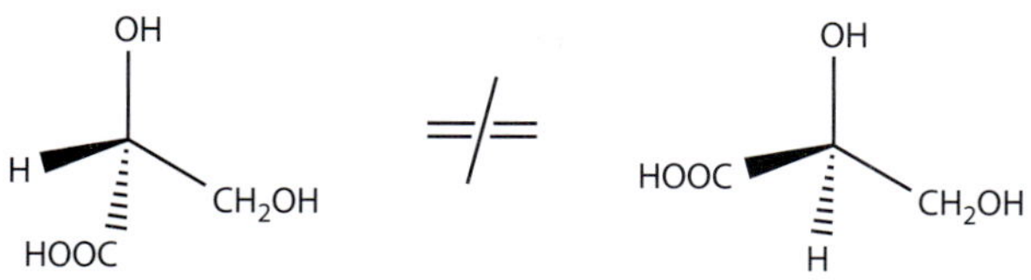

With the Fischer projection, to assign the absolute stereochemistry, we need to put the lowest priority group at the top, by means of a swap:

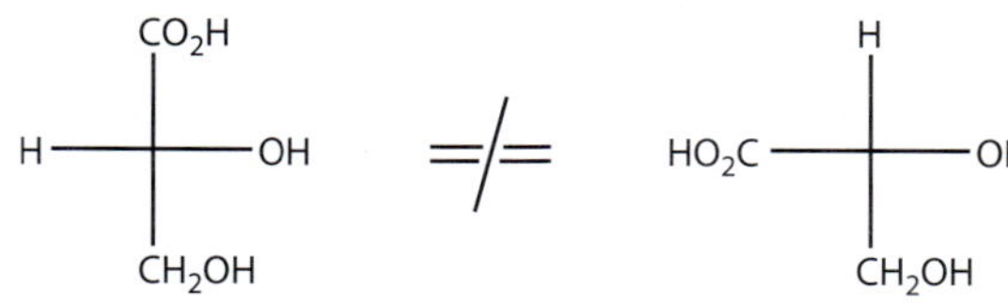

Again, our new molecule is clearly *S*, so the original is *R*. Thus, we conclude that these two molecules are identical.

(b) The right-hand molecule is presented in a reasonably conventional form, as is the lower carbon of the left-hand one. So we need to manipulate the upper carbon of the left-hand structure. We will designate the lower part of the molecule simply as Q:

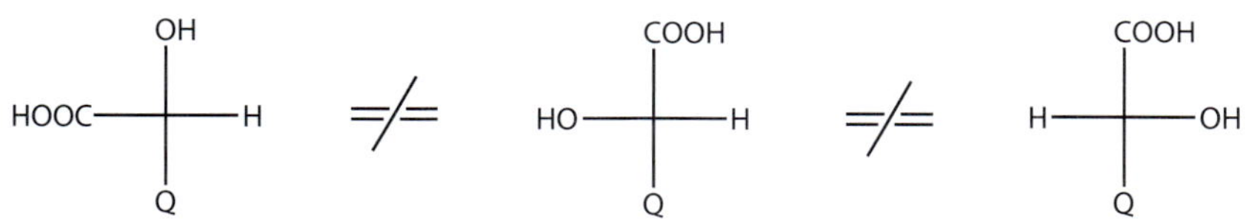

First, we switch the COOH and OH groups; the molecule produced is now of opposite configuration so we need to do another switch to get back to the configuration we started with. This time, we switch H and OH. The first and third structures are identical, so if we now redraw the whole molecule, it is obvious that the two molecules are diastereoisomers; the left-hand structure is a *meso*-compound.

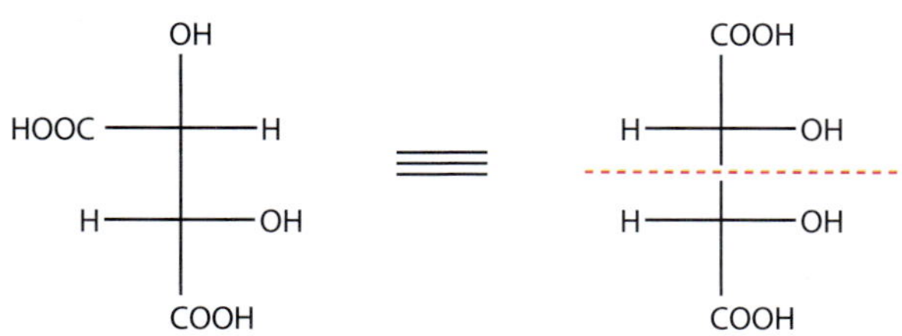

16.3.3 TETROSE SUGARS

There are two naturally occurring aldotetroses, D-erythrose and D-threose (**16.49** and **16.50**), and one ketotetrose, D-erythrulose (**16.51**). Erythrose and threose can be distinguished by nitric acid oxidation (notice the less usual but water-soluble oxidant, selective for the aldehyde and primary alcohol over the secondary alcohols); threose leads to a chiral diacid, while the diacid from erythrose is a *meso*-compound and achiral. Although, in a formal nomenclature sense, these diacids would be threaric acid and erythraric acid, they always go by their more common name of tartaric acid.

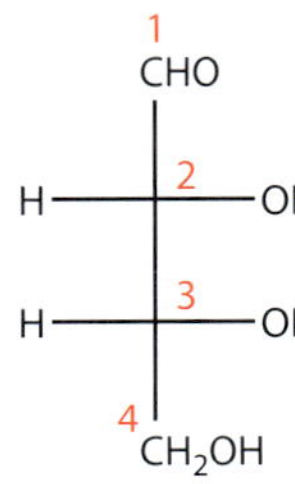

16.49, D-erythrose

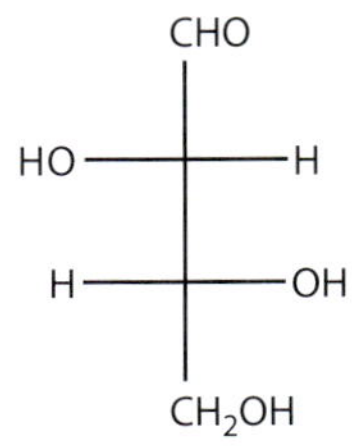

16.50, D-threose

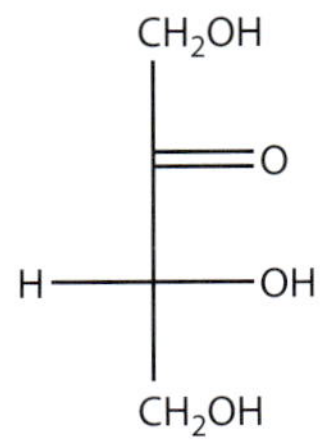

16.51, D-erythulose

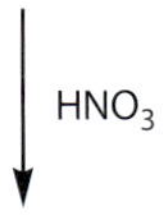

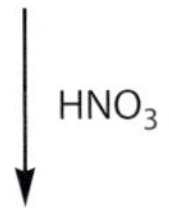

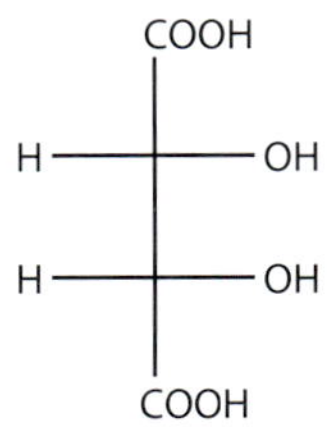

16.52, *meso*-tartaric acid

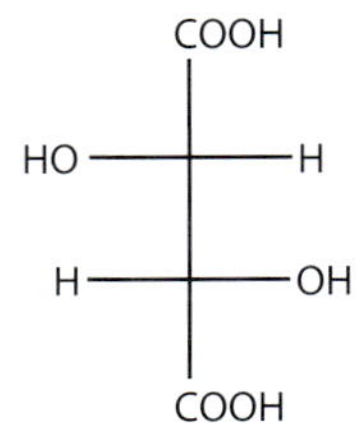

16.53, *S*, *S*-tartaric acid

16.3.4 PENTOSE SUGARS

Every time we add a chiral carbon atom to our sugar chain, there is an additional chiral center, and the number of possible stereoisomers doubles. The four D-aldopentoses are shown in Figure 16.12. All are naturally occurring. Ribose is found in RNA, while arabinose is isolated from the biopolymer pectin. Xylose (Greek ξυλον, wood) is found in the embryos of most edible plants. Lyxose is quite rare and is a component of bacterial glycolipids. An important related structure is 2-deoxyribose, **16.54**, a constituent of DNA.

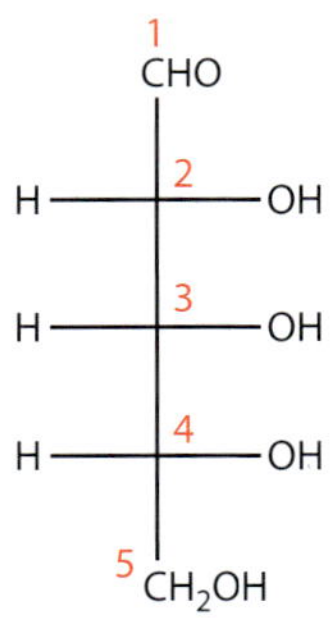

D-ribose

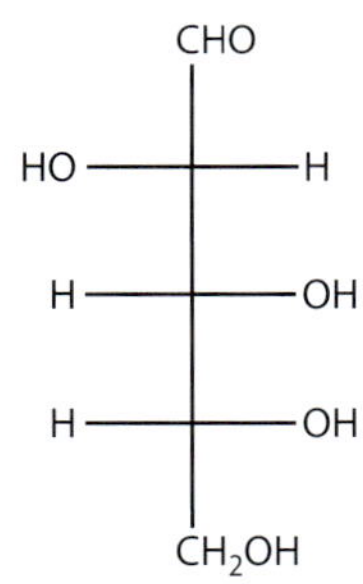

D-arabinose

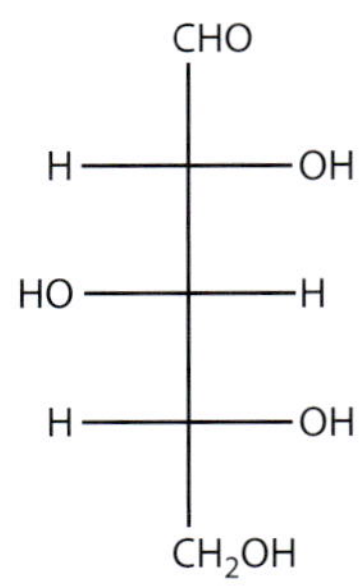

D-xylose

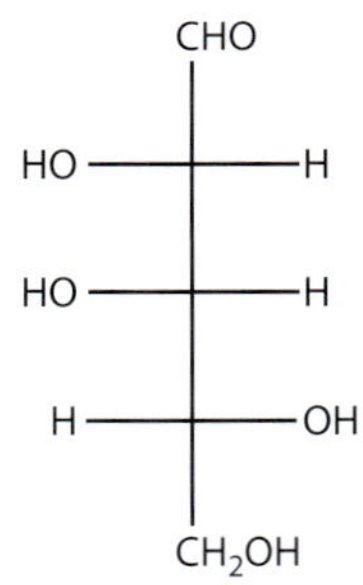

D-lyxose

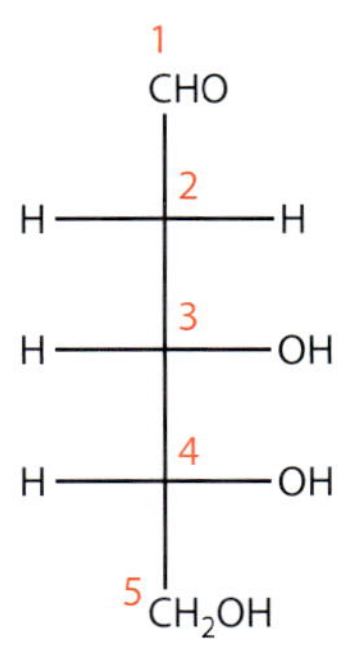

16.54, 2-deoxyribose

FIGURE 16.12 Structures of the D-aldopentoses.

16.3.5 HEXOSE SUGARS

In this family are many of the most important sugars, including glucose (the name comes from the Greek γλυκυς, meaning sweet), mannose, galactose, and fructose. There are eight D-aldohexoses; the family tree of the aldose sugars is shown in Figure 16.13. Allose is isolated from South African proteas. D-Altrose is not naturally occurring, but L-altrose has been isolated from some bacteria. Mannose occurs naturally in cranberries and has been used to treat UTIs; the name is supposedly derived from manna, the food supplied to the Israelites during their wandering in the Sinai desert. Gulose is found naturally only in a few bacteria and archaea, and talose and idose are not naturally occurring, although the oxidation product of idose, iduronic acid, is. Galactose is common and found in sugar beet and dairy products; the name comes from the Greek γαλακτος, meaning of milk. Glucose is ubiquitous and is the main source of metabolic energy in man.

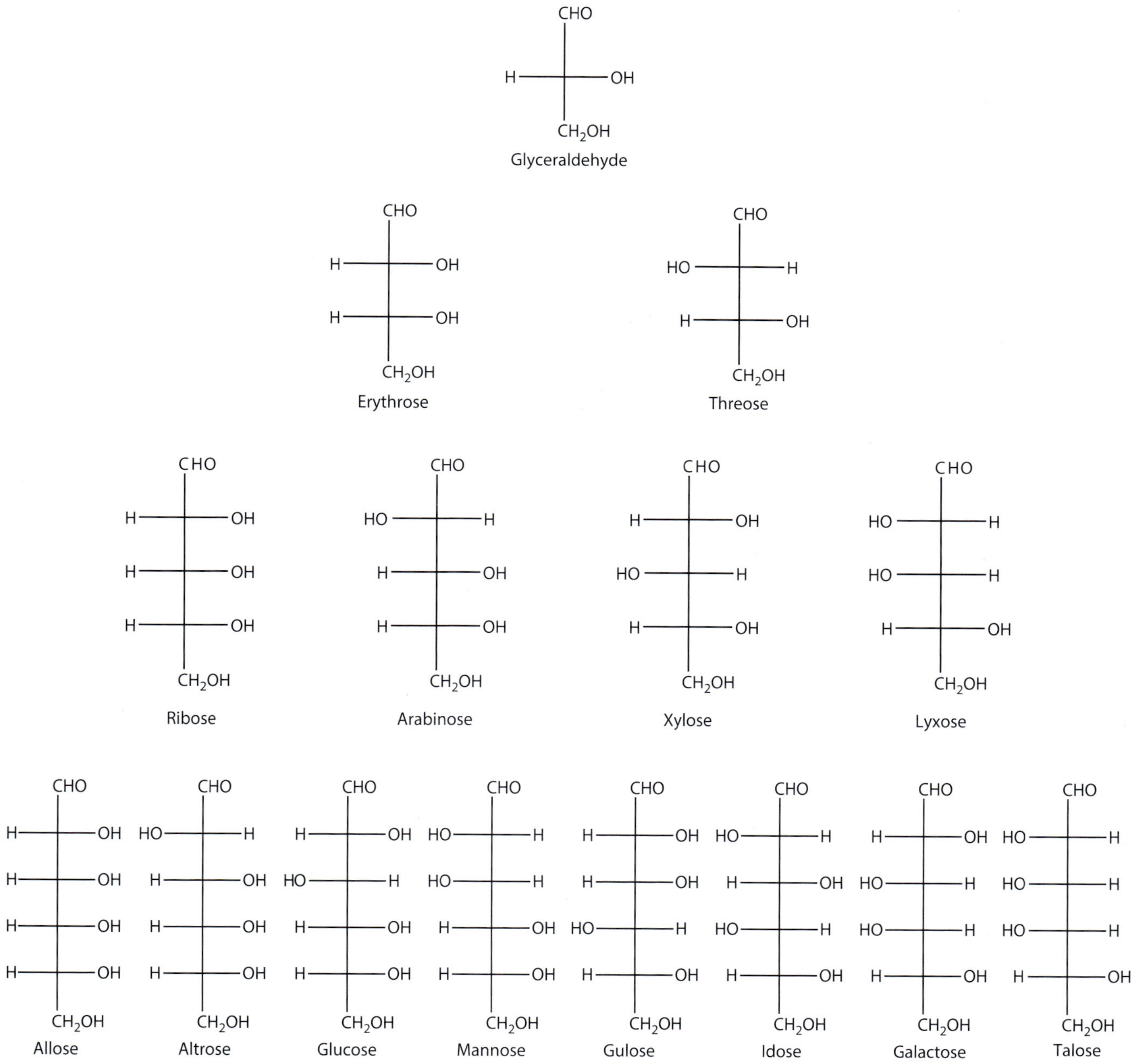

FIGURE 16.13 Sugar family tree.

A similar family tree could be constructed for the less common ketose sugars, but only fructose, **16.43** (*vide supra*), and sorbose are commonly encountered. Fructose is the sweetest of all the naturally occurring monosaccharides, and hence, "high-fructose corn syrup," a mixture of fructose and glucose, is widely used as a sweetening agent for food. Concerns have been raised as it is linked to obesity and diabetes. The natural isomer of sorbose, unusually, is the L-isomer, **16.55**; it can be isolated from mountain ash berries or prepared by oxidation of sorbitol, **16.56**, and it is used in the commercial synthesis of vitamin C.

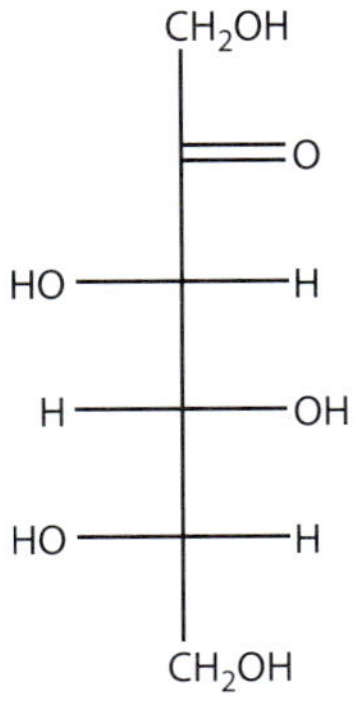

16.55, L-sorbose

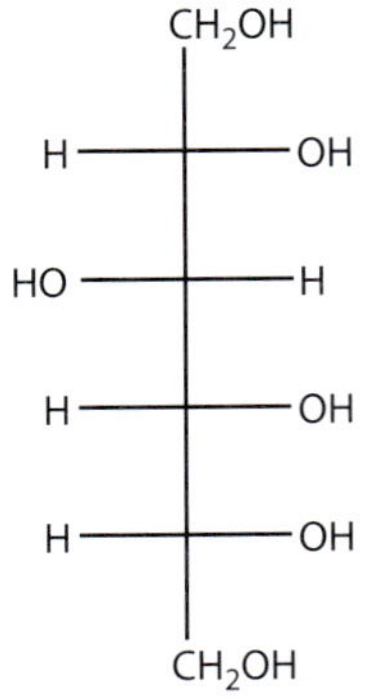

16.56, D-sorbitol, or D-glucitol

PROBLEM 16.6

(a) What is the nature of the structural relationship between **16.57** and **16.58**?

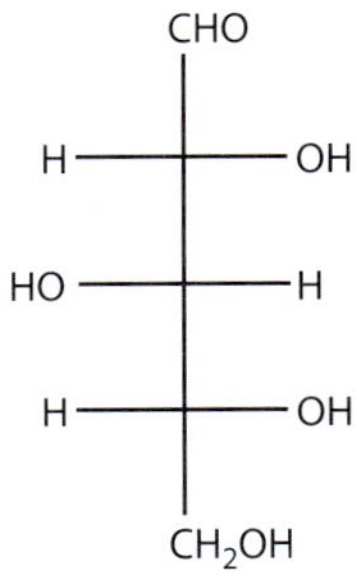

16.57

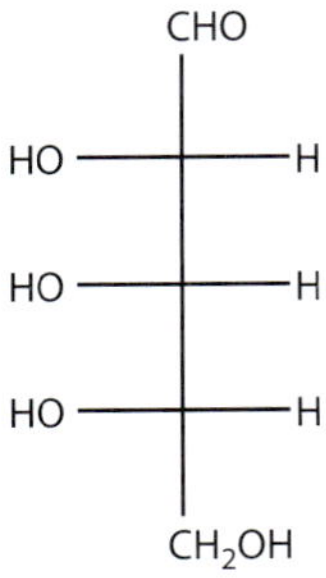

16.58

(b) Give the structure of the molecule **16.59**, which is obtained on nitric acid oxidation of **16.57** (both the aldehyde and the primary alcohol are oxidized).

(c) Is **16.59** chiral?

Solution

(a) The two molecules are not mirror images, nor are they identical. Therefore, they must be diastereoisomers.

(b) Nitric acid oxidizes both aldehydes and primary alcohols to carboxylic acids. So **16.59** is

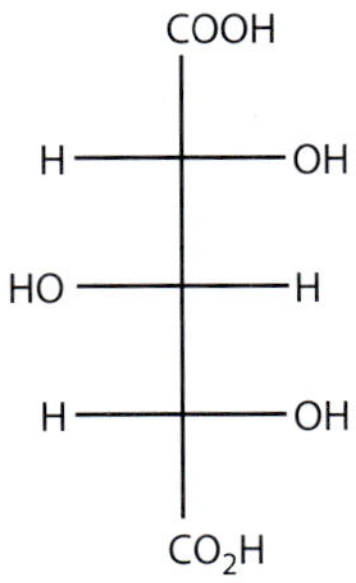

16.59

(c) **16.59** contains a plane of symmetry; hence, it is a *meso*-compound and achiral.

16.3.6 CYCLIC FORMS OF SUGARS

Although we draw sugars as "linear" because this is the simplest way to discern stereochemical relationships, this is not their normal form. We have in these molecules many free OH groups and either an aldehyde or a ketone, and these can react in an intramolecular process to give a cyclic hemiacetal, containing a five- or six-membered ring, a process we already know to be favorable (Section 14.3.2). The reactions of ribose are illustrative (Figure 16.14). In the first process, a five-membered ring is formed by the attack of the OH group at C4 on the aldehyde. Attack on the aldehyde may be on either face, so a new chiral center is formed, and two diastereoisomers, designated α and β, are formed. These two forms are called *anomers*—they are a special class of diastereoisomers specific to sugar chemistry. They differ in configuration only at the carbon atom that was the carbonyl in the noncyclized material. We describe the anomer as β when the absolute configuration at C1 (or C2 for ketose sugars) is the same as that at the furthest chiral center in the ring, the one that defines whether the sugar is D or L, in this case C5. α-Anomers have opposite configurations at these two centers. These five-ring sugars are called furanose forms because of their resemblance to furan, **16.60**. Notice the way in which these are drawn—as if they were flat. This is similar to a Haworth projection (flat but with wedges), which for five-membered rings is a reasonable approximation. Five-membered rings usually adopt an envelope conformation (Chapter 7), but they are very flexible, and a pentagon is a fair average. Six-ring sugars are formed by the attack of the OH at C5 on the aldehyde and are described as pyranose because of their resemblance to pyran, **16.61**. In the chair form of D-sugars, α is axial, hopefully easy to remember. Although Haworth projections were long used also for six-membered rings (like **16.62** for glucose), this is a poor approximation to the real structure. It deprives us of useful information to understand sugar chemistry and is best avoided. In aqueous solution at room temperature, ribose exists in all forms *except* the linear form where the concentration is so low as to be unmeasurable. However, all of the forms interconvert, *via* the linear form. Furanose forms make up 24 % of the total (α 6 %, β 18 %), and pyranose forms 76 % (α 20 %, β 56 %) at equilibrium. We might expect this, given that six-membered rings are generally more stable than five-membered ones. However, in all biological systems, RNA and other, the furanose form is the only one that is used.

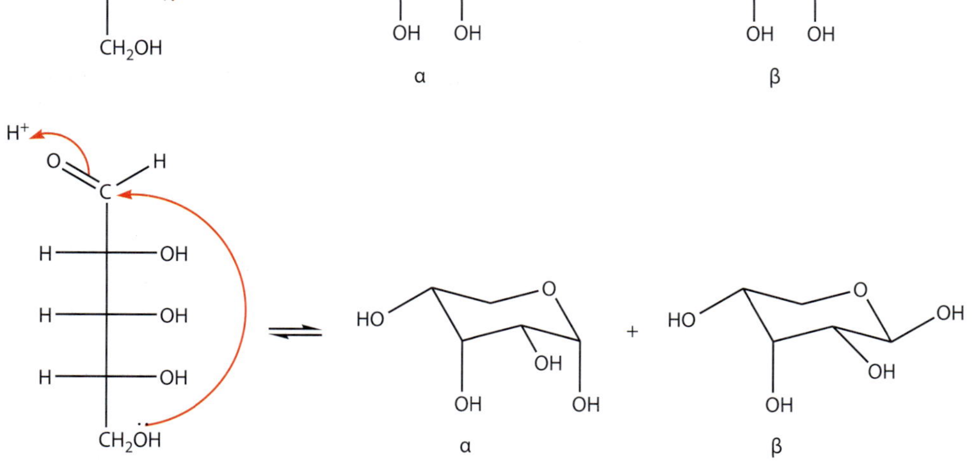

FIGURE 16.14 Cyclization of ribose.

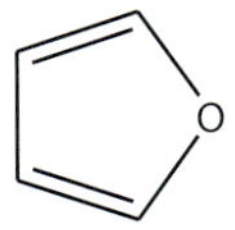

16.60, furan

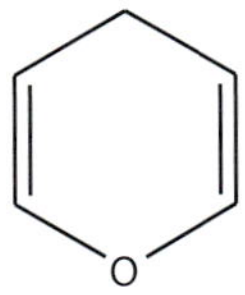

16.61, 4H-pyran

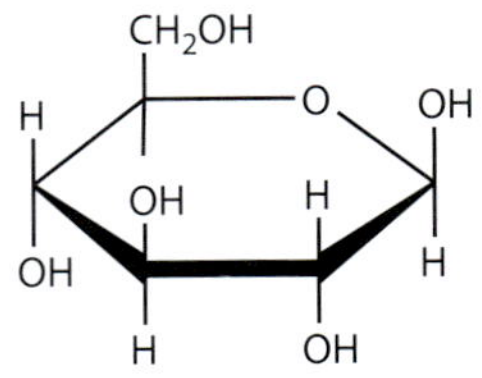

16.62

In solution, glucose exists almost entirely in the cyclic pyranose form (although derivatized furanose forms can be prepared, by some careful synthetic planning), the reaction being shown in Figure 16.15. The fact that glucose has all or almost all of its hydroxyl groups equatorial may be a clue to its biological dominance—it is the most stable of all the aldohexoses. Careful crystallization from water allows the isolation of the α- and β-forms of glucose, which have, respectively, optical rotations of +112 ° and 18.7 °. It is from these data that the position of equilibrium between the two anomers was initially established. If either anomer is redissolved in water, the two forms equilibrate to give a solution with rotation +52.7 °. This phenomenon is called mutarotation, and it is catalyzed by acid or base, like any hemiacetal formation. The mechanism of the acid-catalyzed reaction is shown in Figure 16.16. Ketose sugars such as fructose also exist largely in a cyclic

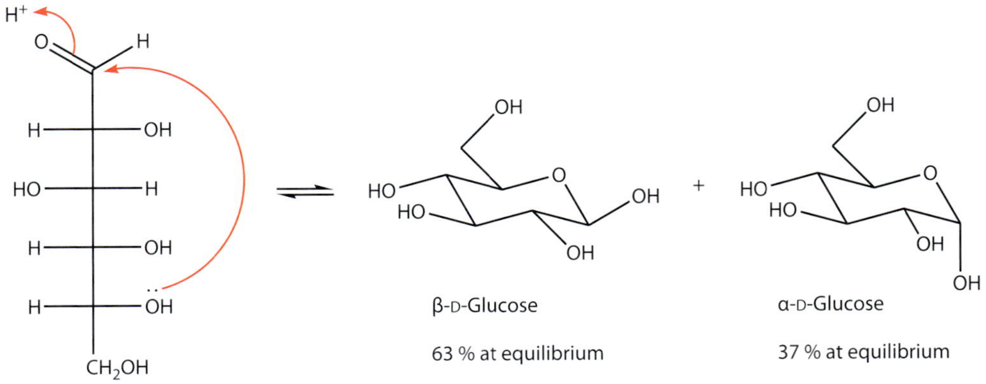

FIGURE 16.15 Pyranose forms of glucose.

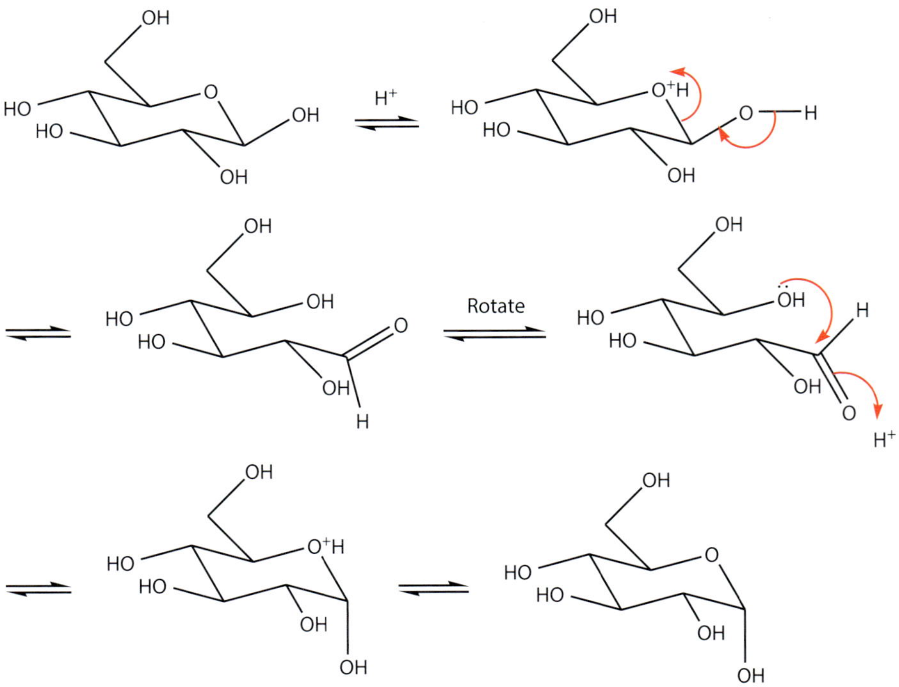

FIGURE 16.16 Mechanism of mutarotation of glucose.

form; although the α-pyranose form, **16.63**, predominates to the extent of 75 % at equilibrium in solution, we often think of fructose as a β-furanose, **16.64**, because this is its form in its most common disaccharide, sucrose or table sugar.

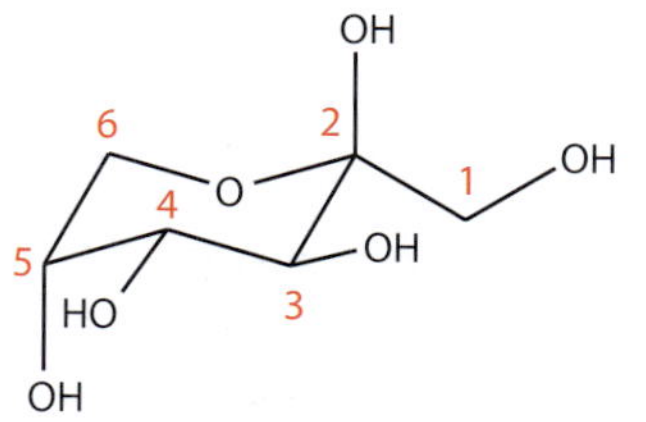

16.63, β-D-fructopyranose

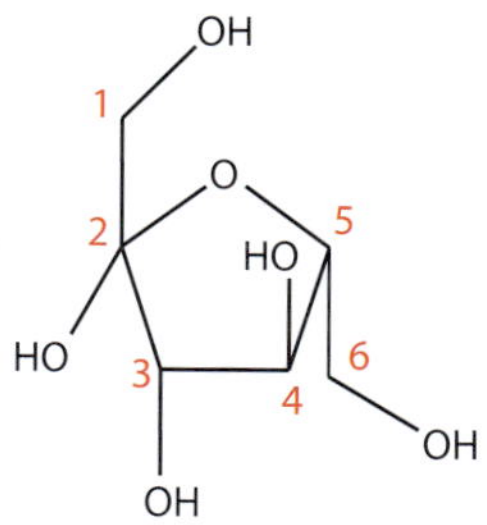

16.64, β-D-fructofuranose

PROBLEM 16.7

Write a mechanism for the mutarotation of glucose catalyzed by base.

Solution

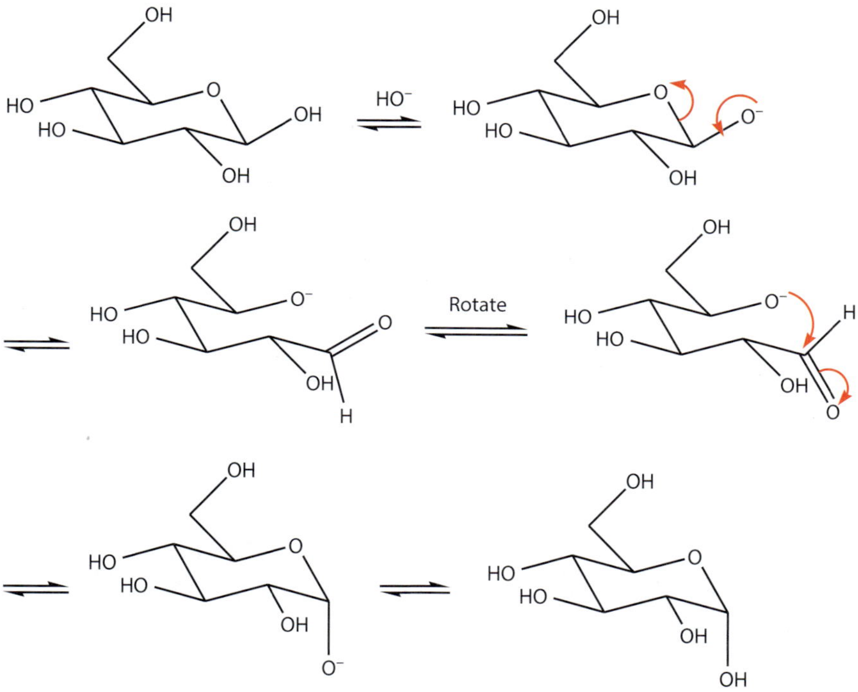

16.3.7 SUBSTITUTION OF THE OH GROUPS OF SUGARS

The hydroxyl groups of sugars undergo many of the same substitution reactions as other alcohols; they react with methylating agents to give *O*-methyl derivatives (**16.65**) and acetic anhydride to give acetates (Figure 16.17). The methylated product is interesting in that while most of the OMe groups are part of ethers, which are resistant to hydrolysis except under extreme conditions, one is part of an acetal, which can be hydrolyzed in aqueous acid. The cyclic form of glucose is a hemiacetal, and it can be converted to an acetal by reaction with an alcohol in the presence of acid (Figure 16.18). The species produced is described as a glycoside, and in this case, methyl β-D-glucopyranoside and methyl α-D-glucopyranoside are produced. The intermediate stabilized cation may be attacked from either the axial or the equatorial direction, and *in vitro* both species are obtained. In a reversal of this process, the acetal, but not the ether, groupings in **16.6** may be hydrolyzed in aqueous acid.

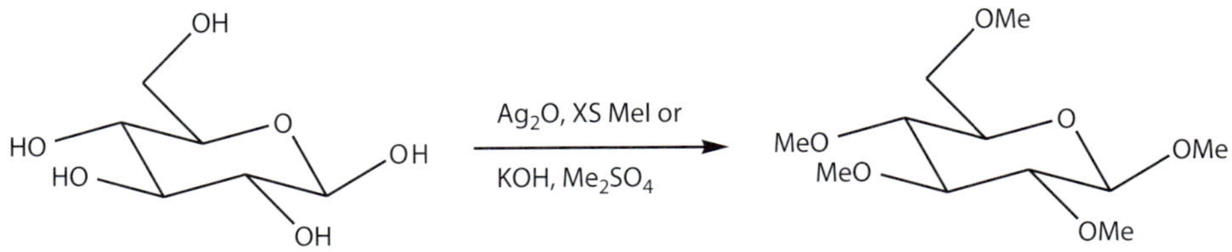

16.65

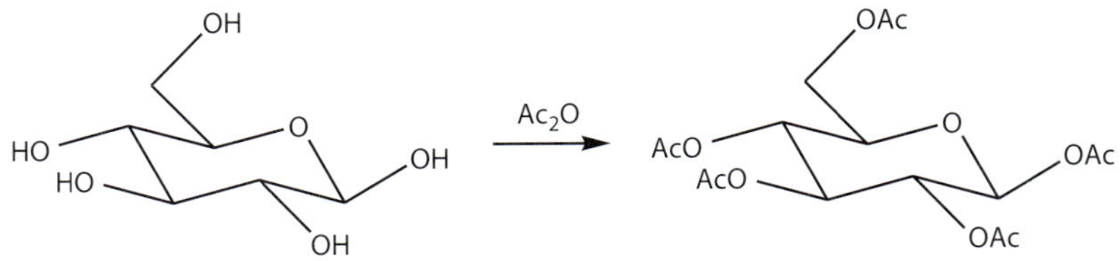

FIGURE 16.17 Substitution reactions by the hydroxyl groups of sugars.

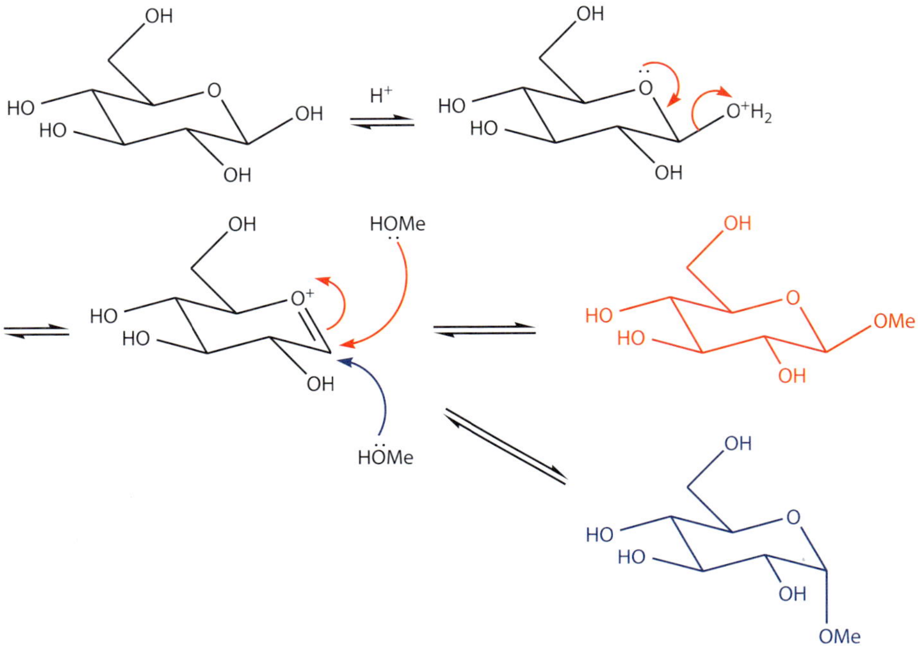

FIGURE 16.18 Mechanism of formation of methyl glycosides.

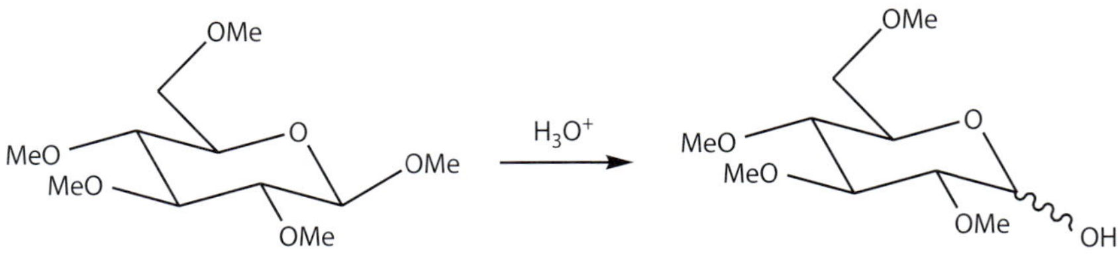

16.66

Monosaccharides are species that cannot be cleaved into simpler carbohydrates. Disaccharides can be cleaved (usually by hydrolysis) into two monosaccharides. Oligosaccharides can be cleaved into more than 3 monosaccharides, and polysaccharides are usually composed of more than 10 monomeric units. Di- and polysaccharides are all glycosides. The formation of glycosides can and does involve a wide range of alcohols—it is the fundamental process by which disaccharides and oligosaccharides are produced. In this case, the methanol is replaced with an OH group that is part of another sugar residue. To prepare these species selectively by means of chemical synthesis is quite challenging—some OH groups must be blocked off in order for the correct one to react. However, enzyme-driven versions of these reactions are ubiquitous. The formation of glycoside derivatives of a wide range of compounds allows for their compatibility with aqueous solutions. For example, the

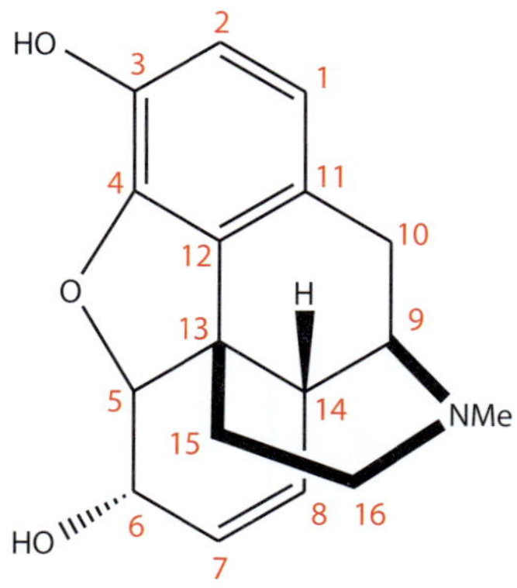

Morphine

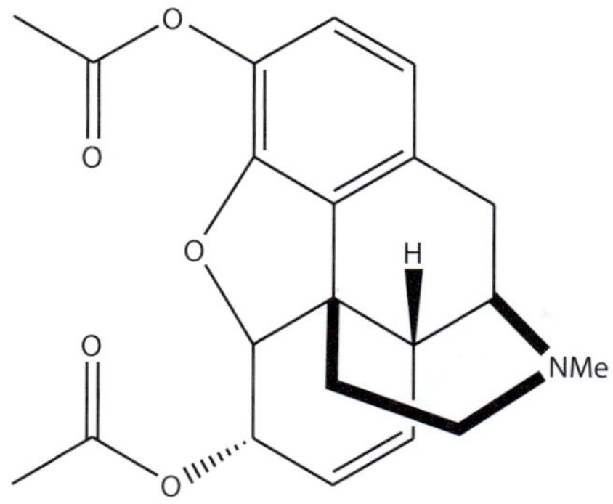

Diamorphine, heroin

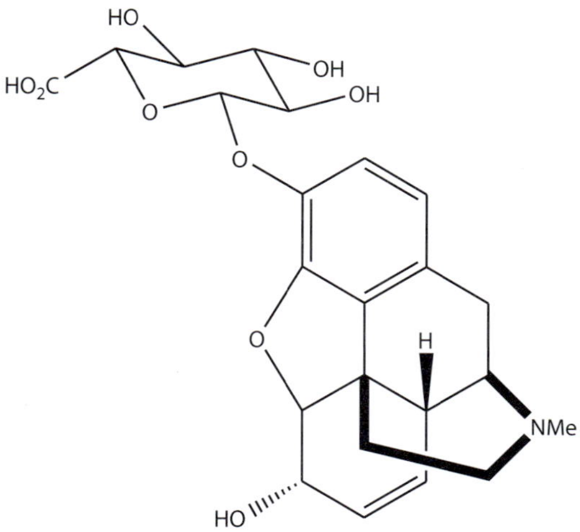

Morphine-3-glucuronide

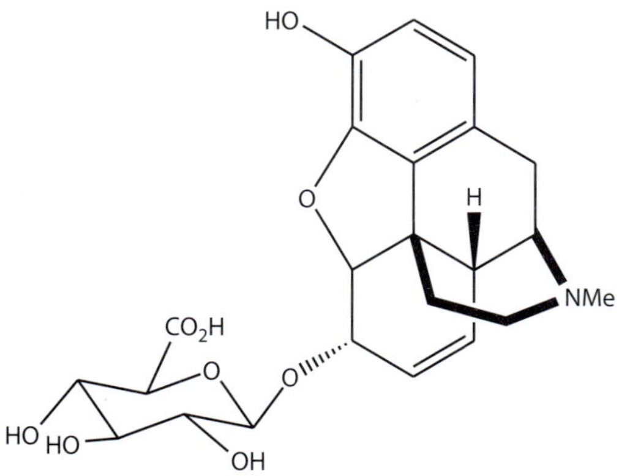

Morphine-6-glucuronide

FIGURE 16.19 Morphine and heroin metabolites.

heart medication, digoxin, **16.67**, is described as a cardiac glycoside. It is also a trisaccharide—with three identical deoxy sugar residues (highlighted in blue). It is the steroidal part of the molecule that has the desired medicinal activity, but without the sugars, it would not be water soluble. Similarly, some drugs must be converted to glycosides to be excreted in urine—again, they would not otherwise be water soluble. Thus, when someone is tested for heroin abuse, it is not the heroin itself that is measured in the urine but morphine 3-glucuronide and morphine 6-glucuronide (Figure 16.19, the numbering refers to the position of the hydroxyl group on the morphine skeleton that has become joined to a sugar), these metabolites being much more water soluble.

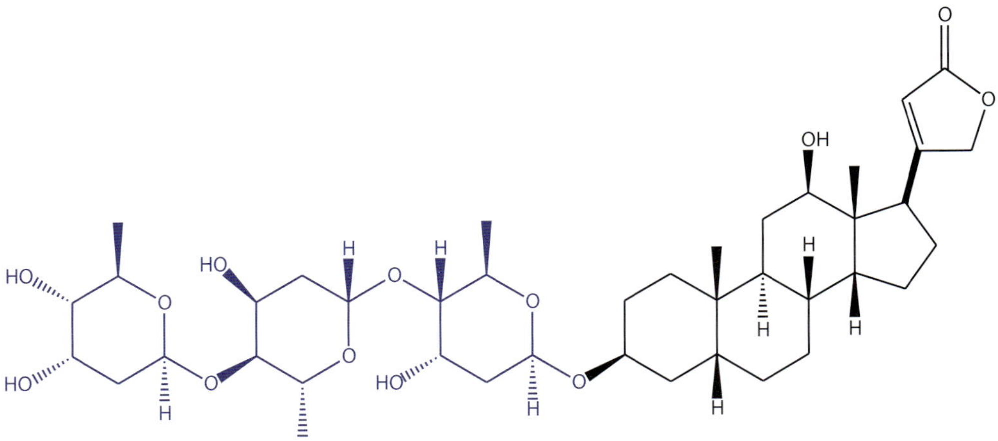

16.67

PROBLEM 16.8

Predict the product(s), including stereochemistry, of each of the following processes and show the mechanism for the reactions:

(a)

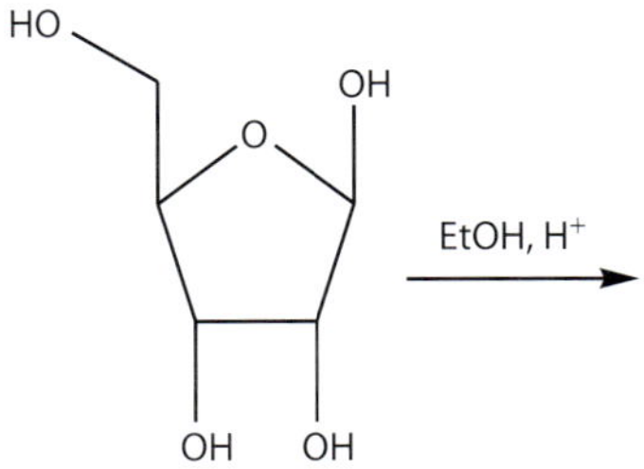

(b)

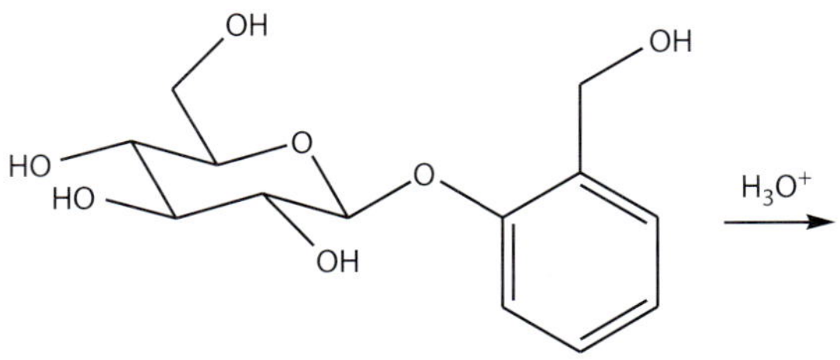

Salicin, isolated from willow bark

Solutions

(a) This is a glycoside-forming reaction; *in vitro* both anomers will be obtained.

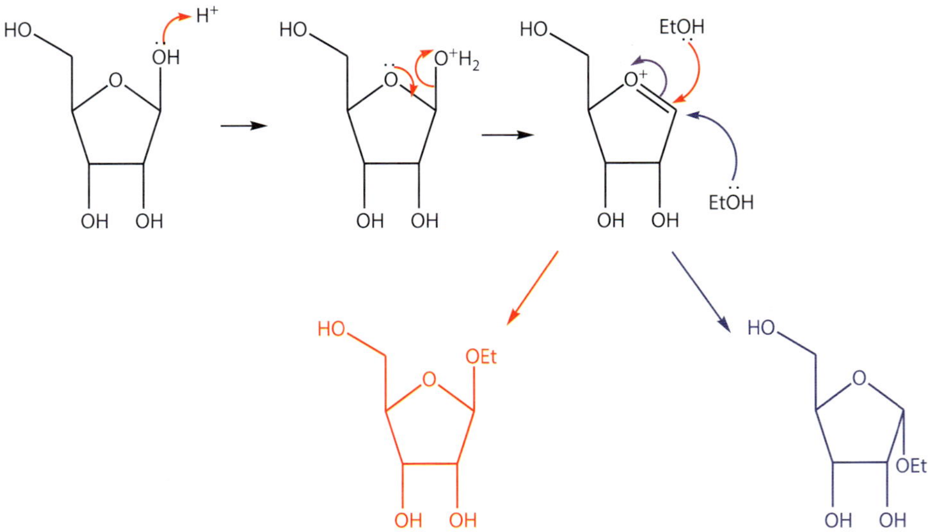

(b) This is the hydrolysis of a glycoside (an acetal) and gives

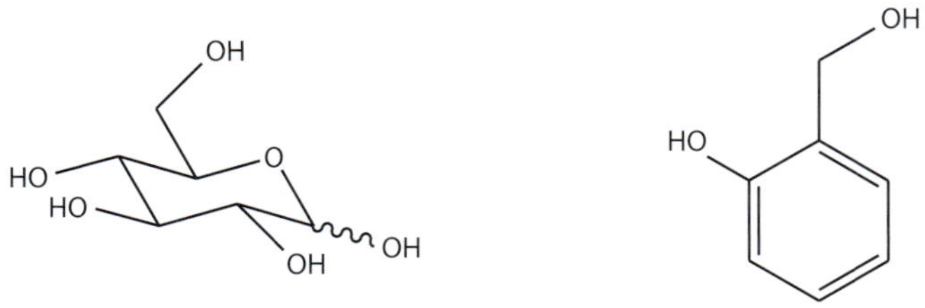

16.3.8 OXIDATION REACTIONS

Mild oxidation conditions will oxidize an aldehyde group to a carboxylic acid, leaving all the alcohol groups unreacted. Typical reagents are silver(I) or copper(II), and the product is an *aldonic* acid (Figure 16.20). These reactions have been used as a test for aldose sugars, both on monosaccharides and disaccharides. Ammoniacal silver nitrate, Tollens' reagent, reacts to give a silver mirror on the side of the flask or tube (Figure 16.21); the redox reaction is given in

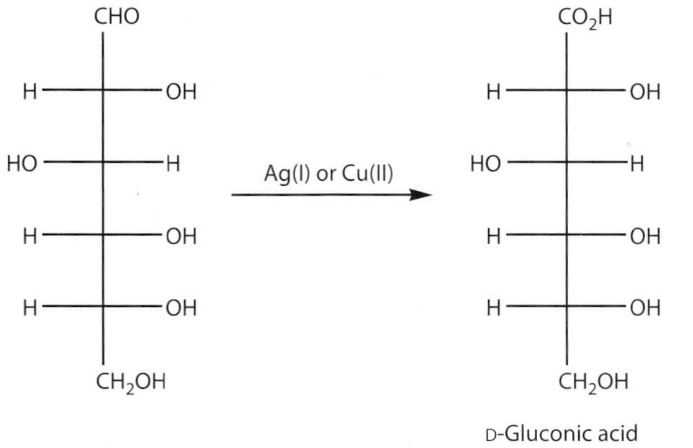

FIGURE 16.20 Mild oxidation of glucose to gluconic acid.

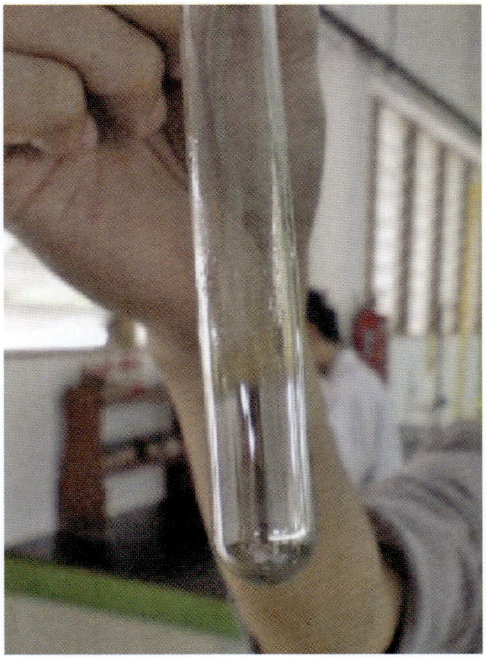

FIGURE 16.21 Tollens' test for a "reducing" sugar. (From http://static.flickr.com/3283/2316913960_4aca252c22_z.jpg.)

$$RCHO + 2HO^- \rightarrow RCOOH + H_2O + 2e^-$$
$$[Ag(NH_3)_2]^+ + e^- \rightarrow Ag(0) + 2NH_3$$
$$RCHO + 2[Ag(NH_3)_2]^+ + 2HO^- \rightarrow RCOOH + 2Ag(0) + 4NH_3 + H_2O$$

$$RCHO + 2HO^- \rightarrow RCOOH + H_2O + 2e^-$$
$$2Cu^{2+} + 2e^- + H_2O \rightarrow Cu_2O + 2H^+$$
$$RCHO + 2HO^- + 2Cu^{2+} \rightarrow RCOOH + Cu_2O + 2H^+$$

FIGURE 16.22 Redox reactions in the oxidation of aldose sugars to aldonic acids.

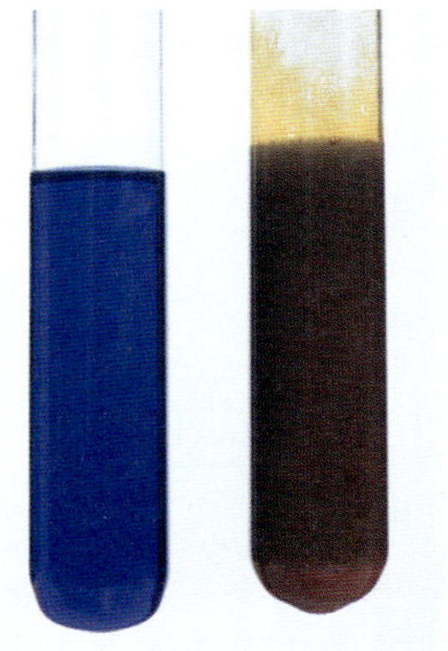

FIGURE 16.23 Fehling's test for a reducing sugar. (From Andrew Lambert/Science.)

Figure 16.22. Fehling's solution, essentially copper(II) sulfate, is initially blue and reacts to give a brown precipitate of copper(I) oxide (Figure 16.23). Sugars that give positive tests with Tollens' or Fehling's solution are termed "reducing sugars." Note that glycosides will NOT give a positive test with these reagents, as there is no "free" aldehyde and no access to the open-chain form of the sugar.

Although it may initially be surprising, fructose and other 2-ketoses do give a silver mirror with Tollens' reagent. To understand the mechanism of this reaction, we need to take a look ahead to the next chapter. We have already met the tautomers of carbonyl compounds, the enols (Section 8.5); **16.68** is the tautomer of acetone. We know that these two species can equilibrate—we will look in detail at the reaction mechanism in Chapter 17. Fructose is also a ketone and therefore capable of forming an enol, **16.69**. Going back to the original ketone is one possibility for this enol (in this case, actually an enediol), but the molecule could equally equilibrate with glucose, particularly under the basic conditions used in the tests for reducing sugars. As soon as this happens, the aldehyde can be oxidized, just as it was for glucose itself (Figure 16.24).

16.68

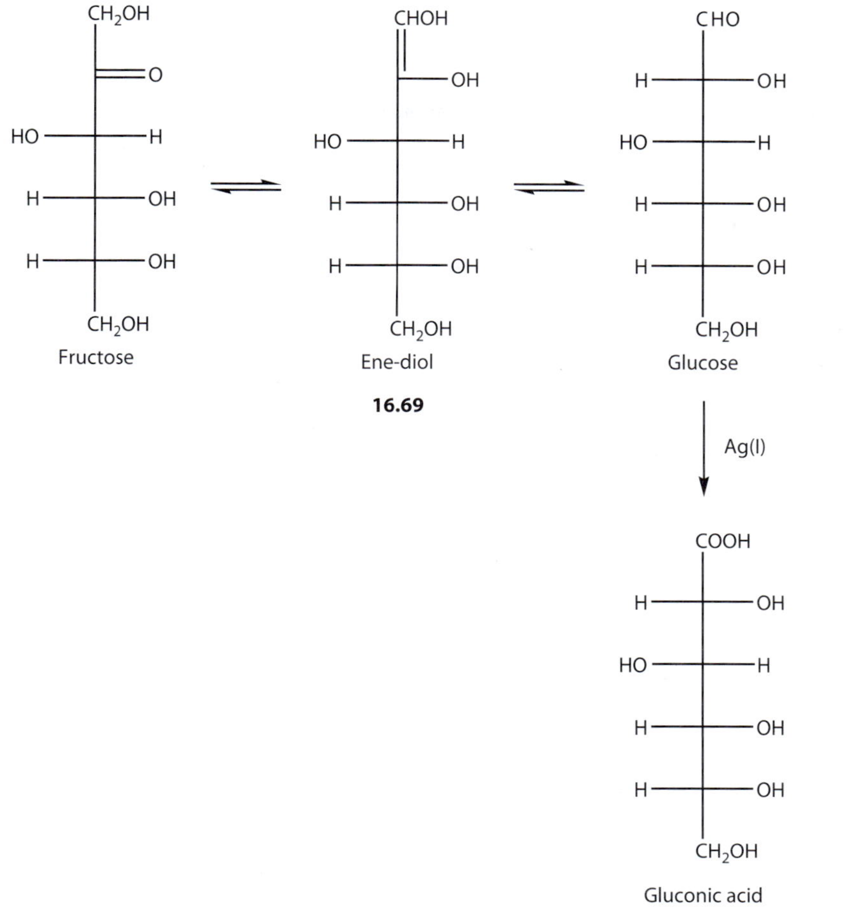

FIGURE 16.24 Silver(I) oxidation of fructose.

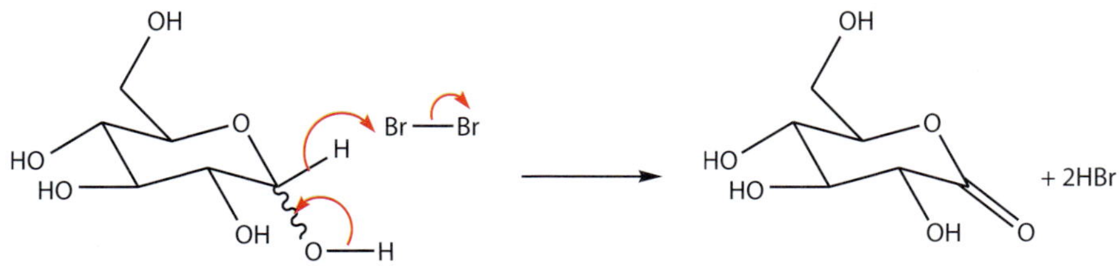

FIGURE 16.25 Oxidation of glucose by bromine.

Fehling's and Tollens' reagents are commonly used as tests for reducing sugars, but for aldonic acid *preparation*, bromine water (Br_2, H_2O) is more commonly used. The reaction proceeds *via* the cyclized hemiacetal form (Figure 16.25). D-Gluconic acid (and other aldonic acids) also exists in a cyclized form, this time as an ester, D-gluconolactone, rather than an acetal (Figure 16.26).

More vigorous oxidation, generally using nitric acid, results in both the aldehyde and the primary alcohol being oxidized to carboxylic acids, called aldaric acids (Figure 16.27). Oxidation of the sugar with the CHO and CH_2OH groups interchanged (L-gulose) gives the *same* diacid as obtained from glucose. This process was crucial in establishing the relative stereochemistries of various sugars. D-Glucaric acid is chiral, but the aldaric acids produced from allose and galactose are *meso*-compounds (Figure 16.28).

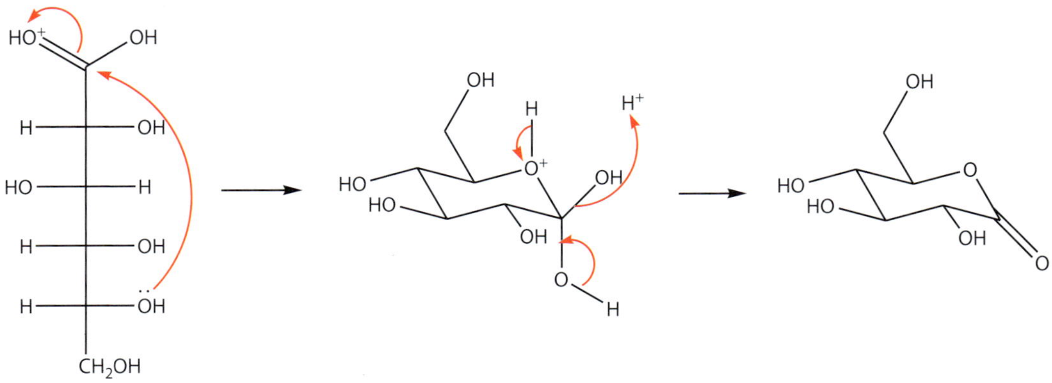

FIGURE 16.26 Cyclization of gluconic acids.

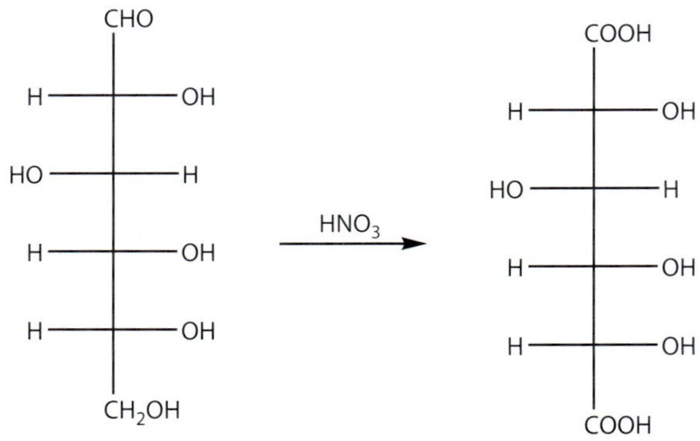

D-Glucaric acid

FIGURE 16.27 Vigorous oxidation of glucose.

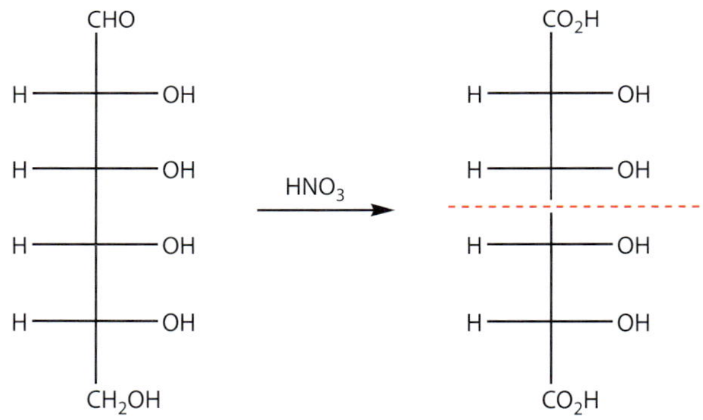

D-Allose

Allaric acid

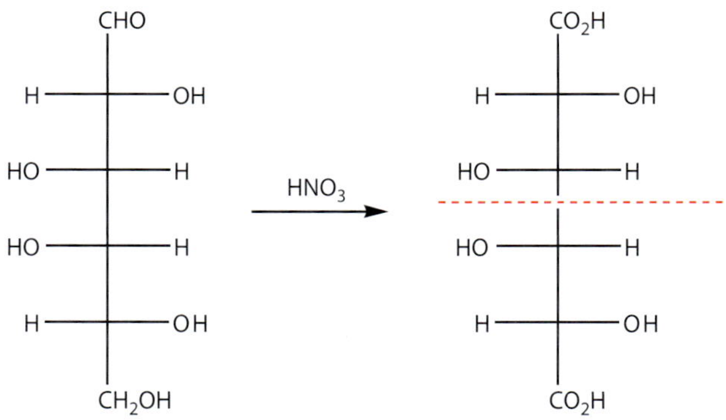

D-Galactose

Galactaric acid

FIGURE 16.28 Oxidation of allose and galactose.

PROBLEM 16.9

Which of the following compounds can be classified as reducing sugars?

(a) D-Glucose
(b) L-Fucose (6-deoxy-L-galactose)

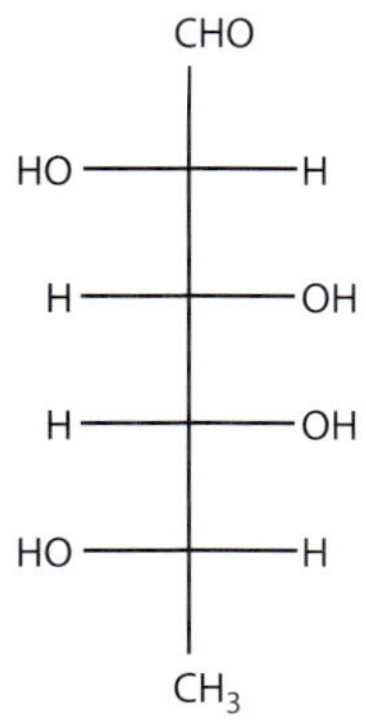

(c) α-D-2,3,4,6-tetra-*O*-Methylglucopyranose
(d) Neotrehalose

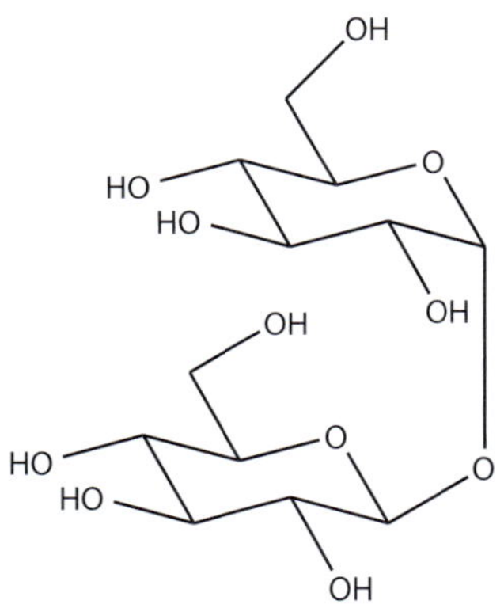

Solutions

(a) and (b) Free aldehydes are obvious, so these are reducing sugars.
(c) The question here is really about whether you can draw the structure correctly—remember in numbering it that the carbon that was the aldehyde in the open chain is number 1, and the CH_2OH is number 6:

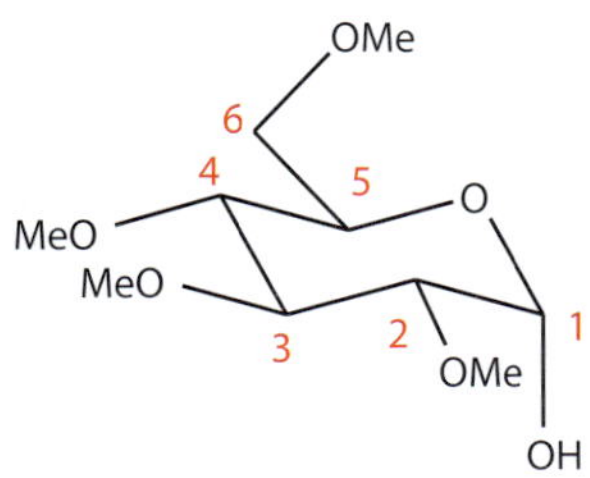

The aldehyde position is not blocked as a glycoside, so there is the possibility of ring opening; hence, this is a reducing sugar.
(d) Both the sugar units are glucose, but they are joined together at the 1-positions, thus blocking either ring opening to give a free aldehyde. This is not a reducing sugar.

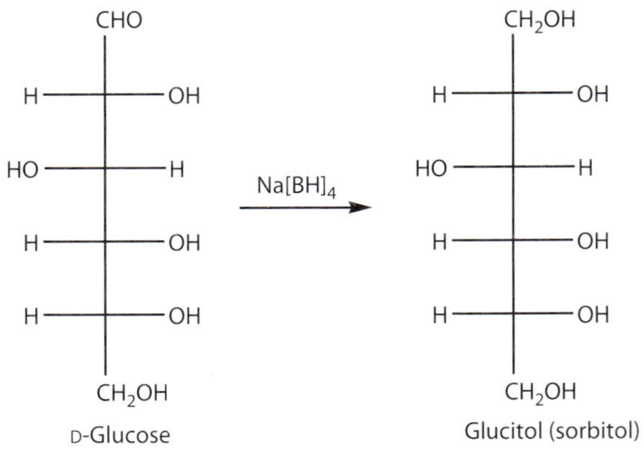

FIGURE 16.29 Borohydride reduction of glucose.

16.3.9 REDUCTION

The reduction of aldose sugars with sodium borohydride gives, as we would expect, a primary alcohol in place of the aldehyde. These diols are called alditols. Thus, glucose is reduced to glucitol (Figure 16.29), more commonly referred to as sorbitol. Sorbitol is less sweet than sucrose and is metabolized slowly (to fructose), so it is often used in "sugar-free" products, especially for diabetics. It occurs naturally in peaches, plums, apples, and pears. Since alditols do not contain a carbonyl group, they cannot cyclize and exist in the open-chain form.

16.3.10 CHAIN LENGTHENING AND SHORTENING REACTIONS

If you look back to the "sugar family tree" in Section 16.3.5, you will see that each sugar has two descendants in the row below, which might be considered to have been formed by adding in an extra chiral carbon atom below the aldehyde. The biosynthesis and metabolism of sugars are beyond the scope of this course, but we can do this chain extension reaction chemically, the Killani–Fischer reaction* (Figure 16.30). In the first step of the reaction, the aldehyde is converted to the cyanohydrin. A new chiral center is created, and as the two products are diastereoisomers, they will be formed in different amounts. Aqueous hydrolysis of the cyanide gives a carboxylic acid (in this case, an aldonic acid), which invariably forms the lactone, and is then reduced to the aldehyde with sodium amalgam. More recently, a procedure has been developed where the cyanide is partially hydrogenated to an imine, using a poisoned catalyst typical of those used to reduce alkynes to alkenes. The imine is produced under conditions in which its hydrolysis is faster than any further reduction. The process was an important one in establishing sugar configurations but has limitations synthetically; yields are low and the reagents toxic. Reduction of the lactone to the hemiacetal with sodium borohydride goes in better yield.

The Wohl degradation is essentially the reverse of the Killani–Fischer chain extension. The classic example is the degradation of glucose to arabinose (Figure 16.31). In the first step, the aldehyde is converted to its oxime. Although glucose does mostly exist in its cyclic form, there is a small amount of free aldehyde present, and, as this is converted, the equilibrium shifts to replenish it. All the hydroxyl groups, including that of the oxime, are then acetylated, and elimination of acetic acid converts the acetylated oxime to the nitrile. The acetates are then removed using methanol (this is a transesterification reaction, driven by the mass action effect

* The spellings Kiliani and Kilani are also found in the literature.

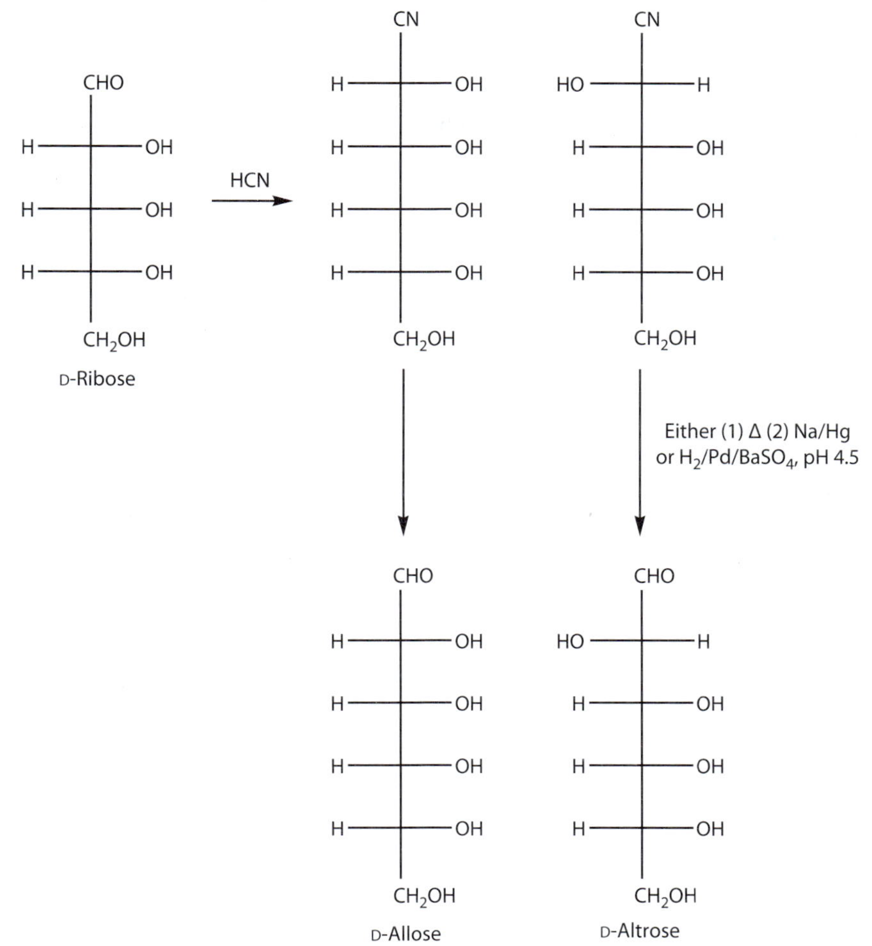

FIGURE 16.30 Killani–Fischer chain extension of ribose.

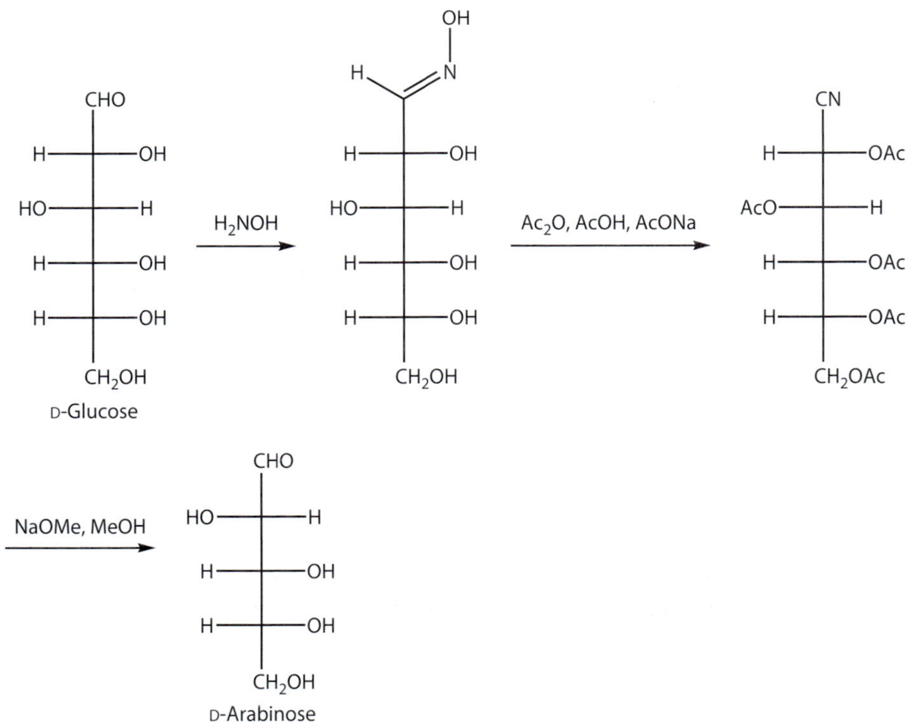

FIGURE 16.31 Wohl degradation of glucose to arabinose.

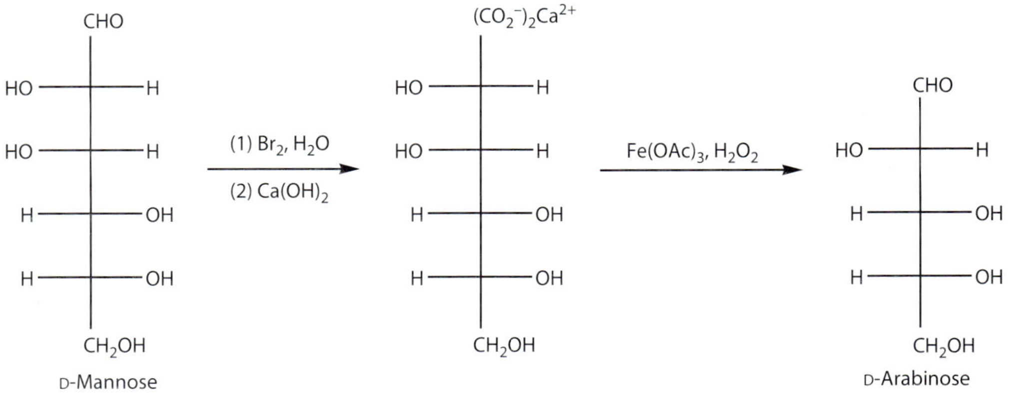

FIGURE 16.32 Ruff degradation of mannose.

of the excess methanol), and the cyanohydrin reverts to the aldehyde. Another method that achieves the same result is the Ruff degradation (Figure 16.32). In the first step, the aldehyde is oxidized to the aldonic acid and then converted to its calcium salt. The mechanism of the final step probably involves a radical process.

PROBLEM 16.10

An unlabeled bottle contains one of the following aldohexoses in pure form:

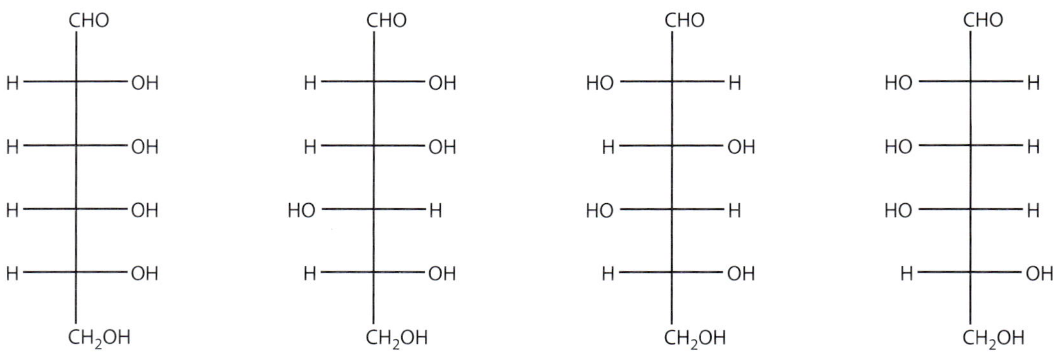

16.70 **16.71** **16.72** **16.73**

To identify the unknown sugar, **16.74**, tests were carried out with the following results:

(i) Oxidation of **16.74** with HNO_3 gave an optically active diacid, **16.75**.
(ii) Ruff degradation of **16.74** (H_2O_2, Fe^{3+}, $CaCO_3$) gave an aldopentose, **16.76**. On treatment with HNO_3, **16.76** gave an optically active diacid, **16.77**.

Deduce the structure of **16.74**. Show each step of your reasoning and give the structures of compounds **16.74–16.77**.

Solution

The first piece of information rules out structure **16.70**, since this would give a *meso*-diacid on oxidation. Ruff degradation of either **16.71** or **16.72** would give structure **16.78**; **16.73** gives **16.79**:

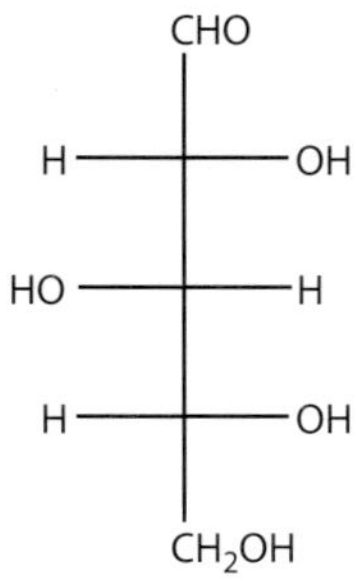

16.78

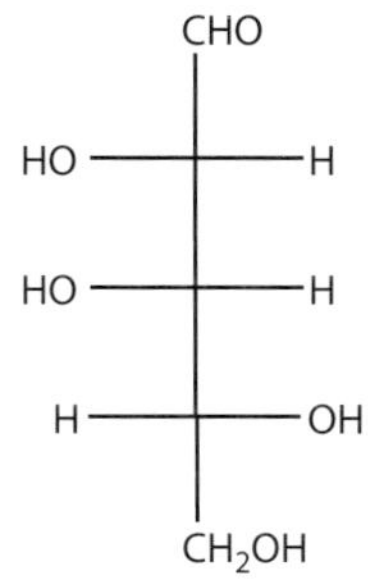

16.79

16.78 would give rise to a *meso*-diacid on nitric acid oxidation. **16.79** would give a chiral diacid. Thus, **16.74** is **16.73**, and the complete sequence is

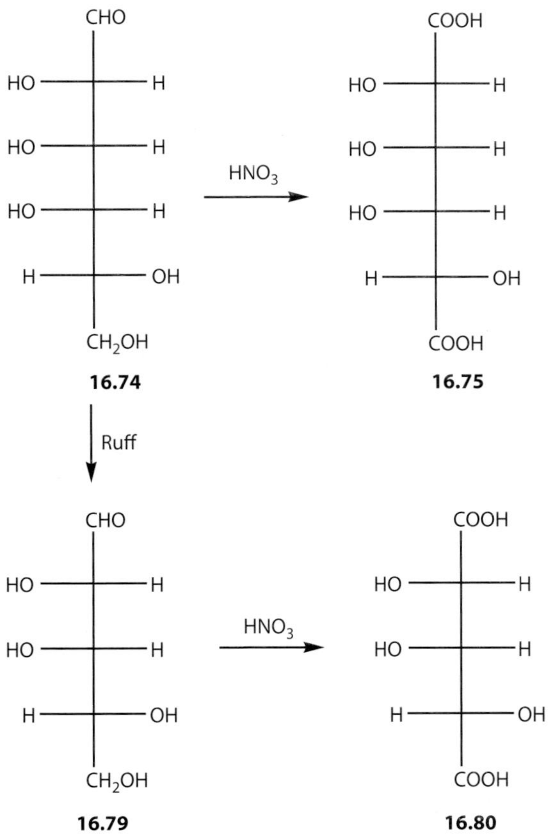

Focus on the Determination of the Configuration of Glucose

The Fischer determination of the configuration of glucose is a *tour de force* in terms of the date at which it was completed—the 1890s. It was not to be until 1951 when Bijvoet used the anomalous dispersion of x-rays in the structure determination of sodium rubidium tartrate (the absolute configuration of which can be reliably related to that of glyceraldehyde) that Fischer's guess as to the absolute configuration of glyceraldehyde was proved to have been correct. In establishing the structure of glucose, Fischer decided to assume the configuration at C5 as *D*. This reduces the number of structures to be considered from 16 to 8—which are shown in the following. Glucose forms an optically active diacid on nitric acid oxidation and an optically active alditol on $Na[BH_4]$ reduction. This rules out structures 1 and 7 in the following, both of which would give *meso*-compounds.

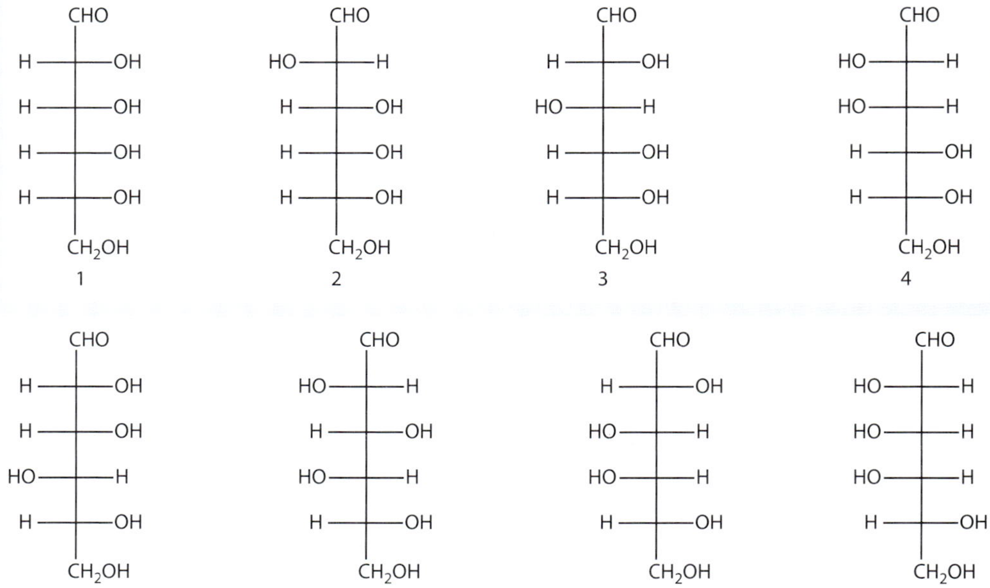

1 2 3 4 5 6 7 8

Glucose and mannose are both obtained from the aldopentose arabinose, by the Killani–Fischer synthesis, and hence, they differ only in their configuration at C2. Arabinose forms an optically active diacid on nitric acid oxidation and an optically active alditol on reduction, so the only two available structures are A and B. On Killani–Fischer chain extension, A gives 3 and 4, and B gives 7 and 8. But we have already excluded 7 as glucose, so arabinose must be A, and 3 and 4 must be glucose and mannose. How do we tell which is which? The hexose gulose forms the same optically active diacid as glucose on nitric acid oxidation. None of our structures can do this for compound 4. Thus, glucose is 3, and mannose is 4. Gulose is 5, and you should prove to yourself that the two acids are identical (it involves quite a lot of manipulation of the Fischer projections, or you could just assign the configurations at each carbon atom—see which way you find quickest).

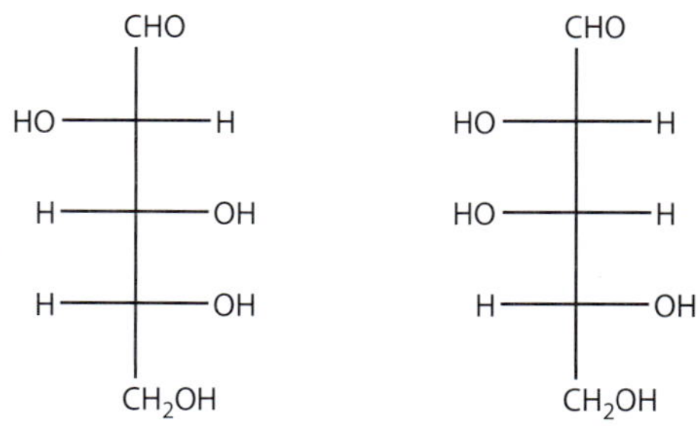

A B

16.3.11 DISACCHARIDES

The most ubiquitous of all the disaccharides is sucrose, table sugar, **16.81**. The glucose is bonded at the anomeric position, with an axial linkage to the anomeric position of fructose, in its furanoside form, with β-orientation with respect to the fructose. Sucrose is not a reducing sugar, as both of the anomeric positions are coupled. This also means that sucrose does not undergo mutarotation. These two observations tell us that it is the anomeric positions that are coupled. Sucrose may be hydrolyzed with dilute aqueous acid or invertase (an enzyme from yeast) to give 1 mol of D-glucose and 1 mol of D-fructose—which tells us what the components of the molecule are. Sucrose can be hydrolyzed by the enzyme α-glucosidase, but not the enzyme β-glucosidase, which implies that the glucose unit is linked at the anomeric position, through an α-linkage. The alternative name for invertase is β-fructofuranosidase, and it is specific for cleavage of fructose units where the fructose is in the furanose form and the linkage is β. Further support for the point of linkage is that when sucrose is reacted with dimethyl sulfate, and the resultant methylated species is hydrolyzed in dilute acid, 2,3,4,6-tetramethylglucose and 1,3,4,6-tetramethyl fructose are obtained (Figure 16.33). The only OH groups that are unmethylated are those that were involved in the linkage between the saccharide units.

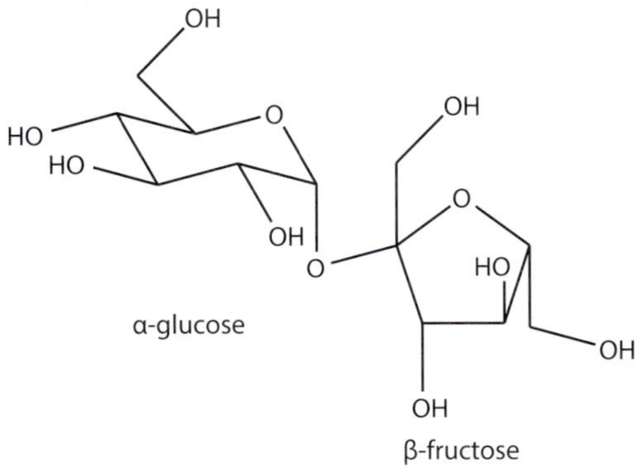

16.81

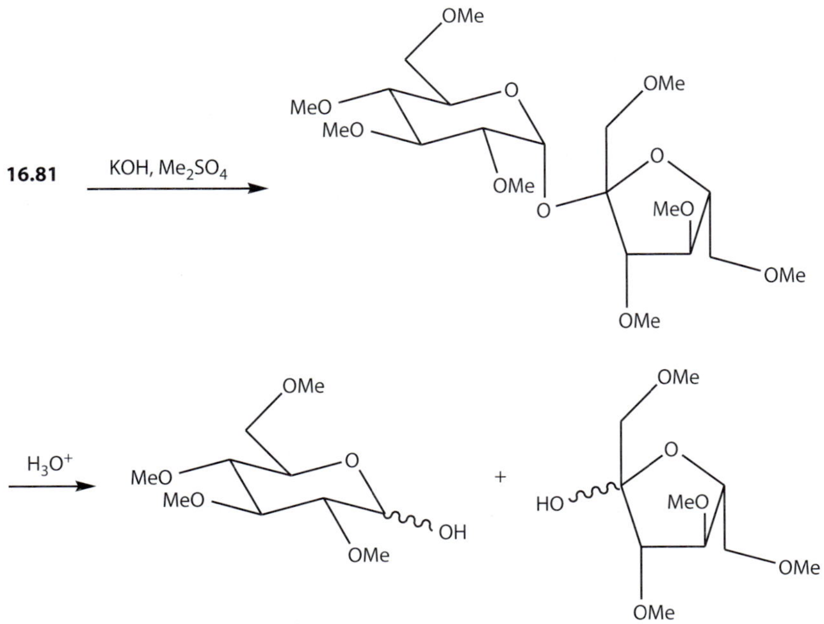

FIGURE 16.33 Methylation and hydrolysis of sucrose.

PROBLEM 16.11

Lactose, milk sugar, is an abundant disaccharide, occurring naturally in human and most animal milk. Crystallization from water gives a single crystalline anomer. Propose a structure of this anomer, based on the information given in the following, showing your reasoning clearly.

(i) Lactose is a reducing sugar and is capable of mutarotation.
(ii) Lactose may be hydrolyzed either in dilute acid or by the enzyme β-galactosidase to give 1 mol of glucose and 1 mol of galactose. (The structure of galactose is shown below.)
(iii) When lactose is treated with dimethyl sulfate, followed by acid hydrolysis, it yields 1 mol of 2,3,6-trimethylglucose and 1 mol of 2,3,4,6-tetramethylgalactose.

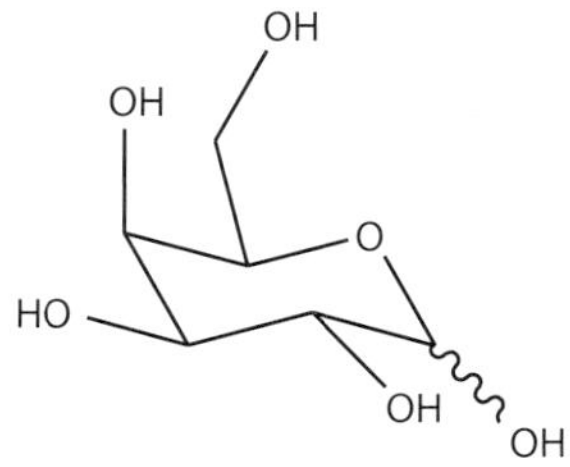

Galactose

Solution

(i) Lactose must contain a "free" aldehyde group in the molecule, one that is not blocked by the linkage.
(ii) Lactose is composed of one glucose and one galactose unit. Since it is hydrolyzed by β-galactosidase, we can conclude that the galactose is coupled at the anomeric carbon and that the linkage is equatorial. In combination with (i), this means that the glucose is *not* coupled at the anomeric carbon; we have a ring that can open and close to mutarotate and for the aldehyde to be oxidized.

So far, we can say we have the units as follows; we know where the galactose (left) is bound, but for the glucose, we only know that it is not at C1.

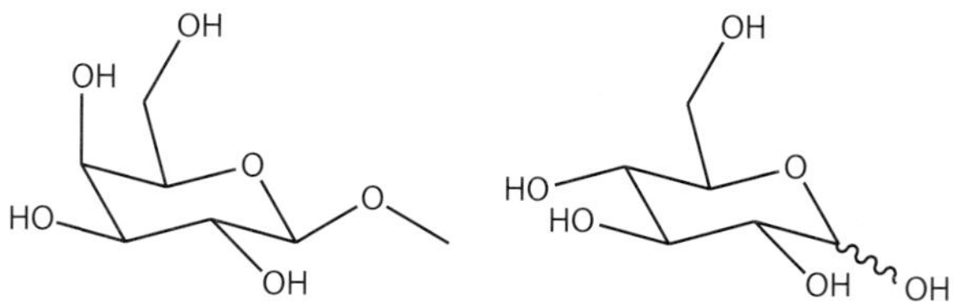

(iii) When lactose is methylated and the glycosidic bond is then hydrolyzed, we get

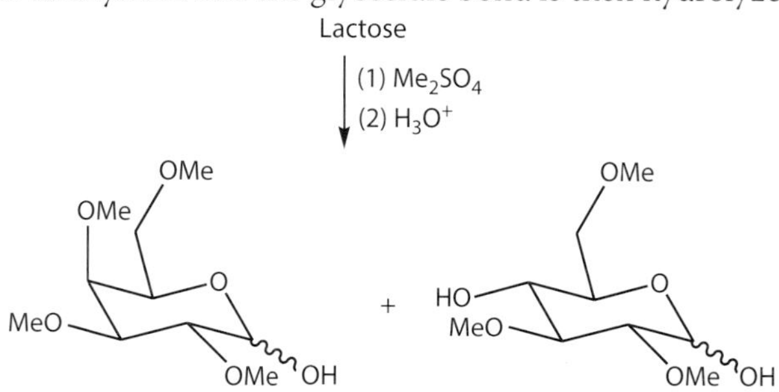

These data allow us to complete the structure—the 4-position of glucose is unmethylated, so this is where the bond to galactose must have been. Note that although the anomeric hydroxyl of glucose will be methylated by dimethyl sulfate, this will be removed during the hydrolysis—it is an acetal, not an ether:

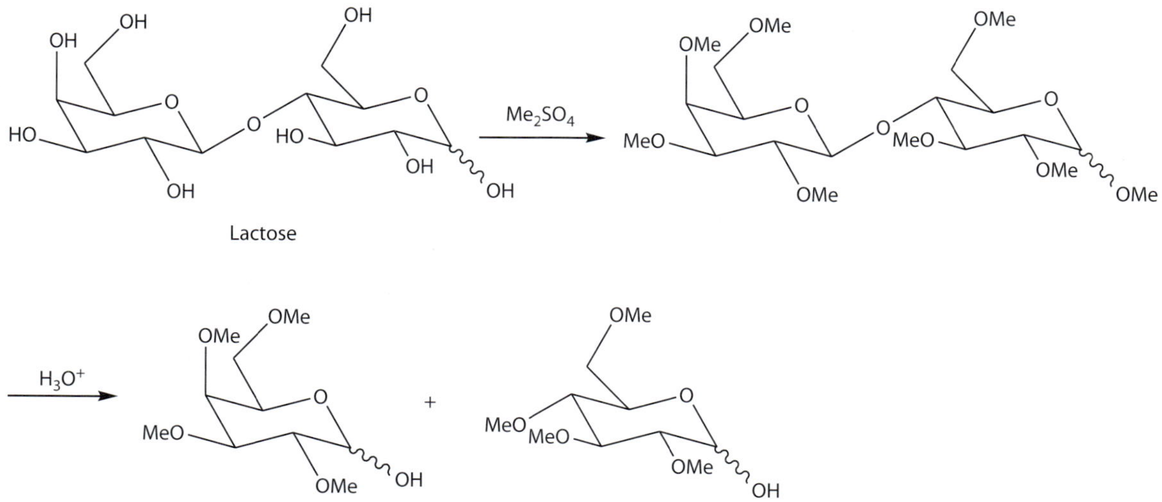

Focus: Sweets for My Sweet, Sugar for My Honey

Honey was the first sweetening agent known to man and has been used for at least 10,000 years. It is stored and used as produced and has a very long shelf life. Bees reduce the water content of plant nectar from 30 %–60 % to 15 %–19 %. The main constituents of honey are glucose (~31 %) and fructose (~38 %), together with small amounts of gluconic acid and higher sugars. Because it is not at this point combined with glucose as sucrose, most of the fructose is in the pyranose form. Honey is as sweet as sucrose but contains fewer nutrient calories per gram (13 kJ g^{-1} as opposed to 17 kJ g^{-1}). Honey has had religious significance—it is one of the elixirs of immortality in Hinduism, and it is used at Rosh Hashanah in Jewish tradition and at the festival of Madhu Purnima in Buddhism. It has been used medicinally as an antibacterial agent, since most microorganisms will not grow in it, in treating sore throats and diabetic ulcers. And Pooh Bear loves hunny.

TABLE 16.6 Relative Sweetness of Various Nutritive and Nonnutritive Sweeteners

Compound	Relative Sweetness	Comments
Sucrose	100	
Glucose	75–80	
Galactose	32	
Fructose	120–175	
Maltose	33–45	
Lactose	16	
Sorbitol	50	Used in "sugar-free" gum, 11 $kJ\ g^{-1}$.
Xylitol	100	Low glycemic index, suitable for diabetics. About two-thirds the calorific value of sucrose. Useful in dental health, so widely used in gum, toothpaste, and mouthwash.
Invert sugar	50	Produced by sucrose hydrolysis; 1:1 mixture of glucose and fructose.
Honey	97	Contains mainly glucose and fructose.
Molasses	74	
High-fructose corn syrup	120–160	
Sodium cyclamate	2,600	Nonnutritive. Banned in the United States in 1970.
Aspartame	18,000	Nonnutritive. Not heat stable. Marketed as *Nutrasweet*, *Aminosweet*. Because of the phenylalanine content, not suitable for sufferers from phenylketonuria.
Saccharin	45,000	Nonnutritive. Banned in Canada.
Sucralose	60,000	Marketed as *Splenda*®, heat stable. Approved in 2002 in the United Kingdom. Essentially nonnutritive. No effect on blood glucose or insulin levels, suitable for diabetics.
Alitame	160,000	Not yet approved in the United States, amide of L-aspartyl-D-alanine. More heat and acid stable than aspartame.
Neotame	60,000	Approved in the United States in 2002, more heat stable than aspartame.
Stevioside	30,000	Extract of stevia leaves, known as sweet leaf or sugar leaf. Marketed as Truvia®, Pure Via®, or Enliten®. All the sugar units are glucose.
Lugduname	22,000,000–30,000,000	Artificial. Name from Lugdunum, the Latin name for Lyons. Not yet approved for food use.

The relative sweetness of various monosaccharides, disaccharides, and other natural and artificial sweeteners is given in Table 16.6. Over the last 90 years, human food consumption in the developed world has shifted from complex carbohydrates (polysaccharides such as starch) toward simple ones such as sugar and syrups. The average percentage of sugar in carbohydrate consumption was 32 % in 1910 and 53 % in 1980; there are obvious consequences in terms of human health. Any artificial, nonnutritive sweetener must satisfy stringent requirements. It must be safe for human consumption, water soluble, fairly stable to heat and light, and at least equal to sucrose in terms of cost per sweetening effect. L-Sugars have attracted interest because they are sweet and they are not metabolized, but they are relatively expensive. Saccharin has a bitter aftertaste, and concerns have been raised about the health effects of both this and cyclamate. Aspartame should be avoided by those with phenylketonuria because of its phenylalanine content (Figure 16.34).

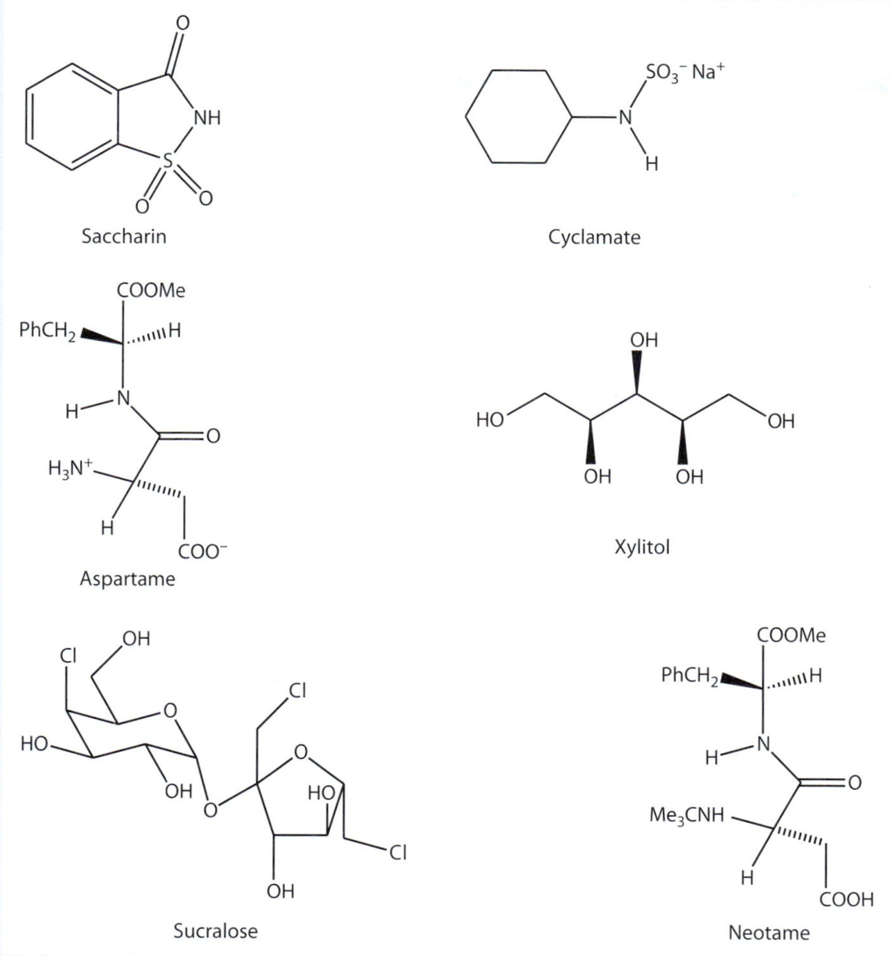

FIGURE 16.34 Structures of some nonsugar sweeteners.

16.3.12 POLYSACCHARIDES

Starch is the carbohydrate that is the most common in a normal diet. It is found abundantly in root vegetables, grains, and rice. About 20 % of starch in plants is the unbranched polymer, amylose, a polymer of glucose with 1,4-α repeating units (**16.82**). The polymer has a helical structure (Figure 16.35) and is readily digestible by all mammals. The other 80 % of starch is the branched species amylopectin—every 25–30 units, there is a branch involving bonding at the 6-position (**16.82**). Carbohydrate storage in humans involves glycogen, a still more branched version, with branches approximately every 10 units.

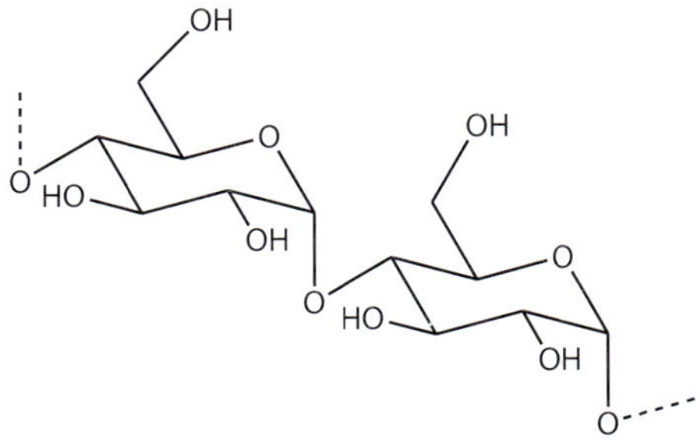

16.82, amylose

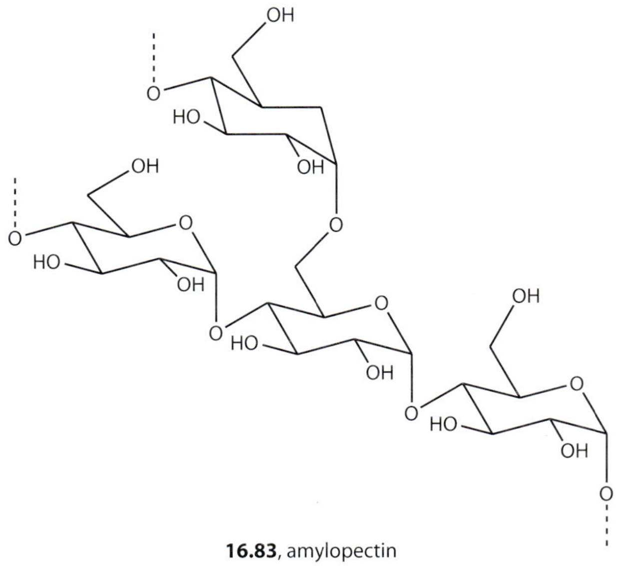

16.83, amylopectin

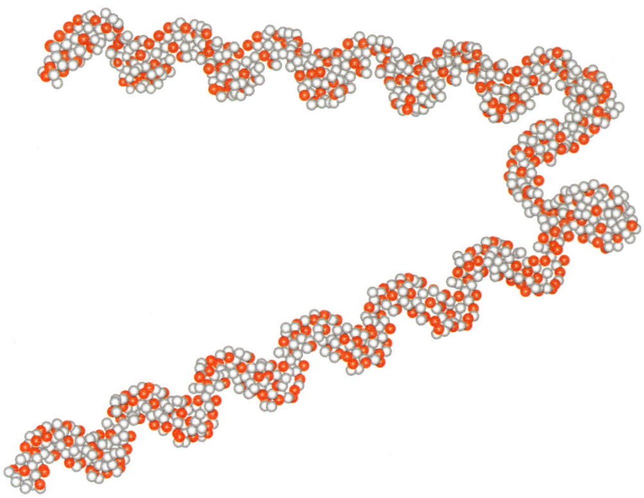

FIGURE 16.35 Three-dimensional helical structure of amylose.

Cellulose is the most abundant biomolecule, indeed the most abundant organic molecule, on the planet. It is used for the cell walls of all green plants and, on average, makes up 30 % of the total mass of plant life. Cellulose is also a 1,4-polymer of glucose, but the linkage is β (**16.84**), giving rise to a linear structure with strong interchain hydrogen bonds. Humans lack the enzyme to break the β-linkages; this is found only in ruminants, where it is provided by symbiotic bacteria. The inability of humans to digest grass has been long known. If we read the Old Testament, Daniel 4.33, we hear

> The same hour was the thing fulfilled on Nebuchadnezzar; and he was driven from men, and did eat grass as oxen, and his body was wet with the dew of heaven, til his hairs were grown like eagles' feathers, and his nails like birds' claws.

It is a matter of historical record that Nebuchadnezzar had a period of madness, but other commentators attribute this to porphyria or tertiary syphilis.

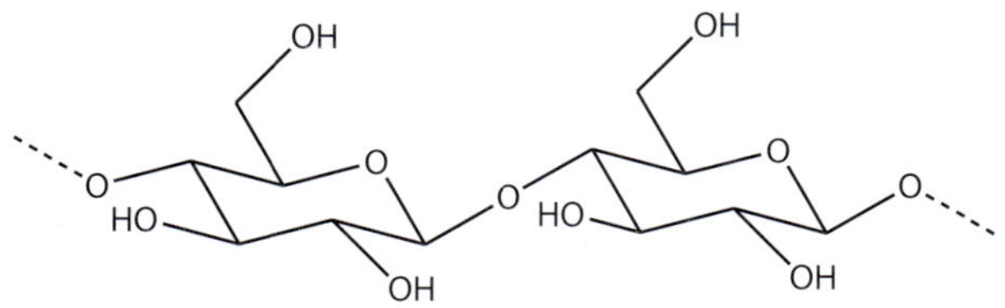

16.84, cellulose

Key Points from Section 16.3

- Sugars are polyhydroxy compounds containing either an aldehyde (aldose sugars) or ketone (ketose sugars) group. Nomenclature is specific to the class, with all names ending in -ose.
- The aldopentose ribose is crucial to the structure of RNA, and 2-deoxyribose takes its place in DNA.
- Glucose, a hexose sugar, is the most common of the hexose sugars and is crucial in many biochemical pathways.
- Both aldose and ketose sugars generally exist in solution in the form of five- or six-membered ring cyclic hemiacetals. These are described as furanose and pyranose forms, respectively. They have one more chiral center than the parent molecule.
- If the cyclic form is in equilibrium with an open-chain aldehyde, the sugar is described as a reducing sugar and undergoes mutarotation.
- The hydroxyl groups of sugars can be methylated or acetylated. Reaction of the hemiacetal forms with an alcohol gives an acetal, called a glycoside.
- Oxidation with mild oxidants such as Ag(I), Cu(II), or Br_2/H_2O results only in aldehyde oxidation to yield an aldonic acid. Vigorous oxidation with HNO_3 results in oxidation of both the aldehyde and the primary alcohol to give a diacid, called an aldaric acid.
- Reduction with $Na[BH_4]$ gives a polyol, which does not exist in a cyclic form.
- The sugar chain can be lengthened (Killani–Fischer reaction) or shortened (Wohl or Ruff degradations.
- Disaccharides such as sucrose can have their structures established by methylation followed by hydrolysis as well as various specific enzymatic reactions.
- Polysaccharides include the structural materials starch (which is digestible) and cellulose (which isn't). These are polymers of glucose, which differ in the stereochemistry of the 1,4-linkage.

William Blake, Nebuchadnezzar, Tate Images

REVIEW PROBLEMS

1. Give systematic names for each of the following compounds:
   (a) $HOOC(CH_2)_7COOH$, produced from ozonolysis of oleic acid
   (b) $CH_3(CH_2)_7C{\equiv}C{-}CH_2{-}CH{=}CH{-}(CH_2)_4COOH$ from nuts of *Ongokea klaineana*
   (c) $CH_3(CH_2)_{10}C{\equiv}C(CH_2)_4CO_2H$ (tariric acid)
   (d) Crepenynic acid from seeds of *Crepis foetida*

2. Give structures for each of the following compounds:
   (a) *n*-Hexadecyl 2,4,6-trimethylnonanoate, produced from the uropygial glands of the sipper duck
   (b) 6-*Z*-6-Hexadecenoic acid (also called sapienic acid, 25 % of human wax sebum)
   (c) Ethyl oleate, a primer pheromone that causes delayed start of foraging in young honey bees
   (d) 2,4,6-Tri-*O*-methylglucopyranoside
3. In analysis of fats, the triglyceride is reacted with acidic methanol to give methyl esters of the fatty acids and glycerol. The various proportions of the fatty acids are then determined by GLC of the esters, using methyl pentadecanoate as an internal standard.
   (a) Write a mechanism for the transesterification reaction.
   (b) Why are the esters, rather than the free acids, analyzed?
   (c) Why is methyl pentadecanoate a suitable internal standard?
4. Explain the following terms and give a specific example of each:
   Disaccharide
   Wax
   Reducing sugar
   Mutarotation
5. Gadoleic acid, $C_{20}H_{38}O_2$, is a fatty acid isolated from cod liver oil. Hydroxylation with OsO_4 followed by cleavage of the diol produced with periodic acid gives the aldehydes $CH_3(CH_2)_9CHO$ and $OHC(CH_2)_7COOH$. What stereoisomeric structures can you propose for gadoleic acid? How might spectroscopy help you to distinguish these stereoisomers?
6. Which of the following triglycerides would be chiral?

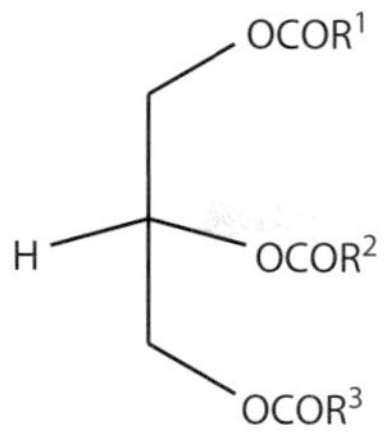

(a) $R^1, R^2, R^3 = C_{17}H_{35}$
(b) $R^1, R^2 = C_{15}H_{31}, R^3 = C_{17}H_{35}$
(c) $R^1, R^3 = C_{17}H_{35}, R^2 = C_{15}H_{31}$

7. Write a mechanism for each of the following reactions:

(a)

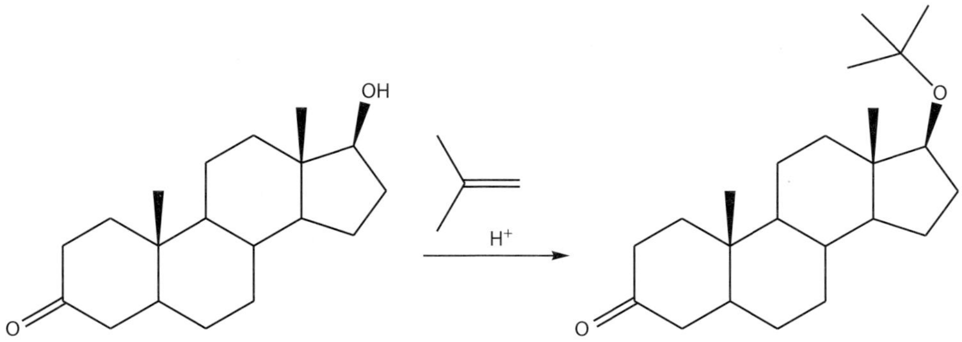

(b)

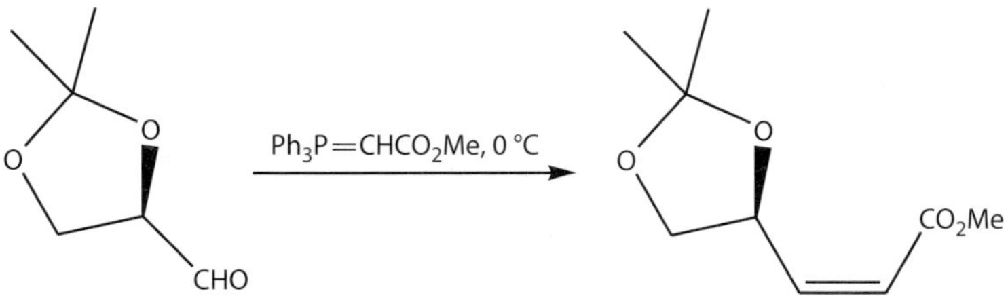

Used in a prostaglandin synthesis

(c)

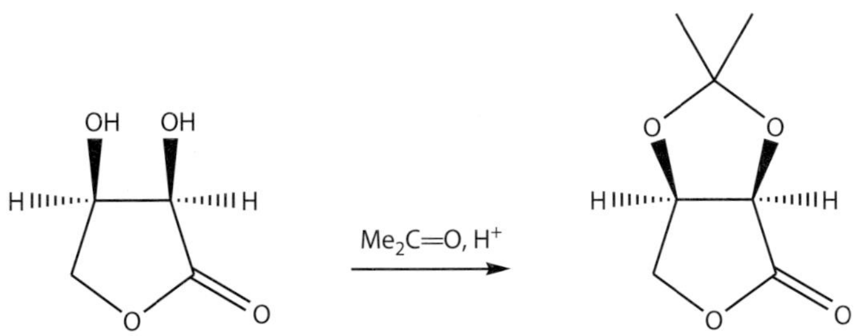

Used in a prostaglandin synthesis

(d)

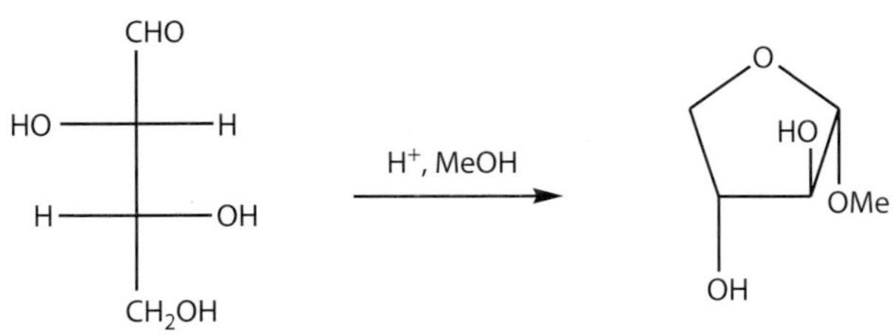

(e)

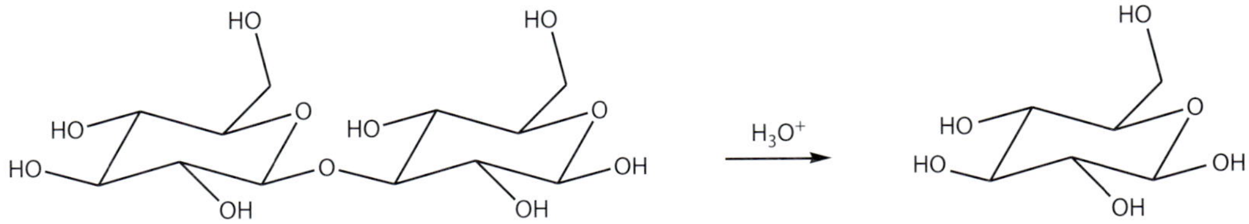

(f)

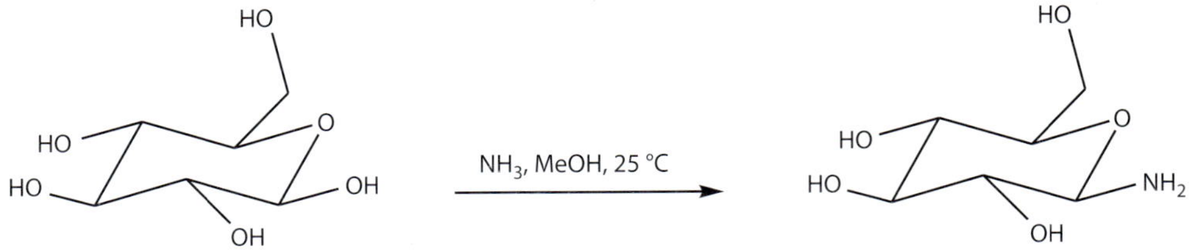

(g)

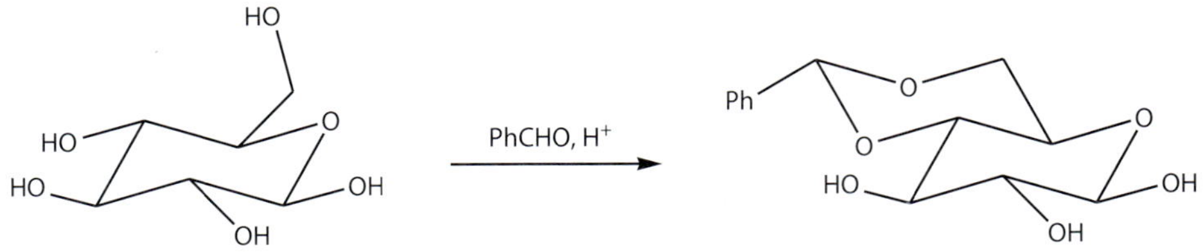

(h)

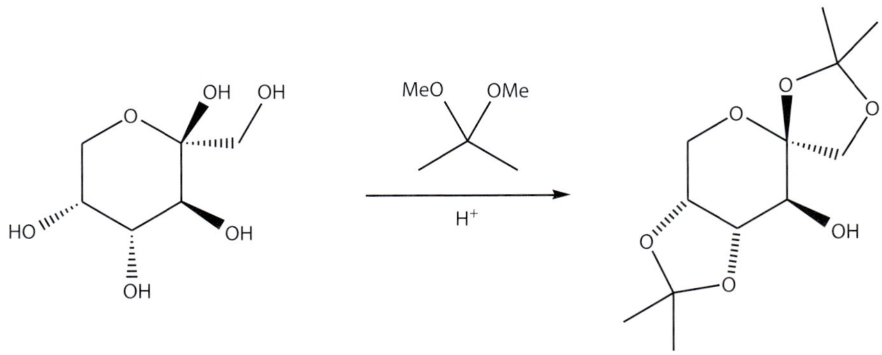

(i)

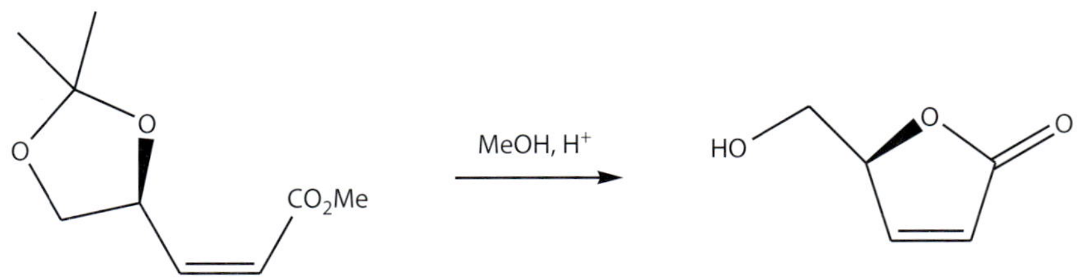

8. Predict the product(s), including stereochemistry as appropriate, of each of the following reactions:

(a)

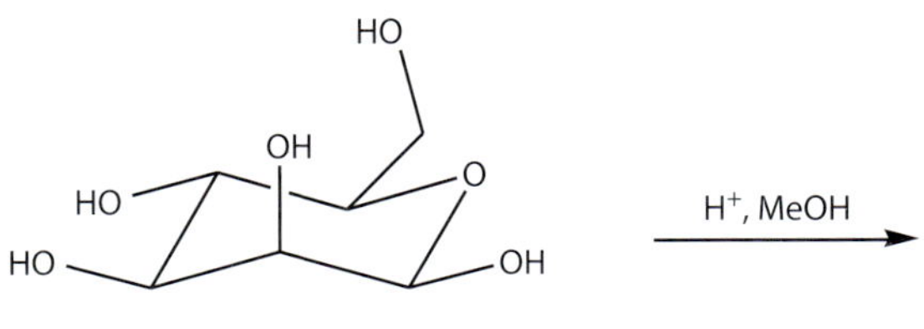

β-D-mannose

(b)

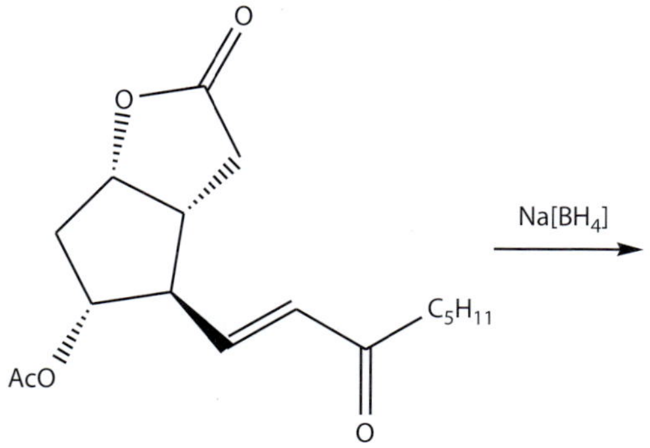

(c)

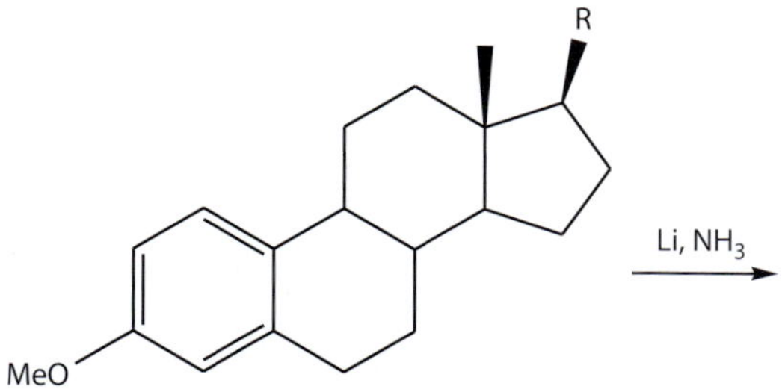

(d)

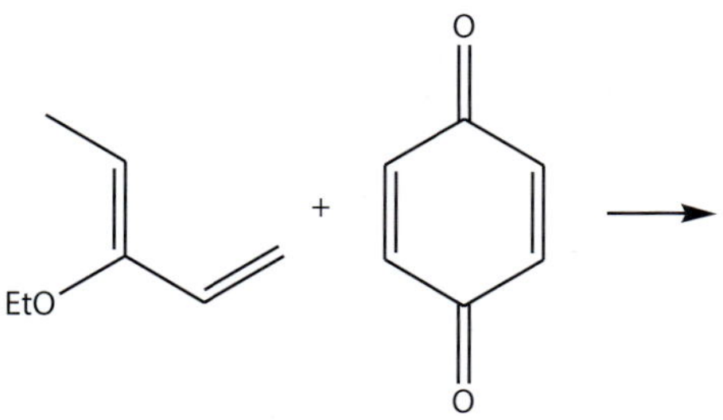

Part of a steroid synthesis

9. How would you carry out each of the following transformations, which may require more than one step?

(a)

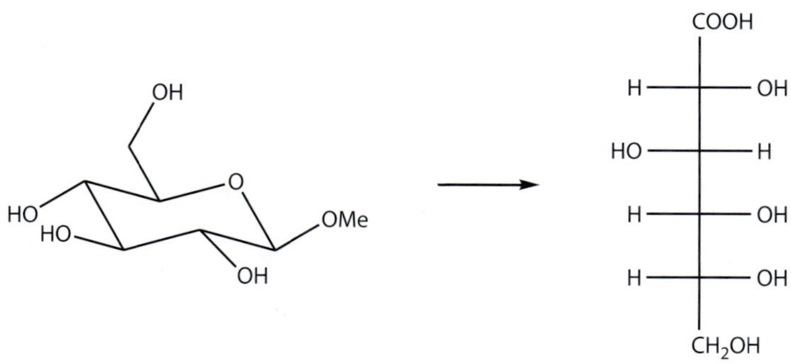

(b)

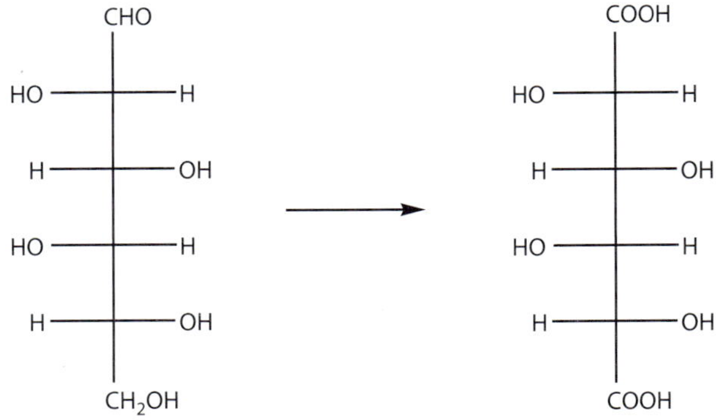

(c)

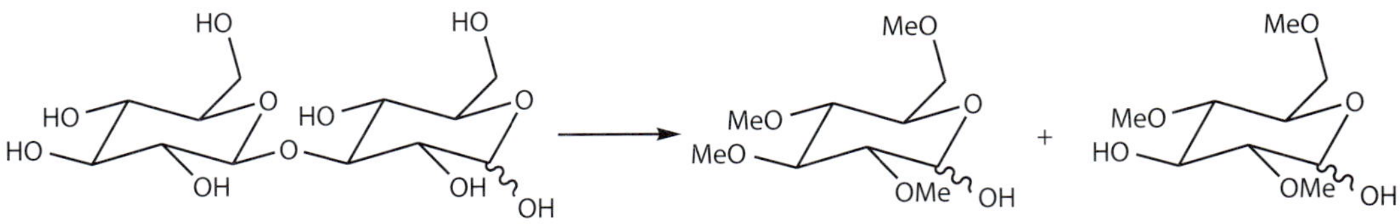

(d)

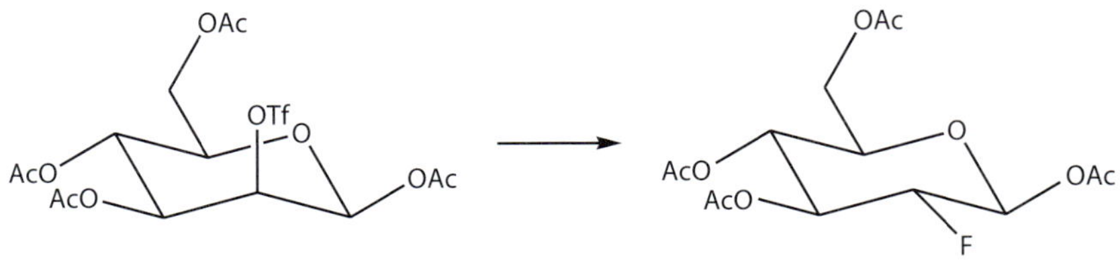

(e)

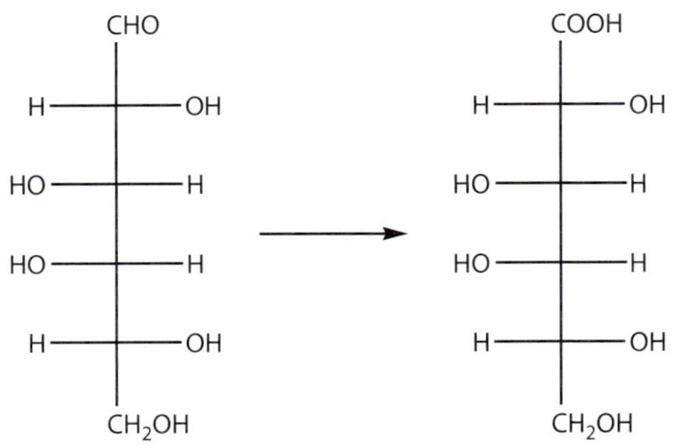

D-(+)-Galactose

(f)

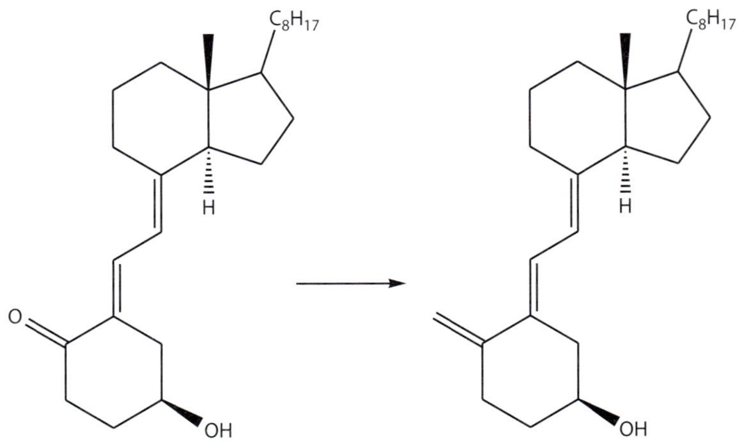

Vitamin D_3

(g)

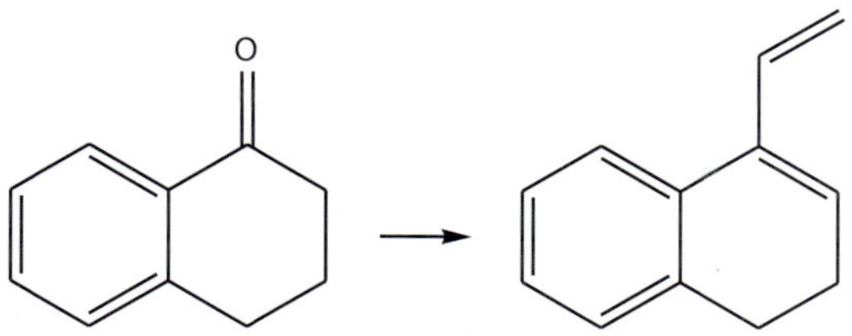

10. Trehalose is a nonreducing disaccharide, which on hydrolysis in aqueous acid gives 2 mol of glucose. Methylation with dimethyl sulfate followed by hydrolysis gives 2 mol of 2,3,4,6-tetra-*O*-methylglucose. Trehalose is hydrolyzed by α- but not β-glucosidase.

    The related compound neotrehalose is hydrolyzed only by β-glucosidase. The related isotrehalose is hydrolyzed by both α- and β-glucosidase. Both neotrehalose and isotrehalose give the same products as trehalose on methylation followed by hydrolysis.

    What are the structures of these three sugars?
11. Gentiobiose, a rare disaccharide found in gentian and saffron, is a reducing sugar and yields only D-glucose on hydrolysis with aqueous acid or β-glucosidase. Reaction of gentiobiose with methyl iodide in the presence of Ag_2O gave an octamethyl derivative, which could be hydrolyzed with aqueous acid to give 2,3,4,6-tetra-*O*-methyl-D-glucopyranose and 2,3,4-tri-*O*-methyl-D-glucopyranose. Deduce the structure of gentiobiose.
12. Isomaltose is a disaccharide obtained by enzymatic hydrolysis of amylopectin and is capable of mutarotation. Hydrolysis of 1 mol of isomaltose by aqueous acid or α-glucosidase gave 2 mol of D-glucose. Methylation of isomaltose followed by acidic hydrolysis gives 2,3,4,6-tetra-*O*-methyl-D-glucose and 2,3,4-tri-*O*-methyl-D-glucose. Deduce the structure of isomaltose, explaining your reasoning clearly.
13. Lactose exists as two anomers, while sucrose doesn't. Why?

MORE CHALLENGING PROBLEMS

1. What is "good cholesterol"? Why is the term misleading?
2. Two important polysaccharides are formed from glucose. What are they and how do their structures determine their function?
3. Write a mechanism for each of the following transformations:

(a)

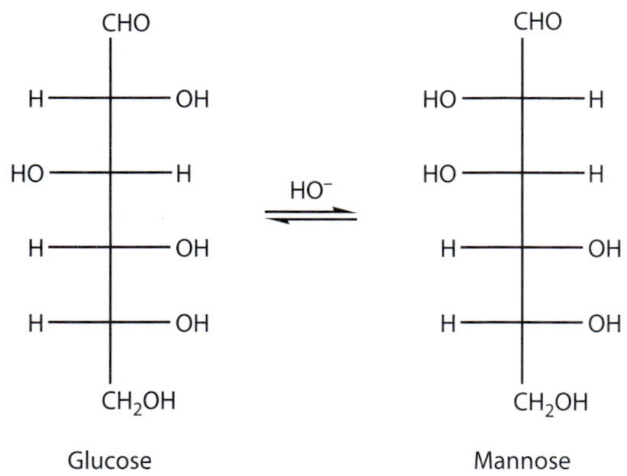

(b)

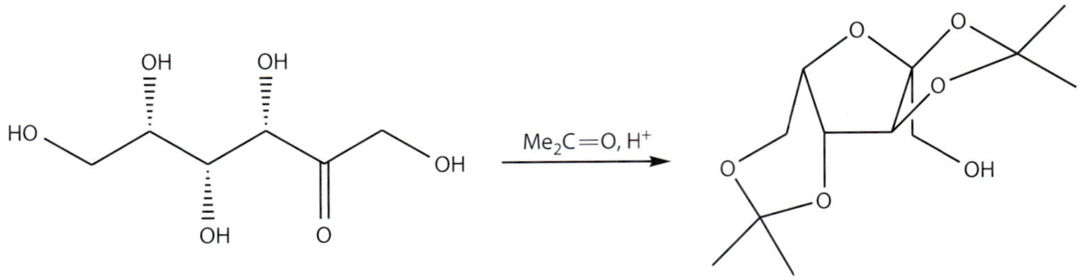

(c)

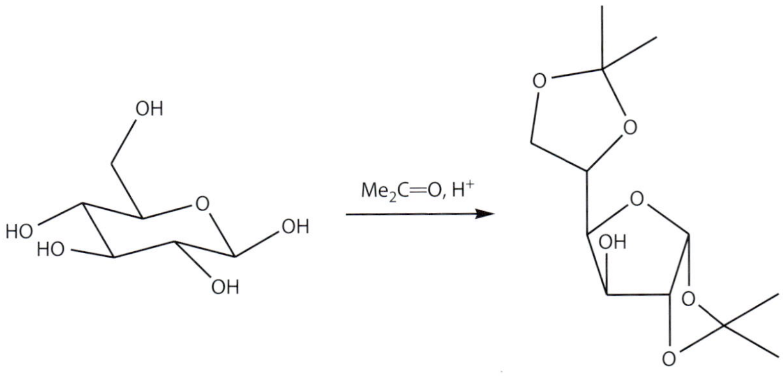

(d)

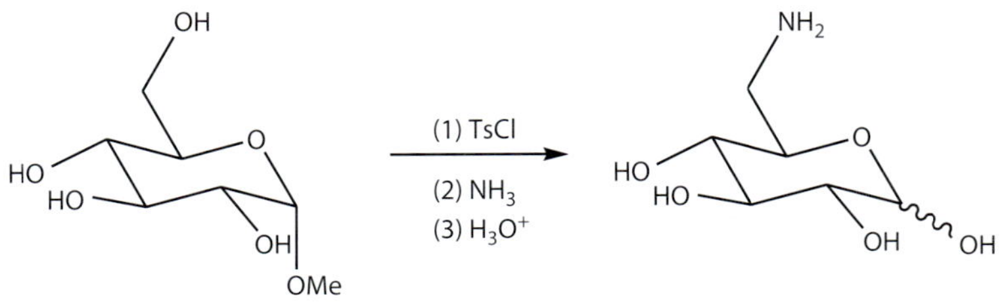

(e)

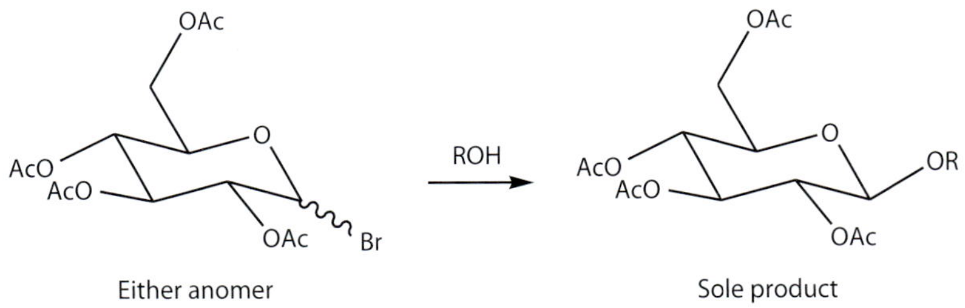

4. When a hexapyranose is in the chair conformation in which the CH_2OH and the C1 OH are both axial, these can form an acetal with the loss of water. This is called the anhydro form of the sugar. The anhydro form of D-idose is shown as follows. In aqueous solution at 100 °C, idose exists as the anhydro form to the extent of 80 %. However, under the same conditions, glucose gives only 0.1 % of the anhydro form. Why?

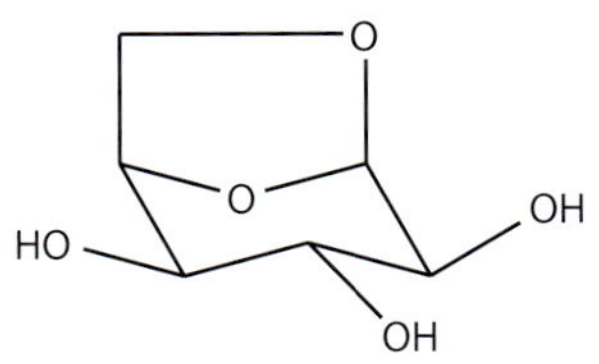

5. Study the reaction scheme shown in the following and answer the questions posed:

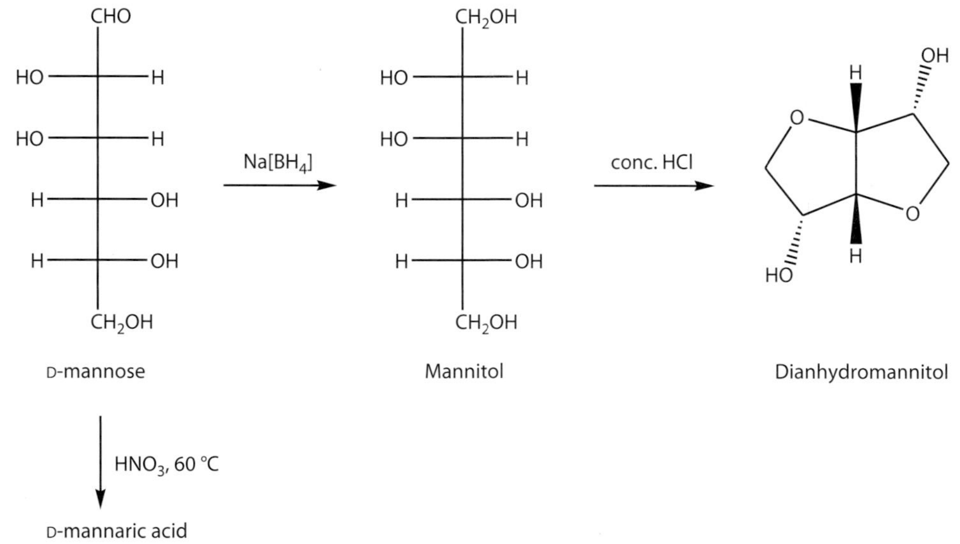

   (a) Give a structure for D-mannaric acid. Is this chiral?
   (b) In the formation of dianhydromannitol, have the stereocenters been inverted?
   (c) Propose a mechanism for the formation of dianhydromannitol.

6. An unknown disaccharide gave Tollens' test with Ag^{+}. A β-glycosidase hydrolyzes it to D-galactose and D-mannose. When it is reacted with MeI/Ag_2O, then subjected to aqueous hydrolysis, 2,3,4,6-tetra-O-Me galactose and 2,3,4-tri-*O*-Me mannose were obtained. Propose a structure for the disaccharide, showing your reasoning clearly:

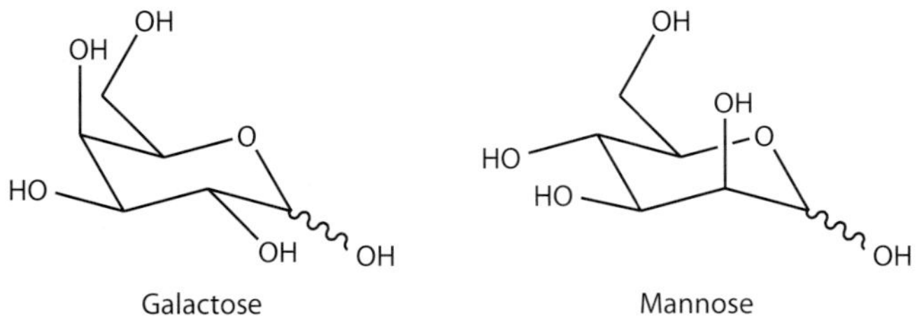

7. Bemisiose is a trisaccharide isolated from the honeydew of the silverleaf whitefly. Hydrolysis by aqueous acid or the enzyme α-glucosidase gives 3 mol of glucose. The enzyme porcine trehalase releases 1 mol of glucose; this enzyme is specific for 1,1-α-linked glucose units. Reaction of bemisiose with dimethyl sulfate, followed by acid-catalyzed hydrolysis, gave 2 mol of 2,3,4,6-tetra-*O*-methylglucose and 1 mol of 2,3,6-tri-*O*-methylglucose. Deduce the structure of bemisiose.
8. Amygdalin is a toxic compound isolable from the stones of almonds, peaches, and apricots. It is not a reducing sugar. Careful aqueous hydrolysis gives benzaldehyde, HCN, and 2 mol of glucose. It is hydrolyzed by β- but not α-glucosidase. Permethylation followed by hydrolysis gives 2,3,4,6-tetra-*O*-methylglucose and 2,3,4-tri-*O*-methylglucose. Deduce the structure.
9. Predict the products **16.85** and **16.86** from the following sequence:

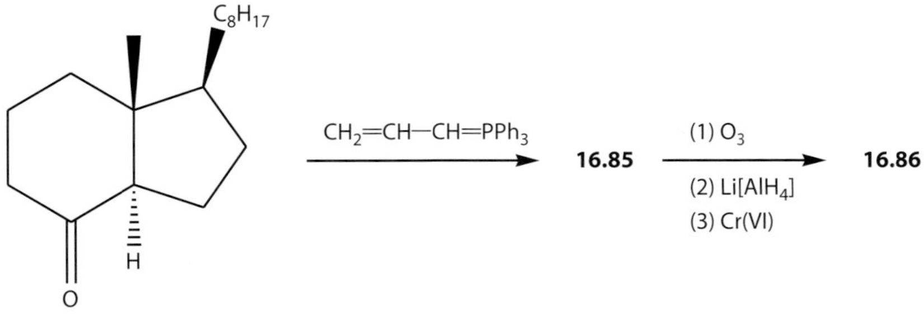

10. Study the following sequence and answer the questions posed:

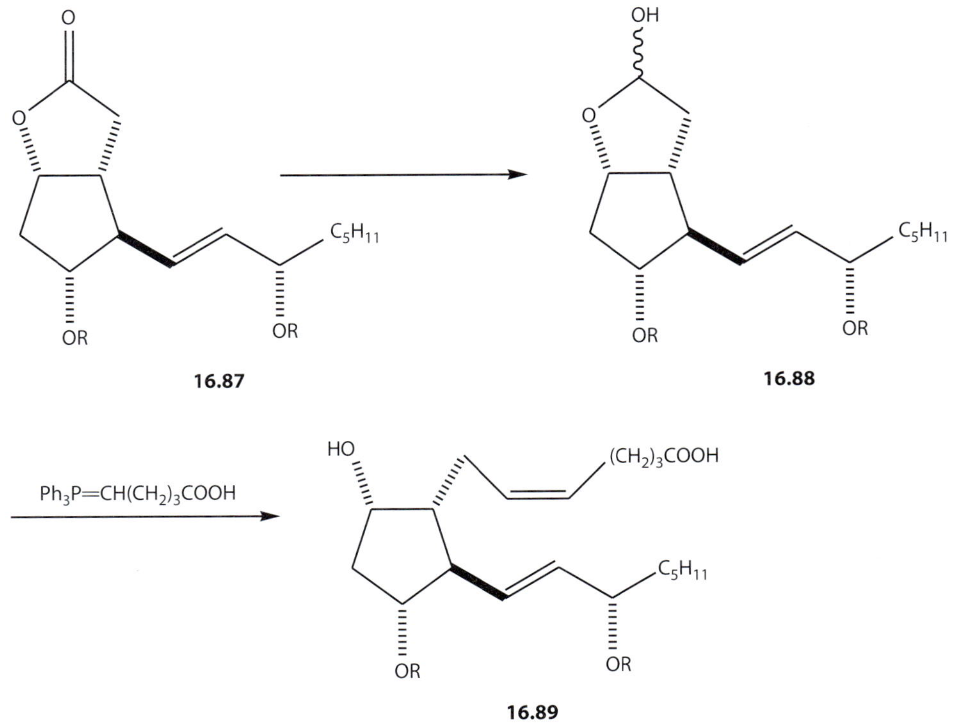

(a) What would be the best reagent to convert **16.87** to **16.88**?
(b) **16.88** is in equilibrium with another compound. Give its structure.
(c) Bearing in mind your answer to (b), explain the conversion of **16.88** to **16.89**.

Substitution α and β to Carbonyl Groups

17

17.1 INTRODUCTION

In Chapters 14 and 15, we saw how carbonyl groups, because of their high polarity, behave as electrophiles toward a wide range of nucleophiles. In this chapter, we will consider reactions not at the carbon of the carbonyl, but at the α and β carbon atoms (**17.1**)—that is, the first and second carbon atoms moving along the chain from the carbonyl carbon. The reactions at the α-position depend on the ability of the carbonyl compound to be converted into its enol tautomer, which behaves as a nucleophile. Reactions at the β-position can occur when the carbonyl is conjugated with a carbon–carbon double bond—attack of nucleophiles can occur at the β-carbon, with the electrons being deposited ultimately on the oxygen of the carbonyl group.

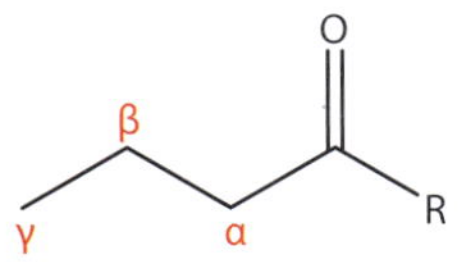

17.1

17.2 FORMATION OF ENOLS AND ENOLATES

We learned previously that carbonyl compounds of most types can interconvert with their enol tautomers. We looked at enol formation as a thought experiment, by removing a proton, writing a resonance form, and then replacing the proton (Section 8.5, Figure 17.1). This is essentially what happens in the base-catalyzed process (Figure 17.2). The acid-catalyzed process is similar, but the order of events is different—the carbonyl group is first protonated, then loses a proton from the α-carbon to give the enol (Figure 17.3).

Any enolizable carbonyl compound is, at least in principle, in equilibrium with its enol tautomer, though equilibrium constants vary widely. The enol is in an acid/base equilibrium with the enolate anion, and the enolate anion has two important resonance forms, with charge residing either on oxygen or on the α-carbon atom (Figure 17.4).

Before we look at the position of the equilibrium between enol and keto forms of carbonyl compounds, it's important to note that not all carbonyl compounds can enolize. Any compound such as those in Figure 17.5, where there is no hydrogen at the α-position, cannot ever form an enol. Although, in principle, you might believe that an α,β-unsaturated carbonyl compound such as **17.2** might enolize, in practice, this is very rare, because the allenyl system, **17.3**, is strongly disfavored. A few other molecules can't form enols even though they have an α-hydrogen atom. For example, in **17.4**, there is an α-hydrogen, at the bridgehead, but any enol formed would have a very strained, nonplanar double bond (**17.5**).

17.2 **17.3**

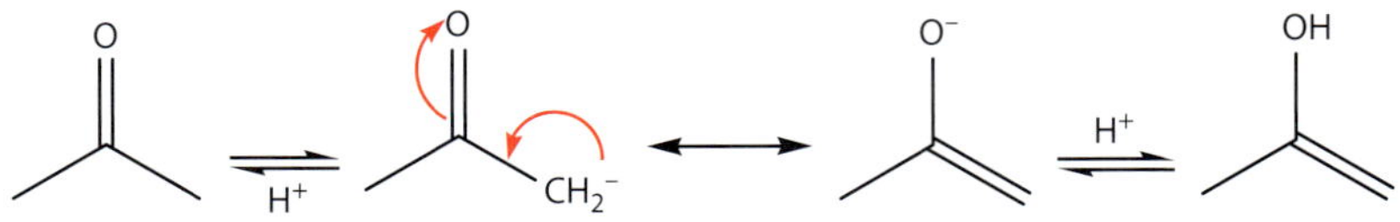

FIGURE 17.1 Thought experiment in making enols.

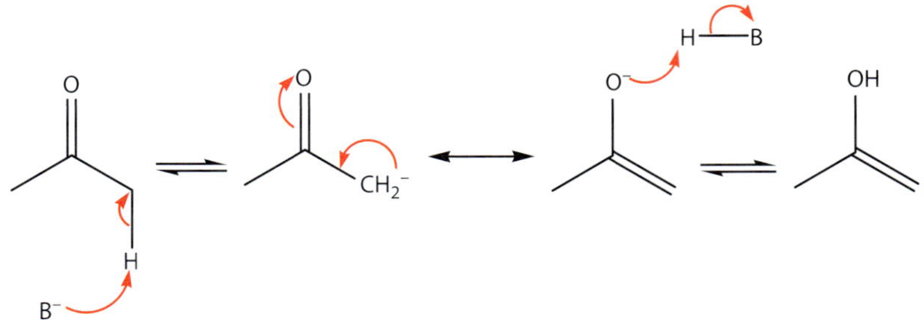

FIGURE 17.2 Base-catalyzed formation of enols.

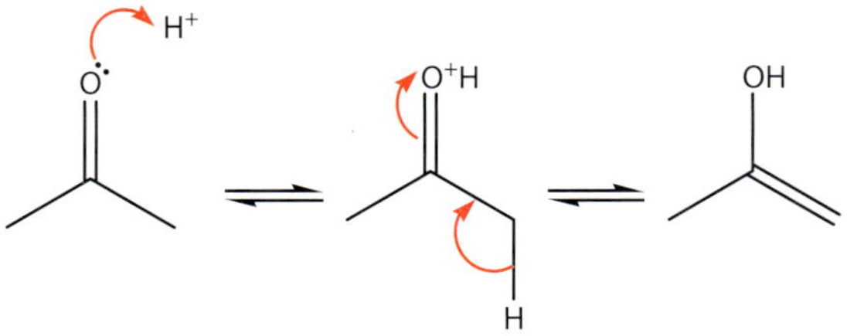

FIGURE 17.3 Acid-catalyzed formation of enols.

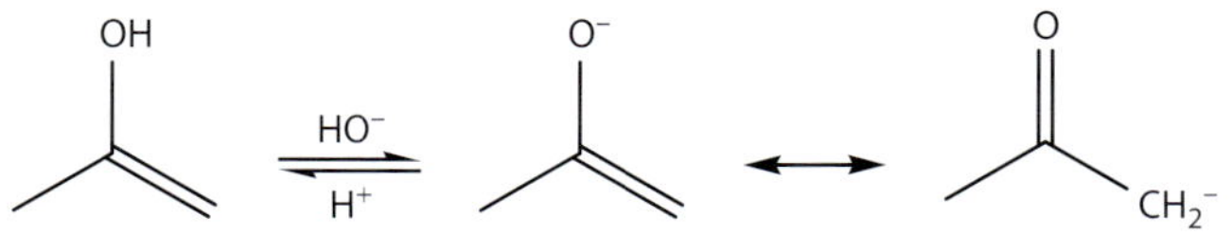

FIGURE 17.4 Acid/base equilibrium of enols and enolates.

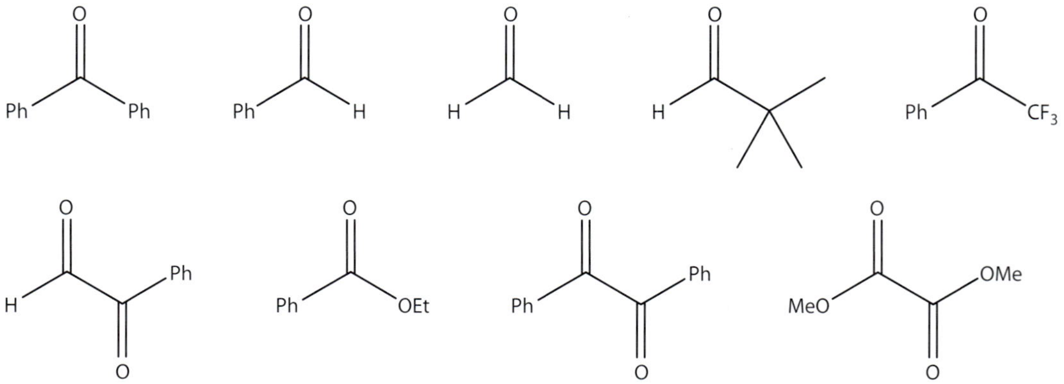

FIGURE 17.5 Nonenolizable carbonyl compounds.

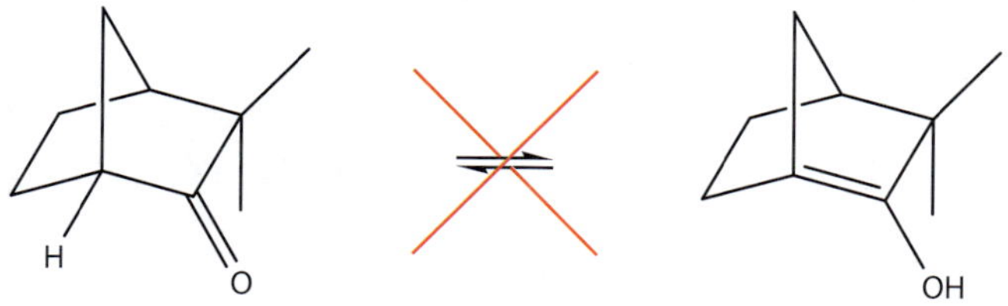

17.4 17.5

In compounds that can enolize, the proportions of enol vary widely at equilibrium. We can identify a number of factors that influence the equilibrium. The most effective is the achievement of aromaticity by enolization. Thus, **17.6** is unknown—the equilibrium is entirely in favor of phenol. That said, the reaction of phenol with electrophiles (**17.7**) has a strong resemblance to many of the reactions of enols with electrophiles that we will meet.

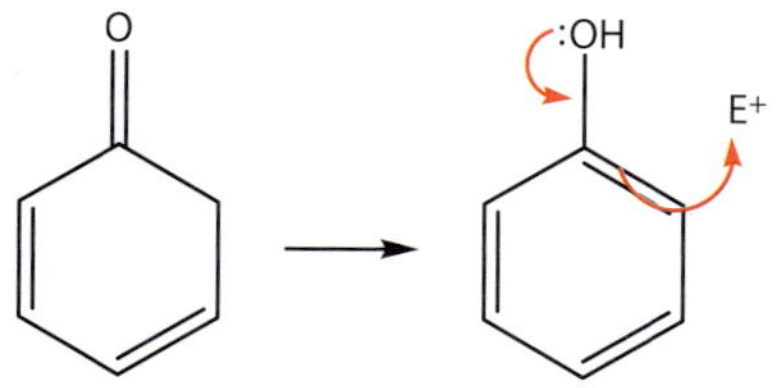

17.6 17.7

Next, enols are stabilized by conjugation; thus, β-dicarbonyl compounds, such as cyclohexane-1,3-dione, **17.8**, contain more enols (33 % in chloroform solution). Vitamin C, **17.9**, is another example where the compound exists entirely in its enediol form at equilibrium. Hydrogen bonding is advantageous; gadusol, **17.10**, where intramolecular hydrogen bonding is possible, exists entirely in the enol form and is a natural product isolated from cod roes. 2,4-Pentanedione (**17.11**, common name acetylacetone and often abbreviated as acac) contains 80 % enol at equilibrium; it is stabilized by both conjugation and hydrogen bonding. Substitution of the double bond in an enol confers some additional stability, just as it does for alkenes. Thus, acetaldehyde has approx. 6 × 10^{-5} % enol at equilibrium, but 2-methylpropanal, **17.12**, has 1.4 × 10^{-2} %. Generally, aldehydes have somewhat more enol content than comparable ketones; thus, acetone has only 6 × 10^{-7} % enol at equilibrium.

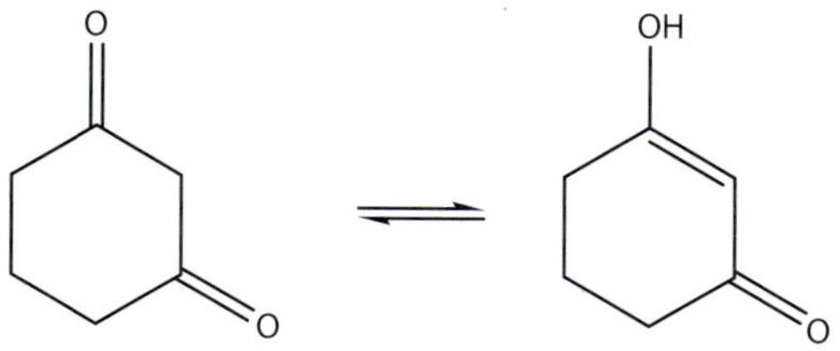

17.8

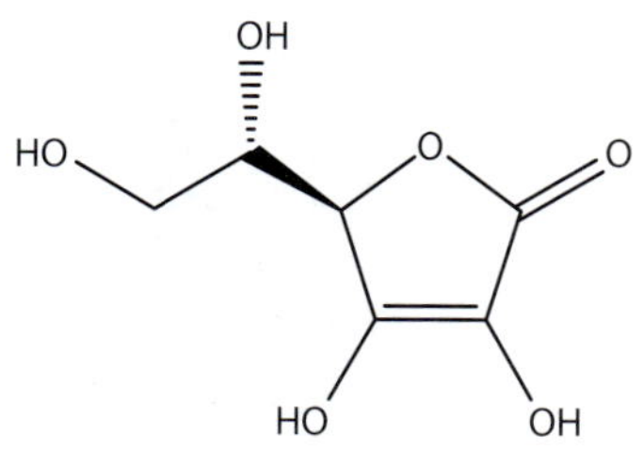

17.9

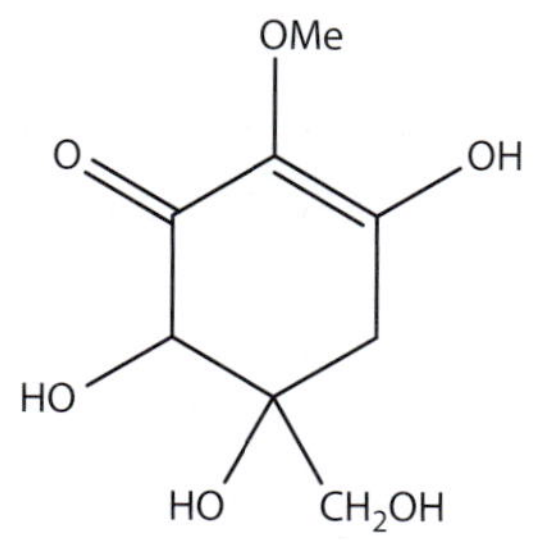

17.10

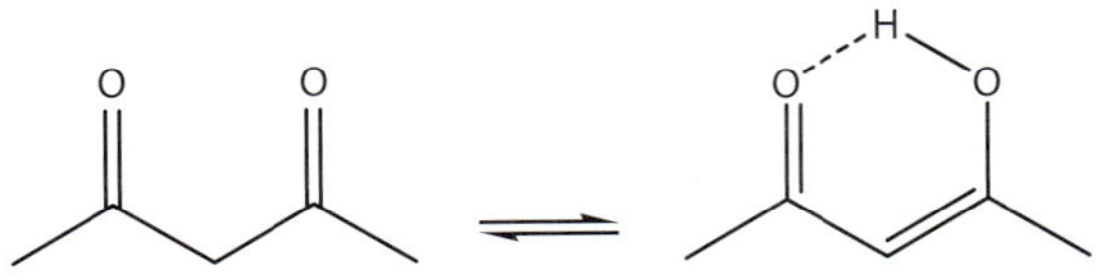

17.11

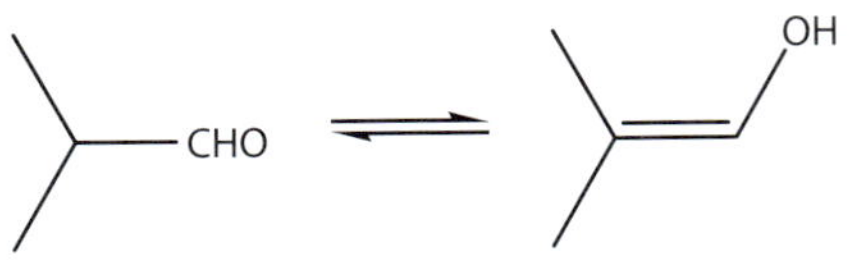

17.12

Solvents and temperature also affect the proportions of enol. Thus pentane-2,4-dione contains 80 % enol as a neat liquid, but in water, it is only 15 % enolized. The equilibrium is shifted because water hydrogen bonds so strongly to the keto form. An increase in temperature also favors the keto form, but this is not a large effect at the temperatures at which ordinary organic reactions are carried out. The rate of ketone/enol interconversion depends strongly on pH, and the individual tautomers of pentane-2,4-dione have been isolated at low temperature, under strictly neutral conditions. Other carbonyl compounds, including esters and amides, can and do enolize, but with much more difficulty. For example, neat diethylmalonate (diethyl propanedioate, **17.13**) contains only 0.01 % enol at equilibrium, in contrast to the 80 % for pentane-2,4-dione.

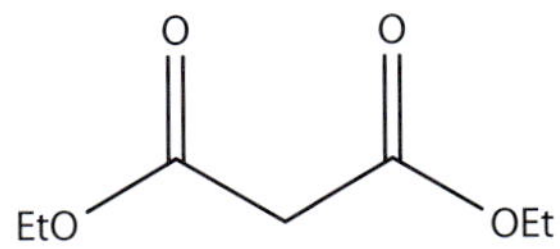

17.13

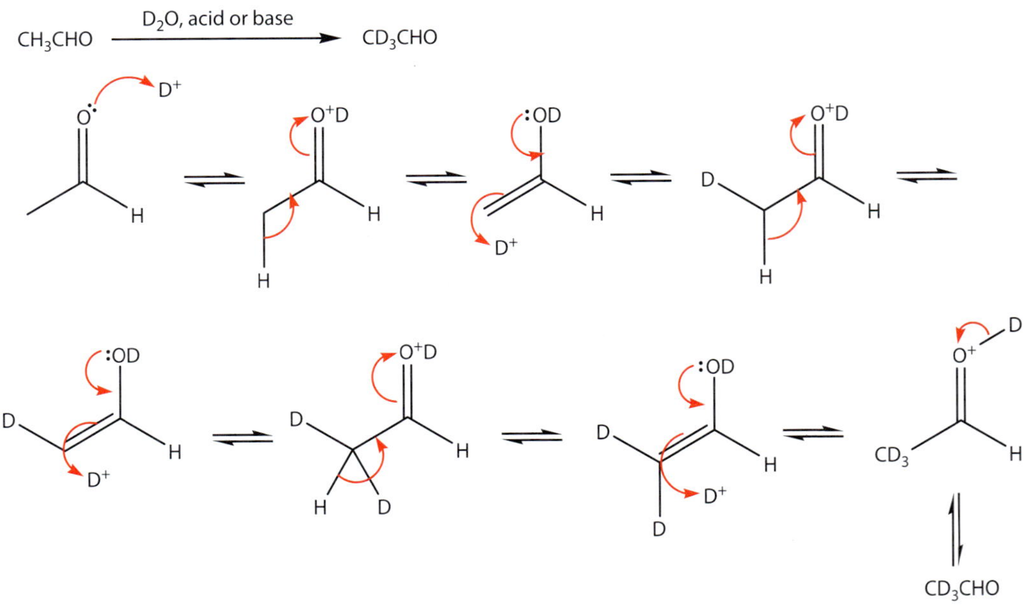

FIGURE 17.6 Mechanism of H/D exchange in acetaldehyde.

In some cases, we can measure the amount of enol at equilibrium using spectroscopic and other techniques, but for many compounds, the amounts of enol present are too low to be detected. So how can we detect that enolization is actually occurring? The usual method relies on the fact that hydrogens that are removed during enolization can be exchanged for deuterium. Thus, if acetaldehyde is dissolved in D_2O in the presence of either acid or base, all of the α-hydrogen atoms are replaced by deuterium (Figure 17.6). Generally, all α-hydrogens in enolizable aldehydes and ketones can be exchanged (mass action and the excess of D_2O solvent ensure that the reaction goes to completion), but the rate varies.

PROBLEM 17.1

If each of the following molecules was treated with DO^-/D_2O, where, if anywhere, would you expect deuterium to be incorporated?

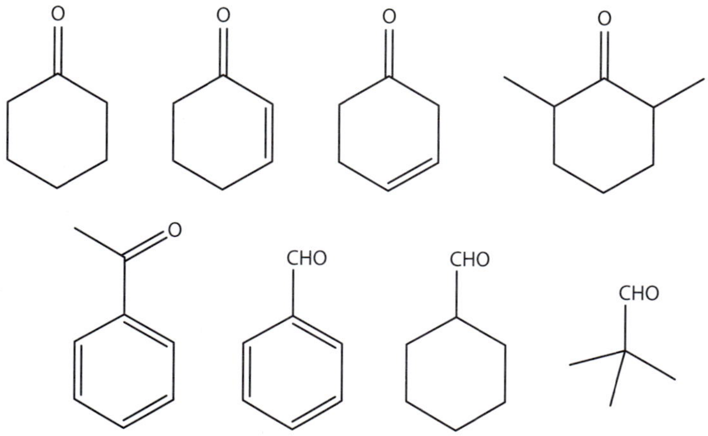

Solution

Two of the compounds can be ruled out of contention immediately as they have no hydrogen atoms at the α-position, *viz.*, PhCHO and Me_3CCHO. In cyclohexanone, all the α-positions are deuterated, and the same is true for 2,5-dimethylcyclohexanone:

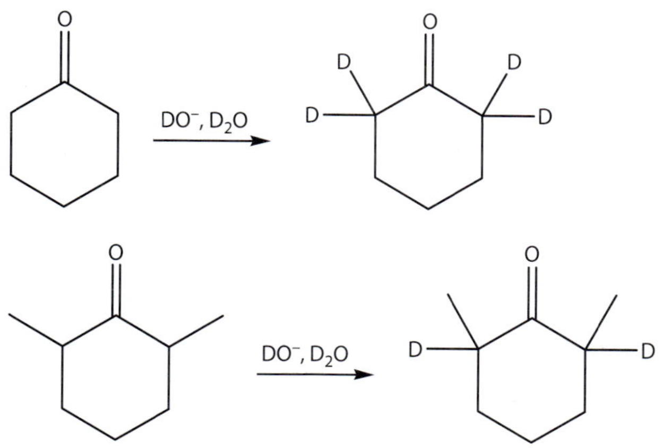

The position is a little more complex with the two cyclohexenones. In both cases, we can easily see that the 6-position will be deuterated. However, for 2-cyclohexenone, the formation of a dienolate, **17.14**, must be considered as a possibility—and this can be deuterated both at the 4- and 6-positions:

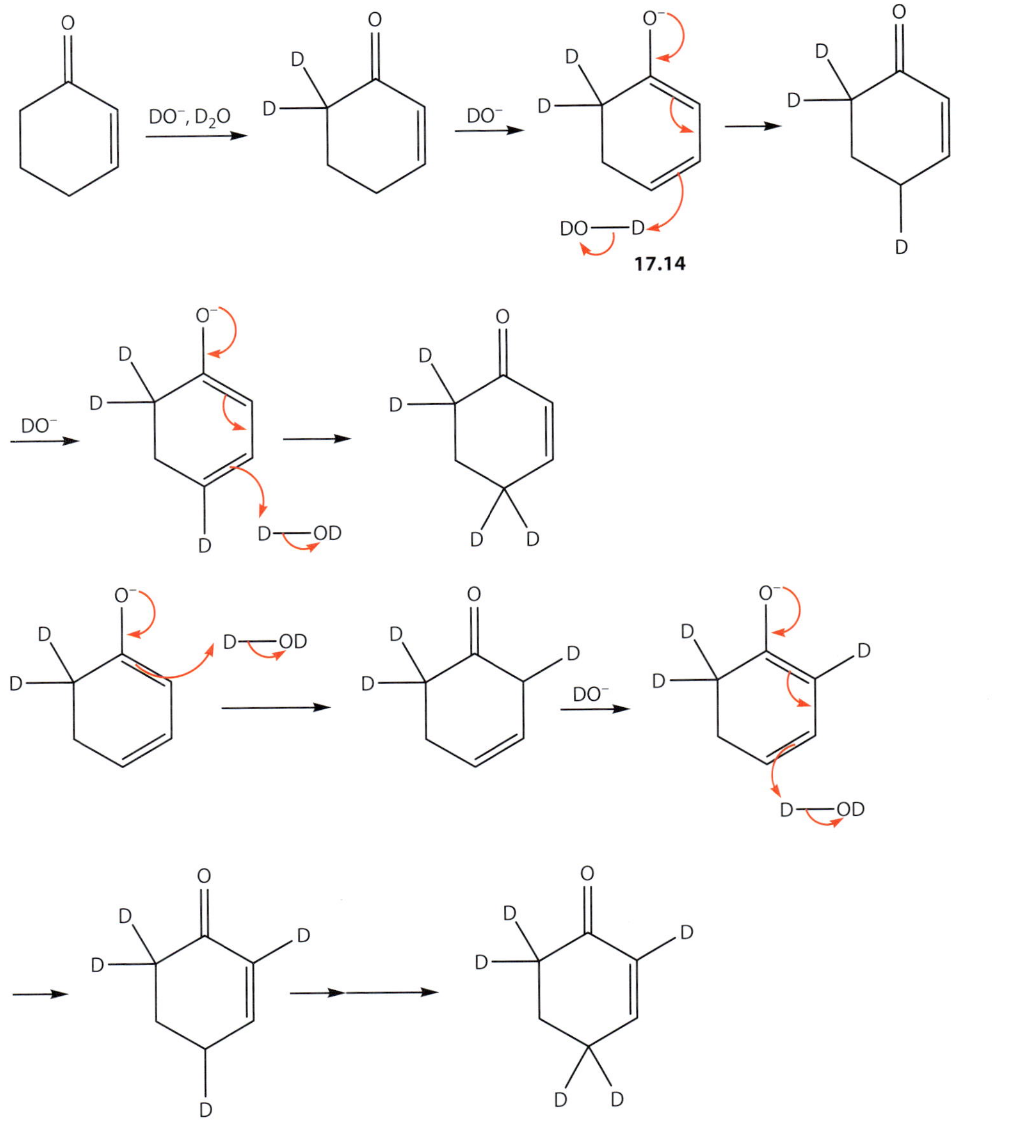

17.14

17.15

While the order of events is impossible to predict, mass action drives the deuteration of every position that can be deuterated, and thermodynamic stability dictates that in an equilibrium process, both 2-cyclohexenone and 3-cyclohexenone will give the same product (the same dienolate is available to both), **17.15**.

Acetophenone and cyclohexane carbaldehyde simply exchange the α-hydrogens:

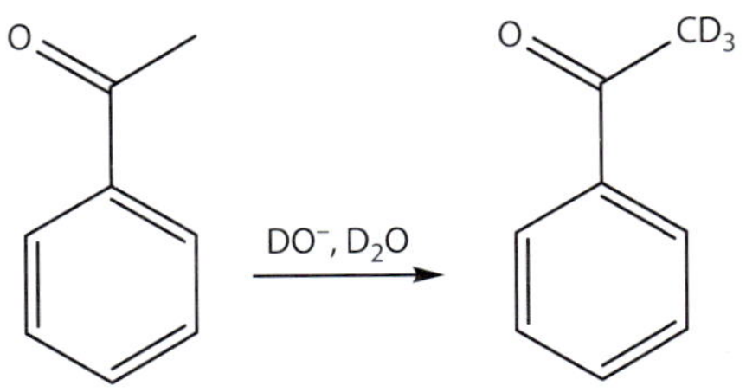

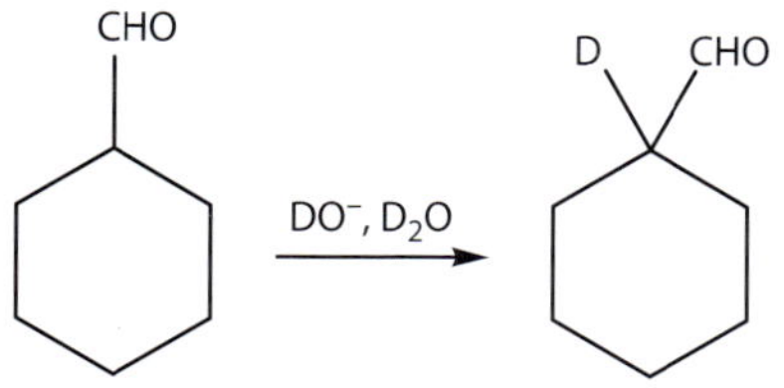

Focus on NMR Spectroscopy of Enols

NMR spectroscopy of enols has been a key tool in determining the relative proportions of keto and enol forms in highly enolized species. For example, in pentane-2,4-dione, the keto and enol forms (Figure 17.7) can be readily identified in the NMR spectrum of the material; the keto–enol tautomerism is slow compared with the timescale of the NMR spectroscopy experiment. This experiment has been extended by changing the temperature, which allows the calculation of thermodynamic parameters for the equilibrium, and the use of a range of solvents to study the effect of polarity. The process has also been studied in the gas phase, where enols predominate for almost all β-dicarbonyl compounds. In solution, the keto forms are generally some 8–9 kJ mol^{-1} more stable than in the gas phase.

The spectroscopic data of Figure 17.7 are instructive, but may also be puzzling. The first unusual feature is the very high value of δ for the enolic proton. As a general rule, the value of δ for the protons of OH groups increases with the extent of hydrogen bonding. For example, the chemical shift for the OH of phenol is δ 4.35 in a 1 % solution, δ 5.95 in a 5 % solution, and δ 7.45 in neat phenol; as the concentration increases, so does the extent of hydrogen bonding. The chemical shift of the OH proton of a carboxylic acid is in the range δ 10–13—and we know that these are extensively hydrogen bonded, even in dilute solutions. But in these enolized β-diketones, there is something quite special happening, over and above a simple hydrogen bond—the chemical shift of the OH proton in **17.16**, which is clearly hydrogen bonded, is δ 3.87. The second puzzling feature of the spectrum of pentane-2,4-dione is that the two possible enolic tautomers give rise to only one signal for a methyl group. If we look at either separately, we would expect to see two signals for the methyl groups—one is attached to the ketone and one to the carbon–carbon double bond. But there is only one—and this is not just a coincidence as the same is true of other comparable β-diketones. We have to consider whether the two forms are in a true equilibrium or what we see is more like a resonance hybrid of the two. Any energy barrier to an equilibrium would be likely to be low—shifting a proton from one oxygen to the other is easy. All the evidence, both theoretical and experimental, however, is that, like other hydrogen-bonded systems, there are two genuine energy minima, but the barrier to interconversion of the two forms is low.

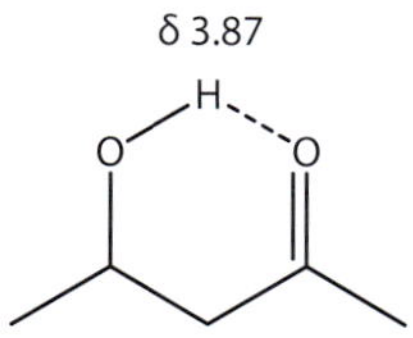

17.16

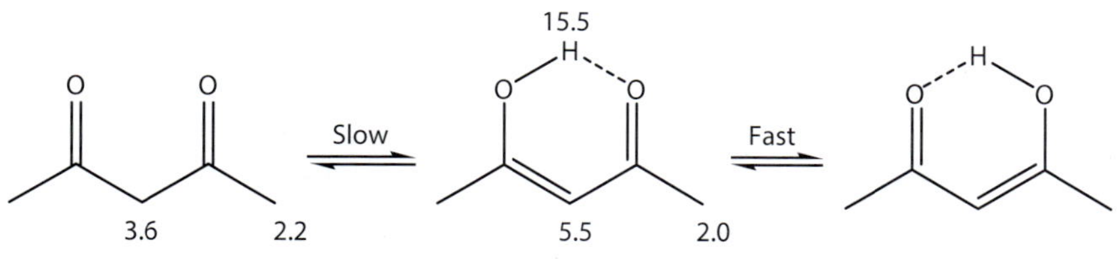

FIGURE 17.7 Proton NMR chemical shifts of keto and enol forms of pentane-2,4-dione.

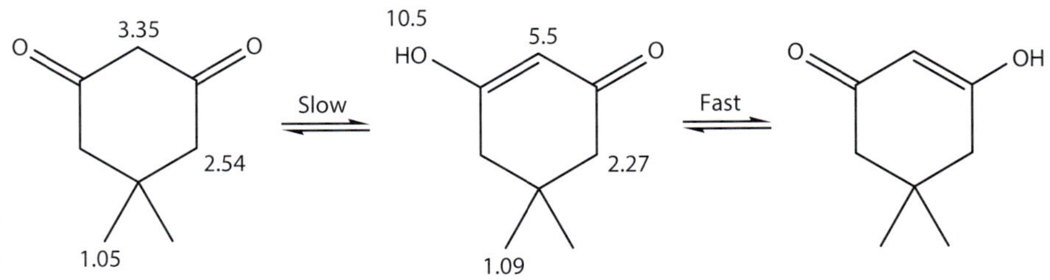

FIGURE 17.8 Proton NMR spectroscopic data for dimedone in $CDCl_3$.

If we now consider the NMR spectrum of the keto and enol tautomers of dimedone (5,5-dimethylcyclohexane-1,3-dione), another puzzle emerges (Figure 17.8). Despite the fact that there can be no intramolecular hydrogen bond, the enol form still appears to be symmetrical—only one type of CH_2. At equilibrium, about 40 % of the material is enolized. It is suggested that under these conditions, the material is part of a dimeric structure, **17.17**, in which rapid proton transfers are possible. However, in very dry dimethylsulfoxide, ^{13}C NMR spectroscopy suggests that dimedone is completely enolized, and monomeric, with distinct CH_2 carbon atoms.

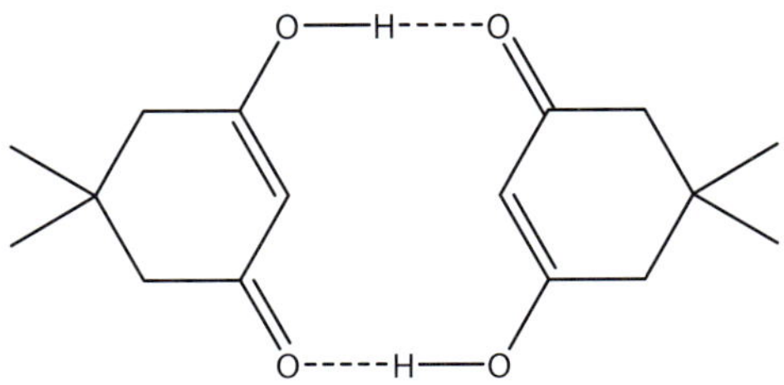

17.17

Other less common NMR active nuclei have also been used to study tautomeric equilibria. For example, ^{19}F NMR spectroscopy was used to study the equilibrium in Figure 17.9; it's an easier experiment as only one tautomer is produced (ketones being easier to enolize than esters), and the keto and enol forms give rise to a single signal in the ^{19}F NMR spectrum. There is 89 % of the enol form in the neat liquid. ^{17}O NMR spectroscopy was used to study **17.18** (R = Me, 8 % enol; R = Ph 17 % enol; R = CF_3 90 % enol).

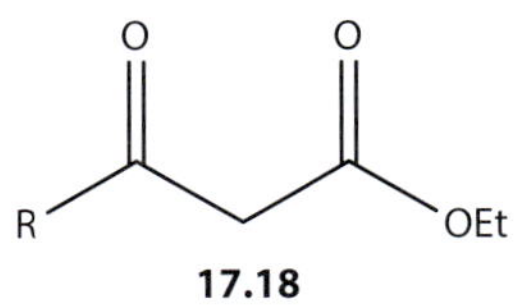

17.18

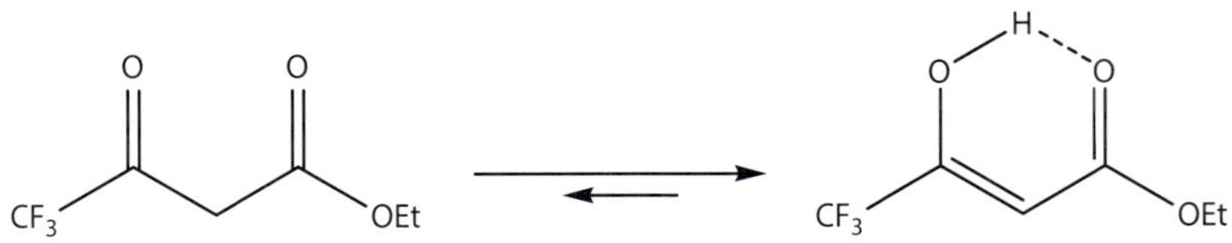

FIGURE 17.9 Enolization of a fluorinated β-dicarbonyl compound.

17.3 ISOMERIZATIONS INVOLVING ENOLS AND ENOLATES

In Problem 17.1, we saw that 3-cyclohexenone was converted to 2-cyclohexenone; in this section, we examine that type of reaction more closely. The formation of an enol, which we describe as tautomerization, is, strictly speaking, an isomerization. However, once we have

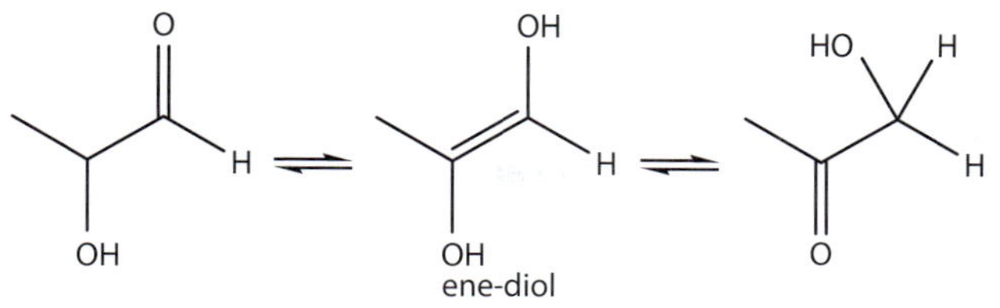

FIGURE 17.10 Isomerization of an α-hydroxyaldehyde.

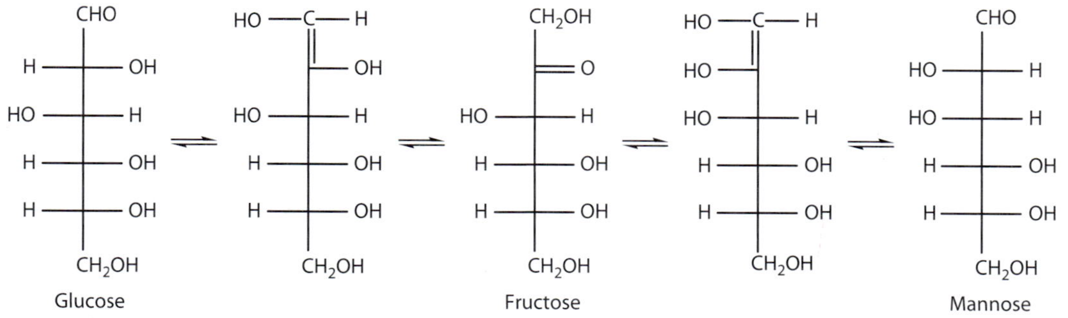

FIGURE 17.11 Interconversion of glucose, fructose, and mannose.

formed an enol, other options may be open to it. An example is provided by the isomerization of α-hydroxyaldehydes, of which a simple example is given in Figure 17.10—when we have two OH groups attached to a double bond, an enediol, this is in equilibrium with both possible keto forms, and in simple cases, thermodynamics will prevail, with the more stable species predominating. Enediol intermediates are proposed in the interconversion of glucose, fructose, and mannose under basic conditions (Figure 17.11). These species are interconverted as their 6-phosphate derivatives by the enzyme phosphohexoseisomerase in biological systems. Biological systems will in general be nonequilibrium systems, as some species are removed from the equilibrium by other metabolic processes.

If there is a chiral center bearing a hydrogen atom adjacent to a carbonyl group, racemization is often easy and rapid. Consider the enolization of compound **17.19**. The enol is planar, and cannot remember which enantiomer it used to be, so it may be reprotonated on either face of the double bond to give either enantiomer. If this compound is subjected to acidic or basic conditions, it will be rapidly racemized. This type of racemization will be of concern to us again when we come to look at peptide synthesis.

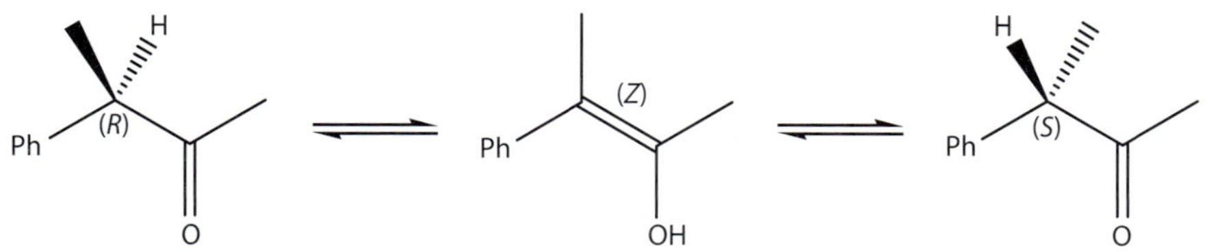

17.19

PROBLEM 17.2

Explain why *R*-2-methylcyclohexanone is racemized in aqueous base, but *R*-3-methylcyclohexanone is not.

Solution

We first need to draw both the molecules and consider how they might enolize. Enolization of 2-methylcyclohexanone toward the 6-position has no effect on stereochemistry, but enolization toward the 2-position produces the planar intermediate, **17.20**, which may be reprotonated from either face resulting in racemization:

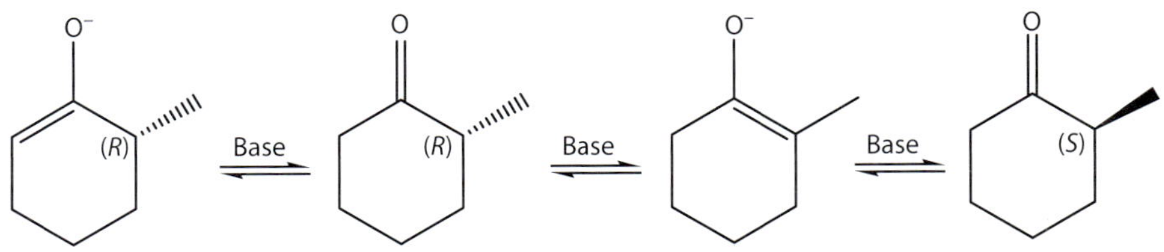

17.20

If we consider the same process for 3-methylcyclohexanone, again, enolization may occur at either site, but in this case, the chiral carbon atom is unaffected, and hence no racemization occurs:

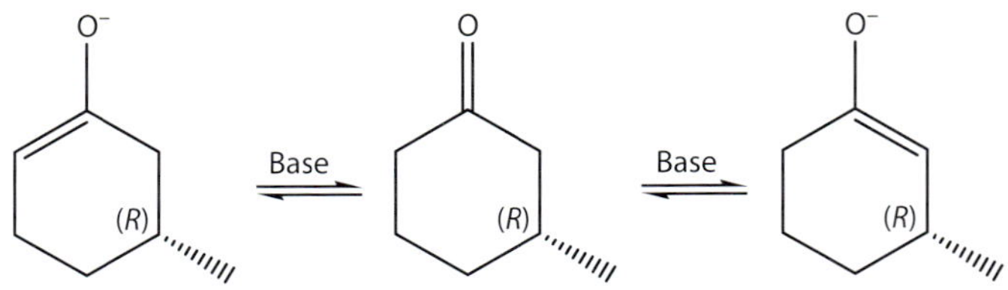

Next, consider the enolization of **17.21**—what is obtained is termed a dienolate/dienol. When the proton is returned to the double bond, where will it go? We can write sensible mechanisms to give both a conjugated and a nonconjugated product (Figure 17.12). Clearly, since these are equilibrating systems, governed by thermodynamic stability, the main product will be the conjugated enone. Thus, treatment of **17.21** with acid or base will result in migration of the carbon–carbon double bond into conjugation with the ketone.

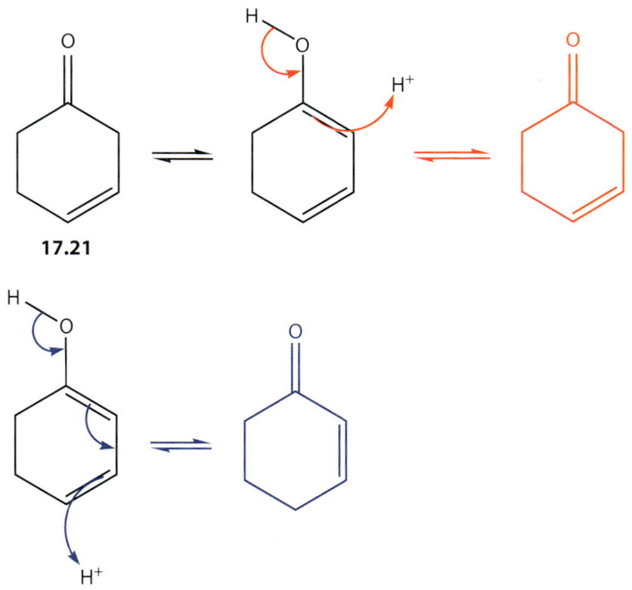

FIGURE 17.12 Reprotonation of a dienolate.

PROBLEM 17.3

Propose a mechanism for the following equilibrium:

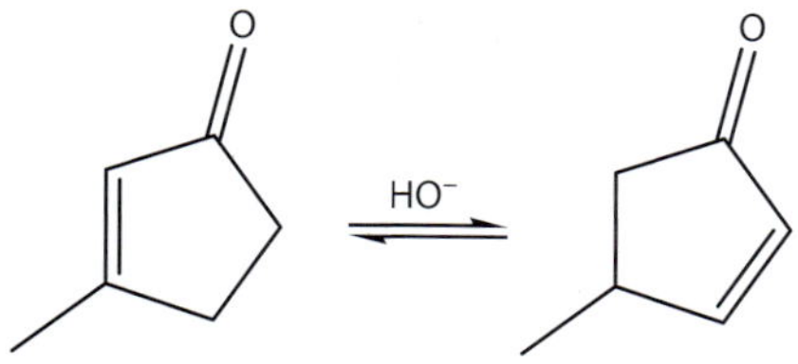

Solution

Both of the enones are in equilibrium, *via* the formation of dienolates, with the disfavored 3-methyl-3-cyclopentenone. Hence, they must be in equilibrium with each other:

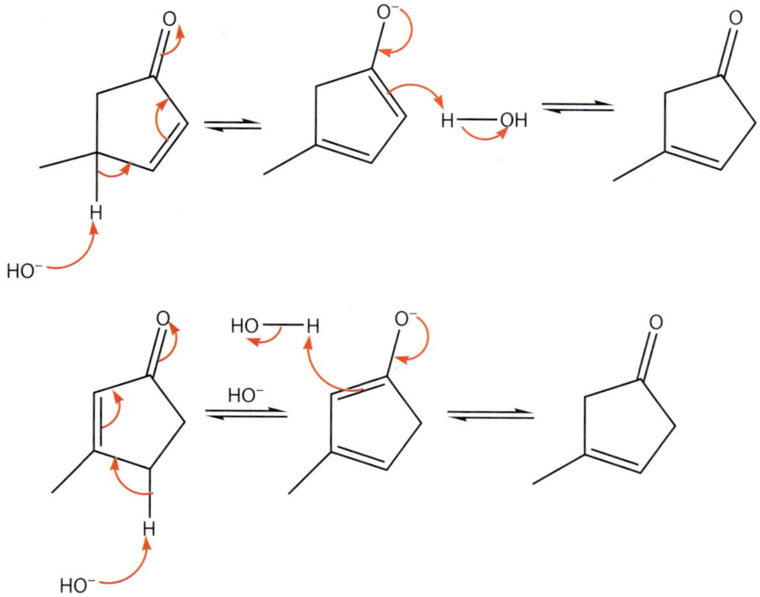

Cis/trans double bond isomerization in enones is also possible if there is a suitable proton available to be removed to give the dienol/dienolate. So if **17.22** is treated with base, a proton is removed from the γ-position to give the dienolate anion (Figure 17.13). In this intermediate, the bond between the α- and β-carbon atoms is a single bond and so may rotate freely; the position of the equilibrium will depend on the relative stabilities of the two species, in this case determined by steric factors.

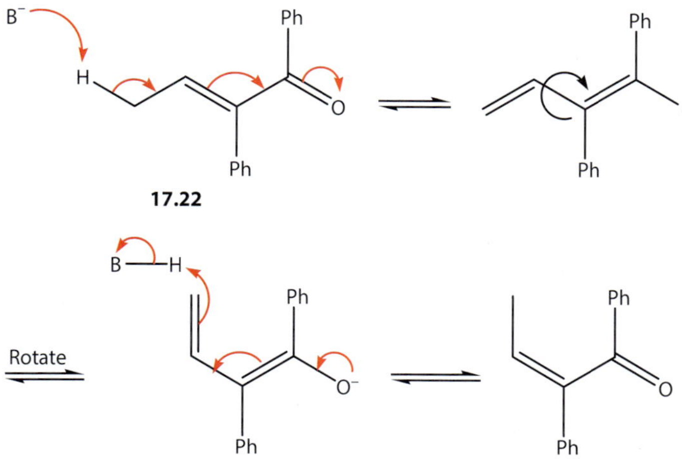

FIGURE 17.13 *cis/trans*-Isomerization as a result of enolization.

Key Points from Sections 17.1 through 17.3

- Most carbonyl compounds possessing hydrogens at the α-position can equilibrate with enol tautomers (unless the enol/enolate would be particularly unstable, for example, if it cannot be planar).
- Enolization may be catalyzed by acid or base; the rate is slowest at neutral pH.
- Enols are formed in acid solution and enolates in basic solution.
- The stability of enols, and hence the position of the carbonyl/enol equilibrium, depends on their structure—conjugation, hydrogen bonding, aromaticity, and solvation are all important.
- The formation of enols may be detected (even where equilibrium constants are very low) by exchange of enolizable protons with D_2O.
- Reversible formation of enols may result in a number of isomerization reactions, including double bond migration, racemization of chiral centers α to a carbonyl group, and *cis*/*trans*-isomerization of double bonds.

17.4 REACTIONS OF ENOLS AND ENOLATES WITH ELECTROPHILES

17.4.1 REACTION WITH HALOGENS

Aldehydes and ketones react with halogens at the α-position *via* the enol or enolate, depending on the solution pH (Figure 17.14). For the acid-catalyzed reaction, the acid is usually HBr or HOAc, and a typical reaction mechanism is shown in Figure 17.15. In acid, the enol reacts with bromine; in base, the enolate is the reacting species. The enolate, as you might expect (remember that the phenolate anion is a better nucleophile than neutral phenol), is the better nucleophile. However, irrespective of pH, the rate of bromination is not dependent on the concentration of molecular bromine. The RDS is enolization, and subsequent reaction of the enol/enolate with bromine is fast.

FIGURE 17.14 Bromination of acetone.

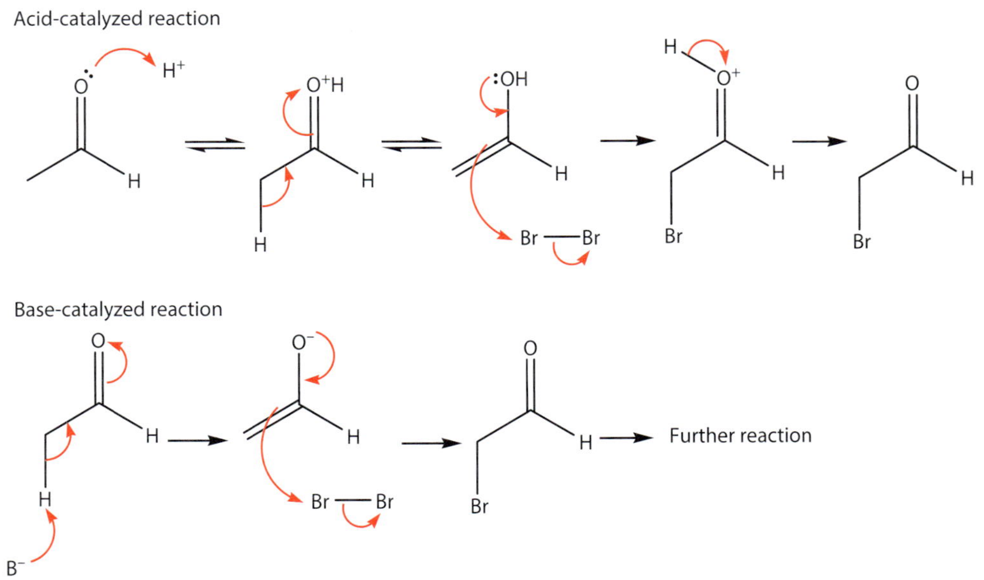

FIGURE 17.15 Mechanism of bromination of acetaldehyde under acidic or basic conditions.

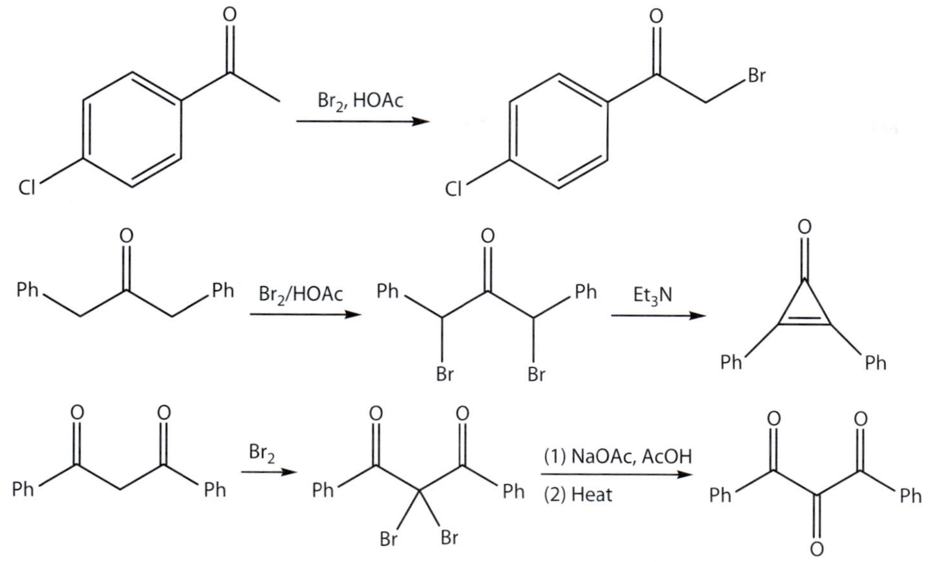

FIGURE 17.16 Bromination of enols/enolates in synthesis.

Some examples of the use of this process in synthesis are shown in Figure 17.16. Most employ acid-catalyzed reactions, as this process is more easily controlled; base-catalyzed process may lead to uncontrolled polyhalogenation. The second example provides us with some useful data—if a second reaction takes place under acid conditions, it occurs preferentially at the carbon that has not already been substituted by halogen. The product here undergoes an intramolecular elimination reaction to give the interesting cyclopropenone—a compound showing significant aromatic properties. In the final example, an acid/base catalyst is not needed as the molecule is inherently significantly enolized. The intermediate in the final step is the hydrate of the ketone.

PROBLEM 17.4

Write a mechanism for the elimination step of the second example in Figure 17.16, and explain why the product should show aromatic character.

Solution

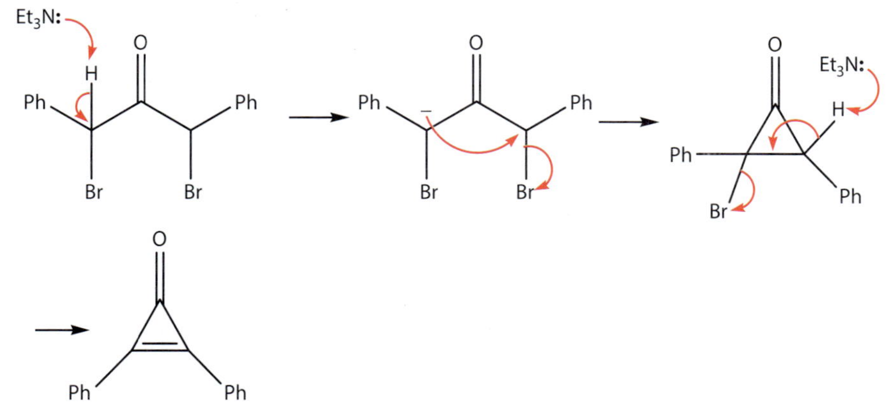

If the carbonyl group in the final product is written in the charge-separated form, then the ring is a two-electron, Hückel-compliant, aromatic ring (see Section 3.2.1).

Methyl ketones react with iodine in the presence of base to give carboxylic acids and iodoform, CHI_3 (Figure 17.17). Writing the mechanism for the reaction is straightforward; successive enolates are formed and halogenated and in the final step $[I_3C]^-$ is expelled. This breaks our general rule that neither hydride nor a carbanion can behave as a leaving group. However, this particular

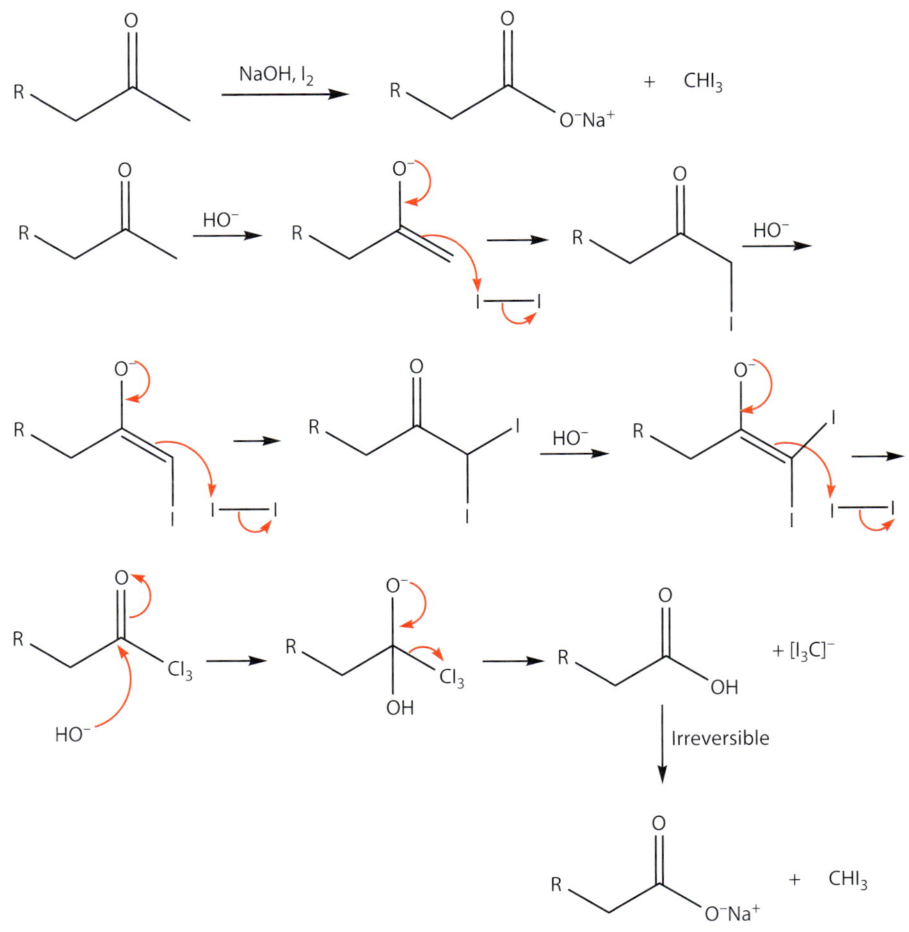

FIGURE 17.17 Iodoform reaction.

FIGURE 17.18 Iodoform reaction.

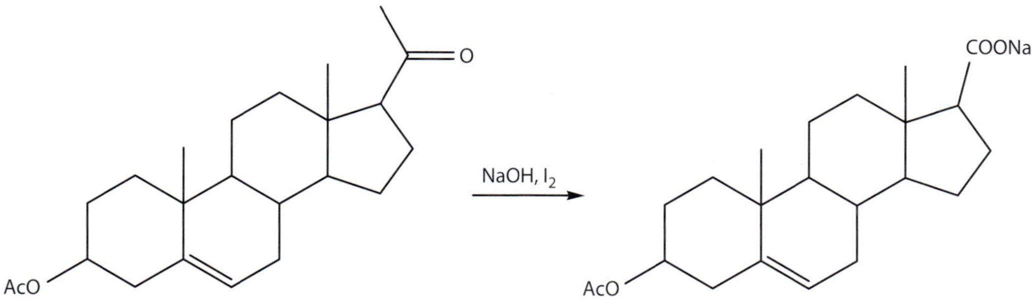

FIGURE 17.19 Iodoform reaction in synthesis.

carbanion is stabilized by the halogen substituents, and its loss is rendered irreversible by its protonation from the carboxylic acid. The iodoform reaction is used as a test for the presence of a methyl ketone, because iodoform (unlike bromoform or chloroform, which are equally easily formed) is insoluble in the reaction mixture, and a visible precipitate is observed (Figure 17.18). The process has also been used synthetically, as shown in Figure 17.19.

PROBLEM 17.5

Explain why bromination and chlorination of acetophenone, PhC(=O)CH_3, occur at the same rate.

Solution

In general, we would not expect bromine and chlorine to react at the same rate with any substrate. This tells us that the RDS in the process is not the reaction with the halogen, but the formation of the enol, in which the halogen is not involved. Note that we do not now show the first slow step as an equilibrium; the forward reaction with the halogen is faster than the reversal to the keto form, so no true equilibrium can be established:

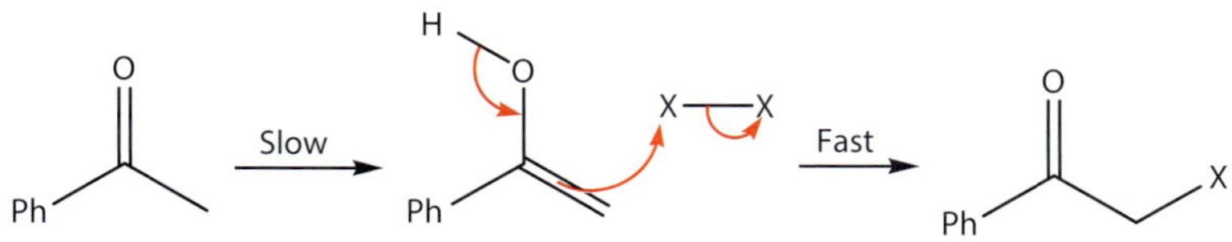

What do these observations tell us about the stability of enols and enolates? The fact that the iodoform reaction works and the methyl group is specifically and persistently iodinated tells us that the terminal enolate, **17.23**, is more stable than the internal one, **17.24**. This implies that there is significant charge density at carbon in an enolate anion—the stability is following that we would expect for a carbanion. If we consider the enolate as a carbanion, then this also explains why successive iodinations occur at the same site—it is getting successively easier to make the enolate anion, because of the electron-withdrawing effect of the iodine.

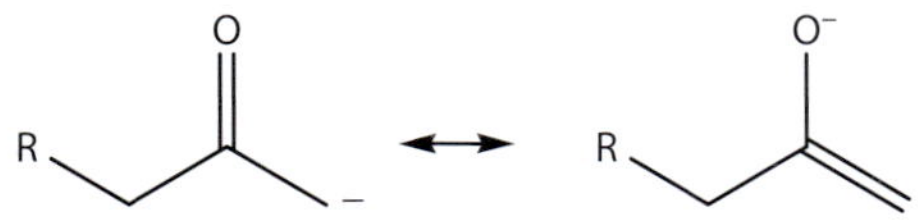

17.23

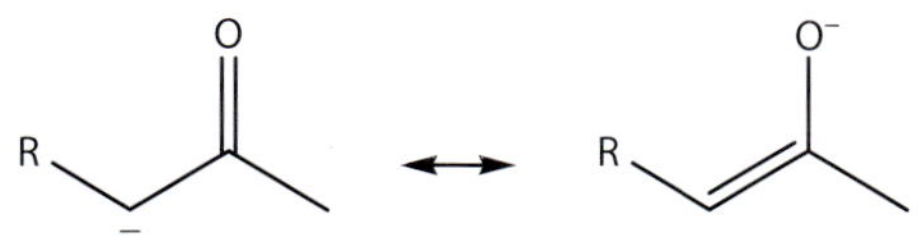

17.24

In acid solution, monohalides are generally formed (the first example of Figure 17.16). We can understand this if we consider the transition state for a second enolization (**17.26**; remember the

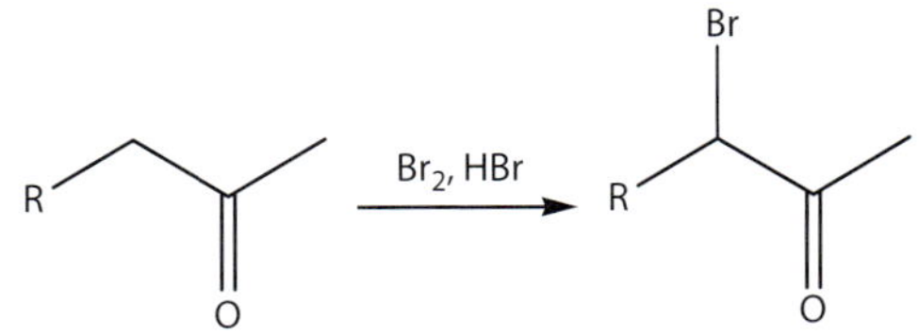

FIGURE 17.20 Halogenation of a nonsymmetric ketone under acid catalysis.

enolization is the RDS) compared with the original one (**17.25**). Bromine is electron withdrawing and hence destabilizes a broadly positive transition state, so that a second halogenation at the same position is discouraged. There is also a difference in the regioselectivity of halogenation under acid conditions; the more substituted site is preferentially halogenated (Figure 17.20). This implies that the more substituted enol, **17.27**, is the more stable one; enols are alkene-like in their stability, so stability increases with substitution.

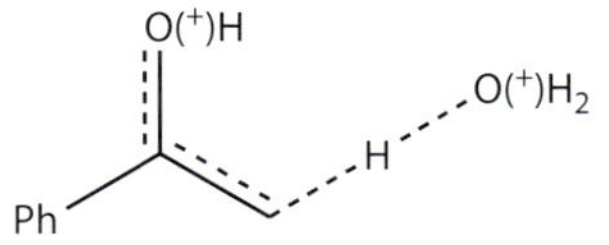

17.25

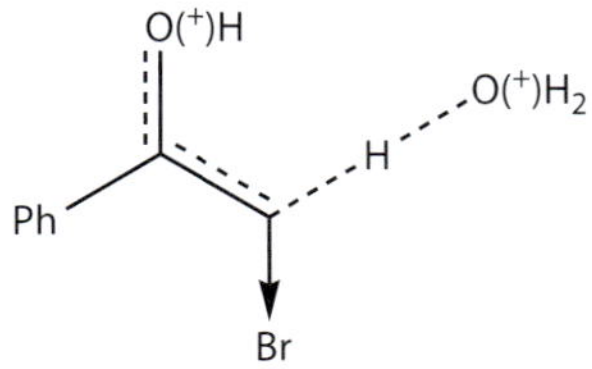

17.26

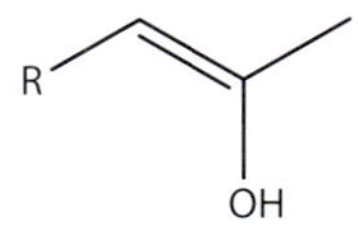

17.27

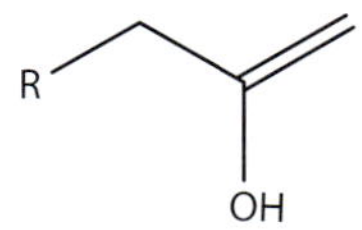

17.28

As we noted earlier, other types of carbonyl compound can also enolize, and this is used in the long-established Hell–Volhard–Zelinsky (HVZ) reaction, to prepare α-halo acids or esters (Figure 17.21). A carboxylic acid is reacted with P/Br_2 to give the acid bromide. This can be enolized and reacts with free bromine to give the α-halo acid bromide. In the absence of any other nucleophile, this exchanges with SM (*via* anhydride formation) to give α-halo acid and acid bromide. Br_2, PBr_3, and PBr_5 are in equilibrium in the reaction mixture, and the phosphorus may be used catalytically, or replaced by PCl_3, which is handled more easily. Synthetic examples are shown in Figure 17.22. The first example demonstrates the value of the reaction in putting a substituent at the α-position of a highly substituted acid, and many amino acids have been prepared analogously to the third example. In the final example, the α-halo acid bromide is subjected to a zinc-mediated debromination to give a ketene.

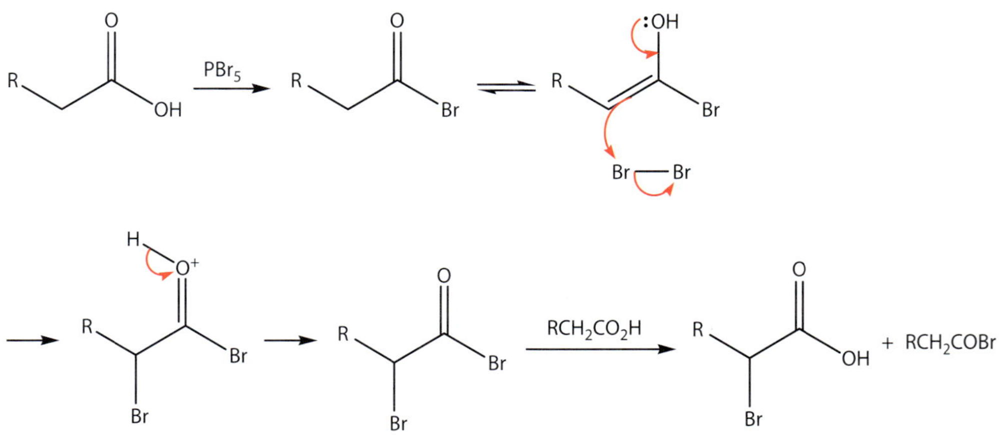

FIGURE 17.21 Hell–Volhard–Zelinsky reaction.

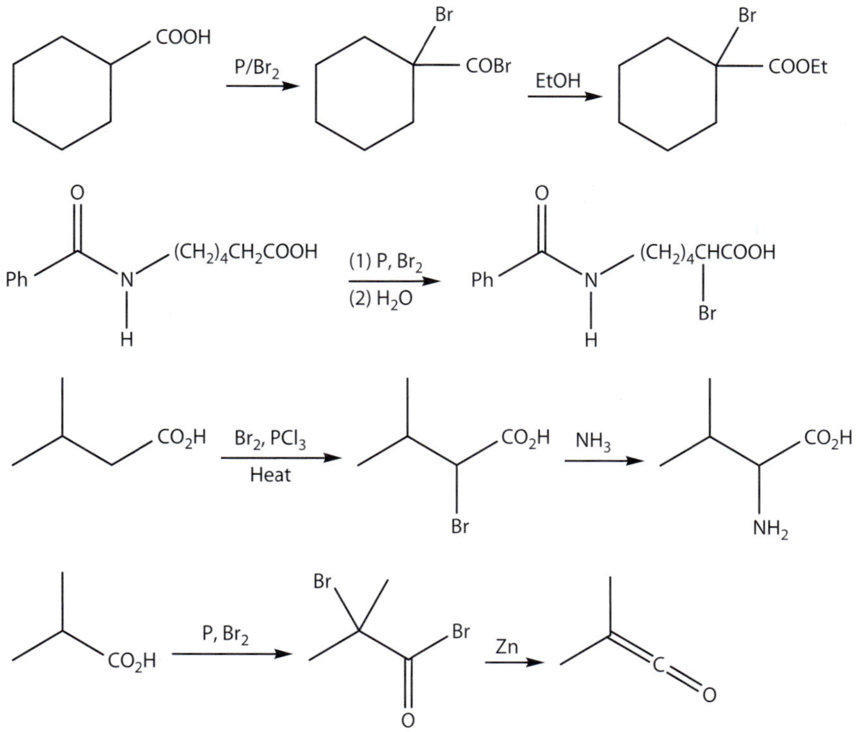

FIGURE 17.22 Examples of the Hell–Volhard–Zelinsky reaction.

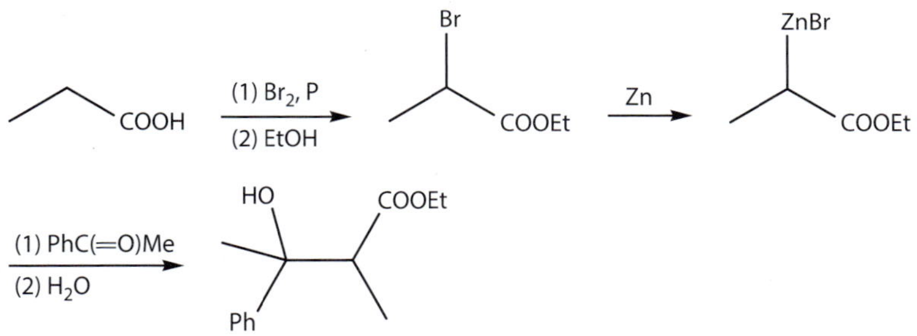

FIGURE 17.23 Hell–Volhard–Zelinsky and Reformatskii reactions.

The HVZ reaction is often combined with the Reformatskii reaction in syntheses—Figure 17.23 illustrates. An α-halo ester is prepared using the HVZ process and is then converted to an organozinc compound. Remember that we cannot make a Grignard reagent here; Grignard reagents react with esters, but organozinc compounds don't. The Reformatskii reagent adds to a ketone to give, after work-up, a highly substituted β-hydroxyester.

PROBLEM 17.6

Predict the product(s) of the following reactions:

(a)

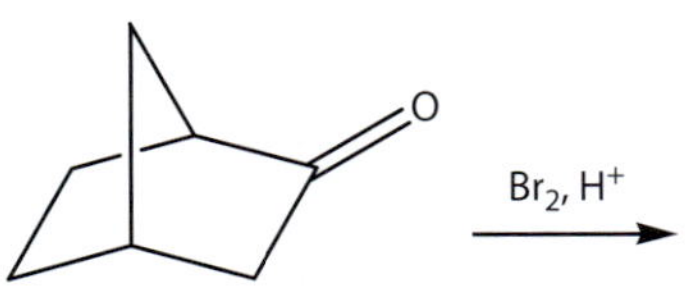

(b)

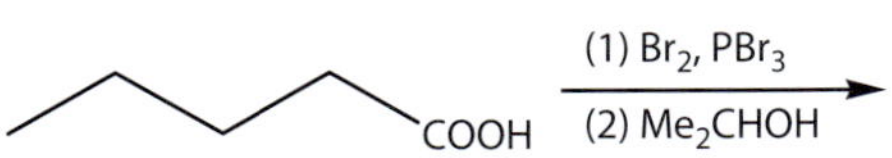

PROBLEM 17.7

How would you accomplish the following conversion?

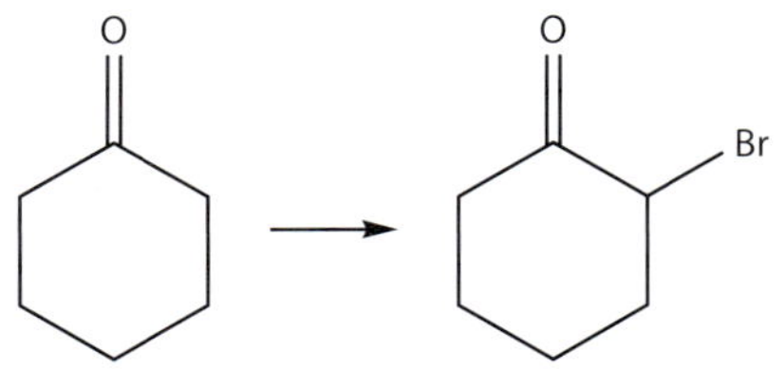

Solutions

17.6 (a) Despite the fact that there is proton at the bridgehead of the bicyclic system, no enolization is possible there. Only one enolate can be formed, and this could in principle react with bromine on either face. Although it might be expected that the less hindered *endo*-product would predominate, there is no simple method to predict the ratio of the products:

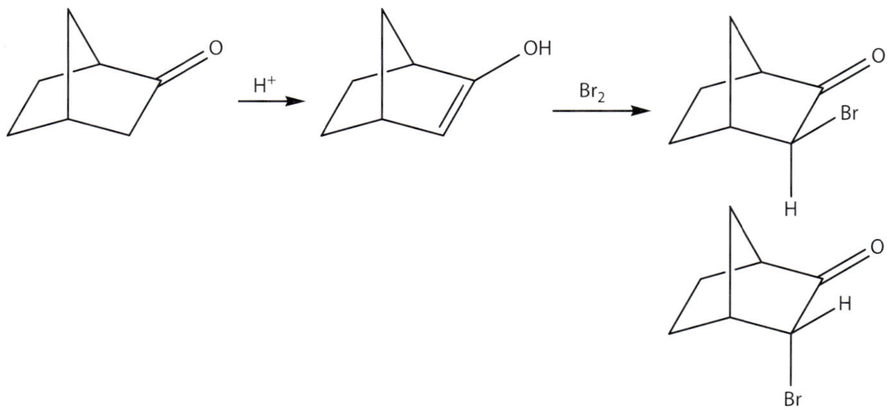

(b) This is a classic Hell–Volhard–Zelinsky reaction with the acid bromide reacting with the alcohol in a final step:

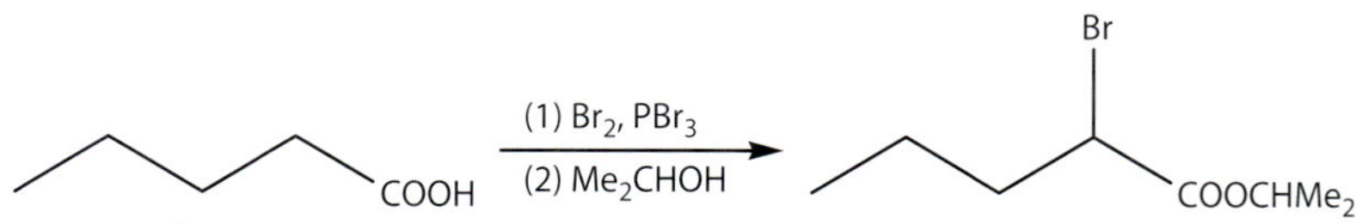

17.7 Since monobromination is required, it will be better to use acid-catalyzed conditions, *via* the enol. Typical conditions would be Br_2/HBr.

Key Points from Section 17.4.1

- Enols and enolates react with halogens to give α-halocarbonyl compounds. Enolization is the RDS.
- Under acidic conditions, monohalogenation is the usual outcome, but in base, polyhalogenation is more common. The iodoform reaction converts methyl ketones to carboxylic acids and iodoform, CHI_3, and can be used as a test for the presence of this functionality.
- Generally, the stability of enols is enhanced by substitution at the double bond. Enolates, by contrast, are most easily formed at unsubstituted positions.
- The Hell–Volhard–Zelinsky reaction converts carboxylic acids to α-halo acids or α-haloesters, *via* the enol of an acyl halide.

17.4.2 ALDOL REACTION

The aldol condensation and related reactions are among the most important, and ubiquitous, construction reactions in organic synthesis. In these condensations, the carbonyl compound acts as both nucleophile and electrophile—the enol or enolate is the nucleophile, and the keto form is the electrophile. The reaction works with enolizable aldehydes (Figure 17.24) or ketones (Figure 17.25) and may be catalyzed by either acid or base. Almost all of the steps we write are equilibria—how can we persuade the reaction to go to completion? In the base-catalyzed reaction, a catalyst such as barium hydroxide is placed inside a Soxhlet thimble, as in Figure 17.26. The reaction mixture is heated so that the acetone refluxes, but the product does not. Thus only the SM, and not the product, comes into contact with the catalyst, ensuring that the reverse reaction does not occur. Note that in the acid-catalyzed processes, it is common for the product to be dehydrated under the reaction conditions—this usually pulls the equilibrium over to the product. The mechanism of the elimination reaction may be E1 or E2 depending on the

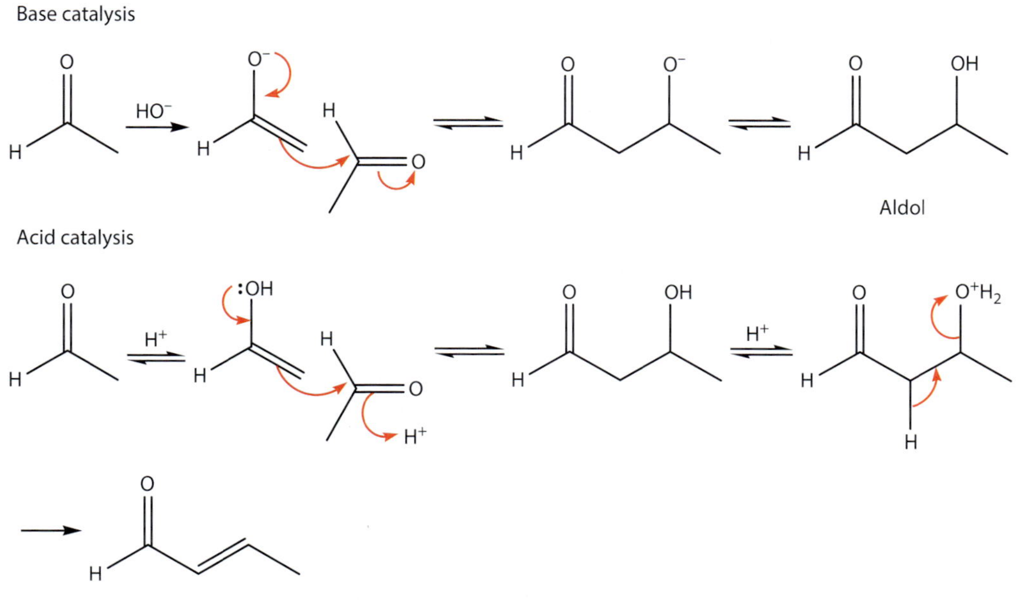

FIGURE 17.24 Aldol reaction of acetaldehyde.

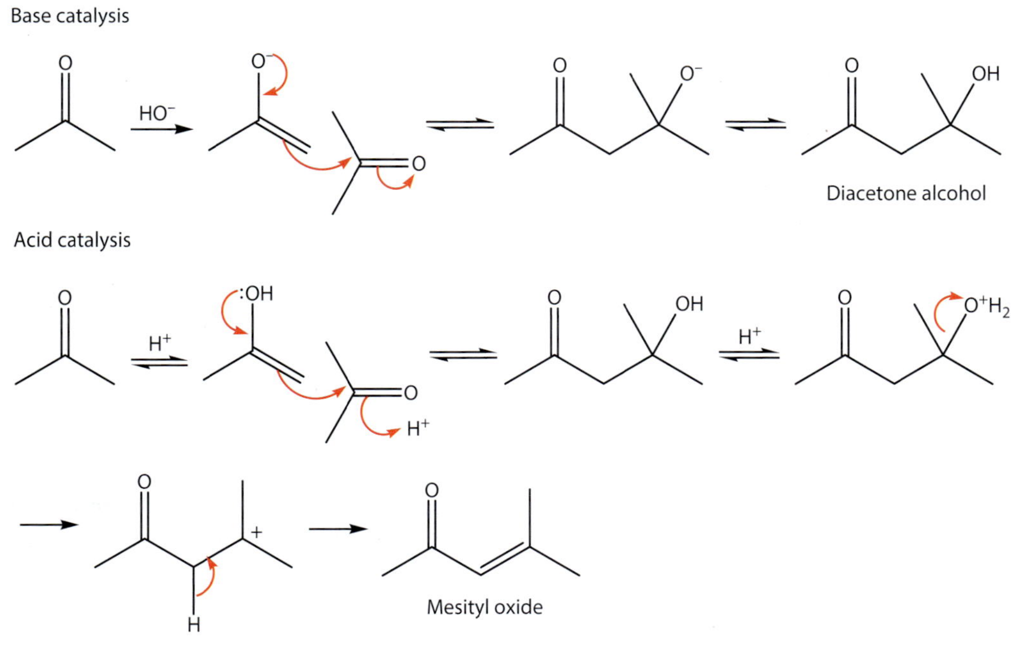

FIGURE 17.25 Aldol reaction of acetone.

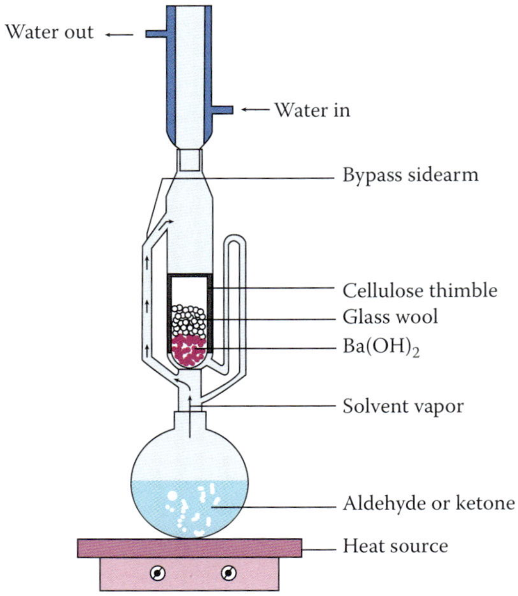

FIGURE 17.26 Soxhlet apparatus for aldol condensation.

degree of substitution. The conjugated product is invariably obtained, and its stability provides an important driving force for the dehydration. Dehydration of the product can also occur in base, by either an E2 or E1cB mechanism.

The reactions earlier, and those in Figure 17.27, are all what we would call self-condensations; the enol and the keto forms of the same molecule react together. Moreover, they have been carefully chosen so that there is only one possible enol/enolate that can be formed. However, if we could use different molecules as nucleophile and electrophile, the scope of the reaction would be greatly increased. Consider the reaction of acetone with acetophenone (Figure 17.28). Clearly, this is not useful—there are four hydroxycarbonyl compounds and four dehydration products that could be obtained. Both acetophenone and acetone can self-condense, and in the cross condensation, either could be nucleophile or electrophile.

The question here is all about control—can we direct the process to give us our target? Although we will develop this theme further in Chapter 20, there are some simple paradigms that work. If we condense two species where one cannot enolize, then this principle reduces our number of primary products to two. If we make sure that the species that cannot enolize is the more electrophilic component, or we use it in large excess, so that the enol is likely to meet this rather than the keto form of itself, then we can get successful reactions. An example is provided by the reaction of 4-chlorobenzaldehyde (can't enolize) with acetophenone

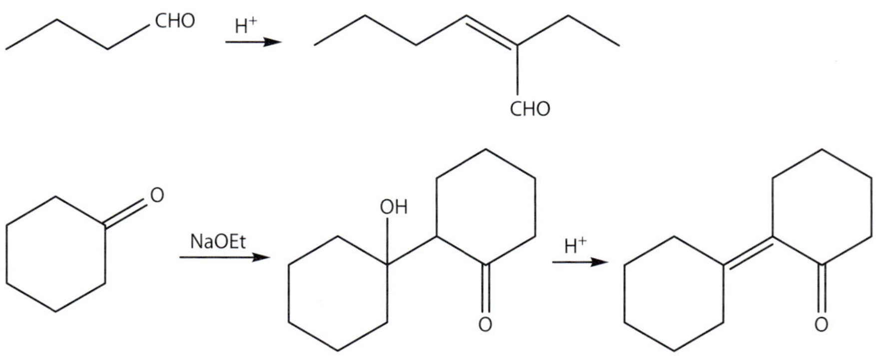

FIGURE 17.27 Self-condensations of aldehydes and ketones.

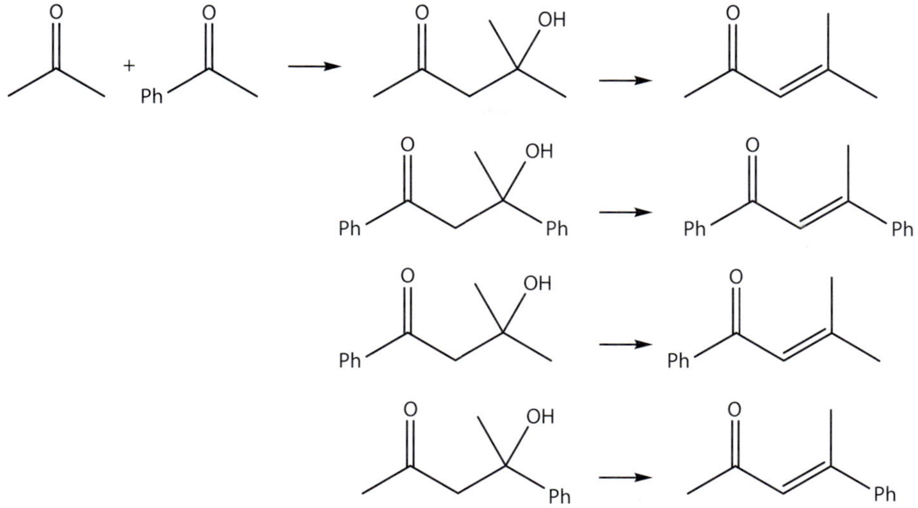

FIGURE 17.28 Cross condensation of acetone and acetophenone.

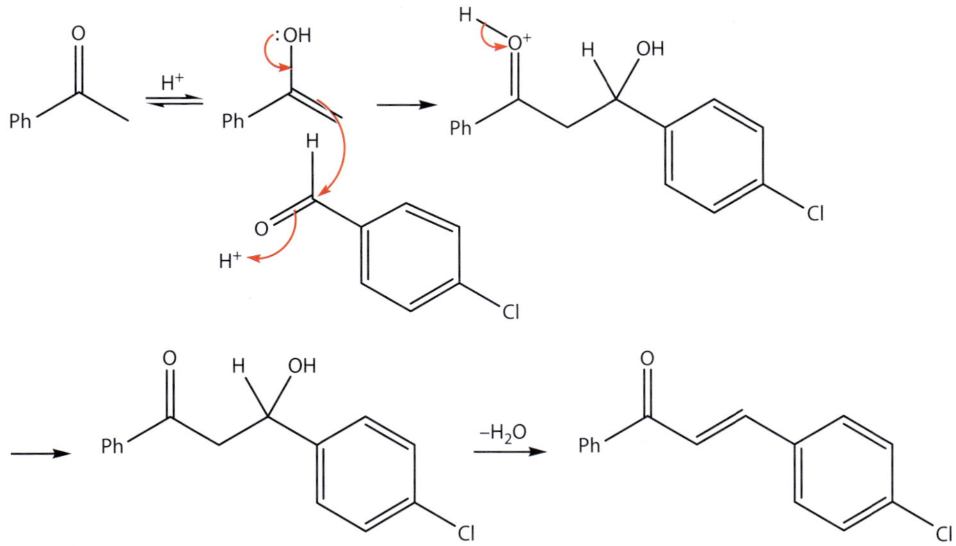

FIGURE 17.29 Cross condensation of acetophenone and 4-chlorobenzaldehyde.

(Figure 17.29). Acetophenone is the only possible source of a nucleophile, but aldehydes are more electrophilic than ketones.

Another way that we can "direct" aldol condensations is when the product is a five- or six-membered ring. Making a ring of this size will always be favored over any alternative process, intra- or intermolecular. So heptane-2,5-dione gives a six-membered ring product, as does nonane-2,8-dione (Figure 17.30). This is not to say that other enols/enolates are not formed or indeed that other cyclizations cannot take place. However, these are equilibrium processes, controlled by thermodynamics, and the five- and six-membered rings are very much more stable than the alternatives. This type of intramolecular process has been widely used in steroid synthesis—for example, in Figure 17.31, showing the last step in the preparation of progesterone.

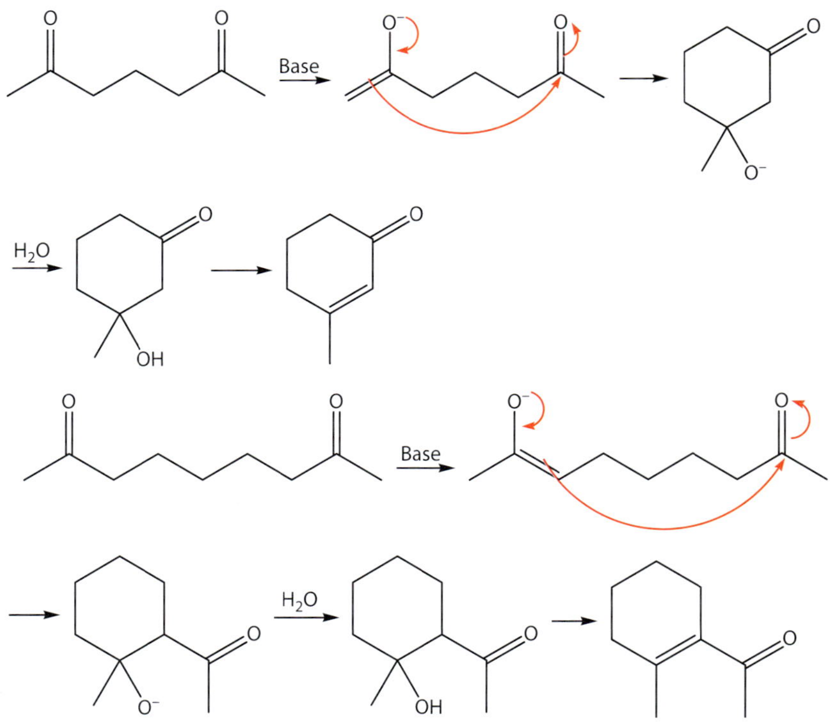

FIGURE 17.30 Intramolecular aldol condensations.

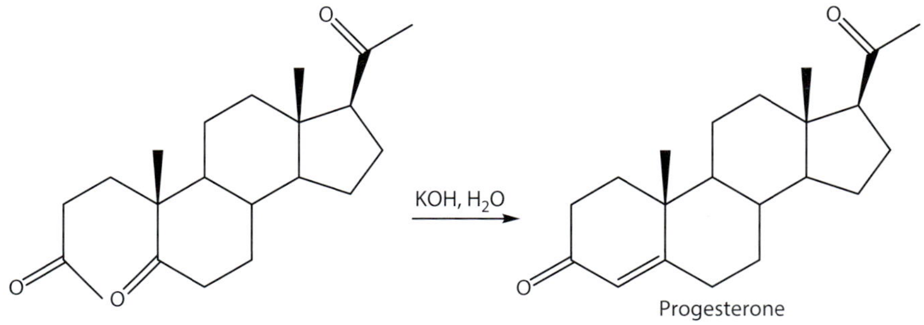

Progesterone

FIGURE 17.31 Intramolecular aldol condensation in steroid synthesis.

PROBLEM 17.8

Write a mechanism for the reaction of Figure 17.29.

PROBLEM 17.9

Study the reaction shown in the following. Write a mechanism, and speculate as to why this stereoisomer might be selectively isolated:

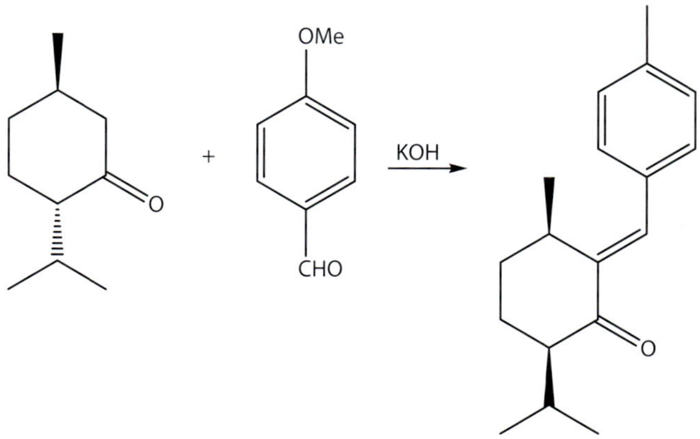

Solutions

17.8

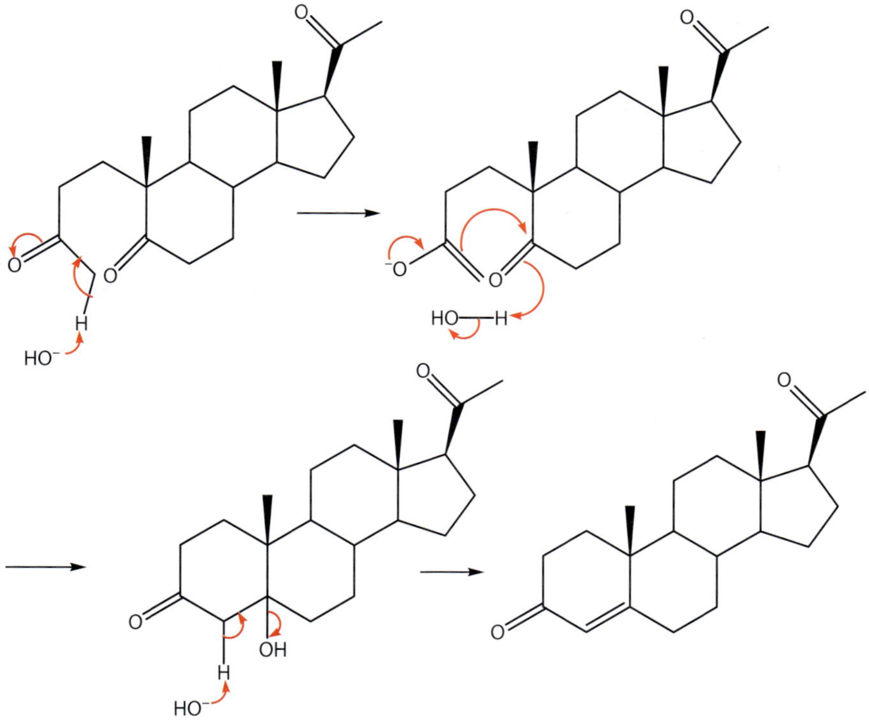

17.9 This is a crossed aldol reaction of the type that we discussed—only one component (the ketone) can enolize, and the second component (the aldehyde) is more electrophilic. Although there are two enolates possible for the ketone, the more stable is the source of the product, and the other might well, in any case, be too hindered for a good reaction:

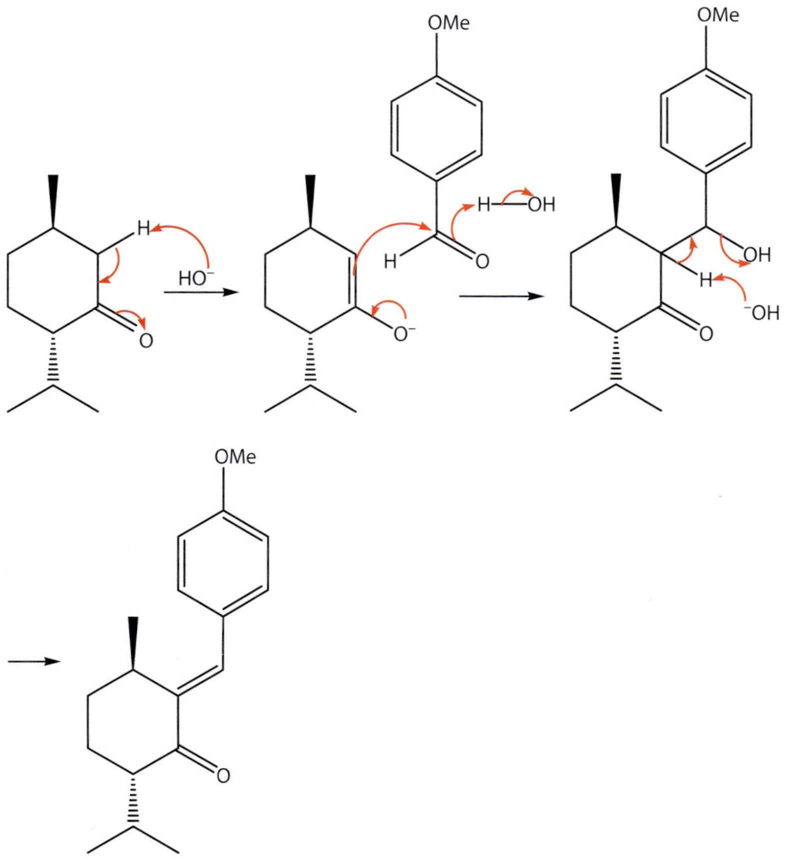

But we have a problem—this is not the product that is obtained! However, it must, through enolate formation, be in equilibrium with the species isolated:

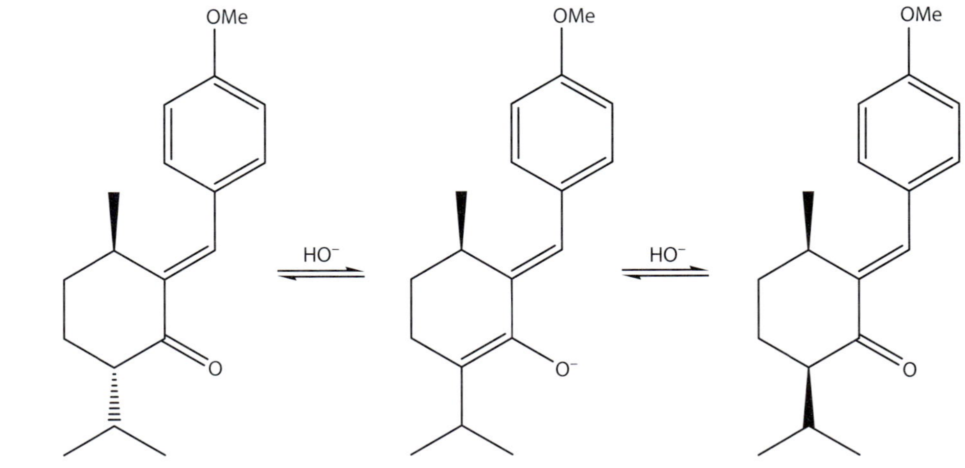

So, why is the *cis*-product isolated? Molecular modeling or the availability of an x-ray crystal structure might tell us that it is more stable for steric reasons, but an equally plausible explanation is more prosaic. If it is less soluble than the *trans*-isomer, then its precipitation from the solution will pull the equilibrium above over to the right.

17.4.3 CLAISEN ESTER CONDENSATION

Other types of carbonyl compounds also undergo condensations of a similar type—but for each type, there is a different name, immortalizing its discoverer. This means that it's harder to look them up in an index or search for examples on a website, but the important thing to remember is that the chemistry is essentially always the same—an enol or enolate is formed, and it attacks another carbonyl compound. A leaving group may be expelled or dehydration of the product may take place, but assuming it does not, the product is always a 1,3-dioxygenated species. The oxidation level of the product will vary depending on the oxidation level of the SMs. The Claisen ester condensation is exemplified by the reaction of ethyl acetate (Figure 17.32). The product is a ketoester, a very useful one, as we shall see later.

What base would be best for this transformation? Hydroxyl would not be useful as there is a risk that the ester would be hydrolyzed. We need a strong base, as esters (pK_a typically 25) are more difficult to convert to their enolates than are ketones (pK_a typically 20). Generally, ethoxide is used as the base—if it attacks either the SM or the product, the only result will be the exchange of one ethoxide for a different one. Since all the steps here are equilibria, how do we ensure that the reaction goes to completion? Fortunately, the reaction is driven by the enolization of the

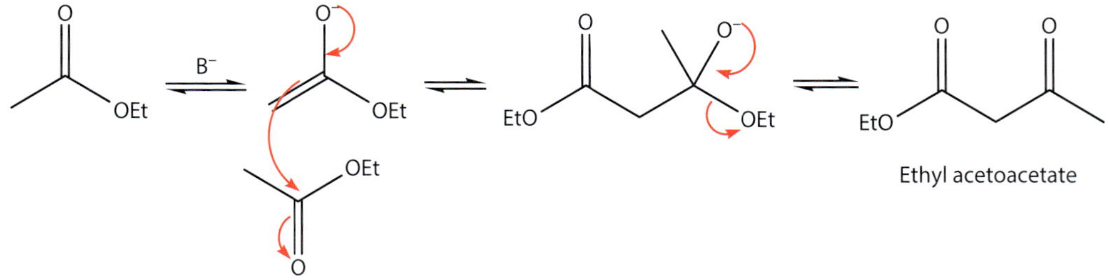

FIGURE 17.32 Claisen ester condensation of ethyl acetate.

product—because the enolate is conjugated, (**17.29**) it is much easier to enolize than the starting ester. The pK_a of ethyl acetoacetate is 11:

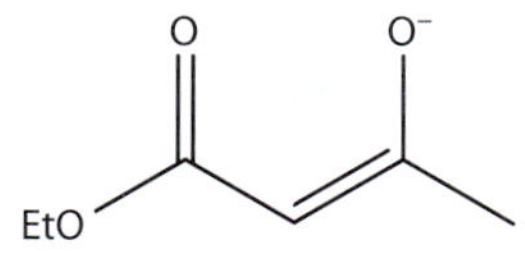

17.29

PROBLEM 17.10

How would you prepare **17.30** using a Claisen condensation?

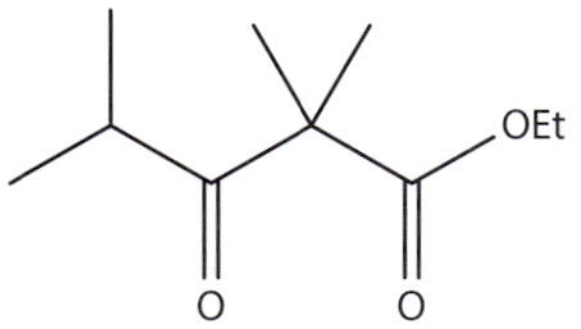

17.30

Solution

The first question we need to answer is which bond is made in the condensation, so that we can dissect the molecule there. This is the bond adjacent to the ketone, marked in red in the following—this ketone used to be an ester, and a leaving group has been displaced:

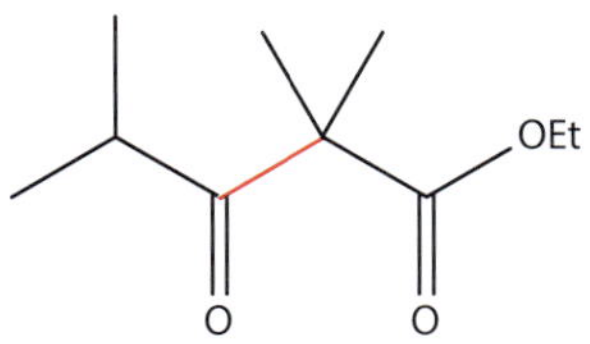

We will assume that the leaving group is OEt, so our two components are the ester and its enolate:

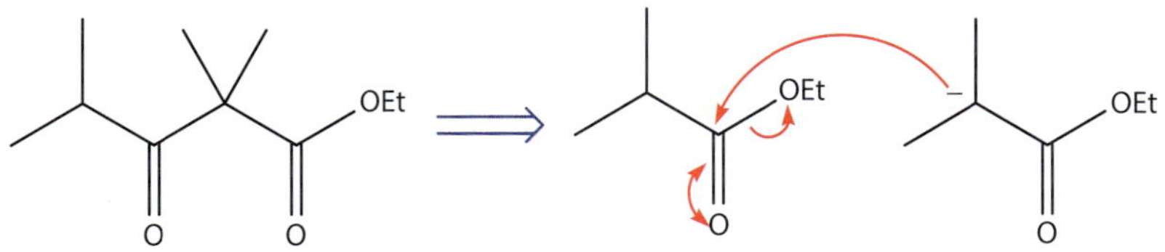

When the Claisen ester condensation is intramolecular, it goes by the name of the Dieckmann condensation. As with the intramolecular aldol reactions, only processes that give rise to five- or six-membered rings are favorable (Figure 17.33). This process has been used in the total synthesis of prostaglandin A2 (Figure 17.34).

Crossed Claisen reactions can be achieved when one component cannot be enolized—confusingly, this name is used whether the nucleophilic component is an ester or a ketone (Figure 17.35). The product of the reaction in the second example was used in a morphine synthesis and is isolated in the enol form shown.

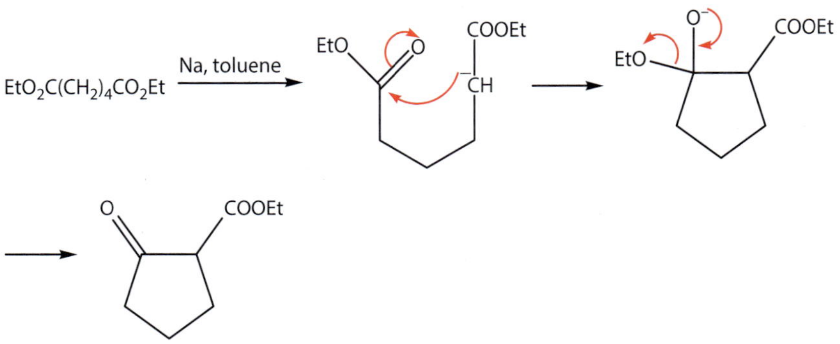

FIGURE 17.33 Dieckmann condensation.

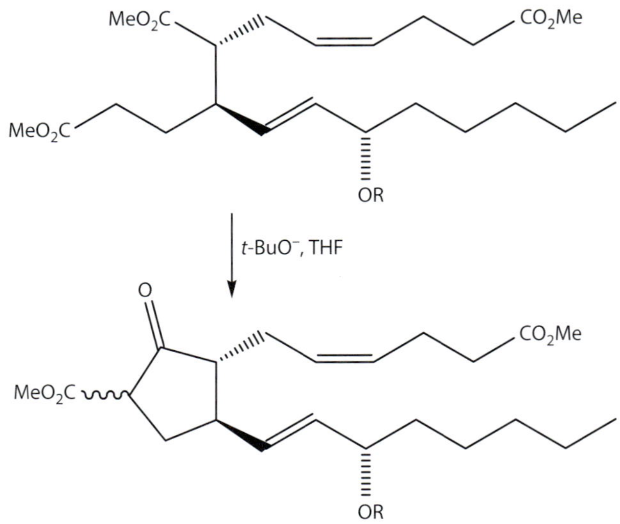

FIGURE 17.34 Application of the Dieckmann condensation.

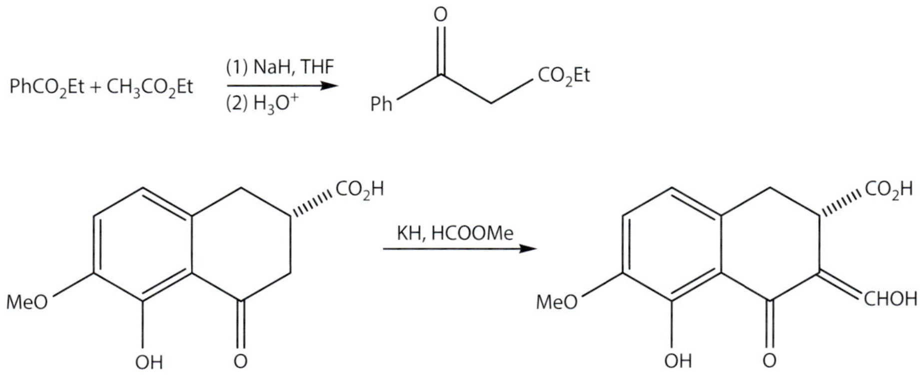

FIGURE 17.35 Crossed Claisen condensations.

Key Points from Sections 17.4.3 and 17.4.4

Students often find it difficult to remember which name goes with which condensation and what each one yields. And of course, there are many more of these reactions, all of whose discoverers wanted to be immortalized to (or cursed by) every new student of organic chemistry. Table 17.1 summarizes the condensations we have seen so far. The names are important only insofar as they make it easier to look up reactions. There are some key pointers that will make the study of these processes easier:

- If there are two components, start by enolizing the one with the lower pK_a; make the most stable enol/enolate.
- Substituents do matter in pK_a, but a good rule of thumb is
  β-Diketone 9; β-ketoester 11; 1,3-diester 13
  ROH, H_2O 15–16
  Aldehyde 17–18; ketone 20; ester 25; amide 30
- React the enol/enolate you have made with the most electrophilic carbonyl compound available.
- Substituents matter, but
  RCOCl > RCHO > $R_2C{=}O$ > RCOOEt ≫ $RCONH_2$
  Formaldehyde and formate esters are particularly reactive, both for electronic and steric reasons

- If you add to an ester, or another species where there is a sensible leaving group, then the leaving group will leave.
- Can your final product enolize more readily than the SM? If it can, this probably is the product that will be formed. Hydrogen bonding may also stabilize your product.
- If an intramolecular reaction producing a five- or six-membered ring is available, this will happen.
- If water can be eliminated to give an α,β-unsaturated carbonyl compound, this is likely to happen under acid conditions and may do so under basic ones.

TABLE 17.1 Important Condensation Reactions

Name	Enol/Enolate Component	"Keto" Component	Reagent/ Conditions	Product(s)	Comment
Aldol homocoupling	Aldehyde or ketone enol or enolate	Same aldehyde or ketone	Acid or base	β-Hydroxyaldehyde or ketone	Product usually dehydrates in acid, sometimes in base. Works best if only one enolate may be formed from the ketone.
Aldol cross coupling	Aldehyde or ketone	Nonenolizable aldehyde or ketone	Acid or base	β-Hydroxyaldehyde or ketone	Product usually dehydrates in acid, sometimes in base. Works best if only one enolate may be formed from the ketone.
Aldol, intramolecular	Dicarbonyl compound. Intramolecular single-component reaction		Acid or base	β-Hydroxyaldehyde or ketone; dehydration is common	Only 5- or 6-membered rings can readily be formed.
Claisen	Ester, RCH_2COOR'	RCH_2COOR'	Usually $R'O^-$ to avoid ester interchange	β-Ketoester	
Dieckmann	$RO_2C(CH_2)_nCO_2R$; intramolecular single-component process		RO^- or Na/ toluene	Cyclic β-ketoester	Only 5- or 6-membered rings can readily be formed.
Crossed Claisen	Enolizable ester	Nonenolizable ester or ketone	RO^- or other base	β-Ketoester or β-hydroxyester	

PROBLEM 17.11

Predict the product(s) of each of the following reactions:

(a)

$$CH_3CO_2Et \xrightarrow{EtO^-}$$

(b)

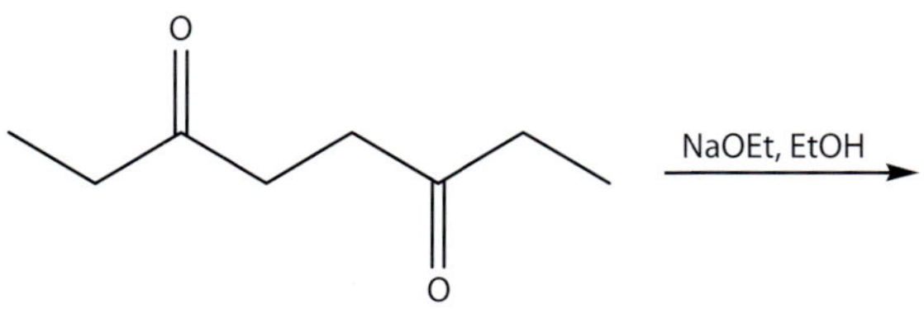

(c)

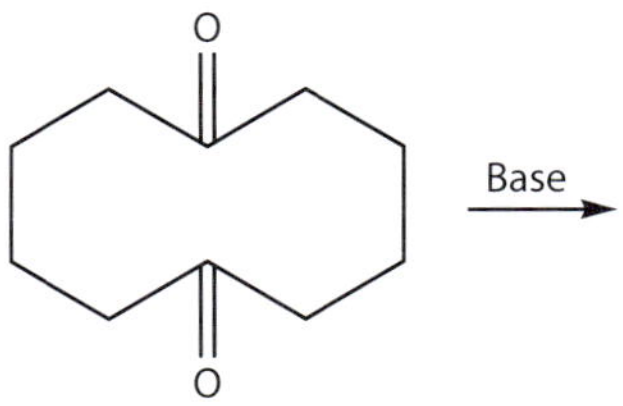

Solutions

(a) This is a classic Claisen condensation, yielding ethyl acetoacetate:

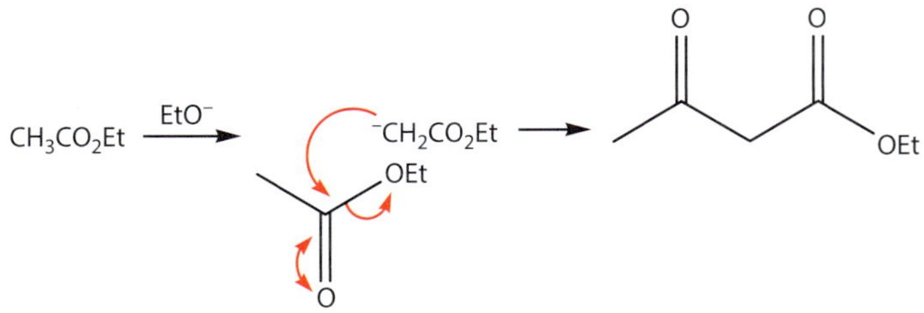

(b) This is an intramolecular aldol reaction; only the reaction giving a five-membered ring occurs:

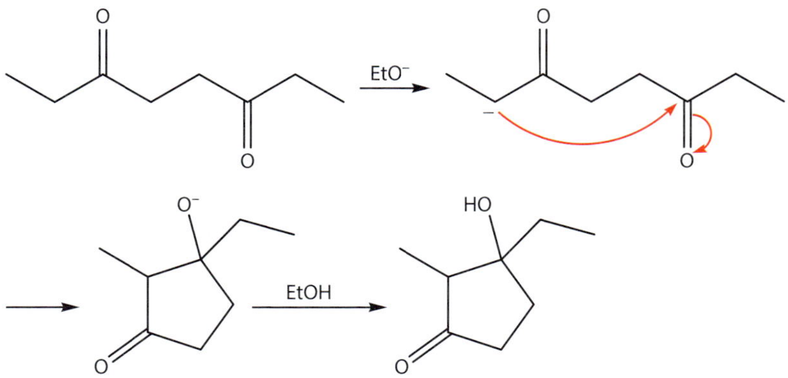

(c) Another intramolecular aldol; the same product is obtained wherever the initial enolization occurs:

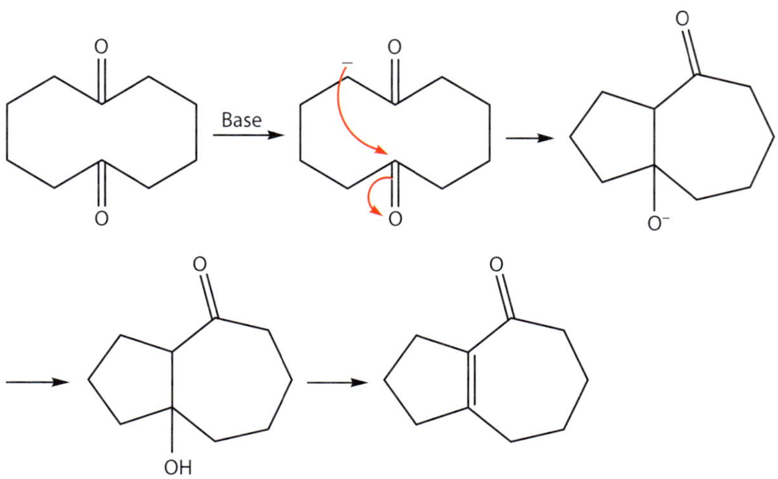

Focus on Biological Claisen Condensations: Synthesis of Fatty Acids and Polyketides

The biological equivalent of acetate is the acetyl derivative (**17.31**) of coenzyme A, **17.30**. This undergoes Claisen condensations, as in Figure 17.36, using a thiolase enzyme. The process is easier with a thioester as these are slightly more acidic than regular oxygen esters. The process can continue, each time building the carbon chain by two atoms, to give a polyketide, of general structure **17.33**. In fatty acid biosynthesis, oxygen is removed at each stage by carbonyl reduction, water elimination, and double bond reduction. The polyketide chain is then elongated using units of malonylCoA (**17.32**), which is subsequently decarboxylated. Polyketides are converted to a wide range of biologically active molecules.

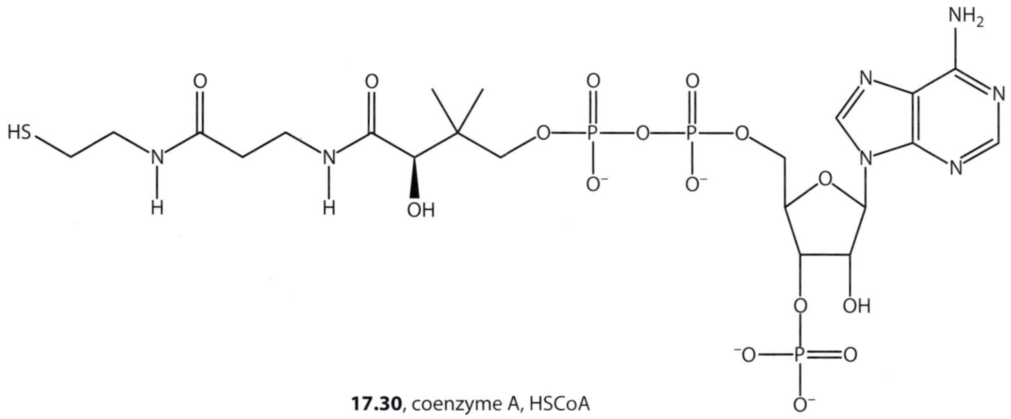

17.30, coenzyme A, HSCoA

17.31, acetylCoA   **17.32**, malonylCoA

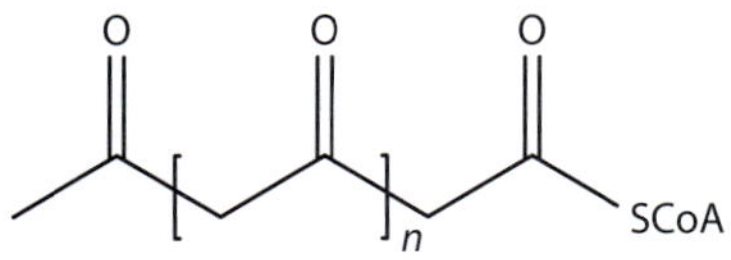

17.33

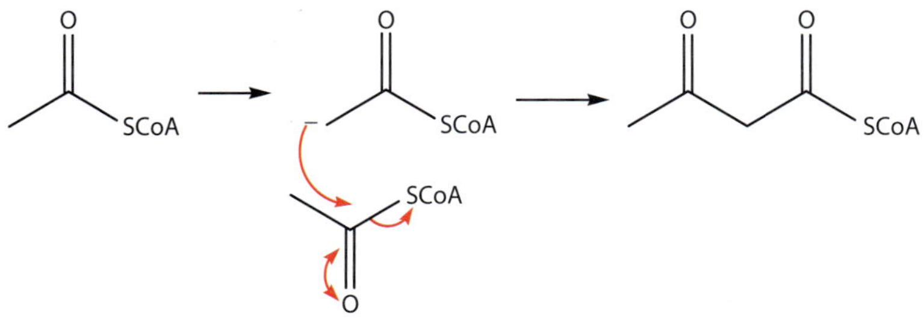

FIGURE 17.36 Claisen condensation of acetyl CoA.

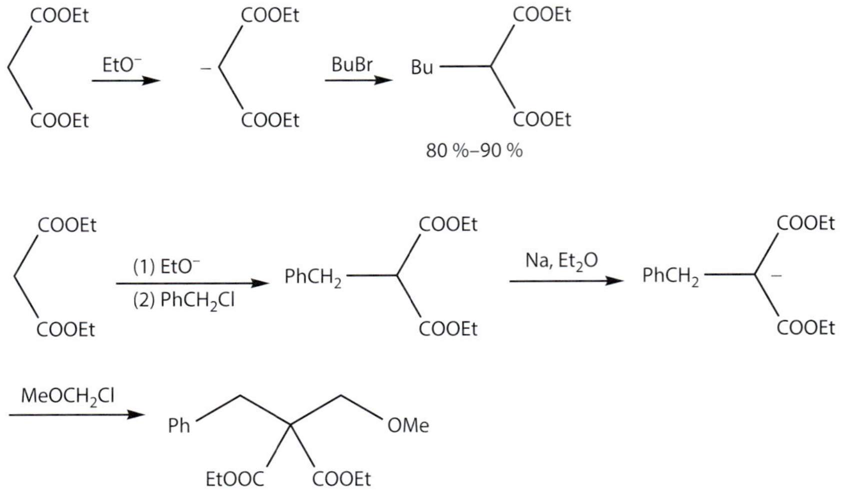

FIGURE 17.37 Alkylation of diethyl malonate.

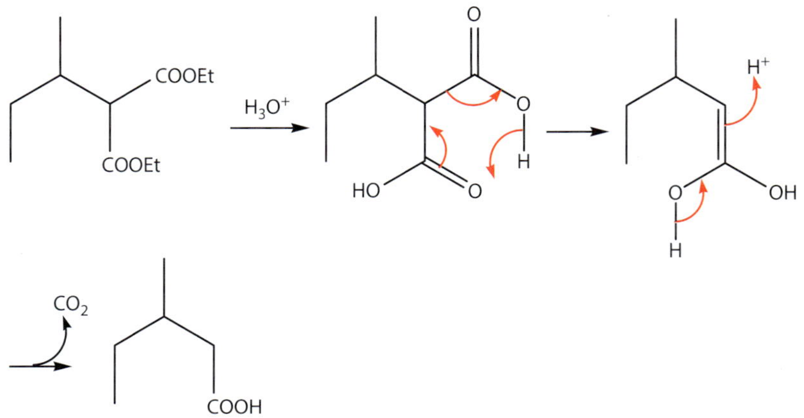

FIGURE 17.38 Decarboxylation of β-carbonyl acids.

17.4.4 ALKYLATION OF β-DICARBONYL COMPOUNDS

We have already met this reaction in its simplest form at the end of Chapter 9 when we studied the nucleophilic substitution of alkyl halides by the anions of dicarbonyl compounds (Figure 17.37). Ethoxide is used as the base in the reaction of diethyl malonate, to avoid any problems with ester interchange. Dialkylation requires a stronger base for the second step—the alkyl group already introduced is electron-donating and destabilizes the carbanion.

Similar reactions are also possible with ethyl acetoacetate. The electrophiles that work well are primary and secondary alkyl halides and sulfonates. Tertiary halides do not react well—elimination is the main reaction. Acyl halides and unhindered epoxides also react well. An important aspect of these processes is that carboxylic acids with a carbonyl group at the β-position are readily decarboxylated (Figure 17.38). We will see in later chapters that we are using the ester as an activating group so that we can make the enolate more easily, but we can eliminate it later. Some examples of the use of this process in synthesis are given in Figure 17.39.

PROBLEM 17.12

Write a reaction mechanism for the final example in Figure 17.39.

Solution

Although this looks complicated, if we start with what we are sure about, the later steps will become obvious. First, the ethyl acetoacetate is deprotonated, and the anion opens the epoxide at the less hindered site (although these anions do react with primary chlorides, this clearly has not

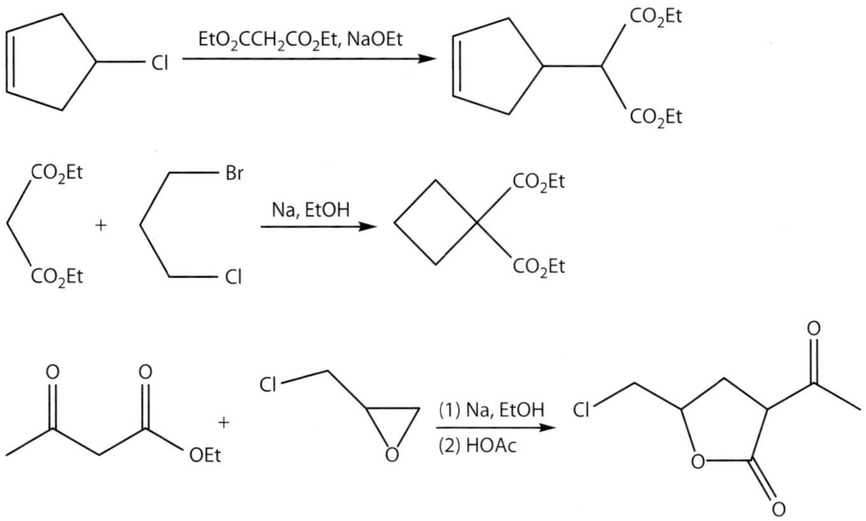

FIGURE 17.39 Ethyl acetoacetate and diethyl malonate in synthesis.

happened here as the chloride is intact). The alkoxide generated is ideally positioned to make the five-membered ring lactone, by an ester exchange process:

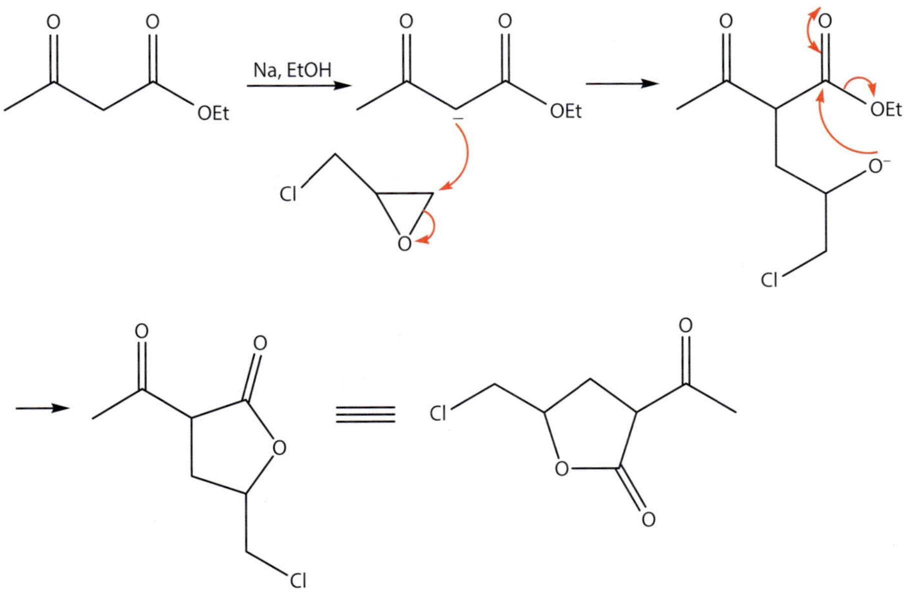

It might be supposed that this technique could be readily extended to alkylation of β-diketones, such as cyclohexane 1,3-dione, **17.34**. These are certainly easy to deprotonate, but the alkylation reaction can present some problems (Figure 17.40). The extent of the *O*-alkylation depends on the base used, the solvent (the alkoxide is "naked" in DMSO, but heavily solvated in methanol) and the electrophile. We describe enolate anions as ambident nucleophiles, since they can react either at carbon or oxygen. RO^- is a hard nucleophile and reacts best with hard electrophiles such as

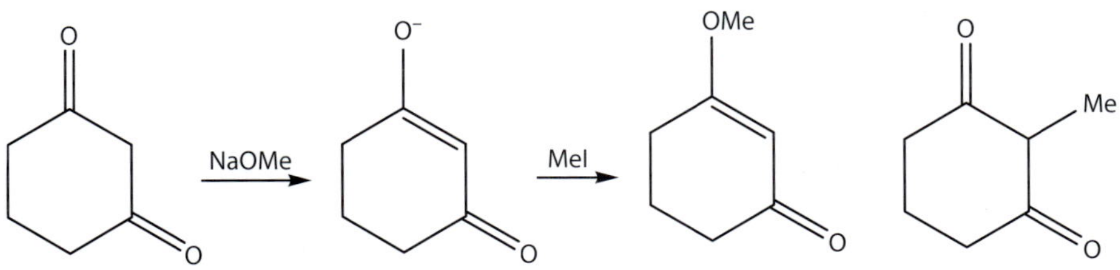

17.34 37 % 15 %

FIGURE 17.40 Alkylation of a β-diketone.

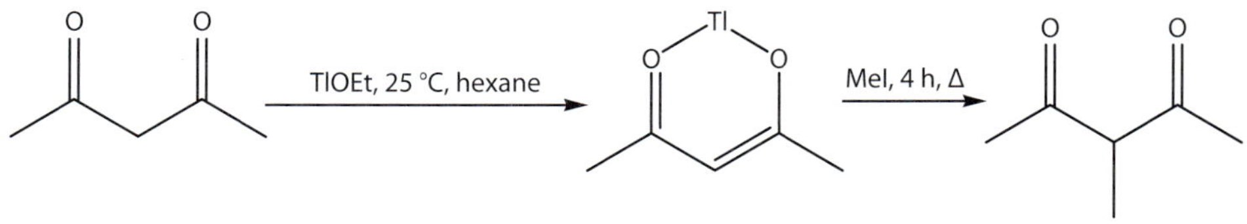

FIGURE 17.41 Thallium ethoxide in alkylation of a β-diketone.

ROTs, RC(=O)Cl. The carbanion form of the enolate is a soft nucleophile (with some character of a carbon–carbon double bond) and reacts best with the soft electrophiles RBr and RI.

Many solutions to this problem have been developed, and some of these will be explored further in Chapters 19 and 20. One effective solution is to use the thallium rather than the sodium enolate. Thallium is a very oxophilic metal and holds on very tightly to the oxygens, so that substitution occurs entirely at carbon (Figure 17.41). However, its toxicity precludes use on a large scale.

PROBLEM 17.13

Write a mechanism for each of the following reactions:

(a)

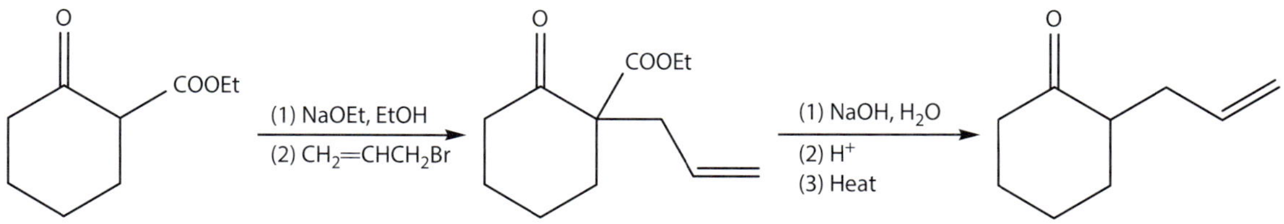

(b)

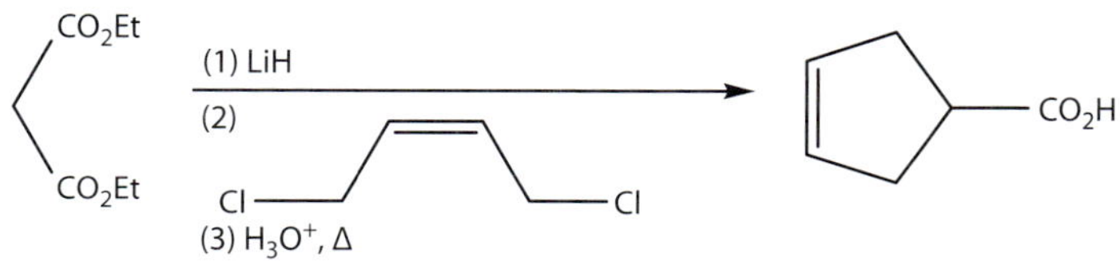

Solutions

(a)

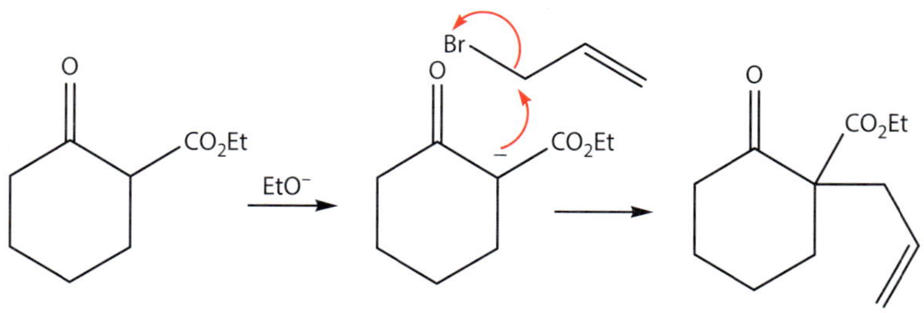

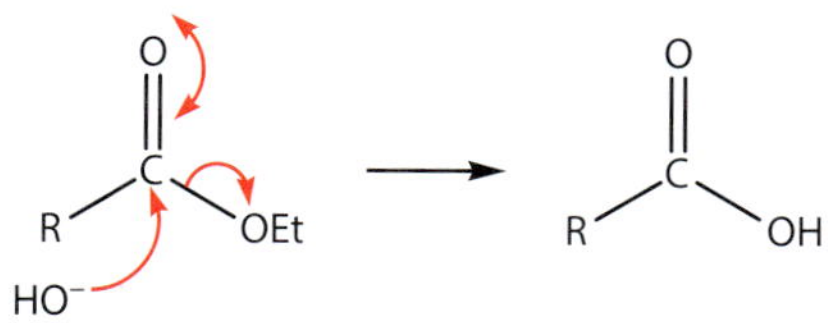

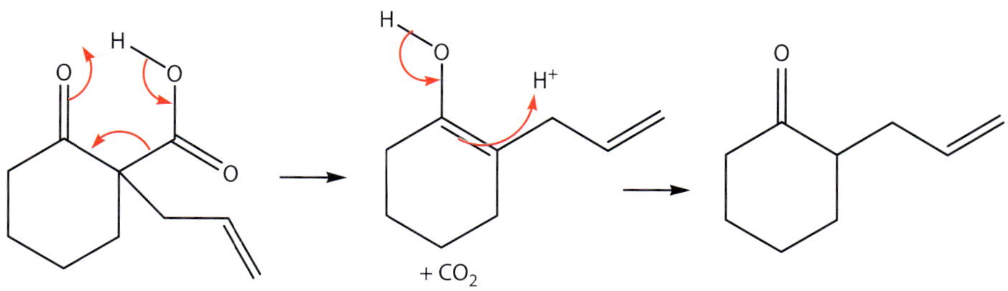

(b)

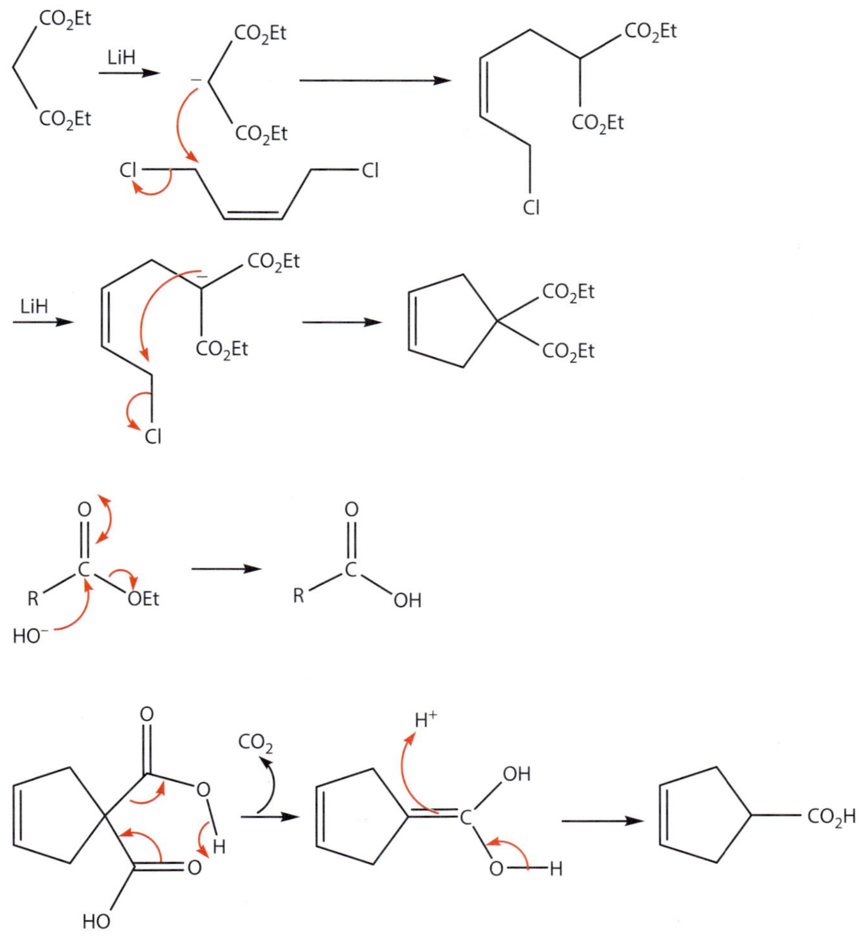

17.4.5 ALKYLATION OF SIMPLE KETONES

Simple ketones have pK_a values of about 20, so they are significantly more difficult to deprotonate than diketones, and a stronger base will be needed to produce an enolate anion. Although *O*-alkylation is rare, polyalkylation does occur, and for nonsymmetric ketones, regioselectivity is often poor. If we consider the deprotonation of 2-methylcyclohexanone (Figure 17.42) under a variety of conditions (Table 17.2), either enolate can be obtained with fair selectivity. Under conditions of kinetic control, the most accessible proton is removed, particularly when the base used is very bulky.

From this, we must conclude that selectivity will not be achieved without considerable care in establishing optimum reaction conditions. In addition, if there is the possibility of proton exchange between enolate and ketone, polyalkylation may occur (Figure 17.43). Numerous approaches to this problem have been developed. The classical solution is to form the enolate completely (so that there is no ketone present to equilibrate) at low temperature (to ensure formation of a kinetic enolate) and then add the enolate to an excess of RX, so that it is all quenched immediately, and

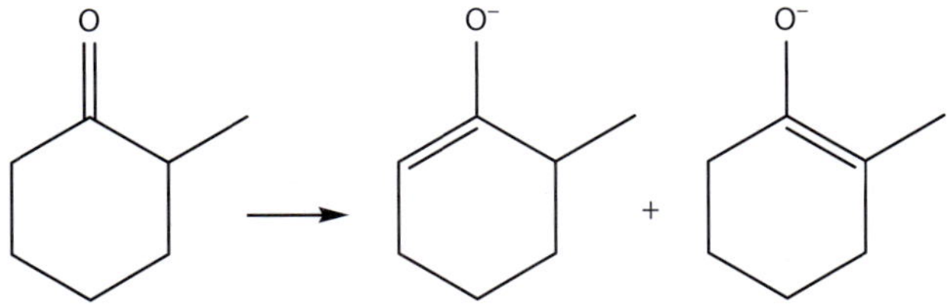

17.35 **17.36**

FIGURE 17.42 Deprotonation of 2-methylcyclohexanone.

TABLE 17.2 Formation of Enolates from 2-Methylcyclohexanone

Reaction Conditions	17.35	17.36
LiN*i*-Pr_2, $MeOCH_2CH_2OMe$, 0 °C	99	1
Ph_3CLi, $MeOCH_2CH_2OMe$, −78 °C	90	10
Et_3N, DMF, 25 °C	22	78
NaH, Et_2O, 25 °C	26	74

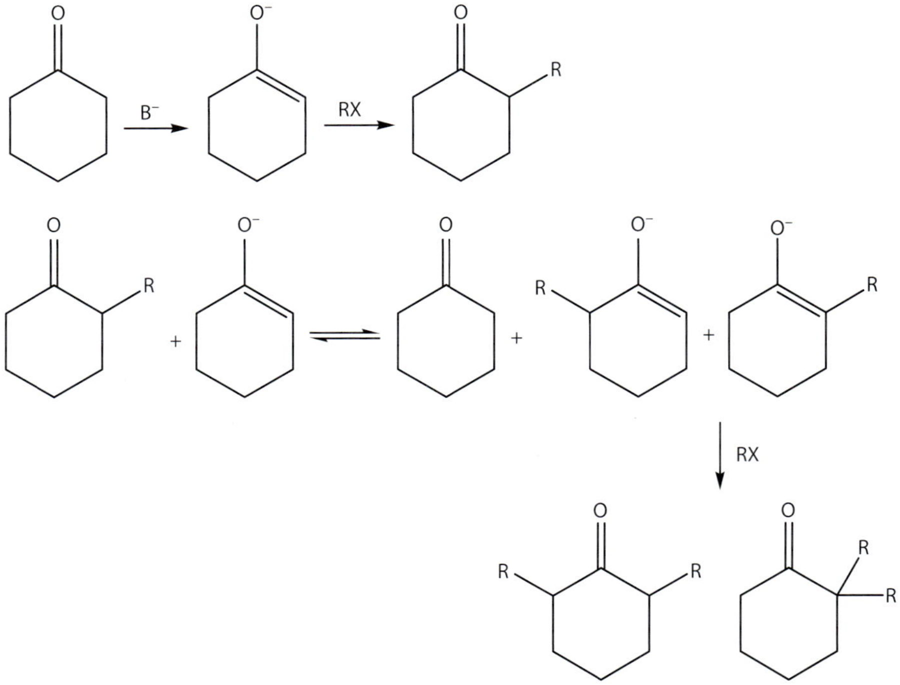

FIGURE 17.43 Formation of polyalkylated products from enolate alkylation.

no equilibration is possible. The base normally chosen is LDA, formed *in situ* from butyl lithium and diisopropylamine (17.1). LDA is a strong enough base completely to deprotonate aldehydes, ketones, esters, amides, and carboxylic acids (where a dianion is formed). Although butyl lithium is an even stronger base, it is also unhindered and hence an excellent nucleophile toward carbonyl compounds.

(17.1)
$$\underset{pK_a=40}{BuLi + HN(CHMe_2)_2} \longrightarrow \underset{pK_a=50}{BuH} + \underset{LDA}{LiN(CHMe_2)_2}$$

A typical process for cyclohexanone is given in Figure 17.44. Primary halides are excellent electrophiles in this process. Many secondary alkyl halides are also useful, but some elimination may occur. The use of a tertiary halide leads mainly to elimination. Comparable reactions of esters, amides, carboxylic acids, nitriles, and nitroalkanes are shown in Figure 17.45. HMPA is hexamethylphosphoramide, which enhances enolate reactivity by complexing the metal cation and hence breaking up the enolate aggregates in solution. In the reaction of the acid, a simple base such as sodamide can be used—there is no issue of addition to the carbonyl group here. In the final example, only a very mild base is needed; the pK_a of aliphatic nitro compounds is typically 9, so they are easily deprotonated, and the reaction is intramolecular leading to a favorable five-membered ring.

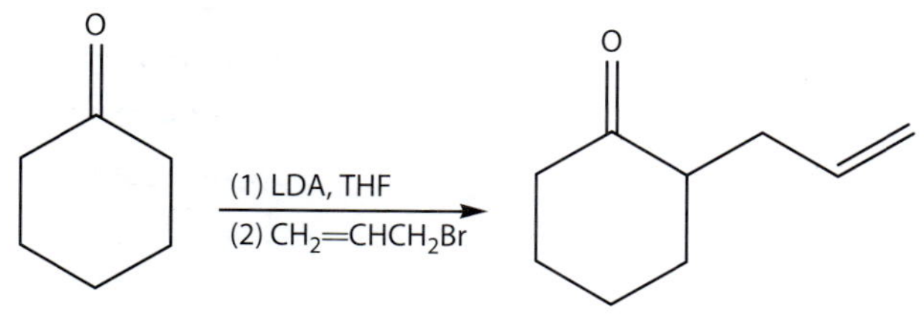

FIGURE 17.44 Selective alkylation of cyclohexanone.

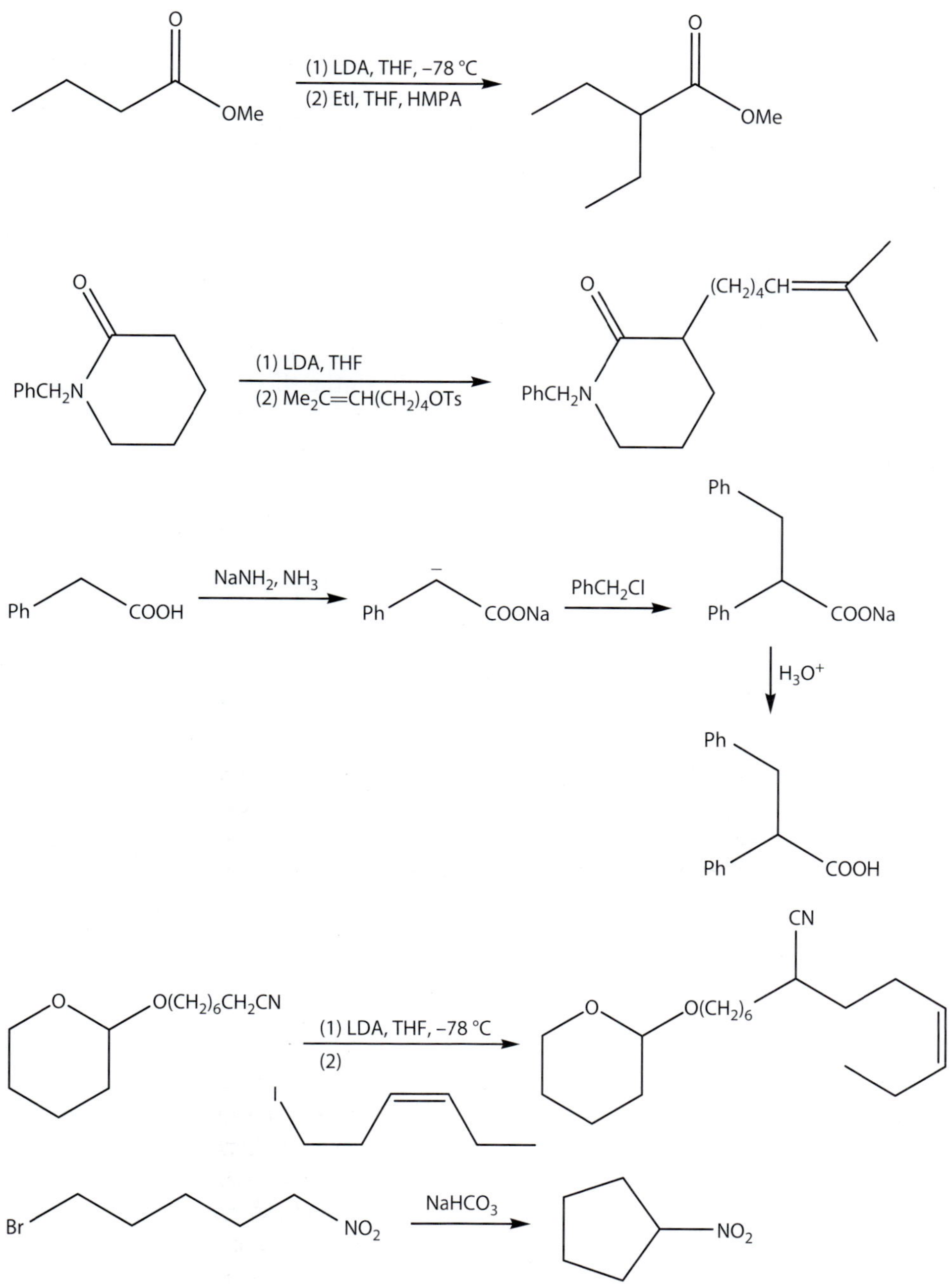

FIGURE 17.45 Alkylation of carbonyl compounds.

PROBLEM 17.14

Write a mechanism for the final reaction of Figure 17.43.

Solution

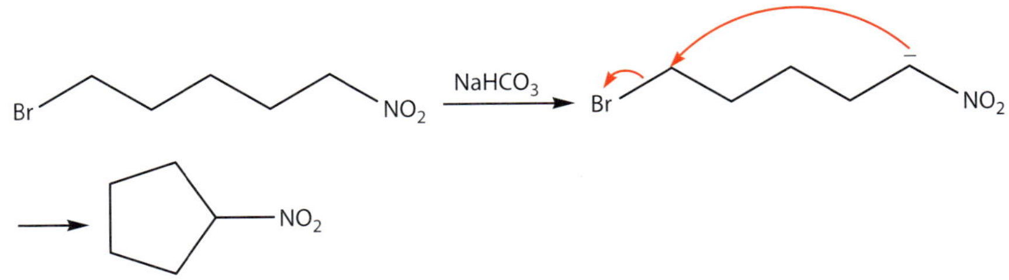

Focus on Enolate Structure

We have discussed how enolates are made and how they react at some length, but what is their structure in solution and in the solid state? Does the nature of the metal matter (yes, but most studies are of lithium enolates), is solvent important (definitely)? X-ray crystal structures of lithium enolates were first published in the 1970s, and it was found that all crystalline lithium enolates are oligomers—dimers, tetramers, or hexamers. Crystalline lithium enolates of ketones, amides, nitriles, and iminoesters have been widely studied. Those of esters are rarer, as they are much less stable in the solid state, even at low temperature. The degree of oligomerization largely does *not* depend on the nature of the carbonyl compound, but on the solvent or any additives. For example, the lithium enolate of 3,3-dimethyl-2-butanone, **17.37**, may be a dimer, a tetramer, or a hexamer, depending on solvent and additive. The more strongly the additive coordinates lithium, the smaller the oligomer. Using LDA in heptane, a solvent-free hexamer is obtained (**17.38**). In THF, a cubic tetramer (**17.39**) is formed and the adduct with $Me_2NCH_2CH_2NHMe$ is a dimer (**17.40**). Numerous other structures have been established (*e.g.*, **17.41**, **17.42**), and it has been shown that sodium, aluminum, magnesium, and zinc enolates also aggregate in the solid state, but these data do not necessarily tell us about solution structures.

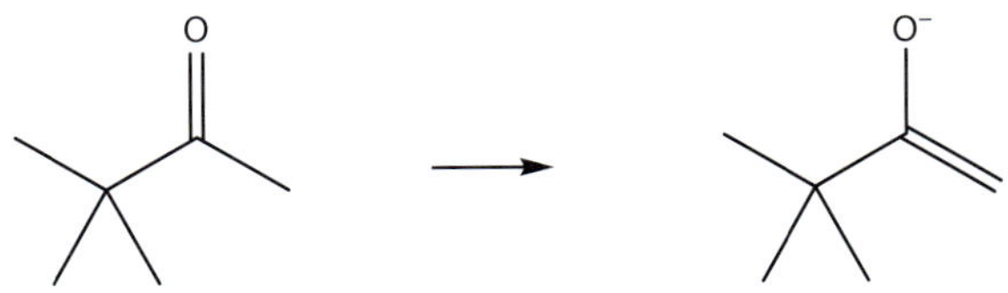

17.37

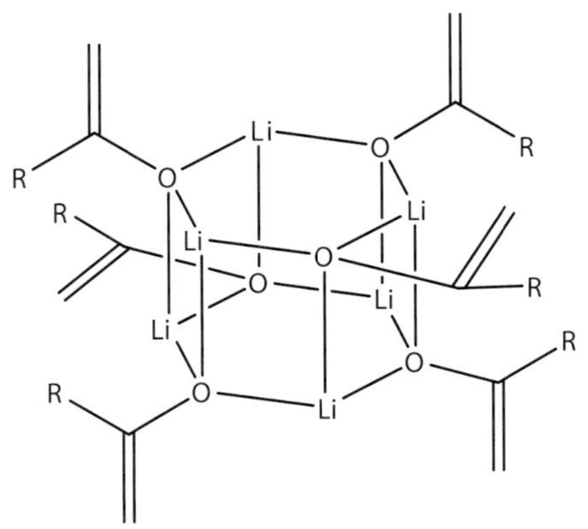

17.38

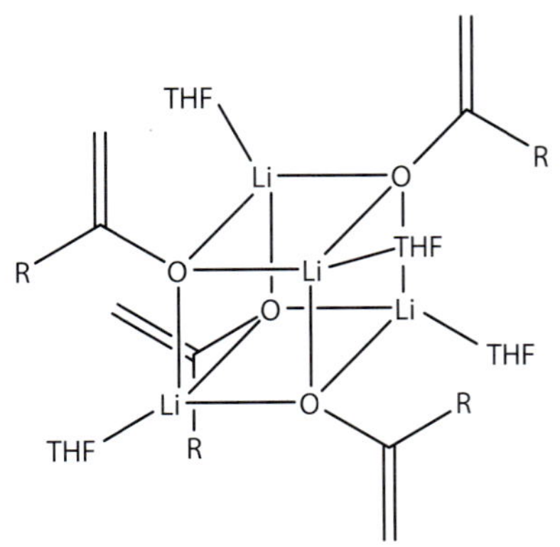

17.39

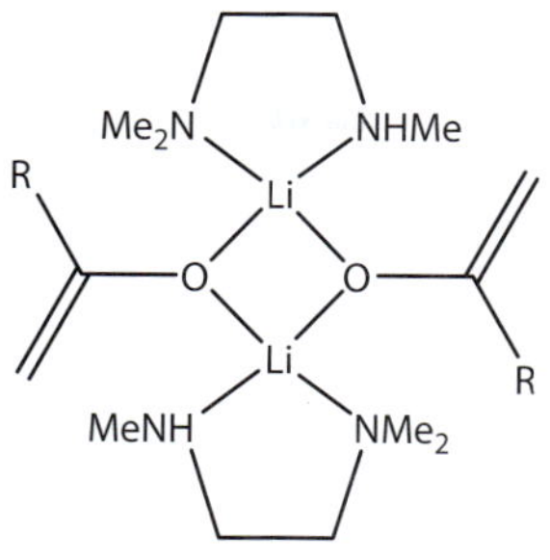

17.40

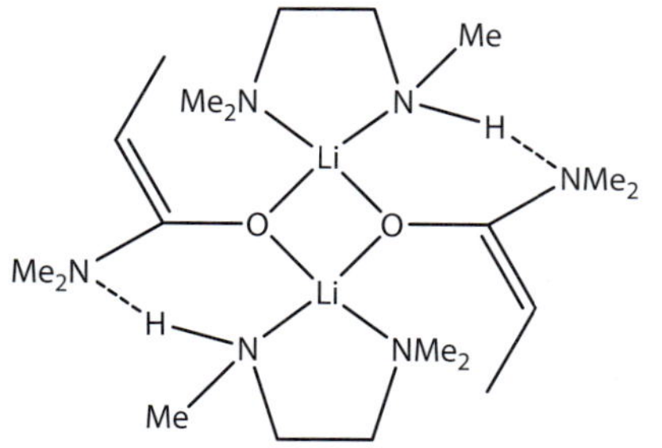

17.41

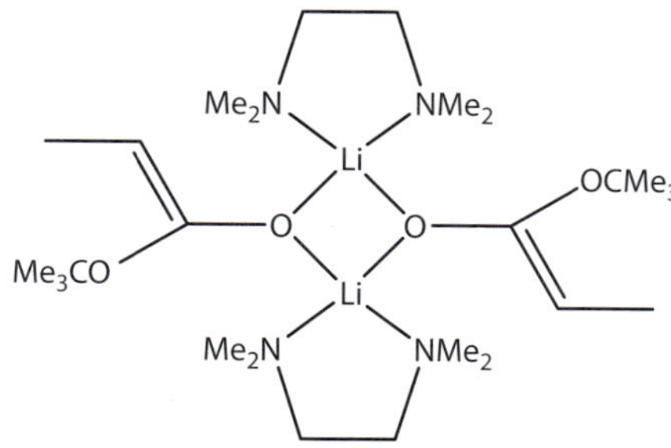

17.42

Recently, ^{6}Li NMR spectroscopy has been used to establish that, in many cases, these oligomers *do* remain intact in solution. Thus, the enolate of cyclohexanone is dimeric in solution when TMEDA, $Me_2NCH_2CH_2NMe_2$, is added, but is a cubic tetramer in THF solutions. The predominance of dimers in TMEDA seems to be fairly general, including the enolates of cyclopentanone, cycloheptanone, 3,3,-dimethyl-2-butanone, acetophenone, and 2,5-dimethylpentanone.

Focus on Silyl Enol Ethers

One rather special enolate, which is very important synthetically, is the silicon enolate. Silicon is not a true metal, and "silicon enolates" such as **17.43** are more usually described as silyl enol ethers. They are prepared by reaction of ketones with base, followed by R_3SiCl. They are stable in the absence of water and can often be purified by distillation. The enolate generated as an intermediate reacts with silicon exclusively at oxygen, because of the strength of the silicon–oxygen bond (452 kJ mol^{-1}). Alone, they are rather unreactive with electrophiles, but a lithium enolate can be regenerated cleanly by reaction with methyl lithium or fluoride ion (because of the strength of the silicon–fluorine bond, 565 kJ mol^{-1}) (Figure 17.46). These are very mild conditions, and enolates generated in this way tend to give few by-products in their reactions with electrophiles, since no isomerization is possible.

Silyl enol ethers have a particular advantage when a specific enolate from a nonsymmetric ketone is required, as both can be generated fairly reliably (Figure 17.47). Many other sets of conditions, and different trialkylsilyl chlorides, are also available.

Another useful attribute of silyl enol ethers is that they react with tertiary halides in the presence of Lewis acids. The reaction mechanism is similar to that of Friedel–Crafts alkylation of a benzene ring with the Lewis acid (usually $TiCl_4$) abstracting the halogen from the tertiary center to leave a carbocation (Figure 17.48). This process is not possible with a lithium enolate, which gives only the products of elimination from the alkyl halide. In the presence of Lewis acids, silyl enol ethers also undergo an aldol process, called the Mukaiyama reaction (Figure 17.49). This process has been very extensively studied and is particularly useful in controlling the stereochemistry of the new stereocenters generated (see also Section 23.3.1).

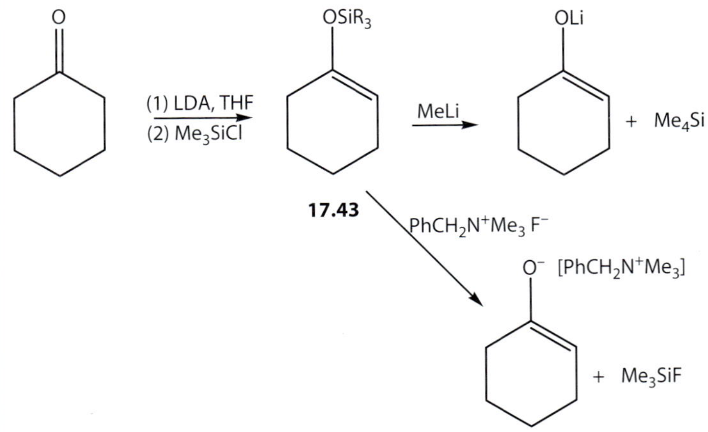

FIGURE 17.46 Formation and reaction of silyl enol ethers.

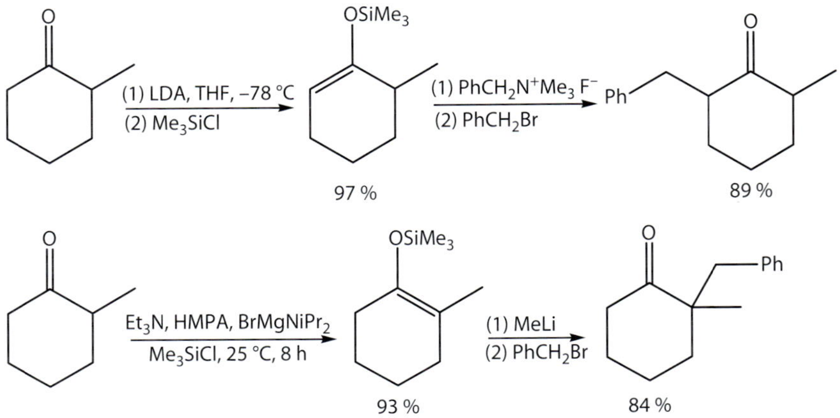

FIGURE 17.47 Regioselective alkylation of methylcyclohexanone, *via* silyl enol ethers.

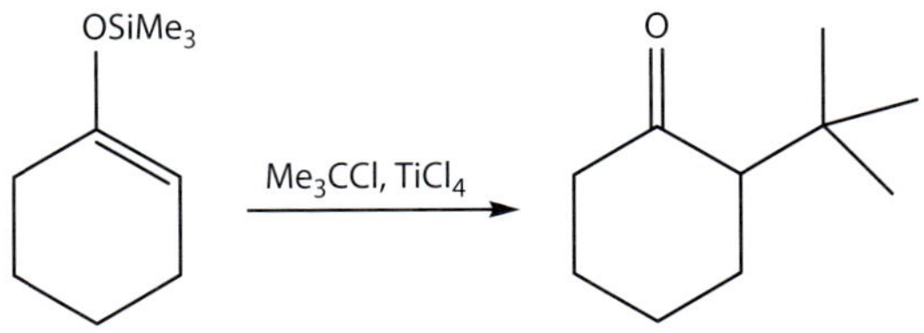

FIGURE 17.48 Reaction of a silyl enol ether with a tertiary halide.

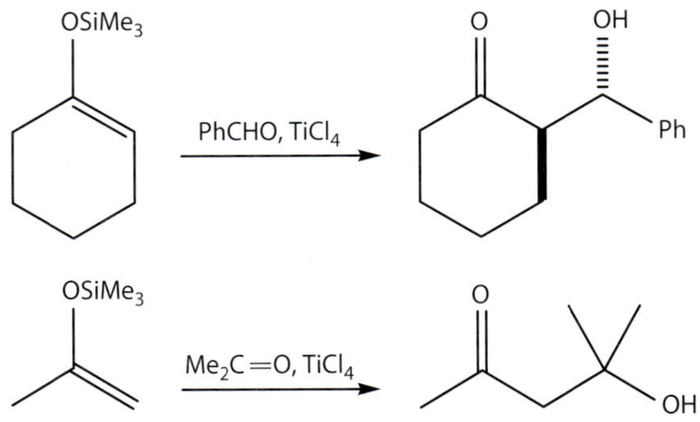

FIGURE 17.49 Examples of the Mukaiyama aldol reaction.

17.4.6 REACTIONS OF ENAMINES

We first met enamines in Chapter 14 when we studied the addition of secondary amines to carbonyl compounds. A typical preparation (Figure 17.50) involves the reaction of a secondary amine (often, but not always cyclic) to a ketone in the presence of acid. In the absence of water, the product enamine is stable and may be purified. The enamine is isoelectronic with an enolate—the lone pair on nitrogen is analogous to the negative charge on the oxygen of an enolate.

PROBLEM 17.15

(Review) Write a complete mechanism for the enamine-forming reaction, Figure 17.50.

Solution

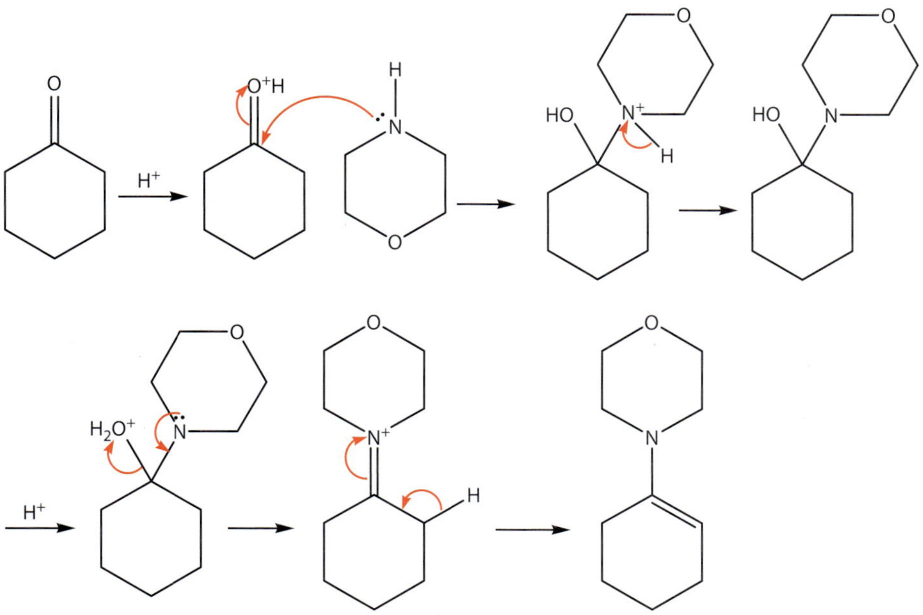

Like enolates, enamines react with electrophiles (Figure 17.51), and the intermediate is hydrolyzed in the work-up to regenerate the ketone. Why use this method rather than the enolate? First, there can be no competing aldol self-condensation of the ketone. Also, this is another way in which the polyalkylation issue can be overcome—the intermediate enamine is purified and the reaction is carried out in the absence of base, so the product cannot be deprotonated. Some interesting and useful selectivity is also possible. In enamine formation from a nonsymmetric ketone, the less substituted enamine is favored, by 90:10 (mainly for steric reasons, so that conjugation between the lone pair of electrons on nitrogen and the double bond is as good as possible), resulting in substitution at the less hindered position. Moreover, the *trans*-product is strongly favored, as the third example demonstrates.

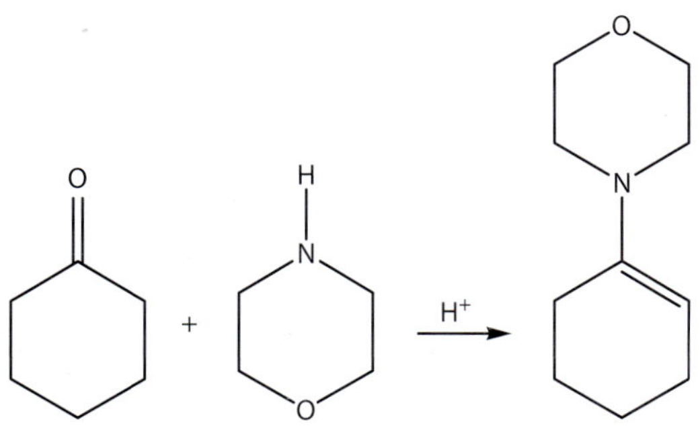

FIGURE 17.50 Enamine preparation.

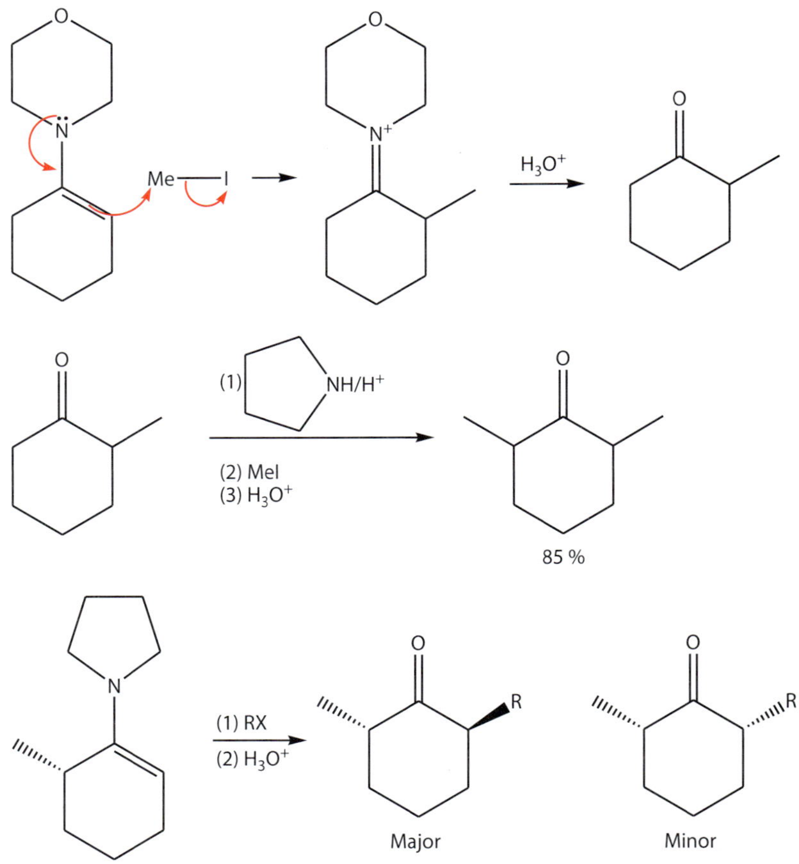

FIGURE 17.51 Selectivity in enamine reactions.

PROBLEM 17.16

Explain the observations in the following:

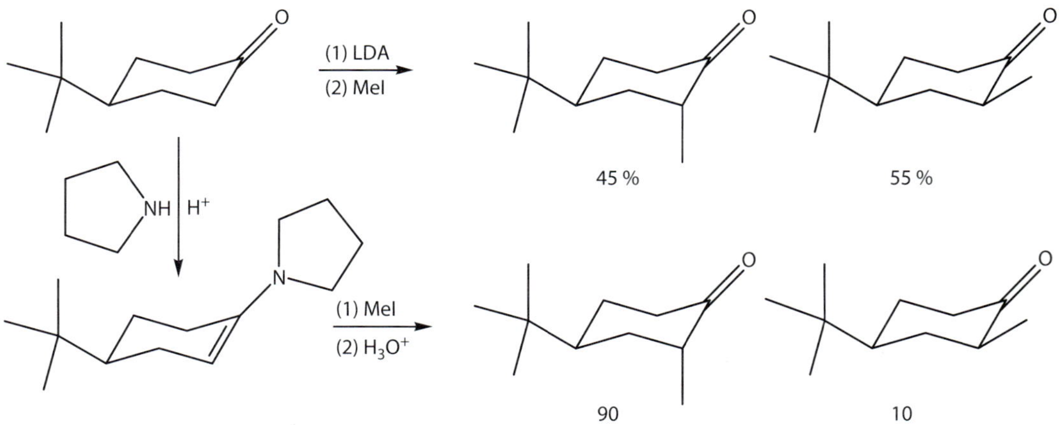

Solution

Both of these reactions result in the alkylation of 4-*t*-butylcyclohexanone, but the selectivity obtained is very different. The lithium enolate produced on reaction with LDA is sterically undemanding, and hence approach of the methyl iodide is possible either axially or equatorially; there is a marginal preference for the more stable equatorial product. However, the pyrrolidine ring of the enamine is much bulkier and inhibits approach to the equatorial position more than it does to the axial position.

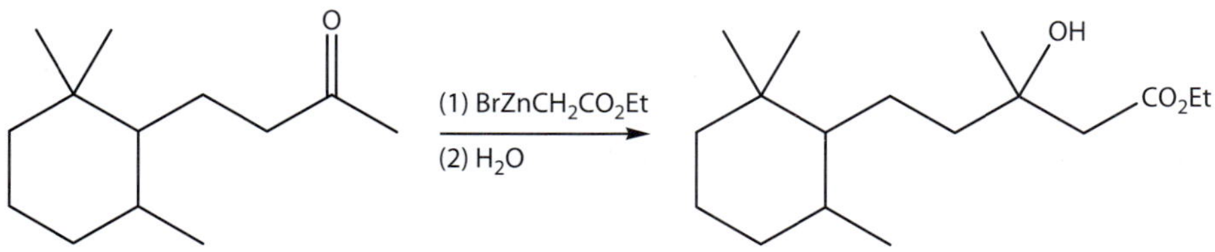

FIGURE 17.52 Reformatskii reaction.

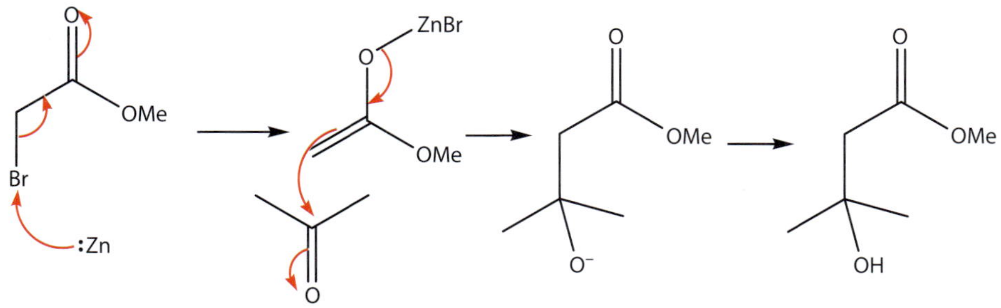

FIGURE 17.53 Reformatskii reagent considered as an ester enolate.

17.4.7 REVISITING THE REFORMATSKII REACTION

When we first met the Reformatskii reaction (Figure 17.52), we considered the reagent as a simple organometallic, analogous to a Grignard reagent. This certainly has merit, and in the solid state, Reformatskii reagents have been shown to have a *C*-metallated structure such as **17.44**. However, we could also see the Reformatskii reagent as a zinc enolate (Figure 17.53), and its addition to a ketone as analogous to a cross condensation reaction, where we have persuaded the less easily enolized component (the ester) to enolize and add to a ketone or aldehyde:

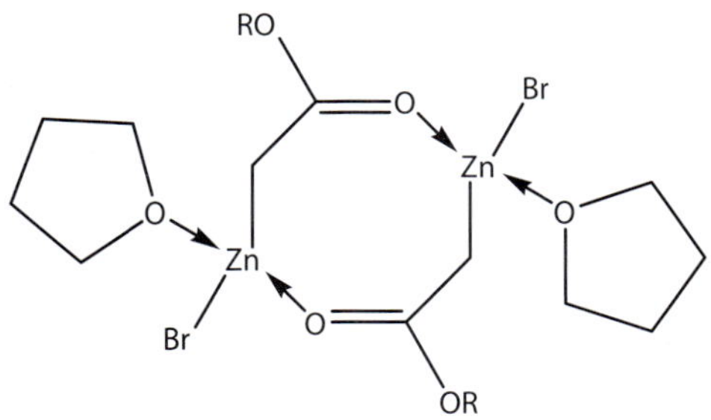

17.44

PROBLEM 17.17

Predict the product of the following reaction:

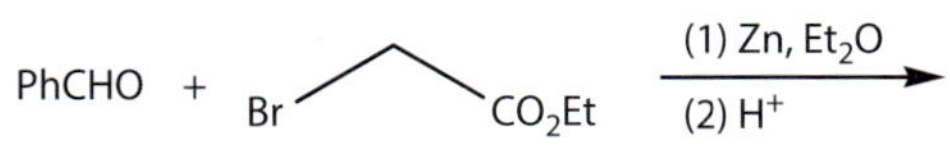

Solution

This is a classic Reformatskii reaction. The bromoester is converted to the organozinc derivative and adds to benzaldehyde. Protonation completes the conversion to **17.45**:

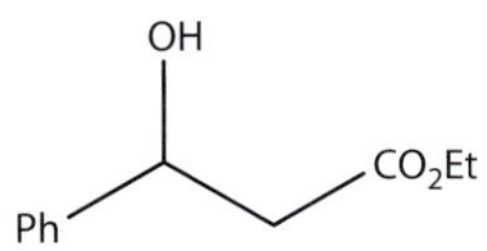

17.45

17.5 TWO USEFUL REACTIONS OF NONENOLIZABLE CARBONYL COMPOUNDS

The benzoin condensation (which really isn't a condensation) and the Cannizzaro reaction are useful processes that may occur when nonenolizable aldehydes and ketones are subjected to conditions that would typically lead to enolization. When benzaldehyde is reacted with a catalytic amount of cyanide ion under basic conditions, the product isolated is benzoin (Figure 17.54). Cyanide attacks the carbonyl group of benzaldehyde to give the anion of a cyanohydrin. This is in equilibrium with a species where the negative charge resides on carbon and which can attack another molecule of benzaldehyde. Finally, the cyanohydrin-forming reaction is reversed, to give benzoin. The reaction depends on the fact that cyanide is a good nucleophile, but cyanohydrin formation is reversible, and that an adjacent cyanide is able to stabilize a negative charge.

When nonenolizable aldehydes are treated with aqueous base, many undergo the Cannizzaro reaction—this is an organic chemistry version of disproportionation, with one molecule of aldehyde reduced to an alcohol and the other oxidized to a carboxylic acid. The reaction mechanism is shown in Figure 17.55. If the reaction is carried out in D_2O, no deuterium is incorporated into the CH_2 of $PhCH_2OH$, implying that the proton never becomes free in solution. All the steps are reversible except the last proton transfer (remember typically an acid has a $pK_a \sim 5$ and an alcohol a $pK_a \sim 15$), which drives the reaction toward products. A cross Cannizzaro reaction is possible if one of the components is formaldehyde—formaldehyde is much more electrophilic than other aldehydes, for both electronic and steric reasons, so it is always attacked preferentially by the hydroxyl ion (Figure 17.56). Effectively, formaldehyde is being used as a reducing agent (an inexpensive one) for another aldehyde.

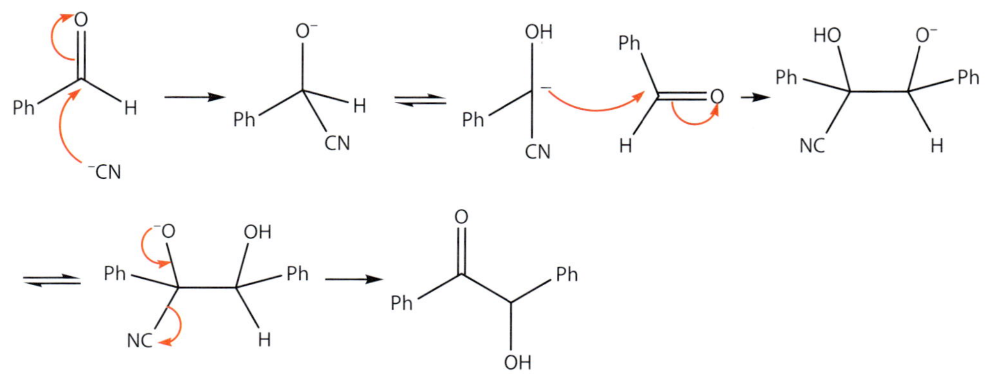

FIGURE 17.54 Benzoin condensation.

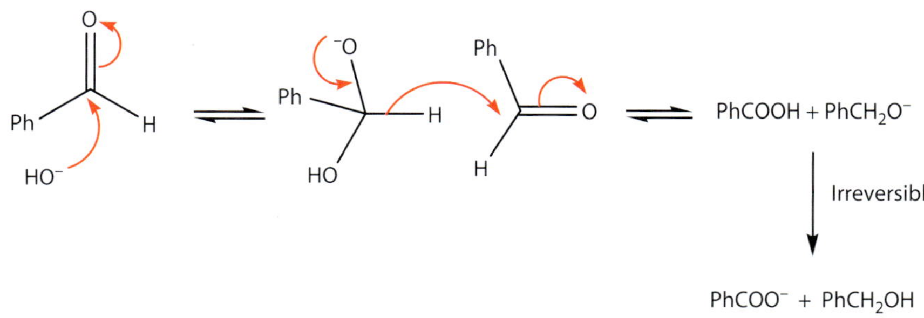

FIGURE 17.55 Cannizzaro reaction of benzaldehyde.

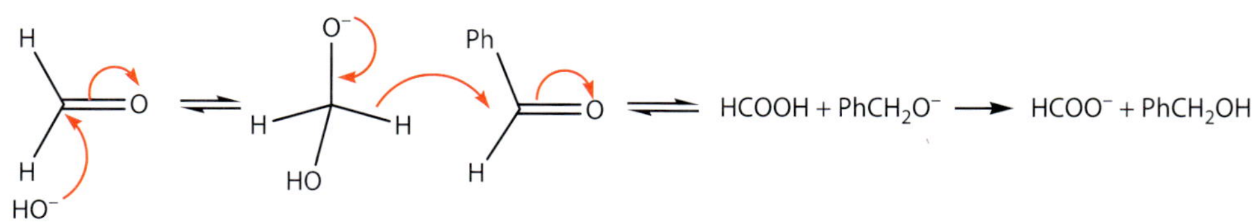

FIGURE 17.56 Cross Cannizzaro reaction using formaldehyde.

Key Points from Sections 17.4.4 through 17.5

- β-Ketoesters and 1,3-diesters are readily alkylated using RO^-, followed by RX.
- β-Carbonyl-substituted carboxylic acids are easily decarboxylated on heating.
- *O*-Alkylation is a serious competing reaction with β-diketone substrates; the use of a thallium rather than a sodium enolate prevents this.
- The regiochemistry of alkylation of simple, non-symmetric ketones may be problematic to control, and polyalkylation can be a competing reaction. The most useful strategy is to ensure complete deprotonation of the carbonyl compound using the strong, non-nucleophilic base LDA, followed by inverse addition of the enolate to an excess of the alkylating agent.
- Enamines can be prepared from carbonyl compounds and secondary amines; they react with electrophiles with predictable regioselectivity.
- A Reformatskii reagent may be regarded as a zinc enolate and is useful for selectivity in cross condensation reactions.
- Non-enolizable aldehydes and ketones may undergo the benzoin or Cannizzaro reactions.

17.6 REACTIONS AT THE β-POSITION OF α,β-UNSATURATED CARBONYL COMPOUNDS AND OTHER ELECTRON-POOR ALKENES

The most important resonance structure for any carbonyl compound is that with a positive charge on carbon and a negative charge on oxygen (**17.46**). This is critical to our understanding that nucleophiles react at carbon and electrophiles at oxygen. If, however, the carbonyl group is conjugated with a double bond, there is an additional important resonance form, with a positive charge β to the carbonyl group (**17.47**). Thus, in an α,β-unsaturated carbonyl compound, a nucleophile may also attack at the β-carbon atom.

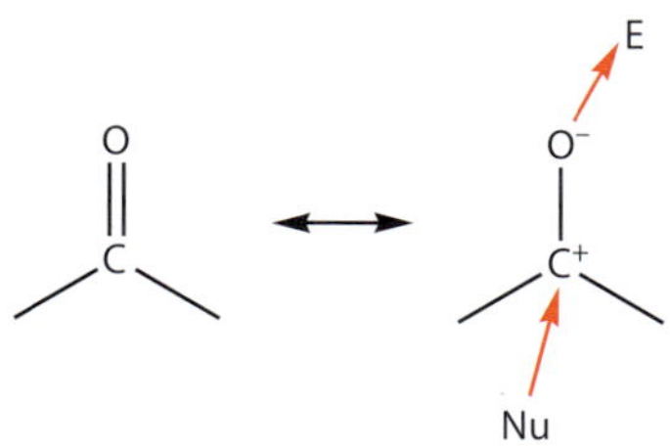

17.46

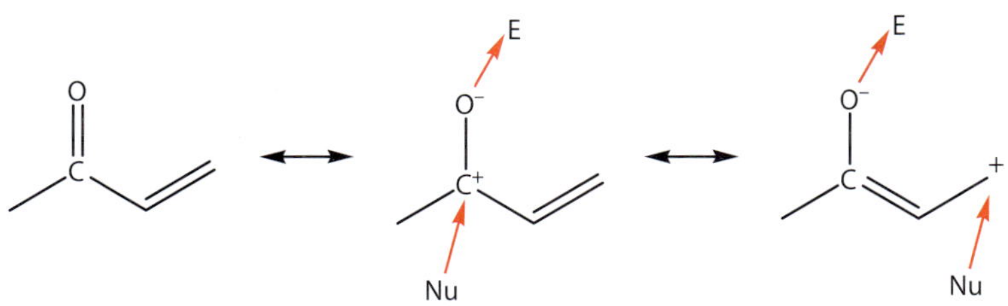

17.47

17.6.1 REACTIONS OF ENONES WITH SIMPLE NUCLEOPHILES

The addition at the β-position of an α,β-unsaturated carbonyl compound is described as conjugate addition, Michael addition, or 1,4-addition. It can be exemplified by the reaction in Figure 17.57, the addition of a cyanide ion to an enone. Students are often puzzled by the term 1,4-addition. Despite the way we usually write this mechanism, with the enolate anion being

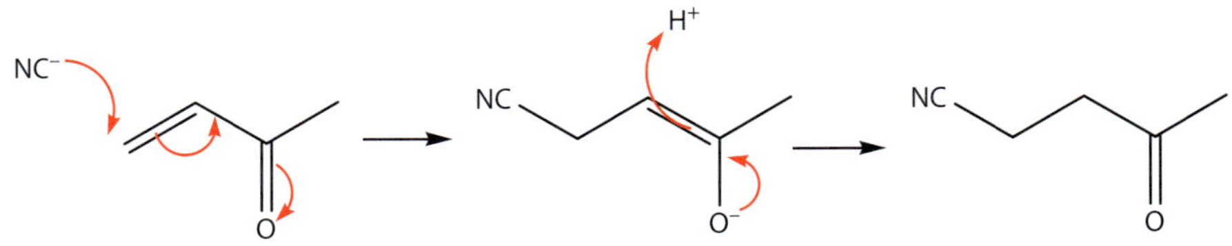

FIGURE 17.57 Addition of cyanide to an enone.

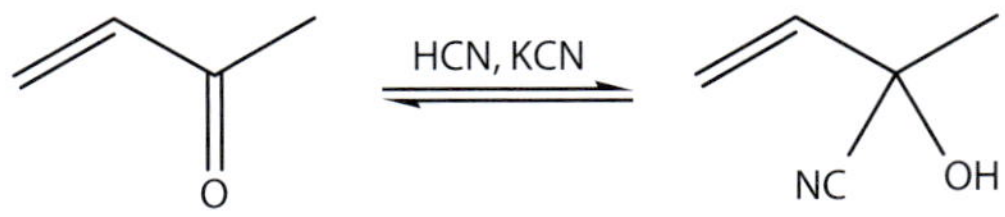

FIGURE 17.58 Reversible cyanohydrin formation from an enone.

immediately protonated on carbon, this is probably not what actually occurs. The protonation of enolate anions occurs on oxygen under conditions of kinetic control. Thus, the first product would be the enol, **17.48**, with the hydrogen added to the oxygen, and the cyanide to the fourth atom, the β-carbon of the original double bond. At −40 °C, the cyanohydrin is formed by the reaction of the enone with HCN/KCN (Figure 17.58); it is the kinetic product, but cyanohydrin formation is reversible. The formation of the 1,4-addition product, which is the thermodynamic product, is irreversible:

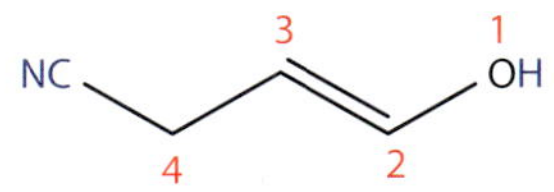

17.48

Additions of oxygen, sulfur, and nitrogen nucleophiles to unsaturated ketones usually give 1,4-addition products; the 1,2-addition reactions are, as we already know, readily reversible. Some examples are given in Figure 17.59.

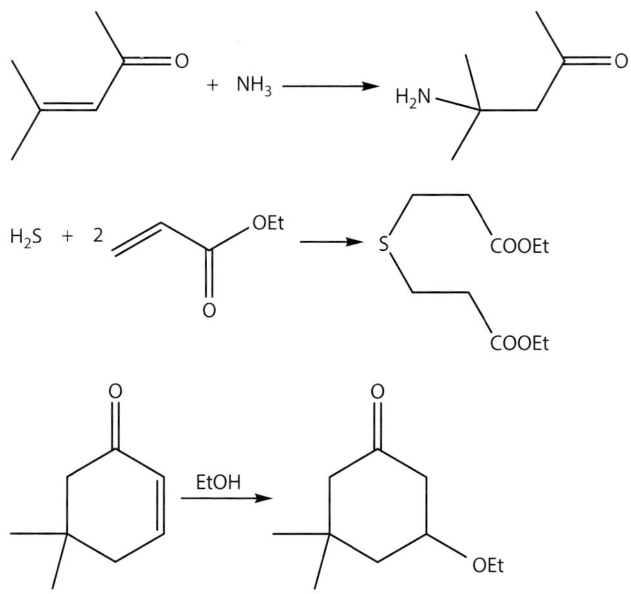

FIGURE 17.59 Addition of simple nucleophiles to enones.

PROBLEM 17.18

Predict the products of the following reactions:

(a)

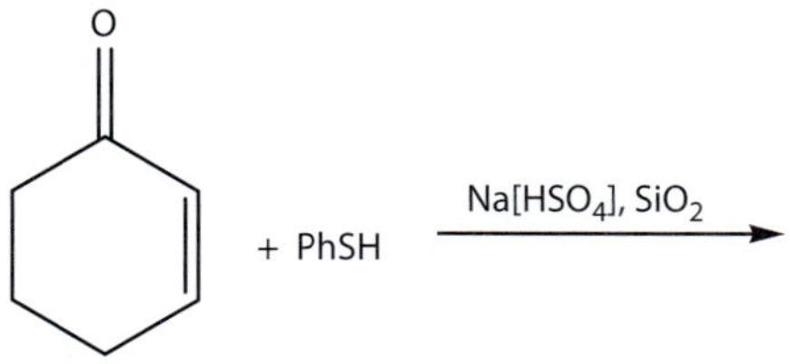

(b)

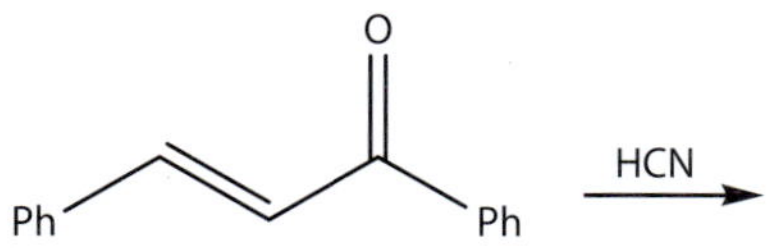

(c)

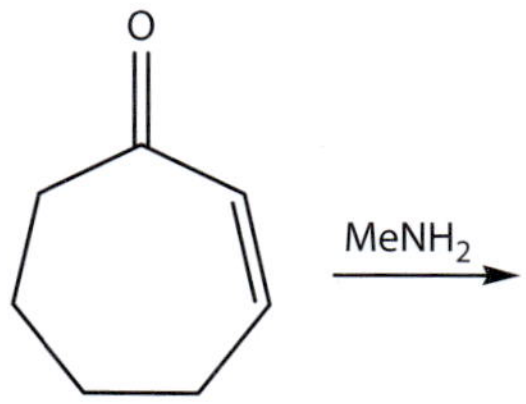

Solutions

(a) Sulfur is an excellent nucleophile and will attack the enone at the conjugate position:

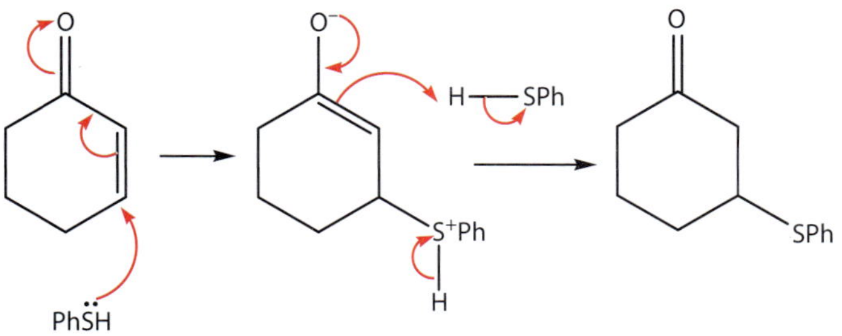

(b) Cyanide attacks at the conjugate position to give **17.49**.

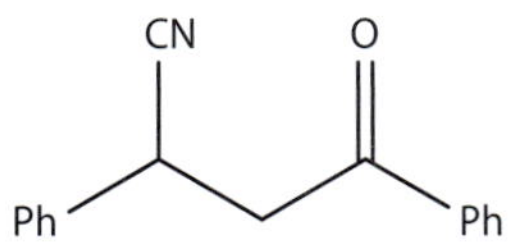

17.49

(c) Again, the mechanism is similar, giving **17.50**.

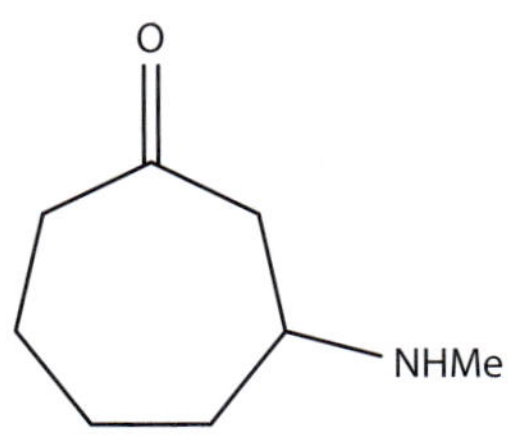

17.50

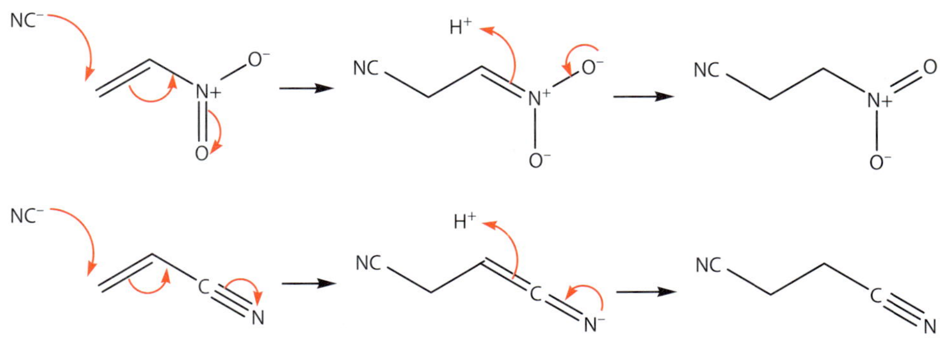

FIGURE 17.60 Conjugate addition to alkenes bearing other electron-accepting groups.

17.6.2 CONJUGATE ADDITION TO OTHER ELECTRON-POOR ALKENES

The nitro and nitrile functional groups can also accept and delocalize electron density (remember the effect of nitro groups in nucleophilic aromatic substitution), so alkenes bearing these functional groups can also undergo conjugate addition (Figure 17.60).

17.6.3 EPOXIDATION OF ELECTRON-POOR ALKENES

The reactions we discussed in Chapter 11 for preparing epoxides involved peracids—effectively electrophilic oxygen. Under these conditions, α,β-unsaturated carbonyl compounds (and other electron-poor alkenes) do not react well. For these, we need a nucleophilic source of oxygen—and the easiest of these is the anion of hydrogen peroxide. We have previously established that this is an excellent nucleophile, and like other oxygen nucleophiles, it attacks enones in the conjugate position (Figure 17.61). But it also contains a weak oxygen–oxygen bond that is easily broken. The enolate anion closes onto the oxygen to form the epoxide and expel hydroxyl ion. With peracids, the most electron-rich double bond is selectively epoxidized, but with alkaline peroxide, the most electron-poor double bond reacts selectively (Figure 17.62).

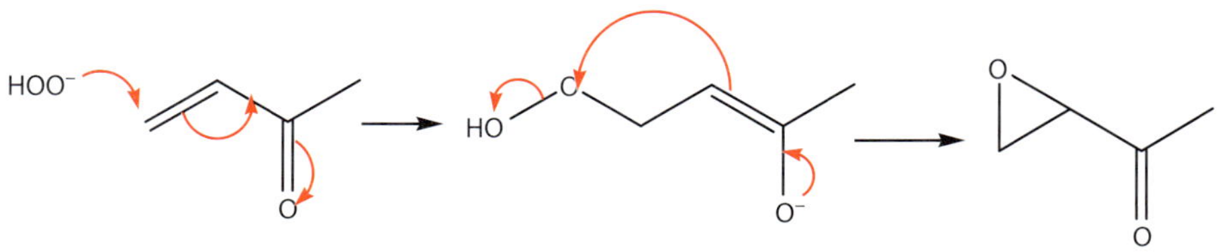

FIGURE 17.61 Epoxidation of enones with alkaline hydrogen peroxide.

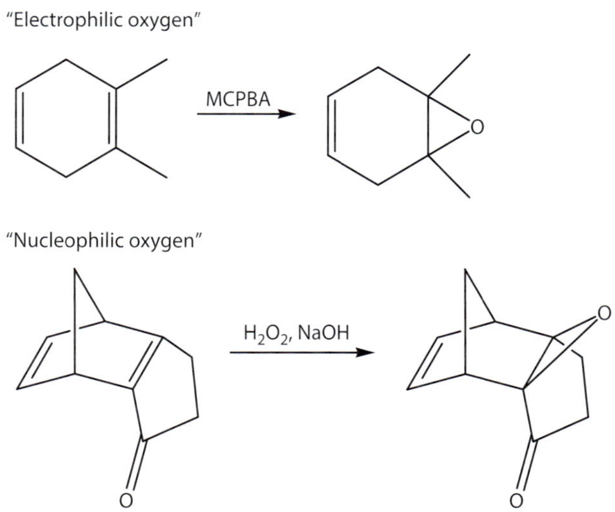

FIGURE 17.62 Selectivity in epoxidation reactions.

PROBLEM 17.19

Predict the product(s) of the following reactions:

(a)

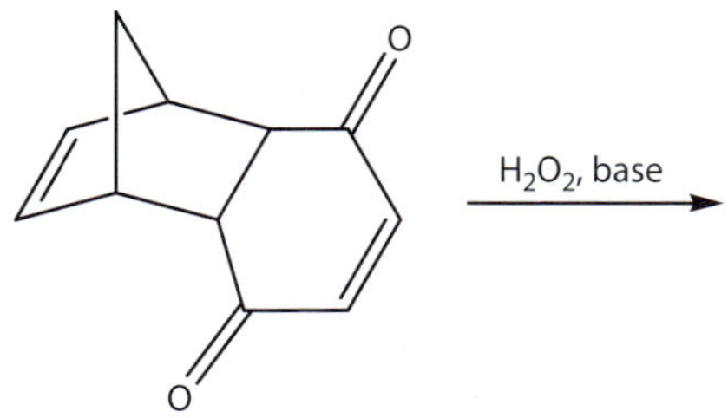

(b) $PhNH_2 + CH_2{=}CHCN \rightarrow$

(c)

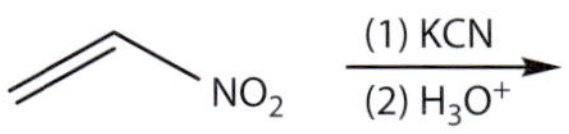

Solutions

(a) The conjugated double bond is selectively epoxidized, in 98 % yield:

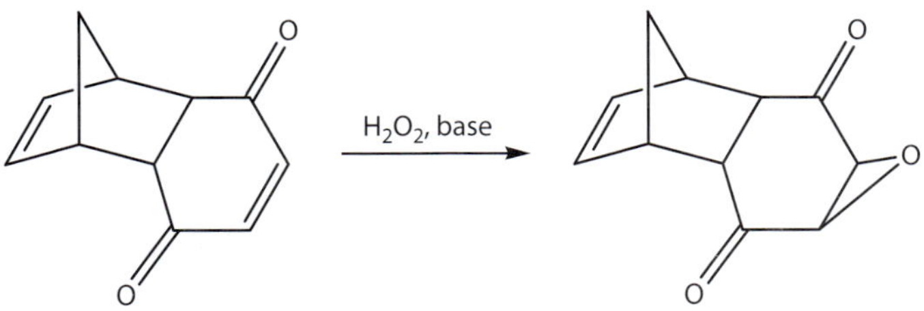

(b) Conjugate addition to the unsaturated nitrile yields $PhNHCH_2CH_2CN$.

(c) The first step of the reaction is the conjugate addition of cyanide to the unsaturated nitro compound to give $NCCH_2CH_2NO_2$. In the second step, the nitrile is hydrolyzed to give the carboxylic acid $O_2NCH_2CH_2COOH$.

17.6.4 ORGANOMETALLIC NUCLEOPHILES

Grignard reagents can add either 1,2- or 1,4- to enones, depending on the reaction conditions. A simple Grignard reagent, in ether or THF, will add directly to the carbonyl, to give an alcohol (Figure 17.63). However, if copper(I) salts are added to the reaction, then addition occurs at the conjugate position (Figure 17.64). Although we can write a simple mechanism for the conjugate addition, and this will serve us adequately to understand what is happening, it's not strictly correct and does not explain the function of the copper salt.

When Grignard reagents react with copper(I), the species formed is initially an alkylcopper reagent, such as MeCu. Solvation of this is very important; the unsolvated reagent polymerizes rapidly. The same results may be obtained by direct preparation of an organocopper reagent—MeLi reacts with CuCl to give Me_2CuLi. Why is copper important here? It's because it has readily accessible, and interconvertible, oxidation states. The mechanism is complex and not fully established in

FIGURE 17.63 1,2-Addition of a Grignard reagent to an unsaturated aldehyde.

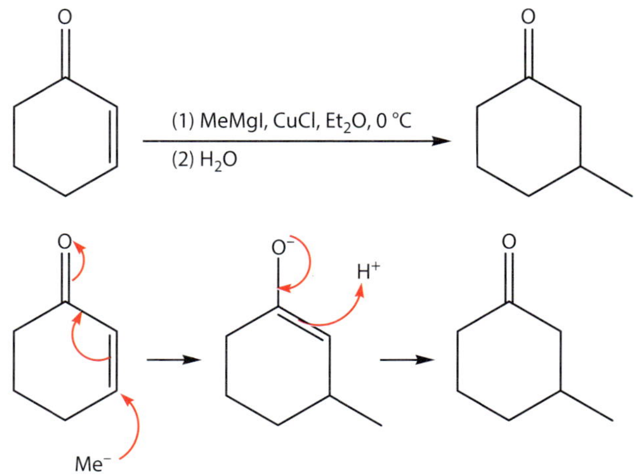

FIGURE 17.64 Conjugate addition of a Grignard reagent to an unsaturated carbonyl compound in the presence of copper(I) salts.

every case. Simple electron transfer processes may be invoked, but with R_2CuLi, a complex of the copper with the enone is formed and converted to a copper(III) intermediate, which collapses to give copper(I) and the enolate (Figure 17.65). A wide range of variously substituted organocopper reagents are now available for this process, and some synthetic examples are given in Figure 17.66. In the second example, we note that α,β-unsaturated esters are also good substrates, and in the final example, the addition reaction is stereospecific, and the enolate is discharged by reaction with an alkyl halide rather than simple protonation.

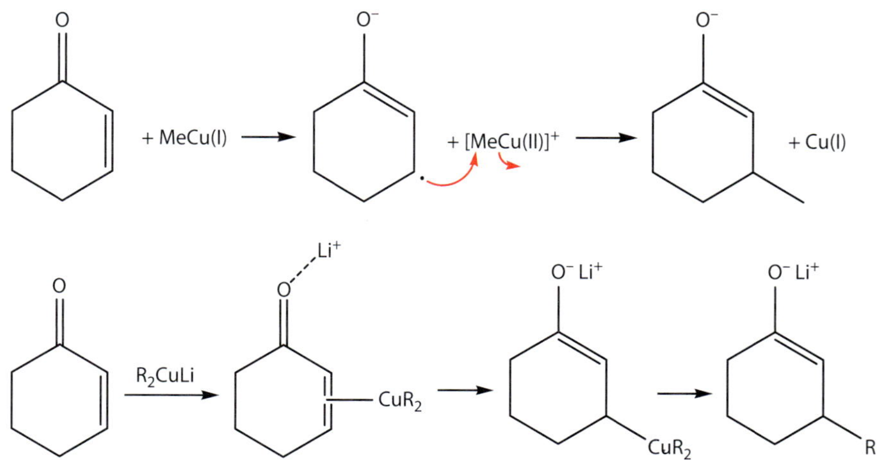

FIGURE 17.65 Mechanisms for addition of organocopper reagents to enones.

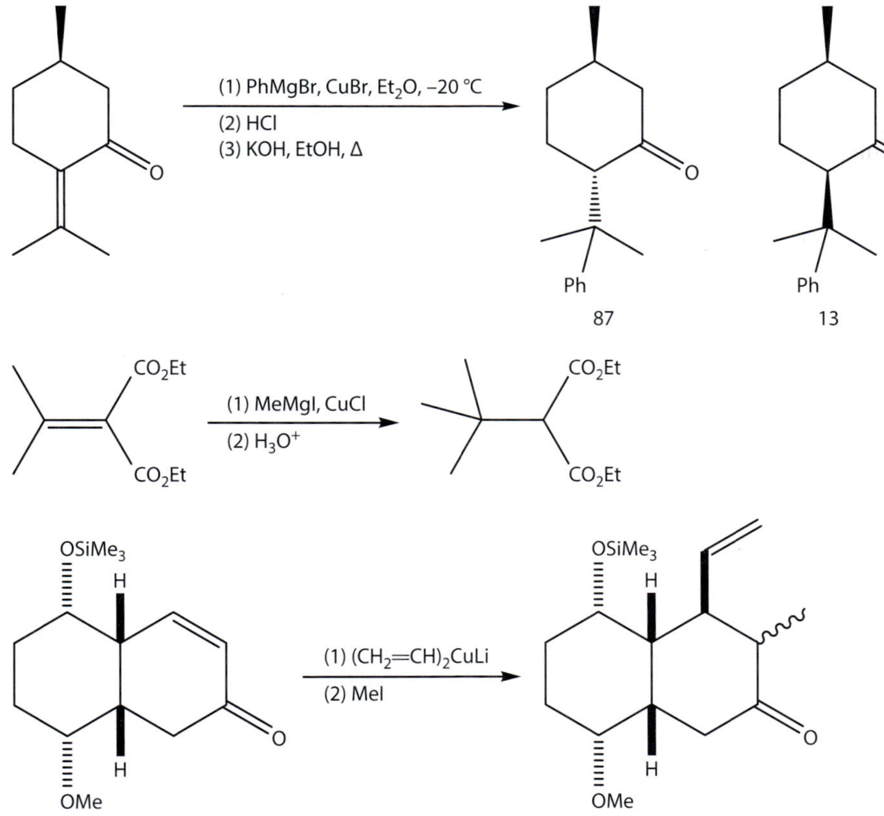

FIGURE 17.66 Conjugate addition of organocopper reagents in synthesis.

PROBLEM 17.20

Predict the products of the following reactions:

(a)

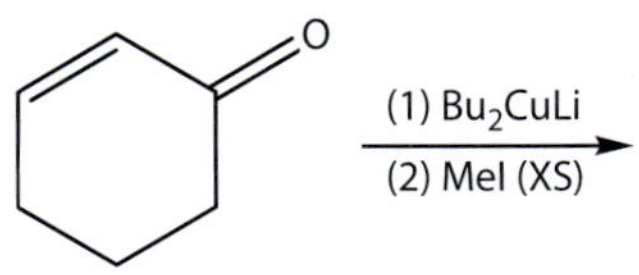

(b)

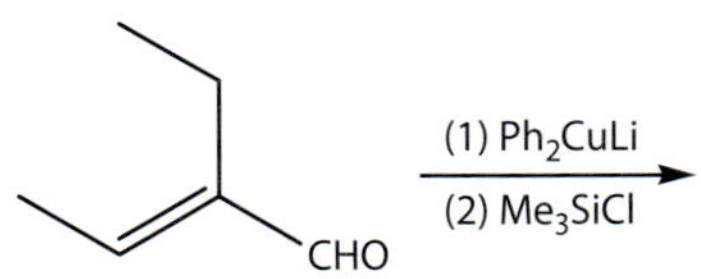

(c)

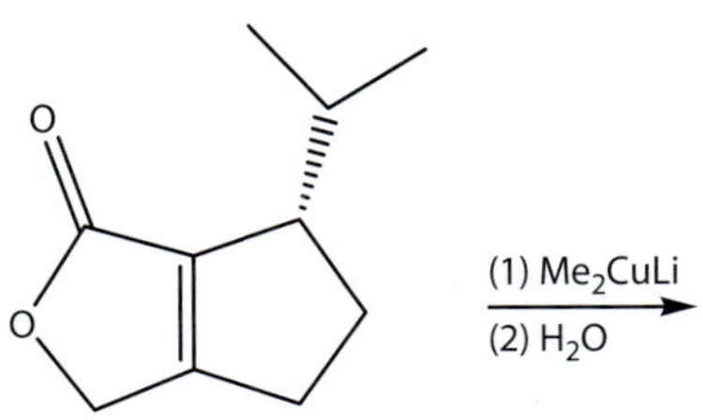

Solutions

The mechanisms given simply use a carbanion as nucleophile; in terms of identifying the product, this will serve.

(a) Conjugate addition gives the enolate, which then substitutes methyl iodide:

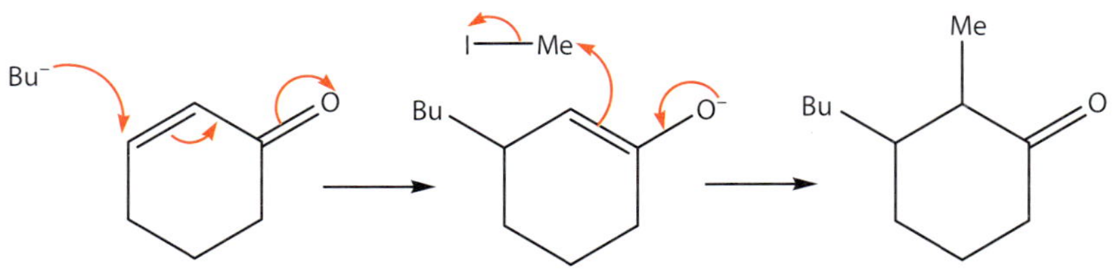

(b) Conjugate addition gives the enolate, which reacts at *oxygen* with trimethylsilylchloride—this is a hard electrophile and almost invariably reacts with enolates at oxygen:

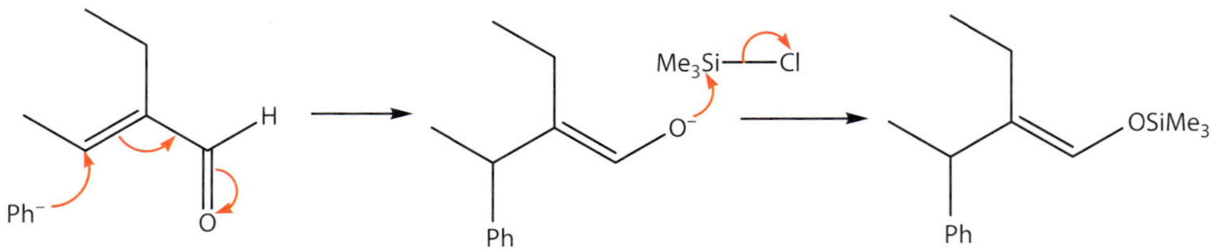

(c) This reaction demonstrates that the same process can occur with unsaturated esters. The nucleophile attacks from the opposite face to the bulky substituent, and the protonation of the enolate yields a *cis*-ring junction, which is more stable in 5,5-systems:

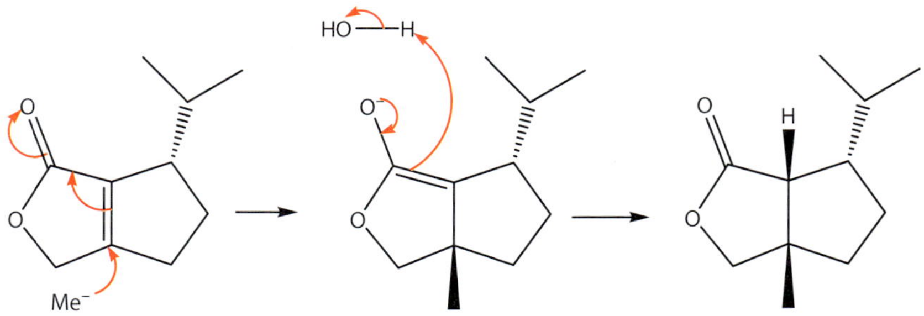

17.6.5 ADDITION OF ENOLATE ANIONS TO ENONES

Enolate anions, whether stabilized (like the anions of diethylmalonate or ethyl acetoacetate) or not, add invariably and predictably at the conjugate position of unsaturated carbonyl compounds. An example is shown in Figure 17.67. The anion of diethylmalonate adds 1,4 to the enone, and the intermediate is reprotonated. The diester can be hydrolyzed and decarboxylated to give the final product. The key feature of this reaction, which is very extensively used in synthesis, is that we have synthesized a 1,5-dicarbonyl compound—and this is the standard synthesis of molecules with this pattern of substitution.

With simpler, less stabilized, enols and enolates, the same selectivity issues arise in this addition process as we saw for aldol and related reactions—these reactions are related to the aldols, but a double bond simply intervenes as an electron relay. Consider the example shown in Figure 17.68. We have chosen a nucleophile that can enolize readily and an α,β-unsaturated aldehyde that can't enolize. The enal is more electrophilic than the ketone, so self-condensation of the ketone is not

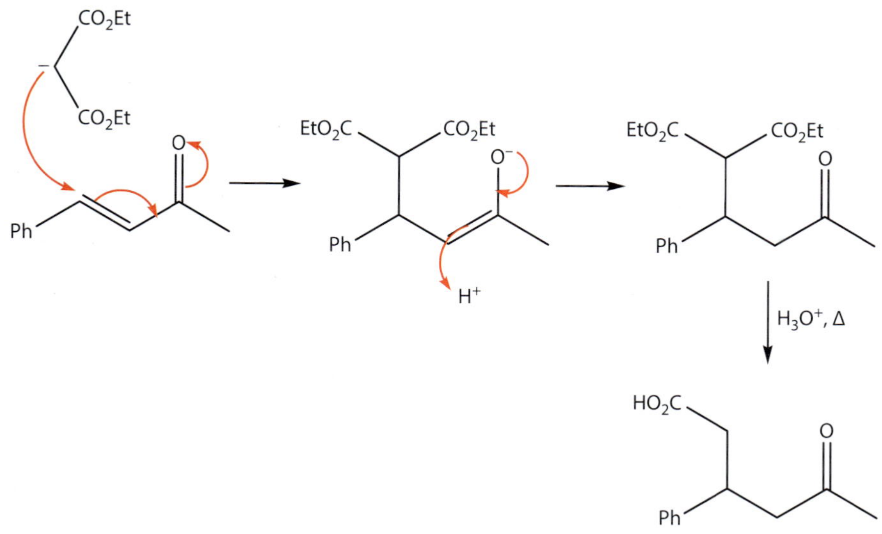

FIGURE 17.67 Addition of diethyl malonate to an enone.

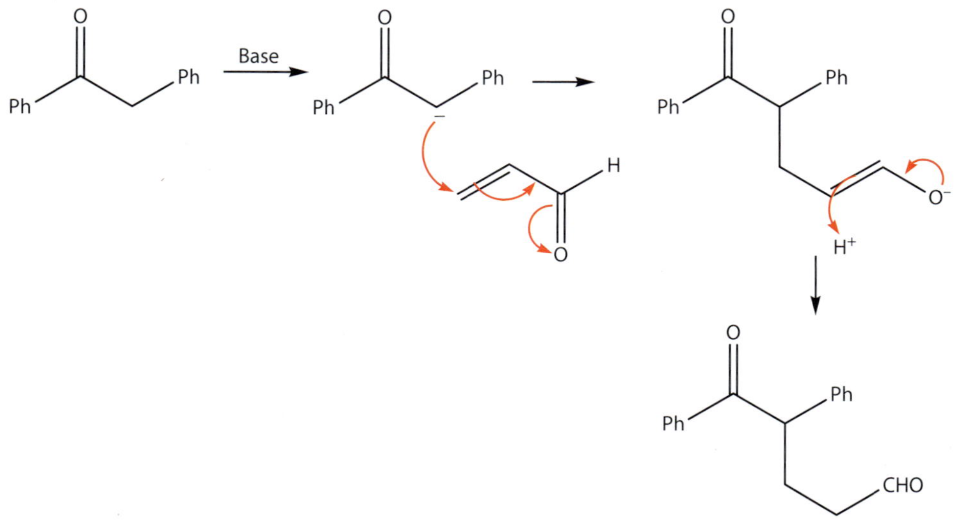

FIGURE 17.68 Reaction of an unstabilized enolate with an enal.

a problem. When writing reactions of this type, it's good practice to count the number of carbon atoms you have at each stage—it's very easy to lose or gain one if you don't.

How could we prepare **17.51**? We can see that it is a 1,5-dicarbonyl compound, so the reaction of an enolate with an enone is a good strategy. There is only one sensible way we could break up the molecule—a stabilized anion is always helpful in establishing selectivity. The SMs are sensible and easy to obtain. The mechanism in the forward direction is shown in Figure 17.69 and some examples of use in synthesis in Figure 17.70. The final example shows that alkenes bearing other electron-accepting groups, such as nitro and nitrile, also undergo this reaction.

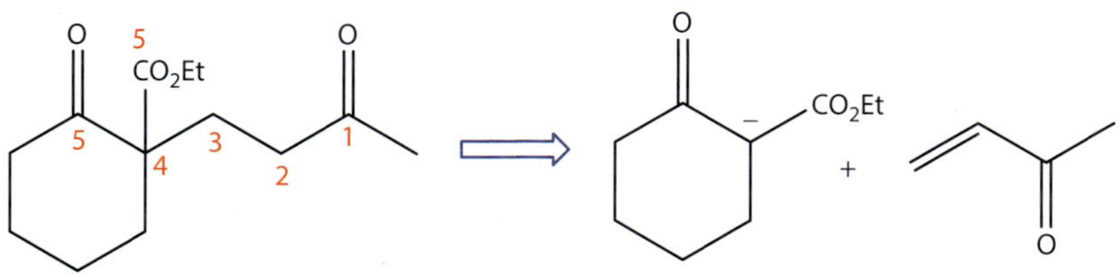

17.51

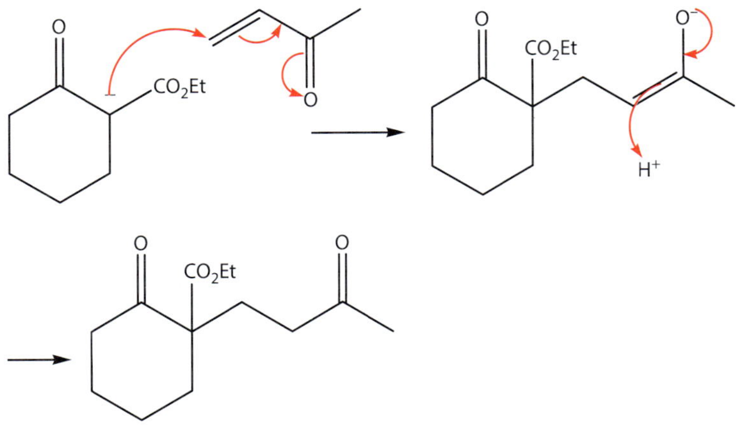

FIGURE 17.69 Mechanism of addition of a stabilized anion to an enone.

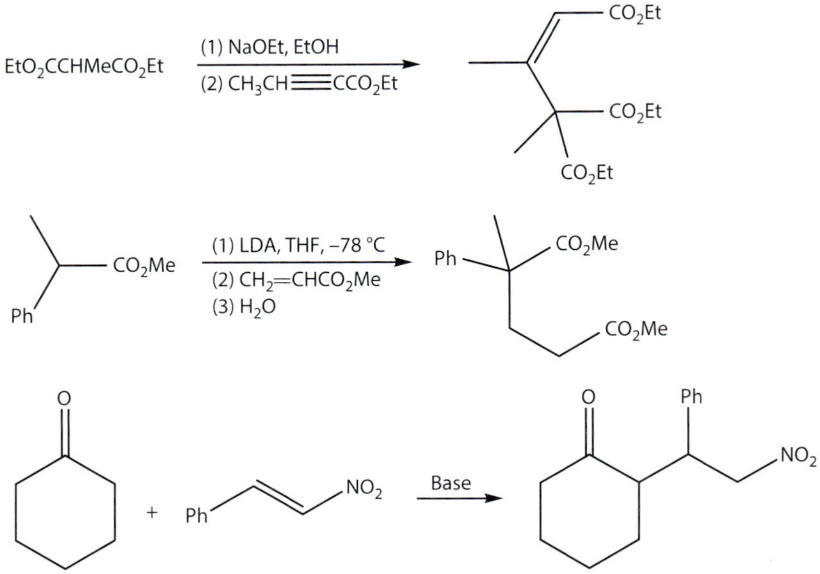

FIGURE 17.70 Addition of enols/enolates to α,β-unsaturated carbonyl compounds and related processes.

One widely used variation to overcome difficulties with selectivity is to transform the enolate into an enamine. These also react predictably at the conjugate position of enones and are hydrolyzed to the carbonyl compound when worked up with aqueous acid. The mechanism is shown in Figure 17.71. In most cases, this product is further transformed, as we will see in the next section.

17.6.6 ROBINSON ANNULATION: A LOOK AHEAD TO SYNTHESIS

The Robinson annulation (or annelation) is a powerful synthesis of six-membered rings involving conjugate addition of an enolate, followed by an aldol condensation. A simple example is shown in Figure 17.72. The diketone is readily converted completely into the enolate anion and then attacks the enone, to give, after reketonization, a 1,5-dicarbonyl compound. This diketone undergoes an

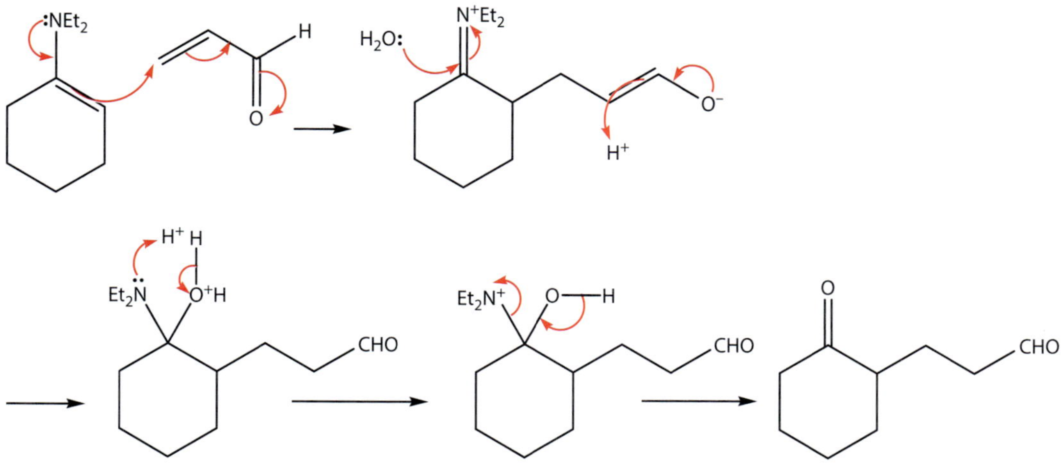

FIGURE 17.71 Mechanism of reaction of enamines with enones.

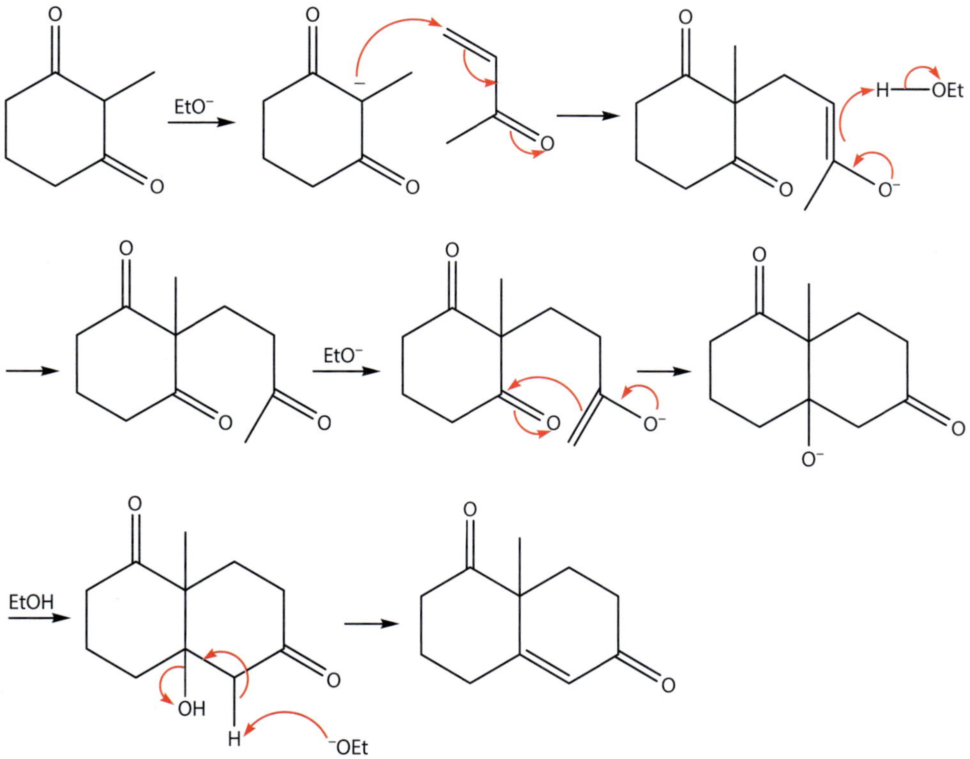

FIGURE 17.72 Mechanism of the Robinson annulation.

intramolecular aldol condensation to form the favorable six-membered ring. Although this is shown as a base-catalyzed reaction, acid-catalyzed processes are also common. The loss of water to form an enone is common in both acid- and base-catalyzed processes. Although the reaction looks to be long and complex mechanistically, it is practically rather simple—the conjugate addition, aldol, and dehydration all proceed in a one-pot process. The reaction is used extensively in synthesis (Figure 17.73 shows some examples) and will be discussed at greater length in Chapter 20.

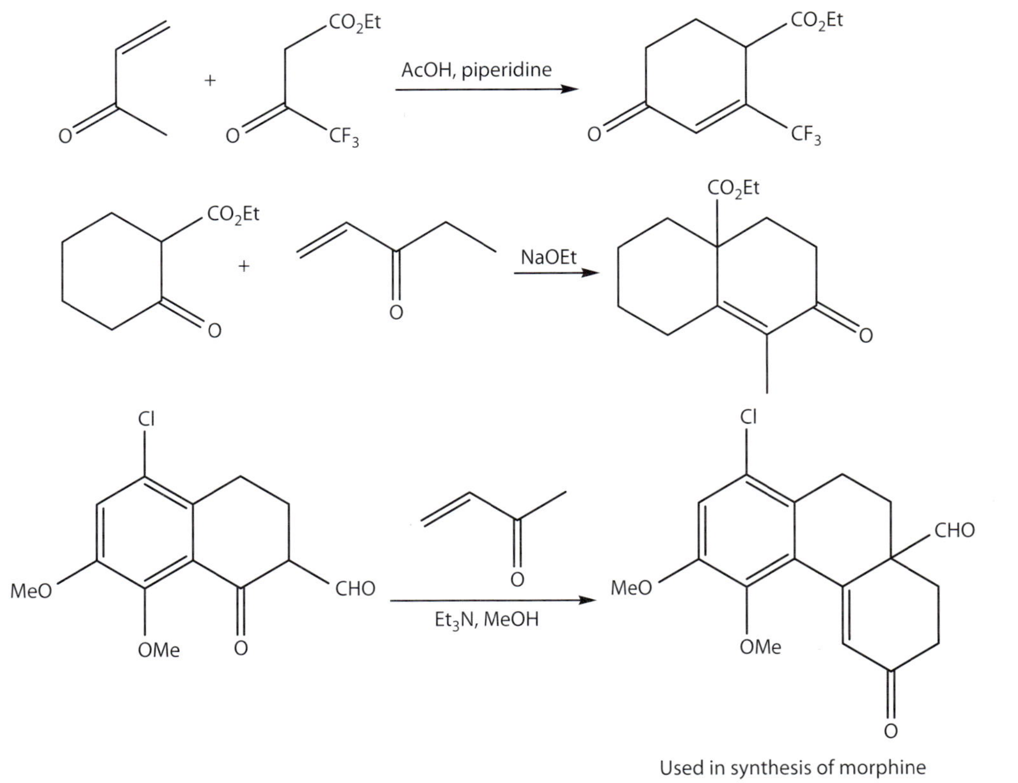

FIGURE 17.73 Robinson annulations in synthesis.

A version of this process involves the 1,5-dicarbonyl compounds that we made using an enamine reaction in the last section. Figure 17.74 shows a typical process. The enamine adds 1,4- to the enone as before, and the 1,5-dicarbonyl compound is generated on work-up. The aldol reaction of the enol gives the favorable six-membered ring, and under the reaction conditions, dehydration to the enone is complete.

Key Points from Section 17.6

- Conjugate, Michael, or 1,4-addition to an α,β-unsaturated carbonyl compound results in a saturated carbonyl compound, substituted at the β-position.
- Alcohols, thiols, amines, and cyanide can add both 1,2- and 1,4- to unsaturated carbonyl compounds, but since 1,2-addition is reversible, and 1,4-addition is not, only the 1,4-addition product is generally isolated.
- Related reactions can occur with other electron-poor alkenes, such as $CH_2{=}CHCN$ or $CH_2{=}CHNO_2$.
- Epoxidation of electron-poor alkenes may be accomplished using the anion of hydrogen peroxide as a source of oxygen.
- While Grignard reagents generally add 1,2- to unsaturated aldehydes and ketones, in the presence of copper(I), or using an organocopper reagent, 1,4-addition occurs.
- Enolate ions also add 1,4- to unsaturated carbonyl compounds, to give 1,5-dioxygenated products. With simple enolates, the same selectivity problems noted with the aldol condensation occur.
- Six-membered rings may be prepared by the Robinson annulation, which involves a conjugate addition followed by an aldol condensation of the product.

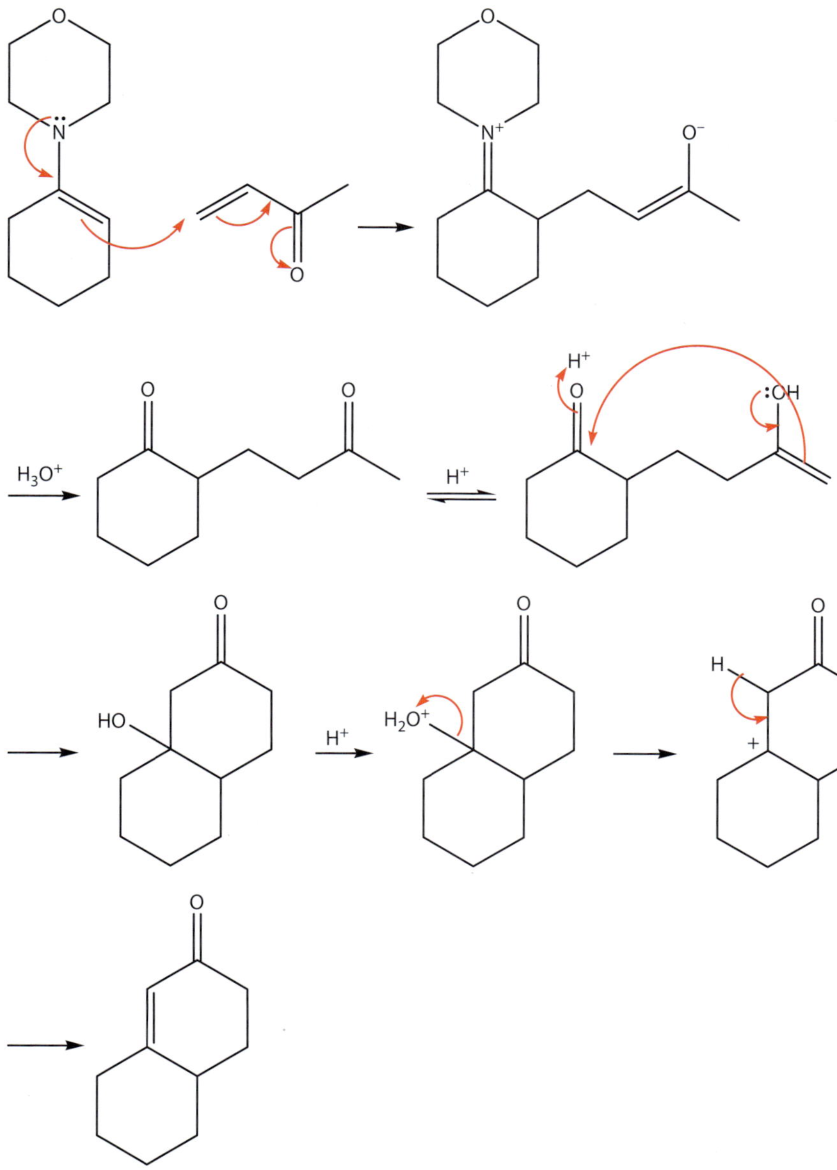

FIGURE 17.74 Robinson annulation using an enamine starting reagent.

REVIEW PROBLEMS

1. Give systematic names for each of the following compounds:

(a)

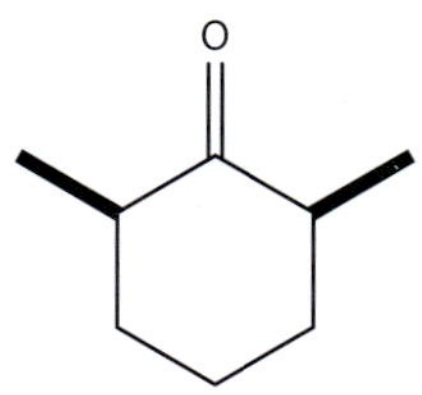

(b)

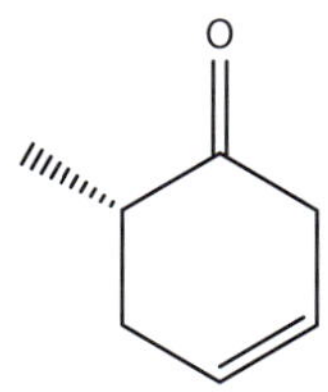

(c)

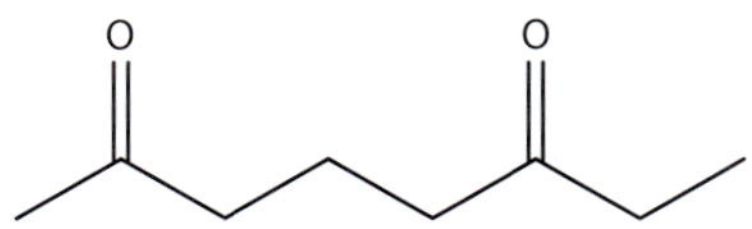

2. Give structures for each of the following compounds:
   (a) Ethyl 3-oxo-4,4,4-trifluorobutanoate
   (b) 1,3-Diphenylpropanone
   (c) Methyl 2-bromo-3-methylbutanoate
3. Predict the equilibrium constant for each of the following processes, giving a brief reason:

   (a)

   (b)

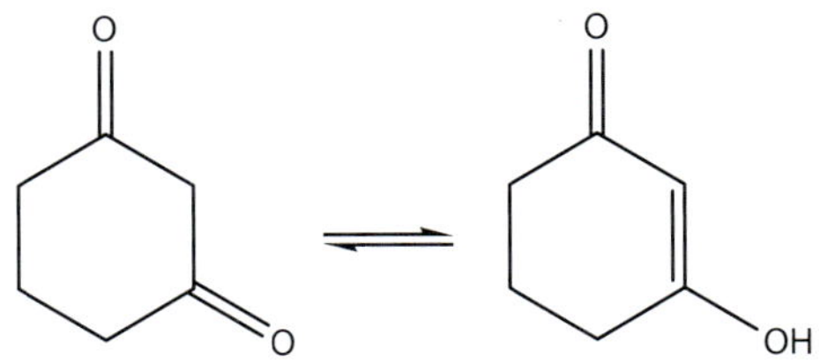

   (c)

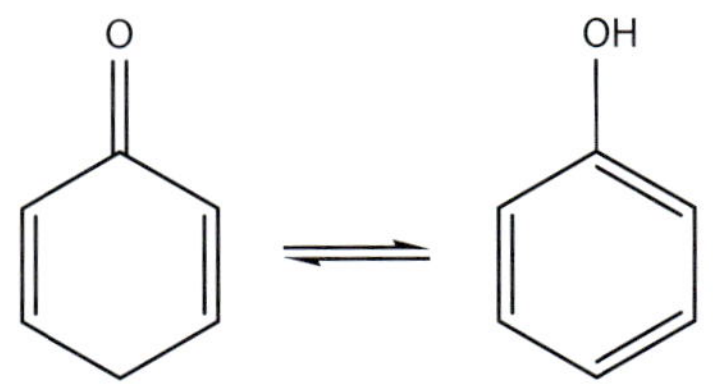

4. Compounds **17.52** and **17.53** have the same molecular formula C_8H_8O. They showed the following IR absorptions (cm^{-1}):

17.52	3030 (m), 2900 (m), 1720 (m), 1600 (m)
17.53	2820 (m), 2730 (m), 1695 (m), 1605 (s)

Their proton NMR spectra showed the following resonances:

17.52	δ	3.66 (d, 2H)
		7.36 (m, 5H)
		9.71 (t, 1H)
17.53	δ	2.41 (s, 3H)
		7.19 (d, 2H)
		7.66 (d, 2H)
		9.81 (s, 1H)

When treated with dilute aqueous alkali, **17.52** undergoes an aldol condensation to give **17.54**, $C_{16}H_{16}O_2$, whereas with conc. aqueous alkali, **17.53** undergoes a Cannizzaro reaction to give **17.55**, $C_8H_{10}O$ and **17.56**, $C_8H_8O_2$.
17.52 and **17.53** both reacted with 2,4-dinitrophenylhydrazine to give a red precipitate. Identify the compounds **17.52–17.56** and explain your reasoning.
5. Predict the product(s) of the following reactions:

   (a)

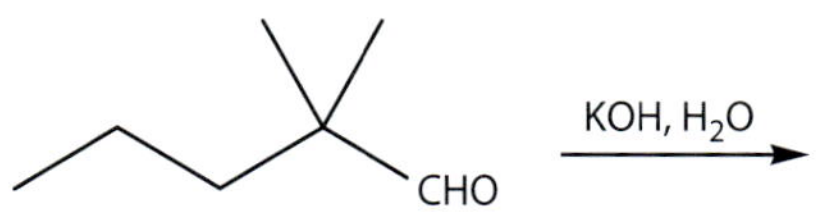

(b)

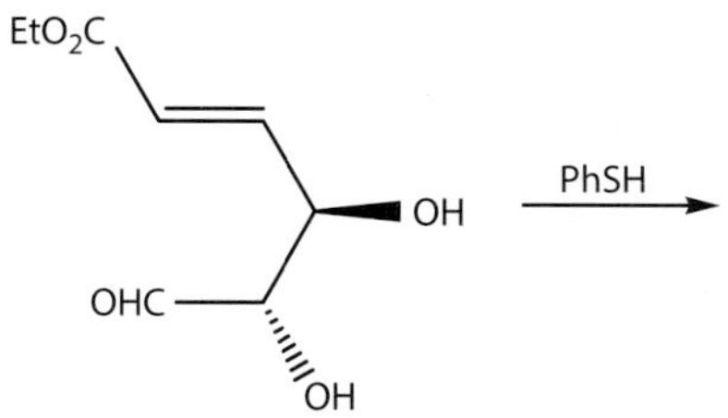

(c)

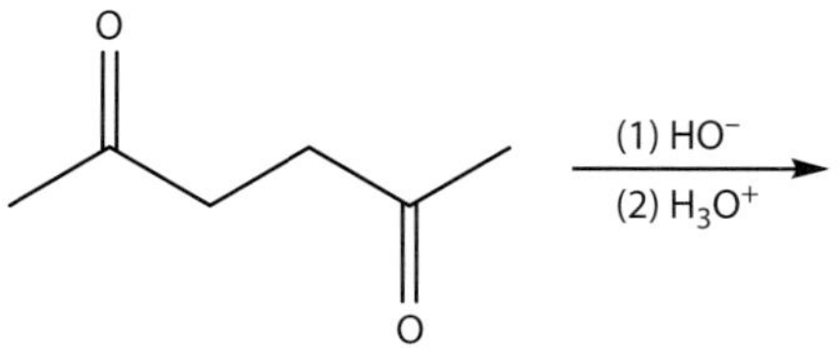

(d)

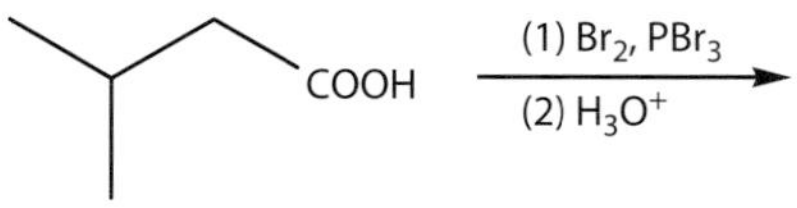

(e)

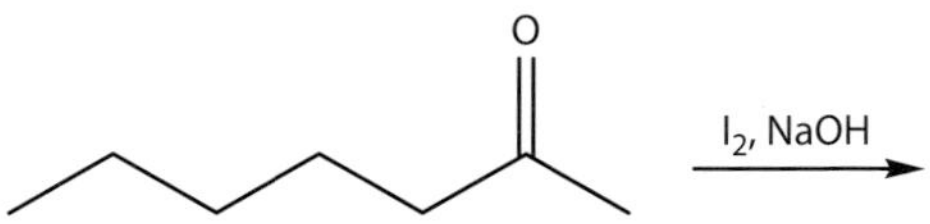

(f)

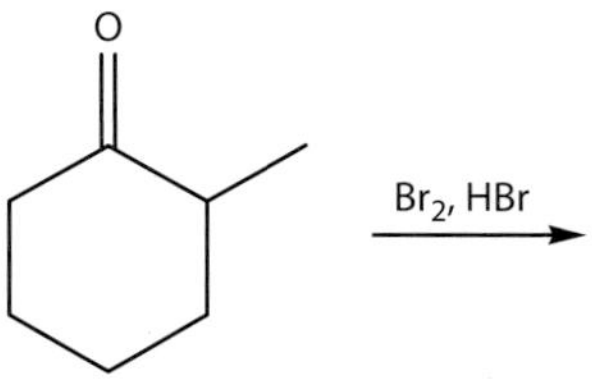

(g)

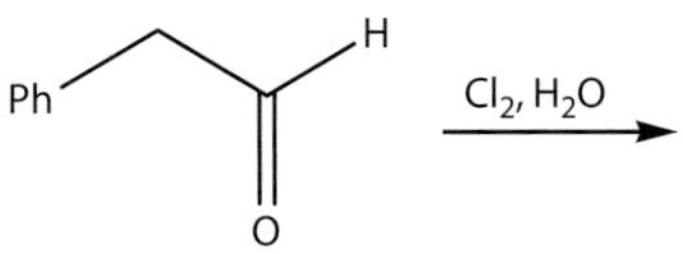

(h)

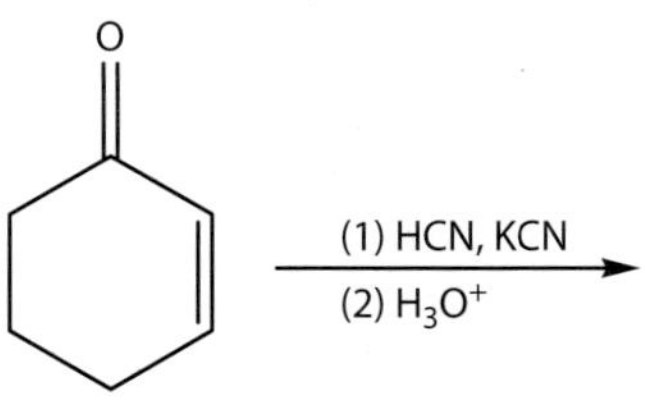

(i) Each of the ketones shown in the following is separately reacted with (1) XS KOBu-*t* and (2) XS MeI

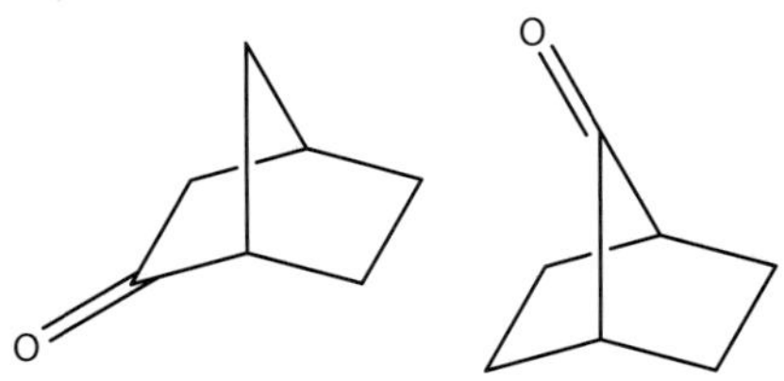

(j)

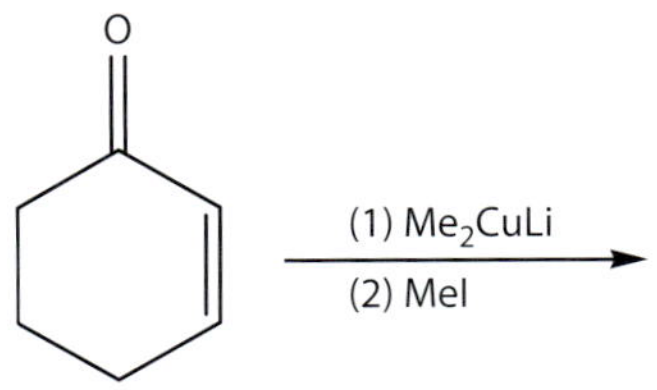

(k)

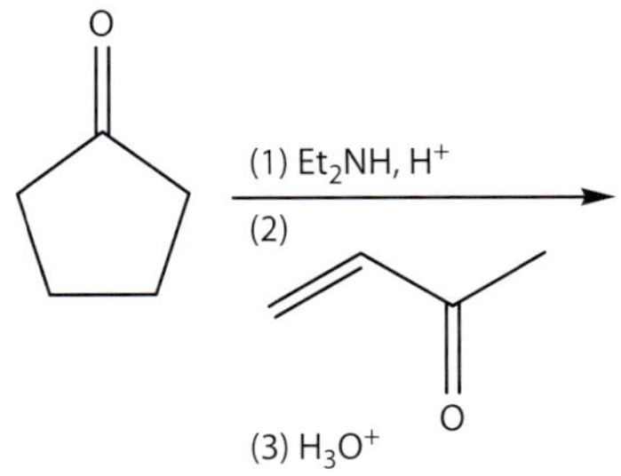

(l)

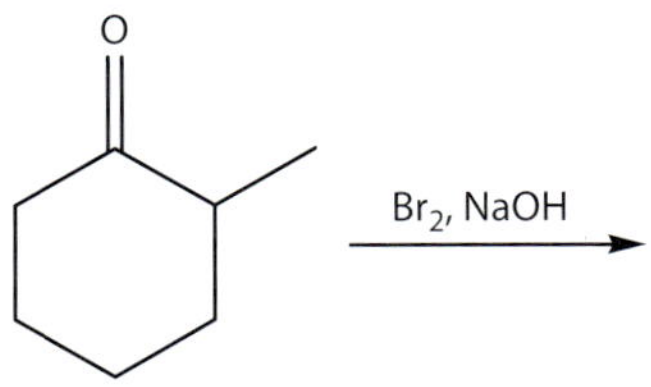

(m) $PhCH_2CO_2Et \xrightarrow{\text{NaOEt, EtOH}}$

(n)

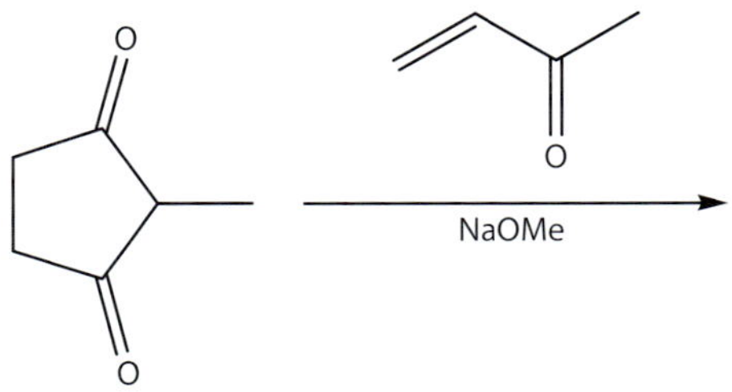

(o)

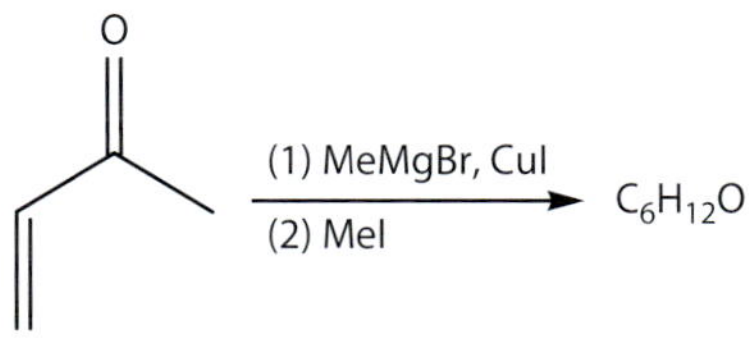

(p) $MeO_2C(CH_2)_5CO_2Me \xrightarrow{\text{NaOMe, MeOH}}$

(q)

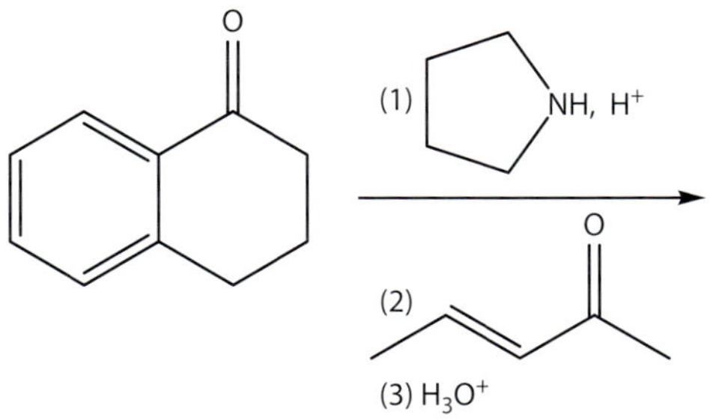

6. Propose a mechanism for each of the following processes:

(a)

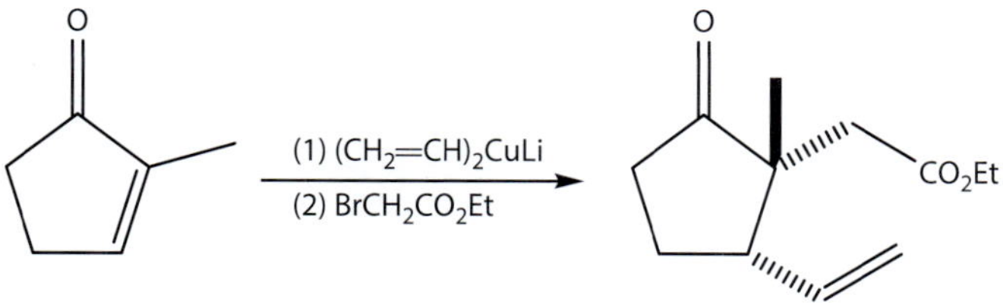

(b)

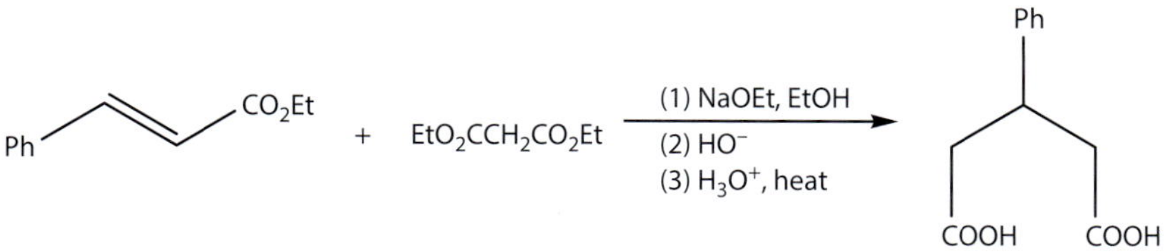

(c)

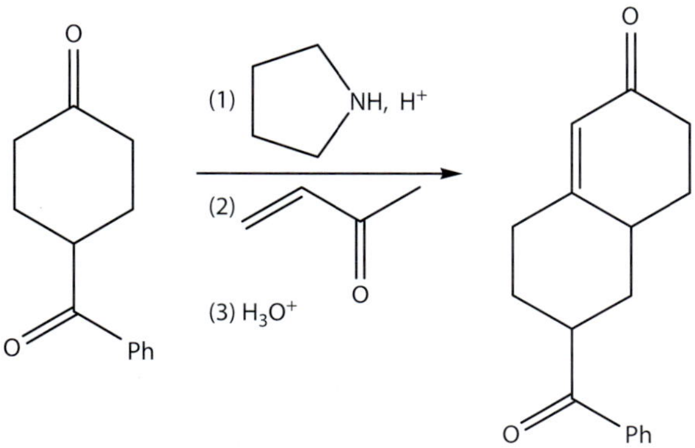

(d)

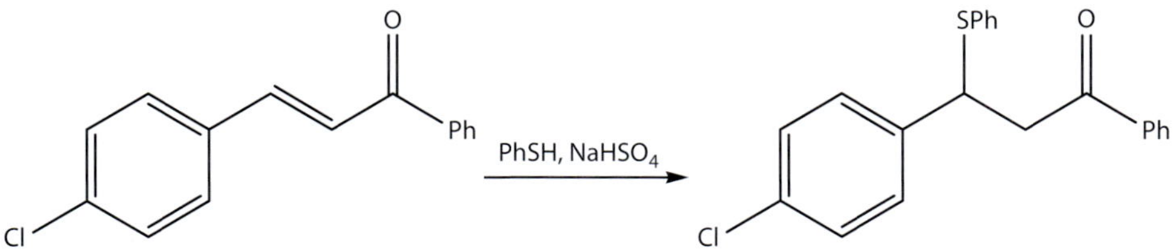

(e)

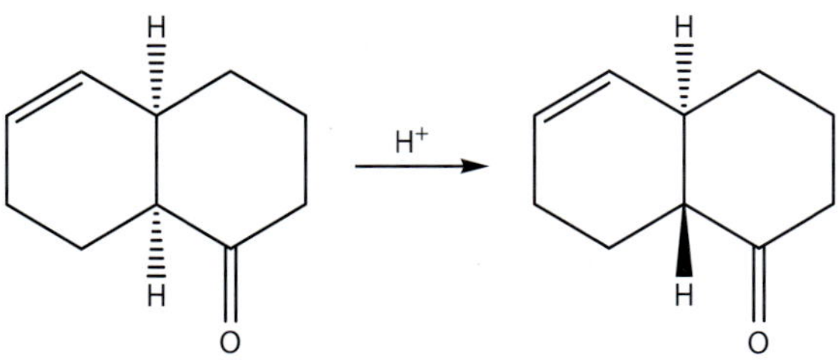

(f)

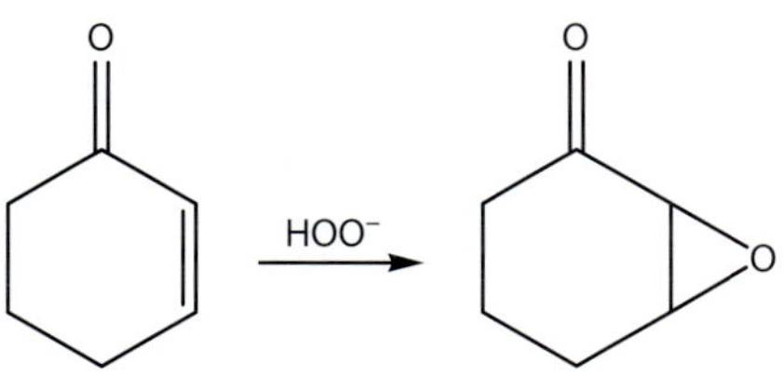

(g)

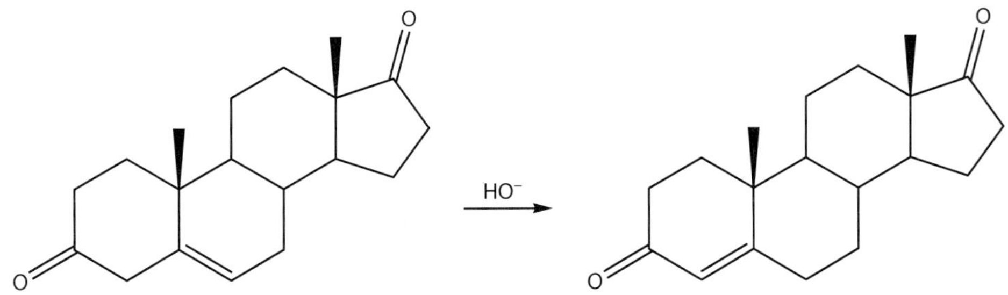
O
O
HO⁻
O
O

(h)

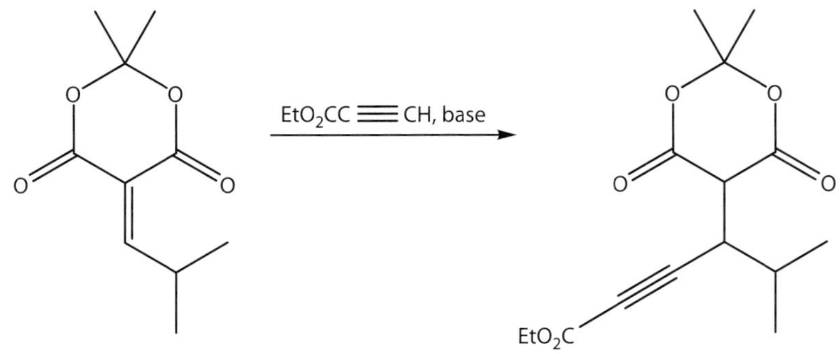
O
O
O
O
EtO2CC≡CH, base
EtO2C

(i)

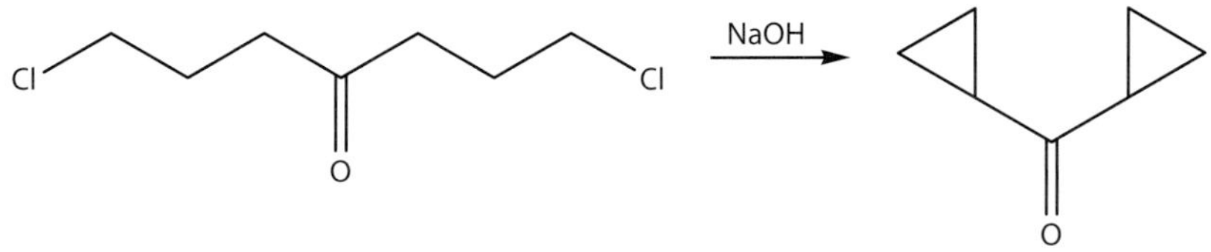
Cl
Cl
O
NaOH
O

(j)

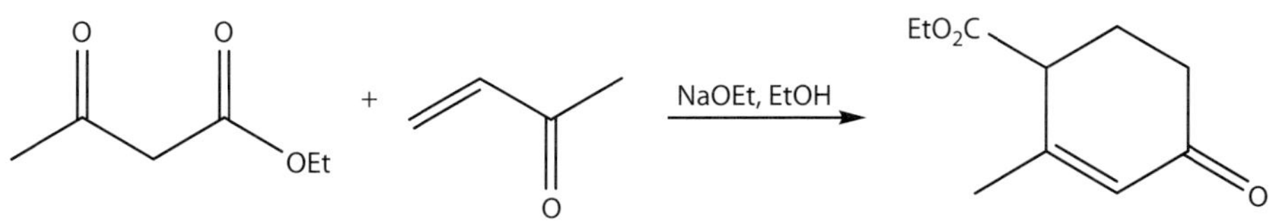
O
O
OEt
+
O
NaOEt, EtOH
EtO2C
O

(k)

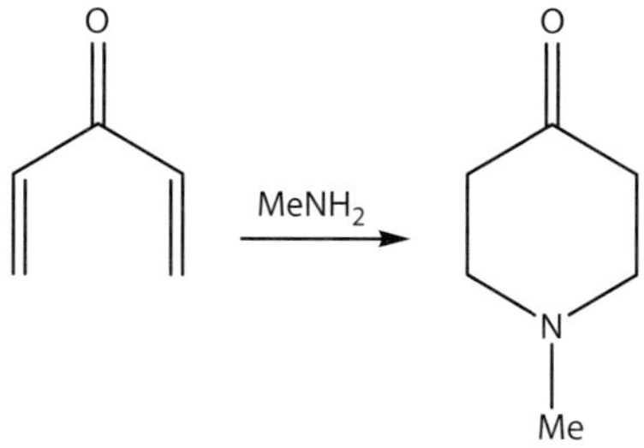
O
MeNH2
O
N
Me

(l)

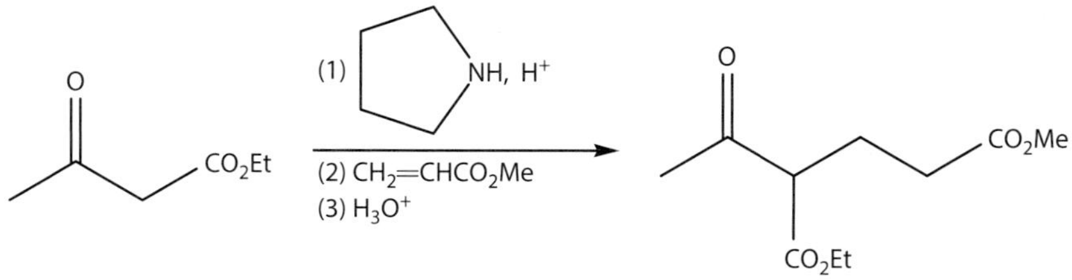
O
CO2Et
(1)
NH, H+
(2) CH2=CHCO2Me
(3) H3O+
O
CO2Me
CO2Et

(m)

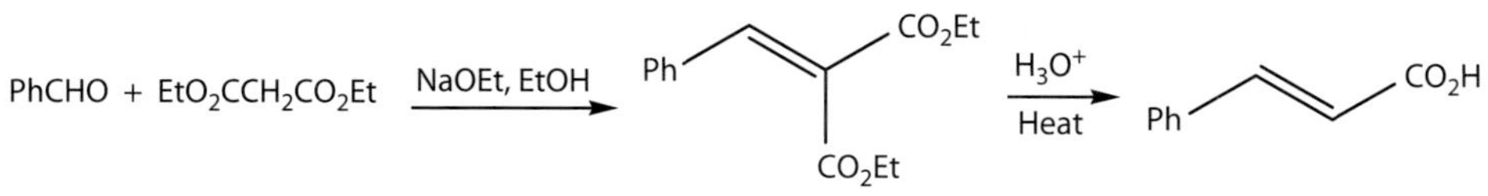
PhCHO + EtO2CCH2CO2Et
NaOEt, EtOH
Ph
CO2Et
CO2Et
H3O+
Heat
Ph
CO2H

(n)

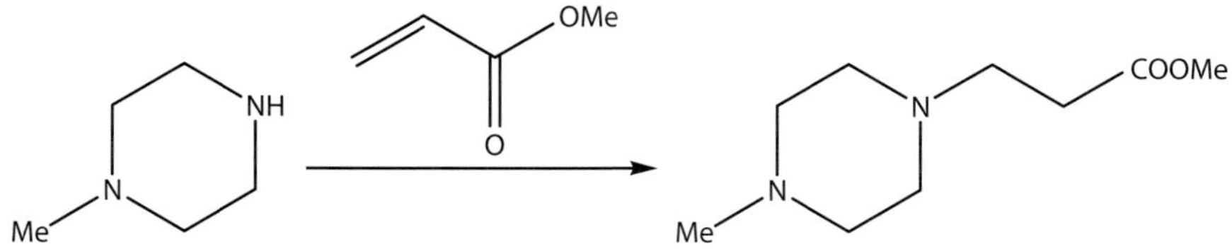
NH
N
Me
OMe
O
N
COOMe
N
Me

(o)

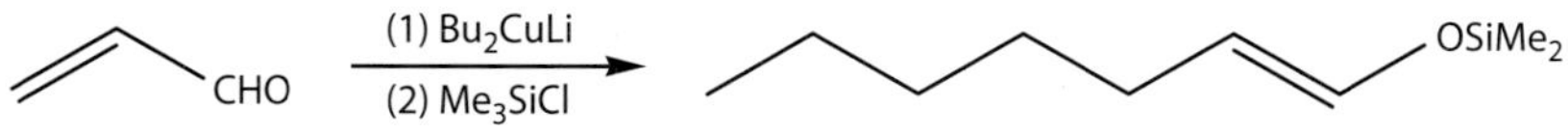
CHO
(1) Bu2CuLi
(2) Me3SiCl
OSiMe2

(p)

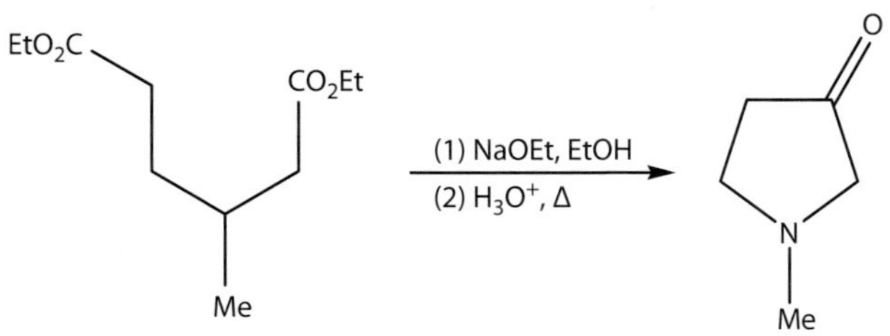
EtO2C
CO2Et
Me
(1) NaOEt, EtOH
(2) H3O+, Δ
O
N
Me

(q)

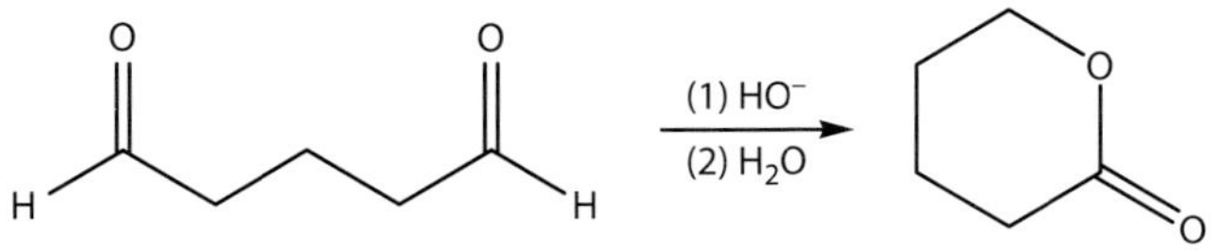
O
O
H
H
(1) HO−
(2) H2O
O
O

(r)

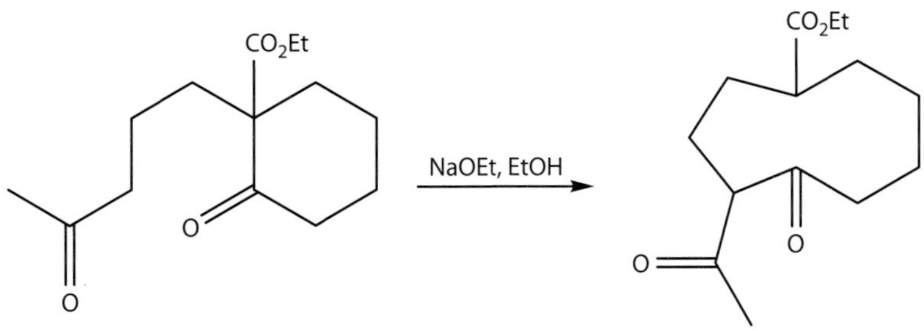
CO2Et
O
O
NaOEt, EtOH
CO2Et
O
O

(s)

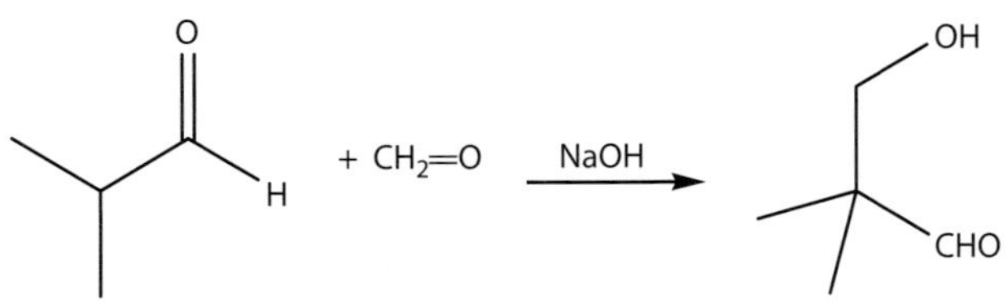
O
H
+ CH2=O
NaOH
OH
CHO

(t)

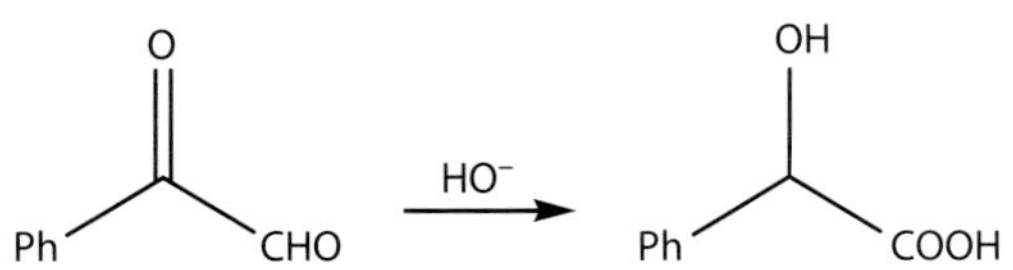
O
Ph
CHO
HO−
OH
Ph
COOH

(u)

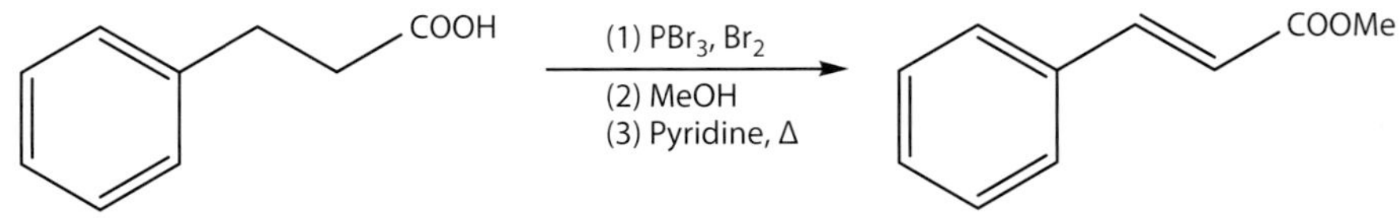

(v)

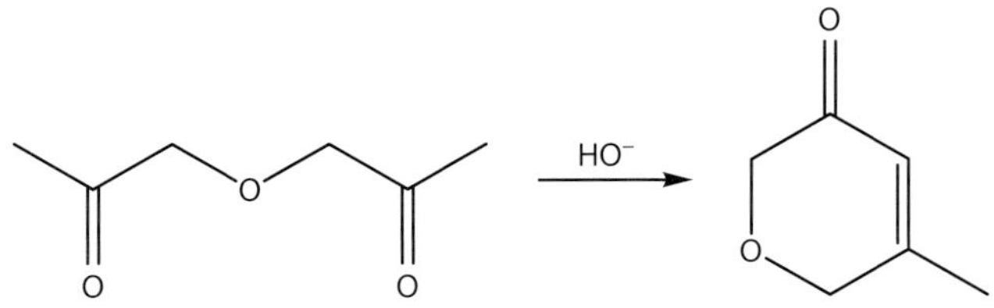

(w)

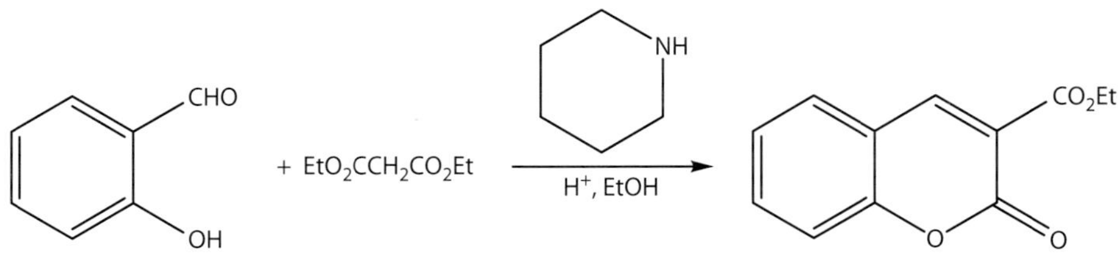

7. How would you accomplish each of the following conversions, which may require more than one step?

(a)

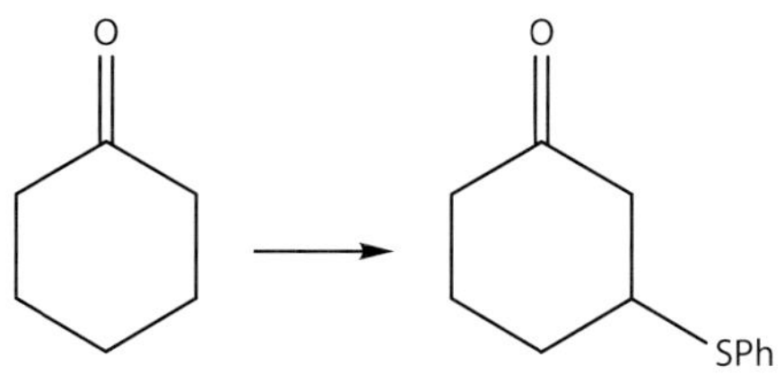

(b)

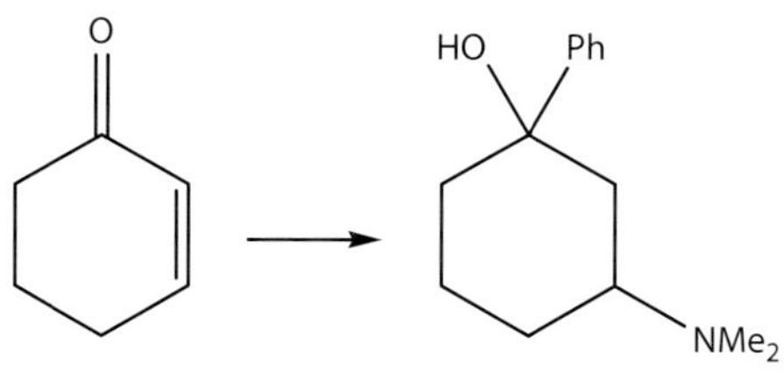

(c)

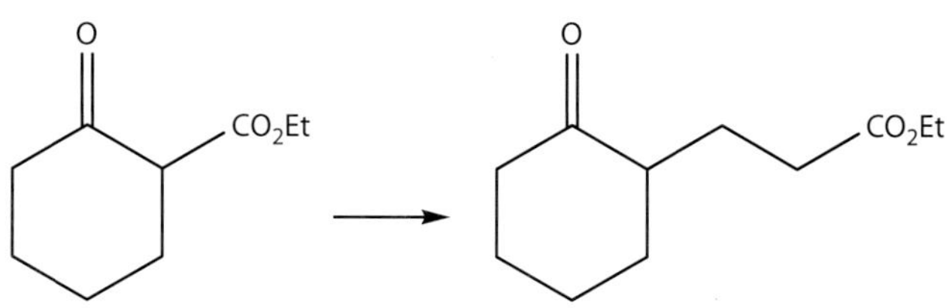

(d)

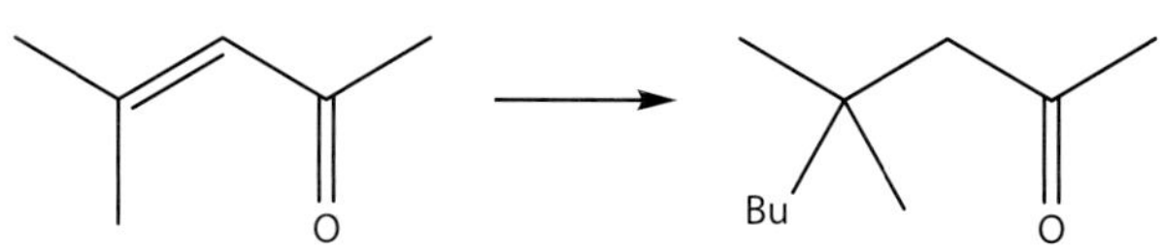

(e)

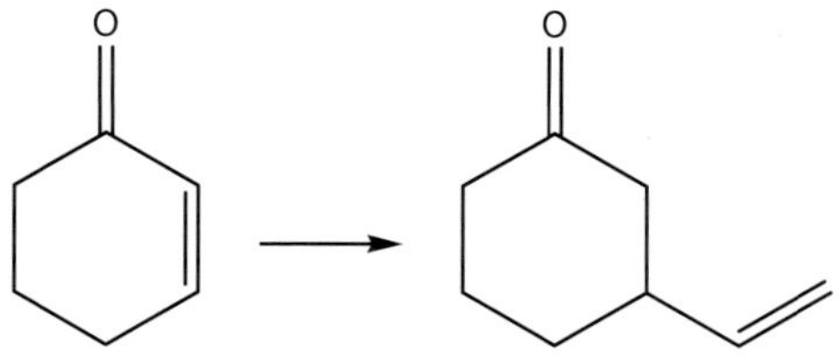

(f)

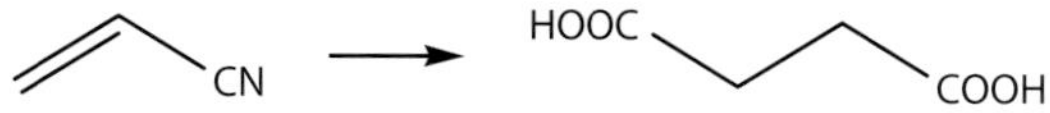

(g)

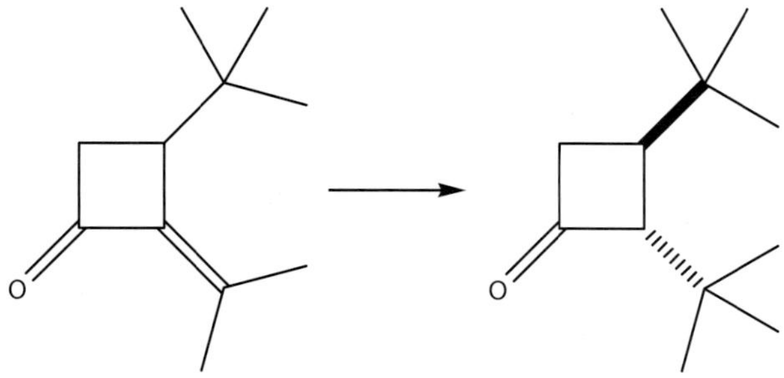

(h)

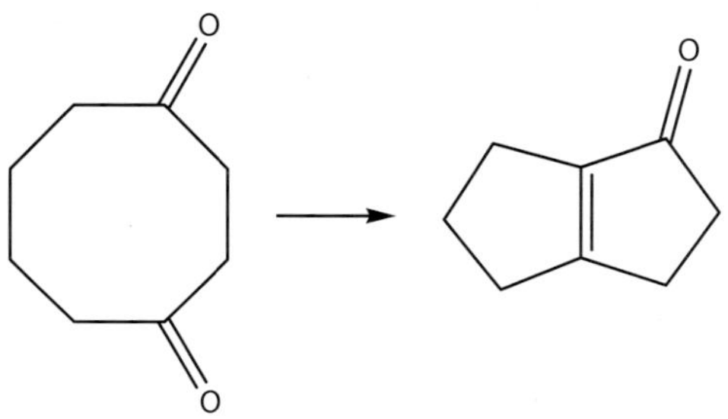

8. How would you synthesize each of the following, using the Robinson annulation? Show both your retrosynthetic analysis and a forward reaction scheme, taking care to use enamines, activating and protecting groups as appropriate:

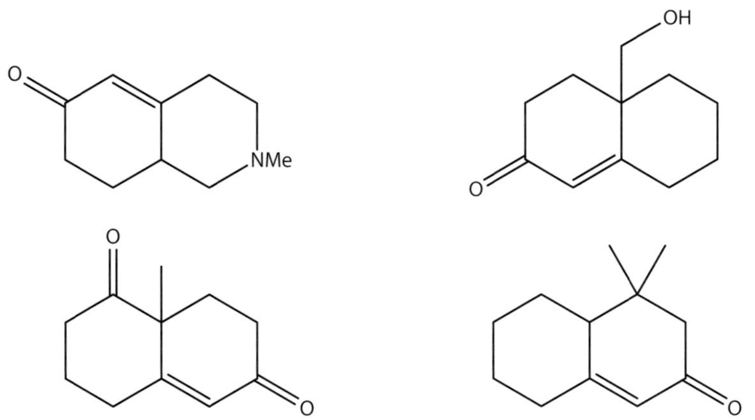

9. Explain the following observations:

(a)

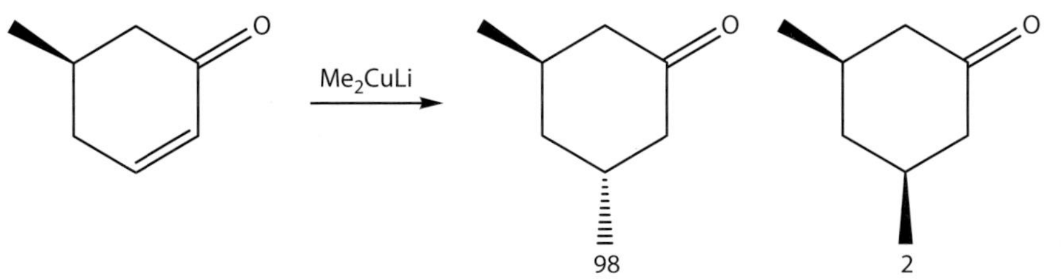

(b) Compound **17.57** racemizes readily in either basic or acidic solution:

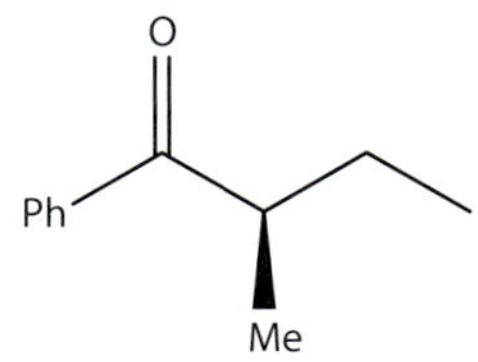

17.57

(c) The rate of bromination of acetone depends on the pH of the solution, but not on the concentration of bromine.

(d) When $PhCH_2CH_2COCH_3$ is reacted with $NaOH/I_2$, a good yield of iodoform is obtained. However, when $PhCH_2COCH_3$ is reacted under the same conditions, very little iodoform is produced.

10. Study the reaction schemes in the following, and answer the questions posed:

(a)

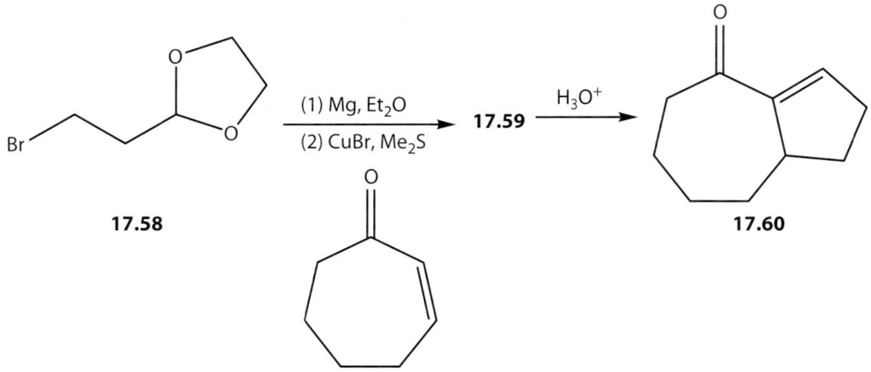

(i) Why is the SM **17.58** protected as an acetal?
(ii) What is the structure of the intermediate **17.59**?
(iii) Write a mechanism for the formation of **17.59**.
(iv) Write a mechanism for the conversion of **17.59** to **17.60**.

(b)

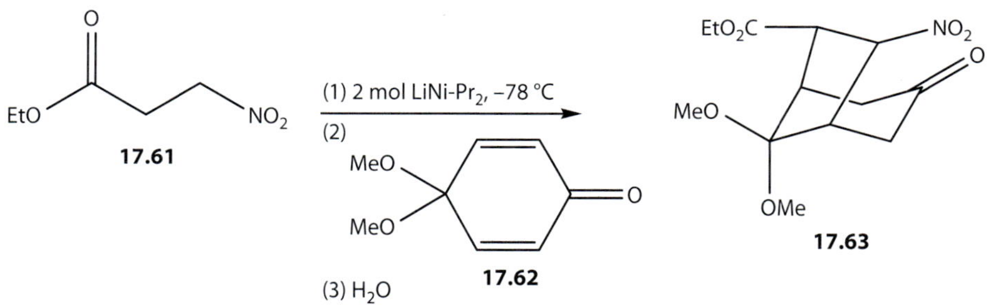

(i) Why do we use 2 mol of LDA?
(ii) Draw the structure of the product of the reaction of 2 mol LDA with **17.61**.
(iii) Write a mechanism for the reaction of this species with **17.62** to give **17.63**.

MORE CHALLENGING PROBLEMS

1. Predict the product(s) of each of the following processes:

(a)

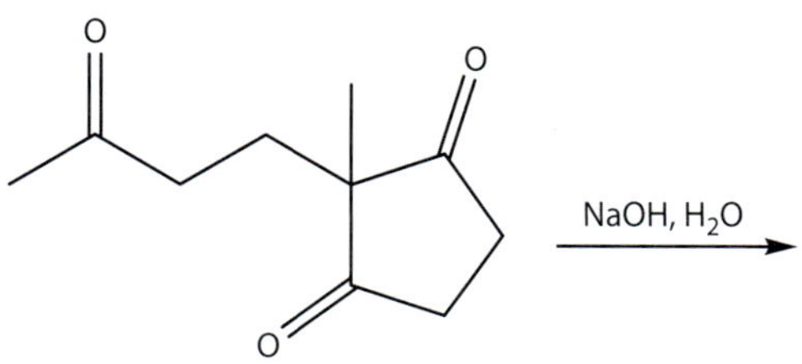

(b)

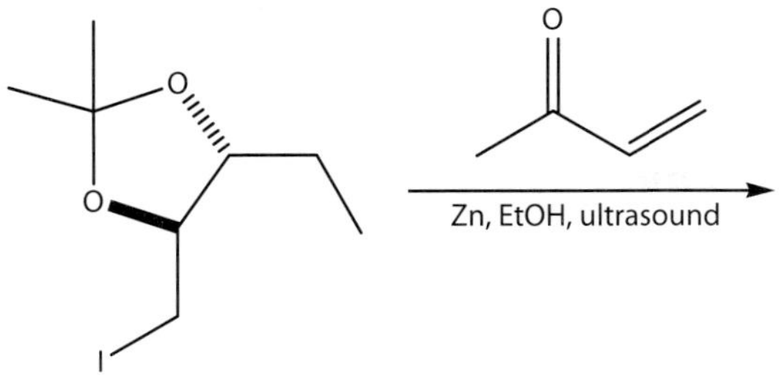

(c)

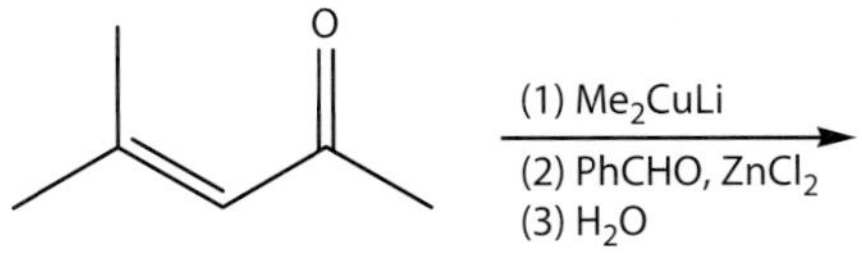

(d)

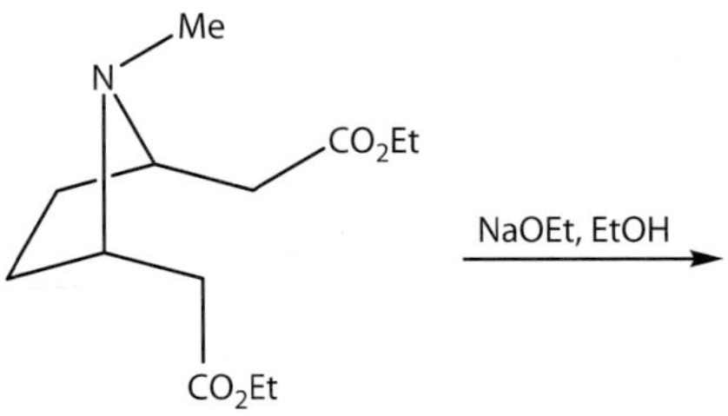

(e)

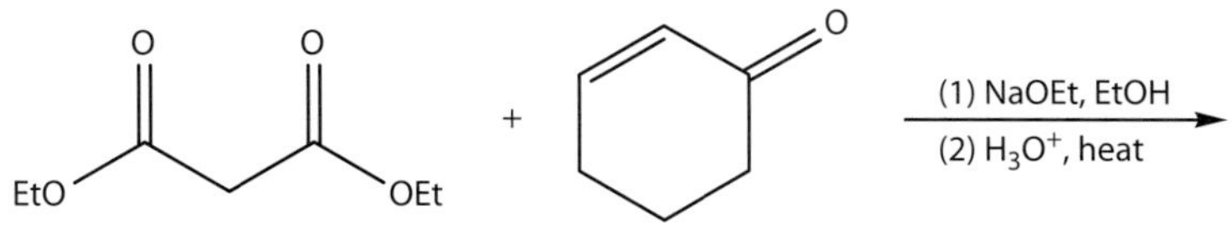

(f)

$EtO_2C(CH_2)_4CO_2Et$ →

(1) NaOEt, EtOH
(2 H^+
(3) KH, MeI
(4) KOH, H_2O
(5) H_3O^+, heat

2. Propose mechanisms for each of the following processes:

(a)

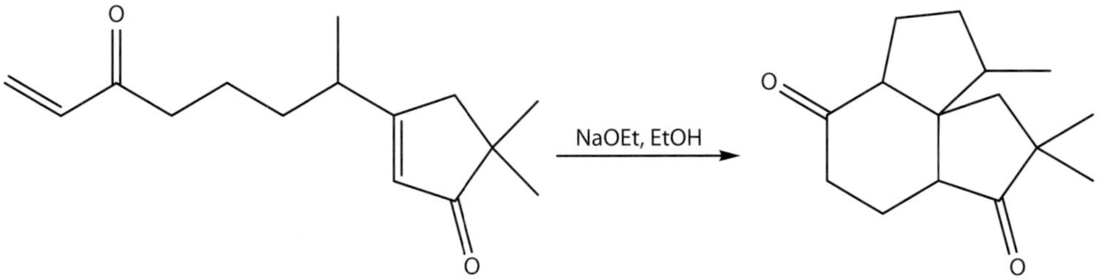

(b)

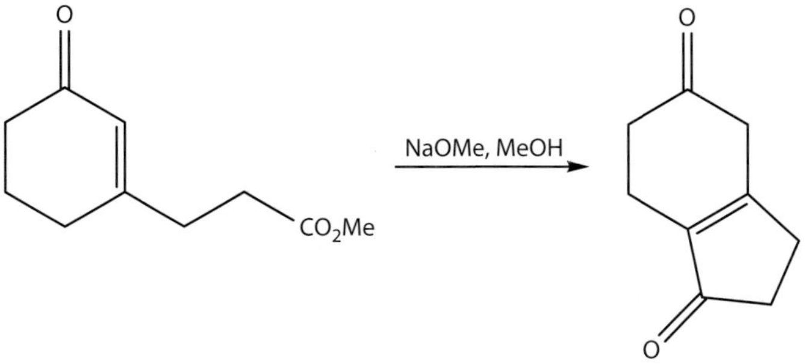

(c)

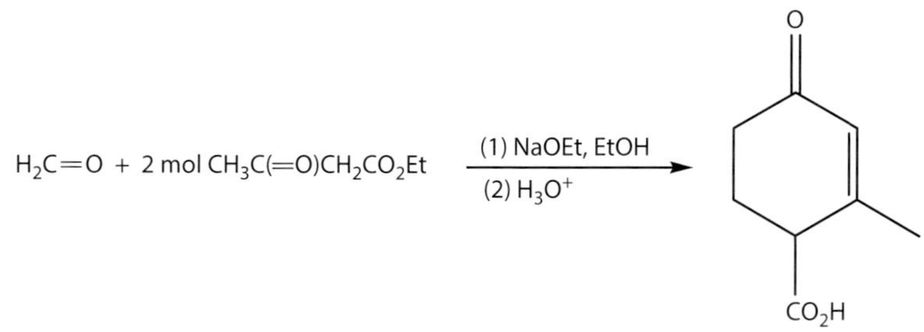
H2C=O + 2 mol CH3C(=O)CH2CO2Et
(1) NaOEt, EtOH
(2) H3O+
O
CO2H

(d)

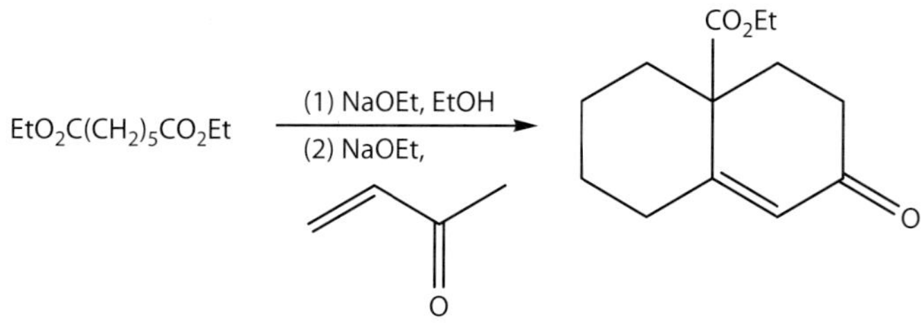
EtO2C(CH2)5CO2Et
(1) NaOEt, EtOH
(2) NaOEt,
O
CO2Et
O

(e)

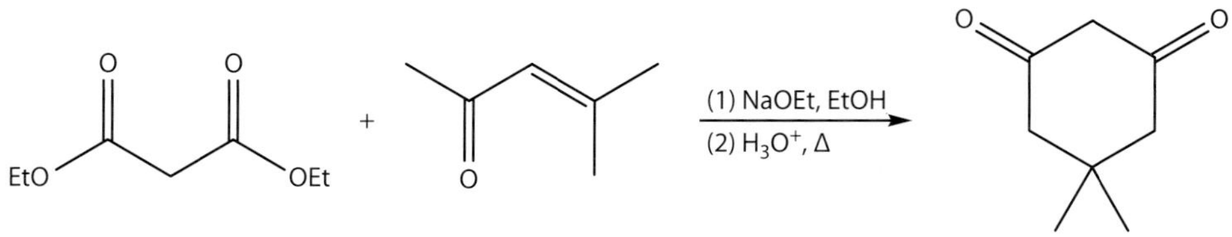
O
O
EtO
OEt
+
O
(1) NaOEt, EtOH
(2) H3O+, Δ
O
O

(f)

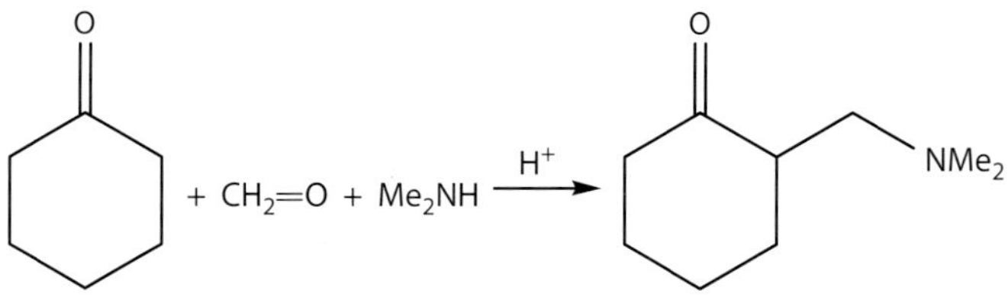
O
+ CH2=O + Me2NH
H+
O
NMe2

(g)

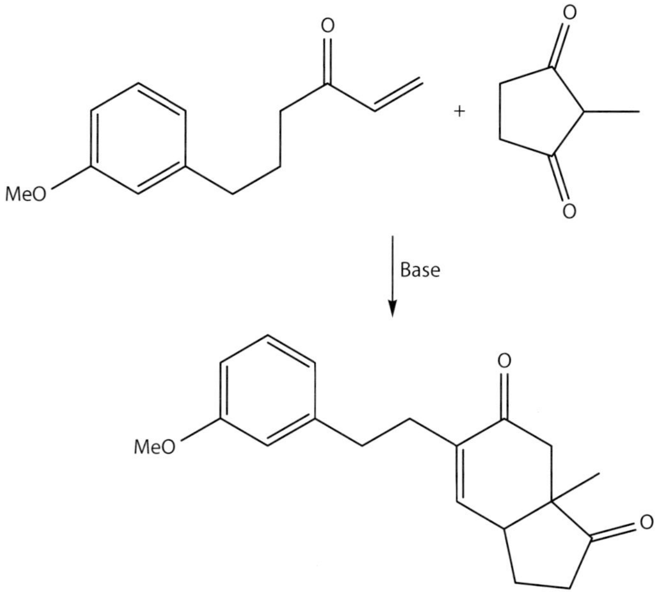
O
MeO
+
O
O
Base
MeO
O
O

(h)

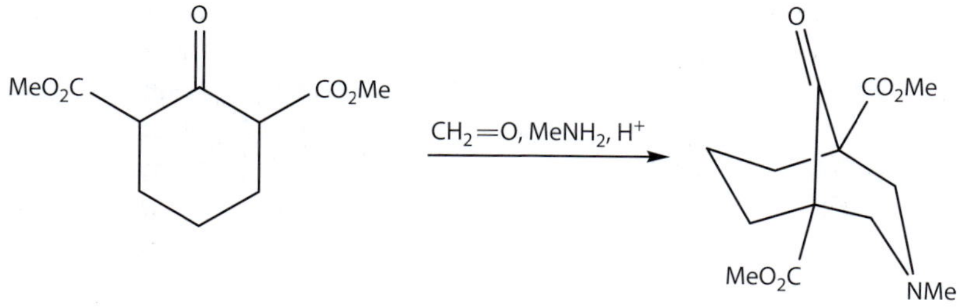
O
MeO2C
CO2Me
CH2=O, MeNH2, H+
O
CO2Me
MeO2C
NMe

(i)

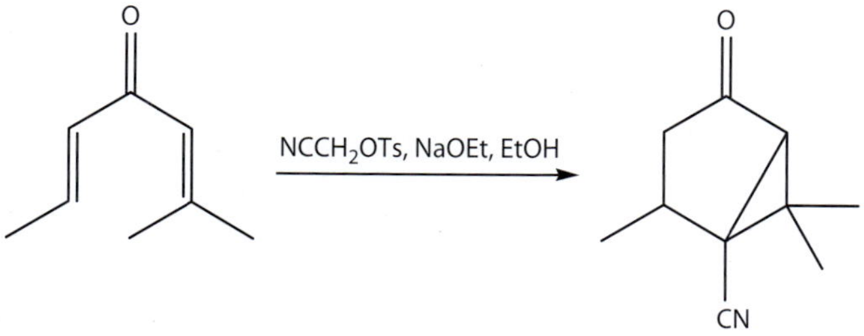
O
NCCH2OTs, NaOEt, EtOH
O
CN

(j)

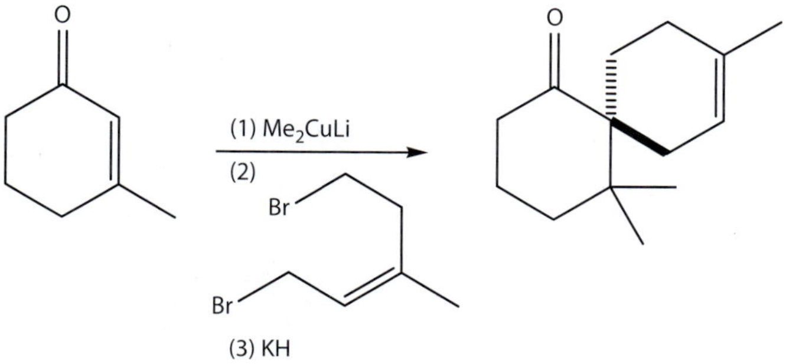
O
(1) Me2CuLi
(2)
Br
Br
(3) KH
O

(k)

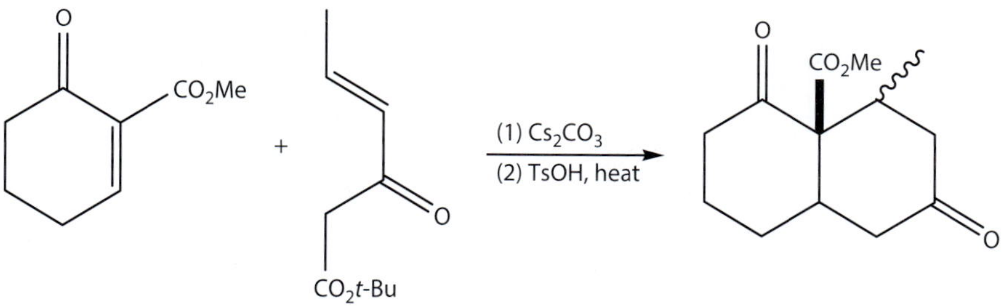
O
CO2Me
+
O
CO2t-Bu
(1) Cs2CO3
(2) TsOH, heat
O
CO2Me
O

(l)

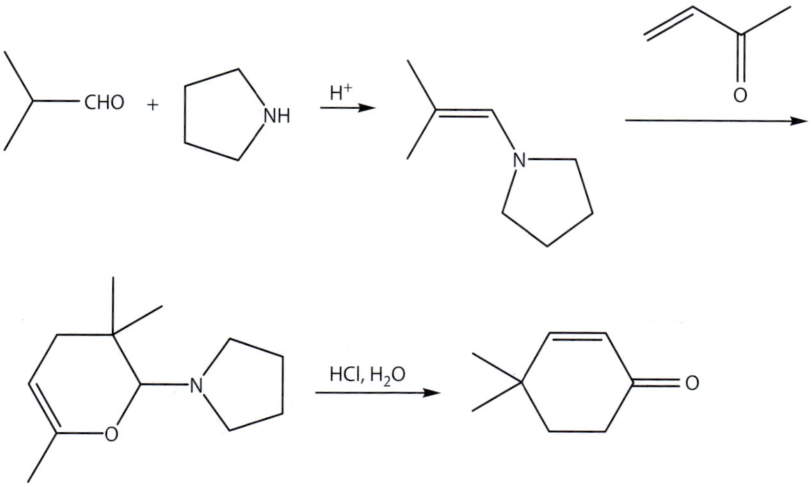
CHO
+
NH
H+
N
O
N
O
HCl, H2O
O

(m)

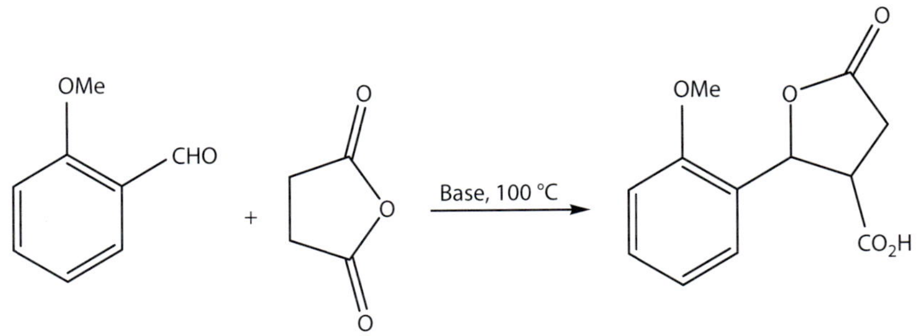
OMe
CHO
+
Base, 100 °C
OMe
CO_2H

(n)

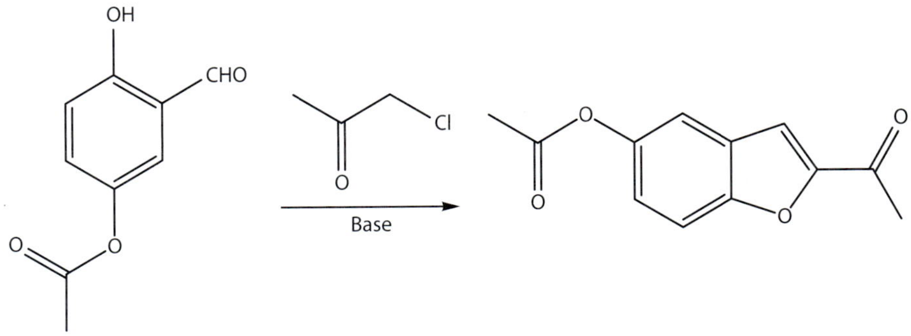
OH
CHO
Cl
Base

(o)

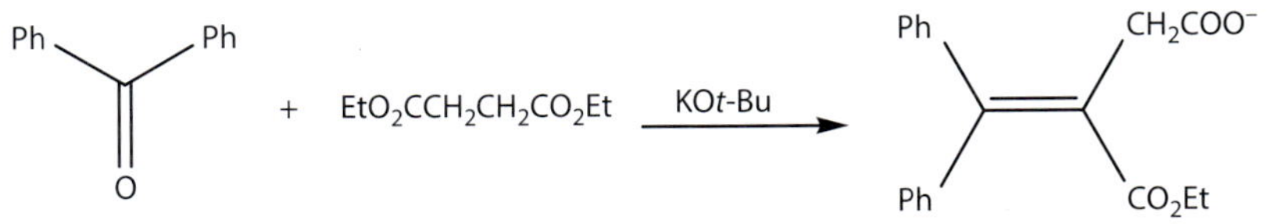
Ph
Ph
+ $EtO_2CCH_2CH_2CO_2Et$
KO*t*-Bu
Ph
CH_2COO^-
Ph
CO_2Et

(p)

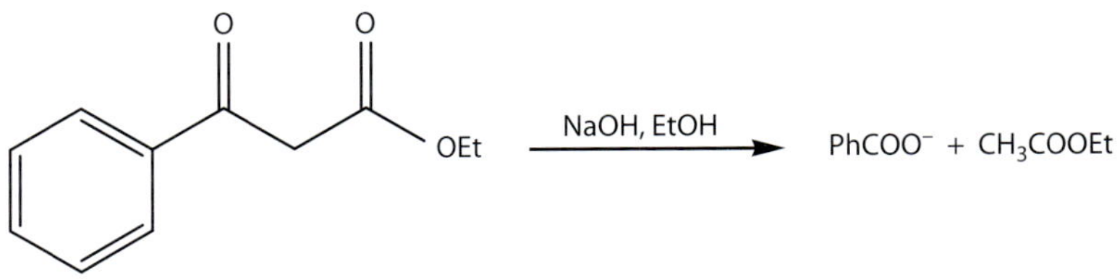
OEt
NaOH, EtOH
$PhCOO^-$ + CH_3COOEt

(q)

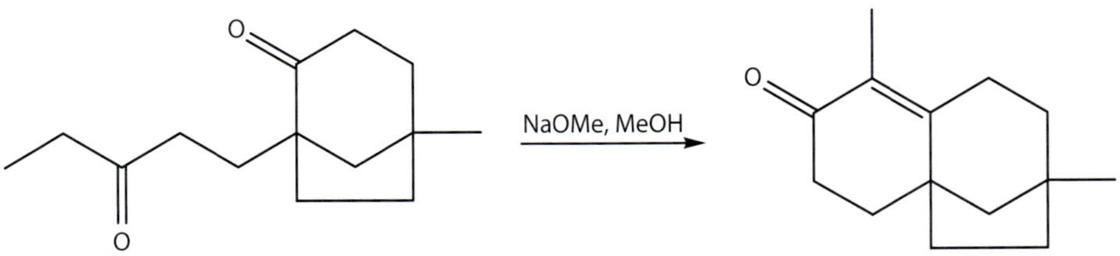
NaOMe, MeOH

(r)

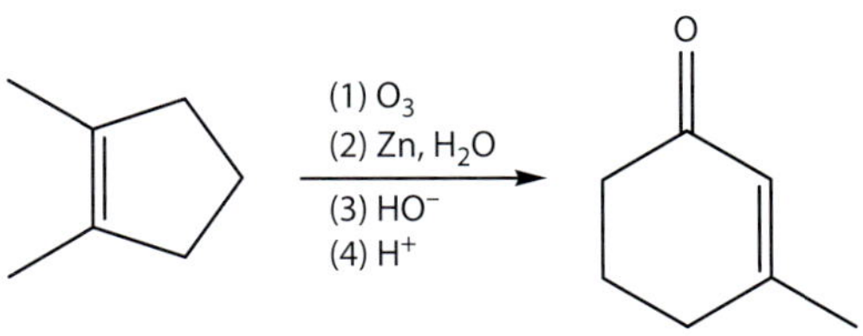
(1) O_3
(2) Zn, H_2O
(3) HO^-
(4) H^+

3. When bromoketone **17.64** is treated with KO*t*-Bu/*t*-BuOH at room temperature, it gives exclusively **17.65**. When **17.64** is treated with LDA at –70 °C, and then warmed, it gives mostly **17.66**. Write the mechanisms for these processes and explain why you get different products:

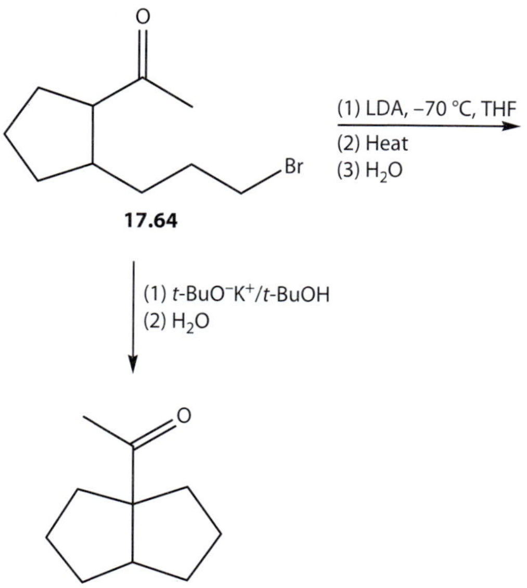

17.65

4. Cycloocta-2,4,6-triene-1-one is unusual in that it is difficult to generate the enolate ion from the ketone, even with strong bases. However, the enolate can be generated by the reaction of cyclooctatetraene monoepoxide with LDA at –78 °C in THF/HMPA. Quenching the enolate with D_2O gives the deuterated compound **17.67** in 85 % yield. The reaction of the enolate with EtI, MeI, or $CH_2{=}CHCH_2Br$ gave the *C*-alkylated products, **17.68**, in 70 %–80 % yield. With EtBr, the yield is less good and mostly of the *O*-alkylated product, **17.69**. $PhCH_2Br$ gave 15 % of *C*-alkylated product and $PhCH_2I$ gave 37 % *C*-alkylation. MeCOCl gave no *C*-alkylated product and 9 % *O*-alkylation; ClCOOEt gave 28 % *O*-alkylation:

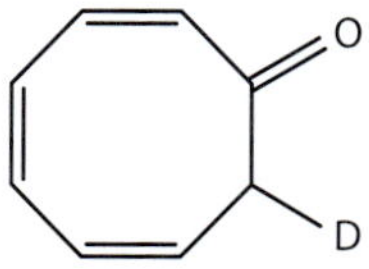

17.67

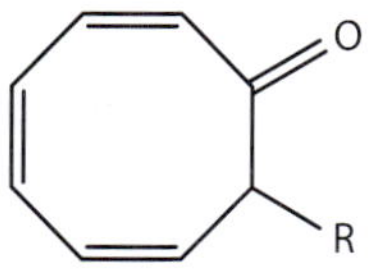

17.68

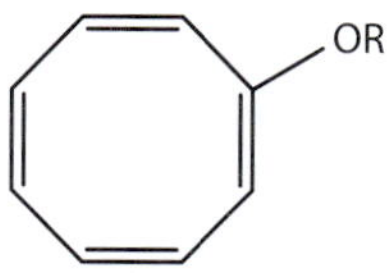

17.69

(a) Why is it difficult to make the enolate directly from the ketone?
(b) Write a mechanism for the LDA-mediated rearrangement of the epoxide to the ketone.
(c) Comment on the amounts of *C*- and *O*-alkylated/acylated products obtained on reaction of the enolate with the various electrophiles.

5. $EtO_2CCH_2CH_2COOEt$ does not undergo a simple Dieckmann condensation, but instead dimerizes to give **17.70**. Explain, writing a full mechanism:

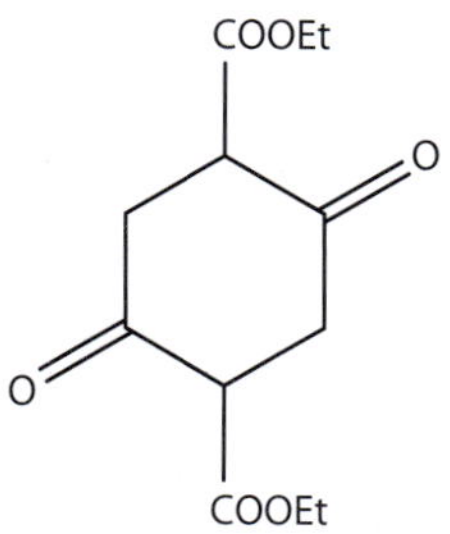

17.70

6. How would you synthesize?

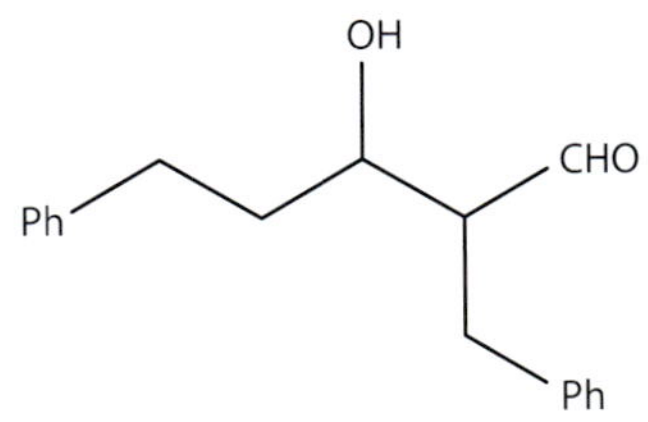

7. Identify the unknown reagents in the following sequence:

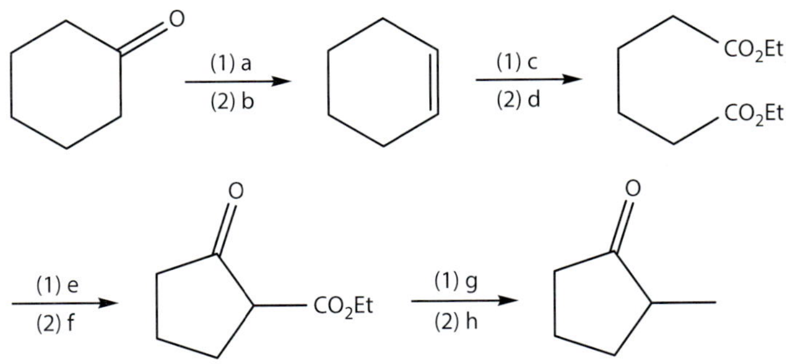

8. Compound **17.71**, $C_{13}H_{18}O$ is not soluble in either aqueous acid or base and decolorizes Br_2/CCl_4. **17.71** reacts with ozone followed by Zn/HOAc to give **17.72** and **17.73**. **17.72** does not give an iodoform test and shows bands in its IR spectrum at 1665 and 850 cm^{-1} and has the ^1H NMR spectrum:

δ	8.0	d, J = 7 Hz, 2H
	6.9	d, J = 7 Hz, 2H
	3.9	s, 3H
	2.9	q, J = 6.5 Hz, 2H
	1.3	t, J = 6.5 Hz, 3H

17.73 shows an IR band at 1710 cm^{-1} and does give a positive iodoform test. Its ^1H NMR spectrum shows only a singlet at δ 2.1. Identify **17.71**, **17.72**, and **17.73**.

9. When benzaldehyde and an excess of acetone react together in the presence of aqueous sodium hydroxide, compound **17.74** is the major product, together with a small amount of compound **17.75**. However, when acetone and an excess of benzaldehyde (at least two equivalents) react together, the main product is **17.75**. Spectroscopic data for **17.74** and **17.75** are given in the following. Deduce structures for **17.74** and **17.75** and rationalize their formation.

17.74
Mass spectrum: m/z = 146 (75 %); 145 (50 %); 131 (100 %); 103 (80 %)
IR: ν_{max} = 1667 (s), 747 (s), 689 (s) cm^{-1}
^1H NMR ($CDCl_3$) δ 2.38 (s, 3H), 6.71 (d, J = 16 Hz, 1H), 7.30–7.66 (m, 5H), 7.54 (d, J = 16 Hz, 1H)

17.75
Mass spectrum: m/z = 234 (75 %); 233 (44 %); 131 (100 %); 128 (26 %); 103 (90 %); 91 (32 %); 77 (100 %)
IR: ν_{max} = 1651 (s), 984 (s), 762 (s) 670 (s) cm^{-1}
^1H NMR ($CDCl_3$) δ 7.10 (d, J = 16 Hz, 1H), 7.30–7.66 (m, 5H), 7.78 (d, J = 16 Hz, 1H)

Rearrangement Reactions

18

18.1 INTRODUCTION

As we have studied a range of organic reactions, we have seen a number of rearrangements, most often as unwanted side reactions of the process we intended. Many of these could be characterized as migration to electron-deficient carbon atoms (carbocations). In this chapter, we will systematize these processes and see how they can be used synthetically. Migrations to other electron-deficient atoms, mainly oxygen and nitrogen, are also useful. There are also rearrangements of anions, but they are less common. We will focus on the well-established, named reactions, but you will encounter rearrangements that are similar, though not identical, to those described. If you remember that energy only runs downhill, this will always point you the right direction. The last category of rearrangements involves neutral molecules—electrocyclic and sigmatropic rearrangements; these have features in common with Diels Alder and other cycloaddition reactions, in that they are controlled by orbital symmetry.

18.2 MIGRATION TO ELECTRON-DEFICIENT CARBON

18.2.1 CARBOCATION REARRANGEMENTS

We have previously met carbocation rearrangements when we studied substitution, elimination, and addition reactions and some Friedel–Crafts reactions. Some examples of these are shown in Figure 18.1. When we write the mechanism for these reactions, in each case, we produce a carbocation, and it rearranges to a more stable carbocation. Thus, in Figure 18.2, a secondary carbocation is produced, and it rearranges to a tertiary cation, before losing a proton to give an alkene. Notice that the methyl group migrates *with* the electron pair of the carbon–carbon bond, so that a cation is left behind at the site from which the methyl came. As with most of these reactions, a less stable secondary carbocation rearranges to a more stable tertiary one. This is almost invariably the case for this type of rearrangement, although we may have to consider other aspects of carbocation stability than the simple tertiary > secondary > primary. These reactions are sometimes known as Wagner–Meerwein rearrangements.

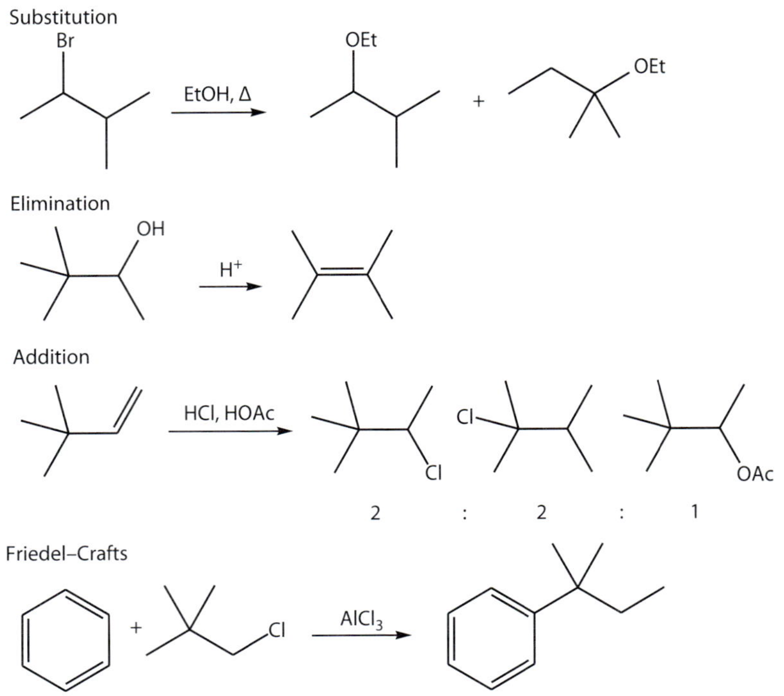

FIGURE 18.1 Review of carbocation rearrangements.

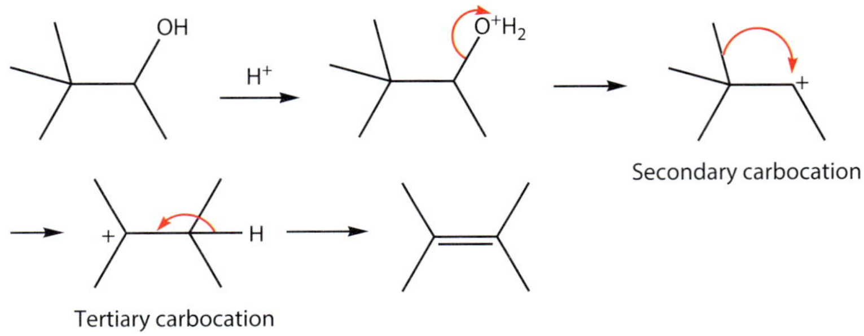

FIGURE 18.2 Mechanism of a carbocation rearrangement.

PROBLEM 18.1

(Review) Draw a mechanism for the rearrangements in the addition and Friedel–Crafts reactions of Figure 18.1.

Solution

Addition reaction:

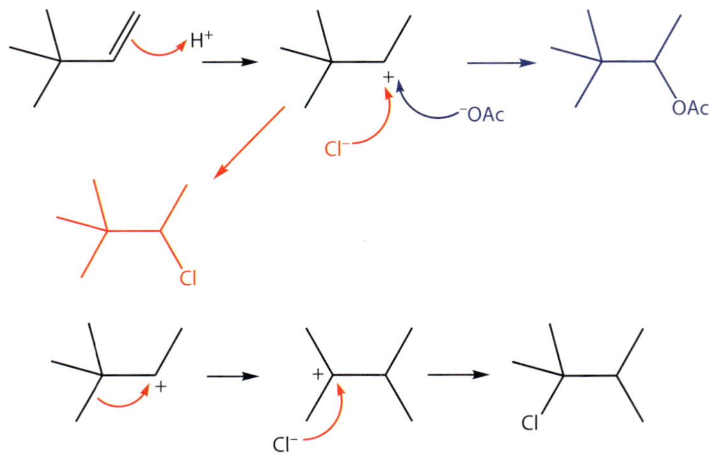

Friedel–Crafts reaction:

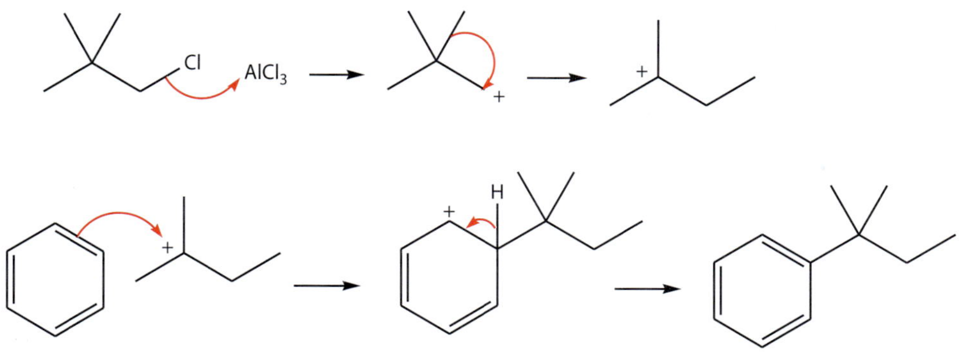

Each of these examples has been carefully chosen so that only one type of group, a methyl group, was available to migrate. What would happen if there was more than one group that could migrate? The group that would be the best carbocation migrates selectively, because at the transition state (**18.1**), the positive charge is partly on the migrating group. Thus, in terms of migratory aptitude, tertiary > secondary > primary > methyl > hydrogen. The process is not generally stereospecific; the leaving group leaves, and then the migration occurs, but there are exceptions. When the migrating group is a phenyl ring, there can be a true intermediate, a phenonium ion, with the same delocalized positive charge on the ring as would be found in a Wheland intermediate in electrophilic substitution. Figure 18.3 illustrates. The phenyl ring acts as a neighboring group, assisting the departure of toluene sulfonic acid, to give a symmetric intermediate. The phenonium ion reacts with acetic acid to give enantiomeric products.

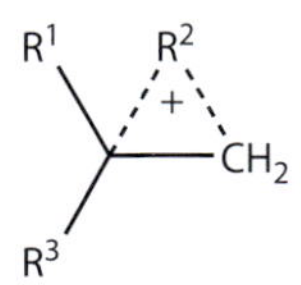

18.1

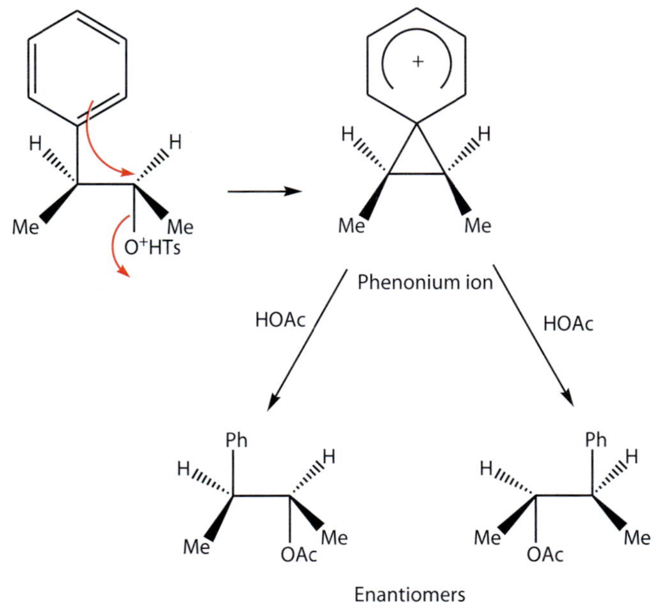

FIGURE 18.3 Migration of phenyl group *via* a phenonium ion.

PROBLEM 18.2

Propose reasonable mechanisms for the following reactions:

(a)

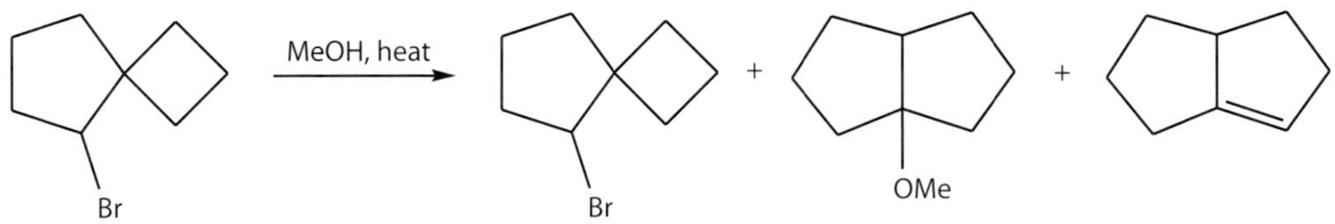

(b)

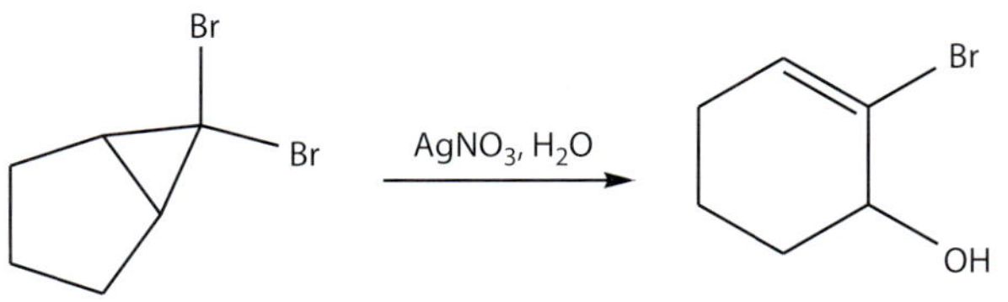

Solutions

In both cases, the reaction will be initiated by the loss of bromide ion to give a carbocation:

(a)

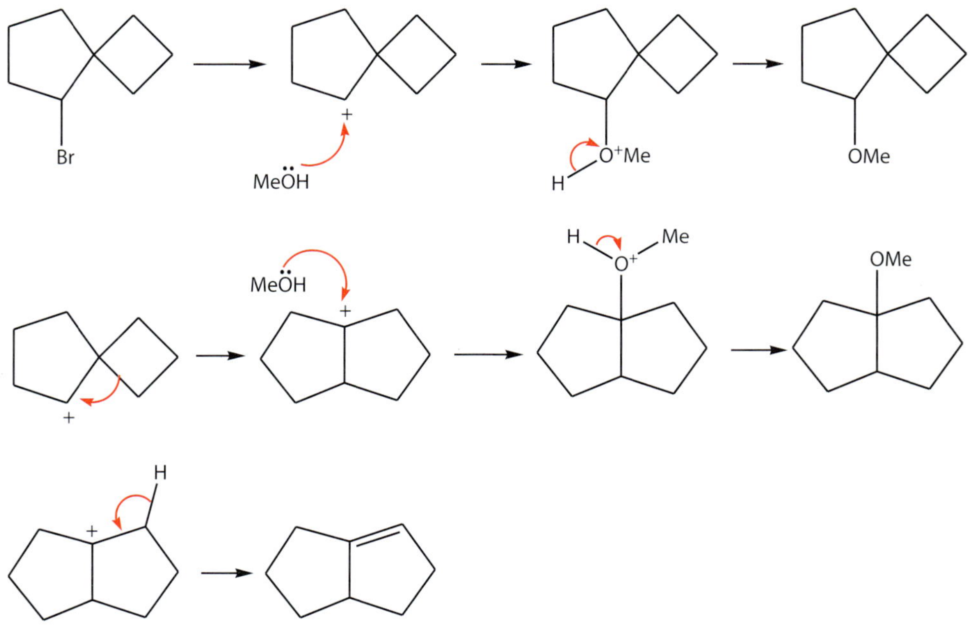

(b) The insolubility of the silver bromide assists the departure of bromide:

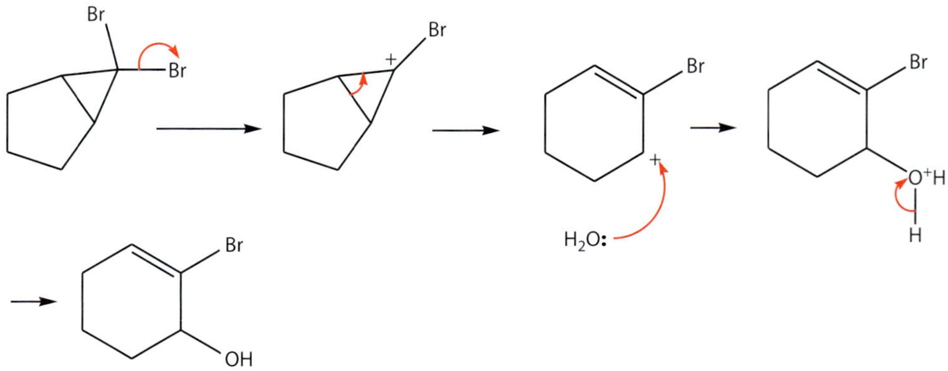

18.2.2 PINACOL AND SEMIPINACOL REARRANGEMENTS

Migratory ability also controls the outcome of the pinacol rearrangement, an acid-catalyzed rearrangement of 1,2-diols. The prototype reaction, Figure 18.4, is of 2,3-dimethylbutane-2,3-diol, which has the common name of pinacol. The product, 3,3-dimethyl-2-butanone, goes by the common name of pinacolone. Although the reaction mechanism is drawn initially as concerted, and the reaction is usually concerted with the migrating group being *trans* and *anti* to the departing water, we could dissect it into three steps. Water departs, leaving a carbocation. The methyl group migrates to leave a carbocation stabilized by its interaction with the lone pair of electrons at oxygen. Finally, a proton is lost to give the ketone.

There have been many studies of relative migratory ability in the pinacol rearrangement, and some consistent data emerge. Phenyl migrates well and for alkyl groups, the migratory aptitude is in the order 3 ° > 2 ° > 1 °. However, the problem is not a simple one—as well as the ability of the migrating group to carry a positive charge, we need to think about the remaining groups' ability to stabilize the new positive center. With cyclic compounds, ring size is also important—will a change in ring size make the molecule more or less stable? Where the two carbon atoms bearing the hydroxyl groups differ, then generally the reaction will start by forming the better carbocation. So in the first example of Figure 18.5, the 4-methoxyphenyl group migrates, because the methoxy group stabilizes the positive transition state. In the second example, the predominant migration is of the $-CH_2R$ rather than the methyl group (note that SM stereochemistry does not matter because the isomers interconvert under the reaction conditions). In the final example, the product derives exclusively from the tertiary carbocation that is stabilized by the two phenyl groups.

The pinacol reaction in its simplest form is not widely used synthetically, because of the possibility of the formation of mixtures of products. Various related reactions exist, where, by modification of one of the hydroxyl groups, we can guarantee that this will be the leaving group. An example is given in Figure 18.6. The less hindered OH group can be selectively converted to the tosylate. Under acid catalysis, the tosylate is a better leaving group than the hydroxyl (both would be protonated), so the direction of the rearrangement is guaranteed.

Another method of addressing this problem of selectivity is through the semipinacol or Tiffeneau–Demjanov reaction. In this reaction, an amino group of a vicinal amino alcohol is converted to a diazo salt using nitrous acid. The reaction mechanism is the same as we wrote for the preparation of aromatic diazo compounds, but when the substituent is an alkyl group, the

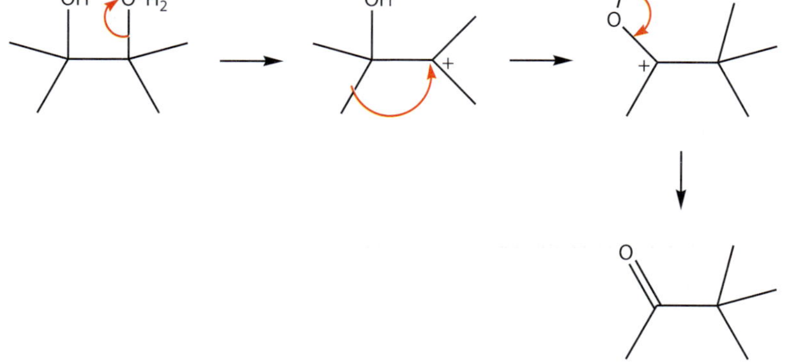

FIGURE 18.4 Pinacol rearrangement.

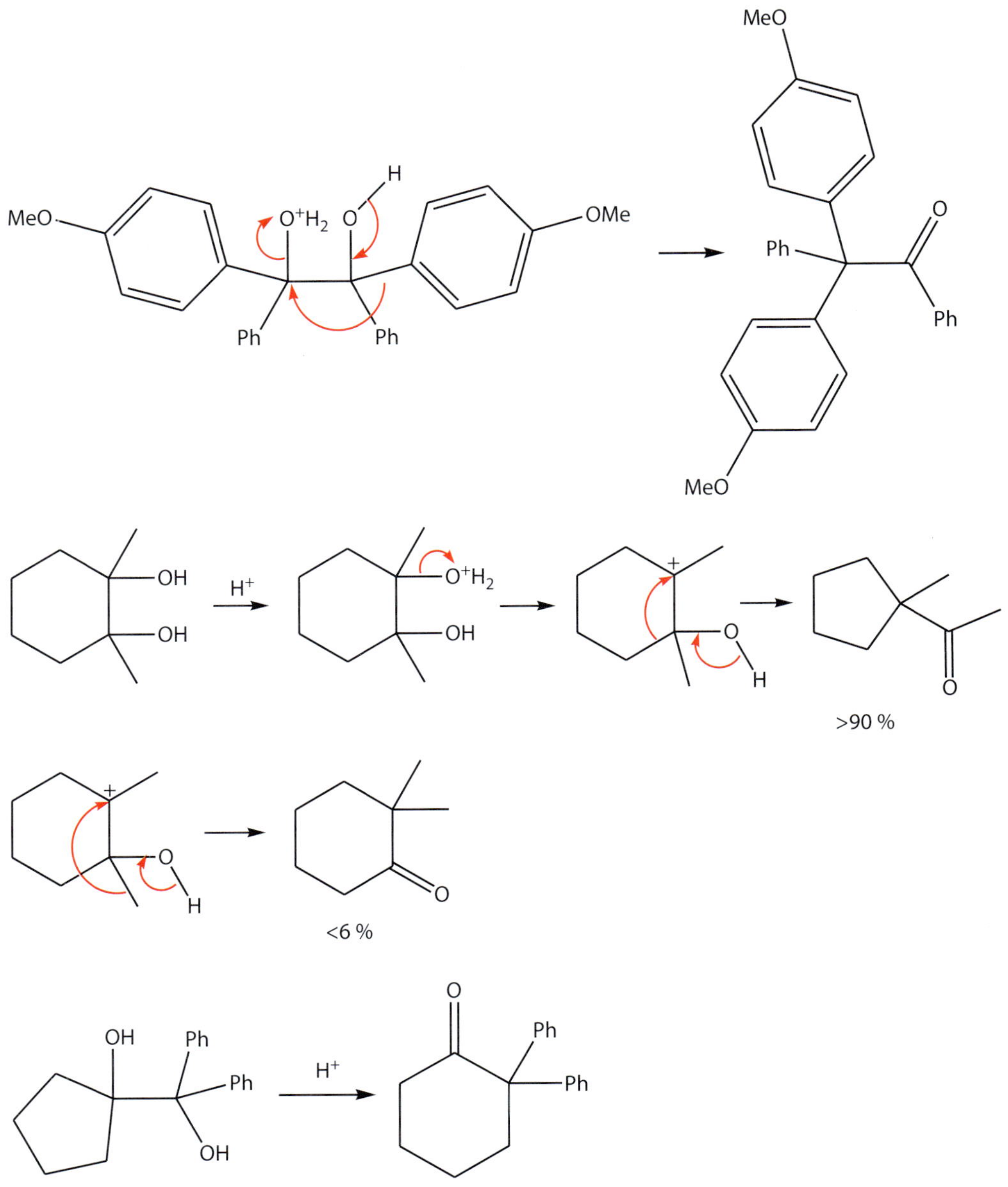

FIGURE 18.5 Selectivity in the pinacol rearrangement.

diazo species are much less stable and decompose even at −50 °C. An alkyl group then migrates, and a proton is lost from oxygen to give a ketone as before (Figure 18.7). Migration invariably involves the group that is *trans* and *anti* to the departing nitrogen; the reaction is concerted, and stereospecific, and is said to be *stereoelectronically* controlled. We can see this by looking that the outcome of the process in a series of cyclohexane derivatives, where the conformation of the ring has been fixed by the presence of a *tert*-butyl group (Figure 18.8). For **18.2** and **18.3**, the ring carbon is *trans* to the departing nitrogen, and these both give the same product, the ring-contracted aldehyde. For **18.4**, the hydrogen migrates, as it is *trans* to the diazo group. In **18.5**, there is no potentially migratory group *trans* to the leaving group, so an S_N2 process closes the epoxide ring.

The semipinacol reaction has been used in ring expansion processes, particularly as an approach to medium rings, which are not easy to prepare by ring closure reactions. In Figure 18.9, cyclohexanone is converted to its cyanohydrin, and this is then reduced to the corresponding amine. A semipinacol rearrangement gives cycloheptanone.

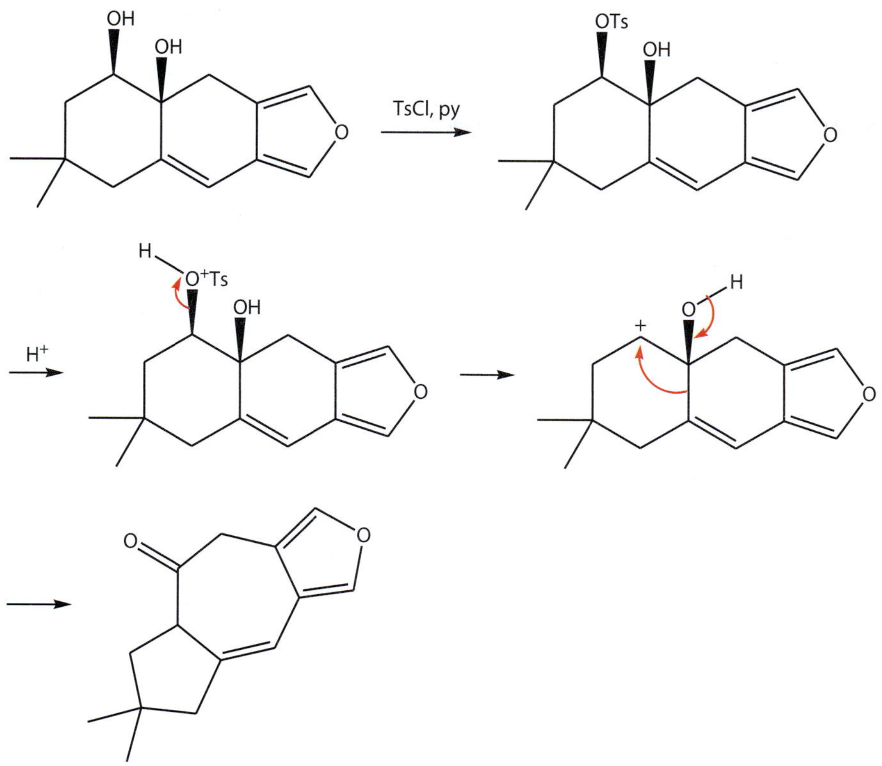

FIGURE 18.6 Selectivity in pinacol rearrangements.

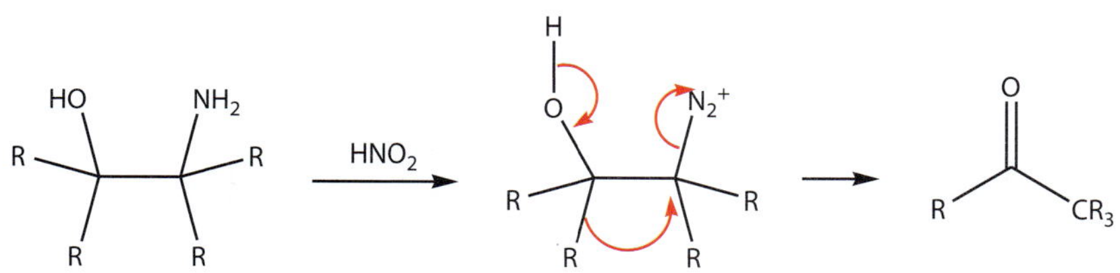

FIGURE 18.7 Semipinacol or Tiffeneau–Demjanov rearrangement.

PROBLEM 18.3

Suggest mechanisms for the following reactions:

(a)

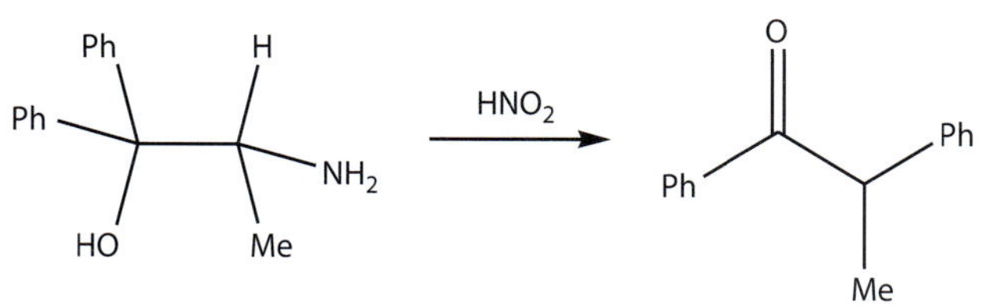

(b)

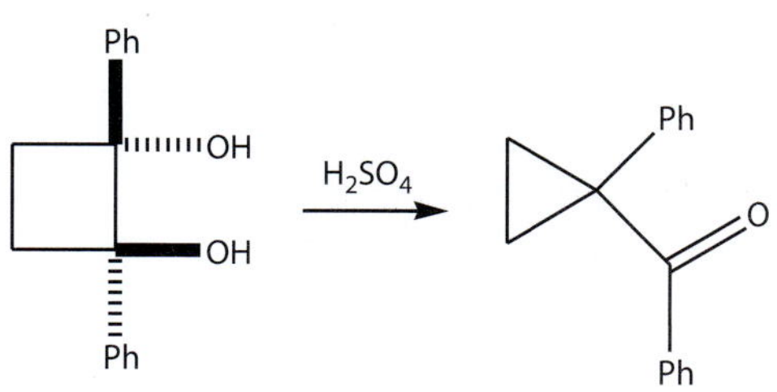

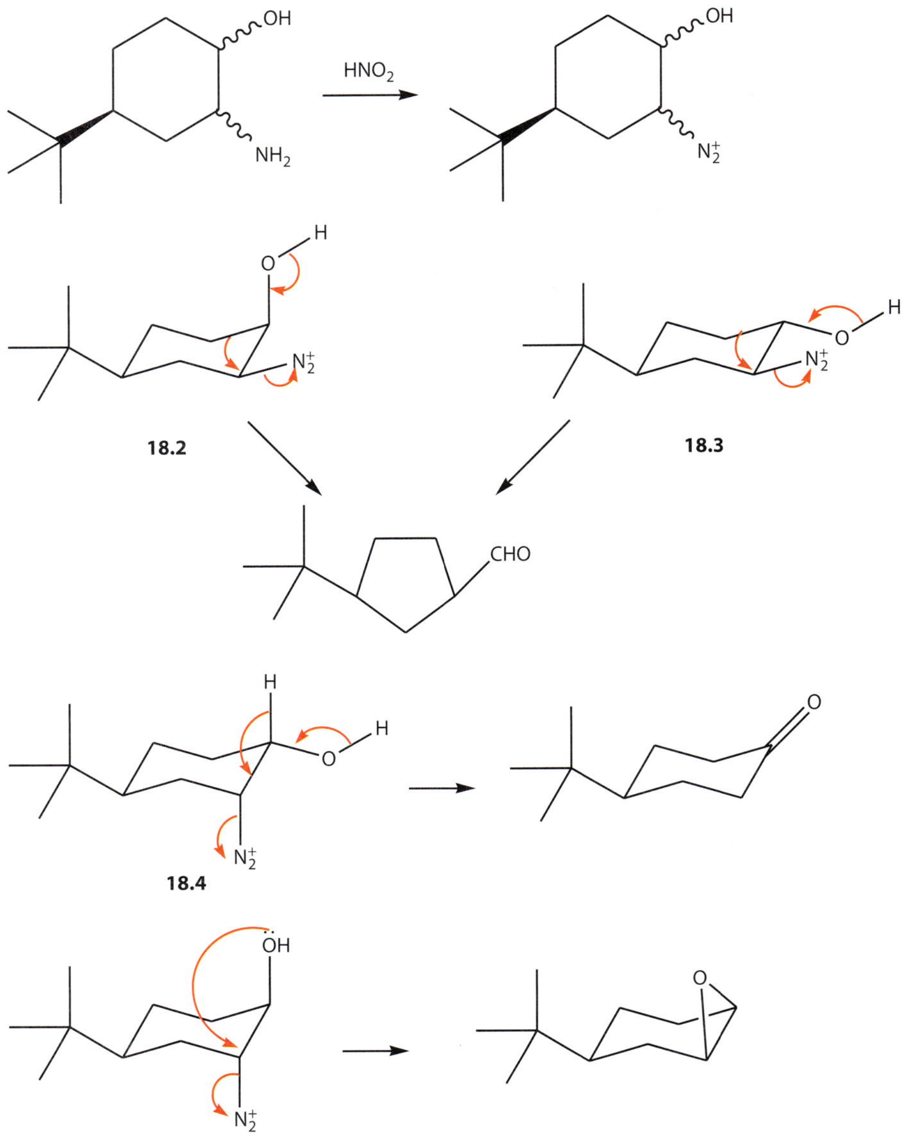

FIGURE 18.8 Stereoelectronic control in the semipinacol rearrangement.

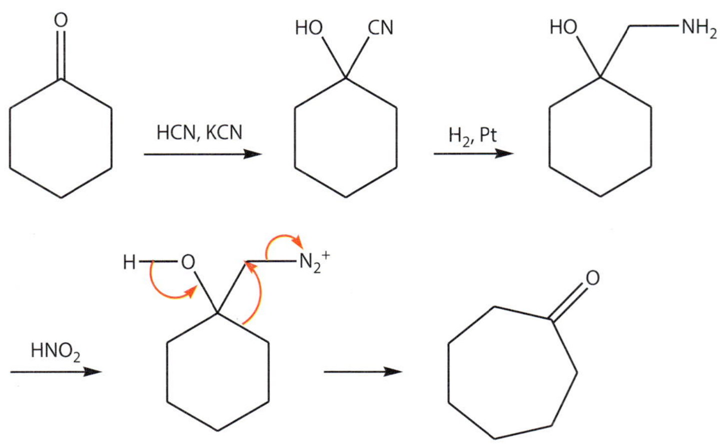

FIGURE 18.9 Ring expansion using the semipinacol reaction.

Solutions to Problem 18.3

(a) This is a classic semipinacol reaction; I have omitted the mechanism for the conversion of the amine to a diazo group, as it is entirely the standard one:

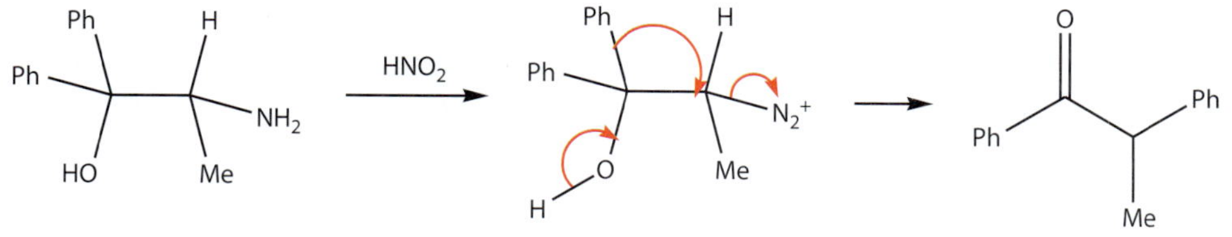

(b) This is a pinacol reaction, and it's reasonably easy to draw the arrows to get to the correct product:

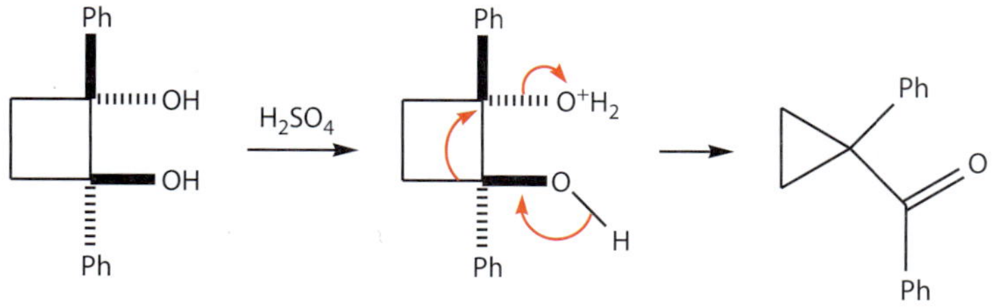

A more interesting question is *why* do we not get the cyclobutanone that would be formed by migration of the phenyl group? Phenyl is a good migrating group, generally much better than CH_2. The reason lies in the stability of the product. Although three-membered rings are inherently more strained than four-membered ones, clearly, a three-membered ring with all sp^3 carbon atoms is less strained than a four-membered one with an sp^2 carbon atom (wanting an angle of 120 °).

PROBLEM 18.4

Predict the product(s) of the following reactions:

(a)

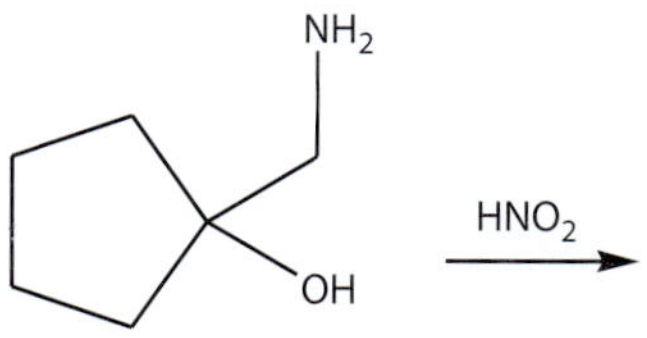

(b)

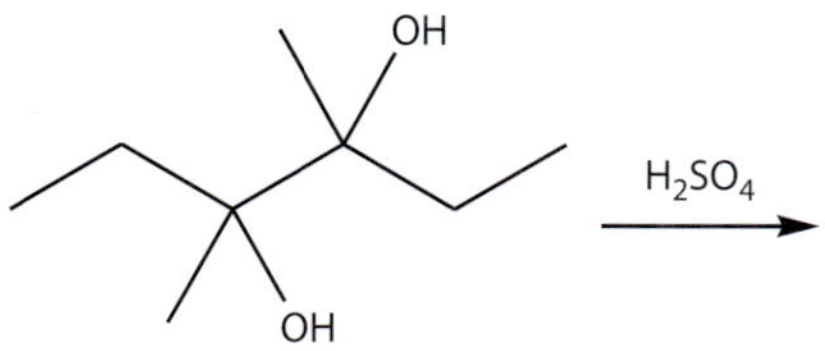

Solutions

(a) This is a classic ring expansion process of the type shown in Figure 18.9. The product is cyclohexanone.

(b) This is a pinacol rearrangement—the predominant product comes from the migration of the ethyl in preference to the methyl group:

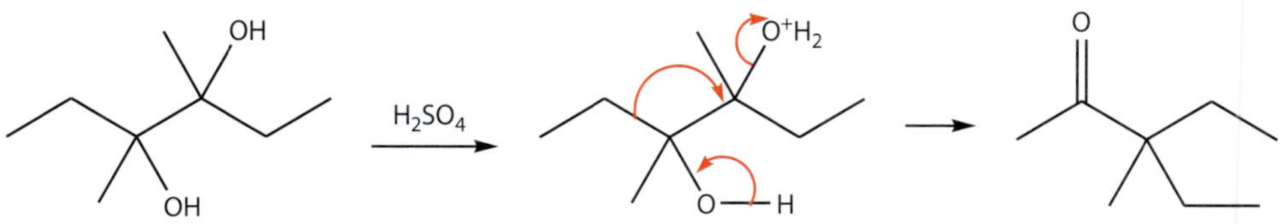

18.2.3 DIENONE–PHENOL REARRANGEMENTS

This type of rearrangement is driven by the stability conferred by the formation of a benzene ring. They occur when we have a cyclohexadienone, with two carbon substituents on one position, so that aromatization cannot occur directly by a proton loss. In acid solution, the oxygen of the carbonyl is protonated, generating a cation that has a resonance form, **18.6**, which is set up for a carbocation rearrangement. The final step of the reaction (Figure 18.10) involves loss of a proton to give the aromatic ring.

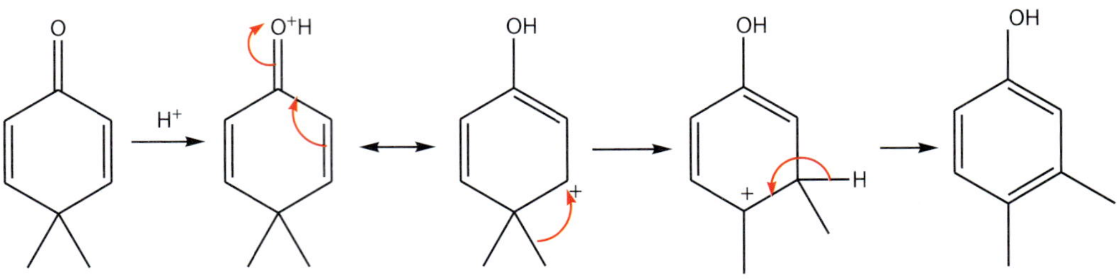

FIGURE 18.10 Mechanism of the dienone–phenol rearrangement.

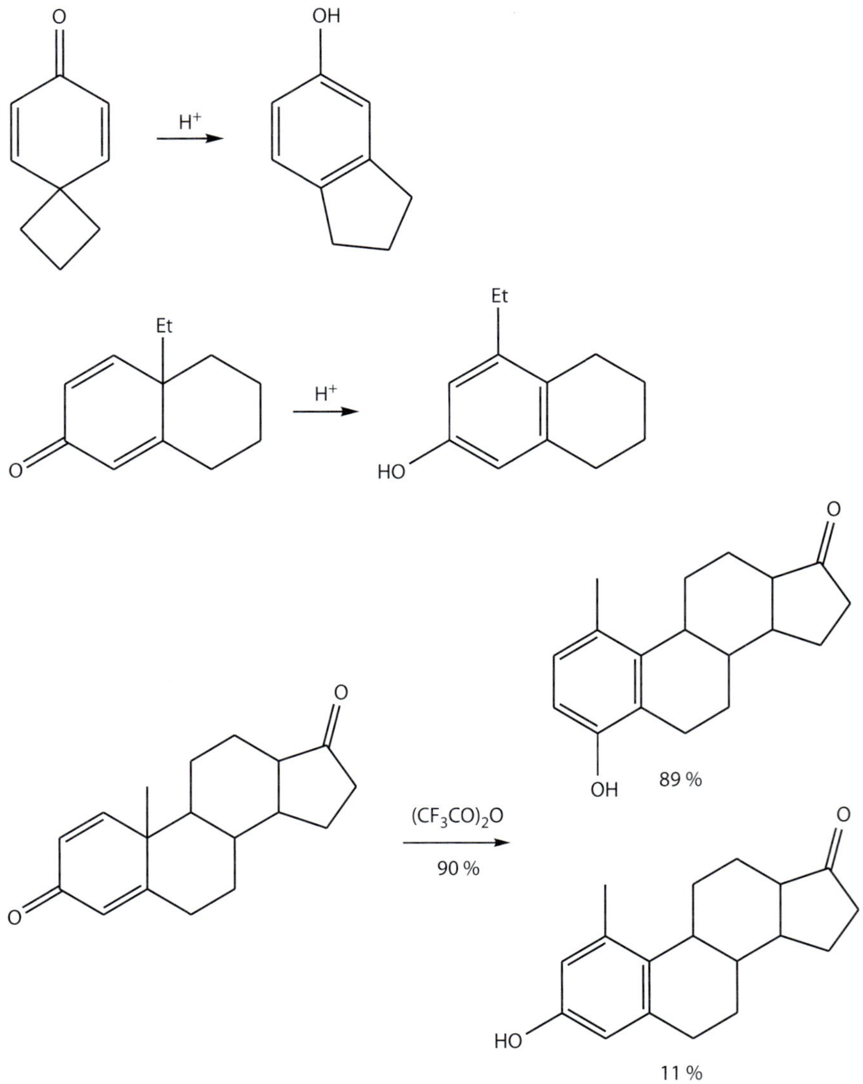

FIGURE 18.11 Dienone–phenol rearrangements.

Numerous versions of this reaction are known, some involving simple alkyl migrations, as shown earlier, while others involve successive movements of bonds in rings, to give more profound restructuring. Examples are shown in Figure 18.11. The final example is particularly interesting. It's quite easy to see how the minor product is formed, but forming the major product requires several more steps. While we can draw a sensible mechanism for both processes, predicting which one would predominate is not so easy.

PROBLEM 18.5

Give reasonable mechanisms for the formation of each of the products in the last example of Figure 18.11.

Solution

Formation of the minor product is an entirely conventional process, with the migration of the methyl group:

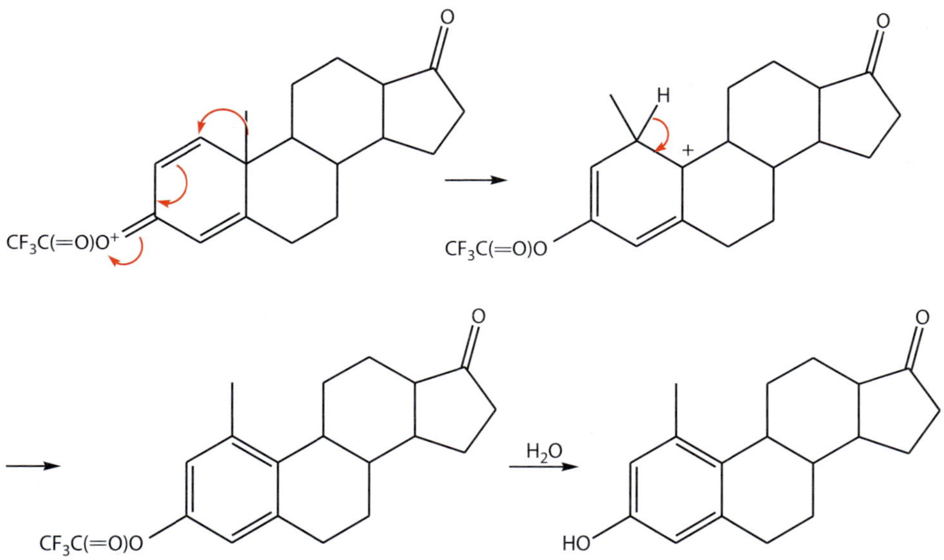

To form the other major product, we should first notice that in this product, the relationship between the methyl group and the oxygen is unchanged, which suggests that the methyl group did not migrate. The other option is to migrate one of the bonds of the adjacent ring. If we try to do this at the top of the molecule, we end up with a rather strained system, with no obvious place to go next—so we will migrate the ring bond downwards, to give a spirocyclic intermediate cation:

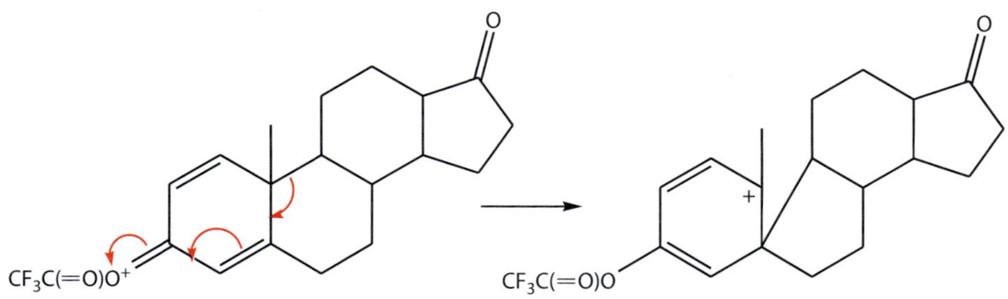

Writing another resonance form of the intermediate allows us to migrate one of the bonds of the spirocycle, reforming a six-membered ring, and able to lose a proton to complete the aromatization. Obviously, as drawn, the molecule looks very odd—in writing the mechanisms for the steps of the rearrangement, bonds and charges have been moved, but atoms haven't. This is deliberate and a good strategy—trying to move the atoms as you go is often confusing and leads to mistakes. "Unwind" the molecule at the end, starting with the parts that have not changed and working out what the actual sizes of the rings you have made are and how it would look best. When we do that here, we find that what we have made is the product required (hydrolysis of the trifluoroacetate ester occurred in the work-up of the reaction):

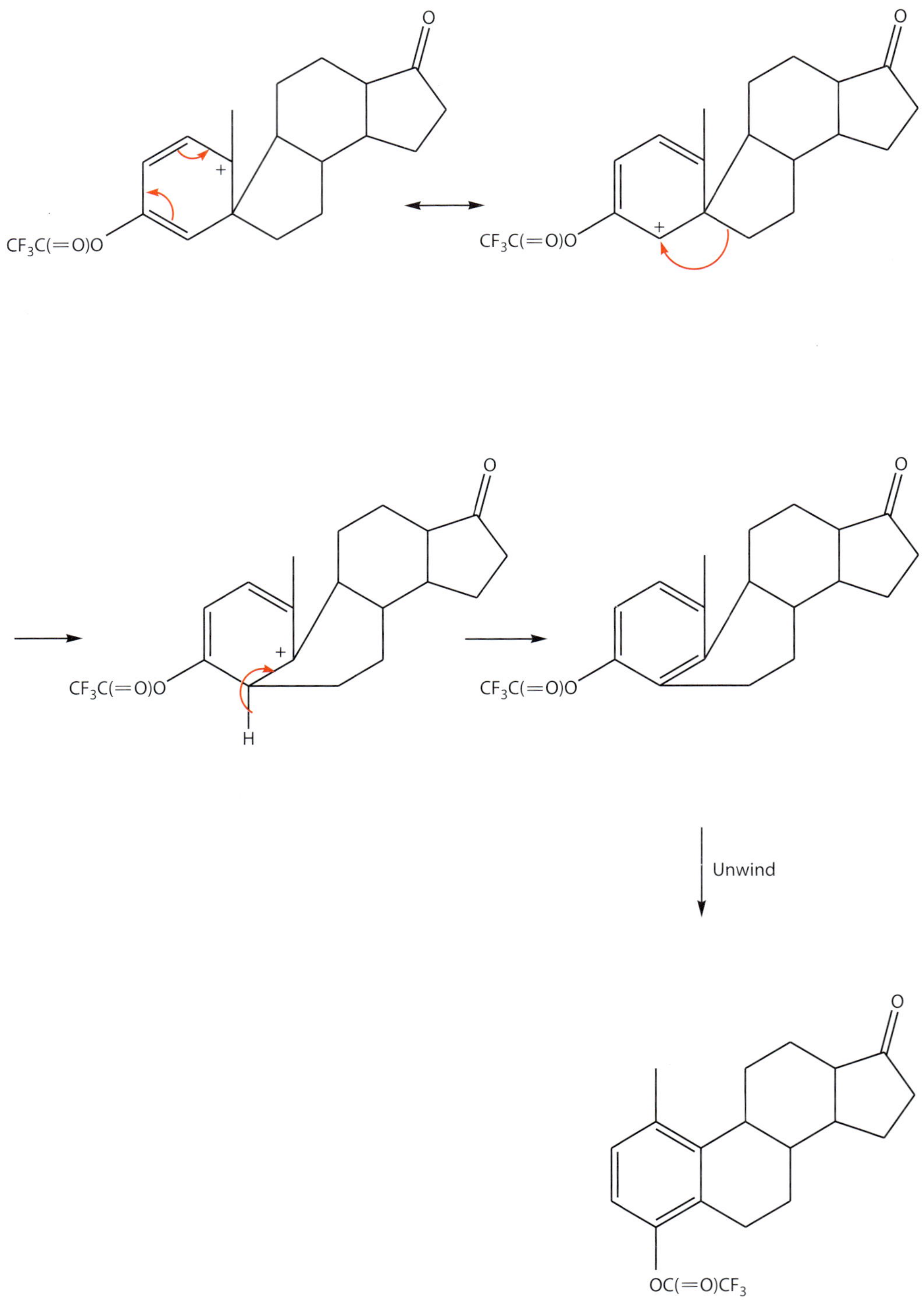

Focus on the Norbornyl Cation

The rearrangement in Figure 18.12, giving *iso*-bornyl chloride, has been known for over a century. The reaction mechanism can be written in classical terms; the right answer is achieved, but should cause us some worries. The migration involves forming a secondary carbocation from a tertiary one, and a single stereoisomer of the product is obtained, for no reason that is readily obvious. There is clearly more to this process than was immediately obvious when it was discovered.

The nature of the norbornyl cation (the unsubstituted version of the rearrangement) was a major controversy of the mid-twentieth century. It had been established that solvolysis of norbornyl brosylates (like tosylates but with bromine rather than methyl at the *para*-position of the benzene ring) depended strongly on stereochemistry. When the leaving group was *exo* to the norbornyl system, the reaction proceeded 350 times faster than for the *endo*-isomer (Figure 18.13). An interesting feature of the process was that the *exo*-isomer gave only an *exo*-product, but the product was racemic.

The structure of the intermediate norbornyl cation was variously proposed as two equilibrating cations (H.C. Brown) or a single nonclassical species (George Olah) (Figure 18.14). Both of these postulates would explain the stability of the norbornyl cation, which is comparable with that of a tertiary carbocation. How can we distinguish something that equilibrates very rapidly, from something that is halfway between the two structures all the time? The controversy raged for many years, incidentally driving the development of physical techniques to study carbocations. Ultimately, Olah was right—one of the key experiments was solid-state NMR spectroscopy at 5 K, which indicated that any barrier to an equilibrium was less than 1 kJ mol^{-1} and that the rate of such a process at 5 K must be faster than a bond vibration. At this point, the case was won—any distinction of the intermediate from a nonclassical, 3-center-2-electron intermediate is pure semantics. The norbornyl

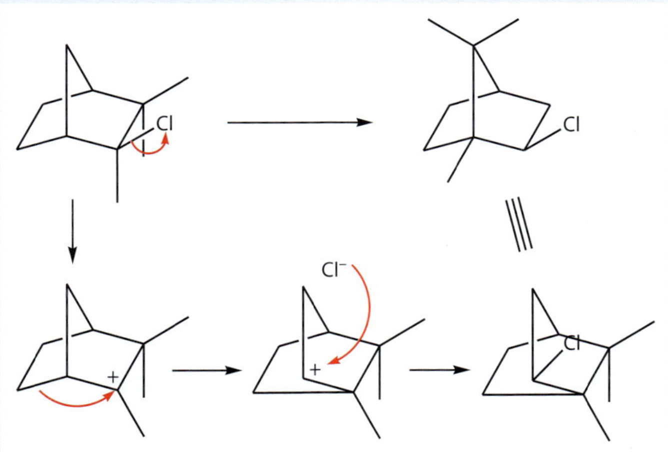

FIGURE 18.12 Rearrangement in the bornyl system.

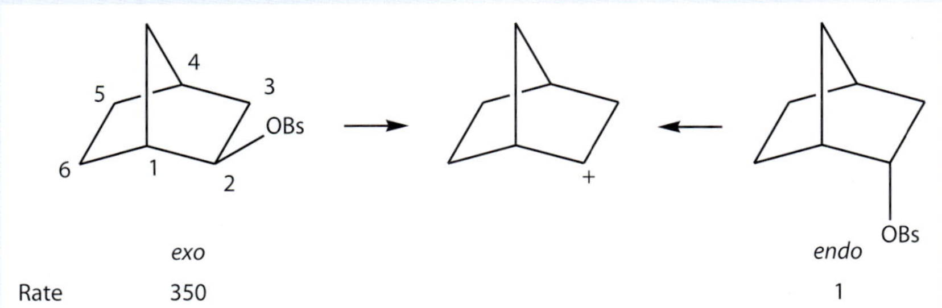

FIGURE 18.13 Rearrangements in the norbornyl system.

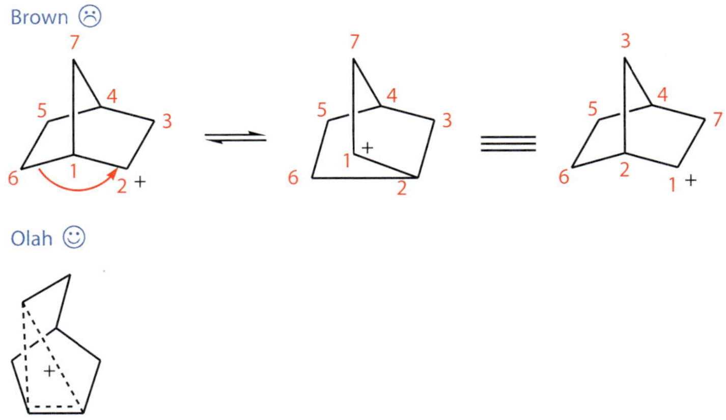

FIGURE 18.14 Proposed structures for the norbornyl cation.

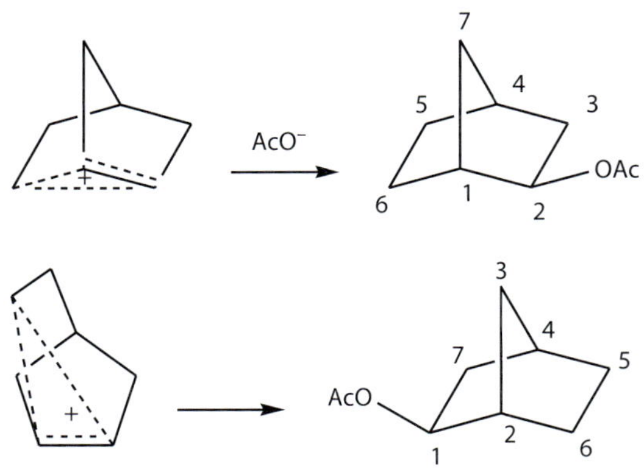

FIGURE 18.15 Reaction of the norbornyl cation with acetate.

cation is a genuine nonclassical species, and the formation of the racemic acetate is readily explained (Figure 18.15). Many other nonclassical cations have been postulated—although few have been subjected to the rigorous physical scrutiny endured by their parent, there is little doubt of their structure.

18.3 MIGRATION TO ELECTRON-DEFICIENT NITROGEN

18.3.1 BECKMANN REARRANGEMENT

This is a rearrangement of oximes (produced from aldehydes and ketones using hydroxylamine) (Figure 18.16). The OH of the oxime is protonated to make it into a leaving group, and migration of the group that is *trans* to the departing water gives an intermediate nitrilium ion. This is attacked by water, and the intermediate tautomerizes to give an amide. The water becomes truly free and exchanges with free solvent; if the reaction is carried out in partially ^{18}O-labeled water, the appropriate percentage of ^{18}O is incorporated into the carbonyl oxygen atom of the product amide. The intermediate nitrilium ion, **18.7**, is stabilized by resonance in the same way as was its oxygen analogue, the acylium ion.

The reaction is strictly stereospecific; only the group that is *trans* to the OH can migrate, so stereoisomeric oximes give distinct products. Any reagent that can convert the OH into a good leaving group will be successful; the reaction of Figure 18.17 used PCl_5, then the halogenated intermediate is hydrolyzed. If the migrating group, R, is chiral, the stereochemistry is retained in the migration. This reaction provides another amide, and hence another amine, synthesis, as we recall that amides can be reduced to amines using $Li[AlH_4]$.

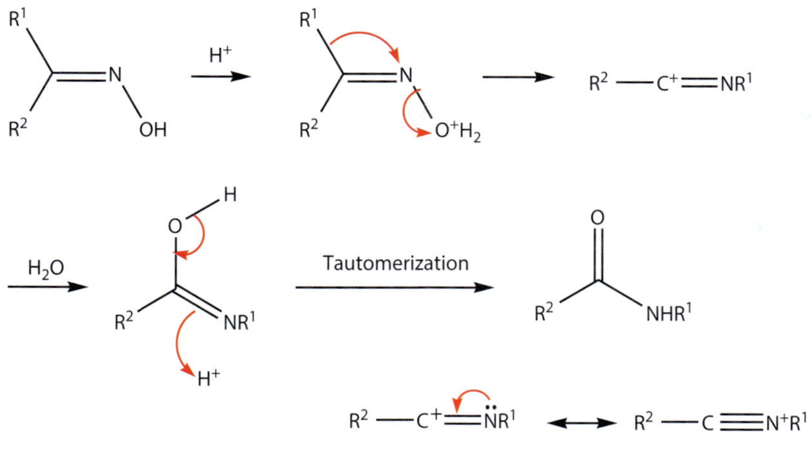

FIGURE 18.16 Mechanism of the Beckmann rearrangement.

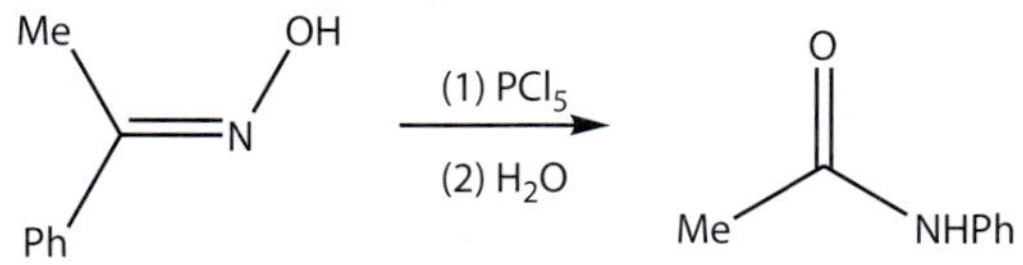

FIGURE 18.17 Beckmann rearrangements are stereospecific.

PROBLEM 18.6

Write a complete reaction mechanism for the reaction of Figure 18.17.

Solution

The first step involves displacement of chloride from PCl_5 by oxygen, making the OH into a good leaving group. Migration of the methyl group as the $POCl_3$ and Cl^- leave gives the nitrilium ion. This reacts initially with chloride. When water is added, this species is hydrolyzed to the amide:

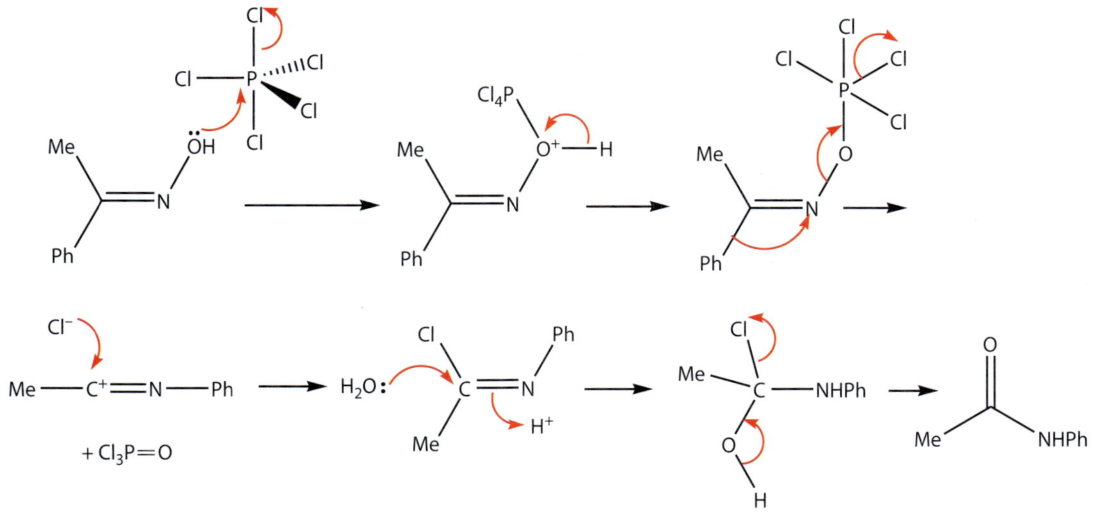

18.3.2 HOFMANN DEGRADATION: A CHAIN-SHORTENING REACTION

We have met a number of reactions that added one or two carbons to a chain; this is the first where one will be removed. The reaction mechanism is quite complex, but it is normally a one-pot process and the outcome straightforward; an amide is converted to an amine,

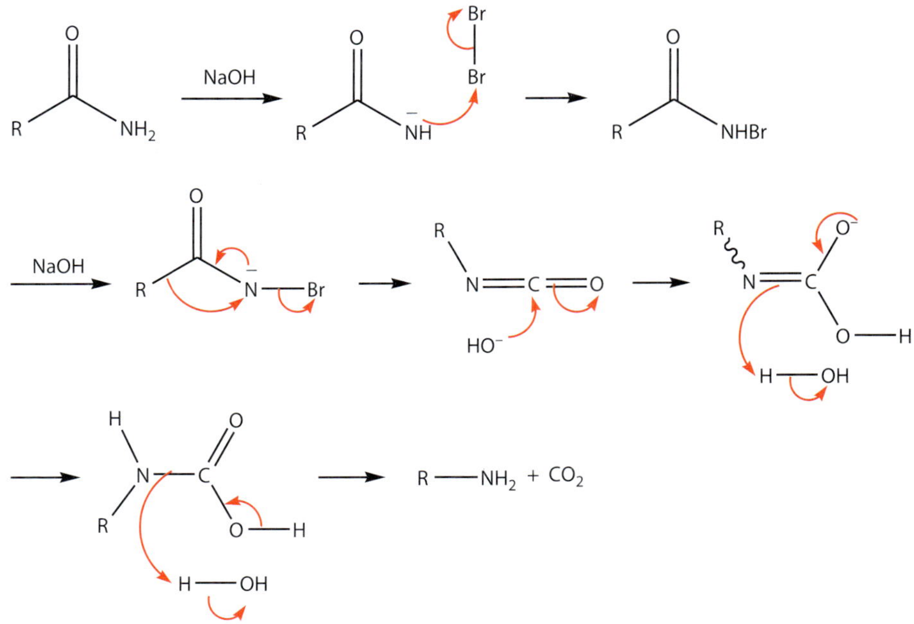

FIGURE 18.18 Mechanism of the Hofmann degradation.

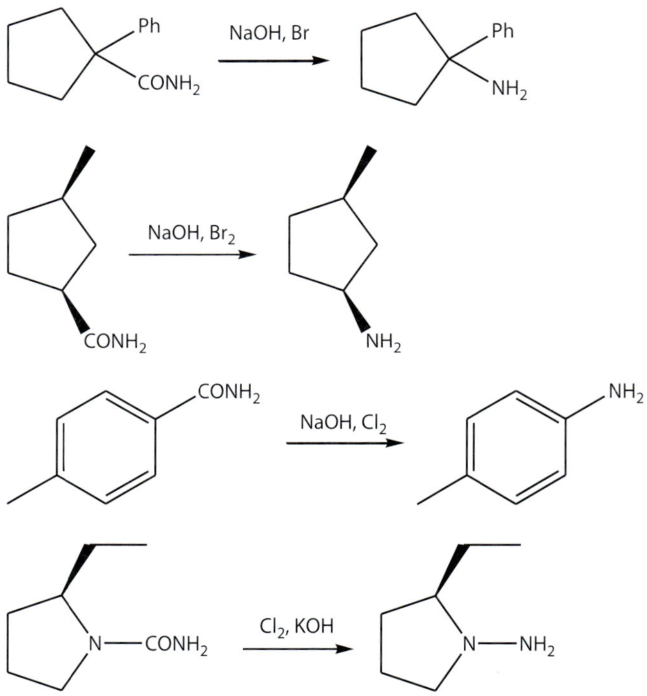

FIGURE 18.19 Examples of the Hofmann degradation.

essentially by removing the carbonyl group (chalk up another amide to amine synthesis). The amide is reacted with bromine and sodium hydroxide. In the first step, the amide is deprotonated and brominated at nitrogen. A further deprotonation gives an anion, which rearranges to RN=C=O, an isocyanate. This is attacked at carbon by the hydroxyl ion, to give a carbamic acid (RNHCOOH), which readily decarboxylates (Figure 18.18). Neither the isocyanate nor the carbamic acid is particularly stable, but they can both be isolated from the reaction, and when they are put back in again, they do convert to product—they are true intermediates. You may be wondering why this reaction is categorized with "migration to electron-deficient nitrogen." Although the mechanism as drawn is correct, at one point, it was thought that loss of bromide ion *preceded* the migration of the alkyl or aryl group to nitrogen. This would give an intermediate containing a nitrogen atom with just one substituent and six electrons—called a *nitrene*, by analogy with carbenes. This clearly is electron deficient! Some examples of the reaction are given in Figure 18.19; alkyl, aryl, and even nitrogen atoms can migrate. The second example shows that the reaction goes with retention of stereochemistry, and from examples 3 and 4, we see that chlorine may be substituted for bromine.

PROBLEM 18.7

Write a mechanism for the following reaction:

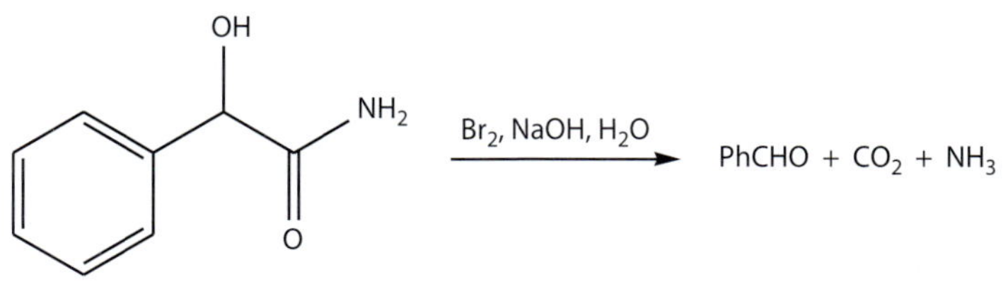

Solution

This is clearly a Hofmann degradation, but something else has happened as well. If we first write out the mechanism of the Hofmann degradation, the last step will become obvious. The loss of CO_2 gives a species that is simply the adduct of ammonia and benzaldehyde—and we know from carbonyl chemistry that this is not stable under aqueous conditions:

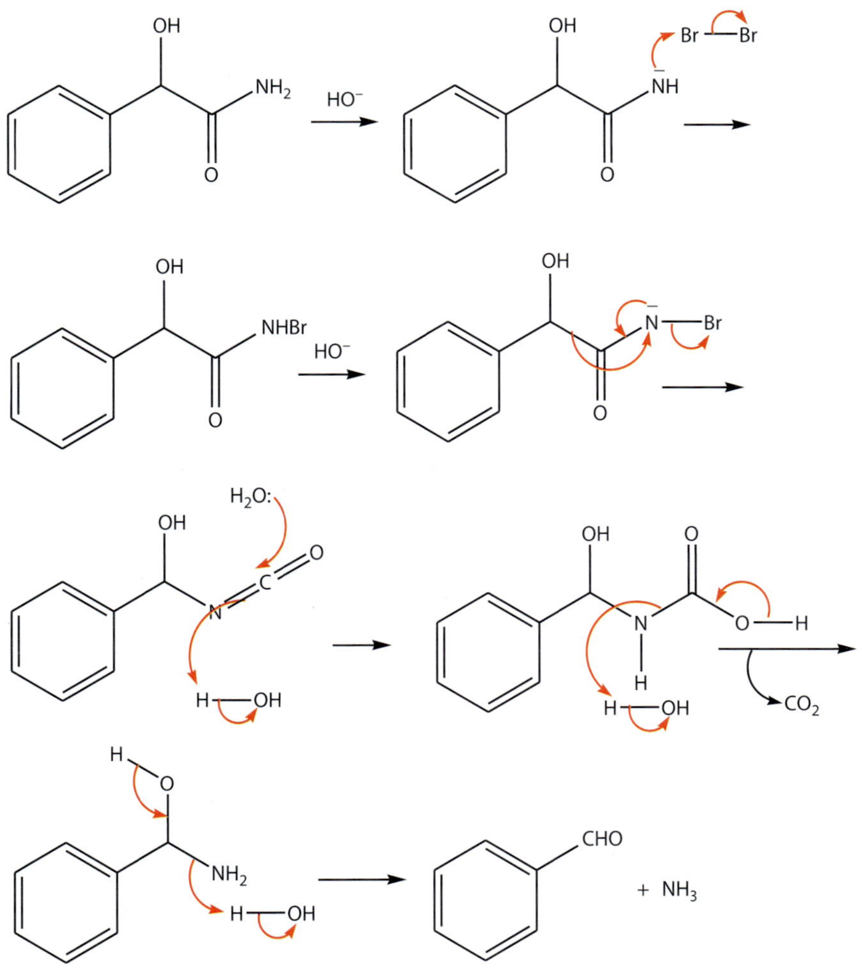

18.4 MIGRATION TO ELECTRON-DEFICIENT OXYGEN: THE BAEYER–VILLIGER OXIDATION

The Baeyer–Villiger oxidation is a method for conversion of a ketone to an ester—although the mechanism is different, the overall *effect* is of inserting an oxygen atom in between one of the substituents on the ketone and the carbonyl group (Figure 18.20). The reagents and reaction conditions vary, but in every case, a compound with a weak oxygen–oxygen bond is involved—and this bond is broken. Figure 18.20 shows a generic mechanism with R–O–OH—this may be a hydrogen peroxide, an alkyl peroxide, or a peracid. (This reaction might remind you of the boron to oxygen migration that we met when we studied the hydroboration of alkenes.) The reaction works well with hydrogen peroxide, but this is rarely the reagent of choice. Both OH groups of hydrogen peroxide can add to ketones, and analogues of explosive cyclic peroxides such as TATP, **18.8** (triacetone triperoxide, colloquially known as "Mother of Satan"), are formed as by-products.

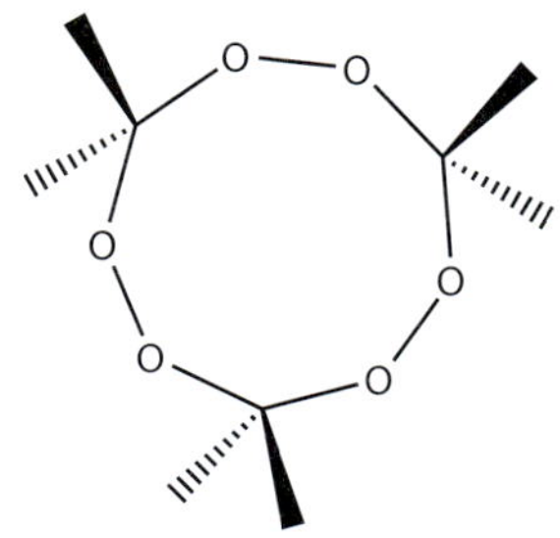

18.8

Although you may again be questioning why this reaction is classed as a migration to an electron-deficient atom, the substituent effects do bear this out. If you were to think of it as having the OR group depart before the migration, the oxygen would indeed be positive, which is obviously unlikely—but the group that migrates is always that which could best bear a positive charge at the transition state. Some examples are given in Figure 18.21. In the first example, the molecule is symmetric, so the same outcome occurs whichever group migrates. In each of the other two, the more substituted group migrates. In the third example, which was used in a prostaglandin synthesis, the Baeyer–Villiger reaction occurs in preference to the epoxidation of the double bond. The final example shows the migration of a chiral center, with retention of configuration.

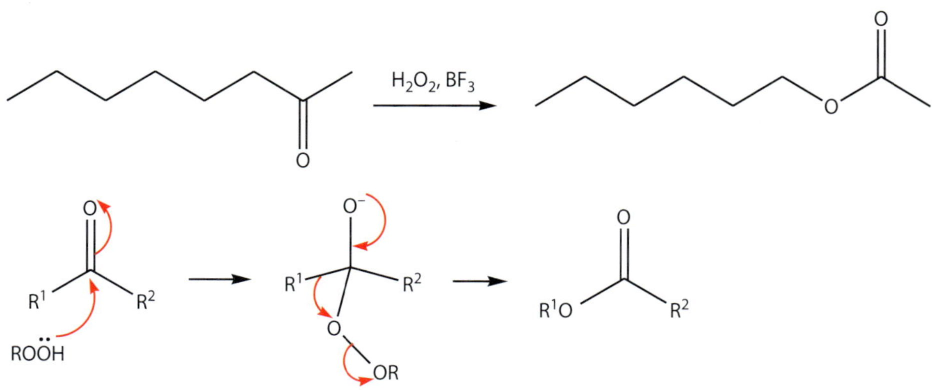

FIGURE 18.20 Baeyer–Villiger oxidation.

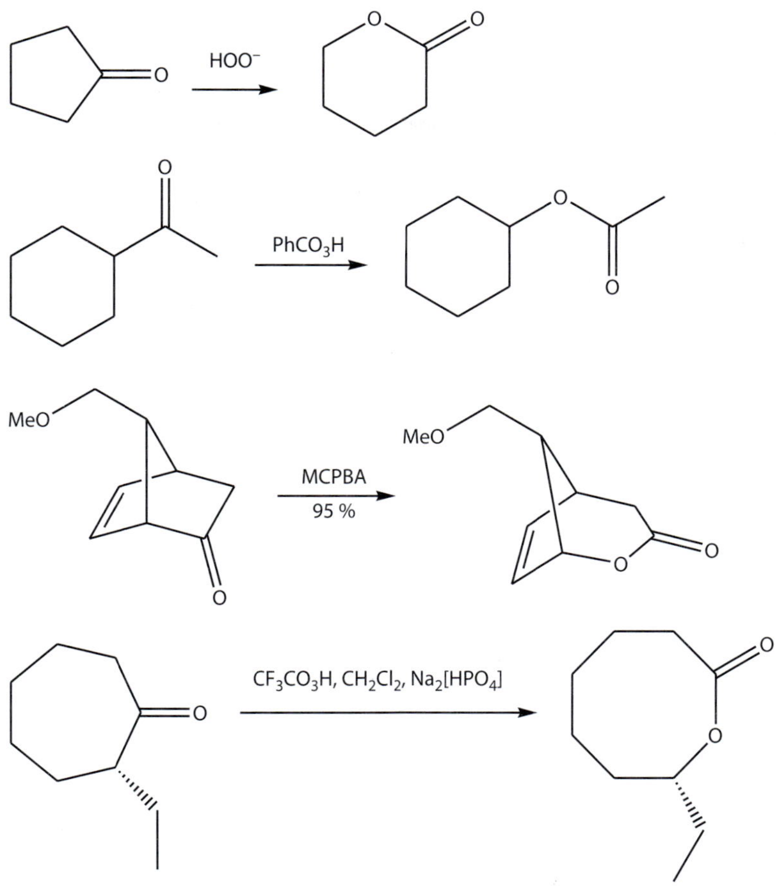

FIGURE 18.21 Examples of the Baeyer–Villiger reaction.

PROBLEM 18.8

Predict the product(s) of each of the following reactions:

(a)

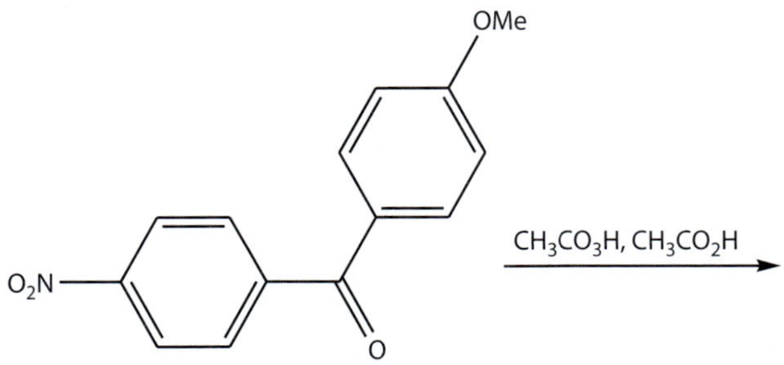

(b)

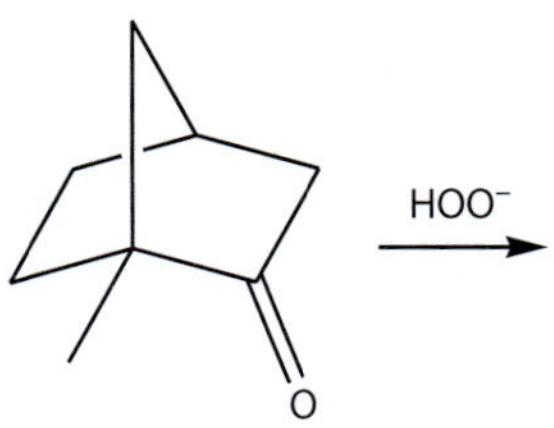

(c)

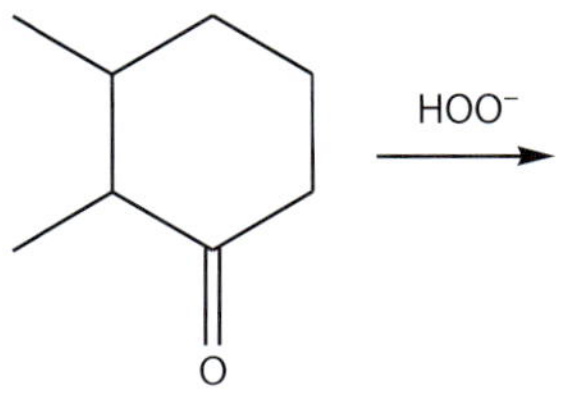

Solutions

(a) The aromatic ring bearing the electron-donating methoxy group is better able to support a positive charge at the transition state, so it is this ring that migrates, and the product is

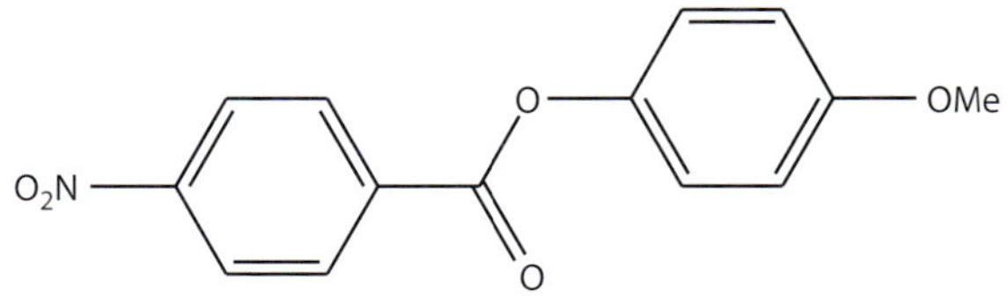

(b) The more substituted carbon atom migrates to give

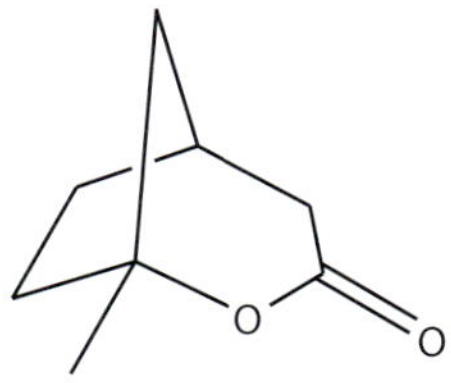

(c) Again, the more substituted carbon atom migrates to give

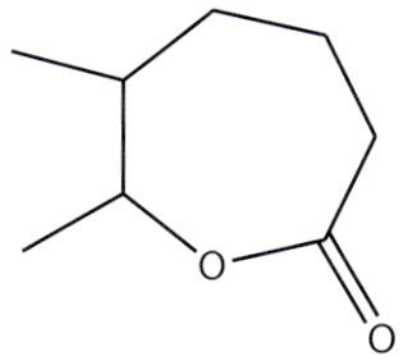

Key Points from Sections 18.2 through 18.4

- Carbocation rearrangements can occur in substitution, addition, elimination, and Friedel–Crafts reactions. The rearrangements involve conversion of a less stable to a more stable carbocation.
- The migratory aptitude of a group in migrations to electron-deficient atoms (C, O, or N) depends on its ability to support a positive charge at the transition state. Stabilization and any irreversible reactions of the product are also important.
- 1,2-Diols rearrange in acid solution to give ketones—pinacol rearrangement.
- 2-Amino alcohols rearrange in the presence of nitrous acid to give ketones. The intermediate is an aliphatic diazonium salt, and the reactions are concerted and stereoelectronically controlled, with the migration of the group *trans–anti* to the departing nitrogen.
- 4,4- or 2,2-disubstituted cyclohexadienones rearrange to give phenols in acid conditions.
- Oximes rearrange stereospecifically in acid solution, with migration of the group *trans* to the OH, to give amides (Beckmann rearrangement).
- Reaction of amides with Br_2/H_2O gives amines with one fewer carbon atom (Hofmann degradation).
- Reaction of ketones with various peroxy reagents yields esters (Baeyer–Villiger oxidation).

18.5 ANIONIC REARRANGEMENTS

Anionic rearrangements are somewhat less common than those involving cations, and we will consider only two here. The first is the benzylic acid rearrangement—the reaction of benzil with strong base (Figure 18.22). Migration of the phenyl group is induced by attack of the base on the carbonyl group, and the process is rendered irreversible by the proton transfer in the final step. In principle, the reaction can occur for any 1,2-diketone (as in Figure 18.23), but if the ketone is enolizable, then other processes are also possible and often predominate.

The Favorskii rearrangement involves α-haloketones. The accepted mechanism (Figure 18.24) in most cases requires the formation of the enolate (albeit *not* the most favorable one) and ring closure to give a cyclopropanone. The cyclopropanone (which is very strained because of the difficulty of accommodating an sp[2] carbon atom in a three-membered ring) is then opened by hydroxide or alkoxide to give a carbanion, which is finally reprotonated. The intermediacy of a symmetric species is proved by ^{13}C-labeling experiments—the label (* in the figure) is equally distributed between the two possible positions of the product.

When enolization is not possible, a similar process may still occur, but the mechanism is different, and it is sometimes described as a semibenzylic acid rearrangement; Figure 18.25 illustrates the process. The carbonyl is attacked by the base, as for the benzylic acid rearrangement, but this time, the migration displaces halide, as in the Favorskii. Some further examples are given in Figure 18.26; the last one is of the semibenzylic acid type.

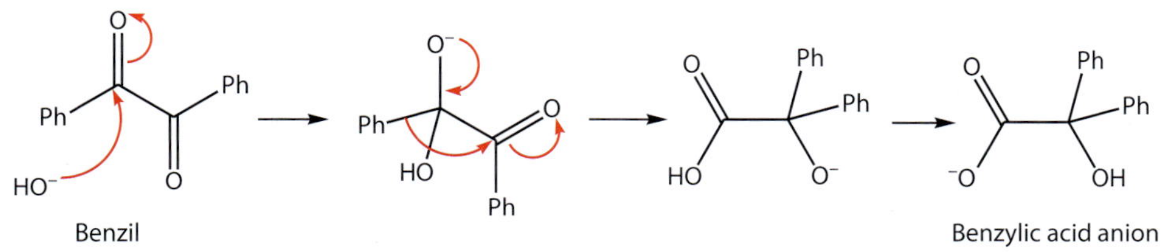

FIGURE 18.22 Mechanism of the benzylic acid rearrangement.

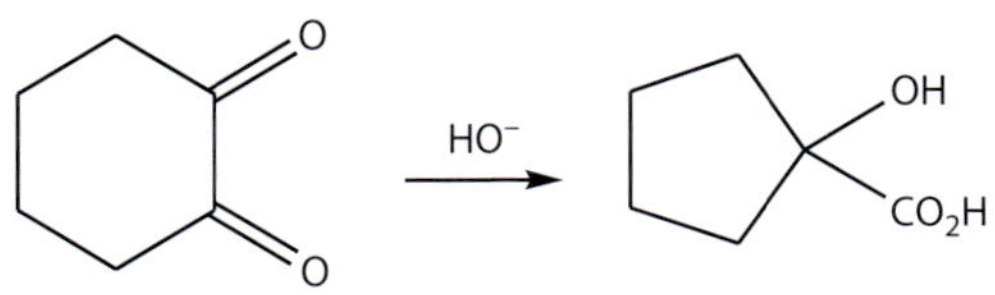

FIGURE 18.23 Benzylic acid-type rearrangement of a 1,2-diketone.

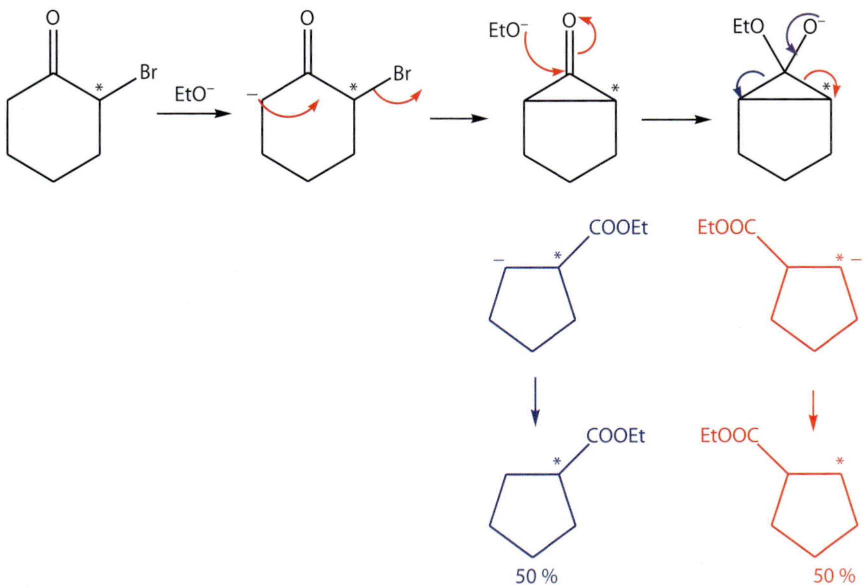

FIGURE 18.24 Mechanism of the Favorskii rearrangement.

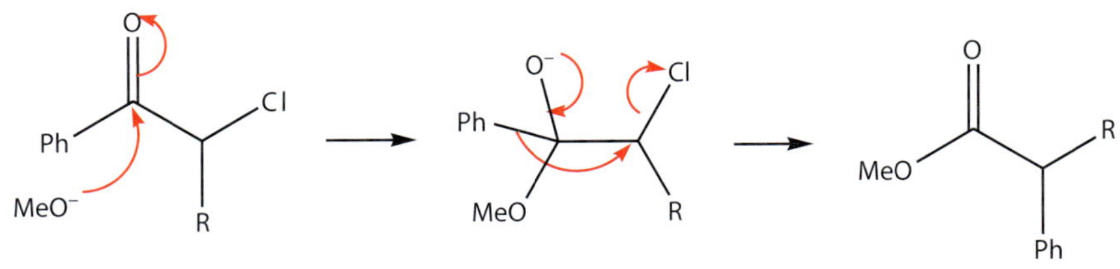

FIGURE 18.25 Favorskii/semibenzylic acid rearrangement.

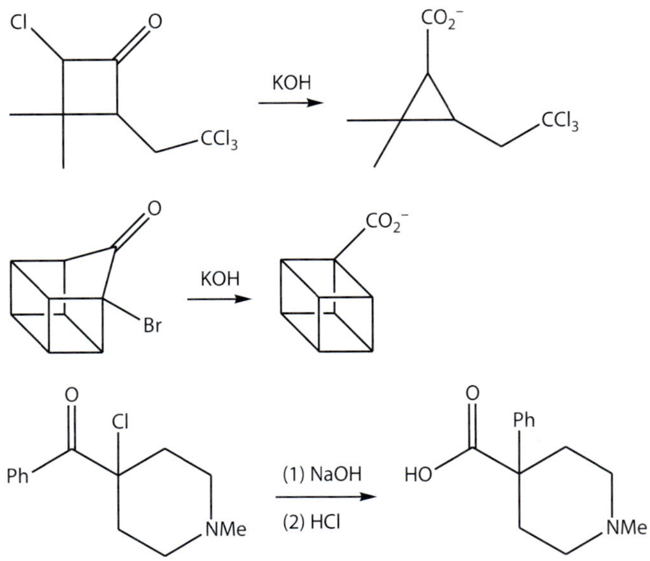

FIGURE 18.26 Examples of the Favorskii rearrangement.

PROBLEM 18.9

Write a mechanism for each of the following reactions:

(a)

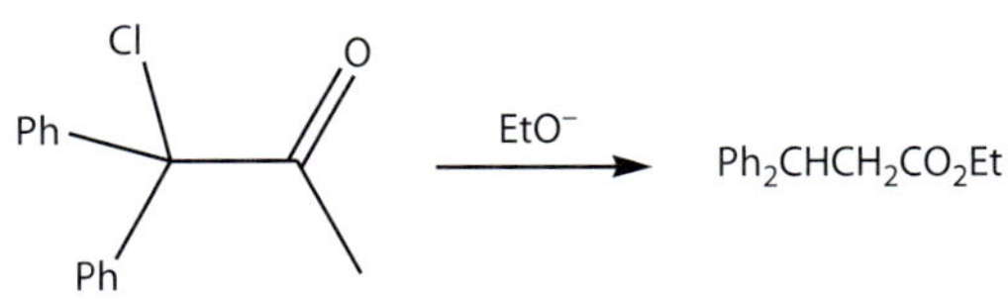

(b)

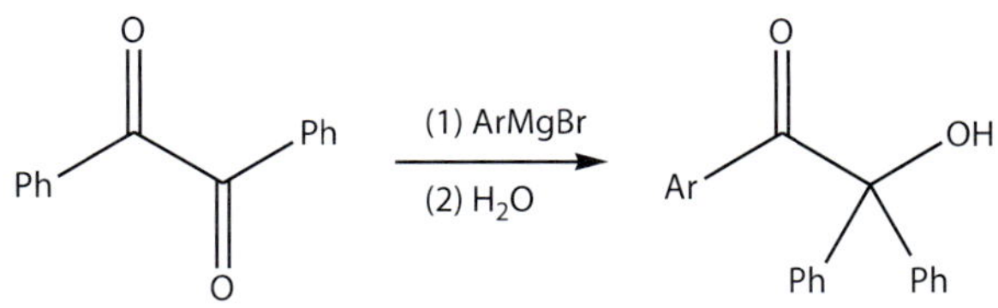

Solutions

(a)

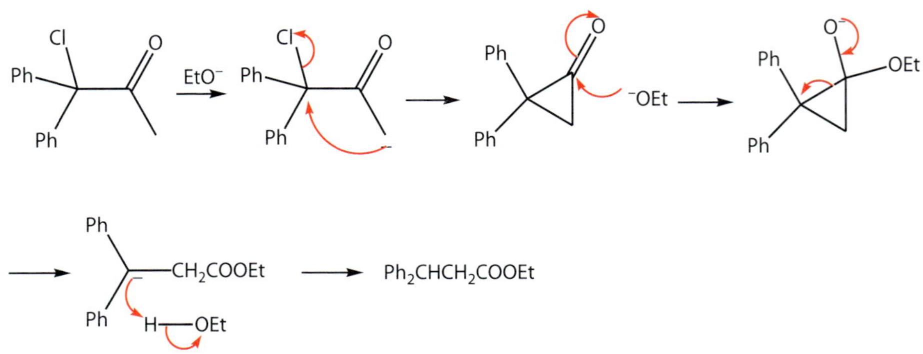

(b) This is clearly related to the benzylic acid rearrangement—the best way to proceed is to do what we are absolutely sure about—the Grignard will attack one of the carbonyls to give a tetrahedral intermediate—this collapses with migration of the phenyl group, and the reaction is completed by work-up:

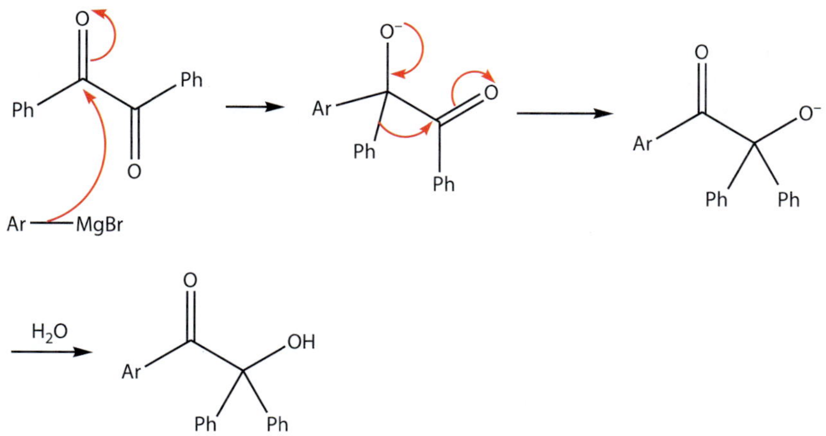

PROBLEM 18.10

Predict the product(s) of the following reaction:

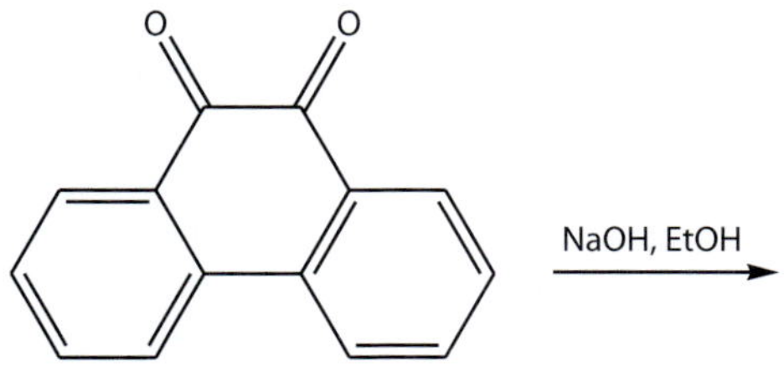

Solution

This is a classic benzylic acid rearrangement, and the reaction proceeds just as it does for benzil—the fact that the two rings are bonded makes no difference:

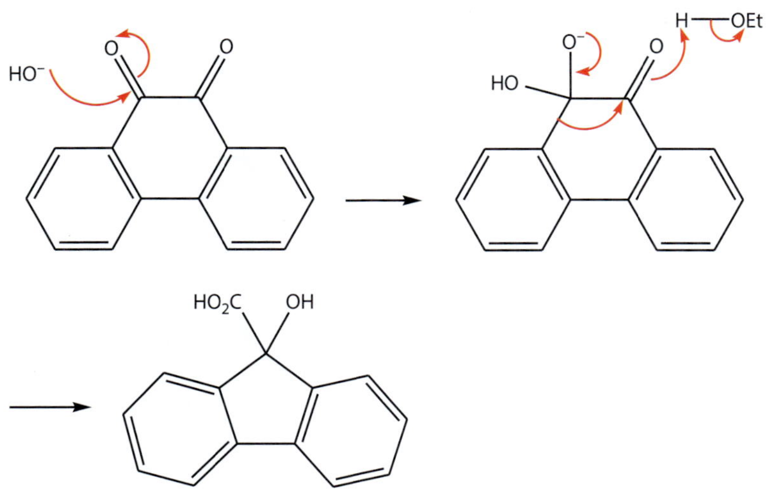

Key Points from Section 18.5

- 1,2-Diketones rearrange in base to 2-hydroxycarboxylic acids (benzylic acid rearrangement). Although possible in principle for all 1,2-diketones, this reaction is only common for nonenolizable species.
- The Favorskii rearrangement of α-haloketones in the presence of base yields carboxylic acids or carboxylate esters. The intermediate in the classic version of the reaction is a cyclopropanone, but an alternative mechanism, related to that of the benzylic acid rearrangement, operates in other cases, with analogous outcome.

18.6 NEUTRAL REARRANGEMENTS

Although most of this section will deal with rearrangements of neutral species, we may see a few anions and cations—the key common feature of this group of reactions is that, like Diels Alder and other cycloaddition processes, they are controlled by orbital symmetry.

18.6.1 ELECTROCYCLIC REACTIONS

Consider the two reactions in Figure 18.27. For each, we can draw reasonable curly arrows, and in these examples, all is well—they both work under thermal conditions. But if the reactions are repeated with some substituents, then differences become apparent. For thermal reactions, we need to consider the HOMO. IF we look at the HOMO of butadiene and consider how we need to rotate the p orbitals to make the new bond, we see that to bring them together in phase, we must rotate them in the same direction, as drawn here, both clockwise (Figure 18.28). This process is described as conrotatory, and conrotatory closure is the norm for all thermal, $4n$, electron processes.

The effect of this conrotatory ring closure becomes clear when we look at a substituted butadiene (Figure 18.29). The reactions are completely stereospecific—no mistakes, ever.

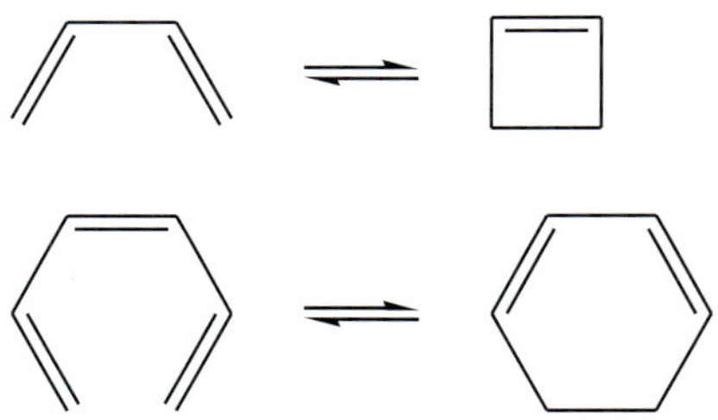

FIGURE 18.27 Electrocyclic reactions.

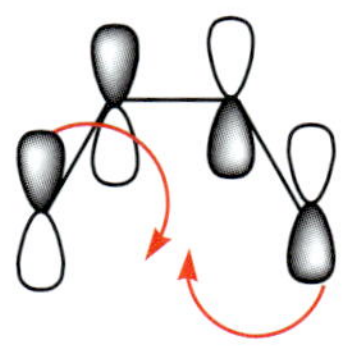

FIGURE 18.28 HOMO of butadiene.

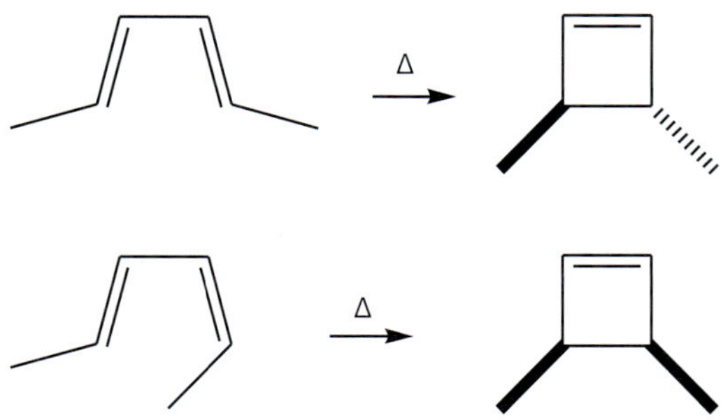

FIGURE 18.29 Thermal electrocyclic ring closures of substituted butadienes.

If we now consider the HOMO of hexatriene (Figure 18.30), where the closure reaction is a six-electron process, to bring the orbitals together to make the new σ-bond, the two orbitals must now rotate in opposite directions, one clockwise and one counterclockwise. This is a ***disrotatory*** process, and this rotation applies to all $4n + 2$ electrocyclic reactions. As always, it is easier to see what is happening, when there are substituents (Figure 18.31).

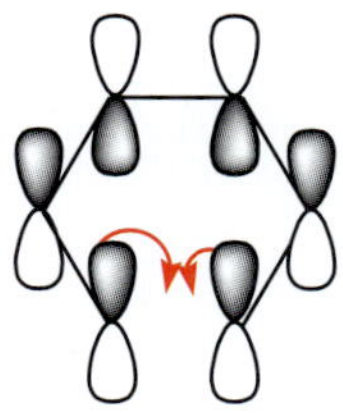

FIGURE 18.30 HOMO of hexatriene.

In photochemical reactions, we consider the LUMO (the light photon promotes electrons into the LUMO); the symmetry of this is invariably opposite to that of the HOMO, so 4e processes are disrotatory, and 6e processes are conrotatory, exactly the opposite of the thermal reaction. Figure 18.32 shows the LUMO of butadiene, and Figure 18.33 shows the photochemical ring closures of substituted butadienes.

Although 2,8, *etc.*, electron electrocyclic reactions are less common, they do exist, so it is useful to summarize the stereochemical outcomes quite generally (Table 18.1). You don't need to memorize the whole of this table—if you remember just one entry, and the fact that $4n$ and $4n + 2$ are different, and thermal and photochemical are different, then you have it all. Some examples of electrocyclic reactions are given in Figure 18.34.

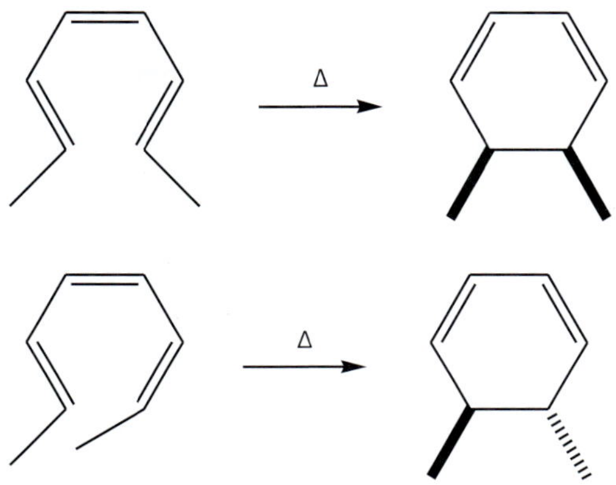

FIGURE 18.31 Thermal electrocyclic ring closure of substituted hexatrienes.

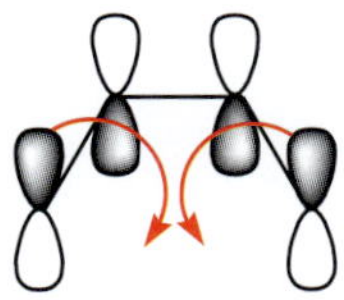

FIGURE 18.32 LUMO of butadiene.

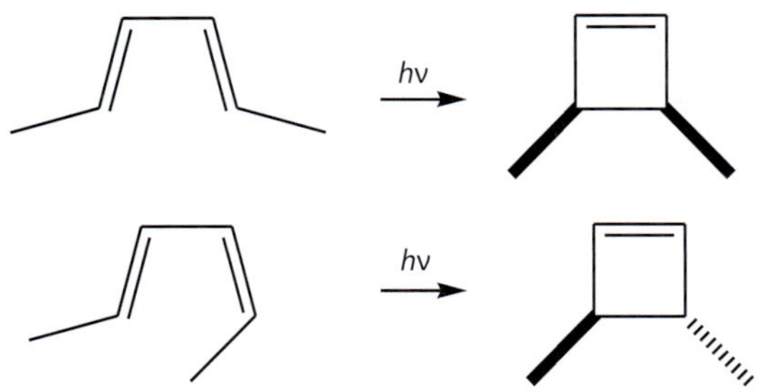

FIGURE 18.33 Photochemical electrocyclic ring closures of substituted butadienes.

TABLE 18.1 Stereochemical Outcomes of Electrocyclic Reactions

Number of Electrons	Conditions	Stereochemical Outcome
$4n + 2$	Δ	Disrotatory
$4n$	Δ	Conrotatory
$4n + 2$	$h\nu$	Conrotatory
$4n$	$h\nu$	Disrotatory

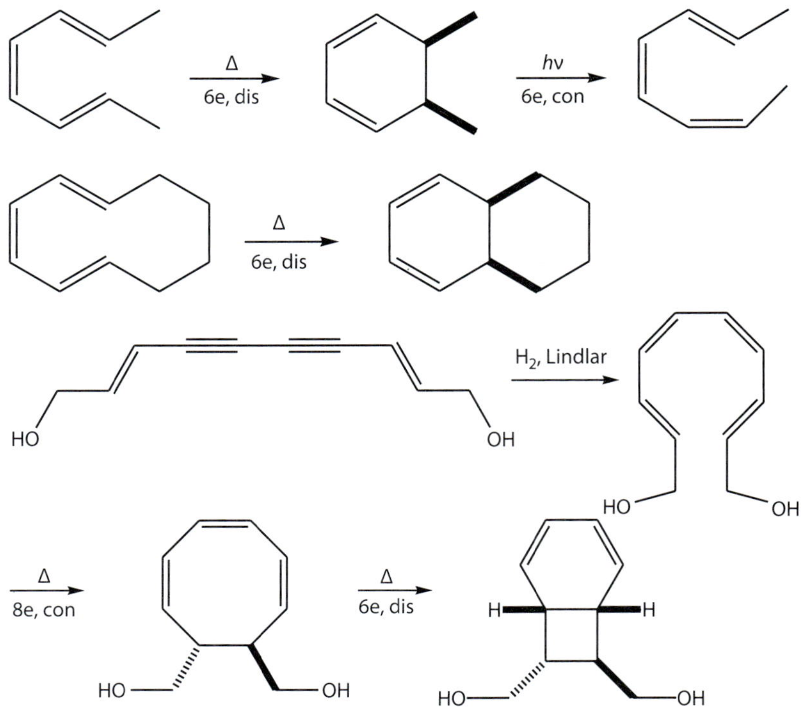

FIGURE 18.34 Examples of electrocyclic reactions.

PROBLEM 18.11

Predict the outcome of the following reactions:

(a)

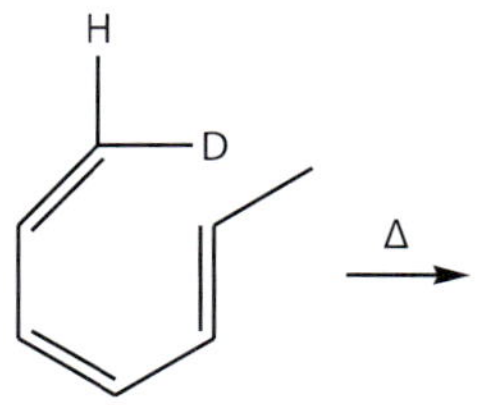

(b)

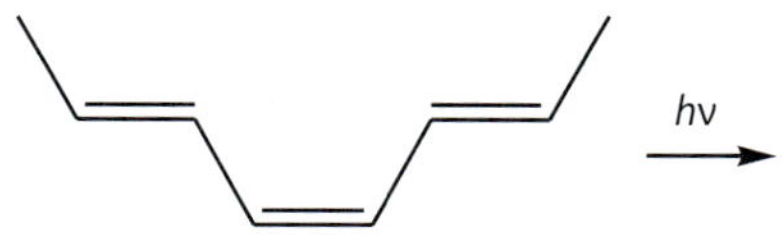

(c)

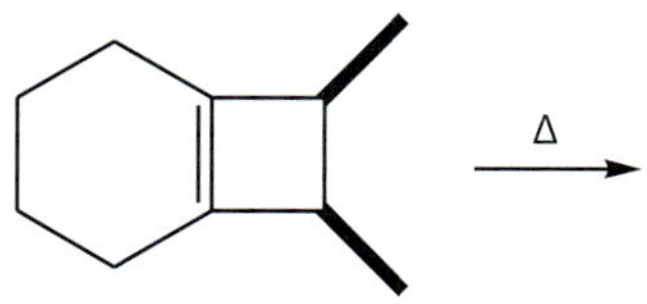

Solutions

(a) A six-electron process, under thermal conditions, so disrotatory:

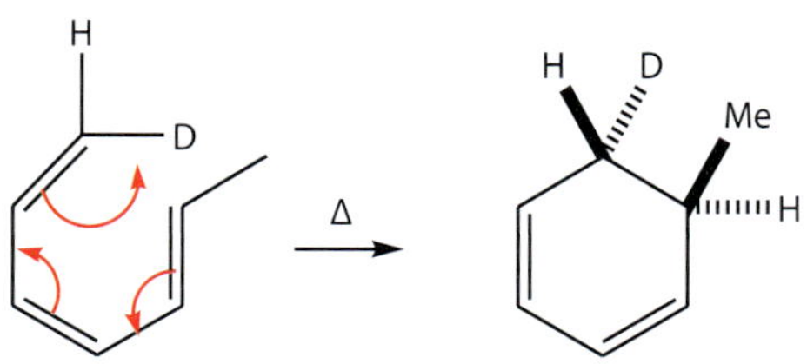

(b) We first need to redraw the molecule in a more suitable conformation for ring closure:

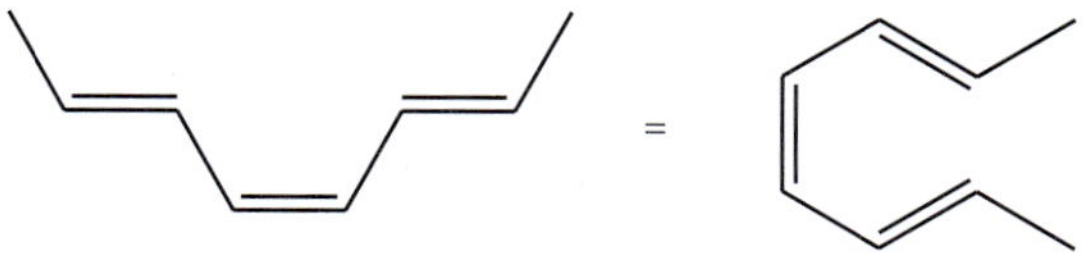

Next, a six-electron process, under photochemical conditions, hence conrotatory:

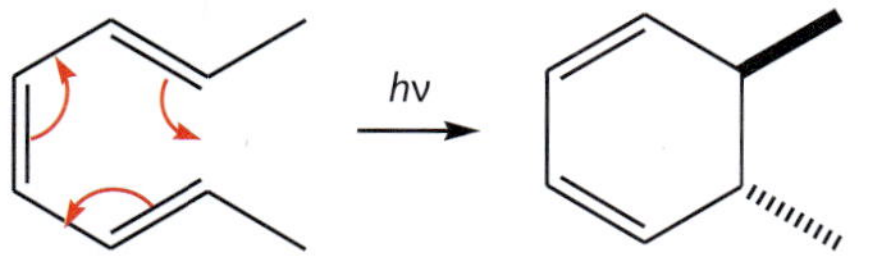

(c) A four-electron process, under thermal conditions, hence conrotatory. Don't be alarmed because the reaction is going in the opposite direction from the one you have seen before—the orbital symmetry considerations are just the same. If you feel less confident opening than closing the rings, that's not unusual; most students do. Check your result by running the reaction the other way, under the same conditions, and make sure that you get back to the SM.

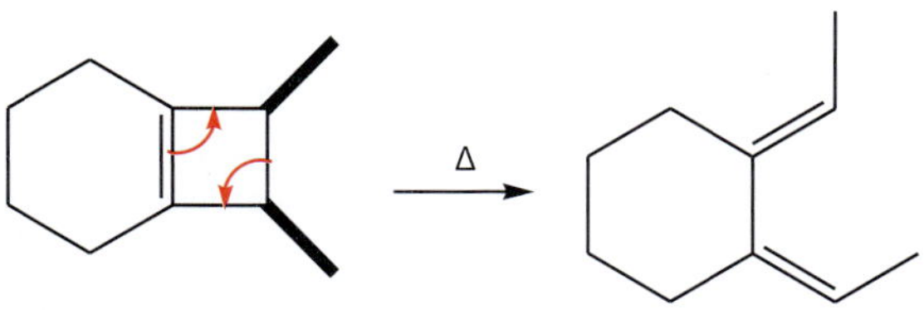

18.6.2 SIGMATROPIC REARRANGEMENTS

Sigmatropic rearrangements are a class of reaction in which an σ-bond is moved effectively "across" a π-system to a new position. Figure 18.35 shows a Cope rearrangement and is a 3,3-sigmatropic process (note this is not an equally balanced equilibrium; the material with the disubstituted double bond is more stable). Figure 18.36 shows a 1,5-sigmatropic shift, of a hydrogen atom, from one end of the molecule to the other. The numbering of these rearrangements often seems confusing. The simplest way to approach the problem is to number the atoms from the σ-bond that is being broken toward its new position (ignore substituents). The numbers at the site of the new σ-bond give the so-called order of the reaction. So for the Cope rearrangement (**18.9**), we see where the 3,3-designation arises. This also works for the 1,5-proton shift—there is only one atom, the hydrogen, to be numbered in the migrating unit, and it moves to atom 5 in the chain (**18.10**).

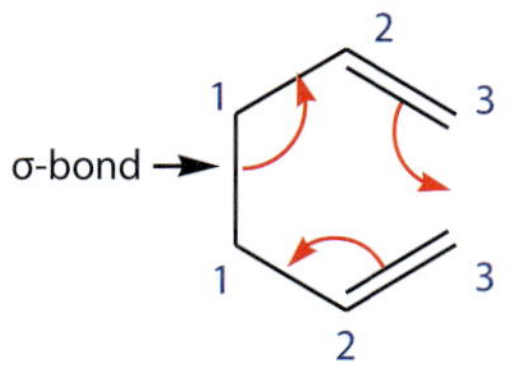

18.9

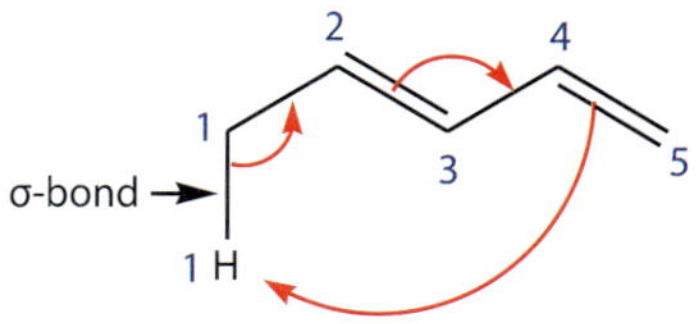

18.10

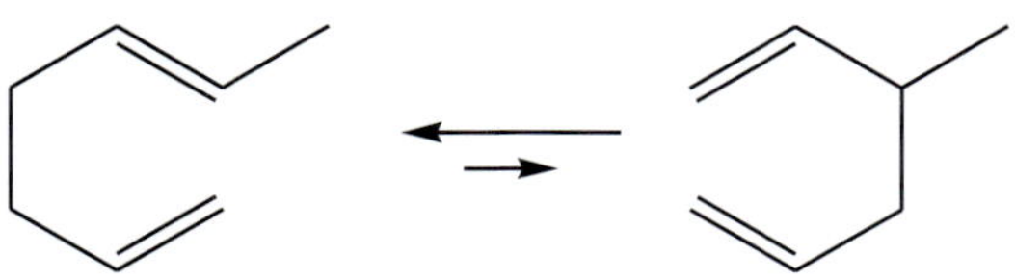

FIGURE 18.35 Cope rearrangement, a 3,3-sigmatropic process.

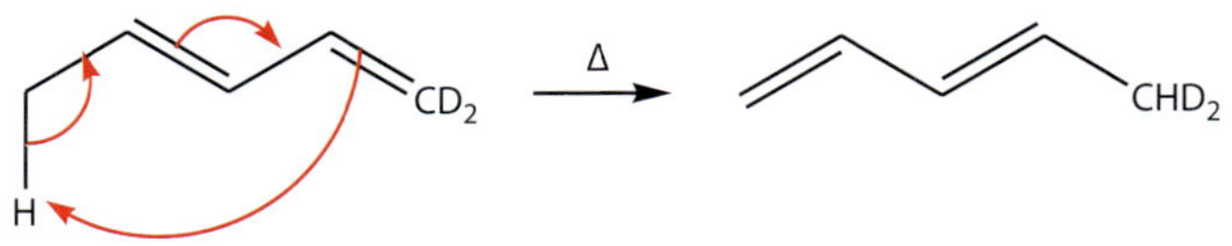

FIGURE 18.36 1,5-Hydride shift.

The first reactions to consider are 1,3- and 1,5-hydrogen shifts. Although the 1,5-shift shown in Figure 18.36 works well, the 1,3-hydogen shift is disallowed thermally. In analyzing these reactions, the convention is that we consider one LUMO and one HOMO, and the HOMO is always the HOMO of the hydride ion, a 1s orbital. The LUMO is the lowest unoccupied molecular orbital of the allyl cation. Drawing these out as in Figure 18.37 shows why this reaction will not work. The hydrogen cannot move across the face of the allyl cation—a suprafacial process, because the orbitals are out of phase, and the "formally allowed" antarafacial process, where it would move from one face to the other, is geometrically impossible.

If we do the same analysis for the 1,5-hydride shift, however, all is well (Figure 18.38). Some examples of 1,5-hydride shifts are shown in Figure 18.39. Notice in the second example that the chirality at the original asymmetric carbon atom is fully transferred to the new chiral carbon, by the suprafacial migration.

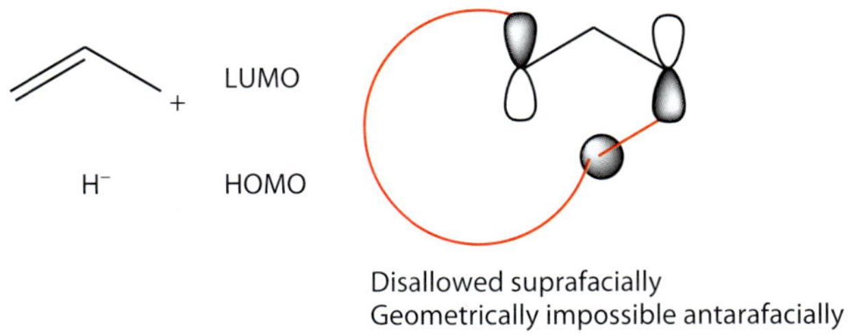

FIGURE 18.37 1,3-Sigmatropic hydride shift.

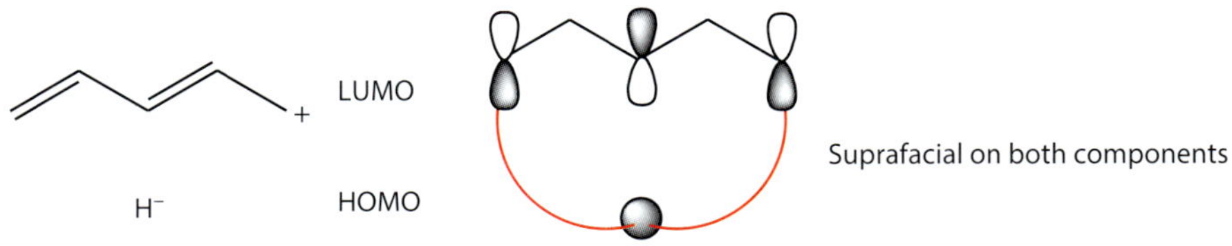

FIGURE 18.38 1,5-Sigmatropic hydride shift.

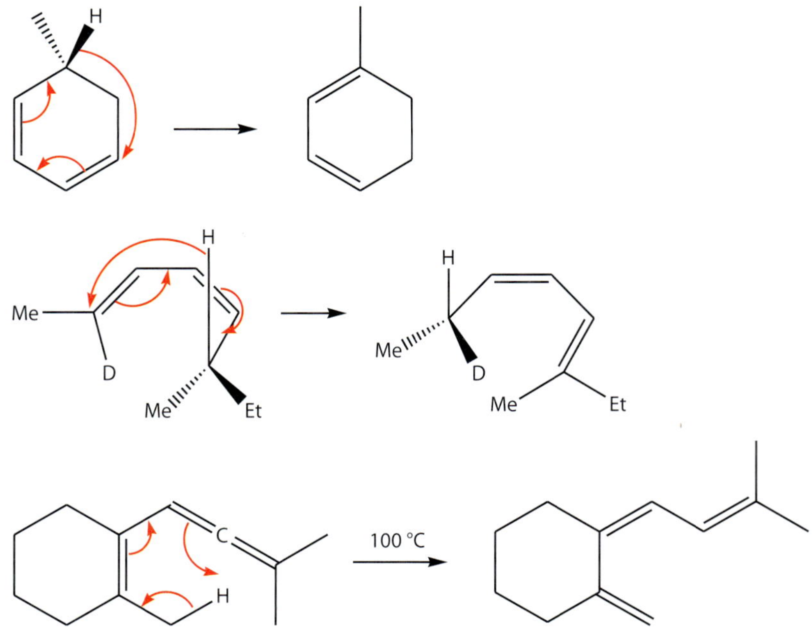

FIGURE 18.39 1,5-Sigmatropic hydride shifts.

Both 1,3- and 1,5-carbon shifts are allowed under thermal conditions, but the stereochemistry of the reaction is different. Considering the 1,3-shift first, we again use the LUMO of the allyl cation, but this time, the HOMO is the "lone pair" of electrons on the carbanion. Although we often don't bother to draw the back lobe of the sp^3 orbital in which this lone pair resides, this does not mean it is not there—indeed, we know that the stereochemistry of carbanions is unstable, and they invert rapidly. In that inversion, one lobe grows and the other shrinks—at the transition state for that inversion, the orbital will look like a simple p orbital. By using this idea, we can show that a 1,3-carbon shift is allowed, with inversion of stereochemistry at the migrating carbon atom (Figure 18.40). An example shows the inversion. To understand the inversion, imagine that the migrating bond is like the edge of a door, and think about swinging it counterclockwise—that gives the wrong stereochemistry (methyl comes forwards; hydrogen goes back), so it is clear that inversion has occurred.

The 1,5-shift of carbon is allowed thermally with retention of configuration at the migrating carbon atom. The orbitals of the LUMO of the pentadienyl cation at the relevant carbon atoms are in phase, so the HOMO of the carbanion must keep the same face toward them as it migrates (Figure 18.41). An example shows the retention—notice that if you simply swing the carbon around this time, the stereochemistry comes out correctly.

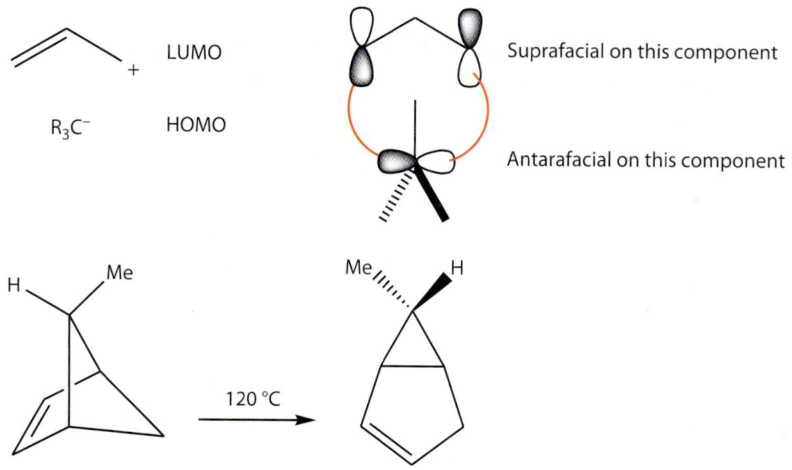

FIGURE 18.40 1,3-Carbon sigmatropic shift with inversion of configuration.

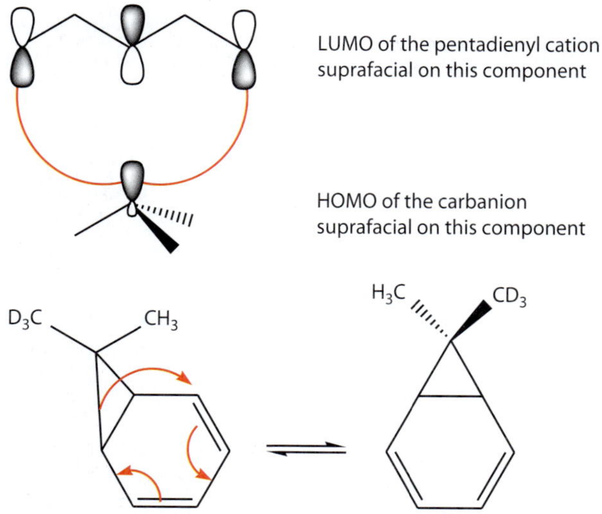

FIGURE 18.41 1,5-Carbon sigmatropic shift with retention of configuration.

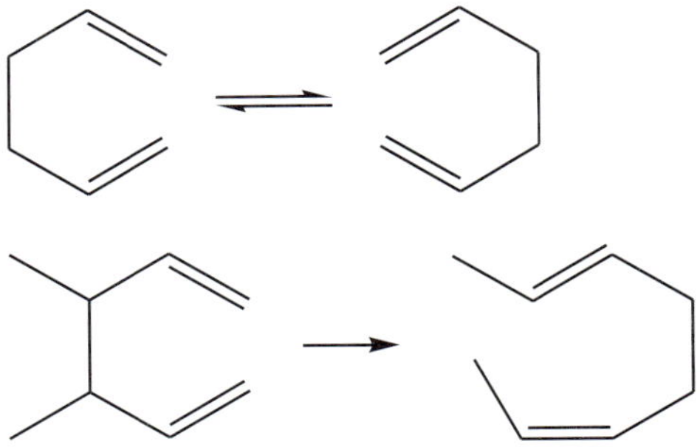

FIGURE 18.42 Cope rearrangements.

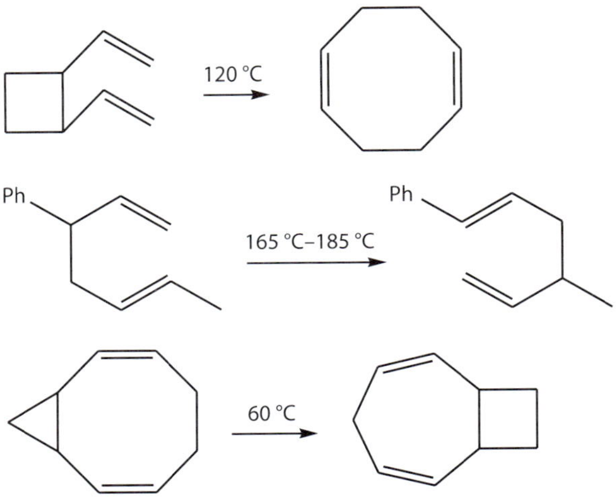

FIGURE 18.43 Examples of the Cope reaction.

The Cope rearrangement (Figure 18.42) is a 3,3-sigmatropic rearrangement. In the first example, the two sides of the equilibrium are identical other than atom labeling, but in the second, the structure with the more substituted double bonds is clearly favored. Although the reaction is generally stereospecific, this is not mandated by any orbital symmetry considerations, but by steric interactions at the transition state. Some examples are shown in Figure 18.43.

Generally, the positions of these equilibria reflect relative energies of the species in equilibrium—thus, in two of the examples of Figure 18.43, they are driven by relief of ring strain and, in the other, by conjugation of the double bond with the aromatic ring. However, if the substrate has an appropriately positioned OH group, the reaction can be rendered irreversible by tautomerization of the enol produced by rearrangement. This is called the oxa-Cope rearrangement, and it is markedly accelerated under basic conditions (up to 10^{17} times faster). Examples are given in Figure 18.44—in each case, the intermediate enol is shown, although it cannot normally be isolated.

When this type of rearrangement occurs with a substrate where there is an oxygen atom in the reacting unit, it is called the Claisen rearrangement, and it was first observed with allyl phenyl ethers (Figure 18.45). In this case, a ketone is produced in the rearrangement and immediately rearomatizes. Following a ^{13}C-labeled carbon atom (shown as the *) shows that the process is a true intramolecular rearrangement, rather than the allyl group becoming free and reacting with the benzene ring. If the *ortho*-positions of the benzene ring are blocked by substituents, then a further Cope rearrangement is needed before the ring can lose a proton and reestablish aromaticity (Figure 18.46)—the whole process is often called the *para*-Claisen rearrangement. However, the reaction is not confined to allyl phenyl ethers and has been widely used in synthesis (Figure 18.47).

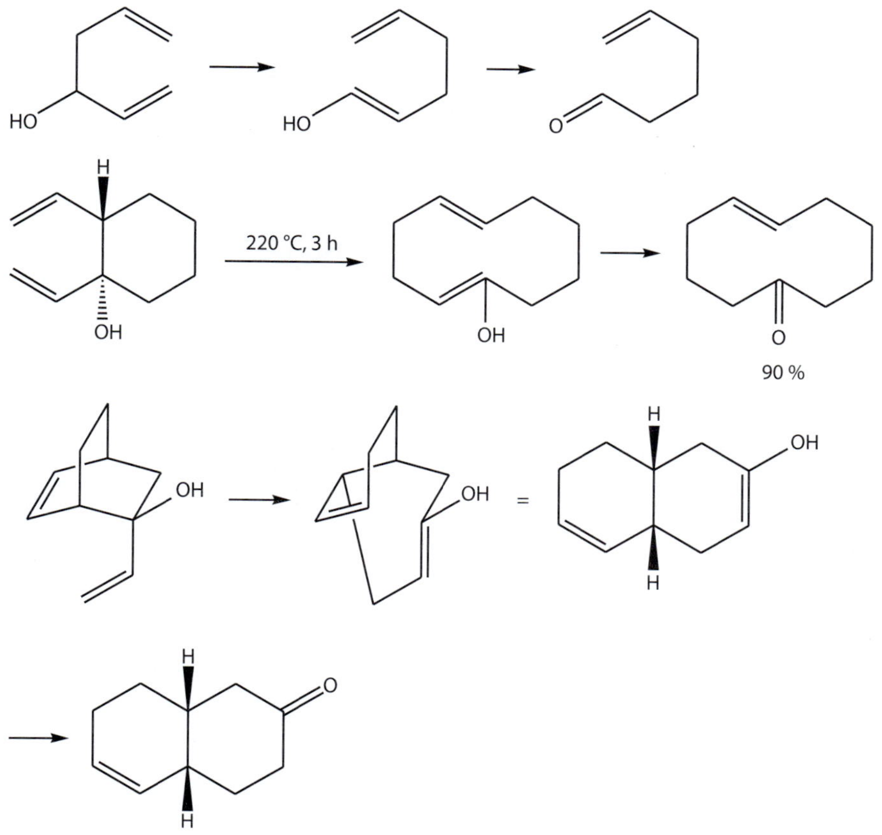

FIGURE 18.44 Examples of the oxa-Cope rearrangement.

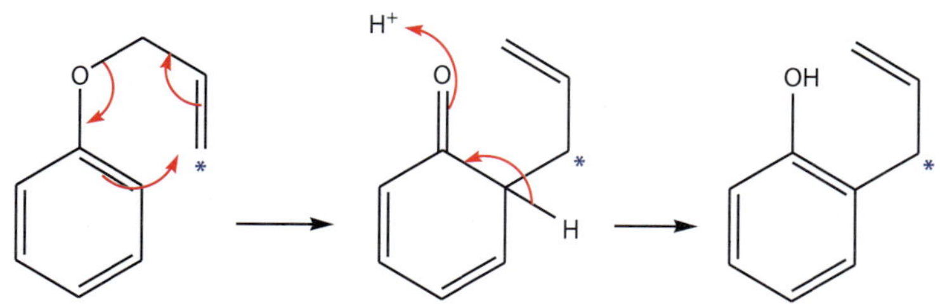

FIGURE 18.45 Claisen rearrangement.

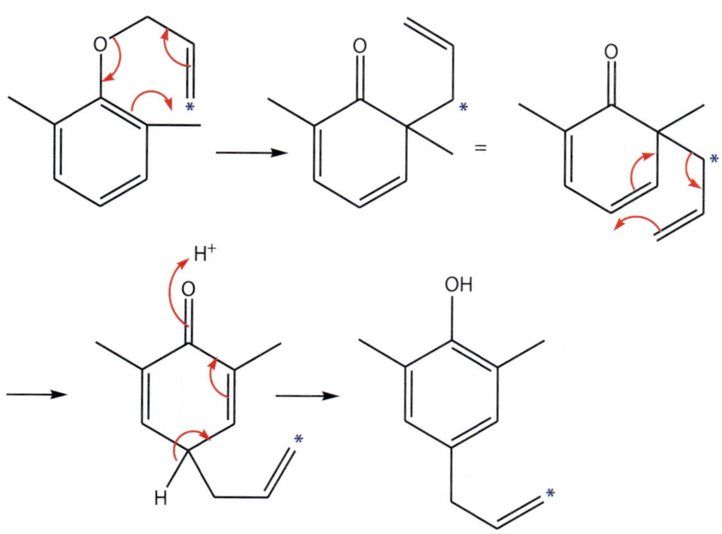

FIGURE 18.46 *Para*-Claisen rearrangement.

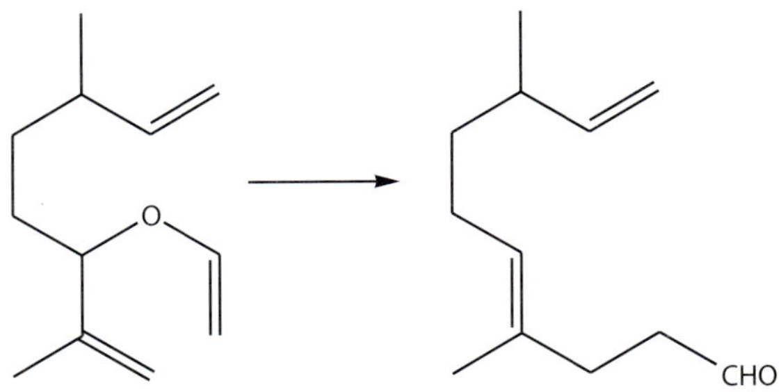

FIGURE 18.47 Claisen rearrangement of an allyl ether.

PROBLEM 18.12

Propose mechanisms for the following reactions:

(a)

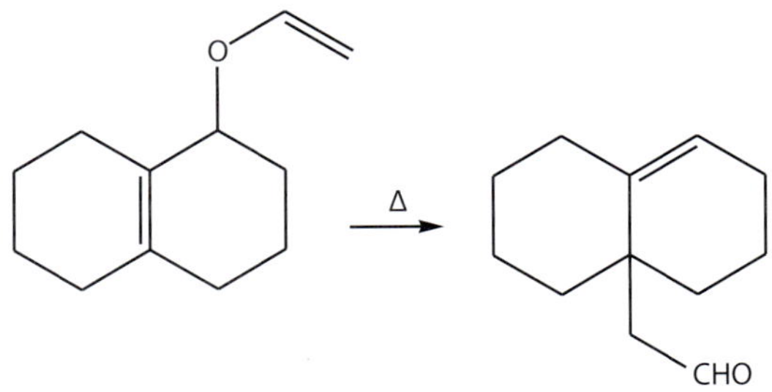

(b)

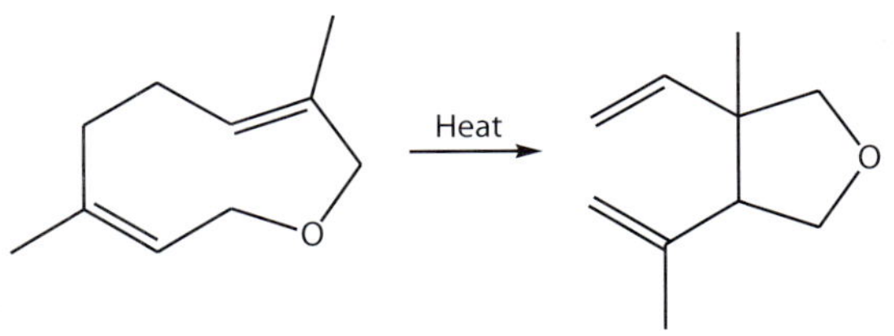

(c)

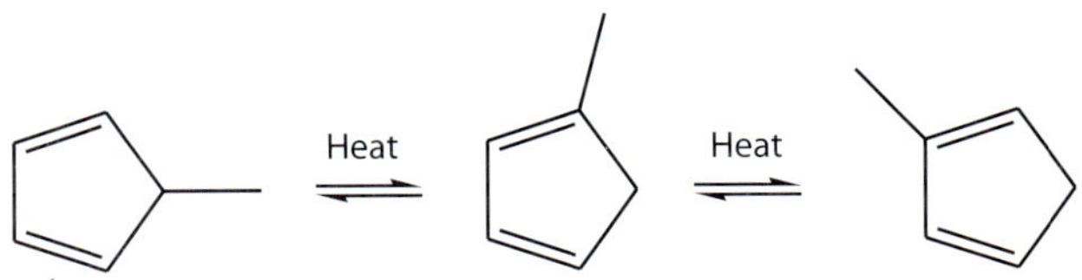

Solutions

(a) This is a Claisen rearrangement:

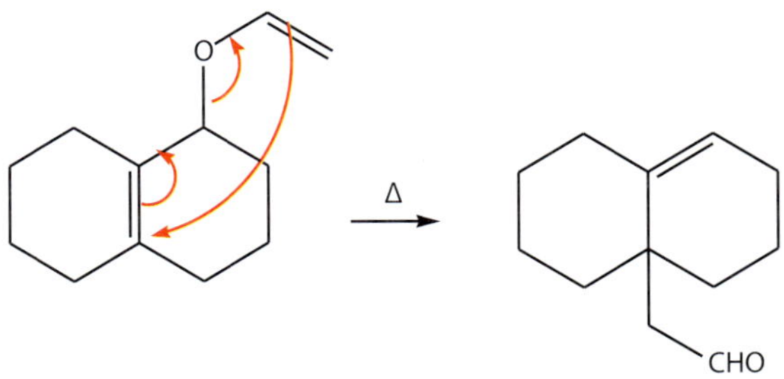

(b) Cope rearrangement—you need to redraw the molecule before it's in the right orientation to do the reaction:

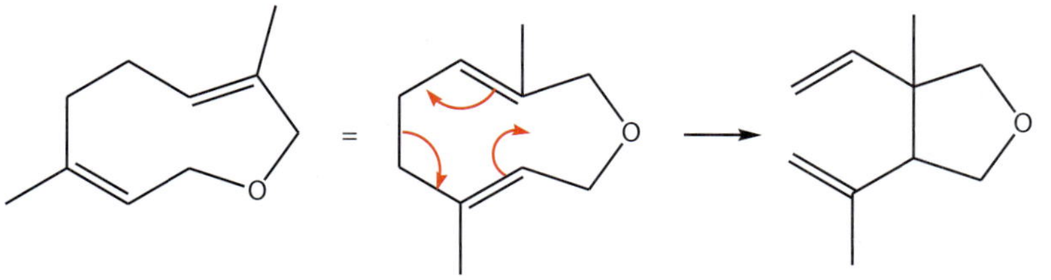

(c) These are all allowed 1,5-sigmatropic hydrogen shifts—although the methyl group may appear to be moving, this is not the case:

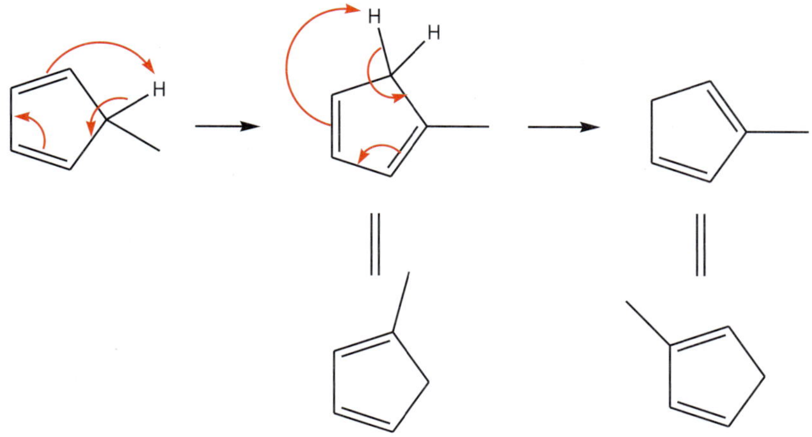

Focus on Vitamin D

You might imagine that the types of reactions discussed in this chapter have limited applications in biology or biochemistry. But that would not be so. Electrocyclic ring openings and sigmatropic hydride shifts are critical to the biosynthesis of the D vitamins. 7-Dehydrocholesterol undergoes a photochemical six-electron conrotatory ring opening to previtamin D3. This undergoes a slow, 1,7-antarafacial hydride shift to give vitamin D3. Vitamin D2 is produced in the same way, but starting from ergosterol (Figure 18.48). The D vitamins are essential to health, and it is difficult to obtain enough from a standard diet. There is now some concern about vitamin D deficiency—fear of skin cancer keeps us covered up or out of the direct sun, and sunlight is essential to the biosynthesis of these vitamins. The most obvious effect of vitamin D deficiency is the bone disease rickets, but there is also evidence that the incidence of diseases such as multiple sclerosis, Parkinson's and Alzheimer's diseases, rheumatoid arthritis, and type 1 diabetes increases with increasing latitude, suggesting a connection with vitamin D deficiency.

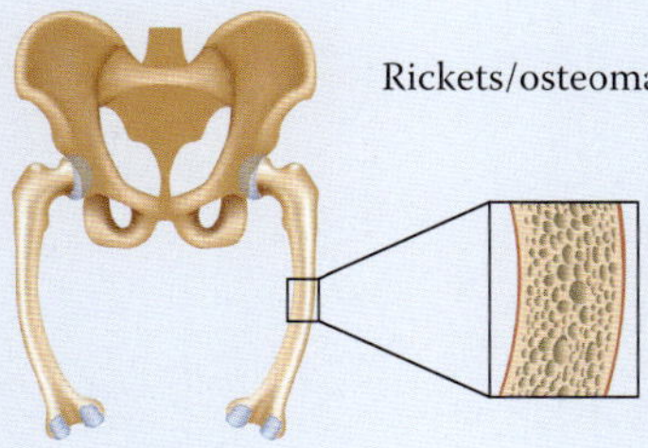

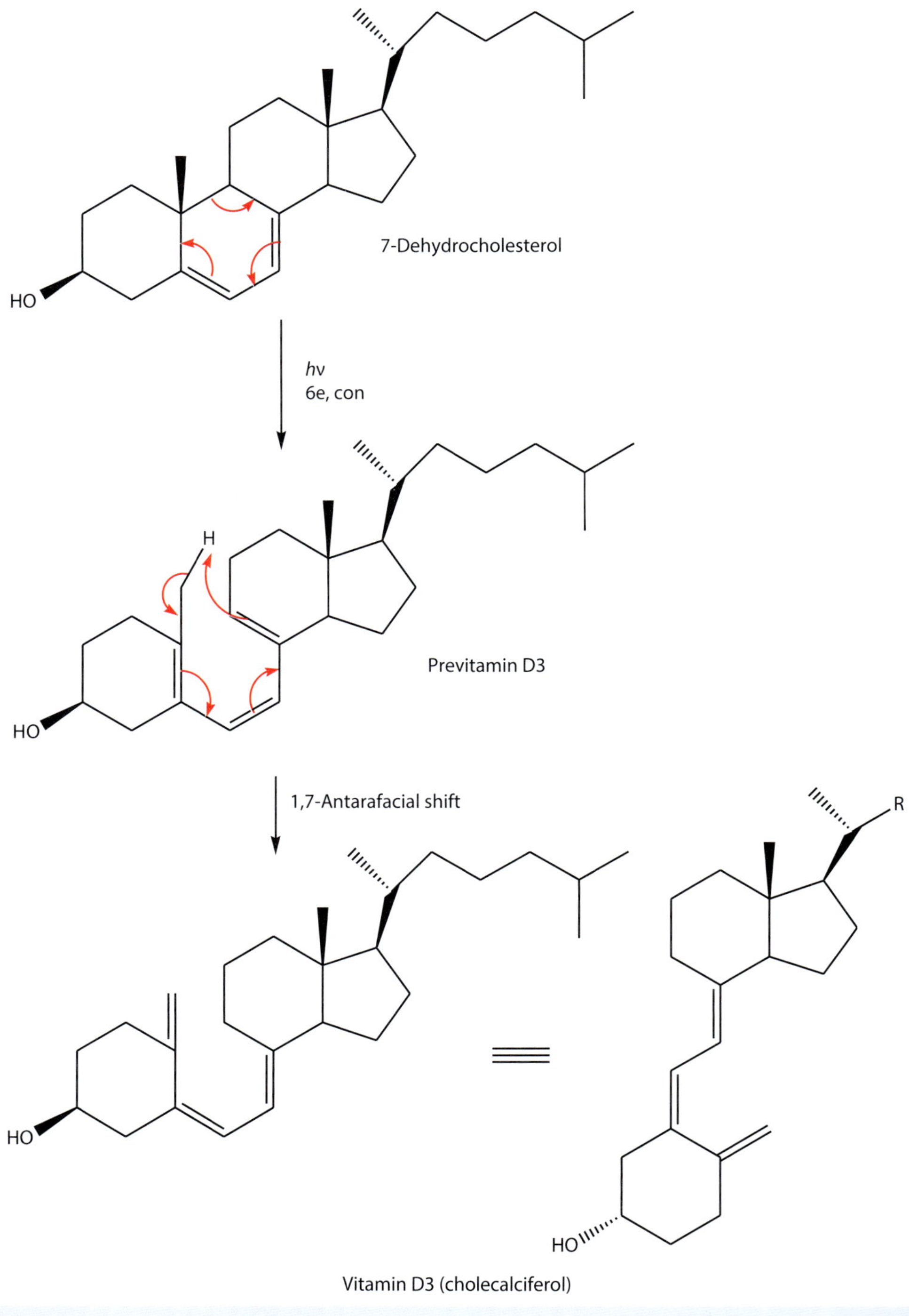

FIGURE 18.48 Biosynthesis of the D vitamins.

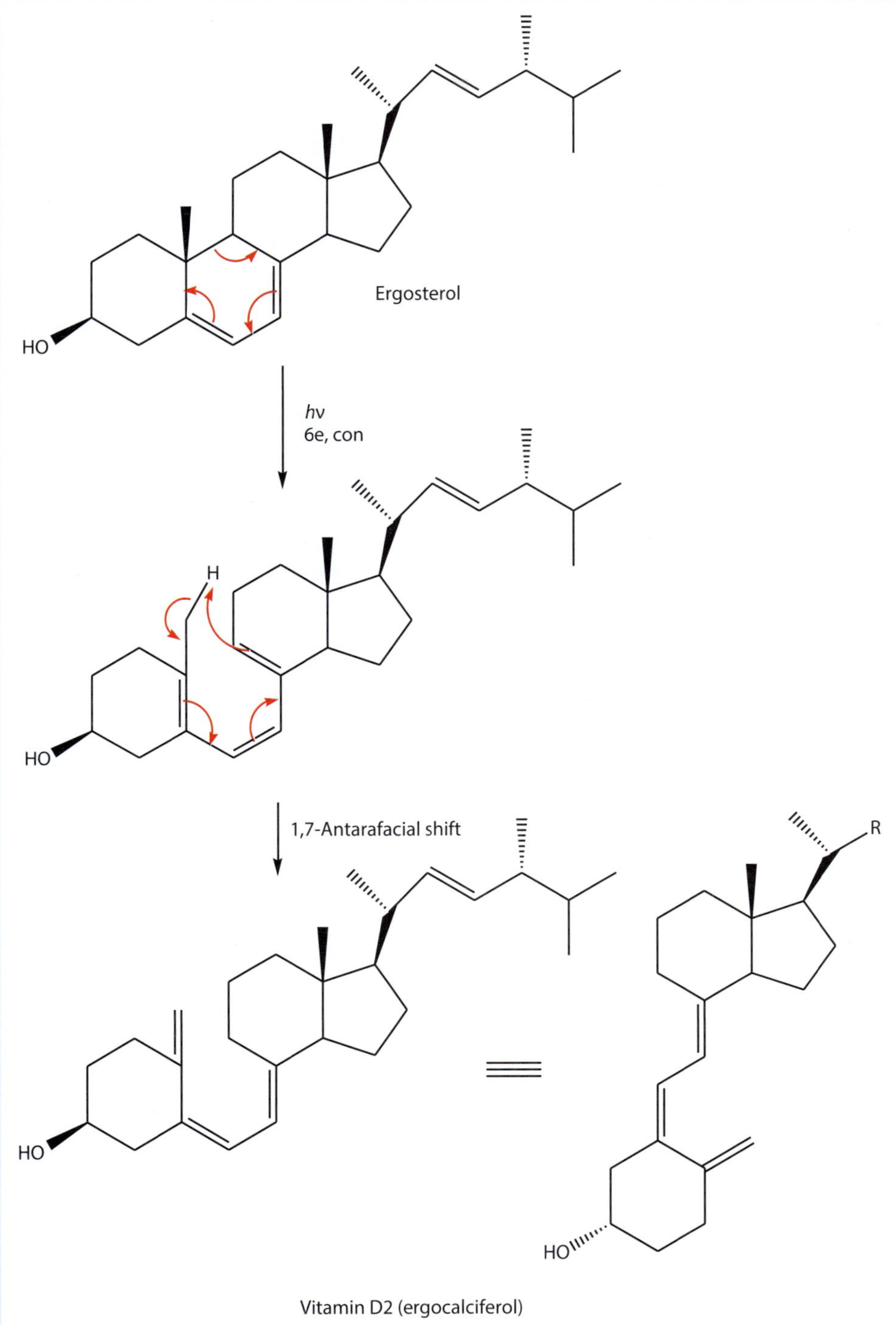

FIGURE 18.48 (continued) Biosynthesis of the D vitamins.

Key Points from Section 18.6

- Electrocyclic ring opening and ring closing reactions of alternating polyenes are controlled by orbital symmetry.
- The symmetry of the HOMO controls the stereochemistry of thermal reactions. The symmetry of the LUMO controls the stereochemistry of photochemical reactions. All of these reactions are strictly stereospecific.
- Sigmatropic rearrangements are also controlled by orbital symmetry considerations—the HOMO and LUMO determine the outcomes of thermal and photochemical reactions, respectively.
- 1,3-Hydride shifts are forbidden thermally, but 1,5-hydride shifts are allowed thermally.
- 1,3-Carbon shifts are allowed thermally with inversion of stereochemistry at the migrating center and 1,5-carbon shifts with retention of stereochemistry at the migrating center.
- The Cope rearrangement is a 3,3-sigmatropic rearrangement of 1,5-hexadienes. Although the rearrangement is degenerate for 1,5-hexadiene itself, substituents will alter the position of the equilibrium in accord with the relative stabilities of the two hexadienes.
- When the product of a Cope reaction is an enol, the process is described as the oxa-Cope reaction. Tautomerization of the enol to the ketone renders the reaction irreversible.
- The related rearrangement of phenyl allyl ethers is called the Claisen rearrangement; this is an irreversible process because the intermediate rearomatizes.

REVIEW PROBLEMS

1. Give a systematic name for each of the following molecules:

(a)

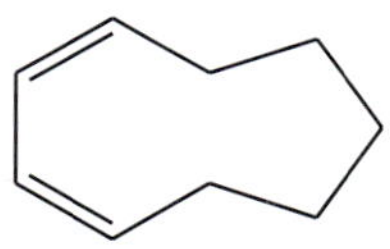

(b)

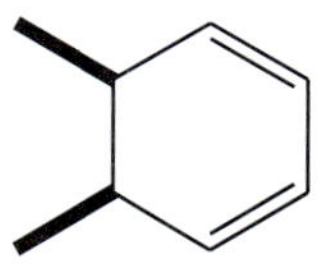

(c)

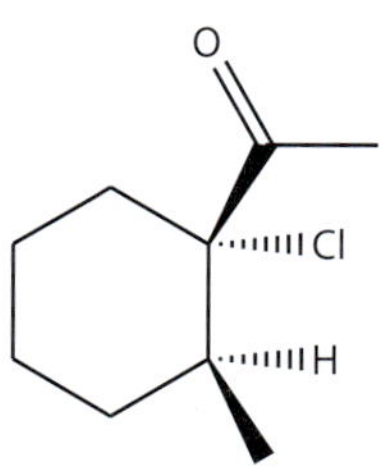

2. Draw a structure for each of the following molecules:
   (a) 2,2-Dimethylpropanamide
   (b) 2,5-Dimethyl-4-methoxybenzaldehyde
   (c) *E,E*-1,5-Cyclodecadiene

3. Predict the products of the following reactions, accounting carefully for the spectroscopic data provided:

(a)

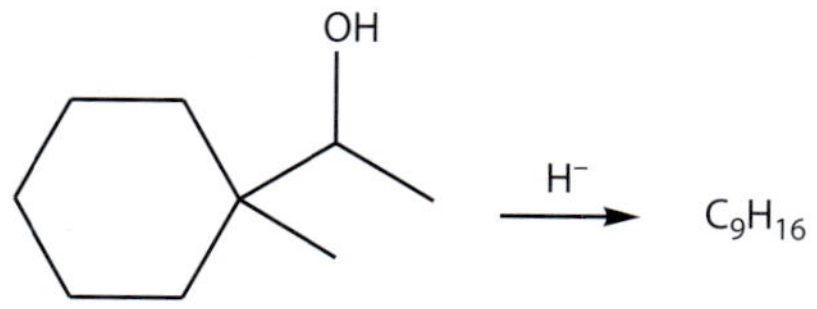

^{1}H NMR spectrum
δ 2.0, 6H, s
2.2, 4H, t
1.3, 4H, m
1.1, 2H, m

(b)

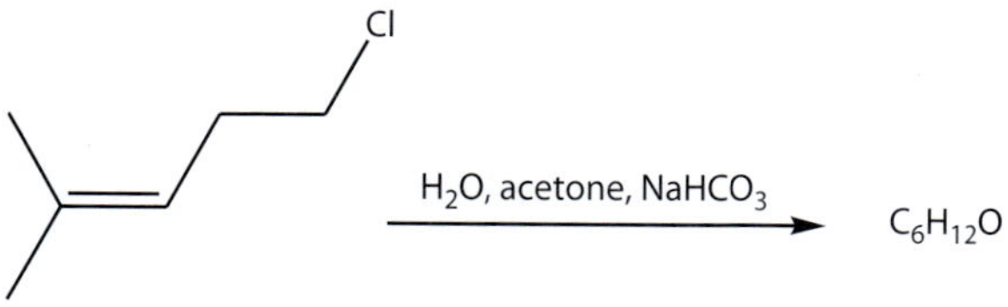

ν_{max} = 3400 cm^{-1}
δ 2.3, 6H, s
2.2, 1H, br, exchanges with D_2O
1.5, 1H, m
0.9, 2H, m
0.7, 2H, m

4. Propose a mechanism for each of the following reactions:

(a)

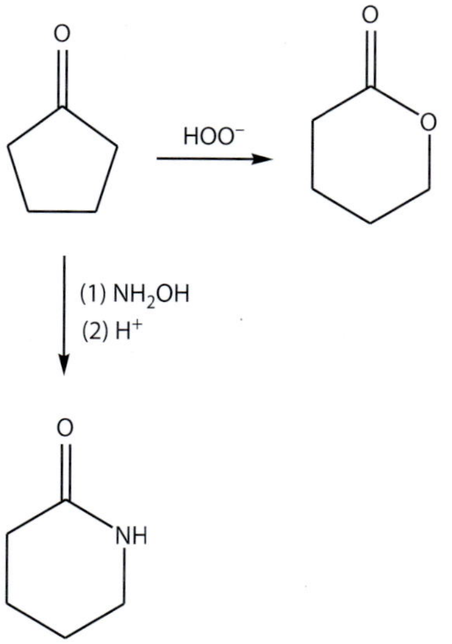

(b)

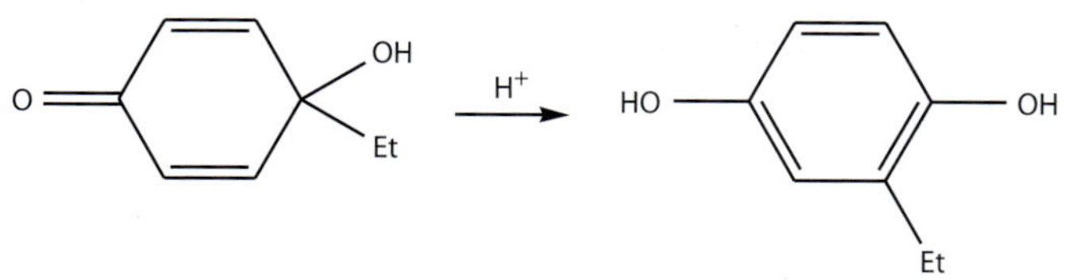

(c)

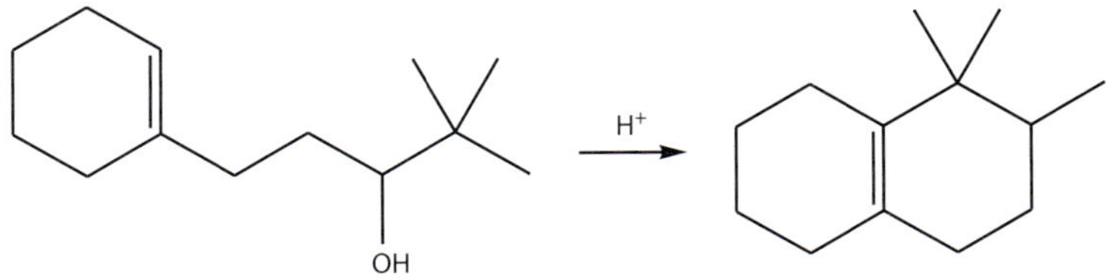
OH
H+

(d)

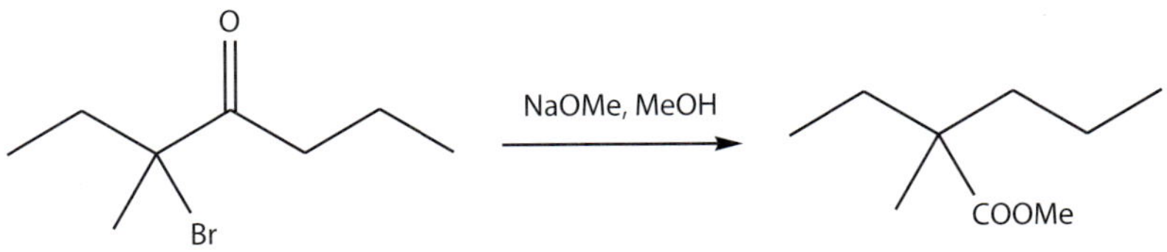
O
Br
NaOMe, MeOH
COOMe

(e)

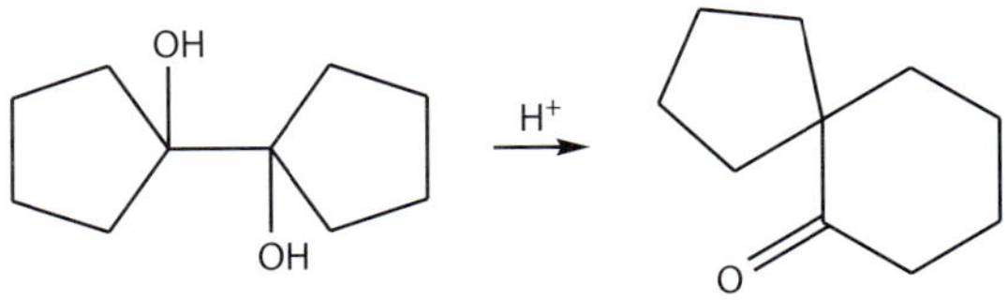
OH
OH
H+
O

(f)

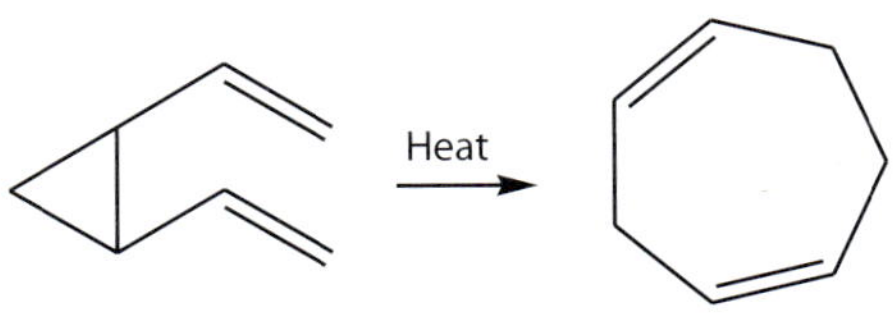
Heat

(g)

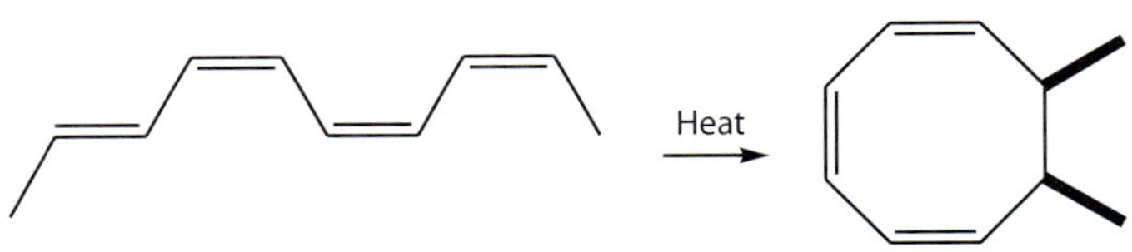
Heat

(h)

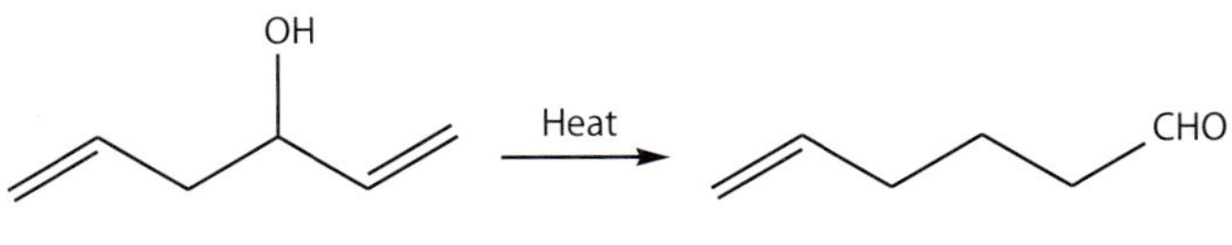
OH
Heat
CHO

(i)

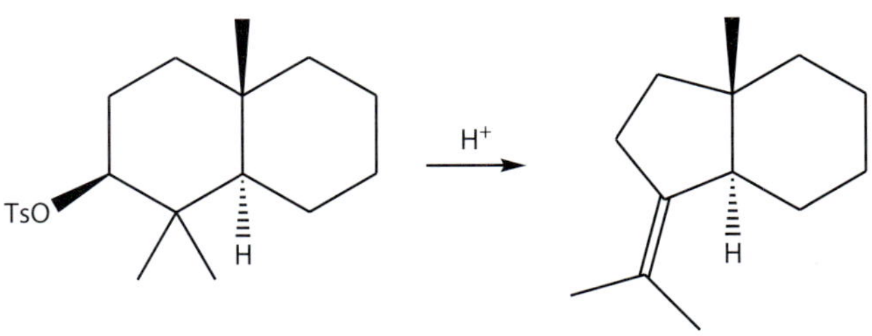
TsO
H
H+
H

(j)

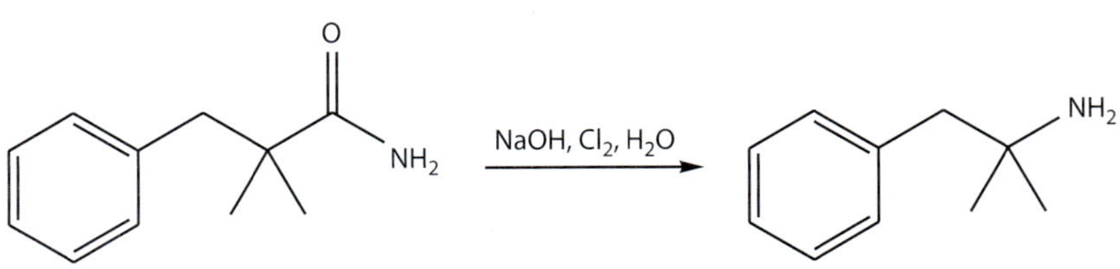
O
NH2
NaOH, Cl2, H2O
NH2

(k)

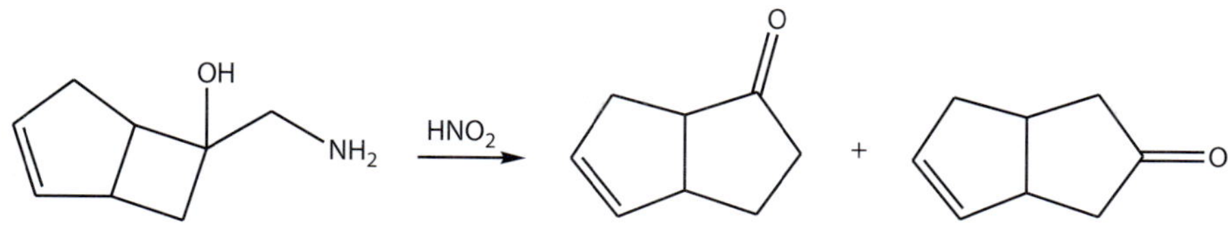

(l)

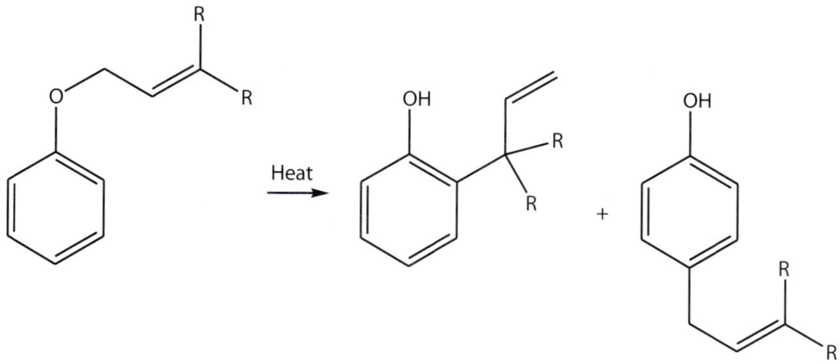

(m)

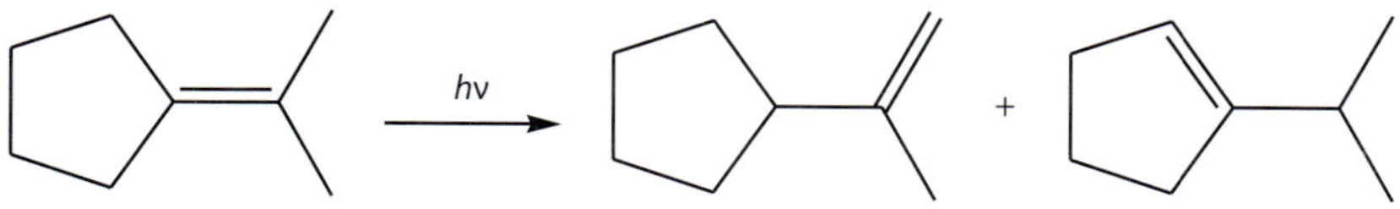

(n)

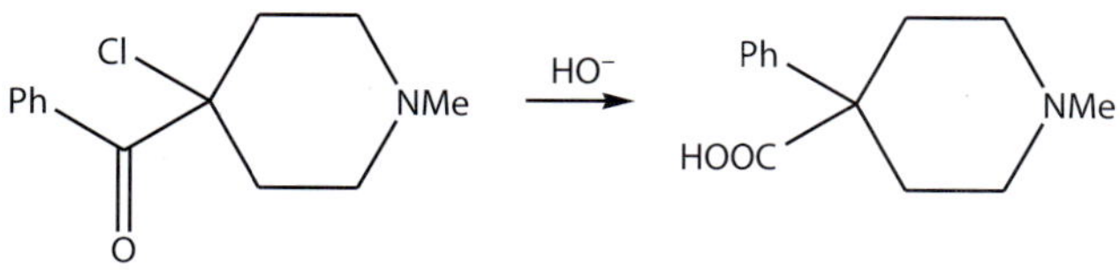

(o)

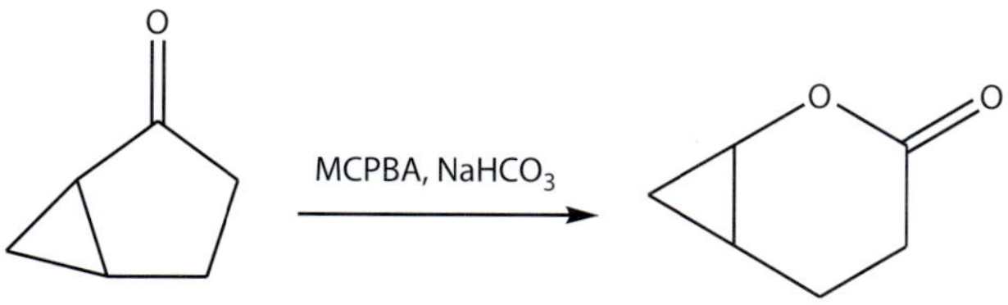

(p)

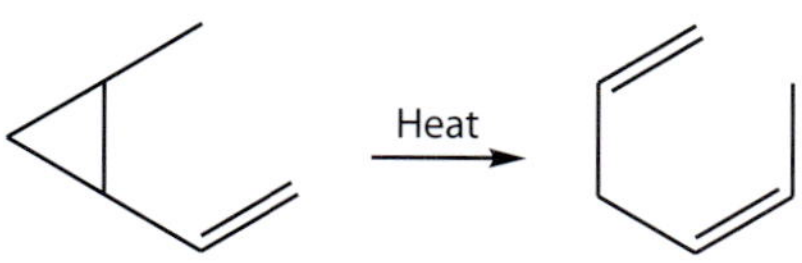

5. Predict the product(s) of each of the following reactions, paying attention to stereochemistry:

(a)

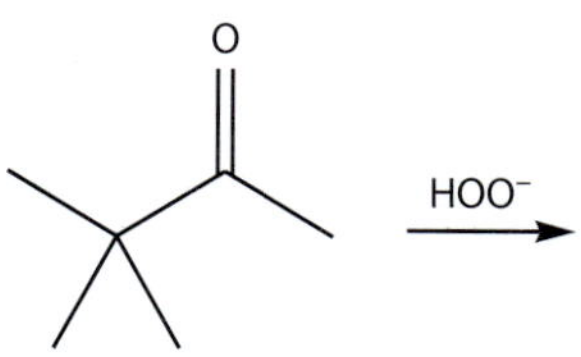

(b)

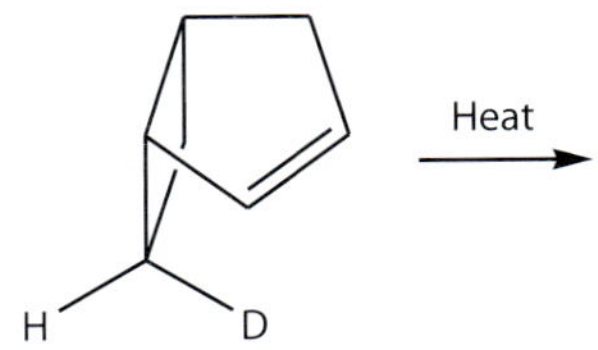
Heat
H
D

(c)

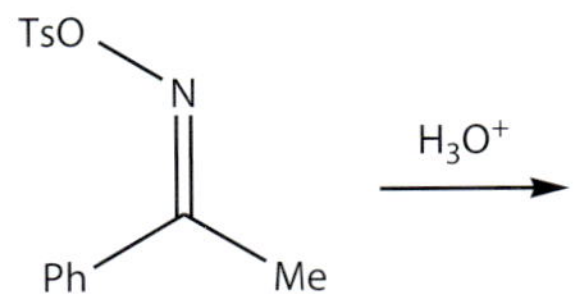
TsO
N
Ph
Me
H_3O^+

(d)

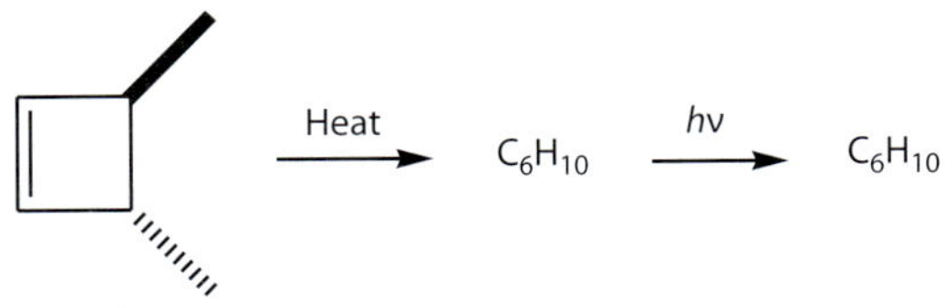
Heat
C_6H_{10}
hv
C_6H_{10}

(e)

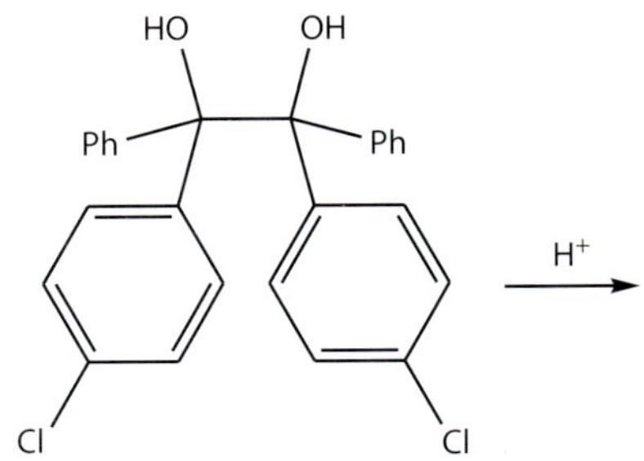
HO
OH
Ph
Ph
Cl
Cl
H^+

(f)

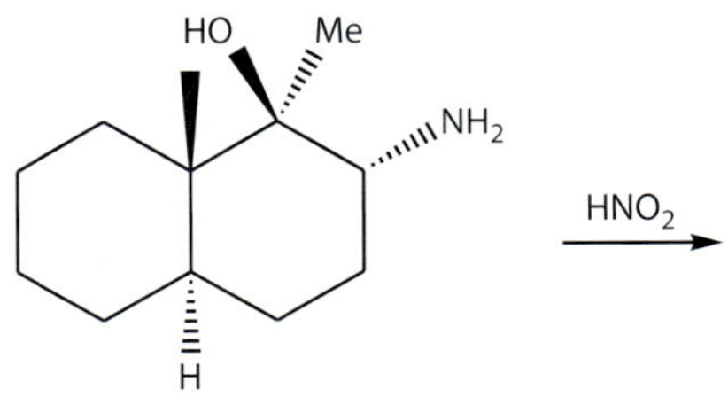
HO
Me
NH_2
H
HNO_2

(g)

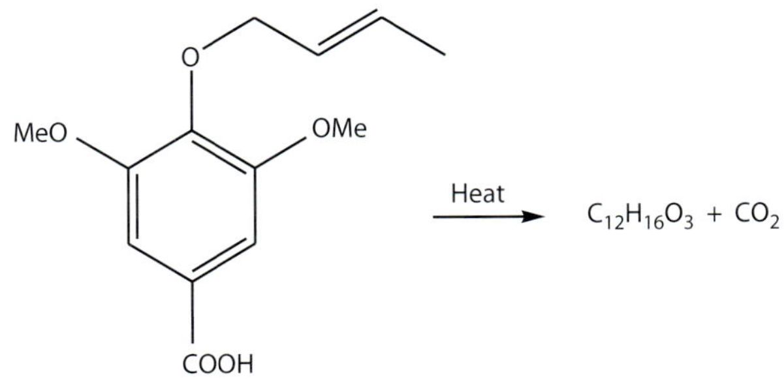
O
MeO
OMe
COOH
Heat
$C_{12}H_{16}O_3 + CO_2$

(h)

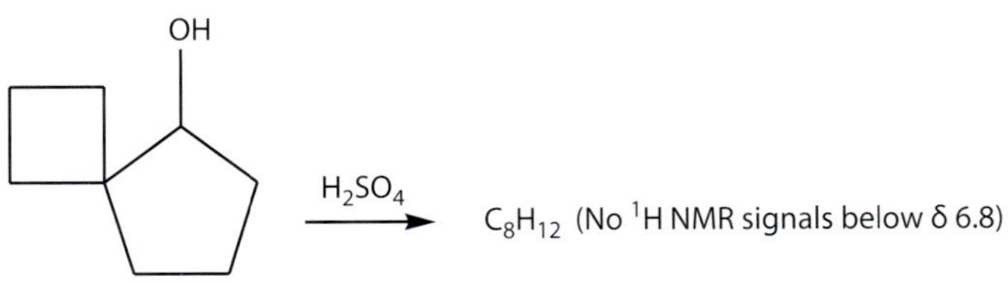
OH
H_2SO_4
C_8H_{12} (No ^1H NMR signals below δ 6.8)

(i)

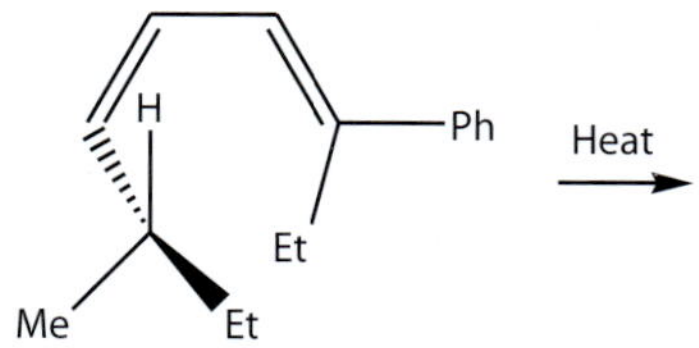
H
Ph
Heat
Et
Me
Et

(j)

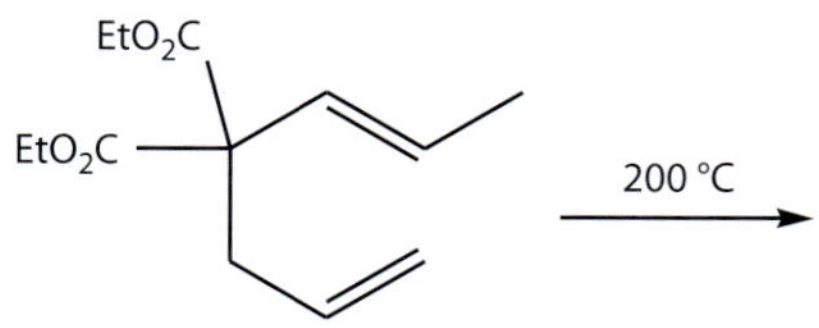
EtO2C
EtO2C
200 °C

(k)

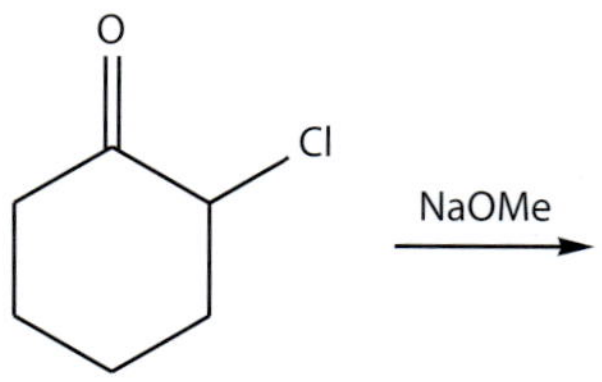
O
Cl
NaOMe

(l)

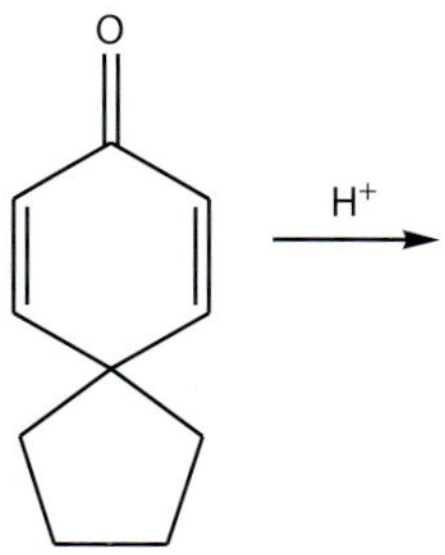
O
H+

(m)

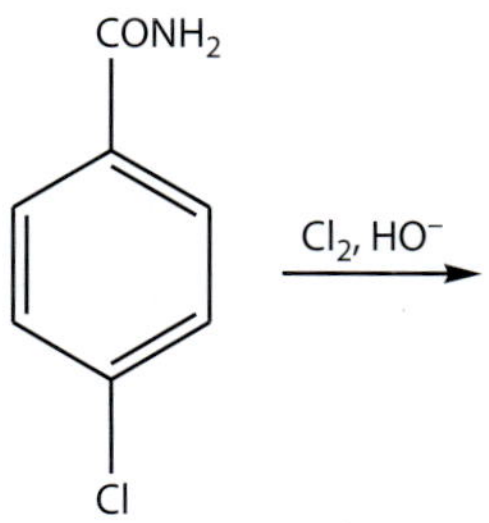
CONH2
Cl2, HO−
Cl

(n)

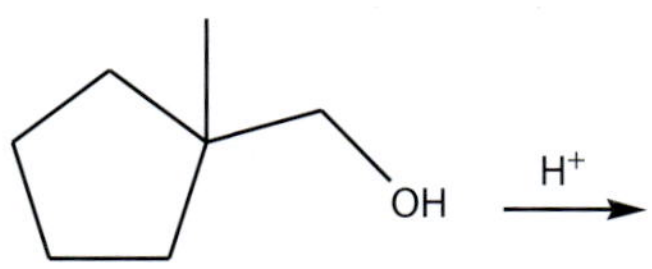
OH
H+

(o)

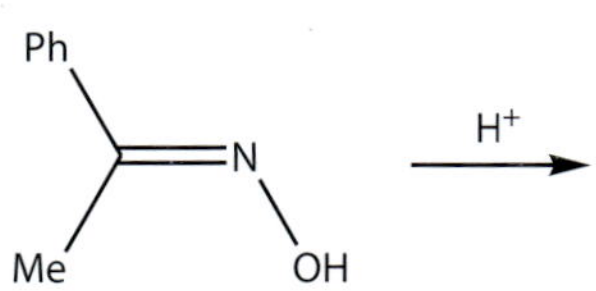
Ph
N
H+
Me
OH

6. Identify the unknowns in the following sequences, justifying your conclusions by means of a mechanism:

(a)

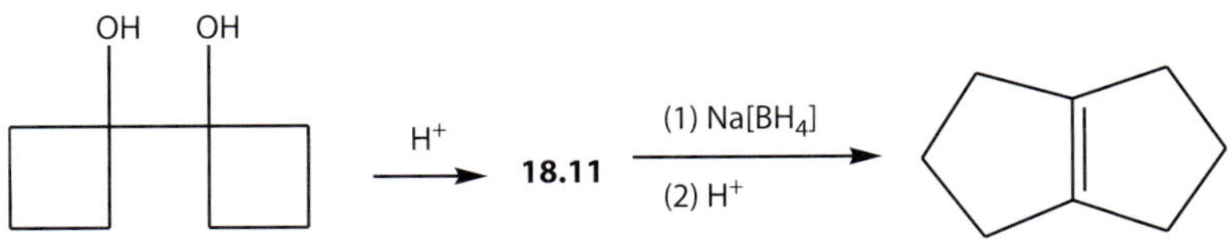

(b)

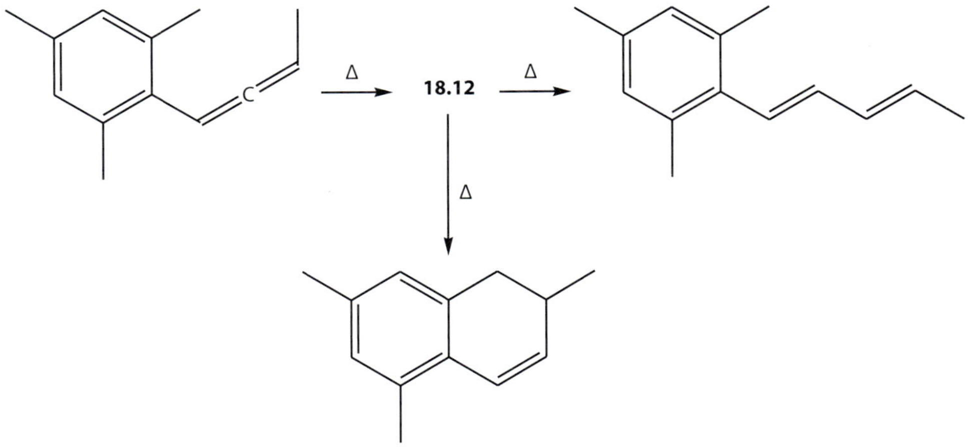

(c)

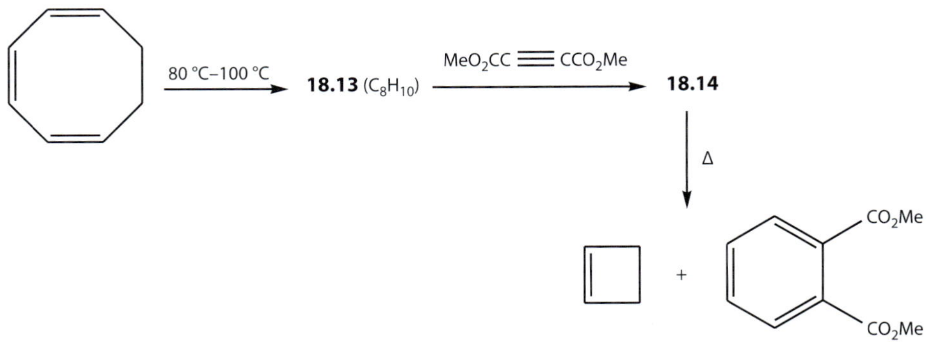

(d)

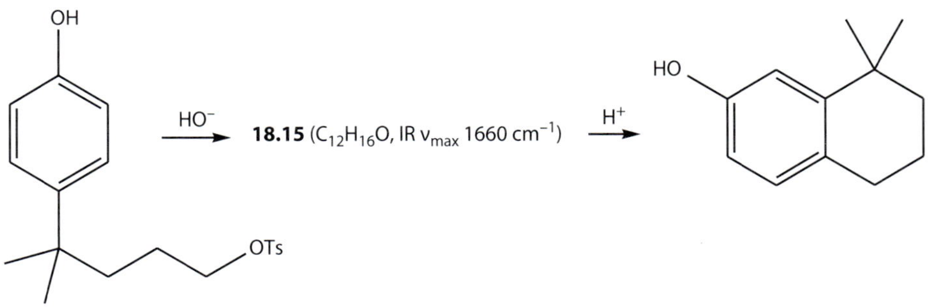

(e)

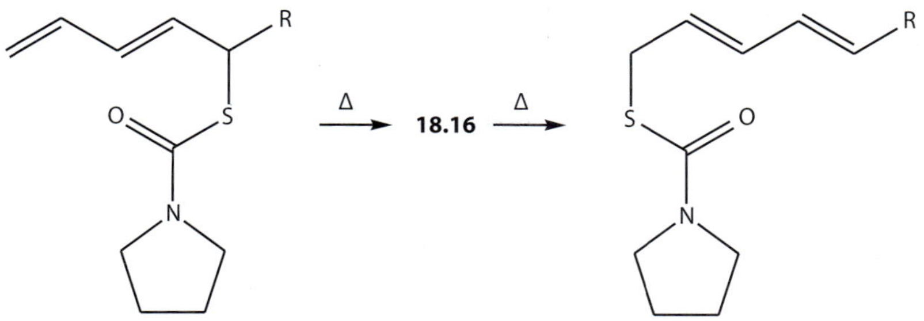

(f)

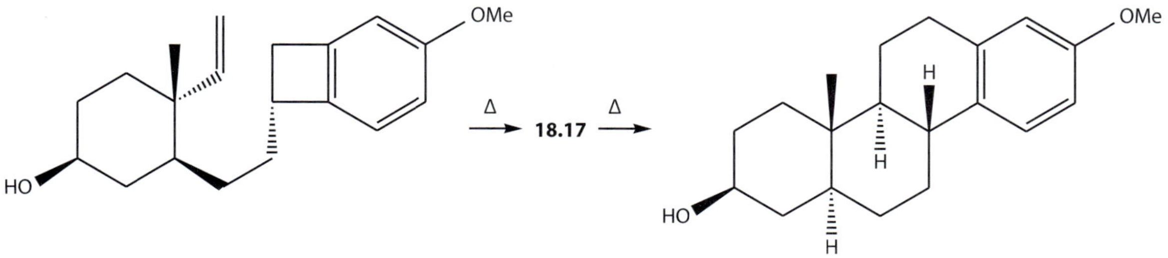

7. How would you accomplish each of the following conversions?

(a)

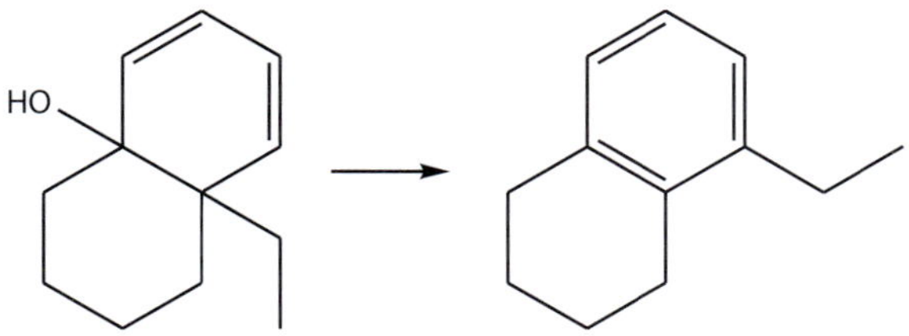

(b)

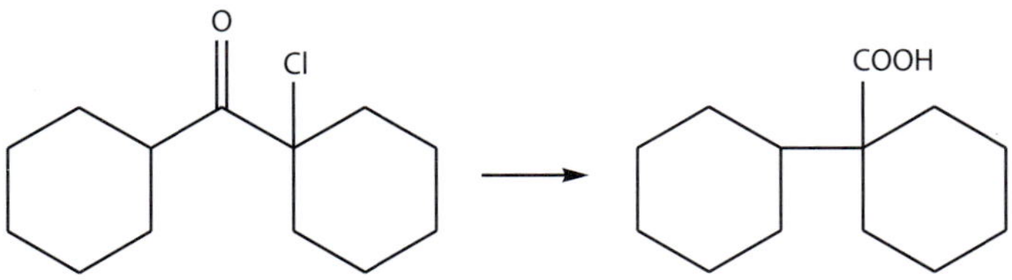

(c)

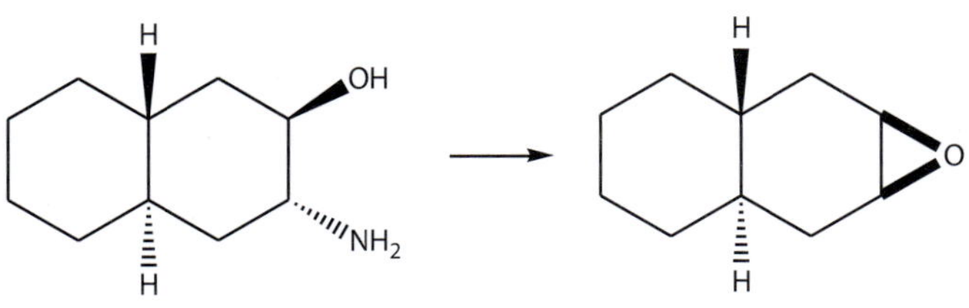

(d)

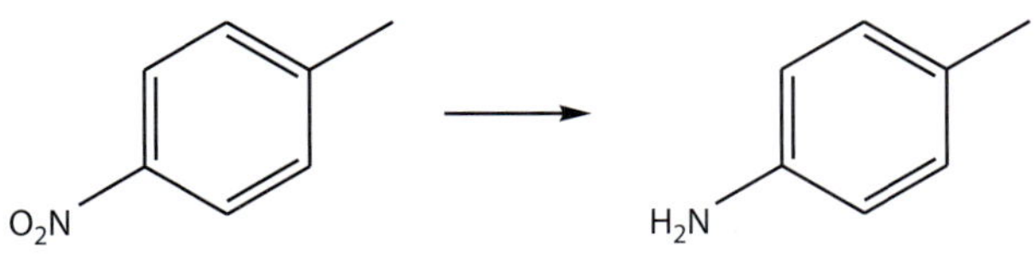

(e)

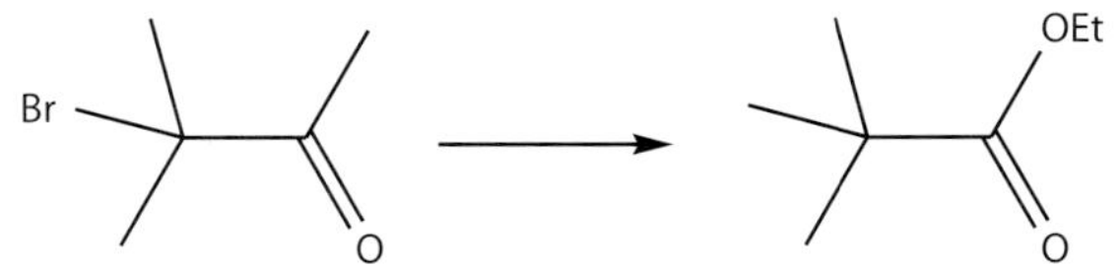

(f)

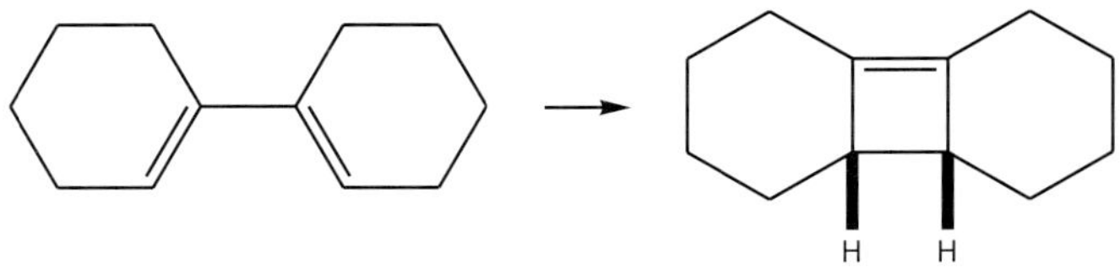

(g)

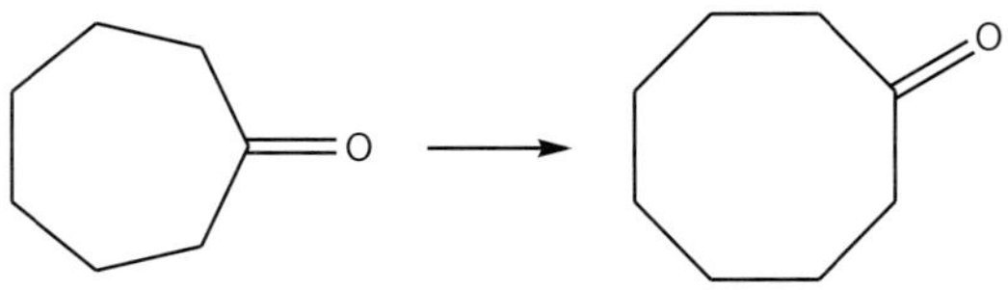

(h)

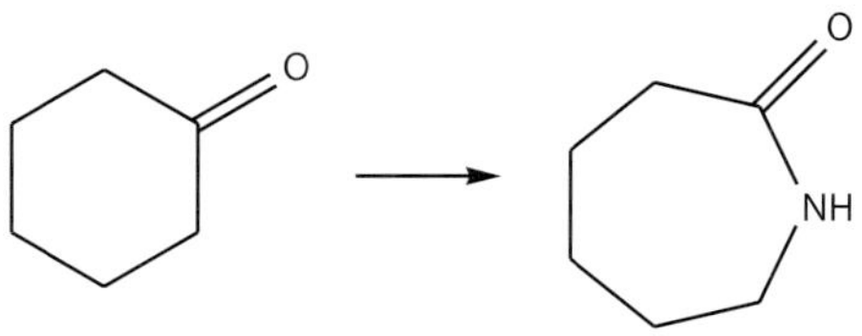

8. Study the following reaction sequence, and answer the questions posed:

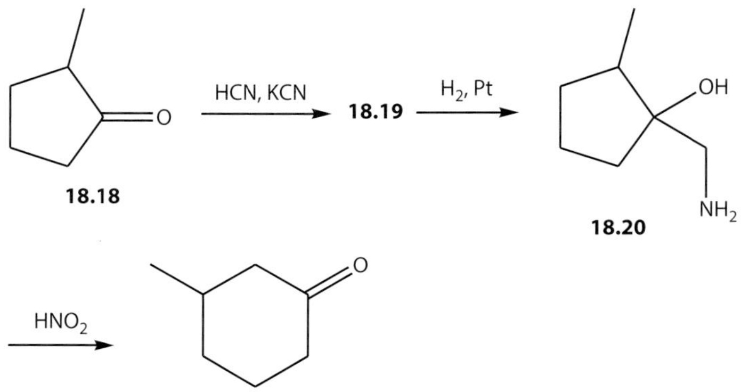

18.21

(a) Identify **18.19** and write a mechanism for its formation.
(b) Propose a mechanism for the conversion of **18.20–18.21**, and comment on the selectivity shown.

9. Explain the following observations:
(a) Dewar benzene is some 250 kJ mol^{-1} higher in energy than its isomer, benzene. However, Dewar benzene can be isolated and in fact rearranges only slowly to benzene:

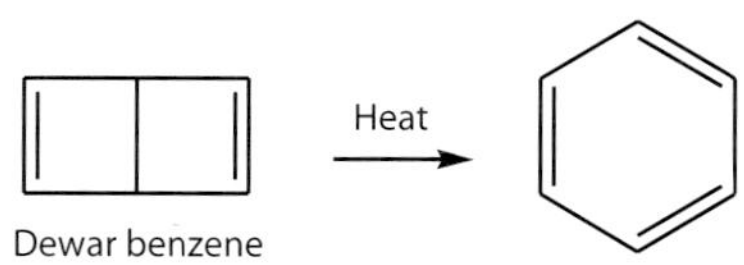

(b) 1-Chloro-1-phenylpropanone and 1-phenyl-3-chloropropanone both react with methoxide ion to give methyl 3-phenylpropanoate.

(c) The Claisen rearrangement of **18.22** gives only **18.23** and not **18.24**:

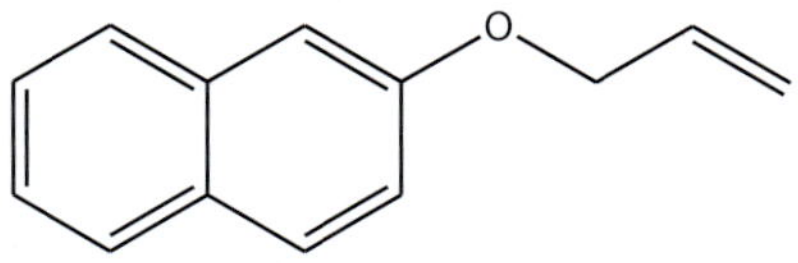

18.22

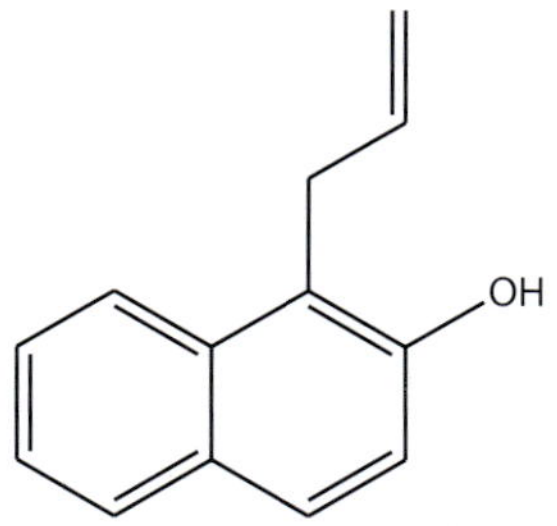

18.23

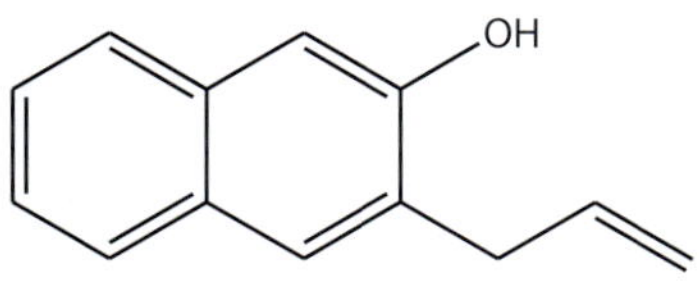

18.24

10. Suggest the best rearrangements for each of the following cations:

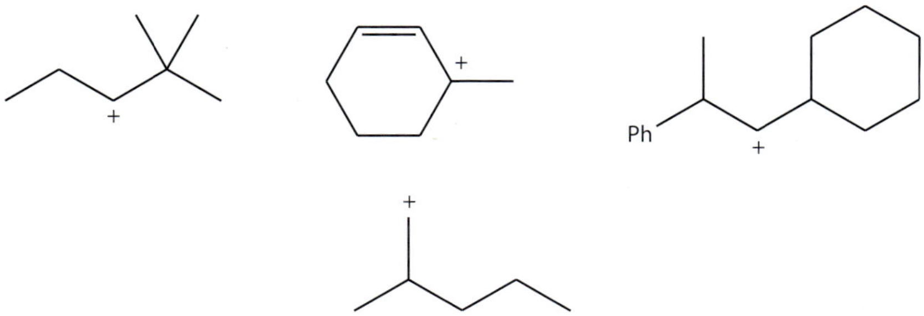

MORE CHALLENGING PROBLEMS

1. Propose a mechanism for each of the following transformations:

(a)

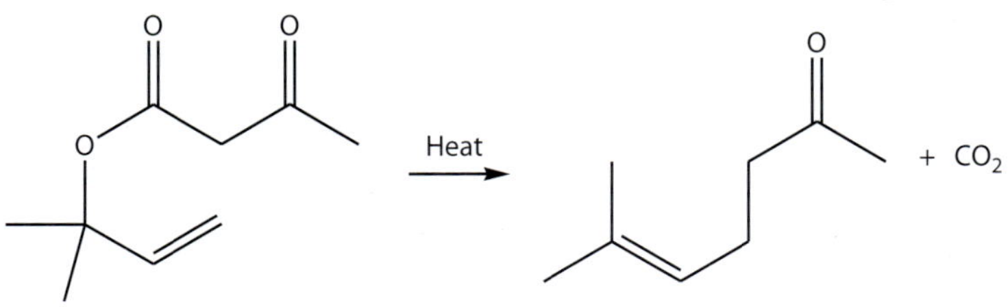

(b)

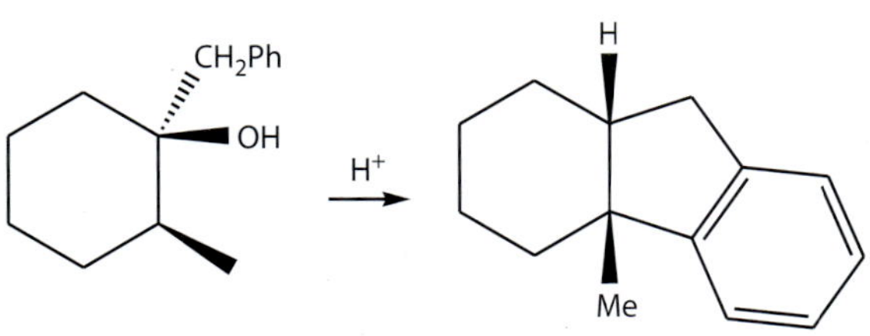

(c)

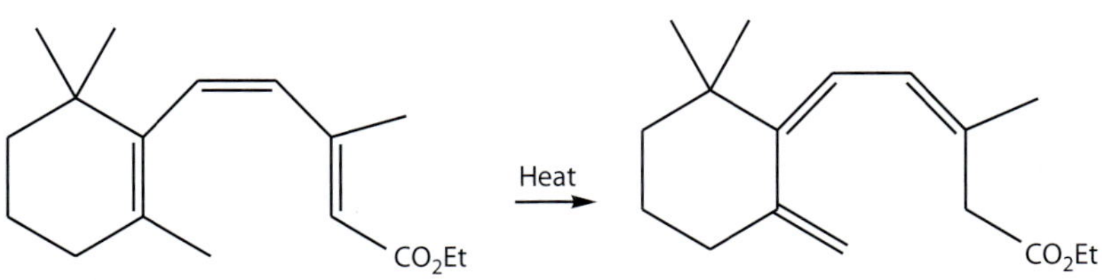

(d)

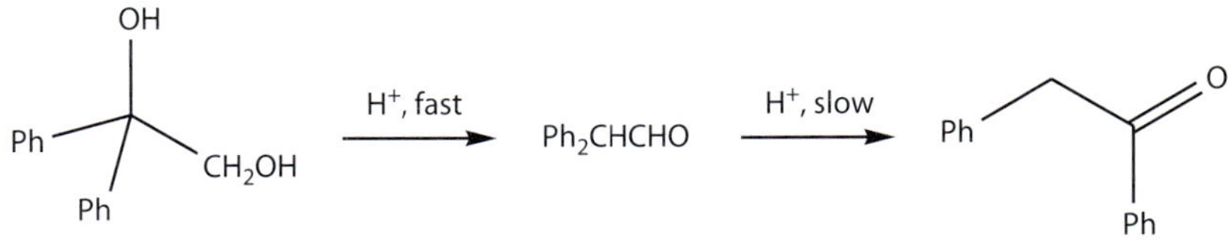
OH
Ph
Ph
CH_2OH
H^+, fast
Ph_2CHCHO
H^+, slow
Ph
O
Ph

(e)

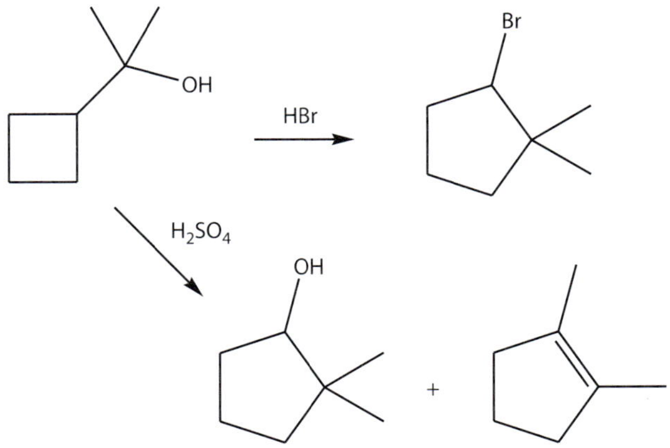
OH
HBr
Br
H_2SO_4
OH
+

(f)

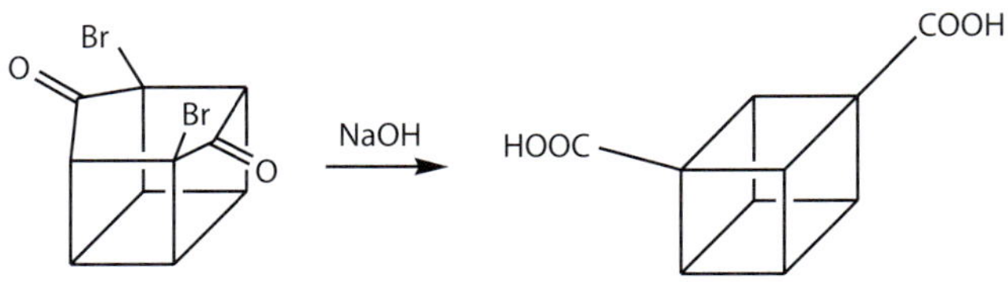
Br
O
Br
O
NaOH
HOOC
COOH

(g)

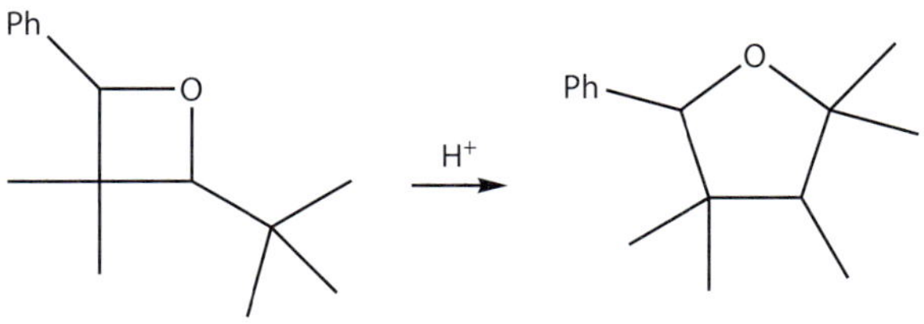
Ph
O
H^+
Ph
O

(h)

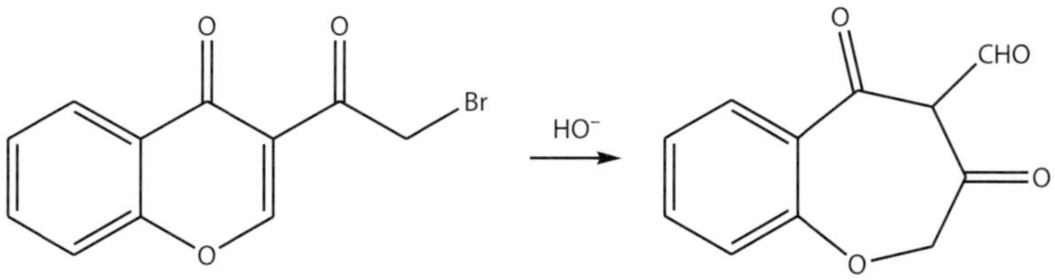
O
O
Br
O
HO^-
O
CHO
O
O

(i)

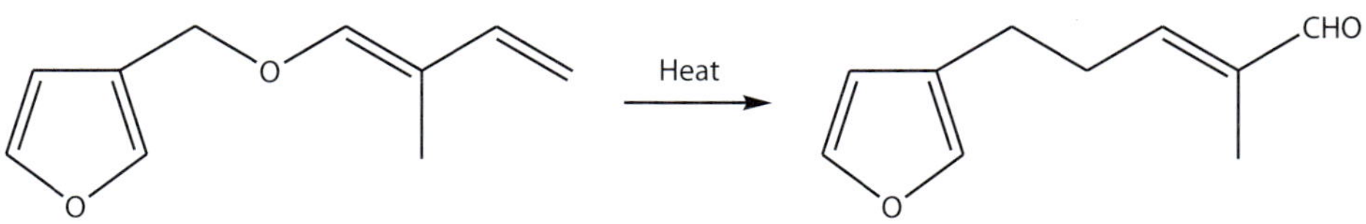
O
O
Heat
CHO
O

(j)

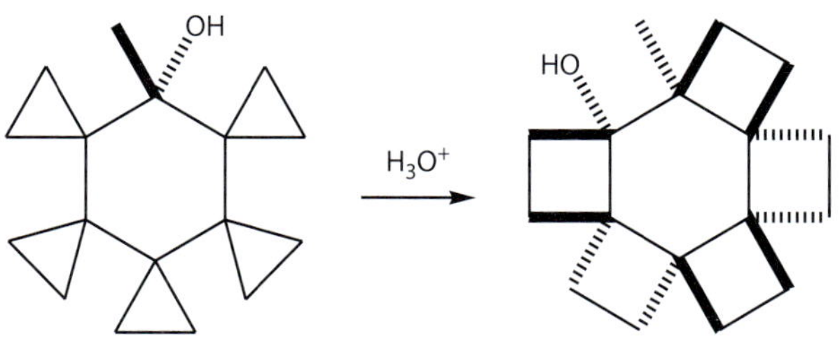
OH
H_3O^+
HO

(k)

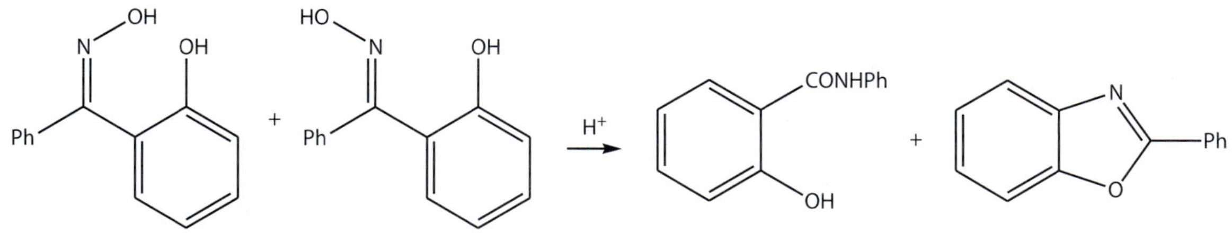

2. Predict the outcome of each of the following processes, paying attention to stereochemistry as appropriate:

(a)

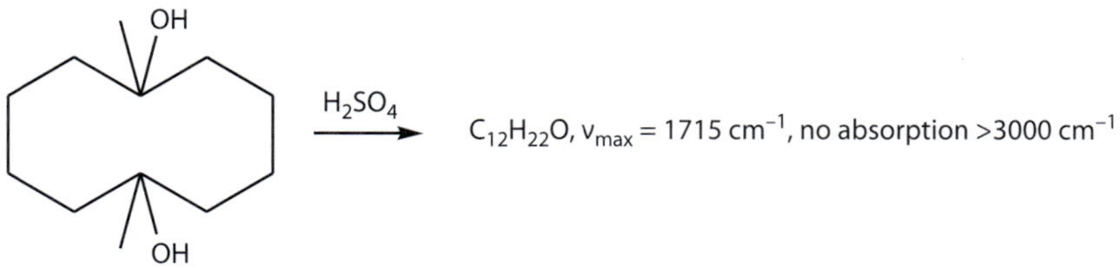

(b)

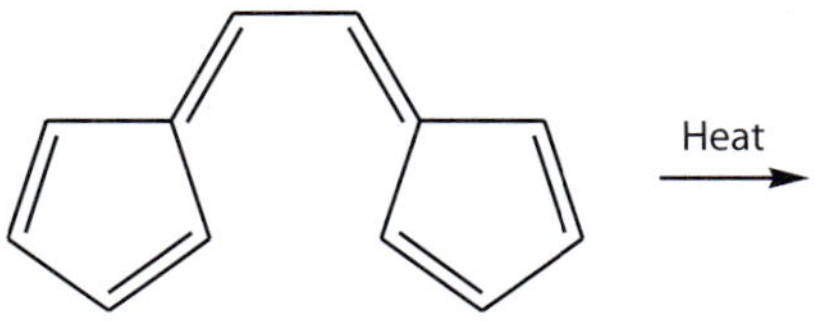

(c)

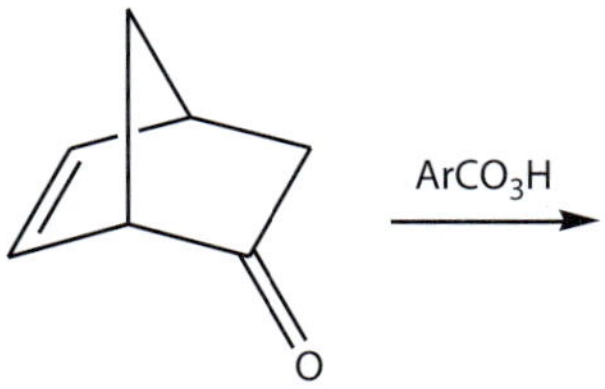

(d)

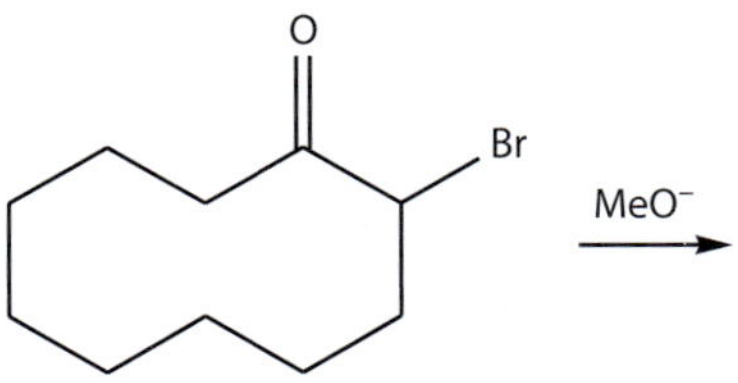

(e)

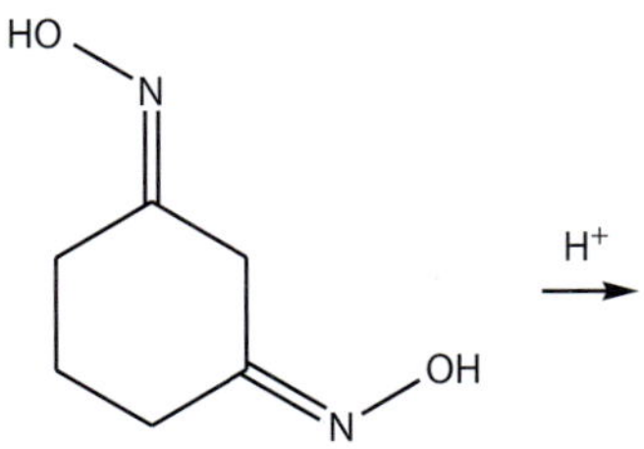

(f)

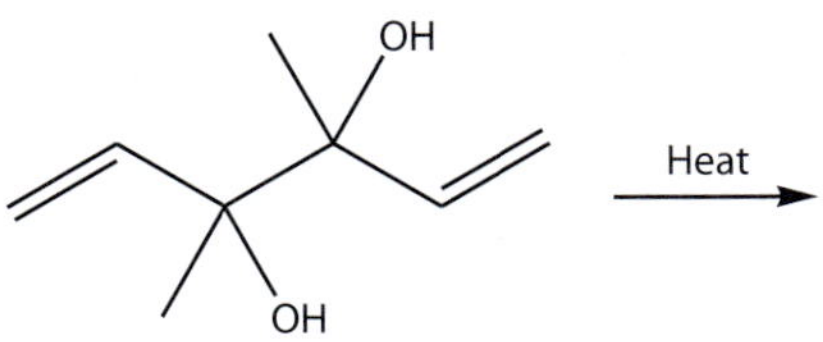

3. Identify the unknown in the following sequence, justifying your conclusion by means of a mechanism:

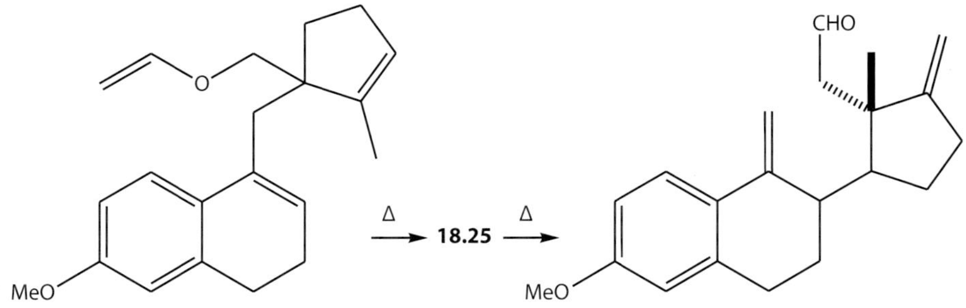

4. Rationalize the following observations, paying particular attention to stereochemistry:

(a)

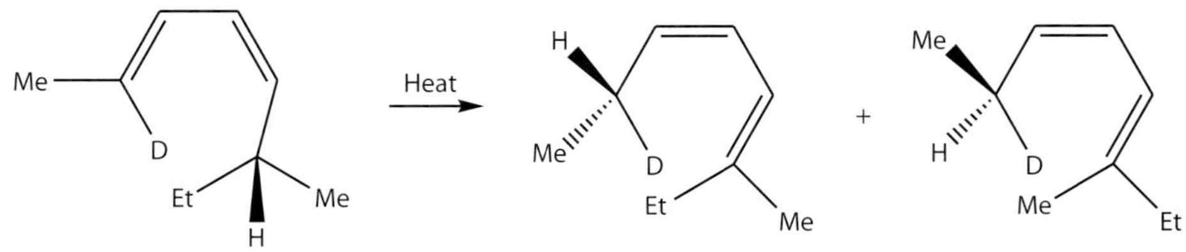

(b)

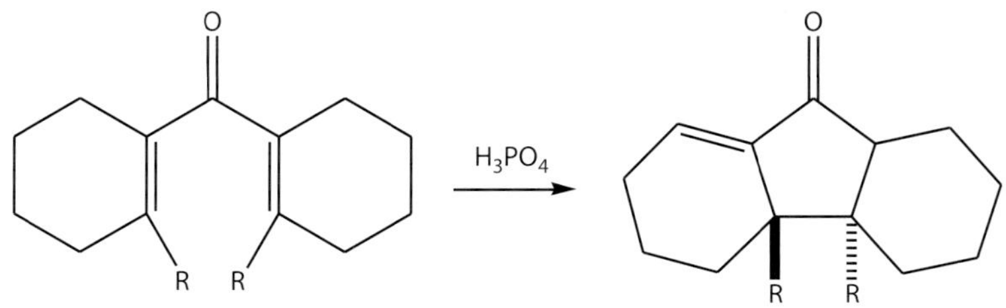

Hint: Start by protonating the carbonyl group:

(c)

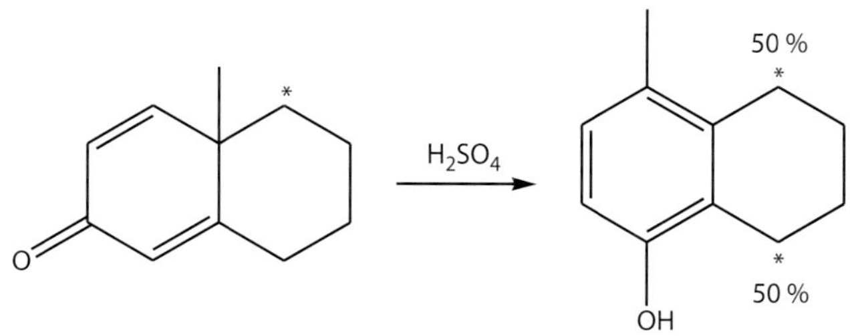

(d)

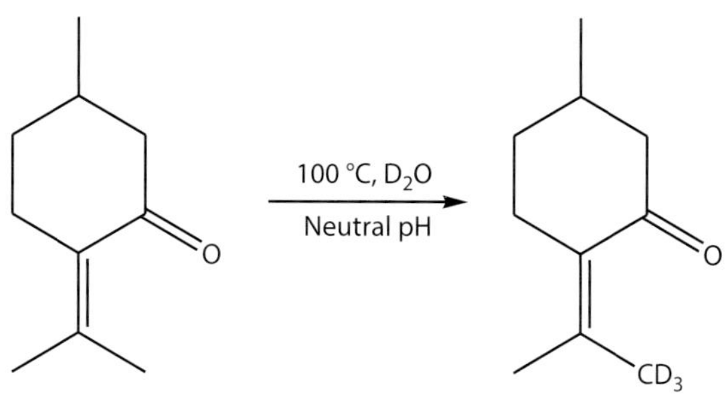

Synthesis 1

One-Group Disconnections

19

19.1 INTRODUCTION

In previous chapters, we have learned a lot about various transformations that will be useful to us in designing syntheses of more complex molecules. Organic synthesis, in all its forms, is the key process of the pharmaceutical, agrochemical, dyestuffs, polymers, and materials industries—if you can't make it efficiently, you can't test its usefulness or sell it. While lead pharmaceutical compounds are frequently natural products, this does not mean that they are available in unlimited quantities (it's also important to consider the environmental damage that might result from over extraction). And the best compound for a particular purpose may be an analog, rather than the naturally occurring material. The next two chapters will be very different—there will be only a few "new" reactions, as this is primarily about strategy. However, these chapters will, more than any others, reinforce the argument that organic chemistry is relentlessly cumulative—you can't simply forget reactions once you have moved on to the next chapter. Look on the bright side—it's an excellent review of much that has gone before.

We need to be able to look at a molecule that we need to synthesize and "pull it apart" into pieces that can be reassembled by reactions that we know to be reliable. A systematic approach to this technique was pioneered by Corey in the late 1960s, with the intention that the process could be computerized. Although programs do exist, it's yet to become clear that they have greater success than an experienced practitioner. The technique is not simple to master and requires both practice and imagination, but it is immensely powerful.

19.2 SIMPLE DISCONNECTIONS AND THE RETROSYNTHETIC APPROACH

The approach is called retrosynthesis—we go backwards from the target molecule (TM) to simpler species. We use a new sort of arrow to show this, the retrosynthetic arrow (**19.1**), which will normally be shown in blue. This does not represent a real reaction but a thought experiment. The other key concept is the "synthon," or synthetic equivalent. When we write a retrosynthesis, this may consist of pulling a bond apart to give an anionic and a cationic species—and clearly these are not usually found in bottles in the stockroom. So we need an equivalent (Figure 19.1 illustrates). In making the amine, I have chosen to break the carbon–carbon bond between C-2 and C-3 (highlighted in blue), to give an anion and a cation. The equivalent of the cation should be easy to guess; it is the benzylic halide. We know that many anions will rapidly displace halide from a primary benzylic position. The anion for which we need an equivalent is $^-[CH_2NH_2]$, at first inspection a most unlikely looking carbanion, as it would be very unstable. However, the cyanide ion contains all the atoms we need and is readily available. Moreover, we already know that cyanides can be hydrogenated or reduced with lithium aluminum hydride to amines. So $[CN]^-$ is the synthon for $^-[CH_2NH_2]$, and the reaction in the forward direction is shown in Figure 19.2.

$$\Longrightarrow$$

19.1

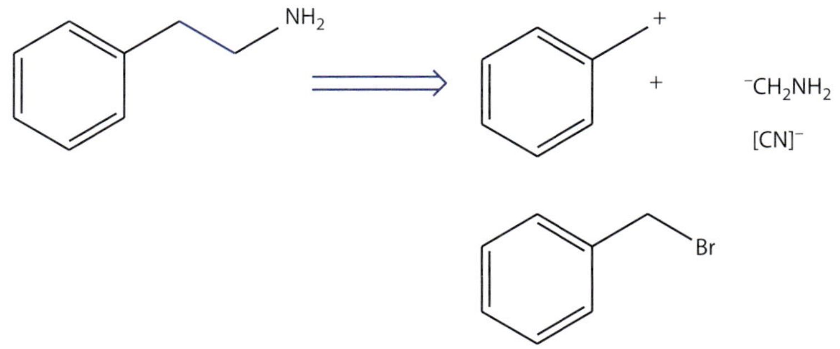

FIGURE 19.1 The concept of a synthon.

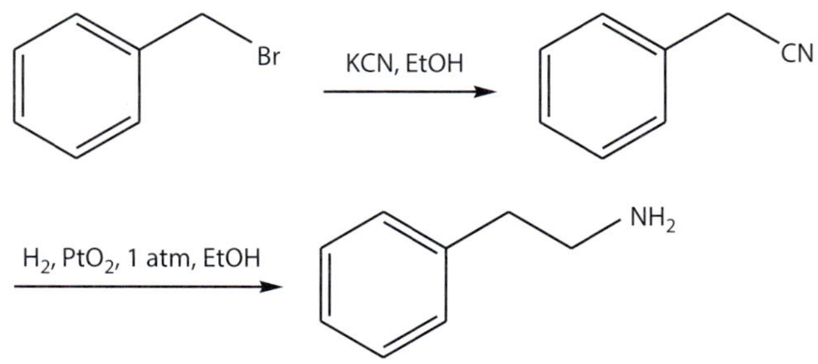

FIGURE 19.2 Synthesis of 2-phenylethylamine.

In approaching synthetic problems, it is not a given that there is a single correct solution—in some cases, there will be more than one plausible approach, and in practice, it may be essential to try each of them and determine which gives the best yield at the lowest cost. To take a very simple example (Figure 19.3), we could envisage two routes to the synthesis of *tert*-butanol. The first is an S_N1 substitution of *tert*-butyl chloride—the disconnection involves breaking the C–O bond to give a *tert*-butyl cation and hydroxyl ion. *tert*-Butyl chloride is the synthon for the cation. Alternatively, we could envisage a Grignard synthesis—the formal disconnection gives a protonated ketone and a carbanion. The synthon for $[Me]^-$ is MeMgBr. How would we choose between the two reactions in practice? Our knowledge of S_N1 reactions suggests that elimination to give 2-methylpropene will be a side reaction of the substitution. So we will lose some of the material. The upside is that 2-methylpropene is a gas at room temperature, so it will be easily removed. There is no obvious side reaction of the Grignard process—but it requires properly dried solvent and a nitrogen atmosphere, and these have costs. The only way to resolve the issue is to try it out and cost it out—not forgetting the waste disposal costs and any environmental issues with waste and by-products.

Those were simple examples, which I hope you would not have found difficult even without a retrosynthetic approach. But how do we choose a disconnection?

- Does the forward reaction have a reasonable mechanism (it does not necessarily need to be one that we have studied before in detail, but it should not contradict what we know about how mechanisms work)?
- Are the intermediate we are drawing sensible ones? For example, are the anions and cations we draw stabilized ones?
- Does the disconnection simplify the problem?

Next, we should use symmetry if we can, and we will need to worry about selectivity in the forward reaction. We should remember that we know many ways to interconvert functional groups

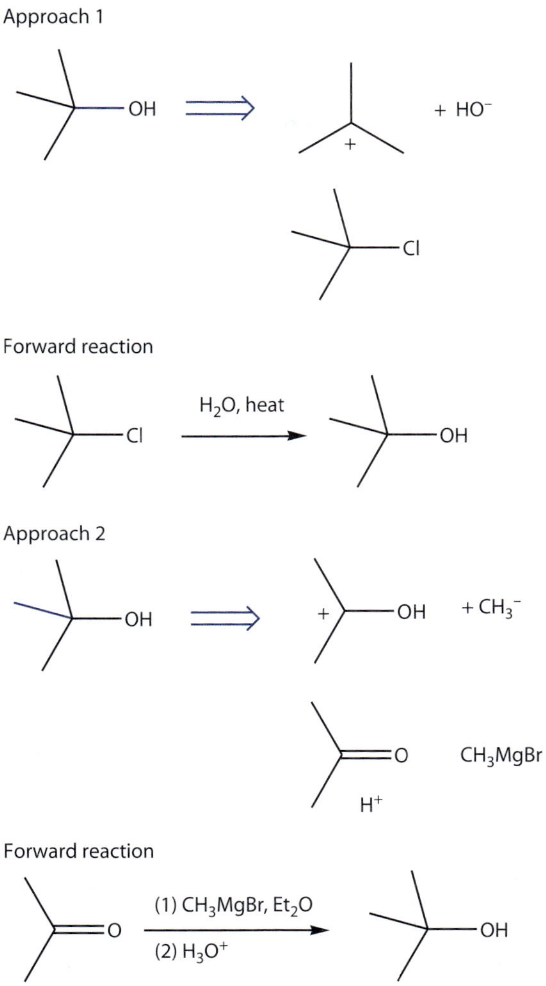

FIGURE 19.3 Disconnection and synthesis of *tert*-butanol.

(functional group interconversion [FGI]), and oxidation level (*e.g.*, primary alcohol, aldehyde, and carboxylic acid) is readily altered. In the next chapter, we will also consider methods of protection and deprotection of functional groups, so that a transformation may be carried out on another part of the molecule.

Returning to the *tert*-butanol synthesis, all the intermediates we have drawn are sensible ones. We know that ketones are available, also Grignard reagents, and that tertiary carbocations are stable. But we could have done something foolish, by cutting the molecule into a methyl cation and $(CH_3)_2C^-OH$. Neither of these intermediates can be accounted as sensible—we know them to be quite unstable. There is another example in Figure 19.4. We might think of disconnecting

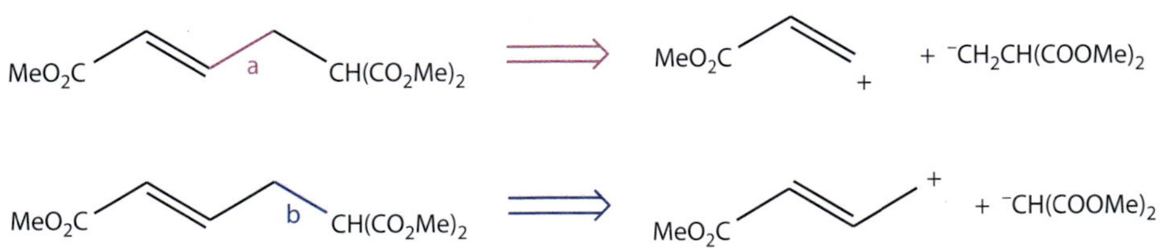

FIGURE 19.4 Not all disconnections give sensible results.

either bond a or bond b, shown, respectively, in pink and blue. Disconnecting bond a gives poor intermediates—vinyl cations are very unstable, and there is no S_N1 or S_N2 reaction at sp^2 carbon atoms. Moreover, the carbanion has no special stability, and using a Grignard reagent or organolithium compound as its synthon is not a possibility because of the presence of the ester groups. Disconnection of bond b would be more successful. The carbocation itself is not particularly stable, but the synthon, the halide, is allylic and reactive. The anion would be stabilized by the electron-withdrawing ester groups.*

All of these would be described as one-group disconnections—the bond breaking is informed by a single functional group in the TM. This does not mean that this only works for molecules with a single functional group, simply that our disconnection only involves one of them. In the next chapter, we will also meet two group disconnections, where the disconnection is directed by the *relationship* between two functional groups and pericyclic disconnections.

There are two stages in designing a synthesis—analysis and synthesis. To summarize, the following applies:

Analysis

- Recognize key functional groups in the TM.
- Disconnect using methods corresponding to reactions with reasonable mechanisms.
- Repeat until you reach "available" SMs.

Students often experience a moment of panic when the phrase "readily available SMs" appears—how are they, with only limited experience of organic chemistry, to judge what is "readily available"? Different instructors will want to tackle this in different ways, but as far as problems in the text are concerned, either you will be given a range of molecules with which you can start, or, if there is no specification, you can have anything in the Aldrich catalog.†

Synthesis

- Write out the synthesis according to your plan from the disconnections, adding relevant reagents, solvents, and reaction conditions.
- Consider any processes that might lead to alternative products or by-products, and modify the plan to take account of these.
- Consider stereochemical problems, and modify the plan accordingly.

19.3 ONE-GROUP DISCONNECTIONS

19.3.1 DISCONNECTION OF ALCOHOLS

In one-group disconnections, a single functional group suggests what disconnection to make. You have already quite a lot of experience of this with alcohols as the TM. For any generic alcohol, **19.2**, there are three possible disconnections to a ketone and a carbanion (Grignard reagent in most cases). The forward reaction is the reaction of a Grignard reagent or other anion, with a ketone, followed by aqueous work-up. But how do we choose which group to disconnect? The general rule is to choose the one that gives the most stable anion. Thus, we would dissect **19.3** to acetone and the anion of phenylethyne. Remember that the pK_a of alkynes is 25, compared with >40 for alkanes; thus, the anions of 1-alkynes can be prepared using either butyl lithium in ether or $Na[NH_2]$ in liquid ammonia. It's also worth remembering that alkynes can be converted to *cis*- or *trans*-alkenes, alkanes, and ketones (Sections 11.2.5 and 11.5).

* If you are worrying about S_N2' substitution of the allylic halide, this is effectively inhibited by the steric hindrance involved; A. Padwa, S.H. Waterson and Z. Ni, *Org. Synth.* (1997) **74**, 147.

† www.sigmaaldrich.com.

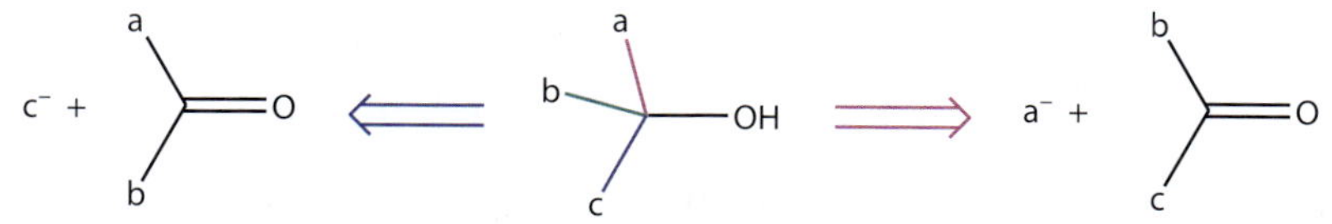

19.2

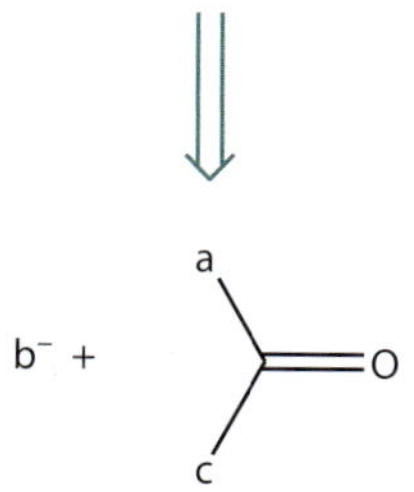

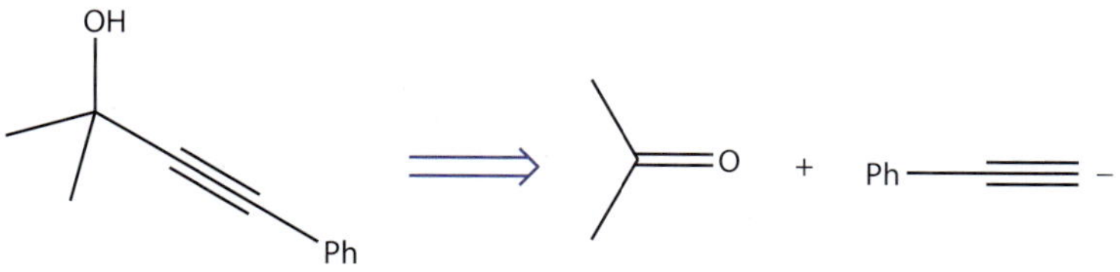

19.3

With substituted cycloalkanols, there is also usually no problem in deciding which bond to break in the disconnection—disconnecting one of the bonds of the ring does not produce any simplification. Examples are shown in Figure 19.5.

In many cases, however, the choice is less obvious (Figure 19.6). All of these disconnections, and reactions between the relevant Grignard reagents and ketones, should work in principle. In practice, a researcher would try them all and check the yields against costs.

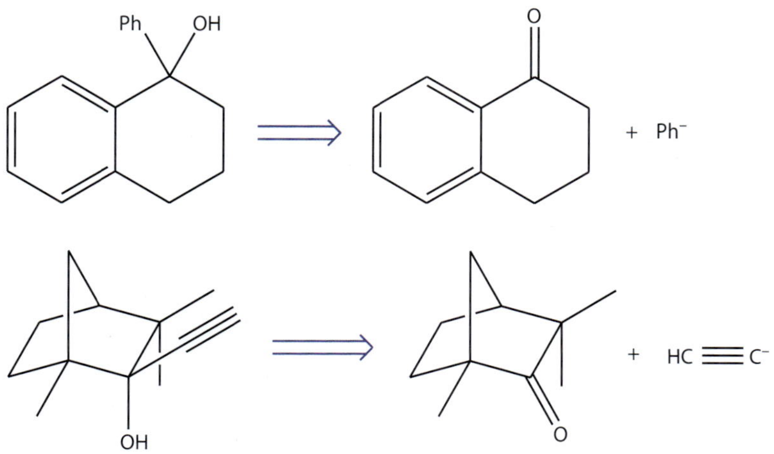

The forward reaction uses HC≡CLi/THF/–78 °C, and the reaction is 97 % selective for the desired isomer. Attack is from the less hindered "top" face of the molecule.

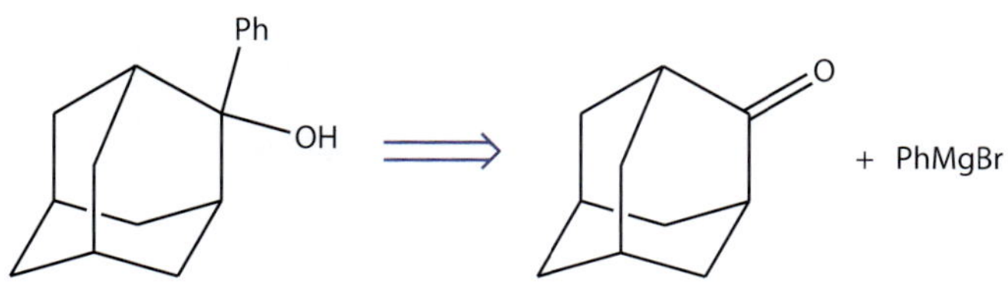

FIGURE 19.5 Disconnections of cycloalkanols.

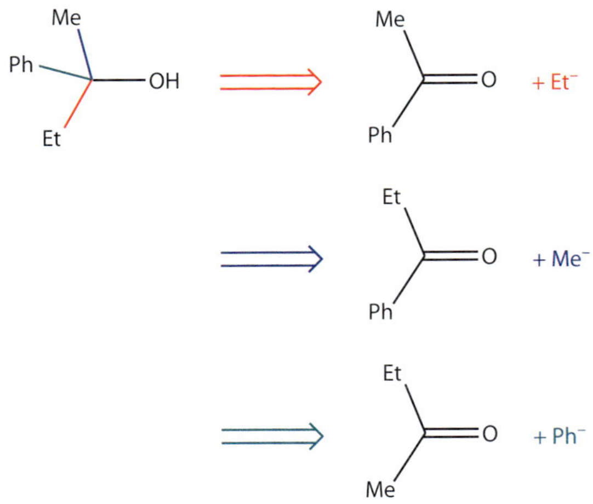

FIGURE 19.6 Disconnections of a tertiary alcohol.

PROBLEM 19.1

Propose a disconnection for each of the following tertiary alcohols:

(a)

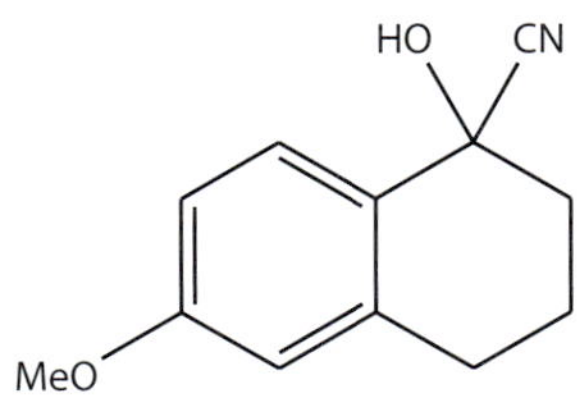

(b)

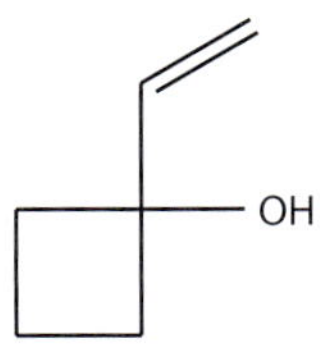

(c)

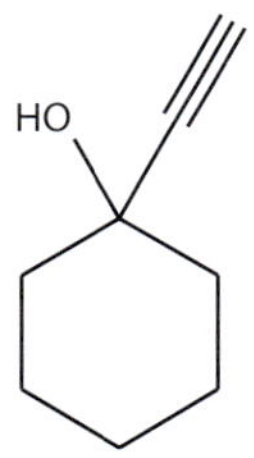

Solutions

(a) The best "carbanion" is $[CN]^-$, and this also gives the greatest simplification. So,

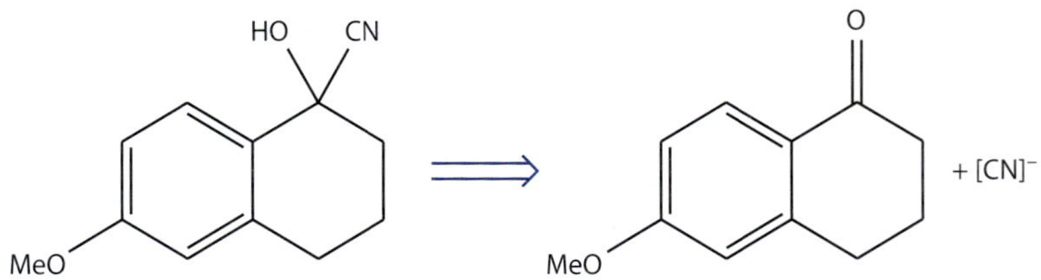

The reagent actually used was Et_2AlCN/toluene.

(b) We disconnect the bond exocyclic to the ring. So,

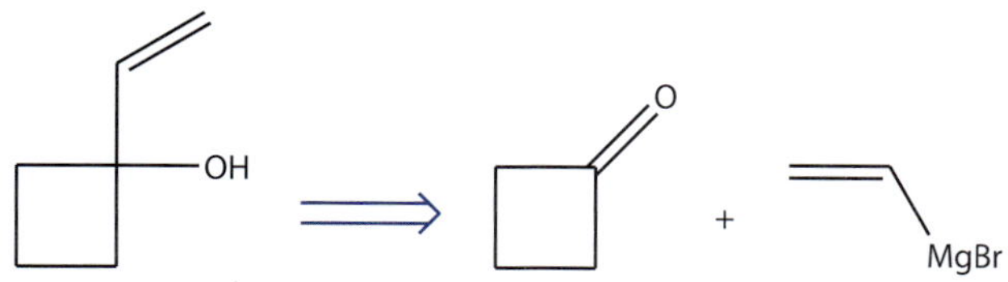

(c) Again the exocyclic bond is disconnected:

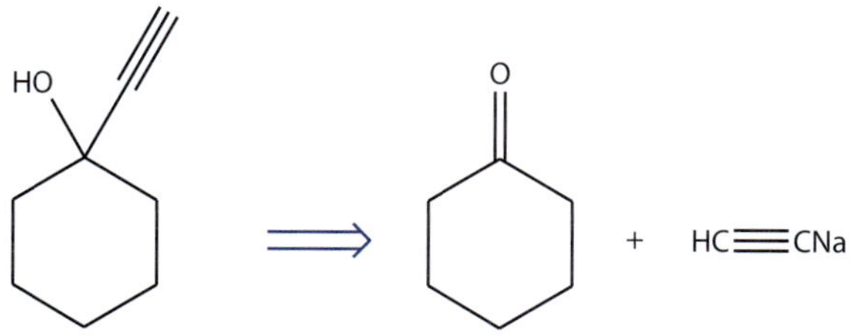

If two of the groups on a tertiary alcohol are identical, then there are two possible approaches. The first is to disconnect the "odd" group and then react a Grignard reagent with a symmetric ketone. Symmetric ketones tend to be more readily available, and usually significantly less expensive, than those with two different groups attached to the carbonyl carbon. Alternatively, the two identical groups can be disconnected simultaneously, to give the ester. Examples are given in Figure 19.7. In the first example, either strategy will be successful, and choice will be made on the basis of yields and costs. The second example is more challenging, and the ester strategy is much more likely to be successful. The preparation of a Grignard reagent from a molecule containing an amine is nontrivial; the amine would need to be deprotonated or protected, or it would protonate the Grignard reagent. Also, compounds with an $-NH_2$ and a bromine on the same carbon atom are inherently unstable. The SM for the ester route, as written, cannot be isolated; the amine would readily displace the alkoxide group from the ester to give a polymer. What is isolated and used for the reaction is the hydrochloride salt of the amine. Thus, extra moles of Grignard reagent are needed first to react with the HCl of the salt and then to deprotonate the amine. In practice, 5 mol of phenyl magnesium chloride was required to obtain a good yield. Fortunately, it is inexpensive, and the by-product, benzene, is easily removed from the product by evaporation.

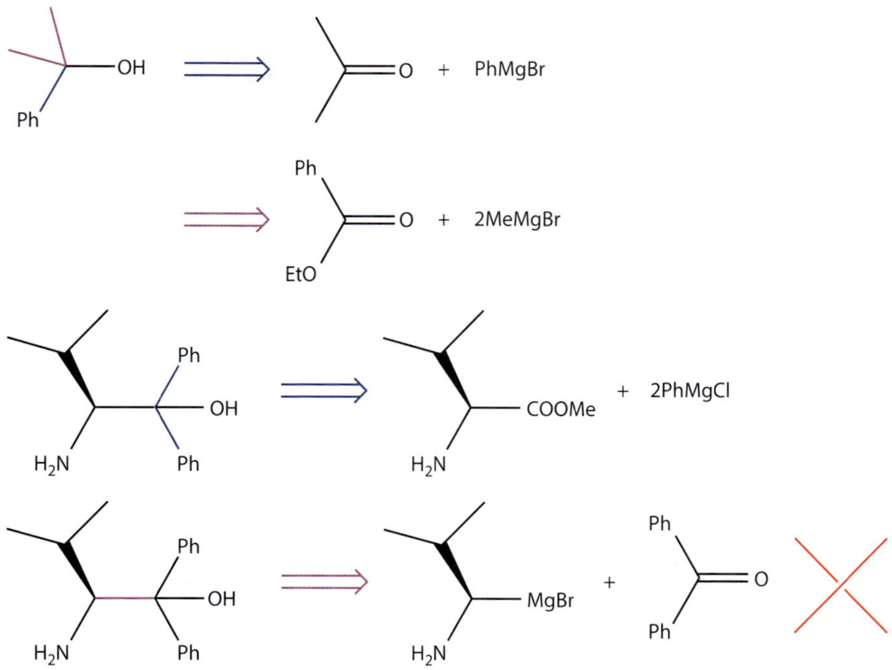

FIGURE 19.7 Disconnection of tertiary alcohols of type $R^1_2R^2COH$.

PROBLEM 19.2

Propose disconnections for each of the following alcohols:

(a) 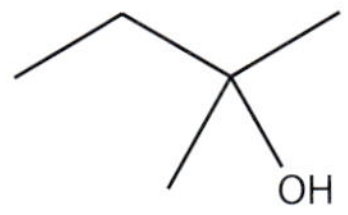

(b) 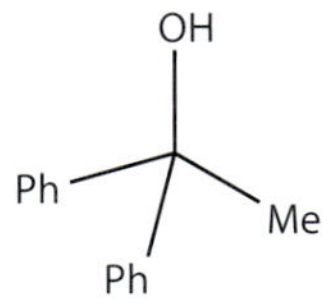

(c) 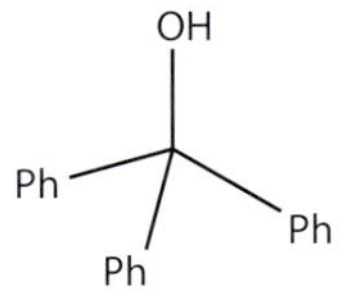

(d) 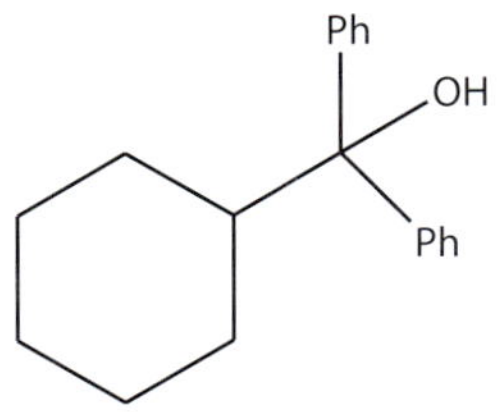

Solutions

(a) Two sensible options here—EtMgBr + $Me_2C{=}O$ and 2 mol MeMgBr + CH_3CH_2COOEt.
(b) In principle, two options: MeMeBr + $Ph_2C{=}O$ and 2 mol PhMgBr + CH_3COOEt. The latter is probably preferred because $Ph_2C{=}O$ is relatively unreactive.
(c) $PhCH_2COOEt$ + PhMgBr.
(d) 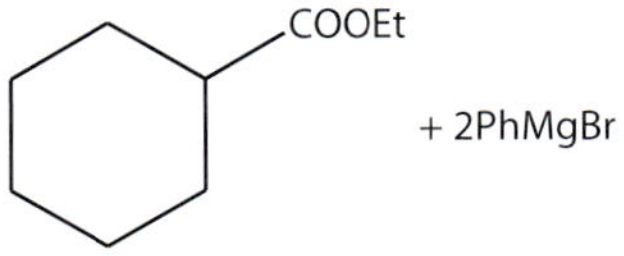

Up to this point, all our target alcohols have been tertiary, but we can also use H^- as a nucleophile with a range of carbonyl compounds. This process would be described as an FGI, a reduction, rather than a disconnection. So cyclopentanol, **19.4**, is "disconnected" to cyclopentanone, **19.5**. The forward reaction is reduction of the ketone with lithium aluminum hydride. The simplest reagents to provide hydride ion are lithium aluminum hydride or sodium borohydride, but there are many other available reagents, including modified versions of $Li[AlH_4]$ or $Na[BH_4]$. Many of these reagents, such as sodium/ethanol, are significantly less expensive, and others such as enzyme systems, or molecular hydrogen with a chiral metal complex as catalyst, may be capable of producing chiral alcohols. Primary alcohols may be prepared by an FGI of either aldehydes or esters (Figure 19.8; although lithium aluminum hydride does reduce carboxylic acids directly, this is rarely the method of choice as more forcing conditions are required, and the intermediates may be rather insoluble). We also recall that the reaction of a Grignard reagent with an aldehyde will give a secondary alcohol, as in the reaction of **19.6**.

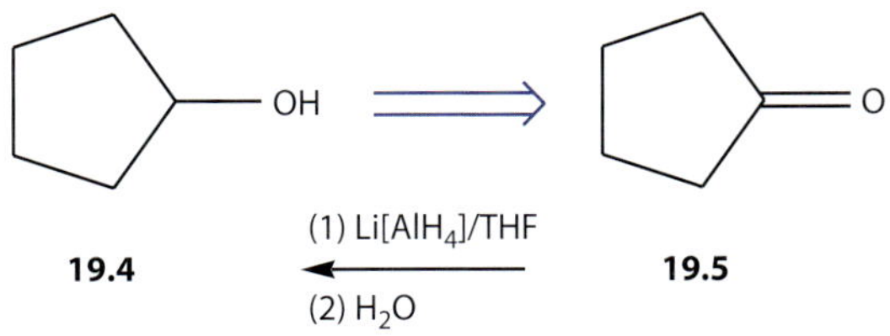

19.4 **19.5**

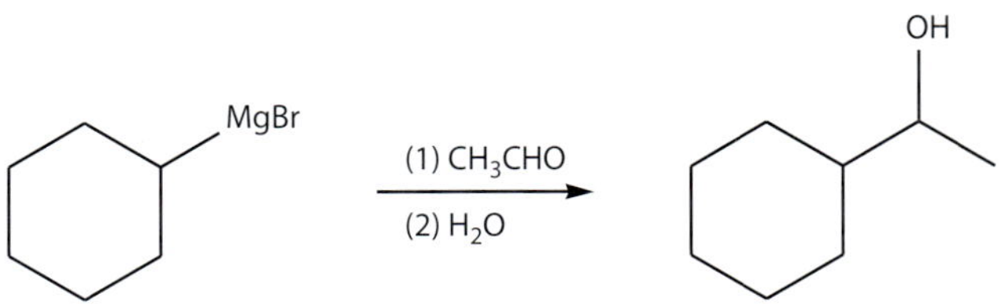

19.6

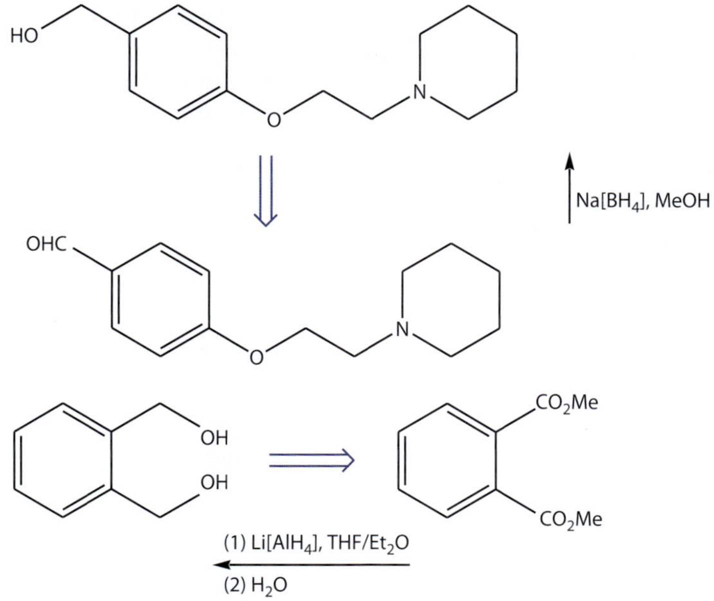

FIGURE 19.8 Disconnection of primary alcohols to aldehydes or esters.

PROBLEM 19.3

Suggest reagents for each of the following transformations:

(a)

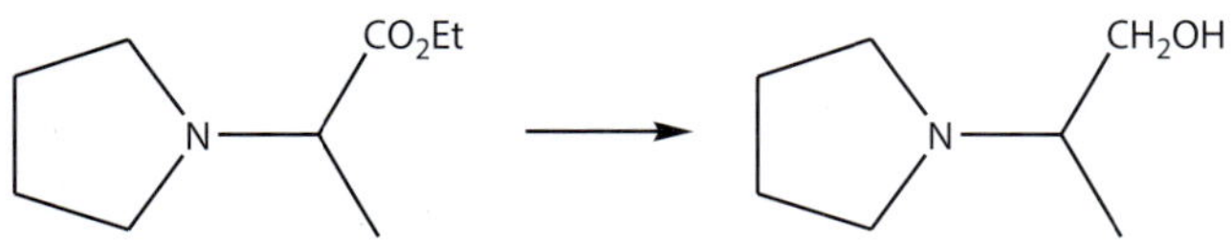

(b)

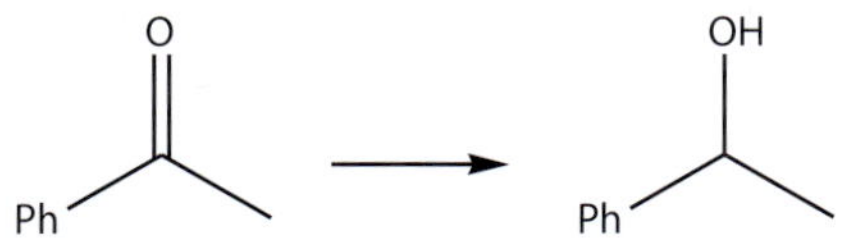

(c)

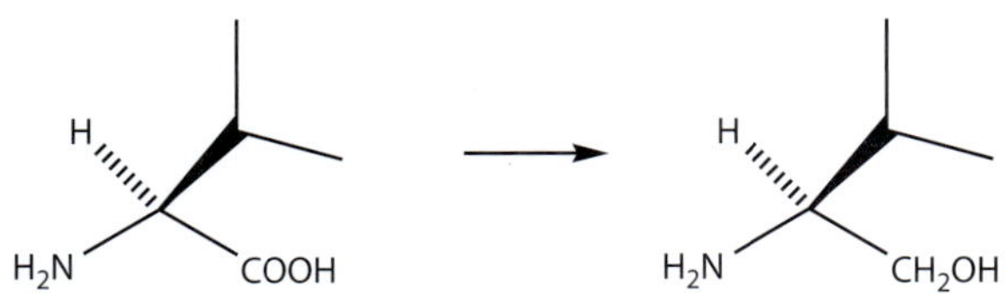

Solutions

(a) (1) Li[AlH$_4$], Et$_2$O, (2) HCl, (3) NaOH
(b) (1) Na[BH$_4$], EtOH, (2) H_3O^+
(c) (1) Li[AlH$_4$], Et$_2$O, (2) H_2O (the SM will be in its zwitterionic form, and an excess of reductant will be needed)

A completely different approach to alcohol synthesis is provided by the ring opening of epoxide with nucleophiles (Section 9.10.4). The bond that is disconnected in this case is not that to the carbon bearing the –OH group but the next one along (as in the reaction of **19.7**). Ethene oxide is a synthon for $[HOCH_2CH_2^+]$, and it may be classified as a 2-carbon electrophile for chain extension. An important aspect of the opening of epoxides by powerful nucleophiles is that the reactions follow the S_N2 mechanism and give *trans*-products. Thus, **19.9** may be disconnected to the epoxide; the forward reaction involves attack of the ethyl Grignard reagent, with inversion, at the less-hindered carbon atom.

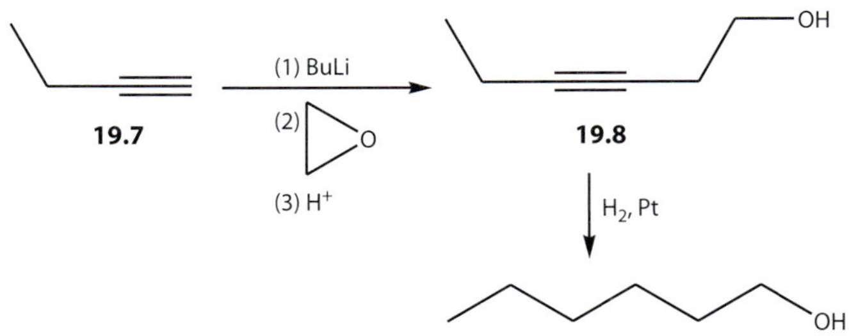

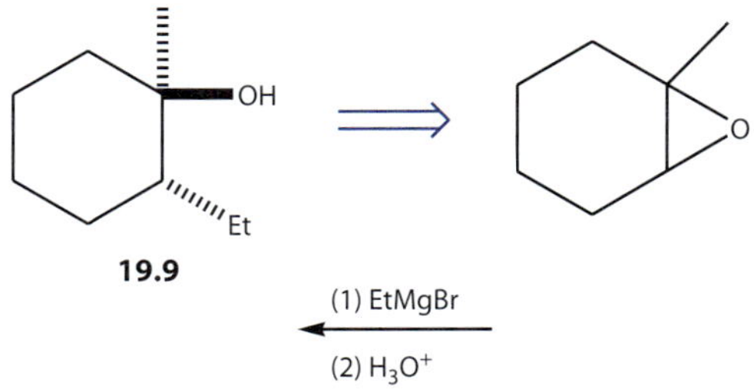

PROBLEM 19.4

Predict the product of each of the following reactions:

(a)

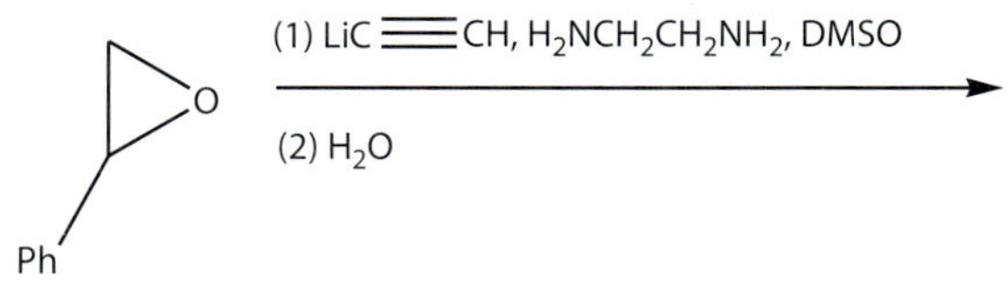

(The ethylenediamine is a complexing agent for lithium; it's needed for the reaction to work well, but you can ignore it for mechanistic purposes.)

(b)

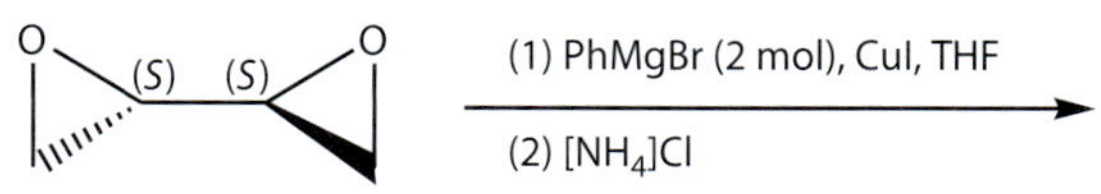

(While organocopper reagents might be involved here, don't worry about this in the mechanism.)

(c)

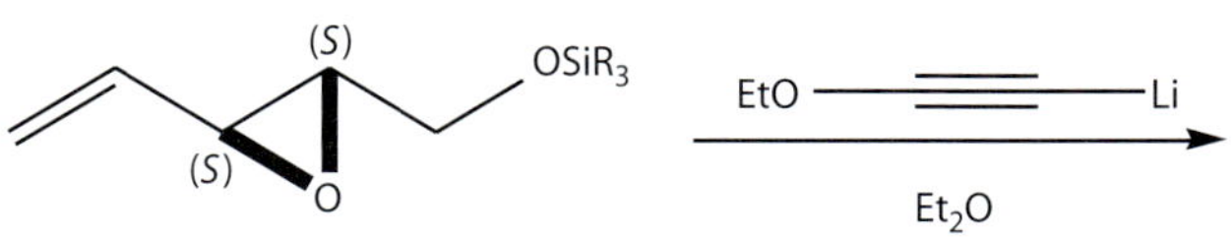

Solutions

(a) Nucleophilic attack takes place at the less substituted center.

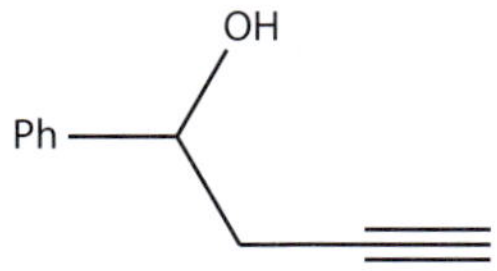

(b) Attack takes place at the less substituted center; note that nothing has happened at the chiral centers, as the attack did not take place there.

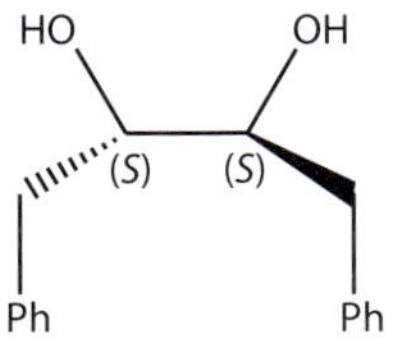

(c) Attack takes place at the less-hindered center with inversion of stereochemistry. It's perhaps not immediately obvious which center is less-hindered, but the substituents at silicon are large, and allylic centers are more reactive in S_N2 processes. In this particular example, no significant S_N2' substitution occurred.

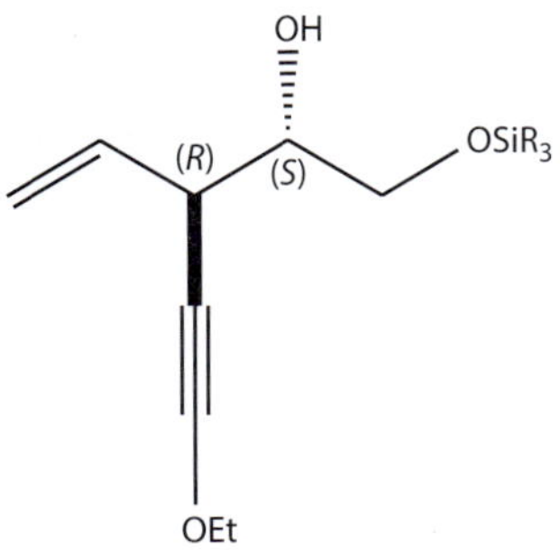

The other synthetic route to alcohols that we discussed in Chapter 9 was the hydration of alkenes. Some reminders are given in Figure 19.9. It's important, however, not to end up in a loop with

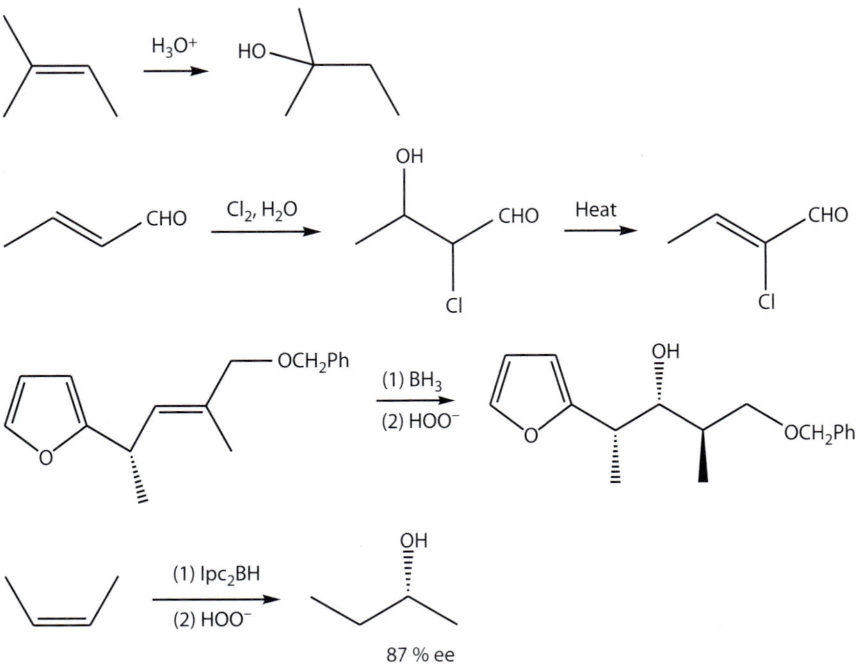

FIGURE 19.9 Additions to alkenes to yield alcohols.

such reactions; many alkenes are prepared by dehydrating alcohols, and simple acid-catalyzed additions of water to alkenes, although providing interesting examples, are rarely seen in modern synthesis. Addition of water through hydroboration/oxidation gives the potential for much more selective reactions. Remember that acid-catalyzed hydration gives Markovnikov products with little or no stereoselectivity, while hydroboration gives stereospecifically *cis*-addition and anti-Markovnikov products.

PROBLEM 19.5

Show how each of the following alcohols could be prepared using a boron reagent:

(a)

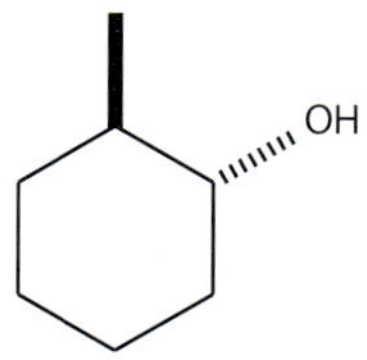

(b)

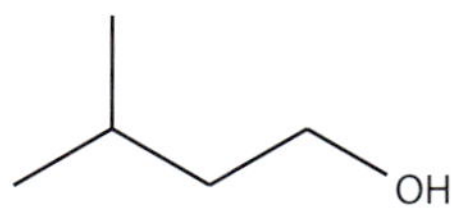

Solutions

(a) 1-Methylcyclohexene with (1) BH_3, (2) H_2O_2/HO^-

(b)

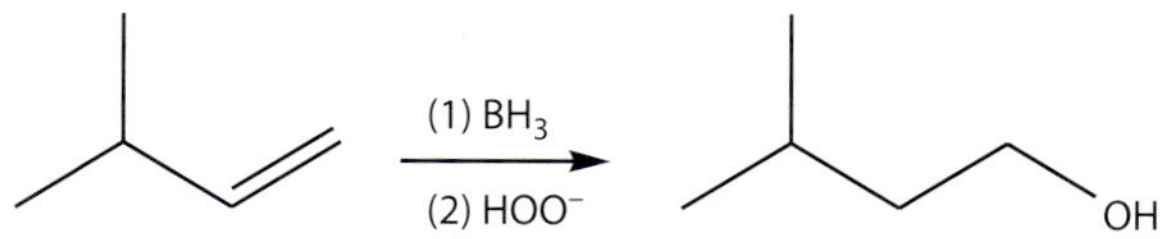

Although many alcohols are of synthetic interest in their own right, their importance to synthetic analysis lies in the fact that they can be converted into so many other classes of compound. Some important examples are in Table 19.1, but neither reaction types nor available reagent lists are intended to be comprehensive.

TABLE 19.1 Simple Conversions of Alcohols

Alcohol	Conversion	Reagent	Product	Comments
RCH_2OH	Oxidation	$K[MnO_4]$, $K_2[Cr_2O_7]$	RCOOH	
RCH_2OH	Oxidation	CrO_3/py; DMSO/$(COCl)_2$/Et_3N	RCHO	
RR′CHOH	Oxidation	CrO_3, $K[MnO_4]$; DMSO/$(COCl)_2$/Et_3N	RR′C=O	
ROH (1 °, 2 °, or 3 °)	Substitution	HX, PX_3 (X = Cl, Br); $SOCl_2$, PCl_5	RX	Elimination competes, especially with 3 ° alcohols. Rearrangement may compete.
ROH (1 °, 2 °)	Esterification	R′COCl, R′COOH/H^+	R′COOR	Some 3 ° alcohols react well, but elimination can be problematic.
ROH	Substitution	R′X (X = halide, tosylate)	ROR′	Works best when R′ is primary.
ROH	Elimination	Acid	Alkenes	Easiest when the alcohol is tertiary. Rearrangement may compete.
ROH	Substitution	4-$MeC_6H_4SO_2Cl$ (TsCl)/py	ROTs	Converts OH into a good leaving group for substitution reactions.
ROH	Substitution	Ethylene oxide/base	$ROCH_2CH_2O^-$	Both intra- and intermolecular versions known.

Key Points from Section 19.3.1

- Alcohols can be prepared by the addition of organometallic reagents to aldehydes (giving secondary alcohols) or ketones (giving tertiary alcohols).
- The best disconnection of alcohols to organometallic and carbonyl compound is normally the one that gives the most stable anion.
- Tertiary alcohols with two identical groups may be prepared by addition of a Grignard reagent to an ester.
- Ring opening of epoxides by organometallics occurs at the less-hindered center and gives reliable inversion of configuration.
- Anti-Markovnikov addition of water to alkenes using boranes is *cis*-stereoselective, and enantioselective versions of the reaction are available.

19.3.2 DISCONNECTIONS OF ALKENES

One of the major synthetic approaches to alkenes is the dehydration of alcohols—which in turn are produced from aldehydes or ketones and nucleophiles (mainly Grignard reagents). Consider **19.10**. In principle, there are two alcohols that could be dehydrated to give this alkene. There are two obvious routes to the tertiary alcohol; $PhCH_2MgBr + Ph_2C{=}O$ and $PhCH_2COOEt$ + 2 mol PhMgBr. Both are successful in making this compound; the only potential problem (which you would not necessarily know until you tried it) is that benzyl Grignards have a tendency to self-couple through radical processes. Synthesis of the secondary alcohol looks straightforward—you would need PhCHO + $Ph_2CHMgBr$. However, the homocoupling of this Grignard reagent is even more facile than that of $PhCH_2MgBr$. It can be overcome, but why make life difficult? Added to this, although there is only one potential product from either dehydration reaction, the elimination is easier from the tertiary alcohol (easier to make the more stable carbocation). This is quite a general observation; tertiary alcohols are easy to make and easy to dehydrate, so "adding" the OH at the most substituted position is generally a good strategy.

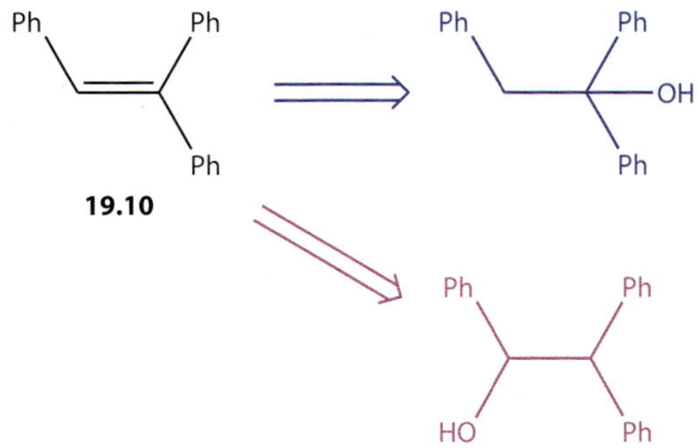

19.10

Another example is provided by **19.11**. Again there are two alcohols that could be dehydrated. If we consider the forward process for the tertiary alcohol, there is no ambiguity about the dehydration; there is only one possible product. There is no reason that the alcohol cannot be prepared from the ketone and PhMgBr. This is the practical process. The ketone is converted to the alcohol with $PhMgBr/Et_2O$ followed by work-up with aqueous acid, and the dehydration occurs under very mild conditions in the presence of acetic anhydride. The secondary alcohol could be prepared by opening an epoxide with a Grignard reagent. The epoxide can be prepared from the alkene, which is available. One extra step is not necessarily critical in overall yield. However, when we consider the dehydration of the alcohol (which will, in practice, have *trans*-stereochemistry, but this should not matter if we choose conditions favorable to an E1 process), there is an ambiguity—two dehydration products are possible. The desired, conjugated product will probably predominate—but why use an ambiguous process when an unambiguous one is available?

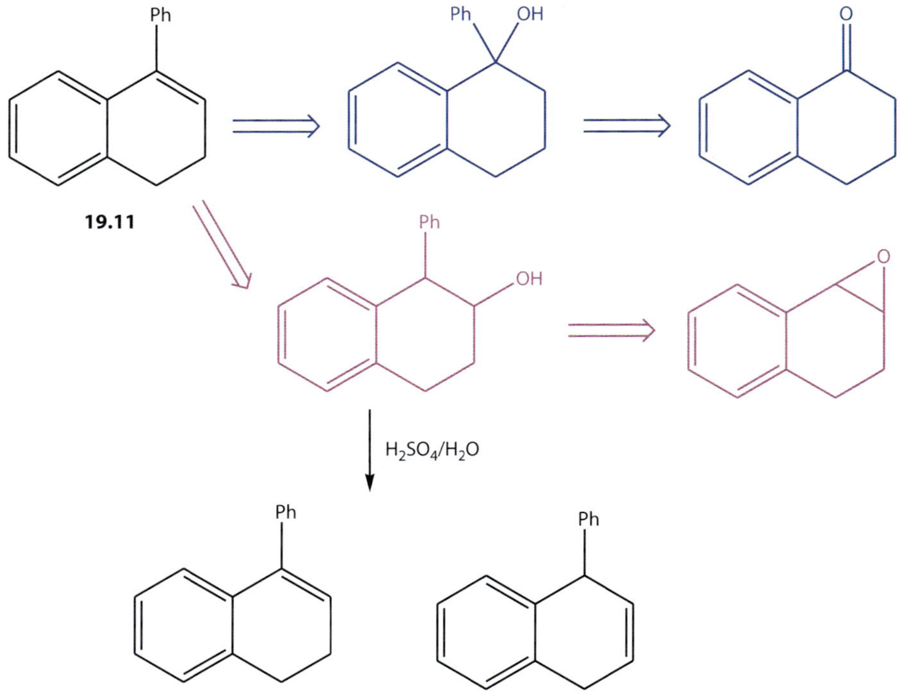

19.11

PROBLEM 19.6

Propose reaction sequences for the following transformations:

(a)

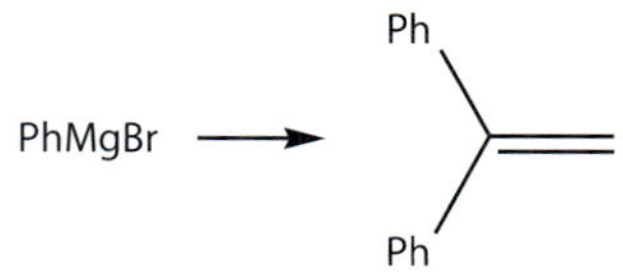

(b)

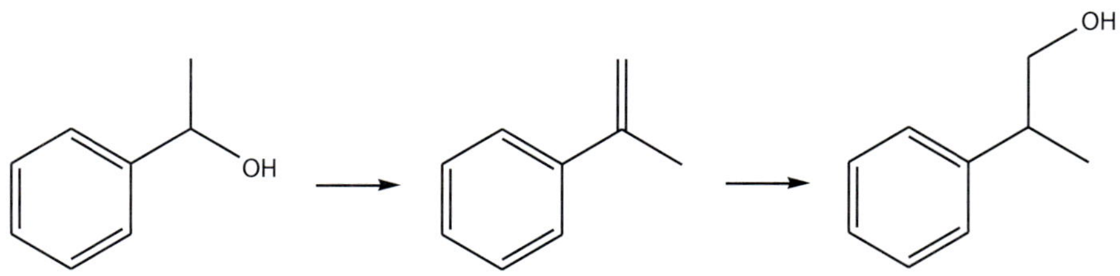

Solutions

(a) The alkene can be prepared by alcohol dehydration:

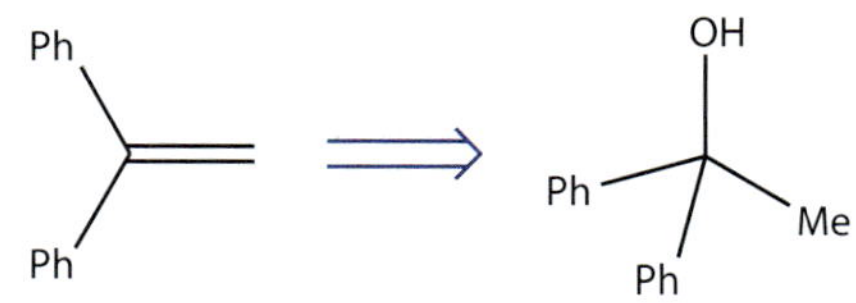

The alcohol is prepared by reaction of PhMgBr with MeCOOEt:

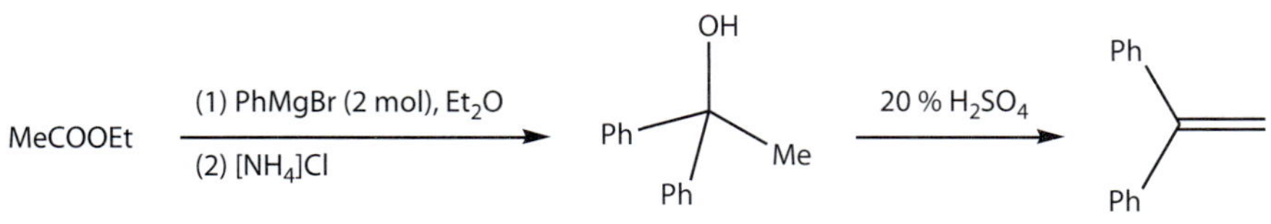

(b) The first step is dehydration to give the alkene; given that it is conjugated, this should be straightforward, using dilute sulfuric acid. The second step is anti-Markovnikov addition of water to the alkene, accomplished using BH_3 followed by H_2O_2/HO^-.

Another useful synthetic approach to alkenes involves starting with an alkyne. This is particularly useful if we need to produce a *cis*- or *trans*-alkene stereospecifically; we have available to us the reactions of Figure 19.10, which accomplish this. The process is more commonly used to produce *cis*-isomers, as they are more difficult to prepare in other ways, being thermodynamically less stable. The alkyne may then be disconnected to give the dianion of ethyne and two alkyl halides or other appropriate electrophile. Practical examples are provided in Figure 19.11.

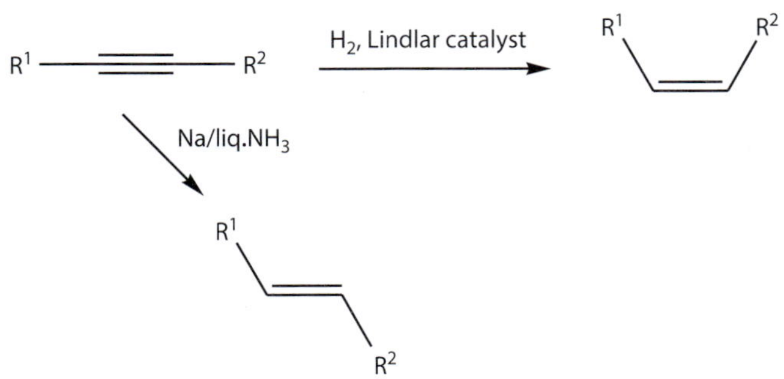

FIGURE 19.10 Stereospecific reductions of alkynes.

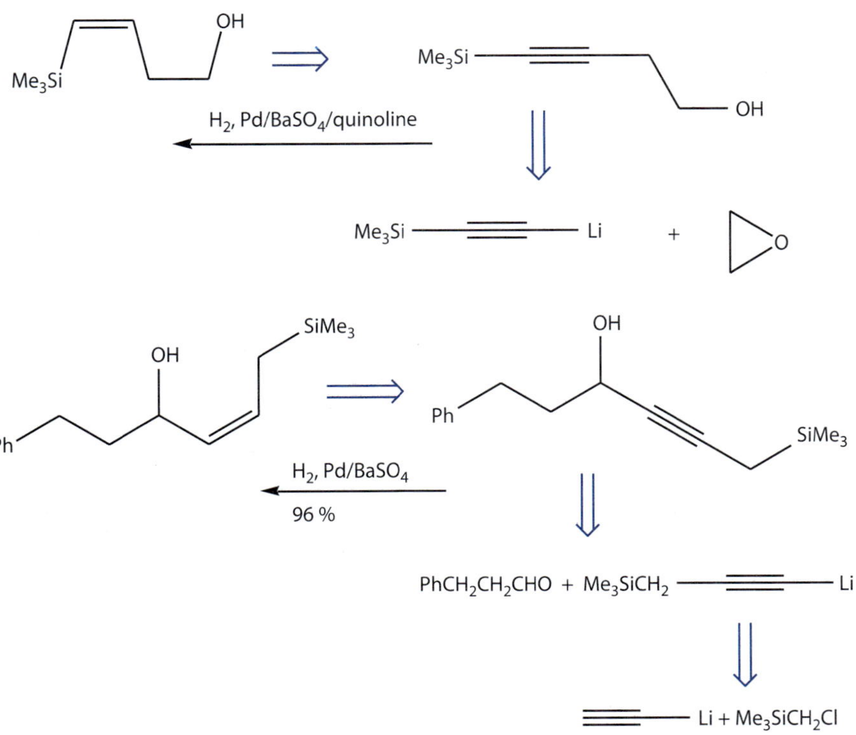

FIGURE 19.11 Disconnection of alkenes *via* an alkyne.

PROBLEM 19.7

Propose reagents for each step of the following reaction sequences:

(a)

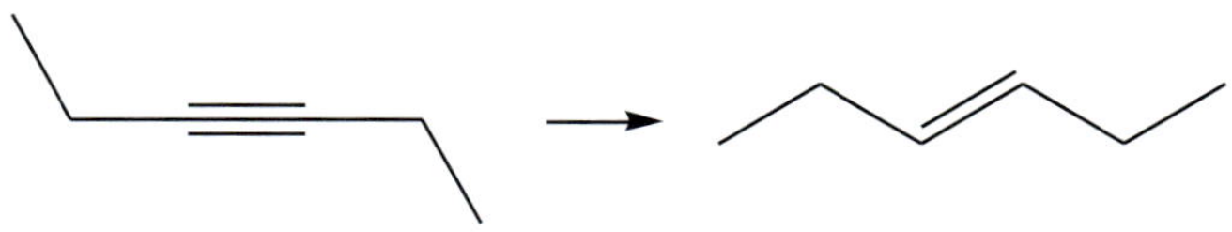

(b)

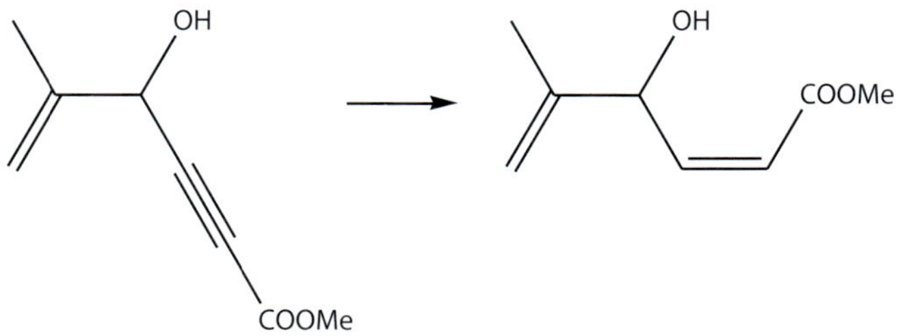

(c)

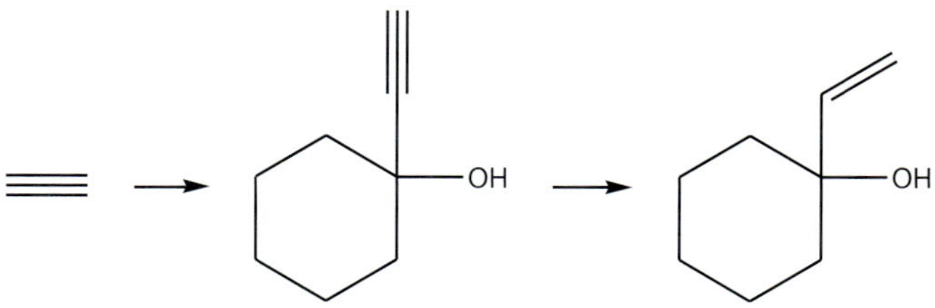

Solutions

(a) Na/liq NH_3.
(b) H_2/Lindlar catalyst would work; the catalyst actually used was Pd/$BaSO_4$/py.
(c) Step 1 (1) BuLi/Et_2O, (2) cyclohexanone, (3) H_3O^+.
Step 2 H_2, Lindlar catalyst.

The reaction that we have previously met that allows us to disconnect *through* a carbon–carbon double bond is the Wittig reaction, reviewed in Figure 19.12. This was developed by Georg Wittig in 1954, and he was awarded the Nobel Prize for this work in 1979. The process is widely used in complex syntheses; aldehydes are significantly more reactive than ketones, but both react so long as the ylid has no additional stabilization from an electron-withdrawing group (Figure 19.13). Some selectivity can also be achieved between carbonyl groups on steric

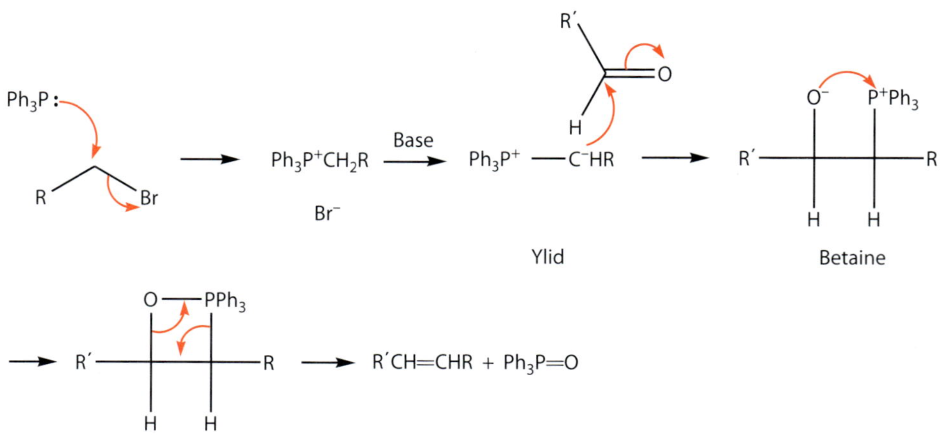

FIGURE 19.12 Mechanism of the Wittig reaction.

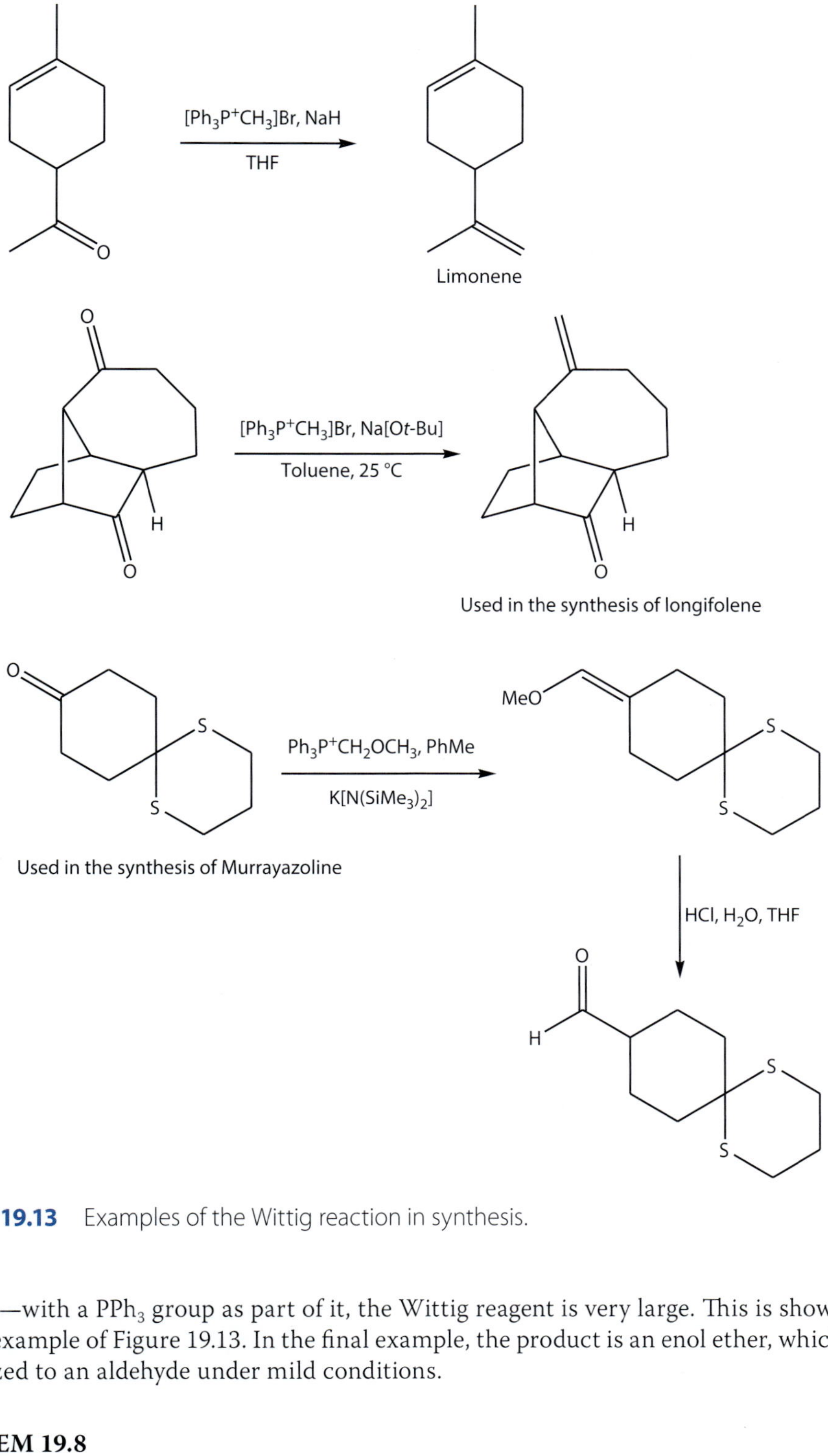

FIGURE 19.13 Examples of the Wittig reaction in synthesis.

grounds—with a PPh_3 group as part of it, the Wittig reagent is very large. This is shown in the second example of Figure 19.13. In the final example, the product is an enol ether, which can be hydrolyzed to an aldehyde under mild conditions.

PROBLEM 19.8

Give the structure of the Wittig reagent required to affect each of the following transformations:

(a)

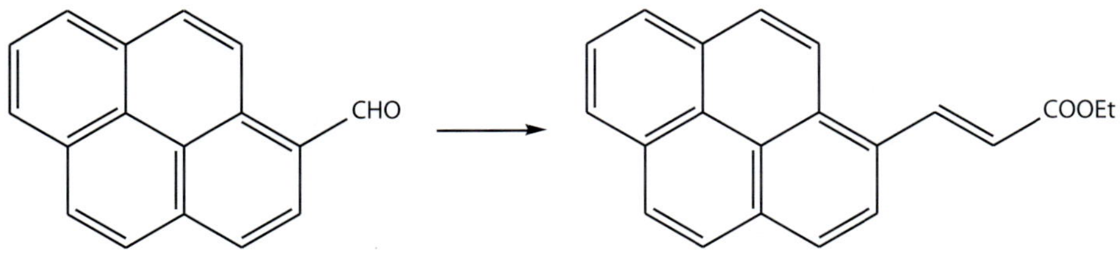

(b)

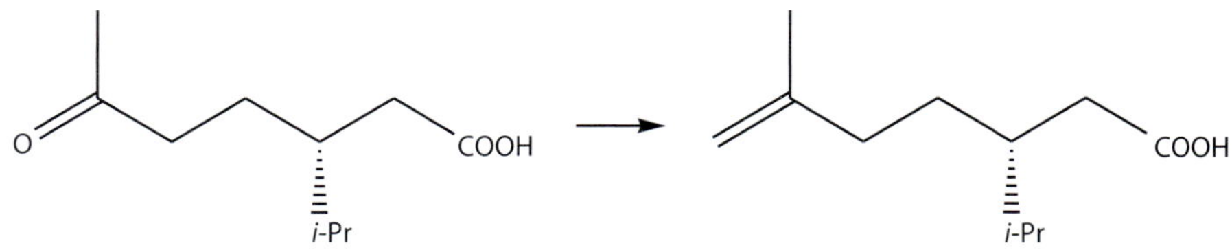

(c)

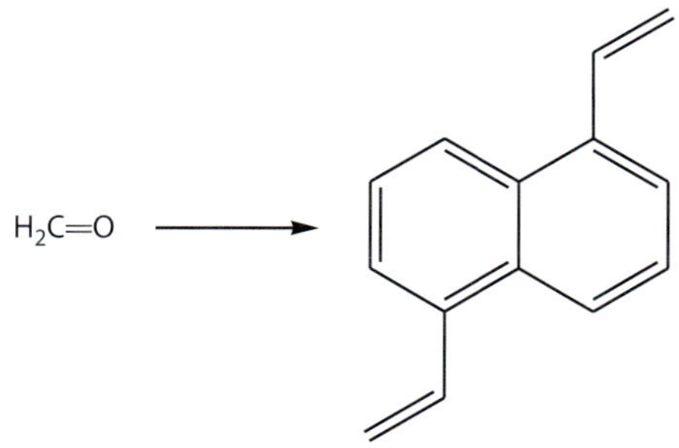

Solutions

(a) $Ph_3P^+C^-HCOOEt$.
(b) $Ph_3P^+{-}CH_2^-$, used in a synthesis of retigeranic acid.
(c) Don't make the assumption that the aldehyde or ketone is always the most valuable part of the molecule.

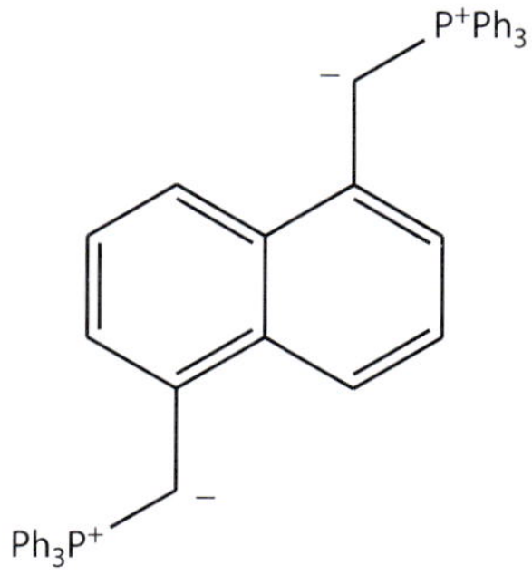

We now extend our knowledge of this process, finding that the stereochemistry of the product is very much dependent on the nature of the Wittig reagent and the exact conditions (nature of the base, kinetic, or thermodynamic conditions) under which it takes place. In general, *cis*-alkenes (must be kinetic control, as these are not the thermodynamic products) are produced when the ylid is unstabilized, as in $Ph_3P^+C^-HCH_2CH_2CH_3$ (Figure 19.14). When the ylid is stabilized by an additional electron-withdrawing group as in $Ph_3P^+C^-HCOR$, a *trans*-alkene is obtained. However, while such ylids react readily with aldehydes, their reactions with ketones are at best sluggish. In practice, there are now enough variants of the reaction in terms of bases, additives, solvents, temperature, *etc.*, that reasonable stereochemical control can usually be achieved.

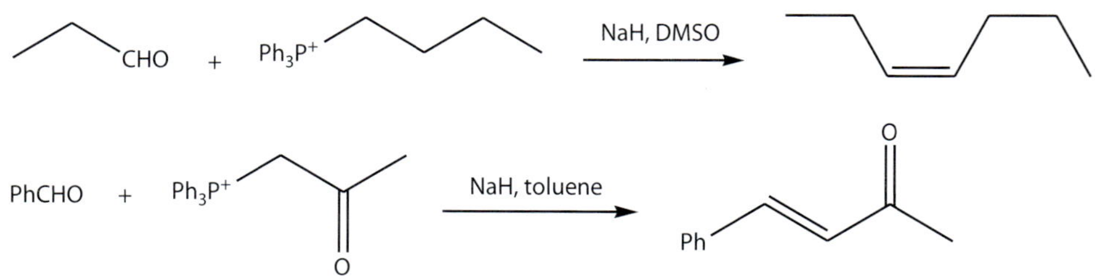

FIGURE 19.14 Stereochemistry of the Wittig reaction.

PROBLEM 19.9

Predict the outcome of the following reactions, paying attention to stereochemistry:

(a) PhCHO + $Ph_3P{=}CHCOOEt$ →

(b)

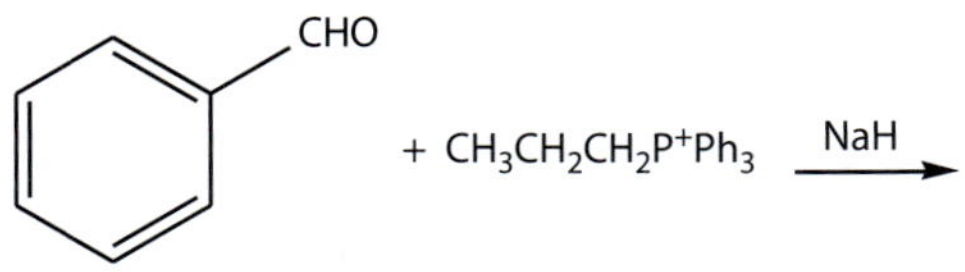

Solutions

(a) Stabilized ylid, *trans*-product:

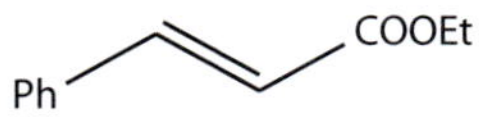

(b) Unstabilized ylid, *cis*-product:

Focus on the Horner–Wadsworth–Emmons Reaction

Two practical problems with the Wittig process were resolved by a modification due to Horner, Wadsworth, and Emmons. The first, common to all Wittig reactions, is that the by-product of the process, $Ph_3P{=}O$, is sometimes quite difficult to separate from the desired alkene product; chromatography is usually required. Secondly, some of the stabilized ylids are relatively unreactive. The modification involves the use of a phosphonate rather than a phosphine as the source of the ylid. $P(OEt)_3$ is reacted with the relevant halide, RCH_2X, and the product is the phosphonate, $RCH_2P({=}O)(OEt)_2$. This is a reaction mechanism we have not met before, the Arbuzov process (Figure 19.15). The ylids formed from these phosphonates are significantly more nucleophilic than simple Wittig reagents and react with either aldehydes or ketones (stereochemical control in reactions of ketones is, however, generally poor). In addition, the by-product, $(EtO)_2P({=}O)O^-$, is water soluble and can be removed from the reaction product by aqueous extraction. Some examples of the process are given in Figure 19.16; the reaction with aldehydes is generally selective for the production of *E*-isomers, and the reaction is thought to proceed under thermodynamic control. The first example shows that ketones do react; the second, that aldehydes react better. This was used in a synthesis of retigeranic acid and the final example in a synthesis of ampelopsin F.

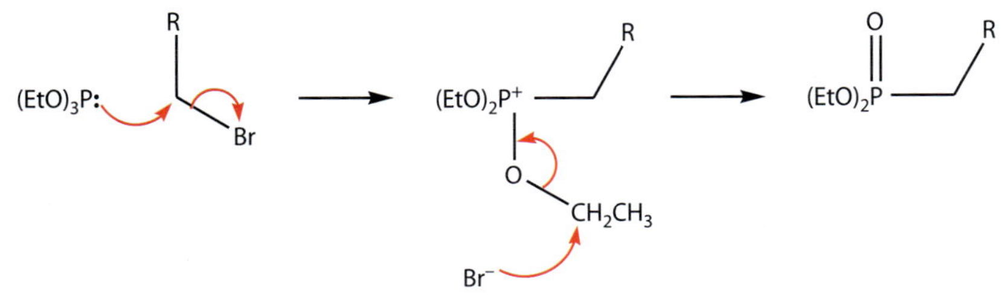

FIGURE 19.15 Mechanism of the Arbuzov reaction.

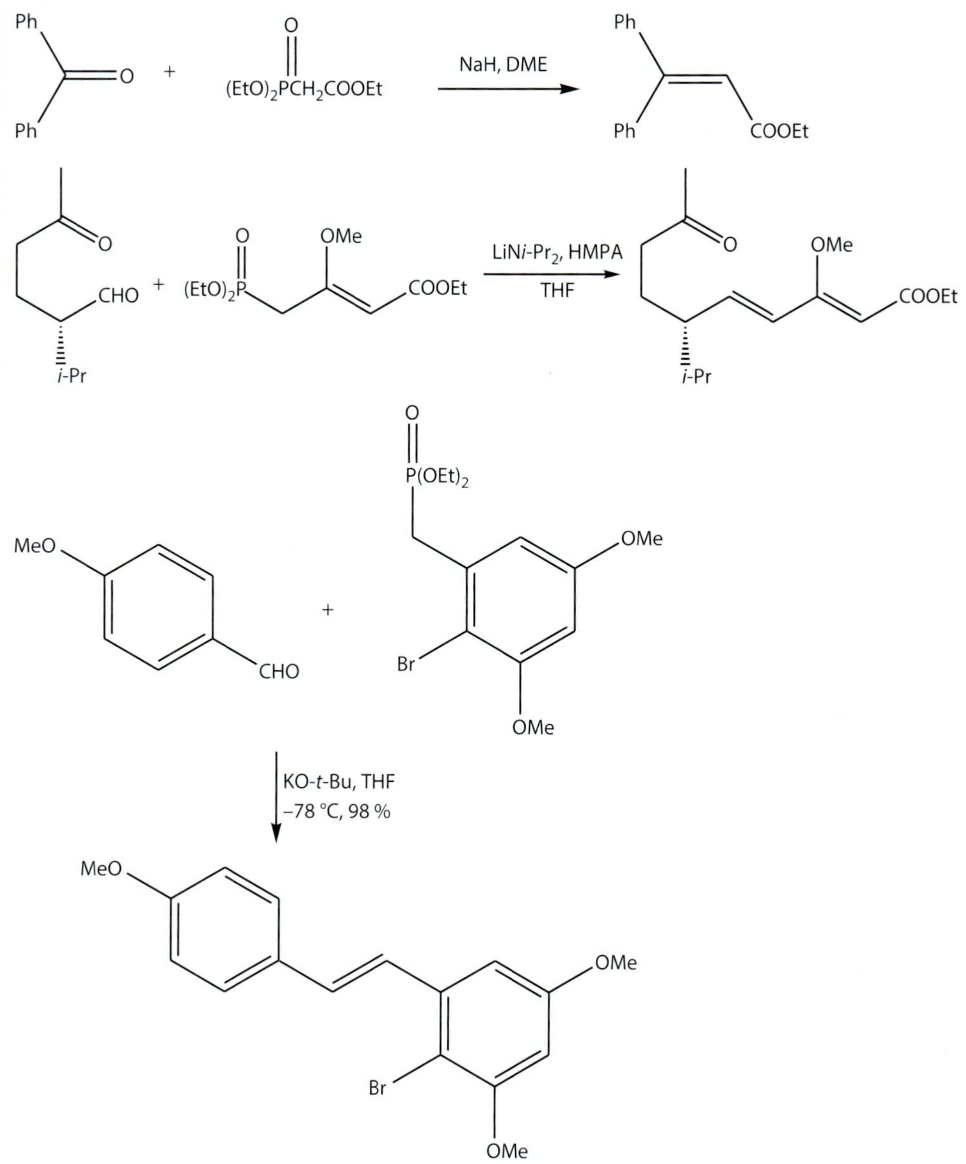

FIGURE 19.16 Uses of the Horner–Wadsworth–Emmons process in synthesis.

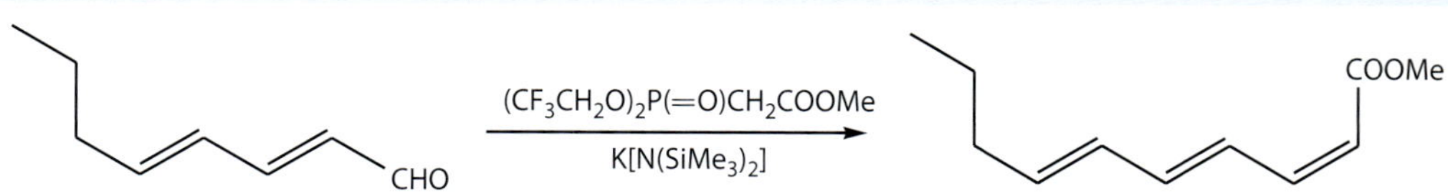

FIGURE 19.17 Still–Gennari modification of the HWE reaction.

But suppose we want the *Z*-alkene. Fortunately another simple modification (Still–Gennari) will oblige. Instead of triethyl phosphite as the SM, we use $(CF_3CH_2O)_3P$ as the SM and make ylids from $(CF_3CH_2O)_2P(=O)CH_2R$. These are deprotonated with the powerful base $K[N(SiMe_3)_2]$. This gives much more reactive ylids, which, under conditions of kinetic control, lead to the production of *cis*-alkenes (Figure 19.17).

Key Points from Section 19.3.2

- Alkenes may be prepared by alcohol dehydration; tertiary alcohols are the easiest to dehydrate.
- Alkynes may be converted stereospecifically to *cis*- or *trans*-alkenes.
- Wittig reactions, and various modifications thereof, convert aldehydes or ketones to alkenes, by one-pot double-bond formation. Chemo- and stereoselective versions of the reaction are available.

19.3.3 FGIs AND DISCONNECTIONS LEADING TO ALDEHYDES AND KETONES

In the earlier section, we saw that aldehydes can be prepared by oxidation of primary alcohols and ketones similarly from secondary alcohols (Figure 19.18). The alcohols in turn are prepared, for example, from Grignard reagents plus carbonyl compounds or epoxides. Chromium trioxide

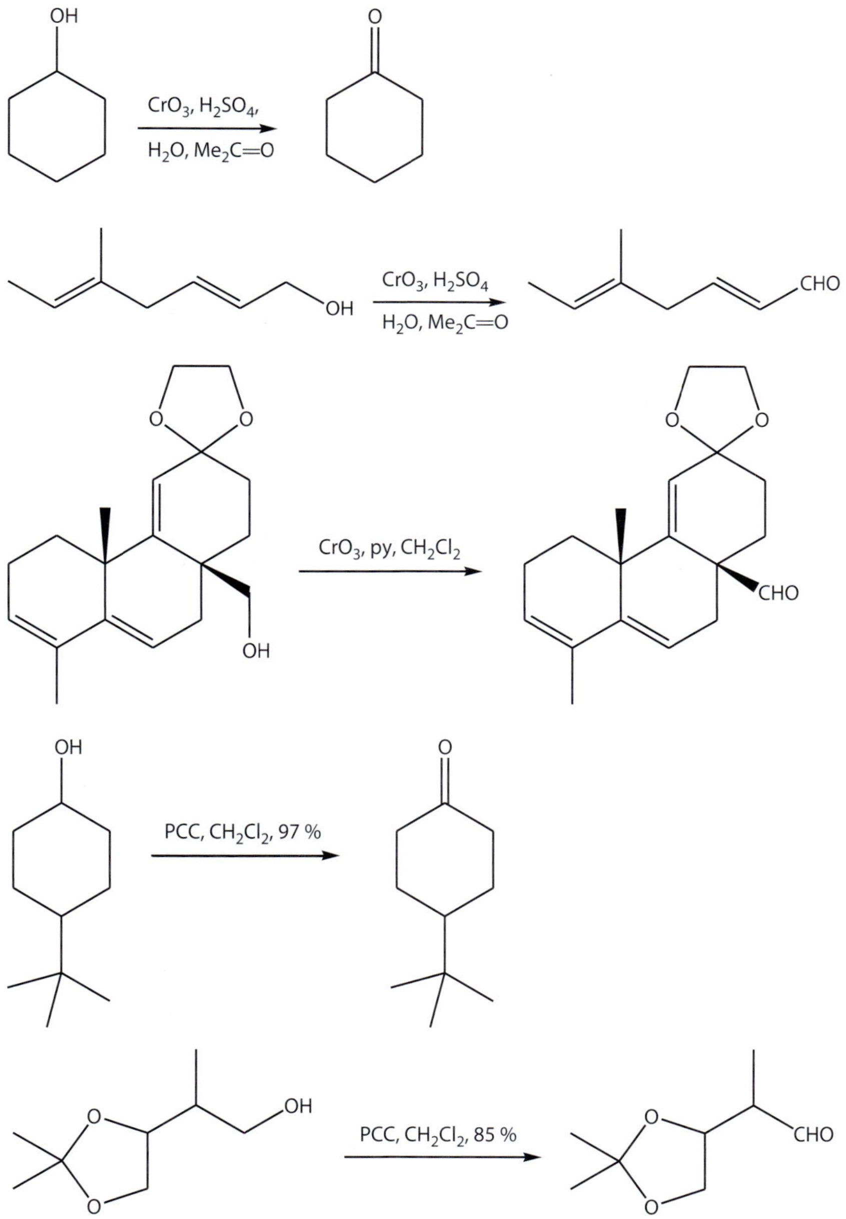

FIGURE 19.18 Aldehydes and ketones by oxidation of alcohols using chromium oxidants.

in dilute sulfuric acid (the Jones reagent) oxidizes both primary and secondary alcohols. Many primary alcohols are easily over oxidized to give carboxylic acids, if the aldehyde produced forms a significant amount of hydrate. The reagent is useful for production of aldehydes only when hydration of the aldehyde is unfavorable, as with allylic or benzylic alcohols. Chromium trioxide in pyridine can be used as an oxidant under essentially anhydrous conditions (the structure of the reagent is **19.12**, the Collins reagent). This oxidizes primary alcohols selectively to aldehydes and secondary alcohols to ketones and is particularly useful for substrates containing acid-sensitive functional groups. PCC, **19.13**, is also used under anhydrous conditions and is selective for the primary alcohol to aldehyde transformation, as well as successfully converting secondary alcohols to ketones. To summarize, most Cr(VI) reagents may be used to oxidize secondary alcohols to ketones, while for primary alcohols to aldehydes, use CrO_3/py or PCC.

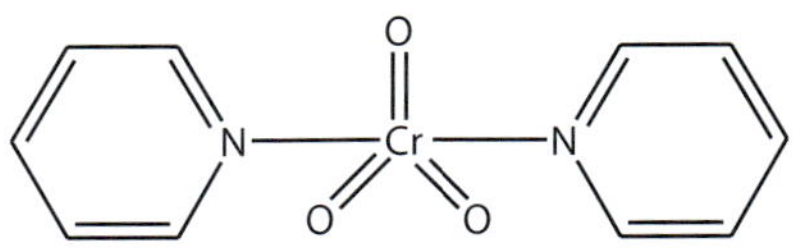

19.12

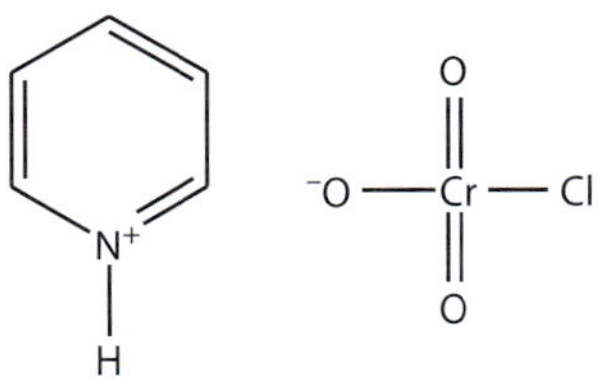

19.13

PROBLEM 19.10

Predict the products of each of the following processes:

(a)

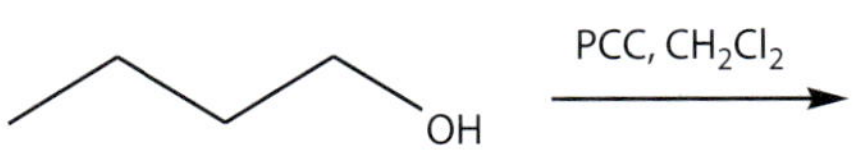

(b)

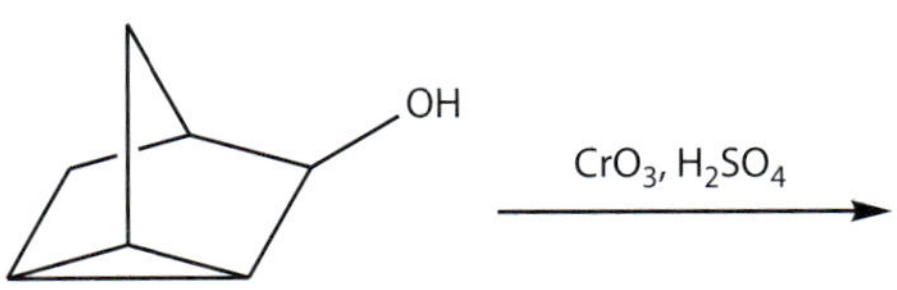

(c)

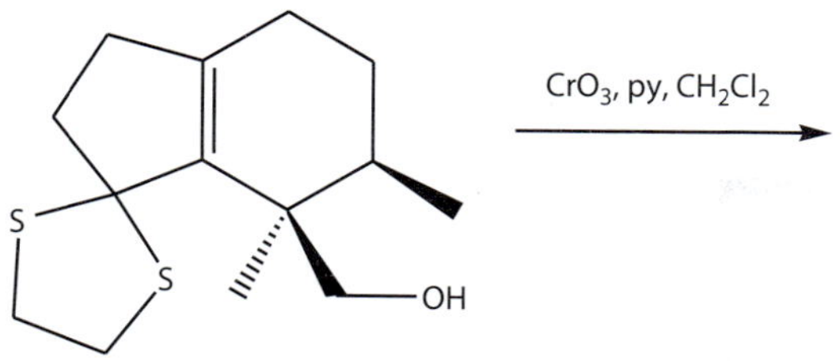

Solutions

(a) PCC is selective for oxidation of primary alcohols to aldehydes:

(b) Jones oxidation:

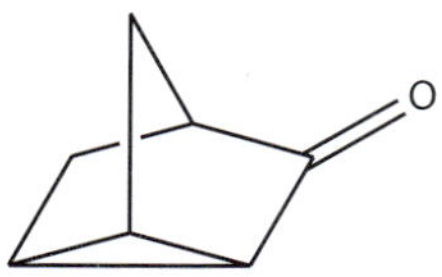

(c) Another reagent selective for primary alcohol to aldehyde:

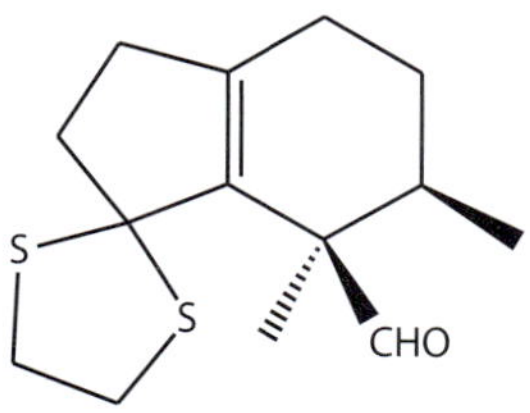

In recent years, chromium reagents have become less popular. Although inexpensive and effective, they are rather toxic and some are carcinogenic. Disposal of waste products also presents an issue. Many researchers now use the Swern oxidation—this uses dimethylsulfoxide as the oxidant, and the alcohol is activated with oxalyl chloride. The mechanism is shown in Figure 19.19. The advantages of the process are the lack of Cr(VI), anhydrous conditions and

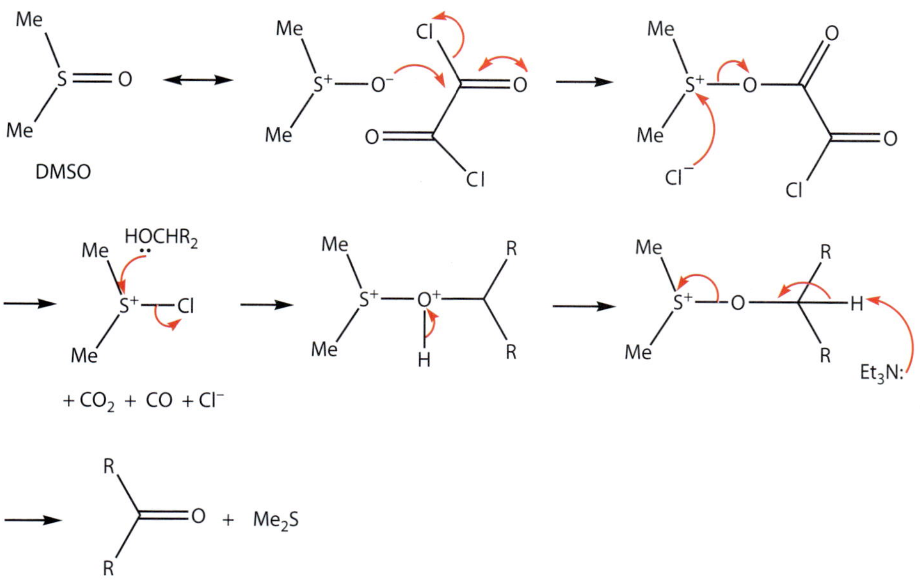

FIGURE 19.19 Mechanism of the Swern oxidation.

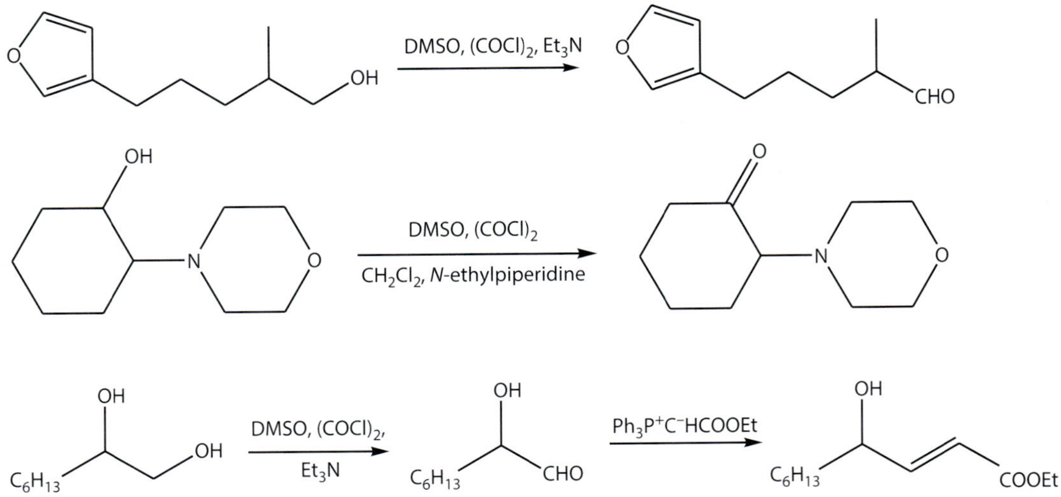

FIGURE 19.20 Examples of the Swern oxidation.

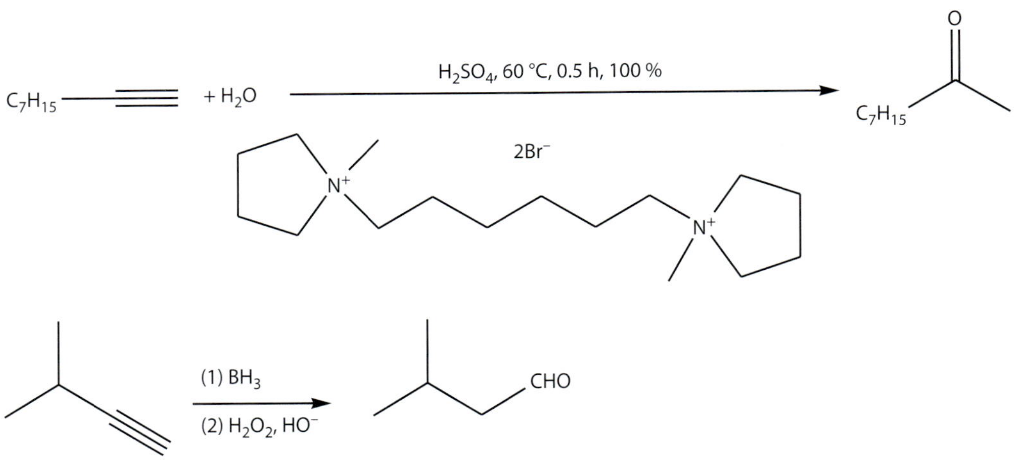

FIGURE 19.21 Preparation of ketones and aldehydes from alkynes.

low temperatures, and selectivity for primary alcohol to aldehyde. The by-products are toxic, but are gaseous, and hence do not contaminate the product. Some examples are shown in Figure 19.20. Although both primary and secondary alcohols can both be oxidized, there is some selectivity for primary centers.

The hydration of alkynes also gives a useful approach to ketones; this may be accomplished either under acid-catalyzed conditions or by hydroboration/oxidation (Figure 19.21). Remember that variously substituted alkynes are easily available.

PROBLEM 19.11

Predict the outcomes of the following reactions:

(a)

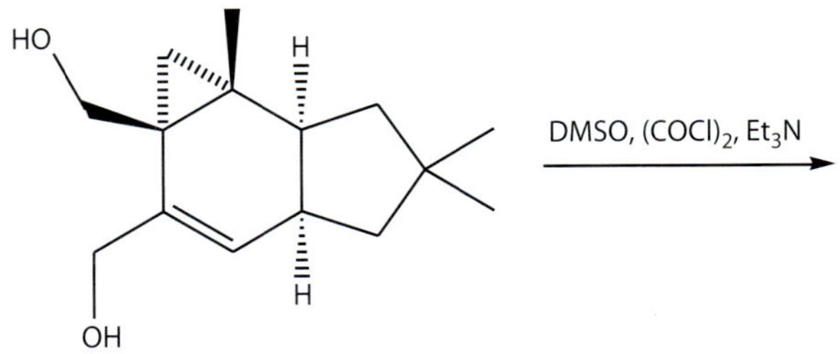

(b)

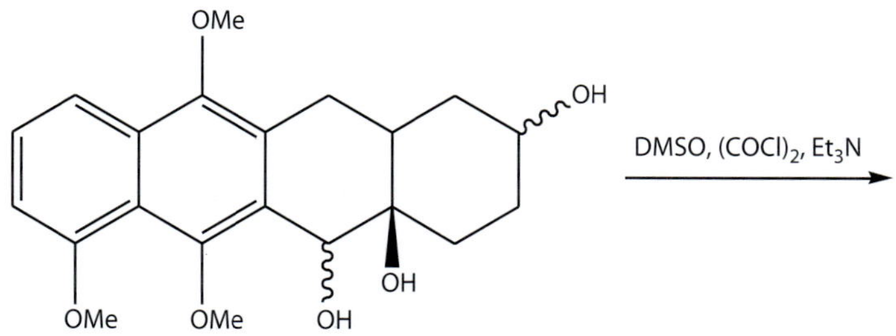

(c)

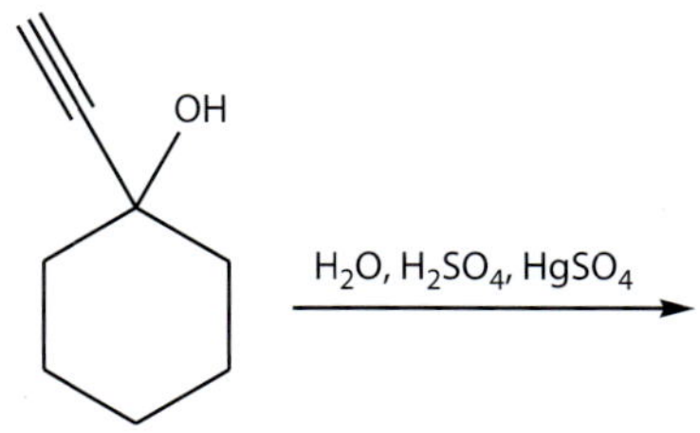

Solutions

(a) Both primary alcohols are oxidized to aldehydes:

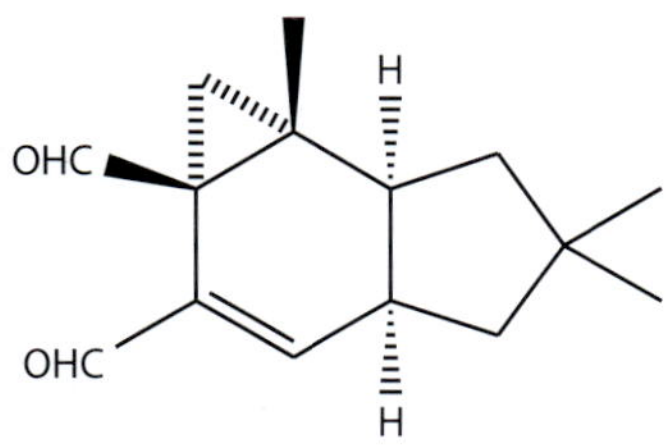

(b) Both the secondary alcohols are oxidized to ketones, but the tertiary alcohol is unaffected:

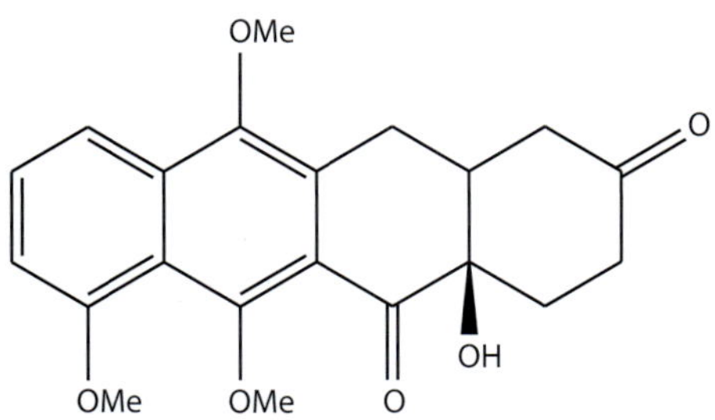

(c) This is a hydration of an alkyne in the presence of mercury (II), followed by tautomerization to give a ketone:

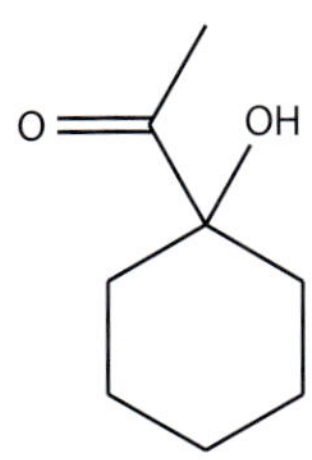

Although there are no obvious reductive approaches to ketones, there are a number for aldehydes. In Section 15.6, we saw that esters could be reduced to aldehydes using the modified hydride-reducing agent DIBAH, i-Bu_2AlH (Figure 19.22). Acid chlorides can be reduced to aldehydes using Li[AlH(O-t-$Bu)_3$] (Figure 19.23). Nitriles are reduced to aldehydes using Li[AlH(OEt)$_3$]; reduction gives the imine, which reacts no further and is hydrolyzed on work-up (Figure 19.24). The oldest, and least expensive, method however uses molecular hydrogen in the presence of a Pd/$BaSO_4$ catalyst with an acid chloride—the Rosenmund reduction (Figure 19.25).

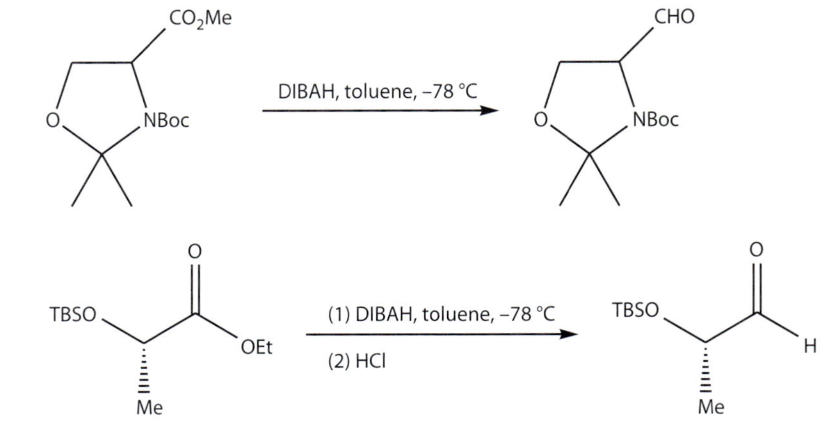

FIGURE 19.22 Reduction of esters to aldehydes.

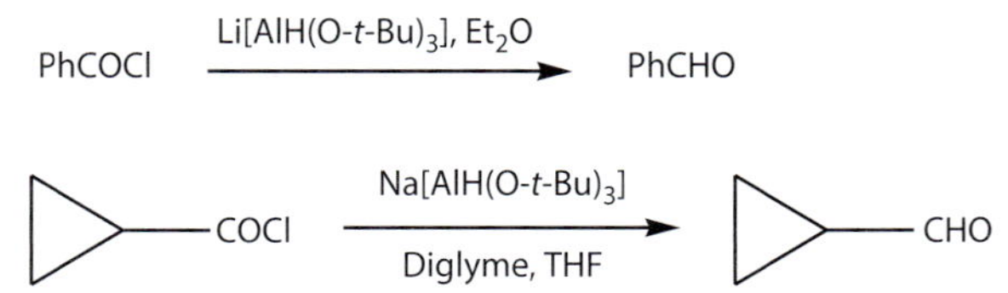

FIGURE 19.23 Reduction of acid chlorides to aldehydes.

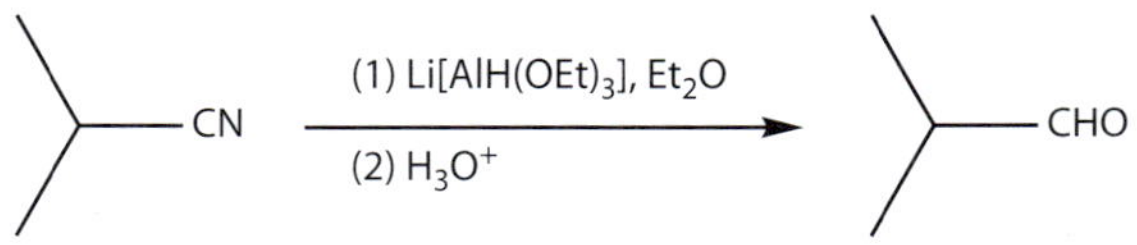

FIGURE 19.24 Reduction of nitriles to aldehydes.

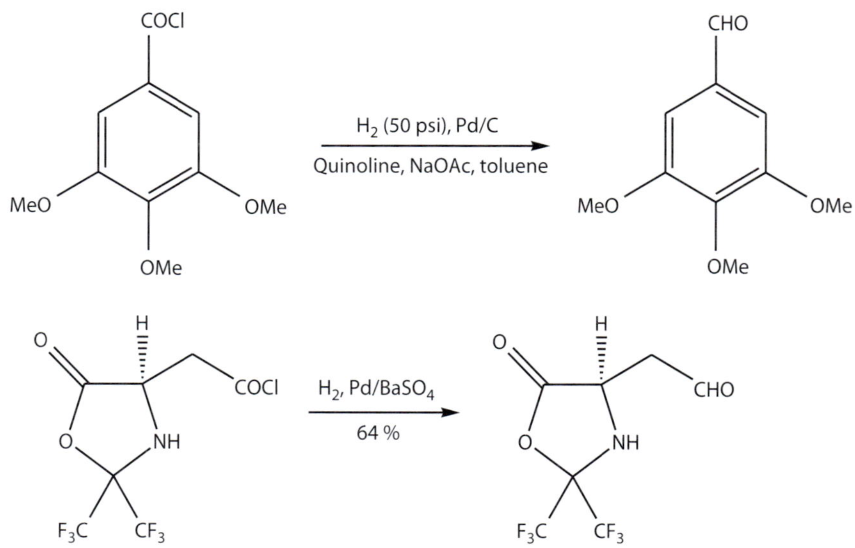

FIGURE 19.25 Rosenmund reduction of acid chlorides.

PROBLEM 19.12

Provide a reagent for each of the following transformations:

(a)

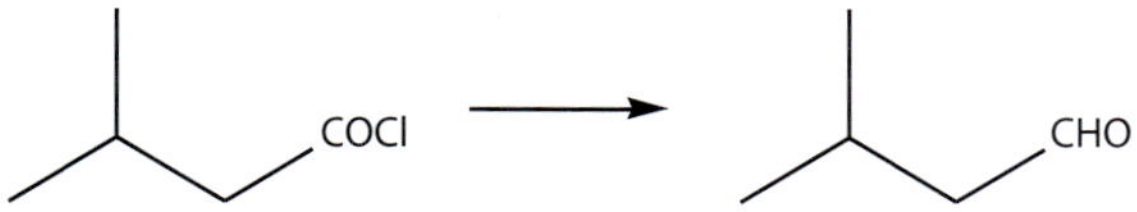

(b)

(c)

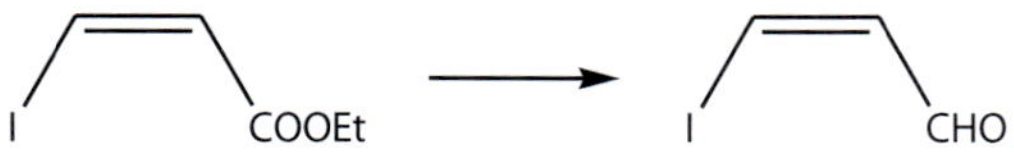

Solutions

(a) $Li[AlH(O\text{-}t\text{-}Bu)_3]$ or Rosenmund reduction ($H_2/Pd/BaSO_4$)
(b) $i\text{-}Bu_2AlH$, toluene, −78 °C, then H_2O or $Li[AlH(OEt)_3]$
(c) $i\text{-}Bu_2AlH$, toluene, −78 °C, then H_2O

The disconnection of a ketone, $R^1C(=O)R^2$, through one of the bonds to the carbonyl group gives us one carbanion and a species $[RC^+{=}O]$, the acylium ion. We know that the synthon for a carbanion is an organometallic compound of some sort (however, the conventional Grignards or organolithium compounds will not work here, as they would also react with the product ketone), but what can we use as a surrogate for the acylium ion? The obvious answer is an acid chloride—reactive and with a good leaving group. Our choice of organometallic nucleophiles is limited—organocadmium compounds or dialkylcopperlithium reagents are the only ones generally suitable, as they don't normally react with ketones. Organocadmium compounds (*e.g.*, reaction 19.1) have become less fashionable, since they are limited to primary alkyl or aryl groups, and cadmium is rather toxic. Most modern syntheses (examples in Figure 19.26) use organocopper reagents. The bonds highlighted in red are those that were disconnected in the TM.

$$(PhCH_2)_2Cd + 2CH_3COCl \longrightarrow 2PhCH_2C(=O)CH_3 \qquad (19.1)$$

Two other acid derivatives may also be converted to ketones by reaction with organometallic compounds. Grignard reagents add to nitriles to give imine salts (Section 14.4.3), which are hydrolyzed to ketones in the reaction work-up. In Figure 19.27, again the disconnected bond is highlighted in red. Remember than many aliphatic nitriles are readily prepared by displacement

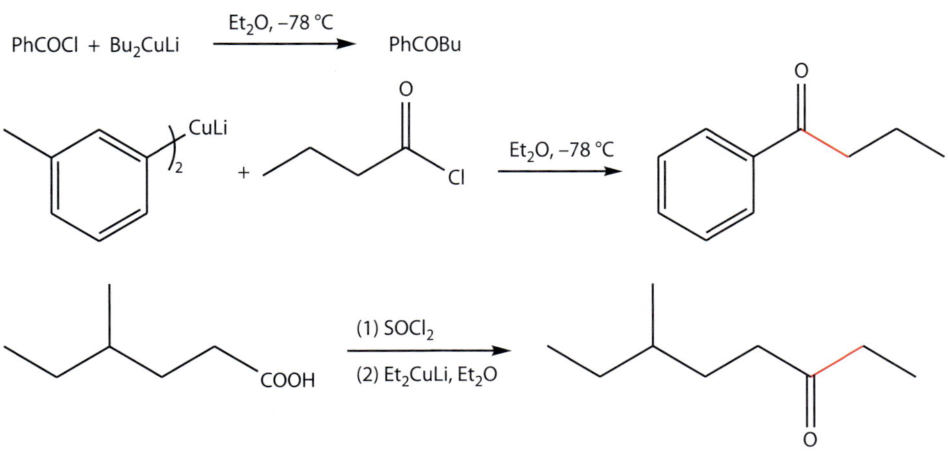

This reaction can also be accomplished using dimethylcadmium on the acid chloride.

FIGURE 19.26 Reactions of dialkylcopper lithium reagents with acyl halides.

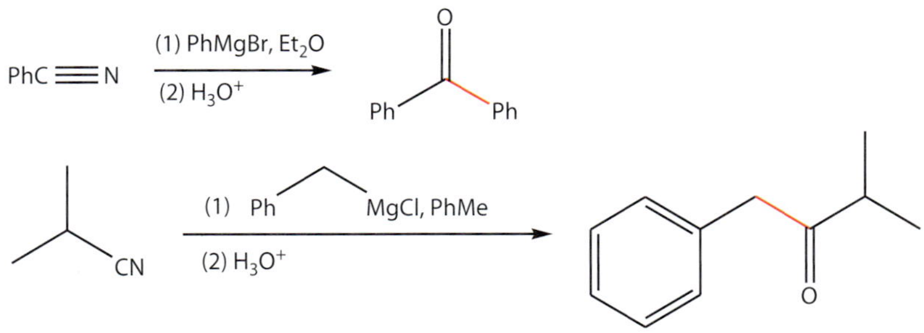

FIGURE 19.27 Synthesis of ketones from nitriles.

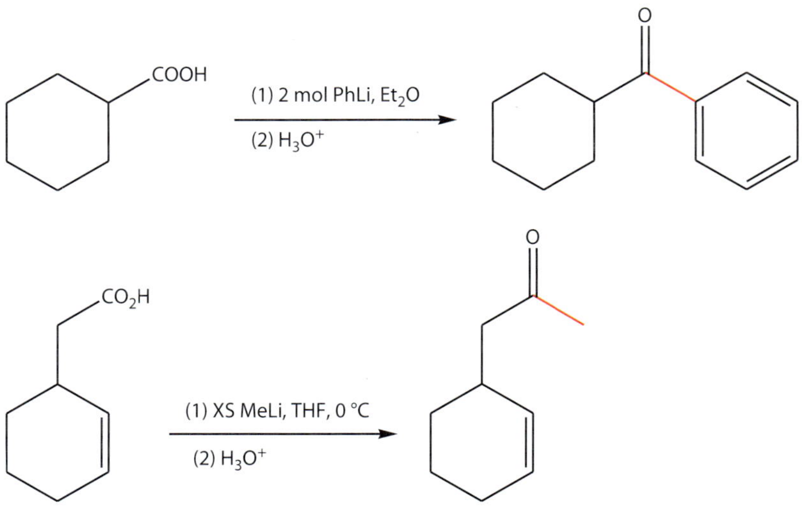

FIGURE 19.28 Ketones from organolithium reagents and carboxylic acids.

of a halide by cyanide ion. Carboxylic acids also react with organolithium reagents to give ketones (Section 15.7.2). An excess of the organometallic is required, since 1 mol is consumed in deprotonation of the acid. Examples are shown in Figure 19.28.

PROBLEM 19.13

Suggest reagents for each of the following transformations:

(a)

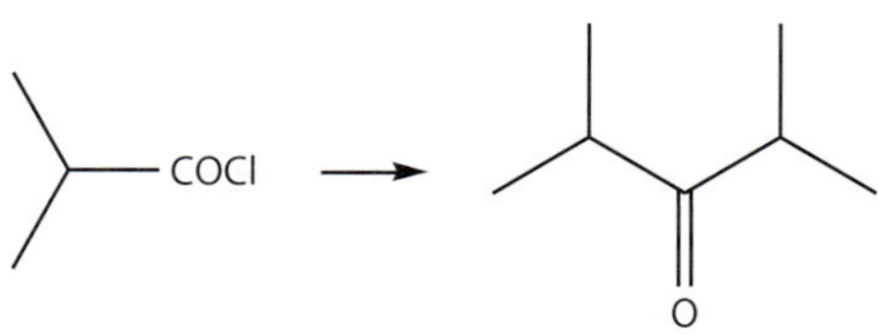

(b)

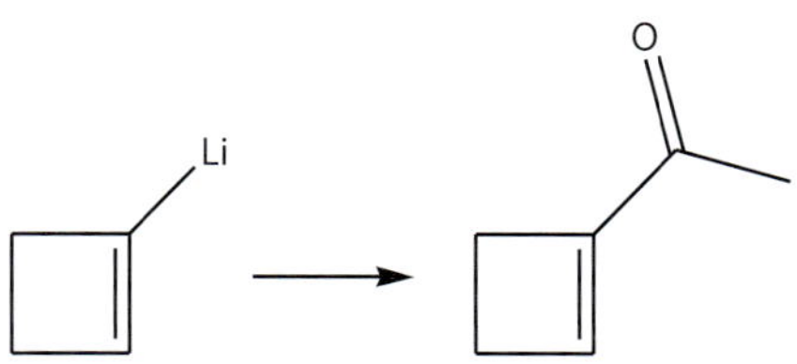

(c)

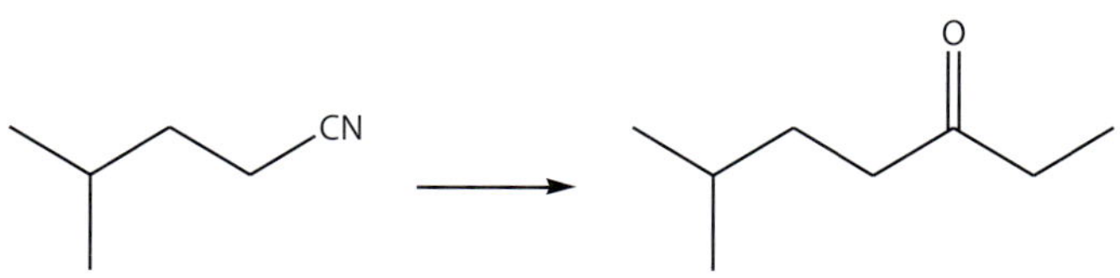

Solutions

(a) $(Me_2CH)_2CuLi$, followed by aqueous work-up
(b) CH_3COOH, then aqueous work-up, or MeCN followed by aqueous work-up
(c) EtMgBr, Et_2O, then aqueous work-up

Aryl ketones are frequently prepared by Friedel–Crafts reactions; an aliphatic version is known but is much less common. Examples are given in Figure 19.29. It is also possible to substitute an aldehyde group directly onto an aryl ring (Figure 19.30). The first reaction is the one that we already know—the Gattermann–Koch process. The other is the Vilsmeier–Haack reaction—which is

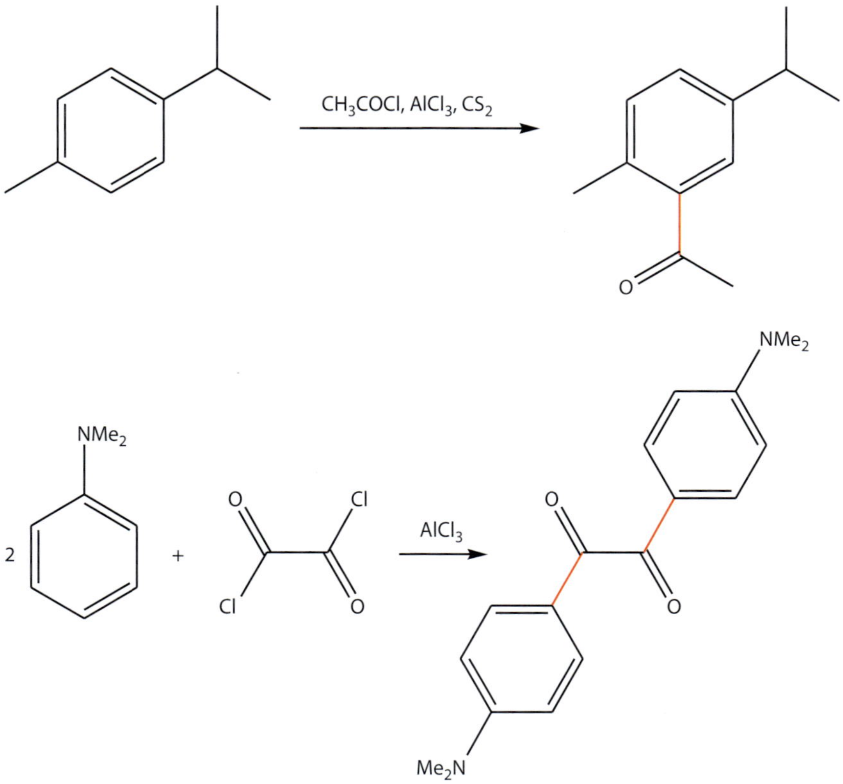

FIGURE 19.29 Synthesis of aryl ketones using the Friedel–Crafts reaction.

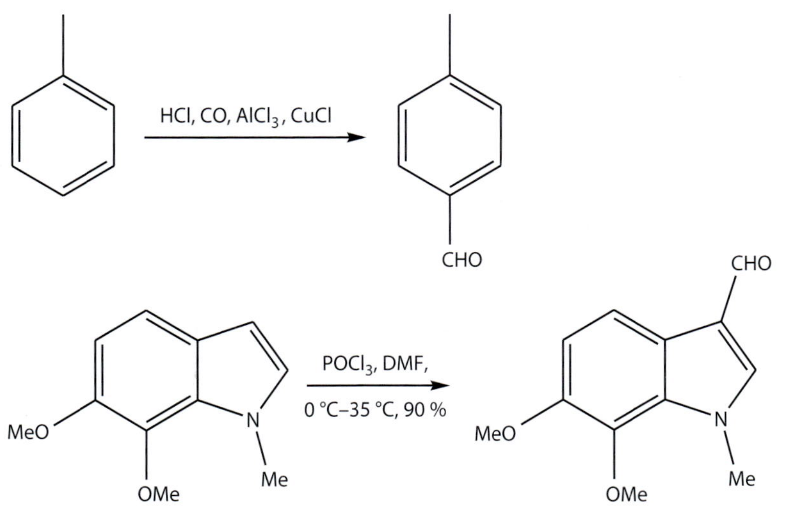

FIGURE 19.30 Formylation of aromatic rings.

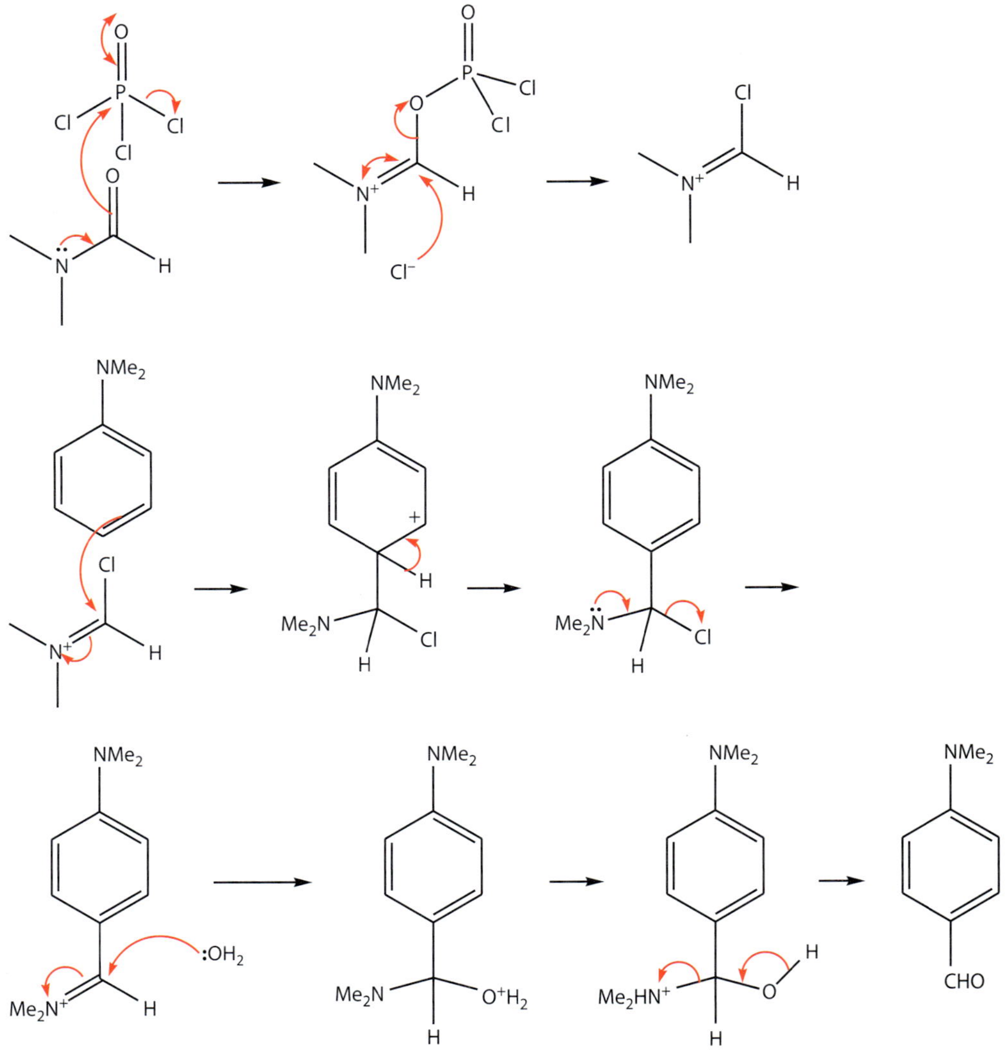

FIGURE 19.31 Mechanism of the Vilsmeier–Haack formylation.

more complicated mechanistically (Figure 19.31) and only for electron-rich arenes but avoids the use of toxic CO. Notice the position of the reaction of indole—refer back to Section 12.5.2 for some clues if you can't work out why the reaction occurs at this position.

PROBLEM 19.14

Predict the outcome of each of the following reactions:

(a)

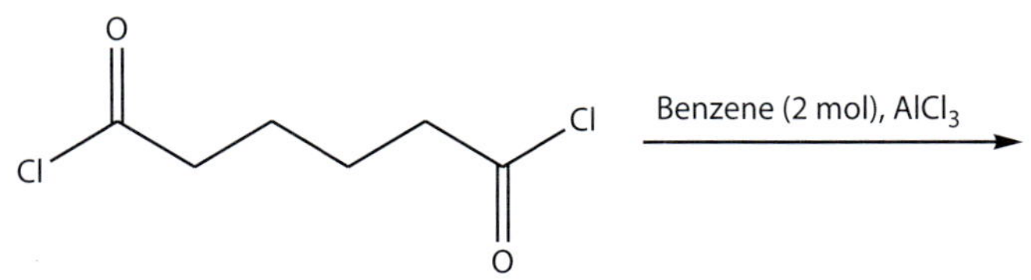

(b)

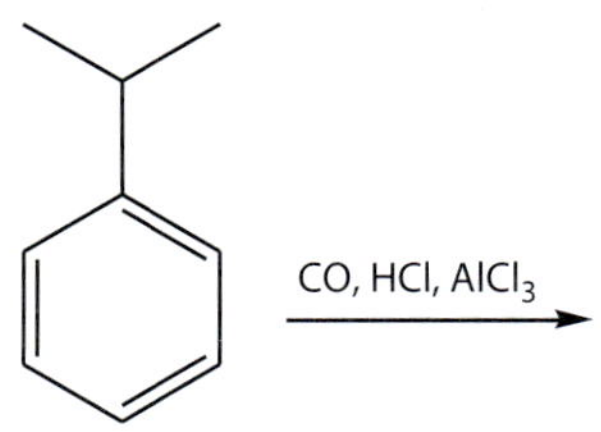

(c)

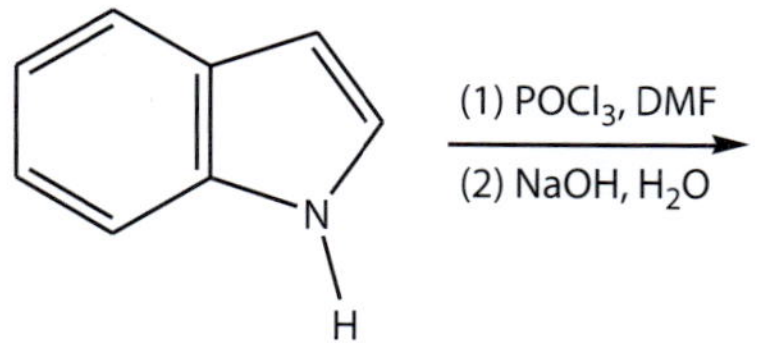

Solutions

(a) A double Friedel–Crafts reaction:

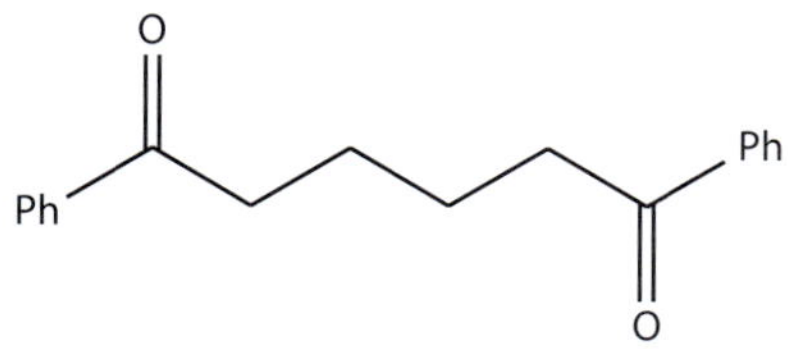

(b) Substitution is almost exclusively *para* because of the size of the substituent:

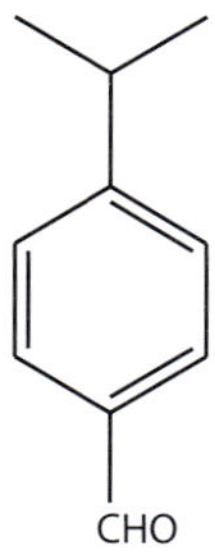

(c) Vilsmeier–Haack formylation occurs in the more electron-rich heterocyclic ring:

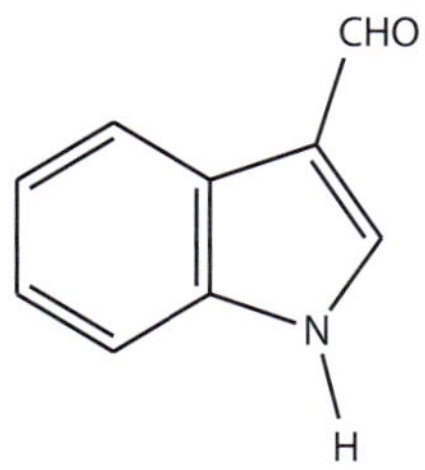

Focus on Dithianes

By now, you should be used to the idea that the carbon atom of a carbonyl group is electrophilic, and to make new bonds to that carbon, we need to attack it with a nucleophile. What if we could turn this around and make the carbon of the carbonyl nucleophilic? This concept is called *umpolung*, the reversal of polarity. For a carbonyl group, one of the ways we can achieve this is by converting the carbonyl to a dithiane (Figure 19.32). Dithiane (**19.14**) itself may be prepared either from $HSCH_2CH_2CH_2SH$ and formaldehyde or the dimethyl acetal of formaldehyde. Dithiane has a pK_a of approx. 31, so it is readily deprotonated by butyl lithium. The lithiated dithiane reacts with a wide range of electrophiles—haloalkanes, epoxides, aldehydes, ketones, acyl halides, and CO_2. Monosubstituted dithianes can also be prepared from aldehydes and can be deprotonated and reacted with a further electrophile. Once the reaction is complete, the dithiane can be hydrolyzed to the carbonyl compound or reduced with H_2/Raney Ni to a hydrocarbon (Figure 19.33).

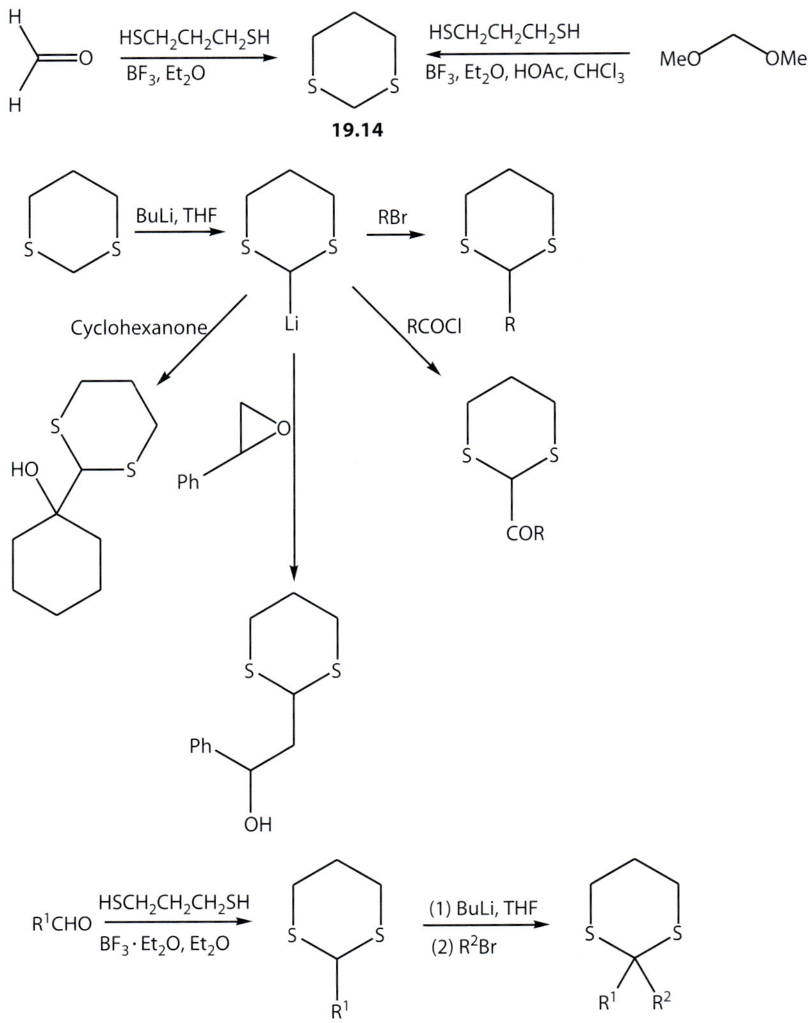

FIGURE 19.32 Preparation and reactions of dithianes.

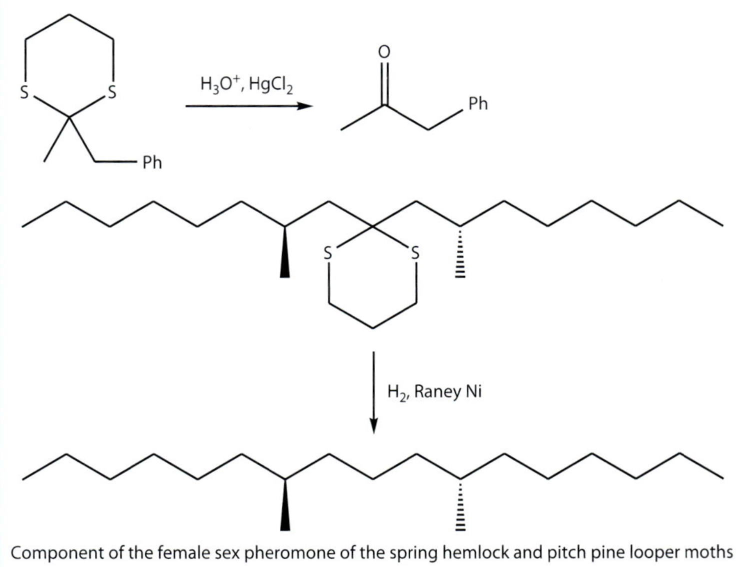

FIGURE 19.33 Removal of sulfur from dithianes.

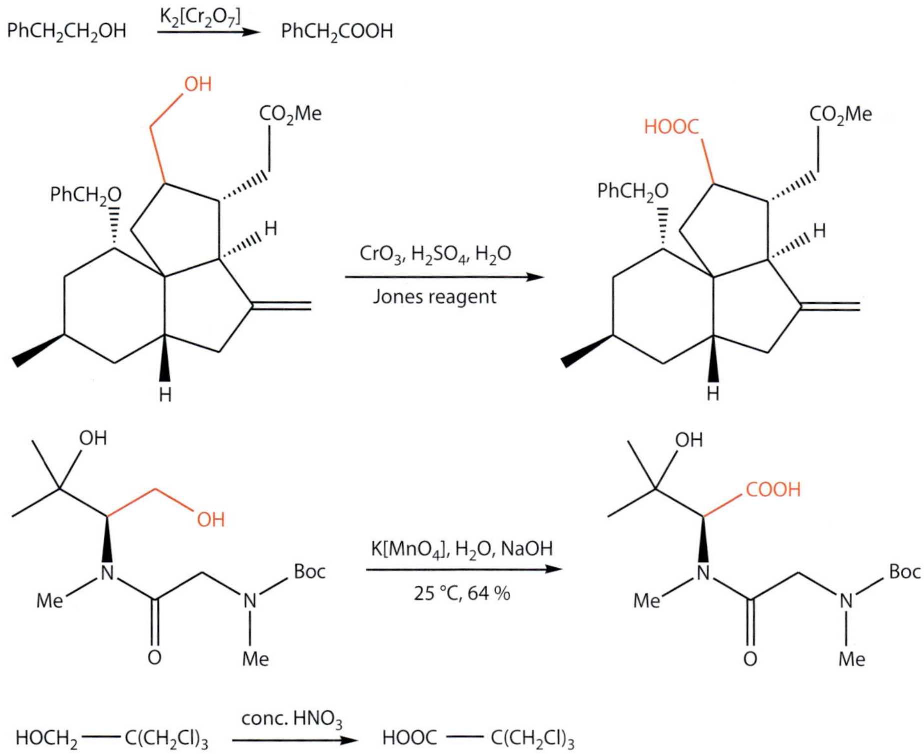

FIGURE 19.34 Oxidation of primary alcohols to carboxylic acids.

19.3.4 FGIs AND DISCONNECTIONS TO PREPARE CARBOXYLIC ACIDS

Primary alcohols are readily oxidized to carboxylic acids; most strong oxidants give good results (Figure 19.34). Note that $K[MnO_4]$ is unsuitable for molecules containing carbon–carbon double bonds, as it reacts to hydroxylate them, in a manner similar to that with OsO_4 (Section 11.4.4). Aldehydes are also readily oxidized to carboxylic acids; this process can be easier than the reaction of primary alcohols, as we saw in the chemistry of sugars (Section 16.3.8, Figure 19.35). Primary and secondary alkyl groups attached to aromatic rings may also be oxidized with $K[MnO_4]$ or $K_2[Cr_2O_7]$ (Figure 19.36); tertiary groups are unaffected.

PROBLEM 19.15

Suggest an appropriate reagent for each of the following reactions:

(a)

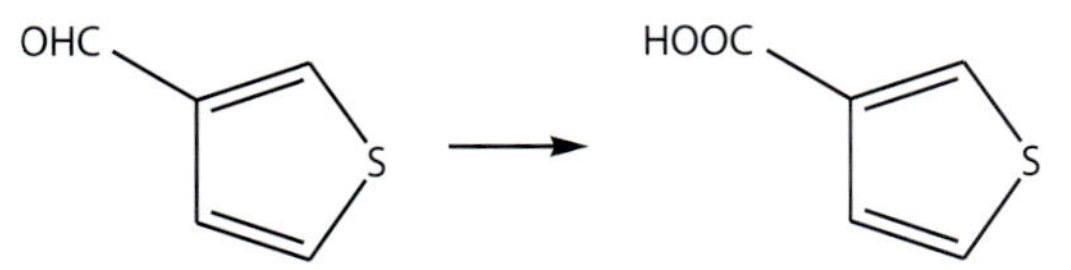

(b)

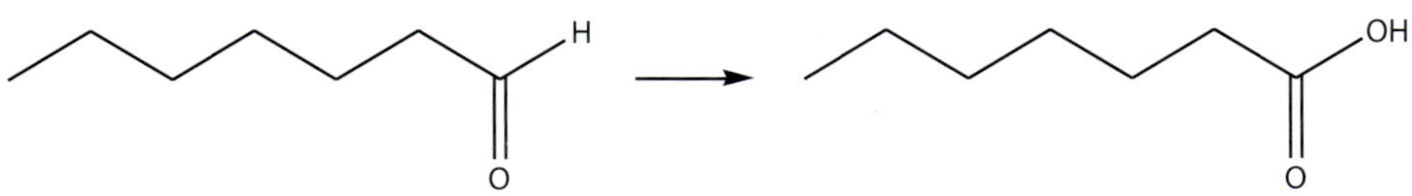

(c)

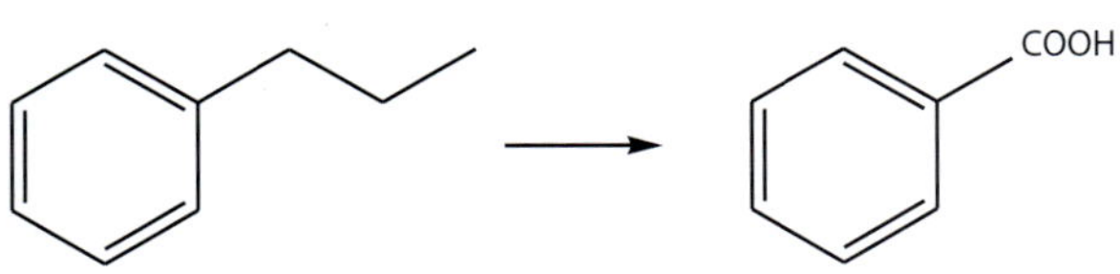

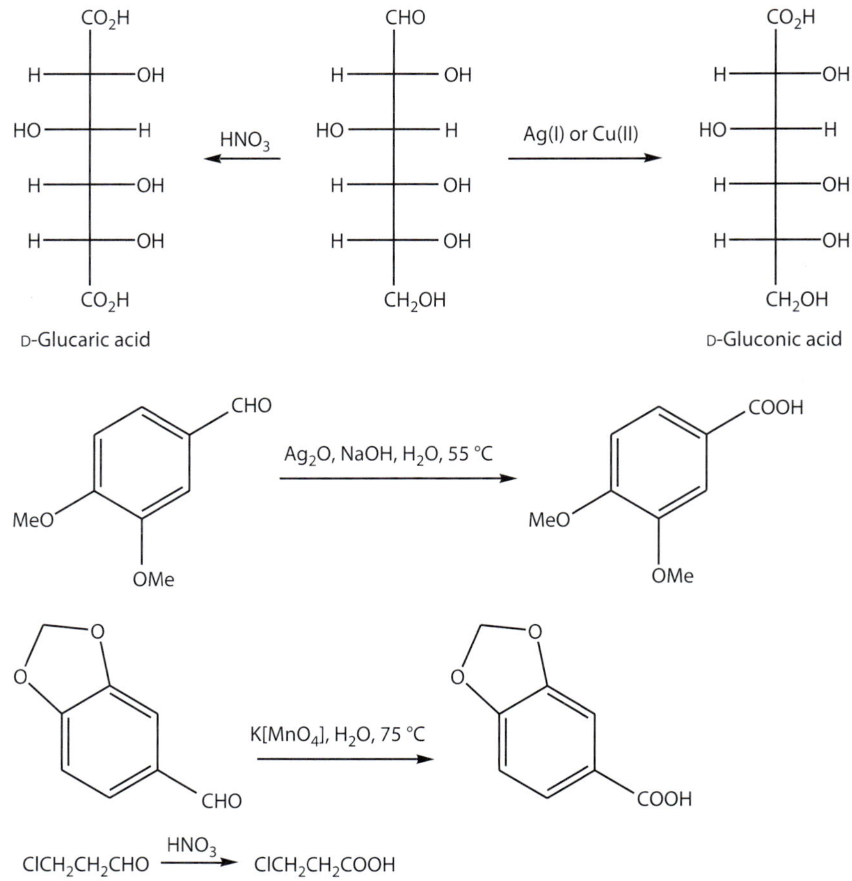

FIGURE 19.35 Oxidation of aldehydes to carboxylic acids.

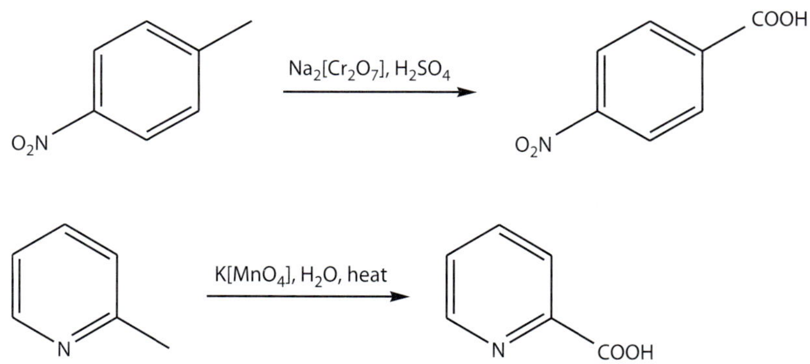

FIGURE 19.36 Oxidation of alkylarenes to aromatic carboxylic acids.

Solutions

(a) Most strong oxidants would oxidize the aldehyde, but sulfur is also oxidizable, so it is sensible to use the mildest available, Ag_2O, NaOH, and H_2O.
(b) Nothing else oxidizable here, so inexpensive $K[MnO_4]$, H_2SO_4, H_2O.
(c) $K[MnO_4]$, H_2O.

Grignard reagents react with CO_2 to give carboxylic acids—thus, an acid can be disconnected to a Grignard reagent and hence a halide. The disconnection in Figure 19.37 was used to prepare benzoic acid labeled with ^{13}C using labeled CO_2 produced from acid on labeled barium carbonate. Other examples are shown in Figure 19.38, and lithium and sodium derivatives react comparably.

Since cyanide groups can be hydrolyzed to carboxylic acids, this provides another route to this class of compounds from alkyl halides (Figure 19.39). The process works well for primary halides but does not succeed for tertiary halides, as elimination then predominates over substitution.

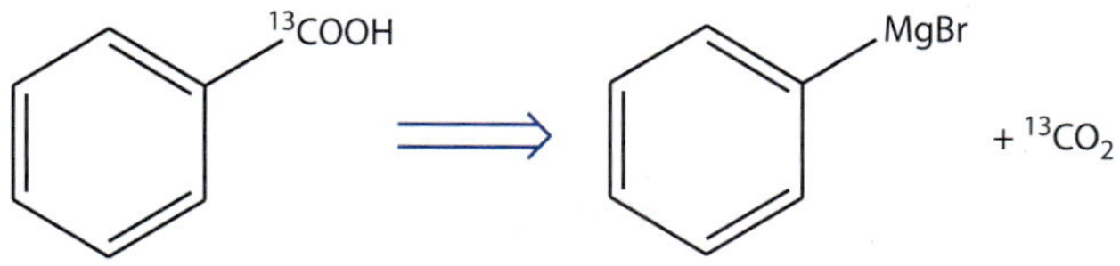

FIGURE 19.37 Preparation of ^{13}C-labeled benzoic acid.

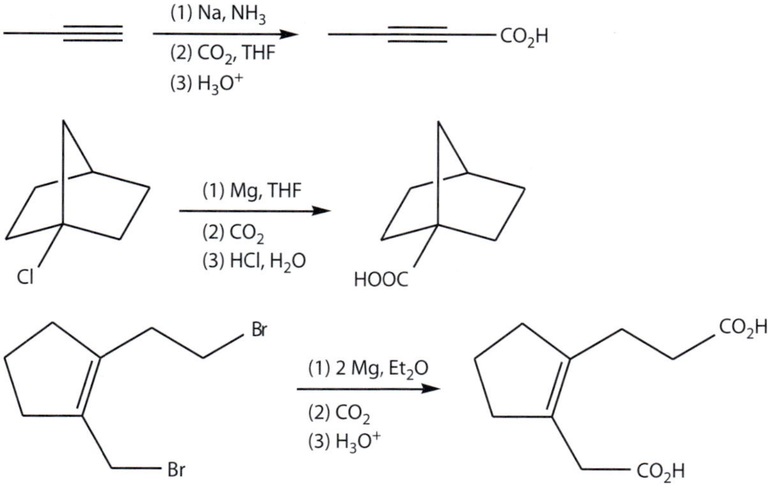

FIGURE 19.38 Preparation of carboxylic acids by carboxylation of Grignard reagents.

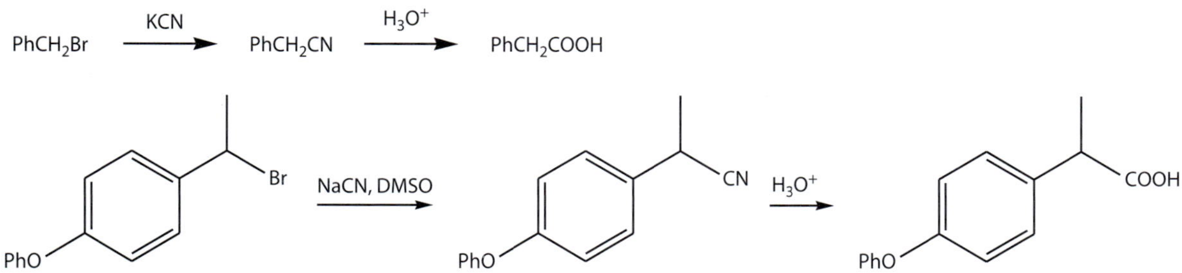

FIGURE 19.39 Synthesis of carboxylic acids *via* cyanides.

Both this and the carboxylation reaction are often described as homologation processes, as a single carbon atom is added to the molecule. CO_2 is a 1-C electrophilic chain extender and $[CN]^-$ a 1-C nucleophilic chain extender.

PROBLEM 19.16

For each of the following molecules, state which homologation pathway (Grignard + CO_2 or cyanide and hydrolysis) you would choose, and explain your reasons:

(a)

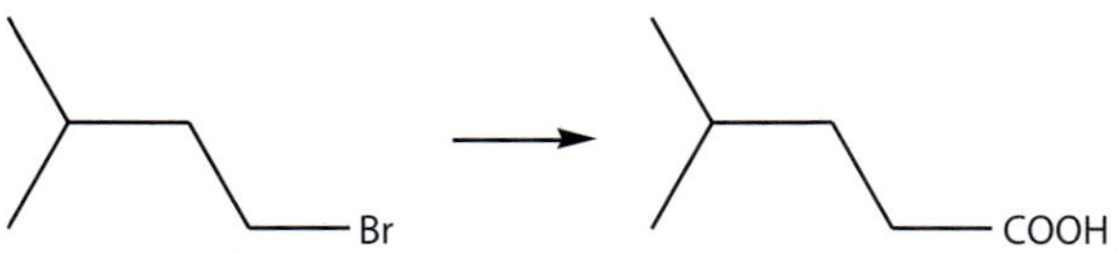

(b)

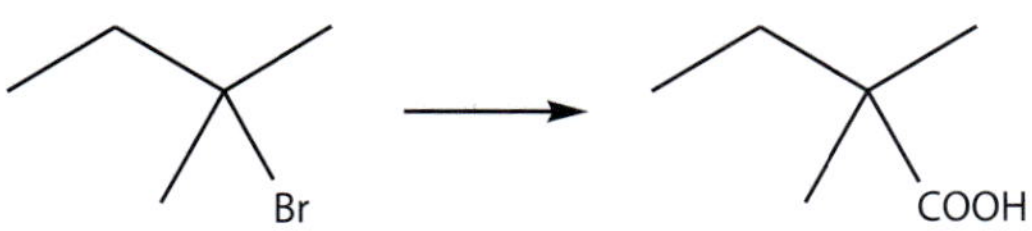

Solutions

(a) Either process would work here; choice should be made on the basis of cost.
(b) This must be the Grignard + CO_2 route; cyanide would not substitute successfully at the tertiary center.

Key FGIs from Sections 19.3.3 and 19.3.4

SM	Reagent	Product	Comments, Limitations
Oxidations			
RCH_2OH	CrO_3, py	RCHO	Cr is toxic.
RCH_2OH	PCC	RCHO	Cr is toxic.
RCH_2OH	DMSO/$(COCl)_2$	RCHO	Swern oxidation; 2 ° alcohols react, but more slowly.
RCH_2OH	CrO_3, H_2SO_4	RCOOH	Jones reagent.
RCH_2OH	$K[MnO_4]$	RCOOH	Not compatible with carbon–carbon double bonds.
RCH_2OH	$K_2[Cr_2O_7]$	RCOOH	Cr is toxic.
R_2CHOH	CrO_3/py, CrO_3/H_2SO_4 or PCC	$R_2C{=}O$	Cr is toxic.
R_2CHOH	DMSO/$(COCl)_2$	$R_2C{=}O$	Swern oxidation; 2 ° alcohols react more slowly than 1 ° alcohols.
RCHO	Ag(I), Cu(I)	RCOOH	More powerful oxidants also give this result, but their use is rarely necessary.
Ar-alkyl	$K[MnO_4]$ or $K_2[Cr_2O_7]$	ArCOOH	Alkyl group must be 1 ° or 2 °; 3 ° alkyl substituents are unreactive.
Reductions			
RCOOalkyl	*i*-Bu_2AlH, DIBAH	RCHO	Sometimes written as DIBAL.
RCOCl	Li[AlH(O-*t*-Bu)$_3$]	RCHO	
RCOCl	H_2/Pd/$BaSO_4$	RCHO	Rosenmund reduction. Inexpensive, but incompatible with carbon–carbon double bonds.
RCN	Li[AlH(OEt)$_3$]	RCHO	
C–C Bond-Forming Reactions			
Ar–H	HCl, CO, $AlCl_3$, CuCl	ArCHO	Vilsmeier–Haack reaction.
Ar–H	$POCl_3$, DMF	ArCHO	Gattermann–Koch reaction.
Ar–H	RCOCl, $AlCl_3$	ArCOR	Friedel–Crafts acylation.
R′COCl	R_2CuLi	R′COR	R_2Cd is also successful, but less commonly used.
R′CN	RMgBr	RCOR′	
R′COOH	RLi	RCOR′	2 mol RLi required.
RMgBr	CO_2	RCOOH	Useful to introduce ^{13}C label.
RX	1. KCN, 2. H_3O^+	RCOOH	1 ° and 2 ° halides only.
Other Reactions			
R–C≡C–R	H_3O^+ *or* BH_3 then HOO^-	RC(=O)CH_2R	Selectivity is opposite for the two reagents.

REVIEW PROBLEMS

1 What is the synthon for each of the following species?
(a) $^-$[COOH]
(b) $^+$[COOH]
(c) $[CH_3]^-$
(d) $[PhCH_2]^+$
(e) $^-[CH_2NH_2]$
(f) $[H]^-$

2. Propose a disconnection, and give reagents for a forward reaction for each of the following:

(a)

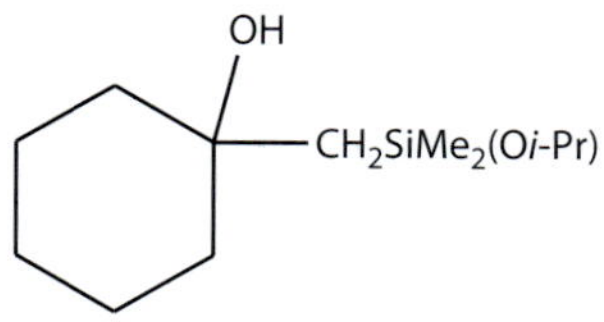

(b)

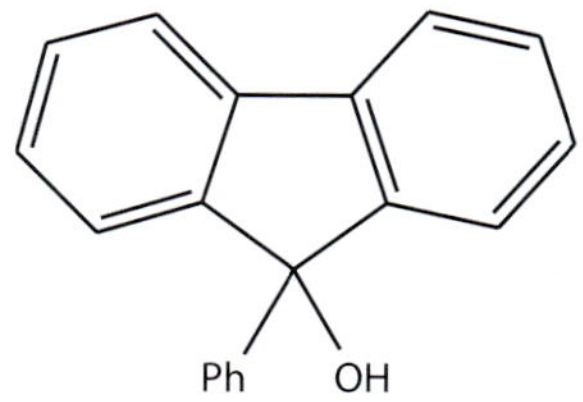

(c)

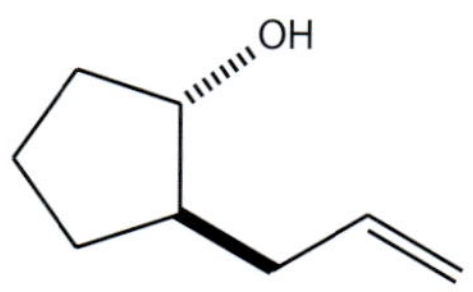

(d)

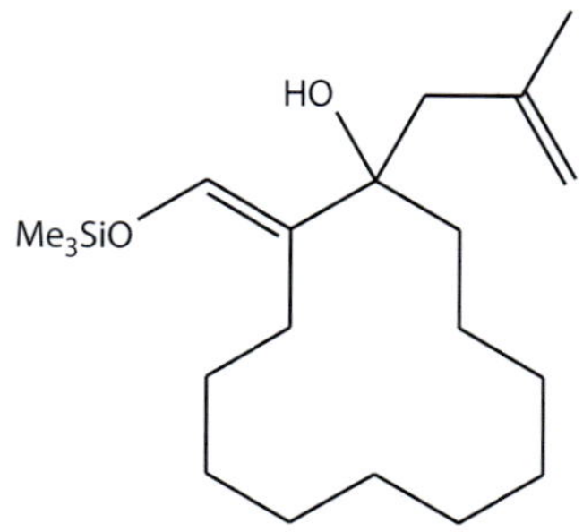

(e)

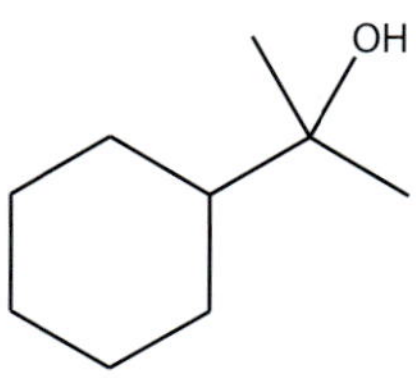

(f)

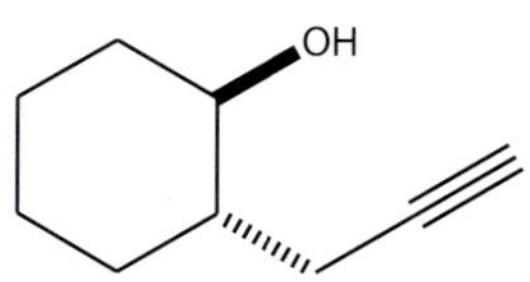

3. Give reagents for each of the following transformations, some of which require more than one step:

(a)

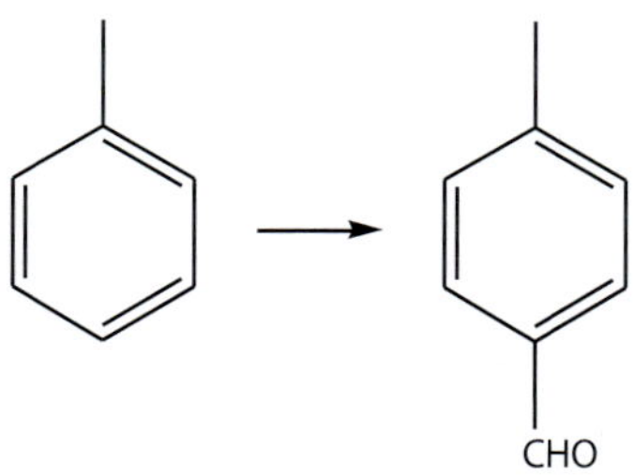

(b)

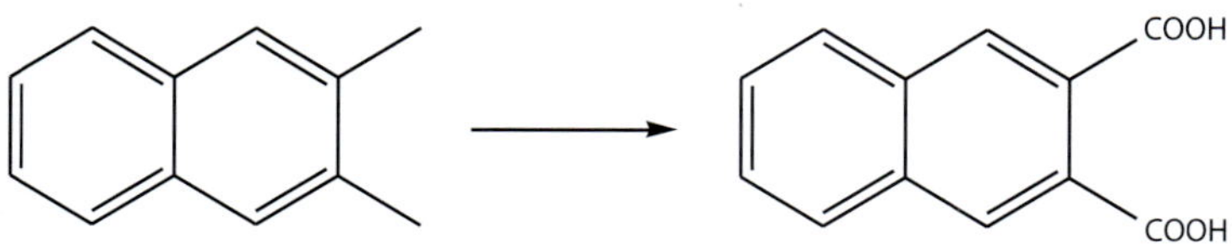

(c)

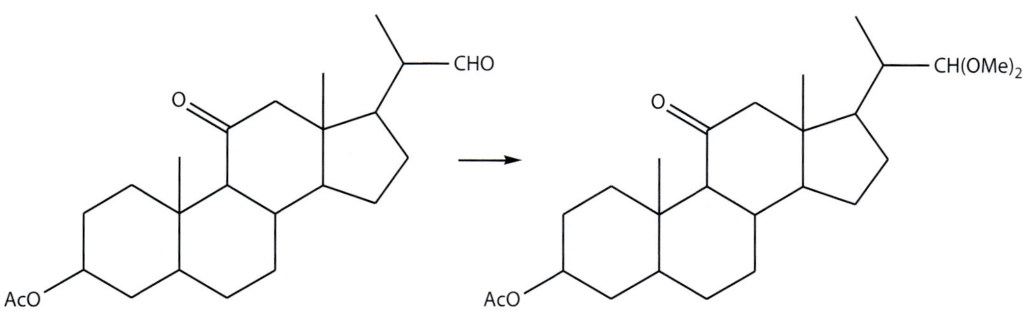

(d)

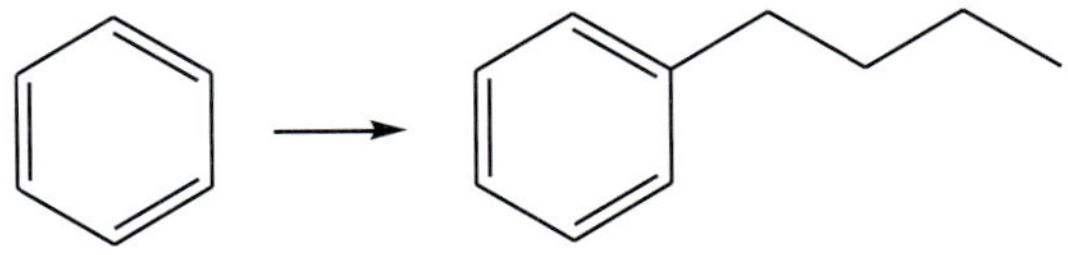

(e)

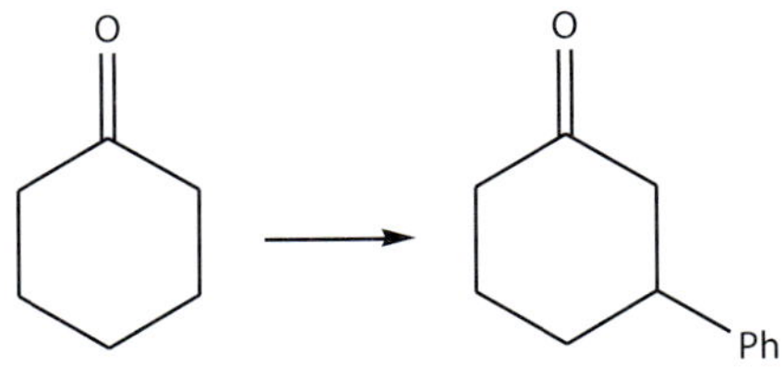

(f)

(g)

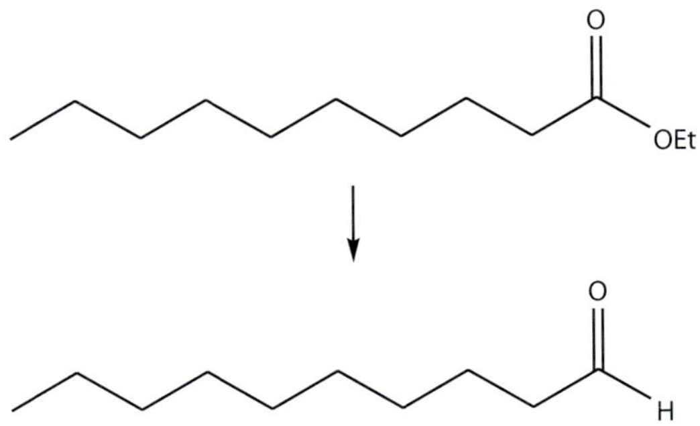

(h)

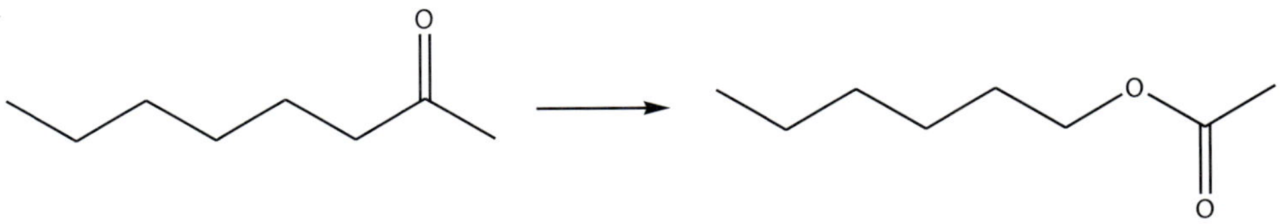

(i)

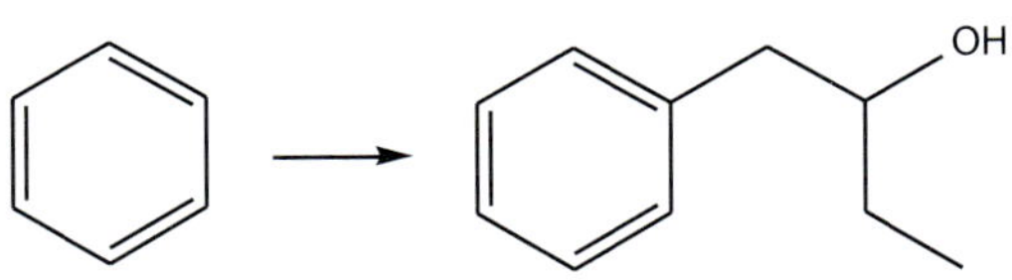

(j)

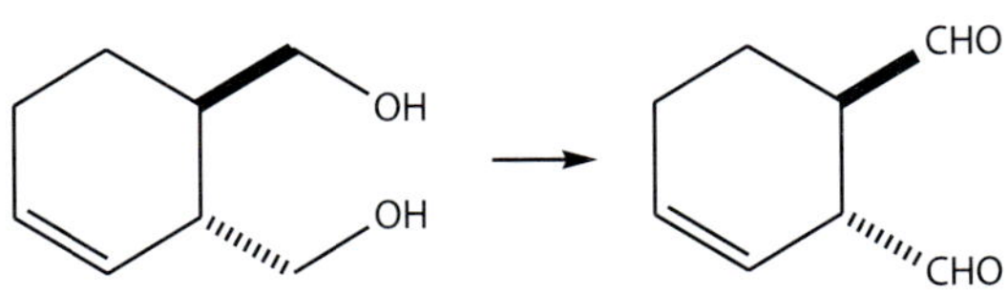

(k)

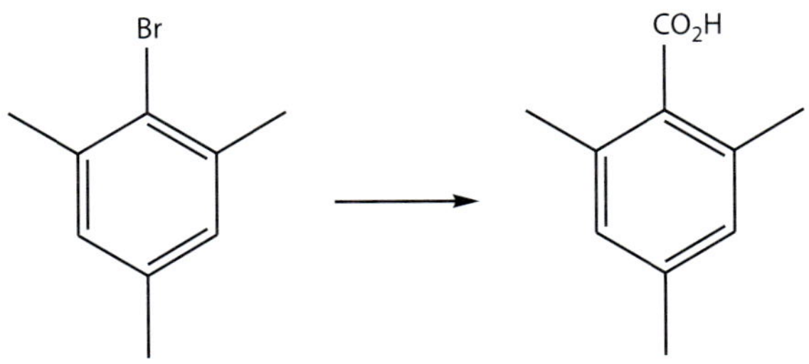

(l) PhCOCl to PhCOPr

(m)

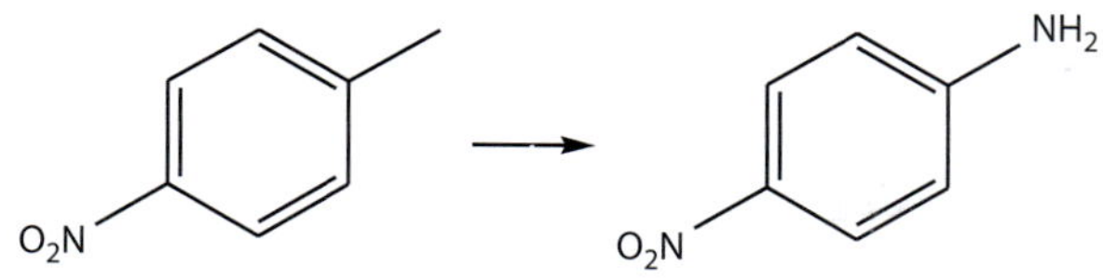

(n)

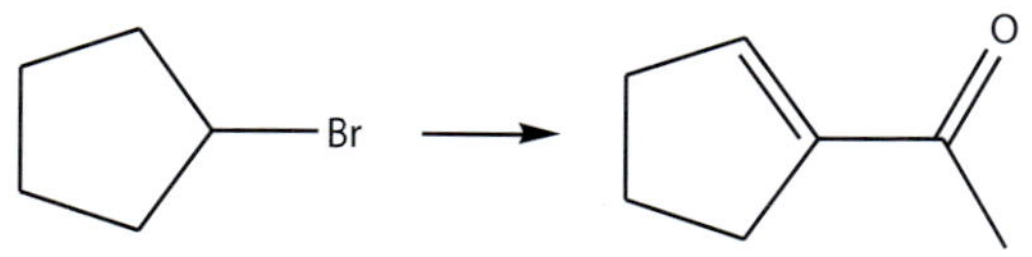

(o) $CH_3CH_2CH_2OH \rightarrow CH_3CH_2CH_2C(CH_3)_2Cl$

(p)

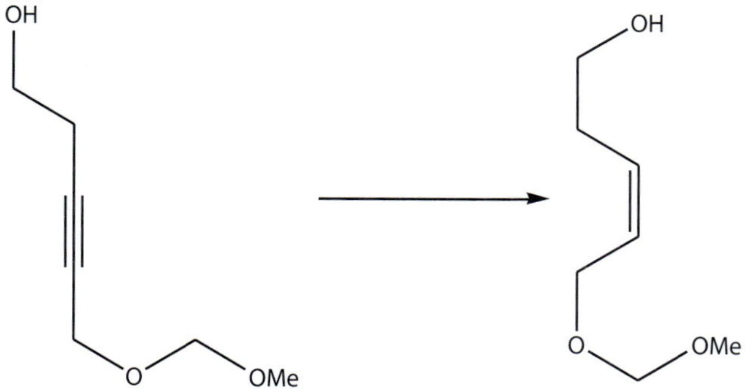

(q)

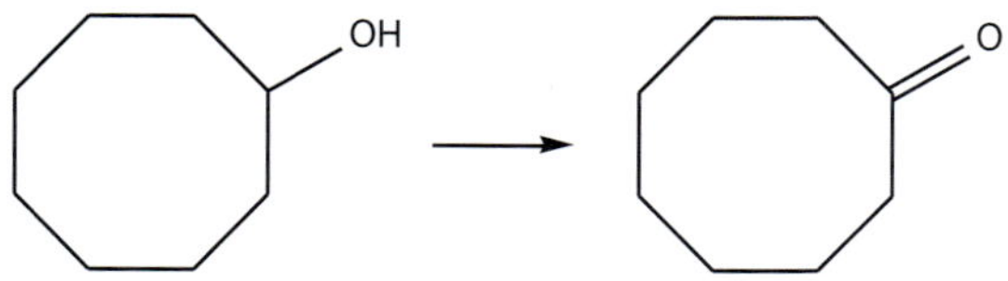

(r)

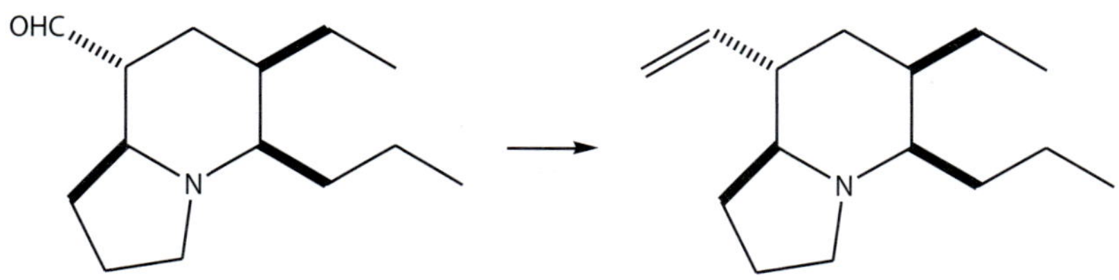

Use in the synthesis of indolizidine 223A

(s)

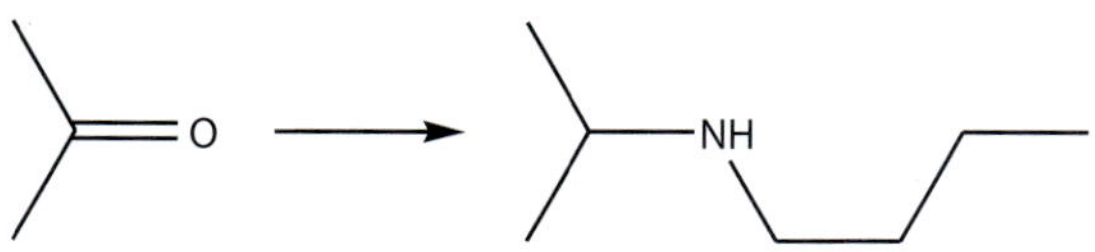

(t)

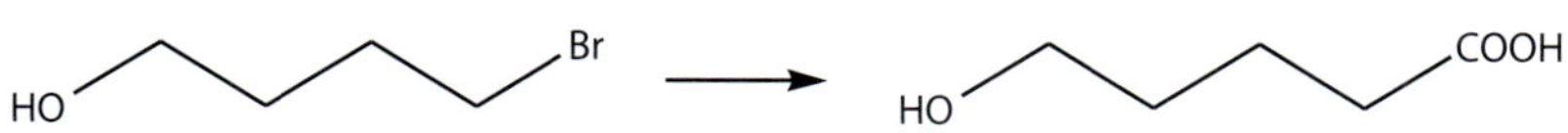

(u) $(PhCH_2)_2Cd + 2CH_3COCl \rightarrow 2PhCH_2C(=O)CH_3$

4. Predict the product(s) of each of the following processes:

(a)

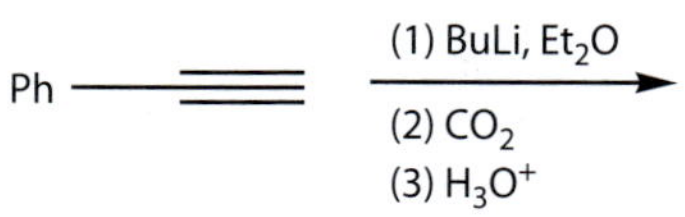

(b)

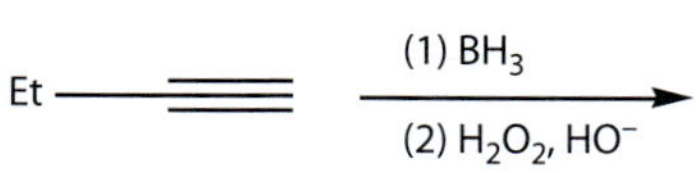

(c)

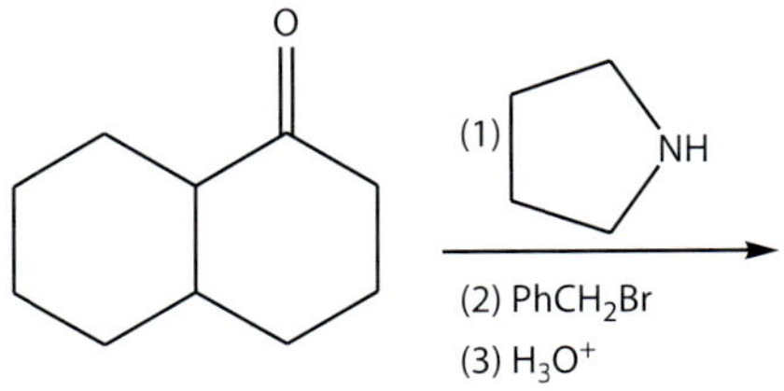
O
(1)
NH
(2) PhCH2Br
(3) H3O+

(d)

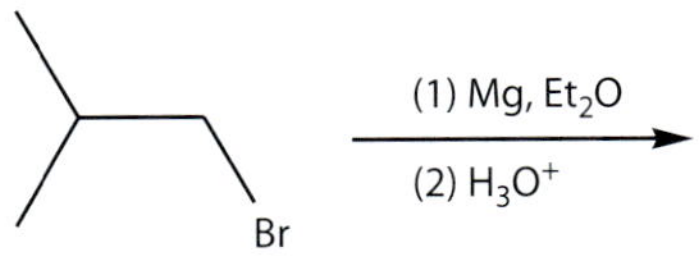
Br
(1) Mg, Et2O
(2) H3O+

(e)

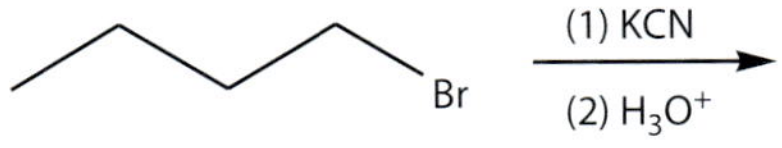
Br
(1) KCN
(2) H3O+

(f)

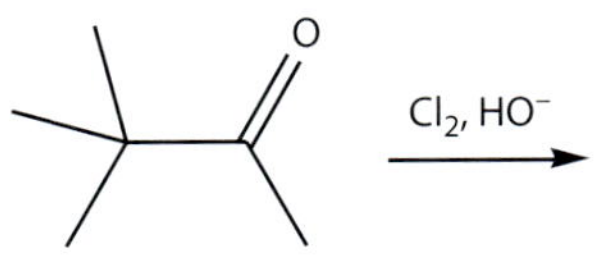
O
Cl2, HO−

(g)

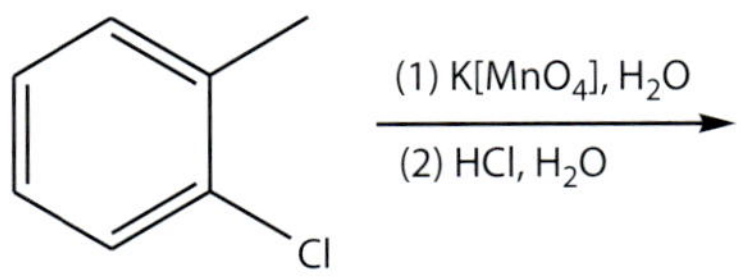
Cl
(1) K[MnO4], H2O
(2) HCl, H2O

(h)

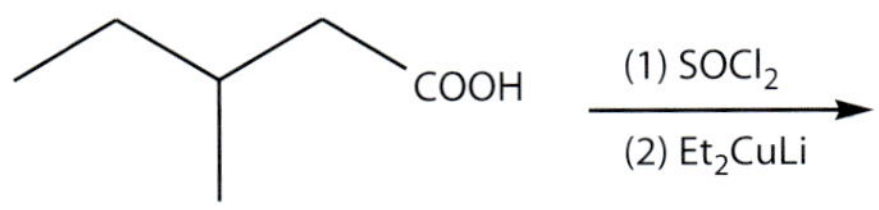
COOH
(1) SOCl2
(2) Et2CuLi

(i)

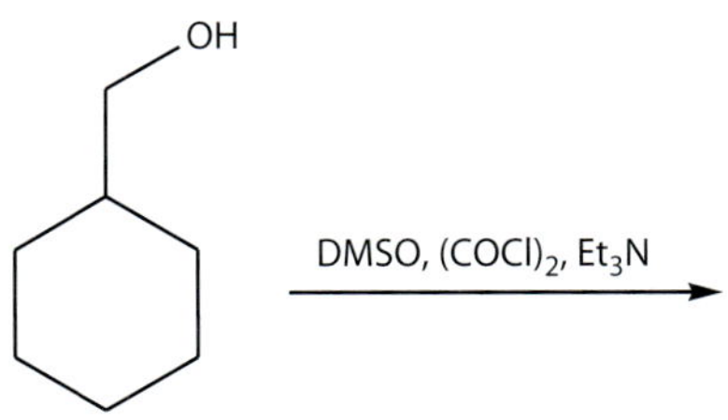
OH
DMSO, (COCl)2, Et3N

(j)

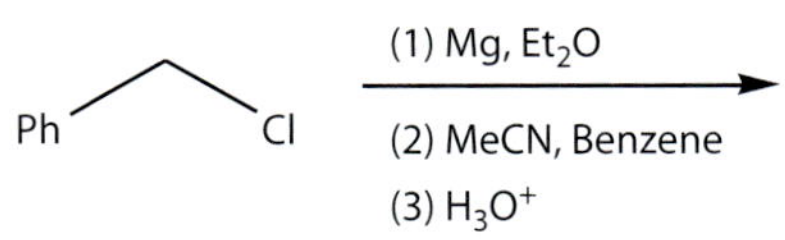
Ph
Cl
(1) Mg, Et2O
(2) MeCN, Benzene
(3) H3O+

(k)

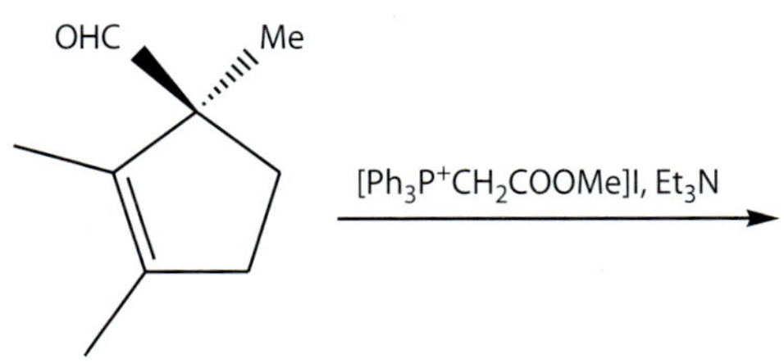
OHC
Me
[Ph3P+CH2COOMe]I, Et3N

(l)

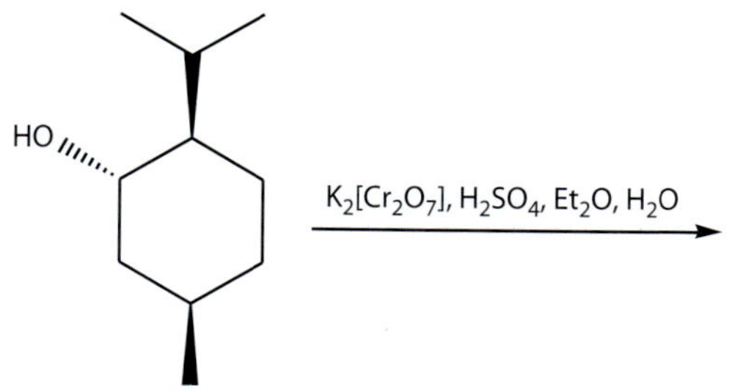

(m)

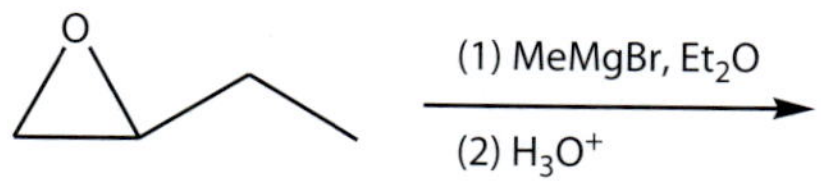

(n)

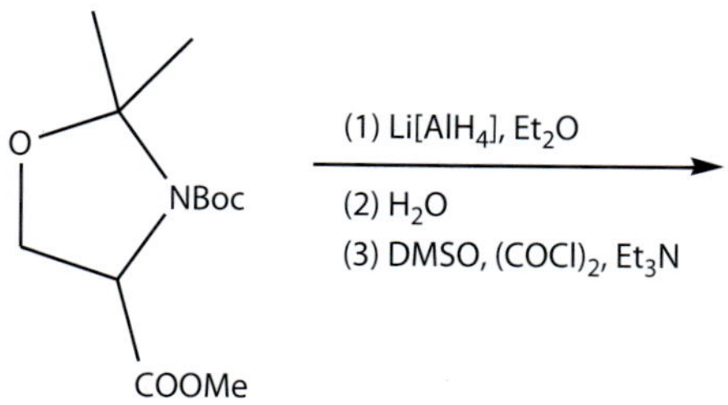

5. Write a mechanism for each of the following reactions:

(a)

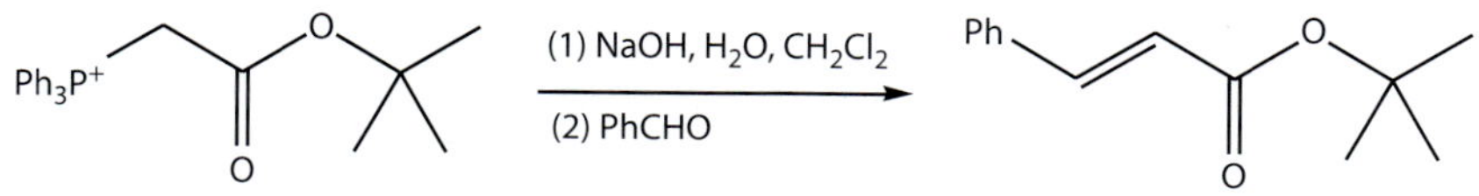

(b)

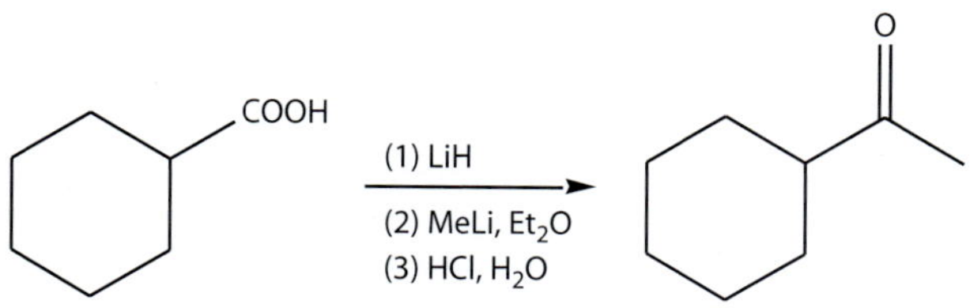

(c)

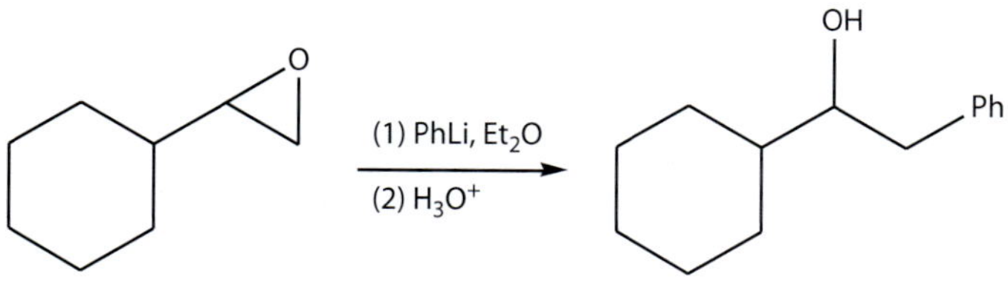

(d)

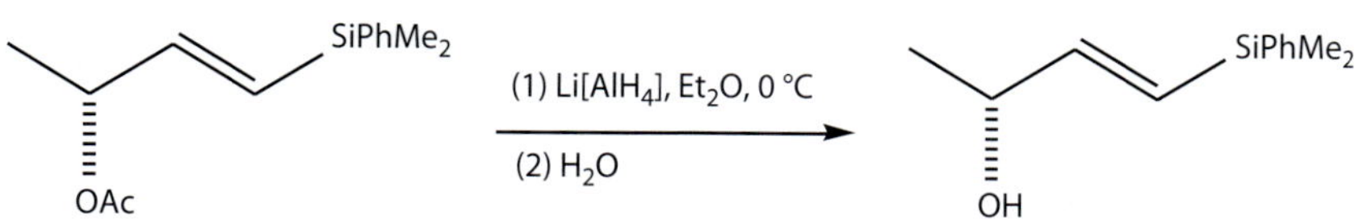

MORE CHALLENGING PROBLEMS

1. What is the synthon for each of the following species?
   (a) $^+[CH_2OH]$
   (b) $[CH_3C{=}O]^-$
   (c) $[C{=}O]^{2-}$
   (d) $[CH_3C{=}O]^+$
   (e) $^+[CH_2CH_2OH]$

2. Explain the following observations:

(a)

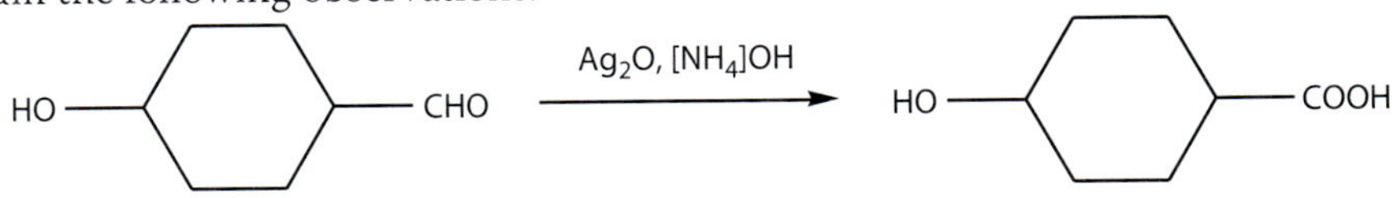

(b) Oxidation of 1-butanol with sodium dichromate gives butyl butanoate as the main product:

(c)

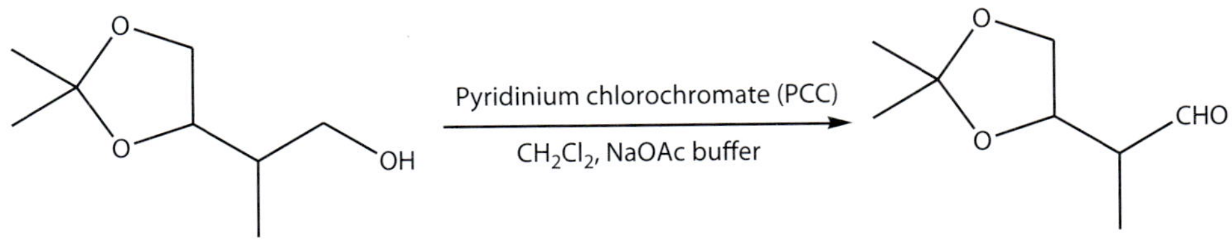

(d) Esters do not normally undergo Wittig reactions, but the reaction sequence below is successful:

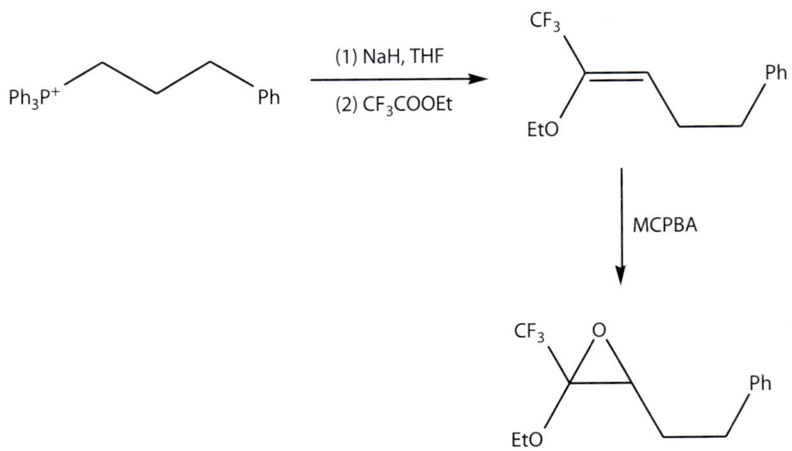

3. How would you carry out the following transformations, which may require more than one step?

(a)

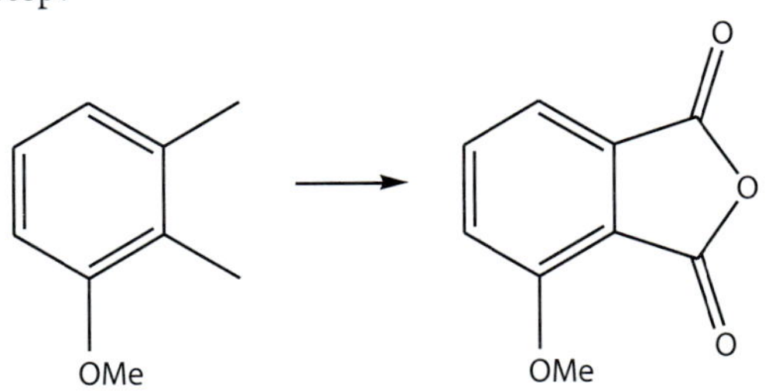

(b)

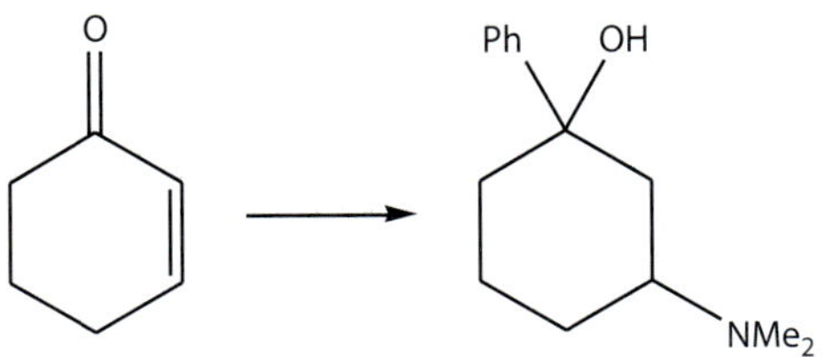

(c) $(CH_3)_2C(OH)CH_2CH_3 \rightarrow CH_3CH_2C(CH_3)_2CHO$

(d)

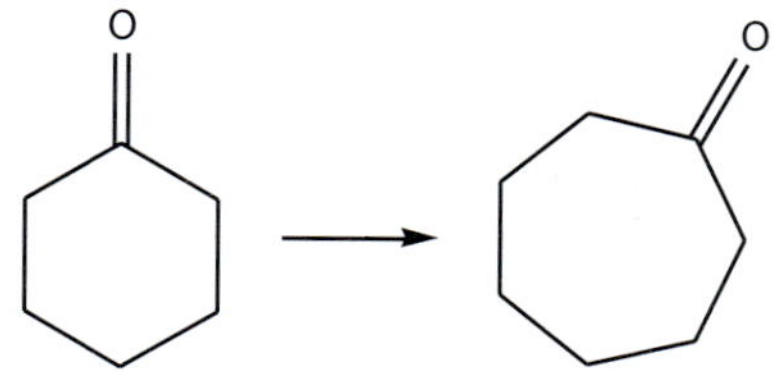

4. Devise a synthesis for each of the following compounds, using the provided SM and other readily available SMs:

(a)

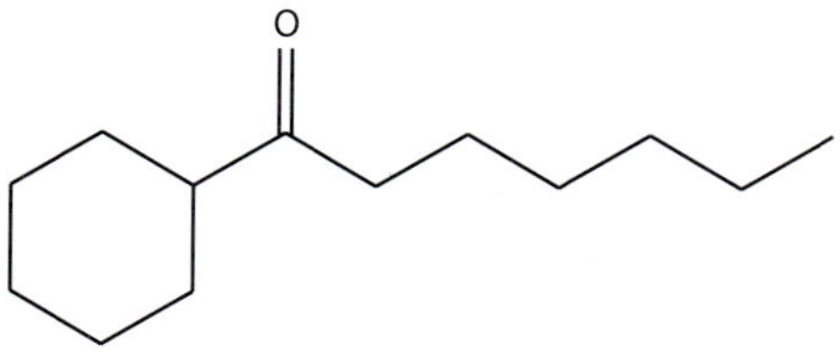 from 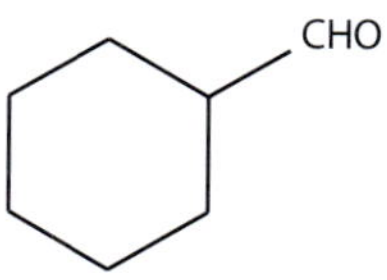

(b)

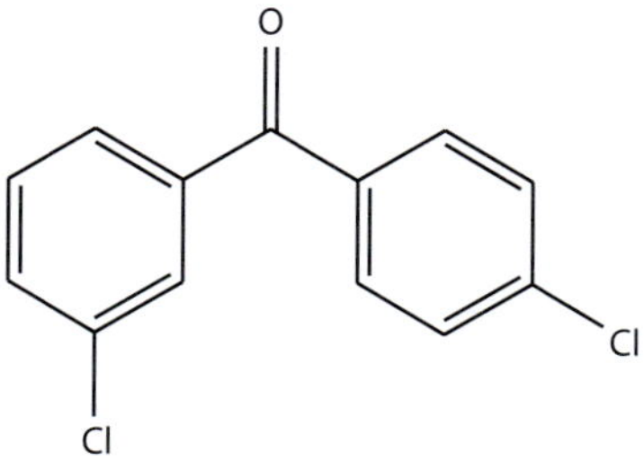 from chlorobenzene

(c)

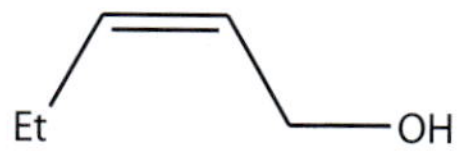 from ethyne

(d)

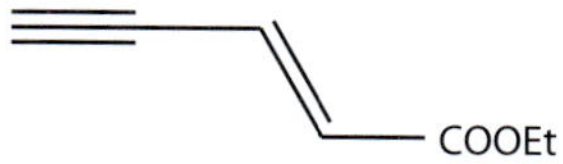 from 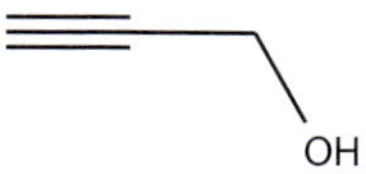

(e)

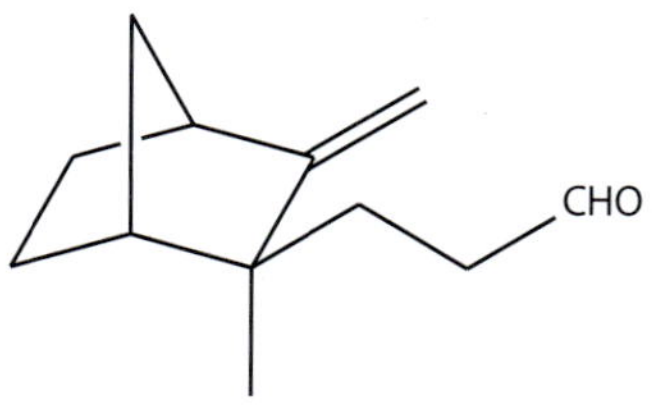 from 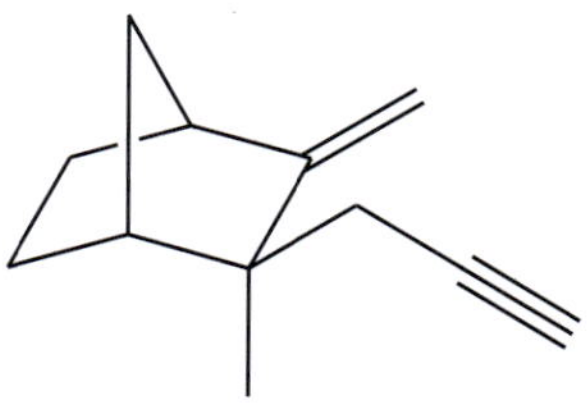

(f)

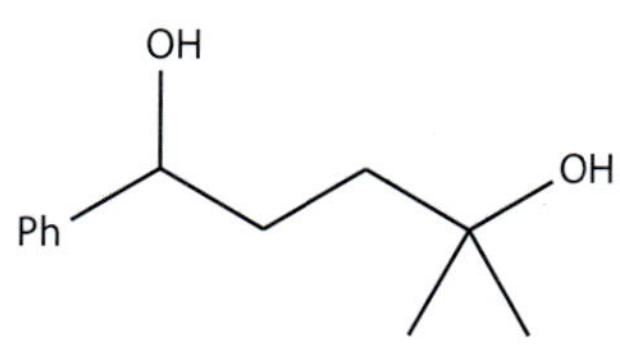 from ethyne

5. Predict the outcomes of each of the following sequences:

(a)

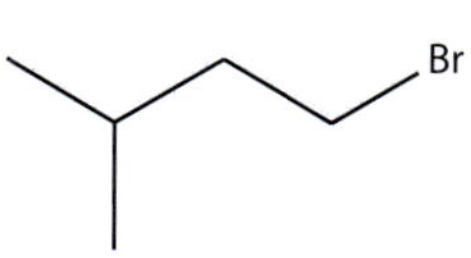

(1) Mg, Et_2O
(2) CO_2
→
(3) H_3O^+
(4) $SOCl_2$
(5) Et_2CuLi

(b)

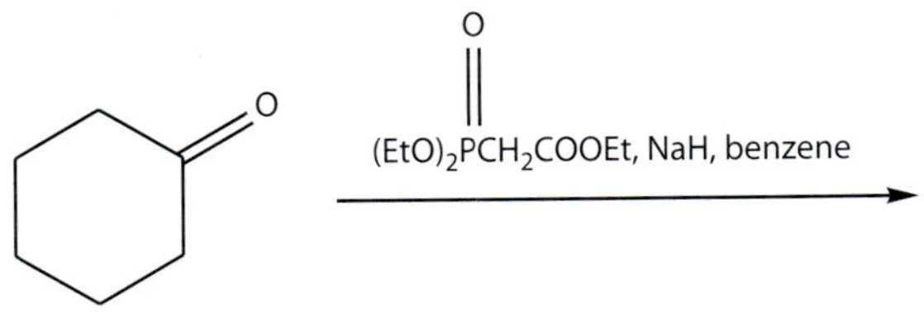
O
O
$(EtO)_2PCH_2COOEt$, NaH, benzene

(c)

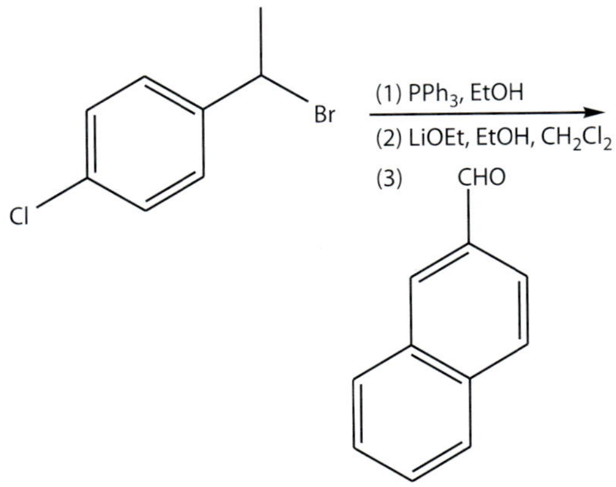
Br
Cl
(1) PPh_3, EtOH
(2) LiOEt, EtOH, CH_2Cl_2
(3)
CHO

(d)

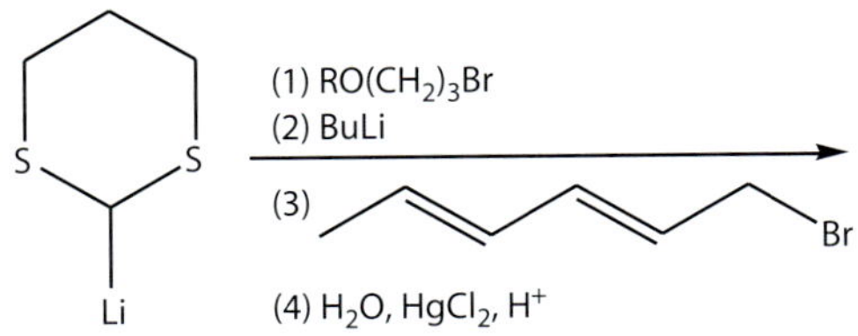
S
S
Li
(1) $RO(CH_2)_3Br$
(2) BuLi
(3)
Br
(4) H_2O, $HgCl_2$, H^+

Synthesis 2

Two-Group Disconnections and Protecting Groups

20

20.1 INTRODUCTION TO TWO-GROUP DISCONNECTIONS

In this chapter, we will be considering how the *relationship* between two functional groups in a molecule can direct us to a possible disconnection. We will study 1,2-, 1,3-, 1,4-, 1,5-, and 1,6-difunctionalized compounds. In general, we will consider how to make 1,*x*-dioxygenated compounds because there are more routes for this—but we remember that these may be readily converted into other functionalities. We will find that 1,3-dioxygenated compounds are readily prepared *via* the carbonyl condensation reactions that we met in Chapter 17 and that 1,5-dioxygenated compounds are prepared by the related conjugate addition reactions. These are sometimes described as *natural* disconnections—they reflect what the molecule "would like to do." The syntheses of 1,2-, 1,4-, and 1,6-difunctionalized molecules require quite different techniques, using what are sometimes referred to as *illogical* disconnections, or the use of *unnatural* nucleophiles. Effectively, we will trick the molecule into a particular reaction.

20.2 1,2-DIFUNCTIONALIZED COMPOUNDS

20.2.1 ADDITION TO DOUBLE BONDS

We already know how to produce 1,2-diols from the reactions of alkenes we studied in Chapter 11. We can make *cis*-diols using osmium tetroxide-mediated dihydroxylation, *trans*-diols by ring opening of epoxides, and chiral diols by Sharpless asymmetric dihydroxylation (Figure 20.1). *trans*-1,2-Amino alcohols can be prepared by ring opening of epoxides, and there is a version of the Sharpless dihydroxylation that leads to chiral amino alcohols.

20.2.2 ACYL ANION EQUIVALENTS

If the equivalent of a negatively charged carbonyl group is added to another carbonyl, then a 1,2-dioxygenated species is produced. The reagents that can do this are called acyl anion equivalents—they are the synthons for $[XC(=O)]^-$. We have already met a number of such species, the cyanide ion, the anions of alkynes, and dithianes.

Cyanohydrins may be used to prepare 1,2-dioxygenated compounds or 1,2-amino alcohols. For example, we could disconnect **20.1** to benzaldehyde and $^-$[COOH], for which the synthon is cyanide ion. Once the cyanohydrin is formed, the –CN group may be hydrolyzed to a carboxylic acid or reduced to an amine. For practical syntheses, it is often preferable to use Me_3SiCN as the reagent for addition of cyanide. Cyanohydrin formation, as you will remember, is reversible (Section 14.3.1), and the equilibrium constant is not always in favor of cyanohydrin formation. The anion of the cyanohydrin may be captured by the trimethylsilyl group, effectively preventing the reverse reaction. The reaction mechanism is shown in Figure 20.2 (trimethylsilyl groups, as we shall see later, are readily removed in dilute aqueous acid) and some practical applications in Figure 20.3.

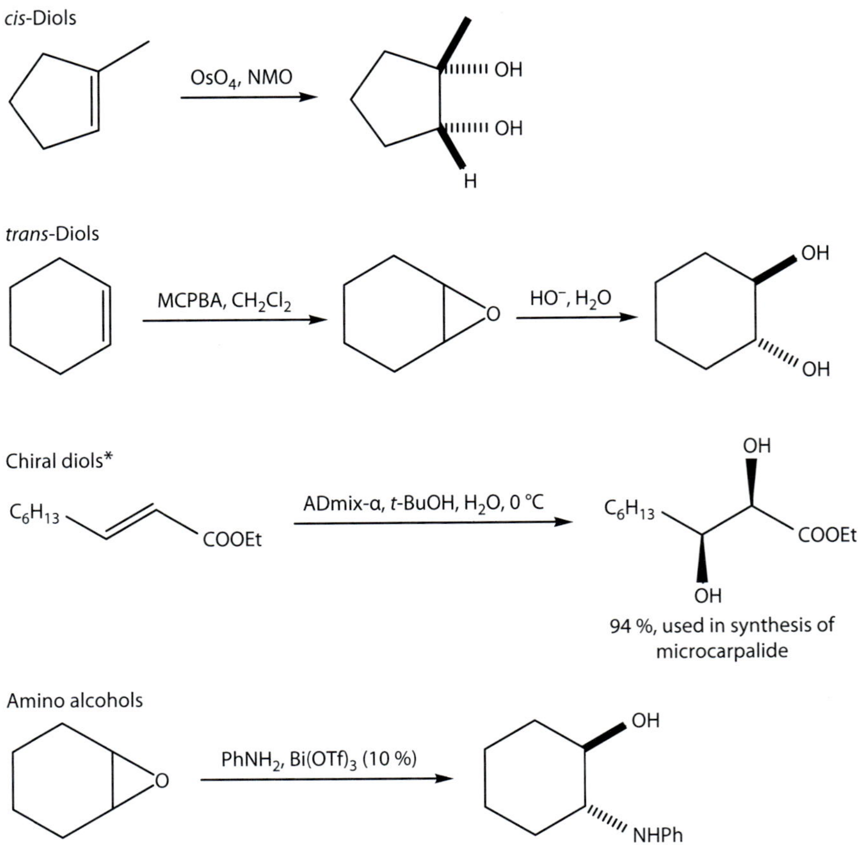

FIGURE 20.1 Synthesis of 1,2-diols and amino alcohols.

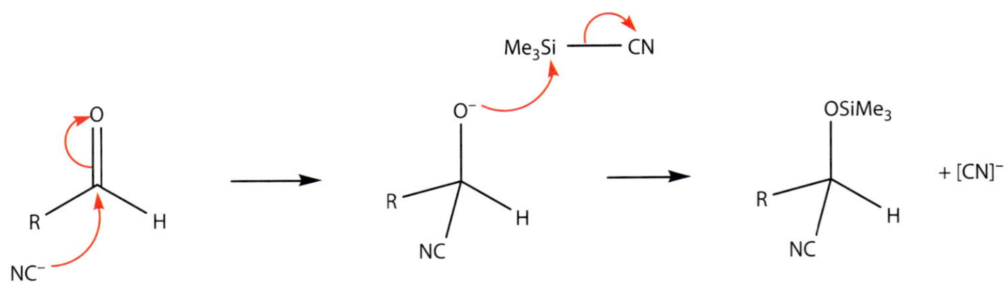

FIGURE 20.2 Mechanism of cyanohydrin formation using Me_3SiCN.

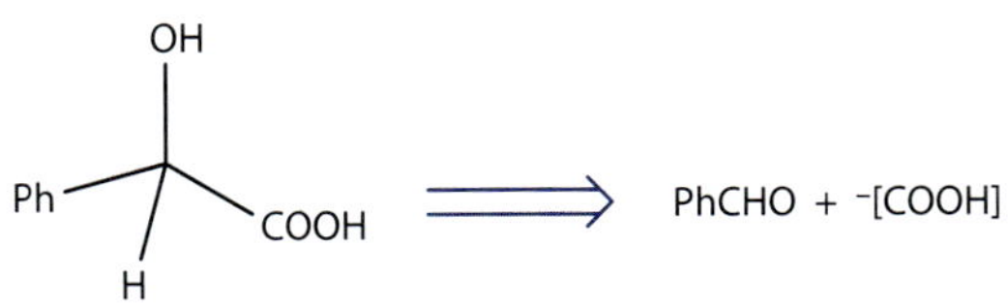

20.1

Related to cyanohydrin formation are the Strecker reaction (see Section 14.3.4) and the benzoin condensation (Figure 20.4, see Section 17.5). Few modern syntheses use this classical form of the benzoin condensation, because of the toxicity of cyanide. Thiamine and related heterocycles (such as **20.2**) are more commonly used and, unlike the cyanide catalyzed reaction (which is restricted to nonenolizable aryl aldehydes) **20.2**, is a good catalysts for coupling of simple aldehydes (Figure 20.5).

---

* K.C. Kolb, M.S. van Nieuwenhze, and K.B. Sharpless, *Chem. Rev.*, **94** (1994) 2483–2547.

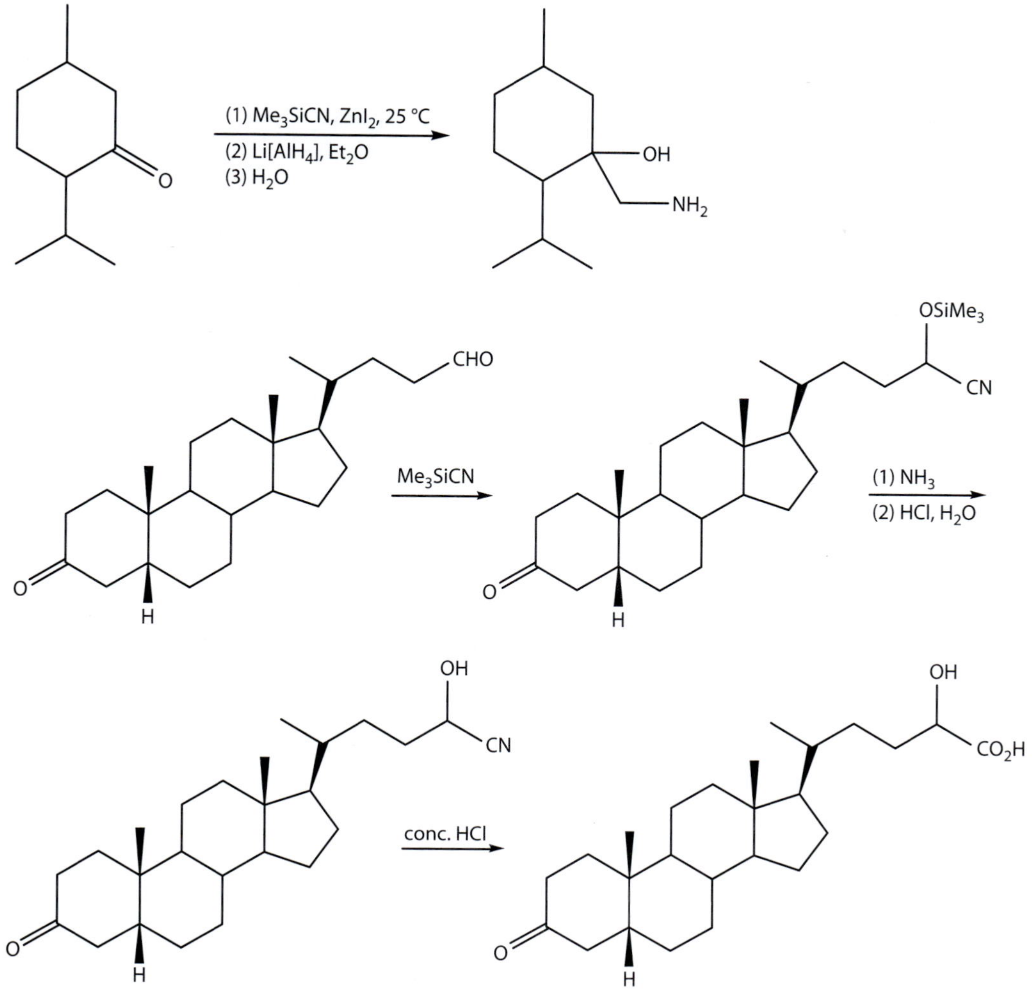

FIGURE 20.3 Uses of cyanohydrins in synthesis of 1,2-difunctionalized compounds.

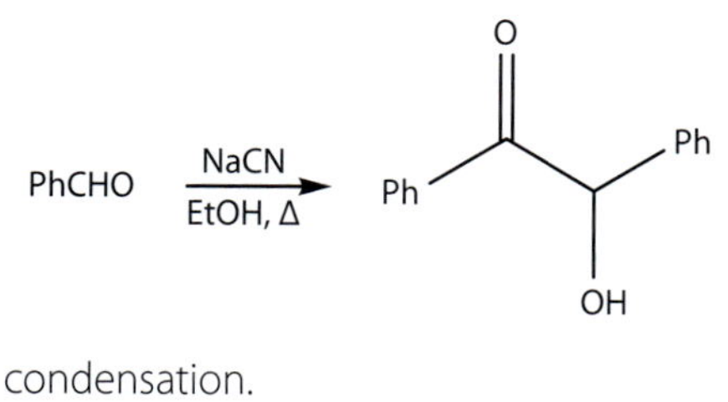

FIGURE 20.4 The Benzoin condensation.

We have already seen the anion of an alkyne as an acyl anion synthon (Figure 20.6). The α-hydroxyketone is disconnected to a ketone and an acyl anion, for which $[HC{\equiv}C]^-$ is the synthon. The final step in the forward reaction is Markovnikov addition of water to the alkyne followed by tautomerization.

Finally, in the last chapter, we noted that lithiated dithianes react with both ketones and acyl halides to give, respectively (after removal of the dithiane), 2-hydroxycarbonyl compounds and 1,2-dicarbonyl compounds (Figure 20.7).

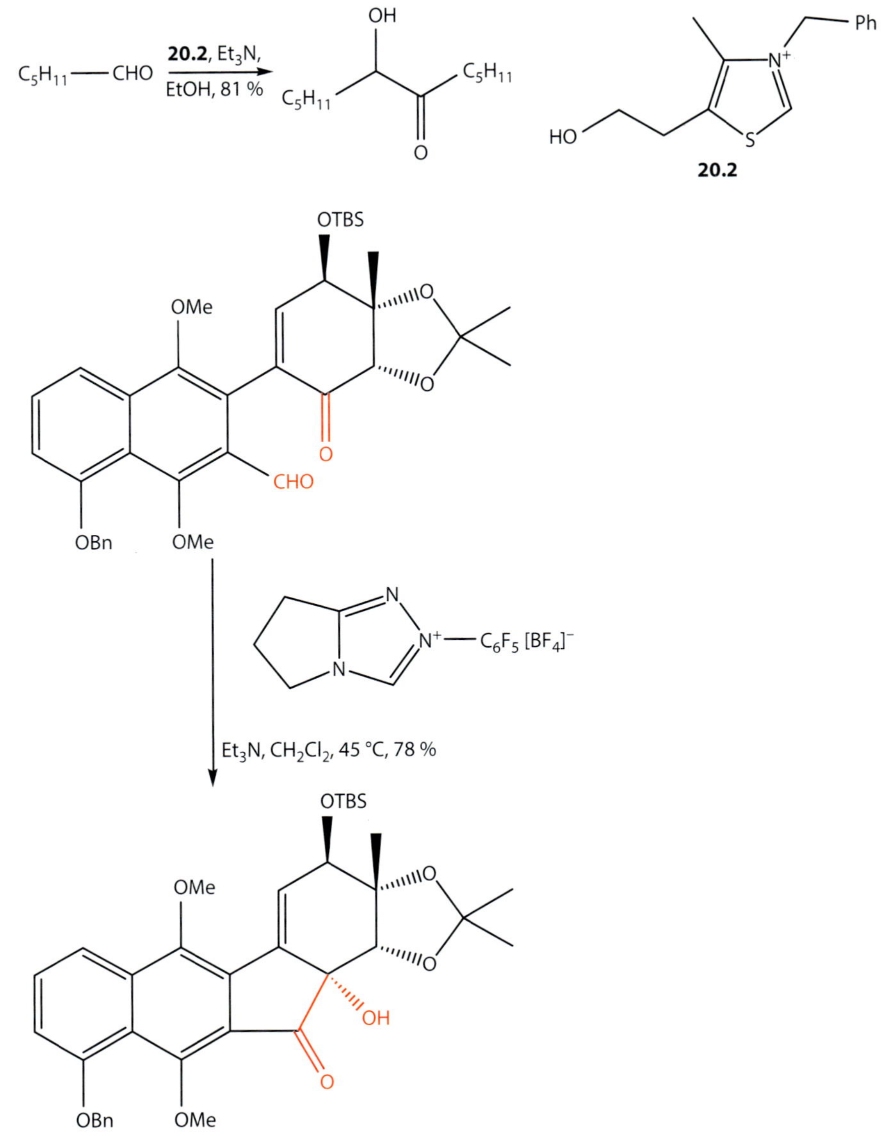

Used in the synthesis of Kinamycin F

FIGURE 20.5 Applications of the benzoin and related reactions in synthesis.

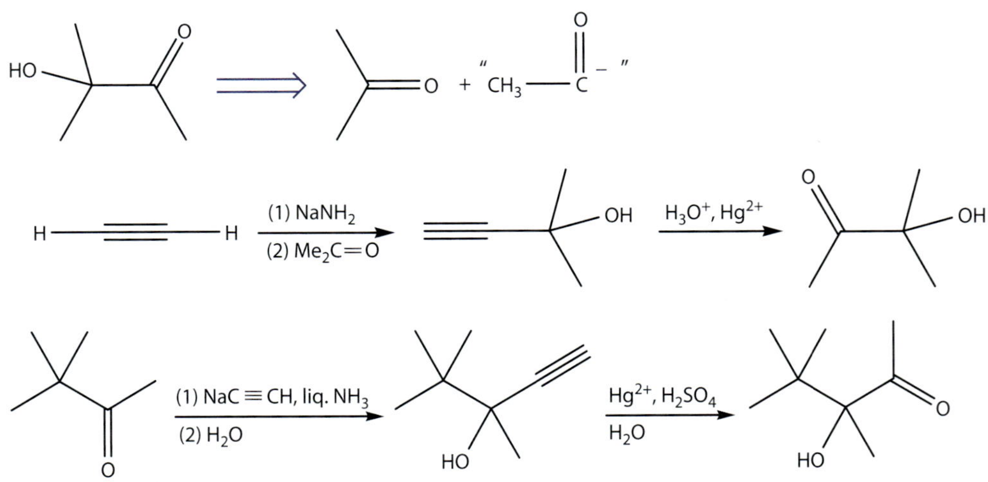

FIGURE 20.6 Alkynyl anions as acyl anion equivalents.

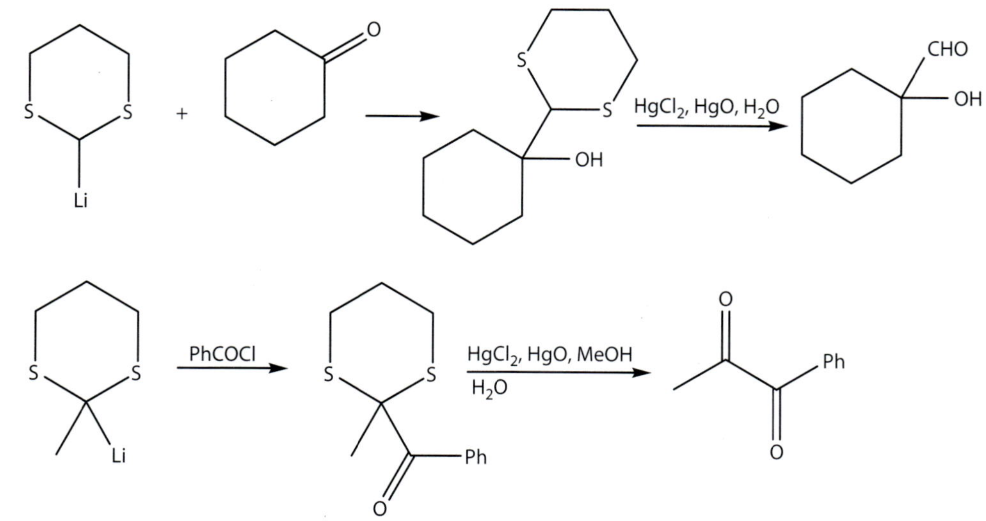

FIGURE 20.7 Use of dithianes as acyl anion equivalents (umpolung).

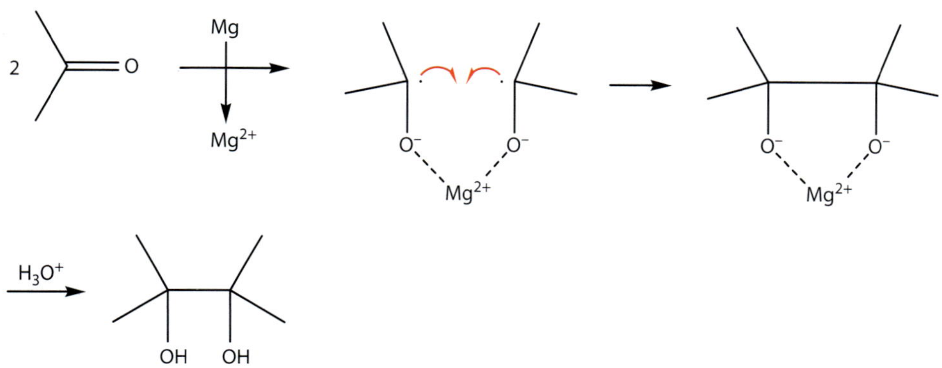

FIGURE 20.8 Mechanism of the pinacol coupling reaction.

20.2.3 RADICAL COUPLING REACTIONS

The two reactions discussed here are new to you—they are reductive processes where an alkali or alkaline earth metal supplies electrons, to convert carbonyl compounds into radical anions, which then couple together on the metal surface. The first is the pinacol coupling reaction of ketones, which, classically, uses magnesium metal as an electron source. The reaction mechanism is shown in Figure 20.8. In recent years, many other metals have been used similarly; titanium and samarium reagents have been particularly popular, and some applications are shown in Figure 20.9.

The acyloin condensation involves essentially the same chemistry as the pinacol reaction, but the substrates are esters, rather than aldehydes or ketones, and most reactions use an alkali metal as the electron source (Figure 20.10); 4 moles of sodium are needed to complete the reaction. The reaction is often modified by addition of trimethylchlorosilane, to capture the dianion. This is called the Rühlman modification and allows the reaction to proceed successfully for intramolecular reactions giving rise to both small and medium rings, mainly by suppressing potential side reactions. Examples are shown in Figure 20.11.

20.2.4 OTHER REACTIONS

Don't forget that halogenation of the α-position of many carbonyl compounds can be accomplished *via* their enolates (Section 17.4, Figure 20.12). Halogens α to carbonyl groups are also particularly easy to displace (Section 9.6.2).

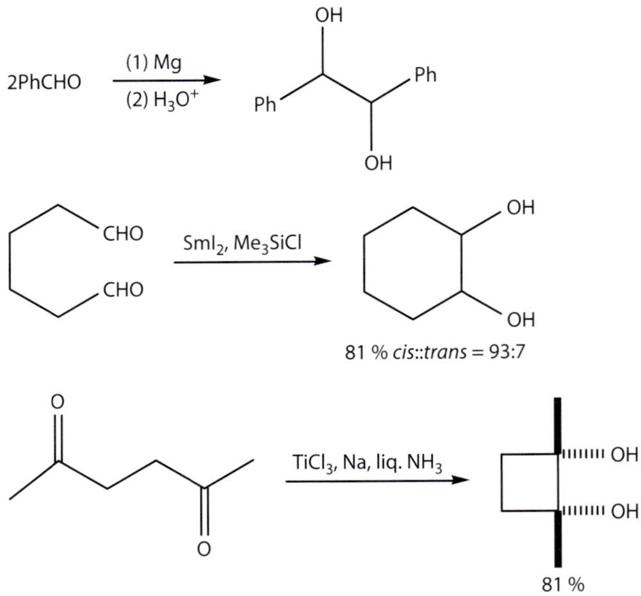

FIGURE 20.9 Applications of the pinacol coupling reaction.

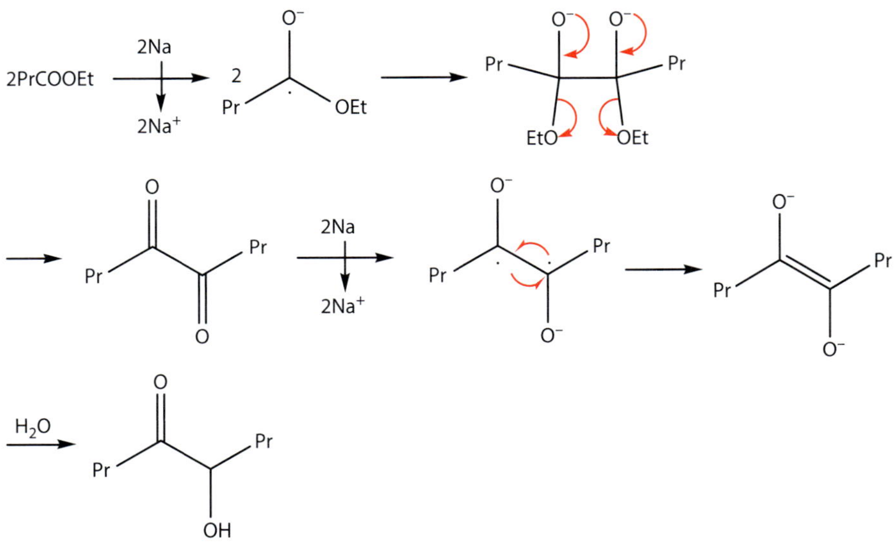

FIGURE 20.10 Mechanism of the acyloin condensation.

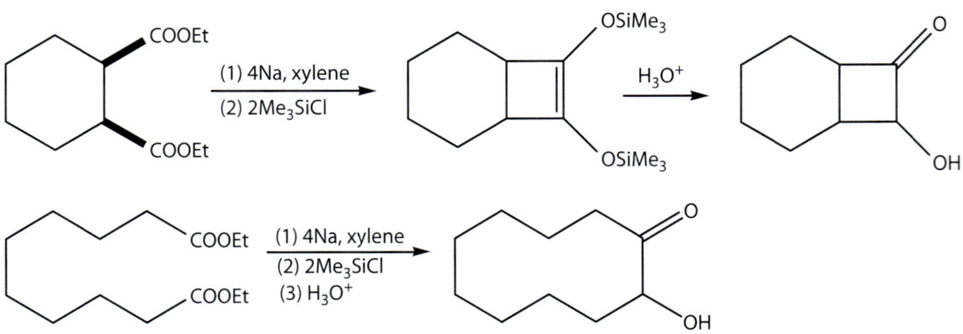

FIGURE 20.11 Applications of the acyloin condensation.

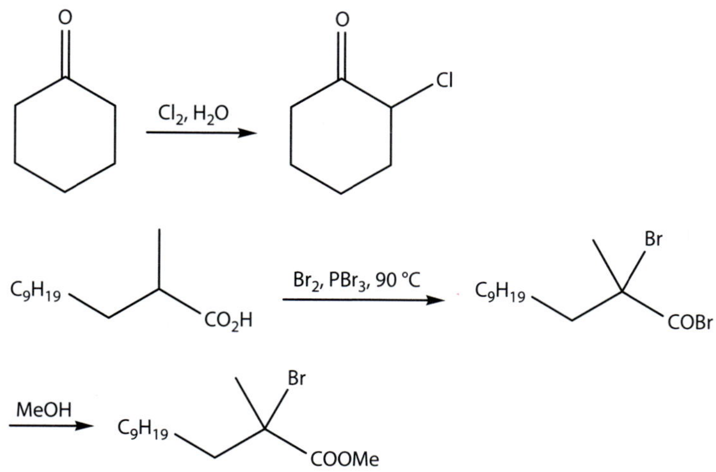

FIGURE 20.12 Halogenation α to carbonyl groups.

PROBLEM 20.1

How would you accomplish each of the following transformations? Some may require more than one step:

(a)

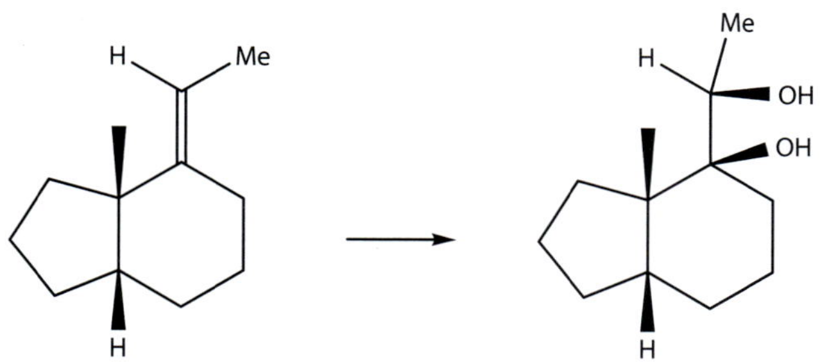

(b)

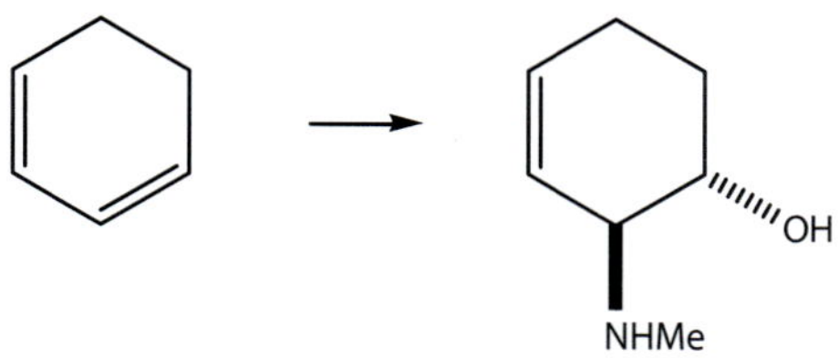

(c)

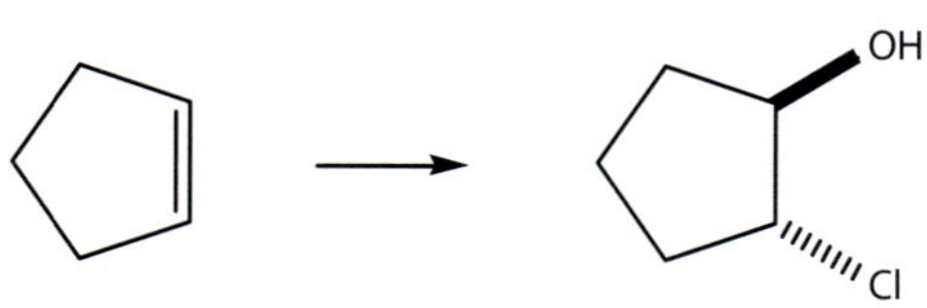

(d)

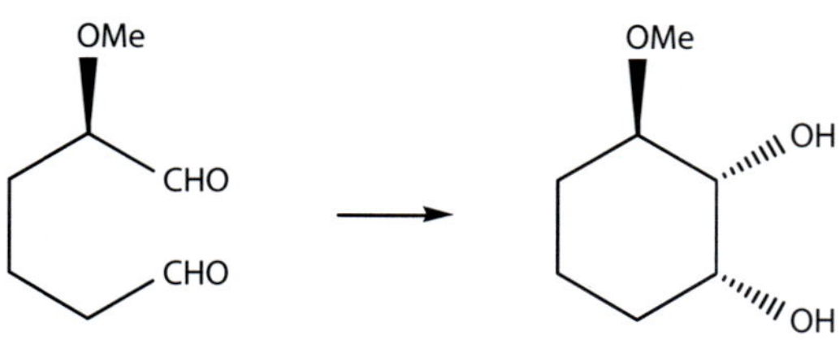

PROBLEM 20.2

Predict the outcome of each of the following reaction sequences:

(a)

$$EtO_2CCH_2CH_2CO_2Et \xrightarrow[\text{(2) } Me_3SiCl \text{ (XS)}]{\text{(1) Na (4 mol), xylene}}$$

(b)

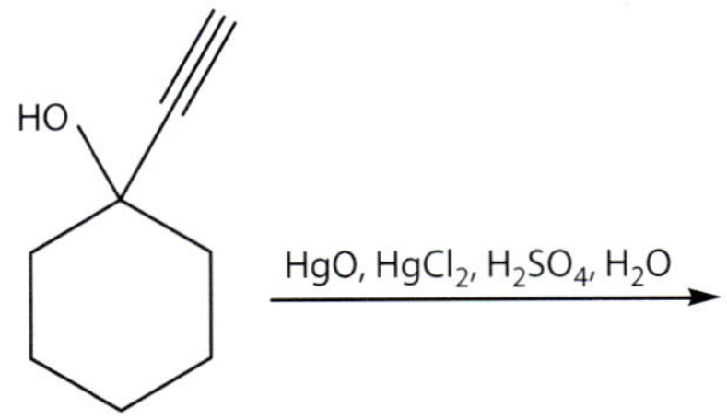

(c)

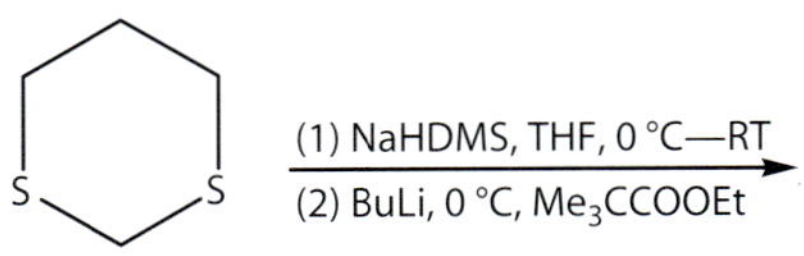

NaHDMS is sodium *bis*(trimethylsilyl)amide, $NaN(SiMe_3)_2$, a strong, nonnucleophilic base.

PROBLEM 20.3

The benzoin condensation is catalyzed by thiamine in the presence of base. Propose a mechanism for the following reaction:

$$PhCHO \xrightarrow{\text{Thiamine}} PhCH(OH)C(=O)Ph$$

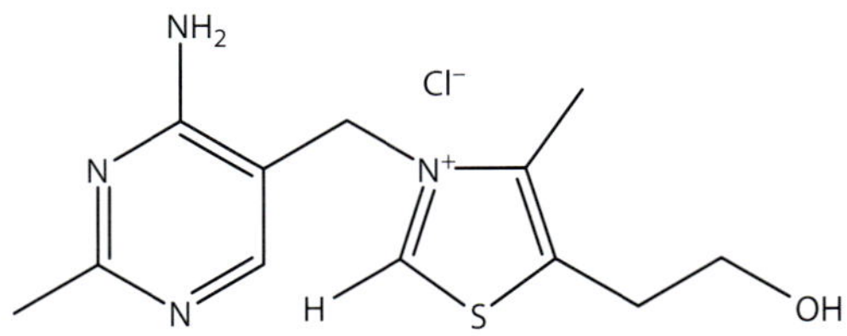

Thiamine hydrochloride

Solutions

20.1(a) OsO_4, NMO.

20.1(b) *trans*-Geometry suggests an epoxide intermediate. So use MCPBA to prepare the epoxide, then open with $MeNH_2$. The regiochemistry of the epoxide opening might be considered problematic, but the allyl position is the more easily substituted.

20.1(c) While the geometry might suggest an epoxide, in fact this reaction is easier—the addition of chlorine water to cyclopentane, involving a chloronium ion. Reagents were aqueous HOCl, HOAc.

20.1(d) Many reagents, especially lanthanide salts, would effect this coupling. The one used in practice was SmI_2 in THF, and the yield was 84 %.

20.2(e) This is an acyloin coupling and gives the protected enediol:

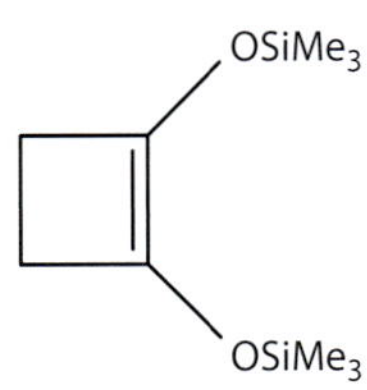

20.2(b) Water is added anti-Markovnikov to the alkyne, and the enol tautomerizes to the ketone:

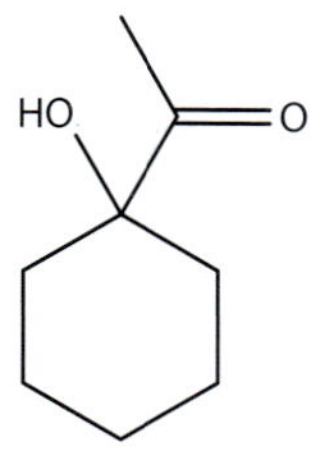

20.2(c) The dithiane is lithiated and then substitutes the ester to give **20.3**. This could be deprotected to the ketoaldehyde with $HgCl_2$, H_3O^+.

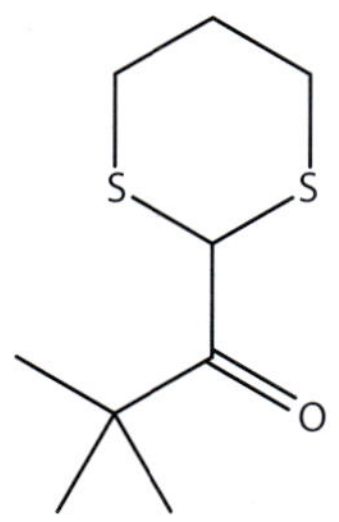

20.3

20.3

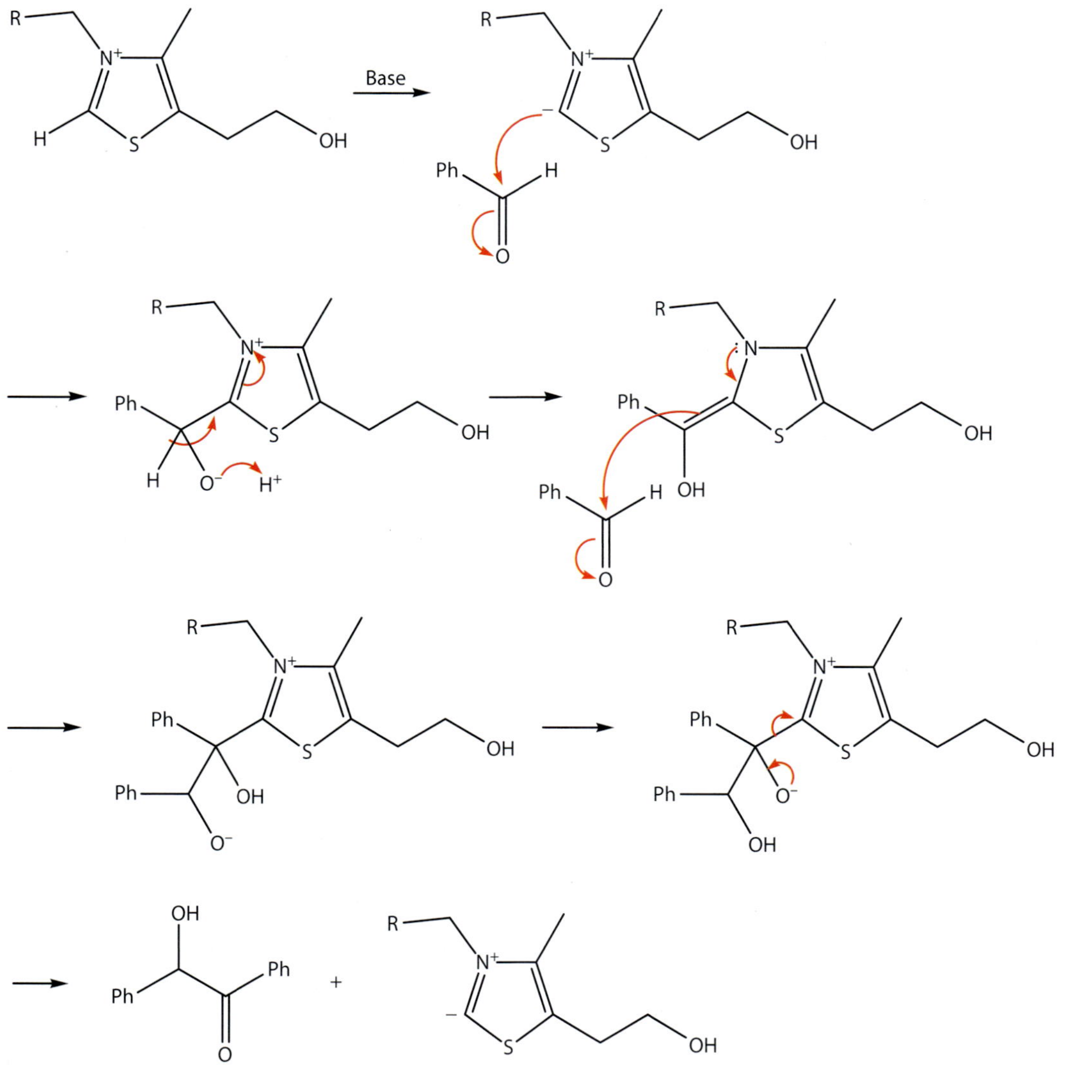

Key Points from Section 20.2

- 1,2-Diols may be prepared using OsO_4-catalyzed processes or *via* epoxides. Sharpless asymmetric dihydroxylation and enantioselective epoxidations allow the preparation of chiral diols.
- Acyl anion equivalents include cyanide ion (makes α-hydroxy acids *via* cyanohydrins and α-amino acids *via* the Strecker reaction), alkynyl anions, and dithianes.
- Electropositive metals may be used to supply electrons to carbonyl compounds to produce radicals, capable of coupling at the metal surfaces in the pinacol and acyloin reactions.

20.3 1,3-DIFUNCTIONALIZED COMPOUNDS

20.3.1 β-HYDROXYKETONES: THE ALDOL CONDENSATION

We met in Chapter 17 the archetypal process for the preparation of β-hydroxyketones, the aldol condensation. This is one of the most important of all construction reactions in organic synthesis. Self-condensation reactions (Figure 20.13) are generally straightforward and work well. The first example is the self-condensation of acetophenone; the product is readily dehydrated. The second example is shown as a disconnection—perhaps not completely obvious that this is a self-condensation until it is disconnected.

The main problem that we face in aldol condensations is control—obviously we do not want to be limited to self-condensations in synthesis. We will discuss three main strategies for control in aldol condensations, which will also be applicable in other types of condensation reactions. These are the use of intramolecular processes forming five- or six-membered rings; the use of nonenolizable, but highly electrophilic species; and activation of the nucleophile by addition of another anion-stabilizing group.

Intramolecular condensation reactions that give rise to five- or six-membered rings are invariably favored over any intermolecular reaction or reactions leading to the formation of rings of other sizes. Thus, **20.4** gives **20.5**, essentially irrespective of which acid/base is used; the formation of the possible eight-membered ring product, **20.6**, would be strongly disfavored (Figure 20.14). The reaction has been widely exploited in synthesis; the examples in Figure 20.15 show the formation of six- and five-membered rings by intramolecular processes, with the new bonds formed highlighted in red. In the first example, the product was enantiomerically enriched as a result of the use of chiral proline as the catalyst (more in Section 23.7); it was used in a synthesis of erythromycin. In the second example, a five-membered ring is formed and the product was used in a triquinacene synthesis.

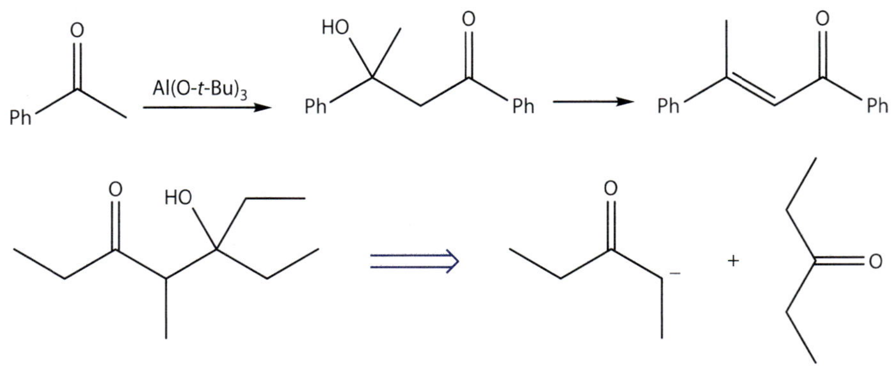

FIGURE 20.13 Aldol self-condensations.

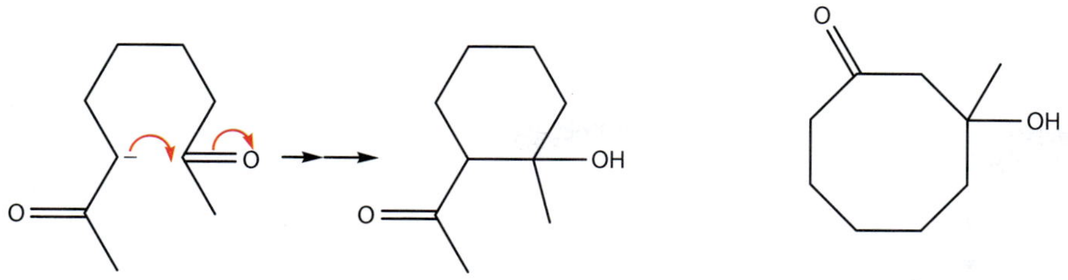

20.4 **20.5** **20.6**, not formed

FIGURE 20.14 Intramolecular aldol reactions are usually selective.

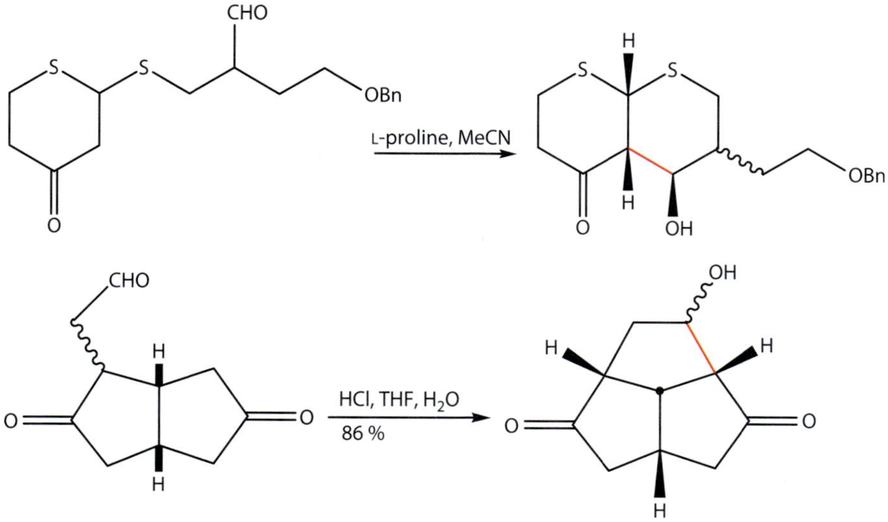

FIGURE 20.15 Intramolecular aldol reactions in synthesis.

PROBLEM 20.4

Predict the product(s) of each of the following processes:

(a)

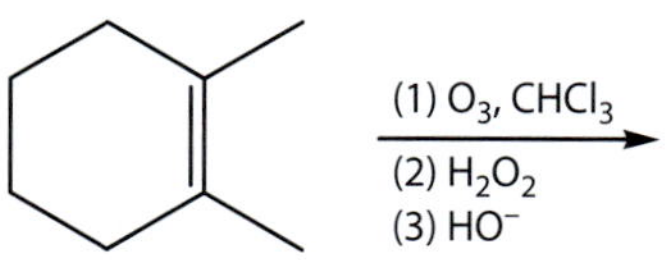

(b)

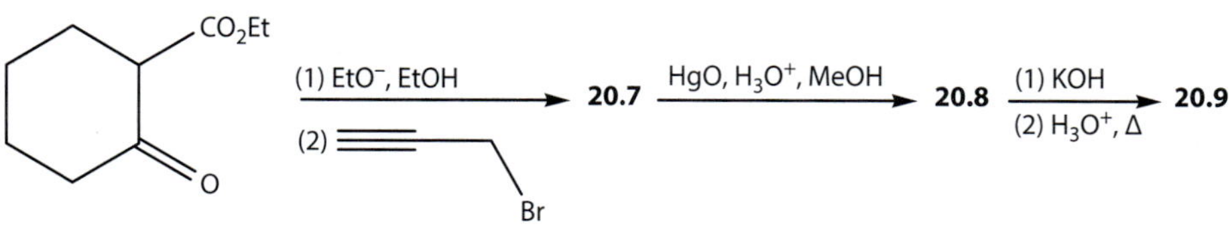

Solutions

20.4(a) The first step is ozonolysis of the double bond to give $CH_3C(=O)(CH_2)_4C(=O)CH_3$. This then undergoes an intramolecular aldol to yield a β-hydroxyketone, which may or may not dehydrate under the reaction conditions. The formation of the five-membered ring is more favorable than the formation of a seven-membered ring.

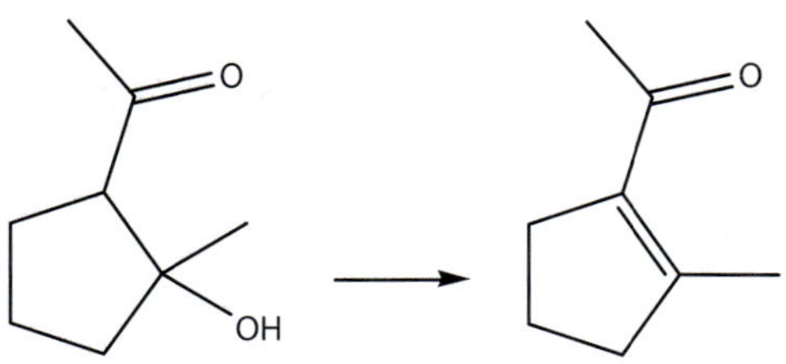

20.4(b) Although this looks a complex problem, none of the reactions should be unfamiliar. The first step is to deprotonate the β-ketoester and then alkylate it with the propargyl bromide to give **20.7**. The triple bond is then hydrated to give the ketone **20.8**. In the presence of base, this undergoes an intramolecular aldol reaction to give a five-membered ring. The ester is hydrolyzed and decarboxylates on heating to give **20.9**.

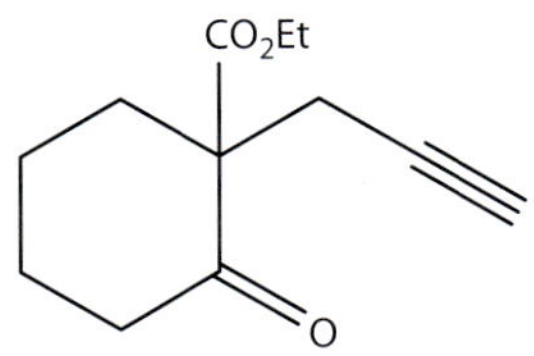

20.7

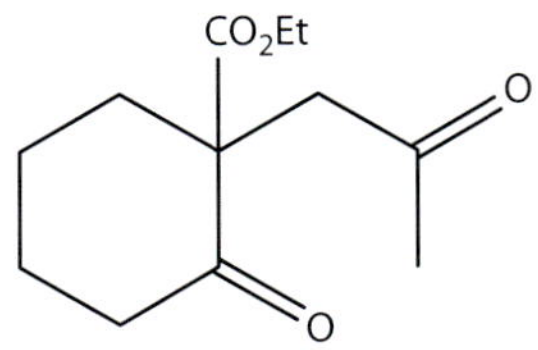

20.8

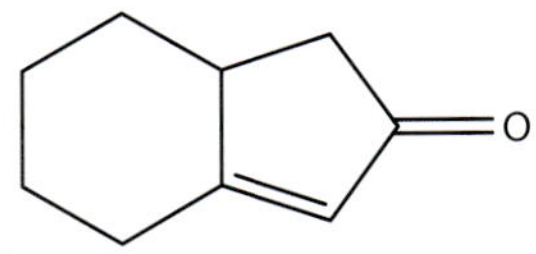

20.9

Our next strategy for selective aldol reactions is the use of very electrophilic but nonenolizable substrates—this effectively ***defines*** which component will be the electrophile. While in principle the other component aldehyde or ketone might still self-condense, this will be suppressed if it has a much better electrophile available. Thus, "highly electrophilic" is a relative term—the species we choose for this must be significantly more electrophilic than the component that we intend as the nucleophile. Useful species in this respect are shown in Figure 20.16. Diethylcarbonate is not an aldehyde or ketone, so reactions with it are not classical aldol processes, but we will find it is a very useful substrate for preparing activated nucleophiles. Examples of these reactions are shown in Figure 20.17, with the carbon–carbon bond formed highlighted in red. In the second example, it did not prove possible to prevent the dehydration; the product of the final example was used in a synthesis of quadrone.

FIGURE 20.16 Nonenolizable carbonyl compounds.

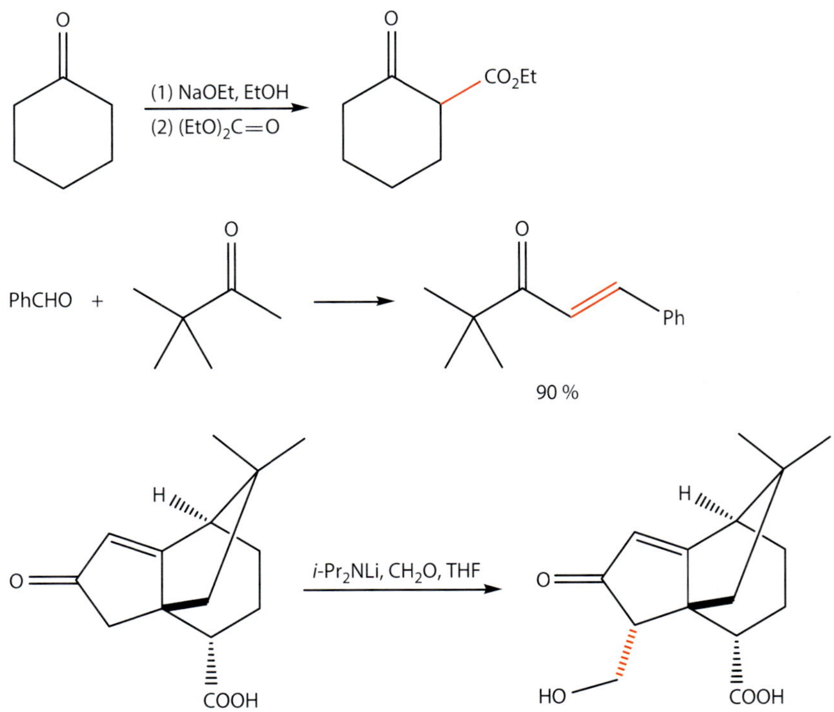

FIGURE 20.17 Aldol reactions with nonenolizable components.

PROBLEM 20.5

How would you carry out the following transformation?

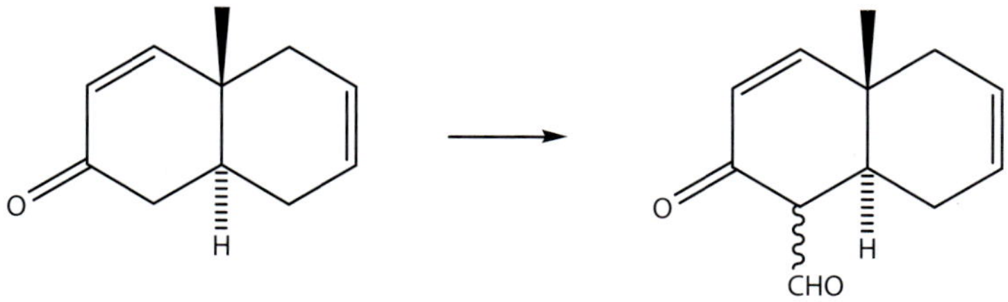

Solution

To do this in a single step and end up at the right oxidation level, you need HCOOMe. NaOMe in benzene was used as the base, and the reaction was complete in 30 min at room temperature.

Finally, we can make sure that we have the correct component as the nucleophile by activation—this means that one component is very much easier to enolize than the other and, if the enolization is essentially complete, means that it cannot act as the electrophile. Thus, instead of using acetone ($pK_a \sim 20$) as a nucleophile, we use ethyl acetoacetate, **20.10**, with a pK_a of 11. This guarantees that using a simple base such as ethoxide (pK_a EtOH 16) ethyl acetoacetate will be fully enolized and hence must be the nucleophilic component. Figure 20.18 shows an example; the product dehydrated readily.

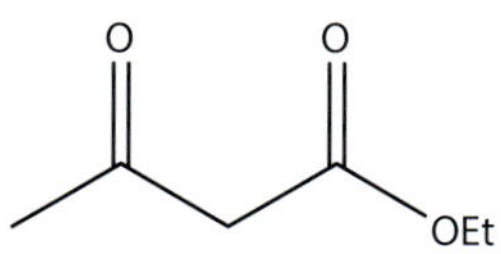

20.10

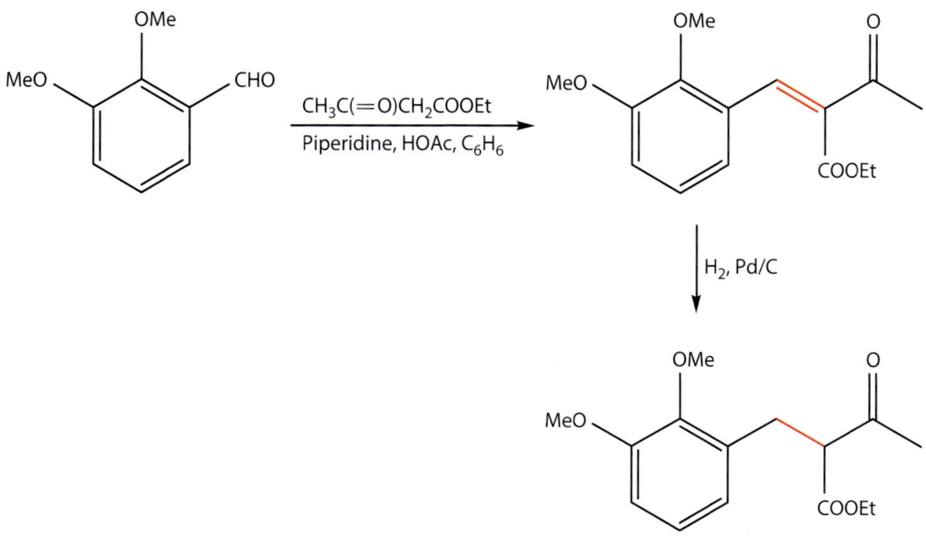

FIGURE 20.18 Reaction of ethyl acetoacetate with an aldehyde.

PROBLEM 20.6

Suggest a mechanism for each of the following reactions:

(a) 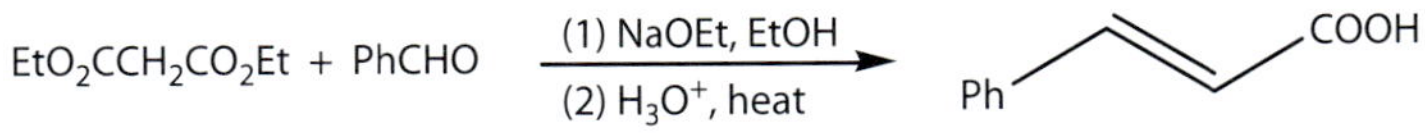

(b) 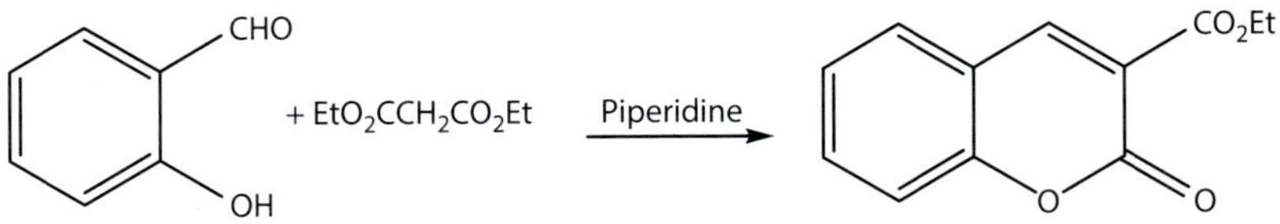

Solutions

(a)

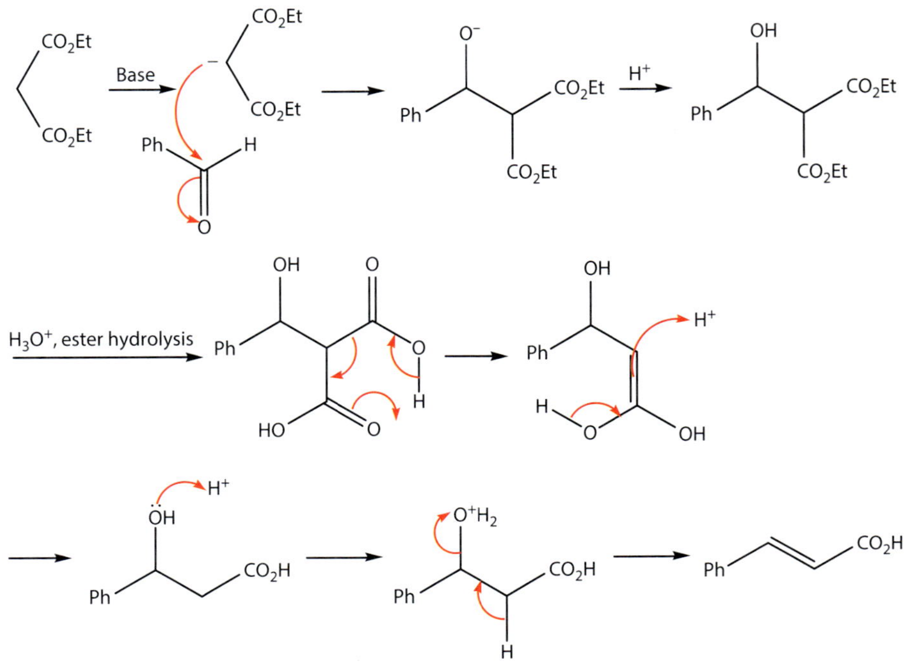

(b) Although it is tempting to dive straight into the condensation reaction, it is more likely that a transesterification is the first step:

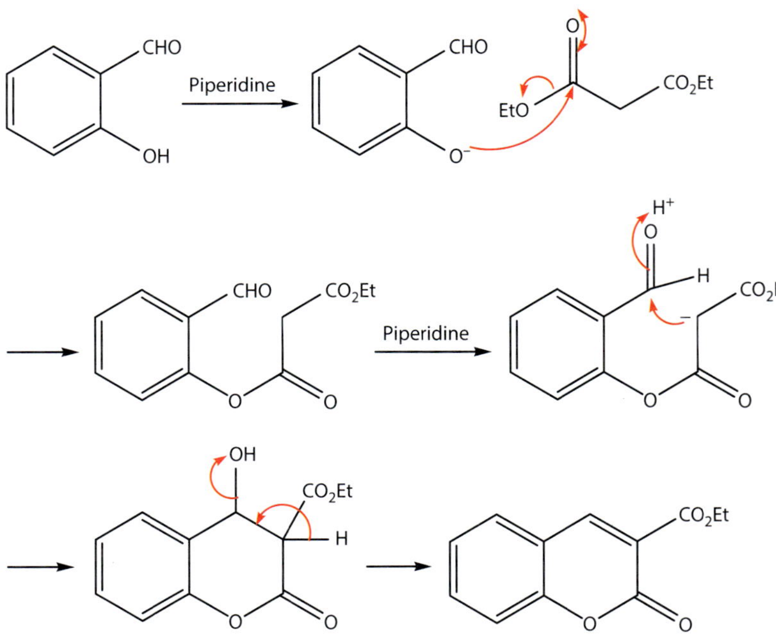

20.3.2 α,β-UNSATURATED ALDEHYDES AND KETONES

In the previous section, we saw several examples of aldol reactions where the dehydration of the β-hydroxyketone was so facile that it was essentially impossible to avoid. Some more examples are shown in Figure 20.19. The first is the self-condensation of a ketone, the second is a cross condensation of an enolizable aldehyde with a nonenolizable aldehyde, and the third is intramolecular reaction forming a five-membered ring.

Now we look at enones from the opposite point of view—how do we disconnect them? If we look at the bond that has been made in each of the previous examples (highlighted in red), it is clear that we must dissect the enone or enal *through* the double bond to give two carbonyl compounds or a dicarbonyl compound. Consider **20.11** (Figure 20.20). We disconnect the double bond so that there is a carbonyl group at one side of it and an enolizable site at the other. If we consider the forward reaction in this case, there should be no problem with it, because only the ketone is able to enolize, but the aldehyde is the more electrophilic component. Most bases will be appropriate here; in practice, sodium hydroxide in aqueous ethanol was used.

Figure 20.21 shows more examples, this time involving intramolecular processes. In the first example, *cis*-jasmone is dissected to a diketone; the real forward reaction is easy, since a

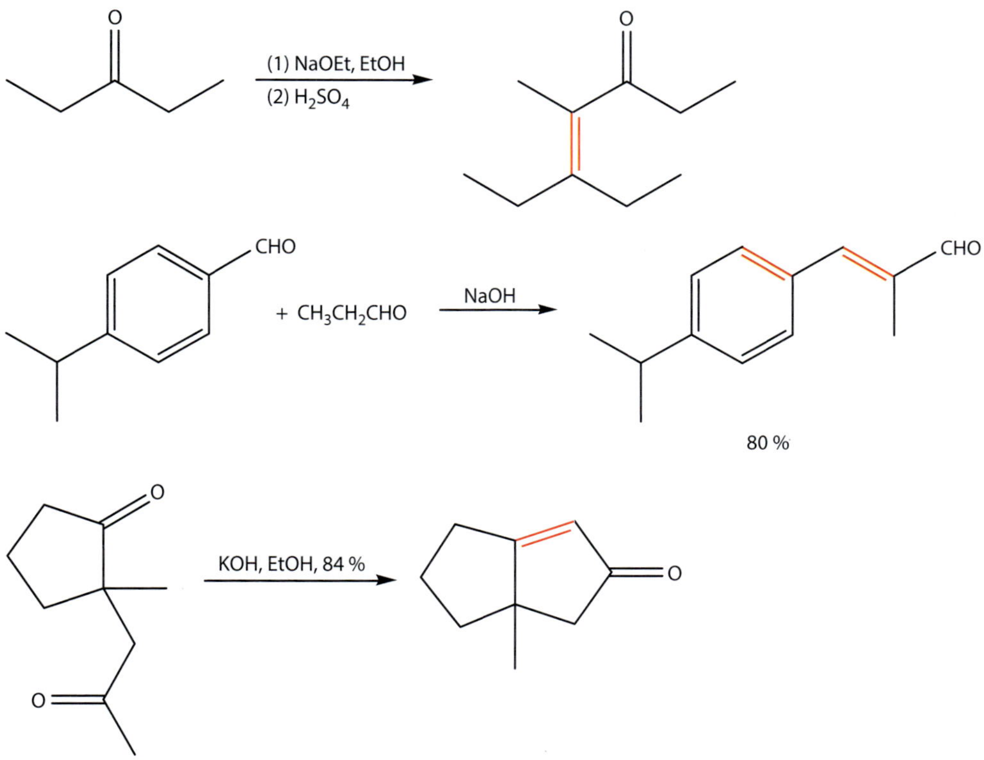

FIGURE 20.19 Synthesis of enals and enones by aldol condensation.

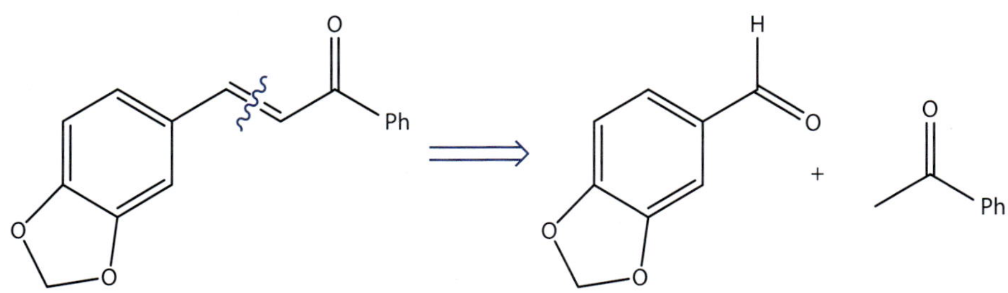

20.11

FIGURE 20.20 Disconnection of an enone.

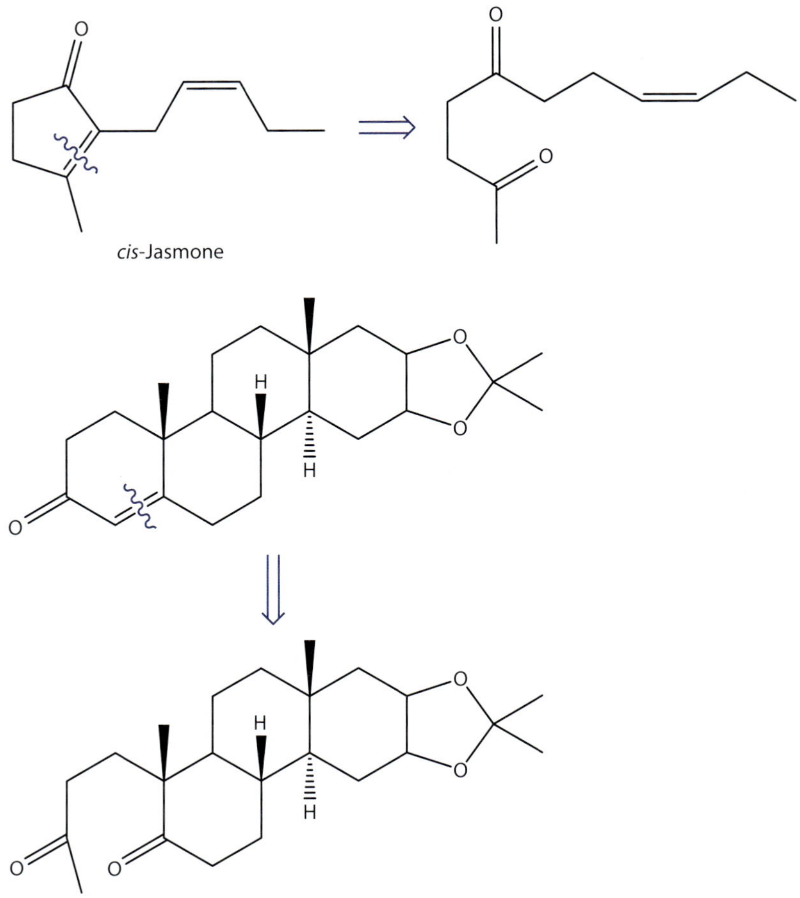

FIGURE 20.21 Disconnection of cyclic enones.

five-membered ring is being formed, and this is accomplished with KOH in aqueous ethanol. In the second example, a six-ring enone is dissected to a diketone; the forward reaction used NaOH/MeOH and the process was part of a synthesis of cholesterol.

PROBLEM 20.7

Show how you would disconnect each of the following enones to a dicarbonyl compound:

(a)

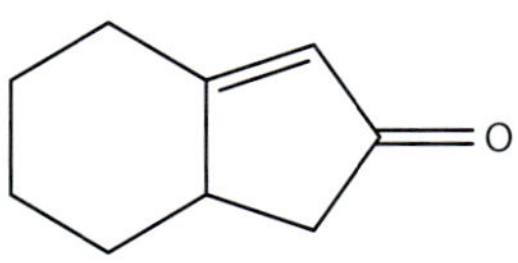

(b)

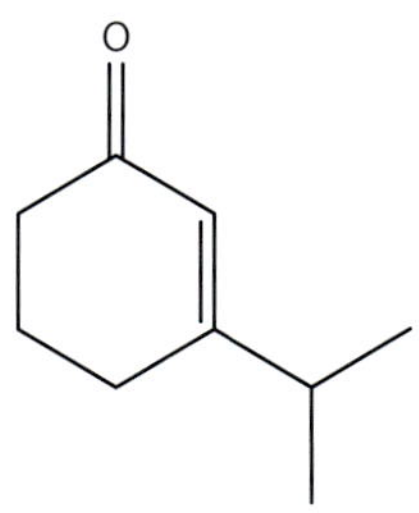

(c)

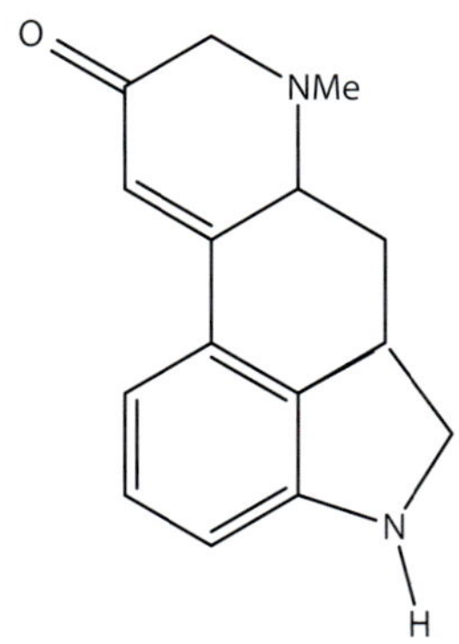

Solutions

(a)

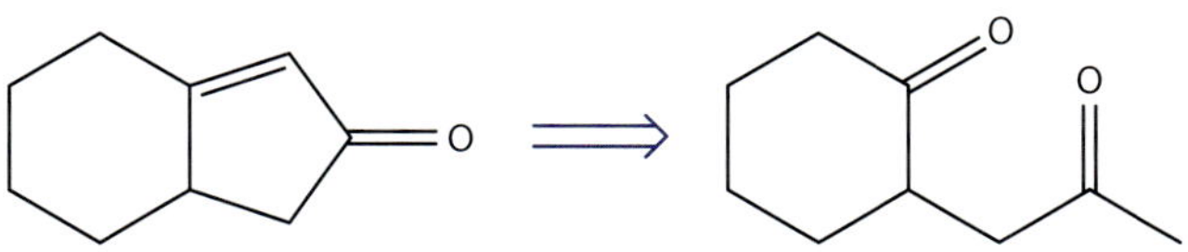

(b)

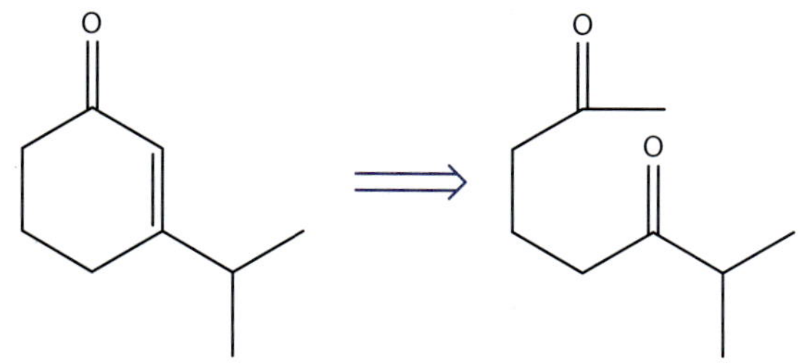

(c)

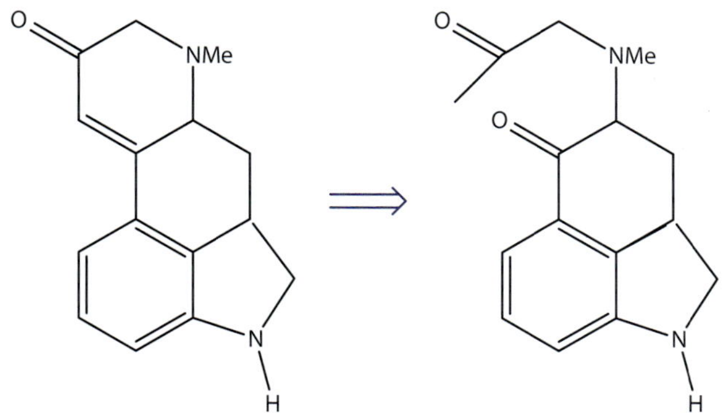

20.3.3 1,3-DIKETONES

1,3-Dicarbonyl compounds could clearly be prepared by oxidation of appropriate β-hydroxyketones; many of the oxidants we discussed in the previous chapter would be suitable. For example, the product of the intramolecular aldol reaction of **20.12** to **20.13** was then oxidized to **20.14** using CrO_3/H_2SO_4.

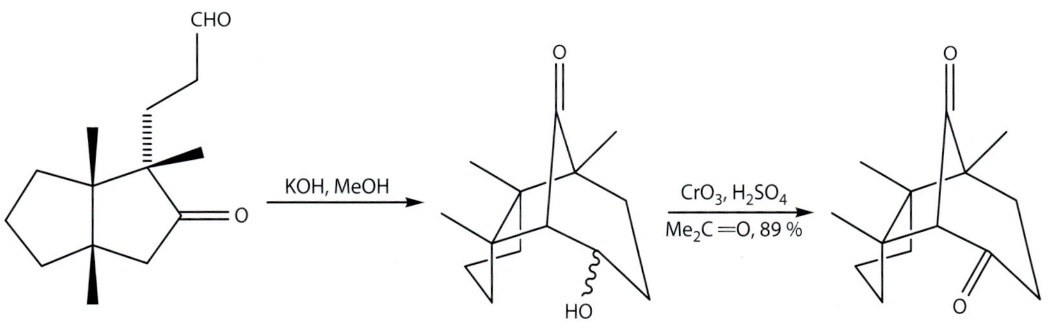

20.12 **20.13** **20.14**

However, 1,3-diketones are more commonly prepared by the crossed Claisen condensation between a ketone or aldehyde enolate and an ester. The disconnection of **20.15** (Figure 20.22) gives an enolate and PhCOX, where X is a leaving group. X is most conveniently OEt, so that we can use EtO^- to deprotonate the ketone. There is no doubt as to which component will be deprotonated as ketones are significantly more acidic (typical pK_a 20) than esters (typical pK_a 25). All of the reaction steps are equilibria except the last one; the enolate is produced irreversibly because it is so stabilized. The irreversible production of this enolate is also the reason that the self-condensation of acetone does not compete significantly with the cross condensation. The acetone self-condensation product cannot be converted to such a stable enolate. A practical example is shown in Figure 20.23, with the key bond highlighted.

20.3.4 β-KETOESTERS

The classic method to synthesize these molecules is the Claisen condensation of two esters. All the same problems of control that we encountered in aldol reactions also exist here and have to be solved in a similar manner. The first reaction of Figure 20.24 is a homocoupling reaction and

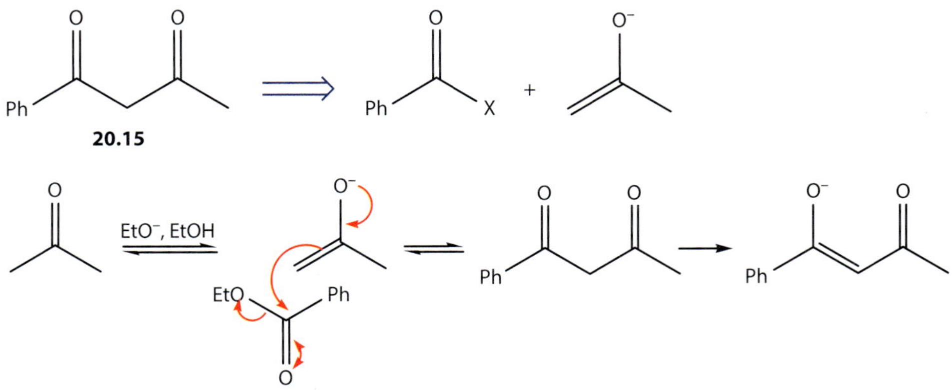

FIGURE 20.22 Crossed Claisen condensation to prepare 1,3-diketones.

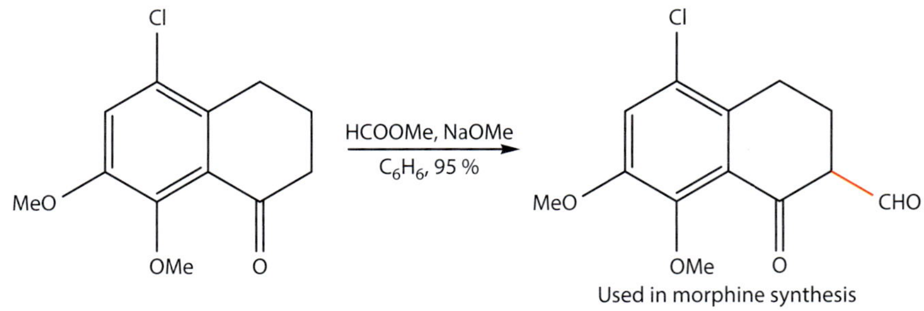

FIGURE 20.23 Crossed Claisen condensation in synthesis.

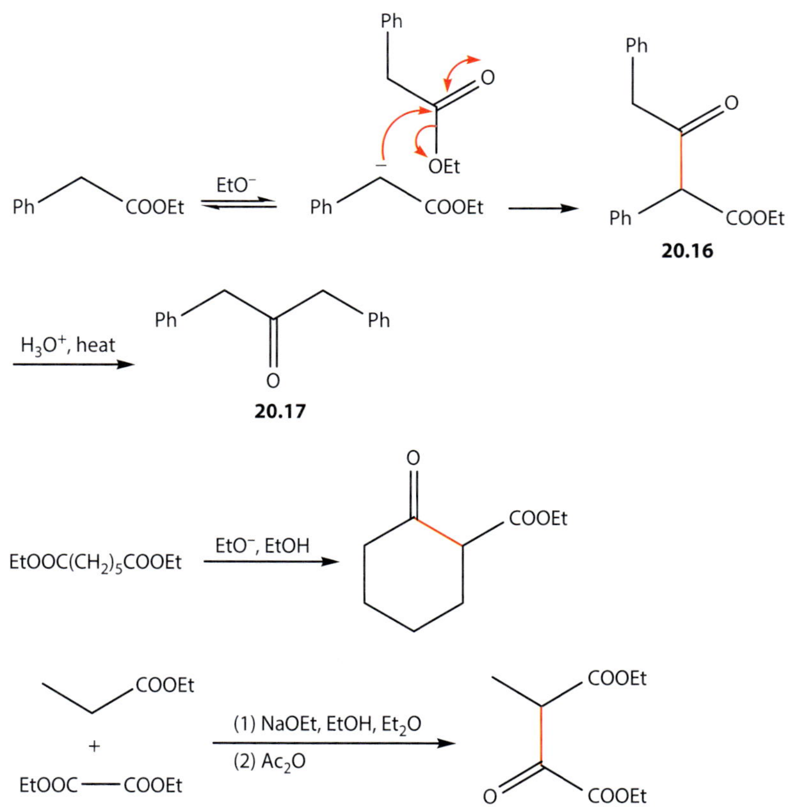

FIGURE 20.24 Synthesis of β-ketoesters.

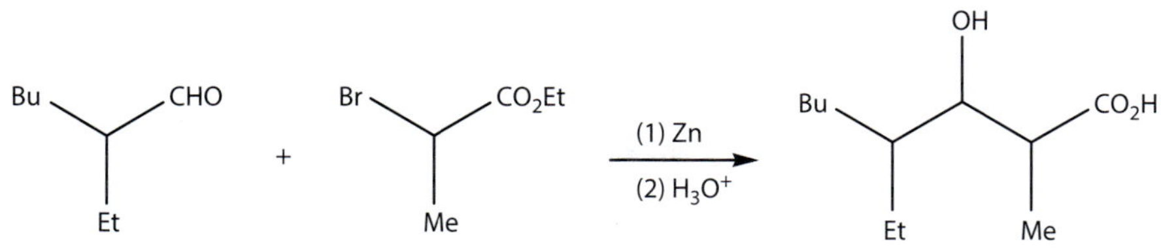

FIGURE 20.25 Reformatskii reaction.

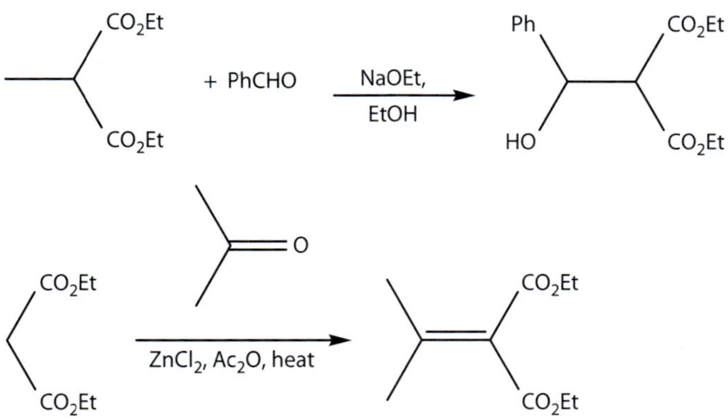

FIGURE 20.26 Use of diethylmalonate in condensation reactions.

the second an intramolecular process, a Dieckmann condensation. The third example is a cross condensation, controlled by the fact that only one of the esters can be enolized, and the other is activated by the adjacent electron-withdrawing group. Remember that the ester group stays there only for as long as you need it—hydrolysis and decarboxylation can remove it (**22.16** to **22.17**), so this strategy provides another approach to ketone synthesis.

20.3.5 TWO IMPORTANT SYNTHONS FOR ESTER ENOLATES

If a synthetic equivalent of the anion of ethyl acetate is needed, then there are two available. The first is the Reformatskii reaction using zinc derivatives of α-haloesters such as [$BrZnCH_2COOEt$] as the attacking reagent (Figure 20.25, see also Sections 17.4.1 and 17.4.7; at this point, it's immaterial as to whether we regard this as a true organometallic or a zinc enolate). The other synthon for the anion of ethyl acetate that we have met is diethylmalonate. After the reaction of the anion, the product is frequently dehydrated or one of the ester groups can be removed by hydrolysis followed by decarboxylation of the diacid (Figure 20.26).

PROBLEM 20.8

How would you carry out each of the following transformations?

(a)

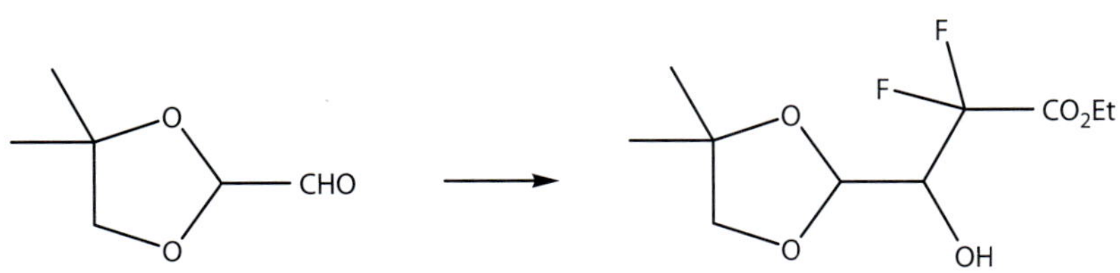

(b)

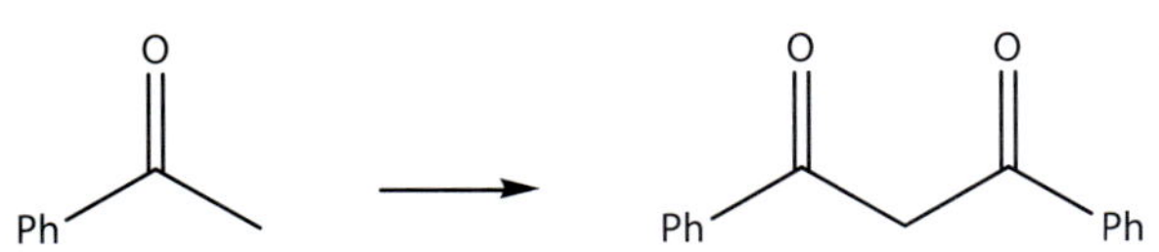

(c)

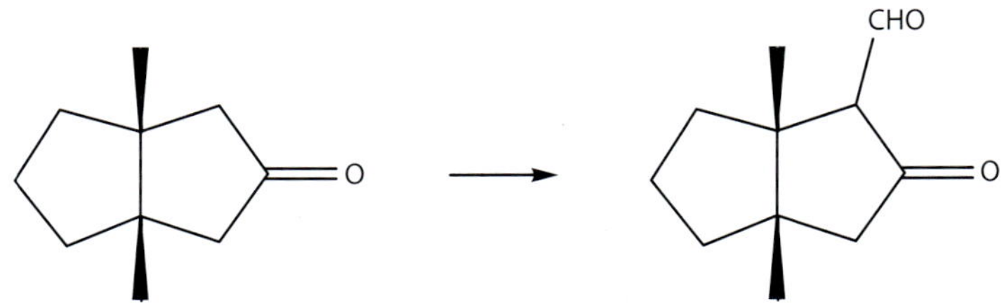

Solutions

(a) This is most easily accomplished using a Reformatskii reagent $BrZnCF_2CO_3Et$.
(b) Crossed Claisen reaction with PhCOOEt. Only the acetophenone can enolize. The equilibrium is shifted toward the dione because it is more easily enolized than either the SM or any other possible product.
(c) Crossed Claisen with HCOOEt, using NaH as base.

Key Points from Section 20.3

- 1,3-Difunctional compounds are generally made by carbonyl compound condensation reactions.
- The aldol condensation uses aldehydes or ketones as substrates and yields β-hydroxycarbonyl compounds.
- Selectivity in crossed condensations is achieved by the use of intramolecular processes yielding five- or six-membered rings, by using a nonenolizable, but very electrophilic substrate, or by activating one component of the reaction by addition of an ester group, thus defining it as the nucleophilic component.
- Enones are produced by dehydration of aldol condensation products.
- 1,3-Diketones are produced by a crossed aldol condensation between an aldehyde or ketone enolate and an ester.
- β-Ketoesters are produced by a Claisen condensation between two esters. The Dieckmann condensation is the intramolecular version of this reaction. The ester group can be removed by hydrolysis and decarboxylation.
- Reformatskii reagents (*e.g.*, $BrZnCH_2COOEt$) and the anion of diethyl malonates are useful equivalents of the anion of ethyl acetate.

20.4 1,4-DIFUNCTIONALIZED COMPOUNDS

As with the 1,2-difunctionalized compounds, there is no single reaction type that will accomplish all these syntheses. We could consider the disconnection in Figure 20.27, which gives an enolate and an α-halocarbonyl compound. Unfortunately, this strategy rarely works—the presence of the electron-withdrawing halogen means that it is generally easier to deprotonate the carbon atom to which it is attached, rather than the other component of the reaction. Thus, as we see in Figure 20.28, the product is not the 1,4-dicarbonyl intended but an epoxide of an α,β-unsaturated ester. This process is called the Darzens condensation—an interesting and useful reaction, but not exactly what we wanted.

We can circumvent this problem and define which species acts as a nucleophile in an unequivocal way, by converting the carbonyl compound that we want as our nucleophile to an enamine

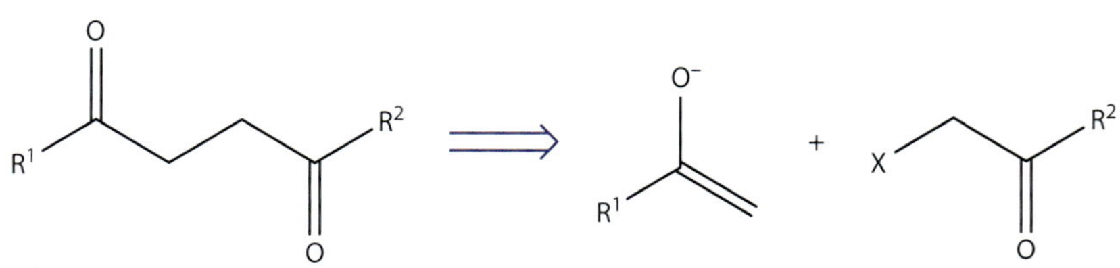

FIGURE 20.27 Disconnection of 1,4-dicarbonyl compounds.

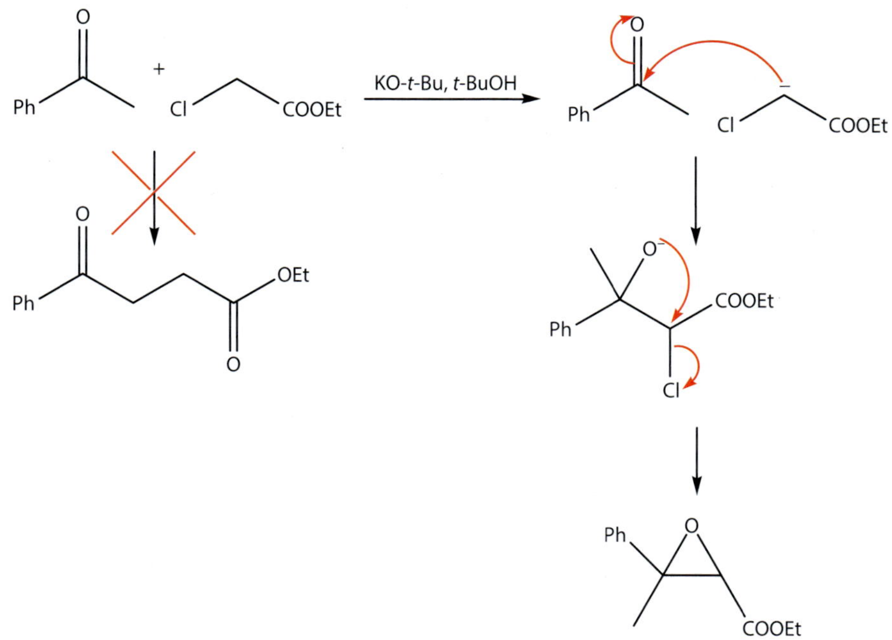

FIGURE 20.28 Mechanism of the Darzens condensation.

(see Section 14.3.4). When a secondary amine such as pyrrolidine reacts with a ketone, it is not possible to form an imine, but nucleophilic attack of the amine on the carbonyl is still favorable. The intermediate iminium salt loses a proton to give an enamine, which is relatively stable and can be purified by distillation. If this then reacts with an α-halocarbonyl compound, the 1,4-dicarbonyl compound is generated after the hydrolytic work-up (Figure 20.29).

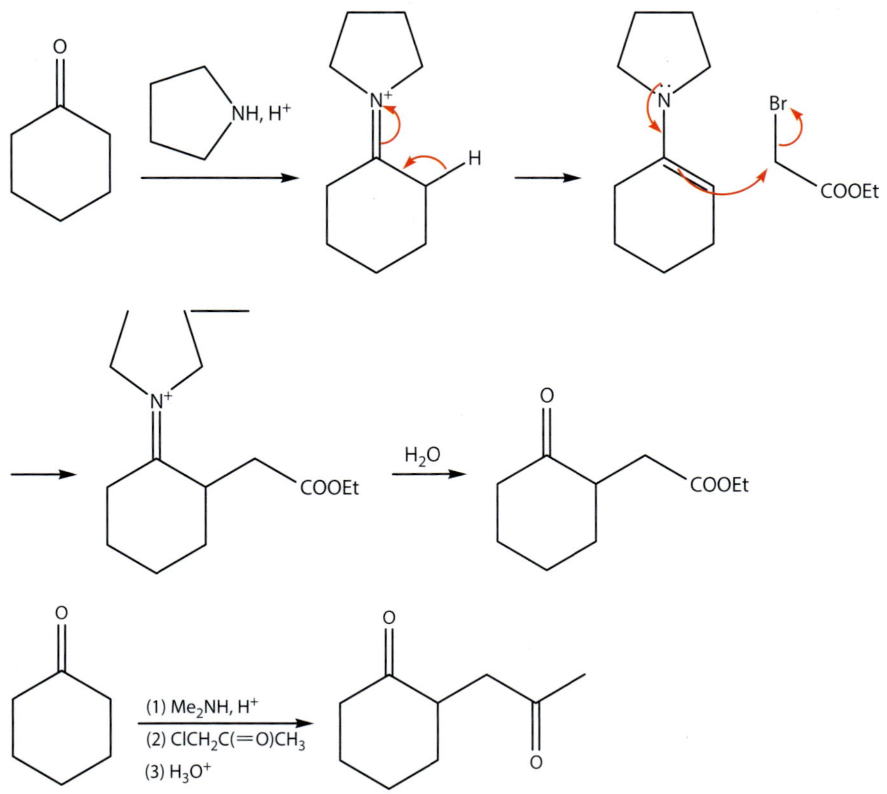

FIGURE 20.29 Enamines in the synthesis of 1,4-difunctionalized compounds.

PROBLEM 20.9

Predict the outcome of each of the following reactions:

(a)

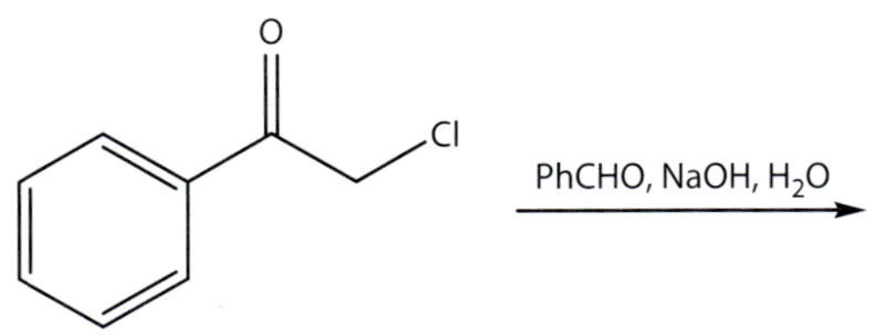

(b)

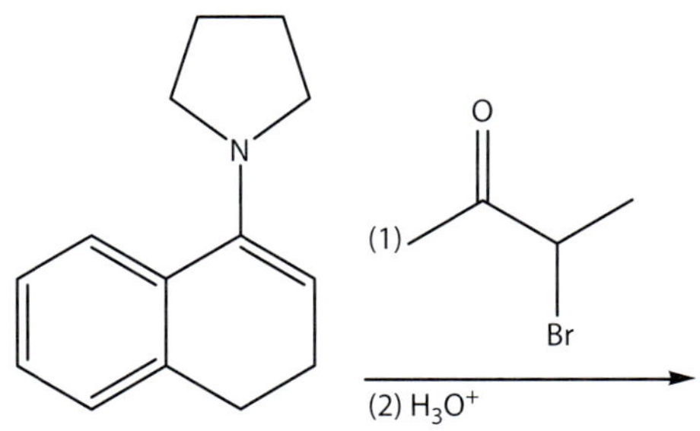

Solutions

(a) This is a Darzens condensation; the *trans*-isomer of the epoxide is the main product:

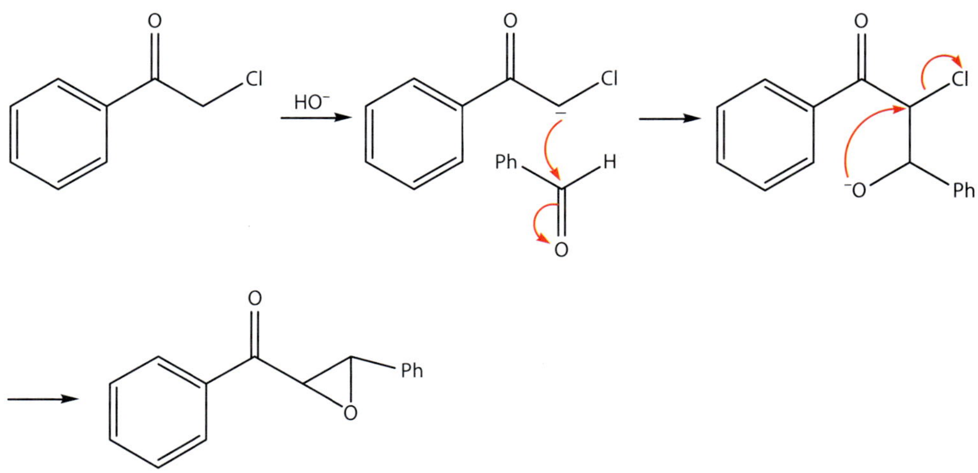

(b) The halide is substituted by the enamine, and the enamine is hydrolyzed in the aqueous acid work-up:

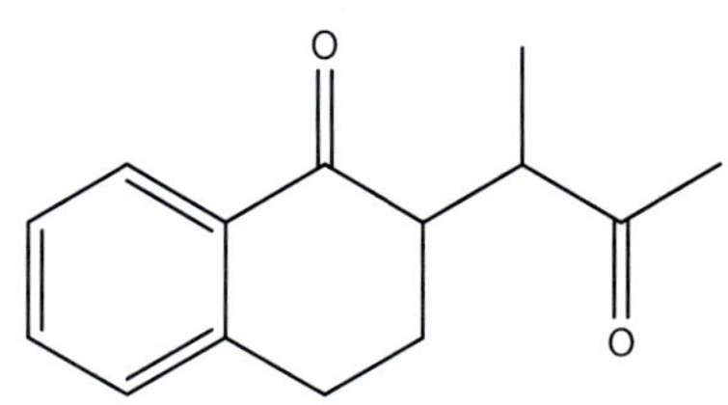

Can we use any of the acyl anion equivalents that we met in Section 20.2.2 to make 1,4-difunctionalized compounds? This should work if, instead of adding them directly to a carbonyl compound, we add them to an α,β-unsaturated species. Thus, we can add cyanide to an enone (Figure 20.30), and the product can be hydrolyzed to a carboxylic acid, reduced to an amine, or otherwise elaborated. Remember that simple nucleophiles such as cyanide undergo conjugate addition to enones under conditions of thermodynamic control.

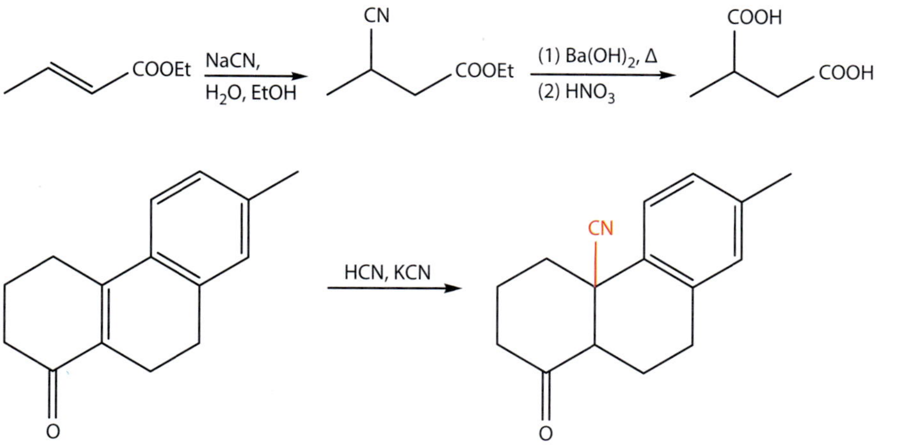

FIGURE 20.30 Addition of cyanide to enones.

The anions of dithianes, which are masked aldehydes, also add to enones and unsaturated esters, though the selectivity for 1,4-addition to enals is poorer. Selectivity for conjugate addition is improved by the addition of HMPA, which acts to complex lithium and separates the ion pairs (Figure 20.31).

Finally, a simple method, albeit one limited in scope, for 1,4-functionalized compounds is the successive reaction of ethynyl anions with carbonyl compounds (Figure 20.32). The triple bond may be partially or wholly reduced, and the usual transformations are available for the alcohols.

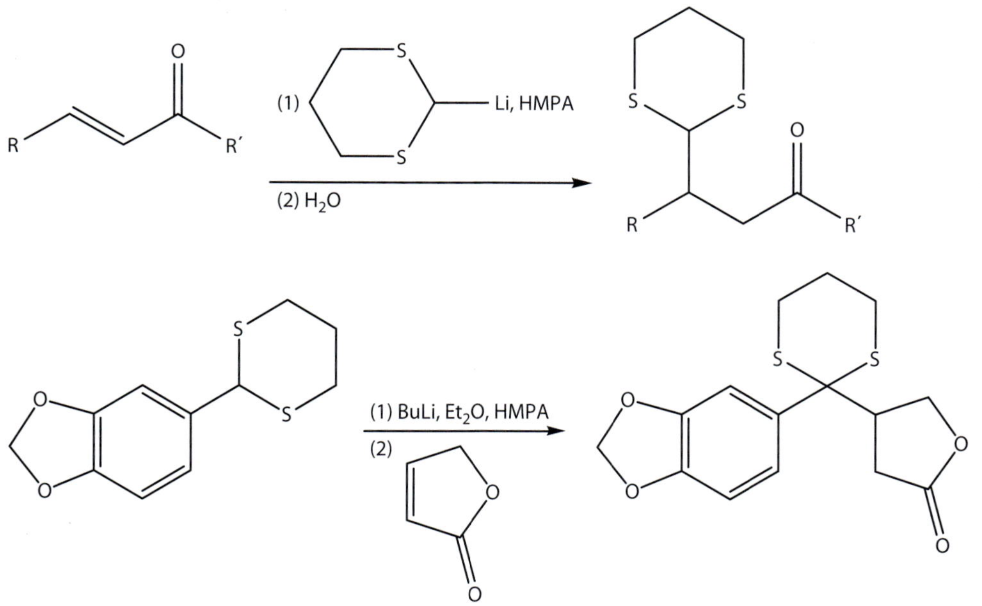

FIGURE 20.31 Conjugate additions of dithianes.

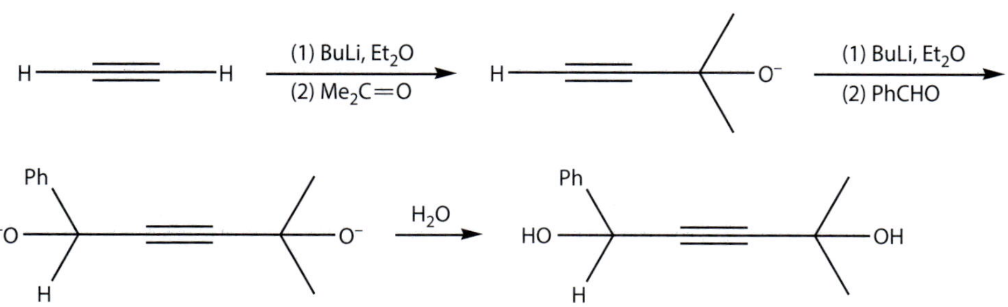

FIGURE 20.32 1,4-Difunctionalized compounds from alkynes.

PROBLEM 20.10

How would you accomplish each of the following transformations?

(a)

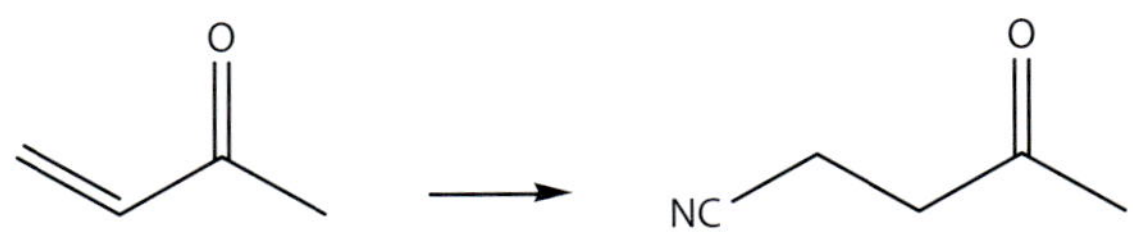

(b)

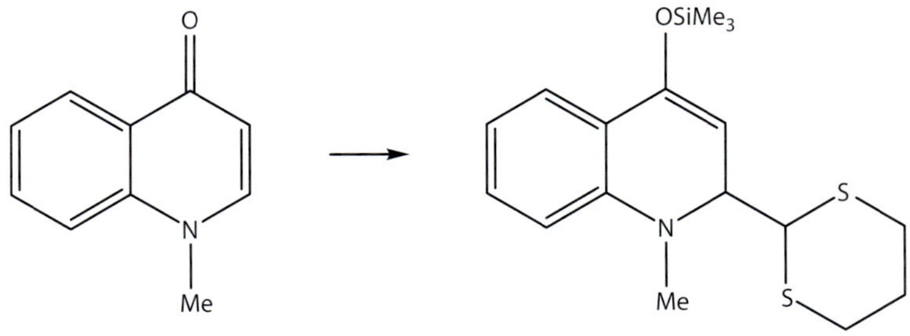

Solutions

(a) HCN, KCN, then H_3O^+.
(b) The anion of dithiane adds in a conjugate manner. Work-up with Me_3SiCl traps the enolate produced as the enol ether.

Key Points from Section 20.4

- 1,4-Dicarbonyl compounds can be synthesized from enamines and α-halocarbonyl compounds. The reaction is not successful with enolates and α-halocarbonyl compounds, which instead yield an epoxide *via* the Darzens condensation.
- Cyanide ion undergoes conjugate addition to most types of α,β-unsaturated carbonyl compounds, and the product may be hydrolyzed to a carboxylic acid.
- Dithianes undergo successful 1,4-addition to enones and unsaturated esters, but selectivity for the 1,4-product is less predictable for enals.
- Anions of ethyne can be reacted with ketones to give 1,4-dihydroxylated compounds.

20.5 1,5-DIFUNCTIONALIZED COMPOUNDS

The normal disconnection of 1,5-dicarbonyl compounds is to an α,β-unsaturated carbonyl compound and an enolate (Figure 20.33). We might sensibly worry about the possibility of direct addition of anions to the carbonyl group, rather than conjugate addition. Generally attack is directly at the carbonyl group when the nucleophile is a strong base, such as PhLi or simple enolates, and at the conjugate position when the nucleophile is a weak base such as cyanide, RSH, or β-dicarbonyl compounds. Hence, in many of the reactions we will discuss here, the nucleophile will be rendered less basic by the addition of an ester group, which may be removed later by decarboxylation. Thus, we will use diethyl malonate rather than ethyl acetate and ethyl acetoacetate rather than acetone. Steric effects

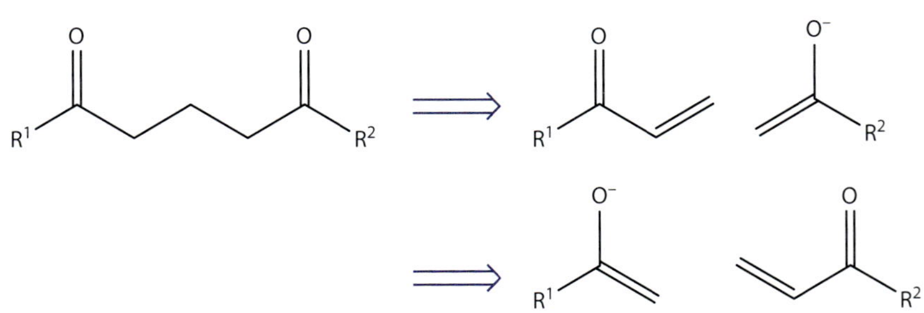

FIGURE 20.33 Disconnection of 1,5-dicarbonyl compounds.

are also important. How do we choose between the two disconnections of Figure 20.33? As in other cases, we will choose on the basis of SM availability and simplification. All the problems of selectivity that we had with aldol and related condensations also exist here—and have similar solutions.

If we consider the synthesis of **20.18**, we first note that it is a 1,5-dicarbonyl compound—the disconnection that we need is suggested by the ring—as in many other examples, we disconnect the bond exocyclic to the ring. The synthon for $^-[CH_2COOH]$ is the anion of diethyl malonate, and the forward reaction is shown in Figure 20.34. Some other synthetic examples are shown in Figure 20.35. In the first example, only one component is able to enolize, and the enolate is stabilized by the phenyl ring. However, the aldehyde is more electrophilic than the ketone. The second example is a reminder that any α,β-unsaturated carbonyl compound will react in this way with any stabilized enolate-type anion. The final example was used in a synthesis of cholesterol; it proceeds *via* the most stable enolate.

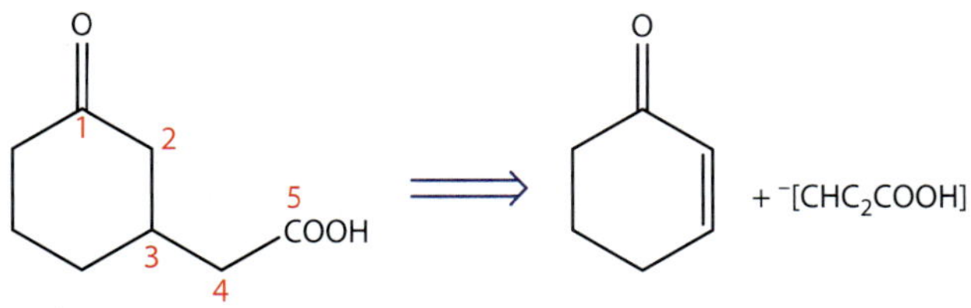

20.18

In these examples, we have chosen α,β-unsaturated carbonyl compounds that are relatively stable, but in some cases, the enone we need is less robust. Consider the disconnection of **20.19**. The enone we require is neither readily available nor very stable. It is readily disconnected to cyclohexanone and formaldehyde through an aldol reaction, but the normal way to make this compound uses a reaction we have not met before, the Mannich reaction (Figure 20.36). In this process, the enol of the ketone reacts with the iminium salt formed from formaldehyde and a secondary amine.

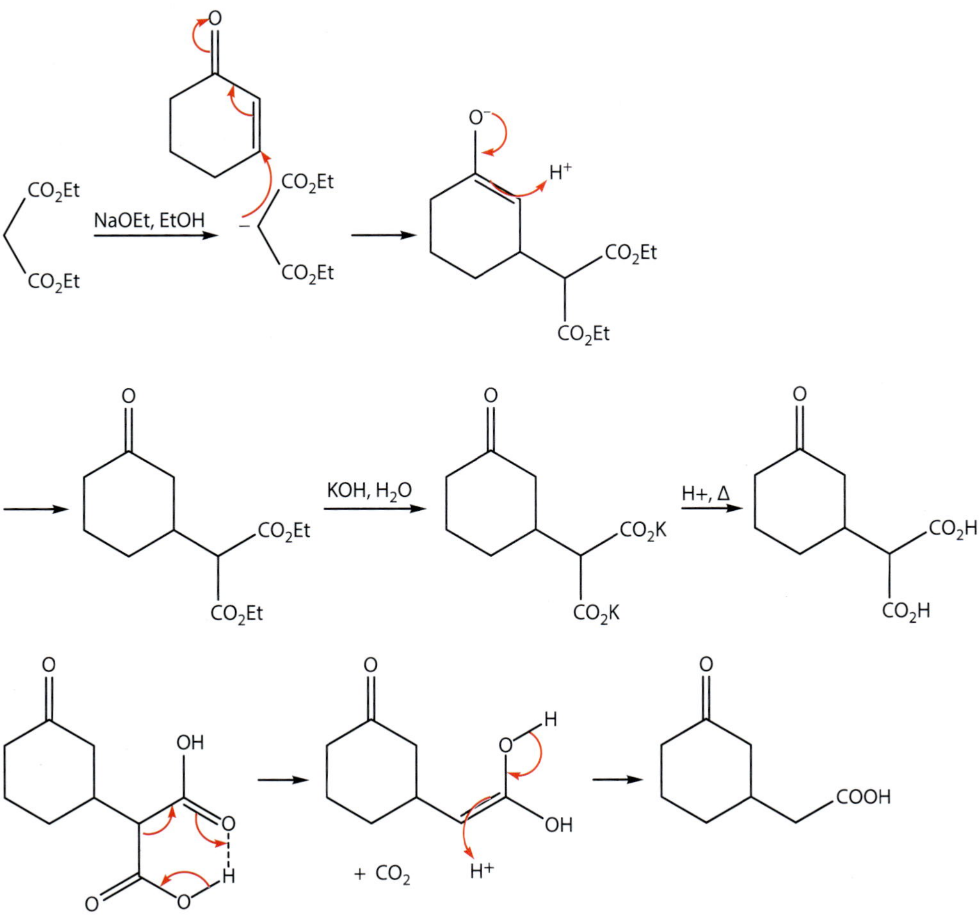

FIGURE 20.34 Reaction of diethyl malonate with cyclohexenone.

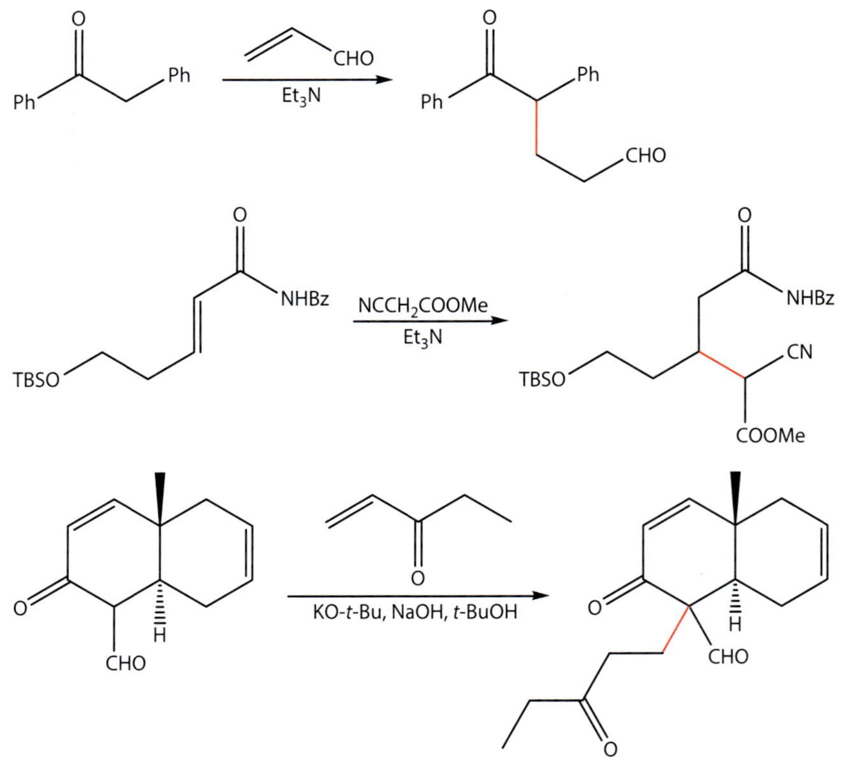

FIGURE 20.35 Preparation of 1,5-dicarbonyl compounds and related molecules.

The amine produced is relatively stable and may be stored, but when it is reacted with methyl iodide, elimination gives the enone, which is reacted with the relevant nucleophile *in situ*:

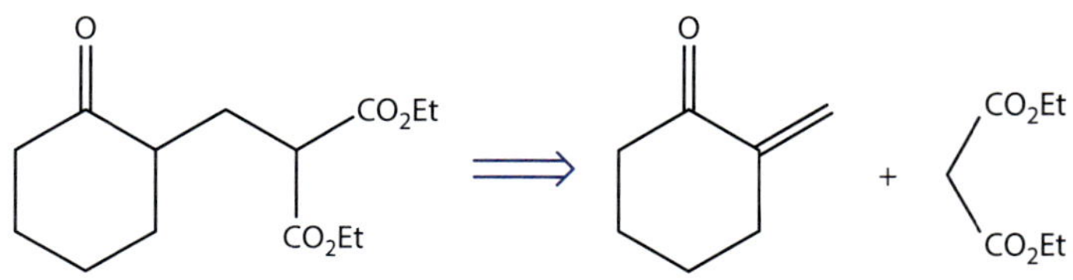

20.19

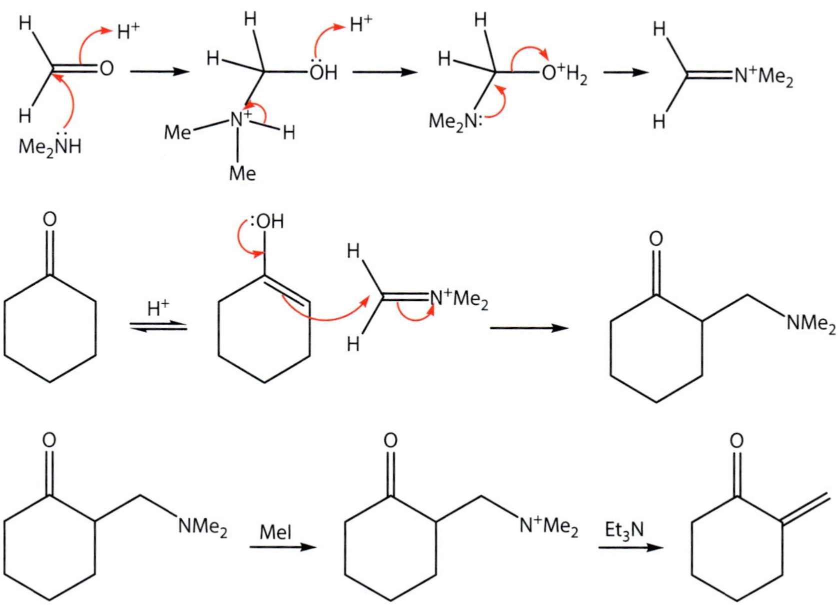

FIGURE 20.36 The Mannich reaction.

PROBLEM 20.11

Propose mechanisms for each of the following reactions:

(a)

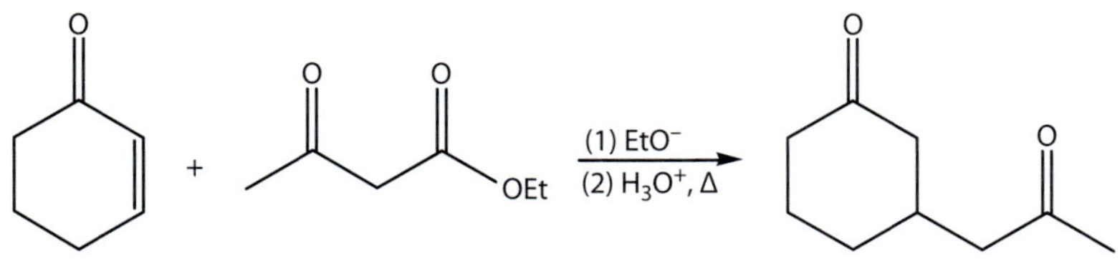

(b)

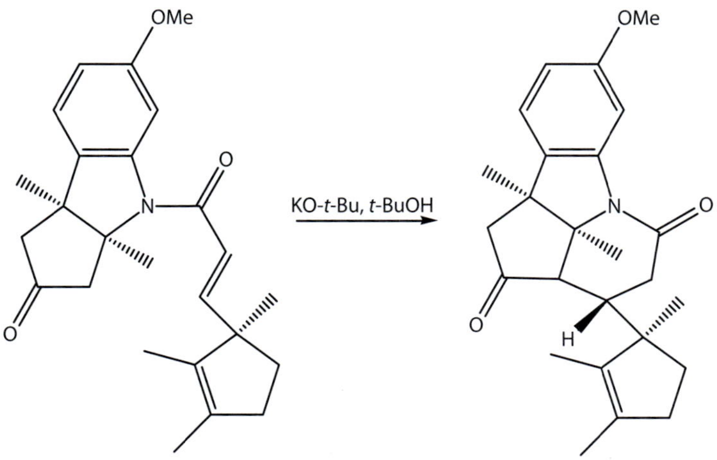

Solutions

(a) This is essentially identical to the process in Figure 20.34, but using ethyl acetoacetate rather than diethyl malonate:

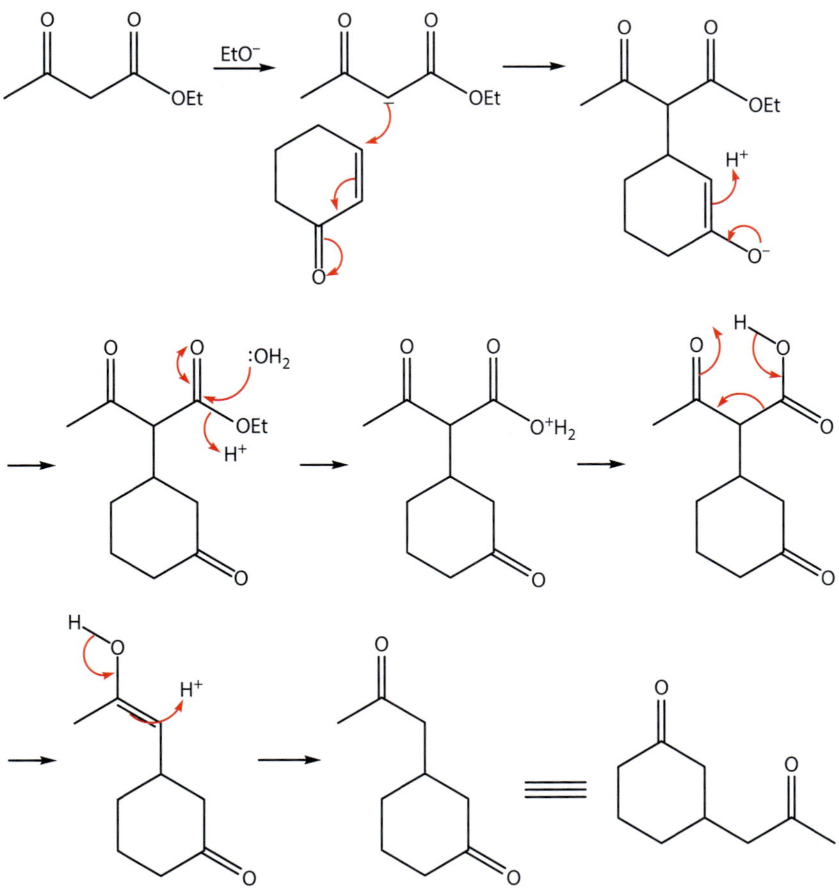

(b) The key point here is to identify the bond that has been made between the ketone enolate and the enone—the mechanism itself is trivial:

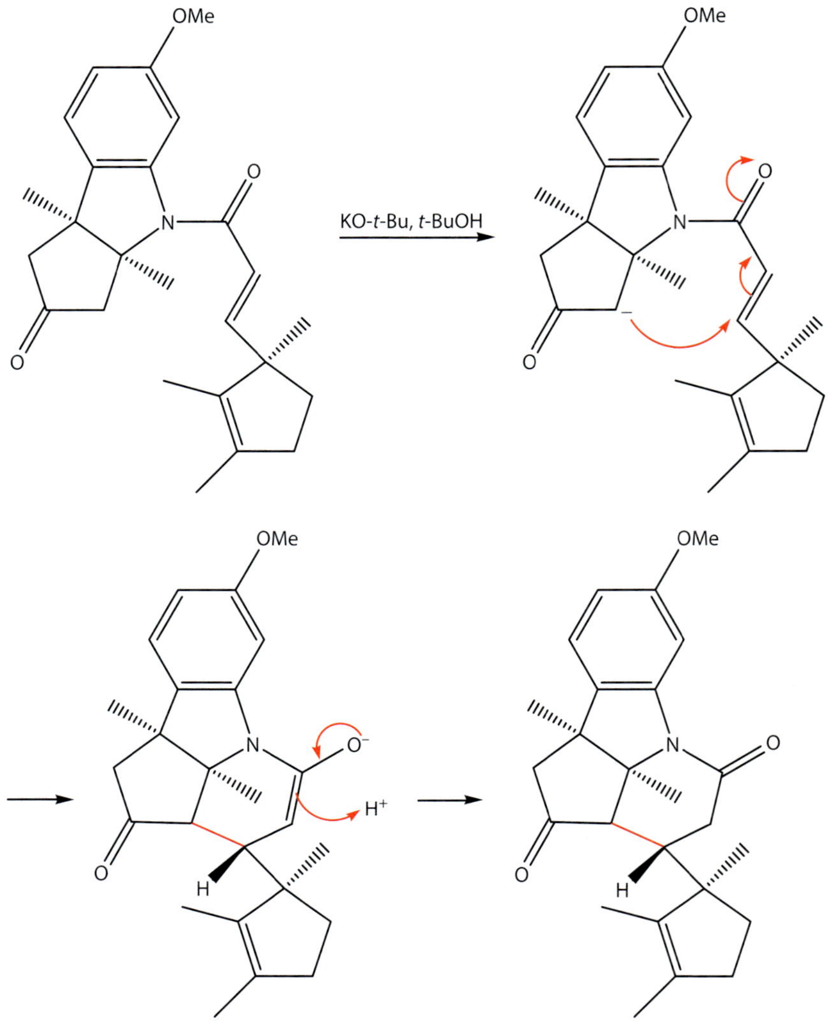

The reaction was used in Woodward's synthesis of cyanocobalamin.

20.6 ROBINSON ANNULATION

In Section 17.5.6, we discussed the Robinson annulation, both with simple enolate nucleophiles and with enamines as the nucleophile. It's a good moment to reread that section and make sure that you are confident with the reaction mechanism, as it is a long and complex one. We will now look at it from the other point of view and see how we can disconnect such six-membered rings. Consider **20.20**. We first disconnect the enone to a dicarbonyl compound—a 1,5-dicarbonyl compound. This is then disconnected to a ketone plus an enone. In practice, the forward reaction is a one-pot process, using KOH/EtOH as the base. Where the initial nucleophile is a β-dicarbonyl compound, no activation is necessary; in other cases, an enamine is sometimes used.

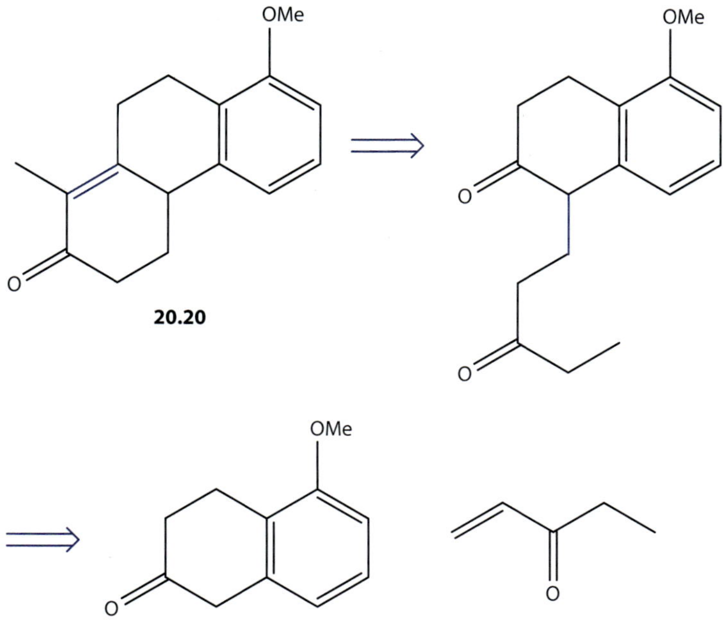

20.20

PROBLEM 20.12

Show how you would disconnect each of the following molecules using a Robinson annulation:

(a)

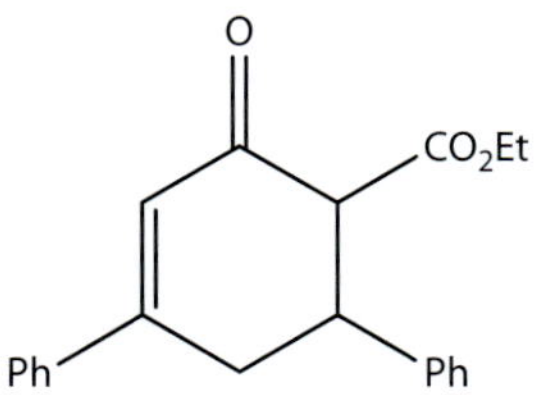

(b)

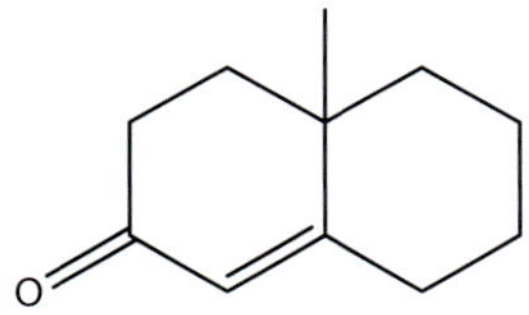

(c)

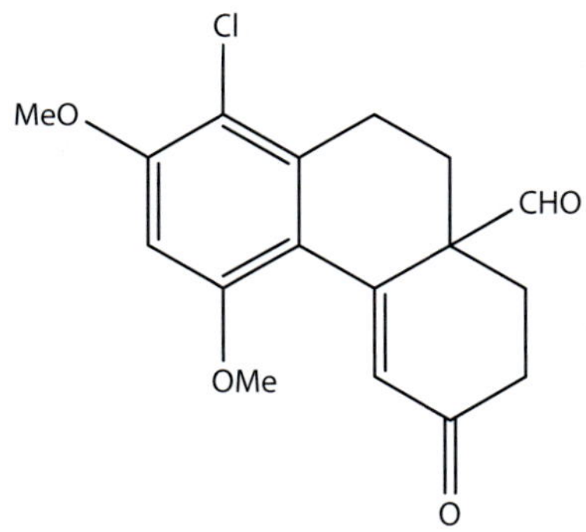

Solutions

(a)

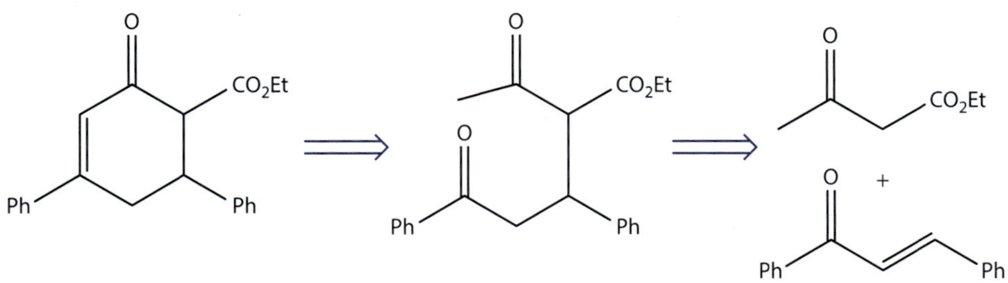

(b)

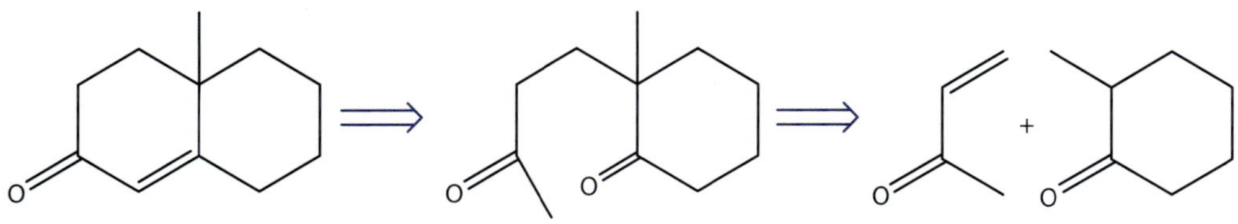

Might need activation

(c)

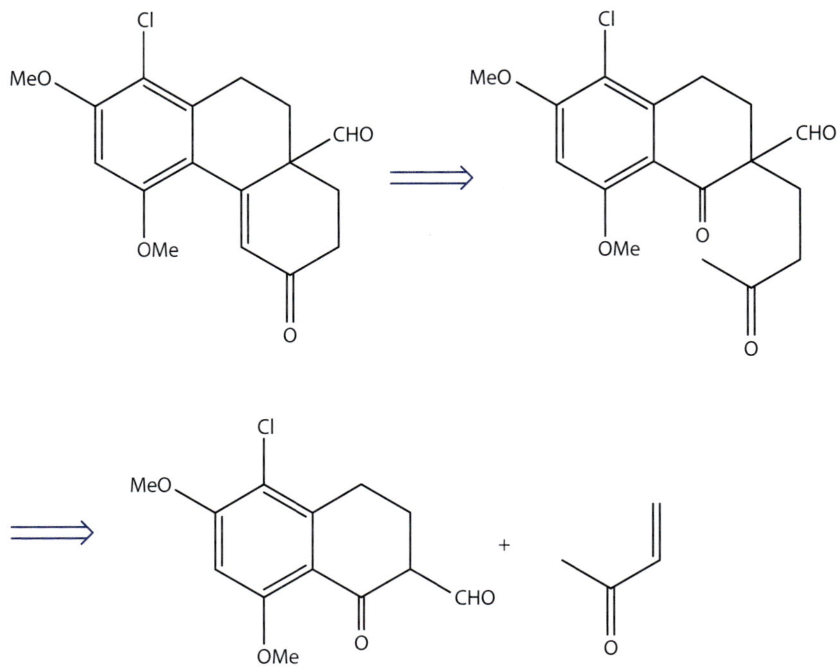

Key Points from Sections 20.5 and 20.6

- 1,5-Dioxygenated compounds are generally prepared by the reaction of an enolate or equivalent with an α,β-unsaturated carbonyl compound. The nucleophile used should be either a β-dicarbonyl compound or an enamine.
- When the enone required for such reactions is unstable, it can be produced *in situ* by amine elimination from a β-amino carbonyl compound, prepared in turn from an enolate and an imine of formaldehyde in the Mannich reaction.
- The Robinson annulation is used to prepare cyclohexenones. The enone is disconnected to a 1,5-dicarbonyl compound, which is synthesized using an enone and an enolate (either of a β-dicarbonyl compound or an enamine). It is a one-pot process.

20.7 1,6-DIFUNCTIONALIZED COMPOUNDS

We might expect the preparation of 1,6-difunctionalized compounds to require the same sort of "unnatural" disconnections and synthons, as we saw for 1,2 and 1,4-difunctionalized compounds. This can be done, but fortunately there is a much simpler approach. We know that the formation of six-membered rings is generally easy (we have seen a number of ring closure reactions, but many are also accessible *via* the cycloaddition reactions in the next section). Cyclohexenes can be ozonized to give dicarbonyl compounds (Figure 20.37). Strictly, the disconnection corresponding to this cleavage process is a *reconnection*. Careful control of reaction and work-up conditions allows the formation of species with the two termini differentiated. Examples of the process are illustrated in Figure 20.38. In the second example, the less-hindered, more electron-rich, double bond is selectively cleaved, and the aldehyde produced is reduced *in situ* to the alcohol.

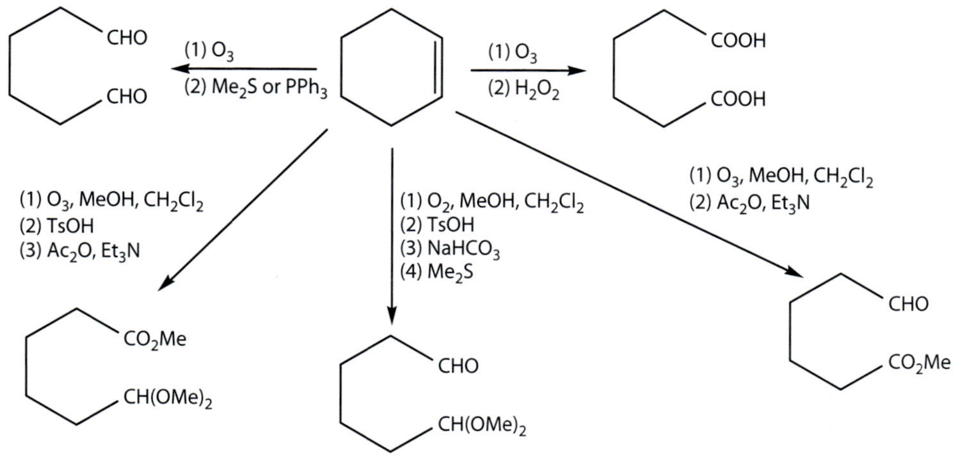

FIGURE 20.37 Ozonolysis of cyclohexene.

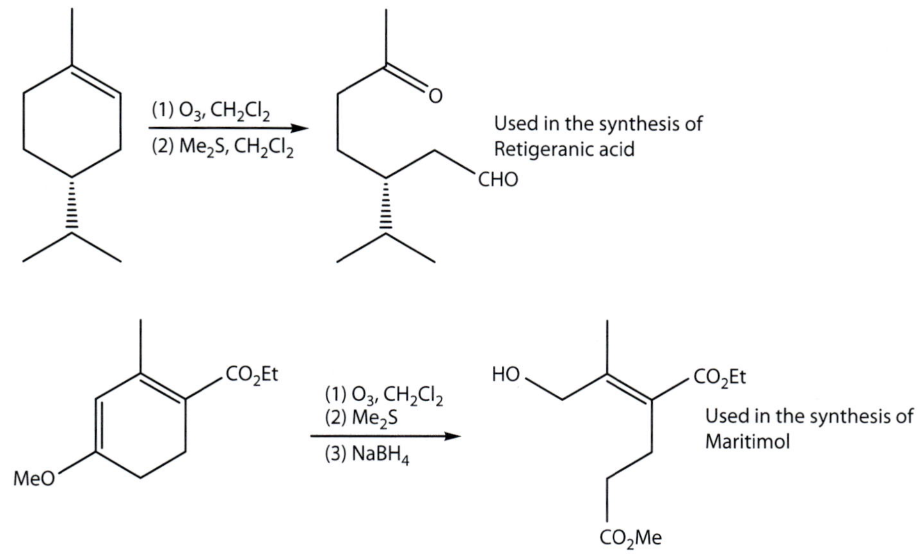

FIGURE 20.38 Ozonolysis in synthesis.

Ozonolysis long required special equipment to generate ozone and stringent safety procedures to use it, making it relatively unpopular. An alternative is provided by the cleavage of 1,2-diols using either periodic acid, HIO_4, or lead tetraacetate, $Pb(OAc)_4$. Since the diols may be readily prepared by osmium tetroxide-catalyzed oxidation of alkenes, this presents another method for alkene cleavage. The reaction is not limited to the preparation of 1,6-difunctionalized compounds; it will work for most alkenes, and the reaction of periodic acid with sugar derivatives is a long-established method of analysis of these species. Figure 20.39 illustrates some uses of this reaction in preparing 1,6-difunctionalized compounds. Both *cis* and *trans*-diols are reactive, but the reactions of *cis*-diols are generally faster because of the more favorable geometry in the cyclic intermediate, **20.21**. In the second example, periodate is used as the reoxidant for osmium as well as for the cleavage of the initially formed diol, in a one-pot process.

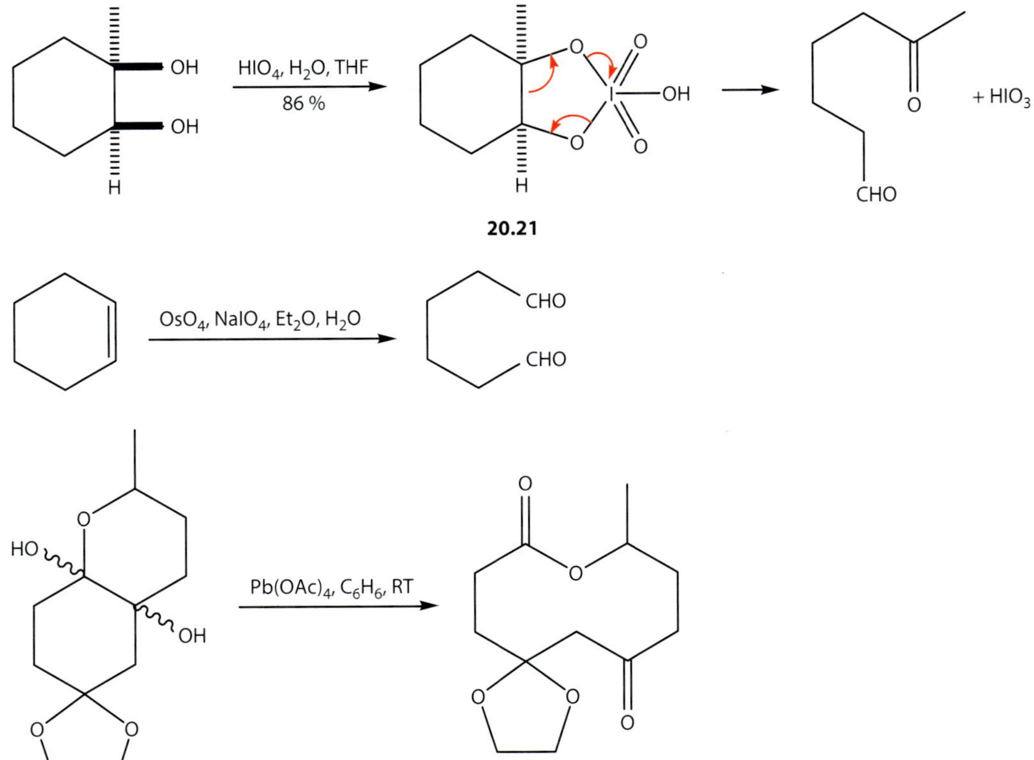

Used in the synthesis of Pyrenolide B

FIGURE 20.39 1,6-Difunctionalized compounds by diol cleavage.

PROBLEM 20.13

Predict the product(s) of each of the following reactions:

(a)

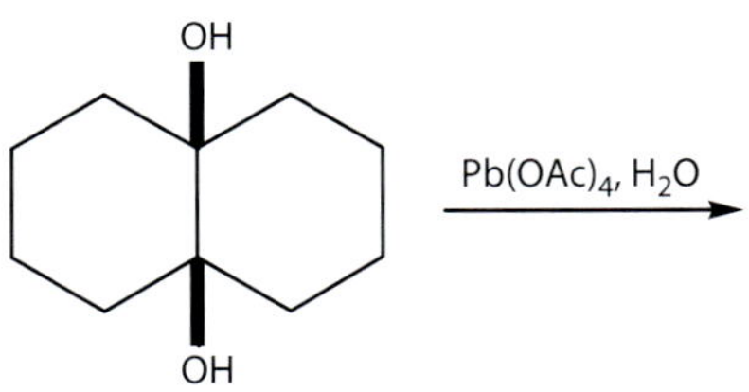

(b)

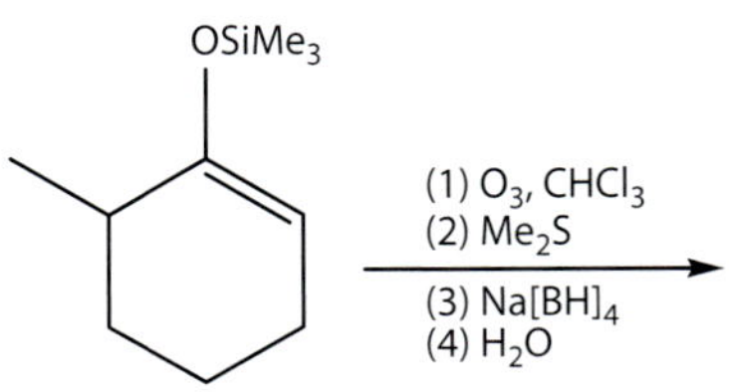

(c)

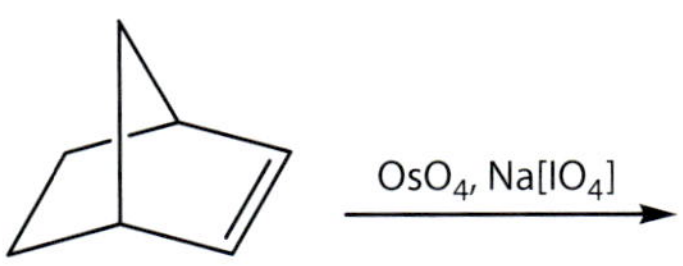

Solutions

(a) The diol is cleaved to a dione:

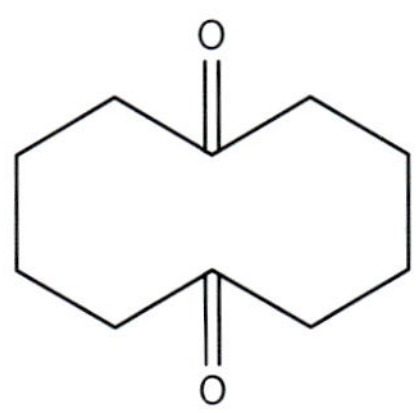

(b) Ozone cleaves the double bond to an aldehyde and a silyl ester. The aldehyde is reduced by $Na[BH_4]$; the ester is not. The ester is, however, hydrolyzed in the reaction work-up:

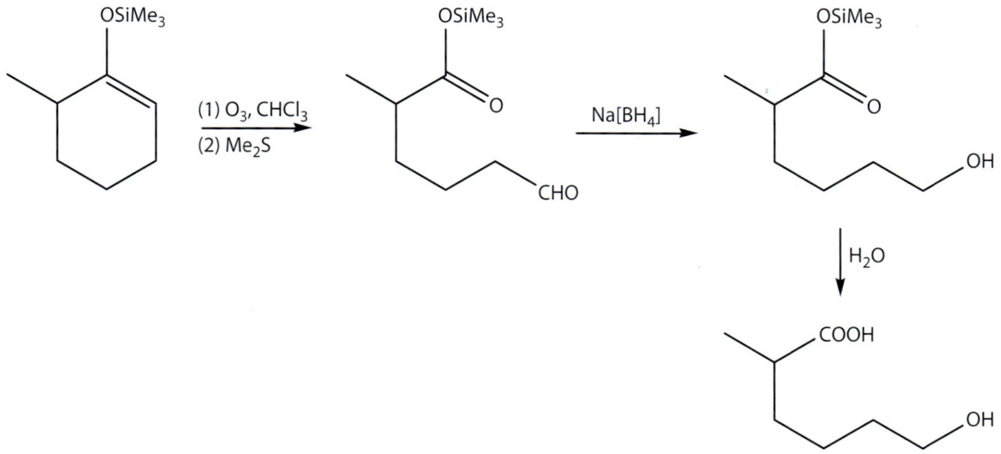

(c) The $OsO_4/NaIO_4$ combination first hydroxylates and then cleaves the double bond to give

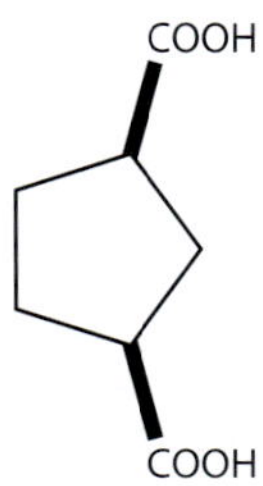

20.8 PERICYCLIC DISCONNECTIONS

Diels–Alder reactions (Section 11.4.6) are a powerful method for the synthesis of cyclohexenes, but normally they only work well when the alkene bears at least one electron-withdrawing group such as COR, COOEt, CN, or NO_2. The reactions are stereospecific, reflecting the stereochemistry of the starting alkene and diene (Figure 20.40; note that relative, not absolute, stereochemistry is implied), and several stereocenters may be generated simultaneously and reliably. When bicyclic systems are prepared, as in the last example, the *endo*-rule applies to reactions that occur under conditions of kinetic control; substituents on the alkenes are on the opposite side of the six-membered ring from the bridge.

When we approach the disconnection of a cyclohexene by a pericyclic process, it's often useful to draw the arrows of the reaction in a backward direction. For example, in Figure 20.41, **20.22** is disconnected to butadiene and quinone. There is no ambiguity in which six-membered ring is disconnected; we know that the electron-withdrawing groups are attached to the alkene component, and if you try the other ring, you'll quickly get to something that you can't make. The forward reaction was accomplished in 93 % yield, at room temperature in benzene over 2–3 weeks. If you have to disconnect a bicyclic target, then the diene was generally a cyclic one. The second example of Figure 20.41 looks complex, but the technique is still valid—draw the arrows backwards, and you will pull out cyclopentadiene, an excellent substrate for Diels–Alder reactions, because of its enforced cisoid geometry. The forward reaction was catalyzed by $BF_3 \cdot Et_2O$, at −78 °C, and was used in the synthesis of Ingenol.

All our examples so far have been carefully chosen so that either the diene or the alkene was symmetric, so that only one product could be produced. This would obviously be a serious limitation in synthesis, if more functionalized alkenes and dienes could not be used with a reliable prediction as to which product would be formed. Fortunately, regioselectivity in such processes is generally good, and the theory of its origins is well understood, though beyond

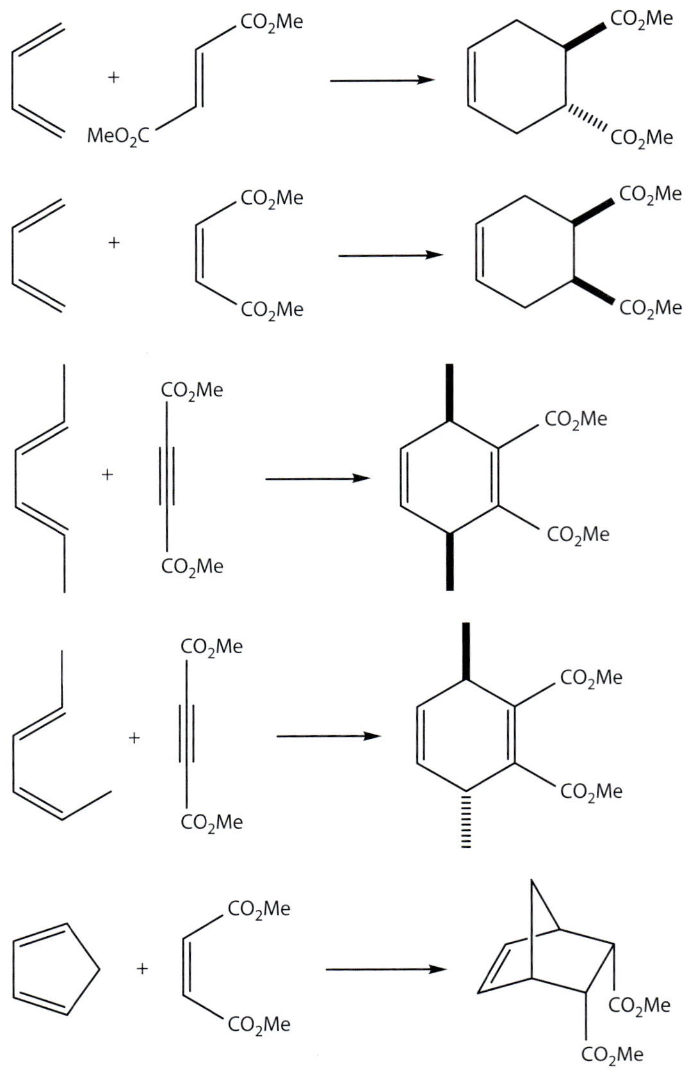

FIGURE 20.40 Stereochemistry of the Diels–Alder reaction.

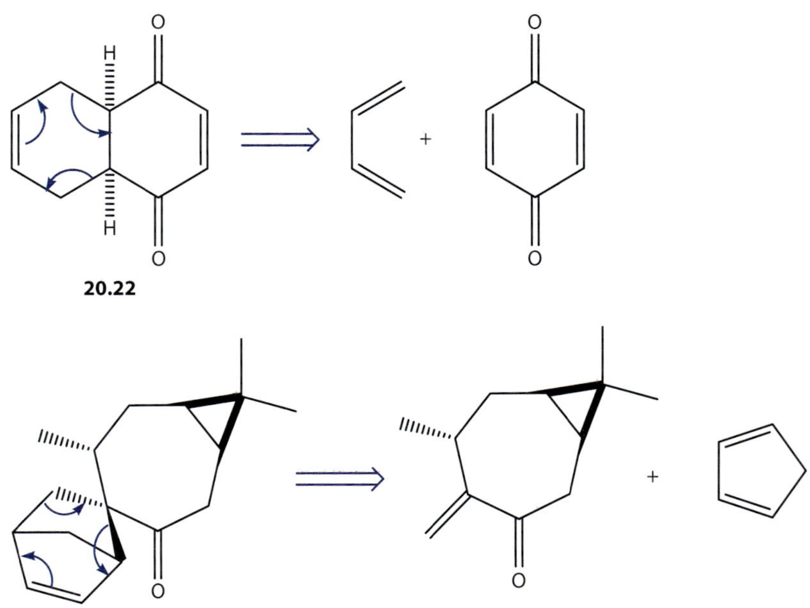

FIGURE 20.41 Pericyclic disconnections.

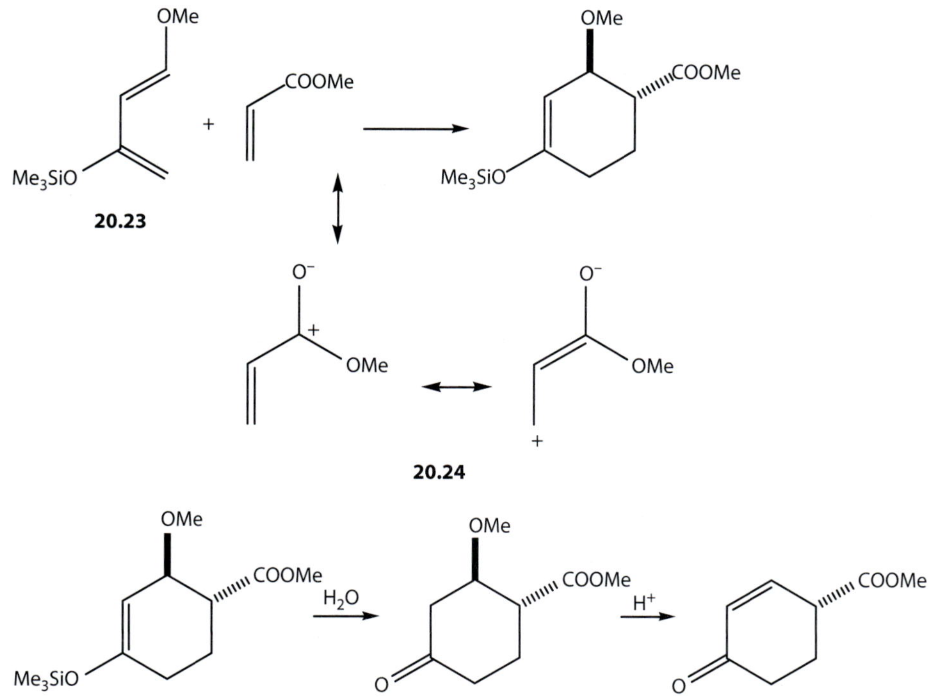

FIGURE 20.42 Reaction of the Danishefsky diene.

the scope of this text.* One very useful diene, which gives good, predictable regioselectivity, is **20.23**, the Danishefsky diene. This diene is very electron rich and hence very reactive. When it reacts with an alkene bearing a single electron-withdrawing group, such as **20.24**, the product regioisomer predominantly formed is that in which the silyl enol ether is next to the more electron-deficient carbon of the alkene. Figure 20.42 illustrates. The other advantage of the use of this diene is that it is readily converted either to a ketone or an enone, by hydrolysis of the silyl enol ether.

PROBLEM 20.14

For each of the following molecules, perform a pericyclic disconnection:

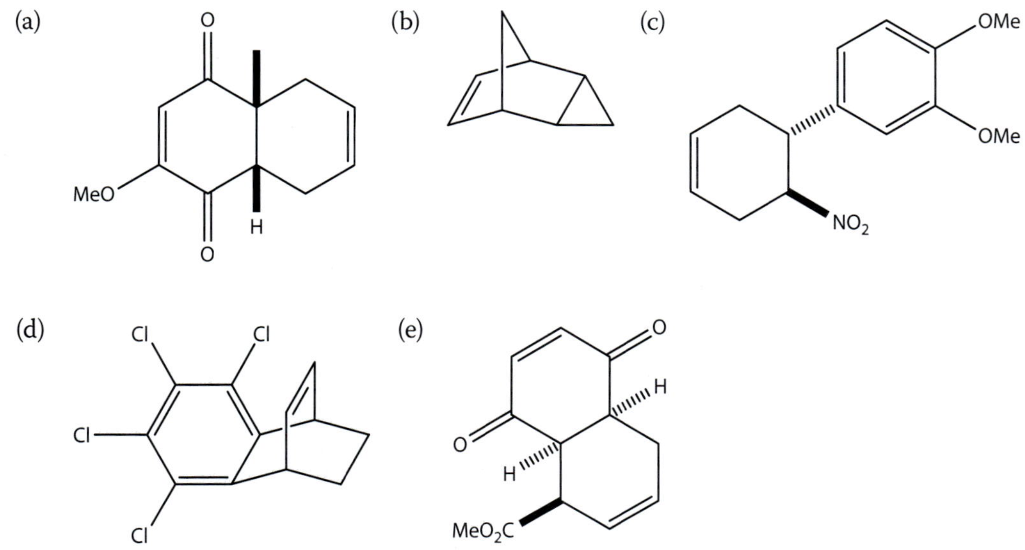

* You can follow this up in *Molecular Orbitals and Organic Chemical Reactions*, Ian Fleming, Wiley, 2010.

Solutions

(a)

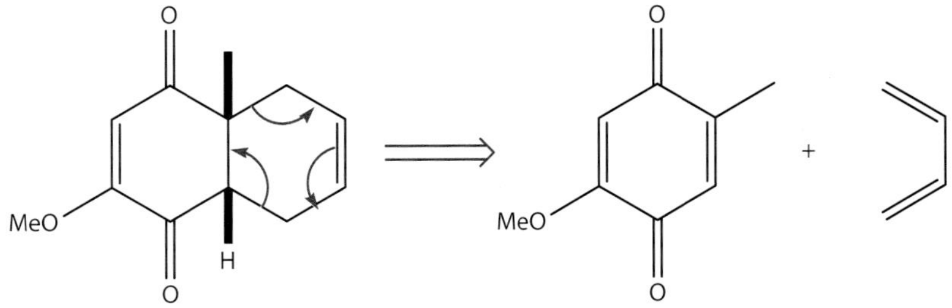

Notice that the more electron poor double bond of the quinone reacts selectively (the methoxy group is a stronger electron donor than the methyl group). The forward reaction was achieved with a benzene solvent, heat, and a long reaction time, in 86 % yield, and the product was used in a synthesis of cholesterol.

(b)

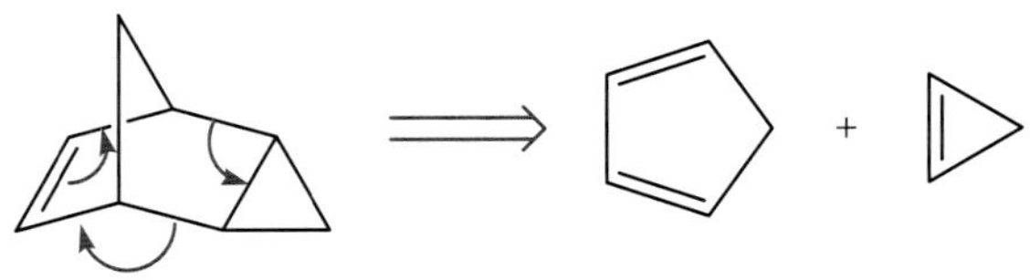

Although the alkene bears no electron-withdrawing groups, the strain makes it highly reactive.

(c)

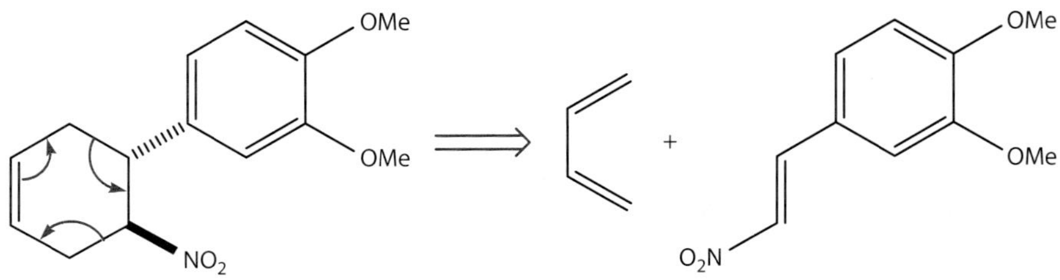

(d)

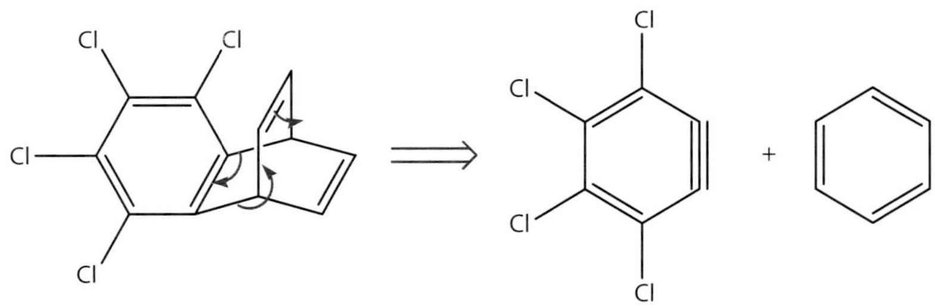

Remember that benzyne is an excellent Diels–Alder substrate, because of its high reactivity; although benzene does not usually react as a diene, in this case, the reaction proceeds below room temperature. The chlorines were removed reductively, to give benzobarrelene. The electron-withdrawing effect of the chlorines is relevant to the ease with which the reaction occurs; without them, the yield is quite low.

(e)

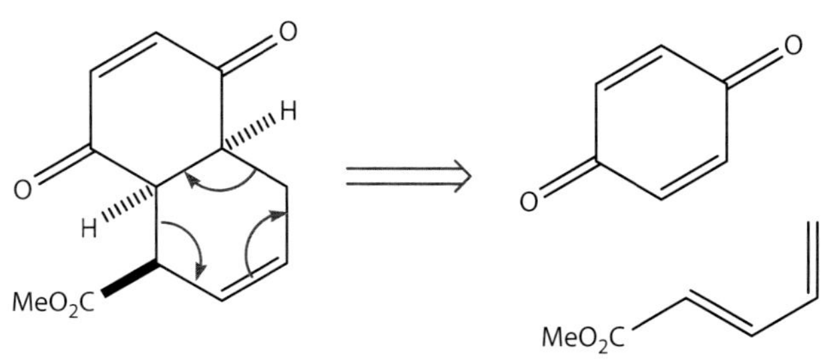

The electron-withdrawing group on the diene means that this is a relatively unfavorable reaction. Only 8 % of the required product was formed after 8 h reflux in benzene. The product was used in the synthesis of reserpine.

Key Points from Sections 20.7 and 20.8

- 1,6-Dioxygenated compounds can be produced by ozonolytic cleavage of cyclohexenes; different work-up procedures allow the differentiation of the termini of the product.
- Alternatively, cyclohexenes can be dihydroxylated and then cleaved with periodic acid or lead tetraacetate.
- 4+2-Cycloaddition reactions (Diels–Alder) between dienes and alkenes yield cyclohexenes. The reaction works well only when the alkene bears at least one electron-withdrawing group.
- Diels–Alder reactions are highly stereospecific. With cyclic dienes, the *endo*-rule operates under conditions of kinetic control. Good regioselectivity is often obtained when both diene and alkenes are nonsymmetrically substituted.

20.9 PROTECTING GROUPS

We have previously met the idea of protecting groups, in the chapters (Chapters 14 and 15) on carbonyl chemistry. We wanted to reduce an ester in the presence of a more reactive functional group, a ketone. It was necessary to protect the ketone. Thus, **20.25** was converted to its acetal, **20.26** (Figure 20.43). The ester could then be reduced by lithium aluminum hydride and the protecting group removed.

What do we expect from a protection/deprotection sequence? The yields for formation of the protected molecule and removal of the protecting group should be high, and the reactions should be free from side products that would complicate purification. It's also best not to introduce any additional stereocenters, as the presence of more than one diastereoisomer, even temporarily, complicates purification. We want the protected functionality to be stable to the reagents/conditions needed for whatever other transformation we wish to carry out elsewhere in the molecule. It's helpful if the protected functionality is stable to as many reagents as possible—we may want to keep it on the molecule for several steps.

20.9.1 PROTECTION OF ALDEHYDES AND KETONES

We are accustomed to protecting aldehydes and ketones as cyclic acetals, as in the sequence of Figure 20.43. In general, acetals are stable toward nucleophiles, bases, and reducing agents but react with electrophiles and oxidizing agents. Acetal formation is generally slower with unsaturated ketones, so some selectivity may be achieved, as in first example of Figure 20.44. However, the reaction is also sensitive to steric effects, as the second example in this figure demonstrates. The examples of Figure 20.45 show that the group is stable to organometallics and two powerful reducing agents.

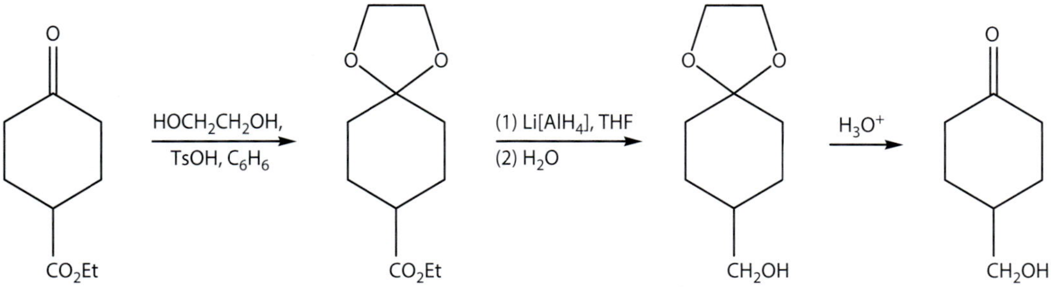

20.25 **20.26**

FIGURE 20.43 Protection of a ketone.

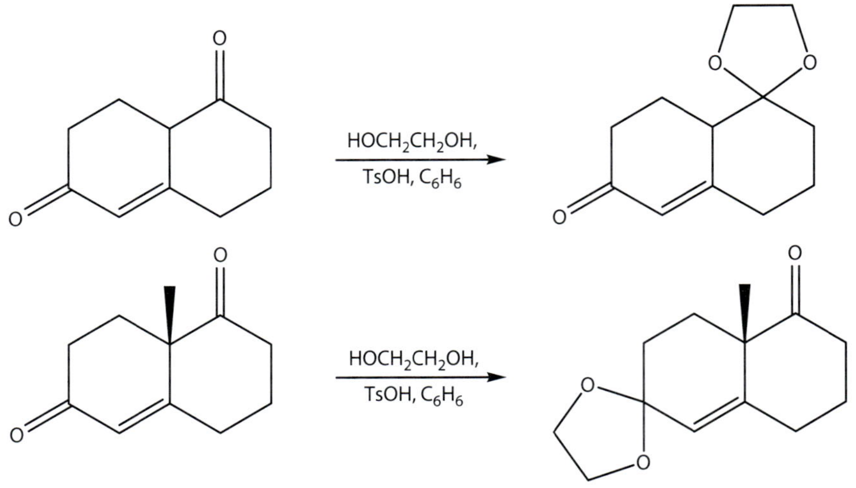

FIGURE 20.44 Selective protection of ketones.

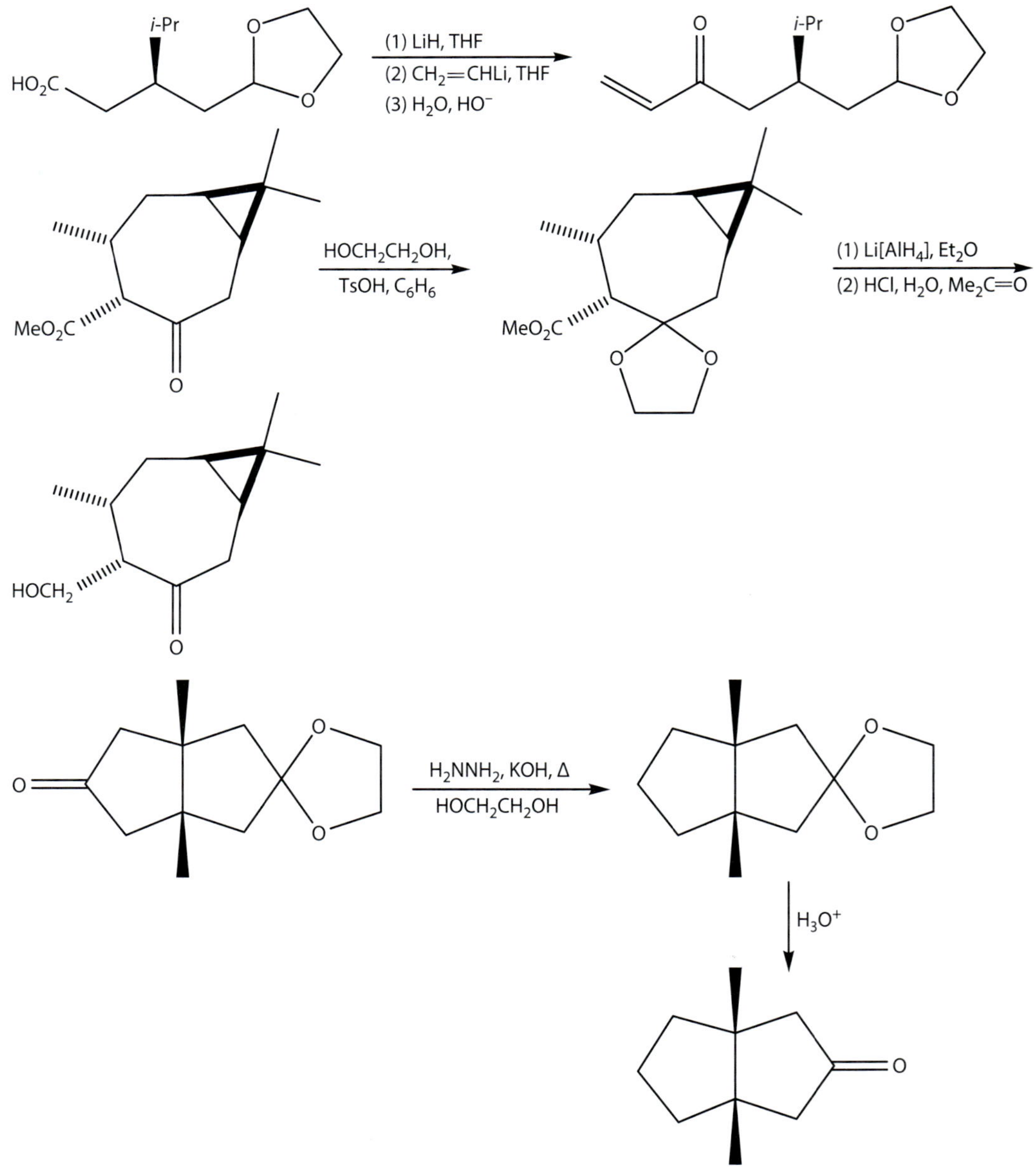

FIGURE 20.45 Reactions of protected aldehydes and ketones.

PROBLEM 20.15

In each of the following reaction sequences, explain *why* a carbonyl group had to be protected:

(a)

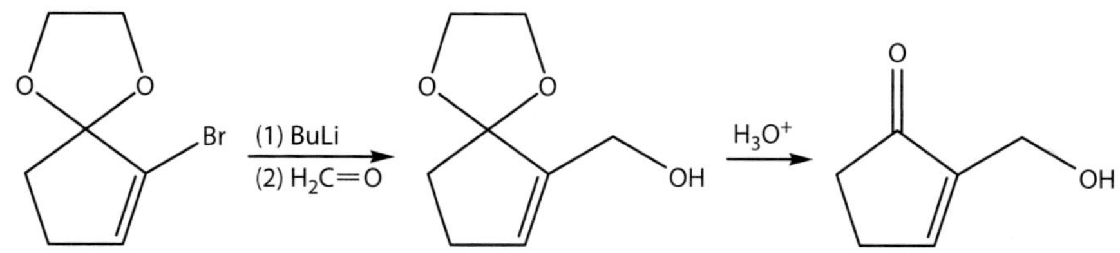

(b)

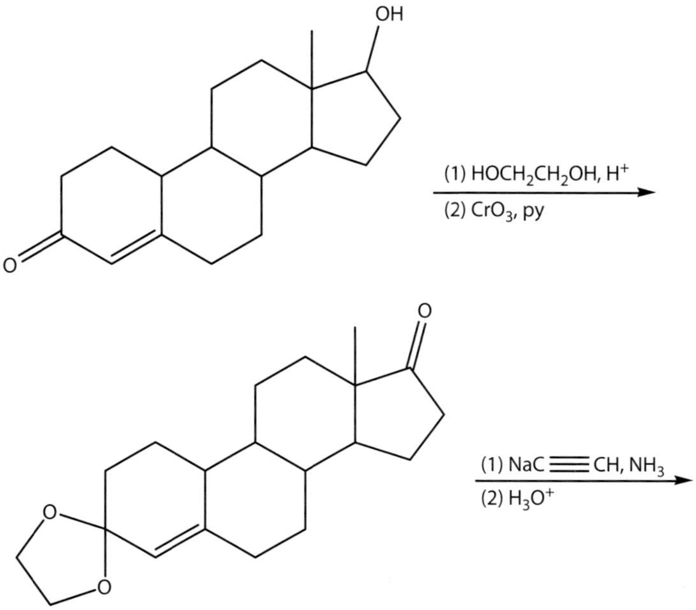

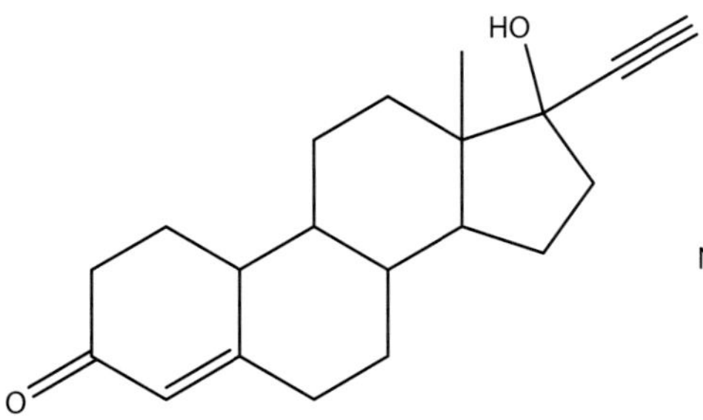

Norethisterone

PROBLEM 20.16

Explain the following observation:

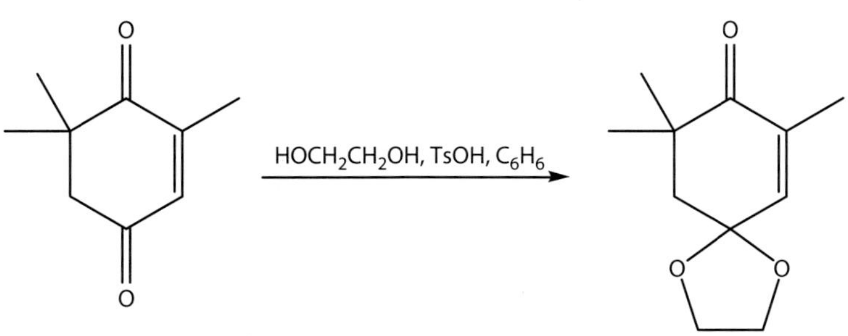

Solutions

20.15(a) If the ketone was not protected, it would react directly with the butyl lithium that is being used to prepare the lithioalkene

20.15(b) The ketone is protected in the first step, so that when the alcohol is oxidized, there is only one "free" ketone to react with the nucleophile; it would otherwise react with both carbonyl groups.

20.16 Both ketones are unsaturated, but one is sterically considerably more hindered than the other.

20.9.2 PROTECTION OF ALCOHOLS

There are many protecting groups for alcohols; the simplest is a benzyl ether, which is usually prepared by a Williamson synthesis, reacting the alkoxide with benzyl chloride or bromide (Equation 20.1). The benzyl group may be removed using HBr if the rest of the substrate is not acid sensitive, but the commonest method is hydrogenolysis using H_2/Pd/C. Benzyl groups are not removed by a simple aqueous acid treatment; the acid must have a strongly nucleophilic counterion.

$$\text{ROH} \xrightarrow{\text{NaH, PhCH}_2\text{Br}} \text{ROCH}_2\text{Ph (ROBn)} \quad (20.1)$$

Alcohols can also be protected as acetals, tetrahydropyranyl (THP), or β-methoxyethoxymethyl (MEM) derivatives (Figure 20.46). These are often preferred to ethers, as they can be removed under very mild conditions. Both of these protecting groups are resistant to bases but react readily with acids. Generally, alcohols are protected to avoid their reaction with reductants such as Li[AlH_4] or organometallic compounds or to avoid their oxidation. Figure 20.47 illustrates. In the first example, the Li[AlH_4] would react with a free alcohol, and in the second one, the system is being protected against overoxidation—and there is an opportunity to protect the alcohol finally produced with a distinct protecting group.

Another common method for alcohol protection is reaction with R_3SiCl to give a silyl ether. Reaction conditions usually involve R_3SiCl, with 4-dimethylaminopyridine (DMAP) as the base. Both the ease of preparation of the silyl ether and the stability of the protected species depend on the nature of the R groups. Trimethylsilyl ethers (ROTMS) are very labile and readily removed with water and dilute acid. The triethylsilyl group (ROTES) is a little more robust but may be removed with fluoride ion (the use of fluoride to cleave silyl groups reflects the strength of the Si–F bond). *t*-BuMe_2SiCl (TBDMSCl) reacts selectively with

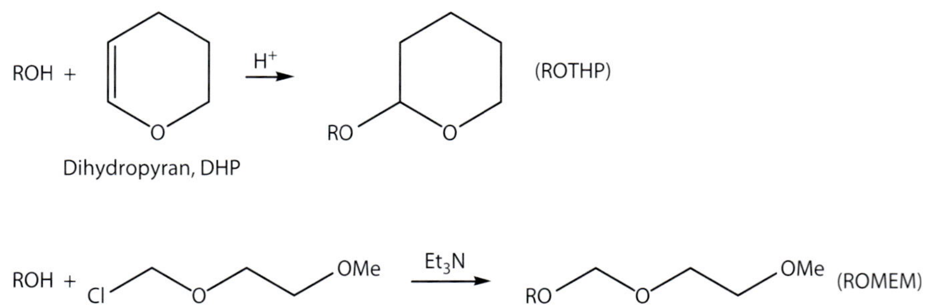

FIGURE 20.46 Protection of alcohols as THP or MEM derivatives.

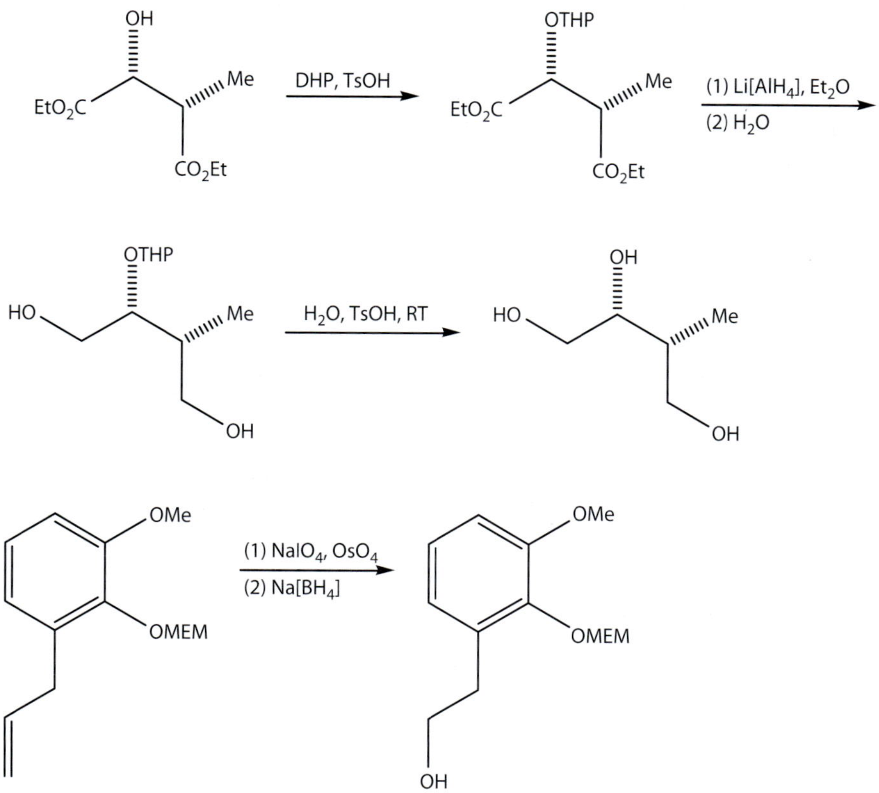

FIGURE 20.47 Uses of protected alcohols.

primary alcohols in the presence of secondary alcohols; the more reactive t-BuMe$_2$SiOTf will silylate secondary alcohols. Protected species ROSiMe$_2t$-Bu are stable to base and dilute acid; they are generally cleaved with fluoride ion. Ph$_2t$-BuSiCl (TBDPSCl) reacts only with primary alcohols, and the protected species are stable to both acid and base and are cleaved only slowly with fluoride ion. This family of protecting groups provides the scope for significant selectivity in forming and cleaving these ethers.* In the first example of Figure 20.48, the trimethylsilyl ether is readily cleaved, leaving the TBDPS ether untouched. The TBDPS ether is also untouched in the second example. In the final example, we see that even when the protecting group is identical, deprotection of primary alcohols is faster than deprotection of secondary alcohols.

20.9.3 PROTECTION OF 1,2- AND 1,3-DIOLS

1,2- and 1,3-diols can be protected as acetals, by reacting them with a simple aldehyde or ketone such as acetone or benzaldehyde.† For example, tartaric acid is converted to its acetone acetal, so that the carboxylate groups can be reduced (Figure 20.49). In the first step, the acetal is prepared, and the acid groups are converted to esters. Next, the esters are reduced to alcohols by lithium aluminum hydride and then protected as benzyl ethers. Finally, the acetal is hydrolyzed to give a partly protected tetraol. These methods are also widely employed in protection of various OH groups of sugars.

* R. Crouch, *Tetrahedron*, **60** (2004); T.D. Nelson and R.D. Crouch, *Synthesis* (1996) 1031–1069.
† F.A.J. Meskens, *Synthesis* (1981) 501–522.

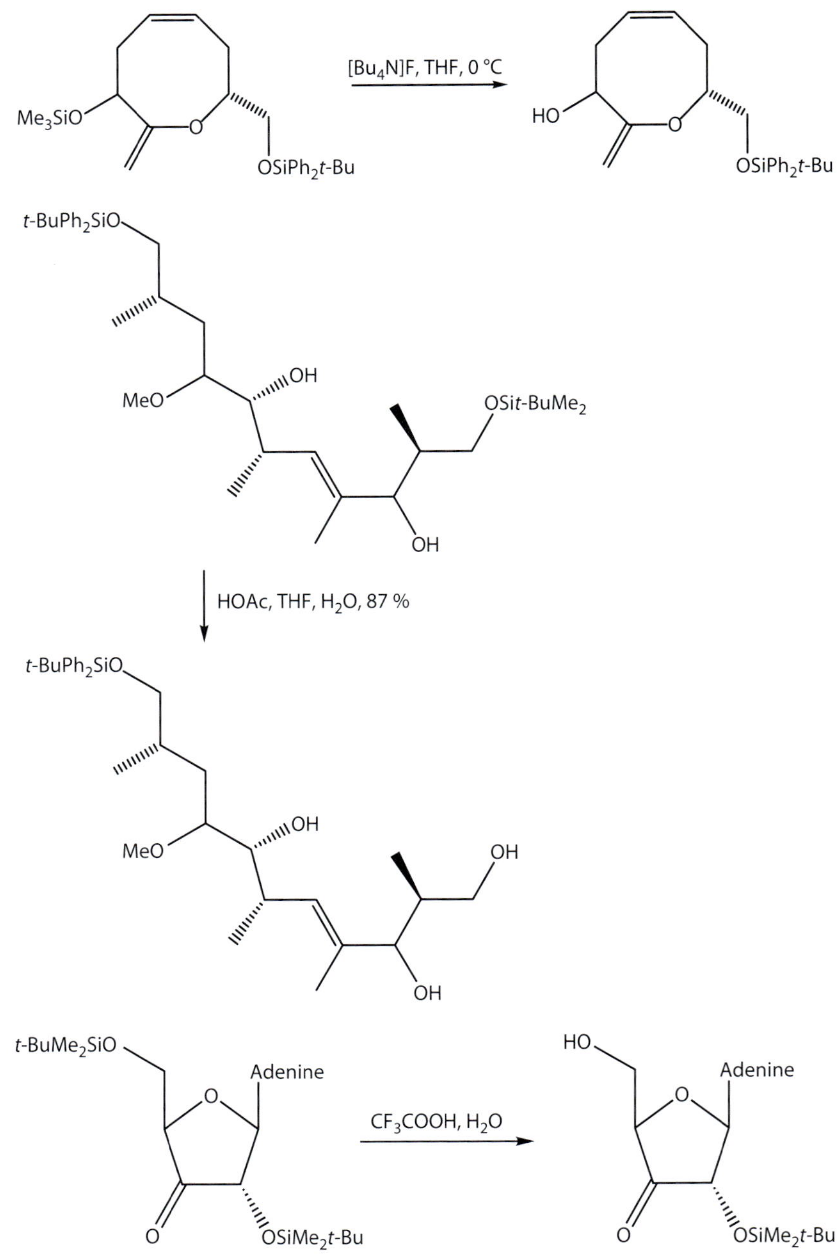

FIGURE 20.48 Selective reactions of silyl ethers.

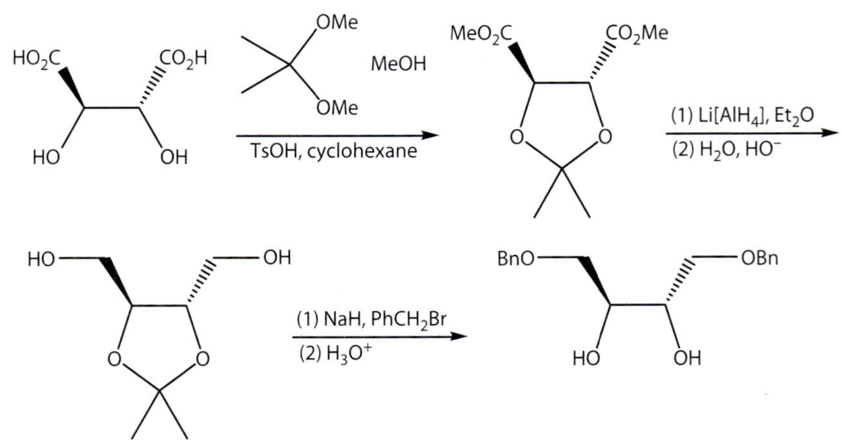

FIGURE 20.49 Protection of diols.

PROBLEM 20.17

Predict the outcome of the following processes:

(a)

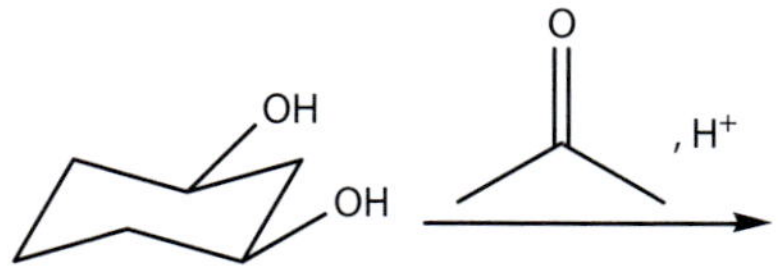

(b)

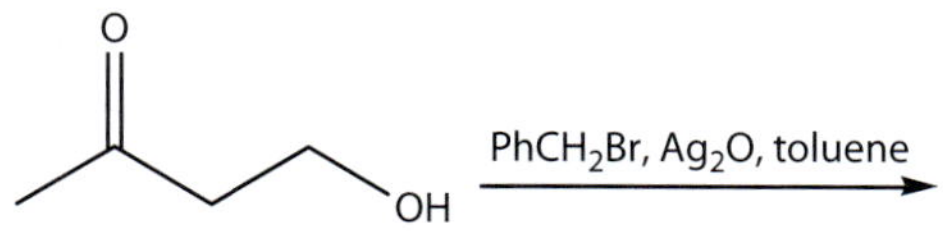

PROBLEM 20.18

Study the following sequence, and answer the questions posed:

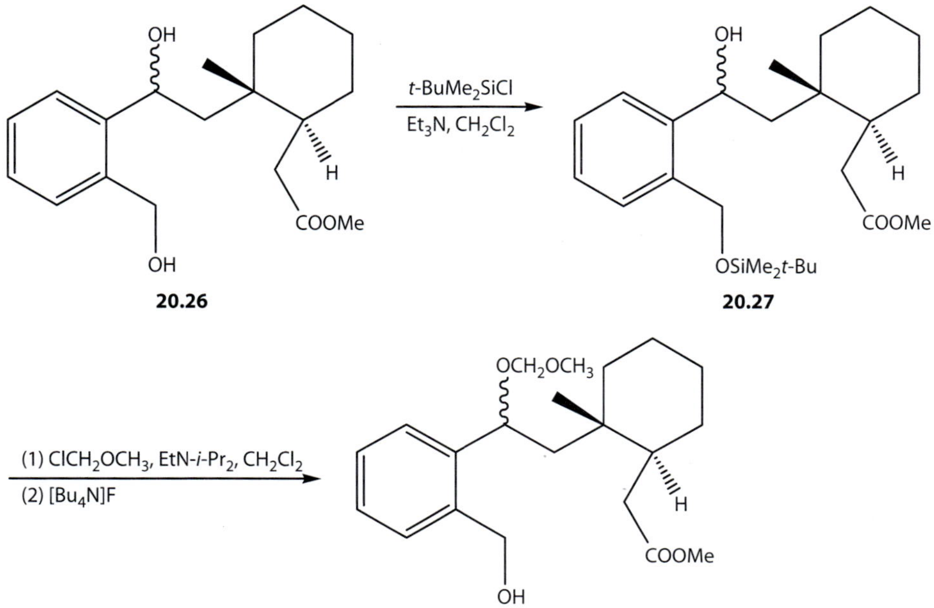

20.26 **20.27** **20.28**

(a) Why does only one OH group become silylated in the conversion of **20.26** to **20.27**?
(b) What is the intermediate formed, but not isolated, in the conversion of **20.27** to **20.28**?
(c) What is the function of the $[Bu_4N]F$?

Solutions

20.17(a) 1,3- and 1,2-diols form rings of a favorable size on reaction with ketones. Thus, the product is

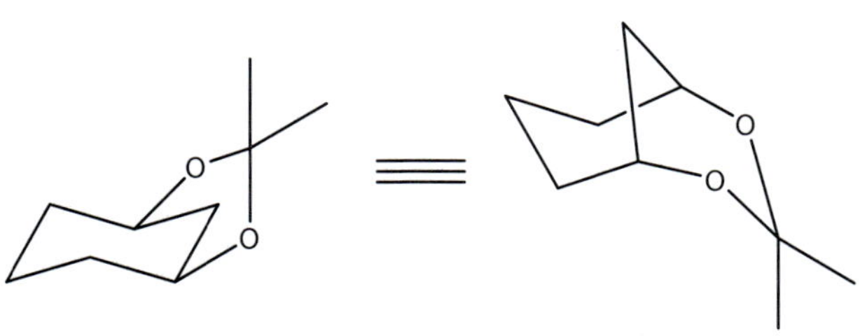

20.17(b) Although the conditions are not quite the usual ones, what is happening here should be obvious. The OH is benzylated. The silver facilitates the reaction by precipitation of AgBr.

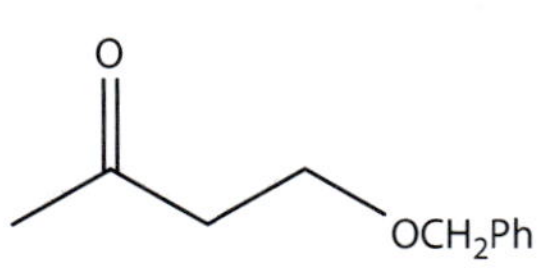

20.18(a) This very hindered silylating reagent only reacts with primary alcohols, and not with secondary alcohols.

20.18(b)

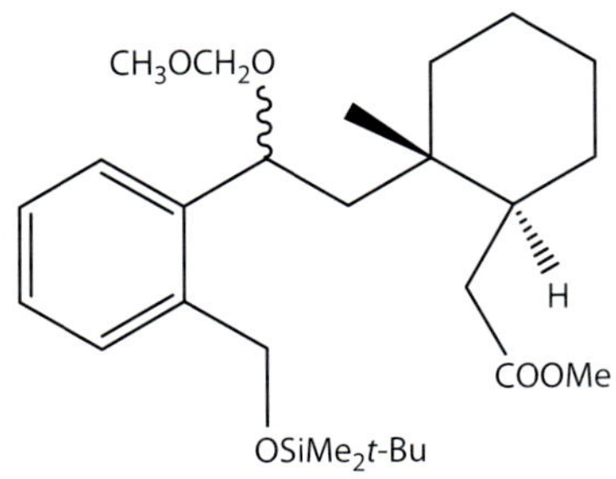

20.18(c) To remove the silicon protecting group.

20.9.4 PROTECTION OF CARBOXYLIC ACIDS

While alcohols are occasionally protected as esters, such protection is much more commonly used for carboxylic acids. The conditions for the preparation and hydrolysis of esters depend on the nature of the alcohol used. "Ordinary" esters such as RCOOEt can be hydrolyzed in either acid or base, but the conditions required are often quite vigorous. *tert*-Butyl esters are hydrolyzed by a different mechanism, *via* the *tert*-butyl cation (Figure 20.50); large amounts of water are not needed. Benzyl esters, like benzyl ethers, can be cleaved by hydrogenolysis* using H_2/Pd/C. If even this is too vigorous, trichloroethyl esters are cleaved by zinc metal (Figure 20.51). Trimethylsilyethyl esters are cleaved with fluoride ion (Figure 20.52).

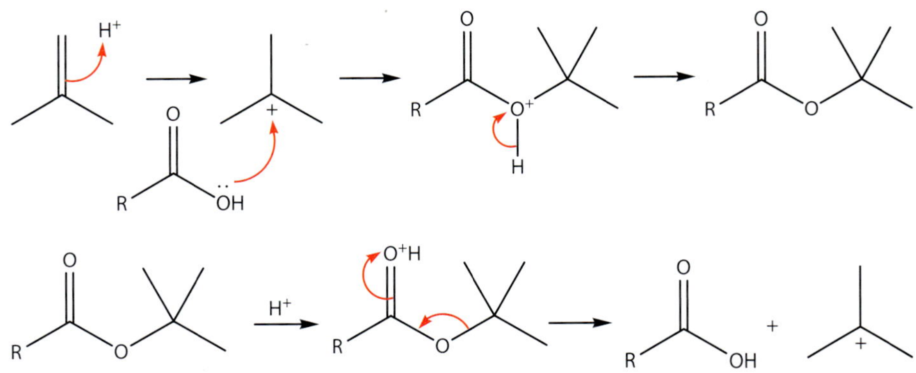

FIGURE 20.50 Formation and hydrolysis of *tert*-butyl esters.

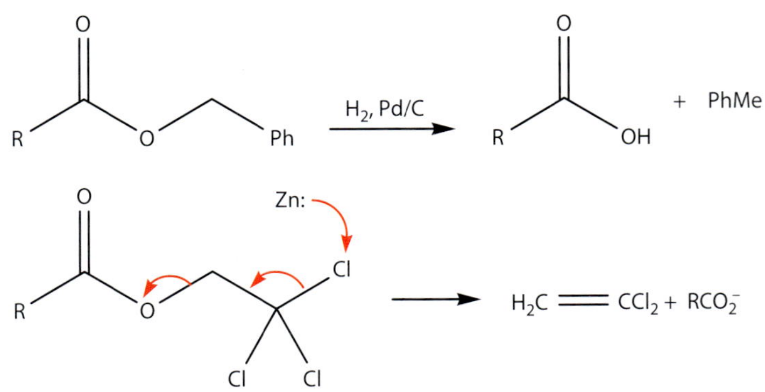

FIGURE 20.51 Cleavage of benzyl and trichloroethyl esters.

* Hydrogenolysis is used to describe the cleavage of a single bond by molecular hydrogen in the presence of a catalyst. It is distinct from hydrogenation, where hydrogen is added across a multiple bond.

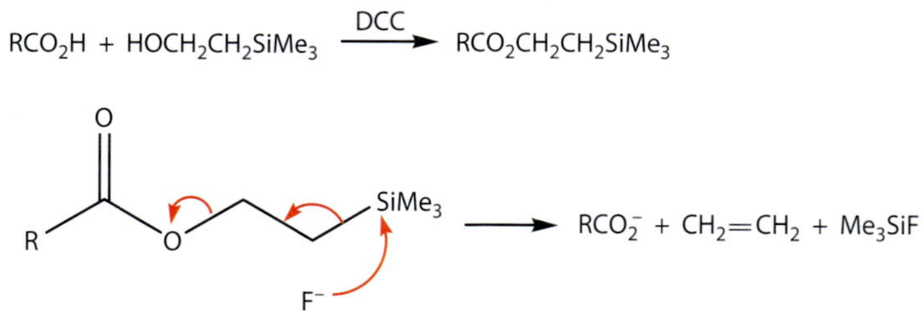

FIGURE 20.52 Formation and cleavage of 2-(trimethylsilyl)ethyl esters.

The mechanism of action of dicyclohexylcarbodiimide (DCC) as a coupling reagent will be discussed in detail in Chapter 22. There, we will also discuss in detail the sort of protection for carboxyl groups that is used for the preparation of peptides.

20.9.5 PROTECTION OF AMINES

This is a topic that we will meet again in more detail in Chapter 22, but for the present, we should just note than we can protect amines as benzyl derivatives, amides, or urethanes. Thus, **20.29** is benzylated so that the labile proton does not interfere with the deprotonation of the ester by LDA; the anion reacts with a source of electrophilic fluorine, and the product is an important precursor to a number of pharmaceutically important materials. Debenzylation may be accomplished by hydrogenolysis.

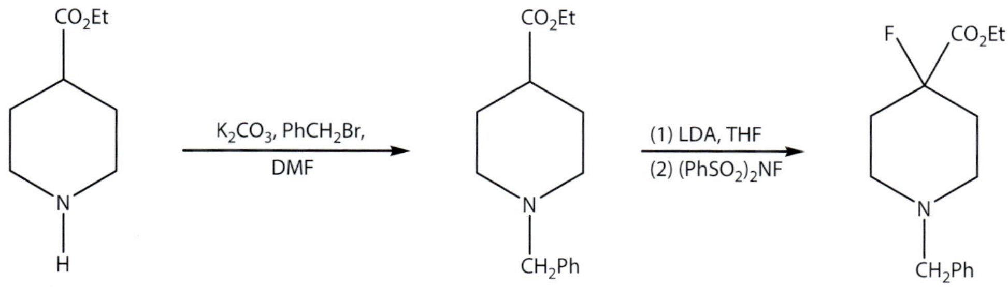

20.29

Simple amides are easy to prepare from either acid chlorides or anhydrides (or for formamides, ethyl formate), but they are quite difficult to hydrolyze and not widely used. Urethanes, RNHCOOR′, are prepared from amines and chloroformates, R′OC(=O)Cl, usually benzylchloroformate. Urethanes can be hydrolyzed with HBr, but the presence of the benzyl group also allows the option of removal by hydrogenolysis. The summary of key points from Section 20.9 is given in Table 20.1.

TABLE 20.1 Key Protecting Groups from Section 20.9

Substrate	Reagent for Protection	Product	Stable to	Removed with	Comments
RCHO, $R_2C{=}O$	$HOCH_2CH_2OH$, H^+	Cyclic acetal	Bases, reducing agents, nucleophiles	H_3O^+	Formed more slowly with enones or hindered substrates
ROH	$PhCH_2Br$, NaH	ROBn	Bases, nucleophiles, oxidants	H_2/Pd/C or HBr	
ROH	Dihydropyran, H^+	ROTHP	Bases, oxidants	Dilute acid	
ROH	$ClCH_2OCH_2CH_2OCH_3$, Et_3N	ROMEM	Bases, oxidants	Dilute acid	
ROH	Me_3SiCl, DMAP	$ROSiMe_3$, ROTMS	Bases, oxidants	Dilute acid	
ROH	*t*-$BuMe_2SiCl$, DMAP	ROTBDMS	Base, oxidants, dilute acid	F^-	Selective for 1° over 2° alcohols
ROH	*t*-$BuMe_2SiOTf$, DMAP	ROTBDMS	Base, oxidants, dilute acid	F^-	Reacts with 1° and 2° alcohols
ROH	Ph_2*t*-BuSiCl	ROTBDPS	Acid, bases, oxidants	F^-, slow reaction	Reacts only with 1° alcohols
1,2-diols	$Me_2C{=}O$ or PhCHO, H^+	Cyclic acetal	Bases, reducing agents, nucleophiles	H_3O^+	Also successful for 1,3-diols
RCOOH	$Me_2C{=}CH_2$, H^+	RCOO*t*-Bu	Base	H^+	
RCOOH	$PhCH_2Br$, base or Ag^+	RCOOBn	Base	H_2/Pd/C	
R_2NH	$PhCH_2Br$, base	R_2NBn	Base	H_2/Pd/C	

PROBLEM 20.19

How would you carry out each of the following transformations?

(a)

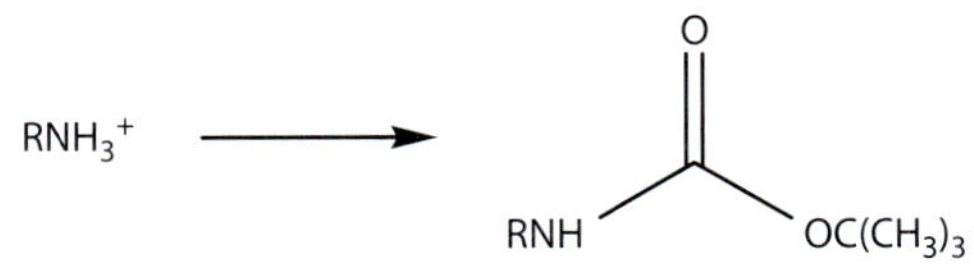

(b)

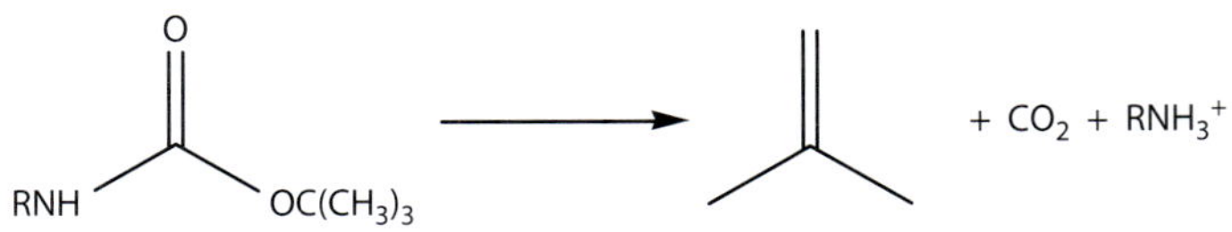

(c) $RCO_2CH_2Ph \longrightarrow RCOOH + PhCH_3$

Solutions

(a) $(CH_3)_3COC(=O)Cl$, base, or $\{(CH_3)_3CO\}_2O$
(b) HBr, H_2O, CF_3COOH, or HCl in dioxane
(c) H_2/Pd/C

FURTHER READING

Kocienski, P.J. *Protecting Groups*, 3rd edn., Georg Thieme Verlag, New York, 2005.
Warren, S. and P. Wyatt. *Organic Synthesis—The Disconnection Approach*, John Wiley & Sons, New York, 2009.
Wuts, P.G.M. and T.W. Greene. *Greene's Protective Groups in Organic Synthesis*, 4th edn., John Wiley & Sons, New York, 2006.

REVIEW PROBLEMS

1. Give reagents for each of the following transformations, some of which require more than one step:

(a)

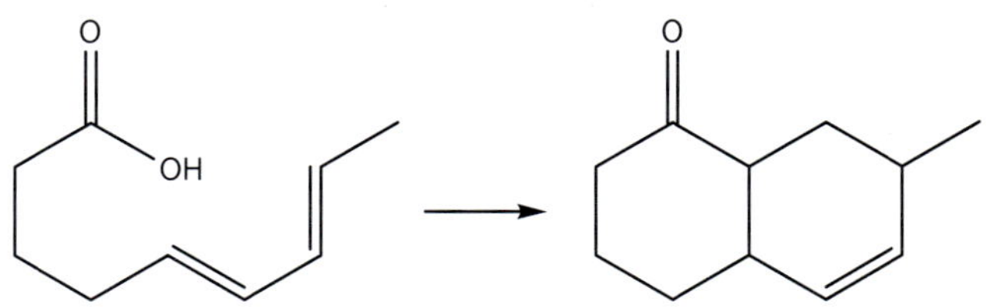

(b)

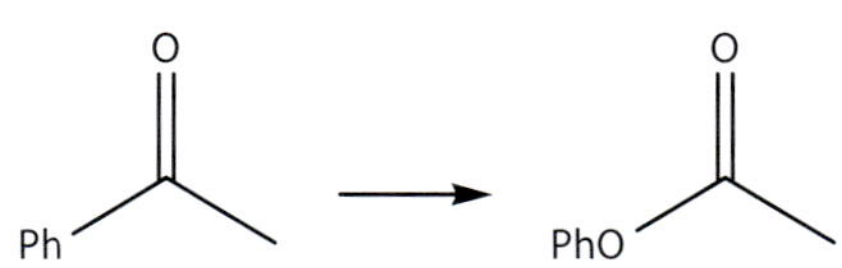

(c)

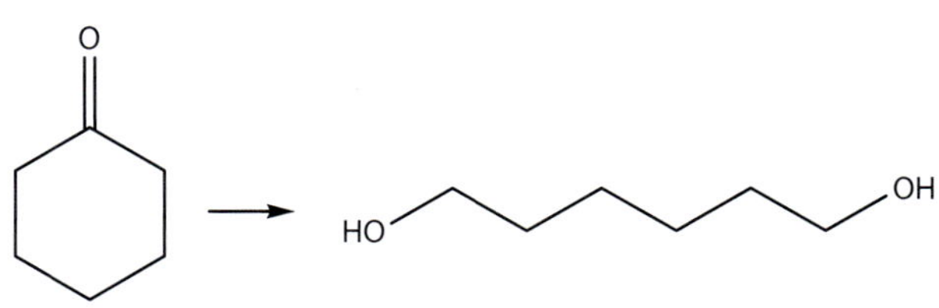

(d)

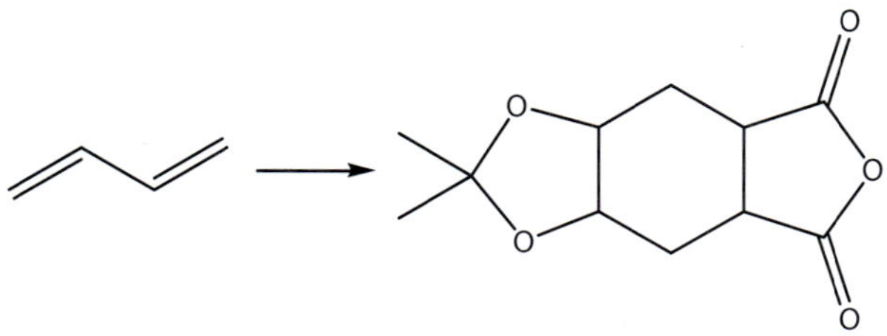

(e) PhCHO to PhCH(OH)CH(OH)Ph

(f)

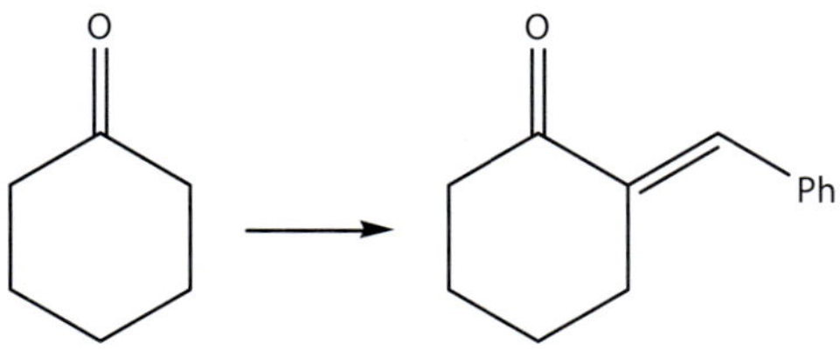

(g)

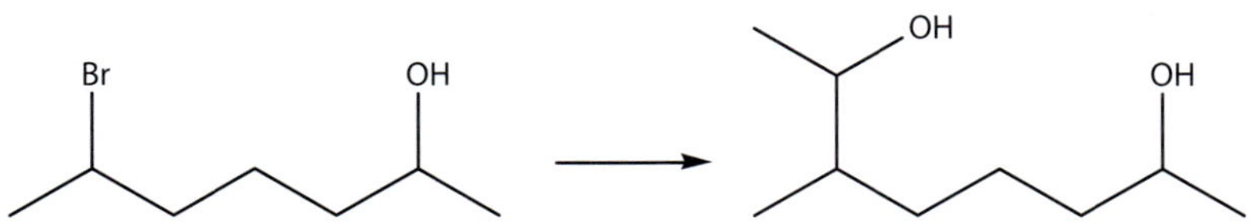

(h)

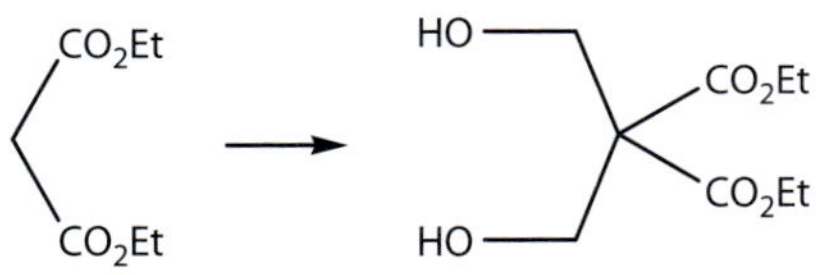

(i)

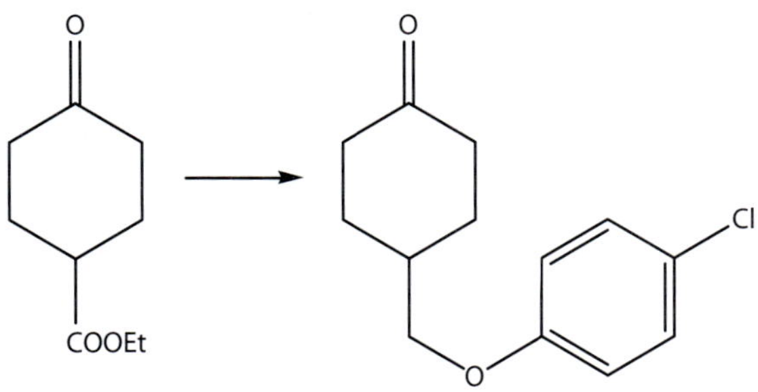

(j)

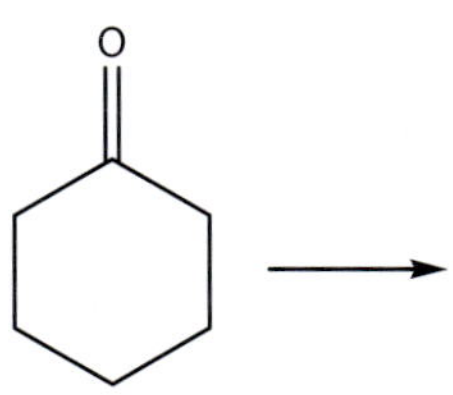

$HOOCCH_2CH_2CH_2CH_2COOH$

(k)

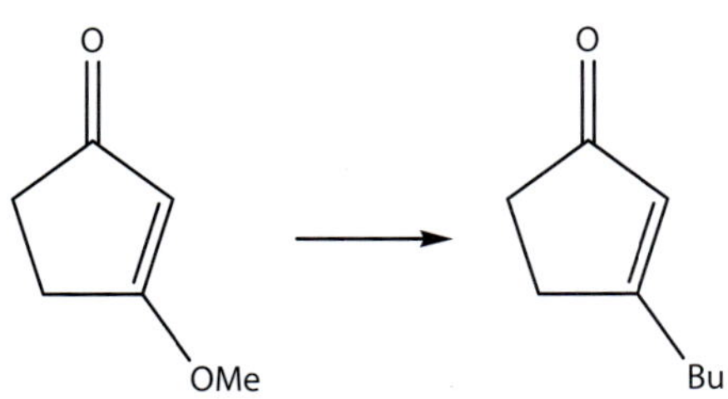

(l)

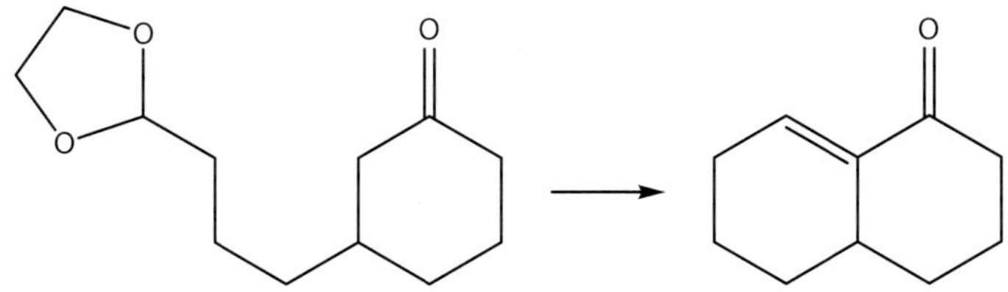
O
O
O
O

(m)

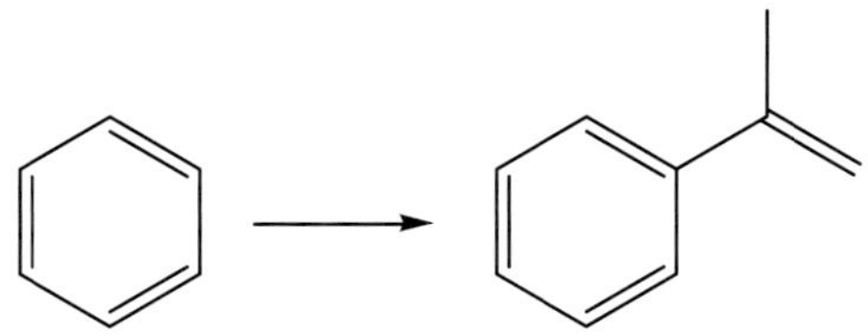

(n)

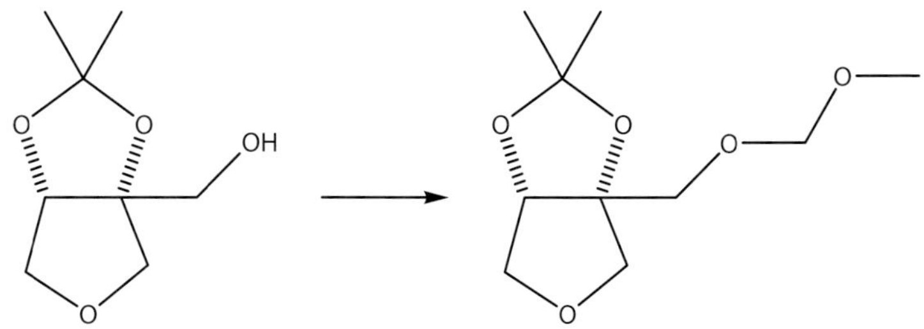
O
O
OH
O
O
O
O
O
O

(o)

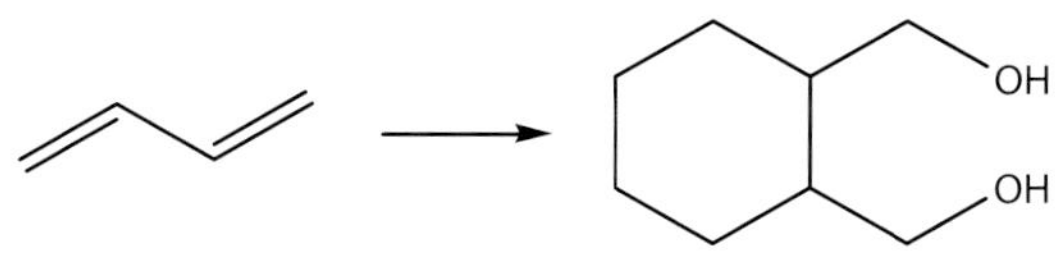
OH
OH

(p)

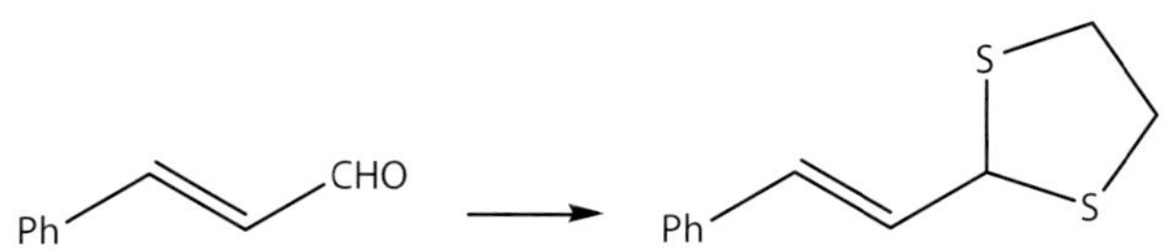
Ph
CHO
Ph
S
S

(q)

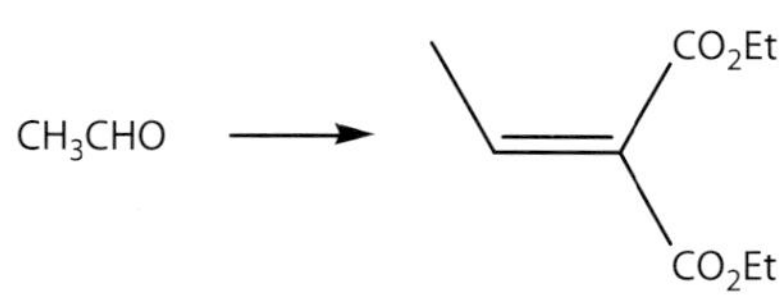
CH$_3$CHO
CO$_2$Et
CO$_2$Et

(r)

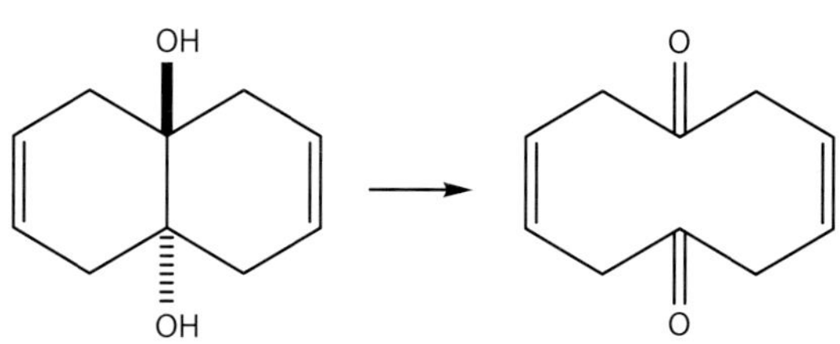
OH
OH
O
O

(s)

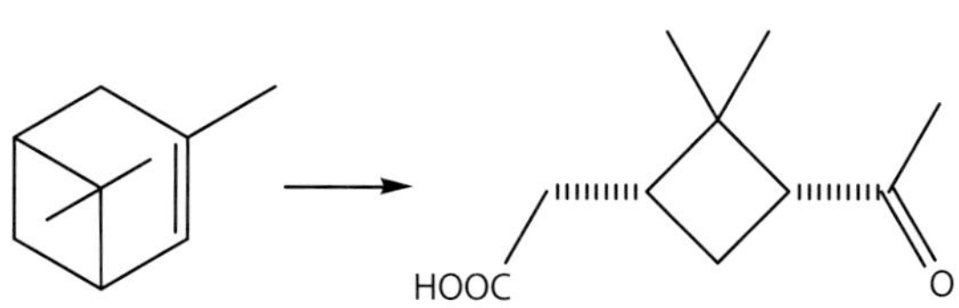
HOOC
O

(t)

2. Predict the product(s) of the following reactions:

(a)

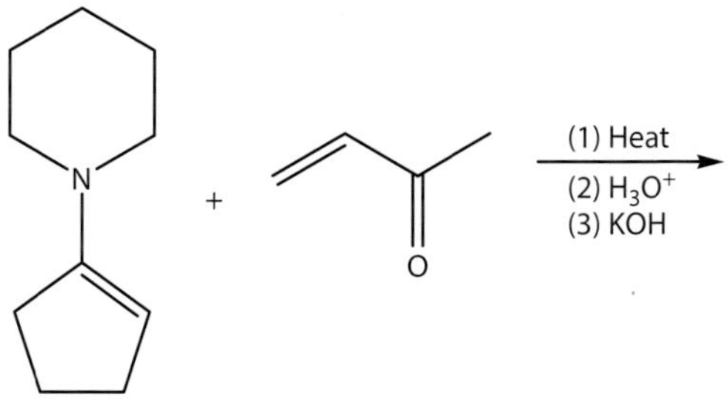

(b)

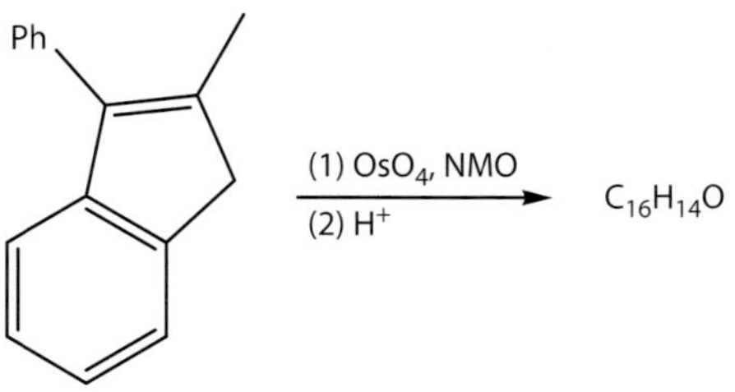

(c)

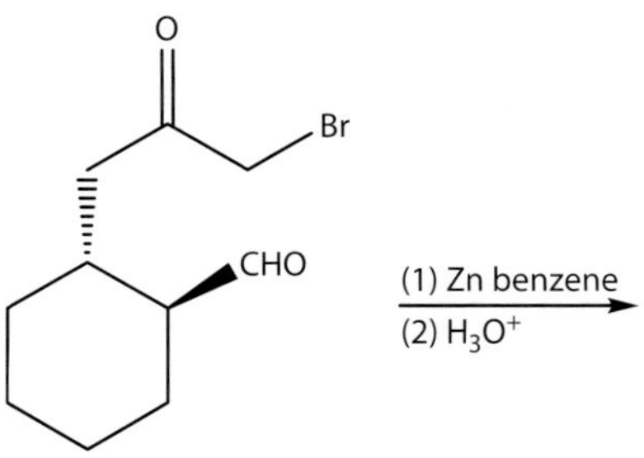

(d)

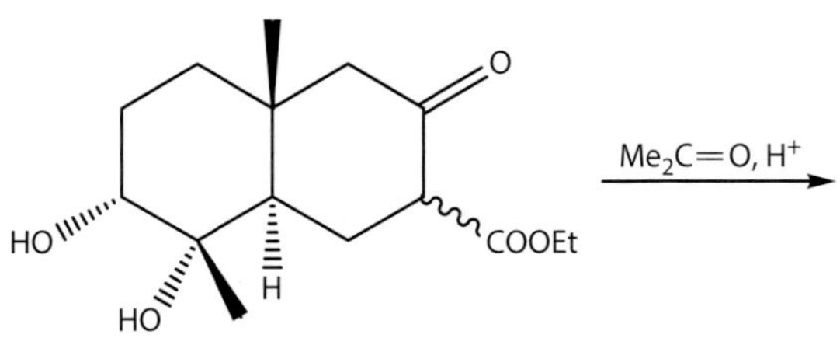

(e)

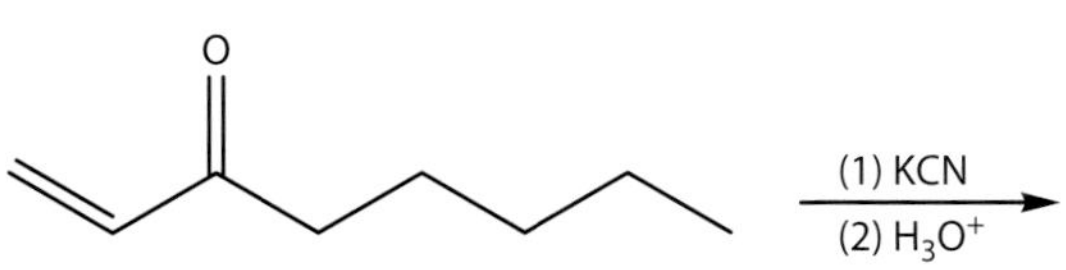

(f)

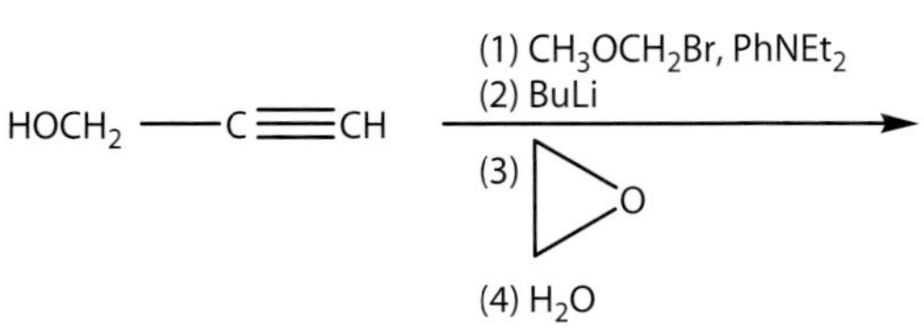

(g)

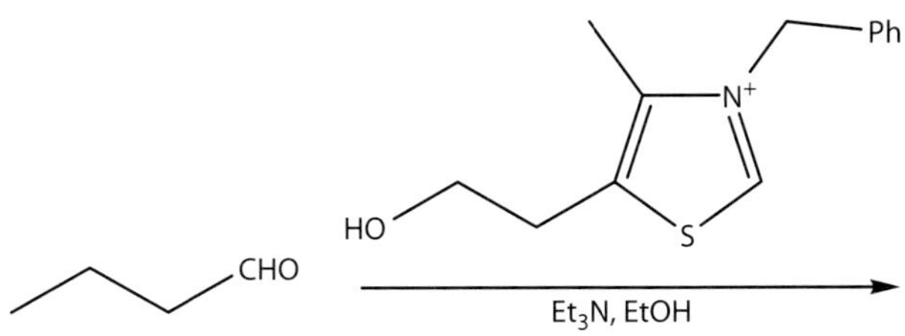
Ph
N+
S
HO
CHO
Et$_3$N, EtOH

(h)

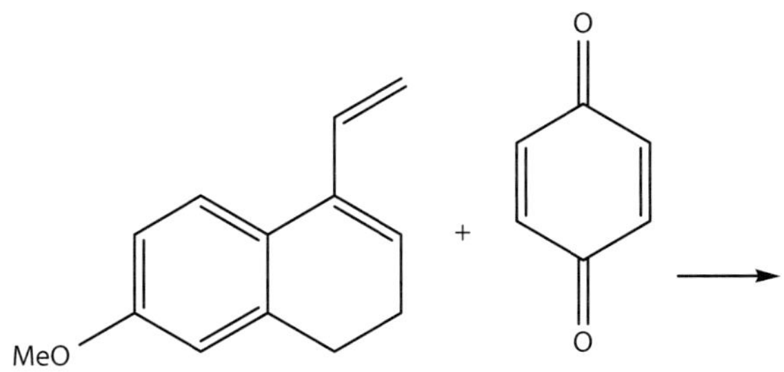
O
+
MeO
O

(i)

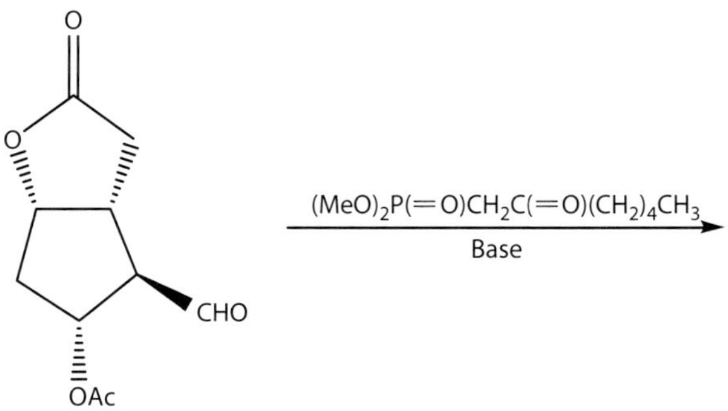
O
O
(MeO)$_2$P(=O)CH$_2$C(=O)(CH$_2$)$_4$CH$_3$
Base
CHO
OAc

(j)

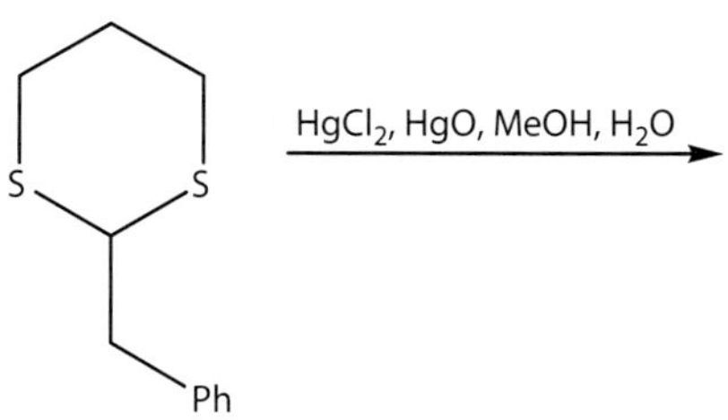
HgCl$_2$, HgO, MeOH, H$_2$O
S
S
Ph

(k)

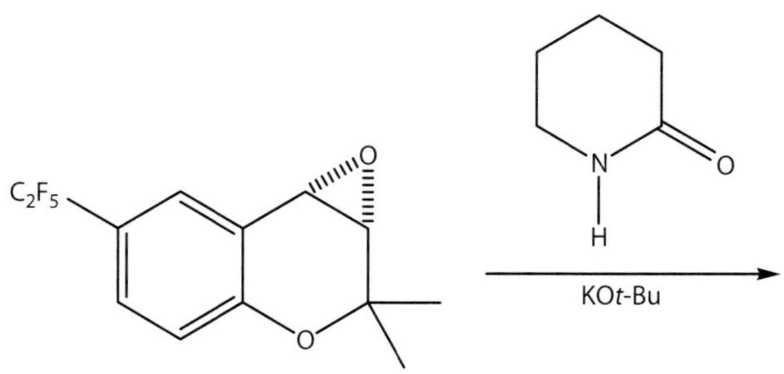
N
O
H
C$_2$F$_5$
O
KO*t*-Bu
O

(l)

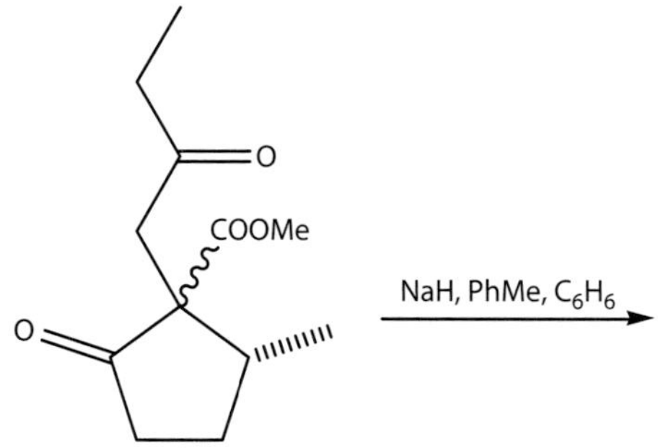

(m)

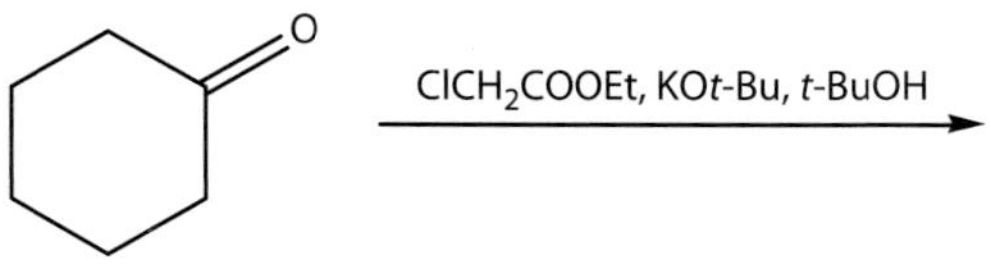

(n)

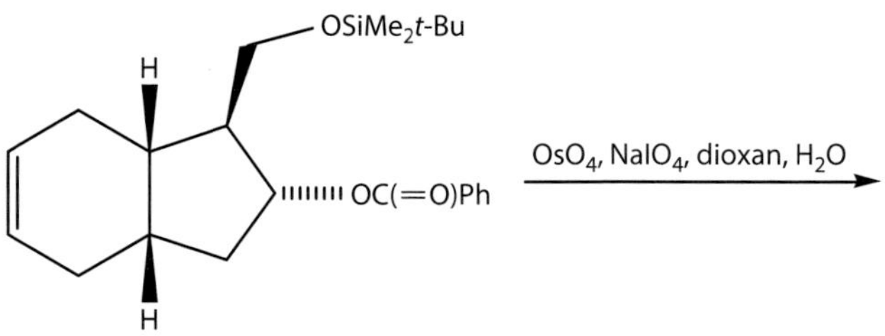

(o)

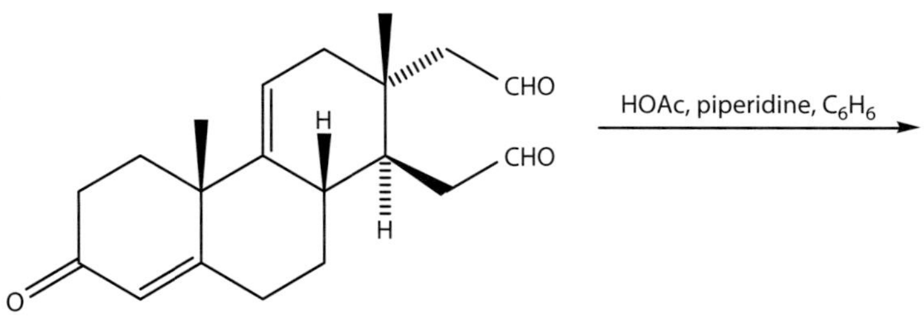

3. Give SMs that would generate each of the following molecules by a Diels–Alder reaction.

(a)

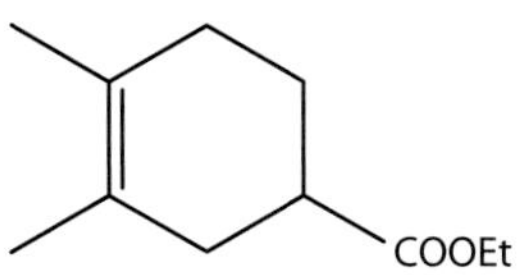

(b)

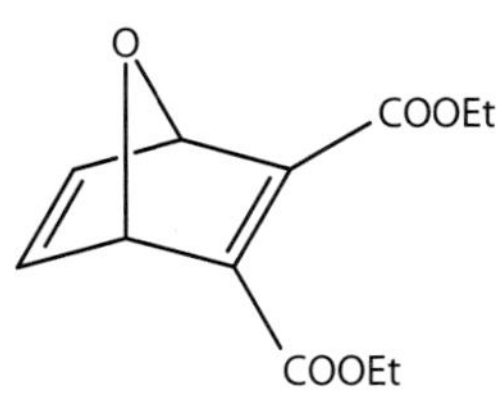

(c)

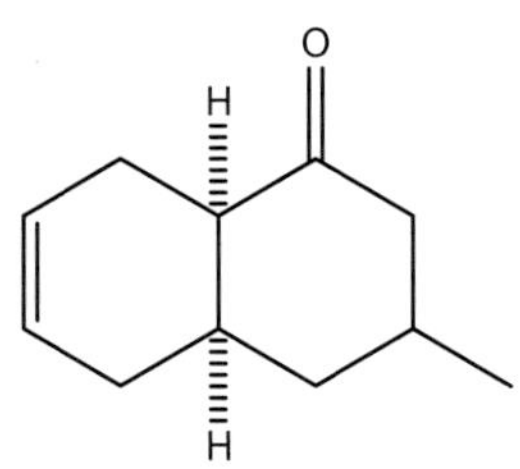

(d)

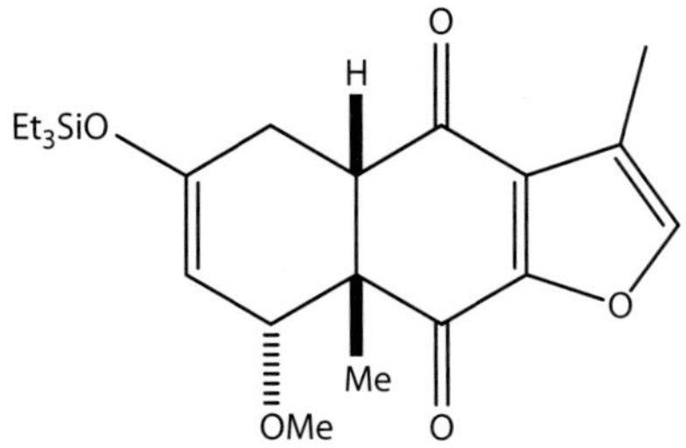

(e)

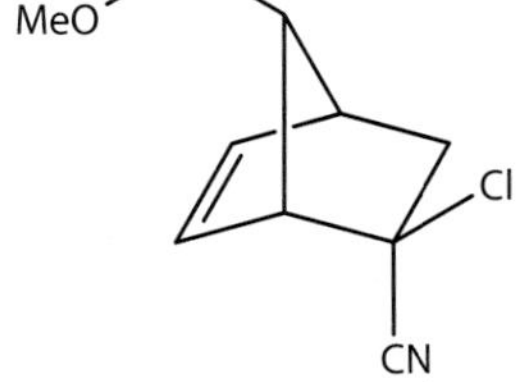

4. Propose a mechanism for each of the following processes:

(a)

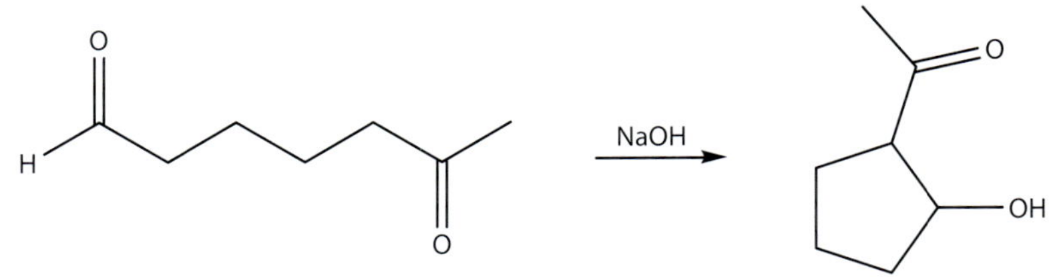

(b)

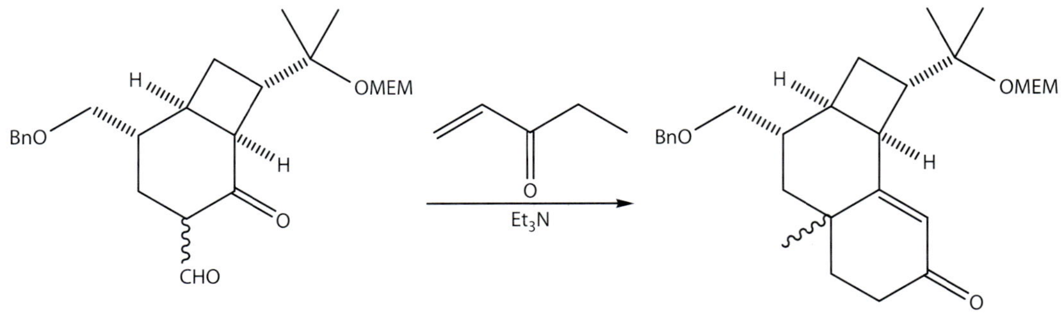

(c)

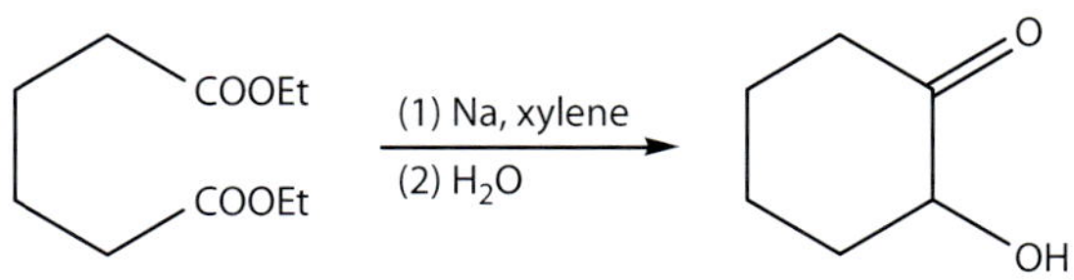

(d)

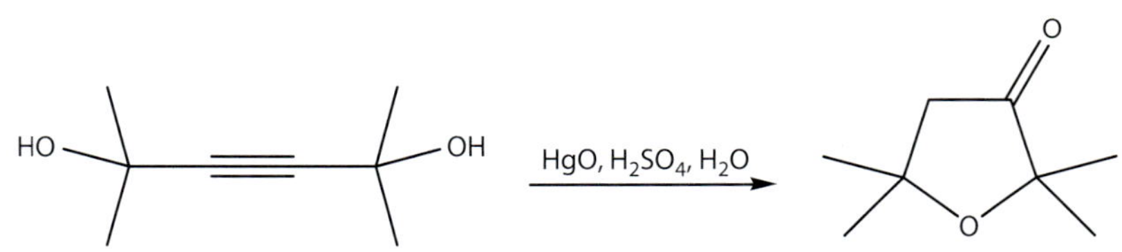

(e)

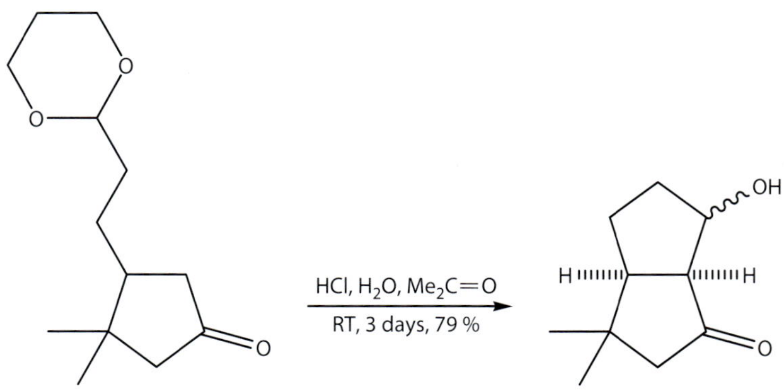

(f)

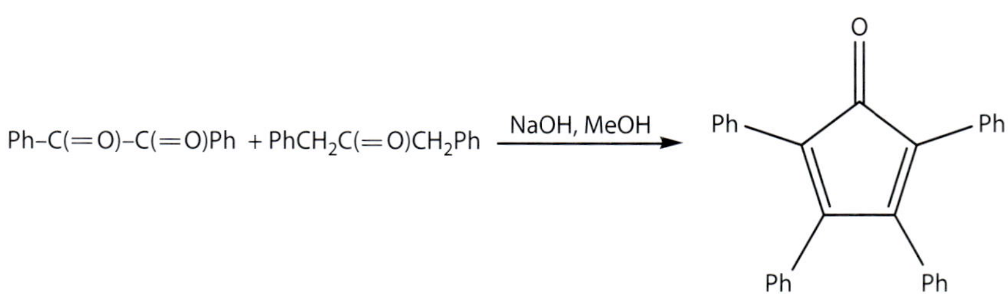

5. Study the reaction sequence shown and answer the questions posed:

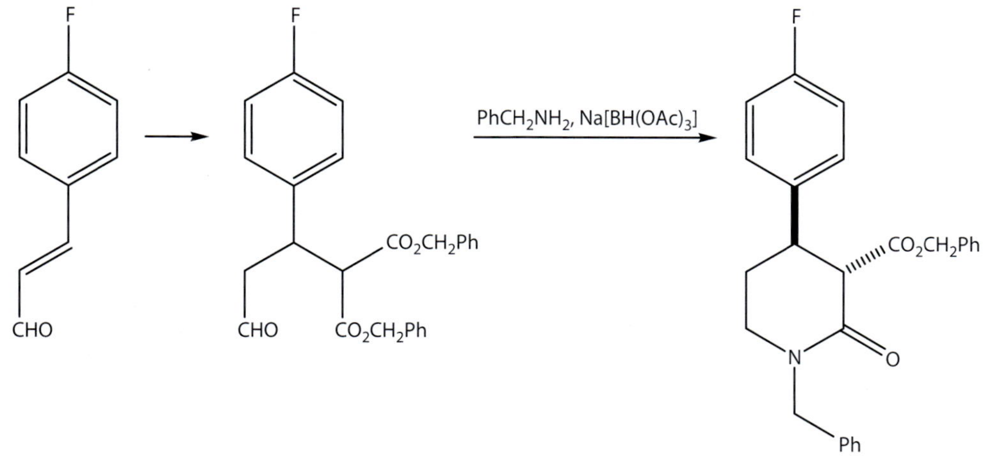

20.30 **20.31** **20.32**

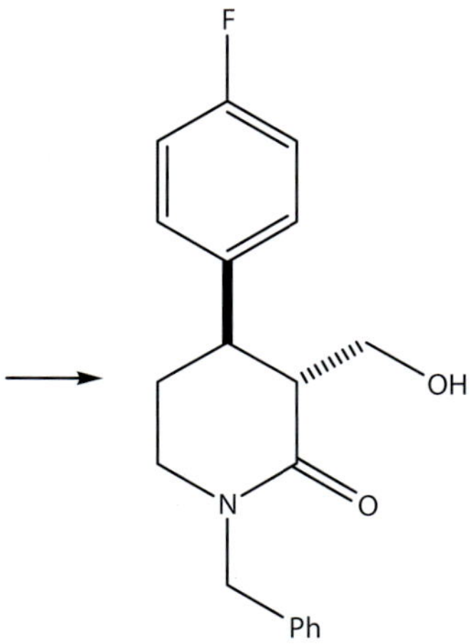

20.33

(a) Give a reagent and conditions for the transformation of **20.30** to **20.31**.
(b) In **20.31**, which carbonyl group is the most electrophilic?
(c) In the light of your answer to (b), propose a mechanism for the transformation of **20.31** to **20.32**. What is the function of the $Na[BH(OAc)_3]$?
(d) What reagent would you use to convert **20.32** to **20.33**?

6. Devise syntheses for each of the following molecules using the SMs shown, compounds with four or fewer carbon atoms and inorganic reagents of choice:

(a)

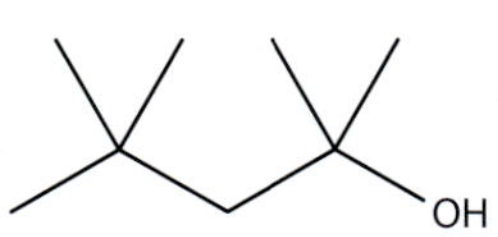

(b) $PhCH_2CH_2OCH_2CH_2C{\equiv}CCH_2OH$

(c)

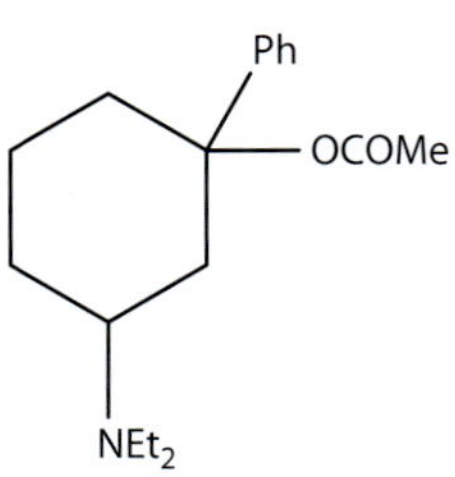

(d)

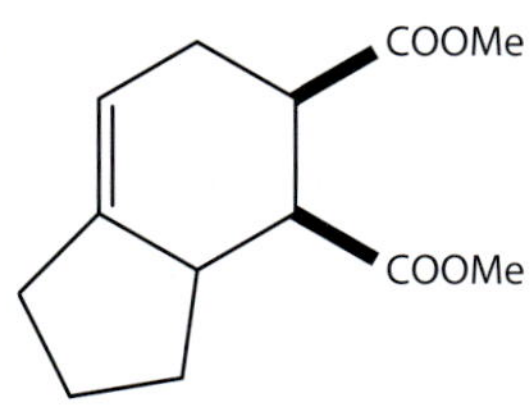

(e)

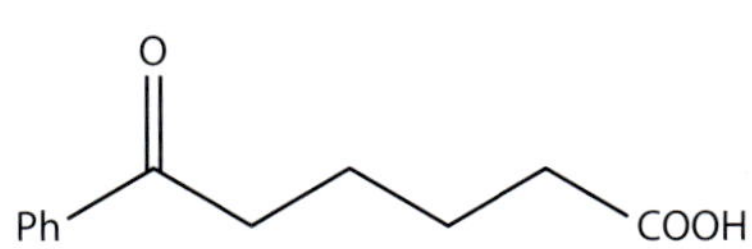

Possible SMs (and anything with four carbons or fewer):

$(EtOOC)_2CH_2$, $EtOOCCH_2COCH_3$ $(EtO)_2C{=}O$, PhMe, PhOMe, PhCOMe, PhC(=O)CH=CH_2, Me_3CCH_2Br, $PhCH_2CH_2Br$, $PhCH_2COOEt$, $PhCH_2CN$, $PhNH_2$, $Br(CH_2)_5Br$, PhCHO, PhCOOH, Me_3CCOOH, PhCl, $Ph_2C{=}O$, PhBr, PhCN, 1,2-$C_6H_4Cl_2$

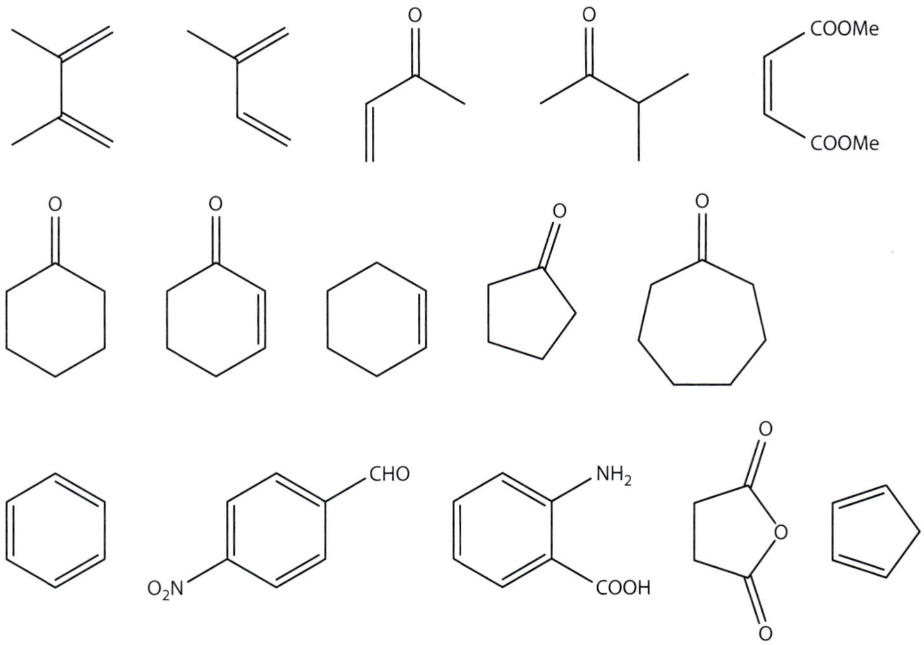

7. Explain the following observations:
   (a) Reaction of **20.34** with acetone/TsOH gives mainly the five-ring acetal **20.35**, with only about 10 % of the six-ring product:

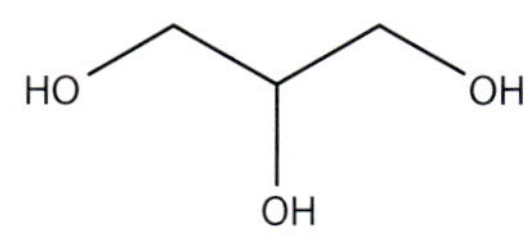

20.34

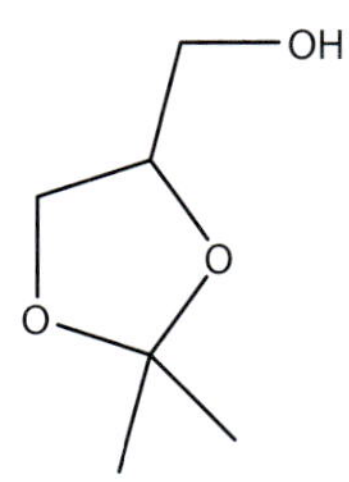

20.35

   (b) Periodate can also cleave α-hydroxyketones as in the reaction of **20.36**:

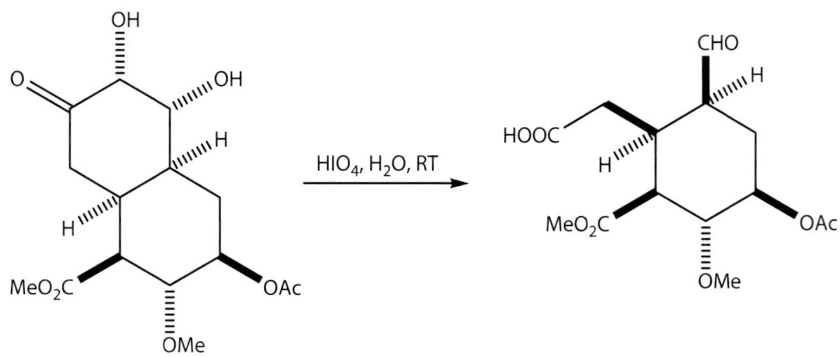

20.36

MORE CHALLENGING PROBLEMS

1. How would you carry out each of the following transformations, which may require more than one step?

(a)

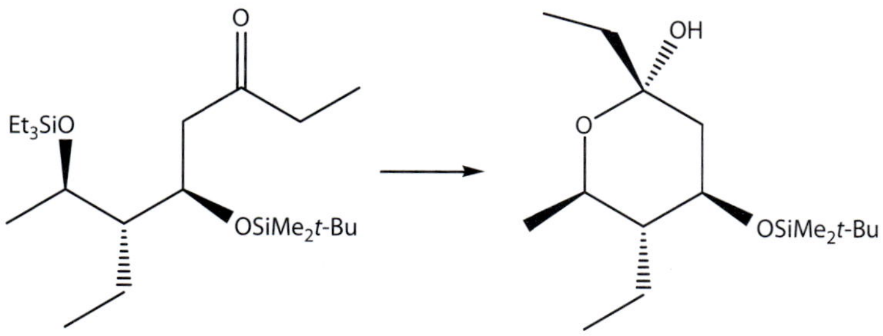

(b)

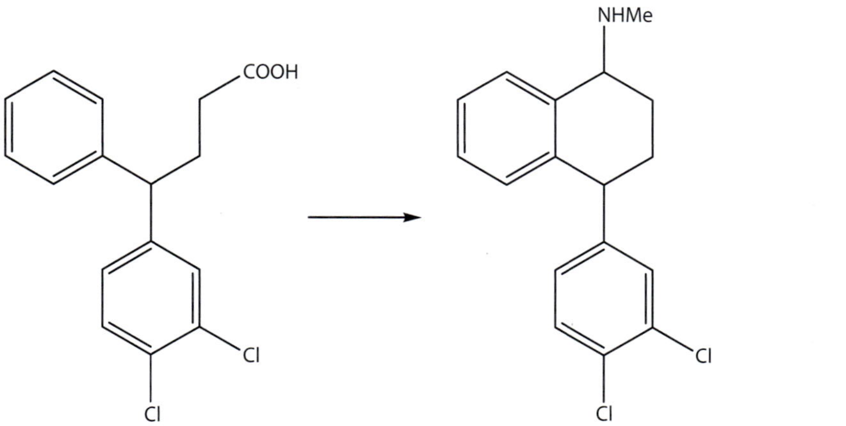

(c)

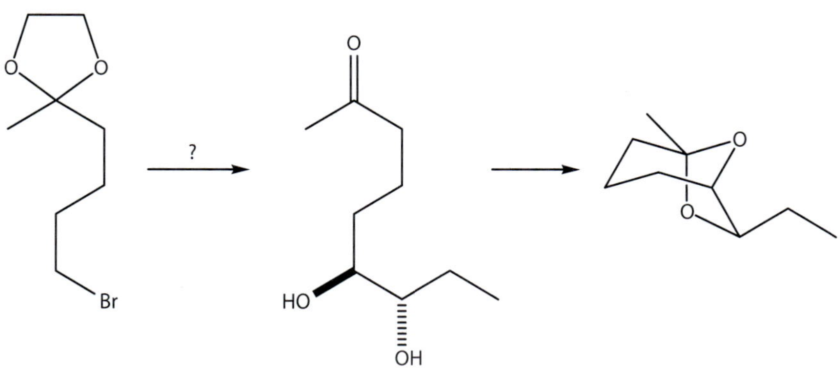

The final step, the cyclization to the acetal, is spontaneous under the reaction conditions.

(d)

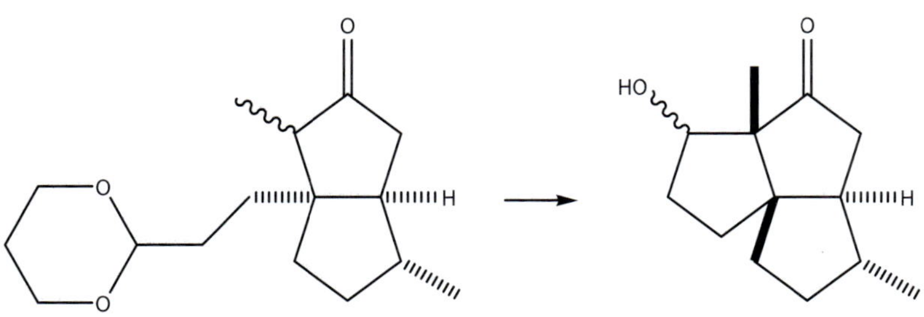

(e)

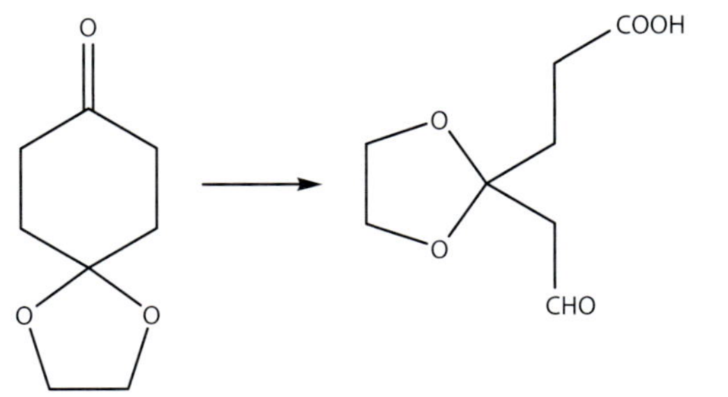

2. Predict the outcome of each of the following processes:

(a)

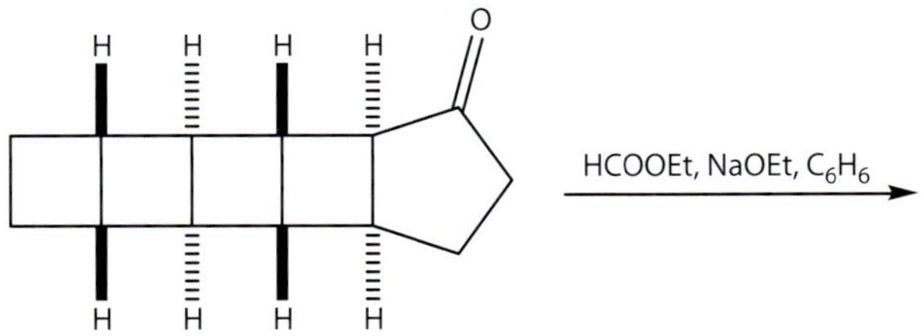

(b)

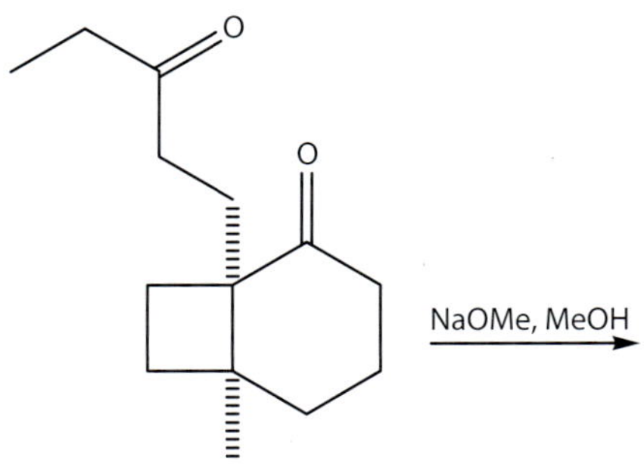

(c)

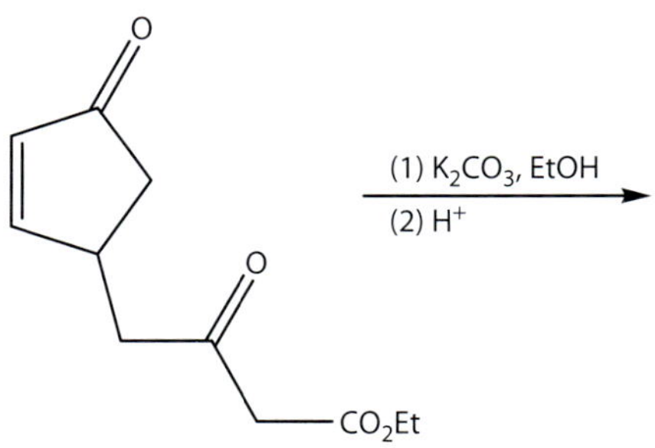

(d)

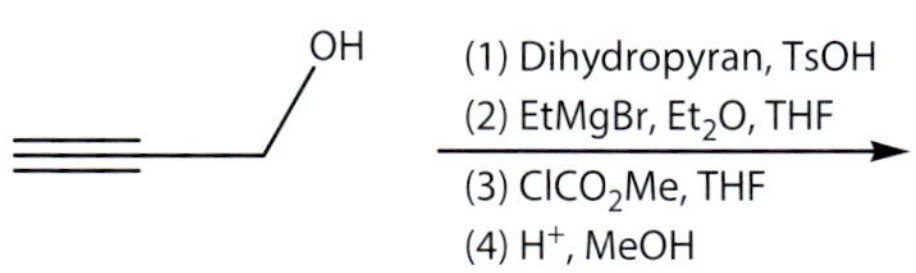

3. Propose a mechanism for each of the following processes:

(a)

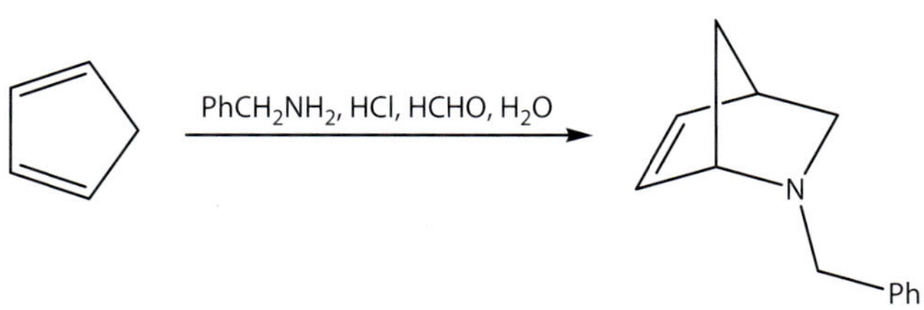

(b)

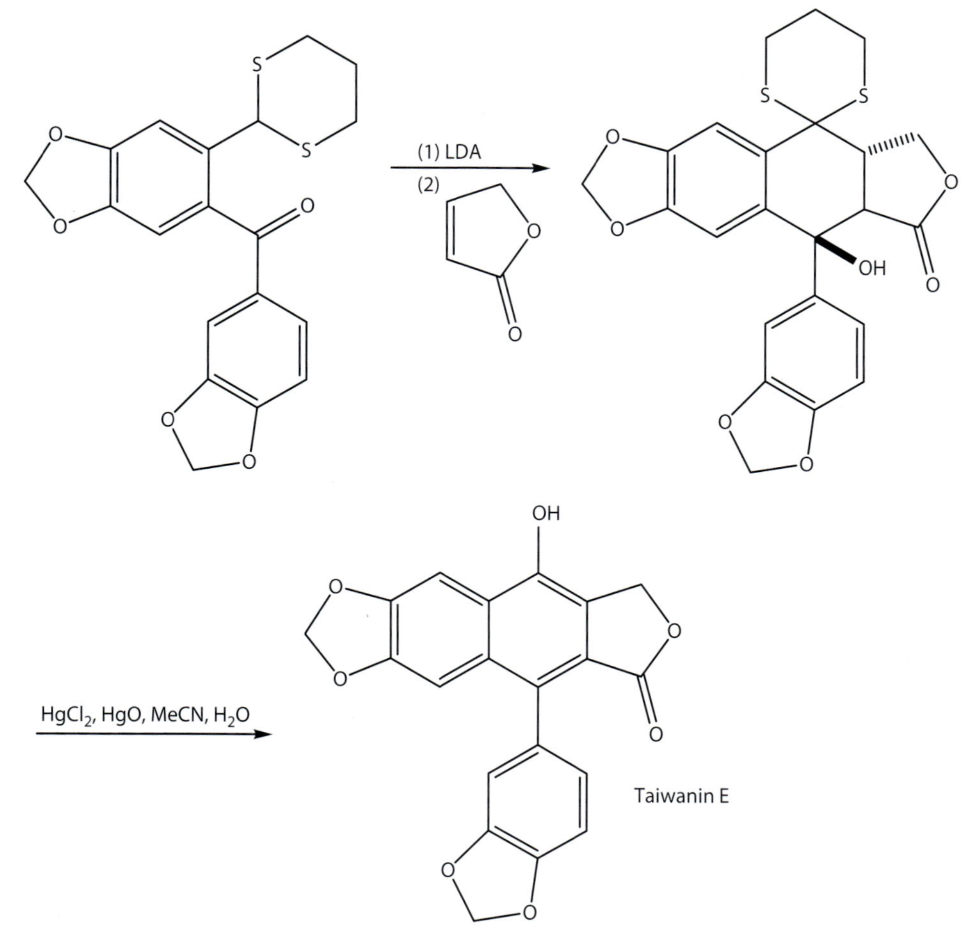
(1) LDA
(2)
OH
HgCl$_2$, HgO, MeCN, H$_2$O
OH
Taiwanin E

(c)

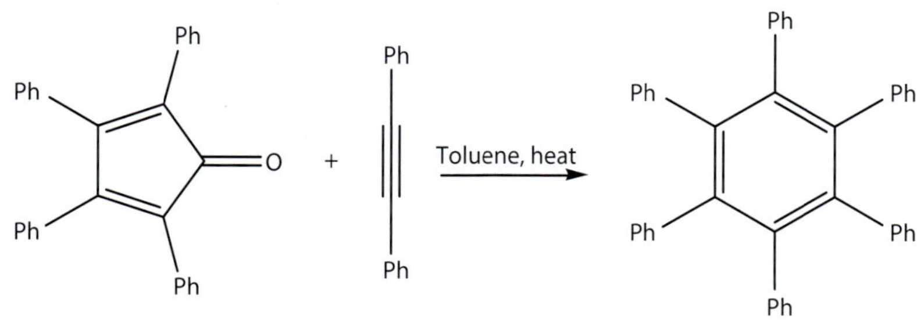
Ph
Ph
Ph
Ph
+
Ph
Ph
Toluene, heat
Ph
Ph
Ph
Ph
Ph
Ph

(d)

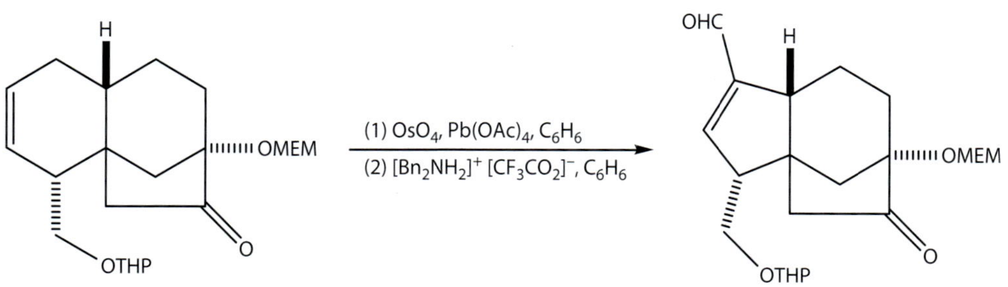
H
OMEM
OTHP
(1) OsO$_4$, Pb(OAc)$_4$, C$_6$H$_6$
(2) [Bn$_2$NH$_2$]$^+$ [CF$_3$CO$_2$]$^-$, C$_6$H$_6$
OHC
H
OMEM
OTHP

(e)

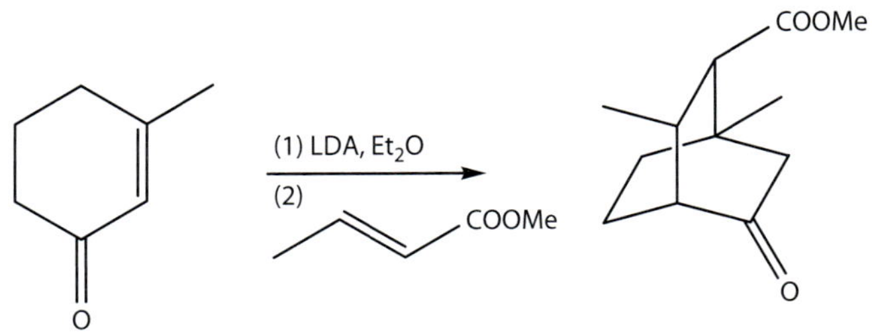

4. Ylids can be formed by deprotonating a carbon atom next to sulfur, in a manner similar to that achieved in the Wittig reaction using phosphorus. Thus, ylids can be formed both from [Me_3S^+] and [$Me_2S(=O)Me]^+$ using *t*-BuOK as the base:

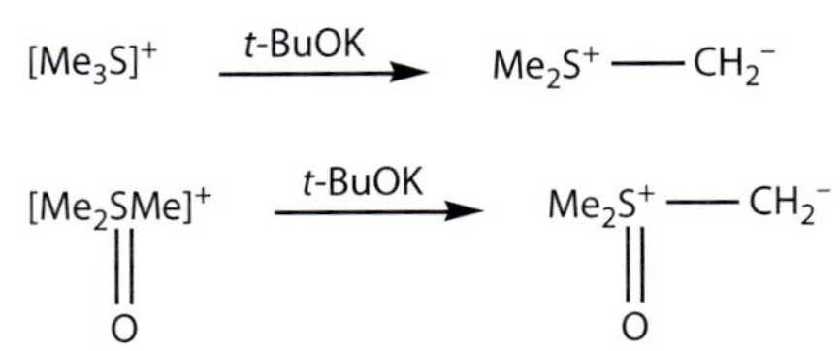

The two ylids react differently with 2,6,6-trimethylcyclo-2,4-heptadienone as shown. Write a mechanism for each process, and suggest a reason for the difference. (You may find it helpful to think about the mechanism of epoxidation of enones with the anion of hydrogen peroxide.)

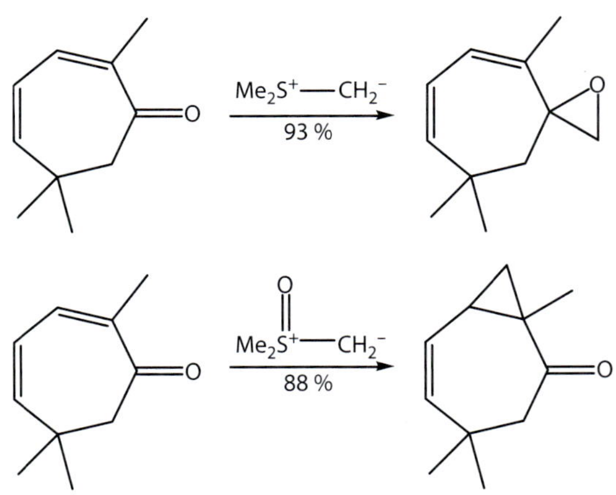

5. Devise syntheses for each of the following molecules using the SMs shown, compounds with four or fewer carbon atoms, and inorganic reagents of choice:

(a)

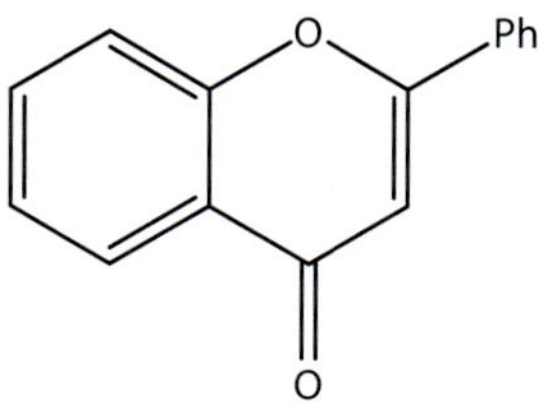

(b)

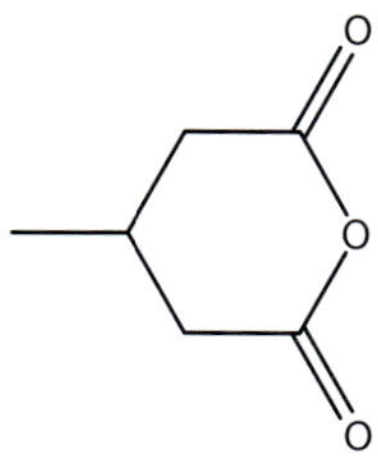

(c)

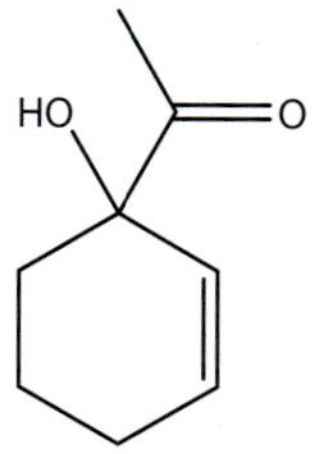

(d)

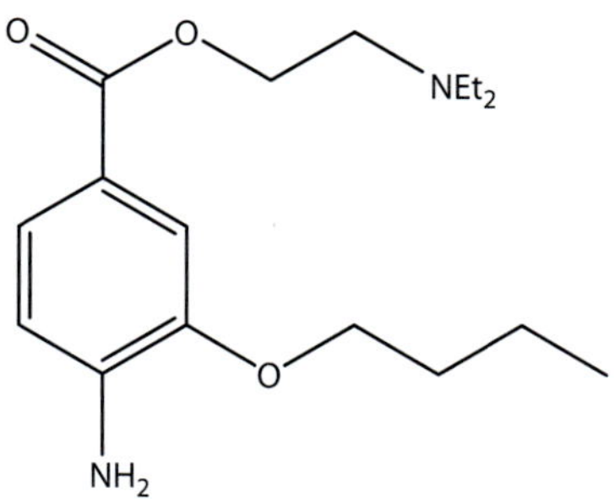

(e)

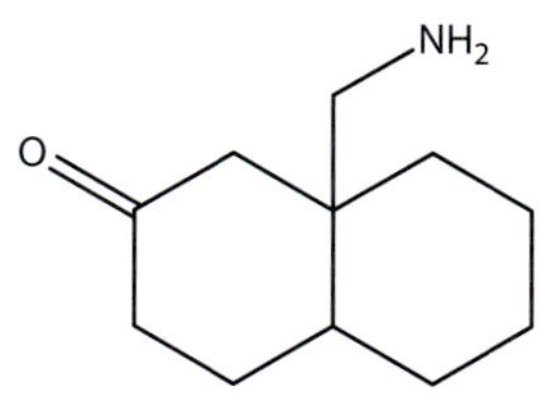

Possible SMs (and anything with four carbons or fewer):

$(EtOOC)_2CH_2$, $EtOOCCH_2COCH_3$ $(EtO)_2C{=}O$, PhMe, PhOMe, PhOH, PhCOMe, PhC(=O)CH=CH_2, Me_3CCH_2Br, $PhCH_2CH_2Br$, $PhCH_2COOEt$, $PhCH_2CN$, $PhNH_2$, $Br(CH_2)_5Br$, PhCHO, PhCOOH, Me_3CCOOH, PhCl, $Ph_2C{=}O$, PhBr, PhCN, 1,2-$C_6H_4Cl_2$, MeCH=CHCOOMe

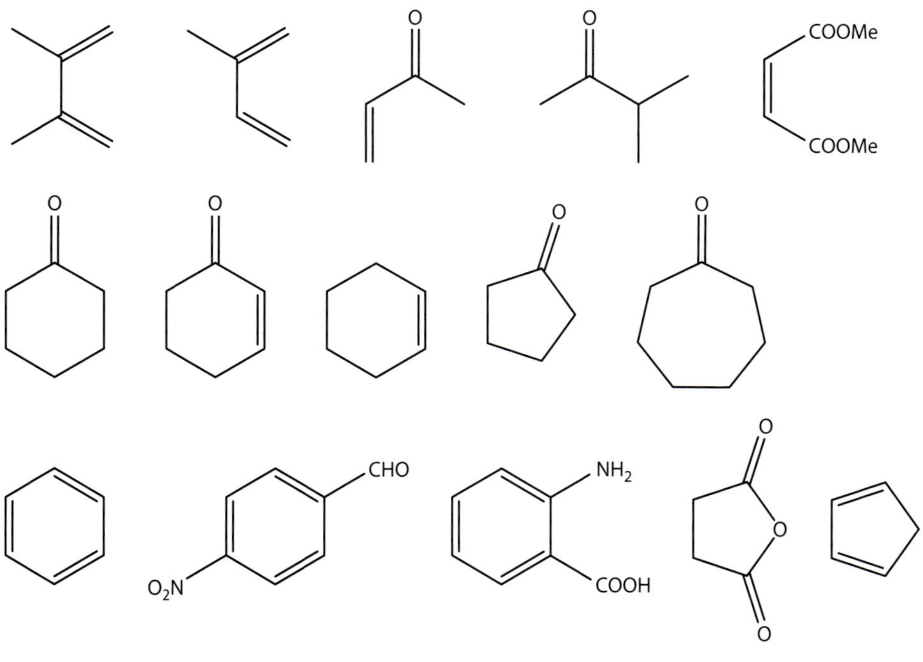

Polymeric Materials

21

21.1 INTRODUCTION

This chapter will focus on the organic aspects of polymer chemistry; the physical chemistry and physics of polymers are fascinating subjects, but must be deferred for a later course. Polymers play a major role in modern everyday life—simply look around your room and at the clothes you are wearing. Polymers in modern artifacts may be simply structural—lighter weight, stronger, and less expensive than natural materials—or they may be the sort of smart materials that make up your computer or cell phone screen. We think of them as a twentieth-century invention, but this is not entirely so; we have been using naturally occurring biopolymers for food, tools, clothing, shelter, and adornment for millennia, and polymers have been made from natural materials for almost as long. Man-made polymers have attracted unfavorable attention from environmentalists, because they degrade too slowly, and from museum conservators because they degrade too quickly. We start with some basic definitions (Table 21.1). Nomenclature for polymers is a problematic topic. We know how various monomers are named under the IUPAC system, but these names are rarely incorporated into the names used by the polymer industry. This chapter will use what is judged to be the "usual" way of describing monomers and polymers, but IUPAC names and/or molecular structures will also be provided.

21.2 POLYMER STRUCTURES

Polymers can have the same composition, but a different structure, and this will profoundly affect their properties. For example, polythene (PE) (also described as polyethylene), the most widely used polymer in the world, may be linear (**21.1**) or branched (**21.2**), and molecular weights can vary greatly. Linear PE is a stiff, tough material because the linear chains can pack at least partially into a crystalline state. Branched PE is less able to crystallize, producing a softer, more flexible material with a lower melting point. Ultra-high MWPE* is a linear polymer with a molecular weight in the range 3–6 × 10^6 and is widely used for joint replacements:

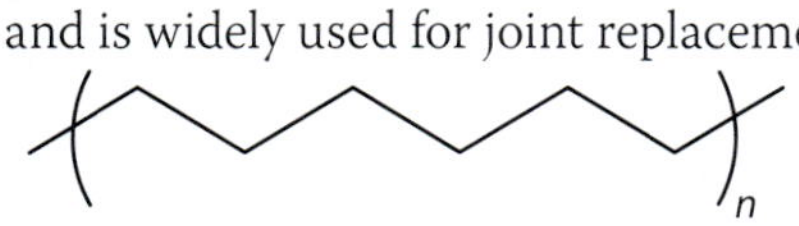

21.1

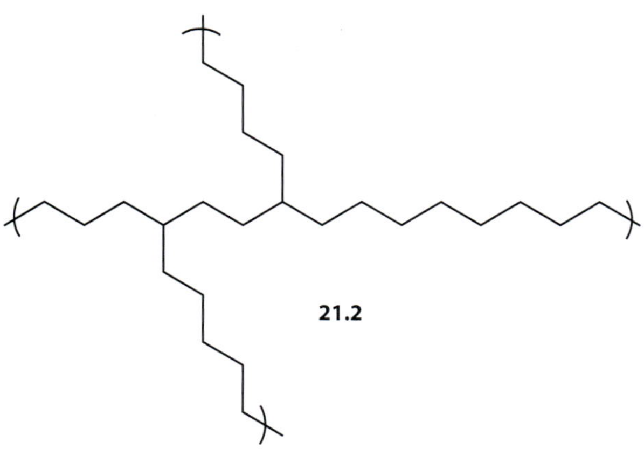

21.2

* Ultra-high molecular weight polyethylene.

TABLE 21.1 Important Definitions in Polymer Chemistry

Term	Definition	Example, Comments
Monomer	A unit that forms a polymer.	Ethene forms PE.
Dimer	Two linked monomers.	1-Butene is a dimer of ethene: 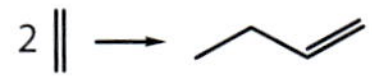
Trimer[a]	Three linked monomers.	1-Hexene is a trimer of ethene: 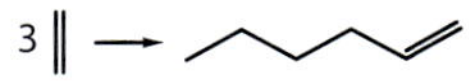
Oligomer	Oligomers contain an intermediate number of monomers, usually up to 20 units.	1-Dodecene is an oligomer of ethene. From the Greek, ὀλίγοι, *oligoi*, few.
Polymer	Many linked monomers.	Polyethene, nylon. From the Greek, πόλυς, *polys*, many.
Addition polymer, or chain-growth polymer	Monomers are added together with no loss of material; all atoms of the monomer are incorporated. Growth occurs at one end of a long chain, and the intermediates are often reactive species such as anions, cations, or radicals.	PE, polystyrene, PEG.
Step-growth polymer	A polymer formed by successive reactions of one or more bifunctional or multifunctional monomers. Characterizable, isolable dimers and oligomers are formed as intermediates.	Nylon. $H_2N(CH_2)_6NH_2$ and $HOOC(CH_2)_4COOH$ form an amide-based polymer, nylon-6,6, with loss of 1 mol of water for each growth step.
Ring-opening polymer	A chain-growth polymer produced by ring opening of a cyclic monomer, usually in an addition polymerization.	PEO, nylon-6, polylactide.
Homopolymer	All the units in the polymer are the same.	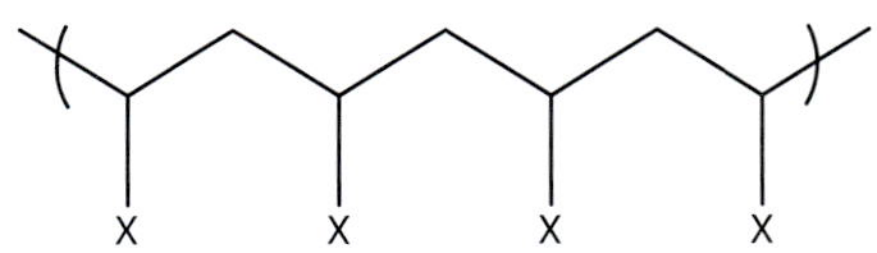 from $CH_2{=}CHX$.
Copolymer	Two or more different types of monomer—may be alternating, in blocks, or statistical.	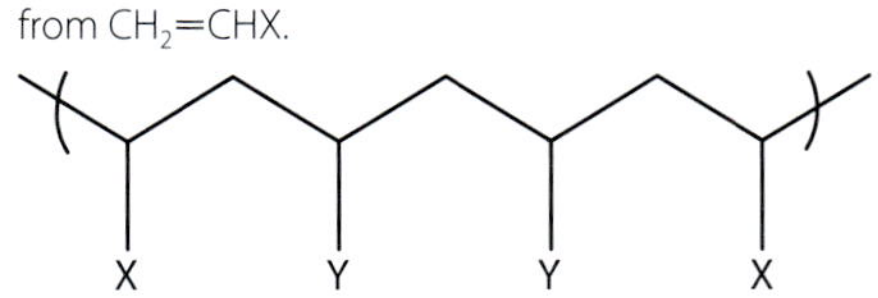 from $CH_2{=}CHX$ and $CH_2{=}CHY$.
Thermoplastics or thermosoftening polymers	Polymers that turn reversibly to liquid on heating, at which point they can be molded. On cooling, they set to a solid glassy or rubbery state.	PE, polystyrene, nylon.
Thermosetting polymers	A polymer, initially soft or liquid (at which point it can be molded), which is cured irreversibly on heating to an infusible, insoluble polymer network. The curing process involves cross-linking, by a chemical reaction. Once hardened, it cannot be remelted.	Bakelite, melamine resin.

[a] The subsequent terms, tetramer, pentamer, hexamer, *etc.*, will be obvious.

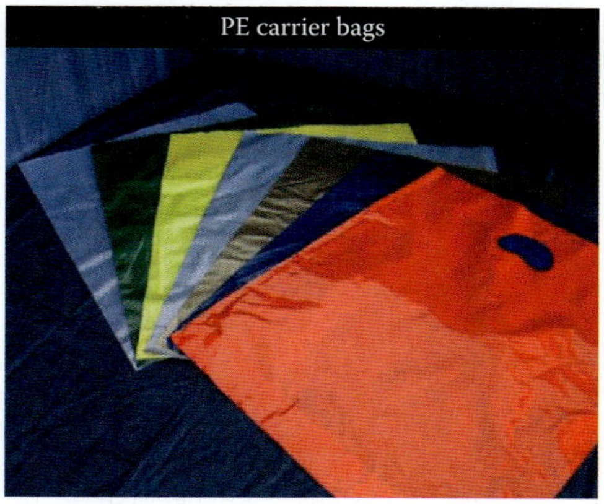

Polymers can be cross-linked by addition of small amounts of a monomer with two possible sites of polymerization. The copolymerization of divinylbenzene and styrene gives cross-linked polystyrene (Figure 21.1). Only 1 %–2 % of the cross-linking agent is needed for the polymer to have very different physical properties; highly cross-linked polymers generally have much greater rigidity, even at high temperatures. However, these materials tend to be less easy to recycle successfully. The most important cross-linking process involves cross-linking of diene rubbers by sulfur atoms, the so-called vulcanization of rubber (Figure 21.2). Both natural rubber (all *cis*-polyisoprene) and its synthetic substitutes are too soft to be used in tires, but vulcanized rubber, although cross-linked, remains flexible and stable over the likely working temperature range of most vehicles. The process, developed in the nineteenth century by Goodyear and Hancock, was named for Vulcan, the Roman god of fires and metalworking. Vulcan was associated with the ancient sulfur mines in Sicily. Neither natural nor vulcanized rubber is black—the black color of tires derives from the added carbon black (20 %–30 %). This acts as a lubricant and increases material lifetime in a tire when it is repeatedly compressed and released.

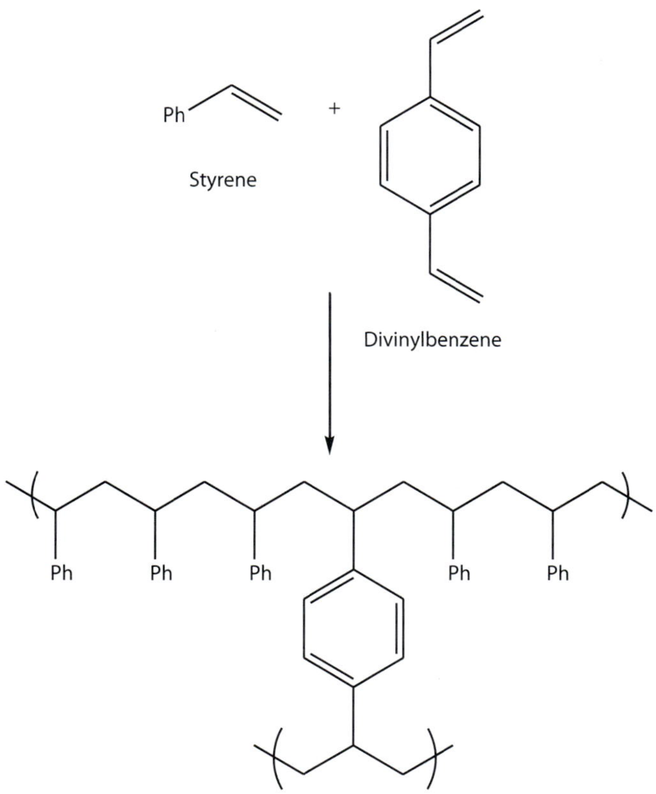

FIGURE 21.1 Cross-linking of polystyrene.

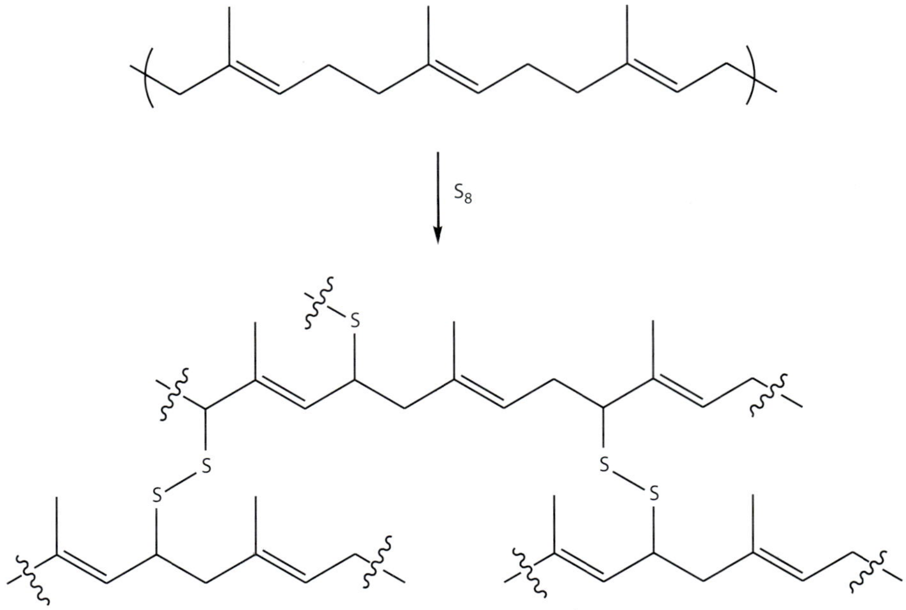

FIGURE 21.2 Vulcanization of rubber.

In the last 25 years, polymers with more exotic structures have been prepared. One group are the "star polymers," **21.3**. In these, there is a central core with a number of linear chains emanating from it. One method of preparation is to start with a polyfunctional core, such as **21.4**, where each functional group can be used to initiate the formation of a polymeric chain. Some of these are used as additives in motor oil. Another group of polymers are the dendrimers (from Greek δένδρον, *dendron*, tree). In a dendrimer, the polymer is perfectly branched at every repeat unit to give a roughly spherical, highly symmetric, molecule (Figure 21.3). Since each cycle (or generation) is identical, the process can in principle be continued indefinitely, though in practice successive generations are increasingly harder to achieve because of steric constraints. Each cycle roughly doubles the molecular weight of the dendrimer and also roughly doubles the number of end groups; all the functionality is on the surface and can be tailored to a range of applications. Both high and low molecular weight species have been prepared, and they are finding promising applications in drug delivery.

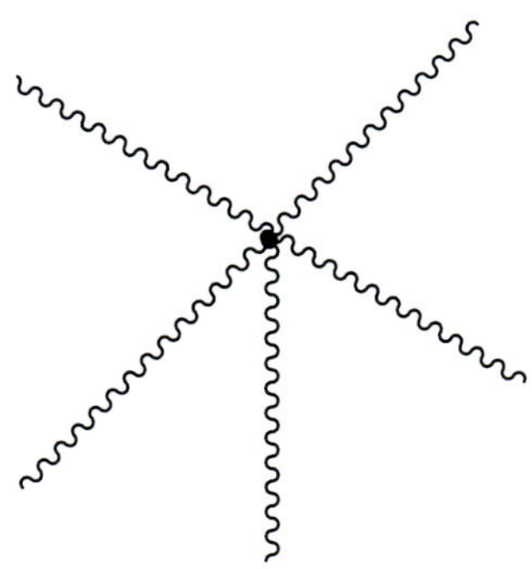

21.3, star polymer

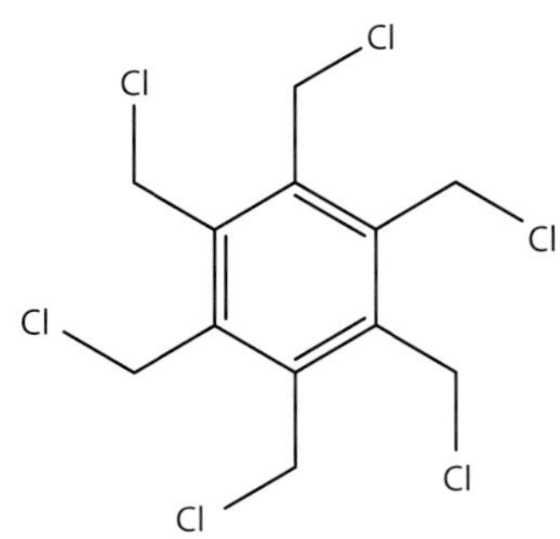

21.4

With chain-substituted polymers, the relative stereochemical orientation of the side chains can profoundly influence the physical properties of the polymer. If we consider the simplest example polypropylene (the old name for propene was propylene), one stereochemical outcome might be that all the methyl groups are oriented in the same way, as in **21.5**, *isotactic* polypropylene.* These stereochemical relationships result in a helical chain conformation for isotactic polypropylene.

* Note that the stereochemistry shown here is relative rather than absolute.

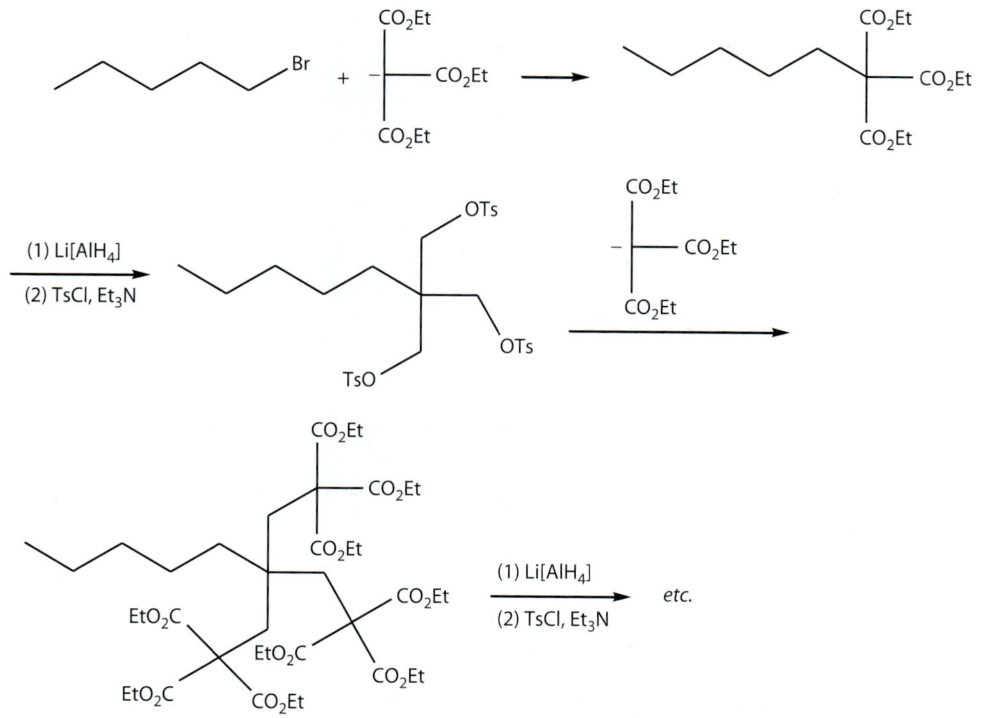

FIGURE 21.3 Synthesis of a simple dendrimer.

Isotactic polypropylene is produced when propene is polymerized using a Ziegler–Natta catalyst such as $R_3Al/TiCl_3$ (see Section 21.3.4). It is hard, rigid, and crystalline, melting at 170 °C. It is resistant to material fatigue and is used in various molded artifacts, hinges, or food containers that will endure repeated flexing. It is also useful in fibers, particularly carpets. *Syndiotactic* polypropylene (**21.6**), prepared using metallocene catalysts, is also partially crystalline, but softer than the isotactic form, melting at 125 °C. Because of the expense of the catalyst, it is not yet produced on a large scale, but it can be used to produce tough, clear moldings. It is said to be more stable toward oxidative degradation than the isotactic material. It may get its first widespread use in medical applications, where gamma-sterilized polypropylene is widely used, but suffers from problems of yellowing and embrittlement due to post-irradiation oxidation. *Atactic* polypropylene has random stereochemistry (**21.7**) and has limited uses in road paint, adhesives, and bitumen asphalt membranes.

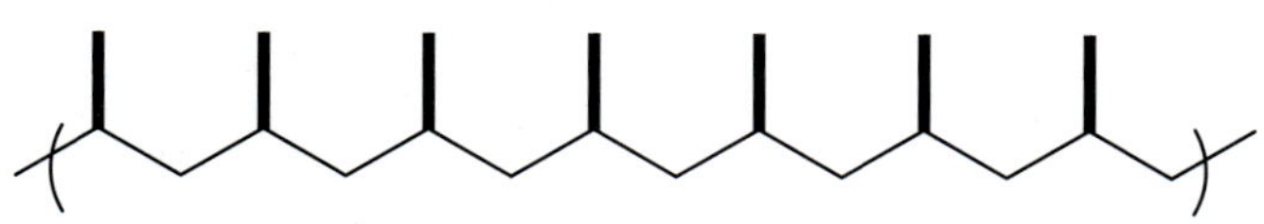

21.5, schematic of a section of isotactic polypropylene

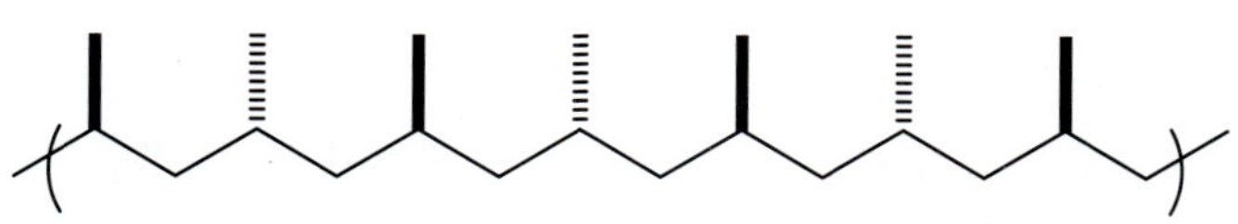

21.6, schematic of a section of syndiotactic polypropylene

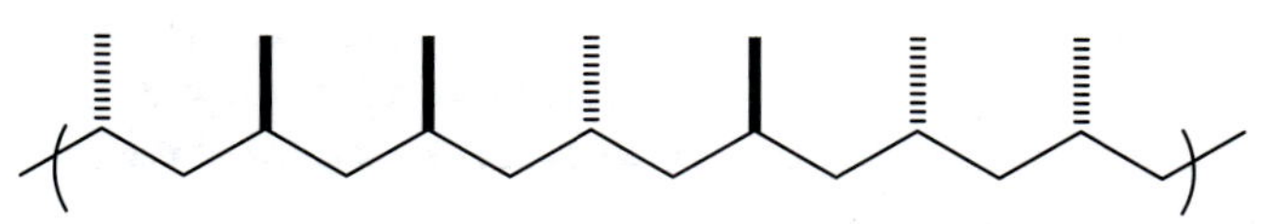

21.7, schematic of a section of atactic polypropylene

PROBLEM 21.1

The reaction shown in the following is the first step in the construction of a star polymer. Propose a mechanism for the process:

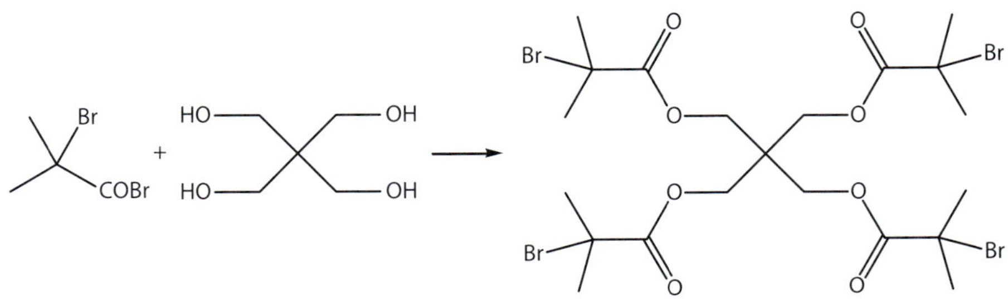

PROBLEM 21.2

Propose a mechanism for the reaction of phenol and formaldehyde in the presence of acid to give bakelite:

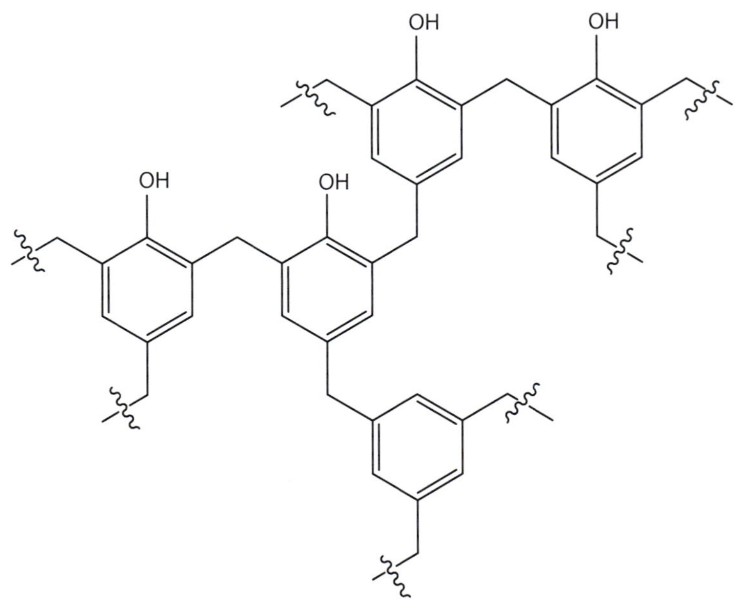

Bakelite

Solutions

21.1 This is the reaction of an alcohol with an acid bromide:

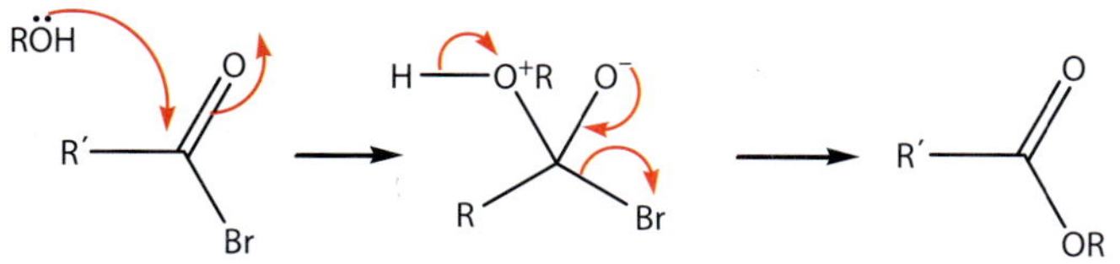

21.2 This process depends on the reaction of phenol with formaldehyde at the *ortho*- and *para*-positions. The OH group activates the aromatic ring toward electrophilic substitution. Once benzylic alcohols have been produced, these too can act as electrophiles, and a rigid, cross-linked structure is produced. Bakelite® was first produced in 1907 and is regarded as the first artificial polymer. It has high thermal and electrical resistivity and is still widely used in electrical fittings, chipboard, and decorative laminates.

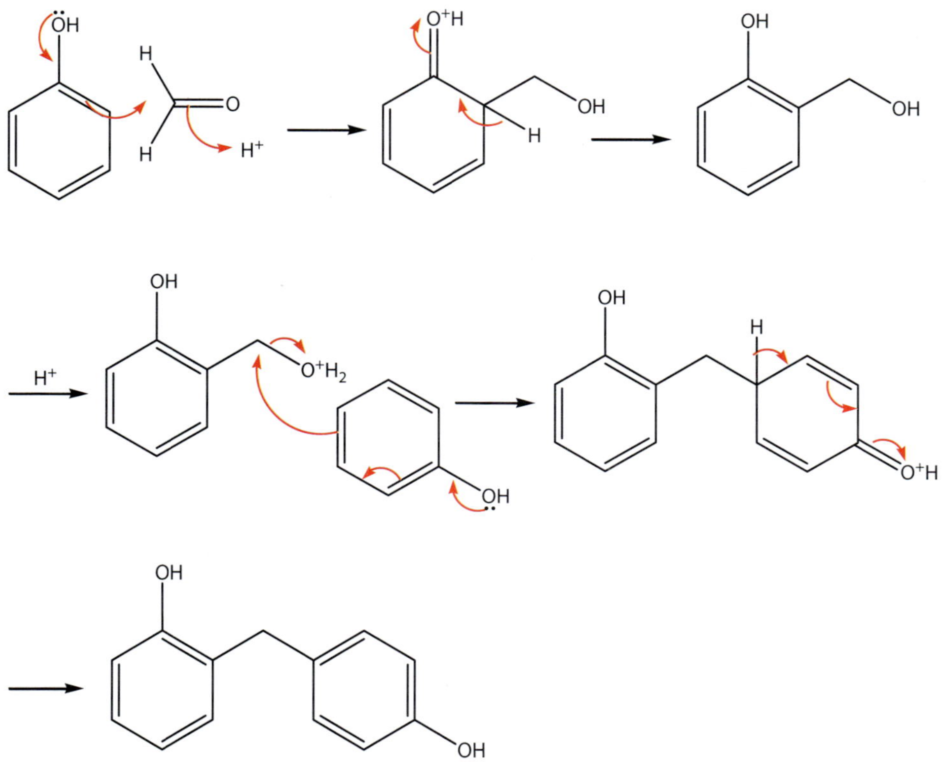

Key Points from Section 21.2

- The structure of a polymer is as important as its composition in determining its properties.
- Polymers may be linear or branched. Cross-linked polymers are prepared using a small amount of the linking molecule, and highly cross-linked materials are generally stronger and more rigid than non-cross-linked alternatives. Thermoset polymers are strong, rigid, and insoluble, but can be molded only prior to the cross-linking curing process that provides the rigidity.
- For the polymers of monosubstituted alkenes, the relative stereochemistry of the side groups is important in determining the properties of the polymer. Polymers where the substituents are all on the "same" side of the chain are called isotactic. Their diastereoisomers, with successive substituents on opposite sides of the chain, are called syndiotactic, and those with random stereochemistry are described as atactic.

21.3 PREPARATION OF ADDITION OR CHAIN-GROWTH POLYMERS

21.3.1 RADICAL POLYMERIZATION

In Chapter 11, we briefly met cationic and radical polymerization of alkenes—something that happened when there was no other reagent present to capture a carbocation or carbon radical. In radical polymerization, an initiator is needed to produce the first radicals. We have previously met dibenzoyl peroxide as an initiator; on heating, the weak oxygen–oxygen bond breaks homolytically (Figure 21.4) to give PhCOO•. This may initiate polymerization or may fragment to give a phenyl radical, which initiates further reactions. Consider the reaction of the phenyl radical with styrene (Figure 21.5). The radical adds to the double bond to give the stabilized benzylic radical. The new radical adds to another molecule of styrene, and the process continues until a polymer is formed. In each step, the more stable benzylic radical is formed, a regiospecific process. This controls the polymer structure—all the units are added head to tail, and the polymer is said to be regioregular.

Other radical initiators include *tert*-butyl peroxide, **21.8**, and azoisobutyronitrile (AIBN), **21.9**. Like all radical chain reactions, these processes have three main steps. The first is initiation, in which a radical is generated. Note the difference between an initiator and a catalyst. Catalysts are involved in every cycle of a reaction, but initiators simply set it going and are consumed by that process. The next steps are those of propagation (this may be a single step or multiple steps—the key feature is that a comparable intermediate is produced after the propagation cycle)—in this case, the successive additions of monomer units to the growing polymer chain. Finally, since reactions do not go on forever, there must be one or more steps that terminate the chain reaction. In radical polymerization of alkenes, there are two processes that commonly terminate the reaction chain. The first is radical recombination, where two radicals meet, and the odd electrons on each are used to form a bond between them. Alternatively, when two chains meet, a hydrogen atom may be transferred from one chain to the other to give one saturated and one unsaturated product, a process known as disproportionation (Figure 21.6). The relative importance of the two processes depends on the monomer. For styrene and acrylonitrile ($CH_2{=}CHCN$), recombination is the main process, whereas for vinyl acetate ($CH_2{=}CHOAc$) and methyl methacrylate

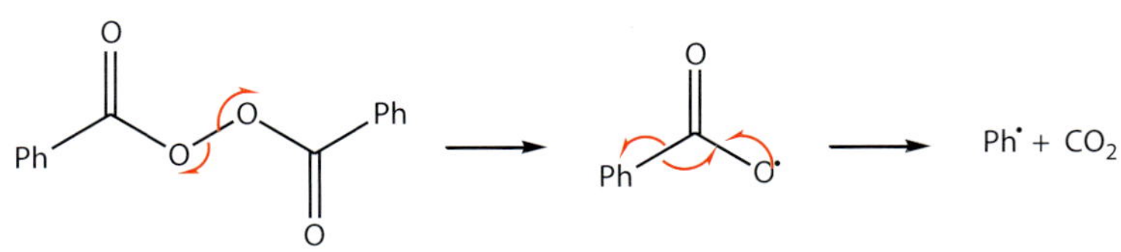

FIGURE 21.4 Thermal decomposition of dibenzoyl peroxide.

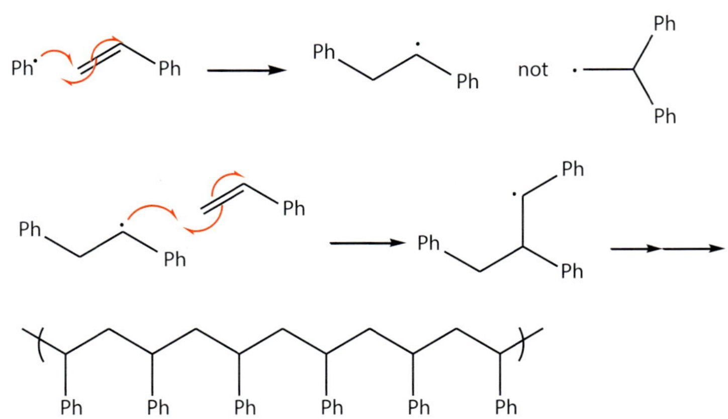

FIGURE 21.5 Radical-initiated polymerization of styrene.

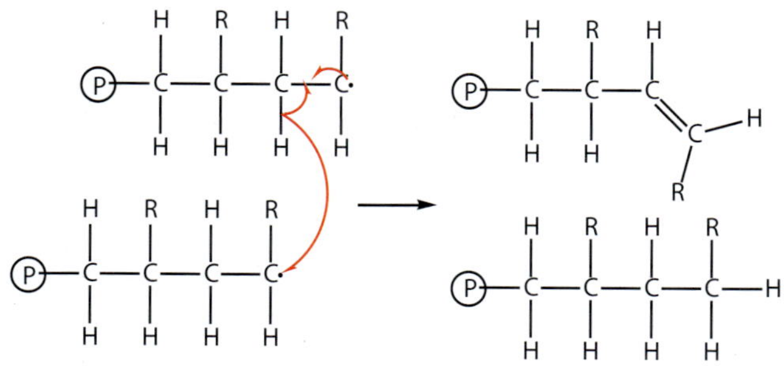

FIGURE 21.6 Termination of radical polymerization by disproportionation.

($CH_2{=}C(Me)COOMe$), disproportionation is more important. The advantage of radical polymerization is that it is easy to perform, tolerant with respect to substrate and solvent, and tolerant of many impurities. The disadvantage is that the product tends to have poor tacticity and a wide molecular weight distribution.

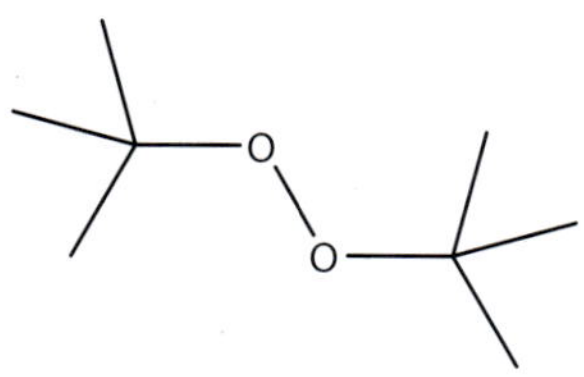

21.8

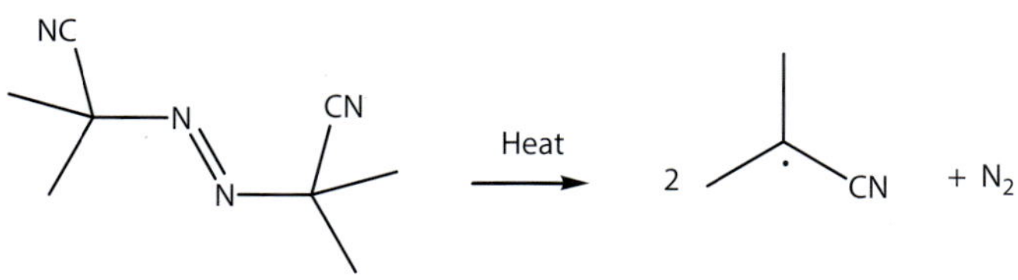

21.9

PROBLEM 21.3

Propose a mechanism for the formation of a segment of poly(vinyl chloride) containing three units of vinyl chloride and initiated by hydroxyl radical.

Solution

The hydroxyl radical will add to the alkene to give the more stable, more substituted radical:

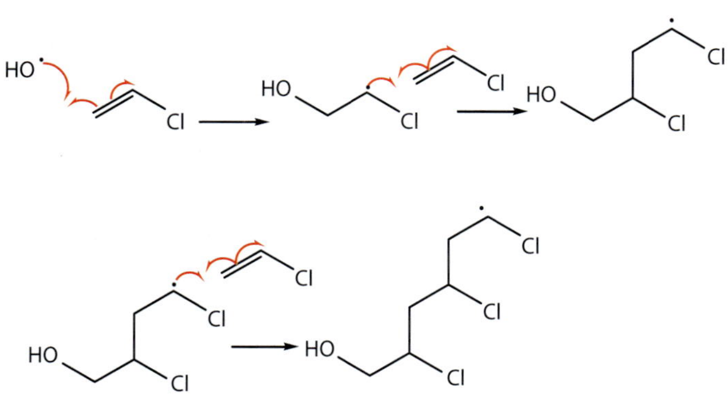

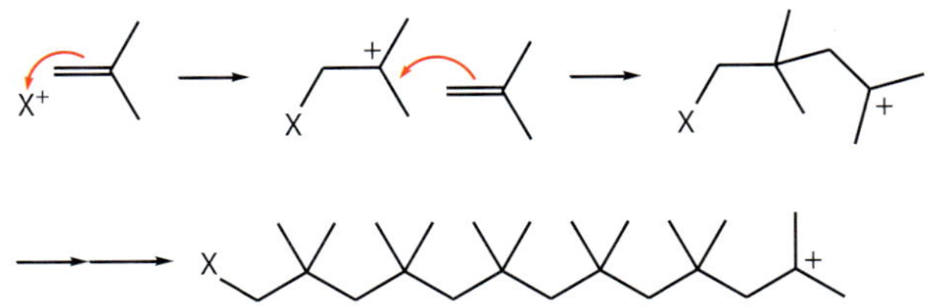

FIGURE 21.7 Cationic polymerization of 2-methylpropene (isobutene).

21.3.2 CATIONIC POLYMERIZATION

Cationic polymerization is initiated by the reaction of an electrophile, usually a proton or a carbocation, with an alkene. The first, and all the subsequent steps, produce the more stable secondary or tertiary (rather than primary) carbocations, so any polymer will once again be regioregular (Figure 21.7). This time, the growing end of the polymer chain is a cation. The growth of the polymer chain is terminated by either loss of a proton from the carbocation to give an alkene or combination of the carbocation with a nucleophile. The process works best when the alkene is substituted with a cation-stabilizing group.

PROBLEM 21.4

Which of each of the following pairs of monomers undergoes cationic polymerization more readily?

(a)

(b)

(c)

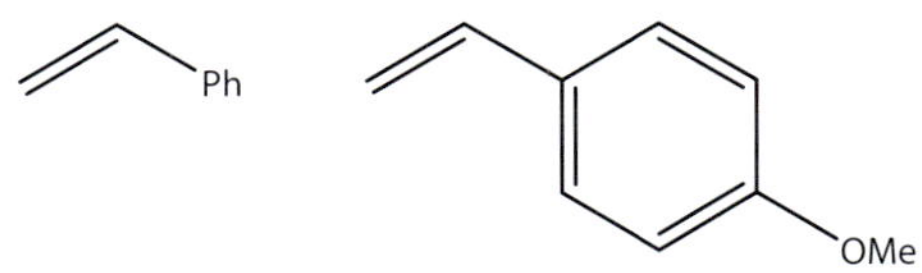

Solution

The question asked here is about the relative stability of the carbocations that are produced on addition of an electrophile to the monomer, which depends on the ability of the substituent to stabilize or destabilize the adjacent carbocation.

(a) $MeOCH{=}CH_2 > MeCH{=}CH_2$; resonance donors are more effective than inductive donors in stabilizing cations, because they are able to donate their lone pair of electrons to the positive center.
(b) $MeOCH{=}CH_2 > MeC({=}O)OCH{=}CH_2$; both cations are stabilized by the resonance donor effect of the oxygen atom. However, in the ester, the oxygen bears an electron-withdrawing carbonyl group, which reduces the electron density at oxygen.
(c) $4\text{-}MeOC_6H_4CH{=}CH_2 > PhCH{=}CH_2$; the cation is stabilized by resonance donation from the methoxy group.

21.3.3 ANIONIC POLYMERIZATION

The most common initiator for anionic polymerization is butyllithium, which adds to an alkene, usually bearing an electron-withdrawing group, to give the most stable carbanion (Figure 21.8) although special conditions are needed with this monomer to avoid branching and attack at the carbonyl group. High-quality, regioregular material is obtained, but the reaction is very sensitive to conditions. Very pure, oxygen-free monomer is required, and the process only works well with monomers bearing anion-stabilizing groups. Chain growth is terminated with water or CO_2.

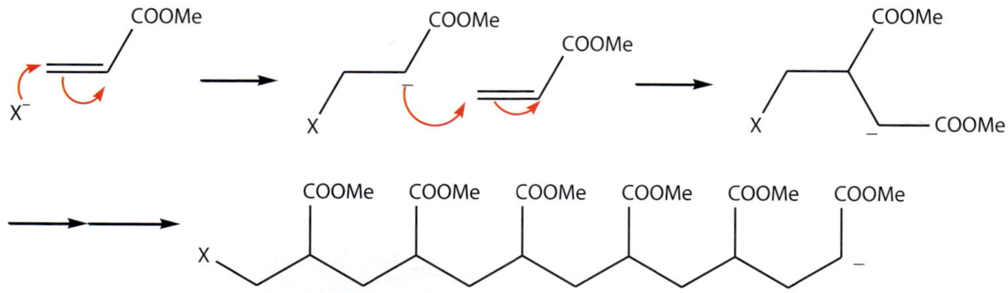

FIGURE 21.8 Anionic polymerization of methyl acrylate.

PROBLEM 21.5

Write a mechanism for the anionic polymerization of 2-vinylpyridine initiated by butyllithium. Why is a regioregular polymer formed?

Solution

In each addition step, the regiochemistry is determined by the ability of the pyridine ring to stabilize the anion formed:

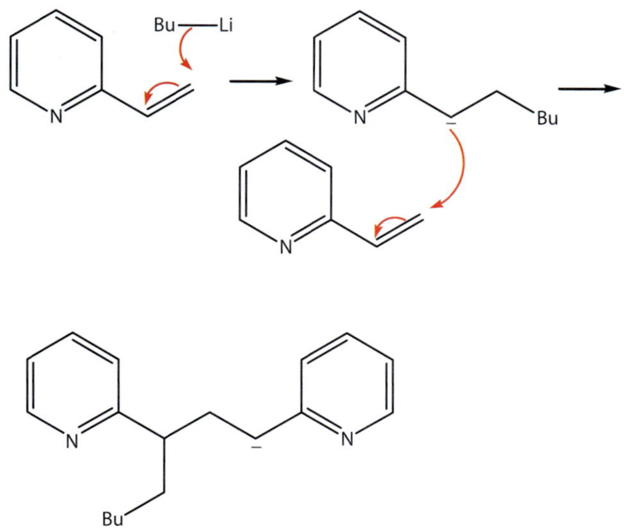

21.3.4 METAL-MEDIATED POLYMERIZATIONS

The mechanistic details of the Ziegler–Natta, metallocene, and supported metal complex catalyzed polymerizations are beyond the scope of this book, but the outline of the Ziegler–Natta process is shown in Figure 21.9. Ziegler and Natta received the Nobel Prize for this work in 1963. A metal catalyst (early versions were prepared from $TiCl_3/Et_3Al$, but many variants and additives have been developed to optimize specific processes) coordinates a molecule of monomer, which is then *inserted* into the bond between the metal center and the growing polymer chain. M in this case is used to represent the metal with all its ligands; the steric interaction between these ligands and the substituent X means that a regioregular and stereoregular polymer is normally produced.

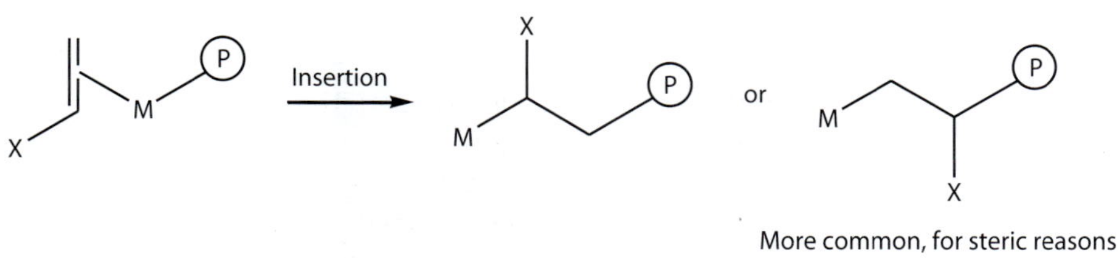

FIGURE 21.9 Ziegler–Natta polymerization.

21.3.5 SCOPE OF POLYMERIZATION OF SUBSTITUTED ALKENES

Alkyl- or aryl-substituted ethenes are the monomers for many of the most widely produced polymers in the world. Tables 21.2 and 21.3 give details of some of the most common ones, how they are made, and what they are used for.

Polystyrene food containers

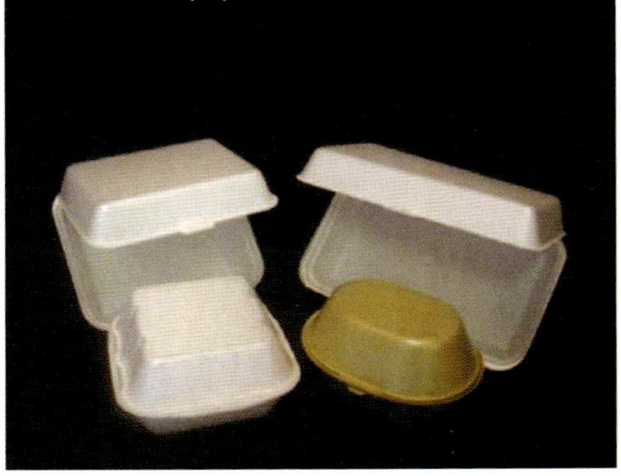

Lucite artifacts. (From http://www.newyorksocialdiary.com/i/shoppingdiary/passionateshopper/rocks/FRAGMENTSyellow _lucite.jpg.)

Ghost chair. (Photo courtesy of Lakeland Furniture, Oldham, U.K.)

Superglue fingerprint. (From http://www.forensicsrus.com/images/SupergluePrint.jpg.)

TABLE 21.2 Polymers of Alkenes without Polar Groups

Monomer	Polymer	Main Method of Synthesis	Uses
Ethene	UHMWPE, ultrahigh molecular weight polyethylene	Chromium complex catalyzed	Artificial joints, personal armor, ultra-stiff ropes, cables
Ethene	HDPE, high-density polyethylene	Ziegler–Natta, metallocene catalysis	Piping, containers, plastic bags
Ethene	MDPE, medium-density polyethylene	Ziegler–Natta, metallocene catalysis	Gas pipes, refuse sacks, plastic bags
Ethene	LDPE, low-density polyethylene	Radical polymerization	Containers, plastic bags, tubing
Ethene	VLDPE, very low-density polyethylene	Metallocene catalysts	Hoses, tubing, food packaging
Propene	Isotactic polypropylene	Ziegler–Natta	Containers, fibers, especially in carpet, car parts
Propene	Syndiotactic polypropylene	Metallocene catalysts	Moldings, medical equipment
Propene	Atactic polypropene	Nonstereoregulating metal catalyst	Road surfacing, road paints
1-Butene	Isotactic polybutylene	Ziegler–Natta	Piping, packaging
4-Methyl-1-pentene	Isotactic polymethylpentene	Ziegler–Natta	Speaker cones, lightweight artifacts, cookware
2-Methylpropene	Polyisobutylene	Cationic polymerization	Adhesives, tire inner tubes
Styrene	Atactic polystyrene	Radical polymerization	Disposable cutlery, expanded polystyrene in cups, packaging. May be functionalized for catalysis of peptide/nucleotide synthesis
Styrene	Syndiotactic polystyrene	Ziegler–Natta	Not yet produced commercially

PVC is commonly used in double-glazed window units

Nonstick pan with Teflon coating

TABLE 21.3 Polymers of Alkenes with Polar Substituents

Monomer	Polymer	Main Method of Synthesis	Uses	Comments
Chloroethene (vinyl chloride)	PVC, polyvinyl chloride	Radical polymerization	Construction. With added plasticizers may be used in clothes, upholstery, fake leather, sewerage pipes.	Third most widely produced plastic.
1,1-Dichloroethene	Polyvinylidene chloride	Radical polymerization	Seat coverings.	Formerly used as a food wrap (Saran); but LDPE has replaced this because of the bad press given to chlorinated materials.
Tetrafluoroethene	Teflon, PTFE, polytetrafluoroethylene	Radical polymerization	Nonstick cookware, containers for corrosive materials, cable insulation, polymer membranes for buildings.	Very high melting point (327 °C). Very low coefficient of friction. Chemically inert because there are no C–H bonds, and the C–F bond is very strong (462 kJ mol^{-1}).
Acrylonitrile, $CH_2{=}CHCN$	PAN, polyacrylonitrile	Radical polymerization	PAN homopolymer is a precursor of high-quality carbon fiber. Also used in yacht sails. Copolymers with up to 15 % of other acrylates are used in fabrics or carpets—described as Orlon, acrylic.	PAN is difficult both to make and to process (the polymer is insoluble in the monomer, and the polymer decomposes before it melts or softens).
Acrylamide, $CH_2{=}CHC({=}O)NH_2$	Polyacrylamide	Radical polymerization	Assists in solid flocculation in liquids; used in the treatment of wastewater. Soil conditioner. Can absorb up to 400 times its own weight of water, slow release watering system for plants.	Cross-linked versions are prepared using $CH_2{=}CHC({=}O)NHCH_2NHC({=}O)CH{=}CH_2$.
Vinyl acetate, $CH_2{=}CHOAc$	PVA, polyvinyl acetate	Radical polymerization	Adhesives, dispersing agents, water-based paints.	Precursor of polyvinyl alcohol.
Methyl methacrylate, $CH_2{=}C(Me)COOMe$	PMMA, polymethyl methacrylate, Lucite, Plexiglass	Radical polymerization	Glass substitute. Car lights, motorcycle helmet visors, aircraft windows, intraocular lenses, hard contact lenses, dentures, acrylic paint, furniture, sculpture, jewelry.	Good environmental stability, but scratches easily. High refractive index, good optical clarity. Soluble in organic solvents.
Methyl cyanoacrylate 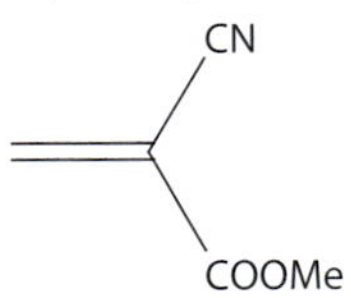	Polymethyl cyanoacrylate, superglue	Anionic, usually with HO^-	Adhesive. Used in wound closures. Fingerprint visualization.	

Orlon fabrics

PROBLEM 21.6

Ethene, vinyl chloride, and acrylonitrile do not undergo cationic polymerization readily. However, 2-methylpropene undergoes cationic polymerization readily. Provide an explanation for this behavior.

PROBLEM 21.7

Methyl cyanoacrylate is readily polymerized, even by relatively weak bases. Draw a mechanism for the process, and explain why this is so easy:

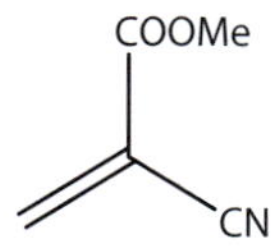

Solutions

21.6 These substrates are poor for cationic polymerization because the carbocations that would need to be produced are either primary ($CH_3CH_2^+$, $ClCH_2CH_2^+$, $NCCH_2CH_2^+$) or destabilized by an electron-withdrawing group at the cationic center (CH_3CH^+Cl or CH_3CH^+CN). The addition of a proton (or other electrophile) to 2-methylpropene gives the very stable, tertiary butyl cation, Me_3C^+.

21.7 This substrate is commonly polymerized using hydroxyl ion as the base; the carbanion produced by addition of a nucleophile to the substrate is stabilized by two electron-withdrawing groups, the ester and the cyanide:

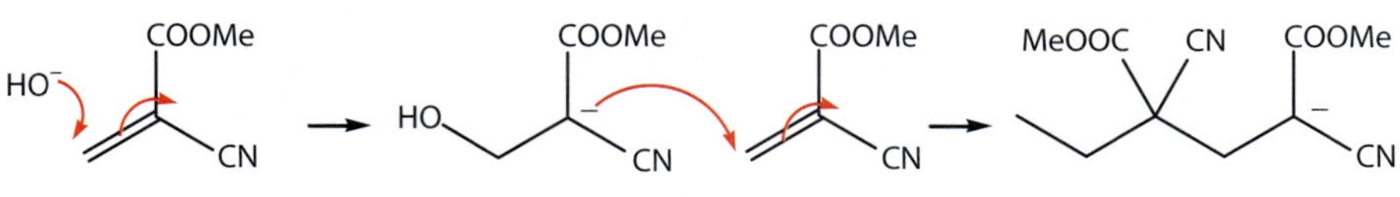

Focus on Polymethyl Methacrylate

The polymerization reaction for methyl methacrylate was discovered in the nineteenth century, but processes to produce practical materials were not developed until about 1930. Polymethyl methacrylate (PMMA) was quickly pressed into service during the Second World War as windshields, gun turrets, and canopies in aircraft, a purpose for which it is still used today. The use in wartime aircraft led to an important observation—PMMA is biocompatible. Examination of injured RAF Spitfire pilots revealed that shards of PMMA in their eyes were not rejected and rarely became infected. This led to the development of PMMA intraocular lenses for use on patients after cataract removal, the first one being implanted in 1949. Rigid PMMA lenses were the norm until relatively recently, but are now being replaced by more flexible materials. PMMA replaced glass in contact lenses in the 1940s but has now been superseded by flexible, oxygen-permeable materials. PMMA is also used in orthopedic surgery to fix prosthetics and as a bone graft template; its elasticity is similar to that of the bone. Recently, it has been used (as Artefill® or Artecoll®) as a permanent "wrinkle filler" in cosmetic surgery, where microscopic PMMA spheres in a collagen filler are injected; the body slowly absorbs the collagen, but PMMA is not absorbed or degraded. It has also been used to make dentures.

PMMA has been marketed as Perspex®, Lucite®, and Plexiglass®. Many of its uses are related to its excellent physical properties. Although it is not as strong as polycarbonate (*vide infra*), it is more scratch resistant and less expensive and does not contain bisphenol A, which is regarded as of concern as a possible endocrine disruptor. It is both less dense (density 1.17–1.2 *vs.* 2.3–3.2 g cm^{-3} for glass) and more transparent than glass; even a 13″ thickness is quite clear, and it has been used in aquaria where a thick wall is needed to withstand high water pressure. This also leads to its use in submersibles and bathyscaphes, as well as aircraft windows and car lights. It found early applications in a range of protective gear, including motorcycle windshields, safety glasses, and protective barriers in ice hockey, but has largely been replaced by polycarbonate. It is used as the core of plastic optical fiber for home and car networks and for illumination; it is not suitable for long-distance data transmission as there is a substantial transmission loss.

Kelp forest in aquarium in Monterey, California

PMMA is generally produced by radical polymerization. Although the polymer is amorphous rather than crystalline and was initially considered to be atactic, it is now known to be significantly syndiotactic. The material is very much like glass—it shows a "glass transition temperature," T_g of 105 °C. This is the temperature at which it becomes sufficiently rubbery to be readily deformed. It is a thermoplastic—it can be softened and melted, molded into a particular shape, and will retain the shape on cooling. This is a reversible process—the changes are physical rather than chemical. As with many established polymers, pure PMMA is becoming a rarity. It has been found that small amounts of various additives improve specific properties for particular applications. For example, acrylate comonomers increase thermal stability, butyl acrylate increases impact strength, and methacrylic acid increases T_g for lighting applications.

21.3.6 POLYMERIZATION OF DIENES

Addition of butadiene to ethene polymerizations gives cross-linked material, but dienes are themselves important substrates for polymerization reactions. Natural rubber is an all-*cis* polymer of isoprene (Figure 21.10), which we encountered in Chapter 11, as an important precursor of the terpenes. Synthetic rubber made by radical polymerization is a mixture of *cis*- and *trans*-polyisoprene, (**21.10**). The material produced by metal-catalyzed polymerization is, however, all-*cis* and essentially identical to natural rubber.

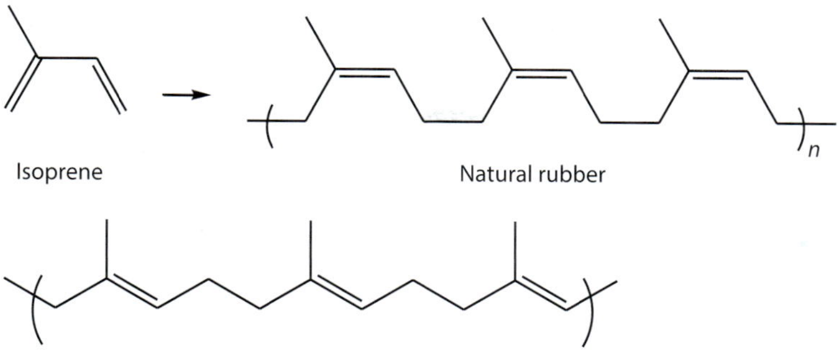

21.10, *trans*-Polyisoprene

FIGURE 21.10 Natural rubber production.

Rubber trees (*Hevea brasiliensis*) are native to South America, and rubber was first used by ancient Mesoamericans to make balls for sport before 1600 BCE. Balls for the modern version of the sport, ulama, are still made by traditional techniques. In the nineteenth century, rubber growing began in Southeast Asia and in Africa, and these are now the most important commercial sources of rubber. Research into the production of synthetic rubber was boosted during the Second World War, when much of Southeast Asia was inaccessible to the Allies. Latex is produced by a number of other plants—probably the most important is gutta-percha, which contains mainly *trans*-polyisoprene, derived from *Palaquium gutta*, a genus of trees native to Southeast Asia. This was widely used to insulate electrical cables, but has now largely been replaced by synthetic materials. A closely related man-made material is neoprene, the polymer of chloroprene (Figure 21.11), which is widely used in car fan belts and electrical insulation. A foamed version of neoprene is used for wetsuits and diving equipment.

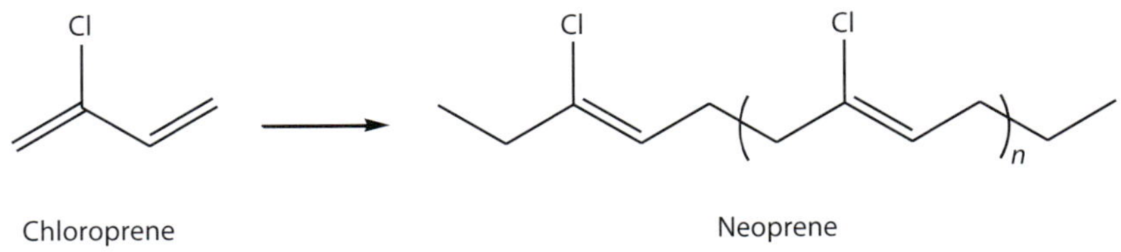

FIGURE 21.11 Polymerization of chloroprene.

Radical, cationic, anionic, and metal catalyzed polymerizations of butadiene are known—and there are three possible products of the polymerization (Figure 21.12). Initial addition of a radical to butadiene gives exclusively the allyl radical, which has two resonance forms. The more stable resonance form adds further butadiene to give 1,2-polybutadiene. The less inherently stable, but also less sterically hindered, form adds further butadiene to give *cis*- or *trans*-1,4-polybutadiene.

Selectivity in producing the various types has been largely achieved. High *cis*-1,4-polybutadiene is formed using $TiCl_3/R_3Al$, or various nickel or neodymium catalysts. High-*trans*-1,4-polybutadiene can be made with VCl_3/AlR_3 or using anionic polymerization with RLi and various additives; it is used to make the outer layer of golf balls. Syndiotactic 1,2-polybutadiene can be prepared using $CrCl_2$·1,2-bis(dimethylphosphino)ethane (DMPE)/methylaluminoxane; this is a soft, pliable material with a low melting point.

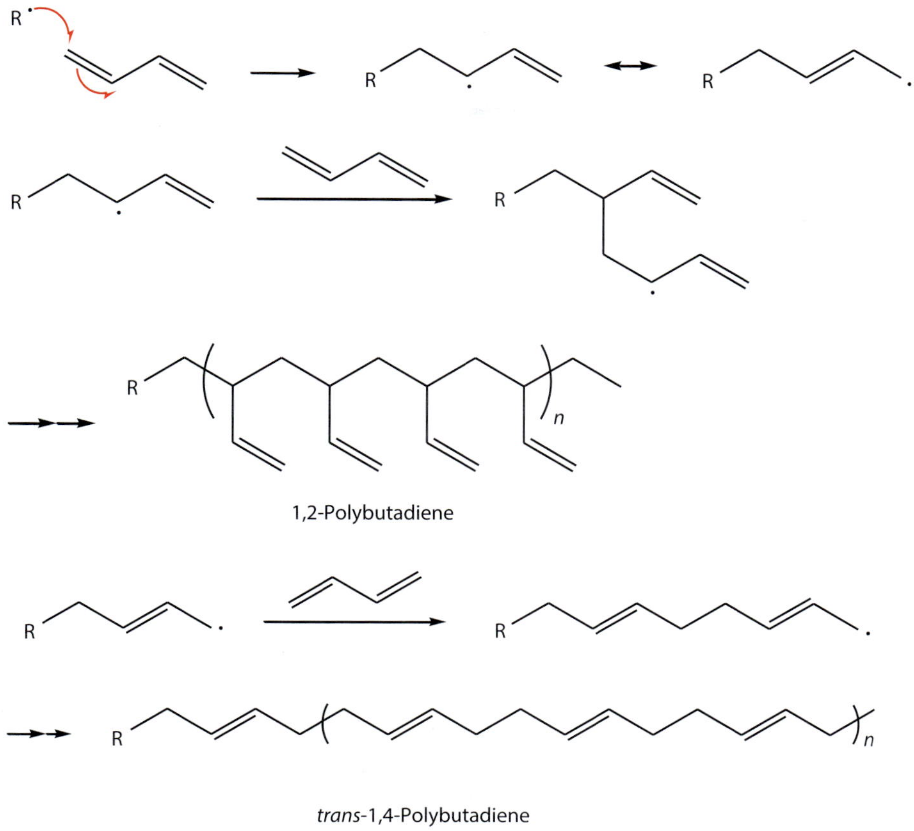

FIGURE 21.12 Radical-initiated polymerization of butadiene.

PROBLEM 21.8

Iso-butene (2-methylpropene) and isoprene can be copolymerized to give "butyl rubber." Only about 2 % of isoprene is needed to give the desired properties (to give sites for vulcanization), and the polymer is prepared by a cationic process.

(a) Draw the structure of pure polyisobutene, and show how the cationic polymerization leads to regioregularity in the chain. You may use a proton to initiate the reaction.
(b) What reactive group is present at the growing end of the polyisobutene chain?
(c) How would isoprene add to this reactive group? Draw both resonance forms of the cationic product formed.
(d) In the final polymer, the isoprene has been incorporated in a 1,4-manner. From the intermediate generated in (c), show how the next isobutylene monomer would react to give this result. How would you rationalize the 1,4-incorporation of isoprene?

Solution

(a) In the initiation step, the most stable tertiary carbocation is formed. Each successive addition step also gives the more stable carbocation:

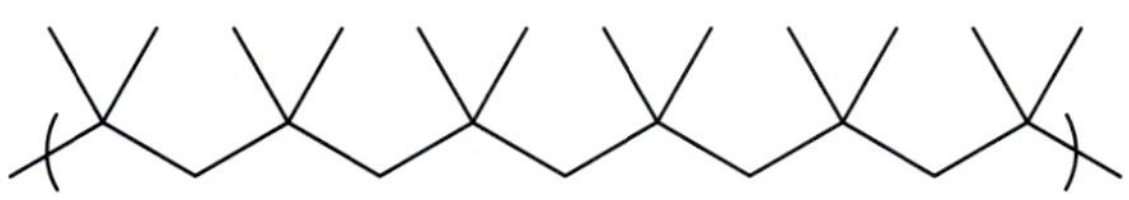

Polyisobutene

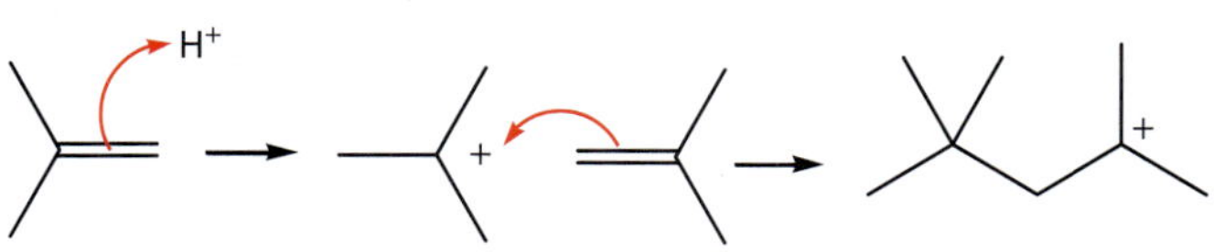

(b)

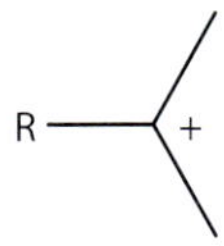

(c) There are two options depending on which "end" of the isoprene reacts:

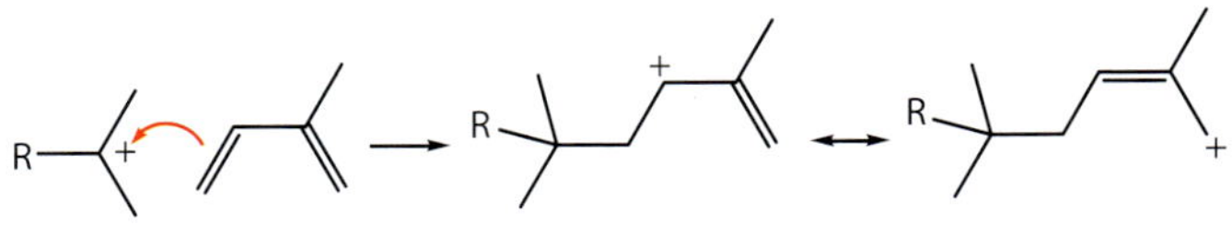

or

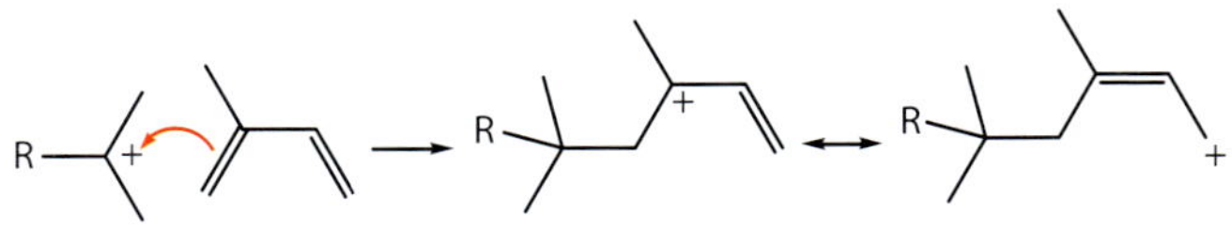

The second option is the better of the two; in this case, the two resonance forms are primary + tertiary rather than primary + secondary carbocations:

(d)

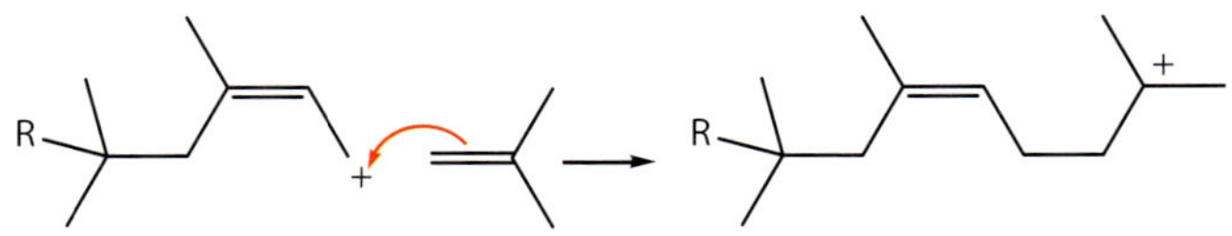

The reason that reaction takes place through the less-favorable resonance form is that the alternative would give two highly substituted carbon atoms next to each other, which would be sterically unfavorable.

21.3.7 RING-OPENING POLYMERIZATION

Unsaturated hydrocarbons are not the only species that can form addition chain-growth polymers. If ethene oxide is treated with a base, the three-membered ring is opened, and the alkoxide produced attacks more epoxide (Figure 21.13). The reaction is driven by the release of strain in the three-membered ring. The product, called polyethylene glycol (PEG) (or polyethylene oxide, PEO), has high water solubility and can be used to transport ions across membranes. The choice of base is tricky—butyllithium in DMSO gives high-quality material but requires that the monomer be very pure and dry, so is rarely used. Using hydroxide or alkoxide is easier, but produces materials with much wider molecular weight distributions.

A number of other oxygen-, nitrogen-, and sulfur-containing heterocycles undergo cationic polymerization (Figure 21.14) (ethene oxide may also be polymerized cationically). It is essential that the initiating cation has a very nonnucleophilic counterion, such as perchlorate or trifluoroacetate, so that these do not compete with substrate for the cationic center on the growing polymer chain. Lewis acids also initiate the process, probably by scavenging adventitious water to give a proton. Formaldehyde (methanal) is also polymerized by a base to give paraformaldehyde (Figure 21.15). This is a useful process in that it allows formaldehyde to be more conveniently and safely stored and transported, but the polymerization reaction is only slightly exothermic and is

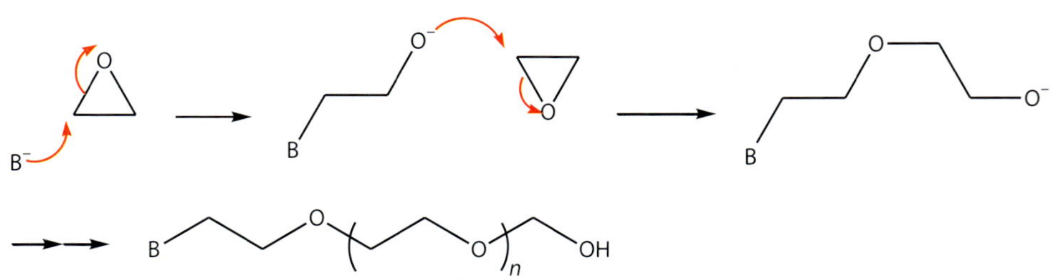

FIGURE 21.13 Polymerization of ethene oxide.

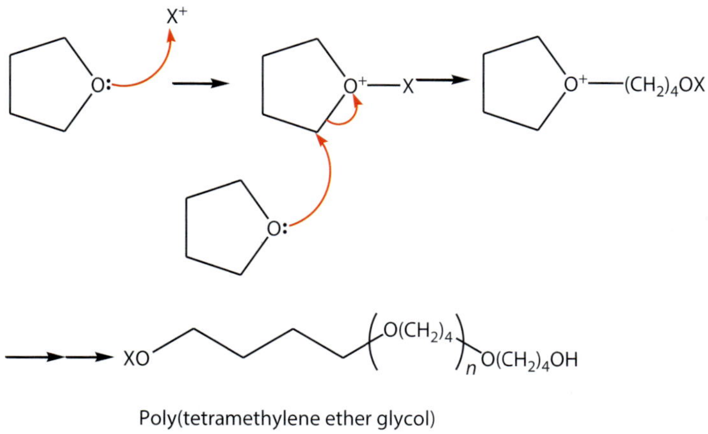

FIGURE 21.14 Cationic polymerization of THF.

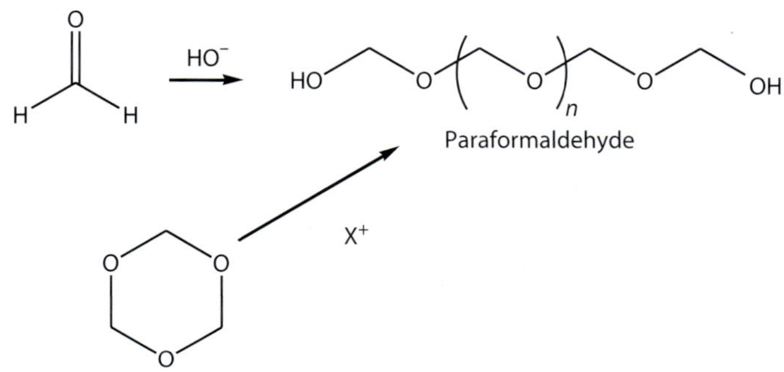

21.11

FIGURE 21.15 Polymerization of formaldehyde (methanal).

reversed in aqueous solution. Trioxan, **21.11**, the cyclic trimer of formaldehyde, gives the same material using a cationic initiator.

PROBLEM 21.9

Predict the product(s) of each of the following reactions:

(a)

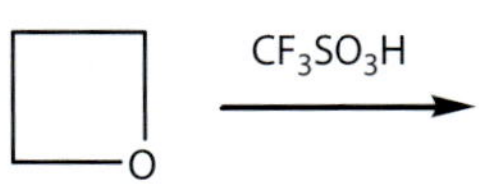

(b)

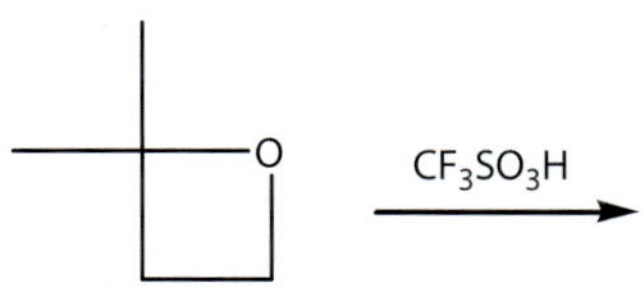

Solutions

(a) This is a ring-opening polymerization, yielding a polyether. The mechanism is that same as that shown for THF in Figure 21.27:

(b) Although similar to the problem earlier, once we have protonated the oxygen, it is reasonable for the four-membered ring to open to give, this time, a rather stable carbocation:

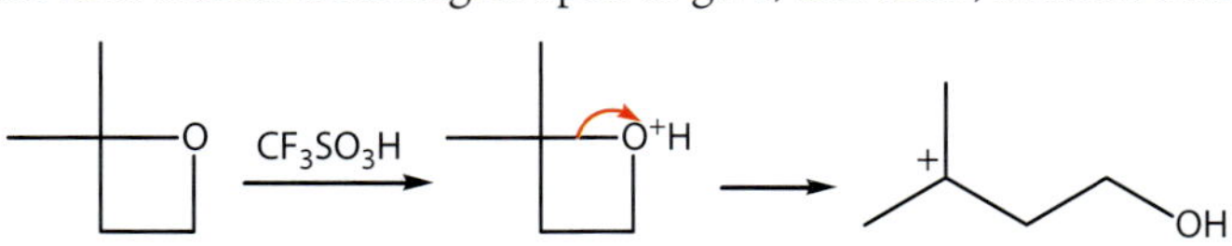

This polymerization should be more favorable, giving

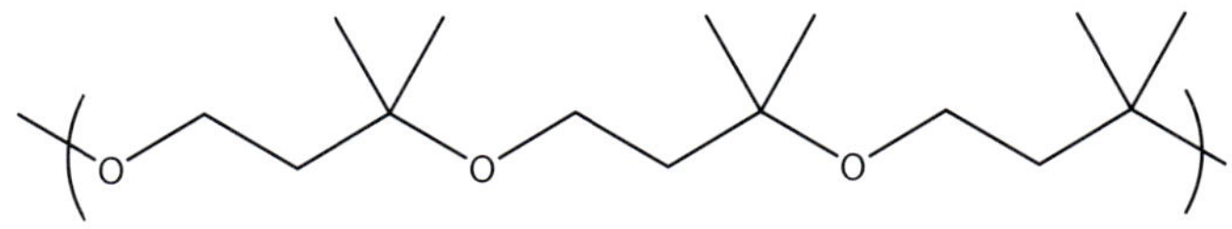

There has been considerable recent interest in polymers formally derived from ω-hydroxy acids. Polycaprolactone is formed from caprolactone, **21.12**, under either anionic or cationic catalysis (Figure 21.16); it is a relatively soft material and has been used to improve the processing characteristics of other materials. It is also slowly biodegradable and is of interest in medical applications. Poly(lactic acid) (which is prepared from the lactic acid dimer, lactide, **21.13**, in turn prepared from starch or sugar) is of particular interest because of its biocompatibility and ability to biodegrade (over a few years). It has found extensive medical uses in sutures, stents, and dialysis equipment and in various disposable clothing, bags, *etc.*, and is being extensively promoted as a biodegradable packaging material for foods. This polymer is particularly interesting as it does not originate in a petroleum feedstock.

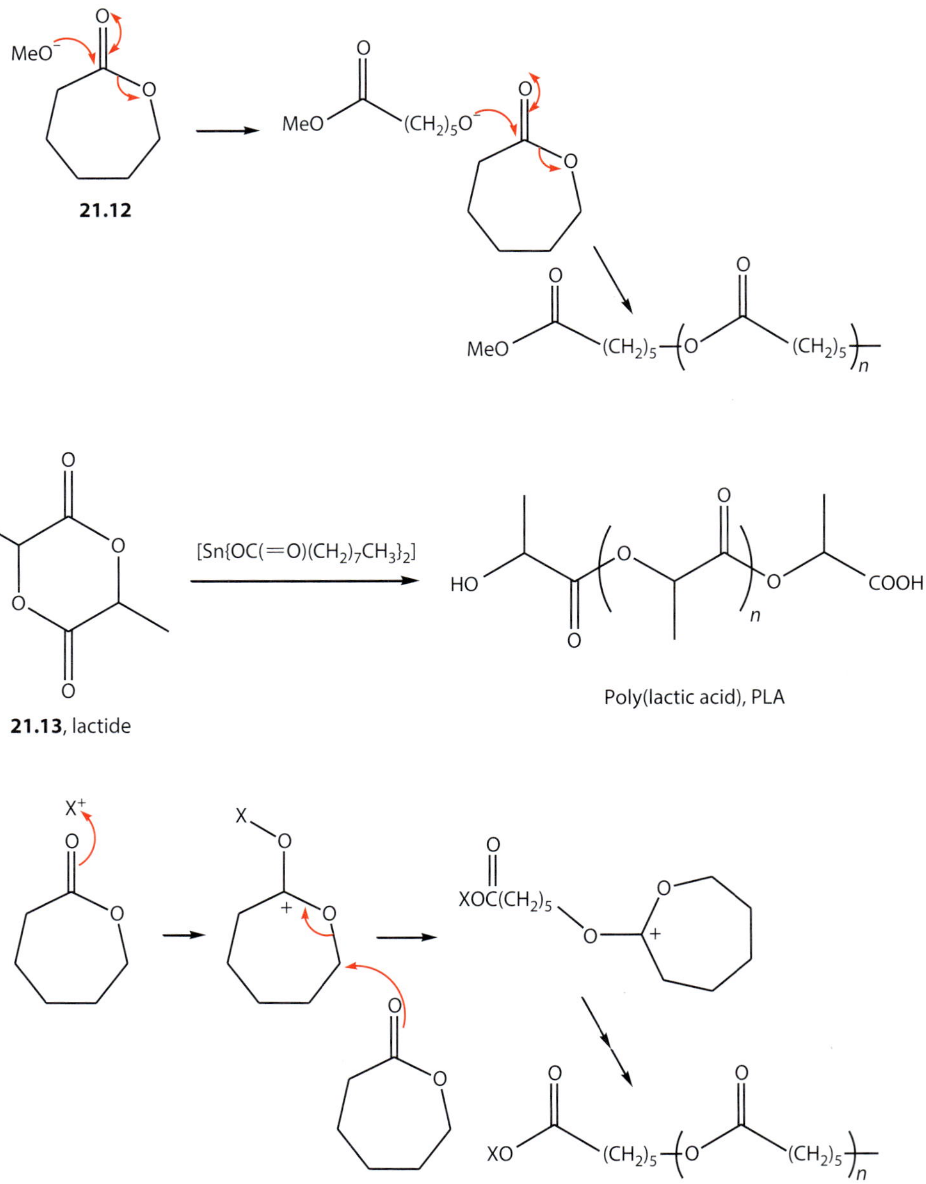

FIGURE 21.16 Polymerization of caprolactone.

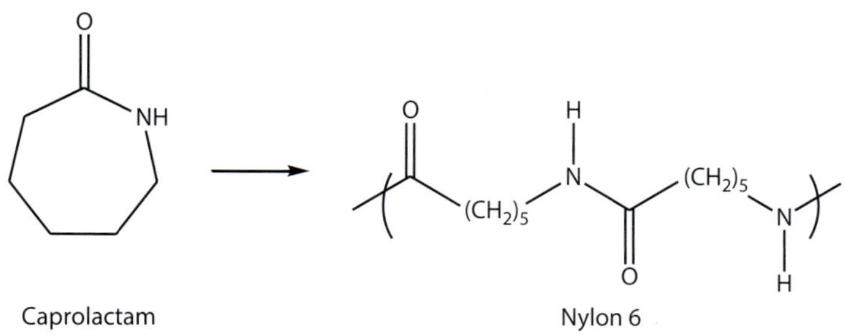

FIGURE 21.17 Preparation of nylon 6.

Caprolactam (Figure 21.17) may be polymerized to nylon-6 under acidic or basic conditions, by mechanisms analogous to those for caprolactone. The name nylon-6 derives from the fact that the material is a polyamide produced from a single monomer, containing six carbon atoms. Its properties are similar to those of other nylons, produced by step-growth polymerization, and will be considered in that section.

Key Points from Section 21.3

- Addition polymerization may be initiated by anions, cations, or radicals, or effected by metal complexes.
- Radical polymerization requires an initiator such as AIBN and proceeds by a radical chain process to give a largely regioregular, but generally atactic or somewhat syndiotactic polymer. Termination involves recombination or disproportionation. The process is tolerant of impurities, but molecular weight distributions are broad.
- Cationic polymerization is initiated by protons, other cations, or Lewis acids. It proceeds by the most stable carbocation to give a regioregular polymer, with variable tacticity. Termination involves reaction with an anion or proton transfer between chains.
- Anionic polymerization is initiated by BuLi or another anion and proceeds *via* the most stable carbanion to give a regioregular polymer. The process is terminated by exposure to oxygen, water, or CO_2. The process requires very pure, dry monomers, and solvents.
- Ziegler–Natta polymerization uses metal complexes as catalysts.
- Heterocyclic compounds, particularly epoxides, can form addition polymers in the presence of acid or base.
- Dienes may be polymerized by radical, cationic, anionic, or metal catalyzed processes. Catalyst choice determines whether the *cis*- or *trans*-1,4-polymers or the 1,2-polymer is formed, though many reactions give mixtures.

21.4 PREPARATION OF STEP-REACTION POLYMERS

Step-growth or step-reaction polymers are formed either by reactions between two functional groups on the same monomer or by reaction of two species, both of which have two or more functional groups that can react together. The easiest way for an organic chemist to distinguish these from the addition polymers that we have discussed is that in these reactions, if they are stopped before completion, ordinary organic molecules (dimers, oligomers) will be present. Unlike the addition polymers, the intermediates are not reactive non-isolable species, and the chemistry involved is that of ordinary organic reactions. In forming such polymers, it is critical that the reactions go to completion. The mathematical treatment of the relationship between monomer consumption and molecular weight of the product is beyond the scope of this text,* but conversion must reach 99 % before molecular weights become significant.

* If you want to pursue this further, look up the Carothers equation in a polymer textbook.

21.4.1 POLYESTERS AND POLYCARBONATES

Polyesters may be formed by the reaction of a diol with a dicarboxylic acid or a dicarboxylate ester. In some older organic textbooks, you will find this referred to as a condensation polymer (because it is formed by a condensation reaction), but the polymer community frowns on this nomenclature. One such polymer is polyethylene terephthalate (PET) (Figure 21.18). Much of this is used in fibers for clothing and home furnishings. As a thin film, Mylar® is used for survival blankets and balloons, as well as thermal and electrical insulation, food and drink packaging, cine films, and musical and electronic equipment. The reaction involved is a transesterification (Figure 21.19).

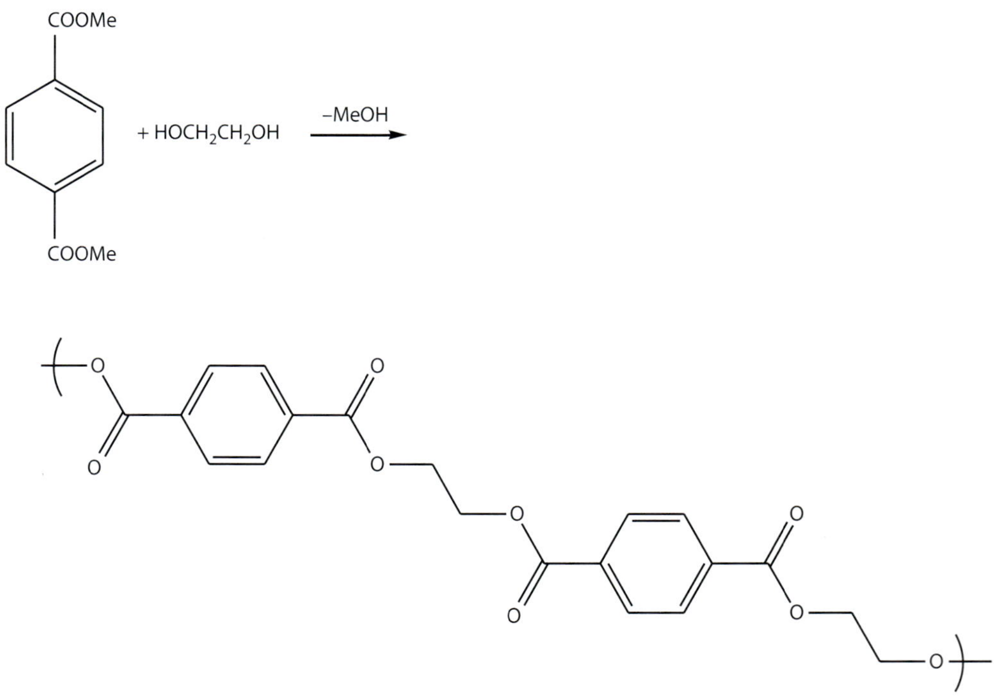

Polyethylene terephthalate, PET

FIGURE 21.18 Formation of PET.

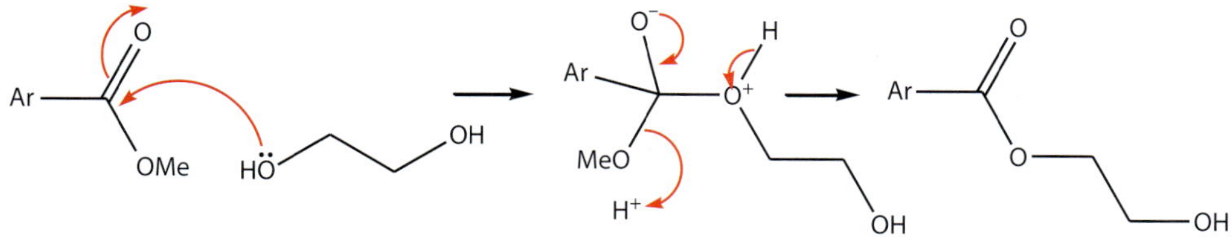

FIGURE 21.19 Transesterification in Mylar production.

Another group of "condensation" polymers that can be made by a transesterification mechanism are the polycarbonates (Figure 21.20). Bisphenol A is prepared from acetone and phenol in the presence of an acid catalyst and reacted with diphenylcarbonate. The polycarbonate produced is a hard, tough material, marketed as Lexan® or Makroclear® and noted for its outstanding resistance to impact. It is used in many "unbreakable" artifacts, CDs and DVDs, automobile parts, and in many safety applications, including riot shields, safety glasses, and bullet-resistant glass. Its use in food containers has become controversial because of the possibility of the leaching of bisphenol A, which has been cited as a possible endocrine disruptor.

"Unbreakable" artifacts made from polycarbonate

Bullet-resistant glass

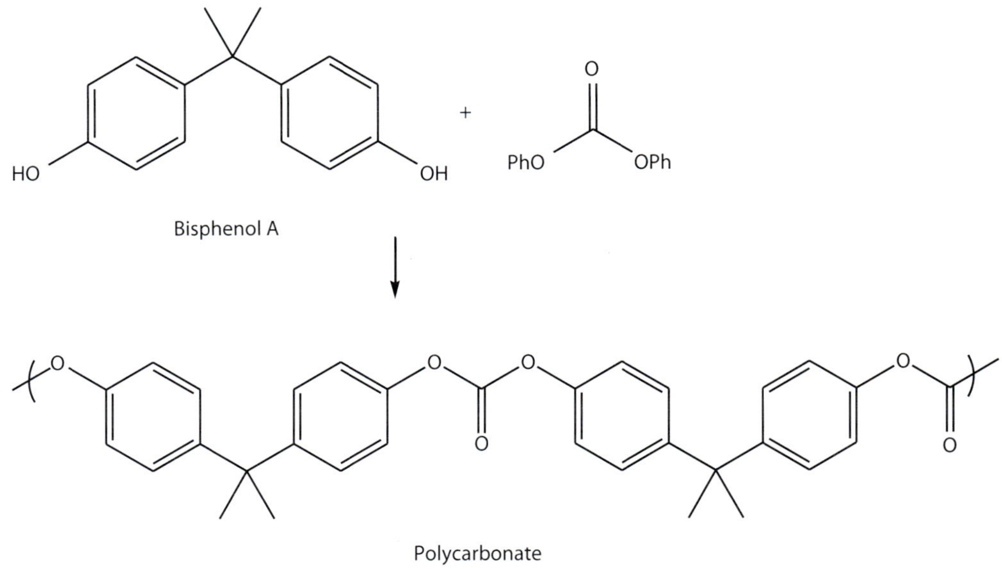

FIGURE 21.20 Synthesis of polycarbonate.

PROBLEM 21.10

Polybutylene terephthalate, shown in the following, is a hydrophobic plastic used in automotive ignition systems. What monomers would you use to make it, and what conditions would you suggest?

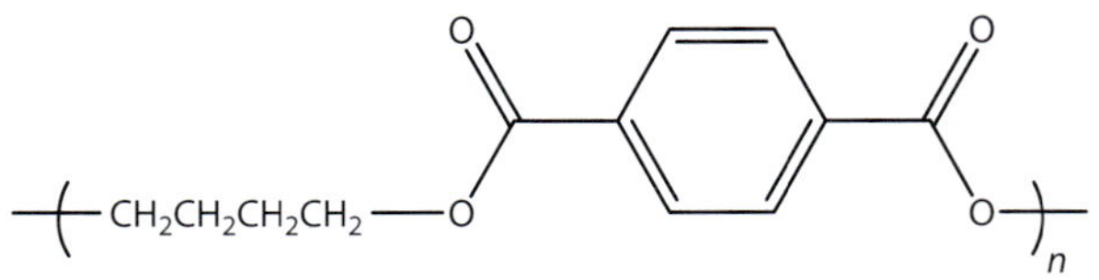

PROBLEM 21.11

Polycarbonates can be formed by nucleophilic aromatic substitution involving aromatic difluorides and carbonate ion. Give a structure for the polymer produced from the monomers shown in the following, and propose a mechanism for its formation:

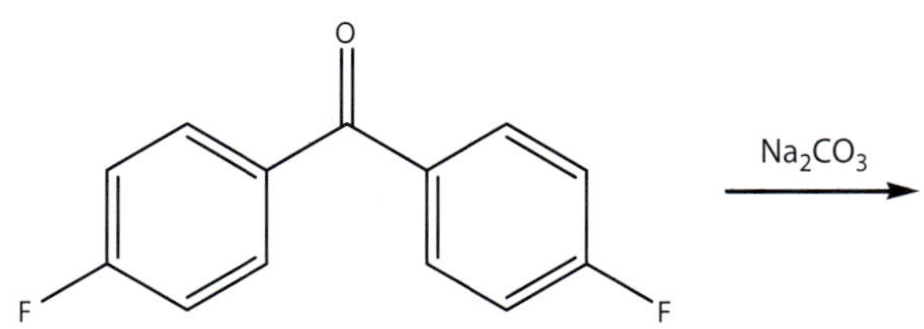

Solutions

21.10 This could be made from terephthalic acid (benzene-1,4-dicarboxylic acid) and butane 1,4-diol. Although the reaction between the diol and the diacid chloride would be more rapid, it would not be commercially viable.

21.11 Because of the electron-accepting nature of the carbonyl group, the fluorines attached to the aromatic rings can be substituted by the carbonate dianion (review Section 13.4 if necessary):

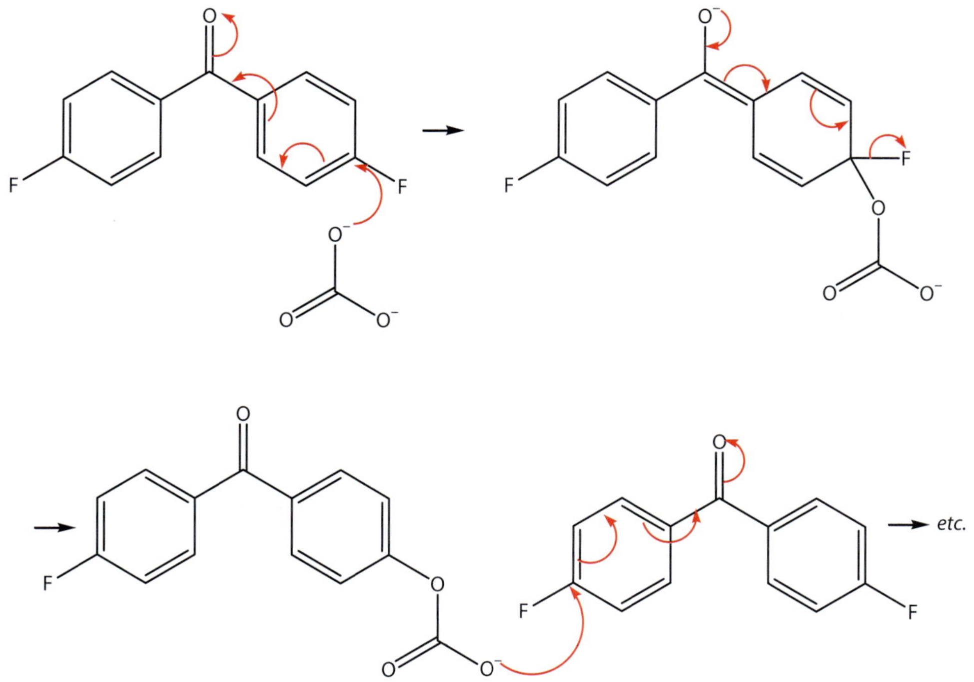

21.4.2 POLYAMIDES

In the previous section, we saw that a polyamide, nylon-6, could be produced by ring-opening polymerization of caprolactam, essentially the equivalent of an ω-amino acid. Other polyamides are prepared from a diacid and a diamine (Figure 21.21). For two-component nylons, the number of carbons in the diamine is followed by the number of carbons in the diacid. For one-component systems, the numbering is simply the number of carbon atoms in the monomer. Careful control of the polymerization process is required as it is difficult to turn the material

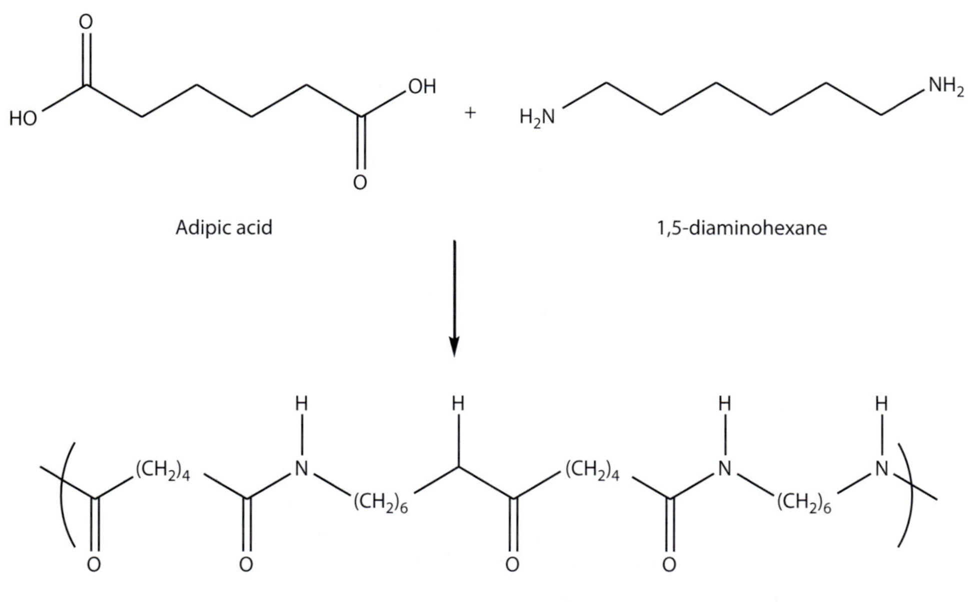

FIGURE 21.21 Preparation of nylon 6,6.

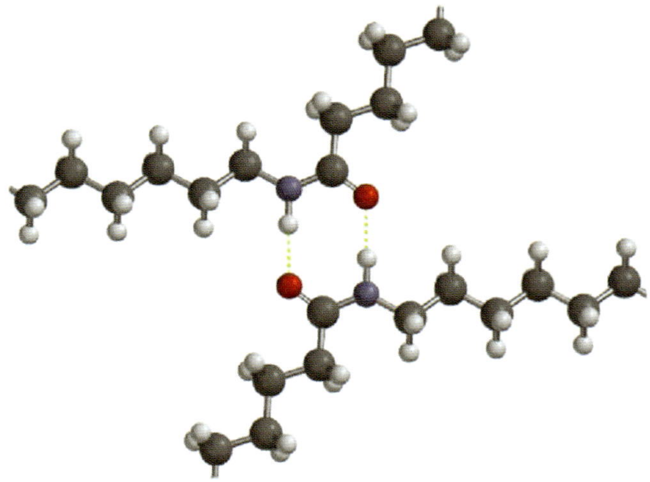

FIGURE 21.22 Hydrogen bonding in nylon 6,6. (Thanks to Prof. Ken Seddon, Queens' University, Belfast for this image.)

into yarn if the molecular weight exceeds 20,000. Although nylon 6 and nylon 6,6* are the most commonly used materials, many others are known, and some have excellent properties—but are more expensive to produce. The strength of nylon derives from the hydrogen bonding between the amide functional groups in adjacent chains—similar to those holding peptides together (Figure 21.22). Nylon is widely used in fabrics for clothing and household fabrics, as well as ropes and small artifacts, and in filled forms as a molding material for power tool casings.

When the diamine and the dicarboxylic acid are attached to aromatic rings, the generic term for the polymers produced is aramide. The reaction of 1,4-diaminobenzene with terephthaloyl chloride gives poly(para-phenylene terephthalamide) (PPTA) marketed as Kevlar® (Figure 21.23). Because of the extensive hydrogen bonding network between the chains (Figure 21.24), this is an extremely stiff material and has found wide applications in body armor and safety equipment.

Kevlar is very stiff because the chain is stiff and highly oriented along the fiber axis, as the fibers are produced by spinning from a liquid crystal solution. The weaker interchain bonds allow some shearing, so that it is like steel along the fiber axis, but like silk at right angles to it.

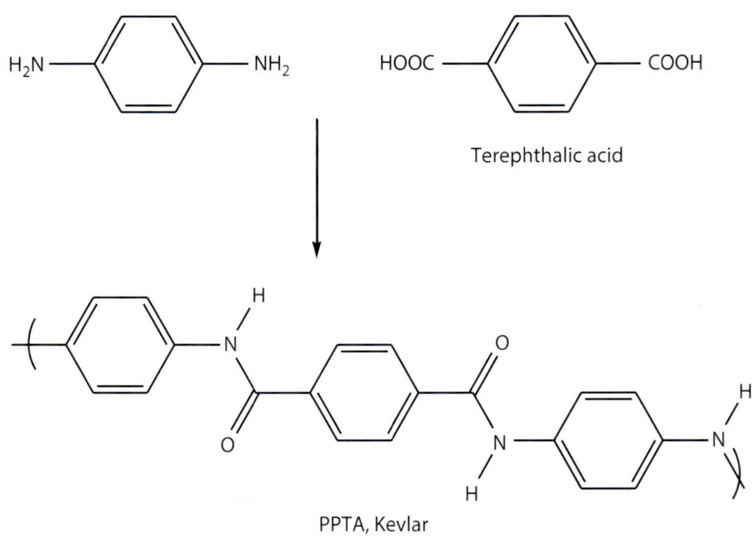

FIGURE 21.23 Preparation of Kevlar.

* So-called because it is formed from two monomers, each with six carbon atoms.

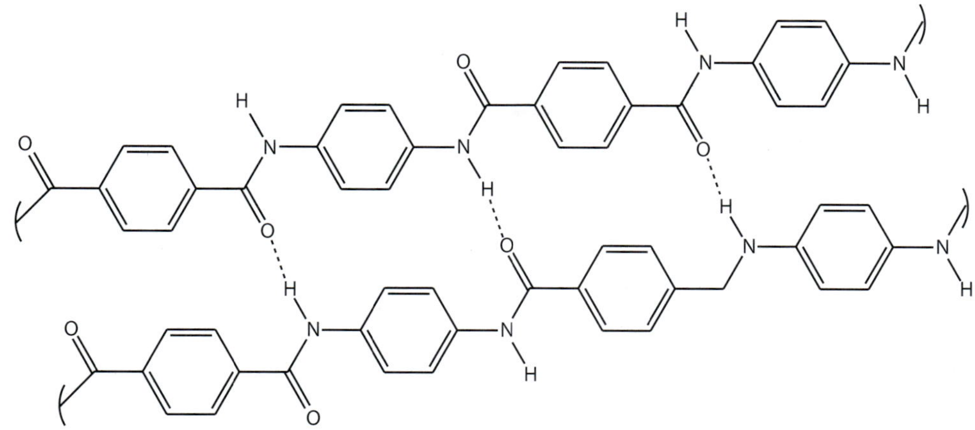

FIGURE 21.24 Interchain hydrogen bonding in PPTA.

PROBLEM 21.12

The polymer Nomex, which is used for parachute cords and jet aircraft tires, is prepared as shown in the following. Give its structure, and write a mechanism for its formation:

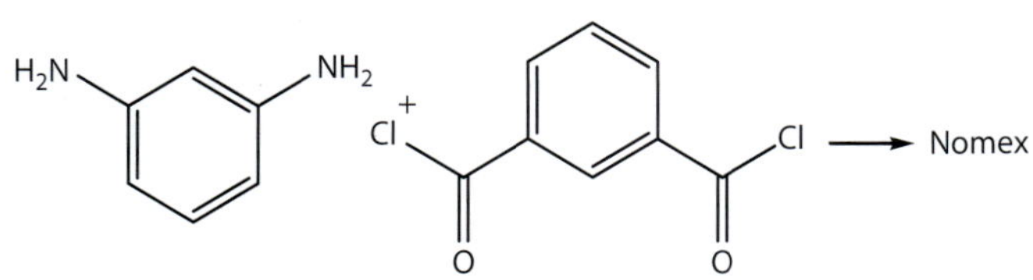

PROBLEM 21.13

Suggest reagents for each of the following steps in the manufacture of 1,6-diaminohexane for nylon production:

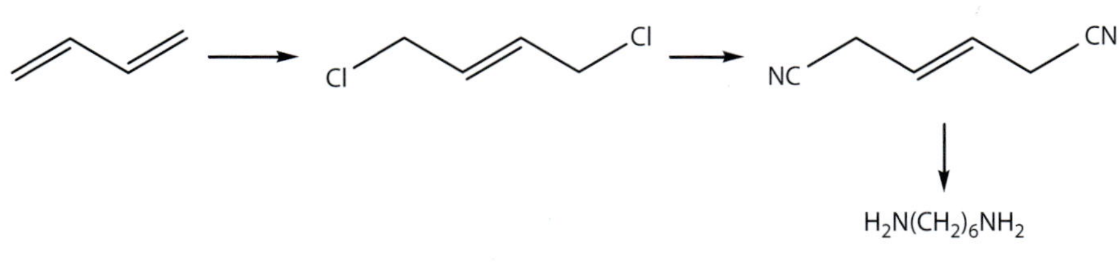

Solutions

21.12 This is the displacement of chloride from an acid chloride by an amine to give an amide:

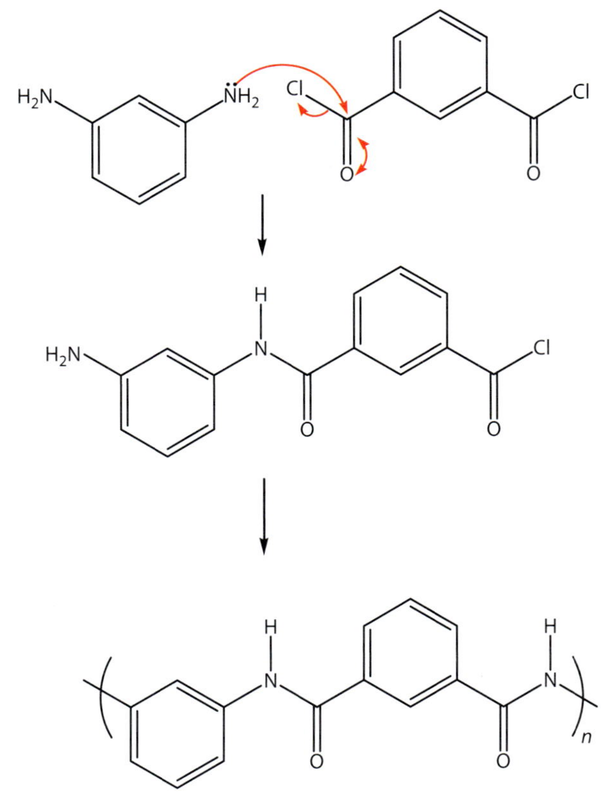

21.13 The first step is 1,4-addition of chlorine to butadiene. In adding chlorine to butadiene, both 1,2- and 1,4-addition products may be produced (Section 11.2.3). The 1,4-product is the thermodynamic product, so the best conditions will be a long period, and some heating. Conversion of the chloride to cyanides involves KCN; S_N2 substitution should predominate over S_N2' reaction (Section 9.8).

In the last step, it is intended that the double bond as well as the cyanide groups should be fully reduced; thus, the best conditions are molecular hydrogen and a platinum catalyst, which reduces both functional groups.

21.4.3 OTHER POLYMERS

Although polyesters and polyamides are by volume the most important of the step-growth polymers, many other types of reaction produce interesting materials. For example, polyurethanes are prepared by the reaction of diisocyanates with diols. The basic reaction mechanism is shown in Figure 21.25, and a common example in Figure 21.26. This is a step reaction, but not a condensation. Depending on the conditions used, both foams and rigid materials for moldings can be produced. For example, if glycerol (propane-1,2,3-triol) is added in small quantities to the reaction in Figure 21.26, a much stiffer foam is obtained by cross-linking. Foaming occurs by reaction of the isocyanate with water to release CO_2; the amine produced can react with further isocyanate to give polyurea structures.

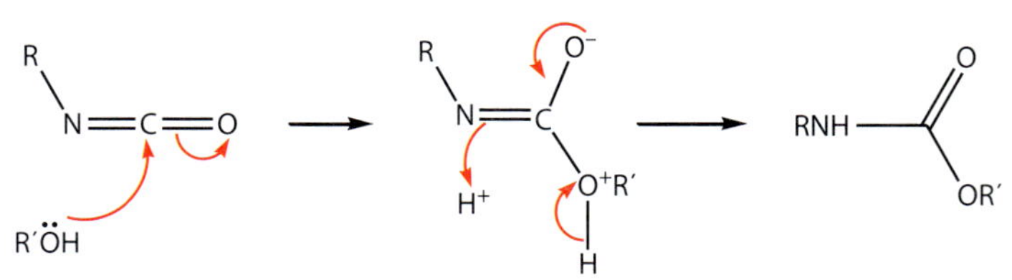

FIGURE 21.25 Mechanism of preparation of polyurethanes.

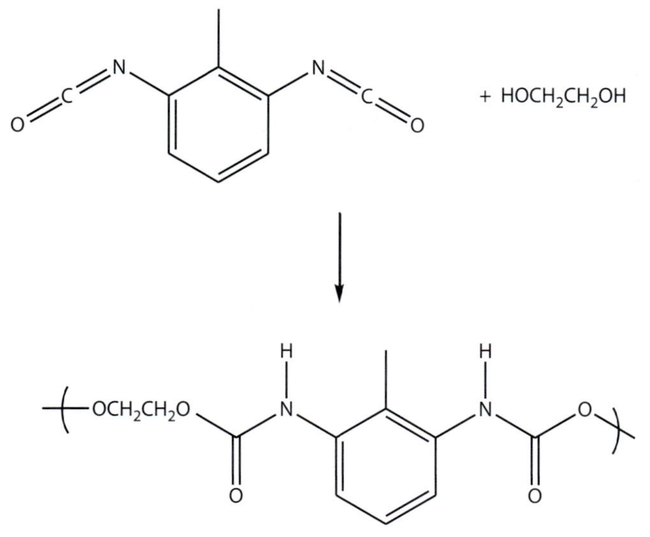

FIGURE 21.26 Preparation of a commonly used polyurethane.

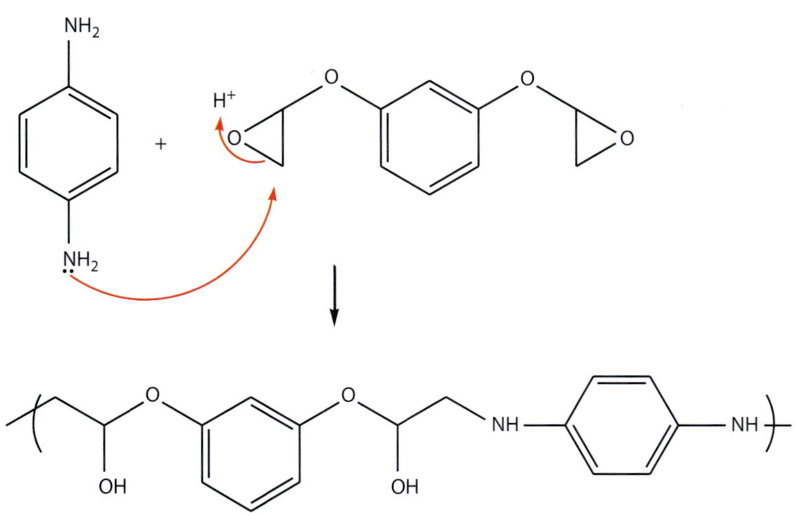

FIGURE 21.27 Polymer synthesis involving opening of strained rings.

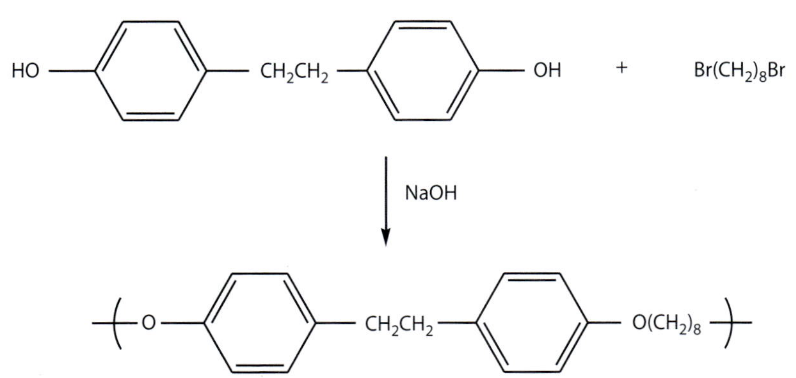

FIGURE 21.28 Polymer synthesis involving S_N2 reactions.

Other examples are shown in Figures 21.27 and 21.28—almost any sort of chemistry could in principle be used to make interesting polymeric materials. However, the reaction needs to go to >99 % conversion to give a polymer with a useful molecular weight, unless high levels of cross-linking agent are used to make a network. Few organic coupling reactions achieve this, so the number used in practice to make linear polymers is relatively small.

PROBLEM 21.14

What would be the repeating unit of the polymer prepared from each of the following pairs of compounds?

(a)

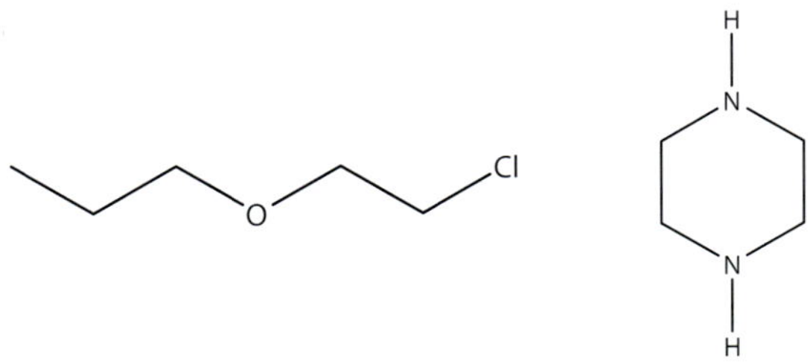

(b)

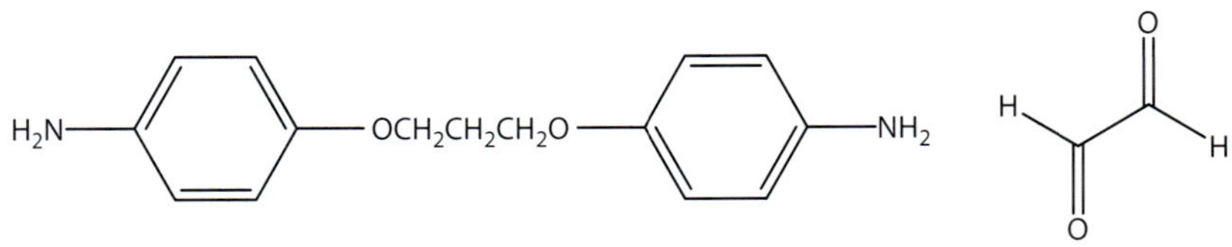

(c)

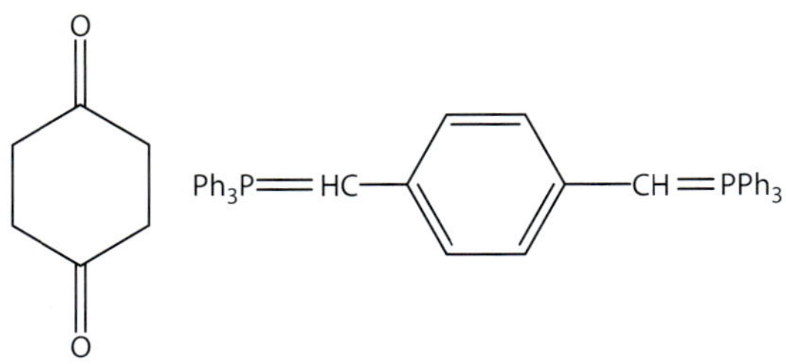

Solutions

(a)

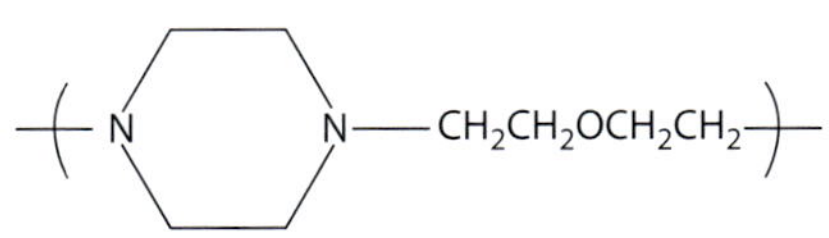

(b)

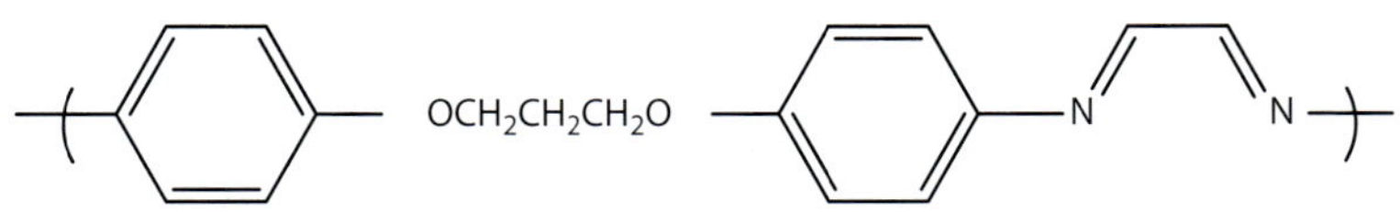

(c)

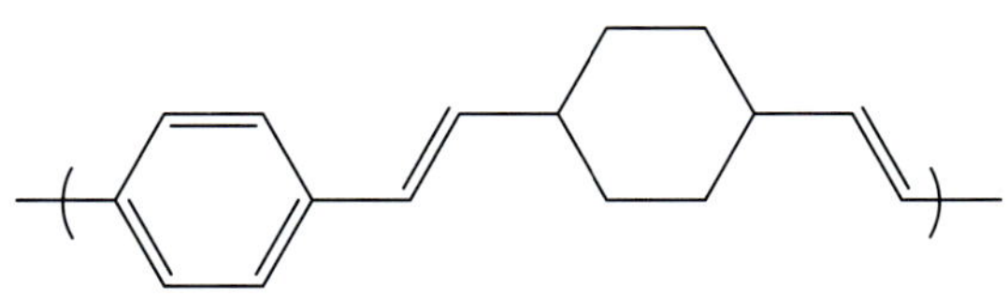

Key Points from Sections 21.4 and 21.5

- Step-growth polymers involve the reaction of a bifunctional monomer, or two bifunctional monomers, to give a polymer. High conversions in the coupling reaction are critical to preparing materials with a useful molecular weight.
- Polyesters and polycarbonates are prepared by esterification or transesterification. Polyamides are prepared from diacids and diamines, usually under acidic conditions.
- The high melting point of polyamides derives from interchain hydrogen bonding.
- Polyurethanes are prepared from diisocyanates and diols.

21.5 UNUSUAL BIOPOLYMERS

We have or will discuss in the relevant chapters the three main classes of biopolymer—polysaccharides, peptides and proteins, and polynucleotides, the study of which makes up the important area of structural molecular biology. In this section, we meet some less obvious aspects of the subject, with a particular emphasis on the uses to which biopolymers have been put, and their relationship to man-made polymers. There will be a discussion of the polymers used in fabrics in Chapter 26.

21.5.1 IT'S A STICKY BUSINESS...

Many modern glues (such as superglue, cyanoacrylate) are artificial, but humans have been using glue throughout history, from the production of the very first tools. There is archaeological evidence for the use of birch bark tar to glue a flint into a wooden spear 200,000 years ago. Animal-based glues have been used at least since 6000 BCE, probably much longer. These are protein colloids formed by partial hydrolysis of animal products including skin, bone, tendon, and fish skins—all essentially formed from the structural protein collagen. Other glues have been prepared from casein, a phosphoprotein from milk, and albumen from denatured protein from egg white.

Plant-based glue can be made from simple starch and water; this type is often used for gluing paper and in other crafts. Wallpaper paste is partially methylated cellulose (**21.14**)—generally between 1.5 and 2.5 of the three hydroxyl groups that could possibly be methylated have been.

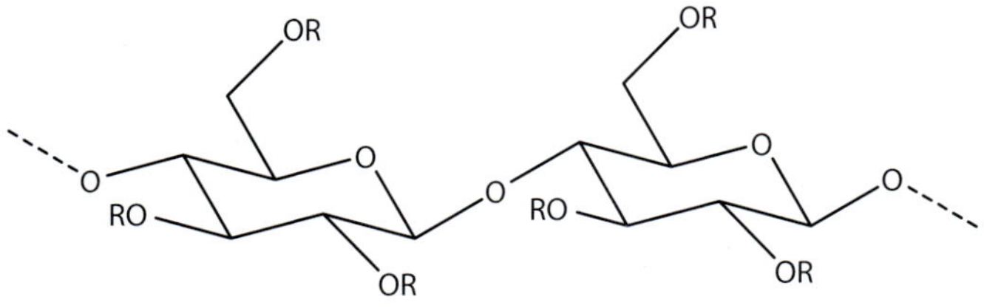

21.14 (R = H or Me)

Gums are usually defined as polysaccharides of natural origin—they are often used in the food industry as thickening agents (*vide infra*), but gum arabic, the hardened sap of acacia trees, was long used as an adhesive for stamps and also as a food stabilizer and a binder in watercolor paints and in lithography. It is a mixture of polysaccharides and glycoproteins, the main sugars being L-arabinose, D-galactose, 4-OMe-L-rhamnose (**21.15**), and D-glucuronic acid. Man-made polymers have been used on stamps since the 1960s and are ubiquitous in pressure-sensitive glues on self-adhesive stamps.

The Penny Black, the world's first postage stamp

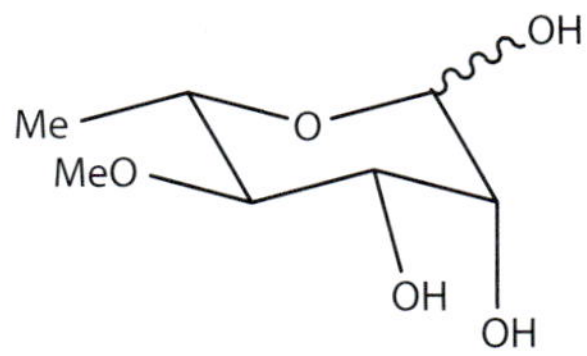

21.15, 4-OMe-L-rhamnose

21.5.2 CHEWING IT OVER...

Gums have been widely used to improve the texture of foods, as emulsifiers, and to prevent the formation of ice crystals. They are distinguished from resins by the fact that they are either water soluble or swell in water. Most are polysaccharides composed of sugars other than glucose, usually arabinose, galactose, mannose, and xylose. They are also widely used in various binders, medicines, and as laxatives. Some gums, such as gum arabic, are exuded by plants in response to injury, while others such as locust bean gum and guar gum are isolated from the plant itself. Both guar gum and locust bean gum are described as galactomannans; these have a 1,4-linked mannose chain substituted at some 6-positions by galactose (**21.16**). They differ in the ratio of galactose to mannose, but both are used in commercial ice creams to improve texture, by reducing ice crystal formation. Guar gum is eight times more effective than cornstarch as a thickening agent. Xanthan gum is a polysaccharide formed by bacterial fermentation of glucose, sucrose, or lactose. It has a β-1,4-glucose backbone, like cellulose, but with some cross-linking and many side chains including mannose and glucuronic acid residues.

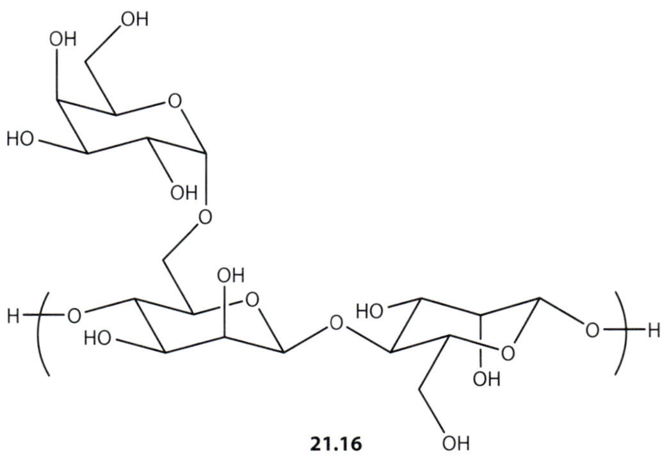

21.16

Agar, a polysaccharide from agarophyte red algae, and carrageenan, from seaweed, are of similar structure but have some sulfonated groups. They are alternating polysaccharides, called agarose with one galactose and one anhydrogalactose unit (agarobiose) as the repeat (**21.17**). Agar is used in desserts in Asia and is widely employed as a culture medium in biology. Both agar and carrageenan are regarded as vegetarian and vegan alternatives to animal-derived gelatin, and are accepted as kosher and hallal.

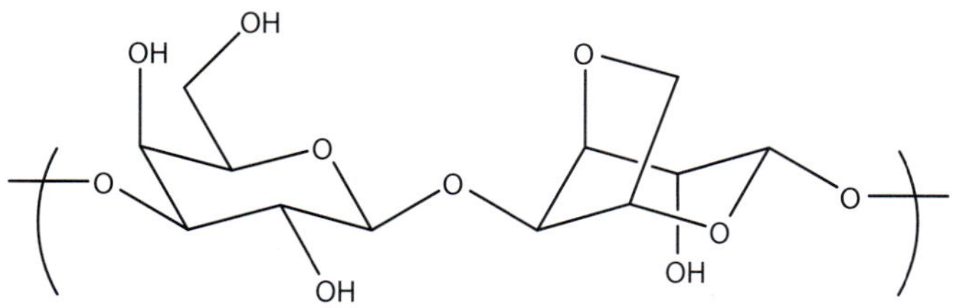

21.17, agarose

Gelatin is prepared from animal skins and bones, just as for glue, and it is essentially partially denatured collagen protein. It is widely used in foodstuffs, especially Jell-O, gummy bears (a kosher, vegan, version using a different gelling agent is also produced), and chewy sweets.

Gummy bears

Pectins, the gelling agents used in making jams and jellies, are ubiquitous in plants and are produced commercially from apple skin or citrus rind. They are structural polysaccharides, with molecular weights between 20,000 and 400,000, containing galactose, galacturonic acid, rhamnose, xylose, and arabinose.

Modern chewing gum is almost all made from polyisobutylene (poly-2-methylpropene). However, chewing gums have been known for millennia. The Greeks chewed mastic gum (μαστιχάω from the Greek verb to gnash the teeth), which contains a polymer (**21.18**) of the terpene myrcene (**21.19**) as well as some volatile oils and various triterpenes. This is essentially a natural rubber, a *cis*-1,4-polybutadiene. Spruce gum was first marketed in the United States in the nineteenth century but was displaced in the twentieth century by chicle gum (a natural rubber from *Manilkara zapota*). None of these materials should strictly be classed as gums—they are resins or latexes, not polysaccharides.

Chewing gum

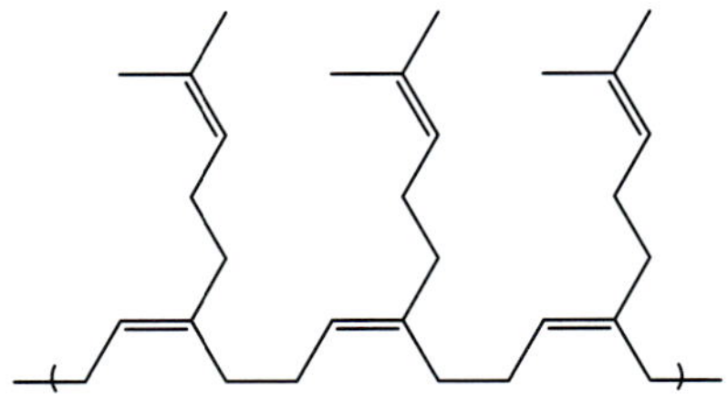

21.18

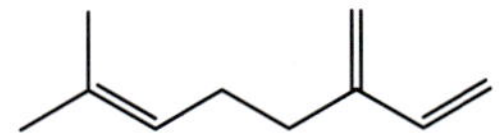

21.19

21.5.3 PRETTY AS A PICTURE...

Polymers are essential to every stage of producing an image and have been throughout history. Paper is a polymer—generally of cellulose—paint and ink use pigments in polymeric binders, and finished pictures were coated with varnish. Egyptian papyrus was prepared before 3000 BCE from the stems of the sedge *Cyperus papyrus,* which were laid across each other, then pounded to release the sap to bind them together. The manufacturing of paper is generally thought to have begun in China around 200 BCE and to have spread slowly to the West, but bark paper was developed independently in South America and Oceania. Modern paper uses wood pulp, pressed into sheets. Wood pulp consists of cellulose fibers embedded in a matrix of the network polymer lignin. For regular paper, the lignin is removed (though not for stiff materials such as cardboard). The finest quality paper, sometimes called rag paper, is made from cotton and linen waste. It is more durable than paper from wood pulp as the fibers are longer (cotton is almost pure cellulose, so less processing is needed) and is therefore used for bank notes (a few countries are now using bank notes made from polypropylene, since they last much longer).

The oldest paper book,
the parable sutra, 256 CE

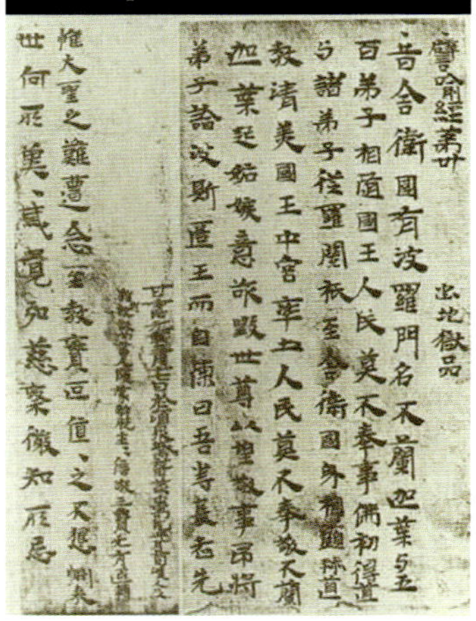

Paints have been used from the earliest history of mankind. Cave art generally only involved pigments (colored materials derived, at that time, from minerals such as iron oxide), but by the bronze age, paints involved both pigment (still mostly mineral in origin) and binders, to hold the particles of pigment together. These include proteinaceous materials such as albumin (from egg white, used since the early middle ages), casein (from milk), tempera (oil and egg yolk, used in ancient Egypt), animal-based glues, and plant-derived materials such as gum arabic. Once applied, paints may either dry, which involves simply allowing evaporation of a volatile solvent, or "cure," where the binding material is polymerized, often with extensive cross-linking. Modern artificial materials fulfilling the same function include acrylics, epoxys, and polyesters.

Leonardo da Vinci's Last Supper mural in Santa Maria delle Grazie, Milan. The paint is egg tempura, and both a chalk/animal glue mixture and mastic gum were used to prepare the surface of the wall.

Varnishes have been used since ancient times to coat pictures, but also for furniture and other wooden artifacts. Originally, these consisted of plant resins dissolved in a solvent for ease of application. The solvent evaporates, and the resin may then be "cured" by exposure to air and heat. The curing process consists of polymerization and cross-linking. Plant resins consist of polymerized terpenes and essential oils. Resins are insoluble in water and generally have a softening, rather than a melting, point. Most are soluble in organic solvents. Copal ("immature amber") dissolves in acetone, though amber itself does not; both have been used in varnishes since ancient times. From copal, communic acid, communol, and biformene (Figure 21.29), all terpenes, can be isolated; in amber, these are polymerized and cross-linked. Intriguing compositions of varnishes for specific purposes were developed, and the reproduction of these for modern restorations has not been easy, for example, the varnishes used for early violins.

Shellac is a varnish of animal origin, and its composition is quite different from that of plant resins. It is secreted by the female lac bug in India and Thailand and has been used as a stain, a varnish, and in artifacts such as 78 rpm record discs. While it has been replaced by man-made polymers for most uses, it is still used in French polishing of furniture and on musical instruments. Hydrolysis gives a mixture of hydroxy acids such as aleuritic acid, *R*-6-hydroxyhexadecanoic acid, jalaric acid, and the colorant laccaic acids (Figure 21.30). The material is composed of low molecular weight, cross-linked polyesters.

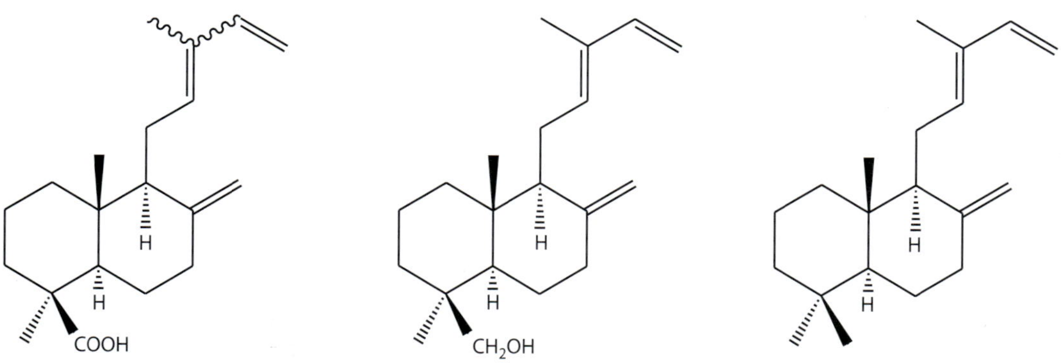

cis and *trans*-Communic acid Communol Biformene

FIGURE 21.29 Terpenes found in copal.

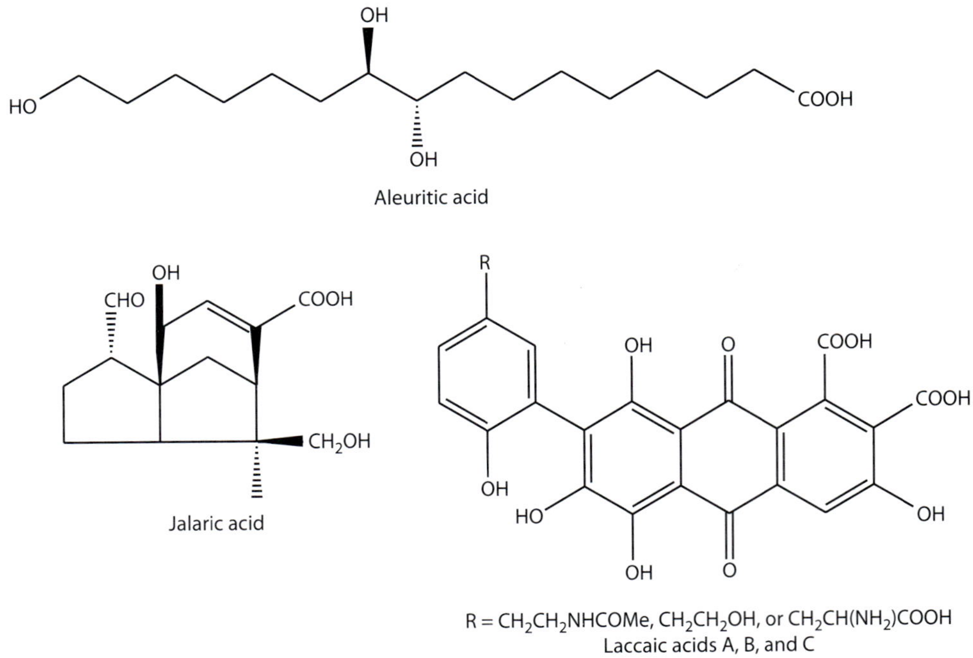

FIGURE 21.30 Some of the hydroxyacids found in shellac.

REVIEW PROBLEMS

1. Explain each of the following terms and give a *specific* example of each:
   (a) Cationic polymerization
   (b) Anionic polymerization
   (c) Radical-initiated polymerization
   (d) Ziegler–Natta polymerization
   (e) Ring-opening polymerization
   (f) Addition polymer
   (g) Isotactic polymer
   (h) Homopolymer
   (i) Atactic polymer
   (j) Syndiotactic polymer
   (k) Random copolymer
   (l) Alternating copolymer
   (m) Cross-linked polymer
   (n) Polyurethane
   (o) Monomer
   (p) Polymer chain branching
   (q) Polyamide
   (r) Polyester
   (s) Polycarbonate
   (t) Radical initiator
   (u) Propagating site
   (v) Head-to-tail addition
   (w) Vulcanization
   (x) Chain termination

2. List each of the following groups of monomers in order of their decreasing ability to undergo cationic polymerization in homogeneous solution. Specifically, you should consider the stability of the carbocation likely to be formed:

(a)

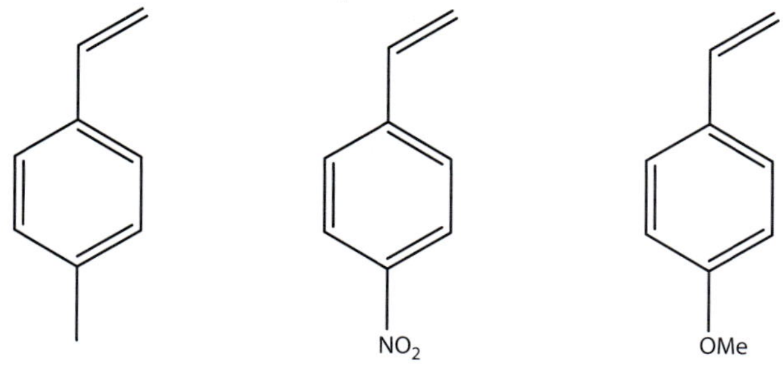

(b)

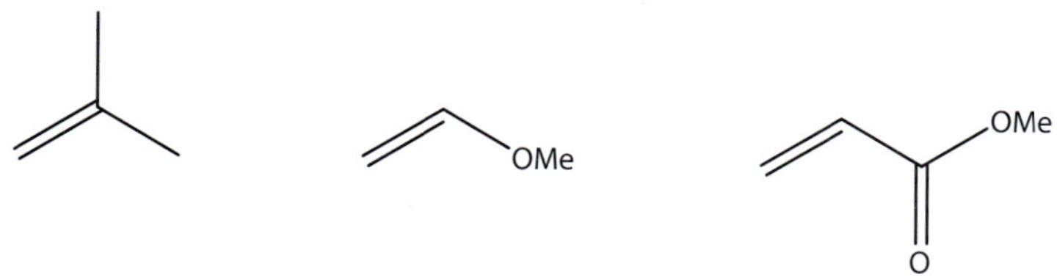

(c)

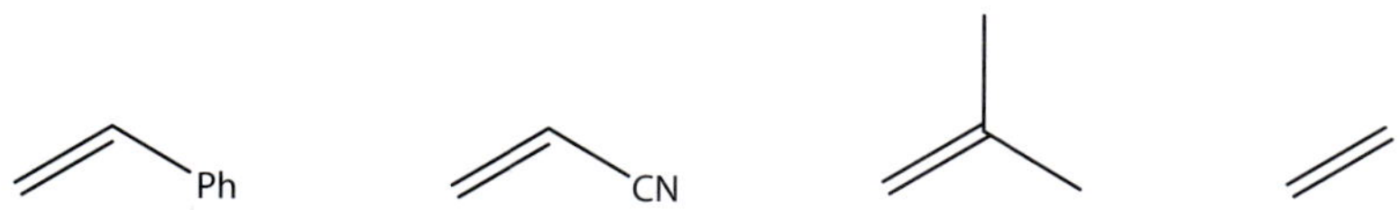

(d)

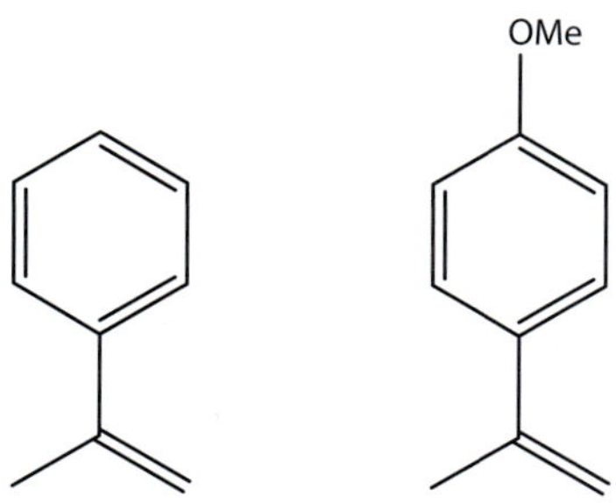

3. List the following groups of monomers in order of decreasing ability to undergo anionic polymerization in homogeneous solution. Consider specifically the likely stability of an anion formed:

(a)

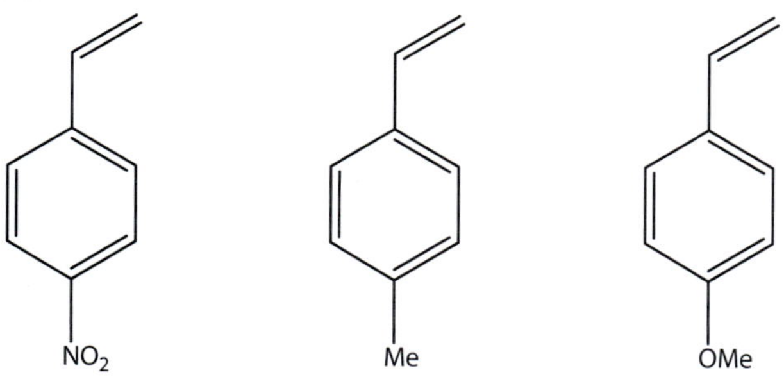

(b)

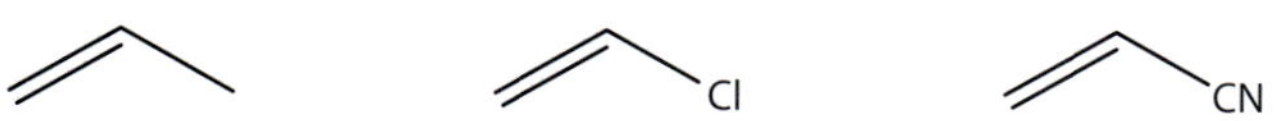

(c)

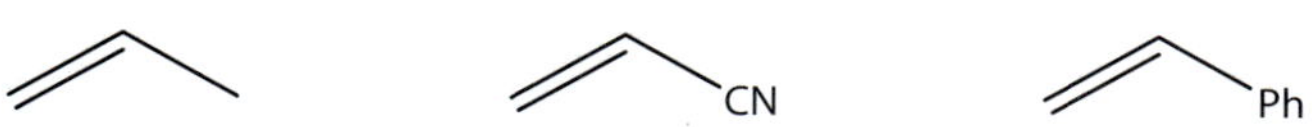

4. $CH_2{=}CHCOOR$ and $CH_2{=}CHCN$ are generally polymerized by radical initiators. However, cationic initiators don't work, and there are a number of side reactions with anionic initiators. Explain.
5. Rationalize the following rate order in radical polymerization:
   $CH_2{=}CHCl > CH_2{=}CHOCOMe > CH_2{=}CHCOOMe > CH_2{=}CHCN > CH_2{=}CHPh$
6. Identify the monomers for each of the following polymers:

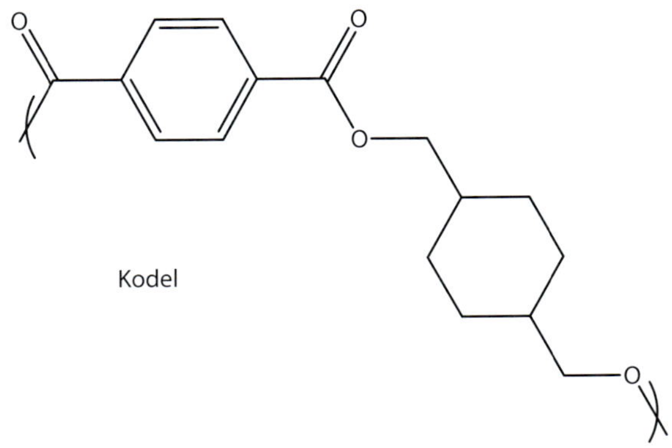

Kodel

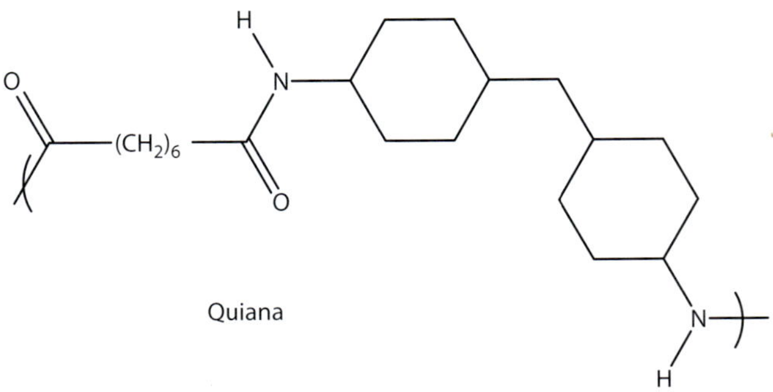

Quiana

7. For the polymers shown in the following,
   (a) Give the structure of the monomer or monomers
   (b) Classify each as a chain-growth or step-growth polymer
   (c) Suggest a synthetic route to their preparation

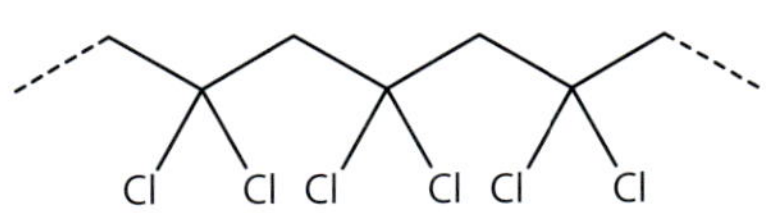

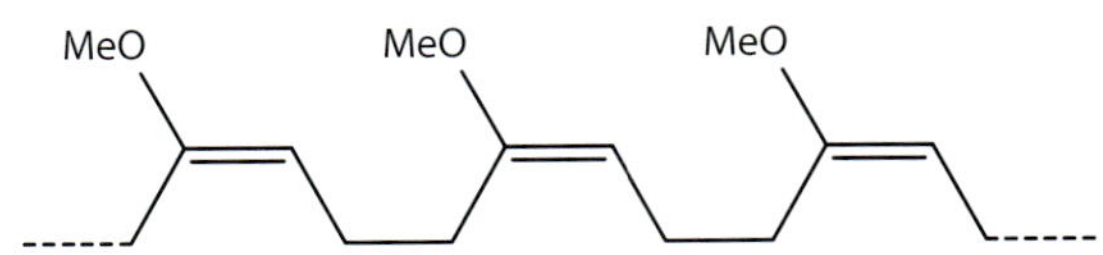

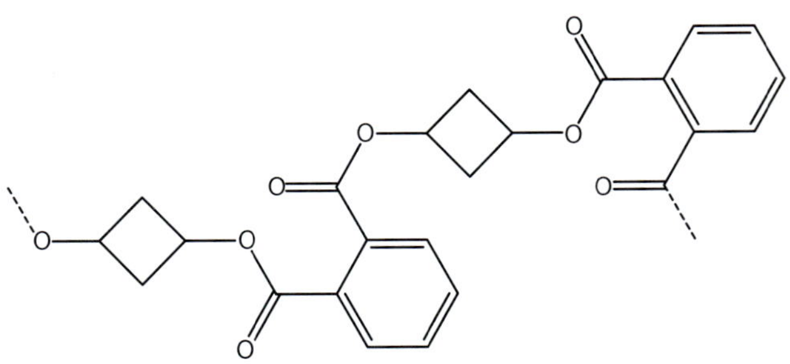

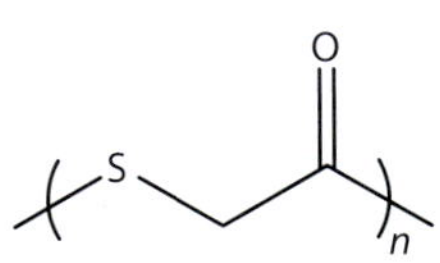

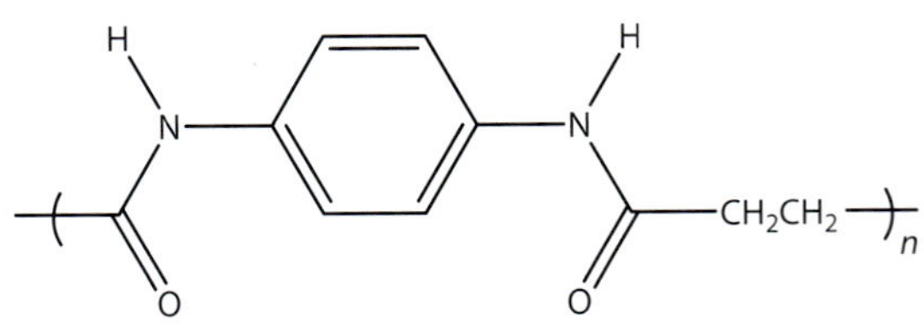

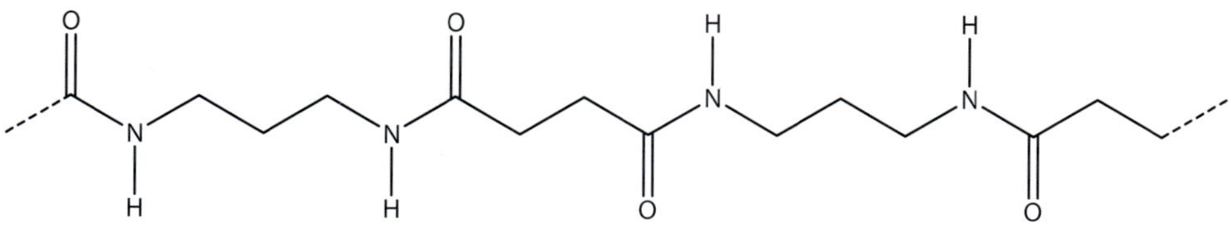

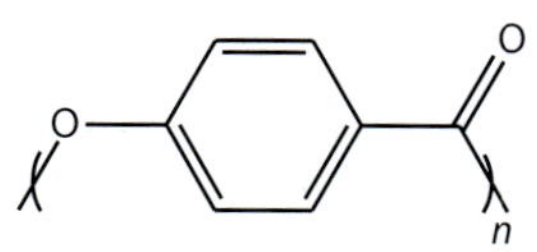

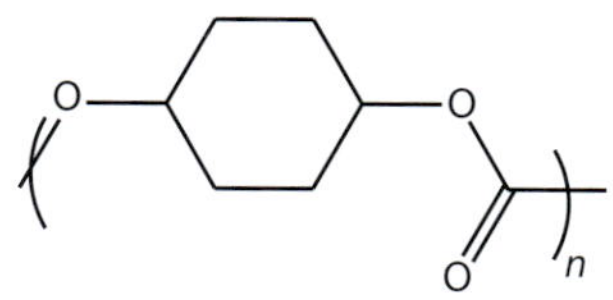

8. What polymers would you expect to be formed in each of the following reactions?

(a)

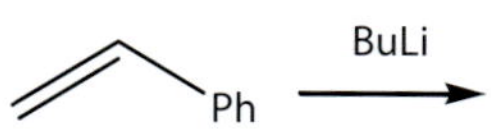

(b)

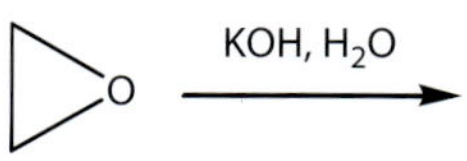

(c)

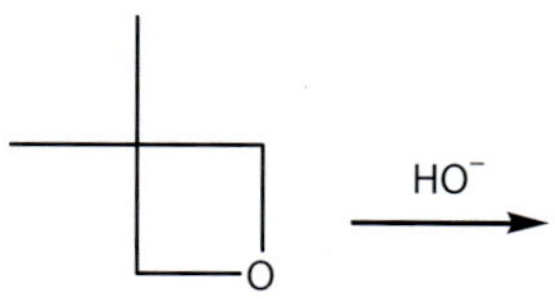

9. Draw a short section of the polymer that would be formed from cationic polymerization of MeOCH=CH_2 with H^+ as initiator.
10. What is the structure of the polymer produced by anionic polymerization of β-propiolactone using NaOH as catalyst?

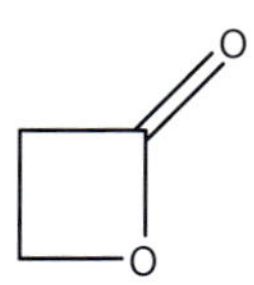

β-Propiolactone

11. Draw a short segment of the polymer formed from the anionic polymerization of

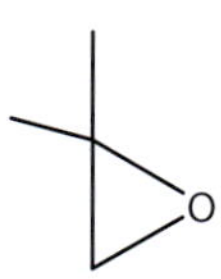

12. For each of the following pairs of monomers, draw the structure of the polymer that they will form, and give a mechanism for the key reaction in polymer formation:
    (a) Dimethyl terephthalate (dimethylbenzene 1,4-dioate) and 1,4-dihydroxycyclohexane (Kodel®)
    (b) Bisphenol A and phosgene (Lexan polycarbonate)

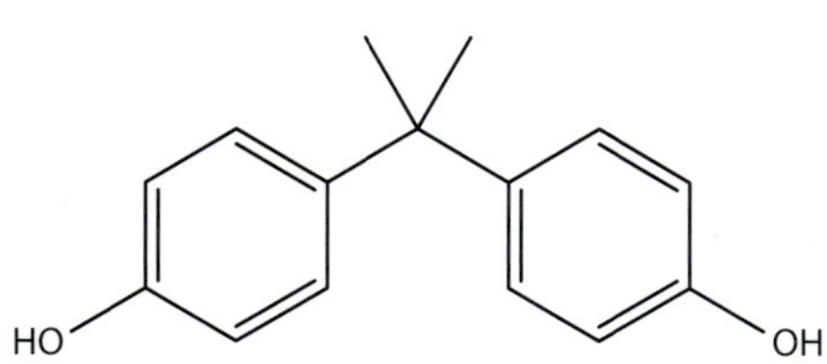

Bisphenol A

(c) Bisphenol A and toluene 2,4-diisocyanate

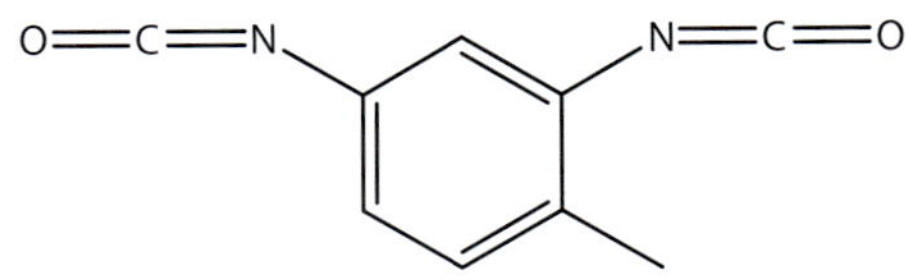

Toluene 2, 4-diisocyanate

(d) Propane-1,3-diol and butanedioic acid

13. Propose a mechanism for the following reaction:

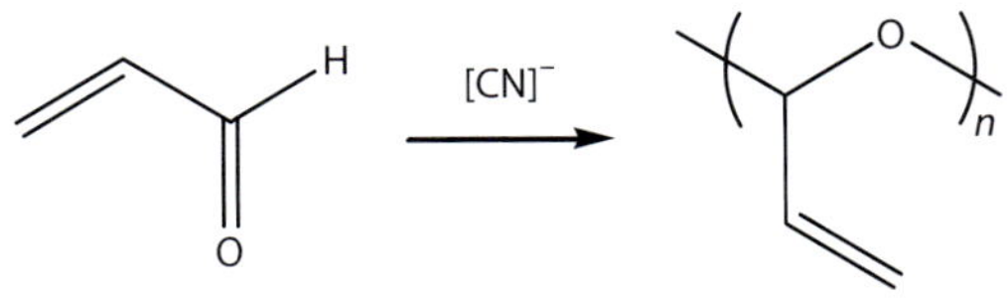

14. Styrene, PhCH=CH_2, may be polymerized in the presence of a cationic initiator, H^+ (it is important to use an acid with a non-nucleophilic counterion, such as H_2SO_4 or H_3PO_4, which will not react with the cation produced on protonation). The product polystyrene is essentially completely regular, as in **21.18**, with the styrene units joined in a head-to-tail fashion. Explain, drawing a reaction mechanism.

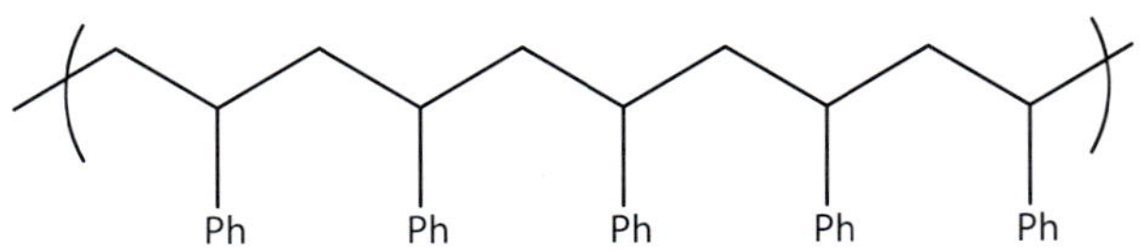

21.18

15. Polymerization of isobutene, $(CH_3)_2C{=}CH_2$, in the presence of Et_2AlCl gives a high molecular weight polymer. A trace of HCl is required, and the reaction works best in polar solvents. What is the mechanism of polymerization, and what structure would you expect for the polymer?

    Hint: Remember that most aluminum compounds are Lewis acids—electron acceptors.

16. Explain why when propylene oxide undergoes anionic polymerization, nucleophilic attack occurs at the less substituted carbon of the epoxide, but when it undergoes cationic polymerization, nucleophilic attack occurs at the more substituted carbon.
17. Polyvinyl alcohol is a polymer used in aqueous adhesives. It is made by polymerization of vinyl acetate, followed by hydrolysis.
    (a) Give the structure of polyvinyl acetate.
    (b) Give the structure of polyvinyl alcohol.
    (c) Why is polyvinyl alcohol made by this roundabout route?
18. Show mechanistically how the reaction of glycerol (propane-1,2,3-triol) with phthalic anhydride leads to a strong cross-linked material, called glyptal:

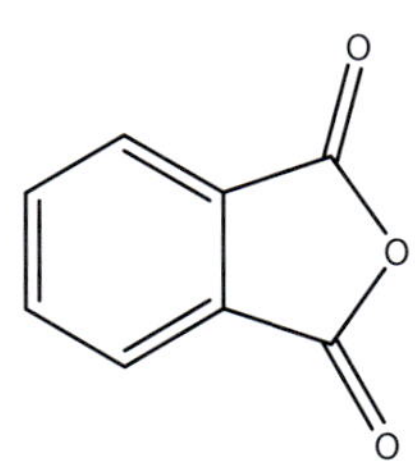

Phthalic anhydride

MORE CHALLENGING PROBLEMS

1. The smoking salons of the Hindenburg and other hydrogen-filled dirigibles of the 1930s were insulated with urea/formaldehyde polymer foams. The structure of the polymer is highly cross-linked. Propose a structure.
2. Write a mechanism for the polymerization of 2-methylpropene initiated by $PhCMe_2Cl/SnCl_4$. Label initiation, propagation, and termination steps, and explain the regioselectivity of the process you propose.
3. Melmac, a thermosetting resin used to make plastic dishes, is prepared by heating melamine with formaldehyde. Propose a structure for Melmac and a mechanism for its formation:

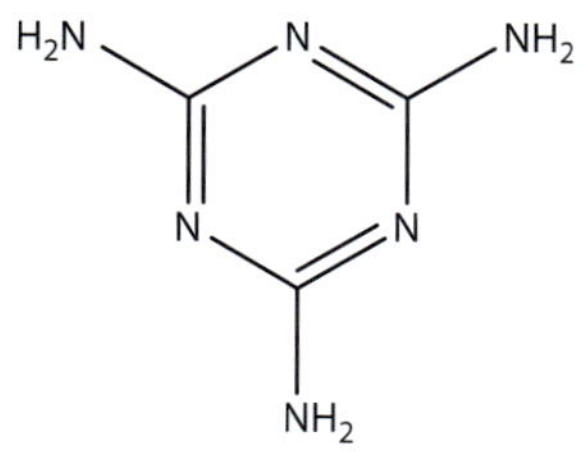

Melamine

4. Write a mechanism for the step-reaction polymerization shown in the following, which gives a high-strength fiber:

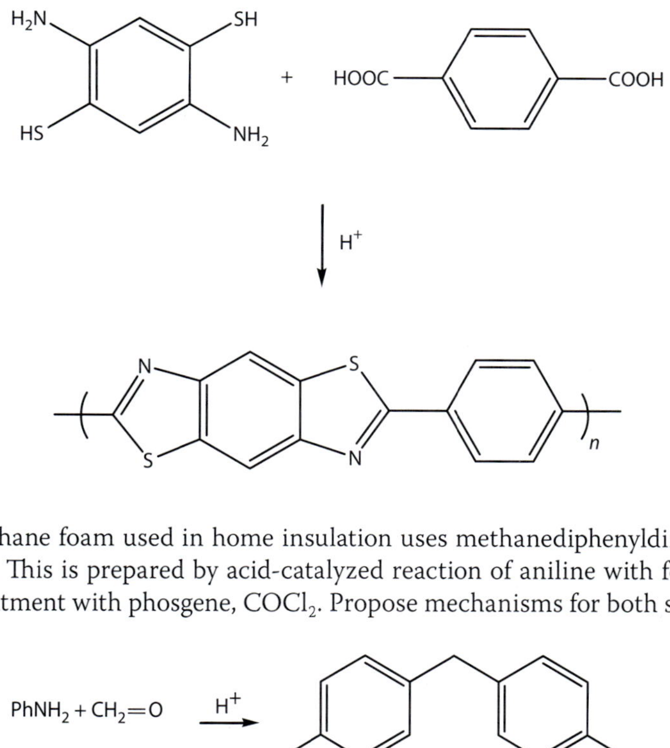

5. The polyurethane foam used in home insulation uses methanediphenyldiisocyanate (MDI) as monomer. This is prepared by acid-catalyzed reaction of aniline with formaldehyde, followed by treatment with phosgene, $COCl_2$. Propose mechanisms for both steps:

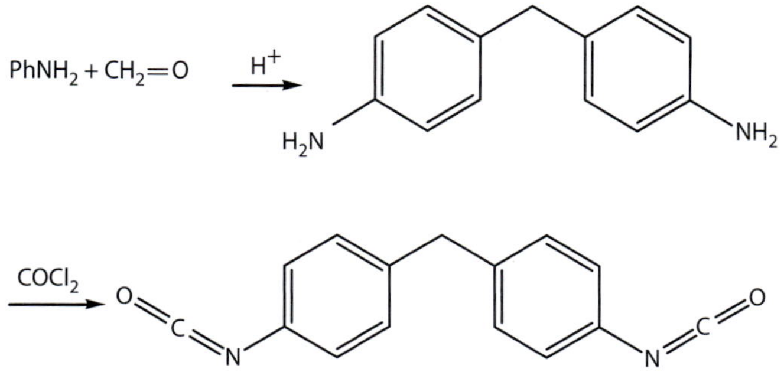

MDI

6. Polyvinyl butyral, shown in the following, is used as the plastic laminate in the preparation of auto windshield safety glass. How would you make this polymer?

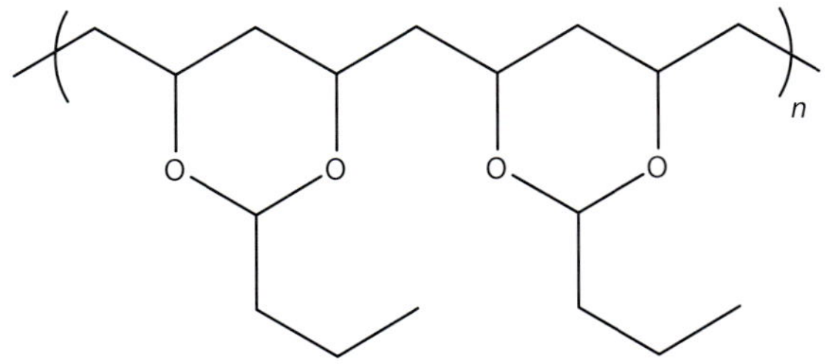

Note: There are about 14 % of OH groups on the final polymer. This should give you a clue as to the synthetic route taken, and you should explain why some free OH groups remain after the last stage of the synthesis.

7. A website on the chemistry of paints and paintings makes the statement that pine resin in North America contains more of the *R*,*R*-isomer of pinene, but that European pines produce more of the *S*,*S*-isomer. This is essentially correct, although there are also other differences in composition. It then states that the *S*,*S*-isomer is less oxygen sensitive, and hence turpentine prepared from French pine oil resinifies less rapidly, and hence is more suitable for artistic uses. Is it plausible that the two enantiomers react at different rates with oxygen? If not, why not? If the observation that French pine oil reacts more slowly with oxygen is correct, what other explanation is possible?

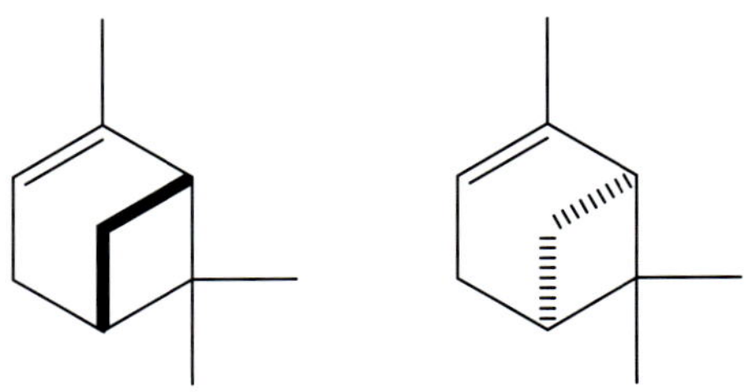

R,*R*-α-pinene *S*,*S*-α-pinene

8. Propose mechanisms for the polymerization of caprolactam under each of acid or base catalysis.
9. Under appropriate conditions, ethene oxide, oxetane, and THF may be polymerized. Dioxane and tetrahydropyran do not undergo polymerization. Explain.

Amines, Alkaloids, Amino Acids, Peptides, and Nucleic Acids

22

The aim in this chapter is to discuss these classes of compounds in chemical terms, while recognizing their biochemical importance. Many of you will go on to take a biochemistry course, and there is no intention here to duplicate that material.

22.1 SYNTHESIS OF AMINES

22.1.1 SYNTHESES INVOLVING NUCLEOPHILIC SUBSTITUTION

In previous chapters, we have encountered a wide range of reactions that lead to amine synthesis, and this section will recap and extend these. Nucleophilic substitution of haloalkanes with ammonia or amines is easy, but does not always give a good synthetic result. If ammonia is reacted with iodoethane, substitution is rapid, and the first product is ethylamine. However, ethylamine is more nucleophilic than ammonia, so the reaction can't easily be halted at this stage. The more alkyl groups are added to the nitrogen, the more nucleophilic it becomes (Figure 22.1). The process can be useful where the initial reaction is fast; one example is provided by an intramolecular substitution to give a five- or six-membered ring (Figure 22.2; DBU* is a strong, non-nucleophilic base that deprotonates the ammonium salt so that it can act as a nucleophile). If the product is very sterically hindered, further reaction may also be inhibited, as in Figure 22.3. We also noted in Chapter 9 that amino acid synthesis is possible by this route; the electron-withdrawing effect of the carboxyl substituent reduces the nucleophilicity of the $-NH_2$ group in the product so that it does not react further (Figure 22.4).

There are also many syntheses that involve nucleophilic substitution using a synthon for ammonia. Thus, the azide anion is an excellent nucleophile, and alkyl azides are readily reduced to amines (Figure 22.5), either by hydrogenation, as in this example, or by reduction with lithium aluminum hydride. Hexamethylenetetramine is also an excellent nucleophile and a synthon for ammonia; in this case, the final product is released by hydrolysis (Figure 22.6). In the Gabriel synthesis, phthalimide acts as a synthon for ammonia (Figure 22.7).

The cyanide ion is also an important synthon, but this time for "$[H_2NCH_2^-]$" rather than ammonia. As well as adding nitrogen, the carbon chain is extended by one unit. The carbon–nitrogen triple bond can be reduced either by catalytic hydrogenation or reduction with lithium aluminum hydride (Figure 22.8).

* DBU—1,8-diazabicyclo [5.4.0] undec-7-ene.

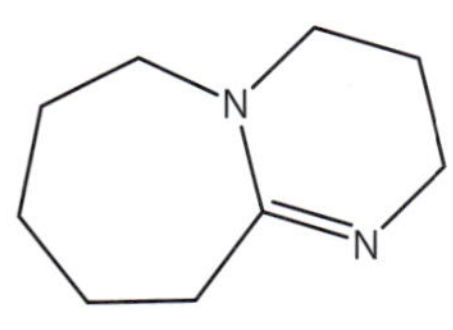

$$NH_3 + EtI \longrightarrow EtNH_2 \xrightarrow{EtI} Et_2NH \xrightarrow{EtI} Et_3N \xrightarrow{EtI} [Et_4N]I$$

FIGURE 22.1 Reaction of ammonia with iodoethane.

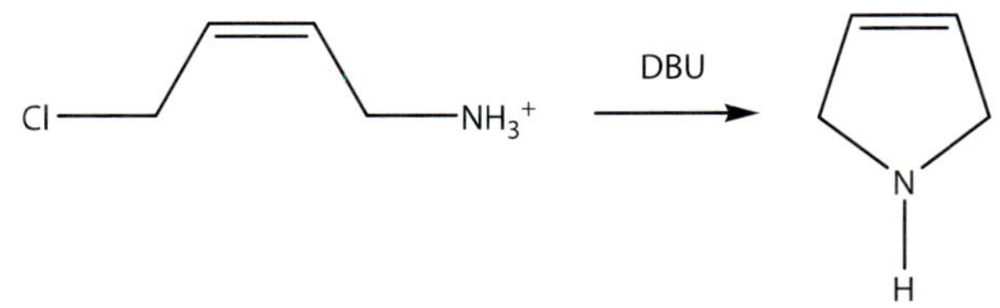

FIGURE 22.2 Nucleophilic substitution in a synthesis of a cyclic amine.

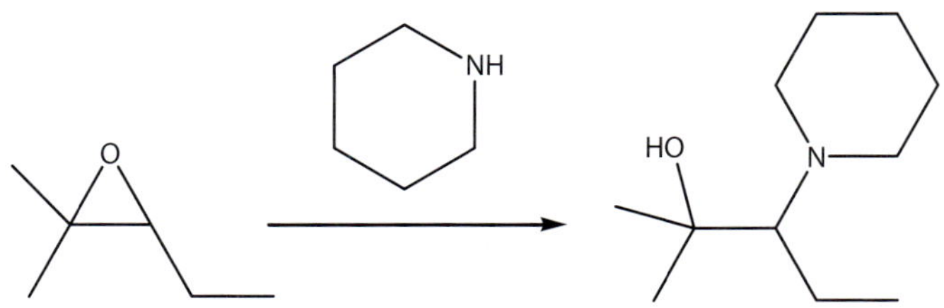

FIGURE 22.3 Amine synthesis by ring opening of a hindered epoxide.

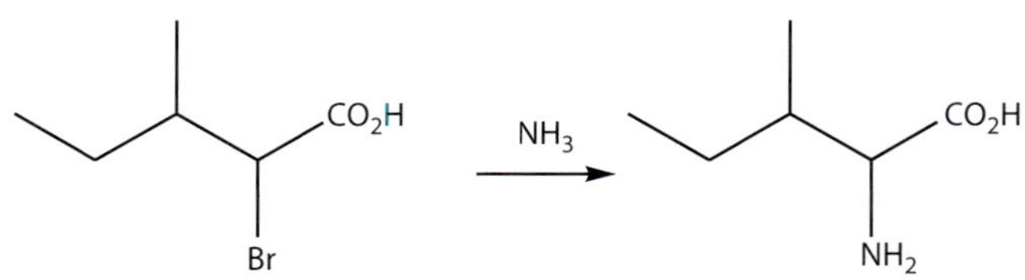

Isoleucine

FIGURE 22.4 Amino acid synthesis by nucleophilic substitution.

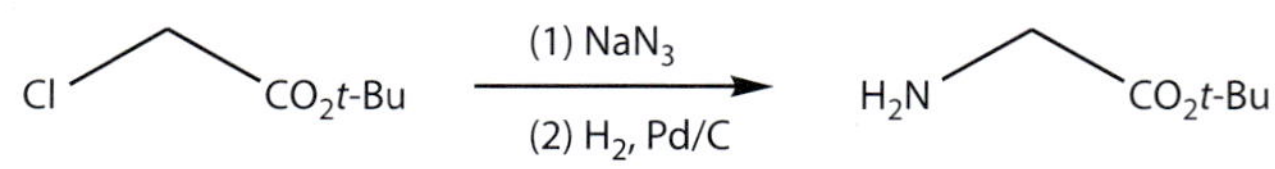

FIGURE 22.5 Amine synthesis by azide reduction.

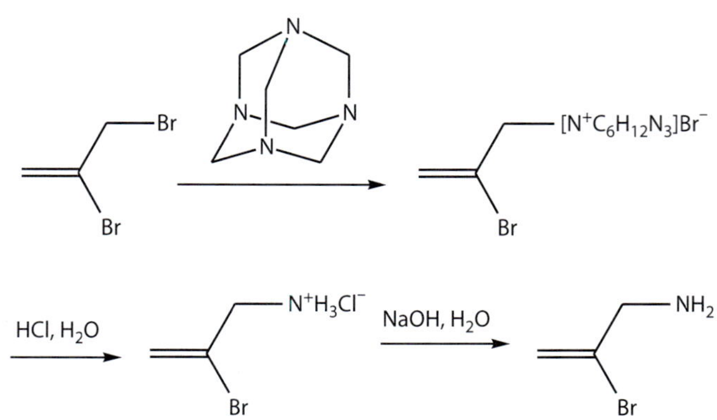

FIGURE 22.6 Amine synthesis using hexamethylenetetramine.

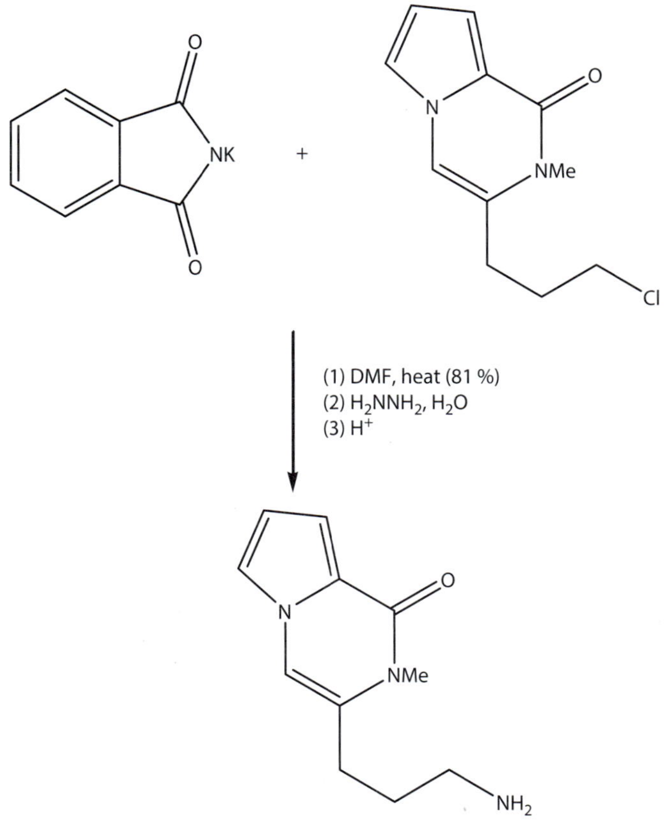

FIGURE 22.7 Gabriel synthesis of amines.

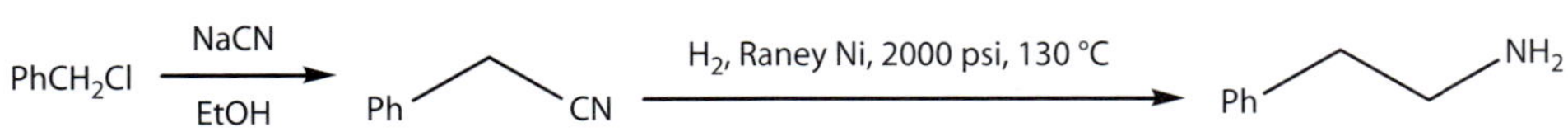

FIGURE 22.8 Amine synthesis using cyanide.

PROBLEM 22.1

Predict the product(s) of each of the following sequences:

(a)

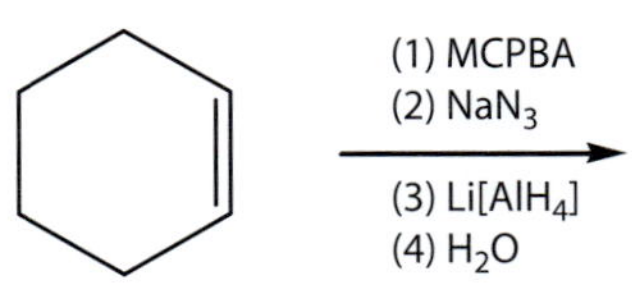

(b)

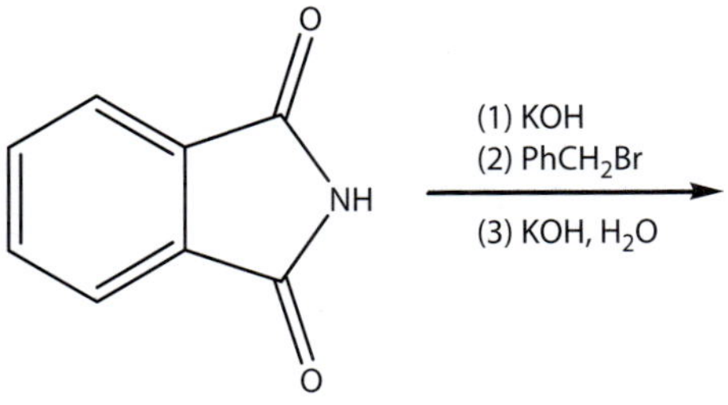

(c)

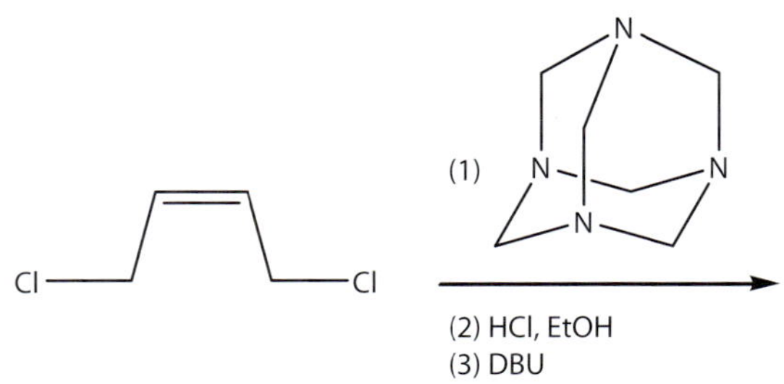

Solutions

22.1(a) The first step here is the epoxidation of the alkene, which is then opened to give the *trans*-azido alcohol. The azide is then reduced to the amine by lithium aluminum hydride:

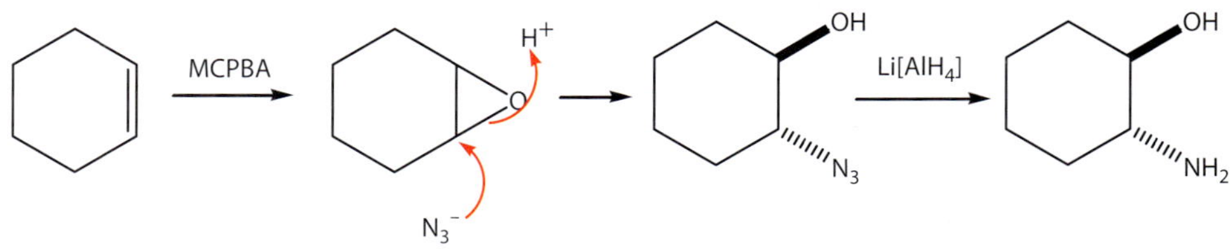

22.1(b) This is a classic phthalimide synthesis:

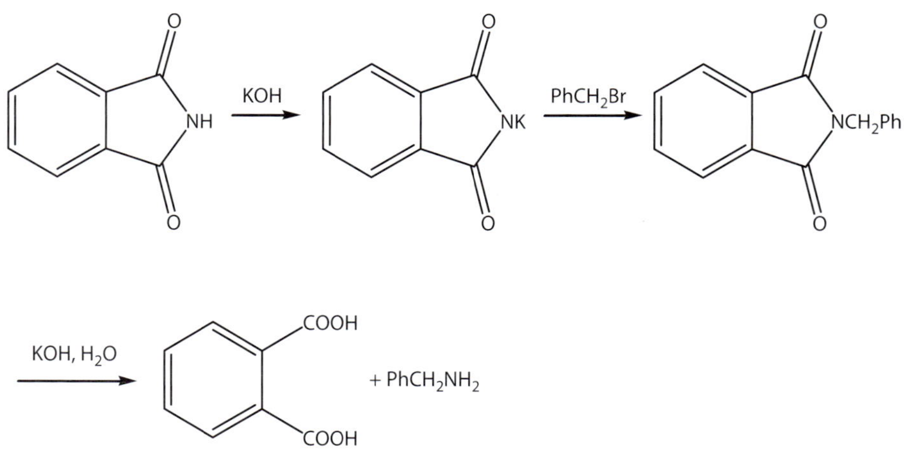

22.1(c) Another classic substitution with a synthon of ammonia—but with a slight twist. Because of the bulk of the reagent, the reaction can be stopped after a single substitution of the allyl dihalide. Hydrolysis of the hexamethylenetetramine gives the ammonium salt; when this is converted to the amine by the strong, non-nucleophilic base DBU, it cyclizes spontaneously to give the five-membered ring.

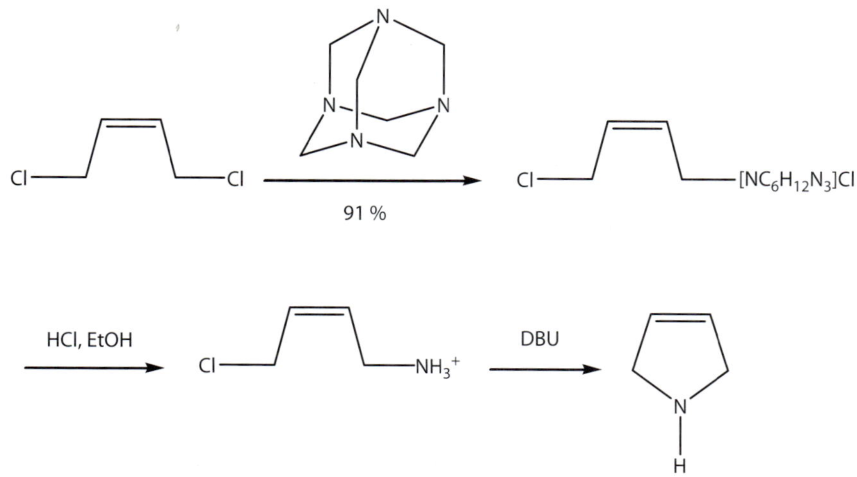

22.1.2 SYNTHESES FROM AMIDES

A key reaction of amides is that they can be reduced, using lithium aluminum hydride, to amines. Amides are readily synthesized from other acid derivatives and are non-nucleophilic because of the electron-withdrawing effect of the carbonyl group, so that there is no further reaction with electrophiles. Figure 22.9 shows an example; in the first step, the esters are converted into amides, then dimethylsulfate methylates the hydroxyl groups. Lithium aluminum hydride then reduces the amides to amines. It's useful to remember that amides can also be prepared by Beckman rearrangements (Figure 22.10).

The Hofmann degradation of amides leads to the formation of amines with one fewer carbon atom than the starting material. This is a particularly useful process in synthesizing aromatic amines or tertiary amines (Figure 22.11) where nucleophilic substitution or amide reduction routes are unavailable. A new reagent has now become popular for effecting this transformation, $PhI\{OC(=O)CF_3\}_2$ (phenyl iodine bis[trifluoroacetate], PIFA). This is a good oxidant, a derivative of iodine(III), and avoids the use of molecular bromine (problems of toxicity and handling) and strong base. The reaction mechanism is shown in Figure 22.12.

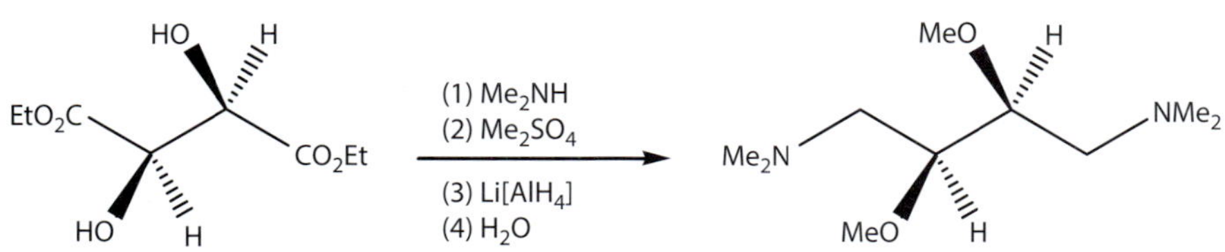

FIGURE 22.9 Synthesis of amines from amides.

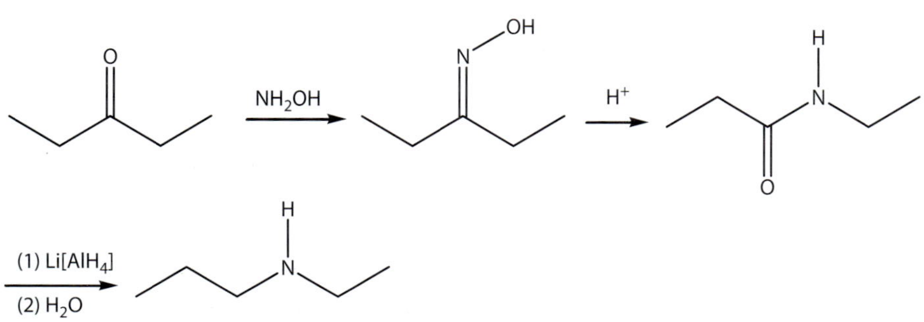

FIGURE 22.10 Synthesis of amides *via* the Beckmann rearrangement.

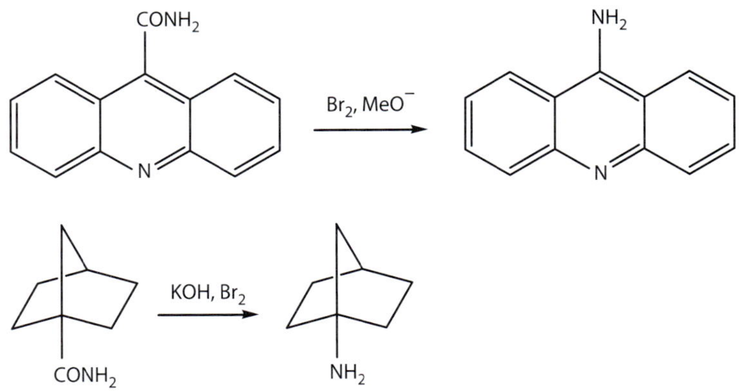

FIGURE 22.11 Synthesis of amines from amides using the Hofmann degradation.

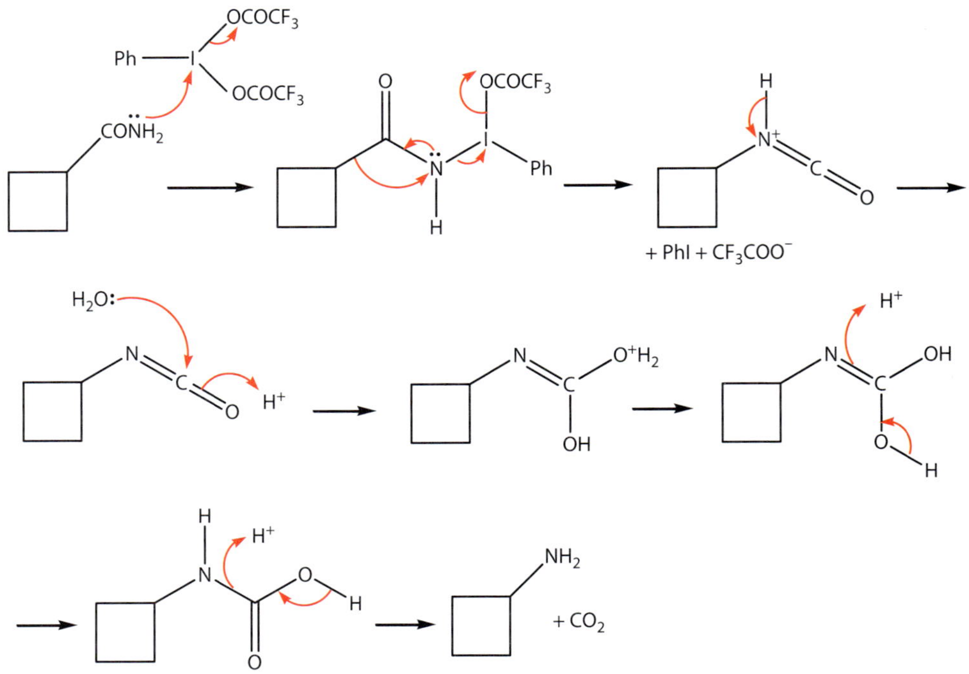

FIGURE 22.12 Mechanism of Hofmann degradation using PhI{OC(=O)CF_3}$_2$.

PROBLEM 22.2

Show how each of the following compounds could be prepared from an amide intermediate:

(a) $CH_3CH_2CH_2CH_2NHCHMe_2$

(b)

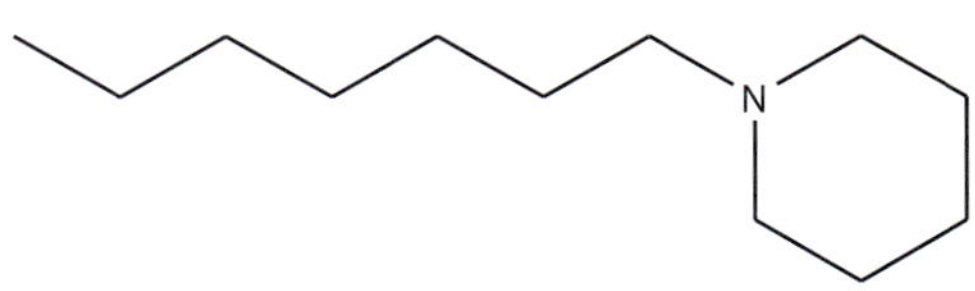

(c)

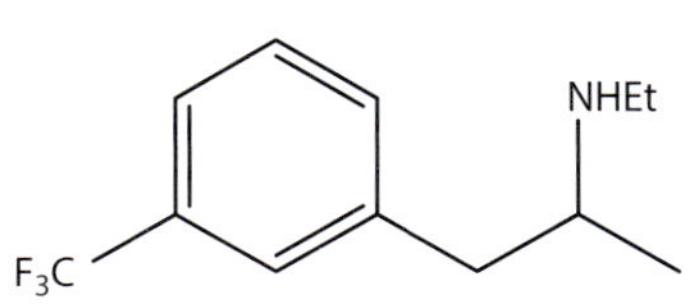

Solutions

(a) There is only one amide, **22.1**, which can easily be reduced to this amine:

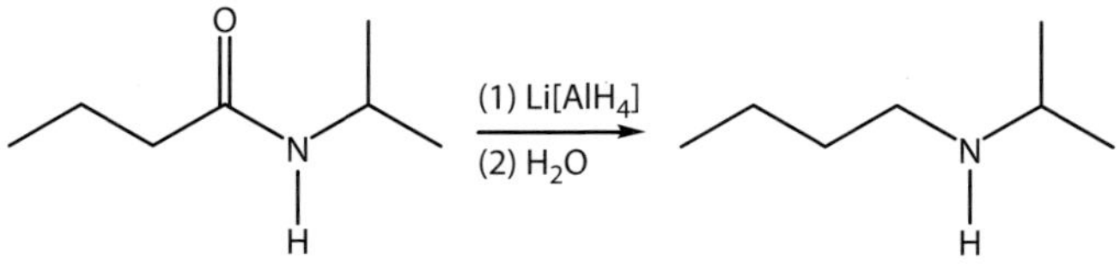

22.1

22.1 can be prepared from $CH_3CH_2CH_2COCl$ and Me_2CHNH_2.

(b) Although the reduction of either **22.2** or **22.3** would give the desired compound, there is no obvious method to synthesize **22.3** from easily available staring materials:

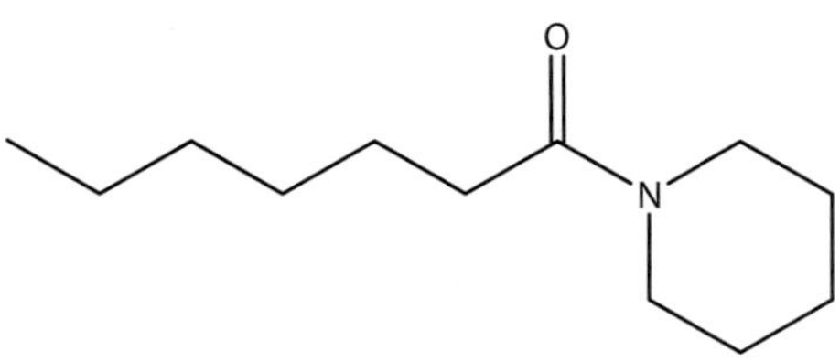

22.2

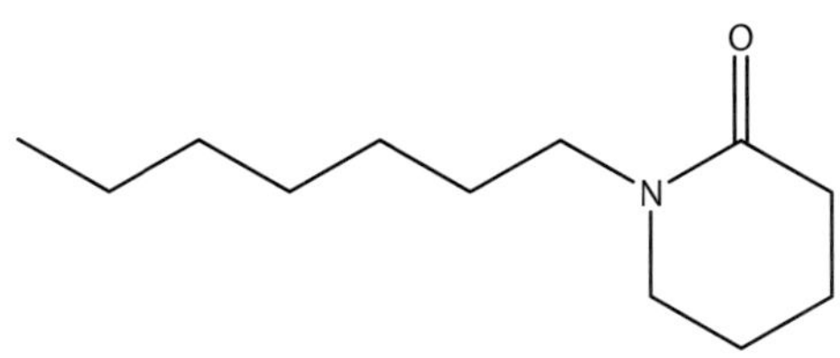

22.3

Preparation of **22.2** is relatively straightforward:

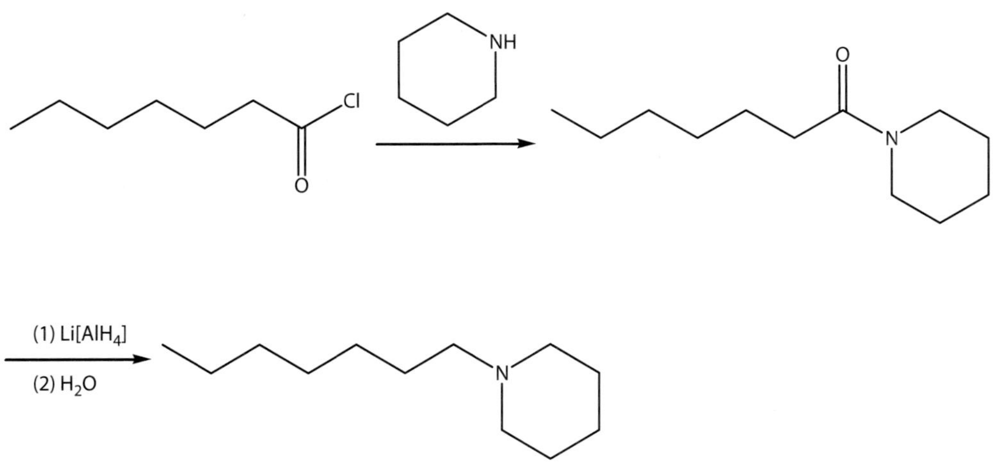

(c) This is a more unusual example in that the only possible amide precursor is

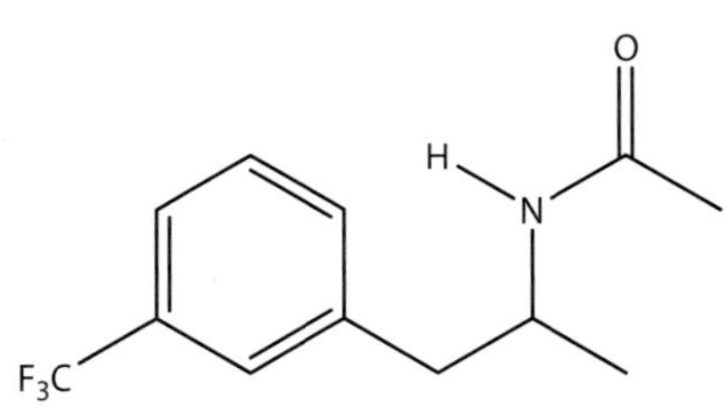

We can now do a retrosynthetic analysis:

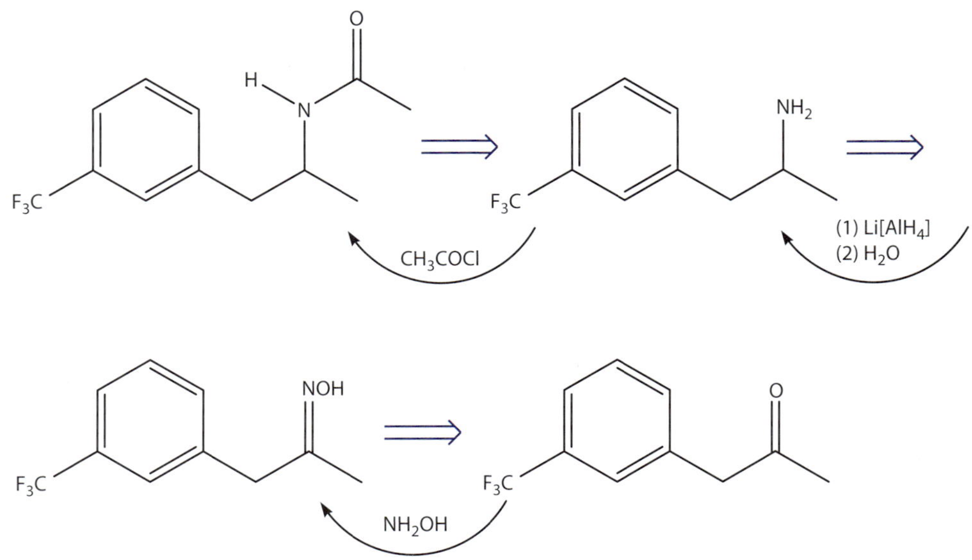

22.1.3 REDUCTION AND ADDITION AT CARBON–NITROGEN DOUBLE BONDS

Imines are readily prepared from aldehydes or ketones and amines. They may be reduced by molecular hydrogen directly to amines (the reaction is also accomplished with Na[BH_4], Na[BH_3CN], or Na[BH(OAc)$_3$]), but also undergo addition with Grignard reagents (Figure 22.13) leading to another synthesis of amines bearing a tertiary group, which are difficult to prepare otherwise. Examples are shown in Figure 22.14. The diastereoselectivity of the addition reaction in the second example allowed the synthesis of a useful chiral ligand.

The process of imine formation and reduction can be accomplished as a one-pot process, described as reductive amination (Figure 22.15). Oximes and hydrazones can also be reduced with lithium aluminum hydride, but these reactions only occasionally have any synthetic advantage over imine reduction.

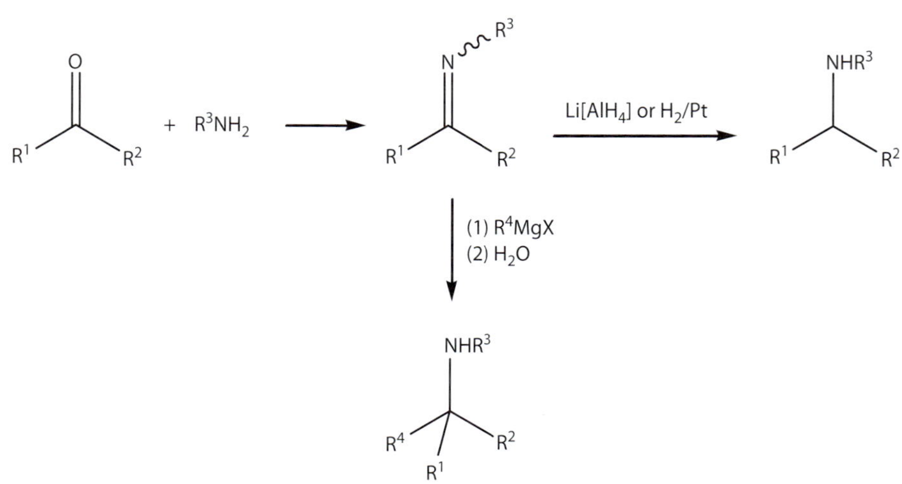

FIGURE 22.13 Syntheses of amines *via* imines.

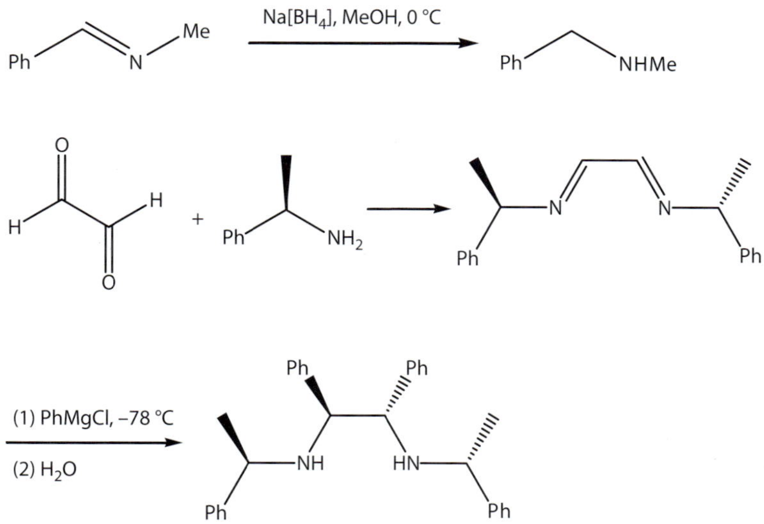

FIGURE 22.14 Examples of amine synthesis from imines.

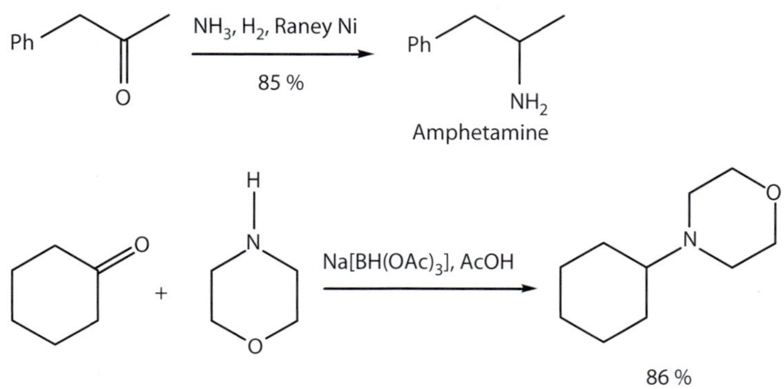

FIGURE 22.15 Reductive amination of ketones.

PROBLEM 22.3

Show how each of the following compounds could be prepared using a reductive amination strategy:

(a) $PhNHCHMe_2$
(b) $CH_3CH_2CH_2CH_2NHCHMe_2$
(c) $CH_3(CH_2)_5NHPh$

Solutions

(a) There is only one possible option here—react aniline, $PhNH_2$, with acetone:

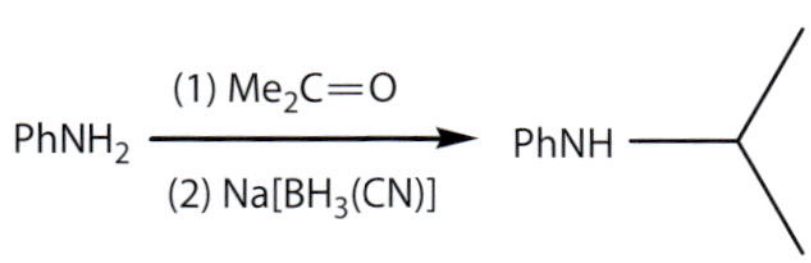

(b) There are two possible approaches here:

$$CH_3CH_2CH_2CH_2NH_2 \xrightarrow[\text{(2) Na[BH}_3\text{(CN)]}]{\text{(1) Me}_2\text{C=O}} CH_3CH_2CH_2CH_2NHCHMe_2$$

$$CH_3CH_2CH_2CHO \xrightarrow[\text{(2) Na[BH}_3\text{(CN)]}]{\text{(1) Me}_2\text{CHNH}_2} CH_3CH_2CH_2CH_2NHCHMe_2$$

(c) A single possible solution here, using aniline and hexanal, conditions as earlier.

22.1.4 REDUCTION OF NITRO COMPOUNDS

In Chapter 13, we frequently used the reduction of aromatic nitro groups to anilines as a part of synthetic strategy (Figure 22.16). There are numerous reagents that will bring about the reduction; the most commonly used are tin or iron in acid, tin(II) chloride, or catalytic hydrogenation. The reaction mechanisms are complex, involving nitroso and hydroxylamine intermediates, and most involve radical processes. Similar reductions occur for aliphatic nitro compounds, mostly using catalytic hydrogenation (Figure 22.17). The use of acidic conditions in most of these reactions results in protonation of the product amine, so that it is no longer nucleophilic and cannot react with any of the intermediates in the process.

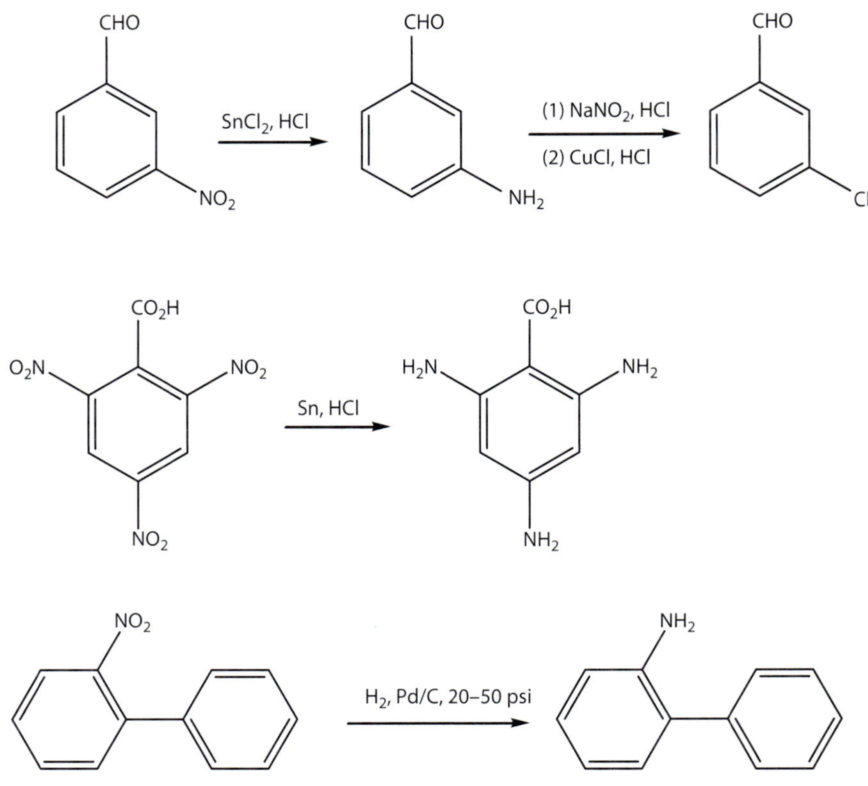

FIGURE 22.16 Synthesis of arylamines by reduction of nitro groups.

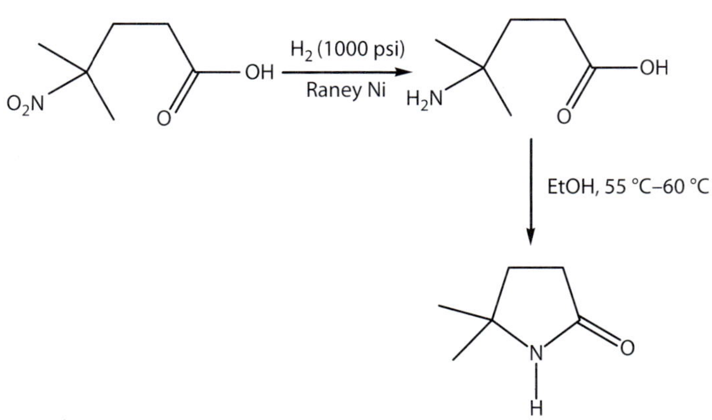

FIGURE 22.17 Reduction of aliphatic nitro compounds to amines.

PROBLEM 22.4

Study the synthetic sequence in the following, and answer the questions posed:

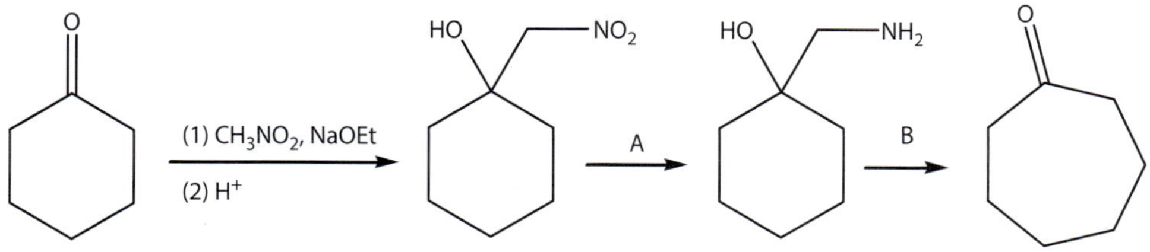

(a) Write a mechanism for the first step of the reaction.
(b) Suggest reagents for the steps marked A and B.
(c) Write a mechanism for the rearrangement in step B.

Solutions

(a)

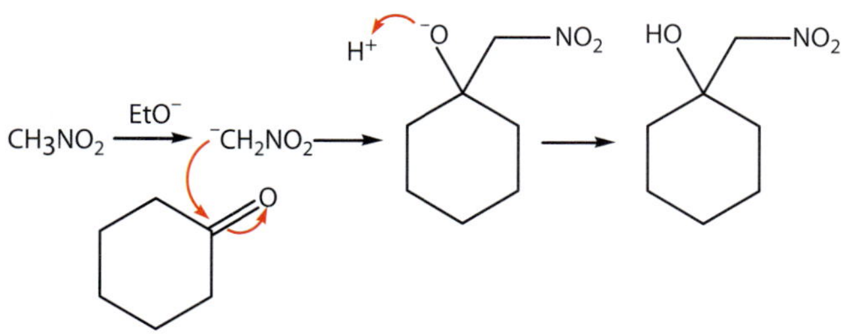

(b) A: H_2, Raney Ni, and AcOH; B: $NaNO_2$ and AcOH

(c)

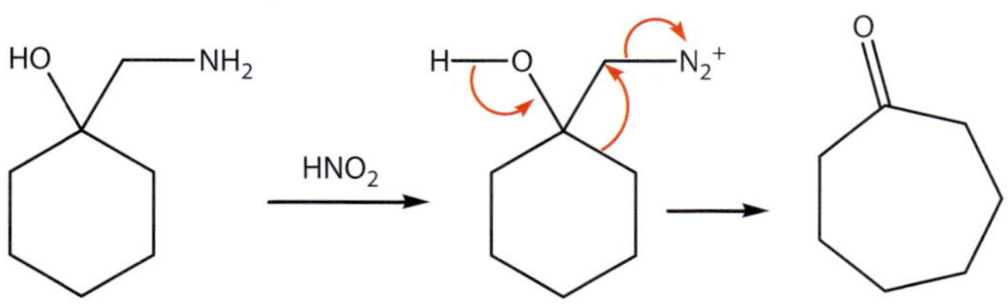

Key Points from Section 22.1

- Synthesis of amines by nucleophilic substitution is only successful with ammonia or amines when steric hindrance restricts multiple reactions, a five- or six-membered ring is being produced, or in the synthesis of amino acids or related compounds.
- Azide ion, hexamethylenetetramine, or phthalimide can be used as synthons for ammonia in amine synthesis by nucleophilic substitution. The cyanide ion is a synthon for $[H_2NCH_2]^-$.
- Amines may be synthesized by reduction or Hofmann degradation of amides.
- Reduction of imines or oximes yields amines. Imines react with Grignard reagents to give amines.
- Reductive amination of aldehydes and ketones using either amine/H_2/metal catalyst or amine followed by $Na[BH_3(CN)]$ gives amines.
- Reduction of nitroarenes to anilines is achieved using metal/acid, Sn(II) reduction. Aliphatic nitro compounds may also be reduced, usually by hydrogen/catalyst.

22.2 ALKALOIDS

The original definition of the class of compounds called alkaloids dates to Ladenburg in the 1880s. They were defined as plant derived, basic (the word alkaloid derives from alkali), and containing a nitrogen atom in a heterocyclic ring. Although most alkaloids are plant derived, it is now known that they are produced in all types of organisms. The definition was also extended to include synthetic analogs of natural materials. Most alkaloids are biosynthesized from amino acids; those that are not are sometimes referred to as pseudoalkaloids. Where the nitrogen atom is not part of a heterocyclic ring, they should be classified as amines rather than alkaloids;

nonetheless, a number of amines are usually referred to as alkaloids. The class of alkaloids is extremely structurally diverse and continues to grow and to present challenges both to structural and synthetic organic chemists. One possible method of subdivision is into monocyclic, bicyclic, and polycyclic structures, exemplified by, **22.4**, coniine (derived from hemlock); **22.5**, cocaine (from coca leaves); and **22.6**, morphine (from the opium poppy) respectively. They may also be categorized according to their mode of biosynthesis.

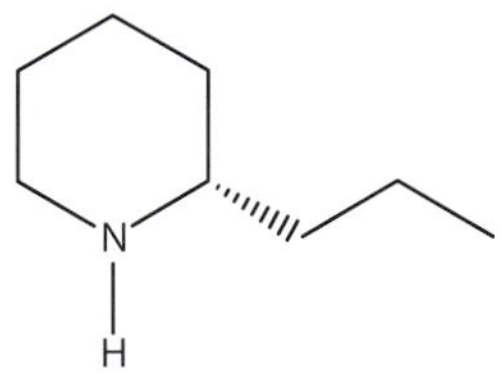

22.4, *S*-(+)-coniine

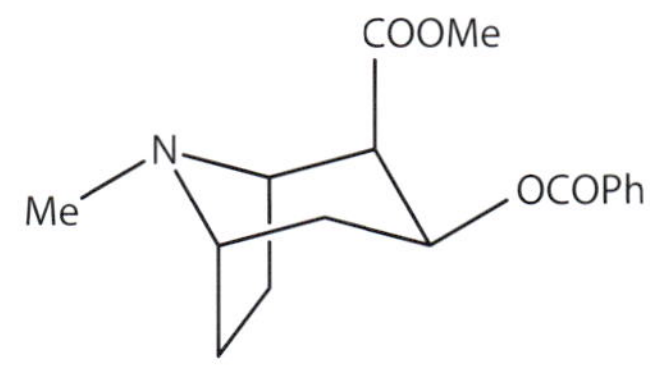

22.5, cocaine

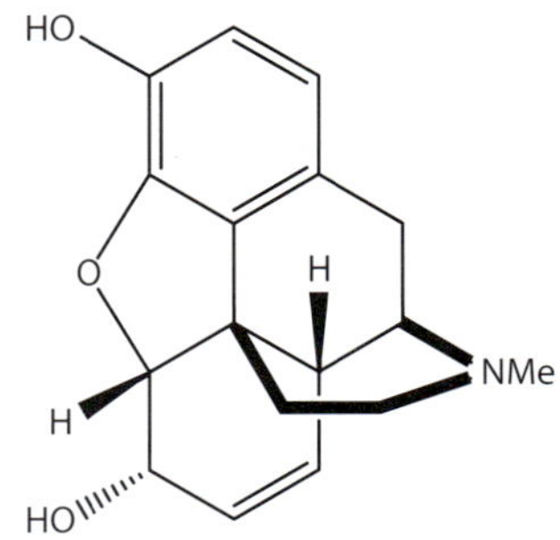

22.6, morphine

Many alkaloids are toxic, some highly so, and most have a bitter taste. The biological purpose of them in the producing organism is probably as an antifeedant compound, but they may also have growth-regulating, and other less well-understood, properties. Many have useful pharmacological attributes. Many names derive from the name of the producing plant, with a suffix –ine: thus, cocaine from *coca* and atropine from *Atropa belladonna*, deadly nightshade. Families of alkaloids with the same basic skeletons but different functionality are often derived from a single plant genus.

The properties of many alkaloids were known in the ancient world. Socrates was executed in 499 BCE using hemlock; Plato gives a detailed description of the progress of the drug and the manner of his death, which could scarcely be bettered today. Opium, derived from *Papaver somniferum* and containing some 50 alkaloids, has been used as an analgesic and narcotic since the Neolithic period. It was cultivated in Mesopotamia by 3400 BCE, in Egypt by 1300 BCE, as well as in Crete, Cyprus, India, Greece, and the Arab and Persian Empires. Morphine is the most abundant of the alkaloids in raw opium, named for Morpheus, the Greek god of dreams. Archaeological evidence indicates that the chewing of coca leaves was common in Peru from 6000 BCE, and alkaloid arrow poisons were used in hunting throughout South America from an early date. Tropane alkaloids, derived from the *Datura* genus, have been used for religious purposes to induce trances across the world, and in India as love potions. A combination of scopolamine and morphine was used to induce sleep until the nineteenth century. Scopolamine (**22.7**) is now used, at very low dose, medicinally, to treat motion sickness, but at higher doses, it has been used as a hallucinogen, a poison, a potential "truth serum" (the Nazis in WWII and the CIA in the 1950s), a date rape drug, and to incapacitate victims of theft. We cannot here discuss every class of alkaloid that exists, and any selection must be arbitrary, but we will return to the alkaloid drugs of abuse in a later chapter on forensic chemistry.

Papaver somniferum, the opium poppy

Datura plant. (Thanks to Camilla Tornoe for this photo.)

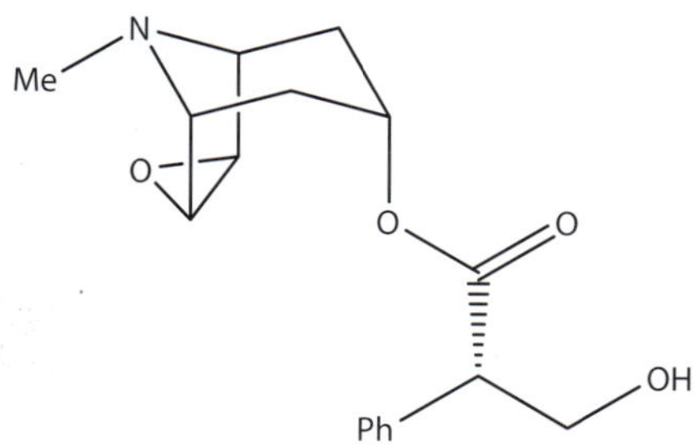

22.7, scopolamine

One of the best known alkaloids to students is caffeine, **22.8**, and the related compounds theobromine (from chocolate), **22.9**, and theophylline, **22.10**—these are described as xanthines (xanthine is the unsubstituted compound, **22.11**). All the xanthines are mild stimulants, and bronchodilators, and caffeine, which particularly increases alertness and ability to work, is often added to analgesics and antihistamines to counteract their soporific effects. Caffeine was isolated from coffee in 1820 and is also present in tea, kola nut, guarana berries, and yerba maté. A single espresso shot contains 80–100 mg caffeine, caffeinated soft drinks 10–50 mg, and Red Bull 80 mg per can. Although it is no longer limited in athletics (World Anti-Doping Agency [WADA] removed the limit in 2004), both caffeine and theobromine are prohibited in horse racing. Both horses and dogs metabolize these substances poorly, and theobromine is toxic to them. Theobromine (does not contain bromine; the name comes from the Greek θεός, *theos*, god, and ἀμβροσία, *ambrosia*, food of the gods, appropriate for something from chocolate ... ☺) is isolated from the cacao plant (genus *Theobroma*), as well as being present in tea and the kola nut. Theophylline is present in trace amounts in tea and in cocoa; it has been used to treat chronic obstructive pulmonary disease and asthma.

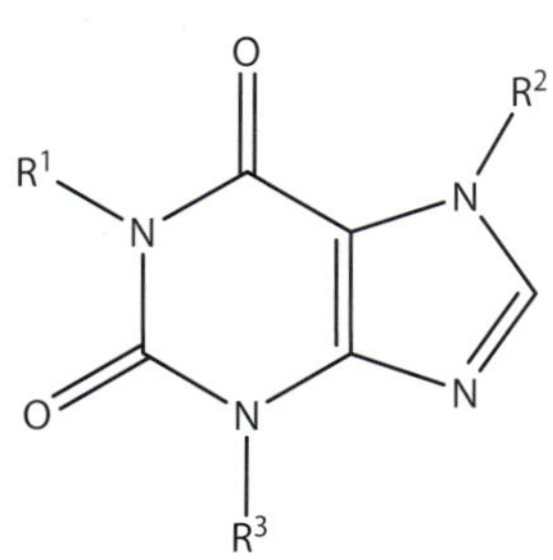

	R^1	R^2	R^3
22.8	Me	Me	Me
22.9	H	Me	Me
22.10	Me	H	Me
22.11	H	H	H

Coffee berries

Cocoa pods

Chocolate fountain

Malaria has been known for at least 10,000 years, with references in the literature of Egypt, India, and China, and archaeological evidence dating to the Stone Age. Until the seventeenth century, there was neither cure nor prophylaxis. Cinchona bark was brought to Europe from Peru by missionaries, in the 1630s, and a relatively crude extract, called Jesuits' bark or Peruvian bark, proved effective as a treatment. Quinine, **22.15**, was purified in 1820. The cinchona tree is supposed to have been named for the Countess of Chinchon, on the basis of a romantic, but untrue, story that she was cured of malaria by the bark. Quinine is named from the Quechua word *quina*, the local name of the cinchona tree. Quinine was used as a prophylactic against malaria until the Second World War; by that time, most production was in Dutch Java and unavailable after the Japanese invasion. Some 700 tons of *cinchona* alkaloids are still produced annually; about half are used medicinally and the rest as food and drink additives. The classic "gin and tonic" began as a palatable way to disguise the very bitter taste of quinine. Although quinine was synthesized in 1944 by Woodward and Doering, no synthesis has proved commercially competitive with extraction of the natural product. Quinidine, **22.13**, a stereoisomer of quinine, is an antiarrhythmic agent.

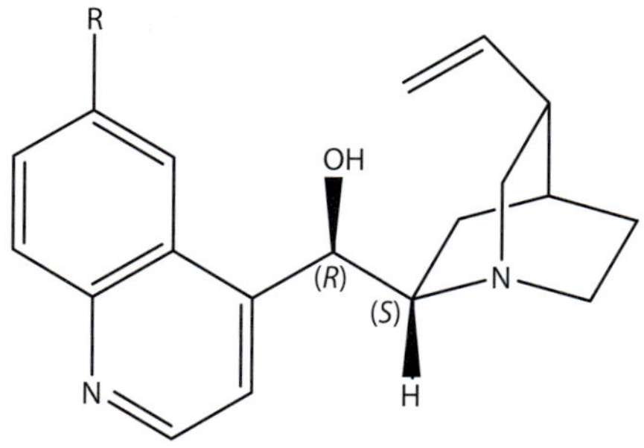

22.12, R = H, cinchonine
22.13, R = OMe, quinidine

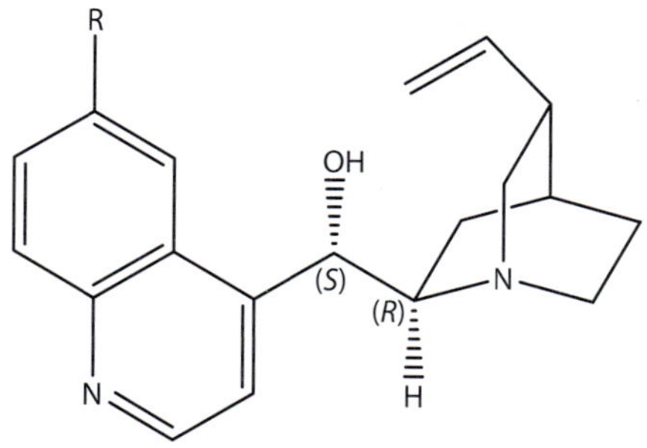

22.14, R = H, cinchonidine
22.15, R = OMe, quinine

Chloroquine, **22.16**, is another related, but synthetic, quinoline alkaloid, developed in the 1940s and widely used prophylactically against malaria until the parasite started to show resistance in the 1970s. Mefloquine (Lariam®), **22.17**, was then used in areas with drug-resistant malaria. Both chloroquine and mefloquine are suppressives and work only once the parasite is in the bloodstream, hence the importance of continuing to take the medication after return from malarial areas. Since the malaria parasite seems to develop resistance to new drugs quickly, drugs with completely different structures and modes of action were sought; one of the best treatments for malaria is now a combination therapy using artemisinin (a terpene, derived from the Chinese herb *Artemisia annua*) or one of its semisynthetic analogs together with one of the quinoline alkaloids. Artemisinin is not itself used in prophylaxis as its biological half-life is too short.

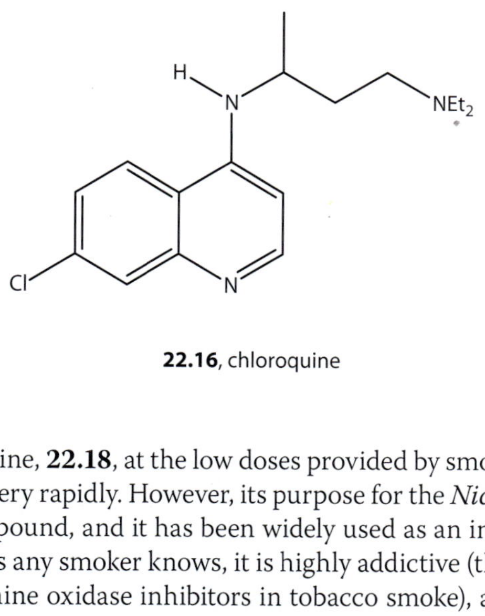

22.16, chloroquine

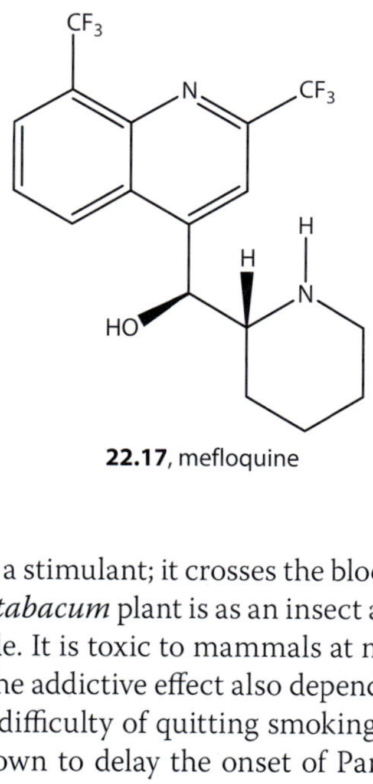

22.17, mefloquine

Nicotine, **22.18**, at the low doses provided by smoking, is a stimulant; it crosses the blood/brain barrier very rapidly. However, its purpose for the *Nicotiana tabacum* plant is as an insect antifeedant compound, and it has been widely used as an insecticide. It is toxic to mammals at moderate doses. As any smoker knows, it is highly addictive (though the addictive effect also depends on the monoamine oxidase inhibitors in tobacco smoke), and the difficulty of quitting smoking is often compared with heroin addiction. Although it has been shown to delay the onset of Parkinson's disease, earlier data indicating benefit for Alzheimer's disease sufferers have not been confirmed. Galanthamine,* **22.19**, originally isolated from snowdrops (*Galanthus caucasicus*) and related species, is a reversible cholinesterase inhibitor and has been successfully used for the treatment of mild to moderate Alzheimer's disease.

* Also spelled galantamine.

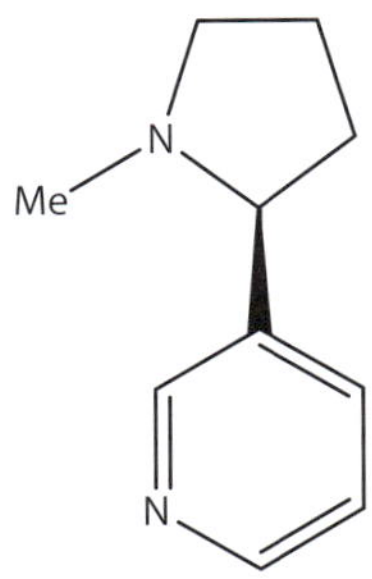

22.18, nicotine

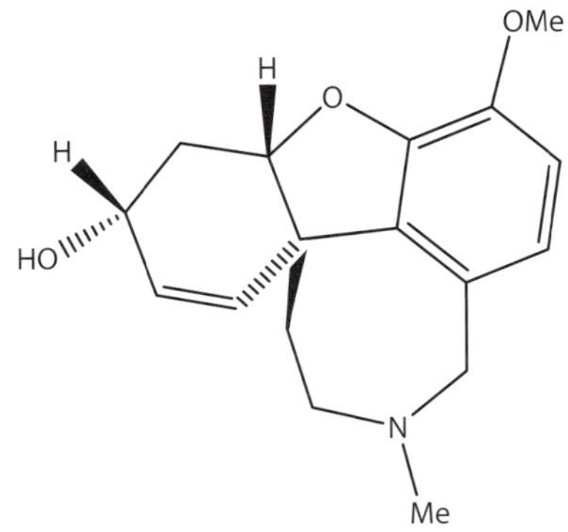

22.19, galanthamine

PROBLEM 22.5

Write the structure of one tautomer of each of the following:

(a)

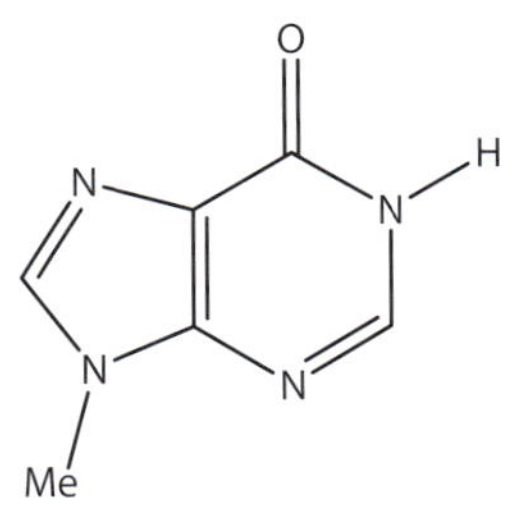

Hypoxanthine

(b)

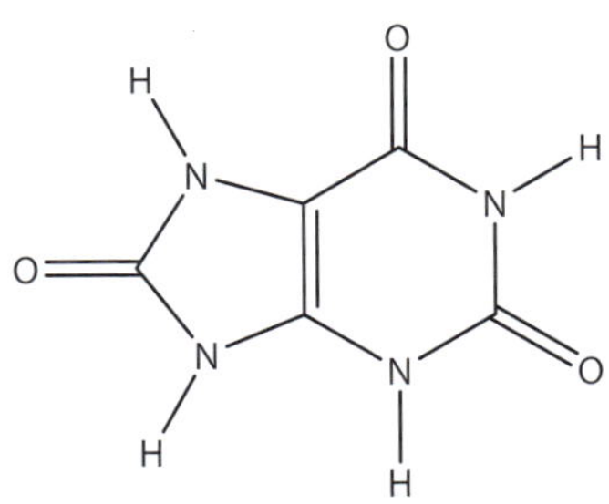

Uric acid

PROBLEM 22.6

The first biomimetic synthesis of the tropane skeleton was reported by Robert Robinson (Nobel Prize, 1947) in 1917; initially, the yield was only of the order of 20 % but it was later improved to 90 %. Write a mechanism for the process:

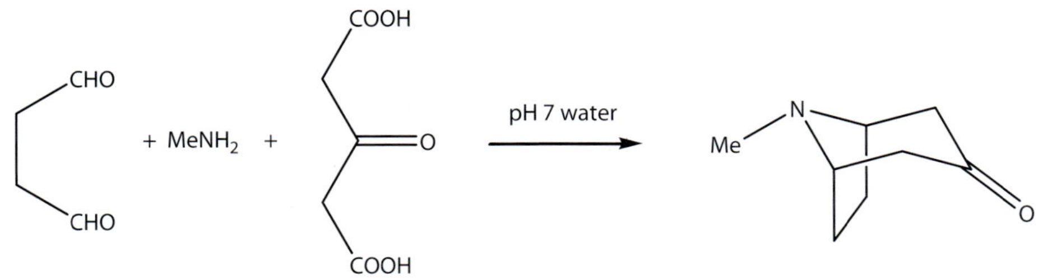

Solutions

22.5(a)

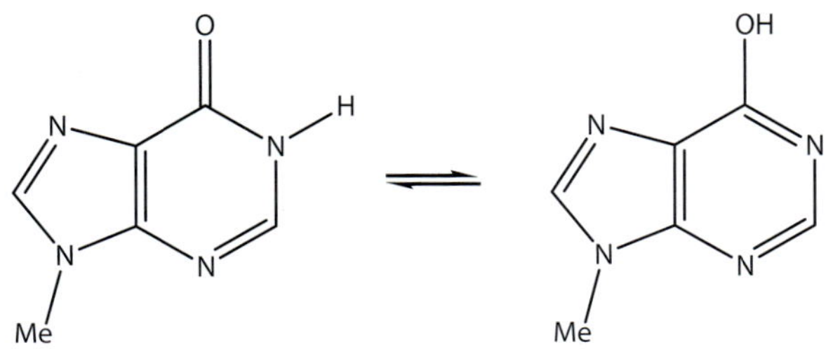

22.5(b)

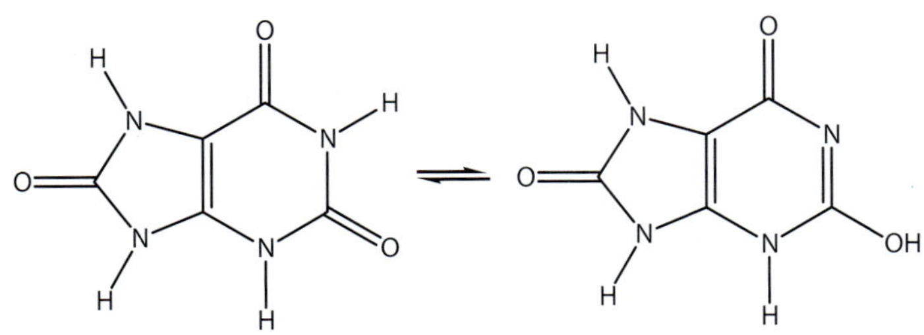

There are other possibilities

22.6

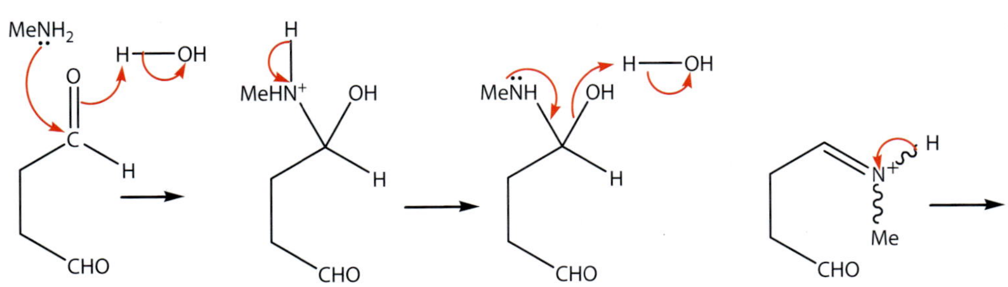

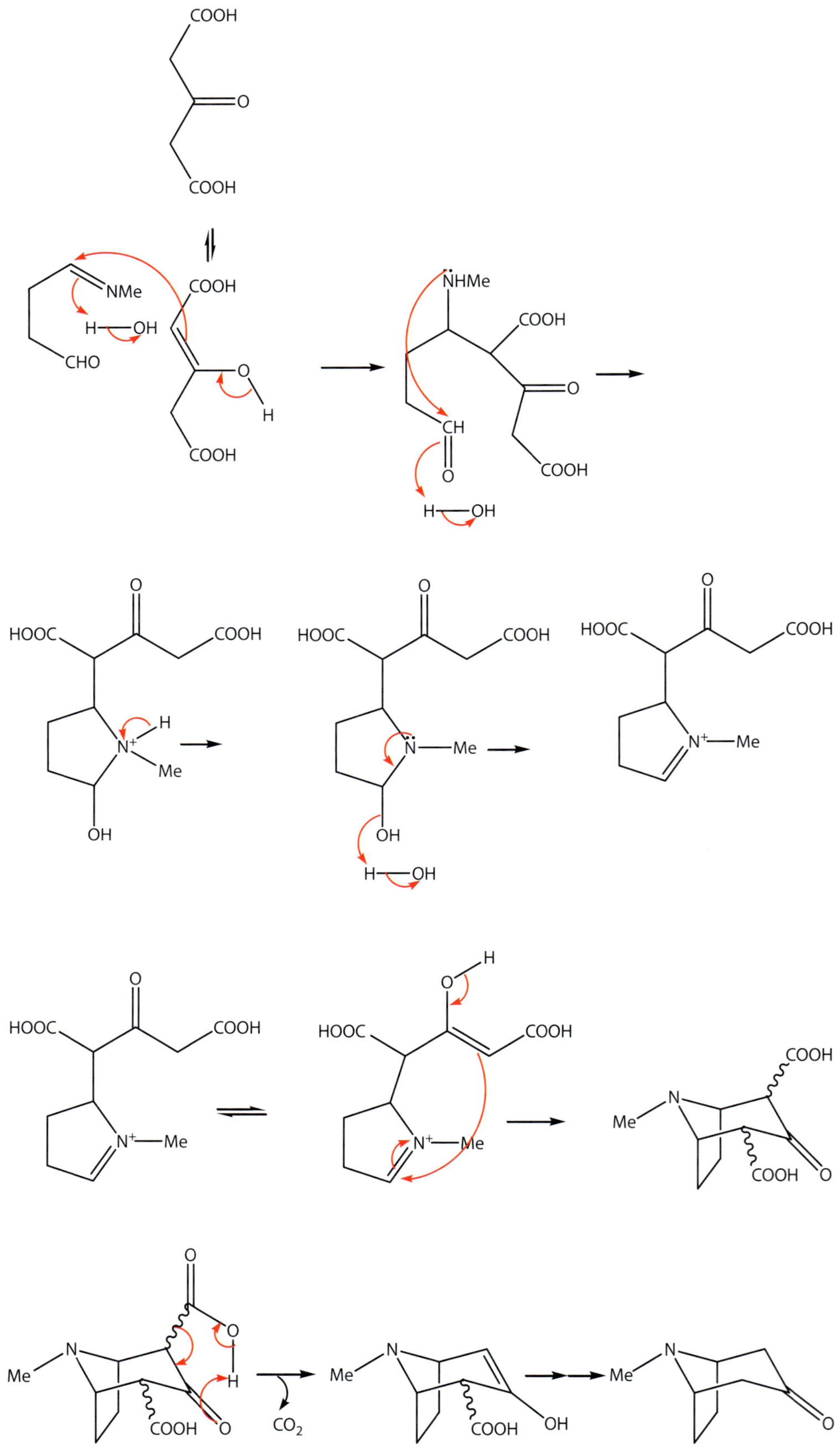

The final steps are not shown in detail—they involve ketonization of the enol and a second, identical decarboxylation.

22.3 AMINO ACIDS

Generally, when chemists or biochemists speak of amino acids, they are referring to the amino acids that make up peptides and proteins in living organisms, having the general structure **22.20**. All of these (except glycine, which is achiral, and cysteine, where the absolute configuration is reversed because sulfur outranks nitrogen and oxygen in the Cahn–Ingold–Prelog system) are chiral and have the *S*-configuration. Biochemists sometimes use an older system and designate them as L. Table 22.1 shows the 20 common amino acids found in proteins. Proline is properly described as an imino acid. Of these amino acids, 11 can be biosynthesized by humans, and 9, the so-called "essential" amino acids (His, Ile, Leu, Lys, Met, Phe, Thr, Try, and Val), must be obtained from dietary sources. Other amino acids may be essential for specific populations, such as small children or individuals with a metabolic abnormality.

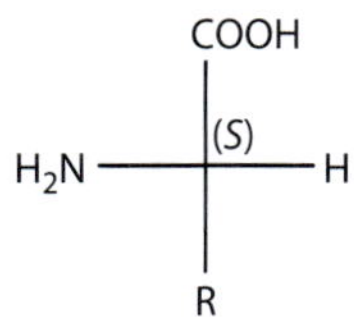

22.20

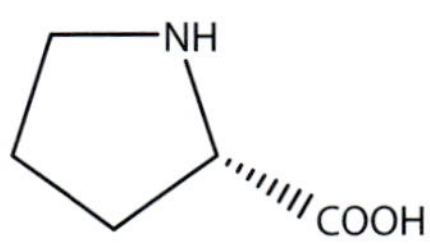

22.21, proline

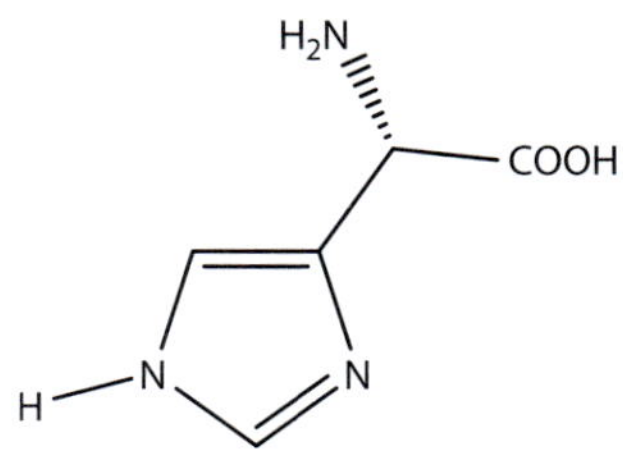

22.22, histidine

TABLE 22.1 Common Proteinogenic Amino Acids

Name	Side Chain	Abbreviations		Hydrophilic	Hydrophobic	Aromatic	Aliphatic	Acidic	Basic
Glycine	H	Gly	G		x		x		
Alanine	CH_3	Ala	A		x		x		
Valine	$CH(CH_3)_2$	Val	V		x		x		
Leucine	$CH_2CH(CH_3)_2$	Leu	L		x		x		
Isoleucine	$CH(CH_3)CH_2CH_3$	Ile	I		x		x		
Phenylalanine	CH_2Ph	Phe	F		x	x			
Tyrosine	$CH_2C_6H_4$-4-OH	Tyr	Y		x	x			
Tryptophan	CH_2-(3-indolyl)	Trp	W		x	x			
Aspartic acid	CH_2COOH	Asp	D	x			x	x	
Glutamic acid	CH_2CH_2COOH	Glu	E	x			x	x	
Arginine	$(CH_2)_3NHC(=NH)NH_2$	Arg	R	x			x		x
Lysine	$(CH_2)_4NH_2$	Lys	K	x			x		x
Serine	CH_2OH	Ser	S	x			x		
Threonine	$CH(CH_3)OH$	Thr	T	x			x		
Cysteine	CH_2SH	Cys	C	(x)			x		
Asparagine	CH_2CONH_2	Asn	N	x			x		
Glutamine	$CH_2CH_2CONH_2$	Gln	Q	x			x		
Methionine	CH_2CH_2SMe	Met	M		x		x		
Histidine[a]		His	H	x		x			x
Proline[a]		Pro	P		x		x		

[a] See structures 22.21 and 22.22.

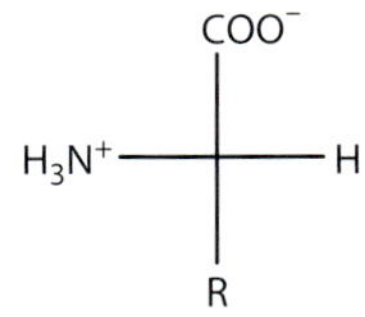

22.23

Amino acids are water-soluble, nonvolatile crystalline solids, with extensive hydrogen bonding in the solid state. In solution, at neutral pH, amino acids exist as zwitterions, **22.23**. For a simple amino acid, such as glycine, in basic water at high pH, the main species is $H_2NCH_2COO^-$. At low pH, in acidic water, both groups are protonated with the main species being $H_3N^+CH_2COOH$. A pH is defined at which the concentrations of these two species are equal, and the remaining material is the zwitterion. This is called the isoelectronic point and designated as pI. For glycine, pK_{a1}, for the ionization of the carboxylate group, is 2.35, and pK_{a2}, for the ionization of the $-N^+H_3$ group, is 9.78. pI is the average of these two, 6.07. Clearly, for the amino acids with additional acidic or basic groups, the pH dependence of the concentrations of various species will be more complex. It is worth noting that at 2.35, the pK_a value for the carboxylate group of glycine is significantly lower than that for acetic acid, at 4.8. This is because the electron-withdrawing effect of the $-N^+H_3$ group stabilizes the carboxylate anion.

22.3.1 SYNTHESIS OF AMINO ACIDS

We have already encountered an important synthesis of amino acids in the sequence of Figure 22.18—a Hell–Volhard–Zelinskii reaction followed by amination of the α-halo acid. There is no multiple substitution of the ammonia because the electron-withdrawing effect of the carboxylic acid makes the amine less nucleophilic. Also, as soon as it is formed, the acid converts to the zwitterionic form. An alternative process uses the Gabriel phthalimide synthesis (Figure 22.19).

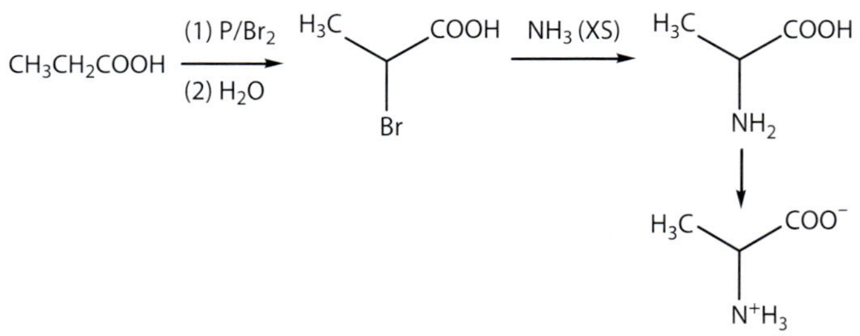

FIGURE 22.18 Synthesis of amino acids.

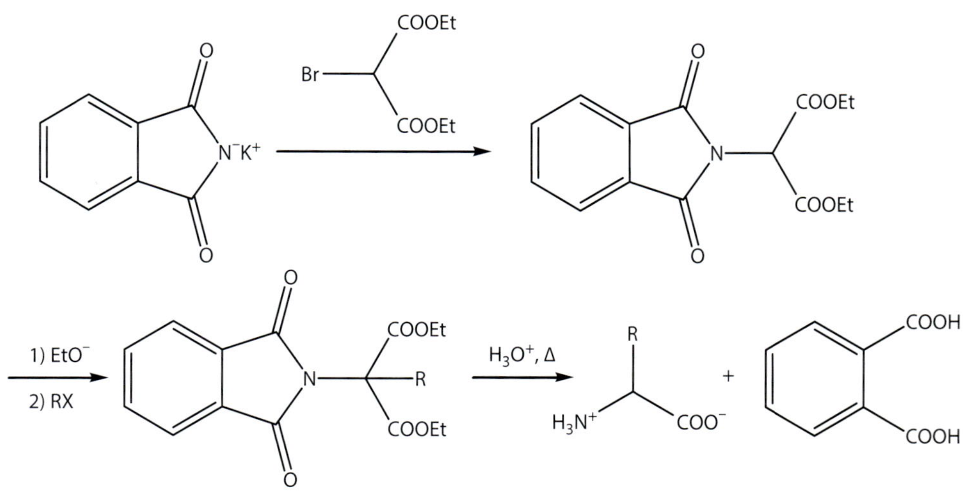

FIGURE 22.19 Gabriel synthesis of amino acids.

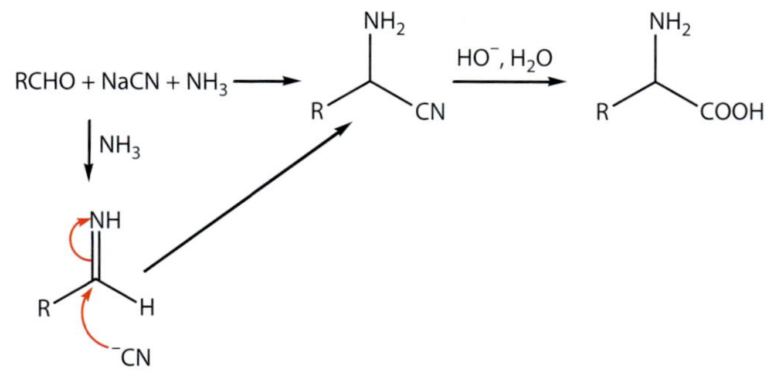

FIGURE 22.20 Strecker synthesis of amino acids.

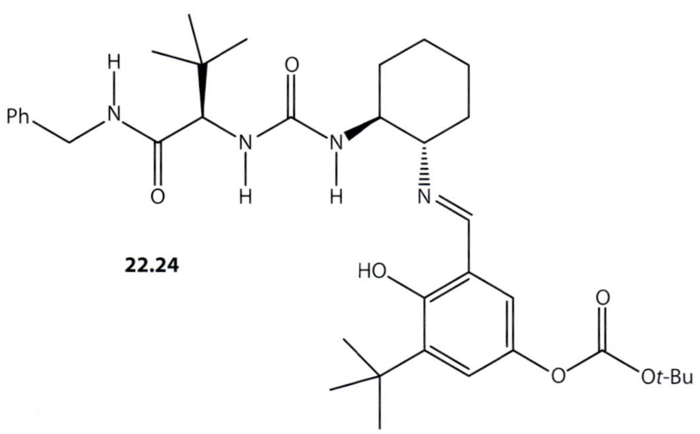

22.24

We have previously met the Strecker synthesis of amino acids (Figure 22.20, Section 14.3.4); like the two previous methods, this normally produces racemic amino acids. However, if the reaction is catalyzed by various additives such as the Jacobsen catalyst, **22.24**, aliphatic α-aminonitriles have been isolated in 80 %–90 % enantiomer excess.

Another fairly direct method for the synthesis of chiral amino acids involves the synthesis of a dehydroamino acid, **22.25**, which is then hydrogenated in the presence of a chiral rhodium complex. The dehydroamino acid is prepared *via* an azlactone, with the full mechanism shown in Figure 22.21. The azlactone is then hydrolyzed to the dehydroamino acid. The dehydroamino acid derivative is readily reduced with a standard palladium catalyst, but the product is then racemic. If the catalyst used is [Rh(COD)(DIPAMP)][BF_4], then the reaction proceeds with up to 99 % enantiomer excess (DIPAMP, **22.26**). This is the basis of the Monsanto process for the production of the anti-Parkinson's drug, L-DOPA, **22.27**, and William Knowles received the Nobel Prize for this work in 2001.

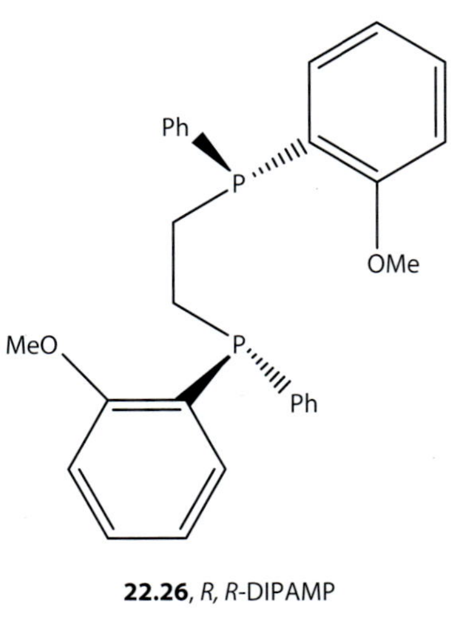

22.26, *R, R*-DIPAMP

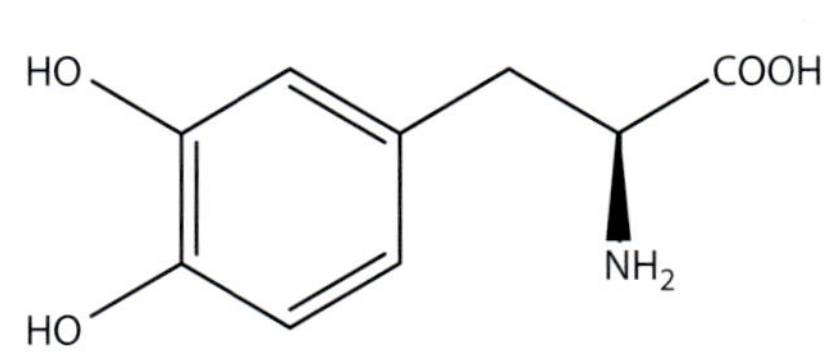

22.27, L-DOPA

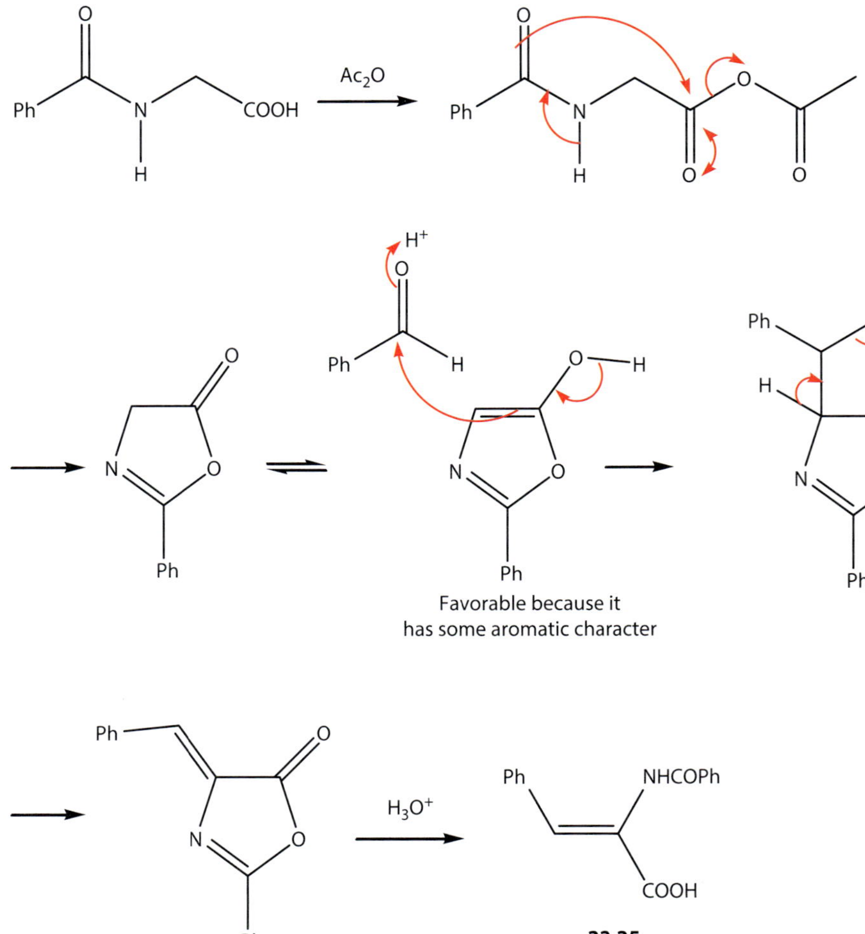

22.25

FIGURE 22.21 Azlactone synthesis of a dehydroamino acid derivative.

PROBLEM 22.7

Predict the products of the following reaction sequences:

(a)

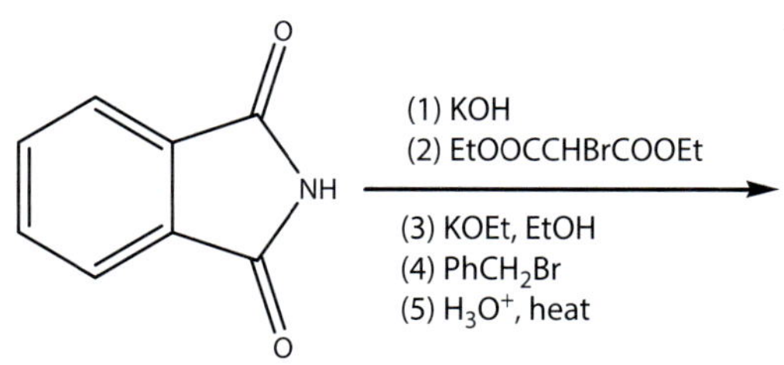

(b)

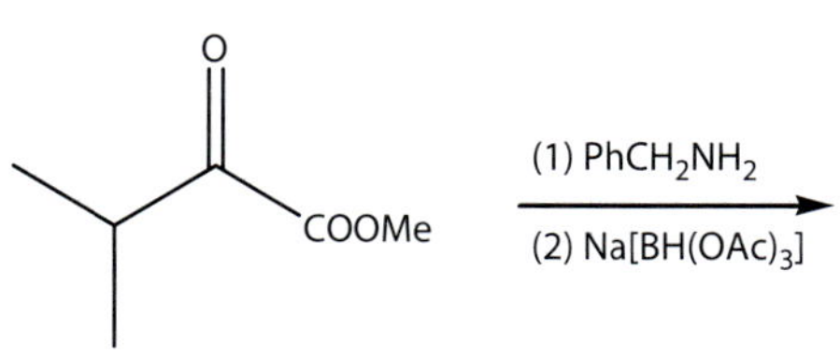

(c)

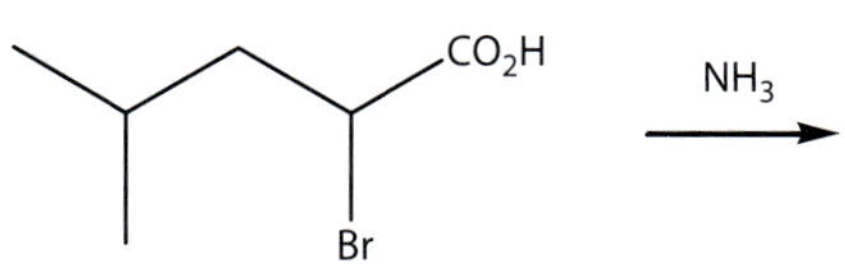

Solutions

(a) This is a standard phthalimide-based synthesis:

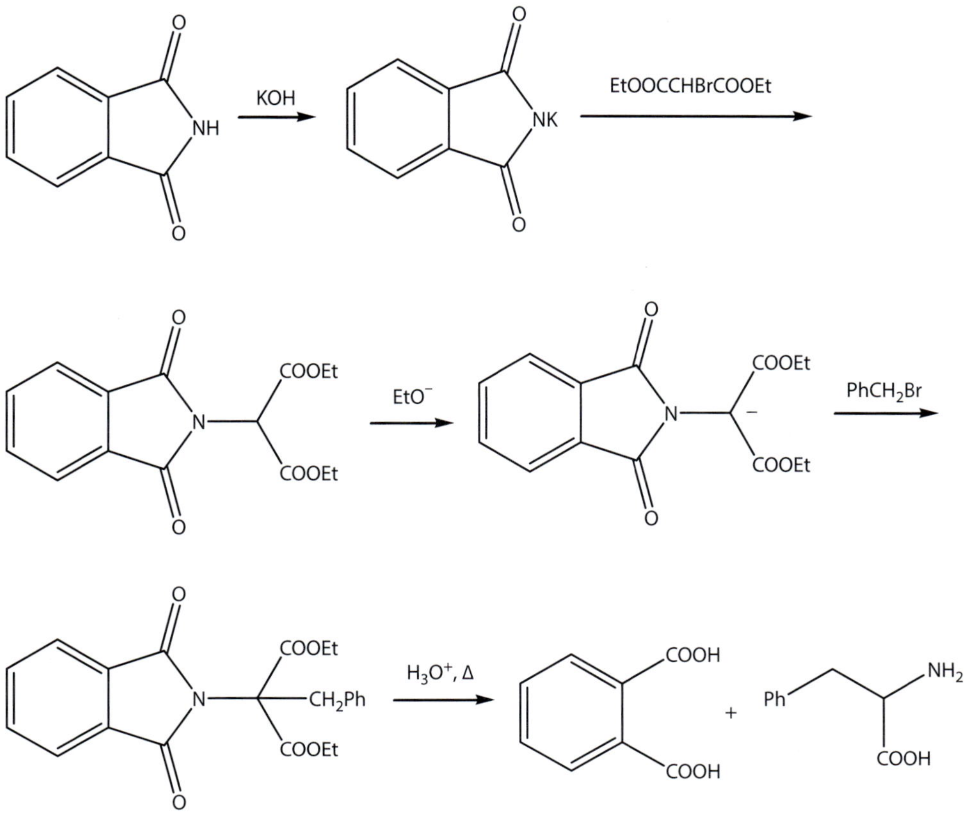

(b) This is a reductive amination, leading to a protected amino acid:

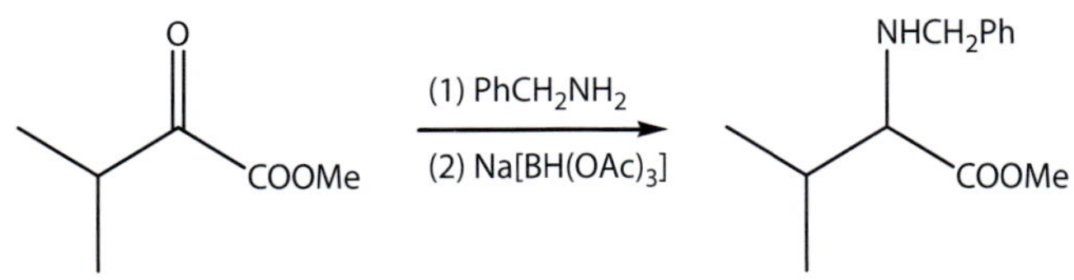

(c) Because this is an α-halo acid, direct substitution of the halide with ammonia is possible:

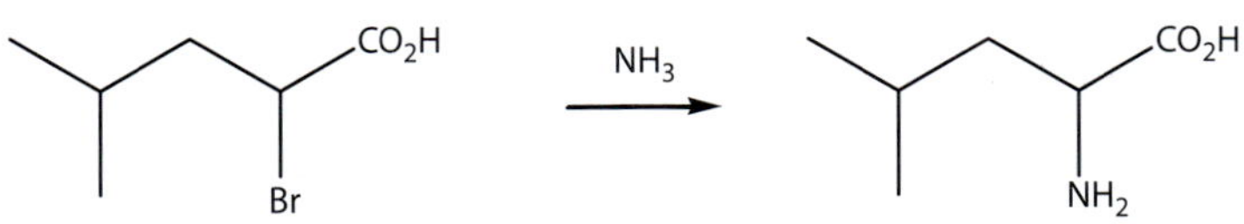

22.3.2 RESOLUTION OF AMINO ACIDS

There are numerous methods for the resolution of amino acids. One of the most straightforward is to convert the amino group into an amide (Figure 22.22), then resolve the amide by fractional crystallization of salt of the acid with an optically active base such as brucine, **22.28**, a toxic alkaloid related to strychnine. Alternatively, an enzyme such as hog kidney acylase may be used to hydrolyze only the natural *S*-enantiomer. However, resolution of amino acids generally means that only half the synthesized material is used, and the unwanted enantiomer is discarded. By using another enzyme from a different microorganism, the unwanted enantiomer can be recycled to give the racemate. In the most favorable cases, the two enzymes can operate together, to convert all of the amide into the required amino acid (Figure 22.23).

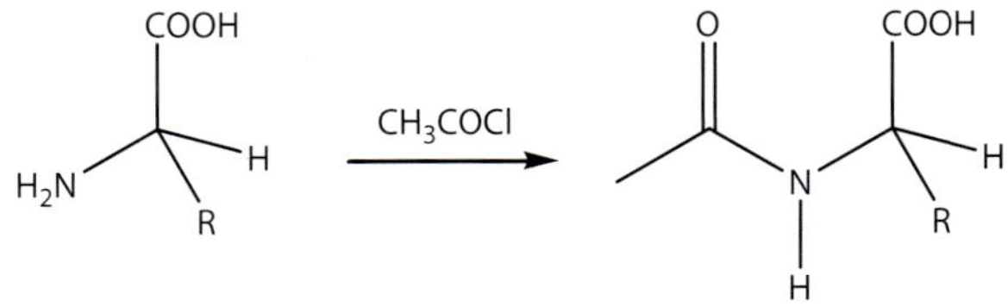

FIGURE 22.22 Amino acid acylation.

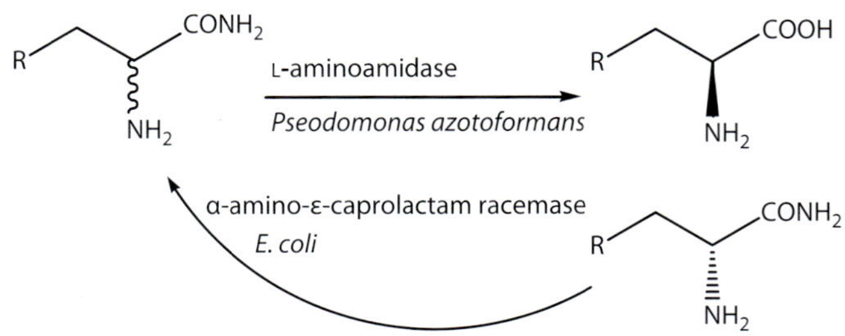

FIGURE 22.23 Enzymatic resolution/recycling of amino acid amides.

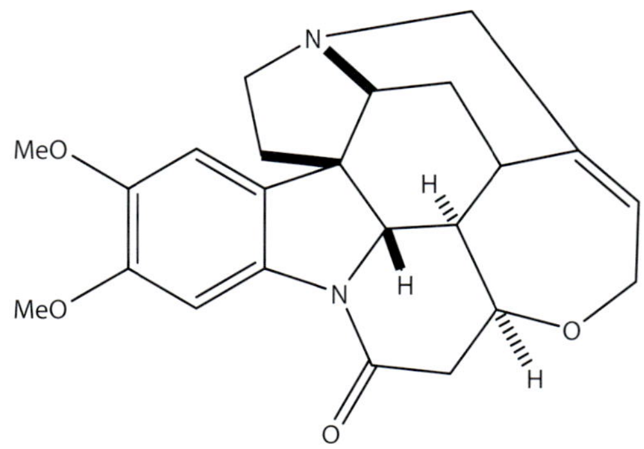

22.28, brucine

22.3.3 REACTIONS OF AMINO ACIDS

In this section, we will consider reactions of amino acids other than those leading to peptide formation—that is much the most significant process and will be covered in the next section. Many of the other reactions are simply those you would expect for primary amines, or carboxylic acids. The amino groups can be acylated by acid chlorides or anhydrides (Figure 22.24). Reaction with nitrous acid leads to diazotization; the reaction in Figure 22.25 goes with a double inversion of configuration, *via* an unstable α-lactone. The product is obtained in better than 95 % enantiomer excess.

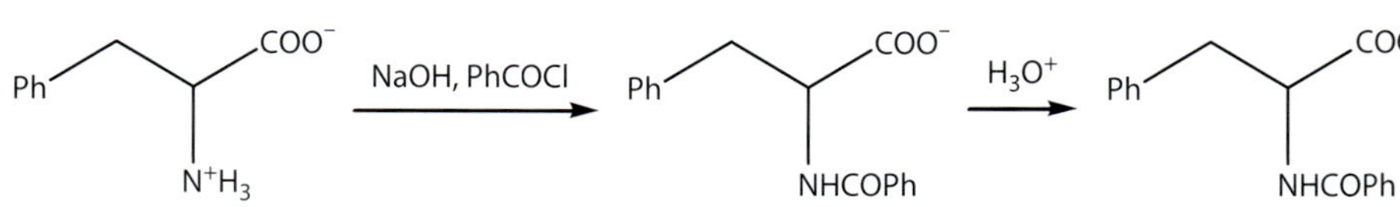

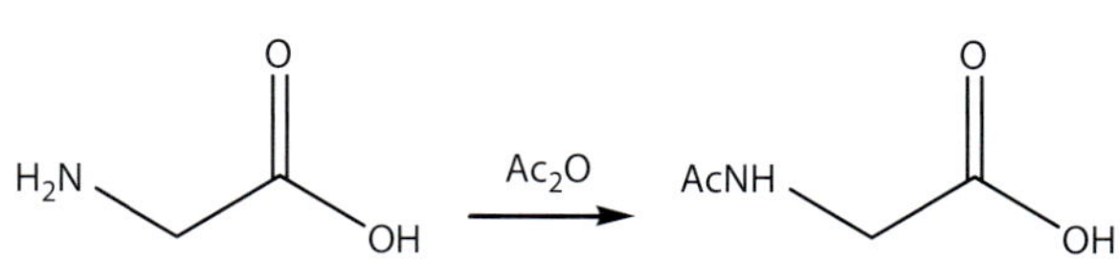

FIGURE 22.24 Acylation of amino acids.

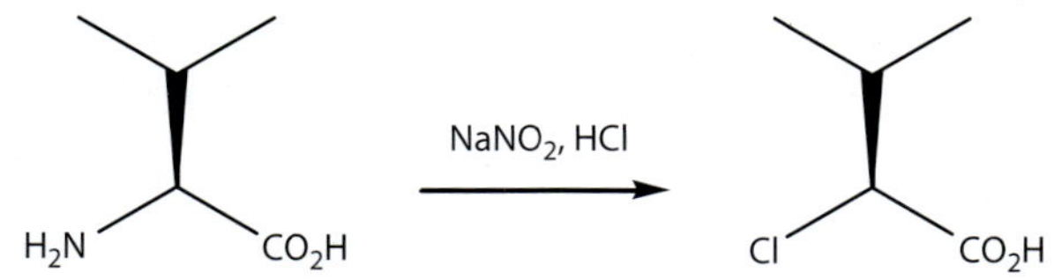

FIGURE 22.25 Diazotization of an amino acid.

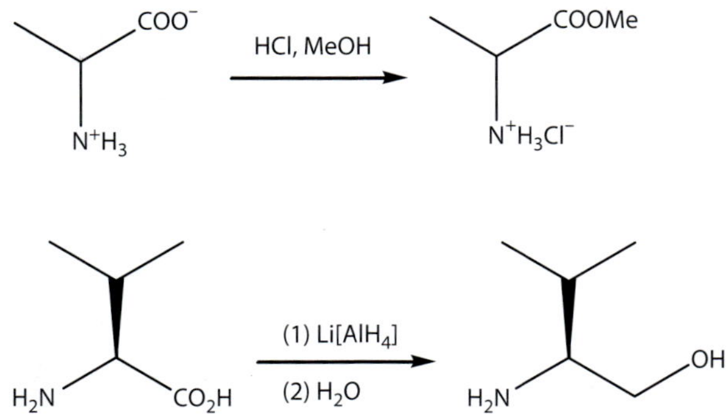

FIGURE 22.26 Reactions of the carboxylate group of amino acids.

The carboxylate group can be esterified using an alcohol and an acid catalyst, or reduced using lithium aluminum hydride (Figure 22.26). The esters are invariably isolated as their hydrochlorides or other salts—the free amino esters can polymerize as the amino group is acylated by the ester. The reduction of amino acids is more commonly carried out with other reagents, however—$BH_3 \cdot Me_2S/BF_3Et_2O$ and $Na[BH_4]$ followed by iodine are less expensive and less hazardous than $Li[AlH_4]$ but give equally good results. The amino alcohols produced have been widely used in the preparation of various chiral ligands for catalytic reactions, or as chiral starting materials in synthesis (Figure 22.27).

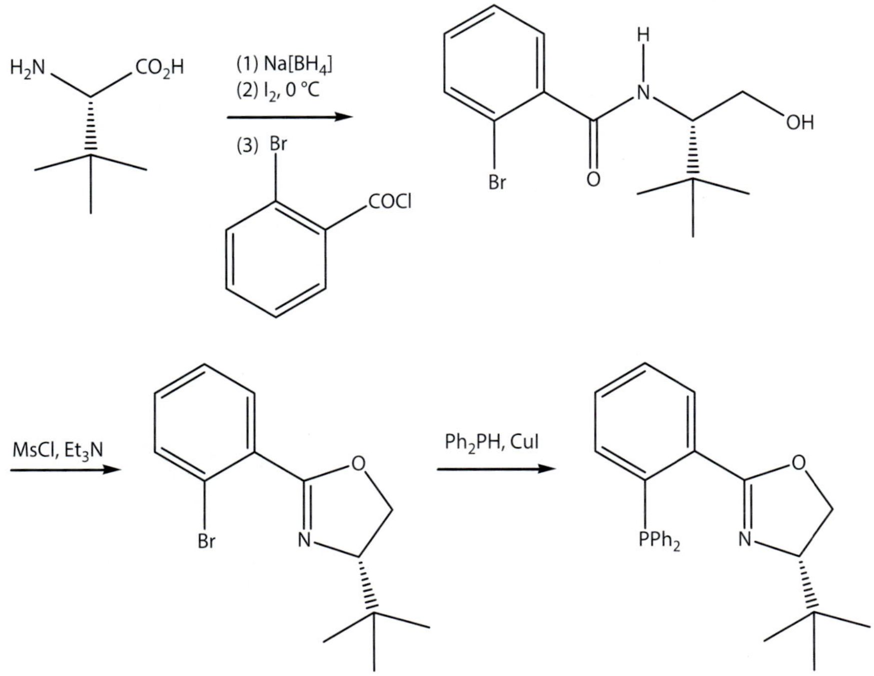

FIGURE 22.27 Amino acid reduction in the synthesis of *S-tert*-BuPHOX. This is a chiral ligand used in a number of enantioselective metal catalysed processes.

One reaction unique to amino acids is that with ninhydrin, **22.29**. Ninhydrin itself is colorless, but the product is an intense blue-purple in color (proline gives a yellow-orange product). The color is exceptionally intense, so that the amounts of amino acids can be determined quantitatively, even if only micrograms are present. Recent applications have included the use of ninhydrin to visualize fingerprints; human sweat contains small amounts of amino acids released from the termini of peptides.

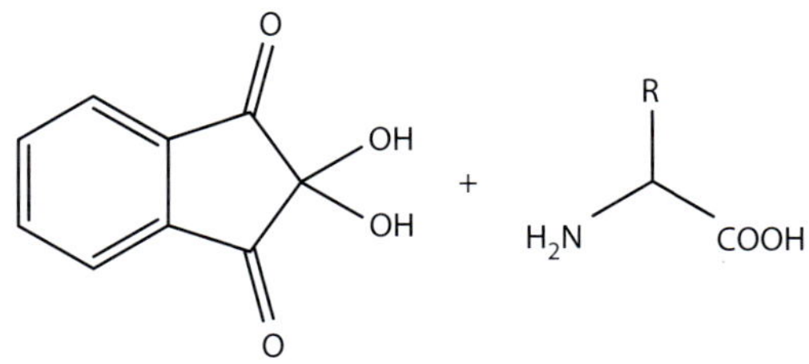

22.29

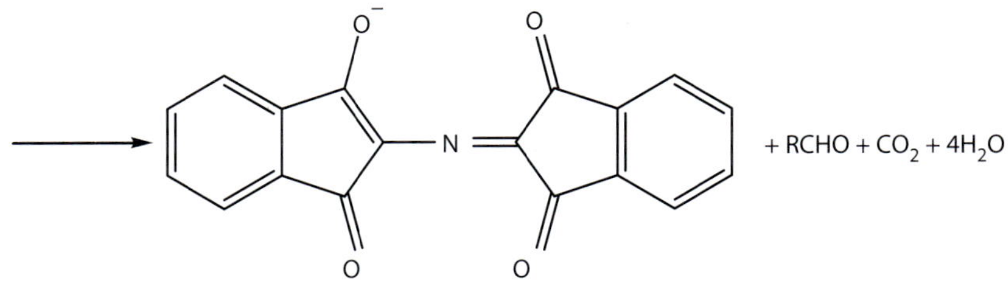

PROBLEM 22.8

Propose a reaction mechanism for the reaction of ninhydrin with amino acids.

PROBLEM 22.9

Predict the product(s) of each of the following reaction sequences:

(a)

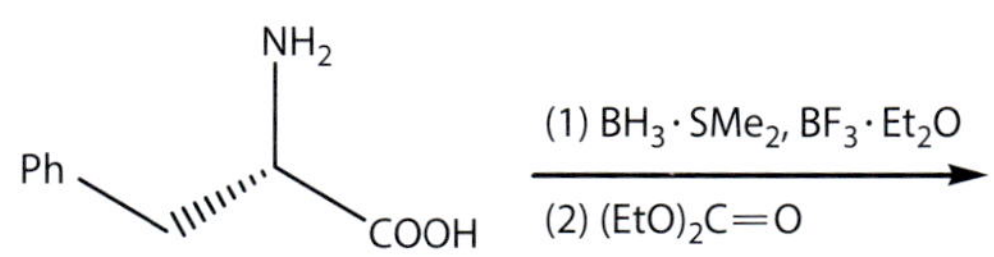

(b)

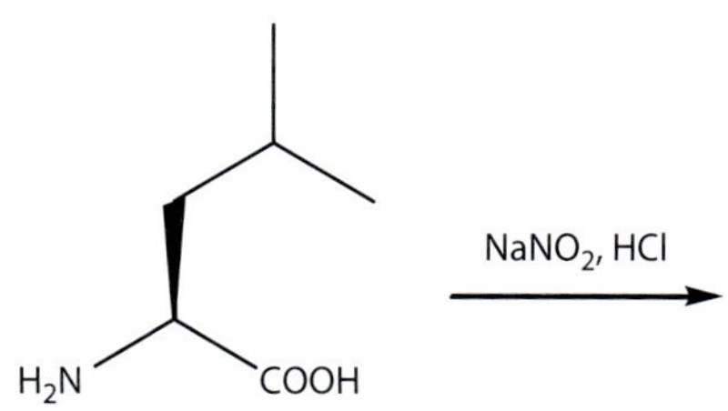

Solutions

22.8

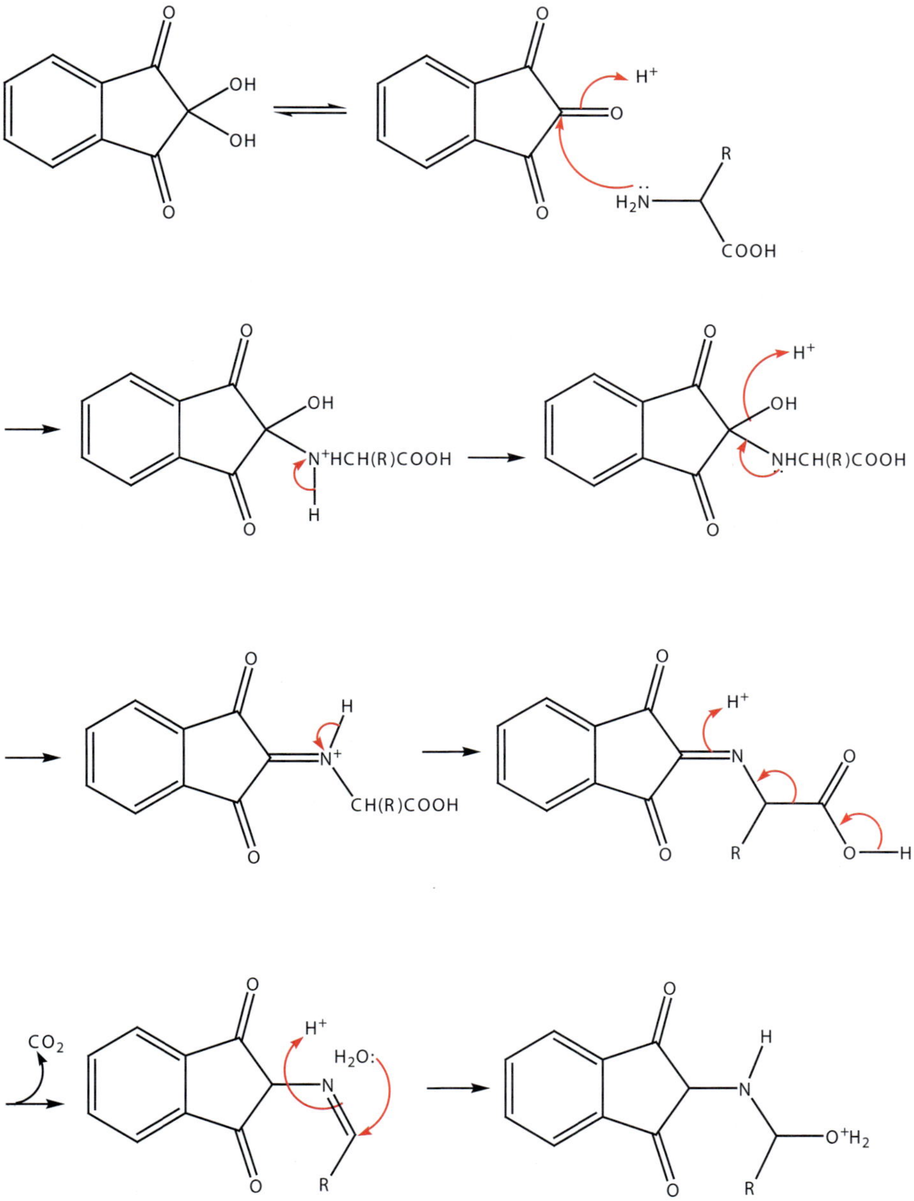

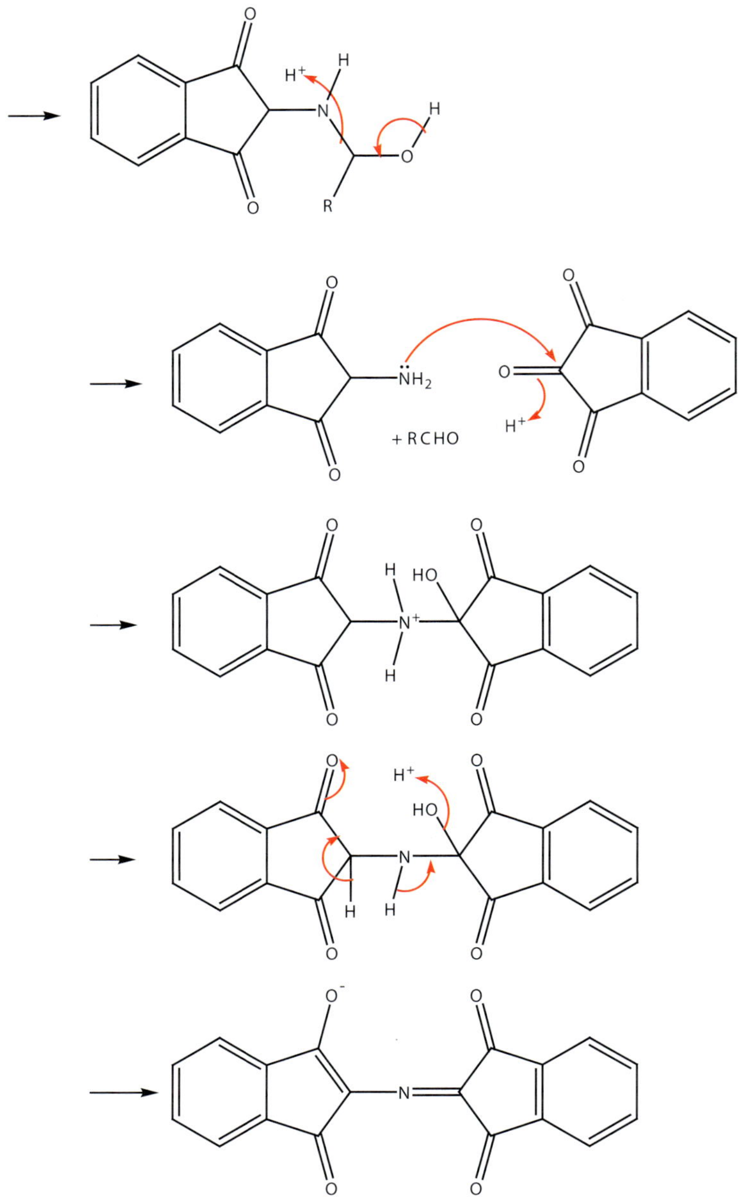

22.9

(a) In the first step, the acid is reduced to the alcohol:

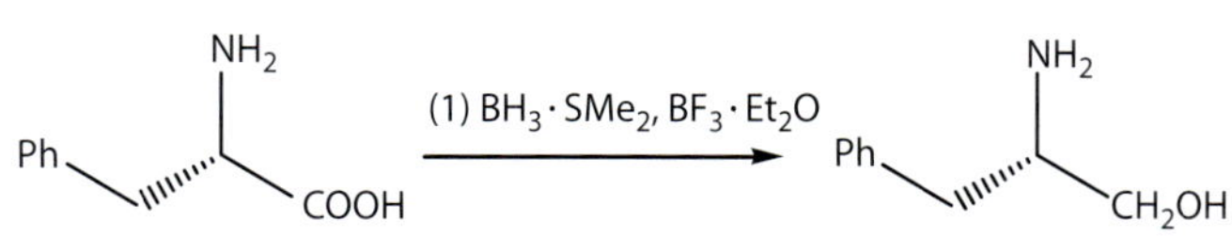

Then, the amino alcohol displaces the two –OEt groups from diethylcarbonate to give a chiral heterocycle:

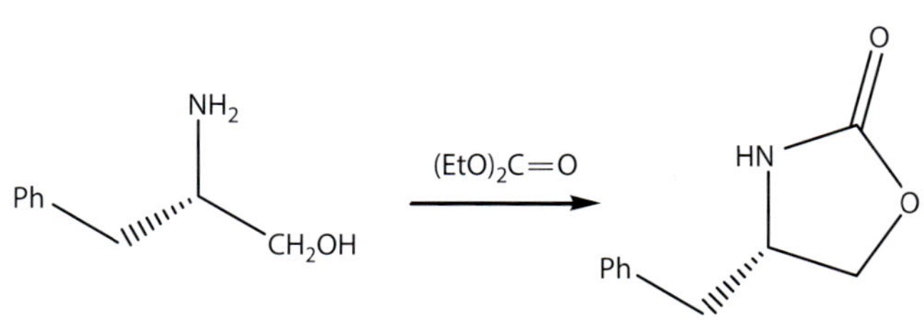

(b) This is a diazotization where the amino group is converted to $-N_2^+$. A double inversion *via* the α-lactone gives the chloride with retention of configuration:

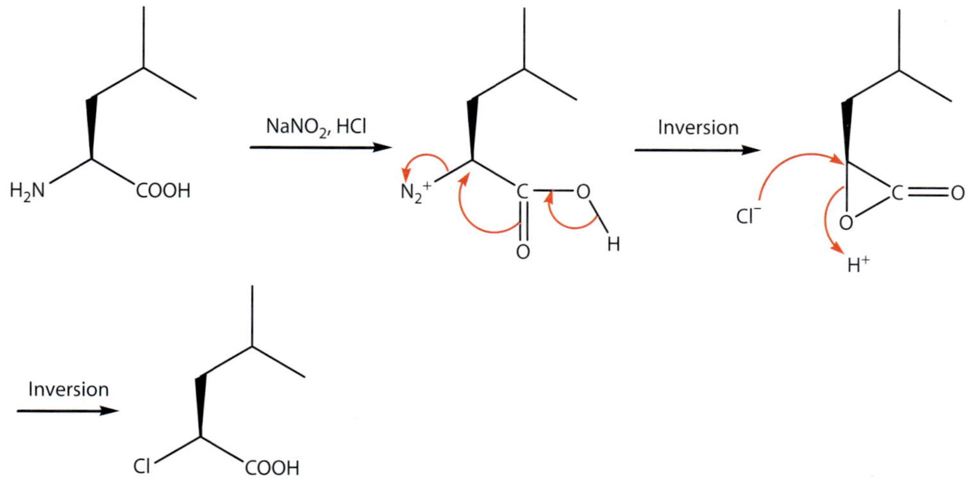

Key Points from Sections 22.2 and 22.3

- Alkaloids are nitrogen heterocyclic compounds, mostly of plant origin, with a wide range of complex structures.
- Many alkaloids are toxic, but the class also includes compounds with narcotic, analgesic, and other medicinal properties, as well as stimulants such as caffeine and theobromine and a range of recreational drugs.
- Twenty common amino acids are found in proteins. They almost (cysteine is *R* and proline is an imino acid) all have the formulation *S*-RCH(NH_2)COOH and have side chains, which may be hydrophilic or hydrophobic and acidic, basic, or neutral.
- Amino acids may be synthesized by displacement of halide from α-halo acids or esters with ammonia, or Gabriel synthesis.
- Racemic amino acids may be resolved by chemical or, more commonly, enzymatic techniques.
- Enantioselective synthesis of amino acids may be achieved by modification of the Strecker synthesis with a chiral additive or hydrogenation of dehydroamino acid derivative (prepared *via* an azlactone synthesis) in the presence of a metal complex of a chiral phosphine.

22.4 PEPTIDES

22.4.1 STRUCTURES OF PEPTIDES AND PROTEINS

We generally refer to structures with <100 amino acids as peptides, and those with more than that as proteins—but the distinction is fairly arbitrary. The key features of peptide structures are dictated by the planarity of the amide functional group, due to the involvement of the lone pair at nitrogen in resonance with the carbonyl group (**22.30**). Hydrogen bonding, both intra- and intermolecular, is important in determining the structure of peptides and proteins (**22.31**); the polarization of the amide because of the contribution of the second resonance form enhances the strength of these bonds.

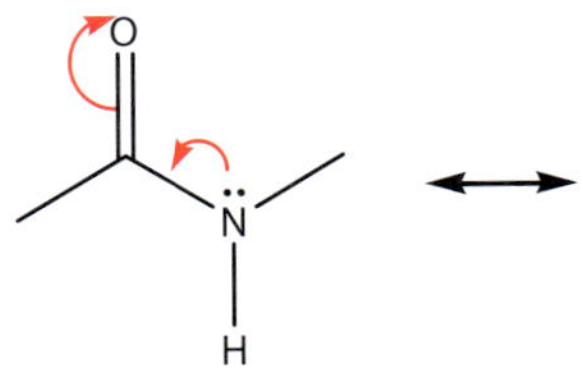

22.30

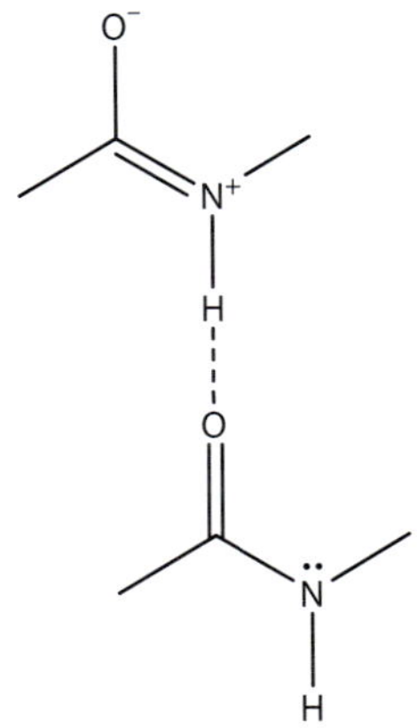

22.31

The primary structure of a peptide or protein is simply the amino acid sequence. The secondary structure is the twists and turns in the chains formed—directed by the planarity of the amide unit and the steric and electronic requirements of the side chains. The tertiary structure is the way in which the chains and sheets of the secondary structure fit together. Many proteins have subunits, which may or may not be identical, and the quaternary structure describes how these are linked together. There is a strong tendency for peptide chains to line up in antiparallel directions (**22.32**) so as to maximize the hydrogen bonding possible. Notice also the forwards-back arrangement of the side chains—this reflects the fact that all naturally occurring amino acids in proteins have the *S*-configuration.

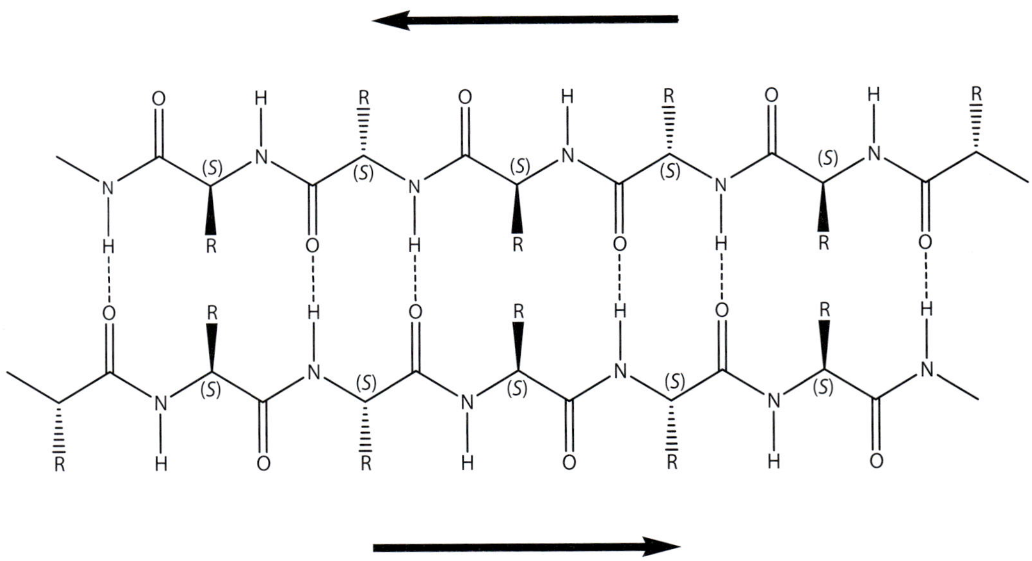

22.32

When the polymer backbone of the peptide is arranged in a sheetlike structure, the side chains (in Figure 22.28, the methyl groups of polyalanine) are directed to the same region of space. Rotation takes place at the α-carbons to reduce the steric interactions, and the structure formed is described as a β-pleated sheet. Another common structure is the α-helix—Figure 22.29a and b, respectively, shows side and top-down views of a polyalanine α-helix. Coiling takes place so that the hydrophobic groups bound to the α-carbon atom are directed toward the outside of the helix (the side chains are shown as gray stubs), with the hydrogen bonding providing the cohesive core of the helix. Two hydrogen bonds are highlighted in magenta. In Figure 22.29b, four carbonyl groups point toward the viewer.

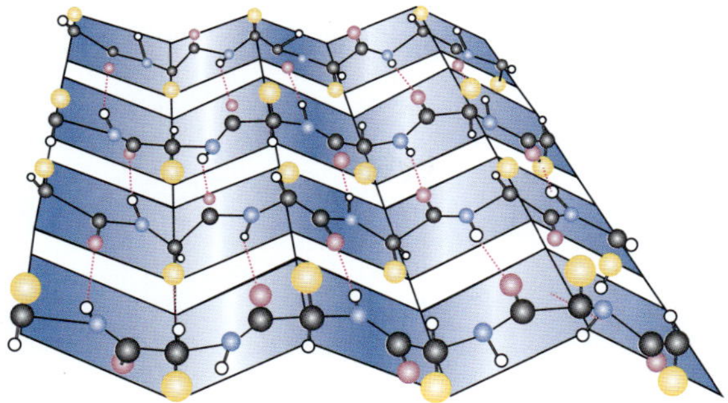

FIGURE 22.28 Schematic of a β-pleated sheet for polyalanine.

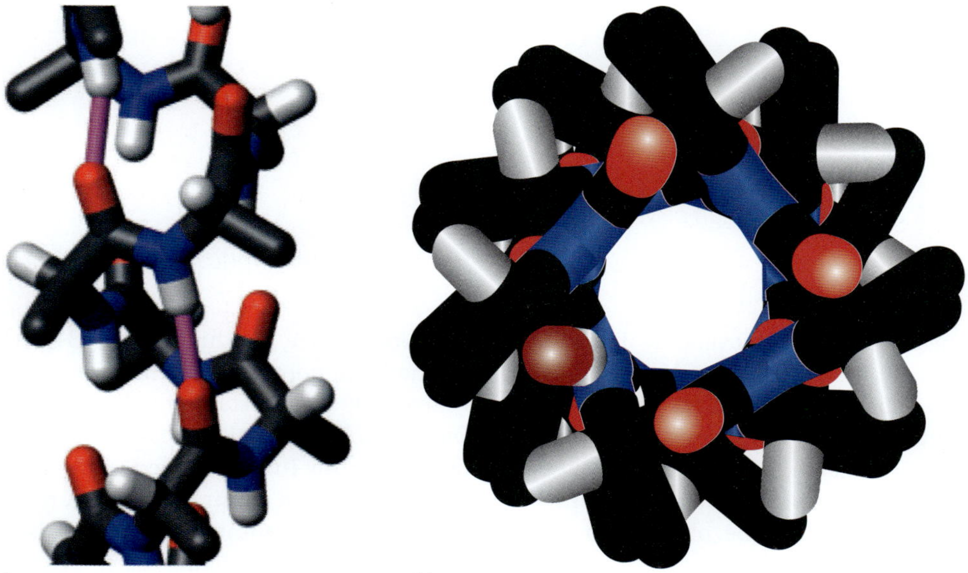

(a) (b)

FIGURE 22.29 (a) Side view of a peptide α-helix. (b) Top down view of a peptide α-helix.

22.4.2 SYNTHESIS

How can chemists imitate the synthesis of peptides and proteins that nature manages to carry out so effectively? Unfortunately, it's not so easy. If we consider the coupling of two amino acids, **22.33** and **22.34**, there are four potential products—each amino acid can couple with an identical molecule, and the coupling between them has two potential directions. Moreover, because they are carbonyl compounds, peptides (amides) epimerize easily. Polypeptides are usually water soluble, but most coupling methods work best in nonaqueous media.

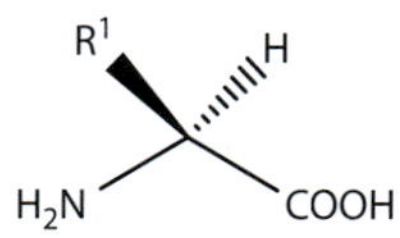

22.33

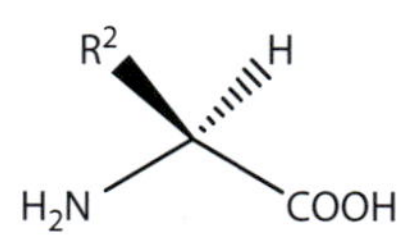

22.34

Many "classical" methods, using protection and deprotection, were developed to ensure that specific couplings took place, but such syntheses were slow and very labor-intensive. Essentially, all modern synthesis uses solid-phase peptide synthesis (SPPS), sometimes called Merrifield

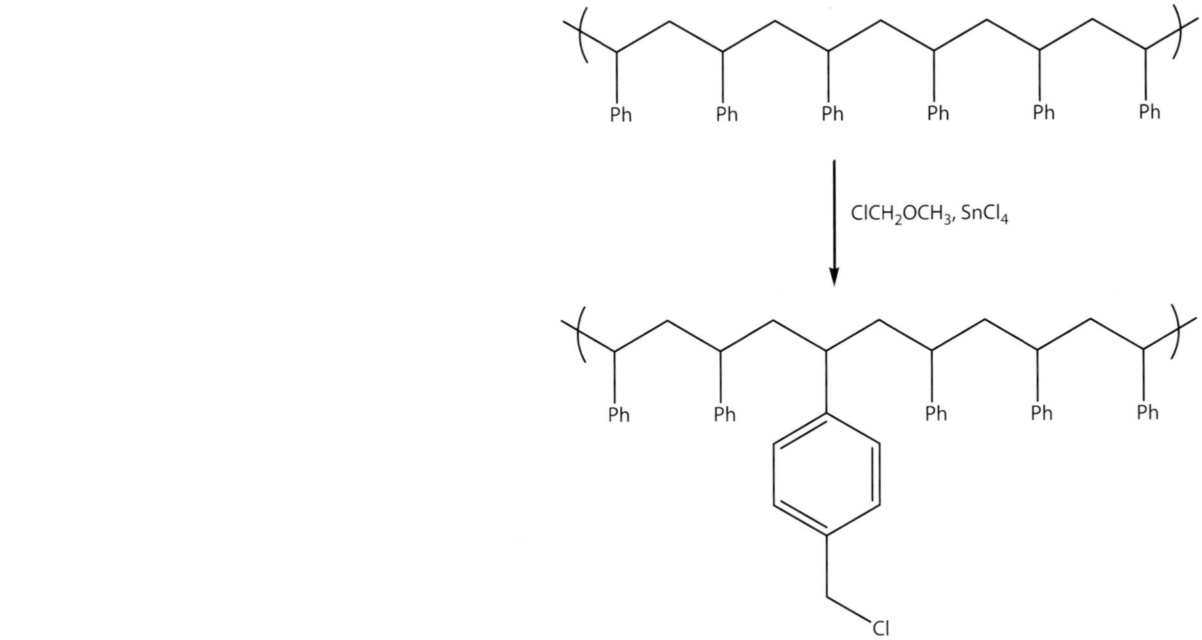

FIGURE 22.30 Preparation of Merrifield resin.

synthesis after its discoverer R. Bruce Merrifield (Nobel Prize for this work, 1984). All the synthesis takes place on a solid support, usually a modified polystyrene with varying degrees of divinylbenzene cross-linking. Polystyrene is physically stable, stable to the solvents and reagents used in the synthesis, and swells sufficiently in solvents to allow reagents to approach the peptide anchored to the resin. Polystyrene may be readily functionalized—not every ring reacts, and materials with different functionality loadings can be prepared (Figure 22.30). Higher loadings can be obtained by copolymerizing styrene with 4-chloromethylstyrene. Benzylic halides are readily displaced, and the synthesis begins by coupling the carboxyl group of a protected amino acid to the resin; this will be the *C*-terminus (the site of the free carboxylic acid group in the peptide) of the finished peptide. Thus, the chemical synthesis works in the opposite direction to the biosynthesis, which begins with the *N*-terminus—the site of the free $-NH_2$ group. An important advantage of the synthesis is the ease of purification of the developing peptide chain, and that an excess of reagents can be used to force reactions to completion, then readily washed out. Racemization of the many chiral centers in a peptide is also reduced with the solid-phase method.

In order to couple the amino acids selectively, we need a way to protect the amino group, while the carboxyl group is coupled to the free amino group of the growing peptide chain. Two protecting groups are commonly used, *t*-Boc (*tert*-butoxycarbonyl) and fMOC (9-fluorenylmethyloxycarbonyl). Figure 22.31 shows the reactions with amino acids. The first protected amino acid is then coupled to the support (Figure 22.32).

In order to couple the next amino acid, the protecting group must now be removed from nitrogen. At this point, the two protecting groups differ in the reagent required. The *t*-Boc group is removed in trifluoroacetic acid, and the fMoc group in base, using piperidine. There is now a free amino group, ready to couple with the carboxyl group of the next amino acid required (Figure 22.33). However, this coupling is not as easy as the reaction of the original carboxyl group with the resin—that was an S_N2 process at a very reactive center. In order for the amine to displace the OH of the carboxylic acid, we need to turn that OH into a much better leaving group. The standard method for this is to react the two in the presence of DCC, **22.35**. The leaving group is then the very stable urea (Figure 22.34). It is important that the coupling is as rapid as possible, to reduce the risk of racemization.

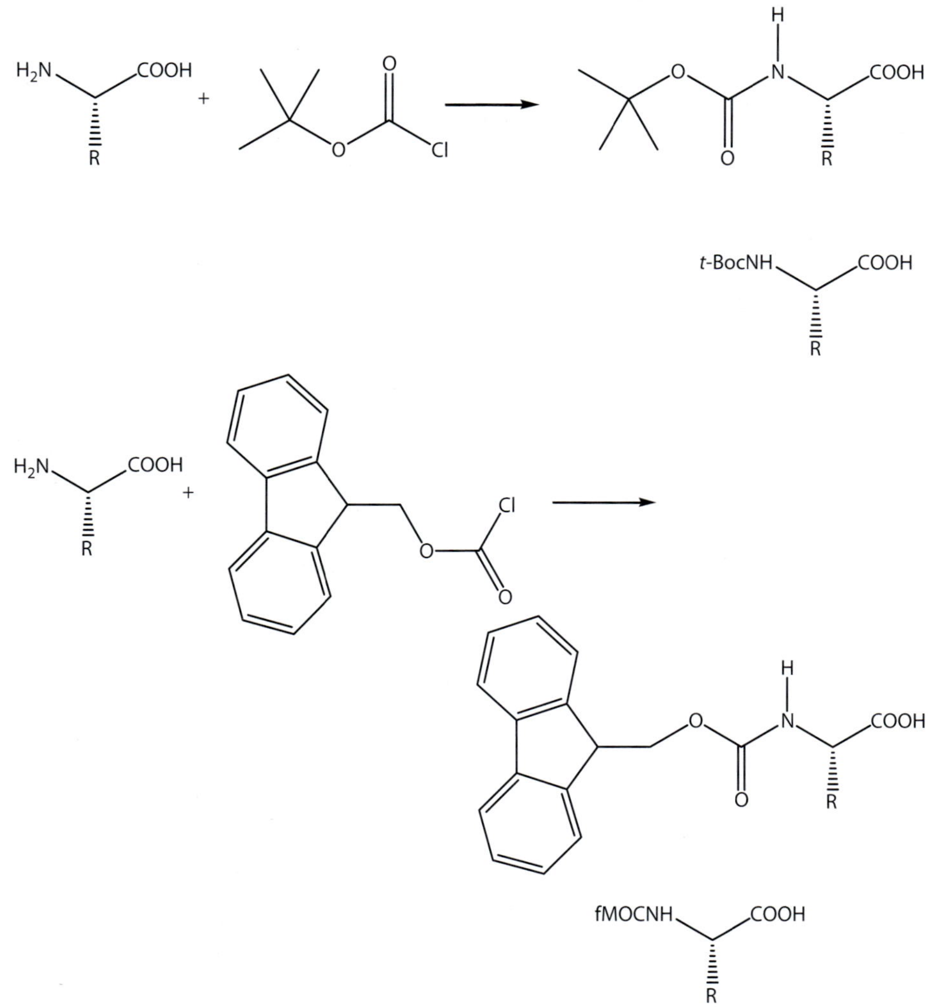

FIGURE 22.31 Protection of the amino groups of amino acids.

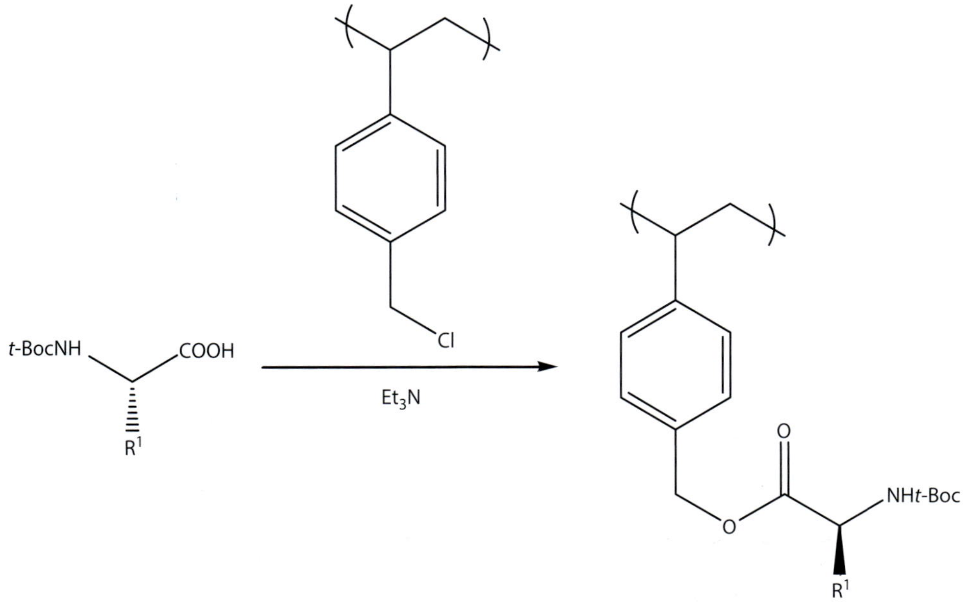

FIGURE 22.32 Coupling of a protected amino acid to the resin support.

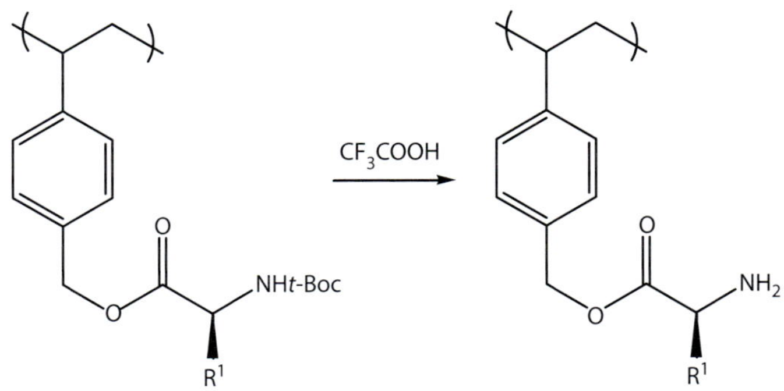

FIGURE 22.33 Deprotection of the amino group.

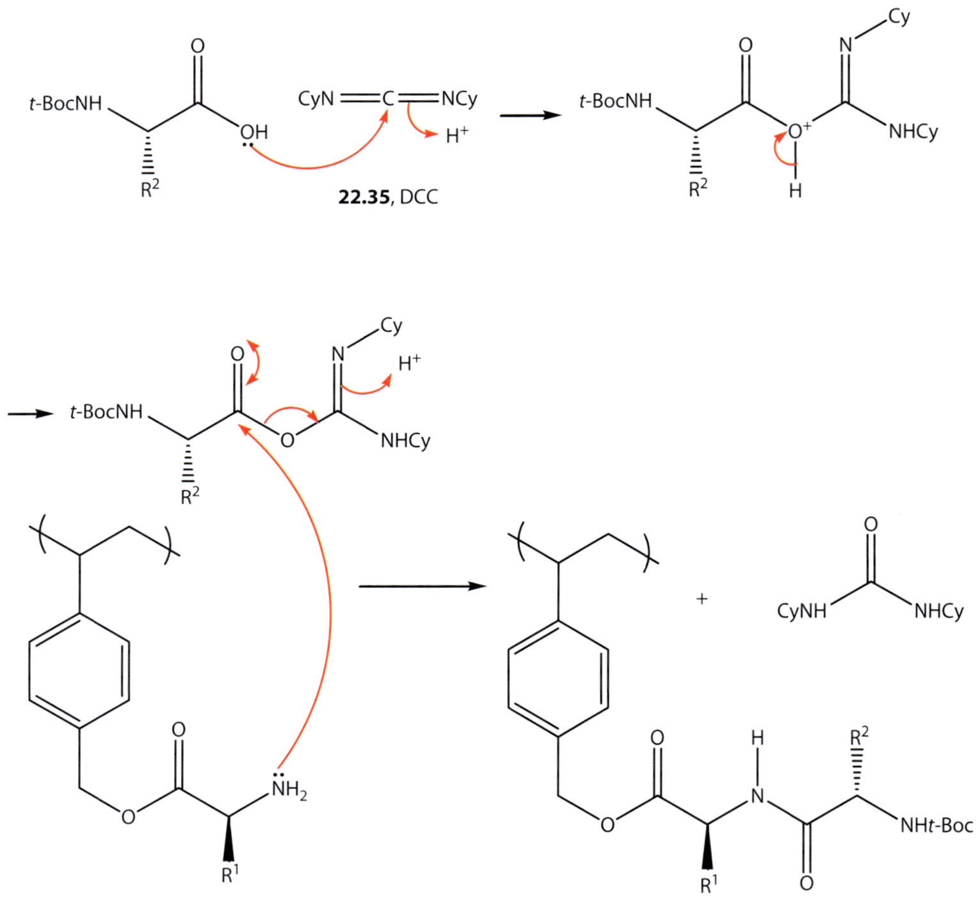

FIGURE 22.34 Coupling of peptides with DCC.

In the final step of the process, the completed peptide chain must be removed from the support; this generally involves HBr in trifluoroacetic acid or anhydrous HF. Although the synthesis can be done manually, automation is now readily available, and synthesis takes from 20 to 60 min per amino acid. In favorable cases, peptides up to 70 units long can be made in this way (but the difficulty of the syntheses also depends on the exact composition of the desired peptide)—beyond that, segments must be made individually and then coupled together.

PROBLEM 22.10

Fill in the blanks in the following sequence:

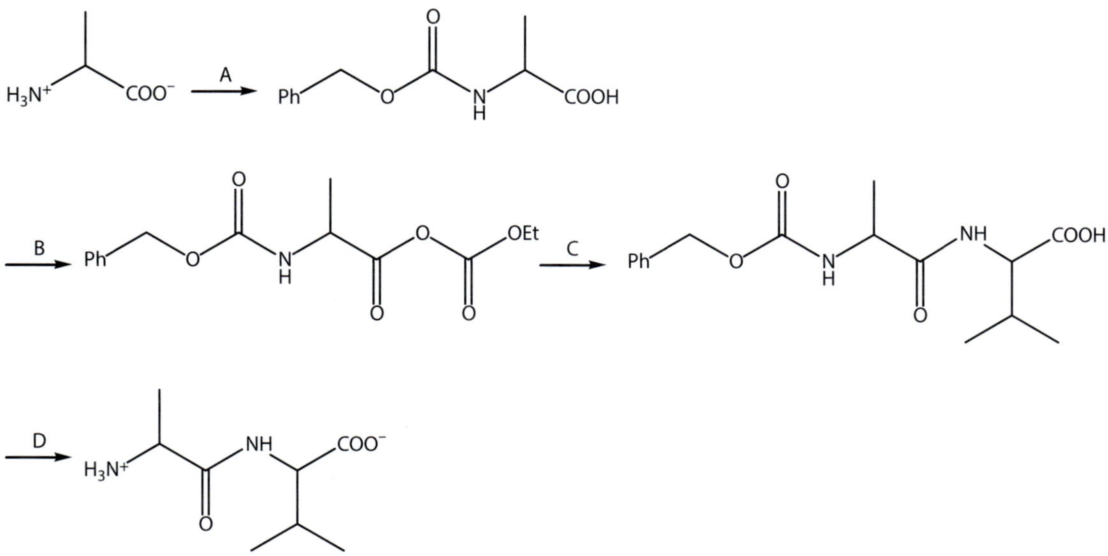

Solution

The first step, A, is an alternative method of protecting an amino group. The required reagent is the relevant acid chloride (or anhydride), $PhCH_2OC(=O)Cl$.

Step B converts the acid into an anhydride, so that there is a better leaving group at the carbonyl. The reagent is ClC(=O)OEt, ethyl chloroformate.

In step C, the activated/protected alanine is coupled with valine, so the reagent required is $Me_2CHCH(NH_2)COOH$.

In step D, we deprotect the amino group on the alanine—we know that the benzyl group can be removed by hydrogenolysis from earlier work on protecting groups. Once this is removed, the –NHCOOH group readily decarboxylates.

22.4.3 ANALYSIS OF PEPTIDES

Many short peptides have biological and medicinal functions, and determining the structures of these shorter sequences proved to be the first step in determining the sequences of proteins. The first step in sequencing any peptide is the determination of its "empirical formula"—the identity and amount of each amino acid that makes it up. Normally, this involves complete hydrolysis in 6 M HCl at 100 °C for 12–24 h. The hydrolysis must be carried out in the absence of oxygen, as the sulfur-containing amino acids, methionine and cysteine, are easily oxidized. Under these conditions, tryptophan is completely destroyed, and the side chain amides in glutamine and asparagine are hydrolyzed to glutamic and aspartic acids, respectively. Hydrolysis under basic conditions (2 M NaOH, 100 °C) destroys arginine, cysteine, serine, and threonine, but tryptophan survives. The amino acids are then separated chromatographically, and the relative quantities determined by reaction with ninhydrin. At this point, the amino acids are listed separated by commas, to indicate composition, but not sequence. When the sequence is known, the three-letter abbreviations are separated by a period, with the *N*-terminus to the left and the *C*-terminus to the right. If the single-letter abbreviations are used, there is no punctuation once the sequence is known. Thus, **22.36** has its composition reported as Ala,Gly,Phe,Ser,Val and its sequence as Val.Ala.Phe.Ser.Gly or VASFG. For the purpose of the discussion here, we will use the three-letter abbreviations.

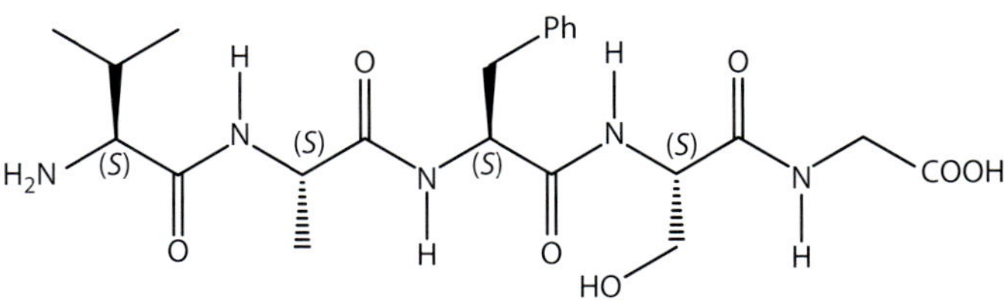

22.36

Once the composition of the peptide is known, then the sequence is generally approached by starting at the "ends." However, we should note that cyclic peptides are known, and these will show none of the characteristic reactions of either the *N*- or *C*-termini. If the peptide is treated with lithium borohydride, then the free carboxylic acid group is reduced to an alcohol. If the peptide is then hydrolyzed, the usual amino acids are obtained, except that the residue at the *C*-terminus now appears as an amino alcohol. There is also an enzyme, carboxypeptidase, that specifically cleaves the *C*-terminal amino acid, which can then be identified. However, this leaves a new *C*-terminus, so cleavage does continue down the chain.

Identification of the *N*-terminus generally depends on methods that "tag" the $-NH_2$ group, so that the tagged species can be identified after hydrolysis. The Sanger method (Figure 22.35) uses 2,4-dinitrofluorobenzene (DNFB) (remember from Section 13.4 that this will undergo ready substitution of the fluoride because of the presence of the electron-accepting nitro groups in an S_N2Ar reaction); the dinitrophenyl substituent gives a bright yellow color to all its derivatives so the tagged material is easily recognized. Note that in lysine and arginine, the free $-NH_2$ group in the side chain will always be substituted by the dinitrophenyl group, but the α-amino group will not, unless the residue is at the *N*-terminus. Thus, chain lysine gives **22.35** and *N*-terminal lysine gives **22.36**.

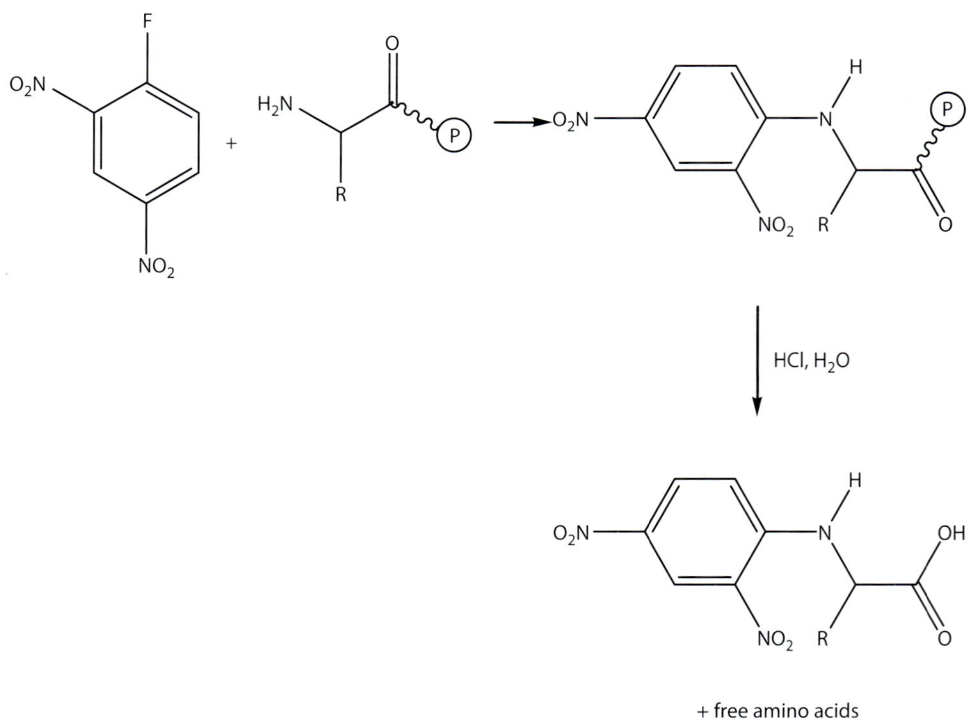

FIGURE 22.35 Sanger method for *N*-terminus determination in peptides.

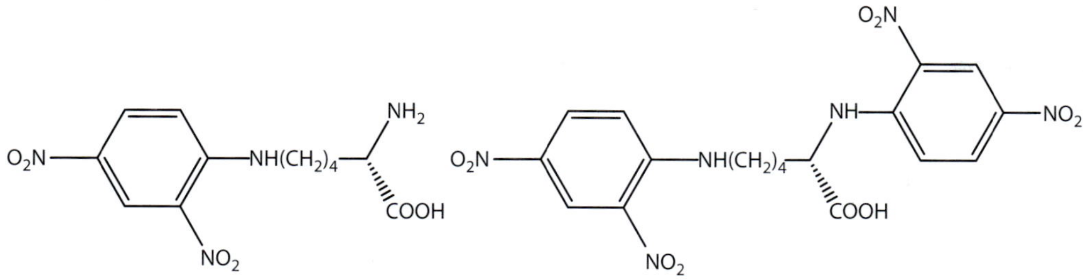

22.37 **22.38**

These methods are, however, completely destructive of the rest of the peptide chain—a single piece of information is gained at the cost of significant material use. When the *N*-terminus of a peptide is reacted with phenyl isothiocyanate, PhN=C=S, the Edman reagent, a phenylthiohydantoin, **22.39**, is obtained and the rest of the peptide chain is left intact (Figure 22.36). So the process can be repeated, and this methodology is the basis for automatic sequenators. The phenylthiohydantoin contains the R group from the *N*-terminal amino acid and can thus be identified.

Having identified the termini of the peptide, it remains to tackle the section in between. One approach to this involves less vigorous hydrolysis—less acidic, shorter time—to give smaller peptides that can be more readily analyzed, and then "fitted" back together to give a whole. There are also a number of reagents and enzymes that cleave the chain at specific points (Table 22.2), again giving useful information.

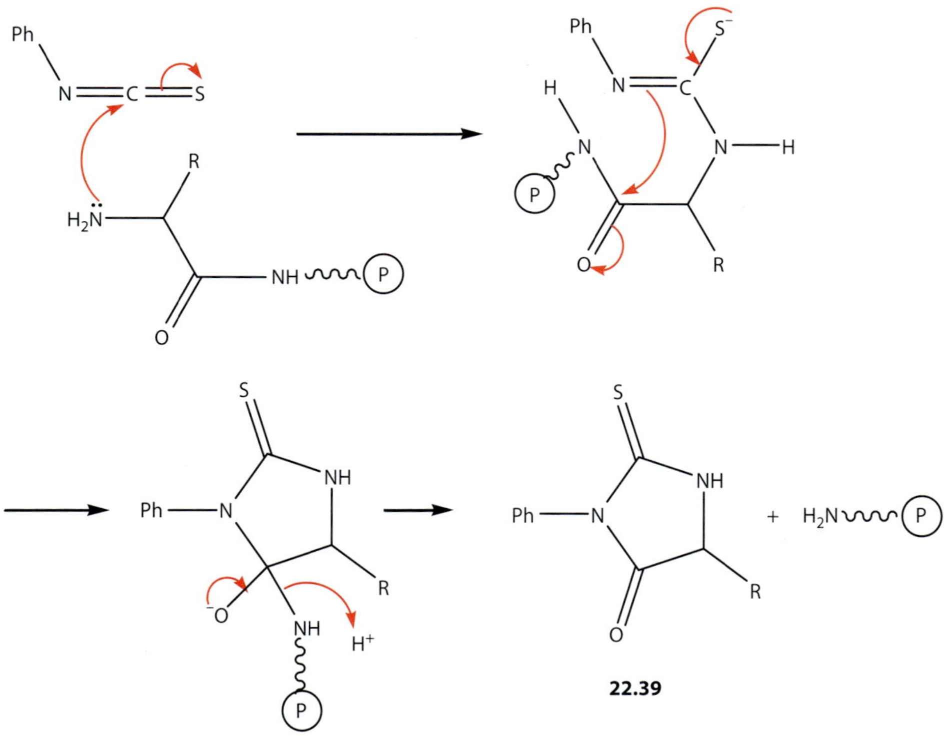

FIGURE 22.36 Reaction of a peptide with the Edman reagent.

TABLE 22.2 Reagents for Selective Cleavage of Peptide Chains

Reagent	Position for Cleavage
CNBr, cyanogen bromide	Carboxyl side of methionine
Trypsin	Carboxyl side of lysine, arginine
Chymotrypsin	Carboxyl side of phenylalanine, tyrosine, tryptophan
Pepsin	Between two hydrophobic amino acids, *e.g.*, Phe.Lys
Elastin	Carboxyl side of alanine and glycine

PROBLEM 22.11

Predict the outcomes of each of the following reactions:

(a) $\text{Phe.Gly.Lys} \xrightarrow{\text{chymotrypsin}}$

(b) $\text{Thr.Phe.Ser} \xrightarrow{\text{chymotrypsin}}$

(c) $\text{Thr.Ser.Phe} \xrightarrow{\text{chymotrypsin}}$

PROBLEM 22.12

A decapeptide has composition Ala,Arg,Gly,Ile,Leu,Lys,Phe,Thr,Trp,Val.
Partial hydrolysis gives two identifiable peptides, composition Ile,Lys and Arg,Leu.
Reaction with 2,4-dinitrophenylbenzene followed by hydrolysis gave Ala–DNP.
Reaction with carboxypeptidase gave Thr.
Reaction with trypsin gave three peptides of composition Ala,Gly,Lys; Arg,Ile,Phe,Val; Leu,Thr,Trp.
Reaction with chymotrypsin gave two peptides Ala,Gly,Ile,Lys,Phe and Arg,Leu,Trp,Val.

Deduce the sequence of the peptide.

Solutions

22.11 Chymotrypsin cleaves at the *C*-side of the aromatic amino acids Phe, Tyr, and Trp. Thus,
(a) Phe + Gly.Lys
(b) Thr.Phe + Ser
(c) No reaction

22.12 These problems are rather like the game, hangman—we have 10 positions to fill. Two of these are easy; the carboxypeptidase data give us the *C*-terminus as Thr, and the DNFB reaction tells us that Ala is at the *N*-terminus. Thus, we have

Ala _ _ _ _ _ _ _ _ Thr

Trypsin cleaves at the *C*-side of arginine and lysine. So for the peptide Ala,Gly,Lys, we know that Lys must be at the *C*-end, and we already know where Ala is. Thus, we have

Ala.Gly.Lys _ _ _ _ _ _ Thr

We also have a peptide of composition Lys,Ile, so we can now position Ile

Ala.Gly.Lys.Ile _ _ _ _ _ Thr

Chymotrypsin cleaves at the *C*-side of aromatic amino acids, so Ala,Gly,Lys,Ile,Phe must have Phe at its *C*-terminus—and we already know where the other amino acids are. Thus, we have

Ala.Gly.Lys.Ile.Phe _ _ _ _ Thr

Trypsin also cleaves the *C*-side of arginine so the peptide Arg,Ile,Phe,Val must have arginine at the *C*-terminal. We already know where Ile.Phe fits in, so we now have

Ala.Gly.Lys.Ile.Phe.Val.Arg _ _ Thr

Chymotrypsin cleaves the *C*-side or aromatic amino acids, so Arg,Leu,Trp,Val must have Trp at the *C*-terminus. We already know where Val.Arg is, so we can now complete the sequence:

Ala.Gly.Lys.Ile.Phe.Val.Arg.Leu.Trp.Thr

An interesting question is—for how much longer will these classical methods of peptide/protein sequencing be relevant to their, or your, study? The last 15 years have seen enormous advances in the analysis of proteins using sophisticated mass spectrometric techniques, coupled with computer analysis of the data obtained. Peptides tend to fragment in the mass spectrometer at the peptide bonds, so it is possible to see individual amino acids being cleaved. Together with advances in protein crystallography and the availability of higher-field NMR spectrometers and multidimensional techniques, it is clear that instrumental and computational techniques represent the future of the subject.

22.5 NUCLEIC ACIDS

The nucleobases are all derivatives of pyrimidine, **22.40**, or purine, **22.41**. Note that as with other heterocycles, the numbering begins on the heteroatom, and where there are two heteroatoms, it goes around the ring to give the second heteroatom the lowest possible number. The key structures for DNA and RNA are shown in Table 22.3. When the nucleobase is attached to a sugar (ribose, **22.42** in RNA, 2-deoxyribose, **22.43** in DNA; both sugars are shown in the furanose form, which they adopt in all common biological molecules), the species is described as a *nucleoside* (comes from *N*-glycoside), and when the nucleoside acquires a phosphate, it is a *nucleotide*.

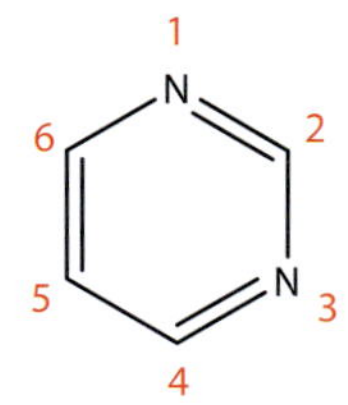

22.40, pyrimidine

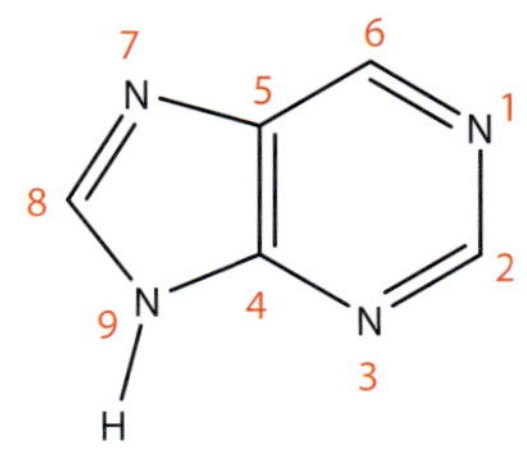

22.41, purine

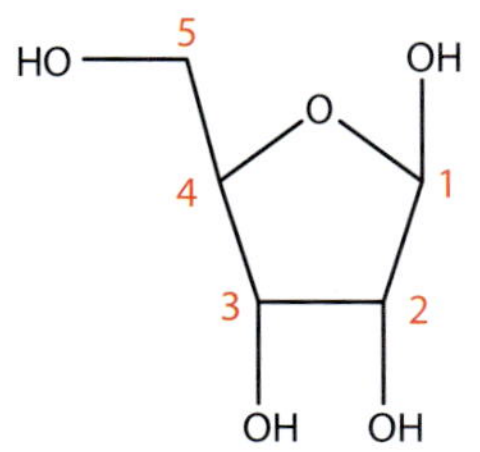

22.42, ribose

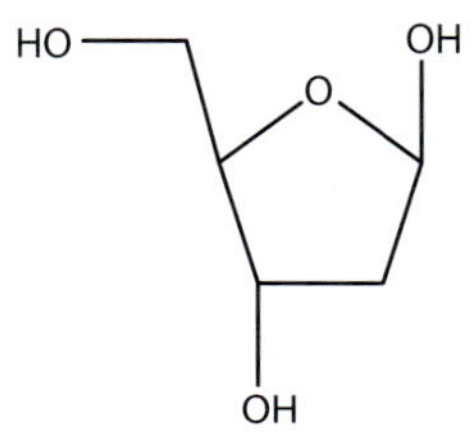

22.43, 2-deoxyribose

Nucleosides are formed by the reaction of the NH group of the base with the sugar at the hemiacetal carbon atom (Figure 22.37). Doing this chemically would give a mixture of stereoisomers, but the biosynthesis is entirely stereospecific.

DNA was first isolated around 1870 by Johan Friedrich Miescher, but its relevance to genetics was not established until the 1940s. The seminal paper on the double helix structure of DNA was published by Watson and Crick in 1953. One important aspect of determining this was to identify which tautomer of the various bases was present. Initially, it had been thought that all of the oxygen-containing molecules were hydroxyaromatic compounds, rather than having ketone functionality—and of course, if they are written in this way, the hydrogen-bonding interactions that hold the double helix together cannot be discerned. Although some x-ray diffraction data for DNA were available, these were not at a resolution that allowed the location of hydrogen atoms in such a large molecule. The problem can be simplified by looking at the structure of 2-hydroxypyridine/2-pyridone, **22.44**. It is now the norm to state that it's obviously 2-pyridone on the basis of the resemblance of the IR and ^{13}C NMR spectra of 2-pyridone to those of the known compound *N*-methyl pyridone, **22.45**, rather than *O*-methylhydroxypyridine, **22.46** (Table 22.4). However, commercial IR spectrometers first appeared in 1957, and the development of NMR spectrometers started around the same period—so these data simply were not available to Watson and Crick. However, commercial UV/visible spectrometers were available, and the data strongly indicated the predominance of the 2-pyridone tautomer. Added to this, the melting point of 2-pyridone is 107 °C–108 °C against a melting point of 40 °C–41 °C for phenol (the amide tautomer would be expected to have more extensive hydrogen bonding), and 2-pyridone is less acidic than phenol (pK_a 11.65 against 9.95 for phenol), so the correct conclusion was reached, despite the lack of modern spectroscopic techniques. The outcome is clear for 2-pyridone, and the related nucleobases, but different results for more highly substituted pyridines have been obtained. More recent studies indicate that the predominance of the pyridone tautomer in solution is strongly dependent on solvation and self-association; in the gas phase, the two tautomers are of comparable energy.

TABLE 22.3 Nucleobases, Nucleosides, and Nucleotides

Base	Nucleoside	5′-Nucleotide	Comment
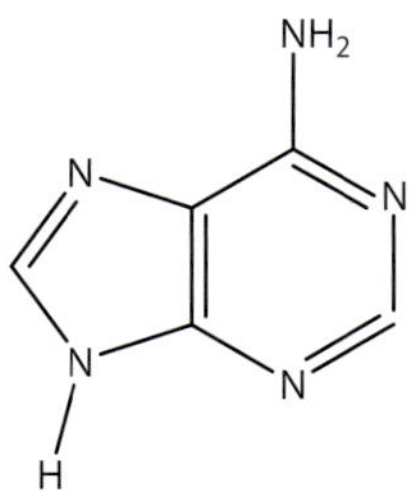 Adenine, A	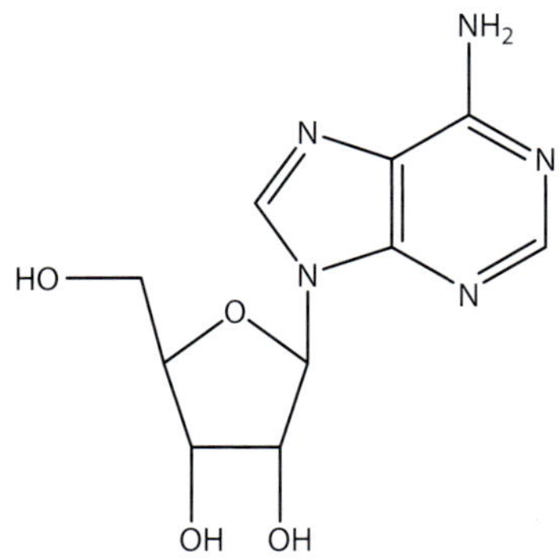 Adenosine	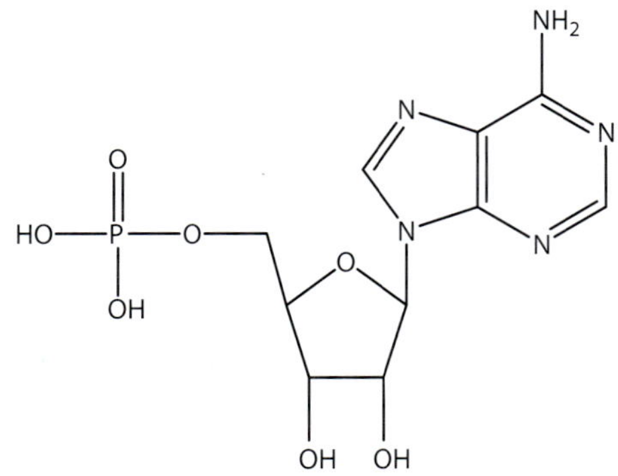 Adenosine 5′-phosphate	The DNA version has no OH at the 2-position of the sugar.
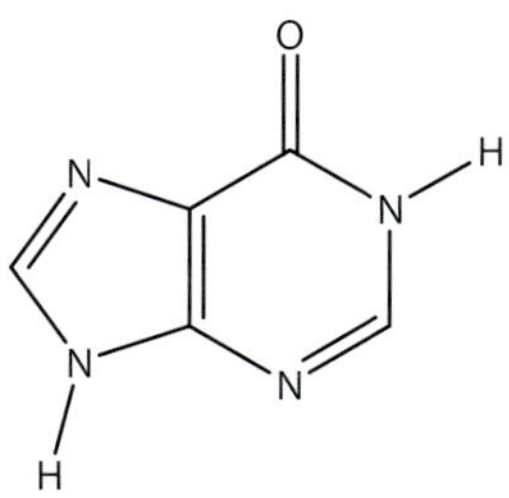 Guanine, G	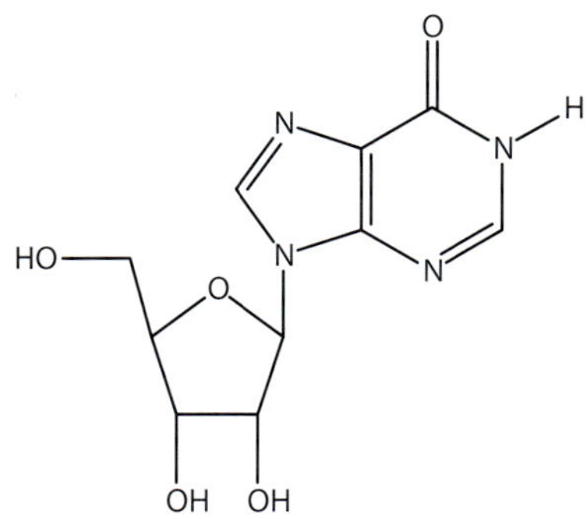 Guanosine	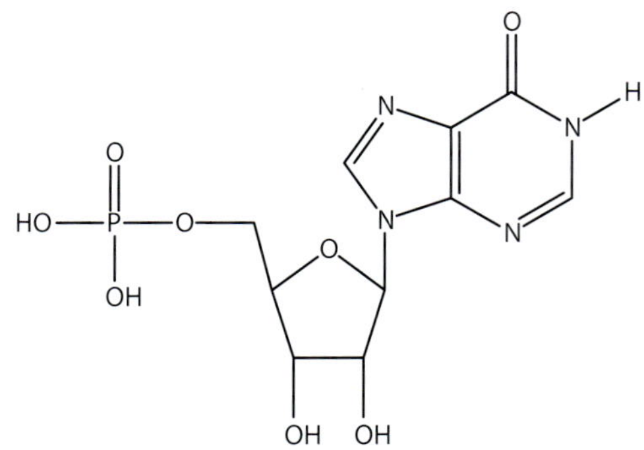 Guanosine-5′-phospate	The DNA version has no OH at the 2-position of the sugar.
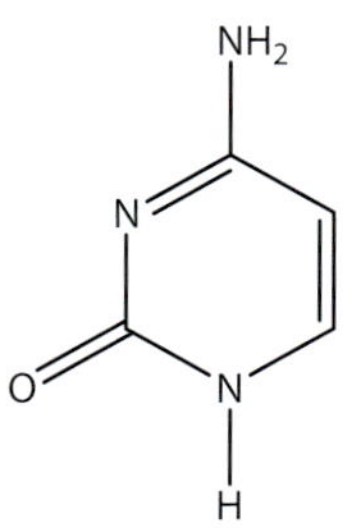 Cytosine, C	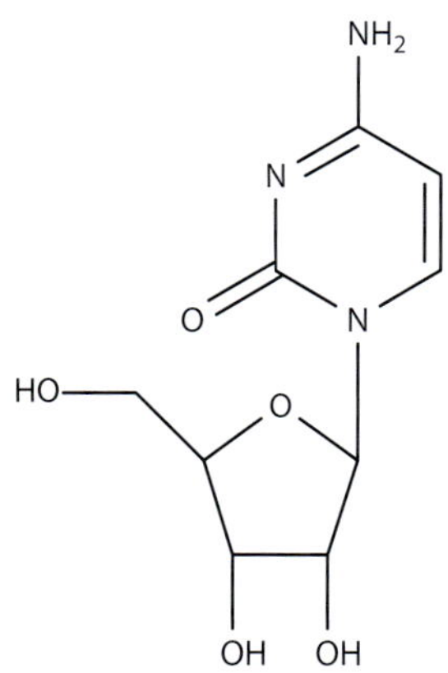 Cytidine	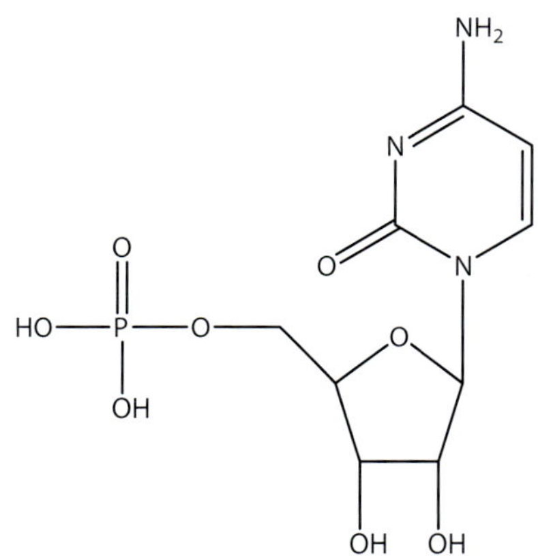 Cytidine-5′-phosphate	The DNA version has no OH at the 2-position of the sugar.

TABLE 22.3 (continued) Nucleobases, Nucleosides, and Nucleotides

Base	Nucleoside	5′-Nucleotide	Comment
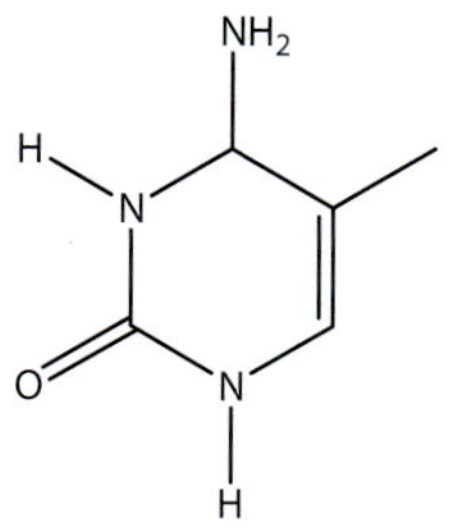 Thymine, T	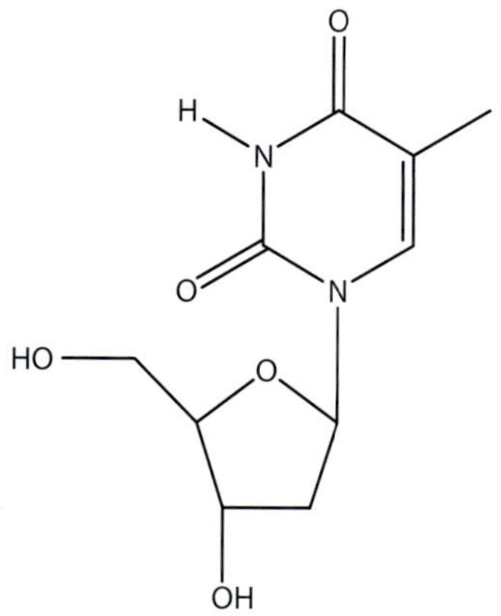 Thymidine	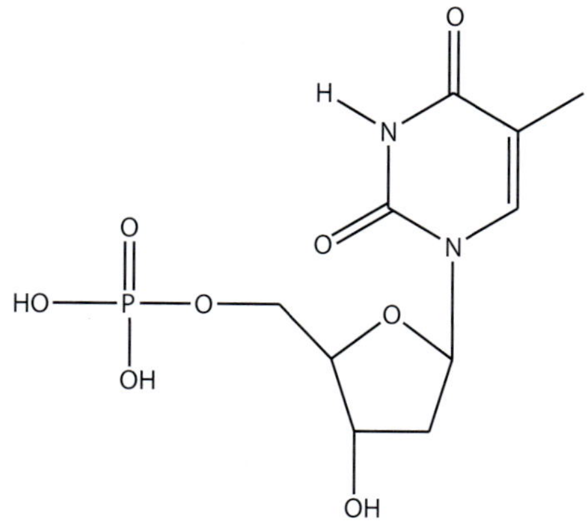 Thymidine-5′-phosphate	Thymine only occurs in DNA, so the sugar is deoxyribose.
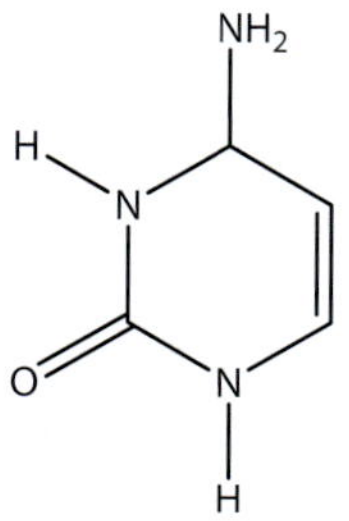 Uracil, U	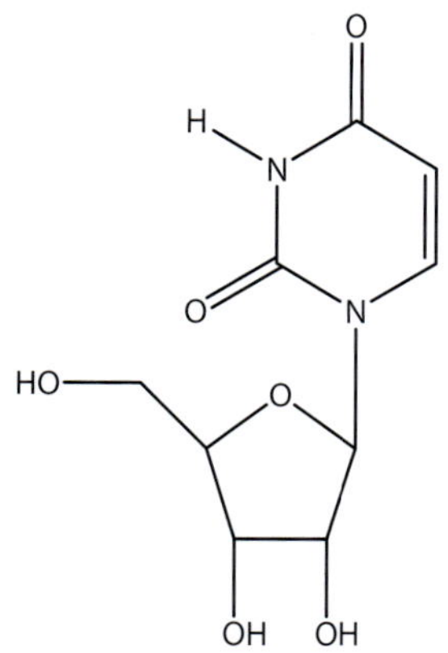 Uridine	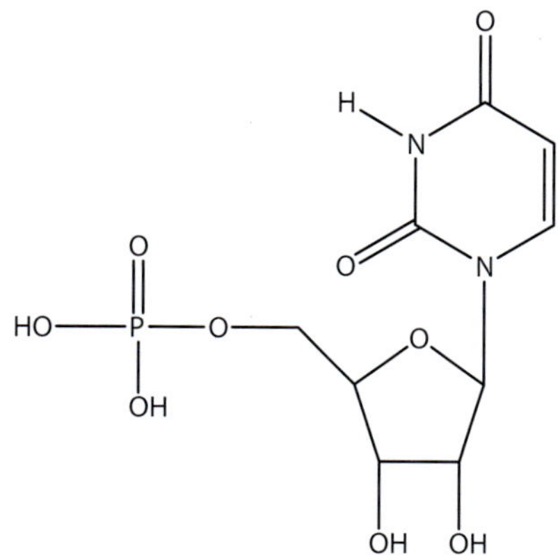 Uridine-5′-phosphate	Uracil is only found in RNA.

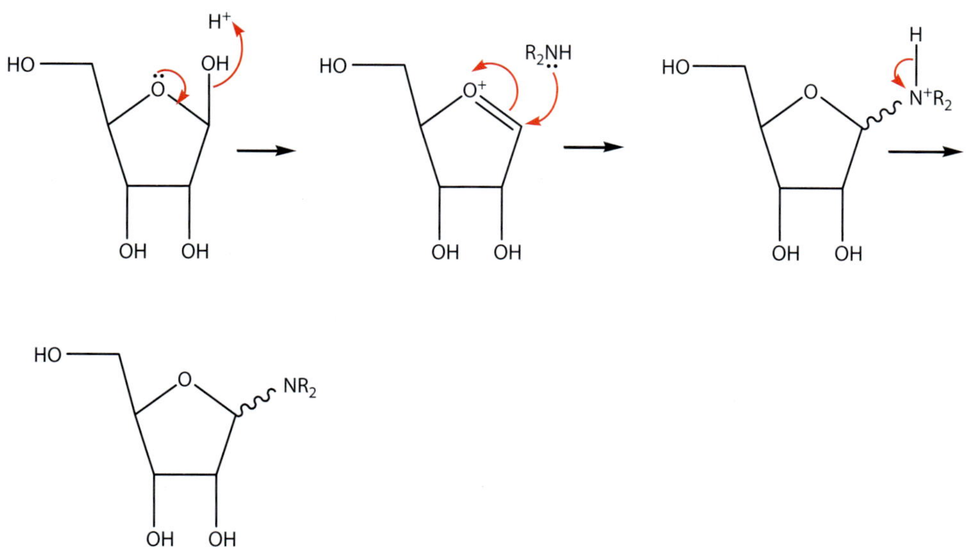

FIGURE 22.37 Nucleoside formation.

TABLE 22.4 Spectroscopic Data for 22.44, 22.45, 22.46

Compound	UV Spectroscopy, (MeOH) λ_{max}, nm (ε)	IR Spectroscopy ν_{max} (cm^{-1})	^{13}C NMR Spectroscopy (δ) C-2, C-3, C-4, C-5, C-6, Others
2-Pyridone, **22.44**	298 (54.1)	3440, 1650, 1610, 1581, 1540	162.3, 119.8, 140.8, 104.8, 135.2
N-Methyl-2-pyridone, **22.45**	301 (51)	1658, 1584, 1539	161.8, 119.1, 139.5, 104.8, 139.5, 41.8
2-Methoxypyridine, **22.46**	271 (42.3)	1603, 1488, 1437	163.1, 110.5, 138.7, 116.7, 146.1, 52.8

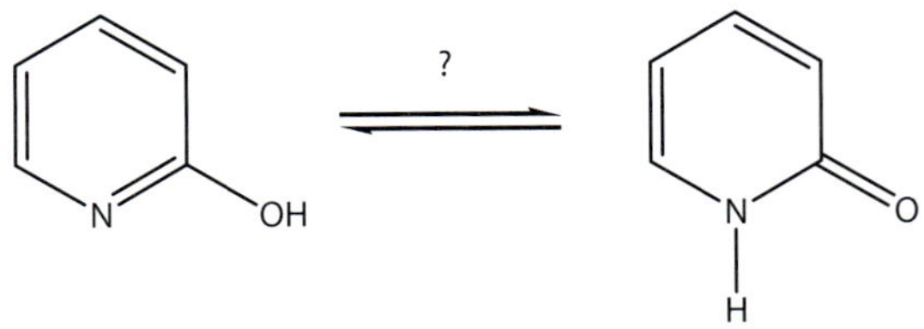

22.44

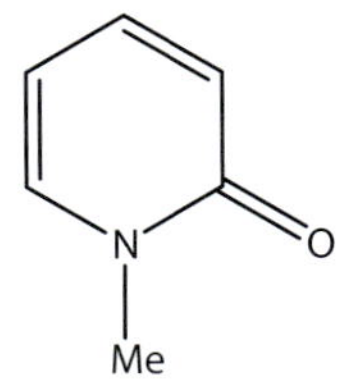

22.45

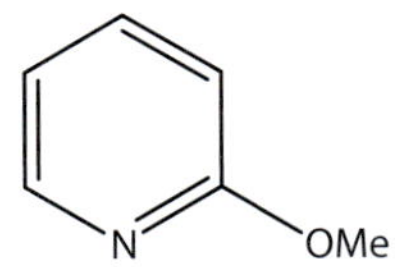

22.46

The base pairing in DNA needs little introduction; Guanine is paired with cytosine (**22.47**) and adenine with thymine (**22.48**). Figure 22.38 shows a standard double helix with the major and minor grooves marked. It is now possible to synthesize short oligonucleotide chains by a solid-phase synthesis method similar to that used to produce peptides. The chemical synthesis proceeds from 3′ to 5′, opposite to the direction of biosynthesis. Because of side reactions, the syntheses are limited to about 200 residues, and in most cases, the best results are obtained in making 20–25 residue units.'

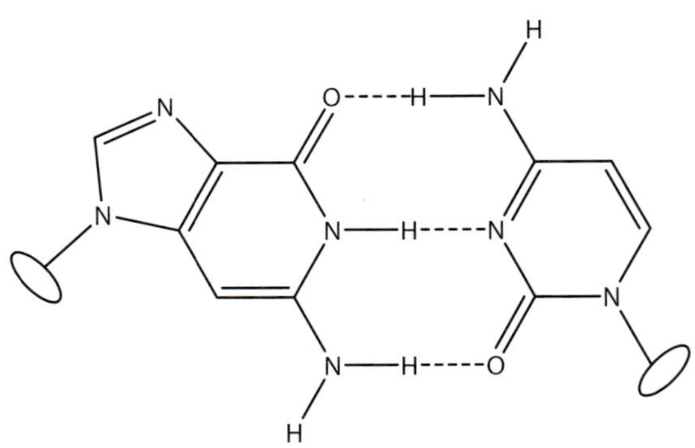

Guanine Cytosine

22.47

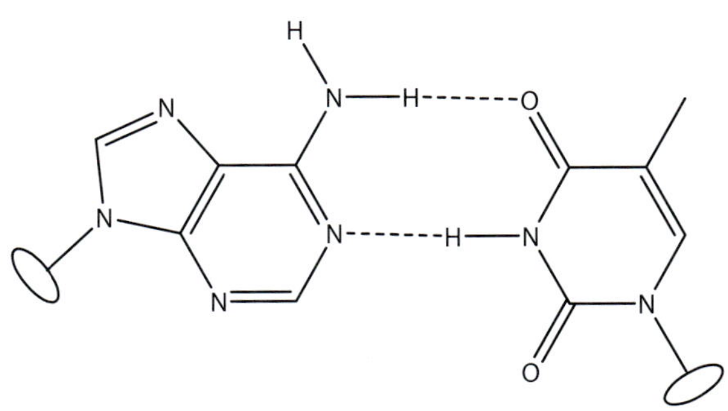

Adenine Thymine

22.48

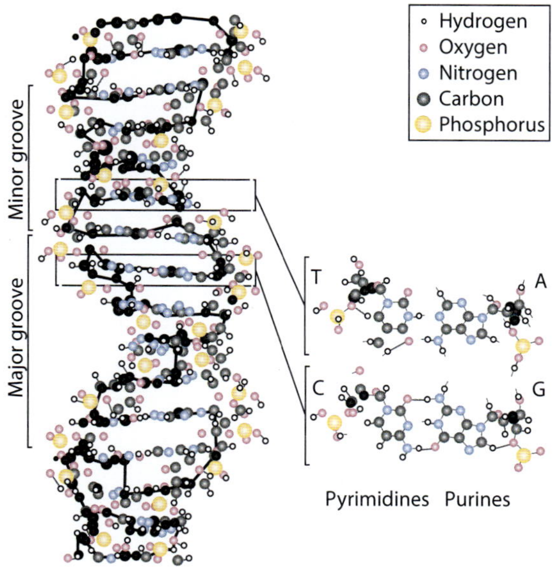

FIGURE 22.38 Double helix of DNA showing major and minor grooves.

Focus on ATP

Adenosine triphosphate (ATP), **22.49**, is the energy battery of the cell. It is also a coenzyme involved in cell signaling, transport of molecules and ions, and various regulatory processes. ATP was discovered in 1929 and synthesized by Todd in 1948. In 1997, the Nobel Prize was awarded to John Walker, Paul Boyer, and Jens Skou for their work on the mode of action of ATP. ATP is hydrolyzed generally to adenosine diphosphate (ADP), but sometimes to adenosine monophosphate (AMP), with considerable release of energy (Equations 22.1 and 22.2; P_i indicates inorganic phosphate). The energy release probably derives from the extra resonance structures available to free phosphate and the absence of interoxygen repulsion in ionized ATP. ATP is synthesized from ADP and inorganic phosphate using the enzyme ATP synthase, and the energy derived from the metabolism for fats and carbohydrates. About one-third of the energy derived from glucose metabolism is used to drive this process, and complete metabolism of glucose to CO_2 produces enough energy to synthesize about 30 molecules of ATP.

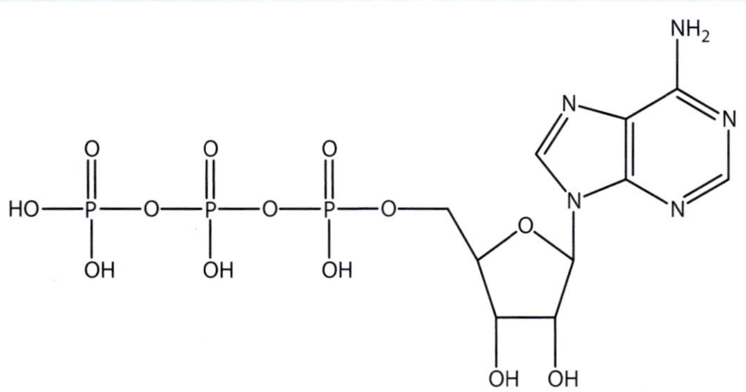

22.49

(22.1) $$ATP + H_2O \rightarrow ADP + P_i \quad \Delta G^{\circ} = -30.5 \text{ kJ mol}^{-1}$$

(22.2) $$ATP + 2H_2O \rightarrow AMP + PP_i \quad \Delta G^{\circ} = -45.6 \text{ kJ mol}^{-1}$$

PROBLEM 22.13

Give a tautomer of each of the following species:

(a)

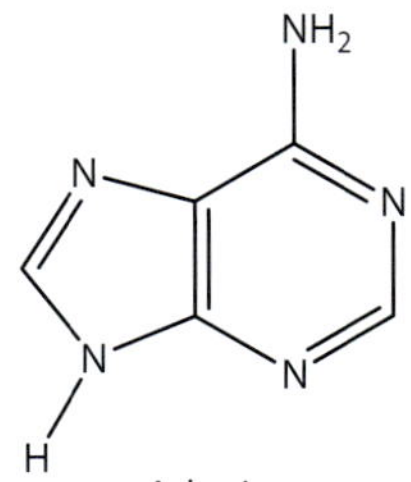

Adenine

(b)

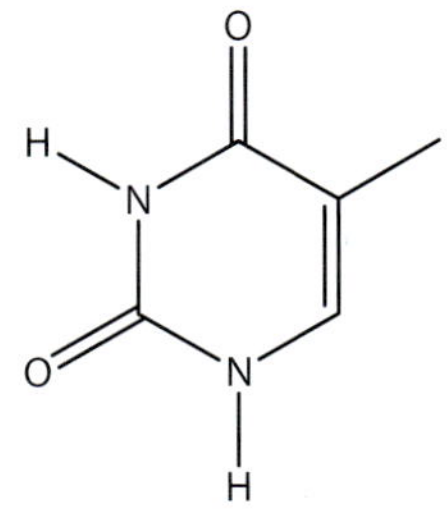

Thymine

(c)

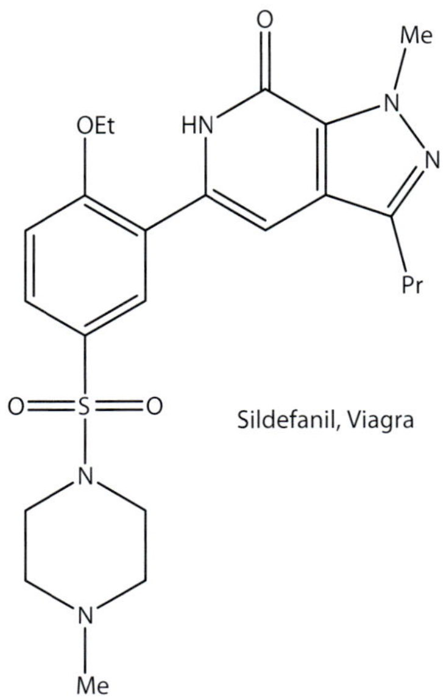

Sildefanil, Viagra

Solution

(a) The simplest possibility here is to shift the proton in the five-membered ring:

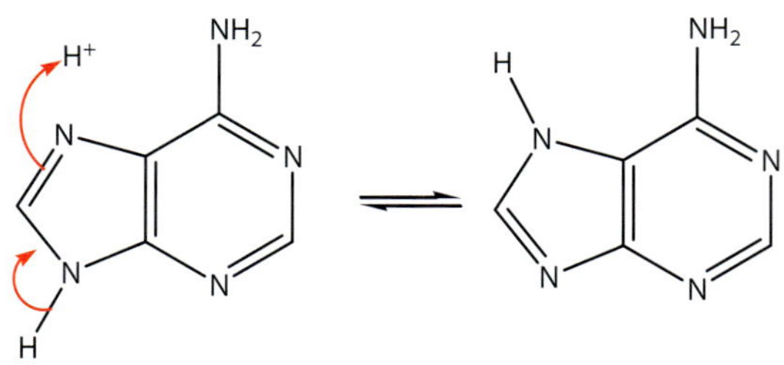

However, it is also possible two write an imino tautomer for the $-NH_2$ group:

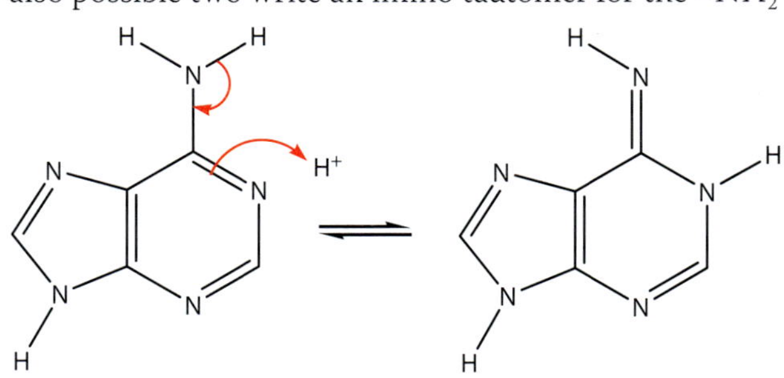

Although all physical measurements indicate that the imino tautomer is not present to any significant extent, it was, at one time, considered seriously as a possibility for DNA structure.

(b)

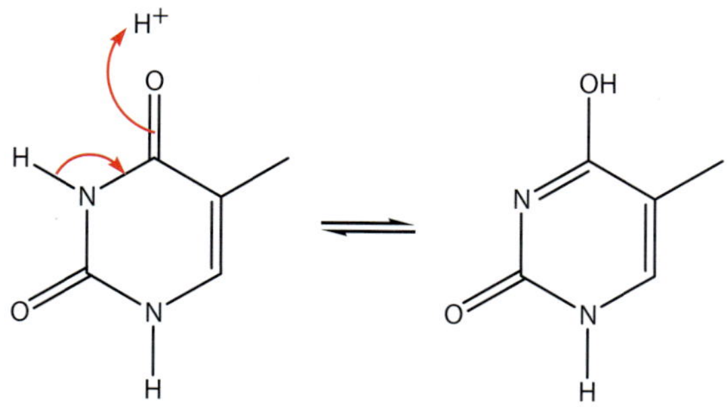

There are various other solutions involving the other carbonyl group:

(c)

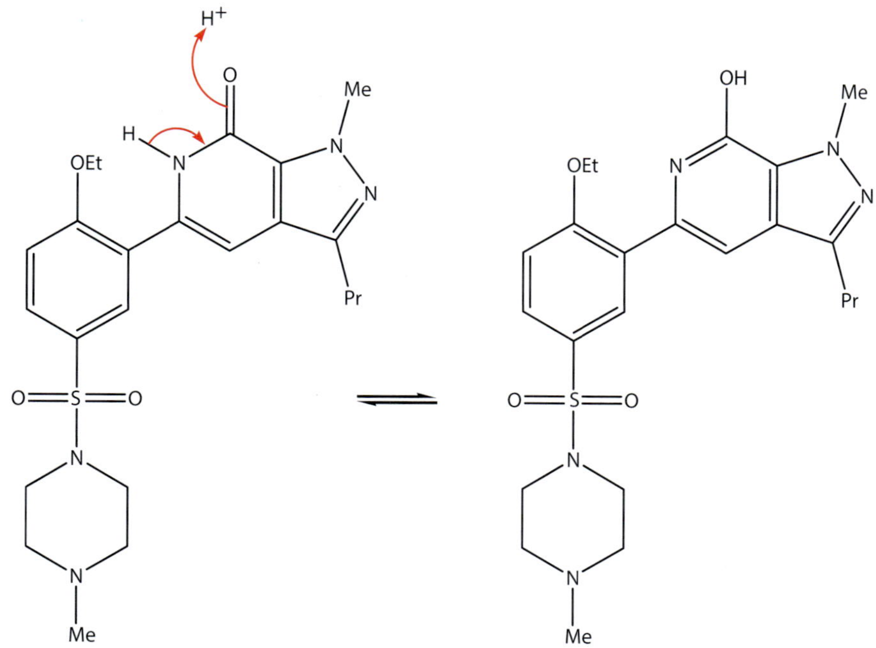

Key Points from Sections 22.4 and 22.5

- Peptides are formed by the condensation of carboxyl and amino groups in amino acids. Their primary structure is the sequence of amino acids. Secondary structure is determined by the required planarity of the amide group and interactions between side chains. α-Helices and β-pleated sheets are common motifs. Tertiary structure is the full three-dimensional structure of the peptide.
- Peptide synthesis involves a sequence of protection, deprotection, and coupling steps, while the growing peptide is attached to a functionalized polystyrene resin. The process is largely automated.
- Classical peptide analysis involves chemical degradation of the peptide to its component amino acids, and specific cleavages of the peptide using chemical reagents or enzymes with predictable selectivities. It is being superseded by instrumental techniques.
- Three nucleobases, adenine, guanine, and cytosine, are common to DNA and RNA. Thymine occurs only in DNA and uracil only in RNA.
- When attached to a sugar (ribose in RNA, 2-deoxyribose in DNA), the bases are converted to nucleosides. Nucleotide is the term used for phosphorylated nucleosides.
- Understanding the hydrogen bonding needed to form the double helix of DNA requires the determination that 2- and 4-hydroxypyridines and their nucleobase analogs exist as the keto tautomer.

REVIEW PROBLEMS

1. Give a systematic name for each of the following compounds:

(a) 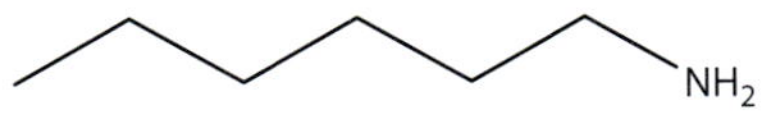 (b) 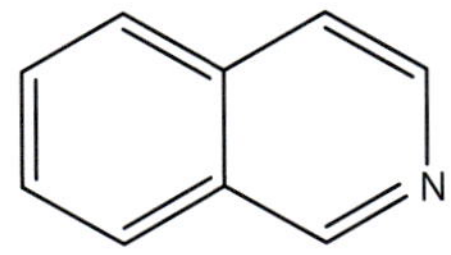

(c) 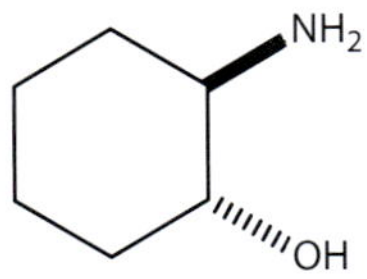

2. Give a structure for each of the following:
   (a) *N*-Methylaniline
   (b) 4-Aminobenzoic acid
   (c) 2-Pentylamine
3. Give a tautomer of each of the following species:

(a) 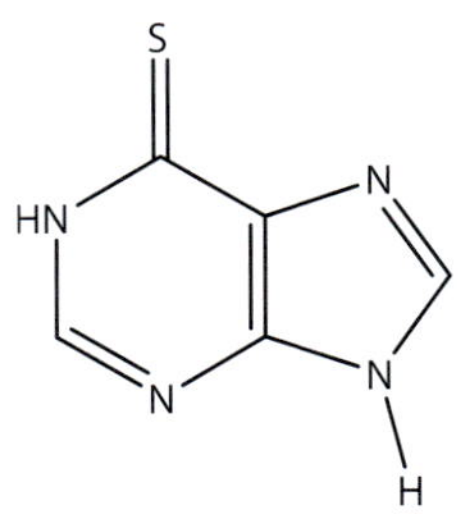

6-Mercaptopurine

(b) 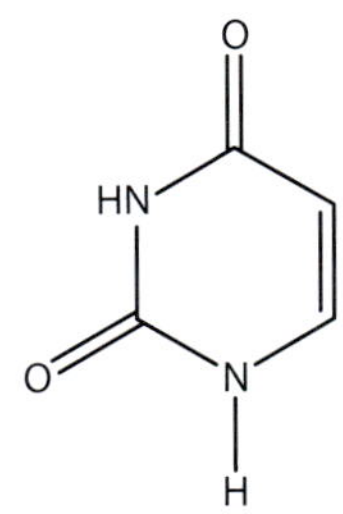

Uracil

(c) 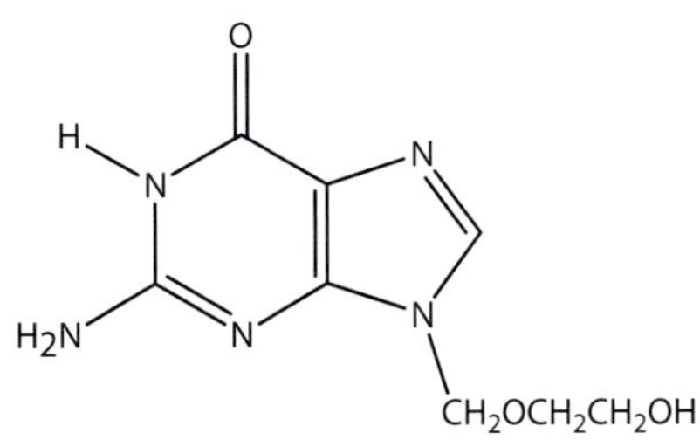

Acyclovir, used against herpes infections

4. A heptapeptide had composition Arg,Gly,Ile,Phe,Pro_2,Val. Trypsin gave Arg and a hexapeptide. Partial hydrolysis gave Gly.Pro, Pro.Pro, and Pro.Phe.Ile.Val. Deduce the structure of the peptide.
5. A peptide, **22.50**, contains 10 residues and has the empirical formula Arg,Cys,Glu,Gly_2,Leu, Phe_2,Tyr,Val. Treatment with the enzyme carboxypeptidase gave Phe. Reaction with DNFB, followed by hydrolysis, gave DNP-labeled Tyr. Reaction with the enzyme chymotrypsin gave free Tyr and Phe, and an octapeptide **22.51**. Treatment of **22.51** with carboxypeptidase gave Phe, and reaction with DNFB followed by hydrolysis gave labeled Leu. Cleavage of **22.50** with trypsin gave a tripeptide of composition Gly,Phe_2 and a heptapeptide **22.52**. Partial chemical hydrolysis of **22.50** gave three identifiable tripeptides of compositions Leu,Tyr,Val, Arg,Gly_2 and Cys,Leu,Val. Identify **22.50**.

6. Predict the product(s) of each of the following reactions:

(a)

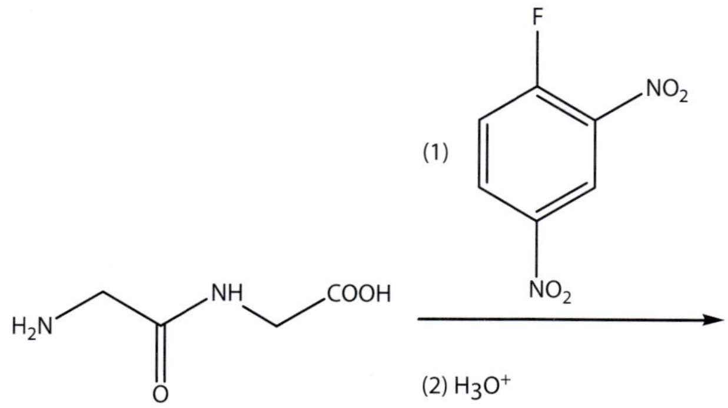

(b) Arg.Leu.Ser $\xrightarrow{\text{trypsin}}$

(c) Tyr.Glu.Val $\xrightarrow{\text{chymotrypsin}}$

(d) PhCHO $\xrightarrow[\text{(2) Li[AlH}_4\text{] (3) H}_2\text{O}]{\text{(1) PhNH}_2\text{, H}^+}$

(e)

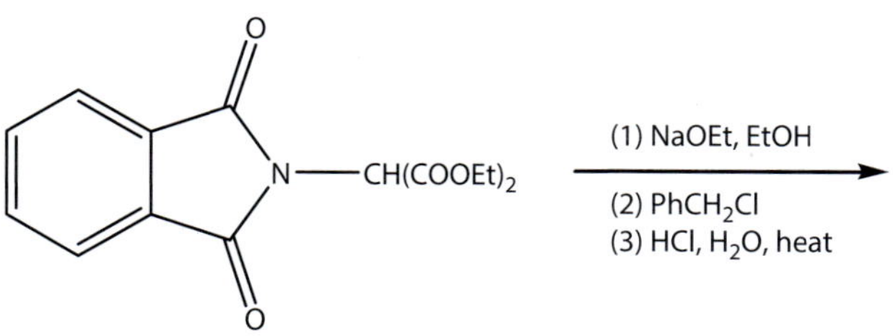

(f)

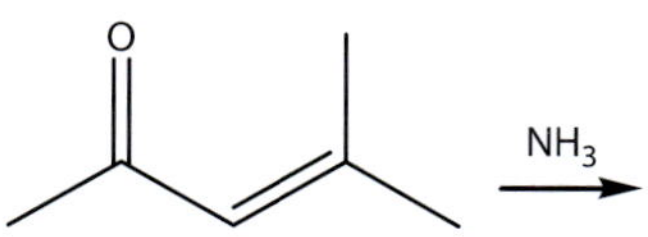

(g)

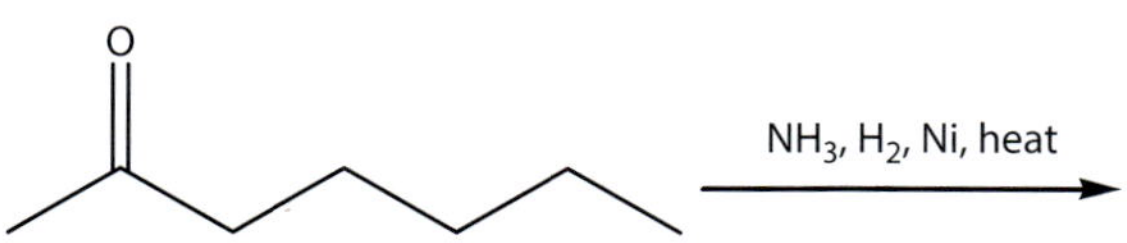

(h)

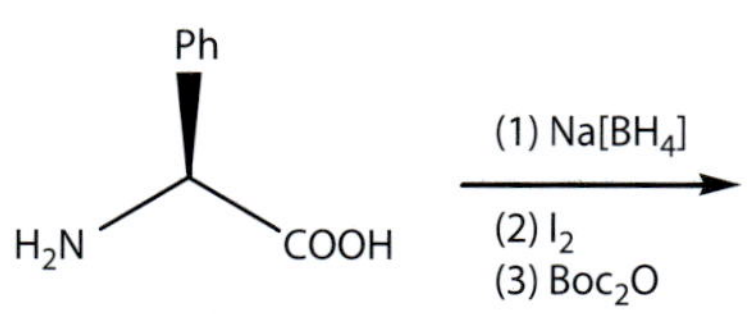

(i)

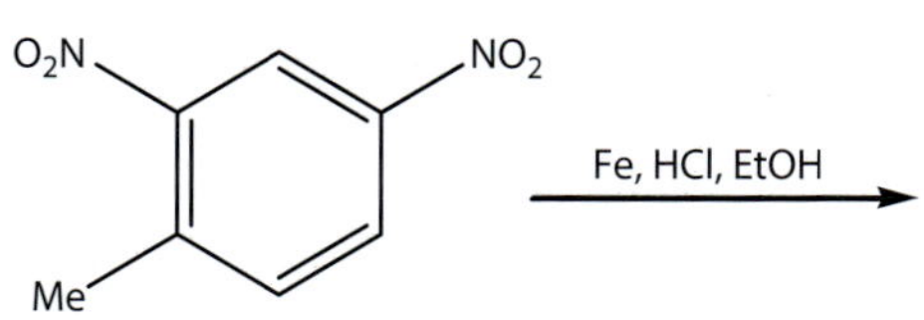

(j)

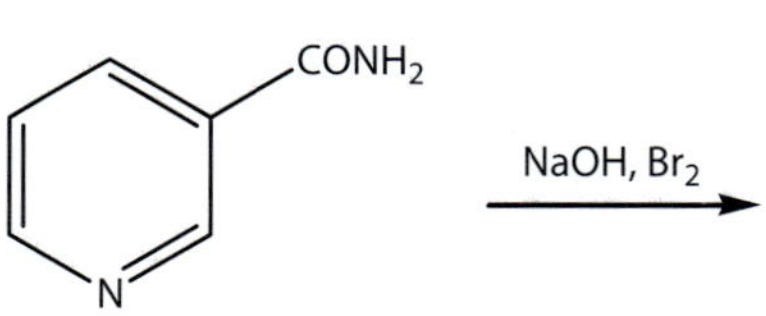

7. Identify reagents A–E in the following sequence:

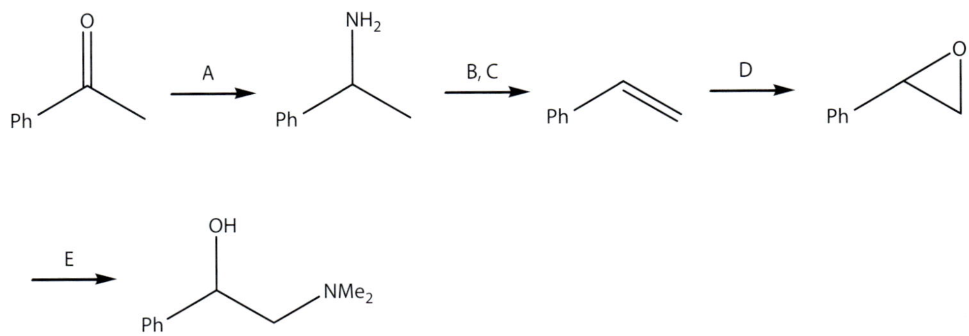

8. Fill in the blanks in the following sequence:

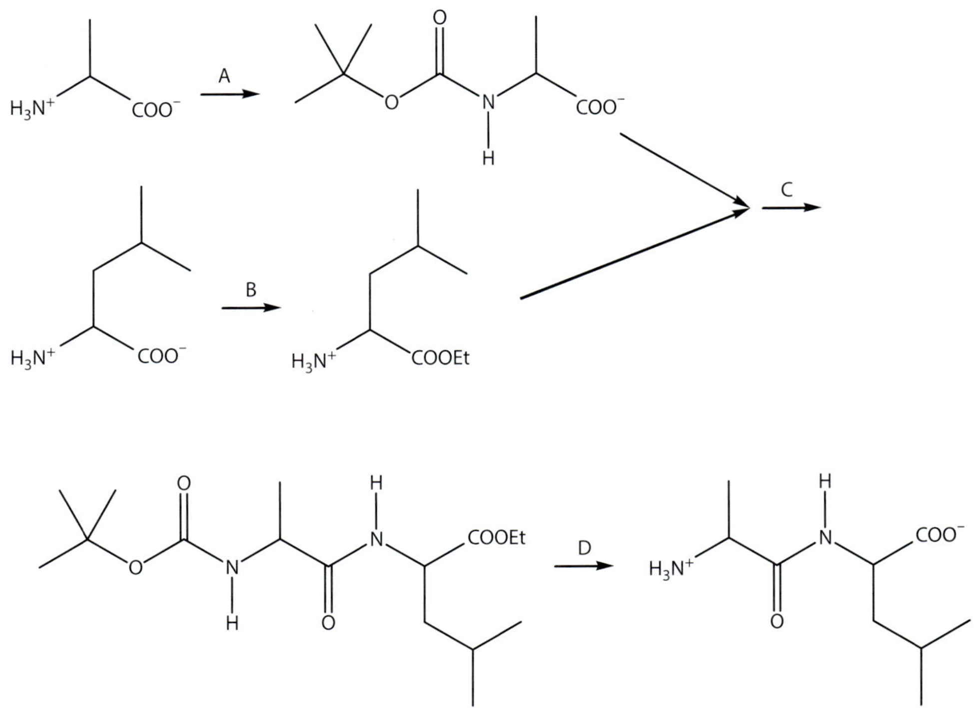

9. Devise a synthesis for Ala–Leu (it's not worth using solid-phase synthesis for a dipeptide).
10. Propose a mechanism for each of the following transformations:

(a)

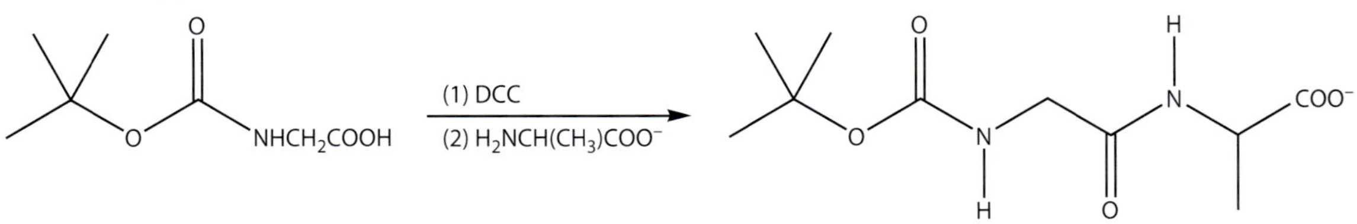

(b)

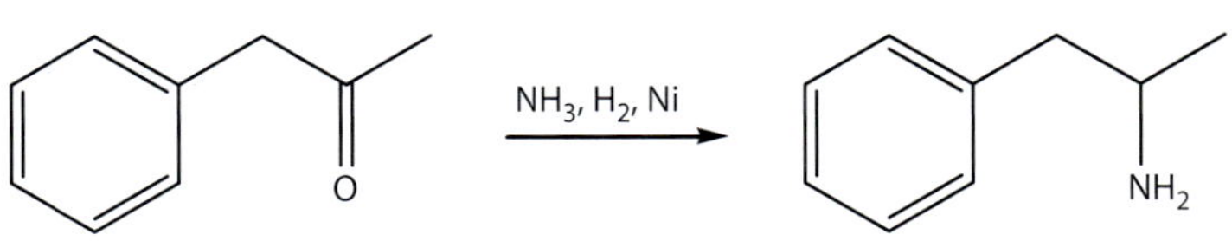

Amphetamine

(c)

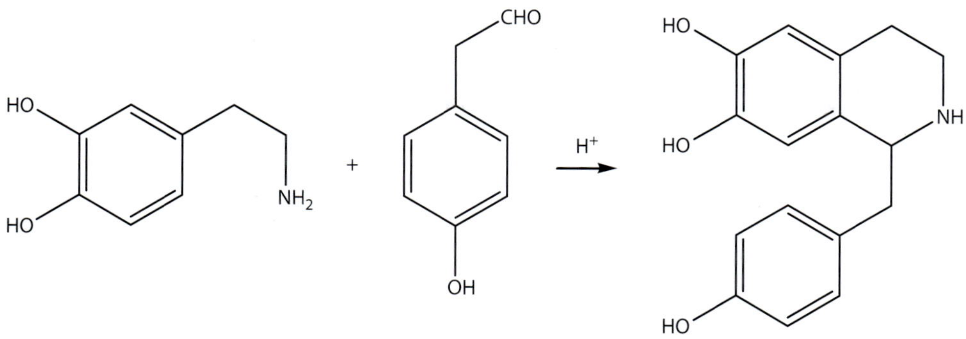

This reaction is a key step in the biosynthesis of the morphine alkaloids.

(d)

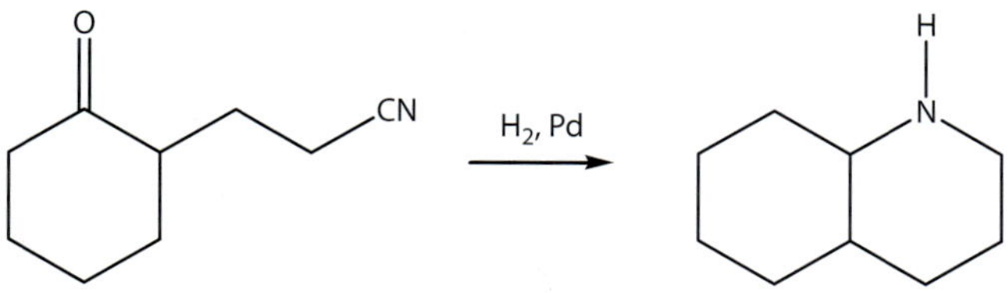

(e)

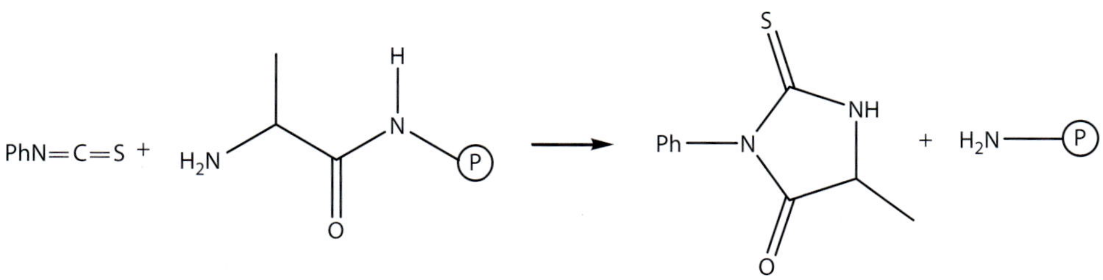

11. How would you carry out each of the following transformations?

(a)

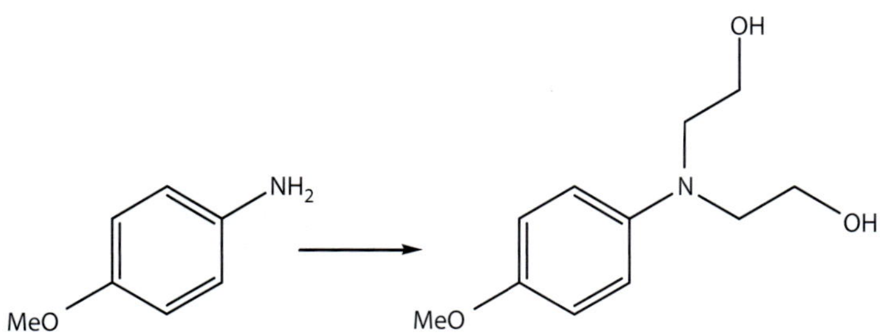

(b)

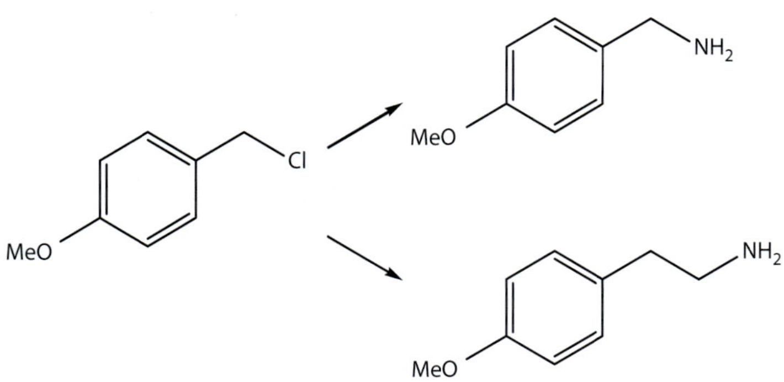

(c)

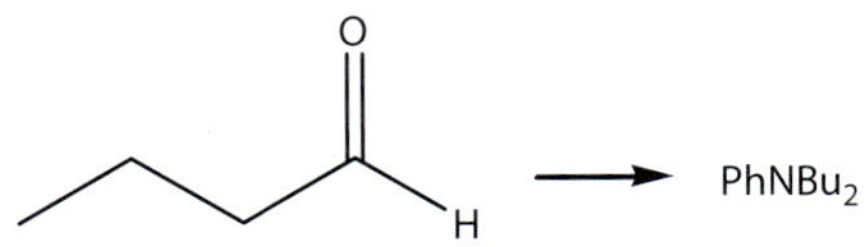

(d) $PhCH_2CH_2Br \rightarrow PhCH_2CH_2NH_2$

(e)

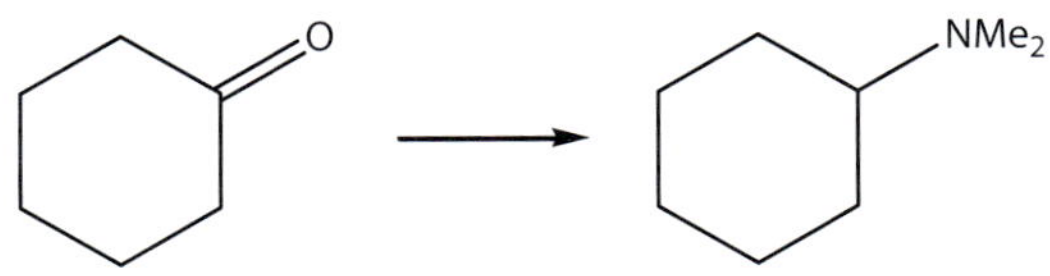

(f) $PhCHO \rightarrow PhCH_2NHCH_2CH_3$

(g)

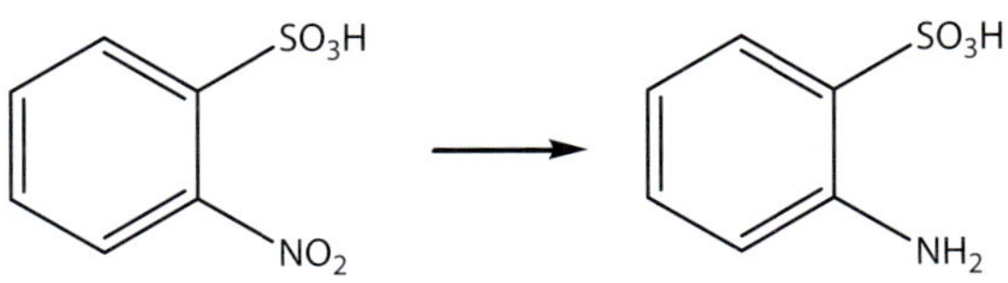

(h)

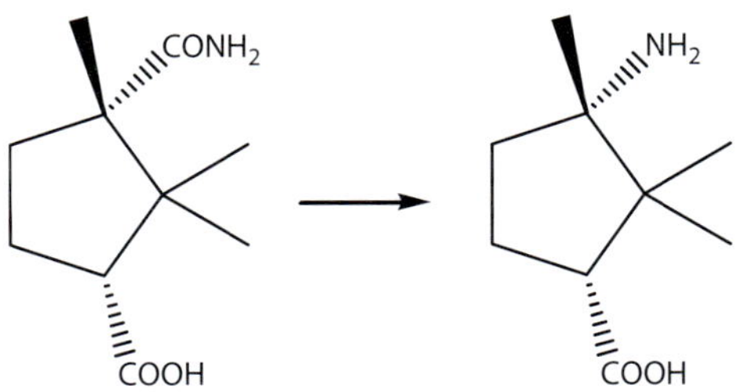

12. How would you distinguish between the peptides **22.53** and **22.54**?

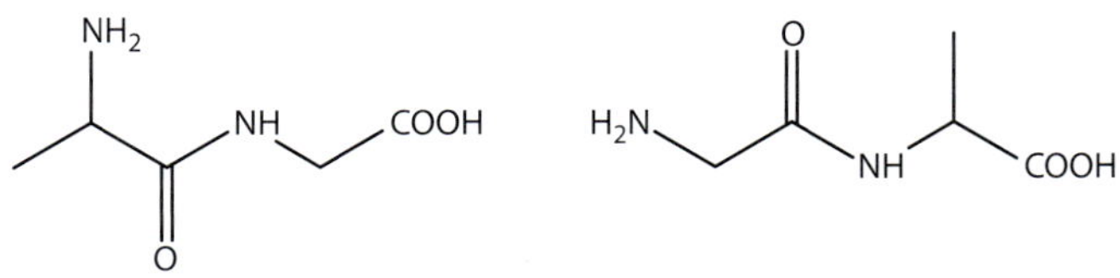

22.53 **22.54**

13. In the sequence shown in the following, identify the intermediate **22.55** and suggest a reagent for its conversion into the final product:

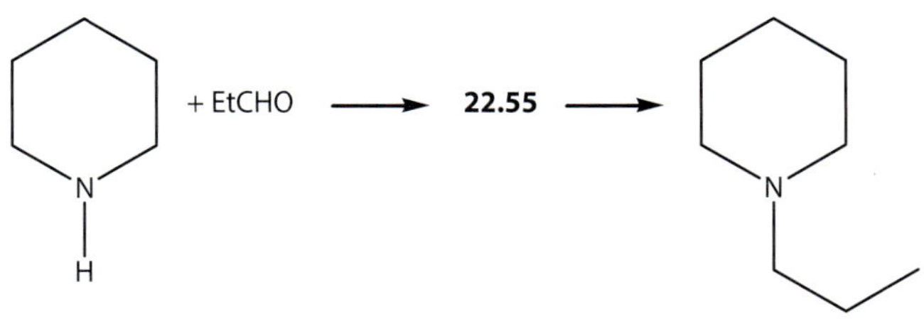

14. Devise a synthesis of each of the following amines, using a DIFFERENT synthetic approach for each one:

$PhCH_2NEt_2$ $CH_3CH_2CH_2CH_2NH_2$ $CH_3CH_2CH(NH_2)CH_3$

15. Identify the intermediates in the following sequences, justifying your conclusion with a full mechanism:

(a)

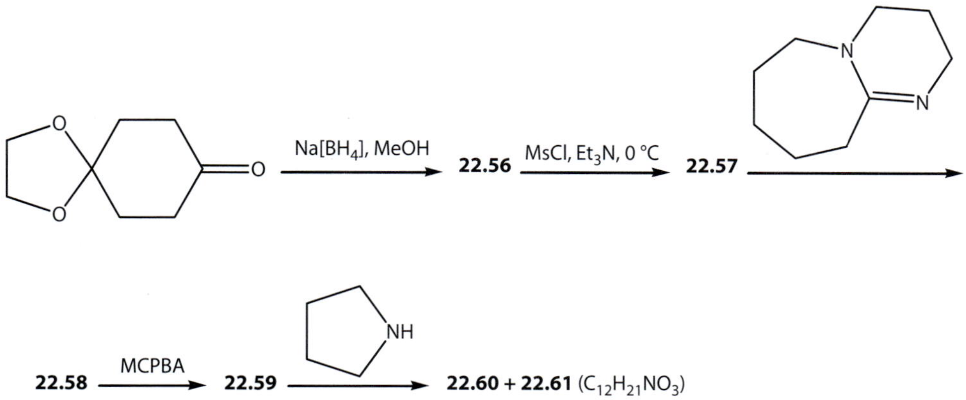

(b)

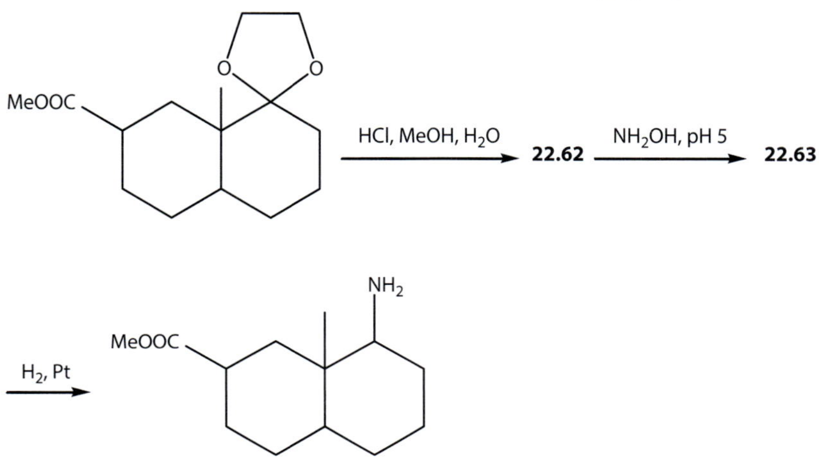

16. Show how you would prepare racemic valine, using three distinct synthetic strategies.
17. Show how each of the following compounds could be prepared using an azide intermediate:
(a) $PhCH_2CH_2NH_2$
(b) Cyclohexylamine
(c) $H_2NCH_2CH_2CH(CH_3)CH_3$
(d)

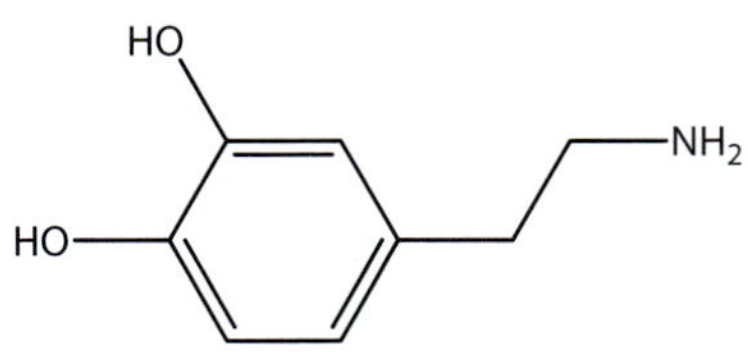

18. Most alkaloids, and many biologically active amines, are biosynthesized from amino acids. Tyrosine is a precursor of each of noradrenaline, mescaline, and mesembrenol:

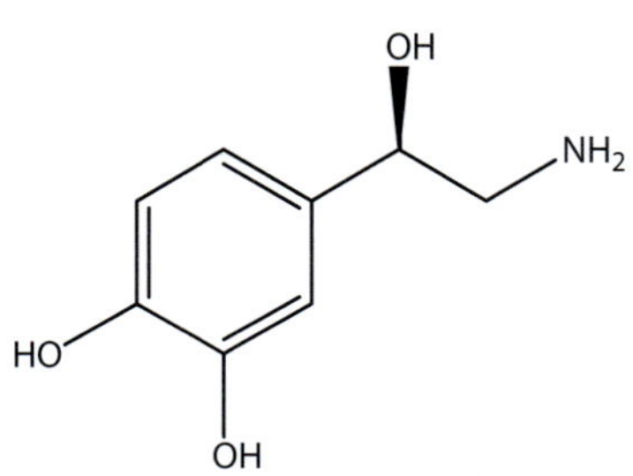

Noradrenaline

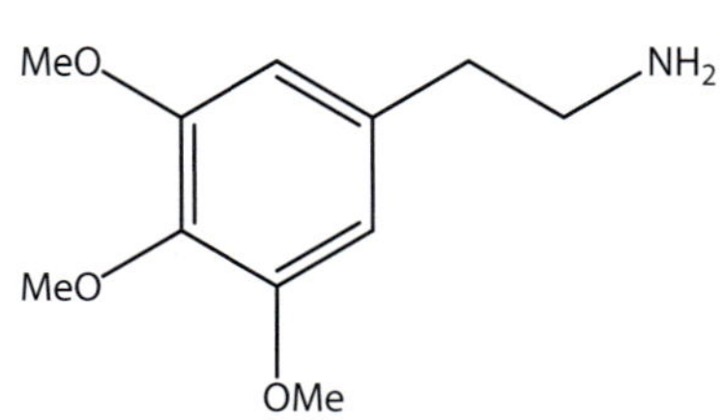

Mescaline

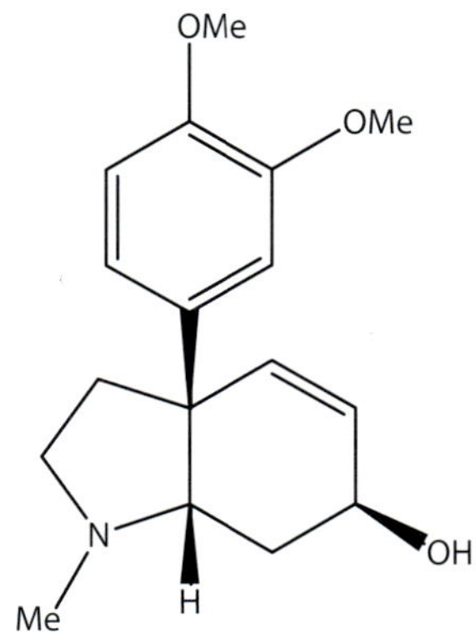

Mesembrenol

In each of these compounds, identify the tyrosine skeleton, bearing in mind that decarboxylation is a common step in such biosynthesis.

19. The sequence in the following shows key steps in the synthesis of the alkaloid yonimbine. Propose mechanisms for steps A and B:

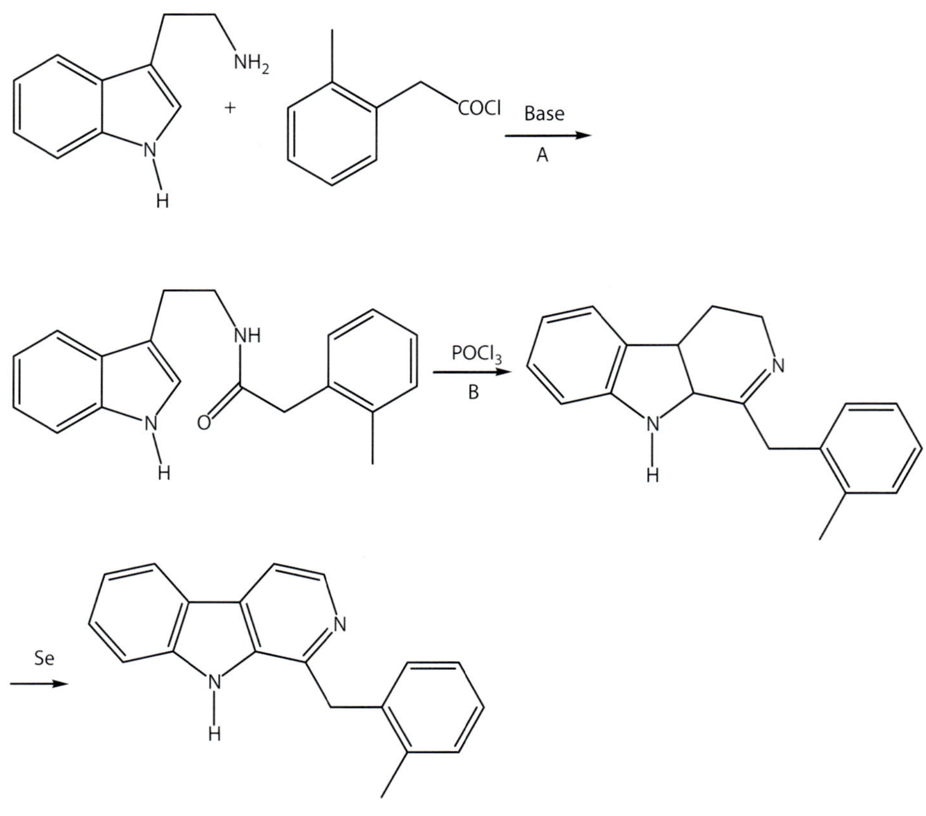

20. The structure of acetyl coenzyme A is shown in the following. Identify the nucleobase and carbohydrate involved in this structure:

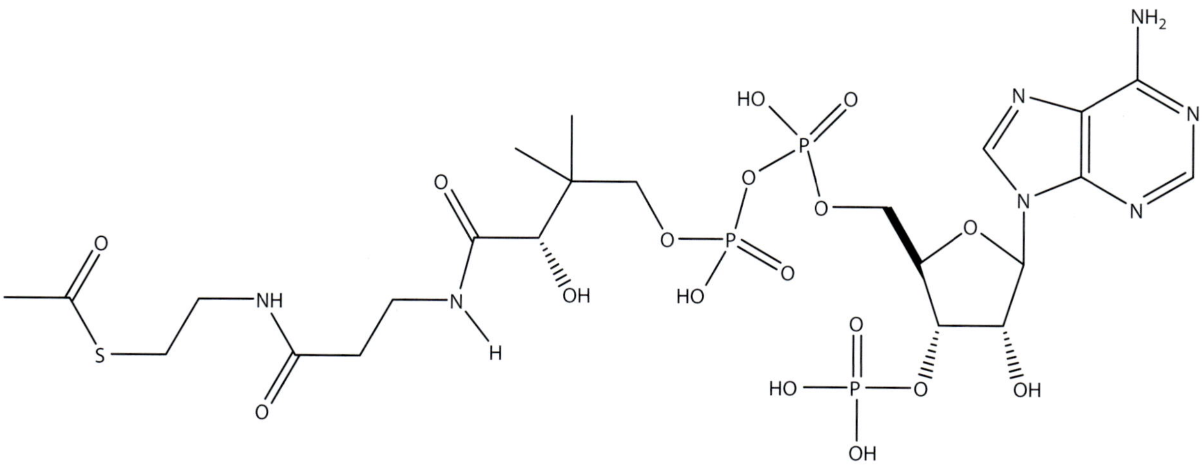

MORE CHALLENGING PROBLEMS

1. Hydrolysis of a polypeptide gave the composition Ala,Arg,Gly,Lys_2,Met,Phe,Pro,Ser_2,Tyr,Val.

   Treatment with the enzyme carboxypeptidase released Ala.
   Treatment with DNFB followed by hydrolysis gave DNP–Val.
   Treatment with cyanogen bromide gave the two peptides:
   Ala,Lys_2,Phe,Pro,Ser,Tyr
   Arg,Gly,Met,Ser,Val
   Treatment with trypsin gave three peptides and free lysine:
   Gly,Lys,Met,Tyr
   Ala,Phe,Pro,Ser
   Arg, Ser,Val
   Treatment with chymotrypsin gave three peptides:
   Lys_2,Pro,Phe
   Arg,Gly,Met,Ser,Tyr,Val
   Ala,Ser

   Use these data to deduce the sequence of the polypeptide. You are not expected to draw out the structures of the individual amino acids, but you should mark the *N*- and *C*-termini clearly.

2. Hypertensin is a polypeptide formed in the kidneys, which causes an increase in blood pressure. Total hydrolysis established the composition of hypertensin to be Arg,Asp,His_2,Leu,Phe, Pro,Tyr,Val_2.

   Treatment with the enzyme carboxypeptidase released Leu.
   Treatment with DNFB followed by hydrolysis gave DNP–Asp and DNP–Asp–Arg.
   Partial enzymatic hydrolysis gave two peptides **22.64** and **22.65**. **22.64** had the composition Arg,Asp,Tyr,Val. **22.65** had the composition His,Phe,Pro,Val. When **22.65** was treated with carboxypeptidase, Phe was released.
   Partial chemical hydrolysis gave the peptides Val.Tyr.Val and Pro.Phe.
   Use these data to deduce the sequence of hypertensin. You are not expected to draw out the structures of the individual amino acids, but you should mark the *N*- and *C*-termini clearly.

3. Propose a mechanism for each of the following transformations:

(a)

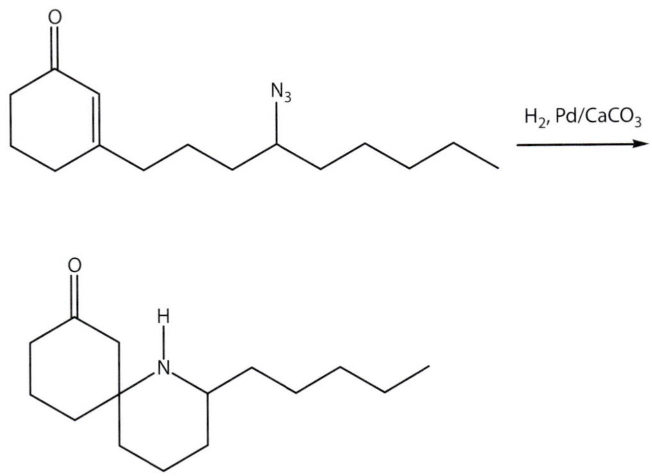

(b)

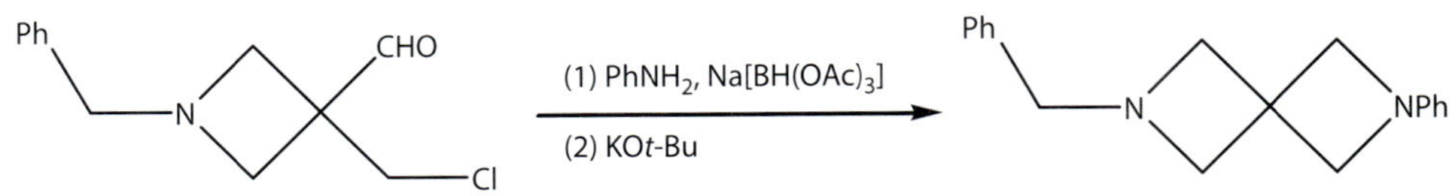

(c)

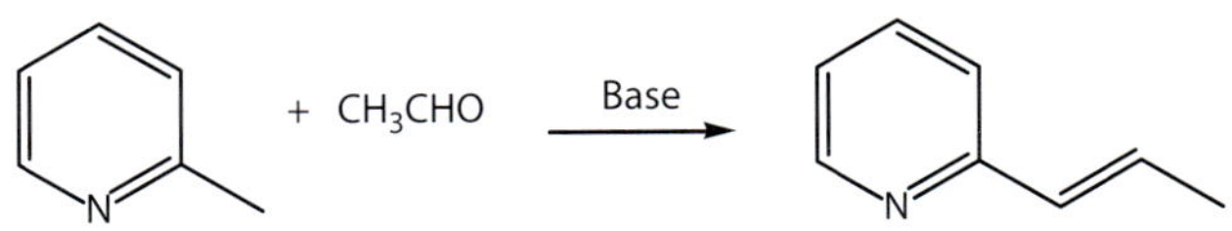

This was a key step in the first synthesis of coniine. How would the product be converted to coniine?

4. How would you carry out each of the following transformations?

(a)

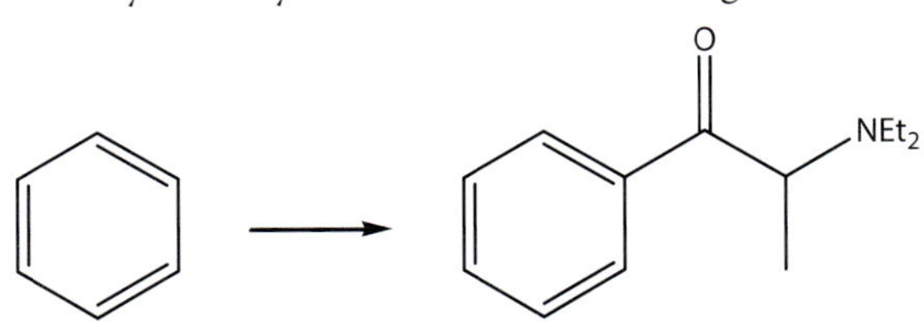

(b)

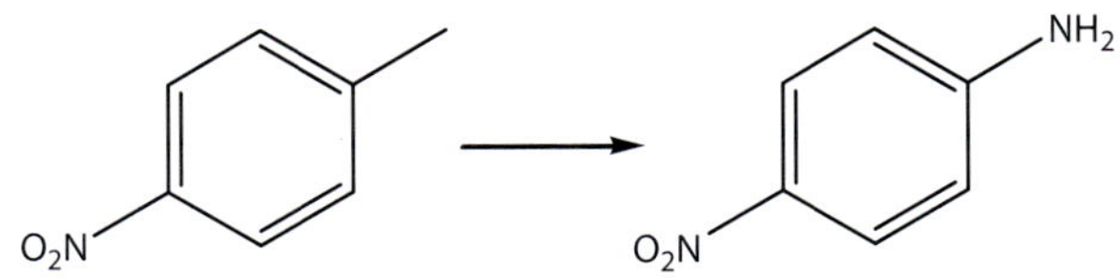

(c)

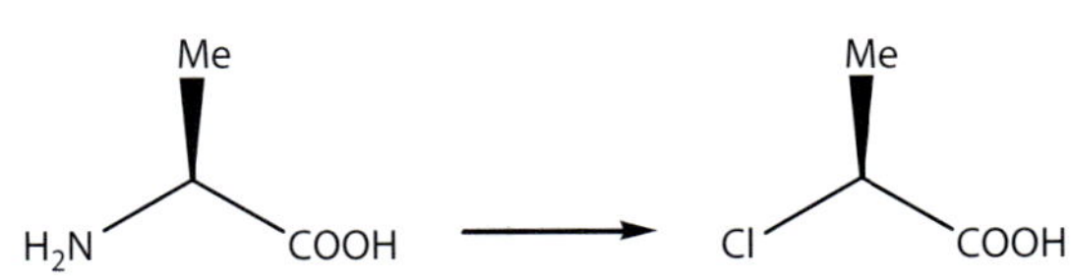

5. Propose three different strategies for the transformation below:

6. A synthesis of pumiliotoxin C, one of the arrow poison alkaloids isolated from the skin of *Dendrobates pumilio*, the strawberry poison dart frog, is shown in the following. Study it, and answer the questions posed:

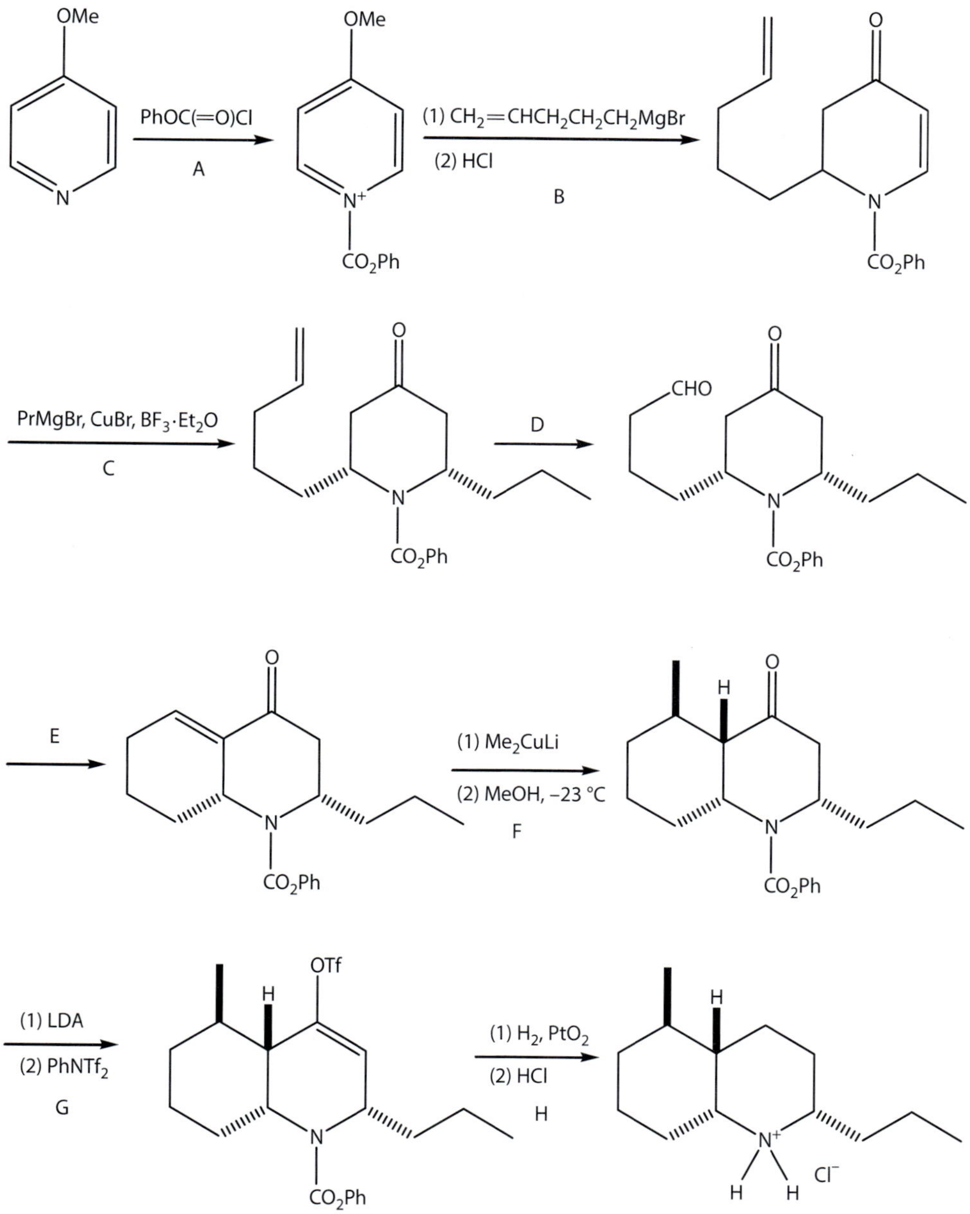

Pumiliotoxin C

(a) Propose mechanisms for steps A–C.
(b) Suggest reagents for step D.
(c) Name the reaction in step E, suggest a reagent, and write a mechanism.
(d) Propose a mechanism for step F.
(e) Although step G is unfamiliar, you should be able to propose a mechanism for it.

7. The normal Watson–Crick structure of DNA is now referred to as the β-form. What are the α and Z forms, and how do they differ from the β-form?
8. RNA is considerably less stable in basic solutions than DNA, due to hydrolysis of the phosphodiester backbone. The 2′-hydroxyl group of ribose in RNA is thought to be responsible for this lability. Propose a mechanism for the hydrolysis accounting for this.

Catalysis of Organic Reactions

23

23.1 INTRODUCTION

You may be familiar with a definition of a catalyst as something that changes the rate of a reaction, but is not consumed by it, and may be recovered unchanged at the end. For many purposes, this is satisfactory. However, things can be a little more complex—some species act as both catalyst and reagent; sometimes it is not possible to recover the catalyst, at least not in the same form as the material you put in; and in some cases, the product of a reaction may also act as a catalyst for it. Note that catalysts are distinct from the initiators that we have used in some polymerization and radical reactions. Catalysts are involved for each molecular event, whereas initiators simply start the process going, and it is then in principle self-sustaining, with the initiator being consumed. Homogeneous catalysts operate in a single homogeneous phase, nearly always a liquid. Heterogeneous catalysts usually involve a solid catalyst with the reactants in either liquid or vapor phase.

We have already met many examples of catalyzed reactions, using acids, bases, solid metals, or metal complexes. There are many types—indeed it's now difficult to find a total synthesis where a catalyzed reaction was *not* used at some stage. The advantages of catalyzed reactions over stoichiometric ones are many. Firstly, the reaction will proceed more rapidly, and this may be translated into an ability to perform the transformation under much milder than usual conditions. For example, acetonitrile, MeCN, is hydrolyzed only slowly by base at 100 °C, but in the complex $[Co(NH_3)_5(MeCN)]^{3+}$, the acetonitrile is hydrolyzed instantaneously at room temperature. Milder reaction conditions, in industrial processes, translate into lower cost, greener processes, and improved safety.

Catalysts may alter or improve the selectivity of a reaction—particular attention has been paid over the last 30 years to chiral species that can induce the formation of enantiomerically enriched products. Only small quantities of an effective catalyst are required, usually 1 % or less. In some cases, recovery and reuse of the catalyst may be possible. This is only rarely the case with homogeneous catalysts,* although every effort is made to recycle those catalysts based on precious metals. Recovery of heterogeneous catalysts is generally easy, requiring only filtration of the reaction mixture.

Catalysts operate by providing an alternative, lower, free-energy pathway for a reaction. They do not allow us to "beat" thermodynamics—they cannot make a thermodynamically forbidden reaction allowed nor can they change the position of an equilibrium. The catalyst operates by introducing a new lower-energy pathway for the reaction, with a lower value of $\Delta G^{\ddagger}$ (Figure 23.1). This may be either similar to, or quite different from, the processes involved in the uncatalyzed process. As with an uncatalyzed system, the slowest step corresponds to the highest-energy transition state. In a catalyzed process, this should technically be referred to as a turnover-limiting step rather than a rate-determining step, but for most purposes, the distinction will not be important. Either the catalyzed or the uncatalyzed pathway may be multistep; the key feature is that $\Delta G^{\ddagger}$ (catalyzed) < $\Delta G^{\ddagger}$ (uncatalyzed).

* The so-called heterogenized homogeneous catalysts, where the catalyst is chemically attached to a solid support such as silica or an ion-exchange resin, are easier to recover, but most eventually leach the catalyst or become inactivated.

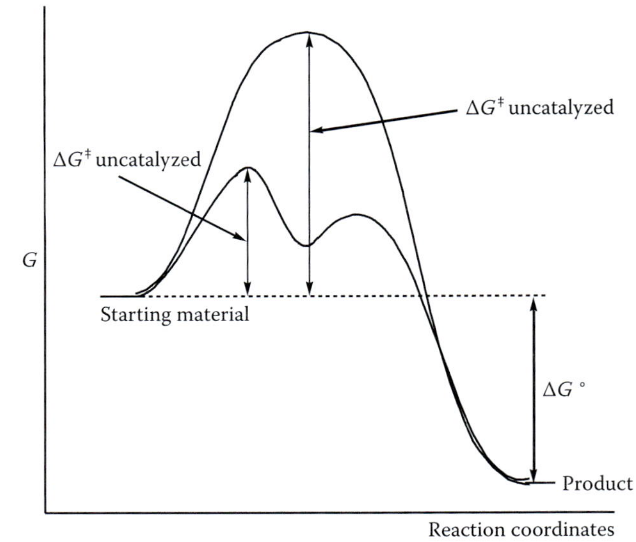

FIGURE 23.1 Reaction profiles of catalyzed and uncatalyzed reactions.

23.2 CATALYSIS BY ACIDS AND BASES

We saw in Chapter 15 that ester hydrolysis can be catalyzed by either acids or bases (Figure 23.2). In the acid-catalyzed process, all the steps are reversible, and the direction of the reaction is effectively controlled by mass action. The rate-determining step is the attack of water on the protonated ester. In the base-catalyzed process, the hydroxyl ion attacks the neutral ester in the rate-determining step. The reaction is rendered irreversible by the final step in which methoxide deprotonates the acid (pK_a MeOH = 15.5, pK_a of a typical acid is 5), and the hydroxyl ion is not regenerated at the end of the reaction. Thus, ester hydrolysis is possible in either acidic or basic solution; if the reaction rate was plotted against pH, it would show a minimum at neutral pH. We have met other reactions of this type, for example, the mutarotation of glucose.

Other reactions show a more complex dependence on pH. In Chapter 14, we studied the reaction of various molecules, RNH_2, with carbonyl compounds to give imines, hydrazones,

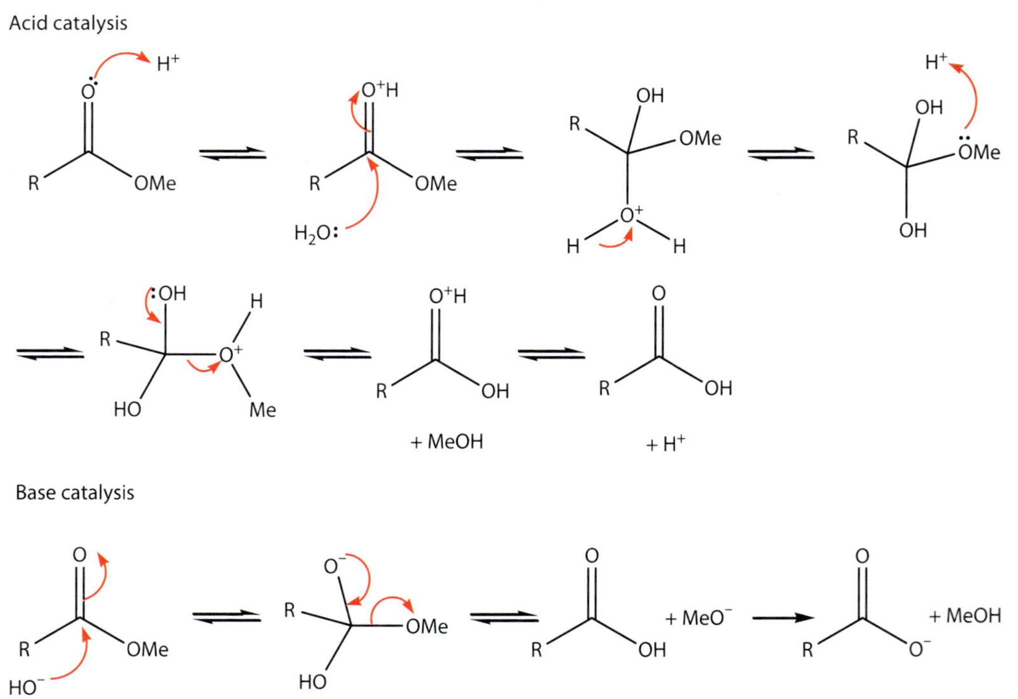

FIGURE 23.2 Catalysis of ester hydrolysis by acids or bases.

semicarbazones, *etc.* Figure 23.3 shows the pH dependence for the formation of acetone oxime, and Figure 23.4 the mechanism of oxime formation. At neutral or basic pH, step 3 is rate limiting; a proton is required for the OH group to leave. However, at low pH in acidic solution, step 1 becomes rate limiting, because the hydroxylamine becomes protonated (pK_a of [H_3N^+OH] is 6.0) and is thus unable to act as a nucleophile.

Some reactions may be catalyzed by any acid or any base present, and the kinetics may then be extremely complex. For example, if the rate of iodination of acetone (Figure 23.5) is measured in

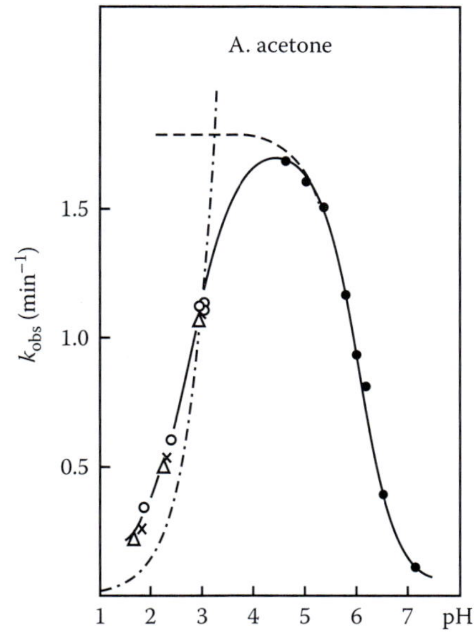

FIGURE 23.3 pH dependendance of the rate of formation of acetone oxime. *Note*: Mechanism remains the same over pH range of interest but still has an upper limit.

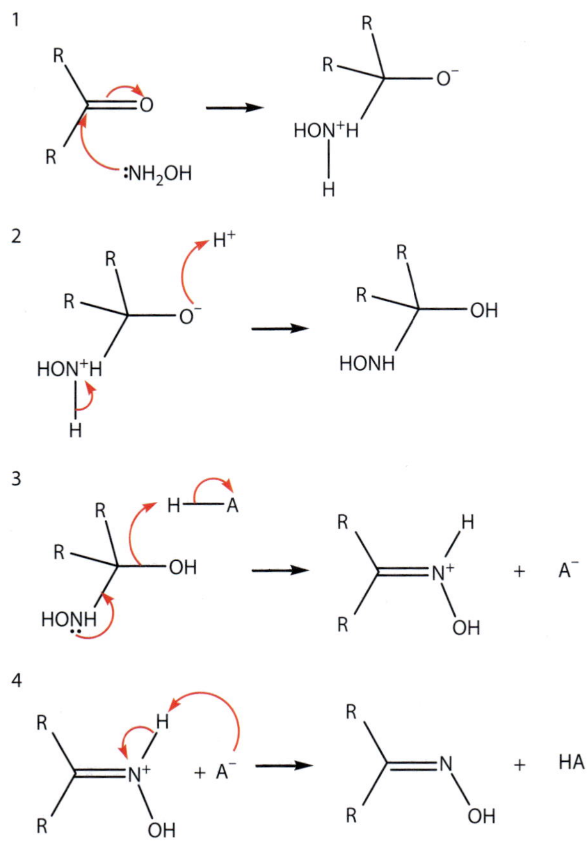

FIGURE 23.4 Mechanism of oxime formation.

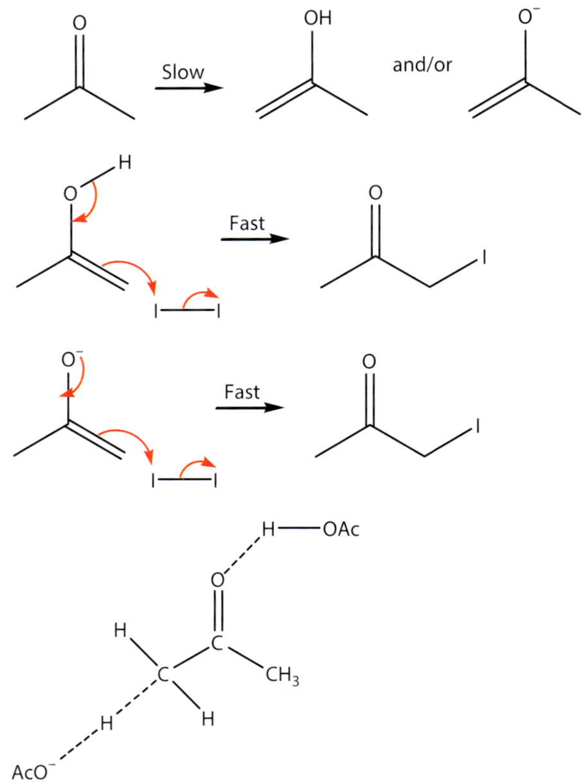

FIGURE 23.5 Iodination of acetone.

an acetic acid/sodium acetate buffer solution of constant ionic strength, then the rate of reaction is given by Equation 23.1. Clearly, acid and base can be involved simultaneously in effecting the enolization of the acetone.

$$\frac{-d[I_2]}{dt} = [\text{acetone}]\{3\times10^{-8} + 1.6\times10^{-3}[H^+] + 15[HO^-] + 5\times10^{-6}[\text{AcOH}] + 1.5\times10^{-6}[\text{AcO}^-] + 2\times10^{-5}[\text{AcOH}][\text{AcO}^-]\} \quad (23.1)$$

PROBLEM 23.1

Write a mechanism for the mutarotation of glucose in acidic, basic, and neutral solution, showing carefully the function of the acid and base in catalysis.

PROBLEM 23.2

The reverse aldol reaction shown below gives a rate equation $k_{obs} = k[\text{starting material}][HO^-]$. Write a mechanism for this process that is consistent with the rate equation, remembering that proton transfers to or from oxygen and nitrogen atoms are rarely rate limiting.

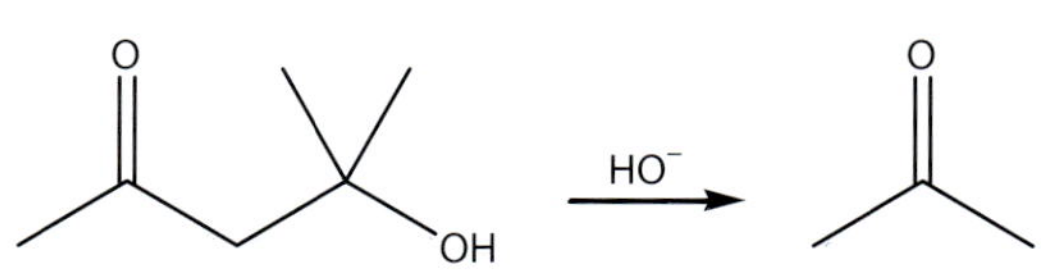

Solutions

23.1 Although we do not often write the reaction in neutral solution, this is the best place to start:

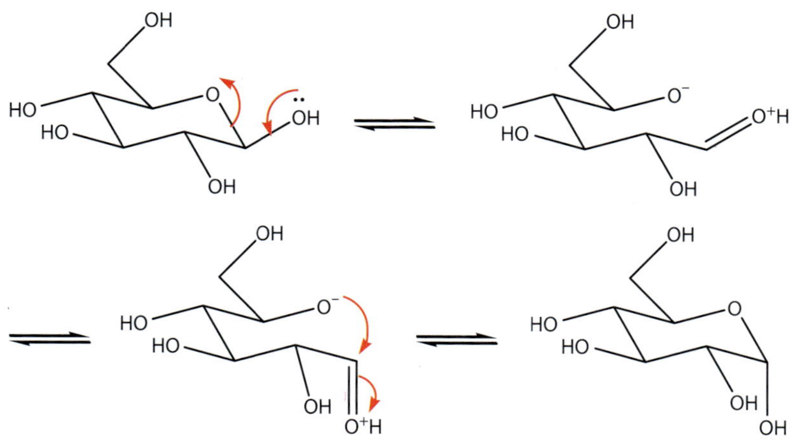

In acid solution, the first step would be to protonate the oxygen that we wish to make into a leaving group. Thus, we would have

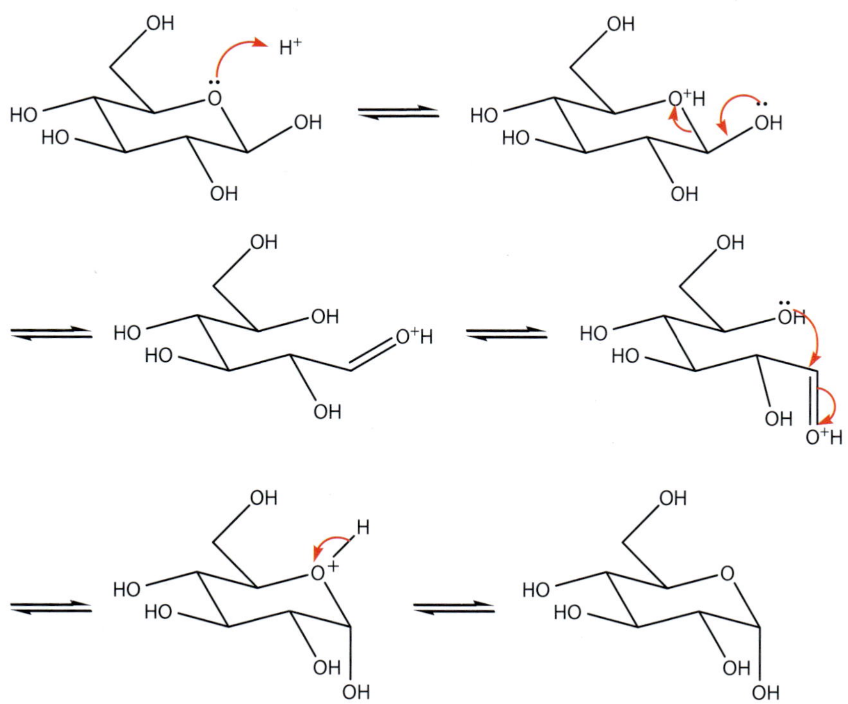

Note that the proton is recovered in the final step of the sequence. In base, we begin by deprotonating the anomeric hydroxyl group. Thus, we have

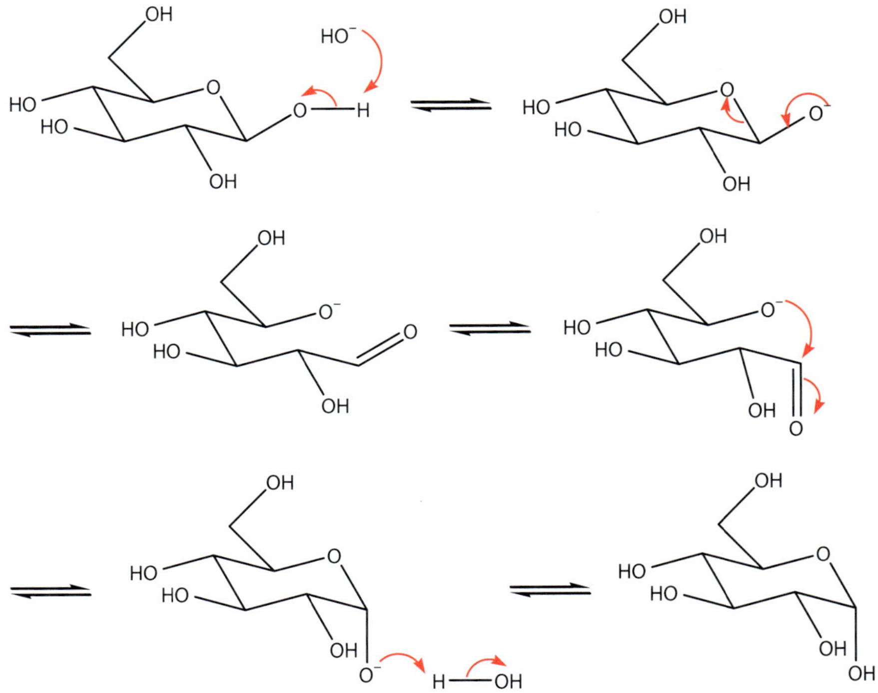

The hydroxyl ion is regenerated in the final step of the process.

23.2 The reaction mechanism involves deprotonation of the OH group and then the reverse of the steps by which an aldol condensation takes place:

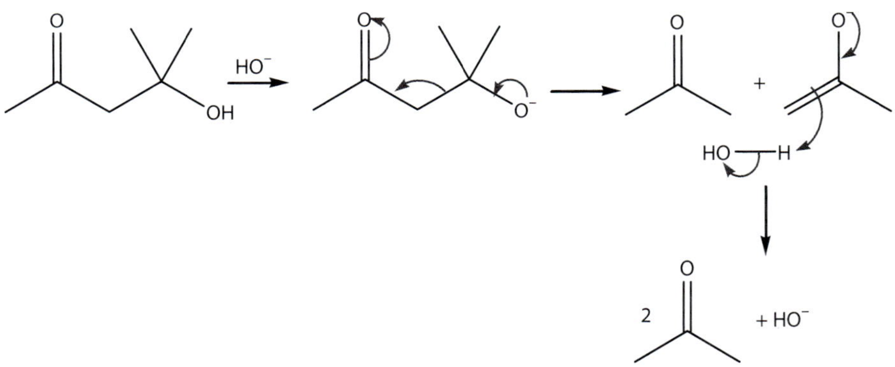

The hydroxyl ion is only involved in the first step of the reaction, and we have already said that this is unlikely to be rate determining. The carbon–carbon bond breaking process of the second step seems a much more likely candidate for a slow step. However, there is something slightly amiss with the mechanism as drawn—deprotonation of an OH group by a hydroxyl ion is likely to be reversible. This means that the concentration of the deprotonated aldol product is determined both by the starting material concentration and hydroxyl ion concentration:

$$[\mathrm{MeC({=}O)CH_2CMe_2O^-}] = K[\mathrm{MeC({=}O)CH_2CMe_2OH}][\mathrm{HO^-}]$$

Now, all is well—the second step is unimolecular and rate determining, with the observed kinetics.

$$\text{Rate} = k[\mathrm{MeC({=}O)CH_2CMe_2O^-}] = k\,\mathrm{K}[\mathrm{MeC({=}O)CH_2CMe_2OH}][\mathrm{HO^-}]$$

Key Points from Sections 23.1 and 23.2

- Catalysts are species that appear in the rate equation at a higher power than in the stoichiometric equation. In some, but not all, cases, they may be recovered at the end of the process.
- Catalysts lower the free energy of activation for a reaction, by providing a different, and lower-energy, pathway for the process to occur. They do not alter $\Delta G°$.
- Many catalyzed reactions show enhanced selectivity over uncatalyzed processes.
- Many reactions are catalyzed by Brønsted acids and/or bases. Since the reaction mechanism and the nature of the rate-determining step may depend on pH, dependence of rate on pH may be quite complex.

23.3 LEWIS ACID CATALYSIS

There is one type of Lewis acid catalysis with which we should be very familiar—a range of Lewis acids such as $AlCl_3$, $SnCl_4$, and $FeCl_3$ catalyze the Friedel–Crafts alkylation and acylation of aromatic compounds (Figure 23.6; see Section 12.2.2).

23.3.1 REACTIONS OF CARBONYL COMPOUNDS

Protic, Brønsted acids activate carbonyl groups toward nucleophilic attack, and coordination of a proton to oxygen (or nitrogen) renders this a better leaving group. The same effect can be achieved using a Lewis acid to coordinate to oxygen (Figure 23.7). Thus, numerous simple transformations of carbonyl groups may be catalyzed by Lewis acids, in some cases, more effectively

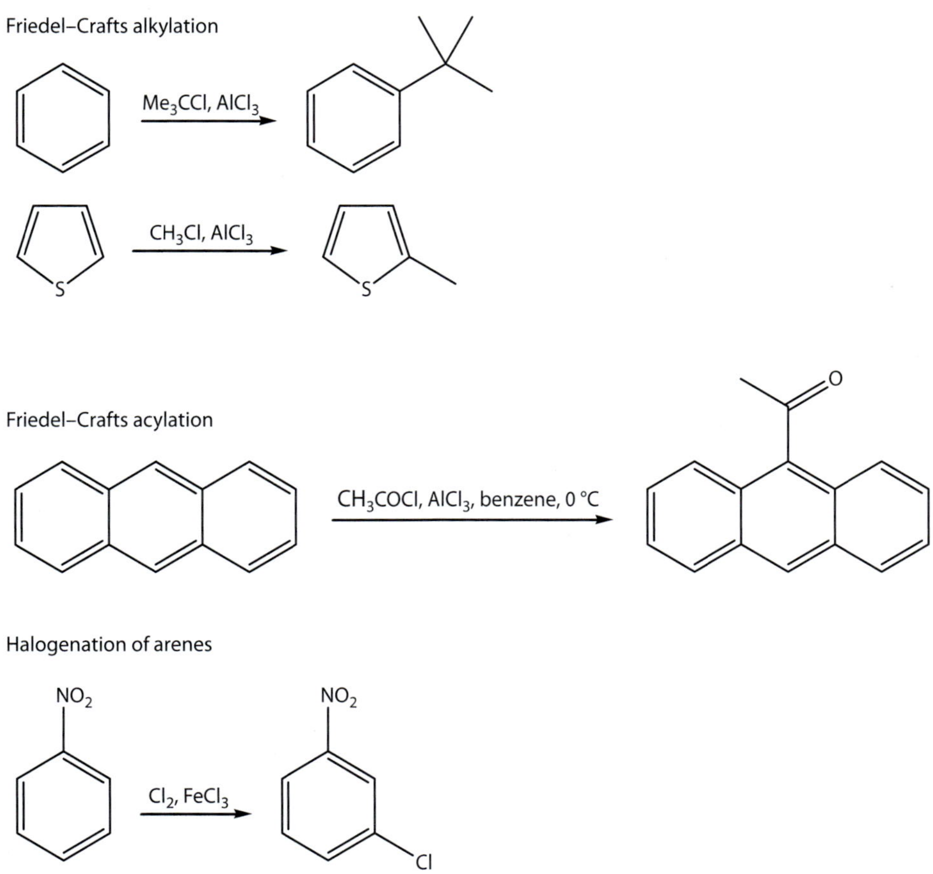

FIGURE 23.6 Lewis acid-catalyzed reactions of aromatic compounds.

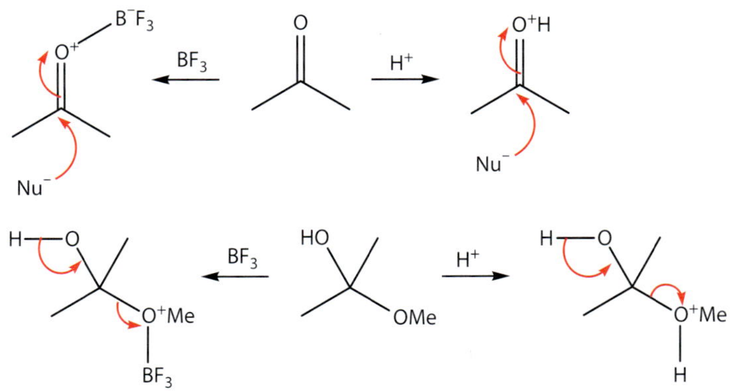

FIGURE 23.7 Coordination of oxygen to a Lewis and a Brønsted acid.

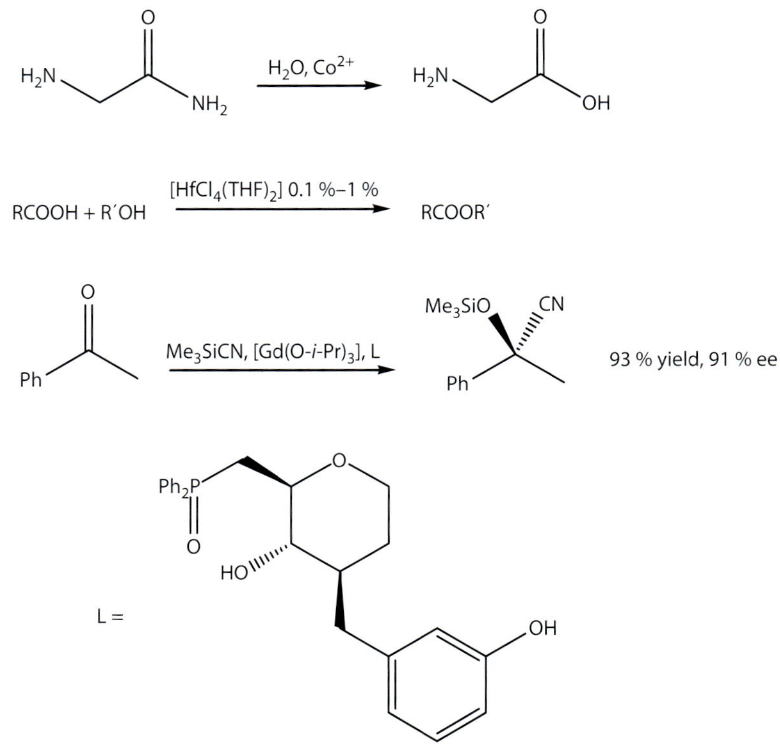

FIGURE 23.8 Lewis acid catalysis of simple carbonyl transformations.

than with protic acids. For example, the preparation of thioacetals is generally catalyzed by BF_3 rather than a protic acid, as better yields are obtained this way. Some other examples are shown in Figure 23.8. The last example is a particularly interesting one in demonstrating that the use of Lewis acids gives the synthetic chemist more opportunity to "tune" the catalyst used to the exact reaction desired. Metals from the main group, the transition series, and the lanthanides have been widely used, and in combination with exotic chiral ligands, many enantioselective transformations can now be achieved.

Carbonyl groups are also activated by Lewis acids to participate in various condensation reactions. One of the most important of these is the Mukaiyama aldol reaction. The first version of this in the 1970s (Figure 23.9) was not catalytic, but catalytic versions and enantioselective variants were quickly developed. The advantage of the process is that a crossed-aldol reaction is achieved without any risk of self-condensation of either component, and reaction conditions are exceptionally mild. However, the starting silyl enol ether does need to be prepared. Some examples are shown in Figure 23.10. The first reaction is one on which many common catalysts

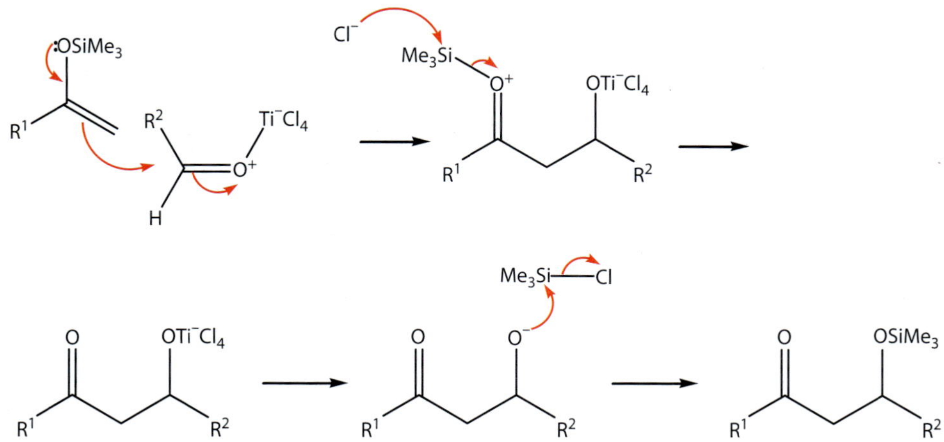

FIGURE 23.9 Mechanism of the Mukaiyama aldol condensation.

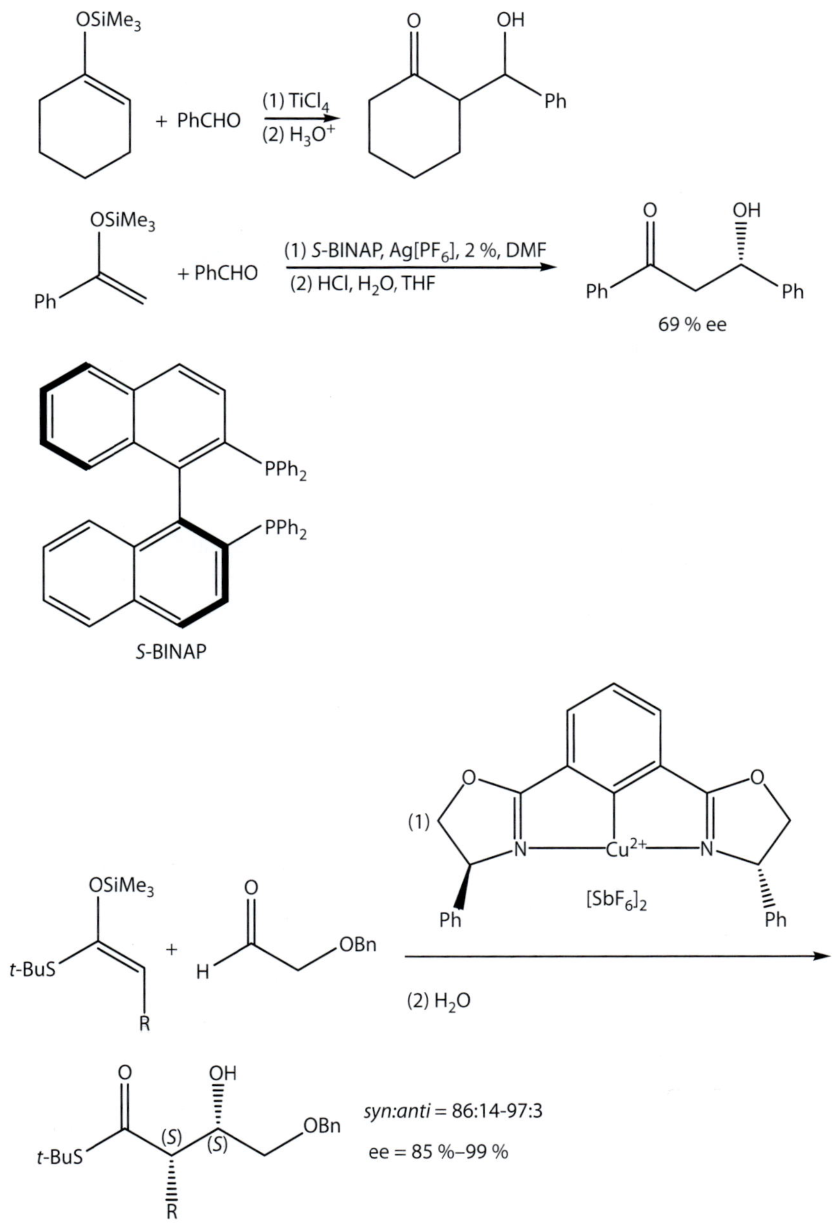

FIGURE 23.10 Examples of the Mukaiyama aldol condensation.

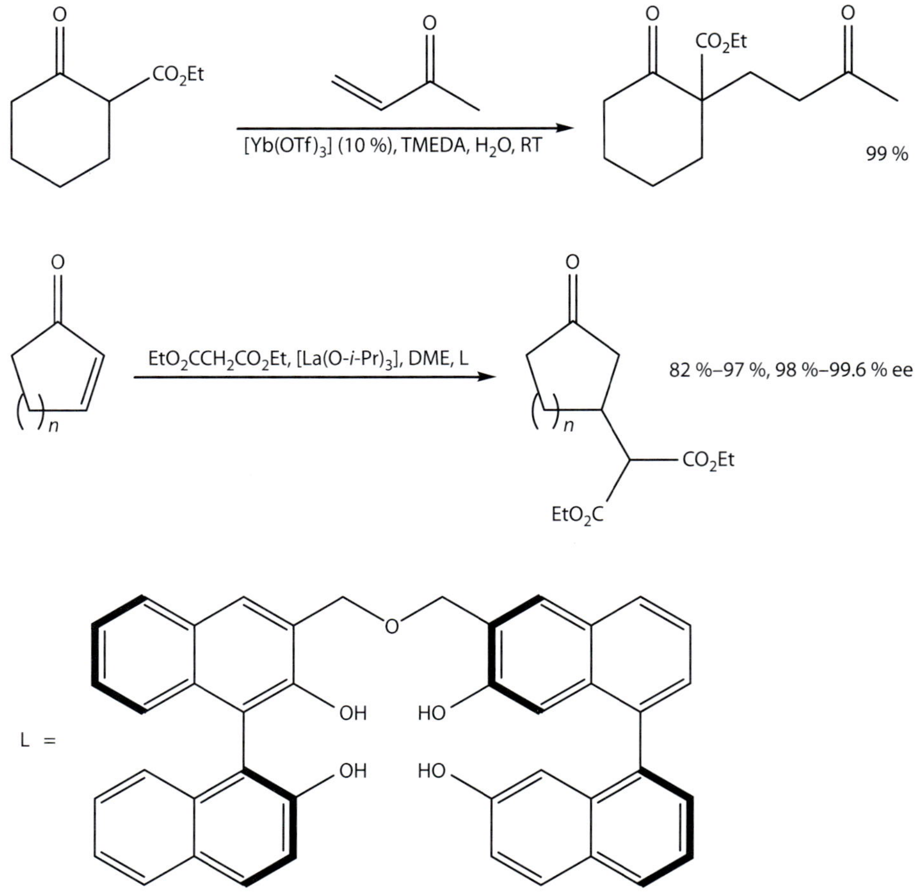

FIGURE 23.11 Conjugate additions catalyzed by Lewis acids.

have been tested, and a wide range is now available. In the second example, a respectable enantioselectivity is achieved. In the final example, the extension of the reaction to silylated enolates of other carbonyl compounds (in this case, a thioester) is demonstrated. This process shows good diastereoselectivity and good enantioselectivity.

Conjugate additions of enolates or their equivalents to enones are also catalyzed by Lewis acids, and again enantioselective versions of the reactions have proved to be very useful (Figure 23.11).

23.3.2 CATALYSIS OF DIELS–ALDER REACTIONS

The rates of Diels–Alder reactions depend, at least in part, on the use of a diene with electron-donor groups and an alkene bearing electron-withdrawing groups. The electron-withdrawing effect of a carbonyl attached to an alkene is enhanced when it is coordinated to a Lewis acid, and many Lewis acids catalyze Diels–Alder reactions. Thus, in the presence of $NbCl_5$, cyclopentadiene reacts with cyclohexenone at −78 °C (Figure 23.12). Diastereoselectivity is often improved, and enantioselective versions of the reaction have been demonstrated. Recently, the use of lanthanide triflates, which may be used in water, rather than organic solvents, allows "greener" versions of the reactions.

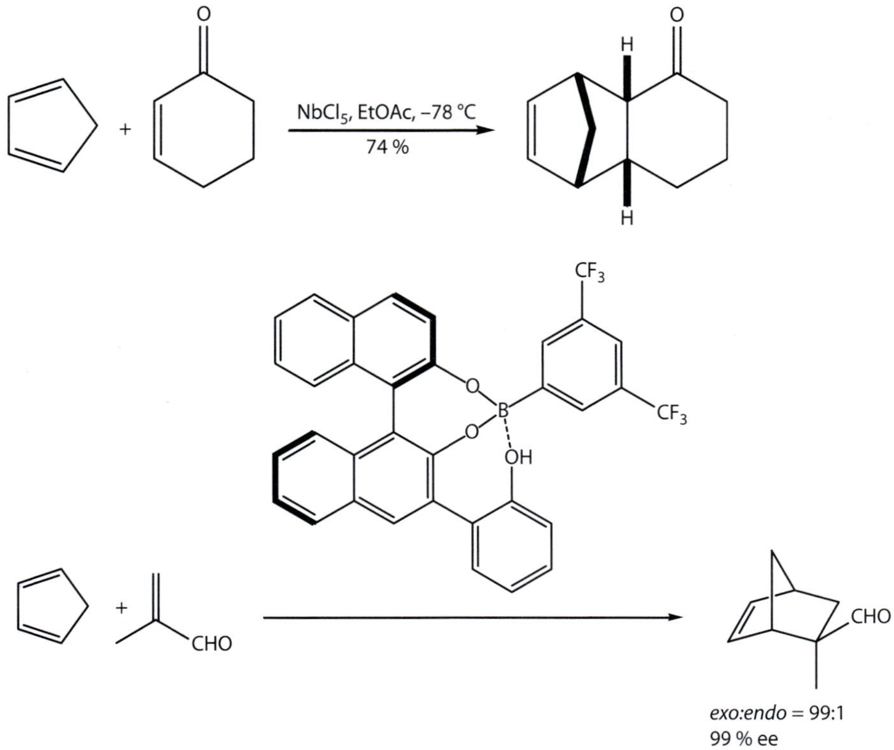

FIGURE 23.12 Lewis acid-catalyzed Diels–Alder reactions.

PROBLEM 23.3

Explain the following observations:

(a) Metal ions increase the rate of decarboxylation of **23.1** but not of **23.2** or **23.3**:

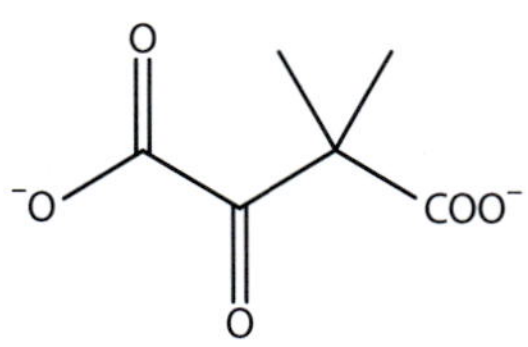

23.1

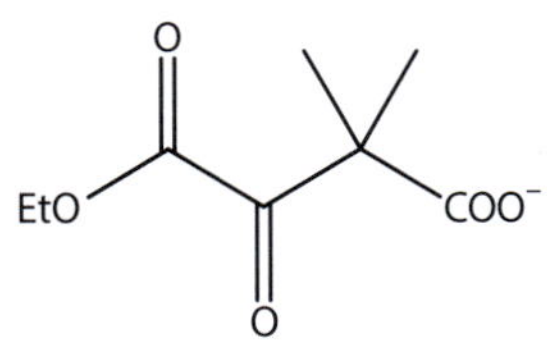

23.2

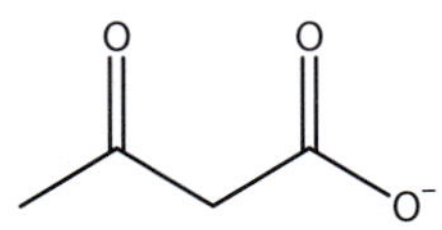

23.3

(b) In the presence of Ni^{2+}, the hydrolysis of **23.4** is up to 10^{10}-fold faster than in pure water:

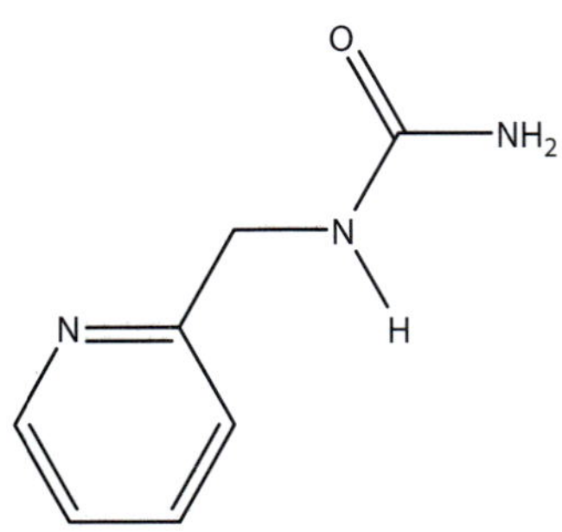

23.4

PROBLEM 23.4

Write a mechanism for each of the following reactions, showing the function of the catalyst:

(a)

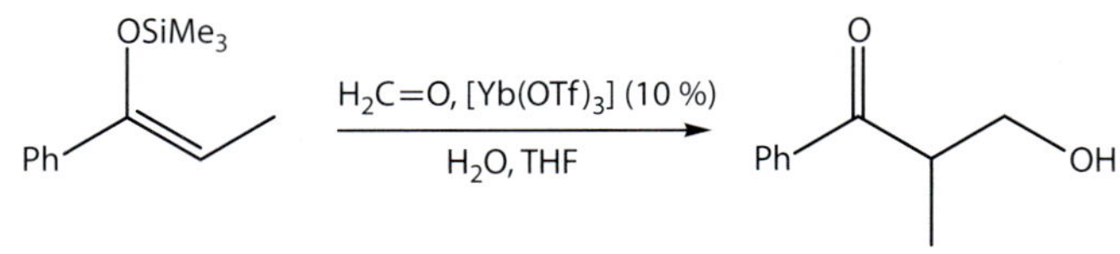

(b)

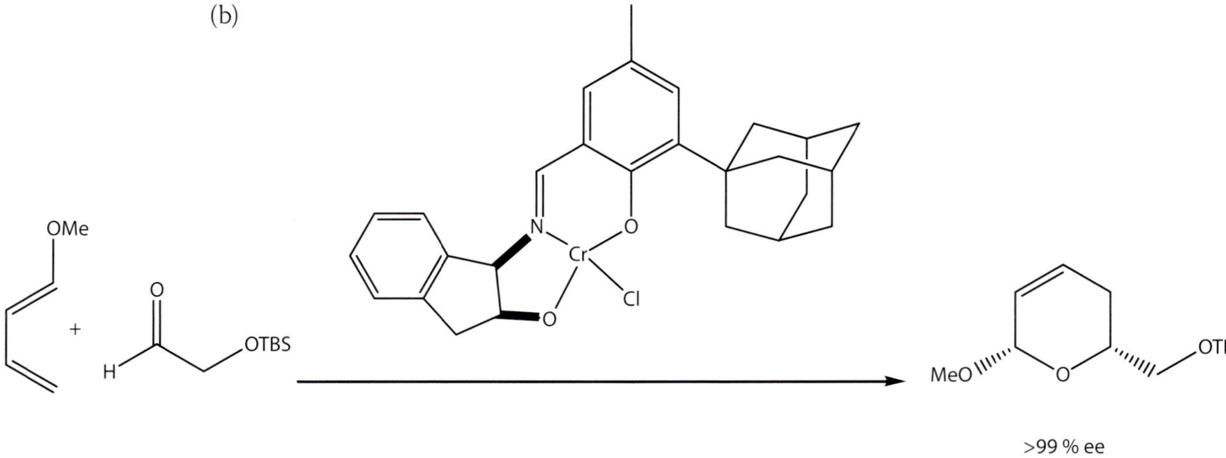

Solutions

23.3(a) In **23.1**, the metal coordinates to the two anionic centers and to the ketone, making the ketone more willing to accept electrons in the decarboxylation process.

23.3(b) The nickel coordinates to both the ring nitrogen and the carbonyl of the urea, enhancing water attack on the carbonyl group.

23.4(a) This is a Mukaiyama aldol reaction; the function of the ytterbium triflate is to enhance the electrophilicity of the formaldehyde by coordination:

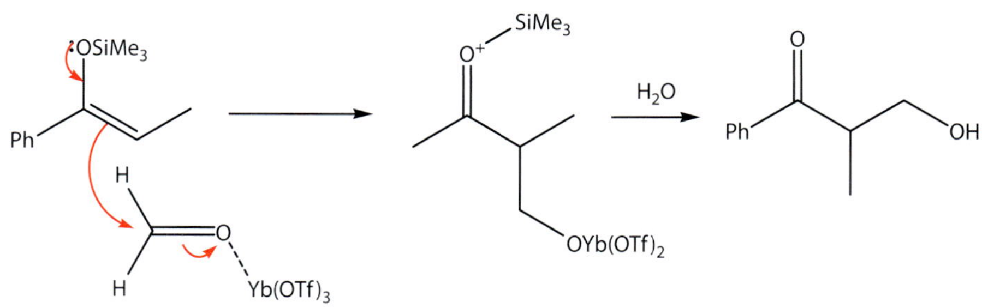

23.4(b) Although not strictly a Diels–Alder reaction, this is a 4+2-cycloaddition and goes by the same mechanism. The function of the Lewis acid is to coordinate to the best available Lewis base, the oxygen of the carbonyl group, to enhance the electron deficiency of this component:

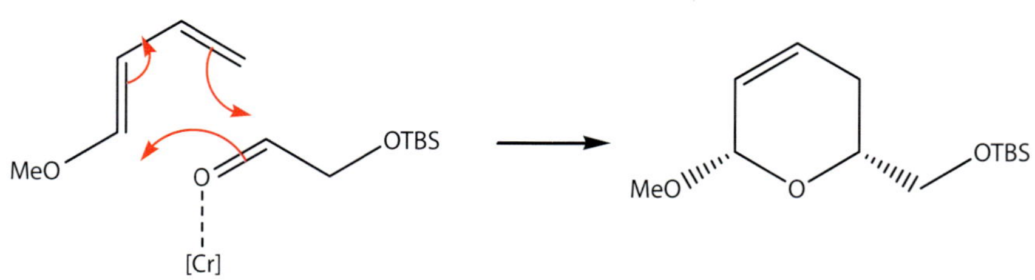

Key Points from Section 23.3

- Coordination of Lewis acids occurs at carbonyl groups and also at sp^3 oxygen and nitrogen atoms. Coordination to a carbonyl group enhances its polarity and alters susceptibility to nucleophilic attack in the same way as does protonation.
- The wide range of Lewis acids available means that the catalyst can be "tuned" to the particular reaction required.
- The Mukaiyama aldol reaction is a highly selective cross aldol condensation using a silyl enol ether as nucleophile and a Lewis acid-coordinated carbonyl compound as electrophile.
- Coordination of carbonyl groups by Lewis acids accelerates simple carbonyl transformations, conjugate additions to unsaturated carbonyl compounds, and Diels–Alder reactions, by polarizing the carbonyl group.

23.4 PHASE-TRANSFER CATALYSIS

Some catalysts, like the ionophores that we saw in Section 4.3.3, work by transporting a reagent or substrate from one medium or solvent to another. These ionophores are examples of the more general catalyst type, phase-transfer (PT) catalysts. These catalysts bring a reactant into a medium in which it would normally not be soluble. Inorganic reagents such as sodium cyanide are usually soluble in water, whereas S_N2 substrates such as alkyl halides are usually soluble in organic solvents. Bringing these together requires a PT catalyst, usually an organic solvent soluble tetraalkylammonium salt, $[R_4N]Br$. A schematic is shown in Figure 23.13. In water, the ammonium bromide is in equilibrium with the ammonium cyanide, and the cyanide is carried into the organic phase as an ion pair with the organic soluble ammonium ion. There, it can substitute the alkyl bromide, and the bromide can be returned to the water layer.

In S_N2 reactions catalyzed in this way, all the normal S_N2 features are retained. Thus, primary substrates react most rapidly, and the usual inversion of stereochemistry occurs. However, the expensive (also difficult to separate from organic products and to dispose of) dipolar aprotic solvents such as DMSO and DMF usually used can be avoided. The order of reactivity is ROMs > RBr > RCl; iodides and tosylates generally give poorer results, as they become too strongly associated with the ammonium ions. In some cases, organic solvents can be completely avoided—the substrate itself is the organic layer. Examples are shown in Figure 23.14. Even highly hindered substrates can give successful results.

Many other types of reactions may be catalyzed by ammonium or phosphonium salt PT catalysts. Figure 23.15 shows some examples. The first is an elimination, and the second the alkylation at the α-position of a nitrile. The third example involves the reaction of an alkynyl anion with a ketone; without the catalyst, the yield is only 7 %. Both this and the previous example will appear surprising—the acid from which the proton is removed has, in both cases, quite a high pK_a

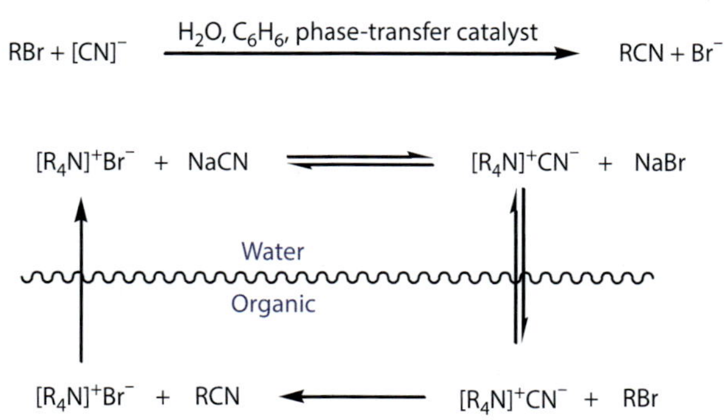

FIGURE 23.13 Mechanism of PT catalysis of an S_N2 reaction.

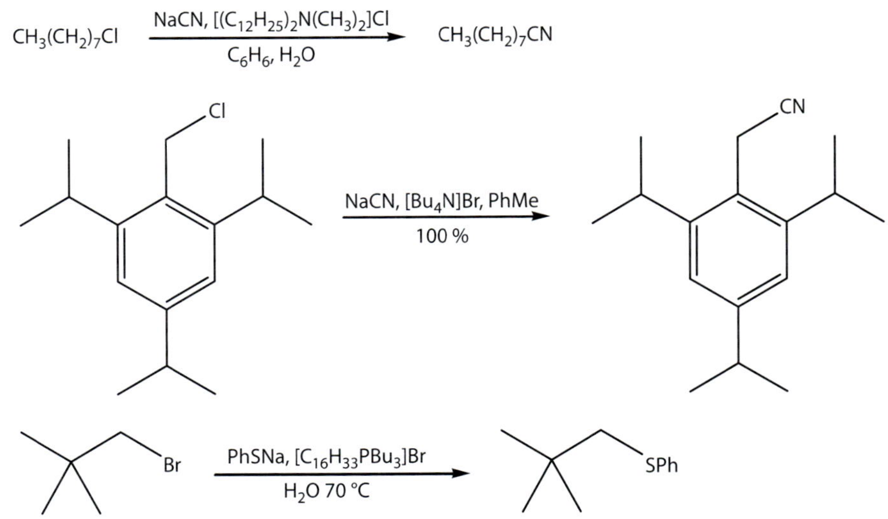

FIGURE 23.14 PT catalysis of S_N2 substitutions.

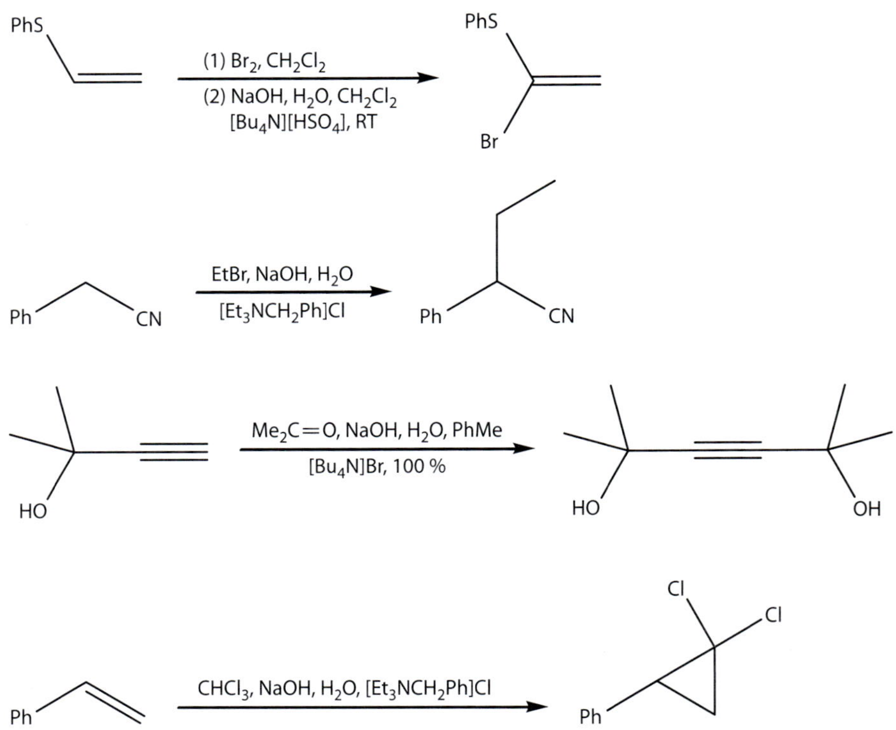

FIGURE 23.15 Examples of PT catalysis.

(25 for the alkyne, about 20 for $PhCH_2CN$, so they do not easily deprotonate using NaOH as base). The process works because any anion generated is rapidly transferred to the aprotic, organic medium. The final example shows how dihalocarbenes can be used to prepare cyclopropanes. The $[CX_3]^-$ anion is generated by deprotonation of the haloform and loses halide to give the carbene, $:CX_2$. However, the carbene is very reactive and is rapidly consumed in the aqueous layer. The transfer of the $[CX_3]^-$ anion to the organic phase isolates it, and hence the carbene, from the aqueous solvent and allows reaction with the alkene.

We met the synthetic ionophores, the crown ethers, in Section 4.3.3; these not only act as PT catalysts to transfer ions into organic media but also encapsulate cations, and hence break up ion pairs. "Naked," unpartnered, ions are much more reactive ☺. Potassium permanganate is insoluble in benzene, so if an aqueous solution of $K[MnO_4]$ is shaken with benzene, there is no color in the organic layer. However, when 18-crown-6 (18-C-6 is the appropriate size to coordinate K^+)

FIGURE 23.16 Permanganate is solubilized in benzene in the presence of 18-C-6.

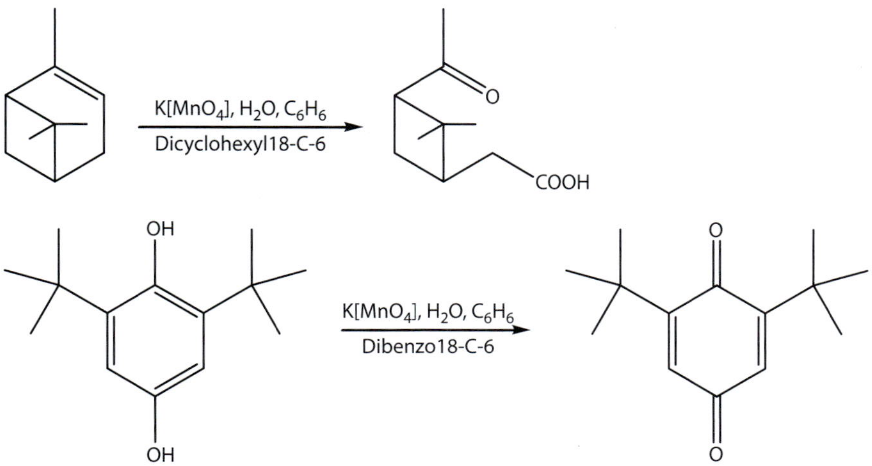

FIGURE 23.17 Permanganate oxidations in two-phase systems containing crown ethers.

is added, then the benzene layer is also colored (concentrations of permanganate up to 0.06 M may be obtained, Figure 23.16). The K^+ is complexed by the crown ether, and the manganate ion crosses the phase boundary as the counterion. Oxidation reactions such as those of Figure 23.17 are thus facilitated.

The ability of crown ethers to sequester metal cations and hence disrupt ion pairing greatly enhances the reactivity of both nucleophiles and bases. For example, **23.5**, when treated with KF/18-crown-6 in MeCN/H_2O, gives a mixture of substitution and elimination products. In water, fluoride is a poor nucleophile and a poor base, because it is very heavily solvated. In acetonitrile, it is only weakly solvated, and hence much more reactive. As shown in Figure 23.18, various substitution and addition reactions are also catalyzed. Note, however, that the use of crown ethers in industry is rare due to their prohibitively high cost—PEGs can often perform the same function almost as well, at much lower cost.

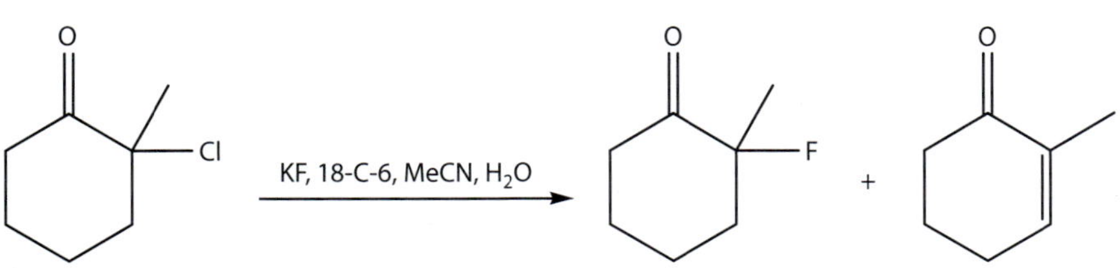

23.5

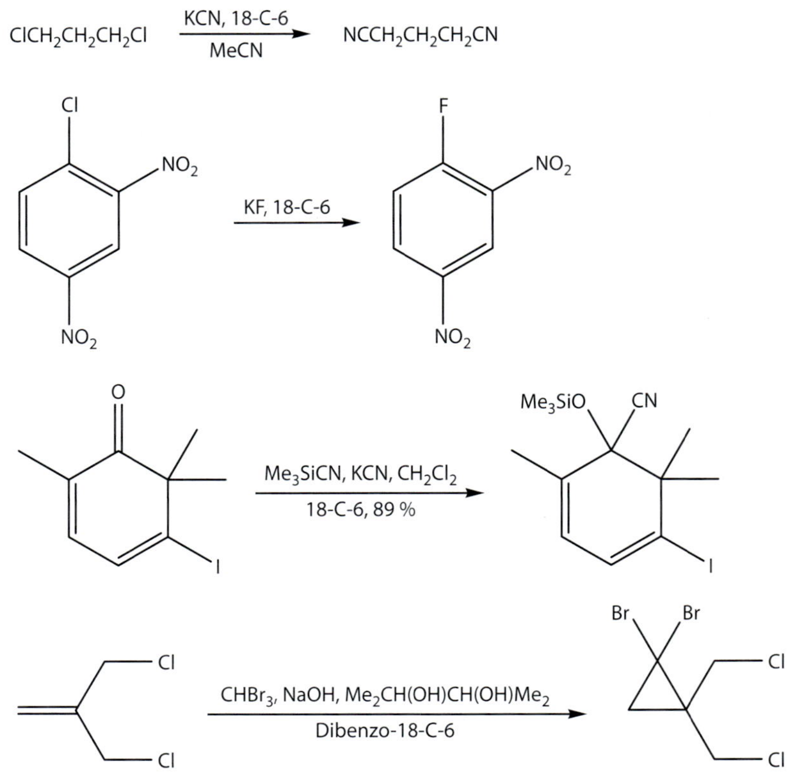

FIGURE 23.18 Use of crown ethers as PT catalysts.

PROBLEM 23.5

Potassium acetate is rather insoluble in MeCN, and acetate is generally considered to be a poor nucleophile. The half-life for the substitution of $PhCH_2Cl$ by potassium acetate in MeCN at room temperature is 685 h. When 18-crown-6 is added to the solution, the rate of the reaction is increased, and the half-life is reduced to 3.5 h. Explain.

PROBLEM 23.6

The condensation of the disodium salt of bisphenol A with phosgene is carried out in a two-phase system as shown in the following. Explain the function of the $[Bu_4N]I$.

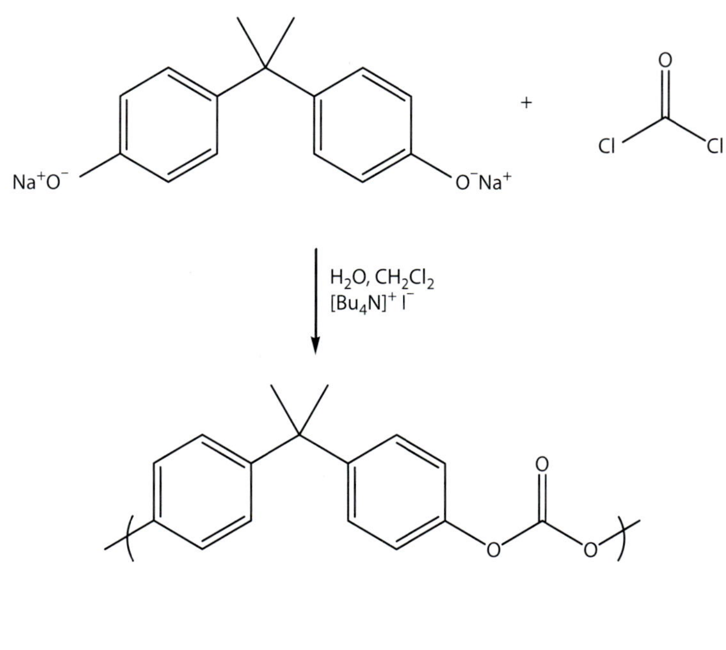

Solutions

23.5 This is a classic crown ether effect. The 18-C-6 sequesters potassium ions, carrying them, and the acetate counterions, into the acetonitrile. Thus, part of the effect derives from the solubilizing of the reagent. Additionally, the 18-C-6 breaks up the potassium acetate ion pairs, leaving the acetate "naked" and hence enhancing its nucleophilicity in the S_N2 substitution reaction.

23.6 Dichloromethane and water are immiscible. The sodium salt of bisphenol A is water soluble; phosgene is soluble in dichloromethane. Phosgene is rapidly decomposed by water. The purpose of the tetraalkylammonium salt is to act as a PT catalyst, to carry the phenolate into the dichloromethane.

Key Points from Section 23.4

- PT catalysts act by bringing a reactant into a medium in which it is not normally soluble. They are usually used in water/immiscible organic solvent mixtures.
- PT catalysts include tetraalkylammonium and tetraalkylphosphonium salts, crown ethers (which also sequester cations and break up ion pairs), and PEGs.
- A wide range of reaction types may be catalyzed using PT catalysts.

23.5 REACTIONS CATALYZED BY METAL SURFACES

We have already met a number of these, particularly in the reduction of carbon–carbon multiple bonds (Section 11.5). The metals most commonly used are platinum, palladium, rhodium, and ruthenium. Adams' catalyst, PtO_2, is widely used for the reduction of alkenes and alkynes to alkanes. It gives more predictable results than finely divided or colloidal platinum and is easier to handle. The dioxide is reduced to the metal *in situ*, and the solid metal recovered (usually by filtration) after completion of the reaction may be pyrophoric and must be treated with care. Examples of its use are shown in Figure 23.19.

Alkynes may be reduced stereospecifically to *cis*-alkenes in the presence of the Lindlar catalyst, or the related system $Pd/BaSO_4$/pyridine (Figure 23.20). More active palladium catalysts can

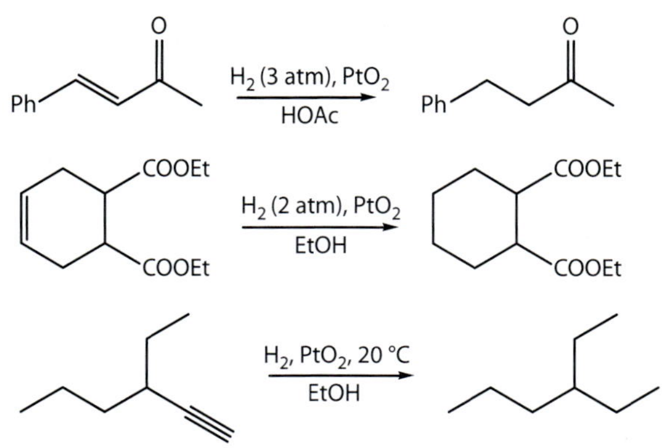

FIGURE 23.19 Reduction of carbon–carbon multiple bonds using Adams' catalyst.

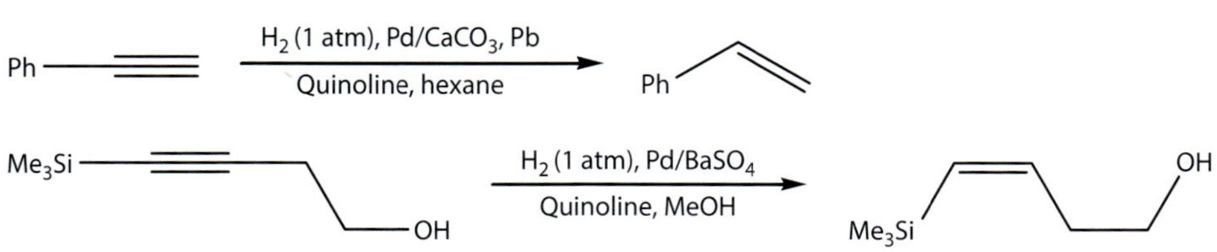

FIGURE 23.20 Stereospecific hydrogenation of alkynes to *cis*-alkenes.

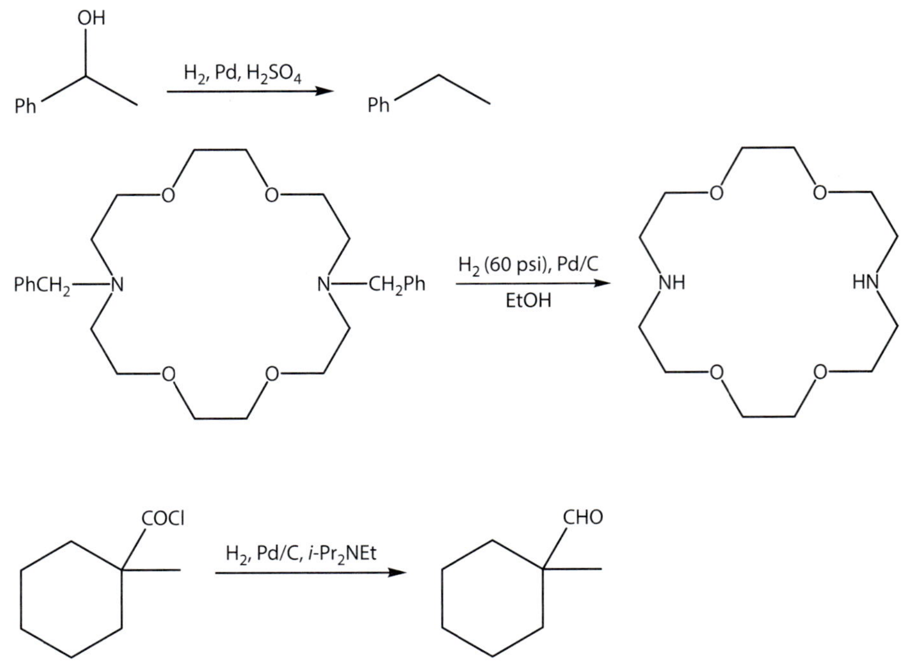

FIGURE 23.21 Hydrogenolysis catalyzed by palladium systems.

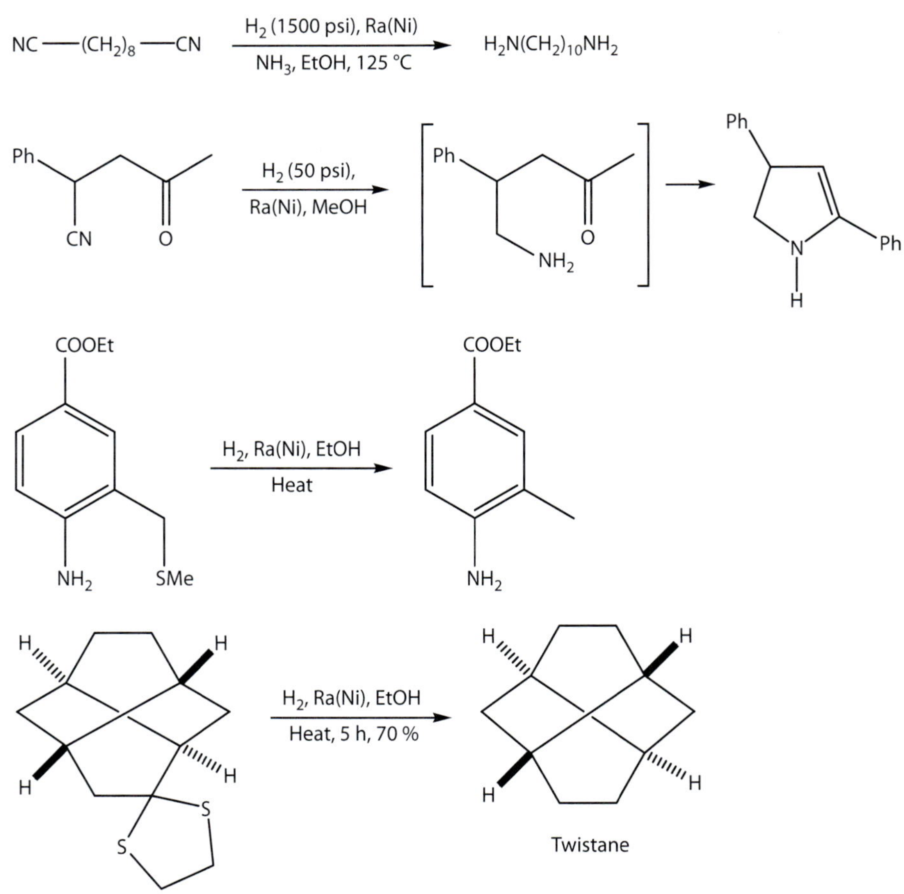

FIGURE 23.22 Raney nickel-catalyzed reactions.

be used for alkene reduction. Palladium catalysts are also widely used for hydrogenolysis in the removal of benzyl-protecting groups, and the Rosenmund reduction of acyl halides to aldehydes (Figure 23.21).

Raney nickel is a highly active, finely divided form of the metal prepared by reaction of a nickel/aluminum alloy with concentrated sodium hydroxide, which removes most of the aluminum as $Na[Al(OH)_4]$. Although active for a very wide range of reductions, it has been particularly widely used for the reduction of nitriles, and the desulfurization of thioacetals, thioethers, and dithianes (see Section 19.3.3) (Figure 23.22). The careful disposal of the catalyst after use is very important; once it has dried out, the metal is highly pyrophoric.

Heterogeneous catalysts predominate in industrial processes, because of their ability to be recovered and recycled relatively easily. Both activity and selectivity may depend significantly on the method of catalyst preparation and the precise conditions of the reaction.

23.6 REACTIONS CATALYZED BY TRANSITION METAL COMPLEXES

The types of reactions that can be catalyzed by transition metal complexes are now very numerous and are very widely used in synthesis. We have already met a number of them—osmium in catalysis of dihydroxylation reactions, titanium in Sharpless epoxidation, various metals in hydrogenation reactions of alkenes, and the Ziegler–Natta process for polymerization. In this section, we will just highlight a few types that have been popular—an oxidation, some hydrogenations, and some coupling reactions. Although outline reaction mechanisms will be given, this is for interest only—they are beyond the scope of this text, and many are more complicated than is shown here.

23.6.1 WACKER OXIDATION AND RELATED REACTIONS

The Wacker oxidation of ethene to ethanal is an important industrial process for the oxygenation of a hydrocarbon feedstock. Essentially, the same process may be used to convert 1-alkenes to methyl ketones. The stoichiometry of the process is shown in Figure 23.23. The reaction is catalytic in both palladium and copper; the ultimate oxidant is (inexpensive) molecular oxygen. A proposed mechanism is shown in Figure 23.24; the main controversy has been as to whether the attack of water on the coordinated alkene is external or *via* prior coordination of the water to palladium. Current thinking is that external attack predominates in high concentrations of chloride ion and internal attack when $[Cl^-]$ is low. Different details of mechanisms under the two conditions are supported by different reaction kinetics.

The Wacker oxidation of ethene is an important commercial process. An adaptation of the process is widely used in total synthesis to convert 1-alkenes to methyl ketones and, in this context, is often referred to as the Wacker–Tsuji reaction.[*,†] Some examples of the reaction are shown in Figure 23.25; note that in the third example, the reaction is selective for the 1-alkene. The use of the process in a steroid synthesis is shown in Figure 23.26.

$$C_2H_4 + PdCl_2 + H_2O \rightarrow CH_3CHO + Pd(0) + 2HCl$$

$$Pd(0) + 2CuCl_2 \rightarrow PdCl_2 + 2CuCl$$

$$2CuCl + 2HCl + \tfrac{1}{2}O_2 \rightarrow 2CuCl_2 + H_2O$$

$$C_2H_4 + \tfrac{1}{2}O_2 \rightarrow CH_3CHO$$

FIGURE 23.23 Stoichiometry of the Wacker oxidation.

* J. Tsuji, *Palladium Reagents and Catalysts*, Wiley 2004.

† *Handbook of Organopalladium Chemistry for Organic Synthesis*, ed. E. Negishi, 2002.

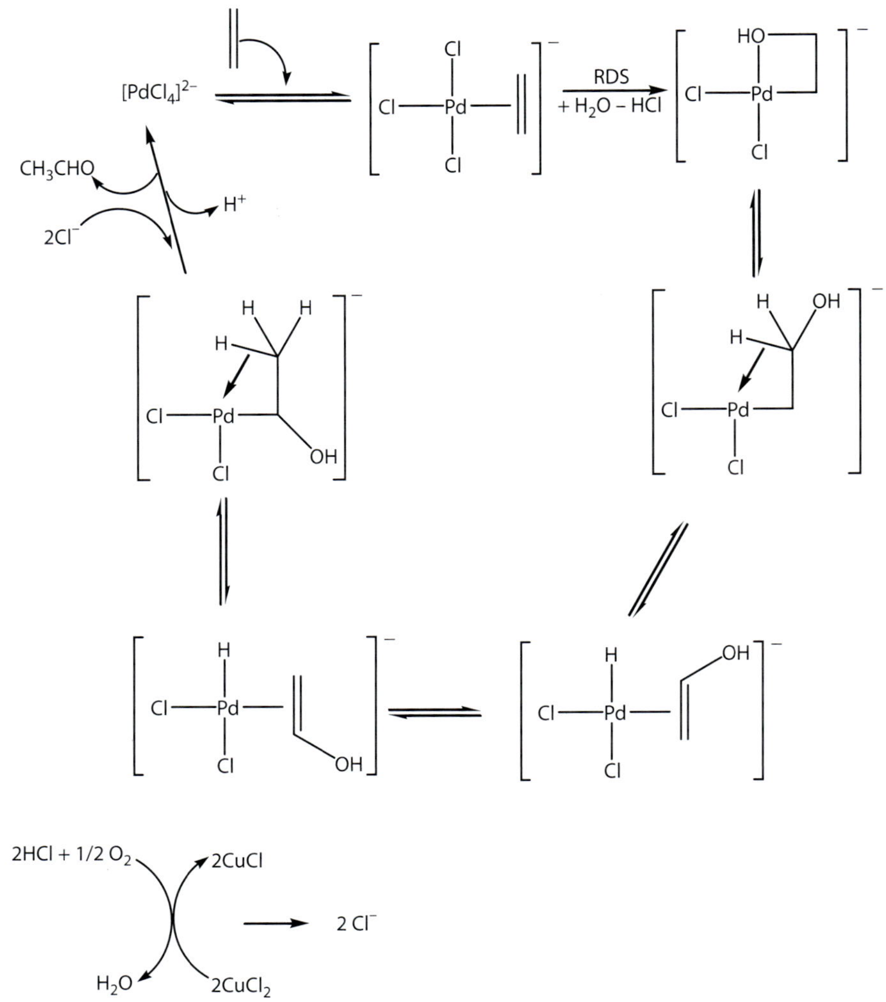

FIGURE 23.24 Mechanism of the Wacker oxidation of ethene.

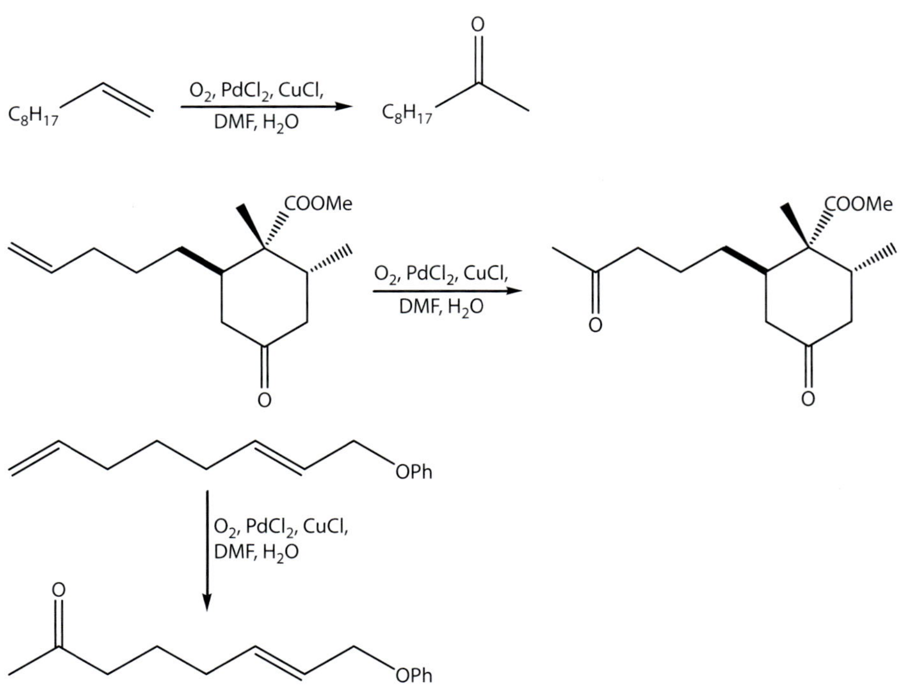

FIGURE 23.25 The Wacker–Tsuji oxidation in synthesis.

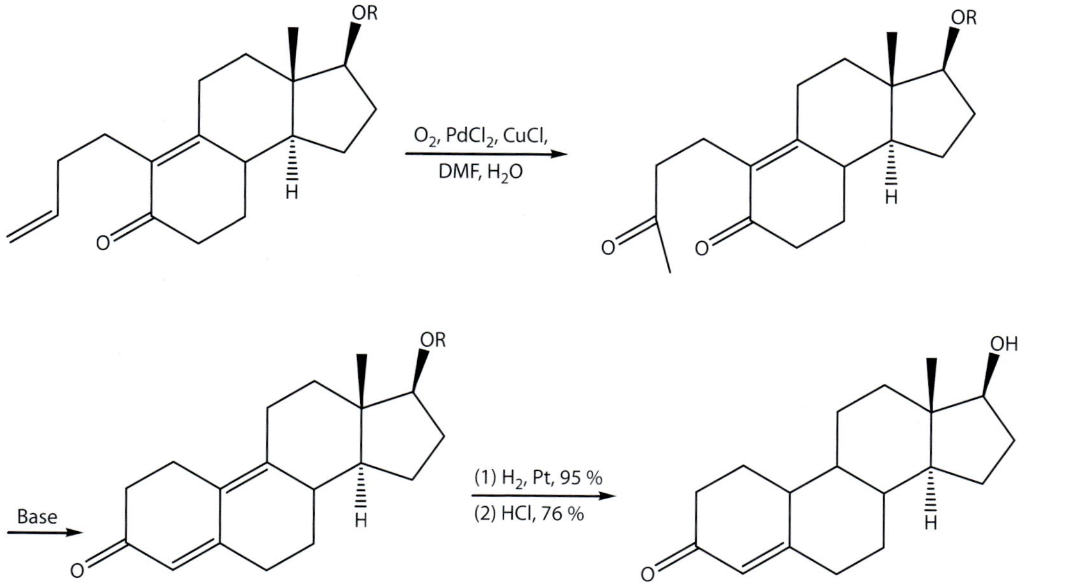

FIGURE 23.26 The Wacker–Tsuji oxidation in steroid synthesis.

PROBLEM 23.7

Study the following sequence:

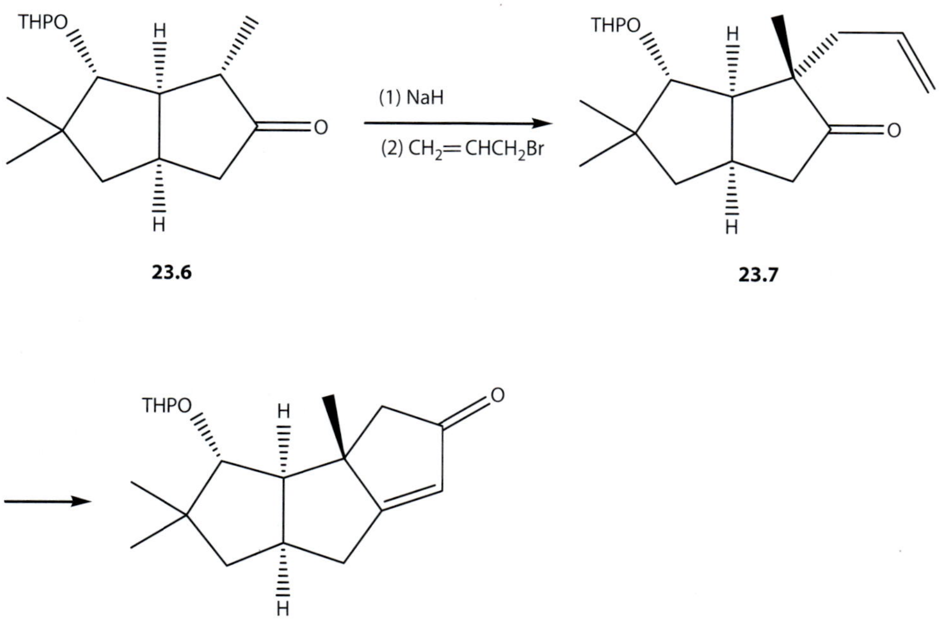

23.6 **23.7** **23.8**

(a) Write a mechanism for the conversion of **23.6** to **23.7**.

(b) Suggest reagents for the conversion of **23.7** to **23.8**, and write a mechanism for the ring closure step of this process.

Solution

(a)

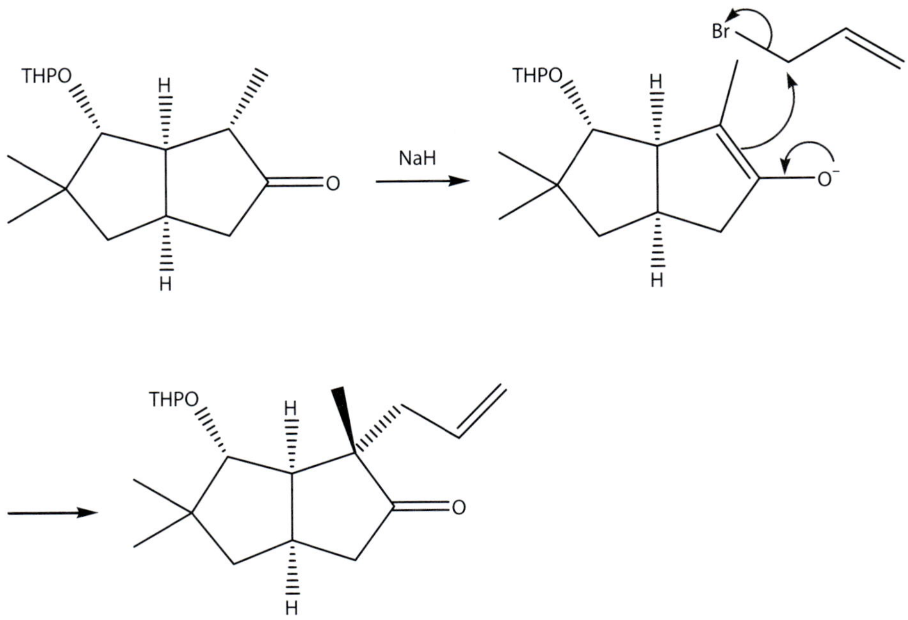

Note that the enolate is alkylated on the less-hindered face.

(b) $O_2/PdCl_2/CuCl_2$ will convert the terminal alkene to a methyl ketone, which can then undergo an aldol reaction in base:

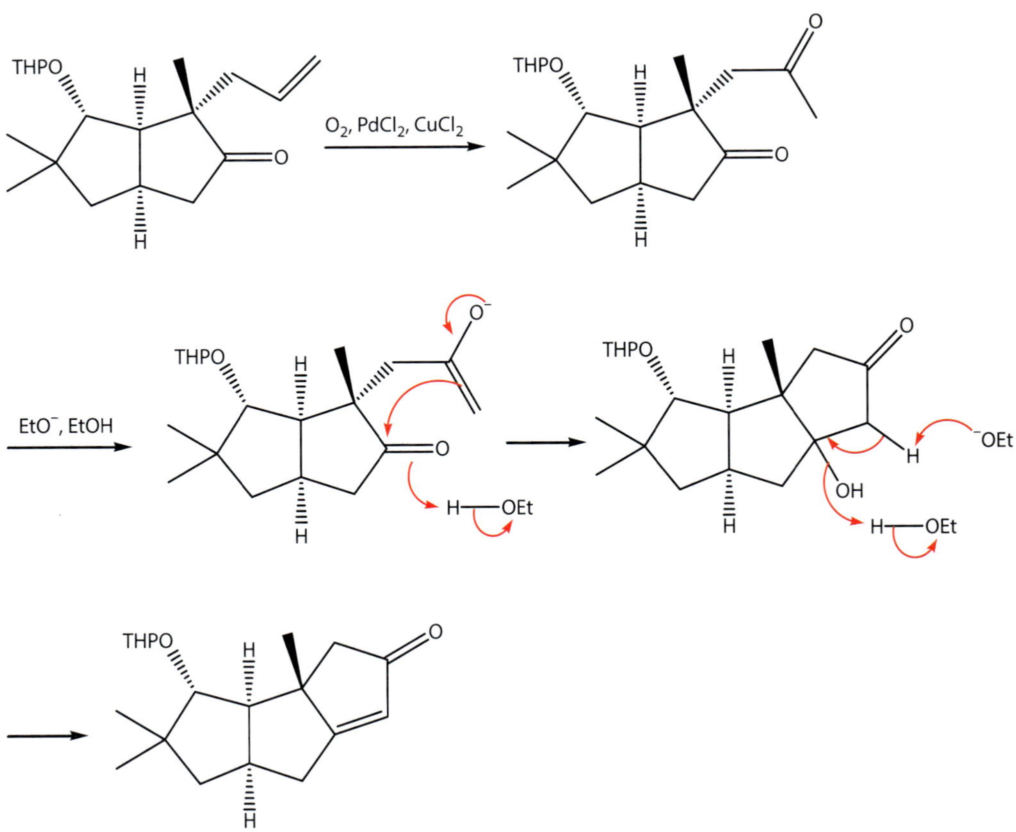

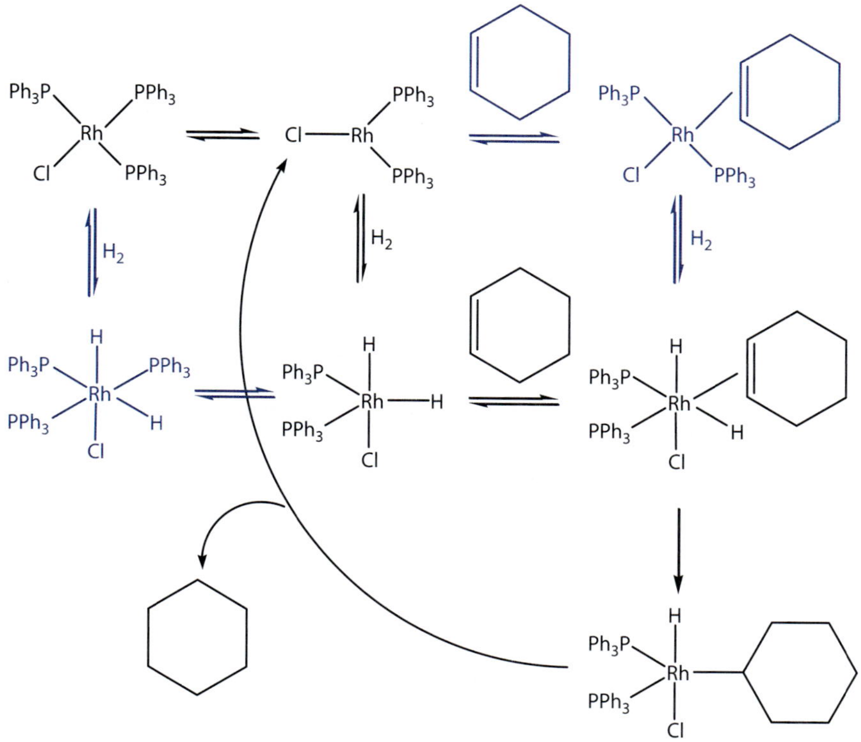

FIGURE 23.27 Mechanism of hydrogenation using Wilkinson's catalyst. The sections in blue are not on the main catalytic pathway; although they occur at measurable rates, these are at least 1000 times slower than the main pathway.

23.6.2 HOMOGENEOUS HYDROGENATION

Modern work on homogeneous catalysis of hydrogenation using metal complexes dates from the work of Wilkinson in the 1960s. Wilkinson's catalyst, *tris*(triphenylphosphine)rhodium(I) chloride, [RhCl(PPh_3)$_3$], remains the granddaddy of the field and is still in wide use. The reaction mechanism is shown in Figure 23.27, with a minor pathway in blue. Addition of hydrogen to double bonds is rigorously *cis*, and the rate of reduction of alkenes depends on the degree of substitution. 1-Alkynes are reduced more slowly than alkenes, and dienes more slowly still. Nitroarenes and benzyl-protecting groups are unaffected. Some examples are shown in Figure 23.28. The first demonstrates the rigorous *cis*-addition, selectivity for the reduction of a disubstituted over a trisubstituted double bond, and diastereoselectivity for the addition to the less-hindered face of the molecule. The second example also shows diastereoselectivity for the less-hindered face of the molecule, as well as the survival of benzyl-protecting groups (which would be removed by hydrogenolysis over a heterogeneous catalyst such as palladium).

In order to reduce tri- and tetrasubstituted double bonds, a more active catalyst is needed. The one most usually chosen is Crabtree's catalyst, [Ir(PCy_3)(py)(COD)][PF_6], **23.9**. This catalyzes the reduction of mono-, di-, tri-, and tetrasubstituted alkenes at sensible rates (Figure 23.29).

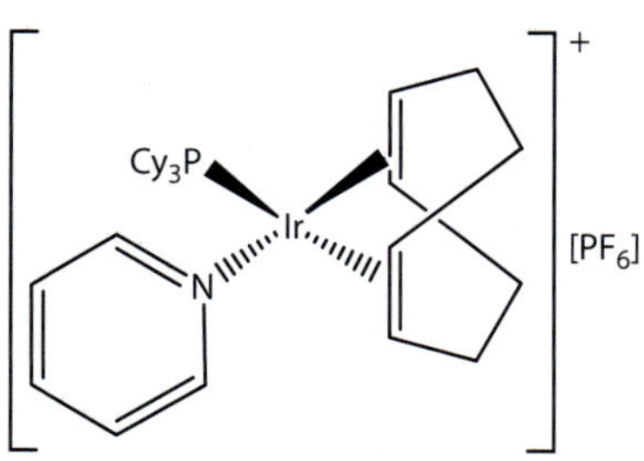

23.9

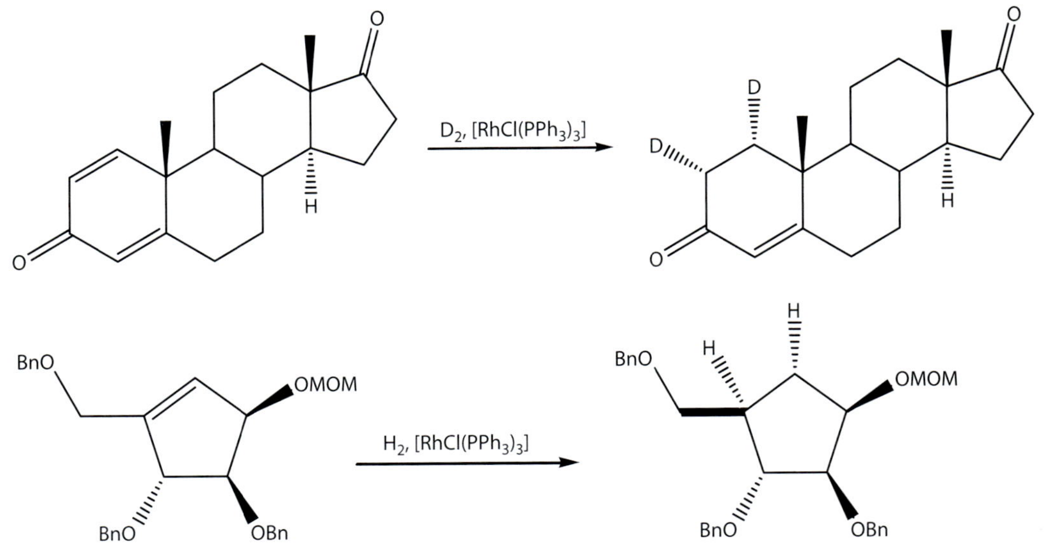

FIGURE 23.28 Uses of Wilkinson's catalyst.

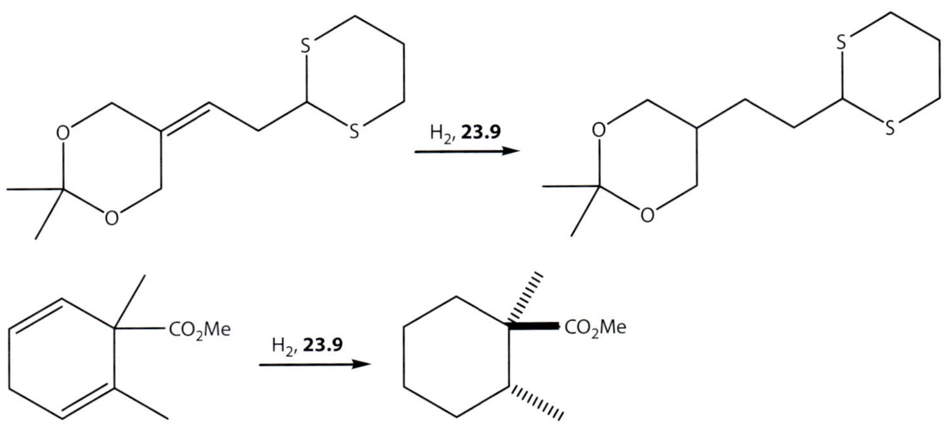

FIGURE 23.29 Applications of Crabtree's catalyst in synthesis.

The great advantage of the Crabtree catalyst is that it is able to coordinate to Lewis basic groups in the substrate, thus directing the reduction to a specific face of the double bond. The diastereospecificity achieved is generally very high (Table 23.1),* and such directed reductions have proved synthetically useful (Figure 23.30).

In Section 22.3.1, we noted that chiral amino acids can be prepared by enantioselective reduction of dehydroamino acids in the presence of rhodium complexes of chiral phosphines (*e.g.*, the reaction of **23.10**). Substrates that do not have the amide functional group are more efficiently reduced using a 2,2′-bis(diphenylphosphino)-1,1′-binapthyl (BINAP) ruthenium complex, and such complexes are also useful for the asymmetric reduction of ketones bearing Lewis base functional groups, although higher pressures of hydrogen are usually required (Figure 23.31).

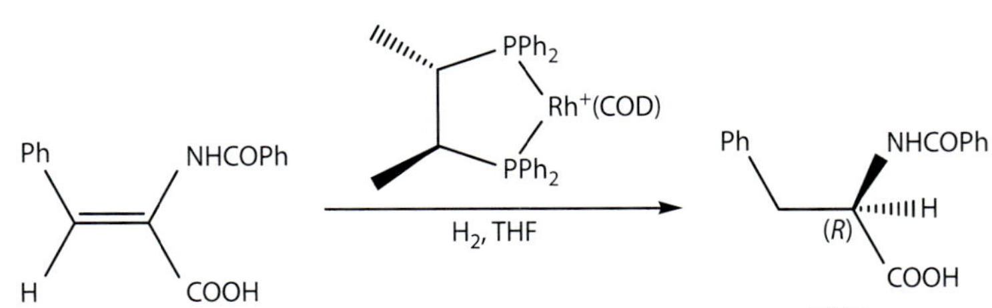

23.10

99 % ee

* J.M. Brown, *Angew. Chem. Int. ed. Engl.*, **26** (1987) 190.

TABLE 23.1 Directed Hydrogenation Using the Crabtree Catalyst

Substrate	Product	Diastereomer Ratio with 23.9 as Catalyst	Diastereomer Ratio with Pd/C as Catalyst
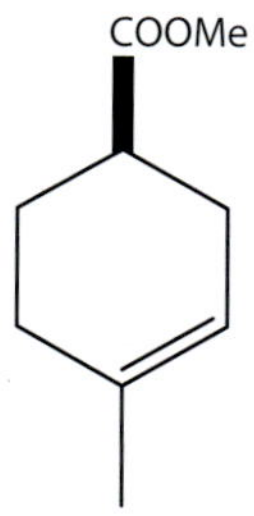	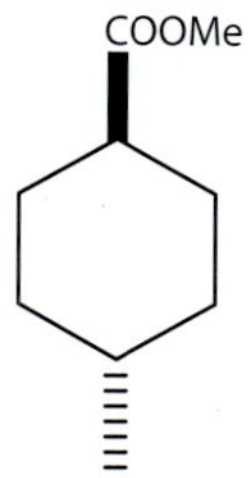	56:1	1.35:1
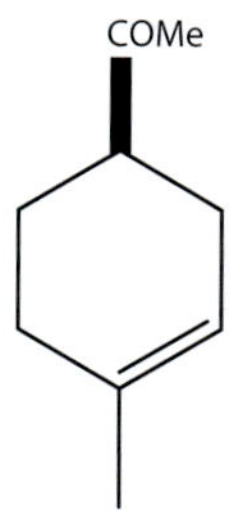	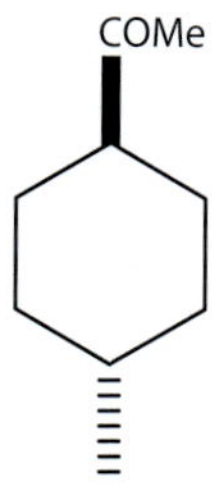	124:1	1.26:1
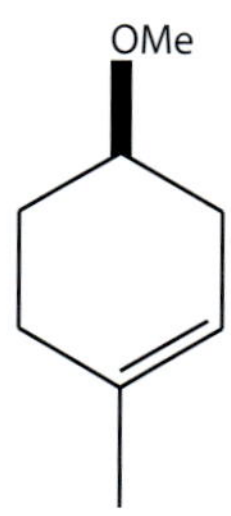	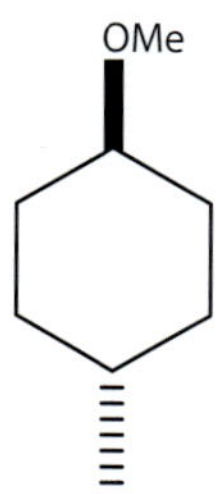	999:1	1.4:1

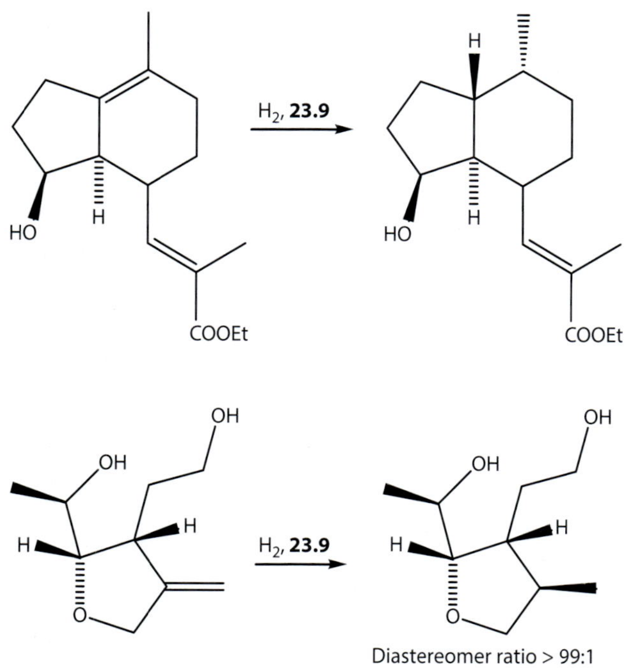

FIGURE 23.30 Applications of directed hydrogenation using the Crabtree catalyst.

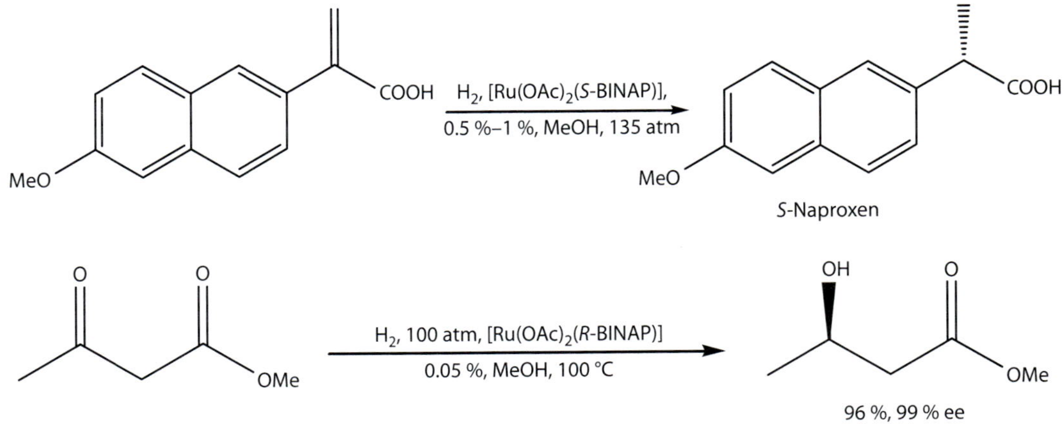

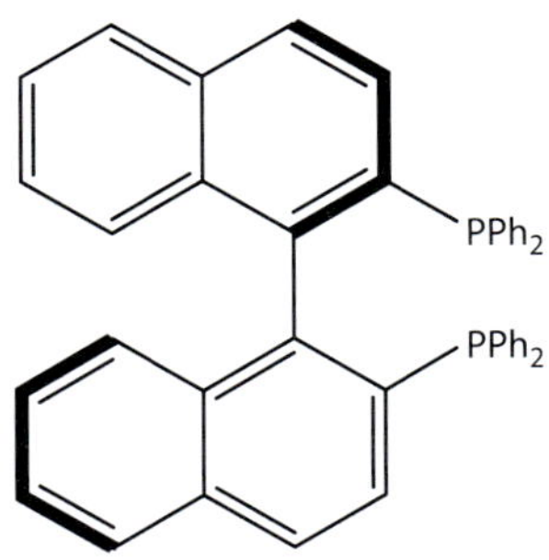

23.11, *R*-BINAP

FIGURE 23.31 BINAP (**23.11**) ruthenium complexes in asymmetric hydrogenation.

PROBLEM 23.8

Predict the product(s) of each of the following reactions:

(a)

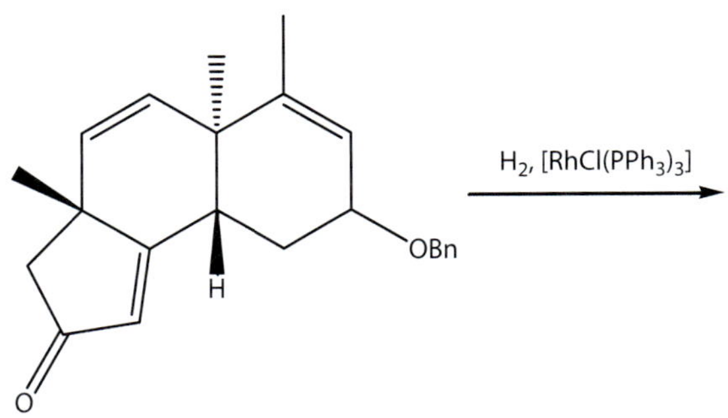

(b)

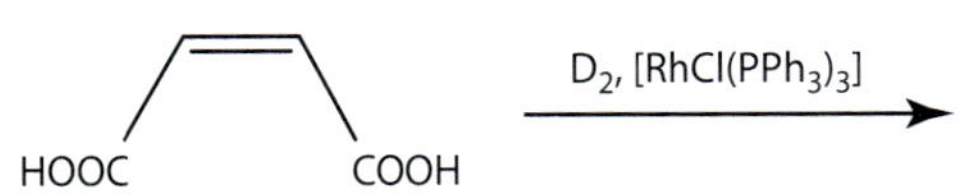

(c)

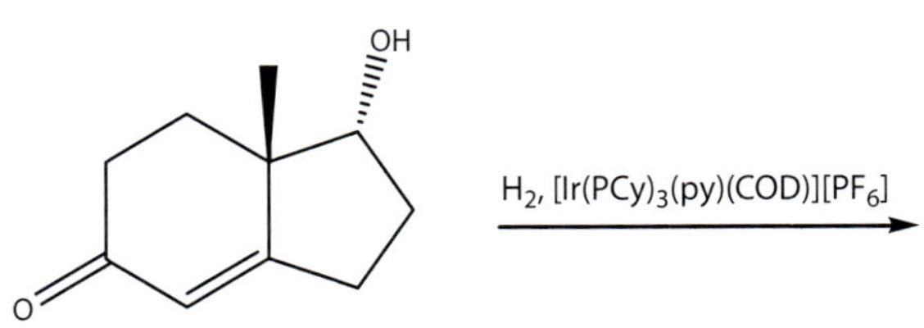

Solutions

(a) Only the disubstituted double bond is reduced using Wilkinson's catalyst. Note that the benzyl-protecting group is not cleaved:

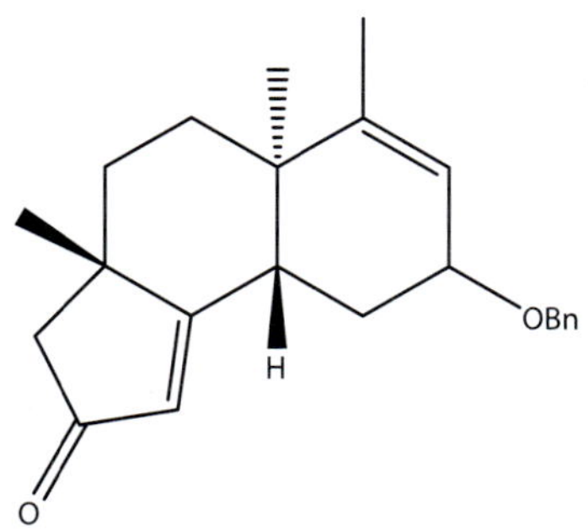

(b) Addition is strictly *cis* to give

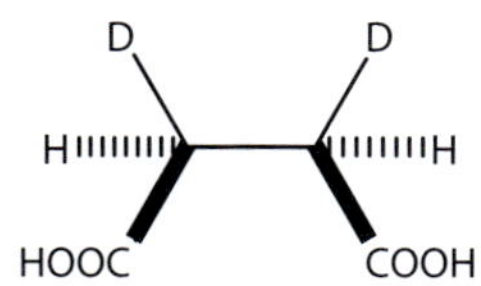

(c) Addition takes place predominantly from the face of the molecule bearing the OH group, using Crabtree's catalyst:

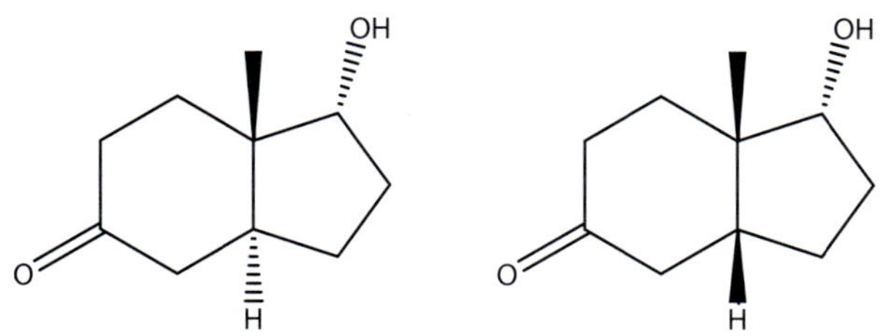

25.6.3 COUPLING REACTIONS

In contemplating synthesis of organic molecules, you might have thought it would be very useful if a carbon–carbon bond could be made simply by taking one positive carbon atom (the synthon would be an alkyl halide or similar) and one negative carbon atom (an organometallic compound such as Grignard reagent) and link them up. But except in a few cases,* this does not work. In particular, we have no simple way to substitute halogens attached to aromatic rings or carbon–carbon double bonds. At least that was so until catalytic reactions began to be developed. Most of these involve the use of a palladium(0) catalyst, and a general mechanism is shown in Figure 23.32. Specific types of coupling may have additional steps, but all of them have those shown here. Palladium(0) reacts with aryl and vinyl halides by a process known as oxidative addition to give a palladium(II) complex. X is displaced from palladium by some type of nucleophilic species. Finally, the coupled product is reductively eliminated from palladium(II) to regenerate palladium(0). The general order of reactivity is RI > ROTf > RBr > RCl. There is now a huge range of such reactions available; most use palladium, but other transition metals are preferred in specific cases. Many different ligands and starting complexes are used, including some that give rise to enantioselective reactions.

* The halide must be attached to an unhindered and very reactive sp^3 carbon atom, and even then, yields are often low.

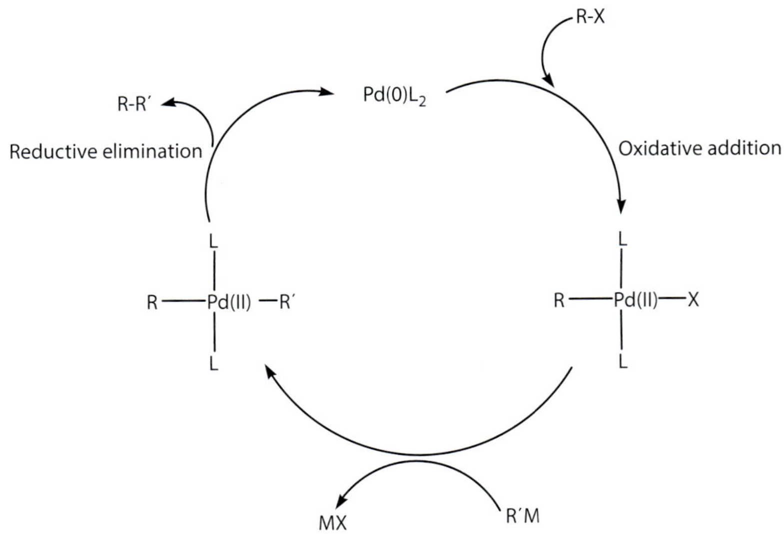

FIGURE 23.32 Outline mechanism of palladium(0)-catalyzed coupling reactions.

The earliest reactions of this type used Grignard reagents as the nucleophiles—the Kumada–Corriu coupling (Figure 23.33). However, the presence of a Grignard reagent is not compatible with unprotected carbonyl groups, so the next development was the Negishi coupling, which used organozinc compounds (Figure 23.34). Although organozinc compounds do not react with esters, they are still incompatible with aldehydes or ketones. More functional group

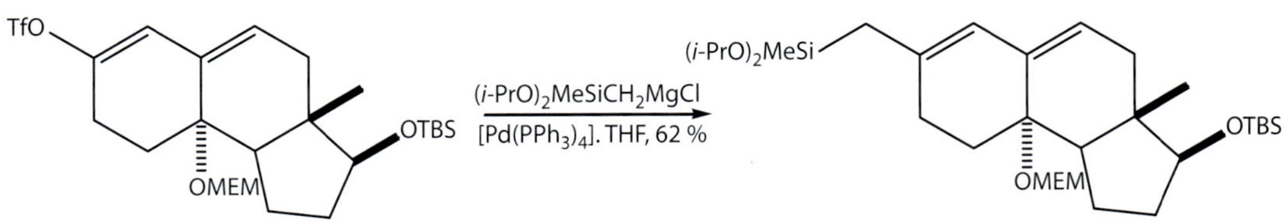

FIGURE 23.33 Kumada–Corriu coupling.

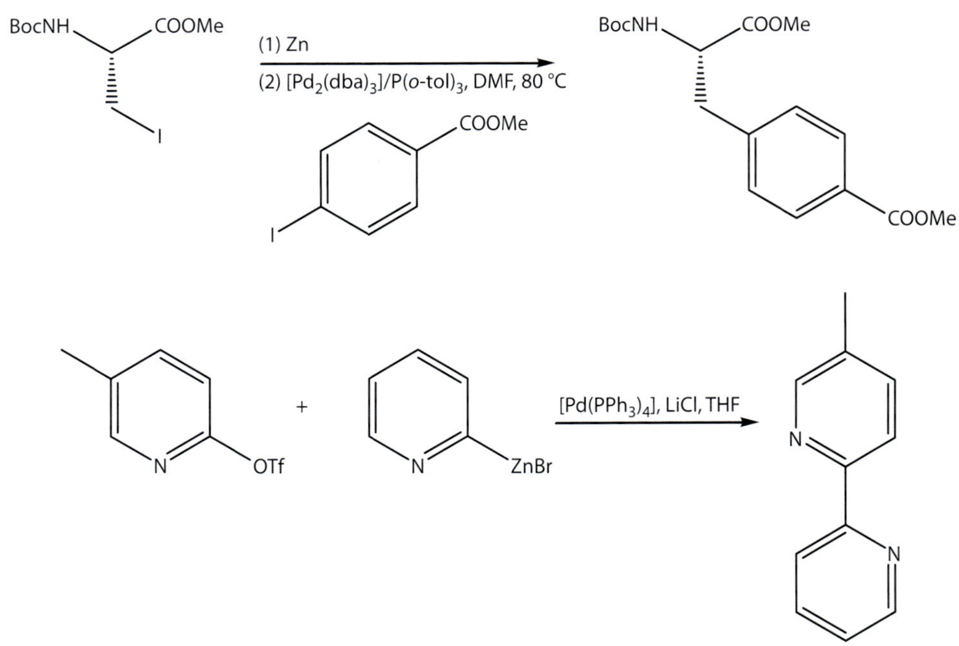

FIGURE 23.34 Negishi couplings.

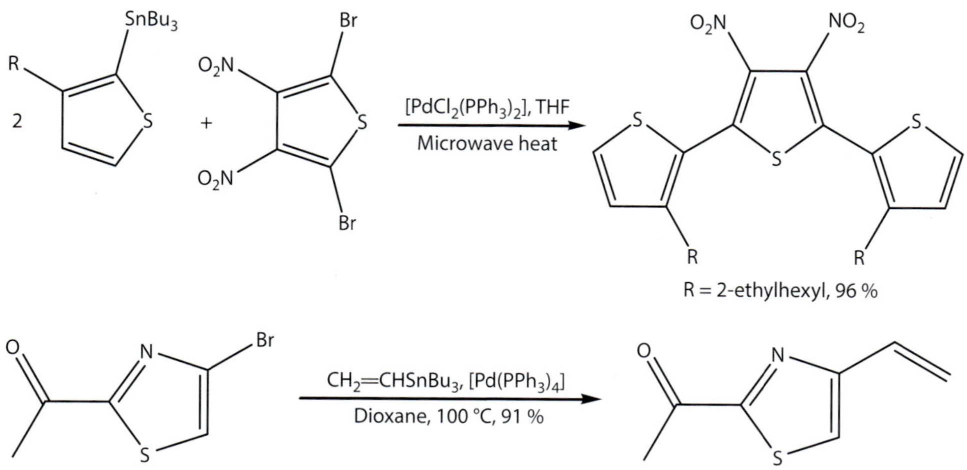

FIGURE 23.35 Stille couplings.

compatibility was achieved by using organotin reagents in the Stille coupling (Figure 23.35). These reagents don't add to aldehydes or ketones, but they are relatively expensive, and some are quite toxic. One of the currently most popular ways to accomplish such couplings is the Suzuki coupling (Figure 23.36). Reagents such as $ArB(OH)_2$ are easy to prepare (the starting material for the first example of Figure 23.36 is prepared by hydroboration of 1-hexyne with catecholborane), stable, and have low toxicity. The reaction is exceptionally efficient, needing in some cases only 0.001 mol % or less of catalyst. The importance of these processes was recognized in 2010 by the award of the Nobel Prize to Suzuki, Negishi, and Heck (the Heck reaction is another, slightly different palladium-catalyzed coupling). Finally, the Sonogashira coupling uses a 1-alkyne as nucleophile and a simple trialkylamine base to generate the copper alkyne that is the effective nucleophile (Figure 23.37).

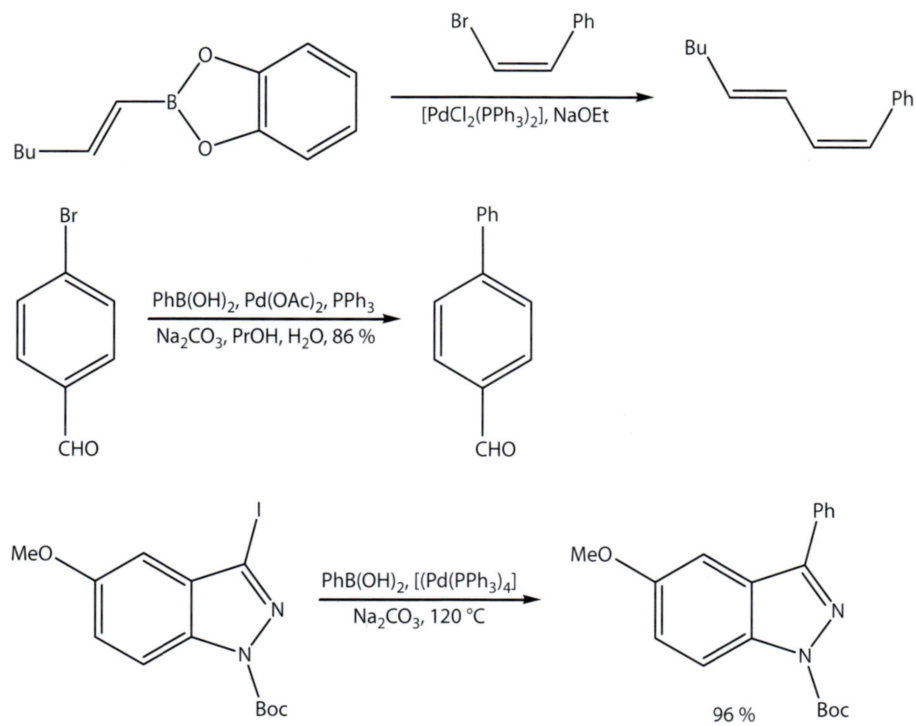

FIGURE 23.36 Suzuki couplings.

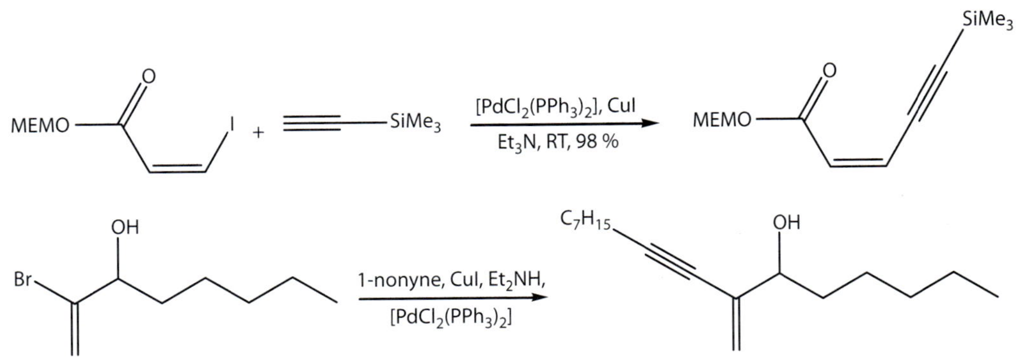

FIGURE 23.37 Sonogashira couplings.

PROBLEM 23.9

Predict the product(s) of each of the following processes:

(a)

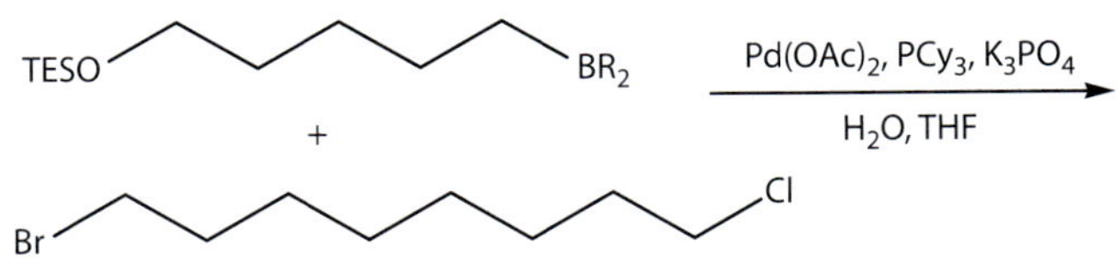

(b)

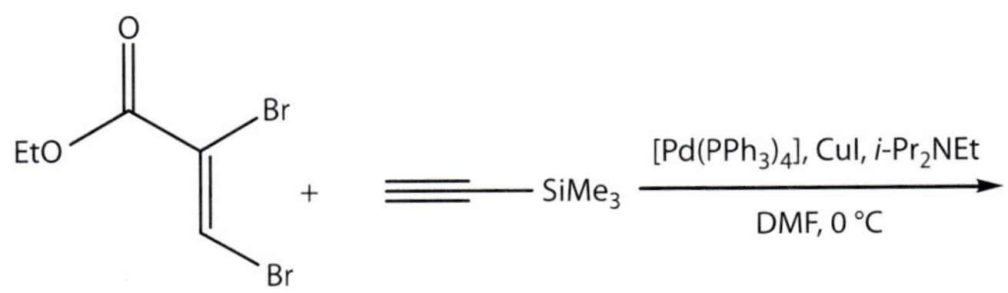

Solutions

(a) This is a Suzuki coupling; the bromide is more reactive than the chloride:

(b) In this Sonogashira coupling, the less hindered site is preferentially substituted to give

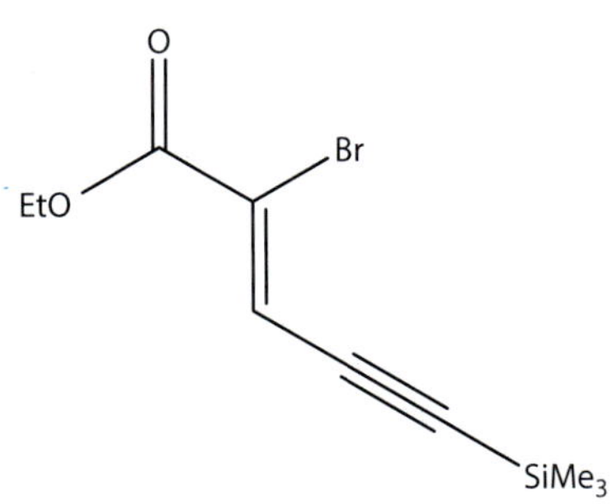

Key Points from Sections 23.5 and 23.6

- Metal surfaces catalyze a wide range of hydrogenation and hydrogenolysis reactions. Activity and selectivity are dependent on both catalyst constitution and the method of preparation.
- The Wacker oxidation is an industrial process using a Pd/Cu catalyst for the air oxidation of ethene to ethanal.
- The Wacker–Tsuji oxidation uses a similar catalyst to convert 1-alkenes to methyl ketones.
- Wilkinson's catalyst, $[RhCl(PPh_3)_3]$, catalyzes *cis*-addition of hydrogen to alkenes; the reaction is sensitive to steric factors, and tri- and tetrasubstituted alkenes are not reduced.
- The Crabtree catalyst, $[Ir(PCy_3)(py)(COD)][PF_6]$, catalyzes the reduction of tri- and tetra-substituted alkenes as well as less substituted ones. The reactions of alkenes bearing a potentially coordinating functional group are highly diastereoselective.
- Rhodium complexes of chiral phosphines catalyze enantioselective hydrogenation of enamides. Ruthenium BINAP complexes give better results for less functionalized alkenes and for the reduction of carbonyl groups, especially in β-ketoesters.
- Organometallics may be coupled with aryl and vinyl halides in the presence of palladium(0) complexes. Organometallics include derivatives of Mg, Zn, Sn, B, and copper derivatives of alkynes.

23.7 ORGANOCATALYSIS*

In the last decade, the topic of organocatalysis has undergone extensive development, particularly directed toward the preparation of chiral compounds. It is a broad term and encompasses processes involving species that form covalent derivatives of reagents, which then undergo reaction and afterwards regenerate the catalyst, as well as species that activate reagents by forming strong directed hydrogen bonds with them. We have already met one excellent example of this in Problem 20.3, where thiamine catalyzed the benzoin condensation, and the aldol reaction of **23.12** was known to be catalyzed by proline in 70 % enantiomer excess some 40 years ago. With the newer catalyst, **23.13**, the enantiomer excess is 91 %, which may be raised to 97 % by a simple recrystallization.

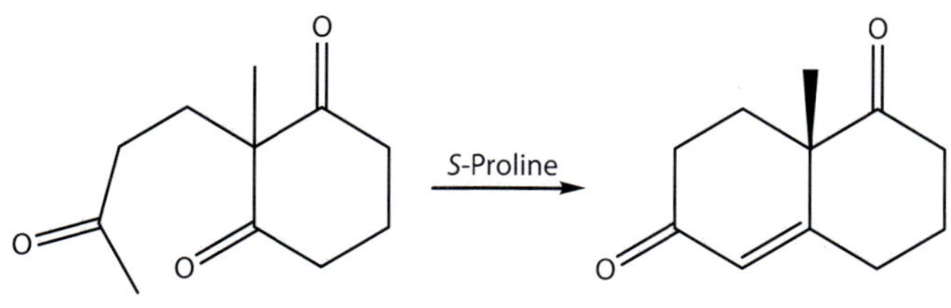

23.12

* A. Berkessel and H. Gröger, *Asymmetric Organocatalysis*, Wiley, 2005.

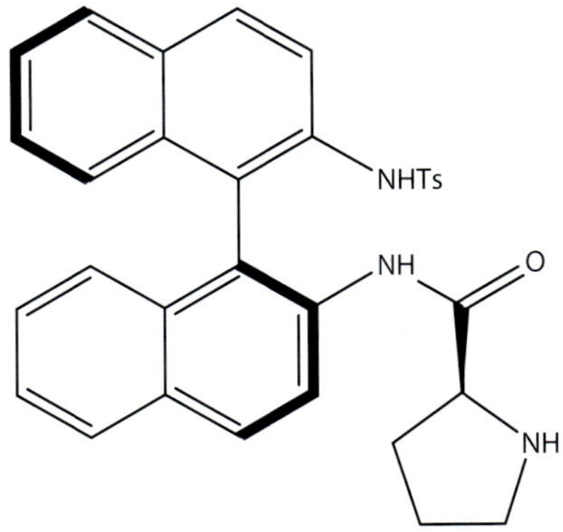

23.13

Proline and related chiral 5-membered ring heterocycles have been used to catalyze a wide range of processes, as shown in Figure 23.38. In almost all cases, the reaction works by converting a carbonyl group, temporarily, to an iminium ion, which is more electrophilic than the original carbonyl compound. At present, the disadvantage of the process is that relatively high (10 %–20 %) catalyst loadings are generally needed, but it is likely that this will change as the field matures.

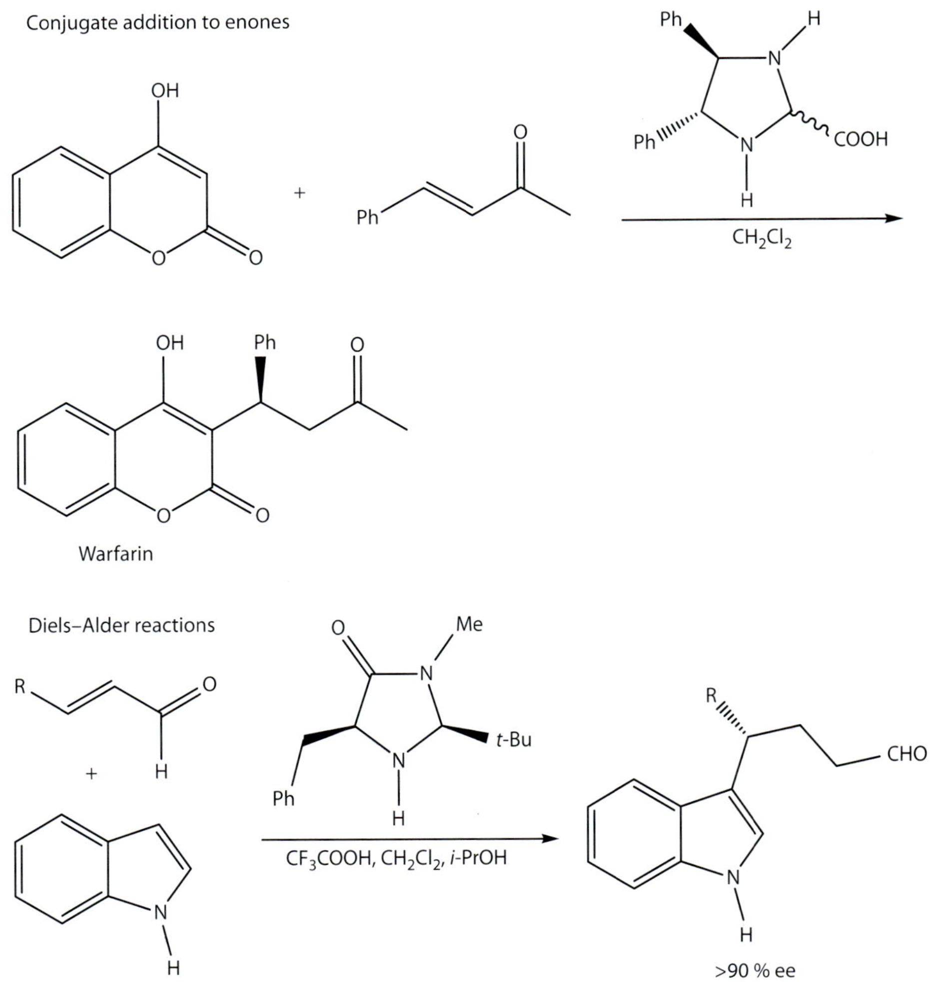

FIGURE 23.38 Organocatalysis in action.

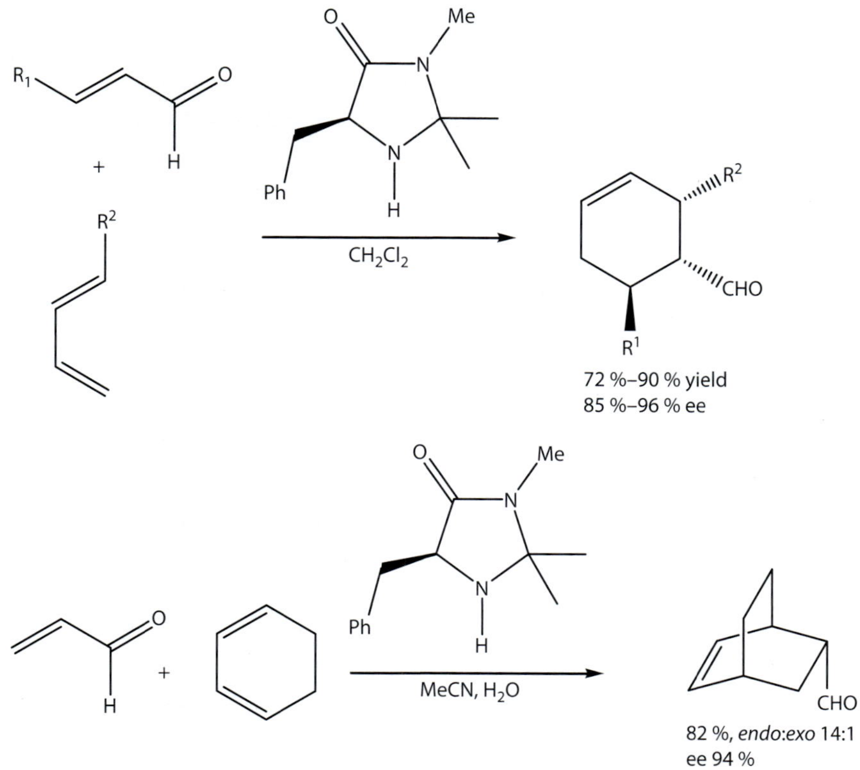

Halogenation

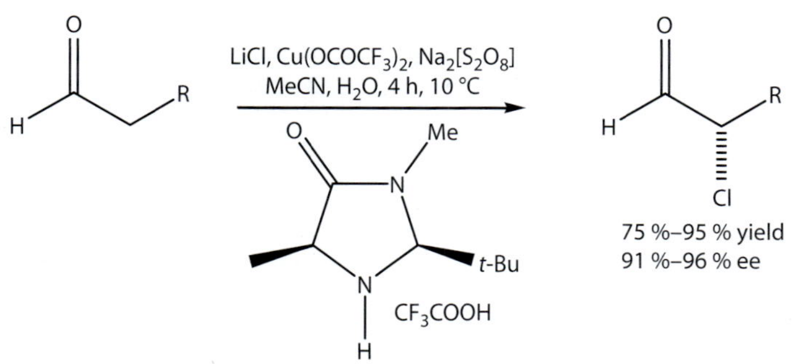

FIGURE 23.38 (continued) Organocatalysis in action.

PROBLEM 23.10

The reaction of chloroethane with hydroxyl ion to give ethanol is accelerated by the presence of I^-. Explain.

PROBLEM 23.11

The Knoevenagel condensation, shown below, is catalyzed by piperidine. Propose a mechanism:

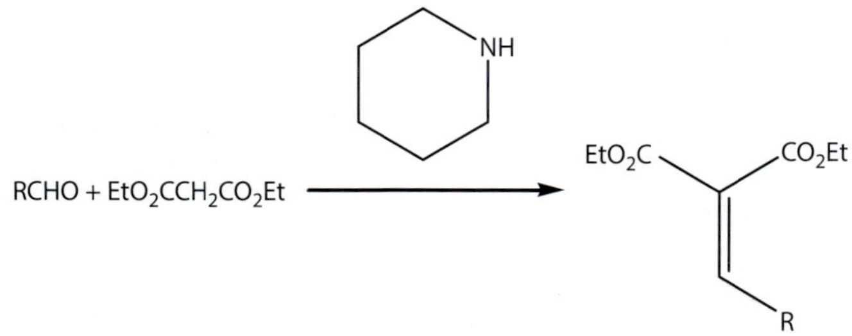

Solutions

23.10 The reaction of alkyl chlorides with nucleophiles is always relatively slow, and hydroxyl is not the best nucleophile. Iodide, however, is an excellent nucleophile. The catalyzed reaction is a two-step process. Iodide displaces chloride, and then hydroxyl displaces iodide. Iodide is a better nucleophile than hydroxyl ion and is also a better leaving group than chloride—hence, the two-step process is faster than the unfavorable single-step reaction:

23.11

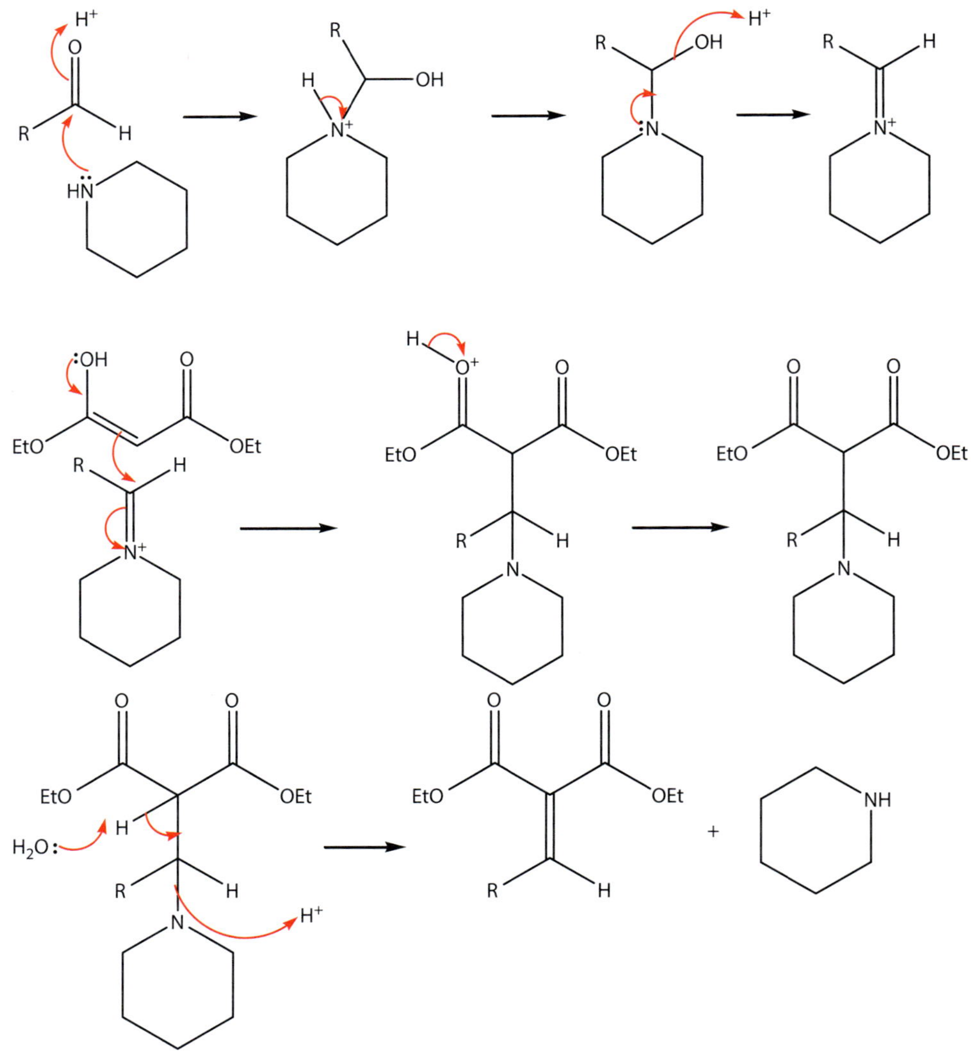

23.8 ENZYME CATALYSIS

Enzymes are natural catalysts *par excellence*—both in their specificity and the extremely high reaction rates that they achieve. Some enzymes are very specific and accept only one or a limited number of substrates. Others are specific for only a limited part of the substrate structure—as we saw in looking at the digestive enzymes that cleave peptides adjacent to particular residues. The specificity arises from the shape of, and the binding sites in, the active site and from the specificity of the reactions available to the bound substrate. Enantiospecificity arises from the fact that the binding site is composed from amino acids of a particular, consistent chirality and hence provides a chiral environment for the reaction.

As with other types of catalysts, the rate acceleration arises from the introduction of a pathway with a lower $\Delta G^{\ddagger}$; there is no change in ΔG°. Accelerations of 10^{10}–10^{12}-fold over the uncatalyzed process are common. The "turnover" frequency of the catalyst, the number of molecules reacting at each active site over time, is usually about 1000 min^{-1}. Turnover, in principle, could reach 10^6 s^{-1}, limited by the rate of diffusion of molecules in and out of the active site.

Binding of the substrate into the active site gives very high local concentrations of catalyst and substrate (compare with the relative rates of intra- and intermolecular processes). Substrate binding may also induce conformational changes, which destabilize the ground state of the molecule relative to the transition state, thus reducing $\Delta G^{\ddagger}$. Acidic, basic, and nucleophilic groups on the enzyme can contribute to the catalysis—they are close to the substrate and correctly positioned for reaction. The active site may also provide a hydrophobic environment, which favors specific kinds of reaction. In this section, we will consider one enzyme reaction mechanism that is very well understood, that of chymotrypsin-catalyzed hydrolysis of peptides; you will meet many more if you take a biochemistry course.

Chymotrypsin-catalyzed hydrolysis of peptides proceeds in four steps. In the first step, the peptide linkage is attacked by a hydroxyl group on serine to form a tetrahedral intermediate (Figure 23.39). Loss of the amine from the tetrahedral intermediate gives an ester involving the

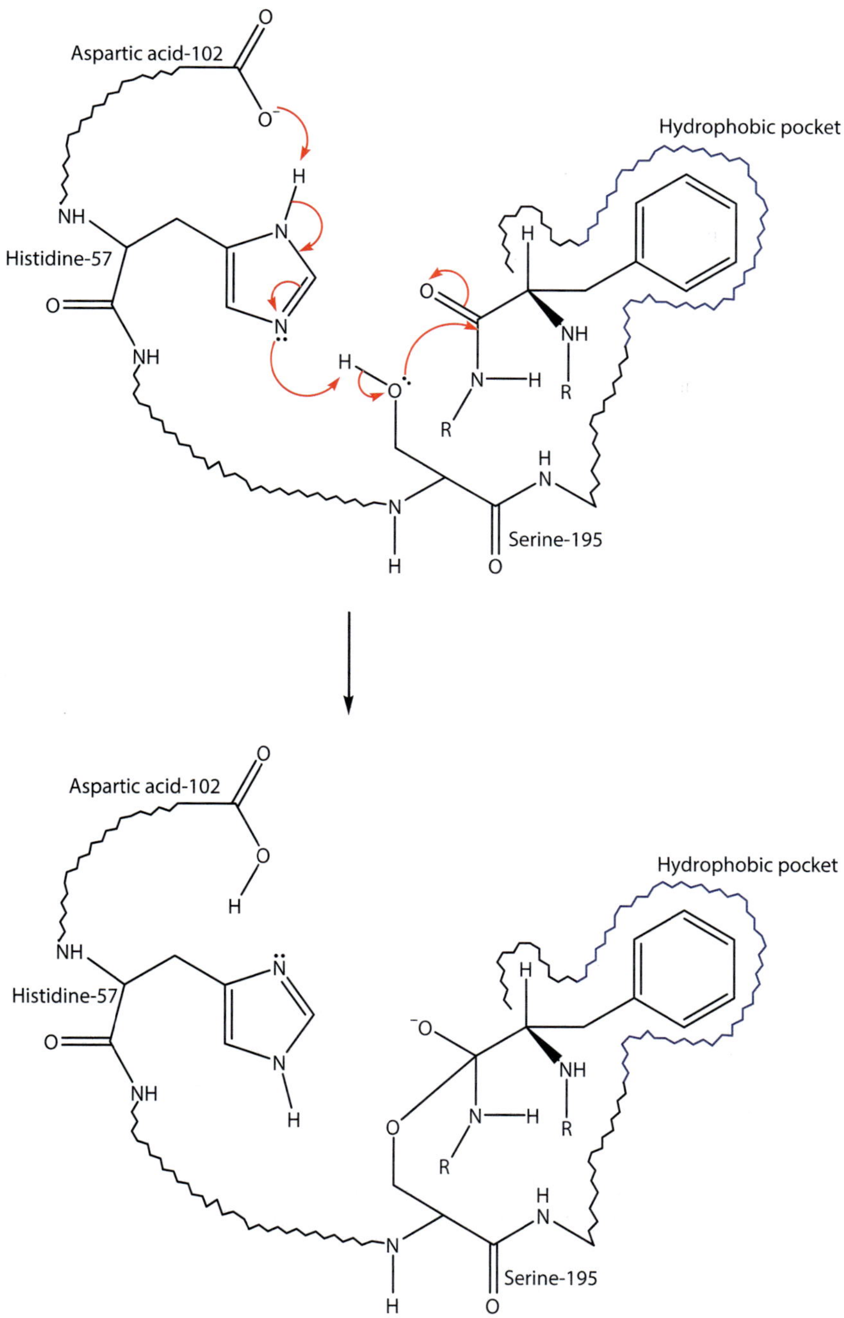

FIGURE 23.39 Step 1 of chymotrypsin-catalyzed hydrolysis of peptides.

serine hydroxyl group. The proton is simultaneously removed from the serine by a shuttle mechanism using a histidine, with the proton being ultimately accepted by an aspartic acid residue. Loss of the amine from the tetrahedral intermediate yields an ester with the serine hydroxyl group. The amine acquires a proton as it departs, again using the histidine shuttle (Figure 23.40).

Nucleophilic attack of water on the serine ester yields a second tetrahedral intermediate; this time, it is the water that is rendered more nucleophilic by the acceptance of one of its proton by histidine (Figure 23.41). Cleavage of the second tetrahedral intermediate frees the carboxylic acid group and regenerates serine (Figure 23.42).

Finally, it's useful to remember what we saw in Section 15.3; enzymes can be used to catalyze useful synthetic transformations, and sometimes only limited purification of the enzyme is necessary for it to be effective. In that section, we studied enantioselective ester hydrolysis and formation reactions, such as that in Figure 23.43. However, both isolated enzymes and microorganisms have the capacity to catalyze a wide range of transformations (Figure 23.44).

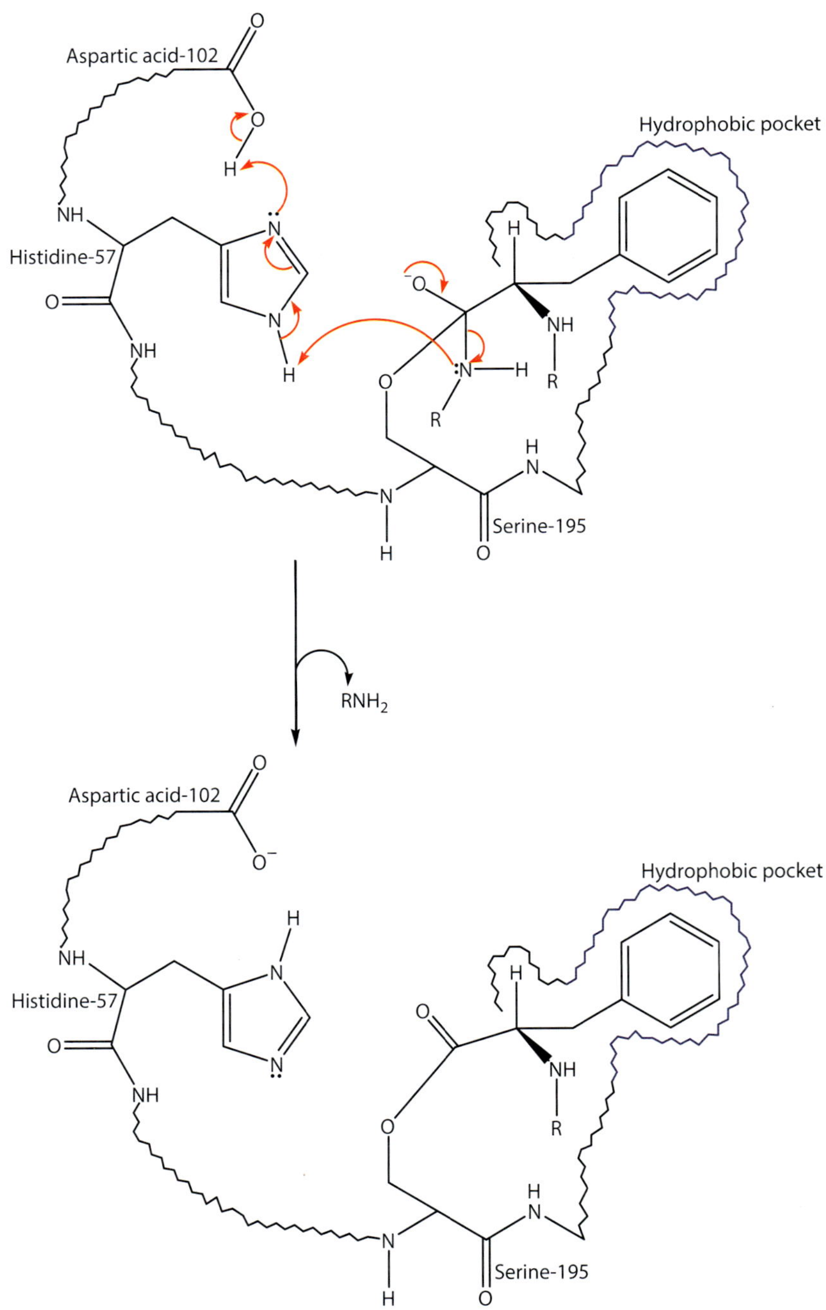

FIGURE 23.40 Step 2 of chymotrypsin-catalyzed hydrolysis of peptides.

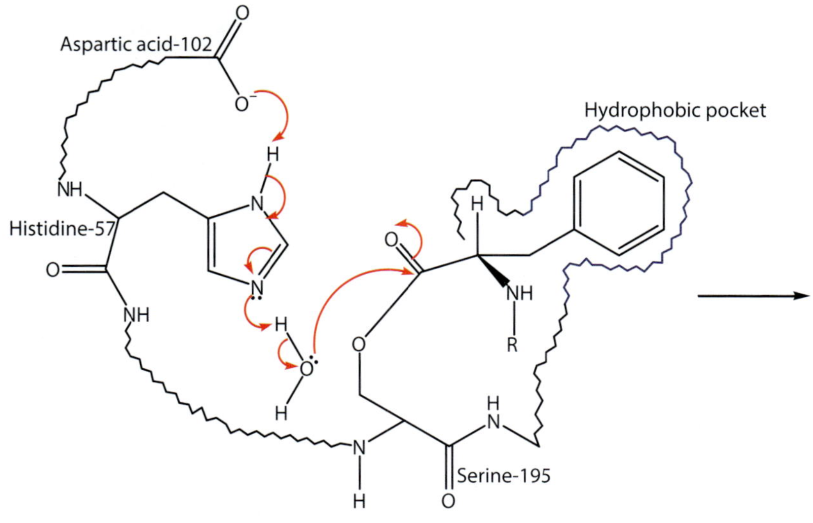

FIGURE 23.41 Step 3 of chymotrypsin-catalyzed hydrolysis of peptides.

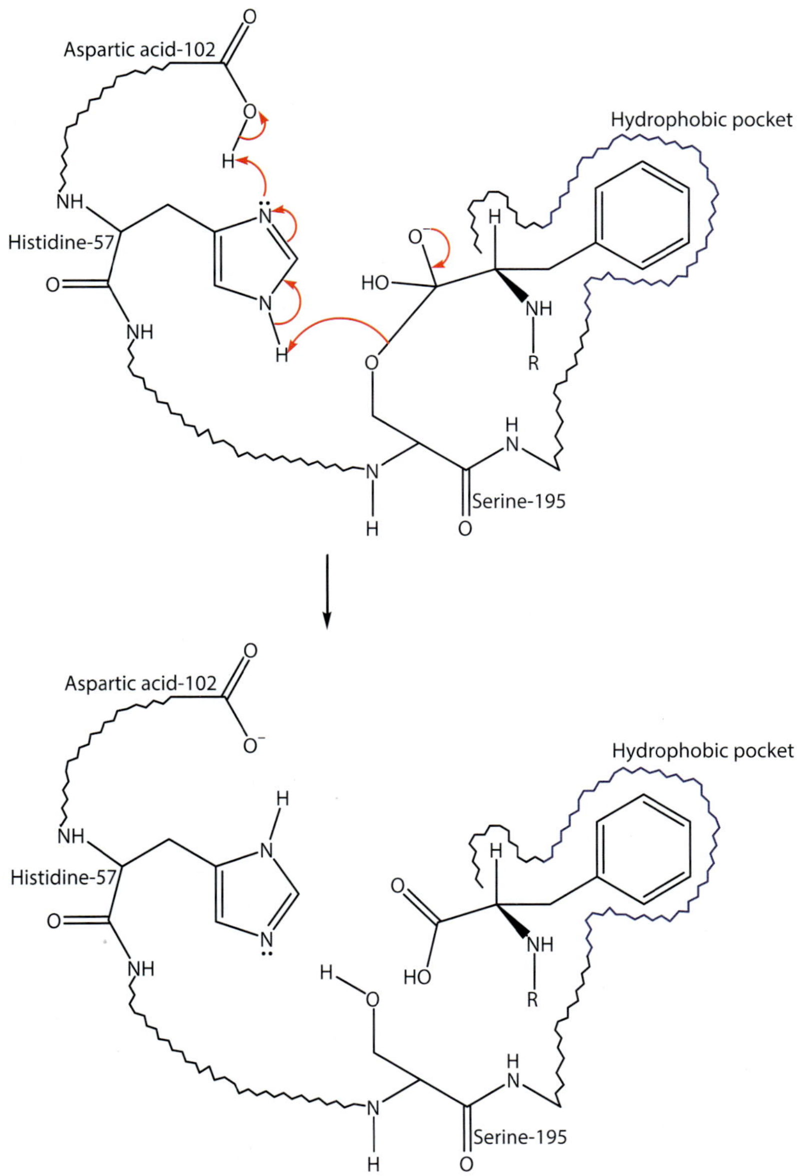

FIGURE 23.42 Step 4 of chymotrypsin-catalyzed hydrolysis of peptides.

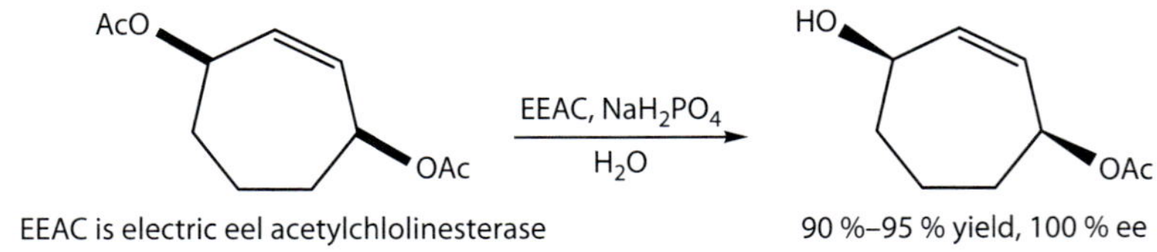

FIGURE 23.43 Enantioselective enzymatic hydrolysis in synthesis.

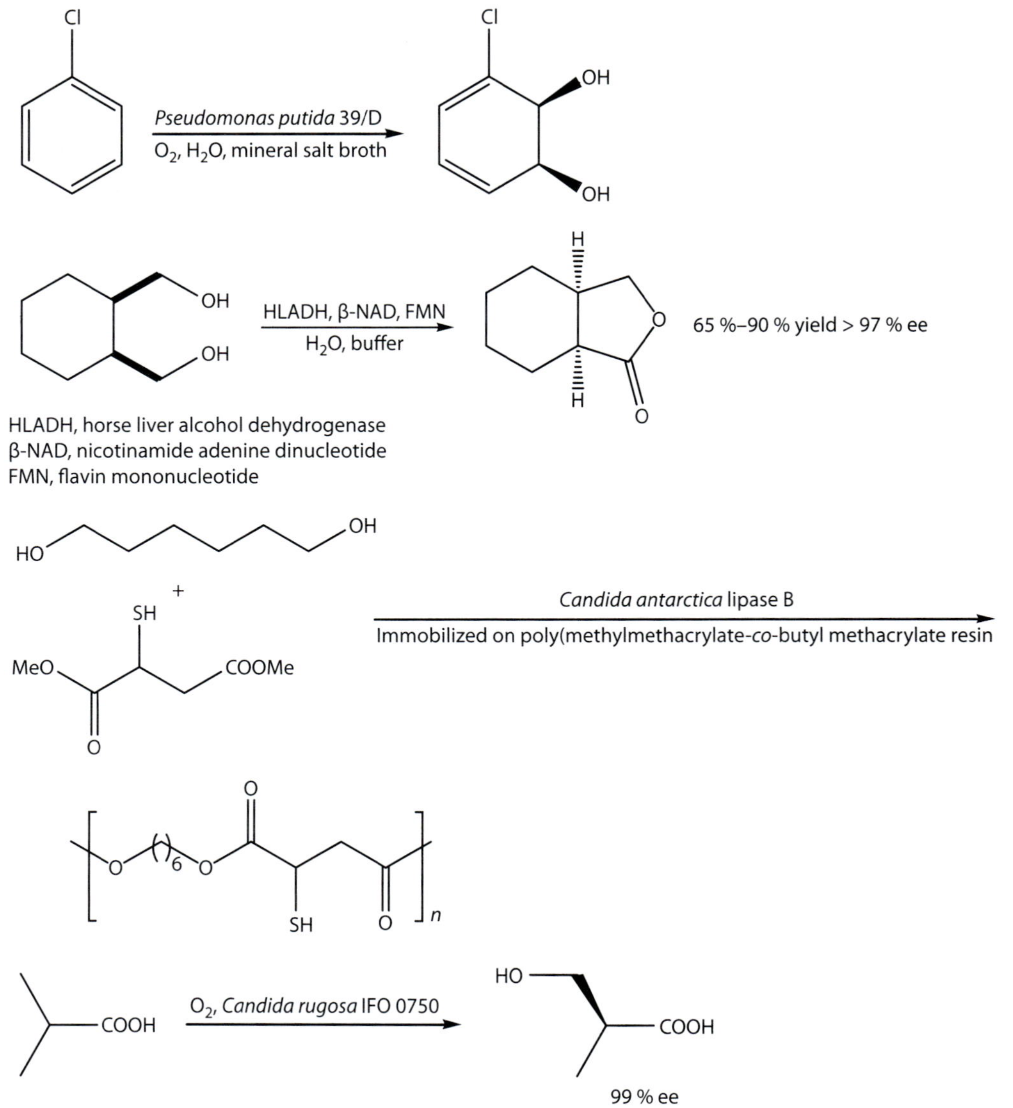

FIGURE 23.44 Other synthetic transformations catalyzed by isolated enzymes or microorganisms.

Key Points from Sections 23.7 and 23.8

- Organocatalysis involves a species that makes a temporary covalent bond with the substrate and is regenerated later in the catalytic cycle.
- Reactions catalyzed in this way include aldol condensations, conjugate additions, halogenations, and Diels–Alder reactions.
- Enzyme catalysis is among the most efficient known and may involve acid and base catalysis, organocatalysis, and processes analogous to PT catalysis. The processes involved in the mechanisms of metalloenzymes may be closely analogous to those of organometallic catalysts.

REVIEW PROBLEMS

1. Below are six catalyzed transformations, showing reagents, but not catalysts. Match each of the following catalysts to the appropriate reaction:

   TsOH; [Pd(PPh$_3$)$_4$], CuI; Pd(OAc)$_2$, K$_2$CO$_3$; *S*-BINAP, [Ru(2-methylallyl)$_2$(COD)], HBr; PdCl$_2$, CuI; [Bu$_4$N]Br

   (a)

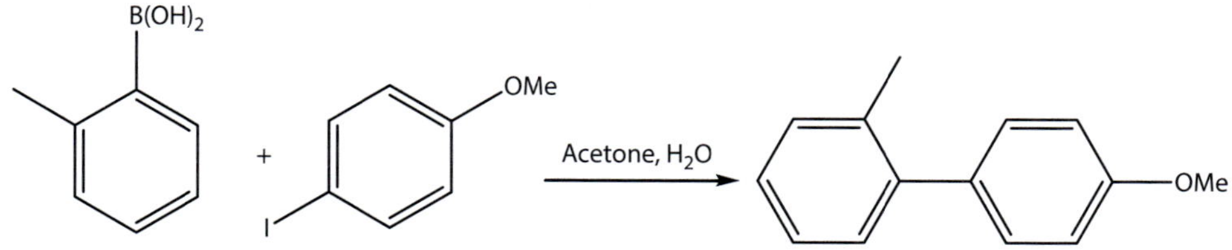

   (b)

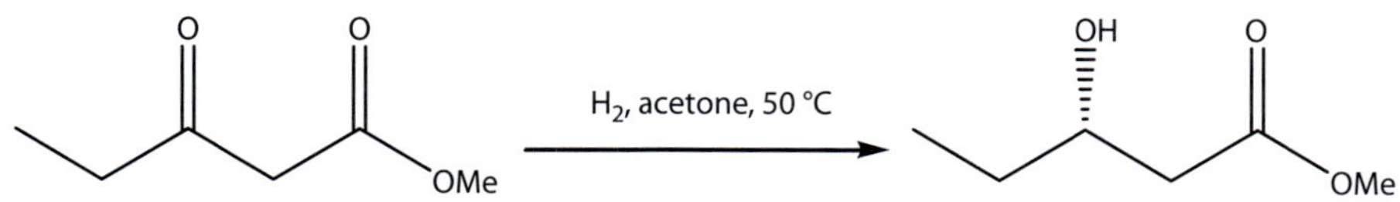

   (c)

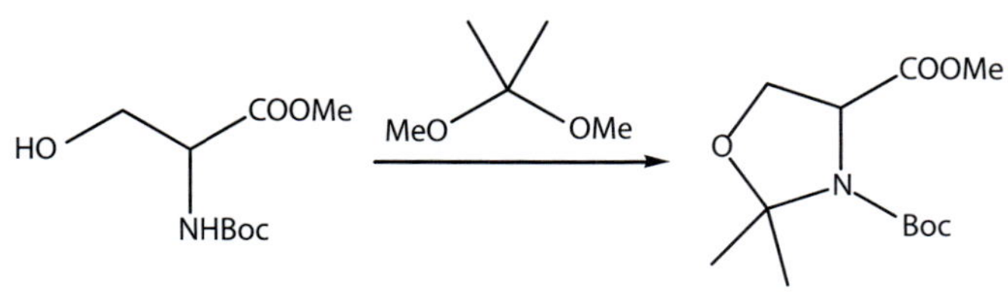

   (d)

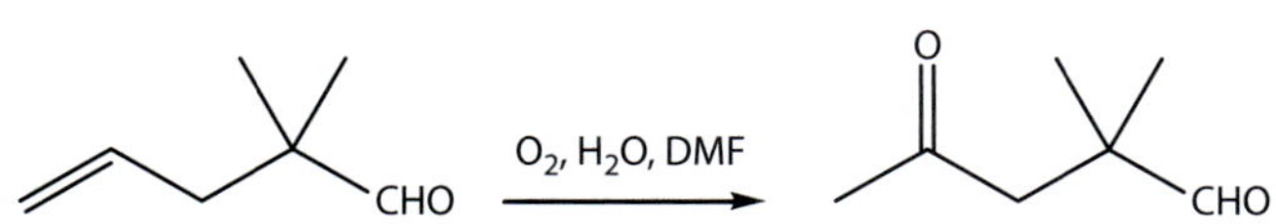

   (e)

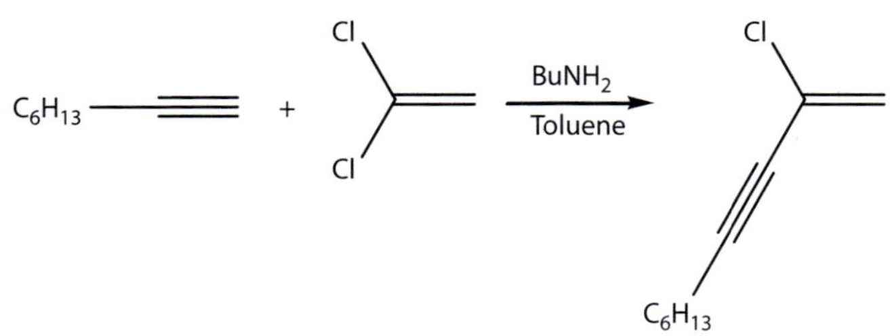

   (f)

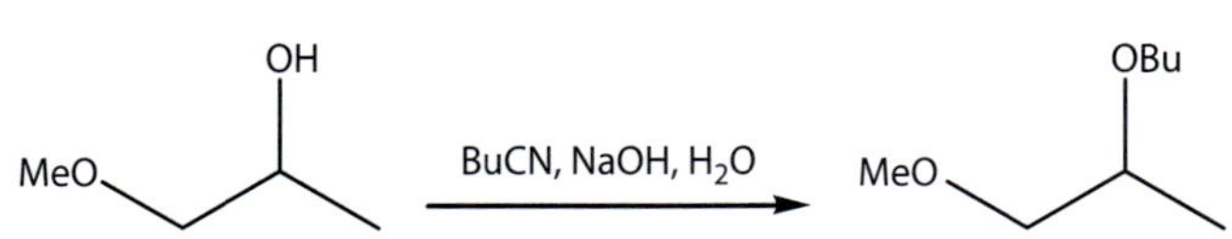

2. Predict the product(s) of the following reactions:

   (a)

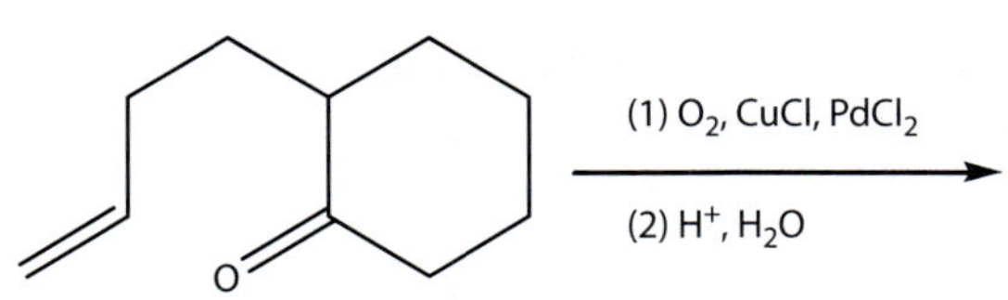

(b)

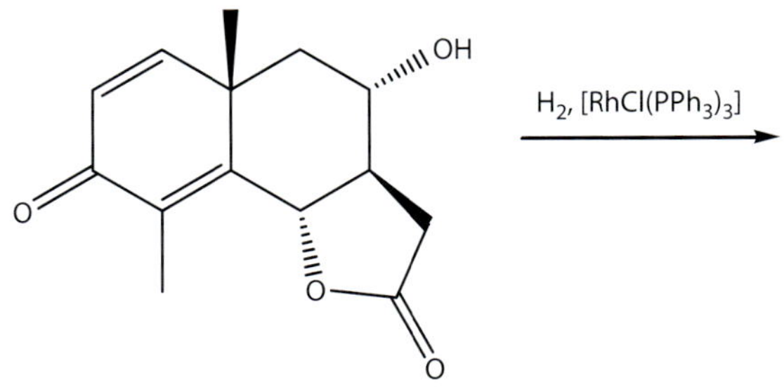

(c)

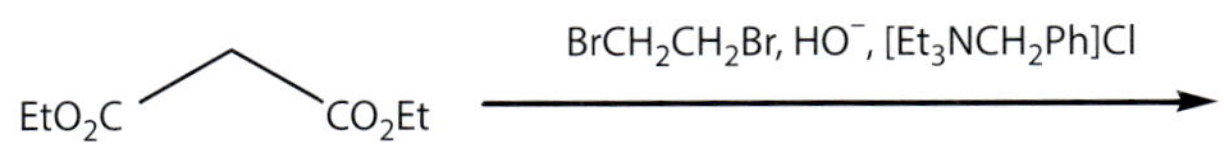

(d)

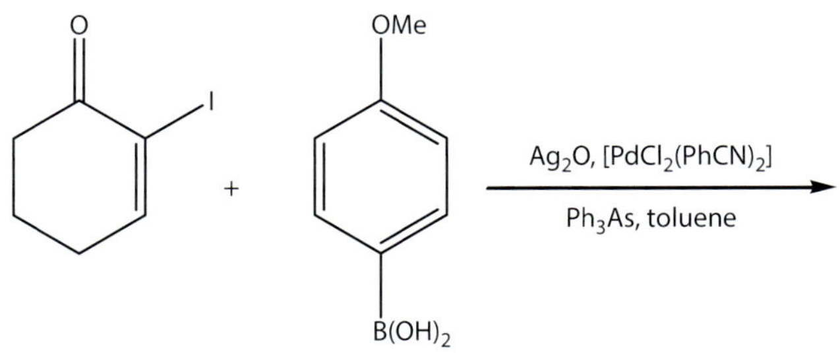

(e)

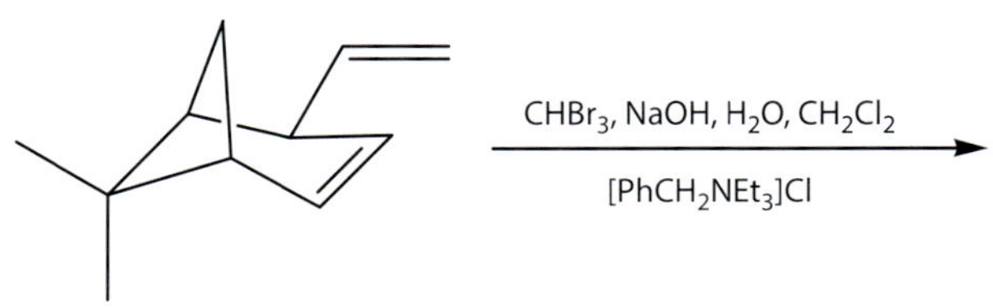

(f)

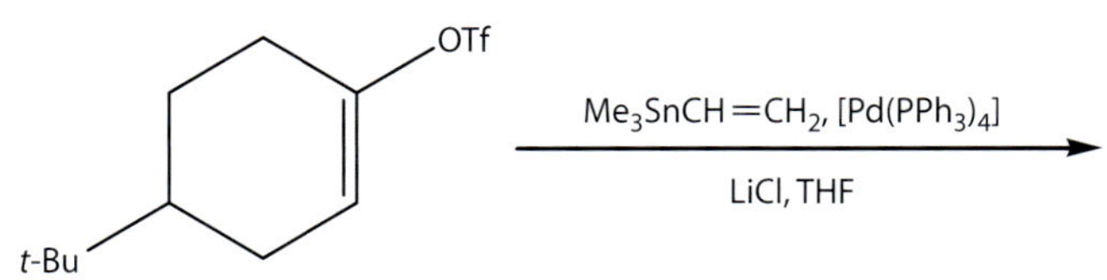

3. Suggest a catalytic system involving at least one transition metal for the following transformations:

(a)

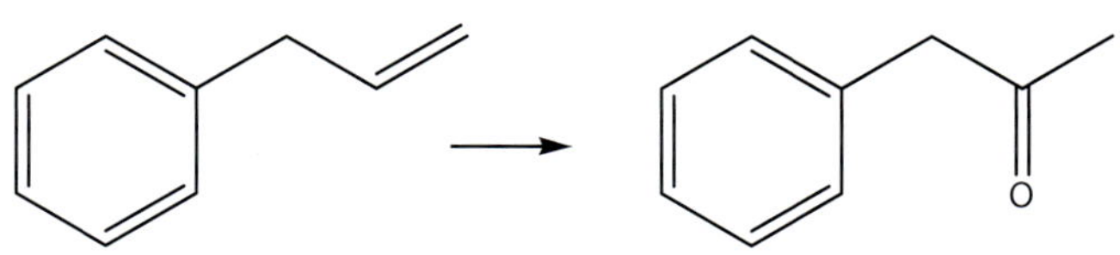

(b)

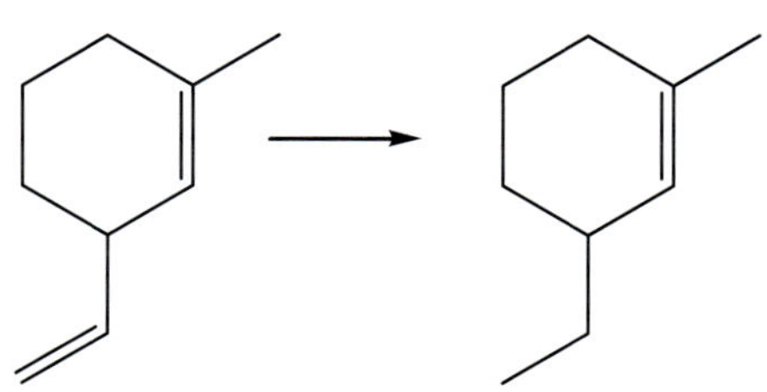

(c)

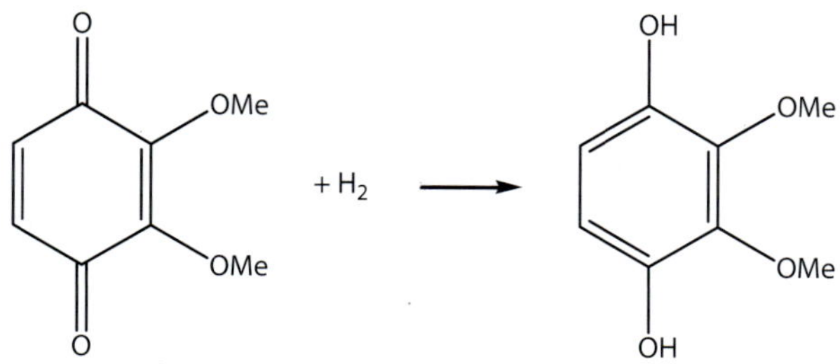

(d)

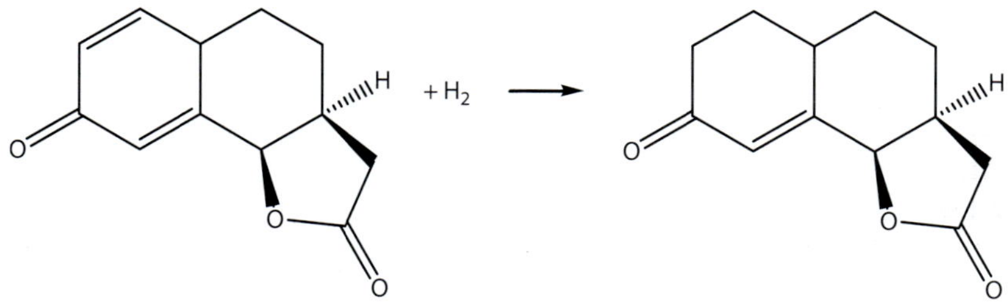

(e)

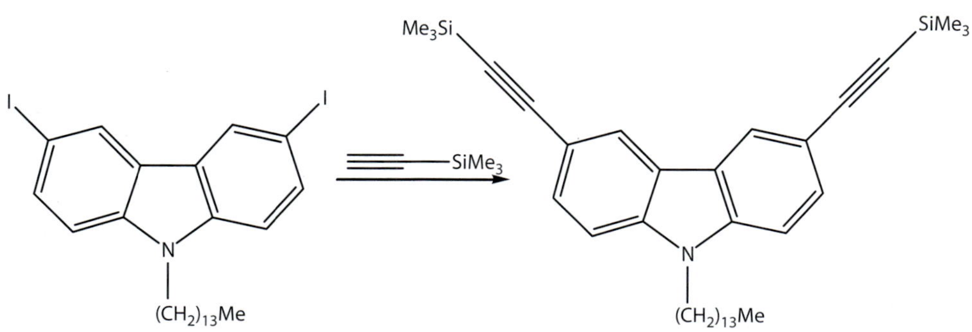

(f)

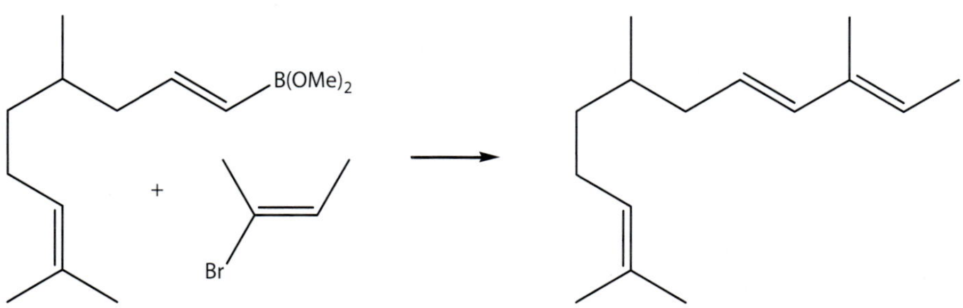

4. Explain the following observations:
   (a) The reaction of benzyl bromide with PPh_3 in xylene, heated under reflux, to give $[PhCH_2PPh_3]Br$ takes several days to go to completion. If the reaction is carried out in the presence of a small amount of lithium iodide, the reaction is complete in 2 h.
   (b) The relative rates of the reactions shown are 1:1000:

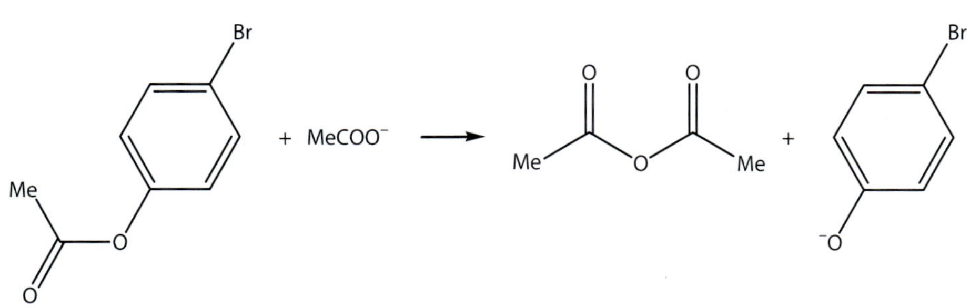

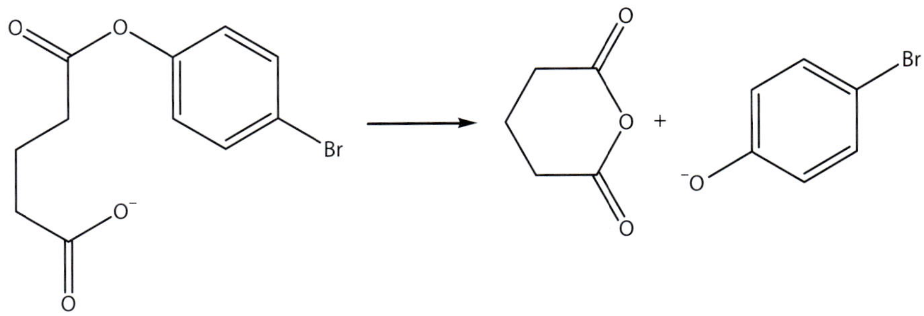

(c)

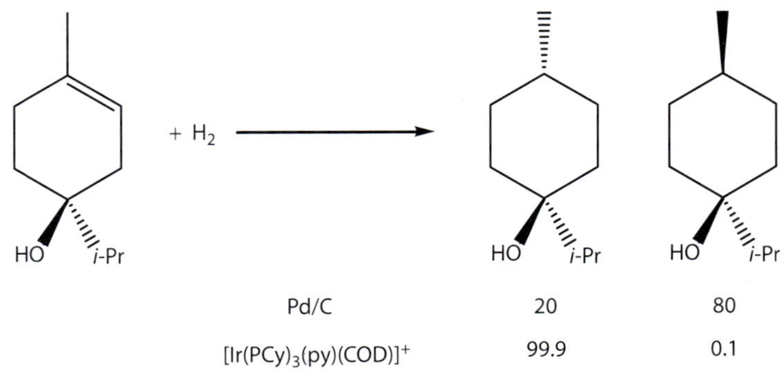

5. Draw a mechanism for each of the following reactions, explaining in each case the function of the crown ether:

(a)

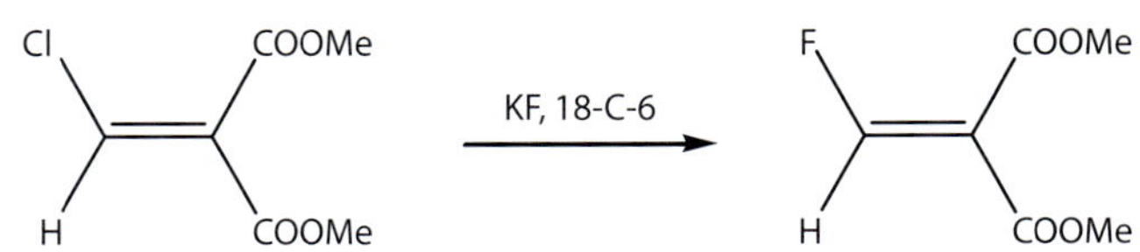

(b)

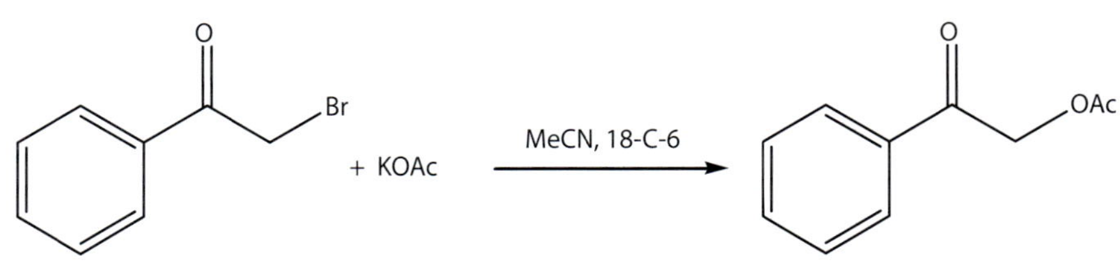

(c)

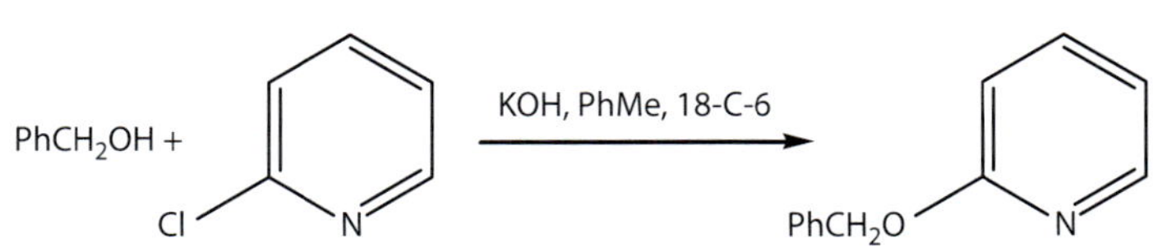

6. Explain what is happening in the reactions shown in the following. You do not need to draw the mechanism for the metal-catalyzed processes, but you do need to identify the product of these reactions and suggest a mechanism for the subsequent transformation(s):

(a)

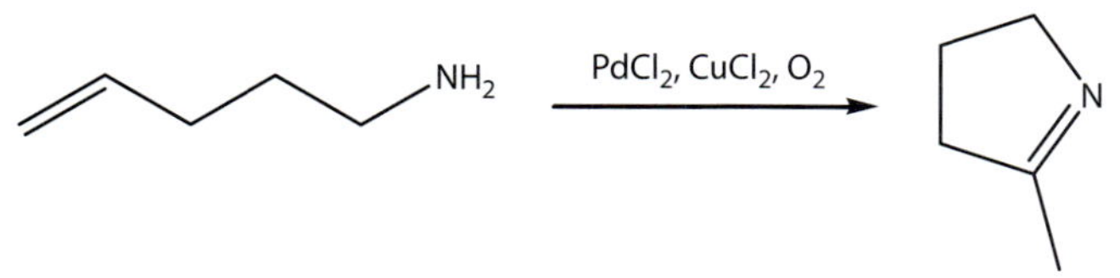

(b)

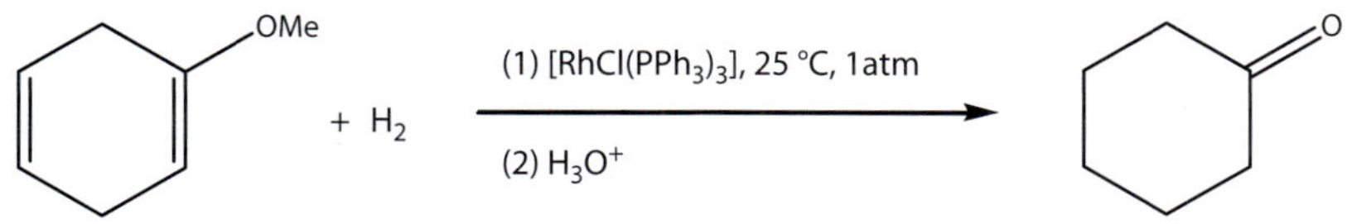

7. The two hydrogenation reactions shown in the following can be effected using different catalysts. Propose a suitable catalyst for each process:

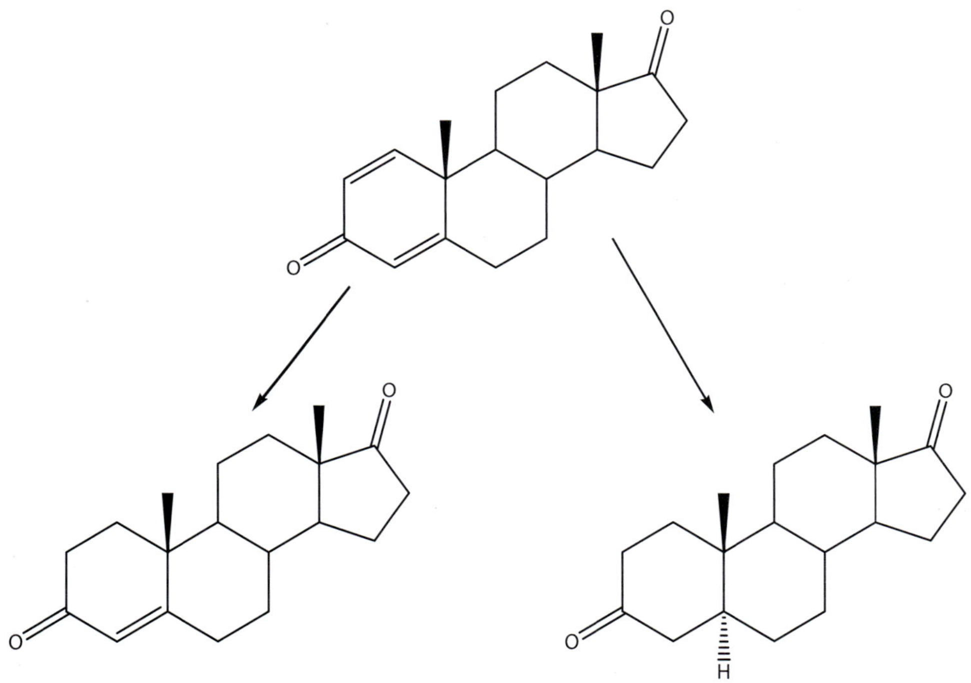

MORE CHALLENGING PROBLEMS

1. Explain the following observations:
   (a) The relative rates of the three reactions shown in the following are 1: 2 × 10^5: 10^7:

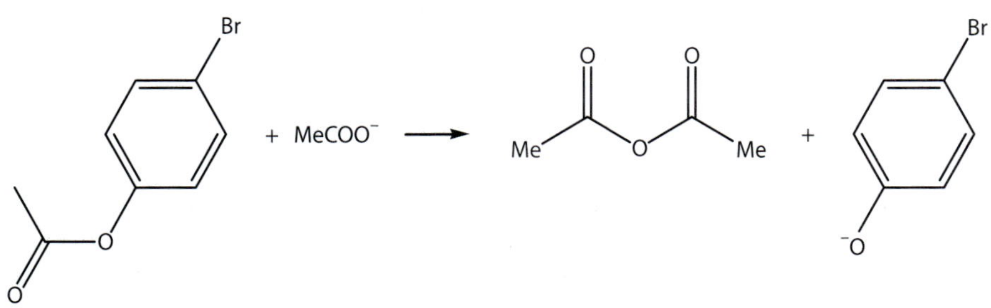

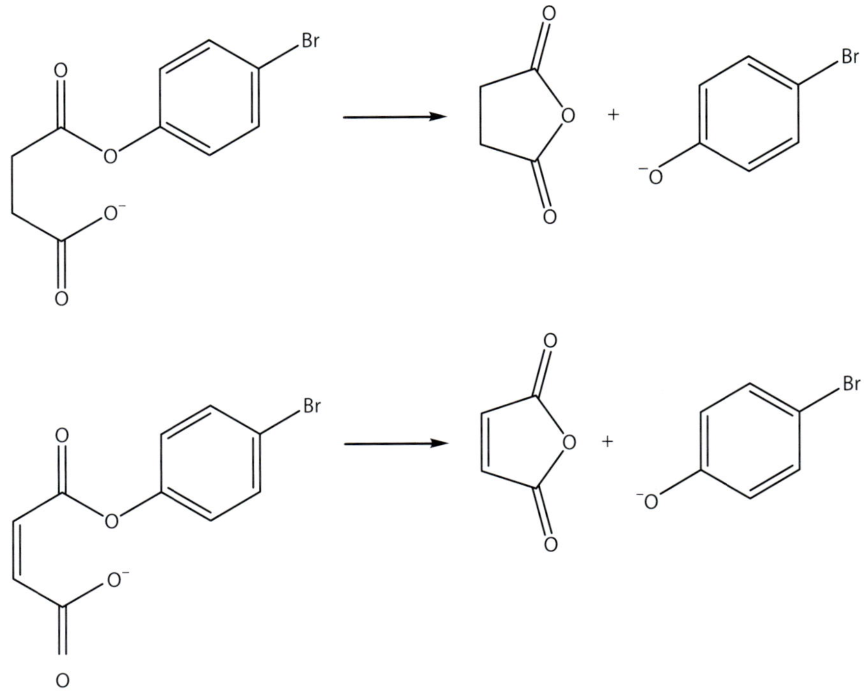

(b) The hydrolysis of acetic anhydride in water is catalyzed by pyridine. Propose mechanisms for the catalyzed and uncatalyzed reactions, showing clearly the function of the pyridine.

(c) The rate of the reaction shown below is directly proportional to [H⁺]. Bear in mind that proton transfers to oxygen or nitrogen are not usually rate determining.

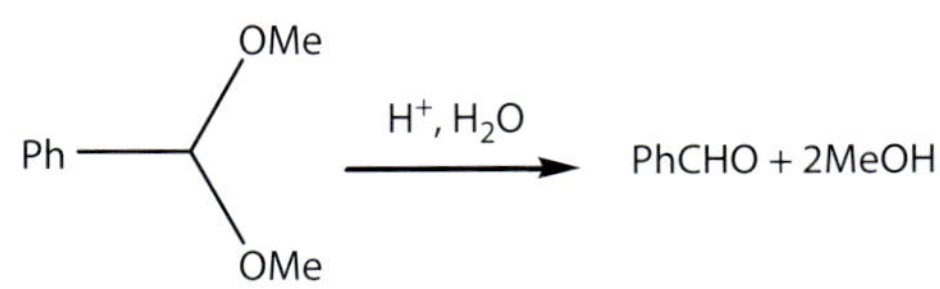

(d)

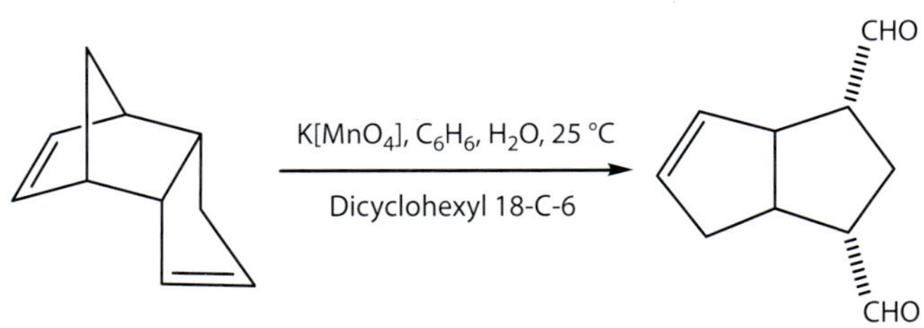

(e)

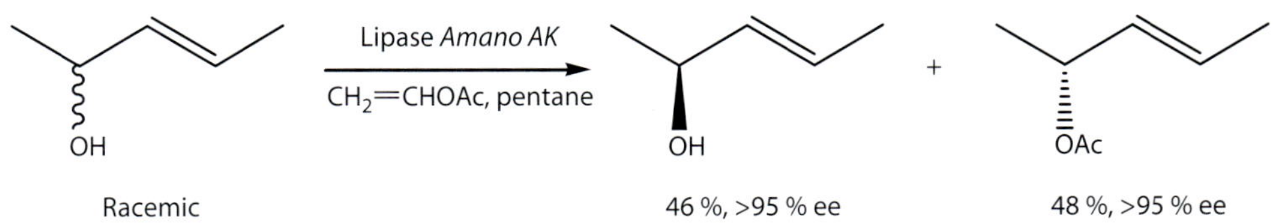

2. Thiamine acts as a catalyst for the decarboxylation of pyruvate. Propose a mechanism for the reaction:

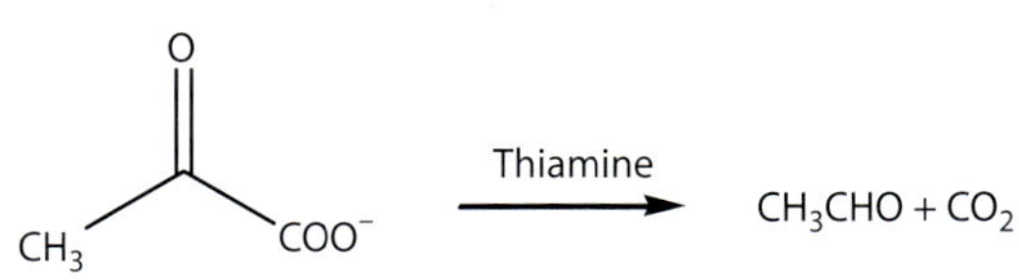

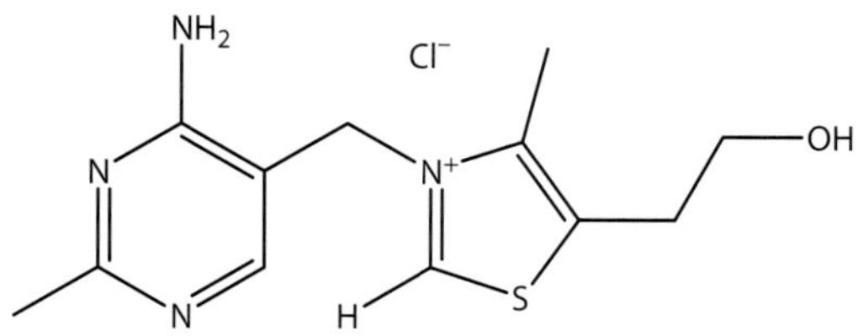

Thiamine hydrochloride

3. The elimination of HX from medium-ring cycloalkylhalides gives both *cis*- and *trans*-products. In general, the *trans*-cycloalkenes are thought to arise from *syn*-elimination and the *cis*-products from *anti*-elimination. The following results were obtained using potassium *tert*-butoxide as base:

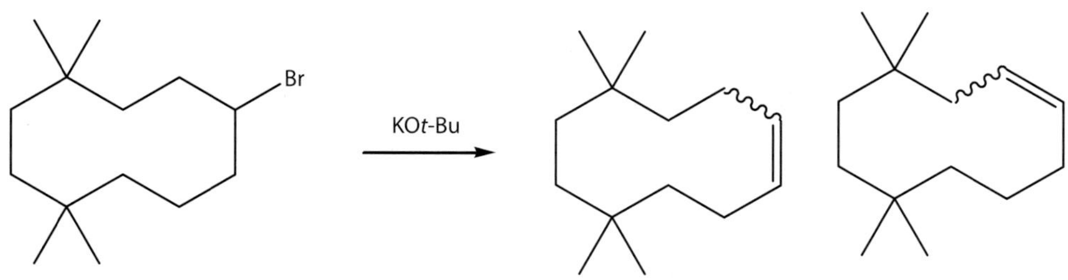

t-BuOH as solvent	*trans*:*cis* = 55	*trans*:*cis* = 9.3
Dicuclohexyl-18-C-6 in *t*-BuOH	*trans*:*cis* = 0.12	*trans*:*cis* = 0.03

Explain the effect of the crown ether.

Organic Chemistry in Industry

24

24.1 INTRODUCTION

Organic chemistry in industry is a vast topic and covers many areas, not all of which we can discuss here. We may first think of it in terms of "big pharma," the drugs and medicinal chemistry industry, but this comprises only about 30 % of the total. Equally important is the manufacture of basic organic chemicals for a wide range of uses, including monomers for the plastics industry. Other large sectors are agrochemicals, dyestuffs for the textile, food, cosmetic, and paper industries, soaps and detergents, perfumes, cosmetics and toiletries, and high-value specialty compounds including adhesives, sealants, catalysts, coatings, paints, smart materials for the electronics industry, and fine chemicals for the chemical research industry.

Large tables of statistics are inappropriate to a textbook—they date far too quickly. Good data (covering all countries) are available every year in the American Chemical Society–produced *Chemical and Engineering News*. However, a few generalizations seem to stand the test of time. In the developed world, the chemical industry is profitable, contributes significantly to the gross domestic product (GDP), and is positive in terms of balance of trade. It has been growing for more than 150 years, and although there is some slowdown during recessions, growth does not stop, especially in the pharmaceutical area. Historically, the largest producers of chemicals have been Western Europe, North America, and Japan, but this is now changing. The outsourcing of chemical manufacture to lower-wage economies (*e.g.*, China, India, Brazil, Vietnam, and Thailand) has developed over the last 25 years; whether this is a permanent feature of the industry as wages in such economies grow and wages in developed economies stagnate remains to be seen.

The chemical industry frequently gets a bad press, as a cause of pollution, using materials later found to be toxic in consumer products and agriculture, and as a contributor to global warming. These criticisms have some justification—but solutions are unlikely to be found by anyone other than the chemical industry. It is also worth remembering that life expectancy is strongly correlated with technological development even in the modern world; there never was a technology-free, chemical-free Utopia. Better chemistry can produce better drugs, reduce pollution, and combat climate change; allocation of resources and setting of priorities are political as well as economic issues. Regulation of the industry by both government and public opinion has increased relentlessly, and issues of intellectual property and compliance costs cannot be ignored. We must also remember that the chemical industry is just that—an industry, which, of its nature, is designed to make a profit. It is not a charitable foundation.

24.2 PROCESS CHEMISTRY *VERSUS* RESEARCH CHEMISTRY

The first steps to a useful new compound in chemistry, be it pharmaceutical, agrochemical, polymer, dye, or fragrance, occur in a research laboratory, and the materials are produced on a small scale. However, commercial products, even high-added-value pharmaceuticals, are produced on a much larger scale and in quite a different environment. The issues are not only ones of a differing scale—they present quite different issues in terms of cost, safety, quality control, sustainability, efficiency, environmental impact, and waste disposal. Process development chemistry has, over the last 25 years or so, emerged as a genuine discipline.

BASF plant in Ludwigshafen, Germany

The relative importance of the factors involved in process development depends, to some extent, on the added value of the material to be produced. For bulk chemicals, cost and efficiency are crucial, but for pharmaceuticals, issues of quality control (robust processes that can repeatedly deliver product with consistent low levels of known impurities, ideally in a closed system) and sustainability (particularly minimization of the use of organic solvents and energy and reduction in waste and water usage) may be more prominent. Safety, however, is always critical. If a process cannot safely be scaled up, then it must be replaced. Although in a research laboratory issues of kinetics and thermodynamics are often sidelined, this cannot be so for a large-scale process. If the kinetics of a process are not understood, then a runaway reaction may occur on a large scale. If a reaction is exothermic, then we must consider the rate of heat loss at the surface of the reactor vessel and methods of mixing (much more difficult in an industrial-size reactor than a round-bottomed flask). The rate of heat loss depends on surface area of the reaction vessel; the ratio of surface area to volume (and hence internal heat generation) is much lower for large than for small reactors. There is also the issue of continuing reliability and quality control when reactions are scaled up.

Some reagents, commonly used in research labs, are party poopers in an industrial context, because of their toxicity or other issues. Stoichiometric osmium tetroxide, organothallium compounds, organomercury compounds (toxicity and toxic waste), benzene, aziridines (carcinogens), dinitroaromatics (explosives), and carbon disulfide (flammable) would be unrealistic for use in large-scale manufacture. Regulatory restrictions are likely to get tougher, not laxer, in the future.

For a process to be commercial, the raw materials must be economically available and of consistent quality, and there must be reasonable prospects of this continuing into the future. Future-proofing is not trivial, and solutions are not always obvious; for example, a material, which is costly because of the high environmental impact of making it, may increase in price because of increased regulation or may reduce in price because a new greener method of manufacture is found or a method of waste recycling discovered. A natural product may increase in price because of depletion of its sources or reduce in price because an inexpensive method, such as yeast fermentation, is found to manufacture it. The availability of some basic chemicals may be constrained by the market for the isomeric by-products in their manufacture. For example, 3,4-dichloronitrobenzene is always produced as a by-product in the nitration of 1,2-dichlorobenzene to obtain 2,4-dichloronitrobenzene, a starting material for a number of agrochemicals. In general, it is best to use expensive reagents/materials toward the end of a synthesis rather than in early stages.

Pacific yew tree in Mount Hood National Forest

The antitumor compound taxol was initially isolated in the 1950s from the bark of the Pacific yew, but yields were low, and extraction on the scale needed would have been environmentally damaging. A semisynthetic route was devised from the needles of the European yew (which could be harvested sustainably as the trees were not killed). Almost all modern production involves plant cell fermentation technology.

The nature of reactor vessels is also important. While glass is almost ubiquitous in a research lab, most industrial synthesis takes place in stainless steel or Hastelloy* vessels. Are the vessels corrosion resistant under reaction conditions, and is there certainty that no metal is leached into the products? Reactor design is a chemical engineering problem—and how are the materials added and recovered, and would a flow system work better? The design of the stirrer and the rate of stirring may also significantly impact product quality, driven by localized pressure around the stirrer blades. The physical method of adding of a reagent to a reactor, such as spraying, dropping, or running down the reactor wall, can also be important, notably in industrial nitration reactions. Current good manufacturing practice rules mean that reactors used to make agrochemical intermediates cannot be used for pharmaceutical intermediates, regardless of cleaning protocols. This means that contract manufacturers have to dedicate reactors to either pharmaceutical or agrochemical/fine chemical use. Agrochemicals are often preferred, because volumes tend to be larger, and there is thus greater plant utilization.

Hastelloy reaction vessel

* A range of corrosion-resistant alloys, mainly nickel and molybdenum based.

At each step of the synthesis in a research lab, the material is generally recovered, and purified, before the next step is attempted. Methods such as chromatography and resolution of a chiral compound are available. Chromatography is generally impracticable on an industrial scale (possible only for low-volume, high-added-value pharmaceuticals and, even then, not desirable). Resolution is wasteful, as half (many new drugs are complex molecules with multiple chiral centers, thereby increasing this challenge by successive powers of 2) the material must be discarded or recycled; the use of chiral starting materials* or enantioselective syntheses are more economic. The additional costs of producing a chiral pharmaceutical are almost invariably lower than the cost of the toxicological and pharmacokinetic work on the "wrong" stereoisomer that would be required for regulatory approval, and few, if any, racemic drugs have been approved in the last decade. Crystallization is the ideal way for material to be purified; in the final stage of pharmaceutical production, it will be important to ensure that the correct (or at least a usable) polymorph† of any crystal is formed. Many products are capable of existing in several different crystal structures—there will always be a preferred one, and the crystal form may influence the solubility in the human body, or "bioavailability," either directly or through the ease, with which the product can be "nanomilled" to aid bioabsorption. If material is amorphous, particle size is also important to pharmacokinetics. Clinical trials of a pharmaceutical can grind to a halt if the process used for producing the 10–100 kg needed for phase II clinical trials results in a new impurity, inconsistent levels of known impurities, or solvent or heavy metal traces.

Environmental impact and waste disposal are increasingly important, for both ethical and economic reasons. Chromium (VI) reagents are very effective oxidants, but chromium is highly toxic, and hence, it's very expensive to dispose of the waste generated. Toxicity to humans, plants, and animals must be considered, as well as the potential of reagents for ozone depletion and climate change. It is only allowed to use Montreal Protocol–listed reagents (ozone depleters) if it can be demonstrated that they are 100 % consumed within the process and cannot escape. Even then, the reagents are often difficult to source, as the move away from such materials has meant traditional suppliers no longer offer them. A classic example is the use of bromofluoromethane in the manufacture of Fluticasone, where it is used to convert a thiol to an $-SCH_2F$ group. Even if materials are not currently heavily restricted, investment in a process that uses something likely to be banned in the future makes no economic sense. The use of energy-efficient processes is also important. Research chemists often believe that good selectivity is most easily achieved at −78 °C. However, cooling on a large scale (just like heating) is expensive, and effective mixing is often much more difficult. The same selectivity can sometimes be achieved by controlled rates of addition of reagents and a proper understanding of reaction kinetics.

In the early years of process development, chemistry, variables were optimized sequentially and through multiple loops. The variables are rarely independent, and this is a slow and inefficient process. More modern work optimizes variables simultaneously, using multiple parallel experiments with computer analysis of the outcomes.

24.3 PHARMACEUTICAL INDUSTRY

24.3.1 INTRODUCTION

In the few pages available here, it's impossible to do justice to the modern pharmaceutical industry; its social and economic importance cannot be overestimated. Until the mid-nineteenth century, most medicinal compounds were herbal remedies. Many of these traditional medicines were successful, and their descendants are still widely used. Aspirin, **24.1**, is derived from salicylic acid (**24.2**) in willow bark (Latin, *salix*, willow tree), and willow bark preparations have been known to have analgesic effects since the Paleolithic. The narcotic effects of

* The so-called chiral pool of materials, generally naturally occurring amino acids or sugars, and some monoterpenes.

† The term polymorph refers to the crystal structure of a molecule; there may be a dozen or more of these including solvates and hydrates. The pharmacodynamics of the different polymorphs may be quite different, especially with respect to absorption.

Datura and poppy juice were well known in the ancient world. Natural materials and their analogs are still important in drug discovery.

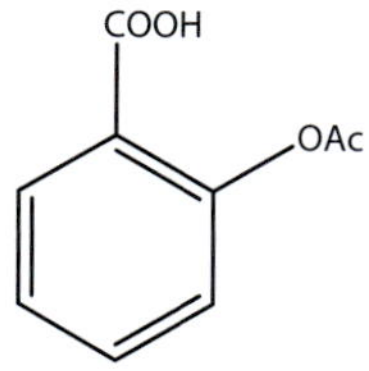

24.1

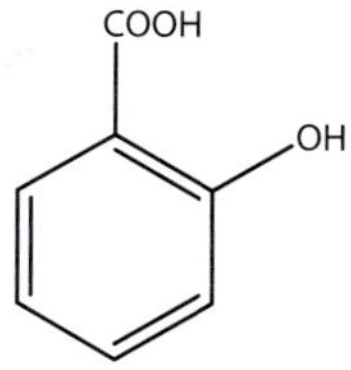

24.2

Most modern pharmaceutical companies have their origins in the late nineteenth and early twentieth century, a period of considerable development in drug discovery. The range of available compounds grew rapidly after the Second World War, with the contraceptive pill, tranquilizers, and blood pressure drugs appearing in the 1960s and novel chemotherapeutic agents in the 1970s. Testing of new medicines was regulated in 1964 by the Helsinki declaration, requiring the informed consent of patients to participate in drug trials. The 1980s saw the development of new drugs for heart disease and antiretrovirals for AIDS treatment, with many new antidepressants appearing in the 1990s. The 1990s also saw an increase in contracting out of development and manufacture (although India has been the largest manufacturer of generic versions of off-patent drugs since 1978), buying up of small start-up companies with one excellent potential drug, and the purchase of drugs through the Internet. Patent protection is a critical feature of the industry; companies cannot spend the resources necessary to develop and test new drugs if they cannot recover those costs by selling it.

The generation of a new drug for clinical use has several stages—discovery and development and then testing and validation. Materials must be patented early on if a company is to proceed. The length of the so-called pipeline for delivery varies with the type of compound; some cancer drugs make it in 7–8 years, but the average for neuropsychological drugs is almost 18 years. Companies have patent protection for 20 years. Of 10,000 compounds that start the process, one will make it through to clinical use. Of 10 new drugs approved, only three will generate enough revenue to cover their development costs (and those of the previous failures), currently averaging between 1 and 2 billion dollars. These are stark, and alarming, figures; new drugs are needed, as there are still many conditions with limited effective treatments, and there is a particular shortage of antibiotics to treat infections that have developed resistance to existing drugs. The slowdown in the development pipeline, and expiration of patents, has made companies more aggressive about leaving areas in which they cannot make a profit, and terminating unpromising materials at an earlier stage of testing, and hence reducing the costs of failed compounds.

24.3.2 DRUG DISCOVERY

Early drug discovery relied on finding the active compounds in traditional remedies. Although this is now a well-mined area, new classes of natural products are discovered each year, both from terrestrial and marine organisms, and most of these will be screened to see if the class is likely to have any medicinal use. Many new compounds are also made for other purposes in research laboratories, and serendipitous discovery still plays its part. The 1980s saw the start of high-throughput screening of compounds and combinatorial chemistry, which generated large libraries of compounds for screening. More recently, research has been more targeted toward interactions with specific proteins. For example, in HIV treatment, nonnucleoside reverse transcriptase inhibitors can bind in a side pocket of HIV's reverse transcriptase enzyme, and this inhibits motions critical in forming infective viral DNA from the virus's genomic DNA. Computer design was used to modify a compound that gave an initial hit in a docking study to give a more potent compound. Although molecular modeling and simulation is not yet a fully enabling technology for drug discovery, the field is advancing rapidly, with serious money being provided by charitable foundations.

24.3.3 DRUG DEVELOPMENT: PRECLINICAL TESTING

Preclinical data on a new compound are needed before any human clinical trial can be contemplated. A safety profile must be developed, using tests on at least two animal species so that safe doses for clinical trials can be estimated. These tests also determine the pharmacodynamics and pharmacokinetics of the material—absorption, distribution, metabolism, and excretion (ADME). Animal testing causes ethical discomfort and is very expensive, so there is a drive to reduce reliance on this. On more practical issues, can the molecule be made in large quantities, with good quality control? Can the availability of high-quality starting material supplies be assured in the long term? How much waste will be produced, and how much will it cost to dispose of it? Is the material stable, how long is its shelf life, and will it require special handing or refrigeration? How is the drug to be administered? There are many options for formulation, and finding the optimum takes significant time. It's frequently found that large-scale production of a material does not give the same crystal polymorph as was initially obtained in a research laboratory. By the end of this stage, the initial group of 10,000 compounds has been reduced to about 250 and, between 6 and 7 years has passed from the original discovery. At this point, the material must have been protected by a patent.

24.3.4 DRUG DEVELOPMENT: CLINICAL TESTING

In the United States, there are three stages to clinical trials, mandated by the Food and Drug Administration (FDA); there are small, but not highly significant, differences in other jurisdictions in the developed world, and this will serve as a generic model. Phase I studies are conducted in a small number (20–100) of healthy volunteers, to determine the human pharmacokinetic and pharmacological effects of the material (ADME, toxicity, appropriate dosage). If these trials are successful, then Phase II studies are designed to ensure that the compound is *effective* in the desired treatment. The number of patients is again small (100–500), and they all have the disease or condition that the material is intended to treat. Of the original 10,000 compounds, 5 will make it through stage II testing. Phase III studies involve a much larger number of patients (1000–5000)—they are intended to test effectiveness and safety, risks, and benefits. What is the incidence and nature of side effects in a large group; are there long-term effects on reproductive health, fertility, and the immune system? These three stages together take an average of 7 years. Finally, there is an FDA review, and one compound of the original 10,000 will make it to the end. The review stage is about 1.5 years, but there are accelerated procedures for drugs that serve an unmet clinical need or show significant benefits relative to treatments currently available. Postapproval surveillance is also often required, so as to pick up long-term effects and rare side effects. This was particularly important in the withdrawal of the NSAID* Rofecoxib, **24.3** (marketed by Merck as Vioxx®), a selective COX-2 inhibitor used in arthritis treatment, in 2004. Long-term use substantially increased the risk of heart attack or stroke.

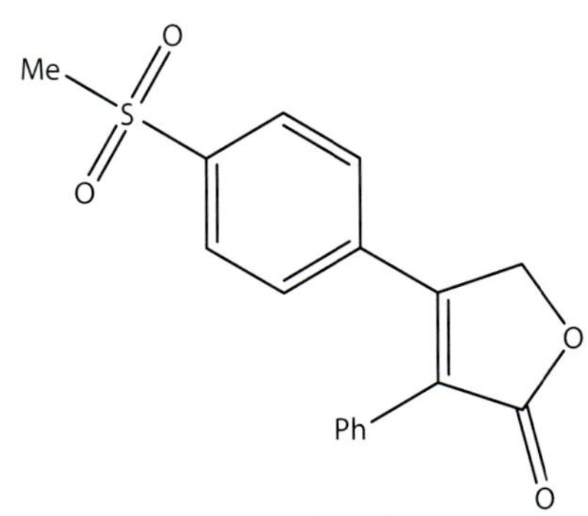

24.3, rofecoxib

* Nonsteroidal anti-inflammatory compound.

However, approval of a drug for use is not the end of the story. In the United Kingdom (and in many other European countries), the drug must also be approved for use within the National Health Service on the basis of cost-effectiveness; it must show that it is efficient as well as effective. This involves the National Institute for Health and Clinical Excellence (NICE) and has been widely criticized. For example, they have benchmarks with respect to new cancer treatments as to the maximum amount an extra month of survival should cost. Brutal, but at least the quality-adjusted life year (QALY) calculation is a transparent way to determine how limited funds for patient care can be most effectively spent. Even in the United States, regulatory approval does not guarantee that real patients will have access to the drugs; some insurers are looking specifically at the costs of new approvals, and some implicit, rather than explicit, rationing is occurring. Companies are increasingly canvassing government healthcare bodies early on in the development cycle of new drugs to determine if they are prepared to pay the projected cost of the treatment, before further investment is made.

24.3.5 PERSONALIZED MEDICINE

For some, while this has been touted as the way of the future, there is little doubt that, where effective, it reduces the chance of adverse effects and maximizes the possibility of a positive outcome. For example, the breast cancer drug tamoxifen, **24.4**, is useful only for the treatment of estrogen-sensitive breast tumors, while the monoclonal antibody Herceptin® is used on patients with tumors that overexpress the HER2 protein.

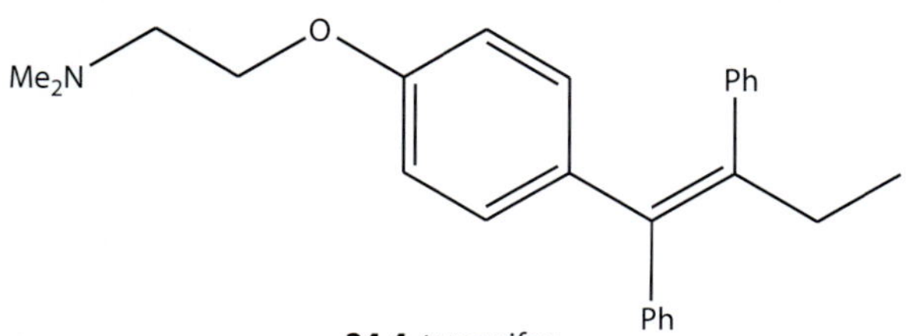

24.4, tamoxifen

Progress is still slow; although numerous tests for biomarkers exist, and appear to work, there is, as yet, less evidence that clinical outcomes generally improve with their use. In some cases, it is more difficult to develop bioreceptor tests. The drug gefitinib, **24.5**, is successful in treating only 10 % of patients with advanced nonsmall cell lung cancer, those with an overactive epidermal growth factor receptor; the FDA withdrew the drug for new patients in 2005, as clinical outcomes did not seem to be improved, but worldwide, biomarker tests do seem to be making progress.

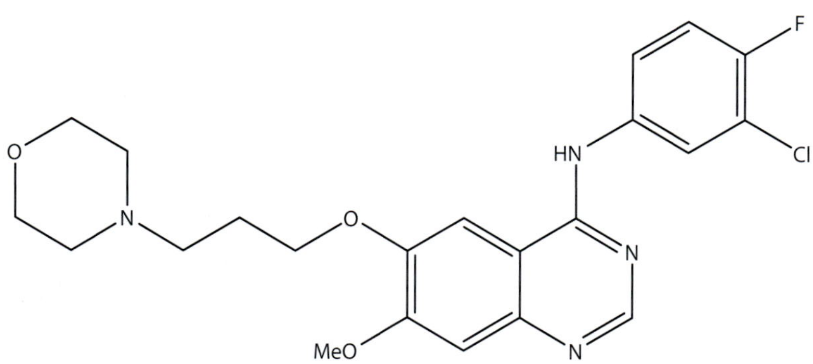

24.5, gefitinib

Patients deficient in activity of the enzyme CYP2C require lower doses of warfarin and more frequent monitoring than patients without the enzyme deficiency. Since 2007, the FDA has recommended genotyping for all patients prescribed with warfarin, but in April 2009, Medicare and Medicaid refused to pay for this; the test is not expensive but the affected population is huge.

24.3.6 THE FUTURE: PROBLEMS AND OPPORTUNITIES

Currently, drug patents are running out at a faster rate than new products are coming to market. Once a patent has expired, rival companies introduce generic versions of the medication, at much lower costs. Development of generics and obtaining approval for the formulation is considerably less expensive (around \$2.5 million *vs.* \$500 million–\$1 billion for the originator), and many large companies are expanding their own generics activities. Commercial decisions about what is likely to be reimbursed by insurers are becoming influential, and there is increasing pressure from both governments and society to deliver value for money. While pharmaceutical companies generally ride out recessions, research and development funding has been reduced, and some companies have axed entire areas of enquiry. For example, in 2011, Pfizer reduced their interests in respiratory medicine, allergy treatment, internal medicine, and antibacterials.

In the last decade, the increased availability of drugs through the Internet has led to a number of new problems. Many drugs can be supplied across borders that would not be permitted in the patient's home country, and appropriate clinical advice and monitoring is frequently inadequate. Moreover, quality control of such materials is limited; the most famous example is the extensive counterfeiting of Pfizer's Viagra®, a treatment for erectile dysfunction.

Sildenafil, marketed as Viagra

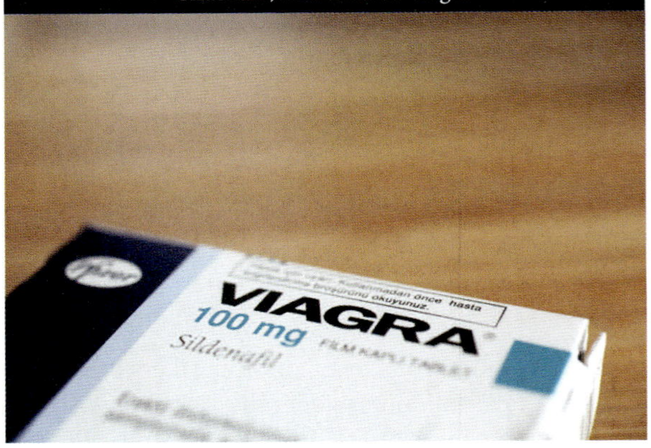

Two areas are of particular concern. The first is new antibiotics for the treatment of resistant "superbugs." MRSA* and VRE† are infectious agents generally acquired and transmitted in hospitals, but strains of *E. coli*, typhoid, tuberculosis, gonorrhea, and salmonella have also become resistant to most antibiotics. Between 1930 and 1968, there were 14 new classes of antibiotic developed. Since 1968, there have only been five new classes, and only two are in development. The development of resistance can probably be attributed to overprescription of antibiotics for minor ailments and overuse in hospitals, but it is now too late to correct this. The area is complex, and not very profitable, so many companies have turned their attention elsewhere.

There is also a dearth of new medications for neuropsychiatric disorders, currently 13 % of the total disease burden according to the World Health Organization (WHO). These include schizophrenia and Parkinson's and Alzheimer's diseases. Clinical trials in these areas take about 2 years longer than average, and review periods by regulatory bodies are 50 % above average. Success rates in clinical trials are lower than average, and failures tend to occur later in the drug pipeline

* Methicillin-resistant *Staphylococcus aureus*.

† Vancomycin-resistant *enterococcus*.

and hence are more costly. It is not always easy to tell whether a reduction in schizoid episodes or cognitive deterioration is a natural fluctuation or as a result of the drug. This is in sharp contrast to anticancer drugs, where the patient either survives or not, or antibacterials, where the infection is cured or it isn't. Moreover, long-term monitoring is essential, as many of these drugs will be taken for a lifetime.

Orphan drugs are defined in the United States as treatments for rare diseases with fewer than 200,000 sufferers in the United States. Many of these are genetic defects, including cystic fibrosis and homozygous familial hypercholesterolemia, and there are many for which there is still no available treatment. The FDA and the EMA* have slightly different definitions (the EMA extends the definition to some conditions common in the developing world but rare in Europe), but have agreed to cooperate in approving drugs for these conditions. Clinical trials must of necessity be much smaller, even in Phase III, because there may simply not be enough potential patients for a full trial. Companies are rewarded for the development of these drugs with an increased length of patent protection/exclusivity and various tax incentives; they are never likely to be profitable.

Another area in which it is difficult for pharmaceutical companies to recover drug development costs is in drugs for diseases of poor countries. If malaria were endemic in Western Europe, it's likely more money would be invested in the development of treatments and vaccinations. The development of drugs for diseases limited primarily to the Third World—malaria, river blindness, lymphatic filiarisis, *etc.*,—is generally only possible through the philanthropic programs of the pharmaceutical industry, since such countries are unable to pay the true cost of such treatments. Charitable foundations and intergovernmental agencies must also pick up the slack here. An area of major controversy has been the costs of supply of antiretrovirals for the treatment of HIV/AIDS in the developing world. In 2001, patented antiretroviral treatment cost some $10,000–$15,000 per annum. A company based in India offered to supply the same drugs for $350 per year, and South Africa's Medicines Act allowed for the use of these despite the patents. In 2001, 40 companies brought action against South Africa for this legislation but were persuaded by international opinion to drop the action. The Doha declaration of 2001 weakens some types of patent protection, requiring companies to license others to produce needed drugs more cheaply. Several patent-protected HIV/AIDS treatments are supplied at cost to sub-Saharan Africa as part of various philanthropic programs, but there is concern that such products are finding their way back into Western markets through corrupt activities.

Costs are rising, patent protection is weakening, regulatory regimes are getting tougher, research and development funding is being reduced, and companies are "derisking" their portfolios. Where is the next generation of developments to come from? This may be good news for research in academic institutions, where it is less expensive to do research, and a number of interesting partnerships between companies, funding bodies, and universities are developing. Rational design using sophisticated IT systems is likely to play a greater role in the future, as is personalized medicine. Future drugs will increasingly be monoclonal antibodies, though these have the disadvantage of needing infusion rather than being available orally. However, they are more difficult to copy, and the cost of bringing a biosimilar (generic monoclonal antibody or other biologic) to market is $100–200 million, much higher than for a small molecule. This also means that market and price erosion postpatent expiry is likely to be much slower, protecting revenue streams for the originator and making the products more appealing to develop.

24.4 AGROCHEMICALS

Rising population and decreasing availability of agricultural land are fast increasing the need for agrochemicals. The problem is compounded as people in the developing world adopt Western-style diets (average consumption of meat in China rose by 250 % between 1990 and 2010), which have higher land requirements. The World Bank estimates that food requirements will rise 50 % by 2030, and climate change is likely to exacerbate the situation. Production of biofuels is thought

* European Medicines Authority.

to have caused some of the recent food price increases, and the food *versus* fuel debate remains unresolved. Other than fertilizers (largely inorganic), the global agricultural agrochemical market was estimated at $38 billion in 2009. The world's top 10 agrochemical companies account for 89 % of the global market. Herbicides account for 46 % of this with about 25 % each for insecticides and fungicides. The remainder comprises specialty products such as plant hormones, including growth regulators (gibberellins, auxins, antigibberellins, and antiauxins). Research spending on agrochemicals has increased significantly in recent years and has become much more sophisticated—away from "spray and pray" and more like the work undertaken in the pharmaceutical industry, involving molecular biology, genomics, proteomics, and informatics. Although less money is available for research than in the pharmaceutical area, regulatory costs are increasing. New molecules must be active, potent, safe with respect to both human and environmental health, and convenient. The pipeline for the launch of a new product is now estimated at about 10 years. However, these are needed, since many older compounds are being legislated out, mainly for environmental reasons, or pests are developing resistance to them. The advantage to agrochemical companies in developing new compounds is the patent protection that is afforded, leading to higher added value, at least for a while.

24.4.1 HERBICIDES

Herbicides may be selective, as for broad-leaved weeds in cereal crops, or unselective, essentially for land clearance. 2,4-Dichlorophenoxyacetic acid (2,4-D), **24.6**, was commercialized in the 1940s; it is inexpensive, is easy to make, and kills broad-leaved weeds in cereal crops. Worldwide, it is the most widely used herbicide. It is a synthetic auxin, or plant hormone, acting only on dicots and not monocots. There is a whole family of related compounds, prepared from chloroacetic acid (or other halo acids) and various chlorinated phenols. Agent Orange, used as a defoliant in Vietnam in the 1970s, was a 1:1 mixture of 2,4-D and 2,4,5-trichlorophenoxyacetic acid (2,4,5-T). The main concerns about its use relate to the possibility that dioxins are formed as contaminants in its manufacture. A few countries ban its use for control of weeds in domestic lawns.

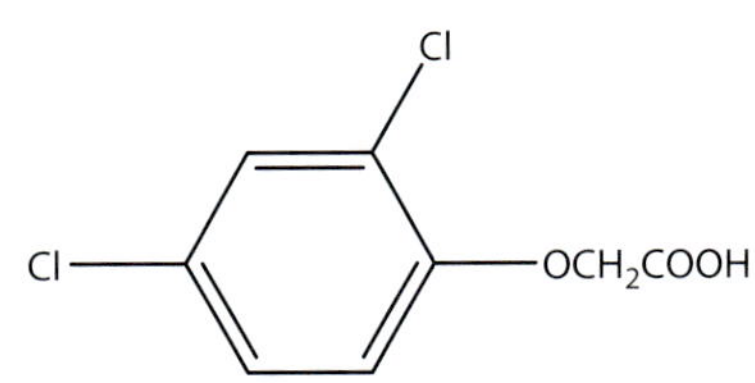

24.6, 2,4-D

The 1950s saw the development of triazine-based herbicides such as atrazine, **24.7**, prepared from amines and the cyanuric chloride, **24.8**. These compounds bind to the plastoquinone-binding protein in photosystem II. They are effective and inexpensive. They are largely banned in the EU (where relatively little corn is grown) as estrogen disruptors but are widely used for corn in the United States.

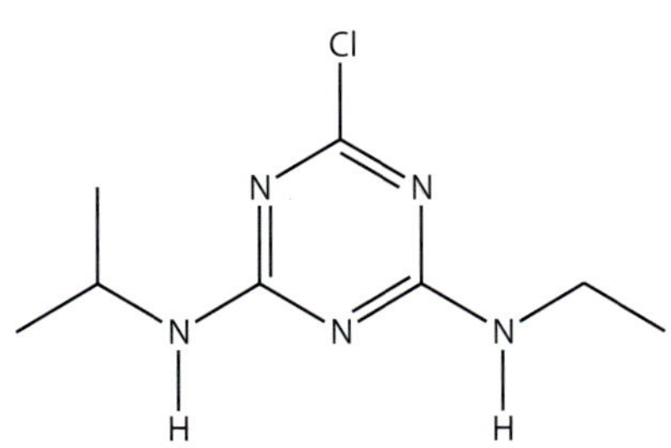

24.7, atrazine

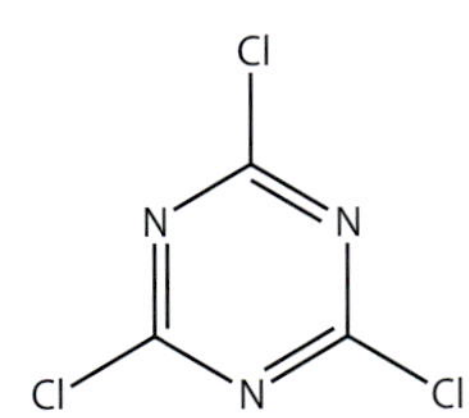

24.8

One of the mainstays of the herbicide market, allowed in all countries, is glyphosate, **24.9**, sold as Roundup® in the United States, where it is the most widely used herbicide. Its mode of action is to inhibit an enzyme involved in amino acid synthesis and it is thus active only for growing plants. It is a systemic herbicide killing grasses, broad-leaved weeds, and woody plants, but it is less environmentally damaging than many other herbicides as it is quickly broken down and does not bioaccumulate. It was off patent in 2000, and its current extensive use owes its origins to its relation with genetically modified crops, described as "Roundup Ready®." More than 85 % of the cotton, corn, soya, and canola grown in the United States are now genetically modified. The rest of the world has been less keen to adopt such crops, and they are largely banned in the EU, India, as well as some countries in Africa, and Latin America. Even if they were to become legally permitted, in Western Europe, at least, there is considerable consumer resistance to such foodstuffs. The arguments for and against genetically modified organisms (GMOs) have been widely rehearsed,* but whatever the conclusions, if consumers resist buying them, they will not succeed in those markets. Meanwhile, there is evidence that some weeds are acquiring resistance to glyphosate in the United States and Australia, and "2,4-D-ready" genetically modified strains are now being developed.

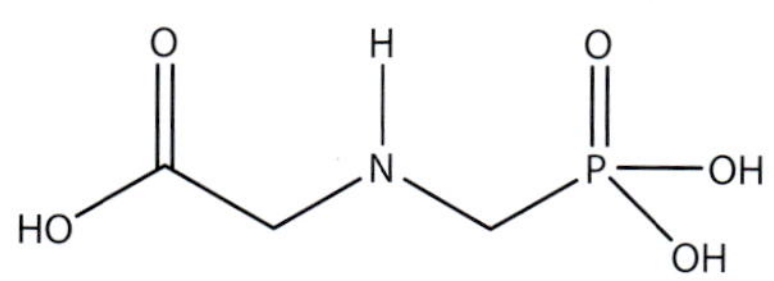

24.9, glyphosate

New herbicides continue to be introduced. Resolve Q® is an herbicide for use on corn, which contains rimsulfuron, **24.10**; thifensulfuron, **24.11**; and isoxadifen ethyl, **24.12**. **24.10** and **24.11** are part of a new family of environmentally compatible sulfonyl urea herbicides, which target the acetolactate synthase enzyme. Alachlor, **24.13**, is used for grasses and broad-leaved weeds in corn, soya, and peanuts but is banned in the EU because of its toxicity. Mesotrione, **24.14**, based on a substance isolated from the Californian bottlebrush plant, is more generally accepted and can be applied both pre- and postemergence; it inhibits the plant enzyme 4-hydroxyphenylpyruvate dioxygenase.

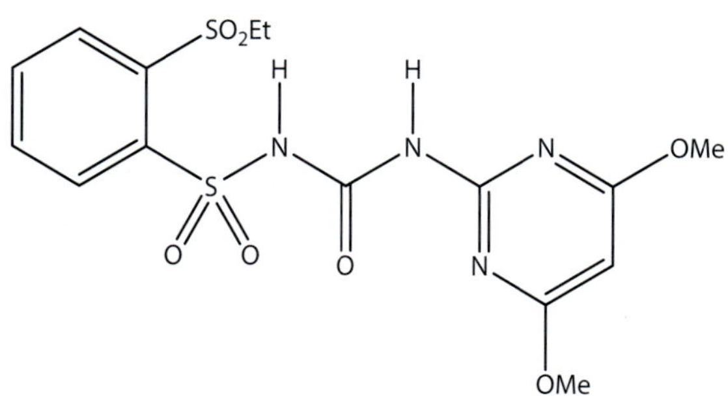

24.10, rimsulfuron

* S. Key, J.K.-C. Ma, and P.M.W. Drake, *J. R. Soc. Med.*, 101 (2008) 290–298; www.organicconsumers.org/; www.trufoodnow.org/.

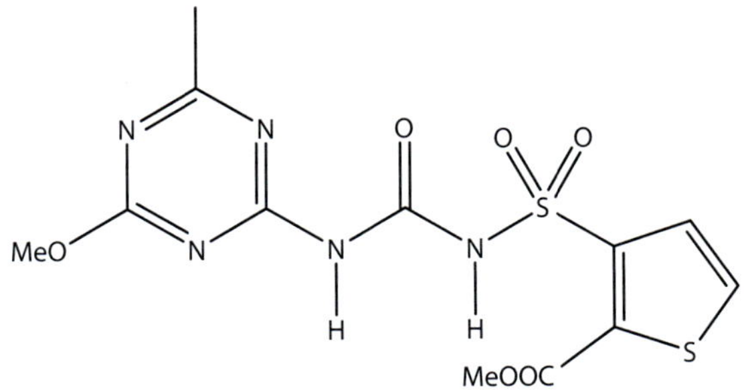

24.11, thifensulfuron methyl

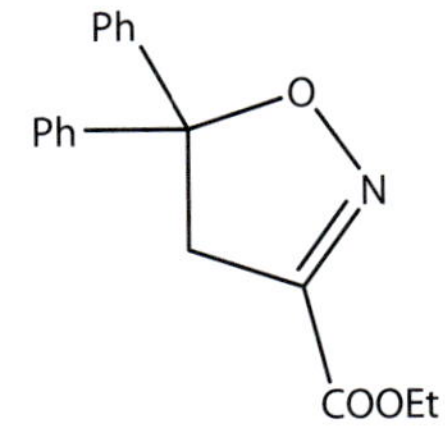

24.12, isoxadifen ethyl

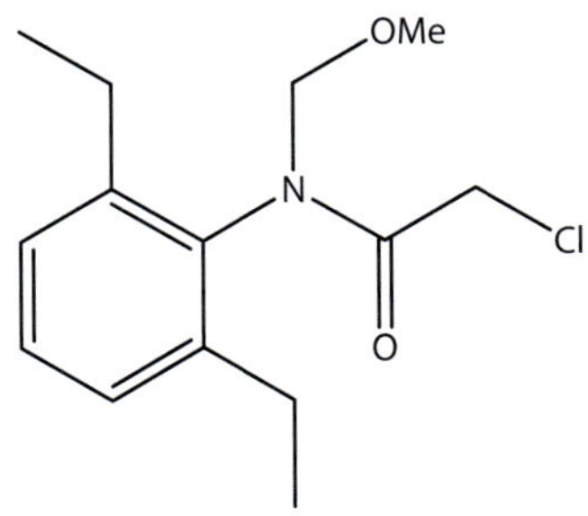

24.13, alachlor

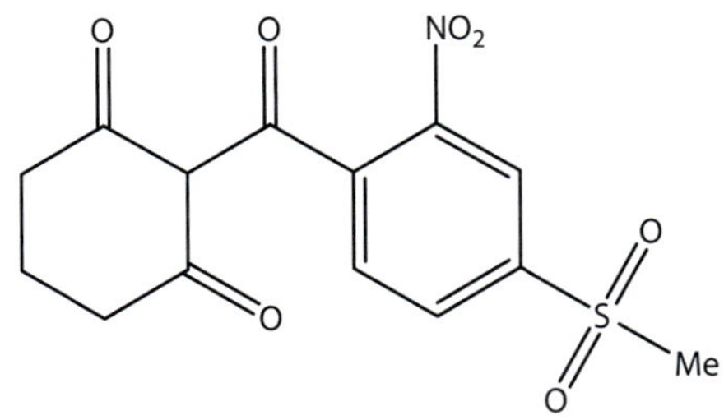

24.14, mesotrione

Focus on the REACH Regulations*

REACH stands for Registration, Evaluation, Authorization and Restriction of Chemicals, and the program applies to all substances manufactured in, or imported into, the EU in greater than 1 ton quantities. It is being implemented in stages between 2007 and 2018. While this does not only apply to agrochemicals (the dyestuffs and food additives industry have also been widely affected, and some 143,000 substances were registered as candidates for approval by the December 2008 deadline), the European Crop Protection Agency mandates reevaluation of all pesticides. Moreover, all existing animal testing data must be shared with other producers. While the aim of reduced animal testing may be laudable, many companies are unhappy at sharing expensively obtained data with rivals. In the agrochemical area, REACH required retesting of more than 800 substances and about 75 % of these were banned in 2008. Of the remaining 268, in early 2011, 210 were still awaiting approval. Many substances have also been restricted to agricultural use and are no longer available for use in domestic gardens.

* http://www.hse.gov.uk/reach/; http://ec.europa.eu/environment/chemicals/reach/reach_intro.htm.

24.4.2 FUNGICIDES

Untreated fungal infections represent at least as serious a threat to crop productivity as insect damage. Almost all types of crop can be infected by fungi; it has been estimated that untreated potato blight may cost up to 95 % of the crop and grape yields can be reduced by 90 % by fungus. Sulfur and Bordeaux mixture (copper sulfate and calcium oxide) are both fungicidal and regarded as "organic" pesticides, despite some concerns about copper buildup in the soil. Although some tin compounds are still used as marine antifoulants, fungicides containing mercury or cadmium have been banned. Most modern fungicides have low mammalian toxicity, and the adverse effects are mainly as allergens and skin irritants.

A number of "families" of fungicides can be identified. Aromatics such as hexachlorobenzene are largely being phased out, as are thiophthalimides such as captan, **24.15** (these latter were mainly used on stored produce). Other groups include triazoles, such as difenoconazole, **24.16** (used for scab on fruit and vegetables and for roses and other ornamental plants), carbamates (*e.g.*, **24.17**, benomyl), thiocarbamates (*e.g.*, Ziram®, $Me_2NCS_2^-$ Zn^{2+} used on apples, peaches, stone fruits, and grapes), and dithiocarbamates (*e.g.*, Propineb, **24.18**). The story is, however, a familiar one; some older compounds have been banned because of their environmental effects, but, more importantly with fungicides, because of the development of resistance. For example, the use of metalaxyl, **24.19**, against potato blight in the 1980s had to be abandoned after a single growing season because of the development of resistance. Fungicides that target multiple sites such as cell membrane permeability or phospholipid biosynthesis are less prone to the development of resistance than those that attack a single site such as RNA polymerase I.

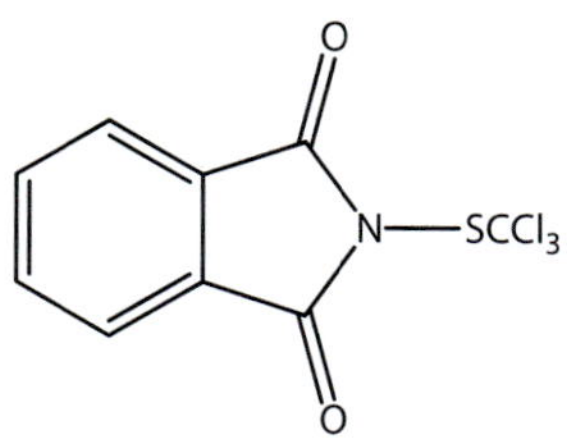

24.15, captan

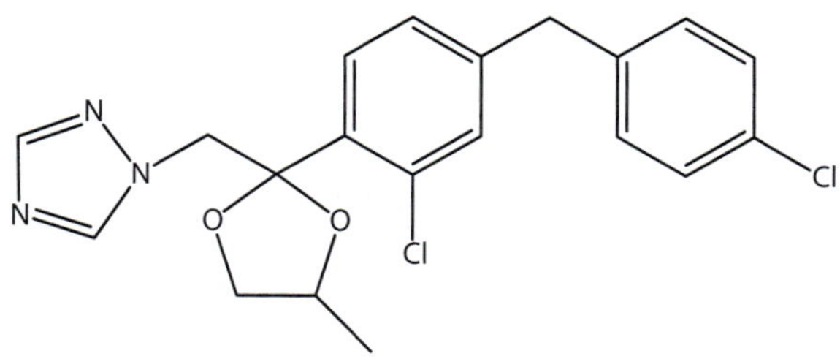

24.16, difenconazole

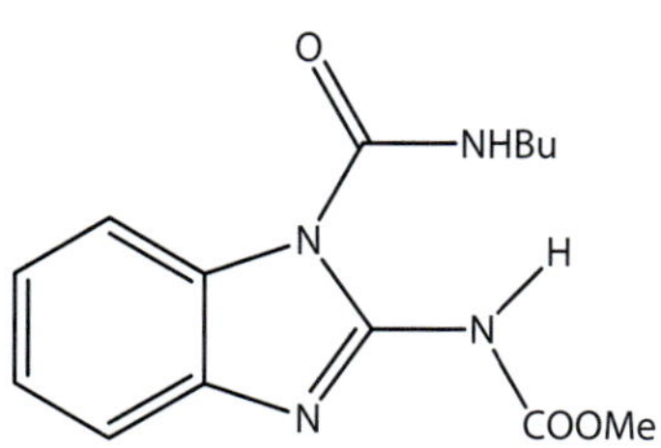

24.17, benomyl

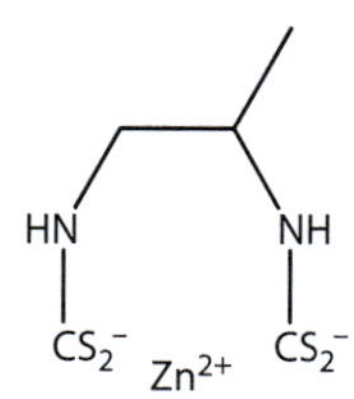

24.18, propineb

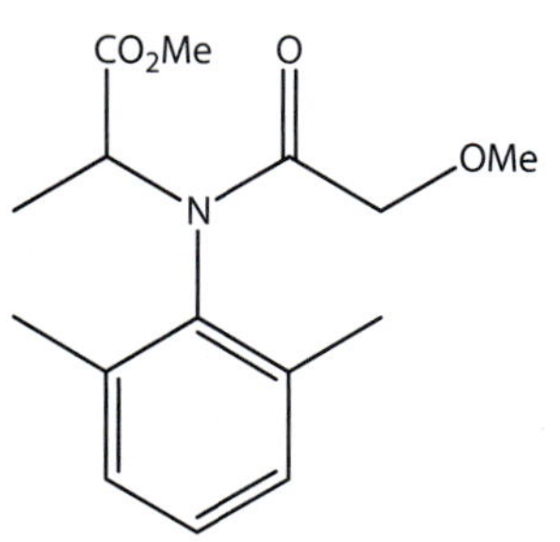

24.19, metalaxyl

New fungicides are in development; one interesting group, the strobilurins, is based on materials isolated from a wood-rotting fungus, *Strobilurus tenacellus*. It destroys the opposition ... F500 contains boscalid, **24.20**, and dimoxystrobin, **24.21**. The first strobilurin to be marketed was azoxystrobin, **24.22**, in the late 1990s, by Zeneca Agrochemicals. A key intermediate in the synthesis, 4,6-dichloropyrimidine, had to be custom made as it was not available in commercial quantities at the time. Over time, fungicides have become increasingly targeted, with a significantly decreased quantity per unit area required for effective use.

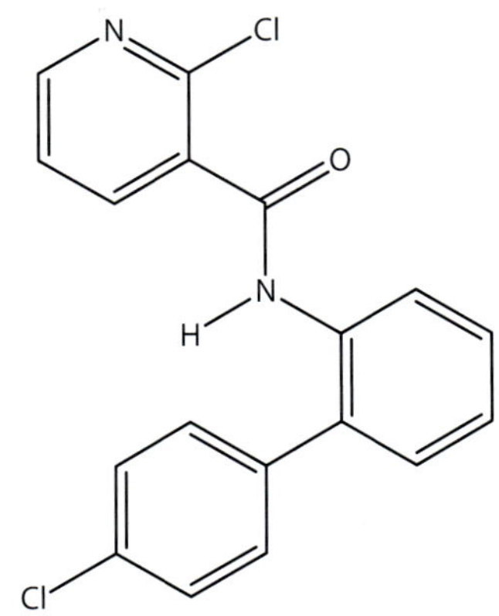

24.20, boscalid

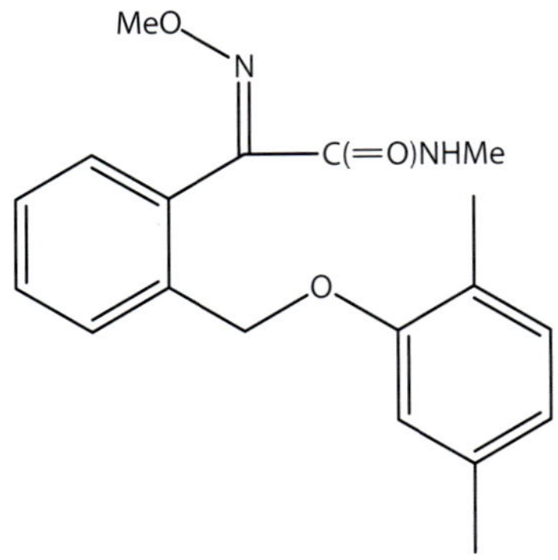

24.21, dimoxystrobin

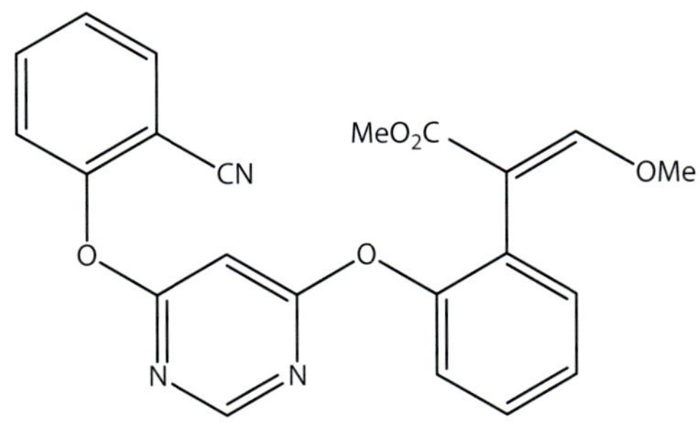

24.22, azoxystrobin

Focus on Potato Blight

Late blight on potatoes, *Phytophthora infestans* (Figure 24.1), is strictly an oomycete infection rather than a fungus, but it responds mainly to fungicidal treatments. These infections caused major famines in the nineteenth century—the 1840s in continental Europe, the Irish potato famine of 1845, and the Highland famines of 1846. The most recent major outbreak was in Estonia in 2010; the genotype was determined as similar to recent Russian and Nordic samples, but some mutations had taken place. Other *Solanaceae* species, such as tomatoes, may also be infected. Potato blight is still extremely difficult to treat and causes losses of $6 billion per annum.

Management of blight is now a very high-tech process. The first consideration is careful monitoring of weather conditions; high humidity and temperatures above 10 °C (50 °F) promote blight development, and growers are advised to spray preventatively after several days of such conditions. Bordeaux mixture, zineb ($[^-S_2CNHCH_2CH_2NHCS_2^-]Zn^{2+}$), and mancozeb are used for this purpose (Figure 24.2), but the use of mancozeb is now restricted and Bordeaux mixture is banned in the EU after 2013. As previously noted, resistance to metalaxyl was developed very rapidly. But this and the related furalaxyl are still used as a "curative" in combination with cymoxanil. The year 2009 saw the introduction of Nando®, fluazinam, to which it is hoped the infection will not develop resistance due to its multisite activity.

FIGURE 24.1 Blighted potato.

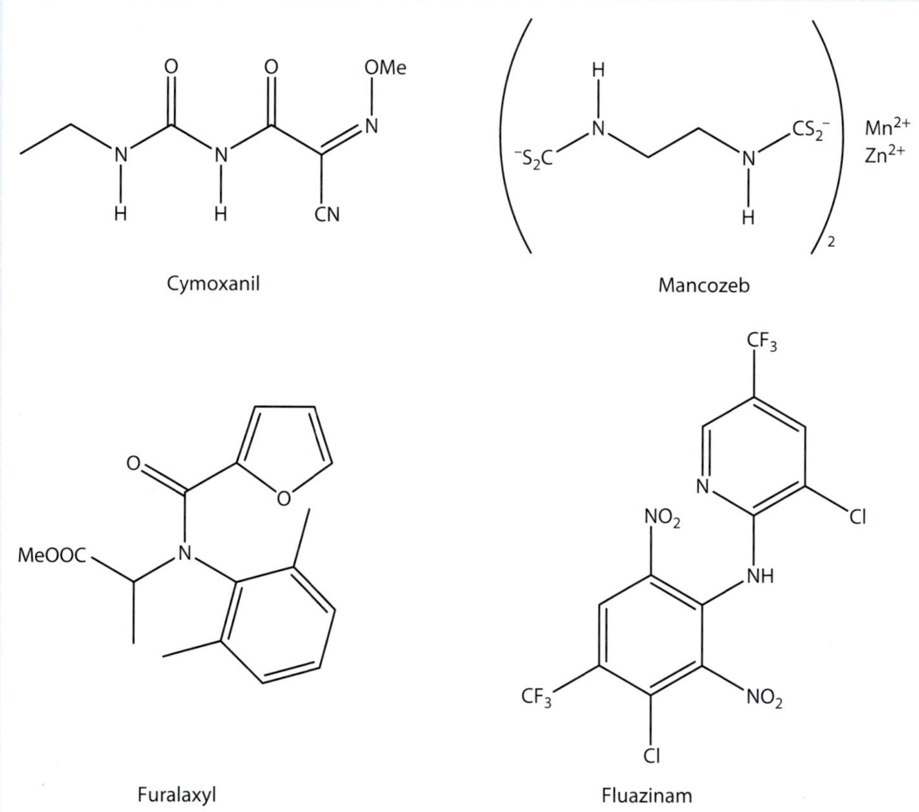

FIGURE 24.2 Fungicides used against potato blight.

Other than Bordeaux mixture, none of these compounds is generally available to the domestic gardener. Much attention has been focused on the development of blight-resistant strains of potato, some generated by selective breeding, but also a range in which a gene from a resistant wild potato relative, *Solanum bulbocastanum*, has been incorporated.

24.4.3 INSECTICIDES

Although not the largest part of the agrochemical sector, insecticides have been much the most controversial in terms of their environmental effects and hence will be discussed more fully in the next chapter. However, the history is similar to other areas; early successful compounds were banned on environmental grounds or abandoned due to the development of resistance, followed by second- and third-generation alternatives, generally based

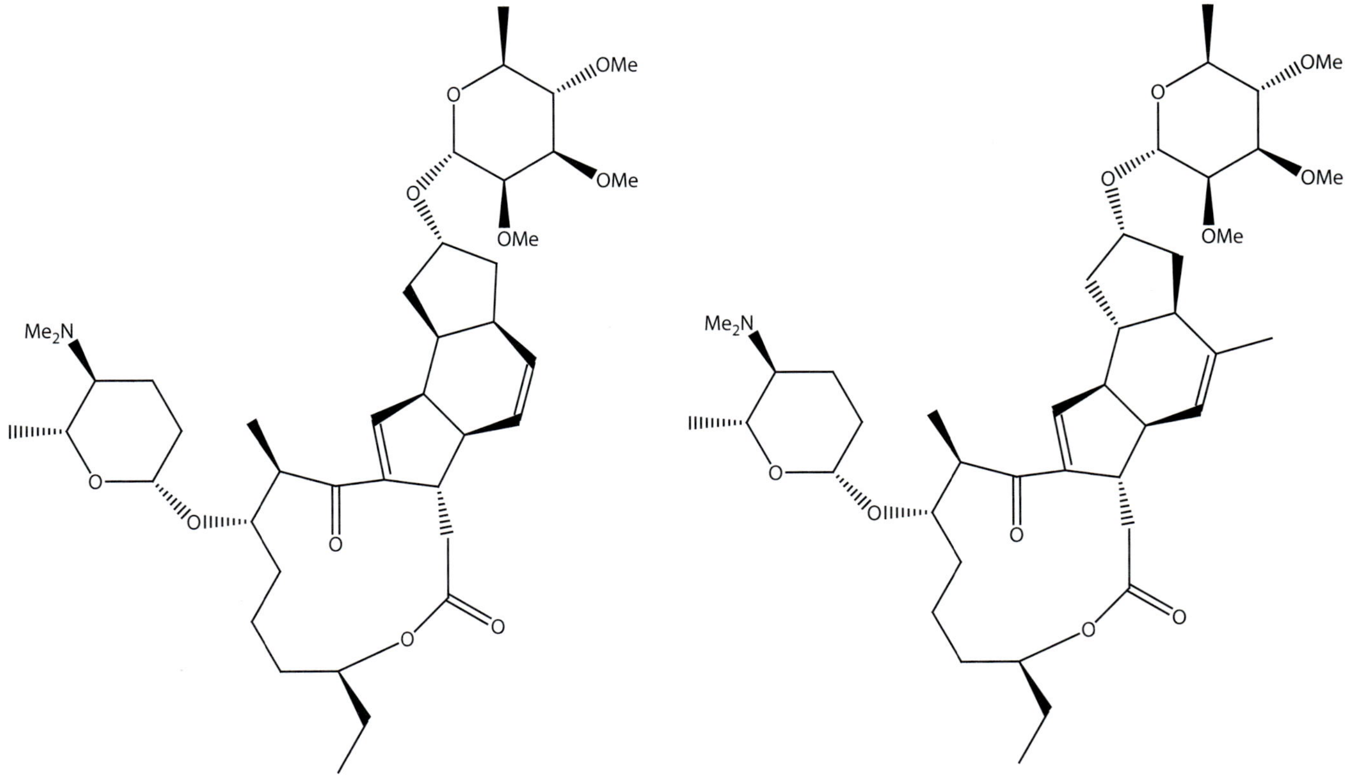

Spinosyn A

Spinosyn D

FIGURE 24.3 Spinosyn structures.

on more detailed biochemical research. Other approaches have included trapping using pheromones (Bag-A-Bug® traps) and control using natural predator insects. An interesting recent development is Entrust Naturalyte®, used against head lice, dog fleas, Colorado potato beetle, fire ants, spider mites, various moths, slugs, and western flower thrip. This is regarded as "organic" because the active ingredients are spinosads, originally isolated from a rare soil dwelling organism, *Saccharopolyspora spinosa*. The discovery was serendipitous; a scientist on holiday collected some soil samples from an abandoned rum distillery in the Caribbean—and it has not been found in nature since. Spinosyn A and D (Figure 24.3) are used in a 5:1 ratio.

Another new class of compounds are the ketonoles, such as spiromesifen, **24.23**, which are insecticides and miticides; resistance is not yet reported, but ...

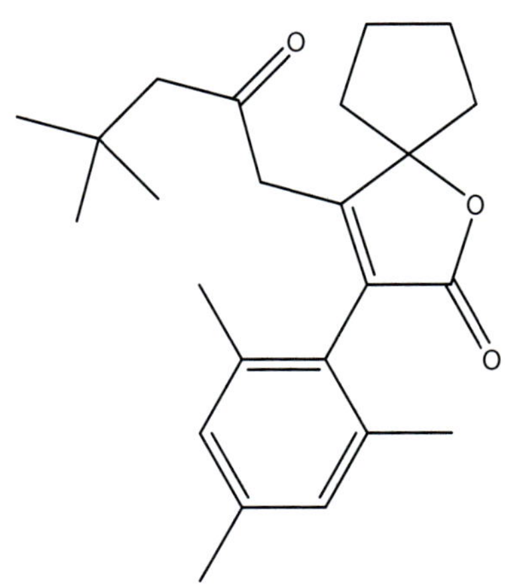

24.23, spiromesifen

24.5 DYES AND COLORANTS

24.5.1 TEXTILE DYES

The history of the dyestuff industry exemplifies the history of the modern chemical industry; along with pharmaceuticals, its development began in the mid-nineteenth century, with serendipitous discoveries, disputes over intellectual property, and protectionism. Until the industrial revolution, all dyes were of natural origin, mostly vegetable, but a few of animal origin (*e.g.*, cochineal). The use of iron oxides in art dates to the Paleolithic era, and many pigments of ancient, medieval, and modern art were of mineral origin, such as ultramarine from lapis lazuli, vermilion from mercury compounds, white lead, and chrome yellow. New dyes arrived in Europe from the Americas in the eighteenth century, including cochineal from Peru and quercetin, a yellow color from the inner bark of the American black oak. These were of course used without knowledge of the molecules that gave them their color, and improvements in technology were generally directed to making dyes on garments more colorfast; the use of alum as a mordant was developed in the thirteenth and fourteenth centuries. Mordants work well for the proteinaceous fibers of wool and silk, but are less successful for cotton. Some examples of natural dyes are shown in Figure 24.4.

Dyed fabrics at a market in Mali.
(Thanks to Dr. Steve Kamin for this photo.)

Aniline was isolated from coal tar in the early nineteenth century; its name comes from ***anil***, Arabic for indigo. Synthetic aniline was prepared in 1845 from benzene.* In the 1850s, Perkin reacted aniline (apparently containing various amounts of ring-methylated anilines) with potassium dichromate in sulfuric acid to give mauveine, which was patented in 1856. In France, Verguin generated fuchsine, or brilliant red, another aniline dye using tin(IV) chloride as oxidant, though arsenic acid was later found to be a better oxidant. Alkylation of aniline red gave a violet dye, known as Hofmann's violet. These dyes could be made soluble in water by sulfonation. With hindsight, little was known about the structures of what was produced, and many were probably mixtures; indeed, a full structural characterization of mauveine was not achieved until the 1990s.† Some of the classic structures are shown in Figure 24.5.

* Benzene was nitrated then the nitrobenzene reduced to aniline.

† O. Meth-Cohn and M. Smith, *J. Chem. Soc., Perkin Trans. I* (1994), 5–7.

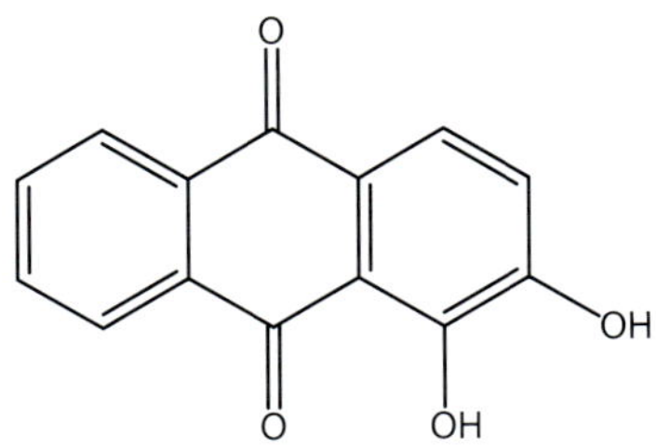

Alizarin, red, *Turkey red,* Mordant red 11®
isolated from the madder plant, name from
al-usara, juice, arabic

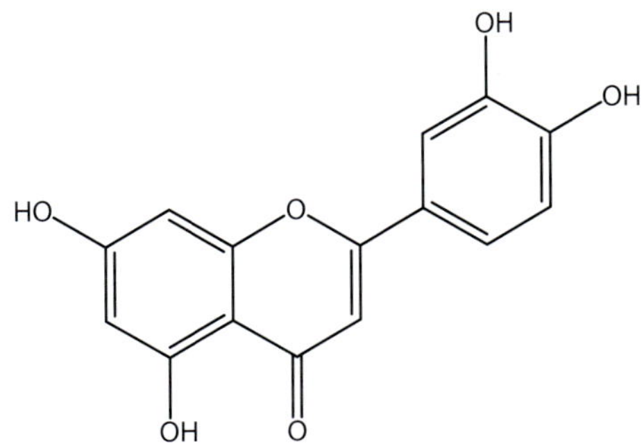

Luteolin, bright yellow, from *Reseda luteola*, Dyer's Rocket, yellow weed, weld

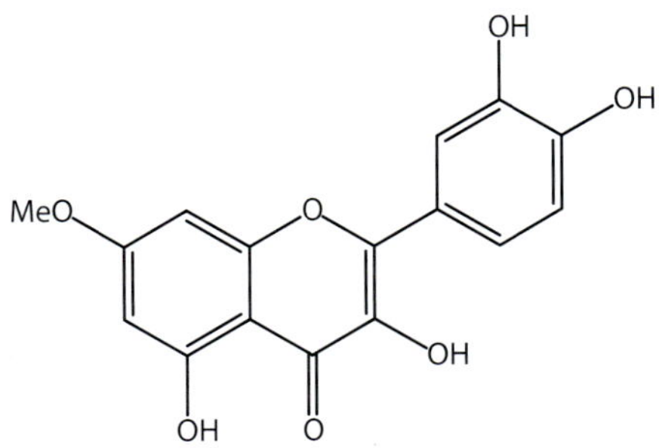

Rhamnetin, yellow, from "Persian Berry," fruit of the Avignon buckthorn, *Rhamnus saxatalis*

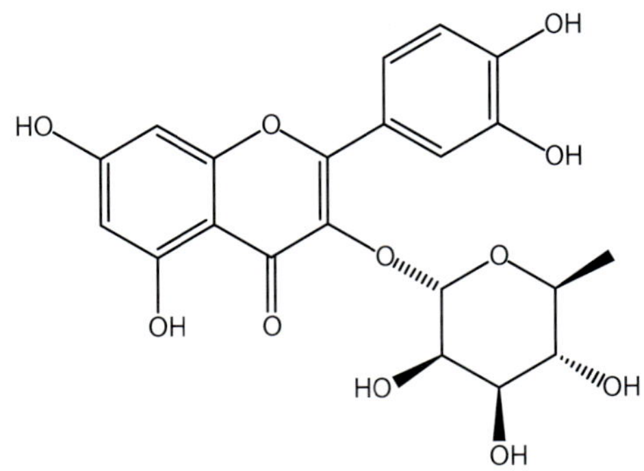

Quercitrin, yellow, from the American black oak

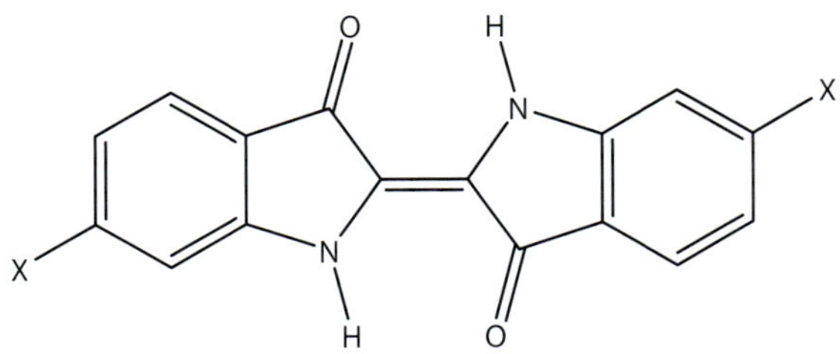

X = H, Indigo, dark blue, from *Indigoferra tinctoria* or woad, *Isatis tinctoria*
X = Br, Tyrian purple, from *Murex brandaris*

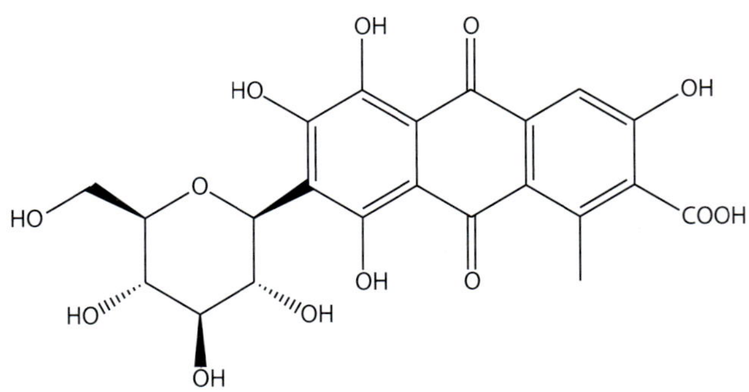

Carminic acid; Aluminum and calcium salts give red/purple dyes called carmine derived from the cochineal insect

FIGURE 24.4 Dyes from natural sources.

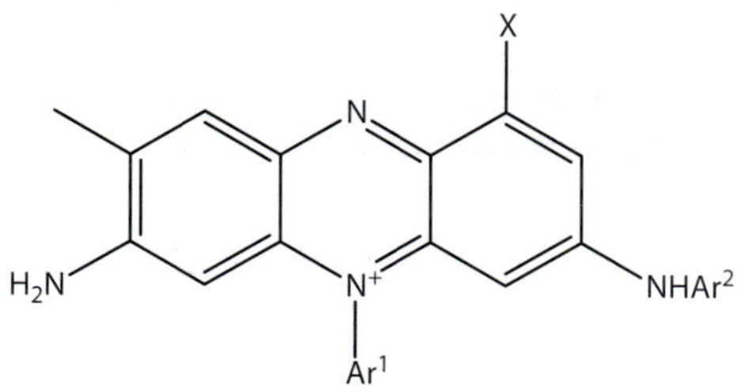

Mauveine A; X = H, Ar[1] = Ph, Ar[2] = C_6H_4-4-Me
Mauveine B; X = Me, Ar[1] = Ph, Ar[2] = C_6H_4-4-Me
Mauveine B2; X = H, Ar[1] = C_6H_4-4-Me, Ar[2] = C_6H_4-4-Me
Mauveine C; X = Me, Ar[1] = C_6H_4-4-Me, Ar[2] = C_6H_4-4-Me

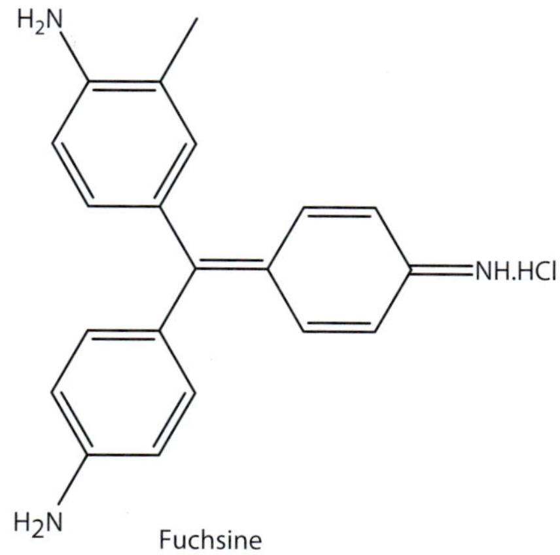

Fuchsine

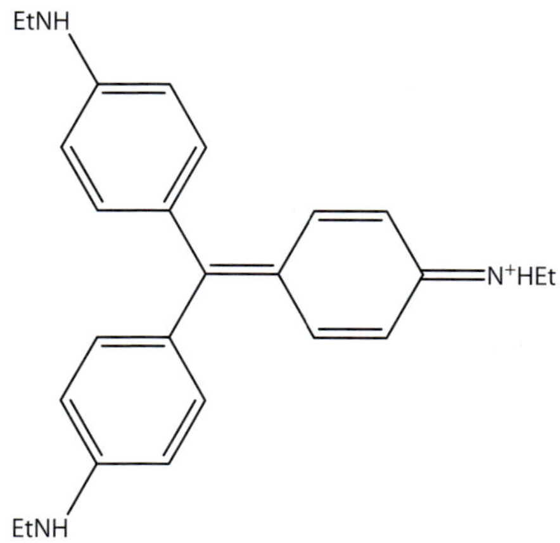

Hoffmann's violet

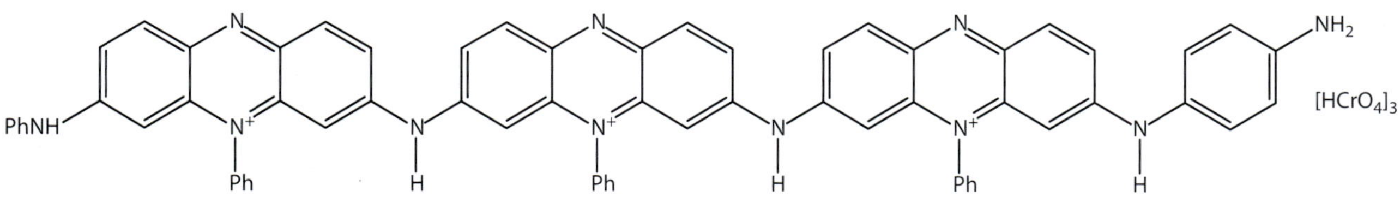

Aniline black

FIGURE 24.5 Structures of some of the early aniline dyes.

By the 1870s, patents were in dispute and the new companies Badische Anilin und Soda Fabrik (BASF) and Bayer became involved in the industry. Synthetic alizarin and other anthraquinone dyes appeared in the 1870s. German academia took a lead in the structural elucidation of the dyes, and phthalein (from phthalic acid and substituted phenols, *e.g.*, **24.24**, a bluish purple dye from 3-methylphenol) and triaryl methane dyes (such as **24.25**, malachite green) emerged. The azo coupling reaction (see Section 13.3) greatly increased the number of variations of dyes and colors that could be produced, and this was systematically investigated. Only a few good dyes emerged, but these were the blockbusters of their time (*e.g.*, **24.26**, aniline yellow and, **24.27**, para red). During the late nineteenth century, theories explaining the colors of molecules started to be developed. Synthetic indigo was produced by BASF in 1897 and, by 1900, had displaced Indian indigo. The industry was dominated by German companies until the outbreak of the First World War, but by 1914, the United States had moved from dye importation to a thriving domestic industry.

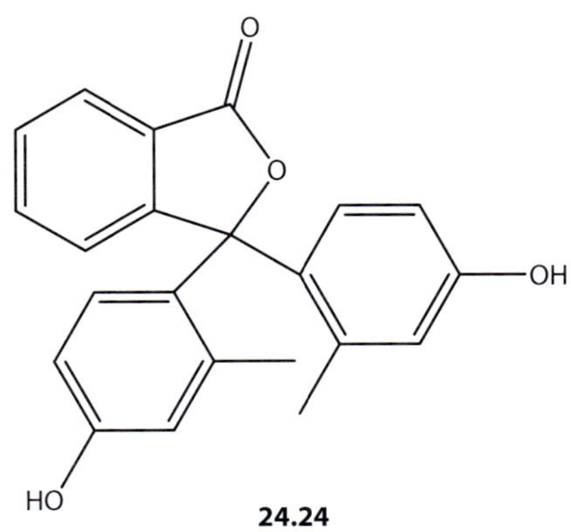

24.24

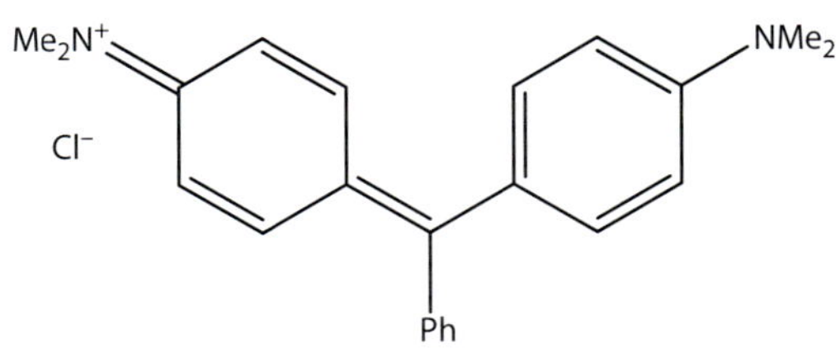

24.25

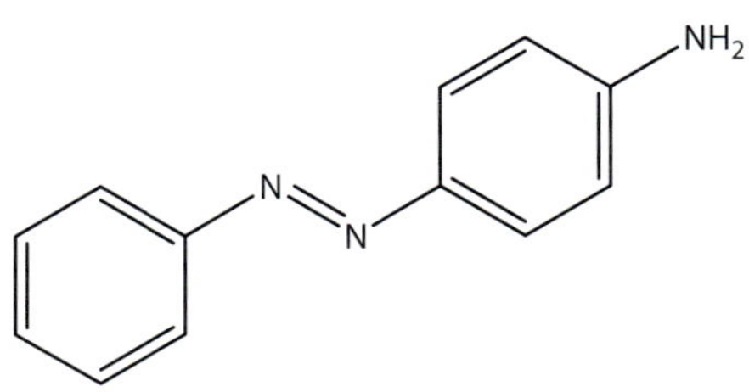

24.26, aniline yellow

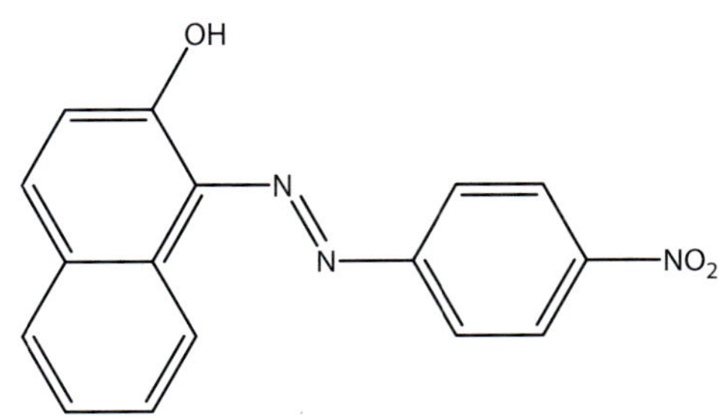

24.27, para-red

The interwar years saw the development of the first dyes for artificial fibers and the founding of ICI in the United Kingdom and IG Farben in Germany. Monastral fast blue, also known as Heliogen blue B (**24.28**), the first of the phthalocyanine dyes, dates from this period. Protectionism and patent disputes continued. In the 1940s and 1950s, more dyes for the new artificial fibers, rayon and nylon, were developed, and countries such as China, India, Korea, Taiwan, Pakistan, Thailand, Mexico, and Indonesia entered the market. The year 1956 saw the development of the first reactive dye, which does not depend on classical techniques to be colorfast, because it becomes chemically bonded to the fiber. These are still marketed in the Procion range, for example, Procion red MX-5B®, **24.29**. Many of these depend on the nucleophilic aromatic substitution reaction of the triazines with an OH group on the fabric and are particularly successful for the fast dyeing of cotton, which is rich in OH groups. Many other companies developed similar products from the 1950s to the 1980s, with improvements such as bifunctional compounds to ensure greater color stability. All of these are now long out of patent and are made worldwide. The production of dyestuff intermediates moved almost entirely out of Europe and the United States to Asia during the 1980s and 1990s leaving many traditional manufacturers with empty reactors. While many simply closed down, others went on to become the custom chemical powerhouses that now supply the specialist needs of both the pharmaceutical and agrochemical industries in Western economies today.

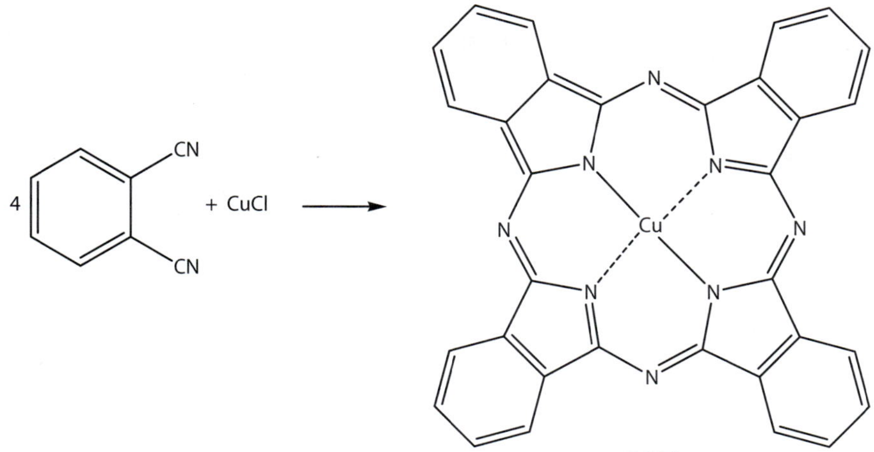

24.28

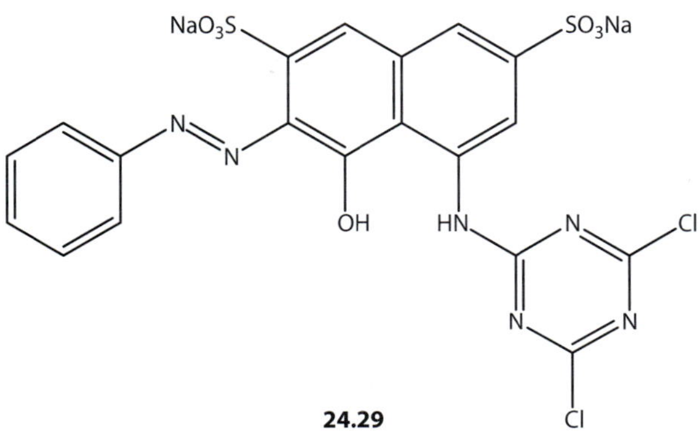

24.29

Yarn dyed with Procion red

PROBLEM 24.1

Write a mechanism for the reactive dyeing of cellulose (designate it as ROH) with **24.30** (X is the chromophore, the part of the molecule that gives rise to the color):

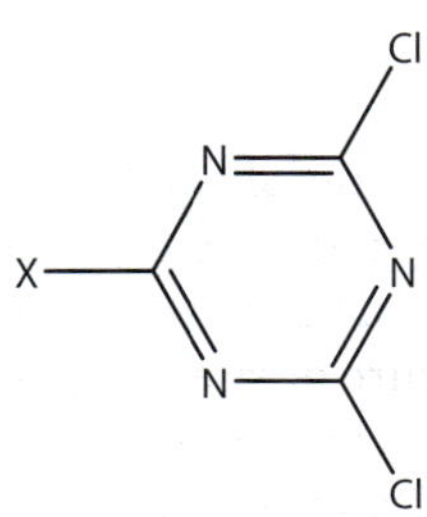

24.30

PROBLEM 24.2

Another class of reactive dyes is vinyl sulfones such as **24.31**. Why do these react readily with alkoxide anions?

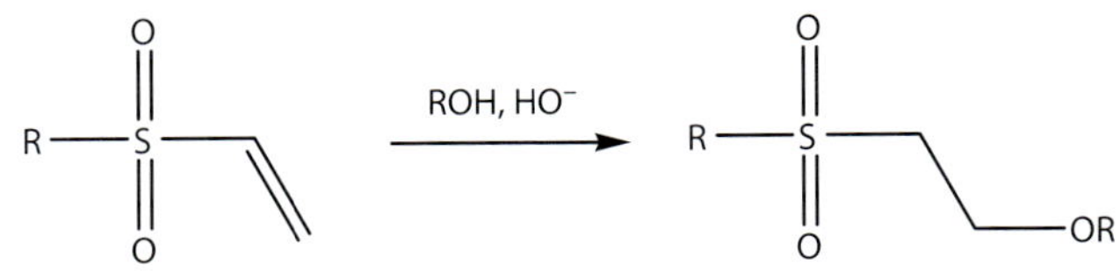

24.31

Solutions

24.1 The substitution is possible because the negative charge in the intermediate is located on any of the three electronegative nitrogen atoms:

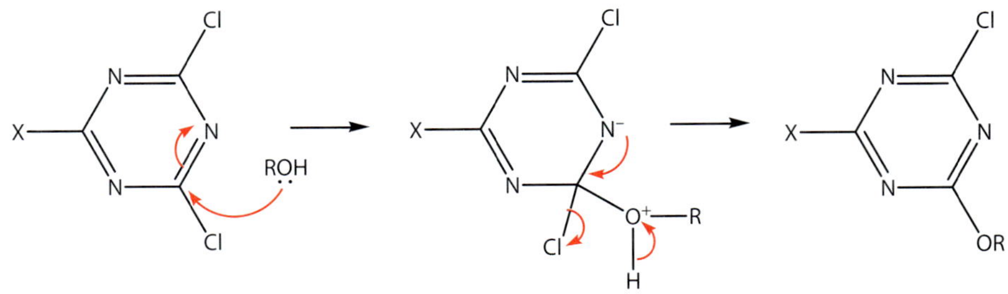

24.2 The sulfur oxygen double bonds of the sulfone are able to accept electron density in the same way that a carbonyl group would:

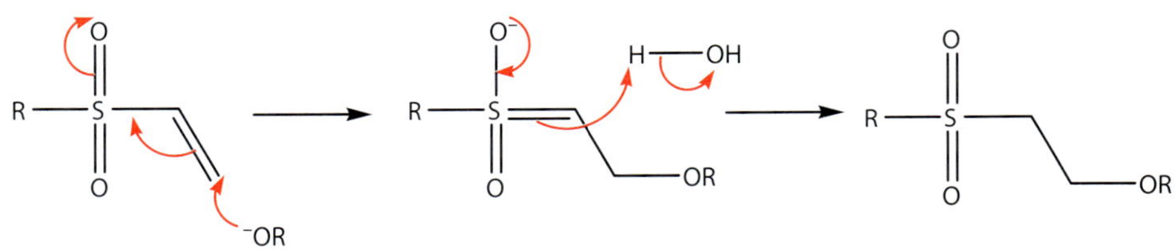

More recent developments have included disperse dyes, designed initially for the dyeing of polyester fibers; these materials do not have functional groups that allow for simple chemical bonding nor can they be dyed in aqueous solutions. These dyes are very finely ground materials, having a large surface area to promote dissolution under the conditions of high temperature and pressure used in the dyeing process. Although a wide range of chemical structural families may be used, for red and blue dyes, the most common are either azo dyes or amino derivatives of anthraquinone. Structures for yellow dyes are more varied (Figure 24.6); yellow is a relatively easy color to achieve—in the nineteenth century, 2,4,6-trinitrophenol, picric acid, was widely used despite its explosive potential. Many materials are sold under a wide variety of trade names and a number are suspected to be allergenic or carcinogenic.

Fabric dyed with Disperse blue 60

The market is now dominated by reactive and dispersive dyes, and the main centers of production are in Asia. World capacity exceeded supply by 2000 and overcapacity and falling margins continue to be problematic for the industry. Many azo dyes either have been or are likely to be banned in Europe.

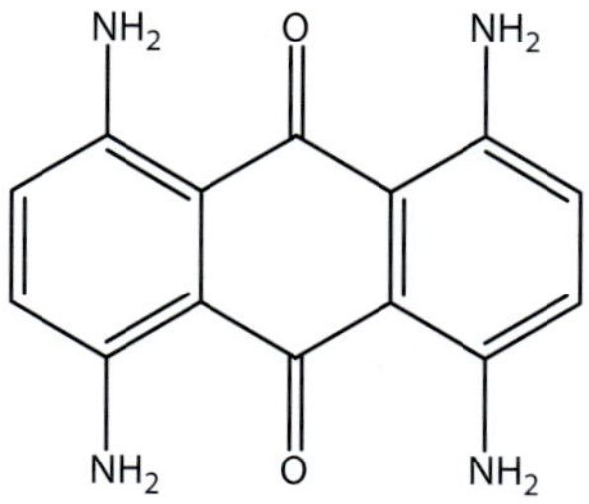

Disperse blue 1, also used to color hair and plastics

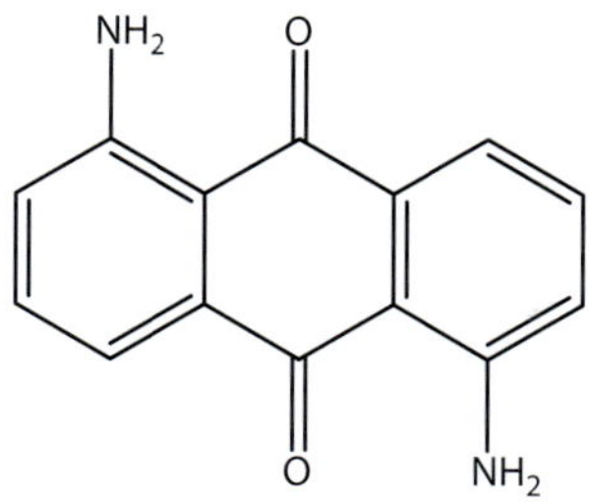

Disperse red 11

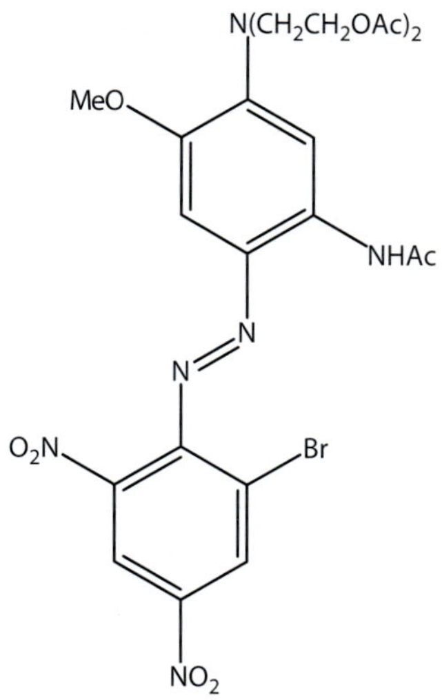

Disperse blue 79

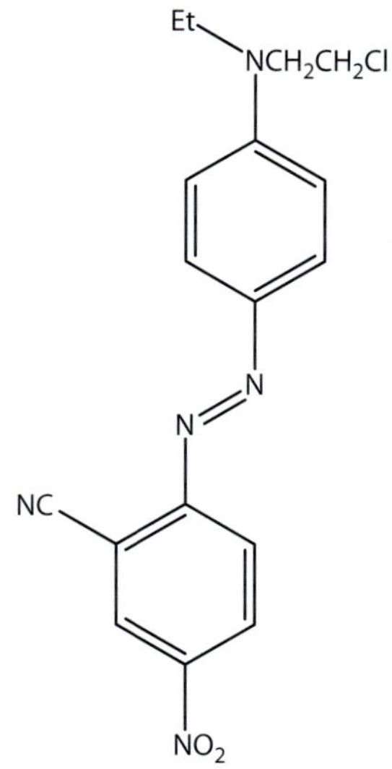

Disperse red 73

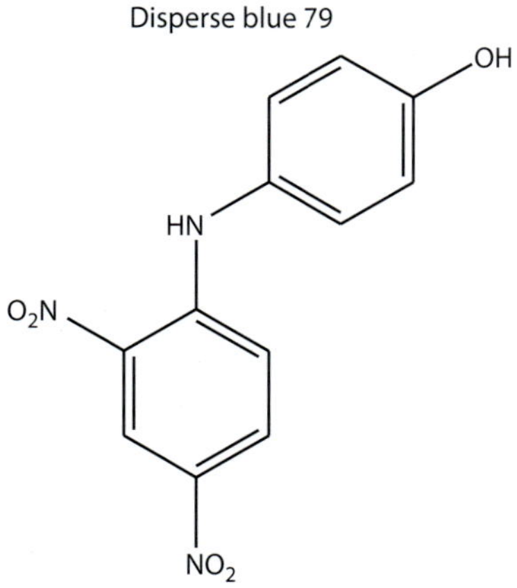

Disperse yellow 1

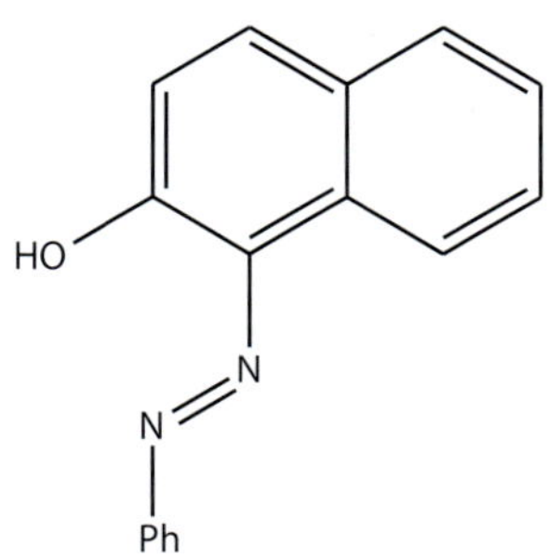

Sudan 1, Disperse yellow 97

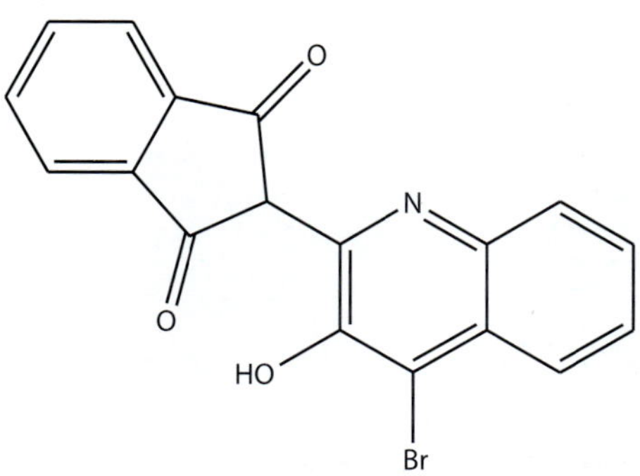

Disperse yellow 64

FIGURE 24.6 Structures of some typical disperse dyes.

24.5.2 FOOD DYES

Food colorants are a niche market within the dyestuff industry and properly and inevitably are subject to much greater regulation than dyes intended for fabrics, plastics, or paper products. Food colorants may be used to simulate what a consumer regards as a natural color. Thus, glacé cherries are colored bright red, to simulate cherry skins, although naturally, they would be beige. Margarine is colored yellow, so that its appearance is similar to that of butter. Dyes may also be used to enhance natural colors, mask natural variation in color, or simply for decoration, as in cake frosting.

Various natural dyes exist; the use of most of these is unrestricted in most countries. Examples include caramel (brown), annatto (red-orange dye from achiote seeds, active ingredient is bixin, **24.32**), chlorophyll (green from *Chlorella* algae), cochineal (red), betanin (red/purple from beets, **24.33**—various tautomers and zwitterions are possible), saffron (yellow; the dye molecule is crocin, **24.34**), turmeric (yellow, active dye is curcumin, **24.35**), paprika (red, zeaxanthin, **24.36**), and elderberry juice. Several of these compounds have been shown to have active benefits for human health. Such compounds are permitted as food dyes even when they are manufactured and synthetic; carotene does not need to be extracted from carrots to be allowed.

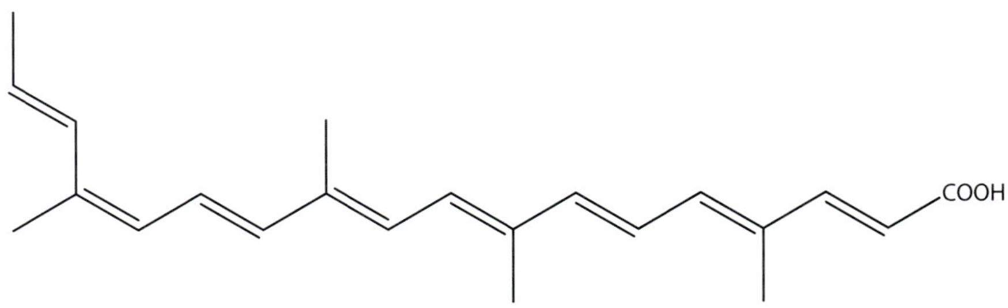

24.32, bixin

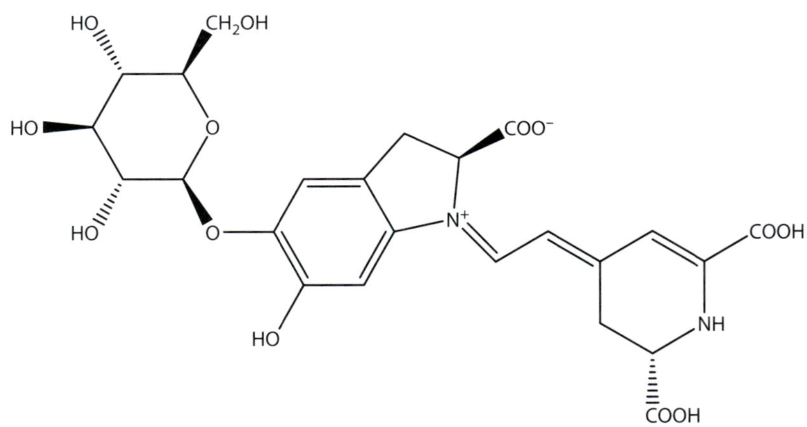

24.33, betanin

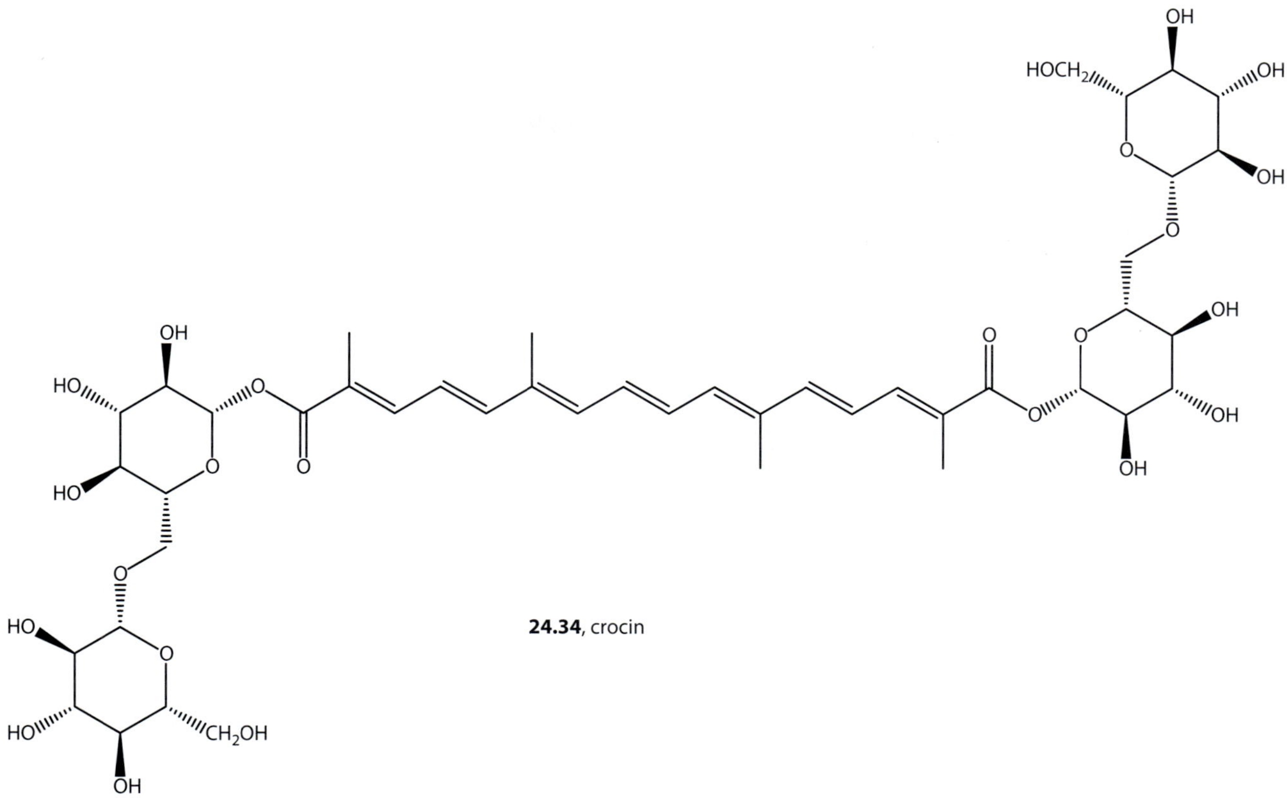

24.34, crocin

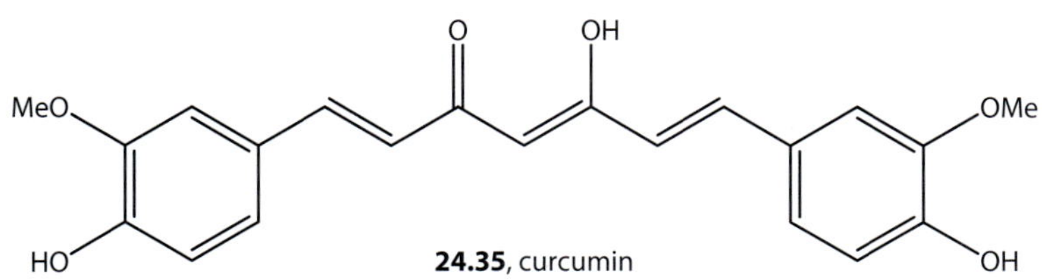

24.35, curcumin

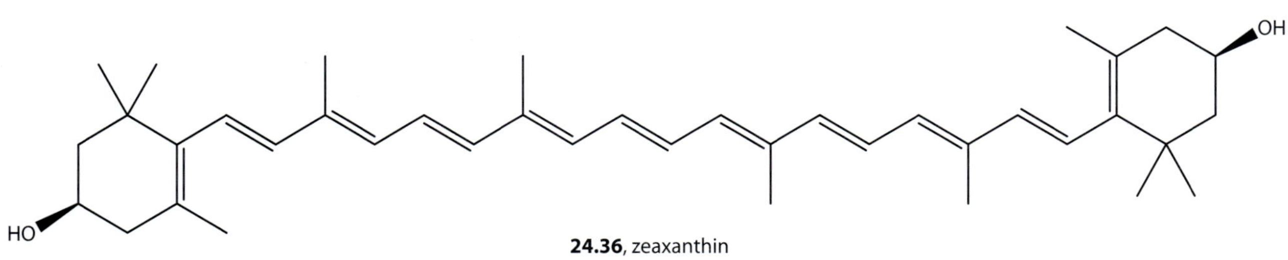

24.36, zeaxanthin

In the United States, only seven artificial food dyes may be used; the restriction also applies imports and the use of the compounds in cosmetics. These are shown in Figure 24.7. Two more dyes are allowed for specific purposes, citrus red 2 to color orange peel and orange B to color hot dog casings. The azo dyes are the most controversial, as they have been connected with attention deficit disorder (ADD) and ADHD in children, and these are being phased out in the EU, and the safe limits for their use lowered in the United States. As with the fabric dyes, the history is one of indiscriminate use followed by health concerns and increasing restrictions, with probably more restrictions to come.

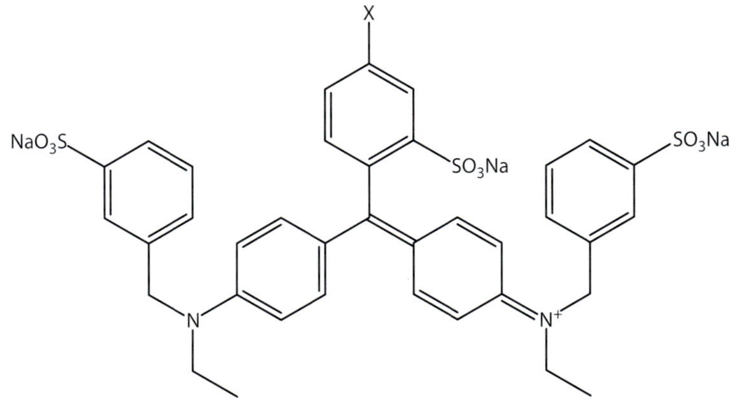

X = H FDC Blue 1, Brilliant blue FCF
X = OH, FDC Green 3, Fast green FCF

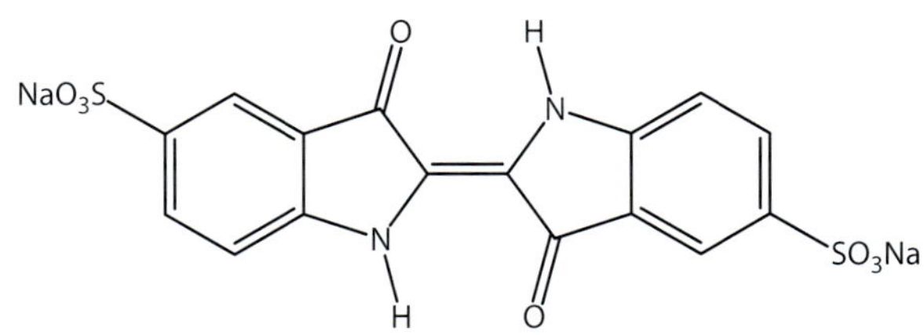

FDC Blue 2 Indigo carmine, blue

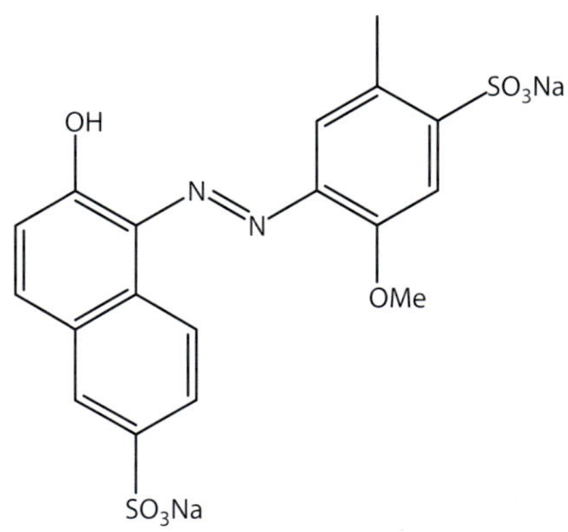

FDC Red 40, Altura red AC

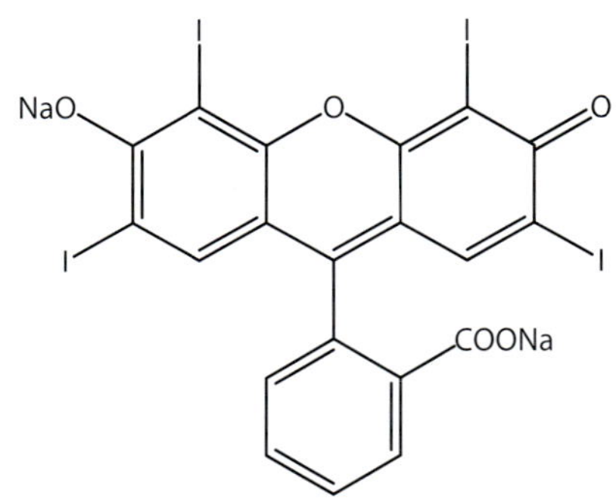

FDC Red 3, Erythrosine, pink

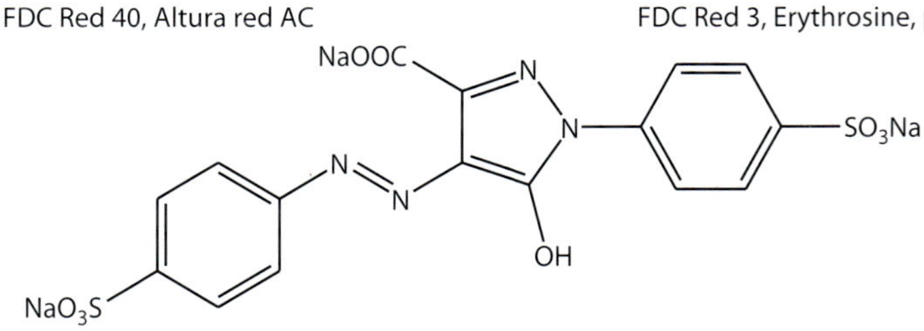

FDC Yellow 5, Tartrazine

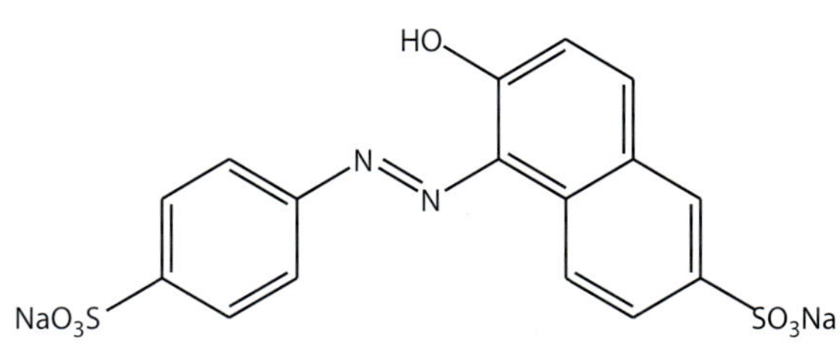

FDC yellow 6, Sunset yellow FCF

FIGURE 24.7 Artificial dyes legally permitted for use in foods and cosmetics in the United States.

REVIEW PROBLEMS

1. In the 1930s, a bright red azo dye, prontosil, was discovered to have antibiotic activities against streptococcal infections. It was off patent before these properties were known, but led to the development of the sulfonamide antibiotics. Discuss the structures and chemistry involved in these developments.
2. Explain the difference between vat dyes, mordant dyes, reactive dyes, and disperse dyes.
3. Identify the sugars involved in the structures of betanin, **24.33**, and crocin, **24.34**.
4. Where might chlorophyll be found with thujone?
5. The synthesis of the COX-2 inhibitor, Celebrex, is shown in the following. Propose a mechanism for each step:

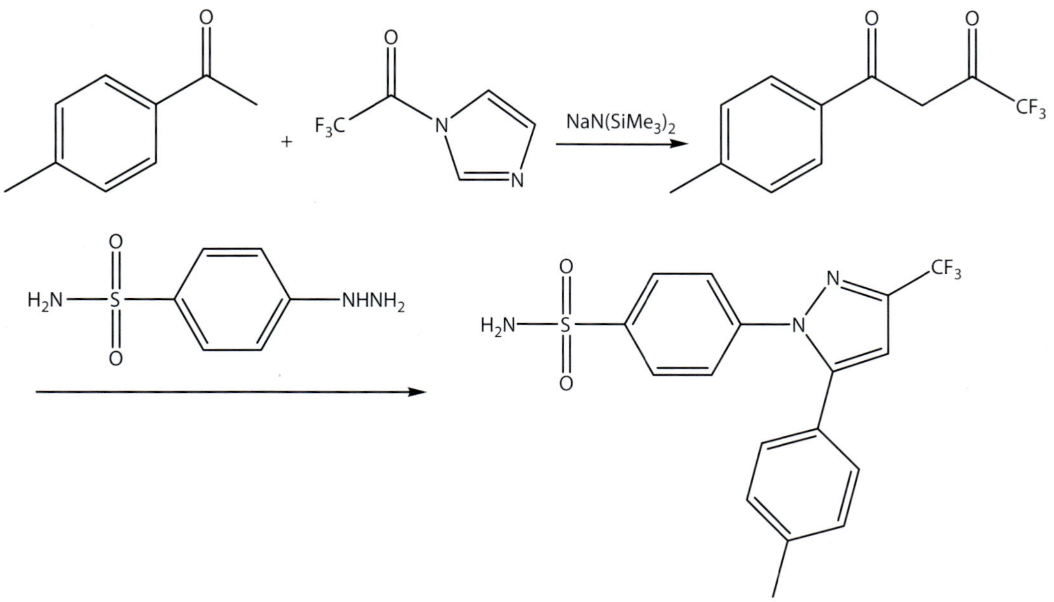

Celebrex

6. The conversion of **24.37** to a cyclopropane in the synthesis of the antibiotic trovafloxacin (a very controversial compound, the use of which is now largely discontinued) was required. The original synthesis was accomplished by a process that was considered to be too hazardous on a large scale, and the process shown in the following was used. Propose a mechanism for this process.

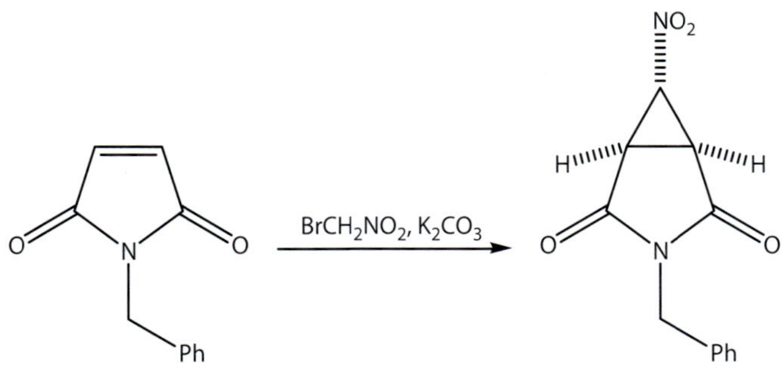

24.37

7. The classic Swern reaction (oxidation of primary alcohols to aldehydes using DMSO/$(COCl)_2$/R_3N) has the drawback on a large scale that large amounts of CO and CO_2 are generated. Many industrial processes have chosen to use the SO_3.pyridine complex as an alternative

activating agent for DMSO. It is a solid and can be used at or near room temperature with good results, as shown in the following. Compare and contrast the mechanisms for the two activating agents:

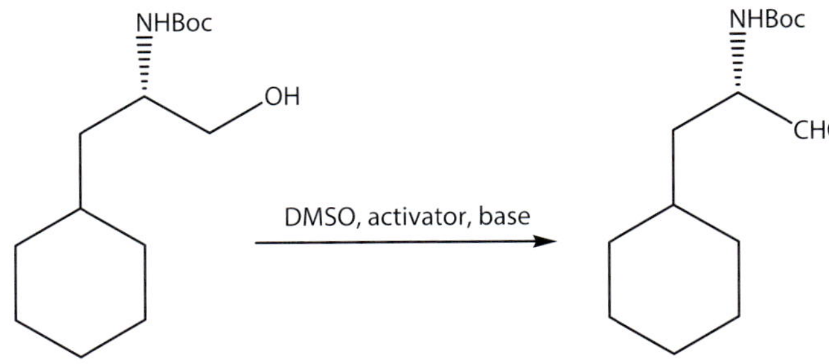

$(COCl)_2$, Et_3N, –70 °C, 90 % yield, 94 % enantiomer excess.
SO_3.py, Et_3N, CH_2Cl_2, 30 °C, 80 % yield, > 95 % enantiomer excess.

FURTHER READING

Blacker, J. and M.T. Williams (eds.). *Pharmaceutical Process Development: Current Chemical and Engineering Challenges*, RSC, Cambridge, U.K., 2011.
Pollak, P. *Fine Chemicals: The Industry and Its Business*, 2nd edn., Wiley, 2011.
Rao, S. *The Chemistry of Process Development in Fine Chemical and Pharmaceutical Industry*, 2nd edn., Wiley, 2007.
Raviña, E. *The Evolution of Drug Discovery: From Traditional Medicines to Modern Drugs*, Wiley, Weinheim, Germany, 2011.
Special issue on process chemistry, *Chemical Reviews*, **106** (2006), 2581–3027.
Tufts Center for the Study of Drug Development. http://csdd.tufts.edu.
Walker, D. *The Management of Chemical Process Development in the Pharmaceutical Industry*, Wiley, 2008.

Organic Chemistry and the Environment

25

25.1 INTRODUCTION

Despite its economic importance, the chemical industry gains limited regard from the general public, being rated below petroleum, paper, and lumber industries (though these are also in some respects "chemical" industries). The pharmaceutical sector gets the least negative reviews, because of its perceived social usefulness. From 1982 to 1999, the DuPont advertising slogan was "Better things for better living through chemistry." It's now been transmuted to "The miracles of science." Foods, cosmetics, and cleaning agents are advertised as "chemical-free," or "natural." But nothing is "chemical-free," and nature is more capable than even the most talented organic chemist of producing toxic materials. The chemical industry has a public relations problem. There are always those who will regard something they don't understand as potentially damaging (insecticides bad, but not necessarily to the 225 million people who suffered from malaria and the 655,000 people who died of it in 2010*), and the tabloid press are quick to jump on any available environmental bandwagon. It's tempting for those knowledgeable about chemistry to decry the ignorant public and the ignorant press, without considering *why* they don't understand—could it be that chemists have failed to communicate?

Environmental chemistry studies the effects of chemical processes on the environment—this includes toxic effects in humans as well as effects on wildlife, water, atmosphere, and soil. Green or sustainable chemistry seeks to achieve chemical transformations with the least possible waste and the lowest impact of what waste is produced. This chapter will consider only the organic chemical aspects of these topics; issues concerning atmospheric chemistry, energy consumption and origins, water usage, and heavy metal pollution are marginal to its scope, although critically important in environmental science. In considering the environmental impact of any process, it's important to look at the whole process, start to finish. It is tempting to say that polystyrene cups are bad and paper cups are good, because polystyrene is not easily degraded, and paper is degraded, and comes from a renewable source. However, if the energy cost of each, and the pollution in the production of each, is factored into the equation, the answer is much more ambiguous. The paper production industry uses large amounts of energy and water, and most paper is bleached.

Another important issue that is often omitted in popular reporting is that correlation does not necessarily imply causation. Many completely unrelated phenomena can be technically correlated. Sometimes the "circumstantial evidence is very strong, as when you find a trout in the milk."† However, in order to establish true causation, some *process* for the causation needs to be established—it is not enough to say that some bad effect has increased along with some pollutant; you need to find out how and why.

Much environmental legislation is informed by the "precautionary principle"—which we all heard as children as "better safe than sorry." This essentially tells us that we should not wait for

* Figures from the World Malaria Report 2011, www.who.int/malaria/world_malaria_report_2011/wmr2011_summary_keypoints.pdf, but see also C.J.L. Murray et al., *The Lancet*, **379** (2011) 413–431, which gives a higher estimate.

† Henry David Thoreau.

absolute proof that something is environmentally damaging before taking steps to mitigate or avoid the damage. The Rio declaration in 1992 stated in Principle 15:

> In order to protect the environment, the precautionary approach shall be widely adopted by States according to their capabilities. Where there are threats of serious or irreversible damage, lack of full scientific certainty shall not be used as a reason for postponing cost-effective measures to prevent environmental degradation.

Note the words "according to their capabilities" and "cost-effective"—there is neither any indication that measures should be taken that a particular state simply cannot effect, nor that there should be no consideration of the costs of the measures. The EU has enshrined the principle in law, but it is not without its critics and has had unintended consequences in some areas. The US Environmental Protection Agency, established in 1970, is responsible for much useful environmental law. In 1988, it eased the restrictions imposed by the Delaney Clause (1958) that forbade the use as a food additive of any substance that had been shown to cause cancer in man or experimental animals, where the risks were very slight.

25.2 PESTICIDES

The Food and Agriculture Organization (FAO) of the United Nations definition of a pesticide is "any substance or mixture of substances intended for preventing, destroying or controlling any pest, including vectors of human or animal disease, unwanted species of plants or animals causing harm during or otherwise interfering in the production, processing, storage, transport or marketing of food, agricultural commodities, wood or wood products, or animal feedstuffs, or substances which may be administered to animals for the control of insects, arachnids or other pests on their bodies." The classes include herbicides, insecticides, fungicides, rodenticides, pediculicides, and biocides. The class that has been of the greatest concern in terms of its environmental effects is the insecticides, consideration of which we deferred from the previous chapter, when we studied agrochemicals more generally.

In the United States, it is estimated that the non-use of insecticides reduces crop yields by at least 10 % and that the return on a farmer's investment in insecticides is fourfold. Losses to pests are greater in tropical regions. There are important insect vectors for a number of serious diseases including malaria, West Nile fever, plague, leishmaniasis, Chagas disease, Lyme disease, yellow fever, toxoplasmosis, human African trypanosomiasis, and typhus. However, it is estimated that 98 % of insecticides and 95 % of herbicides have effects beyond their target species. Groundwater may become contaminated, and persistent soil contaminants reduce soil biodiversity. DDT and methylparathion both inhibit nitrogen fixation. Pesticides may be toxic at high concentrations, but at much lower concentrations, chronic, long-term exposure may result in increases in cancer, neurological deficits, lowered fertility, and birth defects. Bird populations are reduced in areas of pesticide use either from the pesticides themselves but probably more significantly from the reduction in the food supply. *Silent Spring* by Rachel Carson, published in 1962, drew attention to the detrimental effect of pesticides on the environment, especially bird life. The book was widely read and is credited with inspiring the modern environmental movement.

There are three main classes of insecticide. The organochlorine compounds were the first to be developed but have now been almost entirely phased out because of their environmentally damaging effects. DDT was one of the first to be controlled; it suffers biomagnification through the food chain and hence was particularly damaging to higher predators. The thinning of eggshells of birds particularly caught public imagination, and banning DDT brought bald eagles back from the brink of extinction. DDT also affected beneficial insects, fish, and marine invertebrates—and by the time it was banned, most *Anopheles* mosquitoes were in any case resistant to it. In 2001, the so-called dirty dozen (nine of these are pesticides of some sort) of halogenated compounds were banned by the Stockholm Convention as persistent organic pollutants (POPs). Some examples are shown in Figure 25.1; hexachlorobenzene and various polybromodiphenyl ethers, used as fire retardants, were also on the list.

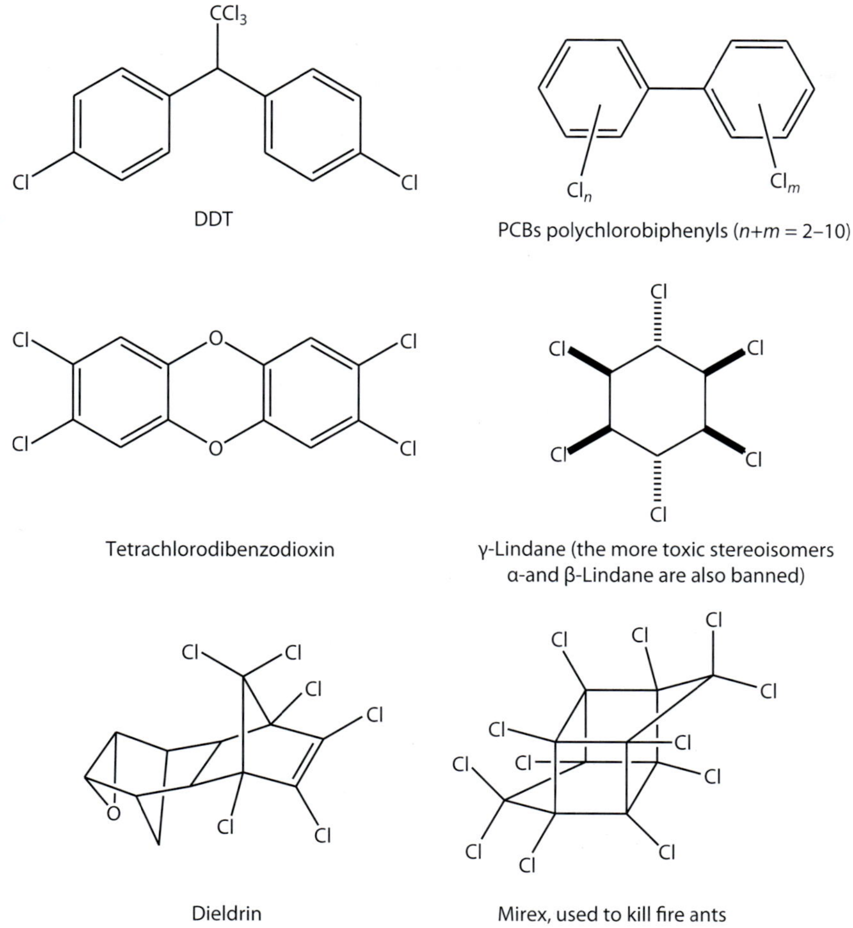

FIGURE 25.1 Compounds banned by the Stockholm convention as POPs.

Focus on POPs and VOCs

POPs have the characteristics that they are resistant to degradation by chemical, biological, or photochemical processes, are moderately volatile, and can be transported long distances in the environment. In general, they have low solubility in water and high solubility in lipids, and they bioaccumulate in humans and other top predators. They generally accumulate in the hydrosphere, soil, and organisms, although some materials can be detected in the atmosphere. Most have molecular weights more than 250 but are still moderately volatile—they tend to volatilize in hot regions and be deposited in colder areas. Lower-molecular-weight species are generally less persistent. POPs should be distinguished from volatile organic compounds (VOCs). VOCs are simply organic materials that have a high vapor pressure at room temperature, with boiling points from 50 °C to 250 °C. Many VOCs are naturally occurring and are responsible for the odor of plants. The most abundant is isoprene, and many related terpenes have pleasant odors (and are harmless). VOCs of anthropogenic origin, and not generally so benign, include solvents from paints, wood preservatives, coatings, aerosols, cleaning materials, adhesives and hobby materials, chlorohydrocarbons and chlorofluorocarbons (CFCs) from dry cleaning, and hydrocarbons and ethers from fuels and fuel additives. These compounds are sometimes, but not always, toxic, and some have been implicated in smog formation.*

* http:www.epa.gov/iaq/voc.html.

PROBLEM 25.1

DDT may be manufactured (DDT is currently only manufactured in India, for the control of the insect vectors of malaria and visceral leishmaniasis) from chlorobenzene and chloral (CCl_3CHO) in the presence of sulfuric acid. Write a mechanism for the process. Only about 77 % of commercial DDT is actually the compound of which we generally write the structure. Commercial samples contain 15 % of **25.1** and smaller amounts of **25.2** and **25.3**. Explain how these compounds could arise from the synthesis.

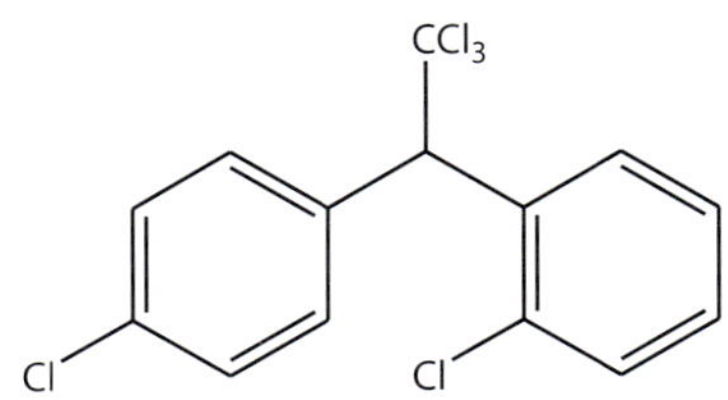

25.1

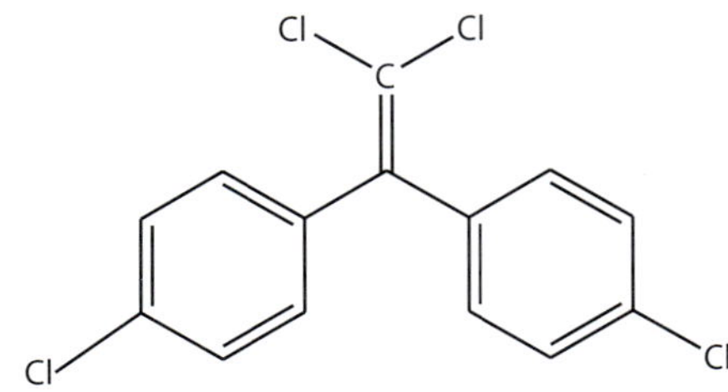

25.2

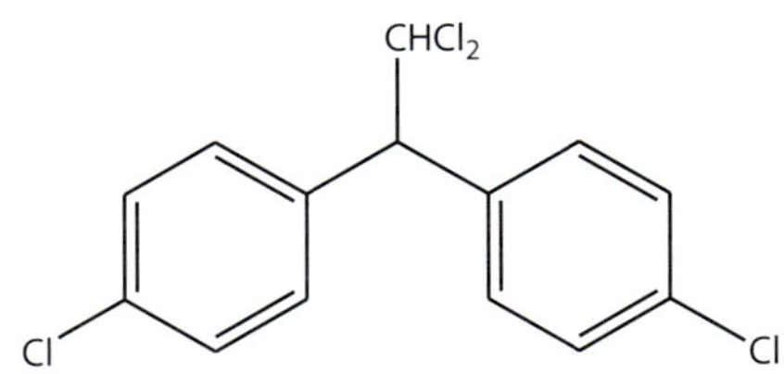

25.3

Solution

The mechanism of the electrophilic substitution is shown below.

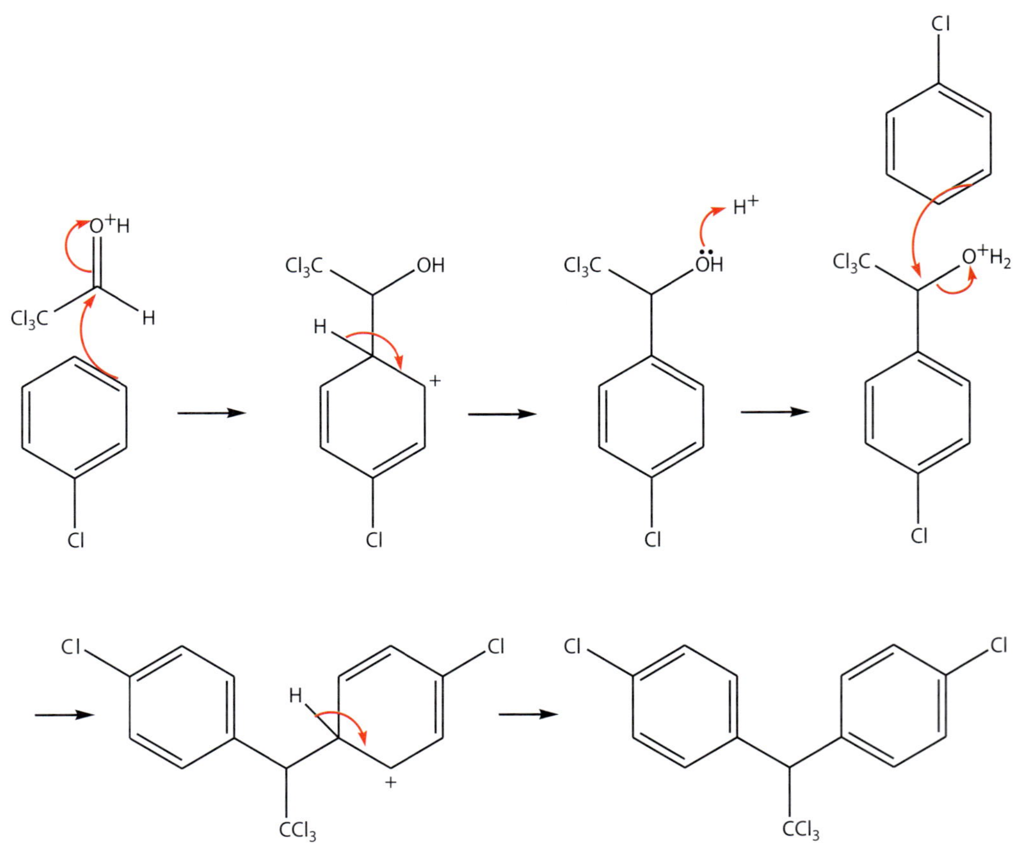

The other isomer of DDT may be formed because chlorine is *ortho-* and *para*-directing in electrophilic aromatic substitution. The *para, para*-isomer is favored for steric reasons but not exclusively. **25.2** is formed by HCl elimination from DDT. There is no obvious source of a reagent for the hydrogenolysis that gives rise to **25.3**; the general consensus is that this is a radical process.

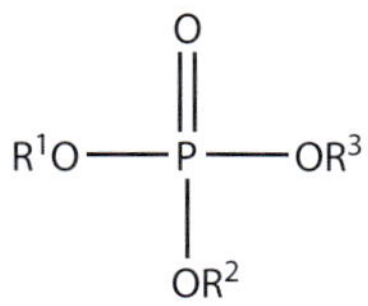

25.4

Even before organochlorine compounds started to be banned as insecticides, various organophosphate-based insecticides (generic structure **25.4**; some oxygen atoms may be replaced by sulfur) were introduced. The term is used to refer to any phosphorus (V) compound. These compounds are not environmentally persistent and are degraded rapidly by hydrolysis. Examples include parathion, methyl parathion, and malathion (Figure 25.2). Malathion was widely used for the control of the Mediterranean fruit fly in California in the 1980s and to combat outbreaks of West Nile fever. It is also used to control head and body lice, although it is not entirely effective (lindane is also permitted to treat head lice, as an exception to the more general agricultural ban, but there is clear development of resistance to both materials). These all have relatively low

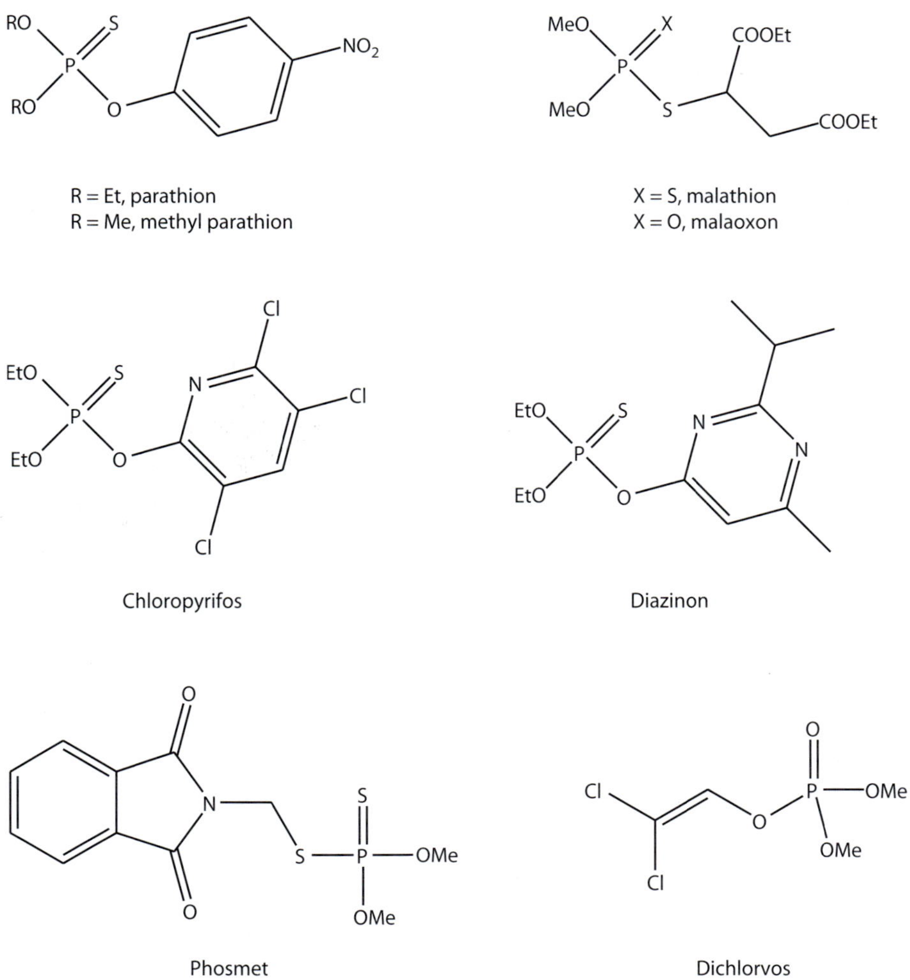

FIGURE 25.2 Organophosphate insecticides.

toxicity in humans. If malathion is ingested, it is converted to the more toxic malaoxon, and there are current concerns that organophosphates may be implicated in development of ADHD. Chlorpyrifos, diazinon, and phosmet are all rather more toxic and are permitted only for agricultural and not domestic use. Dichlorvos is permitted in the United States for use in the home and domestic garden. The mode of action of all the organophosphate insecticides (and their more toxic relatives, the nerve gases) is as acetylcholinesterase inhibitors. This results in the build up of the neurotransmitter acetylcholine, resulting in muscle weakness and paralysis, as well as cardiac problems. Chronic and long-term problems including peripheral neuropathy have been reported to be associated with the use of diazinon in sheep dip.* It is asserted that there are about 1 million organophosphate poisonings worldwide per annum, mainly in the developing world.

PROBLEM 25.2

One of the SMs for the organophosphate insecticide Fenitrothion is 3-methyl-4-nitrophenol. How would you prepare this from a monosubstituted benzene?

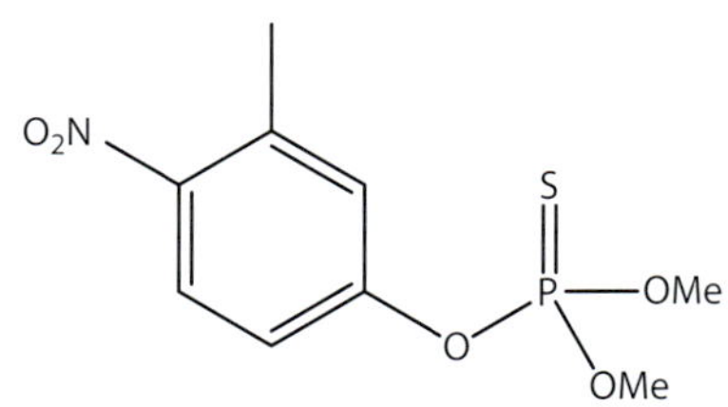

Fenitrothion

Solution

This synthesis may look simple, but it actually isn't. The obvious method is to nitrate 3-methylphenol, but unfortunately, even in conc. sulfuric acid, the yield of the desired isomer is only 52 %.

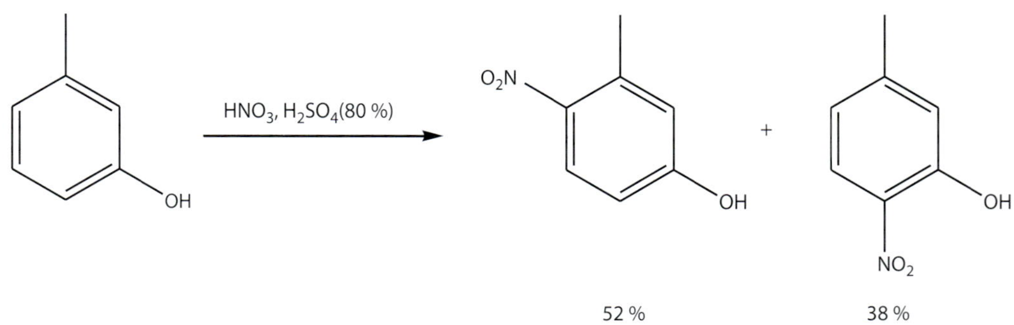

* http://www.iom-world.org/pubs/IOM_TM9902c.pdf.

One industrial synthesis converts the phenol to a phosphate prior to the nitration and then removes the phosphate. This gives a much better isomer ratio, for steric reasons. There's an interesting honors project somewhere here, blocking, temporarily, the OH with some large protecting group, so as to direct the nitration to the right position. The SM, which has a common name *meta*-cresol, is normally obtained from coal tar; the conventional synthesis from toluene is long and tedious.

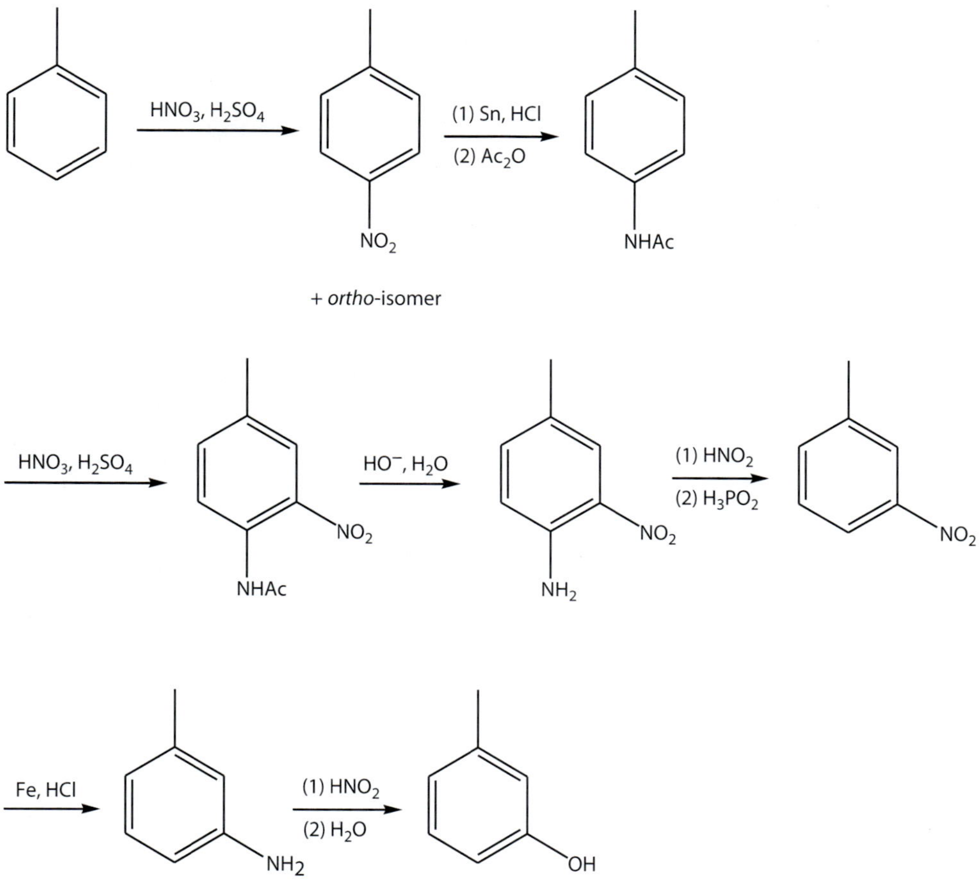

Carbamates are also acetylcholinesterase inhibitors, but their action on the enzyme system generally seems to be more reversible than that of the organophosphates, and they are less toxic to vertebrates. They have the general structure **25.5** and important members of the family are shown in Figure 25.3. Like the organophosphates, these are relatively quickly broken down in the environment.

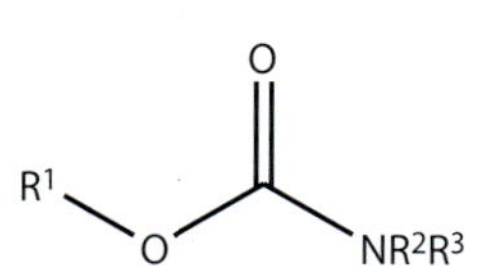

25.5

Pyrethroids are synthetic analogs of derivatives of chrysanthemic acid, **25.6**, naturally occurring in plants of the *Chrysanthemum* genus. The naturally extracted materials, pyrethrin I and II (**25.7**, **25.8**), have been modified for better activity in compounds such as permethrin, **25.9**, and deltamethrin, **25.10**. They are toxic to cats and to aquatic life but have otherwise generally low toxicity to mammals and are broken down rapidly in the environment. They are widely used as domestic insecticides. You should note that although a single stereoisomer

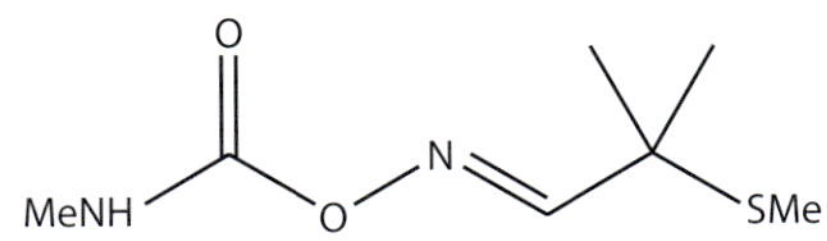

Aldicarb—agricultural use on cotton, beans for thrips, aphids and spider mites

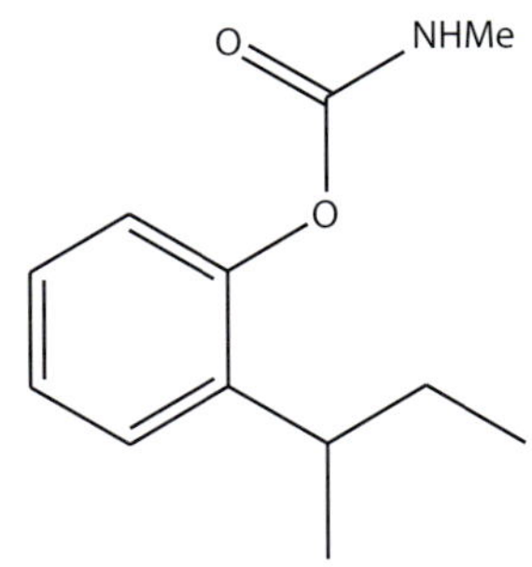

Fenobucarb. Moderately toxic to vertebrates. Used on rice, cotton

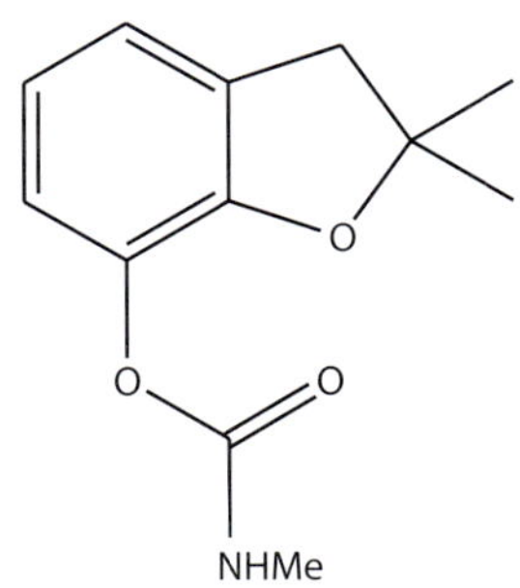

Carbofuran; banned in the EU, likely to be banned in the United States. High toxicity to vertebrates

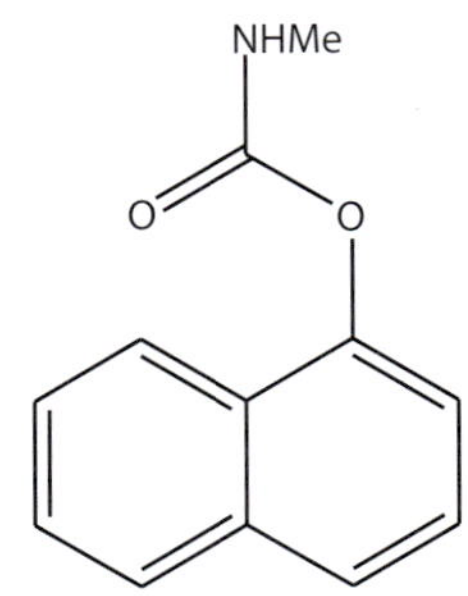

Carbaryl—for agricultural and domestic garden use

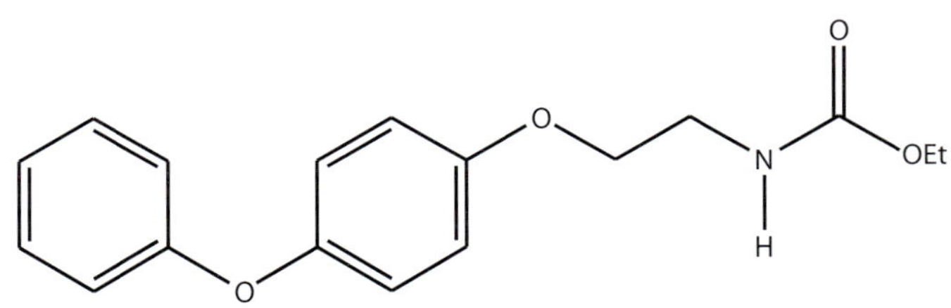

Fenoxycarb; toxic to fish, but only minimally so to mammals, bees. Not a neurotoxin, but a mimic of insect juvenile hormones

FIGURE 25.3 Carbamates used as insecticides.

is shown for each compound below, most commercial preparations are not stereochemically pure, and it is known that some stereoisomers have significantly higher activities than others. One might think that the solution would be to use stereochemically pure material—but this would require significant cost in the purification steps. There is a trade-off between unnecessary environmental burden and cost—and cost generally wins.

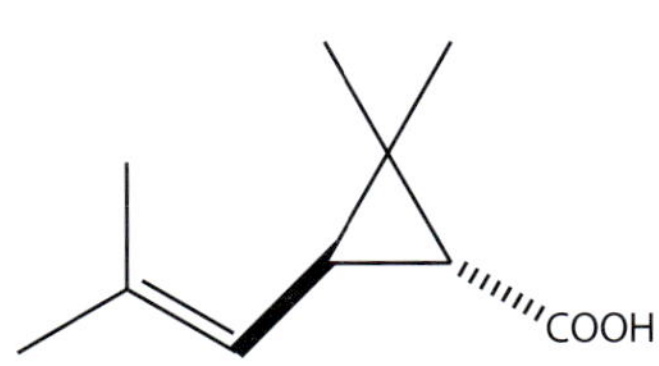

25.6, chrysanthemic acid

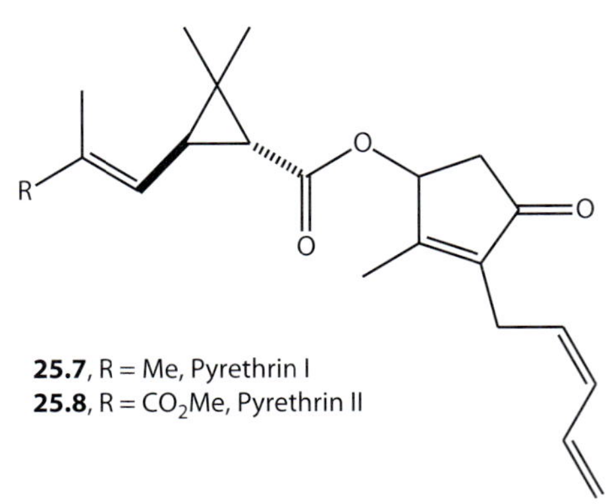

25.7, R = Me, Pyrethrin I
25.8, R = CO_2Me, Pyrethrin II

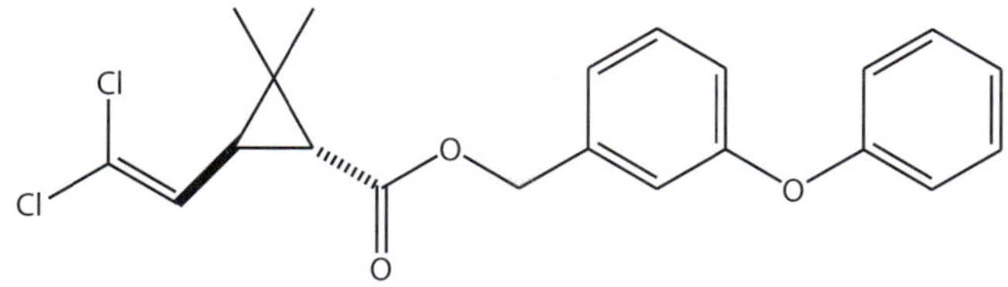

25.9, permethrin

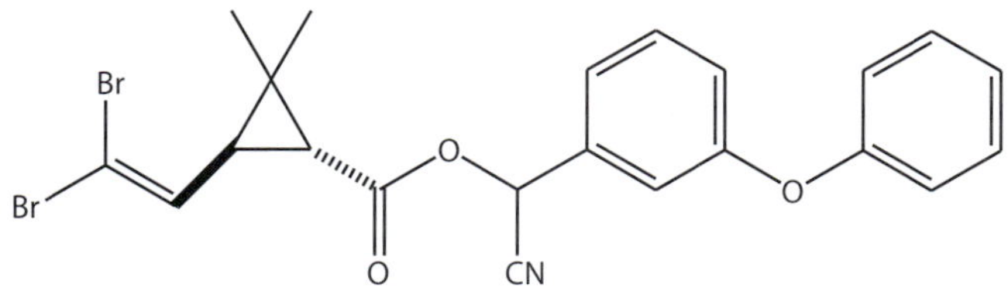

25.10, deltamethrin

Pyrethroid harvest in Ruanda. (Thanks to Dr. Steve Kamin for this photo.)

PROBLEM 25.3

What stereoisomers would you expect for the pyrethroid insecticide allethrin II?

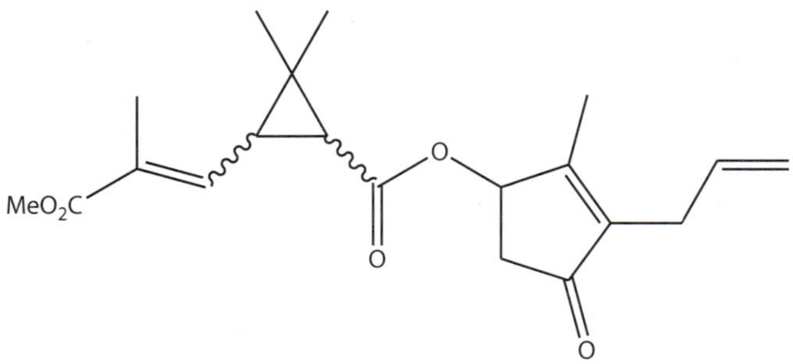

Allethrin II

Solution

The stereochemistry at the three-membered ring is undefined, so *cis*- and *trans*-isomers can be expected. Each of these is also chiral. The double bond may be *E* or *Z*. Thus, there are eight stereoisomers in all; four are shown, and each of these has a stereoisomer with the alternate double-bond geometry.

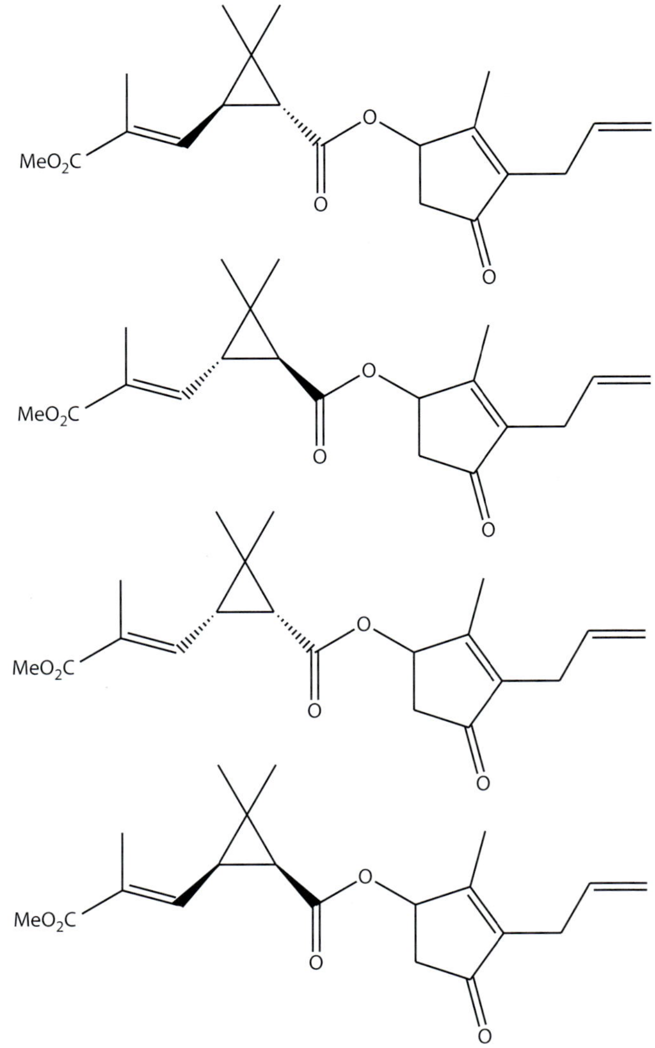

There has been significant pressure for the production of more environmentally friendly methods of insect control. Among these have been the use of juvenile hormone mimics (these prevent the insect developing into its adult form), antifeedant compounds (these prevent the insect from eating and are produced by many plants to defend themselves), bacterial insect diseases, and the use of predator or parasitic insects to attack target species. There is considerable interest in the observation that when tobacco plants are attacked by the tobacco hornworm caterpillar (the saliva of the caterpillar activates the effect), there is a change in the "green leaf volatiles" that are produced (from mostly *Z*-C_6 aldehydes and alcohols to a roughly equal mixture of *E*- and *Z*-compounds). The new mixture is an attractant for *Gecoris* insects, which eat the caterpillar eggs and larvae. There is little doubt that these methods are environmentally more desirable, but it has proved more challenging to make them both as effective, and as cost-effective, as the more traditional pesticides.

25.3 ENDOCRINE DISRUPTORS

The Endocrine Society defines an endocrine disrupting substance as *a compound either natural or synthetic, which, through environmental or inappropriate developmental exposures, alters the hormonal and homeostatic systems that enable the organism to communicate with and respond to its environment.* Effects may include developmental, reproductive, neurological, and immune system defects. The field is a controversial one; most of the data come from wildlife studies, and the dose/response effect is problematic. In laboratory studies, many anthropogenic compounds that are found in the environment can be shown to be endocrine disruptors. There are claims, however, that the concentrations of some of these compounds in the environment are too low to cause any serious effect. Among the most serious effects attributed to endocrine disruptors in the environment are feminizing effects on males and masculinization of females, particularly in aquatic life.

One of the most widely quoted studies of an endocrine disruptor is the case of diethylstilbestrol (DES, **25.11**), which was used medicinally rather than being found environmentally. Prior to its being banned in the 1970s, DES was used to prevent miscarriage (on the basis of poor evidence—later studies showed that it was ineffective)—however, the female children who were exposed to it prenatally had a high incidence of vaginal cancer in later life. DES is one of a class of synthetic estrogens, mimics of estradiol, **25.12**. Many of these, like nonylphenol, **25.13**, resemble the shape of estradiol and hence might be expected to bind to similar biological receptors.* The estrogen receptor has been shown to be rather flexible in the type of molecules it is able to accommodate.

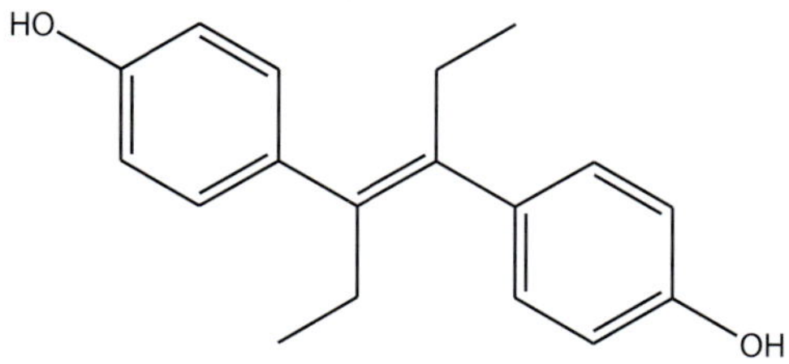

25.11, diethylstilbestrol

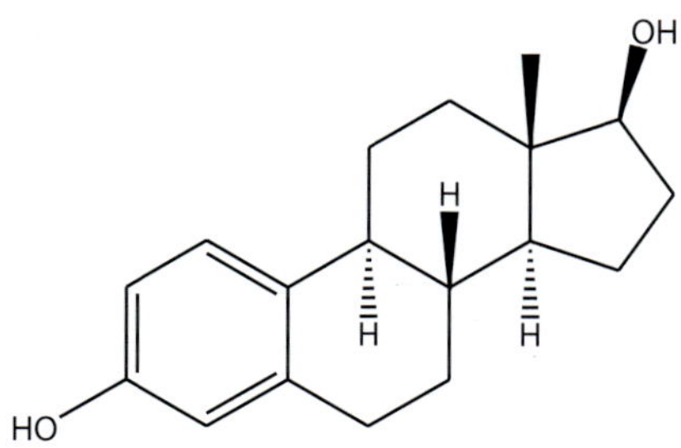

25.12, estradiol

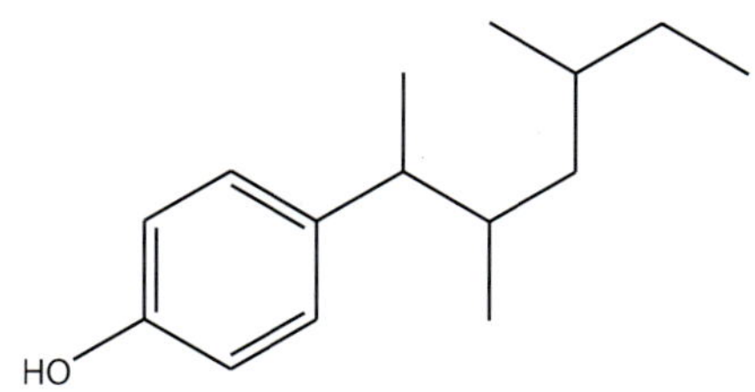

25.13, nonylphenol

Food and water are the main sources of human exposure to endocrine disruptors, though there may be additional exposure from wood floor treatments or monomer leaching from plastic artifacts. Many POPs are also considered to be endocrine disruptors (though not all bind only to the estrogen receptor), including the chlorinated pesticides, PCBs (see Figure 25.1), and polybrominated diphenyl ethers (used in fire retardants for plastics, fabrics, and furniture). PCBs bind to the aryl hydrocarbon receptor, are genotoxic, and have recently been suggested as a possible cause of autism.

PROBLEM 25.4

PCBs that do not have substituents at the *ortho*-positions are described as "coplanar" PCBs and are the most toxic of the PCBs. Given what we learned about the geometry of biphenyl in Chapter 12, comment on this nomenclature.

Solution

Biphenyl itself is not planar in its ground state but has an angle of approx. 44 ° between the rings. It thus seems unlikely that the ground state of more substituted biphenyls is planar either. What should be said is that the barrier to rotation about the interring bond in non-*ortho*-substituted biphenyls is low, 6–8 kJ mol^{-1}, so the molecule is able to become planar for the purpose of interaction with the aryl hydrocarbon receptor.

* The related compound, 4-*tert*-octylphenol (4-(1,1,3,3-tetetramethylbutyl)phenol), has recently been named as a substance of very high concern (SVHC) in the EU as an endocrine disruptor.

Two materials associated with the polymer industry have come under increased scrutiny in recent years—phthalates and bisphenol A. Phthalates are esters of the general structure **25.14**. Their main use is a plasticizer, to soften poly(vinyl chloride) (PVC), which may contain 1 %–40 % phthalate. However, they are also found in pill coatings, stabilizers, dispersants, lubricants, and personal care items. Humans, and most of the environment, are exposed to phthalates, as these are gradually released as the plastic degrades; studies of leaching from new materials are more ambiguous. There is evidence linking feminization of males to fetal and neonatal exposure to phthalates, though other adverse effects have been less conclusively established. In the United States, children's toys may now contain only 0.1 % of phthalates.

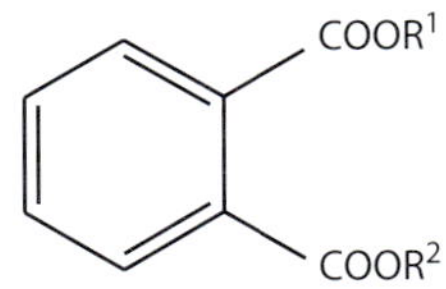

25.14

Bisphenol A, **25.15**, is one of the main components of polycarbonate and of some epoxy resins. It is an estrogen mimic, and it has been blamed at some point, by someone, for a vast array of medical problems. In the environment, the main effects seem to be on fish. As it is covalently bound as a part of the polymer, leaching is always low level, but the use of the material is very widespread, ranging from plastic water bottles to interior coatings of canned foods and drinks. While the material is not environmentally persistent, it is constantly renewed. Regulation of the use of BPA varies widely and is currently in a state of flux. Attention has largely focused on the effects on fetal and neonatal development, and in most countries, polycarbonate is being phased out of use for baby feeding bottles—probably more in response to public concerns and the "precautionary principle" than the hardness of the scientific evidence at present.

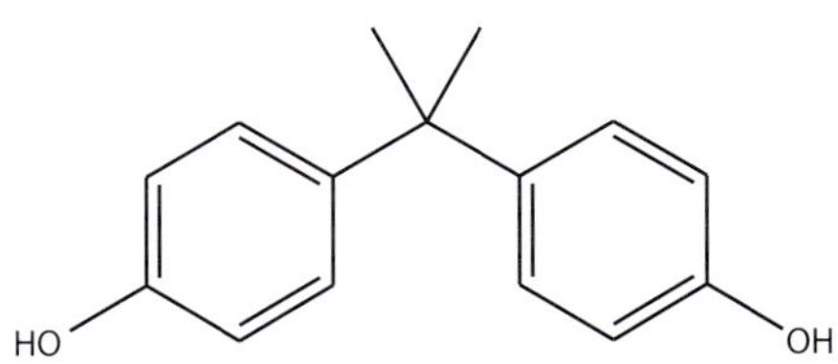

25.15, bisphenol A, BPA

Some of the endocrine disruptors in the aquatic environment are human hormones, from incomplete processing of human waste. Progestins and estrogens are derived from the contraceptive pill, and other materials are derived from treatments for hormone-responsive cancers (such as tamoxifen, an anti-estrogen). It seems that these cause feminization effects in fish, and there have been claims that the incidence of human prostate cancer is higher in countries where the use of the contraceptive pill is higher (correlation, not proven causation).

There are many naturally occurring estrogens in the environment (examples shown in Figure 25.4). These are ubiquitous and bind much more tightly to the estrogen receptor than most synthetics. Most are considered to be benign; daidzein is soya-derived and is sold as a dietary supplement; equol may have benefits in prostate cancer and postmenopausal physiology; coumestrol and genistein are antioxidants and may have antitumor effects and be of benefit in osteoporosis and atherosclerosis.

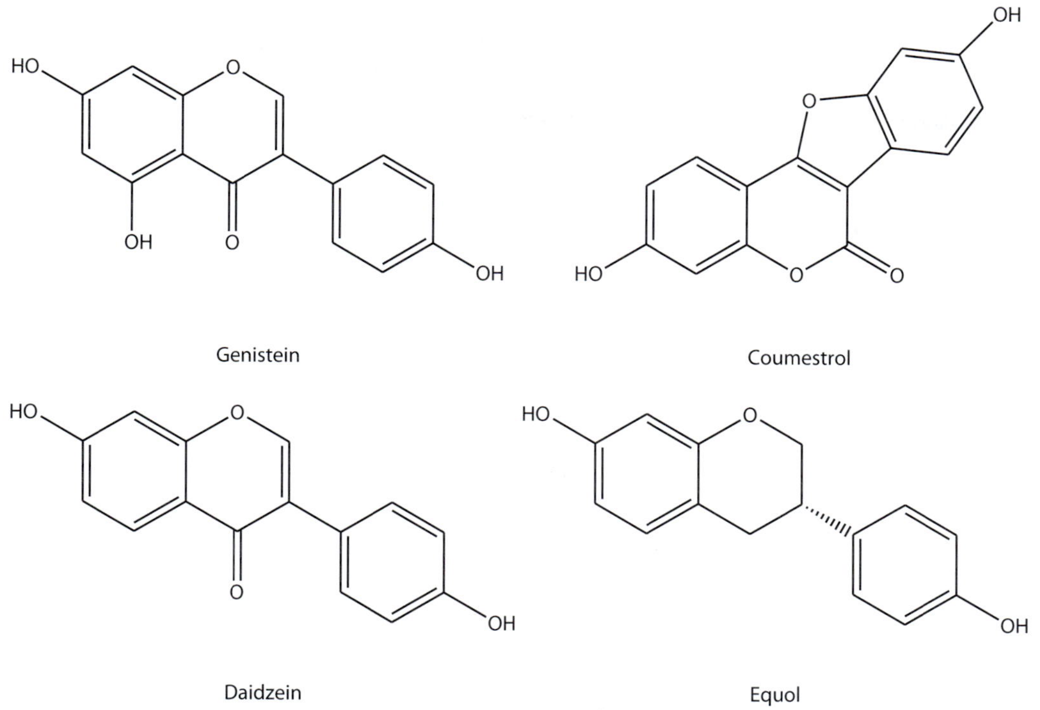

FIGURE 25.4 Phytoestrogens.

25.4 CHLOROFLUOROCARBONS AND THEIR REPLACEMENTS

Refrigeration in the ancient world involved the use of ice or snow, collected in cold regions and seasons and preserved in well-insulated "ice houses." The use of this technique persisted well into the twentieth century, although by then more modern techniques had been developed. Ammonia was used as an effective and low-cost refrigerant from the 1870s, but its toxicity inhibited use in domestic refrigerators. CFCs, sometimes known as Freons, started to be used by the 1930s—they are chemically unreactive, non-toxic, and non-flammable and also found uses as propellants in aerosols and in fire extinguishers. Freon, difluorodichloromethane, was the most popular. By the 1970s, concern was rising about the depletion of ozone in polar regions—the so-called ozone hole. Now the chemistry of chlorocarbons in the upper atmosphere is reasonably well understood; the key reactions involve the homolytic cleavage of a carbon–chlorine bond to give a chlorine radical followed by reaction of the chlorine radical with ozone (Reactions 25.1 through 25.3):

(25.1) $$CCl_3F \xrightarrow{h\nu} Cl^{\bullet} + {}^{\bullet}CCl_2F$$

(25.2) $$Cl^{\bullet} + O_3 \rightarrow ClO + O_2$$

(25.3) $$ClO + O_3 \rightarrow Cl^{\bullet} + 2O_2$$

In 1987, the Montreal protocol mandated a dramatic reduction in the use of CFCs. Interim replacements were hydrochlorofluorohydrocarbons (HCFCs); these do cause ozone depletion, but they are less stable and mostly break down in the lower atmosphere. Hydrofluorocarbons (HFCs) have an even shorter lifetime. All of these compounds are also greenhouse gases; HFCs have 100–1000 times the global warming potential of CO_2. Thus, there is an intention to phase these out, and they were included in the Kyoto protocol of 1997. DuPont is now positioning itself to benefit from a switch to hydrofluoroolefins (HFOs) such as **25.16–25.18**; these have little more global warming potential than CO_2, and HFO 234yf is already sanctioned by the EPA for use in car air-conditioning systems. It is also possible that domestic refrigeration and air-conditioning will ultimately use hydrocarbons, with ammonia being important in industrial applications.

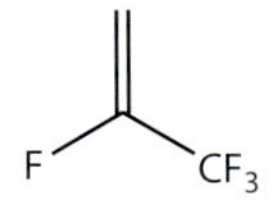

25.16, HFO 1234yf

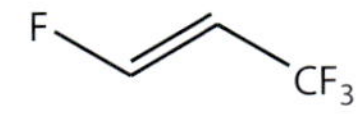

25.17, HFO 1234ze

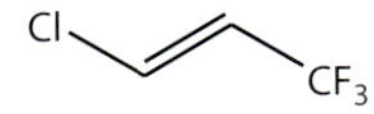

25.18, HCFC 1233t

The Montreal Protocol is generally viewed as having been successful—the ozone holes are now shrinking year on year. It's interesting to consider the reasons for this—there were a relatively small number of manufacturers, replacement chemicals were available/developed, and the financial aspects of the turnaround could be managed. The issues of global warming are, however, many orders of magnitude larger and more difficult to deal with.

The lack of flammability of halogenated hydrocarbons has resulted in their use for fire suppression in high-value areas (aircraft, ships, museums, computer installations) where water is unsuitable, dating back to its use in military aircraft during the Second World War. Most of the compounds used, called halons, contain bromine, such as $CBrF_3$ and CF_2ClBr. These have proved more difficult to replace, and although they are no longer manufactured, they are still recycled, as some countries still require halon fire suppression in aircraft. The nature of the replacement depends on whether toxicity is an issue—thus, is the area occupied or unoccupied? CF_3I or CO_2 may be used in unoccupied areas, but they are toxic. HFCs and fluorocarbons are more suitable for occupied areas. An interesting new development is the use of PFBOCs, perfluorocarbon-based oxygen carriers, as blood substitutes. These have potentially many advantages over donated blood in terms of longer shelf life, no possibility of disease transmission, no need for blood typing, and acceptability to some religious groups who refuse transfused blood. Several formulations had reached Stage II clinical trials as of 2011.

25.5 POLYCYCLIC AROMATIC HYDROCARBONS

In Section 12.4, we saw that PAHs in the environment could be either petrogenic, derived from fossil hydrocarbons, or pyrogenic, derived from various combustion sources. They are regarded as POPs, and the Agency for Toxic Substances and Disease Registry in the United States has identified 17 compounds that are of particular concern. Some of these are shown in Figure 25.5.

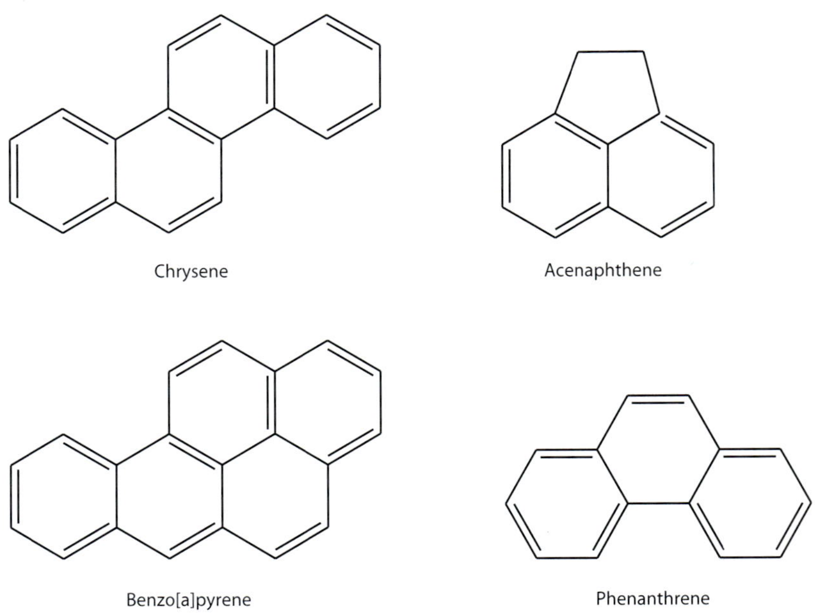

FIGURE 25.5 Toxic PAHs.

Benzo[a]pyrene is among the most toxic and was the first chemical precarcinogen (the actual carcinogen is the epoxide, produced *in vivo*) to be unequivocally identified.

The effect of these compounds on human health has been a major cause of concern. Toxicity to humans depends on exact structure, but exposure is ubiquitous and involves a mixture of compounds, from fossil fuel and natural and anthropogenic burning and from food, mainly grilled or barbecued meat. In the broader environment, PAHs are endocrine disruptors and bind to the aryl hydrocarbon receptor. They have been shown to depress immune function in some wildlife. They may be responsible for reproductive disorders in aquatic organisms, especially in shellfish and sediment dwellers.

25.6 PLASTICS: RECYCLE, DEGRADE, OR BURN?

Plastics are here to stay—although they are villains to some in the environmental lobby, there are also many environmental advantages to their use. Few would quarrel with the replacement of ivory billiard balls or netsuke by plastic ones. Plastics are less dense than metal, so including plastic components in cars and aircraft reduces the use of fossil fuels. Insulation reduces the energy requirements for heating homes and workplaces, and food packaging reduces damage in transit and losses to pests and other spoilage. Cotton may be a renewable resource, but its growth requires significant water and pesticide use. This section will not discuss the at-source recycling, or more properly reprocessing, of industrial waste plastic—this is relatively simple since it is done on the site at which it is produced, and the content of the waste, offcuts, *etc.*, is well established.

How is plastic to be disposed of once it has been used by the consumer for its designated purpose? Currently, plastic represents a significant fraction of trash sent to landfill and is a much higher proportion of domestic refuse. Most plastic household waste comes from packaging, and the proportion that can be, or is, recycled depends on local facilities and collection policy. Improper disposal of such refuse, essentially littering, results in many well-perceived problems. It is very visible and unsightly, mostly it does not degrade, and there are major and well-justified concerns about the effects of plastic debris on coastal and marine ecosystems. Some plastics will always end up in landfill, so we do need to consider what happens to them there—will they degrade, albeit perhaps slowly (everything degrades relatively slowly in the anaerobic dark environment of landfill), and if they do, what are the products? Methane from landfill is a well-known problem (it's a greenhouse gas)/opportunity (if it can be collected).

Another important consideration is the resource and energy cost of plastic relative to other materials in terms of its production. The issue of plastic/paper/cotton bags in supermarkets has been controversial, and some countries are either taxing plastic bags or banning them. However, there can be unintended consequences—in Ireland, plastic bags were taxed, so the use of plastic bags in supermarkets was decreased by about 90 %. However, there was a significant increase in the purchase of plastic bags for kitchen waste bins. It's estimated that some 40 % of supermarket plastic bags are reused in households as bin bags. In order to make the best decisions, the full life cycle of a product must be analyzed. A recent study comparing HDPE, oxobiodegradable HDPE (*vide infra*), starch/polyester blend, paper bags, bags for life (usually LDPE, with some polypropylene), and cotton bags concluded that the total environmental impact for *all* the bags was dominated by the resource and energy use in their production. Transport, secondary packaging, and end-of-life disposal were minor considerations. For all the bags, reuse was critical, and even reuse as a bin bag is better than recycling. Paper bags, starch based bags, bags for life, and cotton bags would have to be reused 3, 4, 11, and 131 times, respectively, to have lower environmental impact than the HDPE bags if they are used only once; the figures are obviously worse if the HDPE bag is reused.

Sourcing material for plastics from renewable sources—so-called bioplastics—was initially embraced as a useful contribution to reduction in use of fossil fuels. These may consist of modified natural polymers, materials from microorganisms or genetically modified crops, or biobased feedstock *via* fermentation. Biobased plastics are not synonymous with biodegradable plastics, although there is a significant area of overlap. Modified cellulose and starch are both biobased and biodegradable. A modified starch-based plastic (modifiers are urea and choline chloride) has been prepared with properties comparable to oil-based plastics and with ability to be recycled or composted. However, there are similar concerns to those now being expressed about biofuels—large areas of land are being diverted from food production, with consequences for food price and availability.

There is, however, general agreement that it is undesirable for plastics to continue to go to landfill—many countries with high population densities are running out of landfill space. So what can be done with plastics after their primary consumer use, if they are not to go to landfill? There are three main choices—recycle, degrade, or burn. It's also important to consider how much we are prepared to pay to feel good about doing this—most established processes are currently only marginally economic. Companies charged with accomplishing these processes do not do so for free—they need to make a profit. Recycling is important but not as simple as the numbers and symbols on our plastic bottles would imply. Some plastics can be returned to approximately the same use as the original material (primary recycling), but this is difficult and accounts for a very small proportion of the total. Almost all recycling is actually downcycling or secondary recycling; the recycled material is used for a lower-grade application than the original. Thus, no. 1, poly(ethylene terephthalate) (PET), used in soft drink bottles, is recycled into polar fleeces, tote bags, furniture, and carpets. Number 2, high-density polyethylene, is initially used for milk containers, shampoo and household cleaner bottles, and butter and yogurt tubs, and is recycled to laundry detergent bottles, oil bottles, pens, floor tiles, fencing, and drainage pipes. Number 5, polypropylene, is used in yogurt cartons, caps, straws, and medicine bottles and is recycled to cables, brooms and brushes, ice scrapers, and trays. Number 6, polystyrene, is used in disposable plates, cups, meat trays, and CD cases and is recycled to egg cartons, takeout containers, insulation, and packaging. Most of the other groups are rarely recycled—cost and energy input are too high to make it economic. The amount of recycled material in the new application may be 50 % or less, and few materials licensed for food packaging are produced. In the waste stream, sorting of plastics to distinguish PET and HDPE may involve near IR spectroscopy or flotation, but neither of these techniques can distinguish materials that were used for food as opposed to nonfood initial applications. In many cases, a final sort requires human input.

There are forms of recycling that are less familiar —in that they do not directly recycle the materials into new artifacts. Tertiary recycling breaks down the plastic to its component monomers, which can be repolymerized. This requires pyrolysis of the materials, sometimes called thermal cracking, and is most easily accomplished with step growth polymers such as polyamides and polyesters but is also possible for poly(methyl methacrylate). The processes require specialized plant, so are capital intensive, and a large-scale operation is necessary for economic viability.

Finally, quaternary recycling converts polymers to various grades of fuel oil. Degradation occurs in the absence of air, at high temperatures, in the presence of catalysts, and the yield of oil is 30 %–40 % by weight, much of the rest being gas that is used to fuel the plant. Mixed, uncleaned, feedstock may be used; both thermoplastic and thermosetting plastics are acceptable, although heteroatoms such as sulfur and chlorine cause some problems. Plant design is important as there is always some solid carbonaceous material produced, which may foul the catalyst. Commercially, the process is only marginally efficient; it's not done to make money but for social, political, and environmental reasons. The recycling of automobile tires is a long-standing problem. They are easily identified, all have similar compositions (butadiene/styrene/isoprene rubber, sulfur, steel, carbon black, zinc salts), there is a lot of material available, and it is easily collected. Cryogenic grinding reduces the tires to a powder that can be used in road surfacing or as supplemental fuel in cement kilns. The EU banned tire waste from landfill in 2006. Pyrolysis is possible but difficult; microwave heating can be induced to give ethene, propene, butene, and higher alkenes, but from an economic point of view, it's better to burn them.

Biological degradation of plastics involves either their being or being converted into something that breaks down in the natural environment, ultimately to CO_2 and water. Thus, the energy and material inputs to production are lost, and a greenhouse gas is generated. For bioplastics, no more CO_2 is usually generated in biodegradation than was originally sequestered by the organism from which they were derived, although significant CO_2 production may have resulted from the manufacture of the plastic and its recovery. Methane, also a greenhouse gas, may be generated if the degradation is anaerobic, as in landfill. Biodegradation of any polymer requires some extracellular process to cleave the chains and release them from the entangled mass as fragments small enough to be transported into the cell and metabolized. In nature, this may be hydrolysis, as in the case of polyesters, or it may be oxidation, as in the case of natural rubber or lignin. In either case, it may be purely chemical or mediated by enzymes released by the cell. Both hydrolytic enzymes (typically esterases) and oxidizing enzymes (*e.g.*, cytochrome systems) are well known. These two well-recognized biodegradation pathways for plastics are distinguished by the mechanism (which may be purely chemical or biologically mediated) by which the chains are cleaved to fragments small enough for bioassimilation. The terms hydro- and oxo-biodegradation are sometimes used to emphasize that biodegradation of a plastic is always a two-stage process; there are hydrolytic and/or oxidative stages leading in the first instance to chain breakdown to materials able to sustain attack by environmentally ubiquitous microorganisms.

Hydrolytically biodegradable plastics may be biological in origin or oil-derived. They include polymers of ω-hydroxyacids, including poly(lactic acid), polyanhydrides, cellulose derivatives, and derivatives of PET modified to make them hydrolytically unstable. Proper degradation may require specific conditions. For example, poly(lactic acid) (in other respects very useful—inexpensive, transparent, and renewable and with mechanical properties similar to PET) only biodegrades at useful rates in an industrial composting plant that operates at a high enough temperature to allow hydrolysis of the ester linkages. Composting, either domestic or industrial, is regarded as biodegradation provided it meets specified standards, and this method of disposal solves the problem of contamination of the materials with food waste. The idea of "plastic-eating" microorganisms was of interest in the late 1960s but has largely been abandoned (what would happen if they "got out"?*), though there is a current interest in trying to isolate bacteria that will efficiently produce, for example, ethanol or butanol from cellulose.

Another approach to degradable materials involves what are described as oxo-biodegradable polymers. These are ordinary hydrocarbon-based polymers, to which is added a prooxidant additive. The mechanism of degradation involves a radical chain oxidation, and the chain is gradually broken down to oxygenated compounds such as acids and alcohols, which are then environmentally degraded to CO_2 and water. The materials have found wide use in agriculture and horticulture, as well as in supermarket bags. Early versions used additives that required exposure to sunlight to degrade the polymer, but more modern systems will also work in the dark. Oxo-biodegradability is only partly a disposal option—it's also insurance against long-term litter in terrestrial and marine environments, but its use is contentious because the timescales for complete degradation are several years.

Hydro- and oxo-biodegradation both convert the carbon in the polymer to CO_2, which is slow incineration without energy recovery. An alternative for some plastics is anaerobic digestion to produce methane recovered as a fuel. Anaerobic bacteria are much less efficient than aerobic ones, and only a few easily hydrolysable plastics are susceptible to this kind of process, though it is being rapidly developed in the EU as an alternative to composting recovery of food and agricultural wastes, which may drive the development of more susceptible plastics for packaging.

Finally, plastics, like most types of waste, can be burned and the energy produced harvested. Many plastics (polyethylene, polypropylene, polystyrene) are simple hydrocarbons, with the same sort of energy content as fossil fuels. Waste-to-energy schemes reduce the solid to be disposed of by 85 % by weight and 95 % by volume. Denmark and Sweden are leaders in such schemes. One of the main objections to incineration as a method of dealing with any type of waste is the perception that large amounts of dioxins are emitted—and this was justifiable in considering early technology. In Germany in 1990, industrial incineration accounted for 30 % of dioxin emissions; by 2010, this had been reduced by 99.9 % and is now well under 10 g nationally per year. Backyard burning of rubbish is now a much more serious problem as such incineration takes place at temperatures too low to destroy dioxins. Also significant amounts of CO_2 are emitted—a greenhouse gas. But landfilled waste generates methane—which is 21 times more damaging as a greenhouse gas.

The most problematic plastic to dispose of is PVC. Although modern formulations have removed the most toxic metals, lead and cadmium, it still contains about 30 % of various additives, as well as having a high chlorine content. It has the highest potential for PCB and dioxin formation on incineration. Mechanical recycling, although difficult, results in lower global warming potential than incineration; the developed world largely exports the problem to developing countries. Fortunately, most PVC is used in long-life materials.

---

* For those of you interested in vintage TV, the first episode of a British series called *Doomwatch* featured just such an eventuality. This was novelized as *Mutant 59: The Plastic Eater*, by Kit Pedler (inventor of the Cybermen, in Dr Who) and Gerry Davis.

25.7 IS "GREEN CHEMISTRY" THE FUTURE?

The term "green chemistry" or sustainable chemistry came into fashion in the 1990s, with the first major books and journals in the field appearing toward the end of the decade. The 12 principles of green chemistry were adumbrated by Paul Anastas of the US Environmental Protection Agency in 1998.*

- It is better to prevent waste than to treat or clean up waste after it is formed.
- Synthetic methods should be designed to maximize incorporation of all the materials used in the process into the final product.
- Wherever practicable, synthetic methodologies should be designed to use and generate substances that possess little or no toxicity to human health and the environment.
- Chemical products should be designed to preserve efficacy of function, while reducing toxicity.
- The use of auxiliary substances (*e.g.*, solvents and separation agents) should be made unnecessary where possible and innocuous where used.
- Energy requirements should be recognized for their environmental and economic impact and should be minimized. Synthetic methods should be conducted at ambient temperature and pressure.
- A raw material or feedstock should be renewable rather than depleting, wherever technically and economically practicable.
- Reduce derivatives—unnecessary derivatization (blocking groups, protection/deprotection/temporary modification) should be avoided wherever possible.
- Catalytic reagents (as selective as possible) are superior to stoichiometric reagents.
- Chemical products should be designed so that at the end of their function, they do not persist in the environment and break down into innocuous degradation products.
- Analytical methodologies need to be further developed to allow for real-time in-process monitoring and control prior to the formation of hazardous substances.
- Substances and the form of a substance used in a chemical process should be chosen to minimize potential for chemical accidents, including releases, explosions, and fire.

The 12 principles are a counsel of perfection, and the chemical industry is far from achieving most of them, let alone all simultaneously. Still less are they likely to be applied in research laboratories, where only small quantities of material are in use, and the aim is generally to synthesize something by any available route and then later find a better one if the compound is valuable. However, UC Berkeley College of Chemistry is rebuilding and redesigning its undergraduate laboratories with a new curriculum based on the principles of green chemistry, with the support of a gift from the Dow Chemical Company Foundation.

Even when it's known how to achieve a reaction in a greener manner, changing a long-established industrial process requires significant investment, which may or may not yield a speedy return. Placed against this, companies must factor in the high costs of compliance with environmental legislation, as well as the fact that most of the principles involved in green chemistry should, in the longer term, drive down costs. If cheap solvents such as water are used and reactions don't need to be heated or cooled, overall costs would be reduced. There are, however, cynics who consider that labeling a process as "green" is a PR device for the chemical industry and a disingenuous method of attracting funding in academia. This is still a new field, with great potential but also much to prove.

An important aspect of green chemistry involves atom "economy." If we consider the oxidation of $PhCH(OH)CH_3$ to $PhC(=O)CH_3$ using chromium trioxide, the atom efficiency, the proportion of the product that is actually something we want, is 42 % (Equations 25.4 and 25.5). If the oxidation could be done using molecular oxygen as the oxidant and a catalyst, then this rises to 86 % (Equations 25.6 and 25.7). In both cases, this is only a theoretical figure; to get a more practical

* www.chemistryexplained.com/Ge-Hy/Green-chemistry.html; P.T. Anastas and J.C. Warner, *Green Chemistry*, OUP, Oxford, U.K., 1998.

estimate, we would need to factor in the chemical yield of the processes. And even this does not take account of temperature and solvent use.

(25.4) $$3PhCH(OH)CH_3 + 2CrO_3 + 3H_2SO_4 \rightarrow 3PhC(=O)CH_3 + Cr_2(SO_4)_3 + 6H_2O$$

(25.5) $$\text{Atom efficiency} = \frac{(3 \times 120)}{\{(3 \times 122) + 392 + (6 \times 18)\}} = 42\,\%$$

(25.6) $$PhCH(OH)CH_3 + \frac{1}{2}O_2 \rightarrow PhC(=O)CH_3 + H_2O$$

(25.7) $$\text{Atom efficiency} = \frac{120}{(122 + 18)} = 86\,\%$$

Considerable effort over the last decade has involved the replacement of organic solvents (particularly chlorinated solvents, implicated in ozone depletion) either ideally, with solvent-free processes, or environmentally more benign solvents. Some changes are quick and simple; for example, benzene is replaced by toluene. Benzene is not easily degraded and is a powerful carcinogen and mutagen. Toluene is oxidized to benzoic acid *in vivo* and then excreted. Although there is no such thing as an intrinsically green solvent, water, supercritical water,* and supercritical CO_2 come closest. Bear in mind, however, that supercritical fluids have to be heated and compressed to be supercritical. Also of interest are the so-called ionic liquids.† which are involatile solvents from which products are recovered by distillation and the solvent can be recycled. Some examples of reactions in water are shown in Figure 25.6. Microwave and ultrasound are economical and effective ways to supply energy.

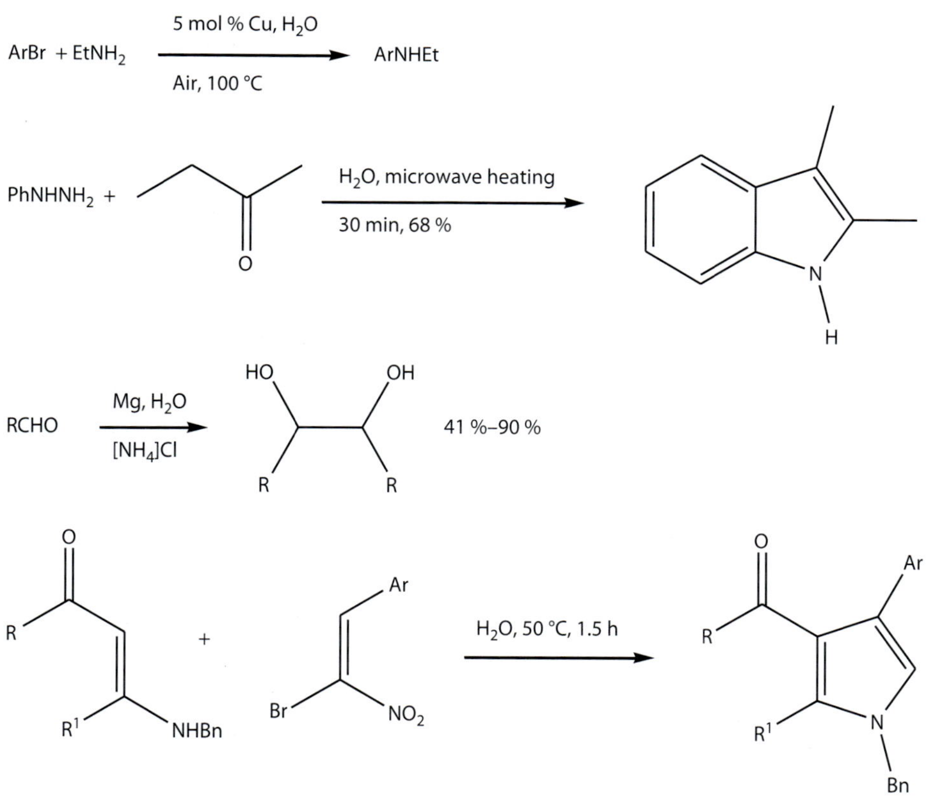

FIGURE 25.6 Green reactions with water as solvent.

* Supercritical fluids are liquids at a temperature and pressure above their critical point, when distinct gas and liquid phases do not exist. They dissolve materials like a liquid but have very low viscosity and surface tension and diffuse through solids like a gas.

† P. Wasserscheid and T. Welton, *Ionic Liquids in Synthesis*, 2nd edn., Wiley-VCH, Weinheim, Germany, 2007; M. Freemantle, *An Introduction to Ionic Liquids*, RSC, Cambridge, MA, 2009.

PROBLEM 25.5

Propose a reaction mechanism for the final example of Figure 25.6.

Solution

The key to solving this is to work out which atoms have been joined up.

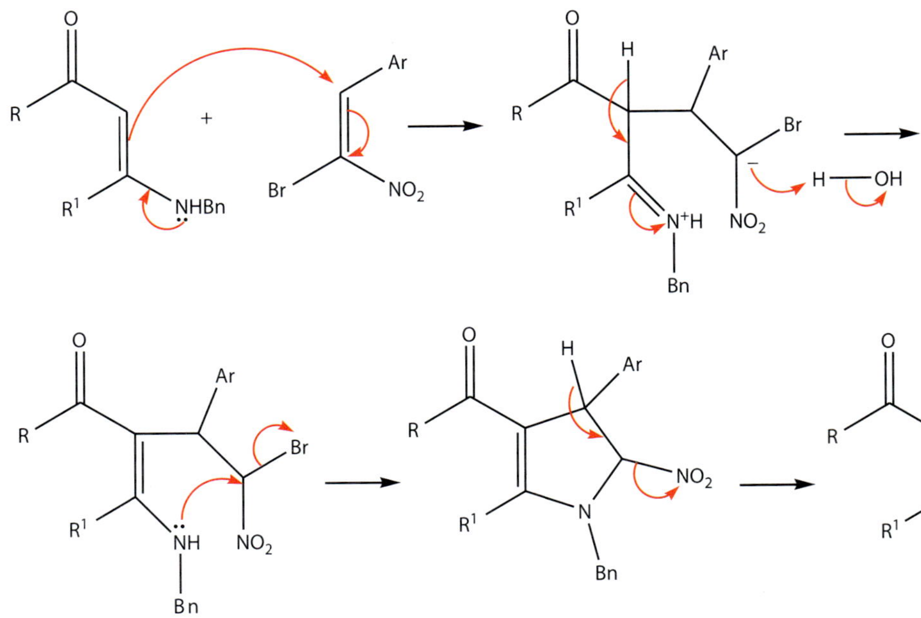

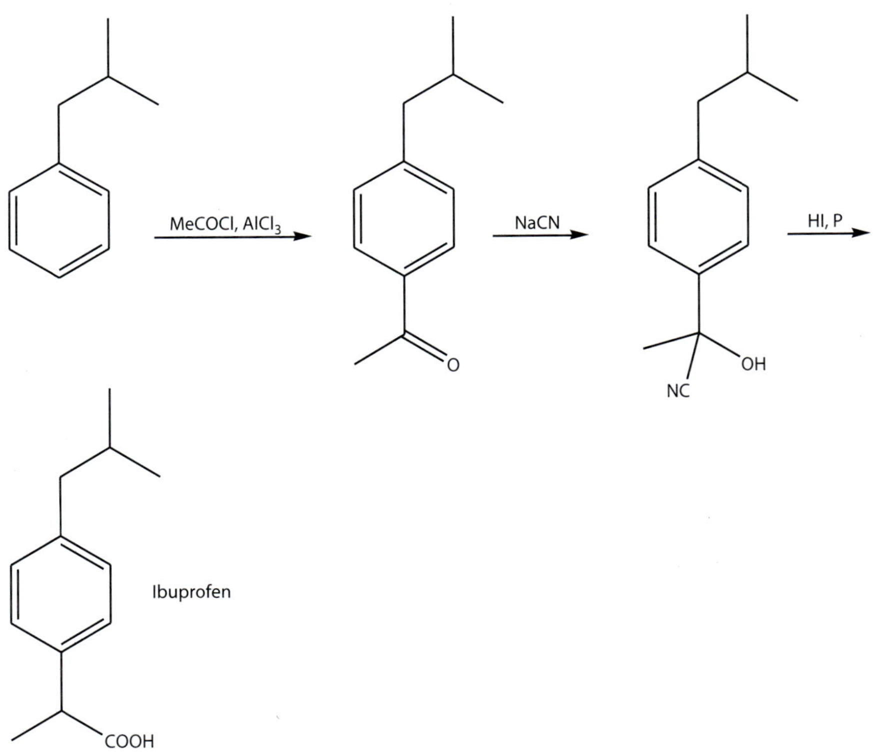

FIGURE 25.7 Traditional synthesis of ibuprofen.

As well as solvent changes, improvements can be achieved by the use of catalytic rather than stoichiometric reagents. The traditional synthesis of ibuprofen is shown in Figure 25.7. Aluminum chloride is not catalytic in this process; it complexes with the product, and 760 tons of $AlCl_3$ are required to produce 1000 tons of ibuprofen. The use of HF as a Friedel–Crafts *catalyst* improves atom efficiency (although HF is very corrosive, it is possible to recycle it), and solid acid catalysts may prove better still.

Catalysts that can be used in water are particularly useful (Figure 25.8). The first reaction is an organocatalyzed process, highly enantioselective, but requiring a relatively large catalyst loading. Diels–Alder reactions are catalyzed by various lanthanides acting as Lewis acids, but scandium gives much the best results for carbocycles. Ytterbium catalysts are often used for 4 + 2-cycloadditions involving heteroatoms. The final example shows that transition metals can also be used as catalysts in water.

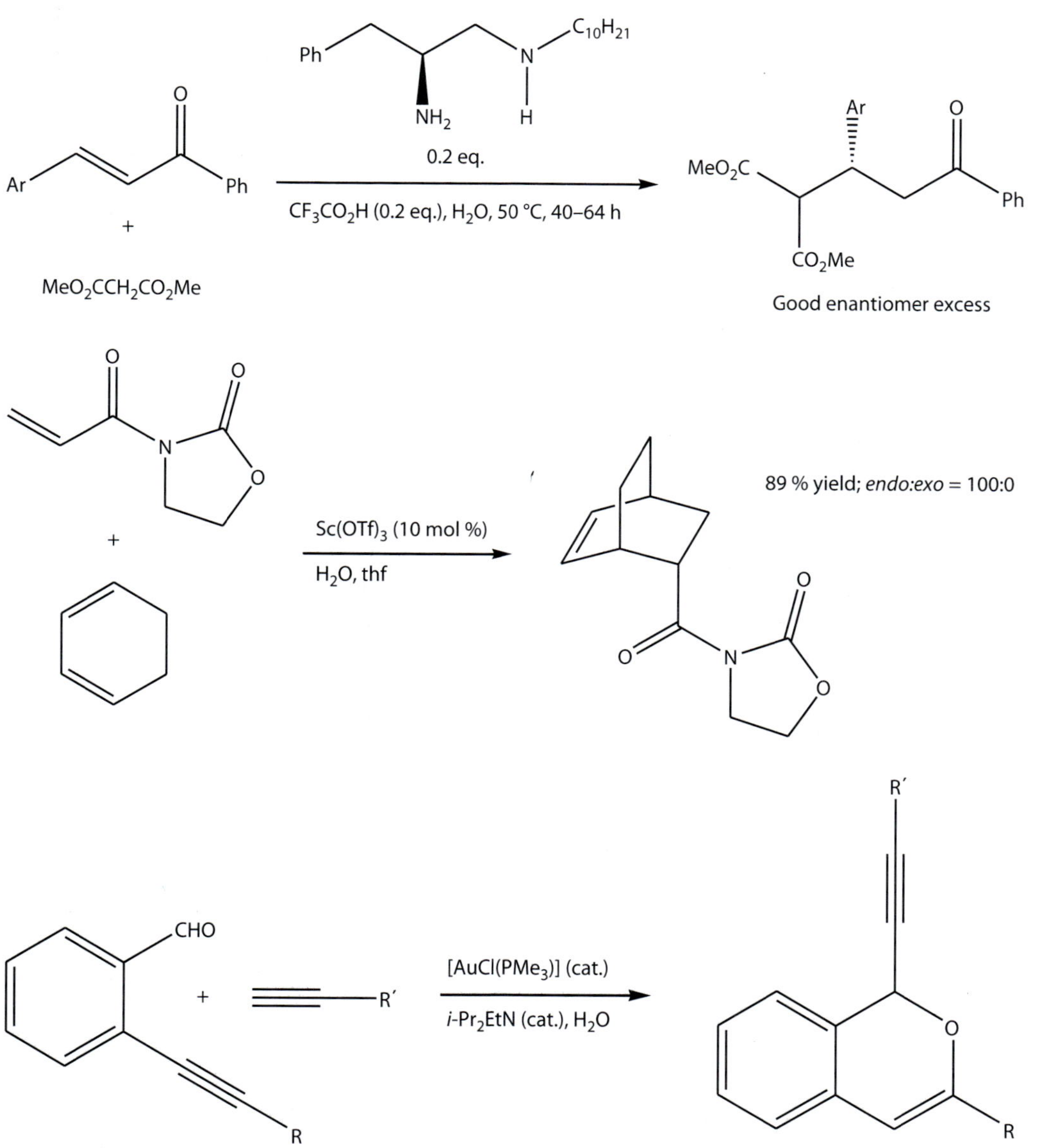

FIGURE 25.8 Catalyzed reactions in water.

REVIEW PROBLEMS

1. The synthesis of dieldrin starts with hexachlorocyclopentadiene, and the intermediate in the synthesis is another insecticide, aldrin. Devise a synthesis and give a mechanism for each step.

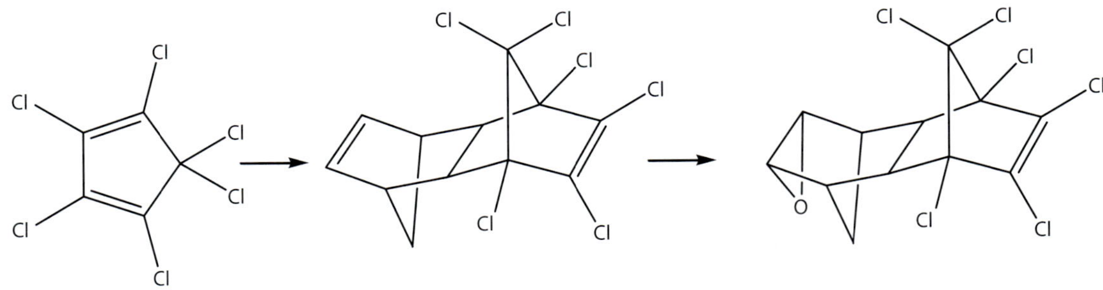

Aldrin Dieldrin

2. Propose reaction mechanisms for the reactions of Figure 25.8.
3. Propose reaction mechanisms for each of the following reactions:

(a)

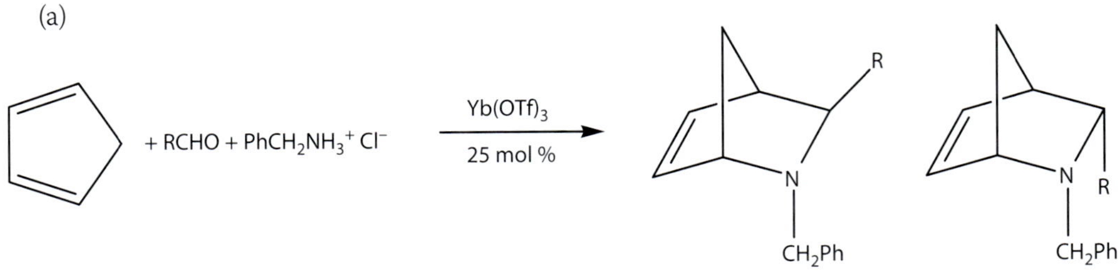

72 % yield, *exo*:*endo* = 1:4

(b)

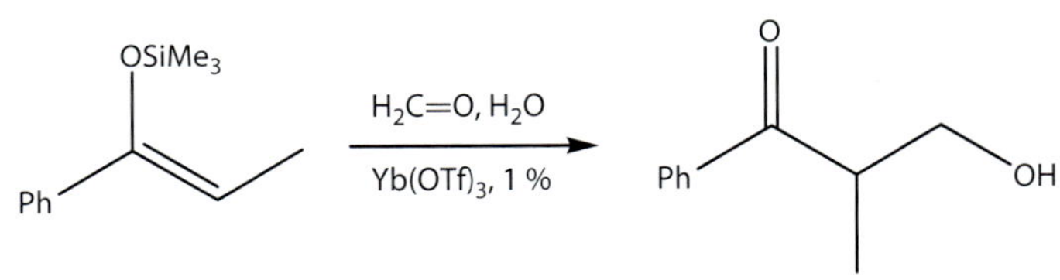

(c)

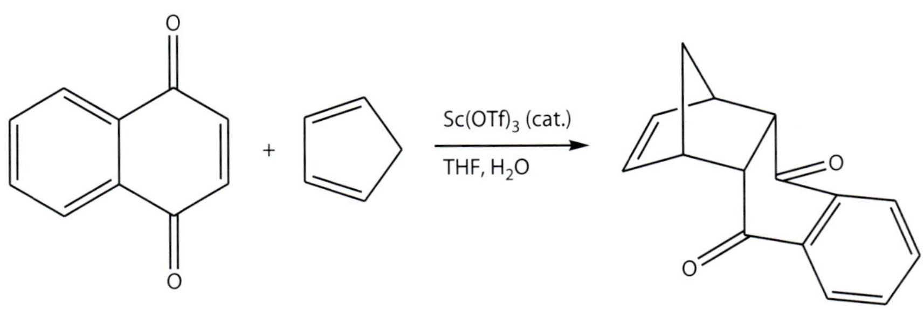

(d)

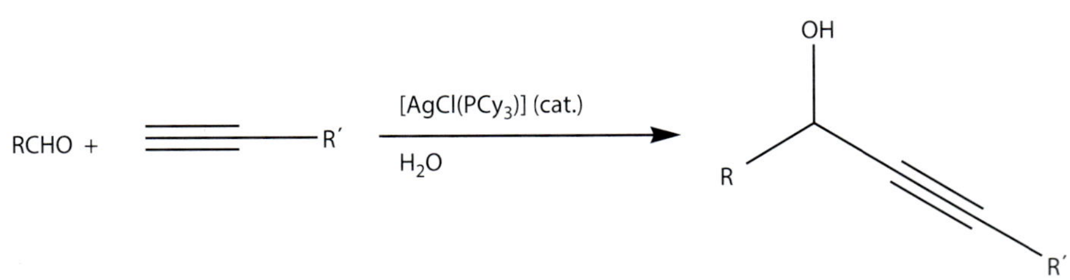

(e)

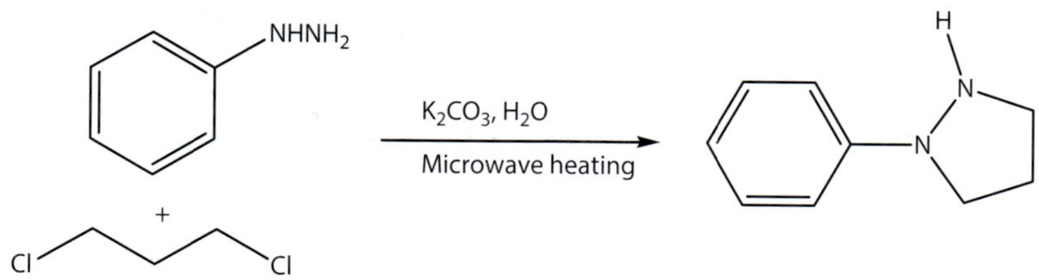

4. For each of the following "crises," trace the origin of the problem, and evaluate whether the measures taken were/are appropriate:
   (a) The Belgian poultry crisis, 1999
   (b) The Irish pork crisis, 2008
   (c) The 2008 Chinese milk scandal
   (d) Honey laundering in 2011
   (e) Di(2-ethylhexyl)phthalate in Taiwanese palm oil in 2011
5. Find out what sort of beverage cups is in use at your home institution. Are these recyclable, and are they recycled? If they are not, why is that? Could they be recycled? What happens to them, and can you suggest, on the basis of a lifecycle analysis, either a better material for the cups or a better way of disposing of them?
6. The Bhopal disaster occurred in a plant designed to manufacture the pesticide carbaryl. Explain the chemistry of the synthesis of carbaryl at Bhopal as it was designed to occur. What went wrong?
7. Explain why methane is a more effective greenhouse gas than carbon dioxide.
8. A statement has been made that in the summer months, 50 % of the PAHs in the atmosphere of Los Angeles are derived from domestic barbecues. How would you verify such a statement? What could or should be done about it, if it's true?
9. Marine oil spillages as a result of ships going aground, leaking, or other accident are sometimes treated with chemical dispersants. What are these, how do they work, and is this the best strategy? How were these applied in the BP Deepwater Horizon oil spillage in the Gulf of Mexico in 2010, and what has been their effect?
10. In 1989, the ship Perenits was wrecked during a storm on the French coast. The cargo included 5.8 tons of lindane, 1 ton of permethrin, and 0.6 tons cypermethrin. If these had been discharged into the marine environment, which would have been the most damaging, and why? Fortunately, only 1 container of lindane was lost. What would be likely its effect on the marine environment?
11. Organochlorine compounds such as lindane, DDT, and PCBs have long been banned from most applications. However, they should still be monitored in the marine environment. Why?

FURTHER READING

http://ec.europa.eu/environment/endocrine/index_en.htm.
http://www.d2w.net/.
http://www.environment-agency.gov.uk/static/documents/resaerch/Carrier_bags_final_18-02-11.pdf.
http://www.epa.gov/scipoly/oscpendo/index.htm.
http://www.epa.gove/endo/pubs/edspoverview/whatare.htm.
http://www.epi-global.com/.
http://www.fda.gov/scienceresearch/bioinformativstools/endocrinedisruptorknowledgebase/default.htm.
http://www.niehs.nih.gov/health/topics/agents/endocrine/index.cfm.
http://www.nrdc.org/health/effects/qendoc.asp.
http://www.organic-chemistry.org/topics/green-chemistry.shtm.
http://www.who.int/ipcs/publications/endocrine_disruptors/endocrine_disruptors.en/.
www.wrap.org.uk.
Scheirs, J. and W. Kaminsky. *Feedstock Recycling and Pyrolysis of Waste Plastics*, Wiley, Chichester, U.K., 2006.
Scott, G. *Plastics in the Environment*, RSC, Cambridge, U.K., 1999.

Organic Chemistry in Forensic Science

26

26.1 INTRODUCTION

The term forensic science covers all aspects of science as applied in the law. The word is derived from *forensis*, Latin, of a public place, and was first used in English around 1650. Every science is now used in the pursuit of justice—from forensic accounting through biochemistry, chemistry, engineering, medicine, psychology, and physics to zoology. A vast subject and one where truly cutting-edge science is being developed. What can chemistry, and more specifically organic chemistry, contribute? In this brief introduction, it's impossible to be comprehensive or to go into great depth; some fascinating topics will receive less coverage than you might wish. I have tried to avoid intrusion from the requirements/regulations of any specific legal system, but inevitably, there will be a few observations on legislation, particularly concerning drugs of abuse.

26.2 DRUGS OF ABUSE

The term "drugs of abuse" is a relatively modern one and is applied to substances considered injurious to the mental or physical health of the individual, and the use of which is therefore controlled, or prohibited, by legislation. Controls on substances were gradually introduced during the late nineteenth and twentieth centuries. While the 1960s have a reputation for the rise of "drug culture," archaeological evidence suggests that many substances have been used for social/medicinal (to allow people to work harder and longer in a challenging and unforgiving environment), recreational, and religious (spirit trances) purposes for millennia. Restrictions are now mandated by UN conventions; interestingly, some of these derive from concerns of governments that drug abuse was reducing the work rate of the population and therefore damaging the country economically, rather than being related to public health. In addition to national and international legislation, sporting bodies impose other restrictions and controls and are vigorous in their testing programs. Most jurisdictions recognize three categories of offense—possession, distribution/supply, and manufacture (or cultivation, for cannabis).

26.2.1 CATEGORIES OF DRUGS OF ABUSE

Categorization of substances as to the seriousness of the harm they do, and hence the seriousness of the offenses associated with them, varies widely around the world. Similarly, legitimate medicinal or research use of materials is permitted differently. However, and more usefully to the organic chemist, the drugs can also be categorized by their effect—hallucinogens, narcotics, stimulants, depressants, analgesics, and human-performance-enhancing drugs (see Section 16.2.5). Hallucinogens distort the perception of time and reality, and most are based on two structural families—derivatives of phenethylamine, **26.1**, or tryptamine, **26.2**. Probably, the most famous of the hallucinogens is lysergic acid diethylamide (LSD), **26.3**, of the tryptamine family. This is a very potent hallucinogen; 25 μg can give a 12 h "trip." It is generally sold as microdots on thin paper or as small pieces of paper that have been soaked in a solution of LSD, so called blotter acid. Doses used recreationally range from 30 to 150 μg. Most production is synthetic from lysergamide, which can be isolated from morning glory seeds or from various ergot alkaloids. Ergotamine, **26.4**, isolated from rye infected with *Claviceps purpurea*, caused outbreaks of *St. Anthony's Fire* documented from the Middle Ages until the nineteenth century (occasionally still occurs in the developing world), and its effects have been suggested as a cause of the hallucinations adduced during the Salem witchcraft trials in 1692/3.* Ergotamine has a legitimate medical use in the treatment of severe migraine.

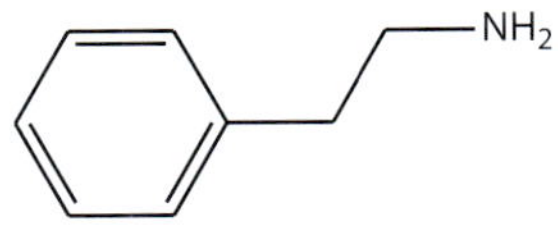

26.1, phenethylamine

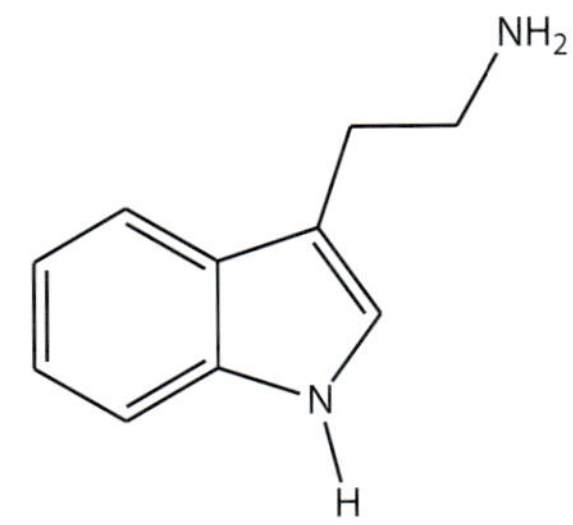

26.2, tryptamine

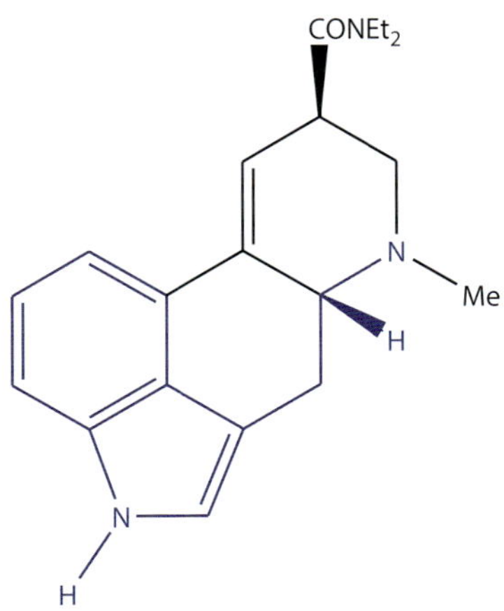

26.3, LSD; the tryptamine section is highlighted in blue

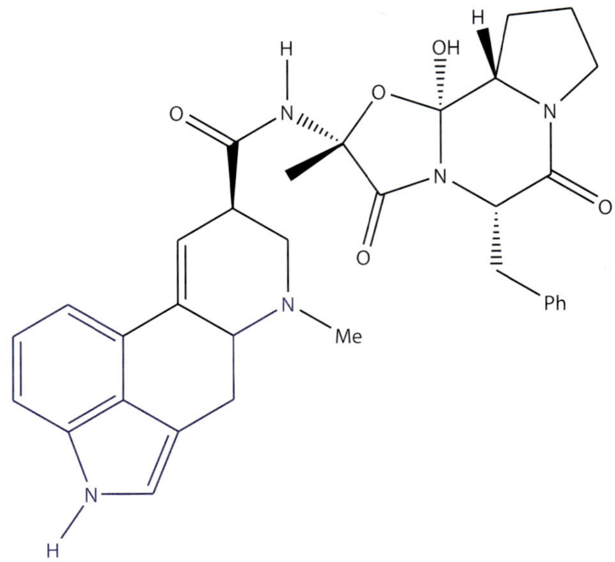

26.4, ergotamine

Blotter acid

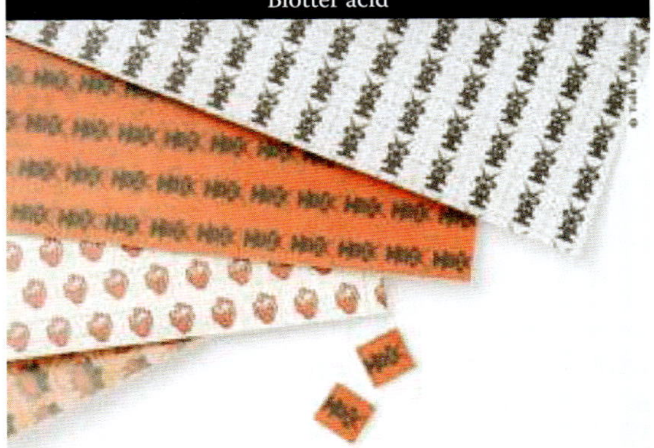

Morning glory flowers. (Thanks to Camilla Tornoe for this photo.)

Another hallucinogen of the tryptamine family is psilocin, **26.5**, isolated from "magic mushrooms." Mescaline, **26.6**, is of the phenethylamine family and is isolated from the Peyote cactus. Both materials have had long use in spiritual practices in Mesoamerica. Methamphetamine, marijuana (where the active constituent has a very different structure), and ecstasy are hallucinogens at high doses.

* H. Sidky, *Witchcraft, Lycanthropy, Drugs and Disease*; an anthropological study of the European Witch Hunts, Peter Lang, 1997.

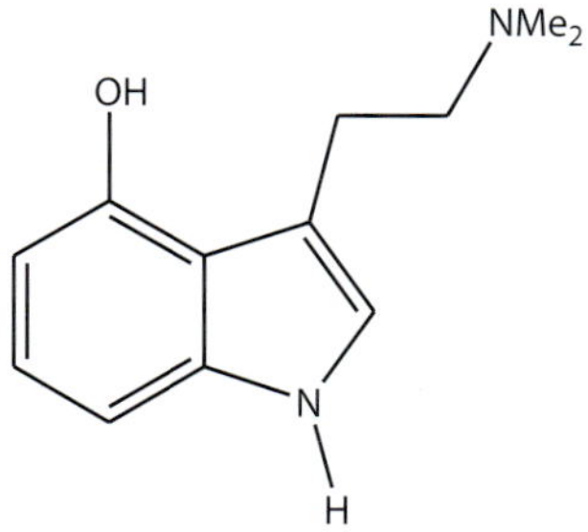

26.5, psylocin

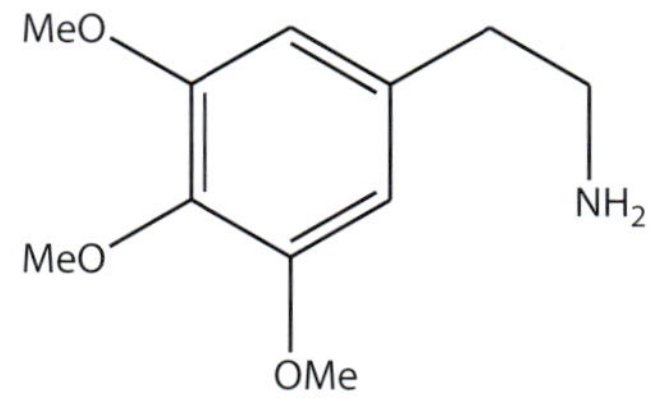

26.6, mescaline

Magic mushrooms, "Shrooms"

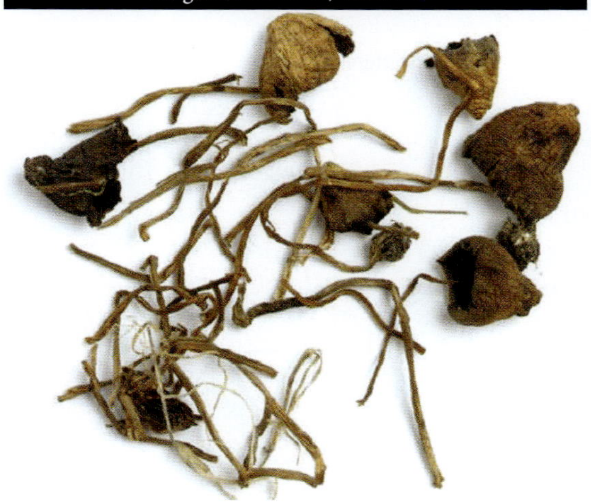

Peyote cactus

PROBLEM 26.1

Identify each of the following hallucinogens as phenethylamine or tryptamine derivatives and highlight the core atoms.

(a)

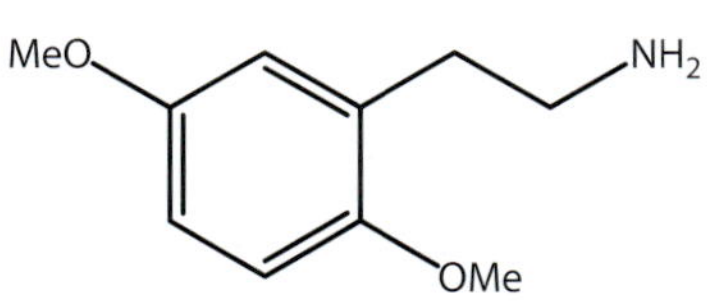

2-CH

(b)

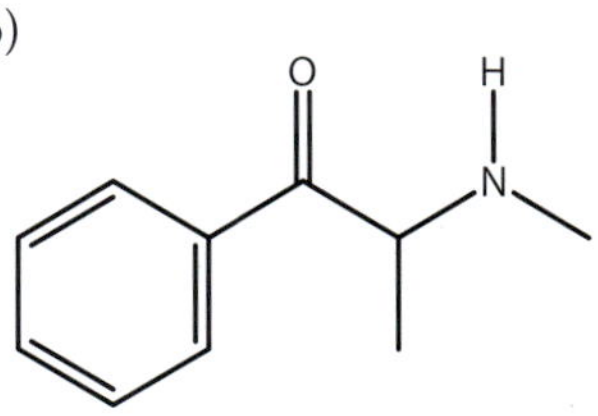

Methcathinone

(c)

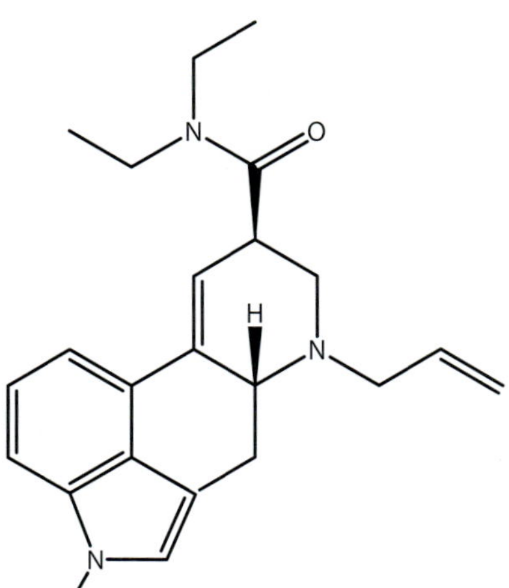

AL-LAD

(d)

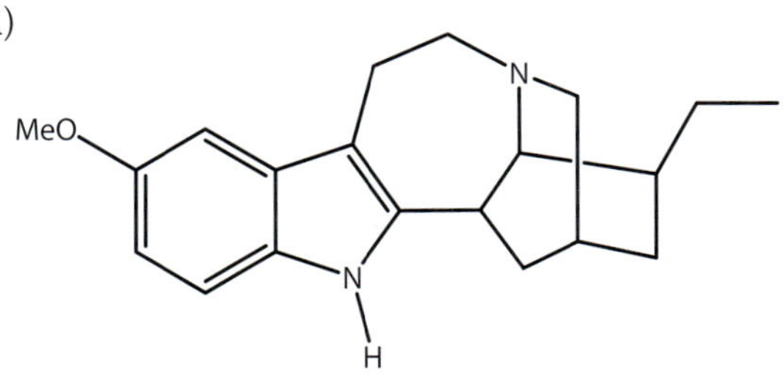

Ibogaine

Solutions

(a) and (b) Phenethylamine
(c) and (d) Tryptamine

(a)

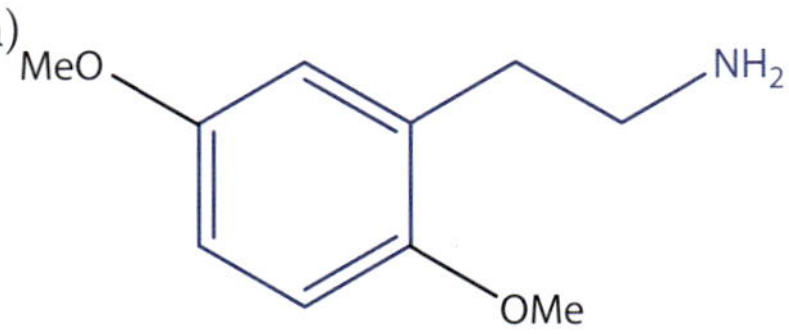

2-CH

(b)

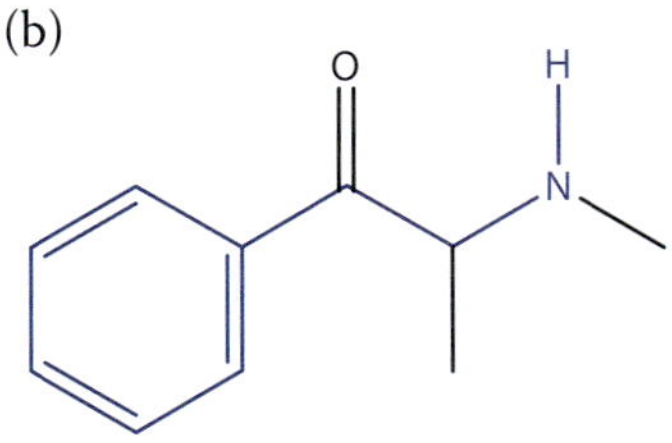

Methcathinone

(c)

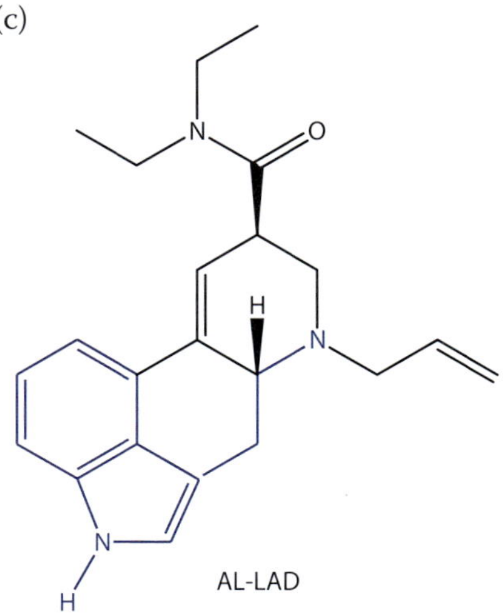

AL-LAD

(d)

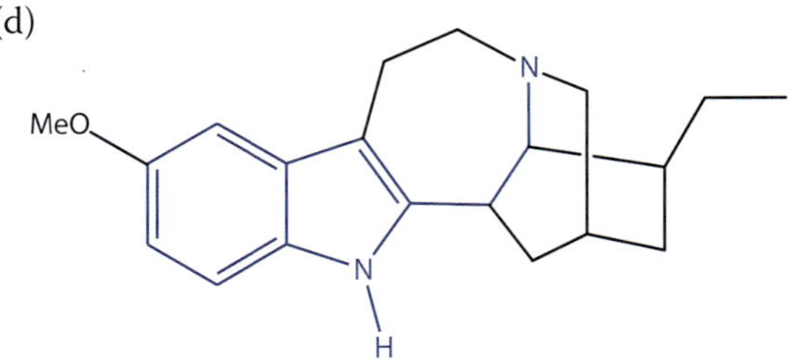

Ibogaine

Narcotics (see also Section 22.2) reduce pain by attaching to the opioid receptors throughout the central nervous system; this blocks the transmission of the nerve impulses that relay the sensation of pain. Other effects include sleepiness and a sense of pleasure and well-being. Morphine, heroin, and codeine, **26.7**, **26.8**, **26.9**, are all derived from the opium poppy, *Papaver somniferum*. Raw opium contains between 4 % and 21 % morphine. Heroin is not a natural product, but is produced during the processing of raw opium. The use of opiates was known to ancient physicians, and they have been used for millennia for pain relief. Street heroin is generally approximately 35 % pure; the additives used to dilute it are referred to as cutting agents and may include quinine, starch, lactose, novocaine, mannitol, or caffeine. The production of heroin from raw opium is a batch process, and impurities may be characteristic of the batch. Small amounts of a number of other morphine alkaloids are also likely to be present; amounts depend on growing conditions. Note that when used medicinally, heroin is called diamorphine, a contraction of diacetylmorphine. Methadone, **26.10**, is used to wean addicts off heroin; it works by blocking the opiate receptors, so that heroin then has little effect.

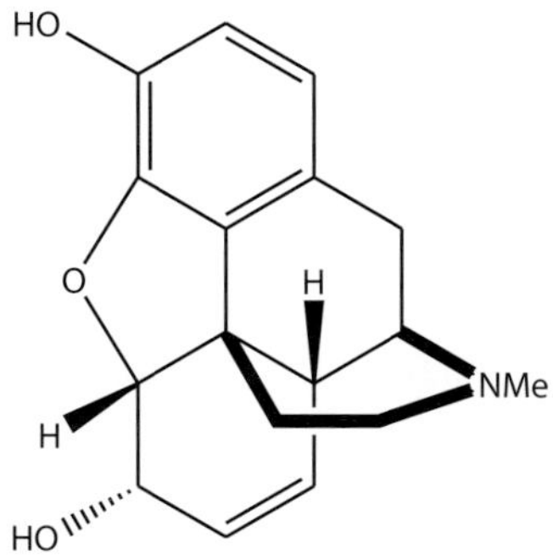

26.7, morphine

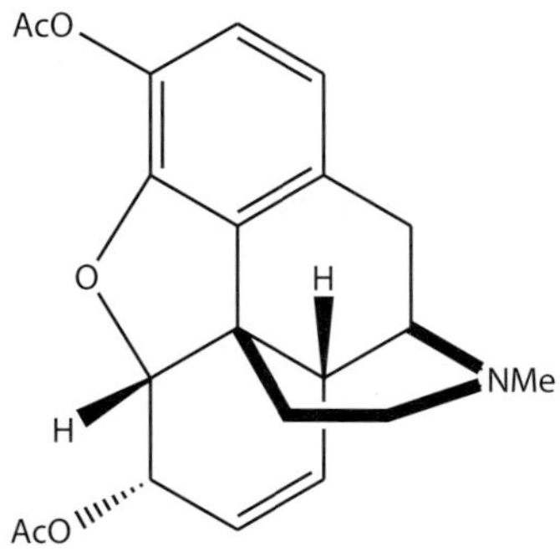

26.8, diacetylmorphine
diamorphine, heroin

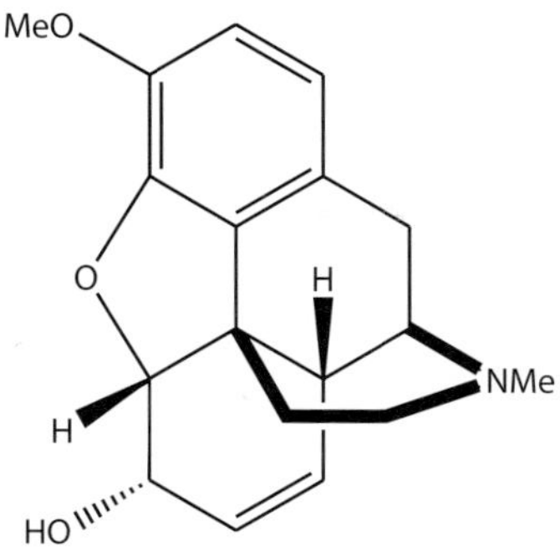

26.9, codeine

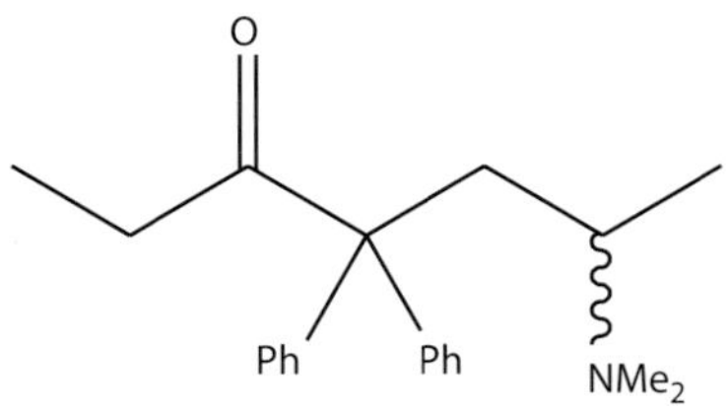

26.10, methadone

PROBLEM 26.2

Testing for drugs *in vivo* depends generally on urinalysis or, less usually, on testing of blood (urine is easily collected; a licensed nurse, physician, or phlebotomist is needed for the collection of blood samples). Testing therefore involves identification of the metabolites of the drugs, rather than the drugs themselves. The main metabolite of both heroin and codeine is morphine 6-glucoronide, **26.11**. What does this tell us about the metabolism of these compounds?

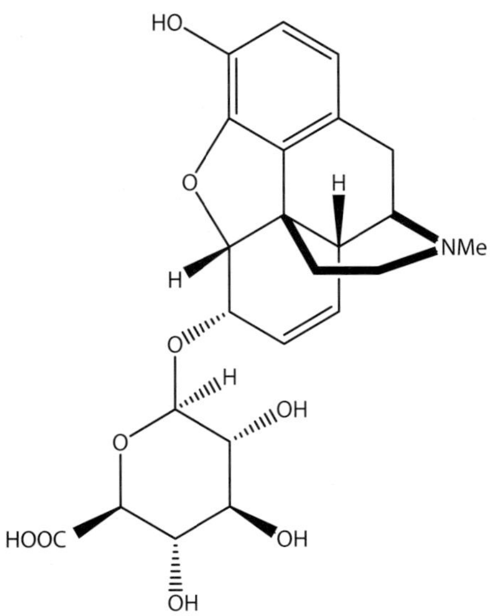

26.11, morphine-6-glucuronide

Solution

During heroin metabolism, the acetyl groups must be cleaved to yield morphine; this might simply be hydrolysis or it may be an enzyme-mediated process. Codeine is also demethylated; this is known to be an enzyme-mediated process (there are no simple chemical processes that would achieve this under physiological conditions), and some 10 %–15 % of the population are deficient in the enzyme, so codeine is not converted to morphine, and codeine has little analgesic effect for this group. Another 10 % of the population has overactivity of the enzyme, so that the codeine is demethylated very rapidly and effectively, and they must be careful not to consume an overdose, which would be less for them than for someone with more standard enzyme activity. Finally, in order to be excreted effectively in urine, the morphine is converted to a sugar derivative, to ensure water solubility.

Stimulants, as the name indicates, stimulate the central nervous system, promote alertness, and interfere with sleep. Many stimulants are hallucinogens at high doses. Cocaine is derived from the South American *Erythroxylum coca*, and coca leaves have been chewed in the Andean region for millennia, to assist the body in coping with hard labor at high altitude. Cocaine is produced by solvent extraction, and batch processing, so that impurities may be detected that are characteristic of the particular batch. Cocaine-containing medicines and beverages were widely sold in the late nineteenth and early twentieth centuries. The stimulant effect of Coca-Cola originally came from cocaine; this was replaced by caffeine in the late 1920s. Most cocaine is supplied as the hydrochloride salt, **26.12**, and this may be converted into the free base, **26.13**, referred to as crack cocaine or crack rocks by simple kitchen chemistry using baking soda, sodium bicarbonate, to deprotonate the salt. In this form, faster euphoria is achieved. In North America, the most common cutting agents are sugars, whereas in Western Europe, cocaine is more commonly cut with local anesthetics such as lignocaine and procaine.

Erythroxylum coca

Cocaine was regarded as a valuable stimulant and tonic in the early twentieth century

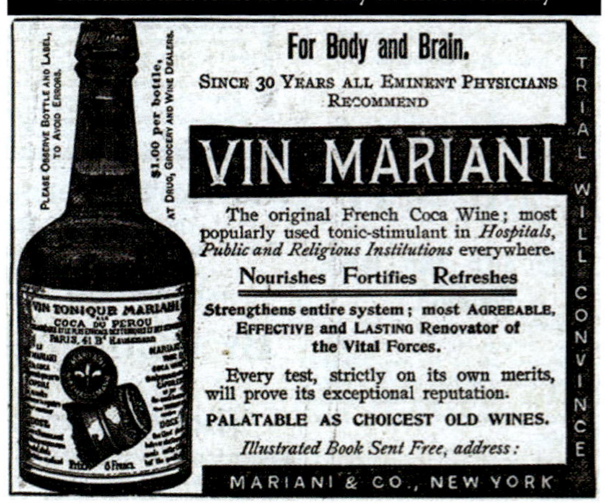

26.12, "snow" **26.13**, "crack"

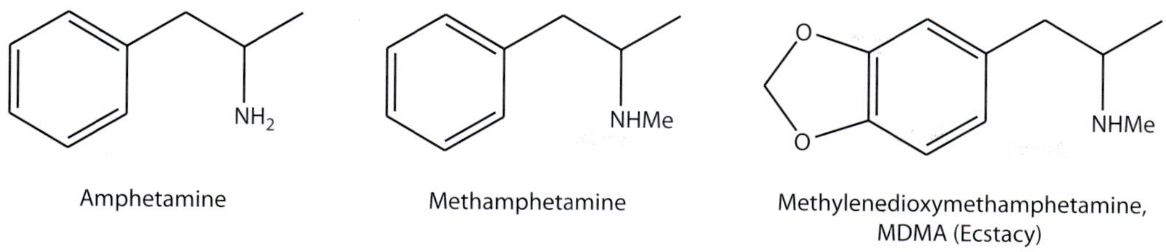

FIGURE 26.1 Amphetamines.

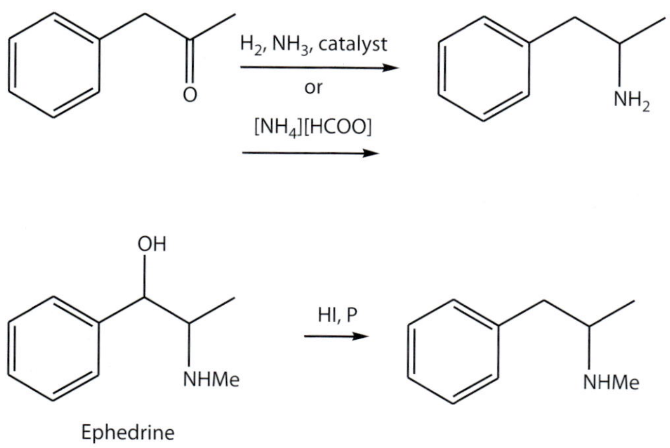

FIGURE 26.2 Syntheses of amphetamines.

The other major group of street drugs that have a stimulant effect are the amphetamines—amphetamine, methamphetamine, and ecstasy (Figure 26.1). These are substituted phenethylamines, explaining their hallucinogenic effects at high doses. Amphetamines are referred to as speed, uppers, or ice, for the smoked version. Ecstasy and most of the other amphetamines were originally patented as appetite suppressants, and their use for recreational purposes was originally designed to circumvent existing legislation. Most jurisdictions now control many substances derived from phenethylamine, though some exceptions are made for medicinal products. In the United States, amphetamine is usually synthesized by reductive amination of phenylacetone (hence, the sale of this is also controlled), and in Europe by the Leuckart synthesis using ammonium formate both for reduction and amination (Figure 26.2). In Japan, reduction of ephedrine, a well-known and legitimate constituent of tonics and cold remedies, is employed. Many of these compounds have legitimate medical uses—amphetamine is marketed as Adderall® for attention deficit disorder and is also used to treat narcolepsy. The use of amphetamines to promote alertness and endurance in the military is more controversial, and accurate information is difficult to obtain, but they have certainly been used since before the Second World War.

PROBLEM 26.3

Estimate the pK_a values for cocaine, amphetamine, and methamphetamine hydrochlorides. Why are the salts more commonly used for oral administration and the free bases for inhalation?

Solution

Cocaine is a tertiary amine, methamphetamine a secondary amine, and amphetamine a primary amine. Typical pK_a values for $[R_3NH]^+$, $[R_2NH_2]^+$, and $[RNH_3]^+$ are, respectively, 9.76, 10.66, and **10.63**. The salts are water soluble and hence more easily ingested, but they are nonvolatile, so difficult to inhale. The free bases are less water soluble but relatively volatile, hence easier to inhale. Since the lungs are richly supplied with blood vessels, inhalation results in higher and faster bioavailability of the drugs.

An observer from the planet Zog might well wonder why anyone would want to take a depressant for recreational purposes. However, it's worth remembering that the most widely abused substance,

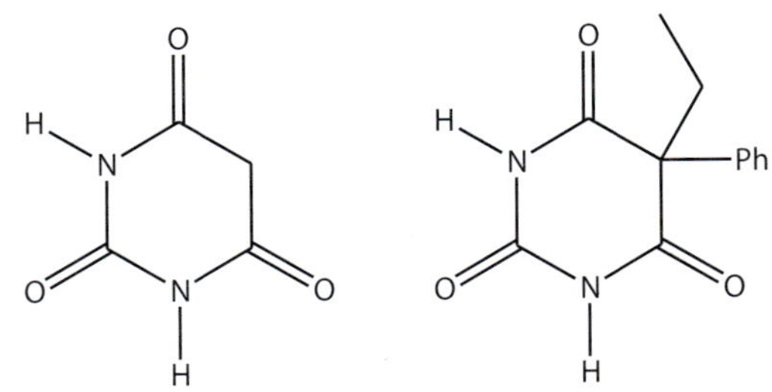

Barbituric acid Phenobarbitone

FIGURE 26.3 Barbiturates.

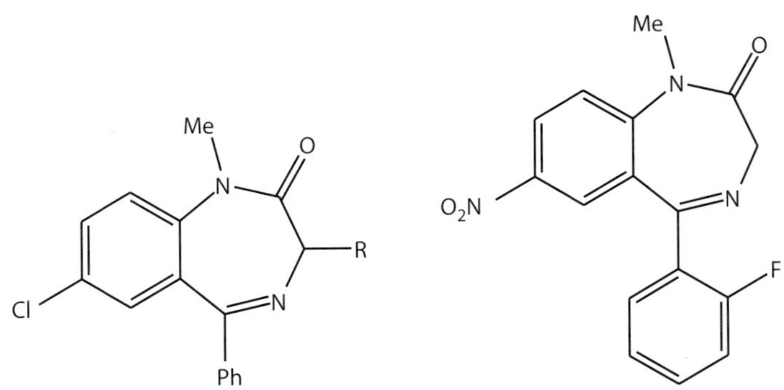

R = H, valium, diazepam
R = OH, temazepam

Rohypnol, flunitazepam, roofie

FIGURE 26.4 Benzodiazepines.

alcohol, is a depressant. Other abused depressants include barbiturates (Figure 26.3) and benzodiazepine tranquilizers (Figure 26.4). Barbiturates have legitimate medical uses for the treatment of sleep disorders and epilepsy, though they have largely been replaced with more modern alternatives. Valium and related compounds interact with the gamma-aminobutyric acid (GABA) receptors in the brain and depress the central nervous system. Many of these have useful medicinal properties in treatment of sleep disorders, management of anxiety, premedication for surgery, and other procedures such as colonoscopy. They may be abused to "come down" from a stimulant high. Rohypnol has a particularly powerful effect, causing anterograde amnesia, and it has become widely associated with drug-assisted rape, although there is more evidence of its use in drug-assisted robbery. It is controlled in most jurisdictions and banned completely in some countries.

Ketamine, **26.14**, and GHB, **26.15**, have also been associated with drug-assisted rape. Ketamine is primarily a veterinary anesthetic but also has uses in human medicine. GHB is controlled in most jurisdictions but gamma-butyrolactone, from which it can be obtained *in vivo* or *in vitro*, is a widely available industrial solvent, and although it has been recently controlled in some countries, it is not yet clear how effective the controls are.

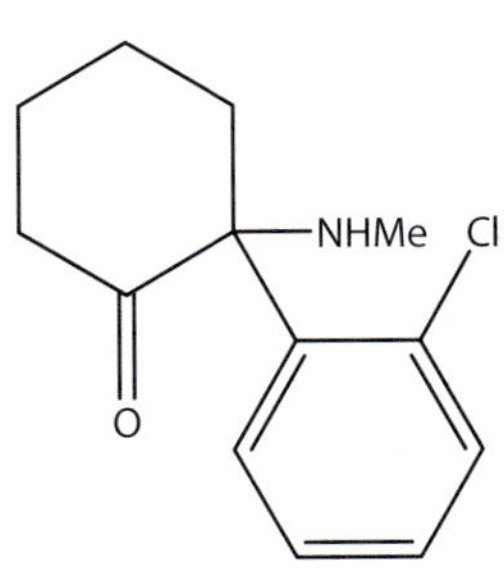

26.14, Ketamine

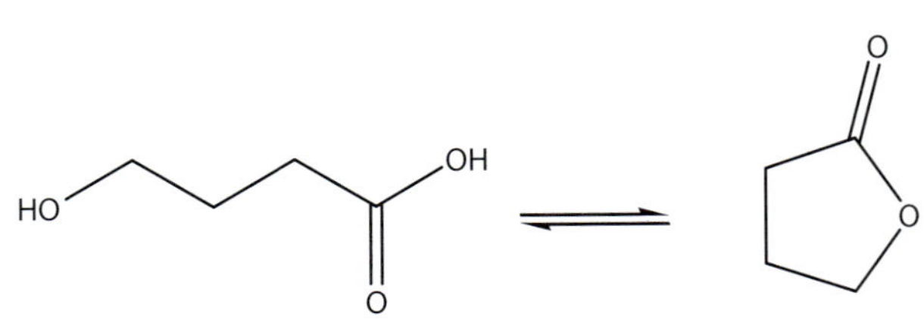

26.15, GHB

γ-butyrolactone

Cannabis, from *Cannabis sativa*, deserves a special mention, although the methods of identification of the material are primarily biological (a study of plant morphology) rather than chemical. It has been in use for 5000 years, with medicinal use recorded in China in 2737 BCE and religious use in India by 1000 BCE. It is most frequently smoked, but oil may be obtained by solvent extraction. The active constituent was identified in 1964 as Δ^9-tetrahydrocannabinol (Δ^9-THC), **26.16**.

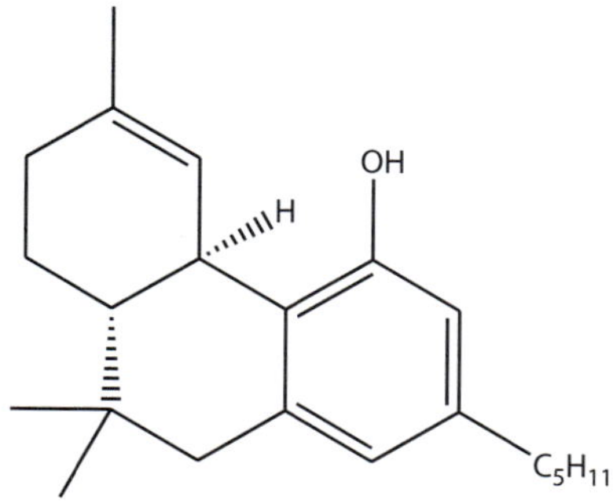

26.16, Δ^9-THC

Cannabis plants

PROBLEM 26.4

Study the synthesis of valium shown in the succeeding text and answer the questions posed:

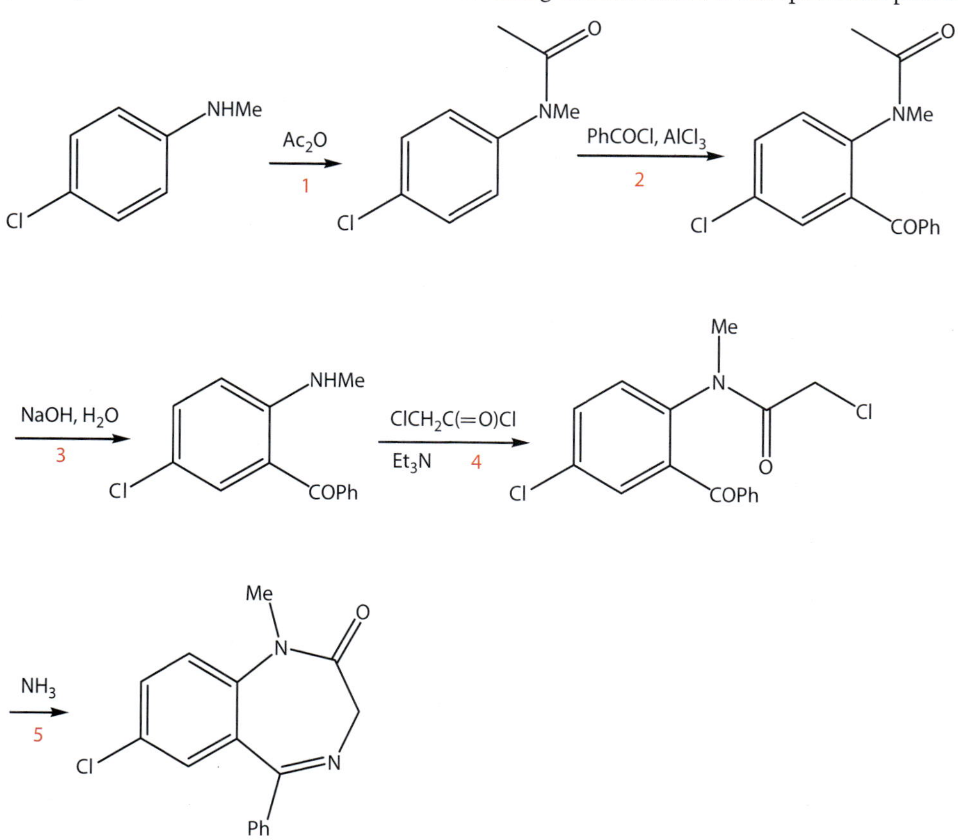

(a) Why is it necessary to protect the amine with an acetyl group (step 1) before the Friedel–Crafts reaction (step 2)?
(b) Explain the regioselectivity of the Friedel–Crafts reaction (step 2).
(c) Suggest mechanisms for steps 4 and 5.

Solutions

(a) The nitrogen would otherwise attack both the acid chloride and the aluminum chloride catalyst.
(b) The amide is *o,p*-directing and activating; chlorine is *o,p*-directing and deactivating. The activating group "wins."

(c)

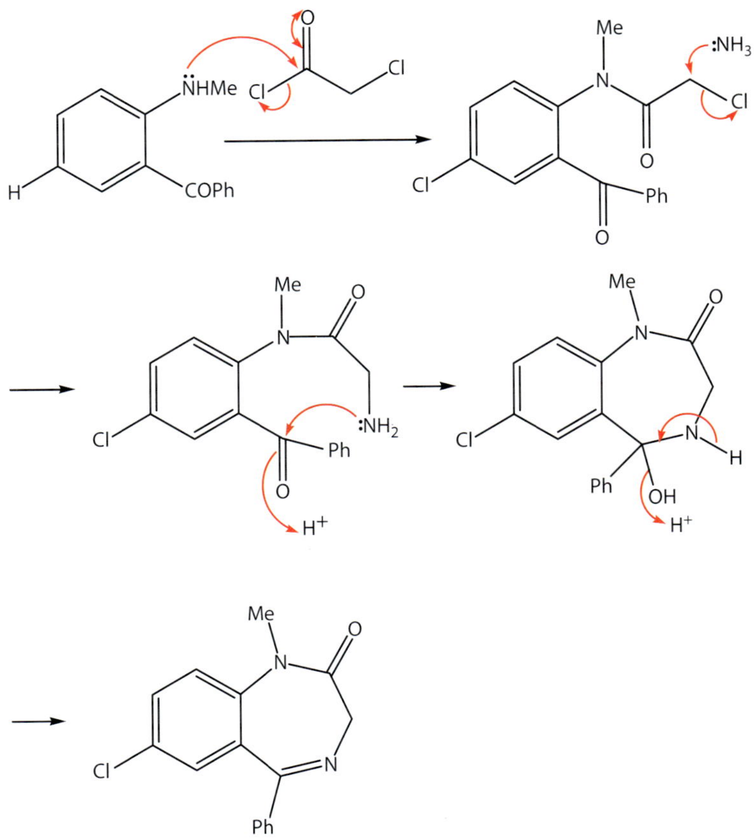

26.2.2 PRESUMPTIVE TESTS FOR DRUGS OF ABUSE

The methods used to test for and analyze drug samples depend on both circumstances and the amount of material available. Where there is ample material, testing will usually begin with presumptive tests in the field. If there is very little, then investigators must proceed directly to instrumental methods. Presumptive tests are designed to be inexpensive, simple, and portable and not to require great knowledge or skill on the part of the investigator. They will give an indication only of the class of material involved, and there will be some false positives with non-controlled substances, so confirmatory tests will also be necessary. Both positive and negative control tests are essential. Negative control tests involve testing the reagent with pure solvent—this ensures that nothing is contaminated with material from previous tests or anything that might interfere with the test. A positive control test involves reacting the testing agent with an authentic sample of the material for which it is testing. This shows that the test is working properly and provides the investigator with a reference color for comparison and a reference time frame in which the color should develop. There are examples where innocent material does give a color test (*e.g.*, some instant coffee when testing for cannabis), but it does so much more slowly than authentic material.

The Marquis test involves a reagent made from 100 cm^3 conc. sulfuric acid and 5 cm^3 40 % formaldehyde solution. Amphetamine and methamphetamine give orange-brown colors, ecstasy and other methylenedioxy-substituted amphetamines give a purple-black color, and opiates give a deep purplish-red color. The test is very sensitive with detection limits from 1 to 20 μg and takes only 1–2 min. However, many other compounds give a color reaction—aspirin gives a dark-red color, and sugar a dark-brown one.*,† The chemistry in these color reactions is complex but is postulated to involve attack of either a benzene or an indole ring on protonated formaldehyde with the ultimate formation of a doubly benzylic cation, which is the source of the color (Figure 26.5). However, other authors suggest that an oligomer or polymer is formed, and most of the materials have not been fully characterized.

Ehrlich's reagent is used to test for LSD. This consists of 4-dimethylaminobenzaldehyde in phosphoric acid. LSD gives a blue-purple color—but all indoles give some color reaction, although the exact color varies. The reaction mechanism is shown in Figure 26.6. A number of other tests depend on the color of inorganic complexes, such as the Mandelin reagent, which uses a 1 % solution of ammonium metavanadate in conc. sulfuric acid to give olive-green products with opiates, or the Scott test for cocaine, which uses cobalt isothiocyanate.

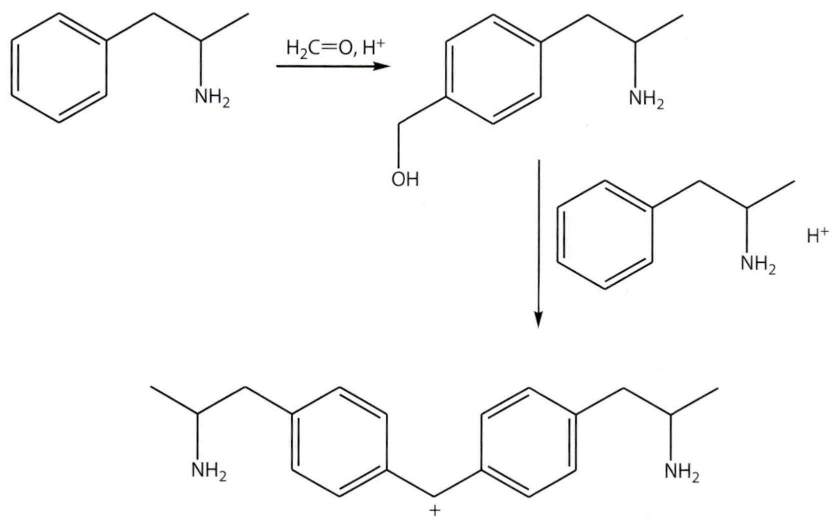

FIGURE 26.5 Marquis reaction of amphetamine.

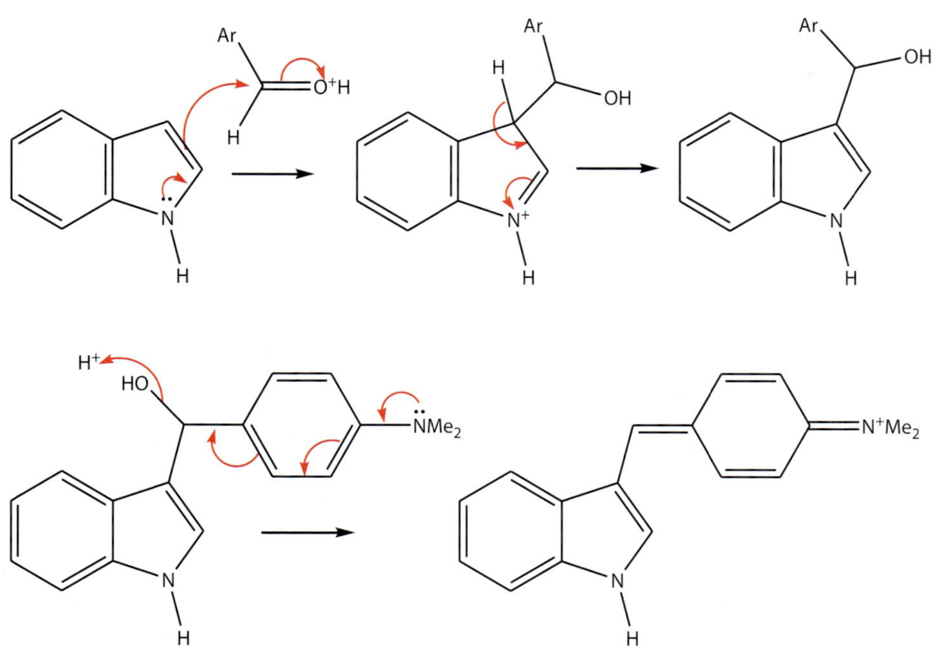

FIGURE 26.6 Mechanism of the reaction of Erlich's reagent with indoles.

* http://www.unodc.org/pdf/publications/st-nar-13-rev1.pdf.
† https://www.ncjrs.gov/pdffiles1/nij/183258.pdf.

PROBLEM 26.5

Suggest a mechanism for the first step of the Marquis test with amphetamine.

Solution

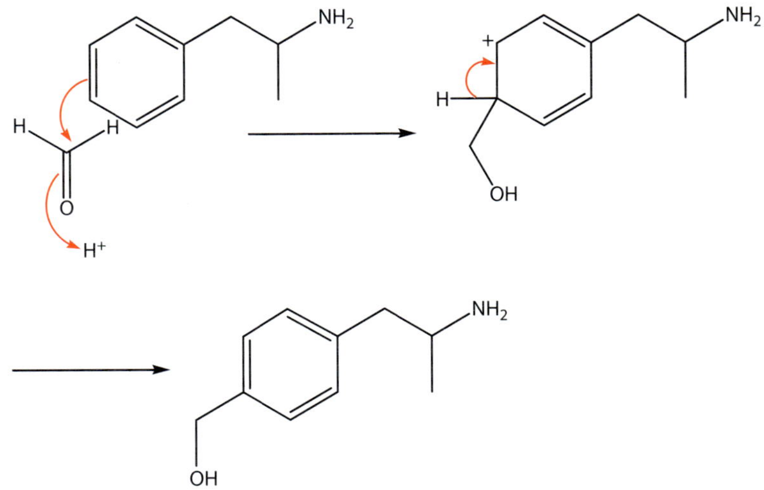

26.2.3 INSTRUMENTAL METHODS

Although there are well-defined protocols for the TLC of all types of controlled substances, these still require a laboratory, and at this point, it is usual to proceed to the more definitive identifications using GCMS and LCMS (see Section 5.2.3). Tailing is often a problem with GCMS (Figure 26.7; note that the molecular ion observed is the protonated amphetamine) and it is sometimes useful to derivatize the materials prior to analysis. Amphetamines may be converted to thiocyanates using CS_2 (Figure 26.8); the most common cutting agent for amphetamine, caffeine, is also successfully identified in this system, giving a good mass spectrum. This process is not, however, useful for secondary amines such as methamphetamine and ecstasy; these are usually reacted with heptafluorobutanoic anhydride. Quantitative determination of amphetamines uses HPLC in methanolic HCl and requires careful calibration. Well-known impurities allow identification of the synthetic route used as well as cutting agents, which frequently allow the determination of the origin of a particular batch of material.

Although heroin itself contains no free OH groups, it is invariably contaminated with a number of other morphine alkaloids including morphine itself. To avoid tailing in GCMS, the material is generally silylated (shown for morphine in Figure 26.9) before analysis. A GCMS trace is shown in Figure 26.10, with the mass spectrum of pure heroin. As with amphetamines, quantitative determinations use HPLC and require careful calibration.

Silylation is also recommended for derivatization of cocaine. Cocaine itself has no free OH groups, but the ecgonines (**26.17**), with which it is invariably found, have free OH and COOH groups. Quantitative determinations use GCMS or UV spectroscopy. Recently, it has become possible to determine where the cocaine was grown from the $^{13}C/^{12}C$ and $^{15}N/^{14}N$ isotope ratios in cocaine and the associated alkaloids.*

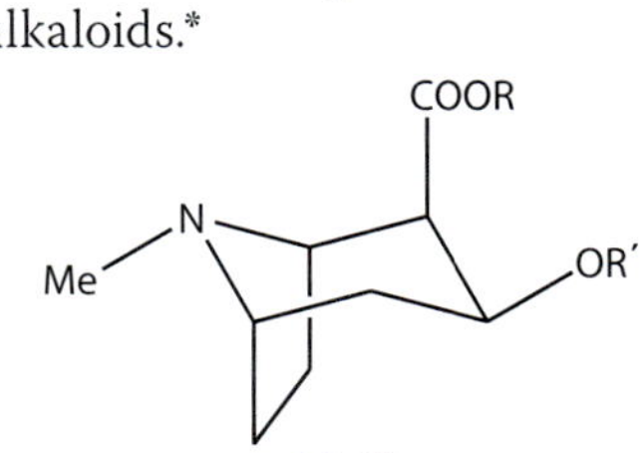

26.17

Ecgonine, R, R′ =H

Benzoyl ecgonine, R=H, R′= C(=O)Ph

Ecgonine methyl ester, R=Me, R′= H

* J.R. Ehleringer, J.F. Casale, M.J. Lott, and V.L. Ford, *Nature*, **408** (2000) 311–312.

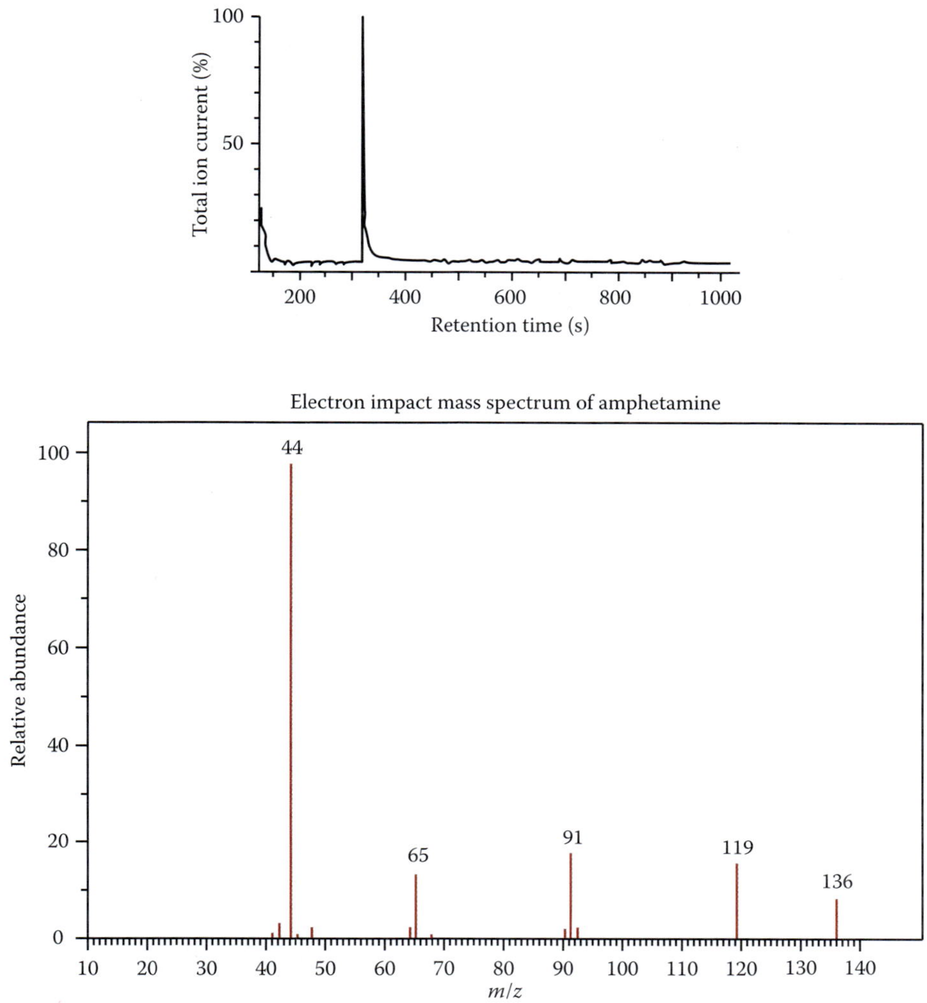

FIGURE 26.7 GC and mass spectrum of amphetamine, showing tailing.

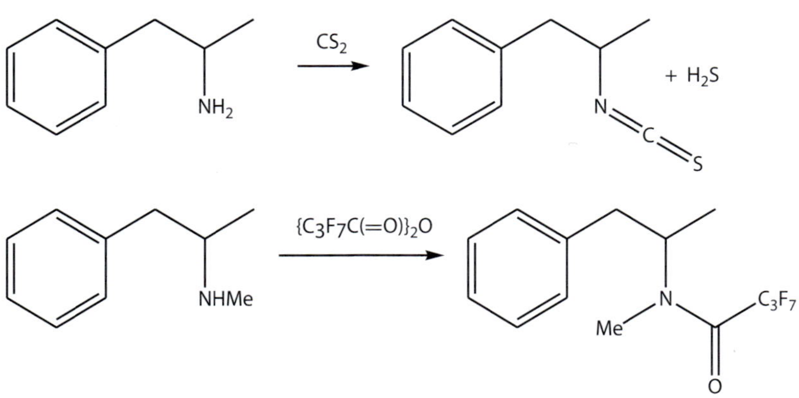

FIGURE 26.8 Derivatization of amphetamines.

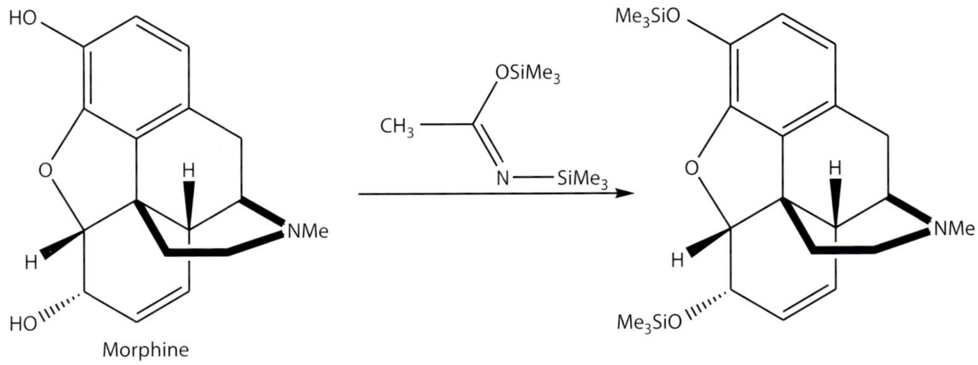

FIGURE 26.9 Derivatization of morphine.

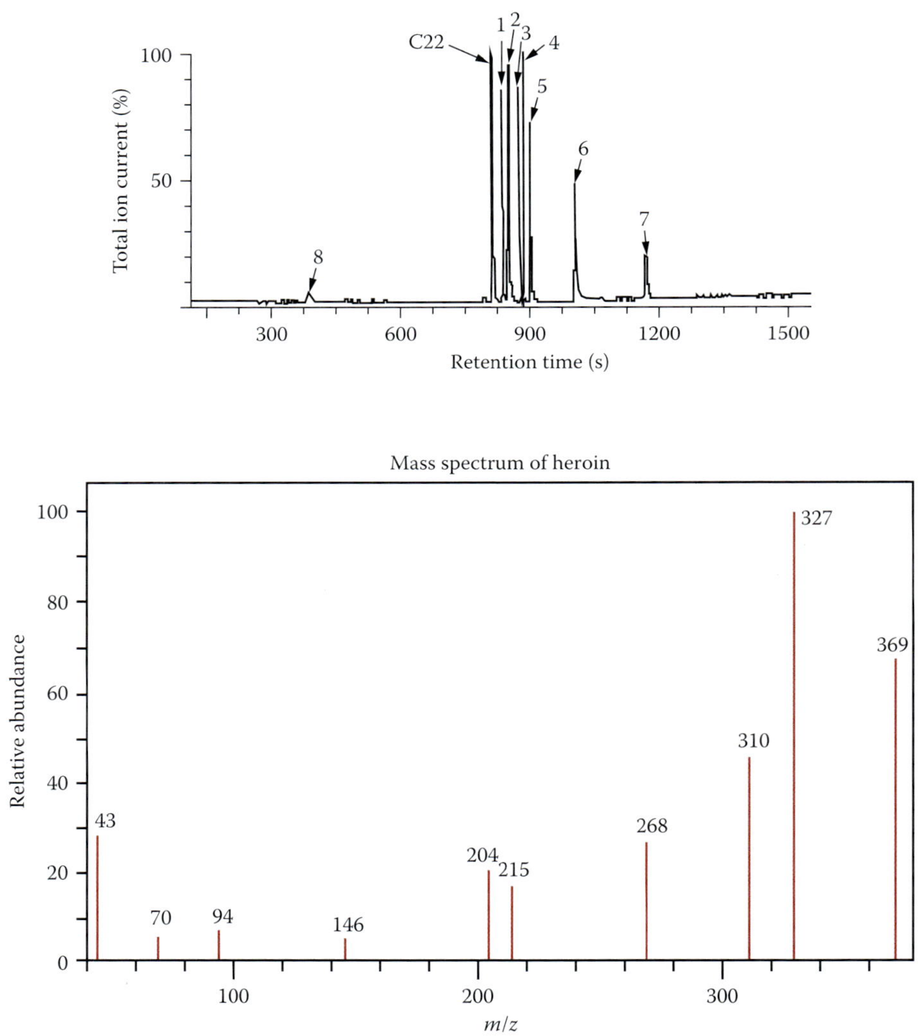

FIGURE 26.10 GCMS of a mixture of derivatized opiates. *Note*: Separation of opiates derivatized with *N,O*-bis(trimethylsilyl)acetamide by GC: (1) codeine, (2) acetyl codeine, (3) morphine, (4) 6-monoacetylmorphine, (5) diamorphine, (6) papaverine, (7) noscapine, and (8) caffeine (C22 is the internal standard).

PROBLEM 26.6

Interpret the mass spectra of amphetamine and heroin.

Solutions

Amphetamine 136 ($M+1^+$), 119 ($M-NH_3$), 91 ($PhCH_2^+$—thought to have particular stability because it rearranges to the tropylium ion), and 44 ($[CH_3CHNH_2]^+$).

Heroin 369 (M^+), 327 (M-42, loss of ketene from one of the acetate groups), and 310 (M-59, $M-CH_3COO$)

26.2.4 DESIGNER DRUGS

"Designer drugs" is the term used to describe materials with psychoactive effects that have been designed to circumvent existing drug legislation. They are often described as legal highs and sold with names such as plant food or bath salts, which circumvents legislation on medicines and foodstuffs that would prevent their being sold for human consumption. In Europe, it is currently estimated that some 40 new compounds are being introduced per year, and both analysis and legislation have a hard time keeping up. Figure 26.11 shows some recent popular introductions,

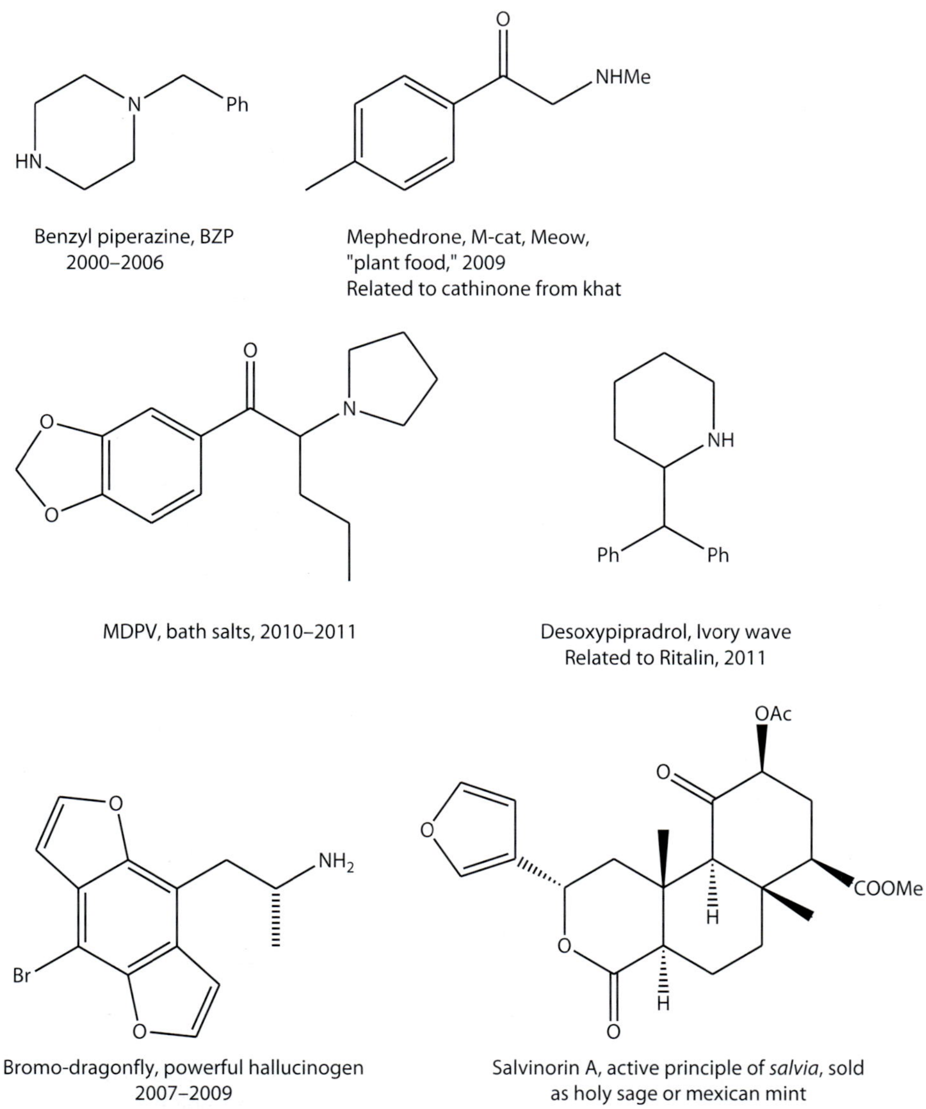

FIGURE 26.11 Designer drugs.

with the years in which they first came to prominence. Additionally, "spice" has been sold as a legal form of cannabis and was said to be a herbal product containing nothing illegal—sometimes true, but it has been found that many samples contained added synthetic analogues of the cannabinoids. Salvia, holy sage or Mexican mint, is the dried leaves of the psychoactive plant *Salvia divinorum*; the active principle, salvinorin A, is of interest as the first established diterpene hallucinogen.

The rate of introduction of these new materials presents many practical as well as legal and ethical problems. Many are simple to prepare, and substituted analogues can be created swiftly. For the analyst, new drugs mean establishing new sets of methods and standards for analysis, which is not trivial, given that the compositions of these materials are frequently different from those claimed by their manufacturers—it's not easy to test complex mixtures when there is little consistency of composition, and some of the materials present may be both unknown and unexpected. Simple reliable roadside tests for the use of *common* drugs of abuse by drivers are only now emerging, and adding new requirements is likely to be a medium-term project at best. A number of governments are now introducing legislation to allow 1-year bans on anything suspicious, to reduce harm while proper pharmacological testing for likely harm is carried out. Another approach that seems useful are the so-called "analog" laws, where any analog of something restricted and known to be harmful is automatically banned unless it can be proven to be harmless.

PROBLEM 26.7

The route for synthesis of a designer drug generally gives rise to some side products/impurities. In a medicinal product, these would of course be removed, but street drugs are rarely of this quality. For each of the following syntheses, predict what impurities might be present.

(a)

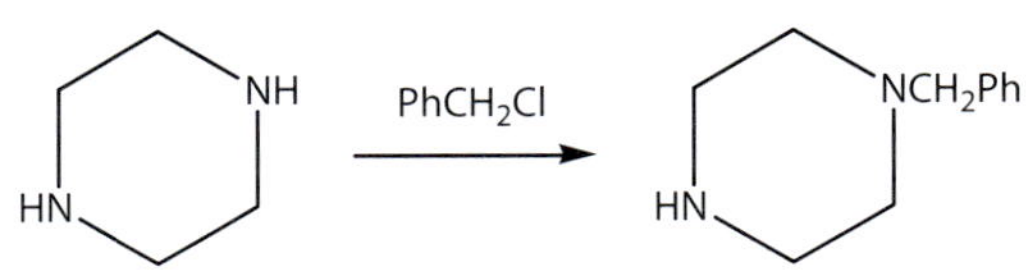

Benzyl piperazine

(b)

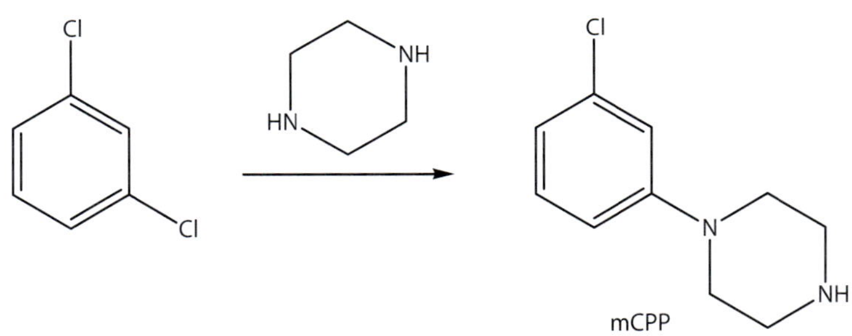

mCPP

(c)

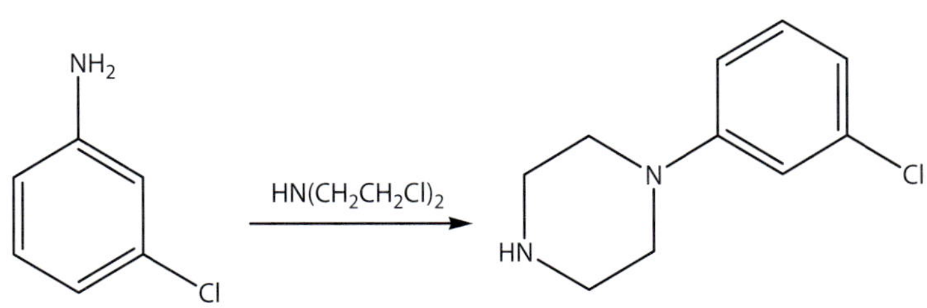

Solutions

(a) The likeliest impurity is dibenzylpiperazine—there is little difference in the reactivity of the two nitrogen atoms. The dibenzyl compound is not currently thought to possess psychoactive properties.

(b) This will be a relatively difficult reaction requiring quite forcing conditions, as the chloro-substituent is not as electron withdrawing as the nitro groups normally used to facilitate nucleophilic aromatic substitution. Any reaction may involve a benzyne intermediate. This makes double substitution of the piperazine unlikely, and the most plausible impurities are unreacted starting materials or other regioisomers of the product.

(c) This reaction produces mCPP by a completely different approach, constructing the piperazine ring by successive S_N2 substitutions on the chlorinated amine. The most plausible impurities would result from reaction of the product with more of the chlorinated amine; the nitrogen of piperazine is considerably more nucleophilic than that of aniline.

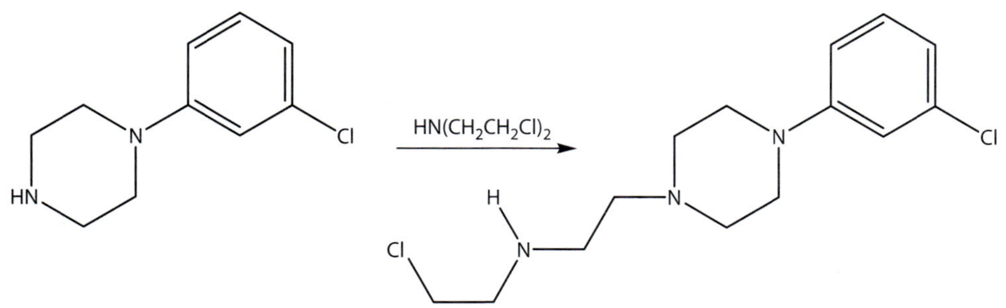

Key Points from Section 26.2

- Governments legislate against the possession, supply, and manufacture of drugs of abuse. The seriousness of the offense is greatest for manufacture/import, and the penalties depend on the perceived harm of the materials involved.
- Classes of drugs include hallucinogens (mostly derivatives of phenethylamine or tryptamine), narcotics (morphine and derivatives), stimulants (cocaine, amphetamines), and depressants (barbiturates, benzodiazepines).
- Although there are chemical tests for cannabis, it is normally identified by microscopic examination of plant morphology.
- Presumptive tests allow materials under suspicion to be tested in the field, by operatives without specific chemical skills. Only the class of compound can be determined in this way, and some false positive do occur. Positive and negative control tests are crucial.
- Instrumental methods (GCMS, LCMS, and various spectroscopic techniques) are used to confirm presumptive tests and analyze small quantities of materials.
- Designer drugs, currently appearing at the rate of 40 new substances per year, present both analytical and legislative challenges to law enforcement.

26.3 POISONS AND POISONING

Toxicology is the term applied to a study of poisons and poisoning—and there are several subdivisions of the topic:

- Clinical toxicology—refers to supervised medical settings
- Environmental toxicology—studies the effects of toxic substances on the broader environment
- Regulatory/animal toxicology—investigations of toxic properties of substances to fulfill the legal requirements for the licensing of commercial products
- Forensic toxicology—investigations in criminal cases, usually provided either by government agencies or university forensic medicine departments

A poison can be defined as any substance that when taken into the body destroys life or impairs health. Almost any substance is capable of being toxic if taken in a high-enough dose, including materials such as water or oxygen, which are also essential for life. Although the words "the dose makes the poison" are not an exact translation, Paracelsus said this in the early sixteenth century. Thus, at therapeutic doses, acetaminophen (paracetamol) will relieve mild pain, but at high doses, there are side effects and, eventually, fatality. For many materials, there is a known dose–response relationship, but for some carcinogenic and mutagenic substances, asbestos, and ionizing radiation, no dose is considered safe. In practice, toxicity may also depend on age and size (the very young and the elderly are more severely affected), state of health, particularly with respect to kidney and liver function (most exogenous materials are metabolized in the liver), history of exposure (which may result in tolerance or sensitization), and some paradoxical effects that are simply unpredictable.

We can divide poisons into a number of categories, not all of which fall into the realm of organic chemistry*:

- Corrosive poisons—acids, alkalis, which cause physical corrosion of body tissue
- Irritant poisons—inorganic and metallic poisons such as lead, mercury, and arsenic compounds
- Systemic poisons—act on biochemical processes within cells; include cyanide, CO, alkaloid poisons, and nerve gases
- Toxins—the term is used to describe naturally occurring substances produced in living organisms; includes snake and insect venoms and substances produced by bacteria and algae

The route of administration and excretion of a poison is important, because it determines its bioavailability and hence its effect. Intravenous administration provides 100 % bioavailability, very rapidly, and inhalation is the next most effective route, since the material is exchanged into the blood in the lungs. Other routes include administration through the mucous membranes (nose, mouth, rectum, vagina) or orally, which is usually the least effective route, as many materials are partially or wholly destroyed by the acidic nature of the digestive tract. It is important to remember that most fatal poisonings are the result of accident (this may be a culpable accident as in some medical or industrial processes, but these are of limited forensic interest since the nature of the toxic material is usually obvious) or suicide. Suicides usually involve overdoses of medicinal drugs, consistent with the principle that people use what is readily available to them. Accidental overdoses are difficult to distinguish from suicide on toxicology alone.

Homicide by poison fascinates us, but in the modern developed world, it is rather rare. In some cases, a toxicologist may be asked to test for a particular substance on the basis of what was found at the crime scene—for example, carbon monoxide in a fire death or a specific drug in a suspected suicide. The toxic material may, however, be partially (a number of possible substances were available in the vicinity of the deceased) or completely unknown. In these cases, it is usual to start with substances locally available and profile the deceased—few pensioners abuse controlled substances or have access to obscure spider venoms.

The sample tested will depend on what is available. Blood is almost always available and can be obtained without an autopsy, if there is a risk of infection or contamination. Urine is available from living patients, but only from some of the deceased (depends on time elapsed after death). Liver samples are almost always available from autopsy, and this is a large organ, so there is always an adequate specimen for analysis. Because the liver is the site where exogenous materials are metabolized, drug concentrations tend to be higher than in blood samples. Concentrations in bile, available from the gall bladder in some autopsies, are also often higher than in a blood sample. Vitreous humor, from the eyeball, can usually be obtained and is useful because few interfering substances are present, but amounts available for testing are inevitably low. Hair and nail clippings are generally used to assess long-term exposure, and

* Many inorganic poisons are fascinating, however—see John Emsley in *Elements of Murder: A History of Poison*, OUP, Oxford, U.K., 2005, and *Molecules of Murder: Criminal Molecules and Classic Cases*, RSC, Cambridge, U.K., 2008.

this is particularly useful for heavy metal poisonings, but this technique has also been used to establish long-term drug abuse or compliance with medical advice or requirement.

Exposure to toxic pesticides may be either accidental or deliberate (in the developing world, agricultural chemicals are more commonly used for homicide and suicide than elsewhere, because of their ready availability) and may be found in an agricultural, horticultural, or home environment. The main symptoms are vomiting and convulsions. Organophosphorus compounds such as parathion (**26.18**) and malathion (**26.19**) were developed as more selective and safer insecticides to replace DDT (**26.20**). Unfortunately, they are much more toxic to humans than is DDT. They act by blocking the nervous tissue enzyme, acetylcholinesterase. When the enzyme is inhibited, the neurotransmitter acetylcholine accumulates at nerve endings, causing continual stimulation of receptors. Related materials may cause the paralytic condition, intermediate syndrome, or neurotoxicity.

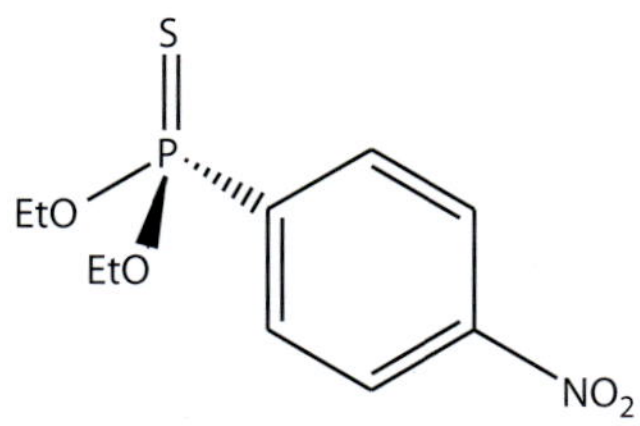

26.18, parathion

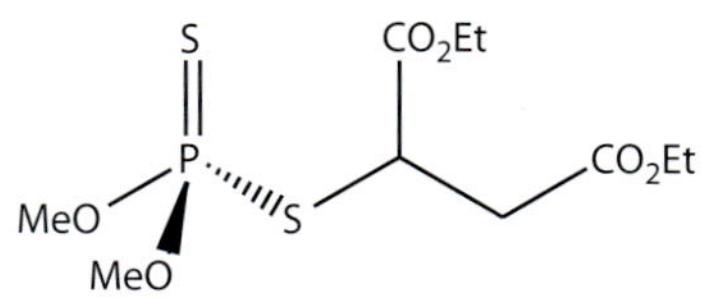

26.19, malathion

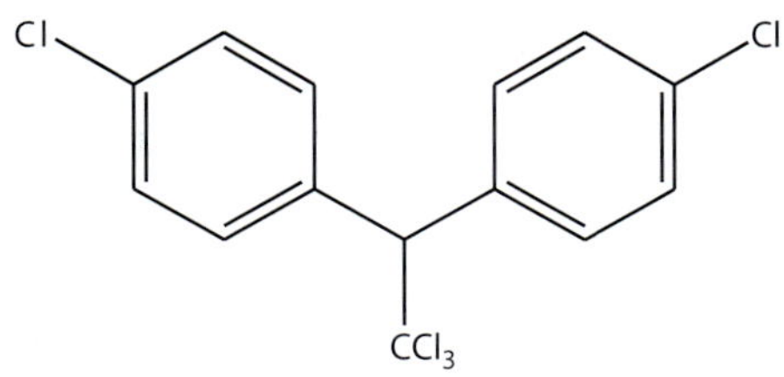

26.20, DDT

The nerve gas sarin, **26.21**, has a related structure and a similar mode of action, though it is much more potent. Sarin is toxic at 0.01 mg kg^{-1} body weight, if no antidote is available. It was discovered in Germany in 1938 and was produced, but not used, for warfare during Second World War. It was produced by both NATO and the Soviet Union during the 1950s. Although it is a liquid, the vapor pressure is high enough for the vapor to contaminate the air at lethal levels. In 1988, Kurdish villagers were the victims of a sarin attack by the Iraqi government of Saddam Hussein. In 1995, the Aum Shinrikyo cult rereleased an impure form of sarin in the Tokyo subway. There were 12 deaths and 50 serious injuries. Seven people had died earlier in the day in a sarin release in Matsumoto, so the symptoms were recognized quickly and the atropine antidote (**26.22**) administered, reducing the level of casualties. In 2013, the use of sarin in the Syrian Civil War was established.

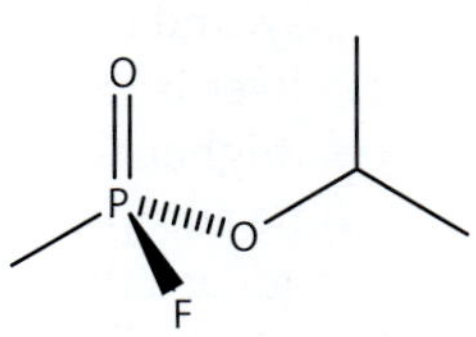

26.21, sarin

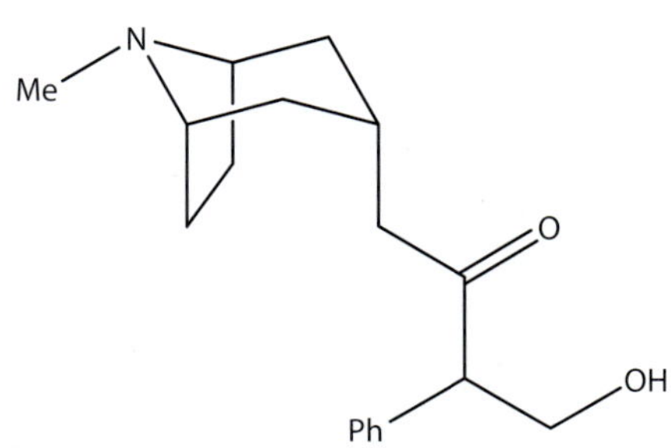

26.22, atropine, from *Atropa belladonna*, deadly nightshade

Paraquat, **26.23** (and other bipyridyl-based herbicides), is now largely banned in the developed world but is still used for murder and suicide in developing countries. Paraquat affects the gastrointestinal tract, kidneys, liver, and heart, and there is no antidote available. In Japan, there were 346 cases of paraquat poisoning recorded between 1998 and 2002; 242 were attempted or successful suicides, 97 were occupational accidents, and 7 were murders. Ninety-seven victims died. In 2002, a bitter substance, bitrex, was added to all paraquat sold, and this effectively ended its misuse. Murders involving paraquat were rare even in the early years of its use. Steven David Caitlin murdered his mother and at least two of his seven wives with paraquat in California between 1977 and 1990, and in Essex, England, in 1981, Susan Barber murdered her husband Michael using paraquat in food.

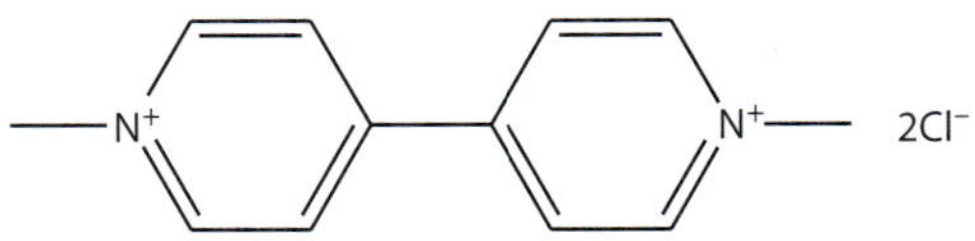

26.23, paraquat

"Dioxin" was first noted as a highly toxic material as a contaminant of agent orange used as a defoliant during the Vietnam War. Vietnam veterans have sued Monsanto and Dow for various medical effects on both them and their children. The most toxic of the dioxin family is 2,3,7,8-tetrachorodibenzo-1,4-dioxin, **26.24**. This is the compound that was formed in the Seveso disaster in 1976, when the reaction of tetrachlorobenzene (**26.25**) with sodium hydroxide went out of control. The best known deliberate use of the material was against Victor Yushenko, who was credited with the "Orange Revolution" in Ukraine. The "before and after" photos (Figure 26.12) show the chloracne that is a major effect of the material; Yushenko was shown to have the second highest dioxin in blood level ever recorded in a living individual.

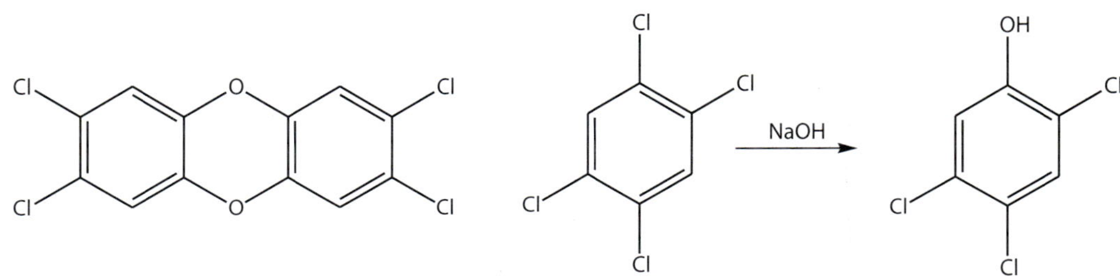

26.24 **26.25** **26.26**

Among the toxins, many plant alkaloids have been used for murder. Although atropine, **26.22**, has legitimate medicinal uses (including its use as an antidote for sarin and other nerve gases and to dilate the pupils for eye examinations), it is poisonous and hallucinogenic in higher doses. The Greek physician Dioscorides knew that deadly nightshade was even more toxic than mandrake (which contains hyoscine). Atropine was used as a poison in ancient Rome and was used in the attempted murder by Paul Agutter of his wife in 1994, when he contaminated bottles of tonic water with atropine. Strychnine, **26.27**, acts by blocking glycine receptors in the central nervous system,

(a) (b)

FIGURE 26.12 (a) Before and (b) after photos of Victor Yushenko who was poisoned with dioxin.

with death occurring in about 20 min at a 5 mg kg^{-1} dosage. Historical strychnine poisoners include the artist and forger Thomas Wainewright, Thomas Neill Cream, the Lambeth poisoner, and Daniel Stott. The Rolling Stone Keith Richards claimed that his worst experience with drugs was when someone adulterated LSD with strychnine, but the idea that street LSD is extensively contaminated with strychnine is now regarded as an urban myth. In the ancient world, Socrates was executed using hemlock (the active principle is coniine, **26.28**) for "corrupting Athenian youth."

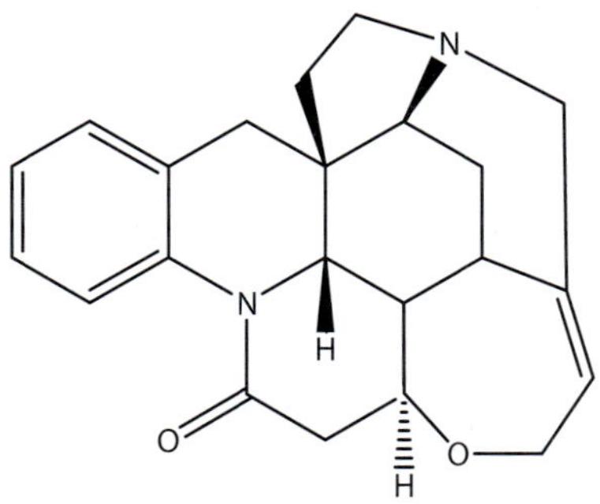

26.27, strychnine

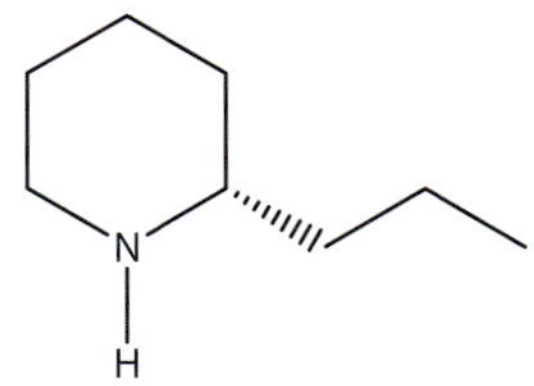

26.28, coniine

Ricin, isolated from castor oil seeds, was used in the assassination of Georgi Markov during the Cold War. A small pellet was extracted from a puncture on his thigh (he had been stabbed with an umbrella in the street), but no toxin could be detected, and it was concluded that ricin was involved by a process of elimination as well as the knowledge that was effective in minute quantities. In August 1978, there was also an attempt to poison Vladimir Kostov in Paris; he survived and the toxin was identified because not all of the ricin escaped from the pellet and his system had time to develop relevant antibodies. Ricin is a glycoprotein and its LD_{50} is 0.1 μg kg^{-1} body weight, and a vaccine is now available. Ricin was said to have been produced in Iraq and to have been found in the Tora Bora caves in Afghanistan. In 2003, a group of six terrorists of North African origin were arrested in London and they were charged with the preparation of ricin with the intention of contaminating a water supply. While they certainly possessed castor beans and some primitive equipment, it was eventually established that they had never managed to produce any ricin. In 2002, however, Kenneth Olsen in Washington State did produce 600 mg of ricin, intending to use it to murder his wife.

Finally, insulin has been used in a number of murders; it is particularly difficult to detect, as it is present naturally. The infamous case of Sunny von Bulow was initially thought to involve insulin poisoning by her husband Claus, but he was acquitted on appeal. The most prolific insulin murderer was the so-called Angel of Death, Beverly Allitt, in the United Kingdom, who murdered 4 children and attempted to harm at least another 9 using insulin; she is currently serving 13 life sentences.

PROBLEM 26.8

Explain mechanistically the conversion of **26.25** to **26.26**, and what happened when the reaction went out of control at Seveso to give **26.24**.

Solution

Although tetrachlorobenzene is not an ideal substrate for nucleophilic aromatic substitution, the chlorines are electron withdrawing and would give some stability to an intermediate anion:

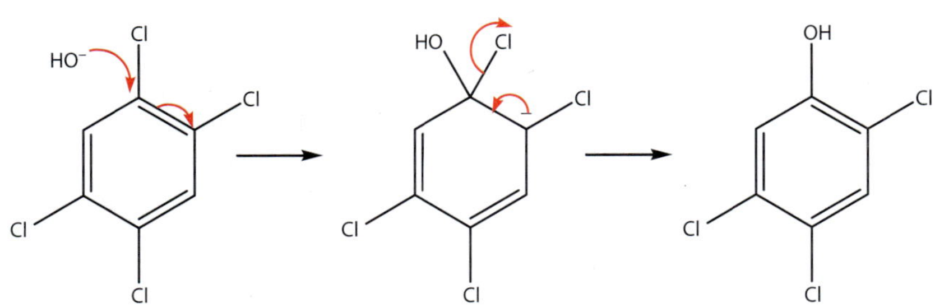

Once the phenol is produced, this can be deprotonated and can attack the ring as an anion:

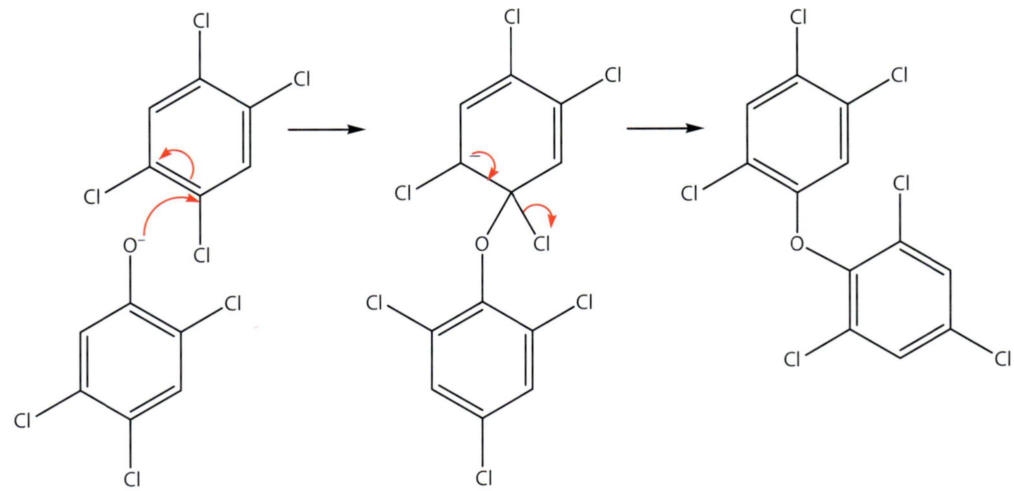

Either ring can then be substituted again by HO^-, and then ring closure achieved.

Key Points from Section 26.3

- Poisons can be divided into corrosive, irritants, systemic poison, and toxins. Most organic compounds fall into the last two categories.
- The route of administration determines the bioavailability of the poison and hence its likely effect.
- Industrial accidents and suicide by poisoning are very much more common than homicide.
- Toxic pesticides including organophosphate insecticides, and the herbicide paraquat, are more commonly encountered in homicide and suicide in the developing world. Nerve gases, such as sarin, which have been used in warfare and terrorism, are structurally related to the organophosphates, and atropine is the antidote for both.
- Toxins include plant alkaloids such as atropine and strychnine, as well as the peptide ricin. Many snake and spider venoms are glycoproteins.

26.4 TESTING FOR BLOOD

Blood is about 7.7 % of body weight and consists of about 55 % plasma, a straw-colored fluid containing dissolved plasma proteins, metal ions, and organic compounds, and about 45 % cells, erythrocytes (red blood cells containing hemoglobin), and leucocytes (white blood cells concerned with the immunological defense of the body). There is some color variation in blood; arterial blood is highly oxygenated, and brighter red than venous blood. Carboxyhemoglobin is formed when CO binds to the iron in heme in CO poisoning; it is cherry red in color. In shed blood, the iron(II) is oxidized to iron(III) and the material appears dark brown. In a forensic sense, however, not all dark-brown stains are blood, so it is necessary to have a method to test whether a particular crime scene stain contains blood.

Almost all the tests for blood depend on the fact that blood contains an enzyme called catalase that, in the presence of hydrogen peroxide, can catalyze the oxidation of other materials. There is a dramatic demonstration of the oxidizing power of catalase. Although the thermodynamics of hydrogen peroxide decomposition are favorable, the material is kinetically stable at room temperature. Addition of a drop or two of mammalian blood will cause foaming and the rapid evolution of oxygen. Most of the materials in the presumptive tests for blood could be described as dyes. They are added to the suspected blood stain in their reduced form, which is usually colorless. When a drop of hydrogen peroxide is then added, the color of the oxidized dye then develops, if blood (and hence catalase) is present. Any mammalian blood will give a positive test

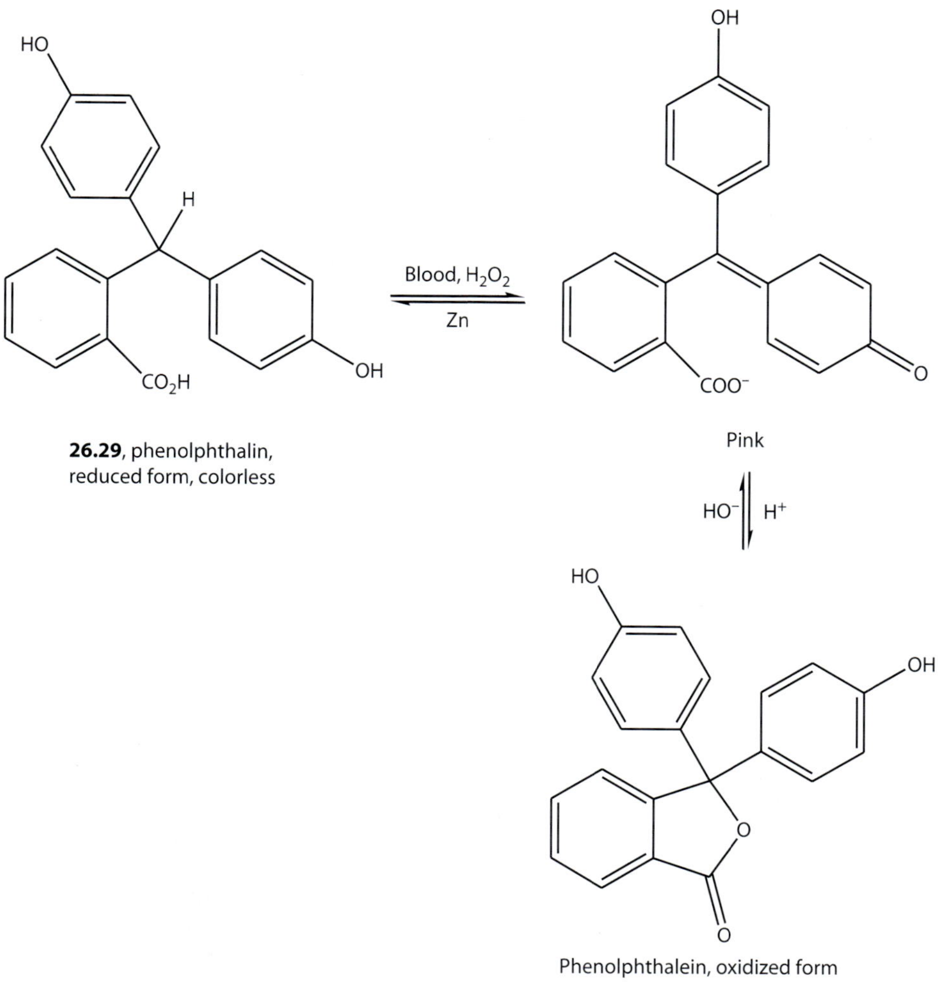

26.29, phenolphthalin, reduced form, colorless

FIGURE 26.13 Phenolphthalein in the KM test.

under these conditions; these methods are not specific for human blood. False positives are also obtained from materials such as potatoes, apples, broccoli, and horseradish, which contain the enzyme peroxidase.

The primary test (and the most popular in TV programs because the color is so distinctive) is the Kastle–Meyer or KM test. You may be familiar with "phenolphthalein" as an acid/base indicator, colorless in acid, bright pink in base. Both the acid and base forms of the indicator are the oxidized form of the dye. For the KM test, the reduced form, **26.29**, (Figure 26.13) is added, followed by a drop of hydrogen peroxide. The oxidized form is generally referred to as phenolphthalein and the reduced form (which needs to be stored cold, in the dark, with zinc metal) as phenolphthalin—but many textbooks and websites are inconsistent in their usage.

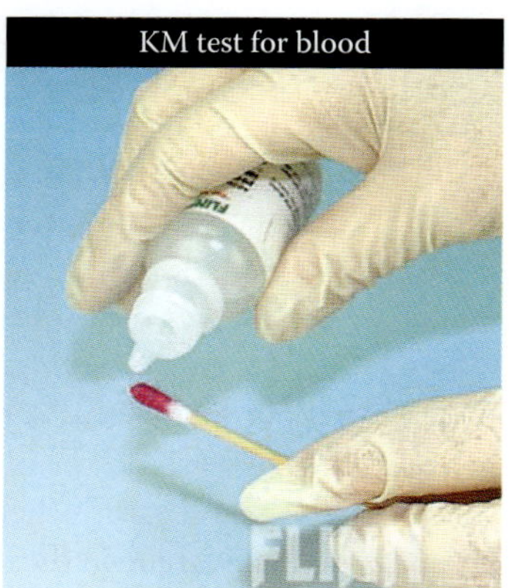

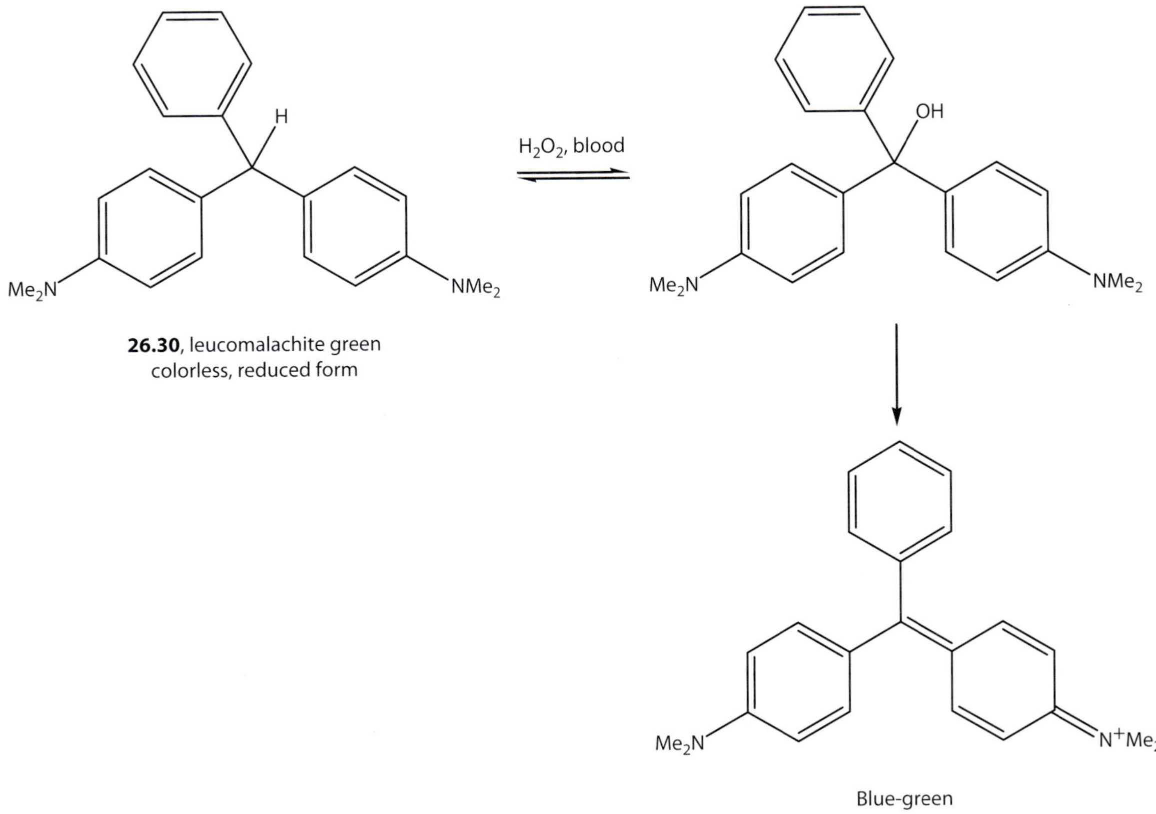

FIGURE 26.14 Oxidation of leucomalachite green.

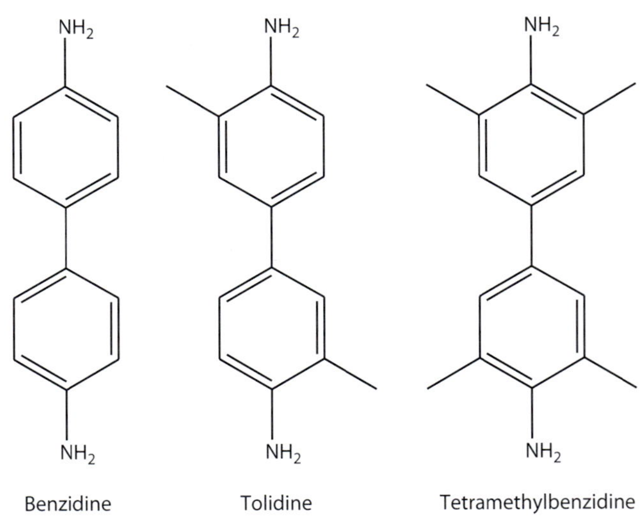

FIGURE 26.15 Benzidine derivatives used in presumptive blood testing.

Leucomalachite green, **26.30**, is also colorless when reduced and blue green when oxidized (Figure 26.14). For greater sensitivity (though lower specificity), benzidine and some substituted benzidines (Figure 26.15) may be used. The oxidation and color changes are shown in Figure 26.16. However, benzidine is a known carcinogen, and the substituted compounds are little better, so their use is not routine.

Another test you may have seen on TV uses luminol—this is more useful when a large area has to be examined or if there has been an attempt to wash off the bloodstains. An alkaline solution of luminol and hydrogen peroxide is sprayed onto the test area. When luminol is oxidized, it undergoes a chemiluminescent reaction (Figure 26.17) and glows blue in the dark, at which point the test area is photographed. False positives can be obtained from vegetable peroxidases, some metals, and bleach.

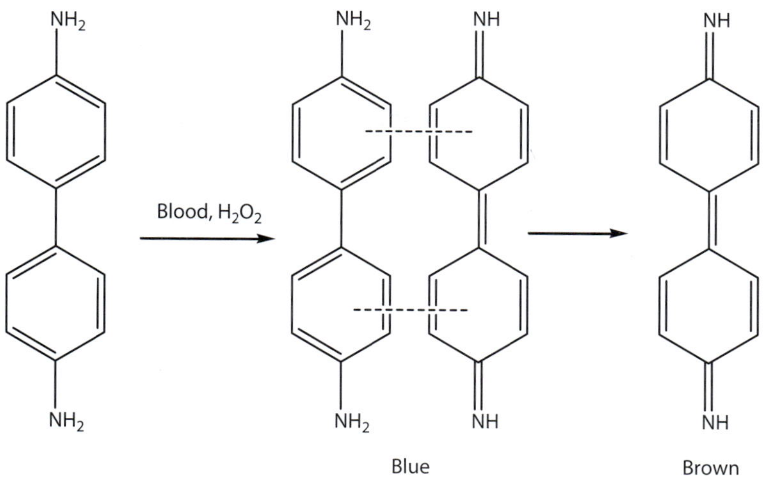

FIGURE 26.16 Oxidation of benzidine in the presence of blood.

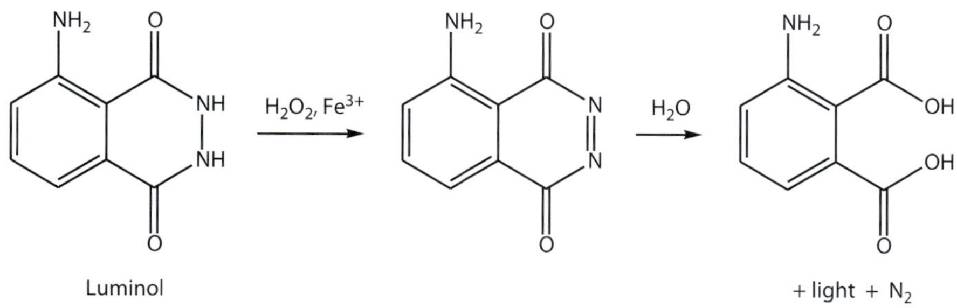

FIGURE 26.17 Oxidation of luminol.

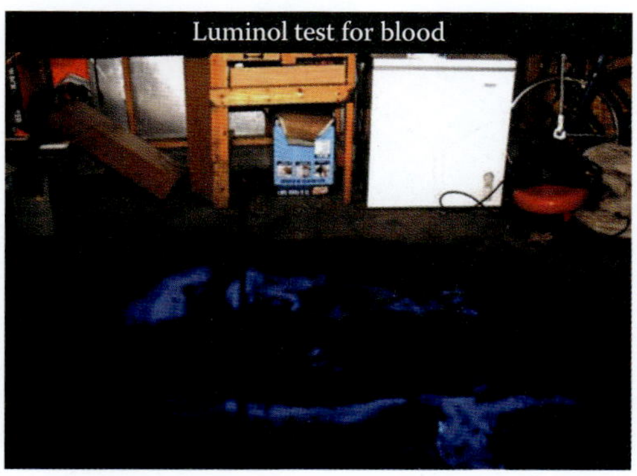

PROBLEM 26.9

Write a mechanism for the acid/base transformation of phenolphthalein (see Figure 26.18 later in this chapter).

Solution

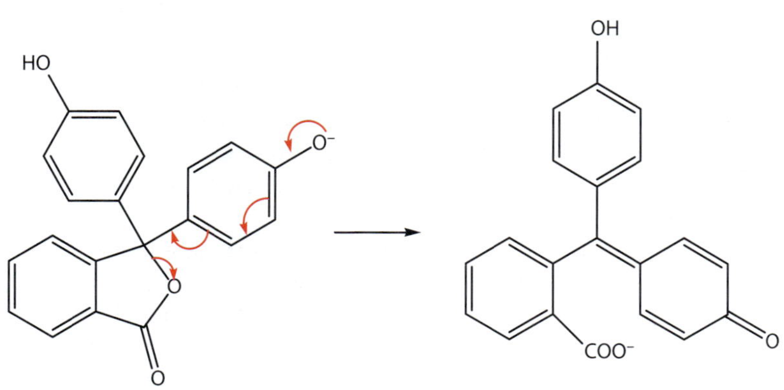

26.5 DYES, INKS, AND PAPER

We describe dyes as materials that are soluble, at least at the point of application, either to fabric or paper. Pigments are insoluble particles, suspended in a medium (something that we also see in paints). Dyes penetrate materials with the solvent in which they are dissolved, but pigments remain on the surface. Dyes can be natural or synthetic, organic or inorganic, and acidic, basic, or neutral. The oldest dyes are of natural origin, such as indigo, **26.31**, and Tyrian purple, **26.32**. Despite their similar structures, they are of very different origin. Indigo was originally isolated by fermentation of the woad plant, *Isatis tinctoria*, and may also be obtained from other plants of the *Indigofera* family. Indigo was known in ancient Egypt and was the woad dye of the ancient Britons. Tyrian purple, famous from its use to signify rank in ancient Rome, is isolated from the mollusk *Murex brandaris*. Alizarin, **26.33**, is isolated from the roots of the madder plant and was used to dye the red coats of the British military in the eighteenth and nineteenth centuries. Dyes with these natural structures are still in wide use, although many are now produced synthetically. Modern synthetic dyes include many of the azo compounds that we discussed in Section 24.5.

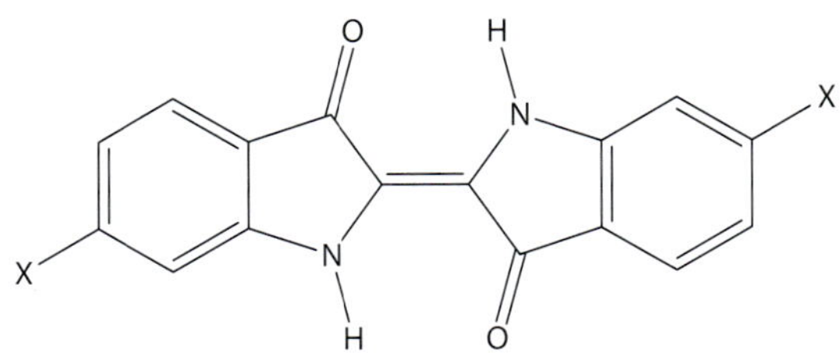

26.31, X = H, indigo
26.32, X = Br, Tyrian purple

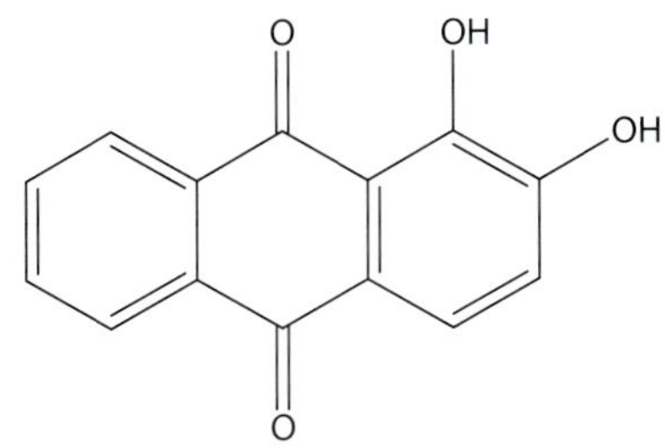

26.33, alizarin

A lump of indigo dye

Murex brandaris

Inks consist of a solvent, sometimes called a vehicle, colorants (dyes or pigments), and additives to control flow, thickness, and appearance. Ink was first developed thousands of years ago in China, and early inks were based on carbon black suspended in glue. This is still in existence, now called Indian ink, and it produces a dense, permanent black line. Iron–tannin inks, mixtures of iron salts and tannin with some glue, were used throughout the medieval and early modern periods. In the nineteenth century, indigo was added to these to give what was called blue-black ink, which darkened over time as the iron–tannin components oxidized. In the nineteenth/twentieth century, dyes have been used, first to replace part of, and later all, the iron–tannin materials. The use of solvents in ink is now common—water, 2-butanone, and glycol ethers. Fountain pens arrived by the mid-nineteenth century, ballpoint pens in the 1940s, and felt tip, fiber tip, and roller ball pens from the 1980s. Ballpoint inks are not solvent based, but are a quick-drying paste; dyes provide the colorant, and resinous material binds it to the paper. Fiber tip and roller ball inks are water based, with alcohols and other solvent to promote quick drying.

Drawing in Indian ink, early nineteenth century

In examining a handwritten document forensically, the examiner wants to find out whether all the writing used the same instrument, when the document was written, and in what order various parts of it were written. More than one instrument or ink on a document might suggest some alteration. The process starts with a nondestructive examination, under different light sources, with different filters. IR and Raman spectroscopy can investigate the components of the ink *in situ*—in particular, they can distinguish "epoxy" and "alkyd" resins in ballpoint ink. These commercial names are unfortunately quite misleading; epoxy resins contain aromatic rings and OH groups, and alkyd resins are essentially polyesters. Many inks contain components that fluoresce in the visible or IR region, when exposed to UV or visible light respectively, and printing inks used in currency printing also have fluorescent security features.

TLC is destructive of some of the material under test, but it has the advantage of being quick and inexpensive. Pyridine is recommended as a solvent for glycol-based ballpoint inks, and ethanol/water for non-ballpoint inks. Since 1968, the US treasury has been compiling a library of commercially used inks, and in the 1990s, there was an attempt to persuade manufacturers to "tag" inks by year of manufacture, but this would need to be a worldwide effort to produce a seriously useful database.

Dating of inks is difficult and frequently not very reliable. GCMS has been used to identify solvents and volatiles in gel pen ink up to 6 months old, and there are some specialized mass spectrometry studies on particular materials—for example, gentian violet, **26.34**, slowly loses methyl groups over time. The only truly reliable determination is to demonstrate that an ink only became commercially available after the date on which the document is purported to be written.

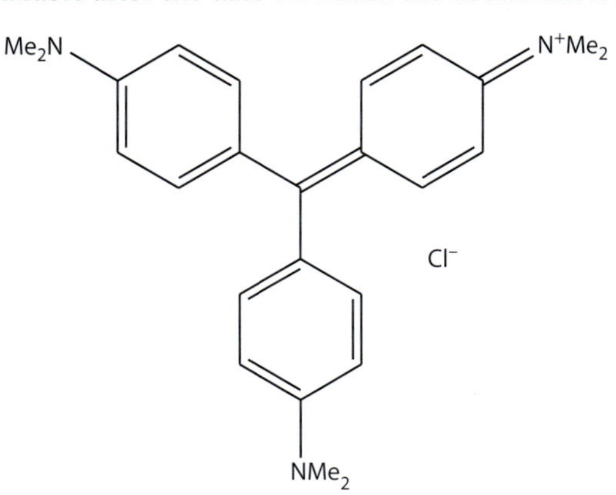

26.34, gentian violet

Paper analysis is difficult, as paper is a mass-produced material, and there is limited variation between samples. Differences derive from the type of raw fiber used, the choice of sizing agent, and the use or otherwise of bleach. Paper is largely composed of cellulose (see Section 16.3.12), a polymer of glucose from plant cell walls, and the quality of the paper is strongly correlated with the amount of cellulose. Cotton, sisal, hemp, and linen have all been used in paper manufacture, but cellulose from wood pulp is the most common source. Wood pulp is bleached and whiteners such as kaolin (china clay) are added. Paper ages as a result of the acid catalyzed breakdown of cellulose to glucose; this gives rise to a yellow coloration and brittleness, so documents intended to have a long lifetime should be printed or written on acid-free paper. The Hitler diary fraud was uncovered when it was found that the paper used contained a whitening agent that had only been introduced in 1954 and that the polyester and viscose threads that had been used to attach the seals were not available until the 1960s.*

26.6 TRACE EVIDENCE

The analysis of trace evidence is based on Locard's principle that every contact leaves a trace—so at every crime scene, the perpetrator will leave something behind and take something from the scene away with them. The materials may be glass, hair, paint, soil, textiles, oil, grease, wax, or vegetable matter. Textiles and paint are of most interest from the organic chemist's point of view. It is important to distinguish between analysis and identification. While analysis may tell us that a fiber came from a particular brand of carpet, or paint from a particular make and model of vehicle, this does not mean that it came from a carpet or a vehicle associated with a particular suspect. Trace evidence is usually supportive, not definitive. That said, the more of it there is, the more definitive it becomes. Transfer of one type of fiber from a suspect to a victim may be a coincidence; transfer of three different types associated with the suspect is much more suspicious. If you are interested in how probabilities stack up in such cases, you should read about Bayes theorem.† Fiber evidence was critical in the case of the murder of Holly Wells and Jessica Chapman in the United Kingdom (2002) and the conviction of Wayne Williams in Atlanta in 1981 for the murders of Nathanial Carter and Ray Payne (although it is likely that he was guilty of more than 20 more homicides).

Fiber traces may come from clothing, curtains, or carpets. The first broad division is between natural and man-made fibers. Natural fibers may come from animal or vegetable sources. Silk and wool are both proteins, fibroin and keratin, but they do have rather different amino acid compositions. Other fur fibers are generally identified microscopically from their morphology. Plant-derived fibers include cotton and linen, both of which are largely cellulose. Conventional wisdom is that IR spectroscopy is not helpful in identifying natural fibers, but this is not strictly true; while the IR spectra of cotton and wool are broad and give limited detail, they are certainly sufficiently different to tell them apart (Figure 26.18). Note that there is only a small carbonyl peak in the cotton spectrum (from oxidized material or impurities) compared with that of wool, but there are strong C–O single-bond stretches.

There are now an enormous number of different man-made fibers available. As with the natural materials, the primary method for identification is microscopy, but IR spectroscopy is also useful in identifying particular functional groups—esters in polyesters, amides in nylon, and, very distinctively, the CN triple bond in anything containing acrylonitrile (Figure 26.19).

Although it is tempting to assume that dyes could be extracted and subjected to TLC or LCMS, this is actually less easy than it might appear. The whole point of dying a fabric is that it is not easy for the dye to be removed, and frequently, attempts to do so result in degradation of both fiber and dye. More usefully, the color can be compared by reflectance microspectrophotometry—a sophisticated method of obtaining a UV visible spectrum.

Finally, it is possible to compare fibers using pyrolysis gas chromatography. The fiber is heated to destruction (Figure 26.20) and then the fragments analyzed by gas chromatography. This technique involves comparison rather than identification and requires reference material for the comparison. However, it does distinguish materials that are quite similar in structure, for example,

* *Selling Hitler: The Extraordinary Story of the Con Job of the Century: The Faking of the Hitler "Diaries,"* Robert Harris, Pantheon, 1986.

† E.W. Weisstein, Bayes Theorem from *Mathworld*—Wolfram Web Resource, http://mathworld.wolfram.com/BayesTheorem.html.

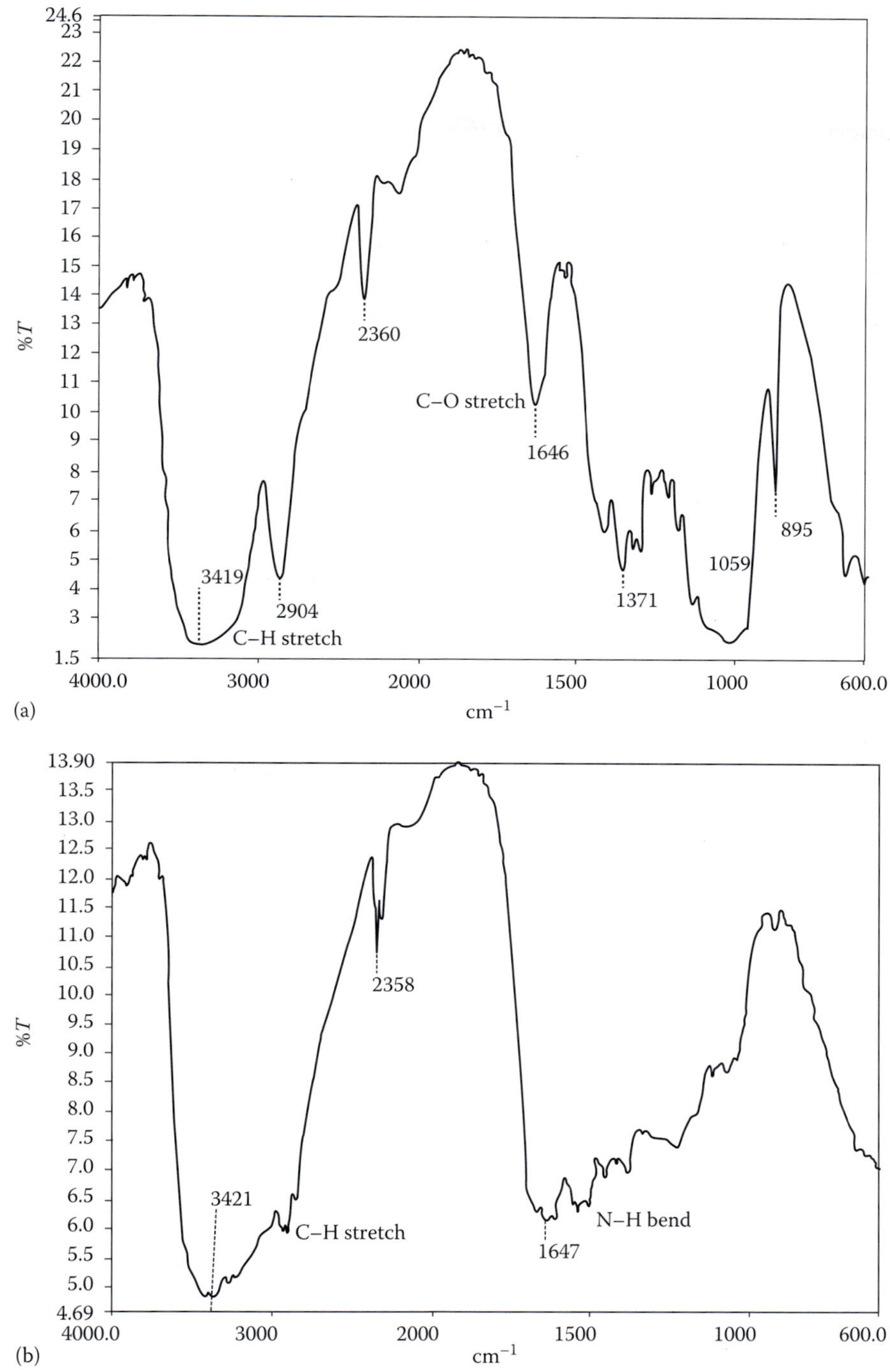

FIGURE 26.18 IR spectra of cotton (a) and wool (b).

nylon-6 and nylon 6,6, which could not be distinguished by IR spectroscopy, since the functional groups are identical (Figure 26.21). Other examples are provided by the pyrograms of some natural materials (Figure 26.22).

Pyrolysis GC is also used in the analysis of the organic components of paint. Paint is a mixture of pigments and additives suspended in a binder/solvent—a complex material for which a full analysis is unlikely to be possible. The pigments impart the color of the paint, and the solvents evaporate (the paint dries) to leave the binders and pigment. In earlier times, most pigments were inorganic, but many are now organic. For vehicle paint (which might be left behind after a hit and run incident), there are many layers—primers, coatings for rust protection, and coatings for appearance. Polyesters, polyurethanes, and acrylics may all be used at various stages of

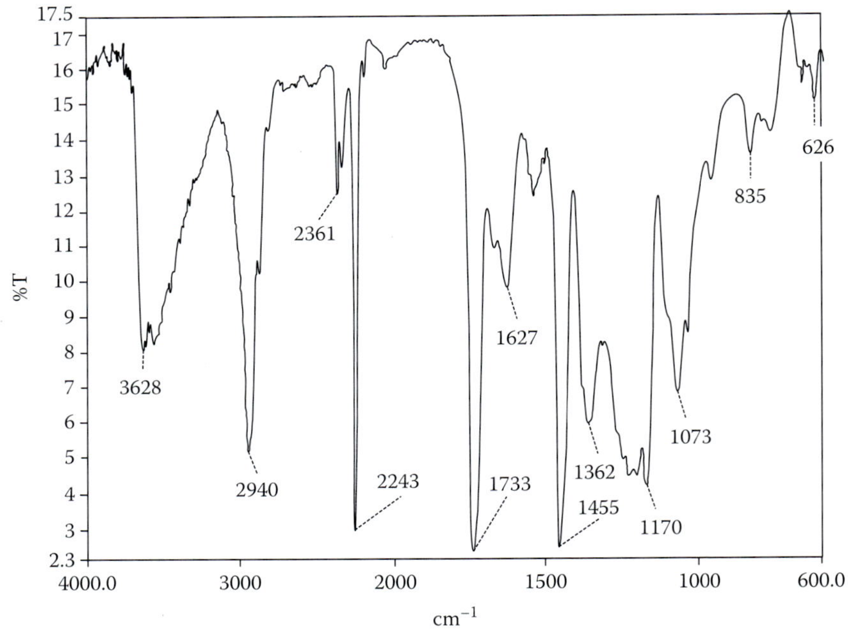

FIGURE 26.19 IR spectrum of an acrylic fiber. Note the CN triple-bond absorption at 2243 cm^{-1} and the carbonyl absorption at 1733 cm^{-1}. If you recall the chapter on polymers, pure polyacrylonitrile is very difficult to process, and usable fibers usually contain other acrylic materials, including esters and amides. Like most fibers, this absorbs some moisture from the atmosphere, so an OH as well as an NH peak is also observed.

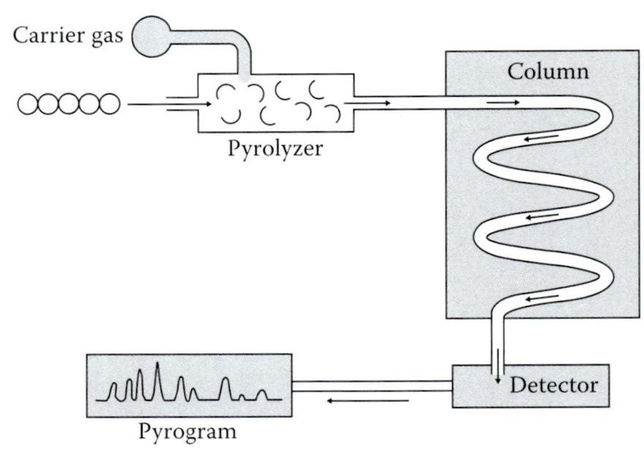

FIGURE 26.20 Pyrolysis GC.

the applications of successive coatings. While quality control of pigments is very high, binder formulations are less well controlled. In the United States, all car manufacturers submit paint samples to a national registry, paint data query (PDQ). Microscopy and microspectrophotometry are important techniques in identifying the layers and the color of the paint. IR spectroscopy is useful in identifying organic binders. Pyrograms of paints are typically very complex, and the technique can only be used for comparison with known samples—but even small differences, which would not show up in an IR spectrum, can be identified. In Figure 26.23, it appears that all three materials are the same by IR spectroscopy, but the pyrograms (Figure 26.24) are distinct.

Many modern vehicles, as well as manufactured goods, contain significant amounts of polymers and IR spectroscopy and pyrolysis GC (old-school polymer chemists tend to put a match to a polymer and then identify it from the odor, but the pyrogram is a little more rigorous ☺) are the main techniques used in their identification (color may also be useful, but this depends on the material involved). Polycarbonate is tough and is frequently used in car reflectors, lenses, headlights, and body panels (Figure 26.25). The acrylonitrile/butadiene/styrene (ABS) copolymer is widely used in car interiors and toys (Figure 26.26—note the very obvious isolated peak for a carbon–nitrogen triple bond). PVC (Figure 26.27) is used in car interiors, as a covering for polyurethane foam, in raincoats, and paddling pools. Polymers may also be identified on adhesive tape and in a range of glues.

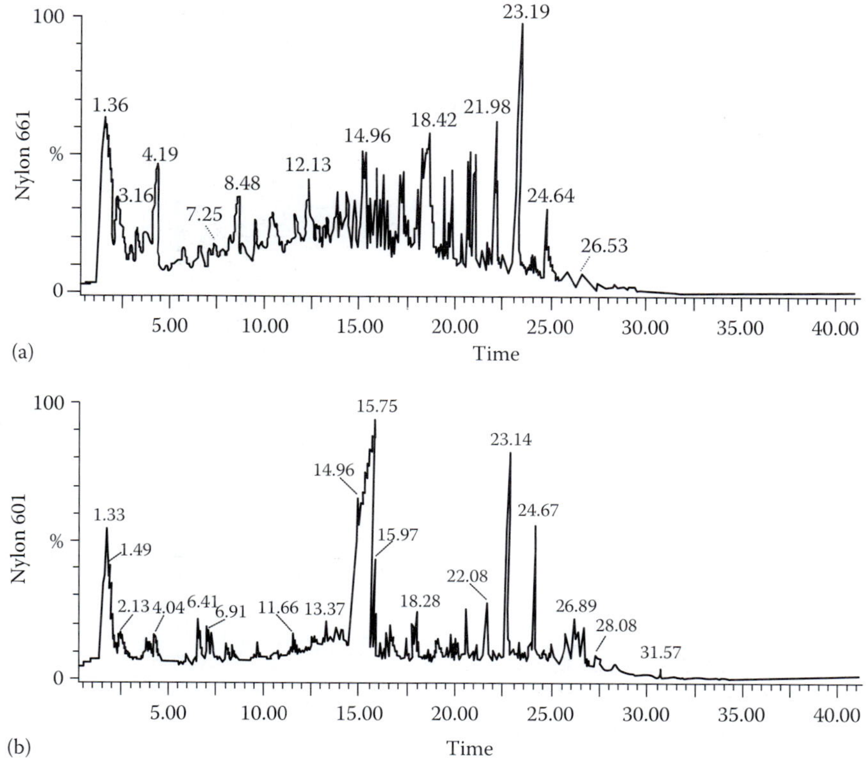

FIGURE 26.21 Pyrolysis GC of (a) nylon-6 and (b) nylon-6,6.

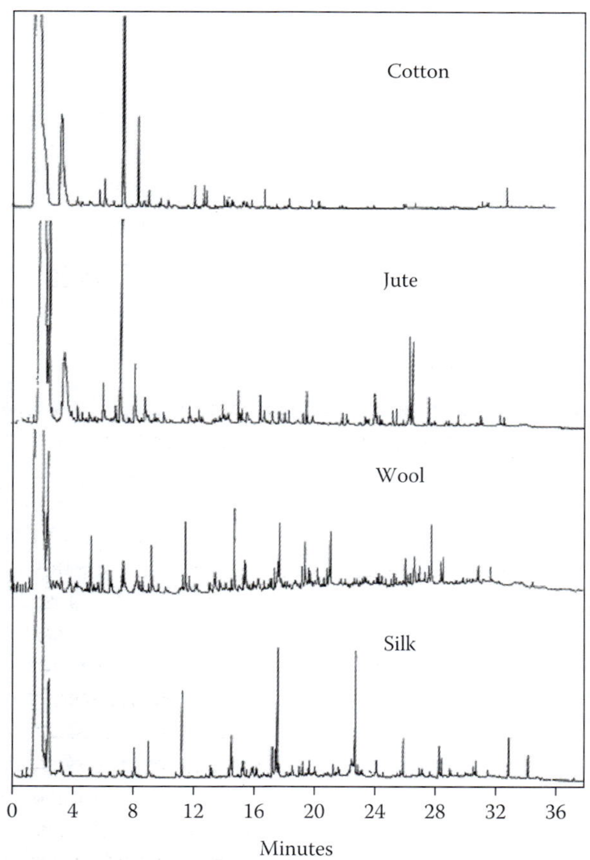

FIGURE 26.22 Pyrograms of some natural materials. (Data from Challinor, J.M., *Forensic Examination of Fibres*, 2nd edn., ed. Robertson, J. and Grieve, M., Boca Raton, FL, CRC Press, 1999, p. 234.)

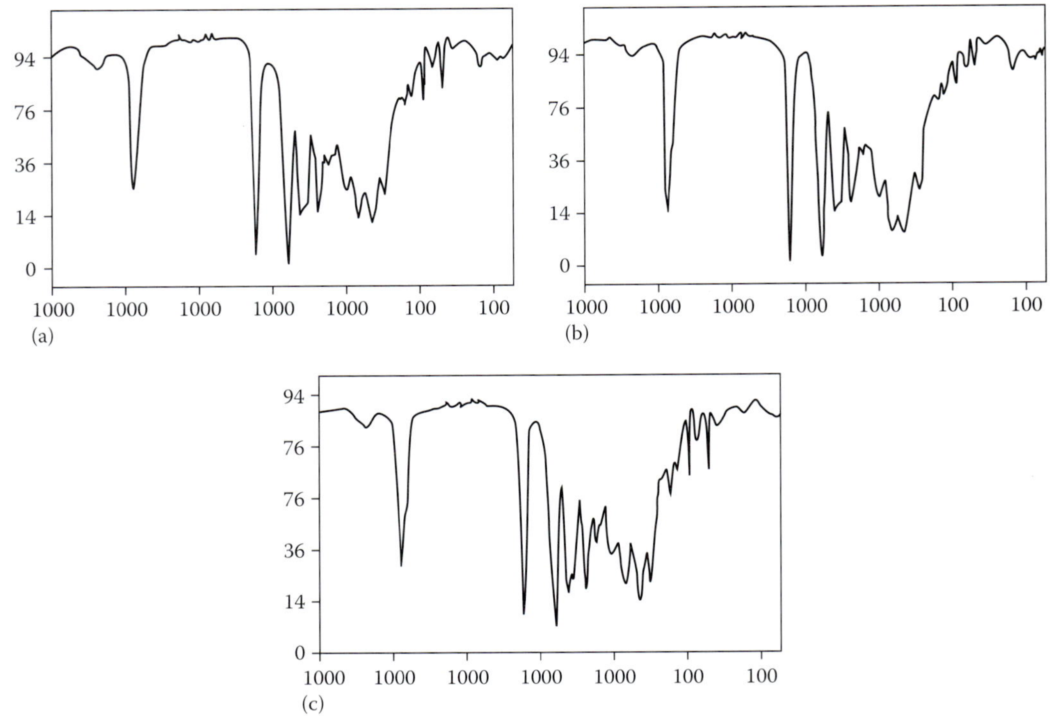

FIGURE 26.23 IR spectra of three clearcoat resins from car paint. (Data from Ryland, S. et al., *J. Forensic Sci.*, 46(1), 31, 2001. Copyright 2001 ASTM International.)

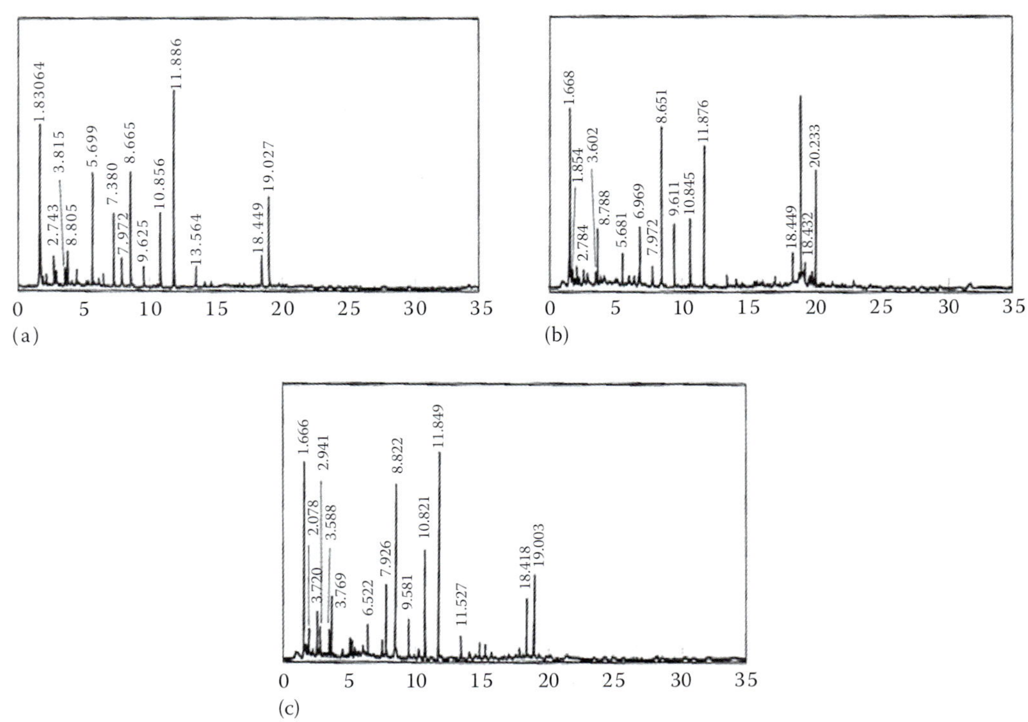

FIGURE 26.24 Pyrograms of the same three clearcoat resins. (Data from Ryland, S. et al., *J. Forensic Sci.*, 46(1), 31, 2001. Copyright 2001 ASTM International.)

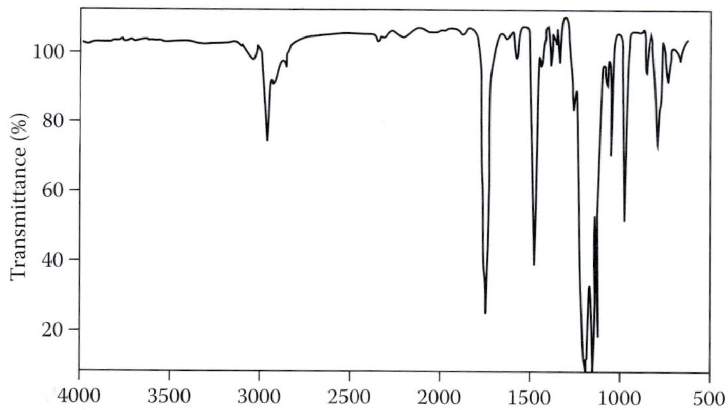

FIGURE 26.25 IR spectrum of polycarbonate.

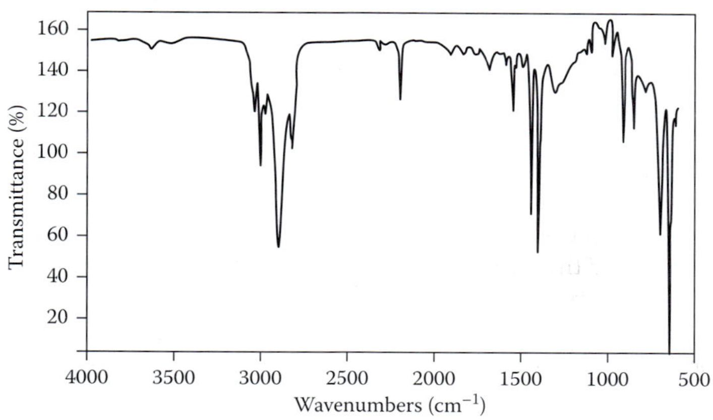

FIGURE 26.26 IR spectrum of ABS copolymer.

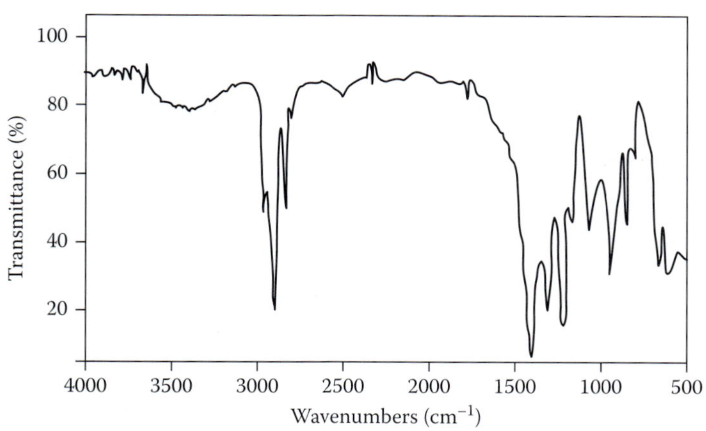

FIGURE 26.27 IR spectrum of PVC.

PROBLEM 26.10

What would you consider to be the most important peaks that can be assigned in the three polymer spectra earlier?

Solution

Polycarbonate—the key features are the carbon–oxygen single- and double-bond stretches.

ABS copolymer—both sp^2 (from the styrene) and sp^3 C–H bonds; carbon–nitrogen triple bond, no carbonyl, OH, or sp^3 C–H.

PVC—the lack of OH, NH, C=O, or sp^2 C–H peaks is significant. The C–Cl bands are difficult to see, around 600–700 cm^{-1}.

26.7 VISUALIZATION OF FINGERPRINTS

Fingerprints are the classic clue of detective fiction. Criminalists have been aware of their use for over a century,* and potters in China and Japan used them to "sign" their work almost 2000 years ago. The modern use of fingerprints dates from the mid-nineteenth century, when law enforcement agencies were attempting to "identify" criminals. Medieval Europe had developed crude forms of identification for criminals, branding and mutilation; these were inexpensive, and visible, but effectively prevented any reintegration of the offender into society. The 1869 Habitual Criminal Act in England provided for longer sentences for repeat offenders—thus, it became essential to be able to identify them. It was a period when there were also various attempts to relate physical features to criminal activity, the so-called criminal type. Before dismissing these activities as offensive superstition, be careful—have you never heard anyone say something about "shifty looking," "eyes too close together," *etc.*? Two systems for fingerprint classification were developed in the late nineteenth century, one by Juan Vucetich in Argentina and one by Sir Edward Richard Henry in the United Kingdom. The latter became much more widely used, although it is probably technically inferior. Fingerprinting was rapidly adopted in Europe and Latin America and became common in the United States in the 1920s. Modern comparison techniques are all computerized, and this type of evidence is accepted as wholly reliable by courts everywhere. However, neither error nor fraud is completely unknown.

Fingerprints are natural skin secretions set down by ridge detail. Most natural fingerprints contain sweat and secretions of the eccine glands. All fingerprints initially contain water, but this is generally lost quickly, while the fatty materials persist. Fingerprints also contain small amounts of amino acids. The quality of the print obtained depends on the elapsed time, the nature of the surface, and the storage of the material. Different techniques are useful for the recovery of prints from different types of surface. For hard, nonabsorbent surfaces, fingerprint powders (various types, but generally inorganic in character) are the most useful. Anthracene, which fluoresces under UV light, is sometimes used for multicolored hard surfaces. All of these powders are "physical" methods of enhancement; they rely on adherence of the material to the moisture and oil in the prints.

On soft materials, chemical treatments are generally used to visualize latent prints. Ninhydrin, **26.35**, reacts irreversibly with amino acids to give a dark-purple product and works well on paper and porous surfaces. It is less useful on wetted items or painted materials. Ninhydrin prints may be treated with zinc(II) chloride, which makes them fluoresce under an argon laser. A related dye is 1,8-diazafluoren-9-one, **26.36**, which is about 2.5 times as sensitive as ninhydrin.

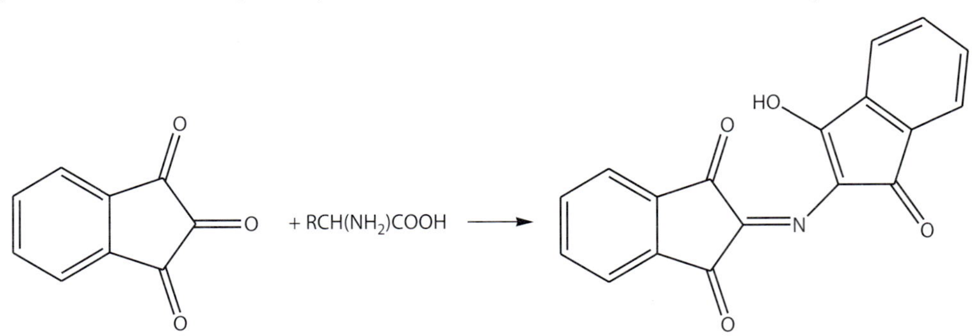

26.35, ninhydrin

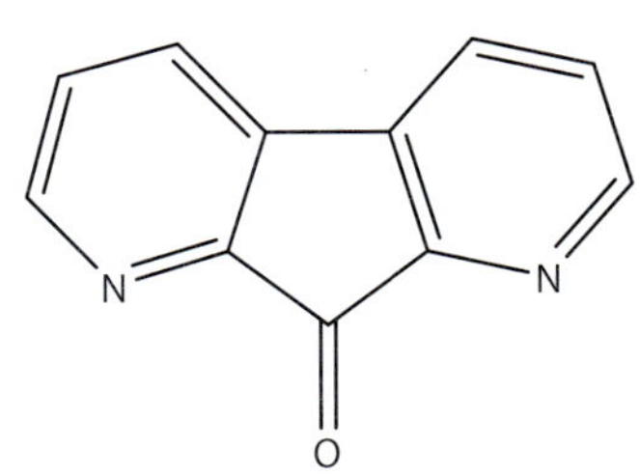

26.36

* Colin Beavan, ***Fingerprints***, 2002, Harper Collins; Simon Cole, ***Suspect Identities***, 2001, Harvard University Press.

Ninhydrin-enhanced fingerprint

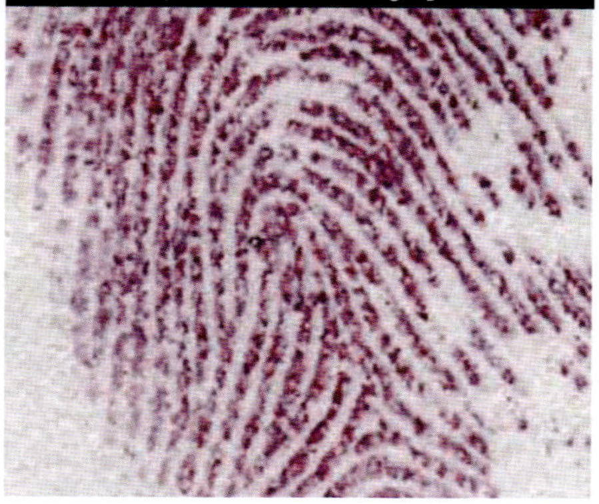

Since 1982, superglue fuming has been added to the armory of techniques for fingerprint visualization. The vapor given off by superglue is ethyl or methyl cyanoacrylate, and this gives a white deposit, an acrylate polymer, on latent prints. Various dyes can be used to enhance color contrast with the background. The formation of the polymer is believed to be catalyzed by water in the fingerprints, so it is only suitable for relatively fresh prints. The technique is useful for metals, electric tape, leather, plastic bags, and even human skin. Normally the "fuming" is done in a lab in a special cabinet, but portable devices are also available, and in some cases, entire rooms can be treated.

Fingerprint developed with superglue fuming and dyed with Basic Yellow 40

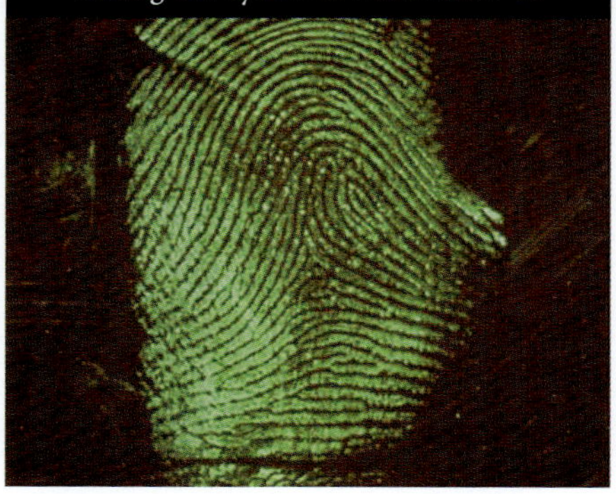

Key Points from Sections 26.4 through 26.7

- Presumptive tests for blood depend on the presence of the enzyme catalase, which is able to oxidize various dye precursors to highly colored materials. Luminol gives a blue color from a chemiluminescent reaction with catalase. Some vegetable materials give false positives.
- Dyes in inks can be characterized using IR and Raman spectroscopy and TLC. The dating of inks and paper is difficult, and the only definitive results are obtained when a particular ink type was not available before the specified date.
- In the area of trace evidence, fibers can be characterized by microscopy, IR spectroscopy, microspectrophotometry, and pyrolysis GC.
- The binders in paints are also identified using IR spectroscopy, microspectrophotometry, and pyrolysis GC.
- Polymers are identified using pyrolysis GC and IR spectroscopy.
- Fingerprints on porous materials may be enhanced using ninhydrin, which reacts with the amino acids in the prints to give a purple color.
- Superglue fuming enhances fingerprints on a variety of surfaces; a polymer of cyano-methacrylate is formed in the presence of water.

REVIEW PROBLEMS

1. (a) Outline one technique that may be used in the identification of fibers. What are the limitations of such analyses as evidence?
   (b) Silver paint traces are found on the clothing of the victim of a hit and run accident. What might be determined from this material, and what techniques should be used?
   (c) How can it be determined whether a stain found at a crime scene is human blood?
2. In the 1980s, a chemistry graduate student attempted to synthesize "synthetic heroin," MPPP, by the sequence shown in the succeeding text (published in the *Journal of Organic Chemistry* in 1947).

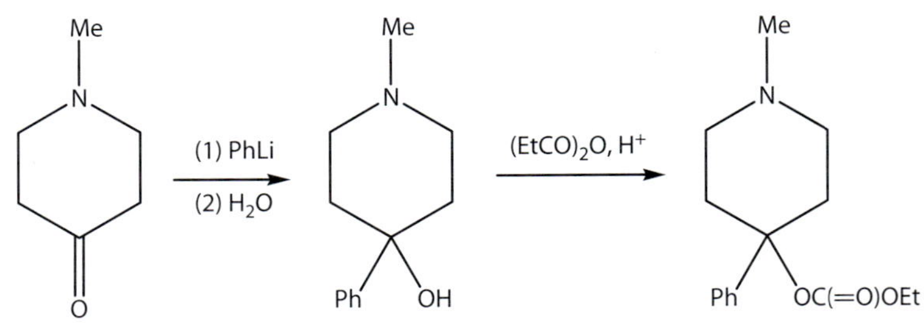

MPPP

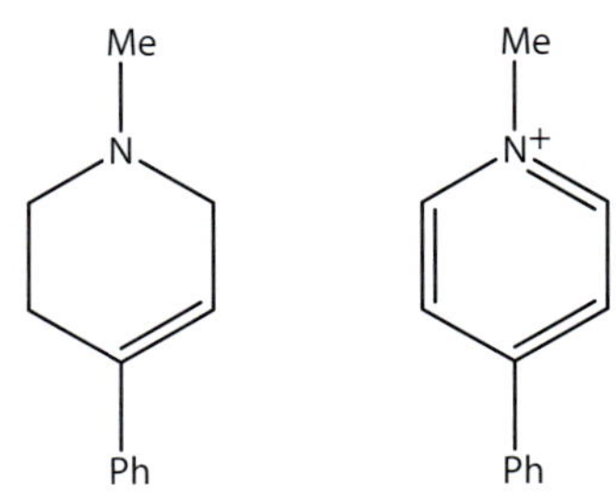

MPTP MPP+

Unfortunately, the preparation went wrong, and significant amounts of MPTP, a neurotoxin, were produced. MPTP can cross the blood–brain carrier and it is then converted to MPP+, which causes the symptoms of Parkinson's disease. This resulted in the so-called case of the frozen addicts (J.W. Langston and J. Palferman, Vintage Books 1996).

Write mechanisms for the steps of the desired synthesis and postulate how the by-product was formed.*
3. Outline the ways in which modern chromatographic techniques have been used in forensic analysis.
4. A plastic bag containing a white powder is confiscated at customs at Heathrow airport. What tests should be done on the material to establish whether an offense has been committed under the Misuse of Drugs Act?
5. How could NMR spectroscopy be used to distinguish the following stimulant molecules?

* Some texts suggest that the reason that the by-product was formed was that the reaction was allowed to overheat—given that the original preparation specifies 3 h at 100 °C, this seems implausible.

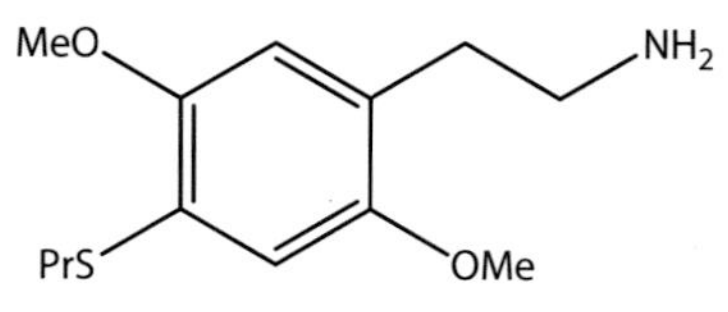

2-CT7

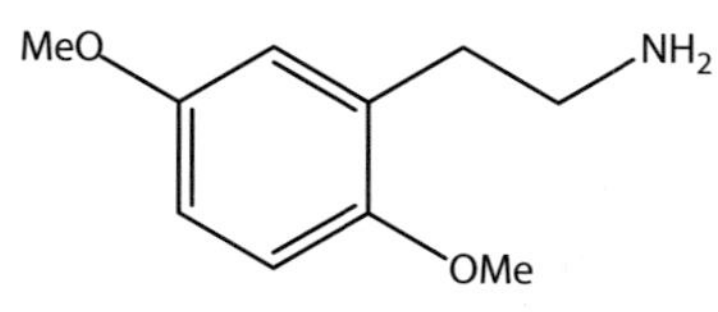

2-CH

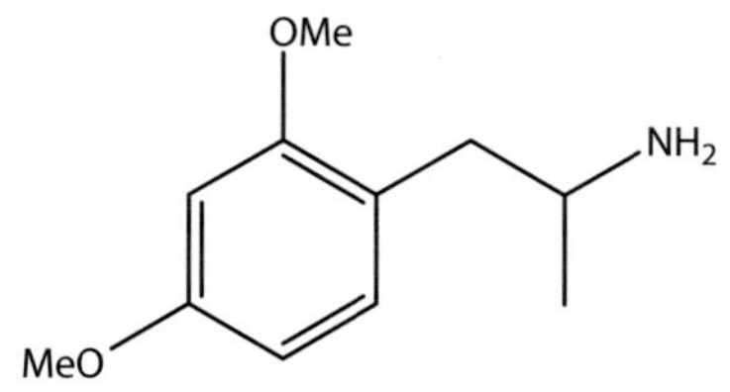

2,4-DMA

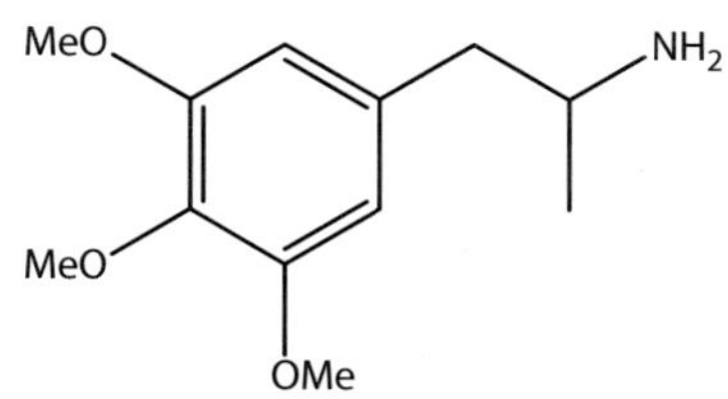

3,4,5-TMA

6. What techniques can be used to establish whether the whole of a questioned document has been written using the same ink?
7. Explain what is meant by presumptive tests for controlled drugs, giving examples. What is meant by positive and negative control experiments and why are they important?
8. What insight is provided by the use of chromatographic techniques in the analysis of illegal drugs?
9. (a) Give advantages and disadvantages of IR *versus* NMR spectroscopies for the analysis of an unknown compound.
   (b) IR stretching frequencies can be predicted using Hooke's law and the following equation:

$$\nu = \frac{1}{2\pi}\sqrt{\frac{k}{\mu}}$$

where
k is the force constant
μ is the reduced mass

Use this information to explain the following IR stretching bands:
S-F ~ 800 cm^{-1}
P-F ~ 1000 cm^{-1}

10. The mushroom fly agaric or *Amanita muscaria* is found through temperate forests in the northern hemisphere. It has been used by shamans in Arctic regions to promote religious experiences (and is supposedly the origin of tales of Santa's flying reindeer), has been implicated in witchcraft, and was said to be used by St. Catherine of Genoa for a spiritual "trip." In fresh materials, the main psychoactive principle is ibotenic acid, but dried material contains mainly muscimol, which is a considerably more potent hallucinogen.
    (a) How would you distinguish ibotenic acid and muscimol spectroscopically?
    (b) What process has occurred in conversion of ibotenic acid to muscimol?
    (c) Can you suggest any reason why muscimol may be more potent than ibotenic acid?

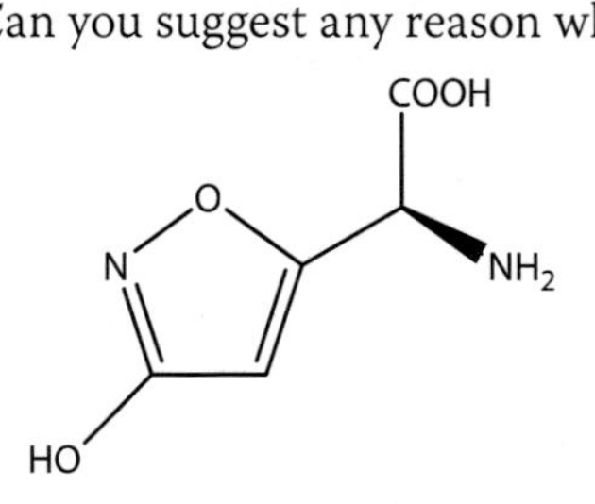

Ibotenic acid

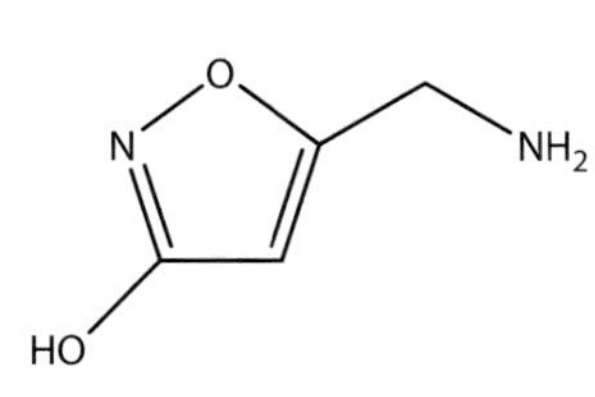

Muscimol

11. What specific features can you note in the spectrum shown below?

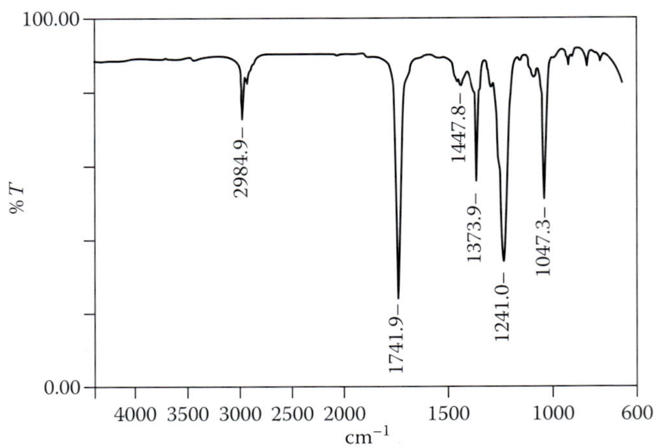

12. How would you interpret the mass spectra of cocaine shown below and heroin in Figure 26.10?

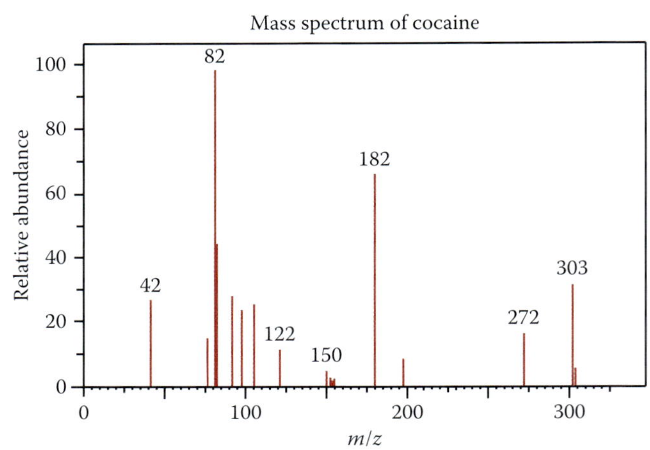

13. Write short notes on
    (a) The Marquis test
    (b) The Scott test
14. Discuss plant and animal derived toxins in the context of their use as poisons. (You may also wish to refer back to Chapter 22.)
15. What methods might be used to investigate the following types of trace evidence?
    (a) Fragments of plastic
    (b) Paint chips
16. Explain how you would distinguish spectroscopically between the controlled substances methamphetamine, **26.36**, methcathinone, **26.37**, and the decongestant ephedrine, **26.38**.

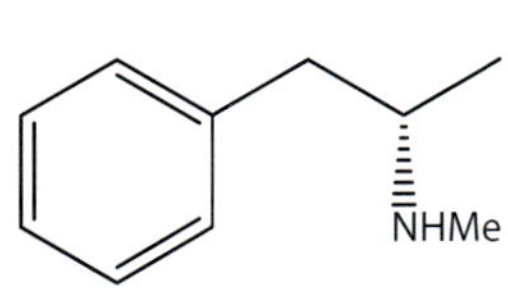

26.36

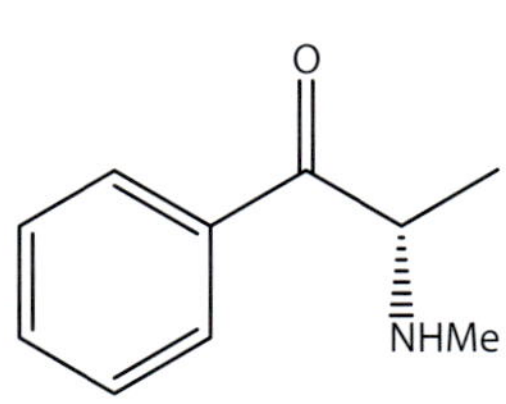

26.37

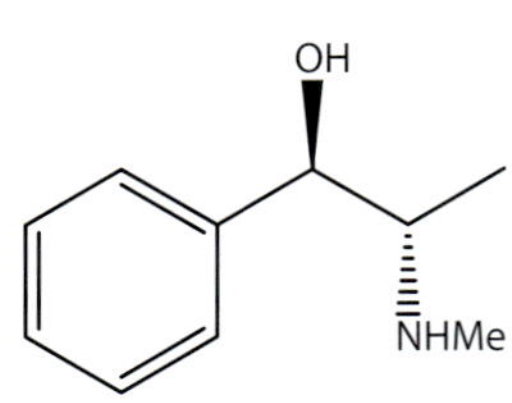

26.38

17. An apparent suicide victim is found in a chair in her living room, with an empty glass and empty bottle of barbiturate sleeping pills and a suicide note on a table by the side of the chair. There is, however, a suspicion that this "suicide" has been staged by the woman's former partner. Describe how the crime scene should be processed to establish the truth about what has happened.
(Note—you are NOT expected to discuss the autopsy or its possible findings)
18. "While testing for the use of illegal substances in the workplace has become relatively common and is ubiquitous in sports, there is, at present, little practical methodology for testing for the use of illegal substances in the context of moving traffic offenses." Discuss.
19. How can analysis of a sample of an illegal drug, confiscated on the street, contribute to knowledge of its origins? You should consider cutting agents as well as impurities in the drug itself.
20. Suggest a mechanism for each step of the following syntheses of benzodiazepines:

(a)

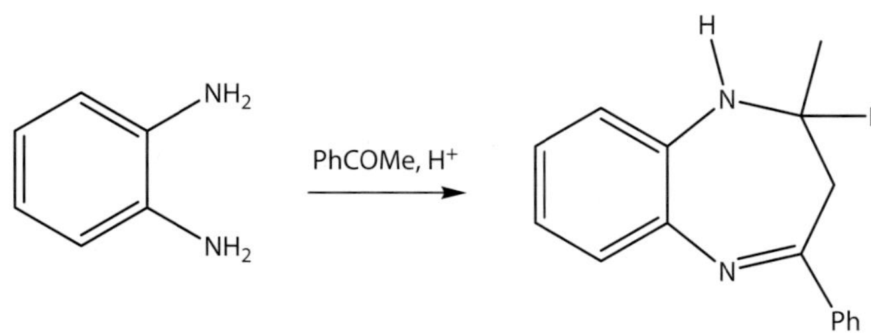

(b)

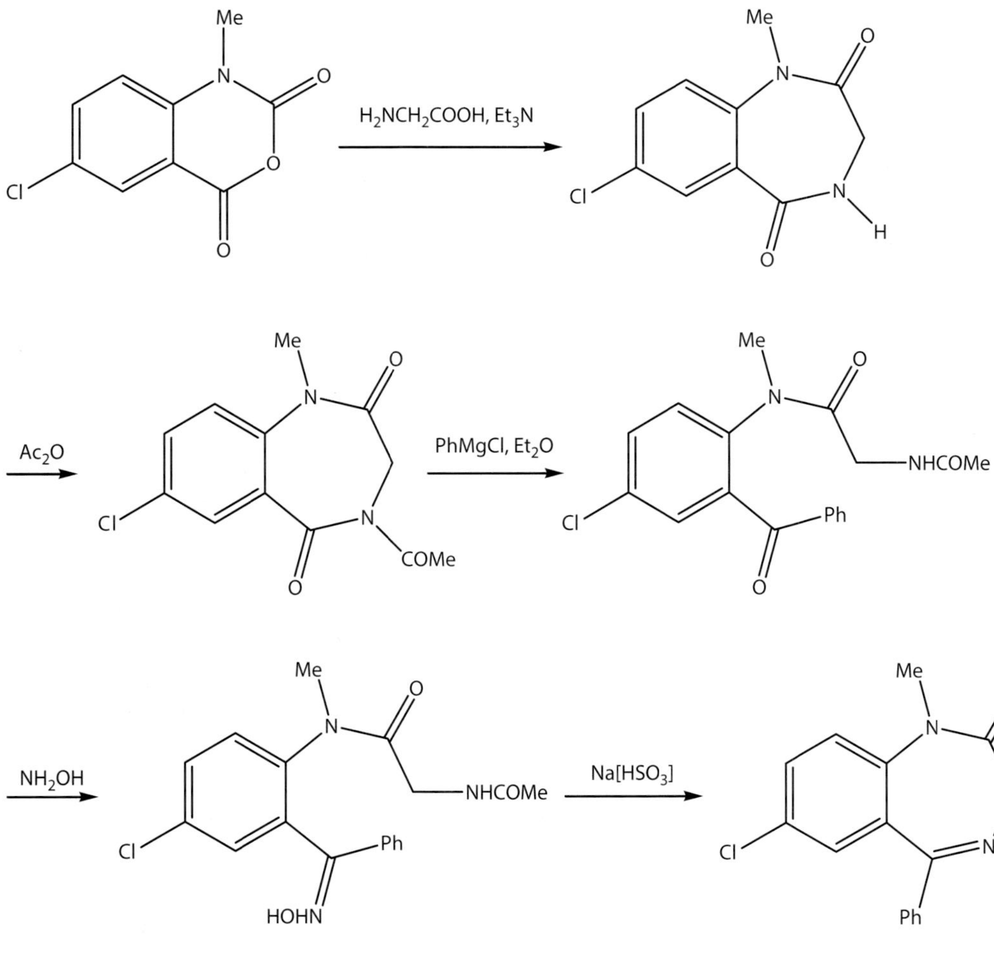

Omit the mechanism of the final step.

21. Suggest mechanisms for each step of the following synthesis of phenobarbitone:

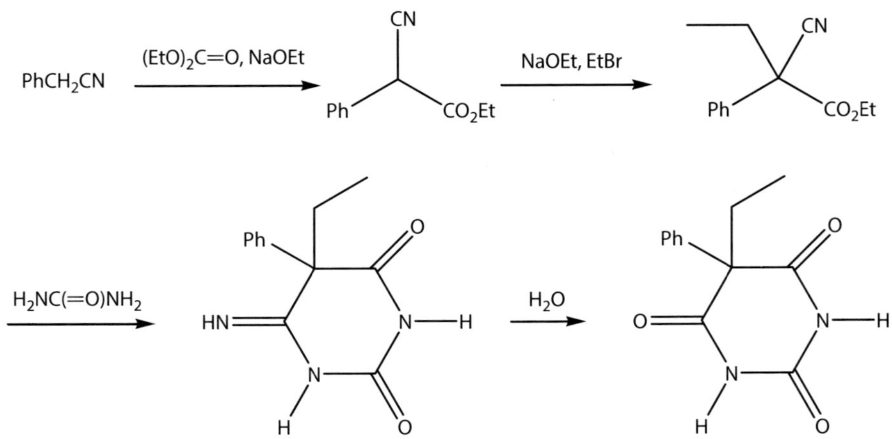

Appendix A: Glossary of Abbreviations and Acronyms

A	Absorbance
A^-	Anion of the generic acid HA
Å	Ångstrom, 10^{-10} m
ABS	Acrylonitrile/butadiene/styrene copolymer
Ac	Acetyl, $CH_3C(=O)-$
acac	Acetyl acetone, pentane-2,4-dione, $CH_3C(=O)CH_2C(=O)CH_3$
acetylCoA	Acetyl coenzyme A
AcOH	Acetic acid, CH_3CO_2H, also HOAc
ADD	Attention deficit disorder
ADHD	Attention deficit hyperactivity disorder
ADME	Absorption, distribution, metabolism, excretion
ADmix	Asymmetric dihydroxylation mixture (α and β variants)
ADP	Adenosine diphosphate
AIBN	Azobisisobutyronitrile, IUPAC 2,2′-[(*E*)-1,2-diazenediyl]*bis*(2-methylpropanenitrile), $NCC(CH_3)_2-N=N-C(CH_3)_2CN$
Ala	Alanine
AMP	Adenosine monophosphate
Ar	Any aryl ring, including phenyl and substituted phenyl groups
Arg	Arginine
Asn	Asparagine
Asp	Aspartic acid
atm	Atmosphere
ATP	Adenosine triphosphate
B	Base
B	Magnetic field strength; used interchangeably with *H*. *B* more commonly used in mass spectrometry
BASF	Badische Anilin und Soda Fabrik
BCE	Before the common era
BINAP	2,2′-*bis*(diphenylphosphino)-1,1′-binapthyl
Bn	benzyl, $-CH_2C_6H_5$
t-Boc	*tert*-Butoxycarbonyl, $Me_3COC(=O)-$
BP	Before present
BPA	Bisphenol A
BPt	Boiling point
br	Broad, refers to width of signals in IR and NMR spectra
i-Bu	*iso*-Butyl, $-CH_2CH_2CH(CH_3)_2$
n-Bu	*n*-Butyl, $-CH_2CH_2CH_2CH_3$
sec-Bu	*sec*-Butyl, $-CH(CH_3)CH_2CH_3$
tert-Bu	*tert*-Butyl, $-C(CH_3)_3$
Bz	Benzoyl, $-C(=O)C_6H_5$
c	Speed of light, 299,792,458 m s^{-1}
c	Concentration
CAS	Chemical Abstracts Services
CFC	Chlorofluorocarbon
cm	Centimeter, 10^{-2} m
CMC	Critical micelle concentration
COD	1,5-Cyclooctadiene

conc.	Concentrated
m-CPBA	*meta*-Chloroperbenzoic acid, also MCPBA
CPK	Corey–Pauling–Koltun—space filling models for the study of stereochemistry
Cy	Cyclohexyl
Cys	Cysteine
d	Doublet—a signal consisting of two lines in an NMR spectrum
D	Debye, the unit for dipole moments
D	An older form of stereochemical designation, still widely used in biochemistry
DBE	Double bond equivalent
DBU	1,8-Diazobicyclo[5.4.0]undece-7-ene, a strong, non-nucleophilic, base
DCC	Dicyclohexylcarbodiimide, CyN=C=NCy
dd	Doublet of doublets in an NMR spectrum
DDT	Dichlorodiphenyltrichloroethane, IUPAC, 1,1,1-trichloro-2,2-di(4-chlorophenyl)ethane
DES	Diethylstilbestrol
DET	Diethyl tartrate
DHP	Dihydropyran
$(DHQ)_2PHAL$	1,4-*bis*(9-*O*-Dihydroquinyl)phthalazine
$(DHQD)_2PHAL$	1,4-*bis*(9-*O*-Dihydroquinidyl)phthalazine
DIBAH	Diisobutylaluminum hydride (sometimes DIBAL)
dil.	Dilute
DIPAMP	Ethane-1,2-diylbis{(2-methoxyphenyl)(phenyl)phosphine}; the acronym is a derivative of the monophosphine, PAMP, phenylanisylmethylphosphine
DIPT	Di-*iso*-propyl tartrate
DMAP	4-Dimethylaminopyridine
DME	1,2-Dimethoxyethane
DMF	Dimethylformamide
DMPE	1,2-*bis*(Dimethylphosphino)ethane
DMSO	Dimethylsulfoxide, $(CH_3)_2S{=}O$
DMT	Dimethyl tartrate
DNA	Deoxyribonucleic acid
DNFB	2,4-Dinitrofluorobenzene
L-DOPA	L-3,4-Dihydroxyphenylalanine
e	Electron
E	Energy
E	Stereochemical designator for a double bond where the two higher priority groups are on opposite sides of the bond
E1	Elimination unimolecular
E2	Elimination bimolecular
E1cB	Elimination, unimolecular, conjugate base
E_a	Activation energy
ee	Enantiomer excess, optical purity
EDTA	Ethylene diamine tetraacetic acid, IUPAC 2,2′,2″,2‴-(1,2-ethanediyldinitrilo)tetraacetic acid
EM	Electromagnetic
EMA	European Medicines Agency (EU)
En	Ethylene diamine, $H_2NCH_2CH_2NH_2$, IUPAC 1,2-diaminoethane
ent	Enantiomeric, used to describe the enantiomer of a structure already designated. Thus, ent-A is the enantiomer of A
EPA	Environmental Protection Agency (United States)
Et	Ethyl, $-CH_2CH_3$
FAO	Food and Agriculture Organization of the United Nations
FDA	Food and drug administration (United States)
FMN	Flavin mononucleotide
fMOC	9-Fluorenylmethoxycarbonyl
g	Gram
G	Gibbs free energy
GABA	Gamma aminobutyric acid, IUPAC 4-aminobutanoic acid, $H_2NCH_2CH_2CH_2CO_2H$

GC/GLC	Gas chromatography
GCMS	Gas chromatography mass spectrometry, sometimes GC-MS
GDP	Gross domestic product
GHB	γ-Hydroxybutyrate, IUPAC 4-hydroxybutanoic acid
GHz	10^9 Hz
Gln	Glutamine
Glu	Glutamic acid
Gly	Glycine
GMO	Genetically modified organisms
h	Planck's constant, 6.626×10^{-34} m^2 kg s^{-1} (J s^{-1})
$h\nu$	Light, may be used for UV/Vis light
H	Enthalpy
H	Magnetic field strength, used interchangeably with *B*. *H* more commonly used in NMR spectroscopy
HA	Generic for an acid
HBcat	Catecholborane
HCFC	Hydrochlorofluorocarbon
HDPE	High density polyethylene
HFC	Hydrofluorocarbon
HFO	Hydrofluoroolefin
His	Histidine
HIV	Human immunodeficiency virus
HLADH	Horse liver alcohol dehydrogenase
HMPA	Hexamethylphosphoramide, $(Me_2N)_3P{=}O$
HOAc	Acetic acid, CH_3CO_2H, also AcOH
HOMO	Highest occupied molecular orbital
HPLC	High-pressure or high-performance liquid chromatography
HRT	Hormone replacement therapy
HSAB	Hard and soft acids and bases
HVZ	Hell–Volhard–Zelinskii
Hz	Hertz, SI unit of frequency, cycles per second
I	Nuclear magnetic moment
Ile	Isoleucine
I_o	Incident light in UV/Vis spectroscopy
IOC	International Olympic Committee
IPA	*iso*-Propanol (*iso*-propyl alcohol)
IPC_2BH	Diisopinocampheylborane, IUPAC *bis*{(1*S*, 2*S*, 3*S*, 5*R*)-2,6,6-trimethylbicyclo[3.3.3[heptan-3-yl}borane
IR	Infrared
iso	Used to denote a branched alkane, isomeric with the normal or linear alkane
I_t	Transmitted light in UV spectroscopy
IUPAC	International Union of Pure and Applied Chemistry
J	Coupling constant in NMR spectroscopy
k	Rate constant
k	Force constant for a bond (or spring) vibration
K	Equilibrium constant
K	Kelvin (temperature relative to absolute zero)
kg	Kilogram
KIE	Kinetic isotope effect
kJ	Kilojoule
KM	Kastle–Meyer—a presumptive test for mammalian blood
l	Angular momentum quantum number
L	A generic ligand
L	An older form of stereochemical designation, still widely used in biochemistry
LAH	Lithium aluminum hydride, $Li[AlH_4]$
LCAO	Linear combination of atomic orbitals
LCMS	Liquid chromatography mass spectrometry, sometimes LC-MS
LD_{50}	Median lethal dose; the dose of a material at which 50 % of the test subjects die after a specified time

LDA	Lithium diisopropylamide, LiN(*i*-Pr)$_2$
LDPE	Low-density polyethylene
Leu	Leucine
LiTMP	Lithium tetramethylpiperidide
LPG	Liquefied petroleum gas
LSD	Lysergic acid diethylamide
LUMO	Lowest unoccupied molecular orbital
Lys	Lysine
m	Meter
m	Mass
m	Medium—refers to absorption strength in IR spectroscopy
m	Multiplet—refers to a group of lines too complex to analyze directly in NMR spectroscopy
m	*meta*, refers to a 1,3-relationship on a benzene ring
M	Used to denote a whole molecule in mass spectrometry, otherwise molecular weight or molarity
m/z	Mass divided by charge in mass spectrometry
max	Maximum, often used with wavelength or frequency in spectroscopy, *e.g.*, λ_{max}
MCPBA	*meta*-Chloroperbenzoic acid, also *m*-CPBA
MDMA	Methylenedioxymethamphetamine, ecstasy
MDPE	Medium-density polyethylene
Me	Methyl, $-CH_3$
Met	Methionine
MEM	β-Methoxyethoxymethyl, ROMEM is $ROCH_2OCH_2CH_2OCH_3$
mg	Milligram
MHz	Megahertz, 10^6 Hz
m_l	Magnetic quantum number
MMPP	Magnesium monoperoxyphthalate
MO	Molecular orbital
mol^{-1}	Per mole
MPa	MegaPascals; the Pascal is the SI unit of pressure, Newton per square meter. MPa is newton per square millimeter
MPt	Melting point
MRI	Magnetic resonance imaging
MRSA	Methicillin-resistant *Staphylococcus aureus*
Ms	Mesyl, methanesulfonate, $-SO_2Me$
n	Principle quantum number
n	Used to denote normal or linear in alkane nomenclature
n	Also used generically for number, *e.g.*, the $(n + 1)$ rule in NMR spectroscopy
NAD^+	Nicotinamide adenine dinucleotide
NADPH	Nicotinamide adenine dinucleotide phosphate
NaHDMS	Sodium *bis*(trimethylsilyl)amide, $NaN(SiMe_3)_2$
NBS	*N*-Bromosuccinimide
nm	Nanometer, 10^{-9} m
NMO	*N*-Methylmorpholine-*N*-oxide
NMR	Nuclear magnetic resonance
NSAID	Non-steroidal anti-inflammatory drug
Nu^-	Nucleophile
o	*ortho*, refers to a 1,2-relationship on a benzene ring
p	*para*, refers to a 1,4-relationship on a benzene ring
PCC	Pyridinium chlorochromate, $[pyH]^+[CrO_3Cl]^-$
ppm	Parts per million
PAH	Polyaromatic hydrocarbon
PAN	Polyacrylonitrile
PCB	Polychlorinated biphenyl
PDB	Pee Dee Belemnite
PE	Polythene, polyethylene
PEG	Poly(ethylene glycol), also called PEO poly(ethylene oxide)

PET	Poly(ethylene terephthalate)
PFBOC	Perfluorocarbon-based oxygen carrier
PFC	Perfluorocarbon
PG	Prostaglandin
Ph	Phenyl, $-C_6H_5$
Phe	Phenylalanine
P_i	Inorganic phosphate
PIFA	Phenyl iodine *bis*(trifluoroacetate)
pK_a	$-\log_{10} K_a$, the equilibrium constant for an acid/base equilibrium; a measure of the acidity of a compound
PLA	Poly(lactic acid)
PMMA	Poly(methyl methacrylate)
POP	Persistent organic pollutant
PP_i	Inorganic diphosphate or pyrophosphate
PPTA	Poly(*para*-phenylene terephthalamide)
i-Pr	*iso*-Propyl, $-CH(CH_3)_2$
n-Pr	*n*-Propyl, $-CH_2CH_2CH_3$
PRF	Partial rate factor
Pro	Proline
psi	Pounds per square inch
PT	Phase transfer
PTFE	Polytetrafluoroethylene, Teflon
PVA	Poly(vinyl acetate)
PVC	Poly(vinyl chloride)
py	Pyridine, C_6H_5N
q	Quartet—a signal consisting of four lines in an NMR spectrum
QALY	Quality-adjusted life year
r	Used for a number of types of distance measurement
R	Stereochemical designator for a chiral molecule
R	Used to denote a substituent in an organic compound, most commonly, but not exclusively, an alkyl group
R	Rydberg constant, 10.974×10^6 m^{-1}
®	Registered trademark
RDS	Rate-determining step
REACH	Registration, Evaluation, Authorization and Restriction of Chemicals (EU)
RF	Radio frequency
R_f	Retardation factor
RNA	Ribonucleic acid
R_s	Measured ratio of a lighter to a heavier isotope
R_{std}	Measured ratio of a lighter to a heavier isotope in a standard
RT	Room temperature—generally between 20 °C and 25 °C
s	Second
s	Strong—refers to absorptions in IR spectroscopy
s	A signal consisting of a single line in an NMR spectrum
s	Spin quantum number
S	Entropy
S	Stereochemical designator for a chiral molecule
SAM	*S*-Adenosylmethionine
Ser	Serine
SFM	Scanning force microscopy
Sia_2BH	Disiamylborane, IUPAC *bis*(1,2-dimethylpropyl)borane, $\{Me_2CHCH(Me)\}_2BH$
SLAP	Standard light Antarctic precipitation
SM	Starting material
SMOW	Standard mean ocean water
S_N1	Substitution, nucleophilic, unimolecular
S_N2	Substitution, nucleophilic, bimolecular
S_Ni	Substitution, nucleophilic, intramolecular

SPPS	Solid-phase peptide synthesis
STM	Scanning tunneling microscopy
SVHC	Substance of very high concern
t	Triplet—a signal consisting of three lines in an NMR spectrum
T	Tesla, SI unit of magnetic flux density, $V\ s\ m^{-2}$
T	Temperature
TATP	Triacetone triperoxide
TBDMS	*tert*-Butyldimethylsilyl; ROTBDMS = ROSiMe$_2$*t*-Bu
TBDPS	*tert*-Butyldiphenylsilyl; ROTBDPS = ROSiPh$_2$*t*-Bu
TES	Triethylsilyl, ROTES = $ROSi(CH_2CH_3)_3$
tert	Tertiary, used to denote a tertiary center in nomenclature as in *tert*-butyl chloride, $(CH_3)_3CCl$
Tf	Triflate, trifluoromethylsulfonate; ROTf = $ROS(=O)_2CF_3$
THF	Tetrahydrofuran
THG	Tetrahydrogestrinone
THP	Tetrahydropyranyl (refers to a group used to protect alcohols)
Thr	Threonine
TLC	Thin layer chromatography
TM	Target molecule
TMEDA	Tetramethylethylene diamine, IUPAC *N,N,N′,N′*-tetramethylethane-1,2-diamine, $Me_2NCH_2CH_2NMe_2$
TMS	Tetramethylsilane
TMS	Trimethylsilyl; ROTMS = $ROSi(CH_3)_3$
TNT	Trinitrotoluene
Tol	Tolyl
TRIS	*tris*(Hydroxymethyl)aminomethane, IUPAC *tris*(hydroxymethyl)methanamine, $H_2NC(CH_2OH)_3$
Trp	Tryptophan
TRT	Testosterone replacement therapy
Ts	Tosyl, toluene sulfonyl, $-SO_2C_6H_4$-4-CH_3
Tyr	Tyrosine
UHMWPE	Ultrahigh molecular weight polythene
UTI	Urinary tract infection
UV	Ultraviolet
UV/Vis	Ultraviolet/visible
V	Voltage
Val	Valine
VLDPE	Very-low-density polyethylene
VOC	Volatile organic compounds
VRE	Vancomycin-resistant *Enterococcus*
vs	Versus
VSEPR	Valence shell electron pair repulsion
w	Weak—refers to strength of absorption in IR spectroscopy
WADA	World Anti-Doping Agency
WHO	World Health Organization
X	Generally used for a halogen substituent but may also be used for a random substituent or a generic leaving group
XS	Excess
Z	Stereochemical designator for a double bond where the two higher priority groups are on the same side of the bond
1 °	Used to denote primary, mainly for alcohols, carbocations and carbanions
12-C-4	14-Crown-4, IUPAC 1,4,7,10-tetraoxacyclododecane
15-C-5	15-Crown-5, IUPAC 1,4,7,10,13-pentaoxacyclopentadecane
18-C-6	18-Crown-6, IUPAC 1,4,7,10,13,16-hexaoxacyclooctadecane
2 °	Used to denote secondary, mainly for alcohols, carbocations and carbanions
2,4-D	2,4-Dichlorophenoxyacetic acid
2,4,5-T	2,4,5-Trichlorophenoxyacetic acid
3 °	Used to denote tertiary, mainly for alcohols, carbocations and carbanions
9-BBNH	9-Borabicyclo[3.3.1]nonane

α	Angle of rotation, used also in designating specific rotation of a chiral compound
Δ	Heat
Δ	Used to express a difference in two values; thus, $\Delta\nu = \nu_1 - \nu_2$
δ	Chemical shift in NMR spectroscopy
ε	Molar extinction coefficient in UV/Vis spectroscopy
γ	Gyromagnetic ratio in NMR spectroscopy
λ	Wavelength
μ	Dipole moment
μm	Micrometer, 10^{-6} m
μM	Micromolar
ν	Frequency
Ψ	Wavefunction in the Schrödinger equation
*	Often used to indicate an atom labeled with a particular isotope such as deuterium or ^{13}C
$	Amounts are in US dollars unless otherwise specified

Appendix B: Some Common/Trivial Names You Need to Know

Many other trivial names will arise in the text, but they are not common enough to be worth learning. The following ones are—if only because they will be on the labels of the reagent bottles in your lab course.

Acetaldehyde	Ethanal
Acetate	Ethanoate
Acetic acid	Ethanoic acid
Acetic anhydride	Ethanoic anhydride
Acetone	Propanone
Acetophenone	1-Phenylethanone ($PhC(=O)CH_3$)
Acetylacetone	Pentane-2-4-dione
Acetyl chloride	Ethanoyl chloride
Acetylene	Ethyne
Acrylamide	Propenamide, $CH_2=CHC(=O)NH_2$
Acrylonitrile	Propenenitrile, $CH_2=CHCN$
Allene	1,2-Propadiene, $H_2C=C=CH_2$
Allyl acetate	2-Propenyl ethanoate
Allyl bromide	3-Bromopropene
Allyl alcohol	2-Propene-1-ol
Aniline	Aminobenzene (IUPAC allows aniline)
Anisole	Methoxybenzene
Benzil	1,2-Diphenyl-1,2-ethanedione
Benzilic acid	2,2-Diphenyl-2-hydroxyethanoic acid
Benzoic acid	Benzene carboxylic acid
Benzoin	1,2-Diphenyl-2-hydroxy-1-ethanone ($PhC(=O)CH(OH)Ph$)
Benzophenone	Diphenylmethanone, $PhC(=O)Ph$
Benzyl chloroformate	Benzyl chloromethanoate, $PhCH_2OC(=O)Cl$
Benzyl cyanide	Phenylacetonitrile, $PhCH_2CN$
Bromoform	Tribromomethane
Butadiene	1,3-Butadiene, unless otherwise specified
iso-Butane	2-Methylpropane
iso-Butene	2-Methylpropene
sec-Butanol	2-Butanol
tert-Butanol	2-Methyl-2-propanol
iso-Butyl	2-Methylpropyl
tert-Butyl	1,1-Dimethylethyl
Butylene	1-Butene
Butyric acid	Butanoic acid
Caprolactam	2-Azepanone
Caprolactone	2-Oxepanone
Carbon tetrachloride	Tetrachloromethane
Chloroform	Trichloromethane
Chloroprene	2-Chloro-1,3-butadiene
Cumene	*iso*-Propylbenzene, 1-methylethylbenzene
Decalin	decahydronaphthalene, two cyclohexyl rings fused along one bond
Diacetone alcohol	4-Hydroxy-4-methyl-2-pentanone
Diethyl malonate	Diethyl propanedioate

Dithiane	1,3-Dithiane
Dry ice	Solid CO_2
Ethyl acetoacetate	Ethyl 3-oxobutanoate
Ethyl alcohol	Ethanol
Ethyl bromide	Bromoethane
Ethyl chloride	Chloroethane
Ethylene	Ethene
Ethylene diamine	1,2-Diaminoethane
Ethylene glycol	1,2-Dihydroxyethane
Formaldehyde	Methanal
Formalin	An aqueous solution of formaldehyde
Formate	Methanoate
Formic acid	Methanoic acid
Freon-12	Dichlorodifluoromethane
Fumaric acid	*trans*-2-Butenedioic acid
Glyceraldehyde	2,3-Dihydroxypropanal
Glyceride	An ester of glycerol
Glycerol	Propane 1,2,3-triol
Iodoform	Triiodomethane
Isoprene	2-Methyl-1,3-butadiene
Lactic acid	2-Hydroxypropanoic acid
Maleic acid	*cis*-2-Butenedioic acid
Maleic anydride	*cis*-2-Butenedioic acid anhydride
Malonic acid	Propanedioic acid
Mesityl oxide	4-Methyl-3-pentene-2-one
Mesyl chloride	Methane sulfonyl chloride
Methyl alcohol	Methanol
Methyl bromide	Bromomethane
Methyl cyanoacrylate	Methyl 2-cyanopropenaote (sometimes cyanomethacrylate)
Methylene chloride	Dichloromethane
Methyl iodide	Iodomethane
Methyl methacrylate	Methyl 2-methylpropenoate
iso-Pentane	2-Methylbutane
neo-Pentane	2,2-Dimethylpropane
Phenol	Hydroxybenzene (IUPAC allows phenol)
Phthalic acid	Benzene-1,2-dicarbocylic acid
Pinacol	2,3-dihyroxy-2,3-dimethylbutane
Pinacolone	3,3-Dimethyl-2-butanone
Propionic acid	Propanoic acid
iso-Propanol	2-Propanol
iso-Propyl	1-Methylethyl
Propyl alcohol	1-Propanol
Propylene	Propene
Salicylic acid	2-Hydroxybenzoic acid
Styrene	Phenyl ethene
Tartaric acid	2,3-Dihydroxybutanedioic acid
Terephthalic acid	Benzene 1,4-dioic acid
Toluene	Methylbenzene (IUPAC allows toluene)
Tosyl chloride	Toluene sulfonyl chloride
Triflate	Trifluoromethylsulfonate
Tropylium ion	Cycloheptatrienyl cation
Vinyl acetate	Ethenyl ethanoate
Vinyl chloride	Chloroethene
Xylene	Dimethylbenzene

Index

B

D

E

F

G

H

I

J

K

L

M

P

Q

R

S

X

Z